# Periodic table

Key: Atomic number, Z · Element symbol · Relative atomic mass, $A_r$

Example:
1
**H**
1.008

| 1 | 2 | 3 | 4 | 5 | 6 | 7 | 8 | 9 | 10 | 11 | 12 | 13 | 14 | 15 | 16 | 17 | 18 |
|---|---|---|---|---|---|---|---|---|---|---|---|---|---|---|---|---|---|
| 1 **H** 1.008 | | | | | | | | | | | | | | | | | 2 **He** 4.00 |
| 3 **Li** 6.94 | 4 **Be** 9.01 | | | | | | | | | | | 5 **B** 10.81 | 6 **C** 12.01 | 7 **N** 14.01 | 8 **O** 16.00 | 9 **F** 19.00 | 10 **Ne** 20.18 |
| 11 **Na** 22.99 | 12 **Mg** 24.31 | | | | | | | | | | | 13 **Al** 26.98 | 14 **Si** 28.09 | 15 **P** 30.97 | 16 **S** 32.06 | 17 **Cl** 35.45 | 18 **Ar** 39.95 |
| 19 **K** 39.10 | 20 **Ca** 40.08 | 21 **Sc** 44.96 | 22 **Ti** 47.90 | 23 **V** 50.94 | 24 **Cr** 52.01 | 25 **Mn** 54.94 | 26 **Fe** 55.85 | 27 **Co** 58.93 | 28 **Ni** 58.69 | 29 **Cu** 63.54 | 30 **Zn** 65.41 | 31 **Ga** 69.72 | 32 **Ge** 72.59 | 33 **As** 74.92 | 34 **Se** 78.96 | 35 **Br** 79.91 | 36 **Kr** 83.80 |
| 37 **Rb** 85.47 | 38 **Sr** 87.62 | 39 **Y** 88.91 | 40 **Zr** 91.22 | 41 **Nb** 92.91 | 42 **Mo** 95.94 | 43 **Tc** 98.91 | 44 **Ru** 101.07 | 45 **Rh** 102.91 | 46 **Pd** 106.42 | 47 **Ag** 107.87 | 48 **Cd** 112.40 | 49 **In** 114.82 | 50 **Sn** 118.71 | 51 **Sb** 121.75 | 52 **Te** 127.60 | 53 **I** 126.90 | 54 **Xe** 131.30 |
| 55 **Cs** 132.91 | 56 **Ba** 137.34 | La–Lu | 72 **Hf** 178.49 | 73 **Ta** 180.95 | 74 **W** 183.85 | 75 **Re** 186.21 | 76 **Os** 190.23 | 77 **Ir** 192.22 | 78 **Pt** 195.08 | 79 **Au** 196.97 | 80 **Hg** 200.59 | 81 **Tl** 204.37 | 82 **Pb** 207.19 | 83 **Bi** 208.98 | 84 **Po** 210 | 85 **At** 210 | 86 **Rn** 222 |
| 87 **Fr** 223 | 88 **Ra** 226.03 | Ac–Lr | 104 **Rf** [261] | 105 **Db** [262] | 106 **Sg** [266] | 107 **Bh** [264] | 108 **Hs** [277] | 109 **Mt** [268] | 110 **Ds** [271] | 111 **Rg** [272] | 112 **Uub** [285] | | | | | | |

**Lanthanoids**

| 57 **La** 138.91 | 58 **Ce** 140.12 | 59 **Pr** 140.91 | 60 **Nd** 144.24 | 61 **Pm** 146.92 | 62 **Sm** 150.35 | 63 **Eu** 151.96 | 64 **Gd** 157.25 | 65 **Tb** 158.92 | 66 **Dy** 162.50 | 67 **Ho** 164.93 | 68 **Er** 167.26 | 69 **Tm** 168.93 | 70 **Yb** 173.04 | 71 **Lu** 174.97 |
|---|---|---|---|---|---|---|---|---|---|---|---|---|---|---|

**Actinoids**

| 89 **Ac** 227.03 | 90 **Th** 232.04 | 91 **Pa** 231.04 | 92 **U** 238.03 | 93 **Np** 237.05 | 94 **Pu** 239.05 | 95 **Am** 241.06 | 96 **Cm** 244.07 | 97 **Bk** 249.08 | 98 **Cf** 252.08 | 99 **Es** 252.09 | 100 **Fm** 257.10 | 101 **Md** 258.10 | 102 **No** 259 | 103 **Lr** 262 |
|---|---|---|---|---|---|---|---|---|---|---|---|---|---|---|

Visit the *Chemistry, Third Edition* Companion Website at
**www.pearsoned.co.uk/housecroft** to find valuable **student**
learning material including:

- Multiple choice questions to help test your learning
- Rotatable 3D structures taken from the book
- Mathematics Tutor
- Annotated links to relevant sites on the web

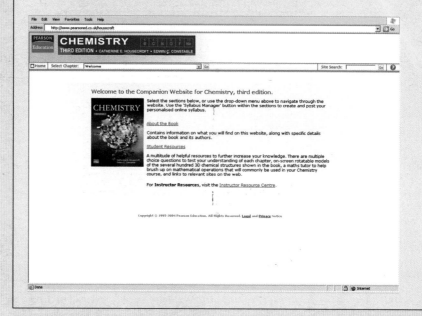

# CHEMISTRY

An Introduction to Organic, Inorganic and Physical Chemistry

**3rd edition**

## Catherine E. Housecroft
## Edwin C. Constable

PEARSON

Prentice
Hall

Harlow, England • London • New York • Boston • San Francisco • Toronto • Sydney • Singapore • Hong Kong
Tokyo • Seoul • Taipei • New Delhi • Cape Town • Madrid • Mexico City • Amsterdam • Munich • Paris • Milan

**Pearson Education Limited**
Edinburgh Gate
Harlow
Essex CM20 2JE
England

and Associated Companies throughout the world

*Visit us on the World Wide Web at:*
www.pearsoned.co.uk

First published under the Longman imprint 1997
Second edition 2002
**Third edition 2006**

ISBN 0 131 27567 4

**British Library Cataloguing-in-Publication Data**
A catalogue record for this book is available from the British Library

**Library of Congress Cataloging-in-Publication Data**
A catalog record for this book is available from the Library of Congress

10 9 8 7 6 5 4 3 2 1
09 08 07 06 05

Typeset in 10pt Times by 60
Printed by Ashford Colour Press Ltd, Gosport

*The publisher's policy is to use paper manufactured from sustainable forests.*

# Summary of contents

# Contents

---

§ Appendices 13 and 14 can be found on the companion website at www.pearsoned.co.uk/housecroft

## Supporting resources

Visit **www.pearsoned.co.uk/housecroft** to find valuable online resources

### Companion Website for students

- Multiple choice questions to help test your learning
- Rotatable 3D structures taken from the book
- Mathematics Tutor
- Annotated links to relevant sites on the web

### For instructors

- Complete, downloadable Instructor's Manual
- PowerPoint slides that can be downloaded and used as OHTs
- Rotatable 3D structures taken from the book

**Also:** The Companion Website provides the following features:

- Search tool to help locate specific items of content
- E-mail results and profile tools to send results of quizzes to instructors
- Online help and support to assist with website usage and troubleshooting

For more information please contact your local Pearson Education sales representative or visit
**www.pearsoned.co.uk/housecroft**

## OneKey: All you and your students need to succeed

OneKey is an exclusive new resource for instructors and students, giving you access to the best online teaching and learning tools 24 hours a day, 7 days a week.

OneKey means all your resources are in one place for maximum convenience, simplicity and success.

A OneKey product is available for *Chemistry, Third Edition* for use with Blackboard™, WebCT and CourseCompass. It contains:

- A test bank of over 5000 questions
- Video clips of experiments
- Simulations and interactions

For more information about the OneKey product please contact your local Pearson Education sales representative or visit
**www.pearsoned.co.uk/onekey**

# Preface

In the third edition of *Chemistry*, our aims continue to be to provide a single book that addresses the requirements of a first year university chemistry course and brings together those areas of chemistry that traditionally have been taught apart from one another. In this new edition, we have tried to take into account some of the feedback from colleagues around the world. Although the core content is based upon attainment expectations in the United Kingdom educational system, we hope that the entry level is now appropriate for a wider cross-section of science students. The structure of the book closely follows that of the second edition. On going to this new edition, we have strengthened the first chapter by including more worked examples that cover basic topics such as masses and moles, concentrations of solutions, reaction stoichiometry and balancing equations. We have expanded the coverage of thermochemistry, and this now appears as a new Chapter 2. Throughout the book, we continue to highlight the relevance of chemistry and related disciplines to everyday life, often through the use of topic boxes. We have tried to introduce examples with which students either will be familiar or will have read about in the popular press. This is an essential way of motivating students who are studying what is often perceived to be a difficult subject. Because of the number of biological science students who are now using *Chemistry*, we have paid increased attention to topics of biological relevance and have introduced more *Environment and biology* topic boxes. With this in mind, a new chapter introduces aromatic heterocyclic compounds to provide a sound chemical basis for subsequent discussion of the structure and function of biological molecules.

Feedback from readers of the second edition has been invaluable as we have considered revisions for the book. Once again, we have agonized about the level of mathematics, especially in chapters dealing with quantum mechanics and thermodynamics. We continue to take the view that, although qualitative discussions are helpful as a way of introducing a student to a topic, a full understanding cannot be achieved without the use of rigorous mathematics. Where necessary, mathematical support material is presented in *Theoretical and chemical background* topic boxes, and students are also guided to the *Mathematics Tutor* that is part of the accompanying website (www.pearsoned.co.uk/housecroft). One of the topics in this tutor deals with significant figures and decimal places. Readers should note that in the worked examples in this book in which several steps are worked in full, we round off the answer at the end of each step.

We make every effort to keep up to date with recommendations put forward by the IUPAC, and a significant change in Chapter 14 arises from the 2001 recommendations of the IUPAC concerning the reporting of chemical shifts in NMR spectroscopy. The recommendation to write, for example, $\delta$ 4.5 ppm, reverses the previous recommendation (1972) to report the value as $\delta$ 4.5.

As in the previous edition of *Chemistry*, the three-dimensional molecular structures in the book have been drawn using atomic coordinates accessed from the Cambridge Crystallographic Data Base and implemented through the ETH in Zürich, or from the Protein Data Bank (http://www.rcsb.org/pdb).

We are grateful to colleagues whom we acknowledged in the preface to the second edition, and to the review panel set up by the publisher. Additional thanks go to Professors Robin Harris, Ken McKendrick, Helma Wennemers and John Nielsen for specific comments about the second edition of the book, to Professor Ron Gillespie for his thoughts on bonding in inorganic molecules, to Dr Egbert Figgemeier for invaluable discussions relating to thermodynamics, and to Professor Neil Connelly for keeping us informed about recent changes to the IUPAC guidelines for inorganic nomenclature.

As always, we are indebted to our colleagues at Pearson Education who have guided the schedule for the project and have developed the accompanying website. We extend warm thanks to Bridget Allen, Kevin Ancient, Melanie Beard, Patrick Bonham, Pauline Gillett, Simon Lake, Mary Lince, Paul Nash and Ros Woodward.

Finally, this book is dedicated to Philby and Isis. Our loyal companions are never far from the computer and have contributed to the project in a unique way that only Siamese cats know how. A little older, certainly; a little wiser, possibly.

Catherine E. Housecroft
Edwin C. Constable
Basel, March 2005

# About the authors

Photo: Pam Marshall

*Chemistry* draws on the experience of this husband and wife author team over 25 years of teaching chemistry in the United Kingdom, North America, Switzerland and South Africa to school, college and university students. After her PhD in Durham, Catherine E. Housecroft began her career as a teacher at Oxford Girls' High School and went on to teach general chemistry to nursing majors at St Mary's College, Notre Dame, USA, and then to undergraduates at the University of New Hampshire. She returned to the UK in 1986 as a teaching Fellow of Newnham College and Lector at Trinity College, Cambridge, later becoming a Royal Society Research Fellow and then university lecturer in Inorganic Chemistry in Cambridge, where she chaired the Teaching Committee in the chemistry department. After his DPhil in Oxford, Edwin C. Constable moved to the University of Cambridge where he became a university lecturer in Inorganic Chemistry. He was a teaching Fellow of Robinson College in Cambridge, and has lectured in and tutored both inorganic and organic chemistry.

After Cambridge, the authors held faculty positions in the Department of Chemistry at the University of Basel, Switzerland. From 2000 to 2002, they held positions as Professors of Chemistry at the University of Birmingham, UK. In 2002, they returned to the University of Basel and are currently Professors in the Department of Chemistry there.

Professor Housecroft has a theoretical and experimental organometallic and supramolecular chemistry research background and has published over 250 research papers and review articles. Professor Constable's research areas span both organic and inorganic chemistry, with particular interest in supramolecular chemistry and nanotechnology; he has published over 350 research papers. Both are established authors and have published both general and inorganic chemistry texts as well as books on specialist topics. Professor Housecroft is an Executive Editor of the international journal *Polyhedron*, and Professor Constable serves on the editorial boards of several international journals.

**Chapter introductions** outline the core content of the chapter.

**Cross-references** in the margins allow students to make links between different areas of chemistry and see it as a more integrated whole.

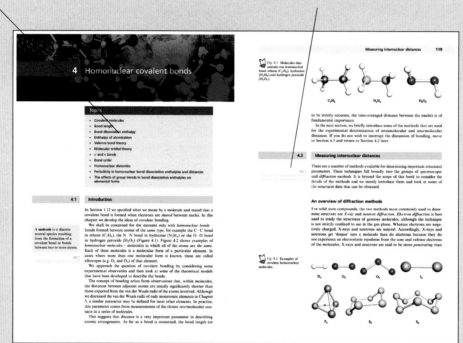

**Key definitions** are highlighted in the text.

**Topic boxes** relate chemistry in each chapter to real life examples from biology, the environment and commercial and laboratory applictions, or provide relevant theoretical background.

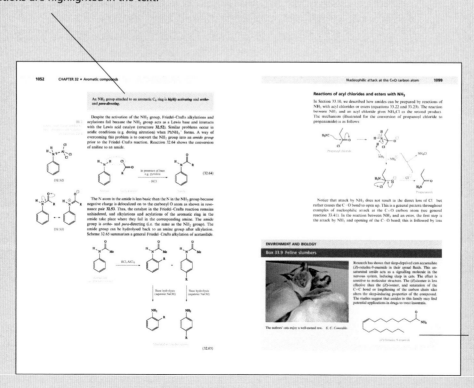

**Molecules** are presented as 3D artwork
supported by Chime-viewable graphics on the website.

**Worked examples** are given
throughout the text.

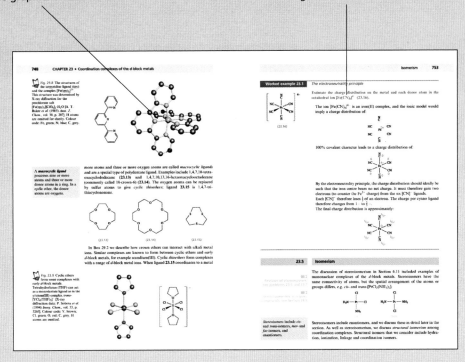

**End-of-chapter problems** test students on
the topics covered in each chapter.
Answers to non-descriptive problems are
provided at the end of the book. Worked
examples are also provided within the text.

**Chapter summary
boxes** help students to
test their understanding
of what they have read.
They also help with
exam revision.

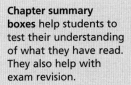

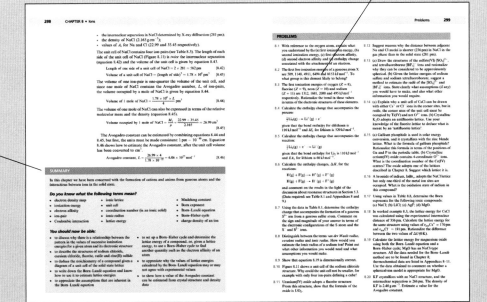

*Chemistry, Third Edition* is supported by a fully interactive Companion Website, available at **www.pearsoned.co.uk/housecroft**, that contains a wealth of additional learning material.

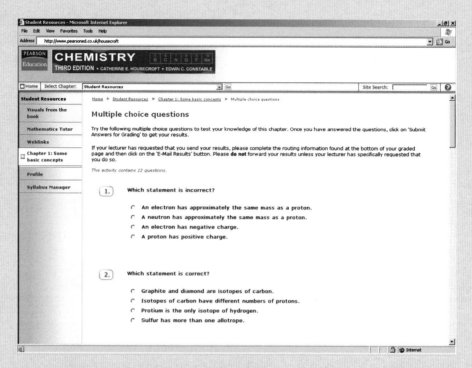

For each chapter **Multiple Choice Questions**, all with feedback on incorrect answers which help develop your subject knowledge and improve your understanding.

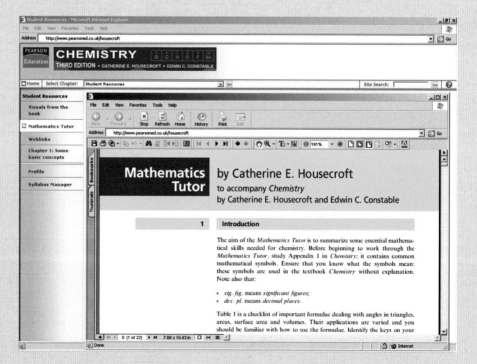

A 22-page *Mathematics Tutor* written by Catherine Housecroft, which summarizes some of the essential mathematical skills needed to study chemistry.

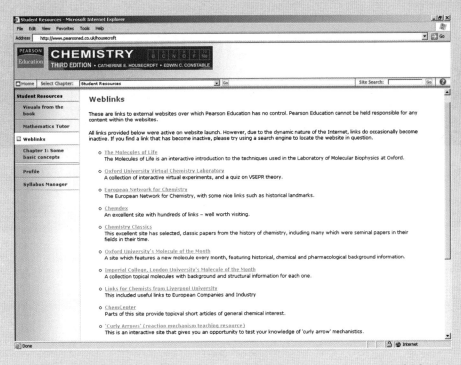

A set of annotated **weblinks** provide guidance and ideas about where to find useful further online resources and information.

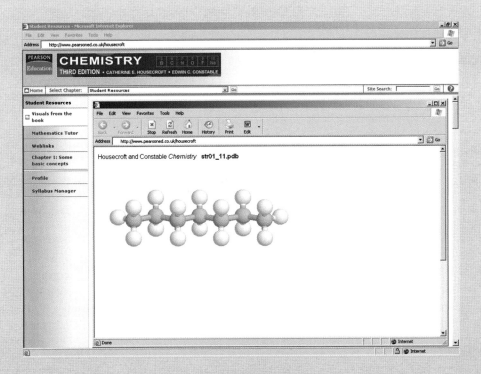

Available here for students are the visuals of the several hundred chemical structures shown in 3D in the book. The visuals are provided in the common web format for chemical structures, Chime. You can view these molecules online or download them to manipulate and explore the structures.

# Acknowledgements

The publishers would like to thank the following people for their advice and feedback during the development of this new edition:

**Dr John Christie**, La Trobe University, Australia
**Dr A. Cooper**, Kingston University, London
**Dr John Crosby**, University of Bristol
**Dr Jason Eames**, Queen Mary College, London
**Dr Bertil Forslund**, Stockholm University, Sweden
**Dr John F. Gallagher**, Dublin City University
**Dr Adrian V. George**, University of Sydney, Australia
**Dr B. J. Howlin**, University of Surrey
**Dr Andrew Hughes**, University of Durham
**Professor Rolf Isrenn**, Bergen University, Norway
**Dr Bob Lauder**, University of Lancaster
**Dr Peter van der Linde**, Hogeschool Leiden, Netherlands
**Professor Ken McKendrick**, Heriot-Watt University
**Dr John Malone**, Queen's University Belfast
**Dr Reijo Suontamo**, Jyvasklya University, Finland
**Professor Anni Vedeler**, Bergen University, Norway
**Dr John Wenger**, University College Cork

We are grateful to the following for permission to reproduce copyright material:

Figure 8.2(a) reprinted with permission from Wahl, A., *Science*, **151**, 961 (1966) © 1966 AAAS; Figure 8.2(b) reprinted with permission from R.F.W. Bader, Ian Keaveny and Paul E. Cade, *Journal of Chemical Physics*, **47**, 3381 (1967) © 1967, American Institute of Physics; Figure 8.2(c) reprinted with permission from R.F.W. Bader and A.D. Bandrauk, *Journal of Chemical Physics*, **49**, 1653 (1968) © 1968, American Institute of Physics; Figure 8.21 reprinted with permission from Schoknecht, G., *Zeitschrift für Naturforschung, Teil A*, **12**, 983 (1957), Verlag der Zeitschrift für Naturforschung.

We are grateful to the following for permission to reproduce photographs:

Frank Greenaway/Dorling Kindersley/University Marine Biological Station, Millport, Scotland; Andy Crawford/Dorling Kindersley; The Nobel Foundation; James King Holmes/OCMS/Science Photo Library; The Library & Information Centre, Royal Society of Chemistry; Brian Cosgrove/Dorling Kindersley; Los Alamos National Laboratory/Science Photo Library; CNRI/Science Photo Library; Alfred Pasieka/Science Photo Library; Crown Copyright/Health & Safety Laboratory/Science Photo Library; VVG/Science Photo Library; Dorling Kindersley; NASA/Science Photo Library; Ken Scripps Laboratory @ Scripps Institute of Oceanography/Oxford Scientific/ photolibrary.com; T Aquilano/GeoScience Features; Colin Keates/Dorling

Kindersley/The Natural History Museum, London; Professor Lothar Beyer, Universität Leipzig; Chris Mason; EMPICS/PA; Arthus Bertrand/Science Photo Library; Steve Gorton/Dorling Kindersley; John Parker/Dorling Kindersley; NASA/Dorling Kindersley; Claude Nuridsany & Marie Perennou/Science Photo Library; The Mary Rose Trust; Argentum/Science Photo Library; Prof. P. Motta/Dept. of Anatomy/University "La Sapienza", Rome/Science Photo Library; Kim Taylor/Dorling Kindersley; Peter Anderson/Dorling Kindersley; Angus Beare; William Ervin/Science Photo Library.

# 1 Some basic concepts

## Topics

- IUPAC
- SI units
- The proton, electron and neutron
- Elements
- Allotropes
- States of matter
- Atomic number and mass number
- Isotopes
- Relative atomic mass
- The mole
- The Avogadro constant
- Gas laws and ideal gases
- The periodic table
- Radicals and ions
- Molecules and compounds
- Relative molecular mass
- Solution concentration
- Stoichiometry
- Oxidation and reduction
- Oxidation states
- Some basic nomenclature

## 1.1 What is chemistry and why is it important?

Matter, be it animal, vegetable or mineral, is composed of chemical elements or combinations thereof. Over a hundred elements are known, although not all are abundant by any means. The vast majority of these elements occur naturally, but some, such as technetium and curium, are artificial.

Chemistry is involved with the understanding of the properties of the elements, how they interact with one another, and how the combination of these elements gives compounds that may undergo chemical changes to generate new compounds.

Life has evolved systems that depend on carbon as a fundamental element; carbon, hydrogen, nitrogen and oxygen are particularly important in biological systems. A true understanding of biology and molecular biology must be based upon a full knowledge of the structures, properties and reactivities of the molecular components of life. This basic knowledge comes from the study of chemistry.

The line between physical and chemical sciences is also a narrow one. Take, for example, the rapidly expanding field of superconducting materials – compounds that possess negligible resistance to the flow of electrons. Typically, this property persists only at very low temperatures but if the superconducting materials are to find general application, they must operate at ambient temperatures. Although a physicist may make the conductivity measurements, it is the preparation and study of the *chemical* composition of the materials that drives the basic research area.

Chemistry plays a pivotal role in the natural sciences. It provides the essential basic knowledge for applied sciences, such as astronomy, materials science, chemical engineering, agriculture, medical sciences and pharmacology. After considering these points, the answer to the question '*why study chemistry as a first year university science?*' should not cause you much problem. Whatever your final career destination within the scientific world, an understanding of chemical concepts is essential.

It is traditional to split chemistry into the three branches of inorganic, organic and physical; theoretical chemistry may be regarded as a division of the physical discipline. However, the overlap between these branches of the subject is significant and real. Let us consider a case study – the Wacker process, in which ethene is oxidized to ethanal (equation 1.1).

■▷
Superconductors:
see Box 9.4

■▷
Oxidation:
see Section 1.16

$$C_2H_4 + [PdCl_4]^{2-} + H_2O \longrightarrow CH_3CHO + 2HCl + 2Cl^- + Pd \qquad (1.1)$$

Ethene              Ethanal

(Acetaldehyde)

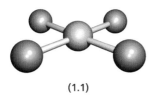

**(1.1)**

■▷
Catalyst:
see Section 15.8

Until recently, this was a significant industrial method for the preparation of ethanal. The reaction occurs in water in the presence of the tetrachloropalladate(II) ion, $[PdCl_4]^{2-}$ **1.1**. The palladium(0) formed in reaction 1.1 is converted back to $[PdCl_4]^{2-}$ by reaction with copper(II) and chloride ions (equation 1.2); the copper(I) so formed (as $[CuCl_2]^-$) is reoxidized by $O_2$ (equation 1.3). In this way, the $[PdCl_4]^{2-}$ and $Cu^{2+}$ function as *catalysts* since they undergo no *overall* change through the cycle of reactions.

$$2Cu^{2+} + Pd + 8Cl^- \longrightarrow [PdCl_4]^{2-} + 2[CuCl_2]^- \qquad (1.2)$$

$$4[CuCl_2]^- + 4H^+ + O_2 \longrightarrow 4Cu^{2+} + 2H_2O + 8Cl^- \qquad (1.3)$$

The basic reaction (ethene to ethanal) is an organic one. The catalyst is inorganic. Physical chemistry is needed to interpret the kinetics and understand the mechanism of the reaction and to work out the most effective conditions under which it should take place in an industrial setting. All three branches of chemistry are used to understand the Wacker process. This is just one of a myriad of cases in which a discussion of a chemical process or concept requires the interplay of two or more areas of chemistry.

In *Chemistry*, topics have been arranged to provide a natural progression through the chemistry that a first year science undergraduate student needs to assimilate. There are no artificial barriers placed between areas of chemistry although the traditional areas of inorganic, organic and physical are distinguishable. By studying chemistry from a single book in which the branches of chemistry are clearly linked, we hope that you will learn to apply the concepts to new situations more easily.

The aim of the rest of this chapter is to review some essential topics and to provide you with a source of reference for basic definitions.

## 1.2 What is the IUPAC?

As chemistry continues to expand as a subject and as the number of known chemical compounds continues to grow at a dramatic rate, it becomes increasingly vital that a set of ground rules be accepted for the naming of compounds. Not only accepted, but, probably more importantly, *used* by chemists.

The International Union of Pure and Applied Chemistry (IUPAC) is, among other things, responsible for making recommendations for the naming of both inorganic and organic compounds, as well as for the numbering system and the collective names that should be in common use for groups of elements in the periodic table.

Throughout this text, we will be using recommended IUPAC nomenclature wherever possible, but old habits die hard and a host of trivial names persists in the chemist's vocabulary. Where use of only the IUPAC nomenclature may cause confusion or may not be generally recognized, we have introduced both the recommended and trivial names. Some basic rules and revision for organic and inorganic nomenclature are dealt with in Section 1.18. As new classes of compounds are introduced in this textbook, we detail the systematic methods of naming them.

*Trivial names in common use: see Appendix 13, www.pearsoned.co.uk/housecroft*

## 1.3 SI units

A system of internationally standardized and recognized units is as important as a system of compound names. The Système International d'Unités (SI units) provides us with the accepted system of measurement.

### Base SI quantities

There are seven base SI quantities (Table 1.1). In addition there are two supplementary units, the radian and the steradian. The *radian* is the SI unit of angle and its symbol is *rad*. For solid geometry, the unit of solid angle is the *steradian* with the unit symbol *sr*.

An important feature of SI units is that they are completely self-consistent. From the seven base units, we can derive all other units as shown in worked examples 1.1, 1.2 and 1.9. Some of the most commonly used derived units

Table 1.1 The base quantities of the SI system.

| Physical quantity | Symbol for quantity | Base unit | Unit symbol |
|---|---|---|---|
| Mass | $m$ | kilogram | kg |
| Length | $l$ | metre | m |
| Time | $t$ | second | s |
| Electrical current | $I$ | ampere | A |
| Temperature (thermodynamic) | $T$ | kelvin | K |
| Amount of substance | $n$ | mole | mol |
| Luminous intensity | $I_v$ | candela | cd |

Degree Celsius, °C (instead of kelvin), is an older unit for temperature which is still in use; note the correct use of the notation °C and K, not °K.

Table 1.2 Some derived units of the SI system with particular names.

| Unit | Name of unit | Symbol | Relation to base units |
|------|--------------|--------|------------------------|
| Energy | joule | J | $kg\,m^2\,s^{-2}$ |
| Frequency | hertz | Hz | $s^{-1}$ |
| Force | newton | N | $kg\,m\,s^{-2}$ |
| Pressure | pascal | Pa | $kg\,m^{-1}\,s^{-2}$ |
| Electric charge | coulomb | C | $A\,s$ |
| Capacitance | farad | F | $A^2\,s^4\,kg^{-1}\,m^{-2}$ |
| Electromotive force (emf) | volt | V | $kg\,m^2\,s^{-3}\,A^{-1}$ |
| Resistance | ohm | $\Omega$ | $kg\,m^2\,s^{-3}\,A^{-2}$ |

have their own names, and selected units of this type, relevant to chemistry, are listed in Table 1.2.

### Derived SI units

Whenever you are using an equation to calculate a physical quantity, the units of the new quantity can be directly determined by including in the same equation the units of each component. Consider the volume of a rectangular box of sides $a$, $b$ and $c$ (equation 1.4). The SI unit of length is the metre (Table 1.1) and so the SI unit of volume can be determined as in equation 1.5. Throughout this book, we combine quantities and units in equations. Thus, equations 1.4 and 1.5 are combined to give the calculation for the volume of the box in the form of equation 1.6.

$$\text{Volume of box} = \text{length} \times \text{width} \times \text{height} = a \times b \times c = abc \qquad (1.4)$$

$$\text{SI unit of volume} = (\text{SI unit of length}) \times (\text{SI unit of width})$$
$$\times (\text{SI unit of height})$$
$$= m \times m \times m = m^3 \qquad (1.5)$$

$$\text{Volume of box} = (\text{length in m}) \times (\text{width in m}) \times (\text{height in m})$$
$$= (a\,m) \times (b\,m) \times (c\,m)$$
$$= abc\,m^3 \qquad (1.6)$$

Some older units still persist, notably atmospheres, atm (instead of pascals), for pressure, and calories, cal (instead of joules), for energy, but on the whole, SI units are well established and are used internationally.

---

**COMMERCIAL AND LABORATORY APPLICATIONS**

### Box 1.1 The importance of being consistent with units

In 1999, NASA lost the spacecraft 'Mars Climate Orbiter'. A preliminary investigation into its loss revealed that one team of scientists had been working in units of inches, feet and pounds, while their international collaborators had been using metric units. Combined information from the two teams was transferred to the Mars Orbiter. The result of providing the flight information in data sets with inconsistent units was that the spacecraft was unable to manoeuvre into the correct orbit about Mars, and was consequently lost in space.

| **Worked example 1.1** | *Defining units* |
|---|---|

**Determine the SI unit of density.**

Density is mass per unit volume:

$$\text{Density} = \frac{\text{Mass}}{\text{Volume}}$$

$$\therefore \text{ the SI unit of density} = \frac{\text{SI unit of mass}}{\text{SI unit of volume}} = \frac{\text{kg}}{\text{m}^3} = \text{kg m}^{-3}$$

## Large and small numbers

Scientific numbers have a habit of being either extremely large or extremely small. Take a typical carbon–carbon (C–C) single bond length. In the standard SI unit of length, this is $0.000\,000\,000\,154$ m. Writing this number of zeros is clumsy and there is always the chance of error when the number is being copied. The distance can be usefully rewritten as $1.54 \times 10^{-10}$ m. This is a much neater and more easily read way of detailing the same information about the bond length. It would be even better if we could eliminate the need to write down the '$\times 10^{-10}$' part of the statement. Since bond lengths are usually of the same order of magnitude, it would be useful simply to be able to write '1.54' for the carbon–carbon distance. This leads us to the unit of the ångström (Å); the C–C bond distance is $1.54$ Å. Unfortunately, the ångström is *not* an SI unit although it *is* in common use. The SI asks us to choose from one of the accepted multipliers listed in Table 1.3. The distance of $1.54 \times 10^{-10}$ m is equal to either $154 \times 10^{-12}$ m or $0.154 \times 10^{-9}$ m, and from Table 1.3 we see that these are equal to 154 pm or 0.154 nm respectively. Both are acceptable within the SI. Throughout this book we have chosen to use picometres (pm) as our unit of bond distance.

## Consistency of units

**A word of warning** about the use of the multipliers in Table 1.3. In a calculation, you must work in a consistent manner. For example, if a density is given in $\text{g cm}^{-3}$ and a volume in $\text{m}^3$, you *cannot* mix these values, e.g. when determining a mass:

$$\text{Mass} = \text{Volume} \times \text{Density}$$

**Table 1.3** Multiplying prefixes for use with SI units; the ones that are most commonly used are given in pink.

| Factor | Name | Symbol | Factor | Name | Symbol |
|---|---|---|---|---|---|
| $10^{12}$ | tera | T | $10^{-2}$ | centi | c |
| $10^{9}$ | giga | G | $10^{-3}$ | milli | m |
| $10^{6}$ | mega | M | $10^{-6}$ | micro | μ |
| $10^{3}$ | kilo | k | $10^{-9}$ | nano | n |
| $10^{2}$ | hecto | h | $10^{-12}$ | pico | p |
| $10$ | deca | da | $10^{-15}$ | femto | f |
| $10^{-1}$ | deci | d | $10^{-18}$ | atto | a |

You must first scale one of the values to be consistent with the other as is shown in worked example 1.2. The most foolproof way of overcoming this frequent source of error is to convert all units into the scale defined by the base units (Table 1.1) *before* you substitute the values into an equation. We shall see many examples of this type of correction in worked examples throughout the book.

---

**Worked example 1.2**     *Units in calculations*

**Calculate the volume occupied by 10.0 g of mercury if the density of mercury is $1.36 \times 10^4 \, \text{kg m}^{-3}$ at 298 K.**

The equation that relates mass ($m$), volume ($V$) and density ($\rho$) is:

$$\rho = \frac{m}{V} \qquad \text{or} \qquad V = \frac{m}{\rho}$$

Before substituting in the numbers provided, we must obtain consistency among the units.

Density is given in $\text{kg m}^{-3}$ but the mass is given in g, and should be converted to kg (or the density converted to $\text{g m}^{-3}$):

$$10.0 \, \text{g} = 10.0 \times 10^{-3} \, \text{kg} = 1.00 \times 10^{-2} \, \text{kg}$$

Now we are a position to calculate the volume occupied by the mercury:

$$V = \frac{m}{\rho} = \frac{1.00 \times 10^{-2} \, \text{kg}}{1.36 \times 10^4 \, \text{kg m}^{-3}} = 7.35 \times 10^{-7} \, \text{m}^3$$

---

## 1.4     The proton, electron and neutron

The basic particles of which atoms are composed are the *proton*, the *electron* and the *neutron*.[§] Some key properties of the proton, electron and neutron are given in Table 1.4. A neutron and a proton have approximately the same mass and, relative to these, the electron has a negligible mass. The charge on a proton is of equal magnitude, but opposite sign, to that on an electron

**Table 1.4** Properties of the proton, electron and neutron.

|  | Proton | Electron | Neutron |
|---|---|---|---|
| Charge / C | $+1.602 \times 10^{-19}$ | $-1.602 \times 10^{-19}$ | 0 |
| Charge number (relative charge) | 1 | $-1$ | 0 |
| Rest mass / kg | $1.673 \times 10^{-27}$ | $9.109 \times 10^{-31}$ | $1.675 \times 10^{-27}$ |
| Relative mass | 1837 | 1 | 1839 |

---

[§] These are the particles considered fundamental by chemists, although particle physicists have demonstrated that there are yet smaller building blocks. This continual subdivision recalls the lines of Jonathan Swift:

'So, naturalists observe, a flea
Hath smaller fleas that on him prey;
And these have smaller fleas to bite 'em,
And so proceed *ad infinitum*.'

and so the combination of *equal numbers* of protons and electrons results in an assembly that is neutral overall. A neutron, as its name suggests, is neutral – it has no charge. The arrangements and energies of electrons in atoms and ions are discussed in Chapter 3.

## 1.5 The elements

An *element* is matter, all of whose atoms are alike in having the same positive charge on the nucleus.

The recommended IUPAC definition of an *element* states that 'an element is matter, all of whose atoms are alike in having the same positive charge on the nucleus'. Each element is given a symbol and these are predominantly internationally accepted even though the names of the elements themselves are subject to linguistic variation (see Box 1.2).

### Metals, non-metals and semi-metals

In the United States, Al is known as *aluminum*.

Elements can be classified as metals, non-metals or semi-metals. The names of *most* metals end in '-ium', e.g. lithium, sodium, magnesium, calcium, aluminium, scandium, chromium, titanium, vanadium, hafnium, ruthenium, rhodium, iridium, osmium and palladium. Common exceptions include tin, lead, iron, cobalt, copper, zinc, tungsten, platinum, silver and gold. Under IUPAC recommendations, the term 'semi-metal' is preferred over 'metalloid'. We look further at what constitutes a 'semi-metal' in Chapter 9.

### Allotropes

Some elements exist in more than one structural form and this property is called *allotropy*. Consider carbon – two commonly quoted *allotropes* of

---

**HISTORY**

### Box 1.2 The elements: names with meanings

The origin of the names of the chemical elements and their variation between languages is fascinating. A wealth of chemical history and fact is hidden within the names, although changing from one language to another may obscure this information. Three examples illustrate this.

#### Mercury (Hg)

The silver-white, liquid metal mercury was known to the ancient Greeks as *liquid silver*. This name evolved into the Latin *hydrargyrum* from which our present-day symbol *Hg* arises. Mercury used to be known as *quicksilver* in English (*quick* means alive) and the element is called *Quecksilber* in German. The word *mercury* refers to the messenger of the Roman gods – anyone who has tried to 'catch' a globule of mercury will appreciate the analogy.

#### Cobalt (Co)

The *d*-block element cobalt (Co) is known as *Kobalt* in German. Nickel and cobalt occur in varying amounts in iron ores. When these ores were processed by early miners, ores containing significant amounts of cobalt gave inferior metallic products. These ores became known as *false ores* and were named after the malicious imps or *Kobold* who supposedly lived in the mines.

#### Tungsten (W)

Tungsten has the unexpected symbol W. In German, tungsten is called *Wolfram* and this refers to its ore *wolframite*. The other common ore is *scheelite*, which was named after Scheele, the discover of tungsten. Scheele was Swedish, and before the ore was called scheelite, it was known as *tungsten* which translates as 'heavy stone'.

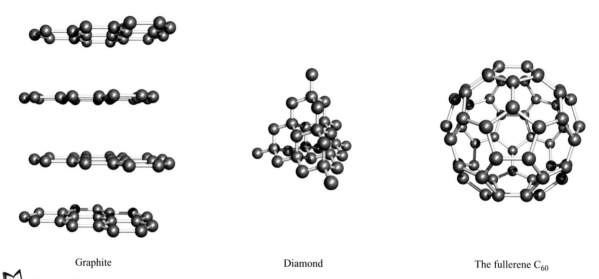

Graphite                    Diamond                    The fullerene $C_{60}$

**Fig. 1.1** The structures of three allotropes of carbon; $C_{60}$ is representative of the group of fullerenes.

***Allotropes*** of an element are different structural modifications of that element.

$C_{60}$: see Section 9.9

carbon are graphite and diamond (Figure 1.1), both of which possess infinite structures. Both allotropes consist only of atoms of carbon and both burn in an excess of $O_2$ to give only carbon dioxide, $CO_2$. The physical appearances of these two allotropes are, however, dramatically different. Diamond is thermodynamically unstable at room temperature and pressure with respect to graphite but, fortunately, the interconversion is extremely slow and diamond is termed *metastable*. In the 1980s, further allotropes of carbon called the fullerenes were discovered. These are present in soot to the extent of a few per cent by mass, and the commonest component consists of discrete $C_{60}$ molecules (Figure 1.1). Other common elements that exhibit allotropy are tin, phosphorus, arsenic, oxygen, sulfur and selenium.

## 1.6    States of matter

### Solids, liquids and gases

At a given temperature, an element is in one of three *states of matter* – solid, liquid or vapour (Figure 1.2). A vapour is called a gas when it is above its *critical temperature* (see Box 1.3).

**Fig. 1.2** The arrangement of particles (atoms or molecules) in a solid, a liquid and a gas. For the solid, the surface is determined by the solid itself. The shape of the liquid is controlled by the shape of the container; the liquid has a definite surface. The atoms or molecules of the gas are free to move throughout the container.

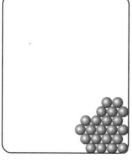

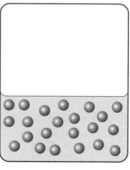

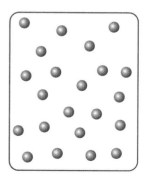

Solid                    Liquid                    Gas

**THEORETICAL AND CHEMICAL BACKGROUND**

## Box 1.3 A simple phase diagram and the distinction between a gas and a vapour

The terms *gas* and *vapour* are often used interchangeably. However, there is a real distinction between them. A vapour, but not a gas, can be *liquefied* by increasing the pressure at a constant temperature. This may at first seem contradictory and to understand it we need to consider a simple one-component phase diagram. One type of *phase diagram* describes the variation in phase of a system as a function of temperature and pressure.

The diagram below shows a simple phase diagram for a compound X. The solid blue lines represent the boundaries between solid X (phase *A*), liquid X (phase *B*), and the vapour or gas state (phase *C*). The broken black line represents phase changes that will occur at a pressure of $10^5$ Pa (1 bar) as the temperature is raised. Solid X first melts ($A \longrightarrow B$) and then

vaporizes ($B \longrightarrow C$). Now, follow the broken blue line from the bottom to the top of the diagram. This represents what happens as you increase the pressure at a constant temperature (298 K). We start in phase *C* and crossing the boundary $C \longrightarrow B$ has the physical effect of liquefying the vapour. Further increase in pressure leads to solidification ($B \longrightarrow A$).

The temperature $T_c$ is the *critical temperature*. Above this temperature, it is no longer possible to liquefy the 'vapour', now called a *gas*; i.e. above temperature $T_c$ it is no longer possible to cross the $C \longrightarrow B$ boundary by increasing the pressure. The *critical point* is shown by the blue dot on the diagram and is defined by both a temperature and pressure.

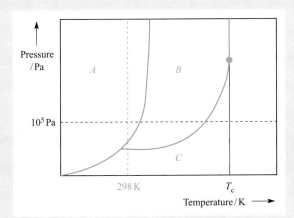

### Exercises

1. Choose any temperature on the temperature axis of the diagram and place a straight edge vertically so you can follow the changes in phase of X as a function of pressure at this particular temperature. At the temperature you have chosen, is phase *C* correctly described as a gas or a vapour? Repeat the exercise at a different temperature.

2. Now choose any pressure on the pressure axis. Follow a horizontal line across the diagram and determine the phase changes that occur as you raise the temperature at your selected (fixed) pressure.

3. For the substance represented in the diagram, in what phase does it exist at 298 K and $1.00 \times 10^5$ Pa?

4. At a constant pressure of $10^2$ kPa, what change in conditions is needed to produce an $A \longrightarrow C$ phase change? Is this a single-step phase transition?

In a *solid*, the atoms are often arranged in a regular fashion. The solid possesses a fixed volume (at a stated temperature and pressure) and shape. A *liquid* also has a fixed volume at a given temperature and pressure, but has no definite shape. It will flow into a container and will adopt the shape of this vessel. The particles (atoms or molecules) in a *gas* move at random and can occupy a large volume. A gas has no fixed shape.

In Figure 1.2, the surface boundary of the solid is determined by the solid itself and is not dictated by the container. On the other hand, unless restricted by a container, the boundary of a gas is continually changing; this permits two (or more) gases to mix in a process called *diffusion* (Figure 1.3). In a liquid, the shape of the liquid mimics the shape of the container, but at the upper surface of the liquid the phenomenon of *surface tension* operates and this defines the surface boundary. When liquids mix they are said to be *miscible*; hexane and octane are miscible, as are water and ethanol. If liquids do not mix (e.g. water and oil, water and hexane), they are *immiscible*.

If two liquids mix, they are *miscible*. *Immiscible* liquids form distinct layers.

**Fig. 1.3** A gas is trapped within the boundaries of the vessel in which it is held. Two gases can mix together (*diffuse*) if the boundary between them is removed.

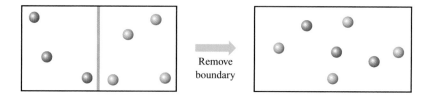

In chemical equations, the states of substance are often included and the standard abbreviations are:

solid        (s)
liquid       (l)
gas          (g)

The use of (aq) refers to an aqueous solution; this is *not* a state of matter.

## Phases

The three states of a substance are *phases*, but within one state there may be more than one phase. For example, each allotrope of carbon is a different phase but each is in the solid state. Each phase exists under specific conditions of temperature and pressure and this information is represented in a *phase diagram*. We explore the use of a simple phase diagram in Box 1.3.

**Table 1.5** Melting and boiling points (at atmospheric pressure) and appearance of selected elements.

| Element (allotrope) | Symbol | Melting point / K | Boiling point / K | Physical appearance at 298 K |
|---|---|---|---|---|
| Aluminium | Al | 933 | 2793 | White metal |
| Bromine | Br | 266 | 332.5 | Brown-orange liquid |
| Calcium | Ca | 1115 | 1757 | Silver-white metal |
| Chlorine | Cl | 172 | 239 | Green-yellow gas |
| Chromium | Cr | 2180 | 2944 | Silver metal |
| Cobalt | Co | 1768 | 3200 | Silver-blue metal |
| Copper | Cu | 1358 | 2835 | Reddish metal |
| Fluorine | F | 53 | 85 | Very pale yellow gas |
| Gold | Au | 1337 | 3081 | Yellow metal |
| Helium | He | – | 4.2 | Colourless gas |
| Hydrogen | H | 14 | 20 | Colourless gas |
| Iodine | I | 387 | 458 | Black solid |
| Iron | Fe | 1811 | 3134 | Silver-grey metal |
| Lead | Pb | 600 | 2022 | Blue-grey metal |
| Lithium | Li | 453.5 | 1615 | Silver-white metal |
| Magnesium | Mg | 923 | 1363 | Silver-white metal |
| Manganese | Mn | 1519 | 2334 | Silver metal |
| Mercury | Hg | 234 | 630 | Silver liquid |
| Nickel | Ni | 1728 | 3186 | Silver metal |
| Nitrogen | N | 63 | 77 | Colourless gas |
| Oxygen | O | 54 | 90 | Colourless gas |
| Phosphorus (white) | P | 317 | 550 | White solid |
| Potassium | K | 336 | 1032 | Silver-white metal |
| Silicon | Si | 1687 | 2638 | Shiny, blue-grey solid |
| Silver | Ag | 1235 | 2435 | Silver-white metal |
| Sodium | Na | 371 | 1156 | Silver-white metal |
| Tin (white) | Sn | 505 | 2533 | Silver metal |
| Zinc | Zn | 692.5 | 1180 | Silver metal |

## Changes of phase

At a given pressure (or temperature), the transformation from one phase to another takes place at a particular temperature (or pressure). Usually, we are concerned with transformations at atmospheric pressure. An element (or a single allotrope if we are dealing with an element that exhibits allotropy) changes from a solid to a liquid at its *melting point* (mp) and from a liquid to a vapour at its *boiling point* (bp). We should refer to the third state as a vapour, and not a gas, until the critical temperature is reached. The melting and boiling points for selected elements are given in Table 1.5. Carbon and sulfur have not been included in this table because neither undergoes a simple change of state. When heated at a particular pressure, some elements transform directly from the solid to the vapour phase and this process is called *sublimation*. Iodine readily sublimes when it is heated.

*Enthalpy changes of fusion vaporization: see Section 2.9*

| 1.7 | **Atoms and isotopes** |

## Atoms and atomic number

An *atom* is the smallest unit quantity of an element that is capable of existence, either alone or in chemical combination with other atoms of the same or another element.

An *atom* is the smallest unit quantity of an element that is capable of existence, either alone or in chemical combination with other atoms of the same or another element. It consists of a positively charged *nucleus* and negatively charged electrons. The simplest atom is hydrogen which is made up of one proton and one electron. The proton of a hydrogen atom is its nucleus, but the nucleus of any other atom consists of protons *and* neutrons.

All atoms are neutral, with the positive charge of the nucleus exactly balanced by the negative charge of a number of electrons equal to the number of protons. The electrons are situated outside the nucleus. Each atom is characterized by its *atomic number*, $Z$. A shorthand method of showing the atomic number and *mass number*, $A$, of an atom along with its symbol, E, is used:

$$\text{mass number} \longrightarrow \quad {}^{A}_{Z}\text{E} \qquad \text{e.g.} \quad {}^{59}_{27}\text{Co}$$

mass number $\longrightarrow$
element symbol $\longrightarrow$
atomic number $\longrightarrow$

Atomic number $= Z =$ number of protons in the nucleus $=$ number of electrons
Mass number $= A =$ number of protons $+$ number of neutrons

## Relative atomic mass

The mass of an atom is concentrated in the nucleus where the protons and neutrons are found. If we were to add together the actual masses of protons and neutrons present, we would always be dealing with very small, non-integral numbers, and for convenience, a system of *relative atomic masses* is used. The *atomic mass unit* (u) has a value of $\approx 1.660 \times 10^{-27}$ kg, and this corresponds closely to the mass of a proton or a neutron (Table 1.4). Effectively, the mass of each proton and neutron is taken to be one atomic mass unit. The scale of relative atomic masses ($A_r$) is based upon measurements taken for carbon with all atomic masses stated *relative to* $^{12}\text{C} = 12.0000$.

## Isotopes

For a given element, there may be more than one type of atom, and these are called *isotopes* of the element. *Do not confuse isotopes with allotropes.* Allotropes are different structural forms of a *bulk* element arising from different spatial arrangements of the atoms (Figure 1.1). Isotopes are atoms of the same element with *different numbers of neutrons*. Some isotopes that do not occur naturally may be produced artificially (see Box 1.4).

An element is characterized by the number of protons and, to keep the atom neutral, this must equal the number of electrons. However, the number of neutrons can vary. For example, hydrogen possesses three isotopes. The most abundant by far (99.985%) is protium, $^1$H, which has one proton and one electron but no neutrons. The second isotope is deuterium $^2$H (also given the symbol D) which has one proton, one electron and one neutron. Deuterium is present naturally at an abundance of 0.015% and is sometimes called 'heavy hydrogen'. Tritium ($^3$H or T) occurs as fewer than 1 in $10^{17}$ atoms in a sample of natural hydrogen and is radioactive. The atomic mass of naturally occurring hydrogen reflects the presence of all three isotopes and is the weighted mean of the masses of the isotopes present. The relative atomic mass of hydrogen is 1.0080, and the value is close to 1 because the isotope $^1$H with mass number 1 makes up 99.985% of the natural mixture of isotopes.

*Isotopes* of an element have the same atomic number, $Z$, but different atomic masses.

Other examples of elements that exist naturally as mixtures of isotopes are lithium ($^6_3$Li and $^7_3$Li), chlorine ($^{35}_{17}$Cl and $^{37}_{17}$Cl) and copper ($^{63}_{29}$Cu and $^{65}_{29}$Cu). Elements that occur naturally as only one type of atom are *monotopic* and include phosphorus ($^{31}_{15}$P) and cobalt ($^{59}_{27}$Co). Isotopes can be separated by *mass spectrometry* and Figure 1.4a shows the isotopic distribution in atomic chlorine. In the mass spectrum of $Cl_2$ (Figure 1.4b), three peaks are

---

**THEORETICAL AND CHEMICAL BACKGROUND**

## Box 1.4 Artificially produced isotopes and β-particle emission

Some isotopes, in particular those of the heaviest elements in the periodic table, are produced by the bombardment of one nucleus by particles which induce nuclear fusion. Typical particles used for the bombardment are neutrons.

An example of an artificially produced isotope is the formation of $^{239}_{94}$Pu, an isotope of plutonium, in a series of nuclear reactions beginning with the isotope of uranium $^{238}_{92}$U. The bombardment of $^{238}_{92}$U with neutrons results in nuclear fusion:

$$^{238}_{92}\text{U} + {}^1_0\text{n} \longrightarrow {}^{239}_{92}\text{U}$$

The relative atomic mass of the product is one atomic mass unit greater than that of the initial isotope because we have added a neutron to it. The isotope $^{239}_{92}$U undergoes spontaneous loss of a β-particle (an electron, see below) to form an isotope of neptunium ($^{239}_{93}$Np) which again undergoes β-decay to give $^{239}_{94}$Pu:

$$^{239}_{92}\text{U} \longrightarrow {}^{239}_{93}\text{Np} + \beta^-$$

$$^{239}_{93}\text{Np} \longrightarrow {}^{239}_{94}\text{Pu} + \beta^-$$

### β-*Decay*

β-Particle emission (β-decay) occurs when an *electron* is lost *from the nucleus* by a complex process which effectively 'turns a neutron into a proton'.[§] The mass of the nucleus undergoing β-decay does not change because the β-particle has negligible mass. The atomic number of the nucleus undergoing the emission increases by one since, effectively, the nucleus has gained a proton:

$$^1_0\text{n} \longrightarrow {}^1_1\text{p} + \beta^-$$

---

[§] This notation is not strictly correct. The decay of a nucleus by the loss of a β-particle is a complex process, the study of which belongs within the remit of the particle physicist. Of interest to us here is the fact that a nucleus that undergoes β-decay retains the same atomic mass but increases its atomic number by one.

**Fig. 1.4** Mass spectrometric traces for (a) atomic chlorine and (b) molecular $Cl_2$.

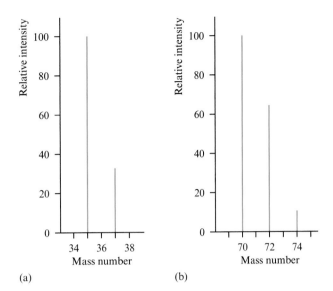

(a)

(b)

observed and are assigned to the possible combinations of the two isotopes of chlorine. We describe the technique of mass spectrometry in Chapter 10.

In the shorthand notation for denoting an isotope, the atomic number is often omitted because it is a constant value for a given element, e.g. the two isotopes of copper can be distinguished by simply writing $^{63}$Cu and $^{65}$Cu.

---

**Worked example 1.3**   *Relative atomic mass*

Calculate the relative atomic mass $A_r$ of naturally occurring magnesium if the distribution of isotopes is 78.7% $^{24}_{12}$Mg, 10.1% $^{25}_{12}$Mg and 11.2% $^{26}_{12}$Mg, and the accurate masses of the isotopes to three significant figures (sig. fig.) are 24.0, 25.0 and 26.0.

The relative atomic mass of magnesium is the weighted mean of the atomic masses of the three isotopes:

$$\text{Relative atomic mass} = \left(\frac{78.7}{100} \times 24.0\right) + \left(\frac{10.1}{100} \times 25.0\right) + \left(\frac{11.2}{100} \times 26.0\right)$$

$$= 24.3 \quad \text{(to 3 sig. fig.)}$$

---

**Worked example 1.4**   *Abundance of isotopes*

Calculate the percentage abundances of $^{12}$C and $^{13}$C if the relative atomic mass, $A_r$, of naturally occurring carbon is 12.011, and the accurate masses of the isotopes are 12.000 and 13.003 (to 5 sig. fig.).

Let the % abundances of the two isotopes be $x$ and $(100 - x)$. The value of $A_r$ is therefore:

$$A_r = 12.011 = \left(\frac{x}{100} \times 12.000\right) + \left(\frac{100 - x}{100} \times 13.003\right)$$

$$12.011 = \left(\frac{12.000x}{100}\right) + \left(\frac{1300.3 - 13.003x}{100}\right)$$

$$12.011 \times 100 = 12.000x + 1300.3 - 13.003x$$

$$1201.1 = 1300.3 - 1.003x$$

$$x = \frac{1300.3 - 1201.1}{1.003} = \frac{99.2}{1.003} = 98.9 \quad \text{(to 3 sig. fig.)}$$

The abundances of the isotopes $^{12}C$ and $^{13}C$ are 98.9% and 1.1%, respectively.

---

## 1.8     The mole and the Avogadro constant

In one mole of substance there are the **Avogadro number**, L, of particles:

$$L \approx 6.022 \times 10^{23} \, \text{mol}^{-1}$$

In Table 1.1 we saw that the SI unit of 'amount of substance' is the *mole*. This unit can apply to any substance and hence it is usual to find the statement 'a mole of *x*' where *x* might be electrons, atoms or molecules. In one mole of substance there are $\approx 6.022 \times 10^{23}$ particles, and this number is called the Avogadro constant or number, L. It is defined as the number of atoms of carbon in exactly 12 g of a sample of isotopically pure $^{12}_{6}C$. Since L is the number of particles in a mole of substance, its units are $\text{mol}^{-1}$.

---

### Worked example 1.5     *The mole and the Avogadro constant (1)*

**Calculate how many atoms there are in 0.200 moles of copper. ($L = 6.022 \times 10^{23} \, \text{mol}^{-1}$)**

The number of atoms in one mole of Cu is equal to the Avogadro constant $= 6.022 \times 10^{23}$.

Number of atoms in 0.200 moles of Cu $= (0.200 \, \text{mol}) \times (6.022 \times 10^{23} \, \text{mol}^{-1})$

$$= 1.20 \times 10^{23} \quad \text{(to 3 sig. fig.)}$$

---

### Worked example 1.6     *The mole and the Avogadro constant (2)*

**Calculate how many molecules of $H_2O$ there are in 12.10 moles of water. ($L = 6.022 \times 10^{23} \, \text{mol}^{-1}$)**

The number of molecules in one mole of water is equal to the Avogadro constant $= 6.022 \times 10^{23}$.

Number of molecules in

12.10 moles of water $= (12.10 \, \text{mol}) \times (6.022 \times 10^{23} \, \text{mol}^{-1})$

$$= 7.287 \times 10^{24} \quad \text{(to 4 sig. fig.)}$$

---

## 1.9     Gas laws and ideal gases

In this section we summarize some of the important laws that apply to *ideal gases*.

### The kinetic theory of gases

The kinetic theory of gases is a model that can explain the observed gas laws in terms of atoms and molecules. In an *ideal gas*, we assume the following (the *postulates of the kinetic theory of gases*):

- the volume occupied by the particles (atoms or molecules) is negligible;
- the particles are in continuous, random motion;
- there are no attractive or repulsive interactions between the particles;
- collisions of the particles with the walls of their container result in the pressure, $P$, exerted by the gas;
- collisions between particles are elastic and no kinetic energy is lost upon collision;
- the average kinetic energy, $E_k$, is directly proportional to the temperature of the gas.

In a *real* gas, there are interactions between particles: see Section 2.10

By treating the kinetic theory of gases in a quantitative manner, we can derive the ideal gas law.

Consider a gaseous particle of mass $m$ moving in a straight-line path within a closed cubic box of side $l$. The motion continues until there is an elastic collision with the wall of the container (Figure 1.5). The velocity, $v$, of the particle can be resolved into three directions coincident with the three Cartesian axes $(x, y, z)$. If the particle is travelling in the $x$ direction, it has a momentum of $mv_x$ (equation 1.7) and on impact, it undergoes a change in momentum (equation 1.8). If the collision is elastic, the new velocity is $-v_x$.

$$\text{Momentum} = \text{mass} \times \text{velocity} \tag{1.7}$$

$$\text{Change in momentum} = (mv_x) - (-mv_x) = 2mv_x \tag{1.8}$$

Every time the particle travels across the container (distance $l$), a collision occurs (Figure 1.5). The number of collisions per second is given by equation 1.9 and it follows that the change in momentum per second is as shown in equation 1.10.

$$\text{Number of collisions per second} = \frac{\text{Velocity}}{\text{Distance}} = \frac{v_x}{l} \tag{1.9}$$

$$\text{Change in momentum per second} = (2mv_x)\left(\frac{v_x}{l}\right) = \frac{2mv_x^2}{l} \tag{1.10}$$

Our particle also has velocity components in the $y$ and $z$ directions. The velocity of the particle, $c$, is given by equation 1.11, and the total change in momentum per second for this particle is given by equation 1.12.

$$c^2 = v_x^2 + v_y^2 + v_z^2 \tag{1.11}$$

$$\text{Total change in momentum} = \frac{2mv_x^2}{l} + \frac{2mv_y^2}{l} + \frac{2mv_z^2}{l} = \frac{2mc^2}{l} \tag{1.12}$$

**Fig. 1.5** A particle in a cubic box of side $l$ collides with a wall at point $A$. The collision is elastic and no kinetic energy is lost. After impact the particle rebounds and travels across the box to the opposite wall where another collision occurs, and so on.

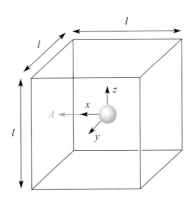

For $a$ particles with velocities $c_i$, the total change in momentum per second is $\dfrac{2m}{l}\sum\limits_{i=1}^{i=a} c_i^2$.

$\sum$ means 'summation of'

Defining the root mean square velocity, $\bar{c}$, such that:

$$\bar{c}^2 = \left(\frac{1}{a}\sum_{i=1}^{i=a} c_i^2\right)$$

leads to the expression for the change in momentum given in equation 1.13.

$$\text{Total change in momentum} = \frac{2ma\bar{c}^2}{l} \tag{1.13}$$

Newton's Laws of Motion: see Box 3.1

By Newton's Second Law of Motion, the force exerted on the walls of the container by the impact of the particles equals the rate of change of momentum. From the force, we can find the pressure since pressure is force per unit area. The total area is equal to that of the six walls of the container (which is $6l^2$) and, so, the pressure is given by equation 1.14, where $V$ is the volume of the cubic container.

$$\text{Pressure, } P = \frac{\text{Force}}{\text{Area}} = \frac{\left(\dfrac{2ma\bar{c}^2}{l}\right)}{6l^2} = \frac{2ma\bar{c}^2}{6l^3} = \frac{ma\bar{c}^2}{3V} \tag{1.14}$$

Let us now move from considering $a$ particles to a molar scale. For $n$ moles of particles, $a = nL$ where $L$ is the Avogadro number. We can therefore rewrite equation 1.14 in the form of equation 1.15, and we shall return to this equation *obtained from a theoretical model* later in this section.

$$PV = \frac{nLm\bar{c}^2}{3} \tag{1.15}$$

## Pressure and Boyle's Law

Pressure ($P$) is defined as force per unit area (equation 1.16) and the derived SI units of pressure are pascals (Pa) (Table 1.2).

$$\text{Pressure} = \frac{\text{Force}}{\text{Area}} \tag{1.16}$$

Although atmospheric pressure is the usual working condition in the laboratory (except when we are specifically working under conditions of reduced or high pressures), a pressure of 1 bar ($1.00 \times 10^5$ Pa) has been defined by the IUPAC as the *standard pressure*.[§] Of course, this may not be the exact pressure of working, and in that case appropriate corrections must be made as we show below.

When a fixed mass of a gas is compressed (the pressure on the gas is increased) at a constant temperature, the volume ($V$) of the gas decreases. The pressure and volume are related by Boyle's Law (equation 1.17). From this inverse relationship we see that doubling the pressure halves the volume, and halving the pressure doubles the volume of a gas.

$$\text{Pressure} \propto \frac{1}{\text{Volume}}$$

$$P \propto \frac{1}{V} \qquad \textit{at constant temperature} \tag{1.17}$$

---

[§] Until 1982, the standard pressure was 1 atmosphere (1 atm = 101 325 Pa) and this pressure remains in use in many textbooks and tables of physical data. The bar is a 'non-standard' unit in the same way that the ångström is defined as 'non-standard'.

## Charles's Law

Boyle's Law is obeyed only under conditions of constant temperature because the volume of a gas is dependent on both pressure and temperature. The volume and temperature of a fixed mass of gas are related by Charles's Law (equation 1.18) and the direct relationship shows that (at constant pressure), the volume doubles if the temperature is doubled.

Volume $\propto$ Temperature

$$V \propto T \qquad \textit{at constant pressure} \qquad (1.18)$$

A combination of Boyle's and Charles's Laws gives a relationship between the pressure, volume and temperature of a fixed mass of gas (equation 1.19), and this can be rewritten in the form of equation 1.20.

$$P \propto \frac{T}{V} \qquad (1.19)$$

$$\frac{PV}{T} = \text{constant} \qquad P = \text{pressure}, V = \text{volume}, T = \text{temperature} \qquad (1.20)$$

Corrections to a volume of gas from one set of pressure and temperature conditions to another can be done using equation 1.21; each side of the equation is equal to the *same* constant.

$$\frac{P_1 V_1}{T_1} = \frac{P_2 V_2}{T_2} \qquad (1.21)$$

---

**Worked example 1.7**  *Dependence of gas volume on temperature and pressure*

If the volume of a sample of helium is $0.0227\,\text{m}^3$ at 273 K and $1.00 \times 10^5\,\text{Pa}$, what is its volume at 293 K and $1.04 \times 10^5\,\text{Pa}$?

The relevant equation is:

$$\frac{P_1 V_1}{T_1} = \frac{P_2 V_2}{T_2}$$

First check that the units are consistent: $V$ in $\text{m}^3$, $T$ in K, $P$ in Pa. (In fact, *in this case* inconsistencies in units of $P$ and $V$ would cancel out – why?)

$$P_1 = 1.00 \times 10^5\,\text{Pa} \qquad V_1 = 0.0227\,\text{m}^3 \qquad T_1 = 273\,\text{K}$$
$$P_2 = 1.04 \times 10^5\,\text{Pa} \qquad V_2 = ? \qquad T_2 = 293\,\text{K}$$
$$\frac{P_1 V_1}{T_1} = \frac{P_2 V_2}{T_2}$$

$$\frac{(1.00 \times 10^5\,\text{Pa}) \times (0.0227\,\text{m}^3)}{(273\,\text{K})} = \frac{(1.04 \times 10^5\,\text{Pa}) \times (V_2\,\text{m}^3)}{(293\,\text{K})}$$

$$V_2 = \frac{(1.00 \times 10^5\,\text{Pa}) \times (0.0227\,\text{m}^3) \times (293\,\text{K})}{(273\,\text{K}) \times (1.04 \times 10^5\,\text{Pa})} = 0.0234\,\text{m}^3$$

---

**Worked example 1.8**  *Dependence of gas pressure on volume and temperature*

The pressure of a $0.0239\,\text{m}^3$ sample of $N_2$ gas is $1.02 \times 10^5\,\text{Pa}$. The gas is compressed to a volume of $0.0210\,\text{m}^3$ while the temperature remains constant at 293 K. What is the new pressure of the gas?

The equation needed for the calculation is:

$$\frac{P_1 V_1}{T_1} = \frac{P_2 V_2}{T_2}$$

but because $T_1 = T_2$, the equation can be simplified to:

$$P_1 V_1 = P_2 V_2$$

The data are as follows:

$$P_1 = 1.02 \times 10^5 \, \text{Pa} \qquad V_1 = 0.0239 \, \text{m}^3$$

$$P_2 = ? \qquad\qquad\qquad V_2 = 0.0210 \, \text{m}^3$$

$$(1.02 \times 10^5 \, \text{Pa}) \times (0.0239 \, \text{m}^3) = (P_2 \, \text{Pa}) \times (0.0210 \, \text{m}^3)$$

$$P_2 = \frac{(1.02 \times 10^5 \, \text{Pa}) \times (0.0239 \, \text{m}^3)}{(0.0210 \, \text{m}^3)} = 1.16 \times 10^5 \, \text{Pa}$$

Check to make sure that the answer is sensible: compression of a gas at constant temperature should lead to an increase in pressure.

## Ideal gases

In real situations, we work with 'real' gases, but it is convenient to assume that most gases behave as though they were *ideal*. *If* a gas is ideal, it obeys the *ideal gas law* (equation 1.22) in which the constant in equation 1.20 is the *molar gas constant*, $R$, and the quantity of gas to which the equation applies is 1 mole.

$$\frac{PV}{T} = R = 8.314 \, \text{J K}^{-1} \, \text{mol}^{-1} \tag{1.22}$$

For $n$ moles of gas, we can rewrite the ideal gas law as equation 1.23.

$$\frac{PV}{T} = nR \quad \text{or} \quad PV = nRT \tag{1.23}$$

■▶
**Van der Waals equation for real gases: see equation 2.20**

The molar gas constant has the same value for *all gases*, whether they are atomic (Ne, He), molecular and elemental ($O_2$, $N_2$) or compounds ($CO_2$, $H_2S$, NO).

We have already seen that the IUPAC has defined *standard pressure* as $1.00 \times 10^5 \, \text{Pa}$ (1 bar) and a value of *standard temperature* has been defined as 273.15 K. For our purposes, we often use 273 K as the standard temperature.[§] The volume of one mole of an ideal gas under conditions of standard pressure and temperature is $0.0227 \, \text{m}^3$ or $22.7 \, \text{dm}^3$ (equation 1.24). We consider the consistency of the units in this equation in worked example 1.9.

$$\text{Volume of 1 mole of ideal gas} = \frac{nRT}{P}$$

$$= \frac{1 \times (8.314 \, \text{J K}^{-1} \, \text{mol}^{-1}) \times (273 \, \text{K})}{(1.00 \times 10^5 \, \text{Pa})}$$

$$= 0.0227 \, \text{m}^3 \tag{1.24}$$

From Table 1.2: $\text{Pa} = \text{kg m}^{-1} \, \text{s}^{-2}$

$$\text{J} = \text{kg m}^2 \, \text{s}^{-2}$$

---

[§] The standard temperature of 273 K is *different* from the *standard state temperature* of 298 K used in thermodynamics; see Section 2.2 and Chapter 17.

and therefore the units work out as follows:

$$\frac{kg\,m^2\,s^{-2}\,K^{-1}\,mol^{-1} \times K}{kg\,m^{-1}\,s^{-2}} = m^3\,mol^{-1}$$

> The **molar volume** of an ideal gas under conditions of standard pressure ($1.00 \times 10^5$ Pa) and temperature (273 K) is 22.7 dm$^3$.

You should notice that this volume differs from the 22.4 dm$^3$ with which you may be familiar! A molar volume of 22.4 dm$^3$ refers to a standard pressure of 1 atm (101 300 Pa), and 22.7 dm$^3$ refers to a standard pressure of 1 bar (100 000 Pa).

We now return to equation 1.15, derived from the kinetic theory of gases. By combining equations 1.15 and 1.23, we can write equation 1.25 for one mole.

$$RT = \frac{Lm\bar{c}^2}{3} \tag{1.25}$$

We can now show the dependence of the kinetic energy of a mole of particles on the temperature. The kinetic energy, $E_k$, is equal to $\frac{1}{2}m\bar{c}^2$, and if we make this substitution into equation 1.25, we can write equation 1.26.

$$RT = \frac{2}{3}LE_k \quad \text{or} \quad E_k = \frac{3}{2}\frac{RT}{L} \tag{1.26}$$

where $E_k$ refers to one mole and $R$ is the molar gas constant. The gas constant per molecule is given by the Boltzmann constant, $k = 1.381 \times 10^{-23}$ J K$^{-1}$. The molar gas constant and the Boltzmann constant are related by equation 1.27, and hence we can express the kinetic energy per molecule in the form of equation 1.28.

$$\text{Boltzmann constant}, k = \frac{R}{L} \tag{1.27}$$

$$E_k = \frac{3}{2}kT \quad \text{where } E_k \text{ refers to 1 molecule} \tag{1.28}$$

---

**Worked example 1.9**  *Finding a derived SI unit*

**Determine the SI unit of the molar gas constant, $R$.**

The ideal gas law is:

$$PV = nRT$$

where $P$ = pressure, $V$ = volume, $n$ = number of moles of gas, $T$ = temperature.

$$R = \frac{PV}{nT}$$

The SI unit of $R = \dfrac{(\text{SI unit of pressure}) \times (\text{SI unit of volume})}{(\text{SI unit of quantity}) \times (\text{SI unit of temperature})}$

The SI unit of pressure is Pa but this is a *derived* unit. In base units, Pa = kg m$^{-1}$ s$^{-2}$ (Tables 1.1 and 1.2):

$$\text{SI unit of } R = \frac{(kg\,m^{-1}\,s^{-2}) \times (m^3)}{(mol) \times (K)}$$
$$= kg\,m^2\,s^{-2}\,mol^{-1}\,K^{-1}$$

While this unit of $R$ is correct in terms of the base SI units, it can be simplified because the joule, J, is defined as kg m$^2$ s$^{-2}$ (Table 1.2).
The SI unit of $R$ = J mol$^{-1}$ K$^{-1}$ or J K$^{-1}$ mol$^{-1}$.

*Exercise*: Look back at equation 1.23 and, by working in SI *base units*, show that a volume in m$^3$ is consistent with using values of pressure in Pa, temperature in K and $R$ in J K$^{-1}$ mol$^{-1}$.

| Worked example 1.10 | *Determining the volume of a gas* |

**Calculate the volume occupied by one mole of $CO_2$ at 300 K and 1 bar pressure. ($R = 8.314\,J\,K^{-1}\,mol^{-1}$)**

Assuming that the gas is ideal, the equation to use is:

$$PV = nRT$$

First ensure that the units are consistent – the pressure needs to be in Pa:

$$1\,bar = 1.00 \times 10^5\,Pa$$

To find the volume, we first rearrange the ideal gas equation:

$$V = \frac{nRT}{P}$$

$$V = \frac{nRT}{P} = \frac{1 \times (8.314\,J\,K^{-1}\,mol^{-1}) \times (300\,K)}{(1.00 \times 10^5\,Pa)}$$

$$= 0.0249\,m^3 = 24.9\,dm^3$$

The cancellation of units is as in equation 1.24.

*NB:* You should develop a habit of stopping and thinking whether a numerical answer to a calculation is sensible. Is its magnitude reasonable? In this example, we know that the molar volume of any ideal gas at $1.00 \times 10^5\,Pa$ and 273 K is $22.7\,dm^3$, and we have calculated that at $1.00 \times 10^5\,Pa$ and 300 K, the volume of one mole of gas is $24.9\,dm^3$. This increased volume appears to be consistent with a relatively small rise in temperature.

## Gay-Lussac's Law

Gay-Lussac's Law states that when gases react, the volumes of the reactants (and of the products if gases) are in a simple ratio, provided that the volumes are measured under the same conditions of temperature and pressure. Gay-Lussac's observations (made at the end of the 18th century) are consistent with the ideal gas equation (equation 1.23). Since $P$, $T$ and $R$ are constants when the volume measurements are made, it follows that the volume of a gaseous component in a reaction is proportional to the number of moles present.

| Worked example 1.11 | *Applying Gay-Lussac's Law* |

**At standard temperature and pressure, it was observed that $22.7\,dm^3$ of $N_2$ reacted with $68.1\,dm^3$ of $H_2$ to give $45.4\,dm^3$ of $NH_3$. Confirm that these volumes are consistent with the equation:**

$$N_2(g) + 3H_2(g) \longrightarrow 2NH_3(g)$$

Work out the ratio of the volumes:

| $N_2$ | | $H_2$ | | $NH_3$ |
|------|---|------|---|------|
| 22.7 | : | 68.1 | : | 45.4 |
| 1 | : | 3 | : | 2 |

This ratio is consistent with the stoichiometry of the equation given.

### Dalton's Law of Partial Pressures

In a mixture of gases, the total pressure exerted by the gas on its surroundings is the sum of the *partial pressures* of the component gases. This is Dalton's Law of Partial Pressures and can be expressed as in equation 1.29.

$$P = P_A + P_B + P_C + \cdots \tag{1.29}$$

where $P$ = total pressure, $P_A$ = partial pressure of gas A, etc.

The partial pressure of each gas is directly related to the number of moles of gas present. Equation 1.30 gives the relationship between the partial pressure of a component gas ($P_X$) and the total pressure $P$ of a gas mixture.

$$\text{Partial pressure of component X} = \frac{\text{Moles of X}}{\text{Total moles of gas}} \times \text{Total pressure} \tag{1.30}$$

---

**ENVIRONMENT AND BIOLOGY**

## Box 1.5 Diving to depths

The gas laws are of great importance to underwater divers. As a diver descends, the pressure increases. At sea level, a diver experiences atmospheric pressure ($\approx 1$ bar). At a depth of 10 m, the pressure is about 2 bar, and when a diver reaches a depth of 40 m, the pressure is 5 bar. Deep-sea divers breathe compressed air. This is composed mainly of $N_2$ (78%) and $O_2$ (21%) (see Figure 21.1). Whereas dioxygen is essential for respiration, $N_2$ is inert with respect to the body's metabolism. As the total pressure increases, the partial pressures of the component gases increase according to Dalton's Law of partial pressures:

$$P_{\text{total}} = P_{O_2} + P_{N_2} + \cdots$$

When air is taken into the lungs, it is in close contact with the blood. At atmospheric pressure, the solubility of $N_2$ in blood is low, but its solubility increases as the pressure increases. This is quantified in Henry's Law, which states that the solubility of a gas is directly proportional to its partial pressure. For a deep-sea diver, the amount of $N_2$ entering the blood and body tissues increases as the depth of the dive increases. Once a diver begins to ascend from a dive, the partial pressure of $N_2$ decreases and $N_2$ that has built up in the bloodstream can escape.

Two medical problems are associated with the accumulation of $N_2$ in the body during a deep-sea dive. The first is *nitrogen narcosis*. The symptoms of nitrogen narcosis depend on depth and pressure, the onset being at a depth of about 25 m. Depending on the depth, the diver may be affected by an impaired ability to make reasoned decisions, or may suffer from drowsiness and confusion. The effects can be reversed by ascending to regions of lower pressure. The ascent can, however, result in another medical problem: decompression sickness. This is caused by bubbles of $N_2$. The amount of $N_2$ escaping as bubbles from body tissue into the bloodstream depends on both the depth of the dive and the

length of time that a diver has been exposed to high partial pressures of $N_2$. Provided that the bubbles are transported in the blood and diffuse back into the lungs, the $N_2$ can be exhaled. However, if large bubbles of $N_2$ are trapped in the body, the diver begins to suffer from decompression sickness. This can be treated by using a decompression (hyperbaric) chamber. The diver is placed in the chamber and subjected to recompression, i.e. the pressure within the chamber in increased to simulate the deep-sea pressures. This assists the break-up of large $N_2$ bubbles which are able to enter the bloodstream. Slow decompression then follows and $N_2$ is released from the body in a controlled manner.

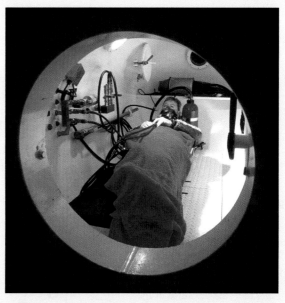

A diver in a decompression chamber. *Frank Greenaway* © *Dorling Kindersley, Courtesy of the University Marine Biological Station, Millport, Scotland.*

| **Worked example 1.12** | *Partial pressures* |
|---|---|

At 290.0 K and $1.000 \times 10^5$ Pa, a 25.00 dm$^3$ sample of gas contains 0.3500 moles of argon and 0.6100 moles of neon. (a) Are these the only components of the gas mixture? (b) What are the partial pressures of the two gases? ($R = 8.314$ J K$^{-1}$ mol$^{-1}$)

Part (a): The ideal gas law can be used to find the *total* number of moles of gas present:

$$PV = nRT$$

The volume must be converted to m$^3$: 25.00 dm$^3$ = $25.00 \times 10^{-3}$ m$^3$

$$n = \frac{PV}{RT} = \frac{(1.000 \times 10^5 \, \text{Pa}) \times (25.00 \times 10^{-3} \, \text{m}^3)}{(8.314 \, \text{J K}^{-1} \, \text{mol}^{-1}) \times (290.0 \, \text{K})} = 1.037 \, \text{moles}$$

From Table 1.2: Pa = kg m$^{-1}$ s$^{-2}$

$$\text{J} = \text{kg m}^2 \, \text{s}^{-2}$$

Total number of moles of gas = 1.037

Moles of argon + moles of neon = $0.3500 + 0.6100 = 0.9600$

Therefore there are $(1.037 - 0.9600) = 0.077$ moles of one or more other gaseous components in the sample.

Part (b): Now that we know the total moles of gas, we can determine the partial pressures of argon ($P_{Ar}$) and neon ($P_{Ne}$):

$$\text{Partial pressure of component X} = \frac{\text{moles of X}}{\text{total moles of gas}} \times \text{total pressure}$$

For argon: $P_{Ar} = \dfrac{(0.3500 \, \text{mol})}{(1.037 \, \text{mol})} \times (1.000 \times 10^5 \, \text{Pa}) = 33\,750 \, \text{Pa}$

For neon: $P_{Ne} = \dfrac{(0.6100 \, \text{mol})}{(1.037 \, \text{mol})} \times (1.000 \times 10^5 \, \text{Pa}) = 58\,820 \, \text{Pa}$

---

One use of partial pressures is in the determination of equilibrium constants for gaseous systems and we look further at this in Chapter 16.

### Graham's Law of Effusion

Consider the two containers shown in Figure 1.6; they are separated by a wall containing a tiny hole. The pressure of the gas in the left-hand container is

**Fig. 1.6** (a) A gas is contained in a chamber that is separated from another chamber at a lower pressure by a wall in which there is a pinhole. (b) Gas molecules or atoms will move from the region of higher to lower pressure, *effusing* until the pressures of the two containers are equal.

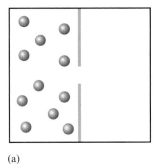

(a)

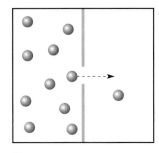

(b)

**Table 1.6** Rates of effusion for a range of gases, determined experimentally in the same piece of apparatus.

| Gaseous compound | Molecular formula | Rate of effusion / $cm^3\,s^{-1}$ |
| --- | --- | --- |
| Methane | $CH_4$ | 1.36 |
| Neon | Ne | 1.22 |
| Dinitrogen | $N_2$ | 1.03 |
| Carbon dioxide | $CO_2$ | 0.83 |
| Sulfur dioxide | $SO_2$ | 0.71 |
| Dichlorine | $Cl_2$ | 0.65 |

higher than that in the right-hand container. If there are no gas molecules at all in the right-hand container, the pressure is zero and the right-hand chamber is said to be *evacuated*. (In practice, the lowest pressures attainable experimentally are of the order of $10^{-8}$ Pa.) Gas molecules (or atoms if it is a monatomic gas such as neon or argon) will pass through the hole from the region of higher pressure to that of lower pressure until the pressures of the gases in each container are equal. This process is called *effusion*.

Graham's Law of Effusion (proven experimentally) is given in equation 1.31 where $M_r$ is the relative molar mass of the gas.

Molar mass: see Section 1.13

$$\text{Rate of effusion of a gas} \propto \frac{1}{\sqrt{M_r}} \qquad (1.31)$$

The law also applies to the process of diffusion, i.e. the rate at which gases, initially held separately from each other, mix (see Figure 1.3). Table 1.6 gives the rates of effusion of several gases; the rates were measured under the same experimental conditions. As an exercise, you should confirm the validity of Graham's Law by plotting a graph of rate of effusion against $1/\sqrt{M_r}$; the plot will be linear.

A classic experiment to illustrate Graham's Law is shown in Figure 1.7. Gaseous $NH_3$ and HCl react to produce ammonium chloride as a fine, white solid (equation 1.32).

$$NH_3(g) + HCl(g) \longrightarrow NH_4Cl(s) \qquad (1.32)$$

Values of $M_r$ for $NH_3$ and HCl are 17.03 and 36.46 respectively, and therefore molecules of HCl diffuse along the tube more slowly than do molecules of $NH_3$. As a result, $NH_4Cl$ forms at a point that is closer to the end of the tube containing the cotton wool soaked in concentrated hydrochloric acid (Figure 1.7). This result can be quantified as shown in equation 1.33.

$$\frac{\text{Rate of diffusion of } NH_3}{\text{Rate of diffusion of HCl}} = \sqrt{\frac{36.46}{17.03}} = 1.463 \qquad (1.33)$$

**Fig. 1.7** A glass tube containing two cotton wool plugs soaked in aqueous $NH_3$ and HCl; the ends of the tube are sealed with stoppers. Solid $NH_4Cl$ is deposited on the inner surface of the glass tube, nearer to the end at which the HCl originates.

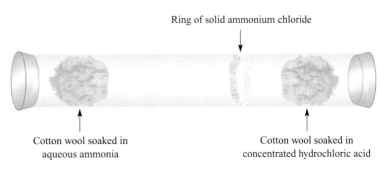

Ring of solid ammonium chloride

Cotton wool soaked in aqueous ammonia

Cotton wool soaked in concentrated hydrochloric acid

| **Worked example 1.13** | *Applying Graham's Law of Effusion* |

If $V$ cm$^3$ of NH$_3$ effuse at a rate of 2.25 cm$^3$ s$^{-1}$, and $V$ cm$^3$ of an unknown gas X effuse at a rate of 1.40 cm$^3$ s$^{-1}$ under the same experimental conditions, what is $M_r$ for X? Suggest a possible identity for X.

By applying Graham's Law, and using values of $A_r$ from the periodic table facing the inside front cover of this book:

$$\frac{\text{Rate of diffusion of NH}_3}{\text{Rate of diffusion of X}} = \sqrt{\frac{(M_r)_X}{(M_r)_{NH_3}}}$$

$$\frac{2.25}{1.40} = \sqrt{\frac{(M_r)_X}{17.03}}$$

$$(M_r)_X = \left(\frac{(2.25 \text{ cm}^3 \text{ s}^{-1})}{(1.40 \text{ cm}^3 \text{ s}^{-1})}\right)^2 \times 17.03 = 44.0$$

The molar mass of X is therefore 44.0, and possible gases include $CO_2$ and $C_3H_8$.

---

## 1.10    The periodic table

### The arrangement of elements in the periodic table

The elements are arranged in the periodic table (Figure 1.8) in numerical order according to the number of protons possessed by each element. The division into *groups* places elements with the same number of valence electrons into vertical columns within the table. Under IUPAC recommendations, the

Valence electrons: see Section 3.19

**Fig. 1.8** The periodic table showing *s*-, *p*- and *d*-block elements and the lanthanoids and actinoids (*f*-block). A complete periodic table with more detailed information is given facing the inside front cover of the book. Strictly, the series of lanthanoid and actinoid metals each consist of 14 elements. However, La (lanthanum) and Ac (actinium) are usually classed with the *f*-block elements.

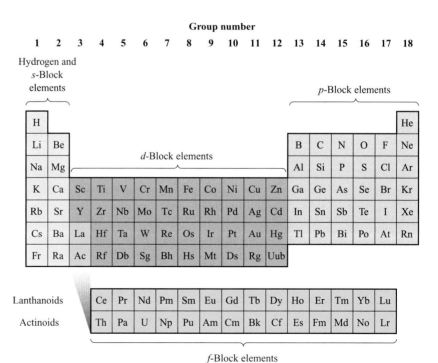

**Table 1.7** IUPAC-recommended names for sections and groups in the periodic table.

| Group number | Recommended name |
|---|---|
| 1 (except H) | Alkali metals |
| 2 | Alkaline earth metals |
| 15 | Pnictogens |
| 16 | Chalcogens |
| 17 | Halogens |
| 18 | Noble gases |
| 1 (except H), 2, 13, 14, 15, 16, 17, 18 | Main group elements |

Inorganic nomenclature:
see Section 1.18

groups are labelled from 1 to 18 (Arabic numbers) and blocks of elements are named as in Figure 1.8.

The *d*-block elements are also referred to as the *transition elements*, although the elements zinc, cadmium and mercury (group 12) are not always included. The vertical groups of three *d*-block elements are called *triads*. The names lanthanoid and actinoid are preferred over the names lanthanide and actinide because the ending '-ide' usually implies a negatively charged ion. However, the terms lanthanide and actinide are still in common use. The lanthanoids and actinoids are collectively termed the *f*-block elements. IUPAC-recommended names for some of the groups are given in Table 1.7.

The last row of elements that are coloured purple in Figure 1.8 consists of artificially produced elements: rutherfordium (Rf), dubrium (Db), seaborgium (Sg), bohrium (Bh), hassium (Hs), meitnerium (Mt), darmstadtium (Ds) and roentgenium (Rg). In mid-2004, the number of elements in the periodic table stood at 112, although the IUPAC had not authenticated element 112. Element 112 is currently known as ununbium ('one-one-two'). This is the usual method of naming newly discovered elements until final names have been approved by the IUPAC.

## Valence electrons

It is important to know the positions of elements in the periodic table. A fundamental property of an element is the *ground state valence electron configuration* (discussed further in Chapter 3) and this can usually be determined for any element by considering its position in the periodic table. For example, elements in groups 1 or 2 have one or two electrons in the outer (valence) shell, respectively. After the *d*-block (which contains 10 groups), the number of valence electrons can be calculated by subtracting 10 from the group number. Nitrogen (N) is in group 15 and has five valence electrons; tellurium (Te) is in group 16 and has six valence electrons. Trends in the behaviour of the elements follow periodic patterns ( *periodicity*).

Periodicity: see Chapter 20

Learning the periodic table by heart is not usually necessary, but it is extremely useful to develop a feeling for the groupings of elements. To know that selenium is in the same group as oxygen will immediately tell you that there will be some similarity between the chemistries of these two elements. Care is needed though – there is a significant change in properties on descending a group as we shall discover in Chapters 21 and 22.

| **Worked example 1.14** | *Relationship between periodic group number and number of valence electrons* |
| --- | --- |

Find the positions of the following elements in the periodic table: **N, K, Si, S, F, Cr, Mg.** For an atom of each element, determine the number of valence electrons.

Use the periodic table in Figure 1.8 to find the group in which each element lies:

| Element | N | K | Si | S | F | Al | Mg |
| --- | --- | --- | --- | --- | --- | --- | --- |
| Group number | 15 | 1 | 14 | 16 | 17 | 13 | 2 |

For elements in groups 1 and 2:

Number of valence electrons = Group number

K has one valence electron.
Mg has two valence electrons.
For elements in groups 13 to 18:

Number of valence electrons = Group number − 10

Si has four valence electrons.
S has six valence electrons.
F has seven valence electrons.
Al has three valence electrons.

---

## 1.11    Radicals and ions

### Radicals, anions and cations

The presence of one or more unpaired electrons in an atom or molecule imparts upon it the property of a *radical*. A superscript • may be used to signify that a species has an unpaired electron and is a radical. The neutral atom $^{19}_{9}F$, with one unpaired electron, is a radical (Figure 1.9).

The fluorine atom readily accepts one electron (Figure 1.9 and equation 1.34) to give an ion with a *noble gas configuration*.

Octet rule: see Section 3.20

$$F^{\bullet}(g) + e^{-} \longrightarrow F^{-}(g) \tag{1.34}$$

There is now a charge imbalance between the positive charge of the fluorine nucleus which has nine protons and the negative charge of the 10 electrons. On gaining an electron, the neutral fluorine radical becomes a negatively charged *fluoride ion*. The change in name from fluor*ine* to fluor*ide* is significant – the ending '*-ide*' signifies the presence of a negatively charged species. A negatively charged ion is called an *anion*.

*A **radical** possesses at least one unpaired electron.*

The loss of one electron from a neutral atom generates a positively charged ion – a *cation*. A sodium atom may lose an electron to give a sodium cation (equation 1.35) and the positive charge arises from the imbalance between the number of protons in the nucleus and the electrons.

$$Na^{\bullet}(g) \longrightarrow Na^{+}(g) + e^{-} \tag{1.35}$$

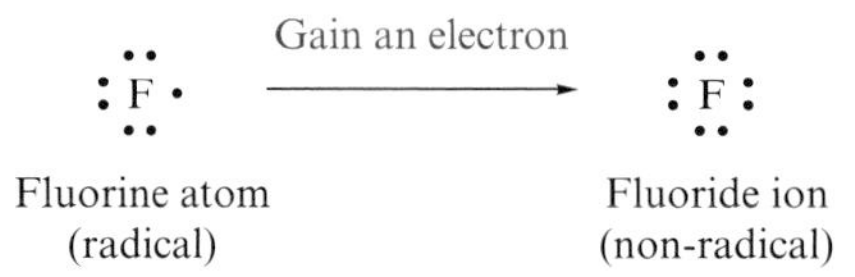

**Fig. 1.9** The fluorine atom (a radical) becomes a negatively charged fluoride ion when it gains one electron. Only the valence electrons are shown. The fluoride ion has a noble gas configuration.

Gain an electron

Fluorine atom (radical) → Fluoride ion (non-radical)

An **anion** is a negatively charged ion, and a **cation** is a positively charged ion.

Although in equations 1.34 and 1.35 we have been careful to indicate that the neutral atom is a radical, it is usually the case that *atoms* of an element are written without specific reference to the radical property. For example, equation 1.36 means exactly the same as equation 1.35.

$$Na(g) \longrightarrow Na^+(g) + e^- \tag{1.36}$$

Elements with fewer than four valence electrons *tend to lose electrons and form cations with a noble gas configuration*, for example:

- Na forms $Na^+$ which has the same electronic configuration as Ne;
- Mg forms $Mg^{2+}$ which has the same electronic configuration as Ne;
- Al forms $Al^{3+}$ which has the same electronic configuration as Ne.

Elements with more than four valence electrons *tend to gain electrons and form anions with a noble gas configuration*, for example:

- N forms $N^{3-}$ which has the same electronic configuration as Ne;
- O forms $O^{2-}$ which has the same electronic configuration as Ne;
- F forms $F^-$ which has the same electronic configuration as Ne.

The terms *dication*, *dianion*, *trication* and *trianion*, etc. are used to indicate that an ion carries a specific charge. A dication carries a 2+ charge (e.g. $Mg^{2+}$, $Co^{2+}$), a dianion has a 2− charge (e.g. $O^{2-}$, $Se^{2-}$), a trication bears a 3+ charge (e.g. $Al^{3+}$, $Fe^{3+}$), and a trianion has a 3− charge (e.g. $N^{3-}$, $PO_4{}^{3-}$).

As we discuss in the next section, ion formation is not the only option for an atom. Electrons may be shared rather than completely gained or lost. For example, in the chemistry of nitrogen and phosphorus, the gain of three electrons by an N or P atom to form $N^{3-}$ or $P^{3-}$ is not as common as the sharing of three electrons to form *covalent compounds*.

| Worked example 1.15 | *Ion formation* |
|---|---|

**With reference to the periodic table (Figure 1.8), predict the most likely ions to be formed by the following elements: Na, Ca, Br.**

Firstly, find the position of each element in the periodic table. Remember that elements with fewer than four valence electrons will tend to *lose* electrons and elements with more than four valence electrons will tend to *gain* electrons so as to form a noble gas configuration.

Na is in group 1.
Na has one valence electron.
Na will easily lose one electron to form an [Ne] configuration.
Na will form an $Na^+$ ion.

Ca is in group 2.
Ca has two valence electrons.
Ca will lose two electrons to form an [Ar] configuration.
Ca will form a $Ca^{2+}$ ion.

Br is in group 17.
Br has seven valence electrons.
Br will tend to gain one electron to give a [Kr] configuration.
Br will form a $Br^-$ ion.

|  1.12  | **Molecules and compounds: bond formation** |

### Covalent bond formation

A *molecule* is a discrete neutral species resulting from *covalent* bond formation between two or more atoms.

When two radicals combine, *pairing of the two electrons* may result in the formation of a covalent bond, and the species produced is molecular. Equation 1.37 shows the formation of molecular dihydrogen from two hydrogen atoms (i.e. radicals).

$$2H^{\bullet} \longrightarrow H_2 \tag{1.37}$$

> A *molecule* is a discrete neutral species containing a covalent bond or bonds between two or more atoms.

### Homonuclear and heteronuclear molecules

The molecule $H_2$ is a *homonuclear diatomic* molecule. *Diatomic* refers to the fact that $H_2$ consists of two atoms, and *homonuclear* indicates that the molecule consists of identical atoms. Molecular hydrogen, $H_2$, should be referred to as *dihydrogen* to distinguish it from atomic hydrogen, H. Other homonuclear molecules include dioxygen ($O_2$), dinitrogen ($N_2$), difluorine ($F_2$), trioxygen ($O_3$) and octasulfur ($S_8$).

A *heteronuclear* molecule consists of more than one type of element. Carbon monoxide, CO, is a *heteronuclear diatomic* molecule. When a molecule contains three or more atoms, it is called a *polyatomic* molecule; carbon dioxide, $CO_2$, methane, $CH_4$, and ethanol, $C_2H_5OH$, are polyatomic, although $CO_2$ may also be called a *triatomic* molecule.

◼▷
Homonuclear diatomics:
see Chapter 4

### Covalent versus ionic bonding

The important difference between covalent and ionic bonding is the average distribution of the bonding electrons between the nuclei. The electrons that form a covalent bond are 'shared' *between* nuclei. This is represented in Figure 1.10a for the $Cl_2$ molecule which contains a Cl–Cl single bond. Since the two atoms are identical, the two bonding electrons are located symmetrically between the two chlorine nuclei.

The complete transfer of the bonding pair of electrons to one of the nuclei results in the formation of *ions*. Figure 1.10b schematically shows the situation for a single pair of ions in a sodium chloride lattice; there is a region between the ions in which the electron density approaches zero. Although Figure 1.10b shows an isolated pair of $Na^+$ and $Cl^-$ ions, this is not a real situation for solid sodium chloride. Positive and negative ions attract one another and the net result is the formation of a 3-dimensional *ionic lattice*.

> In a *covalent bond*, electrons are shared between atoms. When the transfer of one or more electrons from one atom to another occurs, *ions* are formed.

◼▷
Ionic lattices:
see Chapter 8

**Fig. 1.10** Representations of (a) the covalent bond in the $Cl_2$ molecule in which each chlorine atom provides one electron, and (b) the ionic bond formed between a positively charged sodium ion and a negatively charged chloride ion (but see text below).

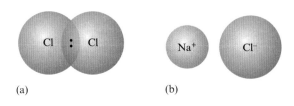

(a)                              (b)

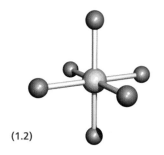

(1.2)

## Molecules and compounds

We must make a careful distinction between the use of the terms *molecule* and *compound*. Compounds are *neutral* and include both covalent and ionic species – NaF (composed of $Na^+$ and $F^-$ ions), CO (a covalent diatomic molecule) and $SF_6$ (**1.2**, a covalent polyatomic molecule) are all compounds. Since a molecule is a *discrete neutral species*, of these three compounds, only CO and $SF_6$ are molecules.

| 1.13 | **Molecules and compounds: relative molecular mass and moles** |

### Relative molecular mass

*$M_r$ is also referred to as the molecular weight of a compound*

The relative molecular mass ($M_r$) of a compound is found by summing the relative atomic masses of the constituent atoms. For example, a mole of carbon monoxide CO has a relative molecular mass of 28.01 (since the relative atomic masses of C and O are 12.01 and 16.00), and $M_r$ for carbon dioxide $CO_2$ is 44.01.

### Moles of compound

The relative molecular mass gives the mass (in g) of *one mole of a compound* and equation 1.38 gives the relationship between the mass (in g) and the number of moles.

$$\text{Amount in moles} = \frac{(\text{Mass in g})}{(M_r \text{ in g mol}^{-1})} \tag{1.38}$$

You should note that the dimensions for $M_r$ in equation 1.38 are not strictly correct. Here, and throughout the book, we implicitly multiply by a standard state $M^\circ$ of 1 g mol$^{-1}$ to adjust $M_r$ to a quantity with units of g mol$^{-1}$. We do not distinguish further between $M_r$ and $(M_r \times M^\circ)$; the usage will be clear, e.g. if we write '$M_r$ (g mol$^{-1}$)' then we really refer to $(M_r \times M^\circ)$.

> The ***molar mass***, $M_r$, is the mass in grams that contains $6.022 \times 10^{23}$ particles and has units of g mol$^{-1}$.

---

**Worked example 1.16**  *Relative molecular mass and moles*

**How many moles of molecules are present in 3.48 g of acetone, $CH_3COCH_3$?**

First, write down the formula of acetone: $C_3H_6O$.

$$M_r = (3 \times 12.01) + (1 \times 16.00) + (6 \times 1.008) = 58.08$$

$$\text{Amount in moles} = \frac{(\text{Mass g})}{(M_r \text{ g mol}^{-1})} = \frac{(3.48 \text{ g})}{(58.08 \text{ g mol}^{-1})} = 0.0599 \text{ moles}$$

Note that in the first line of the calculation, the answer is rounded to two decimal places, but in the second line, it is rounded to three significant figures. For further help with decimal places and significant figures, see the accompanying *Mathematics Tutor* (www.pearsoned.co.uk/housecroft).

---

**1.14    Concentrations of solutions**

### Molarity and molality

Reactions in the laboratory are often carried out in solution, and the *concentration* of the solution tells us how much of a compound or ion is present in a given solution. Two common methods are used for describing the concentration of a solution. The first is to give the amount (in moles) of dissolved substance (the *solute*) per unit volume of solution. The *molarity* of a solution is the amount (in moles) of solute per $dm^3$ of solution and has units of $mol\,dm^{-3}$.

> A solution of volume $1\,dm^3$ containing one mole of solute is called a one molar (1 M) solution.

The second method is to give the amount (in moles) of dissolved substance per mass of *solvent* (not solution). The *molality* of a solution is the amount (in moles) of solute per kg of solvent and has units of $mol\,kg^{-1}$.

> A solution made up of 1 kg of solvent in which one mole of solute is dissolved is called a one molal (1 m) solution.

Although we will usually use molarities in this book, in practice, molality has the significant advantage of being independent of temperature.

### Concentrations and amount of solute

In practice, we do not always work with one molar (1 M) solutions, and the moles of solute in a given solution can be found using equation 1.39 or, if the volume is in $cm^3$ as is often the case in the laboratory, a conversion from $cm^3$ to $dm^3$ must be made (equation 1.40).

$$\text{Amount in moles} = (\text{Volume in } dm^3) \times (\text{Concentration in } mol\,dm^{-3}) \quad (1.39)$$

$$x\,cm^3 = \left(\frac{x}{1000}\right) dm^3 \quad \text{or} \quad x\,cm^3 = (x \times 10^{-3})\,dm^3 \quad (1.40)$$

---

**Worked example 1.17**    *Solution concentrations*

**If 1.17 g of sodium chloride (NaCl) is dissolved in water and made up to 100 cm³ of aqueous solution, what is the concentration of the solution?**

First, we need to find the number of moles of NaCl in 1.17 g:

$A_r$ values are given inside the front cover of the book

$$M_r = 22.99 + 35.45 = 58.44\,g\,mol^{-1}$$

$$\text{Amount of NaCl} = \frac{1.17\,g}{58.44\,g\,mol^{-1}} = 0.0200\,\text{moles (to 3 sig. fig.)}$$

Next, we must ensure that the units for the next part of the calculation are consistent. The volume is given as $100\,cm^3$ and this must be converted to $dm^3$ so that the units of concentration are in $mol\,dm^{-3}$.

$$100\,cm^3 = \frac{100}{1000}\,dm^3 = 0.1\,dm^3$$

$$\text{Concentration in mol dm}^{-3} = \frac{\text{Amount in moles}}{\text{Volume in dm}^3}$$

$$= \frac{0.0200\,mol}{0.1\,dm^3}$$

$$= 0.200\,mol\,dm^{-3} \quad \text{(to 3 sig. fig.)}$$

---

**Worked example 1.18**    *Solution concentrations*

**What mass of potassium iodide (KI) must be dissolved in water and made up to $50.0\,cm^3$ to give a solution of concentration $0.0500\,\text{M}$?**

First we must find the amount of KI in $50.0\,cm^3$ $0.0500\,mol\,dm^{-3}$ solution:

Units conversion for the volume: $50.0\,cm^3 = 50.0 \times 10^{-3}\,dm^3$

$$\text{Amount in moles} = (\text{Volume in dm}^3) \times (\text{Concentration in mol dm}^{-3})$$

$$\text{Amount of KI} = (50.0 \times 10^{-3}\,dm^3) \times (0.0500\,mol\,dm^{-3})$$

$$= 2.50 \times 10^{-3}\,mol \quad \text{(to 3 sig. fig.)}$$

Now convert moles to mass:

$$\text{Amount in moles} = \frac{\text{Mass in g}}{M_r \text{ in g mol}^{-1}}$$

or

$$\text{Mass in g} = (\text{Amount in moles}) \times (M_r \text{ in g mol}^{-1})$$

For KI, $M_r = 39.10 + 126.90 = 166.00$

$$\text{Mass of KI} = (2.50 \times 10^{-3}\,mol) \times (166.00\,g\,mol^{-1}) = 0.415\,g \quad \text{(to 3 sig. fig.)}$$

---

**Worked example 1.19**    *Dilution of a solution*

**The concentration of an aqueous solution of sodium hydroxide, NaOH, is $0.500\,mol\,dm^{-3}$. What volume of this solution must be added to $250\,cm^3$ of water to give a solution of concentration $0.0200\,mol\,dm^{-3}$?**

Let the volume to be added $= x\,cm^3 = x \times 10^{-3}\,dm^3$

Moles of NaOH in $x\,cm^3$ of $0.500\,mol\,dm^{-3}$ solution

$$= (x \times 10^{-3}\,dm^3) \times (0.500\,mol\,dm^{-3})$$

$$= 0.500x \times 10^{-3}\,mol$$

$0.500x \times 10^{-3}$ moles of NaOH are added to $250\,cm^3$ water, and the new volume of the solution is $(250 + x)\,cm^3$.

Unit conversion: $(250 + x)\,cm^3 = (250 + x) \times 10^{-3}\,dm^3$

Concentration of the new solution $= 0.0200\,mol\,dm^{-3}$

$$\text{Amount in moles} = (\text{Volume in dm}^3) \times (\text{Concentration in mol dm}^{-3})$$

$$0.500x \times 10^{-3}\,mol = \{(250 + x) \times 10^{-3}\,dm^3\} \times (0.0200\,mol\,dm^{-3})$$

Solving for $x$:

$$0.500x \times 10^{-3} = \{(250 \times 0.0200) + 0.0200x\} \times 10^{-3}$$

$$0.500x = (250 \times 0.0200) + 0.0200x$$

$$(0.500 - 0.0200)x = (250 \times 0.0200)$$

$$x = \frac{(250 \times 0.0200)}{(0.500 - 0.0200)}$$

$$= \frac{5.00}{0.480} = 10.4 \quad \text{(to 3 sig. fig.)}$$

Therefore the volume to be added is $10.4 \, \text{cm}^3$.

An *aliquot* is a measured fraction of a larger sample, usually a fluid.

In a calculation of this type, it is important not to forget that the addition of an aqueous sample (an *aliquot*) to a larger amount of solution causes the volume of the solution to increase. Ignoring this factor introduces an error into the calculation; the significance of the error depends on the relative magnitudes of the two volumes.

---

**Worked example 1.20**   *Making up a solution to a given volume*

**A stock solution of potassium hydroxide, KOH, is 1.00 M. An $x \, \text{cm}^3$ aliquot of this solution is added to approximately $200 \, \text{cm}^3$ of water in a volumetric flask and the volume of the solution is made up with water to $250 \, \text{cm}^3$. What is the value of $x$ if the concentration of the final solution is $0.0500 \, \text{mol dm}^{-3}$?**

Unit conversion for the volumes:

$$x \, \text{cm}^3 = x \times 10^{-3} \, \text{dm}^3$$

$$250 \, \text{cm}^3 = 250 \times 10^{-3} \, \text{dm}^3$$

In the aliquot, amount of KOH

$$= (\text{Volume in dm}^3) \times (\text{Concentration in mol dm}^{-3})$$

$$= (x \times 10^{-3} \, \text{dm}^3) \times (1.00 \, \text{mol dm}^{-3})$$

$$= x \times 10^{-3} \, \text{mol}$$

In $250 \, \text{cm}^3$ of the final $0.0500 \, \text{mol dm}^{-3}$ solution there are $x \times 10^{-3}$ moles of KOH.

$$x \times 10^{-3} \, \text{mol} = (250 \times 10^{-3} \, \text{dm}^3) \times (0.0500 \, \text{mol dm}^{-3})$$

$$x = 250 \times 0.0500 = 12.5 \quad \text{(to 3 sig. fig.)}$$

Therefore the aliquot needed is of volume $12.5 \, \text{cm}^3$.

---

**1.15**   **Reaction stoichiometry**

When we write a balanced or *stoichiometric* equation, we state the ratio in which the reactants combine and the corresponding ratios of products. In reaction 1.41, one molecule of pentane reacts with eight molecules of dioxygen to give five molecules of carbon dioxide and six molecules of water. This also corresponds to the ratio of *moles* of reactants and products.

$$C_5H_{12}(g) + 8O_2(g) \longrightarrow 5CO_2(g) + 6H_2O(l) \qquad (1.41)$$

Pentane

---

A **balanced** or **stoichiometric** equation for a chemical reaction shows the molecular or the molar ratio in which the reactants combine and in which the products form.

---

**Worked example 1.21** | *Amounts of reactants for complete reaction*

**Zinc reacts with hydrochloric acid according to the equation:**

$$Zn(s) + 2HCl(aq) \longrightarrow ZnCl_2(aq) + H_2(g)$$

**What mass of zinc is required to react completely with 30.0 cm$^3$ 1.00 M hydrochloric acid?**

This is the reaction of a metal with acid to give a salt and $H_2$. First write the stoichiometric equation:

$$Zn(s) + 2HCl(aq) \longrightarrow ZnCl_2(aq) + H_2(g)$$

Now find the amount (in moles) of HCl in the solution:

Unit conversion for the volume: $30.0\,cm^3 = 30.0 \times 10^{-3}\,dm^3$

Amount in moles $= (\text{Volume in dm}^3) \times (\text{Concentration in mol dm}^{-3})$

Amount of HCl $= (30.0 \times 10^{-3}\,dm^3) \times (1.00\,mol\,dm^{-3})$

$$= 3.00 \times 10^{-2}\,mol$$

Now look at the balanced equation. Two moles of HCl react with one mole of Zn, and therefore $3.00 \times 10^{-2}$ moles of HCl react with $1.50 \times 10^{-2}$ moles of Zn.

We can now determine the mass of zinc needed:

Mass of Zn $= (\text{Amount in mol}) \times (A_r \text{ in g mol}^{-1})$

$$= (1.50 \times 10^{-2}\,mol) \times (65.41\,g\,mol^{-1})$$

$$= 0.981\,g \quad \text{(to 3 sig. fig.)}$$

We use $A_r$ in the equation because we are dealing with atoms of an element.

---

**Worked example 1.22** | *Amount of products formed*

**Carbon burns completely in dioxygen according to the equation:**

$$C(s) + O_2(g) \longrightarrow CO_2(g)$$

**What volume of $CO_2$ can be formed under conditions of standard temperature and pressure when 0.300 g of carbon is burnt? (Volume of one mole of gas at $1.00 \times 10^5$ Pa and 273 K = 22.7 dm$^3$)**

First, write the balanced equation:

$$C(s) + O_2(g) \longrightarrow CO_2(g)$$

One mole of carbon atoms gives one mole of $CO_2$ (assuming complete combustion).

Determine the amount (in moles) of carbon:

$$\text{Amount of carbon} = \frac{\text{Mass}}{A_r} = \frac{0.300\,\text{g}}{12.01\,\text{g}\,\text{mol}^{-1}} = 0.0250\,\text{moles}$$

At standard temperature (273 K) and pressure ($1.00 \times 10^5$ Pa), one mole of $CO_2$ occupies $22.7\,\text{dm}^3$.

$$\text{Volume occupied by 0.0250 moles of } CO_2 = (0.0250\,\text{mol}) \times (22.7\,\text{dm}^3\,\text{mol}^{-1})$$

$$= 0.568\,\text{dm}^3 \quad \text{(to 3 sig. fig.)}$$

---

**Worked example 1.23**     *Amount of precipitate formed*

**When aqueous solutions of silver nitrate and sodium iodide react, solid silver iodide is formed according to the equation:**

$$AgNO_3(aq) + NaI(aq) \longrightarrow AgI(s) + NaNO_3(aq)$$

**If $50.0\,\text{cm}^3$ $0.200\,\text{mol}\,\text{dm}^{-3}$ of aqueous $AgNO_3$ is added to $25\,\text{cm}^3$ $0.400\,\text{mol}\,\text{dm}^{-3}$ of aqueous NaI, what mass of AgI is formed?**

The stoichiometric equation:

$$AgNO_3(aq) + NaI(aq) \longrightarrow AgI(s) + NaNO_3(aq)$$

shows that $AgNO_3$ and NaI react in a 1 : 1 molar ratio.

$$\text{Unit conversion for volumes: for } AgNO_3, \; 50.0\,\text{cm}^3 = 50.0 \times 10^{-3}\,\text{dm}^3$$

$$\text{for NaI, } 25.0\,\text{cm}^3 = 25.0 \times 10^{-3}\,\text{dm}^3$$

$$\text{The amount of } AgNO_3 = (\text{Volume in dm}^3) \times (\text{Concentration in mol}\,\text{dm}^{-3})$$

$$= (50.0 \times 10^{-3}\,\text{dm}^3) \times (0.200\,\text{mol}\,\text{dm}^{-3})$$

$$= 0.0100\,\text{mol} \quad \text{(to 3 sig. fig.)}$$

$$\text{The amount of NaI} = (\text{Volume in dm}^3) \times (\text{Concentration in mol}\,\text{dm}^{-3})$$

$$= (25.0 \times 10^{-3}\,\text{dm}^3) \times (0.400\,\text{mol}\,\text{dm}^{-3})$$

$$= 0.0100\,\text{mol} \quad \text{(to 3 sig. fig.)}$$

From the stoichiometric equation, 0.0100 moles of $AgNO_3$ react completely with 0.0100 moles of NaI to give 0.0100 moles of solid AgI. For the mass of AgI, we need to find $M_r$.

Values of $A_r$ are found in the inside cover of the book

$$M_r \text{ for AgI} = 107.87 + 126.90 = 234.77$$

$$\text{Mass of AgI} = (\text{Amount in mol}) \times (M_r \text{ in g}\,\text{mol}^{-1})$$

$$= (0.0100\,\text{mol}) \times (234.77\,\text{g}\,\text{mol}^{-1})$$

$$= 2.35\,\text{g} \quad \text{(to 3 sig. fig.)}$$

---

**Worked example 1.24**     *Excess reagents*

**The reaction of aqueous hydrochloric acid and potassium hydroxide takes place according to the equation:**

$$HCl(aq) + KOH(aq) \longrightarrow KCl(aq) + H_2O(l)$$

$25.0 \, cm^3$ $0.0500 \, mol \, dm^{-3}$ aqueous HCl is added to $30 \, cm^3$ $0.0400 \, mol \, dm^{-3}$ aqueous KOH. The reaction goes to completion. Which reagent was in excess? How many moles of this reagent are left at the end of the reaction?

From the stoichiometric equation:

$$HCl(aq) + KOH(aq) \longrightarrow KCl(aq) + H_2O(l)$$

we see that HCl reacts with KOH in a molar ratio $1:1$.

Unit conversion for volumes: for HCl, $25.0 \, cm^3 = 25.0 \times 10^{-3} \, dm^3$

for KOH, $30.0 \, cm^3 = 30.0 \times 10^{-3} \, dm^3$

Amount of HCl = (Volume in $dm^3$) × (Concentration in $mol \, dm^{-3}$)

$$= (25.0 \times 10^{-3} \, dm^3) \times (0.0500 \, mol \, dm^{-3})$$

$$= 1.25 \times 10^{-3} \, mol \quad \text{(to 3 sig. fig.)}$$

Amount of KOH = (Volume in $dm^3$) × (Concentration in $mol \, dm^{-3}$)

$$= (30.0 \times 10^{-3} \, dm^3) \times (0.0400 \, mol \, dm^{-3})$$

$$= 1.20 \times 10^{-3} \, mol \quad \text{(to 3 sig. fig.)}$$

Since there are more moles of HCl in solution than KOH, the HCl is in excess.

Excess of HCl $= (1.25 \times 10^{-3}) - (1.20 \times 10^{-3}) = 0.05 \times 10^{-3}$ moles

## Balancing equations

In this section, we provide practice in balancing equations. In any *balanced* equation, the number of moles of an element (alone or in a compound) must be the same on the left- and right-hand sides of the equation. An equation such as:

$$HCl + Zn \longrightarrow ZnCl_2 + H_2$$

is unbalanced. By inspection, we can see that $2 \, HCl$ are required on the left-hand side to give the correct number of H and Cl atoms for product formation on the right-hand side, i.e.:

$$2HCl + Zn \longrightarrow ZnCl_2 + H_2$$

If you know the reactants and products of a reaction, you can write an unbalanced equation. However, such an equation is not chemically meaningful until it is balanced. When balancing an equation, it is very important to remember that you must not alter the chemical formulae of the compounds. *The only change that can be made is the addition of coefficients in front of the compound formulae.*

| Worked example 1.25 | *Equation for the combustion of propane* |

Balance the following equation for the complete combustion of propane:

$$C_3H_8 + O_2 \longrightarrow CO_2 + H_2O$$

It is probably best to start with the fact that there are three C atoms and eight H atoms on the left-hand side. Therefore, one mole of propane must produce three moles of $CO_2$ and four moles of $H_2O$:

$$C_3H_8 + O_2 \longrightarrow 3CO_2 + 4H_2O$$

This leaves the O atoms unbalanced. On the right-hand side we have 10 O atoms, and therefore we need five $O_2$ on the left-hand side. This gives:

$$C_3H_8 + 5O_2 \longrightarrow 3CO_2 + 4H_2O$$

Give this a final check through:

| | | | |
|---|---|---|---|
| Left-hand side: | 3 C | 8 H | 10 O |
| Right-hand side: | 3 C | 8 H | 10 O |

---

**Worked example 1.26**    *The reaction of Zn with mineral acids*

**Zinc reacts with $HNO_3$ and $H_2SO_4$ to give metal salts. Balance each of the following equations:**

$$Zn + HNO_3 \longrightarrow Zn(NO_3)_2 + H_2$$
$$Zn + H_2SO_4 \longrightarrow ZnSO_4 + H_2$$

For the reaction:

$$Zn + HNO_3 \longrightarrow Zn(NO_3)_2 + H_2$$

the product $Zn(NO_3)_2$ contains two $NO_3$ groups per Zn. Therefore, the ratio of $Zn : HNO_3$ on the left-hand side must be $1 : 2$. Thus, the initial equation can be rewritten as follows:

$$Zn + 2HNO_3 \longrightarrow Zn(NO_3)_2 + H_2$$

This balances the Zn, N and O atoms on each side of the equation. Now check the numbers of H atoms on the left and right sides of the equation: these are equal and therefore the equation is balanced. Final check:

| | | | | |
|---|---|---|---|---|
| Left-hand side: | 1 Zn | 2 H | 2 N | 6 O |
| Right-hand side: | 1 Zn | 2 H | 2 N | 6 O |

For the reaction:

$$Zn + H_2SO_4 \longrightarrow ZnSO_4 + H_2$$

the product $ZnSO_4$ contains one $SO_4$ group per Zn. Therefore, the ratio of $Zn : H_2SO_4$ on the left-hand side is $1 : 1$:

$$Zn + H_2SO_4 \longrightarrow ZnSO_4 + H_2$$

Now check to see if the H atoms on the left and right sides of the equation are equal; there are two H atoms on each side of the equation, and therefore the equation is balanced. Final check:

| | | | | |
|---|---|---|---|---|
| Left-hand side: | 1 Zn | 2 H | 1 S | 4 O |
| Right-hand side: | 1 Zn | 2 H | 1 S | 4 O |

---

**Worked example 1.27**    *Reactions between acids and bases*

**A reaction of the general type:**

$$\text{Acid} + \text{Base} \longrightarrow \text{Salt} + \text{Water}$$

is a *neutralization reaction.* Balance each of the following equations for neutralization reactions:

$$H_2SO_4 + KOH \longrightarrow K_2SO_4 + H_2O$$

$$HNO_3 + Al(OH)_3 \longrightarrow Al(NO_3)_3 + H_2O$$

For the reaction:

$$H_2SO_4 + KOH \longrightarrow K_2SO_4 + H_2O$$

the product $K_2SO_4$ contains two K per $SO_4$ group. Therefore, the ratio of $KOH : H_2SO_4$ on the left-hand side must be 2:1. The initial equation can be rewritten as follows:

$$H_2SO_4 + 2KOH \longrightarrow K_2SO_4 + H_2O$$

Now look at the formation of $H_2O$. On the left-hand side there are two H (in $H_2SO_4$) and two OH (in 2KOH) and these combine to give two $H_2O$. The balanced equation is:

$$H_2SO_4 + 2KOH \longrightarrow K_2SO_4 + 2H_2O$$

Final check:

| Left-hand side: | 1 S | 2 K | 4 H | 6 O |
|---|---|---|---|---|
| Right-hand side: | 1 S | 2 K | 4 H | 6 O |

For the reaction:

$$HNO_3 + Al(OH)_3 \longrightarrow Al(NO_3)_3 + H_2O$$

the product $Al(NO_3)_3$ contains three $NO_3$ groups per Al. Therefore, the ratio of $HNO_3 : Al(OH)_3$ on the left-hand side of the equation must be 3:1:

$$3HNO_3 + Al(OH)_3 \longrightarrow Al(NO_3)_3 + H_2O$$

Finally look at the formation of $H_2O$. On the left-hand side there are three H (in $3HNO_3$) and three OH (in $Al(OH)_3$) and these combine to give three $H_2O$. The balanced equation is:

$$3HNO_3 + Al(OH)_3 \longrightarrow Al(NO_3)_3 + 3H_2O$$

Final check:

| Left-hand side: | 3 N | 1 Al | 6 H | 12 O |
|---|---|---|---|---|
| Right-hand side: | 3 N | 1 Al | 6 H | 12 O |

---

A *precipitation reaction* is one in which a *sparingly soluble salt* (see Section 17.11) forms from a reaction that occurs in solution. The sparingly soluble salt comes out of solution as a *solid precipitate*.

When aqueous solutions of $AgNO_3$ and NaBr are mixed, a pale yellow *precipitate* of AgBr forms. The reaction is:

$$AgNO_3(aq) + NaBr(aq) \longrightarrow AgBr(s) + NaNO_3(aq)$$

The precipitation reaction could also be represented by the equation:

$$Ag^+(aq) + Br^-(aq) \longrightarrow AgBr(s)$$

Here, the ions that remain in solution and are not involved in the formation of the precipitate are ignored; these ions are called *spectator ions*. When balancing an equation such as the one above, you must make sure that:

*   the numbers of atoms of each element on the left- and right-hand sides of the equation balance, *and*
*   the positive and negative charges balance.

Consider the following equation for the formation of $PbI_2$:

$$Pb^{2+}(aq) + 2I^-(aq) \longrightarrow PbI_2(s)$$

First check the numbers of atoms:

| | | |
|---|---|---|
| Left-hand side: | 1 Pb | 2 I |
| Right-hand side: | 1 Pb | 2 I |

Now check the charges:

| | |
|---|---|
| Left-hand side: | overall charge $= +2 - 2 = 0$ |
| Right-hand side: | no charge |

In the next three worked examples, we look at other examples of reactions involving ions.

---

**Worked example 1.28**    *The precipitation of silver chromate*

Aqueous $Ag^+$ ions react with $[CrO_4]^{2-}$ ions to give a red precipitate of $Ag_2CrO_4$. Write a balanced equation for the precipitation reaction.

First write down an equation that shows the reactants and product:

$$Ag^+(aq) + [CrO_4]^{2-}(aq) \longrightarrow Ag_2CrO_4(s)$$

Since the product contains two Ag, there must be two $Ag^+$ on the left-hand side:

$$2Ag^+(aq) + [CrO_4]^{2-}(aq) \longrightarrow Ag_2CrO_4(s)$$

This equation is now balanced in terms of the numbers of Ag, Cr and O atoms on each side. Check that the charges balance:

| | |
|---|---|
| Left-hand side: | overall charge $= +2 - 2 = 0$ |
| Right-hand side: | no charge |

Therefore, the equation is balanced.

---

**Worked example 1.29**    *The reaction of Zn with $Ag^+$*

Zn reacts with $Ag^+$ to give solid silver:

$$Zn(s) + Ag^+(aq) \longrightarrow Zn^{2+}(aq) + Ag(s)$$

Balance this equation.

In terms of the number of Ag and Zn atoms on the left- and right-hand sides of the equation, the equation is already balanced, *but the charges do not balance*:

| | |
|---|---|
| Left-hand side: | overall charge $= +1$ |
| Right-hand side: | overall charge $= +2$ |

We can balance the charges by having two $Ag^+$ on the right-hand side:

$$Zn(s) + 2Ag^+(aq) \longrightarrow Zn^{2+}(aq) + Ag(s)$$

But, now there are two Ag on the left-hand side and only one Ag on the right-hand side. The final, balanced equation is therefore:

$$Zn(s) + 2Ag^+(aq) \longrightarrow Zn^{2+}(aq) + 2Ag(s)$$

| Worked example 1.30 | *The reaction of Fe²⁺ and [MnO₄]⁻ ions in acidic aqueous solution* |

Balance the following equation:

$$\text{Fe}^{2+} + [\text{MnO}_4]^- + \text{H}^+ \longrightarrow \text{Fe}^{3+} + \text{Mn}^{2+} + \text{H}_2\text{O}$$

The best place to start is with the O atoms: there are four O atoms in $[\text{MnO}_4]^-$ on the left-hand side, but only one O on the right-hand side. Therefore, to balance the O atoms, we need four $\text{H}_2\text{O}$ on the right-hand side:

$$\text{Fe}^{2+} + [\text{MnO}_4]^- + \text{H}^+ \longrightarrow \text{Fe}^{3+} + \text{Mn}^{2+} + 4\text{H}_2\text{O}$$

This change means that there are now eight H on the right-hand side, but only one H on the left-hand side. Therefore, to balance the H atoms, we need eight $\text{H}^+$ on the left-hand side:

$$\text{Fe}^{2+} + [\text{MnO}_4]^- + 8\text{H}^+ \longrightarrow \text{Fe}^{3+} + \text{Mn}^{2+} + 4\text{H}_2\text{O}$$

In terms of the *atoms*, the equation looks balanced, but if you check the *charges*, you find that the equation is not balanced:

Overall charge on left-hand side: $+2 - 1 + 8 = +9$

Overall charge on right-hand side: $+3 + 2 = +5$

The charges can be balanced without upsetting the balancing of the atoms if we have $n$ $\text{Fe}^{2+}$ on the left-hand side and $n$ $\text{Fe}^{3+}$ on the right-hand side, where $n$ can be found as follows:

$$n\text{Fe}^{2+} + [\text{MnO}_4]^- + 8\text{H}^+ \longrightarrow n\text{Fe}^{3+} + \text{Mn}^{2+} + 4\text{H}_2\text{O}$$

Overall charge on left-hand side: $+2n - 1 + 8 = 2n + 7$

Overall charge on right-hand side: $+3n + 2$

The charges on each side of the equation will balance if:

$$2n + 7 = 3n + 2$$
$$3n - 2n = 7 - 2$$
$$n = 5$$

The balanced equation is therefore:

$$5\text{Fe}^{2+} + [\text{MnO}_4]^- + 8\text{H}^+ \longrightarrow 5\text{Fe}^{3+} + \text{Mn}^{2+} + 4\text{H}_2\text{O}$$

Check the charges again:

Left-hand side: $+10 - 1 + 8 = +17$

Right-hand side: $+15 + 2 = +17$

In reality, there are spectator ions in solution which make the whole system neutral.

---

For further practice in balancing equations, try problems 1.24–1.36 at the end of the chapter.

## 1.16 Oxidation and reduction, and oxidation states

### Oxidation and reduction

When an element or compound burns in dioxygen to give an oxide, it is *oxidized* (equation 1.42).

$$2Mg(s) + O_2(g) \longrightarrow 2MgO(s) \tag{1.42}$$

Conversely, if a metal oxide reacts with dihydrogen and is converted to the metal, then the oxide is reduced (equation 1.43).

$$CuO(s) + H_2(g) \xrightarrow{\text{heat}} Cu(s) + H_2O(g) \tag{1.43}$$

In reaction 1.42, $O_2$ is the *oxidizing agent* and in reaction 1.43, $H_2$ is the *reducing agent*.

Although we often think of oxidation in terms of *gaining oxygen* and reduction in terms of *losing oxygen*, we should consider other definitions of these processes. A *reduction* may involve *gaining hydrogen*; in equation 1.44, chlorine is reduced and hydrogen is oxidized.

$$Cl_2(g) + H_2(g) \longrightarrow 2HCl(g) \tag{1.44}$$

*Loss of hydrogen* may correspond to *oxidation* – for example, chlorine is oxidized when HCl is converted to $Cl_2$.

Oxidation and reduction may also be defined in terms of electron transfer – *electrons are gained in reduction* processes and are *lost in oxidation* reactions (equations 1.45 and 1.46):

$$S + 2e^- \longrightarrow S^{2-} \qquad\qquad \textit{Reduction} \tag{1.45}$$

$$Zn \longrightarrow Zn^{2+} + 2e^- \qquad\qquad \textit{Oxidation} \tag{1.46}$$

It is, however, sometimes difficult to apply these simple definitions to a reaction. In order to gain insight into reduction and oxidation processes, it is useful to use the concept of *oxidation states*.

### Oxidation states

The concept of *oxidation states* and the changes in oxidation states that occur during a reaction provide a way of recognizing oxidation and reduction processes. Oxidation states are assigned to each atom of an element in a compound and *are a formalism*, although for ions such as $Na^+$, we can associate the charge of $1+$ with an oxidation state of $+1$.

The oxidation state of an *element* is zero. This applies to both atomic (e.g. He) and molecular elements (e.g. $H_2$, $P_4$, $S_8$).

In order to assign oxidation states to atoms in a *compound*, we must follow a set of rules, but be careful! The basic rules are as follows:

- The sum of the oxidation states of the atoms in a neutral compound is zero.
- The sum of the oxidation states of the atoms in an ion is equal to the charge on the ion, e.g. in the sulfate ion $[SO_4]^{2-}$, the sum of the oxidation states of S and O must be $-2$.
- Bonds between two atoms of the same element (e.g. O–O or S–S) have no influence on the oxidation state of the element (see worked example 1.35).
- The oxidation state of hydrogen is usually $+1$ when it combines with a non-metal and $-1$ if it combines with a metal.
- The oxidation state of fluorine in a compound is always $-1$.
- The oxidation state of chlorine, bromine and iodine is *usually* $-1$ (exceptions are interhalogen compounds and oxides – see Chapter 22).
- The oxidation state of oxygen in a compound is usually $-2$.
- The oxidation state of a group 1 metal in a compound is usually $+1$; exceptions are a relatively small number of compounds which contain $M^-$, with the group 1 metal in a $-1$ oxidation state (see Figure 20.6).

- The oxidation state of a group 2 metal in a compound is +2.
- Metals from the *d*-block will usually have positive oxidation states (exceptions are some low oxidation state compounds – see Section 23.14).

Added to these rules are the facts that most elements in groups 13, 14, 15 and 16 can have variable oxidation states. In reality, it is essential to have a full picture of the structure of a compound before oxidation states can be assigned.

---

**Worked example 1.31**  *Working out oxidation states*

What are the oxidation states of each element in the following: KI, $FeCl_3$, $Na_2SO_4$?

|  | |
|---|---|
| KI: | The group 1 metal is typically in oxidation state +1. This is consistent with the iodine being in oxidation state −1, and the sum of the oxidation states is 0. |
| $FeCl_3$: | Chlorine is usually in oxidation state −1, and since there are 3 Cl atoms, the oxidation state of the iron must be +3 to give a neutral compound. |
| $Na_2SO_4$: | Of the three elements, S can have a variable oxidation state and so we should deal with this element *last*. Na is in group 1 and usually has an oxidation state of +1. Oxygen is usually in an oxidation state of −2. The oxidation state of the sulfur atom is determined by ensuring that the sum of the oxidation states is 0: |

$$(2 \times \text{Oxidation state of Na}) + (\text{Oxidation state of S})$$
$$+ (4 \times \text{Oxidation state of O}) = 0$$
$$(+2) + (\text{Oxidation state of S}) + (-8) = 0$$
$$\text{Oxidation state of S} = 0 + 8 - 2 = +6$$

---

**Worked example 1.32**  *Variable oxidation states of nitrogen*

Determine the oxidation state of N in each of the following species: NO, $[NO_2]^-$, $[NO]^+$, $[NO_3]^-$, $NO_2$, ONF.

Each of the compounds or ions contains oxygen; O is usually in oxidation state −2.

NO:  This is a neutral compound, therefore:

$$(\text{Oxidation state of N}) + (\text{Oxidation state of O}) = 0$$
$$(\text{Oxidation state of N}) + (-2) = 0$$
$$\text{Oxidation state of N} = +2$$

$[NO_2]^-$:  The overall charge is −1, therefore:

$$(\text{Oxidation state of N}) + (2 \times \text{Oxidation state of O}) = -1$$
$$(\text{Oxidation state of N}) + (-4) = -1$$
$$\text{Oxidation state of N} = -1 + 4 = +3$$

[NO]$^+$:   The overall charge is $+1$, therefore:

$$\text{(Oxidation state of N)} + \text{(Oxidation state of O)} = +1$$
$$\text{(Oxidation state of N)} + (-2) = +1$$
$$\text{Oxidation state of N} = +1 + 2 = +3$$

[NO$_3$]$^-$:   The overall charge is $-1$, therefore:

$$\text{(Oxidation state of N)} + (3 \times \text{Oxidation state of O}) = -1$$
$$\text{(Oxidation state of N)} + (-6) = -1$$
$$\text{Oxidation state of N} = -1 + 6 = +5$$

NO$_2$:   This is a neutral compound, therefore:

$$\text{(Oxidation state of N)} + (2 \times \text{Oxidation state of O}) = 0$$
$$\text{(Oxidation state of N)} + (-4) = 0$$
$$\text{Oxidation state of N} = +4$$

ONF:   This is a neutral compound; in a compound, F is always in oxidation state $-1$, therefore:

$$\text{(Oxidation state of N)} + \text{(Oxidation state of O)}$$
$$+\text{(Oxidation state of F)} = 0$$
$$\text{(Oxidation state of N)} + (-2) + (-1) = 0$$
$$\text{Oxidation state of N} = +3$$

## Changes in oxidation states

> An *increase in the oxidation state* of an atom of an element corresponds to an *oxidation process*; a *decrease in the oxidation state* of an atom of an element corresponds to a *reduction process*.

In equation 1.45, the change in oxidation state of the sulfur from 0 to $-2$ is reduction, and in reaction 1.46, the change in oxidation state of the zinc from 0 to $+2$ is oxidation.

In a reduction–oxidation (*redox*) reaction, the change in oxidation number for the oxidation and reduction steps *must balance*. In reaction 1.47, iron is oxidized to iron(III) and chlorine is reduced to chloride ion. The net increase in oxidation state of the iron *for the stoichiometric reaction* must balance the net decrease in oxidation state for the chlorine.

Nomenclature for oxidation states: see Section 1.18

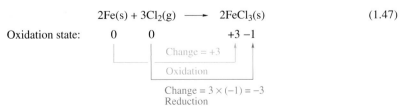

In equation 1.48, Fe$^{2+}$ ions are oxidized by [MnO$_4$]$^-$; at the same time, [MnO$_4$]$^-$ is reduced by Fe$^{2+}$ ions. The [MnO$_4$]$^-$ ion is the *oxidizing agent*, and $Fe^{2+}$ is the *reducing agent*. In the reaction, the oxidation states of H and O remain as $+1$ and $-2$, respectively. The oxidation state of Mn decreases from $+7$ to $+2$, and the oxidation state of Fe increases from $+2$ to $+3$. The balanced equation shows that five $Fe^{2+}$ ions are involved in the reaction and therefore the overall changes in oxidation states balance as shown in equation 1.48.

$$[MnO_4]^- + 5Fe^{2+} + 8H^+ \longrightarrow Mn^{2+} + 5Fe^{3+} + 4H_2O \qquad (1.48)$$

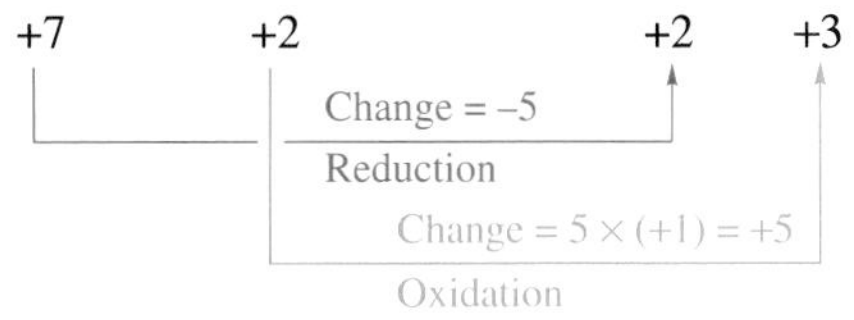

+7          +2                    +2      +3

Change = −5
Reduction
Change = 5 × (+1) = +5
Oxidation

---

| **Worked example 1.33** | *Reduction of iron(III) oxide* |

When heated with carbon, $Fe_2O_3$ is reduced to Fe metal:

$$Fe_2O_3(s) + 3C(s) \xrightarrow{\text{heat}} 2Fe(s) + 3CO(g)$$

**Identify the oxidation and reduction processes. Show that the oxidation state changes in the reaction balance.**

The oxidation state of O is $-2$.
C(s) and Fe(s) are elements in oxidation state 0.
$Fe_2O_3$ is a neutral compound:

$(2 \times \text{Oxidation state of Fe}) + (3 \times \text{Oxidation state of O}) = 0$

$(2 \times \text{Oxidation state of Fe}) - 6 = 0$

$$\text{Oxidation state of Fe} = \frac{+6}{2} = +3$$

In CO, the oxidation state of $C = +2$.
The oxidation process is C going to CO; the reduction process is $Fe_2O_3$ going to Fe. The oxidation state changes balance as shown below:

$$Fe_2O_3 + 3C \xrightarrow{\text{heat}} 2Fe + 3CO$$

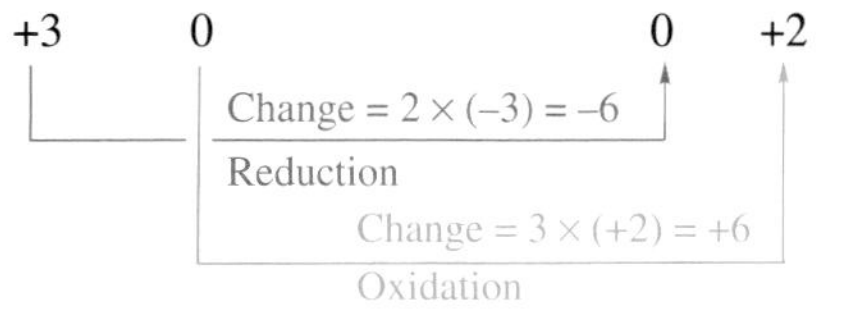

+3      0                              0      +2

Change = 2 × (−3) = −6
Reduction
Change = 3 × (+2) = +6
Oxidation

---

| **Worked example 1.34** | *Reaction of sodium with water* |

When sodium metal is added to water, the following reaction occurs:

$$2Na(s) + 2H_2O(l) \longrightarrow 2NaOH(aq) + H_2(g)$$

**State which species is being oxidized and which is being reduced. Show that the oxidation state changes balance.**

In Na(s), Na is in oxidation state 0.
In $H_2(g)$, H is in oxidation state 0.

$H_2O$ is a neutral compound:

$(2 \times \text{Oxidation state of H}) + (\text{Oxidation state of O}) = 0$

$(2 \times \text{Oxidation state of H}) + (-2) = 0$

$\text{Oxidation state of H} = \dfrac{+2}{2} = +1$

NaOH is a neutral compound; the usual oxidation states of Na and O are $+1$ and $-2$, respectively:

$(\text{Oxidation state of Na}) + (\text{Oxidation state of O}) + (\text{Oxidation state of H}) = 0$

$(+1) + (-2) + (\text{Oxidation state of H}) = 0$

$\text{Oxidation state of H} = 2 - 1 = +1$

Using these oxidation states, we can write the following equation and changes in oxidation states:

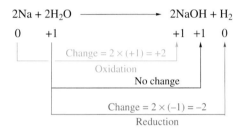

$$2Na + 2H_2O \longrightarrow 2NaOH + H_2$$

The equation shows that Na is oxidized; H is reduced on going from $H_2O$ to $H_2$, but remains in oxidation state $+1$ on going from $H_2O$ to NaOH. The changes in oxidation state balance.

---

**Worked example 1.35**    *Hydrogen peroxide as an oxidizing agent*

**Hydrogen peroxide reacts with iodide ions in the presence of acid according to the following equation:**

$$2I^- + H_2O_2 + 2H^+ \longrightarrow I_2 + 2H_2O$$

**Show that the changes in oxidation states in the equation balance. Confirm that $H_2O_2$ acts as an oxidizing agent.**

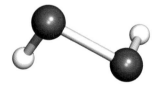

The structure of $H_2O_2$ is shown in the margin. The molecule contains an O−O bond (i.e. a bond between like atoms – a *homonuclear bond*). This bond has no influence on the oxidation state of O. $H_2O_2$ is a neutral compound, therefore:

$(2 \times \text{Oxidation state of O}) + (2 \times \text{Oxidation state of H}) = 0$

$(2 \times \text{Oxidation state of O}) + 2(+1) = 0$

$\text{Oxidation state of O} = \dfrac{-2}{2} = -1$

In $H_2O$, the oxidation states of H and O are $+1$ and $-2$, respectively. In $H^+$, hydrogen is in oxidation state $+1$. In $I^-$ and $I_2$, iodine is oxidation states $-1$ and $0$, respectively. The oxidation state changes in the reaction are shown below; H remains in oxidation state $+1$ and does not undergo a redox reaction.

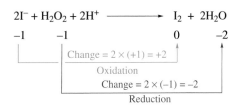

$$2I^- + H_2O_2 + 2H^+ \longrightarrow I_2 + 2H_2O$$

The equation shows that $I^-$ is oxidized by $H_2O_2$, and therefore $H_2O_2$ acts as an oxidizing agent.

---

**1.17**    ## Empirical, molecular and structural formulae

### Empirical and molecular formulae

The *empirical formula* of a compound is the simplest possible formula (with integer subscripts) that gives the composition of the compound.

The *molecular formula* of a compound is the formula consistent with the relative molecular mass.

The empirical formula of a compound gives the ratio of atoms of elements that combine to make the compound. However, this is not necessarily the same as the molecular formula which tells you the number of atoms of the constituent elements in line with the relative molar mass of the compound. The relationship between the *empirical and molecular formulae of a compound* is illustrated using ethane, in which the ratio of carbon : hydrogen atoms is $1:3$. This means that the *empirical formula* of ethane is $CH_3$. The relative molecular mass of ethane is 30, corresponding to two $CH_3$ units per molecule – the molecular formula is $C_2H_6$. In methane, the empirical formula is $CH_4$ and this corresponds directly to the molecular formula.

---

**Worked example 1.36**    *Empirical and molecular formulae*

A compound has the general formula $C_nH_{2n+2}$. This compound belongs to the family of *alkanes*. It contains 83.7% carbon by mass. Suggest the molecular formula of this compound.

Let the formula of the compound be $C_xH_y$. The percentage composition of the compound is 83.7% C and 16.3% H.

$$\% \text{ of C by mass} = \frac{\text{Mass of C in g}}{\text{Total mass in g}} \times 100$$

$$\% \text{ of H by mass} = \frac{\text{Mass of H in g}}{\text{Total mass in g}} \times 100$$

$A_r$ values: see front inside cover of the book

For a mole of the compound, the total mass in g = relative molecular mass $= M_r$.

$$\% \text{ C} = 83.7 = \frac{(12.01 \times x)}{M_r} \times 100$$

$$\% \text{ H} = 16.3 = \frac{(1.008 \times y)}{M_r} \times 100$$

We do not know $M_r$, but we can write down the *ratio* of moles of C : H atoms in the compound – this is the empirical formula. From above:

$$M_r = \frac{12.01x}{83.7} \times 100 = \frac{1.008y}{16.3} \times 100$$

$$\frac{y}{x} = \frac{100 \times 16.3 \times 12.01}{100 \times 83.7 \times 1.008} = 2.32$$

The *empirical formula* of the compound is $CH_{2.32}$ or $C_3H_7$.

The compound must fit into a family of compounds of general formula $C_nH_{2n+2}$, and this suggests that the *molecular formula* is $C_6H_{14}$.

The working above sets the problem out in full; in practice, the empirical formula can be found as follows:

%\, C = 83.7 \qquad $A_r$ C = 12.01

%\, H = 16.3 \qquad $A_r$ H = 1.008

Ratio $C:H = \dfrac{83.7}{12.01} : \dfrac{16.3}{1.008} \approx 6.97:16.2 \approx 1:2.32 = 3:7$

---

**Worked example 1.37**    *Determining the molecular formula of a chromium oxide*

A binary oxide of chromium with $M_r = 152.02$ contains 68.43% Cr. Determine the empirical formula and the molecular formula. [Data: $A_r$ Cr = 52.01; O = 16.00]

The binary chromium oxide contains only Cr and O. The composition by mass of the compound is:

68.43% Cr

$(100 - 68.43)\%$ O = 31.57% O

The ratio of moles of $Cr:O = \dfrac{\%Cr}{A_r(Cr)} : \dfrac{\%O}{A_r(O)}$

$$= \frac{68.43}{52.01} : \frac{31.57}{16.00}$$

$$= 1.316 : 1.973$$

$$= 1 : \frac{1.973}{1.316}$$

$$= 1 : 1.5 \text{ or } 2 : 3$$

The empirical formula of the chromium oxide is therefore $Cr_2O_3$.

To find the molecular formula, we need the molecular mass, $M_r$:

$$M_r = 152.02$$

For the empirical formula $Cr_2O_3$:

$$[2 \times A_r(Cr)] + [3 \times A_r(O)] = (2 \times 52.01) + (3 \times 16.00) = 152.02$$

This value matches the value of $M_r$. Therefore, the molecular formula is the same as the empirical formula, $Cr_2O_3$.

---

**Worked example 1.38**    *Determining the molecular formula of a sulfur chloride*

A binary chloride of sulfur with $M_r = 135.02$ contains 52.51% Cl. What are the empirical and molecular formulae of the compound? [Data: $A_r$ S = 32.06; Cl = 35.45]

The binary sulfur chloride contains only S and Cl. The composition by mass of the compound is:

52.51% Cl

$(100 - 52.51)\%$ S $= 47.49\%$ S

$$\text{The ratio of moles of S : Cl} = \frac{\%S}{A_r(S)} : \frac{\%Cl}{A_r(Cl)}$$

$$= \frac{47.49}{32.06} : \frac{52.51}{35.45}$$

$$= 1.481 : 1.481$$

$$= 1 : 1$$

The empirical formula of the sulfur chloride is therefore SCl.
The molecular mass, $M_r$, of the compound is 135.02.
For the empirical formula SCl: $A_r(S) + A_r(Cl) = 32.06 + 35.45 = 67.51$.
This mass must be doubled to obtain the value of $M_r$. Therefore, the molecular formula is $S_2Cl_2$.

---

### Structural formulae and 'ball-and-stick' models

Covalent bonding and structure: see Chapters 4–7

Neither the empirical nor the molecular formula provides information about the way in which the atoms of a molecule are connected. The molecular formula $H_2S$ does not indicate the arrangement of the three atoms in a molecule of hydrogen sulfide, but structure **1.3** *is* informative. We could also have arrived at this structure by considering the number of valence electrons available for bonding.

Isomers: see Sections 6.11, 23.5 and 24.6–24.8

For some molecular formulae, it is possible to connect the atoms in more than one reasonable way and the molecule is said to possess *isomers*. An example is $C_4H_{10}$ for which two structural formulae **1.4** and **1.5** can be drawn. More detailed structural information can be obtained from 'ball-and-stick' models. Models **1.6–1.8** correspond to formulae **1.3–1.5**. We expand the discussion of drawing structural formulae for organic molecules in Section 24.3.

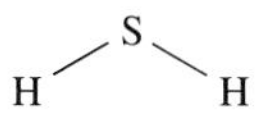

(1.3)

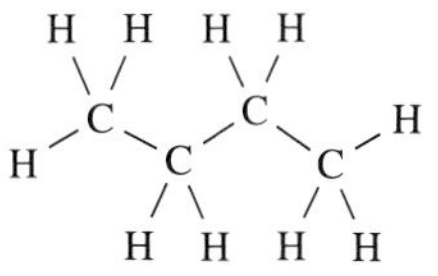

(1.4)

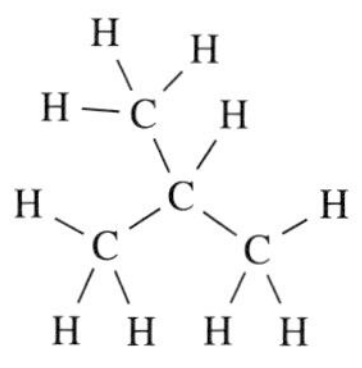

(1.5)

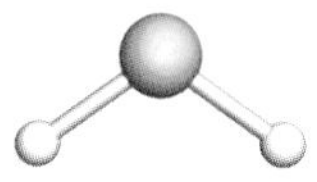

(1.6)

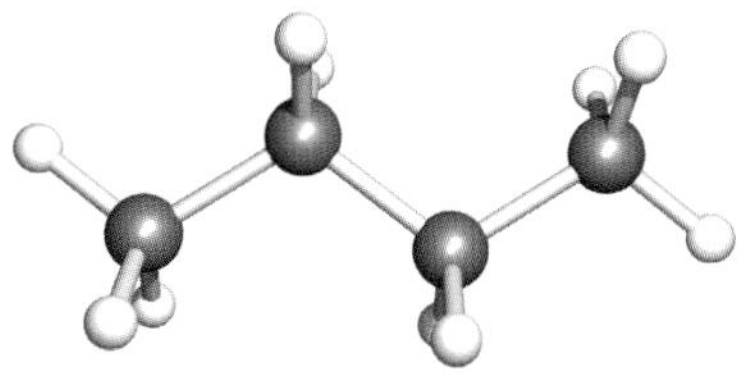

(1.7)

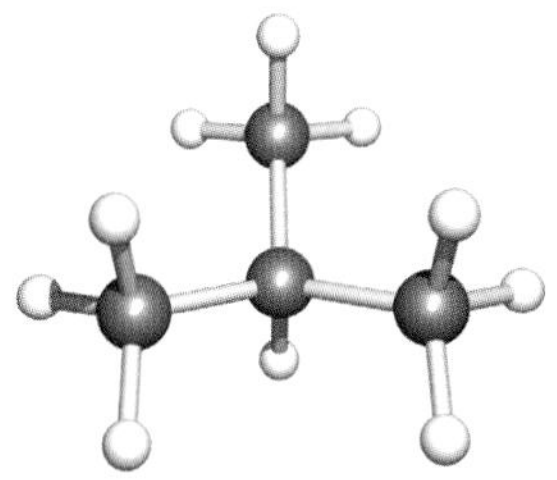

(1.8)

| 1.18 | **Basic nomenclature** |

In this section we outline some fundamental nomenclature and set out some important IUPAC ground rules for organic and inorganic compounds. We summarize some widely used 'trivial' names of compounds and you should become familiar with these as well as with their IUPAC counterparts. Detailed nomenclature rules and the specifics of organic chain numbering are found in Chapters 24–35.

Trivial names for selected organic and inorganic compounds are given in Appendix 13; this can be found on the companion website, www.pearsoned.co.uk/ housecroft

### Basic organic classifications: straight chain alkanes

The general formula for an alkane with a straight chain backbone is $C_nH_{2n+2}$; the molecule is *saturated* and contains only C–C and C–H bonds. The simplest member of the family is *methane* $CH_4$, **1.9**. The name methane carries with it two pieces of information:

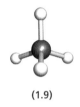

(1.9)

- *meth-* is the prefix that tells us there is one C atom in the carbon chain;
- the ending *-ane* denotes that the compound is an alkane.

The names of organic compounds are composite with the *stem* telling us the number of carbon atoms in the main chain of the molecule. These are listed in the middle column of Table 1.8. For a straight chain *alkane*, the name is completed by adding *-ane* to the prefix in the table. Compounds **1.10** ($C_3H_8$) and **1.11** ($C_7H_{16}$) are both alkanes. Using the stems from Table 1.8, **1.10** with a 3-carbon chain is called propane, and **1.11** with a 7-carbon chain is called heptane.

### Basic organic classifications: functional groups

A *functional group* in a molecule imparts a characteristic reactivity to the compound. The functional group in an alkene is the C=C double bond

**Table 1.8** Numerical descriptors.

| Number | Stem used to give the number of C atoms in an organic carbon chain | Prefix used to describe the number of groups or substituents |
|---|---|---|
| 1 | meth- | mono- |
| 2 | eth- | di- |
| 3 | prop- | tri- |
| 4 | but- | tetra- |
| 5 | pent- | penta- |
| 6 | hex- | hexa- |
| 7 | hept- | hepta- |
| 8 | oct- | octa- |
| 9 | non- | nona- |
| 10 | dec- | deca- |
| 11 | undec- | undeca- |
| 12 | dodec- | dodeca- |
| 13 | tridec- | trideca- |
| 14 | tetradec- | tetradeca- |
| 15 | pentadec- | pentadeca- |
| 16 | hexadec- | hexadeca- |
| 17 | heptadec- | heptadeca- |
| 18 | octadec- | octadeca- |
| 19 | nonadec- | nonadeca- |
| 20 | icos- | icosa- |

and, in an alcohol, the functional group is the $-OH$ unit. The organic functional groups that we will describe in this book are listed in Table 1.9. The presence of most of these groups is recognized by using an instrumental technique such as infrared, electronic or nuclear magnetic resonance spectroscopy – see Chapters 12–14.

(1.10)                                    (1.11)

## Basic inorganic nomenclature

The aim of the IUPAC nomenclature is to provide a compound or an ion with a name that is unambiguous. One problem that we face when dealing with some inorganic elements is the possibility of a variable oxidation state. A simple example is that of distinguishing between the two common oxides of carbon, i.e. carbon monoxide and carbon dioxide. The use of 'mono-' and 'di-' indicates that the compounds are $CO$ and $CO_2$ respectively. The accepted numerical prefixes are listed in Table 1.8. Note that 'di-' should be used in preference to 'bi-'.

**Binary compounds 1**    A binary compound is composed of only *two* types of elements. We deal first with cases where there is no ambiguity over oxidation states of the elements present, e.g. *s*-block metal. Examples of such binary compounds include $NaCl$, $CaO$, $HCl$, $Na_2S$ and $MgBr_2$. The formula should be written with the more electropositive element (often a metal) placed first. The names follow directly:

| | |
|---|---|
| $NaCl$ | sodium chloride |
| $Na_2S$ | sodium sulfide |
| $CaO$ | calcium oxide |
| $MgBr_2$ | magnesium bromide |
| $HCl$ | hydrogen chloride |

A ***binary compound*** is composed of two types of elements.

**Anions**    The endings '-*ide*', '-*ite*' and '-*ate*' generally signify an anionic species. Some examples are listed in Table 1.10. The endings '-ate' and '-ite' *tend* to indicate the presence of oxygen in the anion (i.e. an oxoanion) and are used for anions that are derived from oxoacids; e.g. the oxoanion derived from sulfuric acid is a sulfate.

There is more than one accepted method of distinguishing between the different oxoanions of elements such as sulfur, nitrogen and phosphorus (Table 1.10). Older names such as sulfate, sulfite, nitrate and nitrite are still accepted within the IUPAC guidelines. It is more informative, however, to incorporate the oxidation state of the element that is combining with oxygen, and an alternative name for sulfate is tetraoxosulfate(VI). This shows not only the oxidation state of the sulfur atom, but the number of oxygen atoms as well. A third accepted option is to use the name tetra-oxosulfate(2−). In *Chemistry*, we have made every effort to stay within the

Table **1.9** Selected functional groups for organic molecules.

| Name of functional group | Functional group | Example; where it is in common use, a trivial name is given in pink |
|---|---|---|
| Alcohol | —OH | Ethanol ($CH_3CH_2OH$) |
| Aldehyde | $-C\overset{O}{\underset{H}{\big\backslash}}$ | Ethanal ($CH_3CHO$)<br>Acetaldehyde |
| Ketone | $-C\overset{O}{\underset{R}{\big\backslash}}$   $R \neq H$ | Propanone ($CH_3COCH_3$)<br>Acetone |
| Carboxylic acid | $-C\overset{O}{\underset{O-H}{\big\backslash}}$ | Ethanoic acid ($CH_3CO_2H$)<br>Acetic acid |
| Ester | $-C\overset{O}{\underset{O-R}{\big\backslash}}$<br>e.g. R = alkyl | Ethyl ethanoate ($CH_3CO_2C_2H_5$)<br>Ethyl acetate |
| Ether | $R'-O\overset{R}{\diagup}$  R = R'<br>or R ≠ R' | Diethyl ether ($C_2H_5OC_2H_5$) |
| Amine | —$NH_2$ | Ethylamine ($CH_3CH_2NH_2$) |
| Amide | $-C\overset{O}{\underset{NH_2}{\big\backslash}}$ | Ethanamide ($CH_3CONH_2$)<br>Acetamide |
| Halogenoalkane | —X<br>X = F, Cl, Br, I | Bromoethane ($CH_3CH_2Br$) |
| Acid chloride | $-C\overset{O}{\underset{Cl}{\big\backslash}}$ | Ethanoyl chloride ($CH_3COCl$)<br>Acetyl chloride |
| Nitrile | —C≡N | Ethanenitrile ($CH_3CN$)<br>Acetonitrile |
| Nitro | —$NO_2$ | Nitromethane ($CH_3NO_2$) |
| Thiol | —SH | Ethanethiol ($CH_3CH_2SH$) |

Table 1.10 Names of some common anions. In some cases, more than one name is accepted by the IUPAC.

| Formula of anion | Name of anion |
|---|---|
| $H^-$ | Hydride |
| $[OH]^-$ | Hydroxide |
| $F^-$ | Fluoride |
| $Cl^-$ | Chloride |
| $Br^-$ | Bromide |
| $I^-$ | Iodide |
| $O^{2-}$ | Oxide |
| $S^{2-}$ | Sulfide |
| $Se^{2-}$ | Selenide |
| $N^{3-}$ | Nitride |
| $N_3^-$ | Azide |
| $P^{3-}$ | Phosphide |
| $[CN]^-$ | Cyanide |
| $[NH_2]^-$ | Amide |
| $[OCN]^-$ | Cyanate |
| $[SCN]^-$ | Thiocyanate |
| $[SO_4]^{2-}$ | Sulfate *or* tetraoxosulfate(VI) |
| $[SO_3]^{2-}$ | Sulfite *or* trioxosulfate(IV) |
| $[NO_3]^-$ | Nitrate *or* trioxonitrate(V) |
| $[NO_2]^-$ | Nitrite *or* dioxonitrate(III) |
| $[PO_4]^{3-}$ | Phosphate *or* tetraoxophosphate(V) |
| $[PO_3]^{3-}$ | Phosphite *or* trioxophosphate(III) |
| $[ClO_4]^-$ | Perchlorate *or* tetraoxochlorate(VII) |
| $[CO_3]^{2-}$ | Carbonate *or* trioxocarbonate(IV) |

IUPAC recommendations while retaining the most common alternatives, e.g. sulfate.

**Oxidation states**    The oxidation state is very often indicated by using the Stock system of Roman numerals. The numeral is always an integer and is placed after the name of the element to which it refers; Table 1.10 shows its application to some oxoanions. The oxidation number can be zero, positive or negative.[§] An oxidation state is assumed to be positive unless otherwise indicated by the use of a negative sign. Thus, (III) is taken to read '(+III)'; but for the negative state, write (−III).

In a formula, the oxidation state is written as a superscript (e.g. $[Mn^{VII}O_4]^-$) but in a name, it is written on the line (e.g. iron(II) bromide). Its use is important when the name could be ambiguous (see below).

**Binary compounds 2**    We look now at binary compounds where there could be an ambiguity over the oxidation state of the more electropositive element (often a metal). Examples of such compounds include $FeCl_3$, $SO_2$, $SO_3$, ClF, $ClF_3$ and $SnCl_2$. Simply writing 'iron chloride' does not distinguish between the chlorides of iron(II) and iron(III), and for $FeCl_3$ it is necessary to write iron(III) chloride. Another accepted name is iron trichloride.

The oxidation state of sulfur in $SO_2$ can be seen immediately in the name sulfur(IV) oxide, but also acceptable is the name sulfur dioxide. Similarly, $SO_3$ can be named sulfur(VI) oxide or sulfur trioxide.

[§] A zero oxidation state is signified by 0, although this is not a Roman numeral.

Table 1.11 The names of some common, non-metallic cations.

| Formula of cation | Name of cation |
| --- | --- |
| $H^+$ | Hydrogen ion |
| $[H_3O]^+$ | Oxonium ion |
| $[NH_4]^+$ | Ammonium ion |
| $[NO]^+$ | Nitrosyl ion |
| $[NO_2]^+$ | Nitryl ion |
| $[N_2H_5]^+$ | Hydrazinium ion |

Accepted names for $ClF$, $ClF_3$ and $SnCl_2$ are:

| | |
| --- | --- |
| $ClF$ | chlorine(I) fluoride or chlorine monofluoride |
| $ClF_3$ | chlorine(III) fluoride or chlorine trifluoride |
| $SnCl_2$ | tin(II) chloride or tin dichloride |

**Cations** Cations of metals where the oxidation state does not usually vary, notably the *s*-block elements, may be named by using the name of the metal itself (e.g. sodium ion, barium ion), although the charge may be indicated (e.g. sodium(I) ion or sodium(1+) ion, barium(II) ion or barium(2+) ion).

Where there may be an ambiguity, the charge must be shown (e.g. iron(II) or iron(2+) ion, copper(II) or copper(2+) ion, thallium(I) or thallium(1+) ion).[§]

The names of polyatomic cations are introduced as they appear in the textbook but Table 1.11 lists some of the most common inorganic, non-metallic cations with which you may already be familiar. Look for the ending '-ium'; this often signifies the presence of a cation, although remember that '-ium' is a common ending in the name of elemental metals (see Section 1.5). Many metal ions, in particular those in the *d*-block, occur as complex ions; these are described in Chapter 23.

| 1.19 | **Final comments** |
| --- | --- |

The aim of this first chapter is to provide a point of reference for basic chemical definitions, ones that you have probably encountered before beginning a first year university chemistry course. If you find later in the book that a concept appears to be 'assumed', you should find some revision material to help you in Chapter 1. Section 1.18 gives some basic guidelines for naming organic and inorganic compounds, and more detailed nomenclature appears as the book progresses.

We have deliberately *not* called Chapter 1: 'Introduction'. There is often a tendency to pass through chapters so-labelled without paying attention to them. In this text, Chapter 1 is designed to help you and to remind you of basic issues.

---

[§] An older form of nomenclature which is commonly encountered still uses the suffix '-ous' to describe the lower oxidation state and '-ic' for the higher one. Thus, copper(I) is cuprous and copper(II) is cupric. This system is unambiguous only when the metal exhibits only two oxidation states.

## PROBLEMS

Use values of $A_r$ from the front inside cover of the book.

**1.1** What is 0.0006 m in (a) mm, (b) pm, (c) cm, (d) nm?

**1.2** A typical C=O bond distance in an aldehyde is 122 pm. What is this in nm?

**1.3** The relative molecular mass of NaCl is 58.44 and its density is 2.16 g cm$^{-3}$. What is the volume of 1 mole of NaCl in m$^3$?

**1.4** The equation $E = h\nu$ relates the Planck constant ($h$) to energy and frequency. Determine the SI units of the Planck constant.

**1.5** Kinetic energy is given by the equation: $E = \frac{1}{2}mv^2$. By going back to the base SI units, show that the units on the left- and right-hand sides of this equation are compatible.

**1.6** Calculate the relative atomic mass of a sample of naturally occurring boron which contains 19.9% $^{10}_{5}$B and 80.1% $^{11}_{5}$B. Accurate masses of the isotopes to 3 sig. fig. are 10.0 and 11.0.

**1.7** The mass spectrum of molecular bromine shows three lines for the parent ion, Br$_2$$^+$. The isotopes for bromine are $^{79}_{35}$Br (50%) and $^{81}_{35}$Br (50%). Explain why there are three lines and predict their mass values and relative intensities. Predict what the mass spectrum of HBr would look like; isotopes of hydrogen are given in Section 1.7. (Ignore fragmentation; see Chapter 10.)

**1.8** Convert the volume of each of the following to conditions of standard temperature (273 K) and pressure (1 bar = $1.00 \times 10^5$ Pa) and give your answer in m$^3$ in each case:
(a) 30.0 cm$^3$ of CO$_2$ at 290 K and 101 325 Pa (1 atm)
(b) 5.30 dm$^3$ of H$_2$ at 298 K and 100 kPa (1 bar)
(c) 0.300 m$^3$ of N$_2$ at 263 K and 102 kPa
(d) 222 m$^3$ of CH$_4$ at 298 K and 200 000 Pa (2 bar)

**1.9** The partial pressure of helium in a 50.0 dm$^3$ gas mixture at 285 K and $10^5$ Pa is $4.0 \times 10^4$ Pa. How many moles of helium are present?

**1.10** A 20.0 dm$^3$ sample of gas at 273 K and 2.0 bar pressure contains 0.50 moles N$_2$ and 0.70 moles Ar. What is the partial pressure of each gas, and are there any other gases in the sample? (Volume of one mole of ideal gas at 273 K, $1.00 \times 10^5$ Pa (1 bar) = 22.7 dm$^3$.)

**1.11** Determine the amount (in moles) present in each of the following: (a) 0.44 g PF$_3$, (b) 1.00 dm$^3$ gaseous PF$_3$ at 293 K and $2.00 \times 10^5$ Pa, (c) 3.480 g MnO$_2$, (d) 0.0420 g MgCO$_3$. (Volume of 1 mole of ideal gas at 273 K, $10^5$ Pa = 22.7 dm$^3$.)

**1.12** What mass of solid is required to prepare 100.0 cm$^3$ of each of the following solutions:

(a) 0.0100 mol dm$^{-3}$ KI; (b) 0.200 mol dm$^{-3}$ NaCl; (c) 0.0500 mol dm$^{-3}$ Na$_2$SO$_4$?

**1.13** With reference to the periodic table, write down the likely formulae of compounds formed between: (a) sodium and iodine, (b) magnesium and chlorine, (c) magnesium and oxygen, (d) calcium and fluorine, (e) lithium and nitrogen, (f) calcium and phosphorus, (g) sodium and sulfur and (h) hydrogen and sulfur.

**1.14** Use the information in the periodic table to predict the likely formulae of the oxide, chloride, fluoride and hydride formed by aluminium.

**1.15** Give balanced equations for the formation of each of the compounds in problems 1.13 and 1.14 from their constituent elements.

**1.16** What do you understand by each of the following terms: proton, electron, neutron, nucleus, atom, radical, ion, cation, anion, molecule, covalent bond, compound, isotope, allotrope?

**1.17** Suggest whether you think each of the following species will exhibit covalent or ionic bonding. Which of the species are compounds and which are molecular: (a) NaCl; (b) N$_2$; (c) SO$_3$; (d) KI; (e) NO$_2$; (f) Na$_2$SO$_4$; (g) [MnO$_4$]$^-$; (h) CH$_3$OH; (i) CO$_2$; (j) C$_2$H$_6$; (k) HCl; (l) [SO$_4$]$^{2-}$?

**1.18** Determine the oxidation state of nitrogen in each of the following oxides: (a) N$_2$O; (b) NO; (c) NO$_2$, (d) N$_2$O$_3$; (e) N$_2$O$_4$; (f) N$_2$O$_5$.

**1.19** In each reaction below, assign the oxidation and reduction steps, and, for (b)–(g), show that the changes in oxidation states for the oxidation and reduction processes balance:
(a) Cu$^{2+}$(aq) + 2e$^-$ ⟶ Cu(s)
(b) Mg(s) + H$_2$SO$_4$(aq) ⟶ MgSO$_4$(aq) + H$_2$(g)
(c) 2Ca(s) + O$_2$(g) ⟶ 2CaO(s)
(d) 2Fe(s) + 3Cl$_2$(g) ⟶ 2FeCl$_3$(s)
(e) Cu(s) + 2AgNO$_3$(aq) ⟶ Cu(NO$_3$)$_2$(aq) + 2Ag(s)
(f) CuO(s) + H$_2$(g) ⟶ Cu(s) + H$_2$O(g)
(g) [MnO$_4$]$^-$(aq) + 5Fe$^{2+}$(aq) + 8H$^+$(aq) ⟶ Mn$^{2+}$(aq) + 5Fe$^{3+}$(aq) + 4H$_2$O(l)

**1.20** (a) In a compound, oxygen is usually assigned an oxidation state of $-2$. What is the formal oxidation state in the allotropes O$_2$ and O$_3$, in the compound H$_2$O$_2$, and in the ions [O$_2$]$^{2-}$ and [O$_2$]$^+$?
(b) Determine the oxidation and reduction steps during the decomposition of hydrogen peroxide which occurs by the following reaction:

$$2H_2O_2(l) \longrightarrow 2H_2O(l) + O_2(g)$$

**1.21** Give a systematic name for each of the following compounds: (a) Na$_2$CO$_3$; (b) FeBr$_3$; (c) CoSO$_4$; (d) BaCl$_2$; (e) Fe$_2$O$_3$; (f) Fe(OH)$_2$; (g) LiI; (h) KCN; (i) KSCN; (j) Ca$_3$P$_2$.

**1.22**  Write down the formula of each of the following compounds: (a) nickel(II) iodide; (b) ammonium nitrate; (c) barium hydroxide; (d) iron(III) sulfate; (e) iron(II) sulfite; (f) aluminium hydride; (g) lead(IV) oxide; (h) tin(II) sulfide.

**1.23**  How many atoms make up the carbon chain in (a) octane, (b) hexane, (c) propane, (d) decane, (e) butane?

**1.24**  Balance the following equations:
(a) $C_4H_{10} + O_2 \longrightarrow CO_2 + H_2O$
(b) $SO_2 + O_2 \longrightarrow SO_3$
(c) $HCl + Ca \longrightarrow H_2 + CaCl_2$
(d) $Na_2CO_3 + HCl \longrightarrow NaCl + CO_2 + H_2O$
(e) $HNO_3 + Mg \longrightarrow Mg(NO_3)_2 + H_2$
(f) $H_3PO_4 + NaOH \longrightarrow Na_2HPO_4 + H_2O$

**1.25**  Find $x$, $y$ and $z$ in the following reactions:
(a) $2CO + O_2 \longrightarrow 2CO_x$
(b) $N_2 + xH_2 \longrightarrow yNH_3$
(c) $Mg + 2HNO_3 \longrightarrow Mg(NO_3)_x + H_2$
(d) $xH_2O_2 \longrightarrow yH_2O + zO_2$
(e) $xHCl + CaCO_3 \longrightarrow CaCl_y + CO_2 + H_2O$
(f) $xNaOH + H_2SO_4 \longrightarrow Na_ySO_4 + zH_2O$
(g) $MnO_2 + xHCl \longrightarrow MnCl_2 + Cl_2 + yH_2O$
(h) $xNa_2S_2O_3 + I_2 \longrightarrow yNaI + zNa_2S_4O_6$

**1.26**  Balance the following equations.
(a) $Fe + Cl_2 \longrightarrow FeCl_3$
(b) $SiCl_4 + H_2O \longrightarrow SiO_2 + HCl$
(c) $Al_2O_3 + NaOH + H_2O \longrightarrow Na_3Al(OH)_6$
(d) $K_2CO_3 + HNO_3 \longrightarrow KNO_3 + H_2O + CO_2$
(e) $Fe_2O_3 + CO \xrightarrow{\text{heat}} Fe + CO_2$
(f) $H_2C_2O_4 + KOH \longrightarrow K_2C_2O_4 + H_2O$

**1.27**  Balance the following equations.
(a) $AgNO_3 + MgCl_2 \longrightarrow AgCl + Mg(NO_3)_2$
(b) $Pb(O_2CCH_3)_2 + H_2S \longrightarrow PbS + CH_3CO_2H$
(c) $BaCl_2 + K_2SO_4 \longrightarrow BaSO_4 + KCl$
(d) $Pb(NO_3)_2 + KI \longrightarrow PbI_2 + KNO_3$
(e) $Ca(HCO_3)_2 + Ca(OH)_2 \longrightarrow CaCO_3 + H_2O$

**1.28**  Balance the following equations.
(a) $C_3H_8 + Cl_2 \longrightarrow C_3H_5Cl_3 + HCl$
(b) $C_6H_{14} + O_2 \longrightarrow CO_2 + H_2O$
(c) $C_2H_5OH + Na \longrightarrow C_2H_5ONa + H_2$
(d) $C_2H_2 + Br_2 \longrightarrow C_2H_2Br_4$
(e) $CaC_2 + H_2O \longrightarrow Ca(OH)_2 + C_2H_2$

**1.29**  In each of the following, mixing the aqueous ions shown will produce a precipitate. Write the formula of the neutral product, and then balance the equation.
(a) $Ag^+(aq) + Cl^-(aq) \longrightarrow$
(b) $Mg^{2+}(aq) + [OH]^-(aq) \longrightarrow$
(c) $Pb^{2+}(aq) + S^{2-}(aq) \longrightarrow$
(d) $Fe^{3+}(aq) + [OH]^-(aq) \longrightarrow$
(e) $Ca^{2+}(aq) + [PO_4]^{3-}(aq) \longrightarrow$
(f) $Ag^+(aq) + [SO_4]^{2-}(aq) \longrightarrow$

**1.30**  Balance each of the following equations:
(a) $Fe^{3+} + H_2 \longrightarrow Fe^{2+} + H^+$
(b) $Cl_2 + Br^- \longrightarrow Cl^- + Br_2$
(c) $Fe^{2+} + [Cr_2O_7]^{2-} + H^+ \longrightarrow$
$$Fe^{3+} + Cr^{3+} + H_2O$$
(d) $NH_2OH + Fe^{3+} \longrightarrow N_2O + Fe^{2+} + H_2O + H^+$
(e) $[S_2O_3]^{2-} + I_2 \longrightarrow [S_4O_6]^{2-} + I^-$
(f) $[MoO_4]^{2-} + [PO_4]^{3-} + H^+ \longrightarrow$
$$[PMo_{12}O_{40}]^{3-} + H_2O$$
(g) $HNO_3 + H_2SO_4 \longrightarrow$
$$[H_3O]^+ + [NO_2]^+ + [HSO_4]^-$$

**1.31**  Balance the following equation for the reaction of magnesium with dilute nitric acid:
$$Mg(s) + HNO_3(aq) \longrightarrow Mg(NO_3)_2(aq) + H_2(g)$$
Use the balanced equation to determine the mass of Mg that will completely react with $100.0 \, cm^3$ $0.50 \, M$ nitric acid.

**1.32**  Balance the following equation for the reaction of aqueous phosphoric acid with sodium hydroxide:
$$H_3PO_4(aq) + NaOH(aq) \longrightarrow$$
$$Na_3PO_4(aq) + H_2O(l)$$
$15.0 \, cm^3$ aqueous phosphoric acid of concentration $0.200 \, mol \, dm^{-3}$ is added to $50.0 \, cm^3$ aqueous NaOH of concentration $2.00 \, mol \, dm^{-3}$. Which reagent is in excess, and how many moles of this reagent remain unreacted?

**1.33**  Balance the following equation for the precipitation of silver chromate:
$$AgNO_3(aq) + K_2CrO_4(aq) \longrightarrow$$
$$Ag_2CrO_4(s) + KNO_3(aq)$$
$5.00 \, g$ of $K_2CrO_4$ is dissolved in water and the volume of the solution is made up to $100.0 \, cm^3$. $25.0 \, cm^3$ of a $0.100 \, mol \, dm^{-3}$ solution of $AgNO_3$ is added to the solution of $K_2CrO_4$. Determine the mass of $Ag_2CrO_4$ that is formed.

**1.34**  Balance the following equation:
$$[C_2O_4]^{2-}(aq) + [MnO_4]^-(aq) + H^+(aq) \longrightarrow$$
$$Mn^{2+}(aq) + CO_2(g) + H_2O(l)$$
What volume of $0.200 \, M$ aqueous $KMnO_4$ will react completely with $25.0 \, cm^3$ $0.200 \, M$ aqueous $K_2C_2O_4$ in the presence of excess acid?

**1.35**  $5.00 \, g$ of solid $CaCO_3$ is thermally decomposed in the following reaction:
$$CaCO_3(s) \xrightarrow{\text{heat}} CaO(s) + CO_2(g)$$
What mass of CaO is formed? This oxide reacts with water to give $Ca(OH)_2$. Write a balanced equation for this process.

**1.36** Balance the following equation for the reaction of sodium thiosulfate with diiodine:

$$Na_2S_2O_3 + I_2 \longrightarrow Na_2S_4O_6 + NaI$$

Diiodine is insoluble in water, but dissolves in aqueous potassium iodide solution. $0.0250 \, dm^3$ of a solution of $I_2$ in aqueous KI reacts exactly with $0.0213 \, dm^3$ $0.120 \, M$ sodium thiosulfate solution. What is the concentration of the diiodine solution?

**1.37** An organic compound **A** contains 40.66% C, 23.72% N and 8.53% H. The compound also contains oxygen. Determine the empirical formula of the compound. The molecular mass of **A** is 59.07. What is its molecular formula?

**1.38** (a) A chloride of platinum, $PtCl_x$, contains 26.6% Cl. What is the oxidation state of the platinum in this compound?

(b) Two oxides of iron, **A** and **B**, contain 69.94 and 72.36% Fe, respectively. For **A**, $M_r = 159.70$; for **B**, $M_r = 231.55$. What are the empirical formulae of **A** and **B**?

**1.39** Crystalline copper(II) sulfate contains water. The formula of the crystalline solid is $CuSO_4 \cdot xH_2O$. Determine $x$ if the crystals contain 12.84% S and 4.04% H.

**1.40** (a) Glucose, $C_xH_yO_z$, contains 40.00% C and 6.71% H. What is the empirical formula of glucose? If the molecular mass of glucose is 180.16, determine its molecular formula.

(b) A fluoride of tungsten, $WF_x$, contains 38.27% F. What is the oxidation state of tungsten in this compound?

# 2 Thermochemistry

## Topics

- Enthalpy changes for reactions
- Exothermic and endothermic changes
- Calorimetry
- Standard enthalpy of formation
- Enthalpy of combustion
- Hess's Law of Constant Heat Summation
- Thermodynamic and kinetic stability
- Enthalpies of fusion and vaporization
- An introduction to intermolecular interactions

## 2.1 Factors that control reactions

Much of chemistry is concerned with chemical reactions. The factors that control whether a reaction will or will not take place fall into two categories: *thermodynamic* and *kinetic*. Thermodynamic concepts relate to the energetics of a system, while kinetics deal with the speed at which a reaction occurs. Observations of reaction kinetics are related to the mechanism of the reaction, and this describes the way in which we believe that the atoms and molecules behave during a reaction. We look in detail at kinetics in Chapter 15. We often write equations for chemical reactions with a forward arrow (e.g. equation 2.1) in order to indicate that the reactants take part in a reaction that leads to products, and that the process *goes to completion*.

$$Zn(s) + H_2SO_4(aq) \longrightarrow ZnSO_4(aq) + H_2(g) \tag{2.1}$$

However, many reactions do *not* reach completion. Instead, reactants and products lie in a state of *equilibrium* in which both forward and back reactions take place. Actually, all reactions are equilibria and no reaction under equilibrium conditions goes completely to the right-hand side. We consider equilibria in detail in Chapter 16. The position of an equilibrium is governed by thermodynamic factors. Whether a reaction is favourable, and to what extent it will reach completion, can be assessed from the sign and magnitude of the *change in Gibbs energy*, $\Delta G$, for the overall reaction. Chemical thermodynamics is the topic of Chapter 17. Although, strictly, it is the change in Gibbs energy that gives us information about the favourability of a reaction, we can also gain some insight from *thermochemical data*, i.e. the changes in heat that accompany chemical reactions. The study of heat changes for chemical reactions is called *thermochemistry*. The heat change that accompanies a reaction can be readily determined experimentally by measuring the associated change in temperature. As a consequence, thermo-chemistry is usually the first introduction that a student has to the more detailed subject of thermodynamics. In this chapter, we look at changes in heat (*enthalpy*), not only for chemical reactions, but also for phase transitions. We also give a brief introduction to the enthalpy terms that are associated with interactions between molecules.

| **2.2** | **Change in enthalpy of a reaction** |

The symbol $\Delta$ is used to signify the 'change in' a quantity, e.g. $\Delta H$ means 'change in enthalpy'.

When most chemical reactions occur, heat is either taken in from the surroundings, causing the temperature of the reaction mixture to rise, or is given out to the surroundings. Many common chemical reactions are carried out at constant pressure (e.g. in an open beaker or flask) and under these conditions, the heat transfer, $q$, is equal to the *enthalpy change*, $\Delta H$. The terms *heat* and *enthalpy* are often used interchangeably, although strictly, this is only true under conditions of constant pressure.

> The **enthalpy change**, $\Delta H$, that accompanies a reaction is the amount of heat liberated or absorbed as a reaction proceeds at a given temperature, $T$, at constant pressure.

> The SI units of enthalpy, $H$, are joules, J. Usually, we work with molar quantities and then the units of $H$ and $\Delta H$ are $J\,mol^{-1}$ or $kJ\,mol^{-1}$.

### Standard enthalpy change

The *standard enthalpy change of a reaction* refers to the enthalpy change when all the reactants and products are in their *standard states*. The notation for this thermochemical quantity is $\Delta_r H^{\circ}(T)$ where the subscript 'r' stands for 'reaction', the superscript 'o' means 'standard state conditions', and $(T)$ means 'at temperature $T$'. This type of notation is found for other thermo-dynamic functions that we meet later on.

The **standard state of a substance** is its most stable state under a pressure of 1 bar ($1.00 \times 10^5$ Pa) and at a specified temperature, $T$.

The *standard state of a substance* is its most thermodynamically stable state under a pressure of 1 bar ($1.00 \times 10^5$ Pa) and at a specified temperature, $T$. Most commonly, $T = 298.15\,K$, and the notation for the standard enthalpy change of a reaction at 298.15 K is then $\Delta_r H^{\circ}(298.15\,K)$. It is usually sufficient to write $\Delta_r H^{\circ}(298\,K)$. *Do not confuse standard thermodynamic temperature with the temperature used for the standard temperature and pressure conditions of a gas (Section 1.9).* We return to standard states in Section 2.4.

### Exothermic and endothermic processes

When reactions occur, they may release heat to the surroundings or may absorb heat from the surroundings. By definition, a negative value of $\Delta H$ corresponds to heat given out during a reaction (equation 2.2). Such a

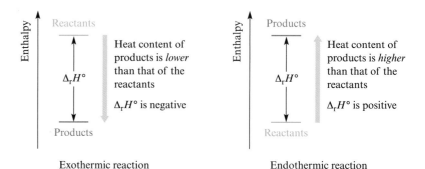

**Fig. 2.1** Enthalpy level diagrams for exothermic and endothermic reactions.

reaction is said to be *exothermic*. Whenever a fuel is burnt, an exothermic reaction occurs.

$$\text{Mg(s)} + \tfrac{1}{2}\text{O}_2(\text{g}) \longrightarrow \text{MgO(s)} \qquad \Delta_r H^\circ(298\,\text{K}) = -602\,\text{kJ}\,\text{mol}^{-1} \qquad (2.2)$$

Although we avoided the use of fractional coefficients when balancing equations in Chapter 1, we now need to use $\tfrac{1}{2}\text{O}_2$ on the left-hand side of equation 2.2 because we are considering the enthalpy change for the formation of *one mole* of MgO. The notation $\text{kJ}\,\text{mol}^{-1}$ refers to the equation as it is written. If we had written equation 2.3 instead of equation 2.2, then $\Delta_r H^\circ(298\,\text{K}) = -1204\,\text{kJ}\,\text{mol}^{-1}$.

$$2\text{Mg(s)} + \text{O}_2(\text{g}) \longrightarrow 2\text{MgO(s)} \qquad (2.3)$$

A positive value of $\Delta H$ corresponds to heat being absorbed from the surroundings and the reaction is said to be *endothermic*. For example, when NaCl dissolves in water, a small amount of heat is absorbed (equation 2.4).

$$\text{NaCl(s)} \xrightarrow{\text{H}_2\text{O}} \text{NaCl(aq)} \qquad \Delta_r H^\circ(298\text{K}) = +3.9\,\text{kJ}\,\text{mol}^{-1} \qquad (2.4)$$

Consider a general reaction in which reactants combine to give products, and for which the standard enthalpy change is $\Delta_r H^\circ(298\,\text{K})$. If the heat content of the reactants is greater than the heat content of the products, heat must be released and the reaction is exothermic. On the other hand, if the heat content of the products is greater than that of the reactants, heat must be absorbed and the reaction is endothermic. Each of these situations is represented schematically in the enthalpy level diagrams in Figure 2.1.

> Heat is given out (liberated) in an *exothermic reaction* ($\Delta H$ is negative).
>
> Heat is taken in (absorbed) in an *endothermic reaction* ($\Delta H$ is positive).

## 2.3    Measuring changes in enthalpy: calorimetry

> A *calorimeter* is used to measure the heat transfer that accompanies a chemical reaction. The technique is called *calorimetry*.

The heat that is given out or taken in when a chemical reaction occurs can be measured using a calorimeter. A simple, constant-pressure calorimeter for measuring heat changes for reactions in solution is shown in Figure 2.2. The container is an expanded polystyrene cup with a lid. This material provides insulation which ensures that heat loss to, or gain from, the surroundings is minimized; the outer cup in Figure 2.2 provides additional insulation. As the reaction takes place, the thermometer records any change in temperature. The relationship between the temperature change

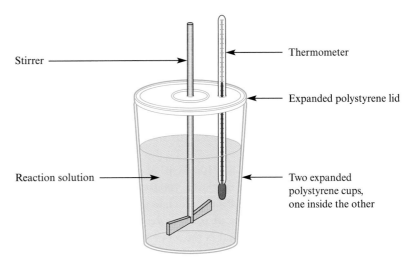

**Fig. 2.2** A simple, constant-pressure calorimeter used for measuring heat changes for reactions in solution. The outer container provides additional insulation.

The *specific heat capacity*, $C$, of a substance is the heat required to raise the temperature of unit mass of the substance by one kelvin.

SI units of $C$ are $J K^{-1} kg^{-1}$, but units of $J K^{-1} g^{-1}$ are often more convenient.

For water, $C = 4.18 \, J K^{-1} g^{-1}$.

and the heat change is given in equation 2.5 where $C$ is the specific heat capacity of the solution. Since the reaction is carried out at constant pressure, the heat change is equal to the enthalpy change. For dilute aqueous solutions, it is usually sufficient to assume that the specific heat capacity of the solution is the same as for water: $C_{water} = 4.18 \, J K^{-1} g^{-1}$. Worked examples 2.1–2.3 illustrate the use of a simple calorimeter to measure enthalpy changes of reaction. In each worked example, we assume that changes in enthalpy of the reaction affect only the temperature of the solution. We assume that no heat is used to change the temperature of the calorimeter itself. Where a calorimeter is made from expanded polystyrene cups, this is a reasonable assumption because the specific heat capacity of the calorimeter material is so small. However, the approximation is not valid for many types of calorimeter and such pieces of apparatus must be calibrated before use. Measurements made in the crude apparatus shown in Figure 2.2 are not accurate, and more specialized calorimeters must be used if accurate results are required.

$$\text{Heat change in J} = (\text{Mass in g})$$
$$\times (\text{Specific heat capacity in } J K^{-1} g^{-1})$$
$$\times (\text{Change in temperature in K})$$

$$\text{Heat change in J} = (m \, g) \times (C \, J K^{-1} g^{-1}) \times (\Delta T \, K) \tag{2.5}$$

Before using equation 2.5, we must emphasize that a *rise* in temperature occurs in an exothermic reaction and corresponds to a negative value of $\Delta H$; a *fall* in temperature occurs in an endothermic reaction and corresponds to a positive value of $\Delta H$.

| Worked example 2.1 | *Heating a known mass of water* |
| --- | --- |

**Calculate the heat required to raise the temperature of 85.0 g of water from 298.0 K to 303.0 K. [Data: $C_{water} = 4.18 \, J K^{-1} g^{-1}$]**

The rise in temperature = $303.0 \, K - 298.0 \, K = 5.0 \, K$

The heat required is given by:

$$\text{Heat in J} = (m\,\text{g}) \times (C\,\text{J K}^{-1}\,\text{g}^{-1}) \times (\Delta T\,\text{K})$$
$$= (85.0\,\text{g}) \times (4.18\,\text{J K}^{-1}\,\text{g}^{-1}) \times (5.0\,\text{K})$$
$$= 1800\,\text{J or } 1.8\,\text{kJ} \quad \text{(to 2 sig. fig.)}$$

---

**Worked example 2.2**   *Estimation of the enthalpy of a reaction*

When 100.0 cm$^3$ of an aqueous solution of nitric acid, HNO$_3$ (1.0 mol dm$^{-3}$), is mixed with 100.0 cm$^3$ of an aqueous solution of sodium hydroxide, NaOH (1.0 mol dm$^{-3}$), in a calorimeter of the type shown in Figure 2.2, a temperature rise of 6.9 K is recorded. (a) Is the reaction exothermic or endothermic? (b) What is the value of $\Delta H$ for this reaction in kJ per mole of HNO$_3$? [Data: density of water = 1.00 g cm$^{-3}$; $C_{\text{water}} = 4.18\,\text{J K}^{-1}\,\text{g}^{-1}$]

(a) A *rise* in temperature is observed. Therefore, the reaction is exothermic.
(b) Total volume of solution = 100.0 + 100.0 = 200.0 cm$^3$.
Assume that the density of the aqueous solution ≈ density of water.

$$\text{Mass of solution in g} = (\text{Volume in cm}^3) \times (\text{Density in g cm}^{-3})$$
$$= (200.0\,\text{cm}^3) \times (1.00\,\text{g cm}^{-3})$$
$$= 200\,\text{g} \quad \text{(to 3 sig. fig.)}$$

The heat change can now be found:

$$\text{Heat change} = (m\,\text{g}) \times (C\,\text{J K}^{-1}\,\text{g}^{-1}) \times (\Delta T\,\text{K})$$
$$= (200\,\text{g}) \times (4.18\,\text{J K}^{-1}\,\text{g}^{-1}) \times (6.9\,\text{K})$$
$$= 5800\,\text{J} \quad \text{(to 2 sig. fig.)}$$

To find $\Delta H$ per mole of HNO$_3$, first determine how many moles of HNO$_3$ are involved in the reaction:

$$\text{Amount of HNO}_3 \text{ in moles} = (\text{Volume in dm}^3)$$
$$\times (\text{Concentration in mol dm}^{-3})$$
$$= (100.0 \times 10^{-3}\,\text{dm}^3) \times (1.0\,\text{mol dm}^{-3})$$
$$= 0.10\,\text{mol}$$

When 0.10 moles of HNO$_3$ react, the heat released is 5800 J. Therefore when 1.0 mole of HNO$_3$ reacts:

$$\text{Heat released} = \frac{5800}{0.10} = 58\,000\,\text{J}$$

The reaction is carried out at constant pressure, and therefore the heat change equals the enthalpy change. The reaction is exothermic and $\Delta H$ is negative.

$$\Delta H = -58\,000\,\text{J mol}^{-1} = -58\,\text{kJ mol}^{-1} \quad \text{(to 2 sig. fig.)}$$

---

**Worked example 2.3**   *Estimation of the enthalpy of dissolution of NH$_4$NO$_3$*

When 2.0 g of ammonium nitrate, NH$_4$NO$_3$, dissolves in 100.0 g of water contained in a simple, constant-pressure calorimeter, a fall in temperature of

1.5 K is recorded. (a) Is the process exothermic or endothermic? (b) Determine the enthalpy change for the dissolution of 1 mole of $NH_4NO_3$.
[Data: $C_{water} = 4.18\,J\,K^{-1}\,g^{-1}$; values of $A_r$ are in the inside cover of the book]

(a) The temperature falls. Therefore, heat is absorbed by the solution. The dissolution of $NH_4NO_3$ is endothermic.
(b) When 2.0 g $NH_4NO_3$ dissolves in 100.0 g of water, we can approximate the heat capacity of the solution to that of 100.0 g of pure water.

$$\text{Heat change} = (m\,g) \times (C\,J\,K^{-1}\,g^{-1}) \times (\Delta T\,K)$$

$$= (100.0\,g) \times (4.18\,J\,K^{-1}\,g^{-1}) \times (1.5\,K)$$

$$= 627\,J$$

$$= 630\,J \quad \text{(to 2 sig. fig.)}$$

For the number of moles of $NH_4NO_3$, we need to find $M_r$ for $NH_4NO_3$:

$$M_r = (2 \times 14.01) + (4 \times 1.008) + (3 \times 16.00) = 80.052\,g\,mol^{-1}$$

$$= 80.05\,g\,mol^{-1} \quad \text{(to 2 dec. pl.)}$$

$$\text{The amount of } NH_4NO_3 = \frac{2.0\,g}{80.05\,g\,mol^{-1}} = 0.025\,mol \quad \text{(to 2 sig. fig.)}$$

630 J of heat are absorbed when 0.025 moles of $NH_4NO_3$ dissolve.
Therefore $\dfrac{630}{0.025}$ J of heat are absorbed when 1.0 mole of $NH_4NO_3$ dissolves.

$$\text{Heat absorbed} = \frac{630}{0.025}\,J\,mol^{-1}$$

$$= 25\,200\,J\,mol^{-1}$$

$$= 25\,kJ\,mol^{-1} \quad \text{(to 2 sig. fig.)}$$

The calorimeter is at constant pressure: heat change = enthalpy change
The dissolution is endothermic, and so $\Delta H$ is positive.

$$\Delta H = +25\,kJ\,mol^{-1}$$

---

Calorimeters are also used to measure the specific heat capacity of solid materials. The solid is heated to a temperature above room temperature. The heated material is then dropped into a known mass of water contained in a well-insulated calorimeter at constant pressure. Assuming that there is no heat loss to the surroundings, all the heat lost by the solid is gained by the water. As a result, the temperature of the water rises. The method shown in worked example 2.4 can be used provided that the solid does not react with or dissolve in water. Table 2.1 lists the specific heat capacities of selected elements and compounds. The high value of $C$ for water is significant for life on Earth: large lakes or seas freeze only slowly because freezing such a large mass of water requires the removal of a huge amount of heat.

---

**Worked example 2.4**     *Determining the specific heat capacity of copper*

100.0 cm³ of water was placed in a constant-pressure calorimeter of the type shown in Figure 2.2. The temperature of the water was recorded as 293.0 K. A 20.0 g block of copper metal was heated to 353.0 K and then dropped into

Table 2.1 Specific heat capacities, $C$, of selected elements and solvents at 298 K and constant pressure.

| Element | $C$ / J K$^{-1}$ g$^{-1}$ | Solvent | $C$ / J K$^{-1}$ g$^{-1}$ |
|---|---|---|---|
| Aluminium | 0.897 | Acetone | 2.17 |
| Carbon (graphite) | 0.709 | Acetonitrile | 2.23 |
| Chromium | 0.449 | Chloroform | 0.96 |
| Copper | 0.385 | Dichloromethane | 1.19 |
| Gold | 0.129 | Diethyl ether | 2.37 |
| Iron | 0.449 | Ethanol | 2.44 |
| Lead | 0.129 | Heptane | 2.25 |
| Magnesium | 1.02 | Hexane | 2.26 |
| Mercury | 0.140 | Methanol | 2.53 |
| Silver | 0.235 | Pentane | 2.32 |
| Sodium | 1.228 | Tetrahydrofuran | 1.72 |
| Sulfur (rhombic) | 0.710 | Toluene | 1.71 |
| Zinc | 0.388 | Water | 4.18 |

the water in the calorimeter. The temperature of the water rose and the maximum temperature attained was 294.1 K. (a) Why must the water be constantly stirred during the experiment? (b) Determine the specific heat capacity of copper, $C_{Cu}$.
[Data: density of water $= 1.00$ g cm$^{-3}$; $C_{water} = 4.18$ J K$^{-1}$ g$^{-1}$]

(a) Constant stirring ensures that the heat lost by the copper is evenly distributed throughout the water. Therefore, the measured temperature rise reflects the true rise for the bulk water and can justifiably be related to the heat loss from the copper.

(b) The heat lost by the copper equals the heat gained by the water. This assumes that the calorimeter is well insulated and that the specific heat capacity of the calorimeter is so small that it can be neglected.

The temperature rise of the water $= 294.1 - 293.0 = 1.1$ K

The temperature fall of the copper $= 353.0 - 294.1 = 58.9$ K

The mass of water $= ($Volume in cm$^3) \times ($Density in g cm$^{-3})$

$$= (100.0 \, \text{cm}^3) \times (1.00 \, \text{g cm}^{-3})$$

$$= 100 \, \text{g}$$

Heat lost by copper $= (m \, \text{g}) \times (C_{Cu} \, \text{J K}^{-1} \text{g}^{-1}) \times (\Delta T \, \text{K})$

$$= (20.0 \, \text{g}) \times (C_{Cu} \, \text{J K}^{-1} \text{g}^{-1}) \times (58.9 \, \text{K})$$

Heat gained by water $= (m \, \text{g}) \times (C_{water} \, \text{J K}^{-1} \text{g}^{-1}) \times (\Delta T \, \text{K})$

$$= (100 \, \text{g}) \times (4.18 \, \text{J K}^{-1} \text{g}^{-1}) \times (1.1 \, \text{K})$$

Heat lost by copper $=$ Heat gained by water

$$(20.0 \, \text{g}) \times (C_{Cu} \, \text{J K}^{-1} \text{g}^{-1}) \times (58.9 \, \text{K}) = (100 \, \text{g}) \times (4.18 \, \text{J K}^{-1} \text{g}^{-1}) \times (1.1 \, \text{K})$$

$$C_{Cu} = \frac{(100 \, \text{g}) \times (4.18 \, \text{J K}^{-1} \text{g}^{-1}) \times (1.1 \, \text{K})}{(20.0 \, \text{g}) \times (58.9 \, \text{K})}$$

$$= 0.39 \, \text{J K}^{-1} \text{g}^{-1} \quad \text{(to 2 sig. fig.)}$$

This value compares with $0.385$ J K$^{-1}$ g$^{-1}$ listed in Table 2.1.

| 2.4 | **Standard enthalpy of formation** |

$\Delta_f H^\circ$(298 K) is the enthalpy change of formation of a compound in its standard state from its constituent elements in their standard states, all at 298 K.

The *standard enthalpy of formation* of a compound, $\Delta_f H^\circ$(298 K), is the enthalpy change at 298 K that accompanies the formation of a compound in its standard state from its constituent elements in their standard states. The standard state of an element at 298 K is the thermodynamically most stable form of the element at 298 K and $1.00 \times 10^5$ Pa. Some examples of the standard states of elements under these conditions are:

- hydrogen: $H_2(g)$
- oxygen: $O_2(g)$
- nitrogen: $N_2(g)$
- bromine: $Br_2(l)$
- iron: $Fe(s)$
- copper: $Cu(s)$
- mercury: $Hg(l)$
- carbon: C(graphite)
- sulfur: $S_8(s)$

$\Delta_f H^\circ$(298 K) for an element in its standard state is defined to be $0 \, kJ \, mol^{-1}$.

The one exception to the definition of standard state of an element given above is phosphorus. The standard state of phosphorus is defined[§] as being white phosphorus, $P_4$(white), rather than the thermodynamically more stable red and black allotropes. *By definition, the standard enthalpy of formation of an element in its standard state is $0 \, kJ \, mol^{-1}$.*

Equations 2.6 and 2.7 describe the formation of carbon monoxide and iron(II) chloride from their constituent elements in their standard states. The values of $\Delta_f H^\circ$ are given 'per mole of compound'. In equation 2.6, the notation '$\Delta_f H^\circ$(CO, g, 298 K)' indicates that the standard enthalpy of formation refers to gaseous CO at 298 K. In equation 2.7, '$\Delta_f H^\circ$($FeCl_2$, s, 298 K)' means 'the standard enthalpy of formation of solid $FeCl_2$ at 298 K'.

$$C(graphite) + \tfrac{1}{2}O_2(g) \longrightarrow CO(g) \qquad \Delta_f H^\circ(CO, \text{ g, } 298 \text{ K}) = -110.5 \, kJ \, mol^{-1}$$
$$(2.6)$$

$$Fe(s) + Cl_2(g) \longrightarrow FeCl_2(s) \qquad \Delta_f H^\circ(FeCl_2, \text{ s, } 298 \text{ K}) = -342 \, kJ \, mol^{-1}$$
$$(2.7)$$

The values of $\Delta_f H^\circ$ for CO(g) and $FeCl_2$(s) show that a significant amount of heat is *liberated* when these compounds are formed from their constituent elements at 298 K and $1.00 \times 10^5$ Pa. Such compounds are described as being exothermic. Under these conditions, CO(g) and $FeCl_2$(s) are both thermodynamically stable with respect to their constituent elements.

**The true guide to thermodynamic stability is the change in Gibbs energy, rather than the change in enthalpy: see Chapter 17**

Not all compounds are formed from their constituent elements in exothermic reactions. Equation 2.8 shows the formation of chlorine dioxide. The relatively large, positive value of $\Delta_f H^\circ$(298 K) indicates that, at 298 K, $ClO_2$ is not stable with respect to its elements. Indeed, $ClO_2$ is explosive, decomposing to $Cl_2$ and $O_2$.

$$\tfrac{1}{2}Cl_2(g) + O_2(g) \longrightarrow ClO_2(g)$$

$$\Delta_f H^\circ(ClO_2, \text{ g, } 298 \text{ K}) = +102.5 \, kJ \, mol^{-1} \quad (2.8)$$

Appendix 11 at the end of the book lists values of $\Delta_f H^\circ$(298 K) for selected organic and inorganic compounds.

[§] The definition of standard state and the exceptional case of phosphorus have been laid down by the National Bureau of Standards.

| 2.5 | **Calculating standard enthalpies of reaction** |

Figure 2.1 showed enthalpy level diagrams for general exothermic and endothermic reactions. The value of the standard enthalpy change for a reaction, $\Delta_r H^\circ(298\,K)$, is the difference between the sum of the standard enthalpies of formation of the products and the sum of the standard enthalpies of formation of the reactants (equation 2.9).

◄►
$\sum$ means 'summation of'

$$\Delta_r H^\circ(298\,K) = \sum \Delta_f H^\circ(\text{products, } 298\,K) - \sum \Delta_f H^\circ(\text{reactants, } 298\,K)$$

(2.9)

If we apply this equation to reaction 2.6, then:

$$\Delta_r H^\circ(298\,K) = \Delta_f H^\circ(\text{CO, g, } 298\,K)$$
$$-[\Delta_f H^\circ(\text{C, graphite, } 298\,K) + \tfrac{1}{2}\Delta_f H^\circ(\text{O}_2\text{, g, } 298\,K)]$$

Since both of the reactants are elements in their standard states, their standard enthalpies of formation are zero. In the special case where a reaction represents the formation of a compound in its standard state, $\Delta_r H^\circ(298\,K) = \Delta_f H^\circ(\text{product, } 298\,K)$.

---

**Worked example 2.5**   *Determination of $\Delta_r H^\circ$(298 K) for the formation of HBr*

Using appropriate values of $\Delta_f H^\circ$(298 K) from Appendix 11, calculate the value of $\Delta_r H^\circ$(298 K) for the following reaction:

$$\text{H}_2(g) + \text{Br}_2(l) \longrightarrow 2\text{HBr}(g)$$

The reaction:

$$\text{H}_2(g) + \text{Br}_2(l) \longrightarrow 2\text{HBr}(g) \qquad (\text{at } 298\,K)$$

refers to the formation of two moles of HBr(g) from its constituent elements in their standard states.

$$\Delta_r H^\circ(298\,K) = 2\Delta_f H^\circ(\text{HBr, g, } 298K)$$
$$-[\Delta_f H^\circ(\text{H}_2\text{, g, } 298\,K) + \Delta_f H^\circ(\text{Br}_2\text{, l, } 298\,K)]$$
$$\qquad\qquad \uparrow \qquad\qquad\qquad\qquad \uparrow$$
$$\qquad = 0\,\text{kJ mol}^{-1} \qquad\qquad = 0\,\text{kJ mol}^{-1}$$
$$= 2\Delta_f H^\circ(\text{HBr, g, } 298\,K)$$

From Appendix 11, $\Delta_f H^\circ(\text{HBr, g, } 298\,K) = -36\,\text{kJ mol}^{-1}$

$$\Delta_r H^\circ(298\,K) = 2(-36) = -72\,\text{kJ per mole of reaction}$$

---

**Worked example 2.6**   *Determination of $\Delta_r H^\circ$(298 K) for the decomposition of NH$_3$ to N$_2$ and H$_2$*

Use data in Appendix 11 to determine $\Delta_r H^\circ$(298 K) for the following reaction:

$$\text{NH}_3(g) \longrightarrow \tfrac{1}{2}\text{N}_2(g) + \tfrac{3}{2}\text{H}_2(g)$$

The reaction:

$$\text{NH}_3(g) \longrightarrow \tfrac{1}{2}\text{N}_2(g) + \tfrac{3}{2}\text{H}_2(g)$$

is the reverse of the formation of one mole of NH$_3$.

From Appendix 11, $\Delta_f H^\circ(\text{NH}_3\text{, g, } 298\,K) = -45.9\,\text{kJ mol}^{-1}$

For a reaction in which one mole of a compound in its standard state *decomposes into its constituent elements in their standard states*:

$$\Delta_r H^\circ(298\,\text{K}) = -\Delta_f H^\circ(298\,\text{K})$$

$$\Delta_r H^\circ(298\,\text{K}) = \sum \Delta_f H^\circ(\text{products, } 298\,\text{K}) - \sum \Delta_f H^\circ(\text{reactants, } 298\,\text{K})$$

$$= [\tfrac{1}{2}\Delta_f H^\circ(N_2,\ \text{g, } 298\,\text{K}) + \tfrac{3}{2}\Delta_f H^\circ(H_2,\ \text{g, } 298\,\text{K})]$$
$$- \Delta_f H^\circ(NH_3, \text{g}, 298\,\text{K})$$

$$= -\Delta_f H^\circ(NH_3,\ \text{g, } 298\,\text{K})$$

$$= -(-45.9)\,\text{kJ mol}^{-1}$$

$$= +45.9\,\text{kJ mol}^{-1}$$

Equation 2.9 can be applied to any reaction in which the standard enthalpies of formation of reactants and products are known. In contrast to worked examples 2.5 and 2.6, the next two worked examples show the application of equation 2.9 to reactions that do not simply have elements in their standard states on one side of the equation.

---

**Worked example 2.7**  *Chlorination of ethene*

Determine the standard enthalpy change for the following reaction at 298 K:

$$C_2H_4(g) + Cl_2(g) \longrightarrow 1,2\text{-}C_2H_4Cl_2(l)$$
Ethene      1,2-Dichloroethane

[Data: $\Delta_f H^\circ(1,2\text{-}C_2H_4Cl_2,\ l,\ 298\,\text{K}) = -167\,\text{kJ mol}^{-1}$; $\Delta_f H^\circ(C_2H_4,\ \text{g},\ 298\,\text{K}) = +53\,\text{kJ mol}^{-1}$]

$$\Delta_r H^\circ(298\,\text{K}) = \sum \Delta_f H^\circ(\text{products, } 298\,\text{K}) - \sum \Delta_f H^\circ(\text{reactants, } 298\,\text{K})$$

$$= \Delta_f H^\circ(1,2\text{-}C_2H_4Cl_2, \text{l}, 298\,\text{K}) - [\Delta_f H^\circ(C_2H_4, \text{g}, 298\,\text{K})$$
$$+ \underset{\uparrow}{\Delta_f H^\circ(Cl_2, \text{g}, 298\,\text{K})}]$$
$$= 0\,\text{kJ mol}^{-1}$$

Substitute the values of $\Delta_f H^\circ(1,2\text{-}C_2H_4Cl_2, \text{l}, 298\,\text{K})$ and $\Delta_f H^\circ(C_2H_4, \text{g}, 298\,\text{K})$ into the above equation:

$$\Delta_r H^\circ(298\,\text{K}) = -167 - (+53)$$

$$= -220\,\text{kJ mol}^{-1}$$

---

**Worked example 2.8**  *The reaction between gaseous NH₃ and HCl*

Use data in Appendix 11 to find the standard enthalpy change for the following reaction at 298 K:

$$NH_3(g) + HCl(g) \longrightarrow NH_4Cl(s)$$

See Figure 1.7

The data needed from Appendix 11 are:

$$\Delta_f H^\circ(NH_3, \text{g}, 298\,\text{K}) = -45.9\,\text{kJ mol}^{-1}$$

$$\Delta_f H^\circ(HCl, \text{g}, 298\,\text{K}) = -92\,\text{kJ mol}^{-1}$$

$$\Delta_f H^\circ(NH_4Cl, \text{s}, 298\,\text{K}) = -314\,\text{kJ mol}^{-1}$$

$$\Delta_r H^\circ(298\,\text{K}) = \sum \Delta_f H^\circ(\text{products, 298 K}) - \sum \Delta_f H^\circ(\text{reactants, 298 K})$$

$$= \Delta_f H^\circ(\text{NH}_4\text{Cl, s, 298 K})$$

$$- [\Delta_f H^\circ(\text{NH}_3, \text{g, 298 K}) + \Delta_f H^\circ(\text{HCl, g, 298 K})]$$

$$= -314 - (-45.9 - 92)$$

$$= -176.1\,\text{kJ mol}^{-1}$$

$$= -176\,\text{kJ mol}^{-1} \quad \text{(rounding to 0 dec. pl.)}$$

---

## 2.6 Enthalpies of combustion

### Combustion of fuels

An everyday example of an exothermic reaction is the burning (combustion) of a fuel such as butane, $C_4H_{10}$. Butane is an example of a *hydrocarbon* and, under standard conditions, complete combustion in $O_2$ gives $CO_2(g)$ and $H_2O(l)$. Reaction 2.10 shows the combustion of one mole of butane.

▶ Hydrocarbons: see Section 24.4

$$C_4H_{10}(g) + \tfrac{13}{2}O_2(g) \longrightarrow 4CO_2(g) + 5H_2O(l) \tag{2.10}$$

For the combustion of a substance in $O_2$, the enthalpy change is called the standard enthalpy of combustion, $\Delta_c H^\circ(298\,\text{K})$. The value of $\Delta_c H^\circ(298\,\text{K})$ for reaction 2.10 can be found from values of $\Delta_f H^\circ(298\,\text{K})$ of the products and reactants:

$$\Delta_c H^\circ(298\,\text{K}) = \sum \Delta_f H^\circ(\text{products, 298 K}) - \sum \Delta_f H^\circ(\text{reactants, 298 K})$$

$$= [4\Delta_f H^\circ(\text{CO}_2, \text{g, 298 K}) + 5\Delta_f H^\circ(\text{H}_2\text{O, l, 298 K})]$$

$$- [\Delta_f H^\circ(\text{C}_4\text{H}_{10}, \text{g, 298 K}) + \tfrac{13}{2}\Delta_f H^\circ(\text{O}_2, \text{g, 298 K})]$$

$$= 4(-393.5) + 5(-286) - (-126) - 0$$

$$= -2878\,\text{kJ mol}^{-1}$$

The reaction is highly exothermic, consistent with the use of butane as a fuel.

If a compound contains C, H and O, the products (under standard conditions) of *complete combustion* are taken to be $CO_2(g)$ and $H_2O(l)$ (e.g. reaction 2.10). If the compound contains C, H and N, the products of *complete combustion* under standard conditions are $CO_2(g)$, $H_2O(l)$ and $N_2(g)$ (e.g. reaction 2.11).

$$2CH_3CH_2NH_2(g) + \tfrac{15}{2}O_2(g) \longrightarrow 4CO_2(g) + 7H_2O(l) + N_2(g) \tag{2.11}$$

When the supply of $O_2$ is limited, *partial combustion* may result in the formation of CO rather than $CO_2$: compare reactions 2.12 and 2.13.

$$CH_4(g) + 2O_2(g) \longrightarrow CO_2(g) + 2H_2O(l) \qquad \textit{Complete combustion} \tag{2.12}$$

$$CH_4(g) + \tfrac{3}{2}O_2(g) \longrightarrow CO(g) + 2H_2O(l) \qquad \textit{Partial combustion} \tag{2.13}$$

Enthalpies of combustion can be measured experimentally by using a *bomb calorimeter* as described in Section 17.3.

**ENVIRONMENT AND BIOLOGY**

## Box 2.1 Energy content of foods: calorific values

In Chapter 35, we look in detail at the structures and properties of biological molecules including carbohydrates and proteins. Carbohydrates constitute a family of compounds consisting of sugars: monosaccharides (e.g. glucose), disaccharides (e.g. lactose) and polysaccharides (e.g. starch). These compounds (along with fats and proteins) are the body's fuels. Examples of natural sugars are glucose, sucrose and lactose. When the body metabolizes glucose, the *overall* reaction is the same as combustion:

$$C_6H_{12}O_6(s) + 6O_2(g) \longrightarrow 6CO_2(g) + 6H_2O(l)$$
Glucose

*Andy Crawford © Dorling Kindersley.*

However, whereas burning glucose in oxygen is rapid, 'burning' glucose in the body is a much slower process and is carried out in a series of steps involving *enzymes* (enzymes are proteins that act as biological catalysts, see Section 15.15). Nonetheless, metabolizing glucose results in the production of energy in just the same way that the combustion of glucose $O_2$ does. The amount of energy that is liberated when a particular food is metabolized is called its *calorific value*. This term is used, not just for foods, but for fuels (e.g. natural gas) more generally. Its origins are in the pre-SI unit *calorie* which is a unit of energy. In the SI system, the calorie is replaced by the joule:

1 calorie (1 cal) = 4.184 joules (4.184 J)

1 kcal = 1000 cal = 4184 J = 4.184 kJ

In the context of foods and nutrition, calorific values are typically given in units of *Calories* (with an upper-case C) where:

1 Calorie = 1000 cal = 1 kcal = 4.184 kJ

The table below lists the calorific values of selected foods, and the percentage of the calories in the food that are obtained from the carbohydrate, fat and protein content.

| Food | Calorific value per 100 g / Calories | % Calories from carbohydrates | % Calories from fats | % Calories from proteins |
|------|--------------------------------------|-------------------------------|----------------------|--------------------------|
| Olive oil | 884 | 0 | 100 | 0 |
| Milk (3.25% milk fat) | 60 | 29 | 48 | 23 |
| Butter (unsalted) | 499 | 0 | 99 | 1 |
| Egg (poached) | 147 | 2 | 61 | 37 |
| Salmon (cooked, dry heat) | 116 | 0 | 27 | 73 |
| Rice (white, cooked) | 130 | 90 | 2 | 8 |
| Potato (boiled, unsalted) | 78 | 89 | 1 | 10 |
| Carrot (boiled, unsalted) | 35 | 90 | 4 | 6 |
| Carrot (raw) | 35 | 92 | 3 | 5 |
| Broccoli (boiled, unsalted) | 35 | 73 | 10 | 17 |
| Broccoli (raw) | 28 | 64 | 10 | 26 |
| Spinach (raw) | 23 | 56 | 14 | 30 |
| Apple (raw) | 52 | 95 | 3 | 2 |
| Orange (raw) | 46 | 91 | 4 | 5 |

Data: Nutritional information provided by NutritionData.com

---

**Worked example 2.9**    *Standard enthalpy of combustion of methane*

Write a balanced equation for the complete combustion of one mole of ethane, $C_2H_6$. Use data from Appendix 11 to determine the value of $\Delta_c H^\circ(C_2H_6, g, 298\,K)$.

The complete combustion of one mole of $C_2H_6$ is given by:

$$C_2H_6(g) + 3\tfrac{1}{2}O_2(g) \longrightarrow 2CO_2(g) + 3H_2O(l)$$

Data needed from Appendix 11 are:

$$\Delta_fH^\circ(CO_2, g, 298\,K) = -393.5\,kJ\,mol^{-1}$$

$$\Delta_fH^\circ(H_2O, l, 298\,K) = -286\,kJ\,mol^{-1}$$

$$\Delta_fH^\circ(C_2H_6, g, 298\,K) = -84\,kJ\,mol^{-1}$$

Since $O_2$ is an element in its standard state, $\Delta_fH^\circ(O_2, g, 298\,K) = 0\,kJ\,mol^{-1}$.

$$\Delta_cH^\circ(298\,K) = \sum \Delta_fH^\circ(\text{products}, 298\,K) - \sum \Delta_fH^\circ(\text{reactants}, 298\,K)$$

$$= [2\Delta_fH^\circ(CO_2, g, 298\,K) + 3\Delta_fH^\circ(H_2O, l, 298\,K)]$$

$$-\Delta_fH^\circ(C_2H_6, g, 298\,K)$$

$$= 2(-393.5) + 3(-286) - (-84)$$

$$= -1561\,kJ\,mol^{-1}$$

---

## 2.7    Hess's Law of Constant Heat Summation

Calculations using equation 2.9 make use of *Hess's Law of Constant Heat Summation*. This states that the enthalpy change on going from reactants to products is independent of the reaction path taken. Consider the reaction of $PCl_3$ and $Cl_2$:

$$PCl_3(l) + Cl_2(g) \longrightarrow PCl_5(s)$$

We can consider this reaction as part of a cycle involving the elements P and Cl:

$$PCl_3(l) + Cl_2(g) \xrightarrow{\quad(1)\quad} PCl_5(s)$$

$$(2)\searrow \qquad \nearrow(3)$$

$$\tfrac{1}{4}P_4(\text{white}) + \tfrac{5}{2}Cl_2(g)$$

Reaction (3) shows the formation of $PCl_5(s)$ from its constituent elements in their standard states. Reactions (2) and (1) describe the formation of $PCl_3(l)$ from its constituent elements, followed by the reaction of $PCl_3(l)$ with $Cl_2(g)$ to give $PCl_5(s)$. Thus, reaction (3) gives a direct route to $PCl_5(s)$ from its constituent elements, while reactions (2) and (1) provide an indirect route. Now consider the enthalpy change for each step:

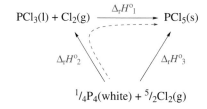

*Hess's Law of Constant Heat Summation* states that the enthalpy change on going from reactants to products is independent of the reaction path taken.

The three enthalpy changes make up a *thermochemical cycle*. By Hess's Law of Constant Heat Summation, the enthalpy change on going from reactants to products is independent of the reaction path taken. In the thermochemical cycle above, the reactants are $P_4$ and $Cl_2$ and the final product is $PCl_5$. There are two routes to $PCl_5$ depending on whether we follow the arrows anticlockwise or clockwise from reactants to product. Application of Hess's Law to this thermochemical cycle leads to equation 2.14.

$$\Delta_r H^\circ_3 = \Delta_r H^\circ_1 + \Delta_r H^\circ_2 \qquad \text{(at 298 K)} \qquad (2.14)$$

$\Delta_r H^\circ_1(298\,\text{K})$ is the standard enthalpy change for the reaction of $PCl_3(l)$ with $Cl_2(g)$, while $\Delta_r H^\circ_2(298\,\text{K})$ and $\Delta_r H^\circ_3(298\,\text{K})$ are the standard enthalpies of formation of $PCl_3(l)$ and $PCl_5(s)$, respectively. We can therefore write an expression for the standard enthalpy change for the reaction of $PCl_3(l)$ with $Cl_2(g)$ in terms of the values of $\Delta H_f^\circ(298\,\text{K})$ for $PCl_3(l)$ and $PCl_5(s)$:

$$\begin{aligned}
\Delta_r H^\circ_1 &= \Delta_r H^\circ_3 - \Delta_r H^\circ_2 \\
&= \Delta_f H^\circ(PCl_5,\,s,\,298\,\text{K}) - \Delta_f H^\circ(PCl_3,\,l,\,298\,\text{K}) \\
&= -444 - (-320) \\
&= -124\,\text{kJ mol}^{-1}
\end{aligned}$$

Hess's Law is particularly useful when we have more complex situations to consider, for example the determination of lattice energies (see Section 8.16) or enthalpy changes associated with the dissolution of salts (see Section 17.11).

---

| **Worked example 2.10** | *Use of Hess's Law: phosphorus oxides* |

**Construct a thermochemical cycle that links the following processes: (i) the combustion of $P_4$(white) to give $P_4O_6(s)$, (ii) the combustion of $P_4$(white) to give $P_4O_{10}(s)$ and (iii) the conversion of $P_4O_6(s)$ to $P_4O_{10}(s)$. Use the cycle to find the standard enthalpy change for the conversion of $P_4O_6(s)$ to $P_4O_{10}(s)$ if values of $\Delta_f H^\circ(298\,\text{K})$ for $P_4O_6(s)$ and $P_4O_{10}(s)$ are $-1640$ and $-2984\,\text{kJ mol}^{-1}$, respectively.**

The thermochemical cycle can be drawn as follows:

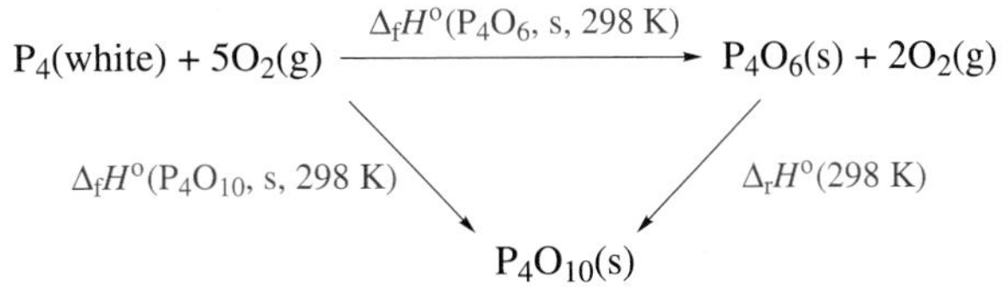

Now apply Hess's Law: the left-hand arrow links $P_4$ to $P_4O_{10}$ directly, while the top and right-hand arrows form an indirect route from $P_4$ to $P_4O_{10}$. Therefore:

$$\Delta_f H^\circ(P_4O_{10},\,s,\,298\,\text{K}) = \Delta_f H^\circ(P_4O_6,\,s,\,298\,\text{K}) + \Delta_r H^\circ(298\,\text{K})$$

Rearranging the equation gives:

$$\begin{aligned}
\Delta_r H^\circ(298\,\text{K}) &= \Delta_f H^\circ(P_4O_{10},\,s,\,298\,\text{K}) - \Delta_f H^\circ(P_4O_6,\,s,\,298\,\text{K}) \\
&= -2984 - (-1640) \\
&= -1344\,\text{kJ mol}^{-1}
\end{aligned}$$

**Worked example 2.11**    *Use of Hess's Law: oxides of nitrogen*

(a) Use data in Appendix 11 to determine a value of $\Delta_r H^\circ(298\,\text{K})$ for the following reaction:

$$2NO_2(g) \longrightarrow N_2O_4(l)$$

(b) Consider the following thermochemical cycle at 298 K:

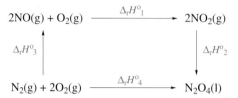

Using data from Appendix 11 and the answer to part (a), determine a value for $\Delta_r H^\circ_1(298\,\text{K})$. (The answer can be checked using values of $\Delta_f H^\circ$ for NO(g) and NO$_2$(g) from Appendix 11.)

(a) To determine the standard enthalpy change for the reaction:

$$2NO_2(g) \longrightarrow N_2O_4(l)$$

we need to look up the standard enthalpies of formation of $N_2O_4(l)$ and $NO_2(g)$. These are $-20$ and $+33\,\text{kJ mol}^{-1}$, respectively. The standard enthalpy change for the reaction is:

$$\Delta_r H^\circ(298\,\text{K}) = \Delta_f H^\circ(N_2O_4, l, 298\,\text{K}) - 2\Delta_f H^\circ(NO_2, g, 298\,\text{K})$$
$$= -20 - 2(+33)$$
$$= -86\,\text{kJ mol}^{-1}$$

(b) Draw out the thermochemical cycle in the question and identify known and unknown values of $\Delta_r H^\circ(298\,\text{K})$:

$$
\begin{array}{ccc}
2NO(g) + O_2(g) & \xrightarrow{\ \Delta_r H^\circ_1\ } & 2NO_2(g) \\
\uparrow \Delta_r H^\circ_3 = 2\Delta_f H^\circ(NO,\,g) & & \downarrow \Delta_r H^\circ_2 = -86\,\text{kJ mol}^{-1} \\
N_2(g) + 2O_2(g) & \xrightarrow[\Delta_r H^\circ_4 = \Delta_f H^\circ(N_2O_4,\,l)]{} & N_2O_4(l)
\end{array}
$$

$\Delta_r H^\circ_2$ was found in part (a) of the question. From Appendix 11, values (at 298 K) of $\Delta_f H^\circ(NO, g)$ and $\Delta_f H^\circ(N_2O_4, l)$ are $+90$ and $-20\,\text{kJ mol}^{-1}$, respectively.

Apply Hess's Law to the thermochemical cycle, noting that three arrows in the cycle follow a clockwise path, while one arrow follows an anticlockwise path:

$$\Delta_r H^\circ_3 + \Delta_r H^\circ_1 + \Delta_r H^\circ_2 = \Delta_r H^\circ_4$$
$$\Delta_r H^\circ_1 = \Delta_r H^\circ_4 - \Delta_r H^\circ_3 - \Delta_r H^\circ_2$$
$$= \Delta_f H^\circ(N_2O_4, l, 298\,\text{K}) - 2\Delta_f H^\circ(NO, g, 298\,\text{K})$$
$$\qquad\qquad\qquad\qquad\qquad - \Delta_r H^\circ_2$$
$$= -20 - 2(+90) - (-86)$$
$$= -114\,\text{kJ mol}^{-1}$$

**2.8**    ## Thermodynamic and kinetic stability

A term that is commonly (and often inconsistently) used is 'stable'. It is meaningless to say that something is stable or unstable unless you specify *'stable or unstable with respect to...'*. Consider hydrogen peroxide, $H_2O_2$. This compound is a liquid at room temperature and a solution can be purchased in a bottle as a hair bleach. Because of this, you may think that $H_2O_2$ is a 'stable' compound. However, the conditions under which the $H_2O_2$ solution is stored are critical. It can decompose to $H_2O$ and $O_2$ (equation 2.15), and the standard enthalpy change for this reaction is $-98.2\,kJ\,mol^{-1}$. The process is slow, but in the presence of some surfaces or alkali, decomposition is rapid, and can even be explosive. Thus, we describe hydrogen peroxide as being *unstable with respect to the formation of $H_2O$ and $O_2$*.

$$H_2O_2(l) \longrightarrow H_2O(l) + \tfrac{1}{2}O_2(g) \qquad (2.15)$$

Hydrogen peroxide is *thermodynamically unstable* with respect to reaction 2.15, but the speed with which the reaction occurs is controlled by kinetic factors. Because the decomposition of $H_2O_2$ is slow, we say that hydrogen peroxide is *kinetically stable* with respect to the formation of $H_2O$ and $O_2$. We return to the kinetics of reactions in Chapter 15.

A notable example of a thermodynamically favourable reaction is the conversion of diamond into graphite (equation 2.16): diamond is thermodynamically unstable with respect to graphite. Fortunately for all owners of diamonds, the transformation takes places *extremely* slowly at room temperature and pressure.

$$C(\text{diamond}) \longrightarrow C(\text{graphite}) \qquad \Delta_r H^\circ(298\,K) = -1.9\,kJ\,mol^{-1} \qquad (2.16)$$

**2.9**    ## Phase changes: enthalpies of fusion and vaporization

### Melting solids and vaporizing liquids

Phases: see Section 1.6

When a solid melts, energy is needed for the phase change from solid to liquid. In a crystalline solid, the atoms or molecules are arranged in a rigid framework and energy is needed to make the structure collapse as the solid transforms to a liquid. In a liquid, the atoms or molecules are not completely separated from one another (see Figure 1.2). If the liquid is heated, heat is initially used to raise the temperature to the boiling point of the liquid. At the boiling point, heat is used to separate the atoms or molecules as the liquid transforms into a vapour. Figure 2.3 illustrates what happens as a constant heat supply provides heat to a sample of $H_2O$ which is initially in the solid phase (ice). The temperature of the solid rises until the melting point is reached. During the process of melting the solid, the temperature remains constant. Once melting is complete, liquid water is heated from the melting point (273 K) to the boiling point (373 K). Heat continues to be supplied to the sample, but at the boiling point, the heat is used to vaporize the sample and the temperature remains constant. After vaporization is complete, the heat supplied is used to raise the temperature of the water vapour. In Figure 2.3, the gradients of the lines representing the heating of

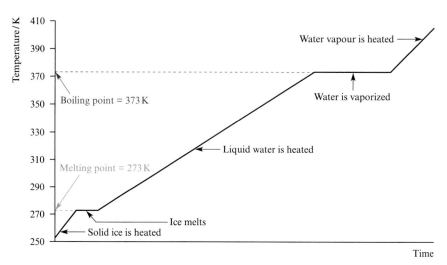

**Fig. 2.3** A heating curve for a constant mass of $H_2O$, initially in the solid state. Heat is supplied at a constant rate. The graph shows both the melting and vaporization of the sample. During phase changes, the temperature remains constant even though heat continues to be supplied to the sample.

solid ice, liquid water and water vapour are different because the specific heat capacity of $H_2O$ in the three phases is different. The specific heat capacity of liquid water $(4.18\,\mathrm{J\,K^{-1}\,g^{-1}})$ is greater than that of ice $(2.03\,\mathrm{J\,K^{-1}\,g^{-1}})$ and water vapour $(1.97\,\mathrm{J\,K^{-1}\,g^{-1}})$; see end-of-chapter problem 2.15.

> The ***molar enthalpy of fusion*** of a substance at its melting point, $\Delta_{fus}H(\mathrm{mp})$, is the enthalpy change for the conversion of one mole of solid to liquid.
>
> The ***molar enthalpy of vaporization*** of a substance at its boiling point, $\Delta_{vap}H(\mathrm{bp})$, is the enthalpy change for the conversion of one mole of liquid to vapour.

The enthalpy change associated with melting one mole of a solid is called the *molar enthalpy of fusion*, and the value refers to the melting point (mp) of the solid. The enthalpy change is written as $\Delta_{fus}H(\mathrm{mp})$. The enthalpy change associated with vaporizing one mole of a liquid is the *molar enthalpy of vaporization* and the value is quoted at the boiling point (bp) of the liquid under specified pressure conditions. The notation is $\Delta_{vap}H(\mathrm{bp})$. Melting a solid and vaporizing a liquid are *endothermic processes*. For $H_2O$, the enthalpy changes for melting and vaporizing are given in equation 2.17.

$$H_2O(s) \xrightarrow{\Delta_{fus}H(\mathrm{mp}) \,=\, +6.0 \text{ kJ mol}^{-1}} H_2O(l) \xrightarrow{\Delta_{vap}H(\mathrm{bp}) \,=\, +40.7 \text{ kJ mol}^{-1}} H_2O(g)$$

$$(2.17)$$

Table 2.2 lists values of $\Delta_{fus}H(\mathrm{mp})$ and $\Delta_{vap}H(\mathrm{bp})$ for selected elements and compounds.

### Solidifying liquids and condensing vapours

When a liquid is cooled to the melting point of the substance, the liquid *solidifies*. This process is also referred to as *freezing*, and, when the temperature is being lowered rather than raised, the melting point is often called the freezing point. When a vapour is cooled to its *condensation point* (the same temperature as the boiling point), the vapour *condenses* to a liquid. As a solid forms from a liquid, heat is released as the atoms or molecules pack more closely together and the system becomes more ordered. The process is exothermic. Similarly, condensing a vapour to form a liquid liberates heat. Suppose we allow a sample of $H_2O$ to cool from 405 K to 253 K. A cooling curve for this process is the mirror image of Figure 2.3: the stages in cooling water vapour to eventually form ice are:

**Table 2.2** Values of melting and boiling points (see also Appendix 11), and enthalpies of fusion, $\Delta_{fus}H$(mp), and vaporization, $\Delta_{vap}H$(bp), for selected elements and compounds; see also Table 3.6.

| | Melting point / K | Boiling point / K | $\Delta_{fus}H$(mp) / kJ mol$^{-1}$ | $\Delta_{vap}H$(bp) / kJ mol$^{-1}$ |
|---|---|---|---|---|
| **Element** | | | | |
| Aluminium (Al) | 933 | 2793 | 10.7 | 294 |
| Bromine (Br$_2$) | 266 | 332 | 10.6 | 30.0 |
| Chlorine (Cl$_2$) | 172 | 239 | 6.4 | 20.4 |
| Fluorine (F$_2$) | 53 | 85 | 0.51 | 6.6 |
| Gold (Au) | 1064 | 2857 | 12.6 | 324 |
| Hydrogen (H$_2$) | 13.7 | 20.1 | 0.12 | 0.90 |
| Iodine (I$_2$) | 387 | 458 | 15.5 | 41.6 |
| Lead (Pb) | 600 | 2022 | 4.8 | 180 |
| Nitrogen (N$_2$) | 63 | 77 | 0.71 | 5.6 |
| Oxygen (O$_2$) | 54 | 90 | 0.44 | 6.8 |
| **Compound** | | | | |
| Acetone (CH$_3$COCH$_3$) | 178 | 329 | 5.7 | 29.1 |
| Ethane (C$_2$H$_6$) | 90 | 184 | 2.9 | 14.7 |
| Ethanol (C$_2$H$_5$OH) | 159 | 351 | 5.0 | 38.6 |
| Hydrogen chloride (HCl) | 159 | 188 | 2.0 | 16.2 |
| Water (H$_2$O) | 273 | 373 | 6.0 | 40.7 |

- the temperature of the vapour falls until the condensation point (the same temperature as the boiling point) is reached;
- the temperature remains constant as water vapour condenses to liquid water;
- the temperature of the liquid falls until the freezing point (the same temperature as the melting point) is reached;
- the temperature remains constant as liquid water freezes (solidifies) to ice;
- the temperature of the ice falls.

The enthalpy change associated with condensation is equal to $-\Delta_{vap}H$(bp), and the enthalpy change associated with solidification is $-\Delta_{fus}H$(mp). For H$_2$O, the enthalpy changes are shown in equation 2.18; compare this with equation 2.17.

$$H_2O(g) \xrightarrow{\Delta_{vap}H(bp) = -40.7 \text{ kJ mol}^{-1}} H_2O(l) \xrightarrow{\Delta_{fus}H(mp) = -6.0 \text{ kJ mol}^{-1}} H_2O(s)$$

(2.18)

**Worked example 2.12** | *Solid and molten copper*

A piece of copper metal of mass 7.94 g was heated to its melting point (1358 K). What is the enthalpy change at 1358 K as the copper melts if $\Delta_{fus}H(mp) = 13 \text{ kJ mol}^{-1}$?

First, look up $A_r$ for Cu in the inside cover of the book: $A_r = 63.54$.

$$\text{Amount of Cu} = \frac{7.94 \text{ g}}{63.54 \text{ g mol}^{-1}} = 0.125 \text{ mol}$$

Melting copper *requires* heat and is an endothermic process.

Therefore, the enthalpy change as the copper melts

$$= (+13\,\text{kJ}\,\text{mol}^{-1}) \times (0.125\,\text{mol})$$

$$= +1.6\,\text{kJ} \quad \text{(to 2 sig. fig.)}$$

---

**Worked example 2.13**   *Liquid acetone and its vapour*

**The boiling point of acetone is 329 K. Use data from Table 2.2 to determine the enthalpy change at 329 K when 14.52 g of acetone, $CH_3COCH_3$, condenses from its vapour.**

Using values of $A_r$ from the inside cover of the book, find $M_r$ for $CH_3COCH_3$: $M_r = 58.08\,\text{g}\,\text{mol}^{-1}$.

$$\text{Amount of acetone condensed} = \frac{14.52\,\text{g}}{58.08\,\text{g}\,\text{mol}^{-1}} = 0.2500\,\text{mol}$$

From Table 2.2: $\Delta_{\text{vap}}H(\text{bp}) = 29.1\,\text{kJ}\,\text{mol}^{-1}$.
Condensation is an *exothermic* process. Therefore, the enthalpy change when 0.2500 moles of acetone condenses

$$= (-29.1\,\text{kJ}\,\text{mol}^{-1}) \times (0.2500\,\text{mol})$$

$$= -7.28\,\text{kJ} \quad \text{(to 3 sig. fig.)}$$

---

In the next section, we look at the types of interactions that exist between molecules and the relative strengths of these interactions.

## 2.10   An introduction to intermolecular interactions

(2.1)          (2.2)

In a *molecular compound* such as methane, $CH_4$ (**2.1**), or dihydrogen, $H_2$ (**2.2**), the atoms in the molecule are held together by covalent bonds. These covalent bonds are present in all phases, i.e. in the solid, the liquid and the vapour states of methane and dihydrogen. In the vapour states of these two compounds, the molecules are well separated and can be regarded as having little effect on one another. This is consistent with one of the postulates of the kinetic theory of gases (Section 1.9), but in reality, the behaviour of a gas is not ideal because the molecules *do* interact with one another. In Section 1.9, we considered gas laws and *ideal* gases. Equation 2.19 holds for $n$ moles of an ideal gas.

See also equation 1.23 and related discussion

$$PV = nRT \qquad R = \text{molar gas constant} = 8.314\,\text{J}\,\text{K}^{-1}\,\text{mol}^{-1} \tag{2.19}$$

For a *real* gas at a given temperature, $PV$ is not a constant because, in contrast to the postulates of the kinetic theory of ideal gases (Section 1.9):

- *real* gas molecules occupy a volume that cannot be ignored; the effective volume of the gas can be corrected from $V$ to $(V - nb)$, where $n$ is the number of moles of the gas and $b$ is a constant;
- *real* gas molecules interact with one another; the pressure has to be corrected from $P$ to $\left(P + \dfrac{an^2}{V^2}\right)$, where $n$ is the number of moles of the gas and $a$ is a constant.

Johannes Diderik van der Waals
(1837–1923).
© *The Nobel Foundation.*

These corrections were first proposed by Johannes van der Waals in 1873, and equation 2.20 gives the van der Waals equation for $n$ moles of a *real* gas. Values of the constants $a$ and $b$ depend on the compound.

$$\left(P + \frac{an^2}{V^2}\right)(V - nb) = RT \qquad \text{van der Waals equation} \qquad (2.20)$$

The strengths of intermolecular interactions (*van der Waals forces or interactions*) vary depending upon their precise nature. The vapour ⟶ liquid and liquid ⟶ solid phase changes are exothermic and this, in part, reflects the enthalpy changes associated with the *formation* of intermolecular interactions. Conversely, solid ⟶ liquid and liquid ⟶ vapour phase changes are endothermic because intermolecular interactions must be overcome before the phase change can occur. When methane gas, for example, is liquefied, the molecules come closer together, and when the liquid is solidified, an ordered structure is formed in which there are intermolecular interactions between the $CH_4$ molecules. In the case of methane, the enthalpy changes associated with fusion and vaporization are small:

$$CH_4(s) \xrightarrow{\Delta_{fus}H(bp) = +1 \text{ kJ mol}^{-1}} CH_4(l) \xrightarrow{\Delta_{vap}H(bp) = +8 \text{ kJ mol}^{-1}} CH_4(g)$$

▮▶
Important! During phase changes, the covalent bonds in $CH_4$ are *not* broken

indicating that the interactions between $CH_4$ molecules in the solid and in the liquid are weak. The interactions between $CH_4$ molecules are called *London dispersion forces* (named after Fritz London) and are the weakest type of intermolecular interactions. They arise from interactions between the electron clouds of adjacent molecules. We will look more closely at the origins of these forces in Section 3.21. Table 2.3 lists four important classes of intermolecular interactions and indicates their relative strengths. We shall have more to say about the origins of these interactions and their effects on physical properties of compounds later in the book. For now, the important point to remember is that values of $\Delta_{fus}H(mp)$ and $\Delta_{vap}H(bp)$, as well as the melting and boiling points of atomic species (e.g. He, Ar) and molecular species (e.g. $CH_4$, $H_2O$, $N_2$ and $C_2H_5OH$) reflect the extent of intermolecular interactions. In an ionic solid (e.g. NaCl) in which ions interact with one another through electrostatic forces, the amount of energy needed to separate the ions is often far greater than that needed to separate covalent molecules. Enthalpies of fusion of ionic solids are significantly higher than those of molecular solids.

**Table 2.3** Types of intermolecular forces.[a]

| Interaction | Acts between: | Typical energy / kJ mol$^{-1}$ |
|---|---|---|
| London dispersion forces | Most molecules | ≤2 |
| Dipole–dipole interactions | Polar molecules | 2 |
| Ion–dipole interactions | Ions and polar molecules | 15 |
| Hydrogen bonds | An electronegative atom (usually N, O or F) and an H atom attached to another electronegative atom | 5–30[b] |

[a] For more detailed discussion, see Sections 3.21 (dispersion forces), 5.11 (dipole moments), 8.6 (electrostatic interactions between ions) and 21.8 (hydrogen bonds).
[b] In $[HF_2]^-$, the H···F hydrogen bond is particularly strong, 165 kJ mol$^{-1}$.

## SUMMARY

This chapter has been concerned with the changes in enthalpy that occur during reactions and during phase changes. We have also introduced different types in intermolecular interactions, and have seen how their differing strengths influence the magnitudes of the enthalpies of fusion and vaporization of elements and compounds.

### Do you know what the following terms mean?

- thermochemistry
- enthalpy change for a reaction
- standard enthalpy change
- standard state of a substance
- exothermic
- endothermic

- calorimetry
- calorimeter
- specific heat capacity
- standard enthalpy of formation
- combustion
- standard enthalpy of combustion

- Hess's Law of Constant Heat Summation
- thermochemical cycle
- molar enthalpy of fusion
- molar enthalpy of vaporization
- van der Waals forces
- intermolecular interactions

### Do you know what the following notations mean?

- $\Delta_r H^\circ(298\,\text{K})$
- $\Delta_f H^\circ(298\,\text{K})$
- $\Delta_c H^\circ(298\,\text{K})$
- $\Delta_{fus} H(\text{mp})$
- $\Delta_{vap} H(\text{bp})$

### You should now be able:

- to define the conditions under which the heat transfer in a reaction is equal to the enthalpy change
- to define what the standard state of an element or compound is, and give examples
- to distinguish between exothermic and endothermic processes, and give an example of each
- to describe the features and operation of a simple, constant-pressure calorimeter
- to describe how to use a constant-pressure calorimeter to measure the specific heat capacity of, for example, copper metal
- to determine the enthalpy of a reaction carried out in a constant-pressure calorimeter, given the change in temperature during the reaction

- to use values of $\Delta_f H^\circ(298\,\text{K})$ to determine standard enthalpies of reactions including combustion reactions
- to construct thermochemical cycles for given situations, and apply Hess's Law of Constant Heat Summation to them
- to distinguish, with examples, between the thermodynamic and kinetic stability of a compound
- to sketch a heating and a cooling curve for a substance (e.g. $H_2O$) undergoing solid/liquid/vapour phase changes
- to explain the origin of the enthalpy changes that accompany phase changes, and to state if a given phase change is exothermic or endothermic

## PROBLEMS

**2.1** From the statements below, say whether the following processes are exothermic or endothermic.
  (a) The addition of caesium to water is explosive.
  (b) The evaporation of a few drops of diethyl ether from the palm of your hand makes your hand feel colder.
  (c) Burning propane gas in $O_2$; this reaction is the basis for the use of propane as a fuel.
  (d) Mixing aqueous solutions of NaOH and HCl causes the temperature of the solution to increase.

**2.2** What are the standard states of the following elements at 298 K: (a) chlorine; (b) nitrogen; (c) phosphorus; (d) carbon; (e) bromine; (f) sodium; (g) fluorine?

**2.3** The standard enthalpy of reaction for the combustion of 1 mole of Ca is $-635\,\text{kJ}\,\text{mol}^{-1}$. Write a balanced equation for the process to which this value refers. Does the reaction give out or absorb heat?

**2.4** $100.0\,\text{cm}^3$ of aqueous hydrochloric acid, HCl $(2.0\,\text{mol}\,\text{dm}^{-3})$ were mixed with $100.0\,\text{cm}^3$ of aqueous NaOH $(2.0\,\text{mol}\,\text{dm}^{-3})$ in a simple, constant-pressure calorimeter. A temperature rise of 13.9 K was recorded. Determine the value of $\Delta H$ for the reaction in kJ per mole of HCl. ($C_{water} = 4.18\,J\,K^{-1}\,g^{-1}$; density of water $= 1.00\,g\,cm^{-3}$)

**2.5** Comment on the fact that the values of $\Delta H$ (quoted per mole of NaOH) for the following reactions are all approximately equal:

$$NaOH(aq) + HCl(aq) \longrightarrow NaCl(aq) + H_2O(l)$$
$$NaOH(aq) + HBr(aq) \longrightarrow NaBr(aq) + H_2O(l)$$
$$NaOH(aq) + HNO_3(aq) \longrightarrow NaNO_3(aq) + H_2O(l)$$

**2.6** When 2.3 g of NaI dissolves in 100.0 g of water contained in a simple, constant-pressure calorimeter, the temperature of the solution rises by 0.28 K. State whether dissolving NaI is an endothermic or exothermic process. Find the enthalpy change for the dissolution of 1 mole of NaI. ($C_{water} = 4.18\,J\,K^{-1}\,g^{-1}$)

**2.7** The specific heat capacity of copper is $0.385\,J\,K^{-1}\,g^{-1}$. A lump of copper weighing 25.00 g is heated to 360.0 K. It is then dropped into $100.0\,cm^3$ of water contained in a constant-pressure calorimeter equipped with a stirrer. If the temperature of the water is initially 295.0 K, determine the maximum temperature attained after the copper has been dropped into the water. What assumptions do you have to make in your calculation? ($C_{water} = 4.18\,J\,K^{-1}\,g^{-1}$; density of water $= 1.00\,g\,cm^{-3}$)

**2.8** Determine $\Delta_r H^\circ(298\,K)$ for each of the following reactions. Data required can be found in Appendix 11.
(a) $H_2(g) + F_2(g) \longrightarrow 2HF(g)$
(b) $4Na(s) + O_2(g) \longrightarrow 2Na_2O(s)$
(c) $2Cl_2O(g) \longrightarrow 2Cl_2(g) + O_2(g)$
(d) $O_2F_2(g) \longrightarrow O_2(g) + F_2(g)$
(e) $3H_2(g) + 2As(grey) \longrightarrow 2AsH_3(g)$
(f) $As(yellow) \longrightarrow As(grey)$
(g) $3O_2(g) \longrightarrow 2O_3(g)$

**2.9** Write a balanced equation for the complete combustion of octane, $C_8H_{18}(l)$. Determine the value for $\Delta_c H^\circ(298\,K)$ using data from Appendix 11.

**2.10** Using data from Appendix 11, determine the standard enthalpy of combustion of propane, $C_3H_8$.

**2.11** Write a balanced equation for the complete combustion of one mole of liquid propan-1-ol, $C_3H_7OH$. Use data from Appendix 11 to find the amount of heat liberated when 3.00 g of propan-1-ol is fully combusted.

**2.12** (a) Sulfur has a number of allotropes. What do you understand by the term *allotrope*? Use data in Appendix 11 to deduce the standard state of sulfur. (b) Determine $\Delta_r H^\circ(298\,K)$ for the conversion of

2.56 g of the orthorhombic form of sulfur to the monoclinic form.

**2.13** Using data from Appendix 11, show by use of an appropriate thermochemical cycle how Hess's Law of Constant Heat Summation can be applied to determine the standard enthalpy change (at 298 K) for the reaction:

$$4LiNO_3(s) \longrightarrow 2Li_2O(s) + 4NO_2(g) + O_2(g)$$

Comment on the fact that LiNO$_3$ does not decompose to Li$_2$O, NO$_2$ and O$_2$ at 298 K.

**2.14** Determine $\Delta_r H^\circ(298\,K)$ for each of the following reactions. For data, see Appendix 11.
(a) $SO_2(g) + \frac{1}{2}O_2(g) \longrightarrow SO_3(s)$
(b) $PCl_5(s) \longrightarrow PCl_3(l) + Cl_2(g)$
(c) $4FeS_2(s) + 11O_2(g) \longrightarrow 2Fe_2O_3(s) + 8SO_2(g)$
(d) $SF_6(g) + 3H_2O(g) \longrightarrow SO_3(g) + 6HF(g)$

**2.15** Figure 2.3 illustrates a heating curve for H$_2$O. Heat is supplied at a constant rate in the experiment. Explain why it takes longer to heat a given mass of liquid water through 1 K than the same mass of solid ice through 1 K.

**2.16** Use data in Table 2.2 to determine the following:
(a) the enthalpy change when 1.60 g of liquid Br$_2$ vaporizes;
(b) the change in enthalpy for the solidification of 2.07 g of molten lead;
(c) the enthalpy change for the condensation of 0.36 g of water.

**2.17** $x$ g of Cl$_2$ are liquefied at 239 K. $\Delta H$ for the process is $-1020\,J$. Use data from Table 2.2 to find $x$.

**2.18** Using data in Table 2.2, determine the enthalpy change for each of the following: (a) melting 4.92 g of gold; (b) liquefying 0.25 moles of N$_2$ gas; (c) vaporizing $150.0\,cm^3$ of water (density of water $= 1.00\,g\,cm^{-3}$).

**2.19** Determine values of $\Delta_r H^\circ(298\,K)$ for the following reactions. For data, see Appendix 11.
(a) $2H_2(g) + CO(g) \longrightarrow CH_3OH(l)$
   methanol
(b) $CuO(s) + H_2(g) \longrightarrow Cu(s) + H_2O(g)$
(c) $4NH_3(g) + 3O_2(g) \longrightarrow 2N_2(g) + 6H_2O(l)$

**2.20** Determine values of $\Delta_c H^\circ(298\,K)$ for (a) the complete combustion of one mole of butane, $C_4H_{10}$, and (b) the partial combustion of one mole of butane in which CO is the only carbon-containing product.

## Additional problems

Data for these problems can be found in Table 2.2 or Appendix 11.

**2.21** (a) Under standard conditions, what products will be formed in the complete combustion of N$_2$H$_4$?

(b) Determine $\Delta_r H^\circ(298\,K)$ for the decomposition of 2.5 g of stibane ($SbH_3$) to its constituent elements.

(c) Find $\Delta_r H^\circ(298\,K)$ for the following reaction:

$$BCl_3(l) + 3H_2O(l) \longrightarrow B(OH)_3(s) + 3HCl(g)$$

**2.22** (a) Write an equation that describes the fusion of silver. Is the process exothermic or endothermic?

(b) Draw out a thermochemical cycle that connects the following interconversions: red to black phosphorus, white to red phosphorus, and white to black phosphorus. Determine $\Delta_r H^\circ(298\,K)$ for $P_4(\text{red}) \longrightarrow P_4(\text{black})$.

**2.23** The conversion of $NO_2(g)$ to $N_2O_4(l)$ is an example of a *dimerization* process. Write a balanced equation for the dimerization of $NO_2$ and determine $\Delta_r H^\circ(298\,K)$ per mole of $NO_2$. Does the value you have calculated indicate that the process is thermodynamically favourable?

**2.24** (a) For crystalline $KMnO_4$, $\Delta_f H^\circ(298\,K) = -837\,kJ\,mol^{-1}$. Write an equation that describes the process to which this value refers.

(b) Cyclohexane, $C_6H_{12}$, is a liquid at 298 K; $\Delta_c H^\circ(C_6H_{12}, l, 298\,K) = -3920\,kJ\,mol^{-1}$. Determine the value of $\Delta_f H^\circ(C_6H_{12}, l, 298\,K)$.

(c) Use your answer to part (b), and the fact that $\Delta_f H^\circ(C_6H_{12}, g, 298\,K) = -123\,kJ\,mol^{-1}$, to determine $\Delta_{vap} H^\circ(C_6H_{12}, 298\,K)$. Why does this value differ from $\Delta_{vap} H(C_6H_{12}, bp) = 30\,kJ\,mol^{-1}$?

**2.25** (a) Hydrogen peroxide decomposes according to the equation:

$$2H_2O_2(l) \longrightarrow 2H_2O(l) + O_2(g)$$

Determine $\Delta_r H^\circ(298\,K)$ per mole of $H_2O_2$.

(b) Write an equation to represent the formation of calcium phosphate, $Ca_3(PO_4)_2$, from its constituent elements under standard conditions.

(c) Find $\Delta_r H^\circ(298\,K)$ for the dehydration of ethanol to give ethene:

$$C_2H_5OH(l) \longrightarrow C_2H_4(g) + H_2O(l)$$

# 3 Atoms and atomic structure

## Topics

- Electrons
- The development of modern atomic theory
- The uncertainty principle
- The Schrödinger wave equation
- Probability density
- Radial distribution functions
- Quantum numbers
- Atomic orbitals
- The hydrogen-like atom
- Atomic spectrum of hydrogen
- The multi-electron atom
- The *aufbau* principle
- Electronic configurations
- The octet rule
- Monatomic gases
- Van der Waals interactions

## 3.1 The importance of electrons

We saw in Chapter 1 that chemistry is concerned with atoms and the species that are formed by their combination. In this chapter, we are concerned with some details of atomic structure. You will learn about a description of atomic structure that is pervasive throughout modern chemistry, and we will extend this description to give an understanding of the structure of multi-atomic ions and molecules.

Atoms consist of nuclei and electrons. The electrons surround the central nucleus which is in turn composed of protons and neutrons; only in the case of protium ($^1$H) is the nucleus devoid of neutrons.

Most chemistry is concerned with the *behaviour of electrons*; the behaviour of nuclei is more properly the realm of the nuclear chemist or physicist. Ions are formed by the gain or loss of electrons; covalent bonds are formed by the sharing of electrons. The number of electrons in an atom of an element controls the *chemical* properties of the element.

Empirical observations led to the grouping together of sets of elements with similar chemical characteristics and to the construction of the periodic table. A periodic table is presented inside the front cover of this book.

Nucleus, proton, electron: see Section 1.4

We now come to the critical question. If the chemical properties of elements are governed by the electrons of the atoms of the elements, why do elements in the same group of the periodic table *but with different total numbers of electrons* behave in a similar manner? This question leads to the concept of *electronic structure* and a need to understand the organization of electrons in atoms.

## 3.2    The classical approach to atomic structure

The road to the present atomic theory has developed from classical mechanics to quantum mechanics. The transition was associated with a crisis in relating theory to experimental observations and with fundamental changes in the understanding of science in the late 19th and early 20th centuries.

The simplest models of atomic structure incorporate no detailed description of the organization of the positively charged nuclei and the negatively charged electrons. The structure merely involves electrostatic attractions between oppositely charged particles. A series of experiments established a model of the atom in which a central positively charged nucleus was surrounded by negatively charged electrons. Consider the consequences of having a positively charged nucleus with static electrons held some distance away. The opposite charges of the electrons and the protons mean that the electrons will be attracted to the nucleus and will be pulled towards it. The only forces opposing this will be the electrostatic *repulsions* between the similarly charged electrons. This model is therefore not consistent with the idea that electrons are found in a region of space *distant* from the nucleus.

This led to attempts to describe the atom in terms of electrons *in motion* about a nucleus. In 1911, Ernest Rutherford proposed an atom consisting of a positively charged nucleus around which electrons move in circular orbits. In a classical model of such an atom, the electrons obey Newton's Laws of Motion, and Rutherford's proposals were flawed because the electron would be attracted towards the nucleus and would plunge towards it. The orbits could not be maintained. Additionally, this classical picture can only describe completely the relative positions and motions of the nucleus and the electron in the hydrogen atom. In atoms with more than one electron it is impossible to solve algebraically the equations describing their motion. Even for hydrogen, however, the Rutherford description was not able to account for some experimentally observed spectroscopic properties.

◼▷
Atomic spectrum of
hydrogen: see Section 3.16

**THEORETICAL AND CHEMICAL BACKGROUND**

### Box 3.1 Newton's Laws of Motion

***The First Law***

A body continues in its state of rest or of uniform motion in a straight line unless acted upon by an external force.

***The Second Law***

The rate of change of momentum of a body is proportional to the applied force and takes place in the direction in which the force acts.

***The Third Law***

For every action, there is an equal and opposite reaction.

## Box 3.2  Atomic theory as a sequence in history

1801  Young demonstrates the **wave properties** of light.

1888  Hertz discovers that radio waves are produced by accelerated electrical charges; this indicates that light is **electromagnetic radiation**.

1897  J. J. Thomson discovers the **electron**.

1900  Rayleigh and Jeans attempt to calculate the energy distribution for **black-body radiation,** but their equation leads to the 'ultraviolet catastrophe' (see Box 3.3).

1900  Planck states that electromagnetic radiation is **quantized**, i.e. the radiation is emitted or absorbed only in discrete amounts ($E = h\nu$).

1905  Einstein considers that light waves also exhibit **particle-like behaviour**; the particles are called **photons** and have energies $h\nu$. (The name *photon* did not appear until 1926.)

1909  Rutherford, Geiger and Marsden show that when $\alpha$-particles strike a piece of gold foil, a small number are deflected (an $\alpha$-particle is an $He^{2+}$ ion). Rutherford suggests that an atom contains a small positively charged **nucleus** surrounded by negatively charged electrons.

1911  Rutherford proposes an atom consisting of a **positively charged nucleus** around which the **electron moves in a circular orbit**; the model is flawed because the orbit would collapse as the electron was attracted towards the nucleus (see Section 3.2).

1913  Bohr proposes a model for the hydrogen atom in which an electron moves around the nucleus in an **orbit with a discrete energy**. Other orbits are possible, also with discrete energies (see Section 3.3).

1924  De Broglie suggests that all particles, including electrons, exhibit both particle and wave properties; this is **wave–particle duality** (see Section 3.5).

1926  The **Schrödinger wave equation** is developed (see Section 3.7).

1927  Davisson and Germer experimentally confirm de Broglie's theory.

1927  Heisenberg states that, because of wave–particle duality, it is impossible to determine simultaneously both the position and momentum of a microscopic particle; this includes an electron. The statement is the **uncertainty principle** (see Section 3.6).

---

### 3.3 The Bohr atom – still a classical picture

In 1913, Niels Bohr developed a *quantized model* for the atom. In the Bohr atom, electrons move in planetary-like orbits about the nucleus. The basic assumption that made the Bohr atom different from previous models was that *the energy of an electron in a particular orbit remains constant and that only certain energies are allowed.* By classical theory, an electron cannot move in a *circular* orbit unless there is a force holding it on this path – if there is no attractive force acting on the electron, it will escape from the orbit (think what happens if you swing a ball tied on the end of a string in circles and then let go of the string). Figure 3.1 shows an electron moving with a velocity $v$ in a circular path about the fixed nucleus. Velocity is a *vector* quantity and for any given point in the path of the electron, the velocity will have a direction tangential to the circle.

A *scalar* quantity possesses magnitude only, e.g. mass.

A *vector* quantity possesses both magnitude and direction, e.g. velocity.

**Fig. 3.1** The Bohr atom. The negatively charged electron is involved in circular motion about a fixed positively charged nucleus. There is an attractive force, but also an outward force due to the electron trying to escape from the orbit.

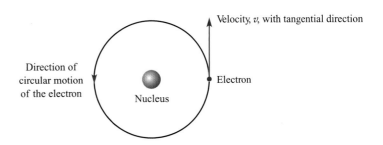

*Quantization* is the packaging of a quantity into units of a discrete size.

In the Bohr atom, the outward force exerted on the electron as it tries to escape from the circular orbit is *exactly* balanced by the inward force of attraction between the negatively charged electron and the positively charged nucleus. Thus, the electron can remain away from the nucleus. Bohr built into his atomic model the idea that different orbits were possible and that the orbits could *only* be of certain energies. This was a very important advance in the development of atomic theory. By allowing the electrons to move *only* in particular orbits, Bohr had *quantized* his model. The model was successful in terms of being consistent with some atomic spectroscopic observations but the Bohr atom fails for other reasons. We return to the Bohr atom in Section 3.16.

## 3.4     Quanta

$E = h\nu$

The units of energy $E$ are joules (J), of frequency $\nu$ are $s^{-1}$ or Hz, and $h$ is the Planck constant, $6.626 \times 10^{-34}$ J s.

In the preceding paragraphs we have considered the organization of the electrons in an atom in terms of classical mechanics and of 'classical' quantum mechanics. In the late 19th century, an important (historical) failure of the classical approach was recognized – the so-called 'ultraviolet catastrophe' (see Box 3.3). This failure led Max Planck to develop the idea of *discrete energies* for electromagnetic radiation, i.e. the energy is *quantized*.

Equation 3.1 gives the relationship between energy $E$ and frequency $\nu$. The proportionality constant is $h$, the Planck constant ($h = 6.626 \times 10^{-34}$ J s).

$$E = h\nu \tag{3.1}$$

## 3.5     Wave–particle duality

Albert Einstein (1879–1955).
© *The Nobel Foundation.*

Planck's model assumed that light was an electromagnetic wave. In 1905, Albert Einstein proposed that electromagnetic radiation could exhibit particle-like behaviour. However, it became apparent that while electromagnetic radiation possessed some properties that were fully explicable in terms of classical particles, other properties could only be explained in terms of waves. These 'particle-like' entities became known as *photons* and from equation 3.1, we see that each photon of light of a given frequency $\nu$ has an energy $h\nu$.

Now let us return to the electrons in an atom. In 1924, Louis de Broglie argued that if radiation could exhibit the properties of both particles *and* waves (something that defeats the laws of classical mechanics), then so too could electrons, and indeed so too could *every* moving particle. This phenomenon is known as *wave–particle duality*.

The de Broglie relationship (equation 3.2) shows that a particle (and this includes an electron) with momentum $mv$ ($m$ = mass and $v$ = velocity of the particle) has an associated wave of wavelength $\lambda$. Thus, equation 3.2 combines the concepts of classical momentum with the idea of wave-like properties.

$$\lambda = \frac{h}{mv} \quad \text{where } h = \text{the Planck constant} \tag{3.2}$$

And so, we have begun to move towards *quantum* (or *wave*) *mechanics* and away from classical mechanics. Let us see what this means for the behaviour of electrons in atoms.

**THEORETICAL AND CHEMICAL BACKGROUND**

## Box 3.3 Black-body radiation and the 'ultraviolet catastrophe'

An ideal *black-body* absorbs all radiation of all wavelengths that falls upon it. When heated, a black-body emits radiation.

If one applies the laws of classical physics to this situation (as was done by Rayleigh and Jeans), the results are rather surprising. One finds that oscillators of very short wavelength, $\lambda$, should radiate strongly at room temperature. The intensity of radiation is predicted to increase continuously, never passing through a maximum. This result disagrees with experimental observations which show a maximum with decreasing intensity to higher or lower wavelengths. Very short wavelengths correspond to ultraviolet light, X-rays and $\gamma$-rays. Hence the term the 'ultraviolet catastrophe'.

The situation can only be rectified if one moves away from classical ideas. This was the approach taken by Max Planck. He proposed that the electromagnetic radiation had associated with it only *discrete energies*. The all-important equation arising from this is:

$$E = h\nu$$

where $E$ is the energy, $\nu$ is the frequency and $h$ is the Planck constant. This equation means that the energy of the emitted or absorbed electromagnetic radiation must have an energy which is a multiple of $h\nu$.

Remember that frequency, $\nu$, and wavelength, $\lambda$, of electromagnetic reaction are related by the equation:

$$c = \lambda\nu$$

where $c$ is the speed of light (see Appendix 4).

Max Karl Ernst Ludwig Planck (1858–1947).
© *The Nobel Foundation.*

---

### 3.6   The uncertainty principle

Werner Karl Heisenberg (1901–1976).
© *The Nobel Foundation.*

For a particle with a relatively large mass, wave properties are unimportant and the position and motion of the particle can be defined and measured almost exactly. However, for an electron with a tiny mass, this is not the case. Treated in a classical sense, an electron can move along a defined path in the same way that a ball moves when it is thrown. However, once we give the electron wave-like properties, it becomes impossible to know exactly both the position and the momentum of the electron at the same instant in time. This is Heisenberg's *uncertainty principle*.

We must now think of the electron in a new and different way and consider the *probability* of finding the electron in a given volume of space, rather than trying to define its exact position and momentum. The probability of finding an electron at a given point in space is determined from the function $\psi^2$ where $\psi$ is the *wavefunction*. A wavefunction is a mathematical function which tells us in detail about the behaviour of an electron-wave.

Now we must look for ways of saying something about values of $\psi$ and $\psi^2$, and this leads us on to the *Schrödinger wave equation*.

The ***probability of finding an electron*** at a given point in space is determined from the function $\psi^2$ where $\psi$ is the ***wavefunction***.

| 3.7 | **The Schrödinger wave equation** |

### The equation

One form of the Schrödinger wave equation is shown in equation 3.3.

$$\mathcal{H}\psi = E\psi \tag{3.3}$$

When first looking at equation 3.3, you might ask: 'Can one divide through by $\psi$ and say that $\mathcal{H} = E$?' This seems a reasonable question, but in order to answer it (and the answer is 'no'), we need to understand what the various components of equation 3.3 mean.

Look first at the right-hand side of equation 3.3. We have already stated that $\psi$ is a wavefunction. $E$ is the total energy associated with the wavefunction $\psi$. The left-hand side of equation 3.3 represents a mathematical operation upon the function $\psi$. $\mathcal{H}$ is called the *Hamiltonian operator* (Box 3.4).

Erwin Schrödinger (1887–1961). © *The Nobel Foundation.*

One form of the *Schrödinger wave equation* is:

$\mathcal{H}\psi = E\psi$

$\mathcal{H}$ is the Hamiltonian operator, the wavefunction $\psi$ is an eigenfunction and the energy $E$ is an eigenvalue.

### Eigenvalues and eigenvectors

Equation 3.3 can be expressed more generally as in equation 3.4. The equation is set up so that, having 'operated' on the function (in our case, $\psi$) the answer comes out in the form of a scalar quantity (in our case, $E$) multiplied by the same function (in our case, $\psi$). For such a relationship, the function is called an *eigenfunction* and the scalar quantity is called the *eigenvalue*. Applied to equation 3.3, this means that $\psi$ is an eigenfunction and $E$ is an eigenvalue, while $\mathcal{H}$ is the operator.

Operator working on a function = (scalar quantity) × (the original function)

$$\tag{3.4}$$

### What information is available from the Schrödinger equation?

It is difficult to grasp the physical meaning of the Schrödinger equation. However, the aim of this discussion is not to study this equation in detail but merely to illustrate that the equation can be set up for a given system and can be solved (either exactly or approximately) to give values of $\psi$ and hence $\psi^2$. From the Schrödinger equation, we can find energy values that

**THEORETICAL AND CHEMICAL BACKGROUND**

### Box 3.4 The Hamiltonian operator

Just as the name suggests, a mathematical operator performs a mathematical operation on a mathematical function.

Differentiation is one type of mathematical operation and is reviewed at a basic level in the *Mathematics Tutor* (on the website accompanying this book; see www.pearsoned.co.uk/housecroft). For example: $\dfrac{d}{dx}$ instructs you to differentiate a variable with respect to $x$.

Equation 3.3 introduces the Hamiltonian operator, $\mathcal{H}$, in the Schrödinger wave equation. The Hamiltonian

operator is a differential operator and takes the form:

$$-\frac{h^2}{8\pi^2 m}\left(\frac{\partial^2}{\partial x^2} + \frac{\partial^2}{\partial y^2} + \frac{\partial^2}{\partial z^2}\right) + V(x, y, z)$$

in which $\dfrac{\partial^2}{\partial x^2}$, $\dfrac{\partial^2}{\partial y^2}$ and $\dfrac{\partial^2}{\partial z^2}$ are *partial differentials* (see Box 17.1) and $V(x, y, z)$ is a potential energy term. The complicated nature of $\mathcal{H}$ arises from the need to describe the position of the electron in three-dimensional space.

are associated with particular wavefunctions. The quantization of the energies is built into the mathematics of the Schrödinger equation.

A very important point to remember is that the Schrödinger equation can *only* be solved exactly for a two-body problem, i.e. for a species that contains a nucleus and only one electron, e.g. H, $He^+$ or $Li^{2+}$; these are all *hydrogen-like* species.

The wavefunction $\psi$ is a solution of the Schrödinger equation and describes the behaviour of an electron in the region in space of the atom called the *atomic orbital*. This result can be extended to give information about a *molecular orbital* in a molecular species.

An atomic orbital is usually described in terms of three integral *quantum numbers*. The principal quantum number, $n$, is a positive integer with values lying between limits 1 and $\infty$. For most chemical purposes, we are within the limits of $n = 1$–7. Two further quantum numbers, $l$ and $m_l$, may be derived (see Section 3.10) and a combination of these three quantum numbers defines a unique *orbital*.

> A **hydrogen-like** species contains one electron.

■▷ Quantum numbers: see Section 3.10

---

A **wavefunction** $\psi$ is a mathematical function that contains detailed information about the behaviour of an electron (electron-wave).

The region of space defined by such a wavefunction is termed an **atomic orbital.**

Each atomic orbital may be uniquely defined by a set of three **quantum numbers**: $n$, $l$ and $m_l$.

---

### Radial and angular parts of the wavefunction

So far, we have described the position of the electron by Cartesian coordinates $(x, y, z)$ with the nucleus at $(0, 0, 0)$. It is very often convenient to use a different coordinate system which separates a radial distance coordinate $(r)$ from the angular components ($\theta$ and $\phi$) (Figure 3.2). These are called

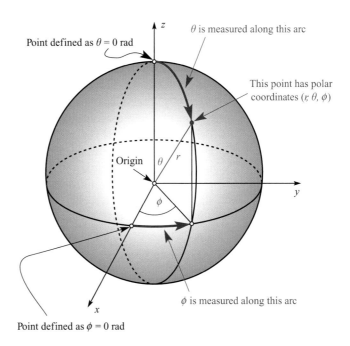

**Fig. 3.2** Definition of the polar coordinates $(r, \theta, \phi)$ for a point shown here in red; $r$ is the radial coordinate and $\theta$ and $\phi$ are angular coordinates. Cartesian axes ($x$, $y$ and $z$) are also shown. The point at the centre of the diagram is the origin.

Point defined as $\theta = 0$ rad

$\theta$ is measured along this arc

This point has polar coordinates $(r, \theta, \phi)$

Origin

$\phi$ is measured along this arc

Point defined as $\phi = 0$ rad

*spherical polar coordinates*. We will not be concerned with the mathematical manipulations used to transform a set of Cartesian coordinates into spherical polar coordinates. However, it is useful to be able to define separate wavefunctions for the radial ($r$) and angular ($\theta, \phi$) components of $\psi$. This is represented in equation 3.5 where $R(r)$ and $A(\theta, \phi)$ are new radial and angular wavefunctions respectively.

$$\psi_{\text{Cartesian}}(x, y, z) \equiv \psi_{\text{radial}}(r)\psi_{\text{angular}}(\theta, \phi) = R(r)A(\theta, \phi) \tag{3.5}$$

An ***atomic wavefunction*** $\psi$ consists of a radial, $R(r)$, and an angular, $A(\theta, \phi)$, component.

The radial component in equation 3.5 is dependent upon the quantum numbers $n$ and $l$, whereas the angular component depends on $l$ and $m_l$. Hence the components should really be written as $R_{n,l}(r)$ and $A_{l,m_l}(\theta, \phi)$. In this book we shall simplify these expressions. The notation $R(r)$ refers to the radial part of the wavefunction, and $A(\theta, \phi)$ refers to the angular part of the wavefunction. It follows that $R(r)$ will give *distance* information, whereas $A(\theta, \phi)$ will give information about shape.

| 3.8 | **Probability density** |
|---|---|

### The functions $\psi^2$, $R(r)^2$ and $A(\theta, \phi)^2$

The function $\psi^{2\S}$ is proportional to the *probability density* of the electron at a point in space. By considering values of $\psi^2$ at points in the volume of space about the nucleus, it is possible to define a surface boundary that encloses the region of space in which the probability of finding the electron is, say, 95%. By doing this, we are effectively drawing out the boundaries of an atomic orbital. Remember that $\psi^2$ may be described in terms of the radial and angular components $R(r)^2$ and $A(\theta, \phi)^2$. Remember, also, that these boundaries are only approximations; in principle, all orbitals are infinitely large.

### The 1s atomic orbital

Atomic orbitals: see Section 3.11

The lowest energy solution to the Schrödinger equation for the hydrogen atom leads to the 1s orbital. The 1s orbital is spherical in shape; it is *spherically symmetric* about the nucleus. If we are interested only in the probability of finding the electron as a function of distance from the nucleus, we can consider only the radial part of the wavefunction $R(r)$. Values of $R(r)^2$ are largest near to the nucleus and then become smaller as we move along a radius centred on the nucleus. Since the chance of the electron being far out from the nucleus is *minutely* small, we draw the boundary surface for the orbital so as to enclose 95% of the probability density of the total wavefunction $\psi^2$. Figure 3.3 gives a plot of $R(r)^2$ against the distance, $r$, from the nucleus. The curve for $R(r)^2$ approaches zero only as the radius approaches $\infty$.

---

§ Although here we use $\psi^2$, we ought really to write $\psi\psi^*$ where $\psi^*$ is the complex conjugate of $\psi$. In one dimension (say $x$), the probability of finding the electron between the limits of $x$ and $(x + \mathrm{d}x)$ (where $\mathrm{d}x$ is an extremely small change in $x$) is proportional to the function $\psi(x)\psi^*(x)\,\mathrm{d}x$. In three dimensions this leads to the use of $\psi\psi^*\,\mathrm{d}\tau$ in which we are considering the probability of finding the electron in a volume element $\mathrm{d}\tau$. Using only the radial part of the wavefunction, the function becomes $R(r)R^*(r)$.

**Fig. 3.3** A plot of $R(r)^2$, as a function of distance, $r$, from the atomic nucleus. This plot refers to the case of a spherically symmetrical $1s$ orbital.

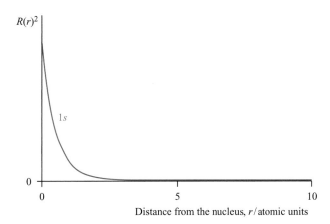

**Normalization**

Wavefunctions are usually *normalized* to unity. This means that the probability of finding the electron somewhere in space is taken to be unity, i.e. 1. In other words, the electron has to be somewhere!

Mathematically, the normalization is represented by equation 3.6. This effectively states that we are integrating ($\int$) over all space ($\mathrm{d}\tau$) and that the total integral of $\psi^2$ (which is a measure of the probability density) must be unity.[§]

$$\int \psi^2 \, \mathrm{d}\tau = 1 \tag{3.6}$$

**3.9**  **The radial distribution function, $4\pi r^2 R(r)^2$**

Another way of representing information about the probability density is to plot a *radial distribution function*. The relationship between $R(r)^2$ and the radial distribution function is given in equation 3.7.

Radial distribution function $= 4\pi r^2 R(r)^2$ (3.7)

The advantage of using this new function is that it represents the probability of finding the electron in a *spherical shell* of radius $r$ and thickness $\mathrm{d}r$, where $(r + \mathrm{d}r)$ may be defined as $(r + $ a small increment of radial distance$)$ (Figure 3.4). Thus, instead of considering the probability of finding the electron as a

**Fig. 3.4** A view into a spherical shell of inner radius $r$ (centred on the nucleus) and thickness $\mathrm{d}r$, where $\mathrm{d}r$ is a very small increment of the radial distance $r$.

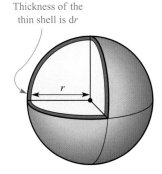

Thickness of the thin shell is $\mathrm{d}r$

$r$

---

[§] Strictly we should write equation 3.6 as $\int \psi\psi^* \mathrm{d}\tau = 1$ where $\psi^*$ is the complex conjugate of $\psi$.

**Fig. 3.5** The radial distribution function $4\pi r^2 R(r)^2$ for the $1s$ atomic orbital. This function describes the probability of finding the electron at a distance $r$ from the nucleus. The maximum value of $4\pi r^2 R(r)^2$ corresponds to the distance from the nucleus where there is the maximum probability of finding the electron.

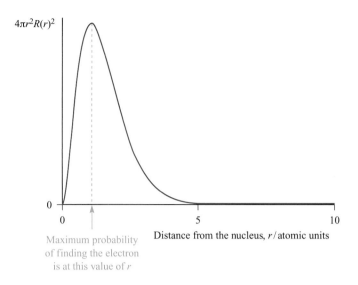

function of distance from the nucleus as illustrated in Figure 3.3, we are now able to consider the complete region in space in which the electron resides.

Figure 3.5 shows the radial distribution function for the $1s$ orbital. Notice that it is zero at the nucleus; this contrasts with the situation for $R(r)^2$ (Figure 3.3). The difference arises from the dependence of the radial distribution function upon $r$; at the nucleus, $r$ is zero, and so $4\pi r^2 R(r)^2$ must equal zero.

> ◼▶
> We return to radial distribution functions in Section 3.13

> The *radial distribution function* is given by the expression $4\pi r^2 R(r)^2$. It gives the probability of finding an electron in a spherical shell of radius $r$ and thickness $dr$. The radius $r$ is measured from the nucleus.

## 3.10    Quantum numbers

An *atomic orbital* is defined by a unique set of three quantum numbers ($n$, $l$ and $m_l$).

An *electron in an atom* is defined by a unique set of four quantum numbers ($n$, $l$, $m_l$ and $m_s$).

In several parts of the preceding discussion, we have mentioned *quantum numbers*. Now we look more closely at these important numbers.

As we have seen, the effects of quantization are that only certain electronic energies, and hence orbital energies, are permitted. Each orbital is described by a set of quantum numbers $n$, $l$ and $m_l$. A fourth quantum number, $m_s$, gives information about the spin of an electron in an orbital. Every orbital in an atom has a unique set of *three* quantum numbers and each electron in an atom has a unique set of *four* quantum numbers.

### The principal quantum number *n*

*Principal quantum number*, $n = 1, 2, 3, 4, 5 \cdots \infty$

The principal quantum number $n$ may have any positive integer value between 1 and $\infty$. The number $n$ corresponds to the orbital energy level or 'shell'. For a particular energy level there may be sub-shells, the number of which is defined by the quantum number $l$. For hydrogen-like species, all orbitals with the same principal quantum number have the same energy; this is not the case for other species where the sub-shells have different energies.

## The orbital quantum number $l$

The quantum number $l$ is also known as the azimuthal quantum number

The quantum number $l$ is called the *orbital quantum number*. For a given value of the principal quantum number, the allowed values of $l$ are positive integers lying between 0 and $(n-1)$. Thus, if $n = 3$, the permitted values of $l$ are 0, 1 and 2. Each value of $l$ corresponds to a particular type of atomic orbital.

Knowing the value of $l$ provides detailed information about the region of space in which an electron may move; it describes the shape of the orbital. The value of $l$ also determines the angular momentum of the electron within the orbital, i.e. its *orbital angular momentum* (see Box 3.6).

*Orbital quantum number*, $l = 0, 1, 2, 3, 4 \ldots (n-1)$.

$n = 1 \qquad l = 0$
$n = 2 \qquad l = 0, 1$
$n = 3 \qquad l = 0, 1, 2$
$n = 4 \qquad l = 0, 1, 2, 3$

## The magnetic quantum number $m_l$

*Magnetic quantum number* $m_l = -l, (-l+1), \ldots 0,$ $\ldots (l-1), l$

The magnetic quantum number $m_l$ relates to the directionality of an orbital and has values which are integers between $+l$ and $-l$. If $l = 1$, the allowed values of $m_l$ are $-1$, 0 and $+1$.

---

| Worked example 3.1 | *Deriving quantum numbers and what they mean* |

**Derive possible sets of quantum numbers for $n = 2$, and explain what these sets of numbers mean.**

Let $n = 2$.
The value of $n$ defines an energy (or principal) level.
Possible values of $l$ lie in the range 0 to $(n-1)$.
Therefore, for $n = 2$, $l = 0$ and 1.
This means that the $n = 2$ level gives rise to two sub-levels, one with $l = 0$ and one with $l = 1$.
Now determine the possible values of the quantum number $m_l$: values of $m_l$ lie in the range $-l \ldots 0 \ldots +l$.
The sub-level $l = 0$ has associated with it one value of $m_l$:

$$m_l = 0 \text{ for } l = 0$$

The sub-level $l = 1$ has associated with it three values of $m_l$:

$$m_l = -1, 0 \text{ or } +1 \text{ for } l = 1$$

The possible sets of quantum numbers for $n = 2$ are:

| $n$ | $l$ | $m_l$ |
|-----|-----|-------|
| 2 | 0 | 0 |
| 2 | 1 | $-1$ |
| 2 | 1 | 0 |
| 2 | 1 | $+1$ |

The physical meaning of these sets of quantum numbers is that for the level $n = 2$, there will be two *types* of orbital (because there are two values of $l$). For $l = 1$, there will be three orbitals *all of a similar type* but with *different directionalities*. (Details of the orbitals are given in Section 3.11.)

### The spin quantum number $s$ and the magnetic spin quantum number $m_s$

In a mechanical picture, an electron may be considered to spin about an axis passing through it and so possesses spin angular momentum in addition to the orbital angular momentum discussed above. The spin quantum number $s$ determines the *magnitude of the spin angular momentum* of an electron and can *only* have a value of $\frac{1}{2}$. The magnetic spin quantum number $m_s$ determines the *direction of the spin angular momentum* of an electron and has values of $+\frac{1}{2}$ or $-\frac{1}{2}$.

An atomic orbital can contain a maximum of two electrons. The two values of $m_s$ correspond to labels for the two electrons that can be accommodated in any one orbital. When two electrons occupy the same orbital, one possesses a value of $m_s = +\frac{1}{2}$ and the other $m_s = -\frac{1}{2}$. We say that this corresponds to the two electrons having *opposite* spins.

> An atomic orbital can contain a maximum of two electrons.

> *Magnetic spin quantum number $m_s = \pm\frac{1}{2}$*

> ◄► We return to quantum numbers in Section 3.18

---

| Worked example 3.2 | *Deriving a set of quantum numbers that uniquely defines a particular electron* |
|---|---|

**Derive a set of quantum numbers that describes an electron in an atomic orbital with $n = 1$.**

First, determine how many orbitals are possible for $n = 1$.
For $n = 1$, possible values for $l$ lie between 0 and $(n - 1)$.
Therefore, for $n = 1$, only $l = 0$ is possible.

Now determine possible values of $m_l$.
Possible values of $m_l$ lie in the range $-l \dots 0 \dots +l$.
Therefore for $l = 0$, the only possible value of $m_l$ is 0.

The orbital that has been defined has the quantum numbers:
$$n = 1, \quad l = 0, \quad m_l = 0$$

This orbital can contain up to two electrons.
Therefore, each electron is uniquely defined by one of the two following sets of quantum numbers:
$$n = 1, \quad l = 0, \quad m_l = 0, \quad m_s = +\frac{1}{2}$$
or
$$n = 1, \quad l = 0, \quad m_l = 0, \quad m_s = -\frac{1}{2}$$

---

**3.11**     ## Atomic orbitals

### Types of atomic orbital

In the previous section we saw that the quantum number $l$ defines a particular type of atomic orbital. The four types of atomic orbital most commonly encountered are the $s$, $p$, $d$ and $f$ orbitals. The letters $s$, $p$, $d$ and $f$ are simply used here as labels. A value of $l = 0$ corresponds to an $s$ orbital, $l = 1$ refers to a $p$ orbital, $l = 2$ refers to a $d$ orbital and $l = 3$ corresponds to an $f$ orbital.

The distinction between these four types of atomic orbital comes from their shapes and symmetries, and the shape of the orbital is governed by the quantum number $l$.

> For an $s$ orbital, $l = 0$
> For a $p$ orbital, $l = 1$
> For a $d$ orbital, $l = 2$
> For an $f$ orbital, $l = 3$

Fig. 3.6 Two ways of representing the phase of a wavefunction. The examples shown are (a) a $1s$ orbital for which there is no phase change across the orbital, and (b) a $2p$ orbital (here the $2p_z$) for which there is one phase change.

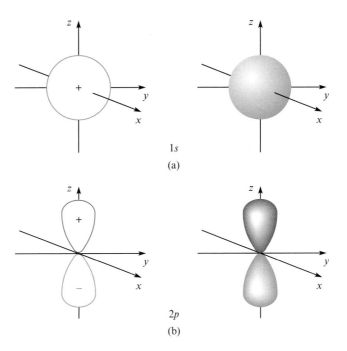

$1s$

(a)

$2p$

(b)

An $s$ orbital is spherically symmetric about the nucleus and the boundary surface of the orbital has a constant *phase*. That is, the amplitude of the wavefunction associated with the boundary surface of the $s$ orbital is always either positive or negative (Figure 3.6a).

For a $p$ orbital, there is *one* phase change with respect to the surface boundary of the orbital, and this occurs at the so-called *nodal plane* (Figure 3.7). Each part of the orbital is called a *lobe*. The phase of a wavefunction is designated as having either a positive or negative amplitude, or by shading an appropriate lobe as indicated in Figure 3.6b. A nodal plane in an orbital corresponds to a node in a transverse wave (Figure 3.8).

> The ***amplitude of the wavefunction*** associated with an $s$ orbital is *either* positive *or* negative. The ***surface boundary*** of an $s$ orbital has no change of phase, and the orbital has *no* nodal plane.
>
> The ***surface boundary*** of a $p$ orbital has *one* change of phase, and the orbital possesses *one* nodal plane.

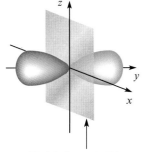

Nodal plane – in this case it lies in the $xz$ plane

Fig. 3.7 The phase change in an atomic $p$ orbital. In this case, the orbital is the $p_y$ orbital and the nodal plane lies in the $xz$ plane.

### The numbers and shapes of orbitals of a given type

For an $s$ orbital, the orbital quantum number, $l$, is 0. This corresponds to only one value of $m_l$: $l = 0$, $m_l = 0$. For a given principal quantum number, $n$, there is only one $s$ orbital, and the $s$ orbital is said to be singly degenerate. The $s$ orbital is spherically symmetric (Figure 3.9).

For a $p$ orbital, the orbital quantum number $l$ has a value of 1. This leads to three possible values of $m_l$, since, for $l = 1$, $m_l = +1, 0, -1$. The physical meaning of this statement is that the Schrödinger equation gives three solutions for $p$ orbitals for a given value of $n$ when $n \geq 2$. (*Exercise*: Why

## Box 3.5  Orbital labels

The labels *s*, *p*, *d* and *f* have their origins in the words 'sharp', 'principal', 'diffuse' and 'fundamental'. These originally referred to the characteristics of the lines observed in the atomic spectrum of hydrogen (see Section 3.16). Nowadays, it is best to think of them simply as labels.

Further types of atomic orbital are labelled *g*, *h*, etc.

| Value of $l$ | 0 | 1 | 2 | 3 | 4 | 5 |
|---|---|---|---|---|---|---|
| Label of atomic orbital | *s* | *p* | *d* | *f* | *g* | *h* |

---

**Fig. 3.8** The definition of a node in a transverse wave. At a node, the amplitude of the wave is zero.

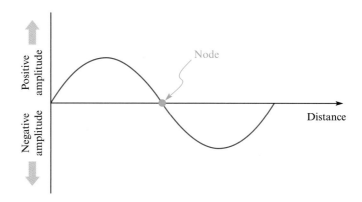

---

## Box 3.6  Angular momentum

Angular momentum = Mass × Angular velocity

Angular velocity is the angle, $\alpha$, turned through per second by a particle travelling on a circular path. The SI units of angular velocity are radians per second.

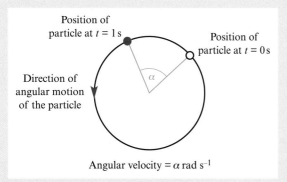

Angular velocity = $\alpha$ rad s$^{-1}$

Consider an electron that is moving in an orbital about the nucleus. The electron has spin *and* orbital angular momenta, both of which are quantized. The *orbital angular momentum* is determined by the quantum number *l*. The amount of orbital angular momentum

that an electron possesses is given by:

$$\frac{h}{2\pi}\sqrt{l(l+1)}$$

where $h$ = Planck constant.

An electron in an *s* orbital (where $l = 0$) has no angular momentum. An electron in a *p* orbital (where $l = 1$) has orbital angular momentum of:

$$\frac{h}{2\pi}\sqrt{l(l+1)} = \sqrt{2}\left(\frac{h}{2\pi}\right)$$

Remember that momentum is a vector property. There are $(2l + 1)$ possible directions for the orbital angular momentum vector, corresponding to the $(2l + 1)$ possible values of $m_l$ for a given value of $l$. Thus, for an electron in a *p* orbital ($l = 1$), there are three possible directions for the orbital angular momentum vector.

*Exercise*

For an electron in a *d* orbital, show that the orbital angular momentum is $\sqrt{6}\left(\frac{h}{2\pi}\right)$ and that there are five possible directions for the orbital angular momentum vector.

**Fig. 3.9** For a given value of the principal quantum number $n$ there is one $s$ orbital. For a given value of $n$ when $n \geq 2$, there are three $p$ orbitals.

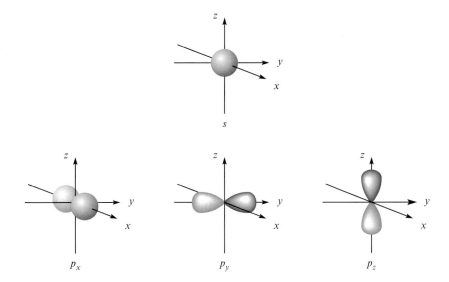

*Degenerate* orbitals possess the same energy.

◼▶
We describe the shapes of $d$ orbitals in Section 23.1

does $n$ have to be greater than 1?) The three solutions are conventionally drawn as shown in Figure 3.9. The surface boundaries of the three $np$ orbitals are identical but the orbitals are mutually orthogonal. A set of $p$ orbitals is said to be *triply* or *three-fold degenerate*. By normal convention, one $p$ orbital points along the $x$ axis and is called the $p_x$ orbital, while the $p_y$ and $p_z$ orbitals point along the $y$ and $z$ axes respectively. This, in effect, defines the $x$, $y$ and $z$ axes!

For a $d$ orbital, the orbital quantum number $l$ is 2. This leads to five possible values of $m_l$ (for $l = 2$, $m_l = +2, +1, 0, -1, -2$). The Schrödinger equation gives five real solutions for $d$ orbitals for a given value of $n$ when $n \geq 3$. (*Exercise:* Why must $n$ be greater than 2?) A set of $d$ orbitals is said to be *five-fold degenerate*.

### Sizes of orbitals

Orbitals that have identical values of $l$ and $m_l$ (e.g. 2s and 4s, or $2p_z$ and $3p_z$) possess the same *orbital symmetry*. A series of orbitals of the same symmetry but with different values of the principal quantum number (e.g. $1s, 2s, 3s, 4s \ldots$) differ in their relative sizes (volumes or spatial extent). The larger the value of $n$, the larger the orbital. This is shown in Figure 3.10 for a series of $ns$ atomic orbitals. The relationship between $n$ and orbital size is not linear. The increase in size corresponds to the orbital being more *diffuse*. In general, the further the maximum electron density is from the nucleus, the higher the energy of the orbital.

**Fig. 3.10** The increase in size of hydrogen $ns$ atomic orbitals exemplified for $n = 1, 2, 3$ and 4. The 4s orbital is *more diffuse* than the 3s, and this in turn is more diffuse than the 2s or the 1s orbitals.

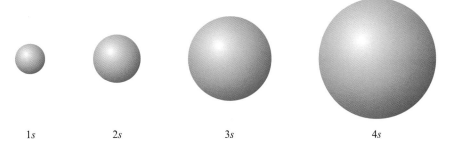

| 3.12 | **Relating orbital types to the principal quantum number** |

For a given value of the principal quantum number $n$, an allowed series of values of $l$ and $m_l$ can be determined as described in Section 3.10. It follows that the number of orbitals and types allowed for a given value of $n$ can therefore be determined. We distinguish between orbitals of the same type but with different principal quantum numbers by writing $ns, np, nd \ldots$ orbitals.

### Consider the case for $n = 1$

For a given value of $n$, the total number of atomic orbitals is $n^2$.

If $n = 1$, then $l = 0$ and $m_l = 0$.
The value of $l = 0$ corresponds to an $s$ orbital which is singly degenerate. There is only one allowed value of $m_l$ ($m_l = 0$).
Thus, for $n = 1$, the allowed atomic orbital is the $1s$.

### Consider the case for $n = 2$

If $n = 2$, then $l = 1$ or 0.
Take each value of $l$ separately.
For $l = 0$, there is a single $s$ orbital, designated as the $2s$ orbital.
For $l = 1$, the allowed values of $m_l$ are $+1, 0$ and $-1$, and this corresponds to a set of three $p$ orbitals.
Thus, for $n = 2$, the allowed atomic orbitals are the $2s$, $2p_x$, $2p_y$ and $2p_z$.

### Consider the case for $n = 3$

If $n = 3$, then $l = 2, 1$ or 0.
Take each value of $l$ separately.
For $l = 0$, there is a single $s$ orbital, designated as the $3s$ orbital.
For $l = 1$, the allowed values of $m_l$ are $+1, 0$ and $-1$, and this corresponds to a set of three $p$ orbitals.
For $l = 2$, the allowed values of $m_l$ are $+2, +1, 0, -1$ and $-2$, and this corresponds to a set of five $d$ orbitals.

▪▶
The labels for the five $3d$ atomic orbitals are given in Section 23.1

Thus, for $n = 3$, the allowed atomic orbitals are the $3s$, $3p_x$, $3p_y$, $3p_z$ and five $3d$ orbitals.

| $n$ | Atomic orbitals allowed | Total number of orbitals, $n^2$ | Total number of electrons |
|---|---|---|---|
| 1 | one $s$ | 1 | 2 |
| 2 | one $s$, three $p$ | 4 | 8 |
| 3 | one $s$, three $p$, five $d$ | 9 | 18 |

| 3.13 | **More about radial distribution functions** |

So far, we have focused on the differences between different types of atomic orbitals in terms of their different quantum numbers, and have stated that the shape of an atomic orbital is governed by the quantum number $l$. For the

**Fig. 3.11** Radial distribution functions, $4\pi r^2 R(r)^2$, for the 1s, 2s and 3s atomic orbitals of hydrogen. Compare the number of radial nodes in these diagrams with the data in Table 3.1.

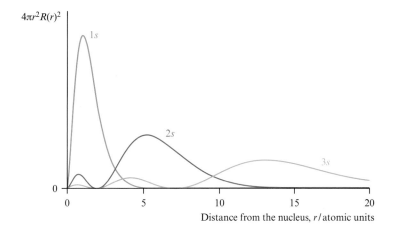

specific case of the 1s orbital, we have also considered the radial part of the wavefunction (Figures 3.3 and 3.5). Now we look at how the *radial distribution function*, $4\pi r^2 R(r)^2$ (see Section 3.9), varies as a function of $r$ for atomic orbitals other than the 1s orbital.

First, we consider the 2s orbital. Figure 3.10 showed that the surface boundaries of the 1s and 2s orbitals are similar. However, a comparison of the radial distribution functions for the two orbitals (Figure 3.11) shows these to be *dissimilar*. Whereas the radial distribution function, $4\pi r^2 R(r)^2$, for the 1s orbital has one maximum, that for the 2s orbital has two maxima and the function falls to zero between them. The point at which $4\pi r^2 R(r)^2 = 0$ is called a *radial node*. Notice that the value of zero at the *radial node* is due to the function $R(r)$ equalling zero, and not $r = 0$ as was true at the nucleus.

*As we are dealing with a function containing $R(r)^2$, the graph can never have negative values. The wavefunction $R(r)$ can have both positive and negative amplitudes.*

Figure 3.11 also shows the radial distribution function for the 3s orbital. The number of radial nodes increases on going from the 2s to the 3s orbital. The trend continues, and thus the 4s and 5s orbitals have three and four

**Fig. 3.12** Radial distribution functions, $4\pi r^2 R(r)^2$, for the 3s, 3p and 3d atomic orbitals of hydrogen. Compare these diagrams with the data in Table 3.1.

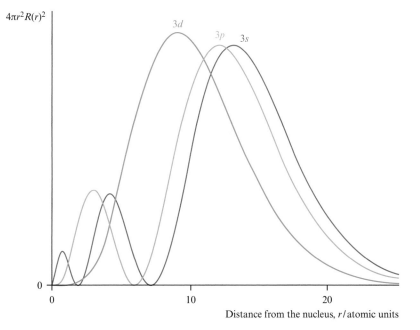

**Table 3.1** Number of radial nodes as a function of orbital type and principal quantum number, $n$. The pattern continues for $n > 4$.

| $n$ | $s\ (l = 0)$ | $p\ (l = 1)$ | $d\ (l = 2)$ | $f\ (l = 3)$ |
|---|---|---|---|---|
| 1 | 0 | | | |
| 2 | 1 | 0 | | |
| 3 | 2 | 1 | 0 | |
| 4 | 3 | 2 | 1 | 0 |

At a **radial node**, the radial distribution function $4\pi r^2 R(r)^2 = 0$.

radial nodes respectively. The presence of the radial nodes in $s$ orbitals with $n > 1$ means that these orbitals are made up, rather like an onion, with concentric layers.

Radial distribution functions for the $3s$, $3p$ and $3d$ orbitals are shown in Figure 3.12. Patterns in numbers of radial nodes as a function of $n$ and $l$ are given in Table 3.1.

---

### 3.14     Applying the Schrödinger equation to the hydrogen atom

In this section we look in more detail at the solutions of the Schrödinger equation relating to the hydrogen atom. This is one of the few systems for which the Schrödinger equation can be solved exactly. You may think that this is restrictive, given that there are over 100 elements in addition to vast numbers of molecules and ions. However, approximations can be made in order to apply quantum mechanical equations to systems larger than hydrogen-like systems.

Schrödinger equation:
see equation 3.3

It is not necessary for us to work through the mathematics of solving the Schrödinger equation. We simply consider the solutions. After all, it is the *results* of this equation that a chemist uses.

### The wavefunctions

Solving the Schrödinger equation for $\psi$ gives information that allows the orbitals for the hydrogen atom to be constructed. Each solution of $\psi$ is in the form of a complicated mathematical expression which contains components that describe the radial part, $R(r)$, and the angular part, $A(\theta, \phi)$. Each solution corresponds to a particular set of $n$, $l$ and $m_l$ quantum numbers.

Some of the solutions for values of $\psi$ are given in Table 3.2. These will give you some idea of the mathematical forms taken by solutions of the Schrödinger equation.

### Energies

The energies for the hydrogen atom that come from the Schrödinger equation represent energies of orbitals (energy levels) and can be expressed according to equation 3.8.

$$E = -\frac{k}{n^2} \tag{3.8}$$

where: $E$ = energy
$n$ = principal quantum number
$k$ = a constant

**Table 3.2** Solutions of the Schrödinger equation for the hydrogen atom that define the $1s$, $2s$ and $2p$ atomic orbitals.

| Atomic orbital | $n$ | $l$ | $m_l$ | Radial part of the wavefunction, $R(r)$ | Angular part of wavefunction, $A(\theta, \phi)$ |
|---|---|---|---|---|---|
| $1s$ | 1 | 0 | 0 | $2e^{-r}$ | $\dfrac{1}{2\sqrt{\pi}}$ |
| $2s$ | 2 | 0 | 0 | $\dfrac{1}{2\sqrt{2}}(2-r)\,e^{-(r/2)}$ | $\dfrac{1}{2\sqrt{\pi}}$ |
| $2p_x$ | 2 | 1 | $+1$ | $\dfrac{1}{2\sqrt{6}}r\,e^{-(r/2)}$ | $\dfrac{\sqrt{3}(\sin\theta\cos\phi)}{2\sqrt{\pi}}$ |
| $2p_z$ | 2 | 1 | 0 | $\dfrac{1}{2\sqrt{6}}r\,e^{-(r/2)}$ | $\dfrac{\sqrt{3}(\cos\phi)}{2\sqrt{\pi}}$ |
| $2p_y$ | 2 | 1 | $-1$ | $\dfrac{1}{2\sqrt{6}}r\,e^{-(r/2)}$ | $\dfrac{\sqrt{3}(\sin\theta\sin\phi)}{2\sqrt{\pi}}$ |

The constant $k$ has a value of $1.312 \times 10^3\,\text{kJ mol}^{-1}$. Thus, we can determine the energy of the $1s$ orbital from equation 3.8 by substituting in a value of $n = 1$. Similarly for $n = 2, 3, 4 \ldots$.

Note that, in this case, only one energy solution will be forthcoming from equation 3.8 for each value of $n$. This means that *for the hydrogen atom, electrons with a particular principal quantum number $n$ have the same energy regardless of the $l$ and $m_l$ values. Such orbitals are degenerate.* For example, the energy is the same when the single electron of a hydrogen atom is in a $2s$ or $2p$ orbital. Similarly, equal energy states are observed when the electron is in any of the $3s$, $3p$ and $3d$ atomic orbitals of the hydrogen atom. (Note that the electron in the hydrogen atom can only occupy an orbital other than the $1s$ orbital when the atom is in an *excited state*; see below.)

Figure 3.13 shows a representation of the energy solutions of the Schrödinger equation for the hydrogen atom. *These solutions are peculiar to the hydrogen atom.* The energy levels get closer together as the value of $n$ increases, and this result is a general one for all other atoms.

### Where does the electron reside in the hydrogen atom?

Once the orbitals and their energies have been determined for the hydrogen atom, the electron can be accommodated in the *orbital in which it has the lowest energy*. This corresponds to the *ground state* for the hydrogen atom and is the most stable state.

Hence, the single electron in the hydrogen atom resides in the $1s$ orbital. This is represented in the notational form $1s^1$ – we say that the *ground state electronic configuration* of the hydrogen atom is $1s^1$.

Other 'hydrogen-like' systems include $He^+$ and $Li^{2+}$. Their electronic configurations are equivalent to that of the H atom, although the nuclear charges of these 'hydrogen-like' systems differ from each other.

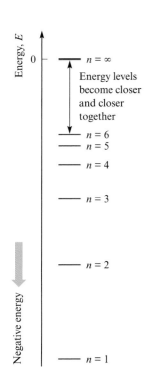

**Fig. 3.13** A schematic (not to scale) representation of the energy solutions of the Schrödinger equation for the hydrogen atom. The energy levels between $n = 6$ and infinity (the continuum) are not shown.

The type and occupancy of an atomic orbital is represented in the form:

$ns^x$ or $np^x$ or $nd^x$, etc.

where $n$ = principal quantum number and $x$ = number of electrons in the orbital or orbital set.

$1s^1$ means that there is one electron in a $1s$ atomic orbital.

▄▶
Occupying atomic orbitals with electrons: see Section 3.18
For clarification: similar notation, see p. 102

The electron in hydrogen can be promoted (raised in energy) to an atomic orbital of higher energy than the $1s$ orbital. For example, it might occupy the $2p$ orbital. Such a state is called an *excited state*. This process is denoted $2p \longleftarrow 1s$.

---

| **Worked example 3.3** | *Hydrogen-like species* |
|---|---|

**Explain why $Li^{2+}$ is termed a 'hydrogen-like' species.**

A hydrogen atom has one electron and a ground state electronic configuration of $1s^1$.
The atomic number of Li = 3.
Therefore, Li has three electrons.
$Li^{2+}$ has $(3 - 2) = 1$ electron.
The ground state electronic configuration of $Li^{2+}$ is therefore $1s^1$, and this is the same as that of a hydrogen atom.

---

## 3.15    Penetration and shielding

### What effects do electrons in an atom have on one another?

So far, we have considered only interactions between a single electron and the nucleus. As we go from the hydrogen atom to atoms with more than one electron, it is necessary to consider the effects that the electrons have on each other.

In Section 3.13 we looked at the radial properties of different atomic orbitals. Now let us consider what happens when we place electrons into these orbitals. There are two types of electrostatic interaction at work:

- electrostatic attraction between the nucleus and an electron;
- electrostatic repulsion between electrons.

In the hydrogen-like atom, the $2s$ and $2p$ orbitals are degenerate, as are all orbitals with the same principal quantum number. The electron can only ever be in one orbital at once. An orbital that is empty is a *virtual orbital* and has no physical significance.

Now take the case of an atom containing two electrons with one electron in each of the $2s$ and $2p$ orbitals. In order to get an idea about the regions of space that these two electrons occupy, we need to consider the radial distribution functions for the two orbitals. These are drawn in Figure 3.14. The presence of the radial node in the $2s$ orbital means that there is a region of space relatively close to the nucleus in which the $2s$ electron is likely to be

*An atom or ion with one electron is called a '**hydrogen-like**' species.*

*An atom or ion with more than one electron is called a '**many-electron**' species.*

**Fig. 3.14** Radial distribution functions, $4\pi r^2 R(r)^2$, for the $2s$ and $2p$ atomic orbitals.

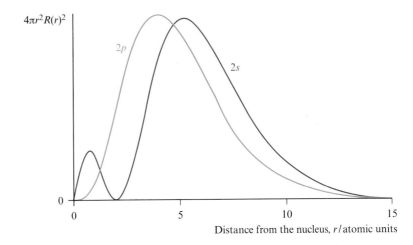

found. In effect, an electron in the $2s$ orbital will spend more time nearer to the nucleus than an electron in a $2p$ orbital. This is described as *penetration*. The $2s$ orbital *penetrates* more than the $2p$ orbital.

In the hydrogen atom, with only one electron, the energy is identical whether the electron is in the $2s$ or the $2p$ orbital. The slightly greater average distance of the electron in the $2s$ orbital from the nucleus than the $2p$ orbital is compensated by the radial node in the $2s$ orbital.

What happens when we have electrons in *both* of the $2s$ and $2p$ orbitals? The presence of the radial node in the $2s$ orbital means that there is a region of negative charge (due to the electron in the $2s$ orbital) between the nucleus and the average position of the electron in the $2p$ orbital. The potential energy of the electron is given by equation 3.9. It is related to the electrostatic attraction between the positively charged nucleus and the negatively charged electron.

$$E \propto \frac{1}{r} \tag{3.9}$$

where $E$ is the Coulombic (potential) energy and $r$ is the distance between two point charges

However, the positive charge (of the nucleus) experienced by the electron in the $2p$ orbital is 'diluted' by the presence of the $2s$ electron density close to the nucleus. The $2s$ electron *screens* or *shields* the $2p$ electron. The nuclear charge experienced by the electron in the $2p$ orbital is less than that experienced by the electron in the $2s$ orbital. Taking into account these effects, it becomes necessary in many-electron atoms to replace the nuclear charge $Z$ by the *effective nuclear charge* $Z_{eff}$. The value of $Z_{eff}$ for a given atom varies for different orbitals and depends on how many electrons there are and in which orbitals they reside. Values of $Z_{eff}$ can be estimated on an empirical basis using Slater's rules.[§]

The *effective nuclear charge*, $Z_{eff}$, is a measure of the positive charge experienced by an electron taking into account the shielding of the other electrons.

The $2s$ orbitals are more *penetrating* and they *shield* the $2p$ orbitals. Electrons in the $2p$ orbitals experience a lower electrostatic attraction and are therefore higher in energy (less stabilized) than electrons in $2s$ orbitals. Similar arguments place the $3d$ orbitals higher in energy than the $3p$ which in turn are higher than the $3s$ (Figure 3.15). An extension of this argument leads to a picture in which all electrons in lower energy orbitals shield

---

[§] We shall not elaborate further on Slater's rules; for further details see: C. E. Housecroft and A. G. Sharpe (2005) *Inorganic Chemistry*, 2nd edn, Prentice Hall, Harlow, Chapter 1.

**Fig. 3.15** A schematic (not to scale) representation of the energy solutions (for $n = 1$, 2 and 3) of the Schrödinger equation for a many-electron atom.

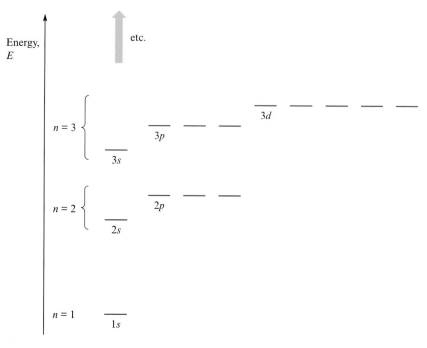

those in higher energy orbitals. For example, in a boron atom $(1s^2 2s^2 2p^1)$, the $2s$ electrons are shielded by the $1s$ pair, whereas the $2p$ electron is shielded by both the $1s$ and $2s$ electron pairs.

### Changes in the energy of an atomic orbital with atomic number

In this section, we briefly consider how the energy of a particular atomic orbital changes as the atomic number increases – for example, the $2s$ atomic orbital is of a different energy in lithium from that in other elements.

As atomic number increases, the nuclear charge must increase, and for an electron in a given $ns$ or $np$ orbital, the *effective nuclear charge* also increases. This results in a decrease in atomic orbital energy – the trend is relatively smooth, although the relationship between $ns$ or $np$ atomic orbital energy and atomic number is *non-linear*. Although the energies of $nd$ and $nf$ atomic orbitals generally decrease with increasing atomic number, the situation is rather complex, and we shall not discuss it further here.[§]

---

**3.16**    ## The atomic spectrum of hydrogen and selection rules

### Spectral lines

When the single electron in the $1s$ orbital of the hydrogen atom is *excited* (that is, it is given energy), it may be *promoted* to a higher energy state. The new state is transient and the electron will fall back to a lower energy state, emitting energy as it does so. Evidence for the discrete nature of orbital energy levels comes from an observation of *spectral lines* in the emission spectrum of hydrogen (Figure 3.16). The emphasis here is on the fact that *single lines are*

---

[§] For a detailed discussion see: K. M. Mackay, R. A. Mackay and W. Henderson (1996) *An Introduction to Modern Inorganic Chemistry*, 5th edn, Chapman and Hall, London, Chapter 8.

**Fig. 3.16** Part of the emission spectrum of atomic hydrogen. Groups of lines have particular names, for example the Balmer and Lyman series.

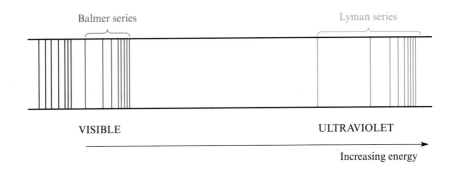

observed; the emission is not continuous over a range of frequencies. Similar single frequencies are also observed in the absorption spectrum.

Some of the electronic transitions which give rise to frequencies observed in the emission spectrum of atomic hydrogen are shown in Figure 3.17. Notice that some transitions are *not* allowed – the *selection rules* given in equations 3.10 and 3.11 must be obeyed.

*Selection rules tell us which spectroscopic transitions are allowed: see also Section 12.3*

In equation 3.10, $\Delta n$ is the change in the value of the principal quantum number $n$, and this selection rule means that there is no restriction on transitions between different principal quantum levels. In equation 3.11, $\Delta l$ is a change in the value of the orbital quantum number $l$ and this selection rule places a restriction on this transition: $l$ can change by only one unit. Equation 3.11 is known as the *Laporte selection rule*, and corresponds to a change of angular momentum by one unit.

$$\Delta n = 0, \pm 1, \pm 2, \pm 3, \pm 4 \dots \tag{3.10}$$

$$\Delta l = +1 \text{ or } -1 \qquad \text{Laporte selection rule} \qquad (3.11)$$

**Fig. 3.17** Some of the allowed transitions that make up the Lyman (shown in blue) and Balmer (shown in pink) series in the emission spectrum of atomic hydrogen. The selection rules in equations 3.10 and 3.11 must be obeyed.

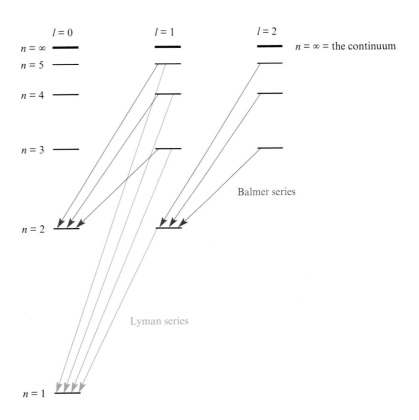

**THEORETICAL AND CHEMICAL BACKGROUND**

## Box 3.7 Absorption and emission spectra

The ground state of an atom (or other species) is one in which the electrons are in the lowest energy arrangement.

When an atom (or other species) absorbs electromagnetic radiation, discrete electronic transitions occur and an *absorption spectrum* is observed.

When energy is provided (e.g. as heat or light) one or more electrons in an atom or other species may be promoted from its ground state level to a higher energy state. This *excited state* is transient and the electron falls back to the ground state. An *emission spectrum* is thereby produced.

Absorption and emission transitions can be distinguished by using the following notation:

Emission: (high energy level) $\longrightarrow$ (low energy level)
Absorption: (high energy level) $\longleftarrow$ (low energy level)

Spectral lines in both absorption and emission spectra may be designated in terms of frequency $\nu$:

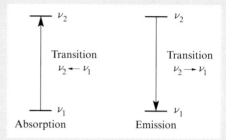

Two useful relationships are:    $E = h\nu$    and    $\nu = \dfrac{c}{\lambda}$

where   $E$ = energy,   $\nu$ = frequency,   $\lambda$ = wavelength,  $c$ = speed of light and $h$ = Planck constant.

---

Basic selection rules for electronic spectra:

$$\Delta n = 0, \pm 1, \pm 2, \pm 3, \pm 4 \ldots$$

$$\Delta l = +1 \text{ or } -1 \qquad \text{Laporte selection rule}$$

◼▷
Degenerate energy levels in
the H atom: see Section 3.14

For example, a transition from a level with $n = 3$ to one with $n = 2$ is only allowed if, at the same time, there is a change in $l$ of $\pm 1$. In terms of atomic orbitals this corresponds to a transition from the $3p \longrightarrow 2s$ or $3d \longrightarrow 2p$ orbitals but *not* $3s \longrightarrow 2s$ or $3d \longrightarrow 2s$. The latter are disallowed because they do not obey equation 3.11; $\Delta l$ values for each of these last transitions are 0 and 2 respectively. Of course, in the hydrogen atom, atomic levels with the same value of $n$ are degenerate and so a transition in which $\Delta n = 0$ (say, $2p \longrightarrow 2s$) does not give rise to a change in energy, and cannot give rise to an observed spectral line.

The lines in the emission spectrum of atomic hydrogen fall into several discrete series, two of which are illustrated (in part) in Figure 3.17. The series are defined in Table 3.3. The relative energies of the transitions mean that each set of spectral lines appears in a different part of the electromagnetic spectrum (see Appendix 4). The Lyman series with lines of type $n' \longrightarrow 1$ ($2 \longrightarrow 1$,

**Table 3.3** Series of lines observed in the emission spectrum of the hydrogen atom.

| Name of series | $n' \longrightarrow n$ | Region in which transitions are observed |
|---|---|---|
| Lyman | $2 \longrightarrow 1, 3 \longrightarrow 1, 4 \longrightarrow 1$, etc. | Ultraviolet |
| Balmer | $3 \longrightarrow 2, 4 \longrightarrow 2, 5 \longrightarrow 2$, etc. | Visible |
| Paschen | $4 \longrightarrow 3, 5 \longrightarrow 3, 6 \longrightarrow 3$, etc. | Infrared |
| Brackett | $5 \longrightarrow 4, 6 \longrightarrow 4, 7 \longrightarrow 4$, etc. | Far infrared |
| Pfund | $6 \longrightarrow 5, 7 \longrightarrow 5, 8 \longrightarrow 5$, etc. | Far infrared |

$3 \longrightarrow 1, 4 \longrightarrow 1 \ldots$) contains energy transitions that correspond to frequencies in the ultraviolet region. The Balmer series with lines of type $n' \longrightarrow 2$ ($3 \longrightarrow 2, 4 \longrightarrow 2, 5 \longrightarrow 2 \ldots$) contains transitions that correspond to frequencies in the visible region. Of course, in the hydrogen atom, the $nd$, $np$ and $ns$ orbitals possess the same energy and the transitions $np \longrightarrow (n-1)s$ and $nd \longrightarrow (n-1)p$ occur at the same frequency for the same value of $n$.

### The Rydberg equation

The frequencies of the lines arising from $n' \longrightarrow n$ transitions where $n' > n$ in the Lyman, Balmer, Paschen and other series obey the general equation 3.12.

$$\nu = R\left(\frac{1}{n^2} - \frac{1}{n'^2}\right) \tag{3.12}$$

where: $\nu$ = frequency in Hz (Hz = s$^{-1}$)
$R$ = Rydberg constant for hydrogen = $3.289 \times 10^{15}$ Hz

*R is used for the Rydberg constant and for the molar gas constant, but the context of its use should be clear*

The frequencies of lines in the atomic spectrum of hydrogen can be calculated using equation 3.12. However, their positions in the spectrum are often quoted in terms of wavenumbers instead of frequencies. Wavenumber is the reciprocal of wavelength (equation 3.13) and the units are 'reciprocal metres' (m$^{-1}$), although 'reciprocal centimetres' (cm$^{-1}$) are often more convenient. The relationship between wavenumber ($\bar{\nu}$) and frequency ($\nu$) follows from the equation relating frequency, wavelength and the speed of light ($c = 2.998 \times 10^{10}$ cm s$^{-1}$) as shown in equation 3.13.

$$\left. \begin{array}{c} \text{Wavenumber } (\bar{\nu} \text{ in cm}^{-1}) = \dfrac{1}{\text{Wavelength } (\lambda \text{ in cm})} \\[2mm] c = \nu\lambda \qquad \lambda = \dfrac{c}{\nu} \qquad \bar{\nu}\,(\text{in cm}^{-1}) = \dfrac{1}{\lambda} = \dfrac{\nu\,(\text{in s}^{-1})}{c\,(\text{in cm s}^{-1})} \end{array} \right\} \tag{3.13}$$

We return to the relationships between frequency, wavelength, wavenumber and energy in Chapter 11, but for now we note that equation 3.12 may be written in the form of equation 3.14 in which the units of the Rydberg constant are cm$^{-1}$.

$$\bar{\nu} = \frac{1}{\lambda} = R\left(\frac{1}{n^2} - \frac{1}{n'^2}\right) \tag{3.14}$$

where: $\bar{\nu}$ = wavenumber in cm$^{-1}$
$R$ = Rydberg constant for hydrogen = $1.097 \times 10^5$ cm$^{-1}$

---

**Worked example 3.4**    *The Rydberg equation*

Determine the frequency of the $7 \longrightarrow 3$ transition in the Paschen series in the emission spectrum of atomic hydrogen.

The notation $7 \longrightarrow 3$ refers to $n' = 7$ and $n = 3$. Use the Rydberg equation in the form:

$$\nu = R\left(\frac{1}{n^2} - \frac{1}{n'^2}\right) \qquad \text{where } R = 3.289 \times 10^{15} \text{ Hz}$$

$$\nu = (3.289 \times 10^{15})\left(\frac{1}{3^2} - \frac{1}{7^2}\right) = (3.289 \times 10^{15})\left(\frac{1}{9} - \frac{1}{49}\right) = 2.983 \times 10^{14} \text{ Hz}$$

### Determining the ionization energy of hydrogen

■▷
*The relationship between energy and frequency is: $E = h\nu$*

The Lyman series of spectral lines is of particular significance because it may be used to determine the ionization energy of hydrogen. This is the energy required to remove the 1*s* electron completely from the atom. The frequencies of lines in the Lyman series give values of $n$ corresponding to the spectral transitions $n' \longrightarrow n = 2 \longrightarrow 1, 3 \longrightarrow 1, 4 \longrightarrow 1$. The series can be extrapolated to find a value of $\nu$ corresponding to the transition $\infty \longrightarrow 1$. The method is shown in worked example 3.5.

> The *first ionization energy* of an atom is the energy required to completely remove the most easily separated electron from the atom in the gaseous state. For an atom X, it is defined for the process:
>
> $$X(g) \longrightarrow X^+(g) + e^-$$

---

**Worked example 3.5**    *Determination of the ionization energy of hydrogen*

**The frequencies of some of the spectral lines in the Lyman series of atomic hydrogen are 2.466, 2.923, 3.083, 3.157, 3.197, 3.221 and $3.237 \times 10^{15}$ Hz. Using these data, calculate the ionization energy of atomic hydrogen.**

The data correspond to spectral transitions, $\nu$. The aim of the question is to find the ionization energy of atomic hydrogen and this is associated with a value of $\nu$ for the transition $n' \longrightarrow n = \infty \longrightarrow 1$.

The level $n' = \infty$ (the continuum, see Figure 3.13) corresponds to the electron being completely removed from the atom. As we approach the level $n = \infty$, differences between *successive* energy levels become progressively smaller until, at the continuum, the difference between levels is zero. At this point, the energy of the electron is independent of the nucleus and it is no longer quantized. Spectral transitions to the level $n = 1$ from levels approaching $n = \infty$ will have virtually identical energies.

In order to determine the ionization energy of the hydrogen atom, we need first to find the point at which the *difference* in energies between successive spectral transitions approaches zero. This corresponds to the convergence limit of the frequencies.

First, calculate the *differences*, $\Delta\nu$, between successive values of $\nu$ for the data given. These are:

$$(2.923 - 2.466) \times 10^{15} = 0.457 \times 10^{15} \text{ Hz}$$

$$(3.083 - 2.923) \times 10^{15} = 0.160 \times 10^{15} \text{ Hz}$$

$$(3.157 - 3.083) \times 10^{15} = 0.074 \times 10^{15} \text{ Hz}$$

$$(3.197 - 3.157) \times 10^{15} = 0.040 \times 10^{15} \text{ Hz}$$

$$(3.221 - 3.197) \times 10^{15} = 0.024 \times 10^{15} \text{ Hz}$$

$$(3.237 - 3.221) \times 10^{15} = 0.016 \times 10^{15} \text{ Hz}$$

Now, plot these differences against *either* the higher *or* the lower value of $\Delta\nu$. Here we use the lower values, and the curve is shown on the next page.

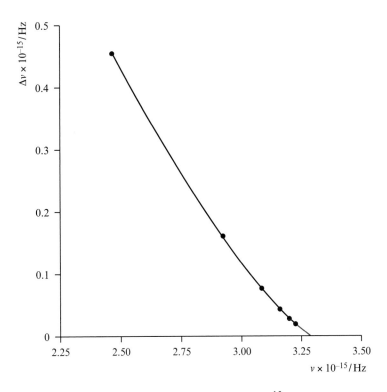

The curve converges at a value of $\nu = 3.275 \times 10^{15}$ Hz. Therefore, the value of $E$ which corresponds to this value of $\nu$ is:

$$E = h\nu = (6.626 \times 10^{-34}\,\text{J s}) \times (3.275 \times 10^{15}\,\text{s}^{-1})$$

But this value is per atom.

The ionization energy should be given in units of energy per mole, which means we need to multiply by the Avogadro number (the number of atoms per mole).

Ionization energy of hydrogen

$$= (6.626 \times 10^{-34}\,\text{J s}) \times (3.275 \times 10^{15}\,\text{s}^{-1}) \times (6.022 \times 10^{23}\,\text{mol}^{-1})$$

$$= 1.307 \times 10^{6}\,\text{J mol}^{-1}$$

$$= 1307\,\text{kJ mol}^{-1}$$

## Bohr's theory of the spectrum of atomic hydrogen

In Section 3.3, we introduced the Bohr model of the atom. The model is quantized, and for the hydrogen atom, the electron can be in one of an infinite number of circular orbits. Lines in the emission or absorption spectrum of atomic hydrogen then arise as energy is emitted or absorbed when the electron moves from one orbit to another. The orbits can be considered to be energy levels having principal quantum numbers $n_1, n_2, \ldots$ and energies $E_1, E_2 \ldots$. The change in energy, $\Delta E$, when the electron changes orbit is given by equation 3.15.

$$\Delta E = E_2 - E_1 = h\nu \tag{3.15}$$

where: $h$ = Planck constant = $6.626 \times 10^{-34}$ J s

$\nu$ = frequency (in Hz = $s^{-1}$) of the transition

The radius, $r_n$, of each allowed orbit in the Bohr atom is given by equation 3.16. The equation arises from a consideration of the centrifugal force acting on the electron as it moves on a circular path.

$$r_n = \frac{\varepsilon_0 h^2 n^2}{\pi m_e e^2} \tag{3.16}$$

where: $\varepsilon_0$ = permittivity of a vacuum = $8.854 \times 10^{-12}$ F m$^{-1}$
$h$ = Planck constant = $6.626 \times 10^{-34}$ J s
$n = 1, 2, 3 \ldots$ describing a given orbit
$m_e$ = electron rest mass = $9.109 \times 10^{-31}$ kg
$e$ = charge on an electron = $1.602 \times 10^{-19}$ C

For the first orbit, $n = 1$, and an evaluation of equation 3.16 for $n = 1$ gives a radius of $5.292 \times 10^{-11}$ m (52.92 pm). This is called the Bohr radius of the hydrogen atom and has the symbol $a_0$. This corresponds to the maximum probability in Figure 3.5.

## 3.17 Many-electron atoms

We have already seen that a neutral atom with more than one electron is called a *many-electron atom* and, clearly, all but the hydrogen atom fall into this category. As we have seen, the electrostatic repulsion between electrons in a many-electron atom has to be considered, and solving the Schrödinger equation exactly for such atoms is impossible. Various approximations have to be made in attempting to generate solutions of $E$ and $\psi$ for a many-electron atom, and for atoms, at least, numerical solutions of high accuracy are possible.

We shall not delve into this problem further, except to stress the important result from Section 3.15. The effects of *penetration* and *shielding* mean that in all atoms with more than one electron, orbitals with the same value of $n$ but different values of $l$ possess different energies (Figure 3.15).

## 3.18 The *aufbau* principle

The word '*aufbau*' is German and means 'building up'. The *aufbau* principle provides a set of rules for determining the ground state electronic structure of atoms. We are concerned for the moment only with the occupancies of atomic (rather than molecular) orbitals.

The *aufbau* principle is used in conjunction with two other principles – Hund's rule and the Pauli exclusion principle. It can be summarized as follows.

Degenerate = same energy

- Orbitals are filled in order of energy with the lowest energy orbitals being filled first.
- If there is a set of degenerate orbitals, pairing of electrons in an orbital cannot begin until *each* orbital in the set contains one electron. Electrons singly occupying orbitals in a degenerate set have the same (*parallel*) spins. This is Hund's rule.
- No two electrons in an atom can have exactly the same set of $n$, $l$, $m_l$ and $m_s$ quantum numbers. This is the Pauli exclusion principle. This means that each orbital can accommodate a maximum of two electrons with different $m_s$ values (different spins).

**Fig. 3.18** Hund's rule: two electrons in a degenerate set of *p* orbitals must, in the ground state, be in separate orbitals and have parallel (the same) spins.

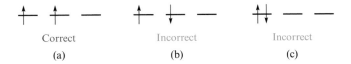

Correct      Incorrect      Incorrect

(a)        (b)        (c)

The first of the above statements needs no further clarification. The second statement can be exemplified by considering two electrons occupying a degenerate set of *p* orbitals. Electrons may be represented by arrows, the orientation of which indicates the direction of spin ($m_s = +\frac{1}{2}$ or $-\frac{1}{2}$). Figure 3.18a shows the *correct* way to arrange the two electrons in the ground state according to Hund's rule. The other possibilities shown in Figures 3.18b and 3.18c are disallowed by Hund's rule (although they represent excited states of the atom).

The Pauli exclusion principle prevents two electrons with the same spins from entering a single orbital. If they were to do so, they would have the same set of four quantum numbers. In Figure 3.19, the allowed and disallowed arrangements of electrons in a 1*s* orbital are depicted.

Wolfgang Pauli (1900–1958).
© *The Nobel Foundation.*

The **Pauli exclusion principle** states that no two electrons in an atom can have the same set of *n*, *l*, $m_l$ and $m_s$ quantum numbers.

This means that every electron in an atom is uniquely defined by its set of four quantum numbers.

**Hund's rule** states that when filling a degenerate set of orbitals in the ground state, pairing of electrons cannot begin until *each* orbital in the set contains one electron. Electrons singly occupying orbitals in a degenerate set have parallel spins.

The **aufbau principle** is a set of rules that must be followed when placing electrons in atomic or molecular orbitals to obtain ground state electronic configurations. The *aufbau* principle combines Hund's rules and the Pauli exclusion principle with the following additional facts:

• Orbitals are filled in order of increasing energy.

• An orbital is fully occupied when it contains two electrons.

**Fig. 3.19** The Pauli exclusion principle: two electrons occupying a 1*s* atomic orbital can only have non-identical sets of the four quantum numbers *n*, *l*, $m_l$ and $m_s$ if their spins are different.

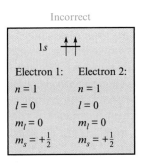

Incorrect

1*s*

| Electron 1: | Electron 2: |
|---|---|
| $n = 1$ | $n = 1$ |
| $l = 0$ | $l = 0$ |
| $m_l = 0$ | $m_l = 0$ |
| $m_s = +\frac{1}{2}$ | $m_s = +\frac{1}{2}$ |

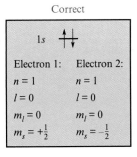

Correct

1*s*

| Electron 1: | Electron 2: |
|---|---|
| $n = 1$ | $n = 1$ |
| $l = 0$ | $l = 0$ |
| $m_l = 0$ | $m_l = 0$ |
| $m_s = +\frac{1}{2}$ | $m_s = -\frac{1}{2}$ |

| Worked example 3.6 | *Using the aufbau principle* |

**Determine the arrangement of the electrons in an atom of nitrogen in its ground state.**

The atomic number of nitrogen is 7, and there are seven electrons to be considered.

The lowest energy orbital is the $1s$ ($n = 1$; $l = 0$; $m_l = 0$).

The maximum occupancy of the $1s$ orbital is two electrons.

The next lowest energy orbital is the $2s$ ($n = 2$; $l = 0$; $m_l = 0$).

The maximum occupancy of the $2s$ orbital is two electrons.

The next lowest energy orbitals are the three making up the degenerate set of $2p$ orbitals ($n = 2$; $l = 1$, $m_l = +1, 0, -1$).

The maximum occupancy of the $2p$ orbitals is six electrons, but only three remain to be accommodated (seven electrons in total for the nitrogen atom; four in the $1s$ and $2s$ orbitals).

The three electrons will occupy the $2p$ orbitals so as to obey Hund's rule. Each electron enters a separate orbital and the three electrons have parallel spins.

The arrangement of electrons in a nitrogen atom in its ground state is represented by:

---

| 3.19 | **Electronic configurations** |

### Ground state electronic configurations

The *ground state electronic configuration* of an atom is its lowest energy state.

The *aufbau* principle provides us with a method of predicting the order of filling atomic orbitals with electrons and gives us the ground state arrangement of the electrons – the *ground state electronic configuration* – of an atom. The result may be represented as in Figure 3.20 for the arrangement of electrons in the ground state of a carbon atom ($Z = 6$).

Another way to present the electronic configuration is given below. For the hydrogen atom, we saw in Section 3.14 that the occupancy of the $1s$ atomic orbital by a single electron could be indicated by the notation $1s^1$. This can be extended to give the electronic configuration of any atom. Thus, for carbon, the configuration shown in Figure 3.20 can be written as $1s^2 2s^2 2p^2$. Similarly, for helium and lithium (Figure 3.21) the configurations are $1s^2$ and $1s^2 2s^1$ respectively.

▣▶
*Z* = atomic number

**Fig. 3.20** The arrangement of electrons in the atomic orbitals of carbon in the ground state.

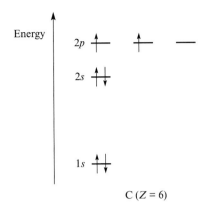

C (*Z* = 6)

The order of occupying atomic orbitals in the ground state of an atom *usually* follows the sequence (lowest energy first):

$$1s < 2s < 2p < 3s < 3p < 4s < 3d < 4p < 5s < 4d < 5p < 6s < 5d$$
$$\approx 4f < 6p < 7s < 6d \approx 5f$$

We emphasize that this series may vary for some atoms because the energy of electrons in orbitals is affected by the nuclear charge, the presence of other electrons in the same orbital or in the same sub-level, and the overall charge. We shall mention this point again when we look at aspects of *d*-block metal chemistry in Chapter 23.

> The *usual* ordering (lowest energy first) of atomic orbitals is:
>
> $$1s < 2s < 2p < 3s < 3p < 4s < 3d < 4p < 5s < 4d < 5p < 6s < 5d$$
> $$\approx 4f < 6p < 7s < 6d \approx 5f$$

By combining this sequence of orbitals with the rules stated in the *aufbau* principle, we can write down the ground state electronic configurations of most atoms. Those for elements with atomic numbers 1 to 20 are given in Table 3.4. As we progress down the table, there is clearly a repetition corresponding to the lowest energy electrons which occupy filled quantum levels. Thus, an abbreviated form of the electronic configuration can be used. Here, only electrons entering new quantum levels (new values of *n*) are emphasized, and the inner levels are indicated by the previous group 18 element (He, Ne or Ar). You should note the relationship between the pattern of repetition of the electronic configurations and the positions of the elements in the periodic table (given inside the front cover of the book). We return to these relationships in Section 3.20.

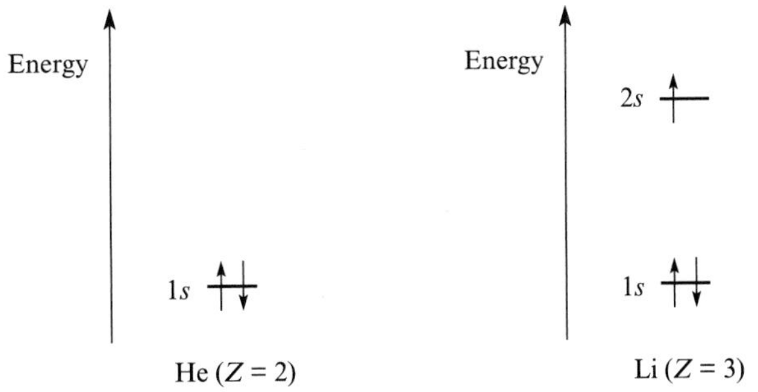

**Fig. 3.21** The arrangement of electrons in the atomic orbitals of helium and lithium in the ground state.

He (*Z* = 2)          Li (*Z* = 3)

**Table 3.4** Ground state electronic configurations for the first 20 elements.

| Atomic number | Element symbol | Electronic configuration | Shortened form of the notation for the electronic configuration |
|---|---|---|---|
| 1 | H | $1s^1$ | $1s^1$ |
| 2 | He | $1s^2$ | $1s^2 = [\text{He}]$ |
| 3 | Li | $1s^2 2s^1$ | $[\text{He}]2s^1$ |
| 4 | Be | $1s^2 2s^2$ | $[\text{He}]2s^2$ |
| 5 | B | $1s^2 2s^2 2p^1$ | $[\text{He}]2s^2 2p^1$ |
| 6 | C | $1s^2 2s^2 2p^2$ | $[\text{He}]2s^2 2p^2$ |
| 7 | N | $1s^2 2s^2 2p^3$ | $[\text{He}]2s^2 2p^3$ |
| 8 | O | $1s^2 2s^2 2p^4$ | $[\text{He}]2s^2 2p^4$ |
| 9 | F | $1s^2 2s^2 2p^5$ | $[\text{He}]2s^2 2p^5$ |
| 10 | Ne | $1s^2 2s^2 2p^6$ | $[\text{He}]2s^2 2p^6 = [\text{Ne}]$ |
| 11 | Na | $1s^2 2s^2 2p^6 3s^1$ | $[\text{Ne}]3s^1$ |
| 12 | Mg | $1s^2 2s^2 2p^6 3s^2$ | $[\text{Ne}]3s^2$ |
| 13 | Al | $1s^2 2s^2 2p^6 3s^2 3p^1$ | $[\text{Ne}]3s^2 3p^1$ |
| 14 | Si | $1s^2 2s^2 2p^6 3s^2 3p^2$ | $[\text{Ne}]3s^2 3p^2$ |
| 15 | P | $1s^2 2s^2 2p^6 3s^2 3p^3$ | $[\text{Ne}]3s^2 3p^3$ |
| 16 | S | $1s^2 2s^2 2p^6 3s^2 3p^4$ | $[\text{Ne}]3s^2 3p^4$ |
| 17 | Cl | $1s^2 2s^2 2p^6 3s^2 3p^5$ | $[\text{Ne}]3s^2 3p^5$ |
| 18 | Ar | $1s^2 2s^2 2p^6 3s^2 3p^6$ | $[\text{Ne}]3s^2 3p^6 = [\text{Ar}]$ |
| 19 | K | $1s^2 2s^2 2p^6 3s^2 3p^6 4s^1$ | $[\text{Ar}]4s^1$ |
| 20 | Ca | $1s^2 2s^2 2p^6 3s^2 3p^6 4s^2$ | $[\text{Ar}]4s^2$ |

For the time being we shall leave ground state electronic configurations for elements with atomic number $>20$. This is the point at which the sequence of orbitals given above begins to become less reliable.

## Valence and core electrons

We now introduce the terms *valence electronic configuration* and *valence electron* which are commonly used and have significant implication. The valence electrons of an atom are those in the outer (highest energy) quantum levels and it is these electrons that are primarily responsible for determining the chemistry of an element. Consider sodium ($Z = 11$) – the ground electronic configuration is $1s^2 2s^2 2p^6 3s^1$. The principal quantum shells with $n = 1$ and 2 are fully occupied and electrons in these shells are referred to as *core electrons*. Sodium has one valence electron (the $3s$ electron) and the chemistry of sodium reflects this.

---

**Worked example 3.7**   *Determining a ground state electronic configuration*

Determine the ground state electronic configuration for sodium ($Z = 11$).

There are 11 electrons to be accommodated.
The sequence of atomic orbitals to be filled is $1s < 2s < 2p < 3s$.
The maximum occupancy of an $s$ level is two electrons.
The maximum occupancy of the $p$ level is six electrons.
Therefore the ground state electronic configuration for sodium is:

$1s^2 2s^2 2p^6 3s^1$   or   $[\text{Ne}]3s^1$

---

| **Worked example 3.8** | *Core and valence electrons* |
|---|---|

**The atomic number of phosphorus is 15. How many core electrons and valence electrons does a phosphorus atom possess? What is its valence electronic configuration in the ground state?**

An atom of phosphorus ($Z = 15$) has a total of 15 electrons.

In the ground state, the electronic configuration is $1s^2 2s^2 2p^6 3s^2 3p^3$.

The core electrons are those in the $1s$, $2s$ and $2p$ orbitals: a phosphorus atom has 10 core electrons.

The valence electrons are those in the $3s$ and $3p$ orbitals: a phosphorus atom has five valence electrons.

The valence electronic configuration in the ground state is $3s^2 3p^3$.

---

## 3.20    The octet rule

Inspection of the ground state electronic configurations listed in Table 3.4 shows a clear pattern, which is emphasized in the right-hand column of the table. This notation makes use of filled principal quantum shells as 'building blocks' within the ground state configurations of later elements. This trend is one of *periodicity* – repetition of the configurations of the outer electrons but for a different value of the principal quantum number.

We see a repetition of the sequences $ns^{(1 \, to \, 2)}$ and $ns^2 np^{(1 \, to \, 6)}$. The total number of electrons that can be accommodated in orbitals with $n = 1$ is only two, but for $n = 2$ this number increases to eight (i.e. $2s^2 2p^6$). The number eight is also a feature of the total number of electrons needed to completely fill the $3s$ and $3p$ orbitals. This is the basis of the *octet rule*. As the name[§] suggests, the octet rule has its origins in the observation that atoms of the s- and p-blocks have a tendency to lose, gain or share electrons in order to end up with eight electrons in their outer shell.

> An atom is obeying the ***octet rule*** when it gains, loses or shares electrons to give an *outer* shell containing eight electrons with the configuration $ns^2 np^6$.

The ground state electronic configuration of fluorine is $[He]2s^2 2p^5$. This is one electron short of the completed configuration $[He]2s^2 2p^6$. Fluorine readily gains an electron to form the fluoride ion $F^-$, the ground state configuration of which is $[He]2s^2 2p^6$. An atom of nitrogen ($[He]2s^2 2p^3$) requires three electrons to give the $[He]2s^2 2p^6$ configuration and $N^{3-}$ may be formed. On the other hand, a carbon atom ($[He]2s^2 2p^2$) would have to form the $C^{4-}$ ion in order to achieve the $[He]2s^2 2p^6$ configuration and this is energetically unfavourable. The answer here is for the carbon atom to *share* four electrons with other atoms, thereby completing its octet without the need for ion formation.

**◼▶**
Formation of ions:
see Chapter 8

Despite its successes, we can see that the concept of the octet rule is rather limited. We can apply it satisfactorily to the principal quantum shell with $n = 2$. Here the quantum shell is fully occupied when it contains eight electrons. But there are exceptions even here: the elements lithium and beryllium which have outer $2s$ electrons (Table 3.4) tend to *lose* electrons so as to possess helium-like ($1s^2$) ground state electronic configurations, and molecules such as $BF_3$ with six electrons in the valence shell are stable.

What about the principal quantum shell with $n = 3$? Here, Na and Mg (Table 3.4) tend to *lose* electrons to possess a neon-like configuration and

---

[§] The word 'octet' is derived from the Latin '*octo*' meaning eight.

in doing so they obey the octet rule. Aluminium with a ground state configuration of $[Ne]3s^2 3p^1$ may lose three electrons to become $Al^{3+}$ (i.e. neon-like) or it may participate in bond formation so as to achieve an octet through sharing electrons. Care is needed though! Aluminium also forms compounds in which it does not formally obey the octet rule, for example $AlCl_3$. Atoms of elements with ground state configurations between $[Ne]3s^2 3p^2$ and $[Ne]3s^2 3p^6$ may also share or gain electrons to achieve octets, but there is a complication. For $n \geq 3$, elements in the later groups of the periodic table form compounds in which the octet of electrons appears to be 'expanded'. For example, phosphorus forms $PCl_5$ in which P is in oxidation state $+5$ and the P atom appears to have 10 electrons in its valence shell. We return to the bonding in such compounds in Section 7.3.

> Aluminium halides: see Section 22.4

The octet rule is very important, and the idea can be extended to an 18-electron rule which takes into account the filling of $ns$, $np$ and $nd$ sublevels. We return to the 18-electron rule in Chapter 23.

---

**Worked example 3.9**     *The octet rule*

**Suggest why fluorine forms only the fluoride $F_2$ (i.e. molecular difluorine), but oxygen can form a difluoride $OF_2$. (For F, $Z = 9$; for O, $Z = 8$.)**

First, write down the ground state electronic configurations of F and O:

F   $1s^2 2s^2 2p^5$

O   $1s^2 2s^2 2p^4$

An F atom can achieve an octet of electrons by sharing one electron with another F atom. In this way, the principal quantum level for $n = 2$ is completed.

Oxygen can complete an octet of outer electrons by sharing two electrons with 2 F atoms in the molecule $OF_2$. The octet of each F atom is also completed by each sharing one electron with the O atom.

---

**3.21**     **Monatomic gases**

### The noble gases

In looking at ground state electronic configurations, we saw that configurations with filled principal quantum shells appeared as 'building blocks' within the ground state configurations of heavier elements. The elements that possess these filled principal quantum shells all belong to one group in the periodic table – group 18. These are the so-called *noble gases*.[§]

The ground state electronic configurations of the noble gases are given in Table 3.5. Note that each configuration (except that for helium) contains an $ns^2 np^6$ configuration for the highest value of $n$ – for example, the configuration for argon ends $3s^2 3p^6$. The $ns^2 np^6$ configuration is often referred

---

[§] Although 'inert gas' is commonly used, 'noble gas' is the IUPAC recommendation; see Section 1.10.

Table 3.5 Ground state electronic configurations of the noble gases.

| Noble gas | Symbol | Atomic number | Ground state electronic configuration |
|---|---|---|---|
| Helium | He | 2 | $1s^2$ |
| Neon | Ne | 10 | $1s^2 2s^2 2p^6$ |
| Argon | Ar | 18 | $1s^2 2s^2 2p^6 3s^2 3p^6$ |
| Krypton | Kr | 36 | $1s^2 2s^2 2p^6 3s^2 3p^6 4s^2 3d^{10} 4p^6$ |
| Xenon | Xe | 54 | $1s^2 2s^2 2p^6 3s^2 3p^6 4s^2 3d^{10} 4p^6 5s^2 4d^{10} 5p^6$ |
| Radon | Rn | 86 | $1s^2 2s^2 2p^6 3s^2 3p^6 4s^2 3d^{10} 4p^6 5s^2 4d^{10} 5p^6 6s^2 4f^{14} 5d^{10} 6p^6$ |

to as an 'inert (noble) gas configuration'. Do not worry that the heaviest elements possess filled $d$ and $f$ shells in addition to filled $s$ and $p$ orbitals.

The fact that each noble gas has a filled outer $np$ shell (He is the exception) means that the group 18 elements exist in the elemental state as monatomic species – there is no driving force for bond formation between atoms.

## Forces between atoms

The noble gases (with the exception of helium, see Table 3.6) solidify only at low temperatures. The melting points of neon and argon are 24.5 K and 84 K respectively. These low values indicate that the *interatomic forces* in the solid state are very weak indeed and are easily overcome when the solid changes into a liquid. The very narrow range of temperatures over which the group 18 elements are in the liquid state is significant. The net interatomic forces in the liquid state are extremely weak and the ease with which vaporization occurs is apparent from the very low values of the enthalpy changes for this process (Table 3.6).

Interatomic interactions: see Section 2.10

$\Delta_{vap}H$: see Section 2.9

What are the nature and form of the forces between atoms? Let us consider two atoms of argon. Each is spherical with a central positively charged nucleus surrounded by negatively charged electrons. At large separations, there will only be very small interactions between the atoms. However, as two argon atoms come closer together, the electron clouds begin to repel each other. At very short internuclear distances, there will be strong repulsive forces as a result of electron–electron repulsions. A good mathematical approximation of the repulsive force shows that it depends on the inter-nuclear distance $d$ as stated in equation 3.17. The *potential energy* is related to the force and varies according to equation 3.18. This repulsive potential

Table 3.6 Some physical properties of the noble gases.

| Element | Melting point / K | Boiling point / K | $\Delta_{fus}H$ / kJ mol$^{-1}$ | $\Delta_{vap}H$ / kJ mol$^{-1}$ | Van der Waals radius $(r_v)$ / pm | First ionization energy / kJ mol$^{-1}$ |
|---|---|---|---|---|---|---|
| Helium | –[a] | 4.2 | – | 0.1 | 99 | 2372 |
| Neon | 24.5 | 27 | 0.3 | 2 | 160 | 2081 |
| Argon | 84 | 87 | 1.1 | 6.5 | 191 | 1521 |
| Krypton | 116 | 120 | 1.4 | 9 | 197 | 1351 |
| Xenon | 161 | 165 | 1.8 | 13 | 214 | 1170 |
| Radon | 202 | 211 | – | 18 | – | 1037 |

[a] Helium cannot be solidified under any conditions of temperature and pressure.

**Fig. 3.22** Potential energy curves due to repulsion (pink line) and attraction (blue line) between two atoms of argon as a function of the internuclear distance, $d$.

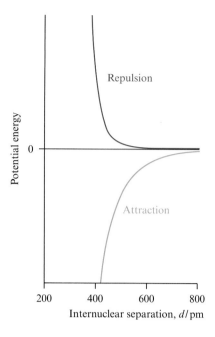

In an atom or molecule, an asymmetrical distribution of charge leads to the formation of a *dipole*. One part of the atom or molecule has a greater share of the negative (or positive) charge than another. The notation $\delta^-$ and $\delta^+$ indicates that there is a partial separation of electronic charge.

(defined with a positive energy) between two argon atoms is shown by the pink curve in Figure 3.22.

$$\text{Repulsive force between atoms} \propto \frac{1}{d^{13}} \tag{3.17}$$

$$\text{Potential energy due to repulsion between atoms} \propto \frac{1}{d^{12}} \tag{3.18}$$

If this repulsive force were the only one operative in the system, then elements such as the noble gases would never form a solid lattice. We need to recognize that there is an attractive force between the atoms that opposes the electronic repulsion.

We may envisage this attractive force as arising in the following way. Within the argon atom, the positions of the nucleus and the electrons are not fixed. At any given time, the centre of negative charge density may not coincide with the centre of the positive charge, i.e. with the nucleus. This results in a net, instantaneous *dipole*. This dipole is transient, but it is sufficient to *induce* a dipole in a neighbouring atom (Figure 3.23) and electrostatic *dipole–dipole interactions* between atoms result. Since the dipoles are transient, each separate interaction will also be transient, but the net result is an attractive force between the argon atoms.

The attractive force due to the dipole–dipole interactions may be represented as shown in equation 3.19. This leads to a potential energy which varies with internuclear distance $d$ (equation 3.20) and this attractive potential (defined with a negative energy) between atoms of argon is shown by the blue curve in Figure 3.22.

Dipole moments: see Section 5.11

**Fig. 3.23** A dipole set up in atom **1** induces a dipole in atom **2** which induces one in **3**, and so on. The distance $r_v$ is the van der Waals radius = half the distance of closest approach of two non-bonded atoms.

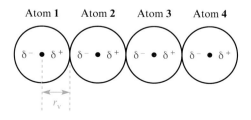

**Fig. 3.24** The overall potential energy of two argon atoms as they are brought together from infinite separation. The value of the internuclear distance, $d$, at the energy minimum ($d_e$) corresponds to *twice* the van der Waals radius of an argon atom.

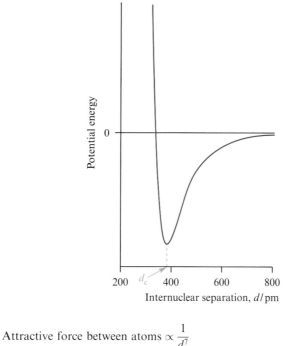

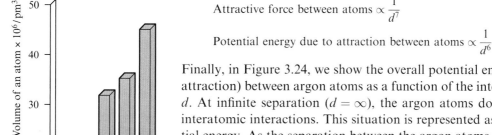

$$\text{Attractive force between atoms} \propto \frac{1}{d^7} \qquad (3.19)$$

$$\text{Potential energy due to attraction between atoms} \propto \frac{1}{d^6} \qquad (3.20)$$

Finally, in Figure 3.24, we show the overall potential energy (repulsion plus attraction) between argon atoms as a function of the internuclear separation $d$. At infinite separation ($d = \infty$), the argon atoms do not experience any interatomic interactions. This situation is represented as having zero potential energy. As the separation between the argon atoms decreases, attractive forces become dominant. As the two argon atoms come together, the attractive potential increases to a given point of maximum stabilization (lowest energy). However, as the atoms come yet closer together, the electronic repulsion between them increases. At very short distances, the electron–electron repulsion is dominant and highly unstable arrangements with *positive* potential energies result.

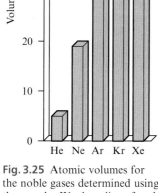

**Fig. 3.25** Atomic volumes for the noble gases determined using the van der Waals radius of each atom. The volume increases as group 18 is descended.

The **van der Waals radius**, $r_v$, of an atom X is measured as half the distance of closest approach of two non-bonded atoms of X.

## Van der Waals radius

The value of $d$ at the point in Figure 3.24 where the potential is at a minimum corresponds to the optimum distance, $d_e$, between two argon atoms. This is a *non-bonded separation*. Note that $d_e$ is *not* the point at which the attractive and repulsive forces are equal. This point is reached when the potential is zero.

It is convenient to consider $d_e$ in terms of an atomic property – the value $d_e/2$ is defined as the van der Waals radius, $r_v$, of an atom of argon. The value of $r_v$ is *half* the distance of closest internuclear separation of adjacent argon atoms. Values of the van der Waals radii for the noble gases are given in Table 3.6. In the particular case of a monatomic noble gas, these radii can be used to give a good idea of atomic size. The atomic volume may be estimated using the van der Waals radius; the volume of the spherical atom is then $\frac{4}{3}\pi r_v^3$. The increase in atomic volume on descending group 18 is represented in Figure 3.25.

## SUMMARY

This chapter has dealt with some difficult concepts, but they are essential to an understanding of the chemistry of the elements. As we introduce other topics in this book, there will be plenty of opportunity to review these concepts and this should provide you with a feeling as to why we have introduced them so early.

### Do you know what the following terms mean?

- vector
- scalar
- quantized
- photon
- wavefunction
- eigenfunction
- eigenvalue
- quantum numbers $n$, $l$, $m_l$ and $m_s$
- radial distribution function
- atomic orbital
- nodal plane

- lobe (of an orbital)
- phase (of an orbital)
- surface boundary of an orbital
- degenerate orbitals
- non-degenerate orbitals
- diffuse (with reference to an atomic orbital)
- penetration
- shielding
- effective nuclear charge
- hydrogen-like atom or ion
- many-electron atom

- first ionization energy
- *aufbau* principle
- Hund's rule
- Pauli exclusion principle
- ground state electronic configuration
- noble (inert) gas configuration
- octet rule
- core electrons
- valence electrons
- van der Waals radius
- induced dipole moment

### You should now be able:

- to describe briefly what is meant by wave–particle duality

- to describe briefly what is meant by the uncertainty principle

- to discuss briefly what is meant by a solution of the Schrödinger wave equation

- to describe the significance of the radial and angular parts of the wavefunction

- to state what is meant by the principal quantum number

- to state what is meant by the orbital and spin quantum numbers and relate their allowed values to a value of the principal quantum number

- to describe an atomic orbital and the electron(s) in it in terms of values of quantum numbers

- to discuss briefly why we tend to use *normalized* wavefunctions

- to sketch graphs for the radial distribution functions of the $1s$, $2s$, $3s$, $4s$, $2p$, $3p$ and $3d$ atomic orbitals; what do these graphs mean?

- to outline the shielding effects that electrons have on one another

- to draw the shapes, indicating phases, of $ns$ and $np$ atomic orbitals

- to distinguish between $np_x$, $np_y$ and $np_z$ atomic orbitals

- to explain why in a hydrogen-like atom, atomic orbitals with the same principal quantum number are degenerate

- to explain why in a many-electron atom, occupied orbitals with the same value of $n$ but different $l$ are non-degenerate

- to estimate the ionization energy of a hydrogen atom given a series of frequency values from the Lyman series of spectral lines

- to describe how to use the *aufbau* principle to determine the ground state electronic configuration of an atom

- to write down the usual ordering of the atomic orbitals from $1s$ to $4p$. Which has the lowest energy? What factors affect this order?

- given the atomic number of an element (up to $Z = 20$), to write down in notational form the ground state electronic configuration

- given the atomic number of an element (up to $Z = 20$), to draw an energy level diagram to show the ground state electronic configuration

- to explain how a potential energy curve such as that in Figure 3.24 arises; what is the significance of the energy minimum?

## PROBLEMS

**3.1** Distinguish between a vector and a scalar quantity and give examples of each.

**3.2** Write down a relationship between wavelength, frequency and speed of light. Calculate the wavelengths of electromagnetic radiation with frequencies of (a) $2.0 \times 10^{13}$ Hz, (b) $4.5 \times 10^{15}$ Hz and (c) $2.1 \times 10^{17}$ Hz. By referring to the spectrum in Appendix 4, assign each wavelength to a particular type of radiation (e.g. X-rays).

**3.3** Determine the possible values of the quantum numbers $l$ and $m_l$ corresponding to $n = 4$. From your answer, deduce the number of possible $4f$ orbitals.

**3.4** Give the sets of four quantum numbers that uniquely define each electron in an atom of (a) He ($Z = 2$) and (b) B ($Z = 5$), each in its ground state.

**3.5** How many orbitals are there in (a) the shell with $n = 3$, (b) the $5f$ sub-shell, (c) the $2p$ sub-shell, (d) the $3d$ sub-shell and (e) the $4d$ sub-shell?

**3.6** Which atomic orbital has the set of quantum numbers $n = 3$, $l = 0$, $m_l = 0$? How do you distinguish between the two electrons that may occupy this orbital?

**3.7** What do the terms *singly degenerate* and *triply degenerate* mean? Give examples of orbitals that are (a) singly and (b) triply degenerate.

**3.8** How does an increase in nuclear charge affect the (a) energy and (b) the spatial extent of an atomic orbital?

**3.9** Determine the energy of the levels of the hydrogen atom for $n = 1$, 2 and 3.

**3.10** Which of the following are hydrogen-like species? (a) $H^+$, (b) $H^-$, (c) He, (d) $He^+$, (e) $Li^+$, (f) $Li^{2+}$.

**3.11** Sketch an energy level diagram showing the approximate relative energies of the $1s$, $2s$ and $2p$ atomic orbitals in lithium. How does this diagram differ from a corresponding diagram showing these levels in the hydrogen atom?

**3.12** For lithium, $Z = 3$. Two electrons occupy the $1s$ orbital. Is it energetically better for the third electron to occupy the $2s$ or $2p$ atomic orbital? Rationalize your choice.

**3.13** What is the difference between the absolute nuclear charge of an atom and its effective nuclear charge. Are these variable or constant quantities?

**3.14** (a) Why does the $2p \longrightarrow 2s$ transition not give rise to a spectral line in the emission spectrum of atomic hydrogen? (b) What is the difference between an absorption and emission spectrum?

**3.15** Identify each of the following $n' \longrightarrow n$ transitions in the emission spectrum of atomic hydrogen as belonging to either the Balmer or Lyman series: (a) $2 \longrightarrow 1$; (b) $3 \longrightarrow 2$; (c) $3 \longrightarrow 1$; (d) $5 \longrightarrow 1$; (e) $5 \longrightarrow 2$.

**3.16** The frequencies of some consecutive spectral lines in the Lyman series of atomic hydrogen are 2.466, 2.923, 3.083, 3.157 and $3.197 \times 10^{15}$ Hz. (a) Use these values to draw a schematic representation of this part of the emission spectrum. (b) How does the spectrum you have drawn relate to the representation of the transitions shown in the figure below? (c) Assign each of the lines in the spectrum to a particular transition.

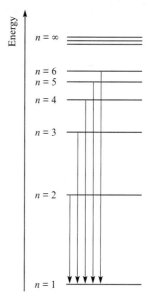

**3.17** Using the frequencies in problem 3.16 and the assignments you have made, plot an appropriate graph to estimate a value for the Rydberg constant.

**3.18** Starting from the Rydberg equation, determine the ionization energy of hydrogen. Write an equation to show to what process this energy refers.

**3.19** Determine the ground state electronic configurations of Be ($Z = 4$), F ($Z = 9$), P ($Z = 15$) and K ($Z = 19$), giving your answers in both notational form and in the form of energy level diagrams.

**3.20** What do you understand by the 'octet rule'?

**3.21** How many electrons are present in the valence shell of the central atom in each of the following species? (a) $SCl_2$; (b) $H_2O$; (c) $NH_3$; (d) $[NH_4]^+$; (e) $CCl_4$; (f) $PF_3$; (g) $BF_3$; (h) $BeCl_2$. Comment on the answers obtained.

## Additional problems

**3.22** Use the data in Table 3.2 to sketch a graph showing the radial wavefunction for a hydrogen $2s$ atomic orbital as a function of $r$. Comment on features of significance of the plot.

**3.23** Use the data in Table 3.4 to organize the elements from $Z = 1$–20 into 'families' with related ground state electronic configurations. Compare the result with the periodic table (see inside the front cover) and comment on any anomalies. On the basis of the ground state electronic configurations, say what you can about the properties of related elements.

# 4 Homonuclear covalent bonds

## Topics

- Covalent molecules
- Bond length
- Bond dissociation enthalpy
- Enthalpy of atomization
- Valence bond theory
- Molecular orbital theory
- $\sigma$ and $\pi$ bonds
- Bond order
- Homonuclear diatomics
- Periodicity in homonuclear bond dissociation enthalpies and distances
- The effects of group trends in bond dissociation enthalpies on elemental forms

## 4.1 Introduction

In Section 1.12 we specified what we mean by a molecule and stated that a covalent bond is formed when electrons are *shared* between nuclei. In this chapter we develop the ideas of covalent bonding.

We shall be concerned for the moment only with *homonuclear bonds* – bonds formed between atoms of the *same type*, for example the C–C bond in ethane ($C_2H_6$), the N–N bond in hydrazine ($N_2H_4$) or the O–O bond in hydrogen peroxide ($H_2O_2$) (Figure 4.1). Figure 4.2 shows examples of *homonuclear molecules* – molecules in which all of the *atoms are the same*. Each of these molecules is a molecular form of a particular element. In cases where more than one molecular form is known, these are called allotropes (e.g. $O_2$ and $O_3$) of that element.

We approach the question of covalent bonding by considering some experimental observables and then look at some of the theoretical models that have been developed to describe the bonds.

The concept of bonding arises from observations that, within molecules, the distances between adjacent atoms are usually significantly shorter than those expected from the van der Waals radii of the atoms involved. Although we discussed the van der Waals radii of only monatomic elements in Chapter 3, a similar parameter may be defined for most other elements. In practice, this parameter comes from measurements of the closest *inter*molecular contacts in a series of molecules.

This suggests that distance is a very important parameter in describing atomic arrangements. As far as a bond is concerned, the bond length (or

A *molecule* is a discrete neutral species resulting from the formation of a covalent bond or bonds between two or more atoms.

Allotropes: see Section 1.5

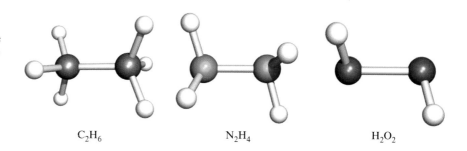

**Fig. 4.1** Molecules that contain one *homonuclear bond*: ethane ($C_2H_6$), hydrazine ($N_2H_4$) and hydrogen peroxide ($H_2O_2$).

$C_2H_6$         $N_2H_4$         $H_2O_2$

to be strictly accurate, the time-averaged distance between the nuclei) is of fundamental importance.

In the next section, we briefly introduce some of the methods that are used for the experimental determination of intramolecular and intermolecular distances. If you do not wish to interrupt the discussion of bonding, move to Section 4.3 and return to Section 4.2 later.

## 4.2      Measuring internuclear distances

▬▶
Vibrational spectroscopy of diatomics: see Section 12.3

There are a number of methods available for determining important structural parameters. These techniques fall broadly into the groups of *spectroscopic* and *diffraction methods*. It is beyond the scope of this book to consider the details of the methods and we merely introduce them and look at some of the structural data that can be obtained.

### An overview of diffraction methods

For solid state compounds, the two methods most commonly used to determine structure are *X-ray* and *neutron diffraction*. *Electron diffraction* is best used to study the structures of gaseous molecules, although the technique is not strictly confined to use in the gas phase. Whereas electrons are negatively charged, X-rays and neutrons are neutral. Accordingly, X-rays and neutrons get 'deeper' into a molecule than do electrons because they do not experience an electrostatic repulsion from the core and valence electrons of the molecules. X-rays and neutrons are said to be more *penetrating* than

**Fig. 4.2** Examples of covalent homonuclear molecules.

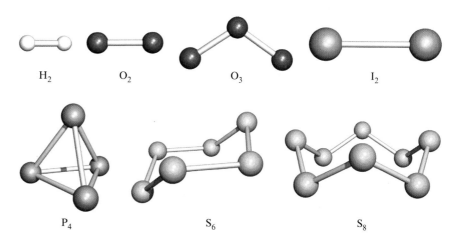

$H_2$         $O_2$         $O_3$         $I_2$

$P_4$         $S_6$         $S_8$

**Table 4.1** A comparison of electron, X-ray and neutron diffraction techniques.

| | **Electron diffraction** | **X-ray diffraction** | **Neutron diffraction** |
|---|---|---|---|
| Approximate wavelength of beam used / m | $10^{-11}$ | $10^{-10}$ | $10^{-10}$ |
| Type of sample studied | Usually gases; also liquid or solid surfaces. | Solid (single crystal or powder); rarely solution. | Usually single crystal or powder. |
| What does the method locate? | Nuclei. | Regions of electron density. | Nuclei. |
| Relative advantages | Allows structural data to be collected for gaseous molecules; neither X-ray nor neutron diffraction methods are suitable for such samples. | Technique is routinely available and relatively straightforward; ionic lattices and molecular solids can be studied; accurate bond parameters can be obtained for non-hydrogen atoms. | Can locate hydrogen atoms; very accurate structural parameters can be determined; both ionic lattices and molecular solids can be studied. |
| Relative disadvantages | Refining data for large molecules is complicated; hydrogen atom location is not accurate. | Hydrogen atom location is not always accurate. | Expensive; refining data for large molecules is complicated. |

electrons. Selected comparative details of the three diffraction methods are listed in Table 4.1.

In devising a method based on diffraction to measure internuclear distances, it is important to note that the distances between bonded atoms are typically in the range of 100 to 300 pm. The wavelength of the radiation that is chosen for the diffraction experiment should be similar in magnitude to the distances of interest; further details are given in Box 4.1.

At this point we stress some relevant facts that affect the type of information obtained from the different diffraction techniques:

- Electrons are negatively charged particles and are predominantly diffracted because they interact with the overall electrostatic field resulting from the negatively charged electrons and the positively charged nuclei of a molecule.
- X-rays are electromagnetic radiation. They are scattered *mainly* by the *electrons* in a molecular or ionic array. It is possible to locate the atomic nuclei approximately because the X-rays are mostly scattered by the *core electrons* which are concentrated around each nucleus. The *scattering power* of an atom is proportional to the atomic number, $Z$, and it is more difficult to locate H atoms ($Z = 1$) than atoms with higher atomic numbers.
- Neutrons are diffracted by the atomic nuclei, and therefore the positions of nuclei can be found accurately from the results of a neutron diffraction study.

Keep in mind that our goal is to measure internuclear distances. Neutron diffraction would appear to be the method of choice for all solid state studies. However, neutron diffraction is a very expensive experimental method and it is not as readily available as an X-ray diffraction facility. Hence, as a routine method for solid samples, X-ray diffraction usually wins over neutron diffraction.

## THEORETICAL AND CHEMICAL BACKGROUND

# Box 4.1 Diffraction methods

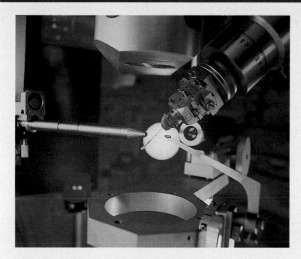

The central part of an X-ray diffractometer. The crystal is encapsulated in a thin glass tube, positioned within the beam of X-rays.
*James King-Holmes/OCMS/Science Photo Library.*

Diffraction methods can be used to measure bond distances but the wavelength of the radiation used must be similar to the internuclear distances.

Internuclear bond distances are usually between 100 and 300 pm.

Since the wavelength of X-rays is about 100 pm, the diffraction of X-rays by a solid (usually crystalline) substance can be used to determine bond lengths.

The wavelength of a beam of electrons can be altered by applying an accelerating voltage, and a wavelength of the order of 10 pm is accessible. This is used for electron diffraction experiments.

Fast neutrons generated in a nuclear reactor can be moderated to give a beam of neutrons with a wavelength of the order of 100 pm. In addition, several neutron sources designed specifically for the purpose of diffraction experiments are available.

For further details of diffraction methods, a suitable book is E. A. V. Ebsworth, D. W. H. Rankin and S. Cradock (1991), *Structural Methods in Inorganic Chemistry*, 2nd edn, Blackwell Scientific Publications, Oxford.

## THEORETICAL AND CHEMICAL BACKGROUND

# Box 4.2 The difficulty of locating a hydrogen nucleus by X-ray diffraction

Consider a C–H bond. The ground state electronic configuration of a carbon atom is $1s^2 2s^2 2p^2$ and that of a hydrogen atom is $1s^1$. The C atom has four valence electrons and two core electrons. The H atom has one valence electron and no core electrons.

When the H atom bonds with the C atom, the single electron from the H atom is used in bond formation. This means that there is little electron density concentrated around the H nucleus:

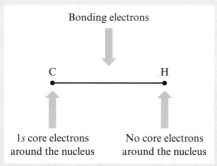

If the length of the C–H bond is being determined by X-ray diffraction, there is a problem. The method locates

regions of electron density. The positions of the nuclei are inferred from the position of the electrons.

The carbon nucleus can be located because it has associated with it the core $1s$ electrons, but the hydrogen nucleus cannot normally be directly located.

However, the region of electron density associated with the C–H bond can be found and hence a C–H distance can be estimated but it is shorter than the true value:

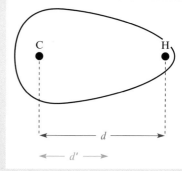

The experimental distance is $d'$ while the true internuclear distance is $d$.

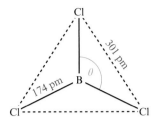

**Fig. 4.3** Intramolecular bonded and non-bonded distances determined from an electron diffraction study of $BCl_3$; the angle $\theta$ is calculated from the distance data; see worked example 4.1.

## Electron diffraction

The gas phase differs from the solid state in that the molecules are continually in motion and there are usually no persistent intermolecular interactions. As a result, diffraction experiments on gases mainly provide information regarding intramolecular distances, and the data collected in an electron diffraction experiment can be analysed in terms of *intra*molecular parameters. The initial diffraction data relate the scattering angle of the electron beam to intensity, and from these results a plot of intensity against distance can be obtained. Peaks in this radial distribution curve correspond to internuclear separations.

Consider the electron diffraction results from the determination of the structure of boron trichloride in the gas phase. They provide the bonded and non-bonded distances shown in Figure 4.3; all the B–Cl bonds are of equal length and all the intramolecular non-bonded Cl···Cl distances are equal. The angle $\theta$ is determined from the B–Cl and Cl–Cl distances by use of trigonometry.

| **Worked example 4.1** | *Determining a bond angle from distance data* |

The results of an electron diffraction experiment for $BCl_3$ show that each B–Cl bond length is 174 pm, and each non-bonded Cl···Cl distance is 301 pm. Use these data to determine the Cl–B–Cl bond angles. Determine whether $BCl_3$ is a planar or non-planar molecule.

For help with the sine rule and cosine rule, visit www.pearsoned.co.uk/housecroft, and consult the accompanying *Mathematics Tutor*

The distance data given in the question are summarized in the diagram below. Since all the B–Cl bond lengths are equal, and all the Cl···Cl non-bonded separations are equal, it follows that the three Cl–B–Cl bond angles are equal. Let $\angle$Cl–B–Cl $= \theta$.

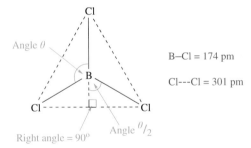

Angle $\theta$ can be found by use of either the cosine rule or the sine rule. For the sine rule:

$$\sin\frac{\theta}{2} = \frac{(301/2)}{174}$$

$$\frac{\theta}{2} = 60°$$

$$\theta = 120°$$

Since the sum of the three angles in $BCl_3$ is 360°, the molecule is planar.

## X-ray diffraction

The diffraction of X-rays by the *core* electrons provides the most useful information regarding the positions of the nuclei in a molecule. The heavier

**Table 4.2** Internuclear S−S distances between pairs of adjacent atoms in molecules present in some allotropes of sulfur.

| Allotrope | Type of molecule | S−S distance / pm |
|-----------|-----------------|-------------------|
| $S_6$ | Ring | 206 |
| $S_8$ (α-form) | Ring | 206 |
| $S_8$ (β-form) | Ring | 205 |
| $S_{10}$ | Ring | 206 |
| $S_{11}$ | Ring | 206 |
| $S_{12}$ | Ring | 205 |
| $S_{18}$ | Ring | 206 |
| $S_{20}$ | Ring | 205 |
| $S_\infty$ | Helical chain | 207 |

an atom is, the more core electrons it possesses, and the greater is its ability to scatter X-rays. As a result, heavier atoms are more easily and accurately located than lighter atoms. We mentioned above the particular difficulty associated with locating hydrogen atoms.

In a solid, the motion of the molecules is restricted. When a compound crystallizes, the molecules form an ordered array and *intermolecular forces* operate between the molecules. The results of an X-ray diffraction study of a crystal provide information about intermolecular as well as intramolecular distances. We illustrate this idea by looking at the solid state structures of various allotropes of sulfur.

▪▷
Allotropes of sulfur: see
Section 9.8

The results of an X-ray diffraction study on crystalline orthorhombic sulfur (α-sulfur, the standard state of sulfur) show that the crystal lattice contains $S_8$ rings (Figure 4.2). The S−S bond distances in each ring are equal (206 pm). Table 4.2 lists some allotropes of sulfur and the internuclear S−S distances between pairs of bonded atoms. The reason that all these S−S distances are close in value is that each represents an S−S single covalent bond length, and for a particular homonuclear single bond, the distance is relatively constant. The shortest distances between sulfur atoms in *adjacent* rings in the solid state (intermolecular distances) are larger than the intramolecular S−S bond lengths; 337 pm for $S_{11}$ and 323 pm for $S_{10}$. These indicate non-bonded interactions.

**4.3**    ## The covalent radius of an atom

In Section 1.12, we saw that, when atoms combine to form molecules, the pairing of two electrons leads to the formation of a covalent bond. When *one pair of electrons* is involved in bond formation, the bond is a *single bond* and it has a *bond order of 1*.

The ***covalent radius***, $r_{cov}$, for an atom X is taken as half the internuclear distance in a homonuclear X−X bond.

The covalent radius is defined for a particular type of bond. Values appropriate for single, double and triple bonds differ.

Consider a gaseous homonuclear diatomic molecule such as $Cl_2$. There is a single bond between the two chlorine nuclei and the bond length is 199 pm. It would be convenient to have an *atomic* parameter that we could use to describe the size of an atom when it participates in covalent bonding. We define the *single bond covalent radius*, $r_{cov}$, of an atom X as half of the internuclear distance in a typical homonuclear X−X single bond. In the case of chlorine, the single bond covalent radius is 99 pm.

As X−X single bonds in molecules are not all exactly the same length, we often use data from a range of compounds to obtain an average $r_{cov}$ value.

**Table 4.3** Van der Waals and covalent radii for hydrogen and atoms in the *p*-block. These values represent the 'best' values from a wide range of compounds containing the elements.

|          | Element | Van der Waals radius / pm | Covalent radius / pm |
|----------|---------|---------------------------|----------------------|
|          | H       | 120                       | 37[a]                |
| Group 13 | B       | 208                       | 88                   |
|          | Al      | –                         | 130                  |
|          | Ga      | –                         | 122                  |
|          | In      | –                         | 150                  |
|          | Tl      | –                         | 155                  |
| Group 14 | C       | 185                       | 77                   |
|          | Si      | 210                       | 118                  |
|          | Ge      | –                         | 122                  |
|          | Sn      | –                         | 140                  |
|          | Pb      | –                         | 154                  |
| Group 15 | N       | 154                       | 75                   |
|          | P       | 190                       | 110                  |
|          | As      | 200                       | 122                  |
|          | Sb      | 220                       | 143                  |
|          | Bi      | 240                       | 152                  |
| Group 16 | O       | 140                       | 73                   |
|          | S       | 185                       | 103                  |
|          | Se      | 200                       | 117                  |
|          | Te      | 220                       | 135                  |
| Group 17 | F       | 135                       | 71                   |
|          | Cl      | 180                       | 99                   |
|          | Br      | 195                       | 114                  |
|          | I       | 215                       | 133                  |

[a] Sometimes it is more appropriate to use a value of 30 pm in organic compounds.

For a given element:
van der Waals radius,
$r_v$ > covalent radius, $r_{cov}$

*p*-Block elements:
see Chapter 22

An estimate for the covalent radius of sulfur can be obtained by taking half the average S−S bond length obtained for various allotropes of sulfur; the average S−S distance from the data in Table 4.2 is 206 pm, which gives a single covalent bond radius for a sulfur atom of 103 pm.

In Table 4.3, values of single bond covalent radii of hydrogen and elements in the *p*-block are compared with those of the van der Waals radii. Since the van der Waals radii are measured from non-bonded distances and the covalent radii are determined from bonded interactions, the former are larger than the latter for a given element. Down any group of the periodic table, the values for both sets of radii generally increase (Figure 4.4).

**Fig. 4.4** Trends in covalent and van der Waals radii upon descending groups 15, 16 and 17.

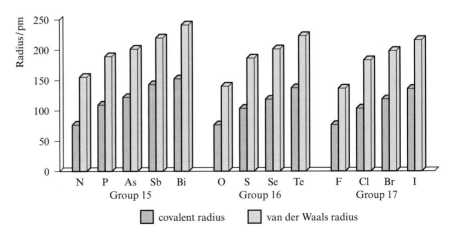

■▷
*Potential energy for two argon atoms: see Section 3.21*

## 4.4    An introduction to bond energy: the formation of the diatomic molecule $H_2$

Let us move now from the lengths of bonds to their energies. Just as we compared covalent (bonded) distances with van der Waals (non-bonded) ones, we begin our discussion of energy by comparing the potential energy associated with bringing together two hydrogen atoms with that of bringing two atoms of argon together. For argon, we considered two opposing factors – the attractive potential due to the induced dipole–dipole interactions and a repulsive term due to interatomic electron–electron interactions. Can we use a similar approach to describe the interaction between two hydrogen atoms?

There is a fundamental difference between the approach of two argon atoms and two hydrogen atoms. Whatever the distance between the argon atoms, the strongest interatomic *attractive* forces are of the induced dipole–dipole type. There is no net covalent bonding resulting from the sharing of electrons between the two argon nuclei. In contrast, as two hydrogen atoms approach each other, electron sharing becomes possible and a covalent $H-H$ bond can be formed.

We can rationalize this by considering the number of electrons in the valence shell of each atom. Each argon atom possesses the noble gas configuration $[Ne]3s^2 3p^6$. There is no driving force for a change of electron configuration.

■▷
*Octet rule: see Section 3.20*

In contrast, each hydrogen atom possesses a $1s^1$ configuration. Just as there is a particular stability associated with a noble gas $ns^2 np^6$ configuration (an octet), the $1s^2$ configuration of helium is also stable (see Box 4.3). There are two ways in which a hydrogen atom could attain the [He] configuration – it could gain an electron and become the $H^-$ anion with a $1s^2$ ground state configuration, or it could *share* an electron with another atom. The simplest atom with which it could share electrons is another hydrogen atom and, in this way, the molecule $H_2$ is formed (Figure 4.5).

**THEORETICAL AND CHEMICAL BACKGROUND**

### Box 4.3 The ground state configuration of helium

Helium is in group 18 of the periodic table and is a noble gas (see Section 3.21). Each group 1 metal readily loses an electron to give a cation possessing a noble gas configuration:

$$K \longrightarrow K^+ + e^-$$
$$[Ar]4s^1 \qquad [Ar]$$

$$Na \longrightarrow Na^+ + e^-$$
$$[Ne]3s^1 \qquad [Ne]$$

$$Li \longrightarrow Li^+ + e^-$$
$$[He]2s^1 \qquad [He]$$

A noble gas atom or an ion with a noble gas configuration has an $ns^2 np^6$ configuration, except for the case where $n = 1$. Here, the noble gas configuration is $1s^2$. This corresponds to the ground state configuration of a helium atom. The $n = 1$ principal quantum level is unique in being fully occupied when it contains only two electrons.

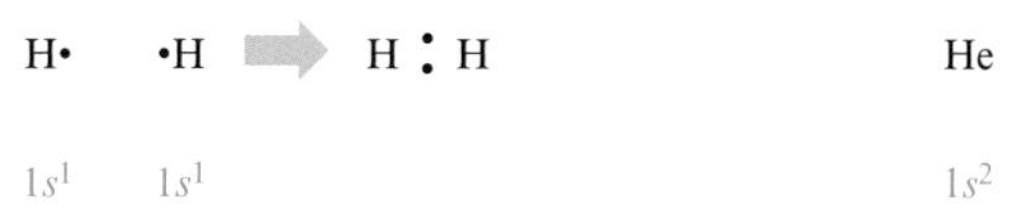

**Fig. 4.5** A hydrogen atom has a valence electronic configuration of $1s^1$ and shares an electron with another hydrogen atom to form $H_2$. In doing so, each H atom becomes like an atom of the noble gas helium which has a valence configuration of $1s^2$.

$$H\cdot \quad \cdot H \implies H \overset{\cdot}{\underset{\cdot}{}} H \qquad\qquad He$$

$$1s^1 \quad 1s^1 \qquad\qquad\qquad\qquad 1s^2$$

**Fig. 4.6** The potential energy curve that describes the approach of two hydrogen atoms to their equilibrium separation.

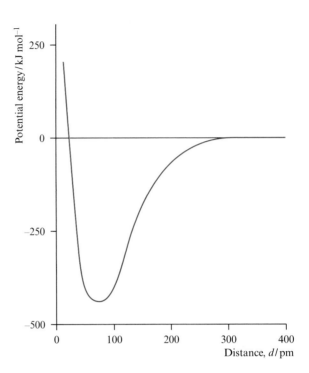

As the two hydrogen atoms approach one another, electron sharing becomes more favourable, and we start to form a covalent bond. Figure 4.6 shows potential energy as a function of internuclear distance. Although this curve is superficially similar to the one in Figure 3.24, the energy terms of which it is composed are rather different and we can characterize four main components:

1. repulsion between electrons;
2. attraction between an electron and a proton in the *same* atom;
3. attraction between an electron and a proton in the *other* atom; and
4. repulsion between the protons.

These interactions are represented diagrammatically in Figure 4.7.

Strictly speaking, terms (2) and (3) are meaningful only at large internuclear distances because, as the atoms come close together and covalent bond formation occurs, it is not reasonable to associate a particular electron with a particular proton.

What does the curve in Figure 4.6 tell us? Experimentally we find that the standard state of the element hydrogen is $H_2$ molecules (dihydrogen) rather than H atoms. The energy minimum should correspond to the equilibrium distance of the two hydrogen nuclei in $H_2$. The equilibrium distance is the position of lowest (most negative) energy for the system, and this

**Fig. 4.7** The approach of two hydrogen atoms. Each contains one proton and one electron. Interactions will be (1) electron–electron (repulsive), (2) electron–proton within the same atom (attractive), (3) electron–proton between different atoms (attractive) and (4) proton–proton (repulsive).

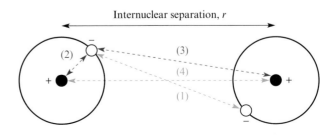

internuclear distance corresponds to the H−H covalent bond distance and is found experimentally to be 74 pm.[§]

## Bond energies and enthalpies

In the preceding section we introduced the H−H covalent bond in the $H_2$ molecule by considering the equilibrium distance and energy of two approaching H atoms. We could draw a curve similar to Figure 4.6 for the formation of any other homonuclear X−X (or heteronuclear X−Y) covalent bond. So far, we have concentrated upon the distance axis – how far apart are the atoms? But we also need to know something about the energy axis – how much energy is involved in the formation of the H−H covalent bond or how much energy does it take to break (*dissociate*) the H−H bond?

### The problem of internal energy and enthalpy

> The *internal energy*, $U$, of a system is its total energy. $\Delta U$ is the change in internal energy.

The energy corresponding to the minimum in Figure 4.6 is 458 kJ mol⁻¹. Unfortunately, there is a problem – the value of 458 kJ mol⁻¹ does *not* correspond exactly to the amount of energy released when the two H atoms form an $H_2$ molecule or, conversely, to the amount of energy required to break the H−H bond in $H_2$.

The curve in Figure 4.6 corresponds to the *internal energy* of the system ($\Delta U$) measured at a temperature of 0 K. However, it is often more convenient to consider enthalpies ($\Delta H$) at 298 K. In practice, the conversion of $\Delta U$ to $\Delta H$ involves only $\approx 3$ kJ mol⁻¹. (The terms $\Delta U$ and $\Delta H$ will be discussed in greater detail in Chapter 17 – at the moment, all that you need to note is that *the difference between $\Delta U$ and $\Delta H$ for a given system is very small*.) The real problem arises from the fact that even at 0 K, the two hydrogen nuclei are not stationary, but are vibrating. We discuss vibrational states in Chapter 12, but for now, note that the lowest vibrational state of the $H_2$ molecule lies about 26 kJ mol⁻¹ higher in energy than the bottom of the curve in Figure 4.6. This residual energy is called the *zero point energy* (see Figure 12.2). Any experimental measurement will relate to the real molecule and to the lowest energy state that this can reach. Thus, experimental measurements do not 'reach' the very bottom of the potential energy well but access only the lowest energy vibrational state.

> The *zero point energy* of a molecule corresponds to the energy of its lowest vibrational level (vibrational ground state); see Figure 12.2.

> ■▶
> $\Delta U$ and $\Delta H$:
> see Section 17.2

$\Delta U$ and $\Delta H$ are related by the equation:

$$\Delta U = \Delta H - P\Delta V \quad \text{at constant pressure}$$

where $P$ = pressure and $V$ = volume.

### Homonuclear bond dissociation in a diatomic molecule

The most convenient parameter to describe the energy associated with the H−H bond in $H_2$ is the energy needed to convert an $H_2$ molecule in its

---

[§] Actually, because of the asymmetry of the curve, the experimentally measured bond length in $H_2$ does not coincide exactly with the distance corresponding to the curve minimum.

**Table 4.4** Internal energy changes ($\Delta U$) and the enthalpy change ($\Delta H$) used to describe the dissociation of $H_2$. The relationship between $\Delta U$ and $\Delta H$ is $\Delta U = \Delta H - P\Delta V$ where $P$ = pressure and $V$ = volume (at constant temperature).

| Quantity | Process | Value / kJ mol$^{-1}$ |
|---|---|---|
| $\Delta U$ (0 K) | $H_2(g) \longrightarrow 2H(g)$ at 0 K | 432 |
| $\Delta U$ (298 K) | $H_2(g) \longrightarrow 2H(g)$ at 298 K | 433 |
| $\Delta H$ (298 K) | $H_2(g) \longrightarrow 2H(g)$ at 298 K | 436 |

The **bond dissociation energy** ($\Delta U$) and the **bond dissociation enthalpy** ($\Delta H$) for an X–X **diatomic molecule** refer to the process:

$$X_2(g) \longrightarrow 2X(g)$$

at a given temperature (often 0 K or 298 K).

lowest vibrational energy level to two hydrogen atoms (equation 4.1). Table 4.4 lists possible internal energy ($\Delta U$) or enthalpy ($\Delta H$) terms that may be used to describe this process.

$$H_2(g) \longrightarrow 2H(g) \tag{4.1}$$

Since the values of $\Delta H$ and $\Delta U$ are so close, here and for other bond dissociation processes, we will not consistently distinguish between them. It is more convenient to deal with enthalpy values (most often at 298 K). Thus, for $H_2$, we associate a value of 436 kJ mol$^{-1}$ with the cleavage of the H–H bond.

### Homonuclear bond cleavage in a polyatomic molecule

In polyatomic molecules, a bond enthalpy term describes the energy needed to break one particular bond. Equations 4.2–4.4 illustrate the processes for the cleavage of the homonuclear bond in each of the molecules shown in Figure 4.1. Figure 4.8 shows this more explicitly for C–C bond cleavage in $C_2H_6$.

$$C_2H_6(g) \longrightarrow 2CH_3^\bullet(g) \tag{4.2}$$
$$N_2H_4(g) \longrightarrow 2NH_2^\bullet(g) \tag{4.3}$$
$$H_2O_2(g) \longrightarrow 2HO^\bullet(g) \tag{4.4}$$

Figure 4.8 also shows the result of homonuclear bond cleavage in a cyclic molecule. When an S–S bond is broken in an $S_6$ ring, the ring opens to form a chain and each terminal sulfur atom has an unpaired electron. Further S–S bond cleavage can occur to give $S_x$ fragments. The enthalpy associated with each S–S bond breakage will be characteristic of that particular process. This introduces the idea that bond enthalpies for chemically equivalent bonds may

**Fig. 4.8** Bond cleavage in a molecule such as ethane will produce two separate radicals. In a cyclic molecule such as $S_6$, cleavage of a bond will cause ring opening.

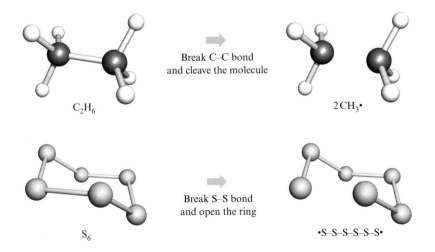

C$_2$H$_6$    Break C–C bond and cleave the molecule    2 CH$_3$•

$S_6$    Break S–S bond and open the ring    •S–S–S–S–S–S•

**THEORETICAL AND CHEMICAL BACKGROUND**

## Box 4.4 Average and individual bond enthalpy terms

Methane is a tetrahedral molecule with four chemically equivalent C–H bonds of equal length.

When methane is completely dissociated:

$$CH_4(g) \longrightarrow C(g) + 4H(g)$$

we can determine an *average* value for $\bar{D}(C{-}H)$ from the enthalpy of atomization (see Section 4.6):

$$\Delta_a H^\circ = 1664 \, kJ \, mol^{-1}$$

$$\bar{D}(C{-}H) = \tfrac{1}{4} \times 1664 = 416 \, kJ \, mol^{-1}$$

However, if we *sequentially* break the C–H bonds, then the enthalpy change for each individual step is not $416 \, kJ \, mol^{-1}$:

$$CH_4(g) \longrightarrow CH_3(g) + H(g) \quad \Delta H = 436 \, kJ \, mol^{-1}$$

$$CH_3(g) \longrightarrow CH_2(g) + H(g) \quad \Delta H = 461 \, kJ \, mol^{-1}$$

$$CH_2(g) \longrightarrow CH(g) + H(g) \quad \Delta H = 428 \, kJ \, mol^{-1}$$

$$CH(g) \longrightarrow C(g) + H(g) \quad \Delta H = 339 \, kJ \, mol^{-1}$$

The conclusion is that, although the bonds in $CH_4$ are chemically equivalent and have equal strengths, their strength will not be the same as the strength of the three C–H bonds in $CH_3$, the two C–H bonds in $CH_2$, or the C–H bond in CH.

---

vary and this is discussed further in Box 4.4. It is not always convenient to use such individual enthalpy values. Instead, we often use *average bond enthalpies*.

$D = \Delta_{diss}H$

We distinguish between a bond enthalpy term which refers to a particular process and a value that is an average of several such enthalpies by using the symbols $D$ and $\bar{D}$ respectively. Thus, for ethane, the bond enthalpy, $D$, for the cleavage of the C–C single bond (equation 4.2) is $376 \, kJ \, mol^{-1}$. However, after considering a wide range of values of $D$ for the cleavage of single C–C bonds in different molecules or in different environments in the same molecule, we arrive at an average value, $\bar{D}$, of about $346 \, kJ \, mol^{-1}$. Values of $\bar{D}$ for a given bond vary slightly between different tables of data, depending upon the number of data averaged and the exact compounds used to obtain them, and separate values must be determined for single, double or triple bonds. Table 4.5 lists some bond enthalpy terms for selected bonds.

A **bond enthalpy**, $\bar{D}$, for a bond X–X represents an average value for the enthalpy change associated with bond cleavage.

Separate values must be determined for single (X–X), double (X=X) or triple (X≡X) bonds.

### 4.6    The standard enthalpy of atomization of an element

#### The standard enthalpy of atomization

Consider the bond dissociation given in equation 4.1. The dissociation of the H–H bond corresponds to the formation of H atoms and this energy is also known as the enthalpy of atomization. The *standard* enthalpy of atomization of an element, $\Delta_a H^\circ(298 \, K)$, is the enthalpy change at 298 K when one mole of gaseous atoms is formed from the element in its standard state. For $H_2$ this is defined for the process given in equation 4.5, while for mercury and nickel it is defined according to equations 4.6 and 4.7 respectively:

Standard state: see Section 2.2

$$\tfrac{1}{2}H_2(g) \longrightarrow H(g) \quad \Delta_a H^\circ = 218 \, kJ \, mol^{-1} \tag{4.5}$$

$$Hg(l) \longrightarrow Hg(g) \quad \Delta_a H^\circ = 61 \, kJ \, mol^{-1} \tag{4.6}$$

$$Ni(s) \longrightarrow Ni(g) \quad \Delta_a H^\circ = 430 \, kJ \, mol^{-1} \tag{4.7}$$

**Table 4.5** Bond enthalpy terms for selected bonds.[a]

| Bond | Bond enthalpy / kJ mol$^{-1}$ | Bond | Bond enthalpy / kJ mol$^{-1}$ | Bond | Bond enthalpy / kJ mol$^{-1}$ |
|------|------|------|------|------|------|
| H–H | 436 | F–F | 159 | C–F | 485 |
| C–C | 346 | Cl–Cl | 242 | C–Cl | 327 |
| C=C | 598 | Br–Br | 193 | C–Br | 285 |
| C≡C | 813 | I–I | 151 | C–I | 213 |
| Si–Si | 226 | C–H | 416[b] | C–O | 359 |
| Ge–Ge | 186 | Si–H | 326 | C=O | 806 |
| Sn–Sn | 152 | Ge–H | 289 | C–N | 285 |
| N–N | 159 | Sn–H | 251 | C≡N | 866 |
| N=N | ≈400 | N–H | 391 | C–S | 272 |
| N≡N | 945 | P–H | 322 | Si–O | 466 |
| P–P | 200 | As–H | 247 | Si=O | 642 |
| P≡P | 490 | O–H | 464 | N–F | 272 |
| As–As | 177 | S–H | 366 | N–Cl | 193 |
| O–O | 146 | Se–H | 276 | N–O | 201 |
| O=O | 498 | F–H | 570 | P–F | 490 |
| S–S | 266 | Cl–H | 432 | P–Cl | 319 |
| S=S | 425 | Br–H | 366 | P–O | 340 |
| Se–Se | 193 | I–H | 298 | S–F | 326 |

[a] Some values can be obtained directly from the dissociation of a gaseous molecule, e.g. $D$(F–F); other values are mean bond enthalpy terms, e.g. $\bar{D}$(C–H).
[b] See the discussion in Section 25.8.

The ***standard enthalpy of atomization***, $\Delta_a H^\circ$, of an element is the enthalpy change (at 298 K) when one mole of gaseous atoms is formed from the element in its standard state. The SI units are J per mole of gaseous atoms formed; kJ mol$^{-1}$ are often more convenient units.

While some elements are solid under standard state conditions, others are liquid or gas. If the standard state of the element is a solid, then the value of the standard enthalpy of atomization (defined at 298 K) includes contributions from:

- the enthalpy needed to raise the temperature of the element from 298 K to the melting point;
- the enthalpy of fusion of the element (i.e. the transition from solid to liquid);
- the enthalpy needed to raise the temperature of the element from the melting point to the boiling point ($T$ K);

**Table 4.6** Selected enthalpies of atomization of the elements ($\Delta_a H^\circ$(298 K)); see also Appendix 10.

| Element | Standard state of element | $\Delta_a H^\circ$(298 K) / kJ mol$^{-1}$ |
|---------|---------------------------|-------------------------------------------|
| As | As(s) | 302 |
| Br | Br$_2$(l) | 112 |
| C | C(graphite) | 717 |
| Cl | Cl$_2$(g) | 121 |
| Cu | Cu(s) | 338 |
| F | F$_2$(g) | 79 |
| H | H$_2$(g) | 218 |
| Hg | Hg(l) | 61 |
| I | I$_2$(s) | 107 |
| K | K(s) | 90 |
| N | N$_2$(g) | 473 |
| Na | Na(s) | 108 |
| O | O$_2$(g) | 249 |
| P | P$_4$(white) | 315 |
| S | S$_8$(rhombic) | 277 |
| Sn | Sn(white) | 302 |

- the enthalpy of vaporization of the element (i.e. the transition from liquid to vapour);
- any enthalpy change associated with the bond breaking of polyatomic gas phase species; and
- the enthalpy change on going from the gas phase monatomic element at $T$ K to 298 K.

■▷
Appendix 10 gives a full list of values of $\Delta_a H^\circ$

Selected values of standard enthalpies of atomization of the elements are listed in Table 4.6.

---

| **Worked example 4.2** | *Contributions to the heat of atomization* |

**Determine the heat of atomization of mercury using Hess's Law of Constant Heat Summation. Information about mercury that is required:**

**bp $= 655$ K**
**molar heat capacity ($C_P$) for liquid mercury $= 27.4$ J K$^{-1}$ mol$^{-1}$**
**molar heat capacity ($C_P$) for gaseous mercury $= 20.8$ J K$^{-1}$ mol$^{-1}$**
**$\Delta_{vap} H(\text{bp}) = 59.11$ kJ mol$^{-1}$.**

The standard enthalpy of atomization is defined for the formation of one mole of gaseous atoms from mercury in its standard state – a liquid.
The molar heat capacity ($C_P$) tells you how much heat energy is needed to raise the temperature of one mole of a substance by one kelvin at constant pressure. We introduced *specific heat capacities* in Section 2.3, and we shall look in detail at *molar heat capacities* in Section 17.3.
Set up a thermochemical cycle that accounts for all the steps in taking mercury from a liquid at 298 K (standard conditions) to gaseous atoms at 298 K:

$$\text{Hg(l) at 298 K} \xrightarrow{\Delta_a H^\circ(298\ \text{K})} \text{Hg(g) at 298 K}$$

$$\downarrow \Delta H_1 \qquad\qquad\qquad\qquad \uparrow \Delta H_3$$

$$\text{Hg(l) at 655 K (boiling point)} \xrightarrow{\Delta H_2 = \Delta_{vap} H(\text{bp})} \text{Hg(g) at 655 K (boiling point)}$$

To find $\Delta H_1$:

Heat required to raise the temperature of one mole Hg(l) from 298 to 655 K

$= $ (Molar heat capacity of liquid) $\times$ (Rise in temperature)

$= (27.4 \text{ J K}^{-1} \text{ mol}^{-1}) \times \{(655 - 298) \text{ K}\}$

$= (27.4 \text{ J K}^{-1} \text{ mol}^{-1}) \times (357 \text{ K})$

$= 9780 \text{ J mol}^{-1}$   (to 3 sig. fig.)

$= 9.78 \text{ kJ mol}^{-1}$

This is an endothermic process. Therefore, $\Delta H_1 = +9.78$ kJ mol$^{-1}$.

To find $\Delta H_2$:
This is the change in enthalpy accompanying the vaporization of one mole of Hg at its boiling point:

$\Delta H_2 = \Delta_{vap} H(\text{bp}) = 59.11$ kJ mol$^{-1}$

To find $\Delta H_3$:

Heat liberated when the temperature of one mole Hg(l) falls from 655 to 298 K

$= $ (Molar heat capacity of gas) $\times$ (Fall in temperature)

$= (20.8\,\mathrm{J\,K^{-1}\,mol^{-1}}) \times \{(655 - 298)\,\mathrm{K}\}$

$= (20.8\,\mathrm{J\,K^{-1}\,mol^{-1}}) \times (357\,\mathrm{K})$

$= 7430\,\mathrm{J\,mol^{-1}}$ (to 3 sig. fig.)

$= 7.43\,\mathrm{kJ\,mol^{-1}}$

This is an exothermic process. Therefore, $\Delta H_1 = -7.43\,\mathrm{kJ\,mol^{-1}}$.
We can determine $\Delta_a H^\circ(298\,\mathrm{K})$ by applying Hess's Law to the thermochemical cycle above:

$\Delta_a H^\circ(298\,\mathrm{K}) = \Delta H_1 + \Delta H_2 + \Delta H_3$

$= (9.78 + 59.11 - 7.43)\,\mathrm{kJ\,mol^{-1}} = 61.46\,\mathrm{kJ\,mol^{-1}}$

## Enthalpy of atomization and bond dissociation enthalpy for a gaseous diatomic molecule

The relationship between the enthalpy of atomization and the bond dissociation enthalpy for a *diatomic* molecule in *the gas phase* is important.

The definition of the bond dissociation enthalpy is in terms of breaking a bond. Thus $D$ is defined per mole of bonds broken (equation 4.8). For dihydrogen, $D = 436\,\mathrm{kJ\,mol^{-1}}$. On the other hand, the enthalpy of atomization is defined per mole of gaseous atoms formed (equation 4.9). This difference in definition is a common cause of error in calculations.

> For a gaseous *diatomic* molecule:
>
> Bond dissociation enthalpy $= 2 \times$ (standard enthalpy of atomization)
>
> $D = 2 \times \Delta_a H^\circ$

$$\mathrm{H_2(g) \longrightarrow 2H(g)} \quad D = 436\,\mathrm{kJ\,mol^{-1}}\ \text{(per mole of bonds broken)} \quad (4.8)$$

$$\tfrac{1}{2}\mathrm{H_2(g) \longrightarrow H(g)} \quad \Delta_a H^\circ = 218\,\mathrm{kJ\,mol^{-1}}\ \text{(per mole of gaseous atoms formed)} \quad (4.9)$$

| Worked example 4.3 | *Bond dissociation enthalpy of $F_2$* |
|---|---|

**Write an equation for the dissociation of gaseous $F_2$ into gaseous atoms. Use data in Appendix 10 to determine the enthalpy change that accompanies this reaction at 298 K.**

The equation for the dissociation of gaseous $F_2$ (i.e. the dissociation of the F−F bond) is:

$$\mathrm{F_2(g) \longrightarrow 2F(g)}$$

or

$$\mathrm{F_2(g) \longrightarrow 2F^{\bullet}(g)}$$

Appendix 10 lists values of $\Delta_a H^\circ(298\,\mathrm{K})$. For fluorine, $\Delta_a H^\circ(298\,\mathrm{K}) = 79\,\mathrm{kJ\,mol^{-1}}$.

For a diatomic molecule:

Bond dissociation enthalpy, $D = 2 \times \Delta_a H^\circ(298\,\mathrm{K})$

For $F_2$:

$$D = 2 \times 79 = 158\,\mathrm{kJ\,mol^{-1}}$$

■▷
Vibrational spectroscopy for
diatomics: see Section 12.3

## 4.7    Determining bond enthalpies from standard heats of formation

The bond dissociation enthalpy of the bond in a diatomic molecule can be measured thermochemically or by using spectroscopic methods. However, for larger molecules, direct measurements of bond enthalpies are not always possible.

■▷
Hess's Law: see Section 2.7

A method that provides values for bond enthalpy terms of a given type uses Hess's Law of Constant Heat Summation. Thermochemical data that *can* be measured experimentally are standard enthalpies of combustion of elements and molecules (see Section 2.6) and from these, standard enthalpies of formation ($\Delta_f H^\circ$) can be obtained. These data may be used to determine bond enthalpies. It is important to keep in mind that values of $D$ obtained this way are *derived values* – they depend upon other bond enthalpy terms as shown below.

Suppose that we wish to determine the bond enthalpy term of an N−N single bond. This bond cannot be studied in isolation because $N_2$ molecules possess N≡N triple bonds (we return to the bonding in $N_2$ later in the chapter). So, we turn to a simple compound containing an N−N bond, $N_2H_4$ (structure **4.1**). This molecule is composed of one N−N and four N−H bonds. If we were to dissociate the molecule completely into gaseous atoms, then all of these bonds would be broken and the enthalpy change would be the sum of the bond enthalpy terms as shown in equation 4.10. Note here the use of $D(N-N)$, but $\bar{D}(N-H)$. This emphasizes the fact that we are dealing with one N−N bond but take an average value for the N−H bond enthalpy.

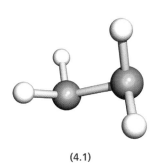

**(4.1)**

$$N_2H_4(g) \longrightarrow 2N(g) + 4H(g) \quad \Delta_a H^\circ = D(N-N) + 4\bar{D}(N-H) \quad (4.10)$$

However, in trying to determine $D(N-N)$, we have introduced another unknown quantity, $\bar{D}(N-H)$. The ammonia molecule (structure **4.2**) is composed only of N−H bonds and the complete dissociation of $NH_3$ provides an *average* value of the N−H bond enthalpy (equation 4.11). The value of $\Delta_a H^\circ$ for the process shown in equation 4.11 can be determined from the standard enthalpy of formation of ammonia (equation 4.12) by using Hess's Law (equation 4.13). The standard enthalpies of atomization for nitrogen and hydrogen are 473 and 218 kJ mol$^{-1}$ respectively.

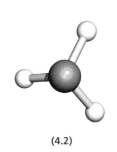

**(4.2)**

$$NH_3(g) \longrightarrow N(g) + 3H(g) \qquad \Delta_a H^\circ = 3\bar{D}(N-H) \quad (4.11)$$

$$\tfrac{1}{2}N_2(g) + \tfrac{3}{2}H_2(g) \longrightarrow NH_3(g) \quad \Delta_f H^\circ = -46 \text{ kJ mol}^{-1} \quad (4.12)$$

$$\begin{array}{ccc} NH_3(g) & \xrightarrow{\quad 3\bar{D}(N-H) \quad} & N(g) + 3H(g) \\ {\scriptstyle \Delta_f H^\circ(NH_3,\, g)} \searrow & & \nearrow {\scriptstyle \Delta_a H^\circ(N) + 3\Delta_a H^\circ(H)} \\ & \tfrac{1}{2}N_2(g) + \tfrac{3}{2}H_2(g) & \end{array} \qquad (4.13)$$

From equation 4.13:

$$3\bar{D}(N-H) = \Delta_a H^\circ(N) + 3\Delta_a H^\circ(H) - \Delta_f H^\circ(NH_3,\, g)$$

$$3\bar{D}(N-H) = 473 + (3 \times 218) - (-46)$$

$$= 1173 \text{ kJ mol}^{-1}$$

$$\bar{D}(N-H) = 391 \text{ kJ mol}^{-1}$$

This result can now be used to find $D(\text{N}-\text{N})$ by setting up a Hess cycle based on equation 4.10:

$$N_2H_4(g) \xrightarrow{\;D(\text{N–N}) + 4\bar{D}(\text{N–H})\;} 2N(g) + 4H(g)$$

$$\Delta_f H^\circ(N_2H_4, g) \qquad\qquad 2\Delta_a H^\circ(N) + 4\Delta_a H^\circ(H)$$

$$N_2(g) \;+\; 2H_2(g)$$

(4.14)

The standard enthalpy of formation of gaseous hydrazine is $95\,\text{kJ mol}^{-1}$.
From equation 4.14:

$$D(\text{N}-\text{N}) + 4\bar{D}(\text{N}-\text{H}) = 2\Delta_a H^\circ(\text{N}) + 4\Delta_a H^\circ(\text{H}) - \Delta_f H^\circ(N_2H_4, g)$$

$$D(\text{N}-\text{N}) + 4\bar{D}(\text{N}-\text{H}) = (2 \times 473) + (4 \times 218) - 95$$

$$= 1723\,\text{kJ mol}^{-1}$$

$$D(\text{N}-\text{N}) = 1723 - 4\bar{D}(\text{N}-\text{H})$$

and substituting in the value we calculated above for $\bar{D}(\text{N}-\text{H})$ gives:

$$D(\text{N}-\text{N}) = 1723 - (4 \times 391)$$

$$= 159\,\text{kJ mol}^{-1}$$

We have made a major assumption in this series of calculations – that the average N−H bond enthalpy term can be *transferred* from ammonia to hydrazine. Ammonia and hydrazine are closely related molecules and the assumption is reasonable. However, the transfer of bond enthalpy contributions between molecules of different types should be treated with caution. Naturally, $\bar{D}$ values for single bonds cannot be used for multiple bonds. Such bonds have very different bond enthalpies.

Some typical values for the bond enthalpies of homonuclear single covalent bonds were listed in Table 4.5. Some, such as $D(\text{H}-\text{H})$, $D(\text{F}-\text{F})$ and $D(\text{Cl}-\text{Cl})$, are directly measured from the dissociation of the appropriate gaseous diatomic molecule. Others can be measured from the dissociation of a small molecule, e.g. $\bar{D}(\text{S}-\text{S})$ from the atomization of gaseous $S_8$ (Figure 4.2). Values such as that for the N−N bond rely upon the method of bond enthalpy transferability described above.

| 4.8 | **The nature of the covalent bond in $H_2$** |

In the previous sections, we have considered two experimental observables by which a bond such as that in $H_2$ can be described: the internuclear distance (which gives a measure of the covalent radius of a hydrogen atom) and the bond dissociation enthalpy. The experimental evidence is that two H atoms come together to form a bond in which the internuclear separation is 74 pm and the bond dissociation enthalpy is $436\,\text{kJ mol}^{-1}$.

We have also seen that the driving force for two H atoms combining to give a molecule of $H_2$ is the tendency of each atom to become 'helium-like'. Each H atom shares one electron with another H atom.

The covalent bond in dihydrogen can be represented in a Lewis structure (see Box 4.5), but is more fully described using other methods: the valence bond (VB) and molecular orbital (MO) approaches are the best known. The following sections deal with these methods.

**THEORETICAL AND CHEMICAL BACKGROUND**

## Box 4.5  Lewis structures

Gilbert N. Lewis (1875–1946). *Reproduced courtesy of the Library and Information Centre, Royal Society of Chemistry.*

In 1916, G. N. Lewis presented a simple, but informative, method of describing the arrangement of valence electrons in molecules. The method uses dots (or dots and crosses) to represent the number of valence electrons associated with the nuclei. The nuclei are designated by the symbol of the element.

If possible, all electrons in a molecule should appear in pairs. Single (unpaired) electrons signify a radical species.

### *Example: Cl₂*

The ground state electronic configuration of Cl is $[Ne]3s^2 3p^5$. Thus, Cl has seven valence electrons.

In forming the $Cl_2$ molecule, a Cl atom shares one electron with another Cl atom to form an argon-like octet.

The Lewis structure for $Cl_2$ is:

$$: \overset{..}{\underset{..}{Cl}} : \overset{..}{\underset{..}{Cl}} :$$

and this can be simplified to show the Cl–Cl covalent bond and the *lone pairs* of electrons:

$$: \overset{..}{\underset{..}{Cl}} — \overset{..}{\underset{..}{Cl}} :$$

Lewis structures are extremely useful for giving connectivity patterns – that is, the bonding arrangement between atoms in a molecule.

Some other examples of Lewis structures are:

$$H — \overset{..}{\underset{..}{F}} : \qquad : \overset{}{\underset{..}{S}} \overset{H}{\underset{H}{|}} \qquad H — \overset{H}{\underset{..}{N}} — H$$

HF        H₂S        NH₃

$$H — \overset{H}{\underset{H}{\overset{|}{\underset{|}{C}}}} — H \qquad H — \overset{H}{\underset{H}{\overset{|}{\underset{|}{C}}}} — \overset{..}{\underset{..}{O}} :$$

CH₄            CH₃OH

$$H — \overset{..}{\underset{\overset{|}{H}}{N}} — \overset{..}{N} — H \qquad : \overset{\overset{: \overset{..}{Cl} :}{|}}{\underset{\underset{: \overset{..}{Cl} :}{|}}{Pb}} — \overset{..}{\underset{..}{Cl}} :$$

N₂H₄              PbCl₄

---

**4.9**

## Lewis structure of H₂

H : H        H——H

(4.3)        (4.4)

Structures **4.3** and **4.4** show two representations of the $H_2$ molecule and are referred to as *Lewis structures*. In structure **4.3**, the electrons are represented by dots[§] but the diagram can be simplified by drawing a line to represent the bond between the two atoms as in **4.4**.

Although these Lewis structures show the connection of one hydrogen atom to the other, they do not provide information about the exact character

---

[§] Dots and crosses are sometimes used to distinguish between electrons from the two adjacent atoms. Remember, however, that *in practice* electrons are *indistinguishable* from each other.

of the bonding pair of electrons, or about the region of space that they occupy.

**4.10**

## The problem of describing electrons in molecules

### The problem

In Section 3.7, we discussed the way in which the Schrödinger equation described the dynamic behaviour of electrons in atoms, and we pointed out that it was possible to obtain exact mathematical solutions to the Schrödinger equation only in the case of hydrogen-like species. If we are to use the Schrödinger equation to describe the behaviour of *electrons in molecules*, we have the same problem – we cannot obtain exact solutions to the wave-equations when we have more than one nucleus and more than one electron. Certainly, we can simplify the problem by assuming that the movement of the nuclei is minimal, but this does not overcome the basic problem.

Our aim is to obtain molecular wavefunctions, $\psi$(molecule), that describe the behaviour of electrons in molecules. What can we do?

### Two methods of approaching the problem

Two main methods are used to make approximations to molecular wavefunctions – *valence bond* and *molecular orbital* theories. These have different starting assumptions but, if successful, they should give similar results. To introduce these methods we need to get a little ahead of ourselves and consider some of the problems that exist with molecules containing more than two atoms. Water makes a good example.

Structures **4.5** and **4.6** give Lewis representations of a water molecule in terms of two O−H bonds and two lone pairs of electrons. The basis of a Lewis structure is the identification of a bond as a pair of electrons shared between two atoms. This is the starting point for the bonding model called *valence bond theory*.

The valence bond (VB) model starts from the chemically familiar viewpoint that a molecule is composed of a number of discrete bonds. To all intents and purposes, the various bonds exist in isolation from one another and each may be described in terms of interactions between two electrons and two nuclei. This is the so-called *2-centre 2-electron* bonding model and the bonding interactions are said to be *localized*. The VB approach attempts to write a separate wavefunction for each of the discrete 2-electron interactions, and then derives a total molecular wavefunction by combining them.

The molecular orbital (MO) approach does *not* start with the assumption that electrons are localized into 2-centre 2-electron bonds. In MO theory, molecular wavefunctions are constructed that encompass any or all of the atomic nuclei in the molecule. An interaction that spreads over more than two nuclei is described as being *delocalized* or *multi-centred*.

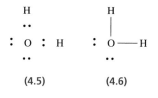

(4.5)        (4.6)

Valence bond theory:
see Section 4.11

Molecular orbital theory:
see Section 4.12

A bonding interaction between two nuclei is **localized**. If it spreads over more than two nuclei, the interaction is **delocalized**.

The MO method is mathematically complex and is not as intuitive as VB theory. In the water molecule, it is far more convenient to think in terms of localized O−H bonds rather than multi-centre H···O···H interactions. However, there are many compounds for which the MO approach provides a more satisfying description of the bonding. In many respects, this is similar to the situation with bond enthalpies. As chemists, we know that the four C−H bonds in methane are equivalent, but as we showed in Box 4.4, if the bonds are broken sequentially, we obtain a different bond enthalpy value for each step.

*Both the VB and MO methods are approximations for obtaining the molecular wavefunction, $\psi$(molecule). Both methods must approach the same 'true' $\psi$(molecule) and should, ultimately, give equivalent wavefunctions.*

> **Valence bond (*VB*) theory** assigns electrons to 2-centre bonds or to atomic-based orbitals.
>
> **Molecular orbital (*MO*) theory** allows electrons to be delocalized over the entire molecule.

### 4.11    Valence bond (VB) theory

#### General overview of valence bond theory

In valence bond theory, a description of the bonding within a diatomic molecule is determined by considering the perturbation that the two atoms have on one another as they penetrate one another's regions of space. In practice this means that we attempt to write an approximation to the 'real' wavefunction for the molecule in terms of a combination of wavefunctions describing individual 2-electron interactions. The details of the mathematics are not relevant to our discussion.[§]

The consideration of *electrons* from the start is an important characteristic of VB models.

#### The bonding in H₂ – an initial approach

Consider the formation of an $H_2$ molecule. Each hydrogen atom consists of a proton and an electron. In Figure 4.9 the two nuclei are labelled $H_A$ and $H_B$ and the two electrons are labelled 1 and 2. When the two H atoms are well separated from one another, electron 1 will be wholly associated with $H_A$, and electron 2 will be fully associated with $H_B$. Let this situation be described by the wavefunction $\psi_1$.

Although we have given each electron a different label, they are actually indistinguishable from one another. Therefore, when the atoms are close together, we cannot tell which electron is associated with which nucleus – electron 2 could be with $H_A$ and electron 1 with $H_B$. Let this situation be described by the wavefunction $\psi_2$.

An overall description of the system containing the two hydrogen atoms, $\psi$(covalent), can be written in the form of equation 4.15.

$$\psi(\text{covalent}) = \psi_1 + \psi_2 \tag{4.15}$$

The wavefunction $\psi$(covalent) defines an energy. When this is calculated as a function of the internuclear separation $H_A$–$H_B$, it provides an energy minimum of $303\,\text{kJ mol}^{-1}$ at an internuclear distance of 87 pm for $H_2$. Using equation 4.15, therefore, does *not* give an answer in good agreement with either the experimental internuclear H−H distance (74 pm) or the bond

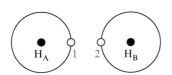

**Fig. 4.9** Labelling scheme used for the nuclei and electrons in a valence bond treatment of the bonding in $H_2$.

---

[§] For more in-depth discussion, see: R. McWeeny (1979) *Coulson's Valence*, 3rd edn, Oxford University Press, Oxford.

dissociation energy ($436 \, \text{kJ mol}^{-1}$) and some refinement of the method is clearly needed.

### The bonding in $H_2$ – a refinement of the initial picture

Although we have so far allowed each of the nuclei $H_A$ and $H_B$ to be associated with *either* of the electrons 1 and 2, we should also allow for a situation in which one or other of the nuclei might be associated with *both* electrons.

There are four situations that can arise as $H_A$ and $H_B$ come close together:

* (nucleus $H_A$ with electron 1) and (nucleus $H_B$ with electron 2);
* (nucleus $H_A$ with electron 2) and (nucleus $H_B$ with electron 1);
* (nucleus $H_A$ with both electrons 1 and 2) and (nucleus $H_B$ with no electrons);
* (nucleus $H_A$ with no electrons) and (nucleus $H_B$ with both electrons 1 and 2).

While the first two situations retain neutrality at each atomic centre, the last two describe the *transfer* of an electron from one nucleus to the other. This has the effect of producing two ions, $H^+$ and $H^-$. If $H_A$ has no electrons, $H_A$ becomes $H_A^+$, and $H_B$ becomes $H_B^-$. If $H_A$ has both electrons, $H_A$ becomes $H_A^-$, and $H_B$ becomes $H_B^+$. There is an equal chance of forming $[H_A^- \ H_B^+]$ or $[H_A^+ \ H_B^-]$ since the two hydrogen atoms are identical.

The effect of allowing for contributions from $[H_A^- \ H_B^+]$ or $[H_A^+ \ H_B^-]$ is to build into the valence bond model the possibility that the wavefunction that describes the bonding region between the two hydrogen atoms may have an *ionic contribution* in addition to the *covalent contribution* that we have already described. This is represented in equation 4.16.

$$\psi(\text{molecule}) = \psi(\text{covalent}) + [c \times \psi(\text{ionic})] \tag{4.16}$$

The coefficient $c$ indicates the relative contribution made by the wavefunction $\psi(\text{ionic})$ to the overall wavefunction $\psi(\text{molecule})$.

With regard to the $H_2$ molecule, equation 4.16 means that we should write three possible structures to represent the covalent and ionic contributions to the bonding description. Structures **4.7a**, **4.7b** and **4.7c** are called *resonance structures* and a *double-headed arrow* is used to represent the resonance between them.

H——H  ⟷  H⁺  H⁻  ⟷  H⁻  H⁺

(4.7a)       (4.7b)       (4.7c)

More about resonance structures: see Section 7.2

It is very important to realize that the purely covalent and the purely ionic forms of the $H_2$ molecule *do not exist separately*. We are merely trying to describe the $H_2$ molecule by the wavefunction $\psi(\text{molecule})$ in terms of contributions from the two extreme bonding models represented by $\psi(\text{covalent})$ and $\psi(\text{ionic})$.

By trying different values of $c$ in equation 4.16, an estimate of the internuclear distance that corresponds to the energy minimum of the system can be obtained which comes close to the experimental bond distance. Normalization: see Section 3.8 If the wavefunctions are normalized, a value of $c \approx 0.25$ gives an H–H internuclear distance of 75 pm, in close agreement with the experimentally determined value. The predicted bond dissociation energy is $398 \, \text{kJ mol}^{-1}$. Further mathematical refinement can greatly improve upon this value at the expense of the simple physical picture.

## 4.12    Molecular orbital (MO) theory

### General overview of molecular orbital theory

Molecular orbital theory differs from the valence bond method in that it does not start from an assumption of localized 2-centre bonding. In the MO approach, we attempt to obtain wavefunctions for the entire molecule. The aim is to calculate regions in space that an electron might occupy which encompass the molecule as a whole – these are *molecular orbitals*. In extreme cases, this procedure permits the electrons to be delocalized over the molecule, while in others, the results approximate to localized 2-centre 2-electron orbitals.

The usual starting point for the MO approach to the bonding in a molecule is a consideration of the atomic orbitals that are available. An approximation which is commonly used within MO theory is known as the *linear combination of atomic orbitals* (LCAO). In this method, wavefunctions ($\psi$) approximately describing the molecular orbitals are constructed from the atomic wavefunctions of the constituent atoms. Remember that the atomic wavefunctions are themselves usually approximations obtained from the hydrogen atom.

Interactions between any two atomic orbitals will be:

- allowed if the symmetries of the atomic orbitals are compatible with one another;
- efficient if the region of overlap of the two atomic orbitals is significant; and
- efficient if the atomic orbitals are relatively close in energy.

The full meaning of the first point will become clearer later in this chapter. With reference to the second point, we should spend a moment considering the meaning of the term 'overlap'. This word is used extensively in MO theory, and the overlap between two orbitals is often expressed in terms of the *overlap integral*, $S$. Figure 4.10 illustrates two $1s$ atomic orbitals. The orbitals in Figure 4.10a are close but there is no common region of space – the probability of finding an electron in one atomic orbital very rarely coincides with the probability of finding an electron in the other orbital. The overlap integral is effectively zero between these two orbitals.[§] In Figure 4.10b, there is a very small common region of space between the two orbitals. The value of $S$ is small, and this situation is not satisfactory for the formation of an effective bonding molecular orbital. In Figure

**Fig. 4.10** Schematic drawing to illustrate the meaning of orbital overlap and the overlap integral $S$: (a) the two $1s$ orbitals effectively do not overlap; (b) there is only a very little overlap; (c) the two $1s$ atomic orbitals overlap efficiently. The two atomic nuclei are represented by the central dots.

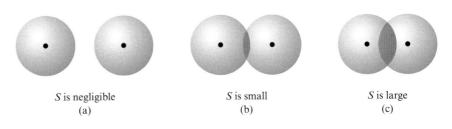

$S$ is negligible    $S$ is small    $S$ is large
(a)                  (b)             (c)

---

[§] The overlap integral $S$ cannot be precisely zero; the surface boundaries of the orbitals do not represent 100% probability of finding the electron as discussed in Chapter 3.

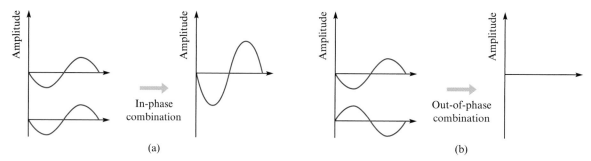

**Fig. 4.11** (a) Constructive and (b) destructive interference between two transverse waves of equal wavelength and amplitude.

4.10c, the overlap region is significant and the overlap integral will have a value $0 < S < 1$; it cannot equal unity. (Keep in mind that the nuclei cannot come too close together because they repel one another.) The situation sketched in Figure 4.10c represents good orbital overlap.

In order to understand the significance of the third point above, look back at Figure 3.10 and think about what would happen if we combined a $1s$ with a $4s$ atomic orbital, as opposed to combining a $1s$ with another $1s$ atomic orbital. The $1s$ and $4s$ orbitals have different energies and different spatial extents. Remember that we need to maximize orbital overlap in order to obtain significant interaction between two atomic orbitals.

The interactions between atomic orbitals may be described by linear combinations of the atomic wavefunctions. Because wavefunctions have wave-like properties, we can think in terms of the interaction between two wavefunctions being similar to the interaction between two transverse waves. Figure 4.11 summarizes what happens when two transverse waves combine either in or out of phase. An in-phase combination leads to constructive interference and, if the initial waves are identical, the amplitude of the wave doubles. An out-of-phase combination causes destructive interference and, if the initial waves are identical, the wave is destroyed.

Now let us move from transverse waves to atomic orbitals – in a similar manner, 'electron waves' (orbitals) may be in or out of phase. Equations 4.17 and 4.18 represent the in- and out-of-phase linear combinations of two wavefunctions, $\psi_1$ and $\psi_2$, which describe two atomic orbitals. Two molecular orbitals result.

$$\psi(\text{in-phase}) = N \times [\psi_1 + \psi_2] \tag{4.17}$$

$$\psi(\text{out-of-phase}) = N \times [\psi_1 - \psi_2] \tag{4.18}$$

**Normalization:**
see Section 3.8

The coefficient $N$ is the normalization factor; this adjusts the equation so that the probability of finding the electron somewhere in space is unity.

A general result that should be remembered is that *the number of molecular orbitals generated will always be equal to the initial number of atomic orbitals –* this becomes important when we construct molecular orbital diagrams in the following sections.

In the construction of a molecular orbital diagram:

$$\begin{pmatrix} \text{The number of molecular} \\ \text{orbitals generated} \end{pmatrix} = \begin{pmatrix} \text{The number of atomic} \\ \text{orbitals used} \end{pmatrix}$$

**Fig. 4.12** The variation in the energy of the molecular wavefunctions $\psi$(in-phase) and $\psi$(out-of-phase) for the combination of two hydrogen $1s$ atomic orbitals as a function of H$\cdots$H separation.

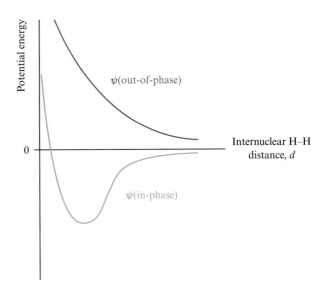

## Constructing the molecular orbitals for H$_2$

Let us now look at the bonding in the H$_2$ molecule using the MO method.

Each hydrogen atom possesses one $1s$ atomic orbital. The MO description of the bonding in the H$_2$ molecule is based on allowing the two $1s$ orbitals to overlap when the two hydrogen nuclei are within bonding distance. The two $1s$ atomic orbitals are allowed to overlap because they have the same symmetries. We shall return to the question of symmetry when we consider the bonding in other diatomic molecules.

Let the wavefunction associated with one H atom be $\psi(1s)_A$ and the wavefunction associated with the second H atom be $\psi(1s)_B$. Using equations 4.17 and 4.18, we can write the new wavefunctions that result from the linear combinations of $\psi(1s)_A$ and $\psi(1s)_B$ and these are given in equations 4.19 and 4.20. Since the H$_2$ molecule is a homonuclear diatomic, each of the atomic wavefunctions will contribute *equally* to each molecular wavefunction.

$$\psi(\text{in-phase}) = N \times [\psi(1s)_A + \psi(1s)_B] \tag{4.19}$$

$$\psi(\text{out-of-phase}) = N \times [\psi(1s)_A - \psi(1s)_B] \tag{4.20}$$

Each of the new molecular wavefunctions, $\psi$(in-phase) and $\psi$(out-of-phase), has an energy associated with it which is dependent on the distance apart of the two hydrogen nuclei. As Figure 4.12 shows, the energy of $\psi$(in-phase) is always lower than that of $\psi$(out-of-phase). Further, if we look at the variation of the energy of $\psi$(in-phase) as a function of internuclear distance, we see that the energy curve has a minimum value corresponding to maximum stability. In contrast, the curve describing the variation in the energy of $\psi$(out-of-phase) never reaches a minimum, and represents an increasingly less stable situation as the two nuclei come closer together.

The wavefunction $\psi$(in-phase) corresponds to a *bonding molecular orbital* and $\psi$(out-of-phase) describes an *antibonding molecular orbital*. The combination of $\psi(1s)_A$ and $\psi(1s)_B$ to give $\psi$(bonding) and $\psi$(antibonding) is represented schematically in Figure 4.13 in which $\psi$(bonding) is stabilized with respect to the two atomic orbitals and $\psi$(antibonding) is destabilized.[§]

---

[§] The amount of energy by which the antibonding MO is destabilized with respect to the atomic orbitals is slightly greater than the amount of energy by which the bonding MO is stabilized. This phenomenon is important but its discussion is beyond the scope of this book.

**Fig. 4.13** In-phase and out-of-phase combinations of two $1s$ atomic orbitals lead to two molecular orbitals at low (stabilized) and high (destabilized) energies respectively.

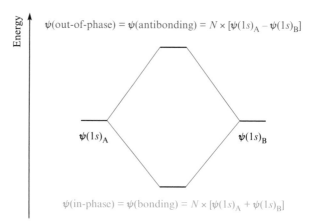

The in-phase combination of atomic orbitals gives a **bonding** molecular orbital (MO).

The out-of-phase combination of atomic orbitals gives an **antibonding** MO that is higher in energy than the bonding MO.

For the combination of two $1s$ atomic orbitals, the normalization factor is $1/\sqrt{2}$ and so we can now rewrite equations 4.19 and 4.20 in their final forms of equations 4.21 and 4.22:

$$\psi(\text{bonding}) = \frac{1}{\sqrt{2}} \times [\psi(1s)_A + \psi(1s)_B] \tag{4.21}$$

$$\psi(\text{antibonding}) = \frac{1}{\sqrt{2}} \times [\psi(1s)_A - \psi(1s)_B] \tag{4.22}$$

### Putting the electrons in the $H_2$ molecule

So far in this MO approach, we have not mentioned the electrons! In simple MO theory, the molecular orbitals are constructed and once their energies are known, the *aufbau* principle is used to place the electrons in them.

*Aufbau* principle: see Section 3.18

The $H_2$ molecule contains two electrons, and by the *aufbau* principle, these occupy the lowest energy molecular orbital and have opposite spins. The MO energy level diagram for $H_2$ is shown in Figure 4.14. Significantly, the two electrons occupy the *bonding MO*, while the antibonding MO remains empty.

**Fig. 4.14** The bonding and antibonding molecular orbitals in $H_2$ formed by the linear combination of two $1s$ atomic orbitals. By the *aufbau* principle, the two electrons in $H_2$ occupy the bonding MO. The labels '$\sigma(1s)$' and '$\sigma^*(1s)$' are explained at the end of Sections 4.12 and 4.15.

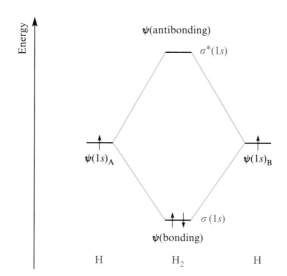

**Fig. 4.15** Schematic representations of (a) the bonding and (b) the antibonding molecular orbitals in the $H_2$ molecule. The H nuclei are represented by black dots.

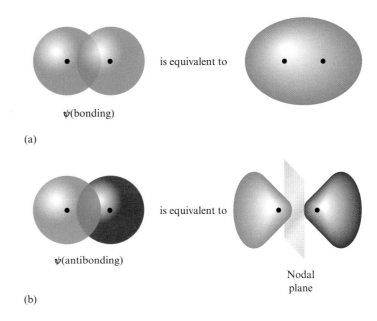

is equivalent to

$\psi$(bonding)

(a)

is equivalent to

$\psi$(antibonding)

Nodal plane

(b)

## What do the molecular orbitals for $H_2$ look like?

The bonding and antibonding molecular orbitals of $H_2$ are drawn schematically in Figure 4.15. The combinations of the two $1s$ orbitals are most simply represented by two overlapping circles, either in or out of phase. This is shown on the left-hand side of Figures 4.15a and 4.15b and is a representation that we shall use elsewhere in this book.

On the right-hand side of Figures 4.15a and 4.15b we show the molecular orbitals in more detail. The diagram representing the bonding orbital for the $H_2$ molecule illustrates that the overlap between the two atomic orbitals gives a region of space in which the two bonding electrons may be found. This corresponds to the $H-H$ bond seen in the VB model. The negatively charged electrons will be found predominantly between the two nuclei, thus reducing internuclear repulsion.

The antibonding orbital for the $H_2$ molecule is shown in Figure 4.15b. The most important feature is the nodal plane passing *between the two H nuclei*. This means that, were this orbital to be occupied, the probability of finding the electrons at any point in this nodal plane would be zero. The outcome is that the probability of finding the electrons between the nuclei in the antibonding MO is less than between two non-interacting hydrogen atoms, and much less than in the bonding orbital. Accordingly, there is an increase in internuclear repulsion.

Nodal plane: see Figure 3.7 and accompanying text

## Labelling the molecular orbitals in $H_2$

'$\sigma$-orbital' is pronounced 'sigma-orbital'; '$\sigma^*$-orbital' is pronounced 'sigma-star-orbital'

Some new notation has appeared in Figure 4.14 – the bonding MO is labelled $\sigma(1s)$ and the antibonding MO is designated $\sigma^*(1s)$. These labels are used to provide information about the nature of the orbital.

A molecular orbital has $\sigma$ symmetry if it is symmetrical with respect to a line joining the two nuclei – there is no phase change when the orbital is rotated about this internuclear axis.

A $\sigma^*$ molecular orbital must meet *two* requirements:

- In order to keep the $\sigma$ label, the MO must satisfy the requirement that if it is rotated about the internuclear axis, there is no phase change.
- In order to take the $^*$ label, there must be a nodal plane *between* the nuclei, and this plane must be orthogonal to the internuclear axis.

Using this new notation, we can write the ground state electronic configuration of the $H_2$ molecule as $\sigma(1s)^2$. This shows that we have formed a $\sigma$-bonding molecular orbital by the overlap of two $1s$ atomic orbitals, and that the MO contains two electrons.

### 4.13    What do the VB and MO theories tell us about the molecular properties of $H_2$?

Now that we have looked at the $H_2$ molecule in several ways, we should consider whether the different bonding models paint the same picture.

- The *Lewis structure* shows that the $H_2$ molecule contains a single covalent $H-H$ bond and the electrons are paired.
- *Valence bond theory* shows that the bond in the $H_2$ molecule can be described by a wavefunction that has both covalent and ionic contributions. The dominant contribution comes from the covalent resonance form. The two electrons in $H_2$ are paired.
- *Molecular orbital theory* models $H_2$ on a predominantly covalent basis. There is a localized $H-H$ bonding MO and the two electrons are paired.

Thus, we find that all the models give us similar insights into the bonding in $H_2$.

### 4.14    Homonuclear diatomic molecules of the first row elements – the *s*-block

In this and later sections in this chapter we construct diagrams like Figure 4.14 for diatomic molecules $X_2$ where X is an element in the first row of the periodic table. For each molecule, we use the molecular orbital approach to predict some of the properties of $X_2$ and compare these with what is known experimentally. We begin by considering the bonding in the diatomic molecules $Li_2$ and $Be_2$. Dilithium, $Li_2$, exists as a gas phase molecule, while there is evidence that $Be_2$ exists only as an extremely unstable species. Can we rationalize the instability of $Be_2$?

#### $Li_2$

A lithium atom has a ground state electronic configuration of $[He]2s^1$. Figure 4.16 shows the combination of the $2s$ atomic orbitals to give the MOs in $Li_2$. There is a clear similarity between this and the formation of $H_2$ (Figure 4.14), although the principal quantum number has changed from 1 to 2.

From Figure 4.16, the electronic configuration of $Li_2$ is $\sigma(2s)^2$. The complete electronic configuration is:

$$\sigma(1s)^2\sigma^*(1s)^2\sigma(2s)^2$$

**Fig. 4.16** The bonding and antibonding molecular orbitals in $Li_2$ formed by the linear combination of two $2s$ atomic orbitals. The $1s$ orbitals (with core electrons) have been omitted.

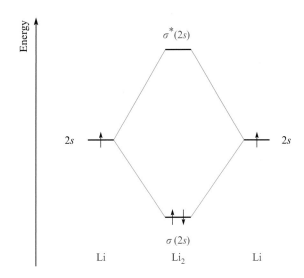

However, of the six electrons in the $Li_2$ molecule, the four in the $\sigma(1s)$ and $\sigma^*(1s)$ orbitals are core electrons. The two electrons in the $\sigma(1s)$ MO 'cancel out' the two electrons in the $\sigma^*(1s)$ MO. Therefore, we need consider only the fate of the valence electrons – those occupying the $\sigma(2s)$ orbital.

Thus, MO theory predicts that the $Li_2$ molecule possesses two electrons which occupy a bonding molecular orbital – the $Li_2$ molecule has a single covalent bond. This is the same answer as the one we obtained by drawing the Lewis structures **4.8** or **4.9**. In VB theory, this corresponds to a localized $Li-Li$ $\sigma$ bond.

Li $\overset{\cdot\cdot}{\phantom{.}}$ Li     Li ——— Li

(4.8)          (4.9)

## Be₂

The ground state configuration of Be is $[He]2s^2$. Figure 4.17 shows the combination of the $2s$ atomic orbitals to give the MOs in $Be_2$. The orbitals themselves are similar to those in $Li_2$, and we can again ignore the contribution of the core $1s$ electrons. There are now four electrons to be accommodated in MOs derived from the $2s$ atomic orbitals. From Figure 4.17, the electron configuration of $Be_2$ is:

$$\sigma(2s)^2\sigma^*(2s)^2$$

**Fig. 4.17** A molecular orbital diagram for the formation of $Be_2$ from two Be atoms. The $1s$ orbitals (with core electrons) have been omitted.

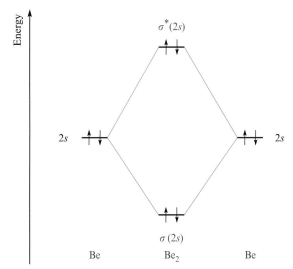

Each of the bonding and antibonding MOs contains two electrons. The result is that there would be no net bonding in a $Be_2$ molecule – this is consistent with its instability.

Before moving on to the other elements in the first row of the periodic table, we need to introduce two further concepts – the orbital overlap of $p$ atomic orbitals, and bond order.

## 4.15    Orbital overlap of *p* atomic orbitals

We saw in Section 4.12 that the overlap of two $s$ atomic orbitals led to two molecular orbitals of $\sigma$ and $\sigma^*$ symmetries. Each of the $s$ atomic orbitals is spherically symmetrical and so the overlap of two such atomic orbitals is *independent* of their relative orientations. To convince yourself of this, look at Figure 4.15 – rotate one of the $1s$ orbitals about an axis passing through the nucleus. Does it affect the overlap with the other $1s$ atomic orbital? Now rotate the first $1s$ orbital about a different axis while keeping the second $1s$ orbital in a fixed position. You should conclude that this operation has no effect on the overlap between the two $1s$ orbitals.

The situation with $p$ atomic orbitals is different. Consider two nuclei placed on the $z$-axis[§] at their equilibrium (bonding) separation. The $2p_z$ atomic orbitals point directly at one another. They can overlap either in or out of phase as shown in Figures 4.18a and 4.18b respectively. The resultant bonding molecular orbital is represented at the right-hand side of Figure 4.18a; this is a $\sigma$ orbital. The corresponding antibonding MO has a nodal plane passing between the nuclei (Figure 4.18b) – this plane is *in addition* to the nodal plane present in each $2p$ atomic orbital at the nucleus. The antibonding MO is a $\sigma^*$ orbital.

> Recall that a $\sigma$ orbital is symmetrical with respect to rotation about the internuclear axis

Now consider the overlap of the $2p_x$ atomic orbitals (Figures 4.18c and 4.18d). Since we have defined the approach of the nuclei to be along the $z$-axis, the two $2p_x$ atomic orbitals can overlap only in a sideways manner. This overlap is less efficient than the direct overlap of the two $2p_z$ atomic orbitals (i.e. the overlap integral $S$ is smaller). The result of an in-phase combination of the $2p_x$ atomic orbitals is the formation of a bonding MO which retains the nodal plane of each individual atomic orbital (Figure 4.18c). This MO is called a $\pi$ orbital. The $\pi$ molecular orbital is asymmetrical with respect to rotation about the internuclear axis – if you rotate the orbital about this axis, there is a change of phase.

The result of an out-of-phase combination of two $2p_x$ atomic orbitals is the formation of an antibonding MO which retains the nodal plane of each individual atomic orbital and has a nodal plane passing between the nuclei (Figure 4.18d). This MO is labelled as a $\pi^*$ orbital.

A molecular orbital has $\pi^*$ symmetry if it meets the following *two* criteria:

- In order to take the $\pi$ label, there must be a change of phase when the orbital is rotated about the internuclear axis; in other words, there is a nodal plane that *contains* the internuclear axis.
- In order to take the $^*$ label, there must be a nodal plane *between* the nuclei; this plane is orthogonal to the internuclear axis.

---

[§] The choice of the $z$-axis is an arbitrary one but it is usual to have it running between the nuclei.

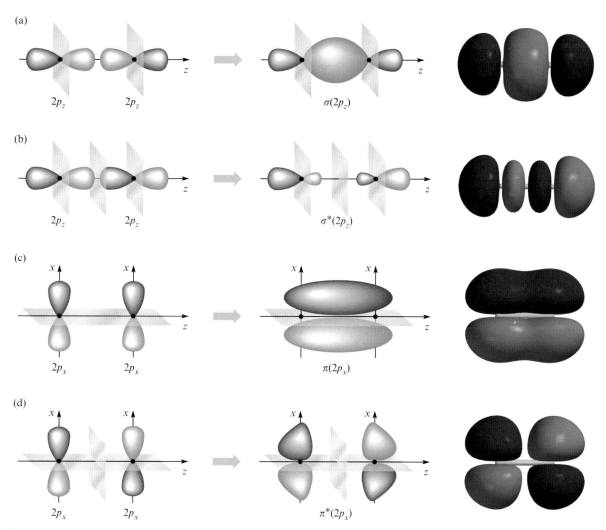

**Fig. 4.18** The overlap of two $2p$ atomic orbitals: (a) direct overlap along the $z$-axis to give a $\sigma$-MO (bonding); (b) the formation of the $\sigma^*$-MO (antibonding); (c) sideways overlap of two $2p_x$ atomic orbitals to give a $\pi$-MO (bonding); (d) the formation of $\pi^*$-MO (antibonding). Atomic nuclei are marked in black, and nodal planes in grey. The diagrams on the right-hand side are more realistic representations of the MOs and have been generated computationally using Spartan '04, © Wavefunction Inc. 2003.

The combination of the two $2p_y$ atomic orbitals is equivalent to the overlap of the two $2p_x$ atomic orbitals but the bonding and antibonding MOs that are formed will lie at right angles to those formed from the combination of the $2p_x$ atomic orbitals. As an exercise, draw diagrams like those in Figure 4.18 to describe the formation of $\pi(2p_y)$ and $\pi^*(2p_y)$ MOs.

In a homonuclear diatomic molecule, the combination of two sets of degenerate $2p$ atomic orbitals (one set per atom) leads to one $\sigma$- and two $\pi$-bonding MOs and one $\sigma^*$- and two $\pi^*$-antibonding MOs. (Remember that six atomic orbitals must give rise to six MOs.) The only difference between the two $\pi$-MOs is their directionality, and consequently, they are degenerate. The same applies to the antibonding MOs. Figure 4.19 shows an energy level diagram that describes the combination of the two sets of degenerate $2p$ atomic orbitals; as we shall see, the relative energies of the MOs may vary.

**Fig. 4.19** A molecular orbital diagram showing the combination of two sets of degenerate $2p_x$, $2p_y$ and $2p_z$ atomic orbitals to give $\sigma$ and $\pi$ bonding and antibonding MOs. The atomic nuclei lie on the $z$-axis.

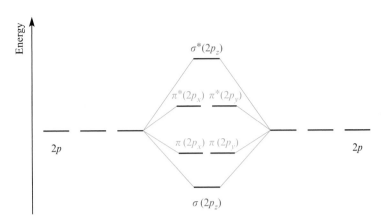

### 4.16     Bond order

▪▶
Fractional bond orders:
see Section 4.22

The bond order of a covalent bond gives a measure of the interaction between the nuclei. The three most common categories are the single (bond order = 1), the double (bond order = 2) and the triple (bond order = 3) bond, although bond orders are not restricted to integral values.

It is easy to understand what is meant by the single bond in $H_2$, but we need to be a little more specific when we talk about the bond order of other molecules. The bond order of a covalent bond can be found by using equation 4.23.[§]

$$\text{Bond order} = \tfrac{1}{2} \times \left[ \left( \begin{array}{c} \text{Number of} \\ \text{bonding electrons} \end{array} \right) - \left( \begin{array}{c} \text{Number of anti-} \\ \text{bonding electrons} \end{array} \right) \right] \quad (4.23)$$

For $H_2$, Figure 4.14 shows that there are two bonding electrons and no antibonding electrons, giving a bond order in $H_2$ of 1. Similarly, from Figure 4.16 the bond order of the bond in $Li_2$ is 1 (equation 4.24). Notice that we only need to consider the valence electrons; you can use equation 4.23 to confirm why in Section 4.14 we said that the core electrons 'cancel out'.

$$\text{Bond order in } Li_2 = \tfrac{1}{2} \times (2 - 0) = 1 \quad (4.24)$$

For $Be_2$, Figure 4.17 illustrates that the number of bonding electrons is two but there are also two electrons in an antibonding orbital. Therefore, from equation 4.23, the bond order is zero. This corresponds to no net bond between the two Be atoms, a result that is consistent with the fact that $Be_2$ is an extremely unstable species.

The bond order of a covalent bond may be found by considering an MO diagram:

$$\text{Bond order} = \tfrac{1}{2} \times \left[ \left( \begin{array}{c} \text{Number of} \\ \text{bonding electrons} \end{array} \right) - \left( \begin{array}{c} \text{Number of anti-} \\ \text{bonding electrons} \end{array} \right) \right]$$

---

[§] Any electrons in non-bonding orbitals are ignored; we return to non-bonding MOs in Section 5.5.

**4.17**    ## Relationships between bond order, bond length and bond enthalpy

In Section 4.5, the discussion of bond enthalpies touched on the fact that the bond enthalpy of a carbon–carbon triple bond (C≡C) is greater than that of a carbon–carbon double (C=C) bond, and this is, in turn, greater than that of a carbon–carbon single (C−C) bond (see Table 4.5). This point is made in Figure 4.20 where we plot bond enthalpy against bond order, and also correlate the bond order and bond enthalpy with bond distance. The following trends emerge:

- bond enthalpy increases as bond order increases;
- bond distance decreases as bond order increases;
- bond enthalpy decreases as bond distance increases.

The graphs also illustrate that these trends apply to nitrogen–nitrogen and oxygen–oxygen bonds, and in fact the trends are general among the *p*-block elements.

**4.18**    ## Homonuclear diatomic molecules of the first row *p*-block elements: $F_2$ and $O_2$

In this section and Section 4.20 we look at the bonding in the diatomic molecules $B_2$, $C_2$, $N_2$, $O_2$ and $F_2$ using the molecular orbital and valence bond approaches, and compare these results with those obtained by drawing a simple Lewis structure for each molecule. The formation of $Li_2$ and $Be_2$ involved the overlap of 2$s$ atomic orbitals, but that of $B_2$, $C_2$, $N_2$, $O_2$ and $F_2$ involves the overlap, not only of 2$s$ atomic orbitals, but also of 2$p$ orbitals.

**Fig. 4.20** Trends in bond order, bond length and bond enthalpies for carbon–carbon, nitrogen–nitrogen and oxygen–oxygen bonds.

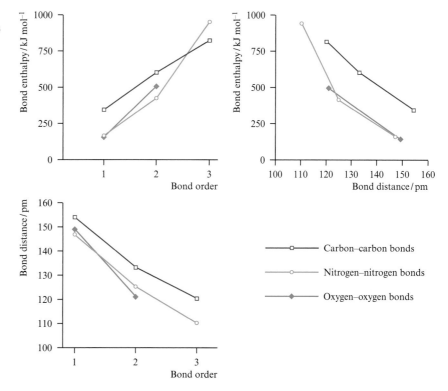

Throughout the following discussion we shall ignore the core ($1s$) electrons of each atom for reasons already discussed.

## $F_2$

*Experimental facts*: The standard state of fluorine is the diamagnetic gas $F_2$.

> A species is **diamagnetic** if all of its electrons are paired.
>
> A species is **paramagnetic** if it contains one or more unpaired electrons.

The ground state electronic configuration of a fluorine atom is $[He]2s^2 2p^5$. Structures **4.10** and **4.11** show two Lewis representations of $F_2$; each fluorine atom has an octet of valence electrons. A single F−F covalent bond is predicted by this approach. The valence bond method also describes the $F_2$ molecule in terms of a single F−F bond.

: F : F :        : F——F :

(4.10)            (4.11)

We can construct an MO diagram for the formation of $F_2$ by considering the linear combination of the atomic orbitals of the two fluorine atoms (Figure 4.21). There are 14 valence electrons and, by the *aufbau* principle,

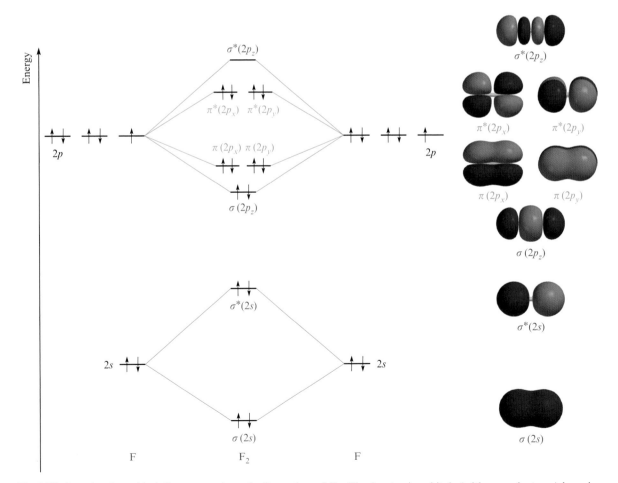

**Fig. 4.21** A molecular orbital diagram to show the formation of $F_2$. The $1s$ atomic orbitals (with core electrons) have been omitted. The F nuclei lie on the $z$-axis. Representations of the MOs (generated using Spartan '04, © Wavefunction Inc. 2003) are shown on the right-hand side of the figure.

these occupy the MOs as shown in Figure 4.21. All the electrons are paired and this is consistent with the observed diamagnetism of $F_2$.

From Figure 4.21, the electronic configuration of $F_2$ may be written as:

$$\sigma(2s)^2\sigma^*(2s)^2\sigma(2p_z)^2\pi(2p_x)^2\pi(2p_y)^2\pi^*(2p_x)^2\pi^*(2p_y)^2$$

▶
**How is this configuration altered if core electrons are included?**

The bond order in $F_2$ can be determined by using equation 4.23 and the single bond (equation 4.25) corresponds to the value obtained from the Lewis structure.

$$\text{Bond order in } F_2 = \tfrac{1}{2} \times (8 - 6) = 1 \qquad (4.25)$$

Thus, the same conclusions about the bonding in the $F_2$ molecule are reached by drawing a Lewis structure or by approaching the bonding using VB or MO theories.

## $O_2$

*Experimental facts*: Dioxygen is a gas at 298 K and condenses to form a blue liquid at 90 K. The $O_2$ molecule is paramagnetic – it is a diradical with two unpaired electrons.

The ground state electronic configuration of an oxygen atom is $[\text{He}]2s^22p^4$. An MO diagram for the formation of $O_2$ from two oxygen atoms is shown in Figure 4.22. There are 12 valence electrons to be accommodated and, by Hund's rule, the last two electrons must singly occupy each of the degenerate $\pi^*(2p_x)$ and $\pi^*(2p_y)$ molecular orbitals. This picture is consistent with experimental observations.

From Figure 4.22, the electronic configuration of $O_2$ may be written as:

$$\sigma(2s)^2\sigma^*(2s)^2\sigma(2p_z)^2\pi(2p_x)^2\pi(2p_y)^2\pi^*(2p_x)^1\pi^*(2p_y)^1$$

$$\text{Bond order in } O_2 = \tfrac{1}{2} \times (8 - 4) = 2 \qquad (4.26)$$

**Fig. 4.22** A molecular orbital diagram to show the formation of $O_2$. The 1*s* atomic orbitals (with core electrons) have been omitted. The O nuclei lie on the *z*-axis. Pictorial representations of the MOs are qualitatively the same as those shown on the right-hand side of Figure 4.21.

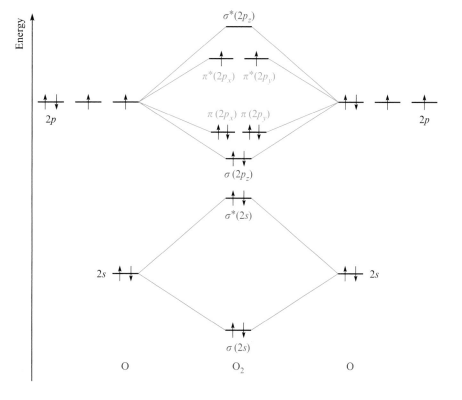

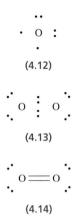

(4.12)

(4.13)

(4.14)

Now let us consider a Lewis structure for $O_2$. The ground state configuration of each oxygen atom is $[He]2s^2 2p^4$ and there are two unpaired electrons per atom as shown in diagram **4.12**. The Lewis picture of the molecule pairs these electrons up to give an O=O double bond in agreement with the bond order calculated in equation 4.26. Each oxygen atom in Lewis structures **4.13** and **4.14** obeys the octet rule.

However, there is a problem – the Lewis structure does *not* predict the presence of two unpaired electrons in the $O_2$ molecule. Simple valence bond theory also gives this result. In contrast, the MO description of $O_2$ gives results about the bond order and the pairing of electrons that are consistent with the experimental data. The explanation of the diradical character of $O_2$ represents one of the classic successes of MO theory. The two unpaired electrons follow naturally from the orbital diagram.

**4.19**    ## Orbital mixing and $\sigma-\pi$ crossover

Before we can progress to the bonding in $B_2$, $C_2$ and $N_2$, we must deal with two new concepts: orbital mixing and $\sigma-\pi$ crossover. However, understanding the reasoning for these phenomena is not critical to a qualitative discussion of the MO diagrams for these diatomics, and you may wish to move on to Section 4.20, returning to this section later.

When we construct an MO diagram using a linear combination of atomic orbitals, our initial approach allows overlap only between like atomic orbitals. Figure 4.23 shows the overlap between the $2s$ atomic orbitals of two identical atoms to give $\sigma(2s)$ and $\sigma^*(2s)$ MOs, and the overlap between the $2p$ atomic orbitals to give $\sigma(2p)$, $\pi(2p)$, $\pi^*(2p)$ and $\sigma^*(2p)$ MOs. The

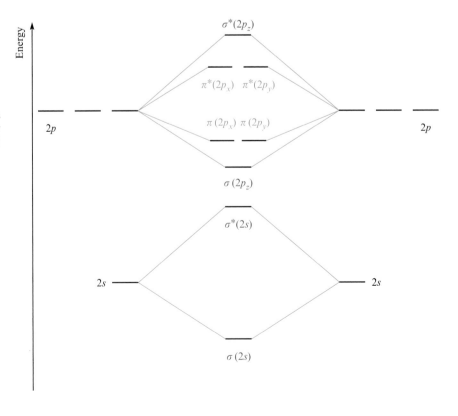

**Fig. 4.23** A molecular orbital diagram showing the approximate ordering of molecular orbitals generated by the combination of $2s$ and $2p$ atomic orbitals. This diagram is applicable to the formation of homonuclear diatomic molecules involving late (O and F) first row $p$-block elements, with the nuclei lying on the $z$-axis.

**Fig. 4.24** The effects of orbital mixing on the ordering of the MOs in a first row diatomic molecule. We consider the effects of the $\sigma$ and $\sigma^*$ mixing sequentially.

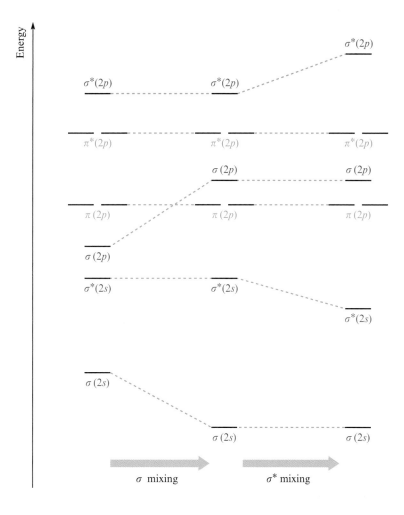

ordering of the molecular orbitals is approximate. Note that the $\sigma(2p)$ orbital lies at lower energy than the $\pi(2p)$ levels.

We can now introduce a new concept. If atomic orbitals of similar symmetry and energy can mix to give MOs, can MOs of similar symmetry and energy also mix? The answer is yes.

In Figure 4.23, we have two $\sigma$ orbitals [$\sigma(2s)$ and $\sigma(2p)$] and two $\sigma^*$ orbitals [$\sigma^*(2s)$ and $\sigma^*(2p)$]. At the beginning of the first row of the $p$-block elements, the $2s$ and $2p$ atomic orbitals are relatively close together in energy ($\approx 550\,\mathrm{kJ\,mol}^{-1}$ for boron), whereas by the end of the row they are very much further apart ($\approx 2100\,\mathrm{kJ\,mol}^{-1}$ for fluorine). It follows that the *molecular* orbitals derived from the $2s$ and $2p$ atomic orbitals are closer together for the earlier $p$-block elements. If the orbitals have similar energies and similar symmetries, they can mix. The results of mixing $\sigma(2s)$ with $\sigma(2p)$, and $\sigma^*(2s)$ with $\sigma^*(2p)$, are shown in Figure 4.24.

Generally, if the $2s$ and $2p$ atomic orbitals are close in energy, mixing of the $\sigma(2s)$ and $\sigma(2p)$ molecular orbitals results in the $\sigma(2p)$ orbital[§] lying higher in energy than the $\pi(2p)$ levels. This is actually the case with the earlier $p$-block elements (B, C and N). Naturally, the mixing of orbitals will also occur with

*s–p separation:*
*see Section 3.15*

---

[§] Strictly, this nomenclature is now incorrect as this orbital contains both $2s$ and $2p$ character. For convenience, we retain the simple notation to indicate the dominant character.

**Fig. 4.25** A molecular orbital diagram showing the ordering of molecular orbitals generated by the combination of $2s$ and $2p$ atomic orbitals *and* allowing for the effects of orbital mixing. This diagram is applicable to the formation of homonuclear diatomic molecules involving early (B, C and N) first row $p$-block elements.

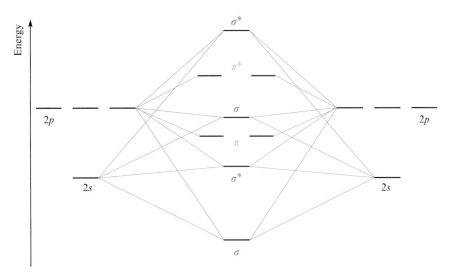

the later elements, but the energy mismatch makes the perturbation less, and the $\sigma(2p)$ orbital lies below the $\pi(2p)$ level.

Finally, in Figure 4.25 we show a full MO diagram for the combination of $2s$ and $2p$ atomic orbitals in which we include the effects of orbital mixing. Notice that the character of the atomic orbitals involved in the mixing is spread into a number of molecular orbitals. We return to this theme in Section 5.14 when we discuss the bonding in carbon monoxide. It is also important to see in Figure 4.25 that although atomic orbitals are now involved in several molecular orbitals, the total number of MOs still equals the total number of atomic orbitals.

Now we are in a position to move on to a discussion of the bonding in $B_2$, $C_2$ and $N_2$ and in it, we use the orbital energy levels obtained from Figure 4.25.

| 4.20 | **Homonuclear diatomic molecules of the first row $p$-block elements: $B_2$, $C_2$ and $N_2$** |
| --- | --- |

### $B_2$

*Experimental fact*: The vapour phase of elemental boron contains paramagnetic $B_2$ molecules.

The ground state electronic configuration of a boron atom is $[He]2s^2 2p^1$. An MO diagram for the formation of a $B_2$ molecule is shown in Figure 4.26. Each boron atom provides three valence electrons. Four electrons occupy the $\sigma(2s)$ and $\sigma^*(2s)$ MOs. This leaves two electrons. For the $B_2$ molecule to be paramagnetic, these electrons must singly occupy each of the degenerate $\pi(2p_x)$ and $\pi(2p_y)$ orbitals. We have experimental evidence from magnetic measurements for the $\sigma$–$\pi$ crossover in $B_2$, as we discuss more fully for $C_2$.

From Figure 4.26, the electronic configuration of $B_2$ may be written as:

$$\sigma(2s)^2 \sigma^*(2s)^2 \pi(2p_x)^1 \pi(2p_y)^1$$

We predict a bond order for the boron–boron bond in $B_2$ of 1 (equation 4.27), i.e. a single bond is predicted.

$$\text{Bond order in } B_2 = \tfrac{1}{2} \times (4 - 2) = 1 \tag{4.27}$$

Now let us draw a Lewis structure for $B_2$. The problem here is that the three electrons of each boron atom will be expected to pair up to give a triple bond as in structures **4.15** and **4.16** – an octet cannot be achieved. This bond order is not in accord either with the molecular orbital result or with the experimental data. An alternative Lewis structure can be drawn in either of the forms **4.17** or **4.18**. Here, a single bond is drawn, leaving a lone pair of electrons per boron atom. Although this result is more acceptable in terms of the B–B bond itself, it still does not predict that the $B_2$ molecule will be paramagnetic – remember that the Lewis approach will always give *paired* electrons whenever possible.

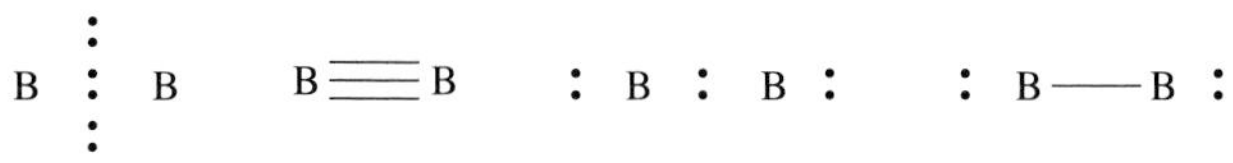

$$B :\!\!: B \qquad B \!\!=\!\!\equiv\!\!= B \qquad : B :\!\!: B : \qquad : B \!\!-\!\! B :$$

$$\textbf{(4.15)} \qquad\qquad \textbf{(4.16)} \qquad\qquad \textbf{(4.17)} \qquad\qquad \textbf{(4.18)}$$

For $B_2$, we see that drawing a Lewis structure does not yield a satisfactory description of the electrons in the molecule. On the other hand, MO theory can be used successfully to rationalize the bond order and the paramagnetism of this diatomic molecule.

## $C_2$

*Experimental facts*: The $C_2$ molecule is a gas phase species and is diamagnetic.

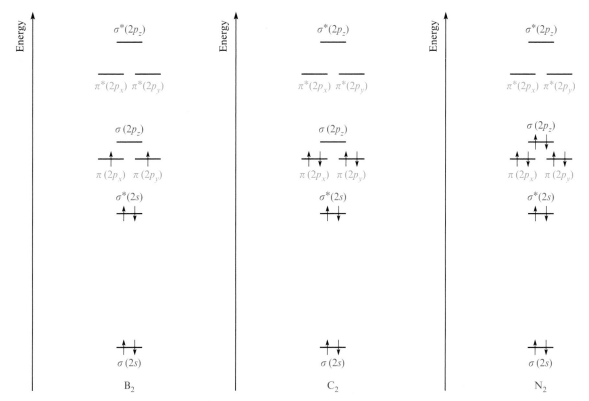

**Fig. 4.26** A molecular orbital diagram for the formation of $B_2$. The $\sigma(1s)$ and $\sigma^*(1s)$ MOs (with core electrons) have been omitted. Refer to Figure 4.25.

**Fig. 4.27** A molecular orbital diagram for the formation of $C_2$. The $\sigma(1s)$ and $\sigma^*(1s)$ MOs (with core electrons) have been omitted. Refer to Figure 4.25.

**Fig. 4.28** A molecular orbital diagram for the formation of $N_2$. The $\sigma(1s)$ and $\sigma^*(1s)$ MOs (with core electrons) have been omitted. Refer to Figure 4.25.

The ground state electronic configuration of a carbon atom is $[He]2s^2 2p^2$. Lewis structures for $C_2$ are given in **4.19** and **4.20**, and show a C=C double bond and a lone pair of electrons per carbon atom. All electrons are paired.

$$: C \;\vdots\; C : \qquad : C =\!\!= C :$$

(4.19)                (4.20)

An MO diagram for the formation of the $C_2$ molecule is shown in Figure 4.27. There are eight valence electrons to be accommodated in the MOs which arise from the combination of the carbon $2s$ and $2p$ atomic orbitals. Four electrons occupy the $\sigma(2s)$ and $\sigma^*(2s)$ MOs. If the $\pi(2p_x)$ and $\pi(2p_y)$ levels are lower in energy than the $\sigma(2p_z)$ level, then the last four electrons will occupy the degenerate set of orbitals as two pairs. This gives rise to a diamagnetic molecule and is in agreement with the experimental observations. Thus, the magnetic data for $C_2$ provide evidence for the $\sigma$–$\pi$ crossover. If the energy of the $\sigma(2p_z)$ level were lower than that of the $\pi(2p_x)$ and $\pi(2p_y)$ levels, the electronic configuration for $C_2$ would be $\sigma(2s)^2\sigma^*(2s)^2\sigma(2p_z)^2\pi(2p_x)^1\pi(2p_y)^1$, predicting (incorrectly) a paramagnetic species.

From Figure 4.27, the electronic configuration of $C_2$ may be written as:

$$\sigma(2s)^2\sigma^*(2s)^2\pi(2p_x)^2\pi(2p_y)^2$$

The bond order of the carbon–carbon bond in $C_2$ may be determined from Figure 4.27 and equation 4.28.

$$\text{Bond order in } C_2 = \tfrac{1}{2} \times (6 - 2) = 2 \qquad (4.28)$$

In conclusion, we see that a C=C bond and complete pairing of electrons in $C_2$ can be rationalized in terms of a Lewis structure, VB and MO theories. We use the term 'rationalized' rather than 'predicted' in regard to the MO approach because of the uncertainties associated with the $\sigma$–$\pi$ crossover. At a simple level of calculation, whether or not this crossover will occur cannot be predicted. However, we *can* rationalize the fact that $C_2$ is diamagnetic by allowing for such a change in energy levels.

## $N_2$

*Experimental facts*: The $N_2$ molecule is relatively unreactive. Dinitrogen is often used to provide an inert atmosphere for experiments in which reagents and/or products react with components in the air, usually $O_2$ or water vapour. The bond dissociation enthalpy of $N_2$ is particularly high and the bond length is short (Table 4.7). The $N_2$ molecule is diamagnetic.

Let us first construct a Lewis structure for $N_2$. The ground state electronic configuration of a nitrogen atom is $[He]2s^2 2p^3$. The three $2p$ electrons are unpaired (as represented in **4.21**) and when two nitrogen atoms come together, a triple bond is required so that each nitrogen atom completes its octet of valence electrons. This is represented in the Lewis structures **4.22** and **4.23**. Similarly, the valence bond model for the $N_2$ molecule shows a localized N≡N triple bond and one lone pair per nitrogen atom.

$$\cdot\; N \;\cdot \qquad : N \;\vdots\; N : \qquad : N =\!\!= N :$$

(4.21)                (4.22)                (4.23)

**Table 4.7** Experimental data for homonuclear diatomic molecules containing elements of the first row of the periodic table.

| Molecule | Bond distance / pm | Bond dissociation enthalpy / kJ mol$^{-1}$ | Magnetic data |
|---|---|---|---|
| $Li_2$ | 267 | 110 | Diamagnetic |
| $Be_2$ (extremely unstable)[a] | – | – | – |
| $B_2$ | 159 | 297 | Paramagnetic |
| $C_2$ | 124 | 607 | Diamagnetic |
| $N_2$ | 110 | 945 | Diamagnetic |
| $O_2$ | 121 | 498 | Paramagnetic |
| $F_2$ | 141 | 159 | Diamagnetic |

[a] There is evidence for an extremely unstable $Be_2$ species with a bond enthalpy of $\sim 10\,kJ\,mol^{-1}$ and a bond distance of $\sim 245\,pm$.

Photoelectron spectroscopy: see Box 4.6

An MO diagram for the formation of $N_2$ is shown in Figure 4.28. There are 10 valence electrons to be accommodated and filling the MOs by the *aufbau* principle gives rise to complete pairing of the electrons. This is consistent with the experimental observation that the $N_2$ molecule is diamagnetic.

For $N_2$, the presence (or not) of the $\sigma$–$\pi$ crossover makes no difference to predictions about the pairing of electrons or to the bond order, but its existence is suggested by the results of *photoelectron spectroscopic* studies of $N_2$.

From Figure 4.28, the electronic configuration of $N_2$ may be written as:

$$\sigma(2s)^2 \sigma^*(2s)^2 \pi(2p_x)^2 \pi(2p_y)^2 \sigma(2p_z)^2$$

and from equation 4.29, the bond order in the $N_2$ molecule is found to be 3. This multiple bond order is in keeping with the high bond dissociation enthalpy of $N_2$ and the short internuclear distance (Table 4.7).

$$\text{Bond order in } N_2 = \tfrac{1}{2} \times (8 - 2) = 3 \tag{4.29}$$

In conclusion, the triple bond in $N_2$ and its diamagnetic character can be correctly predicted by use of a Lewis structure or by the VB or MO approaches.

**THEORETICAL AND CHEMICAL BACKGROUND**

## Box 4.6 Photoelectron spectroscopy (PES)

Photoelectron spectroscopy is a technique used to study the energies of *occupied* atomic and molecular orbitals.

An atom or molecule is irradiated with electromagnetic radiation of energy, $E$, and this causes electrons to be ejected from the sample. Each electron has a characteristic *binding energy* and, to be ejected from the atom or molecule, the electron must absorb an amount of energy that is equal to its binding energy.

If we measure the energy of the electron as it is ejected, then this excess energy will be given by:

Excess energy of electron

$$= E - (\text{binding energy of electron})$$

providing, of course, that $E$ is greater than the binding energy of the electron. Hence, we can measure the binding energy.

Koopmans' theorem relates the binding energy of an electron to the orbital in which it resides. Effectively, it allows the binding energy to be a measure of the orbital energy.

Photoelectron spectroscopy is an important experimental method that permits us to measure the binding energies of both valence and core electrons. From these energies, we can gain insight into the ordering of molecular orbitals for a given atomic or molecular system. We mention photoelectron spectroscopy again in Section 7.9.

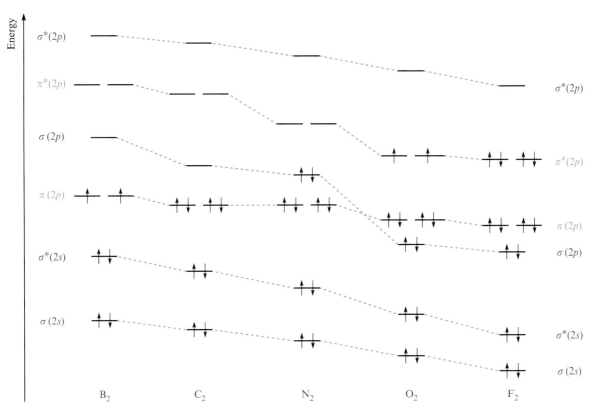

**Fig. 4.29** Changes in the energy levels of the molecular orbitals and the ground state electronic configurations of homonuclear diatomic molecules involving first row $p$-block elements.

### The first row $p$-block elements: a summary

Let us briefly summarize what happens to the bonding in the $X_2$ diatomic molecules for X = B, C, N, O and F. Figure 4.29 illustrates the changes in the energy levels of the molecular $\sigma(2s)$, $\sigma^*(2s)$, $\sigma(2p)$, $\sigma^*(2p)$, $\pi(2p)$ and $\pi^*(2p)$ orbitals in going from $B_2$ to $F_2$.

The main points to note are:

- the effects of molecular orbital mixing which lead to the $\sigma$–$\pi$ crossover;
- the use of the *aufbau* principle to give ground state electronic configurations;
- the use of the MO diagram to determine bond order;
- the use of the MO diagram to explain diamagnetic or paramagnetic behaviour.

---

**4.21**     **Periodic trends in the homonuclear diatomic molecules of the first row elements**

In this section we summarize the information gained about the homonuclear diatomic molecules of the first row elements – $Li_2$ to $F_2$, excluding '$Be_2$'.

Figure 4.30 shows the trends in X–X bond distances and bond dissociation enthalpies in the $X_2$ molecules as we cross the periodic table from left to right. From $B_2$ to $N_2$, there is a decrease in bond length which corresponds to an increase in the bond dissociation enthalpy of the X–X bond. From $N_2$ to $F_2$, the X–X bond lengthens and weakens. These trends follow changes in

**Fig. 4.30** The trends in (a) the X—X bond distances and (b) the X—X bond dissociation enthalpy for $X_2$ molecules containing the first row elements (X = Li to F).

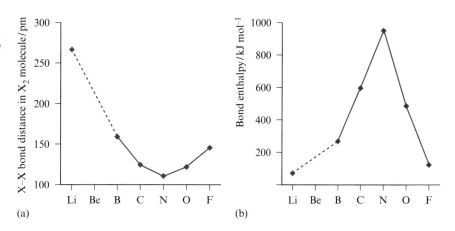

(a)

(b)

the bond order in the sequence $B_2$ (single) $< C_2$ (double) $< N_2$ (triple) $> O_2$ (double) $> F_2$ (single).

Note that we *cannot exactly* correlate particular values of the bond dissociation enthalpy and distance with a given bond order. Factors such as internuclear and inter-electron repulsion play an important part in determining the observed values of these parameters. For example, the $Li_2$ molecule has a single bond and is thus similar to $B_2$ and $F_2$. Yet the bond distance of 267 pm in $Li_2$ is far greater than that of either $B_2$ or $F_2$. Correspondingly, the bond dissociation enthalpy of $Li_2$ is very low. The difference lies in the radial extent of the orbitals involved in bond formation. The $2s$ and $2p$ orbitals of the boron atom experience a greater effective nuclear charge than the $2s$ orbitals of the lithium atom, and are thus more contracted. In order to get an efficient overlap, the atoms in $B_2$ must be closer together than the atoms in $Li_2$.

## 4.22    The diatomic species $O_2$, $[O_2]^+$, $[O_2]^-$ and $[O_2]^{2-}$

Once an MO diagram has been constructed for a homonuclear diatomic molecule, it can be used to predict the properties of (or rationalize experimental data for) species derived from that molecule. We exemplify this by looking at $O_2$, $[O_2]^+$, $[O_2]^-$ and $[O_2]^{2-}$, some physical properties of which are listed in Table 4.8.

Figure 4.22 showed an MO diagram for the formation of the $O_2$ molecule. The configuration $\sigma(2s)^2\sigma^*(2s)^2\sigma(2p_z)^2\pi(2p_x)^2\pi(2p_y)^2\pi^*(2p_x)^1\pi^*(2p_y)^1$ supports the presence of an O=O double bond and the fact that the $O_2$ molecule is paramagnetic.

What happens if we oxidize or reduce $O_2$? The addition of one or more electrons corresponds to *reduction* and the removal of one or more electrons

**Table 4.8** Experimental data for dioxygen and derived diatomic species.

| Diatomic species | Bond distance / pm | Bond enthalpy / kJ mol$^{-1}$ | Magnetic properties |
|---|---|---|---|
| $O_2$ | 121 | 498 | Paramagnetic |
| $[O_2]^+$ | 112 | 644 | Paramagnetic |
| $[O_2]^-$ | 134 | 360 | Paramagnetic |
| $[O_2]^{2-}$ | 149 | 149 | Diamagnetic |

**Oxidation** and **reduction** may be defined in terms of the loss and gain of an electron or electrons.

In an oxidation process, one or more electrons are lost.

In a reduction process, one or more electrons are gained.

is *oxidation*. We can use the MO diagram for $O_2$ to monitor changes (bond order, bond distance, bond dissociation enthalpy and electron pairing) that occur to the diatomic species by adding or removing electrons. The addition of electrons will follow the *aufbau* principle; the removal of electrons follows the same rules but in reverse.

The 1-electron oxidation of the $O_2$ molecule (equation 4.30) corresponds to the removal of one electron from a $\pi^*$ orbital in Figure 4.22. Inspection of the MO diagram shows that the $[O_2]^+$ cation will be paramagnetic with one unpaired electron. It is also a radical species and is known as a *radical cation*.

$$O_2 \longrightarrow [O_2]^+ + e^- \tag{4.30}$$

$$\text{Bond order in } [O_2]^+ = \tfrac{1}{2} \times (8 - 3) = 2.5 \tag{4.31}$$

A **radical cation** is a cation that possesses an unpaired electron.

A **radical anion** is an anion that possesses an unpaired electron.

In going from $O_2$ to $[O_2]^+$, the bond order increases from 2.0 to 2.5 (equations 4.26 and 4.31). This is reflected in an increase in the bond dissociation enthalpy and a decrease in the oxygen–oxygen bond distance (Table 4.8).

The 1- and 2-electron reductions of $O_2$ are shown in equations 4.32 and 4.33. By considering the molecular orbitals in Figure 4.22, we can see that the superoxide ion, $[O_2]^-$, is formed by the addition of one electron to one of the $\pi^*$ orbitals. The $[O_2]^-$ ion is paramagnetic (it is a *radical anion*) and has a bond order of 1.5. The data in Table 4.8 support the weakening and lengthening of the oxygen–oxygen bond on going from $O_2$ to $[O_2]^-$.

$$O_2 + e^- \longrightarrow [O_2]^- \tag{4.32}$$

$$O_2 + 2e^- \longrightarrow [O_2]^{2-} \tag{4.33}$$

Oxides, peroxides and superoxides: see Sections 21.11 and 22.10

The addition of two electrons to $O_2$ to give $[O_2]^{2-}$ (the peroxide ion) results in the formation of a diamagnetic species, and the bond order becomes 1. Again, the experimental data support this. The dianion $[O_2]^{2-}$ has a longer and weaker bond than either of $O_2$ or $[O_2]^-$.

| 4.23 | **Group trends among homonuclear diatomic molecules** |

Now we turn from trends among diatomic molecules in the first row of the periodic table to trends within *groups of elements*, focusing first on the alkali metals and the halogens. We then consider group 15 in which the changes in X–X bond enthalpies in going down the group cause dramatic effects in elemental structure.

### Group 1: alkali metals

Hydrogen is excluded from this discussion

Each group 1 element has a ground state valence electronic configuration of $ns^1$. Homonuclear diatomic molecules of the alkali metals occur *in the gas phase*.

The formation of the $Li_2$ molecule from two lithium atoms was illustrated in Figure 4.16. Similar diagrams can be constructed for the formation of $Na_2$, $K_2$, $Rb_2$, $Cs_2$ and $Fr_2$, although in practice, the experimental data (Table 4.9) on these species become sparse as the group is descended.

Molecular orbital theory predicts that each group 1 diatomic molecule has a bond order of 1. This is also the conclusion when Lewis structures are drawn or when the bonding is considered using VB theory – each alkali metal atom has one valence electron and will share one electron with another group 1 atom.

**Table 4.9** Experimental data for homonuclear diatomic molecules of the group 1 elements.

| Diatomic species, $X_2$ | Ground state electronic configuration of atom X | Bond distance / pm | Bond dissociation enthalpy / kJ mol$^{-1}$ |
|---|---|---|---|
| $Li_2$ | [He]$2s^1$ | 267 | 110 |
| $Na_2$ | [Ne]$3s^1$ | 308 | 74 |
| $K_2$ | [Ar]$4s^1$ | 390 | 55 |
| $Rb_2$ | [Kr]$5s^1$ | – | 49 |
| $Cs_2$ | [Xe]$6s^1$ | – | 44 |

The bond dissociation enthalpies and bond distances for $Li_2$, $Na_2$ and $K_2$ (Table 4.9) indicate that as the group is descended, the M−M bond gets longer and weaker. The trend in enthalpies continues for $Rb_2$ and $Cs_2$. This reflects the change in the principal quantum number of the $ns$ atomic orbital and the spatial properties of the orbitals. We reiterate how the radial distribution function varies among the $ns$ atomic orbitals; remember that this gives the probability of finding the electron in a spherical shell of radius $r$ and thickness $dr$ from the nucleus. Figure 3.11 showed radial distribution functions for $1s$, $2s$ and $3s$ atomic orbitals of hydrogen; the valence orbitals of Li and Na are the $2s$ and $3s$ atomic orbitals.

Spatial properties of orbitals: see Figure 3.10

### Group 17: halogens

Each member of group 17 has an $ns^2np^5$ valence electronic configuration. Lewis structures for $Cl_2$, $Br_2$ and $I_2$ resemble that drawn for $F_2$ in structures **4.10** and **4.11**. Each molecule has a single bond. An MO diagram for each of the $Cl_2$, $Br_2$ and $I_2$ molecules is similar to Figure 4.21, except that the principal quantum numbers are different and this affects the relative energies of the orbitals. In each case, a bond order of 1 is calculated for the $X_2$ molecule, and we find that each diatomic species is diamagnetic.

Table 4.10 gives some data for $F_2$, $Cl_2$, $Br_2$ and $I_2$. The relative values of the distances and enthalpies are in agreement with the notion of single X−X bonds. However, instead of a general trend of decreasing bond dissociation enthalpies and increasing bond distances, the F−F bond is actually much weaker than might be expected. This is emphasized in the graph in Figure 4.31. The additional weakening is usually attributed to repulsion between the lone pairs of electrons on the two fluorine atoms (structure **4.24**). Although each X atom in each $X_2$ molecule has three lone pairs of electrons, those in $F_2$ will experience the greatest repulsion because of the shorter distance between the nuclei.

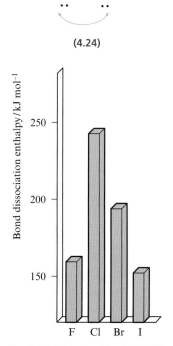

Electron repulsion

(4.24)

**Fig. 4.31** The trend in the X−X bond dissociation enthalpy for $X_2$ molecules where X is a group 17 element.

**Table 4.10** Experimental data for homonuclear diatomic molecules of the group 17 elements.

| Diatomic species, $X_2$ | Ground state electronic configuration of atom X | Bond distance / pm | Bond dissociation enthalpy / kJ mol$^{-1}$ |
|---|---|---|---|
| $F_2$ | [He]$2s^22p^5$ | 141 | 159 |
| $Cl_2$ | [Ne]$3s^23p^5$ | 199 | 242 |
| $Br_2$ | [Ar]$4s^24p^5$ | 228 | 193 |
| $I_2$ | [Kr]$5s^25p^5$ | 267 | 151 |

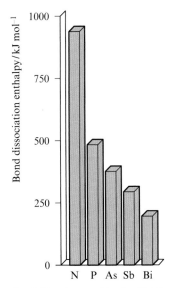

**Fig. 4.32** The trend in the X–X bond dissociation enthalpy for $X_2$ molecules where X is a group 15 element.

## Group 15

The trend in the bond dissociation enthalpies of the molecules $N_2$, $P_2$, $As_2$, $Sb_2$ and $Bi_2$ is shown in Figure 4.32. There is a sharp decrease in the value of the dissociation enthalpy in going from $N_2$ to $P_2$ and then less dramatic decreases in going from $P_2$ through to $Bi_2$.

Let us step back for a moment and consider the choices open to atoms of group 15 elements when they combine with one another. The situation is illustrated in Figure 4.33. The presence of three unpaired electrons suggests that the formation of a diatomic molecule with a triple bond or a tetraatomic molecule with six single bonds are possibilities.

Bond enthalpy is an important factor in determining what degree of molecular aggregation will occur. If we have four nitrogen atoms in the gas phase, they might come together to form two molecules of $N_2$ (equation 4.34) or one molecule of $N_4$ (equation 4.35).

$$4N(g) \longrightarrow 2N_2(g) \tag{4.34}$$
$$4N(g) \longrightarrow N_4(g) \tag{4.35}$$

When two moles of $N_2$ form, two triple bonds are made. Bond formation is exothermic and the associated enthalpy change is:

$$\Delta H^\circ[4N(g) \longrightarrow 2N_2(g)] = -[2 \times D(N{\equiv}N)]$$
$$= -(2 \times 945)$$
$$= -1890 \text{ kJ per 4 moles of N atoms}$$

When one mole of $N_4$ forms, six single bonds are made and the associated enthalpy change is:

> The value of $D(N–N)$ comes from Table 4.5; the value in Figure 4.32 refers to a triple bond

$$\Delta H^\circ[4N(g) \longrightarrow N_4(g)] = -[6 \times D(N–N)]$$
$$= -(6 \times 159)$$
$$= -954 \text{ kJ per 4 moles of N atoms}$$

Thus, in terms of enthalpy it is more favourable to form two moles of $N_2(g)$ than it is to form one mole of $N_4(g)$. In fact, the $N_4$ molecule has not been observed experimentally and the particularly strong $N{\equiv}N$ triple bond is responsible for the $N_2$ molecule being the standard state of this element.

Now let us repeat the exercise for phosphorus.

$$4P(g) \longrightarrow 2P_2(g) \tag{4.36}$$
$$4P(g) \longrightarrow P_4(g) \tag{4.37}$$

For the formation of two moles of $P_2$, the enthalpy change is:

$$\Delta H^\circ[4P(g) \longrightarrow 2P_2(g)] = -[2 \times D(P{\equiv}P)]$$
$$= -(2 \times 490)$$
$$= -980 \text{ kJ per 4 moles of P atoms}$$

**Fig. 4.33** Atoms of a group 15 element may come together to form a diatomic molecule with a triple bond, or a tetraatomic molecule with six single bonds. The $P_4$, $As_4$, $Sb_4$ and $Bi_4$ molecules are formed under drastically different conditions.

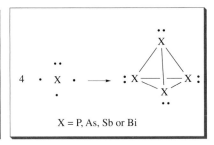

The value of $D$(P–P) comes from Table 4.5

For the formation of one mole of $P_4$, the enthalpy change is:

$$\Delta H^\circ[4P(g) \longrightarrow P_4(g)] = -[6 \times D(P\text{–}P)]$$
$$= -(6 \times 200)$$
$$= -1200 \text{ kJ per 4 moles of P atoms}$$

This time, in terms of enthalpy, it is more favourable for the gaseous $P_4$ molecule to form, although the enthalpy difference between the two forms of phosphorus is not as great as between the two forms of nitrogen. The standard state of phosphorus (white phosphorus) consists of $P_4$ molecules packed in a solid state array.

Above 827 K, $P_4$ dissociates into $P_2$ molecules (equation 4.38) and, from the data calculated above, we would estimate that $\Delta H^\circ$ for this process is 220 kJ mol$^{-1}$ – experimentally, it is found to be 217 kJ mol$^{-1}$.

$$P_4(g) \longrightarrow 2P_2(g) \tag{4.38}$$

In the gas phase, arsenic exists as $As_4$ molecules. At relatively low temperatures, antimony vapour contains mainly $Sb_4$ molecules. $Sb_4$, $Sb_2$, $Bi_4$ and $Bi_2$ molecules have been observed in noble gas matrices. However, all these species are inaccessible at 298 K. The standard states of these elements possess covalent lattice structures similar to that of black phosphorus. We look further at these structures in Chapter 9.

---

**Worked example 4.4**    *Allotropes of oxygen and sulfur*

**Suggest reasons why oxygen forms an $O_2$ molecule rather than a cyclic structure such as $O_6$, whereas sulfur forms $S_6$ rather than an $S_2$ molecule at 298 K and atmospheric pressure.**

**Bond enthalpy data:**

$$D(\text{O–O}) = 146 \text{ kJ mol}^{-1}$$
$$D(\text{S–S}) = 266 \text{ kJ mol}^{-1}$$
$$D(\text{O=O}) = 498 \text{ kJ mol}^{-1}$$
$$D(\text{S=S}) = 425 \text{ kJ mol}^{-1}$$

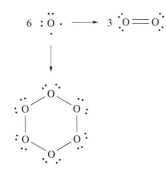

Consider the formation of $O_2$ and $O_6$ molecules from oxygen atoms. In order to make the comparison meaningful, we must consider equal numbers of oxygen atoms:

$$6O(g) \longrightarrow 3O_2(g)$$
$$6O(g) \longrightarrow O_6(g)$$

For the formation of three moles of $O_2(g)$:

$$\Delta H^\circ = -[3 \times D(\text{O=O})]$$
$$= -(3 \times 498)$$
$$= -1494 \text{ kJ per 6 moles of O atoms}$$

For the formation of one mole of $O_6(g)$:

$$\Delta H^\circ = -[6 \times D(\text{O–O})]$$
$$= -(6 \times 146)$$
$$= -876 \text{ kJ per 6 moles of O atoms}$$

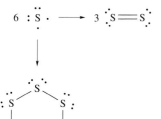

Hence, on enthalpy grounds $O_2$ molecule formation is favoured over $O_6$ ring formation. (Remember that $\Delta H$ is not the only factor involved – see Chapter 17.)

For sulfur, we compare the enthalpies of reaction for the following processes:

$$6S(g) \longrightarrow 3S_2(g)$$

$$6S(g) \longrightarrow S_6(g)$$

For the formation of three moles of $S_2(g)$:

$$\Delta H^\circ = -[3 \times D(S{=}S)]$$
$$= -(3 \times 425)$$
$$= -1275 \text{ kJ per 6 moles of S atoms}$$

For the formation of one mole of $S_6(g)$:

$$\Delta H^\circ = -[6 \times D(S{-}S)]$$
$$= -(6 \times 266)$$
$$= -1596 \text{ kJ per 6 moles of S atoms}$$

Hence, on enthalpy grounds $S_6$ ring formation is favoured over $S_2$ molecule formation. (Other cyclic molecules of type $S_n$ are also possible, as illustrated in Table 4.2.)

## SUMMARY

In this chapter the theme has been bonding in homonuclear molecules.

### Do you know what the following terms mean?

- covalent bond
- homonuclear
- covalent radius
- dissociation
- bond enthalpy
- enthalpy of atomization
- Lewis structure

- valence bond theory
- molecular orbital theory
- linear combination of atomic orbitals
- localized bond
- delocalized bond
- $\sigma$ and $\pi$ bonds

- bonding and antibonding molecular orbitals
- bond order
- diamagnetic
- paramagnetic
- radical cation
- radical anion

### You should now be able:

- to discuss briefly methods of determining bond distances
- to relate bond enthalpy values to enthalpies of atomization
- to distinguish between an individual and an average bond enthalpy value
- to use Hess's Law of Constant Heat Summation to estimate bond enthalpies from enthalpies of formation and atomization
- to draw Lewis structures for simple covalent molecules
- to discuss the principles of MO and VB theories

- to discuss the bonding in $H_2$ in terms of MO and VB theories and compare the two models
- to construct MO diagrams for homonuclear diatomic molecules of the $s$- and $p$-block elements, and make use of experimental data to assess whether your diagram is realistic
- to relate bond order *qualitatively* to bond enthalpy and bond distance for homonuclear bonds
- to discuss trends in homonuclear bond enthalpies and bond distances across the first row of elements in the periodic table and down groups of the $s$- and $p$-block elements

## PROBLEMS

**4.1** State what you understand by each of the following: (a) covalent radius, (b) bond dissociation enthalpy and (c) the standard enthalpy of atomization of an element.

**4.2** Values of van der Waals radii ($r_v$) and covalent radii ($r_{cov}$) for the elements from $Z = 5$ to 9 are:

| $Z$ | 5 | 6 | 7 | 8 | 9 |
|---|---|---|---|---|---|
| $r_v$ / pm | 208 | 185 | 154 | 140 | 135 |
| $r_{cov}$ / pm | 88 | 77 | 75 | 73 | 71 |

(a) In which part of the periodic table are these elements?
(b) Why is $r_v > r_{cov}$ for each element?
(c) Rationalize the trend in $r_v$ values.
(d) For S, $r_v = 185$ pm and $r_{cov} = 103$ pm. How would you estimate these values experimentally?

**4.3** Using data from Appendix 6, estimate bond lengths for the gas phase diatomics $F_2$, $Cl_2$, $Br_2$ and $I_2$. Compare your answers with the values listed in Table 4.10. Suggest reasons for the trend in bond lengths.

**4.4** (a) In an X-ray diffraction experiment, what causes the diffraction of the electromagnetic radiation? (b) Explain why X-ray diffraction results give a value of 103 pm for the length of a localized 2-centre 2-electron terminal B−H bond in the $[B_2H_7]^-$ ion (**4.25**) while the same bond is found to be 118 pm in length from the results of a neutron diffraction experiment.

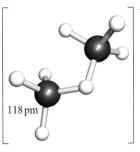

**118 pm**

(4.25)

**4.5** The results of an electron diffraction experiment on a gaseous molecule $XF_3$ give X–F distances of 131.3 pm and F---F separations of 227.4 pm. (a) What are the F–X–F bond angles? (b) Is the molecule planar or non-planar?

**4.6** State, with reasons, whether you can reasonably transfer a value of the stated bond dissociation enthalpy between the following pairs of molecules: (a) the carbon–carbon bond enthalpy between ethane ($C_2H_6$) and ethene ($C_2H_4$), (b) the nitrogen–nitrogen bond enthalpy between hydrazine ($N_2H_4$) and $N_2F_4$, and (c) the oxygen–hydrogen bond enthalpy between water and hydrogen peroxide ($H_2O_2$).

**4.7** Using the bond enthalpy data in Table 4.5, estimate values for the enthalpy of atomization for each of the following gas phase molecules: (a) $P_4$; (b) $S_8$; (c) $As_4$; (d) $I_2$. (The structures of these molecules are given in the chapter.)

**4.8** Using the answer obtained for $P_4(g)$ in question 4.7 and the standard enthalpy of atomization listed in Table 4.6, estimate the standard enthalpy of formation (at 298 K) of *gaseous* $P_4$.

**4.9** The standard enthalpies of formation (at 298 K) for methane ($CH_4$) and ethane ($C_2H_6$) are −75 and −85 kJ mol$^{-1}$ respectively. Determine values for the single C−C and C−H bond enthalpies at 298 K given that the standard enthalpies of atomization of carbon and hydrogen are 717 and 218 kJ mol$^{-1}$ respectively. Compare your answers with the values listed in Table 4.5 and comment on the origins of any differences.

**4.10** Using appropriate data from Appendices 10 and 11, determine the average P–F bond enthalpies in (a) $PF_5$ and (b) $PF_3$. What conclusions can you draw from your answers?

**4.11** (a) Estimate a value for $\bar{D}$(P–H) using appropriate data from Appendices 10 and 11. (b) Using your answer to part (a), estimate a value for $D$(P–P) in $P_2H_4$ which is structurally similar to $N_2H_4$ (Figure 4.1); $\Delta_f H^\circ(P_2H_4, g) = 21$ kJ mol$^{-1}$.

**4.12** By considering only valence electrons in each case, draw Lewis structures for: (a) $Cl_2$, (b) $Na_2$, (c) $S_2$, (d) $NF_3$, (e) HCl, (f) $BH_3$ and (g) $SF_2$.

**4.13** By considering only valence electrons in each case, draw Lewis structures for the following molecules: (a) methane ($CH_4$); (b) bromomethane ($CH_3Br$); (c) ethane ($C_2H_6$); (d) ethanol ($C_2H_5OH$); (e) ethene ($C_2H_4$).

**4.14** Draw Lewis structures for (a) $F_2$, (b) $Br_2$, (c) BrF, (d) $BrF_3$ and (e) $BrF_5$.

**4.15** How many Lewis structures can you draw for $C_2$? In the text, we considered only one structure; suggest why the other structure(s) that you have drawn is (are) unreasonable. [Hint: Think about the distribution of the bonding electrons in space.]

**4.16** (a) By using an MO approach to the bonding, show that the formation of the molecule $He_2$ would not be favourable. (b) Is $[He_2]^+$ likely to be a stable species?

**4.17** Construct a *complete* MO diagram for the formation of $Li_2$ showing the involvement of both the core and valence electrons of the two lithium atoms. Determine the bond order in $Li_2$. Does the inclusion of the core electrons affect the value of the bond order?

**4.18**  When $Li_2$ undergoes a 1-electron oxidation, what species is formed? Use the MO diagram you have drawn in problem 4.17 and the data about $Li_2$ in Table 4.9 to say what you can about the bond order, the bond dissociation enthalpy, the bond length and the magnetic properties of the oxidized product.

**4.19**  Using an MO approach, rationalize the trends in the data given in Table 4.11. Will the $[N_2]^-$ and $[N_2]^+$ ions be diamagnetic or paramagnetic?

**Table 4.11**  Data for problem 4.19.

| Diatomic species | Bond distance / pm | Bond enthalpy / kJ mol$^{-1}$ |
|---|---|---|
| $N_2$ | 110 | 945 |
| $[N_2]^-$ | 119 | 765 |
| $[N_2]^+$ | 112 | 841 |

**4.20**  Table 4.12 lists the bond dissociation enthalpies of diatomic molecules of the group 16 elements. Suggest reasons for the observed trend. Are the values of the bond distances in $O_2$ and $S_2$ consistent with the bond dissociation enthalpies for these molecules?

**Table 4.12**  Data for problem 4.20.

| Diatomic species | Bond distance / pm | Bond enthalpy / kJ mol$^{-1}$ |
|---|---|---|
| $O_2$ | 121 | 498 |
| $S_2$ | 189 | 425 |
| $Se_2$ | 217 | 333 |
| $Te_2$ | 256 | 260 |

**4.21**  The bond dissociation enthalpy of $Na_2$ is $74 \, kJ \, mol^{-1}$. Do you expect the bond enthalpy in $[Na_2]^+$ to be less than, equal to or greater than $74 \, kJ \, mol^{-1}$? Rationalize your answer.

## Additional problems

**4.22**  (a)  Use the LCAO method to construct a general MO diagram (without electrons) to show the interactions between the valence $2s$ and $2p$ atomic orbitals of two X atoms as they form $X_2$. Take the X nuclei to lie on the $z$-axis.

(b)  How does this MO diagram change if you allow mixing between the $\sigma(2s)$ and $\sigma(2p)$, and between the $\sigma^*(2s)$ and $\sigma^*(2p)$ MOs? Draw a revised MO diagram for $X_2$.

(c)  Use each diagram from parts (a) and (b) to deduce the bond order and magnetic properties of $B_2$. What experimental data are available to indicate which of the MO diagrams is the more appropriate to describe the bonding in $B_2$?

**4.23**  Comment on the following observations.

(a)  An O—O bond is weaker than an S—S bond, but O=O is stronger than S=S (see Table 4.5).

(b)  There is evidence for an unstable $Be_2$ species with a bond enthalpy of $10 \, kJ \, mol^{-1}$ and bond distance of 245 pm.

(c)  $P \equiv P$ bonds are weaker than $N \equiv N$ or $C \equiv C$, but stronger than $As \equiv As$.

(d)  The covalent radius for P appropriate for a triple bond is 94 pm, but in $P_4$, the P—P bond distances are 221 pm.

# 5 Heteronuclear diatomic molecules

## Topics

- First row heteronuclear diatomic molecules
- Valence bond theory
- Molecular orbital theory
- Bond enthalpies
- Electronegativity
- Dipole moments
- Isoelectronic molecules
- Multiple bonding

## 5.1 Introduction

In Chapter 4 we discussed *homonuclear* bonds such as those in homonuclear diatomic molecules ($H_2$ or $N_2$) or homonuclear polyatomic molecules ($P_4$ or $S_8$), or homonuclear bonds in *heteronuclear* molecules (the C–C bond in $C_2H_6$ or the O–O bond in $H_2O_2$). In this chapter we are concerned with heteronuclear molecules with an emphasis on diatomic species.

In a homonuclear diatomic molecule $X_2$, the two X atoms contribute equally to the bond. The molecular orbital diagrams drawn in Chapter 4 have a symmetrical appearance, and each molecular orbital contains equal contributions from the constituent atomic orbitals of each atom.

A *heteronuclear bond* is one formed between *different atoms*; examples are the C–H bonds in methane and ethane, the B–F bonds in boron trifluoride and the N–H bonds in ammonia (Figure 5.1). Since the bulk of 'chemistry' deals with molecules and ions containing more than one type of atom, it is clear that an understanding of the interactions and bonding between atoms of different types is crucial to an understanding of such species. Later in this book we

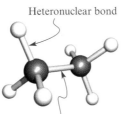

Heteronuclear bond

Homonuclear bond

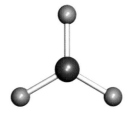

| $CH_4$ | $C_2H_6$ | $BF_3$ | $NH_3$ |

**Fig. 5.1** Molecules with heteronuclear bonds: methane ($CH_4$), ethane ($C_2H_6$), boron trifluoride ($BF_3$) and ammonia ($NH_3$). Colour code: C, grey; B, dark blue; N, blue; F, green; H, white.

shall encounter many heteronuclear *polyatomic* molecules, but we begin our discussion by considering the bonding in the diatomic molecules HF, LiF and LiH.

## 5.2     Lewis structures for HF, LiF and LiH

Lewis structures for hydrogen, lithium and fluorine atoms are shown in Figure 5.2a along with their ground state electronic configurations.

The elements hydrogen and lithium have one valence electron, while fluorine has seven. Just as we formed a covalent bond by sharing electrons between two hydrogen atoms to give the $H_2$ molecule, and between two fluorine atoms to give $F_2$, so we can pair together a hydrogen atom and a fluorine atom to give an HF molecule. Similarly, we can predict the formation of the molecules LiF and LiH and draw Lewis structures for these heterodiatomic molecules as shown in Figure 5.2b. This predicts that each molecule possesses a single bond.

In the gas phase, HF, LiF and LiH can each exist as molecular species. However, as we shall see later, at room temperature and pressure, the bonding is not always as simple as the Lewis approach suggests.

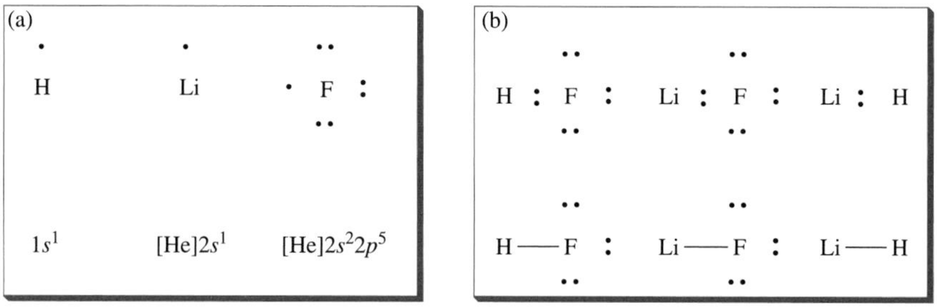

**Fig. 5.2** (a) Lewis diagrams and ground state electronic configurations for hydrogen, lithium and fluorine; (b) Lewis structures for hydrogen fluoride, lithium fluoride and lithium hydride.

## 5.3     The valence bond approach to the bonding in HF, LiF and LiH

### What differences are there between the homonuclear diatomic $H_2$ and a heteronuclear molecule XY?

In Chapter 4 we showed how valence bond theory can be applied to the $H_2$ molecule and how the bonding can be described by resonance structures **5.1**, **5.2** and **5.3**. The resonance forms **5.2** and **5.3** are equivalent in energy.

$$H\!-\!H \quad\longleftrightarrow\quad H_A^+ \quad H_B^- \quad\longleftrightarrow\quad H_A^- \quad H_B^+$$

$$(5.1) \qquad\qquad (5.2) \qquad\qquad (5.3)$$

▶
VB approach to $H_2$: see
Section 4.11

A similar approach can be used to describe the bonding in heteronuclear diatomic molecules. However, there is an important distinction to be made between homo- and heteronuclear diatomics. In a diatomic molecule XY, there will be a resonance structure **5.4** that represents the covalent contribution to the overall bonding picture and *two energetically different* resonance structures, **5.5** and **5.6**, that represent the ionic contributions.

$$X\!-\!Y \quad\longleftrightarrow\quad X^+ \quad Y^- \quad\longleftrightarrow\quad X^- \quad Y^+$$

$$(5.4) \qquad\qquad (5.5) \qquad\qquad (5.6)$$

The relative importance of the three resonance structures depends upon their relative energies, and each of structures **5.4**, **5.5** and **5.6** will make a different contribution to the overall bonding in the molecule XY. Let us apply this idea to HF, LiF and LiH.

### Hydrogen fluoride

The valence bond approach suggests three resonance forms for the hydrogen fluoride molecule, as shown in structures **5.7**, **5.8** and **5.9**.

$$\text{H} \longrightarrow \text{F} \quad \longleftrightarrow \quad \text{H}^+ \quad \text{F}^- \quad \longleftrightarrow \quad \text{H}^- \quad \text{F}^+$$
$$\text{(5.7)} \qquad\qquad \text{(5.8)} \qquad\qquad \text{(5.9)}$$

The covalent form **5.7** implies that the F and H atoms *share* two electrons so that each atom has a noble gas configuration. Forming a covalent single bond between the atoms in the HF molecule allows the hydrogen atom to gain a helium-like outer shell while the fluorine atom becomes neon-like.

In order to assess the relative energies of the ionic forms, consider the ground state configuration of the fluorine atom. Resonance structure **5.8** indicates that an electron has been transferred from the H atom to the F atom (equations 5.1 and 5.2).

$$\text{H} \longrightarrow \text{H}^+ + \text{e}^- \tag{5.1}$$
$$\text{F} + \text{e}^- \longrightarrow \text{F}^- \tag{5.2}$$

The ground state electron configuration of fluorine is $[\text{He}]2s^2 2p^5$ and gaining an electron provides a neon-like configuration $[\text{He}]2s^2 2p^6 \equiv [\text{Ne}]$. The $\text{F}^-$ ion obeys the octet rule. The ionic form **5.8** appears to be a reasonable alternative to the covalent structure **5.7**.

Resonance structure **5.9** indicates that an electron has been transferred from fluorine to hydrogen (equations 5.3 and 5.4).

$$\text{F} \longrightarrow \text{F}^+ + \text{e}^- \tag{5.3}$$
$$\text{H} + \text{e}^- \longrightarrow \text{H}^- \tag{5.4}$$

Although the formation of the $\text{H}^-$ ion gives a helium-like configuration ($1s^2 \equiv [\text{He}]$), the energetic *advantage* of this step is not sufficient to offset the energetic *disadvantage* of forming $\text{F}^+$. The loss of an electron from a fluorine atom to give the $\text{F}^+$ cation means that a species with six valence electrons (a sextet) is generated. This is expected to be *less* stable than the $\text{F}^-$ anion with an octet. The preference for fluorine to *gain* an electron is far, far greater than for it to lose one – reaction 5.2 is favoured over reaction 5.3. (At the end of Chapter 8, problem 8.6 asks you to consider the energetics of reactions 5.1 to 5.4.) Thus, it is not reasonable to expect a resonance structure involving $\text{F}^+$ to feature strongly in a description of the bonding in hydrogen fluoride, and the contribution made by the structure $\text{H}^-\text{F}^+$ is negligible. The bonding in hydrogen fluoride can therefore be effectively described in terms of a covalent structure and an ionic form in which $\text{H}^+$ and $\text{F}^-$ ions are involved.

### Lithium fluoride and lithium hydride

Valence bond theory can be used in a similar manner to that outlined for HF to describe the bonding in the LiF and LiH molecules. Resonance structures for these molecules are shown in Figure 5.3.

In molecular LiF, we can once again disregard the resonance structure involving the $\text{F}^+$ ion. The bonding in lithium fluoride is described in terms of a covalent resonance structure and an ionic resonance form involving $\text{F}^-$.

**Fig. 5.3** Resonance structures for the molecules (a) LiF and (b) LiH. For LiF, the right-hand structure provides a negligible contribution to the overall bonding picture.

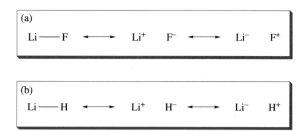

In LiH, three resonance structures can be drawn, all of which contribute to the overall bonding in the compound. The lithium and hydrogen atoms (Figure 5.2) possess $ns^1$ ground state configurations and can readily *lose* an electron to form the $Li^+$ or $H^+$ ions. Each of these ions possesses an $ns^0$ configuration. When hydrogen *gains* an electron, it forms $H^-$ which is a helium-like species.

## VB theory: conclusions

Valence bond theory applied to the molecules HF, LiF and LiH gives bonding descriptions in terms of covalent and ionic resonance structures. However, in a given set of resonance structures, some forms are more important than others. In general, a qualitative treatment in terms of the stability of possible ions proves to be successful, and the method provides a good intuitive description of the bonding in simple molecules. Of course, it *is* possible to approach the problem more rigorously and estimate the contributions that each resonance structure makes to the overall bonding. For us, however, it is sufficient merely to consider their relative merits in the way discussed above.

**5.4**

## The molecular orbital approach to the bonding in a heteronuclear diatomic molecule

In Chapter 4 we used molecular orbital theory to describe the bonding in homonuclear diatomic molecules. The MO diagrams we constructed had a symmetrical appearance; atomic orbitals of the same type but belonging to different atoms had the same energy as seen, for example, for $F_2$ in Figure 4.21. In a molecule $X_2$, the atomic orbitals of each X atom contribute *equally* to a given MO.

Now consider a heteronuclear molecule XY. In this chapter we take the case where the atoms X and Y are different, but with atomic numbers smaller than 10. A fundamental difference between X and Y is that the effective nuclear charge, $Z_{eff}$, experienced by the electrons in the valence shells of the two atoms is not the same. If Y is in a *later* group of the periodic table than X, then $Z_{eff}(Y) > Z_{eff}(X)$ for an electron in a given atomic orbital. As $Z_{eff}$ increases across a row of the periodic table, the energy of a particular atomic orbital is lowered (more negative energy). This is illustrated schematically in Figure 5.4 for the combination of the $2s$ atomic orbitals of atoms X and Y to form the $\sigma(2s)$ and $\sigma^*(2s)$ MOs of the molecule XY.

An important feature of Figure 5.4 is that the energy of the bonding MO is closer to that of the $2s$ atomic orbital of atom Y than to that of the $2s$ atomic orbital of atom X. Conversely, the energy of the antibonding MO is closer to that of the $2s$ atomic orbital of atom X than to that of the $2s$ atomic orbital of atom Y. This has an effect on the nature of the molecular orbitals, as

Effective nuclear charge: see Section 3.15

**Fig. 5.4** Schematic representation of an MO diagram for the combination of the 2s atomic orbitals of atoms X and Y in a heteronuclear diatomic molecule; $Z_{eff}(Y) > Z_{eff}(X)$.

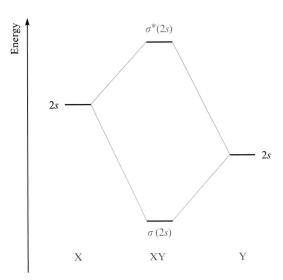

illustrated in Figure 5.5. The 2s atomic orbital of atom Y makes a *larger* contribution to the bonding $\sigma(2s)$ orbital than does the 2s atomic orbital of atom X. For the antibonding MO, the situation is reversed and the 2s atomic orbital of atom Y makes a *smaller* contribution to the $\sigma^*(2s)$ MO than does the atomic orbital of atom X. We say that the $\sigma(2s)$ MO possesses more 2s *character* from Y and the $\sigma^*(2s)$ MO has more 2s *character* from X. In Figures 5.5a and 5.5b, the larger or smaller 2s contributions are represented pictorially. Notice how this affects the final shape of the molecular orbitals and makes them asymmetrical.

The wavefunction for an MO in a heteronuclear diatomic molecule must show the contribution that each constituent atomic orbital makes. This is done by using *coefficients*, $c_n$. In equation 5.5, the coefficients $c_1$ and $c_2$ indicate the composition of the MO in terms of the atomic orbitals $\psi_X$ and $\psi_Y$. The relative values of $c_1$ and $c_2$ are dictated by the effective nuclear charges of the atoms X and Y.

$$\psi_{MO} = \{c_1 \times \psi_X\} + \{c_2 \times \psi_Y\} \tag{5.5}$$

Equation 5.6 gives an expression for the wavefunction that describes the bonding orbital in Figures 5.4 and 5.5.

$$\psi_{MO} = \{c_1 \times \psi_{X(2s)}\} + \{c_2 \times \psi_{Y(2s)}\} \qquad \text{where } c_2 > c_1 \tag{5.6}$$

The fact that coefficient $c_2$ is greater than $c_1$ corresponds to there being more 2s character from Y than X, with the result that the MO (Figure 5.5a) is distorted in favour of atom Y.

One final point remains. In Chapter 4 we noted the importance of *energy matching* between atomic orbitals as a criterion for an efficient orbital interaction. In the case of the homonuclear diatomic molecules, this gave no

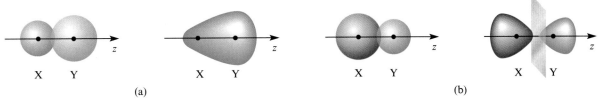

(a)                                                                (b)

**Fig. 5.5** Schematic representations of (a) the $\sigma(2s)$ MO and (b) the $\sigma^*(2s)$ MO of a heteronuclear diatomic molecule XY. The nuclei are shown by black dots.

problems when the discussion was restricted to overlap between like orbitals – in $H_2$, the energies of the two hydrogen $1s$ atomic orbitals were the same. In a heteronuclear diatomic there will always be differences between the orbital energies associated with the two different atoms, and it follows that there may be cases where the energy difference may be too great for effective interaction to be achieved. We illustrate this point later in the discussion of the bonding in LiF and HF.

---

**Worked example 5.1**    *Orbital contributions to molecular orbitals*

**As part of an MO description of the bonding in CO, we consider the interaction between the C $2s$ and O $2s$ orbitals. (a) Which of these atomic orbitals is of lower energy? (b) Will the $\sigma(2s)$ MO have more carbon or more oxygen character? Write an equation that describes the character of the $\sigma(2s)$ MO. (c) Give a qualitative description of the $\sigma^*(2s)$ MO.**

Begin by noting the positions of C and O in the periodic table. Both elements are in the first long period; the $2s$ and $2p$ orbitals are the valence orbitals of C and O. C is in group 14; O is in group 16. Therefore the effective nuclear charge of O is greater than that of C.

(a) Because $Z_{eff}(O) > Z_{eff}(C)$, the energy of the O $2s$ orbital is lower than that of the C $2s$ orbital.

(b) It follows from the answer to part (a) that the $\sigma(2s)$ MO has more O than C character. This can be represented by the equation:

$$\psi_{MO} = \{c_1 \times \psi_{C(2s)}\} + \{c_2 \times \psi_{O(2s)}\} \qquad \text{where } c_2 > c_1$$

(c) The $\sigma^*(2s)$ MO of the CO molecule contains more carbon than oxygen character. The orbital has antibonding character and may be represented pictorially as follows:

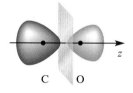

C    O

---

**5.5**    **The molecular orbital approach to the bonding in LiH, LiF and HF**

### Three important preliminary comments

- In using molecular orbital theory to describe the bonding in molecules at a simple level, we need only consider the overlap of the *valence atomic orbitals*.
- The MO treatment of the bonding in LiH, LiF and HF described in this section is at a simple level; we assume that of the lithium $2s$ and $2p$ valence orbitals, only the $2s$ orbital makes significant contributions to the bonding and antibonding MOs. End-of-chapter problem 5.19 looks at the consequences of adding lithium $2p$ orbitals to the valence orbital set.
- Molecular orbital theory describes the bonding in a *covalent* manner. It does not explicitly build into the picture ionic contributions as is done in valence bond theory.

**Fig. 5.6** An MO diagram for the formation of the LiH molecule. Only the valence atomic orbitals and electrons are shown.

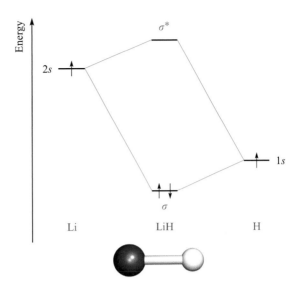

**Fig. 5.6** An MO diagram for the formation of the LiH molecule. Only the valence atomic orbitals and electrons are shown.

### Lithium hydride

The ground state configurations of H and Li are $1s^1$ and $[He]2s^1$ respectively. The H $1s$ orbital is much lower in energy than the Li $2s$ atomic orbital. The symmetries of the $1s$ atomic orbital of H and the $2s$ atomic orbital of Li are compatible and overlap can occur. The combination of these atomic orbitals to give two molecular orbitals is shown in Figure 5.6.

There are two valence electrons and, by the *aufbau* principle, these will have opposite spins and occupy the $\sigma$-bonding MO. The bond order in the LiH molecule is 1 (equation 5.7).

$$\text{Bond order in LiH} = \tfrac{1}{2} \times (2 - 0) = 1 \qquad (5.7)$$

◼▶
Bond order:
see Section 4.16

Figure 5.6 shows that the H $1s$ atomic orbital contributes more to the $\sigma$-bonding MO in LiH than does the Li $2s$ atomic orbital. This means that the $\sigma$-bonding MO exhibits more H than Li character; the orbital has approximately 74% hydrogen character and 26% lithium character, and this is represented pictorially in Figure 5.7. As a result, the electrons spend a greater amount of their time 'associated' with the H than with the Li nucleus.

The wavefunction given in equation 5.8 describes the $\sigma$-bonding MO in LiH.

$$\psi(\text{LiH bonding}) = \{c_1 \times \psi_{\text{Li}(2s)}\} + \{c_2 \times \psi_{\text{H}(1s)}\} \qquad \text{where } c_1 < c_2 \qquad (5.8)$$

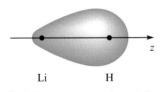

**Fig. 5.7** A representation of the $\sigma$-bonding MO of LiH.

It follows from Figure 5.6 that the $\sigma^*$ MO has more Li than H character. [*Exercise*: Draw an appropriate representation of the $\sigma^*$ MO in LiH.]

### Hydrogen fluoride

The ground state configurations of H and F are $1s^1$ and $[He]2s^2 2p^5$ respectively. Before we consider the overlap of the atomic orbitals, we must think about their relative energies. In going from H to F, the effective nuclear charge increases considerably, and this significantly lowers the energies of the F atomic orbitals with respect to the H $1s$ atomic orbital. Thus, despite the difference in principal quantum number, the F $2s$ and $2p$ orbitals lie at lower energies than the H $1s$ atomic orbital. This is shown in Figure 5.8; the energy axis in the figure is broken in order to signify the fact that the F $2s$ orbital is actually even lower in energy than shown.

**Fig. 5.8** An MO diagram for the formation of HF. Only the valence atomic orbitals and electrons are shown. The break in the vertical (energy) axis signifies that the energy of the fluorine 2s atomic orbital is much lower than is actually shown.

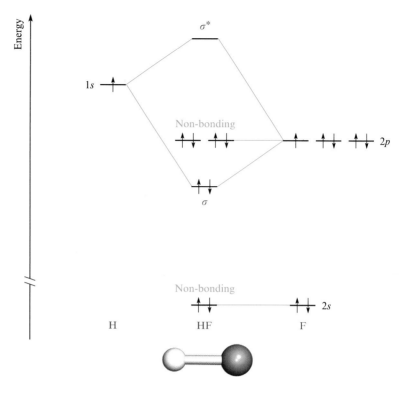

Remember that for efficient overlap, the difference in energy between two atomic orbitals must *not* be too great. We *could* allow the H 1s and the F 2s orbitals to interact with one another just as we did in LiH, but although overlap *is allowed by symmetry*, the energy separation is too large. Thus, overlap is not favourable and this leaves the F 2s atomic orbital unused – it becomes a *non-bonding orbital* in the HF molecule.

A *non-bonding molecular orbital* is one that has neither bonding nor antibonding character overall.

The H and F nuclei are defined to lie on the z-axis, and so the H 1s and F $2p_z$ atomic orbitals can overlap as illustrated in Figure 5.9a. Figure 5.8 shows that there is a satisfactory energy match between these atomic orbitals and overlap occurs to give $\sigma$ and $\sigma^*$ MOs. The contribution made by the fluorine $2p_z$ atomic orbital to the $\sigma$-bonding MO is greater than that made by the hydrogen 1s orbital, and so the bonding orbital contains more fluorine than hydrogen character.

The two remaining F 2p atomic orbitals are orthogonal to the H−F axis and have no net bonding overlap with the H 1s orbital. Figure 5.9b illustrates

**Fig. 5.9** (a) The symmetry-allowed overlap of the H 1s atomic orbital and the F $2p_z$ atomic orbital; note that the H and F nuclei lie on the z-axis. (b) The combination of the H 1s atomic orbital and the F $2p_x$ atomic orbital results in equal in-phase and out-of-phase interactions, and the overall interaction is non-bonding. A similar situation arises for the $2p_y$ orbital.

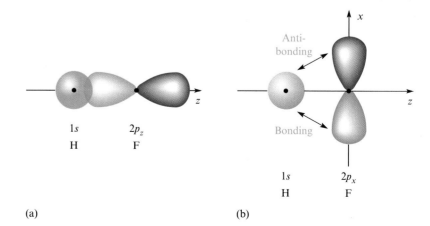

(a)　　　　　　　　　(b)

this for the $2p_x$ atomic orbital. Thus, the F $2p_x$ and $2p_y$ atomic orbitals become non-bonding orbitals in the HF molecule.

There are eight valence electrons in the HF molecule and, by the *aufbau* principle, they occupy the MOs as shown in Figure 5.8. Only two electrons occupy a molecular orbital which has H−F bonding character; the other six are in non-bonding MOs and are localized on the F atom. These three non-bonding pairs of electrons are analogous to the lone pairs drawn in the Lewis structure for the HF molecule in Figure 5.2.

The bond order in hydrogen fluoride (equation 5.9) can be determined in the usual way since electrons in non-bonding MOs have no effect on the bonding:

$$\text{Bond order in HF} = \tfrac{1}{2} \times (2 - 0) = 1 \tag{5.9}$$

> Electrons in **non-bonding** MOs do not influence the bond order.

We can summarize the MO approach to the bonding in the HF molecule as follows, assuming that the H and F nuclei lie on the $z$-axis:

- Overlap occurs between the hydrogen $1s$ and the fluorine $2p_z$ atomic orbitals to give one $\sigma$-bonding MO and one $\sigma^*$-antibonding MO.
- The fluorine $2p_x$ and $2p_y$ orbitals have no net overlap with the hydrogen $1s$ atomic orbital and are non-bonding orbitals.
- The fluorine $2s$ orbital can overlap with the hydrogen $1s$ atomic orbital on symmetry grounds but the mismatch in atomic orbital energies means that the overlap is poor – the fluorine $2s$ orbital is effectively non-bonding.
- Of the eight valence electrons in the HF molecule, two occupy an H−F bonding MO and six occupy non-bonding orbitals localized on the fluorine atom.
- The $\sigma$-bonding MO in HF contains more fluorine than hydrogen character.

### Lithium fluoride

The MO approach to the bonding in LiF closely follows that described for HF. The major difference is that we are dealing with an $[\text{He}]2s^1$ ground state configuration for Li in place of the $1s^1$ configuration for H.

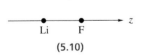

(5.10)

Figure 5.10 gives an MO diagram for lithium fluoride; the Li and F nuclei lie on the $z$-axis, as shown in structure **5.10**. As in HF, the fluorine $2s$ atomic orbital lies at low energy and generates a non-bonding orbital in the LiF molecule. The fluorine $2p_x$ and $2p_y$ atomic orbitals are also non-bonding as they were in HF.

Overlap can occur between the Li $2s$ atomic orbital and the F $2p_z$ atomic orbital. However, the energy separation between these orbitals is significant and the interaction is not particularly effective. This is apparent in Figure 5.10 – the $\sigma$-bonding MO is stabilized *only slightly* with respect to the fluorine $2p_z$ atomic orbital. A comparison of Figures 5.8 and 5.10 shows that the stabilization of the $\sigma$-bonding orbital is smaller in LiF than in HF. What does this mean in terms of the LiF molecule?

The wavefunction that describes the bonding MO in Figure 5.10 is given in equation 5.10.

$$\psi(\text{LiF bonding}) = \{c_1 \times \psi_{\text{Li}(2s)}\} + \{c_2 \times \psi_{\text{F}(2p_z)}\} \qquad \text{where } c_2 \gg c_1 \tag{5.10}$$

If the coefficient $c_1$ is very small, then the wavefunction describes an orbital that is weighted very much in favour of the fluorine atom – in equation 5.10, the term $\{c_2 \times \psi_{\text{F}(2p_z)}\}$ is *much* greater than $\{c_1 \times \psi_{\text{Li}(2s)}\}$. This means that in

**Fig. 5.10** An MO diagram for the formation of the LiF molecule. Only the valence atomic orbitals and electrons are shown. The break in the vertical (energy) axis signifies that the energy of the fluorine 2s atomic orbital is lower than is actually shown.

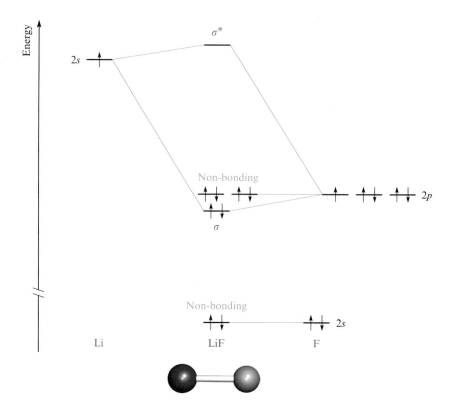

the bonding MO, there is a much greater probability of finding the electrons in the vicinity of F, rather than Li.

In conclusion, molecular orbital theory gives a picture for the bonding in LiF that shows one MO with bonding character; this MO is occupied by two electrons and the bond order in the molecule is 1. There are three filled, nonbonding orbitals which are localized on the F atom; these are analogous to the lone pairs of electrons drawn in the Lewis structure of LiF (Figure 5.2). The bonding MO in the LiF diatomic molecule has predominantly fluorine character ($\approx 97\%$). Thus, although the MO approach is built upon a covalent model, the 'answer' is telling us that the bonding in LiF may be described as *tending towards being ionic*. We return to LiF in Section 8.2.

## MO theory: conclusions

- The MO approach to the bonding in the LiH, HF and LiF molecules shows each possesses a bond order of 1. In each molecule, the bonding MO is a $\sigma$ orbital.
- The bonding MO in LiH is formed by the overlap of the Li 2s and the H 1s atomic orbitals. In HF, the bonding MO is formed by the overlap of the H 1s and one of the F 2p orbitals. In LiF, the character of the bonding MO comes from the Li 2s and one of the F 2p orbitals.
- The use of different atomic orbitals on different nuclei can lead to dramatic differences in the percentage contributions that the atomic orbitals make to the bonding molecular orbitals. In LiH the bonding MO contains more hydrogen than lithium character. In HF, the bonding MO contains more fluorine than hydrogen character. In LiF, the bonding MO possesses almost entirely fluorine character.

Once we know the character of the single bonding MO in each of LiH, HF and LiF, we can say something about the electron distribution in the bonds in these molecules. In contrast to the symmetrical distribution of electron density between the nuclei in a homonuclear diatomic molecule, in LiH, HF and LiF it is shifted in favour of one of the atoms. In LiH, the maximum electron density is closer to the H than the Li nucleus whereas in HF it is closer to the F nucleus. Attempts to rationalize these results lead us eventually to the concept of electronegativity and to polar bonds, but first we look at the bond enthalpies of heteronuclear bonds.

**5.6**

## Bond enthalpies of heteronuclear bonds

$\bar{D}$ = average bond enthalpy term

**(5.11)**

**(5.12)**

In Section 4.7, we discussed the way in which Hess's Law of Constant Heat Summation could be used to estimate values for bond enthalpies. We were primarily concerned with estimating the bond enthalpies of *homonuclear* single bonds, but to do so, it was necessary to involve the bond enthalpies of heteronuclear bonds. An estimate for $D(N-N)$ in $N_2H_4$ depended upon transferring a value of $\bar{D}(N-H)$ from $NH_3$ to $N_2H_4$.

Thus, we are already familiar with the idea that average bond enthalpy values for heteronuclear bonds can be obtained from standard enthalpies of atomization. This is shown for $CCl_4$ (**5.11**), and 1,2-$C_2H_4Br_2$ (**5.12**) in equations 5.11 and 5.12.

$$\Delta_a H^\circ(CCl_4, g) = 4 \times \bar{D}(C-Cl) \tag{5.11}$$

$$\Delta_a H^\circ(C_2H_4Br_2, g) = [4 \times \bar{D}(C-H)] + D(C-C) + [2 \times \bar{D}(C-Br)] \tag{5.12}$$

In Section 4.6, we emphasized the relationship between the bond dissociation enthalpy, $D$, and the standard enthalpy of atomization for a gaseous homonuclear diatomic molecule. This is illustrated for fluorine in equation 5.13.

$$\Delta_a H^\circ(F, g) = \tfrac{1}{2} \times D(F-F) \tag{5.13}$$

Is it feasible to use a *similar* strategy to estimate the bond enthalpies of heteronuclear bonds? For an $X-Y$ single bond, can we estimate values that are 'characteristic' of the contributions made to $D(X-Y)$ by atoms X and Y? A logical approach is to take half of $D(X-X)$ and allow this to be a measure of the contribution made to the bond enthalpy $D(X-Y)$ by atom X. Similarly for Y, let half of $D(Y-Y)$ be a measure of the contribution made to $D(X-Y)$ by atom Y. Estimates for values of $D(X-Y)$ can then be made using equation 5.14.

$$D(X-Y) = [\tfrac{1}{2} \times D(X-X)] + [\tfrac{1}{2} \times D(Y-Y)]$$

or $\tag{5.14}$

$$D(X-Y) = \tfrac{1}{2} \times [D(X-X) + D(Y-Y)]$$

*But*, are we justified in applying an additive rule of this type? This is best answered by comparing some estimates made using equation 5.14 with experimentally determined bond dissociation enthalpies (Table 5.1). For some of the molecules (IBr, BrCl and HI) the model we have used gives good estimates of the bond dissociation enthalpies, but for others (HF and HCl) the discrepancies are large. Why does the model fail in some cases?

The breakdown of this simple model (the non-additivity of bond dissociation enthalpy contributions for heteronuclear bonds) led Linus Pauling (see Box 5.1) to outline the concept of *electronegativity*.

**Table 5.1** Bond dissociation enthalpies, $D$, of homonuclear and heteronuclear diatomic molecules. Experimental values for $D(X-X)$ and $D(Y-Y)$ are used to estimate values of $D(X-Y)$ using equation 5.14. Comparison is then made with experimental data. The value $\Delta D$ in the final column is the difference between the experimental $D(X-Y)$ and the estimated value, i.e. the difference between the value in column 6 in the table and the value in column 5 (see equation 5.15).

| X | Y | $D$ for X–X bond in $X_2$ / kJ mol$^{-1}$ | $D$ for Y–Y bond in $Y_2$ / kJ mol$^{-1}$ | $\frac{1}{2} \times [D(X-X) + D(Y-Y)]$ / kJ mol$^{-1}$ | Experimental $D(X-Y)$ in XY / kJ mol$^{-1}$ | $\Delta D$ (from equation 5.15) |
|---|---|---|---|---|---|---|
| H | F | 436 | 159 | 298 | 570 | 272 |
| H | Cl | 436 | 242 | 339 | 432 | 93 |
| H | Br | 436 | 193 | 315 | 366 | 51 |
| H | I | 436 | 151 | 294 | 298 | 4 |
| F | Cl | 159 | 242 | 201 | 256 | 55 |
| Cl | Br | 242 | 193 | 218 | 218 | 0 |
| Cl | I | 242 | 151 | 197 | 211 | 14 |
| Br | I | 193 | 151 | 172 | 179 | 7 |

---

## HISTORY

### Box 5.1 Linus Carl Pauling (1901–1994)

Linus Carl Pauling (1901–1994). © *The Nobel Foundation.*

You probably associate the name Pauling with electronegativity values but his contributions to chemistry and to science more generally *far* exceed $\chi^P$. His imagination and genius have left a permanent mark on the underlying theories of modern chemistry.

Linus Carl Pauling was born in Oregon in 1901 and received a degree in chemical engineering in 1922. From there, his interests focused on X-ray diffraction methods and the solving of crystal structures. The year 1926 saw Pauling working with Schrödinger and Bohr, trying to get to grips with quantum theory and the structure of the atom. In the 1920s, the structures of silicate minerals were beginning to unfold, but their complexity brought with it challenges to structural perception. In 1929, Pauling set out rules based on ionic radii and charges which helped to rationalize the silicate family. His preoccupation at this time with the chemical bond brought sense to many aspects of bonding that had previously not been understood, and he combined quantum theory with the more qualitative theories of, for example, G. N. Lewis (see Box 4.5). Pauling was also the advocate of resonance structures, realizing that chemical properties could be best understood in terms of a combination of various structural forms. The establishment of Pauling's valence bond theory provided chemists with a new foundation on which to build their chemical models and, in the late 1930s, Pauling's book *The Nature of the Chemical Bond* opened the eyes of many to the world of chemical bonding and structure. This book is still one of the masterpieces of the chemical literature.

Biochemistry has also benefited from Pauling's gift of imagination and it was he who first considered the $\alpha$-helix as a motif in protein structure, and his ideas were confirmed through X-ray diffraction experiments (see Box 35.5). In 1954, for his contributions to chemistry, Pauling received the first of two Nobel prizes – the second was awarded in 1963 for his involvement in world peace efforts. He was an adamant anti-nuclear campaigner and was, for a time, at loggerheads with the US government. From 1966 onwards, Pauling turned his attention to vitamin C and for many of his later years he dedicated himself to it. He was convinced that vitamin C was the cure for all ills including cancer and he took huge doses daily.

## Electronegativity – Pauling values ($\chi^P$)

### Deriving Pauling electronegativity values

When we compare the estimated and experimental values of $D(X-Y)$ listed in Table 5.1, we see that the estimated value is often too small. In 1932, Pauling suggested that these differences (equation 5.15) could be rationalized in terms of his valence bond model. Specifically, he was concerned with the discrepancy, $\Delta D$ (Table 5.1), between the experimentally determined value for the X−Y bond and the average of the values for $X_2$ and $Y_2$.

$$\Delta D = [D(X-Y)_{\text{experimental}}] - \{\tfrac{1}{2} \times [D(X-X) + D(Y-Y)]\} \tag{5.15}$$

Pauling proposed that $\Delta D$ was a measure of the 'ionic' character of the X−Y bond. In valence bond terms, this corresponds to a measure of the importance of resonance structures $X^+Y^-$ and $X^-Y^+$ in the description of the bonding in molecule XY. How much more important are $X^+Y^-$ and $X^-Y^+$ in the bonding picture of XY than are $X^+X^-$ and $Y^-Y^+$ in $X_2$ and $Y_2$ respectively? Pauling then suggested that the resonance forms $X^+Y^-$ and $X^-Y^+$ were not equally important. (We have justified this suggestion for HF in Section 5.3.)

Let us assume that the form $X^+Y^-$ predominates. Remember, this does *not* imply that the molecule XY actually exists as the pair of ions $X^+$ and $Y^-$. All we are saying is that the electrons in the X−Y bond are drawn more towards atom Y than they are towards atom X. That is, in the XY molecule, atom Y has a greater *electron-withdrawing power* than atom X.

Pauling then defined a term that he called *electronegativity* $(\chi^P)^{\S}$ which he described as 'the power of an atom *in a molecule* to attract electrons to itself'. This approach treats Pauling electronegativity as an *atomic* property rather than a property of a bond. It is important to remember that electronegativity is a property of an *atom in a molecule*, and is not a fundamental property of an isolated atom.

*Electronegativity*, $\chi^P$, was defined by Pauling as 'the power of an atom in a molecule to attract electrons to itself'.

Although we approached the idea of electronegativity by considering the energy involved in the formation of the diatomic XY, it is clear that this concept is equally valid when considering X−Y bonds in larger molecules.

Pauling values of electronegativity (Table 5.2) are calculated as follows. The difference $\Delta D$ (equation 5.15) is usually measured in units of kJ mol$^{-1}$. Values of $\chi^P$ relate to thermochemical data given in the non-SI units of electron volts (eV). The value $\Delta D$ was used to estimate the difference in electronegativity values between atoms X and Y (equation 5.16) where $\Delta D$ is found from equation 5.15 and is given in eV.

▬▶
1 eV = 96.485 ≈ 96.5 kJ mol$^{-1}$

$$\chi^P(Y) - \chi^P(X) = \sqrt{\Delta D} \tag{5.16}$$

By solving equation 5.16 for a range of different combinations of atoms X and Y, Pauling estimated a self-consistent set of electronegativity values. Pauling worked with a set of numbers that gave $\chi^P(\text{H}) \approx 2$ and $\chi^P(\text{F}) = 4$. The larger the value of $\chi^P$, the greater the electron-attracting (withdrawing) power of the atom. Fluorine is the most electronegative atom in Table 5.2.

---

$\S$　The superscript in the symbol $\chi^P$ distinguishes Pauling (P) electronegativity values from those of other scales.

**Table 5.2** Pauling electronegativity values ($\chi^P$) for selected elements of the periodic table. For some elements that exhibit a variable oxidation state, $\chi^P$ is given for a specific state. It is conventional (but strictly incorrect) not to assign units to values of electronegativity.

| Group 1 | Group 2 | | Group 13 | Group 14 | Group 15 | Group 16 | Group 17 |
|---|---|---|---|---|---|---|---|
| H 2.2 | | | | | | | |
| Li 1.0 | Be 1.6 | | B 2.0 | C 2.6 | N 3.0 | O 3.4 | F 4.0 |
| Na 0.9 | Mg 1.3 | | Al(III) 1.6 | Si 1.9 | P 2.2 | S 2.6 | Cl 3.2 |
| K 0.8 | Ca 1.0 | (d-block elements) | Ga(III) 1.8 | Ge(IV) 2.0 | As(III) 2.2 | Se 2.6 | Br 3.0 |
| Rb 0.8 | Sr 0.9 | | In(III) 1.8 | Sn(II) 1.8 Sn(IV) 2.0 | Sb 2.1 | Te 2.1 | I 2.7 |
| Cs 0.8 | Ba 0.9 | | Tl(I) 1.6 Tl(III) 2.0 | Pb(II) 1.9 Pb(IV) 2.3 | Bi 2.0 | Po 2.0 | At 2.2 |

Over the years since Pauling introduced this concept, more accurate thermochemical data have become available and values of $\chi^P$ have been updated. Table 5.2 lists values of $\chi^P$ in current use. In the same way that thermochemical data for X–X and Y–Y bonds in the molecules $X_2$ and $Y_2$ may be used, so may values for X–X and Y–Y *bonds in molecules* (e.g. the C–C bond in $C_2H_6$ or the O–O bond in $H_2O_2$) be used to obtain values of $\chi^P$. This last extension allows us to define electronegativity values for elements in various oxidation states.

---

**Worked example 5.2**   *Estimating the bond enthalpy term of a heteronuclear covalent bond*

**Estimate the dissociation enthalpy of the bond in iodine monochloride, ICl, given that:**

$$D(\text{Cl–Cl}) = 242 \text{ kJ mol}^{-1}$$

$$D(\text{I–I}) = 151 \text{ kJ mol}^{-1}$$

$$\chi^P(\text{Cl}) = 3.2$$

$$\chi^P(\text{I}) = 2.7$$

The equations needed are 5.15 and 5.16:

$$\Delta D = D(\text{X–Y}) - \{\tfrac{1}{2} \times [D(\text{X–X}) + D(\text{Y–Y})]\}$$

$$\chi^P(\text{Y}) - \chi^P(\text{X}) = \sqrt{\Delta D}$$

The second equation uses values of $D$ in eV.

First, find $\Delta D$:

$$\sqrt{\Delta D} = \chi^P(\text{Cl}) - \chi^P(\text{I})$$
$$= 3.2 - 2.7$$
$$= 0.5$$
$$\Delta D = (0.5)^2$$
$$= 0.25\,\text{eV}$$

Convert to $\text{kJ}\,\text{mol}^{-1}$:

$$\Delta D = 0.25 \times 96.5$$
$$= 24.1\,\text{kJ}\,\text{mol}^{-1}$$

Now, find $D(\text{I–Cl})$:

$$\Delta D = D(\text{X–Y}) - \{\tfrac{1}{2} \times [D(\text{X–X}) + D(\text{Y–Y})]\}$$
$$24.1 = D(\text{I–Cl}) - \{\tfrac{1}{2} \times [D(\text{I–I}) + D(\text{Cl–Cl})]\}$$
$$D(\text{I–Cl}) = 24.1 + \{\tfrac{1}{2} \times (151 + 242)\}$$
$$= 221\,\text{kJ}\,\text{mol}^{-1}$$

---

## 5.8    The dependence of electronegativity on oxidation state and bond order

### Oxidation state

Some elements exhibit more than one oxidation state, and electronegativity values are dependent on the oxidation state. In Table 5.2, some of the values of $\chi^P$ are given, not just for the element, but for the element in a defined oxidation state, e.g. $\chi^P$ Sn(II) = 1.8 and Sn(IV) = 2.0.

Consider the group 14 element, lead. This is a metal and may use all of its valence electrons to form lead(IV) compounds such as $PbCl_4$. On the other hand, it often uses only two valence electrons to form lead(II) compounds such as $PbCl_2$ and $PbI_2$. The electron-withdrawing power of a lead(IV) centre is greater than that of a lead(II) centre and this is apparent in the electronegativity values.

In general, *electronegativity values for a given element increase as the oxidation state becomes more positive*.

◀▶ Oxidation states: see Section 1.16

### Bond order

The electronegativity of a carbon atom involved in covalent single bonds is not the same as that of a carbon atom involved in multiple bonding.

(5.13)        (5.14)        (5.15)

In the hydrocarbons $C_2H_2$ (**5.13**), $C_2H_4$ (**5.14**) and $C_2H_6$ (**5.15**), the carbon–carbon bond strength varies and so one expects different $\chi^P$

values. On the other hand, Table 5.2 lists only a single value for $\chi^P(C)$! The differences between such electronegativity values are often quite small, but the discrepancy can be sufficient to give some spurious results if the values are used to estimate how electron density is distributed between two atoms. We return to this problem in Section 5.13 and Table 26.1.

| 5.9 | **An overview of the bonding in HF** |

In this section, we review the ideas developed so far in this chapter by considering the bonding in hydrogen fluoride, combining the ideas of the Lewis approach, VB and MO theories, bond enthalpies and electronegativities.

- The resonance structure $H^+F^-$ contributes to the bonding significantly more than does the form $H^-F^+$.
- The MO approach to the bonding in HF indicates that the character of the $\sigma$-bonding MO is weighted in favour of the fluorine atom.
- In both VB and MO theories, the greater electron-withdrawing power of the fluorine atom means that the distribution of electron density along the H$-$F bond-vector is unequal and is significantly displaced towards the fluorine atom.
- An estimate of the bond dissociation enthalpy of the H$-$F bond in HF by a method of simple addition proves to be unsatisfactory (Table 5.1).
- Pauling's concept of electronegativity allows the estimated value of $D(H-F)$ to be corrected for the added strength resulting from a degree of ionic character in the bond. The electronegativity difference between F and H is 1.8 (Table 5.2) and thus, by combining equations 5.15 and 5.16, a corrected value for $D(H-F)$ of $610 \, kJ \, mol^{-1}$ is obtained. This compares with an experimental value of $570 \, kJ \, mol^{-1}$. Typically, bond enthalpies calculated in this way are within $\pm 10\%$ of the experimental values.

| 5.10 | **Other electronegativity scales** |

Although Pauling's electronegativity scale works well, it has a number of problems.

Values of $\chi^P$ are arrived at in an empirical manner. Why should we set the difference in electronegative values between two atoms to the *square root* of the difference in bond dissociation enthalpies? If $\chi^P$ is a property of atoms in molecules, what controls it? How do Pauling electronegativity values relate to electron distributions in molecules?

In an attempt to answer these questions, various other electronegativity scales have been devised based on a variety of atomic properties. We consider two such scales.

### Mulliken electronegativity values ($\chi^M$)

Consider a diatomic molecule XY. The average position of the electron density in the bond depends upon the relative electron-withdrawing powers of the atoms X and Y. There will be a *tendency* for each atom to possess *either* a positive *or* a negative charge. This idea leads to Mulliken's method of estimating electronegativity values. Mulliken suggested that an estimate of the electronegativity of an atom could be obtained from atomic properties relating to the formation of cations and anions. The appropriate properties are the first ionization energy and the first electron affinity of an element. We deal with these quantities in Chapter 8, but for convenience definitions are also given here.

■▷
More about ionization energies and electron affinities in Chapter 8

The first ionization energy ($IE_1$) of an atom is the energy change that is associated with the removal of the first valence electron (equation 5.17); the internal energy change is measured for the *gas phase* process.

$$X(g) \longrightarrow X^+(g) + e^- \tag{5.17}$$

■▷
Internal energy: see Section 4.5

The first electron affinity ($EA_1$) of an element is *minus* the internal energy change ($EA = -\Delta U(0\,\text{K})$) that accompanies the gain of one electron (equation 5.18); the energy change is measured for the *gas phase* process.

$$Y(g) + e^- \longrightarrow Y^-(g) \tag{5.18}$$

Mulliken's proposal followed from Pauling's definition of electronegativity – *the power of an atom in a molecule to attract electrons to itself* – and is consistent with the valence bond concept that the ionic resonance forms $X^+Y^-$ or $X^-Y^+$ may contribute significantly to the bonding in the heteronuclear diatomic molecule XY.

Mulliken electronegativity values are determined using equation 5.19, but numbers calculated from this relationship are *not directly comparable* with Pauling electronegativities, although they may be adjusted to a 'Pauling-compatible' scale.

$$\chi^M = \frac{IE_1 + EA_1}{2} \tag{5.19}$$

where $IE_1$ and $EA_1$ are in eV (1 eV = 96.485 ≈ 96.5 kJ mol$^{-1}$)

### Allred–Rochow electronegativity values ($\chi^{AR}$)

Allred and Rochow devised another electronegativity scale based upon the effective nuclear charge, $Z_{\text{eff}}$, of an atom. Effective nuclear charges may be calculated by using Slater's rules.[§]

The Allred–Rochow electronegativities, $\chi^{AR}$, are determined using equation 5.20 which scales values of $\chi^{AR}$ so that they are *directly comparable* with those of $\chi^P$.

$$\chi^{AR} = \left( 3590 \times \frac{Z_{\text{eff}}}{r_{\text{cov}}^2} \right) + 0.744 \tag{5.20}$$

where $r_{\text{cov}}$ is in pm

One of the difficulties in comparing electronegativity scales lies in their units which we conveniently ignore. Pauling electronegativities have units of eV$^{\frac{1}{2}}$, Mulliken electronegativities have units of eV and Allred–Rochow values have units of pm$^{-2}$.

| 5.11 | **Polar diatomic molecules** |

A homonuclear diatomic molecule such as $H_2$ or $N_2$ possesses a symmetrical distribution of electronic charge, and the two ends of the molecule are indistinguishable. Such molecules are said to be *non-polar*.

---

[§] We shall not discuss details of Slater's rules here; for further details see: C. E. Housecroft and A. G. Sharpe (2005) *Inorganic Chemistry*, 2nd edn, Prentice Hall, Harlow, Chapter 1.

**Fig. 5.11** (a) The polarity of the HBr molecule arises from an asymmetrical charge distribution; (b) this creates a polar bond; (c) the dipole can be denoted by an arrow which, by SI convention, points towards the more *positively* charged end of the molecule.

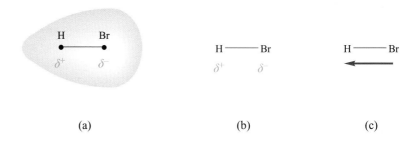

(a)                                    (b)                                    (c)

A heteronuclear diatomic molecule is composed of two different atoms. If these atoms exhibit different electron-withdrawing powers, the electron density will be shifted towards the more electronegative atom. Such molecules are said to be *polar* and possess an *electric dipole moment* ($\mu$). The term 'electric' is needed to distinguish the property from a magnetic dipole, but is not always included.

Magnetic moment: see Section 23.13

> A *polar* diatomic molecule is one in which the time-average charge distribution is shifted towards one end of the molecule. Such molecules possess an *electric dipole moment*, $\mu$. A dipole moment is a vector quantity.
>
> The SI units of $\mu$ are coulomb metres (C m); however, for convenience, debye units (D) are often used (1 D = $3.336 \times 10^{-30}$ C m).

In a polar diatomic molecule, one end of the molecule is negatively charged *with respect to* the other end. This is illustrated in Figure 5.11 for HBr. The notation $\delta^+$ and $\delta^-$ shows that there is a partial charge separation in the molecule but it is not a quantitative measure – it does not indicate the extent to which the separation occurs (see Box 5.2). An arrow is used to indicate the direction of the dipole moment. By SI convention, the arrow points from the $\delta^-$ end of the bond to the $\delta^+$ end (Figure 5.11c). Values of $\mu$ for some heteronuclear diatomic molecules containing single bonds are listed in Table 5.3.

Petrus (Peter) Josephus Wilhelmus Debye (1884–1966). © *The Nobel Foundation.*

**Table 5.3** Dipole moments ($\mu$) of some diatomic molecules in the gas phase. Values are given in debye units (D); these are not SI units but are used for convenience. (1 D = $3.336 \times 10^{-30}$ C m)

| Molecule XY | $\mu$ / D | Polarity ($\delta^+$ and $\delta^-$) |
|---|---|---|
| HF | 1.83 | H ($\delta^+$) F ($\delta^-$) |
| HCl | 1.11 | H ($\delta^+$) Cl ($\delta^-$) |
| HBr | 0.83 | H ($\delta^+$) Br ($\delta^-$) |
| HI | 0.45 | H ($\delta^+$) I ($\delta^-$) |
| ClF | 0.89 | Cl ($\delta^+$) F ($\delta^-$) |
| BrCl | 0.52 | Br ($\delta^+$) Cl ($\delta^-$) |
| BrF | 1.42 | Br ($\delta^+$) F ($\delta^-$) |
| ICl | 1.24 | I ($\delta^+$) Cl ($\delta^-$) |
| IBr | 0.73 | I ($\delta^+$) Br ($\delta^-$) |
| LiH | 5.88 | Li ($\delta^+$) H ($\delta^-$) |
| LiF | 6.33 | Li ($\delta^+$) F ($\delta^-$) |

## THEORETICAL AND CHEMICAL BACKGROUND

### Box 5.2  Dipole moments

Dipole moments arise as a consequence of asymmetrical charge distribution.

In a heteronuclear diatomic molecule XY in which the atoms have different electron withdrawing powers, let the partial charges on the atoms be $+q$ and $-q$:

The molecule has no overall charge and so the *magnitudes* of the charges $q$ are equal and opposite. Let the separation between the two centres of charge be $r$. The dipole moment, $\mu$, of the molecule is given by:

$$\mu = \text{(electronic charge)}$$
$$\times \text{(distance between the charges)}$$

The magnitude of the charge shown in the above diagram is $q$. The *atomic* charge is $(q \times e)$; the factor $e$ ($e = 1.602 \times 10^{-19}$ C) scales the charge to an atomic level.

$$\mu = q \times e \times r$$

Charge is measured in coulombs and the distance, $r$, is in metres. The SI units of $\mu$ are coulomb metres (C m) but, for convenience, $\mu$ is often given in units of debyes (D) where $1\,\text{D} = 3.336 \times 10^{-30}$ C m.

#### Problem

The dipole moment of HF is 1.83 D. The H−F bond

length is 92 pm. Estimate the atomic charge distribution in the HF molecule.

The dipole moment is given by:

$$\mu = q \times e \times r$$

If we assume that the centres of charge coincide with the H and F nuclei, then $r$ is equal to the bond distance.

The H−F bond distance $= 92\,\text{pm} = 9.2 \times 10^{-11}$ m.

$$q = \frac{\mu}{e \times r} \quad \text{where } \mu \text{ is in C m}$$
$$= \frac{(1.83 \times 3.336 \times 10^{-30}\,\text{C m})}{(1.602 \times 10^{-19}\,\text{C}) \times (9.2 \times 10^{-11}\,\text{m})}$$
$$= 0.41$$

This value indicates that the charge in the HF molecule is distributed such that, with respect to the neutral atoms, the fluorine atom has effectively gained 0.41 of an electron and the hydrogen atom has lost 0.41 of an electron.

#### Exercise

The dipole moment of BrF is 1.42 D, and the bond length is 176 pm. Show that $q = 0.17$, and explain what this means in terms of the atomic charge distribution.

Electronegativity values can be used to predict whether a diatomic molecule will be polar and to estimate the magnitude of the dipole moment. Consider hydrogen bromide. The electronegativities, $\chi^P$, of H and Br are 2.2 and 3.0 respectively. This tells us that the Br atom withdraws electrons more than the H atom, and we predict a dipole moment as shown in Figure 5.11.

The difference in electronegativities between the atoms in a heteronuclear diatomic molecule gives some indication of the magnitude of the dipole moment. In the series of hydrogen halides HF, HCl, HBr and HI, $\chi^P$ decreases in the order $\chi^P(\text{F}) > \chi^P(\text{Cl}) > \chi^P(\text{Br}) > \chi^P(\text{I})$. In each case, $\chi^P(\text{halogen}) > \chi^P(\text{H})$. Accordingly, we predict that the most polar molecule in the series is HF and the least polar is HI, but in each hydrogen halide, HX, the dipole moment is in the direction shown in structure **5.16**.

Electronegativity values must be used with caution in molecules containing multiple bonds. Care must be taken to choose values appropriate to the oxidation state and atoms involved.

(5.16)

**5.12**    ## Isoelectric species

Two species are *isoelectronic* if they possess the same number of electrons, that is, {core + valence} electrons.

In this section, we introduce a simple, but valuable, concept: *isoelectronic* species. The term 'isoelectronic' means the 'same number of electrons'. Strictly, this refers to the *total* (core + valence) number of electrons. For example, $N_2$ and CO are isoelectronic because each has 14 electrons:

$$N \quad 1s^2 2s^2 2p^3$$
$$C \quad 1s^2 2s^2 2p^2$$
$$O \quad 1s^2 2s^2 2p^4$$

The term 'isoelectronic' is often used more loosely, however, and molecular species may be considered to be *isoelectronic with respect to their valence electrons*. For example, $CO_2$ and $CS_2$ are isoelectronic if we are considering the valence electrons, since each has 16 valence electrons:

$$C \quad [He]2s^2 2p^2$$
$$O \quad [He]2s^2 2p^4$$
$$S \quad [Ne]3s^2 3p^4$$

Applications of the isoelectronic principle are widespread, for example in discussing the bonding in molecules (see Chapter 7).

Using the periodic table is a quick way of spotting isoelectronic relationships. Consider the part of the *p*-block shown below:

| Group 13 $ns^2 np^1$ | Group 14 $ns^2 np^2$ | Group 15 $ns^2 np^3$ | Group 16 $ns^2 np^4$ | Group 17 $ns^2 np^5$ | Group 18 $ns^2 np^6$ |
|---|---|---|---|---|---|
| B $Z = 5$ | C $Z = 6$ | N $Z = 7$ | O $Z = 8$ | F $Z = 9$ | Ne $Z = 10$ |
| Al $Z = 13$ | Si $Z = 14$ | P $Z = 15$ | S $Z = 16$ | Cl $Z = 17$ | Ar $Z = 18$ |
| Ga $Z = 31$ | Ge $Z = 32$ | As $Z = 33$ | Se $Z = 34$ | Br $Z = 35$ | Kr $Z = 36$ |

From the periodic relationships between these elements, we can see that:

- moving one place to the right in a given row in the table *adds* one valence electron (e.g. O to F);
- moving one place to the left in a given row in the table means there is one valence electron *fewer* (e.g. O to N);
- moving up or down a group does not change the number of valence electrons (e.g. N to P);
- moving up or down a group *does* change the number of core electrons (e.g. N to P).

**Worked example 5.3**    *Isoelectronic atoms and ions*

**Confirm that the Ne atom and the $O^{2-}$ ion are isoelectronic.**

The Ne atom has a ground state electronic configuration $[He]2s^2 2p^6$.
The O atom has a ground state configuration $[He]2s^2 2p^4$.

The $O^{2-}$ ion has a ground state configuration $[He]2s^2 2p^6$.
Therefore, the $O^{2-}$ ion is isoelectronic with the Ne atom.

| **Worked example 5.4** | *Isoelectronic molecules and ions* |
|---|---|

**Show that the species CO, $[CN]^-$ and $N_2$ are isoelectronic.**

An O atom and an $N^-$ ion both possess 8 electrons.
The molecule CO possesses $(6 + 8) = 14$ electrons.
The anion $[CN]^-$ possesses $(6 + 8) = 14$ electrons.
The molecule $N_2$ possesses $(7 + 7) = 14$ electrons.
The species CO, $[CN]^-$ and $N_2$ are all isoelectronic.

## 5.13    The bonding in CO by the Lewis and valence bond approaches

Structures **5.17** and **5.18** show Lewis structures for the homonuclear diatomics $C_2$ and $O_2$. The fact that a double bond is common to both these molecules suggests that we could form a similar molecule between a carbon and an oxygen atom – structure **5.19** shows CO, carbon monoxide.

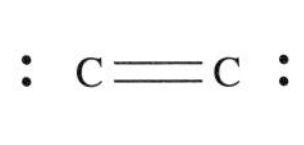

**Bonding in $C_2$ and $O_2$:**
see Sections 4.18 and 4.20

$$: C = C : \qquad O = O \qquad : C = O$$

$$\text{(5.17)} \qquad\qquad \text{(5.18)} \qquad\qquad \text{(5.19)}$$

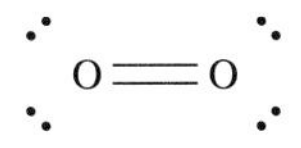

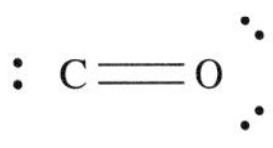

The Lewis structure **5.19** is not the most satisfactory of pictures for the CO molecule. The formation of a C=O double bond gives the oxygen atom an octet, but the carbon atom has only six electrons in its outer shell. The observation that CO is isoelectronic with $N_2$ is an important key in being able to understand how the CO molecule overcomes this problem.

A nitrogen atom possesses five valence electrons. In the $N_2$ molecule **5.20**, two nitrogen atoms combine to form a triple bond and this gives each atom an octet of valence electrons.

$$: N \equiv N :$$

$$\text{(5.20)}$$

A nitrogen atom is isoelectronic with both $C^-$ and $O^+$, and so, by comparing CO with $N_2$, it is possible to write down a Lewis structure for CO by combining $C^-$ and $O^+$. This is done in Figure 5.12; compare this with structure **5.20**.

The two resonance structures that describe carbon monoxide are shown in structures **5.21** and **5.22**. Note that structure **5.21** involves a C=O double

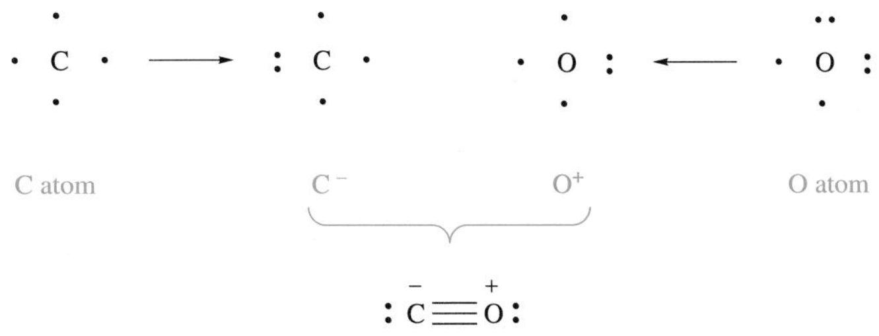

**Fig. 5.12** Using the isoelectronic relationship between $N_2$ and CO, the bonding in CO can be viewed in terms of the combination of $C^-$ and $O^+$.

$$\cdot\, C\, \cdot \longrightarrow : C \cdot \qquad\qquad \cdot\, O : \longleftarrow \cdot\, O :$$

C atom $\qquad\qquad$ $C^-$ $\qquad\qquad$ $O^+$ $\qquad\qquad$ O atom

$$: C \equiv O :$$

bond while structure **5.22** involves a triple bond. In structure **5.22**, each atom has an octet of electrons.

$$: C \!=\!\!=\! O \quad \longleftrightarrow \quad : \overset{-}{C} \!\equiv\!\! \overset{+}{O} :$$

(5.21)                    (5.22)

If structure **5.22** contributes significantly to the overall bonding in CO, it suggests that the molecule should be polar *but with the carbon atom being $\delta^-$.* (This contrasts with what you might have expected simply by looking at the values of the electronegativities of carbon and oxygen in Table 5.2 and emphasizes the dependence of $\chi$ on bond order.) The observed dipole moment of CO is 0.11 D. This is a comparatively low value, and indicates that CO is almost non-polar – the carbon end of the molecule is only slightly negative with respect to the oxygen end. This has important consequences for the chemistry of CO.

Chemistry of CO: see Sections 22.6 and 23.14

---

ENVIRONMENT AND BIOLOGY

## Box 5.3  Carbon monoxide – a toxic gas

Carbon monoxide (CO) is a very poisonous gas. Its toxicity arises because it binds more strongly to the iron centres in haemoglobin than does $O_2$. As a result, CO prevents the uptake and transport of dioxygen in the bloodstream. The structure of one of the chains of the mammalian dioxygen-binding protein haemoglobin is shown in diagram (a) below. The protein chain is shown in a *ribbon representation*. The chain wraps around an iron-containing *haem* (or *heme*) unit. In (a), the haem unit is shown in a space-filling form. Diagram (b) shows a ball-and-stick representation of the haem unit. The iron centre (shown in green) is attached to five nitrogen atoms (shown in blue) in a square-based pyramidal arrangement. Four of the N atoms form part of the haem unit, whereas the fifth N atom is

part of a histidine residue (see Table 35.1) which is connected to the protein backbone. This connection is represented by the broken 'stick' at the right-hand side of diagram (b). When either $O_2$ or CO binds to the Fe centre, the diatomic molecule occupies the vacant site opposite to the histidine residue. A study of the bonding in CO by molecular orbital theory (see Figures 5.13–5.15) shows that the character of the highest occupied molecular orbital (see Figure 5.16) allows CO to bind to an Fe centre in haemoglobin by donating a pair of electrons to the metal centre. The application of MO theory to molecules such as CO and $O_2$ provides valuable information about what happens when they donate or accept electrons from, for example, metal centres in biological systems.

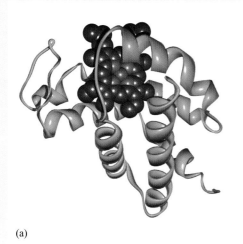

(a)

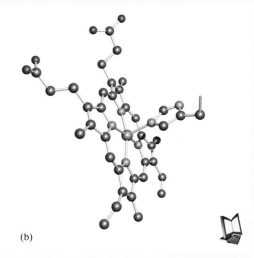

(b)

**Table 5.4** Some properties of CO compared with those of $N_2$.

| Property | CO | $N_2$ |
|---|---|---|
| Bond distance / pm | 113 | 110 |
| Bond dissociation enthalpy / kJ mol$^{-1}$ | 1076 | 945 |
| Dipole moment / D | 0.11 | 0 |
| Magnetic properties | Diamagnetic | Diamagnetic |

If structure **5.21** is important, then the bond dissociation enthalpy of CO and the carbon–oxygen bond distance should be consistent with a double bond. Structure **5.22** suggests a stronger bond with a bond order of 3. Table 5.4 lists some relevant data for CO and also for $N_2$ – remember that $N_2$ has a triple bond. The bond distance in CO is similar to that in $N_2$, and the bond is even stronger. These data support the presence of a triple bond.

In conclusion, valence bond theory describes the bonding in carbon monoxide in terms of resonance structures that show either a double or triple bond. All of the electrons are paired, in keeping with the observed diamagnetism of the compound. The observed properties of CO (the strong, short bond, and the slight polar nature, $C^{\delta-} O^{\delta+}$, of the molecule) can be rationalized in terms of contributions from *both* of the two resonance structures **5.21** and **5.22**. In addition, note that any dipole moment predicted from Pauling electronegativity values opposes the charge separation indicated by resonance structure **5.22**.

Diamagnetism: see Section 4.18

## 5.14    The bonding in carbon monoxide by MO theory

### An initial approach

We now consider the bonding in carbon monoxide using molecular orbital theory and build upon details that we discussed in Chapter 4.

Initial points to note are:

- The effective nuclear charge of an oxygen atom is greater than that of a carbon atom.
- The $2s$ atomic orbital of oxygen lies at lower energy than the $2s$ atomic orbital of carbon (see worked example 5.1).
- The $2p$ atomic orbitals of oxygen lie at lower energy than the $2p$ atomic orbitals of carbon.
- The $2s$–$2p$ energy separation in oxygen is greater than that in carbon.

A linear combination of atomic orbitals of carbon and oxygen gives rise to the MO diagram shown in Figure 5.13. The bonding MOs [$\sigma(2s)$, $\sigma(2p)$ and $\pi(2p)$] all have more oxygen than carbon character. In particular, Figure 5.13 suggests that the $\sigma(2s)$ MO is predominantly oxygen $2s$ in character.

To a first approximation, this MO diagram is satisfactory – it correctly indicates that the bond order in CO is 3. It also predicts that the molecule is diamagnetic and this is consistent with experimental data. However, this simple picture gives an incorrect ordering of the MOs in the molecule and

**Fig. 5.13** An MO diagram for the formation of CO which allows overlap only between 2s(C) and 2s(O) and between 2p(C) and 2p(O) atomic orbitals.

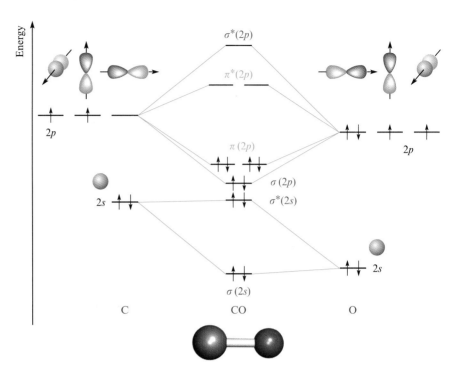

indicates that the $\sigma$ MOs are definitively either bonding or antibonding. You may wish to leave the discussion of the bonding of CO at this point, but for the sake of completeness, the remainder of this section deals with the consequences of orbital mixing.

### The effects of orbital mixing

■▷
$\sigma$–$\pi$ crossover:
see Section 4.19

In Section 4.19, we discussed the consequences of orbital mixing. If the $2s$ and $2p$ atomic orbitals of each atom in a homonuclear diatomic molecule are close enough in energy, then mixing of the $\sigma(2s)$ and $\sigma(2p)$ *molecular* orbitals can occur. Similarly, mixing of the $\sigma^*(2s)$ and $\sigma^*(2p)$ molecular orbitals can occur. This results in a re-ordering of the MOs.

In carbon monoxide, the relative energies of the atomic orbitals are such that some mixing of the molecular orbitals *can* occur. This is illustrated in Figure 5.14. Mixing of the $\sigma(2s)$ and $\sigma(2p)$ orbitals lowers the energy of the $\sigma(2s)$ MO (now labelled $1\sigma$) and raises the energy of $\sigma(2p)$ (now labelled $3\sigma$). The energy of the $3\sigma$ orbital is higher than that of the $\pi(2p)$ orbitals. Similarly, mixing of the $\sigma^*(2s)$ and $\sigma^*(2p)$ orbitals results in a lowering of $\sigma^*(2s)$ (now labelled $2\sigma$) and a raising of $\sigma^*(2p)$ (now labelled $4\sigma$). As a consequence, all four of the $\sigma$-type MOs in the CO molecule (look at the diagrams at the right-hand side of Figure 5.14) contain character from the $2s$ and $2p$ atomic orbitals of each of the carbon and oxygen atoms. The most significant result (seen when comparing Figures 5.13 and 5.14) is the change-over in the highest occupied MO from one of $\pi$ to one of $\sigma$ symmetry. The degree of mixing will control the precise ordering of the $1\sigma$ and $2\sigma$ orbitals.

Unfortunately, even the MO diagram in Figure 5.14 is not fully instructive, because it does not clearly reveal the individual contributions made by the carbon and oxygen atomic orbitals to the four $\sigma$ molecular orbitals. These

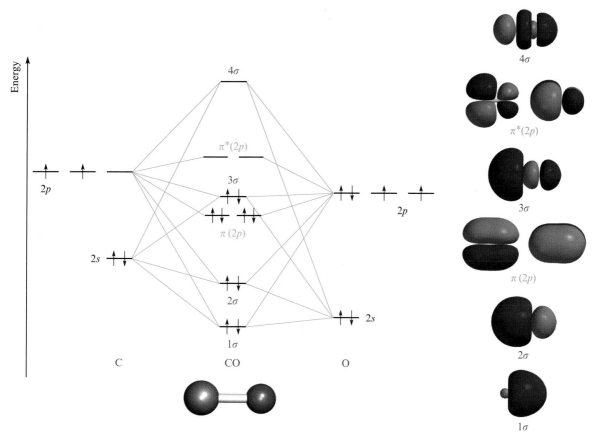

**Fig. 5.14** An MO diagram for the formation of CO which allows for orbital mixing between the $\sigma(2s)$ and $\sigma(2p)$ and between the $\sigma^*(2s)$ and $\sigma^*(2p)$. Compared with Figure 5.13, a major change is the $\sigma$–$\pi$ crossover which results in the highest occupied MO being of $\sigma$-symmetry. The diagrams on the right-hand side show representations of the MOs and have been generated computationally using Spartan '04, © Wavefunction Inc. 2003. These diagrams illustrate that the $1\sigma$ MO has mainly oxygen character; the $2\sigma$, $3\sigma$ and $\pi^*(2p)$ MOs have more carbon than oxygen character.

contributions are unequal, and the consequence of this complication is that the character of the $\sigma$ symmetry MOs is not readily partitioned as being either bonding ($\sigma$) or antibonding ($\sigma^*$). This is the reason these MOs have been labelled as $1\sigma$, $2\sigma$, $3\sigma$ and $4\sigma$ in Figure 5.14.

Two important features emerge when the MO bonding picture for CO is more fully explored:

- The dominant oxygen $2s$ character of the $1\sigma$ MO means that this orbital is essentially non-bonding with respect to the carbon–oxygen interaction.
- The $3\sigma$ MO contains a small percentage of oxygen character – this means that the $3\sigma$ MO is effectively a carbon-centred orbital.

These results lead us to draw a revised MO picture for carbon monoxide – Figure 5.15. Note that this diagram is *over-simplified* but gives a chemically useful view of the bonding in the molecule.

### Summary of the molecular orbital view of the bonding in CO

If we use Figure 5.15 as a basis, then molecular orbital theory suggests the following about the bonding in the CO molecule:

**Fig. 5.15** An MO diagram for CO which allows for the effects of orbital mixing but also recognizes the fact that not all the atomic orbitals will make major contributions to all the $\sigma$ MOs. One change in going from Figure 5.14 to 5.15 is that the $1\sigma$ MO becomes non-bonding. This is an over-simplification.

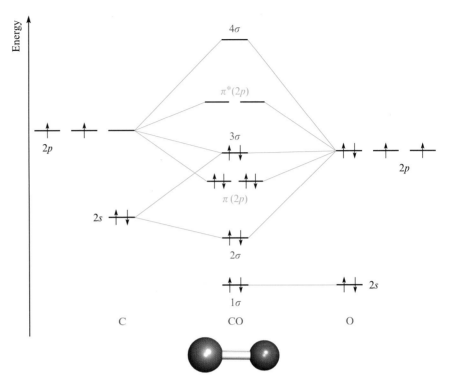

- There are six electrons occupying the $2\sigma$ and $\pi(2p)$ bonding MOs. The other four electrons occupy the non-bonding $1\sigma$ MO, and the essentially carbon-centred $3\sigma$ MO. The $4\sigma$ and $\pi^*(2p)$ MOs are antibonding.
- CO is diamagnetic – this agrees with experimental observations.
- The highest occupied MO (HOMO) in CO has predominantly carbon character and contains contributions from both the carbon $2s$ and $2p$ atomic orbitals, and this results in an outward pointing lobe as shown in Figure 5.16. In Section 23.14 we will discuss some chemical properties of CO that rely, in part, on the nature of this MO and on the fact that the lowest-lying *empty* MOs are of $\pi$ symmetry.
- It is *not* easy to say anything about the polarity of the CO molecule from MO theory at this simple and qualitative level. The charge distribution in a molecule follows from the summed character of the *occupied* molecular orbitals; the empty orbitals cannot affect this molecular property. In the occupied $1\sigma$ MO, the electron density is oxygen centred ($O^{\delta^-}$) while in the $2\sigma$ and $3\sigma$ MOs there is a greater probability of finding the electrons nearer to the carbon nucleus ($C^{\delta^-}$). Both the $\pi(2p)$ MOs will have more oxygen than carbon character ($O^{\delta^-}$).

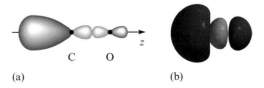

**Fig. 5.16** (a) A schematic representation of the highest occupied MO of carbon monoxide. The black dots show the positions of the C and O nuclei. (b) A more realistic representation of the same orbital generated computationally using Spartan '04, © Wavefunction Inc. 2003.

## [CN]⁻ and [NO]⁺: two ions isoelectronic with CO

The cyanide ion $[CN]^-$ and the nitrosyl cation $[NO]^+$ are well-known chemical species. The valence electron counts in $[CN]^-$ and $[NO]^+$ are as follows:

$[CN]^-$         Valence electrons from C = 4

Valence electrons from N = 5

Electron from negative charge = 1

Total valence electrons = 10

$[NO]^+$         Valence electrons from N = 5

Valence electrons from O = 6

Subtract an electron because of the positive charge = $-1$

Total valence electrons = 10

Carbon, nitrogen and oxygen all have the same number of core electrons. The ions $[CN]^-$ and $[NO]^+$ are therefore isoelectronic. They are also isoelectronic with the molecules CO and $N_2$.

We have already seen that there are some similarities and some differences between the bonding descriptions of the homonuclear diatomic molecule $N_2$ and the heteronuclear CO. How do the ions $[CN]^-$ and $[NO]^+$ fit into this series? Is the bond order the same in each species? Are the magnetic properties the same? What difference does the presence of an external charge make to the bonding in these molecules? Is it possible to predict on which atom the charge in the $[CN]^-$ and $[NO]^+$ ions will be localized?

*Bear in mind throughout this section that C and $N^+$ are isoelectronic with each other, N and $O^+$ are isoelectronic, and $N^-$ and O are isoelectronic.*

### Valence bond theory

Resonance structures for $[NO]^+$ and $[CN]^-$ can readily be drawn by using those of CO as a basis. The C atom can be replaced by the isoelectronic $N^+$, and the O atom by $N^-$ as shown in Figure 5.17. Structure **II** for each species can be compared to the Lewis structure for $N_2$; remember that an N atom is isoelectronic with both $C^-$ and $O^+$.

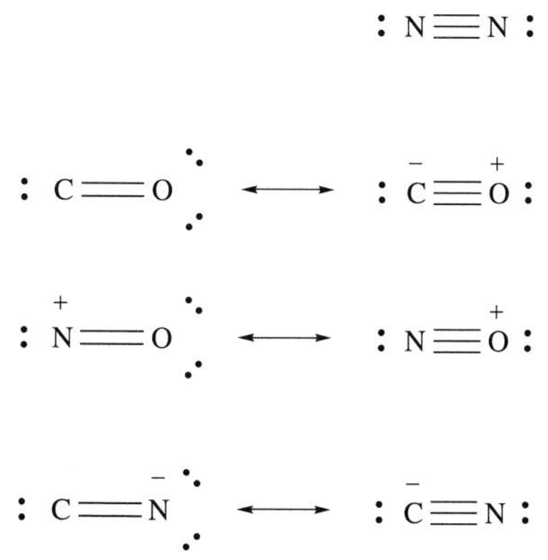

**Fig. 5.17** Resonance structures for $[CN]^-$ and $[NO]^+$ can readily be derived from those of $N_2$ and CO by considering the isoelectronic relationships between C and $N^+$, N and $O^+$, and $N^-$ and O.

                                              **I**                            **II**

We have already seen that the observation of a short, strong bond in CO suggests that resonance structure **II** contributes significantly to the bonding in this molecule and emphasizes the similarity between the bonding in CO and $N_2$. The bond distance in the $[CN]^-$ anion is 114 pm and in the $[NO]^+$ cation is 106 pm. These values compare with distances of 110 pm and 113 pm in $N_2$ and CO respectively, and are consistent with a significant contribution to the bonding in each ion from resonance structure **II**.

The small dipole moment in CO can be rationalized by allowing contributions from both resonance structures **I** and **II** to partly offset one another, making allowances too for the differences in electronegativity between the two atoms. It is, however, difficult to use the valence bond method to *predict* the polarity of the carbon–oxygen bond. Each of the two resonance forms for $[NO]^+$ suggests a different polarity for the ion (Figure 5.17) – **I** implies that the nitrogen atom carries the positive charge while **II** suggests that it is localized on the oxygen atom. The same problem arises if we try to predict whether carbon or nitrogen carries the negative charge in the $[CN]^-$ ion.

## Molecular orbital theory

Despite the differences between the MO diagrams for $N_2$ (Figure 4.28) and CO (Figure 5.15), the results of MO theory are consistent with the observed diamagnetism. The charge distribution in each molecule follows from the character of the occupied molecular orbitals. In $N_2$, the MOs possess equal contributions from both atoms and the bond is non-polar ($\mu = 0$ D). In carbon monoxide, differences in the relative energies of the carbon and oxygen atomic orbitals cause distortions in the MOs and as a result the carbon–oxygen bond is polar ($\mu = 0.11$ D).

The heteronuclear nature of $[NO]^+$ and $[CN]^-$ suggests that it might be better to use CO as a starting point rather than $N_2$ when discussing the bonding in these ions using MO theory. In many regards this viewpoint is correct, but it is still difficult to come up with the 'right' answer when we are using only a qualitative MO picture. As a first approximation it is acceptable to use the rather simple MO picture drawn in Figure 5.13 to depict the bonding in any heteronuclear diatomic species that is isoelectronic with CO. The answers may not be 'correct' in the sense that some predictions may not agree exactly with experimental results. However, with experience, it is possible to see how changes to the atomic orbital energies can influence the energies and the composition of molecular orbitals and, therefore, molecular properties.

Our initial approach is to use the molecular orbital diagram in Figure 5.13, keeping in mind the fact that the electronegativity ordering of O > N > C (Table 5.2) means that the oxygen atomic orbitals will be lower in energy than those of nitrogen, which will, in turn, be lower in energy than those of carbon. Using Figure 5.13, we can make the predictions:

- the bond order in CO, $[CN]^-$ and $[NO]^+$ will be 3;
- the species CO, $[CN]^-$ and $[NO]^+$ will be diamagnetic.

If we use Figure 5.15 as a starting point and couple this with the results of more sophisticated calculations, we discover some subtle differences in the

## ENVIRONMENT AND BIOLOGY

### Box 5.4 Nitrogen monoxide (nitric oxide) in biology

Nitrogen monoxide (NO, also called nitric oxide) plays an important role in biology. Being a small molecule, NO can easily diffuse through cell walls. It behaves as a messenger molecule in biological systems and is active in mammalian functions such as the regulation of blood pressure, muscle relaxation and neurotransmission. The discovery that nitric oxide is involved as a signalling agent in penile erection led to the development and marketing of the drug sildenafil (Viagra) for treatment of male erectile dysfunction.

One of the remarkable properties of NO is that it appears to be cytotoxic (that is, it is able to destroy particular cells) and has an effect on the ability of the body's immune system to kill tumour cells.

In 1998, the Nobel prize in Physiology or Medicine was awarded to Robert Furchgott, Louis Ignarro and Ferid Murad 'for their discoveries concerning nitric oxide as a signalling molecule in the cardiovascular system'. For more information, go to http://www.nobel.se/medicine/laureates/1998/index.html

Viagra

A ball-and-stick model of a molecule of Viagra; structural data from K.Y. J. Zhang *et al.* (2004) *Molecular Cell*, vol. 15, p. 279.

Robert F. Furchgott (1916– ).
© *The Nobel Foundation.*

Louis J. Ignarro (1941– ).
© *The Nobel Foundation.*

Ferid Murad (1936– ).
© *The Nobel Foundation.*

bonding in CO, $[CN]^-$ and $[NO]^+$. Molecular orbital $1\sigma$, which is tending towards being non-bonding in CO, has a greater bonding character in $[NO]^+$ and $[CN]^-$. Molecular orbital $2\sigma$, which is bonding in CO, becomes antibonding in $[NO]^+$ and $[CN]^-$. The orbital $3\sigma$, which is predominantly a carbon-centred lone pair in CO, has more bonding character in $[NO]^+$ and $[CN]^-$ but still possesses an outward-pointing lobe (on carbon in $[CN]^-$ and on nitrogen in $[NO]^+$).

## 5.16    $[NO]^+$, NO and $[NO]^-$

We end this chapter by considering the closely related species $[NO]^+$, NO and $[NO]^-$. Note that NO is isoelectronic with $[O_2]^+$, while $[NO]^-$ is isoelectronic with $O_2$; you may find it useful to compare some of the new results in this section with those detailed in Section 4.22 for the dioxygen species. Some properties of NO and $[O_2]^+$ are listed for comparison in Table 5.5.

**Table 5.5** Some properties of nitrogen monoxide, NO, compared with those of the isoelectronic cation $[O_2]^+$.

| Property | NO | $[O_2]^+$ |
|---|---|---|
| Bond distance / pm | 115 | 112 |
| Bond dissociation enthalpy / kJ mol$^{-1}$ | 631 | 644 |
| Magnetic properties | Paramagnetic | Paramagnetic |
| Dipole moment / D | 0.16 | 0 |

As in Section 5.15 where we considered the ion $[NO]^+$, it is possible to use Figure 5.13 as the basis for a discussion of the bonding in NO and $[NO]^-$. Nitrogen monoxide (NO) has one more valence electron than $[NO]^+$, and by the *aufbau* principle, this will occupy one of the $\pi^*(2p)$ orbitals shown in Figure 5.13. The NO molecule therefore has an unpaired electron and is predicted to be paramagnetic. This agrees with the experimental observation that NO is a radical.

The bond order in NO may be determined using equation 5.21. The increase in bond length in going from $[NO]^+$ (106 pm) to NO (115 pm) is consistent with the decrease in bond order from 3 to 2.5.

$$\text{Bond order in NO} = \tfrac{1}{2} \times (6 - 1) = 2.5 \qquad (5.21)$$

The odd electron occupies an antibonding orbital; this MO will have more nitrogen than oxygen character and therefore our results suggest that the odd electron is more closely associated with the N than the O atom. Of course, the extent to which this statement is true depends on the relative energies of the N and O atomic orbitals.

The anion $[NO]^-$ has two more valence electrons than $[NO]^+$. By the *aufbau* principle, these will singly occupy each of the $\pi^*(2p)$ orbitals shown in Figure 5.13. The $[NO]^-$ ion will be paramagnetic, and will resemble the $O_2$ molecule. The bond order is found to be 2 (equation 5.22).

$$\text{Bond order in } [NO]^- = \tfrac{1}{2} \times (6 - 2) = 2 \qquad (5.22)$$

## SUMMARY

In this chapter the main topic has been bonding in heteronuclear diatomic molecules.

### *Do you know what the following terms mean?*

- heteronuclear
- resonance structure
- orbital character
- coefficient (in respect of a wavefunction)

- orbital energy matching
- non-bonding molecular orbital
- electronegativity
- polar molecule

- (electric) dipole moment
- isoelectronic

### *You should now be able:*

- to summarize the salient differences between homonuclear and heteronuclear diatomic molecules
- to discuss the bonding in HF, LiH, LiF, CO, [NO]$^+$, [CN]$^-$, NO and [NO]$^-$ in terms of VB theory
- to discuss the bonding in HF, LiH, LiF, CO, [NO]$^+$, [CN]$^-$, NO and [NO]$^-$ in terms of MO theory
- to draw schematic representations of the molecular orbitals for heteronuclear diatomic molecules and to interpret these diagrams
- to use bond enthalpies to estimate enthalpies of atomization

- to discuss briefly the concept of electronegativity
- to define the terms (electric) dipole moment and polar molecule
- to use Pauling electronegativity values to determine if a diatomic molecule will be polar
- to write down sets of isoelectronic diatomic molecules and ions
- to use your knowledge about the bonding in one molecule or ion to comment on the bonding in a molecule or ion that is isoelectronic with the first

## PROBLEMS

**5.1** The normalized wavefunction for a bonding MO in a heteronuclear diatomic molecule AB is given by the equation:

$$\psi(MO) = (0.93 \times \psi_A) + (0.37 \times \psi_B)$$

What can you deduce about the relative values of $Z_{eff}$ for atoms A and B?

**5.2** Give resonance structure(s) for the hydroxide ion [OH]$^-$. What description of the bonding in this ion does valence bond theory give?

**5.3** Bearing in mind that [OH]$^-$ is isoelectronic with HF, construct an approximate MO diagram to describe the bonding in [OH]$^-$. What is the bond order in [OH]$^-$? Does the MO picture of the bonding differ from that obtained in the previous question using VB theory?

**5.4** Calculate the bond dissociation enthalpy of chlorine monofluoride, ClF, if the bond dissociation enthalpies of F$_2$ and Cl$_2$ are 159 and 242 kJ mol$^{-1}$ respectively, $\chi^P(F) = 4.0$, and $\chi^P(Cl) = 3.2$ (1 eV = 96.5 kJ mol$^{-1}$). Compare your answer with the experimental value.

**5.5** (a) Are equations 5.15 and 5.16 restricted to use in diatomic XY molecules? (b) Given that $D(O-O)$

and $D(H-H)$ are 146 and 436 kJ mol$^{-1}$ respectively, estimate a bond enthalpy term for an O$-$H bond.

**5.6** (a) Set up a suitable Hess cycle to determine $D(O-H)$ in H$_2$O given that $\Delta_f H^\circ(H_2O, g) = -242$ kJ mol$^{-1}$. Further data are listed in Appendix 10. (b) Compare this value of $D(O-H)$ with that obtained in problem 5.5b and with that in Table 4.5.

**5.7** Rationalize the trends in $\chi^P$ (see Table 5.2) from (a) B to F, and (b) F to I. (c) Why are no $\chi^P$ values listed for the noble gases?

**5.8** (a) Using appropriate *homo*nuclear bond enthalpies from Table 4.5, and $\chi^P$ values from Table 5.2, estimate values of $D(C-Cl)$ and $D(S-F)$. (b) Use these values to estimate $\Delta_a H^\circ(CCl_4, g)$ and $\Delta_a H^\circ(SF_4, g)$.

**5.9** Use values of the first ionization energies and first electron affinities listed in Appendices 8 and 9 to calculate values for the Mulliken electronegativity, $\chi^M$, of N, O, F, Cl, Br and I. Present these values graphically to show the trends in values of $\chi^M$ across the first row of the *p*-block and down group 17. How do these *trends* compare with similar trends in values of $\chi^P$ (Table 5.2)?

**5.10** Write down five ions (cations or anions) that are isoelectronic with neon.

**5.11** Explain whether or not the two species in each of the following pairs are isoelectronic with one another: (a) He and $Li^+$, (b) $[Se_2]^{2-}$ and $Br_2$; (c) NO and $[O_2]^+$; (d) $F_2$ and ClF; (e) $N_2$ and $P_2$; (f) $P^{3-}$ and $Cl^-$.

**5.12** (a) Is a dipole moment a scalar or vector property? (b) Which of the following gas phase molecules are polar: $Cl_2$, HBr, IF, BrCl, $H_2$? For those which are polar, draw a diagram that depicts the direction in which the dipole moment acts.

**5.13** Assuming that you apply 'isoelectronic' in its strictest sense, which of the following species form isoelectronic pairs? Not all species have a partner: $Cl_2$, CO, $[O_2]^{2-}$, $[S_2]^{2-}$, $F_2$, $Br_2$, $N_2$, NO, $[O_2]^+$, $O_2$.

## Additional problems

**5.14** (a) The results of a theoretical study of HCl show that the energies of the valence Cl $3s$ and $3p$ atomic orbitals are $-29$ and $-13$ eV respectively, and the H $1s$ atomic orbital lies at $-14$ eV. Using the data and an LCAO approach, construct an approximate MO diagram for the formation of HCl. (b) Use the MO diagram to find the bond order in HCl. Does the MO diagram indicate that HCl is diamagnetic or paramagnetic? (c) Do these last answers agree with conclusions drawn from a Lewis structure for HCl?

**5.15** Answer 5.14a assumes the simplest of pictures. Figure 5.18 shows an MO diagram for HCl obtained from a computational study. Account for

differences between your answer to problem 5.14a and Figure 5.18.

**5.16** (a) Draw resonance structures for NO that you consider will contribute significantly to the bonding. (b) From (a), what conclusions can you draw about the bond order and magnetic properties in NO? (c) Purely on the basis of $\chi^P$ values, do you think NO is polar? If so, in which direction does the dipole moment act? How does this conclusion fit with the resonance structures drawn in part (a)?

**5.17** (a) Construct an MO diagram using the LCAO approach for the formation of NO. (b) Use the diagram to deduce what you can about the bonding in and magnetic properties of NO. (c) Figure 5.19 shows an MO diagram for NO which allows for some orbital mixing (although the diagram is still oversimplified). Compare Figure 5.19 with your answer to part (a) and comment on the limitations and successes of the simpler of the two bonding schemes.

**5.18** Refer to Figure 5.19. (a) When NO undergoes a 1-electron oxidation, what species is formed and is it paramagnetic or diamagnetic? (b) Do you expect the N−O bond to lengthen, shorten or remain the same length upon 1-electron oxidation?

**5.19** In the bonding descriptions of LiH and LiF (Figures 5.6 and 5.10), we assumed that only the $2s$ orbital of Li was available for bonding. By using Figure 5.6 as a starting point, consider the consequences of adding the lithium $2p$ orbitals to its set of valence orbitals. Construct a revised MO diagram for the formation of LiH with Li $2s$ and $2p$ and H $1s$ orbitals as the valence orbitals.

**Fig. 5.18** An MO diagram for the formation of HCl allowing for some orbital mixing; the diagram is still over-simplified.

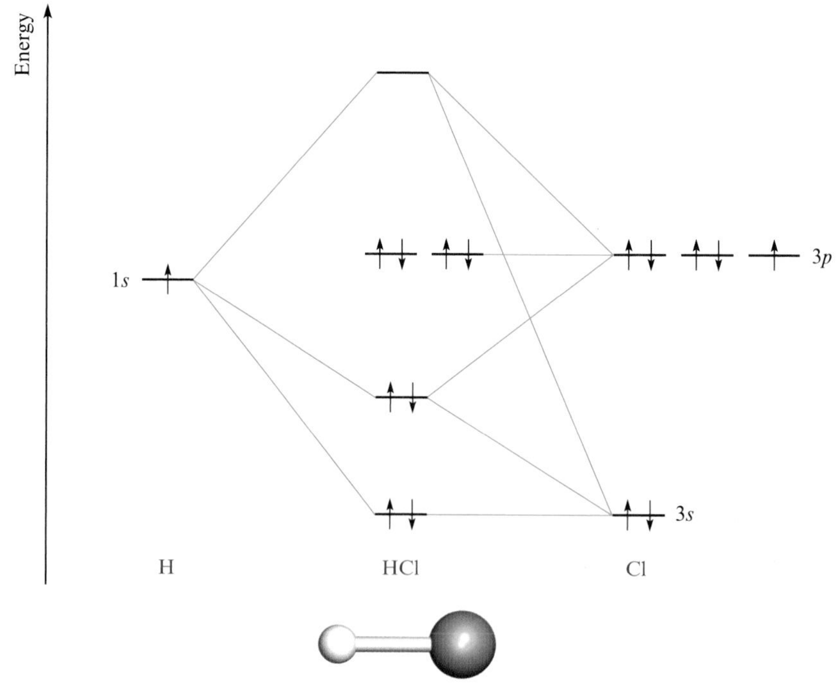

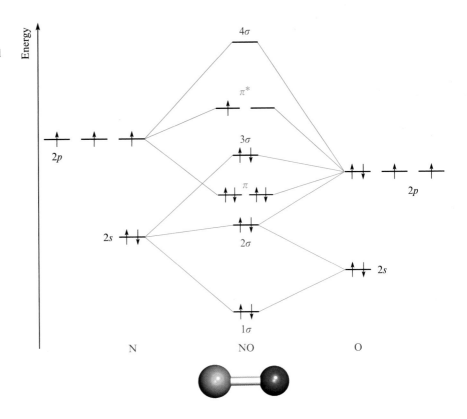

**Fig. 5.19** An MO diagram for the formation of NO. The diagram allows for some orbital mixing, but is still over-simplified.

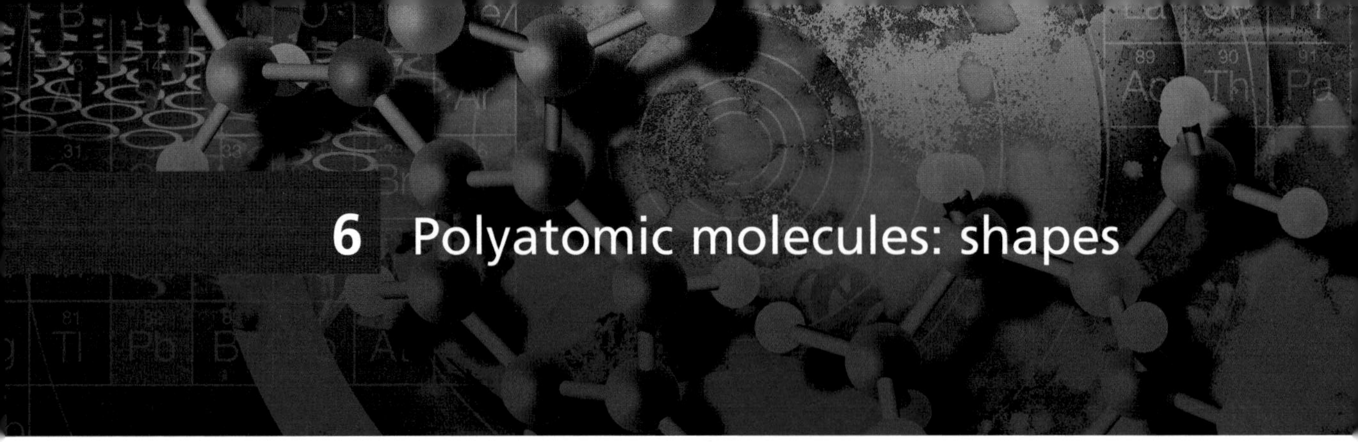

# 6 Polyatomic molecules: shapes

## Topics

- Triatomic molecules: linear or bent
- Shapes of molecules within the *p*-block
- The valence-shell electron-pair model
- The Kepert model
- Stereoisomerism
- Stereochemical non-rigidity
- Molecular dipole moments

## 6.1 Introduction

### Molecular geometry

Our discussion of molecules and molecular ions in Chapters 4 and 5 focused on diatomic species, but descriptions of the shapes of these molecules were omitted. This omission was for a good reason – diatomic molecules *must be linear*! However, only a small number of molecules and molecular ions are diatomic and in this chapter we look at simple *polyatomic* species containing *three or more atoms*. Although a few polyatomic molecules or ions are linear (e.g. $CO_2$, HC≡N, $[I_3]^-$, HC≡CC≡CH), most are not.

It is convenient to describe the 3-dimensional structure of a polyatomic system $XY_n$ in terms of the geometry around the central atom X. Table 6.1 lists some common geometries for molecules of formula $XY_n$. For most, several geometries are possible. We shall describe these geometries in more detail later, but Table 6.1 gives a useful checklist.

A ***polyatomic*** molecule or ion contains three or more atoms.

### Bond angles

In describing a molecular structure, the bond distances and bond angles are of particular significance. The angular relationships between atoms in a molecule conveniently describe the molecular geometry, and a bond angle is defined between three bonded atoms.[§] For example, in a trigonal planar $XY_3$ molecule, each of the angles $\angle$Y–X–Y is 120° (Figure 6.1a).

Molecular shapes are often described in terms of *idealized geometries* and the corresponding *ideal bond angles* are given in the right-hand column of

[§] We are not concerned here with *torsion angles* which define, for four atoms, the degree of twist around the central bond and the deviation from planarity for the four atoms that this generates.

**Table 6.1** Molecular geometries for polyatomic molecules of formula $XY_n$.

| Formula $XY_n$ | Coordination number of atom X | Geometrical descriptor | Spatial representation | Ideal angles ($\angle Y–X–Y$) / degrees |
|---|---|---|---|---|
| $XY_2$ | 2 | Linear | Y — X — Y | 180 |
| $XY_2$ | 2 | Bent or V-shaped | | Variable ($\neq 180$) |
| $XY_3$ | 3 | Trigonal planar | | 120 |
| $XY_3$ | 3 | Trigonal pyramidal | | Variable ($<120$) |
| $XY_3$ | 3 | T-shaped | | $\angle Y_a–X–Y_b \approx 90$<br>$\angle Y_a–X–Y_a \approx 180$<br>(the atoms lie in a plane) |
| $XY_4$ | 4 | Tetrahedral | | 109.5 |
| $XY_4$ | 4 | Square planar | | $\angle Y_1–X–Y_2 = 90^a$<br>$\angle Y_1–X–Y_3 = 180$ |
| $XY_4$ | 4 | Disphenoidal | | $\angle Y_{ax}–X–Y_{eq} \approx 90$<br>$\angle Y_{eq}–X–Y_{eq} \approx 120$ |
| $XY_5$ | 5 | Trigonal bipyramidal | | $\angle Y_{ax}–X–Y_{eq} = 90$<br>$\angle Y_{eq}–X–Y_{eq} = 120$<br>$\angle Y_{ax}–X–Y_{ax} = 180$ |
| $XY_5$ | 5 | Square-based pyramidal | | Variable |
| $XY_6$ | 6 | Octahedral | | $\angle Y_1–X–Y_2 = 90^a$<br>$\angle Y_1–X–Y_3 = 180$ |

[a] Although $Y_1$, $Y_2$ and $Y_3$ are distinguished to define the angle, all Y atoms are equivalent.

**Fig. 6.1** (a) A bond angle is defined as the angle between three bonded atoms. In a trigonal planar molecule $XY_3$, the ideal angle $\angle Y-X-Y$ is 120°. (b) The bond angle is 120° if all the substituents are the same, but if the substituents are different, the bond angles may not be exactly 120°; the geometry is then *approximately trigonal planar.*

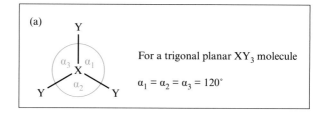

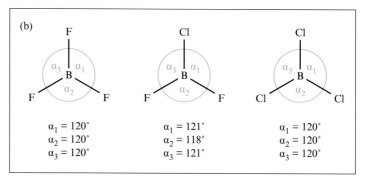

Table 6.1 for common structural types. Note that it is *not* possible to define ideal angles for all the geometries.

In practice, many molecules are described in terms of one of the shapes listed in Table 6.1 *without* actually possessing the ideal bond angles. A common reason for deviation from an ideal geometry is that the substituents in a molecule may not be identical. The molecules $BF_3$ and $BCl_3$ both possess *ideal* trigonal planar structures, but $BCl_2F$ and $BClF_2$ are *approximately* trigonal planar. The chlorine atoms are larger than the fluorine atoms and this difference contributes towards deviations from the ideal 120° bond angles (Figure 6.1b). The bond distances, $d$, also differ: $d(B-Cl) > d(B-F)$.

### Molecules with more than one 'centre'

Table 6.1 focuses on $XY_n$ molecules, each with an atom X at its centre. Many polyatomic molecules possess more than one 'centre'. Ethane, **6.1**, contains two carbon atoms, each of which has a tetrahedral arrangement of atoms about it, and cyclohexene, **6.2**, contains six carbon centres, four of which are tetrahedral and two of which are trigonal planar. The shape of large molecules may more conveniently be described in terms of the geometries about a variety of atomic centres as discussed in Section 6.3.

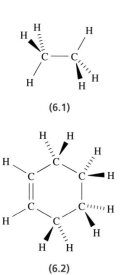

(6.1)

(6.2)

### Coordination number

In molecular compounds, the *coordination number* of an atom defines the number of atoms or groups of atoms that are attached to it. It does *not* rely upon a knowledge of the bond types (single, double or triple) or whether the bonds are localized or delocalized. In later chapters, we meet other ways in which the term 'coordination number' is used.

In molecular compounds, the **coordination number** of an atom defines the number of atoms (or groups of atoms) that are attached to it.

The geometries listed in Table 6.1 give examples of coordination numbers ranging from 2 to 6 for the central atom X in the molecules $XY_n$. In $BF_3$ (**6.3**) and $[CO_3]^{2-}$ (**6.4**), the coordination number of the central atom is 3 even though the bond types are different. Similarly, in $CH_4$ (**6.5**) and $SO_2Cl_2$ (**6.6**), the coordination number of the central atom is 4.

(**6.3**)          (**6.4**)          (**6.5**)          (**6.6**)

### Some aims of this chapter

In this chapter, we describe the shapes of molecules and ions $XY_n$ with a single central atom X, and where Y is an atom or a molecular group. How does the increased molecular complexity involved in going from a diatomic to a polyatomic molecule affect such properties as the molecular dipole moment? We begin by taking a brief 'tour' around the p-block, and summarize the diversity of molecular shapes that is observed, starting with triatomic species.

## 6.2    The geometries of triatomic molecules

A triatomic species may possess a *linear* or *bent* geometry.[§] Examples of linear molecules are carbon dioxide, **6.7**, and hydrogen cyanide, **6.8**, and bent molecules include water, **6.9**, and sulfur dioxide, **6.10**.

(**6.7**)          (**6.8**)          (**6.9**)          (**6.10**)

### Linear triatomic molecules and ions

Figure 6.2 shows the structure of the carbon dioxide molecule, the nitryl cation[†] $[NO_2]^+$, the azide anion $[N_3]^-$ and the cyanate ion $[NCO]^-$. These species are isoelectronic and possess linear structures – they are *isostructural*. Replacing the oxygen atom in the $[NCO]^-$ anion by sulfur gives the thiocyanate anion $[NCS]^-$. *With respect to their valence electrons*, $[NCO]^-$ and $[NCS]^-$ are isoelectronic and possess analogous structures (Figure 6.2). Further examples of species which are isoelectronic with respect to their valence electrons and are isostructural are the linear polyhalide anions $[ICl_2]^-$ (**6.11**) and $[I_3]^-$ (**6.12**).

$$\left[ Cl\!-\!\!-\!I\!-\!\!-\!Cl \right]^-$$

(**6.11**)

$$\left[ I\!-\!\!-\!I\!-\!\!-\!I \right]^-$$

(**6.12**)

Isoelectronic:
see Section 5.12

*Isostructural* means 'having the same structure'.

---

[§] Other terms are often used for the geometry of a bent triatomic molecule; these include non-linear and V-shaped.

[†] Older nomenclature calls $[NO_2]^+$ the nitronium ion; this is still in common use.

**Fig. 6.2** The isoelectronic species $CO_2$, $[NO_2]^+$, $[N_3]^-$ and $[NCO]^-$ all possess linear structures. The cyanate $[NCO]^-$ and thiocyanate $[NCS]^-$ anions have the same number of valence electrons and may be considered to be isoelectronic (*with respect to their valence electrons*). Colour code: C, grey; O, red; N, blue; S, yellow.

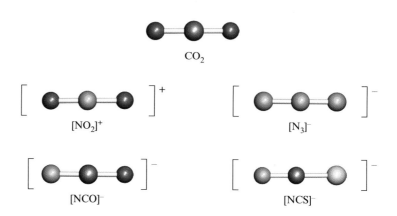

The important message here is that we expect isoelectronic species to possess similar structures; this is true for polyatomic species in general and is not restricted to triatomics. Since the covalent radii of atoms differ, the bond distances in isoelectronic molecules or ions will differ, but the geometry at the central atom will be the same or similar.

However, when comparing molecules and ions that are isoelectronic *only with respect to their valence electrons*, the expectation of isostructural behaviour may *not* hold. A classic example is seen when we compare $CO_2$ with $SiO_2$: at room temperature and pressure, $CO_2$ is a linear molecule but $SiO_2$ has a giant lattice structure containing silicon atoms in tetrahedral environments. An extended solid phase form of $CO_2$ has been made at 1800 K and 40 gigapascal pressure; this has a quartz-like structure, quartz being another polymorph of $SiO_2$. As far as $SiO_2$ is concerned, in the gas phase, we would expect to find linear triatomic molecules.

▭▷
$CO_2$ and $SiO_2$: see Section 22.6

> Molecules or ions that are **isoelectronic** are often isostructural.
>
> However, species that have the same structure are **not necessarily** isoelectronic.

## Bent molecules and ions

**(6.13)**

Figure 6.3 shows some bent molecules. Notice that the bond angle (measured as angle $\alpha$ in structure **6.13**) has no characteristic value. The *exact* geometry of a bent molecule is only known if the bond angle $\alpha$ has been measured experimentally, since the term 'bent' can be used for any triatomic $XY_2$ with $0 < \alpha < 180°$.

Three of the molecules in Figure 6.3 are related to one another: $OF_2$ and $SF_2$ are isoelectronic with respect to their valence electrons, while $H_2O$ and

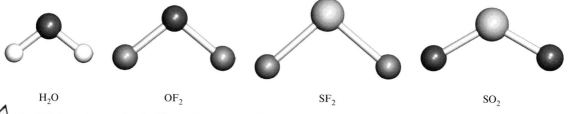

**Fig. 6.3** Some bent molecules illustrating a range of bond angles: $H_2O$ ($\angle H-O-H = 104.5°$), $F_2O$ ($\angle F-O-F = 103°$), $SF_2$ ($\angle F-S-F = 98°$) and $SO_2$ ($\angle O-S-O = 119°$). Colour code: O, red; H, white; S, yellow; F, green.

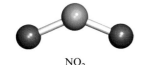

NO$_2$

**Fig. 6.4** The structure of NO$_2$ ($\angle$O$-$N$-$O = 134°). Compare this with the structure of [NO$_2$]$^+$ in Figure 6.2.

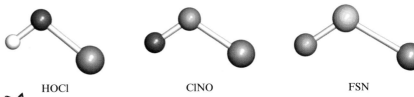

HOCl          ClNO          FSN

**Fig. 6.5** Some asymmetrical bent triatomic molecules: HOCl ($\angle$H$-$O$-$Cl = 102.5°), ClNO ($\angle$O$-$N$-$Cl = 113°) and NSF ($\angle$N$-$S$-$F = 117°). Colour code: O, red; N, blue; Cl, green; F, green; S, yellow; H, white.

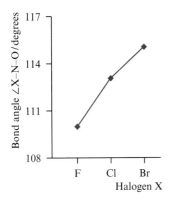

**Fig. 6.6** In the molecules FNO, ClNO and BrNO, the angle $\angle$X$-$N$-$O increases as the halogen atom X becomes larger.

OF$_2$ are related because each possesses a central oxygen atom that forms two single bonds. In addition, SF$_2$ and SO$_2$ are related in that the sulfur atom has a coordination number of 2 in each case. We look at the bonding in SF$_2$ and SO$_2$ in worked example 7.4.

We saw above that the polyhalide anions [ICl$_2$]$^-$ and [I$_3$]$^-$ are linear. The cations [ICl$_2$]$^+$ and [I$_3$]$^+$ are also known but possess bent structures. A change of structure accompanying a change in charge is not unusual. For example, the [NO$_2$]$^+$ cation (Figure 6.2) is linear and yet *both* NO$_2$ and [NO$_2$]$^-$ are bent, although their bond angles are quite different: $\angle$O$-$N$-$O = 134° in NO$_2$ (Figure 6.4) and 115° in the [NO$_2$]$^-$ ion.

Some asymmetrical bent triatomic species are shown in Figure 6.5. The structure of hypochlorous acid (HOCl) is similar to that of H$_2$O, with one chlorine atom replacing a hydrogen atom; another member of this group is Cl$_2$O (related to OF$_2$ in Figure 6.3). Replacing the chlorine atom in ClNO by another group 17 atom gives the structurally similar molecules FNO and BrNO, but with different bond angles (Figure 6.6). Trends in angles can often be correlated with the size and/or the electronegativity of the atoms.

## 6.3    Molecules larger than triatomics described as having linear or bent geometries

Linear and bent geometries are usually associated with triatomic molecules or ions. However, the terms can sometimes be applied to larger molecules in which we describe the molecular shape in terms of the *backbone* of the molecule.

A molecule or ion that contains a backbone of three atoms is readily related to a triatomic species. Figure 6.7a shows that hydrogen cyanide is a linear triatomic molecule, and that acetonitrile (MeCN) is related to hydrogen cyanide. The hydrogen atom in HCN has been replaced by a methyl group, and the methyl group may be regarded as a 'pseudo-atom'. It is useful to consider MeCN as having a CCN backbone as this allows the molecular geometry to be described easily – acetonitrile contains a *linear* CCN backbone.

Me = CH$_3$ = methyl group

Similarly, we can relate the structure of dimethyl sulfide (Me$_2$S) to that of H$_2$S as shown in Figure 6.7b. Me$_2$S retains a bent geometry at the sulfur atom, but the bond angle changes from 92° in H$_2$S to 99° in Me$_2$S. The compound MeSH has a similar structure, but with $\angle$C$-$S$-$H = 96.5°. A related set of compounds is H$_2$O, MeOH and Me$_2$O, each of which is bent with respect to the oxygen centre. Note the way in which this description has been phrased; the emphasis is on the geometry *with respect to each oxygen centre*. The description says nothing about the geometry at each carbon centre.

Alkynes of the general formula RC$\equiv$CR (R is an organic group) possess linear tetraatomic backbones; the first carbon atom in each R group is

**Fig. 6.7** The structural relationships between (a) hydrogen cyanide (HCN) and acetonitrile (MeCN) and (b) $H_2S$ and $Me_2S$. If the methyl group is regarded as a 'pseudo-atom', MeCN can be considered to possess a linear CCN backbone, and $Me_2S$ can be described as a bent molecule. Colour code: C, grey; N, blue; S, yellow; H, white.

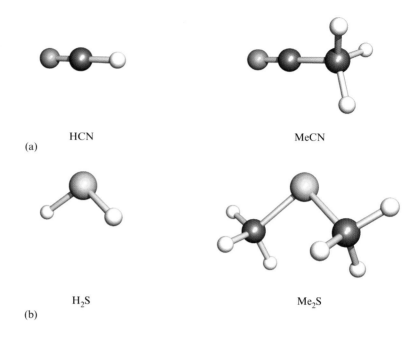

HCN      MeCN

(a)

$H_2S$      $Me_2S$

(b)

counted into the backbone. Figure 6.8 shows the structures of HC≡CH and MeC≡CMe. The change from hydrogen to methyl substituents does not alter the fundamental linear geometry of the central part of the molecule. Remember, though, that describing MeC≡CMe as 'linear' only specifies the geometry with respect to the four central carbon atoms. On the other hand, the HC≡CH molecule *is* truly linear.

The small number of examples in this section make an important and useful point – the geometries of large molecules can often be adequately described by considering the *local geometry* at individual atomic centres.

**Fig. 6.8** The structural relationship between the alkynes HC≡CH and MeC≡CMe; HC≡CH is truly linear, but MeC≡CMe has a linear CCCC backbone.

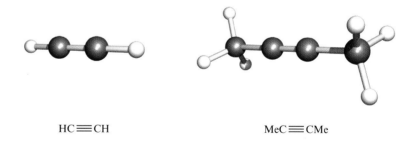

HC≡CH      MeC≡CMe

| 6.4 | **Geometries of molecules within the *p*-block: the first row** |

1–2    3–12    13–18

*s*-block    *d*-block    *p*-block

In this section we consider the shapes of molecules that contain a *p*-block element as the central atom, and confine our survey to compounds of the type $XY_n$ with one central atom and one type of substituent. Figure 6.9 shows the five geometries that are common for such molecules in which X is one of the first row elements boron, carbon, nitrogen or oxygen; Table 6.2 lists some species that adopt these structures. The first row element fluorine is not found at the centre of polyatomic molecules[§] and adopts *terminal* sites, for example in $BF_3$, $CF_4$, $NF_3$ and $OF_2$.

---

[§] Exceptions are some species formed in reactions carried out in liquid HF, e.g. $[H_2F]^+$ and $[H_2F_3]^-$.

**Fig. 6.9** The five common shapes for molecules of type $XY_n$ where X is an element from the first row of the *p*-block.

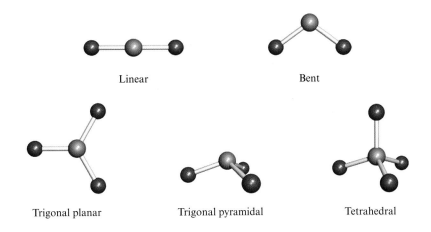

Linear

Bent

Trigonal planar

Trigonal pyramidal

Tetrahedral

## Boron (group 13)

3-Coordinate compounds containing one boron centre are trigonal planar, and examples include $BF_3$, $BCl_3$, $BMe_3$ and $[BO_3]^{3-}$. Tetrahedral boron centres are seen in a range of anions such as $[BH_4]^-$ and $[BF_4]^-$.

See also Section 6.14

O=C, with H and OH

**(6.14)**

## Carbon (group 14)

For molecules containing carbon, one of three geometries is generally observed – linear, trigonal planar or tetrahedral. Linear carbon centres are observed in the triatomic species $CO_2$, $[NCO]^-$ and HCN. A trigonal planar carbon centre is present in $H_2C=O$, $Cl_2C=O$, $HCO_2H$ (**6.14**) and $[CO_3]^{2-}$.

Tetrahedrally coordinated carbon is widely represented, for example in saturated alkanes such as methane, $CH_4$. The isoelectronic relationship between $CH_4$ and $[BH_4]^-$ suggests that they will be isostructural (Figure 6.10).

## Nitrogen (group 15)

Common geometries for nitrogen atoms are linear (Figure 6.2), bent (Figure 6.4), trigonal planar, trigonal pyramidal and tetrahedral.

**Table 6.2** Examples of molecules and ions of general formula $XY_n$, $[XY_n]^{m+}$ or $[XY_n]^{m-}$ in which X is a first row *p*-block element (boron to oxygen). The atom X is shown in pink.

| Shape (refer to Table 6.1 and Figure 6.9) | Examples (Me = methyl = $CH_3$) |
|---|---|
| Linear | $CO_2$, $[NO_2]^+$, $[NCO]^-$, $[N_3]^- = [NNN]^-$ |
| Bent | $H_2O$, $OF_2$, HOF, $NO_2$, $[NO_2]^-$, $[NH_2]^-$ |
| Trigonal planar | $BF_3$, $BCl_3$, $BBr_3$, $BMe_3$, $[BO_3]^{3-}$, $[CO_3]^{2-}$, $[NO_3]^-$ |
| Trigonal pyramidal | $NH_3$, $NF_3$, $Me_3N$, $[H_3O]^+$ |
| Tetrahedral | $[BH_4]^-$, $[BF_4]^-$, $CH_4$, $CMe_4$, $CF_4$, $[NH_4]^+$, $[NMe_4]^+$ |

**Fig. 6.10** The species $[BH_4]^-$, $CH_4$ and $[NH_4]^+$ are isoelectronic and isostructural; $NH_3$ and $[H_3O]^+$ are also isoelectronic and isostructural. Colour code: B, orange; H, white; C, grey; N, blue; O, red.

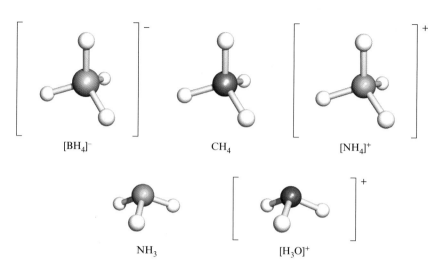

(6.15)

In neutral 3-coordinate compounds, the trigonal pyramidal geometry is usual for nitrogen – for example $NH_3$ (Figure 6.10) and $NF_3$. The family can be extended to include organic amines with the general formulae $RNH_2$, $R_2NH$ and $R_3N$ (**6.15**) where R is an organic group.

The ammonium ion $[NH_4]^+$ is isoelectronic with $CH_4$ and $[BH_4]^-$ (Figure 6.10) and is tetrahedral. Trigonal planar nitrogen is seen in the nitrate ion $[NO_3]^-$, and the isoelectronic principle establishes a structural relationship between $[NO_3]^-$, $[CO_3]^{2-}$ and $[BO_3]^{3-}$.

## Oxygen (group 16)

Simple molecules with oxygen as the central atom are bent (Figure 6.3). In 3-coordinate cations, oxygen is trigonal pyramidal, and the simplest example is the oxonium ion $[H_3O]^+$ which is isoelectronic with $NH_3$ (Figure 6.10).

## Overview of the first row of the *p*-block

- Compounds of the first row *p*-block elements tend to possess one of the five shapes shown in Figure 6.9.
- A greater range of geometries is observed for boron, carbon and nitrogen than for oxygen and fluorine.
- A change in the overall charge often results in a change in geometry, e.g. $NO_2$ to $[NO_2]^-$.

| 6.5 | **Heavier *p*-block elements** |

*p*-Block chemistry: see Chapter 22

The elements that are discussed in this section are shown in Figure 6.11, and the relevant molecular shapes are shown in Figures 6.9 and 6.12; Table 6.3 lists examples of each type of structure.

## Group 13

Temperature-dependent structure of AlCl₃: see Section 22.4

At high temperatures, the aluminium trihalides $AlCl_3$, $AlBr_3$ and $AlI_3$ exist as molecular species with trigonal planar structures. However, a tetrahedral environment is common for aluminium, gallium and indium atoms and is observed in a range of anions including $[AlH_4]^-$, $[AlCl_4]^-$, $[GaBr_4]^-$ and

**Fig. 6.11** The heavier *p*-block elements (excluding the noble gases – group 18).

| | Group 13 | Group 14 | Group 15 | Group 16 | Group 17 |
|---|---|---|---|---|---|
| **2nd row** | **Al** aluminium | **Si** silicon | **P** phosphorus | **S** sulfur | **Cl** chlorine |
| **3rd row** | **Ga** gallium | **Ge** germanium | **As** arsenic | **Se** selenium | **Br** bromine |
| **4th row** | **In** indium | **Sn** tin | **Sb** antimony | **Te** tellurium | **I** iodine |
| **5th row** | **Tl** thallium | **Pb** lead | **Bi** bismuth | **Po** polonium | **At** astatine |

**Fig. 6.12** Molecular shapes (in addition to those shown in Figure 6.9) observed for molecules of type $XY_n$ where X is a heavier *p*-block element.

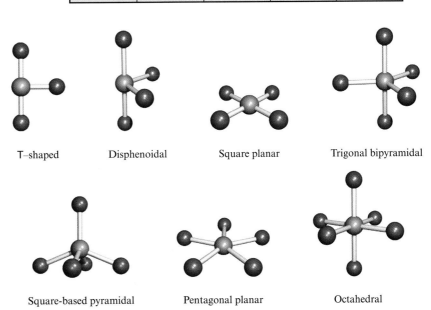

T–shaped          Disphenoidal          Square planar          Trigonal bipyramidal

Square-based pyramidal          Pentagonal planar          Octahedral

**Table 6.3** Examples of molecules of general formula $XY_n$ in which X is a heavy *p*-block element (see Figures 6.9 and 6.12). The atom X is shown in pink.

| Shape | Examples |
|---|---|
| Linear | $[ICl_2]^-$, $[I_3]^-$ |
| Bent | $SnCl_2$, $H_2S$, $SF_2$, $SO_2$, $H_2Se$, $SeCl_2$, $TeCl_2$ |
| Trigonal planar | $AlCl_3$, $AlBr_3$, $SO_3$ |
| Trigonal pyramidal | $PF_3$, $PCl_3$, $AsF_3$, $SbCl_3$, $PMe_3$, $AsMe_3$, $[PO_3]^{3-}$, $[SO_3]^{2-}$, $[SeO_3]^{2-}$, $[ClO_3]^-$ |
| T-shaped | $ClF_3$, $BrF_3$ |
| Tetrahedral | $SiH_4$, $GeH_4$, $SiF_4$, $SiCl_4$, $GeF_4$, $SnCl_4$, $PbCl_4$, $[AlH_4]^-$, $[GaBr_4]^-$, $[InCl_4]^-$, $[TlI_4]^-$, $[PBr_4]^+$, $[PO_4]^{3-}$, $[SO_4]^{2-}$, $[ClO_4]^-$ |
| Disphenoidal | $SF_4$, $[PBr_4]^-$, $[ClF_4]^+$, $[IF_4]^+$ |
| Square planar | $[ClF_4]^-$, $[BrF_4]^-$, $[ICl_4]^-$ |
| Trigonal bipyramidal | $PCl_5$, $AsF_5$, $SbF_5$, $SOF_4$, $[SnCl_5]^-$ |
| Square-based pyramidal | $ClF_5$, $BrF_5$, $IF_5$, $[SF_5]^-$, $[TeF_5]^-$, $[SbF_5]^{2-}$ |
| Pentagonal planar | $[IF_5]^{2-}$ |
| Octahedral | $SF_6$, $SeF_6$, $Te(OH)_6$, $[IF_6]^+$, $[PF_6]^-$, $[AsCl_6]^-$, $[SbF_6]^-$, $[SnF_6]^{2-}$, $[AlF_6]^{3-}$ |

(6.16)

$[InCl_4]^-$. These two shapes resemble those seen in compounds of boron, but simple species containing heavier group 13 elements also exhibit geometries with higher coordination numbers – for example, the ions $[AlF_6]^{3-}$ (6.16) and $[Ga(H_2O)_6]^{3+}$ are octahedral.

## Group 14

CO$_2$ and SiO$_2$:
see Section 22.6

In the gas phase, $SnCl_2$ is bent, but other tin(II) compounds generally have more complex structures. Many compounds of group 14 elements have formulae reminiscent of carbon compounds but caution is needed – $CO_2$ and $SiO_2$ may appear to be similar but their structures are different.

Silane, $SiH_4$, is tetrahedral like $CH_4$, and the same isoelectronic and isostructural relationship exists between $SiH_4$ and $[AlH_4]^-$ as between $CH_4$ and $[BH_4]^-$. Similarly, $GeH_4$ is isoelectronic and isostructural with $[GaH_4]^-$. Other tetrahedral molecules include $SiF_4$, $GeF_4$, $SnCl_4$ and $PbCl_4$.

(6.17)

Within group 14, coordination numbers greater than 4 are exemplified by the $[SnCl_5]^-$ (6.17) and $[Me_2SnCl_3]^-$ anions (which have trigonal bipyramidal structures) and in the octahedral $[SnF_6]^{2-}$ dianion.

## Group 15

Structural diversity is further extended in the compounds formed by the heavier group 15 elements.

The trigonal pyramid is a common shape for species containing phosphorus, arsenic and antimony. Examples include $PF_3$, $AsF_3$, $SbCl_3$, $PMe_3$ (6.18) and $[PO_3]^{3-}$. Although our discussion concentrates on molecules and ions with a single group 15 atomic centre, it is worth remembering that each phosphorus atom in the $P_4$ molecule is in a trigonal pyramidal site (Figure 4.2).

(6.18)

A coordination number of 4 for a group 15 element is usually associated with a tetrahedral geometry but the disphenoidal structure (Figure 6.12) is also observed. The ions $[PCl_4]^+$, $[PBr_4]^+$ and $[PO_4]^{3-}$ are tetrahedral, but the change of charge in going from $[PBr_4]^+$ to $[PBr_4]^-$ is associated with a change in shape and the anion is disphenoidal, as shown in structure 6.19.

The number of 5-coordinate species is large. In the gas phase, the pentahalides $PF_5$, $AsF_5$ and $SbCl_5$ are trigonal bipyramidal, but a change in charge leads to a change in shape – in the solid state, the $[SbF_5]^{2-}$ ion has a square-based pyramidal structure (Figure 6.13).

(6.19)

Octahedral structures are common for anionic species in group 15 and examples include $[PF_6]^-$, $[AsF_6]^-$ and $[SbF_6]^-$.

## Group 16

For oxygen, only bent and trigonal pyramidal shapes are favoured, but for the heavier elements in group 16, the variety of observed geometries is far greater.

**Fig. 6.13** The square-based pyramidal structure of $[SbF_5]^{2-}$: (a) a schematic representation, (b) a 3-dimensional representation and (c) the square-based pyramidal arrangement of the F atoms. Colour code: Sb, orange; F, green.

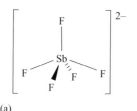

(a)

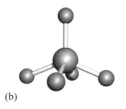

(b)

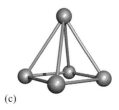

(c)

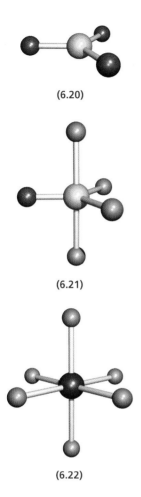

**(6.20)**

**(6.21)**

**(6.22)**

Bent molecules include $H_2S$, $SF_2$, $SO_2$, $SeCl_2$ and $TeCl_2$. Trigonal planar molecules are exemplified by $SO_3$ (**6.20**), while trigonal pyramidal structures are observed for $SOCl_2$, $SeOCl_2$, $[SO_3]^{2-}$ and $[TeO_3]^{2-}$. Once again, a change of shape accompanies a change in charge as is seen in going from $SO_3$ to $[SO_3]^{2-}$.

4-Coordinate sulfur, selenium and tellurium atoms are found in both tetrahedral and disphenoidal environments. The $[SO_4]^{2-}$ ion is tetrahedral, but $SF_4$, $SeF_4$ and $Me_2TeI_2$ possess disphenoidal geometries. For 5-coordinate atoms, both the trigonal bipyramidal (e.g. $SOF_4$, **6.21**) and square-based pyramidal (e.g. $[SF_5]^-$) geometries are observed. Octahedral molecules include $SF_6$, $SeF_6$ (**6.22**) and $TeF_6$.

## Group 17

Unlike fluorine, the heavier halogens are found in a wide range of geometrical environments. Iodine is the central atom in the linear anion $[I_3]^-$ and the bent cation $[I_3]^+$. For 3-coordinate atoms, both T-shaped and trigonal pyramidal species (Figure 6.12) are known; $ClF_3$ is an example of a T-shaped molecule in which the two F–Cl–F bond angles are $87°$. The anion $[ClO_3]^-$ is trigonal pyramidal.

Elements from group 17 may adopt one of *three* different 4-coordinate structures. Tetrahedral anions include $[ClO_4]^-$ and $[IO_4]^-$, while the cations $[ClF_4]^+$ and $[IF_4]^+$ are disphenoidal; $[ClF_4]^+$ is isoelectronic with $SF_4$ (see above). On going from $[ClF_4]^+$ to $[ClF_4]^-$, there is a change in structure from a disphenoidal to a square planar geometry.

5-Coordinate halogen atoms provide examples of trigonal bipyramidal (e.g. $ClO_2F_3$), square-based pyramidal (e.g. $ClF_5$ and $IF_5$) and pentagonal planar ($[IF_5]^{2-}$) shapes. The cation $[IF_6]^+$ is an example of a 6-coordinate, octahedral species.

### Overview of the heavier *p*-block elements

- Compounds of the heavier elements in groups 13 to 17 of the *p*-block show a greater range of shapes than do the first row elements of each respective group.
- The variety of molecular shapes *increases* on moving across the *p*-block from group 13 to 17.
- A change in the overall charge of a species usually causes a change in geometry.

## MID-CHAPTER PROBLEMS

Before moving on to Section 6.6, use the information in Sections 6.1–6.5 to answer these questions.

**1** *n*-Butane, $CH_3CH_2CH_2CH_3$, is a so-called 'straight-chain' alkane and contains four tetrahedral carbon centres. Is the $C_4$ backbone linear or non-linear?

**2** Give examples of three tetrahedral molecular species of type $XY_n$.

**3** Are $Me_3N$ and $[Me_3O]^+$ isoelectronic? Are they isostructural?

**4** Which of the following molecular species are trigonal planar? (a) $PBr_3$; (b) $SO_3$; (c) $AlCl_3$; (d) $NH_3$; (e) $[NO_3]^-$; (f) $[CO_3]^{2-}$; (g) $[SO_3]^{2-}$; (h) $BrF_3$.

**5** Sketch diagrams to show the following shapes for $XY_n$ molecules and check the answers in Figures 6.9 and 6.12: (a) octahedral; (b) tetrahedral; (c) trigonal pyramidal; (d) trigonal bipyramidal; (e) square-based pyramidal; (f) T-shaped; (g) pentagonal planar.

**6** In which of the structures (a) to (g) in problem 5 are the Y atoms *equivalent*?

**7** (a) What shape is $CF_4$? (b) Using isoelectronic relationships to help you, determine which of the following species are likely to have the same structure as $CF_4$: $SF_4$, $[PF_4]^+$, $[IF_4]^-$, $[BF_4]^-$, $[AlF_4]^-$, $SiF_4$.

| 6.6 | **The valence-shell electron-pair repulsion (VSEPR) model** |
|---|---|

In the previous sections, we have discussed the shapes of molecules and molecular ions purely as observables, and we have noted some trends in the geometrical environments found for certain elements within both groups and periods in the $p$-block. In this section, we consider a method for predicting or rationalizing the shape of a molecule or ion.

*The valence-shell electron-pair repulsion* (VSEPR) model was initially proposed by Sidgwick and Powell in 1940, and further developed by Nyholm and Gillespie. The basis of the VSEPR model is the consideration of the *repulsive forces between pairs of valence electrons*, and it is assumed that *only the electrons in the valence shell influence the molecular geometry*.

### Minimizing inter-electron repulsion

The VSEPR model assumes that for maximum stability pairs of valence electrons will be as far apart from one another as possible. In this way, inter-electron repulsions are minimized. Assume that in a molecule $XY_n$, the atoms Y are all bonded to a central atom X and that all the valence electrons of X are involved in bonding. Let each X−Y bond be a single bond; i.e. one pair of electrons is associated with each bond. For a minimum energy system, the relative positions of the $n$ pairs of electrons can be determined by considering each pair of electrons as a point of negative charge, and placing these points on the surface of a sphere.

For $n = 2$, the repulsive interactions between the two point charges are at a minimum if the charges are at opposite sides of the sphere (Figure 6.14a). This corresponds to a *linear* arrangement for the atoms in the molecule $XY_2$.

For $n = 3$, the total inter-electron repulsion between the three point charges is minimized when the charges are placed at the corners of a triangle (Figure 6.14b). This corresponds to a *trigonal planar* arrangement for the atoms in the molecule $XY_3$.

**Fig. 6.14** Pairs of valence electrons can be considered as point charges. They repel one another and move as far away from one another as possible to minimize the energy of the system. (a) Two points of negative charge take up positions at opposite sides of the sphere and this results in a linear arrangement of atoms in the molecule $XY_2$ if all the valence electrons in the molecule are used for bonding. (b) The mutual repulsion between three points of negative charge results in a triangular arrangement for minimum energy; this gives a trigonal planar arrangement of Y atoms in the molecule $XY_3$ if all the valence electrons in the molecule are used for bonding.

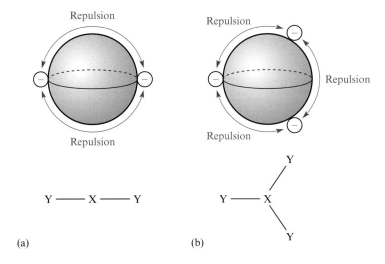

**Fig. 6.15** For four, five or six points of negative charges, repulsions are minimized when they lie at the corners of a tetrahedron, a trigonal bipyramid or an octahedron, respectively. For molecules $XY_n$ in which all the valence electrons are used for bonding, this corresponds to tetrahedral, trigonal bipyramidal or octahedral shapes for the molecules $XY_4$, $XY_5$ and $XY_6$, respectively.

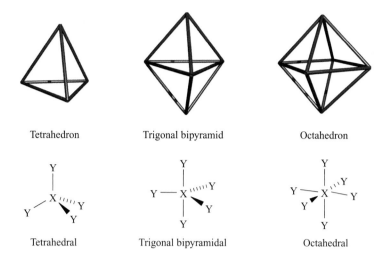

Similarly, for $n = 4$, 5 and 6, the geometries that correspond to minimum inter-electron repulsions are the tetrahedron, the trigonal bipyramid and the octahedron. Figure 6.15 illustrates how these polyhedra relate to the *tetrahedral, trigonal bipyramidal* and *octahedral* shapes of the $XY_4$, $XY_5$ and $XY_6$ molecules.

Notice that, although in Table 6.1 we listed more than one geometrical possibility for $n = 2$, 3, 4 and 5, only one geometry corresponds to the lowest energy structure in each case.

Energy differences between structures: see Section 6.12

### Lone pairs of electrons

Not all valence electrons are necessarily involved in bonding. The Lewis structure for $H_2O$ (**6.23**) reveals that there are two lone pairs in the valence shell of the oxygen atom. On the other hand, in $CH_4$ (**6.24**) all of the valence electrons are used for bonding.

Inter-electron repulsions in the valence shell of the central atom in a molecule involve *both lone and bonding pairs* of electrons, and so the presence of one or more lone pairs influences the molecular shape. Lewis structures for $BH_3$ and $NH_3$ are shown on the left-hand side of Figures 6.16a and 6.16b. The boron atom uses all of its valence electrons to form three B–H single bonds, but the nitrogen atom has five valence electrons, and one lone pair remains after the three N–H bonds have been formed. There are *three* pairs of electrons in the valence shell of the boron atom in $BH_3$, and by the VSEPR model, the molecule will be trigonal planar (Figure 6.16a). In $NH_3$, there are *four* pairs of electrons in the valence shell of the nitrogen

**(6.23)**

**(6.24)**

**Fig. 6.16** The application of the VSEPR model to predict the shapes of the (a) $BH_3$ and (b) $NH_3$ molecules. Note the role of the lone pair of electrons in $NH_3$.

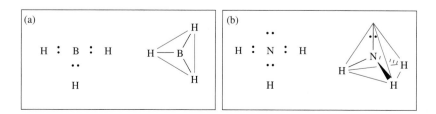

**Fig. 6.17** Multiple bonds contain greater electron density than single bonds, and so inter-electron repulsions involving multiple bonds are greater than those involving single bonds.

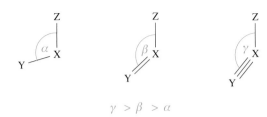

$$\gamma > \beta > \alpha$$

atom, and the VSEPR model predicts that the molecular shape will be based upon a tetrahedron with the lone pair of electrons occupying one of the vertices (Figure 6.16b). The $NH_3$ molecule may be described as being *derived from* a tetrahedron, although the nitrogen and three hydrogen atoms actually define a trigonal pyramid (Figure 6.10).

### Relative magnitudes of inter-electron repulsions

In $XY_n$ compounds that contain *only* bonding pairs of electrons in the valence shell of the central atom, we can assume that the electron–electron repulsions are all equal provided that the bonds are all identical.

- Multiple bonds: multiple bonds contain greater densities of electrons (more negative charge) than single bonds, and therefore, inter-electron repulsions involving multiple bonds are greater than those involving single bonds (Figure 6.17).
- Lone pairs: if one or more lone pairs of electrons are present, the relative magnitudes of the inter-electron repulsions are assumed to follow the sequence: lone pair–lone pair > lone pair–bonding pair > bonding pair–bonding pair.

These differences in repulsions affect the detailed geometry of a molecule as illustrated in the series $CH_4$, $NH_3$ and $H_2O$. The total number of inter-electron repulsive interactions in each molecule is six (Figure 6.18a). In the $CH_4$ molecule, these are all between bonding pairs of electrons, but in the $NH_3$ and $H_2O$ molecules, lone pair–bonding pair or lone pair–lone

**Fig. 6.18** (a) In a tetrahedral arrangement of four electron pairs, there are six repulsive interactions. (b) In the $CH_4$ molecule, there are six bonding pair–bonding pair repulsions and all the H–C–H bond angles are equal. In $NH_3$, there are three lone pair–bonding pair repulsions and three bonding pair–bonding pair repulsions. The H–N–H angle in $NH_3$ is therefore smaller than the H–C–H angle in $CH_4$. In the $H_2O$ molecule, there is one lone pair–lone pair repulsion, one bonding pair–bonding pair repulsion, and four lone pair–bonding pair repulsions. The H–O–H angle is smaller than the H–N–H angle in $NH_3$ or the H–C–H angle in $CH_4$.

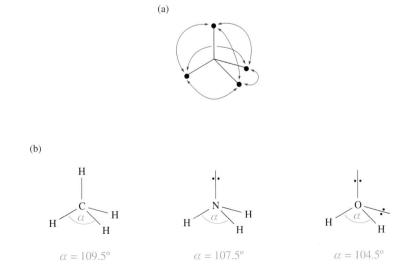

(a)

(b)

H
|
C
H    H
α
H

$\alpha = 109.5°$

N
H    H
α
H

$\alpha = 107.5°$

O
H    H
α

$\alpha = 104.5°$

pair interactions are involved. The result of increased inter-electron repulsion is a decrease in the bond angle as shown in Figure 6.18b.

The *valence-shell electron-pair repulsion* (VSEPR) model is used to predict or rationalize molecular shapes. The model considers only the repulsions between the electrons in the valence shell of the central atom in the molecule or molecular ion.

- Pairs of valence electrons are arranged so as to *minimize* inter-electron repulsions.

- The relative magnitudes of repulsive forces between pairs of electrons follow the order:

  lone pair–lone pair $>$ lone pair–bonding pair $>$ bonding pair–bonding pair

- For $n = 2$ to 6 pairs of valence electrons, the ideal geometries are:
  $n = 2$    linear
  $n = 3$    trigonal planar
  $n = 4$    tetrahedral
  $n = 5$    trigonal bipyramidal
  $n = 6$    octahedral

## Application of the VSEPR model to species involving *p*-block elements

A molecular geometry can be predicted by using the valence-shell electron-pair model by following the procedure set out below.

1. Draw a Lewis structure for the molecule or ion, and determine the number of bonding and lone pairs of electrons in the valence shell of the central atom.
2. The 'parent' geometry (Table 6.4) is determined by the number of points of negative charge – a 'point of charge' is taken to be the electrons involved in a single bond, a multiple bond or a lone pair.
3. If all bonds are single bonds, the amount of distortion away from the ideal bond angles (see Table 6.1) for the arrangement determined in step (2) can be estimated by using the sequence of relative inter-electron repulsions:

lone pair–lone pair $>$ lone pair–bonding pair

$>$ bonding pair–bonding pair

**Table 6.4** Basic arrangements of points of charge (i.e. bonds or lone pairs of electrons) that are used to predict molecular geometries using the VSEPR model. Ideal angles are listed in Table 6.1.

| Number of points of negative charge | Arrangement |
| --- | --- |
| 2 | Linear |
| 3 | Trigonal planar |
| 4 | Tetrahedral |
| 5 | Trigonal bipyramidal |
| 6 | Octahedral |

4. If the electrons are involved in bonds other than single bonds, account must be taken of the fact that:
   repulsion due to the electrons in a triple bond > repulsion due to the electrons in a double bond > repulsion due to the electrons in a single bond

5. In a trigonal bipyramid, lone pairs of electrons usually occupy the equatorial rather than axial sites. A multiple bond also tends to occupy an equatorial site.

6. In an octahedron, two lone pairs are opposite (rather than next) to one another.

**Example 1: H$_2$O**   Oxygen is in group 16 and has six valence electrons.
The hydrogen atom has one valence electron.
The Lewis structure for H$_2$O is:

There are four points of negative charge (two bonding pairs and two lone pairs of electrons).

The parent shape is a tetrahedron. The H$_2$O molecule is therefore derived from a tetrahedron and the three atoms define a bent structure **6.25**.

The ∠H−O−H should be less than 109.5°. (The experimentally determined angle is 104.5°.)

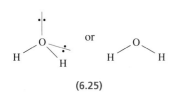

**(6.25)**

**Example 2: [NH$_4$]$^+$**   Nitrogen is in group 15 and has five valence electrons.
The positive charge may be formally assigned to the nitrogen atom; N$^+$ has four valence electrons.

The hydrogen atom has one valence electron.
The Lewis structure for the [NH$_4$]$^+$ ion is:

There are four points of negative charge (four bonding pairs of electrons).
The parent shape is a tetrahedron.

The [NH$_4$]$^+$ ion therefore has the tetrahedral structure **6.26** and the ∠H−N−H will be 109.5°. Compare the predicted result with the structure drawn in Figure 6.10.

**(6.26)**

**Example 3: CO$_2$**   Carbon is in group 14 and has four valence electrons. Oxygen is in group 16 and has six valence electrons.
The Lewis structure for the CO$_2$ molecule is:

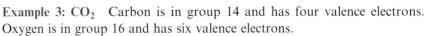

$$O = C = O$$

**(6.27)**

The carbon atom is involved in two double bonds; there are two points of negative charge.

The $CO_2$ molecule (**6.27**) is linear. Compare the predicted result with the structure in Figure 6.2. Why do the lone pairs of electrons on the oxygen atoms not affect the molecular shape?

**Example 4: $SO_2$** Sulfur and oxygen are both in group 16 and each has six valence electrons.

The Lewis structure for the $SO_2$ molecule is:

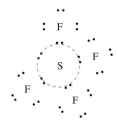

There are three points of negative charge around the sulfur atom consisting of one lone pair of electrons, and two points of charge which are associated with the S=O double bonds.

The geometry of the $SO_2$ molecule **6.28** is derived from a trigonal planar arrangement, the ideal angles for which are 120°. In structure **6.28** it is not easy to assess the relative magnitudes of the lone pair–double bond and double bond–double bond inter-electron repulsions. The experimental value for the angle $\angle O-S-O$ is 119°.

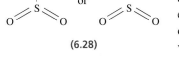

**(6.28)**

**Example 5: $SF_4$** Sulfur is in group 16 and has six valence electrons. Fluorine is in group 17 and has seven valence electrons.

The Lewis structure for the $SF_4$ molecule is:

There are five points of negative charge around the sulfur atom consisting of four bonding pairs and one lone pair of electrons.

The parent geometry is trigonal bipyramidal and the lone pair occupies an equatorial site in order to minimize repulsive interactions. The atoms in the $SF_4$ molecule define the disphenoidal geometry **6.29**. There are two different fluorine environments – the axial ($F_{ax}$) and equatorial ($F_{eq}$) positions. Do the fluorine lone pairs have an effect on the molecular shape?

The $F_{eq}-S-F_{eq}$ bond angle defined by the fluorine atoms in the equatorial plane will be larger than the angles $\angle F_{ax}-S-F_{eq}$. Taking into account the relative lone pair–lone pair, lone pair–bonding pair and bonding pair–bonding pair inter-electron repulsions, we can predict bond angles as follows:

$$\angle F_{ax}-S-F_{ax} \approx 180° \qquad 90° < \angle F_{eq}-S-F_{eq} < 120°$$

**(6.29)**

The experimentally determined bond angles in $SF_4$ are $\angle F_{ax}-S-F_{ax} = 173°$ (measured as shown in **6.29**) and $\angle F_{eq}-S-F_{eq} = 102°$.

**Example 6: $[ICl_4]^-$** Iodine and chlorine are both in group 17 and each has seven valence electrons. The negative charge may be formally assigned to the central iodine atom; $I^-$ has eight valence electrons.

The Lewis structure for the $[ICl_4]^-$ ion is:

There are six points of negative charge around the iodine atom (four bonding pairs of electrons and two lone pairs of electrons).

The geometry of the $[ICl_4]^-$ ion **6.30** is derived from an octahedron. The lone pairs will be opposite one another in order to minimize their mutual repulsions. The five atoms define a square planar structure.

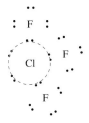

(6.30)

### Apparent expansion of the octet of valence electrons

In examples 4–6 above, the central atom (S in $SO_2$, S in $SF_4$ and I in $[ICl_4]^-$) is shown with more than an octet of electrons in its valence shell. This apparent violation of the octet rule can be ignored when we apply VSEPR theory. The VSEPR model considers the repulsions between pairs of electrons and provides a successful method of rationalizing molecular structures of species containing *p*-block elements. The VSEPR model implies nothing about the actual bonding in a molecule; we return to this in Chapter 7.

---

**6.7**    **The VSEPR model: some ambiguities**

The advantage of the VSEPR model is that it is simple to apply. It is a successful model in *p*-block chemistry and its use is widespread. However, there are some species in which ambiguities can arise, and in this section we look at several such cases.

### Chlorine trifluoride

The structure of the interhalogen compound chlorine trifluoride can be predicted as follows.

The Lewis structure of $ClF_3$ is:

F —— Cl$^{\prime\prime\prime\prime\prime}$F

**(6.31)**

F —— Cl

F

**(6.32)**

F —— Cl$^{\prime\prime\prime\prime\prime}$F

F

**(6.33)**

There are five centres of negative charge around the chlorine atom, and the geometry of $ClF_3$ is based on a trigonal bipyramid.

The problem now is 'which sites do the lone pairs prefer?' The alternatives are given below and the problem comes in assessing the relative merits of each.

1. Place the two lone pairs as far apart as possible as shown in structure **6.31**. This minimizes the lone pair–lone pair interaction, but maximizes the number of 90° lone pair–bonding pair interactions.
2. Place the two lone pairs in the equatorial plane to give structure **6.32**. This places the two lone pairs of electrons at 120° to one another. In addition, there are four lone pair–bonding pair repulsive interactions acting at 90° and two acting at 120°, plus two 90° bonding pair–bonding pair interactions.
3. Place the two lone pairs at 90° to each other, **6.33**. This is the worst possible arrangement in terms of repulsive interactions because the lone pairs of electrons are at 90° to one another. Hence, structure **6.33** can be discounted.

Structures **6.31** and **6.32** are possible but it is extremely difficult to judge between them qualitatively in terms of minimizing the repulsive energy of the system. The experimentally determined structure of $ClF_3$ shows the molecule to be T-shaped, with $F_{ax}$–Cl–$F_{eq}$ bond angles of 87°. This result is consistent with structure **6.32** – the lone pairs are placed in the equatorial plane of the trigonal bipyramid. The reduction in $F_{ax}$–Cl–$F_{eq}$ bond angle from an ideal value of 90° to an experimental one of 87° may be interpreted in terms of lone pair–bonding pair repulsions.

### The anions $[SeCl_6]^{2-}$ and $[TeCl_6]^{2-}$

Selenium and tellurium are in group 16 and each possesses six valence electrons. In the ions $[SeCl_6]^{2-}$ and $[TeCl_6]^{2-}$, the valence shell of the central atom contains 14 electrons (six from the group 6 element, one from each chlorine atom and two from the dinegative charge). The VSEPR model predicts that each ion possesses a structure based upon a 7-coordinate geometry, with the lone pair occupying one site. However, the observed structure of each of the $[SeCl_6]^{2-}$ and $[TeCl_6]^{2-}$ (**6.34**) anions is that of a *regular octahedron*. This experimental result suggests that the lone pair of electrons does not influence the shape in the way that the VSEPR model predicts it should. Such a lone pair is said to be *stereochemically inactive*.

We cannot readily predict the structures of $[SeCl_6]^{2-}$ and $[TeCl_6]^{2-}$, but we can rationalize them in terms of the presence of a stereochemically inactive pair of electrons. Stereochemically inactive lone pairs of electrons are usually observed for the heaviest members of a periodic group, and the tendency for valence shell *s* electrons to adopt a non-bonding role in a molecule is known as the *stereochemical inert pair effect*.

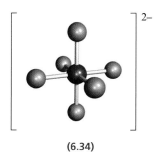

**(6.34)**

When the presence of a lone pair of electrons influences the shape of a molecule or ion, the lone pair is said to be ***stereochemically active***.

If the geometry of a molecule or ion is *not* affected by the presence of a lone pair of electrons, then the lone pair is ***stereochemically inactive***.

The tendency for the pair of valence *s* electrons to adopt a non-bonding role in a molecule or ion is know as the ***stereochemical inert pair effect***.

| | | |
|---|---|---|

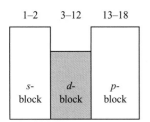

Ligand: see Section 23.3

The ***Kepert model*** predicts the shapes of *d*-block metal compounds $ML_n$ by considering the repulsions between the groups L. Lone pairs of electrons are ignored in the model.

### 6.8 The Kepert model

The VSEPR model is most commonly used for the prediction and rationalization of the shapes of molecules and ions containing central *p*-block atoms. Kepert developed the VSEPR model so that it is applicable to molecules containing a *d*-block metal as the central atom. Consider a molecule of formula $ML_n$ where M is a *d*-block metal and L is a *ligand* – a group attached to the metal centre.

In Kepert's model, the metal lies at the centre of a sphere and the ligands are free to move over the surface of the sphere. The ligands are considered to repel one another in much the same way that the point charges repel one another in the VSEPR model. Kepert's approach predicts the relative positions of the ligands in a molecule $ML_n$ in an analogous manner to the way in which the VSEPR model predicts the relative positions of the groups Y in a molecule $XY_n$. Ions of the type $[ML_n]^{m+}$ or $[ML_n]^{m-}$ can be treated similarly. The polyhedra listed in Table 6.4 apply both to molecules containing *p*-block and *d*-block atomic centres.

The prediction of the shapes of the *d*-block molecules is considered by Kepert to be *independent* of the ground state electronic configuration of the metal centre. The model differs from the VSEPR model in that it ignores the presence of any lone pairs of electrons in the valence shell of the metal centre.

### 6.9 Application of the Kepert model

#### Linear geometry

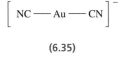

**(6.35)**

In the anion $[Au(CN)_2]^-$, two cyanide groups are bonded to the gold centre. The repulsion between them is minimized if they are 180° apart. The anion is predicted to have the linear NC–Au–CN framework **6.35**, and this agrees with the observed structure in which the bond angle $\angle C-Au-C$ is 180°.

The $[HgCl_2]^-$ and $[AuCl_2]^-$ anions have similar structures to that of $[Au(CN)_2]^-$.

#### Trigonal planar geometry

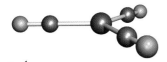

**Fig. 6.19** The trigonal planar structure of the copper(I) ion $[Cu(CN)_3]^{2-}$ [D. S. Marlin *et al.* (2001) *Angew. Chem. Int. Ed.*, vol. 40, p. 4752]. Colour code: Cu, brown; C, grey; N, blue.

When three groups are present around the metal centre as in $[Cu(CN)_3]^{2-}$, the structure is predicted to be trigonal planar so that repulsions between the cyanide groups are minimized. This agrees with the observed structure, shown in Figure 6.19.

#### Tetrahedral geometry

**(6.36)**

Tetrahedrally coordinated *d*-block metals are reasonably common and an example is the $[CoCl_4]^{2-}$ ion in which four chloride groups surround one cobalt(II) centre. In the tetrahedral geometry **6.36**, repulsions between the chlorides are minimized.

Similarly, tetrahedral structures are predicted (and found) for the ions $[MnO_4]^-$, $[CrO_4]^{2-}$, $[NiCl_4]^{2-}$ and $[VCl_4]^-$ and for the neutral compounds $TiCl_4$ and $OsO_4$.

## 5-Coordinate geometries

For the cadmium(II) species $[Cd(CN)_5]^{3-}$, the trigonal bipyramidal structure **6.37** is predicted so that repulsion between the five cyanide groups is minimized. Similarly, structure **6.38** is predicted for the iron(0) compound $[Fe(CO)_5]$.

Me = CH$_3$ = methyl
Ph = C$_6$H$_5$ = phenyl
PMe$_2$Ph =
dimethylphenylphosphine

(6.37)          (6.38)          (6.39)

In the VSEPR model, lone pairs of electrons preferentially occupy the equatorial sites of a trigonal bipyramidal structure because repulsions involving lone pairs of electrons are relatively large. Similarly, in the Kepert model, a sterically demanding (bulky) group is expected to occupy such a site to minimize inter-ligand repulsions.

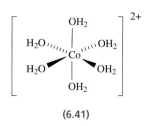

(6.40)

Structure **6.39** is predicted for the nickel(II) compound $[NiCl_2(PMe_2Ph)_3]$ – the PMe$_2$Ph groups are bulky organic phosphines and lie in the three equatorial sites.

The predicted geometry of the compound $[NiBr_3(PMe_2Ph)_2]$ is shown in structure **6.40**, and Figure 6.20a illustrates the structure as determined by X-ray diffraction. Repulsions between the five groups are minimized if the two phosphines are remote from one another. In addition, because the bromides are also sterically demanding, bromide–bromide repulsions are minimized if the Br—Ni—Br bond angles are 120° rather than 90° (see problem 6.8 at the end of this chapter).

(6.41)

## Octahedral geometry

Among molecular species involving the *d*-block elements, the octahedral geometry is extremely common and is the geometry expected from the Kepert model, for example, for the cobalt(II) ion $[Co(H_2O)_6]^{2+}$ (**6.41**).

**Fig. 6.20** The structures (determined by X-ray diffraction) of (a) $[NiBr_3(PMe_2Ph)_2]$ [J. K. Stalick *et al.* (1970) *Inorg. Chem.*, vol. 9, p. 453] and (b) $[Ni(H_2O)_6]^{2+}$ [L. R. Falvello *et al.* (2003) *Acta Crystallogr.*, Sect. C, vol. 59, p. m149]. Colour code: Ni, green; P, orange; Br, brown; C, grey; O, red; H, white.

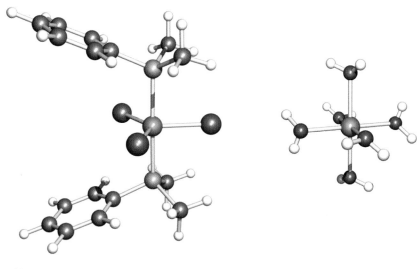

(a)          (b)

Similar structures are predicted (and found) for $[Ni(H_2O)_6]^{2+}$ (Figure 6.20b), $[Fe(CN)_6]^{3-}$, $[Fe(CN)_6]^{4-}$ and $[NiF_6]^{2-}$.

## 6.10    An exception to the Kepert model: the square planar geometry

Although the Kepert model succeeds in predicting the correct geometry for many molecular species involving the $d$-block elements, the simple form of the theory described here can *never* predict the formation of a square planar structure for a 4-coordinate species.

In the VSEPR model, a square planar structure such as $[ICl_4]^-$ (**6.30**) is derived from an octahedral arrangement which contains four bonding pairs and two lone pairs of electrons in the valence shell of the central atom. However, Kepert's model focuses on the repulsions between ligands, and ignores lone pairs of electrons. For a 4-coordinate species, repulsive interactions (i.e. *steric effects*) will *always* favour a tetrahedral geometry.

For $[CoCl_4]^{2-}$, the tetrahedral shape **6.36** gives Cl–Co–Cl bond angles of 109.5°. This structure is both predicted by Kepert and observed in practice. However, the anion $[PtCl_4]^{2-}$ possesses the square planar structure **6.42** with Cl–Pt–Cl bond angles of 90°. Other square planar anions are $[Ni(CN)_4]^{2-}$, $[PdCl_4]^{2-}$ and $[AuCl_4]^-$.

Many square planar molecules and ions, each with a $d$-block metal centre, are known and their structures are clearly *not* controlled by steric effects.[§]

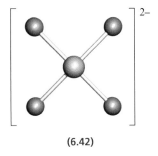

(6.42)

## 6.11    Stereoisomerism

If two compounds have the same molecular formula and the same atom connectivity, but differ in the spatial arrangement of different atoms or groups of atoms, then the compounds are *stereoisomers*.

We illustrate stereoisomerism by looking at the arrangements of atoms in trigonal bipyramidal, square planar and octahedral species and in compounds containing a double bond.

> If two compounds have the same molecular formula and the same atom connectivity, but differ in the spatial arrangement of different atoms or groups about a central atom or a double bond, then the compounds are *stereoisomers*.

### Trigonal bipyramidal structures

In the trigonal bipyramid, there are two different sites that atoms or groups can occupy, the *axial and equatorial sites*, as shown for $[Fe(CO)_5]$ in structure **6.43**.

When a carbonyl group in $[Fe(CO)_5]$ is replaced by another group such as $PMe_3$, two possible structures can be drawn. Structures **6.44** and **6.45** for $[Fe(CO)_4(PMe_3)]$ are *stereoisomers*.

|  |  |  |
|---|---|---|
| CO$_{axial}$ | PMe$_3$ | CO |
| equatorial OC —Fe''''' CO$_{equatorial}$ | OC —Fe''''' CO | OC —Fe''''' PMe$_3$ |
| CO$_{equatorial}$ | CO | CO |
| CO$_{axial}$ | C O | C O |
| (6.43) | (6.44) | (6.45) |

---

§ For a discussion of electronic effects relating to square planar structures, see: C. E. Housecroft and A. G. Sharpe (2005) *Inorganic Chemistry*, 2nd edn, Prentice Hall, Harlow, Chapter 20.

The trigonal bipyramidal molecule $PCl_2F_3$ has three stereoisomers depending upon whether the two chlorine atoms are arranged in the two axial sites (**6.46**), two equatorial positions (**6.47**), or one axial and one equatorial (**6.48**). (The three isomers can also be described in terms of the positions of the fluorine atoms.) As we saw for $[NiCl_2(PMe_2Ph)_3]$ **6.39**, one isomer may be favoured over another and electron diffraction studies in the gas phase confirm that isomer **6.47** is observed.

(**6.46**)   (**6.47**)   (**6.48**)

### Square planar structures

In the anion $[PtCl_4]^{2-}$ (**6.42**), all four chlorides are equivalent. If one chloride is replaced by a new group, only one product can be formed. Thus, $[PtCl_3(NH_3)]^-$ (**6.49**) has no stereoisomers.

If two groups are introduced, they could be positioned next to or opposite one another. This gives two stereoisomers for $[PtCl_2(NH_3)_2]$ as shown in structures **6.50** and **6.51**. Isomer **6.50** is called the *cis*-isomer and **6.51** is the *trans*-isomer of $[PtCl_2(NH_3)_2]$.[§] These structural differences are of more than academic interest. The *cis*-isomer of $[PtCl_2(NH_3)_2]$ is the drug cisplatin (see Box 23.5), which is effective for the treatment of certain forms of cancer; in contrast, *trans*-$[PtCl_2(NH_3)_2]$ is orders of magnitude less active.

The structures of *cis*-$[PtCl_2(PMePh_2)_2]$ and *trans*-$[NiBr_2(PMe_2Ph)_2]$ have been determined by X-ray diffraction and are shown in Figure 6.21.

(6.49)

(6.50)

(6.51)

A square planar species of general formula $XY_2Z_2$ has two stereoisomers.

In *cis*-$XY_2Z_2$, the two Y groups (and thus the two Z groups) are next to each other, and in *trans*-$XY_2Z_2$, the two Y groups (and thus the two Z groups) are in opposite sites.

*cis*   *trans*

### Octahedral structures

In each of the octahedral species $SF_6$, $[Co(H_2O)_6]^{2+}$ and $[MoCl_6]^{3-}$, the six groups around the central atom are the same. If one group is replaced by another, a single product is formed, and, for example, $[MoCl_5(H_2O)]^{2-}$ (**6.52**) does not possess any stereoisomers.

(6.52)

[§] The prefixes *cis* and *trans* are in common use, but the IUPAC has introduced a system of configuration indexes.

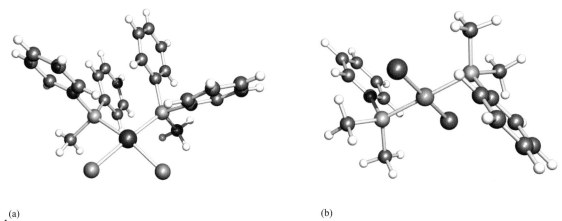

(a)                                                                (b)

**Fig. 6.21** The observed structures (determined by X-ray diffraction) of two square planar compounds: (a) *cis*-[PtCl$_2$(PMePh$_2$)$_2$] [H. Kin-Chee *et al.* (1982) *Acta Crystallogr.*, *Sect. B*, vol. 38, p. 421] and (b) *trans*-[NiBr$_2$(PMe$_2$Ph)$_2$] [S. M. Godfrey *et al.* (1993) *J. Chem. Soc., Dalton Trans.*, p. 2875]. Colour code: Pt, blue; Ni, green; C, grey; Cl, green; Br, brown; P, orange; H, white.

If we introduce *two* groups, two stereoisomers may be formed. The cation [CoCl$_2$(NH$_3$)$_4$]$^+$ can have the chlorides in a *cis* (**6.53**) or *trans* (**6.54**) arrangement.

$$\left[\begin{array}{c} NH_3 \\ H_3N\!\!\!\underset{\phantom{x}}{\overset{\phantom{x}}{\cdots}}\!\!Co\!\cdots\!Cl \\ H_3N\quad\ \ Cl \\ NH_3 \end{array}\right]^+ \qquad \left[\begin{array}{c} Cl \\ H_3N\!\!\!\underset{\phantom{x}}{\overset{\phantom{x}}{\cdots}}\!\!Co\!\cdots\!NH_3 \\ H_3N\quad\ \ NH_3 \\ Cl \end{array}\right]^+$$

(6.53)                                    (6.54)

An octahedral species of general formula XY$_2$Z$_4$ has two stereoisomers.

In *cis*-XY$_2$Z$_4$, the two Y groups are in adjacent sites, whereas in *trans*-XY$_2$Z$_4$, the two Y groups are in opposite sites.

$$\begin{array}{c} Y \\ Y\!\!\!\cdots\!\!X\!\cdots\!Z \\ Z\quad\ \ Z \\ Z \end{array} \qquad \begin{array}{c} Y \\ Z\!\!\!\cdots\!\!X\!\cdots\!Z \\ Z\quad\ \ Z \\ Y \end{array}$$

*cis*                                        *trans*

If an octahedral compound has the general formula XY$_3$Z$_3$, two arrangements of the Y and Z atoms or groups are possible as Figure 6.22 shows. The

**Fig. 6.22** In an octahedral compound of general formula XY$_3$Z$_3$, two arrangements of the Y and Z atoms or groups are possible: the *fac*- and *mer*-isomers.

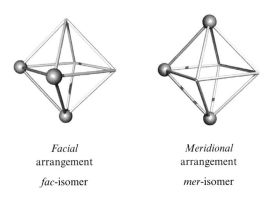

*Facial* arrangement

*fac*-isomer

*Meridional* arrangement

*mer*-isomer

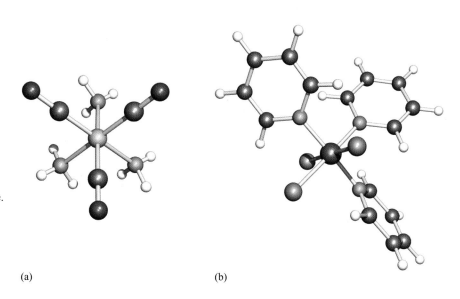

**Fig. 6.23** The octahedral structures (determined by X-ray diffraction) of (a) the cation *fac*-[Mn(CO)$_3$(NH$_3$)$_3$]$^+$ [M. Herberhold *et al.* (1978) *J. Organomet. Chem.*, vol. 152, p. 329] and (b) the compound *mer*-[RhCl$_3$(py)$_3$] (where py = pyridine, C$_5$H$_5$N) [K. R. Acharya *et al.* (1984) *Acta Crystallogr.*, *Sect. C*, vol. 40, p. 1327]. Colour code: Mn, yellow; Rh, brown; N, blue; O, red; Cl, green; C, grey; H, white.

(a)                              (b)

prefix *fac* (or *facial*) means that the three Y (and three Z) groups define one face of an octahedron. The prefix *mer* (or *meridional*) indicates that the three Y (or Z) groups are coplanar with the metal centre.

---

An octahedral species of general formula XY$_3$Z$_3$ has two stereoisomers.

In *fac*-XY$_3$Z$_3$, the three Y groups (and the three Z groups) define a face of the octahedron, whereas in *mer*-XY$_3$Z$_3$, the three Y groups (and the three Z groups) are coplanar with the metal centre.

*fac*                              *mer*

---

The structures of *fac*-[Mn(CO)$_3$(NH$_3$)$_3$]$^+$ and *mer*-[RhCl$_3$(py)$_3$] (where py is the abbreviation for pyridine) have been determined by X-ray diffraction methods and are shown in Figure 6.23. A *fac*-arrangement of the CO groups in the [Mn(CO)$_3$(NH$_3$)$_3$]$^+$ cation necessarily means that the three NH$_3$ groups are also in a *fac*-arrangement. Similarly, if the chloride groups in [RhCl$_3$(py)$_3$] are in a *mer*-arrangement, the pyridine groups must also be in a *mer*-arrangement.

Pyridine: see Section 34.4

## Double bonds

Nomenclature of alkenes: see Sections 24.4 and 26.2

Each carbon atom in ethene (**6.55**) is trigonal planar and the molecule is planar and rigid. In 1,2-dichloroethene, the nomenclature indicates that one chlorine atom is attached to each of the two carbon atoms in the molecule. However, there are two stereoisomers of 1,2-dichloroethene. In the (*Z*)-isomer (**6.56**), the two chlorine atoms are on the *same* side of the double bond, and in (*E*)-1,2-dichloroethene (**6.57**), the two chlorines are attached on *opposite* sides of the double bond.[§]

(6.55)          (6.56)          (6.57)

This type of isomerism is not restricted to organic alkenes. An inorganic compound that shows a similar isomerism is $N_2F_2$. The bent geometry at each nitrogen atom is caused by the presence of a lone pair of electrons and gives rise to the formation of the stereoisomers **6.58** and **6.59**.

(6.58)          (6.59)

### 6.12     Two structures that are close in energy: the trigonal bipyramid and square-based pyramid

#### In the solid state

In the solid state, the structure of the $[Ni(CN)_5]^{3-}$ anion depends upon the cation present. In some salts, the $[Ni(CN)_5]^{3-}$ ion is closer to being a square-based pyramid (**6.60**) than a trigonal bipyramid (**6.61**). This illustrates that these two 5-coordinate structures are often close in energy and this is further emphasized by the observation that in the hydrated salt $[Cr(en)_3][Ni(CN)_5] \cdot 1.5H_2O$, in which the cation is $[Cr(en)_3]^{3+}$, two forms of the $[Ni(CN)_5]^{3-}$ ion are present. Four of the anions, in the arrangement in which they occur in a single crystal of $[Cr(en)_3][Ni(CN)_5] \cdot 1.5H_2O$, are shown in Figure 6.24. Careful inspection of the bond angles reveals just

en is the abbreviation for $H_2NCH_2CH_2NH_2$: see Chapter 23, Table 23.2

**Fig. 6.24** Four of the $[Ni(CN)_5]^{3-}$ anions, shown in the arrangement in which they occur in the crystal lattice of $[Cr(en)_3][Ni(CN)_5] \cdot 1.5H_2O$ (en is the abbreviation for $H_2NCH_2CH_2NH_2$) [K. N. Raymond *et al.* (1968) *Inorg. Chem.*, vol. 7, p. 1362]. The $[Cr(en)_3]^{3+}$ cations and the water molecules of crystallization have been omitted from the diagram for clarity. Evaluation of the bond angles illustrates how similar the structures labelled 'trigonal bipyramid' and 'square-based pyramid' actually are. Colour code: Ni, green; C, grey; N, blue.

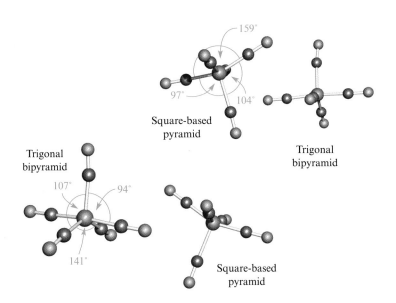

Square-based pyramid

Trigonal bipyramid

Trigonal bipyramid

Square-based pyramid

[§] In an older nomenclature, (*Z*) was termed '*cis*' and (*E*) was termed '*trans*'. These names are still commonly encountered. (*Z*) and (*E*) come from German *zusammen* and *entgegen*, meaning 'together' and 'opposite'.

**Fig. 6.25** The conversion of a trigonal bipyramidal molecule into a square-based pyramidal one is achieved by an angular distortion. This process usually requires only a small amount of energy. The view of the square-based pyramid is not the conventional one; compare with Figure 6.12.

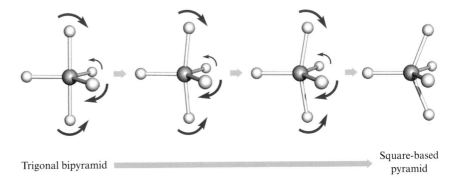

Trigonal bipyramid ⟶ Square-based pyramid

$$\left[ \begin{array}{c} \overset{N}{\underset{C}{|}} \\ NC^{\text{\tiny{IIIII}}}Ni^{\text{\tiny{IIIII}}}CN \\ NC \qquad CN \end{array} \right]^{3-}$$

**(6.60)**

$$\left[ \begin{array}{c} \overset{N}{\underset{C}{|}} \\ NC — Ni^{\text{\tiny{IIIII}}}CN \\ {}^{\blacktriangledown}CN \\ \overset{C}{\underset{N}{|}} \end{array} \right]^{3-}$$

**(6.61)**

how similar the structures labelled 'trigonal bipyramid' and 'square-based pyramid' actually are!

Although ligand–ligand repulsions are minimized in a trigonal bipyramidal arrangement, only a minor change to the structure is needed to turn it into a square-based pyramid. In a trigonal bipyramidal molecule, the two axial groups are related by a bond angle (subtended at the central atom) of 180°. The three equatorial groups are related by bond angles (subtended at the central atom) of 120°. Consider what happens if the angle between the two axial ligands is made a little smaller (<180°), and the angle between two of the equatorial ligands is enlarged a little (>120°). This is shown in Figure 6.25 – the result is to convert the trigonal bipyramid into a square-based pyramid.

The small energy difference between the trigonal bipyramidal and square-based pyramidal structures means that, in the solid state, the nature of a 5-coordinate molecule or ion can be influenced by the forces operating between the species in the crystal. There are many 5-coordinate species with structures that are best described as lying somewhere between the trigonal bipyramid and square-based pyramid.

### In solution: Berry pseudo-rotation

In solution, the small energy difference between the trigonal bipyramidal and square-based pyramidal structures for a given 5-coordinate species can have a dramatic effect. In solutions of $PF_5$ or $[Fe(CO)_5]$ there are fewer restrictions on the molecular motion than in the solid state, and it becomes possible for the process shown in Figure 6.25 to become a *dynamic one* – the process happens in real time. Moreover, further angular distortion can occur to regenerate a trigonal bipyramidal structure. This low-energy process is called *Berry pseudo-rotation*.

An important consequence of Berry pseudo-rotation is the interconversion of axial and equatorial groups in the trigonal bipyramid (Figure 6.26). Repeated structural interconversions of this type allow *all* five of the groups attached to the central atom to exchange places, and the molecule is said to be *fluxional* or *stereochemically non-rigid*.

How do we *observe* fluxional behaviour? Consider the 5-coordinate molecule $XY_5$. If the *rate of exchange* of the axial and equatorial atoms Y is *faster than the timescale* of the experimental technique used, the experiment will never be able to distinguish between the two types of Y atom. Hence, the axial and equatorial atoms appear to be equivalent. If the rate of exchange of

*Berry pseudo-rotation* is the name given to a low-energy process that interconverts trigonal bipyramidal and square-based pyramidal structures with the effect that the axial and equatorial groups of the trigonal bipyramidal structure are exchanged.

**Fig. 6.26** The Berry pseudo-rotation mechanism. This interconverts one trigonal bipyramidal structure into another via a square-based pyramidal intermediate. The atom colour scheme emphasizes that axial and equatorial sites in the trigonal bipyramid are interchanged. The first step in the process is shown in more detail in Figure 6.25.

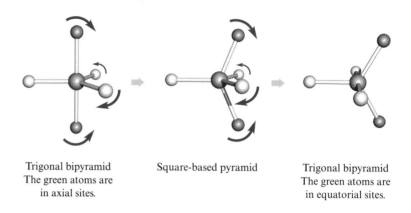

Trigonal bipyramid
The green atoms are
in axial sites.

Square-based pyramid

Trigonal bipyramid
The green atoms are
in equatorial sites.

the axial and equatorial atoms Y is *slower* than the timescale of the experimental technique used, the experiment *should* be able to distinguish between the two types of Y atom, and the axial and equatorial atoms appear to be different.

> Timescale: see Section 11.3

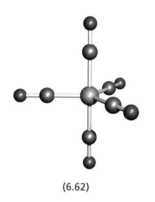

**(6.62)**

> 'NMR active' means that the nucleus can be observed using NMR spectroscopy

> Activation energy: see Section 15.1

---

> A molecule or molecular ion is **fluxional** or **stereochemically non-rigid** if atoms or groups in the system are involved in a dynamic process such that they exchange places.

---

A technique that is often used to study stereochemically non-rigid systems is nuclear magnetic resonance (NMR) spectroscopy. In this section, we deal only with the result of such experiments, and more details of NMR spectroscopy are given in Chapter 14.

Consider the trigonal bipyramidal molecule $[Fe(CO)_5]$ **(6.62)**. The carbon nucleus $^{13}C$ is *NMR active* and *each environment* (axial $^{13}C$ in the two axial CO groups or equatorial $^{13}C$ in the three equatorial CO groups) should give rise to a different spectroscopic signal. At room temperature, sufficient thermal energy is available and interconversion between structures is facile – the $^{13}C$ nuclei all appear to be identical, and only one NMR spectroscopic signal is seen. Each structural interconversion requires a minimum amount of energy called the *activation energy*. This is an 'energy barrier' to interconversion of the 5-coordinate structures and at low temperatures where less energy is available, fewer molecules possess enough energy to overcome the barrier. As the temperature of the molecules is lowered, it may become possible to *freeze out* the fluxional process. However, the energy barriers for the fluxional process in $[Fe(CO)_5]$ are so low that even at 103 K, the axial and equatorial CO groups are still exchanging their positions on the timescale of the NMR spectroscopic experiment.

## 6.13    Shape and molecular dipole moments

In Chapter 5, we discussed polar molecules and (electric) dipole moments ($\mu$). In a polar diatomic, one end of the molecule is partially negatively charged *with respect to* the other end. The situation with polyatomic molecules is more complex, and requires careful consideration.

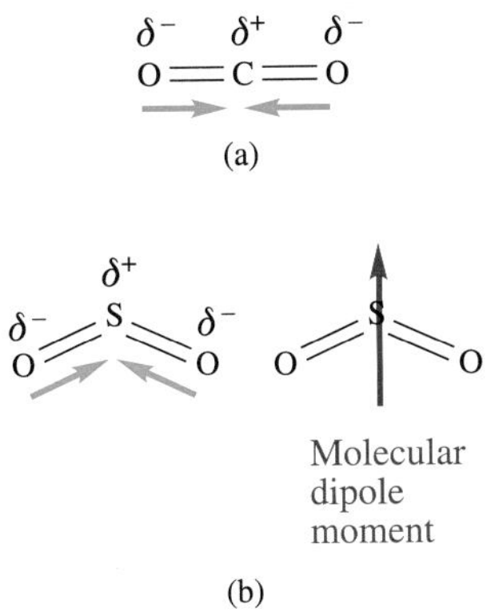

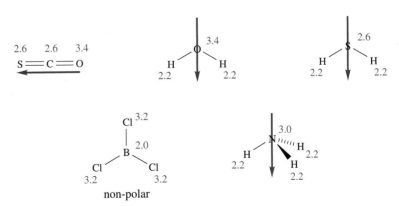

$$\overset{\delta^-}{O}=\overset{\delta^+}{C}=\overset{\delta^-}{O}$$

(a)

$$\overset{\delta^+}{\underset{O}{S}}\overset{\delta^-}{O}$$

Molecular
dipole
moment

(b)

**Fig. 6.27** (a) In a $CO_2$ molecule, each carbon–oxygen bond is polar but the two dipole moments shown in blue (which are vectors) cancel out. (b) In an $SO_2$ molecule, each sulfur–oxygen bond is polar and the bent structure means that the two bond dipoles shown in blue reinforce one another. The direction of the *molecular* dipole moment is shown in pink. By SI convention, the arrow representing the dipole moment points towards the $\delta^+$ part of a bond or molecule.

## Linear molecules

Consider a molecule of $CO_2$. From the Pauling electronegativity values $\chi^P(C) = 2.6$ and $\chi^P(O) = 3.4$, we predict that *each carbon–oxygen bond* in the molecule is polar ($C^{\delta+}O^{\delta-}$). However, the *molecular* dipole moment of $CO_2$ is zero. This is understood if we consider the shape of the molecule *and* recall that a dipole moment is a vector quantity, i.e. it has magnitude *and* direction. Figure 6.27a shows that, although each bond is polar, the two dipole moments act equally in opposite (at 180°) directions. The net result is that the molecule is non-polar ($\mu = 0$ D).

The $CO_2$ (Figure 6.27a) and OCS (Figure 6.28) molecules are both linear, but the former is symmetrical about the carbon atom while the latter is not. The Pauling electronegativity values (Figure 6.28) suggest that OCS will have a net *molecular dipole moment* acting in the direction shown (the oxygen end of the molecule is $\delta^-$). The observed value of $\mu$ for the OCS molecule is 0.72 D (Table 6.5).

## Bent molecules

Dipole moments for several bent molecules are listed in Table 6.5. First, we compare the dipole properties of $CO_2$ and $SO_2$. Although sulfur and carbon have the same Pauling electronegativity values ($\chi^P = 2.6$), the dipole moments of $CO_2$ and $SO_2$ are 0 and 1.63 D respectively. This dramatic difference is due to a difference in molecular shape. Figure 6.27 shows that the *bond dipoles* in $CO_2$ oppose one another and cancel each other out, but in $SO_2$ they reinforce one another, leading to a significant *molecular dipole moment*. This example emphasizes how important it is to differentiate between *bond* and *molecular* polarities.

The ability of water to function as a solvent is partly due to the presence of a molecular dipole. The values listed in Table 6.5 show that the dipole moment for $H_2O$ is relatively large, but in going from $H_2O$ to $H_2S$ the molecular dipole moment decreases from 1.85 D to 0.97 D. Although the shape of the molecule is unchanged (Figure 6.28), the bond polarities decrease and hence the resultant molecular dipole decreases.

**Fig. 6.28** Polar and non-polar polyatomic molecules. The pink numbers correspond to values of the electronegativity ($\chi^P$) for each atom. The pink arrows represent the direction of the molecular dipole moment; by SI convention, the arrow points from $\delta^-$ to $\delta^+$.

**Table 6.5** Dipole moments for selected polyatomic molecules.

| Molecule | Molecular shape | Dipole moment / D | Molecule | Molecular shape | Dipole moment / D |
|---|---|---|---|---|---|
| $BCl_3$ | Trigonal planar | 0 | $OF_2$ | Bent | 0.30 |
| $NCl_3$ | Trigonal pyramidal | 0.39 | $SF_2$ | Bent | 1.05 |
| $NF_3$ | Trigonal pyramidal | 0.24 | $SF_4$ | Disphenoidal | 0.63 |
| $NH_3$ | Trigonal pyramidal | 1.47 | $CF_4$ | Tetrahedral | 0 |
| $PCl_3$ | Trigonal pyramidal | 0.56 | $CO_2$ | Linear | 0 |
| $PF_3$ | Trigonal pyramidal | 1.03 | OCS | Linear | 0.72 |
| $PH_3$ | Trigonal pyramidal | 0.57 | $SO_2$ | Bent | 1.63 |
| $H_2O$ | Bent | 1.85 | $NO_2$ | Bent | 0.32 |
| $H_2S$ | Bent | 0.97 | $N_2O$ | Linear | 0.16 |

## Trigonal planar and trigonal pyramidal molecules

A trigonal planar molecule such as $BCl_3$ (Figure 6.28) has no net molecular dipole moment. Although each B—Cl bond is polar, the three bond dipoles cancel each other out. We look at this in detail in Box 6.1.

The molecules $NCl_3$, $NF_3$, $NH_3$, $PCl_3$, $PF_3$ and $PH_3$ all possess trigonal pyramidal structures, and are all polar (Table 6.5). However, the magnitude and direction of each molecular dipole moment depends on the relative electronegativities of the atoms in each molecule and on the presence of

**THEORETICAL AND CHEMICAL BACKGROUND**

## Box 6.1 Confirmation that the molecule $BCl_3$ is non-polar

The $BCl_3$ molecule has a trigonal planar structure, and each Cl—B—Cl bond angle is 120°.

The Pauling electronegativity values are:

$$\chi^P(B) = 2.0$$

$$\chi^P(Cl) = 3.2$$

Each B—Cl bond is polar, and each bond dipole is a vector quantity.

The three bond dipoles have the same magnitude but act in different directions.

Let each vector be $V$. The three vectors act as follows:

Resolving the vectors into two directions gives:
In a downward direction:

   total vector $= V$

In an upward direction:

   total vector $= 2 \times (V \times \cos 60)$

$$= 2 \times \left(\frac{V}{2}\right) = V$$

Therefore, two equal vectors act in opposite directions and cancel each other out.
The net molecular dipole is zero.

### Conclusion

The $BCl_3$ molecule does not possess a dipole moment.

lone pairs. Consider $NH_3$ ($\mu = 1.47\,D$) (Figure 6.28). Each N–H bond is polar in the sense $N^{\delta-}$–$H^{\delta+}$ and the nitrogen atom also carries a lone pair of electrons. The resultant molecular dipole is such that the nitrogen atom has a partial negative charge ($N^{\delta-}$). Although $NF_3$ is structurally related to $NH_3$, its molecular dipole is smaller ($\mu = 0.24\,D$). Each N–F bond is polar in the sense $N^{\delta+}$–$F^{\delta-}$ but the net dipole due to the three bond dipole moments is almost cancelled by the effect of the lone pair of electrons on the nitrogen atom.

### 4-Coordinate structures

A tetrahedral molecule $XY_4$ with four identical Y atoms or groups is non-polar. In $CF_4$, for example, the four bond dipoles oppose one another, and there is no net molecular dipole moment. Similarly, a square planar molecule $XY_4$ in which the Y atoms or groups are identical has no overall dipole moment.

(6.63)

A 4-coordinate molecule with a disphenoidal structure *does* have a dipole moment – the dipole moment for $SF_4$ (structure **6.63**) is 0.63 D. Each S–F bond is polar and the two bond dipoles that act along the axial directions cancel each other. However, in the equatorial plane there is a net dipole moment.

### 5- and 6-Coordinate species

An octahedral molecule of formula $XY_6$ (in which the Y groups are the same) is non-polar, because, although each X–Y *bond* may be polar, pairs of bond dipoles oppose one another: $\mu = 0\,D$ for $SF_6$.

A trigonal bipyramidal molecule $XY_5$ (in which the Y atoms or groups are identical) does not possess a molecular dipole. In $PF_5$, the two axial P–F bond dipole moments cancel out, and the three equatorial P–F bond dipoles oppose each other with a net cancellation of the dipole. It is more difficult to predict whether a square-based pyramidal molecule $XY_5$ will have a dipole moment, since much depends upon the exact molecular geometry. The molecular dipole moment of $IF_5$ is 2.18 D. Here, resolution of the individual bond dipole moments (vectors) into two directions does not lead to equal and opposite vectors.

> In a polyatomic molecule, the presence of a ***molecular dipole moment*** depends on both the presence of bond dipole moments and the shape of the molecule.

### A summary of molecular dipole moments

- Molecular geometry has an important influence on the property of molecular dipole moments.
- The presence of polar bonds in a molecule does not necessarily mean that the molecule will possess a dipole moment.
- A casual glance at the electronegativity values of the atoms in a molecule may not be enough to establish whether or not a molecule is polar, or in what direction the dipole acts.

### 6.14 Carbon: an element with only three common geometries

In this chapter, we have considered the geometries that are available for 2-, 3-, 4-, 5- and 6-coordinate species. For the heavier group 15, 16 and 17 elements and for *d*-block metals, a larger range of coordination numbers and molecular shapes is possible. However, we have also seen that elements in the first row of the *p*-block have fewer structural choices available. We now briefly consider carbon in order to appreciate the restrictions that its geometrical preferences impose on it. This has an effect on the chemistry of this element, as we shall see in later chapters.

$$R-C\equiv C-R$$

**(6.64)**

### Linear environments

Apart from diatomic and triatomic systems such as CO, $[CN]^-$, $CO_2$, $CS_2$ and HCN, 2-coordinate carbon atoms are most commonly encountered in alkynes (**6.64**) and allenes[§] (**6.65**) (R is any organic substituent). A linear structure is forced on the atom by the presence of a triple bond (in the alkyne) or two adjacent double bonds (in the allene), and the structural restriction imposes *rigidity* in the molecular backbone.

**(6.65)**

### Trigonal planar environments

Two of the carbon atoms in structure **6.65** are 3-coordinate and trigonal planar. This is a characteristic of an alkene as shown in structure **6.66**.

The presence of C=C or C=O bonds in a molecule leads to compounds with trigonal planar carbon centres. One of the fundamental building blocks of aromatic chemistry is the benzene ring (**6.67** and Figure 6.29a) which contains six carbon atoms, each of which is trigonal planar. The effect of the geometrical restriction is that the ring itself is planar. Planar rings are observed in a multitude of molecular systems that are related to benzene, e.g. pyridine (**6.68** and Figure 6.29b) and naphthalene (**6.69** and Figure 6.29c).

**(6.66)**

Aromatic compounds including bonding in benzene: see Chapter 32

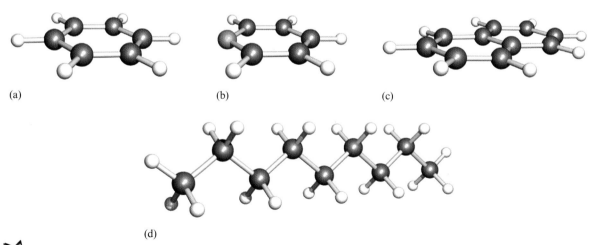

(a)          (b)          (c)

(d)

 **Fig. 6.29** The planar structures of (a) benzene, (b) pyridine, (c) naphthalene. (d) The structure of nonane (an example of an alkane, $C_nH_{2n+2}$) showing the zigzag carbon backbone.

---

[§] An allene is a type of *diene* in which the two double bonds are in adjacent positions in the carbon chain.

(6.67)    (6.68)    (6.69)

Two allotropes of carbon also have bulk structures containing trigonal planar carbon atoms – graphite consists of layers of fused 6-membered rings, while in fullerenes, there is some deviation from strict planarity because the 6-membered rings are fused with 5-membered rings to generate approximately spherical $C_n$ (e.g. $n = 60$) structures.

■▷
$C_{60}$ and graphite: see Sections 9.9 and 9.10

### Tetrahedral environments

Saturated hydrocarbons (alkanes) and their derivatives contain tetrahedral carbon atoms, and in a so-called 'straight chain' alkane their presence leads to a zigzag backbone as illustrated by the structure of nonane, $C_9H_{20}$ (Figure 6.29d). Such an organic chain is far more flexible than one containing 3- or 2-coordinate carbon atoms. This has important consequences in terms of the number of possible arrangements of the chain.

## SUMMARY

The theme of this chapter has been the shapes of polyatomic molecules. We have also looked at how molecular (as opposed to bond) dipole moments arise and seen how molecules that contain polar *bonds* may not be polar overall.

### Do you know what the following terms mean?

- polyatomic
- coordination number
- isostructural
- trigonal planar
- tetrahedral
- trigonal bipyramidal
- square-based pyramidal

- octahedral
- square planar
- disphenoidal
- VSEPR model
- Kepert model
- stereochemically inactive lone pair

- stereochemical inert pair effect
- stereoisomers
- *trans*- and *cis*-isomers
- *mer*- and *fac*-isomers
- stereochemically non-rigid
- Berry pseudo-rotation
- molecular dipole moment

### You should now be able:

- to provide examples of molecular species that possess 2-, 3-, 4-, 5- and 6-coordinate structures
- to discuss why boron, carbon and nitrogen exhibit a greater structural diversity than fluorine and oxygen
- to discuss the range of geometries observed for the *p*-block elements
- to predict the shapes of species containing a central *p*-block or *d*-block element
- to draw out stereoisomers for square planar, trigonal bipyramidal and octahedral species and for molecules containing double bonds

- to discuss the limited range of geometrical environments found for carbon and to give examples of organic molecules with 2-, 3- and 4-coordinate carbon atoms
- to outline a mechanism that exchanges axial and equatorial atoms or groups in a 5-coordinate molecule
- to determine whether a polyatomic molecule is likely to possess a dipole moment

## PROBLEMS

**6.1** Draw diagrams showing Lewis structures for $CO_2$, $[NO_2]^+$, $[N_3]^-$ and $N_2O$.

**6.2** Using VSEPR theory, predict the shape of $[I_3]^-$. Following from this answer and using isoelectronic relationships, suggest the shapes of $[IBr_2]^-$ and $[ClF_2]^-$.

**6.3** (a) What shapes are $CO_2$ and $SO_2$? (b) Using your answer to (a) and the isoelectronic principle, suggest structures for $[NO_2]^+$, $[NO_2]^-$, $CS_2$, $N_2O$ and $[NCS]^-$.

**6.4** Use VSEPR theory to suggest structures for (a) $BCl_3$, (b) $NF_3$, (c) $SCl_2$, (d) $[I_3]^+$, (e) $PCl_5$, (f) $[AsF_6]^-$, (g) $[AlH_4]^-$ and (h) $XeF_4$.

**6.5** In each series of molecular species, pick out the *one* member that is structurally different and specify its shape and the shapes of the remaining members of the series:
(a) $CH_4$, $[BH_4]^-$, $[AlCl_4]^-$, $[SO_4]^{2-}$, $SF_4$;
(b) $AlBr_3$, $PBr_3$, $[H_3O]^+$, $SOCl_2$, $NH_3$;
(c) $[NO_2]^+$, $[N_3]^-$, $[ClO_2]^+$, $[NCO]^-$, $OCS$;
(d) $PF_5$, $SbF_5$, $IF_5$, $SOF_4$, $AsF_5$.

**6.6** Use the Kepert model to predict the shapes of the following molecular species containing *d*-block metal centres: (a) $[AuCl_2]^-$; (b) $[FeF_6]^{3-}$; (c) $[Ag(CN)_2]^-$; (d) $[Ni(H_2O)_6]^{2+}$; (e) $TiCl_4$; (f) $[MoS_4]^{2-}$; (g) $[MnO_4]^-$; (h) $[FeO_4]^{2-}$.

**6.7** (a) Briefly discuss similarities and differences between the VSEPR and Kepert models, and comment on when it is appropriate to apply a particular model. (b) The $[AuCl_4]^-$ ion is square planar. Is this expected by the Kepert model? (c) The $[ICl_4]^-$ ion is square planar. Is this expected by the VSEPR model?

**6.8** (a) What structure does the Kepert model predict for a 5-coordinate compound of general formula $ML_5$? (b) Based on this structure, how many isomers are possible for the compound $[NiBr_3(PMe_2Ph)_2]$?

**6.9** How many stereoisomers are possible for each of the following (not all species possess isomers)?
(a) Square planar $[PdCl_2(PPh_3)_2]$; (b) octahedral $WCl_3F_3$; (c) octahedral $WCl_2F_4$; (d) $CCl_3H$; (e) octahedral $[Cr(H_2O)_5Cl]^{2+}$; (f) $BBrCl_2$; (g) trigonal bipyramidal $PCl_3F_2$; (h) trigonal bipyramidal $PClF_4$. Draw structures of the isomers and give them distinguishing labels where appropriate.

**6.10** Which of the following molecules contain double bonds? (a) $H_2O_2$; (b) $N_2F_2$; (c) $N_2F_4$; (d) $C_2H_4$; (e) $C_2H_2Cl_2$; (f) $COCl_2$.

**6.11** Of those compounds in problem 6.10 that do contain a double bond, which possess stereoisomers? Draw the structures of the isomers and label them appropriately.

**6.12** A representation of the structure of the dye Sudan red B is shown in **6.70**. What is wrong with this structural representation? Redraw the structure so as to give a better representation of the molecule and comment on the result.

**(6.70)**

**6.13** Give structures for the following molecules: (a) $PF_3$; (b) $BBr_2F$; (c) $H_2CO$; (d) $H_2S$; (e) $SO_3$; (f) $PCl_5$; (g) $AlCl_3$. Which do you expect to possess a molecular dipole moment? ($\chi^P$ values are given in Appendix 7.) For those that do, suggest the direction in which it acts.

**6.14** Table 6.6 gives values of the dipole moments of a series of related organic molecules. Rationalize the trends in values; $\chi^P$ values are given in Appendix 7.

**Table 6.6** Data for problem 6.14.

| Molecule | Dipole moment, $\mu$ / D |
|---|---|
| $CH_4$ | 0 |
| $CF_4$ | 0 |
| $CCl_4$ | 0 |
| $CH_3F$ | 1.86 |
| $CH_2F_2$ | 1.98 |
| $CH_3Cl$ | 1.89 |

**6.15** The dipole moment of (Z)-$N_2F_2$ in the gas phase is 0.16 D, but (E)-$N_2F_2$ is non-polar. Explain how this difference arises.

**6.16** (a) How many different fluorine environments are there in $PF_5$ and $SF_4$? By what names are the sites differentiated? (b) At 298 K, NMR spectroscopic investigations of $PF_5$ and $SF_4$ indicate that each possesses only one F environment. Rationalize these data.

**6.17** Describe the Berry pseudo-rotation process using $Fe(CO)_5$ as an example. During the process, are any bonds broken?

**6.18** (a) Discuss, giving examples, the geometrical environments that carbon usually adopts in organic molecules. (b) Look at Figure 1.1 which shows allotropes of carbon. Do these allotropes fit into the general pattern of shapes for 'organic carbon'?

## Additional problems

**6.19** Comment on each of the following observations.

(a) $[IF_5]^{2-}$ and $[XeF_5]^-$ are pentagonal planar.

(b) $[PCl_4][PCl_3F_3]$ contains tetrahedral and octahedral P centres, and the anion possesses isomers.

(c) Members of the series of complexes $[PtCl_4]^{2-}$, $[PtCl_3(PMe_3)]^-$, $[PtCl_2(PMe_3)_2]$, $[PtCl(PMe_3)_3]^+$ do not possess the same number of isomers.

**6.20** Give explanations for each of the following.

(a) $S_n$ rings are non-planar.

(b) In $Br_2O_5$ (which contains a Br−O−Br bridge), the Br atoms are trigonal pyramidal.

(c) In $S_2F_{10}$, each S is octahedral, and the molecule is non-polar.

(d) $H_2O_2$ is non-planar.

# 7 Polyatomic molecules: bonding

## Topics

- The valence octet
- Resonance structures
- Hybridization of atomic orbitals
- Molecular orbital theory: the ligand group orbital approach

## 7.1 Introduction

In Chapters 4 and 5, we considered the bonding in diatomic molecules in terms of Lewis structures, valence bond theory and molecular orbital theory. We concentrated on species containing hydrogen or elements from the period Li to F. In this chapter, we extend our discussion to polyatomic molecules. Bonding schemes are necessarily more complicated than in diatomics because we must consider a greater number of atomic centres.

We begin this chapter with a look at the octet rule, and then move on to ways in which valence bond and molecular orbital theories provide bonding descriptions for polyatomic molecules. Although we shall look at coordination numbers from 3 to 6, we are still limited to relatively simple molecular species. Later in the book, we introduce additional ways of dealing with the bonding in more complex organic and inorganic compounds.

## 7.2 Molecular shape and the octet rule: the first row elements

In Sections 6.4 and 6.5 we saw that:

- in the first row of the $p$-block, boron, carbon and nitrogen show a greater variation in geometrical environment than do oxygen and fluorine;
- heavier elements in groups 13 to 17 of the $p$-block show a greater range of geometries in their compounds than do the first row elements of each group;
- for the heavier $p$-block elements, the range of structures *increases* in going from group 13 to 17.

In this and the following sections we show how the bonding in compounds $XY_n$ (where X is a $p$-block element) can be considered in terms of the central atom X obeying the octet rule.

## Fluorine

The ground state electronic configuration of fluorine is $[He]2s^2 2p^5$. The addition of an electron is required to complete an octet in the valence shell. A fluorine atom forms a single bond and the structural restrictions that we described in Chapter 6 follow from this.

## Oxygen

The ground state electronic configuration of oxygen is $[He]2s^2 2p^4$. An octet is completed if oxygen adds two electrons to its valence shell. This can be done by forming either:

- two single bonds (e.g. $H_2O$, $OF_2$); or
- one double bond (e.g. $O=O$, $O=C=O$)

We can rationalize the limited geometries in which oxygen atoms are found in terms of the octet rule. The formation of two single bonds leads to a bent molecule (e.g. $H_2O$). When an oxygen atom is involved in the formation of a double bond, it is *terminally* attached to another atom (e.g. $CO_2$).

At first glance it may appear that an oxygen atom cannot form three single bonds. However, Figure 6.10 showed the structure of the trigonal pyramidal $[H_3O]^+$ ion. 3-Coordinate species in which oxygen is the central atom have one thing in common – *they are cationic*. If we formally localize the positive charge in $[H_3O]^+$ on the oxygen centre, then the geometry of the ion can be explained in terms of the $O^+$ ion (and not the O atom) obeying the octet rule. Figure 7.1 explains this point. In Figure 7.1a, the *neutral oxygen centre* gains an octet if it forms *two* single bonds. The VSEPR model predicts that the $H_2O$ molecule has a bent structure because the O atom has two bonding pairs and two lone pairs of electrons in its valence shell. Figure 7.1b shows that the *cationic oxygen centre* achieves an octet if it forms *three* single bonds. The VSEPR model predicts that the $[H_3O]^+$ cation is trigonal

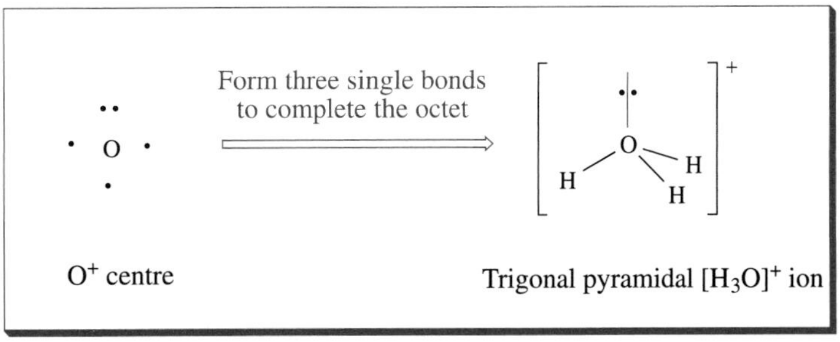

> To achieve an octet of valence electrons, a neutral O atom forms two bonds and an $O^+$ centre forms three bonds.

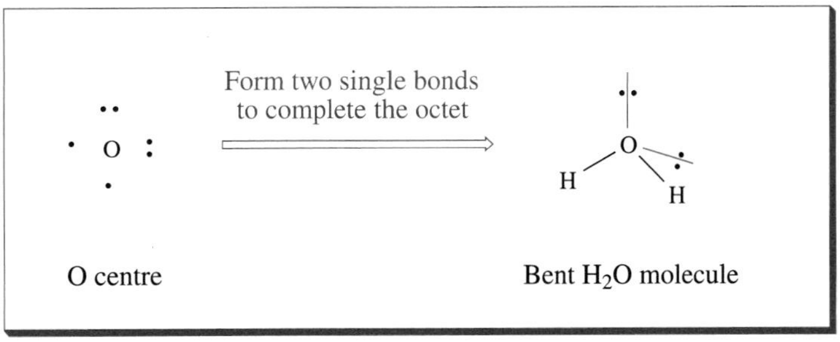

**Fig. 7.1** (a) A *neutral* oxygen atom obeys the octet rule when it forms two single bonds. (b) A *cationic* oxygen centre obeys the octet rule when it forms three single bonds. The localization of the positive charge on the oxygen centre is a formalism used to aid the interpretation of the bonding in the $[H_3O]^+$ ion.

(a)

(b)

pyramidal in shape because the O atom has one lone pair and three bonding pairs of electrons in its valence shell.

### Nitrogen

A nitrogen atom has five valence electrons and can achieve an octet by forming:

- three single bonds (e.g. $NH_3$, $NF_3$, $NMe_3$);
- one double bond and one single bond (e.g. $Cl-N=O$, $F-N=O$);
- one triple bond (e.g. $H-C\equiv N$, $[C\equiv N]^-$).

The geometries then follow as a matter of course. Remembering that there is a lone pair of electrons present on the nitrogen atom, three single bonds lead to a trigonal pyramidal geometry (Figure 7.2a). One double bond and one single bond gives a bent molecule (structure **7.1**). When N forms a triple bond, it is terminally attached.

The number of species in which nitrogen is 4-coordinate is large. We need only consider the wide range of salts containing ammonium $[NH_4]^+$ or tetra-methylammonium $[NMe_4]^+$ ions to realize that 4-coordinate nitrogen is important in the chemistry of this element. Bearing in mind that nitrogen obeys the octet rule, we rationalize the formation of four single bonds by considering an $N^+$ centre rather than a neutral N atom. This is shown in Figure 7.2b and the VSEPR model predicts that the $[NH_4]^+$ cation is tetrahedral.

(7.1)

> To achieve an octet of valence electrons, a neutral N atom forms three bonds, and an $N^+$ centre forms four bonds.

### Carbon

Application of the octet rule shows that carbon ($[He]2s^2 2p^2$) requires four electrons to complete a noble gas configuration, i.e. $[He]2s^2 2p^6 \equiv [Ne]$. The four electrons can be gained by forming:

- four single bonds (e.g. $CH_4$); or
- two single bonds and one double bond (e.g. $O=CCl_2$); or

**Fig. 7.2** (a) A *neutral* nitrogen atom obeys the octet rule when it forms three single bonds. (b) A *cationic* nitrogen centre obeys the octet rule when it forms four single bonds.

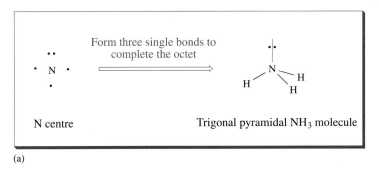

(a)

(b)

- one single and one triple bond (e.g. $H-C{\equiv}N$); or
- two double bonds (e.g. $O{=}C{=}O$).[§]

These bonding options correspond to the three structural preferences we have seen for carbon – tetrahedral, trigonal planar and linear.

### Boron

Neutral molecules that contain a single boron centre are often trigonal planar in shape. This follows directly from the ground state electronic configuration of boron ($[He]2s^2 2p^1$). The three valence electrons can be involved in the formation of three single bonds, but in so doing boron does *not* achieve an octet. Instead, the boron atom has a *sextet* of electrons in its valence shell.

Species such as $[BH_4]^-$ and $[BF_4]^-$ which contain a 4-coordinate boron centre have a feature in common – *they are anionic*. The $B^-$ centre is iso-electronic with a carbon atom and has four valence electrons. It can form four single bonds and so achieve an octet. The tetrahedral shapes of $[BH_4]^-$ and $[BF_4]^-$ follow directly from the VSEPR model.

---

**Worked example 7.1**    *The $[BH_4]^-$ ion*

**Show that the B atom in the $[BH_4]^-$ ion obeys the octet rule. What shape is the $[BH_4]^-$ ion?**

B is in group 13, and the B atom has three valence electrons.
In $[BH_4]^-$, formally assign the negative charge to the B centre.
$B^-$ has four valence electrons, and forms four single B–H bonds.
Using the VSEPR model, a tetrahedral structure is predicted for $[BH_4]^-$.
The bonding in $[BH_4]^-$ is summarized as follows:

$B^-$ centre

Tetrahedral $[BH_4]^-$ ion

---

### Formal localization of charge on a central atom

Assuming that the octet rule is obeyed, the examples above illustrate that in some molecular species, the number of bonds formed by the central atom appears to be inconsistent with the number of valence electrons available. In these cases we can adopt a formalism which involves localizing an overall charge on the central atom. For example, in $[H_3O]^+$, the oxygen centre forms three O–H bonds. This is readily understood in terms of an $O^+$ centre (and not a neutral O atom) obeying the octet rule. In $[BF_4]^-$, there are four B–F single bonds. Their formation can be rationalized in terms of a $B^-$ centre attaining an octet of electrons.

---

[§] The formation of a quadruple bond is, in theory, an option but is not observed.

While this method of working is simple and extremely useful, it is important to realize that *it is only a formalism*. We are not implying that the central atom actually carries all of the overall charge in the ion concerned. In $[H_3O]^+$, the charge distribution is *not* such that the oxygen centre necessarily bears a charge of $+1$. It is also important to realize that the formal charge assignment does *not* imply anything about the oxidation state of the central atom:

- in $[H_3O]^+$, the oxidation state of the oxygen is $-2$; each hydrogen atom has an oxidation state of $+1$;
- in $[BF_4]^-$, the oxidation state of each fluorine is $-1$ and that of the boron is $+3$.

---

**Worked example 7.2**     *Satisfying the octet rule*

**How many single bonds may an $O^-$ centre form while obeying the octet rule?**

First write down the valence electronic configuration of the O atom: $2s^2 2p^4$

$\therefore$ For $O^-$, the ground state configuration is $2s^2 2p^5$.

$\therefore$ Only one electron is needed to form an octet and $O^-$ forms only one single bond, e.g. in $[OH]^-$, a Lewis structure for which is:

$$\overset{\displaystyle ..}{\underset{\displaystyle ..}{: \; \overset{-}{O}}}\!\!-\!\!H$$

It is helpful to use the periodic table to note that $O^-$ is isoelectronic with an F atom and so should behave in a similar manner: HF and $[OH]^-$ are isoelectronic.

---

## Resonance structures

Resonance structures: see
Sections 4.11 and 5.3

The bonding models that we have considered so far in this section have taken into account only *covalent* contributions. In Chapters 4 and 5, we described the bonding in diatomic molecules in terms of both covalent and ionic contributions. This was achieved by representing the bonding by a set of resonance structures (see, for example, structure **4.7**). We can approach the bonding in polyatomic molecules in a similar way, but we must keep in mind that any bonding description must be in accord with experimental data. For example, the O–H bonds in $H_2O$ are equivalent and therefore a description of the bonding in $H_2O$ must reflect this fact.

Resonance structures for the $H_2O$ molecule are drawn in Figure 7.3a. The resonance forms labelled 'covalent' and 'ionic' treat the molecule in a symmetrical manner and therefore the equivalence of the O–H interactions is maintained. '*Partial ionic*' resonance structures for the water molecule can also be drawn but pairs of structures are needed to produce equivalent O–H interactions in the molecule.

Even for a triatomic molecule (which is actually a very 'small' molecule), drawing out a set of resonance structures is tedious. Are all the structures necessary? Do they all contribute to the same extent? In order to assess

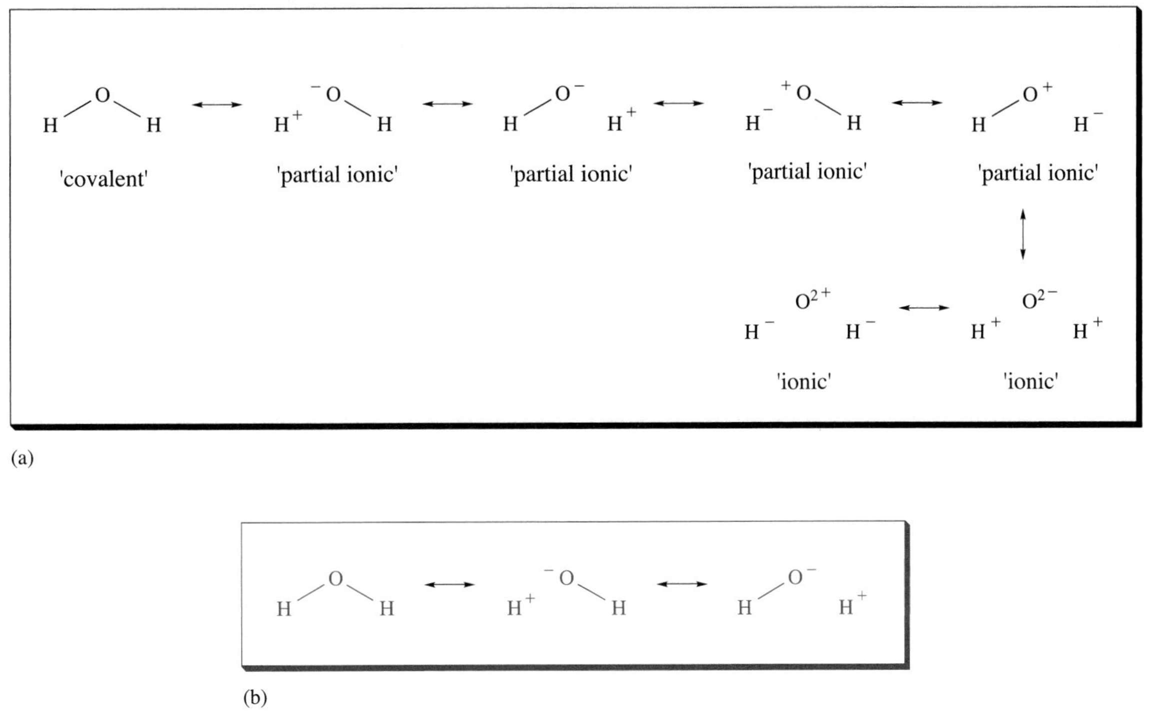

(a)

(b)

**Fig. 7.3** (a) Resonance structures for $H_2O$. (b) The bonding can be adequately described in terms of the three resonance structures shown in pink.

this, we have to use chemical intuition. Three points are crucial:

- Charge separation should reflect the relative electronegativities of the atoms.
- We should avoid building up a large positive or negative charge on any one atom.
- Adjacent charges should be of opposite sign (two adjacent positive or two adjacent negative charges will have a destabilizing effect).

Consider again the resonance structures in Figure 7.3a bearing in mind the electronegativities $\chi^P(O) = 3.4$ and $\chi^P(H) = 2.2$. The ionic resonance structures will contribute negligibly because the charge separation in the molecule is unreasonably large. The three resonance structures shown in Figure 7.3b provide an approximate, but adequate, description of the bonding in $H_2O$.

*Exercise*: In Figure 7.3a, we included resonance forms containing $O^+$ and $H^-$ ions. Why did we ignore them in Figure 7.3b?

The bonding in molecules such as $BF_3$ and $CH_4$ can also be represented by a series of resonance structures. Figure 7.4a illustrates just how tedious the process is becoming! Boron trifluoride is still a 'small' molecule and yet there are 11 possible resonance structures. We can group the resonance structures for $BF_3$ according to the charge on the central atom and eliminate some of them on the grounds that they will contribute only negligibly to the overall bonding description. On this basis, structures with a multiple charge on the boron atom may be neglected. Figure 7.4b shows the resonance structures

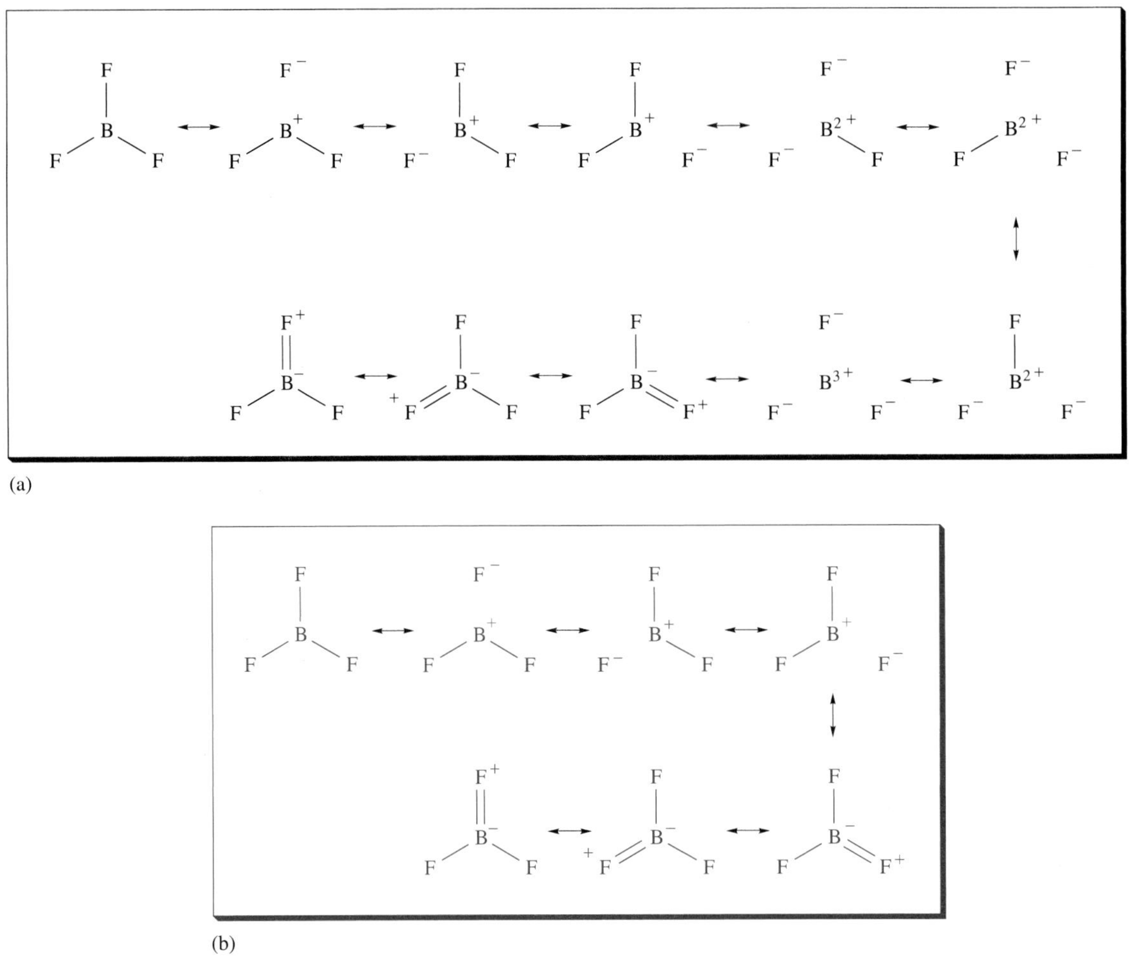

(a)

(b)

**Fig. 7.4** (a) Resonance structures for the $BF_3$ molecule. (b) Only some of the resonance structures contribute significantly and the bonding picture can be approximated to include only these forms.

$BF_3$: see also Sections 7.4 and 7.7

that will contribute the most to the bonding in the $BF_3$ molecule, and this approximate picture is quite satisfactory. Note that we require *all three members of a set* of related partial ionic structures if we are to maintain B–F bond equivalence, an experimentally observed *fact* that must be reproduced by the bonding scheme.

---

**Worked example 7.3**    *Resonance structures for the nitrate ion*

**Draw resonance structures for the nitrate ion and predict its shape.**

The nitrate ion is $[NO_3]^-$.
Nitrogen is in group 15 and has five valence electrons.
Oxygen is in group 16 and has six valence electrons.
The (major contributing) resonance structures for the nitrate ion are:

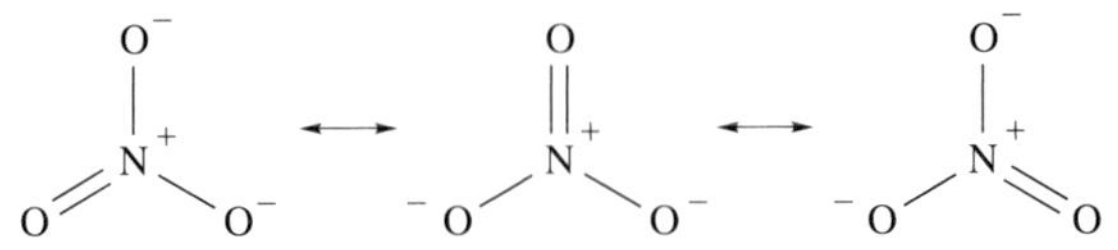

In each resonance structure, each N and O atom obeys the octet rule. The ion has a trigonal planar geometry.

*It is tempting to write down a structure for [NO₃]⁻ involving two N=O bonds and one N−O⁻ bond but this is **incorrect**. Why?*

**7.3**

## Molecular shape and the octet rule: the heavier *p*-block elements

A striking difference as one descends a group of the *p*-block is the tendency for increased coordination numbers (Figure 7.5). For example:

- molecular $PF_5$, $AsF_5$, $SbF_5$ and $BiF_5$ exist, whereas $NF_5$ does not;
- a coordination number of 6 is found in $SF_6$, $[PF_6]^-$ and $[SiF_6]^{2-}$, but is not typical for the analogous first row elements O, N and C;
- the heavier group 17 elements form compounds such as $ClF_3$, $BrF_5$, $IF_7$ in which F is always a terminal atom and forms only one single bond.

It can be argued that a compound such as $NF_5$ does not exist because an N atom is too small to accommodate five F atoms around it. Historically, the differences between, for example, N and P, or O and S, have been attributed to the availability of *d*-orbitals on atoms of the heavier elements which enable the atom to 'expand its octet of valence electrons'. This led to the bonding in, for example, $[SO_4]^{2-}$ **(7.2)** in which all the S−O bond lengths are equal, being represented by a set of resonance structures in which S=O double bonds are important:

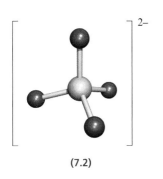

**(7.2)**

This bonding picture for $[SO_4]^{2-}$ results in the S atom having 12 electrons in its valence shell. Current views, however, are that *d*-orbitals on the central atom play no significant role in the bonding in compounds of the *p*-block elements. Even in the presence of highly electronegative atoms such as F or O, the energies of the *d*-orbitals are too high to allow these orbitals to participate in bonding. How, then, can we represent the bonding in species such as $SF_4$, $SF_6$, $PF_5$, $[PF_6]^-$, $[SO_4]^{2-}$ and $[ClO_4]^-$, while allowing the

**Fig. 7.5** The *p*-block elements; the noble gases in group 18 are included, although compounds are only well established for Kr and Xe. The first row elements (shown in beige) tend to exhibit coordination numbers ≤4. The heavier elements (shown in green) exhibit a wider range of coordination numbers.

| | Group number | | | | | |
| --- | --- | --- | --- | --- | --- | --- |
| | 13 | 14 | 15 | 16 | 17 | 18 |
| 1st row | B | C | N | O | F | Ne |
| 2nd row | Al | Si | P | S | Cl | Ar |
| 3rd row | Ga | Ge | As | Se | Br | Kr |
| 4th row | In | Sn | Sb | Te | I | Xe |
| 5th row | Tl | Pb | Bi | Po | At | Rn |

central atom to obey the octet rule? The answer is found by considering *charge-separated species* as the following examples illustrate.

**Example 1: $SF_4$**   S is in group 16 and has six valence electrons. If we simply form four S−F single bonds, the central S atom ends up with 10 electrons in its valence shell (see example 5 in Section 6.6). In order to maintain an octet of electrons around the S atom, we have to apply a similar principle as we described for $[NH_4]^+$ (Figure 7.2b). Instead of considering S as the central atom, consider how many S−F single bonds can be formed by an $S^+$ centre that obeys the octet rule:

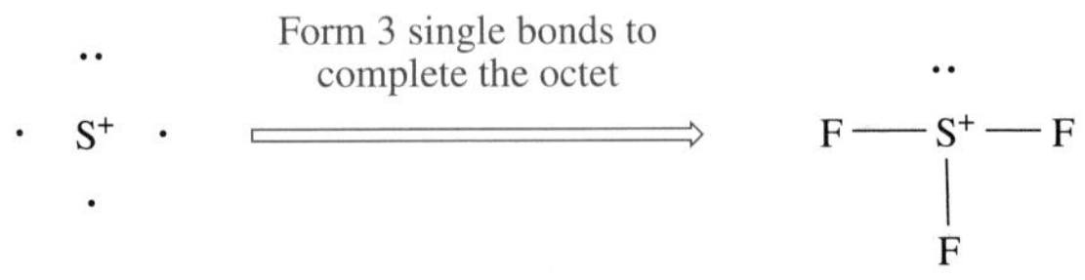

We can now write $SF_4$ as a charge-separated species in which the S centre obeys the octet rule:

Resonance structures: see
Sections 4.11 and 5.3

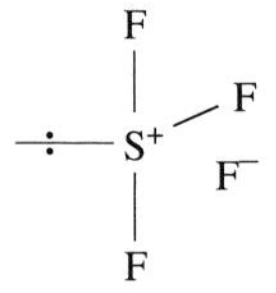

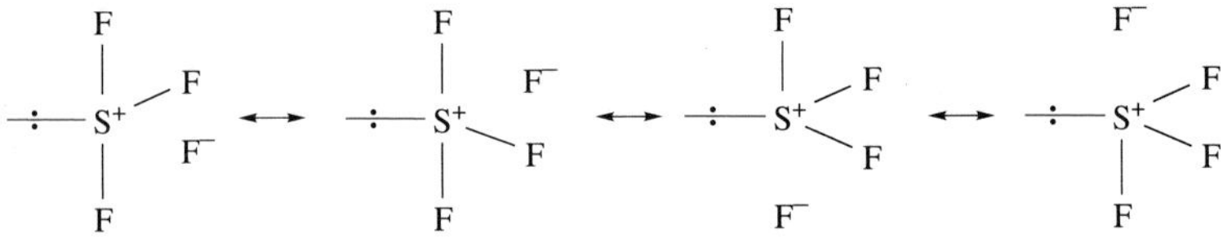

This structure suggests that one S−F interaction is ionic while the other three are covalent. A more realistic representation of the bonding is to draw a set of resonance structures as shown below:

**Example 2: $PF_5$**   P is in group 15 and has five valence electrons. The formation of five P−F bonds would lead to 10 electrons in the valence shell of the P atom. Instead of taking the neutral P atom as the centre of $PF_5$, take a $P^+$ centre:

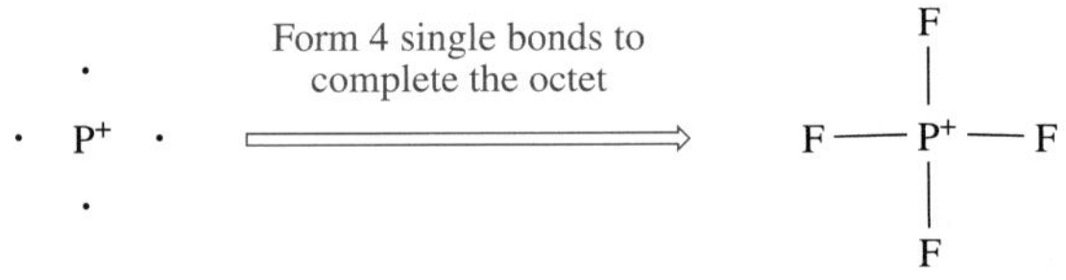

Now write $PF_5$ as a charge-separated species in which the P centre obeys the octet rule:

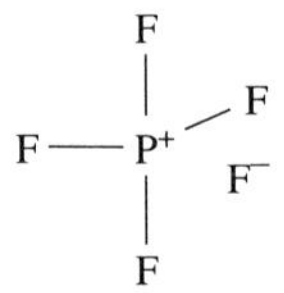

A more realistic view of the bonding is in terms of a set of resonance structures:

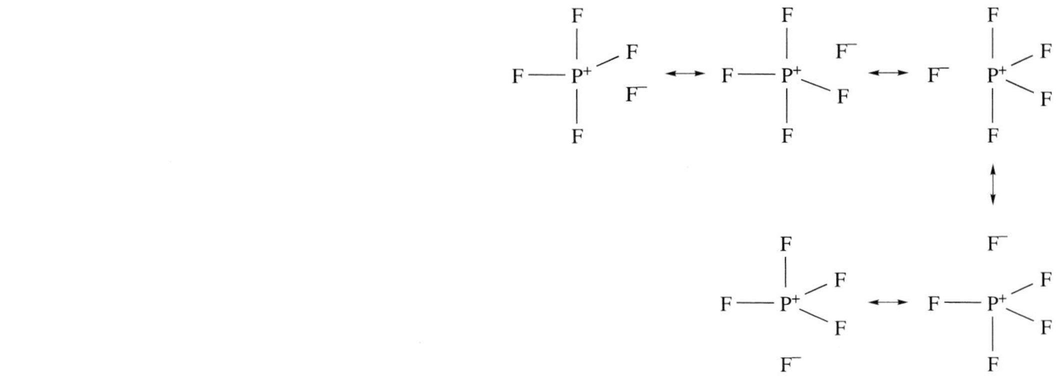

In $PF_5$, the two axial P–F bonds are longer than the three equatorial P–F bonds. The different environments of the axial and equatorial F atoms mean that the resonance forms involving axial $F^-$ need not make the same contributions as those involving equatorial $F^-$.

**Example 3: $SF_6$**  S is in group 16 and has six valence electrons. If we draw a Lewis structure consisting of six S–F covalent bonds, the valence shell of the S atom contains 12 electrons. However, if we consider $SF_6$ in terms of a charge-separated species formally containing an $S^{2+}$ centre, then S obeys the octet rule:

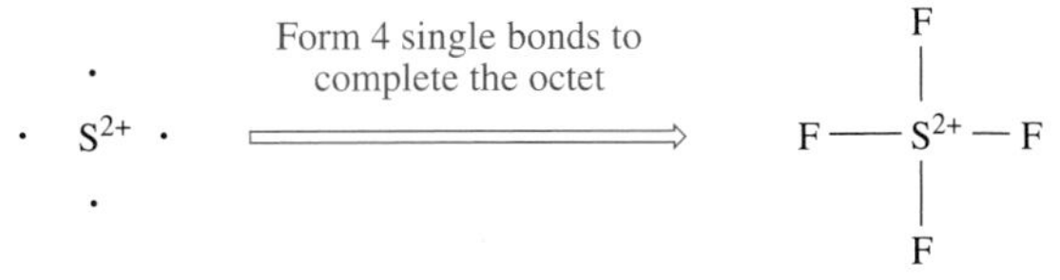

This leads to the following description of the bonding in $SF_6$:

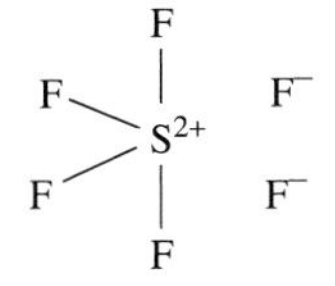

A more realistic description of the bonding involves a set of resonance structures so that all the S–F bonds are equivalent.

**Example 4: $[SO_4]^{2-}$**  We saw earlier that if the bonding in $[SO_4]^{2-}$ involves contributions from two S=O double bonds, the S atom ends up with 12 electrons in its valence shell. A valence bond model for the bonding in $[SO_4]^{2-}$ in which S obeys the octet rule involves bond formation by an $S^{2+}$ centre:

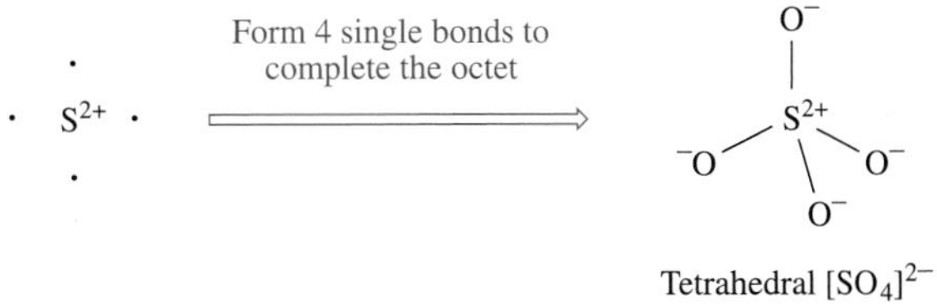

Tetrahedral $[SO_4]^{2-}$

This representation of the bonding results in equivalence of the four S−O interactions, consistent with experimental observations.

| Worked example 7.4 | *Valence bond treatment of the bonding in SF₂ and SO₂* |

*$SF_2$ and $SO_2$ are both bent molecules (see Figure 6.3). Give bonding descriptions for $SF_2$ and $SO_2$ in which the S atom obeys the octet rule.*

**$SF_2$**
S is in group 16 and has six valence electrons. It requires two electrons to complete an octet.
F is in group 17 and has seven valence electrons. It requires one electron to complete an octet.
The bonding in $SF_2$ can be described in terms of two S−F single bonds:

··
· S :       Form 2 single bonds to          ··
·          complete the octet          F — S :
                                       |
                                       F

**$SO_2$**
S is in group 16 and has six valence electrons.
O is in group 16 and has six valence electrons.
If we describe the bonding in $SO_2$ in terms of two S=O double bonds, each O atom obeys the octet rule, but the S atom has 10 electrons in its valence shell:

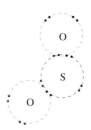

This model is unsatisfactory. For the octet rule to be obeyed, consider $SO_2$ in terms of a charge-separated species formally containing an $S^+$ centre:

··         Form 1 single bond to O⁻
· $S^+$ ·    (see worked example 7.2)        ··  ··
·       and 1 double bond to O       $S^+$ — $O^-$ :
                              //   ··
                            : O
                            ··

Each atom obeys
the octet rule

While this model allows each atom to obey the octet rule, it implies that the two S−O interactions are different. Experimentally, the S−O bonds are equivalent. Therefore, draw a set of resonance structures to take account of this:

··             ··
$S^+$            $S^+$
O ═╱  ╲ O⁻  ⟷  ⁻O  ╱═ O

This now gives a satisfactory bonding picture for $SO_2$ within VB theory.

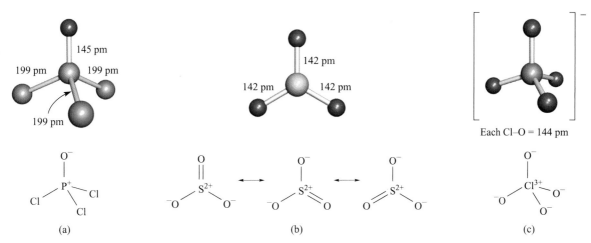

**Fig. 7.6** Experimentally determined structures and bond lengths for (a) $POCl_3$, (b) $SO_3$ and (c) $[ClO_4]^-$. For each, a bonding picture is given in which each atom obeys the octet rule. In (b), the set of resonance structures is needed to account for the equivalence of the S−O bonds. Colour code: P, orange; O, red; Cl, green; S, yellow.

**(7.3)**

**(7.4)**

By using the methods shown above, we can describe the bonding in a wide range of $p$-block compounds so that atoms obey the octet rule. Experimentally determined bond lengths provide information about the equivalence of bonds. Proposed bonding schemes must be consistent with these data as the examples in Figure 7.6 illustrate.

Although the bonding pictures described above are satisfactory in that they show atoms obeying the octet rule, some diagrams (e.g. those for $PF_5$ and $SF_6$ in examples 2 and 3) hide structural information. In addition, when your interest is in emphasizing molecular structure, it is not always convenient to draw sets of resonance structures. Throughout this book, therefore, when we wish to focus on the *structure* of a molecule rather than on its bonding, we shall use representations such as **7.3** and **7.4**. Provided that you remember that a connecting line between two atoms does not *necessarily* mean the presence of a localized pair of bonding electrons, then these representations are perfectly satisfactory.

## 7.4 Valence bond theory and hybridization

In earlier parts of this chapter we used the ground state electronic configuration to determine the number of valence electrons available to an atom, and hence to say how many bonds that atom can form. We also noted that the tendency for an element in the $p$-block to achieve an octet arises from a desire to occupy the $2s$ and $2p$ atomic orbitals completely.

In this section, we consider how VB theory views the participation of the atomic orbitals in bond formation in polyatomic molecules and ions.

### The directional properties of atomic orbitals

Shapes of atomic orbitals: see Section 3.11

With the exception of the spherically symmetric $s$ orbital, an atomic orbital has directional properties. The question is: when atoms are bonded together, are their relative positions restricted by the directional properties of the atomic orbitals involved in the bonding?

**Fig. 7.7** In order to approach the bonding in a trigonal planar molecule such as $BF_3$, we need to consider the valence atomic orbitals of a boron atom, and define a convenient relationship between the relative positions of the atoms in $BF_3$ and the Cartesian axis system. It is not possible for all three of the B–F bond vectors to coincide simultaneously with the directions of the three $2p$ atomic orbitals.

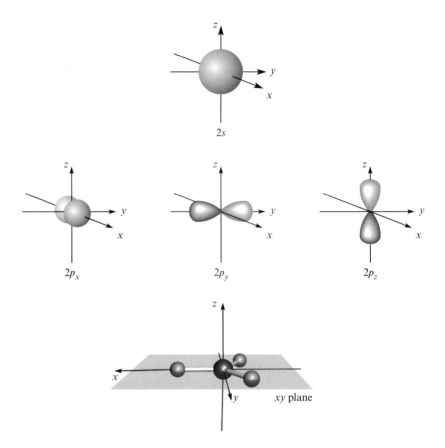

Consider the formation of a trigonal planar molecule such as $BF_3$. The valence shell of the boron atom consists of the $2s$ and three $2p$ atomic orbitals. These are shown in Figure 7.7; while the $2s$ atomic orbital is spherically symmetrical, each $2p$ orbital lies on one of the three axes. Figure 7.7 also shows a convenient relationship between the relative positions of the atoms in a $BF_3$ molecule and the axes. The boron atom has been placed at the origin and one fluorine atom lies on the $+x$-axis. The molecule lies in the $xy$ plane. Although one B–F bond vector coincides with the $2p_x$ atomic orbital, the other two B–F bond vectors do *not* coincide with either of the remaining two $2p$ atomic orbitals. Similar difficulties arise with other molecular shapes.

We can now define the problem. How do we describe the bonding in a molecule in terms of localized orbitals when the directional properties of the valence atomic orbitals do not match the bond vectors defined by the atoms in the molecule?

## Hybridization of atomic orbitals

In the remainder of this section, we describe a method that allows us to generate *spatially directed orbitals* which may be used to produce *localized bonds* in a valence bond scheme. These are known as *hybrid orbitals*, and may be used to construct bonding orbitals in exactly the same way as atomic orbitals. Although hybridization (orbital mixing) is a useful way to describe bonding within the valence bond framework, there is *no* implication that it is an *actual* process.

**Fig. 7.8** The formation of two *sp* hybrid orbitals from in- or out-of-phase combinations of an *s* and a *p* atomic orbital. The effect of altering the phase of the orbital (the *s* atomic orbital) is shown pictorially in the lower part of the figure.

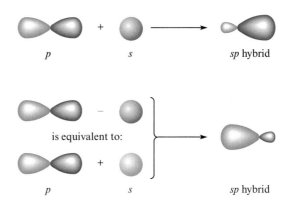

■▷
**Hybridization means 'mixing'**

■▷
**Constructive and destructive interference: see Figure 4.11**

Hybridization of atomic orbitals can be considered *if* the component atomic orbitals are close in energy. The character of the hybrid orbitals depends on the atomic orbitals involved and the percentage contribution that each makes. The number of hybrid orbitals generated is equal to the number of atomic orbitals used; this reiterates the rule in MO theory that equates the number of MOs formed to the initial number of atomic orbitals.

The simplest case is where we have one *p* atomic orbital and one *s* atomic orbital, both from the same principal quantum level (2*s* and 2*p*, or 3*s* and 3*p*, etc.). Figure 7.8 shows that when the *s* and *p* atomic orbitals combine in phase, the orbital will be reinforced (constructive interference). Where the *s* and *p* atomic orbitals combine out of phase, the orbital will be diminished in size (destructive interference). The net result is that two hybrid orbitals are formed which retain some of the directional properties of the atomic *p* orbital. These hybrid orbitals are denoted *sp*, and each *sp* hybrid orbital contains 50% *s* and 50% *p* character.

We can repeat the exercise with a combination of one 2*s* and two 2*p* atomic orbitals. Since we begin with three atomic orbitals, we form three equivalent hybrid orbitals as shown in Figure 7.9. The resultant orbitals are called $sp^2$ hybrids and they lie in the same plane (here, the *xy* plane because we chose to hybridize the *s*, $p_x$ and $p_y$ atomic orbitals) and define a trigonal planar arrangement. We look at this hybridization in detail in Box 7.1.

**Fig. 7.9** The combination of one 2*s* and two 2*p* atomic orbitals to generate three equivalent $sp^2$ hybrid orbitals.

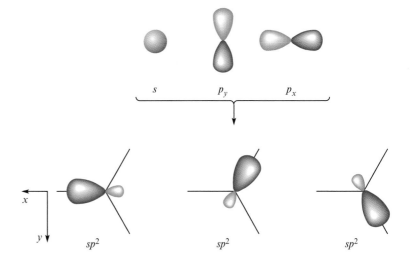

**THEORETICAL AND CHEMICAL BACKGROUND**

## Box 7.1 Directionality of $sp^2$ hybrid orbitals

Figure 7.9 shows that a combination of $s$, $p_x$ and $p_y$ atomic orbitals gives three $sp^2$ hybrid orbitals which are identical except for the directions in which they point. How do the directionalities arise?

First, remember that each $p$ orbital has a directional (vector) property – the $p_x$ orbital lies on the $x$-axis and the $p_y$ orbital lies on the $y$-axis.

The three hybrid orbitals lie in the $xy$ plane. There is no contribution from the $p_z$ atomic orbital and therefore the hybrid orbitals do not include any $z$-component.

The process of combining the $s$, $p_x$ and $p_y$ atomic orbitals can be broken down as follows:

1. The character of the $s$ atomic orbital must be divided equally between the three hybrid orbitals.
2. Each hybrid orbital must end up with the same amount of $p$ character.
3. Let us fix the direction of one of the hybrid orbitals to coincide with one of the initial $p$ atomic orbitals, say the $p_x$ atomic orbital (this is an arbitrary choice). The first $sp^2$ hybrid orbital therefore contains one-third $s$ and two-thirds $p_x$ character:

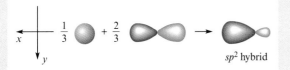

4. The remaining components are one-third of the $p_x$ atomic orbital, all of the $p_y$ atomic orbital, and two-thirds of the $s$ atomic orbital. These must be combined to give two equivalent hybrid orbitals. The directionality of each hybrid is determined by components from both the $p_x$ and $p_y$ atomic orbitals. When $s$ character is added in, two $sp^2$ hybrid orbitals result:

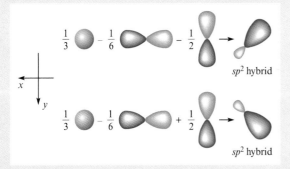

(Remember that changing the sign of a $p$ atomic orbital alters its direction.)

Note that the fact that the $sp^2$ hybrid orbitals lie in the $xy$ plane is wholly arbitrary. They could equally well contain $p_y$ and $p_z$ components (and lie in the $yz$ plane) or $p_x$ and $p_z$ components (and lie in the $xz$ plane).

The approach given above is not rigorous, but does provide a satisfactory and readily visualized method of working out the directionalities and composition of the three $sp^2$ hybrid orbitals.

Consider again an $XY_3$ molecule lying in the $xy$ plane with atom positions as shown below:

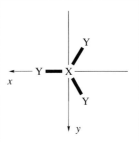

Mathematically, the *normalized* wavefunctions (see Section 3.8) for a set of $sp^2$ hybrid orbitals are written as follows:

$$\psi_{sp^2}(1) = c_1\psi_s + c_2\psi_{p_x} + c_3\psi_{p_y}$$

$$\psi_{sp^2}(2) = c_4\psi_s + c_5\psi_{p_x} + c_6\psi_{p_y}$$

$$\psi_{sp^2}(3) = c_7\psi_s + c_8\psi_{p_x} + c_9\psi_{p_y}$$

We can find the normalized coefficients $c_1$ to $c_9$ as follows:

• The contributions made by $\psi_s$ to each $\psi_{sp^2}$ must be equal:

$$\therefore c_1 = c_4 = c_7$$

• The sum of the squares of the coefficients of $\psi_s$ must equal 1:

$$\therefore c_1{}^2 + c_4{}^2 + c_7{}^2 = 1$$

and since $c_1 = c_4 = c_7$:

$$3c_1{}^2 = 1$$

$$\therefore c_1 = \frac{1}{\sqrt{3}} = c_4 = c_7$$

• Now define the orientation of $\psi_{sp^2}(1)$ to lie along the $x$-axis. This means that there is no contribution from $\psi_{p_y}$:

$$\therefore c_3 = 0$$

- With $c_1$ and $c_3$ known, we can find $c_2$ because for a normalized wavefunction:

$$c_1{}^2 + c_2{}^2 = 1$$

$$\therefore \frac{1}{3} + c_2{}^2 = 1$$

$$c_2 = \sqrt{\frac{2}{3}}$$

- The normalized wavefunction for $\psi_{sp^2}(1)$ can now be written as:

$$\psi_{sp^2}(1) = \frac{1}{\sqrt{3}}\psi_s + \sqrt{\frac{2}{3}}\psi_{p_x}$$

The fractional contribution of each atomic orbital to the hybrid orbital is equal to the square of the normalized coefficient, and so the equation for $\psi_{sp^2}(1)$ indicates that the hybrid orbital is composed of one-third $s$ character and two-thirds $p$ character. This is consistent with the pictorial scheme we derived earlier.

- To find $c_5$, $c_6$, $c_8$ and $c_9$, we again make use of the fact that the sum of the squares of the coefficients of *each atomic orbital* must equal 1, and so must the sum of the squares of the coefficients of *each hybrid orbital*. Therefore:

$$c_2{}^2 + c_5{}^2 + c_8{}^2 = \frac{2}{3} + c_5{}^2 + c_8{}^2 = 1 \qquad (1)$$

$$c_6{}^2 + c_9{}^2 = 1 \qquad (2)$$

$$c_4{}^2 + c_5{}^2 + c_6{}^2 = \frac{1}{3} + c_5{}^2 + c_6{}^2 = 1 \qquad (3)$$

$$c_7{}^2 + c_8{}^2 + c_9{}^2 = \frac{1}{3} + c_8{}^2 + c_9{}^2 = 1 \qquad (4)$$

- Before attempting to solve these equations, look back at the diagram that shows the orientation of the $XY_3$ molecule with respect to the $x$- and $y$-axes. The wavefunction $\psi_{sp^2}(1)$ describes the hybrid orbital which points along the $x$-axis. Wavefunctions $\psi_{sp^2}(2)$ and $\psi_{sp^2}(3)$ are related to

each other in that they contain equal contributions from $\psi_{p_y}$ and equal contributions from $\psi_{p_x}$. Therefore:

$$c_5{}^2 = c_8{}^2 \qquad \text{and} \qquad c_6{}^2 = c_9{}^2$$

Therefore, equation (1) becomes:

$$\frac{2}{3} + 2c_5{}^2 = 1$$

$$\therefore c_5{}^2 = \frac{1}{2}\left(1 - \frac{2}{3}\right) = \frac{1}{6}$$

$$c_5 = \pm\frac{1}{\sqrt{6}}$$

Since $\psi_{sp^2}(2)$ and $\psi_{sp^2}(3)$ point along the $-x$-axis (look again at the diagram opposite):

$$c_5 = -\frac{1}{\sqrt{6}} = c_8$$

Equation (2) becomes:

$$2c_6{}^2 = 1$$

$$\therefore c_6 = \pm\frac{1}{\sqrt{2}} = c_9$$

One hybrid orbital must have a $+y$ contribution and the other a $-y$ contribution, so let:

$$c_6 = +\frac{1}{\sqrt{2}} \qquad \text{and} \qquad c_9 = -\frac{1}{\sqrt{2}}$$

The normalized wavefunctions are therefore:

$$\psi_{sp^2}(2) = \frac{1}{\sqrt{3}}\psi_s - \frac{1}{\sqrt{6}}\psi_{p_x} + \frac{1}{\sqrt{2}}\psi_{p_y}$$

$$\psi_{sp^2}(3) = \frac{1}{\sqrt{3}}\psi_s - \frac{1}{\sqrt{6}}\psi_{p_x} - \frac{1}{\sqrt{2}}\psi_{p_y}$$

These coefficients are consistent with the pictorial representations derived earlier since the contributions of each atomic orbital are given by the squares of the normalized coefficients.

A set of *four* equivalent $sp^3$ hybrid orbitals is formed by combining the $s$, $p_x$, $p_y$ and $p_z$ atomic orbitals as shown in Figure 7.10. Together, they define a tetrahedral arrangement, and each $sp^3$ hybrid orbital possesses 25% $s$ character and 75% $p$ character.

For the principal quantum level with $n = 2$, only $sp$, $sp^2$ and $sp^3$ hybrid orbitals can be formed because only $s$ and $p$ atomic orbitals are available. For higher principal quantum levels, $d$ atomic orbitals may also be available.

The mixing of $s$, $p_x$, $p_y$, $p_z$ and $d_{z^2}$ atomic orbitals gives a set of five $sp^3d$ hybrid orbitals. The spatial disposition of the $sp^3d$ hybrid orbitals corresponds to a trigonal bipyramid (Figure 7.11a). This set of hybrid orbitals is unusual because, unlike the other sets described here, the $sp^3d$ hybrid orbitals are *not* equivalent and divide into two groups: axial and equatorial.

**Fig. 7.10** The combination of one 2s and three 2p atomic orbitals to generate four equivalent $sp^3$ hybrid orbitals.

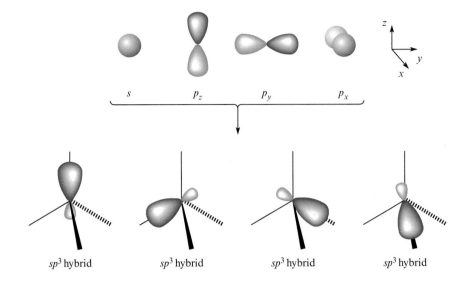

$s$        $p_z$        $p_y$        $p_x$

$sp^3$ hybrid        $sp^3$ hybrid        $sp^3$ hybrid        $sp^3$ hybrid

The characters of atomic orbitals can be mixed to generate hybrid orbitals. Each set of **hybrid** orbitals is associated with a particular shape:

$sp$        linear

$sp^2$       trigonal planar

$sp^3$       tetrahedral

$sp^3d$      trigonal bipyramidal

$sp^3d^2$     octahedral

A combination of $s$, $p_x$, $p_y$, $p_z$, $d_{z^2}$ and $d_{x^2-y^2}$ atomic orbitals generates six $sp^3d^2$ hybrid orbitals, and the spatial arrangement of these hybrids is octahedral (Figure 7.11b).

Other hybridization schemes can be developed in a similar manner to describe other spatial arrangements. As we shall see in the next section, hybridization schemes provide a convenient model within VB theory for rationalizing molecular geometries. However, we must keep in mind current

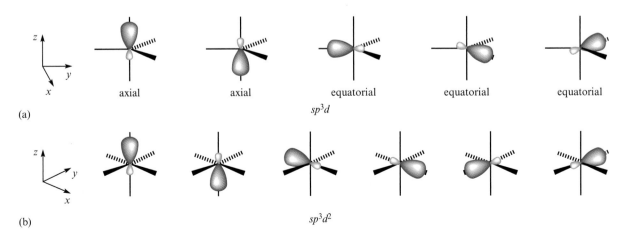

(a)        axial        axial        equatorial        equatorial        equatorial

$sp^3d$

(b)        $sp^3d^2$

**Fig. 7.11** (a) A set of five $sp^3d$ hybrid orbitals and (b) a set of six $sp^3d^2$ hybrid orbitals, shown in relationship to the Cartesian axis set.

**ENVIRONMENT AND BIOLOGY**

## Box 7.2  Hybridization means mixing

A field of oilseed rape.
*Brian Cosgrove © Dorling Kindersley.*

The term 'hybridization' is used not only for orbitals but also for biological mixing. Hybrid plants are often developed to combine the natural properties of different species. For example, oilseed rape (the crop responsible for the bright yellow fields in many parts of Europe) is grown for its edible oil, harvested in high yields. Wild honesty is a plant from which small amounts of long-chain organic acids such as erucic ((*Z*)-13-docosenoic) acid and nervonic ((*Z*)-15-tetracosenoic) acid can be extracted – these are *fatty acids* (see Section 33.9). Their importance lies in the cosmetics industry, other industrial applications and in potential uses as drugs. By producing a hybrid plant from oilseed rape and honesty, it is hoped to produce a plant that will give higher yields of the long-chain fatty acids.

views that in compounds of the *p*-block elements, *d*-orbitals on the central atom do not actually play a significant role in bonding.

| 7.5 | **Hybridization and molecular shape in the *p*-block** |

In the previous section, we saw that each particular set of hybridized orbitals has an associated spatial arrangement. Now we consider the use of hybridization schemes to describe the bonding in some molecules containing *p*-block atoms. By matching the geometries of the hybrid orbital sets with molecular shapes we see that in a linear triatomic molecule (e.g. $CO_2$) the central atom is *sp* hybridized, in a trigonal planar molecule (e.g. $BF_3$) the central atom is $sp^2$ hybridized, and in a tetrahedral molecule (e.g. $CH_4$) the central atom is $sp^3$ hybridized.

A most important point is that *a molecule does not adopt a particular shape because the central atom possesses a particular hybridized set of orbitals.* Hybridization is a convenient model within VB theory which successfully accounts for an observed molecular geometry.

A hybrid orbital can be regarded as a domain in which a pair of electrons resides. In this way, we can see a clear link between hybridization and the VSEPR model. For linear ($[I_3]^-$), trigonal planar ($BBr_3$), tetrahedral ($SiCl_4$), trigonal bipyramidal ($PCl_5$) or octahedral ($SF_6$) molecular species, the hybridization of the central atom may be considered to be *sp*, $sp^2$, $sp^3$, $sp^3d$ or $sp^3d^2$, respectively. However, we have described molecules and ions that possess other shapes – bent, T-shaped, disphenoidal, square planar and square-based pyramidal. Within the VSEPR model, these geometries arise because of lone pairs of electrons within the valence shell of the central atom. In the hybrid orbital approach, a lone pair of electrons occupies an orbital domain in the same way as a bonding pair of electrons.

Consider a molecule of $NH_3$. The VSEPR model predicts that the molecular shape is derived from a tetrahedral arrangement of electron pairs (Figure 6.16b). It follows that the N atom may be $sp^3$ hybridized

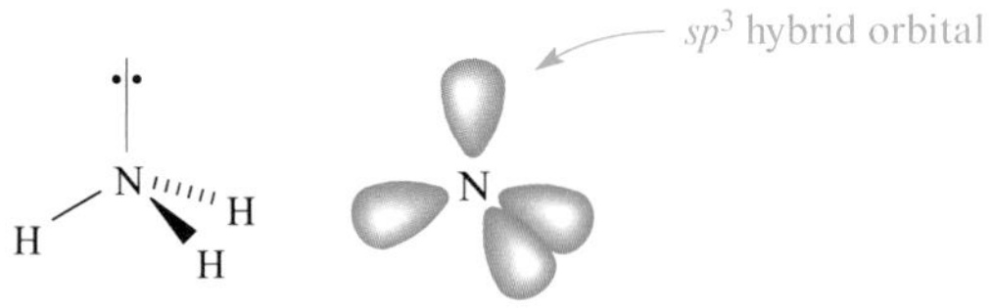

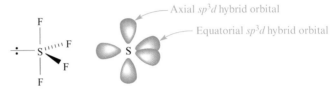

**Fig. 7.12** The VSEPR model predicts that the geometry of the $NH_3$ molecule is based on a tetrahedral arrangement of electron pairs, with the four atoms defining a trigonal pyramidal shape. The N atom is $sp^3$ hybridized.

with the lone pair of electrons occupying one of the $sp^3$ orbitals (Figure 7.12).

Similarly, VSEPR theory predicts that the structure of the $SF_4$ molecule is derived from a trigonal bipyramidal array of one lone and four bonding pairs of electrons. This corresponds to an $sp^3d$ hybridized sulfur atom (Figure 7.13) in which one equatorial orbital is occupied by the lone pair of electrons.

**Fig. 7.13** The VSEPR model predicts that the geometry of the $SF_4$ molecule is based on a trigonal bipyramidal arrangement of electron pairs with the atoms defining a disphenoidal shape. The S atom is $sp^3d$ hybridized.

---

**Worked example 7.5** | *Hybridization schemes*

**Suggest a hybridization scheme that is appropriate for the O atom in $H_2O$.**

First draw a Lewis structure for $H_2O$ and then apply VSEPR theory to suggest a molecular shape:

The tetrahedral arrangement of two bonding and two lone pairs is consistent with an $sp^3$ hybridization scheme for the O atom.

---

### 7.6    Hybridization: the $\sigma$-bonding framework

#### A tetrahedral molecule: $CH_4$

Methane contains four *single* C−H bonds. The $CH_4$ molecule is tetrahedral and the carbon atom is considered to be $sp^3$ hybridized. The mutual orientations of the four $sp^3$ hybrid orbitals coincide with the orientations of the four C−H single bonds (Figure 7.14). Each hybrid orbital overlaps with a hydrogen $1s$ atomic orbital, and two valence electrons (one from carbon and one from hydrogen) occupy each resultant localized bonding orbital. Each C−H single

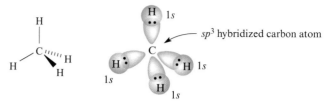

**Fig. 7.14** Valence bond theory in $CH_4$. The orientations of the four $sp^3$ hybrid orbitals centred on the C atom coincide with the orientations of the four C–H single bonds. Each hybrid orbital overlaps with an H $1s$ orbital, to give four localized C–H $\sigma$ bonds.

bond is a *localized $\sigma$-bond*, and the set of four such orbitals describes a $\sigma$-*bonding framework*. This completely describes the bonding in $CH_4$ within the valence bond model. Similarly, VB theory gives a description of the bonding in the $NH_3$ molecule in terms of three N–H $\sigma$-bonds and one lone pair of electrons occupying an $sp^3$ hybrid orbital.

## A trigonal planar molecule: $BH_3$

The boron atom in the $BH_3$ molecule is $sp^2$ hybridized. Valence bond theory describes the bonding in $BH_3$ in terms of the overlap between the B $sp^2$ hybrid orbitals and the H $1s$ orbitals (Figure 7.15). Each resultant bonding orbital contains two electrons (one from B and one from H) and defines a localized B–H $\sigma$-bond. In $BH_3$, the B atom retains an *unhybridized*, empty $2p$ orbital orthogonal to the plane of the four atoms.

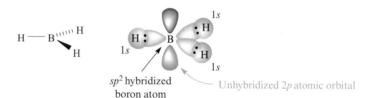

**Fig. 7.15** Valence bond theory in $BH_3$. The orientations of the three $sp^2$ hybrid orbitals centred on the B atom coincide with the orientations of the B–H single bonds. Each hybrid orbital overlaps with an H $1s$ orbital giving three localized B–H $\sigma$-bonds. If the molecule lies in the $xy$ plane, then the unused atomic orbital is a $2p_z$ orbital.

## 7.7    Hybridization: the role of unhybridized atomic orbitals

### 'Left-over' atomic orbitals

A set of one $s$ and three $p$ ($p_x$, $p_y$ and $p_z$) atomic orbitals can be involved in one of three types of hybridization. If an atom is $sp^3$ hybridized, all four of the atomic orbitals are used in the formation of the four hybrid orbitals. However, in the case of an $sp^2$ hybridized atom, the atom retains one pure $p$ atomic orbital, while an $sp$ hybridized atom has two 'left-over' $p$ atomic orbitals. Similarly, when an atom is $sp^3d$ or $sp^3d^2$ hybridized, either four or three (respectively) atomic $d$ orbitals on the central atom remain unhybridized.

In this section we consider the roles that 'left-over' $p$ atomic orbitals play in bonding.

**Fig. 7.16** Valence bond theory in $BF_3$. After the formation of the $\sigma$-bonding framework using an $sp^2$ hybridization scheme, the remaining $2p$ atomic orbital on the B atom is of the correct symmetry to overlap with an occupied $2p$ atomic orbital of an F atom. A $\pi$-bond is formed.

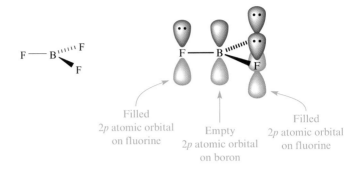

Filled $2p$ atomic orbital on fluorine

Empty $2p$ atomic orbital on boron

Filled $2p$ atomic orbital on fluorine

## A trigonal planar molecule: $BF_3$

In $BF_3$, the $\sigma$-bonding framework is similar to that in $BH_3$. Each localized $B-F$ $\sigma$-bond is formed by the overlap of a B $sp^2$ hybrid orbital and an orbital from an F atom; each resultant $\sigma$-bonding orbital is occupied by two electrons.

As in $BH_3$, once the $\sigma$-bonding framework has been formed in the $BF_3$ molecule, an unhybridized, empty $2p$ atomic orbital remains on the B atom. However, *unlike* the situation in $BH_3$, in $BF_3$ each F atom possesses three lone pairs, and one may be considered to occupy a $2p$ atomic orbital with the same orientation as the unused $2p$ atomic orbital on the B atom (for example, they may be $2p_z$ atomic orbitals). Figure 7.16 illustrates that it is possible for the B atom to form a $\pi$-bond with any one of the F atoms by overlap between the B and F $2p$ atomic orbitals. Look back at the resonance structures for $BF_3$ in Figure 7.4. One group of three resonance structures includes B=F double bond character.

The valence bonding picture for the $BF_3$ molecule illustrates a general point: *unhybridized p orbitals can be involved in $\pi$-bonding*. Related examples are $BCl_3$, $[BO_3]^{3-}$, $[CO_3]^{2-}$ and $[NO_3]^-$. These species all possess trigonal planar structures in which the central atom may be $sp^2$ hybridized. In each case, the central atom possesses an unhybridized $2p$ atomic orbital which can overlap with a $p$ atomic orbital on an adjacent atom to form a $\pi$-bond. *Exercise*: Which atomic orbitals are involved in $\pi$-bond formation in $BCl_3$ and $[CO_3]^{2-}$?

## A linear molecule: $CO_2$

The C atom in the linear $CO_2$ molecule is $sp$ hybridized and retains two $2p$ atomic orbitals (Figure 7.17). The $\sigma$-bonding framework in the molecule involves two $sp$ hybrids on C which overlap with orbitals on the two O atoms. Each localized $\sigma$-bonding orbital contains two electrons (one from C and one from O).

**Fig. 7.17** Valence bond theory in $CO_2$. After forming the $\sigma$-bonding framework using an $sp$ hybridization scheme, the two remaining $2p$ atomic orbitals on the C atom can be used to form two C–O $\pi$-bonds. Each O atom uses a $2p$ atomic orbital for $\pi$-bonding.

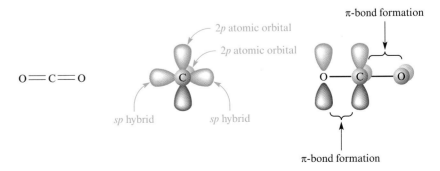

$\pi$-bond formation

$2p$ atomic orbital

$2p$ atomic orbital

$sp$ hybrid    $sp$ hybrid

$\pi$-bond formation

The two remaining $2p$ atomic orbitals on the C atom lie at right angles to one another and each overlaps with a $2p$ atomic orbital on one of the O atoms. This produces two C–O $\pi$-bonds (Figure 7.17) and each $\pi$-bonding orbital is occupied by two electrons (formally, one from C and one from O).

The overall result is the formation of two C=O double bonds, each consisting of a $\sigma$- and a $\pi$-component.

## 7.8    Molecular orbital theory and polyatomic molecules

### Polyatomic molecules create a problem

In this section, we look briefly at the way in which MO theory is used to approach the bonding in polyatomic molecules.[§] A key question is how to represent the problem. If we look back at MO diagrams such as Figure 4.14 (the formation of $H_2$) and Figure 5.8 (the formation of HF), we notice that each side of each diagram represents one of the two atoms of the diatomic molecule. What happens if we wish to represent an MO diagram for a polyatomic molecule? Any diagram that represents the composition of MOs in terms of contributions from various atomic orbitals will become very complicated, and, probably, unreadable!

An approach that is commonly used is to resolve the MO description of a polyatomic molecule into a *two-component* problem. Instead of looking at the bonding in $CH_4$ in terms of the interactions of the $2s$ and $2p$ atomic orbitals of the carbon atom with the individual $1s$ atomic orbitals of the hydrogen atoms (this would be a five-component problem), we consider the way in which the atomic orbitals of the carbon atom interact with the *set* of four hydrogen atoms. This is a two-component problem, and is called a *ligand group orbital* approach.

### Ligand group orbitals

Methane is tetrahedral and to make an MO bonding analysis for $CH_4$ easier, it is useful to recognize that the tetrahedron is related to a cube as shown in Figure 7.18. This conveniently relates the positions of the H atoms to the Cartesian axes, with the C atom at the centre of the cube.

The valence orbitals of C are the $2s$, $2p_x$, $2p_y$ and $2p_z$ atomic orbitals (Figure 7.19a). Remember that the $2s$ atomic orbital is spherically symmetric. The orientations of the $2p$ atomic orbitals are related to a cubic framework; the same axis set is used in Figures 7.18 and 7.19.

Consider now the four $1s$ atomic orbitals that the 4 H atoms contribute. Each $1s$ orbital has two possible phases and, *when the four orbitals are taken as a group*, various combinations of phases are possible. With four atomic orbitals, we can construct four *ligand group orbitals* (LGOs) as shown in Figure 7.19b. The in-phase combination of the four $1s$ atomic orbitals is labelled as LGO(1). We now make use of the relationship between the arrangement of the H atoms and the cubic framework. Each of the $xy$, $xz$ and $yz$ planes bisects the cube in a different direction. In Figure 7.18, mentally

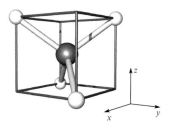

**Fig. 7.18** The relationship between the tetrahedral shape of $CH_4$ and a cubic framework. Each edge of the cube runs parallel to one of the three Cartesian axes.

[§] For a more detailed introduction including the use of group theory, see: C. E. Housecroft and A. G. Sharpe (2005) *Inorganic Chemistry*, 2nd edn, Prentice Hall, Harlow, Chapter 4.

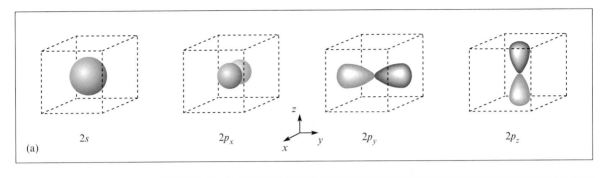

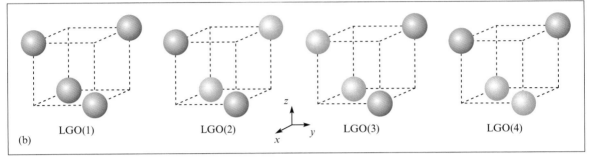

**Fig. 7.19** (a) The $2s$, $2p_x$, $2p_y$ and $2p_z$ atomic orbitals are the valence orbitals of carbon. (b) The four hydrogen $1s$ atomic orbitals combine to generate four ligand group orbitals (LGOs). The rigorous approach to the construction of ligand group orbitals uses group theory.

> The number of ligand group orbitals formed = the number of atomic orbitals used.

sketch the $xy$ plane through the cube – two H atoms lie above this plane, and two lie below. Similarly, the four H atoms are related in pairs across each of the $yz$ and $xz$ planes. Now look at Figure 7.19b. In LGO(2), the four $1s$ orbitals are drawn as two pairs, one pair on each side of the $yz$ plane. The orbitals in one pair are in phase with one another, but are out of phase with the orbitals in the second pair. Ligand group orbitals LGO(3) and LGO(4) can be constructed similarly.

### Combining the atomic orbitals of the central atom with the ligand group orbitals

The next step in the ligand group orbital approach is to 'match up' the valence orbitals of the C atom with LGOs of the four H atoms. The criterion for matching is *symmetry*. Each carbon orbital has to 'find a matching partner' from the set of LGOs.

The carbon $2s$ orbital has the same symmetry as LGO(1). If the $2s$ orbital is placed at the centre of LGO(1), overlap will occur between the $2s$ orbital and *each* of the hydrogen $1s$ orbitals. This interaction leads to one bonding MO in the $CH_4$ molecule: $\sigma(2s)$. This is the lowest-lying MO in Figure 7.20. It possesses C−H bonding character which is *delocalized over all four of the C−H interactions*.

The carbon $2p_x$ orbital has the same symmetry as LGO(2), and if it is placed at the centre of LGO(2), overlap occurs between the $2p_x$ orbital and *each* of the hydrogen $1s$ orbitals. This interaction leads to a second bonding MO in the methane molecule: $\sigma(2p_x)$. This MO contains a nodal plane coincident with the $yz$ plane. Similarly, we can match the carbon $2p_y$

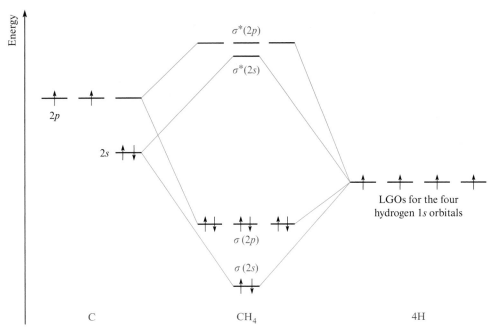

**Fig. 7.20** Molecular orbital diagram for the formation of $CH_4$.

orbital with LGO(3), and the $2p_z$ orbital with LGO(4). These combinations generate two more bonding molecular orbitals. The three MOs that are formed from the combinations of the carbon $2p$ orbitals and LGOs (2) to (4) are equivalent to one another except for their orientations. These MOs are degenerate and are labelled $\sigma(2p)$ in Figure 7.20. Each MO contains C−H bonding character which is *delocalized over all four of the C−H interactions*.

There are therefore four bonding MOs (and hence four antibonding MOs) that describe the bonding in $CH_4$. In each orbital, the C−H bonding character is *delocalized*.

### Putting in the electrons

As usual in MO theory, the last part of the operation is to count the available valence electrons and place them in the molecular orbitals according to the *aufbau* principle.

In $CH_4$ there are eight valence electrons and these occupy the MOs as shown in Figure 7.20. All the valence electrons in the molecule are paired and $CH_4$ is predicted to be diamagnetic, in keeping with experimental data.

|  |  |
|---|---|
| 7.9 | ### How do the VB and MO pictures of the bonding in methane compare? |

### The valence bond and molecular orbital descriptions of methane

Valence bond theory describes the bonding in the $CH_4$ molecule in terms of an $sp^3$ hybridized carbon atom, and four equivalent $\sigma$-bonding orbitals

(Figure 7.14). Each orbital is *localized*, meaning that each C−H bond is described by one σ-bonding orbital.

Molecular orbital theory also produces a description of $CH_4$ which involves four molecular orbitals. However, unlike the valence bond model, the molecular orbital approach gives one unique MO and a set of three degenerate MOs (Figure 7.20). In each MO, the C−H bonding character is *delocalized over all four of the C−H vectors*.

Are these two pictures really different? If so, is one of them wrong? Indeed, is either of them realistic?

## Photoelectron spectroscopy

One of the major differences between the results of the VB and MO treatments of the bonding in the $CH_4$ molecule is that the former suggests that there are four equivalent MOs (equal energy) while the latter is consistent with there being two molecular orbital energy levels, one associated with a unique MO and one with a set of three degenerate orbitals.

In Box 4.6, we briefly mentioned the experimental technique of photoelectron spectroscopy in the context of probing the energies of molecular orbitals. This method can be used to investigate the question of the MO energy levels in $CH_4$ and the photoelectron spectroscopic data support the results of MO theory. But, does this mean that the VB model gives the wrong 'answer'?

## Valence bond versus molecular orbital theory

Molecular orbital theory describes the bonding in the $CH_4$ molecule in terms of four MOs. The unique $\sigma(2s)$ MO is spherically symmetric and provides equal bonding character in all of the four C−H interactions. The $\sigma(2p_x)$, $\sigma(2p_y)$ and $\sigma(2p_z)$ MOs are related to one another by rotations through $90°$. Because they are degenerate, we must consider them *as a set and not as individual orbitals*. Taken together, they describe the four C−H bonds equally. Thus, the MO picture of $CH_4$ is of a molecule with four equivalent C−H bonds. This result arises *despite* the fact that the four MOs are not all identical.

The $sp^3$ hybrid model (VB theory) of $CH_4$ describes four equivalent C−H bonds in terms of four *localized* σ-bonds. The associated bonding orbitals are of equivalent energy.

Both the VB and MO approaches are bonding *models*. Both models achieve a goal of showing that the $CH_4$ molecule contains four equivalent bonds. While the MO model appears to give a more realistic representation of the energy levels associated with the bonding electrons, the VB method is simpler to apply. While MO theory can be applied to small and large molecules alike, it very quickly goes beyond the 'back-of-an-envelope' level of calculation.

## SUMMARY

In this chapter, we have considered the bonding in polyatomic molecules. We have considered the octet rule, and introduced hybridization schemes that can be applied to different spatial arrangements of bonding and lone pairs of electrons around a central atom. Finally, we have introduced the use of MO theory, applying it to a simple example, methane.

### Do you know what the following terms mean?

- the valence shell
- hybrid orbital
- ligand group orbital approach

### You should now be able:

- to work out the number of electrons in the valence shell of atom X in a molecule $XY_n$
- to draw a set of resonance structures for a simple polyatomic molecule or molecular ion
- to discuss what is meant by orbital hybridization
- to relate hybridization schemes to arrangements of bonding and lone pairs of electrons around a central atom

- to relate hybridization schemes to molecular shapes
- to use hybridization schemes to describe the bonding in simple polyatomic molecules containing both single and multiple bonds
- to develop a bonding scheme for $CH_4$ using MO theory and to understand why a 'ligand group orbital approach' is useful

## PROBLEMS

**7.1** In which of the following molecular species does the central atom possess an octet of valence electrons? (a) $BBr_2F$; (b) $OF_2$; (c) $PH_3$; (d) $CO_2$; (e) $NF_3$; (f) $[PCl_4]^+$; (g) $AsF_3$; (h) $BF_3$; (i) $AlCl_3$.

**7.2** In which of the molecular species in problem 7.1 does the central atom possess a sextet of valence electrons?

**7.3** What is the origin of the 'octet rule'?

**7.4** For each of the following, draw Lewis structures that are consistent with the central atom in each molecule obeying the octet rule: (a) $H_2O$, (b) $NH_3$, (c) $AsF_3$, (d) $SF_4$.

**7.5** $N_2O_4$ is a planar molecule. Explain why **7.5** is not a reasonable resonance structure. Suggest what resonance structures do contribute to the bonding in $N_2O_4$. Does this approach indicate why $N_2O_4$ readily decomposes to give $NO_2$?

(7.5)

**7.6** How many single bonds may each of the (a) $N^-$, (b) $B^-$ and (c) $C^-$ centres form while obeying the octet rule?

**7.7** Place charges on the atomic centres in each of the following resonance structures such that each centre obeys the octet rule.

(a) For $[N_3]^-$:

$$N = N = N$$

(b) For $[NCS]^-$:

$$N = C = S$$

(c) For $[NO_2]^+$:

$$O = N = O$$

(d) For $[NO_2]^-$:

$$O - N = O \longleftrightarrow O = N - O$$

Comment on the shapes of these species.

**7.8** The azide ion, $[N_3]^-$, is linear with equal $N-N$ bond lengths. Give a description of the bonding in $[N_3]^-$ in terms of valence bond theory.

**7.9** Outline how the combination of carbon $2s$ and $2p$ atomic orbitals leads to hybrid orbitals that can be used to describe the carbon centres in (a) $CCl_4$, (b) $H_2CO$ (**7.6**), (c) **7.7** and (d) $C_2H_2$ (**7.8**).

(7.6)          (7.7)          (7.8)

**7.10** Write down the hybridization of the central atom in each of the following species: (a) $SiF_4$; (b) $NH_3$; (c) $[NH_4]^+$; (d) $BH_3$; (e) $[CoF_6]^{3-}$; (f) $IF_3$; (g) $H_2S$.

**7.11** What hybridization scheme would you assign to the carbon atoms in each of the following molecules? (a) $CO_2$; (b) $C_2H_6$; (c) $CH_2Cl_2$; (d) $CH_3CH_2OH$; (e) $CH_3CH=CHCH_2CH_3$; (f) $COCl_2$; (g) $[CO_3]^{2-}$.

**7.12** Explain why double bond character in a carbon-containing compound may be described in terms of an $sp^2$ hybridization scheme but is incompatible with $sp^3$ hybridization.

**7.13** Consider the molecule $CO_2$. (a) Use VSEPR theory to rationalize its shape. (b) Draw resonance structures for $CO_2$ and indicate which structure will make the major contribution to the bonding. (c) Describe the bonding in terms of a hybridization scheme, including full descriptions of the formation of $\sigma$- and $\pi$-bonds.

**7.14** Repeat problem 7.13 for HCN.

**7.15** Repeat problem 7.13 for $H_2CO$ (**7.6**).

**7.16** Give appropriate hybridization schemes for the P atom in each of the following species: (a) $PF_5$; (b) $[PF_4]^+$; (c) $[PF_6]^-$; (d) $PF_3$; (e) $POCl_3$; (f) $PMe_3$. For $[PF_6]^-$, draw Lewis structures that are consistent with P obeying the octet rule, and with the P−F bonds being equivalent.

**7.17** What is the 'ligand group orbital approach' and why is it used in MO treatments of polyatomic molecules?

**7.18** $[BH_4]^-$ is isoelectronic with $CH_4$. By using a ligand group orbital approach, construct an MO diagram to show the interactions between $B^-$ and four H atoms for form $[BH_4]^-$.

## Additional problems

**7.19** In Box 7.1, we showed how to solve equations for the normalized wavefunctions for a set of $sp^2$ hybrid orbitals. Follow the same method to derive equations for the normalized wavefunctions to describe (a) $\psi_{sp}$ and (b) $\psi_{sp^3}$ hybrid orbitals.

**7.20** Comment on each of the following observations:
(a) $XeF_6$ has a distorted octahedral structure.
(b) $[N_3]^-$ is linear with equal N−N bond lengths; $[N_5]^+$ is bent at the central N atom.
(c) $NOF_3$ is well established; it has N−O and N−F bond lengths of 116 and 146 pm respectively.
(d) $BH_3$ can accept a pair of electrons to form compounds such as $H_3BNMe_3$ in which the B atom is tetrahedral.

# 8 Ions

## Topics

- Ions
- Electron density maps
- Ionization energy and electron affinity
- Periodic trends in ionization energies
- Trends in values of successive ionization energies
- Electrostatic interactions between ions
- Structural types
- Lattice energy and the Born–Haber cycle

## 8.1 Introduction

In the previous four chapters, we discussed covalent bonding. The electron density in a homonuclear bond is concentrated between the two atoms in the bond, but in a heteronuclear covalent bond, the electron density may be displaced towards one of the atoms according to the relative electronegativities of the atoms.

### A review of the HF molecule

Resonance structures for HF: see Section 5.3

MO approach to HF: see Section 5.5

Within the valence bond model, resonance structures represent different bonding descriptions of a molecule. In a molecule such as HF, the difference between the electronegativities of hydrogen ($\chi^P = 2.2$) and fluorine ($\chi^P = 4.0$) means that the resonance structure describing an ionic form contributes significantly to the overall bonding model. When we used molecular orbital theory to describe the bonding in the HF molecule, the differences in energy between the H and F atomic orbitals meant that the $\sigma$-bonding MO possessed considerably more F than H character. The electron density in the H−F bond is displaced towards the F atom and the bond is polar. In both of these bonding models, the F atom in HF carries a partial negative charge ($\delta^-$) and the H atom bears a partial positive charge ($\delta^+$).

### Approaching the bonding in sodium fluoride by VB and MO theories

In the gas phase, sodium fluoride contains NaF *molecules* (or more correctly, NaF formula units), although in aqueous solution, the liquid or solid state, $Na^+$ and $F^-$ *ions* are present.

**Fig. 8.1** An MO diagram for the formation of an NaF molecule. Only the valence atomic orbitals and electrons are shown. The break in the vertical (energy) axis signifies that the energy of the fluorine 2s atomic orbital is much lower than is actually shown. The σ-bonding MO contains mainly fluorine character.

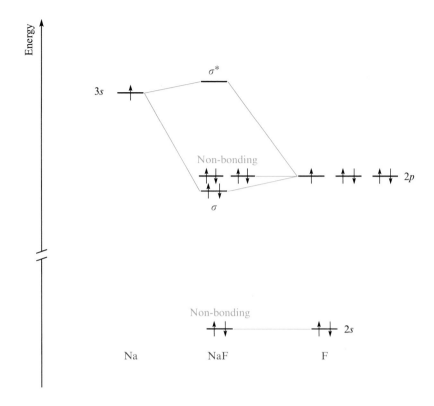

The difference between the electronegativities of Na ($\chi^P = 0.9$) and F ($\chi^P = 4.0$) is larger than the difference between H and F. When we write out resonance structures for sodium fluoride, the *ionic form* ($Na^+F^-$) predominates.

An MO diagram for the formation of an NaF molecule is shown in Figure 8.1. The 3s valence atomic orbital of the Na atom lies at higher energy than the valence 2s and 2p atomic orbitals of F. The energy matching of these orbitals is poor, and the bonding MO contains far more F than Na character. We are *tending towards* a situation in which this MO has so little Na character that it is almost non-bonding.[§] There are eight valence electrons (one from the Na atom and seven from the F atom) and these occupy the orbitals in NaF according to the *aufbau* principle. The important consequence of this is that three of the filled MOs in NaF possess *only fluorine* character and one filled MO possesses *virtually all fluorine* character. It follows that the electron density between the sodium and fluorine nuclei is displaced so far in one direction that we are tending towards the point at which one electron has been transferred from the sodium atom to the fluorine atom. If the transfer is complete, the result is the formation of a fluoride ion ($F^-$) and a sodium ion ($Na^+$).

**Orbital energy matching: see Sections 4.12 and 5.4**

## Ions

In Section 1.11, we reviewed ion formation. An F atom (with a ground state electronic configuration $[He]2s^2 2p^5$) may accept one electron to form

---

[§] We are using 'non-bonding' in the sense usually adopted in discussions of MO diagrams; specifically, there is little sharing of electrons between the nuclei. The charge separation results in 'ionic bonding'.

a fluoride anion ($F^-$) with the noble gas configuration $[He]2s^2 2p^6$ (or [Ne]). The loss of one electron from a neutral Na atom (ground state electronic configuration $[Ne]3s^1$) generates a positively charged cation ($Na^+$) with the noble gas configuration of [Ne].

In the same way that van der Waals forces or dipole–dipole interactions arise between atoms or molecules that possess dipole moments (transient or permanent), so there are strong electrostatic interactions between spherically symmetrical ions of opposite charge. In sodium fluoride, there are attractive interactions between the $Na^+$ cations and *all* the nearby $F^-$ ions. Similarly, there are electrostatic (attractive) interactions between the $F^-$ anions and *all* nearby $Na^+$ cations. An electrostatic interaction depends upon distance and not upon direction. Ultimately this leads to the assembly of an *ionic lattice*.

> An *anion* is a negatively charged ion and a *cation* is a positively charged ion.

This chapter is concerned with *ions* and the enthalpy changes that accompany the processes of cation and anion formation. What forces operate between oppositely charged ions? How much energy is associated with these interactions? How are ions arranged in the solid state?

> A *lattice* is an infinite and regular 3-dimensional array of atoms or ions in a crystalline solid.

Ionic bonding is a *limiting model*, just as a covalent model is. However, just as valence bond theory can be improved by including ionic resonance structures, we see later in this chapter that the ionic model can be adjusted to take account of some covalent character. In fact, very few systems can be considered to be purely ionic or purely covalent. The bonding in many species is more appropriately described as lying somewhere between these two limits.

## 8.2    Electron density maps

The distinction between a fully covalent bond $X-Y$ and an ionic interaction $X^+ Y^-$ is made by considering the regions of electron density around the nuclei X and Y. We are already familiar with the idea that a covalent bond is synonymous with the presence of 'shared' electron density. The distribution of the electron density between the atomic centres in a molecule can be investigated by means of appropriate calculations using forms of the Schrödinger equation. Such results can be represented in the form of a contour map. An *electron density map* is composed of contour lines which connect points of equal electron density. Contours drawn close to the nucleus relate mainly to core electron density.

Schrödinger equation: see Section 3.7

Figure 8.2 shows three electron density maps. Each is drawn as a section through a formula unit and indicates the electron density distribution in that plane. Figure 8.2a shows an electron density map for the homonuclear diatomic molecule $Li_2$. The map has a symmetrical appearance, with the electron density at one Li centre mirrored at the other. There is a region of electron density located symmetrically between the two atomic nuclei and this represents the *bonding electron density*; the $Li-Li$ bond is mainly covalent in character.

An electron density map for HF is shown in Figure 8.2b. The electron density contours in this diagram are typical of those in a polar covalent bond. The map is asymmetrical and the region of electron density is ovate (i.e. 'egg-shaped'). More electron density is associated with the fluorine centre than the hydrogen centre. There is, however, still a region of electron density between the nuclei, and this corresponds to the electron density in the covalent bond.

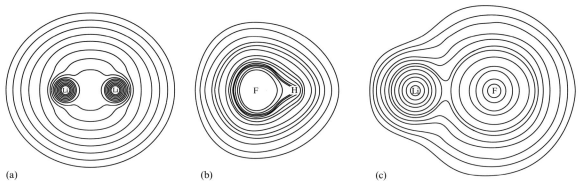

(a)    (b)    (c)

**Fig. 8.2** Electron density maps for (a) $Li_2$, (b) HF and (c) LiF. In $Li_2$, the distribution of electron density is symmetrical and a region of electron density is situated between the two lithium nuclei. This corresponds to a region of Li–Li bonding electron density. Hydrogen fluoride is a polar molecule, and the electron density is displaced towards the fluorine centre, shown on the left-hand side of the figure. A region of bonding electron density between the H and F nuclei is apparent. In LiF, the circular electron density contours correspond to sections through two spherical regions of electron density. The tendency is towards the formation of spherical $Li^+$ and $F^-$ ions. [From A. C. Wahl (1966) *Science*, vol. 151, p. 961 (Figure 8.2a); Reprinted with permission from R. F. W. Bader *et al.* (1967) *J. Chem. Phys.*, vol. 47, p. 3381 (Figure 8.2b); R. F. W. Bader *et al.* (1968) *J. Chem. Phys.*, vol. 49, p. 1653 (Figure 8.2c).]

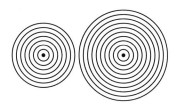

**Fig. 8.3** An idealized electron density map showing sections through a pair of spherical ions.

Figure 8.2c illustrates the electron density distribution in an LiF molecule. Now we can best describe the electron density as being concentrated in two circular regions (that is, spherical in three dimensions). Each region is centred around one of the two nuclei. The peripheral contour lines join points of only low electron density. The bonding in LiF may be described as *tending towards being ionic*.

The electron density map in Figure 8.2c illustrates the same situation that we discussed for sodium fluoride in Section 8.1. As the covalent interaction between two atoms diminishes, the tendency is towards the formation of two *spherical* ions. In the extreme case, the electron density map will appear as shown in Figure 8.3. To a first approximation, we can assume that an ion is spherical, but later in this chapter we discuss to what extent this description is valid.

| 8.3 | **Ionization energy** |

In Section 5.10 we mentioned ionization energies in the context of the Mulliken electronegativity scale, and we defined the first ionization energy ($IE_1$) of an atom as the internal energy change at 0 K ($\Delta U(0\,K)$) associated with the *removal* of the first valence electron (equation 8.1). The energy change is defined for a gas phase process. The units are kJ mol$^{-1}$ or electron volts (eV).

$1\,eV = 96.485 \approx 96.5\,kJ\,mol^{-1}$

$$X(g) \longrightarrow X^+(g) + e^- \tag{8.1}$$

The *first ionization energy* ($IE_1$) of a gaseous atom is the internal energy change at 0 K ($\Delta U(0\,K)$) associated with the removal of the first valence electron:

$$X(g) \longrightarrow X^+(g) + e^-$$

For thermochemical cycles, an associated change in *enthalpy* is used:

$$\Delta H(298\,K) \approx \Delta U(0\,K)$$

**THEORETICAL AND CHEMICAL BACKGROUND**

## Box 8.1 Ionization energies and associated enthalpy values

The first ionization energy ($IE_1$) of an element X is the *internal energy change* at 0 K ($\Delta U(0\,K)$) which is associated with the removal of the first valence electron; it is defined for a gas phase process:

$$X(g) \longrightarrow X^+(g) + e^-$$

Most thermochemical cycles involve enthalpy changes at 298 K ($\Delta H(298\,K)$).

If we assume that the gaseous atoms of element X and the gaseous ions $X^+$ are *ideal* gases, then $\Delta U(0\,K)$ and $\Delta H(298\,K)$ are related as shown on the right; $R$ is the molar gas constant ($8.314 \times 10^{-3}\,kJ\,K^{-1}\,mol^{-1}$) and $\Delta T$ is the difference between the two temperatures (0 K and 298 K):

$$\Delta H(298\,K) = \Delta H(0\,K) + (\tfrac{5}{2} \times R \times \Delta T)$$
$$\Delta H(298\,K) = \Delta H(0\,K)$$
$$+ (\tfrac{5}{2} \times 8.314 \times 10^{-3} \times 298)$$
$$\Delta H(298\,K) = \Delta H(0\,K) + 6.2\,kJ\,mol^{-1}$$

Typically, ionization energies referring to the removal of valence electrons are of the order of $10^3\,kJ\,mol^{-1}$ and the addition of $\approx 6\,kJ\,mol^{-1}$ makes little difference to the value. It is therefore acceptable for most purposes to use tabulated values of ionization energies ($\Delta U(0\,K)$) when an enthalpy of ionization ($\Delta H(298\,K)$) is required:

$$IE = \Delta U(0\,K) \approx \Delta H(298\,K)$$

Hess cycle: see Section 2.7

We often need to incorporate the values of ionization energies into Hess cycles and it is then convenient to use a change in enthalpy ($\Delta H(298\,K)$) rather than an internal energy change ($\Delta U(0\,K)$). We have already come across this problem in Section 4.5, and we discuss the differences between $\Delta U$ and $\Delta H$ more fully in Chapter 17. The difference between an ionization energy and the associated enthalpy change is very small (see Box 8.1) and values of $IE$ can be used in enthalpy cycles provided that a high degree of accuracy is not required.

The second ionization energy ($IE_2$) of an atom refers to the process defined in equation 8.2 – it is also the first ionization step of cation $X^+$. Successive ionizations can also occur; the third ionization energy ($IE_3$) of atom X refers to the loss of the third electron (equation 8.3), and so on. We are generally concerned only with the removal of electrons from the valence shell of an atom. Removing core electrons requires considerably more energy. For example, with an atom in group 2 of the periodic table, the first and second ionization energies refer to the removal of the two valence electrons, while higher values of $IE$ refer to the removal of core electrons.

$$X^+(g) \longrightarrow X^{2+}(g) + e^- \tag{8.2}$$
$$X^{2+}(g) \longrightarrow X^{3+}(g) + e^- \tag{8.3}$$

Values of some ionization energies for the elements are listed in Appendix 8.

| 8.4 | **Trends in ionization energies** |

### First ionization energies across the first period

More about ionization energies in Section 20.4

The first period runs from lithium (group 1) to neon (group 18). The ground state electronic configurations of the two *s*-block elements in the period (lithium and beryllium) are $[He]2s^1$ and $[He]2s^2$ respectively, and of the six *p*-block elements (boron to neon) are $[He]2s^2 2p^1$, $[He]2s^2 2p^2$, $[He]2s^2 2p^3$, $[He]2s^2 2p^4$, $[He]2s^2 2p^5$ and $[He]2s^2 2p^6$ respectively.

**Fig. 8.4** The trend in first
ionization potentials on going
from helium ($Z = 2$) to neon
($Z = 10$). The elements from
lithium to neon make up the first
period.

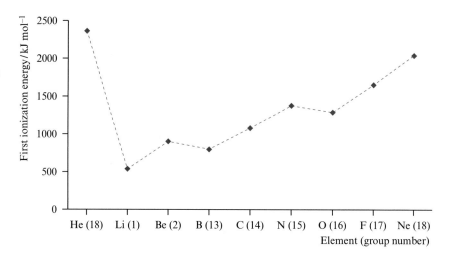

Figure 8.4 shows the trend in the first ionization energies ($IE_1$) as the period is crossed; the value of $IE_1$ for helium (the noble gas which precedes lithium in the periodic table) is also included in the figure. Three features in the graph can be seen:

- there is a sharp decrease in the value of $IE_1$ in going from He to Li;
- there is a *general* increase in the values of $IE_1$ in crossing the period;
- two apparent discontinuities (between Be and B, and between N and O) in the general increase from Li to Ne occur.

These observations can be rationalized in terms of the arrangement of the valence electrons.

Stability of $1s^2$
configuration:
see Section 4.4

**Helium to lithium**    The first ionization energy for He refers to the process in equation 8.4. An electron is removed from a filled $1s$ shell, but as we have already seen there is some special stability associated with the $1s^2$ configuration. Removing an electron from the filled $1s$ orbital requires a relatively large amount of energy.

$$\text{He(g)} \longrightarrow \text{He}^+\text{(g)} + \text{e}^- \qquad (8.4)$$
$$\quad {}_{1s^2} \qquad\qquad {}_{1s^1}$$

The ground state electronic configuration of Li is $[\text{He}]2s^1$ and the first ionization energy is the energy required to remove the $2s$ electron. This is the only electron in the valence shell and removing it is a relatively easy process.

Orbital contraction:
see Section 4.21

**Lithium to neon**    In going from Li to Ne, two significant changes occur: firstly, the valence shell is filled with electrons in a stepwise manner and secondly the effective nuclear charge increases and the atomic orbitals contract. It follows that it becomes increasingly difficult (more energy is needed) to remove an electron from the valence shell on going from an atom with the $[\text{He}]2s^1$ ground state configuration (Li) to one with the $[\text{He}]2s^2 2p^6$ configuration (Ne).

**Beryllium to boron**    In Section 3.15, we noted that the nuclear charge experienced by an electron in the $2p$ orbital is less than that experienced by an electron in the $2s$ orbital *in the same atom*. This is because electrons in the $2s$ atomic orbital shield electrons in the $2p$ orbital. In going from Be to B,

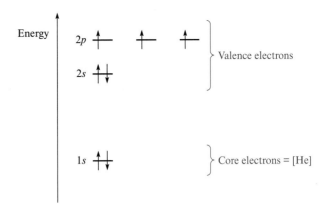

Fig. 8.5 The ground state electronic configuration of nitrogen. Note that the 2p level is half-full.

the ground state electronic configuration changes from $[He]2s^2$ to $[He]2s^22p^1$. But we must be careful. The number of protons also increases by one. The *real* increase in the nuclear charge does *not*, however, outweigh the effect of shielding, and it is easier to remove the $2p$ electron from the B atom than one of the $2s$ electrons from the Be atom.

**Nitrogen to oxygen**   The trend of increasing ionization energies across a row in the *p*-block is interrupted between N and O. The slight fall in values between these two elements is associated with the fact that N possesses the ground state electronic configuration $[He]2s^22p^3$ (Figure 8.5). The $2p$ level is *half full* and this imparts a degree of stability to the N atom.[§] Thus, it is slightly harder to remove an electron from the valence shell of the N atom than might otherwise be expected.

## First ionization energies across the second period

Figure 8.6 shows the trend in first ionization energies on going from neon to argon. The dramatic fall from Ne to Na corresponds to the difference

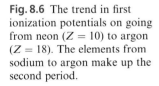

Fig. 8.6 The trend in first ionization potentials on going from neon ($Z = 10$) to argon ($Z = 18$). The elements from sodium to argon make up the second period.

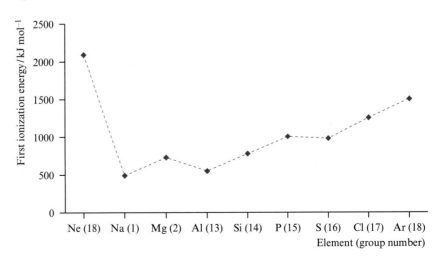

§ The origin of this stability is complex and originates in *exchange energies*. We merely note that there is a particular stability associated with half-filled configurations such as $p^3$ and $d^5$. For further details, see: A. B. Blake (1981) *Journal of Chemical Education*, vol. 58, p. 393; D. M. P. Mingos (1998) *Essential Trends in Inorganic Chemistry*, Oxford University Press, Oxford, p. 14.

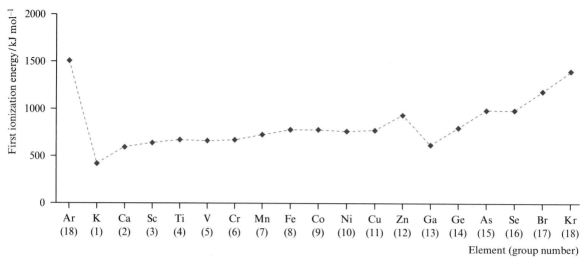

**Fig. 8.7** The trend in first ionization potentials on going from Ar ($Z = 18$) to Kr ($Z = 36$). The elements from K to Kr make up the third period. The elements in groups 1 and 2 are in the *s*-block, elements from groups 3 to 12 inclusive are in the *d*-block, and elements in groups 13 to 18 inclusive are in the *p*-block.

between removing an electron from the full octet in Ne, and removing an electron from a singly occupied 3*s* atomic orbital in Na.

From Na to Ar, the second period is crossed. The same pattern of values in ionization energies is observed in Figure 8.6 as in Figure 8.4 and, not surprisingly, the same arguments explain the trends for the second period as the first, except that now we are dealing with the 3*s* and 3*p* atomic orbitals.

### First ionization energies across the third period

The third period includes elements from the *d*-block as well as from the *s*- and *p*-blocks. The first ionization energies for the elements from Ar (the last element in the second period) to Kr (the last element in the third period) are plotted in Figure 8.7. Compare this figure with Figures 8.4 and 8.6: the same pattern in values of $IE_1$ is observed on crossing the first, second and third rows of the *s*- and *p*-blocks.

The elements from scandium (Sc) to zinc (Zn) possess ground state electronic configurations [Ar]$4s^2 3d^1$ to [Ar]$4s^2 3d^{10}$ respectively. In these elements, the similarity in first ionization energies is associated with the closeness in energy of the 4*s* and 3*d* atomic orbitals. The rise on going from Cu to Zn can be attributed to the particular stability of the Zn atom, the ground state electronic configuration of which is [Ar]$4s^2 3d^{10}$ with the 4*s* and 3*d* levels fully occupied. We return to *d*-block metal ions in Chapter 23.

### First ionization energies down a group

If we look more closely at the graphs in Figures 8.4, 8.6 and 8.7, we observe that, although the *trends* for the first ionization energies of the *s*- and *p*-block elements are similar, the plots are displaced to lower energies on going from Figure 8.4 to 8.6, and from Figure 8.6 to 8.7. This new trend corresponds to a *general decrease in the values of the first ionization energy as a group is descended.*

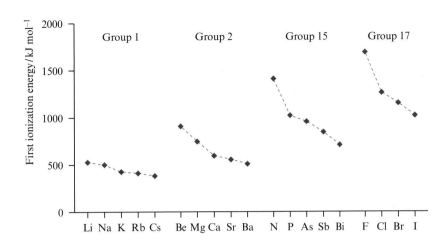

**Fig. 8.8** First ionization potentials decrease as a group in the *s*- or *p*-block is descended.

Consider the elements in group 1. Values of $IE_1$ for lithium, sodium, potassium, rubidium and caesium are plotted in the left-hand graph in Figure 8.8. A decrease in values is observed. The first ionization of Li involves the removal of an electron from the $2s$ orbital; for Na, a $3s$ electron is lost, while for K, Rb or Cs, a $4s$, $5s$ or $6s$ electron is removed, respectively. The decrease in $IE$s reflects an increase in the distance between the nucleus and the valence electron with an increase in the principal quantum number.

We use the same arguments to explain the trends observed in the values of $IE_1$ for the elements in groups 2, 15 and 17 (Figure 8.8).

### Successive ionization energies for an atom

So far, we have been concerned with the formation of singly charged cations (equation 8.1). For an alkali metal with a valence ground state configuration of $ns^1$, the loss of *more* than one electron requires a considerable amount of energy. The first and second ionizations of Na involve the processes shown in equations 8.5 and 8.6.

$$\text{First ionization:} \quad \text{Na(g)} \longrightarrow \text{Na}^+(\text{g}) + \text{e}^- \tag{8.5}$$

$$\text{Second ionization:} \quad \text{Na}^+(\text{g}) \longrightarrow \text{Na}^{2+}(\text{g}) + \text{e}^- \tag{8.6}$$

The value of $IE_1$ for Na is $496 \text{ kJ mol}^{-1}$ but $IE_2$ is $4563 \text{ kJ mol}^{-1}$. This almost 10-fold increase is due to the fact that the second electron must be removed:

1. from a positively charged ion, and
2. from the fully occupied $2p$ level.

Discussions above about neon cover point (2) – Ne and Na$^+$ are isoelectronic. Point (1) addresses the fact that the outer electrons in the Na$^+$ *ion* experience a greater effective nuclear charge than the outer electron in the neutral Na *atom*.

The first five ionization energies for Na are plotted in Figure 8.9. Sharp increases are seen as successive core electrons are removed.

Consider the successive ionizations of magnesium, the ground state electronic configuration of which is $[\text{Ne}]3s^2$. The first five ionization energies

**Fig. 8.9** A plot of the first five ionization potentials for sodium (ground state configuration [Ne]$3s^1$), magnesium (ground state configuration [Ne]$3s^2$) and aluminium (ground state configuration [Ne]$3s^2 3p^1$).

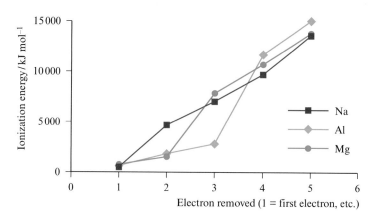

are plotted in Figure 8.9; the first three values refer to the processes given in equations 8.7–8.9.

First ionization:     $Mg(g) \longrightarrow Mg^+(g) + e^-$                     (8.7)

Second ionization:   $Mg^+(g) \longrightarrow Mg^{2+}(g) + e^-$                (8.8)

Third ionization:    $Mg^{2+}(g) \longrightarrow Mg^{3+}(g) + e^-$             (8.9)

The plot of successive ionization energies for Mg in Figure 8.9 shows an initial increase which is followed by a much sharper rise. The discontinuity occurs *after* the second electron has been removed (equation 8.9). Whereas the first two electrons are removed from the valence $3s$ shell, successive ionizations involve the removal of core electrons.

Similarly, a plot (Figure 8.9) of the first five ionization energies for Al (ground state electronic configuration [Ne]$3s^2 3p^1$) shows a gradual increase in energy corresponding to the loss of the three valence electrons, followed by a much sharper rise as we begin to remove core electrons.

In both Mg and Al, it is harder to remove the second valence electron than the first, because the second electron is being removed from a positively charged ion. Similarly, the energy required to remove the third valence electron from Al involves a contribution that reflects the work done in overcoming the attraction between the effective nuclear charge in the $Al^{2+}$ ion and the electron to be removed (equation 8.10).

$$Al^{2+}(g) \longrightarrow Al^{3+}(g) + e^-  \qquad (8.10)$$

Although it is tempting to rationalize the observed oxidation states of Na, Mg and Al (and of other elements) in terms of only the *successive* ionization energies, this is actually dangerous. As an exercise, you should calculate (using *IE* values from Appendix 8) the *sums* of the first two IEs for Na (the *total* energy needed to form the $Na^{2+}$ ion from a gaseous Na atom) and the first three IEs for Al (the *total* energy needed to form the $Al^{3+}$ ion from a gaseous Al atom). Is it still obvious why, in compounds of these elements, the $Al^{3+}$ ion may be present but the $Na^{2+}$ ion is not? And why, for example, are $Al^{2+}$ compounds very rare when the expenditure of energy to form this ion is less than that involved in forming $Al^{3+}$? The answer is that ionization energy is not the only factor governing the stability of a particular oxidation state. This reasoning will become clearer later in the chapter.

## 8.5    Electron affinity

### Definition and sign convention

It is unfortunate (and certainly confusing) that the sign convention used for electron affinity values is the opposite of the normal convention used in thermodynamics. The first electron affinity ($EA_1$) is *minus* the internal energy change ($EA = -\Delta U(0\,\mathrm{K})$) which accompanies the gain of one electron by a *gaseous* atom (equation 8.11). The second electron affinity of atom Y refers to reaction 8.12.

$$Y(g) + e^- \longrightarrow Y^-(g) \tag{8.11}$$

$$Y^-(g) + e^- \longrightarrow Y^{2-}(g) \tag{8.12}$$

In a thermochemical cycle, it is the enthalpy change ($\Delta H(298\,\mathrm{K})$) associated with a process such as reaction 8.11 or 8.12 that is appropriate, rather than the change in internal energy ($\Delta U(0\,\mathrm{K})$). A correction can be made, but the difference between corresponding values of $\Delta H(298\,\mathrm{K})$ and $\Delta U(0\,\mathrm{K})$ is small. Thus, values of electron affinities are usually used directly in thermochemical cycles, remembering of course to ensure that the sign is appropriate: a negative enthalpy ($\Delta H$), but a positive electron affinity ($EA$), corresponds to an exothermic process.[§]

Values of the enthalpy changes ($-EA$) associated with the attachment of an electron to an atom or negative ion are listed in Appendix 9. The units are kJ mol$^{-1}$ or electron volts (eV).

$1\,\mathrm{eV} = 96.485 \approx 96.5\,\mathrm{kJ\,mol}^{-1}$

---

The first *electron affinity* ($EA_1$) of an atom is *minus* the internal energy change at $0\,\mathrm{K}$ ($-\Delta U(0\,\mathrm{K})$) which is associated with the gain of one electron by a gaseous atom:

$$Y(g) + e^- \longrightarrow Y^-(g)$$

In thermochemical cycles, an associated *enthalpy* is used:

$$\Delta H(298\,\mathrm{K}) \approx \Delta U(0\,\mathrm{K}) = -EA$$

---

### Worked example 8.1    *Enthalpy of formation of an anion*

**The bond dissociation enthalpy (at 298 K) of $Cl_2$ is 242 kJ mol$^{-1}$ and the first electron affinity ($EA$) of chlorine is 3.62 eV. Determine the standard enthalpy of formation (at 298 K) of the gas phase chloride ion. ($1\,\mathrm{eV} = 96.5\,\mathrm{kJ\,mol}^{-1}$)**

First, ensure that the standard sign convention is used, and that the units are consistent. The data given refer to the following processes.
Bond dissociation:

$$Cl_2(g) \longrightarrow 2Cl(g) \qquad\qquad D = 242\,\mathrm{kJ\,mol}^{-1}$$

---

[§] Some tables of data list enthalpy values and others, electron affinities. Particular attention must be paid to such data before the numbers are used in calculations.

This is an *enthalpy* value, with a conventional use of sign. Breaking the Cl–Cl bond is an endothermic process.

Gaining an electron:

$$Cl(g) + e^- \longrightarrow Cl^-(g)$$

$$EA = 3.62\,eV = (3.62 \times 96.5)\,kJ\,mol^{-1} = 349\,kJ\,mol^{-1}$$

Firstly, the sign convention for $EA$ is the reverse of the convention used in thermodynamics. When a chlorine atom gains one electron, the process is exothermic.

Secondly, $EA$ is not an enthalpy term, but is an internal energy, specifically:

$$-EA = \Delta U(0\,K) \approx \Delta H(298\,K)$$

$$\Delta H(298\,K) \approx -EA \approx -349\,kJ\,mol^{-1}$$

The aim of the question is to find the *standard enthalpy of formation* ($\Delta_f H^\circ$ at 298 K) of the chloride ion. This is the enthalpy change for the process:

$$\tfrac{1}{2}Cl_2(g) \quad + \quad e^- \longrightarrow Cl^-(g)$$
Standard state

Set up a Hess cycle:

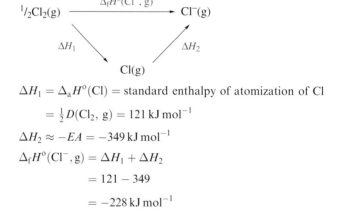

$$\Delta H_1 = \Delta_a H^\circ(Cl) = \text{standard enthalpy of atomization of Cl}$$

$$= \tfrac{1}{2}D(Cl_2,\,g) = 121\,kJ\,mol^{-1}$$

$$\Delta H_2 \approx -EA = -349\,kJ\,mol^{-1}$$

$$\Delta_f H^\circ(Cl^-,\,g) = \Delta H_1 + \Delta H_2$$

$$= 121 - 349$$

$$= -228\,kJ\,mol^{-1}$$

### The enthalpy changes accompanying the attachment of the first and second electrons

Values of the enthalpy changes which accompany the attachment of one electron to a range of *s*- and *p*-block *atoms* are listed in Table 8.1. With one exception, the values are negative indicating that the reactions are exothermic; for nitrogen, the enthalpy change is approximately zero. Enthalpy changes for the attachment of an electron to gaseous $O^-$ and $S^-$ are also given: both are *endothermic* reactions.

When an electron is added to an *atom*, there is repulsion between the incoming electron and the valence shell electrons, but there will also be an attraction between the nucleus and the extra electron. In general, the overall enthalpy change for the process shown in equation 8.11 is negative.

When an electron is added to an *anion*, the repulsive forces are significant and energy must be provided to overcome this repulsion. As a result, the enthalpy change is positive.

**Table 8.1** Approximate enthalpy changes ($\Delta_{EA}H(298\,\mathrm{K})$) associated with the gain of one electron to a gaseous atom or anion.

| Process | $\approx \Delta H\,/\,\mathrm{kJ\,mol^{-1}}$ |
|---|---|
| $H(g) + e^- \longrightarrow H^-(g)$ | $-73$ |
| $Li(g) + e^- \longrightarrow Li^-(g)$ | $-60$ |
| $Na(g) + e^- \longrightarrow Na^-(g)$ | $-53$ |
| $K(g) + e^- \longrightarrow K^-(g)$ | $-48$ |
| $N(g) + e^- \longrightarrow N^-(g)$ | $\approx 0$ |
| $P(g) + e^- \longrightarrow P^-(g)$ | $-72$ |
| $O(g) + e^- \longrightarrow O^-(g)$ | $-141$ |
| $O^-(g) + e^- \longrightarrow O^{2-}(g)$ | $+798$ |
| $S(g) + e^- \longrightarrow S^-(g)$ | $-201$ |
| $S^-(g) + e^- \longrightarrow S^{2-}(g)$ | $+640$ |
| $F(g) + e^- \longrightarrow F^-(g)$ | $-328$ |
| $Cl(g) + e^- \longrightarrow Cl^-(g)$ | $-349$ |
| $Br(g) + e^- \longrightarrow Br^-(g)$ | $-325$ |
| $I(g) + e^- \longrightarrow I^-(g)$ | $-295$ |

Now consider the formation of a gaseous oxide ion $O^{2-}$ from an atom of oxygen (equation 8.13). The enthalpy change for the overall process is the sum of the values of $\Delta H$ for the two individual electron attachments (equation 8.14).

$$O(g) + 2e^- \longrightarrow O^{2-}(g) \tag{8.13}$$

$$\Delta H[O(g) \longrightarrow O^{2-}(g)] = \Delta H[O(g) \longrightarrow O^-(g)] + \Delta H[O^-(g) \longrightarrow O^{2-}(g)]$$

$$= -141 + 798$$

$$= +657\,\mathrm{kJ\,mol^{-1}} \tag{8.14}$$

The formation of the $O^{2-}$ ion from an oxygen atom in the gas phase is an endothermic process and may appear unfavourable. In Sections 8.6 and 8.16 we examine the formation of an ionic salt, and this illustrates how endothermic changes are offset by exothermic ones in order to make salt formation thermodynamically favourable.

### 8.6    Electrostatic interactions between ions

The term *ion-pair* refers to a single pair of oppositely charged ions; in sodium chloride, $Na^+Cl^-$ is an ion-pair.

Consider the formation of a gas phase *ion-pair*, $X^+Y^-$. It is important to remember that in any isolable chemical compound, a *cation cannot occur without an anion.*[§] In a simple process, an electron lost in the formation of a singly charged cation, $X^+$, is transferred to another atom to form an anion, $Y^-$. The overall reaction is given in equation 8.15.

$$\left. \begin{array}{l} X(g) \longrightarrow X^+(g) + e^- \\ Y(g) + e^- \longrightarrow Y^-(g) \\ \hline X(g) + Y(g) \longrightarrow X^+(g) + Y^-(g) \end{array} \right\} \tag{8.15}$$

---

[§] In one or two remarkable cases, the anion is not a *chemical* species, but an electron. An example is a solution of sodium in liquid ammonia – see Section 21.11.

### Electrostatic (Coulombic) attraction

A spherical ion can be treated as a *point charge*. Electrostatic (or *Coulombic*) interactions operate between point charges; oppositely charged ions attract one another, while ions with like charges repel each other. Although isolated pairs of ions are not usually encountered, we initially consider the attraction between a pair of $Na^+$ and $Cl^-$ ions for the sake of simplicity. Equation 8.16 shows the formation of an $Na^+$ ion and a $Cl^-$ ion, and the subsequent formation of an isolated ion-pair in the gas phase.

> A *spherical ion* can be treated as a *point charge*.
>
> *Electrostatic (Coulombic) interactions* operate between ions. Oppositely charged ions attract one another and ions of like charge repel each other.

$$\left.\begin{array}{l} Na(g) \longrightarrow Na^+(g) + e^- \\ \underline{Cl(g) + e^- \longrightarrow Cl^-(g)} \\ Na(g) + Cl(g) \longrightarrow Na^+(g) + Cl^-(g) \end{array}\right\} \qquad (8.16)$$

The change in internal energy, $\Delta U(0\,K)$, associated with the electrostatic attraction between the two isolated ions $Na^+$ and $Cl^-$ is given by equation 8.17, and the units of $\Delta U(0\,K)$ from this equation are joules.

$$\text{For an isolated ion-pair: } \Delta U(0\,K) = -\left(\frac{|z_+| \times |z_-| \times e^2}{4 \times \pi \times \varepsilon_0 \times r}\right) \qquad (8.17)$$

where:[§]

$|z_+|$ = modulus of the positive charge (for $Na^+$, $|z_+| = 1$; for $Ca^{2+}$, $|z_+| = 2$)
$|z_-|$ = modulus of the negative charge (for $Cl^-$, $|z_-| = 1$; for $O^{2-}$, $|z_-| = 2$)
$e$   = charge on the electron = $1.602 \times 10^{-19}\,C$
$\varepsilon_0$   = permittivity of a vacuum = $8.854 \times 10^{-12}\,F\,m^{-1}$
$r$   = internuclear distance between the ions (units = m)

If we have a *mole of sodium chloride* in which each $Na^+$ ion interacts with *only one* $Cl^-$ ion, and each $Cl^-$ ion interacts with *only one* $Na^+$ ion, equation 8.17 is corrected by multiplying by the Avogadro constant ($L = 6.022 \times 10^{23}\,mol^{-1}$) to give equation 8.18, in which the units of $\Delta U(0\,K)$ are $J\,mol^{-1}$.

> Avogadro constant: see Section 1.8

$$\text{For a mole of ion-pairs: } \Delta U(0\,K) = -\left(\frac{L \times |z_+| \times |z_-| \times e^2}{4 \times \pi \times \varepsilon_0 \times r}\right) \qquad (8.18)$$

---

**Worked example 8.2**     *Coulombic attractions between ions*

Calculate the internal energy change, $\Delta U(0\,K)$, which is associated with the attractive forces operating between the cations and anions in a mole of ion-pairs of sodium chloride in the gas phase. The internuclear distance between an $Na^+$ and $Cl^-$ ion in each ion-pair is 236 pm.

Data required: $L = 6.022 \times 10^{23}\,mol^{-1}$, $e = 1.602 \times 10^{-19}\,C$ and $\varepsilon_0 = 8.854 \times 10^{-12}\,F\,m^{-1}$.

$$\text{For a mole of ion-pairs: } \Delta U(0\,K) = -\left(\frac{L \times |z_+| \times |z_-| \times e^2}{4 \times \pi \times \varepsilon_0 \times r}\right) J\,mol^{-1}$$

---

[§] The modulus of a real number is its positive value; although the charges $z_+$ and $z_-$ have positive and negative values respectively, $|z_+|$ and $|z_-|$ are both positive.

but, remember that in this equation, the distance $r$ is in m:

$$r = 236\,\mathrm{pm} = 2.36 \times 10^{-10}\,\mathrm{m}$$

$$\Delta U(0\,\mathrm{K}) = -\left[\frac{(6.022 \times 10^{23}\,\mathrm{mol^{-1}}) \times 1 \times 1 \times (1.602 \times 10^{-19}\,\mathrm{C})^2}{4 \times 3.142 \times (8.854 \times 10^{-12}\,\mathrm{F\,m^{-1}}) \times (2.36 \times 10^{-10}\,\mathrm{m})}\right]$$

$$= -\left[\frac{(6.022 \times 10^{23}\,\mathrm{mol^{-1}}) \times 1 \times 1 \times (1.602 \times 10^{-19}\,\mathrm{A\,s})^2}{4 \times 3.142 \times (8.854 \times 10^{-12}\,\mathrm{A^2\,s^4\,kg^{-1}\,m^{-2}\,m^{-1}}) \times (2.36 \times 10^{-10}\,\mathrm{m})}\right]$$

$$= -589\,000\,\mathrm{kg\,m^2\,s^{-2}\,mol^{-1}} = -589\,000\,\mathrm{J\,mol^{-1}}$$

$$= -589\,\mathrm{kJ\,mol^{-1}}$$

See Table 1.2 for derived units C, F and J

## Coulombic repulsion

If ion-pairs are close to each other, then one cation in one ion-pair will repel cations in other ion-pairs, and anions will repel other anions.

The internal energy associated with Coulombic repulsions can be calculated using a form of equation 8.17 (or equation 8.18 on a molar scale). However, while the internal energy associated with the *attractive* forces has a *negative* value (stabilizing), that associated with the *repulsive* forces has a *positive* value (destabilizing). If we are dealing only with repulsive interactions, equation 8.19 is appropriate.

$$\text{For two similarly charged ions: } \Delta U(0\,\mathrm{K}) = +\left(\frac{|z_1| \times |z_2| \times e^2}{4 \times \pi \times \varepsilon_0 \times r}\right) \qquad (8.19)$$

where $|z_1|$ and $|z_2|$ are the moduli of the charges on the two ions

The repulsive forces between ions in different ion-pairs become important *only* when the ion-pairs are relatively close together, and so for the gas phase ion-pairs that we have been discussing, we can effectively ignore Coulombic repulsions. However, in the solid state their contribution *is* important.

**8.7**    **Ionic lattices**

## Ions in the solid state

In gaseous sodium chloride, each ion-pair consists of one $Na^+$ cation and one $Cl^-$ anion. In a 'thought-experiment', we can condense the gas phase ion-pairs to give solid sodium chloride. The cations and anions will now be close to each other, and if we consider a purely electrostatic model with ions as point charges, then each $Na^+$ cation experiences Coulombic attractive forces from all nearby $Cl^-$ anions. Conversely, each $Cl^-$ anion is attracted to all the nearby $Na^+$ cations. At the same time, any chloride ions that are close will repel one another, and any $Na^+$ ions that approach one another will be repelled.

The net result of the electrostatic attractions and repulsions is the formation of an ordered array of ions called an *ionic lattice* in which the internal energy associated with the electrostatic forces is minimized. This is achieved by the systematic arrangement of ions such that, wherever possible, ions of *opposite* charge are adjacent.

Ionic lattices can be described in terms of the close-packing of spheres: see Chapter 9

**Fig. 8.10** (a) A 6-coordinate ion in an octahedral environment; (b) a 4-coordinate ion in a tetrahedral environment.

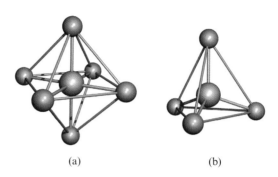

(a)                    (b)

In the solid state, a compound that is composed of ions forms an ordered structure called an *ionic lattice*. There are a number of common structure types which are adopted by compounds containing simple ions.

For many compounds which contain simple ions, the ions are arranged so as to give one of several general *ionic structure types*. Each structure type is described by a generic name – that of a compound (e.g. sodium chloride or caesium chloride) or a mineral that possesses that particular structure (e.g. fluorite, rutile, zinc blende or wurtzite). In this chapter we have chosen some of the more common structure types to illustrate the ways in which ions can be arranged in the solid state.[§] For example, a description of the sodium chloride lattice not only indicates the way in which $Na^+$ and $Cl^-$ ions are arranged in crystalline NaCl, but it also describes the way in which $Ca^{2+}$ and $O^{2-}$ ions are arranged in solid CaO, and the arrangement of $Mg^{2+}$ and $S^{2-}$ ions in crystalline MgS.

### The unit cell

The smallest repeating unit in an ionic lattice is a *unit cell*.

An ionic lattice may be extended indefinitely in three dimensions, and is composed of an infinite number of repeating units. Such a building-block is called a *unit cell* and must be large enough to carry *all* the information necessary to construct *unambiguously* an infinite lattice (see problem 8.10). We illustrate how the concept of a unit cell works by considering several common structure types in the following sections.

### Coordination number

Usage of the term 'coordination number' in molecules: see Section 6.1

The coordination number of an ion in a lattice is the number of closest ions of opposite charge.

In the following sections, we look at the structures of NaCl, CsCl, $CaF_2$, $TiO_2$ and ZnS. Within the 3-dimensional structures of these compounds, ions are found with coordination numbers of 3, 4, 6 and 8. Figure 8.10a shows the octahedral environment of a 6-coordinate ion (shown in orange). Remember that the 'octa' in octahedron refers to the number of faces (eight), not the number of vertices (six). In Figure 8.10b, the orange ion has a coordination number of 4: four ions of opposite charge (the green ions) are arranged tetrahedrally around the central ion.

The *coordination number* of an ion in a lattice is the number of closest ions of opposite charge.

[§] For detailed information about a far wider range of ionic lattices, see: A. F. Wells (1984) *Structural Inorganic Chemistry*, 5th edn, Oxford University Press, Oxford.

## 8.8 The sodium chloride (rock salt) structure type

Inorganic compounds adopt a relatively limited number of *structure types* in the solid state. The structure may be a distorted variant of a particular structure type but the structure is still clearly recognizable as being derived from a certain parent structure. Each structure type is named according to a compound that adopts that structure and in this section we describe the sodium chloride structure type. In Sections 8.10–8.13, other common structure types are introduced.

Figure 8.11a shows a unit cell of the sodium chloride structure. The structure is *cubic*, with alternating $Na^+$ (shown in purple) and $Cl^-$ ions (shown in green). The unit cell has a chloride ion at its centre.[§] The closest neighbours are six $Na^+$ cations, making the $Cl^-$ ion 6-coordinate and octahedrally sited (Figure 8.11b).

Now consider an $Na^+$ ion at the centre of a square face in Figure 8.11a. If we extend the lattice by adding another unit cell to that face (Figure 8.11c), the coordination sphere around that $Na^+$ is completed. It too is octahedral (Figure 8.11d), and the coordination number of $Na^+$ is 6.

Selected compounds that possess the same structure type in the solid state as sodium chloride are listed in Table 8.2. Note that each has a $1:1$ stoichiometry: the ratio of cations to anions is $1:1$.

■▶
Stoichiometry:
see Section 1.15

**Fig. 8.11** (a) A unit cell of the sodium chloride structure type; $Na^+$ ions are shown in purple, $Cl^-$ ions are shown in green. This unit cell is drawn placing a chloride ion at the centre; equally, we could construct a unit cell with a sodium ion at the centre. (b) The coordination number of each chloride ion is 6. (c) Two units cells of sodium chloride with one shared face. (d) The coordination number of each sodium ion is 6.

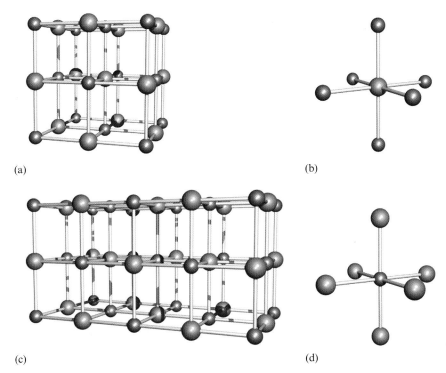

(a)

(b)

(c)

(d)

---

[§] An alternative unit cell could be drawn with a sodium ion at the centre, since all of the sites in the lattice are equivalent.

**Table 8.2** Selected compounds that possess a sodium chloride structure in the solid state (see Figure 8.11) under room temperature conditions.

| Compound | Formula | Cation | Anion |
|---|---|---|---|
| Sodium chloride | NaCl | $Na^+$ | $Cl^-$ |
| Sodium fluoride | NaF | $Na^+$ | $F^-$ |
| Sodium hydride | NaH | $Na^+$ | $H^-$ |
| Lithium chloride | LiCl | $Li^+$ | $Cl^-$ |
| Potassium bromide | KBr | $K^+$ | $Br^-$ |
| Potassium iodide | KI | $K^+$ | $I^-$ |
| Silver fluoride | AgF | $Ag^+$ | $F^-$ |
| Silver chloride | AgCl | $Ag^+$ | $Cl^-$ |
| Magnesium oxide | MgO | $Mg^{2+}$ | $O^{2-}$ |
| Calcium oxide | CaO | $Ca^{2+}$ | $O^{2-}$ |
| Barium oxide | BaO | $Ba^{2+}$ | $O^{2-}$ |
| Iron(II) oxide | FeO | $Fe^{2+}$ | $O^{2-}$ |
| Magnesium sulfide | MgS | $Mg^{2+}$ | $S^{2-}$ |
| Lead(II) sulfide | PbS | $Pb^{2+}$ | $S^{2-}$ |

## 8.9    Determining the stoichiometry of a compound from the solid state structure: NaCl

We stated in Section 8.7 that we must be able to generate an infinite lattice from the information provided in the unit cell. It follows that the stoichiometry of an ionic compound can be determined from its unit cell.

In Figure 8.11a, the number of $Na^+$ ions drawn in the diagram does not equal the number of $Cl^-$ ions. However, since the diagram is of a unit cell, it *must* indicate that the ratio of $Na^+ : Cl^-$ ions in sodium chloride is 1:1. Let us look more closely at the structure. The important feature is that in the NaCl structure, the unit cell shown in Figure 8.11a is immediately surrounded by six other unit cells. One of these adjacent cells is shown in Figure 8.11c.

When unit cells are connected, ions that are at points of connection are *shared* between more than one cell. We can identify three different types of sites of connection and these are illustrated for a cubic unit cell in Figure 8.12.

Figure 8.12a shows two unit cells, with an ion sited in the *centre of the shared face*. Figure 8.12b shows that for an ion sited in the middle of an *edge* of a unit cell, the ion is *shared between four cells*. Figure 8.12c shows that an ion at the *corner* of a unit cell is *shared between eight unit cells*.

Let us now apply this to the unit cell of NaCl shown in Figure 8.13. Firstly, we must identify the different types of site. These are:

- the unique central site;
- 6 sites, one at the centre of each face;
- 12 sites, one in the middle of each edge; and
- 8 sites, one at each corner.

Next, we must determine what fraction of each ion actually 'belongs' to this one unit cell – how does the connection of adjacent unit cells affect the ions in the different sites?

**The unique central site**    This ion is not shared with any other unit cell. In Figure 8.13, the central $Cl^-$ ion belongs wholly to the unit cell shown.

**Fig. 8.12** The sharing of ions between cubic unit cells. (a) An ion at the centre of a face is shared between two unit cells. (b) An ion in the middle of an edge is shared between four unit cells. (c) An ion in a corner site is shared between eight unit cells. In each diagram, the boundaries of a unit cell are shown in yellow.

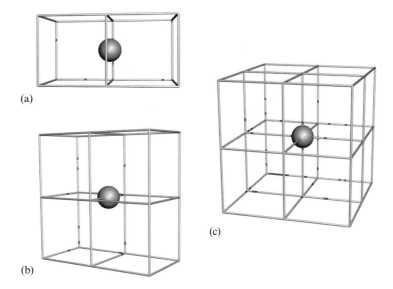

**The sites at the centres of the faces**    An ion in this site is shared between two unit cells. In Figure 8.13, there are six such ions (all $Na^+$ cations) and half of each ion belongs to the unit cell shown.

**The edge sites**    When unit cells are connected together to form an infinite lattice, each edge site is shared between four unit cells. In Figure 8.13, there are 12 such ions, all $Cl^-$ anions. One-quarter of each ion belongs to the unit cell shown.

**The corner sites**    When unit cells are placed together in a lattice, each corner site is shared between eight unit cells. In Figure 8.13, there are eight such $Na^+$ ions, and one-eighth of each ion belongs to the unit cell shown.

The stoichiometry of sodium chloride is determined by summing the fractions of ions that belong to a particular unit cell and Table 8.3 shows how this is done. The total number of $Na^+$ ions belonging to the unit cell is four and this is also the number of $Cl^-$ ions in the unit cell. The ratio of $Na^+ : Cl^-$ ions is $4 : 4$, or $1 : 1$, and it follows that the empirical formula of sodium chloride is NaCl.

**Fig. 8.13** In a unit cell of NaCl, there are four different sites in which an ion can reside. There is one unique site at the centre of the unit cell; this ion (a $Cl^-$ ion shown in green) belongs entirely to the unit cell. There are six sites at the centres of faces, 12 sites situated in the middle of edges and eight corner sites.

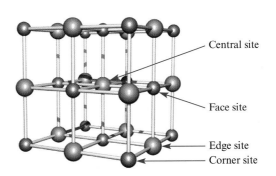

Central site

Face site

Edge site
Corner site

**Table 8.3** Determination of the stoichiometry of sodium chloride from a consideration of the structure of the unit cell.

| Site (see Figure 8.13) | Number of Na$^+$ ions | Number of Cl$^-$ ions |
|---|---|---|
| Central | 0 | 1 |
| Centre of face | $(6 \times \frac{1}{2}) = 3$ | 0 |
| Edge | 0 | $(12 \times \frac{1}{4}) = 3$ |
| Corner | $(8 \times \frac{1}{8}) = 1$ | 0 |
| **Total** | **4** | **4** |

## 8.10   The caesium chloride structure type

A unit cell of caesium chloride, CsCl, is shown in Figure 8.14a. In Figure 8.14b, the structure is extended into parts of adjacent unit cells to illustrate the fact that the extended structure is made up of interpenetrating cubic units of Cs$^+$ ions (shown in yellow) and Cl$^-$ ions (shown in green). Each Cs$^+$ ion is sited at the centre of a cubic arrangement of Cl$^-$ ions, and each Cl$^-$ ion is sited at the centre of a cubic arrangement of Cs$^+$ ions. Each Cs$^+$ ion and each Cl$^-$ ion is therefore 8-coordinate. Figure 8.14b illustrates that a unit cell of CsCl can be drawn either with a Cs$^+$ ion at the centre of a cubic arrangement of Cl$^-$ ions, or with a Cl$^-$ ion at the centre of a cubic arrangement of Cs$^+$ ions.

We can confirm the stoichiometry of caesium chloride from the arrangement of ions in the unit cell as follows. In Figure 8.14a, the Cs$^+$ ion is in the centre of the unit cell and it belongs in its entirety to this cell. There are 8 Cl$^-$ ions, each in a corner site, so the number of Cl$^-$ ions belonging to a unit cell is $(8 \times \frac{1}{8}) = 1$. The ratio of Cs$^+$ : Cl$^-$ ions is 1 : 1.

In the solid state, thallium(I) chloride also adopts a CsCl structural type with Tl$^+$ replacing Cs$^+$.

**Fig. 8.14** (a) A unit cell of the caesium chloride structure; Cs$^+$ ions are shown in yellow and Cl$^-$ ions in green. The central Cs$^+$ ion is 8-coordinate. The edges of the unit cell are shown in yellow. (b) An extension of the unit cell shows the Cl$^-$ ion is also 8-coordinate.

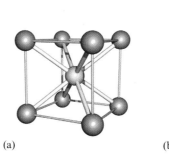   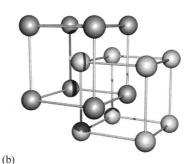

(a)                                    (b)

## 8.11   The fluorite (calcium fluoride) structure type

The mineral fluorite is calcium fluoride, CaF$_2$, a unit cell of which is shown in Figure 8.15a. Note that the arrangement of the Ca$^{2+}$ ions is the same as the arrangement of the Na$^+$ ions in NaCl (Figure 8.11a). However, the relationship between the positions of the F$^-$ and Ca$^{2+}$ ions in fluorite is different from that of the Cl$^-$ and Na$^+$ ions in NaCl.

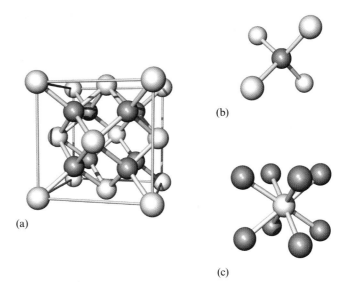

**Fig. 8.15** (a) A unit cell of the fluorite ($CaF_2$) structure; $Ca^{2+}$ ions are shown in silver, $F^-$ ions in green. The edges of the unit cell are shown in yellow. (b) Each $F^-$ ion is 4-coordinate, and (c) each $Ca^{2+}$ ion is 8-coordinate.

(a)

(b)

(c)

Each fluoride ion in $CaF_2$ is in a 4-coordinate, tetrahedral site (Figure 8.15b). The coordination number of the $Ca^{2+}$ ions can only be seen if we extend the structure into the next unit cell, and this is best done by considering one of the $Ca^{2+}$ ions at the centre of a face. It is at the centre of a cubic arrangement of $F^-$ ions and is 8-coordinate (Figure 8.15c).

The number of $Ca^{2+}$ and $F^-$ ions per unit cell in fluorite are calculated in equations 8.20 and 8.21 (see problem 8.11). The ratio of $Ca^{2+} : F^-$ ions is $4 : 8$, or $1 : 2$, and this confirms a formula for calcium fluoride of $CaF_2$.

$$\text{Number of } Ca^{2+} \text{ ions per unit cell of fluorite} = (6 \times \tfrac{1}{2}) + (8 \times \tfrac{1}{8})$$

$$= 4 \tag{8.20}$$

$$\text{Number of } F^- \text{ ions per unit cell of fluorite} = (8 \times 1) = 8 \tag{8.21}$$

Some compounds that adopt a fluorite-type structure in the solid state are listed in Table 8.4.

**Table 8.4** Selected compounds with a fluorite ($CaF_2$) structure in the solid state (see Figure 8.15) under room temperature conditions.

| Compound | Formula | Compound | Formula |
|----------|---------|----------|---------|
| Calcium fluoride | $CaF_2$ | Lead(II) fluoride (β-form) | $\beta\text{-}PbF_2$ |
| Barium fluoride | $BaF_2$ | Zirconium(IV) oxide | $ZrO_2$ |
| Barium chloride | $BaCl_2$ | Hafnium(IV) oxide | $HfO_2$ |
| Mercury(II) fluoride | $HgF_2$ | Uranium(IV) oxide | $UO_2$ |

### 8.12    The rutile (titanium(IV) oxide) structure type

The three structural types discussed so far possess cubic unit cells. In contrast, the solid state structure of the mineral rutile (titanium(IV) oxide, $TiO_2$) has a unit cell which is a rectangular box (Figure 8.16a).

Eight Ti(IV) centres are sited at the corners of the unit cell and one resides within the unit cell. Each of these is 6-coordinate, in an octahedral environment (Figure 8.16b). The structure must be extended in several directions in

**Fig. 8.16** (a) A unit cell of the rutile ($TiO_2$) structure with the titanium centres shown in silver and oxide ions in red. The edges of the unit cell are shown in yellow. (b) The coordination number of each titanium(IV) centre is 6, and (c) each oxide ion is 3.

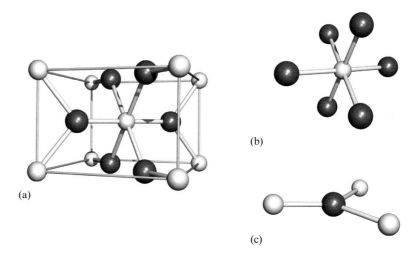

(a)

(b)

(c)

order to see that the Ti centres in the corner sites of the unit cell are octahedrally coordinated in the infinite structure. Each $O^{2-}$ ion is 3-coordinate and is trigonal planar (Figure 8.16c).

The stoichiometry of rutile can be confirmed from the solid state structure. The method of calculating the number of titanium(IV) and oxide ions in the unit cell is shown in equations 8.22 and 8.23. The ratio of Ti:O in rutile is 2:4, or 1:2. This confirms a formula for titanium(IV) oxide of $TiO_2$.

$$\text{Number of Ti(IV) centres per unit cell of rutile} = (1 \times 1) + (8 \times \tfrac{1}{8})$$

$$= 2 \qquad (8.22)$$

$$\text{Number of } O^{2-} \text{ ions per unit cell of rutile} = (4 \times \tfrac{1}{2}) + (2 \times 1) = 4 \qquad (8.23)$$

Table 8.5 lists some compounds that adopt the rutile structural type.

**Table 8.5** Selected compounds that possess a rutile ($TiO_2$) structure in the solid state (see Figure 8.16) under room temperature conditions.

| Compound | Formula | Compound | Formula |
|---|---|---|---|
| Titanium(IV) oxide | $TiO_2$ | Lead(IV) oxide | $PbO_2$ |
| Manganese(IV) oxide (β-form) | $\beta\text{-}MnO_2$ | Magnesium fluoride | $MgF_2$ |
| Tin(IV) oxide | $SnO_2$ | Iron(II) fluoride | $FeF_2$ |

**COMMERCIAL AND LABORATORY APPLICATIONS**

## Box 8.2 Titanium(IV) oxide: brighter than white

Titanium(IV) oxide is a white crystalline, non-toxic compound which is thermally stable. Its applications as a white pigment are widespread – $TiO_2$ is responsible for the brightness of 'brilliant white' paint and white paper. It has now replaced the toxic lead-, antimony- and zinc-containing compounds previously used as pigments. The essential optical properties that make $TiO_2$ so valuable are that it does not absorb visible light and that it has an extremely high refractive index.

**8.13**

## The structures of the polymorphs of zinc(II) sulfide

If a substance exists in more than one crystalline form, it is **polymorphic**. The different crystalline forms of the substance are **polymorphs**.

There are two structural forms (i.e. *polymorphs*) of zinc(II) sulfide (ZnS) in the solid state and both are structure types: *zinc blende* and *wurtzite*.

### The zinc blende structure type

A unit cell of zinc blende is drawn in Figure 8.17a. If we compare this with Figures 8.11a and 8.15a, we see that the $S^{2-}$ ions in zinc blende are arranged in the same way as the $Na^+$ ions in NaCl and the $Ca^{2+}$ ions in $CaF_2$. However, the arrangement of the $Zn^{2+}$ ions in zinc blende differs from those of the counter-ions in either NaCl or $CaF_2$. Each $Zn^{2+}$ ion in zinc blende is 4-coordinate (tetrahedral) (Figure 8.17b).

None of the $S^{2-}$ ions lies within the unit cell, and so the coordination environment can only be seen if the lattice is extended. Each $S^{2-}$ ion is 4-coordinate (tetrahedral) as shown in Figure 8.17c.

The stoichiometry of zinc blende can be confirmed by considering the structure of the unit cell. From equations 8.24 and 8.25, the ratio of $Zn^{2+} : S^{2-}$ ions is $4 : 4$, or $1 : 1$, giving an empirical formula of ZnS.

$$\text{Number of } S^{2-} \text{ ions per unit cell of zinc blende} = (6 \times \tfrac{1}{2}) + (8 \times \tfrac{1}{8})$$

$$= 4 \tag{8.24}$$

$$\text{Number of } Zn^{2+} \text{ ions per unit cell of zinc blende} = (4 \times 1) = 4 \tag{8.25}$$

### The wurtzite structure type

Hexagonal close-packed structure: see Section 9.2

Figure 8.18a shows three unit cells of the mineral wurtzite with the $Zn^{2+}$ ions in grey and the $S^{2-}$ ions in yellow. This compound provides an example of a structure with a *hexagonal* motif, and three unit cells are needed to see clearly the hexagonal unit. Each $Zn^{2+}$ cation and $S^{2-}$ anion in wurtzite is 4-coordinate (tetrahedral) (Figures 8.18b and 8.18c).

The stoichiometry of wurtzite may be confirmed from the solid state structure by using equations 8.26 and 8.27. Note that in a 3-dimensional structure containing hexagonal units, ions in the centre of faces are shared

**Fig. 8.17** (a) A unit cell of the zinc blende structure with the $Zn^{2+}$ ions shown in grey and the $S^{2-}$ ions in yellow. The edges of the unit cell are shown in yellow. (b) Each $Zn^{2+}$ ion is 4-coordinate and (c) each $S^{2-}$ ion is 4-coordinate. Both sites are tetrahedral.

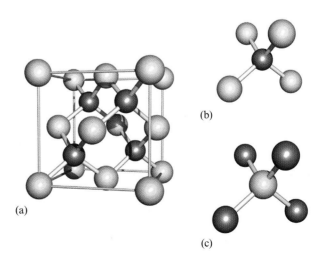

**Fig. 8.18** (a) Three unit cells of the wurtzite structure; the $Zn^{2+}$ ions are shown in grey and the $S^{2-}$ ions in yellow. The yellow lines complete the edges of the hexagonal prism that encloses the three unit cells. (b) Each $Zn^{2+}$ centre is 4-coordinate and (c) each $S^{2-}$ ion is 4-coordinate. Both sites are tetrahedral.

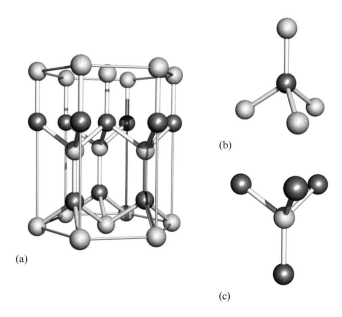

(a)

(b)

(c)

between two units, ions on vertical edges are shared between three units, and ions in corner sites are shared between six units. In this example, we choose to determine the stoichiometry by considering three unit cells for clarity of presentation.

$$\text{Number of } Zn^{2+} \text{ ions per 3 unit cells of wurtzite} = (6 \times \tfrac{1}{3}) + (4 \times 1)$$
$$= 6 \tag{8.26}$$
$$\text{Number of } S^{2-} \text{ ions per 3 unit cells of wurtzite} = (2 \times \tfrac{1}{2}) + (12 \times \tfrac{1}{6}) + (3 \times 1)$$
$$= 6 \tag{8.27}$$

The ratio of $Zn^{2+} : S^{2-}$ ions in wurtzite is $6:6$, or $1:1$, giving an empirical formula of ZnS.

---

**8.14**     ## Sizes of ions

### Determining the internuclear distances in an ionic lattice

X-ray diffraction: see Section 4.2

X-ray diffraction may be used to determine the solid state structure of an ionic compound and gives the distances *between the nuclei*. Thus, the results of an X-ray diffraction experiment give an Na—Cl distance in solid NaCl of 281 pm, and an Na—F distance of 231 pm in crystalline NaF. Each distance corresponds to the equilibrium separation of the closest, oppositely charged ions in the respective crystal lattices.

In the electrostatic model, we assume that the ions are spherical with a finite size. We can also assume that in the solid state structure, ions of opposite charge touch one another. This is shown in Figure 8.19 for the $Na^+F^-$ and $Na^+Cl^-$ ion-pairs. Note that the diagrams of the NaCl, CsCl, $CaF_2$, $TiO_2$ and ZnS structures drawn earlier show the ions with significant gaps between them; this representation is used simply for the purposes of clarity. The ions effectively occupy the space as is illustrated in Figure 8.20 for a unit cell of CsCl.

**Fig. 8.19** The results of X-ray diffraction experiments give internuclear distances. In sodium fluoride, the internuclear distance is 231 pm. In sodium chloride, it is 281 pm. If we assume that the ions are spherical and touch each other, then the sum of the ionic radii equals the internuclear distance in each case.

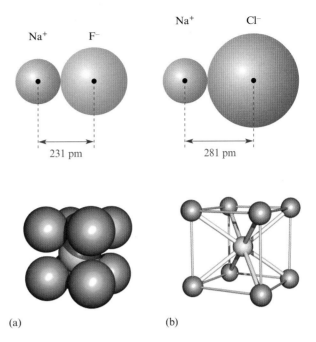

**Fig. 8.20** A space-filling diagram of a unit cell of CsCl is shown in (a). Compare this with the representation of the same unit cell in (b). The space-filling diagram is more realistic, but the lattice structure is easier to see if we use the type of diagram shown in (b).

(a)    (b)

### Ionic radius

We have previously defined the van der Waals and covalent radii of atoms. For an ion, a new parameter, the *ionic radius* ($r_{ion}$), may be derived from X-ray diffraction data. As Figure 8.19 shows, however, there is a problem. The experimental data give only an internuclear distance which is the *sum of the ionic radii of the cation and anion* (equation 8.28).

$$\text{Internuclear distance between a cation and the closest anion in an ionic lattice} = r_{cation} + r_{anion} \tag{8.28}$$

In some cases, electron density contour maps have been obtained. The contour map in Figure 8.21 shows the presence of two different types of ions, $Na^+$ and $Cl^-$ ions. The positions of the nuclei coincide with the centres

**Fig. 8.21** An electron density contour map of part of a sodium chloride lattice showing two $Na^+$ and two $Cl^-$ ions. [From G. Schoknecht (1957) *Z. Naturforsch., Teil A*, vol. 12, p. 983.]

**Table 8.6** Ionic radii for selected ions. The values are given for 6-coordinate ions. The ionic radius varies with the coordination number of the ion (see Section 8.14). A more complete list is given in Appendix 6.

| Anion | Ionic radius ($r_{ion}$) / pm | | Cation | Ionic radius ($r_{ion}$) / pm |
|---|---|---|---|---|
| $F^-$ | 133 | | $Li^+$ | 76 |
| $Cl^-$ | 181 | | $Na^+$ | 102 |
| $Br^-$ | 196 | | $K^+$ | 138 |
| $I^-$ | 220 | | $Rb^+$ | 149 |
| $O^{2-}$ | 140 | | $Cs^+$ | 170 |
| $S^{2-}$ | 184 | | $Mg^{2+}$ | 72 |
| $Se^{2-}$ | 198 | | $Ca^{2+}$ | 100 |
| $N^{3-}$ | 171 | | $Al^{3+}$ | 54 |
| | | | $Ti^{3+}$ | 67 |
| | | | $Cr^{2+}$ | 73 |
| | | | $Cr^{3+}$ | 62 |
| | | | $Mn^{2+}$ | 67 |
| | | | $Fe^{2+}$ | 61 |
| | | | $Fe^{3+}$ | 55 |
| | | | $Co^{2+}$ | 65 |
| | | | $Co^{3+}$ | 55 |
| | | | $Ni^{2+}$ | 69 |
| | | | $Cu^{2+}$ | 73 |
| | | | $Zn^{2+}$ | 74 |

of the spherical[§] regions of electron density. The distance between the nuclei of two adjacent $Na^+$ and $Cl^-$ ions is the internuclear separation (281 pm). Between the two nuclei, the electron density falls to a minimum value as can be seen in Figure 8.21. In a model compound, the minimum electron density would be zero. This point corresponds to the 'boundary' between the $Na^+$ and $Cl^-$ ions, and we can partition the internuclear distance into components due to the radius of the cation ($r_{cation}$) and the radius of the anion ($r_{anion}$). For NaCl, we can then write equation 8.29. If we can accurately measure the electron density between the nuclei, we can use the 'boundary' to find the radius of the two ions.

$$281 \, pm = r_{Na^+} + r_{Cl^-} \tag{8.29}$$

In general, estimated values of ionic radii are not obtained by the method described above. Instead, given a sufficiently large data set of internuclear distances, it is possible to estimate a *self-consistent set of ionic radii* by assuming the relationship given in equation 8.28 and setting up a series of simultaneous equations. We assume that if an ion is in a similar environment in two ionic compounds (e.g. octahedral coordination), then its radius is transferable from one compound to another. We must also assume that, in at least one compound in the set of compounds, the anion–anion distance can be measured experimentally. The need for this assumption is explained in Box 8.3.

Values of ionic radii for selected ions are listed in Table 8.6. The ionic radius of a cation or anion may vary with coordination number; for example,

---

[§] The contour map is a section through the lattice and the circles in such maps correspond to sections through spherical regions of electron density.

**THEORETICAL AND CHEMICAL BACKGROUND**

## Box 8.3  Obtaining a self-consistent set of ionic radii

The following compounds all possess a sodium chloride structure in the solid state. For each, the internuclear distance between a cation and the closest anion has been measured by using X-ray diffraction methods.

| Compound | Formula | Internuclear distance / pm |
|----------|---------|----------------------------|
| Lithium chloride | LiCl | 257 |
| Sodium chloride | NaCl | 281 |
| Lithium fluoride | LiF | 201 |
| Sodium fluoride | NaF | 231 |

Using equation 8.28, we can write down a set of four simultaneous equations:

$$r_{Li} + r_{Cl} = 257$$

$$r_{Na} + r_{Cl} = 281$$

$$r_{Li} + r_{F} = 201$$

$$r_{Na} + r_{F} = 231$$

but there are not enough data to solve these equations and find the four ionic radii. Introducing another compound necessarily introduces another variable and does not improve the situation.

In order to solve the equations, we need to estimate the value of one of the radii by another means. If we choose a lattice that contains a *large anion* and a *small cation* we can make the assumption that, not only do the closest anions and cations touch one another, but so do the closest anions. This is shown below, for LiCl, where the assumption is reasonably valid.

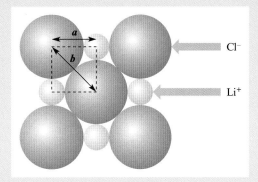

The internuclear distances $a$ and $b$ can both be found experimentally:

$$a = r_{Li} + r_{Cl} = 257 \text{ pm} \tag{i}$$

$$b = 2 \times r_{Cl} = 363 \text{ pm} \tag{ii}$$

From equation (ii):

$$r_{Cl} = 181.5 \text{ pm}$$

and substituting this value into equation (i) gives:

$$r_{Li} = 257 - 181.5 = 75.5 \text{ pm}$$

As an exercise, use these results in the set of four simultaneous equations given above to see what happens when you solve the equations to find the radii of the $Na^+$ and $F^-$ ions. Use all the data to double check the answers. You will find that the answers are not fully consistent with one another and that a larger data set is required to give reliable average values.

$r_{ion}$ ($Ca^{2+}$) is 100 pm for a 6-coordinate ion and 112 pm for an 8-coordinate ion. The values in Table 8.6 all refer to 6-coordinate ions.

For an element that exhibits a variable oxidation state, more than one type of ion is observed. This is commonly seen for the *d*-block metals; for example, both $Fe^{2+}$ and $Fe^{3+}$ are known. A consideration of the effective nuclear charges in these species shows why the $Fe^{3+}$ ion ($r_{ion} = 55$ pm) is smaller than the $Fe^{2+}$ ion ($r_{ion} = 61$ pm). Several other pairs of ions are listed in Table 8.6.

The **ionic radius** ($r_{ion}$) of an ion provides a measure of the size of a spherical ion.

$$\text{Internuclear distance between a cation and the closest anion in a lattice} = r_{cation} + r_{anion}$$

For a given ion, $r_{ion}$ may vary with the coordination number.

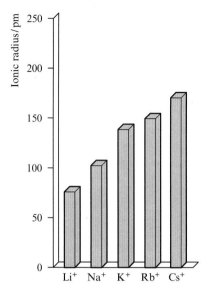

**Fig. 8.22** The increase in ionic radii of the alkali metal ions on descending group 1.

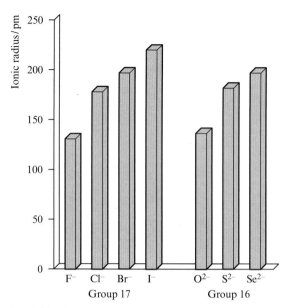

**Fig. 8.23** The increase in the ionic radii of the halide ions on descending group 17, and of the dianions of the group 16 elements.

### Group trends in ionic radii

In Section 3.21 we noted that the van der Waals radii of the noble gas atoms increased on going down group 18 and Figure 4.4 illustrated an increase in values of the covalent radii of atoms on descending groups 15, 16 and 17. Similar trends are observed for series of ionic radii (Table 8.6), provided that we are comparing the radii of ions with a constant charge, for example $M^{2+}$ ions. Figure 8.22 shows that, upon descending group 1, the ionic radius of the $M^+$ ions increases. A similar trend is seen for the halide anions ($X^-$) in group 17, and for the dianions of the group 16 elements (Figure 8.23).

---

**8.15** ### Lattice energy – a purely ionic model

The *lattice energy*, $\Delta U(0\,\mathrm{K})$, of an ionic compound is the change in internal energy that accompanies the formation of one mole of the solid from its constituent gas phase ions at a temperature of $0\,\mathrm{K}$. It can be estimated by assuming that the solid state structure of an ionic compound contains *spherical* ions. We shall examine the extent to which this approximation is true in Sections 8.17 and 8.18. In order to calculate the lattice energy, *all* attractive and repulsive interactions between ions must be considered.

### Coulombic forces in an ionic lattice

In Section 8.6, we discussed the electrostatic interactions between two oppositely charged ions. For a mole of ion-pairs, the change in internal energy associated with the Coulombic attraction between the ions *within* an ion-pair was given by equation 8.18. We assumed that each ion-pair was isolated and that there were no *inter*-ion-pair forces.

Once the ions are arranged in an extended structure, each ion will experience both attractive and repulsive Coulombic forces from surrounding ions. This is apparent if we look at a particular ion in any of the structures shown in Figure 8.11, 8.14, 8.15, 8.16, 8.17 or 8.18. The electrostatic interactions can be categorized as follows:

- attraction between adjacent cations and anions;
- weaker attractions between more distant cations and anions;
- repulsions between ions of like charge that are close to one another;
- weaker repulsions between ions of like charge that are distant from one another.

The number and magnitude of these forces depend upon the arrangement of the ions and the internuclear distance between the ions.

### The change in internal energy associated with Coulombic forces

In order to determine the change in internal energy, $\Delta U(0\,K)$, associated with the formation of an ionic lattice, it is convenient to consider the formation as taking place according to equation 8.30 – that is, the aggregation of *gaseous* ions to form a solid state lattice.

$$X^+(g) + Y^-(g) \longrightarrow X^+Y^-(s) \tag{8.30}$$

In forming the ionic lattice, there are contributions to $\Delta U(0\,K)$ from Coulombic attractive (equation 8.18) and repulsive (equation 8.19) interactions. Equations 8.18 and 8.19 are of the same form and can be readily combined. The total number and relative magnitudes of the Coulombic interactions, and whether these are attractive or repulsive, are taken into account by using a factor known as the *Madelung constant*, $A$. Table 8.7 lists values of $A$ for the structure types we have previously described. The values are calculated by considering the geometrical relationships between the ions as illustrated for NaCl in Box 8.4. Values of $A$ are simply numerical and have no units.

We can now write equation 8.31, where $\Delta U(0\,K)(\text{Coulombic})$ refers to the lowering of internal energy associated with the Coulombic forces that are present when a mole of the ionic solid XY forms from a mole of isolated gaseous $X^+$ ions and a mole of isolated gaseous $Y^-$ ions. As energy is released upon the formation of the lattice, the value of $\Delta U(0\,K)(\text{Coulombic})$ is negative. In equation 8.31, the units of $\Delta U(0\,K)(\text{Coulombic})$ are $J\,mol^{-1}$.

**Table 8.7** Madelung constants, $A$, for some common structure types.

| Structure type | Lattice shown in Figure: | $A$ |
|---|---|---|
| Sodium chloride (NaCl) | 8.11 | 1.7476 |
| Caesium chloride (CsCl) | 8.14 | 1.7627 |
| Fluorite ($CaF_2$) | 8.15 | 2.5194 |
| Rutile ($TiO_2$) | 8.16 | 2.408 |
| Zinc blende ($\beta$-ZnS) | 8.17 | 1.6381 |
| Wurtzite ($\alpha$-ZnS) | 8.18 | 1.6413 |

For an ionic lattice:

$$\Delta U(0\,\text{K})(\text{Coulombic}) = -\left(\frac{L \times A \times |z_+| \times |z_-| \times e^2}{4 \times \pi \times \varepsilon_0 \times r}\right) \tag{8.31}$$

where  $|z_+|$ = modulus of the positive charge

$|z_-|$ = modulus of the negative charge

$e$ = charge on the electron = $1.602 \times 10^{-19}$ C

$\varepsilon_0$ = permittivity of a vacuum = $8.854 \times 10^{-12}$ F m$^{-1}$

$r$ = internuclear distance between the ions (units = m)

$A$ = Madelung constant (see Table 8.7)

$L$ = Avogadro constant = $6.022 \times 10^{23}$ mol$^{-1}$

---

**THEORETICAL AND CHEMICAL BACKGROUND**

## Box 8.4 Determining the value of a Madelung constant

The Madelung constant takes into account the different Coulombic forces, both attractive and repulsive, that act on a particular ion in a lattice.

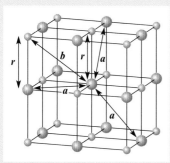

Consider the NaCl structure shown here. We stated in Section 8.8 that each Na$^+$ ion and each Cl$^-$ ion is 6-coordinate. The coordination number gives the number of closest ions of opposite charge. In the diagram above, the central Cl$^-$ ion is surrounded by six Na$^+$ ions, each at a distance $r$, and this is the internuclear distance measured by X-ray diffraction (281 pm). This applies to *all* chloride ions in the lattice since each is in the same environment. Similarly, each Na$^+$ ion is at a distance of 281 pm ($r$) from six Cl$^-$ ions.

Now consider which other ions are close to the central Cl$^-$ ion. Firstly, there are 12 Cl$^-$ ions, each at a distance $a$ from the central ion, and the Cl$^-$ ions repel one another. All 12 Cl$^-$ ions are shown in the diagram, but only three distances $a$ are indicated. Distance $a$ is related to $r$ by the equation:

$$a^2 = r^2 + r^2 = 2r^2$$

$$a = r\sqrt{2}$$

Next, there are eight Na$^+$ ions, each at a distance $b$ from the central Cl$^-$ ion giving rise to attractive

forces. Distance $b$ is related to $r$ by the equation:

$$b^2 = r^2 + a^2$$

$$b^2 = r^2 + (r\sqrt{2})^2$$

$$b^2 = 3r^2$$

$$b = r\sqrt{3}$$

Further attractive and repulsive interactions occur, but as the distances involved increase, the Coulombic interactions decrease. Remember that the energy associated with a Coulombic force is inversely proportional to the distance (see equation 8.17).

The Madelung constant, $A$, contains terms for all the attractive and repulsive interactions experienced by a given ion, and so for the NaCl lattice, the Madelung constant is given by:

$$A = 6 - \left(12 \times \frac{1}{\sqrt{2}}\right) + \left(8 \times \frac{1}{\sqrt{3}}\right) - \cdots$$

where the series will continue with additional terms for interactions at greater distances. Note that $r$ is not included in this equation. The value of $A$ calculated in the equation is for *all* sodium chloride-*type* structures, and $r$ varies depending upon the ions present (see Table 8.2). However, the *ratios* of the distances $r : a : b$ will always be $1 : \sqrt{2} : \sqrt{3}$. The inverse relationships arise, as stated previously, from the fact that the energies associated with the Coulombic forces depend on the inverse of the distance between the point charges.

When all terms are accounted for, we obtain a value of $A = 1.7476$ for the NaCl structure.

*Exercise*: Only the first three terms in the series to calculate $A$ for NaCl are given above. What are the next two terms in the series?

**Table 8.8** Values of the Born exponent, $n$, given for an ionic compound XY in terms of the electronic configuration of the ions $[X^+][Y^-]$. Values of $n$ are numerical and have no units. The value of $n$ for an ionic compound in which the ions have *different* electronic configurations is taken by averaging the component values. For example, for NaF, $n = 7$, but for LiF, $n = \frac{5}{2} + \frac{7}{2} = 6$.

| Electronic configuration of the ions in an ionic compound XY | Examples of ions | $n$ |
|---|---|---|
| [He][He] | $H^-$, $Li^+$ | 5 |
| [Ne][Ne] | $F^-$, $O^{2-}$, $Na^+$, $Mg^{2+}$ | 7 |
| [Ar][Ar], or $[3d^{10}][Ar]$ | $Cl^-$, $S^{2-}$, $K^+$, $Ca^{2+}$, $Cu^+$ | 9 |
| [Kr][Kr] or $[4d^{10}][Kr]$ | $Br^-$, $Rb^+$, $Sr^{2+}$, $Ag^+$ | 10 |
| [Xe][Xe] or $[5d^{10}][Xe]$ | $I^-$, $Cs^+$, $Ba^{2+}$, $Au^+$ | 12 |

## Born forces

In a real ionic lattice, the ions are *not* the point charges assumed in the electrostatic model but have finite size. In addition to the electrostatic forces discussed above, there are also Born (repulsive) forces between adjacent ions. These arise from the electron–electron and nucleus–nucleus repulsions.

The internal energy that is associated with the Born forces is related to the internuclear distance between adjacent ions according to equation 8.32.

$$\Delta U(0\,K)(Born) \propto \frac{1}{r^n} \qquad (8.32)$$

where $r$ = internuclear distance and $n$ = Born exponent

Na$^+$ and Ne are isoelectronic; K$^+$, Cl$^-$ and Ar are isoelectronic

The value of $n$ depends upon the electronic configuration of the ions involved, and values are listed in Table 8.8. Thus, for KCl which consists of argon-like ions, the Born exponent is 9. For NaCl which contains neon- and argon-like ions ($Na^+$ is isoelectronic with Ne; $Cl^-$ is isoelectronic with Ar), $n$ is $(\frac{7}{2} + \frac{9}{2}) = 8$.

## The Born–Landé equation

Equation 8.31 can be modified by including a contribution to $\Delta U(0\,K)$ due to the Born repulsions, and the new expression is given in equation 8.33.

$$\text{Lattice energy} = \Delta U(0\,K) = -\left(\frac{L \times A \times |z_+| \times |z_-| \times e^2}{4 \times \pi \times \varepsilon_0 \times r}\right) \times \left(1 - \frac{1}{n}\right) \qquad (8.33)$$

Given a chemical formula for an ionic compound, equation 8.33 can be readily applied provided we know the internuclear distance, $r$, and the structure type. Both these data can be obtained from the results of X-ray diffraction experiments. Since the Madelung constant is defined for a particular structure type, this piece of information must be available in order to use equation 8.33 (Table 8.7). A value of $r$ can be estimated by summing the radii of the cation and anion (equation 8.28).

The *lattice energy* of an ionic compound is the change in internal energy that accompanies the formation of one mole of a solid from its constituent gas-phase ions at a temperature of 0 K.

*Assuming a model for an ionic lattice that consists of spherical ions*, values of lattice energy can be estimated by using the Born–Landé equation:

$$\text{Lattice energy} = \Delta U(0\,\text{K}) = -\left(\frac{L \times A \times |z_+| \times |z_-| \times e^2}{4 \times \pi \times \varepsilon_0 \times r}\right) \times \left(1 - \frac{1}{n}\right)$$

---

**Worked example 8.3**  | *Lattice energy*

**Estimate the change in internal energy which accompanies the formation of one mole of CsCl from its constituent gaseous ions, assuming an electrostatic model for the solid state lattice.**

**Data required:** $L = 6.022 \times 10^{23}\,\text{mol}^{-1}$; $A = 1.7627$; $e = 1.602 \times 10^{-19}\,\text{C}$; $\varepsilon_0 = 8.854 \times 10^{-12}\,\text{F m}^{-1}$; **Born exponent for CsCl = 10.5; internuclear Cs–Cl distance = 356 pm.**

The change in internal energy (the lattice energy) is given by the Born–Landé equation:

$$\Delta U(0\,\text{K}) = -\left(\frac{L \times A \times |z_+| \times |z_-| \times e^2}{4 \times \pi \times \varepsilon_0 \times r}\right) \times \left(1 - \frac{1}{n}\right)$$

$r$ must be in metres: $356\,\text{pm} = 3.56 \times 10^{-10}\,\text{m}$.
We must also include a conversion factor of $10^{-3}$ for J to kJ.

See worked example 8.2 for a more detailed breakdown of derived units

$$\Delta U(0\,\text{K}) = -\left[\frac{(6.022 \times 10^{23}\,\text{mol}^{-1}) \times 1.7627 \times 1 \times 1 \times (1.602 \times 10^{-19}\,\text{C})^2}{4 \times 3.142 \times (8.854 \times 10^{-12}\,\text{F m}^{-1}) \times (3.56 \times 10^{-10}\,\text{m})}\right]$$
$$\times \left(1 - \frac{1}{10.5}\right) \times 10^{-3}$$

$$= -622\,\text{kJ mol}^{-1}$$

---

**8.16** | **Lattice energy – experimental data**

We have seen how a purely ionic model can be used to *estimate* values of lattice energies and in this section we show how values may be experimentally determined.

### Using Hess's Law: The Born–Haber cycle

Hess's Law: see Section 2.7

Lattice energies are not measured directly, and you can appreciate why if you consider the definition given in Section 8.15. Instead, use is made of Hess's Law. For example, although the lattice energy for KCl is the change in internal energy at 0 K for reaction 8.34, we could also define a *lattice enthalpy change* ($\Delta_{\text{lattice}}H^{\circ}$) as the enthalpy change that accompanies reaction 8.34

under standard conditions. Using Hess's Law, $\Delta_{\text{lattice}}H^{\circ}$ can be determined from equation 8.35.

$$K^+(g) + Cl^-(g) \longrightarrow KCl(s) \tag{8.34}$$

$$\Delta_{\text{lattice}}H^{\circ}(KCl,s) = \Delta_f H^{\circ}(KCl,s) - \Delta_f H^{\circ}(K^+,g) - \Delta_f H^{\circ}(Cl^-,g) \tag{8.35}$$

On the right-hand side of equation 8.35, the standard enthalpy of formation of solid KCl (as for a wide range of inorganic compounds) is readily available from tables of thermochemical data. The standard enthalpies of formation of gaseous ions can be determined by setting up appropriate Hess cycles (equations 8.36 and 8.37); we introduced ionization energies and electron affinities earlier in Chapter 8.

Values of $\Delta_f H^{\circ}$:
see Appendix 11

$$K(s) \xrightarrow{\Delta_f H^{\circ}(K^+,g)} K^+(g)$$

$$\Delta_a H^{\circ}(K,s) \searrow \qquad \nearrow IE_1(K,g)$$

$$K(g)$$

$$\Delta_f H^{\circ}(K^+,g) = \Delta_a H^{\circ}(K,s) + IE_1(K,g) \tag{8.36}$$

$$\tfrac{1}{2}Cl_2(g) \xrightarrow{\Delta_f H^{\circ}(Cl^-,g)} Cl^-(g)$$

$$\Delta_a H^{\circ}(Cl,g) \searrow \qquad \nearrow \Delta_{EA}H(Cl,g)$$

$$Cl(g)$$

$$\Delta_f H^{\circ}(Cl^-,g) = \Delta_a H^{\circ}(Cl,g) + \Delta_{EA}H(Cl,g) \tag{8.37}$$

where $\Delta_{EA}H$ is the *enthalpy change* associated with the attachment of an electron

$\Delta_a H^{\circ}$ and bond enthalpies:
see Section 4.6

Values of standard enthalpies of atomization, $\Delta_a H^{\circ}$, and ionization energies are readily available for the elements, but only a relatively few electron affinities have been measured accurately.

We can now combine equations 8.35, 8.36 and 8.37 to give a single Hess cycle called a *Born–Haber cycle* (Figure 8.24) and from it we can find $\Delta_{\text{lattice}}H^{\circ}$ (equations 8.38 and 8.39). The value determined for $\Delta_{\text{lattice}}H^{\circ}$ in this way represents an *experimental value*, since it is derived from experimentally determined data.

$$\Delta_f H^{\circ}(KCl,s) = \Delta_a H^{\circ}(K,s) + \Delta_a H^{\circ}(Cl,g) + IE_1(K,g)$$
$$+ \Delta_{EA}H(Cl,g) + \Delta_{\text{lattice}}H^{\circ}(KCl,s) \tag{8.38}$$

Equation 8.38 can be rearranged to give equation 8.39 for the lattice enthalpy.

$$\Delta_{\text{lattice}}H^{\circ}(KCl,s) = \Delta_f H^{\circ}(KCl,s) - \Delta_a H^{\circ}(K,s)$$
$$- \Delta_a H^{\circ}(Cl,g) - IE_1(K,g) - \Delta_{EA}H(Cl,g) \tag{8.39}$$

We have seen previously that both ionization energies and electron affinities are changes in internal energy, but, since the necessary corrections are

**Fig. 8.24** A Born–Haber cycle for potassium chloride.

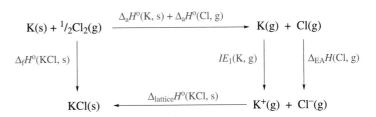

relatively small, they can be approximated to changes in enthalpy. Similarly, $\Delta_{\text{lattice}} H^{\circ} \approx \Delta U(0\,\text{K})$.

Thus for KCl, the lattice energy can be determined (equation 8.40) from equation 8.39 using experimental values for $\Delta_f H^{\circ}(\text{KCl, s}) = -437\,\text{kJ mol}^{-1}$, $\Delta_a H^{\circ}(\text{K, s}) = 90\,\text{kJ mol}^{-1}$, $\Delta_a H^{\circ}(\text{Cl, g}) = 121\,\text{kJ mol}^{-1}$, $IE_1(\text{K, g}) = 418.8\,\text{kJ mol}^{-1}$ and $\Delta_{EA} H(\text{Cl, g}) = -349\,\text{kJ mol}^{-1}$.

$$\Delta U(0\,\text{K}) \approx \Delta_{\text{lattice}} H^{\circ}(\text{KCl, s}) = -437 - 90 - 121 - 418.8 + 349$$

$$= -718\,\text{kJ mol}^{-1} \quad \text{(to 3 sig. fig.)} \tag{8.40}$$

The negative value of $\Delta_{\text{lattice}} H^{\circ}$ indicates that the formation of solid KCl from gaseous $K^+$ and $Cl^-$ ions is an exothermic process. You should note that lattice energy may also be defined as the energy needed to convert an ionic solid into its constituent gaseous ions. In this case, the associated enthalpy change is positive. When quoting a value for a lattice energy or the associated lattice enthalpy change, you should also state the process to which the value refers.

## 8.17 A comparison of lattice energies determined by the Born–Landé equation and the Born–Haber cycle

Table 8.9 gives values for the lattice energies of some ionic compounds, determined both from the Born–Landé equation, $\Delta U(0\,\text{K})$(Born–Landé), and by using a Born–Haber cycle, $\approx \Delta U(0\,\text{K})$(Born–Haber). Compounds have been grouped in the table according to periodic relationships and cover the halides of two group 1 metals (Na and K) and a group 11 metal (Ag). With the exception of silver(I) iodide, each compound in the table has an NaCl structure; AgI has a wurtzite structure at room temperature and pressure.

The difference between $\Delta U(0\,\text{K})$(Born–Haber) and $\Delta U(0\,\text{K})$(Born–Landé) for each compound is expressed as a percentage of the experimental value (the last column of Table 8.9). The agreement between the values of the lattice energies are fairly good for the sodium and potassium halides, and this

**Table 8.9** Values of lattice energies for selected compounds determined using the Born–Landé equation (equation 8.33) and a Born–Haber cycle (Figure 8.24 and equation 8.39). The Born–Haber cycle gives a value of $\Delta_{\text{lattice}} H^{\circ}$ which approximates to $\Delta U(0\,\text{K})$; values in this table have been calculated using thermochemical data from appropriate tables in this book.

| Compound | $\Delta U(0\,\text{K}) \approx \Delta_{\text{lattice}} H^{\circ}$ from a Born–Haber cycle / kJ mol$^{-1}$ | $\Delta U(0\,\text{K})$ from the Born–Landé equation / kJ mol$^{-1}$ | $\Delta U(0\,\text{K})$ (Born–Haber) $-\Delta U(0\,\text{K})$ (Born–Landé) / kJ mol$^{-1}$ | Percentage difference in values of $\Delta U(0\,\text{K})$[a] |
|---|---|---|---|---|
| NaF | −931 | −901 | −30 | 3.2% |
| NaCl | −786 | −756 | −30 | 3.8% |
| NaBr | −736 | −719 | −17 | 2.3% |
| NaI | −671 | −672 | +1 | ≈0 |
| KF | −827 | −798 | −29 | 3.5% |
| KCl | −718 | −687 | −31 | 4.3% |
| KBr | −675 | −660 | −15 | 2.2% |
| KI | −617 | −622 | +5 | 0.8% |
| AgF | −972 | −871 | −101 | 10.4% |
| AgCl | −915 | −784 | −131 | 14.3% |
| AgBr | −888 | −758 | −130 | 14.6% |
| AgI | −858 | −737 | −121 | 14.1% |

[a] The percentage difference is calculated using the equation $\left( \dfrac{\Delta U(0\,\text{K})(\text{Born–Haber}) - \Delta U(0\,\text{K})(\text{Born–Landé})}{\Delta U(0\,\text{K})(\text{Born–Haber})} \right) \times 100$.

suggests that the discrete-ion model assumed for the Born–Landé equation is reasonably appropriate for these compounds.[§]

For each of the silver(I) halides, the Born–Landé equation underestimates the lattice energy by more than 10%. This suggests that the discrete-ion model is *not* appropriate for the silver(I) halides.

We mentioned in Section 8.1 that just as the covalent bonding model can be improved by allowing for ionic contributions to a bond, an ionic model can similarly be improved by allowing for some covalent character. This is exactly the problem that we have encountered with the silver(I) halides.

| 8.18 | **Polarization of ions** |
|---|---|

We now reconsider the assumption that ions are spherical. Is this always true?

Consider a small ion such as $Li^+$, and assume that it is spherical. Since the surface area of the $Li^+$ ion is small, the charge density is relatively large (equation 8.41). For a spherical ion of radius $r$, the surface area is $4\pi r^2$.

$$\text{Charge density of an ion} = \frac{\text{Charge on the ion}}{\text{Surface area of the ion}} \qquad (8.41)$$

Now consider a large anion such as $I^-$ where the charge density is low. If the iodide ion is close to a cation with a high charge density, the charge distribution might be distorted and may no longer be spherical. The distortion means that the centre of electron density in the iodide anion no longer coincides with the position of the nucleus, and the result is a dipole moment, induced by the adjacent, small cation. The electron density is distorted towards the cation. The iodide anion is said to be readily *polarizable*. The cation that causes the polarization to occur is said to be *polarizing*.

▄▶
Induced dipole:
see Section 3.21

$$\text{Charge density of an ion} = \frac{\text{Charge on the ion}}{\text{Surface area of the ion}}$$

Surface area of a spherical ion $= 4\pi(r_{ion})^2$ where $r_{ion}$ is the ionic radius

Small cations such as $Li^+$, $Mg^{2+}$, $Al^{3+}$, $Fe^{3+}$ and $Ti^{4+}$ possess high charge densities, and each has a high *polarizing power*.

Anions such as $I^-$, $H^-$, $N^{3-}$, $Se^{2-}$ and $P^{3-}$ are easily polarizable. As we descend a group, for example from fluoride to iodide where the charge remains constant, the polarizability increases as the size of the ion increases.

When an ion (usually the cation) induces a dipole in an adjacent ion (usually the anion), an ion–dipole interaction results.

| 8.19 | **Determining the Avogadro constant from an ionic lattice** |
|---|---|

The Avogadro number can be estimated from crystal lattice data. The example chosen here is the NaCl structure (Figure 8.11), and the calculation uses the following data:

---

[§] Using the Born–Landé equation is not the only method of calculating lattice energies. More sophisticated equations are available which will improve the answer. However, the pattern of variation noted in Table 8.9 remains the same. For more detailed discussion see: W. E. Dasent (1984) *Inorganic Energetics*, 2nd edn, Cambridge University Press, Cambridge.

- the internuclear separation in NaCl determined by X-ray diffraction (281 pm);
- the density of NaCl ($2.165\,g\,cm^{-3}$);
- values of $A_r$ for Na and Cl (22.99 and 35.45 respectively).

The unit cell of NaCl contains four ion-pairs (see Table 8.3). The length of each side of the unit cell of NaCl (Figure 8.11) is *twice* the internuclear separation (equation 8.42) and the volume of the unit cell is given by equation 8.43.

$$\text{Length of one side of a unit cell of NaCl} = 2 \times 281 = 562\,\text{pm} \qquad (8.42)$$

$$\text{Volume of a unit cell of NaCl} = (\text{length of side})^3 = 1.78 \times 10^8\,\text{pm}^3 \qquad (8.43)$$

The volume of one ion-pair is one-quarter the volume of the unit cell, and since one mole of NaCl contains the Avogadro number, $L$, of ion-pairs, the volume occupied by a mole of NaCl is given by equation 8.44.

$$\text{Volume of 1 mole of NaCl} = \frac{1.78 \times 10^8 \times L}{4}\,\text{pm}^3 \qquad (8.44)$$

The volume of one mole of NaCl can also be expressed in terms of the relative molecular mass and the density (equation 8.45).

$$\text{Volume occupied by 1 mole of NaCl} = \frac{M_r}{\rho} = \frac{22.99 + 35.45}{2.165} = 26.99\,\text{cm}^3 \qquad (8.45)$$

The Avogadro constant can be estimated by combining equations 8.44 and 8.45, but first, the units must be made consistent: $1\,\text{pm} = 10^{-10}\,\text{cm}$. Equation 8.46 shows how to estimate the Avogadro constant, after the unit cell volume has been converted to $cm^3$.

$$\text{Avogadro constant, } L = \frac{26.99 \times 4}{1.78 \times 10^{-22}} = 6.06 \times 10^{23}\,\text{mol}^{-1} \qquad (8.46)$$

## SUMMARY

In this chapter we have been concerned with the formation of cations and anions from gaseous atoms and the interactions between ions in the solid state.

### Do you know what the following terms mean?

- electron density map
- ionization energy
- electron affinity
- ion-pair
- Coulombic interaction

- ionic lattice
- unit cell
- coordination number (in an ionic solid)
- ionic radius
- lattice energy

- Madelung constant
- Born exponent
- Born–Landé equation
- Born–Haber cycle
- charge density of an ion

### You should now be able:

- to discuss why there is a relationship between the pattern in the values of successive ionization energies for a given atom and its electronic structure
- to describe the structures of sodium chloride, caesium chloride, fluorite, rutile and zinc(II) sulfide
- to deduce the stoichiometry of a compound given a diagram of a unit cell of the solid state lattice
- to write down the Born–Landé equation and know how to use it to estimate lattice energies
- to appreciate the assumptions that are inherent in the Born–Landé equation

- to set up a Born–Haber cycle and determine the lattice energy of a compound, or, given a lattice energy, to use a Born–Haber cycle to find another quantity such as the electron affinity of an atom
- to appreciate why the values of lattice energies calculated by the Born–Landé equation may or may not agree with experimental values
- to show how a value of the Avogadro constant can be estimated from crystal structure and density data

## PROBLEMS

**8.1** With reference to the oxygen atom, explain what you understand by the (a) first ionization energy, (b) second ionization energy, (c) first electron affinity, (d) second electron affinity and (e) enthalpy change associated with the attachment of an electron.

**8.2** The first five ionization energies of a gaseous atom X are 589, 1148, 4911, 6494 and 8153 kJ mol$^{-1}$. To what group is this element likely to belong?

**8.3** The first ionization energies of oxygen ($Z = 8$), fluorine ($Z = 9$), neon ($Z = 10$) and sodium ($Z = 11$) are 1312, 1681, 2080 and 495 kJ mol$^{-1}$ respectively. Rationalize the trend in these values in terms of the electronic structures of these elements.

**8.4** Calculate the enthalpy change that accompanies the process:

$$\tfrac{1}{2}Li_2(g) \longrightarrow Li^+(g) + e^-$$

given that the bond enthalpy for dilithium is 110 kJ mol$^{-1}$ and $IE_1$ for lithium is 520 kJ mol$^{-1}$.

**8.5** Calculate the enthalpy change that accompanies the reaction:

$$\tfrac{1}{2}Li_2(g) + e^- \longrightarrow Li^-(g)$$

given that the bond enthalpy for Li$_2$ is 110 kJ mol$^{-1}$ and $EA_1$ for lithium is 60 kJ mol$^{-1}$.

**8.6** Calculate the enthalpy changes, $\Delta H$, for the reactions:

$$H(g) + F(g) \longrightarrow H^+(g) + F^-(g)$$
$$H(g) + F(g) \longrightarrow H^-(g) + F^+(g)$$

and comment on the results in the light of the discussion about resonance structures in Section 5.3. (Data required: see Table 8.1 and Appendices 8 and 9.)

**8.7** Using the data in Table 8.1, determine the enthalpy change that accompanies the formation of a gaseous S$^{2-}$ ion from a gaseous sulfur atom. Comment on the sign and magnitude of your answer in terms of the electronic configurations of the S atom and the S$^-$ and S$^{2-}$ ions.

**8.8** Distinguish between the terms *van der Waals radius*, *covalent radius* and *ionic radius*. How would you estimate the ionic radius of a sodium ion? Point out what other information you would need and what assumptions you would make.

**8.9** Show that equation 8.19 is dimensionally correct.

**8.10** Figure 8.11 shows a unit cell of the sodium chloride structure. Why could the unit cell not be smaller, for example with only four ion-pairs defining a cube?

**8.11** Uranium(IV) oxide adopts a fluorite structure. From this structure, show that the formula of the oxide is UO$_2$.

**8.12** Suggest reasons why the distance between adjacent Na and Cl nuclei is shorter (236 pm) in NaCl in the gas phase than in the solid state (281 pm).

**8.13** (a) Draw the structures of the sulfate(VI) [SO$_4$]$^{2-}$, and tetrafluoroborate [BF$_4$]$^-$ ions and rationalize why they can be considered to be approximately spherical. (b) Given the lattice energies of sodium sulfate and sodium tetrafluoroborate, suggest a method to estimate the radii of the [SO$_4$]$^{2-}$ and [BF$_4$]$^-$ ions. State clearly what assumptions (if any) you would have to make, and also what other information you would require.

**8.14** (a) Explain why a unit cell of CsCl can be drawn with either Cs$^+$ or Cl$^-$ ions in the corner sites, but in rutile, the corner sites of the unit cell must be occupied by Ti(IV) and not O$^{2-}$ ions. (b) Crystalline K$_2$O adopts an antifluorite lattice. Use your knowledge of the fluorite lattice to deduce what is meant by an 'antifluorite lattice'.

**8.15** (a) Gallium phosphide is used is solar energy conversion, and it crystallizes with the zinc blende lattice. What is the formula of gallium phosphide? Rationalize this formula in terms of the positions of Ga and P in the periodic table. (b) Crystalline cerium(IV) oxide contains 4-coordinate O$^{2-}$ ions. What is the coordination number of the Ce(IV) centres? The oxide adopts one of the lattices described in Chapter 8. Suggest which lattice it is.

**8.16** A bromide of indium, InBr$_x$, adopts the NaCl lattice but only one-third of the metal ion sites are occupied. What is the oxidation state of indium in this compound?

**8.17** Using values in Table 8.8, determine the Born exponents for the following ionic compounds: (a) NaCl; (b) LiCl; (c) AgF; (d) MgO.

**8.18** In worked example 8.3, the lattice energy for CsCl was calculated using the experimental internuclear distance of 356 pm. Calculate the lattice energy for the same structure using values of $r_{ion}Cs^+ = 170$ pm and $r_{ion}Cl^- = 181$ pm. Rationalize the difference between the two values of $\Delta U(0\,K)$.

**8.19** Calculate the lattice energy for magnesium oxide using both the Born–Landé equation and a Born–Haber cycle; MgO has an NaCl-type structure. All the data needed for the Born–Landé method are to be found in Chapter 8; thermochemical data are listed in Appendices 8–11. Use the data obtained to comment on whether a spherical-ion model is appropriate for MgO.

**8.20** KF crystallizes with an NaCl structure, and the internuclear separation is 266 pm. The density of KF is 2.48 g cm$^{-3}$. Estimate a value for the Avogadro constant.

**8.21** Using data from worked example 8.3 and Tables 8.6–8.8, estimate the lattice energy of RbF; it crystallizes with an NaCl structure and the internuclear separation is 282 pm.

**8.22** The lattice energy of CaO (which adopts an NaCl structure) is $-3400 \, \text{kJ} \, \text{mol}^{-1}$. Estimate the internuclear separation, and compare your answer with the sum of the ionic radii. (Use data from worked example 8.3 and Tables 8.6–8.8.)

---

**Additional problems: most of these require you to find appropriate data in the Appendices**

---

**8.23** For solid $MgF_2$, $\Delta_f H^\circ(298 \, \text{K}) = -1124 \, \text{kJ} \, \text{mol}^{-1}$. Estimate a value for the lattice energy of $MgF_2$.

**8.24** Determine a value of $\Delta_{\text{lattice}} H^\circ(298 \, \text{K})$ for $K_2O$ given that $\Delta_f H^\circ(K_2O, \text{s}, 298 \, \text{K}) = -362 \, \text{kJ} \, \text{mol}^{-1}$.

**8.25** The enthalpy change associated with the formation of crystalline $CaCl_2$ from gaseous ions is $-2250 \, \text{kJ} \, \text{mol}^{-1}$. What is $\Delta_f H^\circ(CaCl_2, \text{s}, 298 \, \text{K})$?

**8.26** Using information in problem 8.25, determine the energy change for the process:

$$CaCl_2(\text{cryst}) \longrightarrow Ca^{2+}(\text{g}) + 2Cl^-(\text{g})$$

**8.27** (a) Draw a Lewis structure for the cyanide ion. (b) Comment on the fact that sodium cyanide, NaCN, crystallizes with an NaCl structure.

# 9  Elements

## Topics

- Close-packing of spheres
- Solid structures of group 18 elements
- Solids with molecular units
- Solids with infinite covalent lattices
- Metals
- Metallic bonding

## 9.1  Introduction

In Section 3.21, we discussed the van der Waals forces that operate between atoms of a noble gas and noted that the weakness of these forces is reflected in some of the physical properties of the noble gases (Table 3.6). We have already observed that these elements form solids only at very low temperatures, and that helium can only be solidified under pressures greater than atmospheric pressure. When a group 18 element solidifies, the atoms form an ordered structure in which they are *close-packed*. Before considering the specifics of the solid state structures of the noble gases, we discuss what is meant by the *close-packing of spheres*. In Section 9.3, we consider simple and body-centred cubic packing of spheres.

In this chapter we consider not only elemental solids that consist of close-packed atoms, but also the structures of those elements which contain molecular units, small or large, or infinite covalent networks in the solid state.

## 9.2  Close-packing of spheres

### Hexagonal and cubic close-packing

Suppose we have a rectangular box and place in it some spheres of equal size. If we impose a restriction that there must be a *regular arrangement* of the spheres, then the most efficient way in which to cover the floor of the box is to pack the spheres as shown in Figure 9.1. This is part of a *close-packed arrangement*, and spheres not on the edge of the assembly are in contact with six other spheres within the single layer. Figure 9.1 emphasizes that a motif of hexagons is visible.

If we now add a second layer of close-packed spheres to the first, allowing the spheres in the second layer to rest in the hollows between those in the first

**Fig. 9.1** Part of a layer of a close-packed arrangement of equal-sized spheres. It contains hexagonal motifs.

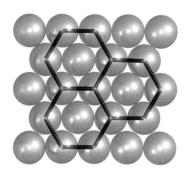

> ***Close-packing of spheres*** represents the most efficient use of space.

layer, there is only enough room for *every other hollow* to be occupied (Figure 9.2 and Figure 9.3a to 9.3b). When a third layer of spheres is added, there is again only room to place them in every other hollow, but now there are *two distinct sets of hollows*. Of the four hollows between the yellow spheres in layer B in Figure 9.3b, one lies over a blue sphere in layer A, and three lie over hollows in layer A. Depending upon which set of hollows is occupied as the third layer of spheres is put in place, one of two close-packed structures results. On going from Figure 9.3b to Figure 9.3c, the red spheres are placed in hollows in layer B that lie over hollows in layer A. This forms a new layer, C, which does not replicate either layer A or layer B. The repetition of layers ABCABC... is called *hexagonal close-packing* (hcp). On going from Figure 9.3b to Figure 9.3d, the blue sphere is placed in a hollow in layer B that lies over a sphere in layer A. This forms a new layer, but it replicates layer A. The repetition of layers ABABAB... is called *cubic close-packing* (ccp).

In both the cubic and hexagonal close-packed arrangements of spheres, each sphere has 12 nearest neighbours, i.e. each sphere is 12-coordinate. Figures 9.4a and 9.4b show how the coordination number of 12 is achieved for the ccp and hcp arrangements, respectively. The two diagrams in Figure 9.4 differ only in the mutual orientations of the two sets of three atoms at the top and bottom of the figures.

Unit cell: see Section 8.7

The unit cells that characterize hexagonal and cubic close-packed arrangements are illustrated in Figures 9.5a and 9.5b. The unit cell shown in Figure 9.5b illustrates the presence of the cubic unit that lends its name to cubic close-packing. It consists of eight spheres at the corners of the cube, with a sphere at the centre of each square face. An alternative name for cubic close-packing is a *face-centred cubic* (fcc) arrangement.

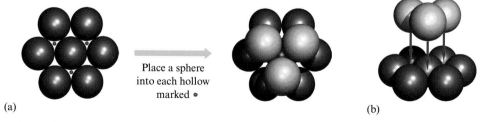

(a)

Place a sphere into each hollow marked •

(b)

**Fig. 9.2** When three spheres are arranged in a triangle and touch one another, there is a hollow at the centre of the triangle. A single layer of close-packed spheres possesses hollows which form a regular pattern. (a) There are six hollows in between the seven close-packed spheres. When a second layer of spheres is placed on top of the first so that the new spheres lie in the hollows in the first layer, there is only room for every other hollow to be occupied. This is emphasized in (b) which gives a side view of the arrangement as the second layer of spheres is added to the first.

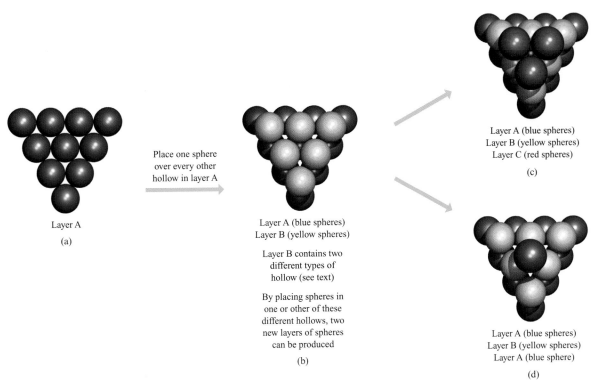

Place one sphere over every other hollow in layer A

Layer A

(a)

Layer A (blue spheres)
Layer B (yellow spheres)

Layer B contains two different types of hollow (see text)

By placing spheres in one or other of these different hollows, two new layers of spheres can be produced

(b)

Layer A (blue spheres)
Layer B (yellow spheres)
Layer C (red spheres)

(c)

Layer A (blue spheres)
Layer B (yellow spheres)
Layer A (blue sphere)

(d)

**Fig. 9.3** (a) One layer (layer A) of close-packed spheres contains hollows arranged in a regular pattern. (b) A second layer (layer B) of close-packed spheres can be formed by occupying every other hollow in layer A. In layer B, there are two types of hollow: one lies over a sphere in layer A, and three lie over hollows in layer A. By stacking spheres over these different types of hollow, two different third layers of spheres can be produced. The red spheres in diagram (c) form a new layer C; this gives an ABC sequence of layers. Diagram (d) shows that the second possible third layer replicates layer A, i.e. the blue sphere in the top layer lies over a blue sphere in layer A to give an ABA sequence.

It is easy to relate Figure 9.5a to the ABABAB... arrangement shown in Figure 9.3 but it is not so easy to recognize the relationship between the unit cell in Figure 9.5b and an ABCABC... layer arrangement. Figures 9.5b and 9.5e show the same ccp unit cell. Whereas Figure 9.5b illustrates the unit cell in an orientation that clearly defines the cubic unit, Figure 9.5e views the cube

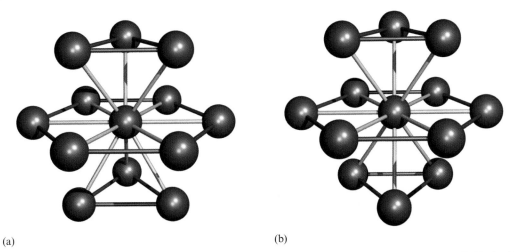

(a)                                    (b)

**Fig. 9.4** In a close-packed assembly of spheres, each sphere has 12 nearest-neighbours. This is true for both (a) the ABA arrangement and (b) the ABC arrangement.

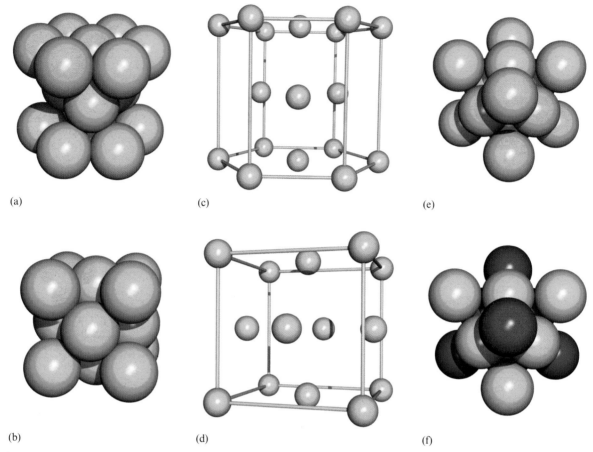

(a)

(c)

(e)

(b)

(d)

(f)

**Fig. 9.5** The ABABAB... and ABCABC... close-packed arrangements of spheres are termed (a) *hexagonal* and (b) *cubic* close-packing respectively. If the spheres are 'pulled apart', (c) the *hexagonal* and (d) the *cubic* units become clear. Cubic close-packing is also called a *face-centred cubic* arrangement. In diagrams (b) and (d), each face of the cubic unit possesses a sphere at its centre. (e) A unit cell of a cubic close-packed arrangement of spheres, viewed along a line that joins two opposite corners of the cube. (f) The same unit cell as shown in diagram (e), but with the ABC layers of spheres highlighted (A, red; B, yellow; C, blue).

down a line that joins two opposite corners of the cube. Now the ABC layers are clearly visible, and in Figure 9.5f, these layers are coloured red, yellow and blue, respectively.

Spheres of equal size may be close-packed in (at least) two ways.

In **hexagonal close-packing** (hcp), layers of close-packed spheres pack in an ABABAB... pattern.

In a **cubic close-packing** (ccp) or face-centred cubic (fcc) arrangement, layers of close-packed spheres are in an ABCABC... pattern.

In both an hcp and a ccp array, each sphere has 12 nearest-neighbours.

## Interstitial holes

Placing one sphere on top of three others in a close-packed array gives a tetrahedral unit (Figure 9.6a) inside which is a cavity called an *interstitial hole* – specifically in this case, a *tetrahedral hole*.

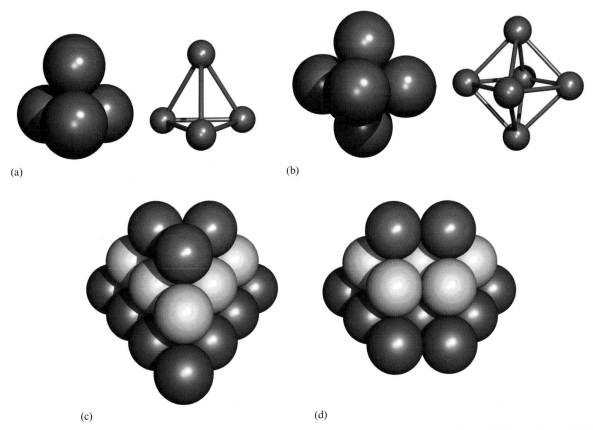

**Fig. 9.6** (a) A tetrahedral hole is formed when one sphere rests in the hollow between three others which are close-packed. (b) An octahedral hole is formed when two spheres stack above and below a square unit of spheres. (c) An ABC . . . arrangement of layers of close-packed spheres (A, red; B, yellow; C, blue). By removing three adjacent spheres (one from each layer) from this arrangement, the arrangement shown in diagram (d) is formed. This possesses two square faces.

A *close-packed* arrangement of spheres contains octahedral and tetrahedral holes.

A close-packed array contains tetrahedral holes but in addition there are *octahedral holes* (Figure 9.6b). Both types of interstitial hole are present in the unit cells shown in Figure 9.5, and this is further explored in Box 9.1.

It is not necessarily obvious why octahedral holes should arise in a close-packed arrangement. Figure 9.6c shows an ABC arrangement of layers of close-packed spheres. On going from Figure 9.6c to 9.6d, three adjacent spheres (one from each layer) have been removed. The result is that two square faces are exposed. As you can see from Figure 9.6b, a square of spheres is a building block within an octahedral arrangement, and such motifs are present in both hexagonal and cubic close-packed assemblies.

## 9.3    Simple cubic and body-centred cubic packing of spheres

In cubic or hexagonal close-packed assemblies, each sphere has 12 nearest-neighbours, and spheres are packed efficiently so that there is the minimum amount of 'wasted' space. It is, however, possible to pack spheres of equal size in an ordered assembly that is *not* close-packed and one such method is that of *simple cubic packing* (Figure 9.7). The unit cell contains eight spheres arranged at the corners of a cube, and when these units are placed next to one another in an extended array, each sphere has six nearest-neighbours and is in an octahedral environment.

THEORETICAL AND CHEMICAL BACKGROUND

## Box 9.1 An alternative description of ionic lattices

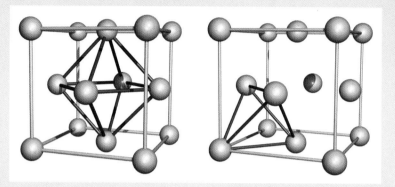

A face-centred cubic (fcc) arrangement of spheres (cubic close-packing) is shown above. The left-hand diagram emphasizes the octahedral hole present at the centre of the fcc unit. The right-hand diagram shows one tetrahedral hole; there are eight within the fcc unit.

An ionic lattice contains spheres (i.e. ions) which are *not* all the same size. In many cases (e.g. NaCl) the anions are larger than the cations, and it is convenient to consider a close-packed array of anions with the smaller cations in the interstitial holes.

Consider a cubic close-packed array of sulfide *anions*. They repel one another when they are in close proximity, but if zinc *cations* are placed in some of the interstitial holes, we can generate an ionic lattice that is stabilized by cation–anion attractive forces. A unit cell of the zinc blende (ZnS) structure is shown opposite.

The $S^{2-}$ ions are in an fcc arrangement. *Half of the tetrahedral holes* are occupied by $Zn^{2+}$ ions. The organization of the cations is such that every other tetrahedral hole is filled with a cation.

This description of the structure of zinc blende relies on the idea that *small cations* can fit into the interstitial

holes which lie between *large anions that are close-packed*. Note that the anions thus define the corners of the unit cell. (You should look back at Chapter 8 and assess to what extent this statement is actually valid.)

Similarly, it is possible to describe the structures of NaCl and $CaF_2$ in terms of cations occupying holes in a close-packed array of anions.

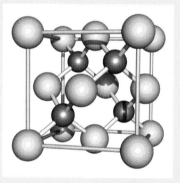

**Fig. 9.7** (a) The arrangement of spheres of equal size in a simple cubic packing. (b) The repeating unit in this arrangement consists of one sphere at each corner of a cube.

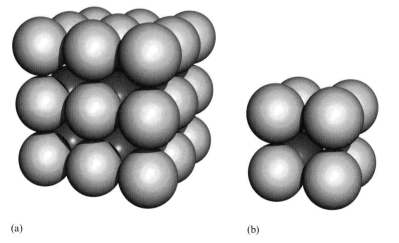

(a)        (b)

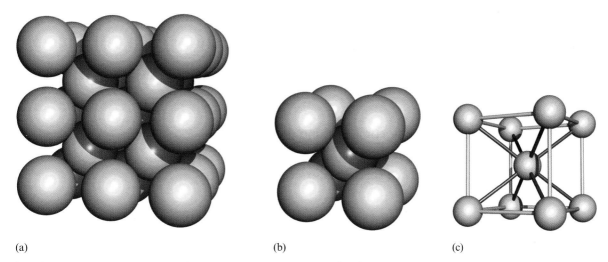

(a)              (b)              (c)

**Fig. 9.8** (a) The packing of spheres of equal size in a body-centred cubic (bcc) arrangement. (b) The unit cell in a bcc arrangement of spheres is a cube with one sphere at the centre. The central sphere touches each corner sphere, but the corner spheres do *not* touch one another. (c) Each sphere in a bcc arrangement has eight nearest-neighbours.

Simple cubic packing of spheres provides an ordered arrangement but there is a significant amount of unused space. Each interstitial hole (a 'cubic' hole) is too small to accommodate another sphere of equal size. If we force a sphere (of equal size) into each interstitial hole in the simple cubic array, the original spheres are pushed apart slightly. The result is the formation of a *body-centred cubic* (bcc) arrangement. The difference between simple and body-centred cubic structures can be seen by comparing Figures 9.7a and 9.8a. Although bcc packing makes better use of the space available than simple cubic packing, it is still less efficient than that in ccp or hcp assemblies.

The repeating unit in a bcc assembly is shown in Figure 9.8b, and the number of nearest-neighbours in the bcc arrangement is eight (Figure 9.8c).

---

The repetition of cubic units of spheres of equal size gives *simple cubic packing*, and each sphere has six nearest-neighbours.

In a *body-centred cubic* (bcc) arrangement, the unit cell consists of a cube of eight, non-touching spheres with one sphere in the centre. Each sphere has eight nearest-neighbours.

The simple and body-centred cubic arrangements are *not* close-packed.

---

**9.4**      **A summary of the similarities and differences between close-packed and non-close-packed arrangements**

### Hexagonal close-packing (Figures 9.5a and 9.5c)

- Layers of close-packed atoms are arranged in an ABABAB... manner.
- Each sphere has 12 nearest-neighbours.
- The arrangement contains tetrahedral and octahedral interstitial sites.

### Cubic close-packing or face-centred cubic (Figures 9.5b and 9.5d)

- Layers of close-packed atoms are arranged in an ABCABC... manner.
- Each sphere has 12 nearest-neighbours.
- The arrangement contains tetrahedral and octahedral interstitial sites.

### Simple cubic packing (Figure 9.7)

- The spheres are *not* close-packed.
- Each sphere has six nearest-neighbours.

### Body-centred cubic packing (Figure 9.8)

- The spheres are *not* close-packed.
- Each sphere has eight nearest-neighbours.

### Relative efficiency of packing

The relative efficiency with which spheres of equal size are packed follows the sequence:

$$\frac{\text{hexagonal}}{\text{close-packing}} = \frac{\text{cubic}}{\text{close-packing}} > \frac{\text{body-centred}}{\text{cubic packing}} > \text{simple cubic packing}$$

## 9.5    Crystalline and amorphous solids

X-ray diffraction:
see Section 4.2

In a **crystalline solid**, atoms, molecules or ions are packed in an ordered lattice with a characteristic unit cell.

In a *crystalline* solid, atoms, molecules or ions are packed in an ordered manner, with a unit cell that is repeated throughout the crystal lattice. For an X-ray diffraction experiment, a single crystal is usually required. If a single crystal shatters, it may cleave along well-defined *cleavage planes*. This leads to particular crystals possessing characteristic shapes.

In an *amorphous* solid, the particles are not arranged in an ordered or repetitive manner. Crushing an amorphous solid leads to the formation of a *powder*, whereas crushing crystals leads to *microcrystalline* materials. However, microcrystals may look like powders to the naked eye!

## 9.6    Solid state structures of the group 18 elements

Enthalpy of fusion:
see Section 2.9

The elements in group 18 are referred to as noble gases, and it is not usual to think of them in other states (but see Box 9.2). The group 18 elements solidify only at low temperatures (Table 3.6) and the enthalpy change that accompanies the fusion of one mole of each element is very small, indicating that the van der Waals forces between the atoms in the solid state are very weak. In the crystalline solid, the atoms of each group 18 element are close-packed. Cubic close-packing is observed for the atoms of each of solid neon, argon, krypton and xenon.

**COMMERCIAL AND LABORATORY APPLICATIONS**

## Box 9.2  Liquid helium: an important coolant

Although the group 18 elements are usually encountered in the gas phase, they have a number of important applications. One such is the use of liquid helium as a coolant.

Liquid nitrogen (bp 77 K) is frequently used as a coolant in the laboratory or in industry. However, it is not possible to reach extremely low temperatures by using liquid $N_2$ alone; liquefaction of gaseous $N_2$ under pressure provides a liquid at a temperature just below the boiling point. In order to reach lower temperatures it is necessary to use a liquid with a boiling point that is much lower than that of $N_2$, and liquid helium is widely used for this purpose. As normally found, helium boils at 4.2 K, and the use of liquid helium is the most important method for reaching temperatures which approach absolute zero. Below 2.2 K, isotopically pure ${}^4$He undergoes a transformation into the so-called He II. This is a remarkable liquid which possesses a viscosity close to zero, and a thermal conductivity which far exceeds that of copper.

In Chapter 14, we look at the technique of nuclear magnetic resonance (NMR) spectroscopy. This is an invaluable tool for probing molecular structure both in solution and in the solid state. The development of the modern generation of NMR spectrometers with high magnetic field strengths (300 to 900 MHz instruments) has allowed scientists to investigate the structures of complex molecules such as proteins, and has led to the development of medical resonance imaging (MRI, see Box 14.1). Highfield NMR spectrometers rely on magnets with superconducting coils that operate only at extremely low temperatures. (Superconductors are also discussed in Box 9.4.) To achieve these temperatures, the superconducting magnetic coil is enclosed within a tank of liquid helium. This in turn is surrounded by liquid nitrogen in order to slow down the rate of helium evaporation. The photograph below shows the magnet for a 600 MHz NMR spectrometer; most of the volume apparently occupied by the magnet is actually occupied by tanks of liquid helium and nitrogen. Spectrometers are conventionally described in terms of the resonance frequency of protons (see Chapter 14).

The magnet for a Bruker 600 MHz NMR spectrometer. *E. C. Constable.*

| 9.7 | **Elemental solids containing diatomic molecules** |

Figure 4.2 showed examples of covalent homonuclear molecules ($H_2$, $O_2$, $O_3$, $I_2$, $P_4$, $S_6$ and $S_8$). Each is a molecular form of an element. In the gas phase, these molecules are separate from one another, but in the solid state, they pack together with van der Waals forces operating between them. In this and the next two sections we consider the solid state structures of $H_2$, elements from groups 17 and 16, an allotrope of phosphorus, and one group of allotropes of carbon – all are *molecular solids* and *non-metals*.

**Fig. 9.9** Molecules of $H_2$ rotate freely in the solid state. (a) Some possible orientations for $H_2$ molecules with respect to the mid-point of the H−H bond are shown. (b) Taking these and all other possible orientations leads to a description of $H_2$ as a sphere.

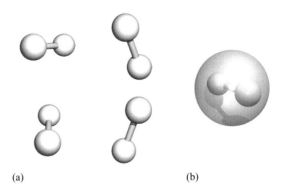

(a)                    (b)

### Dihydrogen and difluorine

When gaseous $H_2$ is cooled to 20.4 K, it liquefies.[§] Further cooling to 14.0 K results in the formation of solid dihydrogen. Even approaching absolute zero (0 K), molecules of $H_2$ possess sufficient energy to rotate about a point in the solid state lattice. Figure 9.9 shows that the result is that each $H_2$ molecule is described by a single sphere, the centre of which coincides with the midpoint of the H−H bond. Solid $H_2$ possesses an hcp arrangement of such spheres, each of which represents one $H_2$ molecule. It is possible to apply the model of close-packed spheres *because the $H_2$ molecules are rotating at the temperature at which the solid state structure has been determined.*

Molecular $F_2$ solidifies at 53 K. Below 45 K, the molecules of $F_2$ can freely rotate, and the structure is described as distorted close-packed, with each $F_2$ molecule being represented by a sphere.[†] The enthalpy of fusion of $F_2$ is $0.5 \, kJ \, mol^{-1}$ and this low value suggests that only van der Waals forces must be overcome in order to melt the solid.

The situation described for crystalline $H_2$ and $F_2$ is unusual. In the solid state, molecules of most elements or compounds are *not* freely rotating. Thermal motion such as the vibration of bonds does occur, but the positions of the atoms in a molecule can often be defined to a reasonable degree of accuracy. This means that the packing of spheres is not an appropriate model for most solid state structures because the component species are not spherical. It is applicable to elements of group 18 because they are monatomic, to $H_2$ and $F_2$ because the molecules are freely rotating, and to metals.

Metal lattices: see Section 9.11

### Dichlorine, dibromine and diiodine (group 17)

Some physical and structural properties of $Cl_2$, $Br_2$ and $I_2$ are given in Table 9.1. At 298 K (1 bar pressure), $I_2$ is a solid, but $Br_2$ is a liquid and $Cl_2$ a gas. Solid $Cl_2$, $Br_2$ and $I_2$ share common structures that differ from those of $F_2$.

In the crystalline state, molecules of $Cl_2$ (or $Br_2$ or $I_2$) are arranged in a zigzag pattern within a layer (Figure 9.10) and these layers of molecules are stacked together. There are three characteristic distances in the structure

---

[§] Here, and throughout the chapter, we consider phase changes at atmospheric pressure, unless otherwise stated.
[†] Above 45 K, a second phase with a more complicated structure exists.

**Table 9.1** Some physical and structural properties of dichlorine, dibromine and diiodine. For details of the solid state structure of these elements, refer to the text and Figure 9.10.

| Element | Melting point / K | $\Delta_{fus} H$ / kJ mol$^{-1}$ | Covalent radius ($r_{cov}$) / pm | Van der Waals radius ($r_v$) / pm | Intramolecular distance, *a* in Figure 9.10 / pm | Intermolecular distance within a layer, *b* in Figure 9.10 / pm | Intermolecular distance between layers / pm | Intramolecular distance for molecule in the gaseous state / pm |
|---|---|---|---|---|---|---|---|---|
| Chlorine | 171.5 | 6.4 | 99 | 180 | 198 | 332 | 374 | 199 |
| Bromine | 265.8 | 10.6 | 114 | 195 | 227 | 331 | 399 | 228 |
| Iodine | 386.7 | 15.5 | 133 | 215 | 272 | 350 | 427 | 267 |

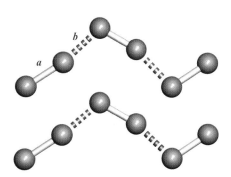

**Fig. 9.10** The solid state structure of $Cl_2$, $Br_2$ and $I_2$ consists of $X_2$ (X = Cl, Br or I) molecules arranged in zigzag chains within layers. Part of one layer is shown. Values of the intramolecular X–X distance, *a*, and the intermolecular distance, *b*, are listed in Table 9.1.

which are particularly informative. Consider the structure of solid $Cl_2$. Within a layer (part of which is shown in Figure 9.10), the *intra*molecular Cl–Cl distance is 198 pm (*a* in Figure 9.10). The measured Cl–Cl bond distance is twice the covalent radius (Table 9.1). Also, within a plane, we can measure *inter*molecular Cl···Cl distances, and the shortest such distance (*b* in Figure 9.10) is 332 pm. This is shorter than twice the van der Waals radius of chlorine and suggests that there is some degree of interaction between the $Cl_2$ molecules in a layer. The shortest *inter*molecular Cl···Cl distance *between* layers of molecules is 374 pm. The degree of intermolecular interaction becomes more pronounced in going from $Cl_2$ to $Br_2$, and from $Br_2$ to $I_2$ as the distances in Table 9.1 indicate. Note also that the I–I bond length in solid $I_2$ is longer than in a gaseous molecule (Table 9.1) although there is little change in either the Cl–Cl or Br–Br bond length in going from gaseous to solid $Cl_2$ or $Br_2$. In solid $I_2$, the bonding interaction between molecules is at the expense of some bonding character within each $I_2$ molecule.

**9.8**    **Elemental molecular solids in groups 15 and 16**

### Sulfur (group 16)

Sulfur forms S–S bonds in a variety of cyclic and chain structures and Table 4.2 listed a range of allotropes of sulfur. One allotrope is $S_6$ which has a cyclic structure with a *chair conformation* (Figure 9.11). When $S_6$ crystallizes, the rings pack together efficiently to give a solid which is the highest density form of elemental sulfur (2.2 g cm$^{-3}$). Only van der Waals forces operate between the rings.

**Fig. 9.11** Two views of an $S_6$ molecule – one view emphasizes the chair conformation of the ring.

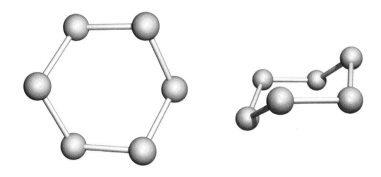

The ***conformation*** of a molecule describes the relative spatial arrangement of the atoms. Two commonly observed conformations of 6-membered rings are the ***chair*** and ***boat*** forms.

        chair                     boat

Chair and boat conformers: see Section 25.2

$r_{cov}(S) = 103\,pm$

Crystalline orthorhombic sulfur (the $\alpha$-form, and the standard state of the element) consists of $S_8$ rings (Figure 9.12) which are packed together with van der Waals interactions between the rings. The average S−S bond length within each ring is 206 pm, consistent with the presence of single bonds. The organization of the $S_8$ rings in the crystalline state is shown in Figure 9.13. The rings do *not* simply stack immediately on top of each other.

Monoclinic sulfur (the $\beta$-form) also contains $S_8$ rings but these are less efficiently packed in the solid state (density $= 1.94\,g\,cm^{-3}$) than are those in orthorhombic sulfur (density $= 2.07\,g\,cm^{-3}$). When orthorhombic sulfur is heated to 368 K, a reorganization of the $S_8$ rings in the lattice occurs and the solid transforms into the monoclinic form. *Single* crystals of orthorhombic sulfur can be rapidly heated to 385 K, when they melt instead of undergoing the orthorhombic to monoclinic transformation. If crystallization takes place at 373 K, the $S_8$ rings adopt the structure of monoclinic

**Fig. 9.12** Two views of an $S_8$ molecule. The shape of the ring is often called a 'crown'. The $Se_8$ molecule also has this geometry.

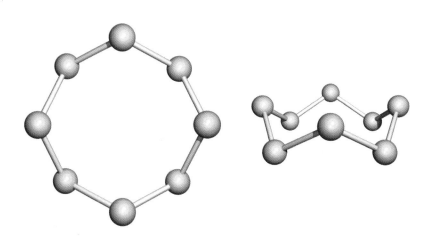

**Fig. 9.13** The arrangement of $S_8$ rings in the solid state of orthorhombic sulfur (the standard state of the element).

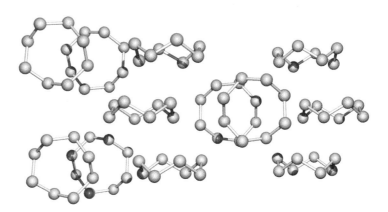

sulfur, but the crystals must be cooled rapidly to 298 K. On standing at room temperature, monoclinic sulfur crystals change into the orthorhombic allotrope within a few weeks.

Most allotropes of sulfur contain cyclic units (Table 4.2) but in some, $S_x$ chains of various lengths are present. Each chain contains S–S single bonds and forms a helix (Figure 9.14). An important property of a helix is its *handedness*. It can turn in either a right-handed or left-handed manner; each form is distinct from the other and they cannot be superimposed.

There are different forms of *polycatenasulfur* which contain mixtures of rings and chains, and these include rubbery and plastic sulfur. Filaments of these can be drawn from molten sulfur; their compositions alter with time, and at 298 K, transformation into orthorhombic sulfur eventually occurs. Two examples of well-characterized allotropes that contain helical chains are fibrous and laminar sulfur. In fibrous sulfur, the chains lie parallel to one another and equal numbers of left- and right-handed helices are present. In laminar sulfur, there is some criss-crossing of the helical chains.

The prefix *catena* is used within the IUPAC nomenclature to mean a chain structure.

### Selenium and tellurium (group 16)

Elemental selenium and tellurium form both rings and helical chains, and selenium possesses several allotropes. Crystalline monoclinic selenium is red and contains $Se_8$ rings with the same crown shape as $S_8$ (Figure 9.12).

**Fig. 9.14** A strand of helical sulfur ($S_\infty$) has a handedness:
(a) a right-handed helix and
(b) a left-handed helix. The two chains are non-superimposable.

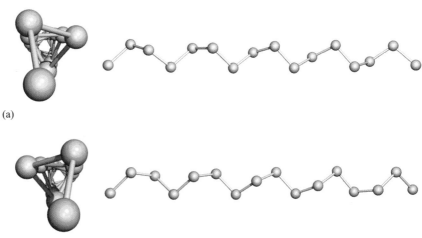

(a)

(b)

➤➤ $Se_x$ or $Te_x$ = chain of infinite length

The standard state of the element is grey (or metallic) selenium, and in the crystalline state it contains helical chains of selenium atoms ($Se_\infty$). Tellurium has one crystalline form and this contains helical $Te_\infty$ chains. In both this and grey selenium, the axes of the chains lie parallel to each other, and a view through each lattice down the axes shows the presence of a hexagonal network.

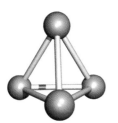

Fig. 9.15 The tetrahedral $P_4$ molecular unit present in white phosphorus. All the P–P distances are equal.

## Phosphorus (group 15)

The standard state of phosphorus is 'white phosphorus'. This allotrope is *not* the thermodynamically most stable state of the element but has been *defined* as the standard state (see Section 2.4). The most stable crystalline form of the element is black phosphorus, and this, and red phosphorus, are described in Section 9.10.

Crystalline white phosphorus contains tetrahedral $P_4$ molecules (Figure 9.15). The intramolecular P–P distance is 221 pm, consistent with the presence of P–P single bonds ($r_{cov} = 110$ pm).

---

**9.9** | **A molecular allotrope of carbon: $C_{60}$**

➤➤ Diamond and graphite: see Section 9.10

The allotropes of carbon that have, in the past, been most commonly cited are diamond and graphite. Since the mid-1980s, new allotropes of carbon – the *fullerenes* – have been recognized.

The fullerenes are discrete molecules, and the most widely studied is $C_{60}$ (Figure 9.16a). The spherical shell of 60 atoms is made up of 5- and 6-membered rings and the carbon atoms are equivalent. Each 5-membered ring (a pentagon) is connected to five 6-membered rings (hexagons). No 5-membered rings are adjacent to each other.

➤➤ Restricted geometry of carbon: see Section 6.14

The geometry about a carbon atom is usually either linear, trigonal planar or tetrahedral, and although apparently complex, the structure of $C_{60}$ complies with this restriction. Each carbon atom in $C_{60}$ is covalently bonded to three others in an approximately trigonal planar arrangement. Since the surface of the $C_{60}$ molecule is relatively large, the deviation from planarity at each carbon centre is small. The C–C bonds in $C_{60}$ fall into two groups – the bonds at the junctions of two hexagonal rings (139 pm) and those at the junctions of a hexagonal and a pentagonal ring (145 pm). Figure 9.16b shows the usual representation of $C_{60}$, with carbon–carbon double and single bonds.

Fig. 9.16 One of the fullerenes – $C_{60}$. (a) The $C_{60}$ molecule is made up of fused 5- and 6-membered rings of carbon atoms which form an approximately spherical molecule. (b) A representation of $C_{60}$ showing only the upper surface (in the same orientation as in (a)) illustrating the localized single and double carbon–carbon bonds.

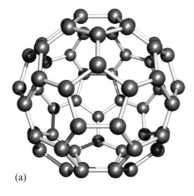

(a)

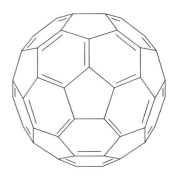

(b)

## THEORETICAL AND CHEMICAL BACKGROUND

### Box 9.3 Why the name fullerene?

The geodesic dome housing Vancouver's interactive Science World. The dome was designed for Expo '86.
*Peter Wilson © Dorling Kindersley.*

The architect Richard Buckminster Fuller has designed geodesic domes such as the one on the left and the one built at EXPO '67 in Montreal. This geodesic dome was constructed of hexagonal motifs, but on its own the network can only lead to a planar sheet. The placement of pentagonal panels at intervals in the structure leads to a curvature of the surface, sufficient to construct a dome. The structure of $C_{60}$ is also represented in a football (soccer ball), which has pentagonal (often black) and hexagonal (often white) panels. $C_{60}$ – buckminsterfullerene – has also been christened 'bucky-ball'.

The name 'fullerene' has been given to the class of near-spherical $C_n$ allotropes which include $C_{60}$, $C_{70}$ and $C_{84}$.

A very readable article that conveys the excitement of the discovery of $C_{60}$ has been written by Harold W. Kroto (1992): '$C_{60}$: Buckminsterfullerene, The Celestial Sphere that Fell to Earth', *Angewandte Chemie, International Edition*, vol. 31, p. 111.

In October 1996, Sir Harry Kroto and Professors Richard Smalley and Robert Curl were awarded the Nobel Prize for Chemistry for their pioneering work on $C_{60}$.

When crystals of a substance are grown from a solution, they may contain *solvent of crystallization*, the presence of which is indicated in the molecular formula.

In the solid state at 298 K, the spherical $C_{60}$ molecules are arranged in a close-packed structure. However, most single crystal X-ray diffraction studies of $C_{60}$ have involved solvated samples rather than the pure solid element. For example, $C_{60}$ is soluble in benzene ($C_6H_6$), and single crystals grown by evaporating solvent from a solution of $C_{60}$ in benzene have the composition $C_{60} \cdot 4C_6H_6$. Figure 9.17 shows part of the crystalline structure of $C_{60} \cdot 4C_6H_6$. The $C_{60}$ molecules are arranged in an ordered manner with the benzene molecules occupying the spaces between them.

**Fig. 9.17** Part of the solid state structure of $C_{60} \cdot 4C_6H_6$. The $C_{60}$ molecules form an ordered array with the benzene molecules in between them. The formula $C_{60} \cdot 4C_6H_6$ indicates that one mole of $C_{60}$ crystallizes with four moles of benzene; this ratio is apparent in the diagram. The structure was determined by X-ray diffraction at 104 K [H. B. Burgi *et al.* (1994) *Chem. Mater.*, vol. 6, p. 1325].

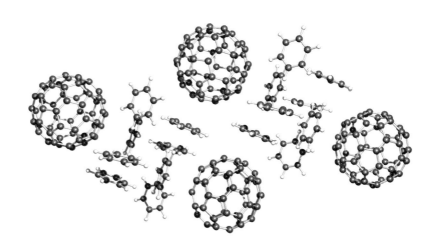

**THEORETICAL AND CHEMICAL BACKGROUND**

## Box 9.4 Superconductivity: alkali metal fullerides $M_3C_{60}$

Alkali metals, M, reduce $C_{60}$ to give fulleride salts of type $[M^+]_3[C_{60}]^{3-}$ and at low temperatures, some of these compounds become *superconducting*. A superconductor is able to conduct electricity without resistance and, until 1986, no compounds were known that were superconductors above 20 K. The temperature at which a material becomes superconducting is called its critical temperature ($T_c$), and in 1987, this barrier was broken – *high-temperature superconductors* were born. Many high-temperature superconductors are metal oxides, for example $YBa_2Cu_4O_8$ ($T_c = 80$ K), $YBa_2Cu_3O_7$ ($T_c = 95$ K), $Ba_2CaCu_2Tl_2O_8$ ($T_c = 110$ K) and $Ba_2Ca_2Cu_3Tl_2O_{10}$ ($T_c = 128$ K).

The $M_3C_{60}$ fulleride superconductors are structurally simpler than the metal oxide systems, and can be described in terms of the alkali metal cations occupying the interstitial holes in a lattice composed of close-packed $C_{60}$ cages. Each $[C_{60}]^{3-}$ anion is approximately spherical and a close-packing of spheres approach is valid. In $K_3C_{60}$ and $Rb_3C_{60}$, the $[C_{60}]^{3-}$ cages are arranged in a face-centred cubic (fcc) arrangement:

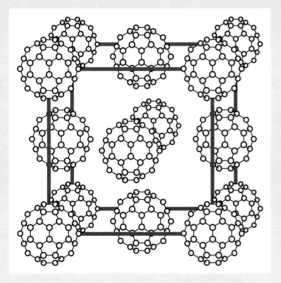

If you look back at Box 9.1, you will see that the fcc unit cell contains an octahedral hole and eight tetrahedral holes. There are also 12 octahedral holes shared between adjacent unit cells. The alkali metal cations in $K_3C_{60}$ and $Rb_3C_{60}$ completely occupy the octahedral (grey) and tetrahedral (red) holes:

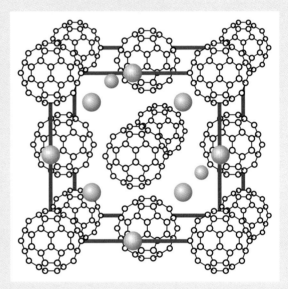

The values of $T_c$ for $K_3C_{60}$ and $Rb_3C_{60}$ are 18 K and 28 K respectively, but for $Cs_3C_{60}$ (in which the $C_{60}$ cages adopt a body-centred cubic lattice), $T_c = 40$ K. $Cs_3C_{60}$ is (at present) the highest temperature superconductor of this family of alkali metal fullerides. [What kind of interstitial holes can the $Cs^+$ ions occupy in the bcc lattice?] $Na_3C_{60}$ is structurally related to its potassium and rubidium analogues, but it is not superconducting. This area of fullerene chemistry is actively being pursued with hopes of further raising the $T_c$ barrier.

A series of well-illustrated articles describing various aspects of superconductivity can be found in *Chemistry in Britain* (1994), vol. 30, pp. 722–748.

---

### 9.10   Solids with infinite covalent structures

Some non-metallic elements in the *p*-block crystallize with *infinite covalent* structures. Diamond and graphite are well-known examples and are described below along with allotropes of boron, silicon, phosphorus, arsenic and antimony. When these elements melt, *covalent bonds* are broken.

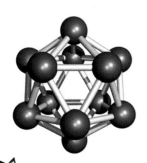

**Fig. 9.18** The $B_{12}$-icosahedral unit is the fundamental building block in both α- and β-rhombohedral boron. These allotropes possess infinite covalent structures in the solid state.

# Boron (group 13)

The standard state of boron is the β-rhombohedral form. The structure of this allotrope is complex and we begin the discussion instead with α-rhombohedral boron. The basic building-block of both α- and β-rhombohedral boron is an icosahedral $B_{12}$-unit (Figure 9.18). Each boron atom is covalently bonded to another five boron atoms within the icosahedron, despite the fact that a boron atom has only three valence electrons. The bonding within each $B_{12}$-unit is *delocalized* and it is important to remember that the B−B connections in Figure 9.18 are *not* 2-centre 2-electron bonds.

The structure of α-rhombohedral boron consists of $B_{12}$-units arranged in an approximately cubic close-packed manner. The boron atoms of the icosahedral unit lie on a spherical surface, and so the close-packing of spheres is an appropriate way in which to describe the solid state structure. However, unlike the close-packed arrays described earlier, the 'spheres' in α-rhombohedral boron are *covalently linked* to each other. Part of the structure (one layer of the infinite lattice) is shown in Figure 9.19, and such layers are arranged in an ABCABC... fashion (Figures 9.3 to 9.5). The presence of B−B covalent bonding interactions between the $B_{12}$-units distinguishes this as an infinite covalent lattice rather than a true close-packed assembly.

The structure of β-rhombohedral boron consists of $B_{84}$-units, linked together by $B_{10}$-units. Each $B_{84}$-unit is conveniently described in terms of three sub-units – $(B_{12})(B_{12})(B_{60})$. At the centre of the $B_{84}$-unit is a $B_{12}$-icosahedron (Figure 9.20a) and radially attached to each boron atom is another boron atom (Figure 9.20b). The term 'radial' is used to signify that the bonds that connect the second set of 12 boron atoms to the central $B_{12}$-unit point outwards from the centre of the unit. The $(B_{12})(B_{12})$ sub-unit so-formed lies inside a $B_{60}$-cage which has the same geometry as a $C_{60}$ molecule (compare Figures 9.20c and 9.16a). The whole $B_{84}$-unit is shown in Figure 9.20d. Notice that each of the boron atoms of the second $B_{12}$-unit is connected to a pentagonal ring of the $B_{60}$-sub-unit. In the 3-dimensional structure of β-rhombohedral boron, the $B_{84}$-units shown in

**Fig. 9.19** Part of a layer of the infinite lattice of α-rhombohedral boron. The building-blocks are $B_{12}$-icosahedra. The overall structure may be considered to consist of spheres in a cubic close-packed arrangement. Delocalized, covalent bonding interactions between the $B_{12}$-units support the framework of the infinite structure, making it rigid.

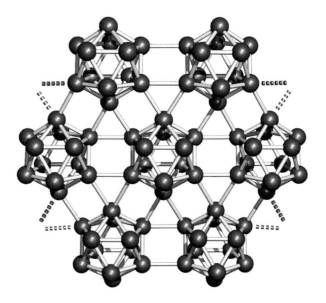

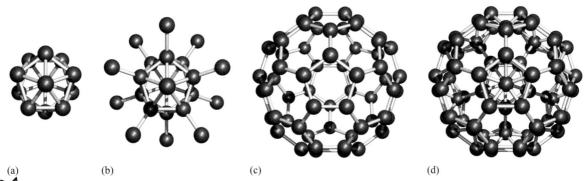

(a)    (b)    (c)    (d)

**Fig. 9.20** The construction of the $B_{84}$-unit that is the main building block in the 3-dimensional structure of β-rhombo-hedral boron. (a) In the centre of the unit is a $B_{12}$-icosahedron. (b) To each boron atom in the central icosahedron, another boron atom is attached. (c) A $B_{60}$-cage is the outer 'skin' of the $B_{84}$-unit. (d) The final $B_{84}$-unit contains three covalently bonded sub-units – $(B_{12})(B_{12})(B_{60})$.

Figure 9.20d are interconnected by $B_{10}$-units. The whole network is extremely rigid, as is that of the α-rhombohedral allotrope, and crystalline boron is very hard, and has a particularly high melting point (2453 K for β-rhombohedral boron).

## Carbon (group 14)

Diamond and graphite are allotropes of carbon and possess 3-dimensional covalent structures in the solid state. They differ remarkably in their physical appearance and properties. Crystals of diamond are transparent, extremely hard and are highly prized for jewellery. Crystals of graphite are black and have a slippery feel.

Figure 9.21 shows part of the crystalline structure of diamond. Each tetrahedral carbon atom forms four single C−C bonds and the overall lattice is very rigid.

The standard state of carbon is graphite. Strictly, this is α-graphite, since there is another allotrope called β-graphite. Both α- and β-graphite have 3-dimensional structures, composed of parallel planes of fused hexagonal rings. In α-graphite, the assembly contains two repeating layers, whereas there are three repeat units in the β-form. Heating β-graphite above 1298 K brings about a change to α-graphite.

The structure of α-graphite ('normal' graphite) is shown in Figure 9.22. The C−C bond distances *within* a layer are equal (142 pm) and the distance between two adjacent layers is 335 pm, indicating that *inter*-layer interactions are weak. The physical nature of graphite reflects this; it cleaves readily *between* the planes of hexagonal rings. This property allows graphite to be used as a lubricant. The carbon atoms in normal graphite are less efficiently packed than in diamond; the densities of α-graphite and diamond are 2.3 and 3.5 g cm$^{-3}$ respectively.

The electrical conductivity of graphite is an important property of this allotrope, and can be explained in terms of the structure and bonding. The valence electronic configuration of carbon is $[He]2s^2 2p^2$, and as each carbon atom forms a covalent bond to each of three other atoms in the same layer, one valence electron remains unused. The odd electrons are

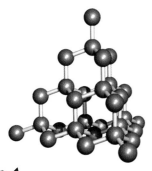

**Fig. 9.21** Part of the infinite covalent lattice of diamond, an allotrope of carbon.

$r_{cov}(C) = 77$ pm
$r_v(C) = 185$ pm

**THEORETICAL AND CHEMICAL BACKGROUND**

## Box 9.5 The relationship between the structure of diamond and zinc blende

The structure of zinc blende (ZnS) was discussed in Section 8.13, and the unit cell was shown in Figure 8.17. If all the zinc(II) and sulfide centres in this structure are replaced by carbon atoms, the unit cell shown below results:

Figure 9.21 becomes apparent:

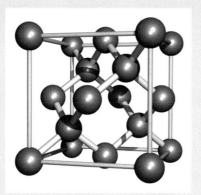

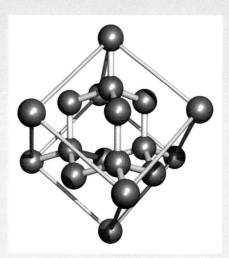

If we view this unit cell from a different angle, the same representation of the diamond lattice that we showed in

We can therefore see that the diamond and zinc blende structures are related.

conducted through the planes of hexagonal rings. The electrical resistivity of α-graphite is $1.3 \times 10^{-5} \, \Omega\,m$ at 293 K in a direction through the planes of hexagonal rings, but is about $1 \, \Omega\,m$ in a direction perpendicular to the planes. The electrical resistivity of diamond is $1 \times 10^{11} \, \Omega\,m$, and diamond is an excellent *insulator*. All valence electrons in diamond are involved in localized C−C single bond formation.

**Fig. 9.22** Part of the layer structure of normal (α) graphite. This allotrope is the standard state of carbon. There are two repeating layers consisting of fused hexagonal rings. The red lines between the layers indicate which carbon atoms lie directly over which other atoms.

Layer I

Layer II

Layer I

Layer II

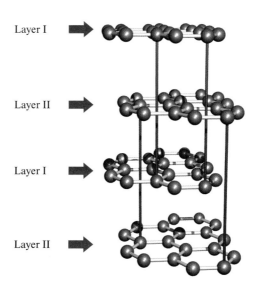

See also Section 9.13

The electrical resistivity of a material measures its resistance (to an electrical current). A good electrical conductor has a very low resistivity, and the reverse is true for an insulator.

For a wire of uniform cross section, the resistivity ($\rho$) is given in units of ohm metre ($\Omega\,m$) where the resistance is in ohms, and the length of the wire is measured in metres.

$$\text{Resistance (in } \Omega) = \frac{\text{resistivity (in } \Omega\,m) \times \text{length of wire (in m)}}{\text{cross section (in m}^2)}$$

$$R = \frac{\rho \times l}{a}$$

## Silicon, germanium and tin (group 14)

In the solid state at 298 K silicon crystallizes with a diamond-type lattice (Figure 9.21). Germanium and the grey allotrope of tin also adopt this infinite structure but the character of these elements puts them into the category of *semi-metals* rather than non-metals. We discuss these allotropes further in Section 9.11, but it is instructive here to compare some of their physical properties (Table 9.2), particularly their electrical resistivities, to understand why a distinction is made between the group 14 elements.

## Phosphorus, arsenic, antimony and bismuth (group 15)

White phosphorus: see Section 9.8

Although it is defined as the standard state of the element, white phosphorus is a metastable state. Other allotropes of phosphorus are either amorphous, or possess infinite covalent lattices, but on melting, all allotropes give a liquid containing $P_4$ molecules.

Amorphous red phosphorus is more dense and less reactive than white phosphorus. Crystallization of red phosphorus in molten lead produces a monoclinic allotrope called Hittorf's (violet) phosphorus. The solid state structure is a complicated 3-dimensional network and two views of the repeat unit are shown in Figure 9.23. Units of this type are connected end-to-end to form chain-like arrays of 3-coordinate phosphorus atoms.

**Table 9.2** Some physical and structural properties of diamond, silicon, germanium and grey tin. All share the same 3-dimensional structure type (Figure 9.21).

| Element | Appearance of crystalline solid | Melting point / K[a] | Enthalpy of fusion / kJ mol$^{-1}$ | Density / g cm$^{-3}$ | Interatomic distance in the crystal lattice / pm | Electrical resistivity / $\Omega\,m$ (temperature) |
|---|---|---|---|---|---|---|
| Carbon (diamond) | Transparent | 3820 | 105 | 3.5 | 154 | $1 \times 10^{11}$ (293 K) |
| Silicon | Blue-grey, lustrous | 1683 | 40 | 2.3 | 235 | $1 \times 10^{-3}$ (273 K) |
| Germanium | Grey-white, lustrous | 1211 | 35 | 5.3 | 244 | 0.46 (295 K) |
| Tin (grey allotrope) | Dull grey | – | – | 5.75 | 280 | $11 \times 10^{-8}$ (273 K) |

[a] The grey allotrope of tin is the low-temperature form of the element. Above 286 K the white form is stable; this melts at 505 K.

**Fig. 9.23** Part of the infinite lattice of Hittorf's phosphorus. (a) Part of the chain-like arrays of atoms; the repeat unit contains 21 atoms, and atoms P′ and P″ are equivalent atoms in adjacent chains; the chains are linked through the P′–P″ bond. (b) The same unit viewed from the end, emphasizing the channels that run through the structure.

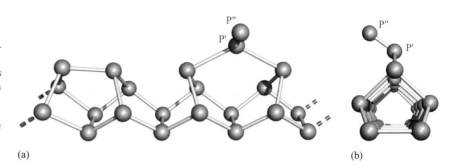

(a)　　　　　　　　　　　　　　　　　　　　　(b)

These chains lie parallel to each other, forming layers, but within a layer the chains are not bonded together. A 3-dimensional network is created by placing the layers one on top of another, such that an atom of the type labelled P′ in Figure 9.23 is covalently bonded to another similar atom (labelled P″) in an adjacent sheet. The chains in one layer lie at right angles to the chains in the next bonded layer, giving a criss-cross network overall. The P–P bond distances in the lattice are all similar ($\approx$222 pm) and are consistent with single covalent bonds.

Black phosphorus is the most thermodynamically stable form of the element. The solid state structure of the rhombohedral form of black phosphorus consists of layers of 6-membered $P_6$-rings each with a chair conformation (Figure 9.24). Hexagonal $P_6$-units are also present in the orthorhombic form of the element.

In the solid state, the rhombohedral allotropes of arsenic, antimony and bismuth are isostructural with the rhombohedral form of black phosphorus. Down group 15, there is a tendency for the coordination number of each atom to change from three (atoms within a layer) to six (three atoms within a layer and three in the next layer). These allotropes of arsenic, antimony and bismuth are known as the 'metallic' forms, and the metallic character of the element increases as group 15 is descended.

**Fig. 9.24** Layers of puckered 6-membered rings are present in the structures of black phosphorus and the rhombohedral allotropes of arsenic, antimony and bismuth, all group 15 elements.

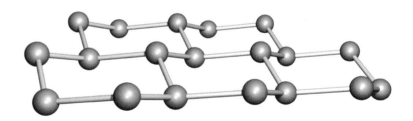

**9.11**　　**The structures of metallic elements at 298 K**

Elements in groups 1 and 2 (the *s*-block) and the *d*-block are metallic. In the *p*-block, a diagonal line *approximately* separates non-metallic from metallic elements (Figure 9.25) although the distinction is *not* clear-cut. The solid state structures of metals are readily described in terms of the packing of their atoms. Table 9.3 lists lattice types for metals of the *s*- and *d*-blocks, and also gives the melting points of these elements.

**Fig. 9.25** A 'diagonal' line is often drawn through the *p*-block to indicate *approximately* the positions of the non-metals (shown in green) and the metals (shown in pink). The distinction is not clear-cut and elements lying along the line may show the characteristics of both, being classed as semi-metals.

| Group 13 | Group 14 | Group 15 | Group 16 | Group 17 | Group 18 |
|---|---|---|---|---|---|
| B | C | N | O | F | Ne |
| Al | Si | P | S | Cl | Ar |
| Ga | Ge | As | Se | Br | Kr |
| In | Sn | Sb | Te | I | Xe |
| Tl | Pb | Bi | Po | At | Rn |

## *s*-Block metals

In the solid state at 298 K, the atoms of each alkali metal (group 1) are packed in a body-centred cubic arrangement (Figure 9.8). All these metals are soft and have relatively low melting points (Table 9.3). The enthalpies of fusion are correspondingly low, decreasing down group 1 from $3.0 \, \text{kJ mol}^{-1}$ for lithium to $2.1 \, \text{kJ mol}^{-1}$ for caesium.

With the exception of barium which has a bcc lattice, the group 2 metals possess hexagonal close-packed structures, and their melting points are higher than those of the group 1 metals.

## *d*-Block metals

At 298 K, the structures of the metals in the *d*-block are either hcp, ccp or bcc, with the exceptions of mercury and manganese (see below). Table 9.3 shows that the structure type adopted depends, in general, on the position of the metal in the periodic table. For most of the *d*-block metals, a close-packed structure is observed at 298 K, and a bcc structure is present as a high-temperature form of the element. Iron is unusual in that it adopts a bcc structure at 298 K, transforms to an fcc lattice at 1179 K, and reverts to a bcc structure at 1674 K.

**Table 9.3** Structures (at 298 K) and melting points (K) of the metallic elements: ▪, hexagonal close-packed; ▪, cubic close-packed (face-centred cubic); ▪, body-centred cubic.

| 1 | 2 | 3 | 4 | 5 | 6 | 7 | 8 | 9 | 10 | 11 | 12 |
|---|---|---|---|---|---|---|---|---|---|---|---|
| Li | Be | | | | | | | | | | |
| 454 | 1560 | | | | | | | | | | |
| Na | Mg | | | | | | | | | | |
| 371 | 923 | | | | | | | | | | |
| K | Ca | Sc | Ti | V | Cr | Mn see text | Fe | Co | Ni | Cu | Zn |
| 337 | 1115 | 1814 | 1941 | 2183 | 2180 | 1519 | 1811 | 1768 | 1728 | 1358 | 693 |
| Rb | Sr | Y | Zr | Nb | Mo | Tc | Ru | Rh | Pd | Ag | Cd |
| 312 | 1050 | 1799 | 2128 | 2750 | 2896 | 2430 | 2607 | 2237 | 1828 | 1235 | 594 |
| Cs | Ba | La | Hf | Ta | W | Re | Os | Ir | Pt | Au | Hg see text |
| 301 | 1000 | 1193 | 2506 | 3290 | 3695 | 3459 | 3306 | 2719 | 2041 | 1337 | 234 |

The melting points of the metals in groups 3 to 11 are far higher than those of the *s*-block metals, and enthalpies of fusion are correspondingly higher. Two *d*-block elements are worthy of special note. The first is mercury. All three metals in group 12 stand out among the *d*-block elements in possessing relatively low melting points, but mercury is well known for the fact that it is a liquid at 298 K. Its enthalpy of fusion is 2.3 kJ mol$^{-1}$, a value that is atypical of the *d*-block elements but is, rather, similar to those of the alkali metals. In the crystalline state, mercury atoms are arranged in a distorted simple cubic lattice (Figure 9.7). The second metal of interest is manganese. Atoms in the solid state are arranged in a complex cubic lattice in such a way that there are four atom types with coordination numbers of 12, 13 or 16. The reasons for this deviation from one of the more common structure types are not simple to understand.

## Metals and semi-metals in the *p*-block

The physical and chemical properties of the heavier elements in groups 13, 14 and 15 indicate that these elements are metallic. Elements intermediate between metals and non-metals are termed semi-metals, for example germanium. The structures described below are those observed at 298 K.

Aluminium possesses a cubic close-packed lattice, typical of a metallic element. The melting point (933 K) is only slightly higher than that of magnesium, the element preceding it in the periodic table, and is dramatically lower than that of boron (2453 K). Atoms of thallium form an hcp structure typical of a metal, and indium possesses a distorted cubic close-packed arrangement of atoms. The solid state structure of gallium is not so easily described; there is one nearest-neighbour (249 pm) and six other close atoms at distances between 270 and 279 pm. Gallium has a low melting point (303 K) which means that it is a liquid metal in some places in the world but a solid in others! The crystalline state of gallium is in between that of a metal and a molecular solid containing Ga$_2$ units.

Elemental boron: see Section 9.10

The 'diagonal' line in Figure 9.25 passes through group 14 between silicon and germanium, suggesting that we might consider silicon to be a non-metal, and germanium a metal. But the distinction is not clear-cut. In the solid state, both elements have the same 3-dimensional structure as diamond, but their electrical resistivities are significantly lower than that of diamond (Table 9.2), indicating metallic behaviour. The heaviest element, lead, possesses a ccp lattice. The intermediate element is tin. White (β) tin is the stable allotrope at 298 K, but at temperatures below 286 K, this transforms into the grey α-form which has a diamond-type lattice (Table 9.2). The structure of white tin is related to that of the grey allotrope by a distortion of the lattice such that each tin atom goes from having four to six nearest-neighbours. The density of white tin (7.31 g cm$^{-3}$) is greater than that of the grey allotrope (5.75 g cm$^{-3}$). Tin is an unusual metal: the density *decreases* on going from β- to α-Sn, whereas it is more usual for there to be an *increase* in density on going from a higher to lower temperature polymorph. The transition from white to grey tin is quite slow, but it can be dramatic. In the 19th century, military uniforms used tin buttons, which crumbled in exceptionally cold winters. Similarly, in 1851, the citizens of Zeitz were alarmed to discover that the tin organ pipes in their church had crumbled to powder!

The structure of bismuth was described in Section 9.10.

| | 9.12 | **Metallic radius** |

▣▷
Van der Waals radius:
see Section 3.21
Covalent radius:
see Section 4.3
Ionic radius:
see Section 8.14

The *metallic radius* is half the distance between the *nearest-neighbour* atoms in a solid state metallic lattice. Table 9.4 lists metallic radii for the *s*- and *d*-block elements. Atom size increases down each of groups 1 and 2. In each of the triads of the *d*-block elements, there is generally an increase in radius in going from the first to second row element, but very little change in size in going from the second to third row metal. This latter observation is due to the presence of a filled 4*f* level, and the so-called *lanthanoid contraction* – the first row of lanthanoid elements lies between lanthanum (La) and hafnium (Hf). The poorly shielded 4*f* electrons are relatively close to the nucleus and have little effect on the observed radius.

> The *metallic radius* is half the distance between the *nearest-neighbour* atoms in a solid state metal lattice.

**Table 9.4** Metallic radii (pm) of the *s*- and *d*-block metals; lanthanum (La) is usually classified with the *f*-block elements.

| 1 | 2 | 3 | 4 | 5 | 6 | 7 | 8 | 9 | 10 | 11 | 12 |
|---|---|---|---|---|---|---|---|---|---|---|---|
| Li | Be | | | | | | | | | | |
| 157 | 112 | | | | | | | | | | |
| Na | Mg | | | | | | | | | | |
| 191 | 160 | | | | | | | | | | |
| K | Ca | Sc | Ti | V | Cr | Mn | Fe | Co | Ni | Cu | Zn |
| 235 | 197 | 164 | 147 | 135 | 129 | 137 | 126 | 125 | 125 | 128 | 137 |
| Rb | Sr | Y | Zr | Nb | Mo | Tc | Ru | Rh | Pd | Ag | Cd |
| 250 | 215 | 182 | 160 | 147 | 140 | 135 | 134 | 134 | 137 | 144 | 152 |
| Cs | Ba | La | Hf | Ta | W | Re | Os | Ir | Pt | Au | Hg |
| 272 | 224 | 187 | 159 | 147 | 141 | 137 | 135 | 136 | 139 | 144 | – |

| | 9.13 | **Metallic bonding** |

### Metals are electrical conductors

An *electrical conductor* offers a low resistance (measured in ohms, $\Omega$) to the flow of an electrical current (measured in amperes, A). An *insulator* offers a high resistance.

One physical property that characterizes a metal is its low electrical resistivity – that is, a metal conducts electricity very well. With the exception of mercury, all elements that are metallic at 298 K are solid. Although the packing of atoms is a convenient means of describing the solid state structure of metals, it gives no feeling for the communication that there must be between the atoms. Communication is implicit in the property of electrical conductivity, since electrons must be able to flow through an assembly of atoms in a metal. In order to understand why metals are such good electrical conductors, we must first consider the bonding between atoms in a metal.

| **Worked example 9.1** | The electrical resistivity of titanium is $4.3 \times 10^{-7}\,\Omega\,\text{m}$. What is the resistance of a 0.50 m strip of titanium wire with cross section $8.0 \times 10^{-7}\,\text{m}^2$? |

Equation needed:

$$\text{Resistance (in } \Omega) = \frac{\text{resistivity (in } \Omega\,\text{m}) \times \text{length of wire (in m)}}{\text{cross section (in m}^2)}$$

Ensure all units are consistent – in this example, they are.

$$\text{Resistance} = \frac{(4.3 \times 10^{-7}\,\Omega\,\text{m})(0.50\,\text{m})}{(8.0 \times 10^{-7}\,\text{m}^2)}$$

$$= 0.27\,\Omega$$

## A 'sea of electrons'

An early approach to metallic bonding (the Drude–Lorentz theory) was to consider a model in which the valence electrons of each metal atom were free to move in the crystal lattice. Thus, instead of simply being composed of neutral atoms, the metal lattice is considered to be an assembly of positive ions (the nuclei surrounded by their core electrons) and electrons (the valence electrons). When a potential difference[§] is applied across a piece of metal, the valence electrons move from high to low potential and a current flows.

In this model, a metallic element is considered to consist of positive ions (arranged, for example, in a close-packed manner) and a 'sea of electrons'. The theory gives a satisfactory general explanation for the conduction of electricity but cannot account for the detailed variation of electrical conductivities amongst the metallic elements. Several other theories have been described, of which *band theory* is the most general.

## Band theory

Band theory follows from a consideration of the energies of the molecular orbitals of an assembly of metal atoms. In constructing the ground state electronic configuration of a molecule, we have previously applied the *aufbau* principle and have arranged the electrons so as to give the lowest energy state. In the case of a degeneracy, this may give singly occupied highest lying molecular orbitals (for example in the $B_2$ and $O_2$ molecules).

A molecular orbital diagram that describes the bonding in a metallic solid is characterized by having a large number of orbitals which are very close in energy. In this case, they form a continuous set of energy states called a *band*. These arise as follows. In an LCAO approach, we consider the interaction

LCAO: see Section 4.12

between similar atomic orbitals, for example 2s with 2s, and 2p with 2p. Figure 9.26 shows the result of the interaction of 2s atomic orbitals for different numbers of lithium atoms. *The energies of these 2s atomic orbitals are the same.* If two Li atoms combine, the overlap of the two 2s atomic orbitals leads to the formation of two MOs. If there are three lithium atoms, three MOs are formed, and if there are four metal atoms, four MOs result. For an assembly containing *n* lithium atoms, there must be *n* molecular orbitals, but

[§] Ohm's Law states: $V = IR$ (potential difference (in V) = current (in A) × resistance (in $\Omega$)).

**Fig. 9.26** The interaction of two $2s$ atomic orbitals in $Li_2$ leads to the formation of two MOs. If there are three lithium atoms, three MOs are formed, and so on. For $Li_n$, there are $n$ molecular orbitals, but because the $2s$ atomic orbitals were all of the same energy, the energies of these MOs are very close together and are described as a *band* of orbitals.

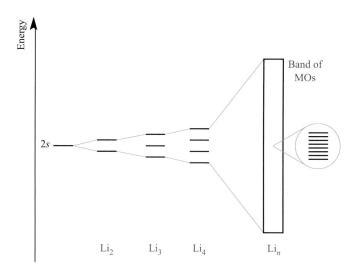

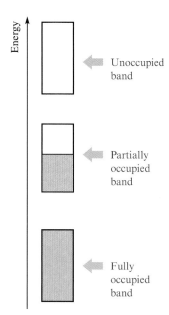

**Fig. 9.27** The molecular orbitals that describe the bonding in a bulk metal are very close in energy and are represented by bands. Bands may be fully occupied with electrons (shown in blue), unoccupied (shown in white) or partially occupied. The figure shows a schematic representation of these bands for a metal.

because the $2s$ atomic orbitals are all of the same energy, the energies of the resultant MOs are very close together and can be described as a *band* of orbitals. The occupation of the band depends upon the number of valence electrons available. Each Li atom provides one valence electron and the band shown in Figure 9.26 is half-occupied. This leads to a delocalized picture of the bonding in the metal, and the metal–metal bonding is *non-directional*.

When metal atoms have more than one type of atomic orbital in the valence shell, correspondingly more bands are formed. If two bands are close together they will overlap, giving a single band in which there is mixed orbital character (for example $s$ and $p$ character). Some bands will be separated from other bands by defined energy gaps, as shown in Figure 9.27. The lowest band is fully occupied with electrons, while the highest band is empty. The central band is partially occupied with electrons, and because the energy states that make up the band are so close, the electrons can move between energy states *in the same band*. In the bulk metal, electrons are therefore mobile. If a potential difference is applied across the metal, the electrons move in a direction from high to low potential and a current flows. The energy gaps *between* bands are relatively large, and it is the presence of a *partially occupied* band that characterizes a metal.

A **band** is a group of MOs that are extremely close in energy. The energy differences are so small that the system behaves as if a continuous, non-quantized variation of energy within the band is possible.

A **band gap** occurs when there is a significant energy difference between two bands. The magnitude of a band gap is typically reported in electron volts (eV); $1\,eV = 96.845\,kJ\,mol^{-1}$.

### Semiconductors

In going down each of groups 13, 14, 15 and 16, there is a transition from a non-metal to a metal (Figure 9.25), passing through some intermediate stage characterized by the *semi-metals*. As we have already seen for the group 14 elements, the metal/non-metal boundary is not well defined.

**Fig. 9.28** The energy difference (the band gap) between occupied (shown in blue) and unoccupied (shown in white) bands of MOs decreases in going down group 14. This allows a change from non-metallic towards metallic character in going down the group.

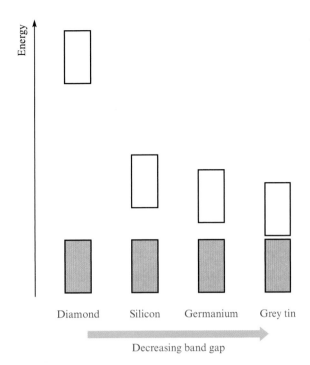

Figure 9.28 gives a representation of the bonding situation for these allotropes. The MO diagram for bulk diamond can be represented in terms of a fully occupied and an unoccupied band. There is a large band gap (5.39 eV) and diamond is an insulator. The situation for silicon and germanium can be similarly represented, but now the band gaps are much smaller (1.10 and 0.66 eV respectively). In grey tin, only 0.08 eV separates the filled and empty bands, and here the situation is approaching that of a single band which is partly occupied. The conduction of electricity in silicon, germanium and grey tin depends upon *thermal population* of the upper band (the *conduction band*) and these allotropes are classed as *semiconductors*. As the temperature is increased, some electrons will have sufficient energy to make the transition from the lower to higher energy band. The smaller the band gap, the greater the number of electrons that will possess sufficient energy to make the transition, and the greater the electrical conductivity.

## SUMMARY

In this chapter, we have discussed the structures of elements in the solid state.

### Do you know what the following terms mean?

- close-packing of spheres
- cubic close-packing
- hexagonal close-packing
- simple cubic lattice
- body-centred cubic lattice
- interstitial hole

- crystalline solid
- amorphous solid
- chair and boat conformers of a 6-membered ring
- handedness of a helical chain
- catena

- electrical resistivity
- insulator
- metallic radius
- metallic bonding
- band gap
- semiconductor

### You should be able:

- to discuss how the close-packing of spheres can give rise to at least two assemblies
- to discuss the relationship between simple and body-centred cubic packing of spheres
- to state how many nearest-neighbours an atom has in ccp, hcp, simple cubic and bcc arrangements
- to compare the efficiencies of packing in ccp, hcp, simple cubic and bcc arrangements
- to appreciate *why* the packing of spheres is an appropriate model for some, but not all, solid state lattices

- to distinguish between intra- and intermolecular bonds in solid state structures (are both types of bonding always present?)
- to describe structural variation among the allotropes of boron, carbon, phosphorus and sulfur
- to describe similarities and differences in the solid state structures of the elements of group 17
- to give examples of metals with different types of lattice structures
- to describe using simple band theory how a metal can conduct electricity

## PROBLEMS

**9.1** When spheres of an equal size are close-packed, what are the features that characterize whether the arrangement is hexagonal or cubic close-packed? Draw a representation of the repeat unit for each arrangement.

**9.2** What is an *interstitial hole*? What types of holes are present in hcp and ccp arrangements?

**9.3** How are spheres organized in a body-centred cubic arrangement? How does the body-centred cubic arrangement differ from a simple cubic one?

**9.4** What is meant by a *nearest-neighbour* in an assembly of spheres? How many nearest-neighbours does each sphere possess in a (a) cubic close-packed, (b) hexagonal close-packed, (c) simple cubic and (d) body-centred cubic arrangement?

**9.5** Write down the ground state electronic configuration of Cl, S and P and use these data to describe the bonding in each of the molecular units present in (a) solid dichlorine, (b) orthorhombic sulfur, (c) fibrous sulfur and (d) white phosphorus.

**9.6** What is meant by electrical resistivity? Use the data in Table 9.5 to discuss the statement: '*All metals are good electrical conductors.*'

**Table 9.5** Table of data for problem 9.6. All the resistivities are measured for pure samples at 273 K unless otherwise stated.

| Element | Electrical resistivity / $\Omega\,m$ |
|---|---|
| Copper | $1.5 \times 10^{-8}$ |
| Silver | $1.5 \times 10^{-8}$ |
| Aluminium | $2.4 \times 10^{-8}$ |
| Iron | $8.6 \times 10^{-8}$ |
| Gallium | $1.4 \times 10^{-7}$ |
| Tin | $3.9 \times 10^{-7}$ |
| Mercury | $9.4 \times 10^{-7}$ |
| Bismuth | $1.1 \times 10^{-6}$ |
| Manganese | $1.4 \times 10^{-6}$ |
| Silicon | $1.0 \times 10^{-3}$ |
| Boron | $1.8 \times 10^{4}$ |
| Phosphorus | $1.0 \times 10^{9}$ (293 K) |

**9.7** Briefly discuss allotropy with respect to carbon. What is the standard state of this element? Explain why α-graphite and diamond show such widely differing electrical resistivities, and suggest why the electrical resistivity in a graphite rod is direction-dependent.

**9.8** What is meant by *solvent of crystallization*? The crystallization of $C_{60}$ from benzene and diiodomethane yields solvated crystals of formula $C_{60}\cdot xC_6H_6\cdot yCH_2I_2$. If the loss of solvent leads to a 32.4% reduction in the molar mass, what is a possible stoichiometry of the solvated compound?

**9.9** What lattice structure is typical of an alkali metal at 298 K? Table 9.6 lists values of the enthalpies of fusion and vaporization for the alkali metals. Describe what is happening in each process in terms of interatomic interactions. Use the data in Table 9.6 to plot graphs which show the trends in melting point, and enthalpies of fusion and atomization down group 1. How do you account for the trends observed, and any relationships that there may be between them?

**Table 9.6** Table of data for problem 9.9.

| Alkali metal | Melting point / K | Enthalpy of fusion / kJ mol$^{-1}$ | Enthalpy of atomization / kJ mol$^{-1}$ |
|---|---|---|---|
| Lithium | 454 | 3.0 | 162 |
| Sodium | 371 | 2.6 | 108 |
| Potassium | 337 | 2.3 | 90 |
| Rubidium | 312 | 2.2 | 82 |
| Caesium | 301 | 2.1 | 78 |

**9.10** Comment on the *relative* values of the metallic (197 pm) and ionic radii (100 pm) of calcium.

**9.11** Account for the following observations:
(a) The density of α-graphite is less than that of diamond.

(b) In group 13, the melting point of β-rhombohedral B is much higher (2453 K) than that of Al (933 K).

(c) The group 1 metals tend to exhibit lower values of $\Delta_a H^\circ(298\,K)$ than metals in the *d*-block.

**9.12** To what processes do the values of (a) $\Delta_{fus}H = 0.7\,kJ\,mol^{-1}$, (b) $\Delta_{vap}H = 5.6\,kJ\,mol^{-1}$ and (c) $\Delta_a H^\circ = 473\,kJ\,mol^{-1}$ for nitrogen refer? Discuss the relative magnitudes of the values.

**9.13** A localized covalent σ-bond is directional, while metallic bonding is non-directional. Discuss the features of covalent and metallic bonding that lead to this difference.

**9.14** Using simple band theory, describe the differences between electrical conduction in a metal such as lithium and in a semiconductor such as germanium.

**9.15** (a) What is a *band gap*? (b) Which of the following would you associate with a metal, a semiconductor and an insulator: (i) a large band gap, (ii) a very small band gap, (iii) a partially occupied band?

## Additional problems

**9.16** $ReO_3$ is a structure type. The structure can be described as a simple cubic array of Re atoms with O atoms located in the middle of each of the cube edges. Construct a unit cell of $ReO_3$. What are the coordination numbers of Re and O? Confirm that your unit cell gives the correct stoichiometry for $ReO_3$.

**9.17** Two types of semiconductors, n- and p-types, are made by *doping* a host such as silicon with a small amount of an element that has more or fewer valence electrons than the host. How do you think doping Si with As to give an n-type semiconductor would change the electronic conductivity of the material?

# 10 Mass spectrometry

## Topics

- Electron impact mass spectrometry
- Isotope abundances
- Parent ions and fragmentations
- Case studies

## 10.1 Introduction

This chapter is the first of five concerned with common laboratory techniques used in the identification of chemical compounds. Our emphasis is on spectroscopic and mass spectrometric methods. The interpretation of data forms a major part of the discussion since this is where many students gain their initial experience of experimental techniques, rather than in the detail of instrument operation. Nonetheless, for a proper understanding of how to interpret spectra, some knowledge of theory is necessary.

We begin our introduction to experimental methods by considering *mass spectrometry*. Most routinely, mass spectrometry is used to determine the molecular weight of a compound. It also provides information about isotopic abundances, and the ways in which molecular ions decompose. A number of different mass spectrometric techniques are available, including electron impact ionization (EI), chemical ionization (CI), fast-atom bombardment (FAB), matrix assisted laser desorption ionization time-of-flight (MALDI-TOF) and electrospray (ES) methods. Different techniques have different advantages over others and a modern chemical laboratory makes use of the range of techniques to tackle different problems. For example, electrospray mass spectrometry is a 'soft' technique, whereas electron impact ionization mass spectrometry often results in fragmentation of the molecular sample under study. Fast-atom bombardment mass spectrometry can be applied to both neutral and ionic compounds, whereas EI mass spectrometry is used to investigate neutral compounds. If a high molecular weight compound (e.g. a high mass polymer) is under study, EI mass spectrometry is not generally suitable, and one of the more modern methods must be applied. Mass spectrometry operates by determining the charge-to-mass ratio for gas phase ions, and the various techniques are different methods of generating these ions.

## 10.2 Recording a mass spectrum

Despite some limitations, EI mass spectrometry remains in routine use and

**Fig. 10.1** Schematic representation of an EI mass spectrometer. The detector is connected to a recorder which outputs a mass spectrum of the type shown in Figure 10.2.

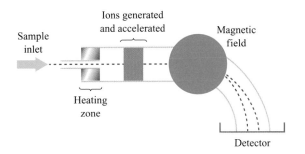

its simplicity of method makes it suitable for an introductory, general discussion of mass spectrometry.

### Electron impact ionization (EI) mass spectrometry

Figure 10.1 shows a schematic representation of an EI mass spectrometer. The sample to be analysed is first vaporized by heating (unless the sample is already in the vapour phase at 298 K) and is then ionized. Ionization takes place with a beam of high-energy ($\approx$70 eV) electrons. Equations 10.1 and 10.2 summarize these first two steps; in equation 10.2, the incoming electron has high energy and the outgoing electrons have a lower energy.

$$E(s) \longrightarrow E(g) \tag{10.1}$$

$$E(g) + e^- \longrightarrow E^+(g) + 2e^- \tag{10.2}$$

The $E^+$ ions pass through a magnetic field where their paths are deflected. The amount of deflection is mass-dependent. The output from the detector (Figure 10.1) is called a *mass spectrum*, and is a plot of signal intensity against mass:charge ($m/z$) ratio. For an ion with $z = 1$, the $m/z$ value corresponds to the molecular mass, $M_r$, of the ion. If $z = 2$, the recorded mass of the ion is half its actual mass, and so on. In EI mass spectrometry, most ions generated in the mass spectrometer have $z = 1$, but in other techniques, molecular ions with $z > 1$ may be common.

In the mass spectrum, the signal intensity is usually plotted in terms of *relative* values, with the most intense signal being arbitrarily assigned a value of 100%. A peak corresponding to the molecular mass of the compound is called the *parent peak*, *parent ion* ($P^+$) or *molecular ion* ($M^+$).

### High- and low-resolution mass spectra

Mass spectrometric measurements may be made at low or high resolution. In a low-resolution mass spectrum, integral $m/z$ values are recorded, and this is satisfactory for most purposes of compound identification. However, it may be necessary to distinguish between two species with very similar masses, and in this case accurate masses (high resolution) are recorded. For example, in a low-resolution mass spectrum, CO and $N_2$ both give rise to peaks at $m/z = 28$. Accurate mass numbers (to five decimal places) for $^{12}C$, $^{16}O$ and $^{14}N$ are 12.000 00 (by definition), 15.994 91 and 14.003 07 respectively, and in a high-resolution mass spectrum, a peak at $m/z = 27.994 91$ can be assigned to CO while one at 28.006 14 is assigned to $N_2$. Accurate mass measurements are important as a means of distinguishing between compound compositions of similar masses, particularly for compounds containing

**Table 10.1** Exact masses (to five decimal places) for selected isotopes. The mass for $^{12}C$ is exactly 12 by definition.

| Isotope | Exact mass | | Isotope | Exact mass |
|---------|-----------|---|---------|-----------|
| $^{1}H$ | 1.007 83 | | $^{28}Si$ | 27.976 93 |
| $^{2}H$ | 2.014 10 | | $^{31}P$ | 30.973 76 |
| $^{7}Li$ | 7.016 00 | | $^{32}S$ | 31.972 07 |
| $^{12}C$ | 12.000 00 | | $^{35}Cl$ | 34.968 85 |
| $^{13}C$ | 13.003 35 | | $^{37}Cl$ | 36.965 90 |
| $^{14}N$ | 14.003 07 | | $^{39}K$ | 38.963 71 |
| $^{16}O$ | 15.994 91 | | $^{79}Br$ | 78.918 34 |
| $^{19}F$ | 18.998 40 | | $^{81}Br$ | 80.916 29 |
| $^{23}Na$ | 22.989 77 | | $^{127}I$ | 126.904 47 |

only C, H, N and O. Measurements need to be to an accuracy of $\approx 7$ significant figures for differences to be unambiguously detected. Exact masses of selected isotopes are listed in Table 10.1. Observed isotope distributions are another method of distinguishing between peaks of the same or very similar $m/z$ value (see below).

See end of chapter problems 10.2 and 10.3

For the rest of this chapter, we work with data from low-resolution mass spectra.

## 10.3    Isotope distributions

### Elements: atoms and molecules

In Section 1.7, we described how some elements occur naturally as mixtures of *isotopes*. Chlorine, for example, occurs as a mixture of 75.77% $^{35}Cl$ and 24.23% $^{37}Cl$. The $3:1$ ratio of $^{35}Cl:^{37}Cl$ is revealed in a mass spectrum of atomized chlorine (Figure 1.4a); the relative intensities of the lines in Figure 1.4a are $100.00:31.98$ (rather than $75.77:24.23$) because the most intense signal (the *base peak*) is arbitrarily assigned a relative intensity of 100.00.

Figure 10.2a shows the mass spectrum of atomized sulfur. The relative intensities of the lines of $100.00:4.44$ correspond to isotopic abundances of 95.02% $^{32}S$ and 4.21% $^{34}S$. The natural abundances of $^{33}S$ and $^{35}S$ are 0.75% and 0.02% respectively, and expansion of the mass spectrum is needed to see the corresponding signals. Monoclinic sulfur consists of $S_8$ rings (**10.1**) and Figure 10.2b shows the parent ion in the mass spectrum of $S_8$. The presence of naturally occurring $^{32}S$ and $^{34}S$ (ignoring for the moment the much smaller amounts of $^{33}S$ and $^{35}S$) leads to a statistical mix of molecules of $(^{32}S)_8$, $(^{32}S)_7(^{34}S)$, $(^{32}S)_6(^{34}S)_2$ and so on. Since $^{32}S$ is by far the most abundant isotope of sulfur, the base peak in the mass spectrum of $S_8$ is at $m/z = 256$ corresponding to $(^{32}S)_8$. The next most abundant peak ($m/z = 258$) corresponds to $(^{32}S)_7(^{34}S)$. The effect of having 0.75% naturally occurring $^{33}S$ present (which is barely visible in the mass spectrum of atomic S) gives rise to the peak at $m/z = 257$, assigned to $(^{32}S)_7(^{33}S)$. Think about the statistics of having any one of the sites in the $S_8$ ring occupied by a $^{33}S$ isotope. The remaining peaks that are visible in Figure 10.2b are assigned to $(^{32}S)_6(^{33}S)(^{34}S)$ and $(^{32}S)_6(^{34}S)_2$; a value of $m/z = 259$ can be assigned to both $(^{32}S)_6(^{33}S)(^{34}S)$ and $(^{32}S)_7(^{35}S)$, but from the relative isotope abundances, the probability of $(^{32}S)_6(^{33}S)(^{34}S)$ is higher than that

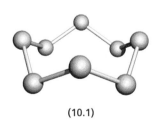

(10.1)

**Fig. 10.2** Low-resolution mass spectrometric traces of (a) atomized S, (b) the parent ion of $S_8$, (c) atomized C and (d) the parent ion of $C_{60}$.

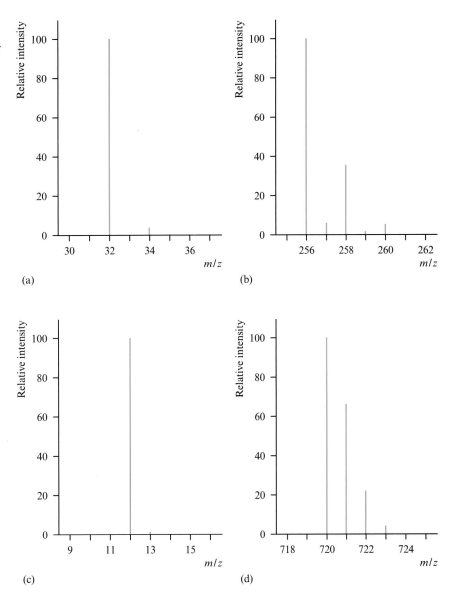

of $(^{32}S)_7(^{35}S)$. Although molecules with other isotopic compositions such as $(^{32}S)_5(^{34}S)_3$ ($m/z = 262$) and $(^{32}S)_5(^{33}S)(^{34}S)_2$ ($m/z = 261$) are possible, their probability is very low.

Naturally occurring phosphorus is *monotopic* ($^{31}P$) and the low-resolution mass spectrum of atomized phosphorus shows a single peak at $m/z = 31$. Similarly, the parent ion in the mass spectrum of white phosphorus, which consists of $P_4$ molecules, appears as a single peak at $m/z = 124$.

The isotopic distribution of naturally occurring carbon is 98.90% $^{12}C$ and 1.10% $^{13}C$, and in the mass spectrum of atomized carbon, the dominant peak is at $m/z = 12$ (Figure 10.2c). Figure 10.2d shows the parent ion in the mass spectrum of $C_{60}$, the molecular structure of which was shown in Figure 1.1c. The most intense peak ($m/z = 720$) can be assigned to $(^{12}C)_{60}$. The effect of having $^{13}C$ present in a sample of *atomic* carbon is negligible, but is significant in a molecule that contains a large number of C atoms. The peak at $m/z = 721$ in Figure 10.2d is assigned to $(^{12}C)_{59}(^{13}C)$, and its relative intensity of 67.35% is consistent with the relatively high probability of finding one

**Fig. 10.3** Low-resolution mass spectrometric traces of (a) atomized Br and (b) the parent ion of IBr.

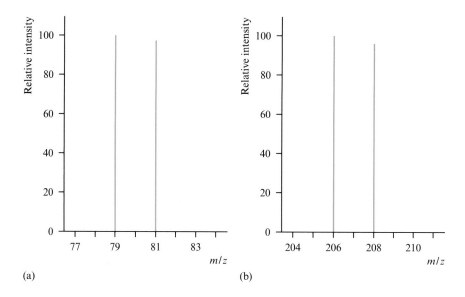

(a)  (b)

$^{13}$C atom in a molecule of $C_{60}$. The remaining two peaks in Figure 10.2d arise from $(^{12}C)_{58}(^{13}C)_2$ and $(^{12}C)_{57}(^{13}C)_3$. There is only a very small chance of there being more than three $^{13}$C in a molecule of $C_{60}$ with a natural isotopic distribution.

## Simple compounds

In a diatomic molecule XY in which X is monotopic (or virtually so) and Y possesses a number of naturally occurring isotopes, the isotopic pattern that is diagnostic of Y appears in the mass spectrum of XY. Consider naturally occurring Br which consists of 50.69% $^{79}$Br and 49.31% $^{81}$Br (Figure 10.3a). Now consider IBr. Iodine is monotopic ($^{127}$I) and Figure 10.3b shows the parent ion in the mass spectrum of IBr. The two peaks can be assigned to $(^{127}I)(^{79}Br)$ and $(^{127}I)(^{81}Br)$, and a comparison of Figures 10.3a and 10.3b shows a replication of the isotopic pattern that is characteristic of naturally occurring bromine. Although this is a simple case, it makes the important point that isotopic patterns can be used to provide evidence for the presence of a certain element in a compound.

---

**Worked example 10.1**   *Parent ions and isotopic distributions*

The parent ion in the mass spectrum of $CCl_4$ contains the following peaks:

| $m/z$ | Relative intensity | $m/z$ | Relative intensity |
|---|---|---|---|
| 152 | 78.18 | 157 | 0.54 |
| 153 | 0.88 | 158 | 10.23 |
| 154 | 100.00 | 159 | 0.11 |
| 155 | 1.12 | 160 | 0.82 |
| 156 | 47.97 | | |

Interpret these data given that naturally occurring C consists of 98.90% $^{12}$C and 1.10% $^{13}$C, and Cl consists of 75.77% $^{35}$Cl and 24.23% $^{37}$Cl.

The most abundant peaks contain $^{12}C$ rather than $^{13}C$. Write down the possibilities for $^{12}C$:

$(^{12}C)(^{35}Cl)_4$        $m/z = 152$

$(^{12}C)(^{35}Cl)_3(^{37}Cl)$     $m/z = 154$

$(^{12}C)(^{35}Cl)_2(^{37}Cl)_2$    $m/z = 156$

$(^{12}C)(^{35}Cl)(^{37}Cl)_3$     $m/z = 158$

$(^{12}C)(^{37}Cl)_4$        $m/z = 160$

Although $^{35}Cl$ is three times more abundant than $^{37}Cl$, the chance of a $CCl_4$ molecule containing *one* $^{37}Cl$ is high – there are four sites that could contain this isotope:

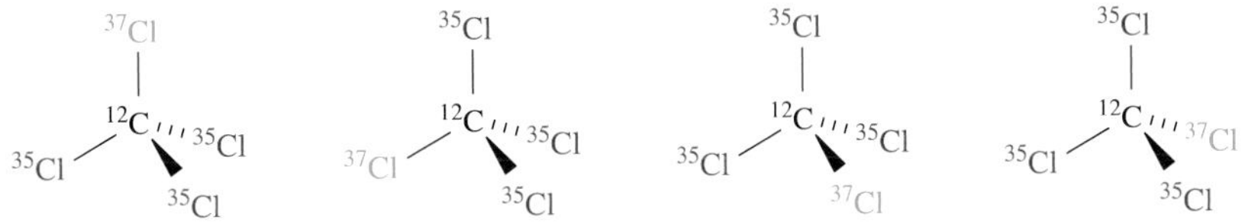

$(^{12}C)(^{35}Cl)_3(^{37}Cl)$ gives rise to the most abundant peak in the spectrum, the next being assigned to $(^{12}C)(^{35}Cl)_4$ and then $(^{12}C)(^{35}Cl)_2(^{37}Cl)_2$. The chance of a $CCl_4$ molecule containing three $^{37}Cl$ is significantly lower, and containing four is very low. The low-intensity peaks at $m/z = 153$, 155, 157 and 159 arise from $(^{13}C)(^{35}Cl)_4$, $(^{13}C)(^{35}Cl)_3(^{37}Cl)$, $(^{13}C)(^{35}Cl)_2(^{37}Cl)_2$ and $(^{13}C)(^{35}Cl)(^{37}Cl)_3$ respectively. No peak is observed for $(^{13}C)(^{37}Cl)_4$ since there is negligible chance of it occurring naturally.

## Carbon-containing compounds: the (P + 1) peak

In carbon-containing compounds, the presence of 98.90% $^{12}C$ and 1.10% $^{13}C$ leads to the $P^+$ peak being accompanied by a low-intensity $(P + 1)^+$ peak. As we saw for $C_{60}$, if the molecule contains a large number of C atoms, a $(P + 2)^+$ peak is also observed. For compounds that contain only C, H and O, the ratio of intensities of the $P^+$ and $(P + 1)^+$ peaks reflects the number of carbon atoms present and this is illustrated for a series of hydrocarbons in Table 10.2. Such characteristic patterns of peaks in a mass spectrum aid the identification of molecular or fragment (see below) ions.

**Table 10.2** Relative intensities of $P^+$ and $(P + 1)^+$ peaks for different numbers of carbon atoms, compared with ratios observed in alkanes, $C_nH_{2n+2}$.

| Number of C atoms | Relative intensities $P^+ : (P + 1)^+$ | Alkane | Relative intensities $P^+ : (P + 1)^+$ |
|---|---|---|---|
| 1 | 100.00 : 1.12 | $CH_4$ | 100.00 : 1.18 |
| 2 | 100.00 : 2.24 | $C_2H_6$ | 100.00 : 2.33 |
| 3 | 100.00 : 3.37 | $C_3H_8$ | 100.00 : 3.49 |
| 4 | 100.00 : 4.49 | $C_4H_{10}$ | 100.00 : 4.64 |
| 5 | 100.00 : 5.61 | $C_5H_{12}$ | 100.00 : 5.79 |
| 6 | 100.00 : 6.73 | $C_6H_{14}$ | 100.00 : 6.94 |

## 10.4    Fragmentation patterns

In a mass spectrum, peaks arising from ions with mass lower than the parent ion are often observed. These arise from bond cleavage, which leads to fragmentation of the molecule. Such fragmentation can greatly complicate the mass spectrum. If hydrogen atoms are lost, fragmentation can also complicate the appearance of the parent ion. For example, based on the abundances of the naturally occurring isotopes of C (98.90% $^{12}C$, 1.10% $^{13}C$), H (99.985% $^{1}H$, 0.015% $^{2}H$) and O (99.76% $^{16}O$, 0.04% $^{17}O$, 0.20% $^{18}O$), we might expect the parent ion of methanol ($CH_3OH$, **10.2**) to contain a base peak at $m/z = 32$ and two low-intensity peaks at $m/z = 33$ and 34. Figure 10.4a shows the observed mass spectrum of methanol. The peak at $m/z = 32$ corresponds to the parent ion $[CH_4O]^+$, with the expected low-intensity peaks at $m/z = 33$ and 34. Fragmentation of the molecular ion through C−H or O−H bond cleavage leads to the fragment ions $[CH_3O]^+$, $[CH_2O]^+$, $[CHO]^+$ and $[CO]^+$ observed at $m/z = 31$, 30, 29 and 28 respectively. Not all

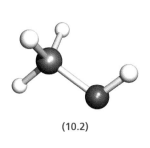

(10.2)

See also Section 30.4

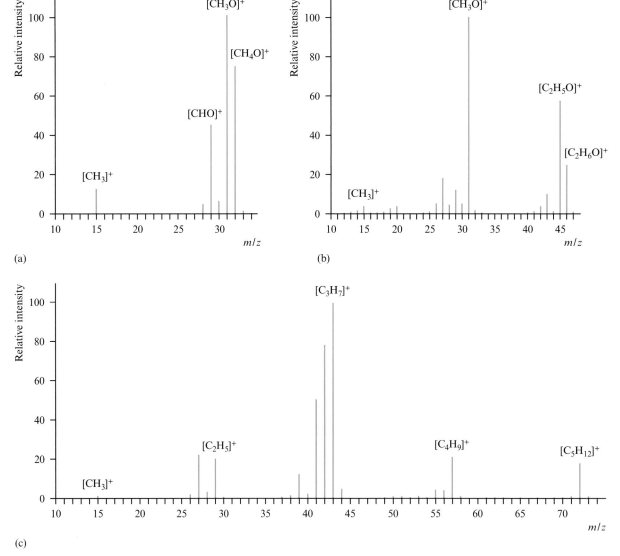

**Fig. 10.4** Mass spectra of (a) methanol ($CH_3OH$), (b) ethanol ($C_2H_5OH$) and (c) pentane ($C_5H_{12}$).

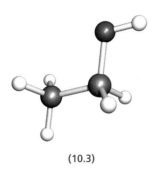

(10.3)

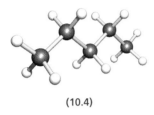

(10.4)

fragmentation ions are equally likely or stable, and predicting the relative intensities of these peaks is not easy. The peak at $m/z = 15$ arises from the methyl fragment $[CH_3]^+$.

Figure 10.4b shows the mass spectrum of ethanol ($C_2H_5OH$, **10.3**). Based only on the natural isotope abundances, a molecular ion for $[C_2H_6O]^+$ at $m/z = 46$ is expected and is observed. Loss of H gives $[C_2H_5O]^+$ at $m/z = 45$, and further fragmentation by loss of H atoms is observed. The most intense peak in the mass spectrum occurs at $m/z = 31$ and corresponds to $[CH_3O]^+$, formed by loss of a $CH_3$ fragment from the parent ion. The peaks with $m/z = 26$ to 29 can be fitted to fragments arising from $[C_2H_6O]^+$ losing an OH group followed by H atom loss, and the $[CH_3]^+$ fragment appears in the spectrum at $m/z = 15$.

The larger the molecule, the more fragments can be generated in the mass spectrometer. The mass spectrum of pentane ($C_5H_{12}$, **10.4**) is shown in Figure 10.4c. The ion at $m/z = 72$ corresponds to the molecular ion $[C_5H_{12}]^+$ (i.e. $[(^{12}C)_5(^1H)_{12}]^+$). The low-intensity peak at $m/z = 73$ arises from $[(^{12}C)_4(^{13}C)(^1H)_{12}]^+$ with a small contribution from $[(^{12}C)_5(^1H)_{11}(^2H)]^+$. The pattern of peaks in Figure 10.4c can be interpreted in terms of the fragmentation of $[C_5H_{12}]^+$ to give $[C_4H_9]^+$, $[C_3H_7]^+$, $[C_2H_5]^+$ and $[CH_3]^+$. In addition, loss of H from these fragments gives ions that include $[C_3H_6]^+$ and $[C_3H_5]^+$.

By assigning identities to fragment peaks in a mass spectrum, it is possible to 'reconstruct' a molecule and hence use fragmentation patterns to aid the identification of unknown compounds.

---

**Worked example 10.2**     *Fragmentation patterns*

**The mass spectrum of a compound A contains major peaks at $m/z = 58, 43$ and 15. Confirm that this fragmentation pattern is consistent with A being acetone (10.5).**

(10.5)

The molecular formula for acetone is $C_3H_6O$. Each of C, H and O possesses a dominant isotope ($^{12}C$, $^1H$ and $^{16}O$), and a peak in the mass spectrum at $m/z = 58$ corresponds to a molecular ion $[C_3H_6O]^+$. From structure **10.5**, loss of $CH_3$ gives a fragment ion $[C_2H_3O]^+$ at $m/z = 43$, and the $[CH_3]^+$ ion appears at $m/z = 15$. The observed fragmentation pattern is therefore consistent with **A** being acetone.

---

**10.5**     ## Case studies

In this section, we present the mass spectra of selected compounds. The examples have been chosen to illustrate the combined effects of isotopic and fragmentation patterns, as well as the fact that, in practice, molecular ions are not always observed.

### Case study 1: tetrachloroethene, $C_2Cl_4$

The mass spectrum of tetrachloroethene is shown in Figure 10.5. The group of peaks at highest masses arise from the parent ion $[C_2Cl_4]^+$ and the pattern of these peaks reflects the statistical distribution of the two isotopes of Cl

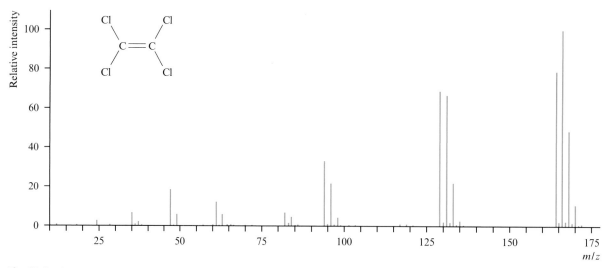

**Fig. 10.5** The mass spectrum of tetrachloroethene, $C_2Cl_4$.

**Fig. 10.6** Simulated isotopic patterns for (a) 4Cl, (b) 3Cl and (c) 2Cl.

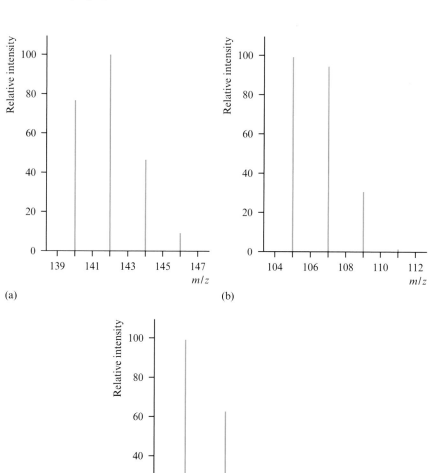

(75.77% $^{35}$Cl and 24.23% $^{37}$Cl) in a compound with four Cl atoms. The simulated pattern for 4Cl is shown in Figure 10.6a. Loss of Cl from $[C_2Cl_4]^+$ gives $[C_2Cl_3]^+$ which gives the group of peaks around $m/z = 130$. The isotopic pattern is dominated by that due to the statistical distribution of the two isotopes of Cl in three sites (Figure 10.6b). The group of peaks around $m/z = 95$ in Figure 10.5 is assigned to $[C_2Cl_2]^+$, and the statistical distribution of $^{35}$Cl and $^{37}$Cl is largely responsible for the pattern of peaks (compare with Figure 10.6c). The remaining groups of peaks in Figure 10.5 are assigned (in order of decreasing $m/z$) to $[CCl_2]^+$, $[C_2Cl]^+$, $[CCl]^+$, $[Cl]^+$ and $[C_2]^+$.

### Case study 2: tetramethyltin, $(CH_3)_4Sn$

The mass spectrum of $(CH_3)_4Sn$ is shown in Figure 10.7a. Figure 10.7b illustrates the isotopic distribution for naturally occurring Sn. A molecular

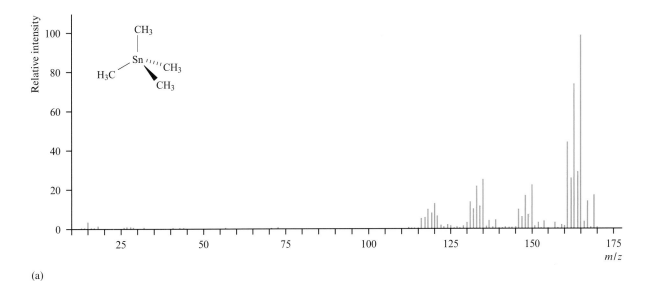

(a)

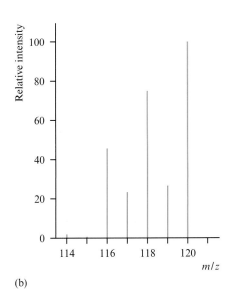

(b)

**Fig. 10.7** (a) The mass spectrum of $(CH_3)_4Sn$ and (b) the isotopic distribution for naturally occurring Sn.

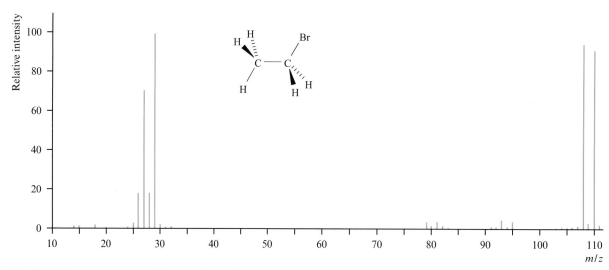

**Fig. 10.8** The mass spectrum of bromoethane, $CH_3CH_2Br$.

ion for $(CH_3)_4Sn$ would appear as a group of peaks, the most intense of which would be at $m/z = 180$ assigned to $(^{120}Sn)(^{12}C)_4(^1H)_{12}$. This is not visible in the mass spectrum of $(CH_3)_4Sn$, and instead the highest mass peak is assigned to the fragment ion $[C_3H_9Sn]^+$, formed after loss of a methyl group from the parent ion. In this example, the parent ion is not sufficiently stable to be carried through the mass spectrometer to the detector. The remaining groups of peaks in Figure 10.7a can be assigned to $[C_2H_6Sn]^+$, $[CH_3Sn]^+$ and $[Sn]^+$. The patterns of peaks for each fragment ion are similar, being dominated by that of the isotopes of Sn (compare Figures 10.7a and 10.7b).

### Case study 3: bromoethane, $CH_3CH_2Br$

Figure 10.8 shows the mass spectrum of bromoethane. Naturally occurring bromine occurs as an approximately $1:1$ mixture of $^{79}Br$ and $^{81}Br$ (Figure 10.3a). This isotopic pattern gives a characteristic appearance to the parent ion $[C_2H_5Br]^+$ in Figure 10.8, with two dominant peaks at $m/z = 108$ and 110. Loss of Br leads to $[C_2H_5]^+$ ($m/z = 29$), and the mass spectrum shows that the fragment ion $[C_2H_3]^+$ ($m/z = 27$) is also stable. Less intense peaks at $m/z = 28$ and 26 are assigned to $[C_2H_4]^+$ and $[C_2H_2]^+$. Loss of $CH_3$ from the parent ion gives $[CH_2Br]^+$ ($m/z = 95$ and 93 with the isotopic pattern characteristic of naturally occurring Br) but the low relative intensity of these peaks indicates that the ion is not particularly stable.

### Case study 4: sulfamic acid, $HOSO_2NH_2$

The mass spectrum and structure of sulfamic acid are shown in Figure 10.9. The parent ion is of relatively low intensity: $m/z = 97$ corresponds to $[SO_3NH_3]^+$. Various fragmentations could occur, but the base peak at $m/z = 80$ indicates that the $[SO_2NH_2]^+$ ion is especially stable and this corresponds to loss of OH from the parent ion. The peaks at $m/z = 64$ and 48 are assigned to the fragment ions $[SO_2]^+$ and $[SO]^+$ respectively. The lowest mass peaks at $m/z = 16$ and 17 could be assigned to $[NH_2]^+$ and $[OH]^+$; the peak at $m/z = 17$ could also arise from $[NH_3]^+$.

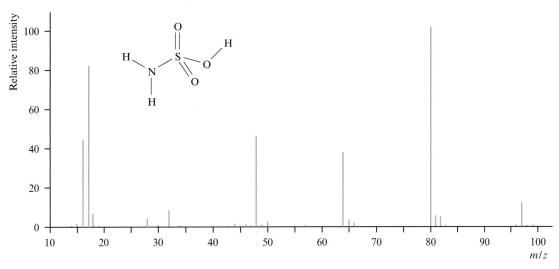

**Fig. 10.9** The mass spectrum of sulfamic acid, $NH_2SO_3H$.

## SUMMARY

In this chapter, we have introduced mass spectrometry, an important and routine technique that allows the molecular weight of a compound to be measured. Isotopic patterns of peaks give information about the elements present. Fragmentation patterns can be used to give supporting evidence for the composition of a compound.

### Do you know what the following terms mean?

- mass spectrometry
- electron impact ionization mass spectrometry
- mass spectrum
- $m/z$ ratio

- parent (or molecular) ion
- high- and low-resolution mass spectra
- base peak
- monotopic

- isotopic distribution of a peak
- (P + 1) peak
- fragmentation pattern
- fragment ion

### You should be able:

- to outline the operation of an electron impact ionization mass spectrometer
- to distinguish between high- and low-resolution mass spectra and recognize their appropriate uses
- to explain why some ions give rise a single peak in a mass spectrum while others give envelopes of peaks
- to explain the varying significance of $^{13}C$ in the mass spectrum species containing different numbers of carbon atoms

- to explain the significance of the parent peak in a mass spectrum
- to describe what information can be obtained from fragmentation peaks
- to assign peaks in the mass spectrum of a known compound
- to determine whether a mass spectrum is consistent with the proposed molecular formula of a compound

## PROBLEMS

Use data from Appendix 5 and Table 10.1 where necessary.

**10.1** Give a brief account of how an EI mass spectrometer functions.

**10.2** A high-resolution mass spectrum is used to give an exact mass determination of 111.055 836 for a compound **X**. Of the possible formulae $C_5H_5NO_2$, $C_5H_7N_2O$, $C_5H_9N_3$ and $C_6H_9NO$, which is supported by the mass spectrometric data?

**10.3** A compound **Y** contains 58.77% C and 27.42% N. From an exact mass determination, the molecular weight of **Y** is found to be 102.11576. Suggest a likely composition for **Y**.

**10.4** The isotopic abundances of naturally occurring gallium are 60.1% $^{69}Ga$ and 39.9% $^{71}Ga$. Sketch the mass spectrum of atomic Ga, setting the most abundant peak to a relative intensity of 100.

**10.5** (a) Why in Figure 10.2b are no peaks at $m/z > 260$ observed? (b) Suggest what peaks might dominate in the parent ion of $S_6$.

**10.6** Suggest what peaks might be present in the mass spectra of (a) $F_2$ and (b) $Cl_2$.

**10.7** Rationalize why the parent ion in the mass spectrum of $CF_2Cl_2$ contains the following peaks:

| $m/z$ | Relative intensity |
|-------|--------------------|
| 120   | 100.00             |
| 121   | 1.12               |
| 122   | 63.96              |
| 123   | 0.72               |
| 124   | 10.23              |
| 125   | 0.11               |

**10.8** In the mass spectra of two samples, **A** and **B**, the appearance of the parent ions are as follows. Which sample is CO, and which $N_2$?

| $m/z$ | Relative intensities for A | Relative intensities for B |
|-------|----------------------------|----------------------------|
| 28    | 100.00                     | 100.00                     |
| 29    | 0.72                       | 1.16                       |
| 30    | –                          | 0.20                       |

**10.9** In its mass spectrum, a compound **A** shows a peak from the molecular ion at $m/z = 78$. What further information from the parent ion allows you to distinguish between **A** being $C_6H_6$ and $C_3H_7Cl$?

**10.10** The mass spectrum of a compound **B** contains major peaks at $m/z = 74, 59, 45, 31, 29$ and 27, and a lower intensity peak at $m/z = 15$. **B** is known to be either butanol (**10.6**) or diethyl ether (**10.7**). Do the mass spectrometric data allow you to unambiguously assign **B** as butanol or diethyl ether?

(10.6)

(10.7)

**10.11** The compounds $CH_3CH_2SH$ and $HOCH_2CH_2OH$ both show parent ions in their mass spectra at $m/z = 62$. How do the fragmentation patterns allow you to distinguish between the two compounds?

**10.12** The mass spectrum of a compound **C** contains intense peaks at $m/z = 60$ (base peak), 45 and 43, and lower-intensity peaks at 61, 46, 44, 42, 41, 40, 31, 29, 28, 18, 17, 16, 15, 14, 13. Show that these data are consistent with **C** having structure **10.8**.

(10.8)

**10.13** The mass spectrum of acetonitrile (**10.9**) contains peaks at $m/z$ (relative intensity) = 42 (3), 41 (100), 40 (51), 39 (18) 38 (9), 26 (2), 15 (1), 14 (7). Account for these peaks.

(10.9)

**10.14** Compound **D** contains 68.13% C, 13.72% H, 18.15% O, and shows a parent ion in its mass spectrum at $m/z = 88$. Determine the molecular formula of **D**.

**10.15** Compound **E** exhibits a parent ion in its mass spectrum at $m/z = 92$, and analysis of **E** gives 91.25% C and 8.75% H. Determine the molecular formula of **E**.

**10.16** The mass spectrum of compound **F** contains a base peak at $m/z = 91$, and a parent ion at 106 (relative intensity 37.5). **F** is known to be a derivative of benzene with structure **10.10** in which R is an unknown group. Identify **F** and rationalize the mass spectrometric data.

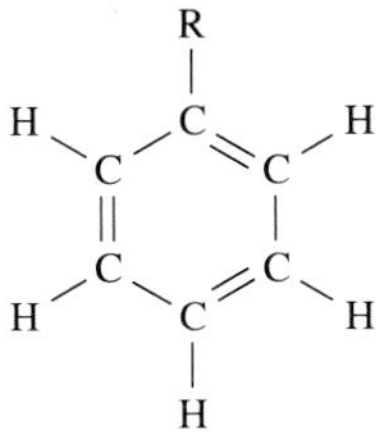

(10.10)

**10.17** Assign *each* of the peaks that make up the group of peaks at highest masses in Figure 10.5.

**10.18** The mass spectrum of nitromethane (**10.11**) contains major peaks as $m/z = 61, 46, 30$ (base peak) and 15. Account for these observations.

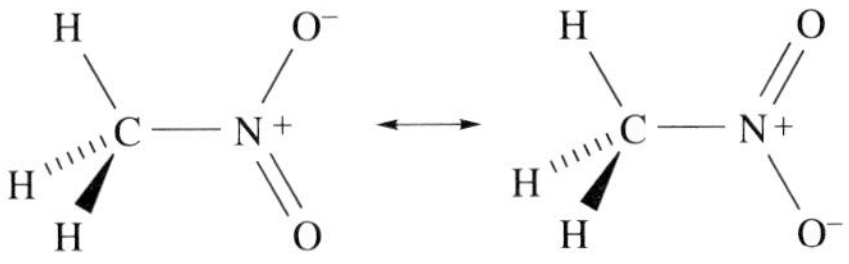

(10.11)

## Additional problems

(*Table 1.9 may be helpful for some problems.*)

**10.19** Naturally occurring Fe consists of 5.8% $^{54}$Fe, 91.7% $^{56}$Fe, 2.2% $^{57}$Fe and 0.3% $^{58}$Fe. Fe(CO)$_5$ (**10.12**) loses CO molecules in a sequential manner in the mass spectrometer. What do you expect to observe in the mass spectrum of Fe(CO)$_5$?

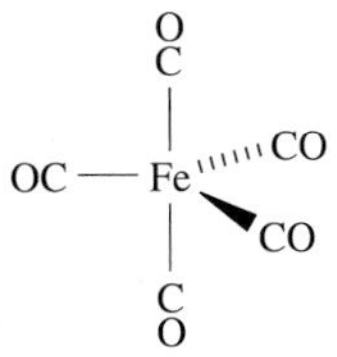

(10.12)

**10.20** C and H elemental analysis of a compound **Z** gives 88.2% C, 11.8% H. An exact mass determination using high-resolution mass spectrometry gives a mass of 68.06260. In the low-resolution mass spectrum, the major peaks that are observed are at $m/z = 68, 67, 53, 41, 40, 39$ and 26. Show that these data are consistent with **Z** being 1,3-pentadiene (**10.13**).

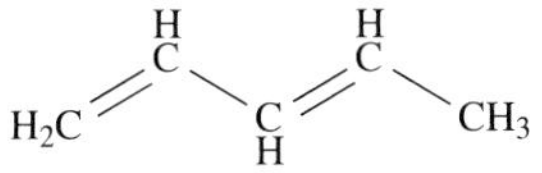

(10.13)

**10.21** (a) Compound **Y** is a solid at 298 K. It has an exact mass of 60.032 37, and elemental analysis shows that it contains 20.0% C, 6.7% H and 46.7% N. Suggest a possible formula for **Y**. (b) Apart from the base peak, the low-resolution mass spectrum of **Y** contains major peaks at 44 and 17, with less intense peaks at 43, 28 and 16. What can you deduce about a possible structure of **Y**?

**10.22** (a) Compound **Z** is a liquid at 298 K. High-resolution mass spectrometry gives an exact mass of 61.016 38, and **Z** contains 19.7% C, 4.9% H and 22.9% N. Suggest a possible formula for **Z**. (b) The low-resolution mass spectrum of **Z** shows intense peaks at 61, 46, 30 and 15. Deduce a possible structure of **Z**.

# 11 Introduction to spectroscopy

## Topics

- Spectroscopic techniques
- Timescales
- Beer–Lambert Law
- Colorimetry

## 11.1 What is spectroscopy?

Atomic spectrum of hydrogen: see Section 3.16
Photoelectron spectroscopy: see Box 4.6

Spectroscopic methods of analysis are an everyday part of modern chemistry, and we have already mentioned the atomic spectrum of hydrogen and photo-electron spectroscopy. There are many different spectroscopic techniques (Table 11.1) and by using different methods, it is possible to investigate many aspects of atomic and molecular structure.

In the next few chapters we consider infrared (IR), rotational, electronic (which includes ultraviolet–visible, UV–VIS) and nuclear magnetic resonance (NMR) spectroscopies – these are the techniques that include those you are most likely to use in the laboratory.

The aim of most of our discussion is *not* to delve deeply into theory, but rather to provide information that will assist in the practical application of these techniques.

### Absorption and emission spectra

Absorption and emission: see Box 3.7

The terms *absorption* and *emission* are fundamental to discussions of spectroscopy; we distinguished between absorption and emission *atomic spectra* in Section 3.16. The absorption of electromagnetic radiation by an atom, molecule or ion causes a transition from a lower to a higher energy level. (We shall deal more explicitly with what is meant by a 'level' later.) Remember (Chapter 3) that only *certain* transitions are allowed. The energy of the radiation absorbed gives the energy difference between the two levels. Figure 11.1 shows a schematic diagram of some components of an *absorption spectrophotometer* used to measure the absorption of electromagnetic radiation. The 'source' provides electromagnetic radiation covering a range of frequencies.

$E = h\nu$

When the radiation passes through a sample, some may be absorbed, but some is transmitted, detected and recorded in the form of a spectrum (Figure 11.1). The spectrum is a plot of the *absorbance* or *transmittance* of the radiation against the energy.

The intensity of an absorption (the reading of absorbance or transmittance in Figure 11.2) depends upon several factors:

**Table 11.1** Some important spectroscopic techniques; note that electron, X-ray and neutron diffraction methods (Chapter 4) and mass spectrometry (Chapter 10) are *not* spectroscopic techniques.

| Name of technique | Comments |
| --- | --- |
| Atomic absorption spectroscopy | Used for elemental analysis; observes the absorption spectra of atoms in the vapour state. |
| Electron spin resonance (ESR) spectroscopy *or* electron paramagnetic resonance (EPR) spectroscopy | Used in the study of species with one or more unpaired electrons. |
| Electronic spectroscopy | Absorption spectroscopy [100–200 nm (vacuum-UV), 200–800 nm (near-UV and visible)] used to study transitions between atomic and molecular electronic energy levels (see Chapter 13). |
| Far infrared spectroscopy | Infrared spectroscopy below $\approx 200\,\mathrm{cm}^{-1}$. |
| Fluorescence spectroscopy | Used to study compounds that *fluoresce, phosphoresce* or *luminesce*; fluorescence is the emission of energy which may follow the absorption of UV or visible radiation, and is a property exhibited only by certain species. The light emitted is at longer wavelength than that absorbed. |
| Infrared (IR) spectroscopy | A form of vibrational spectroscopy; absorptions are usually recorded in the range $200{-}4000\,\mathrm{cm}^{-1}$. Extremely useful as a 'fingerprinting' technique; see Chapter 12. |
| Microwave spectroscopy | Absorption spectroscopy used to study the rotational spectra of gas molecules; see Chapter 12. |
| Mössbauer spectroscopy | Absorption of $\gamma$-radiation by certain nuclei (e.g. $^{57}$Fe and $^{197}$Au); used to study the chemical environment, including oxidation state, of the nuclei. |
| Nuclear magnetic resonance (NMR) spectroscopy | Absorption or emission of radiofrequency radiation used to observe nuclear spin states (see Chapter 14); a very powerful analytical tool which is used to elucidate molecular structures and study dynamic behaviour in solution and the solid state. |
| Photoelectron spectroscopy (PES) | Absorption spectroscopy used to study the energies of occupied atomic and molecular orbitals (see Box 4.6). |
| Raman spectroscopy | A form of vibrational spectroscopy but with different selection rules from IR spectroscopy; some modes that are IR inactive (see Chapter 12) are Raman active. |

- the *probability* of a transition occurring;
- the *populations* of the different energy levels; and
- the amount of the sample.

Whereas an absorption spectrum measures the energy and amount of radiation that has been removed from the initial range of energies, an emission spectrum measures the radiation emitted by the *excited state* of a sample. In an *emission spectrometer*, the sample is *excited* (thermally, electrically or by using electromagnetic radiation) to a short-lived higher energy level. As the transition back to a lower energy level occurs, energy is emitted. The spectrum recorded is a plot of the intensity of emission against energy. The difference in energy between the higher and lower energy levels

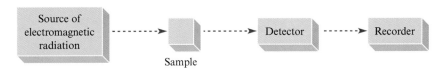

**Fig. 11.1** A schematic representation of parts of a simple absorption spectrometer; the electromagnetic radiation originates from the source and is directed at the sample, where some is absorbed and some transmitted. The resultant radiation passes on to a detector, and the information is then output in the form of a spectrum.

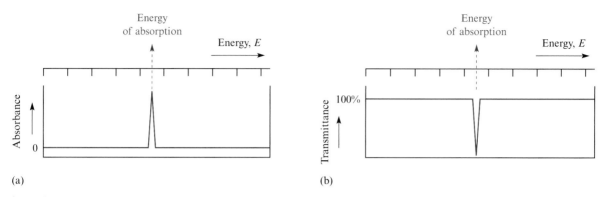

**Fig. 11.2** Schematic representation of an absorption spectrum consisting of a single absorption. The spectrum gives information about the intensity and energy of the absorption; the intensity can be measured in terms of (a) absorbance or (b) transmittance.

corresponds to the energy of the emitted radiation. In an emission spectroscopic experiment, it is important to ensure that sufficient time has elapsed between the initial excitation and the time of recording the emission. If the time delay is too long, the transition from higher to lower level will already have occurred and if it is too short, emission may not have taken place.

## 11.2    The relationship between the electromagnetic spectrum and spectroscopic techniques

The electromagnetic spectrum scale is shown in Appendix 4. Electromagnetic radiation can be described in terms of frequency ($\nu$), wavelength ($\lambda$) or energy ($E$), but another convenient unit is 'wavenumber'. Wavenumbers are the reciprocal of wavelength (equation 11.1) and are used as a convenient quantity which is linearly related to energy. The SI unit of wavenumber is $m^{-1}$ but a more convenient unit is $cm^{-1}$.

$$\text{Wavenumber } (\bar{\nu}) = \frac{1}{\text{Wavelength}} \tag{11.1}$$

Different spectroscopic methods are associated with different parts of the electromagnetic spectrum, since radiation of a particular energy is associated with certain transitions in an atom or molecule. For example, some of the spectral lines in the Lyman series of atomic hydrogen lie between $2.466 \times 10^{15}$ and $3.237 \times 10^{15}$ Hz. This corresponds to the ultraviolet region of the electromagnetic spectrum. In Figure 11.3, the techniques of NMR, rotational, IR and UV–VIS spectroscopies are related to the electromagnetic spectrum.

Lyman series: see worked example 3.5

NMR spectroscopy: see Chapter 14

Nuclear magnetic resonance spectroscopy is concerned with transitions between different nuclear spin states. Such transitions require little energy ($<0.01\,\text{kJ mol}^{-1}$) and can be brought about using radiation from the radio-frequency region of the electromagnetic spectrum.

Rotational spectroscopy: see Chapter 12

Rotational spectroscopy is concerned with transitions between the rotational states of a molecule. Such transitions require between 1 and $0.01\,\text{kJ mol}^{-1}$ and can be brought about by using radiation from the microwave part of the electromagnetic spectrum.

IR spectroscopy: see Chapter 12

Vibrational spectroscopy is concerned with transitions between vibrational states of a molecule. The energy needed to bring about such transitions is typically in the region of 1 to $100\,\text{kJ mol}^{-1}$, corresponding to the infrared

**Fig. 11.3** Part of the electromagnetic spectrum; the radiation can be described in terms of energy, frequency or wavenumber. The relationship to wavelength is shown in Appendix 4. Different regions of the spectrum are associated with different spectroscopic techniques.

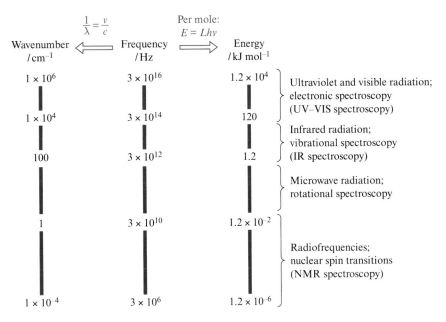

$\dfrac{1}{\lambda} = \dfrac{v}{c}$

Per mole: $E = Lhv$

Wavenumber /cm$^{-1}$ ⟸ Frequency /Hz ⟹ Energy /kJ mol$^{-1}$

| Wavenumber /cm$^{-1}$ | Frequency /Hz | Energy /kJ mol$^{-1}$ | |
|---|---|---|---|
| $1 \times 10^6$ | $3 \times 10^{16}$ | $1.2 \times 10^4$ | Ultraviolet and visible radiation; electronic spectroscopy (UV–VIS spectroscopy) |
| $1 \times 10^4$ | $3 \times 10^{14}$ | 120 | Infrared radiation; vibrational spectroscopy (IR spectroscopy) |
| 100 | $3 \times 10^{12}$ | 1.2 | |
| | | | Microwave radiation; rotational spectroscopy |
| 1 | $3 \times 10^{10}$ | $1.2 \times 10^{-2}$ | Radiofrequencies; nuclear spin transitions (NMR spectroscopy) |
| $1 \times 10^{-4}$ | $3 \times 10^6$ | $1.2 \times 10^{-6}$ | |

Hz = s$^{-1}$ ; $h = 6.26 \times 10^{-34}$ J s, $c = 3 \times 10^8$ m s$^{-1}$, $L$ = Avogadro constant

region of the electromagnetic spectrum. This range corresponds to wavenumbers from 100 to 10 000 cm$^{-1}$. A 'normal' laboratory IR spectrometer operates between 400 and 4000 cm$^{-1}$.

Electronic spectroscopy: see Chapter 13

Electronic spectroscopy is the study of transitions between electronic energy levels in a molecule. The energy differences correspond to radiation from the UV–VIS region of the electromagnetic spectrum – hence the name UV–VIS *spectroscopy*. The normal range of a laboratory UV–VIS spectrophotometer is 190 to 900 nm; the measurement tends to be made as a wavelength.

## 11.3    Timescales

At this point, it is necessary to say something about the spectroscopic *timescales*. This is a complex topic and we are concerned with only one or two aspects. A critical question is: will a particular spectroscopic technique give us a 'snapshot' of the molecule at a given moment or an 'averaged' view?

The conventional 'static' view of a molecule is incorrect – molecules are continually vibrating and rotating, and these motions occur at a rate of $10^{12}$ to $10^{14}$ per second (Figure 11.3). If the spectroscopic technique is *faster* than this, we obtain a 'snapshot' of the event, but if it is *slower*, we see only an averaged view of the molecule undergoing its various motions. Electronic spectroscopy gives a 'snapshot' of a molecule in a given vibrational and rotational state whereas NMR spectroscopy often gives an averaged view, since the timescale of the technique is slower than the molecular vibrational and rotational motions.

A second feature refers to the fact that we are not usually looking at a single molecule but at an assembly of molecules. Although electronic spectroscopy gives a 'snapshot' of a particular molecule in a particular vibrational and rotational state, a typical sample may contain $10^{18}$ molecules, not all of which will be in the same vibrational and rotational states.

A final point is that the molecule may be undergoing a dynamic process (such as Berry pseudo-rotation), and this creates an additional problem of timescales. Lowering the temperature will slow the dynamic behaviour, and *may* make it slower than the spectroscopic timescale although as we have described for $[Fe(CO)_5]$, even at $103\,K$, the axial and equatorial CO ligands are exchanging positions and the $^{13}C$ NMR spectrum shows the presence of only one (on average) $^{13}C$ environment.

■▷
Berry pseudo-rotation:
see Section 6.12

## 11.4    The Beer–Lambert Law

For a given compound, the intensity of absorption depends on the amount of the sample. If a particular frequency of radiation is being absorbed by a molecule, then the more molecules there are, the more radiation of that frequency will be absorbed and less transmitted. The *absorbance* or *optical density* of a sample is related to the *transmittance* (equation 11.2) and this relationship is shown graphically in Figure 11.4.

 $log = log_{10}$    ■▷

$$\text{Absorbance} = -\log(\text{Transmittance}) \qquad A = -\log T \qquad (11.2)$$

Values of transmittance, $T$, lie between 0 and 1, but experimentally we often express $T$ as a percentage. From equation 11.2, 100% transmittance corresponds to zero absorbance, and for zero transmittance, the curve in Figure 11.4 tails off to infinite value of absorbance; for $T = 0.01$ (1%), the absorbance is 2. Typical UV–VIS spectrometers operate up to absorbance values of 4–6.

The transmittance is equal to the ratio of the intensity of the transmitted radiation ($I$) to that of the incident radiation ($I_0$) and combining this with equation 11.2 gives equation 11.3. Figure 11.5 illustrates the relationship between $I$ and $I_0$ and demonstrates how the absorbance of a sample in solution is determined.

$$A = -\log \frac{I}{I_0} \qquad (11.3)$$

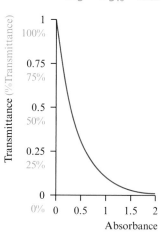

**Fig. 11.4** The relationship between transmittance and absorbance (see equation 11.2). Note that the curve tends to infinity when the transmittance is zero.

In a spectrophotometer, the sample is contained in a *solution cell* of accurately known dimensions. The distance travelled by the radiation through the cell is called the *path length*, $\ell$, and, often, the absorbance is related to the concentration ($c$) and path length by the *Beer–Lambert Law* (equation 11.4), where $\varepsilon$ is the *molar extinction* or *absorption coefficient* of the dissolved compound.

$$A = -\log \frac{I}{I_0} = \varepsilon \times c \times \ell \qquad (11.4)$$

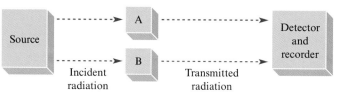

**Fig. 11.5** Solution cell A contains a sample in solution, and cell B contains pure solvent – the *same* solvent as in cell A; the cells have the same path length. If the solvent does *not* absorb any radiation, the intensity of the radiation emerging from cell B is the same as that of the incident radiation ($I_0$); the sample in cell A absorbs some radiation, and the intensity of the radiation emerging from cell A is that of the transmitted radiation ($I$). The transmittance of the sample is the ratio of $I : I_0$, and absorbance can be determined by using equation 11.3.

The concentration is measured in $mol\,dm^{-3}$ and the cell path length in cm, giving the units of $\varepsilon$ as $dm^3\,mol^{-1}\,cm^{-1}$. The extinction coefficient is a *property of the compound* and is independent of concentration for the majority of compounds in *dilute* solution. Values of $\varepsilon$ are often large and it is common to quote $\log\varepsilon$ values.

> The **Beer–Lambert Law** relates the absorbance of a solution sample to the molar extinction coefficient, the concentration and the cell path length:
>
> $$A = \varepsilon \times c \times \ell$$

---

**Worked example 11.1**      *Use of the Beer–Lambert Law*

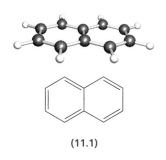

**(11.1)**

Solutions of naphthalene (11.1) absorb light of wavelength 312 nm and the extinction coefficient for this transition is $288\,dm^3\,mol^{-1}\,cm^{-1}$. A solution of naphthalene in ethanol in a cell of path length 1.0 cm gives an absorbance of 0.012. What is the concentration of the solution?

By the Beer–Lambert Law:

$$A = \varepsilon \times c \times \ell$$

Therefore:

$$c = \frac{A}{\varepsilon \times \ell}$$

The absorbance, $A$, is dimensionless (i.e. it has no units).

$$c = \frac{0.012}{(288\,dm^3\,mol^{-1}\,cm^{-1}) \times (1.0\,cm)} = 4.2 \times 10^{-5}\,mol\,dm^{-3}$$

---

## 11.5      Colorimetry

One consequence of the direct relationship between absorbance and sample concentration (equation 11.4) is that the Beer–Lambert Law can be applied analytically. For most compounds, the Beer–Lambert Law is obeyed reasonably well in dilute solution. The technique of *colorimetry* is used to determine the concentration of coloured compounds in solution. Not only can one-off measurements be made, but changes in the concentration of a coloured component during a reaction can be followed by monitoring the change in absorbance.

▬▶
Types of coloured
compounds:
see Section 13.6
and Chapter 23

### Colours

The *colour of a solution* depends upon the wavelength of visible light absorbed by the sample, and the *intensity of colour* depends on the concentration of the solution.

When a coloured compound is dissolved in a solvent, the *intensity* of colour depends on the concentration of the solution, but the *actual* colour depends upon the wavelength of visible light absorbed by the sample.

White light consists of a continuous spectrum of electromagnetic radiation from about 400 to 700 nm. When white light is incident on a solution of a compound that absorbs within the visible region, the transmitted light is coloured, and the observed colour depends upon the 'missing' wavelength(s). Copper(II) sulfate solution absorbs orange light and as a result, the solution

**Table 11.2** The visible part of the electromagnetic spectrum.

| Colour of light absorbed | Approximate wavelength ranges / nm | Colour of light transmitted, i.e. complementary colour of the absorbed light | In a 'colour wheel' representation, complementary colours are opposite to one another |
|---|---|---|---|
| Red | 700–620 | Green | 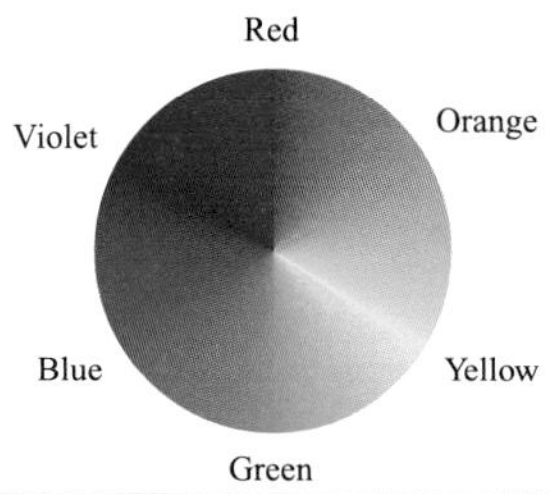 |
| Orange | 620–580 | Blue | |
| Yellow | 580–560 | Violet | |
| Green | 560–490 | Red | |
| Blue | 490–430 | Orange | |
| Violet | 430–380 | Yellow | |

appears blue – blue is the *complementary colour* of orange. Table 11.2 lists the colours of light, corresponding wavelengths and the complementary colours in the visible spectrum.

---

**Worked example 11.2**     *The dependence of absorbance on concentration*

A 0.10 M solution of copper(II) sulfate gives an absorbance of 0.55. What is the absorbance when the concentration is doubled? The same solution cell is used for the two readings.

The relationship between absorbance and concentration is given by the Beer–Lambert Law:

$$A = \varepsilon \times c \times \ell$$

For a constant path length and $\varepsilon$, we can write:

$$\frac{A_1}{A_2} = \frac{c_1}{c_2}$$

where $A_1$ is the absorbance for a concentration $c_1$, and $A_2$ is the absorbance for a concentration $c_2$. Absorbance is dimensionless.

$$A_1 = \frac{c_1 \times A_2}{c_2}$$

$$= \frac{(0.20 \, \mathrm{mol \, dm^{-3}}) \times 0.55}{(0.10 \, \mathrm{mol \, dm^{-3}})}$$

$$= 1.1$$

The Beer–Lambert Law predicts a *linear relationship* between $A$ and $c$ for a given compound *if the path length is constant*.

---

## The colorimeter

Figure 11.5 schematically illustrated how the absorbance of a compound in solution can be measured. In a colorimeter, the 'source' is tuned to provide a particular wavelength – the choice is made based on a knowledge of the wavelength absorbed by the compound or ion under study.

The aim of a colorimetric investigation may be to determine the concentration of a solution, follow a reaction or to determine the stoichiometry of a reaction.

**Fig. 11.6** A plot of absorbance (533 nm) against concentration of a solution of the compound formed between $Fe^{2+}$ ions and 4,7-diphenyl-1,10-phenanthroline. The path length of the solution cell is 1 cm. The plot illustrates the Beer–Lambert Law. Note the way in which the concentrations are given: the values are $0.2 \times 10^{-4}$ mol dm$^{-3}$, etc.

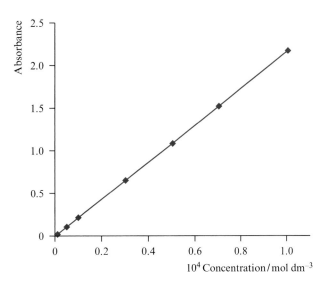

**Case study 1: analysis of iron(II)**    The compound 4,7-diphenyl-1,10-phenanthroline (**11.2**) is used to analyse for $Fe^{2+}$ ions.

When compound **11.2** is added to a solution containing $Fe^{2+}$ ions, an intensely coloured red compound is formed which absorbs light of wavelength 533 nm and has a molar extinction coefficient of 22 000 dm$^3$ mol$^{-1}$ cm$^{-1}$. If the colorimeter is tuned to 533 nm, the concentration of the red compound can be determined by measuring the absorbance. One mole of this red compound contains one mole of $Fe^{2+}$ ions, and if all the iron(II) present in the solution has reacted with compound **11.2**, the concentration of $Fe^{2+}$ can be determined. Figure 11.6 shows how the absorbance (at 533 nm) varies with the concentration of $Fe^{2+}$. Such a plot can be used to analyse a series of samples containing $Fe^{2+}$ (see problem 11.6); since $\varepsilon$ is large, very low concentrations of $Fe^{2+}$ ions can be measured accurately.

(11.2)

**Case study 2: following the reaction between copper metal and potassium dichromate(VI)**    Reactions involving coloured species may be monitored by using colorimetry, and may involve the appearance or disappearance of a particular coloured species. This can provide valuable information about the rate of a reaction (see Chapter 15).

Copper metal reacts with potassium dichromate(VI), $K_2Cr_2O_7$, in the presence of acid (equation 11.5) and during the reaction, the red-orange colour of the $K_2Cr_2O_7$ is lost and the solution becomes blue-green.

$$3Cu(s) + [Cr_2O_7]^{2-}(aq) + 14H^+ \longrightarrow 3Cu^{2+}(aq) + 2Cr^{3+}(aq) + 7H_2O(l)$$

(11.5)

The $[Cr_2O_7]^{2-}$ ion absorbs close to 500 nm, while the products absorb at longer wavelengths. With the colorimeter tuned to 500 nm, the reaction can be monitored by taking readings of absorbance as a function of time as shown in Figure 11.7.

**Case study 3: Job's method**    Colorimetry can be used to determine the stoichiometry of a reaction leading to the formation of a coloured species by using *Job's method*. Consider the reaction between iron(II) ions

**Fig. 11.7** A plot of absorbance (500 nm) against time which follows the disappearance of the $[Cr_2O_7]^{2-}$ ion in its reaction with copper metal in the presence of acid.

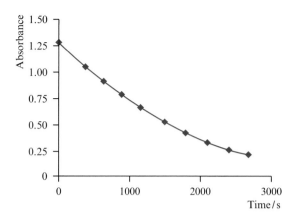

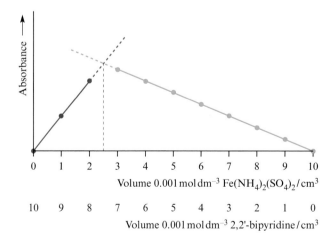

(11.3)

Structure of the 1:3 complex: see Figure 23.7

and 2,2'-bipyridine (**11.3**) which gives a red compound. The aim of the experiment is to determine the ratio of Fe(II) : 2,2'-bipyridine in the coloured product.

Firstly, the wavelength of light absorbed by the complex is found from its absorption spectrum, and the colorimeter is tuned to this wavelength. A series of solution cells (of constant path length) is prepared with solutions containing different ratios of iron(II) to 2,2'-bipyridine, and with a constant total volume as shown on the horizontal axis of the graph in Figure 11.8. A reading of the absorbance for each solution is taken and a graph of absorbance against solution composition is constructed. The maximum absorbance corresponds to the highest concentration of the coloured product in solution. In this case, the maximum absorbance is found by extrapolation and corresponds to an iron(II) : 2,2'-bipyridine ratio of 1 : 3.

**Fig. 11.8** The use of Job's method to determine the stoichiometry of the reaction between iron(II) ions and 2,2'-bipyridine; the iron(II) ammonium sulfate is the source of $Fe^{2+}$.

## SUMMARY

This chapter has given an introduction to spectroscopy with an overview of spectroscopic methods, timescales and differences between absorption and emission spectra. We have discussed the Beer–Lambert Law and its applications.

### *Do you know what the following terms mean?*

- electromagnetic spectrum
- absorption
- emission
- absorbance
- transmittance

- spectroscopic timescale
- wavenumber
- wavelength
- frequency
- molar extinction coefficient

- absorbance (optical density)
- Beer–Lambert Law
- colorimetry
- Job's method

### *Four important equations:*

- $E = h\nu$       $E =$ energy; $h =$ Planck constant; $\nu =$ frequency
- $c = \lambda\nu$       $c =$ speed of light; $\lambda =$ wavelength; $\nu =$ frequency
- $\bar{\nu} = \dfrac{1}{\lambda}$       $\bar{\nu} =$ wavenumber; $\lambda =$ wavelength
- $A = \varepsilon c \ell$       $A =$ absorbance; $\varepsilon =$ molar extinction coefficient; $c =$ concentration; $\ell =$ path length

### *You should be able:*

- to compare the relative energies, frequencies and wavelengths of electromagnetic radiation associated with electronic, rotational, infrared and NMR spectroscopies
- to relate absorbed and transmitted wavelengths of light and understand how this relates to the observed colours of solutions
- to use the relationships between energy, frequency, wavelength and wavenumber
- to describe why one spectroscopic technique 'sees' a molecule as static while another 'sees' a dynamic molecule

- to find the molar extinction coefficient of a compound from absorbance data
- to relate absorbance and transmittance data
- to use the Beer–Lambert Law to determine concentrations from absorbance data
- to use the Beer–Lambert Law to determine the stoichiometry of selected reactions
- to determine the stoichiometry of a coloured compound using Job's method

## PROBLEMS

**11.1** What are the *relative* energies of transitions observed in vibrational, rotational, electronic and nuclear magnetic resonance spectroscopies?

**11.2** Convert the wavelengths (a) 500 nm and (b) 225 nm to cm$^{-1}$ (wavenumbers). (c) Do these wavelengths fall in the visible region?

**11.3** (a) A solution of a compound **X** gives an absorbance reading of 0.446. What is the percentage transmittance (%*T*)? (b) The transmittance for another sample of **X** is 70.9%. To what absorbance reading does this correspond?

**11.4** Aqueous solutions of Ni$^{2+}$ ions appear green. What is the approximate wavelength of light absorbed by aqueous Ni$^{2+}$ ions?

**11.5** The absorption spectrum of benzene dissolved in cyclohexane contains bands at 183, 204 and 256 nm. (a) Which band corresponds to the lowest energy transition? (b) Which band corresponds to the lowest wavenumber? (c) For $\lambda = 256$ nm, $\log \varepsilon = 2.30$. Determine the concentration of a solution (path length $= 1.00$ cm) which gives an absorbance of 0.25.

**11.6** Refer to Figure 11.6. (a) What does the gradient of the graph tell you? (b) Determine the concentration of Fe$^{2+}$ ions in a solution that gives an absorbance of 0.20.

**11.7** Solutions of azulene (**11.4**) in cyclohexane absorb at 357 nm and the value of $\varepsilon$ for this absorption is

$3980 \, dm^3 \, mol^{-1} \, cm^{-1}$. Such a solution contained in a cell of path length 1.0 cm gives an absorbance of 0.58. What is the concentration of the solution?

(11.4)

**11.8** The absorption spectrum of a $5.00 \times 10^{-4} \, mol \, dm^{-3}$ solution of azulene in cyclohexane in a cell of path length 0.50 cm shows an absorption with $A = 0.995$. Calculate the corresponding extinction coefficient.

**11.9** Solutions of naphthalene in ethanol absorb at $\lambda = 312$ nm. A $2.50 \times 10^{-3} \, mol \, dm^{-3}$ solution gives an absorbance of 0.72. Determine the concentration of a solution for which the absorbance is 1.00. The same solution cell was used for both readings.

**11.10** Two solutions, **I** and **II**, of the same compound, in the same solvent and contained in identical solution cells are of concentrations $5.00 \times 10^{-3}$ and $1.75 \times 10^{-3} \, mol \, dm^{-3}$. What is the ratio of their absorbances?

**11.11** The compound $K_3[Fe(CN)_6]$ absorbs at $\lambda = 418$ nm ($\log \varepsilon = 3.01$); experiments are carried out in a cell of path length 1.00 cm. (a) Determine the absorbance of a solution of concentration $6.0 \times 10^{-4} \, mol \, dm^{-3}$. (b) What is the absorbance if the concentration is halved?

**11.12** Azobenzene has two geometrical isomers, **11.5** and **11.6**, although **11.5** rapidly isomerizes to **11.6**. Solutions of **11.5** absorb at $\lambda = 247$ nm ($\log \varepsilon = 4.06$); those of **11.6** absorb at $\lambda = 316$ nm ($\log \varepsilon = 4.34$). Find the relative absorbances of a $5.00 \times 10^{-5} \, mol \, dm^{-3}$ solution of **11.5** and a $2.00 \times 10^{-5} \, mol \, dm^{-3}$ solution of **11.6**. The path length for each is 1.00 cm.

(11.5)          (11.6)

**11.13** Solutions of phenanthrene (**11.7**) in cyclohexane absorb at 357 nm; this is one of four absorptions. The absorbance is measured as a function of solution concentration (path length = 1.00 cm):

| Concentration / mol dm$^{-3}$ | Absorbance |
|---|---|
| 0.0012 | 0.220 |
| 0.0017 | 0.355 |
| 0.0030 | 0.627 |
| 0.0035 | 0.720 |
| 0.0050 | 1.045 |

(a) Use the data to determine the molar extinction coefficient for the absorption band at 357 nm. (b) Why is it more accurate to measure $\varepsilon$ from a series of readings rather than a single reading?

(11.7)

**11.14** Solutions of acridine (**11.8**) in ethanol absorb at $\lambda = 250$ nm ($\log \varepsilon = 5.3$). (a) What is the concentration of acridine in a solution within a cell of path length 0.50 cm for which the absorbance is 0.96? (b) What mass of acridine is required to prepare 250 cm$^3$ of this solution?

(11.8)

## Additional problems

**11.15** The conversion of compound **11.9** to its (*Z*)-isomer (see Section 6.11) is induced by a flash of light (*flash photolysis*). Isomer **11.9** absorbs light at 435 nm, and the (*Z*)-isomer absorbs at 316 nm. Figure 11.9 shows the change in absorbance at $\lambda = 435$ nm for a solution of **11.9** in cyclohexane over a period of 1200 s during which time the solution is subjected to a flash of light. (a) Rationalize the shape of the curve in Figure 11.9. (b) If the path length of the solution cell is 1 cm,

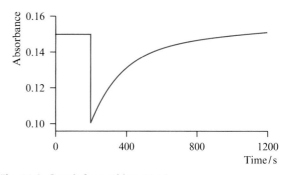

**Fig. 11.9** Graph for problem 11.15.

and the solution concentration is $9.00 \times 10^{-6}\,mol\,dm^{-3}$, determine a value for $\log \varepsilon$ corresponding to the absorption at 435 nm. [Data from: S. R. Hair *et al.* (1990) *J. Chem. Ed.*, vol. 67, p. 709.]

**(11.9)**

**11.16** The reaction between $I^-$ and $[S_2O_8]^{2-}$ occurs as follows:

$$[S_2O_8]^{2-} + 2I^- \longrightarrow 2[SO_4]^{2-} + I_2$$
$$I_2 + I^- \longrightarrow [I_3]^-$$

The second step occurs so long as there is an excess of $I^-$ present in solution; $[I_3]^-$ absorbs at 353 nm ($\log \varepsilon = 4.41$). Explain how measurements of the absorbance at $\lambda = 353$ nm would enable you to measure the change in concentration of $I_2$ during the reaction.

**11.17** An anion $L^{4-}$ reacts with aqueous $Cr^{3+}$ ions to give a coloured complex $[CrL_x]^{n-}$. To find the stoichiometry of this species, $10\,cm^3$ solutions composed of $V\,cm^3$ of a $0.002\,mol\,dm^{-3}$ aqueous solution of $Cr(NO_3)_3$ and $(10 - V)\,cm^3$ $0.002\,mol\,dm^{-3}$ aqueous $H_4L$ were prepared. With a colorimeter tuned to absorb light corresponding to that absorbed by $[CrL_x]^{n-}$, a Job's plot was recorded. Use the following results to determine $x$ and $n$ in $[CrL_x]^{n-}$.

| $V / cm^3$ | Absorbance |
| --- | --- |
| 0 | 0 |
| 1.0 | 0.21 |
| 2.0 | 0.40 |
| 3.0 | 0.59 |
| 4.0 | 0.81 |
| 6.0 | 0.80 |
| 7.0 | 0.60 |
| 8.0 | 0.38 |
| 9.0 | 0.20 |
| 10.0 | 0 |

# 12  Vibrational and rotational spectroscopies

## Topics

- Vibrations of diatomic molecules
- Selection rules
- Vibrational spectroscopy: diatomic and small polyatomic molecules
- Use of IR spectroscopy as an analytical tool
- Rotating molecules and moments of inertia
- Rotational spectroscopy: rigid rotor diatomic molecules

## 12.1  Introduction

Molecules are able to *translate* (i.e. move in space), *vibrate* and *rotate* unless they are restricted, for example by being in a crystalline lattice. In this chapter, we introduce spectroscopic methods that are used to probe vibrational and rotational motions of molecules. One of our main objectives is to consider what information can be derived from vibrational and rotational spectra.

Vibrational spectroscopy is concerned with the study of molecular vibrations and in Table 11.1, two types of vibrational spectroscopy were listed: infrared (IR) and Raman spectroscopies. Of these techniques, the former is more widely available as a routine method in practical laboratories. An IR spectrum measures the energy at which a molecular species absorbs infrared radiation, and the absorption is usually recorded in wavenumbers. Each absorption is associated with a *vibrational mode* of the molecule.

Figure 11.3 showed that molecular rotations require less energy than molecular vibrations. Just as the absorption of *infrared* radiation allows us to study *vibrational* motions of a molecule, so the absorption of *microwave* radiation permits an investigation of molecular *rotational* motion. In a synthetic or analytical laboratory, the primary aim is usually to prepare and identify a compound. In this context, rotational spectroscopy does not play such an important routine role as IR spectroscopy, and the relative coverages of vibrational and rotational spectroscopies in this chapter reflect this.

> Molecular rotations require less energy than molecular vibrations.

## 12.2  The vibration of a diatomic molecule

### The molecule–spring analogy

A diatomic molecule is composed of two bonded atoms and can be likened to a classical model of two objects (corresponding to the atoms) connected by a spring (corresponding to the bonding electrons). Oscillations of the spring correspond to *stretching modes* for the molecule (Figure 12.1). Even at 0 K,

**Fig. 12.1** The stretching mode of a heteronuclear diatomic molecule: (a) the equilibrium position, (b) extension of the bond and (c) compression of the bond.

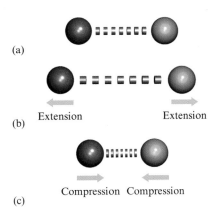

▣▷
Zero point energy:
see Section 4.5

The *zero point energy* of a molecule corresponds to the energy of its lowest vibrational level (***vibrational ground state***).

the $H_2$ molecule is vibrating and possesses an internal energy called the zero point energy. This corresponds to the energy of the *lowest vibrational level* or the *vibrational ground state* of the molecule. By providing the molecule with energy, transitions to higher vibrational levels may be made. *The molecule vibrates more vigorously than in the lowest vibrational level.* If sufficient energy is put into the system, the molecule will dissociate.

The analogy between a spring and a molecule is a common one, but one important difference is easily forgotten. When we talk about a *spring*, the rest state is a static one, but a molecule is never stationary – it is always vibrating, even in the vibrational ground state. When we talk about vibrational modes of a molecule we really mean that we are taking the molecule from one vibrational state (usually its vibrational ground state) to the next vibrational (excited) state. Another important difference between the vibrations of a spring and those of a molecule is that the vibrational energy of a molecule is quantized and *only certain transitions are allowed.*

In general, the energy difference between the vibrational ground state and the first excited state is such that at 298 K, most molecules are in the ground state; the number of molecules in a particular state is called the *population* of that state. At 298 K, the vibrational ground state has a large population, and higher states have only small populations. The distribution of molecules between the vibrational states is given by the *Boltzmann distribution*. If we consider just two states with energies $E_i$ and $E_j$, then the Boltzmann distribution of molecules at a given temperature $T$ is given by equation 12.1, where $N_i/N_j$ is the ratio of the numbers of molecules in the two states.

$$\frac{N_i}{N_j} = e^{\left(-\frac{E_i - E_j}{kT}\right)} = e^{\left(-\frac{\Delta E}{kT}\right)} \qquad (12.1)$$

where: $k$ = Boltzmann constant = $1.38 \times 10^{-23}\,\text{J K}^{-1}$

The populations of states of increasing energies decreases exponentially, the lowest state having the greatest population. In this chapter, we assume that the only transition of importance is from the vibrational ground state to the first excited state, and *whenever we refer to a molecular vibration, we are referring to this vibrational transition.*

### The simple harmonic oscillator

When a spring vibrates, a force must be applied to counter the vibration and bring the spring back to its rest state. This is the *restoring force* and its

**Fig. 12.2** (a) The energy curve of a simple harmonic oscillator. The curve is symmetrical about the point $r_0$, the equilibrium position. (b) The quantized energy levels for a diatomic molecule undergoing simple harmonic motion are equally spaced. (c) The energy curve of an anharmonic oscillator. (d) The quantized energy levels for a diatomic molecule undergoing anharmonic motion; given sufficient vibrational energy, the molecule dissociates.

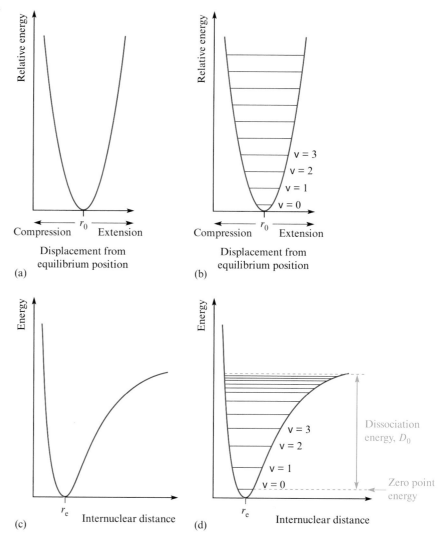

magnitude depends on the magnitude of the oscillation and on a property of the spring called the *force constant*, $k$ (equation 12.2). The units of $k$ are $N\,m^{-1}$ (force per unit distance).

$$\text{Restoring force} = -\left( \begin{array}{c} k \times \text{displacement of particle (in m) from} \\ \text{its equilibrium position} \end{array} \right) \quad (12.2)$$

The energy curve that corresponds to equation 12.2 is shown in Figure 12.2a, and the motion that this energy curve describes is that of a *simple harmonic oscillator*. The vibrational energies, $E_v$, of a molecule undergoing simple harmonic oscillations are quantized and the vibrational energy levels are given by equation 12.3.

$$E_v = (v + \tfrac{1}{2})h\nu \qquad (E_v \text{ in J}) \quad (12.3)$$

where: $v$ is the vibrational quantum number

$h$ = Planck constant

$\nu$ = frequency of vibration

Solving equation 12.3 for $v = 0, 1, 2, 3 \ldots$ gives values of $E_v = \tfrac{1}{2}h\nu, \tfrac{3}{2}h\nu, \tfrac{5}{2}h\nu, \tfrac{7}{2}h\nu \ldots$, showing that the quantized energy levels are *equally spaced* as shown in Figure 12.2b. The lowest energy level ($v = 0$) is the zero point energy of the molecule. It can be shown[§] that for the simple harmonic oscillator, transitions between vibrational states are restricted by the selection rule given in equation 12.4. We return to selection rules in Section 12.3.

$$\Delta v = \pm 1 \tag{12.4}$$

Inspection of Figure 12.2b leads to the conclusion that a diatomic molecule undergoing simple harmonic oscillations can never dissociate, but clearly, for a real molecule, this is not true. Given enough energy, a diatomic *will* dissociate as we discussed in Chapter 4. Nonetheless, the simple harmonic oscillator model is widely used to describe the vibrational motion of diatomic molecules. To understand why this is, we must look at the vibrational motion of real diatomic molecules.

## The anharmonic oscillator

Real molecules undergo *anharmonic oscillations* and Figure 12.2c shows the potential energy curve (a *Morse potential*) that describes this motion. We have already considered such a curve in Figure 4.6. The obvious difference when one compares the curves in Figures 12.2a and 12.2c is that the former is symmetrical, whereas the latter is asymmetrical about the equilibrium position. The quantized energy levels that correspond to the anharmonic oscillator are given by equation 12.5.

$$E_v = (v + \tfrac{1}{2})h\nu - (v + \tfrac{1}{2})^2 h\nu x_e \qquad (E_v \text{ in J}) \tag{12.5}$$

where: $v$ is the vibrational quantum number

$h$ = Planck constant

$\nu$ = frequency of vibration

$x_e$ = anharmonicity constant (a small *positive* number)

Solving equation 12.5 for values of $v = 0, 1, 2, 3 \ldots$ gives energy levels which get closer and closer together and eventually converge as shown in Figure 12.2d. The vibrational energy level for $v = 0$ is the zero point energy of the molecule. The energy difference between this level and the convergence limit equals the dissociation energy of the molecule.

## What are the limitations of the simple harmonic oscillator model?

By comparing Figures 12.2b and 12.2d, we can say something about the limitations of the harmonic oscillator model. Since we are principally concerned with transitions from the vibrational ground state ($v = 0$) to the *first* excited state ($v = 1$, see earlier in this section), we are looking only at the lowest part of the potential energy well. In this region, the curves in Figures 12.2b and 12.2d are similar, and it is valid to apply the simple harmonic oscillator model. For higher excited states, the match between the curves is poor, and the model of the spring is inappropriate. In Figure 12.2d, the flattening of the curve on the right-hand side has no parallel in Figure 12.2b. As this limit (i.e. molecular dissociation) is approached, the model of the simple harmonic oscillator fails completely.

---

[§] For a derivation of this selection rule, see: P. Atkins and J. de Paula (2002) *Atkins' Physical Chemistry*, 7th edn, Oxford University Press, Oxford, p. 513.

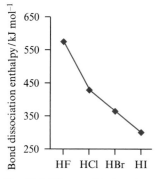

**Fig. 12.3** The trend in bond dissociation enthalpies along the series of the hydrogen halides, HX.

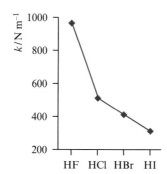

**Fig. 12.4** The trend in force constants for the bonds in the hydrogen halides, HX.

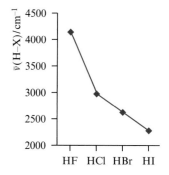

**Fig. 12.5** The trend in values of the fundamental vibrational wavenumbers for the hydrogen halides, HX.

We stated earlier that, at 298 K, the vibrational ground state has a large population and the only transitions of importance are between the vibrational levels $v = 0$ and $v = 1$. For these transitions, therefore, we can consider the diatomic molecule to be undergoing simple harmonic motion, and for the remainder of our discussion, we assume that this model is valid.

### The force constant of a bond

We introduced the force constant, $k$, of a *spring* in equation 12.2. The force constant of a *bond* is related to the bond strength. The stronger the bond, the larger the value of $k$. In Figures 12.3 and 12.4 we plot the dissociation enthalpies and force constants of the bonds in HF, HCl, HBr and HI. The similarity of the trends in these graphs is clear.

### The reduced mass of a diatomic molecule

Let us return to the analogy between a diatomic molecule and a spring. If the two objects connected by the spring have very different masses, then during the oscillation of the spring, the smaller mass will move more freely than the larger one – if you connected a spring between a car and a tennis ball and set the spring in motion, the ball would move *far* more easily than the car, and any oscillation of the spring would effectively describe the movement of *only* the ball. However, if the masses of the two particles are similar, then movement of *both* the particles will be significant as the spring oscillates. The more similar the masses, the more equally the two particles contribute to the overall motion.

A diatomic molecule may contain two nuclei that are of similar mass (e.g. CO) or very different masses (e.g. HI). We need to define a quantity describing the mass in such a way that it reflects the *relative* masses of the nuclei. This is the *reduced mass*, $\mu$,[§] and for two nuclei of masses $m_1$ and $m_2$, the reduced mass is obtained from equation 12.6.

$$\frac{1}{\mu} = \frac{1}{m_1} + \frac{1}{m_2} \tag{12.6}$$

---

[§] The symbol $\mu$ is used for both reduced mass and dipole moment (as well as other things), but the context should minimize confusion.

### The frequency of the vibration of a diatomic molecule

We now consider the frequency of the vibration of the molecule. The analogy with the spring is a good place to start because we can apply the principle of simple harmonic motion, but keep in mind the *relative* motions of the two nuclei in the molecule.

The transition from the vibrational ground state to the first excited state gives rise to a *fundamental absorption* in the vibrational spectrum of a diatomic molecule. The frequency (in Hz) of this absorption is related to the force constant of the bond (in $N\,m^{-1}$) and the reduced mass (in kg) (equation 12.7).

$$\text{Vibrational frequency} = \nu = \frac{1}{2\pi}\sqrt{\frac{k}{\mu}} \tag{12.7}$$

where: $k$ = force constant; $\mu$ = reduced mass

The units of $\nu$ in equation 12.7 are Hz, but the scale on a spectrometer is usually in wavenumbers. Equation 12.8 gives the relationship between *wavenumber* and *stretching frequency* and combining equations 12.7 and 12.8 leads to equation 12.9. Worked example 12.1 illustrates how equation 12.9 may be applied.

$$\bar{\nu} = \frac{\nu}{c} \tag{12.8}$$

where: $\bar{\nu}$ = wavenumber; $\nu$ = frequency; $c$ = speed of light

$$\text{Vibrational wavenumber} = \bar{\nu} = \frac{1}{2\pi \times c}\sqrt{\frac{k}{\mu}} \tag{12.9}$$

It follows that less energy is required to stretch bonds involving heavy elements than light ones, and Figure 12.5 shows the trend in fundamental vibrational wavenumbers of the hydrogen halides. A comparison of Figures 12.3 to 12.5 emphasizes that there is a relationship between the vibrational wavenumber, dissociation enthalpy and force constant of a bond.

---

**Worked example 12.1**        *Determination of the force constant of the bond in carbon monoxide*

The stretching mode of CO gives an absorption in the IR spectrum at $2170\,cm^{-1}$. What is the force constant of the bond in carbon monoxide?

The equation needed is

$$\bar{\nu} = \frac{1}{2\pi \times c}\sqrt{\frac{k}{\mu}}$$

and this can be rearranged to give $k$ as the subject:

$$2\pi \times c \times \bar{\nu} = \sqrt{\frac{k}{\mu}}$$

$$k = 4\pi^2 \times c^2 \times \bar{\nu}^2 \times \mu$$

We must ensure that the units are consistent.
The wavenumber of the absorption is $2170\,cm^{-1}$.
For consistency, $c = 3.00 \times 10^8\,m\,s^{-1} = 3.00 \times 10^{10}\,cm\,s^{-1}$.

The reduced mass, $\mu$, is given by the equation:

$$\frac{1}{\mu} = \frac{1}{m_1} + \frac{1}{m_2} = \left(\frac{1}{12.01 \times (1.66 \times 10^{-27}\,\text{kg})}\right) + \left(\frac{1}{16.00 \times (1.66 \times 10^{-27}\,\text{kg})}\right)$$

$$\mu = \frac{1}{8.78 \times 10^{25}}\,\text{kg}$$

Thus, the force constant is given by:

$$k = 4\pi^2 \times c^2 \times \bar{\nu}^2 \times \mu$$

$$= 4 \times 3.142^2 \times (3.00 \times 10^{10}\,\text{cm s}^{-1})^2 \times (2170\,\text{cm}^{-1})^2 \times \left(\frac{1}{8.78 \times 10^{25}}\,\text{kg}\right)$$

$$= 1906\,\text{N m}^{-1}$$

$$= 1910\,\text{N m}^{-1} \quad \text{(to 3 sig. fig.)}$$

The units of $\text{N m}^{-1}$ follow because the newton (Table 1.2) is a derived unit:

$$\text{N} = \text{kg m s}^{-2}$$

and from the second line of the calculation, the units of $k$ are $\text{kg s}^{-2} = \text{N m}^{-1}$.

---

<table>
<tr><td>**12.3**</td><td>**Infrared spectra of diatomic molecules**</td></tr>
</table>

In the last section, we developed a model in which the vibrations of a diatomic molecule are considered in terms of the simple harmonic oscillations of a spring. However, this model has certain limitations. A molecule will undergo a transition from one vibrational level to another (and usually this is from the vibrational ground state to the first excited state) only if it absorbs radiation of an appropriate frequency. The frequency and wavenumber of the vibration depend on the force constant of the bond and the reduced mass of the system (equations 12.6 and 12.7). The range of frequencies covered corresponds to the infrared (IR) part of the electromagnetic spectrum. An IR spectrum usually records the *wavenumber* of the vibration, but not every vibration gives rise to an observable absorption band in the spectrum. Two *selection rules* must be obeyed.

*Electromagnetic spectrum: see Appendix 4*

### Selection rules

The first selection rule (equation 12.10) restricts the levels between which the transition may occur.

$$\Delta v = \pm 1 \tag{12.10}$$

The vibrational mode of a **homonuclear** diatomic molecule (e.g. $H_2$, $O_2$ or $N_2$) is infrared (IR) inactive. The vibrational mode of a *heteronuclear* diatomic molecule which is polar (e.g. CO, HCl) is infrared (IR) active.

The second selection rule states that for a vibrational mode to be *infrared active* (i.e. observed in an IR spectrum), it must give rise to a change in the molecular dipole moment. Consider the $H_2$ molecule. It is non-polar, and when the $H-H$ bond is stretched or compressed, the molecule does *not* gain a dipole moment. This is true for all *homonuclear* diatomic molecules, e.g. $O_2$, $Br_2$, $N_2$. The vibrational mode of any *homonuclear* diatomic molecule is therefore *IR inactive*. Now consider a *heteronuclear* diatomic molecule such as CO. A CO molecule has a dipole moment $\mu$ of 0.11 D ($C^{\delta+}O^{\delta-}$), with this value corresponding to the equilibrium separation (113 pm) of the nuclei (equation 12.11).

*Dipole moments: see Box 5.2*

$$\mu = (\text{electronic charge}) \times (\text{distance between the charges}) \tag{12.11}$$

As the molecule vibrates, the magnitude of the dipole moment changes, and this leads to the vibrational mode being *IR active*. This is true of any polar diatomic molecule.

---

Vibrational selection rules:

- $\Delta v = \pm 1$

  - a vibrational mode is IR active if it results in a change in molecular dipole moment.

---

**Worked example 12.2**     *IR active and inactive vibrational modes*

**The fundamental vibrational wavenumber of HBr is 2650 cm$^{-1}$. Does this give rise to an absorption in the IR spectrum of HBr?**

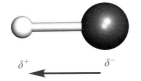

$\delta^+$        $\delta^-$

HBr is a polar molecule: the Pauling electronegativities of H and Br are 2.2 and 3.0 respectively (Table 5.2).

A vibration gives rise to a change in dipole moment and is therefore IR active. An absorption at 2650 cm$^{-1}$ will be observed in the IR spectrum of HBr.

---

**12.4**     **Infrared spectroscopy of triatomic molecules**

### Vibrational modes of freedom

Whereas the vibration of a diatomic molecule is restricted to a single stretching mode (Figure 12.1), the vibrational modes of a polyatomic molecule are more complex.

### CO$_2$: a non-polar linear triatomic molecule

$O{=}C{=}O$

(12.1)

$\chi^P(C) = 2.6, \chi^P(O) = 3.4$

The carbon dioxide molecule (**12.1**) is linear and contains two equivalent C=O bonds. When the molecule vibrates, it can do so in two different ways. Figure 12.6a shows the *symmetric stretch* in which the two oxygen atoms move outwards at the same time and then move in together. Figure 12.6b illustrates the *asymmetric stretch* – here one C=O bond is stretched as the other is compressed. Carbon dioxide does *not* possess a molecular dipole moment, although each C=O bond is polar. During the *symmetric* stretch, there is *no* change in dipole moment, and the symmetric stretch is IR inactive. During the *asymmetric* stretch, a change in dipole moment occurs, and you can see why this happens by studying the right-hand side of Figure 12.6b. The asymmetric stretch is therefore IR active.

A linear triatomic molecule also possesses a *bending mode* (Figure 12.6c); this vibration for CO$_2$ may also be represented in terms of the carbon atom moving up and down with respect to the two oxygen atoms. In Figure 12.6c, the motion is represented in the plane of the paper, but there is an equivalent vibration perpendicular to the plane of the paper. These two vibrational modes require the same amount of energy and are degenerate. The bending

*Degenerate* means 'having the same energy'.

## Box 12.1 Spectroscopy and air-quality monitoring

Monitoring air quality is an important method of keeping a watch on atmospheric pollution, and in many European countries automated monitoring stations are in operation. Spectroscopic methods of analysis are used to monitor levels of CO, $O_3$, $NO_2$ and $SO_2$.

Carbon monoxide has an IR spectroscopic absorption at $2174\,cm^{-1}$, while ozone can be detected by using UV spectroscopy ($\lambda_{max} = 254\,nm$). Monitoring the levels of $NO_2$ is less straightforward: $NO_2$ is first thermally decomposed to NO and then the chemiluminescent reaction between NO and $O_3$ is followed, light from which can be detected and used to quantify the concentration of NO. Fluorescence spectroscopy is used to measure amounts of $SO_2$ in the atmosphere.

Other pollutants are monitored by a range of non-spectroscopic methods including gas chromatography (see Box 24.4).

Dust particles also contribute to atmospheric pollution and samples taken in Leeds, UK, during

A LIDAR air pollution unit uses a laser beam to probe the atmosphere for pollutants. *Los Alamos National Laboratory/Science Photo Library.*

1982–83 illustrate some of the problem particles. (Data illustrated below from *Chemistry in Britain* (1994), vol. 30, p. 987; (1995), vol. 31, p. 131.)

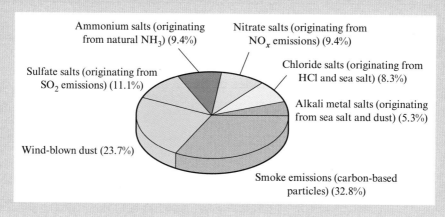

of the $CO_2$ molecule leads to a change in the dipole moment, and this mode is IR active.

The frequencies of the asymmetric stretching and bending vibrations of $CO_2$ are 2349 and $667\,cm^{-1}$ respectively, and this tells us that the molecule requires more energy to undergo an asymmetric stretch than a bend.

### HCN: a polar, linear triatomic molecule

Molecular dipole moment

H —— C ≡≡ N

(12.2)

Hydrogen cyanide (**12.2**) is a polar, linear triatomic molecule ($\mu = 2.98\,D$). Each of the symmetric and asymmetric stretches gives rise to a change in dipole moment, and each is IR active. The bending mode also gives rise to a change in dipole moment and is IR active. Just as in $CO_2$, the bending mode in HCN is doubly degenerate.

**Fig. 12.6** The vibrational modes of carbon dioxide. (a) The symmetric stretch – the diagram on the left-hand side summarizes the motion; follow the figures from top to bottom in the box to see the overall motion of the $CO_2$ molecule during the symmetric stretch. (b) The asymmetric stretch is summarized in the diagram on the left-hand side; follow the figures from top to bottom in the box to see the overall motion of the $CO_2$ molecule during the asymmetric stretch. (c) The bending mode is summarized in the left-hand diagram; in the box, the series of diagrams follows the motion as the molecule bends in the plane of the paper. There is another equivalent bending mode in a plane perpendicular to that of the paper.

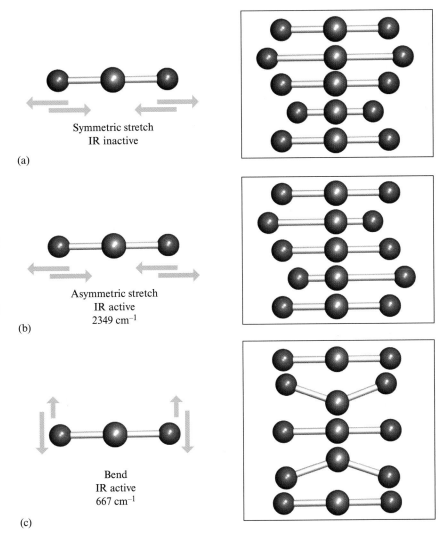

(a)

Symmetric stretch
IR inactive

(b)

Asymmetric stretch
IR active
2349 cm$^{-1}$

(c)

Bend
IR active
667 cm$^{-1}$

We now come to an important general point about linear XYZ molecules, i.e. molecules containing three different atoms. Provided that the atomic masses of X and Z are significantly different, the absorptions observed in the IR spectrum of an XYZ molecule can be assigned to the X−Y stretch, the Y−Z stretch and the XYZ bend. The reason for assigning the stretching modes to individual bond vibrations rather than to the molecule as a whole is that each of the symmetric and asymmetric stretches is essentially dominated by the stretching of one of the two bonds. The IR spectrum of HCN contains three absorptions at 3311, 2097 and 712 cm$^{-1}$ which are assigned to the H−C stretch, the C≡N stretch and the HCN bend respectively.

## H$_2$O: a bent triatomic molecule

The three vibrational modes for H$_2$O are shown in Figure 12.7. Water (**12.3**) possesses a molecular dipole moment ($\mu = 1.85\,\mathrm{D}$) and this changes during the symmetric stretching vibration. This is generally true for bent molecules and the symmetric stretch is IR active. The asymmetric stretching mode is also IR active. If the two hydrogen atoms move towards each other and

Molecular
dipole
moment

H      H

(12.3)

**Fig. 12.7** The vibrational modes of water. (a) In the symmetric stretch, both H atoms move out at the same time, and then in together. (b) In the asymmetric stretch, one H atom moves out as the other moves in. (c) During the symmetric bending or scissoring of the molecule, the H−O−H bond angle increases and decreases; this vibration takes place in the plane of the paper.

then away from one another, this is called a *scissoring* motion or symmetric bending. A change in molecular dipole moment occurs and an additional absorption is observed in the IR spectrum.

---

**Worked example 12.3**    *IR active and inactive vibrational modes*

The IR spectrum of $XeF_2$ exhibits absorptions only at 213 and 555 cm$^{-1}$, while that of $Cl_2O$ has bands only at 296, 639 and 686 cm$^{-1}$. Deduce the shapes of these triatomic molecules.

Triatomics of the general type $YX_2$ can be linear or non-linear. If linear, the $YX_2$ molecule is non-polar. If non-linear, it may be polar.
The vibrational modes are:

Symmetric stretch    IR inactive in linear $YX_2$    IR active in non-linear $YX_2$

Asymmetric stretch    IR active in linear $YX_2$     IR active in non-linear $YX_2$

Bend              IR active in linear $YX_2$     IR active in non-linear $YX_2$

A linear $YX_2$ molecule therefore shows two absorptions in its IR spectrum, but a non-linear $YX_2$ molecule exhibits three.

Conclusion: $XeF_2$ is linear, and $Cl_2O$ is non-linear.

---

## 12.5    Vibrational degrees of freedom

In the discussion above, we appear to have arbitrarily arrived at the *number* of vibrational modes for $CO_2$, HCN and $H_2O$, and certainly when considering larger molecules, it is difficult to write down all the possible vibrational modes. There are, however, some simple rules from which you can work out the number of allowed vibrations.

> A molecule with *n* atoms has 3*n* *degrees of freedom*.

If a molecule has *n* atoms, it possesses 3*n* *degrees of freedom*. The molecule as a whole can move in space and this is described as *translational* motion. Movement is a vector quantity, and translational motion can be described in terms of three degrees of freedom relating to three axes (for example, *x*, *y* and *z*). The molecule possesses *three degrees of translational freedom*.

Having allocated three degrees of freedom for translational motion, the molecule is left with (3*n* − 3) degrees of freedom for other types of motion and these are classified as *rotational* or *vibrational degrees of freedom*. Like translational motion, rotations may be described with respect to an axis set.

We simply state[§] that for a non-linear molecule there are three degrees of rotational freedom, but for a linear molecule, there are only two degrees of rotational freedom. The number of degrees of vibrational freedom is dependent upon the number of atoms in the molecule and on whether the molecule is linear or non-linear. Equations 12.12 and 12.13 summarize these results.

$$\text{Number of degrees of vibrational freedom for a } \textit{non-linear molecule} = 3n - 6 \tag{12.12}$$

$$\text{Number of degrees of vibrational freedom for a } \textit{linear} \text{ molecule} = 3n - 5 \tag{12.13}$$

We now apply equations 12.12 and 12.13 to several small molecules, the shapes of which can be predicted by VSEPR theory.

VSEPR: see Section 6.6

**Example 1: $SO_2$**    $SO_2$ is a bent molecule. $n = 3$
Number of degrees of vibrational freedom $= 3n - 6 = 3$

**Example 2: $CO_2$**    $CO_2$ is a linear molecule. $n = 3$
Number of degrees of vibrational freedom $= 3n - 5 = 4$
Two of the modes of vibration are degenerate (see earlier in this section).

**Example 3: $CH_4$**    $CH_4$ is a tetrahedral molecule. $n = 5$
Number of degrees of vibrational freedom $= 3n - 6 = 9$

In addition to the stretching and scissoring (bending) modes described for $CO_2$ and $H_2O$, larger molecules can undergo *rocking*, *twisting* and *wagging* vibrations. Detailed discussion of these modes is beyond the scope of this book, but Figure 12.8 summarizes the modes of vibration that a methylene group in an alkane chain can undergo.

## 12.6    The use of IR spectroscopy as an analytical tool

We now turn our attention to the everyday use of IR spectroscopy in the laboratory. Compounds may be examined in the solid state, solution or gas phase. There are two ways of interpreting an IR spectrum. It is always correct to interpret the spectrum as a property of the molecule as a whole, but this is extremely time-consuming and possible only for simple molecules. It is common to consider the IR spectrum as a composite of individual absorptions from various components of the molecule. This is related to the functional group model in organic chemistry, which has as a fundamental assumption the fact that a group has similar properties regardless of the precise molecule in which it is found. Similarly, it is convenient to interpret IR spectra in terms of absorptions arising from specific functional groups. *This is an oversimplification*, and is particularly dangerous with 'inorganic' compounds.

Organic functional groups: see Table 1.9

The IR spectra of organic compounds possess several characteristic regions:

- those arising from vibrations and other modes of the molecular framework, particularly C–C single bonds;

---

[§] For further discussion see: P. Atkins and J. de Paula (2002) *Atkins' Physical Chemistry*, 7th edn, Oxford University Press, Oxford.

**Fig. 12.8** The vibrational modes of a methylene (CH$_2$) group in an alkane chain: (a) the symmetric stretching mode; (b) the asymmetric stretching mode; (c) the scissoring mode; (d) the rocking mode (in the plane of the paper); (e) the twisting mode; and (f) the wagging mode (the CH$_2$ unit moves in front of and behind the plane of the paper).

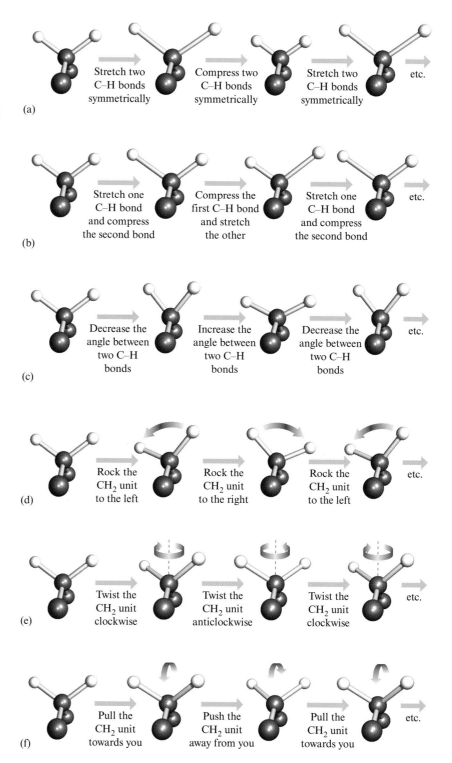

- those arising from vibrations of a functional group, in particular those with multiple bonds.

This effectively divides the spectrum into two regions. Above 1500 cm$^{-1}$, the majority of absorptions are due to stretching modes of multiple bonds (C=C, C≡C, C=O, C=N, C≡N, etc.). Below 1500 cm$^{-1}$, in the *fingerprint region*, absorptions due to C−X single bond stretching modes as well as other

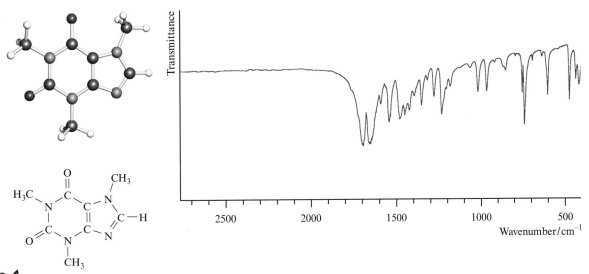

**Fig. 12.9** The IR spectrum of caffeine; the structure consists of two fused rings containing carbon and nitrogen atoms. Colour code in structural diagram: C, grey; O, red; N, blue; H, white.

vibrational modes are found. These distinctions are not absolute, e.g. absorptions due to C–H stretches are found around $3000 \, \text{cm}^{-1}$.

### The fingerprint region of the IR spectrum

Usually the IR spectrum of a compound contains a series of absorptions below about $1500 \, \text{cm}^{-1}$ which *together* are diagnostic of that compound. It is not usual for these absorptions to be individually assigned to particular vibrations within the molecule. Consider the IR spectrum of caffeine (the stimulant found in tea and coffee) shown in Figure 12.9. There are numerous absorptions below $1500 \, \text{cm}^{-1}$, but it is difficult to assign them to individual vibrational modes of the molecule. However, the spectrum is an absolute characteristic of the molecule caffeine – no other molecule exhibits an identical spectrum. This is why this part of the spectrum is known as the *fingerprint region*.

The most effective way to use the fingerprint region of an IR spectrum is to match it with that of an authentic sample of the compound.

### Functional groups and the interpretation of IR spectra

Functional groups in a molecule can give rise to absorptions at particular frequencies in the range $400$–$4000 \, \text{cm}^{-1}$. It is easiest to detect (and assign) functional group absorptions that do not fall within the fingerprint region.

*Carbonyl compounds: see Chapter 33*

We take as an example the carbonyl group C=O which occurs in a range of compounds, including aldehydes (**12.4**), ketones (**12.5**), carboxylic acids (**12.6**), acid chlorides (**12.7**), esters (**12.8**) and carboxylates (**12.9**).

$$O=C\overset{R}{\underset{OR}{\diagdown}} \qquad O=C\overset{R}{\underset{O^{\ominus}}{\diagdown}}$$

**(12.8)**          **(12.9)**

The stretching of the C=O double bond typically gives rise to a *strong* absorption around $1700\,cm^{-1}$ (see Figure 33.4), but the exact frequency depends upon the strength of the bond, and this is in turn affected by other groups in the molecule or ion. Take the aldehyde RCHO as a reference point. On going from this to a ketone, the H is changed for an *electron-releasing alkyl group* and the carbon atom of the C=O group becomes less $\delta^+$. The carbonyl bond polarization ($C^{\delta^+}-O^{\delta^-}$) is therefore smaller in the ketone than in the aldehyde, and as a consequence stretching the C=O bond in the ketone is easier than in the aldehyde. The carbonyl absorptions for acetone (a ketone) and ethanal (an aldehyde) are shown in Figure 12.10.

Now compare the aldehyde RCHO (**12.4**) with the corresponding acid chloride RCOCl (**12.7**). Chlorine is *electron-withdrawing* and the carbon atom of the C=O group carries a greater $\delta^+$ charge in the acid chloride than in the aldehyde. This means that the C=O bond has increased ionic character and is more difficult to stretch – the stretching frequency increases accordingly (Table 12.1).

In an ester, there are two effects. The oxygen atom of the OR group is more electronegative than carbon and this tends to increase the $\delta^+$ charge on the carbonyl C atom. However, the lone pair on the OR oxygen atom can be delocalized (equation 12.14). This effect weakens the carbonyl carbon–oxygen bond. The overall effect is that the absorption due to the

◼▶ **Electron-releasing:** see Section 25.8

◼▶ $\chi^P(Cl) = 3.2,\ \chi^P(H) = 2.2$

**Fig. 12.10** The carbonyl absorptions for acetone, ethanal and ethyl acetate. In the structural diagrams, Me = methyl = $CH_3$, Et = ethyl = $CH_2CH_3$. Colour code: C, grey; O, red; H, white.

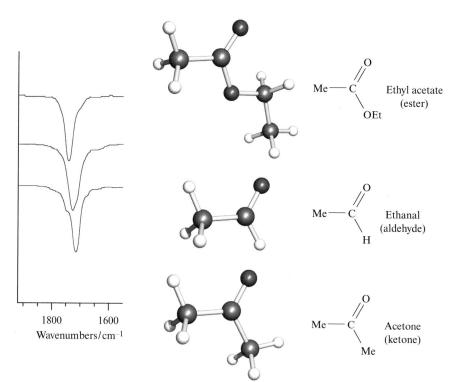

Me—C $\overset{O}{\underset{OEt}{\diagup\diagdown}}$   Ethyl acetate (ester)

Me—C $\overset{O}{\underset{H}{\diagup\diagdown}}$   Ethanal (aldehyde)

Me—C $\overset{O}{\underset{Me}{\diagup\diagdown}}$   Acetone (ketone)

1800    1600
Wavenumbers/cm⁻¹

**Table 12.1** Approximate ranges for carbonyl (C=O) stretching frequencies in compounds in structures **12.4** to **12.9** with R = alkyl.

| Compound type | Structure number | Absorption range / cm$^{-1}$ |
|---|---|---|
| Aldehyde | **12.4** | 1740–1720 |
| Ketone | **12.5** | 1725–1705 |
| Carboxylic acid | **12.6** | 1725–1700 |
| Acid chloride | **12.7** | 1815–1790 |
| Ester | **12.8** | 1750–1735 |
| Carboxylate | **12.9** | 1610–1550 |

carbonyl stretch in an ester may not be much different from that in the corresponding aldehyde (Figure 12.10).

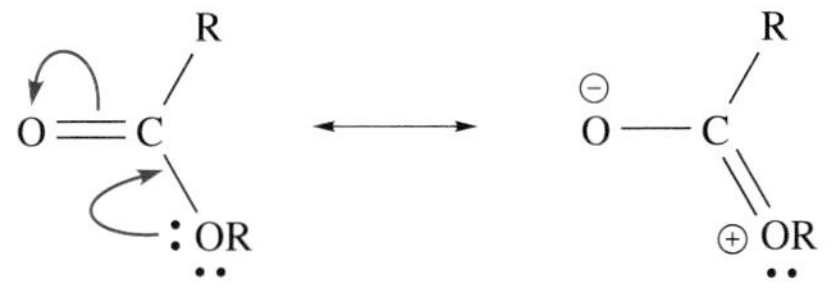

(12.14)

Some other functional groups with multiple bonds to carbon include C≡C, C=C, C≡N, C=N and C=S. Typical frequency ranges for the absorptions due to the stretching of these bonds are listed in Table 12.2.

Many functional groups contain single bonds, and absorptions due to the vibrations of these bonds often fall within the fingerprint region and may not easily be assigned. The stretching frequencies of carbon–halogen bonds are listed in Table 12.3. Whereas the absorptions due to C–Cl stretching are readily observed in the IR spectrum of $CCl_4$ (Figure 12.11a), they disappear into the many bands of the fingerprint region in the spectrum of 1,2-dichloropropane (Figure 12.11b).

See worked example 12.4

**Table 12.2** Approximate ranges for stretching frequencies of some functional groups involving multiple bonds to carbon.

| Functional group | Typical frequency range of absorption band / cm$^{-1}$ |
|---|---|
| $R_2C=CR_2$ | 1680–1620 |
| $RC≡CR$ | 2260–2100 |
| $R_2C=NR$ | 1690–1640 |
| $RC≡N$ | 2260–2200 |
| $R_2C=O$ | 1815–1550 (see Table 12.1) |
| $R_2C=S$ | 1200–1050 |

**Table 12.3** Approximate ranges for stretching frequencies of carbon–halogen bonds.

| Group | Typical frequency range of absorption band / cm$^{-1}$ |
|---|---|
| C–F | 1400–1000 |
| C–Cl | 800–600 |
| C–Br | 750–500 |
| C–I | ≈500 |

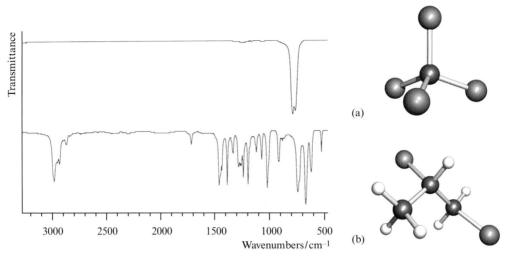

Fig. 12.11 The IR spectra of (a) tetrachloromethane, and (b) 1,2-dichloropropane. Colour code in structural figures: C, grey; Cl, green; H, white.

---

| Worked example 12.4 | *Vibrational modes of CCl₄* |

**The IR spectrum of $CCl_4$ is shown in Figure 12.11a. The absorption at $776 \, cm^{-1}$ is assigned to a vibrational mode. Comment on this fact.**

A molecule of $CCl_4$ has a tetrahedral structure (apply VSEPR theory), and the molecule is non-polar.

Vibrational modes involve stretching the C−Cl bonds.

If the four C−Cl bonds stretch at the same time, the vibrational mode is a *symmetric stretch*:

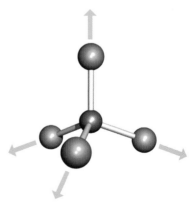

Symmetric stretch
IR inactive

This vibrational mode does not create a change in dipole moment. Therefore, the mode is IR inactive.

The absorption at $776 \, cm^{-1}$ must arise from a stretching mode that generates a change in dipole moment:[§]

---

[§] This asymmetric mode is one of three of equal energy (i.e. the vibrational modes are *degenerate*). The $CCl_4$ molecule has $3n - 6 = 9$ degrees of vibrational freedom: one symmetric stretch, three degenerate asymmetric stretching modes, a set of two degenerate deformation modes, and a set of three degenerate deformation modes, see: C. E. Housecroft and A. G. Sharpe (2005) *Inorganic Chemistry*, 2nd edn, Prentice Hall, Harlow, Chapter 3.

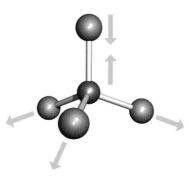

Asymmetric stretch
IR active (776 cm$^{-1}$)

Alcohol nomenclature:
see Section 30.1

Hydrogen bonding in
alcohols: see Section 30.3

Some of the more readily observed IR spectroscopic absorptions are due to stretching modes of O−H, N−H, S−H and C−H bonds (Table 12.4). Figure 12.12 shows the IR spectra of propan-1-ol and propan-2-ol. In each, the band near 3340 cm$^{-1}$ is assigned to the O−H stretch. The broadness of the bands arises from the extensive degree of *hydrogen bonding* that is present between the alcohol molecules and is a common feature of absorptions associated with O−H bonds. Although it would be difficult to assign all the bands in the spectrum of either compound, they exhibit different, but characteristic, IR spectra and can be distinguished provided that authentic samples are

**Table 12.4** Approximate ranges for stretching frequencies of O−H, N−H, S−H and C−H bonds.

| Group | Typical frequency range of absorption band / cm$^{-1}$ |
|---|---|
| O−H | 3600–3200 (broadened by hydrogen bonding) |
| N−H | 3500–3300 |
| S−H | 2600–2350 (weak absorption) |
| C−H ($sp^3$ carbon) | 2950–2850 |
| C−H ($sp^2$ carbon) | 3100–3010 |
| C−H ($sp$ carbon) | ≈3300 |

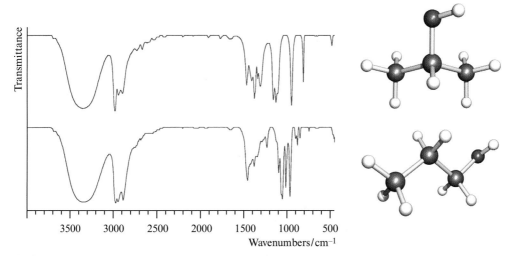

 **Fig. 12.12** The IR spectra of propan-1-ol (bottom) and propan-2-ol (top). Colour code in structural figures: C, grey; H, white; O, red.

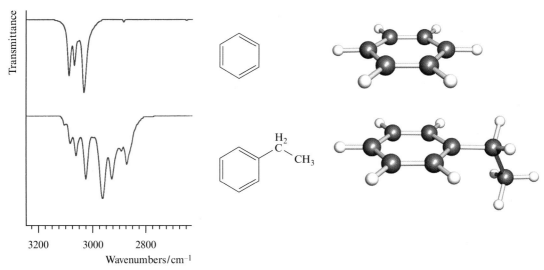

**Fig. 12.13** The IR spectra of benzene (top) and ethylbenzene; only the regions around $3000 \text{ cm}^{-1}$ (the C−H stretching region) are shown. Colour code in structural figures: C, grey; H, white.

available for comparison. Absorptions due to N−H bonds occur in a similar part of the IR spectrum to those of O−H bonds.

In Figures 12.11b and 12.12, the absorptions around $3000 \text{ cm}^{-1}$ are due to C−H stretches. The values in Table 12.4 indicate that the hybridization of the carbon atom influences the stretching frequency of the C−H bond. This is because the polarity (and strength) of a C−H bond decreases in the order $C(sp)−H > C(sp^2)−H > C(sp^3)−H$. Figure 12.13 shows part of the IR spectra of benzene and ethylbenzene. Benzene is a planar molecule and contains only $sp^2$ hybridized carbon atoms – the absorptions due to the C−H stretches fall in the range $\approx 3120–3000 \text{ cm}^{-1}$. In ethylbenzene, absorptions due to *both* $C(sp^2)−H$ and $C(sp^3)−H$ stretches are observed.

We introduce IR spectroscopic absorptions characteristic of other functional groups in the organic chemistry chapters later in the book, and extend the discussion of the use of IR spectroscopy in the laboratory.

▶ **Electronegativities of carbon atoms: see Table 26.1**

---

| **Worked example 12.5** | *The IR spectra of hexane and hex-1-ene* |
|---|---|

**The figure shows the IR spectra of hexane and hex-1-ene. What are the structural features in these molecules that are responsible for the observed absorptions in their spectra?**

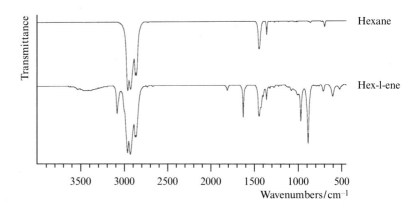

The structures of hexane and hex-1-ene are:

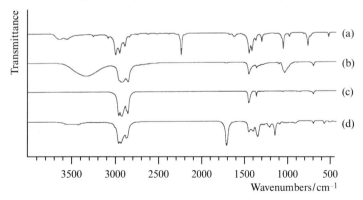

1. The part of each spectrum below 1500 cm$^{-1}$ is the fingerprint region and provides a diagnostic pattern of IR bands for each compound.
2. In the IR spectrum of hexane, the only other feature is the group of absorptions around 3000 cm$^{-1}$ which can be assigned to the C–H stretches; the C atoms are all $sp^3$ hybridized.
3. In the IR spectrum of hex-1-ene, the strong absorptions around 3000 cm$^{-1}$ are due to the C($sp^3$)–H stretches, while the less intense absorption at about 3080 cm$^{-1}$ can be assigned to the C($sp^2$)–H stretches.
4. In the spectrum of hex-1-ene, the sharp absorption at 1643 cm$^{-1}$ is characteristic of a C=C stretch (see Table 12.2).
5. An additional feature in the spectrum of hex-1-ene is the broad band centred around 3400 cm$^{-1}$. This can be assigned to an O–H stretch and suggests that the sample of hex-1-ene is wet!

---

**Worked example 12.6**     *Assignment of some IR spectra*

**The figure below shows the IR spectra of octane, propanenitrile, dodecan-1-ol and heptan-2-one. Assign each spectrum to the correct compound.**

**The structures of the compounds are:**

**Octane**

**Dodecan-1-ol**

**Propanenitrile**

**Heptan-2-one**

Begin the spectral assignments by looking for absorptions that are characteristic of the functional groups present.

Spectrum (a) has a strong absorption at $\approx 2250\,cm^{-1}$. This may be assigned to the $C\equiv N$ group of propanenitrile. The spectrum also contains absorptions around $3000\,cm^{-1}$ assigned to stretches of the aliphatic $C-H$ bonds.

Spectrum (b) exhibits a broad and strong absorption centred at $\approx 3200\,cm^{-1}$ which is characteristic of an alcohol group. The spectrum may be assigned to dodecan-1-ol; $C-H$ stretches around $3000\,cm^{-1}$ are observed and are expected for this compound.

Spectrum (d) shows a strong absorption at $\approx 1700\,cm^{-1}$ indicative of a carbonyl group. This spectrum can be assigned to heptan-2-one. In addition to the fingerprint region, this spectrum shows aliphatic $C-H$ stretches which are expected for this compound.

Above the fingerprint region, spectrum (c) shows only absorptions due to $C-H$ stretches; there appears to be no functional group. This spectrum is assigned to octane.

**Conclusions**

| Spectrum | Assignment |
|----------|------------|
| (a) | Propanenitrile |
| (b) | Dodecan-1-ol |
| (c) | Octane |
| (d) | Heptan-2-one |

## 12.7 Deuteration: the effects on IR spectroscopic absorptions

Naturally occurring hydrogen contains 99.985% $^{1}H$ and 0.015% $^{2}H$. As $^{1}H$ dominates to such a great extent, we can consider, for example, O−H, C−H or N−H bonds in compounds as essentially having the O, C or N atom bonded to $^{1}H$. We saw earlier (equation 12.9) that the fundamental vibrational wavenumber for a diatomic depends on the force constant of the bond and the reduced mass. Similarly, the wavenumber of the stretching mode of, for example, an O−H bond in a compound is *approximately* related to the force constant and reduced mass as shown in equation 12.15.

$$\bar{\nu}_{OH} \propto \sqrt{\frac{k_{OH}}{\mu_{OH}}} \tag{12.15}$$

If the $^{1}H$ atom is replaced by a $^{2}H$ atom (i.e. deuterium, D), then the wavenumber of the stretching mode of the O−D bond is given by equation 12.16.

$$\bar{\nu}_{OD} \propto \sqrt{\frac{k_{OD}}{\mu_{OD}}} \tag{12.16}$$

Exchanging the H in an O−H bond by D has a significant effect on the reduced mass of the system (equations 12.17 and 12.18).

$$\left.\begin{array}{l} \dfrac{1}{\mu_{OH}} = \dfrac{1}{m_O} + \dfrac{1}{m_H} = \dfrac{1}{16.01 \times (1.66 \times 10^{-27}\,kg)} + \dfrac{1}{1.008 \times (1.66 \times 10^{-27}\,kg)} \\[2mm] \mu_{OH} = 1.57 \times 10^{-27}\,kg \end{array}\right\}$$

(12.17)

$$\left.\begin{array}{l} \dfrac{1}{\mu_{OD}} = \dfrac{1}{m_O} + \dfrac{1}{m_D} = \dfrac{1}{16.01 \times (1.66 \times 10^{-27}\,kg)} + \dfrac{1}{2.01 \times (1.66 \times 10^{-27}\,kg)} \\[2mm] \mu_{OD} = 2.96 \times 10^{-27}\,kg \end{array}\right\}$$

(12.18)

We assume that the force constant of the bond is not affected by the isotopic substitution: $k_{OH} = k_{OD}$. We can now combine equations 12.15 and 12.16 to give equation 12.19. This allows us to calculate the ratio of the vibrational wavenumbers of the O−H and O−D bonds.

**(12.10)**

$$\frac{\bar{\nu}_{OH}}{\bar{\nu}_{OD}} = \sqrt{\frac{\mu_{OD}}{\mu_{OH}}} = \sqrt{\frac{2.96 \times 10^{-27}}{1.57 \times 10^{-27}}} = 1.37$$

(12.19)

Equation 12.19 shows that the absorption for an O−D vibration is shifted in the IR spectrum from that assigned to the O−H vibration. The IR spectrum of methanol, **12.10**, exhibits an absorption at $3347\,cm^{-1}$ assigned to the O−H stretch. In $CH_3OD$, the O−D stretch appears at $2443\,cm^{-1}$. Such a dramatic shift can be used to test whether an assignment of an absorption band is correct. For example, if $CH_3OH$ is shaken with $D_2O$, equilibrium 12.20 is established; N−H bonds in amines undergo a similar deuterium exchange.

■▶
*More about equilibria in Chapter 16*

$$CH_3OH + D_2O \rightleftharpoons CH_3OD + HOD$$

(12.20)

Comparisons of the IR spectra of pairs of compounds containing, for example, C−H and C−D, N−H and N−D, or O−H and O−D bonds are valuable in aiding spectroscopic assignments.

---

**Worked example 12.7**     *Deuteration of propan-1-ol*

The IR spectrum of propan-1-ol, $CH_3CH_2CH_2OH$, contains a broad absorption at $3333\,cm^{-1}$. To what wavenumber will this absorption shift if $D_2O$ is shaken with propan-1-ol?

The equations required are:

$$\frac{\bar{\nu}_{OD}}{\bar{\nu}_{OH}} = \sqrt{\frac{\mu_{OH}}{\mu_{OD}}}$$

$$\bar{\nu}_{OD} = \bar{\nu}_{OH} \times \sqrt{\frac{\mu_{OH}}{\mu_{OD}}}$$

The method of calculating $\mu_{OD}$ and $\mu_{OH}$ was shown in equations 12.17 and 12.18. Substitution of these values, and the value of $\bar{\nu}_{OH} = 3333\,cm^{-1}$ gives:

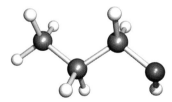

The structure of propan-1-ol

$$\bar{\nu}_{OD} = (3333\,cm^{-1}) \times \sqrt{\frac{(1.57 \times 10^{-27}\,kg)}{(2.96 \times 10^{-27}\,kg)}} = 2430\,cm^{-1} \text{ (to 3 sig. fig.)}$$

| 12.8 | **Rotating molecules and moments of inertia** |
|---|---|

We saw in Section 12.5 that linear molecules have two degrees of *rotational* freedom, while non-linear molecules possess three. In the next two sections, we consider the rotations of diatomic molecules and the spectra that arise from transitions between rotational states. We shall be concerned only with *pure rotational spectra*.

> *Microwaves* are used to study the rotational transitions of gas phase molecules and provide information about internuclear distances and angles.

### Moments of inertia

The rotations of a molecule can be defined with respect to three perpendicular axes $A$, $B$ and $C$ (*principal axes*) illustrated for a diatomic XY in Figure 12.14a. The *moment of inertia* of a rotating molecule is a crucial property in the interpretation of rotational spectra. For a given molecular rotation, the moment of inertia, $I$, of the molecule is given by equation 12.21 where $m$ = mass of an atom (in kg), $r$ = perpendicular distance (in m) of that atom from the axis of rotation, and $\sum$ means summation over all atoms in the molecule.

$$I = \sum mr^2 \quad \text{Units of } I = \text{kg m}^2 \tag{12.21}$$

The *principal moments of inertia* correspond to rotations about the three principal axes $A$, $B$ and $C$ (Figure 12.14a) and are denoted $I_A$, $I_B$ and $I_C$. For a diatomic molecule, $I_A$ corresponds to rotation about the internuclear axis and has no meaning, implying that $I_A = 0$.

The principal axes in Figure 12.14a pass through the *centre of gravity* (*centre of mass*) of molecule XY. This is the point that satisfies equation 12.22, in which the parameters are defined in Figure 12.14b.

$$m_X r_X = m_Y r_Y \tag{12.22}$$

It follows that if X is heavier than Y ($m_X > m_Y$), then the centre of gravity is closer to X than to Y ($r_X < r_Y$).

Now consider rotation about the vertical axis $B$ in Figure 12.14a. The parameters required to determine the moment of inertia $I_B$ are defined in Figure 12.14b, and application of equation 12.21 leads to an expression for

**Fig. 12.14** (a) The principal axes of rotation of a diatomic molecule XY in which X is heavier than Y; the axes (which are perpendicular to each other) pass through the centre of mass (centre of gravity) of the molecule which is closer to X than Y (see equation 12.21). (b) Definitions of the parameters needed to determine the moments of inertia, $I$, of molecule XY.

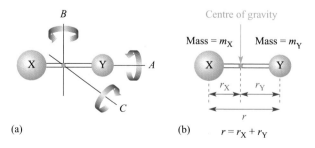

**THEORETICAL AND CHEMICAL BACKGROUND**

## Box 12.2 Moment of inertia of a diatomic molecule, XY

For a diatomic molecule XY, the principal moments of inertia $I_B$ and $I_C$ are given by:

$$I_B = I_C = m_X (r_X)^2 + m_Y (r_Y)^2$$

where $m_X$, $m_Y$, $r_X$ and $r_Y$ are defined in Figure 12.14b. Since the principal moments of inertia are equal, we shall refer to them simply as 'the moment of inertia of XY'. The expression for $I$ can be rewritten in terms of the reduced mass, $\mu$, of the diatomic and the internuclear distance, $r$, and is derived as follows.

$$I = m_X (r_X)^2 + m_Y (r_Y)^2 \qquad (i)$$

$$r = r_X + r_Y \qquad (ii)$$

$$m_X r_X = m_Y r_Y \qquad (iii)$$

Equation (iii) can be rearranged to give:

$$r_X = \frac{m_Y r_Y}{m_X} \quad \text{or} \quad r_Y = \frac{m_X r_X}{m_Y}$$

Substituting these expressions into equation (i) allows the equation to be expanded to:

$$I = \frac{m_X r_X m_Y r_Y}{m_X} + \frac{m_X r_X m_Y r_Y}{m_Y}$$

$$\therefore \quad I = r_X r_Y (m_X + m_Y) \qquad (iv)$$

Now consider equations (ii) and (iii), combination of which gives:

$$m_X r_X = m_Y r_Y = m_Y (r - r_X) = m_Y r - m_Y r_X$$

$$\therefore \quad m_X r_X + m_Y r_X = m_Y r$$

$$r_X (m_X + m_Y) = m_Y r$$

$$r_X = \frac{m_Y r}{m_X + m_Y} \qquad (v)$$

Similarly:

$$r_Y = \frac{m_X r}{m_X + m_Y} \qquad (vi)$$

Substituting equations (v) and (vi) into equation (iv) gives the expression:

$$I = \left( \frac{m_Y r}{m_X + m_Y} \right) \left( \frac{m_X r}{m_X + m_Y} \right) (m_X + m_Y)$$

$$\therefore \quad I = \frac{m_Y m_X r^2}{m_X + m_Y}$$

The reduced mass (Section 12.2) is given by:

$$\frac{1}{\mu} = \frac{1}{m_X} + \frac{1}{m_Y} \quad \text{or} \quad \mu = \frac{m_X m_Y}{m_X + m_Y}$$

and therefore the moment of inertia, $I$, is given by the equation:

$$I = \mu r^2$$

---

$I_B$ (equation 12.23). Rotation of molecule XY about the principal axis $C$ is equivalent to rotation about axis $B$. Therefore, $I_C = I_B$.

$$I_B = m_X (r_X)^2 + m_Y (r_Y)^2 \qquad (12.23)$$

▄▶
Reduced mass:
see Section 12.2

For a diatomic molecule, the moment of inertia ($I = I_B = I_C$) is usually given in terms of the reduced mass, $\mu$, and internuclear distance, $r$, as shown in equation 12.24 (derived in Box 12.2).

$$I = \mu r^2 \qquad (12.24)$$

---

**Worked example 12.8**     *Moments of inertia*

Determine the moments of inertia, $I_B$ and $I_C$, of $H^{35}Cl$ if the bond distance is 127 pm. Atomic masses: $^1H = 1.008$, $^{35}Cl = 34.97\,u$.

For $H^{35}Cl$, the reduced mass is given by:

$$\frac{1}{\mu} = \frac{1}{m_1} + \frac{1}{m_2} \quad \text{or} \quad \mu = \frac{m_1 m_2}{m_1 + m_2}$$

The masses of the atoms must be in kg (see the physical constant table on the inside back cover of the book):

$$m_1 = 1.008 \times 1.66 \times 10^{-27}\,\text{kg}$$

$$m_2 = 34.97 \times 1.66 \times 10^{-27}\,\text{kg}$$

The bond distance must be in m: $127\,\text{pm} = 1.27 \times 10^{-10}\,\text{m}$
For the moment of inertia,

$$I_B = I_C = \mu r^2$$

$$= \left(\frac{m_1 m_2}{m_1 + m_2}\right) r^2$$

$$= \left(\frac{(1.008 \times 1.66 \times 10^{-27}\,\text{kg})(34.97 \times 1.66 \times 10^{-27}\,\text{kg})}{(1.008 + 34.97)(1.66 \times 10^{-27}\,\text{kg})}\right)(1.27 \times 10^{-10}\,\text{m})^2$$

$$= 2.62 \times 10^{-47}\,\text{kg\,m}^2$$

---

## 12.9    Rotational spectroscopy of linear rigid rotor molecules

The *rigid rotor* approximation assumes that the rotating molecule can be considered as a rigid body.

The simplest model of a rotating diatomic molecule is to assume that it is a *linear rigid rotor*, i.e. the internuclear distance remains constant throughout the rotational motion. This is an *approximation* (molecules are clearly not rigid entities) but the model provides a good starting point for a discussion of rotational spectroscopy.

In this section, we deal only with diatomics, and do not distinguish between the moments of inertia $I_B$ and $I_C$, but rather refer to $I$ which equals both $I_B$ and $I_C$.

### Rotational energy levels

A rotational spectrum is made up of lines corresponding to transitions from one rotational energy level to another. The rotational energies are quantized and are denoted by rotational quantum numbers, $J$, for which allowed values are $J = 0, 1, 2, 3 \ldots$. In its rotational ground state ($J = 0$), the molecule is not rotating. Equation 12.25 gives an expression for the allowed rotational energy levels, $E_J$, of a diatomic molecule treated as a rigid rotor.

$$E_J = \frac{h^2}{8\pi^2 I} J(J + 1) \qquad J = 0, 1, 2, 3 \ldots \tag{12.25}$$

where: $E_J$ is in joules

$$h = \text{Planck constant} = 6.626 \times 10^{-34}\,\text{J\,s}$$

$$I = \text{moment of inertia in kg\,m}^2$$

Figure 12.15 shows the relative spacings of the rotational energy levels of a rigid rotor diatomic. Since microwave spectroscopic data are usually recorded in $\text{cm}^{-1}$, it is useful to express the rotational energy levels in terms of $\text{cm}^{-1}$. Wavenumber is reciprocal wavelength, $1/\lambda$. Since $E = h\nu$ and $c = \nu\lambda$, it follows that:

$$E = \frac{hc}{\lambda}$$

$$\frac{1}{\lambda} = \frac{E}{hc}$$

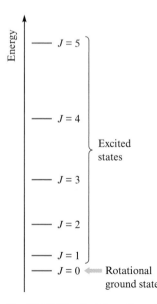

**Fig. 12.15** The first five allowed rotational states (energy levels) of a diatomic molecule, treated as a rigid rotor; $J$ is the rotational quantum number.

We can now write equation 12.26, where values of $E'_J$ are rotational energy levels expressed in $cm^{-1}$.

$$E'_J = \frac{E_J}{hc} = \frac{h}{8\pi^2 cI}J(J+1) \qquad J = 0, 1, 2, 3\ldots \tag{12.26}$$

where: $E'_J$ is in $cm^{-1}$ and $c$ = speed of light = $2.998 \times 10^{10}\,cm\,s^{-1}$

The equation can be further simplified by writing it in terms of the *rotational constant, B*, which is defined in equation 12.27. The value of $B$ depends on the moment of inertia, $I$, and is a molecular property.

$$B = \frac{h}{8\pi^2 cI} \qquad \text{Units of } B = cm^{-1} \tag{12.27}$$

Equations 12.26 and 12.27 give a relationship between the rotational energy levels of a rigid rotor diatomic, the rotational constant and the rotational quantum number (equation 12.28).

$$E'_J = BJ(J+1) \qquad \text{Units of } E'_J = cm^{-1} \tag{12.28}$$

For a given molecule, $B$ is a constant, and so the dependence of $E'_J$ on $J(J+1)$ means that the spacing between the energy levels increases as $J$ increases (Figure 12.15): for $J = 0$, equation 12.28 gives $E'_J = 0$ (rotational ground state), for $J = 1$, $E'_J = 2B$, for $J = 2$, $E'_J = 6B$, etc.

### Selection rules

Rotational transitions and observed rotational spectroscopic lines are governed by selection rules. The first selection rule (equation 12.29) restricts the rotational levels between which a transition can occur.

$$\Delta J = \pm 1 \tag{12.29}$$

Thus, transitions can occur between the rotational ground state ($J = 0$) and the first excited state ($J = 1$) and vice versa, or between levels $J = 1$ and $J = 2$, etc., but *not*, for example, from $J = 0$ to $J = 2$. Using the notation introduced in Box 3.7, a transition from the $J = 0$ to $J = 1$ level is represented as $J = 1 \longleftarrow J = 0$ (absorption of energy), and from the $J = 1$ to $J = 0$ as $J = 1 \longrightarrow J = 0$ (energy emission).

The second selection rule states that, for a diatomic molecule, a pure rotational spectrum is observed only if the molecule is *polar*. Therefore, pure rotational spectra can be observed for gas phase HBr, ClF and CO, but not for $H_2$, $F_2$ and $N_2$.

### The relationship between transitions and spectral lines

Figure 12.15 showed the allowed rotational energy levels of a rigid rotor diatomic. The selection rule $\Delta J = \pm 1$ restricts the possible transitions to $J = 1 \longleftarrow J = 0$, $J = 2 \longleftarrow J = 1$, $J = 3 \longleftarrow J = 2$, etc. The energies of these transitions, $\Delta E'_J$, can be determined in terms of the rotational constant, $B$, from equation 12.28 as shown in worked example 12.9.

| Worked example 12.9 | *Transitions between rotational energy levels* |

For a diatomic molecule (treated as a rigid rotor) with rotational constant $B\,cm^{-1}$, calculate the energies of the rotational ground and first excited states, and determine the value of $\Delta E'_J$ for the transition $J = 1 \longleftarrow J = 0$.

The equation needed is: $E'_J = BJ(J+1)$

For $J = 0$: $\quad\quad\quad\quad\quad E'_J = 0$

For $J = 1$: $\quad\quad\quad\quad\quad E'_J = 2B$

Therefore, for the transition $J = 1 \leftarrow J = 0$, $\Delta E'_J = 2B - 0 = 2B\,\text{cm}^{-1}$

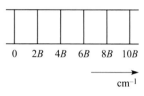

0    2B    4B    6B    8B    10B

$\longrightarrow$

$\text{cm}^{-1}$

**Fig. 12.16** Representation of part of the pure rotational spectrum of a rigid rotor diatomic molecule.

The exercise above illustrates how the lines in the pure rotational spectrum of a rigid rotor diatomic arise. For the successive transitions $J = 1 \leftarrow J = 0$, $J = 2 \leftarrow J = 1$, $J = 3 \leftarrow J = 2$, $J = 4 \leftarrow J = 3$ and $J = 5 \leftarrow J = 4$, values of $\Delta E'_J$ are $2B$, $4B$, $6B$, $8B$ and $10B$. These values correspond to the *relative* spacings between levels in Figure 12.15, and to the equally spaced, discrete spectroscopic lines shown in Figure 12.16. The spacing between each pair of successive lines is $2B$. The application of rotational spectroscopy to the determination of the bond distance of a diatomic molecule now follows: from the spacings of lines in the spectrum (Figure 12.16), the rotational constant, $B$, can be found. Hence we can find the moment of inertia (equation 12.27), and the bond distance (equation 12.24). We reiterate that the *rigid* rotor model is an approximation. In practice, a diatomic molecule is a *non-rigid* rotor and the spacings between the lines in the pure rotational spectrum of a diatomic actually decrease as $J$ increases.

**Worked example 12.10** | *Determination of bond distance from the rotational constant of a diatomic molecule*

Find the H−Br bond length of $^1\text{H}^{79}\text{Br}$ if the rotational constant is $8.58\,\text{cm}^{-1}$; exact atomic masses: $^1\text{H} = 1.008$; $^{79}\text{Br} = 78.92$ u.

Equations needed are:

$$B = \frac{h}{8\pi^2 cI} \quad \text{and} \quad I = \mu r^2$$

Planck constant, $h = 6.626 \times 10^{-34}\,\text{J s}$

$$= 6.626 \times 10^{-34}\,\text{kg m}^2\,\text{s}^{-1} \text{ (see Table 1.2)}$$

Rearrange the first equation to make $I$ the subject:

$$I = \frac{h}{8\pi^2 cB}$$

$$= \frac{(6.626 \times 10^{-34}\,\text{kg m}^2\,\text{s}^{-1})}{8\pi^2 \times (2.998 \times 10^{10}\,\text{cm s}^{-1}) \times (8.58\,\text{cm}^{-1})} = 3.26 \times 10^{-47}\,\text{kg m}^2$$

To find the bond length, $r$:

$$r = \sqrt{\frac{I}{\mu}} = \sqrt{I\left(\frac{m_1 + m_2}{m_1 m_2}\right)}$$

$$= \sqrt{(3.26 \times 10^{-47}\,\text{kg m}^2)\left[\frac{(1.008 \times 1.66 \times 10^{-27}\,\text{kg}) + (78.92 \times 1.66 \times 10^{-27}\,\text{kg})}{(1.008 \times 1.66 \times 10^{-27}\,\text{kg}) \times (78.92 \times 1.66 \times 10^{-27}\,\text{kg})}\right]}$$

$$= \sqrt{(3.26 \times 10^{-47}\,\text{kg m}^2)\left[\frac{(1.008 + 78.92)(1.66 \times 10^{-27}\,\text{kg})}{(1.008 \times 78.92)(1.66 \times 10^{-27}\,\text{kg})^2}\right]}$$

$$= 1.40 \times 10^{-10}\,\text{m} = 140\,\text{pm}$$

## SUMMARY

In this chapter, we introduced vibrational and rotational spectroscopy. Vibrating diatomic molecules can be likened to oscillating springs, and an approximate model is that of the simple harmonic oscillator. In the laboratory, infrared (IR) spectroscopy is a routine method of analysis: functional groups give rise to characteristic absorptions, and for simple molecules, the number of observed absorptions gives information about molecular shape and symmetry. Rotating diatomic molecules can, to a first approximation, be treated as rigid rotors. Transitions between rotational energy levels lead to spectral lines, the spacings of which are related to the rotational constant, the moment of inertia and the bond distance.

### Do you know what the following terms mean?

- vibrational ground state
- vibrational wavenumber
- vibrational mode
- force constant of a bond
- reduced mass
- simple harmonic oscillator
- anharmonic oscillator

- vibrational quantum number
- vibrational energy levels
- degrees of freedom
- IR active mode
- IR inactive mode
- fingerprint region of an IR spectrum

- moment of inertia (of a diatomic)
- centre of gravity (of a diatomic)
- rigid rotor
- rotational ground state
- rotational quantum number
- rotational energy levels
- rotational constant

### You should be able:

- to discuss the vibrating spring model in relation to the vibrations of a diatomic molecule
- to explain the difference between a simple harmonic and an anharmonic oscillator, and the limitations of the simple harmonic oscillator model
- to calculate the reduced mass of a two atom system
- to calculate the fundamental wavenumber of a vibrational mode
- to work out for diatomic and triatomic molecules the allowed modes of vibration, and state if (and why) they are IR active
- to interpret the IR spectra of simple compounds containing, for example, O–H, C=O, C=C, C–Cl and C–H bonds

- to work out the moment of inertia of a diatomic molecule
- to explain the meaning and limitations of the rigid rotor model for a diatomic molecule
- to express the allowed rotational energy levels in terms of the moment of inertia and rotational constant of a rigid rotor diatomic
- to relate the lines in the rotational spectrum to the transitions between rotational energy levels, assuming a rigid rotor model
- to determine the bond length of a diatomic molecule from the rotational constant

## PROBLEMS

**12.1** Give equations for the quantized energy levels of the simple harmonic and anharmonic oscillators and show that the energy levels are equally spaced for the harmonic, but not the anharmonic, oscillator.

**12.2** Calculate the reduced masses of (a) an HD molecule, (b) a $^{12}C$–H unit and (c) a $^{13}C^{16}O$ molecule. Exact masses: $^{1}H$, 1.01; $^{2}H$, 2.01; $^{12}C$, 12.00; $^{13}C$, 13.00; $^{16}O$, 15.99.

**12.3** The force constants for the bonds in ClF, BrF and BrCl are 448, 406 and 282 $N\,m^{-1}$ respectively. Predict the *trend* in their bond enthalpies.

**12.4** The stretching mode of NO gives rise to an absorption in the IR spectrum at 1903 $cm^{-1}$. What is the force constant of the bond?

**12.5** (a) Determine the reduced masses of $H^{35}Cl$ and $H^{37}Cl$; exact masses $^{1}H = 1.01$; $^{35}Cl = 34.97$; $^{37}Cl = 36.97$. (b) If the force constants of the bonds are the same, is the ratio of the reduced masses sufficient to cause a shift in the IR absorption assigned to the H–Cl stretch? (c) Would you expect to see any *chemical* differences between $H^{35}Cl$ and $H^{37}Cl$?

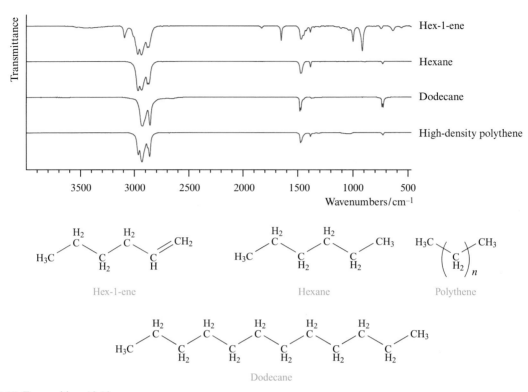

**Fig. 12.17** For problem 12.10.

**12.6** Refer to Figure 12.6. In what direction does the molecular dipole moment act as the $CO_2$ molecule (a) stretches asymmetrically and (b) bends? In either case, is a *permanent* dipole moment generated?

**12.7** What shapes are (a) HCN, (b) $CS_2$, (c) OCS and (d) $XeF_2$? Are the symmetric stretching, asymmetric stretching and bending modes of these molecules IR active?

**12.8** Rationalize why the values for the carbonyl stretching mode given in Table 12.1 for carboxylic acids and carboxylate ions are generally lower than those of the corresponding aldehydes:

Carboxylic acid        Aldehyde

Carboxylate ion

**12.9** For each of the following compounds, state the nature of the functional group present and approximately where in the IR spectrum you would expect to observe a characteristic absorption: (a) acetonitrile; (b) butanol; (c) ethylamine; (d) acetone. Use Tables 1.8 and 1.9 to help you.

**12.10** Figure 12.17 shows the IR spectra of high-density polythene, dodecane, hexane and hex-1-ene.
(a) Why are the spectra of high-density polythene, dodecane and hexane similar to one another?
(b) What *two* features of the spectrum of hex-1-ene allow you to assign this spectrum to an alkene rather than to hexane?

**12.11** Figure 12.18 shows four spectra labelled (a) to (d). Assign each spectrum to one of the compounds cyclohexane, benzene, toluene and phenol shown in Figure 12.18.

**12.12** How could IR spectroscopy be used to distinguish between each of the following pairs of compounds?

(a)

(b)

(c)

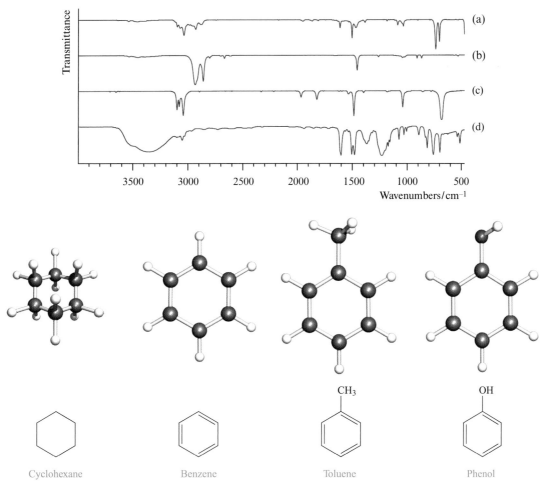

**Fig. 12.18** For problem 12.11.

**12.13** (a) What is the molecular shape of $SO_3$? (b) Draw a diagram to show the symmetric stretching mode of vibration of $SO_3$. (c) Is this mode IR active?

**12.14** (a) Why would you not expect to observe an IR spectroscopic absorption for the $C\equiv C$ stretch of $C_2H_2$? (b) Would you expect to observe such an absorption in the IR spectrum of $CH_3CH_2C\equiv CH$?

**12.15** The IR spectrum of $H_2O$ shows bands at 3756 (asymmetric stretch), 3657 (symmetric stretch) and 1595 $cm^{-1}$ (scissoring). In $D_2O$, the corresponding vibrational modes give rise to absorptions at 2788, 2671 and 1178 $cm^{-1}$. Account for these observations.

**12.16** Methanal, $H_2C{=}O$, exhibits an absorption in its IR spectrum at 2783 $cm^{-1}$ assigned to a $C{-}H$ stretching mode. What will be the shift in vibrational wavenumber for this band on going from $H_2C{=}O$ to $D_2C{=}O$? (Assume that the force constants of the $C{-}H$ and $C{-}D$ bonds are the same; for exact masses, see problem 12.2.)

**12.17** Calculate $I_B$ for $H^{37}Cl$ ($r = 127$ pm) and comment on its value in relation to that of $H^{35}Cl$ (worked example 12.8); accurate masses: $^{1}H = 1.008$; $^{37}Cl = 36.97$.

**12.18** Find the moments of inertia $I_A$, $I_B$ and $I_C$ of $^{79}Br^{19}F$ (bond length $= 176$ pm; accurate masses: $^{79}Br = 78.92$; $^{19}F = 19.00$).

**12.19** The moment of inertia, $I_B$, of $^{127}I^{35}Cl$ is $2.45 \times 10^{-45}$ kg m$^2$. What is the I–Cl bond distance? (Accurate masses: $^{127}I = 126.90$; $^{35}Cl = 34.97$.)

**12.20** Explain why a pure rotational spectrum is observed for HCl but not for $H_2$ or for $Cl_2$.

**12.21** Which of the following transitions between rotational energy levels are allowed?
(a) $J = 2 \leftarrow J = 0$; (b) $J = 3 \leftarrow J = 2$;
(c) $J = 2 \leftarrow J = 1$; (d) $J = 3 \leftarrow J = 1$.

**12.22** Explain what is meant by (a) the vibrational ground state and (b) the rotational ground state of a diatomic molecule.

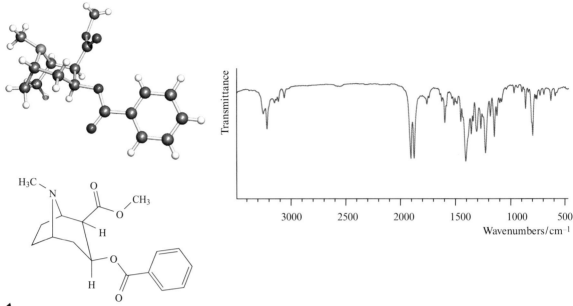

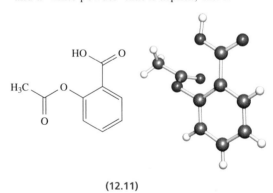

**Fig. 12.19** For problem 12.25. Structure data from: R. J. Hrynchuk *et al.* (1983) *Can. J. Chem.*, vol. 61, p. 481; colour code: C, grey; O, red; N, blue; H, white.

**12.23** (a) To what transition does the first line in a rotational absorption spectrum of a diatomic molecule correspond? (b) Is this line at the low- or high-wavenumber end of the spectrum? Is this the low- or high-energy end of the spectrum?

**12.24** Predict the positions of the lines in the pure rotational spectrum of $^{35}Cl^{19}F$ ($r = 163\,pm$) corresponding to the transitions $J = 1 \leftarrow J = 0$, $J = 2 \leftarrow J = 1$ and $J = 3 \leftarrow J = 2$. What assumption have you made in your calculations? (Accurate masses: $^{35}Cl = 34.97$; $^{19}F = 19.00$.)

### Additional problems

**12.25** Figure 12.19 shows the structure and IR spectrum of the drug cocaine. (a) Suggest assignments for as many spectroscopic absorptions as you can. (b) How useful would IR spectroscopy be as a method for customs officers to verify the content of a seizure of cocaine? (c) Could this

technique assist in distinguishing between cocaine and a 'white powder' that is aspirin, **12.11**?

**(12.11)**

**12.26** Given that the rotational constant of $^{12}C^{16}O$ is $1.92\,cm^{-1}$, determine (a) the bond length of CO, and (b) the positions of the first two lines in the pure rotational spectrum of $^{12}C^{16}O$; accurate masses: $^{12}C = 12.00$; $^{16}O = 15.99$. (c) To what transitions do these lines correspond?

# 13  Electronic spectroscopy

## Topics

- Absorption of UV–VIS radiation
- Electronic transitions in the vacuum-UV
- Choosing a solvent for UV–VIS spectroscopy
- $\pi$-Conjugation in organic molecules
- Compounds that absorb in the visible region

## 13.1    Introduction

Electronic spectroscopy is a branch of spectroscopy that is concerned with transitions of electrons between energy levels, i.e. between atomic or molecular orbitals. We have already examined the atomic spectrum of hydrogen (Section 3.16), and in Box 3.7, we outlined the notation used for distinguishing between transitions in absorption and emission spectra. In this chapter we are concerned with the *electronic spectra of molecules*, and the emphasis of our discussion is on the use of electronic spectroscopy in the laboratory. The energy difference between electronic energy levels is such that electronic spectra are usually observed in the UV or visible regions of the electromagnetic spectrum (see Appendix 4).

## 13.2    Absorption of ultraviolet and visible light

The molecular orbitals of a molecule may be bonding, non-bonding or anti-bonding. If a molecule *absorbs* an appropriate amount of energy, an electron from an occupied orbital may be excited to an unoccupied or partially occupied orbital. As the energies of these orbitals are quantized, it follows that each transition is associated with a specific amount of energy. The energy of the electronic transition from the highest occupied molecular orbital (HOMO) to the lowest unoccupied molecular orbital (LUMO) often corresponds to the ultraviolet (UV) or visible (VIS) region of the electromagnetic spectrum – wavelengths between 100 and 800 nm. Normal laboratory UV–VIS spectrophotometers operate between $\approx$200 and 900 nm (Figure 13.1).

*HOMO* = highest occupied molecular orbital

*LUMO* = lowest unoccupied molecular orbital

Figure 13.2 shows that an electronic transition should occur when light of energy $\Delta E$ is absorbed by a molecule, and the absorption spectrum might be expected to consist of a sharp band (as in Figure 11.2). In practice, however, *molecular* electronic spectra usually consist of *broad absorptions* (Figure 13.3) in contrast to the sharp absorptions of atomic spectra. Why is this? The answer is to do with timescales – the absorption of a photon is fast

**Fig. 13.1** Regions within the UV–VIS range of the electromagnetic spectrum; the vacuum-UV is also called the far-UV.

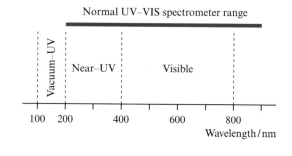

($\approx 10^{-18}$ s) and is more rapid than any molecular vibrations or rotations. Accordingly, the electronic transition is a 'snapshot' of the molecule in a particular vibrational and rotational state at a particular moment in time. As the energies of the molecular orbitals are dependent on the molecular geometry, it follows that a range of $\Delta E$ values (corresponding to the different vibrational and rotational states) will be observed. As the vibrational and rotational energy levels are much more closely spaced than the electronic levels, a broad band is observed in the electronic spectrum. If the rotational and vibrational states are restricted, for example by cooling, sharper spectra are often obtained. Similarly, if transitions are essentially localized on a single atomic centre, the spectrum will also be sharp – this is observed in the $4f$–$4f$ electronic transitions of lanthanoid compounds.

Although the absorption band covers a range of wavelengths, its position may be described by the wavelength corresponding to the absorption maximum – this is written as $\lambda_{max}$ and is measured from the spectrum as indicated in Figure 13.3.

Absorptions in electronic spectra exhibit characteristic intensities and the molecular property describing this is the molar extinction coefficient ($\varepsilon$), determined using the Beer–Lambert Law. Equation 13.1 gives the relationship between $\varepsilon_{max}$ and the maximum absorbance $A_{max}$; compare this with equation 11.4.

$$\varepsilon_{max} = \frac{A_{max}}{c \times \ell} \tag{13.1}$$

where the units are:   $\varepsilon_{max}, \mathrm{dm^3\,mol^{-1}\,cm^{-1}}$

$A_{max}$, dimensionless

$c, \mathrm{mol\,dm^{-3}}$

$\ell, \mathrm{cm}$

Values of $\varepsilon_{max}$ range from close to zero (a very weak absorption) to $>10\,000\,\mathrm{dm^3\,mol^{-1}\,cm^{-1}}$ (an intense absorption). Absorptions in electronic spectra must be described in terms of both $\lambda_{max}$ and $\varepsilon_{max}$.

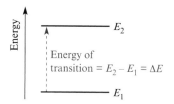

**Fig. 13.2** A transition between two electronic levels may occur when light of energy $\Delta E$ is absorbed by a molecule; the energy is quantized.

Absorption bands in molecular electronic spectra are often broad, and are described in terms of both $\lambda_{max}$ (nm) and $\varepsilon_{max}$ ($\mathrm{dm^3\,mol^{-1}\,cm^{-1}}$).

Beer–Lambert Law: see Section 11.4

**Fig. 13.3** Absorptions in the electronic spectrum of a molecule or molecular ion are often broad, and cover a range of wavelengths. The absorption is characterized by values of $\lambda_{max}$ and $\varepsilon_{max}$ (see equation 13.1).

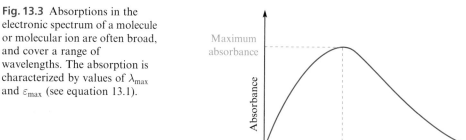

## 13.3    Electronic transitions in the vacuum-UV

Electronic transitions may originate from occupied bonding ($\sigma$ or $\pi$) or non-bonding molecular orbitals. In general, electronic transitions from $\sigma$ MOs to $\sigma^*$ MOs correspond to light in the vacuum-UV region of the spectrum.[§] For example, $C_2H_6$ absorbs light with a wavelength of 135 nm and this transition (written using the notation $\sigma^* \leftarrow \sigma$) is out of the range of the normal UV–VIS spectrophotometer. The energy difference between the $\pi$ and $\pi^*$ MOs in ethyne ($HC\equiv CH$) is smaller than the $\sigma-\sigma^*$ energy gap in ethane, but it still corresponds to light in the vacuum-UV region ($\lambda_{max} = 173$ nm, $\varepsilon_{max} = 6000$ dm$^3$ mol$^{-1}$ cm$^{-1}$). Typically, for an isolated $C=C$ or $C\equiv C$ bond in a molecule, the $\pi^* \leftarrow \pi$ electronic transition lies in the vacuum-UV region.

A third type of molecular electronic transition is designated $\sigma^* \leftarrow n$, where $n$ stands for non-bonding. Such transitions may occur in a molecule in which an atom carries a lone pair of electrons. Examples include water, alcohols ($ROH$), amines ($RNH_2$) and halogenoalkanes ($RF$, $RCl$, $RBr$ and $RI$). (The R in these formulae refers to a general organic group.) In many cases, $\sigma^* \leftarrow n$ electronic transitions occur in the vacuum-UV region. For example, in water, $\lambda_{max} = 167$ nm and $\varepsilon_{max} = 500$ dm$^3$ mol$^{-1}$ cm$^{-1}$, and in methanol, $\lambda_{max} = 177$ nm and $\varepsilon_{max} = 200$ dm$^3$ mol$^{-1}$ cm$^{-1}$.

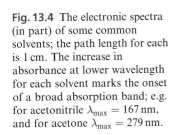

Non-bonding MO:
see Section 5.5

## 13.4    Choosing a solvent for UV–VIS spectroscopy

*Solution samples* are often used for UV–VIS spectroscopy, and the choice of a solvent is important. Ideally, we need a solvent that is *transparent* in the spectral region of interest. Figure 13.4 shows part of the UV–VIS spectra of four solvents. Acetonitrile is 'transparent' over the range from 200 to 800 nm; on moving to a lower wavelength, one observes the onset of a broad absorption ($\lambda_{max} = 167$ nm). A wavelength of 200 nm is the lower limit (or cut-off point) for acetonitrile as a solvent in UV–VIS spectroscopy, and it is a good solvent for laboratory use. Similarly, water is transparent down to 210 nm ($\lambda_{max} = 167$ nm). The lower limit for cyclohexane is also 210 nm, and for chloroform, it is around 240 nm. On the other hand, acetone cuts off at 330 nm and this restricts its use as a solvent. However, this absorption for acetone is relatively weak ($\varepsilon_{max} = 15$ dm$^3$ mol$^{-1}$ cm$^{-1}$) and it may still be used as a solvent for compounds that possess very much larger extinction coefficients.

**Fig. 13.4** The electronic spectra (in part) of some common solvents; the path length for each is 1 cm. The increase in absorbance at lower wavelength for each solvent marks the onset of a broad absorption band; e.g. for acetonitrile $\lambda_{max} = 167$ nm, and for acetone $\lambda_{max} = 279$ nm.

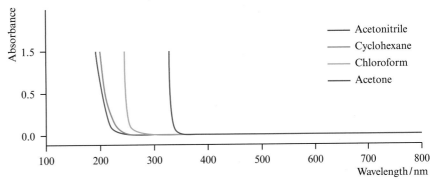

[§] The region below $\approx 200$ nm is referred to as the vacuum-UV or far-UV range. It is necessary to evacuate the sample chamber to make measurements because $O_2$ absorbs in this region.

| 13.5 | $\pi$-Conjugation |
|---|---|

### The dependence of the $\pi^* \leftarrow \pi$ transition on the number and arrangement of C=C bonds

More about alkenes in Chapter 26; for explanation of abbreviated structural formulae, see Section 24.2

Figure 13.5 illustrates the shift in $\lambda_{max}$ for the $\pi^* \leftarrow \pi$ transition along the series ethene, buta-1,3-diene and hexa-1,3,5-triene. The corresponding values of $\varepsilon_{max}$ are 15 000, 21 000 and 34 600 dm$^3$ mol$^{-1}$ cm$^{-1}$. The significant relationship between these molecules is that the carbon chain increases, with an incremental addition of *alternating single and double carbon–carbon bonds*. The shift to longer wavelength with increased chain length is a general phenomenon and indicates that the energy difference between the $\pi$ and $\pi^*$ levels is decreasing. Why does this happen? To answer the question, we consider how carbon–carbon $\pi$-bonding is affected by chain length and the arrangement of single and double bonds, and we approach the problem by sequentially adding carbon atoms to a $C_2$-chain.

### Localized and delocalized $\pi$-bonding

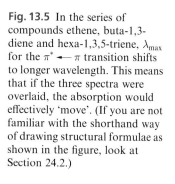

(13.1)

Ethene (**13.1**) is a planar molecule in which each C atom is considered to be $sp^2$ hybridized. The 'left-over' $2p$ atomic orbitals (one per C atom) overlap to give a $\pi$-bond (look back to Section 7.7). The four valence electrons of each C atom are used to form two C−H $\sigma$-bonds, one C−C $\sigma$-bond and one C−C $\pi$-bond. The $\pi$-bond is formed by the in-phase interaction of two $2p$ atomic orbitals that lie perpendicular to the plane of the molecule, and the $\pi^*$ MO results from the out-of-phase interaction of the two $2p$ atomic orbitals (Figure 13.6a). The carbon–carbon $\pi$-bond is *localized*.

For an example of a delocalized $\pi$-system, we consider the allyl carbanion, $[C_3H_5]^-$. The bonding in $[C_3H_5]^-$ can be represented within VB theory by resonance structures (**13.2**). The two resonance forms contribute equally and the net result is that the $\pi$-bonding is delocalized over the three C atoms as represented in structure **13.3**.

(13.2)                    (13.3)

**Fig. 13.5** In the series of compounds ethene, buta-1,3-diene and hexa-1,3,5-triene, $\lambda_{max}$ for the $\pi^* \leftarrow \pi$ transition shifts to longer wavelength. This means that if the three spectra were overlaid, the absorption would effectively 'move'. (If you are not familiar with the shorthand way of drawing structural formulae as shown in the figure, look at Section 24.2.)

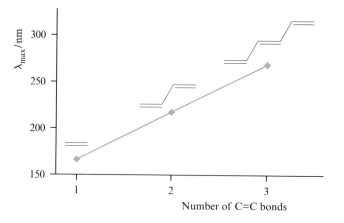

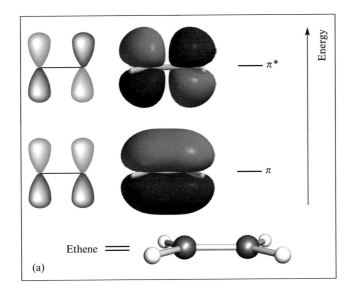

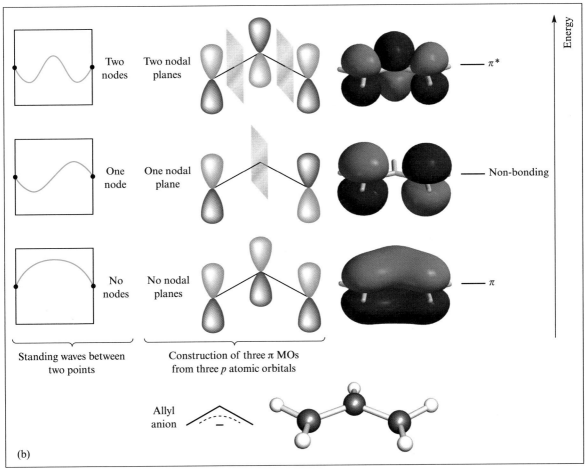

**Fig. 13.6** (a) The π-bonding and antibonding MOs in ethene. (b) The π MOs in the allyl anion can be understood in terms of a standing wave between two fixed points which has no, one or two nodes. Each π-orbital also contains a nodal plane in the plane of the molecule. In addition to schematic diagrams of the MOs, the figure shows more realistic representations that have been generated computationally using Spartan '04, © Wavefunction Inc. 2003.

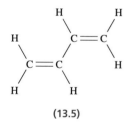

**(13.4)**

The bonding in the allyl anion can also be described using an MO model. The anion is planar **(13.4)**, and after the $\sigma$-bonding framework has been formed, each C atom has one $2p$ atomic orbital remaining lying perpendicular to the plane of the molecule. The $\pi$-bonding in the allyl anion results from the overlap of these atomic orbitals and *since there are three atomic orbitals, we must form three $\pi$ MOs* (Figure 13.6b). The $\pi$-bonding MO arises from the in-phase overlap of all three carbon $2p$ atomic orbitals, and the $\pi$-bonding character is spread over all three carbon centres – the $\pi$-bond is *delocalized* and this result is consistent with structure **13.4** obtained using the VB approach. The $\pi^*$ orbital is readily constructed by taking an out-of-phase combination of the three $2p$ atomic orbitals and is antibonding between each pair of adjacent carbon atoms. At a simple level, the origins of the remaining $\pi$ MO are more difficult to understand – Figure 13.6b shows that this MO has no contribution from the central carbon atom and the MO is labelled 'non-bonding'. One way in which we can understand this result is to take the two terminal carbon atoms in the allyl anion as fixed points and consider a standing wave between the two points. There may be no change of sign in the amplitude of the wave (no node) and this wave corresponds to the $\pi$-bonding MO in Figure 13.6b. The amplitude of the standing wave could change sign once or twice between the fixed points to give one or two nodes, and these situations correspond to the highest two $\pi$ MOs shown in Figure 13.6b. The nodal plane in the second MO passes through the central carbon atom and means that there is *no contribution from this* $2p$ atomic orbital. This MO is therefore *non-bonding* with respect to the C–C–C framework. In the allyl anion the four $\pi$ electrons occupy the two lowest energy MOs.

### $\pi$-Conjugation in buta-1,3-diene

We now extend the carbon chain to four atoms and consider the $\pi$-bonding in buta-1,3-diene **(13.5)**. All the atoms in this molecule are in one plane, i.e. they are coplanar. After the formation of the $\sigma$-framework in buta-1,3-diene, one $2p$ atomic orbital per C atom remains and there must be four $\pi$ MOs (Figure 13.7). As in the allyl anion, there is a pattern in the number of nodal planes. There are no nodal planes in the lowest-lying MO and the $\pi$-character is delocalized over all four C atoms. The next MO has one nodal plane and the orbital possesses $\pi$-character localized as shown in structure **13.5**. The highest-lying MOs have two and three nodal planes respectively. We may label the lower two MOs as $\pi$-bonding (these are fully occupied), and the higher two as $\pi^*$ MOs (which are unoccupied). The LUMO has some $\pi$-bonding character, but is antibonding overall. An important result of this bonding analysis is that, although the name 'buta-1,3-diene' suggests that the $\pi$-bonds are localized in particular positions, the $\pi$-character is actually spread out over the carbon chain. The molecule is *conjugated*.

**(13.5)**

A *polyene* contains more than one C=C bond. If the carbon chain contains alternating C–C and C=C bonds, the molecule is conjugated.

### The effect of conjugation on the $\pi^* \leftarrow \pi$ transition

A consequence of *conjugation* is that the energy difference between the highest-lying $\pi$ MO and the lowest-lying $\pi^*$ MO is lowered, and the associated $\pi^* \leftarrow \pi$ transition shifts to longer wavelength. In buta-1,3-diene, the $\pi^* \leftarrow \pi$ transition corresponds to the HOMO–LUMO separation

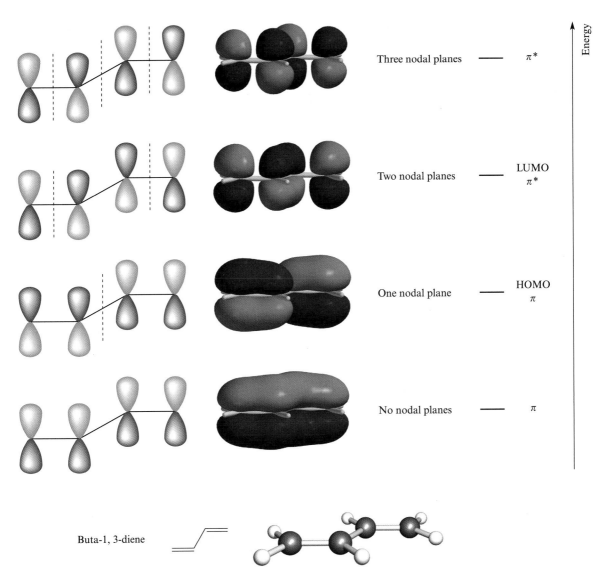

Buta-1, 3-diene

**Fig. 13.7** The π MOs of buta-1,3-diene. Each dotted line represents a nodal plane. Each π-orbital also contains a nodal plane in the plane of the molecule. The figure shows schematic diagrams of the MOs and more realistic representations that have been generated computationally using Spartan '04, © Wavefunction Inc. 2003.

A ***chromophore*** is the group of atoms in a molecule responsible for the absorption of electromagnetic radiation.

(Figure 13.7) and the transition is observed in the near-UV part of the spectrum ($\lambda_{max} = 217$ nm).

The $-C{=}C{-}C{=}C-$ unit is called a *chromophore* – it is the group of atoms in buta-1,3-diene responsible for the absorption of light in the UV–VIS spectrum. The addition of another C=C bond increases the π-conjugation and causes a further shift in the $\pi^* \longleftarrow \pi$ transition to a longer wavelength (Figure 13.5). This is called a *red shift* (because the shift in absorption is towards the red end of the spectrum; see Table 11.2) or *bathochromic effect*. The shift in wavelength for each additional C=C bond added to the alternating $-C{=}C{-}C{=}C-$ chain is $\approx 30$ nm. The absorption also becomes more intense (larger $\varepsilon_{max}$).

The addition of an alkyl substituent to a carbon atom in the chromophore also leads to an increase in conjugation, but as this is due to an interaction

**Table 13.1** Observed values of $\lambda_{max}$ for some conjugated polyenes. Compare these values with those predicted (see text) using buta-1,3-diene as your starting point.

| Compound | $\lambda_{max}$ / nm | $\varepsilon_{max}$ / dm$^3$ mol$^{-1}$ cm$^{-1}$ |
|---|---|---|
| | 217 | 21 000 |
| | 227 | 23 000 |
| | 263 | 30 000 |
| | 352 | 147 000 |

▶ Abbreviated structural formulae are explained in Section 24.2

> In a *polyene*, the $\pi$-conjugation and, therefore, the *chromophore* can be extended by adding another C=C bond or adding an alkyl substituent.

between the $\sigma$-electrons in the alkyl group and the $\pi$-electrons in the alkene chain, the effect is only small. The result is a red shift of $\approx 5$ nm per alkyl substituent. The data in Table 13.1 illustrate the effects of conjugation in some polyenes.

The addition of substituents other than alkyl groups may also affect the position of the absorption maximum in the electronic spectrum. For example, an OR substituent (R = alkyl) causes a red shift of $\approx 6$ nm, an SR group, a red shift of $\approx 30$ nm, and an NR$_2$ group, a red shift of $\approx 60$ nm. Each of these groups can donate electron density from a lone pair into the carbon $\pi$-system, thereby increasing the conjugation in the molecule and extending the chromophore. This is illustrated in resonance structures **13.6**.

**(13.6)**

Conjugation also occurs in alkynes with alternating C−C and C≡C bonds. As the length of the chromophore increases, the absorption in the electronic spectrum undergoes a red shift and becomes more intense.

Conjugation involving $\pi$-electrons is not restricted to carbon–carbon multiple bonds. Other compounds that exhibit absorption spectra in the near-UV region include $\alpha,\beta$-unsaturated ketones and aldehydes. Structure **13.7** shows the group of atoms that must be present; the carbon atoms of the C=C group are labelled $\alpha$ and $\beta$ as shown in the diagram.

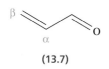

**(13.7)**

▶ $\alpha,\beta$-Unsaturated ketones and aldehydes: see Section 33.15

**(13.8)**

**(13.9)**

The $\pi$-electrons in an $\alpha,\beta$-unsaturated ketone or aldehyde are delocalized in the same way as in an alternating C=C/C−C chain. This leads to a $\pi^* \leftarrow \pi$ transition characterized by an intense absorption around 220 nm. The presence of substituents on the $\alpha$ and $\beta$ carbon atoms may cause significant shifts in the absorption maximum: the $\pi^* \leftarrow \pi$ transition in **13.8** has $\lambda_{max}$ 219 nm ($\varepsilon_{max} = 3600$ dm$^3$ mol$^{-1}$ cm$^{-1}$), but in **13.9**, it is 235 nm ($\varepsilon_{max} = 14\,000$ dm$^3$ mol$^{-1}$ cm$^{-1}$).

In addition to the band assigned to the $\pi^* \leftarrow \pi$ transition, the electronic spectrum of an $\alpha,\beta$-unsaturated ketone or aldehyde contains a less intense absorption that is due to an electronic transition involving an oxygen lone pair of electrons: a $\pi^* \leftarrow n$ transition. Aldehydes, ketones, acid chlorides, carboxylic acids, esters and azo compounds are among those for which $\pi^* \leftarrow n$ transitions can also be observed, and some typical spectroscopic data are listed in Table 13.2.

**Table 13.2** Selected electronic spectroscopic data for compounds exhibiting $\pi^* \leftarrow n$ transitions.

| Compound type (R = alkyl) | $\lambda_{max}$ / nm | $\varepsilon_{max}$/ dm$^3$ mol$^{-1}$ cm$^{-1}$ |
|---|---|---|
| $R_2C=O$ (ketone) | 270–290 | 10–20 |
| RHC=O (aldehyde) | 290 | 15 |
| RCOCl (acid chloride) | 280 | 10–15 |
| RC(O)OR' (ester) or RCO$_2$H (carboxylic acid) | ≤200–210 | 40–100 |
| RN=NR (azo compound) | 350–370 | 10–15 |

### ENVIRONMENT AND BIOLOGY

## Box 13.1  Polyenes as natural colouring agents

Carotenoids are a group of polyenes that are natural colouring agents providing yellow, orange and red hues to a variety of plants and to some animal tissues. Some examples are shown below – notice the structural units that are common to these molecules. Carotenoids fall into two groups: the carotenes and the xanthophylls. Carotenes are hydrocarbons while xanthophylls are derivatives of carotenes with hydroxy or other oxygen-containing functionalities.

The red colour of tomatoes comes from lycopene.
*E. C. Constable.*

The red colour in tomatoes is due largely to the presence of lycopene ($\lambda_{max}$ = 469 nm).

**Lycopene**

β-Carotene gives rise to the orange colour in carrots and mangoes ($\lambda_{max}$ = 452 nm); zeaxanthin is also present in mangoes and contributes to the yellow colour of egg yolks.

**β-Carotene**

**Zeaxanthin**

α-Carotene and violaxanthin are present in oranges.

**α-Carotene**

**Violaxanthin**

The pink pigments of salmon and lobsters are due to astaxanthin.

**Astaxanthin**

## ENVIRONMENT AND BIOLOGY

## Box 13.2 Biological staining in histology and microbiology

Histology is the study of the structure of cells and organisms. The examination of tissue or cells under a microscope is greatly facilitated by staining the materials to selectively dye particular regions of the biological specimen. Two important biological stains are hematoxylin and eosin. Hematoxylin is active as a stain when it is in its oxidized form of hematein. This compound interacts with nucleic acids (see Section 35.7), resulting in the nuclei of cells in animal tissue appearing blue. The stain is most effective when it is combined with a *mordant*. This is a compound that fixes the dye to the substrate, and in the case of hematoxylin, the mordant is usually $KAl(SO_4)_2 \cdot 12H_2O$ (an aluminium-containing *alum*, see Section 22.4).

Hematoxylin

Hematein

Hematoxylin is used to target cell nuclei, but it is often used in combination with eosin Y ($\lambda_{max} = 516$ nm). The net result is that the nuclei are visible as blue regions, and the cytoplasm and collagen appear red.

Eosin Y

Eosin Y is also used as a counter stain in bacterial staining. In the Gram method for bacteriological staining (named after the Danish scientist Hans Christian Gram who first used the technique in 1844), the most commonly used dye is crystal violet.

Crystal violet
(Gentian violet)

After the bacteria have been stained with crystal violet, the specimen is treated with aqueous KI and $I_2$ to fix the stain. Bacteria that appear blue after this treatment are called Gram-positive. Some bacteria (Gram-negative) remain colourless after the initial staining, but appear red when the sample is treated with Eosin Y. The Gram method of staining is important for recognizing the bacteria associated with, for example, whooping cough. The photograph shows a micrograph of Gram-stained *Streptococcus mutans* bacteria.

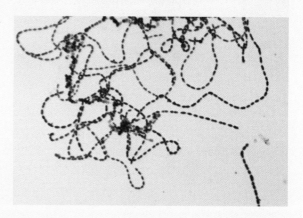

Light micrograph of Gram-stained *Streptococcus mutans* bacteria; these bacteria are found in the mouth cavity and are a major cause of tooth decay.
*CNRI / Science Photo Library.*

The structures, colours and uses of some other biological stains are shown opposite.

Acid fuchsin; red; used for staining plasma
$\lambda_{max} = 540$ nm

Amido black 10B; blue black; used with picric acid to stain
collagen and reticulin
$\lambda_{max} = 618$ nm

Azure B; blue; used to differentiate between RNA
and DNA in plant tissues
$\lambda_{max} = 650$ nm
(RNA and DNA: see Section 35.7)

Sudan IV (Scarlet red); used to stain fats
$\lambda_{max} = 520$ nm

**13.6**     **The visible region of the spectrum**

### The $-N=N-$ chromophore

The electronic spectra of many organic compounds have absorptions *only* in the near-UV region, and the compounds are colourless. However, Table 13.2 indicates that azo compounds absorb light of wavelength approaching the ultraviolet–visible boundary. Just as the absorption maximum for polyenes can be shifted by introducing substituents to the carbon chain, so $\lambda_{max}$ for the $-N=N-$ chromophore can be shifted into the visible region. This has important consequences – most azo compounds are coloured and many are used commercially as dyes. Two examples are shown in Figure 13.8, and the UV–VIS spectrum of methyl orange is shown in Figure 13.9.

The azo chromophore undergoes both $\pi^* \leftarrow n$ and $\pi^* \leftarrow \pi$ transitions but it is the $\pi^* \leftarrow n$ transition that falls close to or in the visible region.

### Ions of the *d*-block metals

*d*-Block metal ions:
see Chapter 23

Compounds of the *d*-block metals are often coloured. In many cases the colours are pale but characteristic. Aqueous solutions of titanium(III) salts often contain the *aqua ion* $[Ti(H_2O)_6]^{3+}$ (ion **13.10**). The electronic spectrum of dilute aqueous $TiCl_3$ shows a band with $\lambda_{max} = 510$ nm due to absorption by the octahedral $[Ti(H_2O)_6]^{3+}$ cation, and the observed colour is violet.

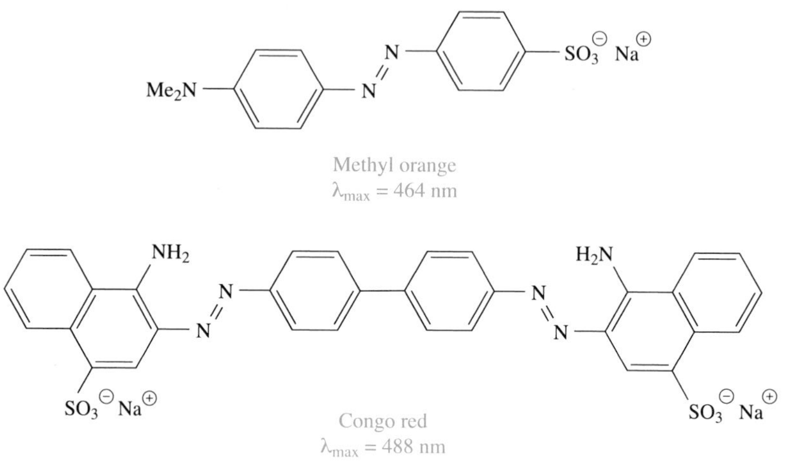

**Fig. 13.8** Examples of azo dyes, all of which contain the −N=N− chromophore. Congo red is used as a biological stain, and methyl orange is an acid–base indicator. Congo red is also used to estimate free mineral acids. See also Box 13.2.

Methyl orange
$\lambda_{max} = 464$ nm

Congo red
$\lambda_{max} = 488$ nm

Other metal ions also form octahedral $[M(H_2O)_6]^{n+}$ ions with characteristic colours (Table 13.3).

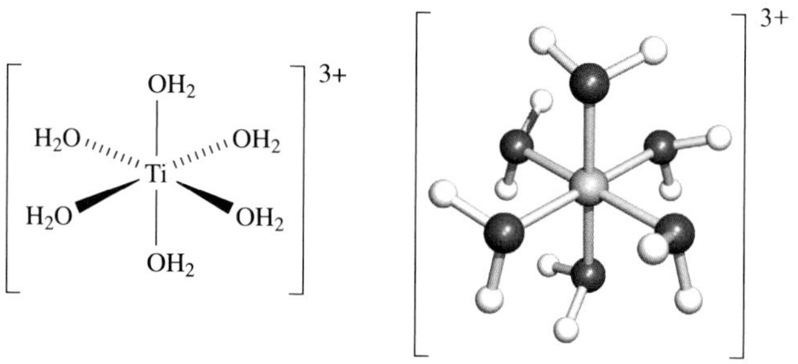

(13.10)

If the water molecules are replaced by other electron-donating groups, the new species may absorb light of a different wavelength. For example, the addition of ammonia to a solution containing pale blue $[Cu(H_2O)_6]^{2+}$ ions results in the formation of a very dark blue solution.

The electronic transitions that occur are known as '*d*–*d*' transitions and we discuss them in greater detail in Section 23.10. A key point is that when an ion of a *d*-block metal possesses *partially filled d orbitals*, electronic

**Fig. 13.9** The UV–VIS spectrum of aqueous methyl orange.

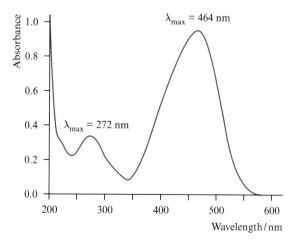

$\lambda_{max} = 464$ nm

$\lambda_{max} = 272$ nm

**Table 13.3** Colours of some hydrated ions of the *d*-block elements; see also Chapter 23.

| Hydrated metal ion | Observed colour | Hydrated metal ion | Observed colour |
|---|---|---|---|
| $[Ti(H_2O)_6]^{3+}$ | Violet | $[Fe(H_2O)_6]^{2+}$ | Very pale green |
| $[V(H_2O)_6]^{3+}$ | Green | $[Co(H_2O)_6]^{2+}$ | Pink |
| $[Cr(H_2O)_6]^{3+}$ | Purple | $[Ni(H_2O)_6]^{2+}$ | Green |
| $[Cr(H_2O)_6]^{2+}$ | Blue | $[Cu(H_2O)_6]^{2+}$ | Blue |

transitions between *d* orbitals may occur. A $Zn^{2+}$ ion has a filled $3d$ shell, and no electronic transitions are possible – solutions of $Zn^{2+}$ compounds are colourless.

## SUMMARY

In this chapter, we have introduced electronic (UV–VIS) spectroscopy with an emphasis on the application of this technique in the laboratory. We have discussed how electronic spectra arise, and the effects that molecular structure has on the wavelength of absorption maxima.

### Do you know what the following terms mean?

- UV–VIS
- HOMO
- LUMO
- $\lambda_{max}$

- $\varepsilon_{max}$
- vacuum-UV
- localized $\pi$-bonding
- delocalized $\pi$-bonding

- conjugated polyene
- chromophore
- bathochromic effect
- red shift

### You should be able:

- to understand how some electronic spectra arise, e.g. the spectrum of a polyene
- to comment on why choice of solvent is important when recording electronic spectra of solutions
- to explain what is meant by notation of the type $\sigma^* \leftarrow \sigma$, $\pi^* \leftarrow \pi$ and $\pi^* \leftarrow n$

- to explain why values of $\lambda_{max}$ may shift in a series of related compounds
- to give examples of compounds that absorb in the near-UV and visible regions

## PROBLEMS

**13.1** Electronic spectra are often reported in terms of wavelength but can also be recorded in wavenumbers. Convert the following absorption maxima given in $cm^{-1}$ into $\lambda$ given in nm: (a) $22\,700\,cm^{-1}$; (b) $35\,000\,cm^{-1}$.

**13.2** Normal laboratory UV–VIS spectrometers operate in the range $\approx 200$–900 nm. Which of the following types of electromagnetic radiation does this range cover: (a) visible light; (b) near-UV radiation; (c) vacuum-UV radiation?

**13.3** Do absorptions in the vacuum-UV involve transitions in which $\Delta E$ (change in energy) is greater or smaller than those observed in the visible region?

**13.4** (a) Why are electronic spectra usually characterized by *broad* absorptions? (b) Absorptions in an electronic spectrum are usually described in terms of $\lambda_{max}$ and $\varepsilon_{max}$ values. Draw a diagram of a typical absorption and indicate how you would measure $\lambda_{max}$ and $\varepsilon_{max}$ for the absorption. (c) Why is it necessary to quote $\lambda_{max}$ and $\varepsilon_{max}$ values rather than $\lambda$ and $\varepsilon$?

**13.5** For each of the following species, the value of $\lambda_{max}$ is given. In which part of the UV–VIS range does each species absorb? (a) $I^-$, 226 nm; (b) $[C_2O_4]^{2-}$, 250 nm; (c) $H_2O$, 167 nm; (d) (Z)-$C_6H_5N{=}NC_6H_5$, 316 nm; (e) $[Ni(H_2O)_6]^{2+}$, 400 nm (one of three bands).

**13.6** (a) What is a polyene? (b) In which of compounds **13.11–13.14** is the $\pi$-bonding delocalized?

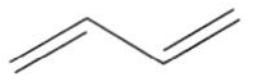

(**13.11**) 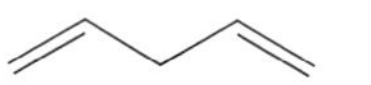　　　(**13.12**)

(**13.13**) 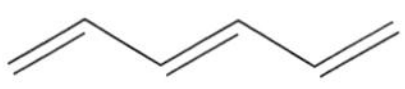　　　(**13.14**)

**13.7** Lycopene ($\lambda_{max} = 469$ nm) is present in ripe tomatoes. What colour of light does lycopene absorb?

**13.8** Suggest which types of transitions ($\sigma^* \leftarrow \sigma$, $\pi^* \leftarrow \pi$, $\sigma^* \leftarrow n$ or $\pi^* \leftarrow n$) give rise to significant features in the electronic spectra of (a) pentane (**13.15**), (b) pent-1-ene (**13.16**), (c) octa-2,4,6-triene (**13.17**) and (d) ethanol (**13.18**).

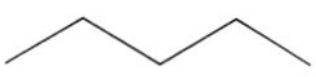

(**13.15**) 　　　(**13.16**)

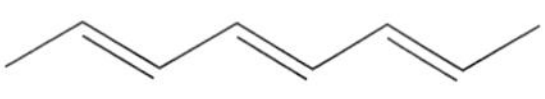

(**13.17**) 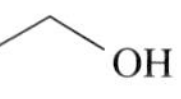　　　(**13.18**)

**13.9** (a) What is a 'red shift'? (b) Suggest why the methoxy (MeO) substituent in the compound $MeOCH{=}CHCH{=}CH_2$ causes a red shift in the UV–VIS spectrum with respect to the spectrum of $CH_2{=}CHCH{=}CH_2$.

**13.10** The following data show the dependence of $\lambda_{max}$ on the number of C=C bonds in the polyenes **13.19**.

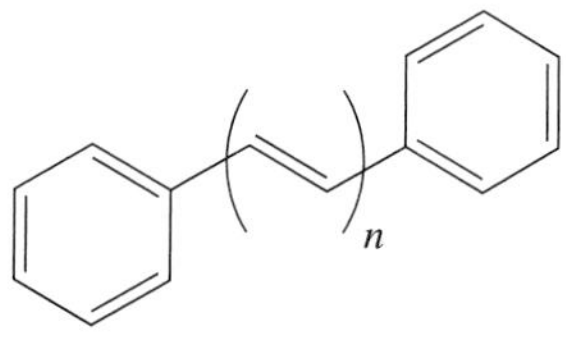

(**13.19**)

| $n$ | 1 | 2 | 3 | 4 |
|---|---|---|---|---|
| $\lambda_{max}$ / nm | 358 | 384 | 403 | 420 |

(a) Rationalize the trend in values of $\lambda_{max}$. (b) Do the compounds absorb in the vacuum-UV, near-UV or visible parts of the spectrum?

## Additional problems

**13.11** The reaction:

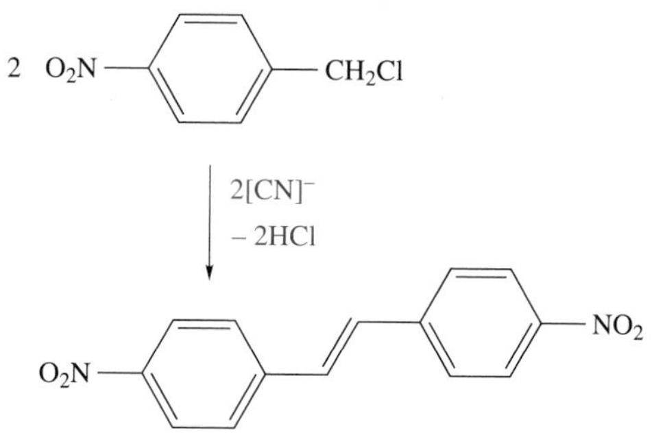

can be followed by UV–VIS spectroscopy. With the spectrometer tuned to $\lambda = 470$ nm, the results show an increase in absorbance over a reaction period of 200 min. What can you deduce from these data?

**13.12** Ketones ($R_2C{=}O$) typically exhibit absorbances in their electronic spectra with $\lambda_{max} \approx 270$–290 nm ($\varepsilon_{max} \approx 10$–$20\,dm^3\,mol^{-1}\,cm^{-1}$) and 190 nm ($\varepsilon_{max} \approx {>}2000\,dm^3\,mol^{-1}\,cm^{-1}$). Suggest an explanation for these observations.

# 14 NMR spectroscopy

## Topics

- Nuclear spin states
- Recording an NMR spectrum
- Chemical shifts and resonance frequencies
- An introduction to analysing $^{13}C$ and $^1H$ NMR spectra
- Homonuclear and heteronuclear coupling between nuclei with $I = \frac{1}{2}$

## 14.1 Introduction

Nuclear magnetic resonance (NMR) spectroscopy is today a routine laboratory technique. Besides its application for the identification of compounds,
NMR spectroscopy can be used to follow reactions, measure rate and
equilibrium constants and study the dynamic behaviour of molecules. It
also has medical application in MRI (magnetic resonance imaging, see Box
14.1) and can be used to determine the structures of large biomolecules
(see Box 14.2). In this chapter, we introduce some basic theory of NMR
spectroscopy but the greater part of our discussion lies with applications of
the technique to compound characterization with an emphasis on $^1H$ and
$^{13}C$ NMR spectroscopies.

We have agonized over the level of presentation in this chapter. Compel-
ling arguments exist for a rigorous (and classical) discussion commencing
with concepts of precession and resonance, as do justifications for a primarily
empirical discussion. We have finally opted for the latter.

In this chapter, we use many examples of organic molecules. More detailed
information about structural diagrams and nomenclature can be found in
Chapter 24.

## 14.2 Nuclear spin states

Many nuclei possess a property described as spin. This is a quantum effect
which may be interpreted classically as the nucleus spinning about an axis.
The nuclear spin (nuclear angular momentum) is quantized and is described
by the nuclear spin quantum number $I$. The nuclear spin quantum number
can have values $I = 0, \frac{1}{2}, 1, \frac{3}{2}, 2, \frac{5}{2}$, etc., but in this chapter we are concerned
with nuclei for which $I = \frac{1}{2}$; examples are $^1H$, $^{13}C$, $^{19}F$ and $^{31}P$.

Nuclei possess positive charges, and, classically, a spinning charge generates a magnetic field, which can interact with an external magnetic field. In
this classical picture, a magnetic spin quantum number $m_I$ describes the

Only certain nuclei are
NMR active. These include
$^1H$, $^{13}C$, $^{19}F$ and $^{31}P$.

**Fig. 14.1** The splitting of nuclear spin states on the application of an external magnetic field.

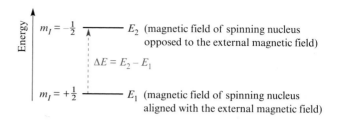

*direction* of spin and $m_I$ has values $+I, +(I-1)\ldots -I$; for nuclei with $I = \frac{1}{2}$, $m_I = +\frac{1}{2}$ or $-\frac{1}{2}$. There is no difference in energy between nuclei spinning in different directions. However, the direction of spin affects the polarity of the associated magnetic field. If an external magnetic field is applied, this could be in the same or opposite direction to that due to the spinning nucleus. The external magnetic field results in an energy difference between the $m_I = +\frac{1}{2}$ and $-\frac{1}{2}$ spin states (Figure 14.1). If splitting of the spin states into different energies occurs upon the application of a magnetic field, it should be possible to use a spectroscopic method to determine the energy difference.

The experiment we have just described leads to *nuclear magnetic resonance* (NMR). The energy difference between the $m_I = +\frac{1}{2}$ and $-\frac{1}{2}$ spin states depends upon the strength of the applied field but is very small at all achievable fields. Typically, the energy difference is less than $0.01\,\text{kJ}\,\text{mol}^{-1}$ and electromagnetic radiation is absorbed in the radiofrequency (RF) region. Absorption of RF energy results in a nucleus being excited from the lower- to higher-energy spin state. *Relaxation* from the upper to lower spin state then occurs.

RF = radiofrequency

## 14.3     Recording an NMR spectrum

An NMR spectrum is a plot of absorbance against frequency. The simplest way to record such a spectrum is to place a sample in a magnetic field and scan through the radiofrequencies until an absorption is observed. Although this method has been widely used, it has an inherent disadvantage. The energy difference between the $m_I = +\frac{1}{2}$ and $-\frac{1}{2}$ spin states is very small, and if we consider a Boltzmann distribution (see Section 12.2) we find that the difference in population between the upper and lower energy levels is about one nucleus in every million! This means that NMR spectroscopy is an inherently insensitive technique. Experimentally, we can assess the sensitivity in terms of the *signal-to-noise ratio* of the recorded spectrum.

One way of increasing the signal-to-noise ratio is to record the same spectrum a number of times, and then add these spectra together. For $n$ spectra, the signal increases according to $n$ and the noise by $n^{\frac{1}{2}}$; the improvement in the signal-to-noise ratio is therefore $n^{\frac{1}{2}}$. However, scanning through a range of frequencies over and over again is time-consuming.

We noted above that the energy difference between the lower and upper energy levels was very small. In the same way that the equilibrium population of these levels is determined by the energy gap, so is the time taken to return to this state after excitation, i.e. *the time for the relaxation process to occur*. The *smaller* the energy gap, the *longer* it takes to return to the equilibrium population. In the case of NMR spectroscopy, the relaxation time may

## COMMERCIAL AND LABORATORY APPLICATIONS

### Box 14.1 Magnetic resonance imaging (MRI)

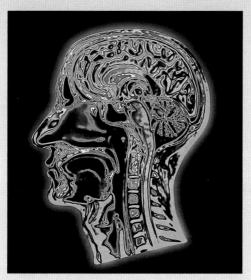

Coloured MRI scan through the human head, showing a healthy brain. *Alfred Pasieka/Science Photo Library*.

Magnetic resonance imaging is a clinical technique to provide images of human organs or tumours. The image is produced from information obtained from the $^1H$ NMR spectroscopic signals of $H_2O$, and the signal intensity depends on the $^1H$ relaxation times and the concentration of $H_2O$. To aid the observation of the organ or tumour, MRI contrast agents are used to alter the relaxation times of the protons. Complexes (see Chapter 23) containing the paramagnetic $Gd^{3+}$ ion appear to be particularly suitable contrasting agents. Introduction of gadolinium ions into the human body must be done with care and it is important that the compound containing the metal does not dissociate and is excreted from the body as rapidly as is practical. Complexes containing *ligands* (see Chapter

23) such as $H_5DTPA$ have proved successful. It is also crucial that a particular contrast agent *targets* an organ or tumour of choice rather than being randomly spread through a patient's body; for example, $Gd^{3+}$ complexes target the liver.

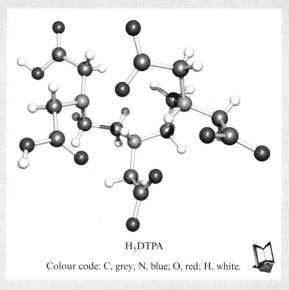

$H_5DTPA$

Colour code: C, grey; N, blue; O, red; H, white.

[Data: L. M. Shkolnikova *et al.* (1984) *Zh. Strukt. Khim.*, vol. 25, p. 103]

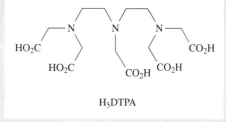

$H_5DTPA$

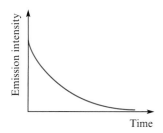

**Fig. 14.2** A plot of energy emission against time from an excited nuclear spin state; the curve is called a *free induction decay* (FID).

vary from milliseconds to many hours. This places another time constraint upon the addition of spectra, because we must wait for the equilibrium population to be re-established between the accumulation of consecutive spectra.

The modern method of recording NMR spectra makes use of some of these effects to provide a partial solution to these problems. We want a plot of intensity of absorption against frequency as an NMR spectrum, but a plot of *emission* against frequency is equivalent, and as the relaxation process is relatively slow, we could just as easily monitor the *emission* of RF after excitation. In itself this would give no advantage – in fact, it would be disadvantageous since the intensity of emission at any given instant will always be lower than any absorption intensity because the absorption is effectively instantaneous and the emission occurs over a period of time (Figure 14.2). However, a plot of *emission intensity against time* contains

■▶
FT = Fourier transform

exactly the same information as a plot of *absorption intensity against frequency* and the mathematical operation of a Fourier transform (FT) interconverts the two sets of data.

Now comes the important point. The precise frequency of the absorption of a nucleus depends upon its chemical environment and if a compound contains several different types of environment, the simple experiment described above involves scanning through the RF frequencies and finding all of the absorptions. By using an FT technique, we can use a mixture of frequencies to excite *all* the nuclei in the different environments at once, and then record the *emission* as a function of time. Such a plot is called a free induction decay (FID). A series of FIDs may be recorded and added together to improve the signal-to-noise ratio, and a Fourier transformation applied to give the conventional absorption versus frequency spectrum.

■▶
FID = free induction decay

## 14.4    Nuclei: resonance frequencies, isotope abundances and chemical shift values

### Resonance frequencies

A particular nucleus ($^1H$, $^{13}C$, etc.) *resonates* at a characteristic frequency, i.e. it absorbs radiofrequencies within a certain range. You can draw an analogy between an NMR spectrometer and a radio. Having tuned a radio to a particular frequency you receive only the station selected because different radio stations broadcast at different frequencies. Similarly, an NMR spectrometer is tuned to a particular resonance frequency to detect a selected NMR active nucleus. The resonance frequencies of some nuclei with $I = \frac{1}{2}$ are listed in Table 14.1. When we record a $^1H$ NMR spectrum (a 'proton spectrum') the *signals* or *resonances* that are observed are due only to the $^1H$ nuclei in the sample. Similarly, a $^{31}P$ NMR spectrum contains resonances due only to $^{31}P$ nuclei.

Richard R. Ernst (1933–) received the Nobel Prize in Chemistry in 1991 for 'his contributions to the development of the methodology of high resolution NMR spectroscopy'.
© *The Nobel Foundation.*

**Table 14.1** Resonance frequencies (referred to 100 MHz for $^1H$)[a] for selected nuclei with $I = \frac{1}{2}$.

| Nucleus | Resonance frequency / MHz | Natural abundance / % | Chemical shift reference |
|---|---|---|---|
| $^1H$ | 100 | ≈99.9 | $Me_4Si$ |
| $^{19}F$ | 94.0 | 100 | $CFCl_3$ |
| $^{31}P$ | 40.48 | 100 | $H_3PO_4$ (85% aqueous) |
| $^{119}Sn$ | 37.29 | 8.6 | $Me_4Sn$ |
| $^{13}C$ | 25.00 | 1.1 | $Me_4Si$ |
| $^{195}Pt$ | 21.46 | 33.8 | $Na_2[PtCl_6]$ |
| $^{29}Si$ | 19.87 | 4.7 | $Me_4Si$ |
| $^{107}Ag$ | 4.05 | 51.8 | Aqueous $Ag^+$ |
| $^{103}Rh$ | 3.19 | 100 | Rh (metal) |

[a] NMR spectrometers can operate at different magnetic field strengths and it is convenient to describe them in terms of the RF needed to detect protons. A 100 MHz instrument records $^1H$ NMR spectra at 100 MHz, while a 250 MHz spectrometer records the $^1H$ NMR spectra at 250 MHz. A 100 MHz instrument operates with a magnetic field of 2.35 tesla and a 250 MHz at 5.875 tesla.

## Natural abundance of a nucleus

We have already seen that a nucleus is only NMR active if it possesses a nuclear spin, but another property that is important is the *natural abundance* of the nucleus (Table 14.1). Fluorine possesses one isotope and when a $^{19}F$ NMR spectrum is recorded, we observe all of the fluorine nuclei present. The situation is similar for hydrogen, because ≈99.9% of natural hydrogen consists of $^1H$ nuclei. When a $^{13}C$ NMR spectrum is recorded, only 1.1% of the carbon nuclei present can be observed, because the remaining 98.9% are $^{12}C$ and are not NMR active ($I = 0$).

Although a low natural abundance can be problematical, it does *not* mean that the nucleus is unsuitable for NMR spectroscopy. A greater number of FIDs may be collected in order to obtain an enhanced signal-to-noise ratio, but alternatively a sample can be isotopically enriched.

## Chemical shifts and NMR spectroscopic references

The resonance frequency of a $^1H$ nucleus depends on its precise chemical environment, and similar effects are found with other NMR active nuclei. Why is this?

We explained earlier that the energy difference between the nuclear spin states arose from the interaction between the magnetic fields of the spinning nuclei and an applied external field. However, the *local field*, **B**, experienced at a nucleus is not precisely the same as that applied. This is because the bonding electron pairs in the molecule are also charges moving in space, and they generate small local magnetic fields. The field **B** is the sum of the applied field, $B_0$, and all these smaller fields. The local fields from the electrons vary with the chemical environment of the nucleus. Typically, the differences in field are small, and variations in absorption frequency of protons in different environments are only a few parts per million. These differences give rise to different signals in the spectrum. The position of a signal is denoted by its *chemical shift value*, $\delta$, relative to the signal of a specified reference compound. The standard reference for $^1H$ and $^{13}C$ NMR spectroscopies is tetramethylsilane (TMS, **14.1**). The 12 hydrogen atoms in TMS are equivalent and there is one proton environment giving rise to one signal in the $^1H$ NMR spectrum of this compound. The chemical shift value of this signal is defined as zero. Similarly, there is only one carbon atom environment in TMS and the signal in the $^{13}C$ NMR spectrum of TMS is also defined as zero.

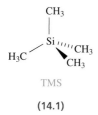

TMS

**(14.1)**

The signals in a particular NMR spectrum are always referenced with respect to a standard compound (Table 14.1), but care should be taken because different references are sometimes used for a given nucleus. Although 85% aqueous phosphoric acid is the standard reference compound in $^{31}P$ NMR spectroscopy, $P(OMe)_3$ (trimethylphosphite) is sometimes used, and it is important to quote the reference when reporting NMR spectra.

Signals due to particular nuclei in a compound that appear in the NMR spectrum are said to be *shifted with respect to the standard reference signal*. A shift to positive $\delta$ is 'shifted to higher frequency' and a shift to negative $\delta$ is 'shifted to lower frequency'. Older terminology that is still used relates a positive $\delta$ value to a 'downfield shift' and a negative $\delta$ to an 'upfield shift'. What exactly is this $\delta$ scale? The problem with NMR spectroscopy is that the energy difference between the $m_I = +\frac{1}{2}$ and $m_I = -\frac{1}{2}$ spin states

The ***chemical shift value*** $\delta$ of a signal in an NMR spectrum is quoted with respect to that of a reference defined at zero.

depends upon the external magnetic field. Similarly, the difference in frequency of absorptions between nuclei in different environments also depends on the applied field. How can we compare data recorded at different magnetic field strengths? The solution is to define a *field-independent parameter* – the chemical shift $\delta$.

We define $\delta$ as follows. The frequency difference ($\Delta\nu$), in Hz, between the signal of interest and some defined reference frequency ($\nu_0$) is divided by the absolute frequency of the reference signal (equation 14.1).

$$\delta = \frac{(\nu - \nu_0)}{\nu_0} = \frac{\Delta\nu}{\nu_0} \tag{14.1}$$

Typically, this leads to a very small number. In order to obtain a more convenient number for $\delta$, it is usual to multiply the ratio in equation 14.1 by $10^6$. This gives $\delta$ in units of parts per million, ppm. The IUPAC defines $\delta$ according to equation 14.1, but equation 14.2 gives a method of calculating $\delta$ in ppm.[§]

$$\delta \text{ in ppm} = \frac{(\nu - \nu_0) \text{ in Hz}}{\nu_0 \text{ in MHz}} \tag{14.2}$$

---

**Worked example 14.1** | *Chemical shift values and frequencies*

**On a 100 MHz spectrometer, what is the shift difference in Hz between signals appearing in the spectrum at $\delta$ 1.5 ppm and $\delta$ 2.6 ppm?**

Equation needed:

$$\delta \text{ in ppm} = \frac{(\nu - \nu_0) \text{ in Hz}}{\nu_0 \text{ in MHz}}$$

$\nu_0 = 100 \text{ MHz}$
$(\nu - \nu_0)$ gives the shift *in Hz* of the signal with respect to a reference signal.
For $\delta$ 1.5 ppm, $(\nu - \nu_0) = 150 \text{ Hz}$
For $\delta$ 2.6 ppm, $(\nu - \nu_0) = 260 \text{ Hz}$
$\therefore$ Shift difference in Hz $= 260 - 150 = 110 \text{ Hz}$
We can combine the working to carry out this calculation as follows:

$$(\Delta\nu \text{ in Hz}) = (\text{Spectrometer frequency in MHz}) \times (\Delta\delta \text{ in ppm})$$

$$\Delta\delta = (2.6 - 1.5) \text{ ppm} = 1.1 \text{ ppm}$$

$$\Delta\nu = (100 \text{ MHz}) \times (1.1 \text{ ppm})$$

$$= 110 \text{ Hz}$$

This method of working is used to calculate coupling constants; see Section 14.9.

---

## Chemical shift ranges

The *chemical shift range* (i.e. the range of chemical shifts over which signals appear in the spectrum) depends on the nucleus. For example, $^1$H NMR

---

[§] R. K. Harris, E. D. Becker, S. M. Cabral de Menezes, R. Goodfellow and P. Granger (2001) *Pure and Applied Chemistry*, vol. 73, p. 1795 – NMR nomenclature. Nuclear spin properties and conventions for chemical shifts (IUPAC recommendations 2001).

resonances usually fall in the range $\delta +15$ to $-35$ ppm, whereas $^{13}C$ NMR spectra are usually recorded over a range of $\delta +250$ to $-50$ ppm, and $^{31}P$ NMR spectra between the approximate limits $\delta +300$ to $-300$ ppm. However, new chemistry can bring surprises in the form of a nucleus in an unprecedented environment, and the spectral 'window' should not be fixed in the experimentalist's mind.

## 14.5    Choosing a solvent for NMR spectroscopy

Although NMR spectra may be recorded in the solid state, such spectra have added complications, and most samples are studied as solutions. Molecules of most solvents contain carbon and hydrogen atoms and so in the $^{1}H$ or $^{13}C$ NMR spectrum of a solution sample, the solvent naturally gives rise to its own spectrum which is superimposed on that of the sample. As the solvent is usually present in a vast excess with respect to the sample, the signals from the solvent are often the largest features observed in the spectrum.

A $^{1}H$ NMR spectrum is easily 'swamped' by the signals from the solvent and those of the sample may be obscured. The problem is solved by using *deuterated solvents* – the $^{1}H$ nuclei in the solvent molecules are replaced by $^{2}H$ (D) nuclei. For example, chloroform is a common solvent and can be isotopically labelled with deuterium to give $CDCl_3$. Although the $^{2}H$ nucleus is NMR active ($I = 1$), its resonance frequency is different from that of $^{1}H$ and therefore the deuterated solvent is effectively 'silent' in a $^{1}H$ NMR spectrum. Such solvents are commercially available and are usually labelled to an extent of $>99.5\%$. Residual molecules containing $^{1}H$ nuclei are present but give rise to relatively low-intensity signals in a $^{1}H$ NMR spectrum. Such solvent signals can serve as internal reference signals, e.g. the $^{1}H$ NMR spectrum of acetone (**14.2**, see p. 410) contains a signal at $\delta 2.05$ ppm.

## 14.6    Molecules with one environment

Nuclear magnetic resonance spectroscopy can be used to 'count' the number of environments of a particular nucleus in a molecule – $^{1}H$ NMR spectroscopy looks only at proton environments and $^{13}C$ NMR spectroscopy can be used to count the different types of carbon atoms in a molecule. The number of signals in the spectrum corresponds to the number of different environments.

In this section we illustrate that more information than simply 'one signal means one environment' can be obtained from an NMR spectrum. The chemical shift can give important information about the environment, for example, the possible geometry of the centre, or the functional groups that are attached to it.

Me = methyl = $CH_3$

Figure 14.3 shows the $^{31}P$ NMR spectra of $PMe_3$ and $PCl_3$. Each spectrum exhibits one signal, consistent with there being one phosphorus environment in each molecule. However, note how the nature of the substituents affects the chemical shift of the signal. The difference in shift between these two trigonal pyramidal $PR_3$ molecules is particularly dramatic. For $PF_3$, $PBr_3$ and $PI_3$ (all trigonal pyramidal) the $^{31}P$ NMR chemical shifts are $\delta = +96$, $+228$ and $+177$ ppm respectively. Each of $CH_4$, $CH_3Cl$, $CH_2Cl_2$ and $CHCl_3$ is tetrahedral, but the presence of the chloro-substituents influences

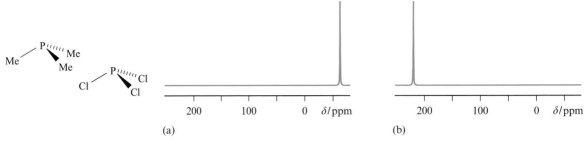

Fig. 14.3 The $^{31}$P NMR spectra of (a) PMe$_3$ and (b) PCl$_3$.

the chemical shift of the $^{13}$C nucleus: CH$_4$ ($\delta$ +2 ppm), CH$_3$Cl ($\delta$ +22 ppm), CH$_2$Cl$_2$ ($\delta$ +54 ppm) and CHCl$_3$ ($\delta$ +77 ppm).

## 14.7     Molecules with more than one environment

### Chemical shifts in $^{13}$C NMR spectra

We can relate the chemical shift of a $^{13}$C NMR signal to the hybridization of the carbon centre, e.g. $sp^2$ carbon atoms resonate at higher frequency (more positive $\delta$) with respect to $sp$ carbon centres. Table 14.2 and Figure 14.4 give *approximate* ranges for carbon nuclei in some environments in organic molecules. Compounds containing a carbonyl group (C=O) are characterized by a $^{13}$C NMR resonance at high frequency. The variety of compounds containing $sp^3$ hybridized carbon atoms is large and includes cyclic and acyclic alkanes, alcohols (X = OH), halogenoalkanes (X = F, Cl, Br or I), amines (X = NH$_2$), thiols (X = SH) and ethers (X = OR).

$^{13}$C NMR spectra of carbonyl compounds: see Section 33.4

### Signal intensities

The *relative intensities* of NMR signals can assist in their assignment because the ratio of intensities gives information about the *ratio* of the number of different environments in the molecule.

The $^{13}$C NMR spectrum of acetone (**14.2**) consists of two signals (Figure 14.5) which can readily be assigned from their relative intensities of 1 : 2. *A word of caution however*. Signal intensities in $^{13}$C NMR spectra can be misleading owing to different relaxation effects of the different $^{13}$C nuclei. The problem arises from our use of the FT method. Because we collect emission data in the time domain, we should wait between acquiring consecutive data sets until the Boltzmann distribution is achieved for nuclei in *all* environments. If this is not done, we observe *rapidly* relaxing nuclei more efficiently than those relaxing slowly. In practice, time constraints may mean that a compromise is made in the time between consecutive excitations of the nuclei and, as a result, the ratios of the signal intensities are not always exactly equal to the ratios of the environments of the nuclei. This is particularly true in $^{13}$C NMR spectra.[§] Fortunately, chemical shift values can come to our aid in assigning signals.

Relaxation: see Section 14.3

---

[§] For more detailed discussion, see: B. K. Hunter and J. K. M. Sanders (1993) *Modern NMR Spectroscopy: A Guide for Chemists*, 2nd edn, Oxford University Press, Oxford.

**Table 14.2** Approximate shift ranges for carbon environments in $^{13}C$ NMR spectra (chemical shifts are with respect to TMS $\delta = 0$ ppm).

| Carbon environment | | Hybridization | $^{13}C$ chemical shift range/ppm | Comments |
|---|---|---|---|---|
| R<br>\\<br>C=O<br>/<br>R | Ketones | $sp^2$ | +230 to +190 | |
| R<br>\\<br>C=O<br>/<br>H | Aldehydes | $sp^2$ | +210 to +185 | |
| R<br>\\<br>C=O<br>/<br>HO | Carboxylic acids | $sp^2$ | +185 to +165 | |
| R<br>\\<br>C=O<br>/<br>X | Amides X = $NH_2$<br>Esters X = OR | $sp^2$ | +180 to +155 | |
| R    R<br>\\  /<br>C=C<br>/  \\<br>R    R | Alkenes | $sp^2$ | +160 to +100 | |
| (benzene ring) | Aromatic compounds | $sp^2$ | +145 to +110 | Note some overlap of the $sp$ and $sp^2$ regions. |
| R—C≡N | Nitriles | $sp$ | +125 to +110 | |
| R—C≡C—R | Alkynes | $sp$ | +110 to +75 | |
| R R<br>R—C—X<br>R | Aliphatic compounds | $sp^3$ | | |
| R R<br>R—C—X<br>H | Aliphatic compounds | $sp^3$ | | Some overlap of regions depending upon X. Overall range +90 to 0. Generally: $CH_3X < RCH_2X < R_2CH < R_3CX$ <br><br>more positive $\delta$ |
| H R<br>R—C—X<br>H | Aliphatic compounds | $sp^3$ | | |
| H H<br>H—C—X<br>H | Aliphatic compounds | $sp^3$ | | |

**Fig. 14.4** Typical $^{13}$C NMR chemical shift ranges for common classes of organic compound. For further details, see Table 14.2.

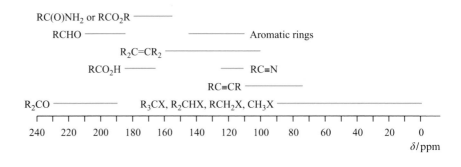

**Fig. 14.5** $^{13}$C NMR spectrum (proton decoupled) of acetone (**14.2**). The spectrum was recorded at 25 MHz.

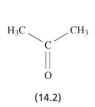

(**14.2**)

### Assigning $^{13}$C NMR spectra on the basis of chemical shifts

**Refer to Table 14.2**

$$\overset{a}{H_3C} - \overset{b}{C} \equiv N$$

Acetonitrile

**Example 1: acetonitrile** The $^{13}$C NMR spectroscopic chemical shifts are $\delta +1.3$ and $+117.7$ ppm (Figure 14.6a). There are two carbon environments:

*a* aliphatic ($sp^3$)
*b* nitrile ($sp$)

and the nitrile carbon should be at higher frequency with respect to the aliphatic carbon.

Assignment: $\delta +1.3$ ppm is assigned to carbon *a*
$\delta +117.7$ ppm is assigned to carbon *b*

Ethanal
(Acetaldehyde)

**Example 2: ethanal** The observed $^{13}$C NMR spectroscopic chemical shifts are $\delta +30.7$ and $+199.7$ ppm (Figure 14.6b). There are two carbon environments:

*a* aliphatic ($sp^3$)
*b* aldehyde ($sp^2$)

**Fig. 14.6** 25 MHz $^{13}$C NMR spectra (proton decoupled) of (a) CH$_3$CN, (b) CH$_3$CHO and (c) Et$_2$NH. The same scale applies to all the spectra.

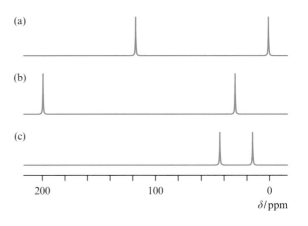

and the aldehyde carbon should be at higher frequency than the aliphatic carbon.

Assignment:  $\delta +30.7$ ppm is assigned to carbon $a$
$\delta +199.7$ ppm is assigned to carbon $b$

**Example 3: diethylamine**    The $^{13}C$ NMR spectroscopic signals are at $\delta +15.4$ and $+44.1$ ppm (Figure 14.6c). There are two $sp^3$ carbon environments:

$a$ terminal $CH_3$ group
$b$ $RCH_2X$ group

and in general $RCH_2X$ carbon atoms are at higher frequency than the terminal $CH_3$ carbons.

Possible assignments:  $\delta +15.4$ ppm is due to carbon $a$
$\delta +44.1$ ppm is due to carbon $b$

Diethylamine

**14.8**    **¹H NMR spectra: chemical environments**

Since the chemical shift range for $^1H$ nuclei is relatively small, there is a significant overlap of the ranges that are characteristic of different proton environments. In Table 14.3 we give some approximate values for $^1H$ NMR spectroscopic signals in selected groups of organic compounds.[§]

Table 14.3 indicates that OH and $NH_2$ protons can give rise to *broad* signals. This phenomenon arises because O−H and N−H protons undergo *exchange* processes on a timescale that is comparable with the NMR spectroscopic time-scale. Equation 14.3 illustrates the exchange of the hydrogen atom in ethanol with a proton from water. The rate of exchange can be slowed down by lowering the temperature and this can cause sharpening and shifting of the signals.

Timescales: see Section 11.3

$$CH_3CH_2OH + HOH \rightleftharpoons CH_3CH_2OH + HOH \qquad (14.3)$$

The chemical shift value of some protons is dependent on solvent and this is particularly true of O−H protons. The $^1H$ NMR spectrum of neat ethanol has a signal at $\delta +5.4$ ppm assigned to the OH proton, but this shifts to lower frequency on dilution in a solvent. This observation is due to *hydrogen bonding*; the intermolecular interactions between ethanol molecules (**14.3**) lead to a reduction in the electron density around the hydrogen atom and causes a shift in the $^1H$ NMR signal to higher frequency.

Hydrogen bonding: see Section 21.8

ıııııııı = Hydrogen bonding interaction

(14.3)                                        (14.4)

---

[§] For greater detail, see: D. H. Williams and I. Fleming (1995) *Spectroscopic Methods in Organic Chemistry*, 5th edn, McGraw-Hill, London.

**Table 14.3** Typical shift ranges for proton environments in $^1$H NMR spectra of organic compounds (chemical shifts are with respect to TMS $\delta = 0$ ppm).

| Proton environment | | $^1$H chemical shift range | Comments |
|---|---|---|---|
| R–C(=O)–OH | Carboxylic acids | +9 to +13 | Often broad and dependent upon solvent |
| R–C(=O)–NH$_2$ | Amides | +5 to +12 | Broad and dependent upon solvent |
| R–C(=O)–H | Aldehydes | +8 to +10 | Sharp (usually no confusion with carboxylic acid, amide or amine signals) |
| X$_3$C–H (X = Cl or Br) | | $\approx$ +7 | CHCl$_3$ $\delta$ +7.25 ppm |
| Aromatic ring | Aromatic compounds | +6 to +10 | C$_6$H$_6$ $\delta$ +7.2 ppm |
| R$_2$C=CHR (alkene) | Alkenes | +4 to +8 | |
| R–C≡C–H | Terminal alkynes | +2.5 to +3 | |
| R–NH$_2$ | Amines | +1 to +6 | Broad and dependent upon solvent |
| R–OH | Alcohols | +0.5 to +8 | Often broad and dependent upon solvent |
| R$_3$C–H | Methine | +1.5 to +4.5 | |
| R$_2$CH$_2$ | Methylene | +1.5 to +4.5 | |
| R–CH$_3$ | Methyl | 0 to +4 | TMS defined as $\delta = 0$ ppm |

These shift ranges are represented diagrammatically below:

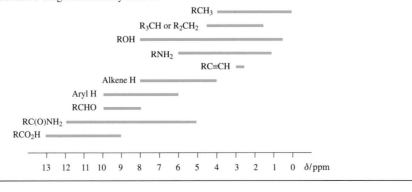

## ENVIRONMENT AND BIOLOGY

### Box 14.2  Biological macromolecular structure determination using NMR spectroscopy

Although this chapter is concerned with the NMR spectra of simple molecules, the technique has now developed to such an extent that the structures of large biomolecules, including proteins, can be determined from their NMR spectroscopic characteristics. Use is made of ¹H, ¹³C and ¹⁵N NMR spectra, and of ¹³C and ¹⁵N isotopically labelled biomolecules, in particular for high molecular weight proteins. A key step forward was the application of two-dimensional NMR spectroscopic methods, originally developed in the 1970s by Jean Jeener. All the NMR spectra that are shown in this book are one-dimensional, but in a two-dimensional spectrum, the signals in a spectrum can be correlated with each other so that nuclei that couple give rise to 'cross peaks'. Both homonuclear (e.g. ¹H–¹H) and heteronuclear (e.g. ¹³C–¹H) two-dimensional spectra can be studied.

atoms, and there are many thousands of distance and angle parameters. Thus, solving a protein structure is time-consuming, and requires extensive computer analysis of the NMR spectroscopic data. Nonetheless, it is now a relatively routine technique. Together with developments in X-ray diffraction methods, structure-solving by NMR spectroscopy has revolutionized our understanding of protein structure. The first protein structure to be determined by NMR spectroscopic methods was that of proteinase inhibitor IIa from bull seminal plasma, and was published in 1985. In 2002, Kurt Wüthrich shared the Nobel Prize in Chemistry for 'his development of nuclear magnetic resonance spectroscopy for determining the three-dimensional structure of biological macromolecules in solution'.

An X-ray diffraction study of a crystalline biomolecule effectively gives a static picture of the molecule. In contrast, NMR spectroscopic data are obtained for solution samples in which the molecule is non-rigid. This is clearly seen in the structure shown below which illustrates the superimposition of 20 NMR spectroscopic structure solutions of a variant human prion protein.

Kurt Wüthrich (1938–).   © *The Nobel Foundation.*

A range of two-dimensional techniques is now available. Two examples are COSY (COrrelated SpectroscopY) and NOESY (Nuclear Overhauser Enhancement SpectroscopY). Using COSY spectra, it is possible to deduce intermolecular connectivity patterns and bond angles using information from spin–spin coupling and the magnitudes of coupling constants; these data depend on through-bond couplings. The results of NOESY spectra provide information about through-space internuclear interactions. Proteins and other large biomolecules consist of thousands of

[Data: R. Zahn *et al.* (2003) *Journal of Molecular Biology*, vol. 326, p. 225.]

For further information, see:

G. Wider (2000) – 'Structure determination of biological macromolecules in solution using NMR spectroscopy'. *BioTechniques*, vol. 29, p. 1278.

M. Pellecchia, D. S. Sem and K. Wüthrich (2002) *Nature Reviews*, vol. 1, p. 211.

**Fig. 14.7** 100 MHz $^1$H NMR spectrum of acetic acid, $CH_3CO_2H$.

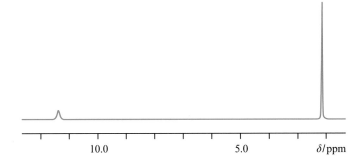

Similarly, hydrogen bonding between carboxylic acid molecules is one of the reasons why these protons are characterized by relatively high frequency signals. The $^1$H NMR spectrum of acetic acid (Figure 14.7) exhibits a broad signal centred at $\delta +11.4$ ppm (assigned to OH) and a sharp, more intense signal at $\delta +2.1$ ppm (assigned to $CH_3$). Acetic acid can form a hydrogen-bonded dimer (**14.4**); we discuss this further in Chapter 33.

$$H - C \equiv C - CH_3$$

(14.5)

(14.6)

Overlapping of the characteristic regions in $^1$H NMR spectra may make spectroscopic assignment difficult. In propyne (**14.5**), the signals for the $C_{alkyne}$–H and the methyl group are coincident ($\delta +1.8$ ppm). The $^1$H NMR spectrum of butan-2-one (**14.6**) has signals at $\delta +1.1$, $+2.1$ and $+2.5$ ppm. Although the relative intensities allow us to distinguish the $CH_2$ resonance from those of the $CH_3$ groups, how can we determine which signal corresponds to which methyl group? Fortunately, a feature of NMR spectroscopy that we have so far ignored now comes to our aid: *nuclear spins that are magnetically inequivalent may couple* to one another and this results in characteristic *splittings* of NMR signals.

## 14.9  Nuclear spin–spin coupling between nuclei with $I = \frac{1}{2}$

### Coupling between two inequivalent $^1$H nuclei

A hydrogen nucleus may be in one of two spin states ($m_I = +\frac{1}{2}$, $m_I = -\frac{1}{2}$) and we stated earlier that the energy difference between these spin states was controlled by the applied magnetic field. Later, we showed that smaller energy differences (giving rise to different chemical shifts) resulted from movement of electrons within bonds. What happens when two *magnetically non-equivalent* hydrogen nuclei in a molecule are close to each other?

Consider a system with two hydrogen atoms $H_A$ and $H_X$ which are in different magnetic environments. Let us introduce a shorthand notation:

$H_A$ with $m_I = +\frac{1}{2}$ is labelled $\alpha(A)$

$H_A$ with $m_I = -\frac{1}{2}$ is labelled $\beta(A)$

$H_X$ with $m_I = +\frac{1}{2}$ is labelled $\alpha(X)$

$H_X$ with $m_I = -\frac{1}{2}$ is labelled $\beta(X)$

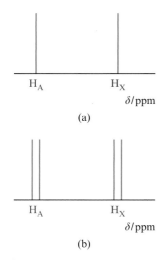

**Fig. 14.8** (a) Two singlet resonances are observed for two protons $H_A$ and $H_X$ if the two nuclei do not couple. (b) If coupling occurs, each resonance is split *by the same amount* to give a doublet.

In the simplest case, the magnetic field experienced by $H_A$ (or $H_X$) is independent of the spin state of $H_X$ (or $H_A$). In other words, the local magnetic field generated by the spin of one nucleus is not detected by the other nucleus. The spectrum will consist of two absorptions corresponding to the $\beta(A) \leftarrow \alpha(A)$ and $\beta(X) \leftarrow \alpha(X)$ transitions (Figure 14.8a). We say that these two resonances are *singlets* (see Box 14.3) and that there is *no coupling* between $H_A$ and $H_X$.

Now let nucleus $H_A$ be affected by the magnetic field associated with $H_X$. The consequence is that the $^1H$ NMR signal for $H_A$ is *split into two equal lines*, depending on whether it is 'seeing' $H_X$ in the $\alpha$ or $\beta$ spin state. It follows that the signal for $H_X$ is also split into two lines in an *exactly equal* manner. We say that $H_A$ and $H_X$ are *coupling* and that the spectrum consists of two *doublets* (Figure 14.8b). This theory is treated more rigorously in Box 14.4.

Coupling can only be detected between *non-equivalent nuclei*, and the magnitude of the coupling is given by the coupling constant, $J$, measured in Hz. The coupling constant in Hz is independent of the field strength of the NMR spectrometer. It is important that the coupling constant is quoted in Hz; if it were given in ppm, then it would vary with the spectrometer field. For example, a coupling constant of 25 Hz corresponds to 0.1 ppm at 250 MHz and 0.25 ppm at 100 MHz (equation 14.4; see also worked example 14.2).

$$(J \text{ in Hz}) = (\text{Spectrometer frequency in MHz}) \times (\Delta\delta \text{ in ppm}) \qquad (14.4)$$

where $\Delta\delta$ = difference in chemical shifts

---

**THEORETICAL AND CHEMICAL BACKGROUND**

## Box 14.3 A more rigorous approach to the observation of two singlets for two nuclei with $I = \frac{1}{2}$ that do not couple

There is in fact a selection rule implicit in the discussion in the main text, and this rule is that $\Delta m_I = \pm 1$.

An alternative way of describing our spin system for nuclei $H_A$ and $H_X$ involves writing the *total* spin states for both nuclei. This gives four energy levels as shown below.

$$\beta(A)\beta(X)$$
$$\beta(A)\alpha(X)$$
$$\alpha(A)\beta(X)$$
$$\alpha(A)\alpha(X)$$

Notice that this refers to the total energy of the system – there is no direct interaction between $H_A$ and $H_X$.

Four transitions are possible when we consider the selection rules $\Delta m_I(H_A) = \pm 1$, and $\Delta m_I(H_X) = \pm 1$, and these are:

1  $\alpha(A)\beta(X) \leftarrow \alpha(A)\alpha(X)$  ⎱ $H_X$ transitions
2  $\beta(A)\beta(X) \leftarrow \beta(A)\alpha(X)$  ⎰

3  $\beta(A)\alpha(X) \leftarrow \alpha(A)\alpha(X)$  ⎱ $H_A$ transitions
4  $\beta(A)\beta(X) \leftarrow \alpha(A)\beta(X)$  ⎰

If there is no interaction between the nuclei, the energy difference between $\alpha(A)\alpha(X)$ and $\alpha(A)\beta(X)$ states must be the same as between the $\beta(A)\alpha(X)$ and $\beta(A)\beta(X)$ states, and so the energy of transitions **1** and **2** must be the same. Similarly, transitions **3** and **4** will occur at the same energy, *but* at a different energy from **1** and **2**. In other words we see two resonances in the NMR spectrum assigned to $H_A$ and $H_X$, and each signal is a singlet.

**THEORETICAL AND CHEMICAL BACKGROUND**

## Box 14.4 A more rigorous approach to the observation of two doublets for two coupled nuclei with $I = \frac{1}{2}$

In Box 14.3, we considered the possible transitions between the four energy levels corresponding to the total spin states of nuclei $H_A$ and $H_X$ which did *not* interact with each other. If there *is* interaction between the nuclei (i.e. if there *is* spin–spin coupling), then each energy level is affected by the same amount, $\Delta E$. This perturbation is represented below, where the left-hand side of the diagram corresponds to that in Box 14.3. The same four transitions are allowed, but now the transition energies are:

1'  $\alpha(A)\beta(X) \leftarrow \alpha(A)\alpha(X)$ $\left.\right\}$ $H_X$ transitions
2'  $\beta(A)\beta(X) \leftarrow \beta(A)\alpha(X)$

3'  $\beta(A)\alpha(X) \leftarrow \alpha(A)\alpha(X)$ $\left.\right\}$ $H_A$ transitions
4'  $\beta(A)\beta(X) \leftarrow \alpha(A)\beta(X)$

where $\begin{cases} \mathbf{1'} = \mathbf{1} - 2\Delta E \\ \mathbf{2'} = \mathbf{2} + 2\Delta E \\ \mathbf{3'} = \mathbf{3} - 2\Delta E \\ \mathbf{4'} = \mathbf{4} + 2\Delta E \end{cases}$

We saw in Box 14.3 that, *without* coupling, transitions **1** and **2** have the same energy, and transitions **3** and **4** are also degenerate. Inspection of the scheme above shows that, *with* coupling, these degeneracies are removed, giving new transition energies of **1'**, **2'**, **3'** and **4'**. In terms of the observed spectrum, this means that the signal for $H_A$ is split into two components, the overall signal being a *doublet*. Similarly, the signal for $H_X$ becomes a doublet with *exactly* the same splitting as is observed for the signal assigned to $H_A$.

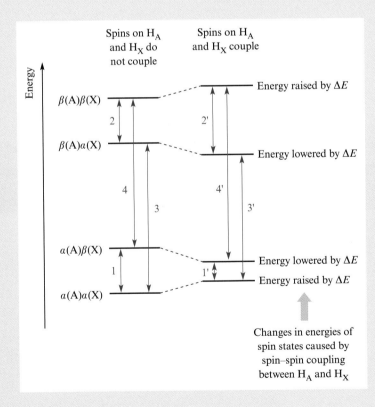

**Worked example 14.2**   *Coupling constants*

A doublet in a $^1H$ NMR spectrum has a coupling constant of 10.0 Hz. To what value of $\Delta\delta$ does this correspond if the spectrum is recorded on (i) a 200 MHz and (ii) a 600 MHz spectrometer?

Equation needed:

$$(\Delta\nu \text{ in Hz}) = (\text{Spectrometer frequency in MHz}) \times (\Delta\delta \text{ in ppm})$$

$$\therefore (\Delta\delta \text{ in ppm}) = \frac{(\Delta\nu \text{ in Hz})}{(\text{Spectrometer frequency in MHz})}$$

(i)   On a 200 MHz spectrometer:   $\Delta\delta = \dfrac{10.0\,\text{Hz}}{200\,\text{MHz}} = 0.050\,\text{ppm}$

(ii)  On a 600 MHz spectrometer:   $\Delta\delta = \dfrac{10.0\,\text{Hz}}{600\,\text{MHz}} = 0.017\,\text{ppm}$

$\Delta\delta$ is *inversely proportional* to the field strength, and therefore $\Delta\delta$ becomes larger if the magnetic field strength of the spectrometer is lower.

The nuclei that are coupling should be indicated by adding subscripts to the symbol for $J$; for example, if the coupling constant between $H_A$ and $H_X$ is 10 Hz, then we write: $J_{H_A H_X}$ 10 Hz, or $J_{HH}$ 10 Hz, or $J_{AX}$ 10 Hz.

The ***coupling constant J*** is measured in Hz:

$J$ in Hz = (Spectrometer frequency in MHz) $\times$ ($\Delta\delta$ in ppm)

The splitting of the resonance for $H_A$ by its coupling with nucleus $H_X$ is summarized in Figure 14.9a, and measurement of $J_{HH}$ is indicated in the diagram.

**Fig. 14.9** (a) The formation of a doublet signal for $H_A$ as it couples to one nucleus $H_X$; the doublet has equal intensity lines, split by an amount equal to the coupling constant, $J_{HH}$. (b) The formation of a triplet signal for $H_A$ as it couples to two equivalent nuclei $H_{X(1)}$ and $H_{X(2)}$; adjacent lines in the triplet are split by an amount equal to the coupling constant, $J_{HH}$.

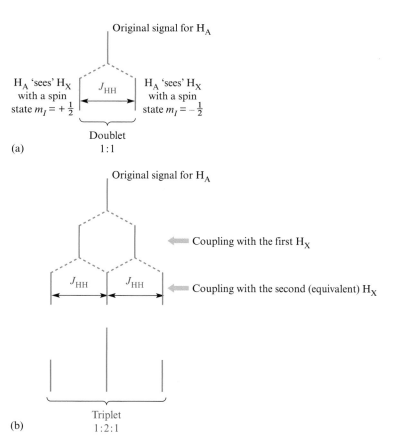

## Coupling between more than two inequivalent $^1$H nuclei

Now consider the case in which $H_A$ can 'see' two nuclei $H_X$, labelled $H_{X(1)}$ and $H_{X(2)}$, which are equivalent to one another, but are inequivalent to $H_A$. Nucleus $H_A$ can couple with both $H_{X(1)}$ and $H_{X(2)}$ and the process is illustrated in a stepwise form in Figure 14.9b. The result is a three-line pattern (a *triplet*) for the $^1$H NMR signal for $H_A$. Each line represents one of the possible combinations of the two spin states of $H_{X(1)}$ and $H_{X(2)}$ as 'seen' by $H_A$ and these are:

$\alpha(X1)\alpha(X2)$

$\left. \begin{array}{l} \alpha(X1)\beta(X2) \\ \beta(X1)\alpha(X2) \end{array} \right\}$ These are degenerate because the energy difference between $\alpha(X1)$ and $\beta(X1)$ must be the same as that between $\alpha(X2)$ and $\beta(X2)$

$\beta(X1)\beta(X2)$

The three lines of the triplet therefore have the relative intensities $1:2:1$ as shown in Figure 14.9b.

In other compounds, $H_A$ could couple to three or four equivalent $H_X$ nuclei to give a quartet $(1:3:3:1)$ or a quintet $(1:4:6:4:1)$ respectively. For coupling to *equivalent* nuclei with $I = \frac{1}{2}$, the number of components in the signal of a simple system (the *multiplicity of the signal*) is given by equation 14.5, and the intensity ratio of the components can be determined by using Pascal's triangle (Figure 14.10). The general name for a doublet, triplet, quartet, etc., is a *multiplet* and this description is often used if the exact nature of the signal cannot be resolved.

$$\text{Multiplicity} = n + 1 \tag{14.5}$$

where $n = $ number of *equivalent* coupled nuclei, each of spin $I = \frac{1}{2}$

A most important point to remember is that *magnetically equivalent nuclei do not exhibit spin–spin coupling.*

> ***Spin–spin coupling*** occurs between non-equivalent nuclei, and if $n$ is the number of *equivalent coupled* nuclei, each of spin $I = \frac{1}{2}$, the multiplicity of the signal is $n + 1$.

Consider the $^1$H NMR spectrum of ethanol (**14.7**). The spectrum shown in Figure 14.11 is of a solution of ethanol in CDCl$_3$ solution. There are three signals: a broad signal at $\delta + 2.6$ ppm, a triplet at $\delta + 1.2$ ppm and a quartet at $\delta + 3.7$ ppm. The broad signal is assigned to the OH group; note that the

**Fig. 14.10** Pascal's triangle gives the relative intensities of the components of multiplet signals when coupling is between nuclei with $I = \frac{1}{2}$. Note that each row begins and ends with 1. The other entries in a given row in the series are obtained by summing pairs of numbers in the previous row – the triplet has intensities $1:(1+1):1$, and the quartet has intensities $1:(1+2):(2+1):1$. The triangle can thereby be extended (see problem 14.9).

| Singlet | | | | | | 1 | | | | |
|---|---|---|---|---|---|---|---|---|---|---|
| Doublet | | | | | 1 | | 1 | | | |
| Triplet | | | | 1 | | 2 | | 1 | | |
| Quartet | | | | 1 | 3 | | 3 | | 1 | |
| Quintet | | | 1 | 4 | | 6 | | 4 | | 1 |
| Sextet | | 1 | 5 | | 10 | | 10 | | 5 | 1 |
| Septet | 1 | 6 | | 15 | | 20 | | 15 | 6 | 1 |
| Octet | 1 | 7 | 21 | | 35 | | 35 | 21 | 7 | 1 |

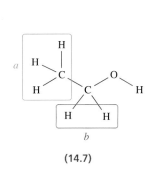

(14.7)

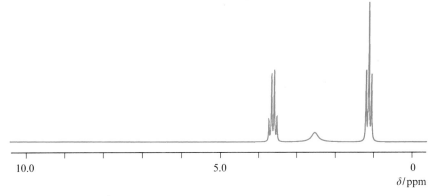

**Fig. 14.11** 100 MHz $^1$H NMR spectrum of ethanol (**14.7**) in CDCl$_3$ solution.

expected coupling to the OH group in ethanol is not observed when the alcohol is dissolved in CDCl$_3$.$^§$ Protons of type *a* couple to the two equivalent protons *b*, and the signal for protons *a* is a triplet. Protons of type *b* couple to the three equivalent protons *a*, and the signal due to protons *b* is a quartet. The spectrum may be assigned on the basis of coupling, but the relative *integrals* of the two signals ($a:b = 3:2$) will confirm the answer. Notice that we have used the word *integral* rather than intensity. Because the signal is split by the coupling, it is the sum of the intensities of the component peaks that we must measure – this is best found by measuring the area under the peak and this is the *integral of the signal*.

The question arises: 'How far along a carbon chain does coupling occur?' It is difficult to give a definitive answer but a good start is to consider $^1$H–$^1$H coupling through one C–C bond (as in ethanol). In practice, more distant coupling may be observed, and this possibility must be kept in mind when interpreting spectra.

If we return to Figure 14.11, both the triplet *and* the quartet have values of $J_{\mathrm{HH}}$ 7 Hz. It is no coincidence that the values are the same – both the triplet *and* the quartet arise from coupling between the same sets of nuclei.

The magnitude of coupling constants depends on a variety of factors, and within this book we can deal with the topic only superficially. However, one type of organic compound that shows a diagnostic relationship between molecular geometry and coupling constants is the family of alkenes. In each of the alkenes **14.8** to **14.10**, substituents X and Y are different from each other, and in each structure, H$_A$ and H$_X$ are magnetically inequivalent and couple to each other. Typically, for **14.8**, $J_{\mathrm{HH}}$ 0–2 Hz, for **14.9**, $J_{\mathrm{HH}}$ 12–18 Hz, and for **14.10**, $J_{\mathrm{HH}}$ 6–12 Hz. These values may allow $^1$H NMR spectroscopic data to be used to distinguish between isomers of an alkene. Coupling between $^1$H nuclei on adjacent C atoms of a C=C bond is called *vicinal* coupling.

Alkenes: see Chapter 26

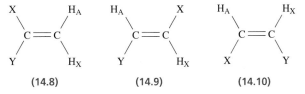

(14.8)     (14.9)     (14.10)

---

$^§$ The $^1$H NMR spectrum of neat ethanol shows a triplet for the OH proton coupling to the CH$_2$ protons; the spectrum is not as simple as in Figure 14.11.

**Fig. 14.12** 100 MHz $^1$H NMR spectrum of 2-chloropropanoic acid.

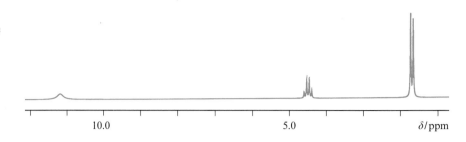

> *Vicinal coupling* is the coupling between $^1$H nuclei on adjacent C atoms of a C=C bond.

## Interpretation of some spectra involving $^1$H–$^1$H coupling

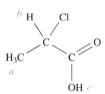

2-Chloropropanoic acid

**Example 1: 2-chloropropanoic acid**   The $^1$H NMR spectrum consists of three signals at $\delta +11.2$ ppm (broad), $+4.5$ ppm (quartet) and $+1.7$ ppm (doublet) (Figure 14.12).

The high-frequency signal is assigned to the OH group.

The three methyl protons $a$ couple with proton $b$, and the signal for $a$ is a doublet.

Proton $b$ couples with three methyl protons and the signal for $b$ is a quartet. Thus the signals at $\delta +4.5$ and $+1.7$ ppm can be assigned to $b$ and $a$, respectively.

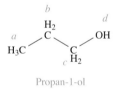

Propan-1-ol

**Example 2: propan-1-ol**   The $^1$H NMR spectrum consists of four signals at $\delta +3.6$ ppm (triplet), $+2.3$ ppm (broad), $+1.6$ ppm (sextet) and $+0.9$ ppm (triplet).

The broad signal can be assigned to the OH group.

The signal for the terminal methyl group should be a triplet owing to coupling to the two equivalent protons $b$. Since terminal $CH_3$ groups are *usually* at less positive $\delta$ than methylene ($CH_2$) groups (see Table 14.3), we can assign the signal at $\delta +0.9$ ppm to protons $a$.

We are now left with a sextet at $\delta +1.6$ ppm and a triplet at $\delta +3.6$ ppm. The best way of tackling the assignment of these two signals is to predict what is expected and try to match the predictions to the observations.

The methylene protons $c$ will couple to the two equivalent protons $b$ to give a triplet. Further coupling to the OH proton is expected but we have already seen in Figure 14.11 that such coupling is not seen in ethanol dissolved in $CDCl_3$, and this is actually a general result. We can assign the triplet at $\delta +3.6$ ppm to methylene group $c$.

Protons $b$ can couple to *both* three equivalent protons of type $a$ and two equivalent protons of type $c$, but the values of $J_{H_a H_b}$ and $J_{H_b H_c}$ will be approximately the same and a sextet (Figure 14.13) due to coupling to five protons is expected.

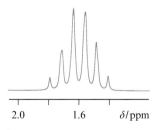

**Fig. 14.13** The signal at $\delta +1.6$ ppm in propan-1-ol is expected to be a sextet because the coupling constants $J_{H_a H_b}$ and $J_{H_b H_c}$ are approximately equal.

*Question*: What would be the effect on the signal shown in Figure 14.13 if $J_{H_a H_b} > J_{H_b H_c}$, or $J_{H_a H_b} < J_{H_b H_c}$?

**THEORETICAL AND CHEMICAL BACKGROUND**

## Box 14.5 First and second order spectra

The description of spin–spin coupling that we give in the main text is a *first order* treatment. This is a valid model if there is a sufficiently large chemical shift difference, $\Delta\delta$, between the signals arising from the nuclei that are coupling. As the signals come closer together, the coupling patterns begin to show *second order* effects. This may involve an effect called 'roofing' in which the relative intensities of the lines making up a pair of doublets, for example, change as $\Delta\delta$ decreases. This is illustrated below in a series of simulated 500 MHz $^1$H NMR spectra for two chemically inequivalent protons which couple with $J_{HH}$ 7.5 Hz:

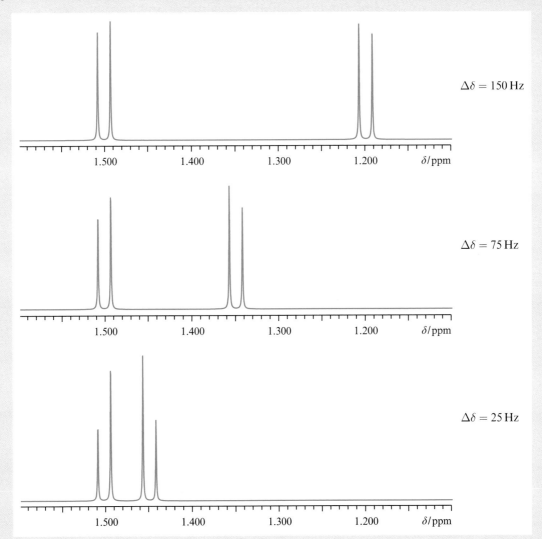

In more complicated cases, additional lines appear as the signals come closer together. As a result, the pair of multiplets can no longer be described in terms of the patterns expected from a Pascal's triangle. Second order effects are particularly noticeable when the chemical shift difference, $\Delta\delta$, between the signals for the coupled nuclei is less than 4–5 times the coupling constant, $J$.

It is conventional to label protons with letters to indicate whether the chemical shift difference between the protons is large (first order spectrum) or small (second order effects). Two well-separated nuclei are labelled as protons A and X and we say that an 'AX pattern is observed'. This corresponds to the labelling that we use in the main text for coupling between protons that leads to first order spectra. If the chemical shift difference between two coupling protons is small, the labels A and B are used. Intermediate cases can be given intermediate labels, e.g. A and M.

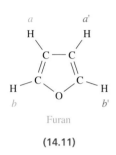

2-Iodopropane

**Example 3: 2-iodopropane**   The $^1$H NMR spectrum of 2-iodopropane contains a doublet at $\delta +1.9$ ppm and a septet at $\delta +4.3$ ppm.

There are two proton environments: $a$ (six protons) and $b$ (one proton).

Protons of type $a$ couple with $b$ and the signal for $a$ is a doublet.

Proton $b$ couples with the six equivalent nuclei $a$, and the signal for $b$ is a septet.

Thus, the signal at $\delta +1.9$ ppm is assigned to the methyl groups and that at $\delta +4.3$ ppm is assigned to $b$.

### Chemical and magnetic equivalence and inequivalence

Furan

**(14.11)**

In the above examples, proton sites have been taken to be magnetically inequivalent on the basis of their *chemical* inequivalence. We now make the distinction between nuclei which are *chemically equivalent but magnetically inequivalent*.

Furan (**14.11**) contains two proton environments ($a$ and $b$). The $^1$H NMR spectrum of furan consists of two signals (as expected for two chemically different $^1$H environments) *but* the signals ($\delta +6.6$ and $+7.4$ ppm) are both multiplets rather than having simple coupling patterns. This can be explained as follows.

Put yourself in the position of one of the protons of type $a$ – you 'see' one proton $b$ attached to the adjacent carbon centre *and* you see another proton $b'$ that *you think* is different from $b$. Although the two protons of type $b$ are *chemically equivalent*, they are (in the eyes of proton $a$) *magnetically inequivalent*. There are two sets of couplings:

- between $a$ and $b$ to give a doublet
- between $a$ and $b'$ to give a doublet

and the signal for $a$ is a *doublet of doublets*.

Similarly, a proton of type $b$ 'thinks' that $a$ is different from $a'$ and so the resonance for $b$ is also a doublet of doublets (Figure 14.14).

### Coupling between $^{13}$C nuclei

When we discussed $^{13}$C NMR spectra, we ignored the possibility of $^{13}$C–$^{13}$C coupling. Why was this? The answer lies in a consideration of the negligible chance of this coupling occurring. Since the natural abundance of $^{13}$C is only 1.1%, it follows that the probability of having two $^{13}$C nuclei attached to one another in a molecule is very low indeed. We are justified in ignoring $J_{CC}$ *unless* a sample has been isotopically enriched with $^{13}$C.

**Fig. 14.14** In furan (**14.11**), proton $a$ couples to $b$ and $b'$ to give a doublet of doublets.

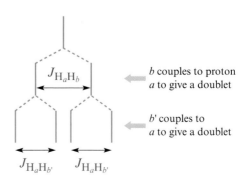

## Coupling between $^1H$ and $^{13}C$ nuclei: heteronuclear coupling

Spin–spin coupling is not restricted to spins of like nuclei (e.g. $^1H–^1H$). Any non-equivalent nuclei that have $I > 0$ may interact with one another, and to illustrate this we consider coupling between $^1H$ and $^{13}C$ nuclei. When different nuclei couple, it is called *heteronuclear coupling*.

The argument as to why we do not observe $^{13}C–^{13}C$ coupling in a $^{13}C$ NMR spectrum also applies to $^1H–^{13}C$ coupling *in a $^1H$ NMR spectrum*.

Consider acetone (**14.12**). The probability of a $^1H$ NMR nucleus being bonded to a $^{13}C$ nucleus is very low. Therefore, in the $^1H$ NMR spectrum, we can ignore the possibility of $J_{CH}$ – the spectrum is a singlet ($\delta +2.1$ ppm). On the other hand, since the natural abundance of $^1H$ is $\approx 99.9\%$, it is approximately true that *every* $^{13}C$ nucleus present in the methyl groups of acetone is attached to $^1H$ nuclei. In the $^{13}C$ spectrum, $^{13}C–^1H$ coupling *is* observed for each methyl $^{13}C$ as Figure 14.15 shows.

---

**THEORETICAL AND CHEMICAL BACKGROUND**

### Box 14.6 Satellite peaks in NMR spectra

We have seen that in a $^1H$ NMR spectrum, coupling between $^{13}C$ and $^1H$ nuclei can be ignored, but that in the $^{13}C$ NMR spectrum of the same sample, $^{13}C–^1H$ coupling is observed. The difference followed from the different natural abundances of the two nuclei.

Consider now the case of the $^1H$ NMR spectrum of $SnMe_4$ which is shown in the simulated spectrum below.

Tin possesses 10 isotopes. Two of these are NMR active: $^{117}Sn$ ($I = \frac{1}{2}$, 7.6%) and $^{119}Sn$ ($I = \frac{1}{2}$, 8.6%). In a sample of $SnMe_4$ there is a statistical distribution of the naturally occurring tin isotopes.

- 83.8% of the $SnMe_4$ molecules possess Sn nuclei with $I = 0$.
- 7.6% of the $SnMe_4$ molecules possess $^{117}Sn$ nuclei.

- 8.6% of the $SnMe_4$ molecules possess $^{119}Sn$ nuclei.

This means that in the $^1H$ NMR spectrum of $SnMe_4$, 83.8% of the $^1H$ nuclei give rise to a singlet (no coupling to tin), while 7.6% of the protons couple to $^{117}Sn$ (to give a doublet) and 8.6% couple to $^{119}Sn$ (also to give a doublet). The $J_{^{119}Sn^1H}$ and $J_{^{117}Sn^1H}$ coupling constants are 54 Hz and 52 Hz respectively and therefore the two doublets are separated from one another. The doublets are called *satellites*.

*Exercises*: In the spectrum below:

- Where is the singlet?
- Where are the two doublets?
- Where would you measure the values of $J_{^{119}Sn^1H}$ and $J_{^{117}Sn^1H}$?

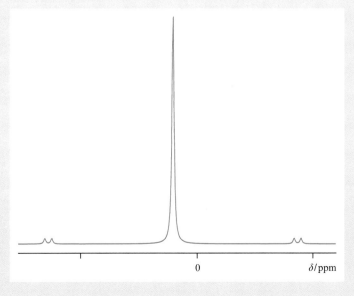

$\delta$/ppm

Acetone

(14.12)

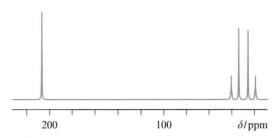

**Fig. 14.15** 25 MHz $^{13}$C NMR spectrum of acetone; for the quartet, $J_{CH} = 172$ Hz. Using the *a* and *b* labelling shown in structure **14.12**, the quartet in the spectrum is assigned to carbon atoms *a*, and the singlet to carbon atom *b*.

Heteronuclear coupling can be suppressed in a spectrum by an instrumental technique called *heteronuclear decoupling* and in the laboratory, you may often encounter *proton-decoupled* $^{13}$C NMR spectra; this is written $^{13}$C{$^1$H}. A comparison of Figures 14.5 and 14.15 shows the differences between proton-decoupled and proton-coupled $^{13}$C NMR spectra of acetone.

| **Worked example 14.3** | *Interpretation of a $^1$H NMR spectrum* |

The 100 MHz $^1$H NMR spectrum of 1-nitropropane (**14.13**) is shown below. (a) Assign the signals in the spectrum. (b) Can $^1$H NMR spectroscopy be used to distinguish between the isomers 1-nitropropane (**14.13**) and 2-nitropropane (**14.14**)? (We introduced isomerism in Section 6.11; 1-nitropropane and 2-nitropropane are examples of *constitutional isomers*. We discuss this further in Chapter 24.)

1-Nitropropane

(14.13)

2-Nitropropane

(14.14)

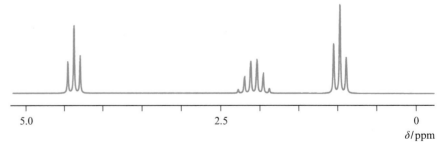

(a) The $^1$H NMR spectrum contains two triplets and a sextet. Look at structure **14.13** and count the number of proton environments: there are three, labelled *a*, *b* and *c* below:

Protons *a* couple to two equivalent protons *b* giving a triplet. Protons *c* couple to two equivalent protons *b* giving a triplet. To assign the two triplets, use the fact that signals for CH$_3$ protons are usually at less positive $\delta$ than signals for CH$_2$ protons (see Table 14.3). Therefore, the triplet at $\delta$ 1.0 ppm can be assigned to protons *a*, and the triplet at $\delta$ 4.4 ppm to protons *c*.

Protons *b* couple to two equivalent protons *c* and to three equivalent protons *a*. The spectrum shows a sextet, indicating that the values of

$J_{H_aH_b}$ and $J_{H_bH_c}$ are virtually the same. This can be confirmed by looking at the two triplets from which values of $J_{H_aH_b}$ and $J_{H_bH_c}$ can be measured. The sextet can unambiguously be assigned to protons $b$.

(b) Consider the structure of the second isomer, 2-nitropropane. There are two proton environments:

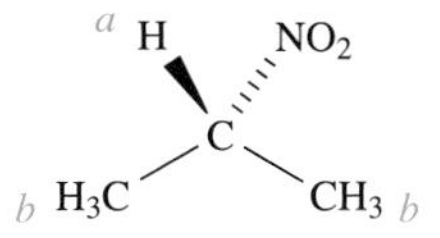

$^1$H NMR spectroscopy can therefore distinguish between the isomers, simply on the basis of the number of signals. The spectrum of 1-nitropropane exhibits three signals (two triplets and a sextet) while that of 2-nitropropane exhibits two signals (a doublet for protons $b$ and a septet for protons $a$).

(In the experimental spectrum of **14.14**, the doublet appears at $\delta$ 1.5 ppm and the septet at $\delta$ 4.7 ppm).

## SUMMARY

Chapter 14 gives an introduction to NMR spectroscopy, with an emphasis on nuclei with $I = \frac{1}{2}$. The level of theory has purposely been kept at a basic level, and the chapter has focused on the interpretation of $^{13}$C and $^1$H NMR spectra.

### Do you know what the following terms mean?

- nuclear spin quantum number, $I$
- magnetic spin quantum number, $m_I$
- nuclear spin states
- NMR active nucleus
- free induction decay, FID

- resonance frequency of a nucleus
- $^1$H NMR spectrum
- $^{13}$C NMR spectrum
- chemical shift value
- chemical shift range
- signal intensity

- signal integral
- spin–spin coupling
- Pascal's triangle
- coupling constant
- proton-decoupled $^{13}$C NMR spectrum

### You should be able:

- to list the properties of a nucleus that make it suitable for NMR spectroscopy
- to discuss whether a high natural abundance of an NMR active nucleus is essential for NMR spectroscopy
- to discuss briefly why FT-NMR spectroscopy made an impact on the development of this technique
- to discuss briefly how an NMR signal arises
- to interpret $^{13}$C NMR spectra in terms of different carbon environments in a molecule

- to understand why a *deuterated* solvent is usually used in $^1$H NMR spectroscopy
- to explain the origin of a simple coupling pattern
- to interpret $^1$H NMR spectra in terms of different proton environments and simple spin–spin couplings
- to exemplify how nuclei may be chemically equivalent but magnetically inequivalent

## PROBLEMS

**14.1** Using the data in Table 14.1, determine the resonance frequency of $^{31}P$ nuclei on a 400 MHz spectrometer, and the resonance frequency of $^{13}C$ nuclei on a 250 MHz instrument.

**14.2** The $^{13}C$ NMR spectrum of acetic acid (**14.15**) has signals at $\delta$ 20.6 and 178.1 ppm. Assign the spectrum.

Acetic acid                2-Methylpropan-2-ol

(**14.15**)                (**14.16**)

**14.3** The $^{13}C$ NMR spectrum of 2-methylpropan-2-ol (**14.16**) has signals at $\delta$ 31.2 and 68.9 ppm. Suggest assignments for these signals. What other feature of the spectrum would confirm the assignments?

**14.4** The $^{13}C$ NMR spectrum of **14.17** possesses signals of approximately equal intensities at $\delta$ 20.5, 52.0, 80.8 and 170.0 ppm. Suggest assignments for the spectrum.

(**14.17**)

**14.5** Compounds **14.18** and **14.19** are isomers of butyne. The $^{13}C$ NMR spectrum of one isomer shows signals at $\delta$ 74.6 and 3.3 ppm, while that of the other shows signals at $\delta$ 13.8, 12.3, 71.9 and 86.0 ppm. (a) Which spectrum belongs to which compound? (b) Assign the signals.

But-2-yne                But-1-yne

(**14.18**)                (**14.19**)

**14.6** Compounds **14.20** and **14.21** both have the molecular formula $C_4H_{10}O$ and are isomers.

2-Methylpropan-2-ol        Diethyl ether

(**14.20**)                (**14.21**)

Peaks in the $^{13}C$ and $^1H$ NMR spectra of these compounds are listed below. Assign the spectra to the correct compounds, and assign the peaks in the spectra.

Isomer I: $^{13}C$ NMR: $\delta$ 69.1, 31.2 ppm; $^1H$ NMR $\delta$ 1.26 (singlet), 2.01 ppm (singlet).

Isomer II: $^{13}C$ NMR: $\delta$ 66.0, 15.4 ppm; $^1H$ NMR $\delta$ 1.14 (triplet), 3.37 ppm (quartet).

**14.7** (a) What is the hybridization of each C atom in $N\equiv CCH_2CH_2CH_2C\equiv N$? (b) The $^{13}C$ NMR spectrum of this compound exhibits signals at $\delta$ 119.3, 24.3 and 16.4 ppm. Which signals can you unambiguously assign?

**14.8** The $^{13}C$ NMR spectrum of compound **14.22** shows signals at $\delta$ 8.4, 28.7 and 170.3 ppm. Assign the spectrum.

(**14.22**)

**14.9** Predict the relative intensities and the coupling patterns of the signals in the $^1H$ NMR spectrum of 2-methylpropane (**14.23**).

2-Methylpropane

(**14.23**)

**14.10** Assign each of the following $^1H$ NMR spectra. What are the expected relative integrals of the signals?

| Compound | Resonances, $\delta$/ppm |
|---|---|
| (a) | 1.3 (doublet), 2.7 (septet) |
| (b) | 1.1 (triplet), 2.2 (quartet), 6.4 (very broad) |
| (c) | 1.1 (triplet), 2.1 (singlet), 2.5 (quartet) |

| Compound | Resonances, $\delta$/ppm |
|---|---|
| (d) 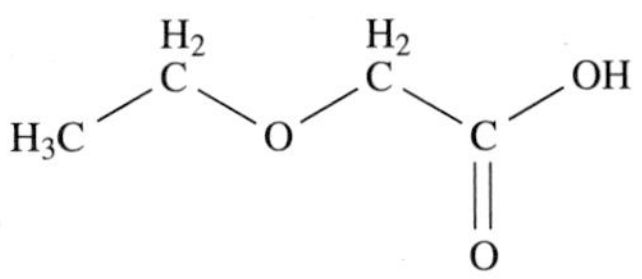 | 1.3 (triplet), 3.7 (quartet), 4.1 (singlet), 10.9 (broadened) |
| (e) 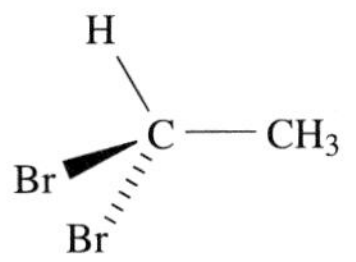 | 2.5 (doublet), 5.9 (quartet) |
| (f) 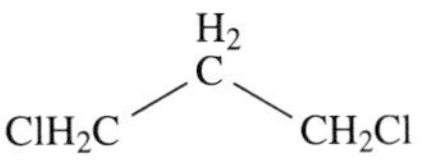 | 2.2 (quintet), 3.7 (triplet) |

**14.11** The $^1$H NMR spectrum of acetaldehyde (**14.24**) consists of signals at $\delta$ 2.21 and 9.79 ppm. (a) Assign the spectrum. (b) The inequivalent protons couple ($J$ 2.9 Hz). What will be the multiplicities of the signals?

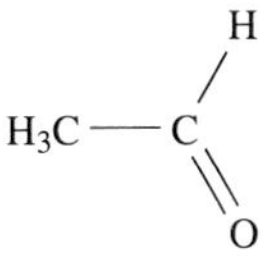

Acetaldehyde
(Ethanal)

(**14.24**)

**14.12** Look back at structures **14.18** and **14.19** (problem 14.5). How would $^1$H NMR spectroscopy allow you to distinguish between these isomers?

**14.13** What would you expect to observe in the $^1$H NMR spectrum of compound **14.25**?

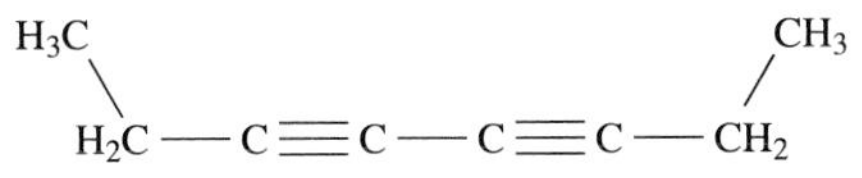

(**14.25**)

**14.14** The $^1$H NMR spectrum of bromoethane, $CH_3CH_2Br$, consists of signals at $\delta$ 1.7 and 3.4 ppm. Assign the signals and predict their coupling patterns.

**14.15** What would you expect to observe in the $^1$H NMR spectrum of triethylamine (**14.26**)?

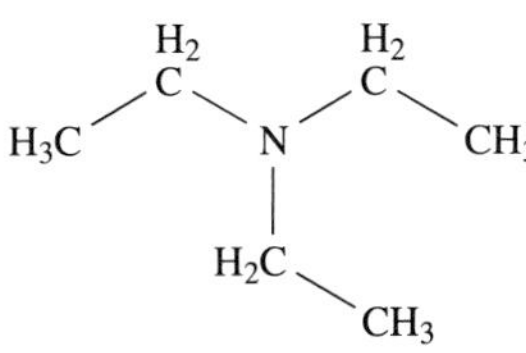

Triethylamine

(**14.26**)

**14.16** Why does the $^1$H NMR spectrum of $CH_2ClCF_2Cl$ appear as a triplet? [Hint: Look at Table 14.1]

**14.17** What would you expect to observe in the $^1$H NMR spectra of the following compounds?

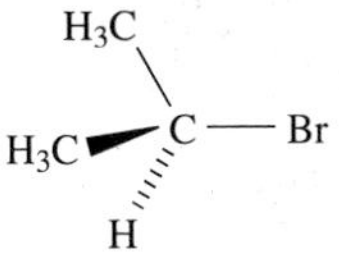

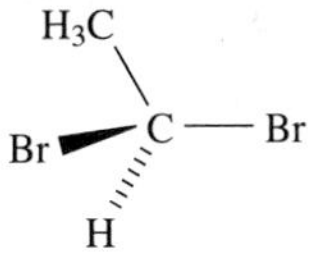

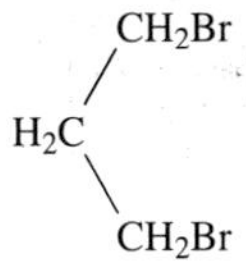

## Additional problems

**14.18** The $^{13}$C NMR spectrum of compound **14.27** shows two quartets at $\delta$ 115 and 163 ppm, with coupling constants of 284 Hz and 44 Hz respectively. Rationalize this observation.

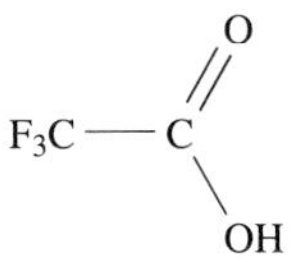

Trifluoroacetic acid

(**14.27**)

**14.19** Two branched-chain alkanes, **A** and **B**, have the molecular formula $C_7H_{16}$ and are isomers (see Section 24.6). Spectroscopic data for the compounds are as follows:

**A**: $^1$H NMR $\delta$ 1.38 (1H, septet, $J$ 6.9 Hz), 0.84 (9H, singlet), 0.83 (6H, doublet, $J$ 6.9 Hz) ppm; $^{13}$C{$^1$H} NMR $\delta$ 37.7, 32.7, 27.1, 17.9 ppm.

**B**: $^1$H NMR $\delta$ 1.62 (2H, multiplet, $J$ 7.2, 6.6 Hz), 1.03 (2H, doublet, $J$ 7.2 Hz), 0.85 (12H, doublet, $J$ 6.6 Hz) ppm; $^{13}$C{$^1$H} NMR $\delta$ 48.9, 25.6, 22.9 ppm.

Suggest possible structures for **A** and **B**, and assign the spectra.

**14.20** (a) Use VSEPR theory to predict the structure of $PF_5$. How many F environments are there? (b) The $^{31}$P NMR spectrum of $PF_5$ at 298 K consists of a binomial sextet. Explain how this arises. (c) Predict what is observed in the $^{19}$F NMR spectrum of $PF_5$ at 298 K. (d) Draw the structure of $[PF_6]^-$, using VSEPR theory to help you. (e) Predict the nature of the $^{31}$P and $^{19}$F NMR spectra of a solution of $[NH_4][PF_6]$. [Hint: Use Table 14.1]

# 15 Reaction kinetics

## Topics

- How fast is a reaction?
- Kinetic and thermodynamic control
- Methods of measuring rate
- The order of a reaction
- Rate equations and rate constants
- Pseudo-first order conditions
- Boltzmann distribution
- The Arrhenius equation
- Catalysts
- Reversible reactions
- Molecularity
- The relationship between a rate equation and reaction mechanism
- The steady-state approximation
- Chain reactions
- Michaelis–Menten kinetics

## 15.1 Introduction

Information about the mechanism of a reaction comes, in part, from a study of the *rate* of the reaction. Such information is concerned with the *reaction kinetics*, and in this chapter we shall discuss this topic in detail. In this introductory section we summarize some basic facts about reaction rates. Throughout this chapter, we make significant use of graphical data. You may wish to refer to the section entitled 'Plotting and interpreting graphs' in the accompanying *Mathematics Tutor* which is available via the website http://www.pearsoned.co.uk/housecroft. The *Mathematics Tutor* also reviews the calculus that is needed to handle some of the rate equations detailed in this chapter.

### Thermodynamics and kinetics

In discussions of the thermodynamics of a reaction, we can talk about various energy terms (see Chapter 17) but for now we restrict our discussion to the enthalpy.

Changes in enthalpy: see Chapter 2

The fact that a reaction is thermodynamically favourable does *not* mean that it necessarily takes place quickly. For example, the reaction of an alkane with $O_2$ is exothermic. Equation 15.1 shows the combustion of pentane. This reaction is not spontaneous – an initiation such as a spark or flame is required. Mixtures of alkanes and $O_2$ may be kept unchanged for long periods.

$$C_5H_{12}(l) + 8O_2(g) \longrightarrow 5CO_2(g) + 6H_2O(l) \qquad (15.1)$$

$$\Delta_r H = -3509 \text{ kJ per mole of } C_5H_{12}$$

Many other reactions which do actually proceed (for example the dissolution of ammonium nitrate in water) are endothermic. Clearly the enthalpy change is not the ultimate arbiter of the spontaneity of a chemical reaction and we will see in Chapter 17 that a new energy term, the change in Gibbs energy $\Delta G$, is an important factor. However, the fact remains that a knowledge of the thermodynamic changes in a reaction gives no knowledge *per se* about the *rate* of the conversion of starting materials to products.

If we consider a reaction such as:

$$A + B \longrightarrow C + D$$

then, in order for compounds A and B to react with one another to give C and D, bonds may be broken or formed. The precise sequence of events is described as a *reaction pathway*. There are many possible pathways by which A and B could react to give C and D, and each of these has a characteristic energy barrier called the *activation energy*, $E_a$. There is a direct relationship between $E_a$ and the rate of a reaction. Reactions usually proceed by the pathway with the lowest activation barrier.

In a reaction with a single step, there is one associated activation energy. However, a reaction may involve several steps, each with a characteristic value of $E_a$. The *reaction profile* in Figure 15.1 describes the reaction between A and B to give C and D by a reaction pathway consisting of two steps. The activation energy for the first step is $E_a(1)$. After this first stage, an *intermediate* species is formed which can then undergo further reaction, with an activation energy $E_a(2)$, to form C and D. Each of the energy *maxima* represents the energy of a *transition state*. In Figure 15.1, $\{TS(1)\}^{\ddagger}$ is the transition state during the first step of the reaction, and $\{TS(2)\}^{\ddagger}$ is the transition state associated with the second step. The braces ('curly brackets') with the superscript 'double-dagger', $^{\ddagger}$, are used to indicate a transition state. The distinction between an intermediate and a transition state is important.

Figure 15.1 shows the first step in the reaction has the higher activation energy: $E_a(1) > E_a(2)$. However, in a reaction pathway with a series of steps, it is not necessarily the first step that has the highest activation energy. The larger the value of $E_a$, the more difficult it is for the reaction to proceed and, in proceeding along a reaction profile, any step with a high activation energy slows the reaction down. We discuss this again in Section 15.7. The step with the highest $E_a$ is called the *slow* or *rate-determining step* (RDS).

Figure 15.2 illustrates reaction profiles for two three-step reactions. In the first profile (Figure 15.2a), step **1** has the highest $E_a$ and is the rate-determining step. In Figure 15.2b, it is the second step that determines how fast the reaction proceeds.

The *activation energy*, $E_a$, of a reaction is the energy in excess of that possessed by the ground state that is required for the reaction to proceed.

A *transition state* occurs at an energy *maximum*, and cannot be isolated. An *intermediate* occurs at a local energy *minimum*, and can be detected (e.g. spectroscopically) and, perhaps, isolated.

## What does 'rate of a reaction' mean?

The progress of a reaction can be monitored in terms of the appearance of a product or the disappearance of a reactant as a function of time as shown in Figure 15.3 for reactions of two different stoichiometries.

**Fig. 15.1** Profile for the two-step reaction of A with B, to give C and D. The first and second transition states are labelled $\{TS(1)\}^{\ddagger}$ and $\{TS(2)\}^{\ddagger}$ respectively. The overall enthalpy change during the reaction is $\Delta_r H$.

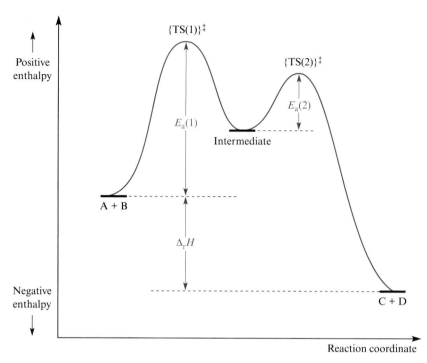

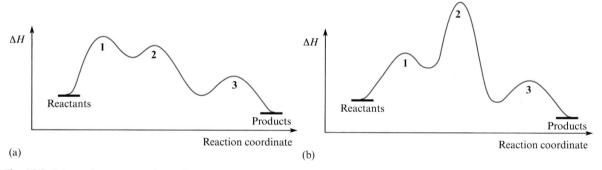

**Fig. 15.2** Schematic representations of reaction profiles for two three-step reactions. In (a), step **1** is the rate-determining step (slow) because it has the highest $E_a$; steps **2** and **3** are relatively fast. In (b), step **2** is the slow step.

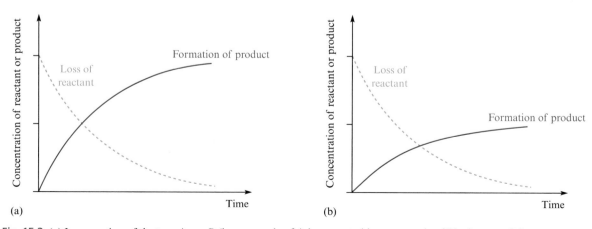

**Fig. 15.3** (a) In a reaction of the type A $\longrightarrow$ B (i.e. one mole of A is converted into one mole of B), the rate of disappearance of A should mirror the rate of appearance of B. (b) In a reaction of type 2A $\longrightarrow$ B, the maximum amount of B formed is half the initial amount of A.

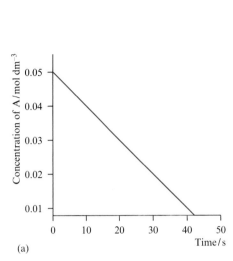

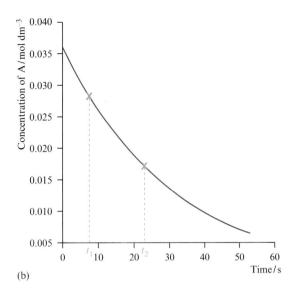

**Fig. 15.4** A plot of the decrease in the concentration of a reactant A against time may be (a) linear or (b) non-linear. The rate of the disappearance of A is constant in graph (a); the rate is equal to the gradient of the line. In graph (b), the rate varies with time, and is found at various times by drawing tangents to the curve and measuring their gradients; rate at $t_1$ > rate at $t_2$.

The change in concentration of a reactant or product with respect to time is a measure of the **rate of a reaction**. The units of rate are often $mol\,dm^{-3}\,s^{-1}$, although other units of time may be used.

*The change in concentration of a reactant or product* (in $mol\,dm^{-3}$) *with respect to time* (in s) is a measure of the rate of a reaction, and this gives units of rate as $mol\,dm^{-3}\,s^{-1}$.

Consider two different reactions (both of the general form given in equation 15.2) involving a reactant A. The rate at which A disappears depends on the reaction concerned and the precise reaction conditions.

$$A \longrightarrow products \tag{15.2}$$

Figure 15.4a describes a reaction in which the concentration of A decreases linearly during the reaction. The rate at which A disappears is equal to the gradient of the graph, and is constant over the course of the reaction, i.e. it does not depend upon the concentration of A.

Figure 15.4b refers to a reaction in which the concentration of A decreases quickly at the beginning of the reaction but then decreases more slowly as time progresses. Here, the rate of reaction changes with time. At time $t_1$, the rate is found by drawing a tangent to the curve at the point $\mathbf{X}(t_1)$ and finding the gradient of the tangent. At a time $t_2$, the rate is determined by drawing a second tangent to the curve at a second point, $\mathbf{X}(t_2)$ as shown in Figure 15.5. In this example, the rate of reaction at $t_2$ is less than that at $t_1$, i.e. the reaction rate decreases as the reaction proceeds.

◼▷ Problem 15.1 asks you to determine rates of reaction from Figure 15.4

### Factors that affect the rate of a reaction

Some of the factors that influence the rate of a reaction are:

- concentration;
- pressure (particularly for reactions involving gases);
- temperature;
- surface area (for reactions involving solids).

Figure 15.6 shows the carbon dioxide evolution from two reaction vessels in which calcium carbonate reacts with dilute hydrochloric acid (equation 15.3).

$$CaCO_3(s) + 2HCl(aq) \longrightarrow CaCl_2(aq) + H_2O(l) + CO_2(g) \tag{15.3}$$

**Fig. 15.5** When a plot of concentration, $c$, against time, $t$, is a curve, the rate of reaction at a particular point can be found by drawing a tangent to the curve at the point and measuring the gradient of this line. For the tangent drawn in the figure, the gradient is $(c_1 - c_2)/(t_1 - t_2)$. This gives a negative gradient (see equation 15.10 and discussion).

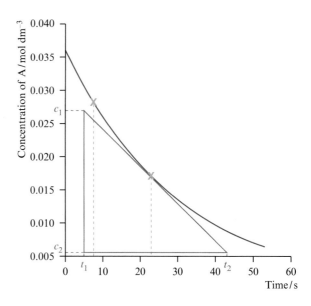

The rate at which $CO_2$ is produced is a measure of the rate of the reaction. The two reactions proceed at different rates; that represented by plot I proceeds approximately twice as fast as that shown in plot II. Clearly, the conditions in the two reaction vessels are different – the rate may be influenced by the concentration of hydrochloric acid, the surface area of the calcium carbonate (e.g. powder instead of granules) or the temperature.

### How do you choose an experimental method for monitoring a reaction?

When choosing a method to follow a reaction, you must look for features that give rise to a *measurable change* between starting materials and products. It is most convenient to use a *non-intrusive* method so that the reaction itself is not perturbed. Conventional spectroscopic methods or some other physical property of the reaction solution such as its refractive index, viscosity or volume may be used. Some reactions proceed too quickly to be followed by *simple* spectroscopic methods.

We are not concerned here with details of particular methods, but in the reactions described in this chapter, we exemplify some possible experimental techniques. Equation 15.3 showed an example of a reaction monitored by the evolution of a gas. In Figure 11.7, a reaction with a change in colour was illustrated.

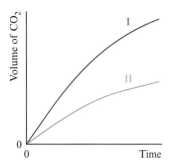

**Fig. 15.6** The rate of reaction between solid $CaCO_3$ and aqueous HCl may be followed by measuring the volume of $CO_2$ evolved. Two reactions, I and II, are carried out under different conditions. The rate of reaction I is approximately twice that of reaction II. What differences might there be between the conditions in the two reaction vessels?

| 15.2 | **Rate equations: the dependence of rate on concentration** |
| --- | --- |

### Rate expressed as a differential equation

Throughout this section, we shall be concerned with the general reaction:

$$A \longrightarrow \text{products}$$

The concentration of A at a given time, $t$, is written as [A] or $[A]_t$. The initial concentration (at $t = 0$) is denoted as $[A]_0$. The rate of reaction can be expressed in terms of the rate of disappearance of A with time, and is written

■►
**More about differential equations** in the *Mathematics Tutor* which can be accessed via the website http://www.pearsoned.co.uk/housecroft
mathematically as the *negative differential coefficent* (*derivative*) of [A] with respect to time, $t$ (equation 15.4). The negative sign is needed to show that A is disappearing.

$$\text{Rate of reaction} = -\frac{d[A]}{dt} \tag{15.4}$$

It is sometimes more convenient experimentally to follow the appearance of a product than the disappearance of a reactant. In a reaction:

$$A \longrightarrow B + C$$

the rate of appearance of B is written as the *positive differential* of [B] with respect to time (equation 15.5). The positive sign shows that B is appearing.

$$\text{Rate of reaction} = +\frac{d[B]}{dt} \tag{15.5}$$

For this particular reaction, the rate at which B and C appear must equal the rate at which A disappears. In addition, the rates at which B and C appear must be the same. It follows that the rate of reaction:

$$A \longrightarrow B + C$$

can be expressed by three different differentials as in equation 15.6.

$$\text{Rate of reaction} = -\frac{d[A]}{dt} = +\frac{d[B]}{dt} = +\frac{d[C]}{dt} \tag{15.6}$$

### The general rate equation

The concentration of A is at a maximum at the beginning of a reaction and decreases during the reaction. Equation 15.7 gives a *general* expression in which *the rate of the reaction depends on the concentration of A*.

$$\text{Rate of reaction} = -\frac{d[A]}{dt} \propto [A]^n \tag{15.7}$$

The power $n$ in equation 15.7 is called the *order of the reaction with respect to A* and may be zero or have an integral or fractional value. The order shows the exact dependence of the rate on [A], but in this chapter we are primarily concerned with values of $n$ of 0, 1 or 2.

Equation 15.7 is often written in the form shown in equation 15.8 where $k$ is the *rate constant* for the reaction. The units of the rate constant vary with the *order of the reaction* as discussed below. Whereas the *rate* of a reaction depends upon the concentration of the reactants, the *rate constant* is independent of concentration. It is important that you distinguish between these two terms. For a given concentration, the larger the value of $k$, the faster the reaction proceeds.

$$\text{Rate of reaction} = -\frac{d[A]}{dt} = k[A]^n \tag{15.8}$$

The **general rate equation** for a reaction A $\longrightarrow$ products is:

$$\text{Rate of reaction} = -\frac{d[A]}{dt} = k[A]^n$$

where [A] is the concentration of A at time $t$, $k$ is the rate constant and $n$ is the order of the reaction with respect to A.

## Zero order with respect to A

The graph in Figure 15.7 corresponds to a reaction in which the rate does *not* depend upon the concentration of A. That is, the power $n$ in equation 15.8 is zero. A number raised to the power zero is equal to unity, and so $[A]^0 = 1$. Thus, the rate of reaction is equal to a constant $k$ (equation 15.9).

$$\text{Rate of reaction} = -\frac{d[A]}{dt} = k[A]^0 = k \tag{15.9}$$

This reaction is *zero* (or *zeroth*) *order with respect to A*. In this case, the units of the rate constant are the same as those of rate: $\text{mol dm}^{-3}\,\text{s}^{-1}$. The rate of the reaction does not vary with time. In Figure 15.7, the rate of reaction can be found by measuring the gradient of the line: from the graph, during the time interval from $t_1$ to $t_2$, the concentration of A has fallen from $[A]_1$ to $[A]_2$. The gradient of the line is determined as shown in equation 15.10.

$$\text{Gradient of line} = \frac{[A]_1 - [A]_2}{t_1 - t_2} = \frac{0.02}{-60}$$
$$= -3.33 \times 10^{-4}\,\text{mol dm}^{-3}\,\text{s}^{-1} \tag{15.10}$$

The gradient is negative because we are looking at the *disappearance* of reactant A. However, we express the rate of disappearance of A as $3.33 \times 10^{-4}\,\text{mol dm}^{-3}\,\text{s}^{-1}$.

---

When a reaction is ***zero order*** with respect to a reactant A, the rate does *not* depend upon the concentration of A and the rate equation is:

$$\text{Rate of reaction} = -\frac{d[A]}{dt} = k$$

The units of $k$ are $\text{mol dm}^{-3}\,\text{s}^{-1}$.

---

## First order with respect to A

If the rate of a reaction is directly proportional to the concentration of reactant A, then the reaction is *first order with respect to A*. The rate equation is given in equation 15.11.

$$\text{Rate of reaction} = -\frac{d[A]}{dt} = k[A] \tag{15.11}$$

This means that as $[A]$ decreases during the reaction, the rate of reaction also decreases. A plot of $[A]$ against time is therefore non-linear, as in Figure 15.8.
The rate constant in equation 15.11 has units of $\text{s}^{-1}$ (or $\text{min}^{-1}$, $\text{h}^{-1}$, etc.).

---

When a reaction is ***first order*** with respect to a reactant A, the rate depends upon the concentration of A according to the rate equation:

$$\text{Rate of reaction} = -\frac{d[A]}{dt} = k[A]$$

The units of $k$ are $\text{s}^{-1}$ (or $\text{min}^{-1}$, $\text{h}^{-1}$, etc.).

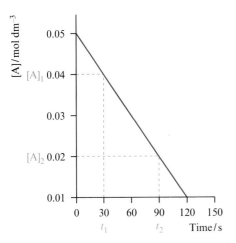

**Fig. 15.7** In a zero order reaction, a plot of [A] against time is linear, and the rate of reaction is constant. See equations 15.9 and 15.10.

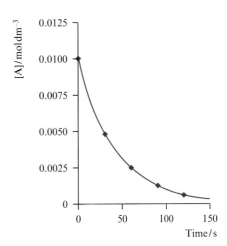

**Fig. 15.8** In a first order reaction, a plot of [A] against time is an exponential curve; the rate of reaction decreases with time. See equation 15.11.

## Second order with respect to A

If the rate of reaction depends upon the square of the concentration of A, then the reaction is *second order with respect to A*. The corresponding rate law is given in equation 15.12:

$$\text{Rate of reaction} = -\frac{d[A]}{dt} = k[A]^2 \tag{15.12}$$

Once again, a plot of [A] against time is non-linear, but is *not* an exponential curve.

The units of the rate constant in equation 15.12 are often $dm^3\,mol^{-1}\,s^{-1}$, but other units of time may be used.

> When a reaction is *second order* with respect to a reactant A, the rate equation is:
>
> $$\text{Rate of reaction} = -\frac{d[A]}{dt} = k[A]^2$$
>
> The units of $k$ are often $dm^3\,mol^{-1}\,s^{-1}$.

## 15.3 Does a reaction show a zero, first or second order dependence on A?

Once again in this section, we refer to the general reaction 15.13.

$$A \longrightarrow \text{products} \tag{15.13}$$

## Graphical methods

If a plot of [A] against time is linear, we may conclude that there is a zero order dependence upon A. On the other hand, if such a plot is non-linear (as in Figure 15.4b), we cannot *immediately* deduce the exact order (other than saying it is non-zero). If we impose a restriction that the order is likely to be first or second, how can we distinguish between them?

**Fig. 15.9** This flowchart summarizes how to use graphical methods to determine whether a reaction of the type:

A ⟶ products

is zero, first or second order with respect to A. It is important that the reaction is followed for a sufficiently long period of time to ensure that the correct plots are obtained.

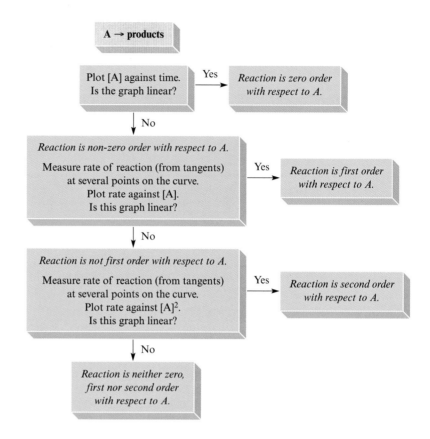

There are several ways of processing experimental results, and one method is summarized in the flow diagram in Figure 15.9. The choices of the graphs to be plotted follow from the rate equations 15.11 and 15.12. The rate of reaction at a given time can be determined from a graph of [A] against time and can be found from the gradient of a tangent drawn to the curve.

---

**Worked example 15.1**    *Thermal decomposition of benzoyl peroxide*

Above 373 K, benzoyl peroxide decomposes. Benzoyl peroxide has a strong absorption in its IR spectrum at $1787\,\text{cm}^{-1}$ and the disappearance of this band gives a measure (using the Beer–Lambert Law) of the concentration of the peroxide. Use the data given (which were recorded at 380 K) to determine the rate equation for the decomposition of benzoyl peroxide.

| Time/min | 0 | 15 | 30 | 45 | 60 | 75 | 90 | 120 | 150 | 180 |
|---|---|---|---|---|---|---|---|---|---|---|
| [Benzoyl peroxide] $\times 10^2/\text{mol dm}^{-3}$ | 2.00 | 1.40 | 1.00 | 0.70 | 0.50 | 0.36 | 0.25 | 0.13 | 0.06 | 0.03 |

(Data from: M. D. Mosher *et al.* (1991), *J. Chem. Educ.*, vol. 68, p. 510.)

Benzoyl peroxide

The initial concentration of the benzoyl peroxide was $2.00 \times 10^{-2}\,\text{mol dm}^{-3}$. (Notice how this information is stated in the table.)

First plot the concentration of benzoyl peroxide as a function of time:

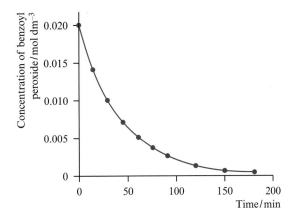

As the graph is non-linear, the decomposition is not zero order, but it may be first or second order with respect to benzoyl peroxide. One method of distinguishing between these possibilities is to see if a plot of rate against [A] is linear.

Next, determine the rate of the decomposition at several points. Five points are sufficient, *but they should be representative of the data as a whole*, and should not be taken from one portion of the curve. The rate of reaction at time $t$ is found by drawing a tangent to the curve at time $t$. The tangents have negative gradients because we have plotted the variation in [benzoyl peroxide]. Since the peroxide is decomposing during the reaction, the negative gradient corresponds to a positive rate of reaction.

Tabulate the new data; you should use the graph above to confirm the values of the rates stated in the table below.

| [Benzoyl peroxide] $\times 10^2$ / mol dm$^{-3}$ | 2.00 | 1.40 | 1.00 | 0.50 | 0.13 |
|---|---|---|---|---|---|
| Rate of reaction $\times 10^4$ / mol dm$^{-3}$ min$^{-1}$ | 4.44 | 3.27 | 2.23 | 1.17 | 0.25 |

Now plot the reaction rate against the concentration of the peroxide:

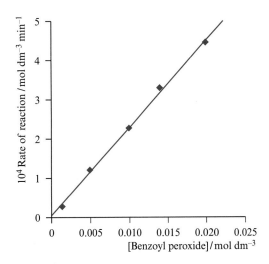

The straight line means that the reaction is first order with respect to benzoyl peroxide and we can write down the rate equation as:

$$\text{Rate of reaction} = -\frac{d[\text{Benzoyl peroxide}]}{dt} = k[\text{Benzoyl peroxide}]$$

The rate constant $k$ can be found from the second graph and equals the gradient of the straight line, $2.25 \times 10^{-2} \, \text{min}^{-1}$, or $3.75 \times 10^{-4} \, \text{s}^{-1}$.

Drawing tangents to a curve is not an accurate method of obtaining your answer, and you will find an alternative method for solving worked example 15.1 later in the chapter.

$$1 \, \text{min}^{-1} = \frac{1}{60} \, \text{s}^{-1}$$

## Initial rates method

In the *initial rates method*, the rate of reaction is measured at a time as close as possible to the point of mixing the reagents. The experiment is repeated several times using different concentrations of reactant. From these data, the dependence of the rate on the initial concentration of a reactant can be determined. Consider a reaction of the general type:

$$A \longrightarrow \text{products}$$

The rate of the reaction is given by equation 15.14.

$$\text{Rate of reaction} = k[A]^n \tag{15.14}$$

**Case 1: $n = 0$**

For a zero order reaction, a plot of [A] against time is linear, and the gradient of this line gives the rate constant, $k$, as in Figure 15.7. It is not necessary to run more than one experiment since both the order and rate constant can be found from one set of experimental data.

**Case 2: $n = 1$ or $2$**

If a reaction is first or second order with respect to A, a plot of [A] against time is a curve. By using different *initial concentrations* of A, $[A]_0$, different curves are obtained (Figure 15.10a). The initial rate of each experiment (i.e. the rate of the reaction at $t = 0$) is found by measuring the gradient of a tangent drawn to each curve at $t = 0$. Figure 15.10b shows this for one of the curves from Figure 15.10a. By comparing the values of the gradients, we can see whether, for example, the rate doubles when the concentration of A is doubled. This particular result would show that the rate depended directly on [A] and that the reaction was first order with respect to A (equation 15.15). As an exercise, draw tangents to the three curves in Figure 15.10a and show that the reaction is first order with respect to A.

$$\text{Rate of reaction} = k[A] \qquad \text{i.e. Rate} \propto [A] \tag{15.15}$$

More generally, the data are treated as follows. By taking logarithms of the left- and right-hand sides of equation 15.14, we can write equation 15.16. Now compare this equation with the general equation for a straight line (equation 15.17).

For practice working with logarithms, look at Section 4 of the *Mathematics Tutor*, accessed via the website http:// www.pearsoned.co.uk/ housecroft

$$\log(\text{Rate}) = \log(k[A]^n) = \log k + n \log[A] \tag{15.16}$$

$$y = mx + c \tag{15.17}$$

where $m$ = gradient

$c$ = value of $y$ when $x = 0$

Using the initial rate values from a series of experiments such as those in Figure 15.10a, a graph of log(Rate) against $\log[A]_0$ is plotted. This is

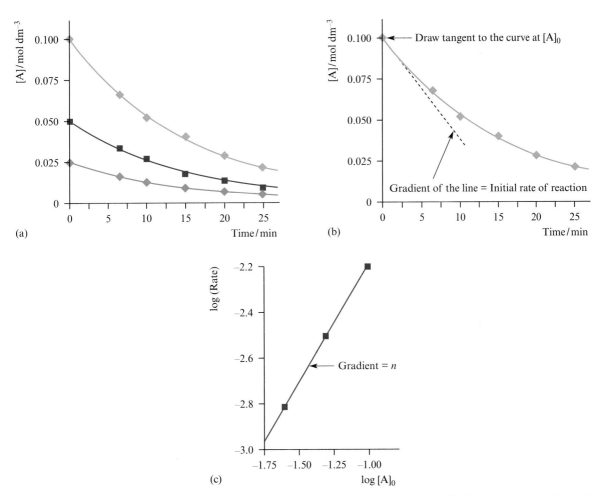

**Fig. 15.10** Initial rates method. (a) For the reaction A ⟶ products, the variation in [A] with time, $t$, is measured for *at least* three different values of $[A]_0$. (b) The gradient of a tangent drawn to the curve at $t = 0$ gives the initial rate of reaction. (c) A plot of the values of log $[A]_0$ against log(Rate) gives the order of the reaction (from the gradient) and the rate constant (from the intercept, see text).

shown in Figure 15.10c for the data from Figure 15.10a. The order of the reaction with respect to A is found directly from the gradient of the line, and log $k$ is found from the intercept on the log(Rate) axis (i.e. the value of log(Rate) when $\log[A]_0 = 0$). As in Figure 15.10c, it may not be possible to obtain this value without extrapolating the line. The data from Figure 15.10c give values of $n = 1$ and $\log k = -1.2$. Therefore, $k = 0.063\ \text{min}^{-1}$. The units of $k$ follow from the fact that the original values of time were given in units of min. As an exercise, you should verify these numbers using the data in the figure.

An advantage of the initial rates method is that the reaction need only be followed for a short period, and not until completion. The disadvantage of the method is that measuring the gradient of a tangent drawn to a curve (Figure 15.10b) is not an accurate means of determining a rate of reaction.

### Half-life method

The time taken for the concentration of reactant A to fall from $[A]_t$ to $[A]_t/2$ (where $[A]_t$ is the concentration of reactant A at a specific time $t$) is called the *half-life* of the reaction.

The **half-life** of a reaction is the time taken for the concentration of reactant A at time $t$, $[A]_t$, to fall to half of its value, $\dfrac{[A]_t}{2}$.

For a first order dependence, half-lives measured for the steps:

$$[A]_t \longrightarrow \frac{[A]_t}{2}, \quad \frac{[A]_t}{2} \longrightarrow \frac{[A]_t}{4}, \quad \frac{[A]_t}{4} \longrightarrow \frac{[A]_t}{8}, \quad \frac{[A]_t}{8} \longrightarrow \frac{[A]_t}{16}, \quad \text{etc.}$$

will be constant. This is not true for a reaction which is second order with respect to A. We return to half-lives in Sections 15.5 and 15.6.

---

**Worked example 15.2**     *The decay of a photogenerated species*

The blue mercury(II) complex **I** shown on the left is formed by irradiating a related mercury(II) compound, but once formed, this photogenerated species converts back to the original compound. The disappearance of **I** ($\lambda_{\max} = 605$ nm, $\varepsilon = 27\,000$ dm$^3$ mol$^{-1}$ cm$^{-1}$) can be followed by UV–VIS spectroscopy, and the concentration of **I** at time $t$ can be found using the Beer–Lambert Law. Use the following table of data to deduce the order of the reaction with respect to **I**.

| Time/min | 0.0 | 0.5 | 1.0 | 1.5 | 2.0 | 2.5 | 3.0 | 3.5 | 4.0 | 4.5 | 5.0 |
|---|---|---|---|---|---|---|---|---|---|---|---|
| $10^5[\mathbf{I}]$ / mol dm$^{-3}$ | 1.447 | 1.061 | 0.826 | 0.612 | 0.449 | 0.336 | 0.249 | 0.200 | 0.139 | 0.106 | 0.083 |

(Data from: R. L. Petersen *et al.* (1985) *J. Chem. Educ.*, vol. 62, p. 802.)

The rate equation for the decay of **I** has the general form:

$$-\frac{d[\mathbf{I}]}{dt} = k[\mathbf{I}]^n$$

First, plot [**I**] against time:

Complex **I**

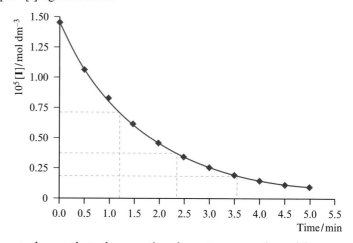

The curve shows that the reaction is not zero order with respect to compound **I**.

There are various ways to proceed, and one is to measure several half-lives.

The initial concentration $[\mathbf{I}]_0 = 1.447 \times 10^{-5}$ mol dm$^{-3}$.

From the graph, the first half-life is the time for the concentration to fall from $[\mathbf{I}]_0$ to $\dfrac{[\mathbf{I}]_0}{2} = 1.2$ min.

The second half-life is the time to go from $\dfrac{[\mathbf{I}]_0}{2}$ to $\dfrac{[\mathbf{I}]_0}{4} = 2.35 - 1.2$ min $= 1.15$ min.

The third half-life is the time to go from $\dfrac{[\mathbf{I}]_0}{4}$ to $\dfrac{[\mathbf{I}]_0}{8} = 3.55 - 2.35\,\text{min} = 1.2\,\text{min}$.

The three half-lives are equal within experimental error and the reaction is therefore first order with respect to compound **I**.

---

**15.4**     **Rate equations for reactions with more than one reactant**

The simple reaction of A that we have used so far is not common in real chemical situations. More often there are two or more reactants and the rate of reaction may depend on the concentration of more than one species. We now consider the implications of this for the rate equation.

### The general rate equation and orders

Consider the reaction in equation 15.18.

$$A + B \longrightarrow \text{products} \tag{15.18}$$

The rate of this reaction may depend upon:

- [A] only; or
- [B] only; or
- both [A] and [B]; or
- neither [A] nor [B]; or
- [A] and/or [B] and some other species.

The rate of the reaction can be measured by monitoring the disappearance of A or B, or the appearance of the products. Since A and B react in a 1 : 1 ratio in reaction 15.18, they will be used up at the same rate as each other. A general rate equation for this situation is written in equation 15.19.

$$\text{Rate of reaction} = -\frac{d[A]}{dt} = -\frac{d[B]}{dt} = k[A]^n[B]^m \tag{15.19}$$

The reaction is $n$th order with respect to A, and $m$th order with respect to B, and the *overall order* of the reaction is $(m + n)$. Compare this with the case of reaction 15.13 where the overall order and the order with respect to A must be identical. The rate constant $k$ refers to the overall rate of the reaction. For example, the rate equation for the hydrolysis of sucrose (reaction 15.20 and Figure 15.11) is given in equation 15.21. The reaction is first order with respect to sucrose and first order with respect to water, and is second order overall.

$$\underset{\text{Sucrose}}{C_{12}H_{22}O_{11}} + H_2O \longrightarrow \underset{\text{Glucose}}{C_6H_{12}O_6} + \underset{\text{Fructose}}{C_6H_{12}O_6} \tag{15.20}$$

$$\text{Rate of reaction} = k[C_{12}H_{22}O_{11}][H_2O] \tag{15.21}$$

The rate of reaction can be written in terms of the disappearance of any reactant or appearance of any product, and relationships between the corresponding differentials follow from the stoichiometry of the reaction. For example, consider reaction 15.22.

$$2A + B \longrightarrow 3C \tag{15.22}$$

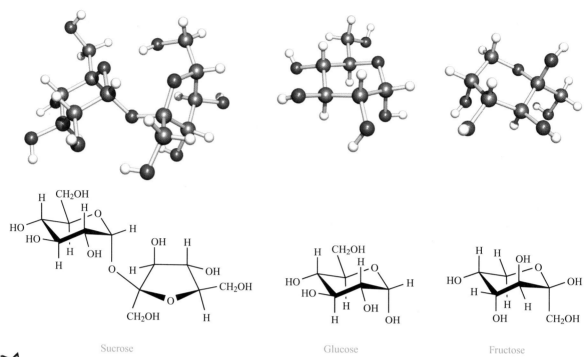

Sucrose                          Glucose                          Fructose

**Fig. 15.11** The structures of sucrose, glucose and fructose. The structures of these sugars have been determined by X-ray diffraction methods [sucrose: M. Bolte *et al.* (2001) private communication to the Cambridge Crystallographic Data Centre; α-D-glucose: A. Mostad (1994) *Acta Chem. Scand.*, vol. 48, p. 276; β-D-fructose: J. A. Kanters *et al.* (1977) *Acta Crystallogr.*, *Sect. B*, vol. 33, p. 665].

The rate of reaction can be written in terms of the disappearance of A or B, or the appearance of C and is given by equation 15.23.

$$\text{Rate of reaction} = -\frac{d[B]}{dt} = -\frac{1}{2}\left(\frac{d[A]}{dt}\right) = +\frac{1}{3}\left(\frac{d[C]}{dt}\right) \tag{15.23}$$

However, it is important to remember that the stoichiometric equation for a reaction tells you *nothing* about the rate equation. In reaction 15.20, both of the reactants *do* occur in the rate equation, but this is not always the case. For example, reaction 15.24 exhibits a rate equation that does *not* follow from the stoichiometry (equation 15.25).

See the discussion of molecularity in Section 15.10

$$[MnO_4]^- + 8H^+ + 5Fe^{2+} \longrightarrow Mn^{2+} + 4H_2O + 5Fe^{3+} \tag{15.24}$$

$$\text{Rate of reaction} \neq k[MnO_4^-][H^+]^8[Fe^{2+}]^5 \tag{15.25}$$

### The problem of investigating the rate when there is more than one reactant

In a system that has more than one reactant, the concentrations of *all* the reactants decrease with time. In order to deduce the rate equation, we should separately observe the effect that each reactant has on the rate of the reaction. This can be achieved by using the *isolation method* in which all but one of the reactants are present in *vast excess* so that as the reaction proceeds, there is no effective change in the concentration of these components.

Consider the following reaction of A with B:

$$A + B \longrightarrow \text{products}$$

Let there be $1 \times 10^{-4}$ moles of A and $1 \times 10^{-2}$ moles of B present initially in $10\,cm^3$ of solution. When half of the initial amount of A has reacted, $9.95 \times 10^{-3}$ moles of B remain unused. Thus the concentration of B has hardly changed, whereas the concentration of A has fallen significantly. (As an exercise, determine the concentration of A and B at the beginning of the reaction and at the point when half of A has reacted. What is the percentage change in each of the two concentrations?) Any effect on the observed rate will be mainly due to changes in the concentration of A. In order to find out how the rate depends upon [B], the reaction must be repeated either with A in vast excess, or with different concentrations of B (always in a large excess) as will be shown in worked example 15.3.

Consider an example in which the overall rate equation for the reaction between A and B is given in equation 15.26. If B is in vast excess, then [B] is assumed to be a constant (known) value, and the rate equation takes the form shown in equation 15.27.

$$\text{Rate of reaction} = -\frac{d[A]}{dt} = k[A][B] \tag{15.26}$$

$$\text{Rate of reaction} = -\frac{d[A]}{dt} = k'[A] \qquad \text{where } k' = k[B] \tag{15.27}$$

See worked examples 15.3–15.5

The reaction is said to show *pseudo-first order* kinetics. It is first order with respect to A, and the condition of a constant concentration of B leads to the reaction *appearing* to be first order overall.

## 15.5  Integrated rate equations

So far we have dealt with rates of reaction in terms of a differential equation, and Figure 15.9 illustrated a graphical strategy for determining simple rate equations. However, determining rates by drawing tangents to curves is not an accurate method.

More about integration in the *Mathematics Tutor*, accessed via the website http:// www.pearsoned.co.uk/ housecroft

The experimental data are often in the form of readings of [A] at various times, $t$, and therefore it would be useful to know exactly how [A] varies with $t$ rather than having to treat the data as outlined in Figure 15.9. Thus, if the reaction is zero, first or second order with respect to A, what is the *actual* relationship between [A] and $t$? For this, we need the *integrated form* of a rate equation.

### Deriving integrated rate equations

In this section we derive integrated rate equations for reactions of the type:

A ⟶ products

or reactions involving reactant A in which reactants other than A are in vast excess. We focus only on reactions in which the rate is zero, first or second order with respect to A.

Equation 15.28 gives the differential form of a zero order rate equation.

$$-\frac{d[A]}{dt} = k \tag{15.28}$$

In order to obtain a relationship between [A] and $t$, we integrate equation 15.28, introducing an integration constant, $c$, because the integration is

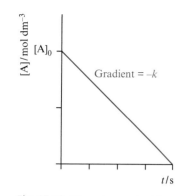

**Fig. 15.12** The rate constant for a reaction A $\longrightarrow$ products which is zero order with respect to A is found from a plot of [A] against time.

without limits. First, separate the variables in equation 15.28, and then perform the integration:

$$-d[A] = k\,dt$$

$$-\int d[A] = \int k\,dt$$

$$c - [A] = kt$$

To find $c$, we know that at $t = 0$, [A] is the initial concentration $[A]_0$. Substituting these values into the integrated equation gives $c = [A]_0$, and therefore the integrated rate equation for a reaction:

A $\longrightarrow$ products

that is zero order with respect to A is given by equation 15.29.

$$[A] = [A]_0 - kt \tag{15.29}$$

Since $[A]_0$ is a constant, equation 15.29 corresponds to a linear relationship between [A] and $t$ (compare equation 15.29 with 15.17). Given experimental data for the variation of [A] with time, the rate constant for the reaction can be found from the gradient of a plot of [A] against $t$ (Figure 15.12).

The differential form of a first order rate equation is given in equation 15.30.

$$-\frac{d[A]}{dt} = k[A] \tag{15.30}$$

In order to obtain a relationship between [A] and $t$, we integrate equation 15.30, introducing an integration constant, $c$:

$$-\frac{d[A]}{[A]} = k\,dt$$

$$-\int \frac{d[A]}{[A]} = \int k\,dt$$

$$c - \ln[A] = kt$$

To find $c$, we apply the condition that when $t = 0$, $[A] = [A]_0$. This gives $c = \ln[A]_0$, and therefore the integrated rate equation for a reaction:

A $\longrightarrow$ products

that is first order with respect to A is given by equation 15.31.

$$\ln[A] = \ln[A]_0 - kt \tag{15.31}$$

This equation corresponds to a linear relationship between $\ln$ [A] and $t$ (compare equation 15.31 with 15.17), and the rate constant for the reaction is determined from the gradient of a plot of $\ln$ [A] against $t$ (Figure 15.13).

Equation 15.32 gives the differential form of a second order rate equation.

$$-\frac{d[A]}{dt} = k[A]^2 \tag{15.32}$$

Integration of this equation leads to a relationship between [A] and $t$, in which $c$ is the integration constant:

$$-\frac{d[A]}{[A]^2} = k\,dt$$

$$-\int \frac{d[A]}{[A]^2} = \int k\,dt$$

$$c + \frac{1}{[A]} = kt$$

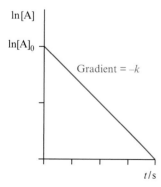

**Fig. 15.13** The rate constant for a reaction A $\longrightarrow$ products which is first order with respect to A is found from a plot of $\ln$ [A] against time.

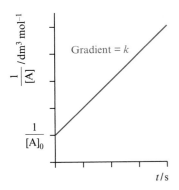

**Fig. 15.14** The rate constant for a reaction A ⟶ products which is second order with respect to A is found from a plot of $\dfrac{1}{[A]}$.

To find $c$, we again apply the condition that when $t = 0$, $[A] = [A]_0$. This gives $c = 1/[A]_0$. Thus, the integrated rate equation for a reaction:

$$A \longrightarrow products$$

that is second order with respect to A is given by equation 15.33.

$$\frac{1}{[A]} = \frac{1}{[A]_0} + kt \tag{15.33}$$

This equation corresponds to a linear relationship between $1/[A]$ and $t$, and the rate constant for the reaction can be found from the gradient of a plot of $1/[A]$ against $t$ (Figure 15.14).

Application of the integrated rate equations 15.29, 15.31 and 15.33 is illustrated in the following three worked examples.

---

**Worked example 15.3**    *The bromination of acetone*

Bromine reacts with acetone in the presence of hydrochloric acid according to the following equation:

$$
\underset{\substack{H_3C \qquad\quad CH_3}}{\overset{\displaystyle O}{\underset{\displaystyle \|}{C}}}
\;+\; Br_2 \;\xrightarrow{\;H^+\;}\;
\underset{\substack{H_3C \qquad\quad CH_2Br}}{\overset{\displaystyle O}{\underset{\displaystyle \|}{C}}}
\;+\; HBr
$$

The reaction can be monitored by using UV–VIS spectroscopy; aqueous $Br_2$ absorbs light of wavelength 395 nm.

The rate equation is given by:

$$\text{Rate of reaction} = -\frac{d[Br_2]}{dt} = k[Br_2]^m[CH_3C(O)CH_3]^n[H^+]^p$$

**(a)** A reaction is carried out in aqueous solution with initial concentrations of acetone, hydrochloric acid and bromine being 1.60, 0.40 and 0.0041 mol dm$^{-3}$ respectively, and the results are tabulated below. Determine the order of the reaction with respect to $Br_2$.

| Time / s | 0 | 10 | 20 | 30 | 40 | 50 | 60 |
|---|---|---|---|---|---|---|---|
| $10^3[Br_2]$ / mol dm$^{-3}$ | | 4.10 | 3.85 | 3.60 | 3.30 | 3.10 | 2.85 | 2.60 |

**(b)** If the acid and acetone are kept in excess but their concentrations are varied, the rate of the reaction alters according to the following results. Use these data along with those in part (a) to derive the overall rate equation.

| Acetone / mol dm$^{-3}$ | 1.60 | 1.60 | 0.80 | 0.80 |
|---|---|---|---|---|
| $[H^+]$ / mol dm$^{-3}$ | 0.40 | 0.20 | 0.40 | 0.20 |
| $[Br_2]_0$ / mol dm$^{-3}$ | 0.0041 | 0.0041 | 0.0041 | 0.0041 |
| Relative rate of reaction | 1.00 | 0.50 | 0.50 | 0.25 |

(Data from: J. P. Birk *et al.* (1992) *J. Chem. Educ.*, vol. 69, p. 585.)

(a) The conditions of the reaction are such that both acid and acetone are in vast excess and so the rate equation can be written as:

$$-\frac{d[Br_2]}{dt} = k'[Br_2]^m$$

We are given no clues about the possible dependence of the rate on $[Br_2]$ and so we may begin by plotting $[Br_2]$ against time:

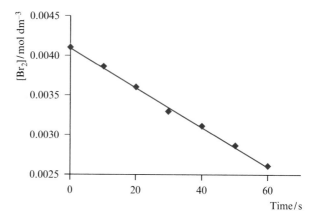

Since this graph is linear, we deduce that the rate is independent of the concentration of $Br_2$. Therefore, the rate is zero order with respect to $Br_2$. The rate equation with respect to $[Br_2]$ is therefore:

$$-\frac{d[Br_2]}{dt} = k'[Br_2]^0 = k'$$

(b) The overall rate equation is given as:

$$-\frac{d[Br_2]}{dt} = k[Br_2]^m[CH_3C(O)CH_3]^n[H^+]^p$$

but since $m = 0$, we can simplify this to:

$$-\frac{d[Br_2]}{dt} = k[CH_3C(O)CH_3]^n[H^+]^p$$

From the table of data, we see that:

- the rate halves when $[H^+]$ halves;
- the rate halves when $[CH_3C(O)CH_3]$ halves;
- the rate falls by a factor of 4 when both $[H^+]$ and $[CH_3C(O)CH_3]$ are halved.

We conclude that the rate depends directly on each of $[CH_3C(O)CH_3]$ and $[H^+]$. The reaction is therefore first order with respect to each component. The rate equation becomes:

$$-\frac{d[Br_2]}{dt} = k[CH_3C(O)CH_3][H^+]$$

*Question*: Work out the percentage change in the concentrations of acetone and $H^+$ when *all* of the $Br_2$ has reacted. Is the assumption of constant concentrations valid?

How can we determine the overall rate constant?
For the reaction in part (a), we know both $[CH_3C(O)CH_3]$ and $[H^+]$, and these are assumed to be constant during the reaction.
Thus, for this particular experiment:

$$-\frac{d[Br_2]}{dt} = k[CH_3C(O)CH_3][H^+]$$

where $[CH_3C(O)CH_3] = 1.6\,mol\,dm^{-3}$ and $[H^+] = 0.40\,mol\,dm^{-3}$

The gradient of the graph in part (a) is $k'$:

$$k' = 2.5 \times 10^{-5} \, \text{mol dm}^{-3} \, \text{s}^{-1}$$

*Question*: Why are the units of $k$ different from those of $k'$?

Therefore, we have two expressions for the rate of disappearance of $Br_2$:

$$-\frac{d[Br_2]}{dt} = k' = 2.5 \times 10^{-5} \, \text{mol dm}^{-3} \, \text{s}^{-1}$$

and

$$-\frac{d[Br_2]}{dt} = k \times (1.6 \, \text{mol dm}^{-3}) \times (0.40 \, \text{mol dm}^{-3})$$

Setting the right-hand sides of these equations equal to each other allows us to find $k$:

$$k \times (1.6 \, \text{mol dm}^{-3}) \times (0.40 \, \text{mol dm}^{-3}) = 2.5 \times 10^{-5} \, \text{mol dm}^{-3} \, \text{s}^{-1}$$

$$k = \frac{(2.5 \times 10^{-5} \, \text{mol dm}^{-3} \, \text{s}^{-1})}{(1.6 \, \text{mol dm}^{-3})(0.40 \, \text{mol dm}^{-3})}$$

$$= 3.9 \times 10^{-5} \, \text{dm}^3 \, \text{mol}^{-1} \, \text{s}^{-1}$$

---

**Worked example 15.4**    *The esterification of trifluoroacetic acid*

**Benzyl alcohol reacts with trifluoroacetic acid according to the following equation:**

$$C_6H_5CH_2OH + \quad CF_3CO_2H \quad \longrightarrow \quad CF_3CO_2CH_2C_6H_5 + H_2O$$

Benzyl alcohol    Trifluoroacetic acid    Benzyl trifluoroacetate    Water

**This is an example of the more general reaction:**

$$\text{Alcohol} + \text{Carboxylic acid} \longrightarrow \text{Ester} + \text{Water}$$

**The disappearance of the alcohol can be followed by $^1$H NMR spectroscopy and monitoring changes in the integral of the signal for the $CH_2$ protons. The data below have been recorded at 310 K for a reaction in which the acid is present in a large excess. Confirm that the reaction is first order with respect to the alcohol, and determine the pseudo-first order rate constant.**

| Time / min | 0 | 6.5 | 10 | 15 | 20 | 25 | 30 | 40 | 50 |
|---|---|---|---|---|---|---|---|---|---|
| [Alcohol] / mol dm$^{-3}$ | 1.00 | 0.67 | 0.57 | 0.41 | 0.30 | 0.23 | 0.14 | 0.07 | 0.04 |

**(Data from: D. E. Minter *et al.* (1985) *J. Chem. Educ.*, vol. 62, p. 911.)**

The rate equation for the reaction may be written as:

$$\text{Rate of reaction} = -\frac{d[C_6H_5CH_2OH]}{dt}$$

$$= k[C_6H_5CH_2OH]^n[CF_3CO_2H]^m$$

but with the acid in vast excess the equation is:

$$-\frac{d[C_6H_5CH_2OH]}{dt} = k'[C_6H_5CH_2OH]^n$$

We need to confirm that $n = 1$, and also find the pseudo-first order rate constant $k'$. This is most efficiently achieved by plotting $\ln[\text{Alcohol}]$ against time. First, tabulate the necessary data:

| Time / min | 0 | 6.5 | 10 | 15 | 20 | 25 | 30 | 40 | 50 |
|---|---|---|---|---|---|---|---|---|---|
| ln [Alcohol] | 0 | −0.400 | −0.562 | −0.892 | −1.204 | −1.470 | −1.966 | −2.659 | −3.219 |

A graph of ln [Alcohol] against time is linear:

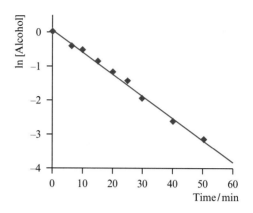

This is consistent with the integrated first order rate equation:

$$\ln[\text{Alcohol}] = \ln[\text{Alcohol}]_0 - k't$$

and the result confirms that the reaction is first order with respect to benzyl alcohol.

The pseudo-first order rate constant $k'$ is found from the slope of the line:

$$k' = 0.066\,\text{min}^{-1} \text{ or } 1.1 \times 10^{-3}\,\text{s}^{-1}$$

---

**Worked example 15.5**  |  *The reaction between $I_2$ and hex-1-ene*

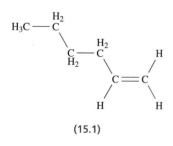

(15.1)

**Hex-1-ene (15.1) reacts with $I_2$ in acetic acid to give 1,2-diodohexane (15.2). The dependence of the rate on $[I_2]$ can be studied if a large excess of hex-1-ene is used. Results recorded at 298 K are tabulated below. Determine the order of the reaction with respect to $I_2$, and write a rate equation for the reaction. Do your results exclude the involvement of acetic acid in the rate-determining step?**

| Time / s | 0 | 1000 | 2000 | 3000 | 4000 | 5000 | 6000 | 7000 | 8000 |
|---|---|---|---|---|---|---|---|---|---|
| $[I_2]$ / mol dm$^{-3}$ | 0.0200 | 0.0156 | 0.0128 | 0.0109 | 0.0094 | 0.0083 | 0.0075 | 0.0068 | 0.0062 |

**(Data from: K. W. Field *et al.* (1987) *J. Chem. Educ.*, vol. 64, p. 269.)**

The overall rate equation can be written as follows, making the assumption that no other species are involved:

$$-\frac{d[I_2]}{dt} = k[I_2]^m[\text{hex-1-ene}]^n$$

but since the hex-1-ene is in excess, we can write:

$$-\frac{d[I_2]}{dt} = k'[I_2]^m$$

(15.2)

Firstly, plot $[I_2]$ as a function of time:

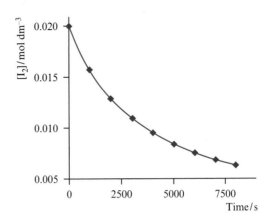

Since this graph is non-linear, we know that $m \neq 0$, but it may be 1 or 2. However, we can rule out a first order dependence by looking at the first two half-lives. The time taken for the initial concentration to halve is $\approx 3500\,\text{s}$. If the reaction were first order with respect to $I_2$, it should take another 3500 s for the concentration of $I_2$ to halve again. This is *not* the case.

*Question*: What is a rough estimate for the second half-life?

The next step is to test for a second order dependence. Using the integrated form of the rate equation, we have:

$$\frac{1}{[I_2]} = \frac{1}{[I_2]_0} + k't$$

and so we need to plot $\dfrac{1}{[I_2]}$ against time.

| Time / s | 0 | 1000 | 2000 | 3000 | 4000 | 5000 | 6000 | 7000 | 8000 |
|---|---|---|---|---|---|---|---|---|---|
| $\dfrac{1}{[I_2]}$ / $\text{dm}^3\,\text{mol}^{-1}$ | 50 | 64 | 78 | 92 | 106 | 120 | 134 | 148 | 162 |

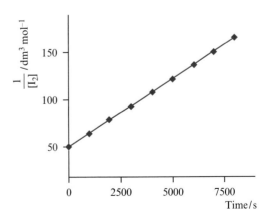

The plot is linear and confirms that the reaction is second order with respect to $I_2$.

The pseudo-second order rate equation is therefore:

$$-\frac{d[I_2]}{dt} = k'[I_2]^2$$

We can determine $k'$ from the gradient of the graph of $\dfrac{1}{[I_2]}$ versus time.

$$\text{Gradient} = k' = 0.014\,\text{dm}^3\,\text{mol}^{-1}\,\text{s}^{-1}$$

The involvement of acetic acid cannot be ruled out even though it is the solvent; it is present in vast excess.

---

### 15.6     Radioactive decay

A nuclide that decomposes into another nuclide is *radioactive*. Decay may involve one of several processes:

- loss of an $\alpha$-particle, i.e. a helium nucleus $[{}^4_2\text{He}]^{2+}$;
- loss of a $\beta$-particle ($\beta^-$), i.e. an electron emitted from the nucleus;
- emission of $\gamma$-radiation which often accompanies loss of $\alpha$- or $\beta$-particles.

When an $\alpha$-particle is emitted by a nuclide, the *daughter nuclide* (the product) has an atomic number two units lower than the parent nuclide and a mass number four units lower (equation 15.34).

$$^{238}_{92}\text{U} \longrightarrow {}^{234}_{90}\text{Th} + {}^4_2\text{He} \tag{15.34}$$

The $\alpha$-particle is represented in equation 15.34 as neutral He because He$^{2+}$ ions readily pick up electrons and helium gas is produced. When a $\beta$-particle is emitted by a nuclide, the daughter nuclide possesses the same mass number as the parent nuclide, but has an atomic number one unit greater (equation 15.35). The decay process in equation 15.35 is the basis of radiocarbon dating (see Box 15.1).

$$^{14}_{6}\text{C} \longrightarrow {}^{14}_{7}\text{N} + \beta^- \tag{15.35}$$

*The **radioactive decay** of any nuclide is a first order process and a radionuclide has a characteristic half-life, $t_{\frac{1}{2}}$.*

The radioactive decay of *any* nuclide follows first order kinetics, although it must be remembered that this statement refers to the decay of a single nuclide. If the daughter nuclide is also radioactive, it will begin to decay as soon as it is formed, thereby complicating the overall process. Consider the decay of a nuclide to a single, stable product. Radioactive decay is often considered in terms of the number of nuclei, $N$, present at a given time. Since the decay is first order with respect to the starting material, equation 15.36 is appropriate.

$$\ln N = \ln N_0 - kt \tag{15.36}$$

where: $N$ = number of nuclei at time $t$

$N_0$ = number of nuclei present at $t = 0$

A plot of $\ln N$ versus time, $t$, is linear and the rate constant, $k$, can be determined from the gradient of the line (as in Figure 15.13). For first order kinetics, the half-life is constant. This leads to a radionuclide having a characteristic half-life, $t_{\frac{1}{2}}$, and a nuclide may be described as being 'short-lived' or 'long-lived'. For example, the half-life of $^{130}_{55}\text{Cs}$ (an artificial radionuclide) is 30.7 minutes, whereas that of naturally occurring $^{235}_{92}\text{U}$ is $7.04 \times 10^8$ years. Radionuclides have many medical applications and the half-life of the nuclide is crucially important. For example, uptake of $^{131}\text{I}$ by the thyroid gland is used as a means of investigating the size

*Artificially made isotopes: see Box 1.4*

## COMMERCIAL AND LABORATORY APPLICATIONS

## Box 15.1 Radiocarbon dating

Radiocarbon dating is a technique used widely by archaeologists to date articles composed of organic material. This includes the dating of ancient textiles, and a much publicized example is the radiocarbon dating of the Turin shroud. The technique was first developed by Willard F. Libby, and in 1960, he was awarded the Nobel Prize in Chemistry for 'his method to use carbon-14 for age determination in archaeology, geology, geophysics and other branches of science'.

Willard F. Libby (1908–1980). © *The Nobel Foundation.*

The method of radiocarbon dating relies on the fact that one isotope of carbon, $^{14}C$, is radioactive with a half-life of 5730 years. Carbon-14 decays to nitrogen-14 by losing a β-particle:

$$^{14}_{6}C \longrightarrow {}^{14}_{7}N + \beta^{-}$$

In a living plant, the ratio of $^{14}C : {}^{12}C$ is constant. Although carbon-14 decays, it is reformed at the same rate by collisions between high-energy neutrons ($^{1}_{0}n$) and atmospheric nitrogen-14:

$$^{14}_{7}N + {}^{1}_{0}n \longrightarrow {}^{14}_{6}C + {}^{1}_{1}H$$

When a plant undergoes photosynthesis, it takes up $CO_2$. Consequently, the uptake of carbon-14 (along with carbon-12 and carbon-13) is a continuous process until the plant dies. At this point, the decaying radioactive carbon-14 is not replaced, and the ratio of $^{14}C : {}^{12}C$ alters with time. If we assume that the $^{14}C : {}^{12}C$ ratio in living species has not altered over an archaeological timescale, then it is possible to date an artefact originating from plant material by measuring the $^{14}C : {}^{12}C$ ratio. Unfortunately, this ratio has altered, but corrections can be made by using data from extremely old, but still living, trees. One example is the American bristlecone pine which grows in the mountains of eastern California.

Samples of organic material for dating are converted to carbon dioxide by combustion. The $CO_2$ gas is then analysed by using mass spectrometry (see Chapter 10). This technique separates $CO_2$ molecules that contain different isotopes of carbon and oxygen, e.g. $(^{14}C)(^{16}O)_2$, $(^{13}C)(^{16}O)_2$ and $(^{12}C)(^{16}O)_2$. In the mid-1970s, accelerator mass spectrometry (AMS) was developed and now provides a highly sensitive technique for determining the ratios of carbon isotopes present in a sample. A crucial point is that, by using AMS, it is possible to distinguish between $^{14}C$ and $^{14}N$. The photograph below shows samples of $CO_2$ being loaded into a linear accelerator for radiocarbon dating.

Linear accelerator used in radiocarbon dating.
*James King-Holmes/Science Photo Library.*

of the gland. Uptake of $^{131}I$ is rapid, allowing diagnostic medicine to be performed efficiently, but equally important is the removal of the radionuclide from the body ($t_{\frac{1}{2}}$ for $^{131}I$ is 8 days, and decay is by $\beta^{-}$ emission).

**Fig. 15.15** Radioactive decay of
any nuclide is a first order process
and a plot of the number of
nuclides against time is an
exponential decay curve. The
curve shown illustrates the decay
of $^{207}_{81}Tl$ to $^{207}_{82}Pb$.

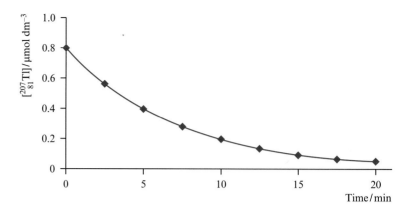

▶
Half-life method:
see worked example 15.2

Figure 15.15 shows the first order decay of $^{207}_{81}Tl$ to the stable isotope $^{207}_{82}Pb$. The half-life of $^{207}_{81}Tl$ can be determined by measuring the time taken for the concentration of $^{207}_{81}Tl$ to fall from 0.8 to 0.4 µmol dm$^{-3}$ (µmol = micromole). From Figure 15.15, this gives $t_{\frac{1}{2}} = 4.8$ min. When using experimental data, at least three consecutive half-lives should be taken from a decay curve, and an average value of $t_{\frac{1}{2}}$ then calculated. As an exercise, measure the first three half-lives from Figure 15.5 and confirm that $t_{\frac{1}{2}} = 4.8$ min.

The relationship between the half-life of a radionuclide and the rate constant for its decay can be derived from equation 15.36. After the first half-life, $t = t_{\frac{1}{2}}$, and $N = N_0/2$. Substitution of these values into equation 15.36 leads to equation 15.37.

$$\left.\begin{array}{c} \ln\left(\dfrac{N_0}{2}\right) - \ln N_0 = -kt_{\frac{1}{2}} \\[2mm] \ln\left(\dfrac{N_0}{2N_0}\right) = -\ln 2 = -kt_{\frac{1}{2}} \\[2mm] t_{\frac{1}{2}} = \dfrac{\ln 2}{k} \end{array}\right\} \qquad (15.37)$$

▶
The SI unit of time is the
second

Since the half-lives of radionuclides vary greatly, units of $t_{\frac{1}{2}}$ can be given as seconds (s), minutes (min), hours (h), days (d) or years (y). Similarly, rate constants may be quoted in s$^{-1}$, min$^{-1}$, h$^{-1}$, d$^{-1}$ or y$^{-1}$. Table 15.1 lists the half-lives of selected radionuclides.

**Table 15.1** Half-lives, $t_{\frac{1}{2}}$, of selected radioactive nuclides. Some nuclides occur naturally, others are artificial.

| Naturally occurring radionuclides | % Natural abundance | $t_{\frac{1}{2}}$ | Artificial radionuclides | $t_{\frac{1}{2}}$ |
|---|---|---|---|---|
| $^{40}K$ | 0.012 | $1.3 \times 10^9$ y | $^{14}C^a$ | 5730 y |
| $^{50}V$ | 0.25 | $>1.4 \times 10^{17}$ y | $^{60}Co$ | 5.27 y |
| $^{87}Rb$ | 27.83 | $4.9 \times 10^{10}$ y | $^{99}Mo$ | 2.75 d |
| $^{115}In$ | 95.7 | $4.4 \times 10^{14}$ y | $^{99m}Tc^b$ | 6.01 h |
| $^{123}Te$ | 0.91 | $1.3 \times 10^{13}$ y | $^{131}I$ | 8.04 d |
| $^{130}Te$ | 33.87 | $2.5 \times 10^{21}$ y | $^{214}Rn$ | 0.27 µs |
| $^{144}Nd$ | 23.80 | $2.1 \times 10^{15}$ y | $^{218}Po$ | 3.04 min |
| $^{235}U$ | 0.72 | $7.0 \times 10^8$ y | $^{222}Rn$ | 3.82 d |
| $^{238}U$ | 99.27 | $4.5 \times 10^9$ y | $^{226}Ra$ | 1599 y |

$^a$ Traces occur naturally;
$^b$ m = metastable.

**Worked example 15.6** | *Radioactive decay*

The following data refer to the decay of $^{222}$Rn:

| Time / days | 0 | 2.0 | 5.0 | 12.0 | 20.0 |
|---|---|---|---|---|---|
| μmoles of $^{222}$Rn | 4.40 | 3.05 | 1.77 | 0.50 | 0.12 |

Determine the half-life of $^{222}$Rn and the rate constant for its decay $(1\,\mu\text{mol} = 10^{-6}\,\text{mol})$.

Two methods could be used to tackle this problem.

*Method 1*: Since radioactive decay is first order with respect to the parent nuclide, use the integrated form of the rate equation and plot ln (moles of $^{222}$Rn) against time. Data to be plotted:

| Time / days | 0 | 2.0 | 5.0 | 12.0 | 20.0 |
|---|---|---|---|---|---|
| ln (moles $^{222}$Rn) | −12.33 | −12.70 | −13.24 | −14.51 | −15.94 |

These data give the following plot:

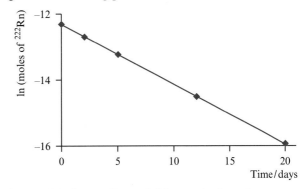

From equation 15.36, the gradient of this graph gives the rate constant, $k$:

$$k = 0.180\,\text{day}^{-1}$$

To find the half-life:

$$t_{\frac{1}{2}} = \frac{\ln 2}{k} = \frac{\ln 2}{(0.180\,\text{day}^{-1})} = 3.85\,\text{days}$$

*Method 2*: Plot a decay curve (moles of $^{222}$Rn against time) and measure three consecutive half-lives. This method is as in worked example 15.2. You should try this method and confirm a value of $t_{\frac{1}{2}} = 3.85$ days. Confirm the value of $k$ found by method 1.

---

**15.7** | ## The dependence of rate on temperature: the Arrhenius equation

### Distribution of the kinetic energies of molecules

In Section 15.1, we stated that the activation energy, $E_a$, of a reaction is the energy in excess of that possessed by the ground state that is required for the

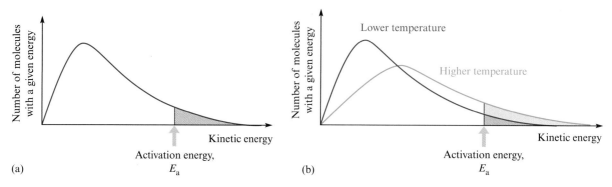

**Fig. 15.16** (a) Boltzmann distribution of kinetic energies. The curve refers to a particular temperature (compare with b). (b) Boltzmann distributions of kinetic energies at two temperatures. For a specified number of molecules, the area under each curve is the same. For a given reaction, the activation energy is *not* dependent on temperature. The blue shaded area overlaps the pink shaded area.

reaction to proceed. Consider a sample of a gas such as $H_2$ contained in a *closed* vessel at a fixed temperature. The molecules are in continual motion, but some molecules are moving faster than others. The *kinetic energy*, $E_k$, of a molecule is related to the molecular velocity, $v$, by equation 15.38.

$$E_k = \tfrac{1}{2}mv^2 \qquad m = \text{molecular mass} \tag{15.38}$$

The distribution of kinetic energies is described by a *Boltzmann distribution curve*. The curve (Figure 15.16a) is asymmetrical and shows that there are more molecules with lower energies than there are with higher energies. When a chemical reaction occurs, the reactant molecules must possess a certain minimum (*activation*) energy (Figure 15.1). The Boltzmann distribution of kinetic energies is temperature-dependent, and Figure 15.16b shows typical distributions at two temperatures. The number of molecules in a given sample stays constant, and so the area under the two graphs must be the same. However, the maximum of the graph shifts to higher energy at higher temperature. The change in shape of the graph corresponds to there being more molecules with higher energies at the higher temperature.

A reaction can proceed only when molecules possess a certain minimum energy, $E_a \approx \Delta H$. For a given reaction, the activation energy (Figure 15.17) essentially has a *fixed* value; $E_a$ is *temperature-independent*. If we mark an arbitrary value on Figure 15.16b as $E_a$, then we can see that the number of molecules that possess this, or a higher, energy increases as the temperature is increased. This means that a greater proportion of molecules can react, and there is an *increase in the rate of reaction*. It is often found that a rise in temperature of 10 K leads to an approximate doubling of the rate of reaction; we consider this statement critically later in this section.

◼▷
**Energy and enthalpy:
see Section 4.5**

**Fig. 15.17** The reaction profile for a single-step reaction showing $E_a$.

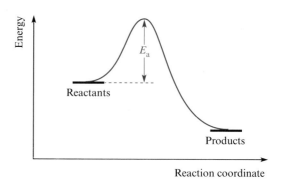

**Table 15.2** Typical values of activation energies. Note the effect that a catalyst has on the value of $E_a$.

| Reaction | Comments | $E_a$ / kJ mol$^{-1}$ |
|---|---|---|
| $2H_2O_2 \longrightarrow 2H_2O + O_2$ | No catalyst | 79 |
| $2H_2O_2 \longrightarrow 2H_2O + O_2$ | Enzyme catalysed | 23 |
| $2NOCl \longrightarrow 2NO + Cl_2$ | | 100 |
| $NO + O_3 \longrightarrow NO_2 + O_2$ | | 10.5 |
| $2HI \longrightarrow H_2 + I_2$ | No catalyst | 185 |
| $2HI \longrightarrow H_2 + I_2$ | With a gold catalyst | 121 |
| $2HI \longrightarrow H_2 + I_2$ | With a platinum catalyst | 59 |
| $2NH_3 \longrightarrow N_2 + 3H_2$ | No catalyst | 335 |
| $2NH_3 \longrightarrow N_2 + 3H_2$ | With a tungsten catalyst | 162 |
| $C_2H_5Br + [OH]^- \longrightarrow C_2H_5OH + Br^-$ | Reaction in aqueous solution | 90 |

## The Arrhenius equation

The discussion above is quantified by the *Arrhenius* relationship (equation 15.39). Notice that the Arrhenius equation involves the *energy* of activation. In Section 4.5 we discussed the difference between an energy and an enthalpy value; here we set $E_a \approx \Delta H$.

$$\ln k = \ln A - \frac{E_a}{RT} \qquad \text{or} \qquad k = A\,e^{-E_a/RT} \qquad (15.39)$$

where the units of $E_a$ are kJ mol$^{-1}$

$k$ = rate constant

$A$ = frequency factor

$R$ = molar gas constant = $8.314 \times 10^{-3}$ kJ K$^{-1}$ mol$^{-1}$

$T$ = temperature (K)

The frequency factor or pre-exponential factor, $A$, may be taken to be a constant for a given reaction. It has the same units as the rate constant and is related to the rate at which collisions occur between reactant molecules and also to the relative orientation of the reactants.

Svante August Arrhenius (1859–1927).
© *The Nobel Foundation.*

The *Arrhenius equation* states:

$$\ln k = \ln A - \frac{E_a}{RT} \qquad \text{or} \qquad k = A\,e^{-E_a/RT}$$

where $k$ = rate constant, $A$ = frequency factor, $R$ = molar gas constant, $T$ = temperature and the units of $E_a$ are kJ mol$^{-1}$.

Some typical values of activation energies are listed in Table 15.2, and the data illustrate the effect that a *catalyst* has on the value of $E_a$ for a given reaction.

Catalyst: see Section 15.8

---

**Worked example 15.7**  |  *Determination of the activation energy for the decomposition of an organic peroxide*

The decomposition of a peroxide ROOR (R = organic substituent) is first order with respect to the peroxide. The rate constant varies with temperature as follows:

| Temperature / K | 410 | 417 | 426 | 436 |
|---|---|---|---|---|
| $k\,/\,\mathrm{s}^{-1}$ | 0.0193 | 0.0398 | 0.0830 | 0.2170 |

**Determine the energy of activation for the reaction.**

The Arrhenius equation gives a relationship between $k$ and $T$:

$$\ln k = \ln A - \frac{E_a}{RT}$$

and so we need to plot $\ln k$ against $\dfrac{1}{T}$.

| $\dfrac{1}{T}\,/\,\mathrm{K}^{-1}$ | 0.002\,44 | 0.002\,40 | 0.002\,35 | 0.002\,29 |
|---|---|---|---|---|
| $\ln k$ | $-3.948$ | $-3.224$ | $-2.489$ | $-1.528$ |

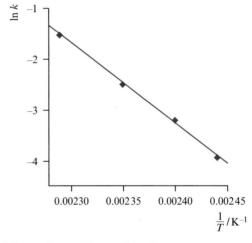

$E_a$ can be found from the gradient of the line:

Physical constants: see inside
back cover of the book

$$\text{Gradient} = -\frac{E_a}{R} = -15\,960\ \mathrm{K}$$

$$E_a = (15\,960\ \mathrm{K}) \times (8.314 \times 10^{-3}\ \mathrm{kJ\,K^{-1}\,mol^{-1}})$$

$$= 133\ \mathrm{kJ\,mol^{-1}} \quad \text{(to 3 sig. figs)}$$

In addition to allowing us to calculate the activation energy of a reaction, the Arrhenius equation also gives an estimate of how the rate of a reaction changes with temperature. For a given reaction, let us assume that $A$ and $E_a$ are constant and thus (using equation 15.39) we can write equation 15.40. This simplifies to equation 15.41 in which $k_n$ is the rate constant at temperature $T_n$.

$$\ln k_2 - \ln k_1 = \left[ \ln A - \frac{E_a}{RT_2} \right] - \left[ \ln A - \frac{E_a}{RT_1} \right]$$

$$= \frac{E_a}{R}\left[ \frac{1}{T_1} - \frac{1}{T_2} \right] \tag{15.40}$$

$$\ln \frac{k_2}{k_1} = \frac{E_a}{R}\left[ \frac{1}{T_1} - \frac{1}{T_2} \right] \tag{15.41}$$

This gives us the *ratio of rate constants* $\dfrac{k_2}{k_1}$, and a ratio of reaction rates can be determined as illustrated in worked example 15.8.

A word of caution: *reactions do not necessarily obey the Arrhenius equation over wide ranges of temperature*, and equation 15.41 must be used with care.

---

| **Worked example 15.8** | *The dependence of rate on temperature* |
| --- | --- |

**The activation energy for the reaction of hydroxide ion with bromoethane to give ethanol is $90.0\,kJ\,mol^{-1}$. How much faster will the reaction proceed if the temperature is raised from 295 to 305 K?**

The larger the value of $k$, the faster the reaction rate.
From the Arrhenius equation:

$$\ln\frac{k_2}{k_1} = \frac{E_a}{R}\left[\frac{1}{T_1} - \frac{1}{T_2}\right]$$

For the reaction in the problem, $E_a = 90.0\,kJ\,mol^{-1}$ and $T_1$ and $T_2$ are 295 and 305 K, respectively.

$$\ln\frac{k_2}{k_1} = \frac{(90.0\,kJ\,mol^{-1})}{(8.314 \times 10^{-3}\,kJ\,K^{-1}\,mol^{-1})}\left(\frac{1}{(295\,K)} - \frac{1}{(305\,K)}\right)$$

$$= 1.20$$

$$\frac{k_2}{k_1} = 3.32$$

Thus, the rate constant increases by a factor of 3.32 when the temperature is raised from 295 to 305 K.
We have been given no information about the rate equation but for *two reactions with equal initial concentrations of reactants*, the ratio of the rates will be equal to the ratio of the rate constants:

$$\frac{\text{Rate 1}}{\text{Rate 2}} = \frac{k_1[C_2H_5Br]^m[OH^-]^n}{k_2[C_2H_5Br]^m[OH^-]^n}$$

Hence the rate of reaction at 305 K is 3.32 times faster than the rate of reaction at 295 K.

---

At the beginning of this section, we commented that a rise in temperature of 10 K leads to an approximate doubling of the reaction rate. However, equation 15.41 shows that there is also a dependence on $E_a$. Figure 15.18 illustrates how an increase in the temperature from 300 to 310 K affects the

**Fig. 15.18** For a rise in temperature from 300 to 310 K, the ratio of the rate constants $\dfrac{k_2}{k_1}$ for a reaction changes as a function of the energy of activation. The approximate doubling of a reaction rate for a 10 degree temperature rise is only valid for values of $E_a$ of $\approx 50\,kJ\,mol^{-1}$.

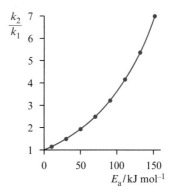

rate of a reaction for values of $E_a$ ranging from 10 to $150 \, \text{kJ} \, \text{mol}^{-1}$. The approximate doubling of rate is valid only for values of $E_a \approx 50 \, \text{kJ} \, \text{mol}^{-1}$. This is a typical value for many reactions in solution.

| **15.8** | **Catalysis and autocatalysis** |

A *catalyst* may speed up or slow down a reaction, and does not appear in the product.

A catalyst is a substance that alters the rate of a reaction without appearing in any of the products of that reaction. Catalysts may speed up or slow down a reaction. The latter is known as *negative catalysis* and is critical in stabilizing many commercially important materials. Examples of some catalysed reactions are given in Table 15.2.

A *catalyst* provides a new pathway for a reaction; if this pathway has a lower activation energy than the uncatalysed pathway, the reaction rate is increased:

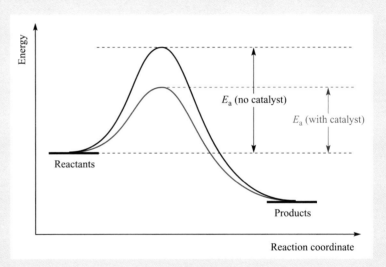

Catalysts fall into two general categories: *homogeneous* and *heterogeneous catalysts*.

Catalytic cycles: see Box 21.7

A *homogeneous catalyst* is in the same phase as the components of the reaction that it is catalysing; a *heterogeneous catalyst* is in a different phase from the components of the reaction.

### Heterogeneous catalysts

Many industrial catalysts are heterogeneous and important examples include bulk metal surfaces and zeolites and involve the reactions of gases. Selected applications are:

Zeolites: see Figure 22.13 and associated text

- in the Haber process, α-Fe catalyses the reaction of $H_2$ and $N_2$ to give ammonia (see text below):

$$N_2 + 3H_2 \longrightarrow 2NH_3$$

- a mixture of Pd, Pt and Rh is used in motor vehicle catalytic converters to catalyse the oxidation of CO and hydrocarbons and reduction of $NO_x$ (see Box 22.9);
- zeolite catalysts are used in the catalytic cracking of heavy petroleum distillates (see Section 25.4).

When molecules of gaseous reactants are passed over the surface of a heterogeneous catalyst, some of the molecules are *adsorbed* on to the surface. If adsorption involves only weak van der Waals interactions between the gas molecules (the *adsorbate*) and the surface atoms, the process is called *physisorption*. If adsorption involves the formation of chemical bonds between surface atoms and adsorbate as in equation 15.42, the process is known as *chemisorption*.

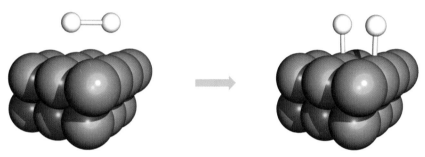

Colour code: metal atom, blue; H, white.

(15.42)

In equation 15.42, the H−H bond of $H_2$ is cleaved, leaving H atoms adsorbed on the metal surface and available for reaction with another adsorbed species. The M−H bond must not be too strong otherwise the atoms remain adsorbed. Consider the reaction between $H_2$ and $N_2$ to give $NH_3$ in more detail. The reaction involves the cleavage of N≡N and H−H bonds, and the formation of N−H bonds. In the absence of a catalyst, the reaction proceeds very slowly because the activation energy for the dissociation of $N_2$ and $H_2$ is very high. In the presence of a metal catalyst, chemisorption of $H_2$ and $N_2$ occurs, and the adsorbed H and N atoms combine to yield $NH_3$ which *desorbs* from the surface. In choosing a suitable metal catalyst for the reaction, early $d$-block metals such as Mo and Re are avoided because, although they chemisorb $N_2$ efficiently, the M−N interaction is so strong that N atoms preferentially remain on the surface rather than reacting with adsorbed H atoms. For Co, Rh, Ir, Ni and Pt, the adsorption of $N_2$ has a large activation energy and this means a slow rate of reaction. The catalyst used industrially is $\alpha$-Fe, produced by reducing $Fe_3O_4$ mixed with $K_2O$ (to enhance catalytic activity), $SiO_2$ and $Al_2O_3$ (to stabilize the structure of the catalyst).

A large surface area is essential for the efficient activity of a heterogeneous catalyst. If a catalyst consists of a piece of bulk metal, the atoms sited within the lump of metal are unavailable for catalytic activity. By using a catalyst that is composed of small metal particles dispersed on, for example, an alumina support, the number of available metal atoms is increased. Very fine powders of alumina have enormous surface areas ($\approx 900 \, m^2 \, g^{-1}$) and depositing metal particles on this surface dramatically increases the surface area of metal with respect to that of a piece of bulk metal, and hence increases the number of catalytic sites. Zeolites are aluminosilicates with structures featuring open lattices with ordered channels and, consequently, huge surface areas. We return to these important catalysts in Section 22.6.

### Homogeneous catalysts

We cannot generalize about the action of a *homogeneous* catalyst. Each catalyst operates in a unique manner and facilitates a pathway from reactants to products in the reaction that it is catalysing. Many industrial homogeneous catalysts involve *organometallic compounds* (see Section 23.14) and examples include:

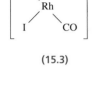

**(15.3)**

- *cis*-$[Rh(CO)_2I_2]^-$ (a square planar complex anion, **15.3**) catalyses the reaction of methanol with CO to form acetic acid in the Monsanto process:

$$CH_3OH + CO \longrightarrow CH_3CO_2H$$

- $HCo(CO)_4$ (which loses CO to give the active catalyst $HCo(CO)_3$) catalyses the conversion of alkenes to aldehydes in the Oxo-process (or *hydroformylation*).

The overall pathway is usually represented in a *catalytic cycle* in which intermediate species are displayed (see Box 21.7). The cycle must be a closed loop because, as products are formed, the catalyst is regenerated and is fed back into the catalytic process.

### Autocatalysis

An interesting phenomenon occurs when one of the products of a reaction is able to act as a catalyst for the process. The reaction is said to be *auto-catalytic*. The reaction in equation 15.43 begins with only A and B present; one of the products, D, is a catalyst for the reaction. Once D is present (equation 15.44), the rate of the reaction increases.

$$\text{\textit{Initially:}} \qquad A + B \longrightarrow C + D \qquad\qquad (15.43)$$

$$\text{\textit{Autocatalysed:}} \quad A + B \xrightarrow{\ D\ } C + D \qquad\qquad (15.44)$$

Figure 15.19 shows how the concentration of $[MnO_4]^-$ varies during reaction with dimethylamine. Initially, the reaction proceeds relatively slowly, but once products are present in solution, the rate at which the $[MnO_4]^-$ ions disappear increases. The reaction is catalysed by the $Mn^{2+}$ ions generated in the reduction of $[MnO_4]^-$.

A biological example is the conversion of trypsinogen to trypsin. Trypsin is a digestive enzyme present in the small intestine. Its function is to cleave protein chains into smaller units. The active site in trypsin involves the amino acid residues histidine, serine and aspartate (see Table 35.1) and these are highlighted in the structural diagram of the enzyme shown in

**Fig. 15.19** A plot of the concentration of manganate(VII) against time for the autocatalytic reaction between $Me_2NH$ and $[MnO_4]^-$ in aqueous solution; the catalyst is Mn(II). (Data from: F. Mata-Perez *et al.* (1987) *J. Chem. Educ.*, vol. 64, p. 925.)

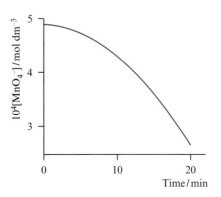

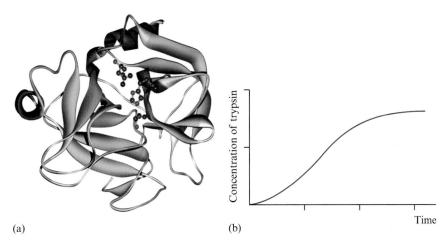

**Fig. 15.20** (a) The structure of the enzyme trypsin (from wild boar, *Sus scrofa*), determined by X-ray crystallography [A. Johnson *et al.* (1999) *Biochim. Biophys. Acta*, vol. 1435, p. 7]. The structure is shown in a ribbon representation, with the amino acid residues in the active site highlighted using ball-and-stick representations (aspartate, top; histidine, middle; serine, bottom; colour code: C, grey; N, blue; O, red). (b) A plot of trypsin concentration against time as trypsinogen is hydrolysed to trypsin. This product catalyses the reaction, and the **S**-shape of the curve is characteristic of an autocatalytic reaction.

Figure 15.20a. Trypsin is formed in the pancreas in the inactive form of trypsinogen. The conversion of trypsinogen to trypsin is catalysed by trypsin. Figure 15.20b shows the appearance of trypsin in the reaction as a function of time. The **S**-shaped plot is characteristic of an autocatalytic reaction – once the catalyst enters the system, the reaction accelerates.

## 15.9    Reversible reactions

So far we have only considered reactions which go in one direction, but in many cases, the forward reaction is opposed by the reverse (or back) reaction. In this section we briefly consider the consequences on the rate equation of having two opposing *first order* processes.

Consider the reactions given in equation 15.45, in which the rate constants for the forward and backward reactions are $k_1$ and $k_{-1}$ respectively.

$$A \underset{k_{-1}}{\overset{k_1}{\rightleftharpoons}} B \tag{15.45}$$

The measured rate of disappearance of A does not obey simple first order kinetics because as B is formed, some of it re-forms A in a first order process. The rate of the forward reaction is given by equation 15.46.

$$\text{Rate of reaction} = -\frac{d[A]}{dt} = k_1[A] - k_{-1}[B] \tag{15.46}$$

Assuming that only A is present initially, then the *initial rate of reaction* is simply $k_1[A]_0$ (from equation 15.11, with the initial concentration of A, $[A]_0$).

The concentration of B at time $t$ is equal to the *difference* in the concentrations of A at times 0 and $t$ (equation 15.47). Note that this relationship is dependent upon the stoichiometry of the reaction; in this case, one mole of A gives one mole of B, and vice versa.

$$[B] = [A]_0 - [A] \tag{15.47}$$

Thus, we can rewrite the rate equation in terms of A as in equations 15.48 and 15.49.

$$\text{Rate of reaction} = -\frac{d[A]}{dt} = k_1[A] - k_{-1}\{[A]_0 - [A]\} \tag{15.48}$$

or

$$\text{Rate of reaction} = -\frac{d[A]}{dt} = (k_1 + k_{-1})[A] - k_{-1}[A]_0 \tag{15.49}$$

Remember that $[A]_0$ is a constant and therefore the second term on the right-hand side of equation 15.49 is also a constant. Thus, the reaction shows a first order dependence upon A. Equation 15.49 has the form:

$$y = mx + c$$

and therefore a plot of rate of reaction against $[A]$ is linear with a gradient equal to $(k_1 + k_{-1})$. The value of $k_{-1}$ can be found from the intercept since the initial concentration of A, $[A]_0$, is known. Thus, both rate constants may be determined as exemplified below.

---

**Worked example 15.9**     *A reversible reaction*

**Experimental data for the reaction**

$$A \underset{k_{-1}}{\overset{k_1}{\rightleftharpoons}} B$$

**are tabulated below. The initial concentration of A is 0.050 mol dm$^{-3}$. Use these results to determine the first order rate constants $k_1$ and $k_{-1}$.**

| $[A]$ / mol dm$^{-3}$ | 0.045 | 0.040 | 0.030 | 0.020 | 0.010 |
|---|---|---|---|---|---|
| $10^4$ Rate / mol dm$^{-3}$ min$^{-1}$ | 8.97 | 7.94 | 5.88 | 3.82 | 1.76 |

The reversible reaction should obey the rate equation:

$$-\frac{d[A]}{dt} = (k_1 + k_{-1})[A] - k_{-1}[A]_0$$

(from equation 15.49) and has the form $y = mx + c$
A plot of the reaction rate against $[A]$ is linear, with a gradient $m = (k_1 + k_{-1})$ and an intercept on the $y$ axis, $c = -k_{-1}[A]_0$.

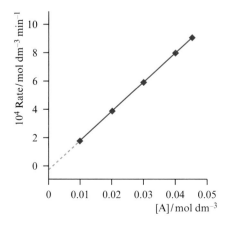

The gradient of the line $= (k_1 + k_{-1}) = 2.1 \times 10^{-2}\,\mathrm{min}^{-1}$

After extrapolation, intercept on the $y$ axis $= -k_{-1}[A]_0$

$$= -0.30 \times 10^{-4}\,\mathrm{mol\,dm}^{-3}\,\mathrm{min}^{-1}$$

From the intercept:

$$k_{-1} = \frac{(0.30 \times 10^{-4}\,\mathrm{mol\,dm}^{-3}\,\mathrm{min}^{-1})}{([A]_0\,\mathrm{mol\,dm}^{-3})}$$

$$= \frac{(0.30 \times 10^{-4}\,\mathrm{mol\,dm}^{-3}\,\mathrm{min}^{-1})}{(0.050\,\mathrm{mol\,dm}^{-3})}$$

$$= 6.0 \times 10^{-4}\,\mathrm{min}^{-1}$$

Substituting the value for $k_{-1}$ into the equation for the gradient allows us to find $k_1$:

$$k_1 = (2.1 \times 10^{-2}\,\mathrm{min}^{-1}) - k_{-1}$$

$$= (2.1 \times 10^{-2}\,\mathrm{min}^{-1}) - (6.0 \times 10^{-4}\,\mathrm{min}^{-1})$$

$$= 0.020\,\mathrm{min}^{-1}$$

(Notice that the rate constant for the back reaction is much smaller than that for the forward reaction. What would be the effect on the reaction if $k_{-1} > k_1$?)

---

## 15.10     Elementary reactions and molecularity

In this section, we move from studying the behaviour of bulk matter to a discussion of reactions at the level of individual molecules. Eventually this will lead us to the microscopic reaction mechanism. We emphasize a crucial point here: the kinetics of a reaction may be determined experimentally, and the *mechanism* of the reaction is inferred from these measurements. *A mechanism cannot be 'proven'* – at best a mechanism is consistent with all of the kinetic data available.

### Elementary reactions

So far, our discussions of rate equations have focused on one-step reactions such as:

A $\longrightarrow$ products

or

A + B $\longrightarrow$ products

However, in many reactions, the complete pathway from reactants to products involves a series of steps, each of which is called an *elementary reaction*. Figure 15.1 showed a profile for a two-step reaction: the intermediate species lies in a local energy minimum along the pathway from reactants to products. Figure 15.21 shows a profile for a three-step reaction in which reactant A is converted to product D through two intermediates, B

**Fig. 15.21** The reaction profile for a reaction A ⟶ D. The mechanism comprises three elementary steps involving two intermediates, B and C.

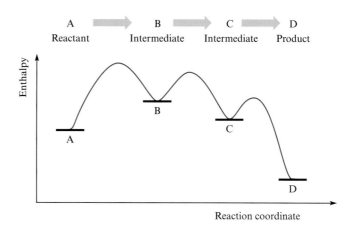

and C. The sequence of elementary reactions is:

$$A \longrightarrow B$$
$$B \longrightarrow C$$
$$C \longrightarrow D$$

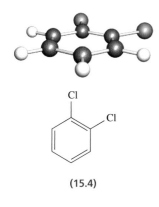

**(15.4)**

and each elementary reaction has a characteristic activation energy and rate constant. The rate-determining step of the complete reaction is the elementary reaction with the highest activation barrier, e.g. in Figure 15.21, this is the conversion of A to B. In some reactions, it is possible to follow the growth and decay of intermediate species, as well as the decay of the reactant(s) and growth of the product(s). Figure 15.22 shows the results of an experiment in which the bromination of 1,2-dichlorobenzene (**15.4**) was followed by using a gas chromatograph (see Box 24.4) coupled to a mass spectrometer. Curve A shows the decay of 1,2-dichlorobenzene and curves B and E illustrate the growth of two brominated products. Curves C and D show that two intermediate species were detected. By finding out information about the intermediates in a reaction and the reaction kinetics, it is possible to build up a picture of the *reaction mechanism* and to propose a sequence of elementary steps. Normally, rates of reactions are reported in terms of the rate of loss of a reactant, or the rate of formation of a product, and do not include terms involving intermediates. Experimentally, it is easier to follow what happens to a reactant or product, rather than an intermediate, during a reaction. Nonetheless, the rate-determining step in a reaction may be one involving an intermediate, and so we must look in detail at the kinetics of elementary steps. In the sections that follow, we see how the overall rate equation for a reaction depends on the rate equations for the elementary steps.

**Fig. 15.22** The variation in concentrations of reactant A (1,2-dichlorobenzene), intermediates C and D, and products B and E during the bromination of 1,2-dichlorobenzene. [Data from: D. A. Annis *et al.* (1995) *J. Chem. Educ.*, vol. 72, p. 460.]

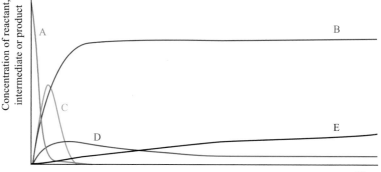

## Molecularity: unimolecular and bimolecular reactions

The *molecularity* of an elementary reaction step is the *number of molecules or atoms of reactant* taking part. The term 'molecularity' only refers to an elementary reaction and must not be confused with the 'order' of an overall reaction. Most elementary reactions involve only one or two species and are known as *unimolecular* or *bimolecular steps* respectively.

Equations 15.50 and 15.51 show examples of unimolecular reactions. The first is a decomposition (by bond cleavage to form radicals, see Section 24.10) and the second is an isomerization in which the relative positions of the two chlorine atoms change.

> The *molecularity of an elementary step* is the number of molecules or atoms of reactant taking part.

$$Br_2 \longrightarrow 2Br^{\bullet} \tag{15.50}$$

$$\begin{array}{ccc} \underset{H}{\overset{Cl}{\diagdown}}C=C\underset{Cl}{\overset{H}{\diagup}} & \longrightarrow & \underset{Cl}{\overset{H}{\diagdown}}C=C\underset{Cl}{\overset{H}{\diagup}} \end{array} \tag{15.51}$$

In a bimolecular step, *two* species combine to give the product(s). The reacting species may be the same or different from each other, as equations 15.52 and 15.53 illustrate:

$$2NO_2 \longrightarrow N_2O_4 \tag{15.52}$$

$$CH_3CH_2Br + [OH]^- \longrightarrow CH_3CH_2OH + Br^- \tag{15.53}$$

▬▶
Mechanisms of reactions are also discussed in Chapters 25–35

In the next section, we develop the relationship between elementary steps and overall reaction mechanisms, and in Sections 15.12 and 15.13 we consider the steady state approximation and its application to multi-step reactions.

For a given reaction, one of the elementary steps must be the slowest (rate-determining) stage. The molecularity of this rate-determining step (RDS) defines the observed kinetics for the overall reaction. This is the reason why the terms appearing in the rate equation may differ from the *overall* stoichiometry of the reaction. This statement is further illustrated in the following sections.

> *RDS* stands for rate-determining step.

| 15.11 | **Microscopic reaction mechanisms: unimolecular and bimolecular elementary steps** |
|---|---|

We have seen that elementary steps play a decisive role in defining the mechanism of a reaction. In this section we look at the problem of writing rate equations in terms of the *experimental* concentrations of starting materials or products.

### Writing rate equations for unimolecular and bimolecular steps

Earlier in this chapter we used experimental data to determine *overall rate equations*. We have also seen that overall reaction mechanisms are composed of elementary reactions, and these are usually unimolecular or bimolecular steps. The aim of studying the kinetics of a reaction is to use the experimental data to write an overall reaction mechanism. The problem is tackled by first suggesting a mechanism, i.e. a series of elementary steps, and then writing down the rate equation in accord with the mechanism. If this rate equation is consistent with that determined experimentally, then it is possible (but not proven) that the proposed mechanism is correct.

We now come to a critical point. Although we *cannot* write down an overall rate equation using the stoichiometry of a reaction, we *can* deduce the rate equation for an elementary step from the stoichiometry of that step.

In a unimolecular step such as the decomposition of $Cl_2$ into chlorine radicals (equation 15.54), the rate of reaction depends directly on the concentration of the reactant – one molecule of reactant is involved in the step. The decomposition of $Cl_2$ occurs when the molecule is excited (thermally or photolytically) and the Cl–Cl bond is broken. The rate of reaction for the fission of $Cl_2$ into two radicals is given in equation 15.55.

◼▷
Overall rate equations:
see Section 15.4

The *rate equation for an elementary step* can be deduced from the stoichiometry of that step, *but* the overall rate equation *cannot* be written down using the stoichiometry of the overall reaction.

$$Cl_2 \longrightarrow 2Cl^{\bullet} \tag{15.54}$$

$$\text{Rate} = k[Cl_2] \tag{15.55}$$

In a bimolecular step, two molecules must collide before a reaction can occur, and therefore the rate of reaction depends on the concentrations of both species. The dimerization of $NO_2$ is a bimolecular process (equation 15.52) and the rate of dimerization is given by equation 15.56. The *squared* dependence tells us that *two* molecules are involved.

$$\text{Rate} = k[NO_2]^2 \tag{15.56}$$

If the bimolecular reaction involves two different molecules, the rate equation reflects this. Dioxygen is formed when an oxygen radical, $O^{\bullet}$, collides and reacts with an ozone molecule (equation 15.57). The rate of the reaction depends on both $[O^{\bullet}]$ and $[O_3]$ as equation 15.58 shows. Although the O atom is a diradical with two unpaired electrons, we represent it here simply as $O^{\bullet}$.

◼▷
Each of O and $O_2$ is a
diradical: see Figure 4.22 and
associated discussion

$$O^{\bullet} + O_3 \longrightarrow 2O_2 \tag{15.57}$$

$$\text{Rate} = k[O^{\bullet}][O_3] \tag{15.58}$$

| | | |
|---|---|---|
| For a unimolecular step: | A ⟶ B | Rate = $k[A]$ |
| For a bimolecular step: | 2A ⟶ B | Rate = $k[A]^2$ |
| For a bimolecular step: | A + B ⟶ C | Rate = $k[A][B]$ |

## 15.12    Combining elementary steps into a reaction mechanism

The rate equations described in the last section involved isolated elementary steps. What happens when we combine a series of such steps into a reaction pathway? It is vital to remember that experimental kinetics data and rate equations refer to overall reactions. A sequence of elementary steps constitutes a proposed mechanism, and if the overall rate equation derived from such elementary steps agrees with the experimentally determined rate equation, then the proposed mechanism *may* be correct, *but it is not proven*. We shall look at some real examples later, but first we consider some general examples.

## A two-step reaction and the steady state approximation

Consider the two-step reaction of A $\longrightarrow$ C proceeding through intermediate B. Each elementary reaction has a characteristic rate constant (equations 15.59 and 15.60).

$$A \xrightarrow{k_1} B \tag{15.59}$$

$$B \xrightarrow{k_2} C \tag{15.60}$$

The smaller the rate constant $k$, the slower the reaction. Let us take the case where $k_1 \ll k_2$, i.e. the rate of formation of B from A is *much slower* than the conversion of B to C. This means that B $\longrightarrow$ C is a *fast step* in the mechanism, and as soon as any B is formed, it is converted to C. There is never a build-up of B, and B cannot be detected by, for example, spectroscopic methods. The rate-determining step of the *overall reaction* is therefore A $\longrightarrow$ B, and the overall rate equation (equation 15.61) is the same as the rate equation for elementary step 15.59.

$$\text{Rate of reaction} = -\frac{d[A]}{dt} = k_1[A] \tag{15.61}$$

In the case where $k_1 \gg k_2$, A is converted to B in a fast step, and the concentration of B increases rapidly. Intermediate B is consumed in the rate-determining step (slow step) and the overall rate equation (equation 15.62) is equal to the rate equation for elementary step 15.60. Since [B] is high, it is very likely that B can be monitored spectroscopically.

$$\text{Rate of reaction} = -\frac{d[B]}{dt} = k_2[B] \tag{15.62}$$

The case where $k_1 \gg k_2$ is an extreme one, and even when $k_1 = k_2$, the concentration of B is still sufficiently high (Figure 15.23a) that it may be monitored and so the expression for the rate of reaction in equation 15.62 is still in terms of an *experimental observable*. Now we must look at the general case where the concentration of intermediate B is very low and effectively constant during most of the reaction period (Figure 15.23b). Because [B] is very low, it cannot be measured experimentally. Therefore, we have a problem because equation 15.62 is expressed in terms of a quantity which is *not* experimentally observable. It would be convenient if we had a way of algebraically expressing [B] in terms of [A] or [C], so that the rate of reaction in equation 15.62 is related to an experimental observable. The *steady state* (or *stationary state*) *approximation* provides such a treatment. We make the assumption that during most of the reaction period, the concentration of B is a constant, steady state value, i.e. there is no increase or decrease in [B]. This leads to equation 15.63.

The ***steady state approximation*** states that the concentration of an intermediate B during most of a reaction is constant:

$$\frac{d[B]}{dt} = -\frac{d[B]}{dt} = 0$$

$$\frac{d[B]}{dt} = -\frac{d[B]}{dt} = 0 \tag{15.63}$$

How can we use equation 15.63? Intermediate B *is generated from A* in a reaction with rate constant $k_1$ (equation 15.59) and is *converted to C* at a rate dependent on $k_2$ (equation 15.60). The rate of formation of B is therefore given by equation 15.64, and rearrangement of this expression gives equation 15.65.

$$\frac{d[B]}{dt} = k_1[A] - k_2[B] = 0 \tag{15.64}$$

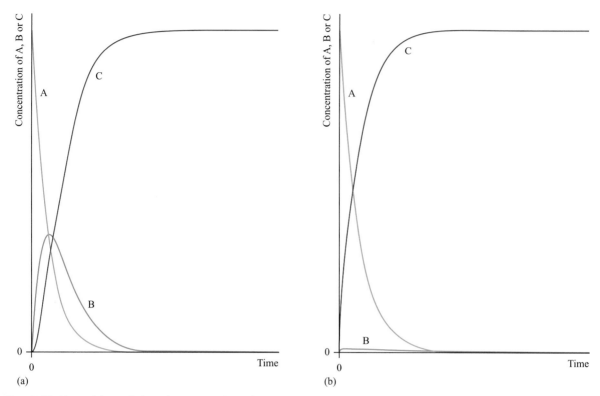

**Fig. 15.23** Plots of the variation of concentrations of A, B and C in the reaction of A ⟶ C through intermediate B (equations 15.59 and 15.60): (a) for the case where $k_1 = k_2$, and (b) for a case where the concentration of B is low and approximately constant for most of the reaction.

$$\left.\begin{aligned} k_1[A] &= k_2[B] \\ [B] &= \frac{k_1[A]}{k_2} \end{aligned}\right\} \tag{15.65}$$

We know from equation 15.62 that the rate of the reaction is related to the concentration of B and so, by combining equations 15.62 and 15.65, we obtain a new expression for the rate of reaction in terms of the *experimentally measurable* concentration of A (equation 15.66).

$$\text{Rate of reaction} = k_2[B] = k_2\left(\frac{k_1[A]}{k_2}\right) = k_1[A] \tag{15.66}$$

The steady state approximation is often applied when a proposed mechanism is complex and goes through several intermediates. In Section 15.13 and Box 15.2, we illustrate applications of the steady state approximation.

### Pre-equilibrium

In this section, we remain focused on the reaction A ⟶ C through intermediate B, but now look at a case in which intermediate B undergoes two competing reactions: reaction to form C, and reaction to re-form A. The three elementary steps can be written as reactions 15.67 and 15.68, where equation 15.67 describes a *pre-equilibrium*.

■▶
More about equilibria in
Chapter 16

$$A \underset{k_{-1}}{\overset{k_1}{\rightleftharpoons}} B \tag{15.67}$$

$$B \overset{k_2}{\longrightarrow} C \tag{15.68}$$

The pre-equilibrium can never be established if $k_2 \gg k_{-1}$ because in this case, B is converted to C too rapidly. A pre-equilibrium is established if $k_2 \ll k_{-1}$, and the conversion of B to C is the rate-determining step. The rate of reaction is given by equation 15.69.

$$\text{Rate of reaction} = \frac{d[C]}{dt} = k_2[B] \tag{15.69}$$

■▷
Equilibrium constants:
see Section 16.3

The equilibrium constant, $K$, for equilibrium 15.67 is given by equation 15.70.

$$K = \frac{[B]}{[A]} \tag{15.70}$$

It follows that $[B] = K[A]$, and substitution of this expression into equation 15.69 allows us to write the rate of reaction in terms of the concentrations of A and C, i.e. *measurable quantities* (equation 15.71).

$$\frac{d[C]}{dt} = k_2[B] = k_2K[A] \tag{15.71}$$

The equilibrium constant, $K$, can also be expressed in terms of the rate constants for the forward and back reactions as shown in equation 15.72.

$$K = \frac{k_1}{k_{-1}} \tag{15.72}$$

By making this substitution into equation 15.71, the rate of reaction can be written in terms of the rate constants $k_1$, $k_{-1}$ and $k_2$ (equation 15.73).

$$\frac{d[C]}{dt} = \left(\frac{k_1 k_2}{k_{-1}}\right)[A] \tag{15.73}$$

This is a first order rate equation, and experimental data provide an observed rate constant, $k_{obs}$ (equation 15.74) where $k_{obs} = \left(\dfrac{k_1 k_2}{k_{-1}}\right)$.

$$\frac{d[C]}{dt} = k_{obs}[A] \tag{15.74}$$

## 15.13    Proposing a mechanism consistent with experimental rate data

Now that we have considered some general cases, let us turn our attention to representative examples. The examples illustrate how experimental data are used to support a proposed mechanism for a reaction.

### Case study 1: decomposition of ozone

Ozone (see Figure 4.2) decomposes according to equation 15.75, and the experimentally determined rate law for the reaction is equation 15.76.

$$2O_3 \longrightarrow 3O_2 \tag{15.75}$$

$$\text{Rate of reaction} = -\frac{d[O_3]}{dt} = k_{obs}\left(\frac{[O_3]^2}{[O_2]}\right) \tag{15.76}$$

Since $O_2$ appears in the rate equation, the reaction mechanism does not involve the simple collision of $O_3$ molecules. In proposing a mechanism, we write a possible series of elementary steps and then deduce from them an overall rate equation. *If the proposed mechanism is plausible, the theoretical rate equation must be the same as that found experimentally*

Radicals: see Section 1.11

under appropriate conditions. The proposed mechanism for the decomposition of $O_3$ is given in scheme 15.77; notice that summation of the elementary steps gives the overall reaction equation 15.75. An oxygen atom is a diradical, but for clarity, we simply write O in the following equations.

$$\left.\begin{array}{l} O_3 \underset{k_{-1}}{\overset{k_1}{\rightleftharpoons}} O_2 + O \\[2ex] O_3 + O \overset{k_2}{\longrightarrow} 2O_2 \end{array}\right\} \tag{15.77}$$

Based on this reaction mechanism, the rate of disappearance of $O_3$ is given by equation 15.78. There are three terms because $O_3$ is consumed in two elementary steps and formed in one step. The negative term on the right-hand side of equation 15.78 shows that this term refers to the appearance (rather than disappearance) of $O_3$.

$$-\frac{d[O_3]}{dt} = k_1[O_3] - k_{-1}[O_2][O] + k_2[O_3][O] \tag{15.78}$$

We now use the steady state approximation to obtain an expression for [O] (equation 15.79), assuming that the overall rate of formation of the intermediate O is zero.

$$\frac{d[O]}{dt} = 0$$

$$k_1[O_3] - k_{-1}[O_2][O] - k_2[O_3][O] = 0$$

$$[O] = \frac{k_1[O_3]}{k_{-1}[O_2] + k_2[O_3]} \tag{15.79}$$

By substituting equation 15.79 into equation 15.78, we obtain an expression for the rate of disappearance of $O_3$ in terms of observable concentrations (equation 15.80).

$$-\frac{d[O_3]}{dt} = k_1[O_3] - \frac{(k_{-1}[O_2] - k_2[O_3])k_1[O_3]}{k_{-1}[O_2] + k_2[O_3]}$$

$$= \frac{k_1 k_{-1}[O_2][O_3] + k_1 k_2[O_3]^2 - k_1 k_{-1}[O_2][O_3] + k_1 k_2[O_3]^2}{k_{-1}[O_2] + k_2[O_3]}$$

$$-\frac{d[O_3]}{dt} = \frac{2k_1 k_2[O_3]^2}{k_{-1}[O_2] + k_2[O_3]} \tag{15.80}$$

Equation 15.80 does not have the form of the observed rate equation 15.76. However, in the proposed reaction mechanism, the $k_2$ step should be much slower than the steps in the pre-equilibrium and therefore $k_2[O_3] \ll k_{-1}[O_2]$. It follows that:

$$k_{-1}[O_2] + k_2[O_3] \approx k_{-1}[O_2]$$

and so the rate of disappearance of $O_3$ can be written as equation 15.81, an expression that does have the same form as the observed rate law, in which $k_{obs} = \frac{2k_1 k_2}{k_{-1}}$.

$$-\frac{d[O_3]}{dt} = \frac{2k_1 k_2[O_3]^2}{k_{-1}[O_2]} \tag{15.81}$$

The agreement between the experimental rate law and that derived from reaction scheme 15.77 means that the proposed mechanism is consistent with the observed reaction kinetics.

It is also possible to express the rate of reaction in terms of the rate of appearance of $O_2$, and from reaction scheme 15.77, we can write equation 15.82.

$$\frac{d[O_2]}{dt} = k_1[O_3] - k_{-1}[O_2][O] + 2k_2[O_3][O] \tag{15.82}$$

The factor of 2 in the last term of equation 15.82 follows from the stoichiometry of the *elementary* step: two moles of $O_2$ are formed in every collision between $O_3$ and O. By making the substitution for [O] from equation 15.79 into rate equation 15.82, we arrive at expression 15.83 for the rate of appearance of $O_2$ in terms of $[O_2]$ and $[O_3]$ (you should verify this derivation).

$$\frac{d[O_2]}{dt} = \frac{3k_1k_2[O_3]^2}{k_{-1}[O_2] + k_2[O_3]} \tag{15.83}$$

If, as above, $k_2[O_3] \ll k_{-1}[O_2]$, then equation 15.83 simplifies to:

$$\frac{d[O_2]}{dt} \approx \frac{3k_1k_2[O_3]^2}{k_{-1}[O_2]}$$

Again, the theoretical rate law (based on the proposed mechanism) is of the same form as that obtained experimentally. Moreover, comparison of equations 15.81 and 15.83 confirms that the rates of formation of $O_2$ and disappearance of $O_3$ are related according to the overall reaction 15.75, i.e. the rate of formation of $O_2$ is $^3/_2$ times the rate of disappearance of $O_3$.

## Case study 2: reaction of NO with $O_2$

The gas phase reaction of NO with $O_2$ (equation 15.84) follows the rate law given in equation 15.85.

$$2NO + O_2 \longrightarrow 2NO_2 \tag{15.84}$$

$$\frac{d[NO_2]}{dt} = k_{obs}[NO]^2[O_2] \tag{15.85}$$

At first glance, this might suggest that the reaction takes place in one step involving a termolecular collision between two molecules of NO and one molecule of $O_2$. However, termolecular elementary steps are uncommon, and a mechanism composed of unimolecular and/or bimolecular steps is more likely. The mechanism proposed for reaction 15.84 is shown in scheme 15.86.

$$\left. \begin{array}{c} 2NO \underset{k_{-1}}{\overset{k_1}{\rightleftharpoons}} N_2O_2 \\[2mm] N_2O_2 + O_2 \overset{k_2}{\longrightarrow} 2NO_2 \end{array} \right\} \tag{15.86}$$

In order to test whether the mechanism is plausible, we must work out the rate law that follows from the mechanism and then see how it relates to the experimental rate equation. The rate of appearance of $NO_2$ is given by equation 15.87; the factor of 2 is needed because two moles of $NO_2$ are formed when one molecule of $N_2O_2$ collides with one molecule of $O_2$ (equation 15.86).

$$\frac{d[NO_2]}{dt} = 2k_2[N_2O_2][O_2] \tag{15.87}$$

We now use the steady state approximation to find an expression for the concentration of the intermediate $N_2O_2$ (equation 15.88).

$$\frac{d[N_2O_2]}{dt} = k_1[NO]^2 - k_{-1}[N_2O_2] - k_2[N_2O_2][O_2] = 0$$

$$k_{-1}[N_2O_2] + k_2[N_2O_2][O_2] = k_1[NO]^2$$

$$[N_2O_2] = \frac{k_1[NO]^2}{k_{-1} + k_2[O_2]} \tag{15.88}$$

Substitution of equation 15.88 into rate equation 15.87 gives an expression (equation 15.89) for the rate of appearance of $NO_2$ in terms of measurable concentrations.

$$\frac{d[NO_2]}{dt} = 2k_2[N_2O_2][O_2] = \frac{2k_1k_2[NO]^2[O_2]}{k_{-1} + k_2[O_2]} \tag{15.89}$$

This rate equation is not consistent with the experimental rate law 15.85. However, if the $k_2$ step in scheme 15.86 is very slow compared to the equilibrium (i.e. if a pre-equilibrium is established), then $k_2[O_2] \ll k_{-1}$, and it follows that:

$$k_{-1} + k_2[O_2] \approx k_{-1}$$

In this case, equation 15.89 can be rewritten as equation 15.90 and now the rate law is of the same form as the experimental equation 15.85 in which $k_{obs} = \frac{2k_1k_2}{k_{-1}}$.

$$\frac{d[NO_2]}{dt} = \frac{2k_1k_2[NO]^2[O_2]}{k_{-1}} \tag{15.90}$$

---

### 15.14     Radical chain reactions

More about radical chain reactions in Section 25.7

The kinetics of radical chain reactions are not simple but the topic is an important one. Flames, atmospheric reactions and explosions are some examples of chain reactions. In this section we illustrate the complex nature of reactions involving radicals.

#### Linear chain processes

A chain reaction involves *initiation* and *propagation* steps in which radicals are generated, and *termination* steps in which radicals are removed from the system. If there is *no net gain in the number of radicals* formed in the propagation steps, such as in reaction 15.91, the chain is called a *linear chain*.

In a *linear chain* there is no increase in the number of radicals present.

$$CH_3{}^{\bullet} + Cl_2 \longrightarrow CH_3Cl + Cl^{\bullet} \tag{15.91}$$

#### Branched chain processes

In a *branched chain*, the number of radicals increases during the reaction.

In some radical reactions, a propagation step results in a net increase in the number of radicals in the system. Equation 15.92 illustrates the reaction between an $H^{\bullet}$ radical and an $O_2$ molecule in which a hydroxyl radical and an oxygen atom (a diradical) are formed.

$$H^{\bullet} + O_2 \longrightarrow OH^{\bullet} + {}^{\bullet}O^{\bullet} \tag{15.92}$$

**Fig. 15.24** The difference between a linear and branched radical chain: (a) a linear chain in which each propagation step uses *and* generates one radical, and (b) a branched chain in which each step (after the initial radical is formed) generates two radicals.

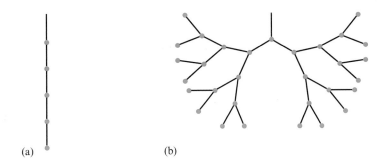

(a)  (b)

Such a process has the effect of *branching* the chain. The increased number of radicals in the system means that more reaction steps involving radicals can now take place. The rate of the reaction increases dramatically and can result in an explosion. Two examples are gas-phase hydrocarbon oxidations (combustion) and the reaction between $H_2$ and $O_2$, initiated by a spark, to give water.

The dramatic difference between a linear and a branched chain is summarized in Figure 15.24.

### The reaction between $H_2$ and $Br_2$

Gaseous $H_2$ and $Br_2$ react at 500 K to form hydrogen bromide; the mechanism is a linear chain reaction. The initiation, propagation and termination steps are given in equations 15.93–15.95 and 15.97. Reaction 15.96 is an *inhibition step* – although a new radical is formed, some of the HBr product is used up. Note that reaction 15.96 is the reverse of propagation step 15.94, and the termination step shown in equation 15.97 is the reverse of the initiation reaction 15.93.

$$Br_2 \xrightarrow{k_1} 2Br^\bullet \qquad\qquad\qquad\qquad \textit{initiation} \quad (15.93)$$

$$Br^\bullet + H_2 \xrightarrow{k_2} HBr + H^\bullet \qquad\qquad \textit{propagation} \quad (15.94)$$

$$H^\bullet + Br_2 \xrightarrow{k_3} HBr + Br^\bullet \qquad\qquad \textit{propagation} \quad (15.95)$$

$$HBr + H^\bullet \xrightarrow{k_{-2}} Br^\bullet + H_2 \qquad\qquad \textit{inhibition} \quad (15.96)$$

$$2Br^\bullet \xrightarrow{k_{-1}} Br_2 \qquad\qquad\qquad\qquad \textit{termination} \quad (15.97)$$

This mechanism is consistent with the observed rate law given in equation 15.98. This is derived in Box 15.2 on the basis of the elementary steps shown above and by using the steady state approximation.

$$\text{Rate} = \frac{d[HBr]}{dt} = \frac{2k_2 k_3 \left(\dfrac{k_1}{k_{-1}}\right)^{\frac{1}{2}} [Br_2]^{\frac{3}{2}} [H_2]}{k_3[Br_2] + k_{-2}[HBr]} \qquad (15.98)$$

Two other elementary steps could have been considered, the initiation step involving the homolytic cleavage of the H−H bond in $H_2$, and the inhibition step involving the reaction between HBr and a $Br^\bullet$ radical. However, the consistency between experimental data and equation 15.98 indicates that neither of these steps is important.

**THEORETICAL AND CHEMICAL BACKGROUND**

## Box 15.2 Use of the steady state approximation in a radical chain reaction

The elementary reactions shown in equations 15.93–15.97 together constitute the linear chain mechanism for the reaction of $H_2$ and $Br_2$ to give HBr. A rate law, which is consistent with experimentally observed kinetic results, can be derived as follows.

HBr is formed in equations 15.94 and 15.95 and consumed in equation 15.96. The rate of formation of HBr is therefore:

$$\frac{d[HBr]}{dt} = k_2[Br^{\bullet}][H_2] + k_3[H^{\bullet}][Br_2] - k_{-2}[H^{\bullet}][HBr] \tag{I}$$

By the steady state approximation:

$$\frac{d[Br^{\bullet}]}{dt} = \frac{d[H^{\bullet}]}{dt} \approx 0 \qquad \frac{d[Br^{\bullet}]}{dt} = 2k_1[Br_2] - k_2[Br^{\bullet}][H_2] + k_3[H^{\bullet}][Br_2] + k_{-2}[H^{\bullet}][HBr] - 2k_{-1}[Br^{\bullet}]^2 \approx 0 \tag{II}$$

and

$$\frac{d[H^{\bullet}]}{dt} = k_2[Br^{\bullet}][H_2] - k_3[H^{\bullet}][Br_2] - k_{-2}[H^{\bullet}][HBr] \approx 0 \tag{III}$$

The addition of equations **(II)** and **(III)** results in the cancellation of three terms, and we have:

$$2k_1[Br_2] - 2k_{-1}[Br^{\bullet}]^2 = 0$$

Therefore:

$$[Br^{\bullet}]^2 = \frac{2k_1[Br_2]}{2k_{-1}} \qquad [Br^{\bullet}] = \left(\frac{k_1}{k_{-1}}\right)^{\frac{1}{2}}[Br_2]^{\frac{1}{2}} \tag{IV}$$

Rearrange equation **(II)** to make $[H^{\bullet}]$ the subject:

$$[H^{\bullet}] = \frac{k_2[Br^{\bullet}][H_2]}{k_3[Br_2] + k_{-2}[HBr]}$$

Substituting in for $[Br^{\bullet}]$ gives:

$$[H^{\bullet}] = \frac{k_2\left(\dfrac{k_1}{k_{-1}}\right)^{\frac{1}{2}}[H_2][Br_2]^{\frac{1}{2}}}{k_3[Br_2] + k_{-2}[HBr]} \tag{V}$$

Now we return to rate equation **(I)** and make substitutions for $[Br^{\bullet}]$ and $[H^{\bullet}]$ using equations **(IV)** and **(V)**. First, note that $[H^{\bullet}]$ is a factor of the last two terms and thus we can write:

$$\frac{d[HBr]}{dt} = k_2[Br^{\bullet}][H_2] + [H^{\bullet}]\{k_3[Br_2] - k_{-2}[HBr]\}$$

$$= k_2\left(\frac{k_1}{k_{-1}}\right)^{\frac{1}{2}}[Br_2]^{\frac{1}{2}}[H_2] + \left(\frac{k_2\left(\dfrac{k_1}{k_{-1}}\right)^{\frac{1}{2}}[H_2][Br_2]^{\frac{1}{2}}}{k_3[Br_2] + k_{-2}[HBr]}\right)\{k_3[Br_2] - k_{-2}[HBr]\}$$

$$= k_2\left(\frac{k_1}{k_{-1}}\right)^{\frac{1}{2}}[Br_2]^{\frac{1}{2}}[H_2]\left(1 + \frac{k_3[Br_2] - k_{-2}[HBr]}{k_3[Br_2] + k_{-2}[HBr]}\right)$$

$$= k_2\left(\frac{k_1}{k_{-1}}\right)^{\frac{1}{2}}[Br_2]^{\frac{1}{2}}[H_2]\left(\frac{k_3[Br_2] + k_{-2}[HBr] + k_3[Br_2] - k_{-2}[HBr]}{k_3[Br_2] + k_{-2}[HBr]}\right)$$

$$= k_2\left(\frac{k_1}{k_{-1}}\right)^{\frac{1}{2}}[Br_2]^{\frac{1}{2}}[H_2]\left(\frac{2k_3[Br_2]}{k_3[Br_2] + k_{-2}[HBr]}\right)$$

$$= \frac{2k_2k_3\left(\dfrac{k_1}{k_{-1}}\right)^{\frac{1}{2}}[Br_2]^{\frac{3}{2}}[H_2]}{k_3[Br_2] + k_{-2}[HBr]}$$

This final rate equation can be written in the simpler form:

$$\frac{d[HBr]}{dt} = \frac{k[Br_2]^{\frac{3}{2}}[H_2]}{[Br_2] + k'[HBr]}$$

where $k = 2k_2\left(\dfrac{k_1}{k_{-1}}\right)^{\frac{1}{2}}$ and $k' = \dfrac{k_{-2}}{k_3}$.

## 15.15    Michaelis–Menten kinetics

*Enzymes* are proteins that act as biological catalysts.

Enzymes are protein molecules (see Chapter 35) that function as catalysts in biological systems. We introduced one example, trypsin, in Figure 15.20. Enzymes interact with specific *substrates*, and the specificity is often rationalized in terms of a 'lock-and-key' model. The enzyme and substrate are complementary in terms of, for example, hydrogen bonding (see Section 21.8) or covalent interactions. This is represented in scheme 15.99 in which enzyme (E) and substrate (S) come together, S is converted to product (P), and finally E and P part company. Equation 15.100 gives the mechanism in a non-pictorial form.

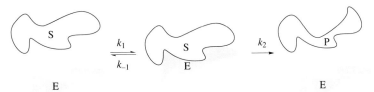

(15.99)

$$E + S \underset{k_{-1}}{\overset{k_1}{\rightleftharpoons}} ES \overset{k_2}{\longrightarrow} E + P \qquad (15.100)$$

In 1913, Leonor Michaelis and Maud Menten observed that:

- at very low substrate concentrations, the rate of enzyme reaction is directly proportional to [S], i.e. first order with respect to [S];
- at very high substrate concentration, the rate is zero order with respect to [S].

These observations are summarized in Figure 15.25. In enzyme kinetics, the rate is often expressed as the *velocity of reaction*, *V*. At the very start of the reaction where [S] is very low, the curve rises steeply and is approximately linear. The flattening of the curve at high [S] corresponds to a constant rate limit being approached. These results are rationalized as follows.

The enzyme–substrate bound state, ES, is an intermediate, and the steady state approximation can be applied to find an expression for [ES] (equation 15.101).

$$\frac{d[ES]}{dt} = k_1[E][S] - k_{-1}[ES] - k_2[ES] = 0$$

$$[ES] = \frac{k_1[E][S]}{k_{-1} + k_2} \qquad (15.101)$$

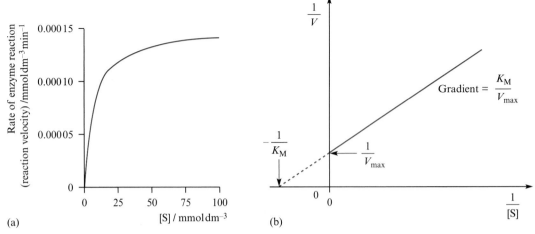

**Fig. 15.25** (a) The dependence of the rate of an enzyme reaction ($V$) on the concentration of substrate (S) illustrated here by experimental data for the urease-catalysed hydrolysis of urea. [Data: K. R. Natarajan (1995) *J. Chem. Educ.*, vol. 72, p. 556]. (b) A plot of $\dfrac{1}{V}$ against $\dfrac{1}{[S]}$ allows the Michaelis constant ($K_M$) and maximum velocity ($V_{max}$) to be determined.

It is not possible to measure [ES], but the *total* concentration of enzyme, $[E]_0$, present can be measured and equals the sum of the concentrations of bound and unbound enzyme:

$$[E]_0 = [E] + [ES] \qquad \text{or} \qquad [E] = [E]_0 - [ES]$$

Substituting for [E] into equation 15.101 gives equation 15.102.

■▷
See problem 15.29

$$[ES] = \frac{k_1([E]_0 - [ES])[S]}{k_{-1} + k_2}$$

$$[ES] = \frac{k_1[E]_0[S]}{k_{-1} + k_2 + k_1[S]} \tag{15.102}$$

The Michaelis constant, $K_M$, is defined in equation 15.103; we see the physical meaning of $K_M$ later.

$$K_M = \frac{k_{-1} + k_2}{k_1} \tag{15.103}$$

Equation 15.102 can be rewritten so that it incorporates the Michaelis constant:

$$[ES] = \frac{[E]_0[S]}{\left(\dfrac{k_{-1} + k_2}{k_1} + [S]\right)} = \frac{[E]_0[S]}{K_M + [S]}$$

According to the proposed mechanism, the rate of the appearance of product P (velocity of reaction) is given by equation 15.104, and substitution of the expression for [ES] gives a rate law in terms of measurable concentrations (equation 15.105).

$$\frac{d[P]}{dt} = V = k_2[ES] \tag{15.104}$$

$$V = \frac{k_2[E]_0[S]}{K_M + [S]} \tag{15.105}$$

The maximum velocity, $V_{max}$, is obtained when [ES] is at a maximum. Ideally, this corresponds to all the enzyme being bound to the substrate, and $[ES] = [E]_0$. From equation 15.104, we can write:

$$V_{max} = k_2[E]_0$$

and substitution of this into equation 15.105 gives the Michaelis–Menten equation in its most commonly quoted form (equation 15.106). The curve in Figure 15.25 is consistent with this equation.

$$V = \frac{V_{max}[S]}{K_M + [S]} \qquad \textit{Michaelis–Menten equation} \quad (15.106)$$

Now we return to the physical meaning of $K_M$. From the Michaelis–Menten equation, we can see that $K_M = [S]$ when $V = V_{max}/2$. (As an exercise, make $K_M$ the subject in equation 15.106, and then substitute in the condition that $V = V_{max}/2$.)

At very low substrate concentrations, $K_M \gg [S]$, and equation 15.106 reduces to equation 15.107, showing that the reaction rate is first order with respect to [S] ($V_{max}$ and $K_M$ are constants).

$$V = \frac{V_{max}[S]}{K_M} \qquad \textit{at very low [S]} \quad (15.107)$$

At very high substrate concentrations, $K_M \ll [S]$, and equation 15.106 reduces to equation 15.108 in which the reaction rate is *independent* of [S] and is a constant value.

$$V = \frac{V_{max}[S]}{[S]} = V_{max} \qquad \textit{at very high [S]} \quad (15.108)$$

In theory, $V_{max}$ can be found from a plot of rate versus [S] (Figure 15.25a); $V_{max}$ is the constant rate corresponding to the horizontal part of the curve. However, in practice, this limit is not usually reached. One of several other approaches uses the Lineweaver–Burk equation (equation 15.109) which is the inverse of the Michaelis–Menten equation.

$$\frac{1}{V} = \frac{K_M + [S]}{V_{max}[S]} = \frac{K_M}{V_{max}}\left(\frac{1}{[S]}\right) + \frac{1}{V_{max}} \qquad \textit{Lineweaver–Burk equation} \quad (15.109)$$

This equation has the form of a straight line graph of $\frac{1}{V}$ plotted against $\frac{1}{[S]}$ (Figure 15.25b). The line has a gradient equal to $\frac{K_M}{V_{max}}$. When $\frac{1}{[S]} = 0$ (i.e. when the line intercepts the $\frac{1}{V}$ axis):

$$\frac{1}{V} = \frac{1}{V_{max}}$$

When $\frac{1}{V} = 0$ (i.e. when the line intercepts the $\frac{1}{[S]}$ axis):

$$0 = \frac{K_M}{V_{max}}\left(\frac{1}{[S]}\right) + \frac{1}{V_{max}}$$

$$\frac{K_M}{V_{max}}\left(\frac{1}{[S]}\right) = -\frac{1}{V_{max}}$$

$$\frac{1}{[S]} = -\frac{1}{K_M}$$

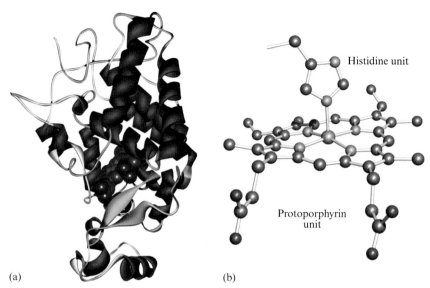

(a)                                                                    (b)

**Fig. 15.26** (a) The structure of the haem-containing enzyme horseradish peroxidase (genetically engineered) determined by X-ray diffraction [G. Berglund *et al.* (2002) *Nature*, vol. 417, p. 463]. In this structure, the enzyme is in a state in which it possesses a 5-coordinate Fe(II) with no bound oxygen-containing species. The protein backbone is shown in a ribbon representation and the iron-containing haem unit (see also Box 5.3) is shown in a space-filling representation. (b) A ball-and-stick representation of the haem unit. The iron centre is bound by a protoporphyrin group through four nitrogen atoms, and by an axially coordinated histidine unit that connects the haem unit to the protein backbone. The point of connection is represented by the 'broken stick'. Colour code: Fe, green; C, grey; N, blue; O, red.

---

**Worked example 15.10**     *Enzymic activity of horseradish peroxidase*

**Horseradish peroxidase (the structure of which is shown in Figure 15.26) is an enzyme found in the root of the horseradish plant. Its function is to catalyse the oxidation of organic substrates that contain aromatic rings (see Chapter 32). The following oxidation reaction can be monitored by using a UV–VIS spectrometer because the product absorbs light of wavelength 450 nm.**

$$\text{(o-phenylenediamine)} \xrightarrow[\text{Horseradish peroxidase (catalyst)}]{\text{H}_2\text{O}_2 \text{ (oxidizing agent)}} \text{(phenazine diamine product)}$$

**The initial velocities, $V$, of the reaction were found to vary with the concentration of the substrate, [S], as follows:**

| $[S] / \text{mmol dm}^{-3}$ | 0.11 | 0.18 | 0.29 | 0.37 | 0.48 |
|---|---|---|---|---|---|
| $10^3 \, V / \text{mmol dm}^{-3} \text{s}^{-1}$ | 1.36 | 1.90 | 2.53 | 2.78 | 3.08 |

**Show that these data are consistent with Michaelis–Menten kinetics. Determine the maximum velocity, and the Michaelis constant for the reaction.**
**[Data from: T. M. Hamilton *et al.* (1999) *J. Chem. Educ.*, vol. 76, p. 642.]**

A plot of $V$ against [S] gives a curve of the type shown in Figure 15.25a. The easiest way to confirm that this curve is consistent with Michaelis–Menten kinetics is to use the method of Lineweaver–Burk. Plot $1/V$ against $1/[S]$ and show that this is linear:

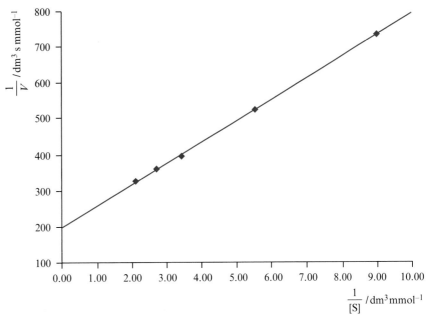

$\dfrac{1}{V_{\max}}$ is found from the intercept on the vertical axis, i.e. the value of $\dfrac{1}{V}$ when $\dfrac{1}{[S]} = 0$.

From the graph: $\qquad \dfrac{1}{V_{\max}} = 195\,\text{dm}^3\,\text{s}\,\text{mmol}^{-1}$

$$V_{\max} = 5.13 \times 10^{-3}\,\text{mmol}\,\text{dm}^{-3}\,\text{s}^{-1}$$

The gradient of the line $= 59.7\,\text{s} = \dfrac{K_{\text{M}}}{V_{\max}}$

$K_{\text{M}} = (59.7\,\text{s}) \times V_{\max}$

$\qquad = (59.7\,\text{s}) \times (5.13 \times 10^{-3}\,\text{mmol}\,\text{dm}^{-3}\,\text{s}^{-1})$

$\qquad = 0.310\,\text{mmol}\,\text{dm}^{-3}$

*Check:* Look at the table of data and confirm that the value of $K_{\text{M}}$ is the substrate concentration when $V = \dfrac{V_{\max}}{2}$.

▣▶
The value of
$195\,\text{dm}^3\,\text{s}\,\text{mmol}^{-1}$, rather
than an approximate value
of $200\,\text{dm}^3\,\text{s}\,\text{mmol}^{-1}$, can be
confirmed by plotting the
graph on a larger scale than
that shown above.

The graph in worked example 15.10 illustrates a general point that, in a Lineweaver–Burk plot, the experimental points tend to be biased towards one part of the line. This can lead to inaccuracies. Another method of treating the data is to use a Eadie–Hofstee plot. Rearrangement of the Michaelis–Menten equation gives a linear relationship between $V/[S]$ and $V$ (equation 15.110):

$$V = \dfrac{V_{\max}[S]}{K_{\text{M}} + [S]} \qquad\qquad \textit{Michaelis–Menten equation}$$

$$VK_{\text{M}} + V[S] = V_{\max}[S]$$

$$VK_{\text{M}} = V_{\max}[S] - V[S]$$

$$\dfrac{V}{[S]} = \dfrac{V_{\max}}{K_{\text{M}}} - V\left(\dfrac{1}{K_{\text{M}}}\right) \qquad \textit{Eadie–Hofstee equation} \qquad (15.110)$$

**Fig. 15.27** Application of the Eadie–Hofstee equation: a plot of $\dfrac{V}{[S]}$ against $V$ for the myosin-catalysed hydrolysis of ATP.

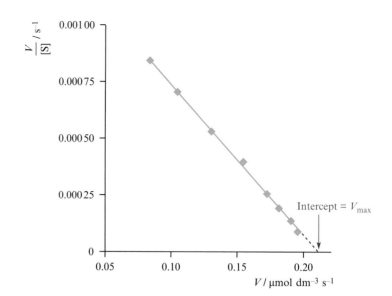

Application of this equation to a set of [S] and $V$ data for the myosin-catalysed hydrolysis of adenosine triphosphate (ATP) gives the linear plot shown in Figure 15.27. The intercept on the $V$ axis when $\dfrac{V}{[S]} = 0$ corresponds to $V_{max}$, and from Figure 15.27, $V_{max} = 0.21\,\mu\text{mol}\,\text{dm}^{-3}\,\text{s}^{-1}$. From equation 15.110, the gradient of the line in Figure 15.27 is $-\dfrac{1}{K_M}$, giving a value of $K_M = 142\,\mu\text{mol}\,\text{dm}^{-3}$. Note that the intercept on the $\dfrac{V}{[S]}$ axis when $V = 0$ corresponds to the value of $\dfrac{V_{max}}{K_M}$.

## SUMMARY

In this chapter, we have discussed some aspects of reaction kinetics and have introduced some important equations. The use of the worked examples has been a critical part of this chapter – learning to deal with experimental data in studies of reaction rates is one of the keys to understanding this topic. You will have noticed that in many cases there is no one way to begin to solve a problem. For example, if you want to confirm that the reaction:

$$\text{A} \longrightarrow \text{B}$$

is first order with respect to A, you could:

- use the half-life method; or
- use the integrated rate equation (is a plot ln [A] against time linear?); or
- measure tangents to a curve of [A] against time, and then see if a plot of the rate (the gradient of a tangent) against [A] is linear.

We have also described how to apply Michaelis–Menten kinetics to enzyme-catalysed reactions.

## Do you know what the following terms mean?

- activation energy
- transition state
- intermediate
- rate of a reaction
- rate-determining step
- rate equation
- rate constant
- order of reaction
- half-life
- pseudo-first order

- integrated form of a rate equation
- radioactive decay
- Boltzmann distribution
- activation energy
- Arrhenius equation
- catalyst
- heterogeneous catalyst
- homogeneous catalyst
- autocatalysis

- elementary step
- molecularity
- reaction mechanism
- unimolecular step
- bimolecular step
- steady state approximation
- chain reaction
- linear chain
- branched chain
- Michaelis–Menten kinetics

## You should be able:

- to explain what is meant by a reaction profile for one- and multi-step reactions
- to explain what is meant by the rate and the rate equation of a reaction
- to distinguish between the order of a reaction with respect to a particular reactant and the overall order
- for a reaction $A \longrightarrow products$, to write down the differential forms of the rate equations for zero, first or second order dependences on A
- to explain why the units of rate constants are not always the same
- for a reaction $A \longrightarrow products$, to write down (or derive) the integrated forms of the rate equations for zero, first or second order dependences on A
- to treat experimental data to determine whether a reaction is zero, first or second order with respect to a particular reactant
- to determine rate constants from experimental data
- to describe how to design an experiment using pseudo-$n$th order conditions and understand when and why these conditions are necessary
- to distinguish between pseudo-$n$th order and overall rate constants and determine them from experimental data
- to derive the differential form of a rate equation for a simple reversible reaction
- to describe the kinetics of radioactive decay
- to discuss why the rate of a reaction depends on temperature

- to write down the Arrhenius equation
- to use the variation of rate constant with temperature to determine the activation energy for a reaction
- to briefly explain why a catalyst alters the rate of a reaction
- to distinguish between homogeneous and heterogeneous catalysts, and give examples of both
- to describe what is meant by autocatalysis
- to discuss the meanings of the terms elementary reaction and unimolecular and bimolecular steps
- to write down rate equations for unimolecular and bimolecular steps
- to relate a series of (two or three) elementary steps to an overall mechanism and use these details to derive an overall rate equation
- to distinguish between a linear and a branched chain mechanism
- to discuss the validity of the steady state approximation and state why it is useful
- to relate the rate equation for a *simple* chain reaction to its mechanism (there is scope here to extend this to the derivation of such rate equations)
- to describe Michaelis–Menten kinetics for an enzyme-catalysed reaction, and know how to treat related experimental data

## PROBLEMS

**15.1** (a) The data plotted in Figure 15.4a refer to a reaction: $A \longrightarrow products$. Determine the rate of this reaction. (b) Figure 15.4b describes a different reaction: $A \longrightarrow products$, in which the rate changes as a function of time. Determine the rates of reaction at times $t_1$ and $t_2$ marked on the graph.

**15.2** For a reaction $A \longrightarrow products$ which is zero order with respect to A, sketch a graph of (a) rate of reaction against time, and (b) [A] against time. Repeat the exercise for reactions that are first and second order with respect to A.

**15.3** For a reaction $A \longrightarrow products$ which is first order with respect to A, show that the units of the rate constant are $s^{-1}$. Similarly, confirm that the units of the second order rate constant are $dm^3 mol^{-1} s^{-1}$.

**15.4** What is meant by each of the following terms: (a) rate of reaction; (b) differential and integrated forms of a rate equation; (c) order of a reaction (both the overall order and the order with respect to a given reactant); (d) rate constant, pseudo-$n$th order rate constant and overall rate constant; (e) activation energy; (f) catalysis and autocatalysis; (g) unimolecular step; (h) bimolecular step?

**15.5** Iron(III) oxidizes iodide according to the following equation:

$$2Fe^{3+} + 2I^- \longrightarrow 2Fe^{2+} + I_2$$

Write down a rate equation for this reaction if doubling the iodide concentration increases the rate by a factor of four, and doubling the $Fe^{3+}$ ion concentration doubles the rate.

**15.6** 2-Chloro-2-methylpropane ($Me_3CCl$) reacts with water according to the equation:

$$Me_3CCl + H_2O \longrightarrow Me_3COH + HCl$$

Changes in concentration of $Me_3CCl$ with time in an experiment carried out at 285 K are as follows:

| Time / s | $10^4[Me_3CCl] / mol\,dm^{-3}$ |
|----------|--------------------------------|
| 20.0 | 2.29 |
| 25.0 | 1.38 |
| 30.0 | 0.90 |
| 34.5 | 0.61 |
| 39.5 | 0.42 |
| 44.5 | 0.26 |

(a) Determine the order of this reaction with respect to $Me_3CCl$.

(b) Calculate a rate constant assuming the rate equation to be of the form:
$Rate = k[Me_3CCl]^n$

(c) Does the rate equation in (b) necessarily mean that only $Me_3CCl$ is involved in the rate-determining step?

(Data from: A. Allen *et al.* (1991) *J. Chem. Educ.*, vol. 68, p. 609.)

**15.7** The kinetics of reactions between alkenes and $I_2$ depend upon the alkene and the solvent. The data below give the results of the reaction of pent-1-ene with $I_2$ in two different solvents; the alkene is always in vast excess. Determine the order with respect to iodine and the pseudo-$n$th order rate constant in each reaction.

Reaction I: Solvent = 1,2-dichloroethane

| Time / s | $[I_2] / mol\,dm^{-3}$ |
|----------|------------------------|
| 0 | 0.0200 |
| 1000 | 0.0152 |
| 2000 | 0.0115 |
| 3000 | 0.0087 |
| 4000 | 0.0066 |
| 5000 | 0.0050 |
| 6000 | 0.0038 |
| 7000 | 0.0029 |
| 8000 | 0.0022 |

Reaction II: Solvent = acetic acid

| Time / s | $[I_2] / mol\,dm^{-3}$ |
|----------|------------------------|
| 0 | 0.0200 |
| 1000 | 0.0163 |
| 2000 | 0.0137 |
| 3000 | 0.0119 |
| 4000 | 0.0105 |
| 5000 | 0.0093 |
| 6000 | 0.0084 |
| 7000 | 0.0077 |
| 8000 | 0.0071 |

(Data from: K. W. Field *et al.* (1987) *J. Chem. Educ.*, vol. 64, p. 269.)

**15.8** Phenyldiazonium chloride is hydrolysed according to the equation:

Phenyldiazonium chloride

The reaction can be followed by the production of dinitrogen. The volume of $N_2$ produced is a direct measure of the amount of starting material consumed in the reaction. Use the following data to deduce the order of the reaction with respect to phenyldiazonium chloride; the reaction has been carried out in aqueous solution. The final volume of $N_2$ collected when the reaction has run to completion is $40.0 \, cm^3$.

| Time / min | Volume $N_2$ / $cm^3$ |
|---|---|
| 0 | 0 |
| 5 | 8.8 |
| 10 | 15.7 |
| 20 | 25.3 |
| 30 | 31.1 |
| 40 | 34.5 |
| 50 | 36.7 |
| 60 | 38.0 |

**15.9** The dimerization of cyclic dienone (**I**) is shown below:

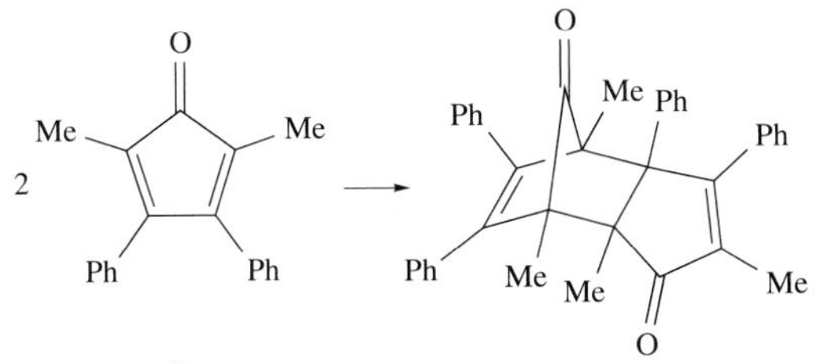

**I**

Me = methyl
Ph = phenyl

Compound **I** absorbs light at 460 nm ($\varepsilon_{max} = 225 \, dm^3 \, mol^{-1} \, cm^{-1}$) and the rate of the reaction has been followed using a UV–VIS spectrometer (path length $= 1$ cm). Use the tabulated absorbance data to determine (a) the order of the reaction with respect to **I** and (b) the rate constant.

| Time / min | Absorbance |
|---|---|
| 0 | 1.20 |
| 5 | 1.10 |
| 10 | 1.00 |
| 20 | 0.81 |
| 30 | 0.72 |
| 35 | 0.67 |
| 45 | 0.59 |
| 55 | 0.54 |
| 75 | 0.45 |
| 120 | 0.33 |

(Data from: H. M. Weiss *et al.* (1990) *J. Chem. Educ.*, vol. 67, p. 707.)

**15.10** Phenolphthalein is an acid–base indicator. Below pH 8 it is colourless and can be represented as $H_2P$, while at higher pH values it is pink and has the form $[P]^{2-}$. In strongly alkaline solution, the colour fades as the following reaction occurs:

$$[P]^{2-} + [OH]^- \longrightarrow [POH]^{3-}$$

An investigation of the kinetics of this reaction using a colorimeter (with a constant path length $= 1$ cm) yielded the following data:

Experiment 1: $[OH^-] = 0.20 \, mol \, dm^{-3}$

| Time / min | Absorbance |
|---|---|
| 0 | 0.560 |
| 1 | 0.454 |
| 2 | 0.361 |
| 3 | 0.292 |
| 4 | 0.228 |

Experiment 2: $[OH^-] = 0.10 \, mol \, dm^{-3}$

| Time / min | Absorbance |
|---|---|
| 0 | 0.560 |
| 2 | 0.449 |
| 4 | 0.361 |
| 6 | 0.301 |
| 8 | 0.247 |

Experiment 3: $[OH^-] = 0.05 \, mol \, dm^{-3}$

| Time / min | Absorbance |
|---|---|
| 0 | 0.560 |
| 3 | 0.468 |
| 6 | 0.415 |
| 9 | 0.368 |
| 12 | 0.313 |

(a) Why was more than one experiment carried out? Why do you think three rather than two experiments were studied?
(b) What is the order of the reaction with respect to $[P]^{2-}$?
(c) What is the order of the reaction with respect to hydroxide ion?
(d) What other data are needed before the overall rate constant can be determined?

(Data from: L. Nicholson (1989) *J. Chem. Educ.*, vol. 66, p. 725.)

**15.11** The $[MnO_4]^-$ ion is an oxidizing agent and the kinetics of the alcohol oxidation:

have been studied. $[MnO_4]^-$ absorbs at 546 nm and the change in absorbance at this wavelength during the reaction was used to monitor the reaction. If [alcohol] $\gg [MnO_4]^-$, use the following data to find the order of the reaction with respect to $[MnO_4]^-$.

| Time / min | Absorbance[a] |
|---|---|
| 1.5 | 0.081 |
| 2.0 | 0.072 |
| 2.5 | 0.062 |
| 3.0 | 0.054 |
| 3.5 | 0.047 |
| 4.0 | 0.040 |
| 4.5 | 0.035 |
| 5.0 | 0.031 |
| 5.5 | 0.027 |

[a]Absorbance values have been corrected to allow for the fact that it does not reach zero when $[MnO_4]^- = 0$.

(Data from: R. D. Crouch (1994) *J. Chem. Educ.*, vol. 71, p. 597.)

**15.12** The rate equation for a reaction of the type: $A + B \longrightarrow$ *products* is of the form:

$$-\frac{d[A]}{dt} = k[A]^x[B]$$

If kinetic runs are carried out with A in vast excess with respect to B, the equation can be rewritten in the form:

$$-\frac{d[A]}{dt} = k_{obs}[B]$$

(a) Use the following kinetic data to determine values of $x$ and $k$.
(b) What is the overall order of the reaction?

| $[A]$ / mol dm$^{-3}$ | $k_{obs}$ / min$^{-1}$ |
|---|---|
| 0.001 | 0.16 |
| 0.002 | 0.31 |
| 0.003 | 0.49 |
| 0.004 | 0.68 |

**15.13** Kinetic data for the reaction:

$$2[MnO_4]^- + 5[C_2O_4]^{2-} + 16H^+ \longrightarrow$$
$$2Mn^{2+} + 10CO_2 + 8H_2O$$

are tabulated below and show the results of an experiment in which the initial concentrations of $[C_2O_4]^{2-}$ and $H^+$ greatly exceed that of $[MnO_4]^-$.

| Time / s | $10^3 [MnO_4^-]$ / mol dm$^{-3}$ |
|---|---|
| 330 | 1.00 |
| 345 | 0.487 |
| 360 | 0.223 |
| 375 | 0.067 |
| 390 | 0.030 |
| 405 | 0.011 |
| 420 | 0.005 |

(a) Which species is the reducing agent in the reaction?
(b) Why were large initial concentrations of two reagents used?
(c) Determine the order of the reaction with respect to $[MnO_4]^-$.

(Data from: B. Miles *et al.* (1990) *J. Chem. Educ.*, vol. 67, p. 269.)

**15.14** The concentration of A during a reaction of the type: $A + 2B \longrightarrow$ *products* changes according to the data in the table.

| Time / min | $[A]$ / mol dm$^{-3}$ |
|---|---|
| 1 | 0.317 |
| 5 | 0.229 |
| 10 | 0.169 |
| 15 | 0.130 |
| 25 | 0.091 |
| 40 | 0.062 |

The reaction is *n*th order with respect to A and zero order with respect to B. (a) Show that $n = 2$.
(b) Write a rate equation for the reaction.
(c) Determine the rate constant.

**15.15** Indium-115 is a β-particle emitter with a half-life of $6 \times 10^{14}$ y. (a) Write an equation for the decay of $^{115}$In. (b) What is the rate constant for the decay?

**15.16** Polonium-211 decays by α-particle emission.
(a) Write an equation for the decay process.
(b) If the half-life of $^{211}$Po is 0.52 s, what is the rate constant for the decay?

**15.17** Americium-241 is an α-particle emitter, and the rate constant for decay is $1.605 \times 10^{-3}$ y$^{-1}$.
(a) Write an equation for the decay of $^{241}$Am.
(b) Determine the half-life of $^{241}$Am.

**15.18** The solvolysis (reaction of a substrate with the solvent) of 2-chloro-2-methylpropane is first order with respect to the substrate. The rate constant, $k$, varies with temperature as shown on the next page. Find $E_a$ for the reaction.

| Temperature / K | $10^5 \, k \, / \, s^{-1}$ |
|---|---|
| 288 | 2.78 |
| 298 | 8.59 |
| 308 | 26.1 |

(Data from: J. A. Duncan and D. J. Pasto (1975) *J. Chem. Educ.*, vol. 52, p. 666.)

**15.19** For the reaction:

$$A \longrightarrow B + C$$

which of the following would you expect to vary over the temperature range 290 to 320 K: (a) reaction rate; (b) rate constant; (c) $E_a$; (d) $[A]_0 - [A]$?

**15.20** Hydrogen peroxide decomposes as follows:

$$2H_2O_2 \longrightarrow 2H_2O + O_2$$

The rate constant for the reaction varies with temperature as follows:

| Temperature / K | $k / s^{-1}$ |
|---|---|
| 295 | $4.93 \times 10^{-4}$ |
| 298 | $6.56 \times 10^{-4}$ |
| 305 | $1.40 \times 10^{-3}$ |
| 310 | $2.36 \times 10^{-3}$ |
| 320 | $6.12 \times 10^{-3}$ |

Determine the activation energy for this reaction.

**15.21** Figure 15.18 showed how the ratio of the rate constants, $\dfrac{k_2}{k_1}$, for a reaction changes as a function of the energy of activation and the result was valid for a rise in temperature from 300 to 310 K. For values of $E_a$ of 10, 30, 50, 70, 90, 110, 130 and 150 kJ mol$^{-1}$, determine $\dfrac{k_2}{k_1}$ for a change in temperature from (a) 320 to 330 K and (b) 420 to 430 K. Plot $\dfrac{k_2}{k_1}$ as a function of $E_a$. What significant differences are there between your graphs and Figure 15.18? Comment critically on the statement that a rise in temperature of 10 K leads to an approximate doubling of the rate of reaction.

**15.22** (a) Explain briefly what you understand by the steady state approximation, and indicate under what circumstances it is valid.

(b) Are the following steps unimolecular or bimolecular?

$$H_2 \longrightarrow 2H^\bullet$$

$$H^\bullet + Br_2 \longrightarrow HBr + Br^\bullet$$

$$2Cl^\bullet \longrightarrow Cl_2$$

$$Cl^\bullet + O_3 \longrightarrow ClO^\bullet + O_2$$

$$O_3 \longrightarrow O_2 + O^\bullet$$

**15.23** Write a rate equation for each of the elementary steps in problem 15.22b.

**15.24** The mechanism of the decomposition: $A \longrightarrow B + C$ can be written as a sequence of the elementary steps:

$$\text{Step I: } A \xrightarrow{k_1} D$$

$$\text{Step II: } D \xrightarrow{k_2} B + C$$

(a) What type of a species is D? (b) For step **I** alone, give an expression that shows the rate of appearance of D. (c) How is your answer to (b) altered if you take into account the sequence of steps **I** and **II**? (d) Use the steady state approximation to show that the rate of formation of B is directly proportional to the concentration of A.

**15.25** The following data were obtained for the urease-catalysed hydrolysis of urea where $V$ is the initial velocity of reaction and $[S]$ is the substrate concentration.

| $[S] \, / \, \text{mmol dm}^{-3}$ | $V \, / \, \text{mmol min}^{-1}$ |
|---|---|
| 2.5 | $3.5 \times 10^{-5}$ |
| 5.0 | $6.2 \times 10^{-5}$ |
| 8.5 | $8.0 \times 10^{-5}$ |
| 10.0 | $8.6 \times 10^{-5}$ |
| 15.0 | $9.4 \times 10^{-5}$ |
| 18.0 | $9.8 \times 10^{-5}$ |
| 40.0 | $1.2 \times 10^{-4}$ |

Show that these data are consistent with Michaelis–Menten kinetics. Determine the maximum velocity and the Michaelis constant for the reaction.

(Data based on: K. R. Natarajan (1995) *J. Chem. Educ.*, vol. 72, p. 556.)

## Additional problems

**15.26** The cobalt(III) complex $[Co(NH_3)_5(NO_2)]^{2+}$ possesses two linkage isomers (see Section 23.5) in which the $[NO_2]^-$ is bonded to $Co^{3+}$ through either the N or O atom. Over a period of days, the O-bonded isomer converts to the N-bonded isomer; let this be represented as:

$$\text{Isomer } \mathbf{I} \longrightarrow \text{Isomer } \mathbf{II}$$

The isomerization can be followed using IR spectroscopy by monitoring the disappearance of an absorption at 1060 cm$^{-1}$. Data are tabulated on the next page and the infinity reading ($t = \infty$) corresponds to the absorbance at 1060 cm$^{-1}$ when isomerization is complete.

| Time / days | Absorbance |
|---|---|
| 0 | 0.480 |
| 0.125 | 0.462 |
| 0.5 | 0.414 |
| 2.0 | 0.349 |
| 4.0 | 0.283 |
| 11.0 | 0.156 |
| 18.0 | 0.118 |
| $\infty$ | 0.108 |

(a) Correct each absorbance value for the fact that the absorption at $1060\,cm^{-1}$ does not completely disappear at $t = \infty$. (b) Plot the corrected absorbance values, $A_{corr}$, against time. What information does the graph give you? (c) Find the order of the reaction with respect to isomer **I**, and determine the rate constant.

(Data from: W. H. Hohman (1974) J. Chem. Educ., vol. 51, p. 553.)

**15.27**  Cerium(IV) ions oxidize $Fe^{2+}$ ions according to the equation:

$$Ce^{4+}(aq) + Fe^{2+}(aq) \longrightarrow Ce^{3+}(aq) + Fe^{3+}(aq)$$

When the reaction is carried out as a redox titration in the presence of the indicator *N*-phenylanthranilic acid, the end point is when the solution becomes purple ($\lambda_{max} = 580\,nm$). If the final solution is left to stand, the purple colour fades to yellow, and the kinetics of the decay can be followed by monitoring the absorbance at $580\,nm$. Typical data are:

| Time / s | Absorbance |
|---|---|
| 30 | 0.65 |
| 60 | 0.60 |
| 90 | 0.55 |
| 120 | 0.52 |
| 180 | 0.46 |
| 240 | 0.40 |
| 300 | 0.32 |
| 360 | 0.28 |
| 420 | 0.24 |
| 480 | 0.21 |

(a) Plot the absorbance against time. How easy is it to state that the decay is *not* zero order? (b) Show that the decay follows first order kinetics. (c) Determine the rate constant.

(Data from: S. K. Mishra *et al.* (1976) J. Chem. Educ., vol. 53, p. 327.)

**15.28**  The overall rate law for the following reaction:

$$2NO(g) + H_2(g) \longrightarrow N_2O(g) + H_2O(g)$$

is:    $-\dfrac{d[H_2]}{dt} = k_{obs}[NO]^2[H_2]$

Confirm that the following mechanism is consistent with the observed rate law:

$$2NO \underset{k_{-1}}{\overset{k_1}{\rightleftharpoons}} N_2O_2$$

$$N_2O_2 + H_2 \xrightarrow{k_2} N_2O + H_2O$$

**15.29**  In the text, the working for the derivation of equation 15.102 was omitted. Show that the substitution for [E] into equation 15.101 gives equation 15.102.

# 16 Equilibria

## Topics

- Le Chatelier's principle
- Equilibrium constants
- Acid–base equilibria
- pH
- Buffers
- Acid–base titrations
- Acid–base indicators

## 16.1 Introduction

Equations for chemical reactions are often written with a single arrow showing reactants going to products, for example, equation 16.1. The assumption is that the reaction *goes to completion*, and at the end of the reaction, no reactants remain.

$$Mg(s) + 2HCl(aq) \longrightarrow MgCl_2(aq) + H_2(g) \tag{16.1}$$

However, many reactions do *not* reach completion. For example, in the reaction of acetic acid with ethanol (equation 16.2), the reaction proceeds to a certain point and gives a mixture of reactants and products. Such a system has both *forward* and *back reactions* occurring concurrently and attains an *equilibrium position*.

$$CH_3CO_2H + CH_3CH_2OH \rightleftharpoons CH_3CO_2CH_2CH_3 + H_2O \tag{16.2}$$

$$\text{Acetic acid} \quad \text{Ethanol} \qquad \text{Ethyl acetate} \quad \text{Water}$$

Carboxylic acids and esters: see Chapter 33

The equilibrium is denoted by the equilibrium sign $\rightleftharpoons$ instead of a full arrow and the *position of the equilibrium*, i.e. to what extent products or reactants predominate, is quantified by the *equilibrium constant*, $K$. All reactions can be considered as equilibria, but in some cases, the forward reaction is so dominant that the back reaction can be ignored.

In this chapter, we consider different types of equilibria, and show how to calculate and use values of $K$. Equilibria are *dynamic* systems with both the forward and back reaction in operation; the rate of the forward reaction equals the rate of the back reaction. In the next section, we discuss how changes in external conditions affect the position of an equilibrium.

A point that often causes confusion is the dual use of square brackets to mean 'concentration of' and to signify a polynuclear ion such as $[NH_4]^+$. In this chapter, therefore, we have, mainly, abandoned the latter usage of square brackets for ions such as $OH^-$, $H_3O^+$, $NH_4^+$ and $CH_3CO_2^-$.

**16.2**

## Le Chatelier's principle

*Le Chatelier's principle* states that when an external change is made to a system in equilibrium, the system will respond so as to oppose the change.

Consider the reaction of dichromate ion with hydroxide ion (equation 16.3).

$$Cr_2O_7^{2-}(aq) + 2OH^-(aq) \rightleftharpoons 2CrO_4^{2-}(aq) + H_2O(l) \qquad (16.3)$$
$$\text{orange} \qquad\qquad\qquad \text{yellow}$$

On reaction with $OH^-$, orange $Cr_2O_7^{2-}$ (dichromate) is converted into yellow $CrO_4^{2-}$ (chromate), but in the presence of acid, $CrO_4^{2-}$ is converted to $Cr_2O_7^{2-}$. By altering the concentration of acid or alkali, the equilibrium can be shifted towards the left- or right-hand side. This is an example of Le Chatelier's principle which states that when an external change is made to a system in equilibrium, the system responds so as to oppose the change. In reaction 16.3, if $OH^-$ is added to the equilibrium mixture, the equilibrium moves towards the right-hand side to consume the excess alkali. Conversely, if acid is added, it neutralizes some $OH^-$ (equation 16.4), and so equilibrium 16.3 moves towards the left-hand side to produce more $OH^-$, thereby restoring equilibrium.

$$OH^-(aq) + H_3O^+(aq) \longrightarrow 2H_2O(l) \qquad (16.4)$$

### Gaseous equilibria: changes in pressure

In an ideal gaseous system at constant volume and temperature, the pressure is proportional to the number of moles of gas present (equation 16.5).

$$P = \frac{nRT}{V} \qquad (16.5)$$

$$P \propto n \qquad\qquad\qquad \text{at constant } T \text{ and } V$$

By Le Chatelier's principle, an increase in external pressure is opposed by a reduction in the number of moles of gas. Conversely, the equilibrium responds to a decrease in pressure by shifting in the direction that increases the number of moles of gas. Consider reaction 16.6.

$$2CO(g) + O_2(g) \rightleftharpoons 2CO_2(g) \qquad (16.6)$$

The forward reaction involves three moles of gaseous reactants going to two moles of gaseous products. In the back reaction, the number of moles of gas increases. In order to encourage the forward reaction, the external pressure could be increased. Equilibrium 16.6 responds to this change by moving towards the right-hand side, decreasing the number of moles of gas and therefore decreasing the pressure. The back reaction is favoured by a decrease in the external pressure.

### Exothermic and endothermic reactions: changes in temperature

Changes to the external temperature of an equilibrium cause it to shift in a direction that opposes the temperature change. For example, an increase in the external temperature encourages an endothermic reaction to occur because this takes in heat and so lowers the external temperature. Consider equilibrium 16.7: the formation of $NH_3$ from its constituent elements in their standard states. The standard enthalpy of formation of $NH_3(g)$ at 298 K is $-45.9\,kJ\,mol^{-1}$. This refers to the forward reaction and to the formation of one mole of $NH_3$.

$$\tfrac{1}{2}N_2(g) + \tfrac{3}{2}H_2(g) \rightleftharpoons NH_3(g)$$

$$\Delta_f H^\circ(NH_3, g, 298\,K) = -45.9\,kJ\,mol^{-1} \qquad (16.7)$$

The data show that the forward reaction is exothermic, and therefore the back reaction is endothermic: if 45.9 kJ are liberated when one mole of $NH_3$ forms from $N_2$ and $H_2$, then 45.9 kJ are required to decompose one mole of $NH_3$ into $N_2$ and $H_2$. If the external temperature is lowered, the equilibrium will move towards the right-hand side because the liberation of heat opposes the external change. If the external temperature is increased, the back reaction is favoured and yields of $NH_3$ are reduced. The effects of changes of temperature and pressure on equilibrium 16.7 have important industrial consequences on the manufacture of $NH_3$ by the Haber process and this is discussed in Box 21.2.

## 16.3        Equilibrium constants

The discussion so far in this chapter has been qualitative. *Quantitative* information about the position of an equilibrium can be obtained from the *equilibrium constant*, *K*, which tells us to what extent the products dominate over the reactants, or vice versa. The *thermodynamic equilibrium constant* is defined in terms of the *activities* of the products and reactants, and before proceeding with the discussion of equilibria, we must define what is meant by activity.

### Activities

> The *activity*, $a_i$, of any pure substance, i, in its standard state is defined as 1; activities are dimensionless (no units).

In dealing with concentrations of solution species, it is common to work in *molarities*: a one molar aqueous solution (1 M or $1\,mol\,dm^{-3}$) contains one mole of solute dissolved in $1\,dm^3$ of *solution*. Another unit may also be used for concentration: a one *molal* aqueous solution contains one mole of solute dissolved in 1 kg of *water* ($1\,mol\,kg^{-1}$).

> A one *molar* aqueous solution contains one mole of solute dissolved in $1\,dm^3$ of *solution* ($1\,mol\,dm^{-3}$).

> A one *molal* aqueous solution contains one mole of solute dissolved in 1 kg of *water* ($1\,mol\,kg^{-1}$).

In the laboratory, we usually deal with *dilute* solutions and we can make some approximations that let us work with concentrations when strictly we should be using the *activities* of solutes. When the concentration of a solute is greater than about $0.1\,mol\,dm^{-3}$, there are significant interactions between the solute molecules or ions. As a result, the *effective* and *real concentrations* are not the same. It is necessary, therefore, to define a new quantity called the *activity*. This is a measure of concentration, but it takes into account interactions between solution species. The relative activity, $a_i$, of a solute i is *dimensionless,* and is related to its molality by equation 16.8 where $\gamma_i$ is the activity coefficient of the solute, and $m_i$ and $m_i{}^o$ are the molality and standard state molality, respectively.

$$a_i = \frac{\gamma_i m_i}{m_i{}^o} \tag{16.8}$$

The standard state of a solute in solution refers to *infinite dilution* at standard molality ($m^o$), 1 bar pressure, where interactions between solute molecules are insignificant. In equation 16.8, $m_i{}^o$ is defined as 1, and so the equation becomes equation 16.9.

$$a_i = \gamma_i m_i \tag{16.9}$$

For *dilute* solutions, we can take $\gamma_i \approx 1$, and from equation 16.9, it follows that the activity is approximately equal to the molality. For most purposes, concentrations are measured in $mol\,dm^{-3}$ rather than molalities, and we can make the approximation that:

> For a **dilute solution**, the activity of the solute is approximately equal to the numerical value of its concentration (measured in $mol\,dm^{-3}$).

For an **ideal gas**, the activity is equal to the numerical value of the partial pressure, the pressure being measured in bar.

The activity, $a_i$, of a gas i is defined in equation 16.10. Note that this definition is similar to that of the activity of a solution species (equation 16.8). Because equation 16.10 contains a ratio of pressures, the units of pressure cancel and the activity is dimensionless.

$$a_i = \gamma_i \frac{P_i}{P_i{}^o} \tag{16.10}$$

where: $\gamma_i$ is the activity coefficient

   $P_i$ is the partial pressure of gas i

   $P_i{}^o$ is the standard state pressure of gas i = 1 bar

In the case of an *ideal gas*, the activity coefficient $\gamma_i = 1$. It therefore follows that the activity of an *ideal gas* is equal to the numerical value of the partial pressure, $P_i$:

$$a_i = \gamma_i \frac{(P_i\,bar)}{(P_i{}^o\,bar)}$$

$$= \frac{(P_i\,bar)}{(1\,bar)} \quad \text{for an ideal gas where } \gamma_i = 1$$

$$= P_i \quad \text{(dimensionless)}$$

The important points to remember from this section are:

- the activity of any *pure substance* in its standard state is defined as 1 (no units);
- the activity of a solute *in a dilute solution* approximately equals the numerical value of its concentration (measured in $mol\,dm^{-3}$);
- the activity of an *ideal gas* approximately equals the numerical value of its partial pressure (measured in bar).

We put these approximations into practice in the following discussion of equilibrium constants. In this and the next chapter, calculations to determine equilibrium constants and reaction quotients involve the use of partial pressures of gases and concentrations of solution species. It is important to remember that a *dimensionless* quantity for concentration or partial pressure is used *because* we approximate activity to the concentration or partial pressure through equation 16.8 or 16.10.

### Thermodynamic equilibrium constants

The **thermodynamic equilibrium constant**, **K**, is defined in terms of the activities of the products and reactants, and is dimensionless.

Consider the general equilibrium 16.11 in which $x$ moles of X and $y$ moles of Y are in equilibrium with $z$ moles of Z.

$$xX + yY \rightleftharpoons zZ \tag{16.11}$$

The thermodynamic equilibrium constant, $K$ (referred to after this point simply as the equilibrium constant), is defined in terms of the activities of the components of the equilibrium as in equation 16.12.

$$K = \frac{(a_Z)^z}{(a_X)^x (a_Y)^y} \tag{16.12}$$

Thus, for reaction 16.13, $K$ is given by the equation shown.

$$2CO(g) + O_2(g) \rightleftharpoons 2CO_2(g) \qquad K = \frac{(a_{CO_2})^2}{(a_{CO})^2(a_{O_2})} \tag{16.13}$$

We now apply the approximation discussed in the previous section and rewrite $K$ (equation 16.14) in terms of the numerical values of the partial pressures of the gases in equilibrium 16.13; we assume that the gases are ideal.

$$K = \frac{(P_{CO_2})^2}{(P_{CO})^2(P_{O_2})} \quad \text{(dimensionless)} \tag{16.14}$$

It is understood that $P$, although indicating a pressure, is included in the expression for $K$ only as a number. Similarly, we can write an expression for equilibrium 16.15 in terms of the activities (equation 16.16) or concentrations (equation 16.17) of the solution species *provided that the solution is dilute*. The expression is simplified by taking the activity of solid Cu as 1.

$$2Cu^+(aq) \rightleftharpoons Cu^{2+}(aq) + Cu(s) \tag{16.15}$$

$$K = \frac{(a_{Cu^{2+}})(a_{Cu})}{(a_{Cu^+})^2} = \frac{(a_{Cu^{2+}})}{(a_{Cu^+})^2} \qquad a_{Cu(s)} = 1 \tag{16.16}$$

$$K = \frac{[Cu^{2+}]}{[Cu^+]^2} \quad \text{(dimensionless)} \tag{16.17}$$

It is understood that the concentrations are included in the expression for $K$ only as numbers.

The next set of worked examples shows how to determine values of thermodynamic equilibrium constants when the composition of an equilibrium mixture is known, and how to use values of $K$ to quantify the position of an equilibrium in terms of its composition.

---

**Worked example 16.1**  |  *Determination of an equilibrium constant*

$N_2O_4$ dissociates according to the following equation:

$$N_2O_4(g) \rightleftharpoons 2NO_2(g)$$

At 350 K, the equilibrium mixture contains 0.13 moles $N_2O_4$ and 0.34 moles $NO_2$. The total pressure is 2.0 bar. Find the value of $K$ at 350 K.

Partial pressures: see worked example 1.12

First find the partial pressures of $NO_2$ and $N_2O_4$.

$$\text{Partial pressure of component X} = \left(\frac{\text{Moles of X}}{\text{Total moles}}\right) \times \text{Total pressure}$$

$$\text{Partial pressure of NO}_2 = \left(\frac{0.34\,\text{mol}}{(0.34+0.13)\,\text{mol}}\right) \times (2.0\,\text{bar}) = 1.45\,\text{bar}$$

$$(1.4\,\text{bar to 2 sig. fig.})$$

$$\text{Partial pressure of N}_2\text{O}_4 = \left(\frac{0.13\,\text{mol}}{(0.34+0.13)\,\text{mol}}\right) \times (2.0\,\text{bar}) = 0.55\,\text{bar}$$

Now we make the assumption that we are dealing with *ideal* gases, and that the activities of $NO_2$ and $N_2O_4$ are equal to the numerical values of the partial pressures.

$$K = \frac{(P_{NO_2})^2}{(P_{N_2O_4})}$$

$$K = \frac{(1.45)^2}{0.55} = 3.8$$

The value of $K$ is dimensionless because only numerical values of $P$ are substituted into the equation.

---

| **Worked example 16.2** | *Determination of an equilibrium constant* |

In an aqueous solution of $CH_3NH_2$, the following equilibrium is established:

$$CH_3NH_2(aq) + H_2O(l) \rightleftharpoons CH_3NH_3^+(aq) + OH^-(aq)$$

Determine $K$ for this equilibrium at 298 K if the initial concentration of $CH_3NH_2$ is $0.10 \, mol \, dm^{-3}$, and at equilibrium, the concentration of $OH^-$ is $0.0066 \, mol \, dm^{-3}$.

Let the total volume be $1 \, dm^{-3}$. Write out the equilibrium composition in terms of moles:

|  | $CH_3NH_2(aq)$ | $+ \; H_2O(l)$ | $\rightleftharpoons CH_3NH_3^+(aq)$ | $+ \; OH^-(aq)$ |
|---|---|---|---|---|
| Moles initially: | 0.10 | excess | 0 | 0 |
| Moles at equilm: | (0.10–0.0066) | excess | 0.0066 | 0.0066 |

Write out an expression for $K$, taking into account that the activity is approximately equal to the numerical value of the concentration in dilute solution, and that the activity of the solvent water is 1.

$$K = \frac{[CH_3NH_3^+][OH^-]}{[CH_3NH_2][H_2O]} = \frac{[CH_3NH_3^+][OH^-]}{[CH_3NH_2]}$$

$$\text{Concentration (in mol dm}^{-3}) = \frac{\text{Number of moles}}{\text{Volume (in dm}^3)}$$

If the volume is $1 \, dm^3$:    Concentration = Number of moles

$$K = \frac{[CH_3NH_3^+][OH^-]}{[CH_3NH_2]}$$

$$K = \frac{(0.0066)(0.0066)}{(0.10 - 0.0066)} = 4.7 \times 10^{-4} \quad \text{(dimensionless)}$$

---

| **Worked example 16.3** | *Finding the composition of an equilibrium mixture* |

At 400 K, the equilibrium constant for the following equilibrium is 40:

$$H_2(g) + I_2(g) \rightleftharpoons 2HI(g)$$

If 2.0 moles of $H_2$ and 2.0 moles of $I_2$ vapour are mixed at 400 K, and the total pressure is 1.0 bar, what is the composition of the equilibrium mixture?

First write down a scheme to show the composition of the equilibrium mixture, using the reaction stoichiometry to work out the ratios of moles of products : reactants.

|  | $H_2(g)$ | $+ \quad I_2(g)$ | $\rightleftharpoons \quad 2HI(g)$ |
|---|---|---|---|
| Moles initially: | 2.0 | 2.0 | 0 |
| Moles at equilibrium: | $2.0 - x$ | $2.0 - x$ | $2x$ |

At equilibrium, the total moles of gas $= (2.0 - x) + (2.0 - x) + 2x = 4.0$

Now find the partial pressures of each gas at equilibrium:

$$\text{Partial pressure of component X} = \left(\frac{\text{Moles of X}}{\text{Total moles}}\right) \times \text{Total pressure}$$

$$\text{Partial pressure of H}_2 = \left(\frac{(2.0 - x)\,\text{mol}}{4.0\,\text{mol}}\right) \times (1.0\,\text{bar}) = \left(\frac{2.0 - x}{4.0}\right) \text{bar}$$

$$\text{Partial pressure of I}_2 = \left(\frac{(2.0 - x)\,\text{mol}}{4.0\,\text{mol}}\right) \times (1.0\,\text{bar}) = \left(\frac{2.0 - x}{4.0}\right) \text{bar}$$

$$\text{Partial pressure of HI} = \left(\frac{2x\,\text{mol}}{4.0\,\text{mol}}\right) \times (1.0\,\text{bar}) = \left(\frac{2x}{4.0}\right) \text{bar}$$

Assume that the system involves *ideal* gases. Write down an expression for $K$ in terms of the numerical values of the partial pressures of the equilibrium components:

$$K = \frac{(P_{HI})^2}{(P_{H_2})(P_{I_2})} = 40$$

$$K = \frac{\left(\dfrac{2x}{4.0}\right)^2}{\left(\dfrac{2.0 - x}{4.0}\right)\left(\dfrac{2.0 - x}{4.0}\right)} = 40$$

$$\frac{(2x)^2}{(2.0 - x)^2} = 40$$

For practice in solving quadratic equations, see Section 5 of the accompanying *Mathematics Tutor*, available via the website www.pearsoned.co.uk/ housecroft

Now solve for $x$:

$$4x^2 = 40(2.0 - x)^2$$

$$4x^2 = 40(4.0 - 4.0x + x^2)$$

$$4x^2 = 160 - 160x + 40x^2$$

$$0 = 160 - 160x + 36x^2$$

The general solution for a quadratic equation of form $ax^2 + bx + c = 0$ is:

$$x = \frac{-b \pm \sqrt{b^2 - 4ac}}{2a}$$

For the quadratic: $36x^2 - 160x + 160 = 0$

$$x = \frac{+160 \pm \sqrt{160^2 - 4(36)(160)}}{2(36)}$$

There are two solutions: $x = 2.9$ or $1.5$, but only $x = 1.5$ is meaningful. (Why?) The equilibrium mixture therefore has the composition:

$$(2.0 - 1.5) = 0.5\,\text{moles H}_2$$

$$(2.0 - 1.5) = 0.5\,\text{moles I}_2$$

$$(2 \times 1.5) = 3.0\,\text{moles HI}$$

*Check:* Total number of moles $= 4.0$

## Values of *K* and extent of reaction

Values of $K$ for different reactions vary from being extremely small (reactants predominate) to extremely large (products predominate). Examples are listed

**Table 16.1** Values of equilibrium constants, $K$, for selected gaseous equilibria at 298 K and 400 K. Values of $\Delta_f H^\circ$ refer to the forward (formation) reactions and are *per mole of compound* formed.

| Equilibrium | $K$ (298 K) | $K$ (400 K) | $\Delta_f H^\circ$(298 K) / kJ mol$^{-1}$ | $\Delta_f H^\circ$(400 K) / kJ mol$^{-1}$ |
|---|---|---|---|---|
| $H_2(g) + Cl_2(g) \rightleftharpoons 2HCl(g)$ | $2.5 \times 10^{33}$ | $1.4 \times 10^{25}$ | $-92.3$ | $-92.6$ |
| $N_2(g) + 3H_2(g) \rightleftharpoons 2NH_3(g)$ | $5.6 \times 10^{5}$ | $36$ | $-45.9$ | $-48.1$ |
| $N_2(g) + 2O_2(g) \rightleftharpoons 2NO_2(g)$ | $4.7 \times 10^{-19}$ | $5.0 \times 10^{-16}$ | $+34.2$ | $+33.6$ |
| $N_2(g) + O_2(g) \rightleftharpoons 2NO(g)$ | $2.0 \times 10^{-31}$ | $2.9 \times 10^{-23}$ | $+91.3$ | $+91.3$ |

in Table 16.1. In the first two equilibria in the table, the large values of $K$ show that products predominate over reactants; in the case of the formation of HCl, the reaction lies so far to the right-hand side that negligible amounts of reactants remain. An increase in temperature from 298 K to 400 K causes a decrease in $K$, i.e. the back reaction is favoured. This is consistent with each of the formations of HCl and NH$_3$ at 298 and 400 K being exothermic (apply Le Chatelier's principle). The last two equilibria listed in Table 16.1 have very small values of $K$, showing that the equilibrium mixtures contain far more reactants than products. In both cases, increasing the temperature from 298 K to 400 K favours the forward reaction, although reactants still predominate at equilibrium. These data are consistent with the formations of NO and NO$_2$ being endothermic.

The dependence of $K$ on temperature is quantified in equation 16.18.

$$\frac{d(\ln K)}{dT} = \frac{\Delta H^\circ}{RT^2} \tag{16.18}$$

Standard integrals: see Table 4 in the accompanying *Mathematics Tutor*, available via the website www.pearsoned.co.uk/housecroft

The integrated form of equation 16.18 (equation 16.19, the derivation of which is given in Section 17.9) shows a linear relationship between $\ln K$ and $1/T$.

$$\ln K = -\frac{\Delta H^\circ}{RT} + c \qquad (c = \text{integration constant}) \tag{16.19}$$

Table 16.2 shows the temperature dependence of $K$ for equilibrium 16.20 over the temperature range 400–700 K.

$$2H_2(g) + O_2(g) \rightleftharpoons 2H_2O(g) \tag{16.20}$$

Figure 16.1 gives a plot of $\ln K$ against $1/T$ using the data from Table 16.2. From equation 16.19, the positive gradient in Figure 16.1 indicates that the forward reaction 16.20 is exothermic (negative $\Delta H^\circ$). An approximate value

**Table 16.2** The temperature dependence of $K$ for the equilibrium:

$$2H_2(g) + O_2(g) \rightleftharpoons 2H_2O(g)$$

| Temperature / K | $K$ |
|---|---|
| 400 | $3.00 \times 10^{58}$ |
| 500 | $5.86 \times 10^{45}$ |
| 600 | $1.83 \times 10^{37}$ |
| 700 | $1.46 \times 10^{31}$ |

**Fig. 16.1** A plot of $\ln K$ against $1/T$ for the equilibrium: $2H_2(g) + O_2(g) \rightleftharpoons 2H_2O(g)$.

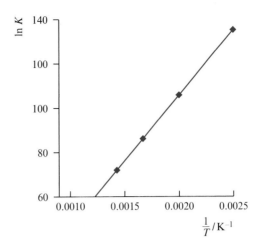

of $\Delta_f H^\circ(H_2O, g)$ can be obtained from the gradient of the line (equation 16.21).

$$\text{Gradient} = 59\,000\,\text{K} = -\frac{\Delta H^\circ}{R} \tag{16.21}$$

$$\Delta H^\circ = -(59\,000\,\text{K}) \times (8.314 \times 10^{-3}\,\text{kJ K}^{-1}\,\text{mol}^{-1})$$

$$= -490\,\text{kJ mol}^{-1} \text{ (per 2 moles of } H_2O(g))$$

The value is only *approximate* because $\Delta H^\circ$ is actually temperature dependent, but the variation over the temperature range 400–700 K is not enormous (about 8 kJ per 2 moles of $H_2O$) and this method of determining enthalpy changes remains a useful experimental approach. The difference in values of $\Delta_f H^\circ(298\,\text{K})$ and $\Delta_f H^\circ(400\,\text{K})$ for selected reactions can be seen from Table 16.1 and these data give an indication of how approximate it is to assume a constant value for $\Delta_f H^\circ$.

**16.4** **Acid–base equilibria**

### Brønsted acids and bases

A *Brønsted acid* is a proton donor, and a *Brønsted base* is a proton acceptor.

A Brønsted acid is a proton donor (e.g. hydrochloric acid) and a Brønsted base is a proton acceptor (e.g. hydroxide ion). Equation 16.22 shows the reaction between a general acid HA and $H_2O$ in which a proton is transferred from the acid to the base: HA acts as a Brønsted acid and $H_2O$ functions as a Brønsted base.

$$\underset{\text{Brønsted acid}}{HA} + \underset{\text{Brønsted base}}{H_2O} \rightleftharpoons H_3O^+ + A^- \tag{16.22}$$

Water can also act as a Brønsted acid and equation 16.23 shows its general reaction with a Brønsted base B.

$$\underset{\text{Brønsted base}}{B} + \underset{\text{Brønsted acid}}{H_2O} \rightleftharpoons BH^+ + OH^- \tag{16.23}$$

If reaction 16.22 goes essentially to completion, then HA is a *strong acid*; an example is HCl which is fully ionized in aqueous solution.

$$HCl(aq) + H_2O(l) \longrightarrow H_3O^+(aq) + Cl^-(aq) \tag{16.24}$$

In aqueous solution, NaOH acts as a strong base because it is fully ionized; the basic species in solution is $OH^-$ (equation 16.25).

$$NaOH(aq) \longrightarrow Na^+(aq) + OH^-(aq) \tag{16.25}$$

## The equilibrium constant, $K_a$

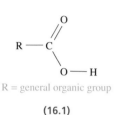

Carboxylic acids: see Chapter 33

In many cases, dissociation of the acid in aqueous solution is *not* complete and an equilibrium is established as shown for aqueous acetic acid in equation 16.26. Acetic acid is an example of a *carboxylic acid* of general formula $RCO_2H$ (**16.1**).

$$CH_3CO_2H(aq) + H_2O(l) \rightleftharpoons H_3O^+(aq) + CH_3CO_2^-(aq) \tag{16.26}$$
Acetic acid                  Acetate ion

R = general organic group

(16.1)

Aqueous acetic acid is a *weak* Brønsted acid and the equilibrium lies to the left-hand side. The equilibrium constant is called the *acid dissociation constant*, $K_a$, and for $CH_3CO_2H$, $K_a = 1.7 \times 10^{-5}$. This is expressed by equation 16.27 where the concentrations are those *at equilibrium*. Remember that the equilibrium constant is strictly expressed in terms of activities, and for the solvent $H_2O$, the activity is 1. Since we approximate the activity (which is dimensionless) to the concentration, and use the numerical values of the concentrations in equation 16.27, the value of $K_a$ is dimensionless.

$$K_a = \frac{[H_3O^+][CH_3CO_2^-]}{[CH_3CO_2H][H_2O]} = \frac{[H_3O^+][CH_3CO_2^-]}{[CH_3CO_2H]} = 1.7 \times 10^{-5} \tag{16.27}$$

Assuming that the contribution to $[H_3O^+]$ from $H_2O$ is negligible, then the stoichiometry of equation 16.26 shows that the concentration of $H_3O^+$ and $CH_3CO_2^-$ ions must be equal and so equation 16.27 can be written in the forms shown in equation 16.28.

$$K_a = \frac{[H_3O^+]^2}{[CH_3CO_2H]} \quad \text{or} \quad K_a = \frac{[CH_3CO_2^-]^2}{[CH_3CO_2H]} \tag{16.28}$$

These equations can be used to calculate the concentration of $H_3O^+$ or $CH_3CO_2^-$ ions as shown in worked example 16.4.

---

**Worked example 16.4**     *The concentration of $H_3O^+$ ions in an aqueous solution of $CH_3CO_2H$ (acetic acid)*

Determine $[H_3O^+]$ in a $1.0 \times 10^{-2} \, mol \, dm^{-3}$ solution of $CH_3CO_2H$ if $K_a = 1.7 \times 10^{-5}$.

The equilibrium in aqueous solution is:

$$CH_3CO_2H(aq) + H_2O(l) \rightleftharpoons H_3O^+(aq) + CH_3CO_2^-(aq)$$

and the acid dissociation constant is given by:

$$K_a = \frac{[H_3O^+][CH_3CO_2^-]}{[CH_3CO_2H]}$$

The stoichiometry of the equilibrium shows that $[CH_3CO_2^-] = [H_3O^+]$, and therefore:

$$K_a = \frac{[H_3O^+]^2}{[CH_3CO_2H]}$$

This equation can be rearranged to find $[H_3O^+]$:

$$[H_3O^+]^2 = K_a \times [CH_3CO_2H]$$

$$[H_3O^+] = \sqrt{K_a \times [CH_3CO_2H]}$$

Strictly, this equation will give us the activity of $H_3O^+$ ions. However, in dilute solution, the activity $\approx$ concentration. The *initial* concentration of $CH_3CO_2H$ is $1.0 \times 10^{-2}$ mol dm$^{-3}$, but we need the *equilibrium* concentration. Since $CH_3CO_2H$ is a *weak* acid, only a very small amount of the acid has dissociated, and therefore we can make the approximation that the concentration of $CH_3CO_2H$ *at equilibrium* is roughly the same as the original concentration:

$$[CH_3CO_2H]_{equilm} \approx [CH_3CO_2H]_{initial} = 1.0 \times 10^{-2}\ \text{mol dm}^{-3}$$

Substituting this value into the equation for $[H_3O^+]$ gives:

$$
\begin{aligned}
[H_3O^+] &= \sqrt{K_a \times [CH_3CO_2H]} \\
&= \sqrt{1.7 \times 10^{-5} \times 1.0 \times 10^{-2}} \\
&= 4.1 \times 10^{-4}\ \text{mol dm}^{-3}
\end{aligned}
$$

*Check* that the assumption that $[CH_3CO_2H]_{equilm} \approx [CH_3CO_2H]_{initial}$ is valid: the initial concentration of acid was $1.0 \times 10^{-2}$ mol dm$^{-3}$, and the equilibrium concentration of $H_3O^+$ is found to be $4.1 \times 10^{-4}$ mol dm$^{-3}$, i.e. the degree of dissociation of the acid is very small:

$$\text{Degree of dissociation} = \frac{(4.1 \times 10^{-4}\ \text{mol dm}^{-3})}{(1.0 \times 10^{-2}\ \text{mol dm}^{-3})} \times 100 = 4.1\%$$

## p$K_a$

For a *weak acid*:
$$pK_a = -\log K_a$$

■▶
$\log = \log_{10}$

Although we often use values of $K_a$, a useful and common way of reporting this number is as the p$K_a$ value. This is defined as the negative logarithm (to the base 10) of the equilibrium constant (equation 16.29).

$$pK_a = -\log K_a \qquad (16.29)$$

If p$K_a$ is known, equation 16.29 can be rearranged to find $K_a$ (equation 16.30).

$$K_a = 10^{-pK_a} \qquad (16.30)$$

From equations 16.29 and 16.30, it follows that as $K_a$ becomes smaller, p$K_a$ becomes larger. For example, $CH_3CO_2H$ is a weaker acid than $HCO_2H$ and the respective values of p$K_a$ and $K_a$ are:

$$CH_3CO_2H \qquad pK_a = 4.77 \qquad K_a = 1.7 \times 10^{-5}$$
$$HCO_2H \qquad pK_a = 3.75 \qquad K_a = 1.8 \times 10^{-4}$$

---

**Worked example 16.5**    *The relationship between $K_a$ and p$K_a$*

p$K_a$ for hydrocyanic acid (HCN) is 9.31. Find the concentration of $H_3O^+$ ions in an aqueous solution of concentration $2.0 \times 10^{-2}$ mol dm$^{-3}$.

The appropriate equilibrium is:

$$HCN(aq) + H_2O(l) \rightleftharpoons H_3O^+(aq) + CN^-(aq)$$

and the equilibrium constant is given by:

$$K_a = \frac{[H_3O^+][CN^-]}{[HCN]}$$

From the stoichiometry of the equilibrium, $[H_3O^+] = [CN^-]$, and so we can write:

$$K_a = \frac{[H_3O^+]^2}{[HCN]}$$

and therefore:

$$[H_3O^+]^2 = K_a \times [HCN]$$

$$[H_3O^+] = \sqrt{K_a \times [HCN]}$$

▣▷
Strictly, this equation gives us the *activity* of $H_3O^+$: see worked example 16.4

$K_a$ can be found from $pK_a$:

$$K_a = 10^{-pK_a} = 10^{-9.31} = 4.9 \times 10^{-10}$$

Using the approximation that $[HCN]_{equilm} \approx [HCN]_{initial} = 2.0 \times 10^{-2} \, mol \, dm^{-3}$, we can calculate $[H_3O^+]$:

$$[H_3O^+] = \sqrt{K_a \times [HCN]}$$

$$= \sqrt{4.9 \times 10^{-10} \times 2.0 \times 10^{-2}}$$

$$= 3.1 \times 10^{-6} \, mol \, dm^{-3}$$

*Check* that the assumption that $[HCN]_{equilm} \approx [HCN]_{initial}$ is valid: the initial concentration of acid was $2.0 \times 10^{-2} \, mol \, dm^{-3}$, and the equilibrium concentration of $H_3O^+$ is found to be $3.1 \times 10^{-6} \, mol \, dm^{-3}$, i.e. the degree of dissociation of the acid is very small.

## Neutralization reactions

A *neutralization* reaction is the reaction of an acid with a base to give a salt and water.

Equations 16.31 and 16.32 summarize general neutralization reactions, and examples are given in reactions 16.33–16.35.

$$H_3O^+(aq) + OH^-(aq) \longrightarrow 2H_2O(l) \tag{16.31}$$

$$\text{Acid} + \text{Base} \longrightarrow \text{Salt} + \text{Water} \tag{16.32}$$

$$\underset{\text{Acid}}{HCl(aq)} + \underset{\text{Base}}{KOH(aq)} \longrightarrow KCl(aq) + H_2O(l) \tag{16.33}$$

$$\underset{\text{Acid}}{H_2SO_4(aq)} + \underset{\text{Base}}{2NaOH(aq)} \longrightarrow Na_2SO_4(aq) + 2H_2O(l) \tag{16.34}$$

$$\underset{\text{Acid}}{CH_3CO_2H(aq)} + \underset{\text{Base}}{NaOH(aq)} \longrightarrow Na[CH_3CO_2](aq) + H_2O(l) \tag{16.35}$$

Even if an acid is weak, the neutralization reaction goes to completion if stoichiometric quantities of reagents are available.

## Conjugate acids and bases

▣▷
Nitrous acid is unstable with respect to disproportionation: see equation 22.72

When a Brønsted acid donates a proton, it forms a species which can, in theory, accept the proton back again. Equation 16.36 shows the dissociation of nitrous acid in water.

$$HNO_2(aq) + H_2O(l) \rightleftharpoons H_3O^+(aq) + NO_2^-(aq) \tag{16.36}$$

**Table 16.3** Conjugate acid and base pairs, and values of $K_a$ for the dilute aqueous solutions of the acids; strong acids are fully dissociated.

| Acid | Formula | $pK_a$ | Conjugate base | Formula |
|---|---|---|---|---|
| Perchloric acid | $HClO_4$ | – | Perchlorate ion | $[ClO_4]^-$ |
| Sulfuric acid | $H_2SO_4$ | – | Hydrogensulfate(1−) ion | $[HSO_4]^-$ |
| Hydrochloric acid | $HCl$ | – | Chloride ion | $Cl^-$ |
| Nitric acid | $HNO_3$ | – | Nitrate ion | $[NO_3]^-$ |
| Sulfurous acid | $H_2SO_3$ | 1.81 | Hydrogensulfite(1−) ion | $[HSO_3]^-$ |
| Hydrogensulfate(1−) ion | $[HSO_4]^-$ | 1.92 | Sulfate ion | $[SO_4]^{2-}$ |
| Phosphoric acid | $H_3PO_4$ | 2.12 | Dihydrogenphosphate(1−) ion | $[H_2PO_4]^-$ |
| Nitrous acid | $HNO_2$ | 3.34 | Nitrite ion | $[NO_2]^-$ |
| Acetic acid | $CH_3CO_2H$ | 4.77 | Acetate ion | $[CH_3CO_2]^-$ |
| Carbonic acid | $H_2CO_3$ | 6.37 | Hydrogencarbonate(1−) ion | $[HCO_3]^-$ |
| Hydrogensulfite(1−) ion | $[HSO_3]^-$ | 6.91 | Sulfite(2−) ion | $[SO_3]^{2-}$ |
| Dihydrogenphosphate(1−) ion | $[H_2PO_4]^-$ | 7.21 | Hydrogenphosphate(2−) ion | $[HPO_4]^{2-}$ |
| Hydrocyanic acid | $HCN$ | 9.31 | Cyanide ion | $[CN]^-$ |
| Hydrogencarbonate(1−) ion | $[HCO_3]^-$ | 10.25 | Carbonate ion | $[CO_3]^{2-}$ |
| Hydrogenphosphate(2−) ion | $[HPO_4]^{2-}$ | 12.67 | Phosphate(3−) ion | $[PO_4]^{3-}$ |

*Increasing strength of acid* (left margin, pointing up)

*Increasing strength of conjugate base* (right margin, pointing down)

In the forward reaction, $HNO_2$ acts as a Brønsted acid and $H_2O$ as a Brønsted base. In the back reaction, the Brønsted acid is $H_3O^+$ and the Brønsted base is $NO_2^-$. The $NO_2^-$ ion is the *conjugate base* of $HNO_2$, and conversely, $HNO_2$ is the *conjugate acid* of $NO_2^-$. Similarly, $H_3O^+$ is the conjugate acid of $H_2O$, and $H_2O$ is the conjugate base of $H_3O^+$. The conjugate acid–base pairs are shown in scheme 16.37.

$$HNO_2(aq) + H_2O(l) \rightleftharpoons NO_2^-(aq) + H_3O^+(aq)$$

Conjugate acid 1    Conjugate base 2    Conjugate base 1    Conjugate acid 2

Conjugate acid–base pair

Conjugate acid–base pair

(16.37)

Table 16.3 lists selected conjugate acid–base pairs and $pK_a$ values for the acids. Note the inverse relationship between the relative strengths of the conjugates acids and bases. We return to this point later in the section (equation 16.54 and the associated discussion).

## The self-ionization of water

The *self-ionization constant*, $K_w$, for water is $1.0 \times 10^{-14}$; $pK_w = 14.00$ (at 298 K).

Water itself is ionized (equation 16.38) to a *very small* extent. The form of the equilibrium sign in equation 16.38 indicates that the reaction lies *far* to the left-hand side. The equilibrium constant for the dissociation is called the *self-ionization constant*, $K_w$ (equation 16.39), and the value reveals how very few ions are present in pure water.

$$2H_2O(l) \rightleftharpoons H_3O^+(aq) + OH^-(aq) \tag{16.38}$$

$$K_w = [H_3O^+][OH^-] = 1.0 \times 10^{-14} \quad \text{(at 298 K)} \tag{16.39}$$

It is convenient to define the term $pK_w$ (equation 16.40) which has the same relationship to $K_w$ that $pK_a$ has to $K_a$.

$$pK_w = -\log K_w = 14.00 \qquad (16.40)$$

We make use of $K_w$ and $pK_w$ later in the section.

### Polybasic acids

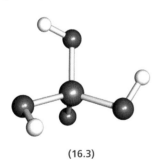

(16.2)

Acids such as HCl, $CH_3CO_2H$, $HNO_2$ and HCN are *monobasic* because they lose only one proton per molecule of acid (e.g. equation 16.36). Some Brønsted acids may lose two, three or more protons. Sulfuric acid (**16.2**) is an example of a *dibasic acid*, and equations 16.41 and 16.42 show the two dissociation steps. The notation $pK_a(2)$ in equation 16.42 shows that the acid dissociation constant refers to the loss of the second proton from the acid, i.e. the second dissociation step of the acid. Similar notation can be used for polybasic acids as is shown in equations 16.45–16.47.

$$H_2SO_4(aq) + H_2O(l) \longrightarrow H_3O^+(aq) + HSO_4^-(aq) \quad \textit{fully dissociated} \quad (16.41)$$

$$HSO_4^-(aq) + H_2O(l) \rightleftharpoons H_3O^+(aq) + SO_4^{2-}(aq) \qquad pK_a(2) = 1.92 \quad (16.42)$$

The first dissociation step goes to completion, because $H_2SO_4$ is a strong acid. The second step is an equilibrium because $HSO_4^-$ is a relatively weak acid. Reactions of sulfuric acid with a base may lead to the formation of a sulfate(2−) or hydrogensulfate(1−) salt as in equations 16.43 and 16.44.

$$H_2SO_4(aq) + KOH(aq) \longrightarrow KHSO_4(aq) + H_2O(l) \qquad (16.43)$$

$$H_2SO_4(aq) + 2KOH(aq) \longrightarrow K_2SO_4(aq) + 2H_2O(l) \qquad (16.44)$$

In general, for a *dibasic acid*:
$K_a(1) > K_a(2)$

It is a general trend that the first dissociation constant, $K_a(1)$, of a dibasic acid is larger than the second, $K_a(2)$, and this is seen in Table 16.3 for $H_2SO_3$ and $H_2CO_3$. Remember that if $K_a(1) > K_a(2)$, then $pK_a(1) < pK_a(2)$. The trend in $K_a$ values extends to tribasic and higher polybasic acids: it is harder to remove a proton from a negatively charged species than from a neutral one. Phosphoric acid, $H_3PO_4$ (**16.3**), is tribasic, and the three dissociation steps are shown in equations 16.45 to 16.47. The acid strength of $H_3PO_4 > H_2PO_4^- > HPO_4^{2-}$.

(16.3)

$$H_3PO_4(aq) + H_2O(l) \rightleftharpoons H_3O^+(aq) + H_2PO_4^-(aq) \quad pK_a(1) = 2.12 \quad (16.45)$$

$$H_2PO_4^-(aq) + H_2O(l) \rightleftharpoons H_3O^+(aq) + HPO_4^{2-}(aq) \quad pK_a(2) = 7.21 \quad (16.46)$$

$$HPO_4^{2-}(aq) + H_2O(l) \rightleftharpoons H_3O^+(aq) + PO_4^{3-}(aq) \qquad pK_a(3) = 12.67 \quad (16.47)$$

### The dissociation constant, $K_b$

In an aqueous solution of a *weak base* **B**, proton transfer is not complete and equation 16.48 shows the general equilibrium and expression for the associated equilibrium constant, $K_b$.

$$B(aq) + H_2O(l) \rightleftharpoons [BH]^+(aq) + OH^-(aq)$$

$$K_b = \frac{[BH^+][OH^-]}{[B][H_2O]} = \frac{[BH^+][OH^-]}{[B]} \qquad (16.48)$$

In the same way that we defined $pK_a$ and $pK_w$, we can also define $pK_b$ (equation 16.49). The weaker the base, the smaller the value of $K_b$ and the larger the value of $pK_b$.

For a *weak base*:
$pK_b = -\log K_b$

$$pK_b = -\log K_b \tag{16.49}$$

---

| **Worked example 16.6** | *Determining the concentration of OH⁻ ions in an aqueous solution of NH₃* |
|---|---|

Calculate the concentration of $OH^-$ ions in an aqueous solution of $NH_3$ of concentration $5.0 \times 10^{-2}\,\mathrm{mol\,dm^{-3}}$ if $pK_b$ for $NH_3 = 4.75$.

The appropriate equilibrium is:

$$NH_3(aq) + H_2O(l) \rightleftharpoons NH_4^+(aq) + OH^-(aq)$$

and the equilibrium constant is given by:

$$K_b = \frac{[NH_4^+][OH^-]}{[NH_3]}$$

where the concentrations are those *at equilibrium*. From the stoichiometry of the equilibrium, it follows that $[NH_4^+] = [OH^-]$ (assuming that the contribution to $[OH^-]$ from $H_2O$ dissociation is negligible) and so we can write:

$$K_b = \frac{[OH^-]^2}{[NH_3]}$$

Rearrangement of the equation gives:

$$[OH^-]^2 = K_b \times [NH_3]$$

$$[OH^-] = \sqrt{K_b \times [NH_3]}$$

Strictly, this equation gives us the activity of OH⁻: see worked example 16.4

We make the assumption that, because $NH_3$ is a *weak* base and is largely undissociated, $[NH_3]_{\text{equilm}} \approx [NH_3]_{\text{initial}} = 5.0 \times 10^{-2}\,\mathrm{mol\,dm^{-3}}$.
We are given the value of $pK_b = 4.75$, and therefore:

$$K_b = 10^{-4.75} = 1.8 \times 10^{-5}$$

Therefore:

$$[OH^-] = \sqrt{1.8 \times 10^{-5} \times 5.0 \times 10^{-2}}$$

$$[OH^-] = 9.5 \times 10^{-4}\,\mathrm{mol\,dm^{-3}}$$

*Check* that the assumption $[NH_3]_{\text{equilm}} \approx [NH_3]_{\text{initial}}$ was valid: the initial concentration of $NH_3$ was $5.0 \times 10^{-2}\,\mathrm{mol\,dm^{-3}}$, and the equilibrium concentration of $OH^-$ is much smaller than this, $9.5 \times 10^{-4}\,\mathrm{mol\,dm^{-3}}$.

---

### The relationship between $K_a$ and $K_b$ for a conjugate acid–base pair

We mentioned earlier that there is an inverse relationship between the relative strengths of a weak acid and its conjugate base (Table 16.3). We can quantify this by considering the relationship between $K_a$ and $K_b$ for a conjugate acid–base pair. Consider the dissociation of acetic acid in aqueous solution (equation 16.50) and the equilibrium established by the acetate ion (i.e. the conjugate base of acetic acid) in aqueous solution (equation 16.51).

$$CH_3CO_2H(aq) + H_2O(l) \rightleftharpoons H_3O^+(aq) + CH_3CO_2^-(aq)$$

Acetic acid

$$K_a = \frac{[H_3O^+][CH_3CO_2^-]}{[CH_3CO_2H]} \quad (16.50)$$

$$CH_3CO_2^-(aq) + H_2O(l) \rightleftharpoons CH_3CO_2H(aq) + OH^-(aq)$$

Acetate ion

$$K_b = \frac{[CH_3CO_2H][OH^-]}{[CH_3CO_2^-]} \quad (16.51)$$

By comparing these equations for $K_a$ and $K_b$ with equation 16.39 for $K_w$, we find that the three equilibrium constants are related by equation 16.52:

$$K_a \times K_b = \frac{[H_3O^+][CH_3CO_2^-]}{[CH_3CO_2H]} \times \frac{[CH_3CO_2H][OH^-]}{[CH_3CO_2^-]}$$

$$= [H_3O^+][OH^-]$$

$$K_a \times K_b = K_w \quad (16.52)$$

If we now take logs of both sides of equation 16.52, we obtain equation 16.53.

$$\log(K_a \times K_b) = \log K_w$$

Therefore:

$$\log K_a + \log K_b = \log K_w \quad (16.53)$$

This equation can be rewritten in terms of $pK_a$, $pK_b$ and $pK_w$ (equation 16.54).

$$-\log K_a - \log K_b = -\log K_w$$

$$pK_a + pK_b = pK_w = 14.00 \quad (16.54)$$

| $pK_a + pK_b = pK_w = 14.00$ |

Thus, for acetic acid for which $pK_a = 4.77$ (Table 16.3), we can use equation 16.54 to determine that $pK_b$ for the acetate ion is $(14.00 - 4.77) = 9.23$. Equation 16.54 allows us to quantify the relationships between the relative strengths of the conjugate acids and bases in Table 16.3. For example, $pK_a$ for $HPO_4^{2-}$ is 12.67, and from equation 16.54, $pK_b$ for the conjugate base $PO_4^{3-}$ is $(14.00 - 12.67) = 1.33$. Thus, while $HPO_4^{2-}$ is the weakest acid listed in Table 16.3, $PO_4^{3-}$ is the strongest base.

## 16.5 Acids and bases in aqueous solution: pH

The concentration of $H_3O^+$ ions in solution is usually denoted by a pH value as defined in equation 16.55.[§] If the pH is known, then the hydrogen ion concentration can be found from equation 16.56.

| $pH = -\log[H_3O^+]$ |

$$pH = -\log[H_3O^+] \quad (16.55)$$

$$[H_3O^+] = 10^{-pH} \quad (16.56)$$

pH meter: see end of
Section 18.8

For strong acids, determining the pH of the solution from the hydrogen ion concentration, or finding the $[H_3O^+]$ from a reading on a pH meter, is straightforward. The method is shown in worked example 16.7. However, for a weak acid, we must take into account that not all the acid is dissociated and this is explained in worked example 16.8.

---

[§] You will also find equation 16.55 written in the form of $pH = -\log[H^+]$; in water, protons combine with $H_2O$ molecules and the predominant species are $H_3O^+$ ions.

**ENVIRONMENT AND BIOLOGY**

## Box 16.1 pH and the gardener

Commercially available kits used to test the pH of soil.
*E. C. Constable.*

Walk around any garden centre, and you will find kits for testing the pH of garden soil. The bedrock and sources of water run-off affect the pH of soil and in the UK, soil pH varies in the approximate range 4.0 to 8.5. Many plants are quite tolerant of the pH conditions in which they grow, and may thrive within pH limits of ±1.0 of an optimum value. Other plants are not at all tolerant, and many agricultural crops grow poorly in acid soils. Rhododendrons and azaleas require acidic soil and grow best where the soil is peat-based. Bilberries and cranberries also need acidic conditions: pH 4.5 is ideal. Anemones (e.g. *Anemone pulsatilla*), clematis and daphne grow best in slightly alkaline conditions (≈pH 7.5) and respond well to the addition of lime. A well-documented example of the effects of pH is the colour of hydrangea flowers: alkaline soil tends to make the flowers pink, while acid soil results in blue flowers, making the plant an indicator of soil pH.

---

**Worked example 16.7**    *pH of an aqueous solution of HCl (hydrochloric acid)*

**What is the pH of an aqueous hydrochloric acid solution of concentration $5.0 \times 10^{-2}\,\text{mol dm}^{-3}$?**

Hydrochloric acid is a strong acid (Table 16.3) and is fully dissociated in water:

$$HCl(aq) + H_2O(l) \longrightarrow H_3O^+(aq) + Cl^-(aq)$$

Therefore, the concentration of $H_3O^+$ ions is the same as the initial concentration of HCl:

$$[H_3O^+] = [HCl] = 5.0 \times 10^{-2}\,\text{mol dm}^{-3}$$

$$pH = -\log[H_3O^+]$$
$$= -\log(5.0 \times 10^{-2})$$
$$= 1.30$$

---

**Worked example 16.8**    *pH of an aqueous solution of HCO$_2$H (formic acid)*

**Formic acid has a p$K_a$ of 3.75. What is the pH of an aqueous solution of concentration $5.0 \times 10^{-3}\,\text{mol dm}^{-3}$?**

Formic acid is a weak acid and is not fully dissociated in aqueous solution:

$$HCO_2H(aq) + H_2O(l) \rightleftharpoons H_3O^+(aq) + HCO_2^-(aq)$$

Therefore, the equilibrium concentration of $H_3O^+$ ions is *not* the same as the initial concentration of $HCO_2H$. To find $[H_3O^+]$, we need the equilibrium constant:

$$K_a = \frac{[H_3O^+][HCO_2^-]}{[HCO_2H]} = 10^{-pK_a}$$

$$\frac{[H_3O^+][HCO_2^-]}{[HCO_2H]} = 10^{-3.75} = 1.8 \times 10^{-4}$$

From the stoichiometry of this equation, $[H_3O^+] = [HCO_2^-]$, therefore:

$$\frac{[H_3O^+]^2}{[HCO_2H]} = 1.8 \times 10^{-4}$$

$$[H_3O^+]^2 = 1.8 \times 10^{-4} \times [HCO_2H]$$

$$[H_3O^+] = \sqrt{1.8 \times 10^{-4} \times [HCO_2H]}$$

where $[HCO_2H]$ is the *equilibrium* concentration of undissociated acid. If only a *very small* amount of the initial acid were dissociated, we could make the assumption that $[HCO_2H]_{equilm} \approx [HCO_2H]_{initial}$. But in this case, *the approximation is not valid* as we can see below. Assume that $[HCO_2H]_{equilm} \approx [HCO_2H]_{initial}$, and so:

$$[H_3O]^+ = \sqrt{1.8 \times 10^{-4} \times 5.0 \times 10^{-3}} = 9.5 \times 10^{-4}\,mol\,dm^{-3}$$

If you compare this concentration to the initial concentration of $HCO_2H$ of $5.0 \times 10^{-3}\,mol\,dm^{-3}$, it shows that 19% of the acid has dissociated, and this is a significant amount:

$$\text{Degree of dissociation} = \frac{(9.5 \times 10^{-4}\,mol\,dm^{-3})}{(5.0 \times 10^{-3}\,mol\,dm^{-3})} \times 100 = 19\%$$

Therefore, we need to solve this problem more rigorously:

$$HCO_2H(aq) \quad + H_2O(l) \quad \rightleftharpoons H_3O^+(aq) \quad + HCO_2^-(aq)$$

$mol\,dm^{-3}$ initially:    $5.0 \times 10^{-3}$      excess    $0$       $0$

$mol\,dm^{-3}$ equilm:    $(5.0 \times 10^{-3}) - x$    excess    $x$       $x$

$$K_a = 1.8 \times 10^{-4} = \frac{[H_3O^+][HCO_2^-]}{[HCO_2H]} = \frac{x^2}{(5.0 \times 10^{-3}) - x}$$

$$x^2 = (1.8 \times 10^{-4})(5.0 \times 10^{-3}) - (1.8 \times 10^{-4})x$$

$$x^2 + (1.8 \times 10^{-4})x - (9.0 \times 10^{-7}) = 0$$

$$x = \frac{-(1.8 \times 10^{-4}) \pm \sqrt{(1.8 \times 10^{-4})^2 + 4(9.0 \times 10^{-7})}}{2}$$

$$= 8.6 \times 10^{-4} \quad \text{or} \quad -1.0 \times 10^{-3}\,mol\,dm^{-3}$$

Clearly, only the positive value is possible and therefore:

$$x = [H_3O]^+ = 8.6 \times 10^{-4}\,mol\,dm^{-3}$$

Now we can find the pH of the solution:

$$pH = -\log[H_3O^+] = -\log(8.6 \times 10^{-4}) = 3.07$$

**Fig. 16.2** The pH scale runs below 0 to 14. The chart gives some examples of aqueous acids and bases covering a range of pH values.

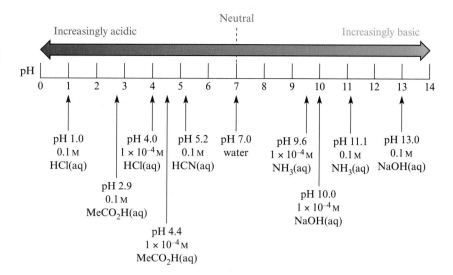

Figure 16.2 illustrates the pH scale with examples of strong and weak acids and bases of different concentrations. The pH of an aqueous solution of a base can be determined by first finding the concentration of hydroxide ions in solution and then using relationship 16.39 to determine the concentration of $H_3O^+$ ions. Worked examples 16.9 and 16.10 consider the cases of a strong (fully dissociated) and weak (partially dissociated) base, respectively.

---

**Worked example 16.9**   *pH of an aqueous solution of NaOH (sodium hydroxide)*

**What is the pH of an aqueous NaOH solution of concentration $5.0 \times 10^{-2}\,\text{mol dm}^{-3}$?**

NaOH is fully dissociated when dissolved in water:

$$NaOH(aq) \longrightarrow Na^+(aq) + OH^-(aq)$$

and the concentration of $OH^-$ ions is equal to the initial concentration of NaOH:

$$[OH^-] = [NaOH] = 5.0 \times 10^{-2}\,\text{mol dm}^{-3}$$

The pH of the solution is given by the equation:

$$pH = -\log[H_3O^+]$$

and so we need to relate $[H_3O^+]$ to the known value of $[OH^-]$:

$$K_w = [H_3O^+][OH^-] = 1.0 \times 10^{-14}$$

$$[H_3O^+] = \frac{1.0 \times 10^{-14}}{[OH^-]} = \frac{1.0 \times 10^{-14}}{5.0 \times 10^{-2}} = 2.0 \times 10^{-13}\,\text{mol dm}^{-3}$$

$$pH = -\log[H_3O^+] = -\log(2.0 \times 10^{-13}) = 12.70$$

---

| Worked example 16.10 | *pH of an aqueous solution of CH$_3$NH$_2$ (methylamine)* |
|---|---|

**CH$_3$NH$_2$ has a p$K_b$ value of 3.34. What is the pH of an aqueous solution of CH$_3$NH$_2$ of concentration 0.10 mol dm$^{-3}$?**

Methylamine is a weak base and is partially dissociated in aqueous solution:

$$CH_3NH_2(aq) + H_2O(l) \rightleftharpoons CH_3NH_3^+(aq) + OH^-(aq)$$

The equilibrium concentration of OH$^-$ ions is *not* the same as the initial concentration of CH$_3$NH$_2$. To find the concentration of OH$^-$ ions, we require the equilibrium constant:

$$K_b = \frac{[CH_3NH_3^+][OH^-]}{[CH_3NH_2]}$$

From the stoichiometry of the equilibrium, $[CH_3NH_3^+] = [OH^-]$, and therefore:

$$K_b = \frac{[OH^-]^2}{[CH_3NH_2]}$$

where these are *equilibrium* concentrations. Make the assumption that $[CH_3NH_2]_{equilm} \approx [CH_3NH_2]_{initial} = 0.10 \text{ mol dm}^{-3}$, and therefore:

$$[OH^-]^2 = K_b \times [CH_3NH_2] = K_b \times 0.10$$

$$[OH^-] = \sqrt{K_b \times 0.10}$$

We are given a value of p$K_b$, and so can find $K_b$:

$$K_b = 10^{-pK_b}$$
$$= 10^{-3.34}$$
$$= 4.6 \times 10^{-4}$$

Therefore:

$$[OH^-] = \sqrt{4.6 \times 10^{-4} \times 0.10}$$
$$= 6.8 \times 10^{-3} \text{ mol dm}^{-3}$$

*Check* that the assumption $[CH_3NH_2]_{equilm} \approx [CH_3NH_2]_{initial}$ was valid: the initial concentration of CH$_3$NH$_2$ was 0.10 mol dm$^{-3}$ and the equilibrium concentration of OH$^-$ is $6.8 \times 10^{-3}$ mol dm$^{-3}$. The degree of dissociation is therefore 6.8%, and this is at the limits of acceptability for the approximation. As an exercise, solve the problem rigorously (see worked example 16.8) and show that the accurate value of $[OH^-] = 6.6 \times 10^{-3}$ mol dm$^{-3}$.

Taking the value of $[OH^-] = 6.8 \times 10^{-3}$ mol dm$^{-3}$, we now have to find $[H_3O^+]$, and this is related to $[OH^-]$ as follows:

$$K_w = [H_3O^+][OH^-] = 1.0 \times 10^{-14} \quad \text{(at 298 K)}$$

$$[H_3O^+] = \frac{1.0 \times 10^{-14}}{[OH^-]} = \frac{1.0 \times 10^{-14}}{6.8 \times 10^{-3}} = 1.5 \times 10^{-12} \text{ mol dm}^{-3}$$

$$pH = -\log[H_3O^+]$$
$$= -\log(1.5 \times 10^{-12})$$
$$= 11.82$$

**16.6**    **Acids and bases in aqueous solution: speciation**

### Dibasic acids

In the last section, we considered the pH of aqueous solutions of *monobasic* acids. Many acids are *polybasic* (see, for example, equations 16.45–16.47 and Table 16.3). To illustrate how to determine the pH of solutions of such acids, we consider the dibasic acids listed in Table 16.4. We have already seen that the first dissociation step lies further to the right-hand side than the second step, i.e. $K_a(1) > K_a(2)$ or $pK_a(1) < pK_a(2)$. However, there is no general way in which to treat the relative importance of the two dissociation steps in terms of estimating the concentration of $H_3O^+$ ions in solution. First, consider $H_2S$ (equations 16.57 and 16.58).

$$H_2S(aq) + H_2O(l) \rightleftharpoons H_3O^+(aq) + HS^-(aq) \qquad (16.57)$$

$$HS^-(aq) + H_2O(l) \rightleftharpoons H_3O^+(aq) + S^{2-}(aq) \qquad (16.58)$$

From Table 16.4, $pK_a(2) \gg pK_a(1)$, and most of the $H_3O^+$ in an aqueous solution of $H_2S$ arises from equilibrium 16.57. In determining the pH of an aqueous solution of $H_2S$, we can effectively ignore equilibrium 16.58 and treat the system as though it were a monobasic acid. Thus, for an $H_2S$ solution of concentration $0.010 \, mol \, dm^{-3}$, the pH is found as follows:

$$K_a(1) = 10^{-7.04} = 9.1 \times 10^{-8} = \frac{[H_3O^+][HS^-]}{[H_2S]} = \frac{[H_3O^+]^2}{[H_2S]}$$

Since $K_a(1)$ is small, $[H_2S]_{equilm} \approx [H_2S]_{initial}$, and therefore:

$$[H_3O^+]^2 = 9.1 \times 10^{-8} \times 0.010$$

$$[H_3O^+] = \sqrt{9.1 \times 10^{-8} \times 0.010} = 3.0 \times 10^{-5} \, mol \, dm^{-3}$$

$$pH = -\log[H_3O^+] = 4.52$$

Now consider $H_2SO_4$ for which the scenario is very different from that of $H_2S$. The first dissociation step for dilute aqueous $H_2SO_4$ goes to completion (equation 16.41). The concentration of $H_3O^+$ ions formed in this step equals the initial concentration of acid. For the second dissociation step, $K_a(2) = 1.2 \times 10^{-2}$ and therefore the amount of $H_3O^+$ ions produced in this step *cannot* be ignored (see problem 16.21). For carbonic and oxalic acids (Table 16.4), the values of $K_a(1)$ are $\approx 10^3$ or $10^4$ times greater than the corresponding value of $K_a(2)$ and in each case, the *total* concentration of $H_3O^+$ can be approximated to that due to the first dissociation step.

The examples in Table 16.4 illustrate that each polybasic acid must be treated as an individual case, although in many instances, the $H_3O^+$ ions produced in the first dissociation step are the major contribution to the total concentration of $H_3O^+$.

**Table 16.4** $pK_a$ values for selected dibasic acids.

| Compound | Formula | $pK_a(1)$ | $pK_a(2)$ |
|---|---|---|---|
| Hydrogen sulfide | $H_2S$ | 7.04 | 19 |
| Carbonic acid | $H_2CO_3$ | 6.37 | 10.25 |
| Oxalic acid (see **16.4**) | $HO_2CCO_2H$ or $H_2C_2O_4$ | 1.23 | 4.19 |
| Sulfuric acid | $H_2SO_4$ | – | 1.92 |

### Speciation in aqueous solutions of acids: the effect of pH

Consider equilibrium 16.59 where HA is a weak acid. How is the equilibrium position affected by a change in pH? The addition of acid shifts the equilibrium to the left-hand side, while the addition of alkali shifts it to the right-hand side (Le Chatelier's principle).

$$HA(aq) + H_2O(l) \rightleftharpoons H_3O^+(aq) + A^-(aq) \qquad (16.59)$$

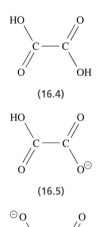

**(16.4)**

**(16.5)**

**(16.6)**

The situation for polybasic acids is more complicated, with the positions of two or more equilibria being altered by a change in pH. We can quantify what happens using speciation curves of the type shown in Figure 16.3 for oxalic acid. Oxalic acid, **16.4**, is a *dicarboxylic acid*, and has two ionizable protons: $pK_a(1) = 1.23$ and $pK_a(2) = 4.19$. The first dissociation step gives **16.5**, and the second step produces **16.6**. At very low pH, the dominant species is $HO_2CCO_2H$, and as the pH is gradually increased, first $HO_2CCO_2^-$ becomes dominant, and then $O_2CCO_2^{2-}$. The concentrations of each species present at a given pH can be determined as follows. Consider a general dibasic acid $H_2A$ (equilibria 16.60 and 16.61) for which the acid dissociation constants are $K_a(1)$ and $K_a(2)$. As you work through the calculations below, remember that, if the pH of the overall system is varied, equilibria 16.60 and 16.61 are interdependent.

$$H_2A(aq) + H_2O(l) \rightleftharpoons H_3O^+(aq) + HA^-(aq)$$

$$K_a(1) = \frac{[H_3O^+][HA^-]}{[H_2A]} \qquad (16.60)$$

$$HA^-(aq) + H_2O(l) \rightleftharpoons H_3O^+(aq) + A^{2-}(aq)$$

$$K_a(2) = \frac{[H_3O^+][A^{2-}]}{[HA^-]} \qquad (16.61)$$

The *total* concentration of $H_2A$, $HA^-$ and $A^{2-}$ *at equilibrium* equals the initial concentration of $H_2A$:

$$[H_2A]_{initial} = [H_2A]_{equilm} + [HA^-]_{equilm} + [A^{2-}]_{equilm}$$

**Fig. 16.3** Speciation curves for aqueous oxalic acid as a function of pH. The red curve corresponds to undissociated $HO_2CCO_2H$ in solution, the green curve to $HO_2CCO_2^-$ and the blue curve to $^-O_2CCO_2^-$ (i.e. $O_2CCO_2^{2-}$).

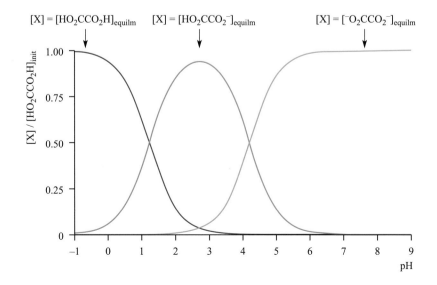

For brevity, we shall write this in the form of equation 16.62, where it is understood that the concentrations on the right-hand side refer to those at equilibrium.

$$[H_2A]_{initial} = [H_2A] + [HA^-] + [A^{2-}] \qquad (16.62)$$

By combining equations 16.60–16.62, we can derive equations for each solution species as a function of $K_a(1)$, $K_a(2)$ and $[H_3O^+]$. To find $[A^{2-}]$:

$$[H_2A]_{initial} = [H_2A] + [HA^-] + [A^{2-}]$$

$$= \left( \frac{[H_3O^+][HA^-]}{K_a(1)} \right) + [HA^-] + [A^{2-}]$$

$$= [HA^-]\left( \frac{[H_3O^+]}{K_a(1)} + 1 \right) + [A^{2-}]$$

$$= \left( \frac{[H_3O^+][A^{2-}]}{K_a(2)} \right)\left( \frac{[H_3O^+]}{K_a(1)} + 1 \right) + [A^{2-}]$$

$$= [A^{2-}]\left\{ \left( \frac{[H_3O^+]}{K_a(2)} \right)\left( \frac{[H_3O^+]}{K_a(1)} + 1 \right) + 1 \right\}$$

$$= [A^{2-}]\left\{ \left( \frac{[H_3O^+]}{K_a(2)} \right)\left( \frac{[H_3O^+] + K_a(1)}{K_a(1)} \right) + 1 \right\}$$

$$= [A^{2-}]\left\{ \left( \frac{[H_3O^+]^2 + K_a(1)[H_3O^+]}{K_a(1)K_a(2)} \right) + 1 \right\}$$

$$= [A^{2-}]\left\{ \frac{[H_3O^+]^2 + K_a(1)[H_3O^+] + K_a(1)K_a(2)}{K_a(1)K_a(2)} \right\}$$

Therefore, as a fraction of the initial concentration of $H_2A$, the concentration of $[A^{2-}]$ is:

$$\frac{[A^{2-}]}{[H_2A]_{initial}} = \frac{K_a(1)K_a(2)}{[H_3O^+]^2 + K_a(1)[H_3O^+] + K_a(1)K_a(2)}$$

We can similarly derive two further expressions (see problem 16.22) to find $\dfrac{[HA^-]}{[H_2A]_{initial}}$ and $\dfrac{[H_2A]}{[H_2A]_{initial}}$:

$$\frac{[HA^-]}{[H_2A]_{initial}} = \frac{K_a(1)[H_3O^+]}{[H_3O^+]^2 + K_a(1)[H_3O^+] + K_a(1)K_a(2)}$$

$$\frac{[H_2A]}{[H_2A]_{initial}} = \frac{[H_3O^+]^2}{[H_3O^+]^2 + K_a(1)[H_3O^+] + K_a(1)K_a(2)}$$

The curves in Figure 16.3 are constructed using the above expressions for $\dfrac{[H_2A]}{[H_2A]_{initial}}$, $\dfrac{[HA^-]}{[H_2A]_{initial}}$ and $\dfrac{[A^{2-}]}{[H_2A]_{initial}}$ for $H_2A = HO_2CCO_2H$ (**16.4**) and $pK_a(1) = 1.23$ and $pK_a(2) = 4.19$. Points to note from the *speciation curves* in Figure 16.3 are:

- the dominance of $HO_2CCO_2H$ under highly acidic conditions;
- the dominance of $O_2CCO_2^{2-}$ below pH 5.5;
- the growth and decay of $HO_2CCO_2^-$ as the pH is gradually raised;
- the crossing point of the $HO_2CCO_2H$ and $HO_2CCO_2^-$ curves is at $pH = pK_a(1) = 1.23$;
- the crossing point of the $HO_2CCO_2^-$ and $O_2CCO_2^{2-}$ curves is at $pH = pK_a(2) = 4.19$.

### What is a buffering effect?

A solution that consists of a weak acid and its salt, or a weak base and its salt, possesses a *buffering effect* and can withstand the addition of small amounts of acid or base without a significant change in pH. The *maximum buffering capacity* is obtained when the concentrations of the weak acid and its salt (or the weak base and its salt) are equal.

A solution that possesses a *buffering effect* is one to which small amounts of acid or base can be added without causing a significant change in pH. A buffer solution usually consists of an aqueous solution of a weak acid and a salt of that acid (e.g. acetic acid and sodium acetate) or of a weak base and its salt. To provide the *maximum buffering capacity*, the relative concentrations of the weak acid and its salt (or the weak base and its salt) must be 1:1; we return to the reasoning behind this later in the section. Buffers are extremely important in living organisms where a constant pH is essential; human blood has a pH of 7.4 and is naturally buffered (see Box 16.2). Consider a solution that is made up of aqueous acetic acid and sodium acetate. Equations 16.63 and 16.64 show that $CH_3CO_2H$ is *partially* dissociated while its sodium salt is *fully* dissociated.

$$CH_3CO_2H(aq) + H_2O(l) \rightleftharpoons H_3O^+(aq) + CH_3CO_2^-(aq) \qquad (16.63)$$

Acetic acid        Acetate ion

$$Na[CH_3CO_2](s) \xrightarrow{\text{Dissolve in water}} Na^+(aq) + CH_3CO_2^-(aq) \qquad (16.64)$$

Sodium acetate        Acetate ion

The combination of these two solutions produces a solution with a buffering effect. Because $CH_3CO_2H$ is a weak acid, most of it is in the undissociated form. The amount of $CH_3CO_2^-$ *from the acid* is negligible with respect to that originating from sodium acetate. If a small amount of acid, $H_3O^+$, is added, it will be consumed by $CH_3CO_2^-$ and form more $CH_3CO_2H$ in the back reaction of equilibrium 16.63. Any $OH^-$ added to the solution will be neutralized by $H_3O^+$, and equilibrium 16.63 will shift to the right-hand side. We now consider why these shifts in the equilibrium do not cause significant changes to the solution pH, provided that appropriate relative concentrations of acetic acid and sodium acetate are used.

### Determining the pH of a buffer solution

Consider $250 \, cm^3$ of a solution of $0.10 \, mol \, dm^{-3}$ $CH_3CO_2H$. From equation 16.63 and the equilibrium constant ($pK_a = 4.77$), we know that the concentrations of $H_3O^+$ and $CH_3CO_2^-$ ions are equal and small. Now let us add $250 \, cm^3$ $0.10 \, mol \, dm^{-3}$ $Na[CH_3CO_2]$ to the same solution. The sodium salt is fully dissociated (equation 16.64) and so the concentration of $CH_3CO_2^-$ ions originating from the salt far exceeds that of $CH_3CO_2^-$ arising from the acid. Thus, we can make the approximation that the *total* concentration of $CH_3CO_2^-$ in solution equals the initial concentration of $Na[CH_3CO_2]$. Let this be designated [base] because $CH_3CO_2^-$ acts as a base in the buffer solution. Equation 16.65 gives the expression for $K_a$ for $CH_3CO_2H$.

$$K_a = \frac{[H_3O^+][CH_3CO_2^-]}{[CH_3CO_2H]} \qquad (16.65)$$

We can now apply the following approximations:

$$[CH_3CO_2^-] \approx [\text{base}]$$

$$[CH_3CO_2H]_{\text{equilm}} \approx [CH_3CO_2H]_{\text{initial}} = [\text{acid}]$$

Substitution of these terms into equation 16.65 leads to equation 16.66 which applies to *a solution of a weak acid and its salt*.

$$K_a = \frac{[H_3O^+][\text{base}]}{[\text{acid}]} \qquad (16.66)$$

If we now take negative logarithms of both sides of the equation, we derive a relationship between pH, $pK_a$ and the initial concentrations of acid and base (equation 16.67).

■▶
For practice working with log functions, see Section 4 in the accompanying *Mathematics Tutor*, available via the website www.pearsoned.co.uk/ housecroft

$$-\log K_a = -\log\left(\frac{[H_3O^+][\text{base}]}{[\text{acid}]}\right)$$

$$= -\log[H_3O^+] - \log\frac{[\text{base}]}{[\text{acid}]}$$

$$pK_a = pH - \log\frac{[\text{base}]}{[\text{acid}]}$$

$$pH = pK_a + \log\frac{[\text{base}]}{[\text{acid}]} \qquad \textit{Henderson–Hasselbalch equation} \quad (16.67)$$

This is the *Henderson–Hasselbalch equation* and can be applied to solutions consisting of a weak acid and its salt, or a weak base and its salt, as is shown in worked examples 16.11 and 16.12.

---

For a solution consisting of a weak acid and a salt of the weak acid:

$$pH = pK_a + \log\frac{[\text{base}]}{[\text{acid}]}$$

where $pK_a$ refers to the weak acid.

For a solution consisting of a weak base and a salt of the weak base:

$$pH = pK_a + \log\frac{[\text{base}]}{[\text{acid}]}$$

where $pK_a$ refers to the conjugate acid of the weak base.

---

**Worked example 16.11**    *pH of a buffer solution consisting of a weak acid and its salt*

Determine the pH of a buffer solution that is $0.050\,\text{mol dm}^{-3}$ with respect to both acetic acid ($pK_a = 4.77$) and sodium acetate.

Equation needed:

$$pH = pK_a + \log\frac{[\text{base}]}{[\text{acid}]}$$

The initial concentrations of both salt and acid are $0.050\,\text{mol dm}^{-3}$. The salt provides $CH_3CO_2{}^-$ which is the base in the buffer solution.

$$pH = 4.77 + \log\frac{(0.050\,\text{mol dm}^{-3})}{(0.050\,\text{mol dm}^{-3})}$$

$$= 4.77$$

This answer illustrates that when the concentrations of acid and base are equal, the pH of a buffer solution is the same as the $pK_a$ of the acid.

**ENVIRONMENT AND BIOLOGY**

## Box 16.2 Buffering human blood

Maintaining a constant pH for a particular human body fluid (blood, intracellular fluids or urine) is critical. We focus here on the ways in which the pH of blood plasma is maintained close to a value of 7.4. The two most important buffering systems in blood involve carbonic acid/hydrogen carbonate, and haemoglobin (see Box 5.3); other blood plasma proteins play a minor role. Whereas phosphate buffer systems are important in controlling the pH in intracellular fluids, they do not have a major function as buffers in blood.

Human metabolism produces $CO_2$ in body tissues. The $CO_2$ diffuses through the capillary walls and into blood plasma. Carbon dioxide is hydrated in aqueous media:

$$CO_2(aq) + H_2O(l) \rightleftharpoons H_2CO_3(aq) \rightleftharpoons$$
$$H^+(aq) + [HCO_3]^-(aq)$$

but the reaction is slow (the rate constant, $k = 0.037 \, s^{-1}$). In aqueous solution, carbonic acid is partially ionized to give protons and hydrogen carbonate ions:

$$H_2CO_3(aq) \rightleftharpoons H^+(aq) + [HCO_3]^-(aq)$$

The $pK_a(1)$ value for $H_2CO_3$ is about 3.6. The $pK_a(1)$ value of 6.37 that is usually quoted refers to the total

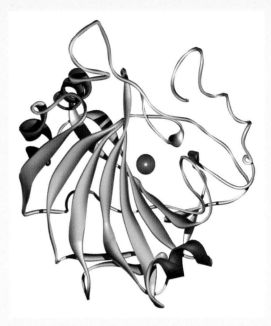

A ribbon representation of the protein chain in the enzyme carbonic anhydrase. The active site is a zinc(II) centre which is shown as the green sphere. [Data: A. E. Eriksson *et al.* (1988) *Proteins*, vol. 4, p. 274.]

amount of $CO_2$ in solution rather than the ionization of carbonic acid; i.e. it refers to the process:

$$CO_2(aq) + H_2O(l) \rightleftharpoons H^+(aq) + [HCO_3]^-(aq)$$

Once $CO_2$ enters the red blood cells (*erythrocytes*), its hydration to $[HCO_3]^-$ is catalysed by *carbonic anhydrase*. This is a zinc(II)-containing enzyme, the structure of which is shown above.

In the presence of carbonic anhydrase, the rate of $CO_2$ hydration increases by a factor of $\approx 10^7$. As a result, in red blood cells, $CO_2$ is transported in the form of $[HCO_3]^-$. As shown above, the hydration of $CO_2$ also produces $H^+$ and, without the action of a buffer, the pH would decrease. Red blood cells also contain haemoglobin. Haemoglobin is a haem-containing protein (see Box 5.3) and has a tetrameric structure. Each haem unit can bind one $O_2$ molecule at the iron centre. As $O_2$ binds, the conformation of the protein chain changes. Protein chains are composed of amino acid residues (see Section 35.6 and Table 35.1), some of which have substituents which may be protonated or deprotonated. In the buffering action of haemoglobin, the role of one particular histidine residue (His-146) is crucial. The imidazole substituent (see

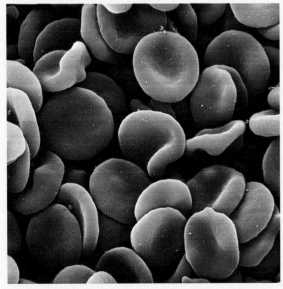

A scanning electron micrograph (SEM) of red blood cells (erythrocytes).
*Science Photo Library.*

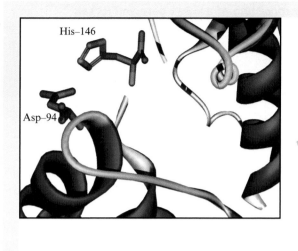

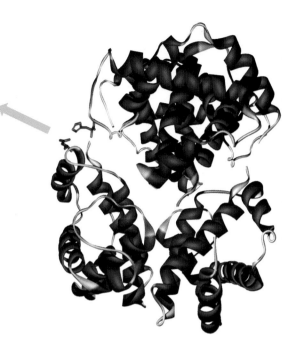

A ribbon representation of the deoxygenated form of human haemoglobin with an enlargement of one pair of His-146 and Asp-94 residues (colour code: N, blue; O, red; C, grey). This illustrates that the imidazole ring of the histidine residue is spatially close to the carboxylate group of the aspartate residue. [Data: F. A. V. Seixas *et al.* (2001), Protein Data Base 1KD2.]

Section 34.11) is protonated and interacts with the carboxylate substituent of the aspartate residue Asp-94. The proximity of residues Asp-94 and His-146 is shown in the figure above. The conformation of the haemoglobin subunits is different in the oxygenated and deoxygenated forms. The favourable interaction between His-146 and Asp-94 can occur only in the *deoxygenated* form of haemoglobin (Hb). It follows that protonation of His-146 promotes the release of $O_2$ and stabilizes the deoxygenated form of the protein:

$$Hb(O_2)_4 + xH^+ \rightleftharpoons [H_xHb(O_2)_{4-n}]^{x+} + nO_2$$

where $n = 1, 2, 3, 4$

This phenomenon is called the Bohr effect and was first observed in 1904 by Christian Bohr.

Now let us look at how $CO_2$ transport in the blood is related to $O_2$ uptake and release in haemoglobin. We have already seen that the catalysed hydration of $CO_2$ to $[HCO_3]^-$ in red blood cells generates $H^+$ which is scavenged by His-146 in haemoglobin. Protonation of haemoglobin has two consequences: it induces release of $O_2$ as explained above, and it also removes $H^+$ ions from solution. The latter effect favours the

conversion of more $CO_2$ to $[HCO_3]^-$ following Le Chatelier's principle:

$$CO_2 + H_2O \rightleftharpoons H^+ + [HCO_3]^-$$

Conversely, transport of $CO_2$ as $[HCO_3]^-$ in red blood cells to the lungs brings it to a region which has a higher partial pressure of $O_2$. Once again, applying Le Chatelier's principle, the deoxygenated form of haemoglobin now binds $O_2$. His-146 can no longer interact with Asp-94 and releases $H^+$:

$$[H_xHb(O_2)_{4-n}]^{x+} + nO_2 \rightleftharpoons Hb(O_2)_4 + xH^+$$

In a final application of Le Chatelier's principle, the lowering of pH results in the formation of $CO_2$ and $H_2O$:

$$H^+ + [HCO_3]^- \rightleftharpoons CO_2 + H_2O$$

The remarkable point is the way in which these processes operate together so that the concentration of $H^+$ in the bloodstream is perfectly controlled and a pH of 7.4 is maintained.

| Worked example 16.12 | *pH of a buffer solution consisting of a weak base and its salt* |

Determine the pH of a buffer solution that is $0.027 \, \text{mol dm}^{-3}$ with respect to aqueous $NH_3$ ($pK_b = 4.75$) and $0.025 \, \text{mol dm}^{-3}$ with respect to $NH_4Cl$.

The equations that describe the system are:

$$NH_3(aq) + H_2O(l) \rightleftharpoons NH_4^+(aq) + OH^-(aq) \qquad pK_b = 4.75$$

$$NH_4Cl(s) \xrightarrow{\text{Dissolve in water}} NH_4^+(aq) + Cl^-(aq)$$

We can assume that $[NH_4^+]_{\text{total}} \approx [NH_4^+]_{\text{salt}} = 0.025 \, \text{mol dm}^{-3}$. In this system, the salt provides $NH_4^+$ which acts as an acid, and so in the Henderson–Hasselbalch equation, $[\text{acid}] = 0.025 \, \text{mol dm}^{-3}$.

The base in the buffer solution is $NH_3$, and $[\text{base}] = 0.027 \, \text{mol dm}^{-3}$. For $NH_3$, $pK_b = 4.75$, and for $pK_a$ of the conjugate acid $NH_4^+$:

$$pK_a = pK_w - pK_b = 14.00 - 4.75 = 9.25$$

The pH of the solution is therefore given by:

$$pH = pK_a + \log \frac{[\text{base}]}{[\text{acid}]}$$

$$= 9.25 + \log \frac{(0.027 \, \text{mol dm}^{-3})}{(0.025 \, \text{mol dm}^{-3})} = 9.28$$

## What happens to the pH of a buffer solution on adding base or acid?

When **acid** is added to a solution with a buffering effect, it is consumed by the weak base present. When **base** is added to a buffer solution, it reacts fully with the weak acid present.

In order to understand why the pH of a solution with a buffering effect is relatively insensitive to the addition of small amounts of acid or base, we consider the addition of $0.20 \, \text{cm}^3$ of $HNO_3$ ($0.50 \, \text{mol dm}^{-3}$) to a buffer solution consisting of $50 \, \text{cm}^3$ $0.045 \, \text{mol dm}^{-3}$ aqueous $CH_3CO_2H$ ($pK_a = 4.77$) and $50 \, \text{cm}^3$ $0.045 \, \text{mol dm}^{-3}$ $Na[CH_3CO_2]$. The total volume before addition is $100 \, \text{cm}^3$, and the addition of the $HNO_3$ causes little increase in volume; let us assume that the total volume remains $100 \, \text{cm}^3$. The initial pH of the buffer solution (before addition of extra acid) is found from the Henderson–Hasselbalch equation, and because the concentrations of acid and base are equal:

$$pH = pK_a + \log \frac{[\text{base}]}{[\text{acid}]}$$

$$= 4.77$$

Now consider the $HNO_3$. It is a strong acid, so fully dissociated in aqueous solution. The $H_3O^+$ from $0.20 \, \text{cm}^3$ of a $0.50 \, \text{mol dm}^{-3}$ solution of $HNO_3$ reacts completely with $CH_3CO_2^-$ in the buffer solution:

$$HNO_3(aq) + CH_3CO_2^-(aq) \longrightarrow CH_3CO_2H(aq) + NO_3^-(aq)$$

Therefore, some $CH_3CO_2^-$ is consumed from the buffer solution, and an equal amount of $CH_3CO_2H$ is produced.

$$\text{Amount of added } H_3O^+ = \text{Amount of added } HNO_3$$
$$= (0.20 \times 10^{-3}\,dm^3) \times (0.50\,mol\,dm^{-3})$$
$$= 1.0 \times 10^{-4}\,moles$$

$x\,cm^3 = x \times 10^{-3}\,dm^3$

$$\text{Before } HNO_3 \text{ addition, amount of } CH_3CO_2^- = (50 \times 10^{-3}\,dm^3)$$
$$\times (0.045\,mol\,dm^{-3})$$
$$= 2.25 \times 10^{-3}\,moles$$

$$\text{After } HNO_3 \text{ addition, amount of } CH_3CO_2^- = (2.25 \times 10^{-3})$$
$$- (1.0 \times 10^{-4})\,moles$$
$$= 2.15 \times 10^{-3}\,moles$$

$$\text{New concentration of } CH_3CO_2^- = \frac{(2.15 \times 10^{-3}\,mol)}{(100 \times 10^{-3}\,dm^3)}$$
$$= 2.15 \times 10^{-2}\,mol\,dm^{-3}$$

$$\text{Before } HNO_3 \text{ addition, amount of } CH_3CO_2H = (50 \times 10^{-3}\,dm^3)$$
$$\times (0.045\,mol\,dm^{-3})$$
$$= 2.25 \times 10^{-3}\,moles$$

$$\text{After } HNO_3 \text{ addition, amount of } CH_3CO_2H = (2.25 \times 10^{-3})$$
$$+ (1.0 \times 10^{-4})\,moles$$
$$= 2.35 \times 10^{-3}\,moles$$

$$\text{New concentration of } CH_3CO_2H = \frac{(2.35 \times 10^{-3}\,mol)}{(100 \times 10^{-3}\,dm^3)}$$
$$= 2.35 \times 10^{-2}\,mol\,dm^{-3}$$

The new pH is therefore:

$$pH = pK_a + \log\frac{[\text{base}]}{[\text{acid}]}$$
$$= 4.77 + \log\frac{(2.15 \times 10^{-2}\,mol\,dm^{-3})}{(2.35 \times 10^{-2}\,mol\,dm^{-3})}$$
$$= 4.73$$

Therefore, there is only a small change in pH (4.77 to 4.73) on adding the $HNO_3$ to the buffer solution. If the nitric acid had simply been added to $100\,cm^3$ of water, the pH would have been 3.00:

$$\text{Concentration of } H_3O^+ = \frac{(1.0 \times 10^{-4}\,mol)}{(100 \times 10^{-3}\,dm^3)} = 1.0 \times 10^{-3}\,mol\,dm^{-3}$$
$$pH = -\log(1.0 \times 10^{-3})$$
$$= 3.00$$

The most effective buffering capacity is obtained when the buffer solution has a $\frac{[\text{base}]}{[\text{acid}]}$ ratio of $1:1$. When acid or base is added to such solutions, minimal changes to the $\frac{[\text{base}]}{[\text{acid}]}$ ratio occur and therefore the pH of the solution is little perturbed. From the Henderson–Hasselbalch equation, it can be seen that when $\dfrac{[\text{base}]}{[\text{acid}]} = 1$, the pH equals the p$K_a$ of the weak acid component of the buffer solution.

## Making up a buffer solution of specified pH

The uses of buffer solutions are widespread, and biological and medical uses are of special importance. For example, experiments with enzymes (biological catalysts) require buffered media because enzyme action is pH-specific. Many buffer solutions are available commercially. To prepare a buffer solution of a given pH, the procedure is as follows:

Enzymes: see Section 15.15

- Choose a weak acid with a p$K_a$ value close to the required pH of the buffer; the weak acid may be the salt of a polybasic acid, e.g. $NaH_2PO_4$.
- Choose an appropriate salt of the weak acid.
- Use the Henderson–Hasselbalch equation to determine the $\dfrac{[\text{base}]}{[\text{acid}]}$ ratio needed to attain the correct pH.
- Remember that for maximum buffering capacity, $\dfrac{[\text{base}]}{[\text{acid}]} = 1$.

---

**Worked example 16.13**     *Making up a buffer solution*

A buffer solution of pH 7.23 is required. Choose a suitable weak acid from Table 16.3, and calculate the ratio of $\dfrac{[\text{base}]}{[\text{acid}]}$ required. Suggest what acid and base combination might be appropriate.

The acid of choice in Table 16.3 is $[H_2PO_4]^-$ (p$K_a = 7.21$). Phosphate buffers are commonly used in the laboratory. From the Henderson–Hasselbalch equation:

$$\text{pH} = \text{p}K_a + \log\frac{[\text{base}]}{[\text{acid}]}$$

$$7.23 = 7.21 + \log\frac{[\text{base}]}{[\text{acid}]}$$

$$\log\frac{[\text{base}]}{[\text{acid}]} = 7.23 - 7.21 = 0.02$$

$$\frac{[\text{base}]}{[\text{acid}]} = 10^{0.02} = 1.05$$

Possible components for the buffer solution are $NaH_2PO_4$ (acid) and $Na_2HPO_4$ (base).

---

**16.8**        ## Acid–base titrations

In this section, we look at the pH changes that accompany neutralization reactions. During an acid–base *titration*, the acid (or base) may be added from a graduated *burette* to the base (or acid) contained in a flask. The reaction can be monitored by measuring the pH of the solution in the flask using a pH meter.

pH meter: see end of Section 18.8

### Strong acid–strong base titration

Figure 16.4 shows the change in pH during the addition of 30.0 cm³ of a 0.10 mol dm⁻³ aqueous solution of NaOH (a strong base) to 25.0 cm³ of a 0.10 mol dm⁻³ solution of hydrochloric acid (a strong acid). Initially, the flask contains a *fully dissociated acid*, and the pH is calculated as follows:

$$[H_3O^+] = [HCl] = 0.10 \, \text{mol dm}^{-3}$$

$$
\begin{aligned}
pH &= -\log[H_3O^+] \\
&= -\log(0.10) \\
&= 1.00
\end{aligned}
$$

This corresponds to the starting point of the titration curve in Figure 16.4. As aqueous NaOH is added, OH⁻ neutralizes $H_3O^+$ (equation 16.68). As the concentration of $H_3O^+$ decreases, the pH value rises.

$$H_3O^+(aq) + OH^-(aq) \longrightarrow 2H_2O(l) \tag{16.68}$$

When 1.0 cm³ of 0.10 mol dm⁻³ aqueous NaOH has been added, the value of the pH can be determined as follows:

$$
\begin{aligned}
\text{Initial amount of } H_3O^+ &= (25.0 \times 10^{-3} \, \text{dm}^3) \times (0.10 \, \text{mol dm}^{-3}) \\
&= 25.0 \times 10^{-4} \, \text{moles}
\end{aligned}
$$

$$
\begin{aligned}
\text{Amount of } OH^- \text{ added} &= (1.0 \times 10^{-3} \, \text{dm}^3) \times (0.10 \, \text{mol dm}^{-3}) \\
&= 1.0 \times 10^{-4} \, \text{moles}
\end{aligned}
$$

$$
\begin{aligned}
\text{Amount of } H_3O^+ \text{ remaining} &= (25.0 \times 10^{-4}) - (1.0 \times 10^{-4}) \\
&= 24.0 \times 10^{-4} \, \text{moles}
\end{aligned}
$$

$$\text{Total volume } after \text{ the addition} = 26.0 \, \text{cm}^3$$

$$
\begin{aligned}
\text{Concentration of } H_3O^+ \text{ after addition} &= \frac{(24.0 \times 10^{-4} \, \text{mol})}{(26.0 \times 10^{-3} \, \text{dm}^3)} \\
&= 9.23 \times 10^{-2} \, \text{mol dm}^{-3}
\end{aligned}
$$

$$
\begin{aligned}
pH &= -\log[H_3O^+] \\
&= -\log(9.23 \times 10^{-2}) \\
&= 1.03
\end{aligned}
$$

**Fig. 16.4** The variation in pH during the addition of 30.0 cm³ of an aqueous solution (0.10 mol dm⁻³) of NaOH to 25.0 cm³ aqueous HCl (0.10 mol dm⁻³).

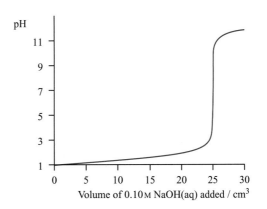

To find the value of the pH after $5.0\,cm^3$ of the $0.10\,mol\,dm^{-3}$ aqueous NaOH has been added:

$$\text{Initial amount of } H_3O^+ = (25.0 \times 10^{-3}\,dm^3) \times (0.10\,mol\,dm^{-3})$$

$$= 25.0 \times 10^{-4}\,moles$$

$$\text{Amount of } OH^- \text{ added} = (5.0 \times 10^{-3}\,dm^3) \times (0.10\,mol\,dm^{-3})$$

$$= 5.0 \times 10^{-4}\,moles$$

$$\text{Amount of } H_3O^+ \text{ remaining} = 20.0 \times 10^{-4}\,moles$$

$$\text{Total volume } after \text{ the addition} = 30.0\,cm^3$$

$$\text{Concentration of } H_3O^+ \text{ after addition} = \frac{(20.0 \times 10^{-4}\,mol)}{(30.0 \times 10^{-3}\,dm^3)}$$

$$= 6.67 \times 10^{-2}\,mol\,dm^{-3}$$

$$pH = -\log[H_3O^+]$$

$$= -\log(6.67 \times 10^{-2})$$

$$= 1.18$$

The pH changes very gradually at first, but then rises sharply (Figure 16.4). The midpoint of the near-vertical section of the curve in Figure 16.4 is the *equivalence point*, i.e. the point at which the alkali has *exactly* neutralized the acid, and neither acid nor alkali is in excess. On either side of the equivalence point, the pH is extremely sensitive to the composition of the solution, with very small amounts of acid or alkali causing large changes in pH. In a strong acid–strong base titration, the equivalence point is at pH 7.00. At this point, the solution contains only a salt (NaCl in the example in Figure 16.4) and water. The salt formed from the reaction of a strong base with a strong acid is neutral. From Figure 16.4, the equivalence point occurs when $25.0\,cm^3$ of a $0.01\,mol\,dm^{-3}$ NaOH solution have reacted with $25.0\,cm^3$ of a $0.01\,mol\,dm^{-3}$ HCl solution, and this is consistent with the stoichiometry of reaction 16.69.

> The equivalence point of a ***strong acid–strong base*** titration is at pH 7.00. The salt formed is neutral.

$$HCl(aq) + NaOH(aq) \longrightarrow NaCl(aq) + H_2O(l) \tag{16.69}$$

### Weak acid–strong base

Figure 16.5 shows the change in pH during the addition of aqueous NaOH $(0.10\,mol\,dm^{-3})$ to $25\,cm^3$ aqueous acetic acid $(0.10\,mol\,dm^{-3})$. Initially the flask contains a weak acid $(pK_a = 4.77)$ and the initial pH is determined

**Fig. 16.5** The variation in pH during the addition of $30.0\,cm^3$ of an aqueous solution $(0.10\,mol\,dm^{-3})$ of NaOH to $25.0\,cm^3$ aqueous $CH_3CO_2H$ $(0.10\,mol\,dm^{-3})$.

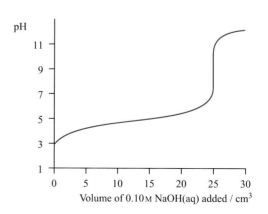

as follows:

$$CH_3CO_2H(aq) + H_2O(l) \rightleftharpoons H_3O^+(aq) + CH_3CO_2^-(aq)$$

$$K_a = 10^{-4.77} = 1.7 \times 10^{-5}$$

$$1.7 \times 10^{-5} = \frac{[H_3O^+][CH_3CO_2^-]}{[CH_3CO_2H]} = \frac{[H_3O^+]^2}{[CH_3CO_2H]}$$

Assume that $[CH_3CO_2H]_{equilm} \approx [CH_3CO_2H]_{initial}$

$$[H_3O^+] = \sqrt{1.7 \times 10^{-5} \times 0.10}$$

$$= 1.3 \times 10^{-3} \, mol \, dm^{-3}$$

$$pH = -\log[H_3O^+] = 2.89$$

This value corresponds to the starting point of the titration curve in Figure 16.5. As alkali is added to the acetic acid, the pH rises gradually, and then steeply as the equivalence point is reached. By comparing the shapes of the titration curves in Figures 16.4 and 16.5, it is clear that the addition of a strong base to a weak acid causes more significant changes in pH early on in the titration than does the addition of a strong base to a strong acid. You can verify this by calculating the pH at several points along the curves and checking the answers against Figures 16.4 and 16.5.

A significant feature of Figure 16.5 is that the equivalence point is *not* at pH 7.00. This is a characteristic result for a weak acid–strong base titration: the equivalence point lies in the range $7.00 < pH < 14.00$, the value depending on the $pK_a$ of the acid. For the $CH_3CO_2H$–NaOH titration, the pH at the end point is found as follows.

At the equivalence point:

$$CH_3CO_2H(aq) + NaOH(aq) \longrightarrow Na[CH_3CO_2](aq) + H_2O(l)$$

$\uparrow$        $\uparrow$        $\uparrow$

All consumed     All consumed     Salt which determines
                                           the pH of the solution

Consider how the $CH_3CO_2^-$ ion from the salt behaves in aqueous solution; $CH_3CO_2^-$ is the conjugate base of $CH_3CO_2H$:

$$CH_3CO_2^-(aq) + H_2O(l) \rightleftharpoons CH_3CO_2H(aq) + OH^-(aq)$$

$$K_b = \frac{[CH_3CO_2H][OH^-]}{[CH_3CO_2^-]} = \frac{[OH^-]^2}{[CH_3CO_2^-]}$$

We need to find the concentration of $CH_3CO_2^-$ at the equivalence point:

Total volume of solution at the equivalence point $= 25.0 + 25.0 = 50.0 \, cm^3$

Amount of $CH_3CO_2^-$ present $=$ Amount of $CH_3CO_2H$ used in the titration

$$= (25.0 \times 10^{-3} \, dm^3) \times (0.10 \, mol \, dm^{-3})$$

$$= 2.5 \times 10^{-3} \, moles$$

Concentration of $CH_3CO_2^-$ at the equivalence point $= \dfrac{(2.5 \times 10^{-3} \, mol)}{(50 \times 10^{-3} \, dm^3)}$

$$= 0.050 \, mol \, dm^{-3}$$

Now, calculate $[OH^-]$, and then $[H_3O^+]$. From above:

$$[OH^-] = \sqrt{K_b \times 0.050}$$

We know $pK_a$ for $CH_3CO_2H$ and so $pK_b$ for the conjugate base is:

$$pK_b = pK_w - pK_a = 14.00 - 4.77 = 9.23$$

$$[OH^-] = \sqrt{10^{-9.23} \times 0.050} = 5.4 \times 10^{-6} \, mol \, dm^{-3}$$

Therefore:

$$[H_3O^+] = \frac{K_w}{[OH^-]} = \frac{1.0 \times 10^{-14}}{5.4 \times 10^{-6}}$$

$$= 1.9 \times 10^{-9} \, mol \, dm^{-3}$$

$$pH = -\log[H_3O^+]$$

$$= 8.72$$

> The equivalence point of a **weak acid–strong base** titration is in the range $7.00 < pH < 14.00$. The salt formed is basic.

This value corresponds to the pH of the midpoint of the near-vertical section of the titration curve in Figure 16.5 and indicates that sodium acetate is a *basic salt*. In solution, it dissociates (equation 16.70), giving the acetate ion which establishes equilibrium 16.71.

$$Na[CH_3CO_2] \xrightarrow{\text{Dissolve in water}} Na^+(aq) + CH_3CO_2^-(aq) \quad (16.70)$$

$$CH_3CO_2^-(aq) + H_2O(l) \rightleftharpoons CH_3CO_2H(aq) + OH^-(aq) \quad (16.71)$$

### Strong acid–weak base

Figure 16.6 illustrates the variation in pH during the addition of aqueous HCl ($0.10 \, mol \, dm^{-3}$) to $25 \, cm^3$ aqueous ammonia ($0.10 \, mol \, dm^{-3}$). Initially the flask contains a weak base ($pK_b = 4.75$) which is partially dissociated (equation 16.72).

$$NH_3(aq) + H_2O(l) \rightleftharpoons NH_4^+(aq) + OH^-(aq) \quad (16.72)$$

The initial pH is determined as follows:

$$K_b = 10^{-4.75} = 1.8 \times 10^{-5}$$

$$1.8 \times 10^{-5} = \frac{[NH_4^+][OH^-]}{[NH_3]} = \frac{[OH^-]^2}{[NH_3]}$$

Assume that $[NH_3]_{equilm} \approx [NH_3]_{initial}$:

$$[OH^-] = \sqrt{1.8 \times 10^{-5} \times 0.10}$$

$$= 1.3 \times 10^{-3} \, mol \, dm^{-3}$$

**Fig. 16.6** The variation in pH during the addition of $30.0 \, cm^3$ of an aqueous solution ($0.10 \, mol \, dm^{-3}$) of HCl to $25.0 \, cm^3$ aqueous $NH_3$ ($0.10 \, mol \, dm^{-3}$).

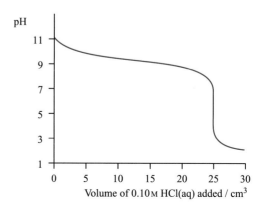

Volume of $0.10 \, M \, HCl(aq)$ added / $cm^3$

$$[H_3O^+] = \frac{K_w}{[OH^-]} = \frac{1.0 \times 10^{-14}}{1.3 \times 10^{-3}}$$

$$= 7.7 \times 10^{-12}\,mol\,dm^{-3}$$

$$pH = -\log[H_3O^+]$$

$$= 11.11$$

The equivalence point of a **strong acid–weak base** titration is in the range $0 < pH < 7.00$. The salt formed is acidic.

This pH value corresponds to the starting point of the titration curve in Figure 16.6. As strong acid is added to the weak base, the pH falls gradually, and then steeply as the equivalence point is reached. The shape of the curve shown in Figure 16.6 is characteristic of a titration involving the addition of a monobasic strong acid to a monobasic weak base. The end point lies in the range $0 < pH < 7.00$, indicating that the salt formed is *acidic*. In the reaction of aqueous $NH_3$ (i.e. $NH_4OH$) with HCl (equation 16.73), the salt formed ionizes to give $NH_4^+$ and $Cl^-$. The salt is acidic because equilibrium 16.74 is established; $pK_a$ for $NH_4^+$ is 9.25.

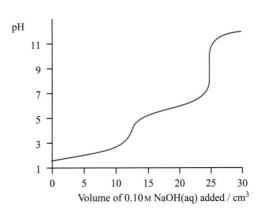

See problem 16.27 for confirmation of the pH of the equivalence point of this titration

$$NH_4OH(aq) + HCl(aq) \longrightarrow NH_4Cl(aq) + H_2O(l) \tag{16.73}$$

$$NH_4^+(aq) + H_2O(l) \rightleftharpoons H_3O^+(aq) + NH_3(aq) \tag{16.74}$$

### Titrations involving polybasic acids

Each of Figures 16.4, 16.5 and 16.6 shows a single equivalence point because each neutralization involves the reaction of a *monobasic* acid with a *monobasic* base. If NaOH is titrated against a *dibasic* acid, two equivalence points are observed. Figure 16.7 shows the change in pH during the addition of a $0.10\,mol\,dm^{-3}$ aqueous solution of NaOH to $12.5\,cm^3$ of a $0.10\,mol\,dm^{-3}$ aqueous solution of maleic acid, **16.7**, which we abbreviate to $H_2A$. The two equivalence points occur after the additions of 12.5 and $25.0\,cm^3$ of alkali and correspond to reactions 16.75 and 16.76 respectively.

$$H_2A(aq) + NaOH(aq) \longrightarrow NaHA(aq) + H_2O(l) \tag{16.75}$$

$$NaHA(aq) + NaOH(aq) \longrightarrow Na_2A(aq) + H_2O(l) \tag{16.76}$$

*Overall*:    $H_2A(aq) + 2NaOH(aq) \longrightarrow Na_2A(aq) + 2H_2O(l)$

The $pK_a$ values for acid **16.7** are $pK_a(1) = 1.92$ and $pK_a(2) = 5.79$, i.e. two weak acids are effectively present in the system. The pH of the first equivalence point can be determined as follows.

**(16.7)**

**Fig. 16.7** The variation in pH during the addition of $30.0\,cm^3$ of an aqueous solution $(0.10\,mol\,dm^{-3})$ of NaOH to $12.5\,cm^3$ aqueous maleic acid **(16.7)** $(0.10\,mol\,dm^{-3})$.

At the first equivalence point, $HA^-$ is present in solution. This can behave as a weak acid:

$$HA^-(aq) + H_2O(l) \rightleftharpoons H_3O^+(aq) + A^{2-}(aq)$$
$\quad$ Acid $\qquad\qquad$ Base

$$HA^-(aq) + HA^-(aq) \rightleftharpoons H_2A(aq) + A^{2-}(aq)$$
$\quad$ Acid $\qquad\qquad$ Base

or as a weak base:

$$HA^-(aq) + H_2O(l) \rightleftharpoons H_2A(aq) + OH^-(aq)$$
$\quad$ Base $\qquad\qquad$ Acid

$$HA^-(aq) + HA^-(aq) \rightleftharpoons H_2A(aq) + A^{2-}(aq)$$
$\quad$ Base $\qquad\qquad$ Acid

Two of these four reactions are identical because they show $HA^-$ acting as both an acid and a base. It is this equilibrium that dominates in the solution. (Think about the relative acid and base strengths of $HA^-$ and $H_2O$.) Our aim is to find the pH of this solution, and so we must find the concentration of $H_3O^+$ ions. But, this term does not appear in the equilibrium:

$$HA^-(aq) + HA^-(aq) \rightleftharpoons H_2A(aq) + A^{2-}(aq)$$

However, since the position of the equilibrium depends upon the concentrations of $H_2A$, $HA^-$ and $A^{2-}$, it must depend upon $K_a(1)$ and $K_a(2)$ (equations 16.77 and 16.78).

$$K_a(1) = \frac{[H_3O^+][HA^-]}{[H_2A]} \qquad\qquad (16.77)$$

$$K_a(2) = \frac{[H_3O^+][A^{2-}]}{[HA^-]} \qquad\qquad (16.78)$$

From these equations, we can write:

$$[H_3O^+] = \frac{K_a(1)[H_2A]}{[HA^-]} = \frac{K_a(2)[HA^-]}{[A^{2-}]}$$

and:

$$\frac{K_a(2)}{K_a(1)} = \frac{[H_2A][A^{2-}]}{[HA^-]^2}$$

From the stoichiometry of the equilibrium above, $[H_2A] = [A^{2-}]$, and so we can write:

$$\frac{K_a(2)}{K_a(1)} = \frac{[H_2A]^2}{[HA^-]^2} \qquad\qquad (16.79)$$

Rearranging equation 16.77 gives:

$$\frac{[H_2A]}{[HA^-]} = \frac{[H_3O^+]}{K_a(1)}$$

and substitution of this expression into equation 16.79 gives:

$$\frac{K_a(2)}{K_a(1)} = \left(\frac{[H_3O^+]}{K_a(1)}\right)^2$$

Therefore:

$$[H_3O^+]^2 = K_a(1)K_a(2)$$

$$[H_3O^+] = \sqrt{K_a(1)K_a(2)} \qquad\qquad (16.80)$$

For maleic acid, $pK_a(1) = 1.92$ and $pK_a(2) = 5.79$:

$$[H_3O^+] = \sqrt{(10^{-1.92})(10^{-5.79})} = 1.40 \times 10^{-4}\,\text{mol dm}^{-3}$$

$$pH = 3.86$$

This value corresponds to the first equivalence point on Figure 16.7. The pH can be found directly from the two $pK_a$ values. Taking the negative logarithms of both sides of equation 16.80 gives equation 16.81.

$$-\log[H_3O^+] = -\log\{K_a(1)K_a(2)\}^{\frac{1}{2}}$$

$$= \tfrac{1}{2}\{-\log K_a(1) - \log K_a(2)\}$$

$$pH = \tfrac{1}{2}\{pK_a(1) + pK_a(2)\} \tag{16.81}$$

The pH of the *first* equivalence point in the titration of a strong base (e.g. NaOH) against a polybasic weak acid is given by:

$$pH = \tfrac{1}{2}\{pK_a(1) + pK_a(2)\}$$

This is a general expression for finding the pH of the *first* equivalence point in the titration of a strong base (e.g. NaOH) against a polybasic weak acid.

Figure 16.7 shows that the second equivalence point in the titration of NaOH against maleic acid, $H_2A$, is at pH $\approx 9$. At this point, all $HA^-$ has been converted to $A^{2-}$ (equation 16.76) and a basic salt is present in solution. To determine the pH of the solution, we must consider equilibrium 16.82.

$$A^{2-}(aq) + H_2O(l) \rightleftharpoons HA^-(aq) + OH^-(aq) \tag{16.82}$$

$$K_b = \frac{[HA^-][OH^-]}{[A^{2-}]} = \frac{[OH^-]^2}{[A^{2-}]}$$

To determine the concentration of $A^{2-}$ at the second equivalence point:

Total volume of solution $= 25.0 + 12.5 = 37.5\,\text{cm}^3$

Amount of $A^{2-}$ present $=$ Amount of $H_2A$ initially

$$= (12.5 \times 10^{-3}\,\text{dm}^3) \times (0.10\,\text{mol dm}^{-3})$$

$$= 1.25 \times 10^{-3}\,\text{moles}$$

Concentration of $A^{2-}$ at the second equivalence point $= \dfrac{(1.25 \times 10^{-3}\,\text{mol})}{(37.5 \times 10^{-3}\,\text{dm}^3)}$

$$= 0.033\,\text{mol dm}^{-3}$$

From above:

$$[OH^-] = \sqrt{K_b \times [A^{2-}]} = \sqrt{K_b \times 0.033}$$

$pK_a$ for $HA^-$ is 5.79, and so $pK_b$ for the conjugate base is:

$$pK_b = pK_w - pK_a = 14.00 - 5.79 = 8.21$$

$$[OH^-] = \sqrt{10^{-8.21} \times 0.033} = 1.4 \times 10^{-5}\,\text{mol dm}^{-3}$$

Therefore:

$$[H_3O^+] = \frac{K_w}{[OH^-]} = \frac{1.0 \times 10^{-14}}{1.4 \times 10^{-5}}$$

$$= 7.1 \times 10^{-10}\,\text{mol dm}^{-3}$$

$$pH = -\log[H_3O^+]$$

$$= 9.15$$

A titration of aqueous NaOH against a tribasic acid $H_3A$ has three equivalence points and the pH value of the first equivalence point (where the predominant solution species is $H_2A^-$) can be found by using equation

16.81. At the second equivalence point, the predominant species is $HA^{2-}$, and the pH can be found using equation 16.83 which can be derived in an analogous manner to equation 16.81.

See problem 16.28

$$pH = \tfrac{1}{2}\{pK_a(2) + pK_a(3)\} \tag{16.83}$$

## 16.9    Acid–base indicators

### Colour changes of acid–base indicators

When it is desirable to monitor the pH changes during an acid–base titration, a pH meter is used as described in the previous section. In many titrations, however, the aim is simply to find the equivalence point and this is detected by adding an *acid–base indicator* to the aqueous solution (usually the base) in the flask and observing a colour change at the *end point*. Clearly, it is important that the end point accurately coincides with the equivalence point, and to ensure this is so, indicators must be chosen carefully as described below. Some common acid–base indicators are listed in Table 16.5.

Acid–base indicators are generally weak acids of the general type HIn (equation 16.84) in which the conjugate base In⁻ has a *different* colour from the undissociated acid HIn.

$$HIn(aq) + H_2O(l) \rightleftharpoons H_3O^+(aq) + In^-(aq) \tag{16.84}$$
Colour **I**                                        Colour **II**

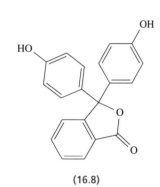

(16.8)

An example is phenolphthalein, the acid form (**16.8**) of which is colourless and the conjugate base of which is pink. By Le Chatelier's principle, equilibrium 16.84 is sensitive to the pH of a solution. It shifts to the left-hand side in acidic media and appears as colour **I**, and moves to the right-hand side under alkaline conditions, thus appearing as colour **II**. The acid dissociation constant for an indicator HIn is given by equation 16.85.

$$K_a = \frac{[H_3O^+][In^-]}{[HIn]} \tag{16.85}$$

The colour of the solution depends on the ratio of concentrations $[In^-]:[HIn]$ (equation 16.86). This expression is usually written in a logarithmic form (equation 16.87).

$$\frac{[In^-]}{[HIn]} = \frac{K_a}{[H_3O^+]} \tag{16.86}$$

**Table 16.5** Selected acid–base indicators in aqueous solution.

| Indicator | $pK_a$ | Colour change from acidic → basic solution | pH range in which indicator changes colour |
|---|---|---|---|
| Phenolphthalein | 9.50 | Colourless → pink | 8.00–10.00 |
| Phenol red | 8.00 | Yellow → red | 6.8–8.2 |
| Bromocresol purple | 6.40 | Yellow → purple | 5.2–6.8 |
| Bromocresol green | 4.90 | Yellow → blue | 3.8–5.4 |
| Methyl orange | 3.46 | Red → yellow | 3.2–4.4 |
| Thymol blue[a] | 1.65 | Red → yellow | 1.2–2.8 |
|  | 9.20 | Yellow → blue | 8.0–9.8 |

[a] Dibasic acid.

For an *acid–base indicator* for which the acid form is HIn, the $[In^-]:[HIn]$ ratio is determined from:

$$\log\frac{[In^-]}{[HIn]} = -pK_a + pH$$

$$\log\frac{[In^-]}{[HIn]} = \log K_a - \log[H_3O^+]$$

$$\log\frac{[In^-]}{[HIn]} = -pK_a + pH \tag{16.87}$$

Application of this equation can be illustrated by determining the pH range in which phenolphthalein ($pK_a = 9.50$) changes colour. The colourless form is HIn, and the pink form is $In^-$. From equation 16.87 we can write:

$$\log\frac{[In^-]}{[HIn]} = -9.50 + pH$$

We could randomly test values of pH in this equation to determine when $In^-$ or HIn dominates as the solution species. However, an important point should be noted:

$$\text{when } pH = pK_a: \qquad \log\frac{[In^-]}{[HIn]} = 0 \qquad \frac{[In^-]}{[HIn]} = 1.00$$

The colour must therefore change around this pH value, so we should test values of the pH on either side of $pH = 9.50$.

At pH 8.00: $\qquad \log\frac{[In^-]}{[HIn]} = -1.50 \qquad \frac{[In^-]}{[HIn]} = 0.03$

At pH 10.00: $\qquad \log\frac{[In^-]}{[HIn]} = 0.50 \qquad \frac{[In^-]}{[HIn]} = 3.16$

The colourless to pink change occurs when the dominant solution species changes from being HIn to $In^-$. At pH 8.00, $[HIn] \approx 33 \times [In^-]$, and the solution is colourless. At pH 10.00, $[In^-] = 3.16 \times [HIn]$ and the solution is pink. The pH range from 8 to 10 may seem rather imprecise, but in a strong acid–strong base titration such as that in Figure 16.4, the pH rises sharply from $\approx 4$ to 10 and, at this point, phenolphthalein changes colour at the addition of a mere drop of base, giving an accurate end point. Conversely, a pink to colourless change is observed if a strong acid is titrated against a strong base.

### Choosing an indicator for a titration

Table 16.5 lists the pH ranges over which selected indicators change colour. If we compare these ranges with the near-vertical sections of the titration curves in Figures 16.4–16.7, we can immediately see that not every indicator is suitable for every titration. In order that the end point is accurate, the indicator *must* undergo a *sharp* colour change, and so the pH range for the change must coincide with the near-vertical part of the titration curve.

For a *strong acid–strong base* titration (Figure 16.4), the equivalence point is at pH 7.00 and is in the middle of a large and rapid pH change. Of the indicators in Table 16.5, phenolphthalein, phenol red, bromocresol purple and bromocresol green are suitable for detecting the end point of this experiment. In a titration in which a *strong base* is added to a *weak acid* (Figure 16.5), the choice of an appropriate indicator is fairly restricted because the change in pH around the equivalence point is quite small. We showed in Section 16.8 that the equivalence point for the titration in Figure 16.5 is at pH 8.72. Of the indicators in Table 16.5, phenolphthalein and thymol blue are suitable; thymol blue is a dibasic acid and it is the second dissociation that is in operation in this weak acid–strong base titration. In the *strong acid–weak base* titration in Figure 16.6, the equivalence point is at pH 5.28, and the near-vertical section of curve is relatively small. Since in Figure 16.6 we are adding acid to a weak base, the pH changes from a higher to lower value (contrast Figures 16.5 and 16.6 with 16.7) through the equivalence point pH of 5.28. A suitable indicator for the titration is therefore bromocresol purple. Many more indicators are available than the few selected in Table 16.5, making the choices less restricted than this discussion might imply.

## SUMMARY

In this chapter we have discussed different types of equilibria. Le Chatelier's principle can be used to assess qualitatively how an equilibrium responds to external changes. Quantitative treatments require the use of equilibrium constants, $K$, which can be applied to gaseous or solution systems; equilibrium constants are temperature-dependent. For equilibria involving acids and bases, the constants $K_a$ and $K_b$ are defined; the self-ionization of water is described by $K_w$. We have shown how to calculate the pH of solutions of acids and bases, illustrated how the solution pH varies during acid–base titrations, and exemplified the use of selected acid–base indicators. Constant pH can be maintained in a buffer solution.

### *Useful equations*

The meanings of all symbols are given in the chapter.

For:    $x\mathrm{X} + y\mathrm{Y} \rightleftharpoons z\mathrm{Z}$    $K = \dfrac{(a_Z)^z}{(a_X)^x(a_Y)^y}$

For any pure substance, i:    $a_i = 1$  (dimensionless)

For a solute, i, *in a dilute solution*:    $a_i = [\mathrm{i}]$  (dimensionless)

For an *ideal gas*, i:    $a_i = P_i$  (dimensionless)

$\dfrac{\mathrm{d}(\ln K)}{\mathrm{d}T} = \dfrac{\Delta H^\circ}{RT^2}$

$\ln K = -\dfrac{\Delta H^\circ}{RT} + c$

For an aqueous solution of a weak acid HA:    $K_a = \dfrac{[\mathrm{H_3O^+}][\mathrm{A^-}]}{[\mathrm{HA}]}$

For an aqueous solution of a weak base B:    $K_b = \dfrac{[\mathrm{BH^+}][\mathrm{OH^-}]}{[\mathrm{B}]}$

$pK_a = -\log K_a$

$K_a = 10^{-pK_a}$

$pK_w = -\log K_w = 14.00$

$pK_b = -\log K_b$

$K_b = 10^{-pK_b}$

$K_a \times K_b = K_w = 1.0 \times 10^{-14}$

$pK_a + pK_b = pK_w = 14.00$

$\mathrm{pH} = -\log[\mathrm{H_3O^+}]$

$$[H_3O^+] = 10^{-pH}$$

$$pH = pK_a + \log\frac{[\text{base}]}{[\text{acid}]} \qquad \text{(Henderson–Hasselbalch equation for a buffer solution)}$$

## Do you know what the following terms mean?

- Le Chatelier's principle
- activity
- molarity and molality
- equilibrium constant (thermodynamic equilibrium constant)
- Brønsted acid
- Brønsted base
- strong acid or base
- weak acid or base
- acid dissociation constant, $K_a$

- $pK_a$
- conjugate acid–base pair
- neutralization reaction
- base dissociation constant, $K_b$
- $pK_b$
- self-ionization constant for water, $K_w$
- $pK_w$
- pH
- speciation in an aqueous solution of a weak, polybasic acid

- buffer solution
- Henderson–Hasselbalch equation
- acid–base titration
- equivalence point (in an acid–base titration)
- acid–base indicator
- end point (in an acid–base titration)

## You should be able:

- to apply Le Chatelier's principle to an equilibrium
- to appreciate the approximations made when writing equilibrium constants in terms of concentrations or partial pressures
- to write down an expression for $K$ given a stoichiometric equation for an equilibrium
- to calculate $K$ given the composition of a system at equilibrium
- to find the composition of a system at equilibrium given $K$ and amounts of initial reagents
- to write down an equation that shows how $K$ depends on temperature, and to apply it to determine an approximate value of $\Delta H^\circ$ for a reaction
- to discuss how the behaviour of a weak acid (or base) differs from that of a strong acid (or base) in solution
- to write equilibria to show how weak acids and weak bases behave in aqueous solutions
- to write an expression for $K_a$ or $K_b$ given a stoichiometric equation for an equilibrium
- to calculate $K_a$ or $K_b$ given the composition of an appropriate system at equilibrium
- to find the extent of dissociation of a weak acid or base given the value of $K_a$ or $K_b$
- to determine $K_a$ from a $pK_a$ value, and vice versa
- to determine $K_b$ from a $pK_b$ value, and vice versa

- to interrelate $pK_w$, $pK_a$ and $pK_b$ for a conjugate acid–base pair
- to write balanced equations for neutralization reactions, and appreciate how different salts can arise from a polybasic acid
- to determine the pH of an acidic or basic solution
- to discuss how the position of an equilibrium involving a weak acid or base is affected by pH
- to discuss what a buffer solution is, its importance and how it functions
- to write down the Henderson–Hasselbalch equation and apply it to calculate the pH of a buffer solution
- to illustrate why the pH of a buffer solution remains almost constant when a small amount of acid or base is added
- to explain how to determine the composition of a buffer solution of a specified pH
- to discuss the differences between strong acid–strong base, strong acid–weak base and weak acid–strong base titrations
- to explain the meaning of the equivalence point in an acid–base titration
- to explain the variation in pH during the titration of a dibasic or tribasic acid against a strong base
- to discuss how acid–base indicators work and illustrate their applications in titrations

## PROBLEMS

**16.1** Consider the equilibrium:

$$2SO_2(g) + O_2(g) \rightleftharpoons 2SO_3(g)$$

$$\Delta_r H^\circ(298\,K) = -96\,kJ \text{ per mole of } SO_3$$

What are the effects of (a) increasing the external pressure, and (b) lowering the external temperature?

**16.2** For $NO_2(g)$, $\Delta_f H^\circ(298\,K) = +34.2\,kJ\,mol^{-1}$.
(a) Write an equation for the reversible formation of $NO_2$ from its constituent elements. (b) What is the effect on this equilibrium of raising the external temperature? (c) If the external pressure is increased, how is the yield of $NO_2$ affected?

**16.3** What is the effect on the following equilibrium of (a) adding propanoic acid, and (b) removing benzyl propanoate by distillation?

$$CH_3CH_2CO_2H + C_6H_5CH_2OH \rightleftharpoons$$
Propanoic acid    Benzyl alcohol

$$CH_3CH_2CO_2CH_2C_6H_5 + H_2O$$
Benzyl propanoate    Water

**16.4** Write down expressions for $K$ in terms of the activities of the components present in the following gaseous equilibria:
(a) $2SO_2 + O_2 \rightleftharpoons 2SO_3$
(b) $N_2 + 3H_2 \rightleftharpoons 2NH_3$
(c) $Al_2Cl_6 \rightleftharpoons 2AlCl_3$
(d) $Cl_2 \rightleftharpoons 2Cl$
(e) $H_2 + I_2 \rightleftharpoons 2HI$

**16.5** Write down expressions for $K$ in terms of the concentrations of the components present in the following equilibria. What are the limitations of using concentrations instead of activities?
(a) $C_6H_5CO_2H(aq) + H_2O(l) \rightleftharpoons$
$$H_3O^+(aq) + C_6H_5CO_2^-(aq)$$
(b) $[Fe(H_2O)_6]^{3+}(aq) + 6CN^-(aq) \rightleftharpoons$
$$[Fe(CN)_6]^{3-}(aq) + 6H_2O(l)$$
(c) $Cr_2O_7^{2-}(aq) + 2OH^-(aq) \rightleftharpoons 2CrO_4^{2-}(aq) +$
$$H_2O(l)$$

**16.6** $I_2$ is very sparingly soluble in water, and laboratory solutions are usually made up in aqueous KI in which the following equilibrium is established:

$$I_2(aq) + I^-(aq) \rightleftharpoons I_3^-(aq)$$

2.54 g of $I_2$ are added to 1 dm$^3$ of a 0.50 mol dm$^{-3}$ aqueous solution of KI, and the solution is allowed to reach equilibrium. At this point, $9.8 \times 10^{-3}$ moles of $[I_3]^-$ are present. Determine the equilibrium constant, assuming no change in solution volume on adding solid $I_2$.

**16.7** Ammonia is manufactured in the Haber process:

$$N_2(g) + 3H_2(g) \rightleftharpoons 2NH_3(g)$$

If 0.50 moles of $N_2$ and 2.0 moles of $H_2$ are

combined at 400 K and 1.00 bar pressure and the system is allowed to reach equilibrium, 0.80 moles of $NH_3$ are present in the system. Determine $K$ under these conditions.

**16.8** The oxidation of $SO_2$ is a stage in the manufacture of sulfuric acid:

$$2SO_2(g) + O_2(g) \rightleftharpoons 2SO_3(g)$$

If 2.00 moles of $SO_2$ react with 0.50 moles of $O_2$ at 1100 K and 1.00 bar pressure and the system is left to establish equilibrium, the final mixture contains 0.24 moles of $SO_3$. Calculate $K$ under these conditions.

**16.9** Consider the equilibrium:

$$H_2(g) + CO_2(g) \rightleftharpoons H_2O(g) + CO(g)$$

At 800 K, $K = 0.29$. If 0.80 moles of $H_2$ and 0.60 moles of $CO_2$ react under a pressure of 1.00 bar, how many moles of $CO_2$ will remain when the reaction mixture has reached equilibrium?

**16.10** The formation of HCl could be considered in terms of the equilibria:

$$H_2(g) + Cl_2(g) \rightleftharpoons 2HCl(g) \qquad K_1$$
or
$$\tfrac{1}{2}H_2(g) + \tfrac{1}{2}Cl_2(g) \rightleftharpoons HCl(g) \qquad K_2$$

What is the relationship between the values of the equilibrium constants for these equilibria?

**16.11** Write equations to show the dissociation in aqueous solution of the following acids: (a) $CH_3CH_2CO_2H$; (b) $HNO_3$; (c) $H_2SO_3$; (d) $H_2SO_4$.

**16.12** Using data from Table 16.3, determine values of $K_a$ for HCN and $HNO_2$.

**16.13** The p$K_a$ values for citric acid (**16.9**) are 3.14, 4.77 and 6.39. Write equations to show the dissociation processes and assign a p$K_a$ value to each step. What are the corresponding $K_a$ values?

$$H_2C - \overset{\overset{\displaystyle OH}{|}}{C} - CH_2$$

with $HO_2C$ and $CO_2H$ groups

(16.9)

**16.14** (a) The p$K_a$ values for acetic acid and chloroacetic acid are 4.77 and 2.85 respectively. Which is the weaker acid in aqueous solution? (b) The values of $K_a$ for HOBr and HOCl are $2.1 \times 10^{-9}$ and $3.0 \times 10^{-5}$ respectively. Which is the stronger acid in aqueous solution?

**16.15** Determine the concentration of $CN^-$ ions in a 0.050 mol dm$^{-3}$ aqueous solution of HCN (p$K_a = 9.31$). How is the concentration of $H_3O^+$ related to that of $CN^-$?

**16.16** (a) Is KOH completely or partially dissociated in aqueous solution? (b) What volume of a $0.20\,mol\,dm^{-3}$ solution of $HNO_3$ is needed to completely neutralize $30\,cm^3$ of a $0.40\,mol\,dm^{-3}$ solution of KOH? (c) Amines of the type $RNH_2$ behave in a similar manner to $NH_3$ in aqueous solution. Write an equation to show what happens when ethylamine ($CH_3CH_2NH_2$) dissolves in water. If $pK_b$ for ethylamine is 3.19, calculate $K_b$ and comment on the position of the equilibrium.

**16.17** (a) To what equilibrium does a value of $pK_a = 9.25$ for $NH_4^+$ refer? (b) To what equilibrium does a value of $pK_b = 4.75$ for $NH_3$ refer? (c) Rationalize why, for $NH_4^+$ and $NH_3$, $(pK_a + pK_b) = 14.00$.

**16.18** (a) Find the pH of a $0.10\,mol\,dm^{-3}$ solution of aqueous HCl. (b) What is the change in pH upon diluting the solution in part (a) by a factor of 10? (c) What is the pH of a $1\,dm^3$ aqueous solution that contains $2.00\,g$ of dissolved NaOH?

**16.19** Calculate the pH of a $0.25\,mol\,dm^{-3}$ aqueous solution of $CH_3CO_2H$; $pK_a = 4.77$.

**16.20** Find the concentration of $OH^-$ ions in a $0.40\,mol\,dm^{-3}$ aqueous solution of $NH_3$ ($pK_b = 4.75$), and hence find the pH of the solution.

**16.21** Determine the pH of an aqueous solution of sulfuric acid of concentration $0.050\,mol\,dm^{-3}$. Data: see Table 16.4.

**16.22** By referring to equations 16.60–16.62, derive the equations:

$$\frac{[HA^-]}{[H_2A]_{initial}} = \frac{K_a(1)[H_3O^+]}{[H_3O^+]^2 + K_a(1)[H_3O^+] + K_a(1)K_a(2)}$$

and

$$\frac{[H_2A]}{[H_2A]_{initial}} = \frac{[H_3O^+]^2}{[H_3O^+]^2 + K_a(1)[H_3O^+] + K_a(1)K_a(2)}$$

given in Section 16.6 for the speciation of a dibasic acid $H_2A$.

**16.23** (a) Rationalize the shapes of the curves in Figure 16.3. (b) Using data from Table 16.3, sketch analogous speciation curves to those in Figure 16.3 to illustrate the behaviour of aqueous $H_3PO_4$ as a function of pH.

**16.24** (a) Briefly explain what a buffer solution is and qualitatively explain how it functions. (b) Calculate the pH of a buffer solution that is $0.50\,mol\,dm^{-3}$ with respect to both $HCO_2H$ ($pK_a = 3.75$) and $Na[HCO_2]$.

**16.25** A buffer solution is prepared by combining $50\,cm^3$ $0.025\,mol\,dm^{-3}$ $Na_2HPO_4$ and $50\,cm^3$ $0.018\,mol\,dm^{-3}$ $NaH_2PO_4$. Find the pH of the solution. Data: see Table 16.3.

**16.26** Sketch a titration curve for the addition of $15\,cm^3$ $0.10\,mol\,dm^{-3}$ aqueous KOH to $20\,cm^3$ $0.050\,mol\,dm^{-3}$ aqueous HCl. Determine the pH values at the start of the titration, at the equivalence point, and at the end of the titration when all the alkali has been added. What species are present in solution at the end of the experiment?

**16.27** Determine the pH of the equivalence point of a titration in which an aqueous solution ($0.10\,mol\,dm^{-3}$) of HCl is added to $25.0\,cm^3$ aqueous $NH_4OH$ ($0.10\,mol\,dm^{-3}$); $pK_a$ for $NH_4^+$ is 9.25.

**16.28** (a) Determine the pH values of the first and second equivalence points during the titration of $20\,cm^3$ aqueous NaOH ($0.020\,mol\,dm^{-3}$) against $20\,cm^3$ $H_3PO_4$ ($0.020\,mol\,dm^{-3}$). $pK_a$ values for $H_3PO_4$ are given in Table 16.3. (b) Malonic acid has the formula $CH_2(CO_2H)_2$ and has $pK_a$ values of 2.85 and 5.67. Sketch a titration curve to illustrate pH changes during the addition of $45.0\,cm^3$ of a $0.10\,mol\,dm^{-3}$ solution of NaOH to $20.0\,cm^3$ of a $0.10\,mol\,dm^{-3}$ solution of malonic acid. Calculate the pH at the equivalence points.

**16.29** Copy Figure 16.5 and mark on it horizontal bands to correspond to the pH ranges over which the indicators in Table 16.5 change colour. Confirm the choices made in the discussion in the last part of Section 16.9. Repeat the exercise for Figures 16.6 and 16.7. What problem is encountered when choosing an indicator for the titration shown in Figure 16.7, and how can this be overcome?

## Additional problems

**16.30** Acetic acid and ethanol react to establish the following equilibrium:

$$CH_3CO_2H + C_2H_5OH \rightleftharpoons CH_3CO_2C_2H_5 + H_2O$$

At 298 K, $K = 4.0$. If 0.30 moles of $CH_3CO_2H$, 0.45 moles of $C_2H_5OH$ and 0.20 moles of $H_2O$ are added together and the mixture is allowed to reach equilibrium, what is the composition of the equilibrium mixture?

**16.31** Look at equation 22.17 in Chapter 22; it shows the dissociation of $[Al(H_2O)_6]^{3+}$ in aqueous solution. Using data from the equation, calculate the pH of a $1.20 \times 10^{-3}\,mol\,dm^{-3}$ aqueous solution of $[Al(H_2O)_6]^{3+}$.

**16.32** Calculate the pH of an aqueous solution of HCl of concentration $1.00 \times 10^{-7}\,mol\,dm^{-3}$.

**16.33** What species are present in an aqueous solution of $Na_2CO_3$?

# 17 Thermodynamics

## Topics

- Internal energy and enthalpy
- Heat capacity
- Reaction quotient
- Gibbs energy and equilibria
- Entropy
- Solubility
- Solubility products

## 17.1 Introduction

In Chapter 15 we dealt with kinetics and considered the question: 'How *fast* is a reaction proceeding?' In Chapter 16, we focused on equilibria, discussing them in terms of equilibrium constants. Now, we discuss chemical *thermodynamics* and ask the questions: 'How *far* will a reaction proceed and how much energy will be consumed or released?' Up to this point, we have dealt with the thermodynamics of reactions in terms of a change in enthalpy, $\Delta H$. This *thermochemical quantity* (see Chapter 2) may provide an *indication* as to whether a reaction is thermodynamically viable or not, *but* some exothermic reactions have a high activation barrier and occur only slowly (or not at all on an observable timescale) and many reactions are endothermic. For example, $N_2H_4$ is kinetically stable with respect to decomposition into its constituent elements even though this process is exothermic (equation 17.1) and when ammonium chloride dissolves in water, the temperature falls indicating that the reaction is *endothermic* (equation 17.2). Why should reaction 17.2 proceed?

$$N_2H_4(g) \longrightarrow N_2(g) + 2H_2(g)$$

$$\Delta H = -50.6 \, \text{kJ per mole of } N_2H_4 \quad (17.1)$$

$$NH_4Cl(s) \xrightarrow{\text{Dissolve in water}} [NH_4]^+(aq) + Cl^-(aq)$$

$$\Delta H = +14.8 \, \text{kJ per mole of } NH_4Cl \quad (17.2)$$

These *thermochemical* data provide only *part* of the story as far as the energetics of a reaction are concerned. The enthalpy change tells us whether the temperature of the *surroundings* or the *system* is raised or lowered as heat is either given out or taken in. Heat is only one form of energy and in Chapter 4 we related the enthalpy change to the change in internal energy, $\Delta U$, of a system. We look again at this relationship in Section 17.2.

In this chapter, we shall see that the *change in Gibbs energy*, $\Delta G$, for a reaction is a better guide as to the favourability of the reaction than is $\Delta H$. Notice that, as for $\Delta U$, we are now using the term *energy* and not enthalpy. The change in Gibbs energy takes into account not only the change in enthalpy, but also the change in the *entropy* of the system. We explore the relationship between changes in Gibbs energy, entropy and enthalpy in Section 17.9, but first we introduce Gibbs energy by considering equilibria. We are already familiar with the physical significance of large and small values of equilibrium constants from Chapter 16. In principle, *all* reactions are equilibria, but a reaction for which the equilibrium constant is very large lies far over towards the right-hand side and to all intents and purposes the reaction has gone to completion. We begin by defining some terms that are fundamental to thermodynamics.

### The system and the surroundings

Two terms used extensively in thermodynamics are the '*system*' and the '*surroundings*'. The system consists of the reaction components (reactants and products) and is distinguishable from the surroundings which make up everything else in the universe except for the system. The system may be contained within the walls of a reaction vessel (Figure 17.1) or there may be no defined boundary, as for example when a piece of solid $CO_2$ (dry-ice) sublimes to give gaseous $CO_2$. In this case the system may seem to disappear into the surroundings but the molecules of $CO_2$ still constitute 'the system'. Thermochemical data, $\Delta H$, refer to the 'heat given out by the system to, or gained by the system from, the surroundings'.

An *open system* is able to exchange matter and energy with the surroundings. For example, if a reaction is carried out in an open flask, gaseous products can escape to the surroundings and energy can be exchanged with it; the system is therefore open. A *closed system* can exchange energy with the surroundings, but matter cannot be exchanged; e.g. if a reaction is carried out in a sealed tube, then the system is closed. An *isolated system* is completely insulated from the surroundings by an *adiabatic* wall and cannot exchange energy or matter.

A system may be *open*, *closed* or *isolated*; an *adiabatic* wall is thermally insulating.

**Fig. 17.1** The distinction between the system and its surroundings may be seen by considering a reaction in which a gas is produced. In this experiment, the gas ($CO_2$) is collected in a syringe which expands into the surroundings during the reaction.

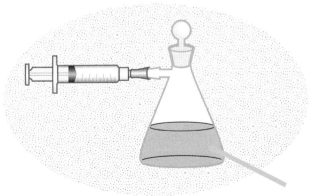

$$2KMnO_4(aq) + 8H_2SO_4(aq) + 5K_2C_2O_4(aq)$$

$$\downarrow$$

$$10CO_2(g) + 2MnSO_4(aq) + 8H_2O(l) + 6K_2SO_4(aq)$$

## State functions

The state of a system may be described by variable quantities such as temperature, pressure, volume, enthalpy, entropy, Gibbs energy and internal energy. These variables are called *state functions* and, provided we know how much material we have in the system, we need to specify only *two* state functions in order to know the other state functions for the system. Consider the ideal gas equation 17.3.

$$PV = nRT \tag{17.3}$$

where $R$ (molar gas constant) $= 8.314 \times 10^{-3} \, \text{kJ K}^{-1} \, \text{mol}^{-1}$

For a particular, fixed system, the amount of material is given by the number of moles, $n$. The interrelationship between pressure $P$, volume $V$ and temperature $T$ means that if we fix *any two* of these variables, then we necessarily know the third. This is the reason why we can specify that the volume of one mole of an ideal gas at 273 K and $10^5$ Pa is 22.7 dm$^3$. Similarly, if we know that at $10^5$ Pa the volume of one mole of the gas is 22.7 dm$^3$, the temperature must be 273 K.

Standard temperature and pressure: see Section 1.9

---

*State functions* are interrelated and knowing any *two state functions* automatically fixes *all the others*.

Some state functions are:

| Temperature | $T$ | Pressure | $P$ |
| Volume | $V$ | Internal energy | $U$ |
| Enthalpy | $H$ | Entropy | $S$ |
| Gibbs energy | $G$ |

---

An important property of a state function is that a change in the function is independent of the manner in which the change is made. We have already come across this in the form of Hess's Law of Constant Heat Summation, where $\Delta H$ for a reaction is the same irrespective of the thermochemical cycle used (Figure 17.2). Equation 17.4 is a general expression that applies to *any* state function.

Hess's Law of Constant Heat Summation: see Section 2.7

$\sum$ = summation of

$$\Delta(\text{State function}) = \sum (\text{State function})_{\text{products}}$$
$$- \sum (\text{State function})_{\text{reactants}} \tag{17.4}$$

## Standard states

Standard states were detailed in Section 2.2.

**Fig. 17.2** FeCl$_3$ may be formed directly from Fe and Cl$_2$, or by first forming FeCl$_2$. The enthalpy change in going from [Fe(s) + $\frac{3}{2}$Cl$_2$(g)] to FeCl$_3$(s) is the same, irrespective of the route: $\Delta H_1 = \Delta H_2 + \Delta H_3$.

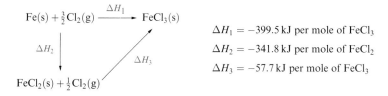

$$\text{Fe(s)} + \tfrac{3}{2}\text{Cl}_2(\text{g}) \xrightarrow{\Delta H_1} \text{FeCl}_3(\text{s})$$

$\Delta H_1 = -399.5 \, \text{kJ per mole of FeCl}_3$

$\Delta H_2 = -341.8 \, \text{kJ per mole of FeCl}_2$

$\Delta H_3 = -57.7 \, \text{kJ per mole of FeCl}_3$

$$\text{FeCl}_2(\text{s}) + \tfrac{1}{2}\text{Cl}_2(\text{g})$$

### Physical constants

Physical constants needed in the chapter are tabulated on the inside back cover of the book.

## 17.2     Internal energy, $U$

### The First Law of Thermodynamics

The ***First Law of Thermodynamics*** states that energy cannot be created or destroyed, merely changed from one form to another.

The internal energy, $U$, of a system is its *total energy*, and is a state function. The total energy is made up of kinetic energy, vibrational energy, rotational energy, etc. In a *perfectly isolated* system, the internal energy remains constant and for any change $\Delta U = 0$. Energy can be converted from one form to another, for example into heat or used to do work, but it cannot be created or destroyed. This is the *First Law of Thermodynamics*.

When the internal energy changes, energy *may* be released in the form of heat, and the temperature of the surroundings increases (provided that the system is not perfectly insulated). This corresponds to a transfer of energy from the system to the surroundings, and while the total energy of the system *plus* the surroundings remains constant (equation 17.5), there is a change in the internal energy of the system itself.

$$\text{Energy given out by system} = \text{Energy gained by surroundings} \qquad (17.5)$$

Alternatively, if the transfer of heat is in the opposite sense (equation 17.6), the internal energy of the system increases.

$$\text{Energy gained by system} = \text{Energy given out by surroundings} \qquad (17.6)$$

Accompanying the heat transfer, $q$, may be energy transfer in the form of 'work done', $w$, and equation 17.7 gives the general relationship for a change in internal energy in the system.

$$\text{Change in internal energy} = \Delta U = q + w \qquad (17.7)$$

where $q$ = change in heat energy (enthalpy) = $\Delta H$, and $w$ = work done

If $q$ is positive, heat energy is added to the system, and if $q$ is negative, heat is given out by the system to the surroundings.

'Work done' *on* the system includes such events as compressing a gas. Conversely, when a gas expands, work is done *by* the system on the surroundings. Consider the expansion of a gas against a frictionless piston upon which an external pressure, $P$, is acting:

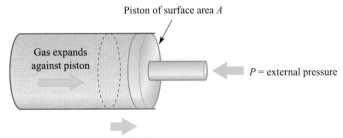

Piston of surface area $A$

Gas expands against piston

$P$ = external pressure

Distance, $d$, travelled by piston

Let the change in volume of the gas be $\Delta V$. From the diagram above:

$$\Delta V = dA$$

Pressure is force per unit area, and therefore:

$$\text{Pressure, } P = \frac{\text{Force}}{\text{Area, } A}$$

$$\text{Force opposing the expansion} = PA = \frac{P\Delta V}{d}$$

Work is given by the equation:

$$\text{Work} = \text{Force} \times \text{Distance}$$

and therefore, the work done by the system as the gas expands can be expressed in terms of the external pressure and change in volume (equation 17.8).

$$\begin{array}{l}\text{Work done at constant pressure} \\ \textbf{\textit{by}} \text{ the system } \textbf{\textit{on}} \text{ the surroundings}\end{array} = -\frac{P\Delta V}{d} \times d$$

$$= -P\Delta V \qquad (17.8)$$

If the pressure is given in Pa and the volume in $m^3$, the units of the work done are J (see Table 1.2). Note the sign convention: *work done* ***by*** *the system* ***on*** the surroundings is defined as negative work.

■▷
$\Delta U \approx \Delta H$: see Sections 4.5, 8.3 and 8.5

The relationship between $\Delta U$ and $\Delta H$ (equation 17.9) that we introduced earlier now follows by combining equations 17.7 and 17.8. In many reactions, the pressure $P$ corresponds to atmospheric pressure (1 atm = 101 300 Pa, or 1 bar = $1.00 \times 10^5$ Pa).

$$\Delta U = \Delta H - P\Delta V \qquad (17.9)$$

For a system at constant pressure doing work on the surroundings:

$$\left( \begin{array}{c} \text{Change in the internal} \\ \text{energy of the system} \end{array} \right) = (\text{Change in enthalpy})$$

$$- (\text{pressure} \times \text{change in volume})$$

$$\Delta U = \Delta H - P\Delta V$$

Figure 17.3 summarizes the changes in enthalpy and volume for a system, and emphasizes the sign convention.

### How large is the term $P\Delta V$ in relation to $\Delta H$?

The work done by or on the system is typically significantly smaller than the enthalpy change and, in many cases $\Delta U$ is very nearly equal to $\Delta H$. This was the approximation that we made in Chapter 4 when we discussed bond dissociation energies, and again in Chapter 8 when we took the lattice energy as being close in value to $\Delta_{\text{lattice}} H$.

Volume changes associated with reactions involving the formation or consumption of gases are always far larger than ones associated with liquids or solids, and volume changes due to the gaseous components of reactions predominate in the $P\Delta V$ term. Assuming that the ideal gas law (equation 17.3) holds for all gaseous components, we can find the work done by, or on, the system at constant pressure and temperature in terms of the change in the number of moles of gas (equation 17.10).

$$\text{Work done by the system} = -P\Delta V = -(\Delta n)RT \qquad (17.10)$$

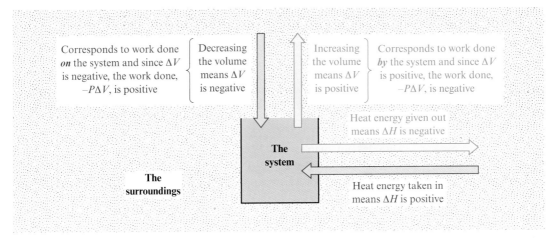

**Fig. 17.3** The system is the term given to the reaction components. The change in enthalpy of a system is given by $\Delta H = H_2 - H_1$ where $H_1$ and $H_2$ are the absolute enthalpies of the starting materials and the products respectively. *Absolute enthalpies cannot be measured; only $\Delta H$ can be found experimentally.* If $H_1 > H_2$, heat energy has been lost to the surroundings; if $H_2 > H_1$, heat energy has been taken in from the surroundings. The change in volume of the system is given by $\Delta V = V_2 - V_1$ where $V_1$ and $V_2$ are the absolute volumes of the starting materials and the products respectively. If the volume of the system increases, for example a gas is evolved in the reaction, then $V_2 > V_1$; if the volume of the system decreases, $V_1 > V_2$.

If the number of moles of gas increases, $\Delta n$ is positive and work is done *by* the system *on* the surroundings (Figure 17.4), i.e. the work done is negative. An example is the reaction between a metal carbonate and acid (equation 17.11).

$$
\left.
\begin{array}{l}
CuCO_3(s) + 2HCl(aq) \longrightarrow CuCl_2(aq) + H_2O(l) + CO_2(g) \\
\text{For 1 mole of } CuCO_3: \\
\Delta n = (\text{moles of gaseous products}) - (\text{moles of gaseous reactants}) \\
\quad = 1 - 0 = 1
\end{array}
\right\} \quad (17.11)
$$

If the number of moles of gas decreases, $\Delta n$ is negative and work is done *on* the system *by* the surroundings, i.e. the work done is positive. Examples are the oxidation of sulfur dioxide (equation 17.12) or the combustion of an alkane (equation 17.13).

$$
\left.
\begin{array}{l}
2SO_2(g) + O_2(g) \longrightarrow 2SO_3(g) \\
\text{For 1 mole of } SO_2: \\
\Delta n = (\text{moles of gaseous products}) - (\text{moles of gaseous reactants}) \\
\quad = 2 - 3 = -1
\end{array}
\right\} \quad (17.12)
$$

**Fig. 17.4** During the reaction of $[C_6H_5N_2]^+Cl^-$ with water, $N_2$ is produced, which fills the balloon. In doing so, the system does work on the surroundings.

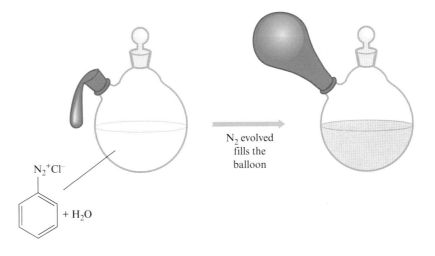

$N_2^+Cl^-$

$+ H_2O$

$N_2$ evolved fills the balloon

$$C_5H_{12}(l) + 8O_2(g) \longrightarrow 5CO_2(g) + 6H_2O(l)$$

For 1 mole of pentane:

$\Delta n =$ (moles of gaseous products) − (moles of gaseous reactants)

$\quad = 5 - 8 = -3$

(17.13)

---

**Worked example 17.1**    *Change in internal energy at standard temperature and pressure*

**By what amount will the change in internal energy differ from the change in enthalpy for the reaction of 0.0500 moles of $MgCO_3$ with an excess of dilute hydrochloric acid under conditions of 1 bar pressure and 273 K in the apparatus shown in Figure 17.1? Assume that the syringe is frictionless.**

**[1 mole of gas at 1 bar pressure and 273 K occupies a volume of 22.7 $dm^3$.]**

The equation needed is:

$$\Delta U = \Delta H - P\Delta V$$

The reaction is at constant pressure (1 bar) and the difference between $\Delta U$ and $\Delta H$ is the term $P\Delta V$.

First write a balanced equation for the reaction:

$$MgCO_3(s) + 2HCl(aq) \longrightarrow MgCl_2(aq) + H_2O(l) + CO_2(g)$$

0.0500 moles of $MgCO_3$ produce 0.0500 moles gaseous $CO_2$ and work is done by the system as the gas produced expands – it 'pushes' against the surrounding atmosphere.

There are two ways of dealing with the problem.

*Method 1:*

Volume occupied by 0.0500 moles of gas at 1 bar and 273 K

$$= 0.0500 \times (22.7\,dm^3) = 1.14\,dm^3$$

Increase in volume, $\Delta V =$ Final volume − initial volume $= 1.14\,dm^3$

*Note that the change in volume is assumed to be entirely due to the gas released.*

Now we must think about the units. By referring to Table 1.2, we see that compatible units of volume, energy and pressure are $m^3$, J and Pa (see also problem 17.1):

Work done by the system $= -P\Delta V$

With consistent units:

$$P = 1.00 \times 10^5\,Pa = 1.00 \times 10^5\,N\,m^{-2}$$

$$\Delta V = 1.14 \times 10^{-3}\,m^3$$

Refer to Table 1.2

Work done by the system $= -(1.00 \times 10^5\,N\,m^{-2}) \times (1.14 \times 10^{-3}\,m^3)$

$$= -114\,N\,m$$

$$= -114\,J$$

Per mole of $MgCO_3$:

Difference between $\Delta U$ and $\Delta H = \dfrac{(114\,J)}{(0.0500\,mol)}$

$$= 2280\,J\,mol^{-1}$$

$$= 2.28\,kJ\,mol^{-1}$$

*Method 2*: The second (and shorter) method of calculation uses the equation for an ideal gas:

$$PV = nRT$$

At constant pressure (1 bar) and temperature (273 K):

$$P\Delta V = (\Delta n)RT$$

where $\Delta n$ is the change in number of moles of gas $= 0.0500$ moles.

$$P\Delta V = (0.0500 \, \text{mol}) \times (8.314 \, \text{J K}^{-1} \, \text{mol}^{-1}) \times (273 \, \text{K}) = 113 \, \text{J}$$

The work is done *by* the system and is therefore negative $= -113 \, \text{J}$, and as above:

$$\text{Difference between } \Delta U \text{ and } \Delta H = \frac{(113 \, \text{J})}{(0.0500 \, \text{mol})}$$

$$= 2260 \, \text{J mol}^{-1}$$

$$= 2.26 \, \text{kJ mol}^{-1}$$

▣▷
*The difference in answers between the methods is due to rounding*

---

**Worked example 17.2**   *Change in internal energy at 1 bar pressure and 298 K*

**The same reaction as in worked example 17.1 is carried out at 1 bar pressure and 298 K. Calculate the work done by the system on the surroundings.**

As in worked example 17.1, there are two methods of calculation. Although both are given here, method 2 is shorter.

*Method 1*: The only difference from worked example 17.1 is that we have to correct the volume of gas for the change in temperature using the relationship:

$$\frac{P_1 V_1}{T_1} = \frac{P_2 V_2}{T_2}$$

The pressure is constant, $P_1 = P_2$, and so:

$$\frac{V_1}{T_1} = \frac{V_2}{T_2}$$

$$V_2 = \frac{V_1 T_2}{T_1} = \frac{(1.14 \times 10^{-3} \, \text{m}^3) \times (298 \, \text{K})}{(273 \, \text{K})} = 1.24 \times 10^{-3} \, \text{m}^3$$

Work done by the system $= -P\Delta V$. With consistent units:

▣▷
*Refer to Table 1.2*

$$\text{Work done } by \text{ the system} = -(1.00 \times 10^5 \, \text{N m}^{-2}) \times (1.24 \times 10^{-3} \, \text{m}^3)$$

$$= -124 \, \text{N m}$$

$$= -124 \, \text{J}$$

Per mole of $MgCO_3$:

$$\text{Difference between } \Delta U \text{ and } \Delta H = \frac{(124 \, \text{J})}{(0.0500 \, \text{mol})}$$

$$= 2480 \, \text{J mol}^{-1}$$

$$= 2.48 \, \text{kJ mol}^{-1}$$

*Method 2*: Assuming an ideal gas at $10^5$ Pa pressure, we can use the equation:

$$PV = nRT$$

and, at constant $T$ and $P$:

$$P\Delta V = (\Delta n)RT$$

where $\Delta n$ is the change in number of moles of gas $= 0.0500$ moles.

$$P\Delta V = (0.0500\,\text{mol}) \times (8.314\,\text{J K}^{-1}\,\text{mol}^{-1}) \times (298\,\text{K}) = 124\,\text{J}$$

The work is done *by* the system and is therefore negative $= -124\,\text{J}$. As above:

$$\text{Difference between } \Delta U \text{ and } \Delta H = \frac{(124\,\text{J})}{(0.0500\,\text{mol})}$$

$$= 2480\,\text{J mol}^{-1}$$

$$= 2.48\,\text{kJ mol}^{-1}$$

## 17.3 The bomb calorimeter

In Section 2.3, we introduced the use of calorimetry to measure heat changes that accompany chemical reactions and to measure the specific heat capacity of a substance. A simple, constant-pressure calorimeter was shown in Figure 2.2. However, such an apparatus is not used for accurate measurements. The *bomb calorimeter* is a constant-volume calorimeter (Figure 17.5) used for the precise determination of the heat change accompanying a reaction. The commonest use is for studying combustion reactions. Heat changes measured at constant volume correspond to changes in internal energy, $\Delta U$ (see equation 17.18). The apparatus consists of a metal bomb in which the reaction takes place. The bomb is a closed system (constant volume) and is immersed in a thermally insulated container filled with water. The combustion is initiated by an electrical heating device. The heat released by the reaction is transferred to the water surrounding the bomb. The water is stirred constantly, and its rise in temperature is recorded. Unlike the simple calorimeter that we described in Figure 2.2, the bomb calorimeter is of known heat capacity, $C_{\text{calorimeter}}$. The calorimeter can be calibrated by using the electrical source to

Specific heat capacity: see Section 2.3

**Fig. 17.5** A schematic representation of a bomb calorimeter used for studying combustion reactions.

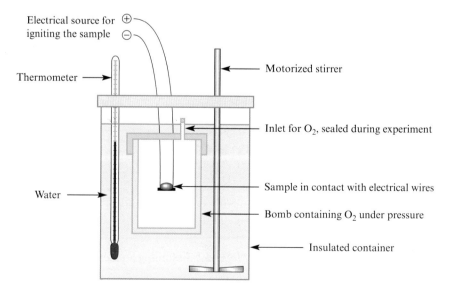

supply a known amount of heat, $q_{calorimeter}$, and measuring the temperature rise, $\Delta T$, of the water surrounding the bomb. The value of $C_{calorimeter}$ is then found from equation 17.14.

$$(q_{calorimeter} \text{ J}) = (C_{calorimeter} \text{ J K}^{-1}) \times (\Delta T \text{ K})$$

$$(C_{calorimeter} \text{ J K}^{-1}) = \frac{(q_{calorimeter} \text{ J})}{(\Delta T \text{ K})} \qquad (17.14)$$

When the sample is combusted, the rise in temperature of the water, $\Delta T_{sample}$, is used to determine $q_{sample}$. The heat gained by the calorimeter, $q_{calorimeter}$, equals the heat liberated on combustion of the sample (equation 17.15).

$$(q_{calorimeter} \text{ J}) = (C_{calorimeter} \text{ J K}^{-1}) \times (\Delta T_{sample} \text{ K})$$

$$(q_{sample} \text{ J}) = -(q_{calorimeter} \text{ J}) \qquad (17.15)$$

---

**Worked example 17.3**    *Calibrating a bomb calorimeter*

A bomb calorimeter is calibrated by using a 12 V electrical supply. A current of 8.0 A is passed through the apparatus for 500 s, and the rise in temperature of the water surrounding the bomb is 6.2 K. Determine the heat capacity of the calorimeter.

$$q_{calorimeter} = (\text{current in A}) \times (\text{potential difference in V}) \times (\text{time in s})$$

$$= (8.0 \text{ A}) \times (12 \text{ V}) \times (500 \text{ s})$$

The derived units (see Table 1.2) can be written in terms of base units:

$$q_{calorimeter} = (8.0 \text{ A}) \times (12 \text{ kg m}^2 \text{ s}^{-3} \text{ A}^{-1}) \times (500 \text{ s})$$

$$= 48\,000 \text{ kg m}^2 \text{ s}^{-2}$$

$$= 48\,000 \text{ J}$$

$$= 48 \text{ kJ}$$

The rise in temperature of the water, $\Delta T$, is 6.2 K. To find the heat capacity of the calorimeter:

$$q_{calorimeter} = (C_{calorimeter} \text{ kJ K}^{-1}) \times (\Delta T \text{ K})$$

$$C_{calorimeter} = \frac{(q_{calorimeter} \text{ kJ})}{(\Delta T \text{ K})}$$

$$= \frac{(48 \text{ kJ})}{(6.2 \text{ K})}$$

$$= 7.7 \text{ kJ K}^{-1}$$

---

**Worked example 17.4**    *The bomb calorimeter*

A calibrated bomb calorimeter has a heat capacity of 8.10 kJ K$^{-1}$. When 0.600 g of graphite is fully combusted in this calorimeter, the temperature of the water surrounding the bomb rises by 2.43 K. What is the heat change per mole of graphite? [Additional data: see the inside front cover of the book.]

Equation needed:

$$(q_{\text{calorimeter}}\,\text{kJ}) = (C_{\text{calorimeter}}\,\text{kJ K}^{-1}) \times (\Delta T_{\text{sample}}\,\text{K})$$

$$= (8.10\,\text{kJ K}^{-1}) \times (2.43\,\text{K})$$

$$= 19.7\,\text{kJ}$$

$$q_{\text{sample}} = -q_{\text{calorimeter}}$$

For 0.600 g of graphite, $q = -19.7\,\text{kJ}$

For 1 mole of graphite, $q = -\dfrac{(19.7\,\text{kJ}) \times (12.01\,\text{g mol}^{-1})}{(0.600\,\text{g})}$

$$= -394\,\text{kJ mol}^{-1}$$

---

## 17.4     Heat capacities

In Chapter 2, we introduced specific heat capacities, and in the previous section, we described how to measure the heat capacity of a bomb calorimeter. Now we look at molar heat capacities.

### Definitions

The *molar* heat capacity, $C$, is a *state function* and is the heat energy required to raise the temperature of one mole of a substance by one kelvin. Equation 17.16 expresses $C$ in a differential form and shows that the heat capacity is defined as the rate of change in $q$ with respect to temperature.

$$C = \frac{\mathrm{d}q}{\mathrm{d}T} \qquad \text{in J K}^{-1}\,\text{mol}^{-1} \tag{17.16}$$

However, we must go a step further and define the two distinct *conditions* under which heat capacity is defined:

- $C_P$ refers to a process at constant pressure.
- $C_V$ refers to a process at constant volume.

From equations 17.7 and 17.9, at constant pressure, $q$ is the change in enthalpy, $\Delta H$ and so we can write equation 17.17. The notation of the differential is examined in Box 17.1; the subscript $P$ in equation 17.17 refers to the fact that the process is occurring at constant pressure.

$$C_P = \left(\frac{\partial H}{\partial T}\right)_P \tag{17.17}$$

A condition of constant pressure refers to experiments such as those carried out at atmospheric pressure where gases are allowed to expand and the pressure remains at atmospheric pressure. In an experiment that is carried out in a *closed container* (for example, in a bomb calorimeter) the pressure may change, but the *volume is constant* and no work is done in expanding the gas. Since the change in volume is zero, equations 17.7 and 17.9 can be simplified to 17.18.

At constant volume: $\Delta U = q = \Delta H$ (17.18)

The heat capacity at constant volume is, therefore, defined in terms of the rate of change of internal energy (equation 17.19).

$$C_V = \left(\frac{\partial U}{\partial T}\right)_V \tag{17.19}$$

**THEORETICAL AND CHEMICAL BACKGROUND**

## Box 17.1 Partial differentials

An introduction to differentiation is given in the accompanying *Mathematics Tutor*, accessible via the website www.pearsoned.co.uk/housecroft. At this level of discussion, we are concerned only with the inter-dependence of two variables, $x$ and $y$. Now, we consider *partial differentiation* using the specific example of internal energy.

The internal energy of a system, $U$, is a function of temperature, $T$, and volume, $V$, i.e. $U$ depends on both these variables. This can be expressed in the form:

$$U = f(T, V)$$

If we differentiate $U$, we must differentiate with respect to each variable, one at a time, with the second variable held constant – this is called *partial differentiation*. The *total differential*, $dU$, is equal to the sum of the *partial differentials*:

$$dU = \left(\frac{\partial U}{\partial T}\right)_V dT + \left(\frac{\partial U}{\partial V}\right)_T dV$$

---

The **molar heat capacity**, $C$, is the heat energy required to raise the temperature of one mole of a substance by one kelvin, and the units are $J\,K^{-1}\,mol^{-1}$.

At constant pressure: $\quad C_P = \left(\dfrac{\partial H}{\partial T}\right)_P$

At constant volume: $\quad C_V = \left(\dfrac{\partial U}{\partial T}\right)_V$

---

### The relationship between $C_P$ and $C_V$

For *one mole* of an ideal gas, the enthalpy and internal energy are related by equation 17.20, which can be rearranged to give equation 17.21.

■▶ *PV = nRT*
For *n = 1, PV = RT*

$$U = H - PV = H - RT \tag{17.20}$$
$$H = U + RT \tag{17.21}$$

We can now write the expression for $C_P$ in terms of $U$ (equation 17.22).

$$C_P = \left(\frac{\partial H}{\partial T}\right)_P = \left(\frac{\partial\{U + RT\}}{\partial T}\right)_P \qquad \text{for 1 mole of ideal gas} \tag{17.22}$$

The differential on the right-hand side can be separated into two terms (remember that $R$ is a constant) to give equation 17.23.

$$C_P = \left(\frac{\partial U}{\partial T}\right)_P + R\left(\frac{\partial T}{\partial T}\right)_P = \left(\frac{\partial U}{\partial T}\right)_P + R \tag{17.23}$$

For an *ideal gas*, $\left(\dfrac{\partial U}{\partial T}\right)_P = \left(\dfrac{\partial U}{\partial T}\right)_V$, and this can be shown as follows. The internal energy, $U$, is a function of $T$ and $V$ and the total differential $dU$ is:

■▶ See Box 17.1

$$dU = \left(\frac{\partial U}{\partial T}\right)_V dT + \left(\frac{\partial U}{\partial V}\right)_T dV$$

Differentiating with respect to $T$ at constant pressure gives equation 17.24.

$$\left(\frac{\partial U}{\partial T}\right)_P = \left(\frac{\partial U}{\partial T}\right)_V\left(\frac{\partial T}{\partial T}\right)_P + \left(\frac{\partial U}{\partial V}\right)_T\left(\frac{\partial V}{\partial T}\right)_P \tag{17.24}$$

**Fig. 17.6** The variation of $C_P$ with temperature is not very great for diatomic molecules, but is significant for polyatomics.

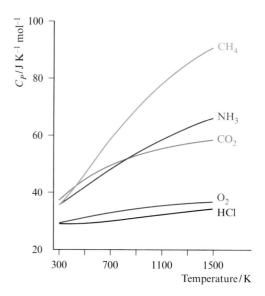

This can be simplified because $\left(\dfrac{\partial T}{\partial T}\right)_P = 1$, and for an *ideal gas*, $\left(\dfrac{\partial U}{\partial V}\right)_T = 0$. Equation 17.24 reduces to equation 17.25, and substitution of this into equation 17.23 gives equation 17.26.

$$\left(\frac{\partial U}{\partial T}\right)_P = \left(\frac{\partial U}{\partial T}\right)_V \tag{17.25}$$

$$C_P = \left(\frac{\partial U}{\partial T}\right)_V + R \tag{17.26}$$

By combining equations 17.26 and 17.19, we derive equation 17.27, which gives a relationship between the molar heat capacities at constant volume and constant pressure.

$$C_P = C_V + R \tag{17.27}$$

For one mole of an ideal gas: $C_P = C_V + R$

## The variation of $C_P$ with temperature

Most of the reactions with which we are concerned in this book are carried out under conditions of constant pressure and so we must look at $C_P$ in more detail.

Over *small* ranges of temperature, $C_P$ is *approximately* constant although Figure 17.6 illustrates that the variation of $C_P$ with temperature over the range 300–1500 K may be considerable. The variation is not very great for diatomic molecules, but is significant for polyatomic molecules. For larger molecules such as $C_2H_6$ and $C_3H_8$, $C_P$ varies from 52.7 to 145.9, and from 73.9 to 205.9 J K$^{-1}$ mol$^{-1}$ respectively between 300 and 1500 K.

### 17.5     The variation of $\Delta H$ with temperature: Kirchhoff's equation

So far we have treated $\Delta H$ as being temperature-independent. However, equation 17.17 shows that enthalpy and heat capacity are related and we have just seen that heat capacities *do* depend on temperature. It follows that $\Delta H$ is also *temperature-dependent*.

Equation 17.28 restates the definition of $C_P$. Both $H$ and $C_P$ are state functions and applying equation 17.4 allows us to write equation 17.29 where $\Delta H$

**THEORETICAL AND CHEMICAL BACKGROUND**

## Box 17.2 Integration of Kirchhoff's equation

Kirchhoff's equation is in the form of a differential:

$$\Delta C_P = \left(\frac{\partial \Delta H}{\partial T}\right)_P$$

Integrated forms are derived as follows.

Separating the variables in Kirchhoff's equation and integrating between the limits of the temperatures $T_1$ and $T_2$ gives:

$$\int_{T_1}^{T_2} \mathrm{d}(\Delta H) = \int_{T_1}^{T_2} \Delta C_P \,\mathrm{d}T$$

Let $\Delta H$ at temperature $T_1$ be $\Delta H_{(T_1)}$, and at $T_2$ be $\Delta H_{(T_2)}$. Thus:

$$\Delta H_{(T_2)} - \Delta H_{(T_1)} = \int_{T_1}^{T_2} \Delta C_P \,\mathrm{d}T$$

There are two ways of treating the right-hand side of the equation.

*Case 1*: Temperature-*independent* $\Delta C_P$. Assume that $\Delta C_P$ is a constant; this is often valid over small temperature ranges (see text).

$$\Delta H_{(T_2)} - \Delta H_{(T_1)} = \int_{T_1}^{T_2} \Delta C_P \,\mathrm{d}T = \Delta C_P \int_{T_1}^{T_2} \mathrm{d}T$$

$$\Delta H_{(T_2)} - \Delta H_{(T_1)} = \Delta C_P (T_2 - T_1)$$

(equation 17.30 in text)

*Case 2*: Temperature-*dependent* $\Delta C_P$. The variation of $C_P$ with $T$ can be empirically expressed as a power series of the type:

$$C_P = a + bT + cT^2 + \ldots$$

where $a$, $b$ and $c$ are constants. $\Delta C_P$ is similarly expressed, and so, if:

$$\Delta C_P = a' + b'T + c'T^2 + \ldots$$

$$\Delta H_{(T_2)} - \Delta H_{(T_1)} = \int_{T_1}^{T_2} \Delta C_P \,\mathrm{d}T$$

$$= \int_{T_1}^{T_2} (a' + b'T + c'T^2) \,\mathrm{d}T$$

$$\Delta H_{(T_2)} - \Delta H_{(T_1)} = \left[a'T + \frac{b'}{2}T^2 + \frac{c'}{3}T^3\right]_{T_1}^{T_2}$$

You are most likely to encounter Case 1.

For revision of integration, see the *Mathematics Tutor*, accessed via the website

www.pearsoned.co.uk/housecroft

Table 4 in the *Mathematics Tutor* lists standard integrals.

---

is the enthalpy change accompanying a reaction and $\Delta C_P$ is the difference in heat capacities (at constant pressure) of products and reactants.

$$C_P = \left(\frac{\partial H}{\partial T}\right)_P \tag{17.28}$$

$$\Delta C_P = \left(\frac{\partial \Delta H}{\partial T}\right)_P \qquad \textit{Kirchhoff's equation} \tag{17.29}$$

This expression is called *Kirchhoff's equation* and in its integrated form (equation 17.30) it can be used to determine how $\Delta H$ for a reaction varies with temperature; the integration steps are shown in Box 17.2.

$$\Delta H_{(T_2)} - \Delta H_{(T_1)} = \Delta C_P (T_2 - T_1) \tag{17.30}$$

where $\Delta H_{(T_2)}$ is the enthalpy change at temperature $T_2$, and $\Delta H_{(T_1)}$ is the enthalpy change at temperature $T_1$

*Kirchhoff's equation* shows how $\Delta H$ for a reaction varies with temperature:

$$\Delta C_P = \left(\frac{\partial \Delta H}{\partial T}\right)_P$$

Equation 17.30 is valid only if we assume that $\Delta C_P$ is independent of temperature. You can test this assumption by determining values of $\Delta C_P$ over the ranges 500–600 K and 1000–1100 K for each gas shown in Figure 17.6. You should be able to demonstrate that equation 17.30 generally works over only *small* temperature ranges. By studying Figure 17.6, deduce how 'small' is defined in this context.

The temperature dependence of the standard enthalpies of formation of selected compounds is illustrated in Figure 17.7. The temperature range is

**Fig. 17.7** The variation of $\Delta_f H^\circ$ with temperature is not very great for diatomic molecules, but is significant for polyatomic molecules; this trend is related to that observed in Figure 17.6 for heat capacities.

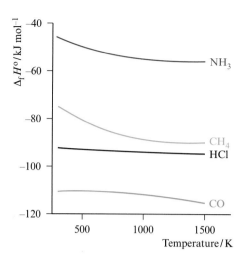

large (300–1500 K) and although $\Delta_f H^\circ(\text{HCl, g})$ varies little, $\Delta_f H^\circ(\text{CH}_4, \text{g})$ changes from $-74.7$ to $-90.2 \text{ kJ mol}^{-1}$. As we expect from Kirchhoff's equation, the *trends* in the variation of $\Delta_f H^\circ$ with temperature (Figure 17.7) are similar to those of $C_P$ with temperature (Figure 17.6).

## 17.6    Equilibrium constants and changes in Gibbs energy

### Equilibrium constants

*K*: see Chapter 16

Apart from $\Delta H$, the other thermodynamic quantities with which you are now familiar are equilibrium constants, $K$. Equilibrium constants can be defined for specific types of system, for example $K_a$ (acid dissociation constant), $K_b$ (base dissociation constant) and $K_w$ (self-ionization constant of water). As we saw in Chapter 16, equilibrium constants are expressed in terms of activities and are therefore dimensionless. However, for *ideal* gases, the activity of a gas is equal to the numerical value of its partial pressure, and for a solution species in *dilute* solution, the activity is equal to the numerical value of the concentration. You must keep these points in mind throughout the discussions and worked examples that follow.

The magnitude of $K$ tells us the position of the equilibrium and provides a measure of the *extent of the reaction* in either a forward or backward direction. Consider the general reaction 17.31. The larger the value of $K$ ($>1$), the more the products dominate over the reactants, and the smaller $K$ is ($<1$), the more the reactants predominate at the equilibrium state.

$$A + B \rightleftharpoons C + D \qquad\qquad K = \frac{[\text{C}][\text{D}]}{[\text{A}][\text{B}]} \qquad (17.31)$$

What do we mean by 'big' and 'small' values of $K$? The acid dissociation constants in Table 16.3 provide some indication of equilibria which lie towards the left-hand side, in favour of the reactants. The range of $K_a$ values for those selected acids was from $1.5 \times 10^{-2}$ for the first dissociation step of $H_2SO_3$, to $2.1 \times 10^{-13}$ for $[\text{HPO}_4]^{2-}$. Very large values of $K$ are exemplified by $3.5 \times 10^{41}$ for the formation of water (equation 17.32) or $1.8 \times 10^{48}$ for the formation of hydrogen fluoride (equation 17.33), both at 298 K. Both these reactions may be explosive and the large values of $K$ reflect the fact that the reactions proceed spontaneously in a forward direction.

◼▶
The form of the equilibrium sign shows that the equilibrium lies predominantly to the right

$$2H_2(g) + O_2(g) \; \longrightarrow\rightleftharpoons \; 2H_2O(l) \tag{17.32}$$

$$H_2(g) + F_2(g) \; \longrightarrow\rightleftharpoons \; 2HF(g) \tag{17.33}$$

The value of $K$ for equation 17.34 is $7.5 \times 10^2$ (at 298 K), and the equilibrium lies towards the right-hand side *but* there are significant amounts of $N_2$ and $H_2$ present in the reaction mixture at equilibrium.

$$N_2(g) + 3H_2(g) \rightleftharpoons 2NH_3(g) \tag{17.34}$$

### Equilibrium constants and $\Delta H$: a lack of consistency?

The standard enthalpies of formation for some selected compounds are given in Table 17.1 and span both exothermic and endothermic reactions. The third column of the table lists values of the equilibrium constants for the corresponding reactions. In most cases, an exothermic reaction is associated with a large value of $K$, and our previous use of $\Delta H$ as an indicator of reaction favourability seems to be fairly well justified. However, the data in the table reveal some inconsistencies. The *trend* in the values of $\Delta_f H^\circ$ is not fully consistent with the trend in the values of $K$. One particular inconsistency is with respect to the formation of $CH_3NH_2$ where the equilibrium constant, $K$, tells us the extent to which equilibrium 17.35 proceeds to the right-hand side – not very far according to a value of $1.8 \times 10^{-6}$, even though the reaction is exothermic.

$$C(gr) + \tfrac{1}{2}N_2(g) + \tfrac{5}{2}H_2(g) \rightleftharpoons CH_3NH_2(g) \tag{17.35}$$

Remember that *the position of equilibrium says nothing about the rate of reaction*. If you mix $N_2$ and $H_2$ (equation 17.34) or graphite, $N_2$ and $H_2$ (equation 17.35) in a reaction vessel in the laboratory, you should not expect to obtain $NH_3$ or $CH_3NH_2$.

Striking inconsistencies between changes in enthalpy and the extent of reaction are also observed in solubility data. Table 17.2 lists the solubilities in water of selected salts and also the standard enthalpy changes that accompany dissolution. The process to which $\Delta_{sol}H^\circ(298\,K)$ refers is given in equation 17.36 for a salt XY; more generally, $\Delta_{sol}H$ refers to dissolution in any solvent (equation 17.37). The *molar enthalpy of solution* refers to the complete dissolution of one mole of XY.

The *standard enthalpy of solution* is denoted by $\Delta_{sol}H^\circ$.

**Table 17.1** Standard enthalpy and Gibbs energy changes, and values of $K$ for the formation of selected compounds at 298 K from their constituent elements in their standard states.

| Compound | $\Delta_f H^\circ$ / kJ mol$^{-1}$ | $K$ | $\Delta_f G^\circ$ / kJ mol$^{-1}$ |
|---|---|---|---|
| $CO_2(g)$ | $-393.5$ | $1.2 \times 10^{69}$ | $-394.4$ |
| $SO_2(g)$ | $-296.8$ | $4.0 \times 10^{52}$ | $-300.1$ |
| $HF(g)$ | $-273.3$ | $1.8 \times 10^{48}$ | $-275.4$ |
| $CO(g)$ | $-110.5$ | $1.1 \times 10^{24}$ | $-137.2$ |
| $HCHO(g)$ | $-108.7$ | $9.8 \times 10^{17}$ | $-102.7$ |
| $HCl(g)$ | $-92.3$ | $5.0 \times 10^{16}$ | $-95.3$ |
| $NH_3(g)$ | $-45.9$ | $7.5 \times 10^{2}$ | $-16.4$ |
| $HBr(g)$ | $-36.3$ | $2.2 \times 10^{9}$ | $-53.4$ |
| $CH_3NH_2(g)$ | $-22.5$ | $1.8 \times 10^{-6}$ | $+32.7$ |
| $BrCl(g)$ | $+14.6$ | $1.5$ | $-1.0$ |
| $HI(g)$ | $+26.5$ | $0.5$ | $+1.7$ |
| $NO_2(g)$ | $+34.2$ | $6.9 \times 10^{-10}$ | $+52.3$ |
| $NO(g)$ | $+91.3$ | $4.5 \times 10^{-16}$ | $+87.6$ |

**Table 17.2** Solubilities in water of selected salts (at the stated temperature) and standard molar enthalpies of solution ($\Delta_{sol}H^\circ$) at 298 K.

| Compound | $\Delta_{sol}H^\circ$ / kJ mol$^{-1}$ | Solubility (at stated temperature) / mol dm$^{-3}$ | Solubility (at stated temperature) / g dm$^{-3}$ |
|---|---|---|---|
| KOH | $-57.6$ | 19.1 (288 K) | 1070 (288 K) |
| LiCl | $-37.0$ | 15.0 (273 K) | 637 (273 K) |
| NaI | $-7.5$ | 12.3 (298 K) | 1845 (298 K) |
| NaCl | $+3.9$ | 6.1 (273 K) | 357 (273 K) |
| NH$_4$Cl | $+14.8$ | 5.6 (273 K) | 295 (273 K) |
| NaI·2H$_2$O | $+16.1$ | 17.1 (273 K) | 3180 (273 K) |
| AgNO$_3$ | $+22.6$ | 7.2 (273 K) | 1224 (273 K) |
| NH$_4$NO$_3$ | $+25.7$ | 14.8 (273 K) | 1184 (273 K) |
| KMnO$_4$ | $+43.6$ | 0.4 (293 K) | 63 (293 K) |

$$XY(s) \xrightarrow{\text{Water}} X^+(aq) + Y^-(aq) \tag{17.36}$$

$$XY(s) \xrightarrow{\text{Solvent}} X^+(solv) + Y^-(solv) \tag{17.37}$$

It seems quite remarkable that salts such as ammonium nitrate that are very soluble in water dissolve in endothermic processes.

## Changes in standard Gibbs energy, $\Delta G^\circ$, and the standard Gibbs energy of formation, $\Delta_f G^\circ$

We must now introduce a new state function called the *Gibbs energy, G*. In Section 17.9 we define *G* fully, but in this and the next section we study some experimental data and illustrate how the change in standard Gibbs energy, $\Delta G^\circ$, for a reaction is related to the equilibrium constant.

Since *G* is a state function we can determine changes in Gibbs energy for reactions by using equation 17.38.

$$\Delta G = \sum G_{\text{products}} - \sum G_{\text{reactants}} \tag{17.38}$$

▣▷
Standard states:
see Section 2.2

The *standard Gibbs energy change* of a reaction, $\Delta_r G^\circ(T)$ (equation 17.39), refers to a system in which the reactants and products are in their standard states at a specified temperature, *T*, and at a pressure of 1 bar. As we see later, $\Delta G^\circ$ *is temperature-dependent* and it is vital to specify *T* when quoting values of $\Delta G^\circ$.

$$\Delta_r G^\circ(298\,\text{K}) = \sum [\Delta_f G^\circ(298\,\text{K})_{\text{products}}] - \sum [\Delta_f G^\circ(298\,\text{K})_{\text{reactants}}] \tag{17.39}$$

The *standard Gibbs energy of formation* of a compound, $\Delta_f G^\circ(T)$, is the Gibbs energy change that accompanies the formation of one mole of a compound in its standard state from its constituent elements in their standard states. By definition, the standard Gibbs energy of formation *and* the standard enthalpy of formation of an *element* in its standard state are zero, at all temperatures. However, $\Delta_f G^\circ$ for a *compound* is temperature-dependent (see Section 17.7).

By definition, $\Delta_f G^\circ$ and $\Delta_f H^\circ$ of an *element* in its standard state are zero, at all temperatures.

The ***standard Gibbs energy of formation*** of a compound, $\Delta_f G^\circ(T)$, is the change in Gibbs energy that accompanies the formation of one mole of a compound in its standard state from its constituent elements in their standard states.

| Worked example 17.5 | *The change in standard Gibbs energy* |

Determine $\Delta_r G^\circ(298\,K)$ for the combustion of one mole of $C_2H_5OH$ given that $\Delta_f G^\circ(298\,K)$ $C_2H_5OH(l)$, $H_2O(l)$ and $CO_2(g)$ are $-175$, $-237$ and $-394\,kJ\,mol^{-1}$ respectively.

First, write a balanced equation for the combustion of $C_2H_5OH$:

$$C_2H_5OH(l) + 3O_2(g) \xrightarrow{\Delta_r G^\circ} 2CO_2(g) + 3H_2O(l)$$

The standard Gibbs energy of formation of $O_2(g) = 0$ (by definition).

$$\Delta_r G^\circ = [2 \times \Delta_f G^\circ(CO_2, g)] + [3 \times \Delta_f G^\circ(H_2O, l)] - \Delta_f G^\circ(C_2H_5OH, l)$$

$$= [2 \times (-394)] + [3 \times (-237)] - (-175)$$

$$= -1324\,kJ \text{ per mole of } C_2H_5OH \text{ (at 298 K)}$$

In order to assess the advantages of using $\Delta G^\circ$ instead of $\Delta H^\circ$, we return to Table 17.1, where standard Gibbs energies of formation are also listed. In contrast to the inconsistencies in the trends of $\Delta_f H^\circ$ and $K$ that we noted earlier, we now see that the trend in values of $\Delta_f G^\circ$ *does* follow that in the values of $K$. A value of $K > 1$ corresponds to a *negative* change in Gibbs energy, and if $K < 1$, $\Delta G^\circ$ is *positive*. The magnitude of the equilibrium constant for the formation of hydrogen iodide (equation 17.40) indicates that the equilibrium lies towards the left-hand side, although there are significant amounts of products in the equilibrium mixture. Although the corresponding change in Gibbs energy is positive, it is small.

▣▷
Strictly, $K$ should be
expressed in terms of
activities: see Section 16.3

$$\tfrac{1}{2}H_2(g) + \tfrac{1}{2}I_2(g) \rightleftharpoons HI(g) \qquad K = \frac{(P_{HI})}{(P_{H_2})^{\frac{1}{2}}(P_{I_2})^{\frac{1}{2}}} = 0.5 \qquad (17.40)$$

The change in Gibbs energy for a reaction is a *reliable* guide to the extent of a reaction *provided that equilibrium is reached* and, in this regard, is a far more useful thermodynamic quantity than $\Delta H$. Remember that $\Delta G^\circ$ and $K$ say nothing about the *rate* of the reaction.

**17.7**

## The temperature dependence of $\Delta G^\circ$ and $K_p$: some experimental data

In this section we consider some experimental data to assess how $K_p$ (the equilibrium constant for a gaseous system) and $\Delta G^\circ$ for a given reaction vary with temperature.

### Case study 1: the formation of NH₃

▣▷
Haber process:
see Box 21.2

▣▷
$\ln = \log_e$

The Haber process is a major industrial process for the production of $NH_3$ from $N_2$ and $H_2$ (reaction 17.41). In selecting optimum conditions for the manufacture of $NH_3$, the variation of $K_p$ with temperature has to be taken into account. Over the temperature range 300–1500 K, $K_p$ varies from $6.7 \times 10^2$ to $5.8 \times 10^{-4}$, and the logarithmic scale ($\ln K_p$) in Figure 17.8a

**Fig. 17.8** (a) The variation with temperature of $\ln K_p$ for the formation of ammonia (equation 17.41), and (b) the variation of $\Delta_f G^\circ(NH_3, g)$ with temperature.

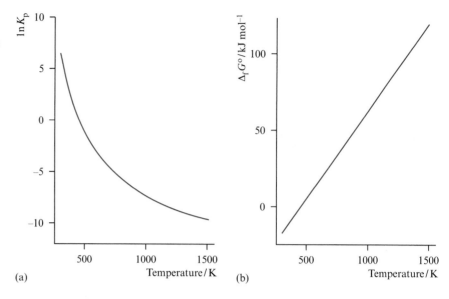

allows this large range of values to be represented more conveniently than plotting $K_p$ itself.

$$\tfrac{1}{2}N_2(g) + \tfrac{3}{2}H_2(g) \rightleftharpoons NH_3(g) \qquad K_p = \frac{(P_{NH_3})}{(P_{N_2})^{\frac{1}{2}}(P_{H_2})^{\frac{3}{2}}} \tag{17.41}$$

Figure 17.8b illustrates that $\Delta_f G^\circ$ per mole of $NH_3$ also varies significantly as a function of temperature; compare this with the much smaller variation in $\Delta_f H^\circ(NH_3, g)$ shown in Figure 17.7. As the temperature is increased, $\Delta_f G^\circ$ becomes increasingly *positive* and the decomposition of $NH_3$ to $N_2$ and $H_2$ is the predominant process. This result is mirrored in the trend in the equilibrium constants where at high temperatures, $\ln K_p$ is large and negative, i.e. $K_p$ is *very* small.

### Case study 2: metal oxide reduction

In Chapter 21, we will see several examples of the reduction of a metal oxide by another metal. Gibbs energy data allow us to approach the problem *quantitatively*.

Figure 17.9 shows the temperature variation of $\Delta_f G^\circ$ for a range of metal oxides. In order that values may be compared to each other, $\Delta_f G^\circ$ refers to the Gibbs energy of formation *per half-mole of* $O_2$.[§]

Thus for CaO, $\Delta_f G^\circ$ refers to reaction 17.42, and for $Al_2O_3$ it corresponds to reaction 17.43; we show both equations with the metal in the solid state, but this is temperature-dependent (see Figure 17.9).

$$Ca(s) + \tfrac{1}{2}O_2(g) \longrightarrow CaO(s) \tag{17.42}$$

$$\tfrac{2}{3}Al(s) + \tfrac{1}{2}O_2(g) \longrightarrow \tfrac{1}{3}Al_2O_3(s) \tag{17.43}$$

**Phase changes: see Section 17.10**

Each plot in Figure 17.9 is either linear (as in the case of NiO, for example) or has two linear sections (e.g. $Al_2O_3$). The change in gradient (which may be small) part way along a line occurs at the melting point of the metal. Two

---

[§] We could equally well have plotted the data per mole of $O_2$.

**Fig. 17.9** An Ellingham diagram showing how the standard Gibbs energies of formation of several metal oxides and carbon monoxide (the pink line) vary as a function of temperature. Values of $\Delta_f G^\circ$ refer to formation reactions involving a half-mole of $O_2$:

$$M + \tfrac{1}{2}O_2 \longrightarrow MO,$$

$$\tfrac{1}{2}M + \tfrac{1}{2}O_2 \longrightarrow \tfrac{1}{2}MO_2, \text{ or}$$

$$\tfrac{2}{3}M + \tfrac{1}{2}O_2 \longrightarrow \tfrac{1}{3}M_2O_3.$$

The points marked ◆ and ◇ are the melting and boiling points, respectively, of the elemental metal.

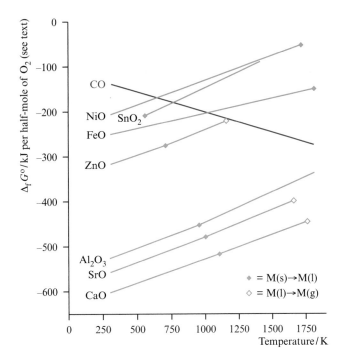

observations from Figure 17.9 are that:

- each metal oxide is *less* thermodynamically stable (less negative $\Delta_f G^\circ$) at higher temperatures, and
- the *relative* stabilities of the oxides at any given temperature can be seen directly from the graph.

THEORETICAL AND CHEMICAL BACKGROUND

## Box 17.3 The thermite reaction

Aluminium reacts with many metal oxides when heated in the solid state; the aluminium is oxidized and the metal oxide reduced to the respective metal. The general name for this reaction is the *thermite reaction* or *thermite process*, although this name is sometimes used for the specific reaction of iron(III) oxide with aluminium:

$$Fe_2O_3 + 2Al \xrightarrow{\text{heat}} 2Fe + Al_2O_3$$

$$3BaO + 2Al \xrightarrow{\text{heat}} 3Ba + Al_2O_3$$

On a small scale, the two powdered reactants may be mixed and placed in a crucible. A 'fuse' of magnesium ribbon is placed partly into the mixture and the free end is ignited. The reactions between aluminium and most metal oxides are violently exothermic and the temperature is raised such that the metallic product is usually formed in the molten state.

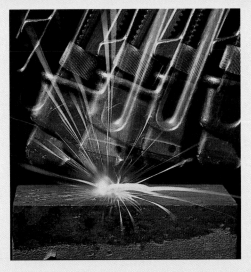

Sparks caused by a thermite reaction as an aluminium wrench strikes a block of rusty iron (iron oxide). *Crown Copyright/ Health & Safety Laboratory/Science Photo Library.*

The second point means that the plots in Figure 17.9 (which constitute an *Ellingham diagram*) can be used to predict which metal(s) will reduce a particular metal oxide, and whether the operating temperature will influence the efficiency of the reduction. The pink line in Figure 17.9 shows the way in which $\Delta_f G^o(CO, g)$ varies with temperature. This plot contrasts with all the others since $CO(g)$ becomes *more* thermodynamically stable at higher temperatures (more negative $\Delta_f G^o$). As the temperature is raised, carbon becomes more effective at reducing a wide variety of metal oxides.[§] Above 1800 K, a greater range of metal oxides than Figure 17.9 shows can be reduced by carbon, although industrially this method of obtaining a metal from its oxide is often not commercially viable.

| **Worked example 17.6** | *Using an Ellingham diagram* |
| --- | --- |

**(a) At an operating temperature of 800 K, which metal oxides in Figure 17.9 will carbon reduce? (b) Estimate the approximate $\Delta_r G^o$ at 800 K for the reduction of NiO by carbon.**

(a) First draw a vertical line on Figure 17.9 at $T = 800$ K. CO is the product when carbon is used as a reducing agent and so as CO is *more* thermodynamically stable than $SnO_2$ or NiO, carbon should reduce these two oxides:

$$C(s) + NiO(s) \longrightarrow CO(g) + Ni(s)$$
$$C(s) + \tfrac{1}{2}SnO_2(s) \longrightarrow CO(g) + \tfrac{1}{2}Sn(l)$$

(b) Since Gibbs energy is a state function, we can apply a Hess-type cycle to find $\Delta_r G^o$:

$$C(s) + NiO(s) \longrightarrow CO(g) + Ni(s)$$
$$\Delta_r G^o = [\Delta_f G^o(Ni, s) + \Delta_f G^o(CO, g)] - [\Delta_f G^o(C, s) + \Delta_f G^o(NiO, s)]$$

where all $\Delta G^o$ values refer to 800 K. For each of the elements in their standard state

$$\Delta_f G^o(Ni, s) = \Delta_f G^o(C, gr) = 0$$

From the graph, read off values of $\Delta_f G^o$ CO and $\Delta_f G^o$ NiO at 800 K:

$$\Delta_f G^o(CO, g) \approx -180 \, kJ \, mol^{-1}$$

$$(per \; \tfrac{1}{2} \; mole \; O_2 \; corresponds \; to \; a \; mole \; of \; CO)$$

$$\Delta_f G^o(NiO, s) \approx -150 \, kJ \, mol^{-1}$$

$$(per \; \tfrac{1}{2} \; mole \; O_2 \; corresponds \; to \; a \; mole \; of \; NiO)$$

$$\Delta_r G^o = \Delta_f G^o(CO, g) - \Delta_f G^o(NiO, s)$$
$$\approx -180 - (-150)$$
$$\approx -30 \, kJ \; per \; mole \; of \; reaction$$

The negative value of $\Delta_r G^o$ is in agreement with the prediction that carbon will reduce nickel oxide.

---

[§] The temperature dependences of $\Delta G$ and of $K_p$ are quantified in the Gibbs–Helmholtz equation and the van't Hoff isochore (see equation 17.58). Details may be found in P. Atkins and J. de Paula (2002) *Atkins' Physical Chemistry*, 7th edn, Oxford University Press.

# The relationship between $\Delta G^\circ$ and $K$: the reaction isotherm

### Deriving the reaction isotherm

Equation 17.44 gives the relationship between the molar Gibbs energy of an ideal gas, $G$, and its partial pressure $P$. In equation 17.44, $G^\circ$ is the standard molar Gibbs energy of the gas, $R$ is the molar gas constant, and $T$ is the temperature in K.

$$G = G^\circ + RT \ln P \tag{17.44}$$

Activities: see Section 16.3

For an ideal gas, the numerical value of the partial pressure is equal to the *activity*, $a$, and so we can rewrite equation 17.44 in the form of equation 17.45.

$$G = G^\circ + RT \ln a \tag{17.45}$$

Application of this equation extends beyond ideal gases to solution species, and for a *dilute solution*, the activity is equal to the numerical value of the solution concentration.

Now consider a general reaction:

$$\text{Reactants} \longrightarrow \text{Products}$$

for which the Gibbs energy change is $\Delta_r G$ and the standard Gibbs energy change is $\Delta_r G^\circ$. From equation 17.45, we can write equation 17.46 for the reaction.

$$\Delta_r G = \Delta_r G^\circ + (RT \ln a_{\text{products}} - RT \ln a_{\text{reactants}})$$

$$\Delta_r G = \Delta_r G^\circ + RT \ln \frac{a_{\text{products}}}{a_{\text{reactants}}} \tag{17.46}$$

The ratio of activities of products to reactants is called the *reaction quotient*, $Q$, and equation 17.46 can be expressed in the form of equation 17.47.

$$\Delta_r G = \Delta_r G^\circ + RT \ln Q \tag{17.47}$$

For a general reaction

$$x\text{X} + y\text{Y} \rightleftharpoons z\text{Z}$$

the value of $Q$ is expressed in terms of the activities of X, Y and Z as shown in equation 17.48.

$$Q = \frac{(a_Z)^z}{(a_X)^x (a_Y)^y} \tag{17.48}$$

This expression is similar to equation 16.12 in which we expressed the equilibrium constant, $K$, in terms of the equilibrium activities of X, Y and Z. There is, however, an important difference. Whereas the equilibrium constant, $K$, refers *only* to a system in equilibrium, the reaction quotient, $Q$, can be calculated at any time during a reaction by substituting into equation 17.48 values of $a_X$, $a_Y$ and $a_Z$ determined at that particular instant. Thus, for a given reaction, the value of $Q$ varies during the reaction, whereas (at a given temperature), $K$ is a constant value.

For a reaction involving ideal gases:

$$x\text{X(g)} + y\text{Y(g)} \rightleftharpoons z\text{Z(g)}$$

we can write an expression for $Q$ in terms of the numerical values of the partial pressures (equation 17.49).

$$Q = \frac{(P_Z)^z}{(P_X)^x (P_Y)^y} \tag{17.49}$$

For a reaction in dilute solution:

$$xX(aq) + yY(aq) \rightleftharpoons zZ(aq)$$

an expression for $Q$ can be written in terms of the numerical values of the concentrations (equation 17.50).

$$Q = \frac{[Z]^z}{[X]^x[Y]^y} \tag{17.50}$$

---

**Worked example 17.7**  *Determination of a reaction quotient*

Gaseous $N_2O_4$ dissociates according to the following equation:

$$N_2O_4(g) \rightleftharpoons 2NO_2(g)$$

The system will eventually come to equilibrium and, at 350 K, $K = 3.8$. At a given moment, the system contains $0.160\,mol\,N_2O_4$ and $0.121\,mol\,NO_2$ and the total pressure is 2.00 bar. Determine the reaction quotient. Has a position of equilibrium been reached?

First, find the partial pressures of $N_2O_4$ and $NO_2$.

$$\text{Partial pressure of component X} = \left(\frac{\text{Moles of X}}{\text{Total moles}}\right) \times \text{Total pressure}$$

Total moles of $N_2O_4$ and $NO_2 = 0.160 + 0.121 = 0.281\,mol$

$$\text{Partial pressure of } N_2O_4 = \left(\frac{0.160\,mol}{0.281\,mol}\right) \times (2.00\,bar) = 1.14\,bar$$

$$\text{Partial pressure of } NO_2 = \left(\frac{0.121\,mol}{0.281\,mol}\right) \times (2.00\,bar) = 0.861\,bar$$

The reaction quotient is the ratio of activities of products to reactants. To find $Q$, use the numerical values of the partial pressures, assuming the gases to be ideal (see Section 16.3):

$$Q = \frac{(P_{NO_2})^2}{(P_{N_2O_4})}$$

$$= \frac{0.861^2}{1.14}$$

$$= 0.650$$

This value is smaller than the equilibrium constant ($K = 3.8$). The system has therefore not yet attained equilibrium and will continue to move to the right-hand side. As it does so, the value of $Q$ increases, approaching the value of $K$.

---

**Worked example 17.8**  *Determination of $\Delta_r G$ for a reaction*

Consider the following reaction:

$$H_2(g) + Cl_2(g) \rightleftharpoons 2HCl(g)$$

for which $\Delta_r G^\circ(298\,K) = -190\,kJ\,mol^{-1}$. A reaction mixture at 298 K contains $0.020\,mol\,H_2$, $0.030\,mol\,Cl_2$ and $0.500\,mol\,HCl$ and the total pressure is 1.00 bar. Determine values of $Q$ and $\Delta_r G$ for the reaction under these conditions. Comment on the direction in which the reaction will proceed.

First, find the partial pressures of the components in the gaseous mixture.

Total moles of $H_2$, $Cl_2$ and HCl $= 0.020 + 0.030 + 0.500 = 0.550\,mol$

Partial pressure of $H_2 = \left(\dfrac{0.020\,mol}{0.550\,mol}\right) \times (1.00\,bar) = 0.036\,bar$

Partial pressure of $Cl_2 = \left(\dfrac{0.030\,mol}{0.550\,mol}\right) \times (1.00\,bar) = 0.055\,bar$

Partial pressure of HCl $= \left(\dfrac{0.500\,mol}{0.550\,mol}\right) \times (1.00\,bar) = 0.909\,bar$

Assuming the gases are ideal, we can determine the reaction quotient, $Q$, from the numerical values of the partial pressures (i.e. the activities):

$$Q = \frac{(P_{HCl})^2}{(P_{H_2})(P_{Cl_2})}$$

$$= \frac{(0.909)^2}{(0.036)(0.055)}$$

$$= 420 \quad \text{(to 2 sig. figs)}$$

The equation needed to find the corresponding value of $\Delta_r G(298\,K)$ is:

$$\Delta_r G = \Delta_r G^\circ + RT \ln Q$$

The value of $\Delta_r G^\circ(298\,K)$ for the reaction is $-190\,kJ\,mol^{-1}$.

$$\Delta_r G = \Delta_r G^\circ + RT \ln Q$$

$$= (-190\,kJ\,mol^{-1}) + \{(8.314 \times 10^{-3}\,kJ\,K^{-1}\,mol^{-1}) \times (298\,K) \times \ln 420\}$$

$$= (-190\,kJ\,mol^{-1}) + (15\,kJ\,mol^{-1})$$

$$= -175\,kJ\,mol^{-1} \quad \text{(i.e. per mole of reaction)}$$

The large negative value of $\Delta_r G$ indicates that the reaction mixture will spontaneously move to the right-hand side until an equilibrium position is attained. The closer the gas mixture comes to equilibrium, the less negative $\Delta_r G$ becomes.

---

As a reaction proceeds, the activities of products and reactants change until, eventually, equilibrium is reached. If at a given time, $Q < K$, the system will move towards the right-hand side. If $Q > K$, the reaction will move to the left-hand side. When $Q = K$, equilibrium has been reached and $\Delta_r G = 0$.

For the specific case of a system *at equilibrium*, equation 17.47 becomes equation 17.51, and this expression (the *reaction isotherm*) is of prime importance in thermodynamics. Note in particular that at equilibrium, $\Delta_r G$ (*not* $\Delta_r G^\circ$) equals zero. The exponential form of the reaction isotherm is given in equation 17.52.

At equilibrium:

$$Q = K$$

and

$$\Delta_r G = 0$$

$$0 = \Delta_r G^\circ + RT \ln K$$

$$\Delta_r G^\circ = -RT \ln K \qquad \text{or} \qquad \Delta G^\circ = -RT \ln K \qquad (17.51)^\S$$

where $\Delta G^\circ$ is in $kJ\,mol^{-1}$, and $R = 8.134 \times 10^{-3}\,kJ\,K^{-1}\,mol^{-1}$

$$K = e^{-\left(\frac{\Delta G^\circ}{RT}\right)} \qquad (17.52)$$

---

$\S$ Equation 17.51 is sometimes expressed in terms of $\log K$ and becomes: $\Delta G^\circ = -2.303 RT \log K$.

**Fig. 17.10** Plot of $\Delta G^{\circ}$ against $\ln K$ at 298 K.

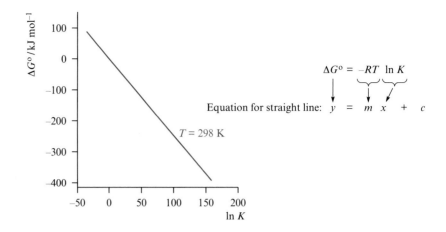

The **reaction isotherm** relates the standard Gibbs energy change for a reaction to the equilibrium constant at a given temperature:

$$\Delta G^{\circ} = -RT \ln K$$

## Using the reaction isotherm

Figure 17.10 illustrates the linear relationship between $\Delta G^{\circ}$ and $\ln K$. The line in Figure 17.10 has a gradient of $-RT$ and passes through the origin.

We are now able to quantify the statements made at the end of Section 17.6. A range of values of $\Delta G^{\circ}$ and corresponding equilibrium constants at 298 K are shown in Figure 17.11; the pairs of values are related by equation 17.51. If $\Delta G^{\circ}$ is negative, $K$ is greater than 1 and the products predominate over reactants. If $\Delta G^{\circ}$ is positive, $K$ is less than 1 and the reactants are dominant. One point on the diagram in Figure 17.11 shows that when $\Delta G^{\circ} = 0$, $K = 1$. This corresponds to a particular case when the standard Gibbs energy of the reactants equals that of the products. This case of $\Delta G^{\circ} = 0$ must not be confused with the condition for equilibrium that $\Delta G = 0$.

**Fig. 17.11** Values of $\Delta G^{\circ}$ and corresponding values of $K$ at 298 K. An indication is also given of the physical significance of the magnitude of such values.

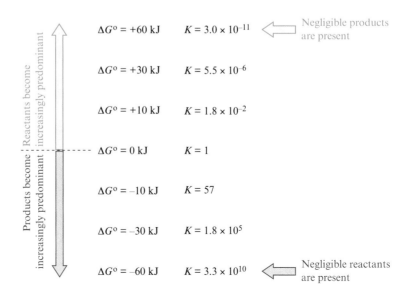

When $\Delta G^\circ = 0$, $K = 1$.

If $\Delta G^\circ$ is negative, $K$ is greater than 1, and the products predominate over the reactants.

If $\Delta G^\circ$ is positive, $K$ is less than 1, and the reactants predominate over the products.

---

**Worked example 17.9**

*Using the reaction isotherm*

Determine $K$ at 298 K for the reaction:

$$C_2H_4(g) + H_2(g) \rightleftharpoons C_2H_6(g)$$

if $\Delta_f G^\circ(C_2H_4, g) = +68.4 \text{ kJ mol}^{-1}$ and $\Delta_f G^\circ(C_2H_6, g) = -32.0 \text{ kJ mol}^{-1}$ at 298 K.

First determine $\Delta_r G^\circ$ for the forward reaction:

$$C_2H_4(g) + H_2(g) \rightleftharpoons C_2H_6(g)$$

$$\Delta_r G^\circ = [\Delta_f G^\circ(C_2H_6, g)] - [\Delta_f G^\circ(C_2H_4, g) + \Delta_f G^\circ(H_2, g)]$$

Since $H_2$ is in its standard state, $\Delta_f G^\circ(H_2, g) = 0$

$$\Delta_r G^\circ = \Delta_f G^\circ(C_2H_6, g) - \Delta_f G^\circ(C_2H_4, g)$$

$$= -32.0 - 68.4$$

$$= -100.4 \text{ kJ mol}^{-1}$$

The equilibrium constant can be found using the equation:

$$\Delta G^\circ = -RT \ln K$$

$$\ln K = -\frac{\Delta G^\circ}{RT}$$

$$= -\frac{(-100.4 \text{ kJ mol}^{-1})}{(8.314 \times 10^{-3} \text{ kJ K}^{-1} \text{ mol}^{-1}) \times (298 \text{ K})}$$

$$= 40.5$$

$$K = 3.9 \times 10^{17} \text{ (at 298 K)}$$

The equilibrium constant is dimensionless.

---

## 17.9    Gibbs energy, enthalpy and entropy

Up to this point we have looked at experimental data and said that there is a state function $G$, changes in which provide a better indicator of reaction spontaneity than do changes in enthalpy. Why is this?

### The relationship between *G* and *H*

Gibbs energy is defined in equation 17.53 where $G$, $H$, $T$ and $S$ are all state functions, and $S$ is the *entropy*.

*Entropy is defined fully in Section 17.10*

$$G = H - TS \tag{17.53}$$

**THEORETICAL AND CHEMICAL BACKGROUND**

**Box 17.4  Deriving $\Delta G = \Delta H - T\Delta S$ from $G = H - TS$**

$G = H - TS$

$dG = dH - (T\,dS + S\,dT)$

$\quad = dH - T\,dS - S\,dT$

At constant temperature, $dT = 0$, and so:

$dG = dH - T\,dS$

or

$\Delta G = \Delta H - T\Delta S$

---

*Gibbs energy* is defined by the equation:

$G = H - TS$

The *change in Gibbs energy*, $\Delta G$, is the maximum work that is available in a reversible process at constant temperature and pressure in *addition* to the work done in expansion (equation 17.54); the derivation of equation 17.54 from 17.53 is outlined in Box 17.4.

$$\Delta G = \Delta H - T\Delta S \qquad (17.54)$$

For standard state conditions, equation 17.55 is appropriate, where $T$ is often (but not necessarily) 298 K.

$$\Delta G^{\circ} = \Delta H^{\circ} - T\Delta S^{\circ} \qquad (17.55)$$

Equations 17.54 and 17.55 show that the change in Gibbs energy for a process at a given temperature has contributions from both the change in enthalpy *and* the change in entropy. From our earlier discussion and the values in Table 17.1, we see that the relationship between the $\Delta G$ and $\Delta H$ terms for a particular reaction depends critically upon the term $T\Delta S$. Consider the formation of carbon dioxide: $\Delta_f H^{\circ}(CO_2, g, 298\,K) = -393.5\,kJ\,mol^{-1}$ and $\Delta_f G^{\circ}(CO_2, g, 298\,K) = -394.4\,kJ\,mol^{-1}$; the closeness of these values means that the $T\Delta S^{\circ}$ term is small (equation 17.56). In this instance, both $\Delta_f H^{\circ}$ and $\Delta_f G^{\circ}$ are good indicators of reaction spontaneity.

For the formation of $CO_2(g)$ from carbon and $O_2$ in their standard states:

$\Delta G^{\circ} = \Delta H^{\circ} - T\Delta S^{\circ}$

$T\Delta S^{\circ} = \Delta H^{\circ} - \Delta G^{\circ} \qquad (17.56)$

$\quad = (-393.5) - (-394.4) = +0.9\,kJ$ per mole of $CO_2$

In contrast, for BrCl, $\Delta_f H^{\circ}(BrCl, g, 298\,K) = +14.6\,kJ\,mol^{-1}$ and $\Delta_f G^{\circ}(BrCl, g, 298\,K) = -1.0\,kJ\,mol^{-1}$, and the $T\Delta S^{\circ}$ term is significant (equation 17.57). Here, the value of $\Delta_f H^{\circ}$ would suggest that the reaction between $Br_2$ and $Cl_2$ may not proceed, but the magnitude of the $T\Delta S^{\circ}$ term compensates for $\Delta_f H^{\circ}$ and gives an equilibrium in which there are slightly more products than reactants.

For the formation of $BrCl(g)$ from $Br_2$ and $Cl_2$ in their standard states:

$T\Delta S^{\circ} = \Delta H^{\circ} - \Delta G^{\circ}$

$\quad = (+14.6) - (-1.0) = +15.6\,kJ$ per mole of BrCl $\qquad (17.57)$

These results reveal the importance of the $T\Delta S$ term. It is the factor that tips the balance with respect to changes in enthalpy so that some *endothermic* reactions occur spontaneously. In the next section we look at entropy in more detail.

At constant temperature and pressure, the **change in Gibbs energy** of a reaction is given by:

$$\Delta G = \Delta H - T\Delta S$$

or for standard state conditions:

$$\Delta G^\circ = \Delta H^\circ - T\Delta S^\circ$$

### The van't Hoff isochore

A combination of equations 17.51 and 17.54 leads to the *van't Hoff isochore* (equation 17.58) from which the temperature dependence of $K$ can be determined.

$$\Delta G^\circ = -RT \ln K$$

$$\Delta G^\circ = \Delta H^\circ - T\Delta S^\circ$$

Therefore: $-RT \ln K = \Delta H^\circ - T\Delta S^\circ$

$$\ln K = -\frac{\Delta H^\circ}{RT} + \frac{\Delta S^\circ}{R} \tag{17.58}$$

Equation 17.58 contains four variables: $K$, $\Delta H^\circ$, $\Delta S^\circ$ and $T$, but the variation in $K$ with temperature is far greater than the variations in $\Delta H^\circ$ and $\Delta S^\circ$. This leads to the form of the *van't Hoff isochore* that we used in Section 16.3 (equation 16.19) to determine an approximate value of $\Delta H^\circ$ for a reaction. That is, if we assume that $\Delta S^\circ$ and $\Delta H^\circ$ have constant values for a given reaction, then equation 17.58 has the form of the equation for a straight line ($y = mx + c$), giving a linear relationship between $\ln K$ and $1/T$ (see Figure 16.1).

Jacobus Henricus van't Hoff (1852–1911).
© *The Nobel Foundation.*

**17.10** **Entropy: the Second and Third Laws of Thermodynamics**

The **entropy**, $S$, of a system *and* surroundings increases during a spontaneous, irreversible process.

The definition above is one way of stating the *Second Law of Thermodynamics*, and in this section we consider entropy in detail. We start by considering thermodynamic and statistical definitions of entropy.

### Change in heat energy for a reversible process

The thermodynamic definition of the change in entropy, $\Delta S$, is given by equation 17.59. This equation relates the change in entropy, $\Delta S$, to the heat transferred, $q_{rev}$, for a *reversible* transfer of thermal energy. Because the process is reversible, the temperature, $T$, is constant, i.e. the process is an *isothermal* one.

An *isothermal* change to a system is one that occurs at *constant temperature*.

$$\Delta S = \frac{q_{rev}}{T} \tag{17.59}$$

We shall return to this equation later when we discuss phase changes.

**Table 17.3** The standard molar entropies $S^o$(298 K) for selected elements and compounds.

| Substance (standard state at 298 K) | $S^o$(298 K) / J K$^{-1}$ mol$^{-1}$ |
| --- | --- |
| Ag(cryst) | 42.6 |
| C(graphite) | 5.7 |
| C(diamond) | 2.4 |
| Fe(cryst) | 27.3 |
| I$_2$(cryst) | 116.1 |
| KI(cryst) | 106.3 |
| Hg(l) | 75.9 |
| PCl$_3$(l) | 217.1 |
| He(g) | 126.2 |
| H$_2$(g) | 130.7 |
| HCl(g) | 186.9 |
| HBr(g) | 198.7 |
| PF$_5$(g) | 300.8 |

## The statistical approach

One way to understand the meaning of the entropy of a system is to use a statistical approach and find the number of possible arrangements of the components (atoms or molecules) of that system. Equation 17.60 gives Boltzmann's definition of entropy, $S$, where $k$ is the Boltzmann constant, and $W$ is the number of ways of arranging the atoms or molecules in the system while maintaining a constant overall energy.[§]

$$S = k \ln W \qquad (17.60)$$

The more arrangements there are, the greater the entropy. Entropy is often expressed in terms of the degree of disorder or randomness of a system: the more randomly arranged are the atoms or molecules in a system, the larger is its entropy.

## Standard molar entropy

The *standard molar entropy*, $S^o$, of an element or compound is the entropy per mole of a pure substance at $10^5$ Pa (1 bar). The units are J K$^{-1}$ mol$^{-1}$.

The *standard molar entropy*, $S^o$, of an element or compound is the entropy per mole of a pure substance at $10^5$ Pa (1 bar). Values of $S^o$(298 K) for some selected elements and compounds are given in Table 17.3. It is tempting to say that there is a general tendency for values of $S^o$ for solid substances to be smaller than those for liquids, while gases possess the largest standard entropies. In the broadest sense this statement is true, but we must exercise caution in comparing entropy values across a wide range of substances – a comparison of the values of $S^o$(298 K) in Table 17.3 for crystalline KI and liquid mercury bears this out.

In principle, as absolute zero is approached, all systems go to their lowest energy states and the components are perfectly ordered in the solid state. This corresponds to the perfect crystalline state, and the *Third Law of Thermodynamics* states that a pure, perfect crystal at 0 K has zero entropy:

The *Third Law of Thermodynamics* states that a pure, perfect crystal at 0 K has zero entropy.

> *For a perfect crystal:*    $S(0 \text{ K}) = 0$

---

[§] The number of ways of arranging the atoms or molecules in the system does not simply refer to the spatial arrangement of identifiable particles, but also to the distribution of energies within the system.

**Fig. 17.12** The variation of $S^o$ with temperature for $CO_2$, $CH_4$, $NH_3$, $O_2$ and HCl, all of which are gases over the temperature range plotted.

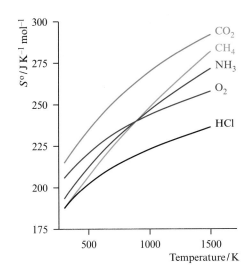

### The dependence of entropy on temperature

If we think of entropy as a measure of the degree of randomness of a system, then an increase in temperature is likely to cause an increase in $S$. Figure 17.12 shows the variation in $S^o$ with temperature for $CO_2$, $CH_4$, $NH_3$, HCl and $O_2$, all of which are gases over the given temperature range.

The change in entropy as a function of temperature is given by equation 17.61, and the integrated form (equation 17.62) can be used to find the variation in entropy between temperatures $T_1$ and $T_2$. The integration steps are set out in Box 17.5.

$$\left(\frac{\partial S}{\partial T}\right)_P = \frac{C_P}{T} \tag{17.61}$$

$$S_{(T_2)} - S_{(T_1)} = C_P \times \ln\left(\frac{T_2}{T_1}\right) \tag{17.62}$$

---

**THEORETICAL AND CHEMICAL BACKGROUND**

### Box 17.5 Integration of $\left(\dfrac{\partial S}{\partial T}\right)_P = \dfrac{C_P}{T}$

The integrated form of the equation:

$$\left(\frac{\partial S}{\partial T}\right)_P = \frac{C_P}{T}$$

is derived as follows.

Separating the variables and integrating between the limits of the temperatures $T_1$ and $T_2$ gives:

$$\int_{T_1}^{T_2} dS = \int_{T_1}^{T_2} \frac{C_P}{T}\, dT$$

Let $S$ at temperature $T_1$ be $S_{(T_1)}$, and at $T_2$ be $S_{(T_2)}$. Integration of the left-hand side of the equation gives:

$$S_{(T_2)} - S_{(T_1)} = \int_{T_1}^{T_2} \frac{C_P}{T}\, dT$$

We now make the assumption that $C_P$ does *not* depend upon temperature; this is valid *only* over small temperature ranges (see Figure 17.6).

With $C_P$ as a constant:

$$S_{(T_2)} - S_{(T_1)} = C_P \int_{T_1}^{T_2} \frac{1}{T}\, dT$$

Integrating the right-hand side of the equation between the limits of $T_2$ and $T_1$ gives:

$$S_{(T_2)} - S_{(T_1)} = C_P \times \ln\left(\frac{T_2}{T_1}\right)$$

**THEORETICAL AND CHEMICAL BACKGROUND**

**Box 17.6 An integral as an area under a curve**

Consider a curve with the general equation $y = f(x)$:

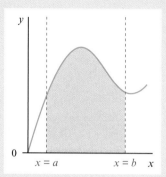

The area under the graph between the limits $x = a$ and $x = b$ is denoted by the definite integral $\int_a^b y \, dx$. We are effectively dividing the shaded area in the graph into an infinite number of vertical strips (drawn parallel to the $y$-axis), finding the area of each strip, and summing the areas together:

$$\int_a^b y \, dx = \text{area under the curve } y = f(x)$$
$$\text{between the limits } x = a \text{ and } x = b$$

Now apply this to the equation:

$$\left(\frac{\partial S}{\partial T}\right)_P = \frac{C_P}{T}$$

The change in entropy between the limits of $T_2$ and $T_1$ is

$$\int_{T_1}^{T_2} dS = \int_{T_1}^{T_2} \frac{C_P}{T} \, dT \quad \text{(from Box 17.5)}$$

or can be found from the area under a plot of $\frac{C_P}{T}$ against $T$ between the limits of $T_2$ and $T_1$, i.e. in this case, the function of $x$ is $\frac{C_P}{T}$.

Equation 17.62 is valid *only* if $C_P$ is approximately constant over the temperature range in question. However, as Figure 17.6 showed, this is not always the case. If we wanted to determine the molar entropy of a substance at 298 K, we would need to evaluate the change in $S^\circ$ between the limits of 0 to 298 K, and over this temperature range, $C_P$ is *not* constant. From equation 17.61, it follows (see Box 17.6) that $S^\circ(298 \text{ K})$ can be determined from the area under a plot of $\frac{C_P}{T}$ against $T$ (Figure 17.13). Measurements of $C_P$ to very low temperatures can be made by calorimetric methods, but the last section of the curve is constructed by an extrapolation to 0 K.

**Fig. 17.13** A plot of $\frac{C_P}{T}$ against temperature, $T$, can be used to find the change in entropy of a substance over a range of temperatures where the heat capacity, $C_P$, is *not* constant. To find $S^\circ(298 \text{ K})$, determine the area under the graph between 0 and 298 K (the shaded area).

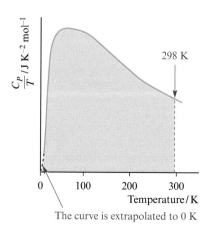

The curve is extrapolated to 0 K

The change in entropy as a function of temperature is given by:

$$\left(\frac{\partial S}{\partial T}\right)_P = \frac{C_P}{T}$$

Assuming that $C_P$ is approximately constant over the temperature range $T_1$ to $T_2$:

$$\Delta S = S_{(T_2)} - S_{(T_1)} = C_P \times \ln\left(\frac{T_2}{T_1}\right)$$

---

**Worked example 17.10**

*Finding the change in entropy of tantalum over a temperature range*

Determine the change in standard entropy per mole of tantalum between 700 and 800 K at constant pressure, if the average value of $C_P$ over this range is $6.53\,\mathrm{J\,K^{-1}\,mol^{-1}}$.

Since we can assume that $C_P$ is constant, the equation needed is:

$$\Delta S^{\circ} = S^{\circ}{}_{(T_2)} - S^{\circ}{}_{(T_1)} = C_P \times \ln\left(\frac{T_2}{T_1}\right)$$

$$\Delta S^{\circ} = (6.53\,\mathrm{J\,K^{-1}\,mol^{-1}}) \times \ln\left(\frac{(800\,\mathrm{K})}{(700\,\mathrm{K})}\right) = 0.87\,\mathrm{J\,K^{-1}\,mol^{-1}}.$$

(For comparison, the *actual* values of $S^{\circ}(700\,\mathrm{K})$ and $S^{\circ}(800\,\mathrm{K})$ are 15.25 and $16.13\,\mathrm{J\,K^{-1}\,mol^{-1}}$, giving a value of $\Delta S^{\circ} = 0.88\,\mathrm{J\,K^{-1}\,mol^{-1}}$.)

---

### The change of entropy associated with a phase change

Now we extend the picture described above to cover different phases of a given substance. Figure 17.14 illustrates that for the *d*-block metal tantalum, there are discontinuities in a plot of $S^{\circ}$ against temperature. These correspond to the solid → liquid and liquid → gas phase transitions. In terms of the statistical model, this makes good sense. Tantalum possesses an ordered crystal structure which is disrupted upon melting. Similarly, the system becomes more random on passing from the liquid to gas phase. Let us consider the phase transitions more carefully. *At the melting point* of tantalum, solid and liquid metal are in equilibrium (equation 17.63).

$$\mathrm{Ta(s)} \rightleftharpoons \mathrm{Ta(l)} \tag{17.63}$$

If an infinitesimally small amount of heat energy is given to the system, a small amount of tantalum melts – equilibrium 17.63 moves to the right-hand side. If an infinitesimally small amount of heat energy is given out by the system, a small amount of tantalum solidifies – equilibrium 17.63 moves to the left-hand side. However, these changes are *so small* that the system remains virtually at equilibrium. Such a situation is one of *thermodynamic reversibility*. When the solid and liquid tantalum are in equilibrium, their Gibbs energies must be equal and $\Delta G$ is zero. It is very important to note: it is $\Delta G$ that is zero, *not* $\Delta G^{\circ}$.

**Fig. 17.14** The variation of $S^\circ$ with temperature for the metal tantalum; the discontinuities in the graph correspond to phase transitions. (Data from *JANAF Tables*, 1972 update, Dow Chemical Company, Midland, Michigan.)

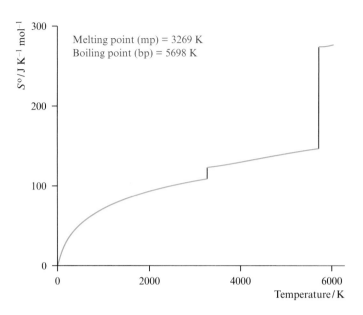

We now apply equation 17.54 to the special case of a phase equilibrium and can write equation 17.64.

At equilibrium: $\Delta G = \Delta H - T\Delta S = 0$

Compare with equation 17.59

$$\Delta H = T\Delta S \qquad \text{or} \qquad \Delta S = \frac{\Delta H}{T} \qquad (17.64)$$

This equation takes two particular forms when we consider the solid $\rightleftharpoons$ liquid and liquid $\rightleftharpoons$ vapour phase changes. The enthalpy change associated with a solid melting is the enthalpy of fusion ($\Delta_{fus}H$). When a liquid vaporizes, the associated enthalpy change is the enthalpy of vaporization ($\Delta_{vap}H$). The entropy change associated with the solid $\rightleftharpoons$ liquid phase change is given by equation 17.65, and equation 17.66 states it for the liquid $\rightleftharpoons$ vapour equilibrium.

Enthalpies of fusion and vaporization: see Section 2.9

*For solid $\rightleftharpoons$ liquid:*     $\Delta_{fus}S = \dfrac{\Delta_{fus}H}{\text{mp}}$     (mp in K)     (17.65)

*For liquid $\rightleftharpoons$ vapour:*     $\Delta_{vap}S = \dfrac{\Delta_{vap}H}{\text{bp}}$     (bp in K)     (17.66)

---

**Worked example 17.11**     *Entropy change of fusion for hexane*

**What is the change in entropy of one mole of hexane, $C_6H_{14}$, when it vaporizes at its boiling point under atmospheric pressure? [$\Delta_{vap}H = 28.9\,\text{kJ mol}^{-1}$, bp = 342 K.]**

$$C_6H_{14}(l) \rightleftharpoons C_6H_{14}(g)$$

First, note carefully the sign of $\Delta_{vap}H$. These data (and those of $\Delta_{fus}H$) are given as positive quantities (heat energy taken in by the system), appropriate for liquid $\longrightarrow$ gas (or solid $\longrightarrow$ liquid). If the phase change is in the other direction, the enthalpy change is negative.

Look back at equations 2.17 and 2.18

$$\Delta_{vap}S = \frac{\Delta_{vap}H}{bp}$$

$$= \frac{(28.9\,\text{kJ mol}^{-1})}{(342\,\text{K})}$$

$$= 0.0845\,\text{kJ K}^{-1}\,\text{mol}^{-1}$$

$$= 84.5\,\text{J K}^{-1}\,\text{mol}^{-1}$$

Trouton's rule and hydrogen bonding: see Section 21.8

A great many liquids possess similar values of $\Delta_{vap}S$ and this is expressed in *Trouton's rule* (equation 17.67).

$$\text{For liquid} \rightleftharpoons \text{vapour:} \quad \Delta_{vap}S \approx 88\,\text{J K}^{-1}\,\text{mol}^{-1} \tag{17.67}$$

Some anomalies can be explained in terms of the presence of extensive hydrogen bonding in the liquid state. For example, values of $\Delta_{vap}S$ for $H_2O$, $H_2S$ and $H_2Se$ are 109, 88 and 85 J K$^{-1}$ mol$^{-1}$ respectively. The hydrogen bonding in liquid water produces a relatively ordered system and lowers its entropy. The *change* in the entropy in going from $H_2O(l)$ to $H_2O(g)$ is larger than would have been the case had hydrogen bonding not been an important effect.

### The change in entropy for a reaction, $\Delta S$

Since entropy is a state function, we can use equation 17.68 to find the change in entropy, $\Delta S$, for a reaction, or equation 17.69 for the change in standard entropy, $\Delta S^o$.

$$\Delta S = \sum S_{products} - \sum S_{reactants} \tag{17.68}$$

$$\Delta S^o = \sum S^o{}_{products} - \sum S^o{}_{reactants} \tag{17.69}$$

Values of the standard entropies for elements and compounds are available in standard tables of data, and can be used to calculate $\sum S^o{}_{products}$ and $\sum S^o{}_{reactants}$ for a reaction.

---

**Worked example 17.12**    *Determination of an entropy change for a reaction*

If the standard entropies (at 298 K) of gaseous $F_2$, $Cl_2$ and ClF are 203, 223 and 218 J K$^{-1}$ mol$^{-1}$ respectively, determine $\Delta S^o(298\,\text{K})$ for the formation of ClF from $F_2$ and $Cl_2$.

First, write a balanced equation:

$$Cl_2(g) + F_2(g) \longrightarrow 2ClF(g)$$

$$\Delta_r S^o = \sum S^o{}_{products} - \sum S^o{}_{reactants}$$

$$= (2 \times 218) - (223 + 203)$$

$$= 10\,\text{J K}^{-1} \text{ per 2 moles of ClF} \qquad or \qquad 5\,\text{J K}^{-1}\,\text{mol}^{-1}$$

---

**Worked example 17.13**    *Determination of an entropy change for a reaction*

Iridium(VI) fluoride is formed by the direct fluorination of iridium metal. Calculate the standard change in entropy (298 K) during this reaction if $S^o(298\,\text{K})$ for Ir(s), $F_2(g)$ and $IrF_6(s)$ are 35, 203 and 248 J K$^{-1}$ mol$^{-1}$ respectively.

First, write a balanced equation for the formation of $IrF_6$.

$$Ir(s) + 3F_2(g) \longrightarrow IrF_6(s)$$

$$\Delta_r S^\circ = \sum S^\circ_{\text{products}} - \sum S^\circ_{\text{reactants}}$$

$$= (248) - [35 + (3 \times 203)]$$

$$= -396 \, \text{J K}^{-1} \, \text{mol}^{-1}$$

---

Changes in entropy for reactions can be either positive (an increase in entropy) or negative (a decrease in entropy) as worked examples 17.12 and 17.13 illustrate – but a word of caution: *each value we have calculated is only the entropy change of the system.* Before the story is complete, we must also consider the *entropy change of the surroundings* during the reaction. If a reaction is exothermic, the heat given out causes an increase in the entropy of the surroundings. Conversely, in an endothermic reaction, heat is taken in by the system, causing a decrease in the entropy of the surroundings (Figure 17.15).

A *spontaneous process* is one that is thermodynamically favourable; it occurs naturally without an external driving force. *But,* the thermodynamics of the process tell us nothing about the rate at which the process occurs.

The balance between the change in entropy of the system and the change in entropy of the surroundings is sometimes critical in determining whether or not a reaction will be thermodynamically viable (i.e. spontaneous). In this context we return to equation 17.54: $\Delta G = \Delta H - T\Delta S$. At a constant temperature, the difference between the terms $\Delta H$ and $T\Delta S$ determines the sign of $\Delta G$.

Consider worked example 17.12: $\Delta S^\circ(298 \, \text{K})$ for the formation of $ClF(g)$ is $+5 \, \text{J K}^{-1} \, \text{mol}^{-1}$, meaning that the system becomes *more disordered* as the reaction takes place. Since $\Delta_f H^\circ(ClF, g, 298 \, \text{K})$ is $-50.3 \, \text{kJ mol}^{-1}$, $\Delta_f G^\circ$ is also negative (equation 17.70) and the formation of $ClF(g)$ is a thermodynamically favoured process.

$$T\Delta S^\circ = (298 \, \text{K}) \times (5 \times 10^{-3} \, \text{kJ K}^{-1} \, \text{mol}^{-1}) \approx 1.5 \, \text{kJ mol}^{-1}$$

$$\Delta G^\circ = \Delta H^\circ - T\Delta S^\circ = -50.3 - 1.5 = -51.8 \, \text{kJ per mole of } ClF(g)$$

$$(17.70)$$

Now consider worked example 17.13: $\Delta S^\circ(298 \, \text{K})$ for the formation of $IrF_6(s)$ is $-396 \, \text{J K}^{-1} \, \text{mol}^{-1}$, meaning that the system becomes *more ordered* as the reaction takes place. On the face of it, this does not seem to be a favourable state of affairs, but $\Delta_f H^\circ(IrF_6, s, 298 \, \text{K})$ is $-579.7$, and this is more than sufficient to compensate for the negative entropy term (equation 17.71).

$$T\Delta S^\circ = (298 \, \text{K}) \times (-396 \times 10^{-3} \, \text{kJ K}^{-1} \, \text{mol}^{-1}) \approx -118 \, \text{kJ mol}^{-1}$$

$$\Delta G^\circ = \Delta H^\circ - T\Delta S^\circ = -579.7 - (-118) \approx -462 \, \text{kJ per mole of } IrF_6(s)$$

$$(17.71)$$

Fig. 17.15 The effect of enthalpy changes in a system on the entropy of the surroundings, emphasizing sign convention.

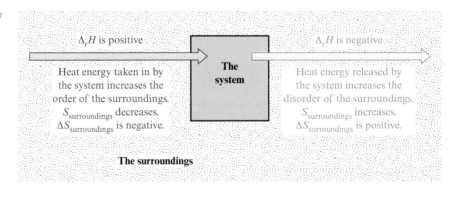

In other words, the high bond energy of the Ir−F bonds compensates for the unfavourable entropy change associated with a gaseous reactant giving a solid product.

This discussion illustrates the *Second Law of Thermodynamics* and brings us back to where we started in Section 17.10. There are various ways in which the Second Law may be stated, but one useful form is that *the total entropy always increases during a spontaneous change* (equation 17.72). Note that it is the *total* entropy with which the law is concerned – the sum of the changes in entropies of the surroundings and the system.

> The *Second Law of Thermodynamics* can be stated in the form: the *total* entropy always increases during a spontaneous change.

$$\Delta S_{total} > 0 \qquad \text{for a spontaneous change} \qquad (17.72)$$

## 17.11    Enthalpy changes and solubility

The data in Table 17.2 illustrated that many salts dissolve in water in endothermic processes. In this section, we show that by considering both the enthalpy *and* entropy changes that accompany the solvation process, it is possible to understand why this phenomenon exists. However, solubility is a complicated topic and our coverage will be relatively superficial.

### The standard enthalpy of solution of potassium chloride

> NaCl structure: see Section 8.8

The $K^+$ and $Cl^-$ ions in solid KCl are arranged in an NaCl structure (Figure 8.11). At 298 K, the solubility of KCl is 0.46 mole (34 g) per 100 g water, indicating that the process of solvation is thermodynamically favourable.

We can initially approach the problem by using Hess's Law, but first we must define the process that occurs as solid KCl dissolves. An ionic lattice in the solid state is changed into solvated cations and anions in the aqueous solution of KCl (equation 17.73), and the standard enthalpy change (at 298 K) that accompanies the dissolution is $\Delta_{sol}H^\circ$.

$$KCl(s) \xrightarrow{\text{Water}} K^+(aq) + Cl^-(aq) \qquad (17.73)$$

The change in equation 17.73 can be considered in two stages:

> $\Delta_{lattice}H^\circ$: see Section 8.16

1. The disruption of the ionic lattice into gaseous ions, i.e. the reverse process to that which defines the lattice energy ($\Delta U(0\,K) \approx \Delta_{lattice}H^\circ$).
2. The solvation (or in this particular case, the hydration) of the gaseous ions which is accompanied by a standard enthalpy change $\Delta_{solv}H^\circ$.

We can now set up the enthalpy cycle shown in equation 17.74 and determine $\Delta_{sol}H^\circ$ using Hess's Law (equation 17.75).

$$KCl(s) \xrightarrow{\Delta_{sol}H^\circ} K^+(aq) + Cl^-(aq)$$

$$-\Delta U(0\,K) \approx -\Delta_{lattice}H^\circ \searrow \qquad \nearrow \sum[\Delta_{solv}H^\circ] = \sum[\Delta_{hyd}H^\circ]$$
$$\text{when the solvent is water}$$

$$K^+(g) + Cl^-(g) \qquad (17.74)$$

$$\Delta_{sol}H^\circ(KCl, s) = \Delta_{hyd}H^\circ(K^+, g) + \Delta_{hyd}H^\circ(Cl^-, g)$$
$$+ [-\Delta_{lattice}H^\circ(KCl, s)] \qquad (17.75)$$

The lattice energy of KCl is $-718\,kJ\,mol^{-1}$ and so if the dissolution of KCl is to be an exothermic process, the sum of the standard enthalpies of solvation of $K^+(g)$ and $Cl^-(g)$ must be *more negative* than $-718\,kJ\,mol^{-1}$. What factors

**Table 17.4** Absolute values of $\Delta_{hyd}H^{\circ}$ (at 298 K) for selected ions.

| Ion | $\Delta_{hyd}H^{\circ}$ / kJ mol$^{-1}$ | Ion | $\Delta_{hyd}H^{\circ}$ / kJ mol$^{-1}$ |
|---|---|---|---|
| $H^{+}$ | $-1091$ | $Sr^{2+}$ | $-1456$ |
| $Li^{+}$ | $-519$ | $Ba^{2+}$ | $-1316$ |
| $Na^{+}$ | $-404$ | $Al^{3+}$ | $-4691$ |
| $K^{+}$ | $-321$ | $F^{-}$ | $-504$ |
| $Rb^{+}$ | $-296$ | $Cl^{-}$ | $-381$ |
| $Cs^{+}$ | $-271$ | $Br^{-}$ | $-330$ |
| $Mg^{2+}$ | $-1931$ | $I^{-}$ | $-285$ |
| $Ca^{2+}$ | $-1586$ | | |

influence the magnitude of $\Delta_{solv}H^{\circ}$ (or $\Delta_{hyd}H^{\circ}$) for a particular ion? The solvent molecules must be able to interact strongly with the ion and this depends partly on the nature of the solvent and partly on the size and charge of the ion. It is impossible to measure values of $\Delta_{hyd}H^{\circ}(298\,K)$ *directly* for individual ions by calorimetric means since each ion must be present in solution with its counter-ion, but values (Table 17.4) can be obtained by indirect methods.[§] By substituting values for the lattice enthalpy and $\Delta_{hyd}H^{\circ}$ into equation 17.75, the standard enthalpy of solution is found to be $+16\,kJ\,mol^{-1}$ (equation 17.76) – the process is *endothermic*.

$$\Delta_{sol}H^{\circ}(KCl, s) = (-321) + (-381) + 718 = +16\,kJ\,mol^{-1} \qquad (17.76)$$

$\Delta_{solv}H^{\circ}(298\,K)$ is the standard enthalpy change that accompanies the *solvation* of a gaseous ion at 298 K; if the solvent is water, the process of solvation is called *hydration*.

$\Delta_{sol}H^{\circ}(298\,K)$ is the enthalpy change that accompanies the *dissolution* of a substance at 298 K.

### The change in entropy when KCl dissolves in water

When solid KCl dissolves in water, it is easy to envisage the destruction of the crystal lattice and you might assume that the entropy of the system necessarily increases. However, other changes occur within the system:

Hydrogen bonding: see Section 21.8

- the crystal lattice breaks into separate ions;
- hydrogen bonds between water molecules in the bulk solvent are broken as KCl enters the solvent;
- ion–dipole bonds form between the ions and water molecules;
- the solvated ions are associated with the bulk solvent molecules through hydrogen bonding involving the coordinated water molecules (see Figure 21.13 and accompanying text).

The entropy change for reaction 17.77 is calculated from values of $S^{\circ}(298\,K)$, available in standard tables.

$$KCl(s) \xrightarrow{\Delta S^{\circ}(298\,K)} K^{+}(aq) + Cl^{-}(aq)$$
$$S^{\circ}(298\,K)\ +83 \qquad\qquad +103 \quad +57 \quad J\,K^{-1}\,mol^{-1} \qquad (17.77)$$

[§] For a more detailed discussion of hydration enthalpies see: W. E. Dasent (1984) *Inorganic Energetics*, 2nd edn, Cambridge University Press, Cambridge.

From equation 17.77, $\Delta S^\circ(298\,K)$ for the dissolution of KCl is $+77\,J\,K^{-1}\,mol^{-1}$ and so an increase in entropy accompanies the process.

### Why does KCl dissolve in water?

By using the values of $\Delta_{sol}H^\circ$ and $\Delta S^\circ$ determined above, we can calculate a value for $\Delta G^\circ(298\,K)$ for the dissolution in water of one mole of KCl (equation 17.78), first making sure to convert the value of $S^\circ$ into appropriate units.

At 298 K:

$$\Delta G^\circ = \Delta H^\circ - T\Delta S^\circ$$
$$= (+16\,kJ\,mol^{-1}) - \{(298\,K) \times (77 \times 10^{-3}\,kJ\,K^{-1}\,mol^{-1})\}$$
$$= -7\,kJ\,mol^{-1} \tag{17.78}$$

The Gibbs energy change is negative, and the dissolution of KCl in water is a thermodynamically favourable process – the solubility at 298 K is $4.6\,mol\,dm^{-3}$.

Over small temperature ranges, it is often the case that both $\Delta H$ and $\Delta S$ are approximately constant. For KCl, we have determined that $\Delta_{sol}G^\circ(298\,K)$ is $-7\,kJ\,mol^{-1}$. We can estimate the change in Gibbs energy for the same process at 320 K from equation 17.79 which assumes that $\Delta H^\circ(298\,K) \approx \Delta H^\circ(320\,K)$ and $\Delta S^\circ(298\,K) \approx \Delta S^\circ(320\,K)$.

At 320 K:

$$\Delta G^\circ = \Delta H^\circ - T\Delta S^\circ$$
$$= (+16\,kJ\,mol^{-1}) - \{(320\,K) \times (77 \times 10^{-3}\,kJ\,K^{-1}\,mol^{-1})\}$$
$$= -9\,kJ\,mol^{-1} \tag{17.79}$$

The more negative value of $\Delta G^\circ$ at 320 K compared with that at 298 K indicates that the solubility of KCl at 320 K is greater than at 298 K. The actual solubilities of 34 g (298 K) and 40 g (320 K) per 100 g of water confirm the calculations.

### Saturated solutions

In a *saturated* solution, the dissolved and undissolved solutes are in equilibrium with one another.

The solubilities of the salts given in Table 17.2 correspond to the *maximum* amount of compound that will dissolve in a given volume of solvent at a stated temperature. Once this limiting value is exceeded, excess solute remains undissolved and the solution is said to be *saturated*. In a saturated solution, the dissolved and undissolved solutes are in equilibrium with each other, and $\Delta G$ for that process is zero.

## 17.12     Solubility product constant, $K_{sp}$

### Sparingly soluble salts

If a salt is *sparingly soluble*, its saturated solution contains only a small number of ions.

Many ionic compounds are only very *sparingly soluble*. This includes many compounds which are commonly described as 'insoluble', e.g. AgCl, $BaSO_4$ and $CaCO_3$. Sparingly soluble means that a saturated solution only contains a small number of ions. When an excess of a sparingly soluble

**Table 17.5** Values of $K_{sp}$(298 K) for selected sparingly soluble salts.

| Compound | Formula | $K_{sp}$(298 K) |
|---|---|---|
| Barium sulfate | $BaSO_4$ | $1.1 \times 10^{-10}$ |
| Calcium carbonate | $CaCO_3$ | $4.9 \times 10^{-9}$ |
| Calcium hydroxide | $Ca(OH)_2$ | $4.7 \times 10^{-6}$ |
| Calcium phosphate | $Ca_3(PO_4)_2$ | $2.1 \times 10^{-33}$ |
| Copper(I) chloride | $CuCl$ | $1.7 \times 10^{-7}$ |
| Copper(I) sulfide | $Cu_2S$ | $2.3 \times 10^{-48}$ |
| Copper(II) sulfide | $CuS$ | $6.0 \times 10^{-37}$ |
| Iron(II) hydroxide | $Fe(OH)_2$ | $4.9 \times 10^{-17}$ |
| Iron(II) sulfide | $FeS$ | $6.0 \times 10^{-19}$ |
| Iron(III) hydroxide | $Fe(OH)_3$ | $2.6 \times 10^{-39}$ |
| Lead(II) iodide | $PbI_2$ | $8.5 \times 10^{-9}$ |
| Lead(II) sulfide | $PbS$ | $3.0 \times 10^{-28}$ |
| Magnesium carbonate | $MgCO_3$ | $6.8 \times 10^{-6}$ |
| Magnesium hydroxide | $Mg(OH)_2$ | $5.6 \times 10^{-12}$ |
| Silver(I) chloride | $AgCl$ | $1.8 \times 10^{-10}$ |
| Silver(I) bromide | $AgBr$ | $5.4 \times 10^{-13}$ |
| Silver(I) iodide | $AgI$ | $8.5 \times 10^{-17}$ |
| Silver(I) chromate | $Ag_2CrO_4$ | $1.1 \times 10^{-12}$ |
| Silver(I) sulfate | $Ag_2SO_4$ | $1.2 \times 10^{-5}$ |
| Zinc(II) sulfide | $ZnS$ | $2.0 \times 10^{-25}$ |

compound is added to a solvent, a system is obtained in which the solid is in equilibrium with ions in the saturated solution. Equation 17.80 exemplifies this for silver chloride, the solubility of which is $1.35 \times 10^{-5}\ mol\,dm^{-3}$ (at 298 K).

$$AgCl(s) \rightleftharpoons Ag^+(aq) + Cl^-(aq) \qquad K = \frac{[Ag^+][Cl^-]}{[AgCl]} \qquad (17.80)$$

In the expression for the equilibrium constant, $[Ag^+]$ and $[Cl^-]$ are the *equilibrium* concentrations of the aqueous ions and $[AgCl]$ is the equilibrium concentration of the undissolved solid. From Le Chatelier's principle, the addition of more chloride ion to this saturated solution will result in the precipitation of AgCl. It is more convenient to describe the equilibrium position in terms of the *solubility product* (or *solubility constant*) $K_{sp}$ (equation 17.81).[§]

$$K_{sp} = [Ag^+][Cl^-] \qquad (17.81)$$

The expressions for the solubility products for two other salts which are sparingly soluble in water are given in equations 17.82 and 17.83, and values of $K_{sp}$ for selected salts are listed in Table 17.5.

$$CaF_2(s) \rightleftharpoons Ca^{2+}(aq) + 2F^-(aq) \qquad K_{sp} = [Ca^{2+}][F^-]^2 \qquad (17.82)$$

$$Ca_3(PO_4)_2(s) \rightleftharpoons 3Ca^{2+}(aq) + 2[PO_4]^{3-}(aq)$$

$$K_{sp} = [Ca^{2+}]^3[PO_4{}^{3-}]^2 \qquad (17.83)$$

See also Section 18.8

### Determining values of $K_{sp}$ from the solubilities of salts

The solubility of silver chloride in water at 298 K is $1.35 \times 10^{-5}\ mol\,dm^{-3}$. When $1.35 \times 10^{-5}$ moles of AgCl dissolve in $1\,dm^3$ of solution, $1.35 \times 10^{-5}$

---

[§] The substitution of $[AgCl] = 1$ is done on the basis that the activity of a solid is unity and is constant; activities are discussed in Section 16.3.

**ENVIRONMENT AND BIOLOGY**

## Box 17.7 Biomineralization: the assembly of biological architectures from inorganic minerals

Some inorganic compounds which are sparingly soluble (typically referred to as being insoluble) play an important role in biology. Some living organisms have the ability to engineer parts of their body or shell from inorganic minerals such as silica ($SiO_2$) or calcium carbonate ($CaCO_3$). The process is called *biomineralization* and examples include gravity sensors in a variety of animals, sea urchin spines, corals, mother of pearl in sea shells and the exoskeletons of some phytoplankton. The photograph shows the marine phytoplankton *Emiliana huxleyi*. This is a single-celled alga which possesses a calcareous exoskeleton made up of an intricate assembly of calcium carbonate plates called coccoliths. The structural features of the plates have been reproduced for millions of years; when the alga dies, the plates are deposited as ocean sediments and become the building blocks of sedimentary rocks such as chalk.

The crystallization processes associated with the formation of the coccoliths in the exoskeleton of *Emiliana huxleyi* are clearly not the same as those that lead to natural crystals of pure calcite or aragonite. Left to their own devices, these polymorphs of $CaCO_3$ crystallize with characteristic crystal habits, quite unlike the crystal shapes observed as a result of biomineralization. The role of organic materials in controlling the assembly of inorganic components in biomineralization processes is crucial. Each coccolith in *Emiliana huxleyi* consists of 30–40 single crystals of calcite arranged in an oval disc. The single crystals are clearly seen as the 'spokes' in

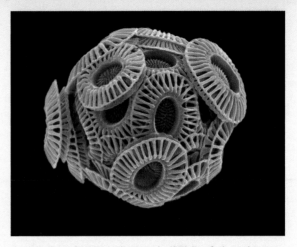

A scanning electron micrograph (SEM) of the calcareous phytoplankton *Emiliana huxleyi*. The exoskeleton consists of oval plates called coccoliths. The diameter of the organism is <10 μm. *VVG/Science Photo Library.*

the coccoliths in the photograph. The biomineralization process that produces this unit depends on calcite crystallization taking place within membrane-bound cavities of specific shape and size. These tube-shaped cavities are called *vesicles*. Complex biological encoding results in the crystals being aligned to produce oval discs, and in each crystal being terminated in a structure that connects it to its neighbours.

moles of $Ag^+$ and $1.35 \times 10^{-5}$ moles of $Cl^-$ ions are formed, and $[Ag^+]$ and $[Cl^-]$ are both equal to $1.35 \times 10^{-5}$ mol dm$^{-3}$. The solubility product is given by equation 17.84.

$$K_{sp} = [Ag^+][Cl^-] = (1.35 \times 10^{-5})(1.35 \times 10^{-5}) = 1.82 \times 10^{-10} \qquad (17.84)$$

Now consider iron(II) hydroxide with a solubility of $2.3 \times 10^{-6}$ mol dm$^{-3}$ at 298 K. Dissolution of a mole of $Fe(OH)_2$ produces one mole of $Fe^{2+}$ ions and *two* moles of hydroxide ions (equation 17.85).

$$Fe(OH)_2(s) \rightleftharpoons Fe^{2+}(aq) + 2[OH]^-(aq) \qquad (17.85)$$

In a saturated solution of $Fe(OH)_2$ at 298 K, the concentration of $Fe^{2+}$ ions is equal to the solubility of the salt ($2.3 \times 10^{-6}$ mol dm$^{-3}$), but $[OH^-]$ is double this ($4.6 \times 10^{-6}$ mol dm$^{-3}$). Equation 17.86 gives the expression for $K_{sp}$.

$$K_{sp} = [Fe^{2+}][OH^-]^2$$
$$= (2.3 \times 10^{-6})(4.6 \times 10^{-6})^2 = 4.9 \times 10^{-17} \qquad (17.86)$$

| **Worked example 17.14** | *Determination of the solubility of a salt from $K_{sp}$* |

**What is the solubility of $Ca(OH)_2$ in water at 298 K if $K_{sp} = 4.7 \times 10^{-6}$?**

First, write an equilibrium to show the dissolution of the salt:

$$Ca(OH)_2(s) \rightleftharpoons Ca^{2+}(aq) + 2[OH]^-(aq)$$

The equation for $K_{sp}$ is:

$$K_{sp} = [Ca^{2+}][OH^-]^2$$

In solution, the concentration of $[OH]^-$ is twice that of $Ca^{2+}$, and we can rewrite the above equation in the form:

$$K_{sp} = [Ca^{2+}](2[Ca^{2+}])^2 = 4[Ca^{2+}]^3$$

We choose to write this equation in terms of $[Ca^{2+}]$ rather than the concentration of hydroxide ions because the solubility of the salt (the ultimate aim of the exercise) is equal to $[Ca^{2+}]$.

$$K_{sp} = 4.7 \times 10^{-6} = 4[Ca^{2+}]^3$$

$$[Ca^{2+}] = \sqrt[3]{\frac{4.7 \times 10^{-6}}{4}}$$

$$= 1.1 \times 10^{-2} \, \text{mol dm}^{-3}$$

The number of moles of $Ca^{2+}$ ions in solution is equal to the number of moles of $Ca(OH)_2$ that have dissolved, and so the solubility of $Ca(OH)_2$ is $1.1 \times 10^{-2} \, \text{mol dm}^{-3}$.

## SUMMARY

Thermodynamics is not the easiest of subjects and in this chapter we have combined experimental observations with theory so that you might see *why* the theory is necessary. Our emphasis has purposely not been on the derivation of equations – this material can be found in more advanced physical chemistry texts.

An understanding of thermodynamics is central to an understanding of why (and in what direction) chemical processes occur. Some of the most important equations that we have encountered in this chapter are listed below. You should be familiar with them all, and *be able to use them*; the meanings of the symbols are given in the chapter.

$$\Delta U = q + w$$

$$\Delta U = \Delta H - P\Delta V$$

$$C = \frac{dq}{dT}$$

$$C_P = \left(\frac{\partial H}{\partial T}\right)_P$$

$$C_V = \left(\frac{\partial U}{\partial T}\right)_V$$

$$C_P = C_V + R \qquad \text{(for 1 mole of an ideal gas)}$$

$$\Delta(\text{State function}) = \sum(\text{State function})_{\text{products}}$$
$$- \sum(\text{State function})_{\text{reactants}}$$

$$\Delta C_P = \left(\frac{\partial \Delta H}{\partial T}\right)_P \qquad \text{Kirchhoff's equation}$$

$$\Delta_r G = \Delta_r G^\circ + RT \ln Q$$

$$\Delta G^\circ = -RT \ln K \qquad \text{Reaction isotherm}$$

$$K = e^{-\left(\frac{\Delta G^\circ}{RT}\right)}$$

$$\Delta G = \Delta H - T\Delta S$$

$$\ln K = -\frac{\Delta H^\circ}{RT} + \frac{\Delta S^\circ}{R} \qquad \text{van't Hoff isochore}$$

$$\Delta S = \frac{q_{\text{rev}}}{T}$$

$$\left(\frac{\partial S}{\partial T}\right)_P = \frac{C_P}{T}$$

$$S_{(T_2)} - S_{(T_1)} = C_P \times \ln\left(\frac{T_2}{T_1}\right)$$
$$\text{(for a constant } C_P\text{)}$$

$$\Delta_{\text{fus}} S = \frac{\Delta_{\text{fus}} H}{\text{mp}}$$

$$\Delta_{\text{vap}} S = \frac{\Delta_{\text{vap}} H}{\text{bp}}$$

$$\Delta S_{\text{total}} > 0 \qquad \text{for a spontaneous change}$$

$$S(0\,\text{K}) = 0 \qquad \text{for a perfect crystal}$$

$$K_{\text{sp}} = [X^+][Y^-] \qquad \text{for a salt XY}$$

$$K_{\text{sp}} = [X^{n+}][Y^-]^n \qquad \text{for a salt XY}_n$$

### Do you know what the following terms mean?

- enthalpy and $\Delta H$
- internal energy and $\Delta U$
- system and surroundings
- state function
- work done (on or by the system)
- heat capacities ($C_P$ and $C_V$)
- Gibbs energy and $\Delta G$

- spontaneous reaction
- entropy and $\Delta S$
- standard molar entropy
- the First, Second and Third Laws of Thermodynamics
- the reaction isotherm
- Ellingham diagram

- enthalpy of solution
- enthalpy of solvation
- enthalpy of hydration
- saturated solution
- sparingly soluble salt
- solubility product constant

### You should now be able:

- to distinguish between the system and the surroundings
- to distinguish between an isothermal and an adiabatic change
- to define and give examples of state functions
- to recognize the limitations of assuming that $C_P$ and $\Delta H$ are temperature-invariant
- to discuss the reasons why values of $\Delta G$ rather than $\Delta H$ are reliable guides to the extent of reactions

- to understand the meaning of a 'reaction quotient' and know how it differs from an equilibrium constant
- to understand the physical significance of the magnitudes of values of $K$
- to understand the physical significance of the magnitudes and signs of values of $\Delta G$
- to understand the distinction between $\Delta G$ and $\Delta G^\circ$

- to determine $\Delta_r G^\circ$ from values of $\Delta_f G^\circ$
- to determine $\Delta S^\circ$ from values of $S^\circ$
- to set up and use a Hess cycle to determine the enthalpy change of solution for an ionic salt
- to discuss the relationship between $\Delta_{sol} H$, $\Delta_{sol} G$ and $\Delta_{sol} S$ for the dissolution of an ionic salt at a given temperature

- to justify why some salts dissolve in endothermic processes while others dissolve in exothermic processes
- to describe what is meant by a sparingly soluble salt
- to calculate $K_{sp}$ given the solubility of a sparingly soluble salt
- to calculate the solubility of a sparingly soluble salt given $K_{sp}$

## PROBLEMS

**17.1** In equation 17.9, show that units of Pa for pressure and $m^3$ for volume are compatible with J for $\Delta H$. [Hint: refer to Table 1.2.]

**17.2** (a) Calculate the work done on the surroundings at 1 bar pressure and 298 K when 0.200 mole $H_2O_2$ decomposes (volume occupied by 1 mole of gas at 1 bar and 273 K = 22.7 $dm^3$). (b) If the standard enthalpy of reaction is −98.2 kJ per mole of $H_2O_2$, what is the corresponding change in the internal energy of the system?

**17.3** In equation 17.13, for 1 mole of pentane, $\Delta n = -3$. What would happen if the reaction were defined so as to produce water in the gas phase?

**17.4** How does the *molar* heat capacity of a substance differ from the *specific* heat capacity?

**17.5** Estimate the enthalpy change at 400 K that accompanies the reaction:

$$Li(s) + \tfrac{1}{2}Cl_2(g) \longrightarrow LiCl(s)$$

if $\Delta_f H^\circ (298\,K)$ LiCl(s) = −408.6 kJ mol$^{-1}$, and values of $C_P$ (298 K) are Li 24.8, $Cl_2$ 33.9 and LiCl 48.0 J K$^{-1}$ mol$^{-1}$.

**17.6** Consider the reaction: $H_2(g) \longrightarrow 2H(g)$
At 298 K, $\Delta_f H^\circ (H, g) = 218.0$ kJ mol$^{-1}$. Determine the value of $\Delta_f H^\circ (H, g)$ at 400 K if values of $C_P$ for $H_2(g)$ and H(g) over the temperature range 298–400 K are 29.0 and 20.8 J K$^{-1}$ mol$^{-1}$ respectively.

**17.7** How valid is it to assume that $C_P$ remains constant for the following gaseous species over the temperature range 298–500 K: (a) $H_2$; (b) $CClF_3$; (c) HI; (d) $NO_2$; (e) NO; (f) $C_2H_4$?

**17.8** (a) Determine $\Delta_r H^\circ$ for the reaction:

$$CO(g) + \tfrac{1}{2}O_2(g) \longrightarrow CO_2(g)$$

given that, at 298 K,
$\Delta_f H^\circ (CO, g) = -110.5$ kJ mol$^{-1}$ and
$\Delta_f H^\circ (CO_2, g) = -393.5$ kJ mol$^{-1}$.
(b) Calculate $\Delta_r H^\circ (320\,K)$ for the above reaction

given that $C_P$ (298–320 K) for CO(g), $O_2(g)$ and $CO_2(g) = 29.2$, 29.4 and 37.2 J K$^{-1}$ mol$^{-1}$ respectively.

**17.9** Determine the Gibbs energy change at 298 K that accompanies the reaction:

$$B_2H_6(g) + 6H_2O(l) \longrightarrow 2B(OH)_3(s) + 6H_2(g)$$

if values of $\Delta_f G^\circ (298\,K)$ are $B_2H_6(g)$ +87, $B(OH)_3(s)$ −969, $H_2O(l)$ −237 kJ mol$^{-1}$.

**17.10** Using data from Appendix 11, determine $\Delta_r G^\circ (298\,K)$ for the following reactions:
(a) $C_2H_2(g) + 2H_2(g) \longrightarrow C_2H_6(g)$
(b) $CaCO_3(s) \longrightarrow CaO(s) + CO_2(g)$
(c) $Mg(s) + 2HCl(g) \longrightarrow MgCl_2(s) + H_2(g)$

**17.11** Are all the reactions in problem 17.10 thermodynamically favourable at 298 K?

**17.12** What are the values of (a) $\Delta_f G^\circ (O_2, g, 298\,K)$, (b) $\Delta_f G^\circ (O_2, g, 450\,K)$, (c) $\Delta_f H^\circ (O_2, g, 298\,K)$ and (d) $\Delta_f H^\circ (O_2, g, 500\,K)$?

**17.13** (a) At 298 K, $\Delta_f G^\circ (OF_2, g) = +42$ kJ mol$^{-1}$. From this value, can you say whether $OF_2$ will form when $F_2$ and $O_2$ combine at 298 K? (b) For the reaction:

$$H_2O_2(l) \longrightarrow H_2O(l) + \tfrac{1}{2}O_2(g)$$

$\Delta_r G^\circ (298\,K) = -117$ kJ mol$^{-1}$. Suggest why $H_2O_2$ does not decompose spontaneously on standing. Under what circumstances might $H_2O_2$ rapidly decompose at 298 K?

**17.14** Figure 17.16 shows how $\Delta_f G^\circ$ for four gases varies with temperature. (a) Which compounds become less thermodynamically favoured as the temperature increases from 300 to 1500 K? (b) For which compounds is the formation from the constituent elements favourable between 300 and 1500 K? (c) For which compound is $\Delta_f G^\circ$ approximately temperature-independent between 300 and 800 K? (d) What can you say about the

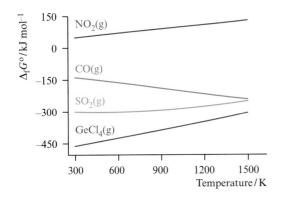

Fig. 17.16 The dependence of $\Delta_f G^\circ$ on temperature for gaseous $NO_2$, CO, $SO_2$ and $GeCl_4$.

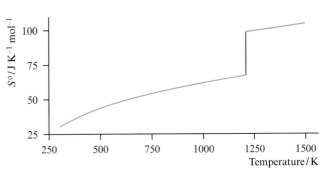

Fig. 17.17 The variation of $S^\circ$ of germanium with temperature.

*relative* thermodynamic stabilities of $SO_2(g)$ and CO(g) with respect to their constituent elements as a function of temperature? (e) Write an equation for the reaction to which $\Delta_f G^\circ(GeCl_4, g)$ refers.

**17.15** By using the Ellingham diagram in Figure 17.9, (a) predict which metals will reduce FeO at 500 K, (b) estimate $\Delta G^\circ$ for the reaction between FeO and carbon at 1200 K, and (c) assess to what extent the reaction between Sn and NiO is affected by changing the temperature from 600 to 1000 K.

**17.16** Use the data in Table 17.1 to verify the reaction isotherm and to calculate a value for the molar gas constant.

**17.17** The value of $\Delta_f G^\circ(C_2H_4, g, 298 K)$ is $+68 \text{ kJ mol}^{-1}$. What is $K$ for this formation reaction? Write an equation for the formation of $C_2H_4$ from its elements in their standard states and indicate in which direction the equilibrium lies.

**17.18** (a) Determine $K$ for the formation of NO from $O_2$ and $N_2$ at 298 K if $\Delta_f G^\circ(NO, g, 298 K) = +87 \text{ kJ mol}^{-1}$. (b) What does the answer to part (a) tell you about the position of the equilibrium:
$\frac{1}{2}N_2(g) + \frac{1}{2}O_2(g) \rightleftharpoons NO(g)$?

**17.19** The value of $K$ for the equilibrium:

$$\frac{1}{2}H_2(g) + \frac{1}{2}I_2(g) \rightleftharpoons HI(g)$$

is 0.5 at 298 K and 6.4 at 800 K. Calculate $\Delta_r G^\circ$ at each temperature and comment on how raising the temperature from 298 to 800 K affects the formation of HI from its constituent elements.

**17.20** The value of $\log K$ for the formation of $SO_2(g)$ from $S_8$ and $O_2$ at 400 K is 39.3. Calculate $\Delta_f G^\circ(SO_2, g)$ at this temperature. Comment on the value.

**17.21** (a) Write an equation for the formation of HBr from its constituent elements. (b) At 298 K, $\Delta_f G^\circ(HBr, g) = -53.4 \text{ kJ mol}^{-1}$. What is the corresponding value of $K$? (c) Is the formation of HBr from $H_2$ and $Br_2$ thermodynamically favoured at 298 K?

**17.22** Predict, with reasoning, whether $\Delta S^\circ$ will be positive, negative or zero for each of the following:
(a) $Na(s) \longrightarrow Na(g)$
(b) $2Cl(g) \longrightarrow Cl_2(g)$
(c) $2CO(g) + O_2(g) \longrightarrow 2CO_2(g)$

**17.23** Using the following data, quantify your answers to problem 17.22. $S^\circ(298 K) / \text{J K}^{-1} \text{mol}^{-1}$: Na(s), 5.3; Na(g), 153.7; Cl(g), 165.2; $Cl_2(g)$, 223.0; $O_2(g)$, 205.1; CO(g), 197.7; $CO_2(g)$, 213.8.

**17.24** The molar entropy, $S^\circ$, of HCl at 298 K is $186.9 \text{ J K}^{-1} \text{mol}^{-1}$. Find $S^\circ(350 K)$ if $C_P(298–350 K)$ is $29.1 \text{ J K}^{-1} \text{mol}^{-1}$.

**17.25** Values of $S^\circ(Cu, s)$ at 400 K and 500 K are 40.5 and $46.2 \text{ J K}^{-1} \text{mol}^{-1}$ respectively. Estimate the molar heat capacity of copper over this temperature range.

**17.26** Figure 17.17 shows how $S^\circ$ for germanium varies between 300 and 1500 K. Rationalize the shape of the curve.

**17.27** Using data in Appendix 11, determine $\Delta S^\circ(298 K)$ for the following reactions:
(a) $2Fe(s) + 3Cl_2(g) \longrightarrow 2FeCl_3(s)$
(b) $Fe(s) + 2HCl(g) \longrightarrow FeCl_2(s) + H_2(g)$
(c) $CaCO_3(s) \longrightarrow CaO(s) + CO_2(g)$
Which reactions are entropically favoured?

**17.28** The enthalpies of fusion and vaporization of $F_2$ are 0.5 and $6.6 \text{ kJ mol}^{-1}$, and the melting and boiling points are 53 and 85 K respectively. Determine values (per mole of $F_2$) of $\Delta_{fus}S$ and $\Delta_{vap}S$, and comment on their relative magnitudes.

**17.29** Calculate the standard entropy change (298 K) when one mole of sulfur is oxidized to $SO_3$; $S^\circ(298 K)$ for S(s), $O_2(g)$ and $SO_3(g)$ are 32.1, 205.2, $256.8 \text{ J K}^{-1} \text{mol}^{-1}$ respectively. Given that the standard enthalpy of formation (298 K) of $SO_3(g)$ is $-395.7 \text{ kJ mol}^{-1}$, determine $\Delta_f G^\circ(298 K)$.

**17.30** The solubility of NaBr at 298 K is 0.88 moles per 100 g of water. The value of $\Delta_{sol}H^{\circ}$ is $-0.60$ kJ mol$^{-1}$. If $\Delta_{sol}S^{\circ} = +57.3$ J K$^{-1}$ mol$^{-1}$, calculate $\Delta_{sol}G^{\circ}(298$ K$)$ and comment on the value in view of the stated solubility.

**17.31** Determine the solubility of $BaSO_4$ in water if $K_{sp} = 1.1 \times 10^{-10}$.

**17.32** (a) If the solubility of $Ag_2Cr_2O_4$ is $6.5 \times 10^{-5}$ mol dm$^{-3}$, determine the solubility product for this salt.
(b) $K_{sp}$ for AgCl is $1.8 \times 10^{-10}$. Why can silver chromate (which is red) be used to indicate the end point in titrations involving the precipitation of AgCl?

## Additional problems

**17.33** The table shows the variation of ln $K$ with $T$ for the equilibrium:

$$S + O_2 \rightleftharpoons SO_2$$

| $T/$K | 298 | 400 | 500 | 700 | 900 |
|-------|-----|-----|-----|-----|-----|
| ln $K$ | 121.1 | 90.49 | 72.36 | 51.44 | 39.55 |

(a) Write down an expression for $K$ indicating any assumptions made.
(b) Use the data to estimate values of $K$ at 300 and 600 K.
(c) Determine $\Delta_f G^{\circ}(SO_2$, g$)$ at 300 K.
(d) Values of $S^{\circ}$ for S(s), $O_2$(g) and $SO_2$(g) at 300 K are 32.2, 205.3 and 248.5 J K$^{-1}$ mol$^{-1}$.

Determine $\Delta_f H(SO_2$, g$)$ at 300 K.
(e) Given that $S^{\circ}(SO_2$, g, 400 K$) = 260.4$ J K$^{-1}$ mol$^{-1}$, determine $C_P$ for $SO_2$(g). Comment on the validity of your answer.

**17.34** $NO_2$ can be prepared by heating $Pb(NO_3)_2$. Passing the gas formed through a U-tube contained in a beaker of ice-water results in the formation of liquid $N_2O_4$ (bp 294 K). At 298 K, $NO_2$ and $N_2O_4$ are in equilibrium and at 298 K, $K = 8.7$.
(a) Find $\Delta G^{\circ}(298$ K$)$ for the dimerization process.
(b) Is the formation of $NO_2$ entropically favoured? Rationalize your answer.
(c) How would the magnetic properties of $NO_2$ and $N_2O_4$ assist in determining the composition of an equilibrium mixture?
(d) The table shows the variation of $K$ with $T$ for the gas phase dissociation of $N_2O_4$ into $NO_2$:

| $T/$K | 298 | 350 | 400 | 450 | 500 |
|-------|-----|-----|-----|-----|-----|
| $K$ | 0.115 | 3.89 | 47.9 | 346 | 1700 |

Write down an expression for $K$ to which the data refer, and state how the values are related to those for the gas phase dimerization of $NO_2$. Discuss the data in the table.
(e) Determine values of $\Delta G^{\circ}(350$ K$)$ for the dissociation of $N_2O_4$ and for the dimerization of $NO_2$.
(f) If $\Delta H^{\circ}(350$ K$)$ for the dissociation of $N_2O_4$ is 58 kJ mol$^{-1}$, find $\Delta S^{\circ}(350$ K$)$ and comment on your answer in terms of the process taking place.

# 18 Electrochemistry

## Topics

- Galvanic cells
- Electrode potentials
- Nernst equation
- Electrolytic cells
- Faraday Law
- Some applications of electrochemistry

## 18.1 Introduction

Electricity is a part of everyday life. In this chapter, we focus on electro-chemical cells: reactions in which electrical work is done, and reactions that are driven by the input of electrical energy. In a photovoltaic cell, solar energy is converted directly to electrical energy and this type of cell is described in Box 18.1.

There are two types of electrochemical cell: the *electrolytic cell* and the *galvanic cell*. In the electrolytic cell, the passage of an electrical current through an electrolyte causes a chemical reaction to occur. Examples of such electrochemical cells are seen in many industrial extractions such as the Downs process for the production of Na and $Cl_2$ from molten NaCl (equation 18.1 and Figure 21.16).

$$NaCl(l) \xrightarrow{\text{electrolyse}} Na(l) + \tfrac{1}{2}Cl_2(g) \qquad (18.1)$$

Redox = *red*uction-*ox*idation

Without the input of electrical current, reaction 18.1 does not occur. Reaction 18.1 is a *redox reaction* in which one species is reduced and one is oxidized, and the overall process can be separated into two half-equations 18.2 and 18.3 in which the addition or removal of electrons can be seen.

$$Na^+(l) + e^- \longrightarrow Na(l) \qquad\qquad \textit{Reduction} \quad (18.2)$$
$$Cl^-(l) \longrightarrow \tfrac{1}{2}Cl_2(g) + e^- \qquad\qquad \textit{Oxidation} \quad (18.3)$$

In a galvanic cell, a *spontaneous redox reaction* occurs and generates an electrical current. In such a cell, *electrical work* is done by the system. Dropping a piece of magnesium ribbon into dilute mineral acid (e.g. HCl or $H_2SO_4$) results in a spontaneous redox reaction (equation 18.4), i.e. a galvanic cell is created. The two half-reactions that make up the electrochemical cell are given in equations 18.5 and 18.6.

$$Mg(s) + 2H^+(aq) \longrightarrow Mg^{2+}(aq) + H_2(g) \qquad (18.4)$$
$$2H^+(aq) + 2e^- \longrightarrow H_2(g) \qquad\qquad \textit{Reduction} \quad (18.5)$$
$$Mg(s) \longrightarrow Mg^{2+}(aq) + 2e^- \qquad\qquad \textit{Oxidation} \quad (18.6)$$

**COMMERCIAL AND LABORATORY APPLICATIONS**

## Box 18.1 Solar cells: electricity from solar energy

The reserves of energy in the Sun are enormous, and harnessing solar energy is an environmentally acceptable method of producing power. The sunlight falling on the Earth's surface each day provides about $1.4 \times 10^{19}$ kJ of energy. However, collecting solar energy and converting it into other forms of energy is not trivial. Conversion using heat exchange units (*solar panels*) gives thermal energy to raise the temperature of swimming pools and to provide domestic hot water or power lighting in gardens. Conversion via photovoltaic cells produces electricity and involves the use of semiconductors. The initial development of solar cells was linked to NASA's space programme. Although applications in satellites and space probes are still at the cutting edge of design technology, the use of solar cells is a now part of our everyday life. The photograph shows the silicon solar cell and printed circuit components of a solar-powered calculator. Semiconductors in such cells are usually fabricated from silicon, and the thickness of a typical cell is 200–350 μm. In Section 9.13, we discussed the bonding in silicon in terms of simple band theory, and saw that the band gap in pure silicon is 1.1 eV. The semiconducting properties of silicon (a group 14 element) can be altered by *doping* the bulk material with, for example, gallium (a group 13 element) or arsenic (a group 15 element). It is essential that only very small amounts of the dopant are introduced. Substituting a Ga for Si atom produces an electron deficient site. This in turn introduces a discrete, *unoccupied* level (an *acceptor level*) into the band structure; the acceptor levels remain discrete provided that the concentration of Ga atoms is low. Diagram (a) below illustrates that the band gap between the acceptor level and the lower-lying occupied band (the *valence band*) is small ($\approx 0.1$ eV). As a result, thermal population of the acceptor level by electrons is possible. The positive holes left behind in the valence band act as *charge carriers*. You can think in terms of an electron moving into the hole, thereby leaving another hole into which another electron can move and so on. The Ga-doped silicon is called a p-type (p stands for positive) semiconductor. In arsenic-doped silicon, As atoms substitute for Si atoms, thereby introducing electron-rich sites. Diagram (b) below shows that the extra electrons occupy a discrete level and there is a small band gap ($\approx 0.1$ eV) between this band (a *donor level*) and the high-lying unoccupied band (the *conduction band*). Electrons from the donor level thermally populate the conduction band where they are free to move. Electrical conduction can be described in terms of the movement of negatively charged electrons and this generates an n-type (n stands for negative) semiconductor.

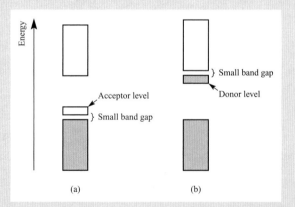

A solar cell is constructed from an n-doped layer which faces the sun, a p-doped layer and a metal-contact grid on the top and bottom surfaces. The latter are connected by a conducting wire. When light falls on the solar cell, electrons move from the p-type to the n-type silicon at the p–n junction. This leads to a flow of electricity around the circuit. Power output per cell is small, and a large number of cells must operate together to produce a viable voltage supply.

Other semiconductors in use in solar cells include GaAs (e.g. in space satellites), CdTe (a newcomer to solar cell development) and $TiO_2$ (used in the Grätzel cell which involves a $TiO_2$ film coated with an organic dye).

A solar cell inside a solar-powered calculator; the multimeter illustrates the electrical resistance of the cell.
© *Dorling Kindersley.*

We begin the chapter with a detailed look at galvanic cells, and then move on to discuss electrolytic cells. Finally, we discuss some applications of electrochemistry.

## 18.2    The Daniell cell and some definitions

Figure 18.1 illustrates a simple electrochemical cell called the *Daniell cell* after its inventor, John Daniell. It consists of two *half-cells*, one containing a strip of copper metal immersed in an aqueous solution of copper(II) sulfate, and the other containing a zinc foil immersed in an aqueous solution of zinc(II) sulfate. The two solutions are connected via a *salt bridge*. This may consist of gelatine containing $KCl$ or $KNO_3$ solution which permits ions to pass between the two half-cells but does not allow the copper(II) and zinc(II) solutions to mix too quickly. The metal strips are connected by an electrically conducting wire, and a voltmeter may be placed in the circuit.

When a metal is placed in a solution containing its ions (e.g. copper in copper(II) sulfate), a state of equilibrium is attained (equation 18.7) and an electrical potential exists which depends on the position of the equilibrium.

$$Cu^{2+}(aq) + 2e^- \rightleftharpoons Cu(s) \qquad (18.7)$$

The combination of Cu metal and $Cu^{2+}$ ions is called a *redox couple*; there is the possibility of either the reduction of $Cu^{2+}$ to Cu or the oxidation of Cu to $Cu^{2+}$.

The tendency for $Cu^{2+}$ to be reduced is greater than that of $Zn^{2+}$. This is readily demonstrated by placing a piece of Cu metal in a solution containing $Zn^{2+}$ ions, and in a separate beaker, placing a strip of Zn metal in a solution containing $Cu^{2+}$ ions. In the first beaker, no reaction occurs, but in the second, zinc is oxidized and copper metal is deposited as $Cu^{2+}$ ions are reduced. Reaction 18.8 is the *spontaneous process*.

$$Zn(s) + Cu^{2+}(aq) \longrightarrow Zn^{2+}(aq) + Cu(s) \qquad (18.8)$$

When the two half-cells in the Daniell cell are connected as in Figure 18.1, reaction 18.8 occurs even though the reactants are not in immediate contact. The difference in the electrical potentials of the two half-cells results in a *potential difference* (which can be measured on the voltmeter), and electrons flow *along*

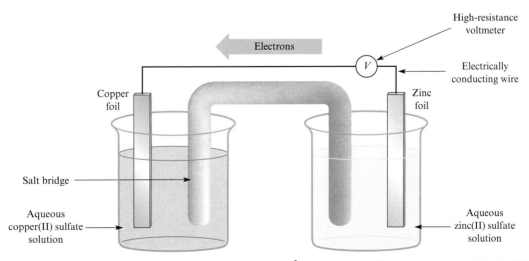

**Fig. 18.1** A representation of the Daniell cell. In the left-hand cell, $Cu^{2+}$ ions are reduced to copper metal, and in the right-hand cell, zinc metal is oxidized to $Zn^{2+}$ ions. The cell diagram is: $Zn(s)|Zn^{2+}(aq) \vdots Cu^{2+}(aq)|Cu(s)$.

*the wire* from one cell to the other as the reaction proceeds. The direction of electron flow is from the zinc to the copper foil (right to left along the wire in Figure 18.1). This corresponds to the production of electrons in the zinc half-cell (equation 18.9) and their consumption in the copper half-cell (equation 18.10).

$$Zn(s) \longrightarrow Zn^{2+}(aq) + 2e^- \tag{18.9}$$

$$Cu^{2+}(aq) + 2e^- \longrightarrow Cu(s) \tag{18.10}$$

The electrical circuit is completed by the passage of ions across the salt bridge. The overall cell reaction (equation 18.8) is obtained by combining the two half-cell reactions, ensuring that the number of electrons used in the reduction step balances the number of electrons produced in the oxidation step. In this case, each half-cell equation involves two electrons.

The overall electrochemical cell is denoted by the *cell diagram*:

$$Zn(s)|Zn^{2+}(aq) \vdots Cu^{2+}(aq)|Cu(s)$$

where the solid vertical lines represent boundaries between phases, and the dotted line stands for the salt bridge. *The spontaneous process reads from left to right across the cell diagram* – Zn metal is converted to $Zn^{2+}$ ions as $Cu^{2+}$ ions transform to Cu metal. *The reduction process is always drawn on the right-hand side of the cell diagram.*

The potential difference between the two half-cells is denoted by $E_{cell}$ and its value (in volts) is directly related to the change in Gibbs energy for the overall reaction. Equation 18.11 relates $E_{cell}$ to $\Delta G$, and equation 18.12 refers to the *standard cell potential*. The temperature must always be stated, because $\Delta G$ (and hence $E_{cell}$) is temperature-dependent (see Section 17.7).

$$\Delta G = -zFE_{cell} \tag{18.11}$$

$$\Delta G^o = -zFE^o_{cell} \tag{18.12}$$

where: $z$ = number of moles of electrons transferred in the reaction
$F$ = Faraday constant = $96\,485\,C\,mol^{-1}$

**Physical constants: see inside back cover of this book**

The *standard cell potential* refers to the conditions:

- the concentration of any solution is $1\,mol\,dm^{-3}$;[§]
- the pressure of any gaseous component is 1 bar ($1.00 \times 10^5\,Pa$);[†]
- the temperature is 298 K;
- a solid component is in its standard state.

For a spontaneous reaction, $E^o_{cell}$ is always positive, and this corresponds to a negative value of $\Delta G^o$.

---

The change in standard Gibbs energy of an **electrochemical cell** is:

$$\Delta G^o = -zFE^o_{cell}$$

where: $z$ = number of moles of electrons transferred **per mole of reaction**,

$F$ = Faraday constant = $96\,485\,C\,mol^{-1}$.

For a **spontaneous** reaction, $E^o_{cell}$ is always **positive**.

---

[§] Strictly, the standard conditions refer to unit activity, but for a *dilute* solution, the concentration is approximately equal to the activity. Solutions of $1\,mol\,dm^{-3}$ would not be used in the laboratory; this is discussed later in this section.
[†] The standard pressure may in some tables of data be taken as 1 atm (101 300 Pa); see Table 18.1.

| Worked example 18.1 | *The Gibbs energy change in a Daniell cell* |

If the solutions in the Daniell cell are both of concentration $1\,mol\,dm^{-3}$, $E^{\circ}{}_{cell}$(298 K) is 1.10 V. What is $\Delta G^{\circ}$ for the cell reaction?

The equation needed is:

$$\Delta G^{\circ} = -zFE^{\circ}{}_{cell}$$

The cell reaction is:

$$Zn(s) + Cu^{2+}(aq) \longrightarrow Zn^{2+}(aq) + Cu(s)$$

In the reaction, two moles of electrons are transferred: $z = 2$.

$$\Delta G^{\circ}(298\,K) = -zFE^{\circ}{}_{cell}$$
$$= -2 \times (96\,485\,C\,mol^{-1}) \times (1.10\,V)$$
$$= -212\,000\,J\,mol^{-1}$$
$$= -212\,kJ\,mol^{-1}$$

Refer to Table 1.2:

$C = A\,s$

$V = kg\,m^2\,s^{-3}\,A^{-1}$

$J = kg\,m^2\,s^{-2}$

This large negative value of $\Delta G^{\circ}$ is consistent with a spontaneous reaction.

## 18.3 The standard hydrogen electrode and standard reduction potentials

In the Daniell cell, we could determine $E^{\circ}{}_{cell}$ experimentally but it is not possible to determine the potentials of the half-cells, although we know qualitatively that the tendency for zinc to lose electrons exceeds that of copper. It would be useful if we knew this information *quantitatively*, and for this purpose, we may use the *standard hydrogen electrode* as a *reference electrode* for which the cell potential, $E^{\circ}{}_{cell}$, has been defined as 0 V.

Equation 18.13 gives the half-cell reaction that occurs in a standard hydrogen electrode – either $H_2$ is oxidized or $H^+$ ions are reduced.

$$2H^+(aq, 1.0\,mol\,dm^{-3}) + 2e^- \rightleftharpoons H_2(g, 1\,bar) \qquad (18.13)$$

Neither component can be physically connected into the electrical circuit, and when the electrode is constructed, a platinum wire is placed in contact with both the $H_2$ gas and the aqueous $H^+$ ions. Platinum metal is inert and is not involved in any redox process. By constructing the electrochemical cell shown in Figure 18.2, the *standard reduction potential* or *standard electrode potential*, $E^{\circ}$, for the $Zn^{2+}/Zn$ couple could, in theory, be determined. In practice, a more dilute solution of aqueous metal ions would be used (see below). The cell diagram for this cell reaction is denoted as follows with the reduction process on the right-hand side:

$$Zn(s)|Zn^{2+}(aq, 1.0\,mol\,dm^{-3}) \vdots 2H^+(aq, 1.0\,mol\,dm^{-3})|[H_2(g, 1\,bar)]Pt$$

The spontaneous reaction, reading from left to right, is the oxidation of zinc (equation 18.14) and the reduction of hydrogen ions (equation 18.15).

$$Zn(s) \longrightarrow Zn^{2+}(aq) + 2e^- \qquad (18.14)$$
$$2H^+(aq) + 2e^- \longrightarrow H_2(g) \qquad (18.15)$$

**Fig. 18.2** The standard hydrogen electrode consists of a platinum wire immersed in a $1.0\,mol\,dm^{-3}$ solution of aqueous hydrochloric acid through which $H_2$ (at 1 bar pressure) is bubbled. When the standard hydrogen electrode is combined with a standard $Zn^{2+}/Zn$ couple, the reading on the voltmeter, $E°_{cell}$, is equal to the standard reduction potential, $E°$, of the $Zn^{2+}/Zn$ couple.

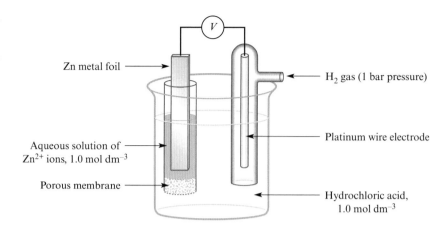

The cell potential (measured on the voltmeter shown in Figure 18.2) gives a direct measure of the potential of the $Zn^{2+}/Zn$ couple with respect to the standard hydrogen electrode, i.e. $E°$ for the $Zn^{2+}/Zn$ couple. We must, however, pay attention to the sign convention. *After writing the cell diagram such that the spontaneous reaction is read from left to right*, the standard reduction potential of the $Zn^{2+}/Zn$ couple can be determined using equation 18.16. In this instance, the electrode at which the reduction occurs is the standard hydrogen electrode and the electrode at which oxidation occurs is the $Zn^{2+}/Zn$ electrode.

$$E°_{cell} = [E°_{reduction\ process}] - [E°_{oxidation\ process}] \tag{18.16}$$

For the cell in Figure 18.2, $E°_{cell}$ is 0.76 V. It is not necessary to include a sign with the value of $E°_{cell}$ because it is always positive for a spontaneous reaction. Equation 18.17 then shows how the standard reduction potential of the $Zn^{2+}/Zn$ couple is determined. Remember that $E° = 0\,V$ for the standard hydrogen electrode (at 298 K).

$$0.76\,V = 0 - E°_{Zn^{2+}/Zn} \qquad E°_{Zn^{2+}/Zn} = -0.76\,V \tag{18.17}$$

Selected standard reduction potentials for half-cells are given in Table 18.1; a wider range is listed in Appendix 12. *The half-cell is always quoted in terms of a reduction reaction.*

In an electrochemical cell under standard conditions, the concentration of each aqueous solution species is $1.0\,mol\,dm^{-3}$, and so in Table 18.1, each half-cell contains the specified solution species at a concentration of $1.0\,mol\,dm^{-3}$. As a result, the reduction of $O_2$ appears twice in the table, once under acidic conditions:

$$O_2(g) + 4H^+(aq) + 4e^- \rightleftharpoons 2H_2O(l)$$

$$E° = +1.23\,V \text{ when } [H^+] = 1.0\,mol\,dm^{-3}\ (pH = 0)$$

and once under alkaline conditions:

$$O_2(g) + 2H_2O(l) + 4e^- \rightleftharpoons 4[OH]^-(aq)$$

$$E° = +0.40\,V \text{ when } [OH^-] = 1.0\,mol\,dm^{-3}\ (pH = 14)$$

Similar situations arise for other species, e.g. the reduction of $[BrO_3]^-$ (see Appendix 12). To ensure that values of $E°$ are unambiguous, we have adopted the following notation:

**Table 18.1** Selected standard reduction potentials (298 K); see Appendix 12 for a more detailed list. The concentration of each aqueous solution is $1.0\,mol\,dm^{-3}$, and the pressure of a gaseous component is 1 bar ($10^5\,Pa$). (Changing the standard pressure to 1 atm (101 300 Pa) makes no difference to the values of $E^o$ at this level of accuracy.)

| Reduction half-equation | $E^o$ or $E^o_{[OH^-]=1}$ / V |
|---|---|
| $Li^+(aq) + e^- \rightleftharpoons Li(s)$ | −3.04 |
| $K^+(aq) + e^- \rightleftharpoons K(s)$ | −2.93 |
| $Ca^{2+}(aq) + 2e^- \rightleftharpoons Ca(s)$ | −2.87 |
| $Na^+(aq) + e^- \rightleftharpoons Na(s)$ | −2.71 |
| $Mg^{2+}(aq) + 2e^- \rightleftharpoons Mg(s)$ | −2.37 |
| $Al^{3+}(aq) + 3e^- \rightleftharpoons Al(s)$ | −1.66 |
| $Zn^{2+}(aq) + 2e^- \rightleftharpoons Zn(s)$ | −0.76 |
| $S(s) + 2e^- \rightleftharpoons S^{2-}(aq)$ | −0.48 |
| $Fe^{2+}(aq) + 2e^- \rightleftharpoons Fe(s)$ | −0.44 |
| $Cr^{3+}(aq) + e^- \rightleftharpoons Cr^{2+}(aq)$ | −0.41 |
| $Co^{2+}(aq) + 2e^- \rightleftharpoons Co(s)$ | −0.28 |
| $Pb^{2+}(aq) + 2e^- \rightleftharpoons Pb(s)$ | −0.13 |
| $Fe^{3+}(aq) + 3e^- \rightleftharpoons Fe(s)$ | −0.04 |
| $2H^+(aq) + 2e^- \rightleftharpoons H_2(g, 1\,bar)$ | 0.00 |
| $2H^+(aq) + S(s) + 2e^- \rightleftharpoons H_2S(aq)$ | +0.14 |
| $Cu^{2+}(aq) + e^- \rightleftharpoons Cu^+(aq)$ | +0.15 |
| $Cu^{2+}(aq) + 2e^- \rightleftharpoons Cu(s)$ | +0.34 |
| $O_2(g) + 2H_2O(l) + 4e^- \rightleftharpoons 4[OH]^-(aq)$ | +0.40 |
| $Cu^+(aq) + e^- \rightleftharpoons Cu(s)$ | +0.52 |
| $I_2(aq) + 2e^- \rightleftharpoons 2I^-(aq)$ | +0.54 |
| $Fe^{3+}(aq) + e^- \rightleftharpoons Fe^{2+}(aq)$ | +0.77 |
| $Ag^+(aq) + e^- \rightleftharpoons Ag(s)$ | +0.80 |
| $Br_2(aq) + 2e^- \rightleftharpoons 2Br^-(aq)$ | +1.09 |
| $O_2(g) + 4H^+(aq) + 4e^- \rightleftharpoons 2H_2O(l)$ | +1.23 |
| $[Cr_2O_7]^{2-}(aq) + 14H^+(aq) + 6e^- \rightleftharpoons 2Cr^{3+}(aq) + 7H_2O(l)$ | +1.33 |
| $Cl_2(aq) + 2e^- \rightleftharpoons 2Cl^-(aq)$ | +1.36 |
| $[MnO_4]^-(aq) + 8H^+(aq) + 5e^- \rightleftharpoons Mn^{2+}(aq) + 4H_2O(l)$ | +1.51 |
| $Ce^{4+}(aq) + e^- \rightleftharpoons Ce^{3+}(aq)$ | +1.72 |
| $F_2(aq) + 2e^- \rightleftharpoons 2F^-(aq)$ | +2.87 |

- for half-cells that involve $H^+$ and for which the reduction potential is therefore pH-dependent (see Section 18.4), $E^o$ refers to $[H^+] = 1.0\,mol\,dm^{-3}$ (pH = 0);
- for the specific case of $[OH^-] = 1.0\,mol\,dm^{-3}$, we use the notation $E^o_{[OH^-]=1}$.

---

**Worked example 18.2**    *Standard reduction potential*

The value of $E^o{}_{cell}$ for the cell shown below is 0.80 V.

$$Pt[H_2(g, 1\,bar)]|2H^+(aq, 1.0\,mol\,dm^{-3}) \vdots Ag^+(aq, 1.0\,mol\,dm^{-3})|Ag(s)$$

What is the spontaneous cell reaction, and what is the standard reduction potential for the $Ag^+(aq)/Ag(s)$ electrode?

The spontaneous cell reaction is obtained by reading the cell diagram from left to right. The two half-cell reactions are:

$$H_2(g) \longrightarrow 2H^+(aq) + 2e^-$$

$$Ag^+(aq) + e^- \longrightarrow Ag(s)$$

The overall cell reaction can be written by combining the two half-cell reactions. In order that the number of electrons used for reduction balances the number produced during oxidation, the silver half-cell reaction must be multiplied by 2. The two equations to be combined are therefore:

$$H_2(g) \longrightarrow 2H^+(aq) + 2e^-$$

$$2Ag^+(aq) + 2e^- \longrightarrow 2Ag(s)$$

The overall reaction is:

$$H_2(g) + 2Ag^+(aq) \longrightarrow 2H^+(aq) + 2Ag(s)$$

For the cell in the cell diagram in the question, the standard cell potential is given by the equation:

$$E^o{}_{cell} = [E^o{}_{reduction\ process}] - [E^o{}_{oxidation\ process}]$$

$$0.80\,V = E^o{}_{Ag^+/Ag} - 0$$

$$E^o{}_{Ag^+/Ag} = +0.80\,V$$

## Other half-cells

Inspection of Table 18.1 or Appendix 12 shows that by no means all half-cells are of the type $M^{n+}/M$, i.e. they do not all contain a metal strip immersed in an aqueous solution of the metal ions. For example, half-equations 18.18–18.20 involve aqueous species.

$$Ce^{4+}(aq) + e^- \rightleftharpoons Ce^{3+}(aq) \tag{18.18}$$

$$Br_2(aq) + 2e^- \rightleftharpoons 2Br^-(aq) \tag{18.19}$$

$$[MnO_4]^-(aq) + 8H^+(aq) + 5e^- \rightleftharpoons Mn^{2+}(aq) + 4H_2O(l) \tag{18.20}$$

In these cases, the half-cells are assembled in a similar way to the standard hydrogen electrode, with the reduced and oxidized components in contact with each other and with a platinum wire to facilitate the flow of electrons. The cell notation for systems involving more complex half-cells follows the same principle as that described above. For the combination of a standard hydrogen electrode and the $Fe^{3+}/Fe^{2+}$ couple, the cell diagram is:

$$Pt[H_2(g, 1\,bar)]|2H^+(aq, 1.0\,mol\,dm^{-3}) \vdots Fe^{3+}(aq), Fe^{2+}(aq)|Pt$$

and reaction 18.21 is the spontaneous cell process:

$$H_2(g) + 2Fe^{3+}(aq) \longrightarrow 2H^+(aq) + 2Fe^{2+}(aq) \tag{18.21}$$

## Using standard reduction potentials to find $E^o{}_{cell}$

Standard reduction potentials can be used to determine values of $E^o{}_{cell}$ for a reaction. Since we can find the related standard Gibbs energy change from $E^o{}_{cell}$ (equation 18.12), standard reduction potentials are useful means of predicting the *thermodynamic* viability of a reaction.

The half-cell reactions in Table 18.1 are all written as *reduction reactions* and the name *standard reduction potentials* for the associated values of $E^o$ follows accordingly. A positive $E^o$ means that the reduction is thermodynamically favourable with respect to the reduction of $H^+$ ions to $H_2$. A negative $E^o$ means that the preferential reduction is that of the hydrogen ions. Consider the following three half-equations and their corresponding standard reduction potentials:

$$Cr^{3+}(aq) + e^- \rightleftharpoons Cr^{2+}(aq) \qquad E^o = -0.41\,V$$

$$2H^+(aq) + 2e^- \rightleftharpoons H_2(g) \qquad E^o = 0.00\,V$$

$$Ce^{4+}(aq) + e^- \rightleftharpoons Ce^{3+}(aq) \qquad E^o = +1.72\,V$$

In an electrochemical cell consisting of a standard hydrogen electrode and a $Ce^{4+}/Ce^{3+}$ couple, reaction 18.22 is the spontaneous process, but if the hydrogen electrode is combined with the $Cr^{3+}/Cr^{2+}$ couple, reaction 18.23 occurs.

$$2Ce^{4+}(aq) + H_2(g) \longrightarrow 2Ce^{3+}(aq) + 2H^+(aq) \qquad (18.22)$$

reduction | oxidation

$$2Cr^{2+}(aq) + 2H^+(aq) \longrightarrow 2Cr^{3+}(aq) + H_2(g) \qquad (18.23)$$

oxidation | reduction

These results show that the $Ce^{4+}$ ion is a more powerful oxidizing agent than $H^+$, but that $H^+$ is a better oxidizing agent than $Cr^{3+}$ ions. This means that $Ce^{4+}$ ions will oxidize $Cr^{2+}$ ions (equation 18.24).

$$Ce^{4+}(aq) + Cr^{2+}(aq) \longrightarrow Ce^{3+}(aq) + Cr^{3+}(aq) \qquad (18.24)$$

reduction | oxidation

The thermodynamic favourability of each reaction can be assessed by calculating $E^o_{cell}$ and $\Delta G^o(298\,K)$. For reaction 18.22, the cell diagram (showing the spontaneous process from left to right) is:

$$Pt[H_2(g, 1\,bar)]|2H^+(aq, 1.0\,mol\,dm^{-3}) \vdots Ce^{4+}(aq), Ce^{3+}(aq)|Pt$$

and $E^o_{cell}$ (equation 18.25) and $\Delta G^o$ (equation 18.26) can be determined.

$$E^o_{cell} = [E^o_{reduction\ process}] - [E^o_{oxidation\ process}]$$

$$= +1.72 - 0.00 = 1.72\,V \qquad (18.25)$$

▶
Details of units are given in worked example 18.1

$$\Delta G^o(298\,K) = -zFE^o_{cell} = -2 \times (96\,485\,C\,mol^{-1}) \times (1.72\,V)$$

$$= -332\,000\,J\,mol^{-1} = -332\,kJ\,mol^{-1} \qquad (18.26)$$

You must note carefully to what this value of $\Delta G^o$ refers: from equation 18.22, $\Delta G^o$ is $-332\,kJ$ per 2 moles of $Ce^{4+}$ or per 1 mole of $H_2$. You should be able to show that $\Delta G^o(298\,K)$ for reactions 18.23 and 18.24 are $-79$ and $-206\,kJ\,mol^{-1}$ respectively.

It is important to remember that $E^o_{cell}$ is independent of the amount of material present. However, whereas $E^o_{cell}$ has identical values for reactions 18.22 *and* 18.27, the value of $z$, and therefore of $\Delta G^o$, varies.

$$Ce^{4+}(aq) + \tfrac{1}{2}H_2(g) \rightleftharpoons Ce^{3+}(aq) + H^+(aq) \qquad (18.27)$$

For reaction 18.22, we showed above that $\Delta G^\circ(298\,\text{K}) = -332\,\text{kJ}$ per mole of $H_2$ or per two moles of $Ce^{4+}$. For reaction 18.27, $\Delta G^\circ(298\,\text{K})$ is given by:

$$\Delta G^\circ(298\,\text{K}) = -zFE^\circ{}_{\text{cell}}$$
$$= -1 \times (96\,485\,\text{C}\,\text{mol}^{-1}) \times (1.72\,\text{V})$$
$$= -166\,000\,\text{J}\,\text{mol}^{-1} = -166\,\text{kJ}\,\text{mol}^{-1}$$

i.e. $-166\,\text{kJ}$ per mole of $Ce^{4+}$ or per half-mole of $H_2$.

In Table 18.1, the half-reactions are arranged in order of the values of $E^\circ$ with the most negative value at the top of the table, and the most positive at the bottom. This is a common format for such tables, and allows you to see immediately the relative oxidizing and reducing abilities of each species. Consider the $Ag^+/Ag$ couple: $E^\circ = +0.80\,\text{V}$. Silver(I) ions will oxidize any of the reduced species on the list which lie *above* the $Ag^+/Ag$ couple – for example, Cu metal. On the other hand, any oxidized species *below* the $Ag^+/Ag$ couple is able to oxidize Ag to $Ag^+$. The extent of the reaction depends upon the magnitude of $\Delta G^\circ$. For reaction 18.28, $E^\circ{}_{\text{cell}}$ is 0.46 V, corresponding to a value of $\Delta G^\circ(298\,\text{K})$ of $-89\,\text{kJ}$ per mole of reaction. This is sufficiently large and negative that few reactants remain in the system.

$$2Ag^+(aq) + Cu(s) \rightleftharpoons 2Ag(s) + Cu^{2+}(aq)$$
$$E^\circ{}_{\text{cell}} = +0.80 - (+0.34) = 0.46\,\text{V} \tag{18.28}$$

In contrast, consider the reaction between $Cr^{3+}$ ions and Fe (equation 18.29). The corresponding value of $\Delta G^\circ(298\,\text{K})$ is $-6\,\text{kJ}$ per mole of reaction, and this indicates that there are only slightly more products than reactants in the system.

$$2Cr^{3+}(aq) + Fe(s) \rightleftharpoons 2Cr^{2+}(aq) + Fe^{2+}(aq)$$
$$E^\circ{}_{\text{cell}} = -0.41 - (-0.44) = 0.03\,\text{V} \tag{18.29}$$

A useful (if somewhat artificial) 'rule' to remember is the *anticlockwise cycle* which can be used with a table of standard reduction potentials that are arranged as in Table 18.1. Consider any pair of redox couples in the table; each couple is made up of an oxidized and a reduced species. The spontaneous reaction will be between the lower oxidized species and the upper reduced species. For example, $Ag^+$ (*an oxidized species*) oxidizes Cu (*a reduced species*) and the direction of the spontaneous reaction may be deduced by constructing an anticlockwise cycle:

$$Cu^{2+}(aq) + 2e^- \rightleftharpoons Cu(s) \qquad\qquad E^\circ = +0.34\,\text{V}$$

Direction of

$$Ag^+(aq) + e^- \rightleftharpoons Ag(s) \qquad\qquad E^\circ = +0.80\,\text{V}$$

spontaneous

reaction

---

**Worked example 18.3**     *Redox equilibria*

**Use the standard reduction potentials in Table 18.1 to determine the products of the reaction between iron and aqueous $Br_2$.**

From Table 18.1, there are three redox couples that are relevant to this question:

$$Fe^{2+}(aq) + 2e^- \rightleftharpoons Fe(s) \qquad E^\circ = -0.44\,\text{V}$$
$$Fe^{3+}(aq) + 3e^- \rightleftharpoons Fe(s) \qquad E^\circ = -0.04\,\text{V}$$
$$Br_2(aq) + 2e^- \rightleftharpoons 2Br^-(aq) \qquad E^\circ = +1.09\,\text{V}$$

This suggests two possible reactions:

**1.** $Fe(s) + Br_2(aq) \rightleftharpoons Fe^{2+}(aq) + 2Br^-(aq)$

or

**2.** $2Fe(s) + 3Br_2(aq) \rightleftharpoons 2Fe^{3+}(aq) + 6Br^-(aq)$

The value of $E^o_{cell}$ for each reaction is:

**1.** $E^o_{cell} = 1.09 - (-0.44) = 1.53\,V$

**2.** $E^o_{cell} = 1.09 - (-0.04) = 1.13\,V$

The larger (more positive) $E^o_{cell}$ might initially suggest that reaction **1** is the more thermodynamically favourable process, but the true arbiter is $\Delta G^o$, not $E^o_{cell}$. In reaction **1**, $z = 2$, but in reaction **2**, $z = 6$. $\Delta G^o(298\,K)$ for each reaction is found using the equation:

$$\Delta G^o(298\,K) = -zFE^o_{cell}$$

**1.** $\Delta G^o(298\,K) = -2 \times (96\,485\,C\,mol^{-1}) \times (1.53\,V)$

$$= -295\,000\,J\,mol^{-1}$$

$$= -295\,kJ\,mol^{-1}$$

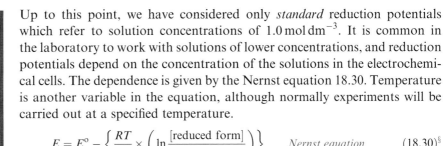

This corresponds to $-295\,kJ$ *per mole of iron*.

**2.** $\Delta G^o(298\,K) = -6 \times (96\,485\,C\,mol^{-1}) \times (1.13\,V)$

$$= -654\,000\,J\,mol^{-1}$$

$$= -654\,kJ\,mol^{-1}$$

This corresponds to $-327\,kJ$ *per mole of iron*.

Reaction **2** is therefore thermodynamically more favourable than **1**. This example emphasizes that although a positive value of $E^o_{cell}$ certainly indicates spontaneity, it is $\Delta G^o$ that is the true indicator of the extent of reaction.

---

## 18.4 The effect of solution concentration on $E^o$: the Nernst equation

Up to this point, we have considered only *standard* reduction potentials which refer to solution concentrations of $1.0\,mol\,dm^{-3}$. It is common in the laboratory to work with solutions of lower concentrations, and reduction potentials depend on the concentration of the solutions in the electrochemical cells. The dependence is given by the Nernst equation 18.30. Temperature is another variable in the equation, although normally experiments will be carried out at a specified temperature.

$$E = E^o - \left\{ \frac{RT}{zF} \times \left( \ln \frac{[\text{reduced form}]}{[\text{oxidized form}]} \right) \right\} \qquad \textit{Nernst equation} \qquad (18.30)^\S$$

where:  $R$ = molar gas constant = $8.314\,J\,K^{-1}\,mol^{-1}$

$T$ = temperature in K

$z$ = number of moles of electrons transferred *per mole of reaction*

$F$ = Faraday constant = $96\,485\,C\,mol^{-1}$

Walther Hermann Nernst (1864–1941).
© *The Nobel Foundation.*

For treatment of units see worked example 18.1

---

$\S$ The Nernst equation may also be written in the form $E = E^o - \left\{ \frac{RT}{zF} \times \ln Q \right\}$ where $Q$ is the *reaction quotient*; see Section 17.8.

In Table 18.1, the oxidized forms of the half-cells are on the left-hand side and the reduced forms on the right. If the half-cell involves a metal ion/metal couple, the activity of the solid metal is taken to be 1 and equation 18.30 simplifies to 18.31.

For an $M^{n+}/M$ couple:
$$E = E^{\circ} - \left( \frac{RT}{zF} \ln \frac{1}{[M^{n+}]} \right)$$

or

$$E = E^{\circ} + \left( \frac{RT}{zF} \ln [M^{n+}] \right)$$

(18.31)

The **Nernst equation** states:

$$E = E^{\circ} - \left\{ \frac{RT}{zF} \times \left( \ln \frac{[\text{reduced form}]}{[\text{oxidized form}]} \right) \right\}$$

---

**Worked example 18.4**     *The dependence of E on solution concentration*

The standard reduction potential of the $Ag^+/Ag$ couple is $+0.80$ V. What is the value of $E$ at 298 K when the concentration of $Ag^+$ ions is $0.020 \, \text{mol dm}^{-3}$?

The electrode is of the type $M^{n+}/M$, and so $E$ can be found using the equation:

$$E = E^{\circ} + \left( \frac{RT}{zF} \ln [M^{n+}] \right)$$

The half-equation for the couple is:

$$Ag^+(aq) + e^- \rightleftharpoons Ag(s)$$

and one mole of electrons is transferred per mole of reaction: $z = 1$.

Refer to Table 1.2:

$J = kg \, m^2 \, s^{-2}$

$C = A \, s$

$V = kg \, m^2 \, s^{-3} \, A^{-1}$

$$E = (0.80 \, \text{V}) + \left\{ \frac{(8.314 \, \text{J K}^{-1} \, \text{mol}^{-1}) \times (298 \, \text{K}) \times \ln 0.020}{1 \times (96\,485 \, \text{C mol}^{-1})} \right\}$$

$$= (0.80 \, \text{V}) - (0.10 \, \text{V})$$

$$= 0.70 \, \text{V}$$

---

Equation 18.31 is of the form:

$$y = c + mx$$

and a plot of $E$ against $\ln [M^{n+}]$ is linear. The gradient of the line is $\dfrac{RT}{zF}$ and the value on the vertical axis corresponding to $\ln [M^{n+}] = 0$ is equal to $E^{\circ}$ for the $M^{n+}/M$ couple. Figure 18.3 shows a plot of $E_{Ag^+/Ag}$ against $\ln [Ag^+]$ using experimental data collected for dilute solutions. The half-cell potentials may be obtained from the electrochemical cell:

$$Pt[H_2(g, 1 \, bar)] \| 2H^+(aq, 1.0 \, \text{mol dm}^{-3}) \vdots Ag^+(aq)|Ag(s)$$

for which $E_{cell}$ is equivalent to $E_{Ag^+/Ag}$ (equation 18.32).

$$E_{cell} = E_{Ag^+/Ag} - E^{\circ}_{\text{standard hydrogen electrode}} = E_{Ag^+/Ag}$$

(18.32)

In equation 18.32, notice the use of $E^{\circ}$ for the hydrogen electrode denoting standard conditions, but $E$ for the $Ag^+/Ag$ electrode because the concentration

**Fig. 18.3** The dependence of the reduction potential of the $Ag^+/Ag$ couple on the concentration of $Ag^+$ ions. When $[Ag^+] = 1.0\,mol\,dm^{-3}$, the cell is under standard conditions; this corresponds to $\ln[Ag^+] = 0$, and the reduction potential is the standard value, $E^o = +0.80\,V$.

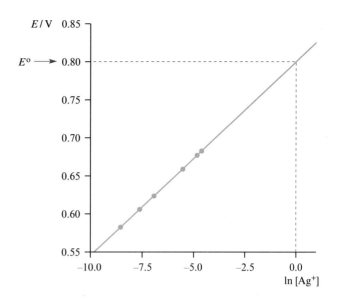

of $Ag^+$ is variable. The *cell potential* is therefore non-standard ($E_{cell}$ rather than $E^o_{cell}$). The value of $E^o_{Ag^+/Ag}$ can be found by extrapolation; when $[Ag^+] = 1.0\,mol\,dm^{-3}$ (standard conditions), $\ln[Ag^+] = 0$, and $E^o_{Ag^+/Ag}$ can be determined from Figure 18.3 as shown.

### The effect of pH on reduction potential

If the half-cell of interest involves either $H^+$ or $[OH]^-$ ions, then the value of $E_{half-cell}$ will depend upon the pH of the solution. Consider half-cell 18.33.

$$\underbrace{[MnO_4]^-(aq) + 8H^+(aq)}_{\text{oxidized form}} + 5e^- \rightleftharpoons \underbrace{Mn^{2+}(aq) + 4H_2O(l)}_{\text{reduced form}} \qquad (18.33)$$
$$E^o = +1.51\,V$$

Using the Nernst equation at 298 K, the value of $E$ can be found (equation 18.34) and we immediately see the dependence of $E$ upon hydrogen ion concentration.

For treatment of units, see worked example 18.4

$$E = E^o - \left(\frac{RT}{zF}\ln\frac{[Mn^{2+}]}{[MnO_4^-][H^+]^8}\right)$$

$$E = (1.51\,V) - \left(\frac{(8.314\,J\,K^{-1}\,mol^{-1}) \times (298\,K)}{5 \times (96\,485\,C\,mol^{-1})}\ln\frac{[Mn^{2+}]}{[MnO_4^-][H^+]^8}\right) \quad (18.34)$$

$pH = -\log[H^+]$

For a solution in which the ratio $[Mn^{2+}]:[MnO_4^-]$ is $1:100$, equation 18.34 shows that at $pH = 0$, $E = +1.53\,V$, but at $pH = 3$, $E = +1.25\,V$. You should consider what effect this difference will have, for example, on the ability of $KMnO_4$ to oxidize chloride ions to $Cl_2$ (equation 18.35).

$$Cl_2(aq) + 2e^- \rightleftharpoons 2Cl^-(aq) \qquad\qquad E^o = +1.36\,V \quad (18.35)$$

---

**18.5**    ### Reference electrodes

Throughout this section, we have emphasized the use of the standard hydrogen electrode as a reference electrode in electrochemical cells. While this is useful because $E^o_{2H^+/H_2} = 0$, the cell itself is not very convenient to use. A

**Fig. 18.4** A schematic
representation of the silver/silver
chloride reference electrode.

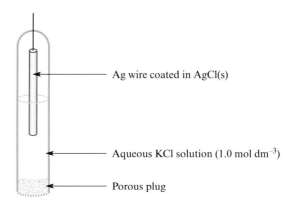

**Fig. 18.4** A schematic
representation of the silver/silver
chloride reference electrode.

cylinder of $H_2$ is needed and the pressure must be maintained at exactly 1 bar. A range of other reference electrodes is available, including the silver/silver chloride electrode. Figure 18.4 shows the silver/silver chloride electrode. An Ag wire coated with AgCl dips into an aqueous solution of KCl of concentration $1.0 \, mol \, dm^{-3}$. The half-cell reaction is given in equation 18.36, and the reduction potential depends on the chloride concentration; for $[Cl^-] = 1.0 \, mol \, dm^{-3}$, $E^o$ is $+0.22 \, V$.

$$AgCl(s) + e^- \rightleftharpoons Ag(s) + Cl^-(aq) \tag{18.36}$$

Other reduction potentials may be measured *with respect to the silver/silver chloride electrode*, and one use of this reference electrode is in combination with a pH electrode in a pH meter (see Section 18.8).

Another reference electrode is the calomel electrode, the half-cell reaction for which is given in equation 18.37.

$$Hg_2Cl_2(s) + 2e^- \rightleftharpoons 2Hg(l) + 2Cl^-(aq) \qquad E^o = +0.27 \, V \tag{18.37}$$

Because a variety of reference electrodes are in use, it is always necessary to quote reduction potentials *with respect to a specific reference*.

## 18.6    Electrolytic cells

### Electrolytic cells and reactions at the anode and cathode

We have seen that in a galvanic cell, the combination of two half-cells can lead to a spontaneous reaction which generates a current. In contrast, in an *electrolytic cell*, the passage of a current through an *electrolyte* drives a redox reaction which is otherwise non-spontaneous. A typical electrolytic cell consists of two electrodes dipping into an electrolyte. The electrodes are connected to the positive and negative terminals of a battery as shown in Figure 18.5. The role of the electrodes is to provide an interface at which electrons can be transferred into the electrolyte. It is important that the material from which the electrodes are made is inert with respect to the electrolysis reaction. Graphite rods are often used in simple cells. The power supply in the electrolytic cell supplies electrons which flow around the circuit to the negative electrode, the *cathode* (Figure 18.5). Positively charged ions in the electrolyte migrate to the cathode and a reduction reaction occurs (e.g. half-reaction 18.2). The other electrode (the *anode*) is positively charged. Anions in the electrolyte migrate to the anode (Figure 18.5) and an oxidation reaction takes place (e.g. half-reaction 18.3).

▪▷
Electrical conductivity of
graphite: see Section 9.10

**Fig. 18.5** A schematic representation of an electrolytic cell. The passage of electrical current through the electrolyte causes cations to migrate to the cathode and anions to migrate to the anode.

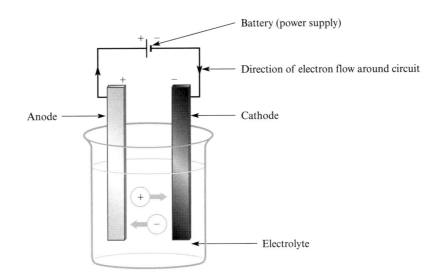

In an *electrolytic cell*, the passage of a current through an *electrolyte* drives a redox reaction; reduction occurs at the *cathode* and oxidation takes place at the *anode*. The overall process is *electrolysis*.

If an electrolyte consists of a molten ionic salt such as NaCl, NaBr or $MgCl_2$, there is only one type of cation and one type of anion. Equations 18.2 and 18.3 showed the cell half-reactions for the electrolysis of molten NaCl. The electrolysis of molten $MgCl_2$ is shown in scheme 18.38.

$$\left.\begin{array}{ll} Mg^{2+}(l) + 2e^- \longrightarrow Mg(l) & \textit{Reduction} \\ 2Cl^- \longrightarrow Cl_2(g) + 2e^- & \textit{Oxidation} \end{array}\right\} \tag{18.38}$$

If the electrolyte is an aqueous solution of a salt, then a competitive situation arises because water itself can be electrolysed (equation 18.39).

$$2H_2O(l) \xrightarrow{\text{electrolysis}} 2H_2(g) + O_2(g) \tag{18.39}$$

In order to drive reaction 18.39 (at pH 7.0), the battery (see Figure 18.5) must supply a minimum of 1.23 V. This is because the *reverse* of reaction 18.39 is the spontaneous process and its inherent 'driving force' corresponds to a value of $E_{cell} = 1.23\,V$ at pH 7.0 (see Box 18.2). In practice, this potential is insufficient to bring about the electrolysis of $H_2O$ and an additional potential called the *overpotential* is required. If the electrodes are platinum, the overpotential is $\approx 0.60\,V$. The origins of overpotentials are not simple, but a basic explanation can be given in terms of the problems involved in carrying out the reduction or oxidation at a solid surface and releasing a gaseous product.

Now consider what happens when we electrolyse an aqueous solution of NaCl. The possible cathode half-reactions are given in equations 18.40 and 18.41. Of course, if Na were formed it would immediately react with the water. From Table 18.1, it is clear that at pH 0, $H^+$ is more readily reduced than $Na^+$. At pH 7 (see Box 18.2), this preference still appertains, and so $H_2$ is preferentially liberated at the cathode.

$$2H^+(aq) + 2e^- \longrightarrow H_2(g) \tag{18.40}$$

$$Na^+(aq) + e^- \longrightarrow Na(s) \tag{18.41}$$

At the anode, the competition is between the oxidation of $Cl^-$ or $H_2O$ (equations 18.42 and 18.43).

$$2Cl^-(aq) \longrightarrow Cl_2(g) + 2e^- \tag{18.42}$$

$$2H_2O(l) \longrightarrow O_2(g) + 4H^+(aq) + 4e^- \tag{18.43}$$

**THEORETICAL AND CHEMICAL BACKGROUND**

**Box 18.2 Calculating $E_{cell}$ for the reaction: $2H_2(g) + O_2(g) \longrightarrow 2H_2O(l)$**

The reaction:

$$2H_2(g) + O_2(g) \longrightarrow 2H_2O(l)$$

can be considered in terms of the two half-reactions:

*Oxidation:*   $H_2(g) \longrightarrow 2H^+(aq) + 2e^-$

*Reduction:*   $O_2(g) + 4H^+(aq) + 4e^- \longrightarrow 2H_2O(l)$

Tables of standard reduction potentials give the following data:

$$2H^+(aq) + 2e^- \rightleftharpoons H_2(g) \qquad E^\circ = 0.00\,V$$

$$O_2(g) + 4H^+(aq) + 4e^- \rightleftharpoons 2H_2O(l)$$
$$E^\circ = +1.23\,V$$

The $E^\circ$ values refer to solutions under standard conditions, i.e. $[H^+] = 1.0\,mol\,dm^{-3}$ (pH 0). However, in pure water, pH = 7.0 and $[H^+] = 10^{-7}\,mol\,dm^{-3}$. By applying the Nernst equation, values of $E$ for the half-cells shown above can be determined.

For:   $2H^+(aq) + 2e^- \rightleftharpoons H_2(g)$

$$E = E^\circ - \left( \frac{RT}{zF} \ln \frac{1}{[H^+]^2} \right)$$

$$= (0.00\,V) - \left( \frac{(8.314\,J\,K^{-1}\,mol^{-1}) \times (298\,K)}{2 \times (96\,485\,C\,mol^{-1})} \right.$$

$$\left. \times \ln \frac{1}{[10^{-7}]^2} \right)$$

$$= -0.41\,V$$

For:   $O_2(g) + 4H^+(aq) + 4e^- \rightleftharpoons 2H_2O(l)$

$$E = E^\circ - \left( \frac{RT}{zF} \ln \frac{1}{[H^+]^4} \right)$$

$$= (1.23\,V) - \left( \frac{(8.314\,J\,K^{-1}\,mol^{-1}) \times (298\,K)}{4 \times (96\,485\,C\,mol^{-1})} \right.$$

$$\left. \times \ln \frac{1}{[10^{-7}]^4} \right)$$

$$= +0.82\,V$$

For the overall reaction:

$$2H_2(g) + O_2(g) \longrightarrow 2H_2O(l)$$

at pH 7.0:

$$E_{cell} = E_{\text{reduction process}} - E_{\text{oxidation process}}$$

$$= +0.82 - (-0.41)$$

$$= 1.23\,V$$

Standard reduction potentials in Table 18.1 show that at pH 0, $H_2O$ is more readily oxidized than $Cl^-$, and at pH 7 (see Box 18.2) this preference is also true. However, caution is needed in interpreting these data. In practice, the electrolysis of aqueous NaCl yields $Cl_2$ (or a mixture of $Cl_2$ and $O_2$) at the anode because of the higher overpotential needed for the liberation of $O_2$.

## Commercial applications of electrolysis

The application of electrolysis to the extraction of elements is of great commercial significance and is illustrated in Chapters 21 and 22 by the following processes:

- the Downs process for the extraction of Na and $Cl_2$ from NaCl (Section 21.11 and Figure 21.16);
- extraction of Be from $BeCl_2$, and Ca from $CaCl_2$ (Section 21.12);
- extraction of Mg from $MgCl_2$ (Section 21.12 and Figure 21.18);
- extraction of Al from a mixture of cryolite and alumina (Section 22.4).

**Fig. 18.6** A schematic representation of an electrochemical cell in which Cu can be transferred from anode to cathode during electrolysis. In the commercial purification of copper, the anode is composed of impure ('blister') copper and the electrolyte is a mixture of aqueous $CuSO_4$ and $H_2SO_4$.

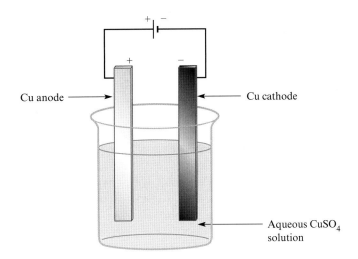

The effective transfer of Cu from anode to cathode is the basis for the industrial purification of copper. An electrolysis cell similar to that represented in Figure 18.6 is constructed. Impure copper is used as the anode and Cu metal is oxidized during electrolysis (equation 18.44) while pure Cu metal is deposited at the cathode (equation 18.45).

*At the anode*: $\quad Cu(s) \longrightarrow Cu^{2+}(aq) + 2e^-$ （18.44)

*At the cathode*: $\quad Cu^{2+}(aq) + 2e^- \longrightarrow Cu(s)$ （18.45)

This last application is a convenient means of leading us to Faraday's Laws of Electrolysis.

**18.7**　**Faraday's Laws of Electrolysis**

If the electrolysis cell shown in Figure 18.6 is constructed with copper electrodes of known mass, and current flows around the circuit, it is found that the mass of the anode *decreases* by a mass $x$ g, and the mass of the cathode *increases* by a mass $x$ g.

*Faraday's First Law of Electrolysis*: the mass of substance liberated at an electrode during electrolysis is proportional to the quantity of charge (in coulombs) passing through the electrolyte.

*Faraday's Second Law of Electrolysis*: the numbers of coulombs needed to liberate one mole of different products are in whole number ratios.

Michael Faraday (1791–1867). *Reproduced courtesy of the Library and Information Centre, Royal Society of Chemistry.*

The best way to understand these laws is to put them into practice. Equation 18.46 defines a coulomb (a measure of the amount of electricity).

Electrical charge (coulombs) = Current (amps) × Time (seconds) （18.46)

Equation 18.47 follows from Faraday's First Law.

Mass of product deposited at electrode $\propto$ Electrical charge (coulombs)

（18.47)

■▶
Faraday constant:
see Section 18.2

Experimentally, it is found that 96 485 C are required to liberate one mole of product from a singly charged ion (e.g. Ag from $Ag^+$, $\frac{1}{2}Cl_2$ from $Cl^-$, Na from $Na^+$). This quantity of electricity is known as a Faraday: $1\,F = 96\,485\,C$. For a doubly charged ion, $2\,F$ are needed to liberate one mole of product (e.g. Cu from $Cu^{2+}$, Pb from $Pb^{2+}$), and for a triply charged ion, $3\,F$ are needed to deposit one mole of product (e.g. Al from $Al^{3+}$).

| Worked example 18.5 | *Application of Faraday's Laws* |

**What mass of Cu metal is deposited at the cathode during the electrolysis of aqueous $CuSO_4$ solution if a current of 0.10 A passes through the solution for 30 min?**

■▶
Refer to Table 1.2:
$C = A\,s$

The amount of electrical charge (C) = Current (A) × Time (s)
Convert the time to SI units: $30\,min = 30 \times 60\,s = 1800\,s$
Electrical charge $= (0.10\,A) \times (1800\,s) = 180\,C$
For the reaction at the cathode:

$$Cu^{2+}(aq) + 2e^- \longrightarrow Cu(s)$$

Therefore, deposition of one mole of Cu requires $2\,F = 2 \times 96\,485\,C$
$$= 192\,970\,C$$

Number of moles of Cu deposited by $180\,C = \dfrac{(180\,C)}{(192\,970\,C\,mol^{-1})}$
$$= 9.33 \times 10^{-4}\,mol$$

Mass of one mole of $Cu = 63.54\,g$ (see periodic table inside the front cover of the book).
Mass of Cu deposited in the experiment
$$= (9.33 \times 10^{-4}\,mol) \times (63.54\,g\,mol^{-1})$$
$$= 5.93 \times 10^{-2}\,g$$

| 18.8 | **Selected applications of electrochemical cells** |

In addition to the industrial applications that we have already exemplified and their uses in batteries (e.g. see Box 22.7), electrochemical cells have widespread applications. In this section, we consider two uses in the chemical laboratory.

### Determination of solubility products, $K_{sp}$

The dependence of the reduction potential, $E$, on solution concentration can be used to measure very low concentrations of ions, and from these data, solubilities and values of $K_{sp}$ for sparingly soluble salts can be determined. Suppose that we wish to find $K_{sp}$ for AgCl. A suitable electrochemical half-cell for this determination must contain a saturated solution of AgCl (equation 18.48) in addition to an Ag electrode to provide an $Ag^+/Ag$ redox couple (equation 18.49).

$$AgCl(s) \rightleftharpoons Ag^+(aq) + Cl^-(aq) \qquad \text{equilibrium constant} = K_{sp} \qquad (18.48)$$

$$Ag^+(aq) + e^- \rightleftharpoons Ag(s) \qquad\qquad\qquad\qquad\qquad\qquad\qquad (18.49)$$

The concentration of $Ag^+$ ions in solution depends on the position of equilibrium 18.48 and therefore on $K_{sp}$, and $[Ag^+]$ affects the value of $E_{Ag^+/Ag}$. By combining the half-cell described above with a standard hydrogen electrode, $E_{Ag^+/Ag}$ is found to be $+0.512\,V$. Using the Nernst equation, the concentration of $Ag^+$ ions in the saturated solution of AgCl can be found:

$$E = E^\circ - \left( \frac{RT}{zF} \ln \frac{1}{[Ag^+]} \right)$$

$$E = E^\circ + \left( \frac{RT}{zF} \ln[Ag^+] \right)$$

For treatment of units, look back at worked example 18.4

$$(0.512\,V) = (0.80\,V) + \left( \frac{(8.314\,J\,K^{-1}\,mol^{-1}) \times (298\,K)}{1 \times (96\,485\,C\,mol^{-1})} \ln[Ag^+] \right)$$

$$\ln[Ag^+] = (0.512\,V - 0.80\,V)\left( \frac{(96\,485\,C\,mol^{-1})}{(8.314\,J\,K^{-1}\,mol^{-1}) \times (298\,K)} \right) = -11.22$$

$$[Ag^+] = e^{-11.22} = 1.34 \times 10^{-5}\,mol\,dm^{-3}$$

The equilibrium concentrations of $Ag^+$ and $Cl^-$ ions in the saturated solution of AgCl are equal, and $K_{sp}$ (which is dimensionless) is found from equation 18.50.

$$K_{sp} = [Ag^+][Cl^-] = [Ag^+]^2$$

$$K_{sp} = (1.34 \times 10^{-5})^2 = 1.80 \times 10^{-10} \tag{18.50}$$

### Ion-selective electrodes: the pH meter

An electrode that is sensitive to the concentration of a specific ion is called an *ion-selective electrode*. One example is the *glass electrode* which is sensitive to $[H_3O]^+$ ions and is used in pH meters. The glass electrode consists of a thin glass-walled membrane containing an aqueous HCl solution of known concentration (e.g. $0.10\,mol\,dm^{-3}$). A reference electrode such as an Ag/AgCl electrode is often built within the assembly and allows the pH meter to operate with a single probe (Figure 18.7). When the glass electrode is dipped into a solution of unknown pH, the electrical potential of the glass electrode depends on the difference in concentrations of $[H_3O]^+$ ions on each side of the glass membrane.

Other ion-selective electrodes include those sensitive to $Na^+$, $K^+$, $Cl^-$, $F^-$, $[CN]^-$ and $[NH_4]^+$ ions.

**Fig. 18.7** A schematic representation of a glass electrode.

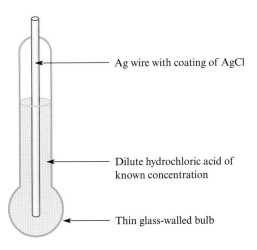

Ag wire with coating of AgCl

Dilute hydrochloric acid of known concentration

Thin glass-walled bulb

## SUMMARY

In this chapter, we have introduced electrochemistry by looking at galvanic and electrolytic cells. In a galvanic cell, a spontaneous redox reaction occurs, generating an current; in an electrolytic cell, a current is used to drive a thermodynamically, non-spontaneous reaction. Values of $E^\circ$ for electrochemical half-cells refer only to *standard conditions*, and the dependence of the reduction potential on the concentration of solution species is given by the Nernst equation. Applications of electrochemical cells range from commercial extraction of elements from ores to pH meters.

### Do you know what the following terms mean?

- redox reaction
- reduction
- oxidation
- galvanic cell
- electrode
- electrolyte
- salt bridge
- potential difference

- half-cell
- cell diagram
- standard hydrogen electrode
- standard reduction potential, $E^\circ$
- standard cell potential, $E^\circ_{cell}$
- Nernst equation
- reference electrode
- electrolytic cell

- anode
- cathode
- electrolysis
- overpotential
- Faraday's Laws of Electrolysis
- ion-selective electrode
- glass electrode

### Some useful equations

$$\Delta G^\circ = -zFE^\circ_{cell}$$

$$\Delta G = -zFE_{cell}$$

$$E^\circ_{cell} = [E^\circ_{reduction\ process}] - [E^\circ_{oxidation\ process}]$$

$$E = E^\circ - \left( \frac{RT}{zF} \ln \frac{[\text{reduced form}]}{[\text{oxidized form}]} \right) \qquad (Nernst\ equation)$$

### You should be able:

- to distinguish between a galvanic cell and an electrolytic cell

- to describe the construction of, and reaction within, the Daniell cell

- to apply a knowledge of the Daniell cell to illustrate galvanic cells in a wider sense

- to define what is meant by a *reference electrode*

- to describe the construction and operation of the standard hydrogen electrode

- to use tables of standard reduction potentials (e.g. Table 18.1 or Appendix 12) to determine values of $E^\circ_{cell}$ for given redox reactions

- to determine values of $\Delta G^\circ$ from $E^\circ_{cell}$, and to use the values to judge whether a reaction is spontaneous

- to use tables of $E^\circ$ values to predict possible reactions

- to use the Nernst equation to find the dependence of $E$ on the concentration of solution species

- to show why $E$ values may depend on the pH of the solution

- to illustrate the construction of a typical electrolytic cell

- to describe what happens during the electrolysis of a molten salt

- to explain which ions may be preferentially discharged during the electrolysis of an aqueous solution of a salt

- to describe briefly what is meant by an overpotential

- to give selected examples of commercial and laboratory applications of electrochemical cells

## PROBLEMS

Data for these problems can be found in Table 18.1, Appendix 12 and the inside back cover of the book.

**18.1**  Balance the following half-equations by finding the number, $z$, of electrons:
(a) $Cr^{2+} + ze^- \rightleftharpoons Cr$
(b) $[Cr_2O_7]^{2-} + 14H^+ + ze^- \rightleftharpoons 2Cr^{3+} + 7H_2O$
(c) $[S_4O_6]^{2-} + ze^- \rightleftharpoons 2[S_2O_3]^{2-}$
(d) $O_2 + 4H^+ + ze^- \rightleftharpoons 2H_2O$
(e) $[BrO_3]^- + 3H_2O + ze^- \rightleftharpoons Br^- + 6[OH]^-$
(f) $PbO_2 + 4H^+ + [SO_4]^{2-} + ze^- \rightleftharpoons PbSO_4 + 2H_2O$

**18.2**  For the reaction:

$$2Ag^+(aq) + Zn(s) \longrightarrow 2Ag(s) + Zn^{2+}(aq)$$

$E^o_{cell} = 1.56\,V$. Calculate $\Delta G^o$ per mole of $Ag^+$.

**18.3**  What information can you obtain from the following cell diagram?

$$Pt[H_2(g, 1\,bar)] \,|\, 2H^+(aq, 1.0\,mol\,dm^{-3})$$
$$\vdots\, Ce^{4+}(aq), Ce^{3+}(aq) \,|\, Pt$$

**18.4**  Consider the following cell diagram:

$$Mg(s) \,|\, Mg^{2+}(aq)$$
$$\vdots\, 2H^+(aq, 1.0\,mol\,dm^{-3}) \,|\, [H_2(g, 1\,bar)]Pt$$

(a) Write down the spontaneous cell reaction.
(b) Calculate $E^o_{cell}$ for the reaction. (c) Determine $\Delta G^o(298\,K)$ for the reaction.

**18.5**  Determine $E^o_{cell}$ and $\Delta G^o(298\,K)$ for the following reactions:
(a) $Co^{2+}(aq) + 2Cr^{2+}(aq) \rightleftharpoons Co(s) + 2Cr^{3+}(aq)$
(b) $\frac{1}{2}Co^{2+}(aq) + Cr^{2+}(aq) \rightleftharpoons \frac{1}{2}Co(s) + Cr^{3+}(aq)$
(c) $[Cr_2O_7]^{2-}(aq) + 14H^+(aq) + 6Fe^{2+}(aq) \rightleftharpoons$
$$2Cr^{3+}(aq) + 7H_2O(l) + 6Fe^{3+}(aq)$$

**18.6**  Write balanced equations for the spontaneous reactions that should occur when the following pairs of half-cells are combined:
(a) $Zn^{2+}(aq) + 2e^- \rightleftharpoons Zn(s)$ with
$Cl_2(aq) + 2e^- \rightleftharpoons 2Cl^-(aq)$
(b) $2H^+(aq) + 2e^- \rightleftharpoons H_2(g)$ with
$Zn^{2+}(aq) + 2e^- \rightleftharpoons Zn(s)$
(c) $Cl_2(aq) + 2e^- \rightleftharpoons 2Cl^-(aq)$ with
$Br_2(aq) + 2e^- \rightleftharpoons 2Br^-(aq)$
(d) $[IO_3]^-(aq) + 6H^+ + 6e^- \rightleftharpoons I^-(aq) + 3H_2O(l)$
with $I_2(aq) + 2e^- \rightleftharpoons 2I^-(aq)$

**18.7**  A half-cell containing a Zn metal foil dipping into an aqueous solution of $Zn(NO_3)_2$ ($1.0\,mol\,dm^{-3}$) is connected by a salt bridge to a half-cell containing a Pb metal foil dipping into an aqueous solution of $Pb(NO_3)_2$ ($1.0\,mol\,dm^{-3}$). (a) Write down a cell diagram for the overall electrochemical cell. (b) Give an equation for the spontaneous redox process and specify which species act as reducing

and oxidizing agents. (c) Calculate $\Delta G^o(298\,K)$ for the reaction.

**18.8**  Find the value of $E$ (at 298 K) for the $Cu^{2+}/Cu$ couple when $[Cu^{2+}] = 0.020\,mol\,dm^{-3}$.

**18.9**  If the concentration of $Ag^+(aq)$ in an $Ag^+/Ag$ half-cell is $0.10\,mol\,dm^{-3}$, what is $E$ (at 298 K)?

**18.10**  Write down four half-cells in Table 18.1 for which $E$ is pH-dependent.

**18.11**  Determine the reduction potential for the half-cell:

$$[MnO_4]^-(aq) + 8H^+(aq) + 5e^- \rightleftharpoons$$
$$Mn^{2+}(aq) + 5H_2O(l)$$

at 298 K and at pH 5.0 when the ratio of $[MnO_4^-]:[Mn^{2+}]$ is $75:1$.

**18.12**  In order that the $Sn^{2+}/Sn$ couple has a value of $E = -0.16\,V$ at 298 K, what concentration of $Sn^{2+}$ ions should be used in the half-cell?

**18.13**  Find $E^o(Cu^{2+}/Cu)$ using the following experimental data.

| $[Cu^{2+}]\,/\,mol\,dm^{-3}$ | $E\,/\,V$ |
|---|---|
| 0.0010 | 0.250 |
| 0.0020 | 0.259 |
| 0.0025 | 0.262 |
| 0.0040 | 0.268 |
| 0.010 | 0.280 |

**18.14**  Use the data in Appendix 12 to suggest which of the following species could be oxidized by aqueous $Fe^{3+}$ ions. What are the limitations of the data? (a) $Co^{2+}(aq)$; (b) $Cl^-(aq)$; (c) $I^-(aq)$; (d) $Sn^{2+}(aq)$.

**18.15**  Consider the two half-equations:

$$O_2(g) + 2H^+(aq) + 2e^- \rightleftharpoons H_2O_2(aq)$$
$$E^o = +0.70\,V$$

$$H_2O_2(aq) + 2H^+(aq) + 2e^- \rightleftharpoons 2H_2O(l)$$
$$E^o = +1.78\,V$$

(a) If aqueous $H_2O_2$ were mixed with aqueous $Fe^{2+}$ in the presence of acid, would $H_2O_2$ act as a reducing or oxidizing agent? Write an equation for the reaction that takes place.
(b) Suggest what happens when acidified aqueous $H_2O_2$ and aqueous $KMnO_4$ are mixed.
(c) What do the two half-equations above suggest may happen if $H_2O_2$ is left to stand at 298 K? Write an equation for the reaction that you suggest may occur and calculate $\Delta G^o(298\,K)$ for the reaction. Comment on the result, and explain how your answer fits in with the discussion associated with equation 21.53 (Chapter 21).

**18.16** Suggest what products are discharged at the anode and cathode during the electrolysis of (a) molten KBr; (b) fused $CaCl_2$; (c) dilute aqueous NaCl; (d) concentrated aqueous NaCl (brine); (e) aqueous $CuSO_4$ using Cu electrodes; (f) dilute $H_2SO_4$ using Pt electrodes.

**18.17** What mass of Mg is deposited at the cathode if a current of 5.0 A is passed through molten $MgCl_2$ for 60 min?

**18.18** The manufacture of Al by electrolysis of fused $Al_2O_3$ and $K_3[AlF_6]$ demands especially large amounts of electricity, and aluminium refineries are often associated with hydroelectric schemes. Rationalize why this is so.

**18.19** What volumes of $H_2$ and $O_2$ are produced when acidified water is electrolysed for 30 min by a current of 1.50 A? (Volume of one mole of an ideal gas at 1 bar $= 22.7\,dm^3$.)

**18.20** The reduction potential for a half-cell consisting of an Ag wire dipping into a saturated solution of AgI is +0.325 V at 298 K. Calculate the solubility of AgI and the value of $K_{sp}$.

## Additional problems

*For some of these problems, you may need data from Table 18.1.*

**18.21** The half-cell:

$$[Cr_2O_7]^{2-}(aq) + 14H^+(aq) + 6e^- \rightleftharpoons$$
$$2Cr^{3+}(aq) + 7H_2O(l)$$

is prepared in aqueous solution at pH 5.0 at 298 K. Determine $E$ if the concentrations of $[Cr_2O_7]^{2-}$ and $Cr^{3+}$ are 0.50 and 0.0050 $mol\,dm^{-3}$ respectively. Comment on the ability of $[Cr_2O_7]^{2-}$ in this half-cell to act as an oxidizing agent with respect to the half-cell under standard conditions.

**18.22** Discuss the following statements and data.
(a) In dilute aqueous solution, $[ClO_4]^-$ is very difficult to reduce despite the values of $E^o$ for the following half-cells at pH 0:

$$[ClO_4]^-(aq) + 2H^+(aq) + 2e^- \rightleftharpoons$$
$$[ClO_3]^-(aq) + H_2O(l)\quad E^o = +1.19\,V$$
$$[ClO_4]^-(aq) + 8H^+(aq) + 8e^- \rightleftharpoons$$
$$Cl^-(aq) + 4H_2O(l)\quad E^o = +1.24\,V$$

(b) Lead–acid batteries involve the half-cells:

$$PbSO_4(s) + 2e^- \rightleftharpoons$$
$$Pb(s) + [SO_4]^{2-}(aq)\quad E^o = -0.36\,V$$
$$PbO_2(s) + 4H^+(aq) + [SO_4]^{2-}(aq) + 2e^- \rightleftharpoons$$
$$PbSO_4(s) + 2H_2O(l)\quad E^o = +1.69\,V$$

Batteries can be recharged by reversing the cell reaction.

**18.23** Determine $K$ for the equilibrium:

$$Ag^+(aq) + 3I^-(aq) \rightleftharpoons [AgI_3]^{2-}$$

given the following half-cell reduction potential:

$$[AgI_3]^{2-}(aq) + e^- \rightleftharpoons Ag(s) + 3I^-(aq)$$
$$E^o = -0.02\,V$$

# 19 The conductivity of ions in solution

## Topics

- Conductivity and conductance
- Strong and weak electrolytes
- Kohlrausch's Law
- Kohlrausch's Law of Independent Migration of Ions
- Ostwald's Dilution Law
- Conductance changes during reactions

## 19.1 Some definitions and units

Conductivity: see Section 9.10 under 'carbon' and Section 9.13

We have already discussed *electrical conductivity* and the *resistivity* of a wire (equation 19.1). *Conductivity* is the inverse of resistivity (equation 19.2).

$$\text{Resistance (in } \Omega) = \frac{\text{Resistivity (in } \Omega \text{ m)} \times \text{Length of wire (in m)}}{\text{Cross section (in m}^2)} \quad (19.1)$$

$$\text{Conductivity (in S m}^{-1}) = \frac{1}{\text{Resistivity (in } \Omega \text{ m)}} \quad (19.2)$$

The units of conductivity may be quoted as siemens per metre ($S\,m^{-1}$) or reciprocal ohms per metre ($\Omega^{-1}m^{-1}$); $1\,S = 1\,\Omega^{-1}$.

In Chapter 18, we described electrolysis cells in which a current passes through an electrolyte causing ions to migrate to either the cathode or anode. In this chapter, we are concerned with the conductivities and conductances of solutions containing ions.

Two terms are frequently encountered in discussions of ions in solution and should not be confused with each other or used interchangeably. The *conductance*, $G$, of a solution is measured in siemens, and is given by equation 19.3.

$$\text{Conductance (in S)} = \frac{1}{\text{Resistance (in } \Omega)} \quad (19.3)$$

The *conductivity*, $\kappa$, of a solution is given by equation 19.2 and is measured in units of $S\,m^{-1}$. The conductivity of a solution depends on the number of ions present and is usually expressed in terms of the *molar conductivity*, $\Lambda_m$ (equation 19.4), the units of which are $S\,m^2\,mol^{-1}$.

$$\text{Molar conductivity } (\Lambda_m) = \frac{\text{Conductivity}}{\text{Concentration}} = \frac{\kappa}{c} \quad (19.4)$$

Since concentration is usually quoted in $mol\,dm^{-3}$, a conversion to $mol\,m^{-3}$ is needed before substitution into equation 19.4:

$$1\,dm^3 = 10^{-3}\,m^3 \qquad 1\,mol\,dm^{-3} = 10^3\,mol\,m^{-3}$$

$$\text{Molar conductivity (S m}^2\,mol^{-1}) = \frac{\text{Conductivity (S m}^{-1})}{\text{Concentration (mol m}^{-3})}$$

**Fig. 19.1** The variation in molar conductivities with concentration for aqueous KCl and NaCl and $CH_3CO_2H$ at 298 K. Values of $\Lambda_m^\infty$ for KCl and NaCl are $1.498 \times 10^{-2}$ and $1.264 \times 10^{-2}\,S\,m^2\,mol^{-1}$ respectively. For $CH_3CO_2H$, $\Lambda_m^\infty = 3.905 \times 10^{-2}\,S\,m^2\,mol^{-1}$ and the figure shows the curve tending towards this point.

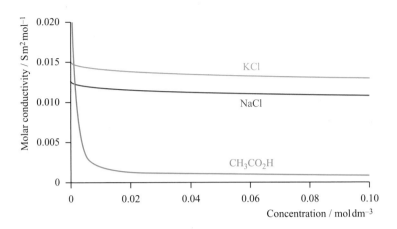

## 19.2 Molar conductivities of strong and weak electrolytes

### Strong electrolytes

*A **strong electrolyte** is fully ionized in aqueous solution.*

The variation in molar conductivities with concentration ($c$) for aqueous KCl and NaCl is shown in Figure 19.1. Values of $\Lambda_m$ vary to only a small extent as the concentration increases. This is reasonable because the conductivity should bear some relationship to the number of ions present, and, for a fully ionized compound, as the concentration doubles, the number of ions in solution doubles. Therefore, the *molar* conductivity (equation 19.4) is expected to remain constant. However, interactions between ions are only negligible at very low concentrations, and so the maximum molar conductivity is observed when the ions are free of each other's influence. This value is the *molar conductivity at infinite dilution*, $\Lambda_m^\infty$, and can be read from Figure 19.1 as the points of intercept of the curves for KCl and NaCl with the molar conductivity scale, i.e. when $c = 0$. This corresponds to a situation in which the ions are infinitely far apart, and is a hypothetical state. Thus, values of $\Lambda_m^\infty$ cannot be measured directly. Values of $\Lambda_m$ can be determined experimentally, and $\Lambda_m^\infty$ is then found by extrapolation.

Sodium and potassium chlorides are examples of *strong electrolytes*. A strong electrolyte is *fully ionized* in aqueous solution, and the behaviour shown for NaCl and KCl in Figure 19.1 is typical. Further examples of strong electrolytes are listed in Table 19.1.

Studies of conductivity were first rigorously investigated by Kohlrausch at the end of the 19th century. He showed that each ion in solution has a characteristic molar conductivity at infinite dilution, $\lambda^\infty$, and that the molar conductivity at infinite dilution of the electrolyte was equal to the

**Table 19.1** Selected strong electrolytes and values of molar conductivities at infinite dilution, $\Lambda_m^\infty$, at 298 K.

| Ionic salt | $\Lambda_m^\infty$ / $S\,m^2\,mol^{-1}$ | Ionic salt | $\Lambda_m^\infty$ / $S\,m^2\,mol^{-1}$ |
|---|---|---|---|
| NaCl | $1.264 \times 10^{-2}$ | KCl | $1.498 \times 10^{-2}$ |
| NaI | $1.269 \times 10^{-2}$ | KBr | $1.516 \times 10^{-2}$ |
| $NaClO_4$ | $1.174 \times 10^{-2}$ | $KNO_3$ | $1.449 \times 10^{-2}$ |
| $MgCl_2$ | $2.586 \times 10^{-2}$ | HCl | $4.259 \times 10^{-2}$ |

**Table 19.2** The molar conductivities at infinite dilution, $\lambda^{\infty}$ (298 K) for selected cations and anions.

| Cation | $\lambda^{\infty}$ / S m$^2$ mol$^{-1}$ | Anion | $\lambda^{\infty}$ / S m$^2$ mol$^{-1}$ |
|---|---|---|---|
| $H^+$ | $3.496 \times 10^{-2}$ | $Cl^-$ | $0.763 \times 10^{-2}$ |
| $Na^+$ | $0.501 \times 10^{-2}$ | $Br^-$ | $0.781 \times 10^{-2}$ |
| $K^+$ | $0.735 \times 10^{-2}$ | $I^-$ | $0.768 \times 10^{-2}$ |
| $Cs^+$ | $0.772 \times 10^{-2}$ | $[OH]^-$ | $1.980 \times 10^{-2}$ |
| $[NH_4]^+$ | $0.735 \times 10^{-2}$ | $[NO_3]^-$ | $0.714 \times 10^{-2}$ |
| $Ag^+$ | $0.619 \times 10^{-2}$ | $[ClO_4]^-$ | $0.673 \times 10^{-2}$ |
| $Mg^{2+}$ | $1.060 \times 10^{-2}$ | $[CH_3CO_2]^-$ | $0.409 \times 10^{-2}$ |
| $Ca^{2+}$ | $1.189 \times 10^{-2}$ | $[SO_4]^{2-}$ | $1.600 \times 10^{-2}$ |

sum of the values of $\lambda^{\infty}$ for the cations and anions. For a 1:1 electrolyte such as KCl, this is expressed by equation 19.5.

$$\Lambda_m^{\infty} = \lambda_+^{\infty} + \lambda_-^{\infty} \qquad \textit{for a 1:1 electrolyte} \quad (19.5)$$

More generally, *Kohlrausch's Law of Independent Migration of Ions* is stated in the form of equation 19.6 where $\nu_+$ and $\nu_-$ are the numbers of cations and anions per formula unit, e.g. for $Cu(NO_3)_2$, $\nu_+ = 1$ and $\nu_- = 2$.

$$\Lambda_m^{\infty} = \nu_+\lambda_+^{\infty} + \nu_-\lambda_-^{\infty} \qquad \begin{array}{c}\textit{Kohlrausch's Law of}\\ \textit{Independent Migration of Ions}\end{array} \quad (19.6)$$

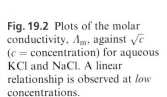

**Kohlrausch's Law of Independent Migration of Ions** states:
$$\Lambda_m^{\infty} = \nu_+\lambda_+^{\infty} + \nu_-\lambda_-^{\infty}$$

Table 19.2 lists values of $\lambda^{\infty}$ (at 298 K) for selected cations and anions. You should show that these values are consistent with the values of $\Lambda^{\infty}$ listed in Table 19.1.

Kohlrausch further showed that, at *low concentrations*, the molar conductivity of a strong electrolyte has a linear dependence on the square root of the concentration (equation 19.7).

$$\Lambda_m = \Lambda_m^{\infty} - k\sqrt{c} \qquad \textit{Kohlrausch's Law} \quad (19.7)$$

where $k$ is a constant and $c$ is the solution concentration

Plots of $\Lambda_m$ against $\sqrt{c}$ for aqueous KCl and NaCl are shown in Figure 19.2. As the concentration increases, the plots deviate from a linear relationship. The dashed lines show the linear relationships determined from the low concentration parts of each plot. The value of $\Lambda_m$ when $\sqrt{c} = 0$ corresponds to $\Lambda_m^{\infty}$, and the constant $k$ is equal to the gradient of the line. Variations in $k$ are observed with variation in compound stoichiometry.

**Fig. 19.2** Plots of the molar conductivity, $\Lambda_m$, against $\sqrt{c}$ ($c$ = concentration) for aqueous KCl and NaCl. A linear relationship is observed at *low* concentrations.

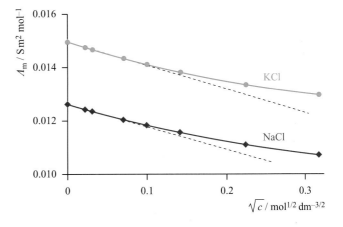

### Weak electrolytes and Ostwald's Dilution Law

A *weak electrolyte* is partially dissociated in aqueous solution.

Figure 19.1 illustrates that the variation in $\Lambda_m$ with concentration for $CH_3CO_2H$ contrasts with those of NaCl and KCl. At infinite dilution, the molar conductivity is high; $\Lambda_m^\infty$ for $CH_3CO_2H$ at 298 K is $3.905 \times 10^{-2}\,S\,m^2\,mol^{-1}$ and is off the scale in the figure. As the concentration increases to about $0.01\,mol\,dm^{-3}$, $\Lambda_m$ drops dramatically and then remains fairly constant as the concentration is raised to $0.10\,mol\,dm^{-3}$. This behaviour is typical of a *weak electrolyte*, that is, one that is *partially dissociated*. There is a hypothetical state at infinite dilution where the compound is fully ionized. As we saw in Chapter 16, acetic acid is a weak acid and, in aqueous solution, establishes equilibrium 19.8.

$$CH_3CO_2H(aq) + H_2O(l) \rightleftharpoons [CH_3CO_2]^-(aq) + [H_3O]^+(aq) \qquad (19.8)$$

Finding $\Lambda_m^\infty$ by extrapolation from experimental values of $\Lambda_m$ is not possible. However, we can make use of Kohlrausch's Law of Independent Migration of Ions. To determine $\Lambda_m^\infty(CH_3CO_2H)$, we must choose a group of *strong* electrolytes which together include the $[CH_3CO_2]^-$ and $[H_3O]^+$ ions. A suitable group would be NaCl, HCl and $Na[CH_3CO_2]$, all of which are fully ionized in aqueous solution. By Kohlrausch's Law of Independent Migration of Ions, we can write equations 19.9–19.12.

H⁺ in H₂O forms [H₃O]⁺

$$\Lambda_m^\infty(NaCl) = \lambda_+^\infty(Na^+) + \lambda_-^\infty(Cl^-) \qquad (19.9)$$

$$\Lambda_m^\infty(HCl) = \lambda_+^\infty(H^+) + \lambda_-^\infty(Cl^-) \qquad (19.10)$$

$$\Lambda_m^\infty(Na[CH_3CO_2]) = \lambda_+^\infty(Na^+) + \lambda_-^\infty([CH_3CO_2]^-) \qquad (19.11)$$

$$\Lambda_m^\infty(CH_3CO_2H) = \lambda_+^\infty(H^+) + \lambda_-^\infty([CH_3CO_2]^-) \qquad (19.12)$$

Values of $\Lambda_m^\infty(NaCl)$, $\Lambda_m^\infty(HCl)$ and $\Lambda_m^\infty(Na[CH_3CO_2])$ may be determined as described above for strong electrolytes. Appropriate combination of equations 19.9–19.11 allows us to write:

$$\Lambda_m^\infty(HCl) + \Lambda_m^\infty(Na[CH_3CO_2]) - \Lambda_m^\infty(NaCl)$$

$$= \lambda_+^\infty(H^+) + \lambda_-^\infty(Cl^-) + \lambda_+^\infty(Na^+)$$

$$+ \lambda_-^\infty([CH_3CO_2]^-) - \lambda_+^\infty(Na^+) - \lambda_-^\infty(Cl^-)$$

$$= \lambda_+^\infty(H^+) + \lambda_-^\infty([CH_3CO_2]^-)$$

Combining this expression with equation 19.12 gives equation 19.13.

$$\Lambda_m^\infty(HCl) + \Lambda_m^\infty(Na[CH_3CO_2]) - \Lambda_m^\infty(NaCl) = \Lambda_m^\infty(CH_3CO_2H) \quad (19.13)$$

Experimental values of $\Lambda_m^\infty(NaCl)$, $\Lambda_m^\infty(HCl)$ and $\Lambda_m^\infty(Na[CH_3CO_2])$ are $1.264 \times 10^{-2}$, $4.259 \times 10^{-2}$ and $0.910 \times 10^{-2}\,S\,m^2\,mol^{-1}$ respectively, giving:

$$\Lambda_m^\infty(CH_3CO_2H) = (4.259 \times 10^{-2}) + (0.910 \times 10^{-2}) - (1.264 \times 10^{-2})$$

$$= 3.905 \times 10^{-2}\,S\,m^2\,mol^{-1}$$

This is the value of $\Lambda_m^\infty$ that the curve for $CH_3CO_2H$ tends towards in Figure 19.1.

Approximation of concentrations to activities: see Section 16.3

The behaviour of a weak electrolyte is described by Ostwald's Dilution Law which relates the equilibrium constant for the weak electrolyte to the concentration and molar conductivity. Consider an aqueous solution of $CH_3CO_2H$ of concentration $c\,mol\,dm^{-3}$, for which we can write the following scheme to describe the composition at equilibrium where $\alpha$ is the *degree of dissociation* of the acid:

$$CH_3CO_2H(aq) + H_2O(l) \rightleftharpoons [CH_3CO_2]^-(aq) + [H_3O]^+(aq)$$

Concentrations at equilm:    $c(1 - \alpha)$    excess    $c\alpha$    $c\alpha$

**Degree of dissociation:** see also worked examples in Chapter 16

Values of $\alpha$ may be quoted as percentage or fractional degrees of dissociation, e.g. $\alpha = 20\%$ corresponds to $\alpha = 0.20$. The equilibrium constant, $K$, is therefore given by equation 19.14.

$$K = \frac{(c\alpha)(c\alpha)}{c(1 - \alpha)} = \frac{c\alpha^2}{(1 - \alpha)} \tag{19.14}$$

At infinite dilution, the molar conductivity is $\Lambda_m^\infty$, and Arrhenius proposed relationship 19.15 where $\Lambda_m$ is the molar conductivity at a given concentration. This relationship can be confirmed from experimental data.

**See problem 19.9**

$$\alpha = \frac{\Lambda_m}{\Lambda_m^\infty} \tag{19.15}$$

Substitution of equation 19.15 into equation 19.14 leads to Ostwald's Dilution Law (equation 19.16).

$$K = \frac{c\alpha^2}{(1 - \alpha)} = \frac{c\left(\dfrac{\Lambda_m}{\Lambda_m^\infty}\right)^2}{\left(1 - \dfrac{\Lambda_m}{\Lambda_m^\infty}\right)}$$

$$= \frac{c\left(\dfrac{\Lambda_m}{\Lambda_m^\infty}\right)^2}{\left(\dfrac{\Lambda_m^\infty - \Lambda_m}{\Lambda_m^\infty}\right)}$$

$$K = \frac{c\Lambda_m^2}{\Lambda_m^\infty(\Lambda_m^\infty - \Lambda_m)} \qquad \textit{Ostwald's Dilution Law} \tag{19.16}$$

Wilhelm Ostwald (1853–1932). © *The Nobel Foundation.*

Ostwald's Dilution Law is obeyed by weak electrolytes at concentrations of $\leq 0.10 \, \text{mol dm}^{-3}$. This is illustrated in worked example 19.1.

---

**Worked example 19.1**    *Ostwald's Dilution Law*

**Use the following data to show that $CH_3CO_2H$ obeys Ostwald's Dilution Law, and determine $K_a$ for $CH_3CO_2H$. ($\Lambda_m^\infty = 3.905 \times 10^{-2} \, \text{S m}^2 \, \text{mol}^{-1}$)**

| $[CH_3CO_2H]$ / $\text{mol dm}^{-3}$ | $\Lambda_m$ / $\text{S m}^2 \, \text{mol}^{-1}$ |
| --- | --- |
| 0.0752 | $6.021 \times 10^{-4}$ |
| 0.0354 | $8.750 \times 10^{-4}$ |
| 0.0180 | $1.221 \times 10^{-3}$ |
| 0.0067 | $1.987 \times 10^{-3}$ |

Ostwald's Dilution Law states:

$$K = \frac{c\Lambda_m^2}{\Lambda_m^\infty(\Lambda_m^\infty - \Lambda_m)}$$

where $c$ and $\Lambda_m$ are variables, and $K$ and $\Lambda_m^\infty$ are constants.

Rearrange the equation into the form of a linear relationship, $y = mx + c$:

$$c\Lambda_m^2 = K(\Lambda_m^\infty)^2 - K\Lambda_m^\infty \Lambda_m$$

$$c\Lambda_m = \frac{K(\Lambda_m^\infty)^2}{\Lambda_m} - K\Lambda_m^\infty$$

Tabulate the data to be plotted using the data from the table given in the problem:

| $c$ / mol dm$^{-3}$ | $\Lambda_m$ / S m$^2$ mol$^{-1}$ | $\dfrac{1}{\Lambda_m}$ / mol S$^{-1}$ m$^{-2}$ | $c\Lambda_m$ / S m$^2$ dm$^{-3}$ |
|---|---|---|---|
| 0.0752 | $6.021 \times 10^{-4}$ | 1661 | $4.528 \times 10^{-5}$ |
| 0.0354 | $8.750 \times 10^{-4}$ | 1143 | $3.098 \times 10^{-5}$ |
| 0.0180 | $1.221 \times 10^{-3}$ | 819.0 | $2.198 \times 10^{-5}$ |
| 0.0067 | $1.987 \times 10^{-3}$ | 503.3 | $1.331 \times 10^{-5}$ |

Now plot $c\Lambda_m$ against $\dfrac{1}{\Lambda_m}$.

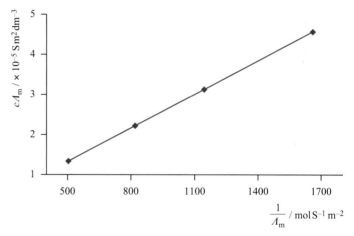

The linear plot confirms that $CH_3CO_2H$ obeys Ostwald's Dilution Law over the range of concentrations studied. The acid dissociation constant, $K_a$, can be found from the gradient of the graph. The equation for the straight line is:

$$c\Lambda_m = \frac{K(\Lambda_m^\infty)^2}{\Lambda_m} - K\Lambda_m^\infty$$

and therefore:

$$\text{Gradient} = K(\Lambda_m^\infty)^2$$
$$= 2.77 \times 10^{-8} \, S^2 \, m^4 \, mol^{-1} \, dm^{-3}$$

$\Lambda_m^\infty$ can also be found from the graph, but is given in the problem: $\Lambda_m^\infty = 3.905 \times 10^{-2} \, S \, m^2 \, mol^{-1}$.

$$K_a = \frac{(2.77 \times 10^{-8} \, S^2 \, m^4 \, mol^{-1} \, dm^{-3})}{(\Lambda_m^\infty)^2}$$

$$= \frac{(2.77 \times 10^{-8} \, S^2 \, m^4 \, mol^{-1} \, dm^{-3})}{(3.905 \times 10^{-2} \, S \, m^2 \, mol^{-1})^2}$$

$$= 1.82 \times 10^{-5} \, mol \, dm^{-3}$$

Note that the method gives an answer for $K_a$ in units of $mol\,dm^{-3}$, consistent with an expression for $K_a$ in terms of concentrations:

$$K_a = \frac{[H_3O^+][CH_3CO_2^-]}{[CH_3CO_2H]}$$

but since, strictly, $K$ is expressed in terms of activities, the thermodynamic equilibrium constant is dimensionless (see Chapter 16).

## 19.3    Conductivity changes during reactions

### Acid–base conductometric titrations

Acid–base titrations:
see Sections 16.8 and 16.9

In Chapter 16, we discussed two ways of following an acid–base titration: monitoring the pH of the solution, and using an acid–base indicator. A third option is to monitor the conductivity of the solution by carrying out a *conductometric titration*.

Suppose we wish to find the end point of a *strong acid–strong base* conductometric titration such as that between aqueous HCl and NaOH. The experiment may be set up with the base being added from a burette to the acid contained in a beaker into which dip electrodes connected to a conductivity meter. To minimize the changes in conductivity caused by dilution effects, the concentration of the reagent being added from the burette must be significantly greater than that of the reagent in the beaker. For example, let us titrate a $0.50\,mol\,dm^{-3}$ solution of NaOH against $50\,cm^3$ of $0.050\,mol\,dm^{-3}$ hydrochloric acid. Initially the flask contains only aqueous HCl which is fully ionized (equation 19.17). As aqueous NaOH is added, neutralization reaction 19.18 occurs.

$$HCl(aq) + H_2O(l) \longrightarrow [H_3O]^+(aq) + Cl^-(aq) \tag{19.17}$$

$$HCl(aq) + NaOH(aq) \longrightarrow NaCl(aq) + H_2O(l) \tag{19.18}$$

Aqueous NaCl is fully ionized and, as the neutralization reaction progresses during the titration, $[H_3O]^+$ ions are replaced by $Na^+$ ions (equation 19.19).

$$[H_3O]^+(aq) + NaOH(aq) \longrightarrow Na^+(aq) + 2H_2O(l) \tag{19.19}$$

Ions possess characteristic *mobilities* in solution, and Table 19.3 lists the mobilities of selected ions. The ion mobility, $\mu$, is related to the molar conductivity at infinite dilution by equation 19.20. Thus, values in Tables 19.2 and 19.3 are related by equation 19.20.

$$\mu = \frac{\lambda^\infty}{zF} \tag{19.20}$$

where:    $z$ = numerical charge on the ion (e.g. $z = 1$ for $Na^+$ or $Cl^-$)

$F$ = Faraday constant = $96\,485\,C\,mol^{-1}$

---

**Worked example 19.2**    *The relationship between the molar conductivity at infinite dilution and ion mobility*

At 298 K, the molar conductivity at infinite dilution, $\lambda^\infty$, of $Mg^{2+}$ is $1.060 \times 10^{-2}\,S\,m^2\,mol^{-1}$. What is the ion mobility, $\mu$, of $Mg^{2+}$ in aqueous solution at 298 K? Confirm that the units of $\mu$ are $m^2\,s^{-1}\,V^{-1}$.

The equation needed is:

$$\mu = \frac{\lambda^\infty}{zF}$$

For $Mg^{2+}$, $z = 2$

$$\mu = \frac{(1.060 \times 10^{-2}\, S\, m^2\, mol^{-1})}{2 \times (96\,485\, C\, mol^{-1})}$$

$$= 5.493 \times 10^{-8}\, S\, m^2\, C^{-1}$$

The units for $\mu$ are given as $m^2\, s^{-1}\, V^{-1}$. Refer to Table 1.2 and write the units of $S\, m^2\, C^{-1}$ in base units, remembering that $1\, S = 1\, \Omega^{-1}$:

$$S = \Omega^{-1} = s^3\, A^2\, kg^{-1}\, m^{-2}$$

$$C^{-1} = A^{-1}\, s^{-1}$$

Therefore:

$$S\, m^2\, C^{-1} = s^3\, A^2\, kg^{-1}\, m^{-2}\, m^2\, A^{-1}\, s^{-1}$$

$$= s^2\, A\, kg^{-1}$$

The base units of the volt are:

$$V = kg\, m^2\, s^{-3}\, A^{-1}$$

Therefore $s^2\, A\, kg^{-1}$ can be rewritten as $m^2\, s^{-1}\, V^{-1}$

Check: $m^2\, s^{-1}\, V^{-1} = m^2\, s^{-1}\, kg^{-1}\, m^{-2}\, s^3\, A$

$$= s^2\, A\, kg^{-1}$$

---

Among the ions listed in Table 19.3, $H^+$ and $[OH]^-$ stand out as possessing significantly higher mobilities than other ions. Proton mobility is especially facile. The movement of protons is considered to be a type of 'proton hopping' between $[H_3O]^+$ ions and $H_2O$ molecules, and probably also involves $[H_5O_2]^+$ and $[H_9O_4]^+$ ions.

If we now return to equation 19.19, the replacement of $[H_3O]^+$ by less mobile $Na^+$ ions results in a decrease in the conductivity as shown in Figure 19.3. At the end point, the conductivity reaches a minimum and then increases as excess NaOH is added to the solution. *After* the end point, the conductivity is due to $Na^+$ and $Cl^-$ from the salt *and* $Na^+$ and $[OH]^-$ ions from the excess alkali.

As an example of a *weak acid–strong base* conductometric titration, consider the addition of $0.50\, mol\, dm^{-3}$ NaOH to $50\, cm^3$ $0.050\, mol\, dm^{-3}$ $CH_3CO_2H$.

**Table 19.3** Mobilities of selected ions in aqueous solution at 298 K.

| Cation | Ion mobility, $\mu\, /\, m^2\, s^{-1}\, V^{-1}$ | Anion | Ion mobility, $\mu\, /\, m^2\, s^{-1}\, V^{-1}$ |
|---|---|---|---|
| $H^+$ | $36.23 \times 10^{-8}$ | $[OH]^-$ | $20.52 \times 10^{-8}$ |
| $Na^+$ | $5.19 \times 10^{-8}$ | $Cl^-$ | $7.91 \times 10^{-8}$ |
| $K^+$ | $7.62 \times 10^{-8}$ | $Br^-$ | $8.09 \times 10^{-8}$ |
| $Ag^+$ | $6.42 \times 10^{-8}$ | $[NO_3]^-$ | $7.40 \times 10^{-8}$ |
| $Mg^{2+}$ | $5.49 \times 10^{-8}$ | $[CH_3CO_2]^-$ | $4.24 \times 10^{-8}$ |
| $Zn^{2+}$ | $5.47 \times 10^{-8}$ | $[SO_4]^{2-}$ | $8.29 \times 10^{-8}$ |

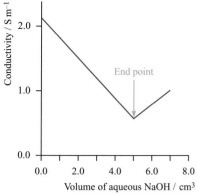

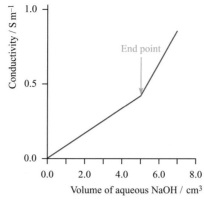

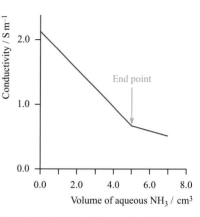

**Fig. 19.3** Variation in conductivity during the titration of a $0.50 \, mol \, dm^{-3}$ solution of NaOH against $50 \, cm^3$ of $0.050 \, mol \, dm^{-3}$ HCl.

**Fig. 19.4** Variation in conductivity during the titration of a $0.50 \, mol \, dm^{-3}$ solution of NaOH against $50 \, cm^3$ of $0.050 \, mol \, dm^{-3}$ $CH_3CO_2H$.

**Fig. 19.5** Variation in conductivity during the titration of a $0.50 \, mol \, dm^{-3}$ solution of aqueous $NH_3$ against $50 \, cm^3$ of $0.050 \, mol \, dm^{-3}$ HCl.

Equation 19.21 gives the neutralization reaction, and the conductivity changes during the titration are shown in Figure 19.4.

$$CH_3CO_2H(aq) + NaOH(aq) \longrightarrow Na[CH_3CO_2](aq) + H_2O(l) \qquad (19.21)$$

Initially, the conductivity is due to the partially dissociated $CH_3CO_2H$. The molar conductivity of $CH_3CO_2H$ is low (Figure 19.1) and so the measured conductivity (equation 19.4) is also low. As NaOH is added, $Na[CH_3CO_2]$ forms and is fully ionized in solution; as a result, the conductivity increases. *After* the end point, the solution contains $Na^+$ and $[CH_3CO_2]^-$ ions from the salt, *and* $Na^+$ and $[OH]^-$ ions from the excess NaOH. The high mobility of the $[OH]^-$ contributes to a significant increase in the conductivity (Figure 19.4).

Figure 19.5 shows the change in conductivity during a *strong acid–weak base* titration in which a $0.50 \, mol \, dm^{-3}$ solution of $NH_3$ is added to $50 \, cm^3$ of $0.050 \, mol \, dm^{-3}$ aqueous HCl. The neutralization reaction is given in equation 19.22.

**Remember:** aqueous $NH_3 = NH_4OH$

$$HCl(aq) + NH_4OH(aq) \longrightarrow NH_4Cl(aq) + H_2O(l) \qquad (19.22)$$

The initial high conductivity (Figure 19.5) arises from the presence of a fully dissociated acid and the high proton mobility. As protons are replaced by $[NH_4]^+$ ions (equation 19.23), the conductivity falls.

$$[H_3O]^+(aq) + NH_4OH(aq) \longrightarrow [NH_4]^+(aq) + 2H_2O(l) \qquad (19.23)$$

At the end point, the solution contains $[NH_4]^+$ and $Cl^-$ ions. *After* the end point, additional aqueous $NH_3$ enters the solution (equilibrium 19.24). However, because the solution already contains $[NH_4]^+$ ions, equilibrium 19.24 is shifted to the left-hand side (apply Le Chatelier's principle) and the conductivity of the solution decreases slightly (Figure 19.5).

$$NH_3(aq) + H_2O(l) \rightleftharpoons [NH_4]^+(aq) + [OH]^-(aq) \qquad (19.24)$$

### Conductometric titrations: precipitation reactions

**Coordination and ionization isomers:** see Section 23.5

The use of *quantitative* precipitation reactions is important for compound analysis. For example, two isomers of a compound might differ in having one or two moles of free $Cl^-$ per mole of compound, and the free $Cl^-$

**Fig. 19.6** Variation in conductivity during the titration of a 0.20 mol dm$^{-3}$ solution of AgNO$_3$ against 50 cm$^3$ of 0.020 mol dm$^{-3}$ aqueous MgCl$_2$.

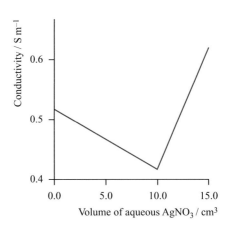

Sparingly soluble salts: see Section 17.12

could be precipitated using Ag$^+$ ions (equation 19.25). Other precipitations include reactions 19.26–19.28. All nitrate salts are soluble in water and therefore suitable reagents are often metal nitrates such as AgNO$_3$, Ba(NO$_3$)$_2$ and Pb(NO$_3$)$_2$.

$$Ag^+(aq) + Cl^-(aq) \longrightarrow AgCl(s) \tag{19.25}$$

$$Ag^+(aq) + I^-(aq) \longrightarrow AgI(s) \tag{19.26}$$

$$Ba^{2+}(aq) + [SO_4]^{2-}(aq) \longrightarrow BaSO_4(s) \tag{19.27}$$

$$Pb^{2+}(aq) + 2I^-(aq) \longrightarrow PbI_2(s) \tag{19.28}$$

Consider a conductometric titration in which aqueous AgNO$_3$ is added to an aqueous solution of a compound such as MgCl$_2$. This contains ionizable chloride ion, and the initial conductivity is due to the fully dissociated salt (equation 19.29).

$$MgCl_2(s) \xrightarrow{\text{dissolve in water}} Mg^{2+}(aq) + 2Cl^-(aq) \tag{19.29}$$

On addition of aqueous AgNO$_3$ (which is fully dissociated), Cl$^-$ is removed from the solution as AgCl and is replaced by [NO$_3$]$^-$. At the end point, the *solution* contains dissolved Mg(NO$_3$)$_2$ which is fully dissociated (equation 19.30).

$$MgCl_2(aq) + 2AgNO_3(aq) \longrightarrow Mg(NO_3)_2(aq) + 2AgCl(s) \tag{19.30}$$

After the end point, additional AgNO$_3$ provides further ions in the solution and the conductivity increases. In Figure 19.6, the end point coincides with the minimum reading of conductivity; the conductivity of aqueous Mg(NO$_3$)$_2$ is less than that of MgCl$_2$ (see Tables 19.2 and 19.3).

## SUMMARY

In this chapter, we have been concerned with the conductivity of ions in solution. Electrolytes are classified as being strong (fully ionized) or weak (partially ionized), and the behaviours of these types of electrolytes are very different. The molar conductivity of an ionic *compound* is related to the molar conductivities of the independent ions. Weak electrolytes obey Ostwald's Dilution Law, and application of this law can be used to determine values of acid dissociation constants. Changes in conductivity provide a means of monitoring, for example, acid–base titrations and precipitation reactions.

### Do you know what the following terms mean?

- conductivity
- conductance
- molar conductivity, $\Lambda_m$
- strong electrolyte
- weak electrolyte

- molar conductivity at infinite dilution, $\Lambda_m^\infty$
- molar conductivity of an ion, $\lambda_+$ or $\lambda_-$
- Kohlrausch's Law of Independent Migration of Ions

- Kohlrausch's Law
- degree of dissociation, $\alpha$
- Ostwald's Dilution Law
- ion mobility, $\mu$
- conductometric titration

### You should be able:

- to distinguish between conductivity and conductance
- to write down a relationship between the molar conductivity of a compound and the conductivity of an aqueous solution of known concentration
- to distinguish between a strong and weak electrolyte
- to explain what is mean by infinite dilution, and explain why $\Lambda_m$ is not identical to $\Lambda_m^\infty$
- to apply Kohlrausch's Law of Independent Migration of Ions to calculate the molar conductivity of a compound from values of $\lambda_+$ and $\lambda_-$

- to give an expression that shows how $\Lambda_m$ for a solution varies with its concentration
- to describe how $\Lambda_m^\infty$ could be determined for strong and weak electrolytes
- to show how conductivity measurements can be used to determine $K_a$ of a weak acid
- to explain how conductivity measurements can be used to find the end points of strong acid–strong base, weak acid–strong base and strong acid–weak base titrations
- to rationalize the changes in conductivity during a titration involving a precipitation reaction.

## PROBLEMS

**19.1** Write down expressions that define (a) resistivity, (b) conductivity and (c) conductance.

**19.2** The variation of $\Lambda_m$ with the concentration of aqueous $AgNO_3$ was determined at 298 K, and the results are tabulated below. Use the data to estimate $\Lambda_m^\infty$(298 K) for $AgNO_3$. Compare your answer with one calculated using data in Table 19.2.

| [AgNO₃] / mol dm⁻³ | $\Lambda_m$ / S m² mol⁻¹ |
|---|---|
| 0.000 50 | $1.313 \times 10^{-2}$ |
| 0.001 0 | $1.305 \times 10^{-2}$ |
| 0.005 0 | $1.271 \times 10^{-2}$ |
| 0.010 | $1.247 \times 10^{-2}$ |
| 0.020 | $1.214 \times 10^{-2}$ |

**19.3** Use data in Table 19.2 to calculate the molar conductivities at infinite dilution (at 298 K) of (a) $MgCl_2$; (b) $CsOH$; (c) $Ca(NO_3)_2$; (d) $Na_2SO_4$.

**19.4** Classify solutions of the following compounds as strong or weak electrolytes: (a) $AgNO_3$, (b) $HNO_3$; (c) $CH_3CH_2CO_2H$; (d) $K[CH_3CO_2]$; (e) $NH_4Cl$.

**19.5** Which of the following compounds are sparingly soluble in water (a) $NH_4Br$; (b) $AgBr$; (c) $KOH$; (d) $BaSO_4$? How does this affect conductivity measurements?

**19.6** The value of $\Lambda_m^\infty$(298 K) for $Na[C_3H_7CO_2]$ is $8.27 \times 10^{-3}$ S m² mol⁻¹. Using appropriate data from tables in the chapter, calculate $\Lambda_m^\infty$(298 K) for $C_3H_7CO_2H$ (butanoic acid).

**19.7** $Mg(OH)_2$ is sparingly soluble. Determine $\Lambda_m^\infty$ for $Mg(OH)_2$ given that $\Lambda_m^\infty$ for $MgCl_2$, NaOH and NaCl are $2.586 \times 10^{-2}, 2.477 \times 10^{-2}$ and $1.264 \times 10^{-2}\,S\,m^2\,mol^{-1}$. Compare the answer with a value of $\Lambda_m^\infty$ calculated from data in Table 19.2.

**19.8** Explain why the molar conductivity at infinite dilution of a weak acid such as $CH_3CO_2H$ is high, but then decreases dramatically as the concentration is raised even slightly.

**19.9** A $0.050\,mol\,dm^{-3}$ aqueous solution of $CH_3CO_2H$ has molar conductivity of $7.35 \times 10^{-4}\,S\,m^2\,mol^{-1}$. Using data from Table 19.2, calculate the degree of dissociation of the acid, and a value of $K_a$.

**19.10** The data in the table below were recorded to show the dependence of $\Lambda_m(CH_3CO_2H, 298\,K)$ on concentration. Confirm that $CH_3CO_2H$ obeys Ostwald's Dilution Law, and hence find $pK_a$ for the acid. Additional data: Table 19.2.

| $[CH_3CO_2H]$ / mol dm$^{-3}$ | $\Lambda_m$ / S m$^2$ mol$^{-1}$ |
|---|---|
| $5.45 \times 10^{-2}$ | $7.029 \times 10^{-4}$ |
| $1.95 \times 10^{-2}$ | $1.171 \times 10^{-3}$ |
| $8.50 \times 10^{-3}$ | $1.757 \times 10^{-3}$ |
| $4.70 \times 10^{-3}$ | $2.343 \times 10^{-3}$ |
| $2.30 \times 10^{-3}$ | $3.319 \times 10^{-3}$ |

**19.11** Using values from Table 19.2, calculate the ionic mobilities of (a) $Ca^{2+}$, (b) $[OH]^-$, (c) $[SO_4]^{2-}$, (d) $[NH_4]^+$ and (e) $I^-$.

**19.12** The ionic mobilities of $K^+$ and $[MnO_4]^-$ at 298 K are $7.62 \times 10^{-8}$ and $6.35 \times 10^{-8}\,m^2\,s^{-1}\,V^{-1}$. Determine $\Lambda_m^\infty(298\,K)$ for $KMnO_4$.

**19.13** The ionic mobilities of $Sr^{2+}$ and $Cl^-$ at 298 K are $6.16 \times 10^{-8}$ and $7.91 \times 10^{-8}\,m^2\,s^{-1}\,V^{-1}$. Determine $\Lambda_m^\infty(298\,K)$ for $SrCl_2$.

**19.14** As a $0.20\,mol\,dm^{-3}$ aqueous solution of KOH is titrated into $0.010\,mol\,dm^{-3}$ aqueous HCl, the conductivity decreases to a minimum value at the end point and then increases again. Explain this observation with reference to appropriate tabulated data in the chapter.

**19.15** The precipitation reaction:

$$Pb(NO_3)_2(aq) + 2NaI(aq) \longrightarrow$$
$$PbI_2(s) + 2NaNO_3(aq)$$

was monitored by measuring the conductivity of the solution and the data are shown in Figure 19.7; $0.10\,mol\,dm^{-3}$ NaI was added to $50\,cm^3$ $0.010\,mol\,dm^{-3}$ $Pb(NO_3)_2$. Account for the results.

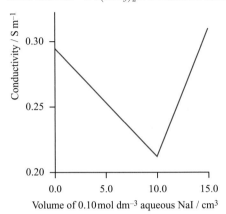

**Fig. 19.7** Graph for problem 19.15.

### Additional problems

**19.16** Discuss the following statements.
(a) The value of $\Lambda_m^\infty$ for HCl can be estimated directly from readings of $\Lambda_m$ at different concentrations, but $\Lambda_m^\infty$ for $CH_3CO_2H$ cannot be found by this method.
(b) Proton mobility is more facile than that of other ions.
(c) The kinetics of reactions between RBr and $R_3N$ or pyridine can be followed by taking conductivity measurements.

**19.17** Discuss how a conductometric titration could be used to find the end point of the reaction between a weak acid $H_xA$ ($0.010\,mol\,dm^{-3}$) and NaOH ($0.10\,mol\,dm^{-3}$). How do the data allow you to determine $x$?

# 20 Periodicity

## Topics

- Ground state electronic configurations
- Melting and boiling points
- Standard enthalpies of atomization
- First ionization energies
- Electron affinities
- Metallic, covalent and ionic radii
- Periodicity in groups 1 and 18

## 20.1 Introduction

In 1869 and 1870, Dmitri Mendeléev and Lothar Meyer stated that the 'properties of elements can be represented as periodic functions of their atomic weights'. This laid the foundation of the periodic table, but since the end of the 19th century, many new elements have been discovered and the form of the periodic table has been significantly modified. As we discussed in Section 1.10, the modern periodic table places elements in the *s-*, *p-*, *d-* and *f-blocks* and in *groups 1–18*. What is now known as *periodicity* is a consequence of the variation in ground state electronic configurations, and many physical and chemical properties follow periodic trends.

This short chapter presents a summary of trends in important physical properties of the elements, and sets the scene for the descriptive inorganic chemistry that follows in Chapters 21–23. In addition to ground state electronic configurations, properties such as melting and boiling points, ionization energies, electron affinities, electronegativities (see Section 5.7), and atom and ion sizes influence much of the chemistry that we discuss later.[§]

First, we recap on the metallic, non-metallic or semi-metallic characters of the elements. In Section 9.11, we looked at the structures of metallic elements and described the so-called 'diagonal line' that separates metals from non-metals in the periodic table (Figure 9.25). The distinction between metals and non-metals is not clear-cut, and elements close to the line may exhibit properties characteristic of both metals *and* non-metals. Such elements are classed as semi-metals. Figure 20.1 shows the periodic table, colour coded to distinguish metals, non-metals and semi-metals. Elements in the *s*-block, *d*-block and *f*-block have metallic character. Elements at the far right of the *p*-block (groups 17 and 18) are non-metallic. Elements in groups 13–16

■▶
Blocks of elements in the periodic table: see Section 1.10

---

[§] For more detailed discussions of trends within the periodic table, see: D. M. P. Mingos (1998) *Essential Trends in Inorganic Chemistry*, Oxford University Press, Oxford.

*s*-block elements ◄——————————— *d*-block elements ———————————► ◄——————— *p*-block elements ———————►

| Group 1 | Group 2 | Group 3 | Group 4 | Group 5 | Group 6 | Group 7 | Group 8 | Group 9 | Group 10 | Group 11 | Group 12 | Group 13 | Group 14 | Group 15 | Group 16 | Group 17 | Group 18 |
|---|---|---|---|---|---|---|---|---|---|---|---|---|---|---|---|---|---|
| 1 H | | | | | | | | | | | | | | | | | 2 He |
| 3 Li | 4 Be | | | | | | | | | | | 5 B | 6 C | 7 N | 8 O | 9 F | 10 Ne |
| 11 Na | 12 Mg | | | | | | | | | | | 13 Al | 14 Si | 15 P | 16 S | 17 Cl | 18 Ar |
| 19 K | 20 Ca | 21 Sc | 22 Ti | 23 V | 24 Cr | 25 Mn | 26 Fe | 27 Co | 28 Ni | 29 Cu | 30 Zn | 31 Ga | 32 Ge | 33 As | 34 Se | 35 Br | 36 Kr |
| 37 Rb | 38 Sr | 39 Y | 40 Zr | 41 Nb | 42 Mo | 43 Tc | 44 Ru | 45 Rh | 46 Pd | 47 Ag | 48 Cd | 49 In | 50 Sn | 51 Sb | 52 Te | 53 I | 54 Xe |
| 55 Cs | 56 Ba | 57–71 La–Lu | 72 Hf | 73 Ta | 74 W | 75 Re | 76 Os | 77 Ir | 78 Pt | 79 Au | 80 Hg | 81 Tl | 82 Pb | 83 Bi | 84 Po | 85 At | 86 Rn |
| 87 Fr | 88 Ra | 89–103 Ac–Lr | 104 Rf | 105 Db | 106 Sg | 107 Bh | 108 Hs | 109 Mt | 110 Ds | 111 Rg | 112 Uub | | | | | | |

*f*-block elements

| | 58 Ce | 59 Pr | 60 Nd | 61 Pm | 62 Sm | 63 Eu | 64 Gd | 65 Tb | 66 Dy | 67 Ho | 68 Er | 69 Tm | 70 Yb | 71 Lu |
|---|---|---|---|---|---|---|---|---|---|---|---|---|---|---|
| Lanthanoids | 58 Ce | 59 Pr | 60 Nd | 61 Pm | 62 Sm | 63 Eu | 64 Gd | 65 Tb | 66 Dy | 67 Ho | 68 Er | 69 Tm | 70 Yb | 71 Lu |
| Actinoids | 90 Th | 91 Pa | 92 U | 93 Np | 94 Pu | 95 Am | 96 Cm | 97 Bk | 98 Cf | 99 Es | 100 Fm | 101 Md | 102 No | 103 Lr |

**Fig. 20.1** The classification of elements in the periodic table as metals (purple), non-metals (green) and semi-metals (blue). The semi-metals lie along the so-called 'diagonal line'.

may be metallic, non-metallic or semi-metallic; metallic character predominates among the elements in group 13, while non-metallic character is typical of most elements in group 16 (Figure 20.1).

## 20.2 Ground state electronic configurations

We introduced *ground state electronic configurations* in Section 3.19, and focused on the first 20 elements (Table 3.4). Now we extend the discussion, and Table 20.1 lists the ground state electronic configurations for elements from $Z = 1$ to 103. These configurations are derived from experimental data, usually from studies of atomic spectra. If we take Table 20.1 in combination with the periodic table (Figure 20.1), then periodic patterns are clear and these influence, for example, values of the first ionization energies discussed in Section 20.4.

Elements in the periodic table are grouped according to the number of valence electrons that they possess. Elements in the same group typically possess related outer electronic configurations. For example, Li, Na, K, Rb, Cs and Fr are in group 1 and have an $ns^1$ configuration (one valence electron). N, P, As, Sb and Bi are in group 15 and possess an $ns^2 np^3$ ground state outer configuration (five valence electrons). F, Cl, Br, I and At lie in group 17 and have an $ns^2 np^5$ ground state configuration (seven valence electrons). These relationships work well for the *s*- and *p*-block elements. A general pattern for *d*-block elements is that the three metals in a particular group (a *triad*) possess the same number of valence electrons. Thus, Cr, Mo and W in group 6 have six valence electrons, Fe, Ru and Os

**Table 20.1** Ground state electronic configurations of the elements up to $Z = 103$.

| Atomic number | Element | Ground state electronic configuration | Atomic number | Element | Ground state electronic configuration |
|---|---|---|---|---|---|
| 1 | H | $1s^1$ | 53 | I | $[Kr]5s^2 4d^{10} 5p^5$ |
| 2 | He | $1s^2 = [He]$ | 54 | Xe | $[Kr]5s^2 4d^{10} 5p^6 = [Xe]$ |
| 3 | Li | $[He]2s^1$ | 55 | Cs | $[Xe]6s^1$ |
| 4 | Be | $[He]2s^2$ | 56 | Ba | $[Xe]6s^2$ |
| 5 | B | $[He]2s^2 2p^1$ | 57 | La | $[Xe]6s^2 5d^1$ |
| 6 | C | $[He]2s^2 2p^2$ | 58 | Ce | $[Xe]4f^1 6s^2 5d^1$ |
| 7 | N | $[He]2s^2 2p^3$ | 59 | Pr | $[Xe]4f^3 6s^2$ |
| 8 | O | $[He]2s^2 2p^4$ | 60 | Nd | $[Xe]4f^4 6s^2$ |
| 9 | F | $[He]2s^2 2p^5$ | 61 | Pm | $[Xe]4f^5 6s^2$ |
| 10 | Ne | $[He]2s^2 2p^6 = [Ne]$ | 62 | Sm | $[Xe]4f^6 6s^2$ |
| 11 | Na | $[Ne]3s^1$ | 63 | Eu | $[Xe]4f^7 6s^2$ |
| 12 | Mg | $[Ne]3s^2$ | 64 | Gd | $[Xe]4f^7 6s^2 5d^1$ |
| 13 | Al | $[Ne]3s^2 3p^1$ | 65 | Tb | $[Xe]4f^9 6s^2$ |
| 14 | Si | $[Ne]3s^2 3p^2$ | 66 | Dy | $[Xe]4f^{10} 6s^2$ |
| 15 | P | $[Ne]3s^2 3p^3$ | 67 | Ho | $[Xe]4f^{11} 6s^2$ |
| 16 | S | $[Ne]3s^2 3p^4$ | 68 | Er | $[Xe]4f^{12} 6s^2$ |
| 17 | Cl | $[Ne]3s^2 3p^5$ | 69 | Tm | $[Xe]4f^{13} 6s^2$ |
| 18 | Ar | $[Ne]3s^2 3p^6 = [Ar]$ | 70 | Yb | $[Xe]4f^{14} 6s^2$ |
| 19 | K | $[Ar]4s^1$ | 71 | Lu | $[Xe]4f^{14} 6s^2 5d^1$ |
| 20 | Ca | $[Ar]4s^2$ | 72 | Hf | $[Xe]4f^{14} 6s^2 5d^2$ |
| 21 | Sc | $[Ar]4s^2 3d^1$ | 73 | Ta | $[Xe]4f^{14} 6s^2 5d^3$ |
| 22 | Ti | $[Ar]4s^2 3d^2$ | 74 | W | $[Xe]4f^{14} 6s^2 5d^4$ |
| 23 | V | $[Ar]4s^2 3d^3$ | 75 | Re | $[Xe]4f^{14} 6s^2 5d^5$ |
| 24 | Cr | $[Ar]4s^1 3d^5$ | 76 | Os | $[Xe]4f^{14} 6s^2 5d^6$ |
| 25 | Mn | $[Ar]4s^2 3d^5$ | 77 | Ir | $[Xe]4f^{14} 6s^2 5d^7$ |
| 26 | Fe | $[Ar]4s^2 3d^6$ | 78 | Pt | $[Xe]4f^{14} 6s^1 5d^9$ |
| 27 | Co | $[Ar]4s^2 3d^7$ | 79 | Au | $[Xe]4f^{14} 6s^1 5d^{10}$ |
| 28 | Ni | $[Ar]4s^2 3d^8$ | 80 | Hg | $[Xe]4f^{14} 6s^2 5d^{10}$ |
| 29 | Cu | $[Ar]4s^1 3d^{10}$ | 81 | Tl | $[Xe]4f^{14} 6s^2 5d^{10} 6p^1$ |
| 30 | Zn | $[Ar]4s^2 3d^{10}$ | 82 | Pb | $[Xe]4f^{14} 6s^2 5d^{10} 6p^2$ |
| 31 | Ga | $[Ar]4s^2 3d^{10} 4p^1$ | 83 | Bi | $[Xe]4f^{14} 6s^2 5d^{10} 6p^3$ |
| 32 | Ge | $[Ar]4s^2 3d^{10} 4p^2$ | 84 | Po | $[Xe]4f^{14} 6s^2 5d^{10} 6p^4$ |
| 33 | As | $[Ar]4s^2 3d^{10} 4p^3$ | 85 | At | $[Xe]4f^{14} 6s^2 5d^{10} 6p^5$ |
| 34 | Se | $[Ar]4s^2 3d^{10} 4p^4$ | 86 | Rn | $[Xe]4f^{14} 6s^2 5d^{10} 6p^6 = [Rn]$ |
| 35 | Br | $[Ar]4s^2 3d^{10} 4p^5$ | 87 | Fr | $[Rn]7s^1$ |
| 36 | Kr | $[Ar]4s^2 3d^{10} 4p^6 = [Kr]$ | 88 | Ra | $[Rn]7s^2$ |
| 37 | Rb | $[Kr]5s^1$ | 89 | Ac | $[Rn]6d^1 7s^2$ |
| 38 | Sr | $[Kr]5s^2$ | 90 | Th | $[Rn]6d^2 7s^2$ |
| 39 | Y | $[Kr]5s^2 4d^1$ | 91 | Pa | $[Rn]5f^2 7s^2 6d^1$ |
| 40 | Zr | $[Kr]5s^2 4d^2$ | 92 | U | $[Rn]5f^3 7s^2 6d^1$ |
| 41 | Nb | $[Kr]5s^1 4d^4$ | 93 | Np | $[Rn]5f^4 7s^2 6d^1$ |
| 42 | Mo | $[Kr]5s^1 4d^5$ | 94 | Pu | $[Rn]5f^6 7s^2$ |
| 43 | Tc | $[Kr]5s^2 4d^5$ | 95 | Am | $[Rn]5f^7 7s^2$ |
| 44 | Ru | $[Kr]5s^1 4d^7$ | 96 | Cm | $[Rn]5f^7 7s^2 6d^1$ |
| 45 | Rh | $[Kr]5s^1 4d^8$ | 97 | Bk | $[Rn]5f^9 7s^2$ |
| 46 | Pd | $[Kr]5s^0 4d^{10}$ | 98 | Cf | $[Rn]5f^{10} 7s^2$ |
| 47 | Ag | $[Kr]5s^1 4d^{10}$ | 99 | Es | $[Rn]5f^{11} 7s^2$ |
| 48 | Cd | $[Kr]5s^2 4d^{10}$ | 100 | Fm | $[Rn]5f^{12} 7s^2$ |
| 49 | In | $[Kr]5s^2 4d^{10} 5p^1$ | 101 | Md | $[Rn]5f^{13} 7s^2$ |
| 50 | Sn | $[Kr]5s^2 4d^{10} 5p^2$ | 102 | No | $[Rn]5f^{14} 7s^2$ |
| 51 | Sb | $[Kr]5s^2 4d^{10} 5p^3$ | 103 | Lr | $[Rn]5f^{14} 7s^2 6d^1$ |
| 52 | Te | $[Kr]5s^2 4d^{10} 5p^4$ | | | |

in group 8 possess eight valence electrons, and Ni, Pd and Pt in group 10 have 10 valence electrons. However, details of the ground state electronic configurations (Table 20.1) do not necessarily follow consistent patterns. For example, group 5 consists of V, Nb and Ta and each element has five valence electrons, but the outer electronic configurations of V, Nb and Ta are $4s^2 3d^3$, $5s^1 4d^4$ and $6s^2 5d^3$ respectively. Such irregularities arise because the ordering of orbital energy levels that we discussed in Section 3.19 is only approximate. The energies of different atomic orbitals are close together for high values of the principal quantum number. We shall not be concerned in this book with detailed discussion of the $f$-block elements,[§] although their ground state electronic configurations are included in Table 20.1 for completeness. An $nf$ level is fully occupied when it contains 14 electrons.

As we shall see in Chapters 21 and 22, chemical properties of the $s$- and $p$-block elements in a particular group have certain similarities (although there are differences too as the group is descended) and the relationships between electronic configurations and group number are crucial to these observations.

## 20.3 Melting and boiling points, and enthalpies of atomization

### Melting and boiling points

Figure 20.2 shows the variation of melting and boiling points of the elements with atomic number; elements can be identified with reference to Figure 20.1

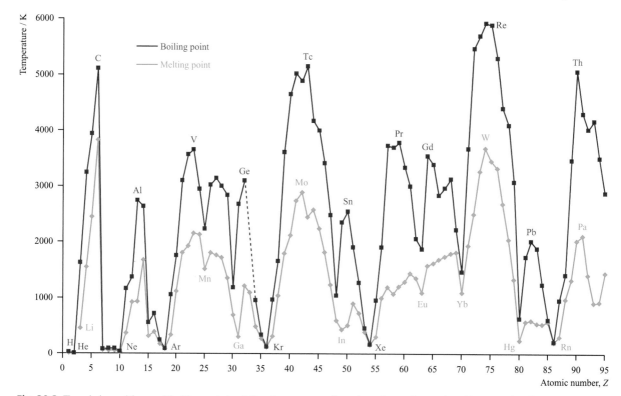

**Fig. 20.2** Trends in melting and boiling points of the elements as a function of atomic number. For $Z > 95$, data available are sparse.

---

[§] For an introduction to the $f$-block, see: C. E. Housecroft and A. G. Sharpe (2005) *Inorganic Chemistry*, 2nd edn, Prentice Hall, Harlow, Chapter 24.

and the periodic table facing the inside front cover of the book. Data for two elements are missing: it is not possible to solidify helium under any conditions of pressure and temperature and no melting point is therefore recorded, and arsenic ($Z = 33$) *sublimes* at 889 K. Some points to note from Figure 20.2 are:

Sublimation: see Section 1.6

- the extremely low values of melting and boiling points of hydrogen;
- the very low melting and boiling points of the group 18 elements (the noble gases);
- the relatively short liquid ranges of H, He, N, O, F, Ne, Cl, Ar, Kr, Xe and Rn, i.e. the elements that are gases at 298 K;
- the extremely high melting and boiling points of carbon;
- the sharp drop in boiling points after the group 14 elements C, Si, Ge, Sn;
- the similar trends across the three rows of *d*-block elements, $Z = 21$–30, 39–48 and 72–80, and the low melting points of the last (group 12) metal of each row of the *d*-block;
- the roughly repeating patterns of the melting and boiling points of the *s*-, *p*- and *d*-block elements, and the insertion of the first row of the *f*-block metals ($Z = 57$–71) at the beginning of the third row of the *d*-block;
- the high melting points of metals in the middle of the third row of the *d*-block metals;
- the rather similar melting points of the first row *f*-block metals.

Critical inspection of Figure 20.2 reveals a clear periodic pattern in both melting and boiling points.

## Standard enthalpies of atomization

Standard enthalpy of atomization: see Section 4.6

Values of the standard enthalpies of atomization, $\Delta_a H^\circ$, of the elements are listed in Appendix 10, and the variation in values is illustrated in Figure 20.3. No values are listed for the noble gases because they are monatomic in the standard state. Values of $\Delta_a H^\circ$ refer to reaction 20.1 where $E_n$ refers to the element in the standard state and E(g) to gaseous atoms.

$$\frac{1}{n} E_n(\text{standard state}) \longrightarrow E(g) \qquad (20.1)$$

Inspection of Figure 20.3 in conjunction with Figure 20.1 reveals the following points:

- the group 1 elements (alkali metals) and group 17 elements (halogens) have the lowest values of $\Delta_a H^\circ$, and the group 12 metals (Zn, Cd and Hg) are also readily atomized;
- across the first two rows of *s*- and *p*-block elements (Li to F, Na to Cl), values of $\Delta_a H^\circ$ peak at group 14 (C and Si);
- the third and fourth rows (K to Br, Rb to I) include *d*-block metals, and across each row of *d*-block metals, the trend is for the highest values of $\Delta_a H^\circ$ to be associated with the middle elements (Mn is an exception);

Structure of Mn: see Section 9.11

- the row of elements from Cs to At runs through the *s*-, *f*-, *d*- and *p*-blocks, although values for the lanthanoids (first row of the *f*-block) are omitted because data are sparse; values of $\Delta_a H^\circ$ reach a maximum in the middle of the row of *d*-block metals with W having the highest atomization enthalpy of any element in the periodic table;
- if we focus *only* on the *s*- and *p*-block elements, then the pattern noted for the first two rows (Li to F, Na to Cl) is similar for the heavier elements (K to Br, Rb to I, Cs to At).

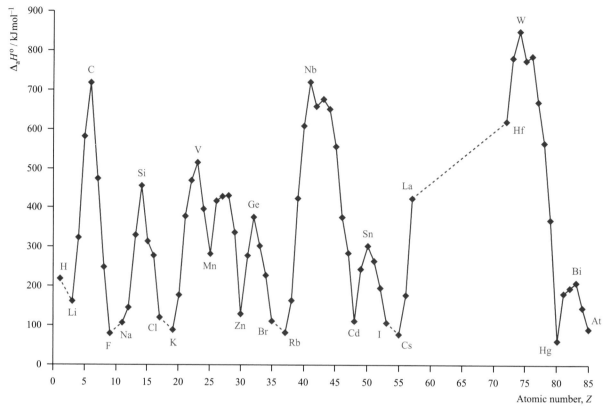

**Fig. 20.3** The trend in standard enthalpies of atomization, $\Delta_a H^\circ$, of the elements (up to $Z = 85$) as a function of atomic number; the first row $f$-block elements and noble gases (see text) are omitted.

See problem 20.7

Finally, a comparison of Figures 20.2 and 20.3 illustrates that trends in melting and boiling points are in many ways mirrored by those in values of $\Delta_a H^\circ$.

---

**20.4**    **First ionization energies and electron affinities**

### First ionization energies, $IE_1$

Values of ionization energies: see Appendix 8

We have already discussed ionization energies in detail in Section 8.4, and have related trends across a row of the periodic table to valence electronic configurations. Now we give an overview of the variation in first ionization energies, $IE_1$, throughout the periodic table. Values of $IE_1$ refer to process 20.2 and are plotted in Figure 20.4 as a function of atomic number.

$$X(g) \longrightarrow X^+(g) + e^- \tag{20.2}$$

The extremely high values for the noble gases (group 18) are immediately apparent from Figure 20.4, and are associated with the removal of an electron from the filled quantum level. Also of note are the low values of the alkali metals (group 1), associated with the removal of an $ns^1$ electron. Trends in values of $IE_1$ down groups 1, 2, 15 and 17 were illustrated in Figure 8.8, and the decrease in values for each of these groups is typical. The same trends can be observed for groups 13, 14, 16 and 18 by finding related points in Figure 20.4, e.g. B, Al, Ga, In and Tl in group 13.

A general trend across a row of the $p$-block (B to Ne, Al to Ar, Ga to Kr, In to Xe, and Tl to Rn) is the steady increase in the energy needed to remove the

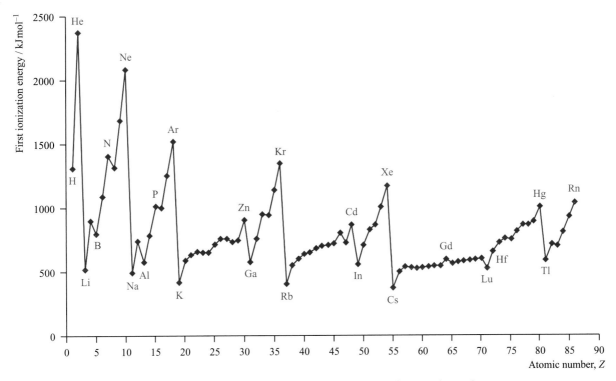

**Fig. 20.4** The variation of values of the first ionization energy ($IE_1$) with increasing atomic number.

first electron from the steadily filling $np$ shell. Discontinuities at the group 15 elements N, P and As have already been discussed in Section 8.4. The general increase in values of $IE_1$ across a row of the $p$-block extends to a longer row of the periodic table (e.g. K to Kr, or Rb to Xe) and reflects an increase in the effective nuclear charge, $Z_{eff}$.

$Z_{eff}$: see Section 3.15

The three rows of $d$-block metals can be picked out easily in Figure 20.4. There is relatively little variation in $IE_1$ on going from Sc to Zn ($Z = 12$–30), Y to Cd ($Z = 39$–48), or Hf to Hg ($Z = 72$–80). In each row, the group 12 metal (Zn, Cd and Hg) has the highest ionization energy. Strictly, La ($Z = 57$) is in the third row of the $d$-block metals but it is often classed with the lanthanoid metals in the first row of the $f$-block. Values of $IE_1$ for La and the 14 lanthanoid metals (Ce to Lu) are very similar to one another.

### Electron affinities

Electron affinities: see Appendix 9

We introduced electron affinities (electron attachment energies) in Section 8.5, and discussed why it is useful to consider the enthalpy change associated with the attachment of an electron (process 20.3) rather than the actual electron affinity: $\Delta_{EA}H(298\,\text{K}) = -EA$.

$$X(g) + e^- \longrightarrow X^-(g) \tag{20.3}$$

Figure 20.5 shows the variation in $\Delta_{EA}H$ for hydrogen and the $s$- and $p$-block elements for the first four periods. The two gaps in the figure correspond to the first and second rows of $d$-block metals. The horizontal dashed line in Figure 20.5 corresponds to $\Delta_{EA}H = 0\,\text{kJ}\,\text{mol}^{-1}$. Thus, points above the line mean that process 20.3 is endothermic, and points below the line indicate

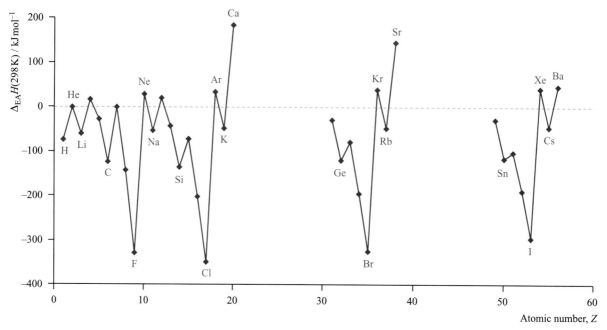

**Fig. 20.5** Trends in values of the changes in enthalpy associated with the attachment of the first electron to a gaseous atom (equation 20.3). Only theoretical values are available for the noble gases.

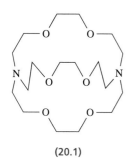

(20.1)

Ligands and complexes: see Section 23.3

$\Delta G$ and $\Delta H$: see Chapter 17

an exothermic electron attachment. The highly exothermic electron attachments for the group 17 elements (F, Cl, Br, I) follow from the fact that gaining 1 electron gives a filled quantum shell and is therefore favourable:

$$F(g) + e^- \longrightarrow F^-(g)$$
$$[He]2s^2 2s^5 \qquad\qquad [He]2s^2 2s^6 = [Ne]$$

There is a clear periodic pattern that repeats across Figure 20.5: start from any group 1 metal and follow the trend to the noble gas at the end of the period. Interestingly, values of $\Delta_{EA}H$ for the group 1 metals are *negative*. Normally, one associates metals with the formation of *positive* ions, but the alkali metals do form $M^-$ ions in a group of compounds called *alkalides*. For example, compound **20.1** is able to shift equilibrium 20.4 to the right-hand side by forming a complex with $Na^+$ (Figure 20.6).

$$2Na \rightleftharpoons Na^+ + Na^- \tag{20.4}$$

The group 2 metals Ca, Sr and Ba have highly positive values of $\Delta_{EA}H$, and the lighter group 2 metals (Be and Mg) have small positive values. This is as expected for the addition of an electron to an atom with a ground state $ns^2$ outer electronic configuration. However, one has to be careful about necessarily interpreting positive values of $\Delta_{EA}H$ as being indicative of a thermodynamically unfavourable process. Apart from the fact that $\Delta G$, and not $\Delta H$, is the true guide to spontaneity, the attachment of an electron *in a chemical reaction* is not an isolated event. The formation of gaseous $O^{2-}$ from gaseous, atomic O is highly endothermic (see equation 8.14) but in the formation of a metal oxide such as MgO (containing $Mg^{2+}$ and $O^{2-}$ ions), other factors contribute to the overall process, making the formation of MgO from its constituent elements highly exothermic.

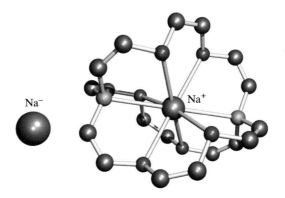

**Fig. 20.6** The structure (determined by X-ray diffraction) of $[Na(crypt-[222])]^+Na^-$. The ligand 'crypt-[222]' is a *cryptand* (see structure **20.1**) that encapsulates the $Na^+$ ion [F. J. Tehan *et al.* (1974) *J. Am. Chem. Soc.*, vol. 96, p. 7203]. Colour code: Na, purple; C, grey; N, blue; O, red.

## 20.5    Metallic, covalent and ionic radii

### Metallic radii

Metallic, covalent and ionic radii: see Sections 4.3, 9.12 and 8.14 respectively

Figure 20.1 showed that the majority of elements are metals. The size of a metal atom is characterized by its *metallic radius*, $r_{metal}$, and values are plotted in Figure 20.7 as a function of atomic number up to $Z = 83$. Several trends emerge from Figure 20.7:

- on descending a group of the *s*- or *p*-blocks, values of $r_{metal}$ increase; for the *p*-block elements, series of elements are incomplete because of the non-metal–metal divide (see Figure 20.1);
- across a row, later elements are non-metallic and, in the case of the first two rows, trends can only be based on two or three metals. Nonetheless, the trend of decreasing atomic size observed on going from Li to Be, can be regarded more generally because it is repeated in the later rows;
- crossing each row of *d*-block metals is accompanied by a small variation in $r_{metal}$, with the largest radii being associated with the earliest and latest metals; Mn is as exception, and this may be traced to its unique structure (see Section 9.11);

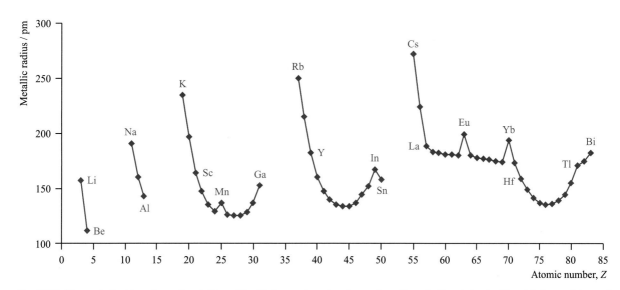

**Fig. 20.7** The variation in values of metallic radii with increasing atomic number. Gaps in the plot coincide with the positions of non-metals or semi-metals.

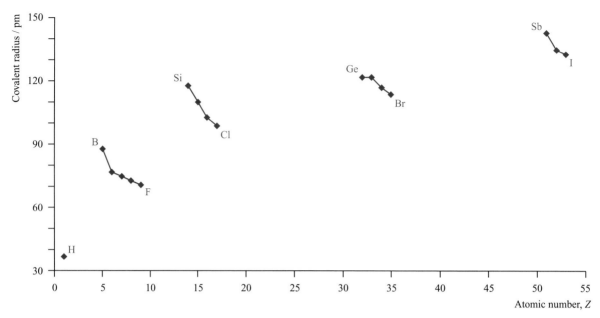

**Fig. 20.8** The variation in values of covalent radii with increasing atomic number for non-metallic or semi-metallic elements. Gaps in the plot coincide with the positions of metals.

- metallic radii for the first row $d$-block metals are smaller than for second and third row metals in the same group, *but* second and third row metals in the same group have similar values of $r_{metal}$;
- the lanthanoid metals, with the exceptions of Eu and Yb, are almost of constant atomic size.

### Covalent radii

The gaps in Figure 20.7 correspond to the non-metallic and semi-metallic elements in the $p$-block (see Figure 20.1). For a comparison of their atomic sizes, we consider values of *covalent radii*, $r_{cov}$, trends in which are presented in Figure 20.8. The covalent radius of H is significantly smaller than that of any other element. Across each row of the $p$-block elements (B to F, Si to Cl, Ge to Br, Sb to I) there is a general decrease in atomic size, consistent with an increase in $Z_{eff}$. Radii tend to increase down a group, e.g. $r_{cov}$ for I > Br > Cl > F.

### Ionic radii

Meaningful comparisons of the sizes of ions must be made between ions of the same charge, and so an assessment of trends throughout the periodic table is not possible. Some metals such as Fe exhibit more than one cation ($Fe^{2+}$ and $Fe^{3+}$), and to complicate matters, the electronic configuration of a given $d$-block ion may be affected by its coordination environment giving so-called *high-spin* and *low-spin* ions with different ionic radii. Values of $r_{ion}$ are listed in Appendix 6. Some important trends that emerge are:

High- and low-spin: see Section 23.11

- $r_{ion}$ for group 1 $M^+$ ions increases down the group (see Figure 8.22);
- $r_{ion}$ for group 2 $M^{2+}$ ions increases down the group;

- $r_{ion}$ for group 17 $X^-$ ions increases down the group (see Figure 8.23);
- $r_{ion}$ for group 16 $X^{2-}$ ions increases down the group (see Figure 8.23);
- increasing the charge on a *cation* leads to a reduction in $r_{ion}$ (other factors being constant).

## 20.6     Periodicity in groups 1 and 18

So far in this chapter, we have considered periodicity by looking at variations in given physical properties. Now we take a closer look at the elements in groups 1 and 18, and illustrate how the related valence electronic configurations of elements within a group influence physical and chemical properties.

### Group 18: the noble gases

More about the noble gases in Section 22.13

In Section 3.21, we introduced the *noble gases* as a *group* of elements possessing fully occupied atomic orbitals in their outer quantum level. Helium is the odd one out with a $1s^2$ configuration; each of the heavier noble gases has an $ns^2np^6$ ground state electronic configuration and obeys the octet rule. The physical and chemical properties of the noble gases reflect the fact that they are stable entities. All are stable, monatomic gases at 298 K. Their melting points are exceptionally low (Table 3.6) and the temperature ranges over which the group 18 elements are liquids are very small (Table 3.6 and Figure 20.2). The first ionization energy of each noble gas is extremely high (Figure 20.4), precluding the easy formation of a positive ion. However, for Xe, and to a lesser extent for Kr, the presence of the completed octet does *not* result in a total lack of chemical reactivity. Some reactions involving Xe are given in equations 20.5–20.7; notice that Xe forms compounds in which it exhibits different oxidation states. Reasons for the ability of Xe and Kr to undergo such reactions are discussed in Section 22.13.

$$Xe + F_2 \longrightarrow XeF_2 \qquad\qquad (20.5)$$

$$3XeF_4 + 6H_2O \longrightarrow XeO_3 + 2Xe + \tfrac{3}{2}O_2 + 12HF \qquad\qquad (20.6)$$

$$XeF_6 + H_2O \longrightarrow XeOF_4 + 2HF \qquad\qquad (20.7)$$

### Group 1: the alkali metals

The metals in group 1 are characterized by having an $ns^1$ ground state valence electronic configuration. Selected physical and chemical data are listed in Table 20.2. All the alkali metals are solid at 298 K, but their melting points are relatively low, indicating that the interatomic forces in the solid state are overcome quite easily. In some countries, caesium is a liquid at ambient temperatures! The ranges of temperature over which the alkali metals remain in the liquid state are much greater than those of the noble gases (Tables 20.2 and 3.6 and Figure 20.2). The values of the first ionization energies of the group 1 elements are low relative to elements in other groups (Figure 20.4), and indicate that ionization occurs easily. However, in a chemical reaction involving the metal at 298 K, formation of the ion

**Table 20.2** Physical and chemical data for the group 1 elements (alkali metals).

| Element | Symbol | Melting point / K | Boiling point / K | Ground state electronic configuration | First ionization energy / kJ mol$^{-1}$ | $\Delta_a H^\circ$ / kJ mol$^{-1}$ | Ion formed | Formula of chloride |
|---|---|---|---|---|---|---|---|---|
| Lithium | Li | 453.5 | 1615 | [He]$2s^1$ | 520.2 | 161 | Li$^+$ | LiCl |
| Sodium | Na | 371 | 1154 | [Ne]$3s^1$ | 495.8 | 108 | Na$^+$ | NaCl |
| Potassium | K | 336 | 1032 | [Ar]$4s^1$ | 418.8 | 90 | K$^+$ | KCl |
| Rubidium | Rb | 312 | 961 | [Kr]$5s^1$ | 403.0 | 82 | Rb$^+$ | RbCl |
| Caesium[a] | Cs | 301.5 | 978 | [Xe]$6s^1$ | 375.7 | 78 | Cs$^+$ | CsCl |

[a] In US-speaking countries, Cs is *cesium*.

*formally* requires input of energy to atomize the metal and *then* to ionize the gaseous atoms (scheme 20.8).

$$M(s) \xrightarrow{\Delta_a H^\circ} M(g) \xrightarrow{IE_1} M^+(g) + e^- \tag{20.8}$$

Table 20.2 and Figure 20.3 show that the group 1 metals possess low values of $\Delta_a H^\circ(298\,\text{K})$, and therefore the energetics of the steps in scheme 20.8 are (relative to other elements) quite favourable. The formation of $M^+$ ions is observed throughout the chemistry of the alkali metals, as exemplified in reactions 20.9–20.12.

$$2Na + Cl_2 \longrightarrow 2NaCl \tag{20.9}$$

$$2K + 2H_2O \longrightarrow 2KOH + H_2 \tag{20.10}$$

$$4Na + O_2 \longrightarrow 2Na_2O \tag{20.11}$$

$$Li_2CO_3 \xrightarrow{heat} Li_2O + CO_2 \tag{20.12}$$

## SUMMARY

In this chapter, we have looked at periodic trends in a number of physical properties of the elements, and have also overviewed trends within two groups of elements: the noble gases and the alkali metals. Application of periodic trends to the detailed chemistry of the elements follows in Chapters 21–23.

### *You should be able:*

- to discuss the relationship between the ground state electronic configuration of an element and its position in the periodic table
- to discuss trends in melting and boiling points and $\Delta_a H^\circ$ within the periodic table
- to discuss variations in the values of the first ionization energies across rows and down groups in the periodic table
- to illustrate how values of $\Delta_{EA} H$ for elements in the s- and p-blocks vary, and the

consequences of this variation on monoanion formation

- to discuss trends in metallic and covalent radii within the periodic table
- to illustrate variations in the values of $r_{ion}$ down groups in the periodic table, and explain why discussions of more general trends are less meaningful
- to give summaries of trends in physical and chemical properties for elements in groups 1 and 18.

## PROBLEMS

**20.1** Identify the groups of the periodic table to which the elements with the following general ground state electronic configurations belong:
(a) $ns^2np^4$;
(b) $ns^1$;
(c) $ns^2np^6$;
(d) $ns^2$;
(e) $ns^2(n-1)d^{10}$;
(f) $ns^2np^1$.

**20.2** An element, X, has the ground state electronic configuration $1s^22s^22p^63s^23p^64s^23d^5$.
(a) How many valence electrons does X possess?
(b) What is the atomic number of X?
(c) To which block of elements does X belong?
(d) To which group does X belong?
(e) Is X a metal, semi-metal or non-metal?
(f) Write out the ground state electronic configuration of the element *after* X in the periodic table.

**20.3** An element, Z, has the ground state electronic configuration $1s^22s^22p^63s^23p^4$.
(a) How many valence electrons does Z possess?
(b) To which block of elements does Z belong?
(c) To which group does Z belong?
(d) Does Z readily form a cation or anion? If so, suggest its formula.
(e) Is Z a metal, semi-metal or non-metal?
(f) Write out the ground state electronic configuration of the element *above* Z in the periodic table.

**20.4** An element, E, has an outer ground state electronic configuration of $ns^2np^2$.
(a) How many valence electrons does E possess?
(b) How many unpaired electrons does E possess in the ground state?
(c) To which block of elements does E belong?
(d) To which group does E belong?
(e) Is E likely to be a metal, semi-metal or non-metal?
(f) What will be the likely formula of the compound that E forms with hydrogen?
(g) Write out the outer ground state electronic configuration of the element three places to the right of E in the periodic table.

**20.5** In terms of electronic configurations, why is He unique among the group 18 elements?

**20.6** Refer to Figure 20.2. Suggest reasons why:
(a) the group 1 metals have relatively low melting points;
(b) Ne, Ar, Kr and Xe possess extremely low melting and boiling points;

(c) there is a dramatic fall in melting and boiling points on going from B and C to N, O and F.

**20.7** Although the trends in melting points in Figure 20.2 roughly parallel those in values of $\Delta_aH^\circ(298\,K)$ in Figure 20.3, the processes to which these physical quantities refer are different. Comment on this statement.

**20.8** Write equations to define the processes to which $\Delta_aH^\circ(298\,K)$ refers for each of the following elements: (a) F; (b) Rb; (c) Br; (d) V; (e) Si.

**20.9** Values of the first five ionization energies of an element, A, are 737.7, 1451, 7733, 10 540 and 13 630 kJ mol$^{-1}$. Suggest (a) to which group A belongs, and (b) whether A is a metal or non-metal.

**20.10** Give explanations for the trends in values of $IE_1$ (Figure 20.4) on going from (a) Xe to Cs, (b) Zn to Ga, (c) H to He, (d) B to N, (e) Na to Al, and (f) P to K.

**20.11** Suggest reasons why (a) Li, Na, K, Rb and Cs appear at the lowest points in Figure 20.4, and (b) He has the highest first ionization energy of any element.

**20.12** Refer to Figure 20.5. If $\Delta_{EA}H$ refers to the attachment of one electron to a gaseous atom, suggest reasons why:
(a) F, Cl, Br and I have the most negative values of $\Delta_{EA}H$;
(b) C, Si, Ge and Sn have significantly negative values of $\Delta_{EA}H$;
(c) the formation of Ca$^-$ is highly endothermic;
(d) an endothermic electron attachment is not necessarily an indication of unfavourable ion formation.

**20.13** Rationalize why the metallic radii of the group 1 and 2 metals increase down each group.

**20.14** Give explanations for the trends in values of $r_{metal}$ (Figure 20.7) on going from: (a) K to Sc; (b) In to Sn; (c) Al to K.

**20.15** (a) Atoms of Li, Mg and Sc are essentially the same size. What is their relationship in the periodic table?
(b) Atoms of Fe, Co, Ni and Cu have similar metallic radii. How are these metals related in the periodic table?
(c) Values of $r_{metal}$ for Ru, Os, Rh and Ir are virtually identical. Find the positions of these elements in Figure 20.7 and comment on their periodic relationships.

**20.16** How will the atomic or ionic radius change on going from (a) Cr to Cr$^{3+}$, (b) Cl to Cl$^-$, (c) Cs to Cs$^+$, (d) Na$^+$ to Na$^-$, and (e) Fe$^{3+}$ to Fe$^{2+}$?

**20.17** How do you expect the van der Waals radii of the group 18 elements to vary down the group? Rationalize your answer.

**20.18** Use data in Table 5.2 to discuss periodic trends in Pauling electronegativity values.

## Additional problems

**20.19** Less is known about francium, Fr, than the other alkali metals. Use data in this chapter to say what you can about the expected properties of Fr.

**20.20** Suggest why $\Delta_{EA}H$ is more negative for Cl than for F, but less negative for Br than for Cl (see Figure 20.5).

**20.21** Comment on the following values of $r_{ion}$ for vanadium:

| Oxidation state | Coordination number | $r_{ion}$ / pm |
|---|---|---|
| +2 | 6 | 79 |
| +3 | 6 | 64 |
| +4 | 6 | 58 |
| +4 | 5 | 53 |
| +5 | 6 | 54 |
| +5 | 5 | 46 |

**20.22** Why can metallic radii not be directly compared with covalent radii (Figures 20.7 and 20.8)?

# 21 Hydrogen and the s-block elements

## Topics

- The element hydrogen
- Hydrides of the s-, d- and p-block elements
- Hydrogen bonding
- Elements in group 1 (the alkali metals)
- Elements in group 2
- The diagonal relationship between lithium and magnesium

| 1 | 2 | | 13 | 14 | 15 | 16 | 17 | 18 |
|---|---|---|---|---|---|---|---|---|
| H | | | | | | | | He |
| Li | Be | | B | C | N | O | F | Ne |
| Na | Mg | | Al | Si | P | S | Cl | Ar |
| K | Ca | d-block | Ga | Ge | As | Se | Br | Kr |
| Rb | Sr | | In | Sn | Sb | Te | I | Xe |
| Cs | Ba | | Tl | Pb | Bi | Po | At | Rn |
| Fr | Ra | | | | | | | |

## 21.1 Introduction

Earlier, we discussed periodicity and arranged the elements into families or groups in the periodic table in which properties are related to the ground state atomic electronic configurations. In Chapters 21 and 22 we look at the chemistries of hydrogen and the s- and p-block elements,[§] and in addition to describing physical and chemical properties we relate some of the trends in properties.

The title of this chapter 'Hydrogen and the s-block elements' emphasizes the fact that, although hydrogen has a ground state electronic configuration

[§] For more detailed coverage see: N. N. Greenwood and A. Earnshaw (1997) *Chemistry of the Elements*, 2nd edn, Butterworth-Heinemann, Oxford; C. E. Housecroft and A. G. Sharpe (2005) *Inorganic Chemistry*, 2nd edn, Prentice Hall, Harlow.

of $1s^1$, it is *not* an alkali metal. Some versions of the periodic table place hydrogen at the head of group 1, while others do not associate it with any particular group.

<table>
<tr><td>21.2</td><td></td></tr>
</table>

## The element hydrogen

Hydrogen is the most abundant element in the universe, and the third most abundant on Earth (after oxygen and silicon). On Earth, it occurs mainly in the form of water or combined with carbon in organic molecules – hydrocarbons, plant and animal material. Dihydrogen is *not* a major constituent of the Earth's atmosphere (Figure 21.1), occurring to an extent of less than one part per million by volume. Light gases such as $H_2$ and He are readily lost from the atmosphere.

### Isotopes of hydrogen

Isotopes: see Section 1.7

Hydrogen possesses three isotopes but $^1H$ is by far the most abundant (99.984%). Although the natural abundance of deuterium, $^2H$ or D, is

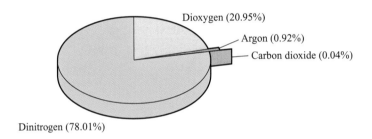

**Fig. 21.1** The principal components (by percentage volume) of Earth's atmosphere. Note that $H_2$ is not among these constituents.

Dioxygen (20.95%)
Argon (0.92%)
Carbon dioxide (0.04%)
Dinitrogen (78.01%)

### ENVIRONMENT AND BIOLOGY

### Box 21.1 Inside Jupiter and Saturn

The cores of Saturn and Jupiter are probably composed of metallic hydrogen although, until recently, the metallic state of this element had not been observed on the Earth. Early in 1996, researchers from the Livermore Laboratory in the USA reported that they had subjected a thin layer of liquid $H_2$ to tremendous pressure and observed changes in conductivity that were consistent with the formation of metallic hydrogen. Now that more is known about the conditions needed for the liquid–solid phase transition, it should be possible to gain a better picture of the composition of the cores of Saturn and Jupiter.

Computer-enhanced Voyager 2 photograph of Saturn.
*NASA/Science Photo Library.*

only 0.0156%, deuterium-labelled compounds are important, e.g. as solvents in NMR spectroscopy. Of particular significance is their use in following what happens to hydrogen atoms in reactions.

We have seen that the stretching frequency and vibrational wavenumber of an IR spectroscopic absorption are related to the reduced mass of the system (equations 12.7 and 12.9). Exchanging the hydrogen atom in an X−H bond for a deuterium atom alters the reduced mass and shifts the position of the absorption in the IR spectrum. Consider a C−H bond. The vibrational wavenumber is inversely proportional to the square root of the reduced mass (equation 21.1), and we can write a similar equation for a C−D bond. Combining the two expressions gives equation 21.2.

$$\bar{\nu}_{C-H} \propto \frac{1}{\sqrt{\mu_{C-H}}} \tag{21.1}$$

$$\frac{\bar{\nu}_{C-D}}{\bar{\nu}_{C-H}} = \sqrt{\frac{\mu_{C-H}}{\mu_{C-D}}} \tag{21.2}$$

The relative atomic masses of C, H and D are 12.01, 1.01 and 2.01 respectively. The reduced mass of a C−H bond is calculated in equation 21.3 and similarly we find that $\mu_{C-D}$ is $2.86 \times 10^{-27}$ kg.

$$\frac{1}{\mu_{C-H}} = \frac{1}{m_1} + \frac{1}{m_2} = \frac{1}{(12.01 \times 1.66 \times 10^{-27}\,\text{kg})} + \frac{1}{(1.01 \times 1.66 \times 10^{-27}\,\text{kg})}$$

$$= 6.47 \times 10^{26}\,\text{kg}^{-1}$$

$$\mu_{C-H} = \frac{1}{6.47 \times 10^{26}\,\text{kg}^{-1}} = 1.55 \times 10^{-27}\,\text{kg} \tag{21.3}$$

An absorption at $3000\,\text{cm}^{-1}$ due to a C−H vibration will shift to $2209\,\text{cm}^{-1}$ (equation 21.4) upon exchanging the hydrogen for a deuterium atom. (This assumes that the force constants of C−H and C−D bonds are the same.)

*Force constant: see Section 12.2*

$$\bar{\nu}_{C-D} = \bar{\nu}_{C-H} \times \sqrt{\frac{\mu_{C-H}}{\mu_{C-D}}} = (3000\,\text{cm}^{-1}) \times \sqrt{\frac{(1.55 \times 10^{-27}\,\text{kg})}{(2.86 \times 10^{-27}\,\text{kg})}}$$

$$= 2209\,\text{cm}^{-1} \tag{21.4}$$

This aids the assignment of bands in the IR spectra of organic compounds. For example, N−H, O−H and C−H bonds all absorb around $3000-3500\,\text{cm}^{-1}$. If the organic compound is shaken with $D_2O$, usually only the OH and NH groups undergo a *deuterium exchange reaction* (equation 21.5).

$$R-XH + D_2O \rightleftharpoons R-XD + HOD \tag{21.5}$$

By seeing which IR spectroscopic bands shift (and by how much) and which remain unchanged, it is possible to separate N−H, O−H and C−H absorptions (see problem 21.1).

Deuterium labelling may also be used to probe the mechanism of a reaction. Consider the C−H bond again. Let us assume that in the rate-determining step of a reaction a particular C−H bond is broken. If this is the case, then labelling the compound with deuterium at that site will result in cleavage of a C−D rather than a C−H bond. The bond dissociation enthalpy of a C−D bond is higher than that of a C−H bond because the zero point energy is lowered when the reduced mass of a bond is increased $[\mu(C-D) > \mu(C-H)]$, as is shown in Figure 21.2. The fact that it requires more energy to break the C−D than the C−H bond should cause the rate-determining step to proceed more slowly – this is called a *kinetic isotope effect* and is quantified by comparing the rate constants ($k_H$ and $k_D$) for

*Zero point energy: see Sections 4.5 and 12.2*

**Fig. 21.2** The zero point energy (the lowest vibrational state) of a C−D bond is lower than that of a C−H bond and this results in the bond dissociation enthalpy of the C−D bond being greater than that of the C−H bond.

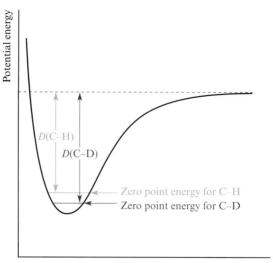

the reactions that involve the non-deuterated and deuterated compounds. A value of the ratio $k_H/k_D$ greater than 1 corresponds to the observation of a kinetic isotope effect.

## Physical properties of $H_2$

Dihydrogen[§] is a colourless and odourless gas (mp $= 13.7\,\text{K}$, bp $= 21.1\,\text{K}$), of particularly low density ($0.09\,\text{g}\,\text{dm}^{-3}$) which is almost insoluble in water.

Dihydrogen is quite unreactive at 298 K largely because of the high H−H bond dissociation enthalpy ($436\,\text{kJ}\,\text{mol}^{-1}$) and many reactions of $H_2$ require high temperatures, high pressures or catalysts as we describe for the addition of $H_2$ to ethene in Section 26.4.

## Sources of $H_2$

Industrially, $H_2$ is prepared by the reaction of carbon or a hydrocarbon (e.g. $CH_4$) with steam followed by the reaction of the carbon monoxide so-formed with more water vapour (equation 21.6). The mixture of CO and $H_2$ produced in the first reaction is called *synthesis gas* and the sequence of reactions is called the *water–gas shift reaction*.

$$\left.\begin{array}{c} CH_4(g) + H_2O(g) \xrightarrow{\;1200\,\text{K, nickel catalyst}\;} CO(g) + 3H_2(g) \\[1em] CO(g) + H_2O(g) \xrightarrow{\;700\,\text{K, iron oxide catalyst}\;} CO_2(g) + H_2(g) \end{array}\right\} \quad (21.6)$$

*The electrochemical series is quantified in Chapter 18*

On a laboratory scale, $H_2$ can be formed by reactions between acids and *electropositive* metals. The success of a particular reaction depends upon both the thermodynamics and kinetics of the process. The electrochemical series can be used *qualitatively* to predict whether the metal should be oxidized by $H^+$ ions. Table 21.1 lists some reduction half-reactions. At the top of the table, oxidation of the metal to the corresponding metal ion is thermodynamically *favourable* with respect to the reduction of $H^+$ to $H_2$.

---

[§] Each H nucleus in an $H_2$ molecule could have a nuclear spin of $+\frac{1}{2}$ or $-\frac{1}{2}$. The forms of $H_2$ with the spin combinations $(+\frac{1}{2},+\frac{1}{2})$ [or $(-\frac{1}{2},-\frac{1}{2})$] and $(+\frac{1}{2},-\frac{1}{2})$ are known as *ortho-* and *para*-dihydrogen respectively.

**Table 21.1** The electrochemical series: a qualitative approach to predicting reactions between $H^+$ and metals, and between $H_2$ and metal ions. See also Appendix 12.

$$\longleftarrow \text{Oxidation}$$

$$Li^+ + e^- \rightleftharpoons Li$$
$$K^+ + e^- \rightleftharpoons K$$
$$Ca^{2+} + 2e^- \rightleftharpoons Ca$$
$$Na^+ + e^- \rightleftharpoons Na$$
$$Mg^{2+} + 2e^- \rightleftharpoons Mg$$
$$Al^{3+} + 3e^- \rightleftharpoons Al$$
$$Mn^{2+} + 2e^- \rightleftharpoons Mn$$
$$Cr^{2+} + 2e^- \rightleftharpoons Cr$$
$$Zn^{2+} + 2e^- \rightleftharpoons Zn$$
$$Cr^{3+} + 3e^- \rightleftharpoons Cr$$
$$Fe^{2+} + 2e^- \rightleftharpoons Fe$$
$$Cr^{3+} + e^- \rightleftharpoons Cr^{2+}$$
$$Co^{2+} + 2e^- \rightleftharpoons Co$$
$$Ni^{2+} + 2e^- \rightleftharpoons Ni$$
$$Sn^{2+} + 2e^- \rightleftharpoons Sn$$
$$Pb^{2+} + 2e^- \rightleftharpoons Pb$$
$$2H^+ + 2e^- \rightleftharpoons H_2$$
$$Cu^{2+} + 2e^- \rightleftharpoons Cu$$
$$Cu^+ + e^- \rightleftharpoons Cu$$
$$Fe^{3+} + e^- \rightleftharpoons Fe^{2+}$$
$$[Hg_2]^{2+} + 2e^- \rightleftharpoons 2Hg$$
$$Ag^+ + e^- \rightleftharpoons Ag$$
$$Hg^{2+} + 2e^- \rightleftharpoons Hg$$
$$Ce^{4+} + e^- \rightleftharpoons Ce^{3+}$$
$$Co^{3+} + e^- \rightleftharpoons Co^{2+}$$

$$\text{Reduction} \longrightarrow$$

Equations 21.7 and 21.8 illustrate two reactions suitable for the preparation of $H_2$ in the laboratory.

$$Mg(s) + 2H^+(aq) \longrightarrow Mg^{2+}(aq) + H_2(g) \tag{21.7}$$

oxidation

reduction

*Electropositive metals* release $H_2$ from acids.

$$Zn(s) + 2H^+(aq) \longrightarrow Zn^{2+}(aq) + H_2(g) \tag{21.8}$$

oxidation

reduction

## Some uses of $H_2$

An important use of $H_2$ is in the industrial *fixation of dinitrogen*, a process by which $N_2$ is removed from the atmosphere and converted into commercially useful compounds including ammonia. The reversible reaction 21.9 is achieved in the Haber process, and Box 21.2 illustrates the application of Le Chatelier's principle to this system.

$$3H_2(g) + N_2(g) \rightleftharpoons 2NH_3(g) \tag{21.9}$$

A second major use of $H_2$ is in the manufacture of methanol (equation 21.10). Reaction between $H_2$ and CO takes place at high pressure (25 300 kPa), high

**THEORETICAL AND CHEMICAL BACKGROUND**

## Box 21.2 Chemical equilibria and Le Chatelier's principle: application to the Haber process

Fritz Haber (1868–1934). ©*The Nobel Foundation.*

Industrially, dinitrogen may be fixed in the following reversible reaction:

$$3H_2(g) + N_2(g) \rightleftharpoons 2NH_3(g)$$

The forward reaction is exothermic; at 298 K, $\Delta_f H^\circ(NH_3, g) = -45.9 \, kJ \, mol^{-1}$. Unfortunately the reaction is extremely slow at room temperature – mixtures of $N_2$ and $H_2$ are indefinitely stable. We saw in Chapter 15 that the *rate* of a reaction increases with increasing temperature.

Le Chatelier's principle (Chapter 16) states that when a change (e.g. pressure or temperature) is made to a system in equilibrium, the equilibrium will tend to change to counteract the external change. Thus, in the Haber process, lowering the temperature will cause the equilibrium to shift to the right-hand side. Heat is produced and this raises the temperature and opposes the external change. The result can be seen by looking at the temperature dependence of the equilibrium constant, $K_p$:

$$K_p = \frac{(P_{NH_3})^2}{(P_{H_2})^3 (P_{N_2})}$$

where $P_{NH_3}$, $P_{H_2}$ and $P_{N_2}$ are the numerical values of the partial pressures of the three gases in the system. Remember that, strictly, $K$ should be expressed in terms of activities (see Chapter 16).

For this reaction, $K_p$ varies as follows:

$$\left.\begin{array}{ll} 298 \, K & K_p = 5.6 \times 10^5 \\ 500 \, K & K_p = 0.1 \\ 800 \, K & K_p = 9.0 \times 10^{-6} \end{array}\right\} \Rightarrow \begin{array}{l} NH_3 \text{ production} \\ \text{is increased} \\ \text{at lower} \\ \text{temperature} \end{array}$$

For the process to be industrially viable, ammonia should be formed both in good yield *and* at a reasonable rate. A low temperature may favour the formation of ammonia but the *rate* at which it is formed would be enhanced by a higher temperature. These two results conflict: at higher temperatures, the reaction is faster but the conversion to $NH_3$ is low. This problem can be resolved in part by considering Le Chatelier's principle again.

In the gas phase reaction, four moles of reactants produce two moles of products. If the pressure of the system is increased, the equilibrium will counter the change by lowering the pressure, i.e. it will tend to move to the right-hand side and more $NH_3$ will be produced. Remember that the pressure of a gas mixture depends upon the number of molecules present. Thus, working at a higher temperature *and* a higher pressure can produce favourable amounts of $NH_3$. However, the rate of the reaction is still rather slow, and a catalyst is needed. The final reaction conditions are a temperature of 723 K, a pressure of 20 260 kPa, and $Fe_3O_4$ mixed with KOH, $SiO_2$ and $Al_2O_3$ as the catalyst.

temperature ($\approx$600 K) and in the presence of a catalyst such as $Al_2O_3$. Methanol is industrially important as an additive in unleaded motor fuel, a precursor to organic compounds such as methanal (HCHO) and acetic acid, and in the synthesis of plastics and fibres.

$$2H_2(g) + CO(g) \longrightarrow CH_3OH(g) \tag{21.10}$$

The very low density of $H_2$ has, in the past, meant that it was used in balloons, but the high risk of an explosive reaction with $O_2$ means that helium (which is approximately twice as dense as $H_2$) is favoured nowadays. A relatively new use is as a rocket fuel (see Box 21.3), but the storage of $H_2$ is not simple. It can be stored as a *liquid* but the low boiling point makes this impractical for many purposes.

## COMMERCIAL AND LABORATORY APPLICATIONS

### Box 21.3 The space shuttle

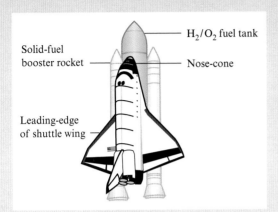

Solid-fuel booster rocket

$H_2/O_2$ fuel tank

Nose-cone

Leading-edge of shuttle wing

When the space shuttle lifts off, it is attached to the back of a huge fuel tank which contains liquid $H_2$ (about 1 457 000 litres) and liquid $O_2$ (about 530 000 litres). The two elements are kept in separate storage compartments inside the tank. The fuel is used up within the initial combustion and the fuel tank is jettisoned, leaving the shuttle to fly on into space. The

fuel for the booster rockets of the shuttle is composed of ammonium perchlorate $[NH_4][ClO_4]$ and aluminium – the $[ClO_4]^-$ anions are the source of oxygen for the oxidation of the aluminium. (Perchlorates are discussed in Section 22.12.)

In January 1986, the shuttle *Challenger* began an ill-fated mission. Exhaust gases escaped from a joint in one of the solid-fuel boosters and caused a fire which broke open the $H_2/O_2$ fuel tank. The resulting explosion and fires destroyed *Challenger* and its crew. The leak is thought to have been caused by a rubber O-ring which failed to keep a seal in the cold January weather.

Lifting off is one thing, but returning to the Earth is another. The space shuttle must be able to withstand the stress and thermal shock of re-entry into the Earth's atmosphere. The nose-cone and leading edges of the wings are constructed of *carbon–carbon composites*, which are fibre-reinforced materials with properties that include high strength, rigidity, chemical inertness, thermal stability, high resistance to thermal shock and retention of mechanical properties at high temperatures.

In 1839, William Grove observed that when the current was switched off in an electrolysis cell in which water was being electrolysed to give $O_2$ and $H_2$ using Pt electrodes, a small current continued to flow, but in the opposite direction to the current that had driven the electrolysis cell. The observation constituted the first *fuel cell* in which chemical energy produced from the reaction:

$$2H_2 + O_2 \longrightarrow 2H_2O \quad \textit{catalysed by Pt}$$

is converted into electrical energy. The combustion of $H_2$ produces *only* $H_2O$ and, therefore, $H_2$ is an environmentally clean fuel. From the end of the 20th century, the motor industry has been increasingly interested in fuel cell development. This research is driven by environmental legislation for pollution control. The motor industry's current strategy is for fuel cells to be powering millions of vehicles by 2020. Again, we come back to the problem of hydrogen storage, something that vehicle manufacturers must overcome if $H_2$ is to become a viable fuel. Potential methods of storing hydrogen include the use of some metal hydrides as 'hydrogen storage vessels'.

**▧▷**
Interstitial metal hydrides: see Section 21.4

**▧▷**
Saturated and unsaturated fats: see Box 33.5

Dihydrogen is widely used as a reducing agent, and the reduction of C=C double bonds by $H_2$ is the basis of the formation of saturated oils and fats for human consumption. $H_2$ can reduce some metal ions to lower oxidation state ions (equation 21.11) or to the elemental state (equation 21.12), and this is applied to the extraction of certain metals from their ores. The electrochemical series (Table 21.1) can be used to predict whether a metal ion may be reduced by $H_2$. At the top of the table, reduction is thermodynamically *unfavourable* with respect to the oxidation of $H_2$ to $H^+$. At the bottom of the table, reduction of the metal ion is thermodynamically *favourable* with respect to the oxidation

■▷
Thermodynamics of these
reactions: see Chapter 18

of $H_2$ to $H^+$. Thus, $H_2$ should reduce $Ag^+$ salts to metallic Ag, but will not reduce $Mg^{2+}$ to Mg. In fact, $H_2$ gas reduces hot silver(I) oxide and copper(II) oxide but does not reduce the ions when in solution under ambient conditions.

$$2Ce^{4+}(aq) + H_2(g) \longrightarrow 2Ce^{3+}(aq) + 2H^+(aq) \qquad (21.11)$$

$$CuO(s) + H_2(g) \xrightarrow{\Delta} Cu(s) + H_2O(g) \qquad (21.12)$$

## Oxidation of $H_2$

The 'pop' that is heard when a flame is placed into the mouth of a test-tube containing $H_2$ is a familiar sound in a chemical teaching laboratory, and is a qualitative test for the gas. In the reaction, $H_2$ is oxidized to water by the reaction with $O_2$ in the air (equation 21.13).

$$2H_2 + O_2 \longrightarrow 2H_2O \qquad (21.13)$$

■▷
Radical chain reactions:
see Section 15.14

When larger amounts of $H_2$ are involved, the 'pop' becomes an explosion. The mechanism is a complicated radical branched chain reaction and we give only a simplified form. One initiation step, brought about by a spark, is the homolytic cleavage of the $H-H$ bond (equation 21.14) and another occurs when $H_2$ and $O_2$ molecules collide (equation 21.15).

$$H_2 \longrightarrow 2H^\bullet \qquad \qquad Initiation \quad (21.14)$$
$$H_2 + O_2 \longrightarrow 2OH^\bullet \qquad \qquad Initiation \quad (21.15)$$

Branching of the chain then takes place (see Figure 15.24), increasing the number of radicals (equations 21.16 and 21.17). Efficient branching results in a rapid reaction and an explosion.

$$H^\bullet + O_2 \longrightarrow OH^\bullet + {}^\bullet O^\bullet \qquad \qquad Branching \quad (21.16)$$
$${}^\bullet O^\bullet + H_2 \longrightarrow OH^\bullet + H^\bullet \qquad \qquad Branching \quad (21.17)$$

Water is produced (for example) in a propagation step (equation 21.18).

$$OH^\bullet + H_2 \longrightarrow H_2O + H^\bullet \qquad \qquad Propagation \quad (21.18)$$

| 21.3 | ## What does *hydride* imply? |

The term 'hydride' conjures up the notion of the hydride ion, $H^-$. For compounds of the type $MH_x$ in which M is a *metal,* the hydrogen will either be in the form of $H^-$ or will carry a $\delta^-$ charge.

How realistic is the term 'hydride' for a compound formed between hydrogen and a *p*-block element? Table 21.2 lists the Pauling electronegativity values for elements in groups 13 to 17 and highlights whether $\chi^P$ for the element is less than, greater than, or equal to $\chi^P(H) = 2.2$. The green boxes show those elements that are more electronegative than hydrogen. A $B-H$ bond is polar in the sense $B^{\delta+}-H^{\delta-}$ but an $N-H$ bond is polar in the opposite sense $N^{\delta-}-H^{\delta+}$. In $PH_3$ and $AsH_3$, the $P-H$ and $As-H$ bonds are effectively

**Table 21.2** Pauling electronegativity values, $\chi^P$, for elements in groups 13 to 17. A more complete list is given in Table 5.2. For hydrogen $\chi^P = 2.2$. The shading in the table codes the elements according to whether they are more electronegative than H (▨), less electronegative than H (▨) or have the same value of $\chi^P$ (▨).

| 13 | 14 | 15 | 16 | 17 |
|---|---|---|---|---|
| B | C | N | O | F |
| 2.0 | 2.6 | 3.0 | 3.4 | 4.0 |
| Al(III) | Si | P | S | Cl |
| 1.6 | 1.9 | 2.2 | 2.6 | 3.2 |
| Ga(III) | Ge(IV) | As(III) | Se | Br |
| 1.8 | 2.0 | 2.2 | 2.6 | 3.0 |
| In(III) | Sn(IV) | Sb | Te | I |
| 1.8 | 2.0 | 2.1 | 2.1 | 2.7 |
| Tl(III) | Pb(IV) | Bi | Po | At |
| 2.0 | 2.3 | 2.0 | 2.0 | 2.2 |

non-polar. In groups 13 to 15, with a few exceptions, the name *hydride* correctly implies that the hydrogen atom is in an oxidation state of $-1$, although in $PH_3$ and $AsH_3$ the situation is rather ambiguous.

As we move to groups 16 and 17, the H atom becomes *less* electronegative than the atom, E, to which it is attached, and the term *hydride* (although correct nomenclature) may appear somewhat misleading for these binary $EH_x$ compounds. For this reason we treat the hydrides of groups 13 to 15 (Sections 21.5–21.7) separately from our discussion of those of the group 16 and 17 elements (Sections 21.9 and 21.10).

◼▷
Binary compound: see
Section 1.18

## 21.4    Binary hydrides of the *s*- and *d*-block metals

Many millions of compounds contain hydrogen and it is impossible to discuss them all here. It is convenient to think about hydrogen-containing compounds in terms of the other components of the molecules, and this is adequately illustrated by thinking about organic compounds where C–H bonds are major building blocks. In the next few sections, we consider binary hydrides of a range of elements.

### Hydrides of the *s*-block metals

The reactions of group 1 or group 2 elements with $H_2$ at high temperatures lead to the formation of the binary hydrides MH or $MH_2$ respectively (equations 21.19 and 21.20).

$$\textit{Group 1} \qquad 2K + H_2 \xrightarrow{\Delta} 2KH \tag{21.19}$$

$$\textit{Group 2} \qquad Ba + H_2 \xrightarrow{\Delta} BaH_2 \tag{21.20}$$

The electropositive *s*-block metals form ionic compounds, and the group 1 hydrides $M^+H^-$ possess sodium chloride structures in the solid state. The

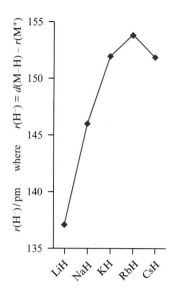

**Fig. 21.3** The apparent ionic radius of the $H^-$ ion is variable and depends on the metal ion present. The graph shows the variation in the radius of the $H^-$ ion in the alkali metal hydrides.

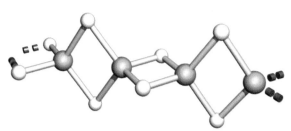

**Fig. 21.4** Part of the chain structure of $BeH_2$. Each $Be-H-Be$ bridge is a delocalized 3-centre 2-electron (3c–2e) interaction. Colour code: Be, yellow; H, white.

radius of the hydride ion varies with the metal cation (Figure 21.3). This is attributed to the relatively weak interaction between the proton and the two $1s$ electrons of the $H^-$ anion and, consequently, the $H^-$ ion is easily deformed. Extreme distortion results in a covalent interaction and the sharing of electrons between H and M.

The hydrides of Li, Be and Mg show considerable covalent character. Beryllium hydride, $BeH_2$, has a polymeric structure in which each Be centre is 4-coordinate (Figure 21.4) and is connected to the next Be atom by two 2-coordinate or *bridging* hydrogen atoms. Each H atom possesses only one valence electron and each Be atom has only two valence electrons, and yet the connectivities of H and Be in the polymeric beryllium hydride appear to exceed the bonding capabilities of the two elements. This can be rationalized by describing the bonding in the $Be-H-Be$ bridges as *delocalized*. Each Be atom may be considered to be $sp^3$ hybridized. The $1s$ atomic orbital of each H atom can overlap with *two $sp^3$* hybrid orbitals from different Be atoms (Figure 21.5). The distribution of electrons means that each bridge is associated with only two electrons and we have a *3-centre 2-electron (3c–2e) interaction*. Beryllium hydride is usually prepared by the thermal decomposition of beryllium alkyls such as $^tBu_2Be$ rather than by the direct reaction of Be with $H_2$.

Reactions of the *s*-block metal hydrides often involve formation of $H_2$ (equation 21.21) and in many cases the source of $H^+$ is a weak acid such as $H_2O$ (equation 21.22).

$$H^- + H^+ \longrightarrow H_2 \qquad (21.21)$$

$$H^-(aq) + H_2O(l) \longrightarrow H_2(g) + [OH]^-(aq) \qquad (21.22)$$

Beryllium hydride is fairly stable in water, but the other *s*-block metal hydrides react rapidly, releasing $H_2$. NaH reacts violently (equation 21.23). $CaH_2$ reacts more slowly and is routinely used as a drying agent for organic solvents (hydrocarbons, ethers, amines and higher alcohols) and also as a means of 'storing' $H_2$. The controlled release of $H_2$ can be achieved by contact between $CaH_2$ and limited amounts of water (equation 21.24).

$$NaH(s) + H_2O(l) \longrightarrow NaOH(aq) + H_2(g) \qquad (21.23)$$

$$CaH_2(s) + 2H_2O(l) \longrightarrow Ca(OH)_2(aq) + 2H_2(g) \qquad (21.24)$$

$^tBu$ = *tert*-butyl: see Section 24.7

**Fig. 21.5** The formation of two $Be-H-Be$ bridging interactions by the overlap of Be $sp^3$ and H $1s$ orbitals. Each H atom has one valence electron and each Be atom has two. The electrons are omitted from the diagram, but a pair of electrons is associated with each 3-centre $Be-H-Be$ bridge.

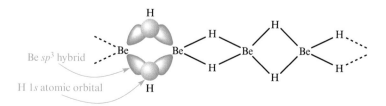

**Fig. 21.6** (a) The tricapped trigonal prismatic structure of the $[ReH_9]^{2-}$ anion in the dipotassium salt; six H atoms define a trigonal prism and each of the three remaining H atoms caps one square face of the prism. The trigonal prism is defined by the yellow lines. (b) The structure of the $[Pt_2H_9]^{5-}$ anion in the pentalithium salt. Colour code: Re, red; Pt, brown; H, white.

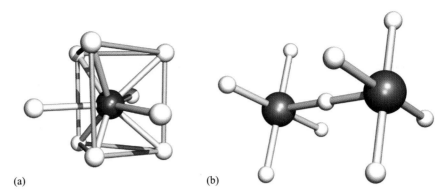

(a)    (b)

In general, the reactivity towards water increases down each group but also depends on the purity of the metal hydride. In moist air, RbH, CsH and $BaH_2$ spontaneously ignite.

### Hydrides of the *d*-block metals

The binary hydrides of the early *d*-block metals include $ScH_2$, $YH_2$, $TiH_2$ and $HfH_2$. Hafnium dihydride is a lustrous metallic solid, formed according to equation 21.25.

$$Hf \xrightarrow{1000\,K} \xrightarrow{\text{cool under } H_2\ (300\,kPa)} HfH_2 \qquad (21.25)$$

*Non-stoichiometric* compounds are also formed when titanium, zirconium, hafnium and niobium react with $H_2$. The unusual stoichiometry of compounds such as $TiH_{1.7}$, $HfH_{1.98}$ and $HfH_{2.10}$ arises because the hydrogen atoms are small enough to enter the metal lattice and occupy the interstitial holes. Niobium forms a series of non-stoichiometric hydrides of formula $NbH_x$ ($0 < x \leq 1$) and at low hydrogen content, the body-centred cubic structure of metallic niobium is retained. When these *d*-block metal hydrides are heated, $H_2$ is released and this property means that they are convenient storage 'vessels' for dihydrogen.

A ***non-stoichiometric compound*** is one that does not have the exact stoichiometric composition expected from the electronic structure, and this is often associated with a *defect* in the crystal lattice.

Anionic *d*-block hydrides include $[ReH_9]^{2-}$, $[TcH_9]^{2-}$ and $[Pt_2H_9]^{5-}$. A neutron diffraction study of $K_2[ReH_9]$ has shown that the $[ReH_9]^{2-}$ dianion has an unusual structure in the solid state (Figure 21.6a) in which there are two hydrogen environments – six prism-corner sites and three capping sites. In solution, the $^1H$ NMR spectrum shows one signal indicating that the dianion is stereochemically non-rigid on the NMR spectroscopic timescale. The salt $Li_5[Pt_2H_9]$ contains the anion $[Pt_2H_9]^{5-}$ (Figure 21.6b) and although a high pressure of $H_2$ is needed to prepare it, the compound is stable with respect to loss of $H_2$ at 298 K.

Interstitial holes: see Section 9.2

Neutron diffraction: see Section 4.2

Non-rigidity and timescales: see Sections 6.12 and 11.3

**21.5**    ## Binary hydrides of group 13 elements

Boron forms a range of hydrides (*boranes*) including neutral compounds of general formulae $B_nH_{n+4}$ and $B_nH_{n+6}$ and hydroborate dianions of formula

**THEORETICAL AND CHEMICAL BACKGROUND**

## Box 21.4 Boron hydride (borane) clusters

Boron forms a large group of neutral boron hydrides (boranes) which possess 3-dimensional *cluster* structures. In each, the bonding is considered to be delocalized, since the connectivity of each boron atom is typically between five and seven, despite the fact that a boron atom has only three valence electrons.

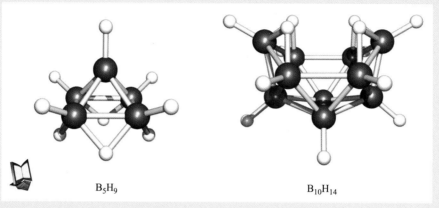

$B_5H_9$                    $B_{10}H_{14}$

A full discussion of these novel compounds is beyond the scope of this text but further details of these and related species may be found in: N. N. Greenwood and A. Earnshaw (1997) *Chemistry of the Elements*, 2nd edn, Butterworth-Heinemann, Oxford, Ch. 6; C. E. Housecroft (1994) *Boranes and Metallaboranes: Structure, Bonding and Reactivity*, 2nd edn, Ellis Horwood, Hemel Hempstead; C. E. Housecroft (1994) *Cluster Molecules of the p-Block Elements*, OUP, Oxford.

$[B_nH_n]^{2-}$. These possess cage-like structures (see Box 21.4) and *delocalized* bonding schemes are usually needed to describe the bonding.

Two small molecular hydrides of boron are diborane[§] $B_2H_6$ (Figure 21.7a) and the tetrahydroborate(1−) anion $[BH_4]^-$ (Figure 21.7b).

Boron has a ground state electronic configuration of $[He]2s^2 2p^1$ and is expected to form a hydride of formula $BH_3$. This compound is known in the gas phase but is very reactive. The planar $BH_3$ molecule possesses an empty $2p$ atomic orbital which readily accepts a pair of electrons from a *Lewis base* such as tetrahydrofuran, THF, and in this way boron completes its valence octet of electrons. A Lewis base compound such as $THF \cdot BH_3$ (Figure 21.7c) is called an *adduct* and can be represented either by the valence bond structure **21.1** or by showing a *dative* or *coordinate bond* from the donor atom of the Lewis base to the boron atom (structure **21.2**). The $BH_3$ molecule acts as a *Lewis acid* by accepting a pair of electrons.

$[BH_4]^-$ is also called borohydride

Bonding in $BH_3$: see Section 7.6

Donor–acceptor compounds: see Chapter 23

**(21.1)**                    **(21.2)**

[§] Strictly the name should be diborane(6) to indicate the presence of *both* two B atoms and six H atoms.

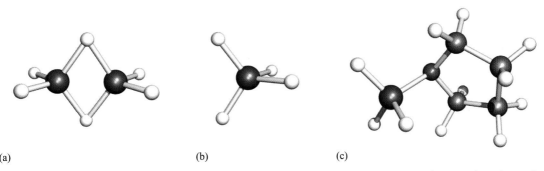

(a)                    (b)                    (c)

**Fig. 21.7** The structures of (a) $B_2H_6$, (b) $[BH_4]^-$ and (c) THF·$BH_3$. In each, the boron atom is approximately tetrahedral. Colour code: B, blue; O, red; C, grey; H, white.

A **Lewis base** can donate a pair of electrons; a **Lewis acid** can accept a pair of electrons.

A **coordinate bond** forms when a Lewis base donates a pair of electrons to a Lewis acid and the resulting compound is an **adduct**.

The $[BH_4]^-$ anion can be considered to be a Lewis base–acid adduct, formed by the donation of a pair of electrons from an $H^-$ ion to $BH_3$. Diborane is a dimer of $BH_3$ and the formation of the B−H−B bridges (Figure 21.7a) completes the octet of each boron atom as scheme **21.3** shows. Each bridge in $B_2H_6$ consists of a delocalized 3c–2e bonding interaction and is similar to a Be−H−Be bridge in $BeH_2$ (Figure 21.5). Each boron atom in $B_2H_6$, $[BH_4]^-$ and THF·$BH_3$ is tetrahedral.

**3c–2e = 3-centre 2-electron**

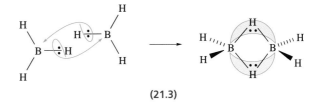

**(21.3)**

Diborane and adducts such as THF·$BH_3$ are used extensively as reducing agents and hydroborating agents. THF·$BH_3$ is a convenient reagent and is purchased as a solution in THF. Both $B_2H_6$ and THF·$BH_3$ are water-sensitive and hydrolyse rapidly (equation 21.26) and the product is trioxo-boric acid, commonly called boric acid, $B(OH)_3$ or $H_3BO_3$ (**21.4**).

**Hydroboration: see Section 26.10**

$$B_2H_6(g) + 6H_2O(l) \longrightarrow 2B(OH)_3(aq) + 6H_2(g) \qquad (21.26)$$

Boric acid

HO — B with OH, OH groups

**(21.4)**

Tetrahydroborate(1−) salts are also important reducing agents. The sodium salt Na[$BH_4$] is a white, reasonably air-stable crystalline solid which is available commercially. It can be prepared from sodium hydride (equations 21.27 and 21.28) under water-free conditions. The $[BH_4]^-$ anion is *kinetically* stable towards hydrolysis, but pre-formed, cobalt-doped pellets of Na[$BH_4$] can be used as a convenient source of $H_2$ – simply add to water.

**See problem 21.5**

$$4NaH + BCl_3 \longrightarrow Na[BH_4] + 3NaCl \qquad (21.27)$$

$$4NaH + B(OCH_3)_3 \xrightarrow{\;520\,K\;} Na[BH_4] + 3NaOCH_3 \qquad (21.28)$$

**COMMERCIAL AND LABORATORY APPLICATIONS**

## Box 21.5 Lithium aluminium hydride as a reducing agent and a source of hydride

The use of Li[AlH$_4$] as a *reducing agent* is widespread. Organic reductions include the following:

| Starting material | General reaction | | Product |
|---|---|---|---|
| Halogenoalkane | RX $\xrightarrow{\text{Li[AlH}_4]}$ RH | | Alkane |
| Aldehyde | $\begin{array}{c}R\\ \backslash\\ C=O\\ /\\ H\end{array}$ $\xrightarrow{\text{Li[AlH}_4]}$ RCH$_2$OH | | Primary alcohol |
| Ketone | $\begin{array}{c}R\\ \backslash\\ C=O\\ /\\ R\end{array}$ $\xrightarrow{\text{Li[AlH}_4]}$ R$_2$CHOH | | Secondary alcohol |
| Carboxylic acid | $R-C\begin{array}{c}O\\ \\ OH\end{array}$ $\xrightarrow{\text{Li[AlH}_4]}$ RCH$_2$OH | | Primary alcohol |
| Acid chloride | $R-C\begin{array}{c}O\\ \\ Cl\end{array}$ $\xrightarrow{\text{Li[AlH}_4]}$ RCH$_2$OH | | Primary alcohol |
| Amide | $R-C\begin{array}{c}O\\ \\ NH_2\end{array}$ $\xrightarrow{\text{Li[AlH}_4]}$ RCH$_2$NH$_2$ | | Primary amine |
| Azide | $R-N\overset{+}{=}N\overset{-}{=}N$ $\xrightarrow{\text{Li[AlH}_4]}$ RNH$_2$ | | Amine |
| Nitrile | $R-C\equiv N$ $\xrightarrow{\text{Li[AlH}_4]}$ RCH$_2$NH$_2$ | | Primary amine |
| Hydroperoxide | $R-O\begin{array}{c}O-H\\ \end{array}$ $\xrightarrow{\text{Li[AlH}_4]}$ ROH | | Alcohol |
| Peroxide | $R-O\begin{array}{c}O-R\\ \end{array}$ $\xrightarrow{\text{Li[AlH}_4]}$ 2ROH | | O–O bond cleavage and formation of alcohols |

Lithium aluminium hydride may be used to *convert some halides to hydrides*:

Metal or non-metal halide $\xrightarrow{\text{Li[AlH}_4]}$ Metal or non-metal hydride

e.g.

SiCl$_4$ $\xrightarrow{\text{Li[AlH}_4]}$ SiH$_4$    *similarly with Ge and Sn halides*

PCl$_3$ $\xrightarrow{\text{Li[AlH}_4]}$ PH$_3$    *similarly with As and Sb halides*

Hydrides may be formed from organometallic compounds:

$[\text{ZnMe}_4]^{2-}$ $\xrightarrow{\text{Li[AlH}_4]}$ $[\text{ZnH}_4]^{2-}$

In some cases reduction occurs instead of hydride formation:

AgCl $\xrightarrow{\text{Li[AlH}_4]}$ Ag

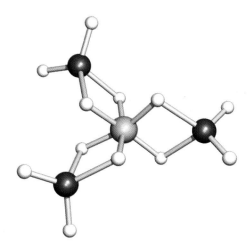

**Fig. 21.8** The structure of $[Al(BH_4)_3]$. The aluminium centre is octahedral and is associated with six 3c–2e Al−H−B bridges. Colour code: Al, yellow; B, blue; H, white.

Although $Na[BH_4]$ is an ionic salt, some metal tetrahydroborates such as $[Al(BH_4)_3]$ are covalent with the hydrogen atoms taking part in 3c–2e bridges between the boron and the metal centres (Figure 21.8).

In comparison with boron, other members of group 13 form fewer hydrides. Aluminium trihydride is a colourless solid. It is unstable if heated above 420 K, and reacts violently with water to produce $H_2$. In the solid state, $AlH_3$ units are connected through Al−H−Al bridges. An aluminium analogue of $B_2H_6$ has not yet been established.

Like $Na[BH_4]$, $[AlH_4]^-$ salts are widely used as reducing agents (see Box 21.5); $Li[AlH_4]$ and $Na[AlH_4]$ are available commercially. Lithium tetrahydridoaluminate(1−) (also called lithium aluminium hydride or lithal) can be prepared from lithium hydride (equation 21.29) and is a white, crystalline solid. It is very reactive and must be handled in moisture-free conditions; it is soluble in a range of ethers.

$$4LiH + AlCl_3 \longrightarrow Li[AlH_4] + 3LiCl \tag{21.29}$$

Gallium hydride was first fully characterized in the early 1990s. It is prepared from $GaCl_3$ as shown in scheme 21.30. Digallane, $Ga_2H_6$, condenses at low temperature as a white solid which melts at 223 K to give a colourless, viscous liquid; it decomposes above 253 K (equation 21.31). Electron diffraction studies have confirmed that the gas phase structure of $Ga_2H_6$ resembles that of $B_2H_6$ (Figure 21.7a).

Electron diffraction: see Section 4.2

$$GaCl_3 \xrightarrow{Me_3SiH} \quad \overset{H\,\text{\tiny''''}}{\underset{H}{}}Ga\overset{Cl}{\underset{Cl}{\diamondsuit}}Ga\overset{\text{\tiny''''}H}{\underset{H}{}} \xrightarrow[240\,K]{Li[AlH_4]} \quad \overset{H\,\text{\tiny''''}}{\underset{H}{}}Ga\overset{H}{\underset{H}{\diamondsuit}}Ga\overset{\text{\tiny''''}H}{\underset{H}{}}$$

$$\tag{21.30}$$

$$Ga_2H_6 \longrightarrow 2Ga + 3H_2 \tag{21.31}$$

Both $B_2H_6$ and $Ga_2H_6$ react with Lewis bases, and scheme 21.32 summarizes the reactions with $NH_3$ and $NMe_3$. Whereas two of the small $NH_3$ molecules can attack the *same* boron or gallium centre, reaction with the more sterically demanding $NMe_3$ tends to follow a different pathway. Competition between these routes (known as asymmetric and symmetric cleavage of the $E_2H_6$ molecule) may be observed.

$$[EH_2(NH_3)_2]^+[EH_4]^- \xleftarrow[\textit{asymmetric cleavage}]{NH_3} \quad \text{(structure)} \quad \xrightarrow[\textit{symmetric cleavage}]{NMe_3} 2Me_3N\cdot EH_3$$

$$E = B \text{ or } Ga \tag{21.32}$$

Neutral, binary hydrides of indium and thallium are not known. The compounds $Li[EH_4]$ for $E = Ga$, In and Tl are all thermally unstable. $Li[GaH_4]$ is prepared by reaction 21.33 and decomposes at 320 K. $Li[InH_4]$ and $Li[TlH_4]$ both decompose around 273 K.

$$4LiH + GaCl_3 \longrightarrow Li[GaH_4] + 3LiCl \tag{21.33}$$

**21.6**        **Binary hydrides of group 14 elements**

Hydrocarbons:
see Chapters 25 and 26

Hydrocarbons might be considered to be 'carbon hydrides' although they are rarely described as such. The contrast between the wealth of hydrocarbons and the small number of silicon hydrides (*silanes*) is dramatic. The silicon analogue of methane is $SiH_4$ and it is formed by reacting $SiCl_4$ or $SiF_4$ with $Li[AlH_4]$. It is an important source of pure silicon (equation 21.34) for use in semiconductors. It is a colourless gas (bp 161 K), insoluble in water (although reaction with alkali is violent), and spontaneously inflammable in air. Mixtures of $SiH_4$ and $O_2$ are explosive (equation 21.35).

Semiconductors:
see Section 9.13

$$SiH_4 \xrightarrow{\Delta} Si + 2H_2 \tag{21.34}$$
$$SiH_4 + 2O_2 \longrightarrow SiO_2 + 2H_2O \tag{21.35}$$

Further members of the family of saturated silanes with the general formula $Si_nH_{2n+2}$ are known up to $n = 10$ with both straight and branched chains. A mixture of $SiH_4$ and higher silanes is produced when magnesium silicide ($Mg_2Si$) reacts with aqueous acid. The structures of the $Si_nH_{2n+2}$ molecules are directly related to those of their carbon (alkane) counterparts. Potassium reacts with $SiH_4$ (equation 21.36) to form the white, crystalline salt $K[SiH_3]$ which is a useful synthetic reagent.

$$2SiH_4 + 2K \longrightarrow 2K[SiH_3] + H_2 \tag{21.36}$$

The heavier members of group 14 form few binary hydrides. $GeH_4$ as well as several higher germanes are known; $GeH_4$ is a colourless gas (bp 184 K) and is insoluble in water and inflammable in air. The reaction of $SnCl_4$ with $Li[AlH_4]$ gives $SnH_4$ (bp 221 K) but at 298 K it decomposes to its constituent elements. It is unlikely that the lead analogue $PbH_4$ has been prepared.

**21.7**        **Binary hydrides of group 15 elements**

### Ammonia

Ammonia, $NH_3$, is prepared on an industrial scale by the Haber process (equation 21.9 and Box 21.2). The scale of production is huge, and $NH_3$ is used extensively for the manufacture of fertilizers, nitric acid, explosives and synthetic fibres. It is a colourless gas (bp 239 K) with a characteristic

**Fig. 21.9** The structures of (a) ammonia, $NH_3$, and (b) hydrazine, $N_2H_4$. Colour code: N, blue; H, white.

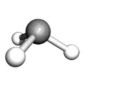
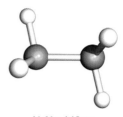

$N–H = 101\,pm$
(a)   $\angle H–N–H = 107.5°$         (b)     $N–N = 145\,pm$
                                                $N–H = 102\,pm$

smell and is corrosive and an irritant. The trigonal pyramidal structure (Figure 21.9a) is consistent with VSEPR theory. Combustion in air takes place according to equation 21.37.

$$4NH_3 + 3O_2 \longrightarrow 2N_2 + 6H_2O \tag{21.37}$$

oxidation

reduction

Ammonia is very soluble in water and a concentrated solution (density $0.88\,g\,cm^{-3}$) is commercially available. The high solubility is due to the extensive hydrogen bonding that occurs between $H_2O$ and $NH_3$ molecules. Aqueous solutions are alkaline because of the hydroxide ions that are formed in equilibrium 21.38, and may be neutralized by acids. In the expression for the corresponding equilibrium constant (equation 21.39), the concentration of water (the bulk solvent) is defined as unity.[§]

Hydrogen bonding: see Section 21.8

$$NH_3(aq) + H_2O(l) \rightleftharpoons [NH_4]^+(aq) + [OH]^-(aq) \tag{21.38}$$

$$K_b = \frac{[NH_4^+][OH^-]}{[NH_3][H_2O]} = \frac{[NH_4^+][OH^-]}{[NH_3]} = 1.8 \times 10^{-5} \text{ (at 298 K)} \tag{21.39}$$

This value of $K$ is 'small' and means that the concentration of $NH_3$ in aqueous solution vastly exceeds the concentration of ions. Although you may find aqueous solutions of ammonia referred to as 'ammonium hydroxide', no such compound exists in the solid state.

Anhydrous liquid ammonia is *self-ionizing* (equation 21.40) and is used extensively as a *non-aqueous solvent*. An interesting property of liquid ammonia is the fact that it dissolves alkali metals to give electrically conducting solutions.

Alkali metals in liquid $NH_3$: see Section 21.11

$$2NH_3(l) \rightleftharpoons [NH_4]^+(solv) + [NH_2]^-(solv) \tag{21.40}$$

Ammonium cation     Amide anion

If a pure liquid partially dissociates into ions, it is said to be *self-ionizing*.

### Hydrazine

Hydrazine, $N_2H_4$, is a colourless liquid (mp 275 K, bp 386 K) which is miscible with water and with a range of organic solvents. It is corrosive and toxic, and its vapour forms explosive mixtures with air. Although the formation of $N_2H_4$ from its constituent elements is an endothermic process (equation 21.41), hydrazine is *kinetically* stable with respect to $N_2$ and $H_2$.

$$N_2(g) + 2H_2(g) \longrightarrow N_2H_4(l) \qquad \Delta_fH^\circ(298\,K) = +50.6\,kJ\,mol^{-1} \tag{21.41}$$

[§] $[H_2O] = \dfrac{\text{Molar concentration of } H_2O}{\text{Molar concentration of pure water}} \approx 1$; strictly $K$ is expressed in terms of activities (see Section 16.3).

Conformer and Newman
projection: see Section 24.9

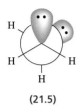

**(21.5)**

Figure 21.9b shows the molecular structure of $N_2H_4$. It is related to that of $C_2H_6$ but the presence of two lone pairs (one per N atom) increases the number of possible conformers. The most favourable conformation for the gas phase molecule is the *gauche* form, shown as a Newman projection in structure **21.5**. You might expect a staggered conformer would be preferred but this is not the case; this observation is not readily explained.

Hydrazine can be prepared from ammonia (equation 21.42) by the *Raschig reaction* upon which its industrial synthesis is based. Hydrazine is used in the agricultural and plastics industries, and also as a rocket fuel. It is a powerful reducing agent, and one application is in the removal of $O_2$ from industrial water boilers to minimize their corrosion (equation 21.43).

$$2NH_3 + \underset{\text{Sodium hypochlorite}}{NaOCl} \longrightarrow N_2H_4 + NaCl + H_2O \qquad (21.42)$$

$$N_2H_4 + O_2 \longrightarrow N_2 + 2H_2O \qquad (21.43)$$

oxidation

reduction

## Phosphane

Phosphane is also called
phosphine

Phosphane, $PH_3$, is produced by the action of water on calcium phosphide (equation 21.44) or magnesium phosphide, or by the reaction of $PCl_3$ with lithium hydride (equation 21.45) or $Li[AlH_4]$. Phosphane is a colourless gas (bp 185 K) with a garlic-like odour, and is extremely poisonous. It has been detected in the atmospheres of Saturn, Jupiter and Uranus.

$$Ca_3P_2 + 6H_2O \longrightarrow 2PH_3 + 3Ca(OH)_2 \qquad (21.44)$$

$$PCl_3 + 3LiH \longrightarrow PH_3 + 3LiCl \qquad (21.45)$$

Phosphane ignites in dry air at about 420 K (equation 21.46), or spontaneously if $P_2H_4$ (formed as a by-product in its synthesis) is present. It is a strong reducing reagent.

$$PH_3 + 2O_2 \longrightarrow \underset{\text{Phosphoric acid}}{H_3PO_4} \qquad (21.46)$$

In a number of respects, $PH_3$ is rather different from $NH_3$. It is less soluble than $NH_3$ in water, and dissolves to give neutral solutions. Like $NH_3$, the molecular structure of $PH_3$ is trigonal pyramidal, but whereas the H−N−H bond angle is 107.5°, the H−P−H angle is only 93°. The former is consistent with $sp^3$ hybridization and the latter suggests that $p$ orbitals play a major role in bonding. This difference has an important chemical consequence. The lone pair in the $sp^3$ hybridized $NH_3$ molecule is readily available and ammonia is a *reasonably* strong base. In contrast, the lone pair in $PH_3$ is less available and $PH_3$ is a very weak base. Although phosphonium, $[PH_4]^+$, salts can be formed, they are decomposed by water (equation 21.47).

$$[PH_4]^+(aq) + H_2O(l) \longrightarrow PH_3(g) + [H_3O]^+(aq) \qquad (21.47)$$

## Diphosphane

Diphosphane, $P_2H_4$, is a colourless liquid at room temperature (mp 174 K, bp 329 K), is toxic and spontaneously inflammable. The gas phase structure of the $P_2H_4$ molecule has a *gauche* conformation like $N_2H_4$ (Figure 21.9b and structure **21.5**) with bond distances of P−P = 222 pm and P−H = 145 pm.

**Fig. 21.10** The differing thermal stabilities of arsine and stibine can be used to distinguish between them. $SbH_3$ is *less* stable than $AsH_3$ and decomposes in region **A** while $AsH_3$ decomposes in region **B**.

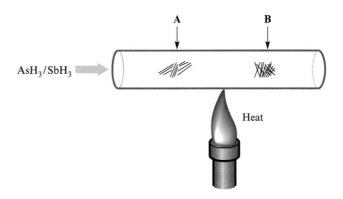

### Hydrides of arsenic, antimony and bismuth

Arsane, stibane and bismuthane (with the more common names of arsine, stibine and bismuthine) are all unstable gases at 298 K and are toxic. They can be prepared by the reaction of the respective trichlorides with $Na[BH_4]$. Arsane, $AsH_3$, has a particularly repulsive smell. It has been detected in the atmospheres of Jupiter and Saturn. $AsH_3$ and $SbH_3$ are readily oxidized to $As_2O_3$ and $Sb_2O_3$ respectively.

One of the old tests (the Marsh test) for the presence of arsenic or antimony involved the production of the hydrides. The material of interest was treated with zinc dust and acid, and the gas evolved, containing $H_2$ and either $AsH_3$ or $SbH_3$, was passed through a tube that was heated at one point. The thermal instabilities of $SbH_3$ and $AsH_3$ meant that they decomposed – $SbH_3$ is *less* stable than $AsH_3$ and decomposes in the warm region *before* the flame (Figure 21.10). The result is that Sb-containing materials give a black deposit (which appears as a mirror) on the wall of the glass tube in region **A** in Figure 21.10 (equation 21.48) while As-containing compounds produce a deposit of arsenic in region **B** (equation 21.49).

$$2SbH_3(g) \xrightarrow{\Delta} 2Sb(s) + 3H_2(g) \tag{21.48}$$

$$2AsH_3(g) \xrightarrow{\Delta} 2As(s) + 3H_2(g) \tag{21.49}$$

## 21.8    Hydrogen bonding

Before we consider the compounds formed between hydrogen and the group 16 and 17 elements, we must mention two important consequences of having a hydrogen atom bonded to an electronegative atom.

### The hydrogen bond

A *hydrogen bond* is an interaction between a hydrogen atom attached to an electronegative atom, and an electronegative atom that possesses a lone pair of electrons. The strongest hydrogen bonds involve the first row elements F, O or N, and structure **21.6** shows the formation of a hydrogen bond between two water molecules.

(21.6)

**ENVIRONMENT AND BIOLOGY**

## Box 21.6 Hydrogen bonding and DNA

DNA (deoxyribonucleic acid) is the key to life – when a cell divides, the genetic information from the original cell is passed to the new cell by the DNA molecule. The second major role of DNA is in the synthesis of proteins.

DNA is a nucleic acid polymer and its structure consists of two helical chains which interact with one another through hydrogen bonding to form a *double* helix. The backbone of each chain consists of sugar units (each containing a 5-membered $C_4O$-ring) linked by phosphate groups. Attached to each sugar is an organic base. In DNA, there are four different bases: adenine, guanine, cytosine and thymine. The left-hand structure below represents part of the polymeric backbone of DNA and shows the positions of attachment of the bases (shown on the right).

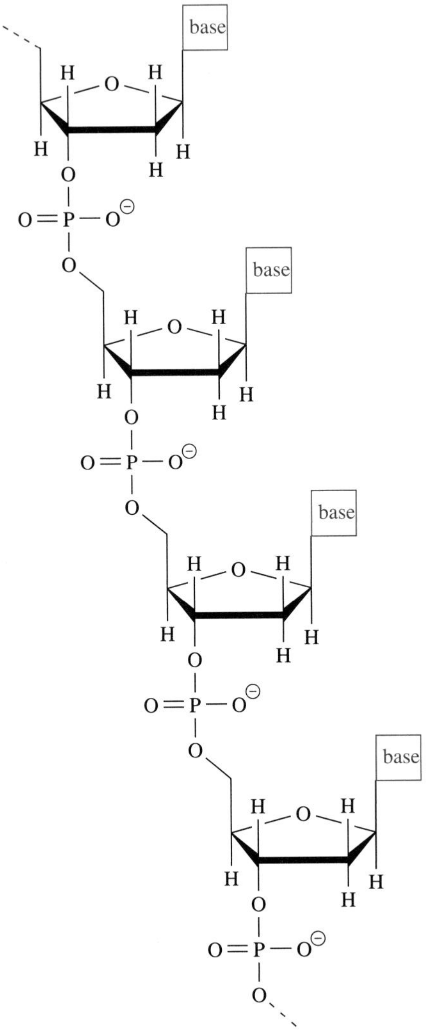

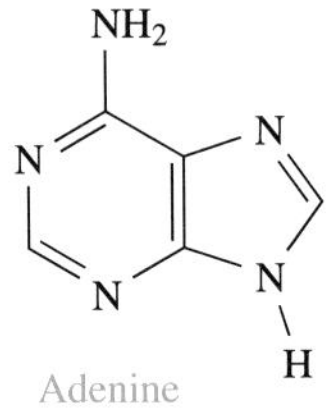

Adenine

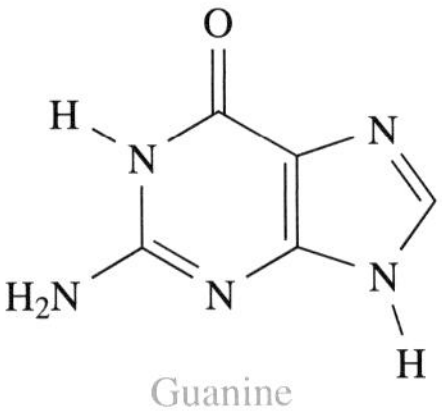

Guanine

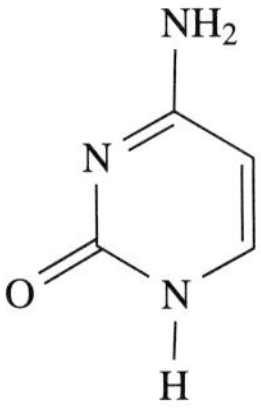

Cytosine

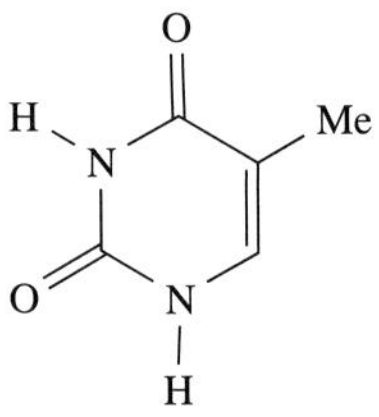

Thymine

The structures of adenine and thymine are *exactly* matched to permit hydrogen bonding interactions between these bases – adenine and thymine are *complementary bases*. Guanine and cytosine are similarly matched.

Adenine–thymine base pair

Guanine–cytosine base pair

The result of this base-pairing is that the sequence of the bases in one chain of DNA is *complemented* by a sequence in a second chain, and the two hydrogen-bonded chains combine to form a *double helix*.

Proteins consist of a sequence of amino acids, and the sequence is specific to a particular protein. During protein synthesis, DNA is able to *code* the sequence correctly by transmitting information in the form of its own base sequence. Each amino acid in the protein chain has an associated code of three adjacent DNA bases (a *codon*).

For related information in this book see: ATP (structure **22.38**), heterocyclic organic molecules (Chapter 34), anti-cancer drugs and DNA (Box 23.5) and nucleobases, nucleotides and nucleic acids (Section 35.7).

*Further reading*: C. K. Mathews, K. E. van Holde and K. G. Ahern (2000) *Biochemistry*, 3rd edn, Benjamin/Cummings, New York, Chapter 4.

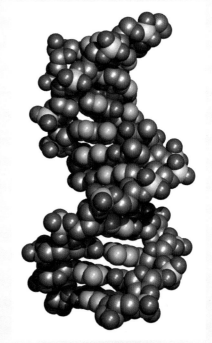

 Colour code: C, grey; N, Blue; O, red; P, yellow.

**Fig. 21.11** The trends in (a) boiling points and (b) $\Delta_{vap}H$ for the group 15 hydrides, and (c) the trends in boiling points for the group 16 and 17 hydrides.

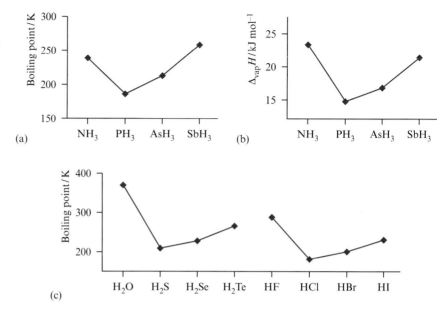

> A **hydrogen bond** is an interaction between a hydrogen atom attached to an electronegative atom, and an electronegative atom that possesses a lone pair of electrons.

The physical properties of compounds may indicate the presence of hydrogen bonds because such interactions result in the association of molecules. In this section we restrict our discussion to binary hydrides but further examples of the effects of hydrogen bonding are given in later chapters.

### Boiling points and enthalpies of vaporization

Figure 21.11a shows the trend in the boiling points of the compounds $EH_3$ where E is a group 15 element. The trend in enthalpies of vaporization (Figure 21.11b) is the same and this is *not* a coincidence! Trouton's rule (equation 21.50) gives an approximate relationship between $\Delta_{vap}H$ (measured at the boiling point of a liquid) and the boiling point (in K) of the liquid.

**Trouton's rule:**
**see Section 17.10**

$$\frac{\Delta_{vap}H}{bp} \approx 0.088 \, kJ \, K^{-1} \, mol^{-1} \qquad \textit{Trouton's rule} \quad (21.50)$$

Figure 21.11c illustrates that similar plots are obtained for the boiling points of hydrides of the group 16 and 17 elements. Although the general trend is an increase down a group, $NH_3$, $H_2O$ and HF stand out as having anomalously high boiling points.

The transition from a liquid to a vapour involves the cleavage of some (but not necessarily all) of the *intermolecular* interactions. The strength of a normal[§] hydrogen bond is $\approx 25 \, kJ \, mol^{-1}$ and although this is considerably weaker than a covalent bond, it is strong enough to affect the enthalpy of

---

[§] For a more detailed discussion of different types of hydrogen bond, see: C. E. Housecroft and A. G. Sharpe (2005) *Inorganic Chemistry*, 2nd edn, Prentice Hall, Harlow, Chapter 9.

vaporization and the boiling point of a compound significantly. The unusually high values for $NH_3$, $H_2O$ and HF are due to the additional energy needed to break the N···H, O···H or F···H hydrogen bonds which are present in the liquid state. Note, however, that while $H_2O$ is hydrogen bonded in the liquid but not in the vapour state, HF is hydrogen bonded in both states.

### Association in the solid state

The association of molecules through hydrogen bonding in the solid state often leads to well-organized assemblies of molecules. In solid hydrogen fluoride, the HF molecules are arranged in zigzag chains (Figure 21.12a) in which each H atom is in a linear environment.

Intermolecular hydrogen bonding occurs in both water and ammonia in the solid state. In water, the relative orientations of the two lone pairs and the two hydrogen atoms (Figure 21.12b) are set up so that when approximately *linear* O−H···O interactions occur, a 3-dimensional structure results. In ice, each oxygen atom is in an approximately tetrahedral environment (Figure 21.12c) with two O−H distances of 101 pm and two of 175 pm. The network structure is a very open one and when ice melts at 273 K, the density *increases* as some (but not all) of the hydrogen bonds are destroyed. The structure is similar to that of diamond (Figure 9.21).

**Fig. 21.12** (a) In the solid state of HF, the molecules are hydrogen bonded together to form zigzag chains; (b) the relative orientations of the two H atoms and the two lone pairs in a molecule of water are perfectly set up to generate a 3-dimensional network of hydrogen-bonded molecules in ice as shown in diagram (c). From: L. Pauling (1960) *The Nature of the Chemical Bond*, Cornell University Press, Ithaca, NY.

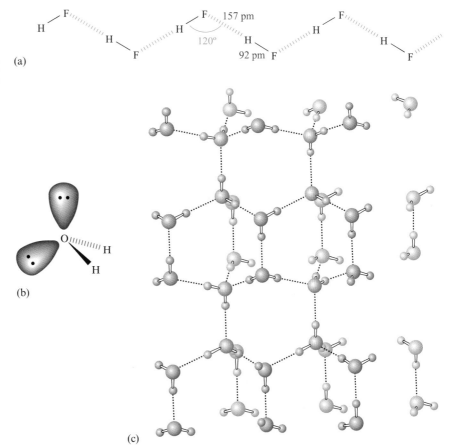

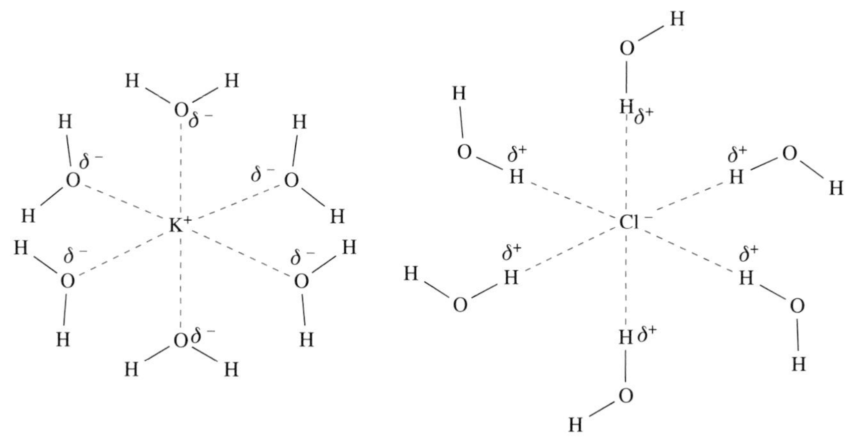

**Fig. 21.13** When solid KCl dissolves in water, the ions are freed from the crystal lattice and are hydrated. Ion–dipole interactions are responsible for the formation of hydration shells around each ion.

### Hydrogen bonding and solvation

*Solvation* is the intermolecular interaction between a *solute* and a *solvent* that leads to the solute dissolving. *Hydration* is the specific case in which the solvent is water.

Just as $H_2O$ or $NH_3$ molecules can associate with one another in liquid water or ammonia, so too can $H_2O$ *and* $NH_3$ molecules when ammonia dissolves in water. Some of the ammonia ionizes in the aqueous solution (equation 21.38) but most is present as *solvated* $NH_3$ molecules. Solvation is the intermolecular interaction between a solute and a solvent that leads to the solute dissolving, and *hydration* is the specific case in which the solvent is water. Many other liquids can be used as solvents, and non-aqueous solvents such as $NH_3$, HF, $N_2O_4$ and $SO_2$ have been widely investigated as they dissolve ionic solids.

The use of water as a solvent is widespread – think how many reactions are carried out in *aqueous solution*! When water behaves as a solvent, some hydrogen bonds between the $H_2O$ molecules must be broken, but new ion–dipole or dipole–dipole (electrostatic) interactions are generated as the solute is solvated. To illustrate this, we consider the solvation of potassium chloride in water. Figure 21.13 schematically illustrates the hydrated $K^+$ and $Cl^-$ ions in which ion–dipole interactions are indicated by the dotted pink lines. The water molecules shown make up the first hydration shell and hydrogen bonds are present between these $H_2O$ molecules and others in the bulk solvent.

Solubility of KCl: see Section 17.12

## 21.9    Hydrides of group 16 elements

### Water

Water is the most important hydride of oxygen and its properties colour our description of many other chemical properties. When we talk about substances being soluble or insoluble without qualification, we are usually referring to solubility in water. Water is the commonest liquid on Earth and possesses unique and remarkable properties. The strong hydrogen bonding network within liquid water ensures the large liquid range (mp 273 K, bp 373 K), and ultimately is responsible for the existence and maintenance of life on Earth.

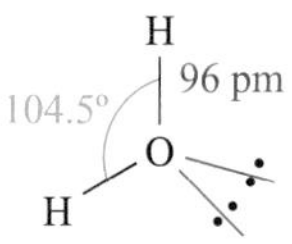

The structure of gas phase $H_2O$ is shown in **21.7**. The following topics about water have been covered earlier:

(21.7)

* structure and bonding (Sections 6.2, 6.6, 7.2, worked example 7.5);
* molecular dipole moment (Section 6.13);
* solution equilibria in aqueous solution and pH (Chapter 16).

## Hydrogen peroxide

Hydrogen peroxide, $H_2O_2$, is a blue, viscous liquid (mp 272 K, bp 425 K) but is usually encountered as an aqueous solution where the concentration is given in terms of a 'volume' or percentage. A 10 vol. aqueous solution of $H_2O_2$ means that $1 \, cm^3$ of the solution will liberate $10 \, cm^3$ of $O_2$ upon decomposition (see below); this corresponds to a 3% solution. Extensive hydrogen bonding occurs both between $H_2O_2$ molecules in the pure liquid, and between $H_2O_2$ and $H_2O$ molecules in aqueous solutions. A molecule of $H_2O_2$ (Figure 21.14) possesses a *skew conformation*, with a *torsion angle* (the angle between the planes as illustrated in Figure 21.14) of 90° in the solid state; the barrier to rotation about the O−O bond is low and in the gas phase the torsion angle increases to 111°.

Hydrogen peroxide is produced commercially by using compound **21.8** (a derivative of anthraquinone) in a two-step process involving reaction with $H_2$ (equation 21.51) followed by oxidation with $O_2$ (equation 21.52, see Box 21.7). Common uses of $H_2O_2$ are as an oxidizing agent, and include its roles as a bleach, antiseptic and in pollution control. The use of $H_2O_2$ as a bleach for hair has led to the expression 'peroxide blonde'.

**(21.8)**      $+ \; H_2$    $\xrightarrow{\text{Pd or Ni catalyst}}$    **(21.51)**

$+ \; O_2 \longrightarrow H_2O_2 \; +$    **(21.8)**    **(21.52)**

Hydrogen peroxide is *thermodynamically* unstable with respect to decomposition to water and $O_2$ (equation 21.53) but is *kinetically* stable. The reaction is catalysed by traces of $MnO_2$, $[OH]^-$ or some metal surfaces. Blood also catalyses this decomposition and this may be the basis of $H_2O_2$

**Fig. 21.14** The structure of $H_2O_2$. In the gas phase, the O−O bond distance is 147 pm, and the angles O−O−H are 95°; the angle shown is the *torsion angle*. In the solid state, the torsion angle is 90°; in the gas phase it is 111°.

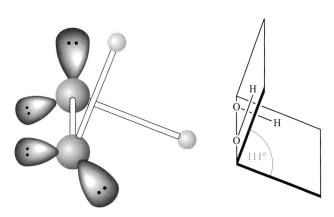

THEORETICAL AND CHEMICAL BACKGROUND

## Box 21.7 Catalytic cycles

Equations 21.51 and 21.52 together show the conversion of $H_2$ to $H_2O_2$ in the presence of compound **21.8** which acts as a catalyst. The two-step reaction can be represented in the *catalytic cycle* shown on the right.

Catalytic cycles of this type are a convenient way of illustrating the key steps of reactions involving a sequence of steps. They are commonly used to describe both industrial and biological processes.

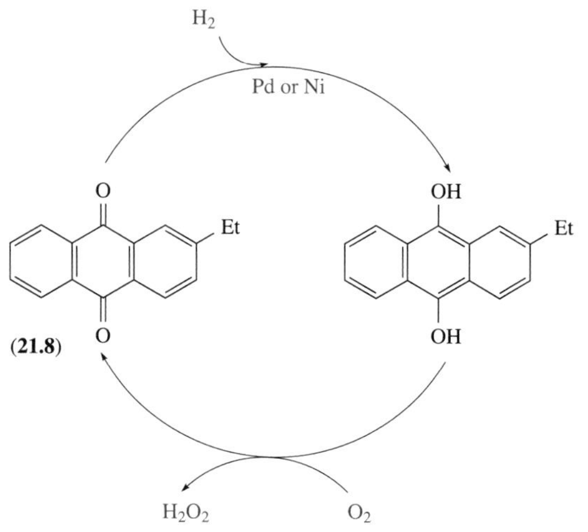

(**21.8**)

A species ***disproportionates*** if it undergoes simultaneous oxidation and reduction.

being used as an antiseptic as the $O_2$ released effectively kills anaerobic bacteria. Reaction 21.53 is a *disproportionation*. The oxidation state of oxygen in $H_2O_2$ is −1; in water, it is −2 and in $O_2$ it is zero. In reaction 21.53 the oxygen is simultaneously oxidized and reduced.

$$2H_2O_2(l) \longrightarrow 2H_2O(l) + O_2(g) \qquad (21.53)$$

reduction: $H_2O_2 \rightarrow H_2O$; oxidation: $H_2O_2 \rightarrow O_2$

Hydrogen peroxide can act as an oxidizing agent in both acidic and alkaline solutions (reactions 21.54 and 21.55). As soon as any $MnO_2$ is formed in reaction 21.55, it catalyses the decomposition of $H_2O_2$, and so the reaction shown is not very efficient.

$$2Fe^{2+} + H_2O_2 + 2H^+ \longrightarrow 2Fe^{3+} + 2H_2O \qquad \textit{Acidic solution} \qquad (21.54)$$

oxidation: $Fe^{2+} \rightarrow Fe^{3+}$; reduction: $H_2O_2 \rightarrow H_2O$

$$Mn(OH)_2 + H_2O_2 \longrightarrow MnO_2 + 2H_2O \qquad \textit{Alkaline solution} \qquad (21.55)$$

oxidation: $Mn(OH)_2 \rightarrow MnO_2$; reduction: $H_2O_2 \rightarrow H_2O$

Hydrogen peroxide can also act as a reducing agent in both acidic and alkaline solutions, but it will only reduce species that are themselves strong oxidizing agents (equations 21.56 and 21.57).

$$2[MnO_4]^- + 5H_2O_2 + 6H^+ \longrightarrow 2Mn^{2+} + 8H_2O + 5O_2 \qquad \textit{Acidic solution} \qquad (21.56)$$

reduction: $[MnO_4]^- \rightarrow Mn^{2+}$; oxidation: $H_2O_2 \rightarrow O_2$

$$Cl_2 + H_2O_2 + 2OH^- \longrightarrow 2Cl^- + 2H_2O + O_2 \qquad \textit{Alkaline solution} \qquad (21.57)$$

reduction

oxidation

The dissociation of aqueous $H_2O_2$ (equation 21.58) occurs to a greater extent than does the self-ionization of water, and $H_2O_2$ is a slightly stronger acid than $H_2O$ (equation 21.59).

The $[HO_2]^-$ ion is also called hydrogendioxide(1–)

$$H_2O_2(aq) + H_2O(l) \rightleftharpoons [H_3O]^+(aq) + [HO_2]^-(aq) \qquad (21.58)$$

Hydroperoxide ion

$$K_a = \frac{[H_3O^+][HO_2^-]}{[H_2O_2]} = 1.78 \times 10^{-12} \qquad pK_a = 11.75 \qquad (21.59)$$

## Hydrogen sulfide

The hydrides of sulfur are called *sulfanes* but for $H_2S$, the name hydrogen sulfide is usually used. Hydrogen sulfide is a colourless gas (bp 214 K) with a characteristic smell of bad eggs. It is extremely toxic. The fact that it anaesthetizes the sense of smell means that the intensity of smell is not a reliable measure of the amount of $H_2S$ present.

Hydrogen sulfide is a natural product of decaying sulfur-containing matter, and it occurs in coal pits, gas wells and sulfur springs. Historically, laboratory preparations by reaction 21.60 used a Kipp's apparatus, which was designed so as to minimize the escape of gaseous $H_2S$. The hydrolysis of calcium sulfide (equation 21.61) produces purer $H_2S$.

$$FeS(s) + 2HCl(aq) \longrightarrow H_2S(g) + FeCl_2(aq) \qquad (21.60)$$

$$CaS + 2H_2O \longrightarrow H_2S + Ca(OH)_2 \qquad (21.61)$$

---

**ENVIRONMENT AND BIOLOGY**

### Box 21.8 $H_2S$: Toxic or life-supporting?

Anemone *Actinauge abyssorum* on glass sponge stalk, 4100 m deep off California. © *Ken Smith Laboratory @ Scripps Institute of Oceanography/Oxford Scientific/photolibrary.com.*

Hydrothermal fluids are discharged from volcanic vents on the ocean floor, and around these vents there are large quantities of $H_2S$ but virtually no $O_2$.

Hydrogen sulfide is normally considered too toxic to support life, and the lack of $O_2$ should also pose a critical problem. Nature, however, has adapted to the environment and many new species of animals, including certain clams and mussels, have been discovered only at these volcanic vent sites. Light does not penetrate to such ocean depths, and food chains that depend upon photosynthesis cannot operate. Research has shown that the vent communities use geothermal energy (rather than solar energy), and the reaction that starts the food chain may be represented as:

$$H_2S + CO_2 \longrightarrow [carbohydrate] + H_2SO_4$$

from the    from
volcanic    seawater
vent

This is a case of one animal's poison being the stuff of life to another!

The presence of $H_2S$ can be confirmed by its reaction with lead acetate. Filter paper dipped in aqueous $Pb(O_2CCH_3)_2$, and then dried, turns black when it comes into contact with $H_2S$ (equation 21.62).

*Qualitative test for gaseous $H_2S$:*

$$H_2S(g) + Pb(O_2CCH_3)_2 \longrightarrow PbS(s) + 2CH_3CO_2H \qquad (21.62)$$
Colourless        Black

**(21.9)**

The bent structure of $H_2S$ **(21.9)** is consistent with VSEPR theory, *but* the H–S–H bond angle is much smaller than would be predicted. In the liquid state, hydrogen bonding is not very important as we saw from the trends in boiling points in Figure 21.11.

Hydrogen sulfide is slightly soluble in water; a saturated solution at 298 K and atmospheric pressure has a concentration of $\approx 0.1 \, mol \, dm^{-3}$. Aqueous solutions of $H_2S$ are weakly acidic (equation 21.63) but the extremely small value of the second dissociation constant (equation 21.64) means that this second equilibrium lies to the left-hand side. A consequence of this is that *soluble* metal sulfides ($Na_2S$, $K_2S$, $[NH_4]_2S$ and $CaS$) are readily hydrolysed (equation 21.65).

> Only a few metal sulfides are soluble in water: $Na_2S$, $K_2S$, $[NH_4]_2S$, $CaS$.

$$H_2S + H_2O \rightleftharpoons [H_3O]^+ + [SH]^- \qquad pK_a = 7.04 \qquad (21.63)$$

$$[SH]^- + H_2O \rightleftharpoons [H_3O]^+ + S^{2-} \qquad pK_a \approx 19 \qquad (21.64)$$

$$Na_2S + H_2O \rightleftharpoons NaSH + NaOH \qquad (21.65)$$

Hydrogen sulfide burns in air with a blue flame and is oxidized to sulfur dioxide or sulfur depending upon the supply of $O_2$ (equations 21.66 and 21.67).

*Excess air/$O_2$:*        $$2H_2S + 3O_2 \longrightarrow 2SO_2 + 2H_2O \qquad (21.66)$$

*Limited supply of air/$O_2$:*    $$2H_2S + O_2 \longrightarrow 2S + 2H_2O \qquad (21.67)$$

In acidic conditions, $H_2S$ is a mild reducing agent and may be oxidized to sulfur(0) (equation 21.68) or to higher oxidation states.

$$2Fe^{3+} + H_2S \longrightarrow 2Fe^{2+} + 2H^+ + \tfrac{1}{8}S_8 \qquad (21.68)$$

### Hydrogen selenide and hydrogen telluride

| | E = Se | E = Te |
|---|---|---|
| $d$ / pm | 146 | 169 |
| $\alpha$ / deg | 91 | 90 |

**(21.10)**

Hydrogen selenide ($H_2Se$) and hydrogen telluride ($H_2Te$) are colourless, foul smelling and extremely toxic gases. $H_2Se$ is a dangerous fire hazard – on exposure to $O_2$ it rapidly decomposes to give red selenium. $H_2Te$ decomposes in air or water or when heated. Molecules of $H_2Se$ and $H_2Te$ are bent but, like $H_2S$, have smaller bond angles than expected from VSEPR theory. Bond parameters for the gas phase molecules are given in **21.10**.

Hydrogen selenide may be prepared from its constituent elements (equation 21.69) but the thermal instability of $H_2Te$ means that it cannot be prepared by an analogous reaction. Both $H_2Se$ and $H_2Te$ can be prepared by hydrolysing the appropriate aluminium chalcogenide (equation 21.70).

$$H_2 + Se \xrightarrow{630 \, K} H_2Se \qquad (21.69)$$

$$Al_2E_3 + 6H_2O \longrightarrow 3H_2E + 2Al(OH)_3 \qquad (21.70)$$
E = Se or Te

<table>
<tr><td>**21.10**</td><td>## Binary compounds containing hydrogen and group 17 elements: hydrogen halides</td></tr>
</table>

Elements in group 17 form *hydrogen halides* of general formula HX (X = F, Cl, Br or I) which dissolve in water to give acidic solutions called *hydrohalic acids*.

### Hydrogen fluoride

Radical chain reactions: see Section 15.14

Hydrogen fluoride, HF, is produced in an explosive radical chain reaction when $H_2$ and $F_2$ are mixed (equation 21.71). Reaction 21.72 is a more convenient method of preparation.

$$H_2(g) + F_2(g) \longrightarrow 2HF(g) \tag{21.71}$$

$$CaF_2(s) + 2H_2SO_4(aq) \longrightarrow 2HF(g) + Ca(HSO_4)_2(aq) \tag{21.72}$$
$$\text{Conc}$$

A problem of working with HF in the laboratory is its ability to etch silica glass (equation 21.73) which corrodes glass reaction vessels. On the other hand, the same reaction is used commercially for etching patterns on glass. Monel metal (a nickel alloy) or polytetrafluoroethene (PTFE) containers are suitable for storing and handling HF.

$$4HF + SiO_2 \longrightarrow SiF_4 + 2H_2O$$

$$\bigg\downarrow 2HF(aq) \tag{21.73}$$

$$H_2SiF_6$$

We have already described the role of hydrogen bonding in the solid state structure of hydrogen fluoride (Figure 21.12). Hydrogen fluoride has a long liquid range (mp 190 K, bp 293 K) and at room temperature is a colourless, fuming and corrosive liquid or a gas depending upon the ambient conditions. Extensive hydrogen bonding exists in the liquid and this contributes to its low volatility; even in the vapour state, some intermolecular interactions persist, giving species of formulae $(HF)_x$ ($x \leq 6$). Although the ability of HF to react with glass colours our view of the compound, it is in many respects a hydrogen-bonded liquid like $H_2O$. For example, the protein insulin may be recovered unchanged after dissolution in pure liquid HF.

Hydrogen fluoride is completely miscible with water but its solution chemistry is complex. For the dissociation in equation 21.74, $pK_a$ is 3.45 (at 298 K). This indicates that HF is a weaker acid than HCl, HBr or HI; a contributing factor is the high bond dissociation enthalpy of the H−F bond (Figure 21.15). Equilibrium 21.74 is complicated by the interaction of fluoride ions and hydrogen fluoride (equation 21.75). By applying Le Chatelier's principle, we see that if $F^-$ produced in reaction 21.74 is removed from the system by reaction with HF, more $F^-$ (and, necessarily, more $[H_3O]^+$) forms.

$$HF(l) + H_2O(l) \rightleftharpoons [H_3O]^+(aq) + F^-(aq) \tag{21.74}$$

$$F^-(aq) + HF(aq) \rightleftharpoons [HF_2]^-(aq) \qquad K = \frac{[HF_2^-]}{[HF][F^-]} = 0.2 \tag{21.75}$$

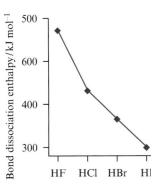

**Fig. 21.15** The trend in bond dissociation enthalpies for the group 17 hydrides.

X-ray and neutron diffraction: see Section 4.2

$$\left[ F\!-\!\!-\!\!-H\!-\!\!-\!\!-F \right]^{-}$$

**(21.11)**

The formation of $[HF_2]^-$ depends on the ability of H and F to be involved in strong hydrogen bonding. Diffraction (X-ray and neutron) studies of $K[HF_2]$ confirm that the $[HF_2]^-$ anion is linear (**21.11**) with an F⋯F distance of 226 pm. Compare this with twice the H−F bond length in HF ($2 \times 92$ pm), and we see that the hydrogen bonding in $[HF_2]^-$ is very strong. Note that an H⋯F hydrogen bond will always be weaker and longer than a 2-centre covalent H−F bond.

Anhydrous liquid HF is self-ionizing (equation 21.76) and is used as a non-aqueous solvent.

$$3HF(l) \rightleftharpoons [H_2F]^+(solv) + [HF_2]^-(solv) \tag{21.76}$$

## Hydrogen chloride

See Figure 1.7

Hydrogen chloride, HCl, is a poisonous, colourless gas (bp 188 K) and can be detected by contact with $NH_3$ (equation 21.77). It is conveniently prepared by the action of concentrated $H_2SO_4$ on solid NaCl (equation 21.78).

$$\underset{\text{Colourless gas}}{HCl(g)} + \underset{\text{Colourless gas}}{NH_3(g)} \longrightarrow \underset{\text{Fine, white powder}}{NH_4Cl(s)} \tag{21.77}$$

$$NaCl(s) + \underset{\text{Conc}}{H_2SO_4} \longrightarrow HCl(g) + NaHSO_4(s) \tag{21.78}$$

Hydrogen chloride is also formed from the reaction of $H_2$ and $Cl_2$. This reaction is by no means as vigorous as that between $H_2$ and $F_2$; the standard enthalpies of formation of gaseous HF and HCl are $-273$ and $-92\,\text{kJ mol}^{-1}$ respectively. On a commercial scale, reaction 21.78 may be used to make HCl, but the reaction of $H_2$ with $Cl_2$ is preferred. Hydrogen chloride has a wide range of uses, not only in the chemical laboratory as a gas or an aqueous acid (see below), but also in the manufacture of a wide range of inorganic and organic chemicals.

In liquid HCl, the intermolecular association through hydrogen bonding is less than in HF, and the degree of self-ionization is small. The zigzag chain structure of solid HCl is similar to that of HF (see Figure 21.12) and is due to hydrogen bonding.

Some reactions involving HCl (e.g. additions to unsaturated hydrocarbons) are described in Chapter 26. Hydrogen chloride oxidizes some metals and the electrochemical series (Table 21.1) can be used to predict which metal chlorides can be formed in this way (equations 21.79 and 21.80).

$$Mg(s) + 2HCl(g) \xrightarrow{\Delta} MgCl_2(s) + H_2(g) \tag{21.79}$$

$$Fe(s) + 2HCl(g) \xrightarrow{\Delta} FeCl_2(s) + H_2(g) \tag{21.80}$$

Hydrogen chloride fumes in moist air and dissolves in water to give a strongly acidic solution (hydrochloric acid) in which it is fully ionized. The chloride ion is precipitated as white AgCl when silver(I) nitrate is added (equation 21.81) and this is a standard qualitative test for the presence of free chloride ions in aqueous solution (equation 21.82).

$$AgNO_3(aq) + HCl(aq) \longrightarrow AgCl(s) + HNO_3(aq) \tag{21.81}$$

*Qualitative test for Cl⁻ ions:*

$$AgNO_3(aq) + Cl^-(aq) \longrightarrow \underset{\text{White precipitate}}{AgCl(s)} + [NO_3]^-(aq) \tag{21.82}$$

Displacement of $H_2$ from hydrochloric acid occurs with metals above hydrogen in the electrochemical series. These reactions are analogous to the oxidations shown in equations 21.79 and 21.80 but involve aqueous HCl (equation 21.83).

$$Mg(s) + 2HCl(aq) \longrightarrow MgCl_2(aq) + H_2(g) \tag{21.83}$$

Hydrochloric acid neutralizes bases (equations 21.84 and 21.85) and reacts with metal carbonates (equation 21.86) – *these are typical reactions of dilute aqueous acids.*

$$HCl(aq) + NaOH(aq) \longrightarrow NaCl(aq) + H_2O(l) \tag{21.84}$$

$$2HCl(aq) + CuO(s) \longrightarrow CuCl_2(aq) + H_2O(l) \tag{21.85}$$

$$2HCl(aq) + MgCO_3(s) \longrightarrow MgCl_2(aq) + H_2O(l) + CO_2(g) \tag{21.86}$$

The above reactions all occur with *dilute* hydrochloric acid, typically $2 \, mol \, dm^{-3}$. *Concentrated* hydrochloric acid is oxidized by $KMnO_4$ or $MnO_2$, to produce $Cl_2$ (equations 21.87 and 21.88) and the former is a convenient synthesis of $Cl_2$.

$$\underset{\text{Conc}}{16HCl} + 2KMnO_4(s) \longrightarrow 5Cl_2(g) + 2KCl(aq) + 2MnCl_2(aq) + 8H_2O(l) \tag{21.87}$$

oxidation

reduction

$$\underset{\text{Conc}}{4HCl} + MnO_2(s) \xrightarrow{\Delta} Cl_2(g) + MnCl_2(aq) + 2H_2O(l) \tag{21.88}$$

oxidation

reduction

### Hydrogen bromide and hydrogen iodide

Both HBr and HI are choking colourless gases (bp 206 and 238 K respectively). They cannot be prepared in reactions analogous to reaction 21.78 because both HBr and HI are oxidized by $H_2SO_4$. Instead, phosphoric acid may be used (equation 21.89) or the reaction of red phosphorus with the respective halogen and water (equation 21.90). $H_2$ reacts with $Br_2$ or $I_2$ to give HBr or HI respectively, but high temperatures and a metal catalyst are needed; compare this with the explosive combination of $H_2$ and $F_2$.

$$KX + H_3PO_4 \xrightarrow{\Delta} HX + KH_2PO_4 \tag{21.89}$$
X = Br or I

$$3X_2 + 2P + 6H_2O \longrightarrow 6HX + 2H_3PO_3 \tag{21.90}$$
X = Br or I

Both HBr and HI are soluble in water and are fully ionized; hydrobromic and hydroiodic acids are strong acids. The bromide or iodide can be precipitated by adding silver nitrate. Solid AgBr is cream-coloured and AgI is pale yellow. These precipitations are used in qualitative analysis for $Br^-$ and $I^-$ ions. Both acids undergo similar reactions to hydrochloric acid; HBr and HI are easily oxidized as exemplified by reactions 21.91–21.93.

$$2Cu^{2+} + 4I^- \longrightarrow 2CuI + I_2 \tag{21.91}$$

$$2X^- + Cl_2 \longrightarrow X_2 + 2Cl^- \tag{21.92}$$
X = I or Br

$$2I^- + Br_2 \longrightarrow I_2 + 2Br^- \tag{21.93}$$

| 21.11 | **Group 1: the alkali metals** |
|---|---|

We have already discussed several aspects of the group 1 metals:

- trends in ionization energies (Section 8.4);
- NaCl structure (Sections 8.8 and 8.9);
- CsCl structure (Section 8.10);
- ionic radii (Table 8.6);
- solid state structures (Section 9.11);
- melting points (Table 9.3);
- metallic radii (Table 9.4);
- metallic bonding (Section 9.13).

The chemistry of the group 1 metals is dominated by the formation of singly charged cations. The ionization energies relating to equation 21.94 are given in Appendix 8.

$$M(g) \longrightarrow M^+(g) + e^- \tag{21.94}$$

Diagonal relationship: see Section 21.13

In many of its properties, lithium is atypical and bears a resemblance to magnesium to which it is *diagonally* related in the periodic table.

### Appearance, physical properties, sources and uses

'Caesium' is spelt 'cesium' in the US and American-speaking countries

Each of the elements lithium, sodium, potassium, rubidium and caesium is a soft, silver-grey solid metal at 298 K. However, the particularly low melting point of caesium (301.5 K, Table 9.3) means that at ambient temperatures, it may be a liquid. Francium (named after France, the country in which it was discovered in 1939) is a radioactive element and only minute quantities of it have ever been handled. Their high reactivities mean that the group 1 metals do not occur *native*, i.e. in the elemental state.

Lithium is produced by electrolysing LiCl in a manner analogous to the Downs process (see below). It has the lowest density ($0.53 \, \text{g cm}^{-3}$) of all the metals in the periodic table.

Sodium is the sixth most abundant element on Earth (in the form of compounds such as NaCl) and the most abundant of the group 1 metals. Compounds of sodium have many applications, including uses in the paper, glass, detergent, chemical and metal industries. Both sodium and potassium are biologically important. They are involved, for example, in osmotic control and the body's nervous system. Sodium is manufactured in the Downs process in which molten NaCl is electrolysed (Figure 21.16). $CaCl_2$ is added to reduce the operating temperature to about 870 K; pure NaCl melts at 1073 K. Molten NaCl is composed of free $Na^+$ and $Cl^-$ ions. Reduction of the $Na^+$ ions to form Na (equation 21.95) occurs at the cathode while $Cl^-$ ions are oxidized at the anode (equation 21.96). The net result is reaction 21.97, and the molten Na passes out of the cell where it is cooled and solidifies. The design of the electrolysis cell is critical; NaCl will reform if the metallic Na and gaseous $Cl_2$ come into contact with each other.

$$\textit{At the cathode:} \quad Na^+(l) + e^- \longrightarrow Na(l) \tag{21.95}$$

$$\textit{At the anode:} \quad 2Cl^-(l) \longrightarrow Cl_2(g) + 2e^- \tag{21.96}$$

**Fig. 21.16** A schematic representation of the electrolysis cell used in the Downs process to produce sodium commercially from NaCl. The CaCl$_2$ is present to lower the working temperature from 1073 K (the mp of NaCl) to $\approx$870 K. The Na and Cl$_2$ must be kept separate to prevent reformation of NaCl.

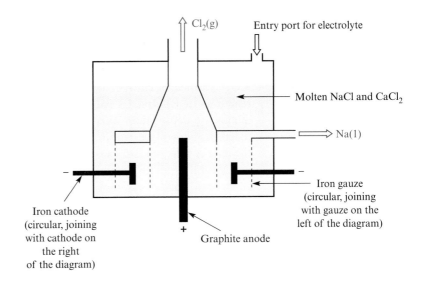

$$\textit{Overall reaction}:\quad 2Na^+(l) + 2Cl^-(l) \longrightarrow 2Na(l) + Cl_2(g) \tag{21.97}$$

Potassium is almost as abundant as sodium in the Earth's crust (2.4% versus 2.6%). Potassium salts are widely used as fertilizers, potassium being an essential plant nutrient. The industrial production of potassium follows reaction 21.98. Rubidium can be prepared similarly by reducing RbCl (equation 21.99).

$$KCl(l) + Na(g) \xrightarrow{1000\,K} K(g) + NaCl(s) \tag{21.98}$$

$$2RbCl + Ca \xrightarrow{\Delta} 2Rb + CaCl_2 \tag{21.99}$$

> Almost all salts of the group 1 metals are soluble in water.

*Almost all salts of the group 1 metals are soluble in water.* Caesium salts have found some applications in organic syntheses, and the large Cs$^+$ cation is often used to give good crystals with large anions.

## Alkali metal hydroxides

Sodium hydroxide is an important industrial chemical and Figure 21.17 shows some of its uses.

The alkali metals react exothermically with water (equation 21.100) and the reactivity *increases* significantly as the group is descended.

$$2M + 2H_2O \longrightarrow 2MOH + H_2 \qquad M = Li, Na, K, Rb \text{ or } Cs \tag{21.100}$$

**Fig. 21.17** Industrial uses of sodium hydroxide in Western Europe in 1994. (Data from *Chemistry & Industry* (1995), p. 832.)

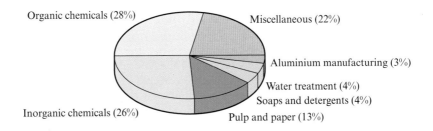

Lithium, sodium and potassium are less dense than water and therefore reaction 21.100 takes place on the surface of the liquid. In the case of potassium, the reaction is violent enough to ignite the $H_2$ produced. This ignition is an explosive radical chain process, and coupled with the fact that the aqueous KOH formed is caustic, the reaction must be conducted with care. The reactions of Rb and Cs with water are even more spectacular. Both these metals are denser than water (1.53 and $1.84\,g\,cm^{-3}$ respectively) and so the reaction takes place within the bulk liquid. As ignition of the $H_2$ occurs, the energy released can cause violent explosions.

Lithium, sodium, potassium and rubidium hydroxides are white, crystalline solids; caesium hydroxide is pale yellow. The crystals are *deliquescent*: they absorb water from the surrounding air and finally become liquid. The solubility of the group 1 hydroxides in water tends to increase as the group is descended. The hydroxides of the metals sodium to caesium dissociate fully (equation 21.101) and the presence of free $[OH]^-$ ions means that these aqueous solutions are strongly basic.

**◼▸**
Radical chain reaction: see Section 15.14

A *deliquescent* substance absorbs water from the surrounding air and eventually forms a liquid.

$$MOH(s) \xrightarrow{\text{Water}} M^+(aq) + [OH]^-(aq) \qquad (21.101)$$

## Alkali metal oxides, peroxides and superoxides

Alkali metals are often stored in paraffin oil to prevent reaction with atmospheric $O_2$ and water vapour. A piece of freshly cut Na or K is shiny but quickly tarnishes as it comes into contact with the air.

The combustion of the group 1 metals in $O_2$ leads to various oxides depending on the metal and conditions. Lithium reacts to give lithium oxide, $Li_2O$ (equation 21.102), but Na usually forms sodium peroxide, $Na_2O_2$, which contains the diamagnetic $[O_2]^{2-}$ ion (equation 21.103). By increasing the pressure or temperature, $NaO_2$ may be produced (equation 21.104) and this contains the paramagnetic superoxide anion, $[O_2]^-$.

$$4Li + O_2 \longrightarrow 2Li_2O \qquad (21.102)$$
$$\text{Lithium oxide}$$

---

**THEORETICAL AND CHEMICAL BACKGROUND**

### Box 21.9 Suboxides of rubidium and caesium

When oxidized under *controlled* conditions, rubidium and caesium form *suboxides* such as $Rb_6O$, $Cs_7O$, $Rb_9O_2$ and $Cs_{11}O_3$.

An interesting question arises: 'Do the usual oxidation states of +1 for the alkali metal and −2 for the oxygen apply in these compounds?' If we assign an oxidation state of −2 to the oxygen in $Rb_6O$, then the *formal* oxidation state of rubidium is $+\frac{1}{3}$. However, the formula is more usefully considered in terms of $Rb^+$ and $O^{2-}$ ions and $Rb_6O$ may be written as $(Rb^+)_6(O^{2-})\cdot4e^-$.

The structures of $Rb_6O$, $Cs_7O$, $Rb_9O_2$ and $Cs_{11}O_3$ all feature octahedral units of metal ions with an oxide ion at the centre of each octahedron. The octahedra are fused together by sharing faces. The structure of $Cs_{11}O_3$ is shown on the right.

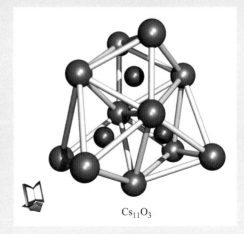

$Cs_{11}O_3$

$$2Na + O_2 \longrightarrow \underset{\text{Sodium peroxide}}{Na_2O_2} \tag{21.103}$$

$$Na + O_2 \longrightarrow \underset{\text{Sodium superoxide}}{NaO_2} \tag{21.104}$$

Although under varying conditions all three oxides can be prepared for each of the five metals, the tendency to form the peroxide and the superoxide increases down group 1 with the $[O_2]^{2-}$ and $[O_2]^-$ anions being increasingly stabilized by the larger metal cations $K^+$, $Rb^+$ and $Cs^+$. All the superoxides are, however, relatively unstable. The colours of the compounds vary from white to orange, following a general trend as the group is descended: $Li_2O$ and $Na_2O$ form white crystals, $K_2O$ is pale yellow, $Rb_2O$ is yellow and $Cs_2O$ is orange.

*Trend in ionic radii: see Figure 8.22*

Potassium superoxide is used in breathing masks; $KO_2$ absorbs water and in so doing produces both $O_2$ for respiration and KOH which absorbs exhaled $CO_2$. All the potassium-containing components in equations 21.105 and 21.106 remain in the solid phase and are contained effectively in the breathing mask.

$$4KO_2(s) + 2H_2O(l) \longrightarrow 4KOH(s) + 3O_2(g) \tag{21.105}$$

$$KOH(s) + CO_2(g) \longrightarrow KHCO_3(s) \tag{21.106}$$

### Alkali metal halides

Each group 1 metal reacts with $X_2$ ($X = F$, Cl, Br or I) when heated (equation 21.107).

$$2M + X_2 \xrightarrow{\Delta} 2MX \tag{21.107}$$

The alkali metal halides are white, crystalline high-melting solids with ionic lattices. Sodium chloride, an industrially important chemical, occurs naturally as rock salt and in seawater.

### Alkali metals as reducing agents

When an alkali metal reacts with water or $O_2$, it is oxidized. Conversely, the metal behaves as a reducing agent, reducing the hydrogen in $H_2O$ to $H_2$, or reducing the oxygen in $O_2$ to $O^{2-}$, $[O_2]^-$ or $[O_2]^{2-}$. Alkali metals are powerful reducing agents and this property is associated with the ease with which each metal loses an electron: the more thermodynamically favourable is this loss, the more powerful the reducing ability of the metal. Equations 21.108 to 21.111 illustrate the role of alkali metals as reducing agents, and in Chapter 26 we describe how alkali metals can be used to reduce alkynes. The reducing power can be increased by dissolving the alkali metal in liquid ammonia as described below.

$$2Na + Cl_2 \longrightarrow 2NaCl \tag{21.108}$$

$$2Cs + F_2 \longrightarrow 2CsF \tag{21.109}$$

$$4Na + ZrCl_4 \longrightarrow 4NaCl + Zr \tag{21.110}$$

$$2Na + 2C_2H_5OH \longrightarrow \underset{\text{Sodium ethoxide}}{2Na^+[C_2H_5O]^-} + H_2 \tag{21.111}$$

## Alkali metals in liquid ammonia

Alkali metals dissolve in dry liquid ammonia to give coloured solutions. At low concentrations, these are blue and paramagnetic and have high electrical conductivities, but at higher concentrations, bronze coloured 'metallic' solutions are formed in which a variety of aggregates are present. These properties are caused by the ionization of the metal (equation 21.112) and a factor that contributes to this unusual phenomenon is the low value of the first ionization potential of each group 1 metal.

$$Na(s) + NH_3(l) \longrightarrow Na^+ \text{ (solvated by liquid } NH_3)$$
$$+ e^- \text{(solvated by liquid } NH_3) \qquad (21.112)$$

The electron removed from the Na atom is solvated by the $NH_3$ molecules but can be considered to be a 'free electron'. Hence, solutions of alkali metals in liquid $NH_3$ are powerful reducing agents (equations 21.113 to 21.115). The alkali metal is oxidized to the corresponding metal ion and this becomes the counter-ion for the reduced partner in the reaction, e.g. in reaction 21.115, the product is isolated as $Na_2[Fe(CO)_4]$.

$$O_2 \xrightarrow{\text{M in liquid } NH_3} [O_2]^- \qquad (21.113)$$

$$[MnO_4]^- \xrightarrow{\text{M in liquid } NH_3} [MnO_4]^{2-} \qquad (21.114)$$

$$[Fe(CO)_5] \xrightarrow{\text{Na in liquid } NH_3} [Fe(CO)_4]^{2-} \qquad (21.115)$$

Solutions of an alkali metal in ammonia are unstable with respect to the formation of a metal amide and $H_2$ (equation 21.116). The reaction is slow but is catalysed by some metals (e.g. iron).

$$2Na + 2NH_3 \longrightarrow \underset{\text{Sodium amide}}{2NaNH_2} + H_2 \qquad (21.116)$$

**21.12**    ## The group 2 metals

Each of the group 2 metals has two electrons in the valence shell and most of their chemistry is that of the $M^{2+}$ ion. Beryllium stands out from the group because it exhibits significant covalency in its compounds. Calcium, strontium, barium and radium are collectively called the *alkaline earth metals*.

### Appearance, physical properties, sources and uses

Some physical properties of the group 2 metals are listed in Table 21.3; with the exception of barium, these metallic elements have close-packed lattices in the solid state (see Table 9.3).

Metal lattices:
see Section 9.11

Beryllium is relatively rare but it is one of the lightest metals known and possesses one of the highest melting points. It is non-magnetic, has a high thermal conductivity, and is resistant to both attack by concentrated nitric acid and oxidation in air at 298 K. These properties contribute to its being of great industrial importance, and beryllium is used in the manufacture of body-parts in high-speed aircraft and missiles, and in communication satellites; it is also used in nuclear reactors as a moderator and a reflector.

**Table 21.3** Selected physical properties of the alkaline earth metals.

| Element (symbol) | Physical appearance | Melting point / K | Boiling point / K | Density / g cm$^{-3}$ | Notes |
|---|---|---|---|---|---|
| Beryllium (Be) | Steel-grey metal | 1560 | 2744 | 1.85 | Beryllium is extremely toxic |
| Magnesium (Mg) | Silver-white metal | 923 | 1363 | 1.74 | |
| Calcium (Ca) | Silver coloured metal | 1115 | 1757 | 1.55 | |
| Strontium (Sr) | Silver coloured metal | 1050 | 1655 | 2.58 | |
| Barium (Ba) | Silver-white metal | 1000 | 2170 | 3.50 | Soluble compounds are toxic |
| Radium (Ra) | White metal | 973 | 1413 | 5.00 | Radioactive |

X-rays: see Section 4.2

Since each beryllium atom has only four electrons, it allows X-rays to pass through virtually unperturbed and is used in the windows of X-ray tubes. Beryllium is found in many natural minerals; emerald and aquamarine, which are two precious forms of the mineral *beryl* (the mixed oxide $3BeO \cdot Al_2O_3 \cdot 6SiO_2$), are used in jewellery. Despite the widespread technical uses of beryllium, care must be taken when handling the metal as it is extremely toxic. The element can be prepared by reducing $BeF_2$ (equation 21.117) or by the electrolysis of molten $BeCl_2$.

$$BeF_2 + Mg \longrightarrow MgF_2 + Be \qquad (21.117)$$

Magnesium is the eighth most abundant element in the Earth's crust but it is not present in the elemental state. Two of the major sources are the minerals dolomite (the mixed carbonate $CaCO_3 \cdot MgCO_3$) and magnesite ($MgCO_3$). It is also present in seawater as magnesium(II) salts. Figure 21.18 summarizes the extraction of magnesium from seawater. It begins with the precipitation of $Mg^{2+}$ ions in the form of magnesium hydroxide, the hydroxide ions being provided by adding $Ca(OH)_2$ (slaked lime) which itself is formed from calcium carbonate, which is widely available as limestone, sea shells and

**Fig. 21.18** A summary of the industrial process used to extract magnesium from seawater.

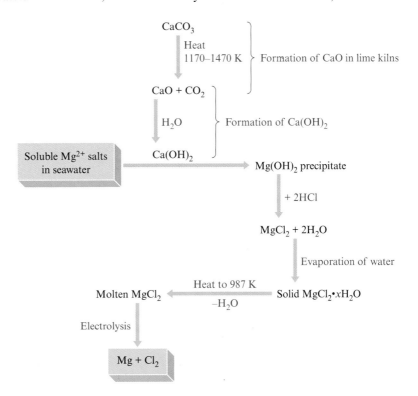

other calcareous deposits. The $Mg(OH)_2$ is neutralized with hydrochloric acid (equation 21.118) and the solid $MgCl_2$ deposited after the water has been evaporated is melted and electrolysed (equations 21.119 and 21.120).

$$2HCl(aq) + Mg(OH)_2(s) \longrightarrow MgCl_2(aq) + 2H_2O(l) \qquad (21.118)$$

*At the cathode:* $\qquad Mg^{2+}(l) + 2e^- \longrightarrow Mg(l) \qquad (21.119)$

*At the anode:* $\qquad 2Cl^-(l) \longrightarrow Cl_2(g) + 2e^- \qquad (21.120)$

Large-scale production of Mg is required to meet the commercial demands of this metal. Its high reactivity is responsible for its uses in photographic flashlights, flares and fireworks, and its low density (Table 21.3) makes it invaluable as a component in alloys. The presence of Mg in an Mg/Al alloy improves the mechanical strength, welding and fabrication properties of the material, and increases its resistance to corrosion. The properties of a particular alloy depend upon the ratio of Mg : Al and alloys have wide uses in aircraft body-parts, missiles and lightweight tools and containers.

> An *alloy* is an intimate mixture of two or more metals, or metals and non-metals. The aim of alloying is to improve the physical properties and resistance to corrosion, heat, etc. of the material.

Magnesium plays an important biological role: it is involved in phosphate metabolism and is present in chlorophyll. Medical uses of magnesium salts include indigestion powders ('milk of magnesia', $Mg(OH)_2$) and medication for constipation ('Epsom salts', $MgSO_4$).

Chlorophyll: see Box 23.3

Calcium is the fifth most abundant element in the Earth's crust and occurs in limestone ($CaCO_3$), gypsum ($CaSO_4 \cdot 2H_2O$) and fluorite ($CaF_2$), but not in the element state. The metal may be produced by electrolysing molten $CaCl_2$ (reduction of $Ca^{2+}$ occurs at the cathode) or by the reduction of CaO with Al (equation 21.121). The uses of Ca metal are not as widespread as those of its compounds.

$$3CaO + 2Al \xrightarrow{\Delta} 3Ca + Al_2O_3 \qquad (21.121)$$

Strontium and barium both occur as the sulfate and carbonate: $SrSO_4$ (celestite), $SrCO_3$ (strontianite), $BaSO_4$ (barite) and $BaCO_3$ (witherite). The oxides of these elements are reduced by Al to produce Sr and Ba respectively in reactions analogous to equation 21.121. Barium is used as a 'getter' in vacuum tubes; the high reactivity of the metal means that it combines with gaseous impurities such as $O_2$ and $N_2$ (equations 21.122 and 21.123) and the effect is to remove residual traces of such gases to give a good vacuum.

$$2Ba + O_2 \longrightarrow 2BaO \qquad (21.122)$$

$$3Ba + N_2 \longrightarrow Ba_3N_2 \qquad (21.123)$$

Barium sulfate is a white compound and is insoluble in water; it is used in paints, in X-ray diagnostic work ('barium meals') and in the glass industry. The addition of barium chloride to aqueous solutions containing sulfate ions results in the formation of a white precipitate of $BaSO_4$ (equation 21.124). All water- or acid-soluble compounds of barium are highly toxic and one past use of $BaCO_3$ was as a rat poison.

*Qualitative test for $[SO_4]^{2-}$ ions:*

$$BaCl_2(aq) + [SO_4]^{2-}(aq) \longrightarrow \underset{\text{White precipitate}}{BaSO_4(s)} + 2Cl^-(aq) \qquad (21.124)$$

## Group 2 metal oxides

With the exception of beryllium, the group 2 oxides (MO) are usually formed by the thermal decomposition of the respective carbonate ($MCO_3$) as shown for calcium in Figure 21.18. In air, the shiny surface of a group 2 metal quickly tarnishes due to the formation of a thin coat of metal oxide. This protects the metal from further reaction under ambient conditions and the metal is said to be *passivated*. The property is particularly noticeable for beryllium which is resistant to reaction with water and acids because of the presence of the protective oxide covering.

A metal is *passivated* if it has a surface coating of the metal oxide which protects it from reaction with e.g. water.

$$2Be(s) + O_2 \longrightarrow 2BeO(s) \tag{21.125}$$

The complete oxidation of beryllium takes place if the powdered metal is ignited in $O_2$ (equation 21.125). BeO is a white solid that crystallizes with the wurtzite structure. When heated in air, magnesium burns with a brilliant white flame, reacting not only with $O_2$ but also with $N_2$ in reactions analogous to those for barium. The oxides MgO, CaO, SrO and BaO are white crystalline solids with ionic structures of the sodium chloride type. Barium oxide reacts further with $O_2$ when heated at 900 K (equation 21.126), and strontium peroxide can be similarly produced (620 K and under a high pressure of $O_2$).

Wurtzite structure: see Section 8.13

NaCl structure: see Section 8.8

$$2BaO(s) + O_2(g) \xrightarrow{900\,K} 2BaO_2(s) \tag{21.126}$$

Barium peroxide

Calcium oxide is commonly called quicklime or lime. The addition of a molar equivalent of water to solid CaO produces solid $Ca(OH)_2$ (slaked lime) in a highly exothermic reaction. In contrast, BeO does not react with water, MgO reacts slowly, and BaO quickly; BaO is used as a drying agent for liquid alcohols and amines. Calcium hydroxide is used as an industrial alkali and has a vital role in the building trade as a component in mortar. Dry mixtures of sand and CaO can be stored and transported without problem. Once water is added, the mortar quickly sets, forming solid calcium carbonate as $CO_2$ is absorbed (equation 21.127). The sand from the mortar acts as a binding agent between the particles of $CaCO_3$.

$$CaO(s) \xrightarrow{H_2O(l)} Ca(OH)_2(s) \xrightarrow{CO_2(g)} CaCO_3(s) + H_2O(l) \tag{21.127}$$

A solution of calcium hydroxide is called limewater and is used to test for $CO_2$. Bubbling $CO_2$ through limewater turns the latter cloudy owing to the precipitation of $CaCO_3$ (equation 21.128).

*Qualitative test for CO₂:*

$$Ca(OH)_2(aq) + CO_2(g) \longrightarrow \underset{\text{White precipitate}}{CaCO_3(s)} + H_2O(l) \tag{21.128}$$

Limewater

Large amounts of CaO are used in the chemical industry in the manufacture of $Na_2CO_3$ and NaOH. However, the importance of the Solvay process which has been used to produce $Na_2CO_3$ and $NaHCO_3$ is now significantly less than it was. Calcium oxide is also used in the manufacture of calcium carbide, $CaC_2$ (equation 21.129), which contains the anion $[C{\equiv}C]^{2-}$.

$$CaO(s) + 3C(s) \xrightarrow{2200\,K} CaC_2(s) + CO(g) \tag{21.129}$$

The oxides have very high melting points (all >2000 K) and BeO, MgO and CaO are used as *refractory materials*; such materials are not decomposed at high temperatures. Magnesium oxide is a good thermal conductor but is an electrical insulator, and coupled with its thermal stability, these properties make it suitable for use as an electrical insulator in heating and cooking appliances.

COMMERCIAL AND LABORATORY APPLICATIONS

## Box 21.10 Flame tests and fireworks

When the salt of an *s*-block metal is treated with concentrated hydrochloric acid (to produce a volatile metal chloride) and is heated strongly in the non-luminous flame of a Bunsen burner, the flame appears with a characteristic colour – this *flame test* is used in qualitative analysis to determine the identity of the metal ion (see table below).

The principal requirement of a firework (apart from the noise!) is to produce bright, coloured lights, and the pyrotechnics industry capitalizes on the emission of coloured light by *s*-block (and other) elements. Typically, white light is the result of the oxidation of magnesium or aluminium (e.g. in sparklers), and yellow, red and green are due to sodium, strontium and barium respectively. Producing fireworks that emit a blue colour is not so easy; copper(I) chloride is used but the intensity of colour is not particularly good. The explosion in a firework is initiated by an oxidizing agent such as potassium chlorate, $KClO_3$, or potassium perchlorate, $KClO_4$. In rockets, an exothermic reaction that will produce the energy to launch the firework is also needed. Fireworks are stored for quite long time periods and must not absorb water – as we saw in the main discussion, many compounds of the *s*-block elements are deliquescent or hygroscopic and thus the component compounds in the fireworks must be carefully chosen.

A display of fireworks from a floating barge.
*T. Aquilano/GeoScience Features.*

| Metal ion | Colour of flame | Additional confirmatory tests |
|---|---|---|
| Lithium | Red | |
| Sodium | Intense yellow | Cobalt-blue glass completely absorbs sodium light; thus, the flame appears colourless though a piece of blue glass. |
| Potassium | Lilac (difficult to observe) | The flame appears crimson when viewed through a piece of blue glass. |
| Calcium | Brick-red | The flame appears pale green when viewed through a piece of blue glass. |
| Strontium | Crimson | The flame appears purple when viewed through a piece of blue glass. |
| Barium | Apple-green | |

## Group 2 metal hydroxides

Beryllium does not react with water even when heated, since it is passivated. To a certain extent magnesium is also passivated, but it reacts with steam or when heated in water (equation 21.130). Calcium, strontium and barium all react with water in a similar manner, with the reactivity increasing down the group. Calcium reacts with cold water and a steady flow of $H_2$ is produced; with hot water the reaction is rapid. Beryllium hydroxide may be prepared by precipitation from the reaction of $BeCl_2$ and NaOH (but see below).

$$Mg(s) + 2H_2O(l) \xrightarrow{\Delta} Mg(OH)_2(aq) + H_2(g) \qquad (21.130)$$

The properties of beryllium hydroxide differ from those of the later hydroxides. In the presence of an excess of $[OH]^-$, $Be(OH)_2$ forms a tetrahedral *complex anion* **21.12** (equation 21.131) and this exemplifies its behaviour as a Lewis acid.

$$Be(OH)_2(s) + 2[OH]^-(aq) \longrightarrow [Be(OH)_4]^{2-}(aq) \qquad (21.131)$$

Beryllium hydroxide is neutralized by acid and can also function as a Brønsted base (equation 21.132).

$$Be(OH)_2(s) + H_2SO_4(aq) \longrightarrow BeSO_4(s) + 2H_2O(l) \qquad (21.132)$$

When a compound can act as either an acid or a base, it is said to be *amphoteric*. Whereas $Be(OH)_2$ is amphoteric, the later group 2 hydroxides act only as bases with the base strength increasing down the group.

►
Complex ions:
see Chapter 23

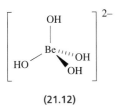

**(21.12)**

A compound is *amphoteric* if it can act as either an acid or a base.

## Group 2 metal halides

Equation 21.133 shows the general reaction for a group 2 metal, M, with a halogen $X_2$, but this is not always a convenient route to the metal halides.

$$M + X_2 \longrightarrow MX_2 \qquad (21.133)$$

$$BeO + Cl_2 + C \xrightarrow{900-1100\,K} BeCl_2 + CO \qquad (21.134)$$

Beryllium chloride is prepared from the oxide (equation 21.134) and possesses a polymeric structure similar to that of $BeH_2$ (Figure 21.4). We described the bonding in $BeH_2$ in terms of 3c–2e Be−H−Be bridges, but in $BeCl_2$ (**21.13**), there are sufficient electrons for all the Be−Cl interactions to be localized 2-centre 2-electron bonds. Each Cl atom has three lone pairs of electrons, and one lone pair per Cl atom is used to form a coordinate bond.

**(21.13)**

►
Fluorite and rutile structures:
see Sections 8.11 and 8.12

The halides of the later metals form ionic lattices: $MgF_2$ has a rutile structure and $CaF_2$, $SrF_2$ and $BaF_2$ crystallize with the fluorite structure. Calcium fluoride occurs naturally as the mineral fluorite and is a major source of $F_2$. With the exception of $BeF_2$ which forms a glass, the group 2 halides are white or colourless crystalline solids and the fluorides tend to have considerably higher melting points than the chlorides, bromides and iodides. There is a tendency for crystals of the chlorides, bromides and iodides to be deliquescent, and $SrBr_2$ crystals are *hygroscopic*. A hygroscopic solid absorbs water from the surrounding air but, unlike a deliquescent crystal, does not become a liquid. The deliquescent nature of calcium chloride renders it an excellent drying agent, for example in desiccators.

A *hygroscopic* solid absorbs water from the surrounding air but does not become a liquid.

## Anomalous behaviour of beryllium

After reading about the chemistry of Be, you should have gained the impression that beryllium is not a typical group 2 metal. Magnesium,

**Fig. 21.19** The structure of $[Be_4O(O_2CMe)_6]$ determined by X-ray diffraction; the central $\{Be_4O\}^{6+}$ unit is a common structural motif in beryllium chemistry. The H atoms of the methyl groups have been omitted. Colour code: Be, yellow; O, red; C, grey. [Data: A. Tulinsky *et al.* (1959) *Acta Cryst.*, vol. 12, p. 623.]

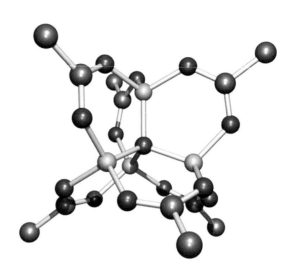

calcium, strontium and barium have many properties in common, although the degree of reactivity changes as the group is descended. Beryllium (the first member of the group) shows anomalous chemical behaviour and many of its compounds exhibit covalent character.

The crucial property that makes beryllium distinct in group 2 is the high charge density of the $Be^{2+}$ ion. If it existed as a naked ion, it would be extremely polarizing. The consequence is that beryllium compounds either exhibit partial ionic character or contain solvated ions. In water, $Be^{2+}$ is hydrated, forming $[Be(H_2O)_4]^{2+}$ in which the positive charge is spread out over the beryllium centre and the four water molecules. The $[Be(H_2O)_4]^{2+}$ ion readily loses a proton because the $Be^{2+}$ centre significantly polarizes the already polar O—H bonds (structure **21.14**).

We could envisage that loss of $H^+$ (as $[H_3O]^+$) might lead to the formation of $[Be(H_2O)_3(OH)]^+$ but the process does not stop here. Various solution species are present and equation 21.135 gives one example. By applying Le Chatelier's principle to this equilibrium, we can understand why $[Be(H_2O)_4]^{2+}$ exists only in strongly acidic solution; the presence of excess $[H_3O]^+$ ions drives the equilibrium to the left-hand side. A common structural motif is the $\{Be_4O\}^{6+}$ unit and Figure 21.19 shows this in $[Be_4O(O_2CMe)_6]$.

$$4[Be(H_2O)_4]^{2+}(aq) + 2H_2O(l) \rightleftharpoons$$
$$2[(H_2O)_3Be-O-Be(H_2O)_3]^{2+}(aq) + 4[H_3O]^+(aq) \qquad (21.135)$$

In Chapter 22, we see that aluminium behaves in a similar way to beryllium, the two elements being *diagonally related* in the periodic table.

**▄▶**
**Charge density:**
**see Section 8.18**

$$\delta^+$$
$$H$$
$$\diagdown$$
$$O \longrightarrow Be^{2+}$$
$$\diagup$$
$$H$$
$$\delta^+$$

**(21.14)**

## 21.13    Diagonal relationships in the periodic table: lithium and magnesium

Although lithium does not stand out so distinctly from group 1 as beryllium does from group 2, there is no doubt that lithium is atypical of the alkali metals in many respects. The $Li^+$ ion is small (see Table 8.6 and Figure 8.22) and highly polarizing, and this results in a degree of covalency in some of its compounds. On going from $Li^+$ to $Be^{2+}$, the ionic radius *decreases*, but on going down group 2 from $Be^{2+}$ to $Mg^{2+}$, the ionic radius *increases*. The net result is that the sizes of $Li^+$ and $Mg^{2+}$ are similar (see Table 8.6) and we observe similar patterns in behaviour between lithium and magnesium despite the fact that they are in different groups. The $Li^+$ and $Mg^{2+}$ ions are *diagonally related* in the periodic table, as are $Be^{2+}$ and $Al^{3+}$, and the three ions $Na^+$, $Ca^{2+}$ and $Y^{3+}$.

Lithium and magnesium ions are more strongly hydrated in aqueous solution than are the ions of the heavier group 1 and 2 metals. When the nitrates of lithium and magnesium are heated, they decompose to give oxides (equation 21.136) whereas thermal decomposition of the remaining group 1 and 2 nitrates gives a nitrite (equation 21.137) which further decomposes to an oxide.

$$4LiNO_3(s) \xrightarrow{\Delta} 2Li_2O(s) + O_2(g) + 2N_2O_4(g) \quad (21.136)$$

$$2MNO_3(s) \xrightarrow{\Delta} 2MNO_2(s) + O_2(g) \quad (21.137)$$
$$M = Na, K, Rb, Cs$$

## SUMMARY

In this chapter we have discussed some descriptive inorganic chemistry of hydrogen and the *s*-block elements and have introduced hydrogen bonding.

### Do you know what the following terms mean?

- kinetic isotope effect
- electropositive
- non-stoichiometric compound
- 3-centre 2-electron interaction
- Lewis acid and base
- coordinate bond

- adduct
- non-aqueous solvent
- self-ionization
- hydrogen bond
- alkali metal
- alkaline earth metal
- solvation

- hydration
- disproportionation
- deliquescent
- hygroscopic
- amphoteric
- alloy
- diagonal relationship

### You should now be able to discuss aspects of the chemistry, including sources (or syntheses) and uses, of:

- the element hydrogen
- selected hydrides of the *s*-, *p*- and *d*-block elements
- the alkali metals

- halides, oxides and hydroxides of the alkali metals
- the group 2 metals
- halides, oxides and hydroxides of the group 2 metals

### Do you know how to test qualitatively for the following?

- $H_2(g)$
- $CO_2(g)$

- $H_2S(g)$
- $HCl(g)$

- $Cl^-$, $Br^-$ and $I^-$ ions
- $[SO_4]^{2-}$ ions

## PROBLEMS

**21.1** An absorption in an IR spectrum of an organic compound comes at $3100\,cm^{-1}$ and is assigned to an X–H bond where X = C, O or N. Determine the shift in this band upon deuteration of the X–H bond for (a) X = $^{12}C$, (b) X = $^{16}O$ and (c) X = $^{14}N$.

**21.2** Why are kinetic isotope effects useful for reactions involving C–H bonds but are not likely to be of significant use in studying reactions involving $^{12}C=O$ and $^{13}C=O$ labelled compounds?

**21.3** Which of the following metal ions might be reduced to the respective metal by $H_2$: (a) $Cu^{2+}$; (b) $Zn^{2+}$; (c) $Fe^{2+}$; (d) $Ni^{2+}$; (e) $Mn^{2+}$? Conversely, which of the following metals would react with dilute sulfuric acid to generate $H_2$: (a) Mg; (b) Ag; (c) Ca; (d) Zn; (e) Cr?

**21.4** The standard enthalpies of formation (298 K) of methanol and octane are $-239$ and $-250\,kJ\,mol^{-1}$. Determine the enthalpy change when a mole of each is fully combusted, and assess the relative merits of these compounds as fuels. ($\Delta_f H^\circ$ $CO_2(g) = -394$, $H_2O(l) = -286\,kJ\,mol^{-1}$.)

**21.5** In Section 21.5 we stated that 'the tetrahydroborate anion is *kinetically* stable towards hydrolysis'. What does this mean? How does kinetic stability in this case differ from thermodynamic stability? Suggest why the cobalt is present in the 'cobalt-doped pellets of $Na[BH_4]$' mentioned in Section 21.5.

**21.6** Figure 21.20 shows the trends in the boiling points for the group 14 hydrides $E_nH_{2n+2}$ (E = C, Si or Ge) for $n = 1$, 2 and 3. Comment briefly on the observed trends and the factors that govern the boiling points of these compounds.

**21.7** What is the oxidation state of nitrogen in (a) $NH_3$, (b) $[NH_4]^+$, (c) $N_2H_4$ and (d) $[NH_2]^-$?

**21.8** Use the data in Table 21.4 to verify Trouton's rule.

**21.9** Discuss what is meant by hydrogen bonding and how it may influence (a) the boiling points and (b) the solid state structures of hydrides of some of the p-block elements.

**21.10** Comment on the following data which relate to different hydrogen bonded species:

| Hydrogen bond (···) | Approximate enthalpy term / $kJ\,mol^{-1}$ |
| --- | --- |
| F···H···F in $[HF_2]^-$ | 163 |
| O–H···O in $H_2O···H_2O$ | 20 |
| S–H···S in $H_2S···H_2S$ | 5 |

**21.11** For each of the following molecules $EH_n$: state whether the E–H bond is polar and if so, show in which direction the dipole moment acts; and state whether the molecule is polar and if so, indicate the direction of the resultant molecular dipole moment: (a) $CH_4$; (b) $H_2O$; (c) $BH_3$; (d) $NH_3$; (e) $SiH_4$; (f) HCl; (g) $H_2Se$.

**21.12** Suggest products for the following reactions:
(a) electrolysis of molten KBr;
(b) electrolysis of aqueous NaCl;
(c) $BaO_2$ with aqueous $H_2SO_4$;
(d) passage of $H_2$ over hot PbO;
(e) addition of a small amount of $MnO_2$ to $H_2O_2$;
(f) reaction of $H_2O_2$ with acidified KI solution.

**21.13** Construct an MO diagram for $O_2$. Assuming the same ordering of energy levels in $O_2$, $[O_2]^{2-}$ and $[O_2]^-$, show that $O_2$ and $[O_2]^-$ are paramagnetic, but $[O_2]^{2-}$ is diamagnetic.

**21.14** According to the electrochemical series, K is a better reducing agent than Na, and yet in equation 21.98 we showed that Na reduces $K^+$ ions in the production of K metal. Using Le Chatelier's

**Table 21.4** Data for problem 21.8.

| Compound | bp / K | $\Delta_{vap}H$ / $kJ\,mol^{-1}$ |
| --- | --- | --- |
| $PCl_3$ | 349 | 30.5 |
| HI | 237 | 19.8 |
| $CHCl_3$ | 334 | 29.2 |
| $C_7H_{16}$ (heptane) | 371 | 31.8 |

**Fig. 21.20** For problem 21.6.

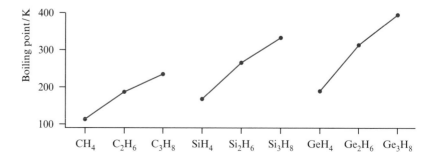

principle, suggest why the following equilibrium might lie to the right-hand side (at 1120 K) rather than to the left:

$$K^+(l) + Na(g) \rightleftharpoons K(g) + Na^+(l)$$

**21.15** Suggest how lattice energy effects might be used to explain why the stabilities of potassium, rubidium and caesium superoxides are greater than that of sodium superoxide, and why lithium usually forms the oxide $Li_2O$ rather than the peroxide or superoxide.

**21.16** Suggest products for the following reactions which are not necessarily balanced on the left-hand side:
(a) $Na_2O_2(s) + CO_2(g) \longrightarrow$
(b) $Na_2O + H_2O \longrightarrow$
(c) $KOH(aq) + H_2SO_4(aq) \longrightarrow$
(d) $KOH(aq) + HI(aq) \longrightarrow$
(e) $O_2 \xrightarrow{\text{Na in liquid NH}_3}$
(f) $Na + Cl_2 \xrightarrow{\Delta}$

**21.17** Suggest products for the following reactions which are not necessarily balanced on the left-hand side:
(a) $SrCO_3(s) \xrightarrow{1560\,K}$
(b) $Ca(OH)_2(s) + H_2O(l) \longrightarrow$
(c) $Be(OH)_2(s) + \text{excess } [OH]^-(aq) \longrightarrow$
(d) $Mg(OH)_2(s) + HCl(aq) \longrightarrow$
(e) $CaF_2(s) + H_2SO_4(conc) \longrightarrow$
(f) $CaH_2(s) + H_2O(l) \longrightarrow$

**21.18** The first members of periodic groups are often noted for their 'anomalous behaviour'. Discuss some of the chemical properties of (a) lithium and (b) beryllium that would support this statement.

**21.19** Distinguish between the terms *deliquescent* and *hygroscopic*.

**21.20** What do you understand by the term *disproportionation*? Give two examples of disproportionation reactions.

**21.21** In this chapter we have highlighted several tests used in qualitative inorganic analysis. How would you test for: (a) $Cl^-$, (b) $Br^-$, (c) $I^-$, (d) $[SO_4]^{2-}$, (e) $CO_2$, (f) $H_2$, (g) gaseous HCl and (h) $H_2S$?

## Additional problems

**21.22** 'Hard water' is a characteristic of areas where limestone ($CaCO_3$) is present and contains $Ca^{2+}$ ions. The following equilibrium describes how calcium ions are freed into the tap-water:

$$CaCO_3(s) + H_2O(l) + CO_2(g) \rightleftharpoons$$
$$Ca^{2+}(aq) + 2[HCO_3]^-(aq)$$

Carbon dioxide is less soluble in hot water than cold water. Use these data to explain why scales of calcium carbonate form inside a kettle when the tap-water is hard.

**21.23** Discuss how hydrogen bonding affects the properties of HF and $H_2O$ with respect to HCl and $H_2S$.

**21.24** 'Binary hydrides can be classified as saline ("salt-like"), molecular, polymeric and metallic.' Discuss this statement.

**21.25** Rationalize the following observations.
(a) In liquid ammonia, $NaNH_2$ acts as a base.
(b) Liquid water is approximately 55 molar.
(c) The superoxide ion is paramagnetic.
(d) Caesium metal has a low melting point.
(e) In the solid state, $BeH_2$ and $BeCl_2$ are structurally similar but a common bonding scheme in these compounds is not appropriate.
(f) The hydride ion is similar in size to $F^-$.

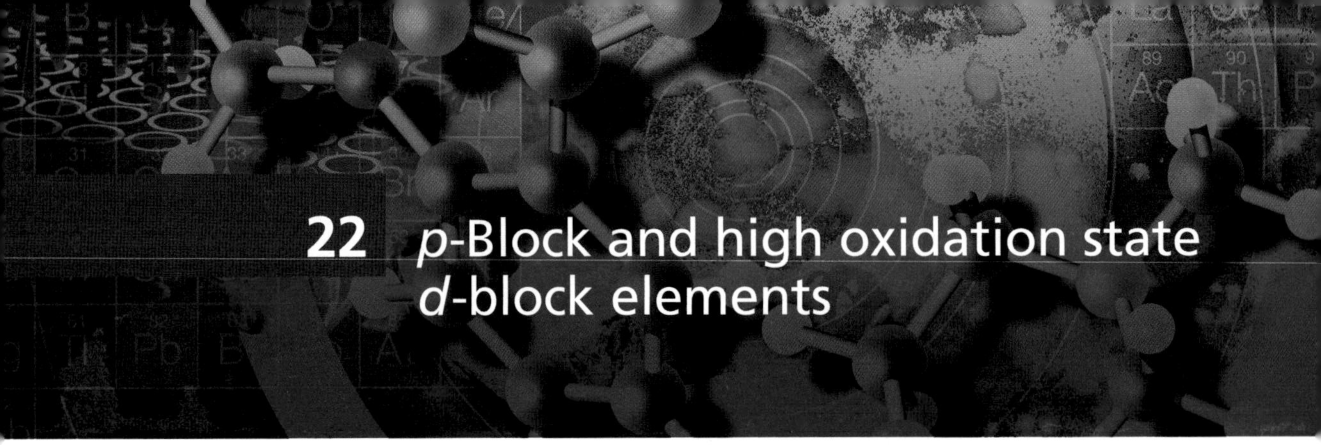

# 22 *p*-Block and high oxidation state *d*-block elements

## Topics

- Oxidation states
- Boron
- Aluminium
- Indium, gallium and thallium
- Carbon and silicon
- Germanium, tin and lead
- Nitrogen and phosphorus
- Arsenic, antimony and bismuth
- Oxygen and sulfur
- Selenium and tellurium
- The halogens
- The noble gases
- High oxidation state *d*-block metals

| 1 | 2 | 3 | 4 | 5 | 6 | 7 | 8 | 9 | 10 | 11 | 12 | 13 | 14 | 15 | 16 | 17 | 18 |
|---|---|---|---|---|---|---|---|---|----|----|----|----|----|----|----|----|----|
| H |   |   |   |   |   |   |   |   |    |    |    |    |    |    |    |    | He |
| Li | Be |   |   |   |   |   |   |   |    |    |    | B | C | N | O | F | Ne |
| Na | Mg |   |   |   |   |   |   |   |    |    |    | Al | Si | P | S | Cl | Ar |
| K | Ca | Sc | Ti | V | Cr | Mn | Fe | Co | Ni | Cu | Zn | Ga | Ge | As | Se | Br | Kr |
| Rb | Sr | Y | Zr | Nb | Mo | Tc | Ru | Rh | Pd | Ag | Cd | In | Sn | Sb | Te | I | Xe |
| Cs | Ba | La | Hf | Ta | W | Re | Os | Ir | Pt | Au | Hg | Tl | Pb | Bi | Po | At | Rn |
| Fr | Ra |   |   |   |   |   |   |   |    |    |    |    |    |    |    |    |    |

## 22.1 Introduction

In this chapter, we continue some descriptive chemistry and apply the thermodynamic principles introduced in Chapters 17 and 18 to understand why certain reactions take place.

'Diagonal line': see
Section 9.11

Groups 13 to 18 of the periodic table make up the *p*-block and the 'diagonal line' that runs through the block *approximately* partitions the elements into metals and non-metals. In group 13, only the first member, boron, is a non-metal, but by groups 17 and 18, all the members of the group are essentially non-metallic.

The *d*-block is composed of 10 triads of metallic elements in groups 3 to 12. Although lanthanum (La) is in group 3, it is also the first member of the series of elements called the lanthanoids, which with the actinoids are known as the *f*-block elements. The group number gives the total number of valence $(n + 1)s$ and $nd$ electrons. For Fe in group 8, the ground state, outer electronic configuration is $4s^2 3d^6$, and for Ir at the bottom of group 9, the configuration is $6s^2 5d^7$. Ground state electronic configurations for all the elements up to $Z = 103$ are listed in Table 20.1.

Hydrides of the *p*- and
*d*-blocks: see Chapter 21

The chemistry in this chapter is arranged by group for the *p*-block, with selected *d*-block elements described in Section 22.14. Our purpose in combining some chemistry of the *d*-block elements with that of the *p*-block is to illustrate the extent to which halides, oxides and related anions of the *high oxidation state* *d*-block elements may be considered as typical covalent species.

## 22.2　Oxidation states

By way of introduction, we look at the relationship between the oxidation state of an element in a compound and the periodic group. In going from group 13 to group 18, the number of electrons in the valence shell of an element increases from three to eight. For a group 13 element, a maximum oxidation state of +3 is expected, but on going down the group, an oxidation state of +1 is also observed as the *thermodynamic 6s inert pair effect* becomes significant. In group 14, the maximum oxidation state is +4, but the heavier elements form compounds in both the +2 and +4 oxidation states. The prevalence of *positive* oxidation states for these elements arises from the fact that they are most likely to combine with elements that are more electronegative than themselves (Table 21.2).

A group 15 element has five valence electrons ($ns^2 np^3$), and a range of oxidation states from +5 (e.g. $[NO_3]^-$, $PF_5$) to –3 (e.g. $Li_3N$, $NH_3$) is observed. As we discussed in Chapter 7, the bonding in species such as $PF_5$, $[PF_6]^-$ and $[NO_3]^-$ can be described in terms of the octet rule. Thus, structure **22.1** (one of a set of three resonance structures) describes the bonding in $[NO_3]^-$. Structure **22.2** is incorrect because it implies an 'expansion' of the octet of electrons around the N centre. The bonding in $PF_5$ may be described in terms of a set

Example 2 in Section 7.3
provides a fuller description
of the bonding in $PF_5$

of resonance structures of which the charge-separated species **22.3** represents one contribution. Such charge-separated species are not always helpful when it comes to providing 3-dimensional information about the structure of a compound. Thus, if you are interested in illustrating the *structure* of $PF_5$ rather than describing its bonding, it is more useful to use diagram **22.4** from which the trigonal bipyramidal structure is clear.

(22.1)　　　　　(22.2)　　　　　(22.3)　　　　　(22.4)

**Fig. 22.1** The variation in oxidation states exhibited by the first row *d*-block metals. The most common oxidation states are marked in pink. Note the pattern: a single oxidation state (other than zero) is observed for scandium (Sc) and zinc (Zn), and the maximum number of oxidation states is attained for manganese (Mn) in the centre of the row. The +8 state is not attained by Fe (see text).

| Oxidation state | 3 Sc | 4 Ti | 5 V | 6 Cr | 7 Mn | 8 Fe | 9 Co | 10 Ni | 11 Cu | 12 Zn |
|---|---|---|---|---|---|---|---|---|---|---|
| 0 | * | * | * | * | * | * | * | * | * | * |
| +1 |  | * | * | * |  |  | * | * | * |  |
| +2 |  | * | * | * | * | * | * | * | * | * |
| +3 | * | * | * | * | * | * | * | * | * |  |
| +4 |  | * | * | * | * | * | * | * |  |  |
| +5 |  |  | * | * | * | * |  |  |  |  |
| +6 |  |  |  | * | * | * |  |  |  |  |
| +7 |  |  |  |  | * |  |  |  |  |  |
| +8 |  |  |  |  |  |  |  |  |  |  |

A group 16 element has six valence electrons ($ns^2 np^4$), and oxidation states range from +6 (e.g. $[SO_4]^{2-}$, $SO_3$) to −2 (e.g. $H_2S$, $H_2Se$). Bonding descriptions can be formulated so that the group 16 element obeys the octet rule. Detailed examples were given in Section 7.3.

Each member of group 17 shows an oxidation state of −1. In addition, positive oxidation states are observed for the heavier elements (e.g. in $[ClO_4]^-$, $[BrO_4]^-$ and $[IO_4]^-$). Of the noble gases, xenon (and to a lesser extent krypton) forms compounds, but only with very electronegative elements (fluorine and oxygen) and oxidation states of +2, +4, +6 and +8 are observed.

One characteristic of a *d*-block metal is the observation of a range of oxidation states. This is true for most of the metals, but in group 3 (the first triad in the *d*-block), the +3 state predominates and this corresponds to the availability of only three electrons in the valence shell. In group 12, the +2 oxidation state is usual for Zn and Cd. Figure 22.1 shows the range of oxidation states found for the first row metals. The pattern in the second and third rows is similar, although Ru and Os show higher oxidation states than Fe, as do Rh and Ir compared with Co. Group 11 should be singled out in particular: Cu(I) and Cu(II) are observed, in contrast to a predominance of the +1 state for Ag, and the +1 and +3 states for Au.

It is important to remember that the highest oxidation states in the *d*-block occur when the atom is in a *covalent environment* – structures **22.5** to **22.7** contain manganese(VII), iridium(VI) and gold(III).

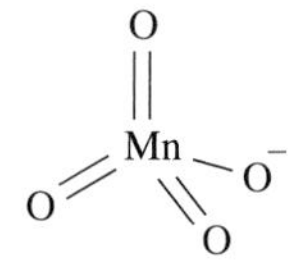

(22.5)

(22.6)

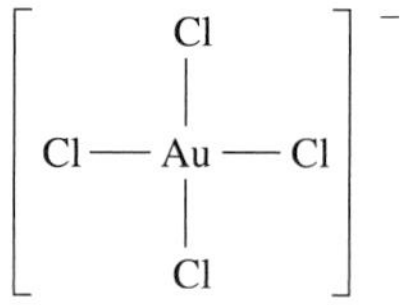

(22.7)

---

## 22.3    Group 13: boron

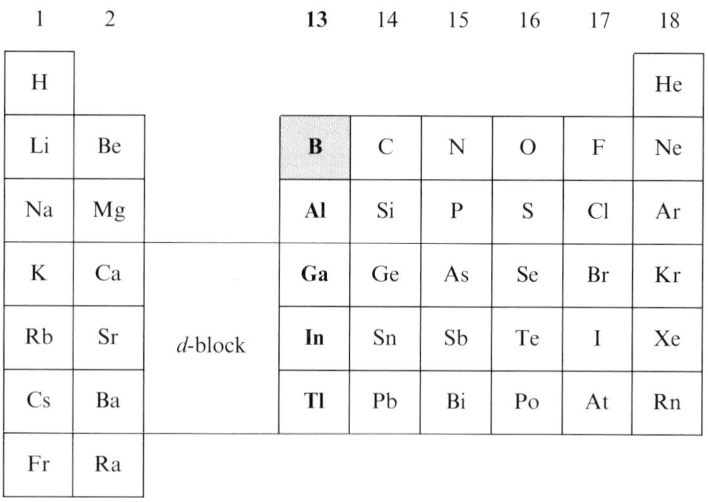

| 1 | 2 |  | 13 | 14 | 15 | 16 | 17 | 18 |
|---|---|---|---|---|---|---|---|---|
| H |  |  |  |  |  |  |  | He |
| Li | Be |  | **B** | C | N | O | F | Ne |
| Na | Mg |  | **Al** | Si | P | S | Cl | Ar |
| K | Ca | | **Ga** | Ge | As | Se | Br | Kr |
| Rb | Sr | *d*-block | **In** | Sn | Sb | Te | I | Xe |
| Cs | Ba |  | **Tl** | Pb | Bi | Po | At | Rn |
| Fr | Ra |  |  |  |  |  |  |  |

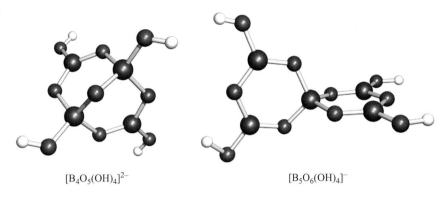

**Fig. 22.2** The structures of
the anions $[B_4O_5(OH)_4]^{2-}$
and $[B_5O_6(OH)_4]^-$, determined
by X-ray diffraction studies of
the salts
$[H_3NCH_2CH_2NH_3][B_4O_5(OH)_4]$
and $[Bu_3NH][B_5O_6(OH)_4]$.
[Data: A. S. Batsanov *et al.*
(1982) *Cryst. Struct. Commun.*,
vol. 11, p. 1629; K. M.
Turdybekov *et al.* (1992) *Zh.
Neorg. Khim.*, vol. 37, p. 1250.]
Both anions contain trigonal
planar and tetrahedral boron
centres. Colour code: B, blue; O,
red; H, white.

$[B_4O_5(OH)_4]^{2-}$          $[B_5O_6(OH)_4]^-$

In Chapter 9, we discussed the complex structures of the allotropes of boron. Among the group 13 elements, boron is unique in forming a wide variety of hydrides, some of which were described in Section 21.5.

**(22.8)**

## Boron–oxygen compounds

Boron occurs naturally in *borax*, $Na_2[B_4O_5(OH)_4]\cdot 10H_2O$, and *kernite*, $Na_2[B_4O_5(OH)_4]\cdot 2H_2O$, which are mined in the Mojave Desert, California. The borate anion present in these ores is shown in structure **22.8** and Figure 22.2.

The oxide $B_2O_3$ is obtained by dehydrating boric acid $B(OH)_3$ (equation 22.1), which is prepared commercially from borax by treatment with aqueous sulfuric acid. After conversion to $B_2O_3$, boron is produced as a brown, amorphous powder by reaction 22.2.

The formula for boric acid, structure 21.4, can be written as $H_3BO_3$

$$2B(OH)_3(s) \xrightarrow{\Delta} B_2O_3(s) + 3H_2O(g) \tag{22.1}$$

$$B_2O_3(s) + 3Mg(s) \xrightarrow{\Delta} 2B(s) + 3MgO(s) \tag{22.2}$$

At room temperature, $B_2O_3$ is usually encountered in the form of a vitreous (glassy) solid and is a typical non-metal oxide. It reacts slowly with water to give $B(OH)_3$. *Molten* $B_2O_3$ (>1270 K) reacts rapidly with steam to give $HBO_2$ (metaboric acid). The main use of $B_2O_3$ is in the manufacture of borosilicate glass including Pyrex. The low thermal expansion allows the glass to be heated and cooled rapidly, making it invaluable for laboratory and kitchen glassware.

Boric acid has a layered structure in the solid state with $B(OH)_3$ units linked by hydrogen bonds (Figure 22.3) within each layer. Inter-layer interactions are very weak and as a consequence, boric acid crystals feel slippery and the compound is a good lubricant. This structure-related property is reminiscent of graphite. Boric acid is soluble in water ($63.5\,g\,dm^{-3}$ at 303 K), and aqueous solutions are weakly acidic ($pK_a = 9.1$). In aqueous solution it acts as a *Lewis*, rather than a Brønsted, acid (equation 22.3) – the protons in solution are *not directly* provided by the $B(OH)_3$ molecule, rather $B(OH)_3$ accepts $[OH]^-$.

Graphite: see Section 9.10

$$B(OH)_3(aq) + 2H_2O(l) \rightleftharpoons [B(OH)_4]^-(aq) + [H_3O]^+(aq) \tag{22.3}$$

In addition to its use as a precursor to $B_2O_3$ and a range of inorganic borates and related compounds, boric acid is used as a flame retardant, a preservative

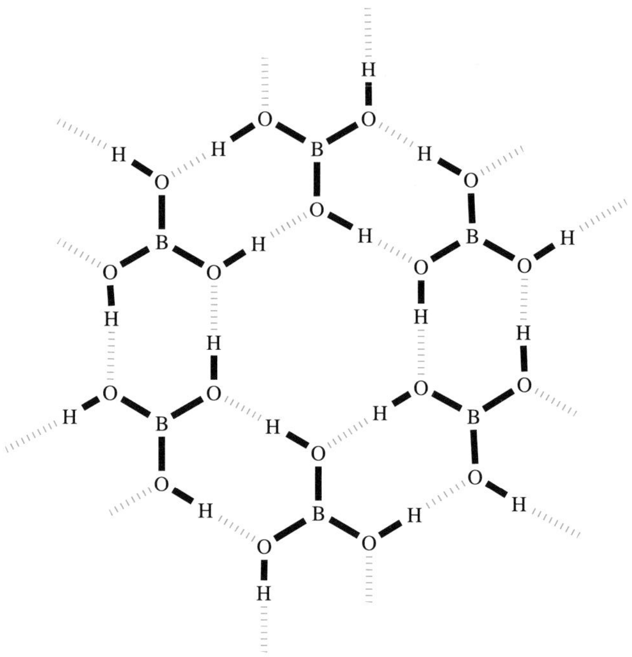

**Fig. 22.3** Boric acid possesses a layer structure. The diagram shows part of one layer in which B(OH)$_3$ molecules are hydrogen bonded together. The bonds within each molecule are highlighted in bold, and the hydrogen bonds are shown by the hashed lines. The hydrogen bonds are asymmetrical with O···H longer than O−H.

■▷ Borane hydrolysis: see equation 21.26

for wood and leather, and as a mild antiseptic. It is also the product from the hydrolysis of many boron-containing compounds. There are numerous borate anions, some salts of which occur in nature. Examples are [BO$_3$]$^{3-}$ (**22.9**), [B$_3$O$_6$]$^{3-}$ (**22.10**), [B$_4$O$_5$(OH)$_4$]$^{2-}$ (**22.8**) (Figure 22.2) and [B$_5$O$_6$(OH)$_4$]$^-$ (**22.11**) (Figure 22.2). Their structures fall into general types in which the boron is in either a trigonal planar BO$_3$ or a tetrahedral BO$_4$ unit. The B−O bond lengths are generally shorter ($\approx$136 pm) in a trigonal planar unit than in a tetrahedral one ($\approx$148 pm). This is due to the donation of $\pi$-electrons from an oxygen lone pair to the vacant $2p$ atomic orbital on the trigonal planar boron atom (Figure 22.4a). This is called $(p-p)\pi$-bonding, and can occur in all three B−O interactions, giving *partial* $\pi$-character to each. In a tetrahedral geometry ($sp^3$ hybridized), there is no vacant atomic orbital available on the boron atom and B−O $\pi$-bonding cannot occur (Figure 22.4b).

The cyclic structures that are characteristic of the larger borate anions are also observed in some organic derivatives. The dehydration of boronic acids of the general form RB(OH)$_2$ where R is an organic substituent, leads to the formation of cyclic trimers (RBO)$_3$ (equation 22.4). Figure 22.5 shows the

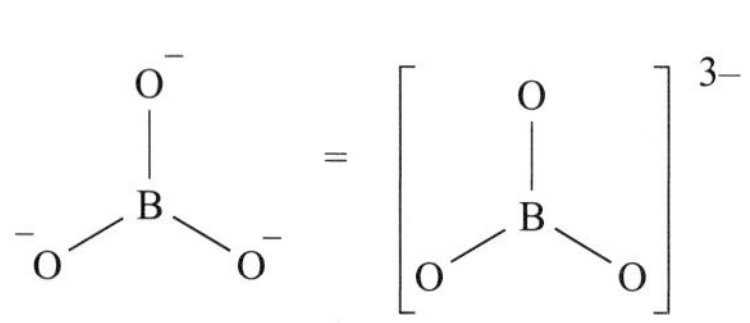

(22.9) 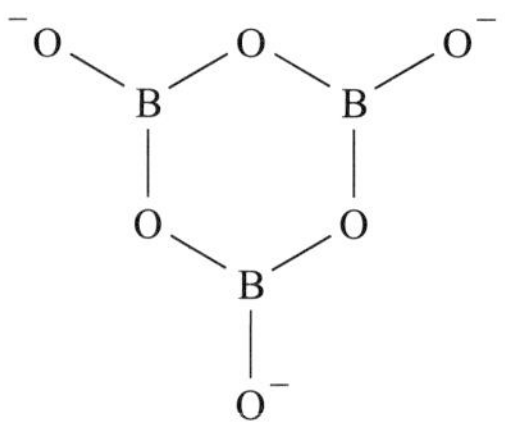  (22.10) 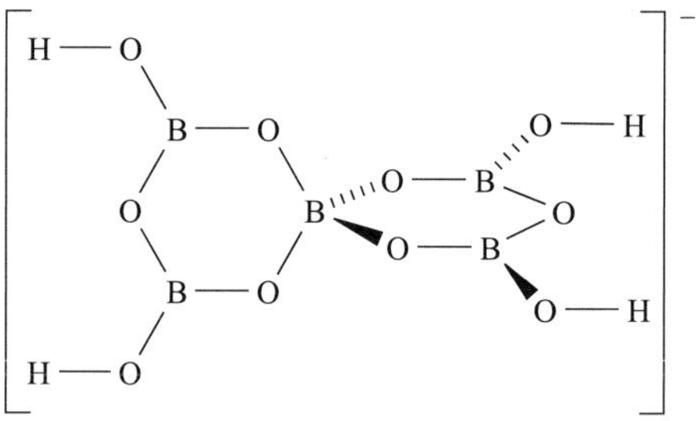  (22.11)

**Fig. 22.4** (a) In a trigonal planar $BO_3$ unit, the B atom is $sp^2$ hybridized and the vacant $2p$ orbital can accept lone pair electrons from *each* oxygen atom; only one such $\pi$-interaction is shown. If all three oxygen atoms are involved, there can only be a *partial $\pi$-bond* along each B–O vector. The resultant $(p–p)\pi$-interactions strengthen the B–O bonds. (b) In a tetrahedral $BO_4$ unit, the B atom is $sp^3$ hybridized, and forms four B–O $\sigma$-bonds.

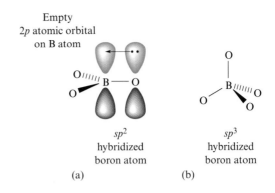

$(p–p)\pi$-**Bonding** arises from the overlap of $p$-atomic orbitals.

structure of $(EtBO)_3$. The boron–oxygen ring is planar as a result of $(p–p)\pi$ B–O interactions.

(22.4)

### Boron halides

The boron trihalides $BF_3$, $BCl_3$, $BBr_3$ and $BI_3$ are all monomeric and at room temperature, $BF_3$ is a colourless gas (bp 172 K), $BCl_3$ a fuming liquid (bp 285 K), $BBr_3$ a colourless liquid (mp 227 K, bp 364 K) and $BI_3$ a colourless solid (mp 323 K). Boron trifluoride may be prepared from $B_2O_3$ by reaction 22.5. $BCl_3$ and $BBr_3$ are formed by heating boron and $Cl_2$ or $Br_2$ together. Boron triiodide can be prepared from $BCl_3$ by reaction 22.6.

$$B_2O_3 + 3CaF_2 + 3H_2SO_4 \longrightarrow 2BF_3 + 3CaSO_4 + 3H_2O \tag{22.5}$$

$$BCl_3 + 3HI \xrightarrow{\Delta} BI_3 + 3HCl \tag{22.6}$$

Each $BX_3$ molecule is trigonal planar, consistent with an $sp^2$ hybridized boron atom. Donation of $\pi$-electrons from each halogen atom to the empty $2p$ orbital on the boron atom gives partial B–X $\pi$-character and stabilizes the monomer. The importance of B–X $(p–p)\pi$-bonding decreases along the series $F > Cl > Br > I$; the energy of the $2p$ orbital of boron is poorly matched with that of the $4p$ atomic orbital of bromine, and even less well matched

▣▷ Electron diffraction data for $BCl_3$: see Section 4.2

📖 **Fig. 22.5** The structure of the trimer $(EtBO)_3$ determined by X-ray diffraction [R. Boese *et al.* (1987) *Angew. Chem., Int. Ed.*, vol. 26, p. 245]. The $B_3O_3$ ring is planar and each B–O bonding interaction consists of a B–O $\sigma$-bond, strengthened with partial $\pi$-character. Colour code: B, blue; O, red; C, grey; H, white.

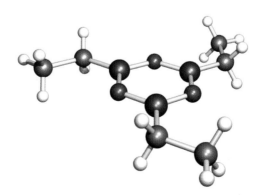

**Fig. 22.6** When a Lewis base, L, uses a lone pair of electrons to coordinate to the Lewis acid $BX_3$ (X = F, Cl, Br or I), an adduct is formed. During the reaction, the geometry of the boron centre changes with the loss of B−X $(p−p)\pi$-bonding.

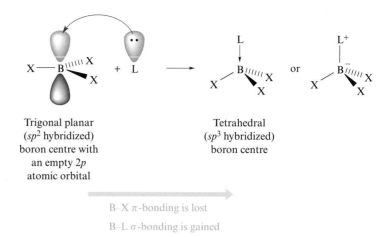

Trigonal planar ($sp^2$ hybridized) boron centre with an empty $2p$ atomic orbital

Tetrahedral ($sp^3$ hybridized) boron centre

B−X $\pi$-bonding is lost

B−L $\sigma$-bonding is gained

with the iodine $5p$ orbital. As the amount of B−X $(p−p)\pi$-bonding *decreases*, the availability of the $2p$ acceptor orbital on the boron centre *increases* and this affects the relative Lewis acid strengths of the trihalides. Consider the formation of a donor–acceptor complex between $BX_3$ and a Lewis base, L. Figure 22.6 shows the attack by the lone pair of the Lewis base on the boron atom and the formation of a B−L bond. This causes a change from trigonal planar to tetrahedral boron. Concomitant with this change is a *loss* of the bond enthalpy associated with the B−F $(p−p)\pi$-bonding, but a *gain* of the bond enthalpy due to the formation of the new B−L $\sigma$-bond. The thermodynamic stability of the adduct with respect to the separate molecules $BX_3$ and L depends, in part, upon the difference in these two enthalpy terms. When the Lewis base is pyridine (py), the experimentally determined order of stabilities of the adducts **22.12** is py·$BF_3$ < py·$BCl_3$ < py·$BBr_3$, in keeping with the degree of B−X $(p−p)\pi$-bonding in $BX_3$ following the order F > Cl > Br.

Boron trifluoride plays a part in facilitating some organic transformations by acting as a Lewis acid. It is relatively resistant to hydrolysis, but in an excess of water, reaction 22.7 occurs, giving boric acid and aqueous tetrafluoroboric acid.

$$4BF_3 + 6H_2O \longrightarrow \underbrace{3[H_3O]^+ + 3[BF_4]^-}_{\text{Aqueous tetrafluoroboric acid}} + B(OH)_3 \qquad (22.7)$$

Pure tetrafluoroboric acid, $HBF_4$, is not isolable but is commercially available in diethyl ether solution, or as solutions formulated as $[H_3O][BF_4]\cdot 4H_2O$. The acid is fully ionized in aqueous solution and behaves as a strong 'mineral' acid.[§]

The halides $BCl_3$, $BBr_3$ and $BI_3$ react rapidly with water (equation 22.8).

$$BX_3 + 3H_2O \longrightarrow B(OH)_3 + 3HX \qquad \text{X = Cl, Br, I} \qquad (22.8)$$

The reaction of $BF_3$ with alkali metal fluorides, MF, in a non-aqueous solvent such as HF, leads to the formation of the salts $M[BF_4]$. The tetrahedral anion $[BF_4]^-$ can be viewed as a donor–acceptor complex in which the Lewis base is the $F^-$ ion. The B−F bond length increases from 130 pm in $BF_3$ to 145 pm in $[BF_4]^-$, consistent with the loss of the $\pi$-contribution to each B−F bond. All the B−F bonds are equivalent. The corresponding

X = F, Cl or Br

**(22.12)**

Lewis acid catalysts: see Chapter 32

Non-aqueous solvent: see Sections 21.7 and 21.10

---

[§] The $[BF_4]^-$ ion only coordinates (see Chapter 23) very weakly to metal centres and is often used as an 'innocent' anion to precipitate cationic species; $[PF_6]^-$ behaves similarly.

**Fig. 22.7** The solid state structure of $B_2Cl_4$ is planar, but in the gas phase, a staggered conformation is favoured. In each, the boron atoms are trigonal planar. Colour code: B, blue; Cl, green.

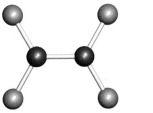

Solid state structure

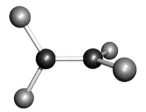

Gas phase structure

Bu = $C_4H_9$ ▮▷

anions $[BCl_4]^-$, $[BBr_4]^-$ and $[BI_4]^-$ are stabilized only in the presence of large cations such as the tetrabutylammonium ion, $[NBu_4]^+$.

A second series of boron halides has the formula $B_2X_4$. $B_2Cl_4$ can be prepared by co-condensing $BCl_3$ and Cu vapours on a surface cooled by liquid $N_2$. In the solid state, $B_2Cl_4$ is planar, but in the gas phase the most stable conformation is a staggered one in which steric interactions between the $BCl_2$ units are minimized (Figure 22.7). When it reacts with $SbF_3$, $B_2Cl_4$ is converted to $B_2F_4$ which is planar in both the solid state *and* the gas phase. The diboron tetrahalides are inflammable in air. $B_2Cl_4$ can be reduced to $B_2H_6$ (equation 22.9).

▮▷
$SbF_3$ as a fluorinating agent: see Section 22.9

$$B_2Cl_4 \xrightarrow{H_2} B_2H_6 \tag{22.9}$$

## 22.4    Group 13: aluminium

| 1 | 2 | | 13 | 14 | 15 | 16 | 17 | 18 |
|---|---|---|---|---|---|---|---|---|
| H | | | | | | | | He |
| Li | Be | | **B** | C | N | O | F | Ne |
| Na | Mg | | **Al** | Si | P | S | Cl | Ar |
| K | Ca | *d*-block | **Ga** | Ge | As | Se | Br | Kr |
| Rb | Sr | | **In** | Sn | Sb | Te | I | Xe |
| Cs | Ba | | **Tl** | Pb | Bi | Po | At | Rn |
| Fr | Ra | | | | | | | |

▮▷
In the US and American speaking areas, the element is known as 'aluminum'

Aluminium is the most abundant metal in the Earth's crust and occurs to an extent of 8.1%. The major source is the hydrated oxide *bauxite*, but it is also present in minerals such as *feldspars*. Bauxite is refined by the Bayer process to give aluminium oxide $Al_2O_3$ (*alumina*), and the metal is then extracted electrolytically. Whereas alumina melts at 2345 K, a mixture of cryolite, $Na_3[AlF_6]$, and alumina melts at 1220 K, making this mixture more convenient to use than pure alumina. Electrolysis of the melt produces aluminium at the cathode (equation 22.10).

*At the cathode:*   $Al^{3+}(l) + 3e^- \longrightarrow Al(l)$ (22.10)

Aluminium is a silvery-white, non-magnetic metal (mp 933 K) which is both *ductile* and *malleable*. It has a low density ($2.7\,g\,cm^{-3}$), a high thermal conductivity and is very resistant to corrosion, being passivated by a coating of oxide. These properties make Al industrially important. Its strength can be increased by alloying with Cu or Mg. Thin sheets of Al are used as household aluminium foil, confectionery wrappers and emergency (space) blankets.

Passivate/alloy: see
Section 21.12

> When a metal can be extruded into wires or other shapes without a reduction in strength or the appearance of cracks, the metal is ***ductile***. When a metal can be beaten or rolled into sheets, it is ***malleable***.

## Oxides and hydroxides

Electrochemical series:
see Tables 18.1 and 21.1

Aluminium oxide can be produced by burning aluminium in $O_2$ (equation 22.11) or by heating with oxides of metals lower than Al in the electrochemical series (equation 22.12). There are several different structural forms (polymorphs) of $Al_2O_3$, in addition to a large number of hydrated species.

$$4Al + 3O_2 \xrightarrow{\Delta} 2Al_2O_3 \tag{22.11}$$

$$Fe_2O_3 + 2Al \xrightarrow{\Delta} 2Fe + Al_2O_3 \tag{22.12}$$

The α-form of $Al_2O_3$, *corundum*, is extremely hard and unreactive and is used industrially as an abrasive. The γ-form (formed by heating α-$Al_2O_3$ to 2300 K) absorbs water and has amphoteric properties as illustrated in reactions 22.13 and 22.14.

$$\gamma\text{-}Al_2O_3(s) + 3H_2O(l) + 2[OH]^-(aq) \longrightarrow 2[Al(OH)_4]^- \tag{22.13}$$
$$\mathbf{(22.13)}$$

$$\gamma\text{-}Al_2O_3(s) + 3H_2O(l) + 6[H_3O]^+(aq) \longrightarrow 2[Al(H_2O)_6]^{3+} \tag{22.14}$$
$$\mathbf{(22.14)}$$

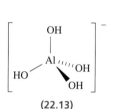

**(22.13)**

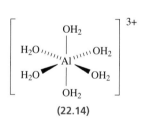

**(22.14)**

Ions **22.13** and **22.14** are also formed when aluminium metal reacts with alkali or hot hydrochloric acid, respectively (equations 22.15 and 22.16).

$$2Al(s) + 2KOH(aq) + 6H_2O(l) \longrightarrow 2K^+(aq) + 2[Al(OH)_4]^-(aq) + 3H_2(g) \tag{22.15}$$

$$2Al(s) + 6HCl(aq) + 12H_2O(l) \xrightarrow{\Delta} 2[Al(H_2O)_6]^{3+} + 6Cl^-(aq) + 3H_2(g) \tag{22.16}$$

Diagonal relationships:
see Section 21.13

The charge density on the Al(III) centre is high, making the $[Al(H_2O)_6]^{3+}$ cation acidic (equation 22.17). This behaviour resembles that of beryllium, and exemplifies the periodic *diagonal relationship* between beryllium and aluminium.

$$[Al(H_2O)_6]^{3+}(aq) + H_2O(l) \rightleftharpoons [Al(H_2O)_5(OH)]^{2+}(aq) + [H_3O]^+(aq)$$

$$pK_a = 5.0 \quad (22.17)$$

Aluminium hydroxide, $Al(OH)_3$, is a white solid and is insoluble in water. It is prepared by the reaction of $CO_2$ with $[Al(OH)_4]^-$. Reactions 22.18 and 22.19 illustrate the amphoteric behaviour of $Al(OH)_3$. Contrast this with the

## MINERAL AND CHEMICAL RESOURCES

## Box 22.1 Aluminium oxide and gemstones

The mineral *corundum* is aluminium oxide $Al_2O_3$; crystals are often barrel-shaped:

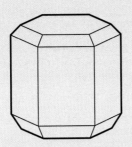

Different coloured corundum. *Colin Keates © Dorling Kindersley, courtesy of the Natural History Museum, London.*

The only naturally occurring mineral harder than corundum is diamond. The presence of trace metals leads to the formation of corundum gemstones and these possess characteristic colours: ruby (red), sapphire (blue), oriental amethyst (purple), oriental emerald (green) and oriental topaz (yellow).

The red colour of ruby is due to the presence of chromium(III), while titanium(IV) causes sapphires to be blue, and iron(III) is present in topaz.

Synthetic corundum crystals are made from bauxite by melting it in furnaces. Artificial rubies and sapphires can also be manufactured, and large crystals of these are made by similar methods to corundum. Ruby (and other) crystals are the critical components in many lasers.

acidic nature of $B(OH)_3$, equation 22.3. One use of $Al(OH)_3$ is as a *mordant* – it absorbs dyes and is used to fix dyes to fabrics.

$$Al(OH)_3(s) + [OH]^-(aq) \longrightarrow [Al(OH)_4]^-(aq) \tag{22.18}$$

$$Al(OH)_3(s) + 3[H_3O]^+(aq) \longrightarrow [Al(H_2O)_6]^{3+}(aq) \tag{22.19}$$

### Aluminium halides

The aluminium trihalides $AlX_3$ (X = Cl, Br or I) may be prepared by reactions between aluminium and the respective halogen, while reaction 22.20 is a route to $AlF_3$.

$$Al_2O_3 + 6HF \longrightarrow 2AlF_3 + 3H_2O \tag{22.20}$$

Aluminium trifluoride is a colourless solid and sublimes at 1564 K, while $AlCl_3$ is a deliquescent, white solid (sublimation temp. 450 K) which hydrolyses rapidly in moist air, releasing HCl. The tribromide (mp 529 K) and triiodide (mp 655 K) are deliquescent solids.

**Fig. 22.8** (a) The gas phase dimeric structure of $Al_2Cl_6$. (b) The bonding scheme for $Al_2Cl_6$ involves $Cl \rightarrow Al$ coordinate bonds within the Al–Cl–Al bridges. Colour code: Al, pale grey; Cl, green.

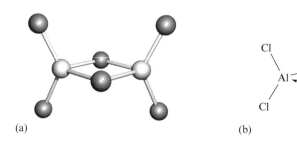

(a)                    (b)

In the solid state, $AlF_3$ possesses an extended structure in which each Al is bonded to six F atoms and each F is connected to two Al atoms. An octahedral geometry is also observed in $[AlF_6]^{3-}$ in *cryolite*, $Na_3[AlF_6]$. In crystalline aluminium trichloride, each Al centre is also 6-coordinate but on vaporizing, the structure collapses to give $Al_2Cl_6$ molecules (Figure 22.8a). Each Al atom is approximately tetrahedral, and the terminal Al−Cl bonds are slightly shorter (206 pm) than the Al−Cl bonds involved in the bridges (220 pm). Although this structure looks reminiscent of that of $B_2H_6$, the bonding schemes in the two molecules are *not* the same. In $B_2H_6$, 3c–2e bridge bonds are invoked because there are insufficient electrons for localized 2c–2e bonds. In $AlCl_3$, each Cl atom has three lone pairs of electrons and, on forming a dimer, each bridging Cl donates a pair of electrons into the empty orbital on an adjacent Al atom (Figure 22.8b).

At higher temperatures in the gas phase, the dimers dissociate into trigonal planar monomers (equation 22.21). Aluminium tribromide and triiodide are dimeric in the solid and liquid states, and in the gas phase at lower temperatures. Dissociation into monomers occurs at higher temperatures and the associated enthalpy changes are lower than for $Al_2Cl_6$.

> ■▶ 2c–2e = 2-centre 2-electron, 3c–2e = 3-centre 2-electron

$$Al_2Cl_6(g) \rightleftharpoons 2AlCl_3(g) \tag{22.21}$$

> ■▶ $AlCl_3$ in Friedel–Crafts reactions: see Section 32.8

The trihalides are strong Lewis acids because the planar $AlX_3$ molecule possesses a vacant $3p$ atomic orbital on the Al atom. The trihalides form adducts such as $Cl_3Al \cdot OEt_2$ and $[AlCl_4]^-$ with Lewis bases. Conducting (ionic) liquids are formed at 298 K when *N*-butylpyridinium chloride (**22.15**) reacts with $AlCl_3$ (equation 22.22). These ionic liquids are good solvents for a variety of compounds, although they are very water-sensitive.

[pyBu]Cl

(**22.15**)

$$Al_2Cl_6 + 2[pyBu]Cl \rightleftharpoons 2[pyBu][AlCl_4] \tag{22.22}$$

### Aluminium sulfate and alums

An alum has the general formula $M^IM^{III}(SO_4)_2 \cdot 12H_2O$. The formula may also be written as $(M^I)_2(M^{III})_2(SO_4)_4 \cdot 24H_2O$ to emphasize that these are *double salts* containing, for example, aluminium sulfate, $Al_2(SO_4)_3$.

Aluminium sulfate forms a series of double salts with sulfates of singly charged metal and ammonium cations; the general formula of an *alum* is $M_2Al_2(SO_4)_4 \cdot 24H_2O$ where $M^+$ is an alkali metal cation or $[NH_4]^+$. These include $K_2Al_2(SO_4)_4 \cdot 24H_2O$ (potash alum or 'alum') which is used in water treatment and as a mordant. Alums are often beautifully crystalline. Members of the series of compounds in which the aluminium is replaced by another $M^{3+}$ ion such as $Fe^{3+}$, $Cr^{3+}$ or $Co^{3+}$ are also alums, despite the absence of aluminium. These include $[NH_4]_2Fe_2(SO_4)_4 \cdot 24H_2O$ (commonly called *ferric or iron(III) alum*) and $KCr(SO_4)_2 \cdot 12H_2O$ (*chrome alum*), often used in the laboratory for crystal growing experiments.

### Aluminium alkyls

Compounds containing Al–C bonds (*organoaluminium compounds*) are important industrial chemicals and are used principally as catalysts. Ziegler–Natta catalysts are mixtures of alkyl aluminium compounds (such as triethylaluminium AlEt$_3$ (**22.16**)) and titanium(IV) chloride and are used to catalyse the polymerization of alkenes at low temperatures and pressures. The catalysed processes produce stereoregular polymers (isotactic or syndiotactic, see Box 26.3) which are characteristically high-density materials with crystalline structures.

**Addition polymers: see Section 26.8**

$$\begin{array}{c} H_3C \\ \diagdown \\ \quad CH_2 \\ \quad | \\ H_2C \diagup Al \diagdown \diagup CH_3 \\ \quad \quad \quad \quad C \\ \quad | \quad \quad \quad H_2 \\ CH_3 \end{array}$$

(22.16)

## 22.5    Group 13: gallium, indium and thallium

| 1 | 2 | | 13 | 14 | 15 | 16 | 17 | 18 |
|---|---|---|---|---|---|---|---|---|
| H | | | | | | | | He |
| Li | Be | | **B** | C | N | O | F | Ne |
| Na | Mg | | **Al** | Si | P | S | Cl | Ar |
| K | Ca | *d*-block | **Ga** | Ge | As | Se | Br | Kr |
| Rb | Sr | | **In** | Sn | Sb | Te | I | Xe |
| Cs | Ba | | **Tl** | Pb | Bi | Po | At | Rn |
| Fr | Ra | | | | | | | |

Gallium is a silvery-coloured metal (mp 303 K) with an exceptionally long liquid range (303–2477 K). It is a trace element in several naturally occurring minerals, and has important applications in the semiconductor industry and in the manufacture of solid state devices. Indium is a soft, silvery-coloured metal (mp 429 K) and when the pure metal is bent, it emits a high-pitched noise. It occurs naturally in association with zinc-containing minerals. Thallium is a soft metal (mp 577 K) and is found as a minor component in some zinc- and lead-containing minerals. Soluble thallium compounds are extremely toxic. Thallium sulfate was used to kill ants and rats, but it is too toxic to other animals (including humans) for general use as a method of pest control.

### Oxides and hydroxides

Gallium hydroxide, $Ga(OH)_3$, is amphoteric and behaves in a similar manner to $Al(OH)_3$. It forms $[Ga(OH)_4]^-$ in basic solution and $[Ga(H_2O)_6]^{3+}$ in the presence of acid. The $[Ga(H_2O)_6]^{3+}$ ion is a *stronger* acid than $[Al(H_2O)_6]^{3+}$ (compare equations 22.23 and 22.17).

$$[Ga(H_2O)_6]^{3+}(aq) + H_2O(l) \rightleftharpoons [Ga(H_2O)_5(OH)]^{2+}(aq) + [H_3O]^+(aq)$$

$$pK_a = 2.6 \quad (22.23)$$

Indium and thallium(III) oxides and hydroxides are basic (typical of a metal). Thallium(III) oxide loses $O_2$ on heating (equation 22.24), thallium(III) being reduced to thallium(I).

$$Tl_2O_3(s) \xrightarrow{370\,K} Tl_2O(s) + O_2(g) \quad (22.24)$$

### Halides

Gallium and indium trifluorides are similar to $AlF_3$ and have high melting points. $GaF_3$ and $InF_3$ are prepared by the thermal decomposition of $[NH_4]_3[MF_6]$ (M = Ga or In). $GaCl_3$, $GaBr_3$ and $GaI_3$ and their indium analogues are formed by reactions of the appropriate metal and halogen. Each gallium compound is a low-melting, deliquescent solid at room temperature. In the solid and liquid states, dimers exist and in the gas phase, dissociation to monomers occurs at higher temperature.

The thallium(III) halides are less stable than those of the earlier group 13 elements. Thallium(III) chloride and bromide are thermally unstable with respect to conversion to the respective thallium(I) compounds (equation 22.25).

$$TlBr_3 \longrightarrow TlBr + Br_2 \quad (22.25)$$

In the case of $TlI_3$, the formula is deceptive. It is *not* a thallium(III) compound but contains thallium(I) and the $[I_3]^-$ anion (**22.17**). In the presence of a large excess of iodide ion, the thallium(III) anion $[TlI_4]^-$ (**22.18**) is formed.

The Lewis acidity of the $MX_3$ molecules permits the formation of the ions $[MCl_6]^{3-}$, $[MBr_6]^{3-}$, $[MCl_5]^{2-}$, $[MCl_4]^-$ and $[MBr_4]^-$ (M = Ga or In) and

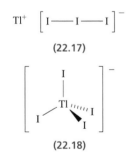

$$Tl^+ \quad \left[ I\!-\!\!-\!I\!-\!\!-\!I \right]^-$$

**(22.17)**

$$\left[ \begin{array}{c} I \\ | \\ I\!-\!Tl\!\cdots\!I \\ {}^{\backslash}I \end{array} \right]^-$$

**(22.18)**

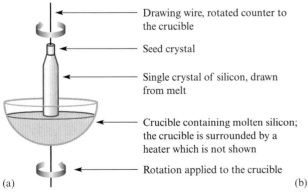

Drawing wire, rotated counter to the crucible

Seed crystal

Single crystal of silicon, drawn from melt

Crucible containing molten silicon; the crucible is surrounded by a heater which is not shown

Rotation applied to the crucible

(a)                                                            (b)

**Fig. 22.9** (a) In the Czochralski process, a seed crystal of silicon is first lowered into molten silicon. It is withdrawn in a specifically controlled manner, and a single crystal of silicon grows. The same technique is used to grow other single crystals, such as gallium arsenide for semiconductor use. (b) A photograph of crystalline silicon produced from the Czochralski process. (The film case indicates size.)  *Photo courtesy Professor Lothar Beyer, University of Leipzig.*

▶ See problem 22.5

$[TlCl_5]^{2-}$. Reactions occur with neutral Lewis bases, L, to yield adducts of the type $L \cdot GaX_3$ or $L \cdot InX_3$.

When $GaCl_3$ is heated with gallium metal, the product appears to be '$GaCl_2$'. However, crystallographic and magnetic data (the compound is diamagnetic) show that it should be formulated as $Ga^I[Ga^{III}Cl_4]$, i.e. a mixed Ga(I)–Ga(III) compound.

## 22.6   Group 14: carbon and silicon

| 1 | 2 | | | 13 | **14** | 15 | 16 | 17 | 18 |
|---|---|---|---|---|---|---|---|---|---|
| H | | | | | | | | | He |
| Li | Be | | | B | **C** | N | O | F | Ne |
| Na | Mg | | | Al | **Si** | P | S | Cl | Ar |
| K | Ca | | | Ga | **Ge** | As | Se | Br | Kr |
| Rb | Sr | *d*-block | | In | **Sn** | Sb | Te | I | Xe |
| Cs | Ba | | | Tl | **Pb** | Bi | Po | At | Rn |
| Fr | Ra | | | | | | | | |

▶ For further chemistry of carbon: see Chapters 24 to 35

The allotropes of carbon were described in Chapter 9. Silicon does not occur naturally in the elemental state, but is still the second most abundant element on Earth. Silicon-based minerals include quartz, feldspars, micas, amphiboles and pyroxenes. Silicon may be produced by reducing silica ($SiO_2$) in an electric furnace. The Czochralski process (Figure 22.9) is used to draw single crystals of silicon from melts of the element for use in solid state devices or the semiconductor industry.

### Oxides

Carbon dioxide is a minor component of the atmosphere of Earth (Figure 21.1). It is a product of respiration, but is removed from the atmosphere during photosynthesis (see Box 22.2). When carbon is burned in an excess of $O_2$, $CO_2$ forms, but when less $O_2$ is present, carbon monoxide can also be produced (equations 22.26 and 22.27). Carbon monoxide and carbon dioxide are also produced in the water–gas shift reaction (Section 21.2) and we have already seen the importance of carbon as a reducing agent for metal oxides (Section 17.7). Carbon monoxide is also used industrially to reduce metal and non-metal oxides and is oxidized to $CO_2$ during the process.

▶ Carbon cycle: see Box 24.1

*With an excess of $O_2$:*   $C(s) + O_2(g) \xrightarrow{\Delta} CO_2(g)$    (22.26)

*With limited $O_2$:*   $2C(s) + O_2(g) \xrightarrow{\Delta} 2CO(g)$    (22.27)

**ENVIRONMENT AND BIOLOGY**

## Box 22.2 The 'greenhouse effect'

The level of carbon dioxide in the Earth's atmosphere is steadily rising owing to the combustion of fossil fuels in industry and the burning of forests (e.g. tropical rainforests) when land is being cleared. Not only does the latter produce *more* $CO_2$, but it removes from the Earth vegetation that would otherwise be undergoing photosynthesis and removing $CO_2$ *from* the atmosphere.

The increased amount of $CO_2$ may lead to a rise in the atmospheric temperature. Short-wave radiation from the Sun heats the Earth, but energy that is radiated back from the Earth into the atmosphere has a longer wavelength. This is coincident with the range of the electromagnetic spectrum in which $CO_2$ absorbs,

and the result would be an increase in the temperature of the atmosphere around the Earth. The phenomenon is known as the 'enhanced greenhouse effect'. Consequences of a rise in the Earth's temperature include partial melting of the polar ice caps and of glaciers, and general changes in climate.

Industrialized countries that signed the 1997 Kyoto Protocol are committed to reducing their 'greenhouse gas' emissions. With respect to 1990 emission levels, a target of $\approx 5\%$ reduction (averaged over all participating countries) must be achieved by 2008–2012.

Other 'greenhouse gases' include $CH_4$ (see Box 25.1) and $N_2O$.

Both CO (bp 82 K) and $CO_2$ are colourless, odourless gases at room temperature. Solid $CO_2$ ('dry ice', see Box 22.3) sublimes at 195 K but may be kept in insulated containers in the laboratory.

Carbon monoxide is almost insoluble in water. It is a very poisonous gas, and its toxicity arises because, acting as a Lewis base, it binds more strongly to the iron centres in haemoglobin than does $O_2$ and prevents the uptake and transport of $O_2$ in the bloodstream. The ability of CO to coordinate to low oxidation state *d*-block metals leads to the formation of a wide range of metal carbonyl compounds.

■▷
Haemoglobin: see Box 5.3

■▷
Metal carbonyls:
see Section 23.14

**COMMERCIAL AND LABORATORY APPLICATIONS**

## Box 22.3 Low-temperature baths

In the laboratory, it is often necessary to carry out reactions at constant, low temperatures. Dry ice and liquid $N_2$ may be used in conjunction with organic solvents to produce baths covering a range of temperatures. Dry ice (solid $CO_2$) sublimes at 195 K, and liquid $N_2$ boils at 77 K.

In liquid $N_2$ baths, the cold liquid is poured from a Dewar flask into a solvent that is constantly stirred. As the two are mixed, a *slush* is formed, the temperature of which can be maintained by occasionally adding more liquid $N_2$.

In a bath that is cooled by dry ice, *small* pieces of solid $CO_2$ are carefully added to the solvent. Initially, the solid carbon dioxide sublimes, but eventually the temperature of the bath decreases to a point at which solid dry ice is present. The low temperature of the bath is kept constant by adding small pieces of dry ice at intervals.

Typical ranges of temperatures that may be achieved can be seen from the selected bath systems listed below.

| Bath contents | Temperature / K | Bath contents | Temperature / K |
|---|---|---|---|
| Liquid $N_2$ + cyclohexane | 279 | Dry ice + ethane-1,2-diol | 258 |
| Liquid $N_2$ + cycloheptane | 261 | Dry ice + heptan-3-one | 235 |
| Liquid $N_2$ + acetonitrile | 232 | Dry ice + acetonitrile | 231 |
| Liquid $N_2$ + octane | 217 | Dry ice + ethanol | 201 |
| Liquid $N_2$ + heptane | 182 | Dry ice + acetone | 195 |
| Liquid $N_2$ + hexa-1,5-diene | 132 | Dry ice + diethyl ether | 173 |

[ethylene glycol = ethane-1,2-diol]

**(22.19)**

The reaction of CO and $Cl_2$ in the presence of an activated carbon catalyst produces carbonyl dichloride (or phosgene), $COCl_2$ (**22.19**). This is a very toxic, colourless gas (bp 281 K) with a choking smell, and was used in the First World War in chemical warfare.

When $CO_2$ dissolves in water, equilibrium 22.28, which lies to the left-hand side, is established. When we refer to 'carbonic acid', we actually mean a solution of $CO_2$ in water. Pure carbonic acid, $H_2CO_3$, is not isolable under ambient conditions.

$$CO_2(aq) + H_2O(l) \rightleftharpoons H_2CO_3(aq) \qquad K \approx 1.7 \times 10^{-3} \qquad (22.28)$$

Carbonic acid is a weak acid, although carbonate and hydrogencarbonate salts are commonly encountered in the laboratory. The $pK_a$ values for equilibria 22.29 and 22.30 indicate that the $[HCO_3]^-$ ion is the dominant species in aqueous solutions of $CO_2$, and that the $[CO_3]^{2-}$ ion will form only in basic solution. The *rate* of forward reaction 22.28 is very slow, and fresh solutions contain significant amounts of $CO_2(aq)$. You should consider how this affects equilibria 22.29 and 22.30. (See also Box 16.2.)

$$H_2CO_3(aq) + H_2O(l) \rightleftharpoons [H_3O]^+(aq) + [HCO_3]^-(aq) \qquad pK_a(1) = 6.4 \quad (22.29)$$

$$[HCO_3]^-(aq) + H_2O(l) \rightleftharpoons [H_3O]^+(aq) + [CO_3]^{2-}(aq) \quad pK_a(2) = 10.3 \quad (22.30)$$

There are remarkable differences between the oxides of carbon and silicon. Silicon monoxide can be prepared by reducing $SiO_2$ but it is the dioxide that is the more important. Silica, $SiO_2$, is a crystalline solid with an extended structure. There are many different structural forms of $SiO_2$, and the most thermodynamically stable is α-quartz (equation 22.31).

$$\text{α-quartz} \underset{845\,K}{\rightleftharpoons} \text{β-quartz} \underset{1145\,K}{\rightleftharpoons} \text{β-tridymite} \underset{1745\,K}{\rightleftharpoons} \text{β-cristobalite} \qquad (22.31)$$

α-Quartz is present in granite and sandstone and occurs as rock crystal and as a range of coloured, crystalline forms such as purple amethyst. The structural building block is an $SiO_4$ unit (**22.20**), which is often represented as a tetrahedron (**22.21**). Such units are interconnected through the oxygen atoms to form a 3-dimensional structure, and the zeolite structure described later (Figure 22.13b) is represented in the form of connected tetrahedra.

In the various structural modifications of silica, the $SiO_4$ units are connected in different ways and the simplest to visualize is *cristobalite*. Its structure is related to the diamond-type lattice (Figure 9.21) of elemental silicon but with an oxygen atom bonded between each pair of adjacent silicon atoms giving Si–O–Si bridges.

The critical difference between carbon–oxygen and silicon–oxygen bond formation lies in their relative bond enthalpies. The bond dissociation enthalpy of the C≡O triple bond in CO is 1076 kJ mol$^{-1}$, and in $CO_2$, each C=O bond has a dissociation enthalpy of 532 kJ mol$^{-1}$. In both molecules, the π-contributions to the carbon–oxygen bond are significant. However, if we look at the difference in bond enthalpies between two Si=O bonds in a hypothetical $SiO_2$ *molecule*, O=Si=O, and four Si–O bonds in an extended structure, then the latter are energetically favourable. The opposite is true for carbon. However, in 1999, an extended solid phase form of $CO_2$ with a quartz-like structure was made under the extreme conditions of 1800 K and 40 gigaPa pressure.

Silica is resistant to attack by acids, with the exception of HF.

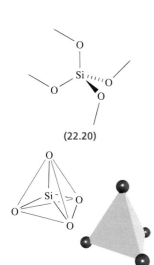

**(22.20)**

**(22.21)**

Etching of silica glass: see Section 21.10

$$SiO_2 + 2Na_2CO_3 \xrightarrow{1770\,K} \underset{\text{Sodium silicate}}{Na_4[SiO_4]} + 2CO_2 \qquad (22.32)$$

**Fig. 22.10** Three-dimensional representations of the silicate anions $[Si_3O_9]^{6-}$ (**22.22**) and $[Si_6O_{18}]^{12-}$ (**22.23**). Colour code: Si, pale grey; O, red.

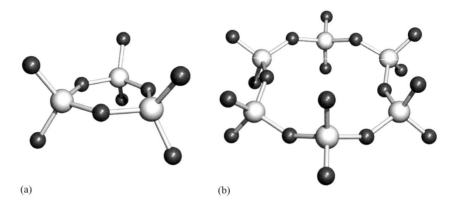

(a)    (b)

Silica reacts slowly with molten $Na_2CO_3$ (equation 22.32), and after purification, sodium silicate (commonly known as 'water glass') is obtained; it is used commercially in detergents, where it maintains a constant pH and can degrade fats by hydrolysis. Silicon dioxide (and glass) react with concentrated aqueous MOH (M = Na or K) and rapidly with molten hydroxides. *Silica gel* is an amorphous form of $SiO_2$, produced by treating aqueous sodium silicate with acid. After careful dehydration, silica gel is used as a drying agent. Small sachets of silica gel are placed in the cases of optical equipment (e.g. cameras) to keep them moisture-free. The structure of silica gel is an open network into which water penetrates and is trapped. Self-indicating silica gel is doped with cobalt(II) chloride which turns from blue to pink as water is absorbed. Silica gel is used as a stationary phase in chromatography and as a solid state (*heterogeneous*) catalyst.

■▶
Heterogeneous catalysts:
see Section 15.8

Sodium silicate, $Na_4[SiO_4]$, is an example of a large family of silicates, the simplest of which (the *orthosilicates*) contain tetrahedral $[SiO_4]^{4-}$ ions. Members of another group of silicates contain *cyclic silicate anions*, such as $[Si_3O_9]^{6-}$ (**22.22**) and $[Si_6O_{18}]^{12-}$ (**22.23**), which occurs in the mineral *beryl*, $Be_3Al_2Si_6O_{18}$. Three-dimensional representations of these silicate anions are shown in Figure 22.10.

The *pyroxenes* are a group of rock-forming silicate minerals in which the silicon and oxygen atoms are arranged in chain structures with a repeat unit of formula $\{Si_2O_6\}_n$ (**22.24**). In the mineral *diopside* $CaMgSi_2O_6$, the silicate chains lie parallel to one another and the cations lie between the chains. In the *amphibole* family of minerals (which includes the fibrous

(22.22)    (22.23)    (22.24)

**Box 22.4 Asbestos: fire protection, but a health hazard**

In the past, fibrous asbestos found extensive use in the manufacture of heat- and fire-resistant material, and was routinely used in chemical laboratories and in building materials. However, inhalation of asbestos fibres can result in chronic lung disease. Its uses have been drastically curtailed, and it is being removed from many older buildings in which asbestos dust continues to pose a health risk.

*asbestos*, see Box 22.4), alternate pairs of adjacent silicate chains are linked together by Si–O–Si bridges and the basic building block is the $[Si_4O_{11}]^{6-}$ unit.

Silicate minerals with sheet structures (the *phyllosilicates*) are produced when $SiO_4$ units are linked together by Si–O–Si bridges to give a layer of fused $Si_4O_4$ rings (Figure 22.11). These silicates tend to be flaky, and an example is *talc*, $Mg_3(OH)_2Si_4O_{10}$. *Micas* constitute a related mineral group in which one in four of the Si atoms is replaced by Al; *muscovite mica* has the formula $KAl_2(OH)_2[AlSi_3O_{10}]$. The replacement of Si by Al to give *aluminosilicates* also occurs in some discrete anions and two related species are shown in Figure 22.12. Since $Si^{4+}$ and $Al^{3+}$ are about the same size, the substitution of one for the other results in little perturbation of the structural unit. Note that although we can consider the aluminosilicate in terms of possessing $Si^{4+}$ ions, this is a formalism; an $Si^{4+}$ ion is unlikely on ionization energy grounds.

Members of another group of silicate minerals including quartz possess 3-dimensional lattices in which each Si is connected to another Si atom by an Si–O–Si bridge. The Si:O ratio in these compounds is 1:2.

*Zeolites* represent a large group of natural and synthetic compounds based on aluminosilicates and have a number of unique and useful properties. They

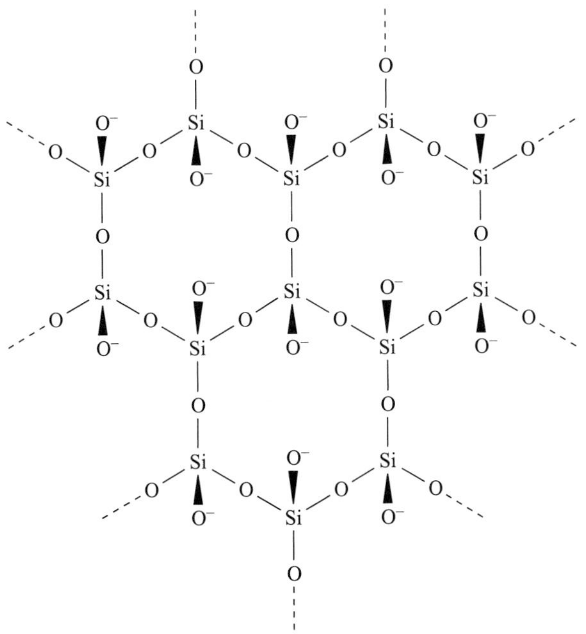

**Fig. 22.11** Part of the layer structure of a sheet silicate. Note that the diagram does *not* show a single repeat unit of the layer. Each Si atom is tetrahedrally sited, and is attached to one terminal and three bridging oxygen atoms. In the 3-dimensional structure, the environment at each O atom is bent.

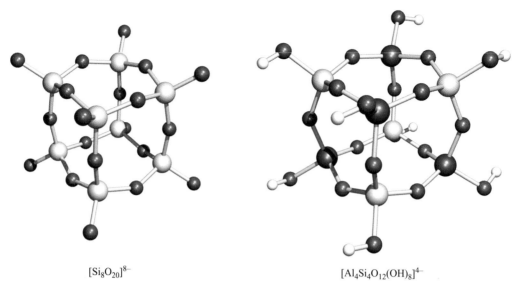

[Si$_8$O$_{20}$]$^{8-}$                    [Al$_4$Si$_4$O$_{12}$(OH)$_8$]$^{4-}$

**Fig. 22.12** The structures of the silicate anion [Si$_8$O$_{20}$]$^{8-}$ and of the alumino-silicate anion [Al$_4$Si$_4$O$_{12}$(OH)$_8$]$^{4-}$. The [Si$_8$O$_{20}$]$^{8-}$ anion contains a cubic arrangement of silicon atoms supported by bridging oxygen atoms; each Si atom also has a terminal oxygen atom attached to it. Replacement of a tetrahedral Si by Al causes little structural perturbation. Colour code: Si, pale grey; O, red; Al, blue; H, white.

contain structured channels (Figure 22.13) of various sizes which small molecules can enter. As different zeolites contain different sized channels, it follows that different sizes of molecules can enter them. This is the basis for the use of zeolites as catalysts in many industrial processes. An example is the zeolite ZSM-5 with a composition Na$_n$[Al$_n$Si$_{96-n}$O$_{192}$]· $\approx$ 16H$_2$O ($n < 27$) which is used as a catalyst in benzene alkylation, xylene isomerization and methanol to hydrocarbon (for motor fuels) conversions. Another important property of zeolites is related to the OH groups on the aluminium centres; these are strongly acidic and the protons may be replaced by a variety of other cations. In practice, this means that the zeolites

Benzene and xylene:
see Chapter 32

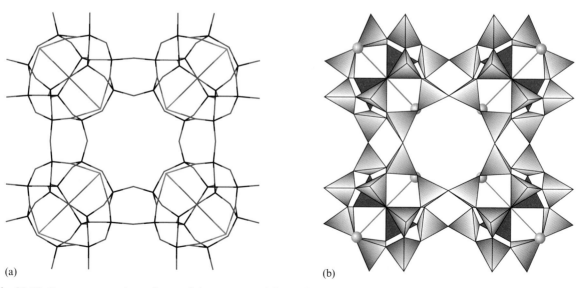

(a)                                                                  (b)

**Fig. 22.13** Two representations of part of the structure of the zeolite ZK5 showing one of the large cavities within the lattice. (a) A conventional 'chemical' representation of the structure. (b) A representation in which the SiO$_4$ or AlO$_3$(OH) units are represented by tetrahedra (see structures **22.20** and **22.21**). The small spheres represent Na$^+$ ions in the lattice.

are excellent *ion-exchange* materials and find widespread uses in water purification, washing powder manufacture and other applications. In the laboratory, zeolites are used as drying agents in the form of *molecular sieves*.

## Halides

In Chapter 25 we look at the halogenation of alkanes, which ultimately gives compounds of the type $C_nX_{2n+2}$ where X = F, Cl, Br or I. In this section we focus on $CX_4$ molecules and their silicon analogues.

Tetrafluoromethane (or carbon tetrafluoride) $CF_4$ is a stable gas (bp 145 K) which may be prepared from silicon carbide (SiC) and $F_2$, or by reactions 22.33. $CF_4$ is extremely resistant to chemical attack by acids, alkalis and oxidizing and reducing agents.

$$\left.\begin{array}{c} CO \xrightarrow{\;SF_4\;} CF_4 \\[1ex] CO_2 \xrightarrow{\;SF_4\;} CF_4 \end{array}\right\} \tag{22.33}$$

$\chi^P(C) = 2.6; \chi^P(F) = 4.0$

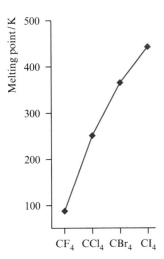

The Montreal Protocol: see Box 25.3

Although each C−F *bond* is polar, the tetrahedral $CF_4$ *molecule* is non-polar and $CF_4$ dissolves only in non-polar organic solvents. Fluorocarbons and chlorofluorocarbons (CFCs) have had widespread applications as high-temperature lubricants and as refrigerants, but they are not 'environmentally friendly' and their use is being phased out.

Tetrachloromethane is a colourless liquid at 298 K with a characteristic odour. The trend in melting points of the carbon halides $CX_4$ is shown in Figure 22.14 and as expected, the melting point rises with increasing molecular weight. $CCl_4$ is prepared by the chlorination of $CH_4$ or by reaction 22.34. It is non-polar, immiscible with water and kinetically stable with respect to hydrolysis.

$$CS_2 + 3Cl_2 \xrightarrow{\;Fe\ catalyst\;} CCl_4 + S_2Cl_2 \tag{22.34}$$

Tetrachloromethane was widely used as a solvent in the past (e.g. for dry cleaning) but its potential carcinogenic properties mean that its use is now severely restricted.

The pale yellow $CBr_4$ is a solid at 298 K, and is insoluble in water but soluble in non-polar organic solvents. It is less stable with respect to decomposition than $CF_4$ and $CCl_4$, and is usually prepared from $CCl_4$ (reaction 22.35), or from $CH_4$ and $Br_2$.

$$3CCl_4 + 2Al_2Br_6 \xrightarrow{\;373\ K\;} 3CBr_4 + 2Al_2Cl_6 \tag{22.35}$$

**Fig. 22.14** As the molecular weight of the tetrahalomethane molecules increases, the melting point increases.

The C−X bond enthalpy decreases in the order $D(C-F) > D(C-Cl) > D(C-Br) > D(C-I)$, and $CI_4$ is even less stable than $CBr_4$. Red crystals of $CI_4$ may be prepared from $CCl_4$ by reaction 22.36 in which $AlCl_3$ acts as a Lewis acid catalyst. $CI_4$ decomposes in the presence of light or when heated (equation 22.37).

$$CCl_4 + 4C_2H_5I \xrightarrow{\;AlCl_3\;} CI_4 + 4C_2H_5Cl \tag{22.36}$$

$$2CI_4 \xrightarrow{\;h\nu\ or\ \Delta\;} C_2I_4 + 2I_2 \tag{22.37}$$

The molecular silicon tetrahalides $SiX_4$ (X = F, Cl, Br or I) can be prepared from their constituent elements (equation 22.38). Whereas the reaction between silicon and $F_2$ is spontaneous, heat is needed for the other three halogens.

$$Si + 2X_2 \longrightarrow SiX_4 \qquad X = F, Cl, Br \text{ or } I \tag{22.38}$$

The halides are colourless, and at 298 K, $SiF_4$ is a gas, $SiCl_4$ and $SiBr_4$ are fuming liquids and $SiI_4$ is a solid. The fuming of the liquids is due to rapid hydrolysis by moisture in the air and the formation of HCl or HBr (equation 22.39).

$$SiX_4 + 2H_2O \longrightarrow SiO_2 + 4HX \qquad X = Cl \text{ or } Br \tag{22.39}$$

The ease of hydrolysis of the silicon halides is in marked contrast to the behaviour of the carbon analogues and has traditionally been attributed to the availability of low-lying 3*d* atomic orbitals on Si which may stabilize a 5-coordinate transition state. However, such arguments are no longer favoured (see Section 7.3). The difference in reactivities is better rationalized in terms of steric effects: the smaller size of C versus Si means that the four Cl atoms in $CCl_4$ are closer together than in $SiCl_4$, and this protects the C atom from attack by $H_2O$. $SiF_4$ undergoes a more complex hydrolysis reaction than $SiCl_4$ and $SiBr_4$, and forms the hexafluorosilicate ion in addition to silica (equation 22.40). Aqueous fluorosilicic acid is a strong acid and is commercially available as an aqueous solution – pure $H_2SiF_6$ is not isolable.

$$2SiF_4 + 4H_2O \longrightarrow SiO_2 + 2HF + \underbrace{2[H_3O]^+ + [SiF_6]^{2-}}_{\text{Fluorosilicic acid}} \tag{22.40}$$

---

**COMMERCIAL AND LABORATORY APPLICATIONS**

## Box 22.5 Silicones in personal care products

The manufacture of personal care products (shampoos, conditioners, toothpastes, anti-perspirants, cosmetics, etc.) is a large component of the chemical industry, and *silicone* products play an important role as the ingredients in shampoos and conditioners which improve the softness and silkiness of hair, making it more manageable and easy to style. Silicones are polymeric organosilicon compounds containing Si−O−Si bridges and have many uses other than in the personal care industry – e.g. silicone grease, sealants, varnishes, synthetic rubbers and hydraulic fluids. In the personal product industry in particular, there are manufacturing challenges to overcome – silicones tend to be viscous oils and are immiscible with water. Silicones may be produced by one company for use by another in the manufacture of shampoos and other products – the second manufacturer may require that the silicone product is in a form that is easy to combine with the basic shampoo which is water-based. One way of overcoming the immiscibility problem is for the silicone to be dispersed in water to give an emulsion. In this form, the silicone and water layers do not separate out from each other, and the silicone can be readily incorporated into a shampoo or conditioner.

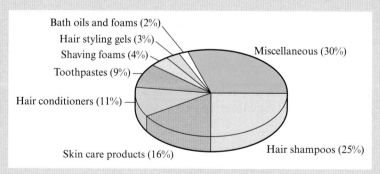

Bath oils and foams (2%)
Hair styling gels (3%)
Shaving foams (4%)
Toothpastes (9%)
Hair conditioners (11%)
Skin care products (16%)
Miscellaneous (30%)
Hair shampoos (25%)

(Data from *Chemical & Engineering News* (1995), p. 34.)

Silicon tetrachloride and related organosilicon halides such as $RSiCl_3$ (R = alkyl) are used widely as synthetic reagents, e.g. in the formation of silicon alkoxides $Si(OR)_4$ (R = alkyl) and siloxanes (equation 22.41). In the siloxane cage produced in this reaction, each silicon atom is tetrahedral, being bonded to one terminal R group and three Si−O−Si bridges. Compare this structure with that of the $[Si_8O_{20}]^{4-}$ anion in Figure 22.12.

$$8RSiCl_3 + 24H_2O \xrightarrow[\text{}]{-24\,HCl} 8RSi(OH)_3 \xrightarrow{\text{dehydrate}}$$

Colour code: Si, pale grey; O, red; R group, dark grey    (22.41)

## 22.7    Group 14: germanium, tin and lead

|  | 1 | 2 |  | 13 | **14** | 15 | 16 | 17 | 18 |
|---|---|---|---|---|---|---|---|---|---|
| | H | | | | | | | | He |
| | Li | Be | | B | C | N | O | F | Ne |
| | Na | Mg | | Al | **Si** | P | S | Cl | Ar |
| | K | Ca | *d*-block | Ga | **Ge** | As | Se | Br | Kr |
| | Rb | Sr | | In | **Sn** | Sb | Te | I | Xe |
| | Cs | Ba | | Tl | **Pb** | Bi | Po | At | Rn |
| | Fr | Ra | | | | | | | |

Elemental structures: see Sections 9.10 and 9.11

Germanium is a grey-white semi-metal which does not tarnish in air. It occurs naturally as the mineral *argyrodite* (a mixed germanium–silver sulfide) and is also obtained from smelters used in processing zinc ores. Crystals of germanium are used in the semiconductor industry and, when doped with gallium or arsenic, act as transistors for use in electronic equipment.

Tin occurs naturally as *cassiterite* ($SnO_2$) from which the metal is extracted by reduction with carbon (equation 22.42).

$$SnO_2 + C \xrightarrow{\Delta} Sn + CO_2 \qquad (22.42)$$

One use of tin is in tin-plating metal cans, but because tin is so soft, many applications require that it is alloyed with other elements; alloys include pewter, soldering metal, bronze and die-casting alloy. High-quality window

**ENVIRONMENT AND BIOLOGY**

## Box 22.6 Snails get heavy!

Since Roman times, lead mining has been carried out in parts of Britain. This brings lead to the surface – spoil heaps containing lead-bearing materials are open and may be a health risk. But sometimes, evolution can provide startling solutions. Researchers have found that snails from populations in areas where lead mining has been carried out for many centuries have been able to concentrate lead into their shells where it cannot harm them. Snails from other populations are not able to do this and suffer lead accumulation in body tissue.

Garden snail, seen from below through glass.
*Frank Greenaway © Dorling Kindersley.*

glass is generally manufactured by the Pilkington process which involves floating molten glass on molten tin to generate a flat surface.

The mineral *galena* (PbS) is a major source of lead, and the metal is extracted by converting the sulfide to the oxide, followed by reduction (equation 22.43).

$$PbS \xrightarrow{O_2, \Delta} PbO \xrightarrow{C \text{ or } CO, \Delta} Pb \tag{22.43}$$

Lead is a very soft, malleable and ductile, blue-white metal. Its resistance to corrosion leads to its use for storing corrosive liquids such as sulfuric acid. Major uses of lead include storage batteries, radiation protection (sheet lead) and plumbing materials, although applications for household pipes are now limited owing to the toxicity of this metal (see Box 22.6).

### Oxides and hydroxides

Germanium(IV) oxide can be prepared by hydrolysing $GeCl_4$ and is stable in hot water. It resembles silica, possessing several crystalline forms with similar structures to those of $SiO_2$. The commercially available form has a β-quartz structure. When heated with germanium, $GeO_2$ is reduced to GeO which is stable in air and water at 298 K. At high temperatures, GeO is oxidized in air back to $GeO_2$. There are several oxoanions of germanium(IV) (*germanates*) but, structurally, they do not resemble analogous silicates.

◼▶
Rutile structure:
see Section 8.12

Crystalline $SnO_2$ has a rutile structure; the 6-coordinate Sn atoms in the structure illustrate a geometry that is quite common for Sn(IV). Tin(IV) oxide is amphoteric and dissolves in acidic and basic solutions. In aqueous hydrochloric acid, the solution species is octahedral $[SnCl_6]^{2-}$ (equation 22.44); in contrast, lead(IV) oxide does *not* dissolve in acids. In strongly alkaline solution, $[Sn(OH)_6]^{2-}$ forms (equation 22.45), but in less basic media, $[SnO_3]^{2-}$ is present.

$$SnO_2(s) + 6HCl(aq) \longrightarrow 2[H_3O]^+(aq) + [SnCl_6]^{2-}(aq) \tag{22.44}$$

$$SnO_2(s) + 2KOH(aq) + 2H_2O(l) \longrightarrow 2K^+(aq) + [Sn(OH)_6]^{2-}(aq) \tag{22.45}$$

**COMMERCIAL AND LABORATORY APPLICATIONS**

## Box 22.7 Lead storage batteries

A lead-acid battery in a motor vehicle consists of a series of electrochemical cells. Each cell contains several lead–antimony electrodes in the form of grid-like plates, the design of which ensures a large surface area. Alternate electrode-plates possess a coating of lead(IV) oxide or lead(II) sulfate which renders them as either a cathode or an anode respectively. Each electrochemical cell is filled with sulfuric acid which functions as the electrolyte.

The two half reactions that are involved in the lead-acid cell are:

$$PbSO_4(s) + 2e^- \rightleftharpoons Pb(s) + [SO_4]^{2-}(aq)$$
$$E^\circ = -0.36\,V$$

$$PbO_2(s) + 4H^+(aq) + [SO_4]^{2-}(aq) + 2e^- \rightleftharpoons$$
$$PbSO_4(s) + 2H_2O(l) \quad E^\circ = +1.69\,V$$

giving (under standard conditions) a value of $E^\circ_{cell}$ of $+2.05\,V$.

A normal automobile battery runs at $12\,V$ and contains six cells connected *in series*; when cells are connected in series, the overall cell potential equals the sum of the separate cell potentials.

By combining the two half-cells above (as detailed in Chapter 18), the overall cell reaction can be written as:

$$PbO_2(s) + 4H^+(aq) + 2[SO_4]^{2-}(aq) + Pb(s) \rightleftharpoons$$
$$2PbSO_4(s) + 2H_2O(l)$$

Sulfuric acid is constantly consumed by the electrolysis cell. Recharging a cell drives the cell-reaction in the opposite direction to its spontaneous route, thereby regenerating aqueous $H_2SO_4$.

Tin(II) oxide, SnO, is prepared by the reaction of any tin(II) salt with dilute alkali. The oxide is amphoteric, dissolving in both acids and alkalis. In strongly alkaline solution, the anion $[Sn(OH)_3]^-$ is formed but more complex species may exist, depending upon the pH.

The red form of lead(II) oxide is called *litharge*, but above 760 K, a yellow form exists. Lead(II) oxide has wide application in lead-acid batteries (see Box 22.7). It is prepared by heating lead and $O_2$, and further reaction in air at 730 K yields $Pb_3O_4$ (*red lead*). This diamagnetic oxide is a mixed lead(II)–lead(IV) compound ($2PbO\cdot PbO_2$), and is used as a pigment and a corrosion-resistant coating for steel and iron. Lead oxides have important uses in the manufacture of 'lead crystal' glass. Strong oxidizing agents are needed to prepare lead(IV) oxide because $PbO_2$ is readily reduced as the standard reduction potential for half-equation 22.46 indicates.

$$PbO_2(s) + 4H^+(aq) + 2e^- \rightleftharpoons Pb^{2+}(aq) + 2H_2O(l) \quad E^\circ = +1.46\,V \quad (22.46)$$

$PbO_2$ reacts with concentrated hydrochloric acid to give solutions of lead(IV) halo-compounds which evolve $Cl_2$.

### Halides

Both metal(IV) and metal(II) halides are known, with the +2 oxidation state becoming more stable towards the bottom of group 14. Thus $PbX_2$ compounds are more stable than $PbX_4$, but $GeX_4$ compounds are more stable than $GeX_2$.

Compounds in the series $GeX_4$ (X = F, Cl, Br or I) can all be prepared from Ge and the relevant halogen. Under ambient conditions, $GeF_4$ is a colourless gas, $GeCl_4$ a colourless liquid, while $GeBr_4$ melts at 299 K; $GeI_4$ is a red-orange solid (mp 417 K). The tetrahalides are tetrahedral monomers. Hydrolysis occurs with the formation of HX. The Lewis acidity of the

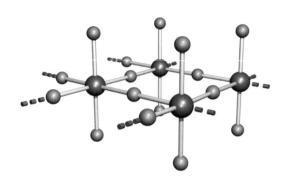

**Fig. 22.15** In SnF$_4$, bridging fluorine atoms between tin centres results in the formation of a 'sheet' structure in the solid state. The tin atom is 6-coordinate but the stoichiometry remains Sn : F = 1 : 4. Colour code: Sn, brown; F, green.

**(22.25)**

In a ***conproportionation*** reaction, oxidation and reduction occur to give the same species. It is the reverse of a ***disproportionation*** reaction.

compounds is illustrated by their ability to form ions such as $[GeCl_6]^{2-}$ (from $GeCl_4$ and $Cl^-$) and adducts such as *trans*-$[Ge(py)_2Cl_4]$ (**22.25**) (from $GeCl_4$ and pyridine).

Germanium reduces $GeF_4$ in a *conproportionation* reaction (equation 22.47) to $GeF_2$ (a white solid), and $GeCl_2$ can be prepared similarly. The reduction of $GeBr_4$ with zinc gives $GeBr_2$. Heating $GeX_2$ (X = Cl, Br) causes disproportionation (equation 22.48).

$$Ge + GeF_4 \longrightarrow 2GeF_2$$

oxidation

reduction

(22.47)

$$2GeX_2 \xrightarrow{\Delta} GeX_4 + Ge \qquad \begin{array}{l} X = Cl,\ T = 1270\ K \text{ in a vacuum} \\ X = Br,\ T = 420\ K \end{array}$$

(22.48)

The Lewis acidity of $GeF_2$ is illustrated in its reaction with $F^-$ to give $[GeF_3]^-$ (equation 22.49), and a similar reaction occurs between $GeCl_2$ and $Cl^-$. The solid state structure of the products may contain discrete pyramidal $[GeX_3]^-$ anions (as in $Rb[GeCl_3]$), but the nature of the cation has a significant influence on the structure and the presence of separated $[GeX_3]^-$ anions cannot be assumed.

$$\underset{\text{Lewis acid}}{GeF_2} + \underset{\text{Fluoride ion donor}}{CsF} \longrightarrow Cs[GeF_3]$$

(22.49)

**Hygroscopic: see Section 21.12**

Tin(IV) fluoride may be prepared from $SnCl_4$ and HF. Crystals of $SnF_4$ are hygroscopic, and $SnF_4$ sublimes at 978 K to give a vapour containing tetrahedral molecules. At 298 K, $SnF_4$ is a white solid and possesses an extended structure consisting of sheets containing 6-coordinate Sn atoms (Figure 22.15). These physical and structural properties contrast greatly with those of $CF_4$, $SiF_4$ and $GeF_4$.

Tin(IV) chloride, bromide and iodide are prepared from the constituent elements and resemble their silicon and germanium analogues. Hydrolysis occurs with loss of HX, but hydrates can also be isolated, for example $SnCl_4 \cdot 4H_2O$ which forms opaque crystals. The tin(IV) halides are Lewis acids; $SnCl_4$ reacts with chloride ion to give $[SnCl_6]^{2-}$ (**22.26**) (equation 22.50) or $[SnCl_5]^-$ (**22.27**), and adducts with other Lewis bases include $THF \cdot SnF_4$ and *cis*-$[Sn(MeCN)_2Cl_4]$ (**22.28**). The Lewis acid strengths of the $SnX_4$ molecules decrease in the order $SnF_4 > SnCl_4 > SnBr_4 > SnI_4$.

$$2KCl + SnCl_4 \xrightarrow{\text{In presence of HCl(aq)}} K_2[SnCl_6]$$

(22.50)

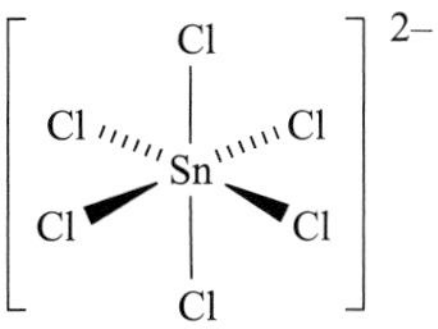

(22.26) 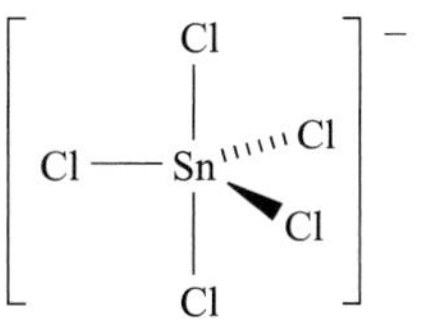      (22.27) 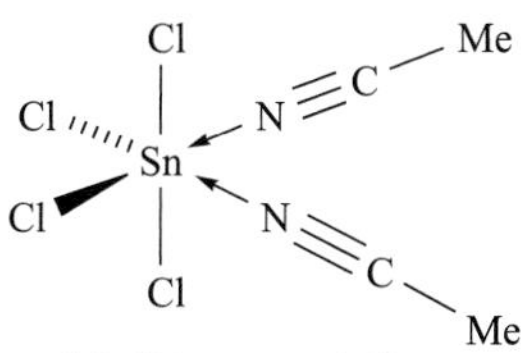      MeCN = acetonitrile   (22.28)

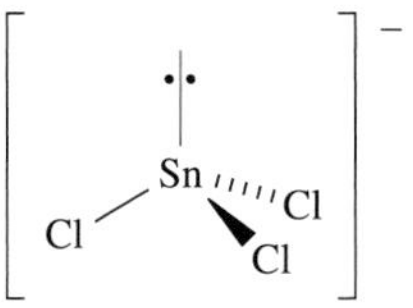

(22.29)

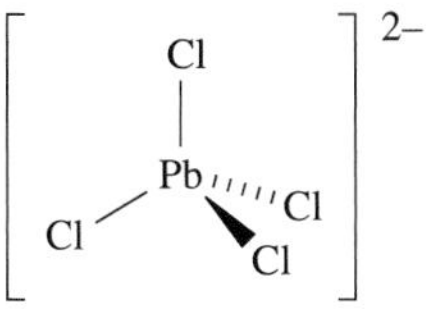

(22.30)

The reaction of tin and hydrogen chloride gives tin(II) chloride, a white solid at room temperature. $SnCl_2$ reacts slowly in alkaline solutions to give a range of oxo and hydroxy species, but the white hydrate $SnCl_2 \cdot 2H_2O$, used as a reducing agent, can be isolated and is commercially available. $SnCl_2$ is a Lewis acid and reacts with chloride ion to give $[SnCl_3]^-$ (**22.29**).

The lead(IV) halides are far less stable than the lead(II) compounds. These are crystalline solids at 298 K, and can be precipitated by mixing aqueous solutions of a soluble halide salt and a soluble lead(II) salt (equation 22.51). The choice of lead(II) salt is critical since few lead(II) compounds are very soluble in water.

$$Pb(NO_3)_2(aq) + 2NaCl(aq) \longrightarrow PbCl_2(s) + 2NaNO_3(aq) \qquad (22.51)$$

Lead(II) chloride is sparingly soluble in water ($K_{sp} = 1.2 \times 10^{-5}$) but the solubility in hydrochloric acid is greater because $PbCl_2$ acts as a Lewis acid and forms $[PbCl_4]^{2-}$ (**22.30**).

## 22.8    Group 15: nitrogen and phosphorus

| 1 | 2 | | 13 | 14 | **15** | 16 | 17 | 18 |
|---|---|---|----|----|----|----|----|----|
| H | | | | | | | | He |
| Li | Be | | B | C | **N** | O | F | Ne |
| Na | Mg | | Al | Si | **P** | S | Cl | Ar |
| K | Ca | | Ga | Ge | **As** | Se | Br | Kr |
| Rb | Sr | *d*-block | In | Sn | **Sb** | Te | I | Xe |
| Cs | Ba | | Tl | Pb | **Bi** | Po | At | Rn |
| Fr | Ra | | | | | | | |

The group 15 elements are called the *pnictogens*. Dinitrogen makes up 78% of the Earth's atmosphere (Figure 21.1) and can be obtained by liquefaction and fractional distillation of air. Reaction 22.52 is a convenient laboratory synthesis, but since ammonium nitrite is potentially explosive, $N_2$ is better prepared as required by mixing aqueous solutions of sodium nitrite and an ammonium salt. $N_2$ is also formed when $NH_4NO_3$ is heated *strongly* (equation 22.53) but this can be explosive. Ammonium nitrate is a powerful

◼▶
TNT: See Box 30.3 oxidizing agent and a component of amatol (a mixture of TNT and $NH_4NO_3$).

$$NH_4NO_2(aq) \xrightarrow{\Delta} N_2(g) + 2H_2O(l) \qquad (22.52)$$

$$\underset{\text{Ammonium nitrite}}{}$$

$$2NH_4NO_3(s) \xrightarrow{>570\,K} 2N_2(g) + O_2(g) + 4H_2O(g) \qquad (22.53)$$

$$\underset{\text{Ammonium nitrate}}{}$$

Dinitrogen is a colourless, odourless gas, and is generally unreactive. The $N\equiv N$ bond is strong (bond dissociation enthalpy = $945\,kJ\,mol^{-1}$) due to ◼▶
Sizes of orbitals:
see Section 3.11 the effective overlap of nitrogen 2*p* atomic orbitals. Nitrogen is the only member of group 15 to be able to form effective $\pi$-interactions in a homo-nuclear bond; for the heavier elements, the *np* atomic orbitals are more diffuse. The lack of reactivity means that $N_2$ is used to provide inert

---

**ENVIRONMENT AND BIOLOGY**

## Box 22.8 The nitrogen cycle

The nitrogen cycle illustrates how $N_2$ is removed from the atmosphere and converted into useful sources. Ultimately, the element is returned to the atmosphere. The action of lightning during thunderstorms converts $N_2$ to nitrogen monoxide:

$$N_2(g) + O_2(g) \longrightarrow 2NO(g)$$

Subsequent reaction with $O_2$ occurs:

$$2NO(g) + O_2(g) \longrightarrow 2NO_2(g)$$

Nitric acid is produced by the action of rain:

$$4NO_2(g) + 2H_2O(l) + O_2(g) \longrightarrow 4HNO_3(aq)$$

Soluble nitrates are formed when the nitric acid reacts with, for example, calcium carbonate:

$$2HNO_3(aq) + CaCO_3(s) \longrightarrow$$
$$Ca(NO_3)_2(aq) + H_2O(l) + CO_2(g)$$

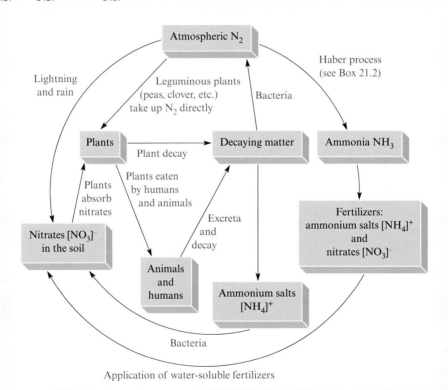

$N_2$ as a coolant:
see Box 22.3

Reaction of $N_2$ with $H_2$:
see Section 21.2

atmospheres in the laboratory. In the electronics industry, it is used as a 'blanket gas' during the production of transistors and other components. Liquid $N_2$ (bp 77 K) is used widely as a coolant.

When heated with Li, Mg or Ca, $N_2$ reacts to give a nitride (equation 22.54).

$$N_2(g) + 3Mg(s) \xrightarrow{\Delta} Mg_3N_2(s) \tag{22.54}$$

In contrast to nitrogen, phosphorus does not occur in the elemental state, but is found in minerals such as *apatites* ($Ca_5X(PO_4)_3$, X = OH, F, Cl) and phosphate rock ($Ca_3(PO_4)_2$). One method of extraction involves heating phosphate rock with silica and carbon in a furnace (equation 22.55). Phosphorus is formed as a vapour and can be collected in water or oxidized to phosphoric acid which is used for the manufacture of fertilizers.

$$2Ca_3(PO_4)_2 + 10C + 6SiO_2 \xrightarrow{\Delta} P_4 + 6CaSiO_3 + 10CO \tag{22.55}$$

In addition to its vital role in the fertilizer industry, phosphorus finds application in the production of steels and phosphor-bronze. The salt $Na_3PO_4$ is used as a water softener (see later).

Allotropes of phosphorus:
see Sections 9.8 and 9.10

In sharp contrast to the inactivity of $N_2$ at 298 K, $P_4$ is stored under water because it ignites spontaneously in air (equation 22.56). The product, $P_4O_{10}$, is often referred to as phosphorus pentoxide, '$P_2O_5$' (see later).

$$P_4(\text{white}) + 5O_2(g) \xrightarrow{\Delta} P_4O_{10}(s) \tag{22.56}$$

## Oxides of nitrogen

$$\overset{-}{N}=\overset{+}{N}=O$$
113 pm   119 pm
**(22.31)**

$$\overset{\bullet}{N}=O$$
115 pm
**(22.32)**

Some common oxides of nitrogen are listed in Table 22.1. In $N_2O$ and NO, the nitrogen atom is in a formal oxidation state of +1 and +2 respectively. $N_2O$ is linear and structure **22.31** gives a valence bond representation of the molecule. Nitrogen monoxide is a radical, and although the valence bond structure **22.32** shows the odd electron on nitrogen, a molecular orbital description places the electron in an NO $\pi^*$ orbital (see Section 5.16).

Dinitrogen oxide $N_2O$ is a colourless gas at 298 K, and is quite unreactive. It is prepared by the potentially explosive conproportionation reaction 22.57 but the conditions must be carefully controlled – relatively mild heating is required (compare reactions 22.53 and 22.57).

$$NH_4NO_3 \xrightarrow{520\text{ K}} N_2O + 2H_2O \tag{22.57}$$

oxidation

reduction

**Table 22.1** Oxides of nitrogen.

| Formula | Name |
| --- | --- |
| $N_2O$ | Dinitrogen oxide; nitrous oxide |
| NO | Nitrogen oxide; nitric oxide |
| $N_2O_3$ | Dinitrogen trioxide |
| $N_2O_4$ | Dinitrogen tetraoxide |
| $NO_2$ | Nitrogen dioxide |
| $N_2O_5$ | Dinitrogen pentaoxide |

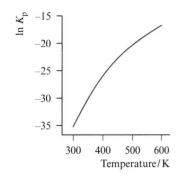

**Fig. 22.16** A graph showing the temperature-dependence of $\ln K_p$ for the equilibrium: $\frac{1}{2}N_2(g) + \frac{1}{2}O_2(g) \rightleftharpoons NO(g)$. At the high temperatures of fuel combustion in aviation and motor engines, the equilibrium is shifted towards the formation of NO. Nitrogen oxides ($NO_x$) are major pollutants.

Dinitrogen oxide is probably best known for its use as a general anaesthetic (laughing gas), but it is used commercially in the synthesis of sodium azide (equation 22.58), and as a precursor to other azides such as $Pb(N_3)_2$ (equation 22.59) which is used as a detonator.

$$N_2O + NaNH_2 \xrightarrow{470\,K} \underset{\text{Sodium azide}}{NaN_3} + H_2O \qquad (22.58)$$

$$2NaN_3 + Pb(NO_3)_2 \longrightarrow Pb(N_3)_2 + 2NaNO_3 \qquad (22.59)$$

Nitrogen monoxide NO is prepared on an industrial scale from $NH_3$ (equation 22.60). It is also one of several oxides formed when $N_2$ is oxidized; generally, these combustion products are written as $NO_x$ (pronounced 'NOX').

$$4NH_3 + 5O_2 \xrightarrow{1300\,K,\ Pt\ catalyst} 4NO + 6H_2O \qquad (22.60)$$

$$\tfrac{1}{2}N_2(g) + \tfrac{1}{2}O_2(g) \rightleftharpoons NO(g) \quad \Delta H^\circ = +91\ kJ\ per\ mole\ of\ NO \qquad (22.61)$$

Equilibrium 22.61 shows the formation of gaseous NO, and Figure 22.16 illustrates how this endothermic reaction shifts to the right-hand side as the temperature is raised – a situation relevant to the combustion of motor and aircraft fuels. *Catalytic converters* in motor vehicles decrease $NO_x$, CO and hydrocarbon emissions (see Box 22.9). $NO_x$ contributes to the formation of smogs over major cities, notably in Los Angeles, but its role as a pollutant is in sharp contrast to the essential role of NO in biological systems.

Even though NO is a radical, it does not dimerize unless it is cooled to low temperatures. The reactions of NO with halogens lead to FNO, ClNO and BrNO (equation 22.62). With respect to decomposition into NO and halogen, FNO is the most stable of the three compounds. The halides FNO and ClNO are used widely as sources of the nitrosyl cation, $[NO]^+$, and

NO in biology: see Box 5.4

FNO, ClNO and BrNO are bent molecules: see Figures 6.5 and 6.6

## COMMERCIAL AND LABORATORY APPLICATIONS

### Box 22.9 Catalytic converters

A catalytic converter consists of a honeycomb ceramic structure coated in finely divided aluminium oxide in which particles of catalytically active metals including platinum, palladium and rhodium are dispersed. The assembly is enclosed in a stainless steel vessel. Exhaust gases from a motor vehicle pass through the heated catalytic chamber in which CO and hydrocarbons (e.g. $C_3H_8$) are oxidized and $NO_x$ is reduced:

$$2CO(g) + O_2(g) \longrightarrow 2CO_2(g)$$

$$C_3H_8(g) + 5O_2(g) \longrightarrow 3CO_2(g) + 4H_2O(g)$$

$$2NO + 2CO \longrightarrow N_2 + 2CO_2$$

$$2NO(g) + 2H_2(g) \longrightarrow N_2(g) + 2H_2O(g)$$

$$C_3H_8(g) + 10NO(g) \longrightarrow 3CO_2(g) + 4H_2O(g)$$

$$+5N_2(g)$$

Originally, catalytic converters concentrated on the oxidation processes, but they are now designed to catalyse both oxidation and reduction reactions, so enhancing their environmental impact. Under legislation, $CO_2$, $N_2$ and $H_2O$ are the only acceptable emission gases. The design of the converter provides a large surface area and facilitates the efficient conversion of the toxic gases into non-pollutants.

A problem, however, is that the catalytic converter must be hot ($\geq 600\,K$) before it becomes effective, and this precludes its being effective during the starting of a car engine and its initial cold-running. Electrically heated converters have been designed but they can drain car-battery resources.

**Fig. 22.17** The molecular structures of $NO_2$, $N_2O_4$, $N_2O_3$ and $N_2O_5$, and resonance structures for $N_2O_3$ indicating that two of the N−O bonds have partial $\pi$-character. Colour code: N, blue; O, red.

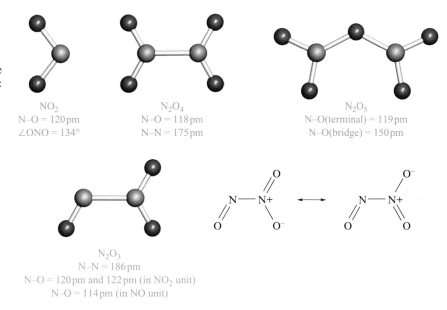

$NO_2$
N−O = 120 pm
$\angle$ONO = 134°

$N_2O_4$
N−O = 118 pm
N−N = 175 pm

$N_2O_5$
N−O(terminal) = 119 pm
N−O(bridge) = 150 pm

$N_2O_3$
N−N = 186 pm
N−O = 120 pm and 122 pm (in $NO_2$ unit)
N−O = 114 pm (in NO unit)

reactions with suitable Lewis acids yield nitrosyl salts (equations 22.63 and 22.64).

$$2NO + Cl_2 \xrightarrow{190\,K} 2ClNO \qquad (22.62)$$

$$ClNO \xrightarrow{AlCl_3} [NO]^+[AlCl_4]^- \qquad (22.63)$$

$$FNO \xrightarrow{PF_5} [NO]^+[PF_6]^- \qquad (22.64)$$

Oxidation of NO with $O_2$ gives $NO_2$ (equation 22.65), and $NO_2$, like NO, has an unpaired electron.

$$\underset{\text{Colourless gas}}{2NO(g)} + O_2(g) \rightleftharpoons \underset{\text{Brown gas}}{2NO_2(g)} \qquad (22.65)$$

$$2NO_2(g) \rightleftharpoons N_2O_4(g) \qquad (22.66)$$

See worked example 17.7

Nitrogen dioxide (Figure 22.17) readily dimerizes and exists in the gas phase in equilibrium with $N_2O_4$ (equation 22.66), although above 420 K the system essentially contains only $NO_2$. In the liquid phase, the equilibrium lies well over to the right-hand side, and in the solid state (mp 262 K), only $N_2O_4$ molecules are present. The $N_2O_4$ molecule is planar (Figure 22.17) with a particularly long N−N bond (175 pm). The N−O bond distances in $NO_2$ and $N_2O_4$ are consistent with the presence of a degree of nitrogen–oxygen $(p–p)\pi$-bonding (Figure 22.18).

A convenient synthesis of $NO_2$ or $N_2O_4$ is the thermal decomposition of *dry* lead(II) nitrate (equation 22.67). In the presence of water, $NO_2$ is hydrolysed to nitrous and nitric acids (see below). By cooling the brown gaseous $NO_2$ to about 273 K, the equilibrium in equation 22.67 is shifted towards the dimer and $N_2O_4$ condenses as a yellow liquid.

$$2Pb(NO_3)_2(s) \xrightarrow{\Delta} 2PbO(s) + 4NO_2(g) + O_2(g)$$
$$\updownarrow \text{cool to} \approx 273\,K \qquad (22.67)$$
$$2N_2O_4(l) \text{ as the major component}$$

**Fig. 22.18** In $NO_2$, the nitrogen atom can be considered to be *sp*² hybridized with a lone pair of electrons occupying one hybrid orbital. Overlap can occur between the nitrogen 2*p* orbital (which is singly occupied) and the two oxygen 2*p* atomic orbitals, giving *partial* π-character to each N−O bond.

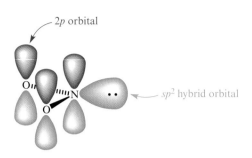

Hydrazine: see Section 21.7

Dinitrogen tetraoxide is a powerful oxidizing agent and the reaction with hydrazine has been used to fuel rockets. Pure liquid $N_2O_4$ is self-ionizing (equation 22.68) and is used as a non-aqueous solvent. The equilibrium lies towards the left-hand side but can be pushed to the right if nitromethane ($MeNO_2$) or dimethyl sulfoxide ($Me_2SO$) is added as a co-solvent.

$$N_2O_4(l) \rightleftharpoons [NO]^+(solv) + [NO_3]^-(solv) \tag{22.68}$$

Dry liquid $N_2O_4$ is used to prepare anhydrous metal nitrates such as $Cu(NO_3)_2$ which cannot be obtained by the evaporation of an aqueous solution or by heating hydrated copper(II) nitrate. $NaNO_3$ and $Zn(NO_3)_2$ may be prepared by reactions 22.69 and 22.70. As Na is oxidized in reaction 22.69, $[NO]^+$ undergoes a 1-electron reduction.

$$Na + N_2O_4 \xrightarrow{\text{N}_2\text{O}_4 \text{ solvent}} NaNO_3 + NO \tag{22.69}$$

$$ZnCl_2 + 2N_2O_4 \xrightarrow{\text{N}_2\text{O}_4 \text{ solvent}} Zn(NO_3)_2 + 2ClNO \tag{22.70}$$

The reaction of $NO_2$ or $N_2O_4$ with water gives a 1:1 mixture of nitrous and nitric acids (equation 22.71), although nitrous acid disproportionates to give NO and $HNO_3$ (equation 22.72). Because of the formation of these acids, atmospheric $NO_2$ is corrosive and contributes to the pollution problem called 'acid rain'.

Acid rain: see Box 22.13

$$2NO_2(g) + H_2O(l) \longrightarrow \underset{\text{Nitric acid}}{HNO_3(aq)} + \underset{\text{Nitrous acid}}{HNO_2(aq)} \tag{22.71}$$

$$3HNO_2(aq) \longrightarrow 2NO(g) + [H_3O]^+(aq) + [NO_3]^-(aq) \tag{22.72}$$

reduction

oxidation

When equal molar amounts of NO and $NO_2$ are mixed at low temperature, $N_2O_3$ forms as a blue solid. The molecule is planar (Figure 22.17) with a long N−N single bond. The different N−O bond lengths suggest that the unique bond is of bond order 2, but the other two nitrogen–oxygen interactions possess *partial* π-character. Above its melting point (173 K), $N_2O_3$ dissociates into NO and $NO_2$. $N_2O_3$ dissolves in water to give nitrous acid (equation 22.73) which may then disproportionate as shown above.

$$N_2O_3 + H_2O \longrightarrow 2HNO_2 \tag{22.73}$$

Dinitrogen pentaoxide is prepared by dehydrating nitric acid (equation 22.74) and forms colourless deliquescent crystals.

Deliquescent: see Section 21.11

$$2HNO_3 \xrightarrow[\text{(dehydrating agent)}]{P_4O_{10}} N_2O_5 + H_2O \tag{22.74}$$

In the solid state, crystalline $N_2O_5$ is composed of $[NO_2]^+$ and $[NO_3]^-$ ions, but it sublimes at 305 K to a give a vapour containing planar molecules (Figure 22.17). $N_2O_5$ is a powerful oxidizing agent (e.g. reaction 22.75).

$$N_2O_5 + I_2 \longrightarrow I_2O_5 + N_2 \tag{22.75}$$

## Oxoacids of nitrogen

An **oxoacid** is a compound that contains oxygen, and at least one hydrogen atom attached to oxygen, and which produces a conjugate base by proton loss.

The two most important *oxoacids* of nitrogen are $HNO_3$ (nitric acid) and $HNO_2$ (nitrous acid). Nitric acid is a strong acid, but for $HNO_2$, $pK_a = 3.37$. The conjugate bases are the nitrate and nitrite ions respectively. In the gas phase, molecules of $HNO_2$ (**22.33**) are planar with a *trans*-configuration.

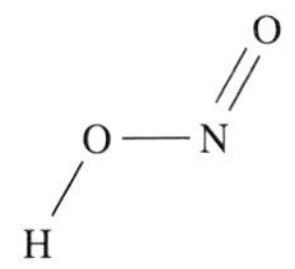

Pure $HNO_2$ is unstable, but dilute aqueous solutions can be prepared from *s*-block metal nitrites (equation 22.76). On standing in aqueous solution, $HNO_2$ disproportionates (equation 22.72).

O
‖
O — N
/
H

**(22.33)**

$$NaNO_2(aq) + HCl(aq) + H_2O(l) \xrightarrow{<273\,K}$$
$$NaCl(aq) + [H_3O]^+(aq) + [NO_2]^-(aq) \tag{22.76}$$

Diazonium salts: Section 32.13

Reaction 22.76 is used to prepare $HNO_2$ *in situ* for the formation of diazonium salts; we describe these salts in Chapter 32.

Most nitrite salts are water-soluble. Although slightly toxic, nitrites are used in the curing of meat because they inhibit the oxidation of blood and prevent the meat from turning brown.

Nitric acid is manufactured on a large scale by the Ostwald process (equations 22.77 and 22.78).

$$4NH_3(g) + 5O_2(g) \xrightarrow[\text{Platinum/rhodium catalyst}]{1120\,K,\,500\,kPa} 4NO(g) + 6H_2O(g) \tag{22.77}$$

$$2NO(g) \xrightarrow{O_2} 2NO_2(g) \xrightarrow{H_2O} HNO_3(aq) + NO(g) \tag{22.78}$$

Pure nitric acid is a colourless liquid (bp 356 K) and the brown-yellow coloration that concentrated aqueous solutions often develop is due to the decomposition reaction 22.79.

$$4HNO_3(aq) \rightleftharpoons 4NO_2(aq) + 2H_2O(l) + O_2(g) \tag{22.79}$$
Colourless      Brown

In the gas phase, $HNO_3$ is monomeric with a planar $NO_3$ framework (Figure 22.19a). The resonance structures shown account for the observed equivalence of two of the N–O bonds (121 pm) and the distance of 141 pm for the third N–O bond indicates a bond order of 1. In the planar nitrate ion, all the N–O bond distances are equal (Figure 22.19b). One of *three* contributing resonance structures is shown in the figure, but the bonding can also be described in terms of three N–O σ-bonds and additional $(p–p)\pi$-bonding delocalized over the $NO_3$ framework.

As well as being a strong acid, $HNO_3$ is a good oxidizing agent and some half-equations involving $HNO_2$, $[NO_2]^-$, $HNO_3$ and $[NO_3]^-$ are listed in Table 22.2 along with the corresponding $E^\circ$ values. By combining these data with those in Appendix 12, you should be able to suggest redox reactions in which $HNO_2$, $[NO_2]^-$, $HNO_3$ and $[NO_3]^-$ are involved. Nitric acid oxidizes most metals, but gold, rhodium, platinum and iridium are

**Fig. 22.19** (a) The molecular and resonance structures of $HNO_3$. (b) The molecular structure of the planar $[NO_3]^-$ anion; the equivalence of the three N–O bonds can be rationalized by valence bond theory (one of three resonance structures is shown) or by MO theory (partial $\pi$-bonds are formed by overlap of N and O $2p$ atomic orbitals and the $\pi$-bonding is delocalized over the $NO_3$ framework). Colour code: N, blue; O, red; H, white.

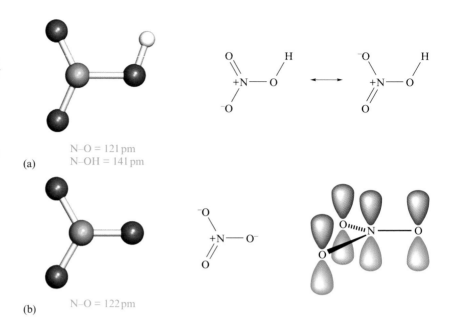

(a) N–O = 121 pm
N–OH = 141 pm

(b) N–O = 122 pm

exceptions. With *very* dilute acid, the oxidations may proceed as in reaction 22.80, but at higher concentrations, the reactions become more complex (equations 22.81 and 22.82).

$$Mg(s) + 2HNO_3(aq) \longrightarrow Mg(NO_3)_2(aq) + H_2(g) \qquad (22.80)$$
Very dilute

$$3Cu(s) + 8HNO_3(aq) \longrightarrow 3Cu(NO_3)_2(aq) + 4H_2O(l) + 2NO(g) \qquad (22.81)$$
Dilute

$$Cu(s) + 4HNO_3(aq) \longrightarrow Cu(NO_3)_2(aq) + 2H_2O(l) + 2NO_2(g) \qquad (22.82)$$
Conc

Passivate: see Section 21.12

Some metals such as aluminium, iron and chromium are passivated when placed in concentrated $HNO_3$. Although concentrated $HNO_3$ does not attack gold, a 3:1 (by volume) combination of concentrated HCl and $HNO_3$ solutions oxidizes gold to gold(III), the latter being formed as $[AuCl_4]^-$. This mixture of acids is called *aqua regia* and is used in cleaning baths for laboratory glassware.

**Table 22.2** Standard reduction potentials involving the nitrate and nitrite ions. Some reactions occur in acidic solution, and some in basic medium. For acidic solution, $E^o$ refers to pH 0; for alkaline solution, $E^o$ refers to pH 14.

| Half-equation | $E^o$ or $E^o_{[OH^-]=1}$ / V |
|---|---|
| $2[NO_3]^-(aq) + 2H_2O(l) + 2e^- \rightleftharpoons N_2O_4(g) + 4[OH]^-(aq)$ | $-0.85$ |
| $[NO_2]^-(aq) + H_2O(l) + e^- \rightleftharpoons NO(g) + 2[OH]^-(aq)$ | $-0.46$ |
| $[NO_3]^-(aq) + H_2O(l) + 2e^- \rightleftharpoons [NO_2]^-(aq) + 2[OH]^-(aq)$ | $+0.01$ |
| $2[NO_2]^-(aq) + 3H_2O(l) + 4e^- \rightleftharpoons N_2O(g) + 6[OH]^-(aq)$ | $+0.15$ |
| $2HNO_2(aq) + 4H^+(aq) + 4e^- \rightleftharpoons H_2N_2O_2(aq) + 2H_2O(l)$ | $+0.86$ |
| $[NO_3]^-(aq) + 3H^+(aq) + 2e^- \rightleftharpoons HNO_2(aq) + H_2O(l)$ | $+0.93$ |
| $[NO_3]^-(aq) + 4H^+(aq) + 3e^- \rightleftharpoons NO(g) + 2H_2O(l)$ | $+0.96$ |
| $HNO_2(aq) + H^+(aq) + e^- \rightleftharpoons NO(g) + H_2O(l)$ | $+0.98$ |
| $2HNO_2(aq) + 4H^+(aq) + 4e^- \rightleftharpoons N_2O(g) + 3H_2O(l)$ | $+1.30$ |

**Fig. 22.20** The 'brown ring' test is used in qualitative analysis to confirm the presence of nitrate ions.

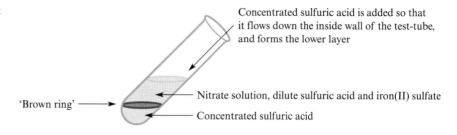

Concentrated sulfuric acid is added so that it flows down the inside wall of the test-tube, and forms the lower layer

'Brown ring'

Nitrate solution, dilute sulfuric acid and iron(II) sulfate

Concentrated sulfuric acid

▣▷
*Nitryl ion = nitronium ion; Nitration: see equations 32.16–32.17 and discussion*

A mixture of concentrated sulfuric and nitric acids produces the nitryl ion, $[NO_2]^+$ (equation 22.83), and is used widely as a nitrating agent in organic chemistry. In reaction 22.83, $HNO_3$ acts *as a Brønsted base*, since $H_2SO_4$ is a stronger acid than $HNO_3$.

$$2H_2SO_4(aq) + HNO_3(aq) \longrightarrow 2[HSO_4]^-(aq) + [NO_2]^+(aq) + [H_3O]^+(aq)$$
$$\text{Conc} \qquad\qquad \text{Conc}$$

$$(22.83)$$

Most nitrate salts are water-soluble and a qualitative test for the $[NO_3]^-$ ion is the *brown ring test*. This is carried out by acidifying the test solution with a few drops of *dilute* sulfuric acid, boiling the solution and allowing it to cool. After the addition of an equal volume of aqueous iron(II) sulfate solution, cold, concentrated sulfuric acid is added *carefully* (Figure 22.20). If $[NO_3]^-$ is present, a brown ring forms at the interface of the concentrated acid and aqueous layers. The origin of the brown colour is the formation of an octahedral ion containing iron and a nitrosyl ligand (equation 22.84).

▣▷
*Complex ions: see Chapter 23*

$$Fe^{2+}(aq) + [NO_3]^-(aq) \xrightarrow{\text{Conc } H_2SO_4} [Fe(NO)(H_2O)_5]^{2+}(aq) \qquad (22.84)$$

We conclude this section on oxoacids of nitrogen by comparing the results of heating some nitrite and nitrate salts (equations 22.85 and 22.86).

$$\left.\begin{array}{l} 2LiNO_2 \xrightarrow{\Delta} Li_2O + NO + NO_2 \\[6pt] 4LiNO_3 \xrightarrow{\Delta} 2Li_2O + 4NO_2 + O_2 \end{array}\right\} \qquad (22.85)$$

$$\left.\begin{array}{l} Pb(NO_2)_2 \xrightarrow{\Delta} PbO + NO + NO_2 \\[6pt] 2Pb(NO_3)_2 \xrightarrow{\Delta} 2PbO + 4NO_2 + O_2 \end{array}\right\} \qquad (22.86)$$

### Oxides of phosphorus

When white phosphorus is exposed to air, it burns spontaneously to give phosphorus(V) oxide $P_4O_{10}$, but if the supply of $O_2$ is limited, phosphorus(III) oxide, $P_4O_6$ (**22.34**) can be isolated. It is often convenient to represent the structure of $P_4O_{10}$ as in diagram **22.35**. However, structure **22.36** is more realistic; it shows a charge-separated species in which all the atoms obey the octet rule.

(22.34)

(22.35)

(22.36)

Fig. 22.21 The molecular structures of (a) $P_4$, (b) $P_4O_6$ and (c) $P_4O_{10}$. The P atoms in $P_4O_6$ and $P_4O_{10}$ are still in a tetrahedral shape but are not within bonding distance of one another. Colour code: P, brown; O, red.

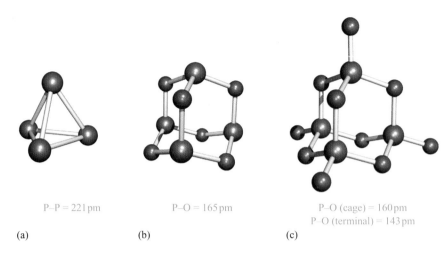

P–P = 221 pm          P–O = 165 pm          P–O (cage) = 160 pm
                                             P–O (terminal) = 143 pm

(a)                    (b)                   (c)

Figure 22.21 illustrates the relationship between the structures of $P_4$, $P_4O_6$ and $P_4O_{10}$. The bonding in $P_4$ is described in terms of six localized P–P single bonds. In the first step of oxidation, each P–P bond is converted into a P–O–P unit; P–O bond distances of 165 pm are consistent with single bonds. Although a tetrahedral arrangement of P atoms is retained in $P_4O_6$, the P–P distances increase from 221 pm (bonding) in $P_4$ to 295 pm in $P_4O_6$ (non-bonding). Further oxidation occurs to give $P_4O_{10}$. The molecular units are retained in the gas phase.

Hydrolysis of $P_4O_6$ yields $H_3PO_3$, while $P_4O_{10}$ gives $H_3PO_4$ (see below). $P_4O_{10}$ is available commercially as a white powder but is extremely hygroscopic. Phosphorus(V) oxide also exists in polymeric forms, '$P_2O_5$'. The oxide is used as a drying and dehydrating agent; if a dish containing phosphorus(V) oxide is exposed to moist air, a brown, viscous layer forms on the surface of the white powder preventing it from absorbing more water.

## Oxoacids of phosphorus

Phosphorus possesses far more oxoacids than does nitrogen, and Table 22.3 lists selected oxoacids and gives their structures. Each phosphorus atom is tetrahedrally bonded. The basicity of each acid corresponds to the number of OH groups, *and not simply to the number of hydrogen atoms*. Phosphinic acid, $H_3PO_2$, is monobasic (equation 22.87) because the two *hydrogen atoms which are bonded directly to the phosphorus atom are not lost as protons.*

$$H_3PO_2(aq) + H_2O(l) \rightleftharpoons [H_3O]^+(aq) + [H_2PO_2]^-(aq) \tag{22.87}$$

An *acid anhydride* is formed when one or more molecules of acid lose one or more molecules of water.

Phosphoric acid, $H_3PO_4$, is formed by reaction 22.88 – the oxide is the *anhydride* of the acid – or when calcium phosphate ('phosphate rock') reacts with concentrated sulfuric acid (equation 22.89). In its pure state, $H_3PO_4$ forms deliquescent, colourless crystals (mp 315 K) which quickly turn into a viscous liquid. The acid is available commercially as an 85% aqueous solution; even this is viscous, owing to extensive hydrogen bonding.

$$P_4O_{10}(s) + 6H_2O(l) \longrightarrow 4H_3PO_4(aq) \tag{22.88}$$

$$Ca_3(PO_4)_2 + 3H_2SO_4 \longrightarrow 2H_3PO_4 + 3CaSO_4 \tag{22.89}$$
                          Conc

**Table 22.3** Selected oxoacids of phosphorus. The names in parentheses are older names, still in use. The number of OH groups determines the basicity of the acid; this is *not necessarily* the same as the total number of hydrogen atoms, e.g. $H_3PO_3$.

| Formula | Name | Structure |
|---|---|---|
| $H_3PO_4$ | Phosphoric acid (orthophosphoric acid) | |
| $H_3PO_3$ or $HPO(OH)_2$ | Phosphonic acid (phosphorous acid) | |
| $H_3PO_2$ or $H_2PO(OH)$ | Phosphinic acid | |
| $H_4P_2O_7$ | Diphosphoric acid | |
| $H_4P_2O_5$ | Diphosphonic acid (diphosphorous acid) | |
| $H_5P_3O_{10}$ | Triphosphoric acid | |
| $H_4P_2O_6$ | Hypophosphoric acid | |

Industrially, phosphoric acid is very important and is used on a large scale in the production of fertilizers, detergents and food additives; it is responsible for the sharp taste of many soft drinks. It is also used to remove oxide and scale from the surfaces of iron and steel.

Aqueous $H_3PO_4$ is a tribasic acid, but the values of $pK_a$ for equilibria 22.90 to 22.92 show that only the first proton is readily lost. With NaOH, $H_3PO_4$ can react to give three salts $NaH_2PO_4$, $Na_2HPO_4$ and $Na_3PO_4$.

$$H_3PO_4(aq) + H_2O(l) \rightleftharpoons [H_3O]^+(aq) + [H_2PO_4]^-(aq) \quad pK_a = 2.12 \quad (22.90)$$

$$[H_2PO_4]^-(aq) + H_2O(l) \rightleftharpoons [H_3O]^+(aq) + [HPO_4]^{2-}(aq) \quad pK_a = 7.21 \quad (22.91)$$

$$[HPO_4]^{2-}(aq) + H_2O(l) \rightleftharpoons [H_3O]^+(aq) + [PO_4]^{3-}(aq) \quad pK_a = 12.67 \quad (22.92)$$

## ENVIRONMENT AND BIOLOGY

### Box 22.10 Phosphates in lakes: friend or foe?

Phosphates from detergents and fertilizers are a factor in the increase of algae in lakes – excessive growth of algae is called *eutrophication* and its presence depletes the lakes of a supply of $O_2$ and affects fish and other water-life. The algae population can increase as zooplankton (algae are their natural food) are killed off. But the phosphate problem is not clear cut.

Eutrophic lake in Lithuania. *Reproduced courtesy of Chris Mason.*

Firstly, it has been shown that phosphates are not the only pollutant to blame for this upset in the natural balance – zooplankton are also affected by heavy metal wastes, oil and insecticides.

Secondly, the presence of phosphates in a lake may be able to offset some of the problems caused by acid rain (see Box 22.13). Acid rain lowers the pH of lakes (i.e. the water becomes more acidic). Although neutralization can be achieved by adding lime, this leads to higher levels of $Ca^{2+}$ ions which in turn affect the variety of life supported by the lake's waters. A study carried out in the Lake District of Britain has shown the presence of phosphates encourages water-plant growth (i.e. the phosphate acts as a fertilizer) and the plants absorb nitrates (a natural part of the $N_2$ cycle, Box 22.8). Hydroxide ions are produced as nitrate ions are taken up, and neutralization of the lake's water can occur. The report of this research (*Nature* (1994), vol. 377, p. 504) brings into question whether phosphates are as bad for our water-courses as other reports may suggest.

Phosphoric acid is not a strong oxidizing agent, as indicated by the value of $E^\circ$ for half-equation 22.93. Compare this with the values listed in Table 22.2 for nitric and nitrous acids.

$$H_3PO_4(aq) + 2H^+(aq) + 2e^- \rightleftharpoons H_3PO_3(aq) + H_2O(l)$$

$$E^\circ = -0.28\,\text{V} \quad (22.93)$$

When $H_3PO_4$ is heated above 480 K, it is dehydrated to diphosphoric acid (equation 22.94).

(22.94)

Further heating results in further $H_2O$ elimination and the formation of polyphosphoric acids such as triphosphoric acid. The sodium salt of this pentabasic acid (Table 22.3) $Na_5P_3O_{10}$ is used in detergents where it acts as a water softener. 'Hard water' is due to the presence of $Ca^{2+}$ and $Mg^{2+}$ ions, and the $[P_3O_{10}]^{5-}$ anion (**22.37**) is a *sequestering agent* which forms stable water-soluble complexes with the group 2 metal ions, preventing them from forming insoluble complexes with the stearate ion in soaps. Such precipitates appear as 'scum' in household sinks and baths.

*Stearic acid and soap: see Section 33.9*

**(22.37)**

Phosphates are biologically important, and the structural material of bones and teeth is *apatite* $Ca_5(OH)(PO_4)_3$. Tooth decay involves acid attack on the phosphate, but reaction with fluoride ion (which is commonly added to water supplies) can lead to the formation of fluoroapatite as a coating on teeth (equation 22.95), making them more resistant to decay.

$$Ca_5(OH)(PO_4)_3 + F^- \longrightarrow Ca_5F(PO_4)_3 + [OH]^- \qquad (22.95)$$
$$\text{Apatite} \qquad\qquad\qquad \text{Fluoroapatite}$$

All living cells contain *adenosine triphosphate* (ATP, **22.38**), and its conversion to adenosine diphosphate (ADP) by the cleavage of one phosphate group during hydrolysis (equation 22.96) releases energy used for functions such as cell growth and muscle movement. The conversion of ATP to ADP occurs at pH 7.4 (biological pH) and $\Delta G$ for reaction 22.96 (which is an over-simplification of this essential biological process) is approximately $-40\,\text{kJ}$ per mole of reaction. ATP is continually being reformed using energy from the burning of organic compounds, so that there is a constant supply of stored energy.

**(22.38)**

$$[ATP]^{4-} + 2H_2O \longrightarrow [ADP]^{3-} + [HPO_4]^{2-} + [H_3O]^+ \qquad (22.96)$$

Phosphonic acid is formed when $P_4O_6$ or $PCl_3$ is hydrolysed (equations 22.97 and 22.98) and the latter method is used industrially.

$$P_4O_6(s) + 6H_2O(l) \longrightarrow 4H_3PO_3(aq) \qquad (22.97)$$

$$PCl_3(l) + 3H_2O(l) \longrightarrow H_3PO_3(aq) + 3HCl(aq) \qquad (22.98)$$

Pure $H_3PO_3$ forms colourless, deliquescent crystals (mp 343 K). It contains *two* OH groups (Table 22.3) and in aqueous solution acts as a dibasic acid (equations 22.99 and 22.100), forming salts such as $NaH_2PO_3$ and $Na_2HPO_3$.

$$H_3PO_3(aq) + H_2O(l) \rightleftharpoons [H_3O]^+(aq) + [H_2PO_3]^-(aq) \qquad pK_a = 2.00 \quad (22.99)$$

$$[H_2PO_3]^-(aq) + H_2O(l) \rightleftharpoons [H_3O]^+(aq) + [HPO_3]^{2-}(aq)$$

$$pK_a = 6.59 \quad (22.100)$$

Salts containing the $[HPO_3]^{2-}$ ion are called *phosphonates* but the name 'phosphite' is still commonly used. This can cause confusion since 'phosphite'

## Box 22.11 ATP and fireflies: bioluminescence

Fireflies (visible as faint streaks of green light) gathered over bean fields in Iowa, USA.    *Keith Kent/Science Photo Library.*

The firefly is a common sight at dusk in North America. This remarkable insect attracts its mate by emitting

flashes of light from its rear body. This is called *bioluminescence* and is caused by a light-emitting chemical reaction. A compound called luciferin is oxidized in this process. The overall reaction uses an enzyme catalyst (luciferase) and the biological fuel ATP which is reduced to adenosine monophosphate (AMP) and inorganic phosphate. Many bioluminescent reactions are known and a walk through the fields or woods at night in summer or autumn may reveal many spectacular light-emitting insects and fungi.

Luciferin

---

**(22.39)**

**PH$_3$ is also called phosphane**

is also used for organic compounds of formula $P(OR)_3$, e.g. $P(OMe)_3$ **(22.39)** is trimethylphosphite. There is *no inorganic acid* of formula $P(OH)_3$. It is a common mistake to think that the formula $H_3PO_3$ (see Table 22.3) refers to a tribasic acid of this structural type.

Pure phosphinic acid $H_3PO_2$ is difficult to isolate as it readily decomposes (reaction 22.101). It forms deliquescent, white crystals which rapidly absorb water to give an oily liquid. In aqueous solution, the acid is monobasic ($pK_a = 1.24$) (equation 22.102).

$$2H_3PO_2 \longrightarrow PH_3 + H_3PO_4 \qquad (22.101)$$
$$\text{Phosphine}$$

$$H_3PO_2(aq) + H_2O(l) \rightleftharpoons [H_3O]^+(aq) + [H_2PO_2]^-(aq) \qquad (22.102)$$
$$\text{Phosphinate ion}$$

Phosphinate salts are produced when white phosphorus reacts with aqueous alkali (equation 22.103).

$$P_4 + 4[OH]^- + 4H_2O \longrightarrow 4[H_2PO_2]^- + 2H_2 \qquad (22.103)$$

### Halides of nitrogen and phosphorus

**Dipole moment of NX$_3$: see Section 6.13**

Nitrogen forms the trihalides $NF_3$ and $NCl_3$ **(22.40)**. Although $NBr_3$ and $NI_3$ exist, they are less well characterized than $NF_3$ and $NCl_3$. At room temperature, $NCl_3$ is an explosive liquid. It hydrolyses rapidly in water (equation 22.104).

**Hypochlorous acid: see Section 22.12**

$$NCl_3 + 3H_2O \longrightarrow NH_3 + 3HOCl \qquad (22.104)$$
$$\text{Hypochlorous acid}$$

Nitrogen trifluoride can be produced by reaction 22.105. It is a colourless gas at 298 K and is thermodynamically more stable than $NCl_3$. $NF_3$ resists attack

X = F or Cl

**(22.40)**

by water and dilute acids and alkalis, but is an effective fluorinating agent, converting some elements into their respective fluorides.

$$4NH_3 + 3F_2 \xrightarrow{\text{copper catalyst}} NF_3 + 3NH_4F \qquad (22.105)$$

When $NF_3$ is heated with copper (equation 22.106), dinitrogen tetrafluoride is formed. This colourless gas exists in equilibrium with $NF_2$ radicals (equation 22.107).

$$2NF_3 + Cu \xrightarrow{\Delta} N_2F_4 + CuF_2 \qquad (22.106)$$

$$\underset{\text{Colourless}}{N_2F_4(g)} \rightleftharpoons \underset{\text{Blue}}{2NF_2(g)} \qquad\qquad K(420\,K) = 0.03 \quad (22.107)$$

Dinitrogen tetrafluoride (Figure 22.22) adopts both staggered and *gauche* conformers at 298 K; compare this with hydrazine (**21.5**). $N_2F_4$ is very reactive and is a powerful fluorinating agent (reactions 22.108 and 22.109). It reacts with *fluoride acceptors* such as $AsF_5$ to give salts (equation 22.110).

Fluoride acceptors:
see Section 22.9

$$N_2F_4 + 10Li \xrightarrow{\Delta} 4LiF + 2Li_3N \qquad (22.108)$$

$$N_2F_4 + SiH_4 \longrightarrow SiF_4 + N_2 + 2H_2 \qquad (22.109)$$

$$N_2F_4 + AsF_5 \longrightarrow [N_2F_3]^+[AsF_6]^- \qquad (22.110)$$

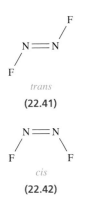

*trans*

**(22.41)**

*cis*

**(22.42)**

The difluoride $N_2F_2$ is planar and exists as both the *cis*- and *trans*-isomers (**22.41** and **22.42**), *cis*-$N_2F_2$ being more thermodynamically stable, but also more reactive.[§] The *trans*-isomer can be selectively prepared by reaction 22.111, but on heating to 370 K, it isomerizes to give an equilibrium mixture of isomers in which *cis*-$N_2F_2$ predominates.

$$2N_2F_4 + 2AlCl_3 \xrightarrow{200\,K} \textit{trans-}N_2F_2 + N_2 + 3Cl_2 + 2AlF_3 \qquad (22.111)$$

In comparing the halides of nitrogen with those of phosphorus, we must bear in mind that:

- N–N $2p$–$2p$ $\pi$-interactions are stronger than P–P $3p$–$3p$ $\pi$-interactions, and as a consequence, there is no phosphorus analogue of $N_2F_2$;
- phosphorus forms halides of the type $PX_5$ (see also Section 7.3).

The phosphorus trihalides $PX_3$ (X = F, Cl, Br or I) have trigonal pyramidal structures. At 298 K, $PF_3$ is a colourless gas, $PCl_3$ and $PBr_3$ are colourless liquids and $PI_3$ is a red crystalline solid. $PF_3$ is highly toxic and, like CO, reacts with the iron centre in haemoglobin. In contrast to the later trihalides, $PF_3$ hydrolyses slowly in moist air (equation 22.112).

Haemoglobin: see Box 5.3

$$PF_3 + 3H_2O \longrightarrow 3HF + H_3PO_3 \qquad (22.112)$$

The compounds $PCl_3$, $PBr_3$ and $PI_3$ may be prepared from their respective elements. $PF_3$ is made from $PCl_3$ using $CaF_2$ as the fluorinating agent. $PCl_3$ is manufactured on a large scale, and is a precursor to many *organophosphorus compounds* used as flame retardants and fuel additives, and in the formation of insecticides and nerve gases for chemical warfare (see Box 22.12).

---

[§] Inorganic nomenclature generally retains *cis*- and *trans*- rather than (*Z*)- and (*E*)-isomers; see Chapter 26.

## COMMERCIAL AND LABORATORY APPLICATIONS

### Box 22.12 Chemical weapons

In the First and Second World Wars, chemical weapons based on simple toxic chemicals such as phosgene and mustard gas were available for active use and also as a deterrent. During and after the Second World War, the development of chemical weapons moved to nerve gases, for example Sarin and Soman. These agents work by enzyme inhibition in the nervous system and inhalation of as little as 1 mg is fatal. Sarin was the nerve gas used in the attacks in Japan that brought chaos to commuters in March 1995.

Ways must be found to destroy chemical weapons that are stockpiled, since current policy in most countries is towards chemical weapon disarmament. For example, Sarin may be chemically destroyed by hydrolysis:

$$(Me_2HCO)P(O)(Me)F + H_2O \longrightarrow$$
$$(Me_2HCO)P(O)(Me)OH + HF$$
$$\downarrow H_2O$$
$$Me_2HCOH + MeP(O)(OH)_2$$

and if aqueous sodium hydroxide is used for the reaction, effectively harmless sodium salts are formed.

The Sarin gas attack carried out on the Tokyo subway in March 1995. *EMPICS/PA*.

Phosgene    Mustard gas    Sarin    Soman

---

(22.43)

The reaction between water and $PCl_3$ is vigorous (equation 22.113), while oxidation with $O_2$ gives the oxychloride $POCl_3$ (**22.43**) (equation 22.114).

$$PCl_3 + 3H_2O \longrightarrow 3HCl + H_3PO_3 \tag{22.113}$$
$$2PCl_3 + O_2 \longrightarrow 2POCl_3 \tag{22.114}$$

All four $P_2X_4$ compounds (X = F, Cl, Br or I) are known, but $P_2I_4$ is the most stable member of the series. It can be prepared by heating red phosphorus with $I_2$, and is a red crystalline solid at 298 K. It consists of $P_2I_4$ molecules with a staggered conformation (as in Figure 22.22).

The phosphorus(V) halides exist as trigonal bipyramidal molecules in the gas phase, consistent with VSEPR theory, but only $PF_5$ is a gas at 298 K. $PCl_5$, $PBr_5$ and $PI_5$ are *ionic* crystalline solids. For $PCl_5$, chloride transfer occurs to give a solid consisting of tetrahedral $[PCl_4]^+$ cations and octahedral $[PCl_6]^-$ anions. In the solid bromide and iodide, free halide ions are present and the compounds are formulated as $[PBr_4]^+Br^-$ and $[PI_4]^+I^-$. The molecular structure of $PF_5$ (**22.44**) possesses two distinct fluorine sites (axial and equatorial). However, the $^{19}F$ NMR spectrum of a

F ←—— Axial

F —— P<sup>\\\\\</sup>F
    F } Equatorial

F ←—— Axial

(22.44)

**Fig. 22.22** $N_2F_4$ exists in both the staggered and *gauche* conformers, shown here as molecular structures and Newman projections (see Section 24.8). Colour code: N, blue; F, green.

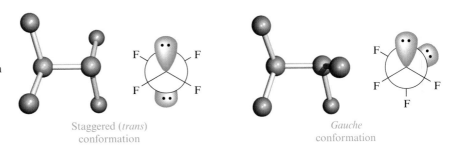

Staggered (*trans*)
conformation

*Gauche*
conformation

Stereochemically non-rigid: see Section 6.12

$$\begin{bmatrix} F \\ F''''' \overset{|}{\underset{|}{P}} ''''' F \\ F \overset{}{\diagup} \overset{|}{\underset{F}{\diagdown}} F \end{bmatrix}^{-}$$

**(22.45)**

solution of $PF_5$ shows only one signal, because $PF_5$ is stereochemically non-rigid and the fluorine nuclei are equivalent on the NMR spectroscopic timescale.

Of the phosphorus(V) halides, $PCl_5$ is the most important, and its synthetic use is widespread. It is made industrially from $PCl_3$ (equation 22.115) and may be converted to $PF_5$ by reaction with a fluorinating agent such as $CaF_2$. However, $PF_5$ itself is a good fluoride acceptor and fluorination of $PCl_5$ with KF gives $K[PF_6]$ containing the $[PF_6]^-$ ion (**22.45**). $PF_5$ and $PCl_5$ are moisture-sensitive (equation 22.116) but if the conditions of $PCl_5$ hydrolysis are controlled, $POCl_3$ can be produced (equation 22.117).

$$PCl_3 + Cl_2 \longrightarrow PCl_5 \tag{22.115}$$

$$PCl_5 + 4H_2O \longrightarrow H_3PO_4 + 5HCl \tag{22.116}$$

$$PCl_5 + H_2O \longrightarrow POCl_3 + 2HCl \tag{22.117}$$

**22.9**

## Group 15: arsenic, antimony and bismuth

| 1 | 2 | | 13 | 14 | **15** | 16 | 17 | 18 |
|---|---|---|----|----|----|----|----|----|
| H | | | | | | | | He |
| Li | Be | | B | C | **N** | O | F | Ne |
| Na | Mg | | Al | Si | **P** | S | Cl | Ar |
| K | Ca | | Ga | Ge | **As** | Se | Br | Kr |
| Rb | Sr | *d*-block | In | Sn | **Sb** | Te | I | Xe |
| Cs | Ba | | Tl | Pb | **Bi** | Po | At | Rn |
| Fr | Ra | | | | | | | |

Arsenic is a steel-grey, brittle semi-metal and occurs in the elemental state and in the sulfide ores *arsenopyrite* (FeAsS), *realgar* ($As_4S_4$) and *orpiment* ($As_2S_3$). When arsenopyrite is heated, it decomposes to give arsenic and FeS. Compounds of arsenic are toxic – much beloved by crime-writers! Gallium arsenide has widespread uses in solid state devices and semiconductors.

Antimony is not an abundant element and is found in various minerals, of which *stibnite* $Sb_2S_3$ is the most important. Antimony can be extracted by reduction using scrap iron (equation 22.118).

$$Sb_2S_3 + 3Fe \longrightarrow 2Sb + 3FeS$$

reduction

oxidation

(22.118)

Antimony is a toxic, brittle, blue-white metal, which is a poor electrical and thermal conductor. It resists oxidation in air at 298 K, but is converted to oxides at higher temperatures (see below). Some of its uses are in the semiconductor industry, in alloys (e.g. it increases the strength of lead) and in batteries.

Bismuth is a soft, grey metal, occurring in its elemental state and in *bismite* ($Bi_2O_3$), *bismuthinite* ($Bi_2S_3$) and in some copper-, lead- and silver-containing ores. It is a component of alloys (e.g. with tin) and compounds of bismuth are used in the cosmetics industry, as oxidation catalysts and in high-temperature superconductors.

## Oxides

When heated with $O_2$, arsenic forms the highly toxic $As_2O_3$ (equation 22.119) which has a smell resembling garlic. This oxide is an important precursor to many arsenic compounds and is manufactured by several routes, including roasting the sulfide ores realgar or orpiment in air.

$$4As + 3O_2 \longrightarrow 2As_2O_3 \qquad (22.119)$$

At 298 K, $As_2O_3$ is a crystalline solid containing $As_4O_6$ units, and in the vapour state, $As_4O_6$ molecules, similar to $P_4O_6$ (Figure 22.21), are present. Unlike $P_4O_6$, $As_4O_6$ does not readily oxidize to arsenic(V) oxide. Instead, $As_4O_{10}$ can be formed by first making $H_3AsO_4$ (equation 22.120). As this reaction suggests, $As_4O_{10}$ dissolves in water to give arsenic acid, $H_3AsO_4$.

$$2As_2O_3 \xrightarrow{\text{Conc HNO}_3} 4H_3AsO_4 \xrightarrow{\text{Dehydration}} As_4O_{10} + 6H_2O \qquad (22.120)$$

Antimony(III) oxide, $Sb_2O_3$, forms when antimony burns in $O_2$ and, structurally, it is similar to $As_2O_3$. It is used as a flame retardant and in paints, adhesives and plastics. Oxidation to antimony(V) oxide occurs when $Sb_2O_3$ reacts with high pressures of $O_2$ at elevated temperatures. Neither arsenic(V) nor antimony(V) oxide resembles $P_4O_{10}$ in structure. Bismuth(III) oxide (the mineral *bismite*) has many uses in the glass and ceramic industry, and for catalysts and magnets. Bismuth(V) oxide is very unstable.

The properties of the group 15 oxides are summarized as follows:

- each element, E, from P to Bi forms two oxides containing E(III) or E(V);
- the stabilities of the E(V) oxides decrease down the group;
- the E(V) oxides and $P_4O_6$ are acidic;
- the As(III) and Sb(III) oxides are amphoteric;
- $Bi_2O_3$ is basic.

## Oxoacids and related salts

The later members of group 15 form fewer oxoacids than phosphorus. Arsenic acid, $H_3AsO_4$ is tribasic (equations 22.121–22.123).

$$H_3AsO_4(aq) + H_2O(l) \rightleftharpoons [H_3O]^+(aq) + [H_2AsO_4]^-(aq)$$

$$pK_a = 2.25 \quad (22.121)$$

$$[H_2AsO_4]^-(aq) + H_2O(l) \rightleftharpoons [H_3O]^+(aq) + [HAsO_4]^{2-}(aq)$$

$$pK_a = 6.77 \quad (22.122)$$

$$[HAsO_4]^{2-}(aq) + H_2O(l) \rightleftharpoons [H_3O]^+(aq) + [AsO_4]^{3-}(aq)$$

$$pK_a = 11.60 \quad (22.123)$$

In contrast to $H_3PO_4$, arsenic acid is a good oxidizing agent as the $E^o$ value for half-reaction 22.124 indicates.

$$H_3AsO_4(aq) + 2H^+(aq) + 2e^- \rightleftharpoons HAsO_2(aq) + 2H_2O(l)$$

$$E^o = +0.56\,V \quad (22.124)$$

The pure acid '$H_3AsO_3$' has not been isolated, but *arsenite* salts containing $[AsO_3]^{3-}$ are known. Although antimonites are well-characterized salts, oxoacids of antimony(III) are not stable. The trend continues for bismuth, and no oxoacids are known, although some bismuthate salts are. Sodium bismuthate is an orange solid which is insoluble in water and is a powerful oxidizing agent.

## Halides

Each of arsenic, antimony and bismuth forms a halide $EX_3$ ($X = F$, Cl, Br or I). We focus here on the fluorides and chlorides. Arsenic(III) fluoride is a colourless liquid (mp 267 K, bp 330 K) and may be prepared from arsenic and $F_2$, but glass containers should be avoided for storage, because, in the presence of moisture, $AsF_3$ reacts with silica. The trichloride can be prepared by the reaction of $As_2O_3$ with HCl (equation 22.125) and $AsCl_3$ is a colourless liquid at 298 K. Both $AsF_3$ and $AsCl_3$ have molecular structures (**22.46**) in the solid, liquid and gas phases. Each of the arsenic trihalides is hydrolysed by water, liberating HX.

$$X = F \text{ or } Cl$$

**(22.46)**

$$As_2O_3 + 6HCl \longrightarrow 2AsCl_3 + 3H_2O \quad (22.125)$$

Arsenic(V) fluoride is the only halide of arsenic to be stable in the +5 oxidation state. It is a colourless gas at 298 K, with a trigonal bipyramidal structure (**22.47**). It can be made by reaction 22.126 in which $Br_2$ acts as an oxidizing agent. $AsF_5$ is an excellent *fluoride acceptor* as we discuss below.

$$AsF_3 + 2SbF_5 + Br_2 \longrightarrow AsF_5 + 2SbBrF_4 \quad (22.126)$$

In contrast to gaseous $AsF_3$, $SbF_3$ is a colourless crystalline solid (mp 563 K). Trigonal pyramidal molecular units are present in the solid, but fluorine atoms form bridges between Sb atoms such that each Sb is 6- (rather than 3-) coordinate. $SbF_3$ is prepared by reacting $Sb_2O_3$ with HF, and is widely used as a fluorinating agent (e.g. reaction 22.127). However, reactions may be complicated by $SbF_3$ acting as an oxidizing agent *in addition to* donating fluorine.

$$SiCl_4 \xrightarrow{\ SbF_3\ } SiCl_3F + SiCl_2F_2 + SiClF_3 + SiF_4 \quad (22.127)$$

The reaction of $Sb_2O_3$ with HCl gives $SbCl_3$. This is a solid at 298 K, forming white, deliquescent crystals.

$$SbF_3 + F_2 \longrightarrow SbF_5 \quad (22.128)$$

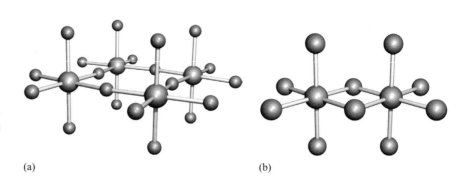

**Fig. 22.23** (a) Solid antimony(V) fluoride contains tetrameric molecules. Each Sb atom is 6-coordinate, but the stoichiometry is Sb : F = 1 : 5. (b) Solid antimony(V) chloride consists of dimers, $Sb_2Cl_{10}$; the coordination number of Sb is 6 but the stoichiometry Sb : Cl = 1 : 5. Colour code: Sb, orange; F and Cl, green.

(a)                                     (b)

Difluorine oxidizes $SbF_3$ to $SbF_5$ (equation 22.128). This is a viscous liquid (mp 281 K) due to the formation of Sb−F−Sb bridges between $SbF_5$ units. In the solid state, $SbF_5$ is tetrameric (Figure 22.23a). The oxidation of $SbCl_3$ with $Cl_2$ gives $SbCl_5$. In the liquid, trigonal bipyramidal molecules are present, but the solid (mp 276 K) consists of dimers with Sb−Cl−Sb bridges (Figure 22.23b).

### The Lewis acidity of arsenic and antimony halides

An important property of arsenic and antimony trihalides and pentahalides is their Lewis acidity. The solid state structures of $SbF_5$ and $SbCl_5$ illustrate the tendency of Sb to accept a lone pair of electrons from a halide with the formation of $\{SbF_5\}_4$ and $\{SbCl_5\}_2$ (Figure 22.23). Halide bridge formation involves the donation of a lone pair of electrons, but *not* the complete transfer of a halide ion. When complete halide *transfer* occurs, ionic products are obtained as in reactions 22.129 and 22.130. $SbCl_5$ is a particularly strong chloride acceptor.

$$\underset{\text{Chloride acceptor}}{SbCl_5} + \underset{\text{Chloride donor}}{AlCl_3} \longrightarrow [AlCl_2]^+[SbCl_6]^- \qquad (22.129)$$

$$\underset{\text{Chloride acceptor}}{SbCl_5} + \underset{\text{Chloride donor}}{PCl_5} \longrightarrow [PCl_4]^+[SbCl_6]^- \qquad (22.130)$$

In anhydrous hydrogen fluoride, $SbF_5$ acts as a Lewis acid and accepts $F^-$ as shown in equation 22.131.

$$2HF + SbF_5 \rightleftharpoons [H_2F]^+ + [SbF_6]^- \qquad (22.131)$$

Antimony(V) and arsenic(V) fluorides react with alkali metal fluorides or with FNO to give salts (scheme 22.132).

$$\left.\begin{array}{c} MF \\ \text{e.g. M = Na, K, Cs} \\ \text{or} \\ FNO \end{array}\right\} \begin{array}{c} \xrightarrow{AsF_5} M^+[AsF_6]^- \text{ or } [NO]^+[AsF_6]^- \\ \\ \xrightarrow{SbF_5} M^+[SbF_6]^- \text{ or } [NO]^+[SbF_6]^- \end{array} \qquad (22.132)$$

The $[AsF_6]^-$ anion is octahedral, but $[SbF_6]^-$ shows a tendency to associate with Lewis bases; for example, with $SbF_5$, F⟶Sb coordinate bond formation leads to the formation of $[Sb_2F_{11}]^-$ (Figure 22.24).

Coordinate bond formation is not restricted to halides, and adducts with oxygen and nitrogen donors are known. When $SbF_5$ reacts with anhydrous $HSO_3F$ (equation 22.133), the product is a *super-acid* because it is capable of protonating exceptionally weak bases.

$$2HSO_3F + SbF_5 \rightleftharpoons [H_2SO_3F]^+ + [SbF_5OSO_2F]^- \qquad (22.133)$$

**(22.48)**

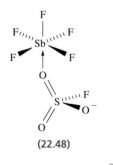

(22.48)

HSO$_3$F: see Section 22.10

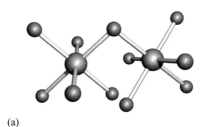

**Fig. 22.24** (a) The structure of the $[Sb_2F_{11}]^-$ anion, and (b) a representation of the structure indicating coordinate bond formation between $[SbF_6]^-$ and $[SbF_5]$. Colour code: Sb, orange; F, green.

(a)                                    (b)

**22.10**     **Group 16: oxygen and sulfur**

| 1 | 2 | | 13 | 14 | 15 | **16** | 17 | 18 |
|---|---|---|---|---|---|---|---|---|
| H | | | | | | | | He |
| Li | Be | | B | C | N | **O** | F | Ne |
| Na | Mg | | Al | Si | P | **S** | Cl | Ar |
| K | Ca | *d*-block | Ga | Ge | As | **Se** | Br | Kr |
| Rb | Sr | | In | Sn | Sb | **Te** | I | Xe |
| Cs | Ba | | Tl | Pb | Bi | **Po** | At | Rn |
| Fr | Ra | | | | | | | |

The group 16 elements are called the *chalcogens*. Dioxygen makes up 21% of the Earth's atmosphere (see Figure 21.1). Oxygen-containing compounds (e.g. water, sand ($SiO_2$), limestone, silicates, bauxite and haematite ($Fe_2O_3$)) make up 49% of the Earth's crust. It is a component of many hundreds of thousands of compounds and is essential to life, being taken in and converted to $CO_2$ during respiration. Reactions 22.134 and 22.135 are convenient laboratory preparations of $O_2$, and a mixture of $KClO_3$ and $MnO_2$ used to be sold as 'oxygen mixture'. It can be produced by the electrolysis of water (equation 22.136), and on an industrial scale, $O_2$ is obtained by the liquefaction and fractional distillation of air.

$$2H_2O_2 \xrightarrow{\text{MnO}_2 \text{ or Pt catalyst}} O_2 + 2H_2O \tag{22.134}$$

■▷
Caution! Chlorates are
potentially explosive

$$2KClO_3 \xrightarrow{\Delta, \text{ MnO}_2 \text{ catalyst}} 3O_2 + 2KCl \tag{22.135}$$
Potassium chlorate

$$\left.\begin{array}{ll} \textit{At the anode}: & 4[OH]^-(aq) \longrightarrow O_2(g) + 2H_2O(l) + 4e^- \\ \textit{At the cathode}: & 2[H_3O]^+ + 2e^- \longrightarrow H_2(g) + 2H_2O(l) \end{array}\right\} \tag{22.136}$$

■▷
The diradical $O_2$:
see Section 4.18

Dioxygen is a colourless, odourless gas (bp 90 K), but liquid $O_2$ is blue. It is paramagnetic owing to the presence of two unpaired electrons and is very reactive. Most elements combine with $O_2$ to form oxides. Electric discharges (for example in thunderstorms) or UV light convert $O_2$ into ozone (equation 22.137). Ozone is a perceptibly blue, toxic gas with a characteristic 'electric'

**Fig. 22.25** The structure of the ozone molecule $O_3$, and valence bond representations of the molecule. The contribution made by the cyclic resonance structure means that the average $O-O$ bond order will be less than 1.5.

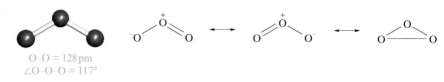

$O-O = 128\,pm$
$\angle O-O-O = 117°$

The ozone layer: see Box 25.3

smell. Molecules of $O_3$ are bent (Figure 22.25) with an $O-O$ bond distance (128 pm) longer than in $O_2$ (121 pm) consistent with a lower bond order. The layer of ozone in the Earth's atmosphere is essential in preventing harmful UV radiation reaching the planet.

$$3O_2 \xrightarrow[\text{or } h\nu]{\text{Electrical discharge}} \underset{\text{Ozone}}{2O_3} \tag{22.137}$$

Allotropes of sulfur: see Section 9.8

Elemental sulfur occurs in deposits around volcanoes and hot springs, as well as in minerals including *iron pyrites* ($FeS_2$, also called *fool's gold*, and which contains the $[S_2]^{2-}$ anion), *galena* (PbS), *sphalerite* (or zinc blende ZnS), *cinnabar* (HgS), *realgar* ($As_4S_4$) and *stibnite* ($Sb_2S_3$). Sulfur has, traditionally, been produced using the Frasch process. Superheated water (440 K under pressure) is used to melt sulfur deposits, and compressed air then forces it to the surface. For environmental reasons, the Frasch process is being replaced by methods that recover sulfur from crude petroleum refining and natural gas production. In natural gas, the source of sulfur is $H_2S$. It occurs in concentrations of up to 30%, and sulfur is recovered using reaction 22.138.

$$2H_2S + O_2 \xrightarrow{\text{Activated carbon or alumina catalyst}} 2S + 2H_2O \tag{22.138}$$

A major use of sulfur is in the production of sulfuric acid, ranked number one in manufactured chemical products in the US. On an industrial level, it is also incorporated into a wide range of sulfur-containing organic compounds and inorganic compounds. These include $CaSO_4$, $[NH_4]_2SO_4$, $CS_2$, $H_2S$ and $SO_2$.

## Oxides of sulfur

The two most important oxides of sulfur are sulfur dioxide and sulfur trioxide. $SO_2$ is formed when sulfur burns in air or $O_2$ (equation 22.139), or by roasting sulfide ores with $O_2$ (equation 22.140). This is the first step in the manufacture of sulfuric acid (see below).[§]

$$S(s) + O_2(g) \xrightarrow{\Delta} SO_2(g) \tag{22.139}$$

$$4FeS_2 + 11O_2 \xrightarrow{\Delta} 2Fe_2O_3 + 8SO_2 \tag{22.140}$$

Reactions 22.141 or 22.142 are convenient laboratory preparations of $SO_2$ and the gas is best dried by passage through concentrated sulfuric acid.

$$\underset{\text{Conc}}{Na_2SO_3(s) + 2HCl(aq)} \longrightarrow SO_2(g) + 2NaCl(aq) + H_2O(l) \tag{22.141}$$

$$\underset{\text{Conc}}{Cu(s) + 2H_2SO_4(aq)} \longrightarrow SO_2(g) + CuSO_4(aq) + 2H_2O(l) \tag{22.142}$$

Sulfur dioxide is a colourless, dense and toxic gas with a choking smell. It is easily liquefied (bp 263 K) and liquid $SO_2$ is widely used as a non-aqueous

---

[§] In equations, we represent elemental sulfur simply as 'S' and do not specify whether $S_8$ rings or other allotropes are present.

**Fig. 22.26** A graph illustrating the temperature-dependence of the equilibrium:

$$2SO_2(g) + O_2(g) \rightleftharpoons 2SO_3(g)$$

As the temperature is raised, $\ln K$ *decreases*, meaning that there is less $SO_3$ in the equilibrium mixture. In industrial processes in which $SO_3$ is produced, a temperature of about 700 K is used.

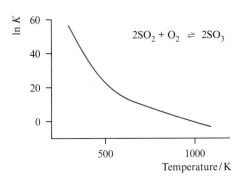

solvent. It dissolves in water to give sulfurous acid, but equilibrium 22.143 lies well to the left-hand side. Hence, solutions contain significant amounts of dissolved $SO_2$. This situation parallels that of $CO_2$ in water.

$$SO_2(aq) + H_2O(l) \rightleftharpoons H_2SO_3(aq) \qquad\qquad K < 10^{-9} \quad (22.143)$$

The $SO_2$ molecule is bent (**22.49**). Although structure **22.49** is a convenient representation of the $SO_2$ molecule, the pair of resonance structures **22.50** gives a more realistic picture of the bonding in the molecule. In each contributing resonance structure, each atom obeys the octet rule.

Sulfur dioxide reacts with $COCl_2$ to give *thionyl dichloride* (equation 22.144) – commonly called 'thionyl chloride' – and it is prepared industrially by reaction 22.145. It is used commercially in the synthesis of anhydrous metal chlorides and for the conversion of organic $-OH$ to $-Cl$ groups. It is a colourless liquid at 298 K, consisting of trigonal pyramidal molecules (**22.51**), and is rapidly hydrolysed by water (equation 22.146).

$$SO_2 + COCl_2 \longrightarrow SOCl_2 + CO_2 \qquad\qquad (22.144)$$

$$2SO_2 + S_2Cl_2 + 3Cl_2 \longrightarrow 4SOCl_2 \qquad\qquad (22.145)$$

$$SOCl_2 + H_2O \longrightarrow SO_2 + 2HCl \qquad\qquad (22.146)$$

Sulfur dioxide is oxidized to $SO_3$ by $O_2$ in the presence of a catalyst such as vanadium(V) oxide. Figure 22.26 shows that equilibrium 22.147 shifts further towards the left-hand side as the temperature is raised, and industrial processes in which $SO_3$ is produced use a relatively low operating temperature ($\approx 700$ K). The importance of this equilibrium lies in its role in the formation of sulfuric acid (see below).

$$2SO_2(g) + O_2(g) \rightleftharpoons 2SO_3(g) \qquad\qquad (22.147)$$

Sulfur dioxide is a weak reducing agent in acidic solution, and a rather stronger one in basic aqueous solution (when $[SO_3]^{2-}$ and $[HSO_3]^-$ ions are formed) as indicated by the $E^o$ values for half-equations 22.148 and 22.149. The *oxidation* product in each case is the *sulfate ion*. Although we are considering the reactivity of $SO_2$, these equations show the presence of $H_2SO_3$ or $[SO_3]^{2-}$ in acidic or alkaline aqueous conditions rather than indicating a direct involvement by $SO_2$. This is because the aqueous solutions of $SO_2$ behave as though they contain these species as we discuss later in this section.

$$[SO_4]^{2-}(aq) + 4H^+(aq) + 2e^- \rightleftharpoons H_2SO_3(aq) + H_2O(l)$$
$$E^o = +0.17\,V \quad (22.148)$$

$$[SO_4]^{2-}(aq) + H_2O(l) + 2e^- \rightleftharpoons [SO_3]^{2-}(aq) + 2[OH]^-(aq)$$
$$E^o = -0.93\,V \quad (22.149)$$

S  143 pm
O   O
119°

**(22.49)**

**(22.50)**

**Use of $SOCl_2$: see Section 28.2**

**(22.51)**

## ENVIRONMENT AND BIOLOGY

### Box 22.13 Acid rain

Sulfur dioxide is produced naturally when volcanoes erupt, but the combustion of fossil fuels (e.g. coal and oil) releases large quantities of $SO_2$ into the atmosphere and these emissions are concentrated in the industrial regions of the world. The $SO_2$ dissolves in atmospheric water vapour and forms $H_2SO_3$ and $H_2SO_4$. The formation of the acids takes time – the reactions leading to the formation of $H_2SO_4$ involve radicals:

$$SO_2 + O_2 + OH^{\bullet} \longrightarrow SO_3 + HO_2^{\bullet}$$

$$SO_3 + H_2O \rightleftharpoons H_2SO_4$$

The acids are transported in the atmosphere in directions that are dictated by the prevailing winds. In Europe, $SO_2$ from the industrialized areas of the United Kingdom, France and Germany finds its way to Scandinavia and is deposited on the landscape as *acid rain*. One of the consequences of acid rain is that trees die, and when the acid rain enters natural water courses, fish die and this in turn affects other wildlife which depend upon the fish for food. Acid rain falling on stone buildings causes corrosion – the ever-disappearing faces of gargoyles on ancient churches are reminders of the problems of pollution. The photograph opposite shows one of the stone lions outside Leeds Town Hall in the north of England. The loss of clearly defined features

on the lion is a consequence of acid rain damage.

Increasing efforts to control environmental pollution have led to the development of methods to 'scrub' industrial waste gases. In *desulfurization* processes, $SO_2$ is removed from emissions into the atmosphere. This can be achieved by neutralizing $H_2SO_4$ using $Ca(OH)_2$, $CaCO_3$ or $NH_3$.

$NO_x$ emissions, discussed in Section 22.8, also contribute to the problem of acid rain.

The effects of acid rain damage.
*Adam Hart-Davis/Science Photo Library.*

---

O
120°  || 143 pm
S
O       O

**(22.52)**

At 298 K, $SO_3$ is a volatile white solid which is usually stored in sealed vessels. There are three polymorphs of which α-$SO_3$ (mp 290 K) and β-$SO_3$ (mp 335 K) crystallize as silky, fibrous needles. Both α- and β-$SO_3$ possess extended structures. In the gas phase, trigonal planar $SO_3$ molecules (**22.52**) exist. In the liquid state, monomers and some trimers (Figure 22.27) are present. The bonding in monomeric $SO_3$ is best described in terms of a set of resonance structures **22.53**. Structure **22.52** is, however, a convenient way of representing the molecule.

(22.53)

**Fig. 22.27** The structure of the trimer $S_3O_9$, present (along with monomeric $SO_3$) in liquid sulfur trioxide. In the monomer, each sulfur atom is in a trigonal planar environment, but in the trimer, each is tetrahedrally sited. Colour code: S, yellow; O, red.

$SO_3$ reacts extremely vigorously with water (equation 22.150) but concentrated $H_2SO_4$ dissolves $SO_3$ to give *oleum*, consisting of a mixture of polysulfuric acids (see below). The reaction of oleum with water gives concentrated $H_2SO_4$ in a less vigorous reaction.

$$SO_3(g) + H_2O(l) \longrightarrow H_2SO_4(aq) \tag{22.150}$$

With HF and HCl, $SO_3$ reacts to give fluoro- or chlorosulfuric acid (equation 22.151). Anhydrous $HSO_3F$ undergoes self-ionization (equation 22.152) –

**(22.54)**

the equilibrium lies well over to the left-hand side, but is pushed towards the right in the presence of $SbF_5$ (equation 22.133). The combination of $HSO_3F$ and $SbF_5$ generates one of the strongest acids known (*magic acid*).

$$SO_3 + HF \longrightarrow HSO_3F \qquad (22.151)$$
$$\text{(22.54)}$$

$$2HSO_3F \rightleftharpoons [H_2SO_3F]^+ + [SO_3F]^- \qquad K \approx 10^{-8} \quad (22.152)$$

## Oxoacids of sulfur and their salts

Selected oxoacids of sulfur, of which sulfuric acid is by far the most important, are listed in Table 22.4. Sulfuric acid is manufactured from $SO_2$ by the *Contact process*. $SO_2$ is first converted to $SO_3$ (equilibrium 22.147), the $SO_3$ is dissolved in concentrated $H_2SO_4$ to give oleum, dilution of which with water yields $H_2SO_4$. Uses of sulfur and sulfuric acid are summarized in Figure 22.28).

Pure $H_2SO_4$ is a viscous liquid due to extensive intermolecular hydrogen bonding. Self-ionization occurs (equation 22.153), but is complicated by dehydration steps such as that in equation 22.154.

$$\underset{\text{Pure acid}}{2H_2SO_4(l)} \rightleftharpoons [H_3SO_4]^+(\text{solv}) + [HSO_4]^-(\text{solv}) \qquad K \approx 1.7 \times 10^{-4} \quad (22.153)$$

$$H_2SO_4 + [HSO_4]^- \rightleftharpoons H_2O + [HS_2O_7]^- \qquad (22.154)$$

**Table 22.4** Selected oxoacids of sulfur. The number of OH groups determines the basicity of the acid; notice that *unlike* the case for phosphorus where some acids contain direct P–H bonds (Table 22.3), there are no oxoacids of sulfur containing direct S–H bonds.

| Formula | Name | Structure |
|---------|------|-----------|
| $H_2SO_4$ | Sulfuric acid | |
| $H_2S_2O_3$ | Thiosulfuric acid | |
| $H_2S_2O_7$ | Disulfuric acid | |
| $H_2S_2O_8$ | Peroxydisulfuric acid | |
| $H_2SO_3$ | Sulfurous acid | |

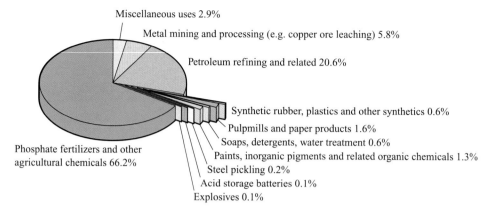

Phosphate fertilizers and other agricultural chemicals 66.2%

Miscellaneous uses 2.9%
Metal mining and processing (e.g. copper ore leaching) 5.8%
Petroleum refining and related 20.6%
Synthetic rubber, plastics and other synthetics 0.6%
Pulpmills and paper products 1.6%
Soaps, detergents, water treatment 0.6%
Paints, inorganic pigments and related organic chemicals 1.3%
Steel pickling 0.2%
Acid storage batteries 0.1%
Explosives 0.1%

**Fig. 22.28**  Uses of sulfur and sulfuric acid (by sulfur content) in the US in 2001. [Data: US Geological Survey.]

Concentrated $H_2SO_4$ is a powerful dehydrating agent. If it is poured carefully onto a pile of sugar crystals, hydrogen and oxygen are removed from the sugar as water (equation 22.155) and a sticky, black deposit of carbon is left. The black product appears to 'grow' rather like a volcano erupting as gaseous CO and $CO_2$ are produced by oxidation processes that accompany the reaction.

$$C_{12}H_{22}O_{11}(s) \xrightarrow{\text{Conc } H_2SO_4} 11H_2O(l) + 12C(s) \qquad (22.155)$$
Sucrose

In aqueous solution, $H_2SO_4$ acts as a strong acid (equation 22.156) but the $[HSO_4]^-$ ion is a fairly weak acid (equation 22.157).

$$H_2SO_4(aq) + H_2O(l) \longrightarrow [H_3O]^+(aq) + [HSO_4]^-(aq)$$
Hydrogensulfate(1−) ion

*fully dissociated*   (22.156)

$$[HSO_4]^-(aq) + H_2O(l) \rightleftharpoons [H_3O]^+(aq) + [SO_4]^{2-}(aq) \qquad pK_a = 1.92 \quad (22.157)$$
Sulfate ion

▇▶
Sulfate(2−) is usually known simply as *sulfate*

Two series of salts are formed, for example $NaHSO_4$ and $Na_2SO_4$. *Dilute* aqueous sulfuric acid (typically 2 M) neutralizes bases (equations 22.158 and 22.159), and reacts with metal carbonates (equation 22.160) and some metals (equation 22.161). By looking at the standard reduction potentials in Table 18.1, you should be able to predict which metals will react with 0.5 M $H_2SO_4$; remember that the standard conditions for the redox cells is $[H^+] = 1 \text{ mol dm}^{-3}$, and one mole of sulfuric acid provides two moles of $H^+$ ions.

$$H_2SO_4(aq) + NaOH(aq) \longrightarrow NaHSO_4(aq) + H_2O(l) \qquad (22.158)$$

$$H_2SO_4(aq) + 2NaOH(aq) \longrightarrow Na_2SO_4(aq) + 2H_2O(l) \qquad (22.159)$$

$$H_2SO_4(aq) + CuCO_3(s) \longrightarrow CuSO_4(aq) + H_2O(l) + CO_2(g) \qquad (22.160)$$

$$H_2SO_4(aq) + Mg(s) \longrightarrow MgSO_4(aq) + H_2(g) \qquad (22.161)$$

Many sulfate salts have commercial applications and an important one is in the fertilizer industry. Other uses include lead(II) sulfate in lead storage batteries (see Box 22.7), copper(II) sulfate in fungicides, magnesium sulfate as a laxative and hydrated calcium sulfate ('plaster of Paris') in casts for broken bones.

Concentrated sulfuric acid is a good oxidizing agent (reaction 22.162).

$$Cu(s) + 2H_2SO_4(aq) \xrightarrow[\text{Conc}]{\Delta} CuSO_4(aq) + SO_2(g) + 2H_2O(l) \qquad (22.162)$$

When we refer to 'sulfurous acid' (equations 22.163 and 22.164) we mean a *solution of sulfur dioxide in water* (equilibrium 22.143). Pure $H_2SO_3$ is *not* isolable. Nonetheless, salts containing the $[HSO_3]^-$ and $[SO_3]^{2-}$ ions are common. Equilibria 22.163 and 22.164 are complicated by the fact that any aqueous solution of '$H_2SO_3$' contains significant amounts of $SO_2$. Earlier in this section, we described the use of $SO_2$ as a reducing agent, and it should now be clear why half-equations 22.148 and 22.149 showed the presence of $H_2SO_3$ or $[SO_3]^{2-}$, rather than $SO_2$, in acidic or alkaline aqueous conditions respectively. Sodium sulfite is prepared by dissolving $SO_2$ in aqueous sodium hydroxide solution (reaction 22.165).

$$H_2SO_3(aq) + H_2O(l) \rightleftharpoons [H_3O]^+(aq) + [HSO_3]^-(aq) \qquad pK_a = 1.82 \quad (22.163)$$

$$[HSO_3]^-(aq) + H_2O(l) \rightleftharpoons [H_3O]^+(aq) + [SO_3]^{2-}(aq) \qquad pK_a = 6.92 \quad (22.164)$$

$$SO_2(aq) + 2NaOH(aq) \longrightarrow Na_2SO_3(aq) + H_2O(l) \qquad (22.165)$$

Although the anhydrous thiosulfuric acid $H_2S_2O_3$ exists, it decomposes in aqueous solutions. However, salts containing the thiosulfate anion, $[S_2O_3]^{2-}$, are stable and sodium thiosulfate is commercially important. It is used in the traditional (i.e. non-digital) photographic industry as a reducing agent and to remove excess $Cl_2$ in bleaching processes. It is also commonly encountered in the laboratory as a reagent in $I_2$ analysis (reactions 22.166 and 22.167).

$I_2$ is insoluble in water: see Section 22.12

$$I_2(s) + \underset{\text{Aqueous KI}}{I^-(aq)} \longrightarrow [I_3]^-(aq) \qquad (22.166)$$

$$2Na_2S_2O_3(aq) + [I_3]^-(aq) \longrightarrow 2NaI(aq) + Na_2S_4O_6(aq) + I^-(aq) \qquad (22.167)$$

## Fluorides and chlorides of oxygen and sulfur

$H_2O_2$: see Section 21.9

Oxygen forms halides of the type $OX_2$ and $O_2X_2$ (X = halogen)[§] and these compounds are structurally similar to $H_2O$ and $H_2O_2$. At 298 K, oxygen difluoride, $OF_2$, is a highly toxic, pale yellow gas. It is formed by reaction 22.168, although this preparative route may be complicated by the fact that $OF_2$ reacts with hydroxide ion.

$$2NaOH(aq) + 2F_2(g) \longrightarrow OF_2(g) + 2NaF(aq) + H_2O(l) \qquad (22.168)$$

The role of the ClO• radical in the depletion of the ozone layer: see Box 25.3

Oxygen difluoride is hydrolysed by water only slowly, but with steam it explodes. When irradiated, $OF_2$ decomposes, forming FO• radicals which combine to give $O_2F_2$ (equation 22.169). Above 220 K, this yellow gas is unstable with respect to decomposition into $O_2$ and $F_2$. When $O_2F_2$ combines with $BF_3$ or $SbF_5$, fluoride transfer occurs to give salts of the $[O_2]^+$ ion (reaction 22.170).

---

[§] In formulae for binary compounds between oxygen, fluorine and chlorine, the symbols should be placed in the order Cl, O, F, i.e. $OF_2$ and $O_2F_2$, but $Cl_2O$ and $ClO_2$, are correct.

$$2OF_2 \xrightarrow{h\nu} 2FO^{\bullet} + 2F^{\bullet}$$

$$\downarrow$$

$$O_2F_2 \qquad\qquad (22.169)$$

$$2O_2F_2 + 2SbF_5 \longrightarrow 2[O_2][SbF_6] + F_2 \qquad (22.170)$$

Oxygen dichloride $Cl_2O$ is explosively unstable, and reacts with water to give hypochlorous acid (equation 22.171).

$$Cl_2O + H_2O \longrightarrow 2HOCl \qquad (22.171)$$

Chlorine dioxide $ClO_2$ is a paramagnetic, yellow gas at room temperature and is prepared in a hazardous reaction between concentrated $H_2SO_4$ and potassium chlorate ($KClO_3$), or by reaction 22.172 which is marginally less hazardous. An industrial application of $ClO_2$ is as a bleaching agent.

$$\underset{\substack{\text{Potassium} \\ \text{chlorate}}}{2KClO_3} + \underset{\substack{\text{Ethanedioic acid} \\ \text{(oxalic acid)}}}{2H_2C_2O_4} \longrightarrow 2ClO_2 + K_2C_2O_4 + 2CO_2 + 2H_2O \qquad (22.172)$$

Disulfur difluoride

**(22.55)**

**(22.56)**

We consider some further compounds formed between oxygen and the halogens in Section 22.12.

The number of sulfur fluorides is greater than the number of other halides of sulfur. Two structural isomers of $S_2F_2$ exist, **22.55** and **22.56**. Structure **22.55** has a *gauche* conformation similar to that of $H_2O_2$ (Figure 21.14).

Sulfur difluoride $SF_2$ is not a well-known species, unlike its chloride analogue $SCl_2$, but the higher fluorides $SF_4$ and $SF_6$ are stable gases at room temperature.

$$3SCl_2 + 4NaF \longrightarrow S_2Cl_2 + SF_4 + 4NaCl \qquad (22.173)$$

Sulfur tetrafluoride can be prepared from $SCl_2$ (equation 22.173) and is useful as a selective fluorinating agent. $SF_4$ acts both as a fluoride acceptor (reaction 22.174) and as a fluoride donor (reaction 22.175). Anhydrous conditions are essential in reactions involving $SF_4$, because it rapidly hydrolyses to give HF and $SOF_2$.

$$SF_4 + KF \longrightarrow K[SF_5] \qquad (22.174)$$

$$SF_4 + BF_3 \longrightarrow [SF_3]^+ + [BF_4]^- \qquad (22.175)$$

Although $SF_4$ can be formed when sulfur reacts with $F_2$, $SF_6$ is the dominant product when an excess of $F_2$ is used. Sulfur hexafluoride consists of octahedral molecules and is *kinetically* stable. Reaction 22.176 is *thermodynamically* favoured, but $SF_6$ is resistant to attack even by steam.

$$SF_6(g) + 3H_2O(g) \longrightarrow SO_3(g) + 6HF(g)$$

$$\Delta G^{\circ} = -221 \text{ kJ per mole of } SF_6 \qquad (22.176)$$

The lower chlorides of sulfur are $S_2Cl_2$ and $SCl_2$. $S_2Cl_2$ is a toxic, foul-smelling yellow liquid and is prepared by reacting molten sulfur with $Cl_2$. Structurally it resembles isomer **22.55** of disulfur difluoride. $S_2Cl_2$ reacts further with $Cl_2$ to give $SCl_2$. This is a poisonous, red liquid at 298 K, but tends to decompose according to equilibrium 22.177.

$$2SCl_2 \rightleftharpoons S_2Cl_2 + Cl_2 \qquad (22.177)$$

Both $SCl_2$ and $S_2Cl_2$ are reagents in synthetic chemistry for the preparation of organosulfur compounds.

## 22.11      Group 16: selenium and tellurium

| 1 | 2 | | 13 | 14 | 15 | **16** | 17 | 18 |
|---|---|---|----|----|----|----|----|----|
| H | | | | | | | | He |
| Li | Be | | B | C | N | **O** | F | Ne |
| Na | Mg | | Al | Si | P | **S** | Cl | Ar |
| K | Ca | *d*-block | Ga | Ge | As | **Se** | Br | Kr |
| Rb | Sr | | In | Sn | Sb | **Te** | I | Xe |
| Cs | Ba | | Tl | Pb | Bi | **Po** | At | Rn |
| Fr | Ra | | | | | | | |

Allotropes of selenium: see Section 9.8

Selenium is found in only a few minerals. A commercial source is flue dusts deposited during the refining of copper sulfide ores. An important property of selenium is its ability to convert light into electricity, and the element is used in photoelectric cells and in photographic exposure meters. Below its melting point (490 K), selenium is a semiconductor.

Tellurium is usually found combined with other metals, e.g. with gold in the mineral *calaverite*. Pure metallic tellurium has a silver-white appearance and both it and its compounds are toxic; if you are exposed to tellurium, your breath develops a garlic-like smell. Some of the applications of the element arise from its semiconducting properties.

The heaviest element in group 16 is polonium, but only milligram quantities of this radioactive element are available.

### Compounds with selenium–oxygen or tellurium–oxygen bonds

In contrast to $SO_2$, $SeO_2$ and $TeO_2$ (formed from the element and $O_2$) are crystalline solids at 298 K with lattice structures. Selenium dioxide dissolves in water to give selenous acid, $H_2SeO_3$, and its ease of reduction makes it a suitable oxidizing agent for some organic reactions. There are two structural forms of solid $TeO_2$; white crystalline $\alpha$-$TeO_2$ is sparingly soluble in water giving $H_2TeO_3$. Whereas both $SO_2$ and $TeO_2$ are readily oxidized to trioxides, $SeO_3$ is thermodynamically *unstable* with respect to $SeO_2$ and $O_2$.

In contrast to sulfur, selenium and tellurium do not form a wide range of oxoacids. Both $H_2SeO_3$ and $H_2TeO_3$ are white solids at 298 K, and can be dehydrated to the respective metal dioxide. The acids are dibasic, and two sets of salts containing the $[HSeO_3]^-$ and $[SeO_3]^{2-}$, or $[HTeO_3]^-$ and $[TeO_3]^{2-}$ anions are formed.

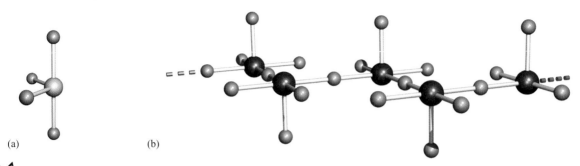

(a)                                    (b)

**Fig. 22.29** The structures of (a) $SeF_4$ and (b) $TeF_4$ in the solid state. Each tellurium atom in (b) is 5-coordinate but the stoichiometry remains Te : F = 1 : 4. Colour code: Se, yellow; Te, red; F, green.

### Fluorides and chlorides

A mixture of $SeF_2$, $Se_2F_2$ and $SeF_4$ is obtained when selenium vapour and $F_2$ react, but only $SeF_4$ is well characterized. Selenium tetrafluoride is a good fluorinating agent, and since it is a liquid at room temperature (mp 259 K), it is more convenient to use than gaseous $SF_4$. Liquid $SeF_4$ contains discrete molecules (Figure 22.29a) but in the solid state, they form an ordered array with significant intermolecular interactions. In solid $TeF_4$, association between the $TeF_4$ units is more extensive owing to the formation of Te−F−Te bridges (Figure 22.29b).

Like $SF_4$, $SeF_4$ acts as a *fluoride donor* but equilibrium 22.178 lies further to the left than the corresponding equilibrium involving $SF_4$ for which $K = 4 \times 10^{-2}$. $SeF_4$ can also act as a *fluoride acceptor* as in reaction 22.179. Similarly, $TeF_4$ behaves as both a fluoride donor and acceptor.

$$SeF_4 + HF \rightleftharpoons [SeF_3]^+ + [HF_2]^- \qquad\qquad K = 4 \times 10^{-4} \quad (22.178)$$

$$SeF_4 + CsF \rightleftharpoons Cs^+[SeF_5]^- \qquad\qquad (22.179)$$

**Fig. 22.30** The structure of the molecular $Se_4Cl_{16}$ unit present in the crystal lattice of $SeCl_4$. Similar molecular units of formula $Te_4Cl_{16}$ are present in $TeCl_4$. Colour code: Se, yellow; Cl, green.

**▶** Stereochemically inactive lone pair: see Section 6.7

Whereas $SCl_4$ is unstable, $SeCl_4$ is produced by reacting Se with $Cl_2$, but it decomposes when heated (equation 22.180). $TeCl_4$ is prepared similarly and is the only stable chloride of tellurium. Both $SeCl_4$ and $TeCl_4$ are crystalline solids at 298 K, and contain tetrameric $Se_4Cl_{16}$ or $Te_4Cl_{16}$ units (Figure 22.30).

$$SeCl_4 \xrightarrow{\Delta} SeCl_2 + Cl_2 \qquad\qquad (22.180)$$

Selenium and tellurium tetrachlorides *accept* chloride ions to give $[SeCl_6]^{2-}$ and $[TeCl_6]^{2-}$ (equation 22.181). Each has a *regular octahedral* structure, despite having 14 electrons in the valence shell; the lone pair is stereochemically inactive.

$$SeCl_4 + 2KCl \longrightarrow K_2[SeCl_6] \qquad\qquad (22.181)$$

*Donation* of chloride ion by $SeCl_4$ is observed in its reaction with $AlCl_3$. An X-ray diffraction study of the crystalline product confirms the presence of discrete $[SeCl_3]^+$ and $[AlCl_4]^-$ ions.

Selenium hexafluoride (produced from Se and $F_2$) is a gas at 298 K. It is quite stable, resisting hydrolysis even in alkaline aqueous solution. $TeF_6$ is well characterized, but binary *chlorides* of selenium(VI) and tellurium(VI) are not known. It is the oxidizing ability of $F_2$ (emphasized by the value of

$E^o$ for half-equation 22.182) and the small size of fluorine that make the formation of $SF_6$, $SeF_6$ and $TeF_6$ feasible.

$$F_2(aq) + 2e^- \rightleftharpoons 2F^-(aq) \qquad\qquad E^o = +2.87\,V \quad (22.182)$$

## 22.12    Group 17: the halogens

| 1 | 2 | | 13 | 14 | 15 | 16 | **17** | 18 |
|---|---|---|---|---|---|---|---|---|
| H | | | | | | | | He |
| Li | Be | | B | C | N | O | **F** | Ne |
| Na | Mg | | Al | Si | P | S | **Cl** | Ar |
| K | Ca | *d*-block | Ga | Ge | As | Se | **Br** | Kr |
| Rb | Sr | | In | Sn | Sb | Te | **I** | Xe |
| Cs | Ba | | Tl | Pb | Bi | Po | **At** | Rn |
| Fr | Ra | | | | | | | |

Fluorine is not found native, but occurs as the fluoride ion in *fluorite* $CaF_2$, *cryolite* $Na_3[AlF_6]$ and a range of other minerals. Because of its strongly oxidizing nature, $F_2$ must be prepared by electrolytic oxidation of $F^-$, and in the industrial route, a mixture of anhydrous molten KF and HF is electrolysed (equation 22.183).

*At a carbon anode*:   $2F^-(l) \longrightarrow F_2(g) + 2e^-$ \qquad (22.183)

Difluorine is a pale yellow gas (bp 85 K) consisting of discrete $F_2$ molecules. Fluorine is the most electronegative of all the elements, and $F_2$ reacts with most elements and compounds. It is a dangerous gas to use. It is extremely corrosive and must be handled in special steel containers. The developments of the atomic bomb and the nuclear energy industry have demanded large quantities of $F_2$. A major use is in the production of uranium hexafluoride during nuclear fuel enrichment processes (see Box 22.14).

**The structures of the halogens: see Section 9.7**

Dichlorine is among the most important industrial chemicals and is extracted from NaCl using the Downs process (Figure 21.16). On a laboratory scale, reaction 22.184 is a convenient synthesis of $Cl_2$.

$$MnO_2(s) + 4HCl(aq) \longrightarrow MnCl_2(aq) + Cl_2(g) + 2H_2O(l) \qquad (22.184)$$

Dichlorine is a pale green-yellow gas (bp 239 K) with a characteristic smell. When inhaled, it causes irritation of the respiratory system and liquid $Cl_2$ burns the skin. Drinking-water supplies are often chlorinated in order to make the water safer for human consumption, and $Cl_2$ is used widely as a disinfectant and a bleach. A disadvantage of using $Cl_2$ in the paper industry is the production of toxic effluents; as a result, the use of alternative bleaching agents is encouraged on environmental grounds. Its conversion into chlorine-

## COMMERCIAL AND LABORATORY APPLICATIONS

### Box 22.14 Nuclear reactors and the reprocessing of nuclear fuels

Uranium is a nuclear fuel, and consists naturally of the isotopes $^{234}_{92}U$ (<0.01%), $^{235}_{92}U$ (≈0.72%) and $^{238}_{92}U$ (≈99.28%). Of these, $^{235}_{92}U$ has the ability to *capture a thermal neutron* to yield the unstable isotope $^{236}_{92}U$. This is followed by nuclear fission:

$$^{235}_{92}U + ^{1}_{0}n \longrightarrow ^{236}_{92}U \xrightarrow{\text{Nuclear fission}} ^{141}_{56}Ba + ^{92}_{36}Kr + 3^{1}_{0}n$$

and release of about $2 \times 10^{10}$ kJ of energy per mole of uranium. The production of extra neutrons means that there is the potential for a *chain reaction* (see Figure 15.24) and this is the basis for the atomic bomb. In a nuclear reactor, nuclear energy is harnessed for electricity among other uses, but the nuclear reaction must be efficiently *controlled*. This is done by inserting boron or cadmium rods (which absorb neutrons) between the uranium fuel rods.

Discharging fuel from a reactor core to cooling ponds at a pressurized water reactor nuclear power station.
*Arthus Bertrand/Science Photo Library.*

As $^{235}_{92}U$ is used up in the nuclear reactor, the fuel becomes spent. Spent nuclear fuels are not disposed of but are *reprocessed* – this recovers uranium and separates $^{235}_{92}U$ from fission products. After preliminary processes including *pond storage* (which allows time for short-lived radioactive products to decay), reprocessing begins by converting uranium metal to soluble uranyl nitrate:

$$U(s) + 8HNO_3(aq) \longrightarrow$$
$$[UO_2][NO_3]_2(aq) + 4H_2O(l) + 6NO_2(g)$$

This is followed by conversion to uranium(VI) oxide, reduction to uranium(IV) oxide, and fluorination to give $UF_6$:

$$[UO_2][NO_3]_2(s) \xrightarrow{570\,K}$$
$$UO_3(s) + NO(g) + NO_2(g) + O_2(g)$$

$$UO_3(s) + H_2(g) \xrightarrow{970\,K} UO_2(s) + H_2O(g)$$

$$UO_2(s) + 4HF(aq) \longrightarrow UF_4(s) + 2H_2O(l)$$

$$UF_4(s) + F_2(g) \xrightarrow{720\,K} UF_6(g)$$

The final stage of reprocessing is the separation of $^{235}_{92}UF_6$ from $^{238}_{92}UF_6$. This is achieved by using a special centrifuge. *Graham's Law* states that the rate of gas effusion (or diffusion) depends upon the molecular mass:

$$\text{Rate of effusion} \propto \frac{1}{\sqrt{\text{Molecular mass}}}$$

Application of Graham's Law (see Chapter 1) shows that $^{235}_{92}UF_6$ and $^{238}_{92}UF_6$ can be separated by subjecting them to a centrifugal force which moves the molecules to the outer wall of their container, but at different rates. In this way it is possible to obtain $^{235}_{92}U$ *enriched* $UF_6$. After this process, the hexafluoride is converted back to uranium metal before the fuel is ready for reuse in the nuclear reactor:

$$UF_6(g) + H_2(g) + 2H_2O(g) \xrightarrow{870\,K}$$
$$UO_2(s) + 6HF(g)$$

$$UO_2(s) + 2Mg(s) \xrightarrow{\Delta} U(s) + 2MgO(s)$$

**Fig. 22.31** Uses of $Cl_2$ in Western Europe in 1994. The percentage used for pulp and paper bleaching and water chlorination is decreasing owing to environmental concern. Applications for CFCs have also been phased out (see Box 25.3). (Data from *Chemistry & Industry* (1995), p. 832.)

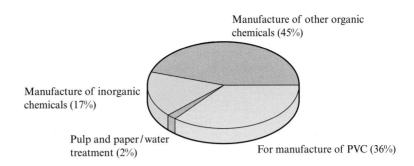

Manufacture of other organic chemicals (45%)

Manufacture of inorganic chemicals (17%)

Pulp and paper / water treatment (2%)

For manufacture of PVC (36%)

containing compounds, including HCl, is a major industrial use (Figure 22.31). It reacts with many elements to give the respective chlorides.

Dibromine is a dark orange, volatile liquid (mp 266 K, bp 332 K). It is the only non-metallic element that is a liquid at 298 K. In the laboratory it is often encountered as the aqueous solution 'bromine water'. $Br_2$ is toxic – the liquid produces skin burns and the vapour has an unpleasant odour and causes eye and respiratory irritations. Bromine occurs as the bromide ion in brine wells in parts of the US, and can be extracted using reaction 22.185. Some uses of $Br_2$ include its role in flame-proofing materials, dyes, bromide compounds for the non-digital photographic industry and organic bromides.

$$2Br^-(aq) + Cl_2(g) \longrightarrow Br_2(aq) + 2Cl^-(aq) \qquad (22.185)$$

oxidation

reduction

At 298 K, $I_2$ forms dark purple crystals (mp 387 K, bp 458 K) containing molecules (see Table 9.1 and Figure 9.10). The solid sublimes readily at atmospheric pressure into a purple vapour which irritates the eyes. Iodine occurs as the $I^-$ ion in seawater and is taken up by seaweed from which $I_2$ can be extracted. On an industrial scale, $I_2$ is extracted from oil and brine wells by using $Cl_2$ to oxidize $I^-$ to $I_2$ (analogous to reaction 22.185). Unlike $F_2$, $Cl_2$ and $Br_2$, $I_2$ is only very sparingly soluble in water, and laboratory solutions are usually made up in aqueous potassium iodide in which $I_2$ is soluble giving $[I_3]^-$ salts (equation 22.166). Iodine is an essential element for life, and a deficiency results in a swollen thyroid gland; 'iodized salt' (NaCl with added $I^-$) provides supplemental iodine to the body. Iodide salts such as KI have applications in the non-digital photographic industry. Solutions of $I_2$ in aqueous KI are used as disinfectants for wounds.

The trend in reactivities of the halogens reflects the decrease in electronegativity values from F ($\chi^P = 4.0$) to I ($\chi^P = 2.7$) and the decreasing oxidizing strength of the element (Table 22.5). Fluorine is the most electronegative

**Table 22.5** Standard reduction potentials for the reduction of the halogens. $F_2$ is the most readily reduced and is the most powerful oxidizing agent in the series.

| Reduction half-reaction | $E^o$ / V |
|---|---|
| $I_2(aq) + 2e^- \rightleftharpoons 2I^-(aq)$ | +0.54 |
| $Br_2(aq) + 2e^- \rightleftharpoons 2Br^-(aq)$ | +1.09 |
| $Cl_2(aq) + 2e^- \rightleftharpoons 2Cl^-(aq)$ | +1.36 |
| $F_2(aq) + 2e^- \rightleftharpoons 2F^-(aq)$ | +2.87 |

element in the periodic table, and in its compounds it always has a formal oxidation state of $-1$. The later halogens form compounds with a range of oxidation states (see below). The relative sizes of the halogen atoms and ions (Figure 22.32) also contribute to some of the differences between the chemistries of the group 17 elements. The small covalent radius of fluorine permits elements with which fluorine combines to attain high coordination numbers. For example, six F atoms fit around an S atom, but six I atoms will not. In ionic metal fluorides, the small size of the $F^-$ ion contributes to the high values of lattice enthalpies and to the thermodynamic stability of the compounds.

▶
Dependence of lattice enthalpy on ion size: see Section 8.15

The relative strength of $F_2$ as an oxidizing agent with respect to the other halogens is seen in the formation of *interhalogen compounds*. These are compounds formed by combining two *different* halogens. In each of ClF, $ClF_3$, $BrF_5$ and $IF_7$, the formal oxidation state of fluorine is $-1$, while the oxidation states of the other halogens range from $+1$ to $+7$. In *none* of the interhalogen compounds is fluorine in a positive oxidation state – $F_2$ is able to oxidize the lower halogens (equation 22.190) but $Cl_2$, $Br_2$ and $I_2$ are *not* able to oxidize $F_2$.

$$F_2 \begin{array}{l} \xrightarrow{Cl_2,\ 470\ K} ClF_3 \\ \xrightarrow{Br_2,\ 420\ K} BrF_5 \\ \xrightarrow{I_2,\ 530\ K} IF_7 \end{array} \qquad (22.186)$$

▶
Hydrogen halides: see Section 21.10

Examples of reactions of halogens with metals and non-metals are considered in the relevant sections of Chapters 21 and 22.

### Oxides

We have already described some of the lower oxides of fluorine and chlorine ($OF_2$, $O_2F_2$, $Cl_2O$ and $ClO_2$). Fluorine does not form higher oxides because it is a stronger oxidizing agent than oxygen, but the other halogens form compounds such as $Cl_2O_6$, $Cl_2O_7$, $Br_2O_3$, $Br_2O_5$ and $I_2O_5$.

Dichlorine hexaoxide $Cl_2O_6$ is a dark red liquid at 298 K and is prepared by the reaction between ozone and $ClO_2$ (equation 22.187). It must be handled with extreme care as it explodes when in contact with organic material.

$$2ClO_2 + 2O_3 \longrightarrow Cl_2O_6 + 2O_2 \qquad (22.187)$$

There has been much speculation as to the exact structure of $Cl_2O_6$, but the probable structure in the liquid state is shown in Figure 22.33a. In the solid state (mp 277 K), $[ClO_2]^+$ and $[ClO_4]^-$ ions are present. The oxide is unstable with respect to decomposition into $ClO_2$ and $O_2$, and with water, reaction 22.188 occurs.

$$Cl_2O_6 + H_2O \longrightarrow \underset{\text{Perchloric acid}}{HClO_4} + \underset{\text{Chloric acid}}{HClO_3} \qquad (22.188)$$

The oxide $Cl_2O_7$ is a colourless oily liquid (mp 182 K, bp 355 K) which is shock-sensitive and potentially explosive. It is formed by dehydrating perchloric acid (equation 22.189), but the reaction is reversed when $Cl_2O_7$ contacts water. The structure of $Cl_2O_7$ is shown in Figure 22.33b.

$$2HClO_4 \xrightarrow{H_3PO_4,\ 263\ K} H_2O + Cl_2O_7 \qquad (22.189)$$

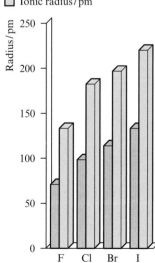

☐ Covalent radius/pm
☐ Ionic radius/pm

**Fig. 22.32** The trend in the covalent and ionic radii of the halogens. Blue bars = covalent radius; orange bars = ionic radius.

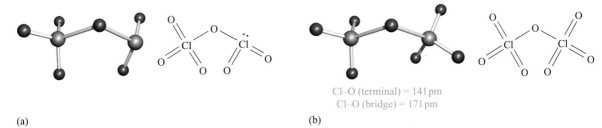

Cl–O (terminal) = 141 pm
Cl–O (bridge) = 171 pm

(a)                                                                (b)

**Fig. 22.33** (a) The proposed structure of $Cl_2O_6$ in the liquid state and (b) the confirmed structure of $Cl_2O_7$. Colour code: Cl, green; O, red. The schematic diagrams of the structures are a convenient way of illustrating these molecules. However, it is more realistic to represent the bonding using charge-separated species so that each Cl atom obeys the octet rule (see Section 7.3).

The bromine oxides $Br_2O_3$ and $Br_2O_5$ are unstable, but iodine forms $I_2O_5$ as white, hygroscopic crystals which decompose above 570 K. The solid state structure consists of an extended array. The dehydration of $HIO_3$ is one method of preparing $I_2O_5$, but the compound readily takes up water again to regenerate the acid. $I_2O_5$ is a powerful oxidizing agent and oxidizes CO to $CO_2$. This reaction may be used in the quantitative analysis of carbon monoxide. The $I_2$ produced is analysed by titration with sodium thiosulfate using a starch indicator (equation 22.190 and Box 22.15). The $I_2$ in equation 22.190 is in the form of $[I_3]^-$ (see equation 22.166).

$$\left.\begin{array}{ll}\textit{Quantitative} & 5CO + I_2O_5 \longrightarrow 5CO_2 + I_2 \\ \textit{analysis for } CO: & I_2 + 2Na_2S_2O_3 \longrightarrow 2NaI + Na_2S_4O_6\end{array}\right\} \quad (22.190)$$

### Oxoacids and the oxoanions

The oxides $Cl_2O$, $Cl_2O_7$ and $I_2O_5$ are the *anhydrides* of HOCl, $HClO_4$ and $HIO_3$ respectively. The acids and their oxoanions are all good oxidizing agents, and Table 22.6 lists selected half-equations and the corresponding standard reduction potentials. Remember that the more positive the value

**THEORETICAL AND CHEMICAL BACKGROUND**

## Box 22.15 Starch test for $I_2$

Starch is a widespread component of plant material and is obtained from potatoes, wheat and rice. One of the components of starch is amylose, which consists of α-D(+)glucopyranose units linked together to produce helical coils.

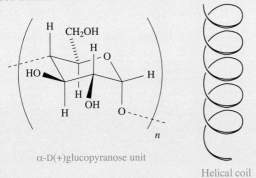

α-D(+)glucopyranose unit

Helical coil

Solutions of $I_2$ in aqueous KI contain a series of polyiodide ions including $[I_3]^-$ and $[I_5]^-$:

$$\left[I - I - I\right]^- \qquad \left[\begin{array}{c} I \\ I \diagdown I \diagup I \diagdown I \end{array}\right]^-$$

The $[I_5]^-$ ion is trapped within the channels that run through the helical structures of α-D(+)glucopyranose units. The adduct formed has a characteristic blue-black colour. It is the formation of this coloured adduct that accounts for the use of starch as an 'indicator' in titrations involving iodine.

**Table 22.6** Standard reduction potentials for selected reduction reactions involving oxoacids of chlorine, bromine and iodine.

| Reduction half-reaction | $E^o$ or $E^o_{[OH^-]=1}$ / V |
| --- | --- |
| $[ClO_4]^-(aq) + H_2O(l) + 2e^- \rightleftharpoons [ClO_3]^-(aq) + 2[OH]^-(aq)$ | +0.36 |
| $[BrO_3]^-(aq) + 3H_2O(l) + 6e^- \rightleftharpoons Br^-(aq) + 6[OH]^-(aq)$ | +0.61 |
| $[BrO]^-(aq) + H_2O(l) + 2e^- \rightleftharpoons Br^-(aq) + 2[OH]^-(aq)$ | +0.76 |
| $[ClO]^-(aq) + H_2O(l) + 2e^- \rightleftharpoons Cl^-(aq) + 2[OH]^-(aq)$ | +0.84 |
| $[IO_3]^-(aq) + 6H^+(aq) + 6e^- \rightleftharpoons I^-(aq) + 3H_2O(l)$ | +1.09 |
| $2[IO_3]^-(aq) + 12H^+(aq) + 10e^- \rightleftharpoons I_2(aq) + 6H_2O(l)$ | +1.20 |
| $[ClO_4]^-(aq) + 8H^+(aq) + 8e^- \rightleftharpoons Cl^-(aq) + 4H_2O(l)$ | +1.39 |
| $2[ClO_4]^-(aq) + 16H^+(aq) + 14e^- \rightleftharpoons Cl_2(aq) + 8H_2O(l)$ | +1.39 |
| $[BrO_3]^-(aq) + 6H^+(aq) + 6e^- \rightleftharpoons Br^-(aq) + 3H_2O(l)$ | +1.42 |
| $[ClO_3]^-(aq) + 6H^+(aq) + 6e^- \rightleftharpoons Cl^-(aq) + 3H_2O(l)$ | +1.45 |
| $2[ClO_3]^-(aq) + 12H^+(aq) + 10e^- \rightleftharpoons Cl_2(aq) + 6H_2O(l)$ | +1.47 |
| $2[BrO_3]^-(aq) + 12H^+(aq) + 10e^- \rightleftharpoons Br_2(aq) + 6H_2O(l)$ | +1.48 |
| $HOCl(aq) + H^+(aq) + 2e^- \rightleftharpoons Cl^-(aq) + H_2O(l)$ | +1.48 |
| $2HOCl(aq) + 2H^+(aq) + 2e^- \rightleftharpoons Cl_2(aq) + 2H_2O(l)$ | +1.61 |

of $E^o$, the more readily the reduction process occurs. Furthermore, the values of $E^o$ give only *thermodynamic* data, and provide no indication of *kinetic* factors; we return to this point when discussing $HClO_4$.

The only oxoacid of fluorine is HOF, and it spontaneously decomposes to HF and $O_2$. Hypochlorous acid, HOCl, and its heavier congeners HOBr and HOI, cannot be isolated as pure compounds, but are encountered as aqueous solutions. They are weak, monobasic acids (equilibrium 22.191), but the $[OX]^-$ anions are unstable with respect to disproportionation (equation 22.192). The reaction is slow for $[OCl]^-$, fast for $[OBr]^-$ and very rapid for $[OI]^-$.

$$HOX(aq) + H_2O(l) \rightleftharpoons [H_3O]^+(aq) + [OX]^-(aq) \qquad (22.191)$$

$$X = Cl \quad pK_a = 4.53$$
$$X = Br \quad pK_a = 8.69$$
$$X = I \quad pK_a = 10.64$$

$$3[OX]^-(aq) \longrightarrow [XO_3]^-(aq) + 2X^-(aq) \qquad (22.192)$$

Salts such as NaOCl, KOCl and $Ca(OCl)_2$ can be isolated (e.g. reaction 22.193) and are valuable oxidizing agents. $Ca(OCl)_2$ is a component of bleaching powder, and NaOCl is a bleaching agent and disinfectant.

$$2CaO(s) + 2Cl_2(g) \longrightarrow Ca(OCl)_2(s) + CaCl_2(s) \qquad (22.193)$$

Chloric and bromic acids, $HClO_3$ and $HBrO_3$, are both strong acids but cannot be isolated as pure compounds. Aqueous solutions of $HClO_3$ can be made by reaction 22.194, and bromic acid is prepared similarly.

$$Ba(ClO_3)_2(aq) + H_2SO_4(aq) \longrightarrow BaSO_4(s) + 2HClO_3(aq) \qquad (22.194)$$

Iodic acid $HIO_3$ is a stable, white solid at room temperature, and is produced by reacting $I_2O_5$ with water. In aqueous solution it behaves as a monobasic acid ($pK_a = 0.77$).

The three halic acids and the oxoanions derived from them are strong oxidizing agents (see Table 22.6). A conproportionation reaction used in volumetric analysis combines the reduction of the $[XO_3]^-$ ion to $X_2$ with

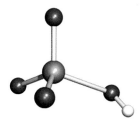

Cl–O = 141 pm
Cl–OH = 164 pm

**Fig. 22.34** The structure of perchloric acid, $HClO_4$. Colour code: Cl, green; O, red; H, white.

Fireworks:
see Box 21.10

The space shuttle:
see Box 21.3

the oxidation of $X^-$ to $X_2$ (e.g. reaction 22.195). Reactions 22.196 and 22.197 illustrate the uses of $[ClO_3]^-$ and $[IO_3]^-$ as oxidizing agents.

$$[IO_3]^-(aq) + 5I^-(aq) + 6H^+(aq) \longrightarrow 3I_2(aq) + 3H_2O(l) \qquad (22.195)$$

$$[ClO_3]^-(aq) + 6Fe^{2+}(aq) + 6H^+(aq) \longrightarrow Cl^-(aq) + 6Fe^{3+}(aq) + 3H_2O(l) \qquad (22.196)$$

$$[IO_3]^-(aq) + 3[SO_3]^{2-}(aq) \longrightarrow I^-(aq) + 3[SO_4]^{2-}(aq) \qquad (22.197)$$

The disproportionation of the chlorate ion (reaction 22.198) is thermodynamically favourable but takes place only slowly. Alkali metal chlorates decompose by this route when they are heated above their melting points. In the presence of a suitable catalyst, $KClO_3$ decomposes with loss of $O_2$ (equation 22.199) and, as we saw earlier, this method can be used for small-scale preparations of $O_2$. *Caution is required as there is a risk of explosion.* Potassium chlorate is used as a component of fireworks and in safety matches where its ready reduction by, for example, sulfur provides a source of $O_2$ for the combustion of the match.

$$\underset{\text{Chlorate ion}}{4[ClO_3]^-} \longrightarrow \underset{\text{Perchlorate ion}}{3[ClO_4]^-} + \underset{\text{Chloride ion}}{Cl^-} \qquad (22.198)$$

$$2KClO_3(s) \xrightarrow{\Delta,\ MnO_2\ catalyst} 2KCl(s) + 3O_2(g) \qquad (22.199)$$

Of salts containing $[XO_4]^-$, perchlorates are most commonly encountered in the laboratory, *but they should always be regarded as hazardous explosives and avoided whenever possible.* Mixtures of ammonium perchlorate and aluminium are used in missile propellants. The parent acid $HClO_4$ (perchloric acid) can be produced by reacting $NaClO_4$ with concentrated HCl, and is a colourless, hygroscopic liquid when pure. The chlorine atom is tetrahedrally sited (Figure 22.34) and is in its highest oxidation state of +7. In the tetrahedral $[ClO_4]^-$ anion, all the Cl–O bond distances are equal (144 pm).

Anhydrous $HClO_4$ is a strong oxidizing agent, but aqueous solutions of the acid or its salts are *kinetically* stable; remember that the values of $E^o$ in Table 22.6 for reductions involving $[ClO_4]^-$ provide *thermodynamic*, but not kinetic, information. In aqueous solution, $HClO_4$ behaves as an oxidizing agent and a strong, monobasic acid and most metals form perchlorate salts. Alkali metal perchlorates are usually prepared by the disproportionation of the corresponding chlorate salts (reaction 22.198).

## ENVIRONMENT AND BIOLOGY

### Box 22.16 Radon levels in our homes

The radioactive decay of uranium leads to the formation of radon. Uranium occurs naturally in low abundance in many rock deposits, and the radon gas released may be trapped under the Earth's surface or may find a route to above the ground – much depends upon the geological rock formations. In some areas of Britain, levels of radon in houses are higher than in surrounding areas, and this appears to be related to the bedrock formations in each region. Research points towards a link between exposure to radon and lung cancer, and there is some concern over the health risks that may be involved in homes in which radon levels are relatively high.

| 22.13 | **Group 18: the noble gases** |

| 1 | 2 | | 13 | 14 | 15 | 16 | 17 | **18** |
|---|---|---|----|----|----|----|----|--------|
| H | | | | | | | | **He** |
| Li | Be | | B | C | N | O | F | **Ne** |
| Na | Mg | | Al | Si | P | S | Cl | **Ar** |
| K | Ca | | Ga | Ge | As | Se | Br | **Kr** |
| Rb | Sr | *d*-block | In | Sn | Sb | Te | I | **Xe** |
| Cs | Ba | | Tl | Pb | Bi | Po | At | **Rn** |
| Fr | Ra | | | | | | | |

▣▶
*Physical and structural properties of noble gases: see Table 3.6 and Section 9.6*

Helium is the second most abundant element in the universe (after hydrogen) and is used widely wherever inert atmospheres are needed (e.g. as a protective environment during the growth of single silicon crystals for the semiconductor industry) and in balloons and advertising 'blimps'. Liquid helium is an important coolant and is used in highfield NMR spectrometers (see Box 9.2), including those used for medical imaging.

Neon is a rare element obtained by the liquefaction of air followed by distillation. It is used in electric discharge tubes in which it produces a red glow, and a major application is in advertising signs. Argon constitutes 0.94% of the atmosphere and is obtained by the liquefaction of air. It is used to provide inert environments including those in glove or dry boxes in the laboratory, and as an atmosphere in fluorescent lighting tubes. Krypton too is rare, being present in the Earth's atmosphere to an extent of only one part per million (1 ppm). Xenon is even less abundant. Nonetheless, it is xenon that is the focus of attention of this section, because it is the only noble gas for which a reasonably extensive chemistry is known. There is little evidence that helium, neon and argon form compounds, and the chemistry of krypton is essentially limited to that of $KrF_2$ and its derivatives. The heaviest noble gas is radon; it is radioactive and has been a serious health hazard in uranium mines where people exposed to it have developed lung cancer (see Box 22.16).

Bartlett prepared the first compound of xenon in 1962 by reacting Xe with $PtF_6$ which is a very strong oxidizing agent (equation 22.200). Even after over 40 years, the *exact* nature of this product is unclear.

$$Xe + PtF_6 \longrightarrow \text{`}XePtF_6\text{'} \tag{22.200}$$

A noble gas is characterized by having a *complete* valence shell of electrons, and it is reasonable to assume that the reactivity of the group 18 elements will, as a consequence, be low or even non-existent. Xenon (and to a lesser extent, krypton), however, combines with fluorine and oxygen to form a wide range of compounds and this restriction is because:

- $F_2$ and $O_2$ are strong oxidizing agents;
- F and O are highly electronegative elements.

### Fluorides of xenon

Xenon forms the neutral fluorides $XeF_2$, $XeF_4$ and $XeF_6$ which are prepared from Xe and $F_2$ under different conditions (scheme 22.201). All the fluorides are crystalline solids at room temperature.

$$Xe + F_2 \begin{cases} \xrightarrow{\text{670 K, 2 : 1 ratio}} XeF_2 \\ \xrightarrow{\text{670 K, 6 bar, 1 : 5 ratio}} XeF_4 \\ \xrightarrow{\text{570 K, 50 bar, 1 : 20 ratio}} XeF_6 \end{cases} \qquad (22.201)$$

The molecular structures of $XeF_2$ (linear) and $XeF_4$ (square planar) are as expected by VSEPR theory. Xenon hexafluoride is isoelectronic with $[TeCl_6]^{2-}$ but they are *not* isostructural. The *regular* octahedral structure of $[TeCl_6]^{2-}$ is explained in terms of a stereochemically inactive lone pair. In the vapour phase, $XeF_6$ is stereochemically non-rigid with the lone pair of electrons occupying different sites and causing the overall structure to be *distorted octahedral* (Figure 22.35). In the solid state, $[XeF_5]^+$ and $F^-$ ions are present, and the fluoride ions bridge between the square pyramidal cations.

Xenon(II) fluoride is commercially available and is used widely as a fluorinating and oxidizing agent. Often the solvent for the reaction is anhydrous hydrogen fluoride (equations 22.202 and 22.203). Using $XeF_2$ as a fluorinating agent is advantageous because the only other product is inert, gaseous xenon.

$$S + 3XeF_2 \xrightarrow{\text{Anhydrous HF}} SF_6 + 3Xe \qquad (22.202)$$

$$2Ir + 5XeF_2 \xrightarrow{\text{Anhydrous HF}} 2IrF_5 + 5Xe \qquad (22.203)$$

Xenon(II) fluoride is reduced by water to xenon (equation 22.204) but the reaction is slow unless a base is present. In the presence of water, xenon(IV) fluoride disproportionates (equation 22.205), while $XeF_6$ may be partially (equation 22.206) or completely (equation 22.207) hydrolysed without a change in oxidation state. In each hydrolysis, HF is produced.

$$2XeF_2 + 2H_2O \longrightarrow 2Xe + 4HF + O_2 \qquad (22.204)$$

$$6XeF_4 + 12H_2O \longrightarrow 4Xe + 2XeO_3 + 24HF + 3O_2 \qquad (22.205)$$

$$XeF_6 + H_2O \longrightarrow XeOF_4 + 2HF \qquad (22.206)$$

$$\textbf{(22.57)}$$

$$XeF_6 + 3H_2O \longrightarrow XeO_3 + 6HF \qquad (22.207)$$

A problem of working with $XeF_6$ is that it attacks silica glass. Initially, $XeOF_4$ is formed (equation 22.208) but further reaction leads to $XeO_3$. Copper or platinum apparatus is commonly used in place of glass.

$$2XeF_6 + SiO_2 \longrightarrow 2XeOF_4 + SiF_4 \qquad (22.208)$$

$[TeCl_6]^{2-}$ : see Section 6.7

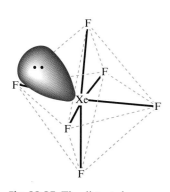

**Fig. 22.35** The distorted octahedral structure of $XeF_6$ in the gas phase is caused by the presence of a lone pair of electrons, but the molecule is stereochemically non-rigid and exchanges between several possible arrangements.

**(22.57)**

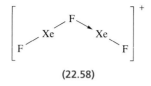

(22.58)

The donation and acceptance of fluoride ions by $XeF_2$, $XeF_4$ and $XeF_6$ are illustrated by the formation of the ions $[XeF]^+$, $[XeF_5]^+$, $[XeF_5]^-$ and $[XeF_8]^{2-}$. The $[XeF]^+$ ion acts as a Lewis acid and forms an adduct with $XeF_2$ (**22.58**).

## Oxides of xenon

Xenon(VI) oxide can be prepared by the hydrolysis of $XeF_4$ or $XeF_6$ (see above) but is highly explosive. It dissolves in water without change, but in strongly alkaline solution, equilibrium 22.209 is set up involving the xenate anion $[HXeO_4]^-$ which slowly disproportionates giving the perxenate ion and xenon (equation 22.210).

$$Xe^{VI}O_3 + [OH]^- \rightleftharpoons [HXe^{VI}O_4]^- \qquad (22.209)$$

$$2[HXe^{VI}O_4]^- + 2[OH]^- \longrightarrow [Xe^{VIII}O_6]^{4-} + Xe^0 + 2H_2O + O_2 \qquad (22.210)$$

In acidic solution, both $XeO_3$ and $[XeO_6]^{4-}$ are very powerful oxidizing agents as the standard reduction potentials for equilibria 22.211 and 22.212 show. By comparing these potentials with those in Appendix 12, it can be seen that $[XeO_6]^{4-}$ is capable of oxidizing aqueous $Cr^{3+}$ to $[Cr_2O_7]^{2-}$, $Br_2$ to $[BrO_3]^-$ and $Mn^{2+}$ to $[MnO_4]^-$.

$$XeO_3(aq) + 6H^+(aq) + 6e^- \rightleftharpoons Xe(g) + 3H_2O(l) \qquad E^o = +2.10\,V \quad (22.211)$$

$$H_4XeO_6(aq) + 2H^+(aq) + 2e^- \rightleftharpoons XeO_3(aq) + 3H_2O(l)$$
$$E^o = +2.42\,V \quad (22.212)$$

## 22.14    Some compounds of high oxidation state *d*-block elements

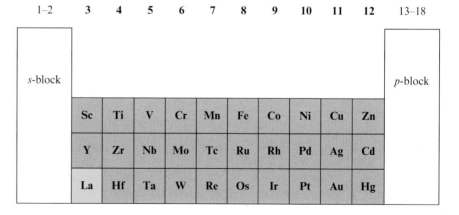

Older nomenclature labelled the groups in the *d*-block so that each was related to a group in the *s*- or *p*-block (Figure 22.36). Although this notation has been largely superseded by current IUPAC recommendations, the reasons behind the choice of labels remind us that, in their highest oxidation states, the chemistry of the transition metals is reminiscent of that of related elements in the *s*- and *p*-blocks. We illustrate this point by briefly considering some oxides, oxoanions and halides of the *d*-block metals.

| IUPAC group number | 3 | 4 | 5 | 6 | 7 | 8 | 9 | 10 | 11 | 12 |
|---|---|---|---|---|---|---|---|---|---|---|
| Older group number | IIIA | IVA | VA | VIA | VIIA | ← | VIII | → | IB | IIB |

| Related *s*- or *p*-block group | 13 | 14 | 15 | 16 | 17 | ← | 18 | → | 1 | 2 |

**Fig. 22.36** The IUPAC group numbering for the *d*-block elements, and the corresponding older numbering system which emphasized the relationship between these metals and elements in the *s*- and *p*-blocks.

Pyridine (py)

The highest oxidation state of titanium (group 4) is +4, and the properties of the titanium(IV) halides are, in many respects, similar to those of the *heavier* group 14 halides. For example, $TiX_4$ compounds are all readily hydrolysed by water (equation 22.213). Titanium(IV) halides are Lewis acids and in concentrated hydrochloric acid, $TiCl_4$ forms the octahedral anion $[TiCl_6]^{2-}$. The $[TiCl_5]^-$ anion can also be formed and this is isostructural with $[SnCl_5]^-$. $TiCl_4$ also forms a complex with pyridine (equation 22.214) with a structure reminiscent of that of $[Ge(py)_2Cl_4]$ (**22.25**). The oxide $TiO_2$ possesses the rutile structure, as does $SnO_2$.

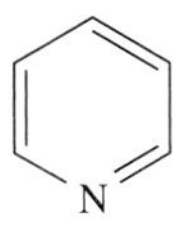

$$TiX_4 + 2H_2O \longrightarrow TiO_2 + 4HX \qquad X = F, Cl, Br \text{ or } I \quad (22.213)$$

$$TiCl_4 + 2py \longrightarrow [Ti(py)_2Cl_4] \qquad (22.214)$$

In group 5, vanadium forms compounds in the +5 oxidation state that find analogues in phosphorus chemistry. For example, vanadium reacts with $O_2$ to give $V_2O_5$ (as well as lower oxides) and with $F_2$ to yield $VF_5$. However, the product of heating vanadium metal with $Cl_2$ is $VCl_4$, and $Br_2$ and $I_2$ only oxidize the metal as far as the +3 state, giving $VBr_3$ and $VI_3$, respectively. The structure of $V_2O_5$ is not like that of phosphorus(V) oxide; in the solid state it consists of a complicated extended structure. In strongly basic solution, $V_2O_5$ reacts to give oxoanions including $[VO_4]^{3-}$ which is tetrahedral like $[PO_4]^{3-}$. If the pH of the solution is lowered (i.e. the solution becomes less basic), condensation of the $[VO_4]^{3-}$ anions occurs, which is reminiscent of (but more complicated than) the condensation of phosphate anions.

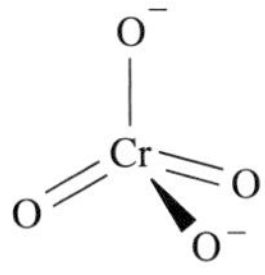

The highest oxidation state of chromium (group 6) is +6 as observed in the oxide $CrO_3$ and the oxoanions $[CrO_4]^{2-}$ (chromate) and $[Cr_2O_7]^{2-}$ (dichromate). Solutions of chromium(VI) oxide are acidic like those of $SO_3$, and in basic solution it forms the tetrahedral chromate ion (**22.59**). Solutions of chromate salts are yellow, but turn orange as the pH is lowered, because in acidic solution, the dichromate ion (**22.60**) is favoured. Equilibrium 22.215 can be pushed to the left- or the right-hand side by altering the pH of the solution; chromate is favoured in basic solution and dichromate in acidic solution.

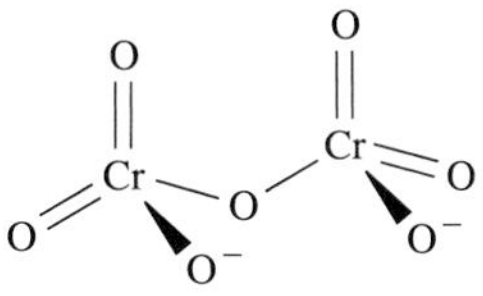

(**22.59**)

(**22.60**)

$$[Cr_2O_7]^{2-}(aq) + 2[OH]^-(aq) \rightleftharpoons 2[CrO_4]^{2-}(aq) + H_2O(l) \qquad (22.215)$$
Orange                                                  Yellow

Acidic solutions of dichromate salts are strongly oxidizing as the standard reduction potential in equation 22.216 indicates.

$$[Cr_2O_7]^{2-}(aq) + 14H^+(aq) + 6e^- \rightleftharpoons 2Cr^{3+}(aq) + 7H_2O(l)$$

$$E^\circ = +1.33 \text{ V} \quad (22.216)$$

Manganese is in group 7 and the maximum oxidation state is +7. The oxide $Mn_2O_7$ is unstable and explosive; $Cl_2O_7$ has similar properties. The manganate(VII) anion $[MnO_4]^-$ (commonly known as permanganate) is most usually encountered as the purple potassium salt, $KMnO_4$, and is a powerful oxidizing agent as indicated by the standard reduction potential for equation 22.217. Compare this with the reduction potentials for the perchlorate ion given in Table 22.6. Equation 22.217 refers to an acidic solution; in basic aqueous solution, $[MnO_4]^-$ is reduced to manganese(IV) oxide (equation 22.218). The reduction potential of manganate(VII) is such that it can oxidize water; solutions of $KMnO_4$ are *thermodynamically* unstable with respect to equation 22.218, but are *kinetically* stable and may be kept for short periods in the dark.

$$[MnO_4]^-(aq) + 8H^+(aq) + 5e^- \rightleftharpoons Mn^{2+}(aq) + 4H_2O(l)$$

$$E^\circ = +1.51 \text{ V} \quad (22.217)$$

$$[MnO_4]^-(aq) + 2H_2O(aq) + 3e^- \rightleftharpoons MnO_2(s) + 4[OH]^-(l)$$

$$E^\circ = +0.59 \text{ V} \quad (22.218)$$

Although iron has eight valence electrons, the maximum oxidation state is +6, and this is only commonly observed in the tetrahedral ferrate ion $[FeO_4]^{2-}$. The red-purple salts $Na_2FeO_4$ and $K_2FeO_4$ are strong oxidizing agents (equation 22.219).

$$[FeO_4]^{2-}(aq) + 8H^+(aq) + 3e^- \rightleftharpoons Fe^{3+}(aq) + 4H_2O(l)$$

$$E^\circ = +2.20 \text{ V} \quad (22.219)$$

As with the noble gases, the highest oxidation state for the group 8 metals is only observed for the heaviest elements in the group. Osmium forms the white crystalline oxide $OsO_4$ and the hexafluoride $OsF_6$, neither of which has an iron analogue. Similarly, in group 9, cobalt exhibits a maximum oxidation state of +4, whereas iridium forms iridium(VI) fluoride. A similar trend is seen in group 10, where nickel forms compounds in highest oxidation state +4, but platinum, at the bottom of the triad, forms the octahedral $PtF_6$. As expected, $PtF_6$ is an extremely powerful oxidizing agent, and the first example of a noble gas compound was obtained from the reaction of $PtF_6$ with xenon (equation 22.200). Bartlett was encouraged to perform this reaction after his observation that $PtF_6$ oxidized $O_2$ to $[O_2]^+[PtF_6]^-$; the ionization potential of $O_2$ is similar to that of Xe.

## SUMMARY

This chapter has been concerned with the descriptive chemistry of elements in the *p*-block and of higher oxidation state compounds of the *d*-block metals.

### Do you know what the following terms mean?

- (*p–p*)π-bonding
- conproportionation

- ductile
- malleable

- oxoacid
- acid anhydride

### You should now be able

- to discuss trends in chemical reactivities of elements within the *p*-block groups
- to comment on the differences between the first and second members of a group
- to state the characteristic oxidation states of *p*-block elements
- to give selected and characteristic reactions of *p*-block elements
- to illustrate the diversity of oxoacids and oxoanions formed by nitrogen, phosphorus and sulfur and discuss some of their reactions

- to discuss some of the properties of halides of the *p*-block elements
- to understand the reasons why compounds of fluorine contain the element in the −1 oxidation state, but the other group 17 elements can attain positive oxidation states
- to discuss the reactivity of xenon
- to give representative chemistry of the high oxidation state *d*-block elements

## PROBLEMS

**22.1** What is the oxidation state of the central atom (in pink) in each of the following compounds or ions: (a) $SCl_2$; (b) $POCl_3$; (c) $[XeF_5]^-$; (d) $[Sb_2F_{11}]^-$; (e) $[HPO_4]^{2-}$; (f) $Al_2O_3$; (g) $Me_2SnF_2$; (h) $[Cr_2O_7]^{2-}$; (i) $[MnO_4]^-$; (j) $[IO_4]^-$, (k) $[S_2]^{2-}$?

**22.2** Calculate $\Delta G^\circ(298\,K)$ for the reaction given in equation 22.2 if (at 298 K) $\Delta_f G^\circ\ B_2O_3(s)$ and $MgO(s) = -1194$ and $-569\,kJ\,mol^{-1}$ respectively.

**22.3** The Pauling electronegativities of B, F, Cl and Br are listed in Table 5.2. (a) What are the *relative* dipole moments of B−F, B−Cl and B−Br bonds? (b) When pyridine reacts with $BF_3$, $BCl_3$ and $BBr_3$, which trihalide might you expect to attract the Lewis acid the most, based solely on your answer to part (a)? (c) Rationalize why the formation of $Br_3B\cdot py$ (from pyridine and $BBr_3$) is more favourable than the formation of $Cl_3B\cdot py$, which in turn is more favourable than the formation of $F_3B\cdot py$. Does this ordering support or contradict your answer to (b)? What conclusions can you draw?

**22.4** How do magnetic data assist in the formulation of '$GaCl_2$' as $Ga[GaCl_4]$?

**22.5** Use VSEPR theory to predict the structures of (a) $[InBr_6]^{3-}$, (b) $[GaCl_5]^{2-}$ and (c) $[GaCl_4]^-$. In the

salt $[(C_2H_5)_4N]_2[InCl_5]$, the dianion has a square-based pyramidal structure. Suggest possible reasons for this geometry.

**22.6** By considering the atomic orbitals that are available on carbon and silicon, suggest why two C=O double bonds are energetically preferred to four C−O single bonds, but four Si−O single bonds are energetically preferred to two Si=O double bonds.

**22.7** Describe, with structural diagrams, representative groups of silicates.

**22.8** In equations 22.52 and 22.53, what are the oxidation states of the N atoms in the reactants and products? What general class of reaction is each?

**22.9** Draw the structure of the azide ion. With which oxide of nitrogen is it isoelectronic?

**22.10** Which of the following oxides of nitrogen are paramagnetic: (a) NO; (b) $N_2O$; (c) $NO_2$; (d) $N_2O_4$; (e) $N_2O_3$?

**22.11** Write down an expression for the equilibrium constant for reaction 22.61 and, using Figure 22.16, estimate the value of *K* at 400 K.

**22.12** Draw resonance structures to describe the bonding in $N_2O_4$, bearing in mind the bond lengths shown in Figure 22.17.

**22.13** Suggest likely products for the following reactions which *are balanced* on the left-hand side:

$$SbF_5 + BrF_3 \longrightarrow$$

$$KF + AsF_5 \longrightarrow$$

$$PF_5 + 2SbF_5 \longrightarrow$$

**22.14** (a) In equation 22.57, what are the oxidation state changes for N?
(b) Give the oxidation states of N in the oxides in Table 22.1.
(c) What resonance structures contribute to the bonding in $NO_2$, $N_2O_4$ and $N_2O_5$ (see Figure 22.17)?
(d) What are the reduction and oxidation steps in reactions 22.80–22.82?

**22.15** The structure of molecular $SeF_4$ is shown in Figure 22.29. How many fluorine environments are there? The $^{19}F$ NMR spectrum liquid $SeF_4$ shows only one signal at 298 K. Suggest a reason for this observation.

**22.16** Suggest why some *p*-block elements occur in the elemental state while others occur only as compounds. Give examples.

**22.17** Write down possible products from the reaction of (a) $I_2$ with $F_2$ and (b) $Cl_2$ with $Br_2$. In each case, also give the expected reactions of the products you have written with water.

**22.18** Suggest why compounds of the noble gases appear to be essentially restricted to xenon. Your arguments might lead you to suggest that radon should form compounds similar to those of xenon – why do you think such compounds have not been included in our discussions?

**22.19** If xenon oxo-compounds are such good oxidizing agents, and only give gaseous xenon as a product, why do you think they are not more commonly used in the laboratory?

**22.20** What do you understand by the term *acid anhydride*? What are the anhydrides of (a) $HNO_2$, (b) $HNO_3$, (c) $H_3PO_4$, (d) $H_3PO_3$, (e) $HOCl$ and (f) $HIO_3$?

**22.21** (a) Why is the $Mn^{7+}$ ion not thermodynamically stable?
(b) What shapes are consistent with VSEPR theory for $[AlCl_2]^+$, $[SbCl_6]^-$, $[PCl_4]^+$ and $[I_3]^-$?

**(c)** Consider reaction 22.162. Why does *dilute* aqueous $H_2SO_4$ *not* react with Cu to give $CuSO_4$?

**22.22** (a) In reaction 22.175, why are the products not $[BF_2]^+$ and $[SF_5]^-$?
(b) Look at Figure 22.29. Why is the geometry of $SeF_4$ not tetrahedral, and why are the centres in $TeF_4$ not trigonal bipyramidal?
(c) How might $XeF_6$ react with NaF?

## Additional problems

**22.23** Rationalize the following observations.
(a) The p$K_a$ value of $[Ga(H_2O)_6]^{3+}$ is 2.6.
(b) The $^{19}F$ NMR spectrum of $PF_5$ at 298 K consists of a doublet.
(c) $N_2F_2$ possesses geometrical isomers, while $S_2F_2$ has structural isomers.
(d) $H_3PO_3$ is a dibasic acid, but $H_3PO_4$ is tribasic.
(e) Zeolites are used as drying agents (molecular sieves), as catalysts and as ion-exchange materials.

**22.24** (a) Tin(IV) oxide has a rutile structure. Draw a unit cell of this structure and use your diagram to confirm the stoichiometry as $SnO_2$. (b) When $SnO_2$ reacts with HCl(aq), what is the tin-containing product? (c) Give representative reactions that illustrate the Lewis acidity of $SnF_4$.

**22.25** $N_2F_4$ exists in both staggered and *gauche* conformations, while $H_2O_2$ favours a *gauche* conformation and a staggered conformation is observed for $P_2I_4$. Draw representations of these structures. Comment on factors that contribute to the preference of a particular conformation, paying attention to other possible conformations.

**22.26** Discuss each of the following statements, giving examples (where appropriate) to support the statements.
(a) Liquid $H_2SO_4$ is self-ionizing.
(b) $SbF_5$ is a fluoride acceptor.
(c) $XeF_2$ is used as a combined oxidizing and fluorinating agent.
(d) $Al_2O_3$ is amphoteric.
(e) $GeCl_4$ is a Lewis acid.
(f) $[IF_5]^{2-}$ and $[XeF_5]^-$ are isostructural.
(g) At 298 K and 1 bar pressure, elemental nitrogen occurs as $N_2$ but elemental phosphorus occurs as $P_4$ molecules.

# 23 Coordination complexes of the *d*-block metals

## Topics

- Electronic configurations
- Ligands and donor atoms
- Electroneutrality principle
- Isomerism
- Formation of complex ions of the first row *d*-block metals
- Ligand exchange and stability constants
- Hard and soft metals and donor atoms
- Electrode potentials
- Colours of complexes
- Crystal field theory
- Spectrochemical series
- High- and low-spin
- Magnetism: the spin-only formula
- π-Acceptor ligands and metal carbonyl compounds

| 1 | 2 | 3 | 4 | 5 | 6 | 7 | 8 | 9 | 10 | 11 | 12 | 13 | 14 | 15 | 16 | 17 | 18 |
|---|---|---|---|---|---|---|---|---|----|----|----|----|----|----|----|----|----|
| H | | | | | | | | | | | | | | | | | He |
| Li | Be | | | | | | | | | | | B | C | N | O | F | Ne |
| Na | Mg | | | | | | | | | | | Al | Si | P | S | Cl | Ar |
| K | Ca | Sc | Ti | V | Cr | Mn | Fe | Co | Ni | Cu | Zn | Ga | Ge | As | Se | Br | Kr |
| Rb | Sr | Y | Zr | Nb | Mo | Tc | Ru | Rh | Pd | Ag | Cd | In | Sn | Sb | Te | I | Xe |
| Cs | Ba | La | Hf | Ta | W | Re | Os | Ir | Pt | Au | Hg | Tl | Pb | Bi | Po | At | Rn |
| Fr | Ra | | | | | | | | | | | | | | | | |

## 23.1  Introduction: some terminology

### Coordination complexes and coordinate bonds

This chapter is concerned with some aspects of the chemistry of the *d*-block metals. In Section 22.14, we considered some *d*-block metal halides, oxides

$$\left[ \begin{array}{c} NH_3 \\ H_3N_{\prime\prime\prime\prime\prime} \; Ni \; \prime\prime\prime\prime\prime NH_3 \\ H_3N \quad \quad NH_3 \\ NH_3 \end{array} \right]^{2+}$$

**(23.1)**

$$\left[ \begin{array}{c} NH_3 \\ H_3N \; \diagdown \; Ni \; \diagup \; NH_3 \\ H_3N \; \diagup \quad \diagdown \; NH_3 \\ NH_3 \end{array} \right]^{2+}$$

**(23.2)**

Alfred Werner (1866–1919).
© *The Nobel Foundation.*

*trans-* and *cis-*isomers: see
Section 6.11

and oxyanions (e.g. $[VO_4]^{3-}$, $CrO_3$, $[MnO_4]^-$ and $OsF_6$) which contained metals in high oxidation states. In each species, the metal is essentially *covalently bonded* to the surrounding atoms. Now we turn our attention to *metal complexes* in which the *d*-block elements are in lower oxidation states and are surrounded by molecules or ions which function as Lewis bases and form coordinate bonds with the central metal ion. We have already discussed related complexes when we described the formation of *hydrated ions* in Section 21.8, but here, *ion–dipole interactions* were primarily responsible for the formation of the first hydration shell (Figure 21.13).

A great deal of the chemistry of the *d*-block metals is concerned with *coordination compounds or complexes* and these include species such as $[Ni(NH_3)_6]^{2+}$ (represented by **23.1** or **23.2**) and $[PtCl_4]^{2-}$ (represented by **23.3** or **23.4**) in which the *coordination numbers* of Ni(II) and Pt(II) are 6 and 4 respectively. Two representations of each complex are shown; one emphasizes the *stereochemistry* while the other emphasizes the nature of the *bonding*. Each of the chloride ions or ammonia molecules forms a bond with the metal centre by donating a lone pair of electrons. The molecules or ions surrounding the metal ions are called *ligands*; this term is derived from the Latin verb '*ligare*' meaning 'to bind'. As an extension of the formalisms introduced in structures **21.1** and **21.2**, we denote a bond between a metal ion and a *neutral ligand* by an arrow (as in **23.2**), and between a metal ion and an *anionic ligand* by a line (as in **23.4**). In a coordination compound, the *ligands coordinate to the metal ion.*

$$\left[ \begin{array}{c} Cl_{\prime\prime\prime\prime\prime} \; Pt \; \prime\prime\prime\prime\prime Cl \\ Cl \quad \quad Cl \end{array} \right]^{2-} \qquad \left[ \begin{array}{c} Cl \diagdown \quad \diagup Cl \\ \quad Pt \\ Cl \diagup \quad \diagdown Cl \end{array} \right]^{2-}$$

**(23.3)**            **(23.4)**

The existence of coordination complexes was first recognized by Alfred Werner, who was awarded a Nobel prize for chemistry in 1913 for his pioneering studies. A typical example of one of the problems encountered by Werner was in the structural formulation of a compound with the molecular formula $CoCl_3(NH_3)_6$. This compound was found to be ionic and it possesses a *complex cation* $[Co(NH_3)_6]^{3+}$ in which the cobalt(III) ion is sited at the centre of an octahedral arrangement of six ammonia molecules (structure **23.5**); the three chlorides are present as free ions. Further members of this group of compounds include $CoCl_3(NH_3)_5$ and $CoCl_3(NH_3)_4$. In each, Co(III) is octahedrally sited with one or two chloride ions directly bonded to the metal ion (structures **23.6** and **23.7**); the remaining chloride ions are non-coordinated. Structure **23.7** shows one of two possible *stereoisomers* of the $[Co(NH_3)_4Cl_2]^+$ cation. The chloride ions are drawn in a *trans*-arrangement, but a *cis*-isomer is also possible.

$$\left[ \begin{array}{c} NH_3 \\ H_3N_{\prime\prime\prime\prime\prime} \; Co \; \prime\prime\prime\prime\prime NH_3 \\ H_3N \quad \quad NH_3 \\ NH_3 \end{array} \right]^{3+} \; 3Cl^-$$

**(23.5)**

$$\left[ \begin{array}{c} NH_3 \\ H_3N_{\prime\prime\prime\prime\prime} \; Co \; \prime\prime\prime\prime\prime NH_3 \\ H_3N \quad \quad NH_3 \\ Cl \end{array} \right]^{2+} \; 2Cl^-$$

**(23.6)**

$$\left[ \begin{array}{c} Cl \\ H_3N_{\prime\prime\prime\prime\prime} \; Co \; \prime\prime\prime\prime\prime NH_3 \\ H_3N \quad \quad NH_3 \\ Cl \end{array} \right]^{+} \; Cl^-$$

**(23.7)**

**Fig. 23.1** The five $3d$ atomic orbitals. Each has two nodal planes.

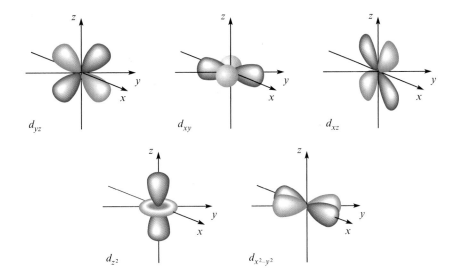

## Primary valence and coordination number

In a ***complex***, the **coordination number** of the metal centre is *not* related to the oxidation state of the metal ion.

In each of $[Co(NH_3)_6]Cl_3$, $[Co(NH_3)_5Cl]Cl_2$ and $[Co(NH_3)_4Cl_2]Cl$, the *primary valence* or *oxidation state* of Co is $+3$, but the coordination number is 6. Once recognized, the concept of complex ion formation opened up a new and exciting area of chemistry. A crucial principle is that *the coordination number of the metal centre is not related to the primary valence (the oxidation state) of the metal ion.*

## Atomic *d* orbitals

We have already mentioned $d$ orbitals in discussions of atomic structure (Chapter 3), molecular shapes (Chapter 7) and the chemistry of the $p$-block and high oxidation states $d$-block elements (Chapter 22). Now we look at the spatial properties of these atomic orbitals.

Quantum numbers: see Section 3.10

For the principal quantum number $n = 3$, there are three possible values for the quantum number $l$ ($l = 0, 1, 2$). An orbital quantum number $l = 2$ is associated with a $d$ orbital. For $l = 2$, there are five possible values of $m_l$ (for $l = 2$, $m_l = +2, +1, 0, -1, -2$) and this means that there are five real solutions to the Schrödinger equation. A set of $d$ orbitals is *five-fold degenerate*. The labels given to these five atomic orbitals are $d_{xy}$, $d_{xz}$, $d_{yz}$, $d_{z^2}$ and $d_{x^2-y^2}$ and their spatial properties (shapes) are shown in Figure 23.1. Notice that, like $p$ orbitals, $d$ orbitals are *directional*. Each $d$ orbital possesses *two nodal planes* and this is shown for the $d_{xy}$ orbital in Figure 23.2.

**Fig. 23.2** The two sign changes in an atomic $d_{xy}$ orbital. The nodal planes lie in the $xz$ and $yz$ planes.

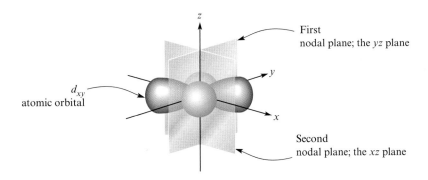

## THEORETICAL AND CHEMICAL BACKGROUND

### Box 23.1 The shapes of *d* atomic orbitals

Figure 23.1 shows that the lobes of the $d_{xy}$, $d_{xz}$ and $d_{yz}$ orbitals point *between* the Cartesian axes and each orbital lies in one of the three planes defined by these axes.

The $d_{x^2-y^2}$ orbital is related to $d_{xy}$ but its lobes point *along* (rather than between) the $x$ and $y$ axes.

We could envisage being able to draw two more atomic orbitals which are related to the $d_{x^2-y^2}$ orbital, namely the $d_{z^2-x^2}$ and $d_{z^2-y^2}$ orbitals. This would give a total of *six* orbitals. But, only *five* real solutions (i.e. five values of $m_l$) of the Schrödinger equation are allowed for $l = 2$. The problem is solved by taking a *linear combination* (see Section 4.12) of the $d_{z^2-x^2}$ and $d_{z^2-y^2}$ orbitals. This means that the two are combined, with the result that the fifth real solution to the Schrödinger equation corresponds to the $d_{z^2}$ orbital:

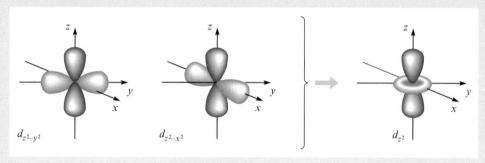

The radial distribution function of a $3d$ atomic orbital possesses *no radial node* (Figure 3.12) and we look further at the dependence of $d$ orbital energy on effective nuclear charge in Section 23.2.

### Geometries of complexes: review of Kepert theory

◼▷
Kepert theory:
see Sections 6.8–6.10

Kepert theory allows us to rationalize the shape of a complex by considering the repulsive forces between ligands. To recap, the donor atoms lie at the vertices of a polyhedral shell that surrounds the metal ion and the basic structural types are based upon:

- a linear arrangement for 2 donor atoms;
- a trigonal planar arrangement for 3 donor atoms;
- a tetrahedral arrangement for 4 donor atoms;
- a trigonal bipyramidal or square-based pyramidal arrangement for 5 donor atoms;
- an octahedral arrangement for 6 donor atoms.

◼▷
Shapes and orbital
hybridization:
see Section 7.4

An exception to these rules is that some 4-coordinate complexes (e.g. $[PtCl_4]^{2-}$ and $[AuCl_4]^-$) are square planar.

### Nomenclature

The nomenclature of coordination complexes is beyond the scope of this text and we shall refer to complexes only by formulae. The formula of a coordination complex, even when it is a neutral molecule, is written *within square brackets*. For an ionic species, the charge is written outside the square brackets. Examples are: $[PtCl_2(NH_3)_2]$, $[Co(NH_3)_6]^{2+}$ and $[ZnCl_4]^{2-}$.

### Coordination complexes of the *d*-block metals: chapter aims

In the discussion that follows, we consider some terms and definitions that are fundamental to coordination chemistry. Then we look at some descriptive chemistry of the *d*-block metals, and rationalize some of the observations in terms of the electronic structure of the metal ions. You should bear in mind several properties that typify transition metal ions:

- a *d*-block metal usually possesses several stable oxidation states (Figure 22.1);
- compounds of *d*-block metals are often coloured, although the colour may not be intense;
- compounds of *d*-block metals may be diamagnetic or paramagnetic;
- the formation of complexes is common, but not all *d*-block compounds are complexes.

> ■▷
> **Diamagnetism and paramagnetism: see Section 4.18**

## 23.2 Electronic configurations of the *d*-block metals and their ions

Figure 23.3 shows the *d*-block metals and Table 23.1 lists the names and ground state electronic configurations of each element. The group number gives the *total number of valence electrons* for each member of a particular triad. For example, in group 4, each of Ti, Zr and Hf possesses a configuration $(n + 1)s^2 nd^2$. In Section 3.19, we stated that the *approximate* order in which atomic orbitals are occupied in the ground state of an atom follows the sequence:

$$1s < 2s < 2p < 3s < 3p < 4s < 3d < 4p < 5s < 4d < 5p < 6s < 5d$$

$$\approx 4f < 6p < 7s < 6d \approx 5f$$

However, we emphasized that the energy of electrons in orbitals is affected by the nuclear charge, the presence of other electrons, and the overall charge. By applying the *aufbau* principle, we predict that for a Ti atom, the 4*s* level is occupied before the 3*d* level because the energy of the 4*s* level is lower than that of the 3*d* level. Similarly, for Zr, the 5*s* level is filled before the 4*d* level, and for hafnium, the 6*s* level is filled before the 5*d* level. We might expect this pattern to continue for other metal atoms, and that the $(n + 1)s$ orbital will always be filled before the corresponding *nd* level. However, the ground state electronic configurations in Table 23.1 show that this is not always the case. Some apparent irregularities may be rationalized (for example, Cr and Cu which are considered below), but others are not easily explained and these irregularities are noted in the table.

> ■▷
> *aufbau* **principle: see Sections 3.18 and 3.19**

**Fig. 23.3** The periodic table highlighting the *d*-block elements. Each group of three metals is called a *triad*. The 14 *lanthanoid* metals come in between La and Hf, but are members of the *f*-block. La is usually classified with the lanthanoids. Refer also to the periodic table inside the front cover of the book.

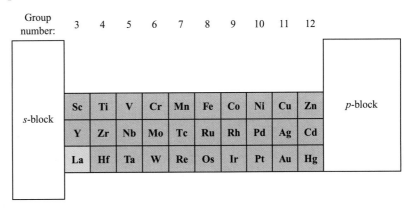

**Table 23.1** The *d*-block metals and their ground state electronic configurations.

| Element symbol | Element name | Group number | Ground state electronic configuration | Comment |
|---|---|---|---|---|
| Sc | Scandium | 3 | $[Ar]4s^2 3d^1$ | |
| Ti | Titanium | 4 | $[Ar]4s^2 3d^2$ | |
| V | Vanadium | 5 | $[Ar]4s^2 3d^3$ | |
| Cr | Chromium | 6 | $[Ar]4s^1 3d^5$ | Half-filled *d*-level |
| Mn | Manganese | 7 | $[Ar]4s^2 3d^5$ | Half-filled *d*-level |
| Fe | Iron | 8 | $[Ar]4s^2 3d^6$ | |
| Co | Cobalt | 9 | $[Ar]4s^2 3d^7$ | |
| Ni | Nickel | 10 | $[Ar]4s^2 3d^8$ | |
| Cu | Copper | 11 | $[Ar]4s^1 3d^{10}$ | Filled *d*-level |
| Zn | Zinc | 12 | $[Ar]4s^2 3d^{10}$ | Filled *d*-level |
| Y | Yttrium | 3 | $[Kr]5s^2 4d^1$ | |
| Zr | Zirconium | 4 | $[Kr]5s^2 4d^2$ | |
| Nb | Niobium | 5 | $[Kr]5s^1 4d^4$ | Irregular configuration |
| Mo | Molybdenum | 6 | $[Kr]5s^1 4d^5$ | Half-filled *d*-level |
| Tc | Technetium | 7 | $[Kr]5s^2 4d^5$ | Half-filled *d*-level |
| Ru | Ruthenium | 8 | $[Kr]5s^1 4d^7$ | Irregular configuration |
| Rh | Rhodium | 9 | $[Kr]5s^1 4d^8$ | Irregular configuration |
| Pd | Palladium | 10 | $[Kr]5s^0 4d^{10}$ | Filled *d*-level; irregular |
| Ag | Silver | 11 | $[Kr]5s^1 4d^{10}$ | Filled *d*-level |
| Cd | Cadmium | 12 | $[Kr]5s^2 4d^{10}$ | Filled *d*-level |
| La | Lanthanum | 3 | $[Xe]6s^2 5d^1$ | |
| Hf | Hafnium | 4 | $[Xe]4f^{14}6s^2 5d^2$ | |
| Ta | Tantalum | 5 | $[Xe]4f^{14}6s^2 5d^3$ | |
| W | Tungsten | 6 | $[Xe]4f^{14}6s^2 5d^4$ | |
| Re | Rhenium | 7 | $[Xe]4f^{14}6s^2 5d^5$ | Half-filled *d*-level |
| Os | Osmium | 8 | $[Xe]4f^{14}6s^2 5d^6$ | |
| Ir | Iridium | 9 | $[Xe]4f^{14}6s^2 5d^7$ | |
| Pt | Platinum | 10 | $[Xe]4f^{14}6s^1 5d^9$ | Irregular configuration |
| Au | Gold | 11 | $[Xe]4f^{14}6s^1 5d^{10}$ | Filled *d*-level |
| Hg | Mercury | 12 | $[Xe]4f^{14}6s^2 5d^{10}$ | Filled *d*-level |

Copper is in group 11 and has 11 valence electrons. We might predict that the ground state electronic configuration is $[Ar]4s^2 3d^9$ but in practice the electrons occupy the orbitals so as to give a configuration $[Ar]4s^1 3d^{10}$. This provides the Cu atom with a completely filled $3d$ shell, a situation that is energetically favourable. Similarly, Cr with six valence electrons has a ground state electronic configuration of $[Ar]4s^1 3d^5$ in preference to $[Ar]4s^2 3d^4$, and has a half-filled $3d$ level.[§]

In *d*-block *metal ions*, the configurations are $(n + 1)s^0 nd^m$. For example, the electronic configuration of $Fe^{2+}$ is $[Ar]3d^6$, and that of $Ti^{3+}$ is $[Ar]3d^1$. The origins of this effect are rather complex, but the end result is that the energies of the $nd$ orbitals drop *below* those of the $(n + 1)s$ orbital ($3d$ is lower than $4s$). In part this is due to the different ways in which these orbitals behave as the effective nuclear charge increases as we move from the metal atom to metal ion. The effective nuclear charge experienced by an electron is dependent upon the orbital in which it resides. The overall effect is a

*Effective nuclear charge, penetration and shielding: see Section 3.15*

---

[§] An electronic arrangement in which a set of orbitals is half or fully occupied is particularly favoured. For discussion, see: A. B. Blake (1981) *Journal of Chemical Education*, vol. 58, p. 393; B. J. Duke (1978) *Education in Chemistry*, vol. 15, p. 186.

*contraction* of the *d* orbitals in the metal *ions* such that the electrons are closer to the nucleus. This means that the electrostatic attraction between the negatively charged electrons and the positively charged nucleus increases and the *nd* orbitals are stabilized with respect to the $(n + 1)s$ orbital.[§]

## 23.3          Ligands

### Ligand denticity and monodentate ligands

A ligand is a Lewis base and has at least one pair of electrons with which it can form a coordinate bond with a metal centre. Most ligands are neutral or anionic, and all can donate one or more electrons through one or more atoms to a metal ion. An atom on which the electrons reside is called a *donor atom* of the ligand – when $NH_3$ acts as a ligand as in structure **23.1**, the donor atom is nitrogen.

The number of donor atoms through which a ligand coordinates to a metal ion is defined as the *denticity of the ligand*, and a ligand that coordinates to a metal centre through one donor atom is *monodentate*. Examples of monodentate ligands are given in Table 23.2.

> In a **coordination complex**, a metal ion is surrounded by molecules or ions (the ligands) which act as Lewis bases and form coordinate bonds with the metal centre which acts as a Lewis acid. The atoms in the ligands that are bonded to the metal centre are the **donor atoms**.

### Polydentate ligands and chelate rings

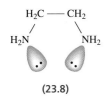

**(23.8)**

If a ligand binds to a metal ion through two donor atoms, the ligand is termed *bidentate* (Table 23.2). Ethane-1,2-diamine (**23.8**, usually abbreviated to en) can act as a bidentate ligand as in the complex ion $[Co(en)_3]^{3+}$ (Figure 23.4a). However, en may also function as a monodentate ligand as in Figure 23.4b.

> The **denticity** of a ligand is the number of donor atoms that bind to the metal centre in a complex. Collectively, ligands that bond through more than one donor atom are called **polydentate**.

| Number of donor atoms bonded in the metal centre | Denticity |
|---|---|
| 1 | Monodentate |
| 2 | Bidentate |
| 3 | Tridentate |
| 4 | Tetradentate |
| 5 | Pentadentate |
| 6 | Hexadentate |

[§] For a more in-depth discussion of this topic, see M. Gerloch and E. C. Constable (1994) *Transition Metal Chemistry: The valence shell in d-block chemistry*, VCH, Weinheim.

Table 23.2 Some typical monodentate and polydentate ligands and representative *d*-block metal complexes.

| Name of free ligand | Name of coordinated ligand | Formula or structure of ligand | Common abbreviation | Donor atoms | Representative complex |
|---|---|---|---|---|---|
| **Monodentate ligands** | | | | | |
| Water | Aqua | $H_2O$ | | *O* | $[Mn(H_2O)_6]^{2+}$ |
| Hydroxide | Hydroxo | $[OH]^-$ | | *O* | $[Zn(OH)_4]^{2-}$ |
| Ammonia | Ammine | $NH_3$ | | *N* | $[Co(NH_3)_6]^{3+}$ |
| Fluoride | Fluoro | $F^-$ | | *F* | $[CrF_6]^{3-}$ |
| Chloride | Chloro | $Cl^-$ | | *Cl* | $[ZnCl_4]^{2-}$ |
| Bromide | Bromo | $Br^-$ | | *Br* | $[FeBr_4]^{2-}$ |
| Iodide | Iodo | $I^-$ | | *I* | $[AuI_2]^-$ |
| Acetonitrile | Acetonitrile | MeCN | | *N* | $[Fe(NCMe)_6]^{2+}$ |
| Cyanide | Cyano | $[CN]^-$ | | *C* | $[Fe(CN)_6]^{3-}$ |
| Thiocyanate | Thiocyanato | $[NCS]^-$ | | *N* or *S* | $[Co(CN)_5(SCN)]^{3-}$ In solution, linkage isomers are present, $[Co(CN)_5(SCN)]^{3-}$ and $[Co(CN)_5(SCN)]^{3-}$ |
| Pyridine | Pyridine | | py | *N* | $[Fe(py)_6]^{2+}$ |
| Tetrahydrofuran | Tetrahydrofuran | | THF or thf | *O* | *trans*-$[YCl_4(THF)_2]^-$ |
| **Bidentate ligands** | | | | | |
| Ethane-1,2-diamine[a] | Ethane-1,2-diamine | | en | *N,N'* | $[Co(en)_3]^{3+}$ |
| Acetylacetonate ion | Acetylacetonato | | [acac]$^-$ | *O,O'* | $[Cr(acac)_3]$ |

| | | | | |
|---|---|---|---|---|
| Oxalate or ethanedioate ion | Oxalato | $[ox]^{2-}$ | $O,O'$ | $[Fe(ox)_3]^{3-}$ |
| Glycinate ion | Glycinato | $[gly]^{-}$ | $N,O$ | $[Ni(gly)_3]^{-}$ |
| 2,2'-Bipyridine | 2,2'-Bipyridine | bpy (or bipy) | $N,N'$ | $[Ru(bpy)_3]^{2+}$ |
| 1,10-Phenanthroline | 1,10-Phenanthroline | phen | $N,N'$ | $[Fe(phen)_3]^{2+}$ |
| **Tridentate ligands** | | | | |
| 1,4,7-Triazaheptane[a] | 1,4,7-Triazaheptane | dien | $N,N',N''$ | $[Co(dien)_2]^{3+}$ |
| 2,2':6',2''-Terpyridine | 2,2':6',2''-Terpyridine | tpy (or terpy) | $N,N',N''$ | $[Ru(tpy)_2]^{2+}$ |
| **Hexadentate ligands** | | | | |
| $N,N,N',N'$-Ethylenediaminetetraacetate ion[b] | $N,N,N',N'$-Ethylenediaminetetraacetato | $[edta]^{4-}$ | $N,N',O,O',O'',O'''$ | $[Co(edta)]^{-}$ |

[a] Older names (still in common use) are 1,2-diaminoethane or ethylenediamine, and diethylenetriamine.

[b] Although not systematic by IUPAC, this is the commonly accepted name for this anion. The prefix $N,N,N',N'$ indicates the positions of substitution of the $CH_2CO_2^{-}$ groups.

**Fig. 23.4** The solid state structures (determined by X-ray diffraction methods) and diagrammatic representations of (a) $[Co(en)_3]^{3+}$ (en = ethane-1,2-diamine) determined for the salt $[Co(en)_3][NO_3]_3$ [S. Haussuhl *et al.* (1998) *Z. Kristallogr.*, vol. 213, p. 161] and (b) *mer*-$[IrCl_3(en-N)(en-N,N')]$ in which one en ligand is monodentate and one is bidentate [F. Galsbol *et al.* (1990) *Acta Chem. Scand.*, vol. 44, p. 31]. Hydrogen atoms in the structural figures have been omitted for simplicity. Colour code: Co, yellow; Ir, red; N, blue; C, grey; Cl, green.

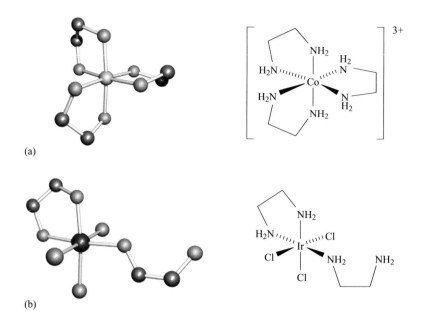

(a)

(b)

The coordination to a metal ion of the *bidentate* $H_2NCH_2CH_2NH_2$ ligand results in the formation of a *non-planar, 5-membered chelate ring* (**23.9**) (Figure 23.4a). The $[acac]^-$ ligand forms a *planar 6-membered* chelate ring (**23.10**) as in $[Cr(acac)_3]$ (Figure 23.5). When a ligand coordinates in a mode such as **23.9** or **23.10**, it is said to be *chelating*. Usually, complexes containing chelating ligands are more stable than complexes with similar monodentate ligands.

Bonding in $[acac]^-$ : see Section 33.12

The attachment of two donor atoms from one ligand to a single metal ion forms a ***chelate ring***; 5- and 6-membered chelate rings are common.

(23.9)

(23.10)

Conformers: see Section 24.9

The ability of ethane-1,2-diamine to act as a bidentate or monodentate ligand arises because the N–C–C–N chain can alter its conformation by rotation about the C–C bond (Figure 23.6) or the C–N bonds. Some

**Fig. 23.5** The solid state structure of $[Cr(acac)_3]$ and a schematic representation of the complex. Each of the chelate rings is planar. Colour code: Cr, yellow; O, red; C, grey. H atoms are omitted. [Data: S. G. Bott *et al.* (2001) *J. Chem. Soc., Dalton Trans.*, p. 2148.]

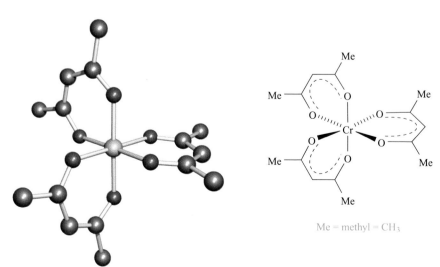

Me = methyl = $CH_3$

**Fig. 23.6** The ligand ethane-1,2-diamine (en) can undergo free bond rotation about C–C and C–N single bonds, and the result is a change in conformation. The diagram illustrates bond rotation about the C–C bond. Colour code: N, blue; C, grey; H, white.

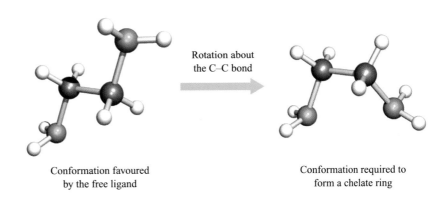

Rotation about the C–C bond

Conformation favoured by the free ligand

Conformation required to form a chelate ring

**(23.11)**

**(23.12)**

bidentate ligands such as 1,10-phenanthroline (**23.11**) (abbreviated to phen) have a rigid framework and a chelating mode is usually forced upon the ligand. In Section 11.5, we saw how a related ligand (4,7-diphenyl-1,10-phenanthroline, **11.2**) was used to analyse for $Fe^{2+}$ ions. The complex formed in this case is the octahedral dication $[FeL_3]^{2+}$, where L is ligand **11.2**. The ligand phen forms a similar complex $[Fe(phen)_3]^{2+}$.

A ligand such as 2,2'-bipyridine (bpy) (**23.12**) may undergo free bond rotation about the C–C bond connecting the two pyridine rings. However, in most of its complexes, this ligand acts as a bidentate ligand and forms a planar, 5-membered chelate ring (Figure 23.7).

A tridentate ligand possesses three donor atoms. An example is 2,2':6',2''-terpyridine (abbreviated to tpy) and Figure 23.8 shows its structure and that of the complex $[Fe(tpy)_2]^{3+}$. Ligands with four, five and six donor atoms are tetradentate, pentadentate and hexadentate, although collectively any ligand with more than one donor atom may be referred to as being *polydentate*. Some common ligands and examples of complexes involving them are listed in Table 23.2.

## Cyclic ethers and thioethers

Cyclic ethers and crowns: see Chapter 29

The oxygen atoms in a cyclic ether are potential donor atoms. An ether such as THF (Table 23.2) can function as a monodentate ligand as in the complex anion *trans*-$[YCl_4(THF)_2]^-$ (Figure 23.9). Larger cyclic ethers with nine or

**Fig. 23.7** The structure of $[Fe(bpy)_3]^{3+}$, determined by X-ray diffraction for the perchlorate salt [ B. N. Figgis *et al.* (1978) *Aust. J. Chem.*, vol. 31, p. 57]. A schematic representation of the complex cation is also shown. Colour code: Fe, green; N, blue; C, grey; H atoms are omitted.

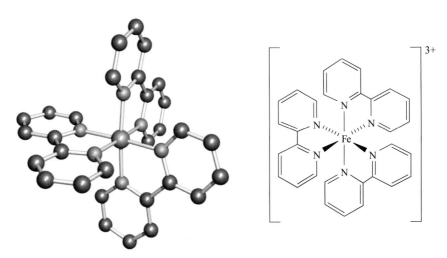

$3+$

**Fig. 23.8** The structures of the terpyridine ligand (tpy) and the complex $[Fe(tpy)_2]^{2+}$. This structure was determined by X-ray diffraction for the perchlorate salt $[Fe(tpy)_2][ClO_4]_2 \cdot H_2O$ [A. T. Baker *et al.* (1985) *Aust. J. Chem.*, vol. 38, p. 207]; H atoms are omitted for clarity. Colour code: Fe, green; N, blue; C, grey.

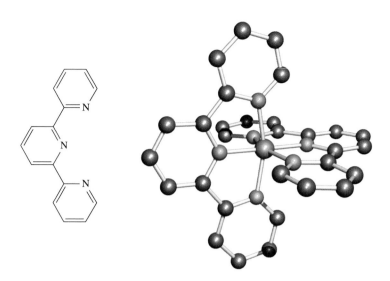

A *macrocyclic ligand* possesses nine or more atoms and three or more donor atoms in a ring. In a cyclic ether, the donor atoms are oxygens.

more atoms and three or more oxygen atoms are called *macrocyclic ligands* and are a special type of polydentate ligand. Examples include 1,4,7,10-tetra-oxacyclododecane **(23.13)** and 1,4,7,10,13,16-hexaoxacyclooctadecane (commonly called 18-crown-6) **(23.14)**. The oxygen atoms can be replaced by sulfur atoms to give *cyclic thioethers*; ligand **23.15** is 1,4,7-tri-thiacyclononane.

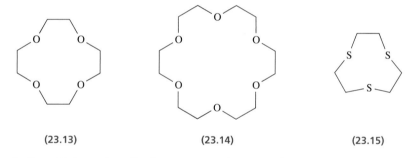

(23.13)                    (23.14)                    (23.15)

In Box 29.2 we describe how crown ethers can interact with alkali metal ions. Similar complexes are known to form between cyclic ethers and *early d*-block metals, for example scandium(III). Cyclic *thioethers* form complexes with a range of *d*-block metal ions. When ligand **23.15** coordinates to a metal

**Fig. 23.9** Cyclic ethers form some complexes with *early d*-block metals. Tetrahydrofuran (THF) can act as a monodentate ligand as in the yttrium(III) complex *trans*-$[YCl_4(THF)_2]^-$ [X-ray diffraction data: P. Sobota *et al.* (1994) *Inorg. Chem.*, vol. 33, p. 5203]. Colour code: Y, brown; Cl, green; O, red; C, grey. H atoms are omitted.

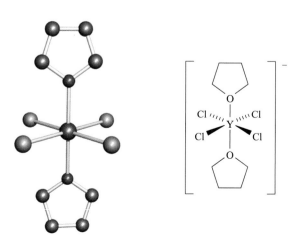

## THEORETICAL AND CHEMICAL BACKGROUND

### Box 23.2 Molecular architecture by ligand design

The emphasis in this chapter is on coordination compounds that contain a single *d*-block metal centre – *mononuclear* complexes. Ligands such as 2,2'-bipyridine and 1,10-phenanthroline usually act in a bidentate mode, binding to one metal centre, although bpy is more flexible than phen. By designing ligands with donor atoms in particular sites, it is possible to encourage the formation of *polynuclear* complexes. An example is the ligand 4,4'-bipyridine, in which the two nitrogen donor atoms are remote from each other. The only flexibility that the ligand has is rotation about the inter-ring C–C bond.

4,4'-Bipyridine cannot act as a chelating ligand *but* it can coordinate to two different metal atoms and has the

4,4'-bipyridine

potential to be a building-block in a polymeric structure. This is seen in the silver(I) complex [Ag(4,4'-bpy)$_2$][CF$_3$SO$_3$] which, in the solid state, possesses a network structure with the 4,4'-bipyridine ligands connecting pairs of silver(I) centres. The [CF$_3$SO$_3$]$^-$ anions are distributed within the network and are not shown in the structural diagram below:

Part of the polymeric structure of [{Ag(4,4'-bpy)$_2$}$_n$][CF$_3$SO$_3$]$_n$ with the anions omitted; colour code: Ag, red; C, grey; N, blue. H atoms are omitted. [X-ray diffraction data: L. Carlucci *et al.* (1994) *J. Chem. Soc., Chem. Commun.*, p. 2755.]

3,6-Di(2-pyridyl)pyridazine is related to 2,2'-bipyridine and is tetradentate:

3,6-di(2-pyridyl)pyridazine

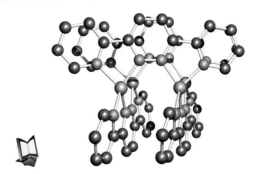

With copper(I) ions, it forms a tetranuclear complex, the structure of which is shown on the right. The structure is described as a [2 × 2]-grid, and arises because of the combined effects of the arrangement of the *N*-donor atoms and the tetrahedral coordination of the Cu(I) centres.

Colour code: Cu, orange; C, grey; N, blue. H atoms are omitted. [X-ray diffraction data: M.-T. Youinou *et al.* (1992) *Angew. Chem., Int. Ed.*, vol. 31, p. 733.]

**ENVIRONMENT AND BIOLOGY**

## Box 23.3 Porphyrin ligands

Some biologically important macrocyclic ligands contain porphyrin rings. The parent porphyrin is of the type shown on the left-hand side below – the R groups may be, for example, phenyl (Ph).

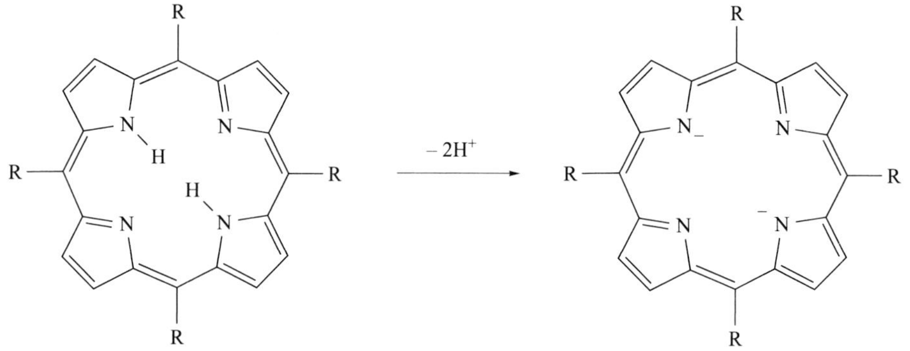

The central cavity of a porphyrin molecule contains two *N*-donor atoms and two NH groups. When these are deprotonated, the *porphyrinato ligand* is formed. The four *N*-donors lie in a plane and can coordinate to a variety of metal centres. In Box 5.3, we illustrated coordination to iron in haemoglobin. Chlorophyll (the green pigment in plants which is responsible for photosynthesis) contains a porphyrin coordinated to $Mg^{2+}$.

Photosynthesis involves the enzyme Photosystem II which converts $H_2O$ to $O_2$. Photosystem II operates in conjunction with cytochrome $b_6 f$. The crystal structure of cytochrome $b_6 f$ from the alga *Chlamydomonas reinhardtii* confirms the presence of chlorophyll A units, one of which is shown below. [X-ray diffraction data: D. Stroebel *et al.* (2003) *Nature*, vol. 426, p. 413.]

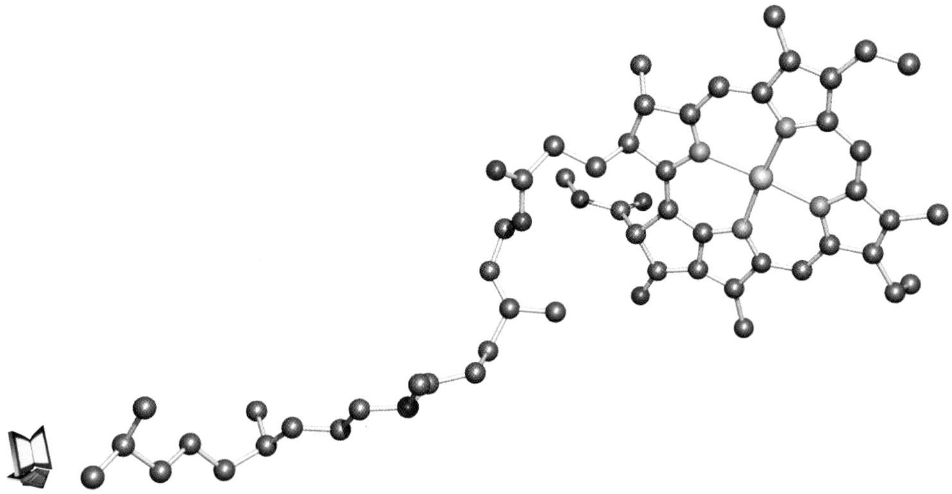

Colour code: Mg, yellow; C, grey; N, blue; O, red. H atoms are omitted.

centre, the ion lies out of the plane containing the three sulfur atoms (Figure 23.10a) but larger cyclic ethers provide a cavity that can accommodate the metal ion within the ring as the structure of $[Sc(benzo-15-crown-5)Cl_2]^+$ illustrates (Figure 23.10b).

**Fig. 23.10** (a) When a macrocyclic ligand coordinates to a metal ion, the metal may lie above the plane containing the donor atoms as in the iron(III) complex [Fe(L)Cl₃] where L is 1,4,7-trithiacyclononane (ligand **23.15**) [X-ray diffraction data: J. Ballester *et al.* (1994) *Acta Crystallogr., Sect. C*, vol. 50, p. 712]. (b) Larger macrocyclic ligands may be able to accommodate the metal ion within the ring as in the scandium(III) complex *trans*-[ScCl₂(benzo-15-crown-5)]⁺ [X-ray diffraction data: G. R. Willey *et al.* (1993) *J. Chem. Soc., Dalton Trans.*, p. 3407]. Colour code: Fe, blue; S, yellow; Cl, green; C, grey; Sc, brown; O, red. H atoms are omitted.

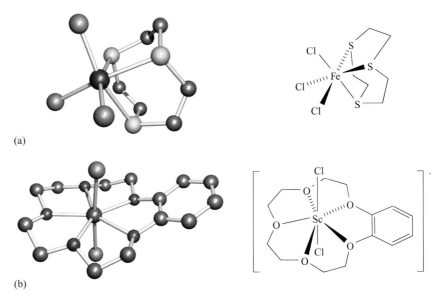

(a)

(b)

### Donor atom notation

When we write the formula of a complex ion such as $[Co(en)_3]^{2+}$, there is an in-built assumption that the en ligand coordinates through both the nitrogen donor atoms. However, we have also seen that the en ligand can coordinate in a monodentate fashion. In order to distinguish between these two situations, suffixes are added to the ligand name to denote the donor atoms involved in coordination. In a monodentate mode, the ligand may be specified as en-*N*, and in a bidentate mode, as en-*N*,*N'*. Notice two features of this notation:

- the symbols for the donor atoms are written in italic;
- the two *different* nitrogen donor atoms are distinguished by using a prime mark on the second atom.

The structure shown in Figure 23.4b is of the compound *mer*-[IrCl₃(en-*N*)(en-*N*,*N'*)]. Further examples of the use of this notation are shown in the donor atom list in Table 23.2; additional prime marks are added for additional like-atoms, for example in the [edta]⁴⁻ ligand (Table 23.2).

---

**23.4**  ### The electroneutrality principle

In Chapter 5 we discussed polar covalent bonds, and in Chapter 8 we considered ionic compounds but emphasized that the ionic model may be adjusted to take account of some covalent character. The coordinate bond formed between the donor atom of a ligand and a metal ion is a particular type of bond that requires some thought. What we are implying by the representation shown in Figure 23.11a for $[Co(NH_3)_6]^{3+}$ is that a total of six *pairs of electrons* are donated from the ligands to the metal centre. In the *covalently* bonded representation, a single net unit of charge is transferred from each ligand to $Co^{3+}$ and Figure 23.11b shows the resulting charge distribution. The differences between Figures 23.11a and 23.11b are equivalent to the differences between the two representations of the bonding in THF·BH₃ in structures **21.1** and **21.2**.

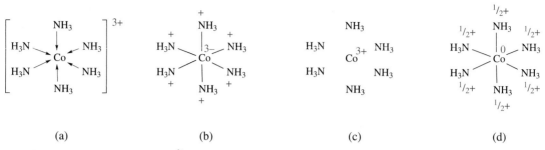

(a)                          (b)                          (c)                          (d)

**Fig. 23.11** The complex cation $[Co(NH_3)_6]^{3+}$ (a) showing the presence of coordinate bonds, (b) showing the charge distribution that results from a 100% covalent model, (c) showing the charge distribution that results from a 100% ionic model, and (d) showing the charge distribution that results from applying the electroneutrality principle.

Linus Pauling: see Box 5.1

Now consider the consequence of assuming a wholly *ionic* model for the bonding in $[Co(NH_3)_6]^{3+}$. The tripositive charge remains localized on the cobalt ion and the six $NH_3$ molecules remain neutral (Figure 23.11c). Which of these two models is correct, or is an intermediate situation realistic?

The solution and solid state properties of coordination compounds suggest that the ionic model is not satisfactory. For example, the compound $[Co(NH_3)_6]Cl_3$ dissolves in water to give $[Co(NH_3)_6]^{3+}$ and $Cl^-$ ions, rather than $Co^{3+}$ ions, $Cl^-$ ions and $NH_3$ molecules. Simple electrostatic interactions as indicated by the ionic model are not likely to be sufficiently strong to allow the $[Co(NH_3)_6]^{3+}$ complex ion to be maintained in aqueous solution. If we return to the covalent model for $[Co(NH_3)_6]^{3+}$, the charge distribution in Figure 23.11b causes a problem in terms of the electron-withdrawing capabilities of the cobalt(III) and nitrogen centres. The metal centre is *electropositive* whereas the nitrogen is *electronegative* and we expect the Co–N bond to be polarized in the sense $Co^{\delta+}–N^{\delta-}$. This suggests that the charge distribution in Figure 23.11b is unrealistic.

Neither a wholly ionic nor a wholly covalent model is appropriate to describe the interactions between ligands and metal ions. The problem was addressed by Pauling who put forward the *electroneutrality principle*:

> Pauling's ***electroneutrality principle*** states that the distribution of charge in a molecule or ion is such that the charge on any single atom is within the range $+1$ to $-1$ (ideally close to zero).

If we apply the electroneutrality principle to the $[Co(NH_3)_6]^{3+}$ ion, then, ideally, the net charge on the cobalt centre should be zero. That is, the $Co^{3+}$ ion may accept a total of only three electrons from the six $NH_3$ ligands, and this leads to the charge distribution shown in Figure 23.11d. The charge transferred from each N to Co is only half an electron and this is midway between the limiting covalent (one electron transferred) and ionic (no electrons transferred) models. The electroneutrality principle results in a bonding description for the $[Co(NH_3)_6]^{3+}$ ion which is 50% ionic (or, 50% covalent). **Note:** the electroneutrality principle is only an *approximate method* of estimating the charge distribution in molecules and complex ions.

| Worked example 23.1 | *The electroneutrality principle* |

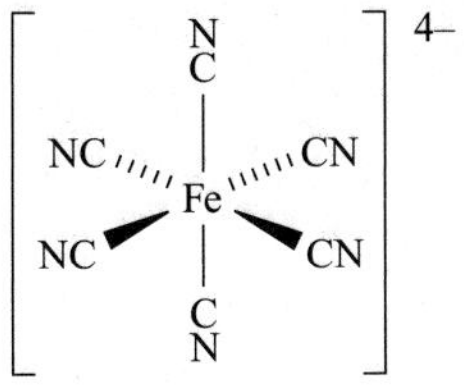

(23.16)

**Estimate the charge distribution on the metal and each donor atom in the octahedral ion $[Fe(CN)_6]^{4-}$ (23.16).**

The ion $[Fe(CN)_6]^{4-}$ is an iron(II) complex, and the ionic model would imply a charge distribution of:

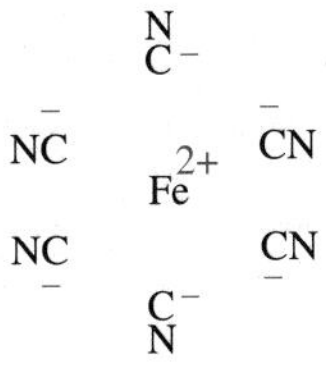

100% covalent character leads to a charge distribution of:

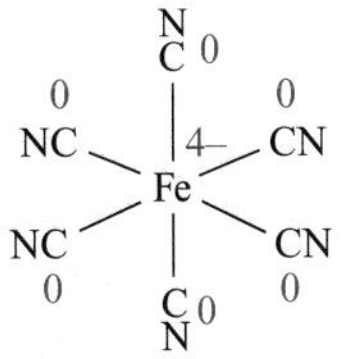

By the electroneutrality principle, the charge distribution should ideally be such that the iron centre bears no net charge. It must therefore gain two electrons (to counter the $Fe^{2+}$ charge) from the six $[CN]^-$ ligands.
Each $[CN]^-$ therefore loses $\frac{1}{3}$ of an electron. The charge per cyano ligand therefore changes from $1-$ to $\frac{2}{3}-$.
The final charge distribution is approximately:

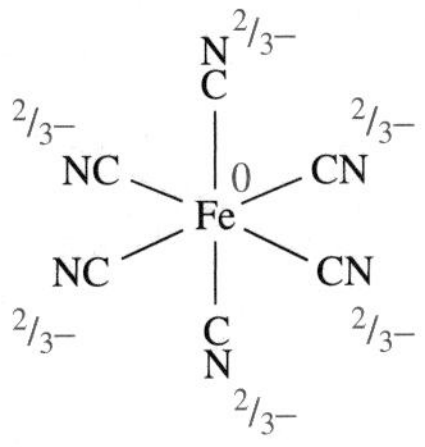

---

## 23.5    Isomerism

▢▶
Revision of stereoisomers: see problems 23.5 and 23.7

▢▶
Stereoisomerism in organic compounds: see Section 24.8

The discussion of stereoisomerism in Section 6.11 included examples of mononuclear complexes of the *d*-block metals. Stereoisomers have the same connectivity of atoms, but the spatial arrangement of the atoms or groups differs, e.g. *cis-* and *trans-*$[PtCl_2(NH_3)_2]$:

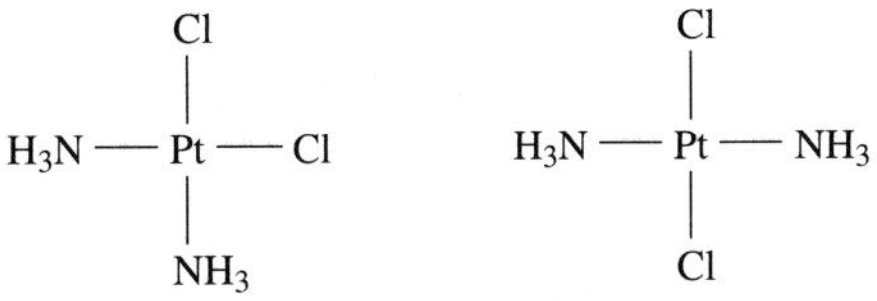

*Stereoisomers* include *cis-* and *trans-*isomers, *mer-* and *fac-*isomers, and enantiomers.

Stereoisomers include enantiomers, and we discuss these in detail later in the section. As well as stereoisomerism, we discuss *structural isomerism* among coordination complexes. Structural isomers that we consider include hydration, ionization, linkage and coordination isomers.

## Hydration and ionization isomers

Hydration and ionization isomers result from the interchange of ligands within the first coordination sphere with those outside it. The compound $CrCl_3(H_2O)_6$ is obtained commercially as a green solid and has the structural formula $[CrCl_2(H_2O)_4]Cl \cdot 2H_2O$ – two $Cl^-$ ions are coordinated to the $Cr^{3+}$ centre and there are one non-coordinated $Cl^-$ and two water molecules of crystallization. When this solid dissolves in water, the solution slowly turns blue-green as a result of an isomerization in which one $Cl^-$ ligand is replaced by $H_2O$; crystallization yields $[CrCl(H_2O)_5]Cl_2 \cdot H_2O$. The compound consists of the complex cation $[CrCl(H_2O)_5]^{2+}$ (**23.17**), two $Cl^-$ ions and a water molecule of crystallization.

Ionization isomers result from an exchange involving an anionic ligand. The purple complex $[Co(NH_3)_5(SO_4)]Br$ and the pink complex $[Co(NH_3)_5Br][SO_4]$ are ionization isomers, and differ in having either a $Br^-$ or $[SO_4]^{2-}$ ion coordinated (or not) to the cobalt(III) centre.

> *Hydration* and *ionization isomers* of a compound are the result of the exchange of ligands within the first coordination sphere with those outside it. *Hydration isomers* involve the exchange of water, and *ionization isomers* involve the exchange of anionic ligands.

## Linkage isomerism

If a ligand possesses two or more different donor sites, there is more than one way in which it can bond to a metal ion. One resonance structure of the thiocyanate ion $[NCS]^-$ is shown in **23.18**. Both the nitrogen and sulfur atoms are potential donor atoms, and *linkage isomers* of a complex may exist. Structures **23.19** and **23.20** illustrate linkage isomers of the complex cation $[Co(NH_3)_5(NCS)]^{2+}$ in which the thiocyanato ligand is either *S*- or *N*-bonded. The formulae of the two isomers are written as $[Co(NH_3)_5(NCS-N)]^{2+}$ (*N*-bonded thiocyanato ligand) and $[Co(NH_3)_5(NCS-S)]^{2+}$ (*S*-bonded thiocyanato ligand).

> If a ligand possesses two or more different donor sites, *linkage isomers* may be formed because the ligand can attach to the metal centre in more than one way.

## Coordination isomerism

If a coordination compound contains both a complex cation *and* a complex anion, it may be possible to isolate isomers of the compound in which ligands are exchanged between the two metal centres. Such complexes are examples of *coordination isomers*. Consider $[Co(bpy)_3][Fe(CN)_6]$ which contains the cation $[Co(bpy)_3]^{3+}$ and the anion $[Fe(CN)_6]^{3-}$. This is not the only arrangement of the ligands, and other possibilities are:

$$[Co(bpy)_2(CN)_2]^+[Fe(bpy)(CN)_4]^-$$

$$[Fe(bpy)_2(CN)_2]^+[Co(bpy)(CN)_4]^-$$

$$[Fe(bpy)_3]^{3+}[Co(CN)_6]^{3-}$$

(23.17)

(23.18)

(23.19)

(23.20)

**Fig. 23.12** The complex cation *cis*-[Co(en)$_2$Cl$_2$]$^+$ possesses enantiomers.
(a) Isomers A and B are related by a reflection through a mirror plane.
(b) Rotating isomer B shows that it is *non-superposable* on isomer A. A and B are therefore enantiomers. Colour code: Co, yellow; Cl, green; N, blue; C, grey. H atoms are omitted.
[X-ray diffraction data: K. Matsumoto *et al.* (1970) *Bull. Chem. Soc. Jp.*, vol. 43, p. 3801.]

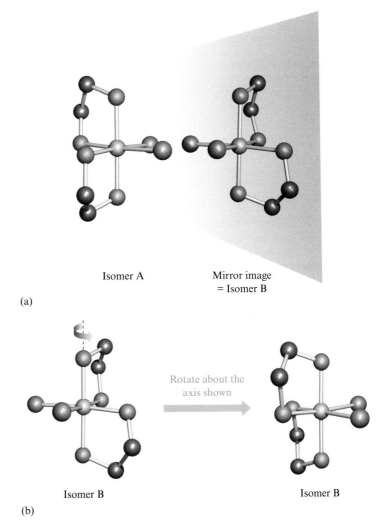

Isomer A          Mirror image
                  = Isomer B

(a)

Rotate about the
axis shown

Isomer B          Isomer B

(b)

Each of these is a distinct chemical compound and may be isolated and characterized. For example, [Co(bpy)$_3$][Fe(CN)$_6$] is prepared from the reaction of K$_3$[Fe(CN)$_6$] and [Co(bpy)$_3$]Cl$_3$.

If a coordination compound contains both a complex cation *and* a complex anion, it may be possible to isolate *coordination isomers* in which ligands are exchanged between the two different metal centres.

### Enantiomerism (optical isomerism)

More about enantiomers in Section 24.8

A pair of *enantiomers* consists of two molecular species which are mirror images of each other and are non-superposable.[§] Enantiomers of a coordination compound most often occur when chelating ligands are involved. The cation *cis*-[Co(en)$_2$Cl$_2$]$^+$ and its mirror image are shown in Figure 23.12a. Figure 23.12b shows that the mirror image (labelled isomer B) is non-superposable

[§] This definition is taken from Basic Terminology of Stereochemistry: IUPAC Recommendations (1996) *Pure and Applied Chemistry*, vol. 68, p. 2193.

**Fig. 23.13** Enantiomers of the coordination complex [Cr(acac)$_3$]. See Table 23.2 for the structure of the ligand [acac]$^-$. Colour code: Cr, green; O, red; C, grey. H atoms are omitted. [X-ray diffraction data: S. G. Bott *et al.* (2001) *J. Chem. Soc., Dalton Trans.*, p. 2148.]

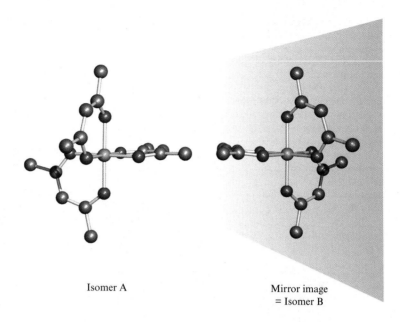

Isomer A

Mirror image
= Isomer B

upon the original structure (labelled isomer A). Isomers A and B are enantiomers and *cis*-[Co(en)$_2$Cl$_2$]$^+$ is a *chiral* cation. Similarly, [Cr(acac)$_3$] possesses enantiomers because it has a non-superposable mirror image (Figure 23.13). On the other hand, you should be able to show that the octahedral complexes *trans*-[Co(en)$_2$Cl$_2$]$^+$ and [Mn(en)(H$_2$O)$_4$]$^{2+}$ do *not* possess enantiomers. Further complications arise if the ligand itself possesses enantiomers (see Section 24.8).

> A pair of molecular species which are mirror images of each other and are non-superposable are *enantiomers*. They may also be referred to as *optical isomers*.[§]
>
> When an object (e.g. a molecule) is non-superposable on its mirror image, it is said to be *chiral*. If an object is superposable on its mirror image, the object is *achiral*.

| | 23.6 | **The formation of first row *d*-block metal coordination complexes** |

| Group number | 3 | 4 | 5 | 6 | 7 | 8 | 9 | 10 | 11 | 12 |
|---|---|---|---|---|---|---|---|---|---|---|
| First row *d*-block metal | Sc | Ti | V | Cr | Mn | Fe | Co | Ni | Cu | Zn |

### Aqua complexes and complexes with hydroxide ligands

The simplest method of forming a complex ion is to dissolve a *d*-block metal salt such as anhydrous CoCl$_2$ in water. On dissolution, the Co$^{2+}$ ion is present as the *hexaaqua ion* [Co(H$_2$O)$_6$]$^{2+}$ (Figure 23.14), and solutions containing this ion are pale pink. Although 'naked' Co$^{2+}$ ions (and other *d*-block metal ions) are *not present in aqueous solution*, you will often find

---

[§] The IUPAC recommends the use of the term *enantiomer* in preference to *optical isomer*.

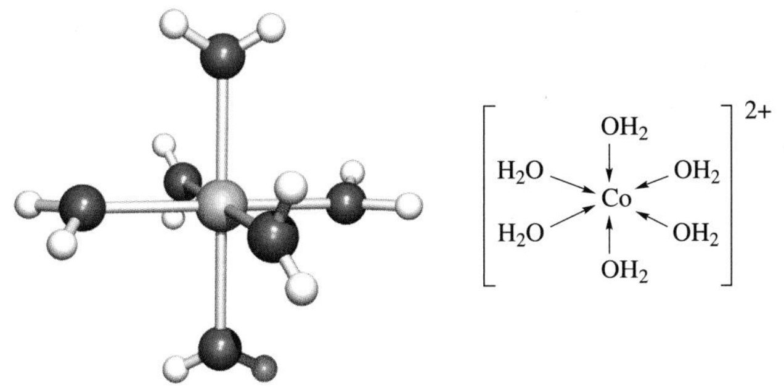

**Fig. 23.14** The octahedral structure of the complex cation $[Co(H_2O)_6]^{2+}$ and a schematic representation showing the coordinate bonds. Colour code: Co, yellow; O, red; H, white. [X-ray diffraction data: B. Viossat *et al.* (1981) *Bull. Soc. Chim. Fr.*, p. 69.]

$Co^{2+}(aq)$ written in equations. This is strictly incorrect because water molecules are present in the *first coordination sphere* of the metal ion.

Hexaaqua ions such as $[Co(H_2O)_6]^{2+}$ are found for all of the first row *d*-block *metal(II) ions*, although the $[M(H_2O)_6]^{2+}$ ions are not always stable in aqueous solution. For example, $[Ti(H_2O)_6]^{2+}$ is not stable and solutions readily oxidize to purple $[Ti(H_2O)_6]^{3+}$. However, since the charge density on $Ti^{3+}$ is high, it is strongly polarizing and $[Ti(H_2O)_6]^{3+}$ readily loses a proton (equation 23.1). In dilute hydrochloric acid, equilibrium 23.1 is pushed towards the left-hand side, and $[Ti(H_2O)_6]^{3+}$ is stabilized. This behaviour resembles that of $[Al(H_2O)_6]^{3+}$.

Polarizing power: see Section 8.18

$[Al(H_2O)_6]^{3+}$: see equation 22.17

$$[Ti(H_2O)_6]^{3+}(aq) + H_2O(l) \rightleftharpoons [Ti(H_2O)_5(OH)]^{2+}(aq) + [H_3O]^+(aq)$$

$$pK_a = 3.9 \quad (23.1)$$

The next metal in the first row of the *d*-block is vanadium. Pale purple solutions containing $[V(H_2O)_6]^{2+}$ are readily oxidized and solutions turn green as $[V(H_2O)_6]^{3+}$ is formed. The $V^{3+}$ ion is strongly polarizing and $[V(H_2O)_6]^{3+}$ is acidic (equation 23.2); the $pK_a$ value shows it to be a stronger acid than $[Ti(H_2O)_6]^{3+}$, and $[V(H_2O)_6]^{3+}$ is stabilized only in acidic solutions.

$$[V(H_2O)_6]^{3+}(aq) + H_2O(l) \rightleftharpoons [V(H_2O)_5(OH)]^{2+}(aq) + [H_3O]^+(aq)$$

$$pK_a = 2.9 \quad (23.2)$$

The *nuclearity* of a metal complex refers to the number of metal centres. A *mononuclear* complex has one metal centre, a *dinuclear* complex contains two and a *trinuclear* complex possesses three.

The formation of a *hydroxo ligand*, $[OH]^-$, is often associated with a change in the *nuclearity* of the solution species. Equation 23.2 shows the formation of a *mononuclear* complex cation (i.e. one vanadium centre), but a *dinuclear* solution species is also present and this is possible because of the formation of *hydroxo bridges* between two vanadium centres. The presence of a ligand in a bridging mode is denoted in the formula by the Greek letter μ (pronounced 'mu'), written as a prefix. The structure of $[(H_2O)_4V(\mu\text{-OH})_2V(H_2O)_4]^{4+}$ is shown in **23.21**. It is the hydroxo and not the aqua ligands that adopt the bridging sites.

In a polynuclear coordination compound, the use of the prefix μ- means that the ligand is in a *bridging* position between two metal centres.

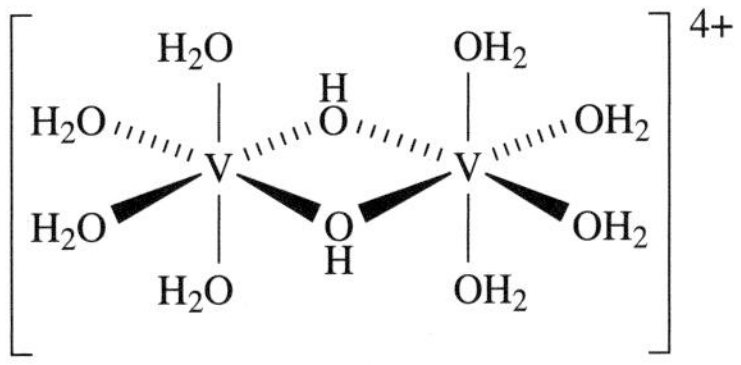

**(23.21)**

Vanadium(IV) is even more polarizing than $V^{3+}$ and $[V(H_2O)_6]^{4+}$ does not exist because it readily loses two protons *from the same aqua ligand* to give

**(23.22)**

▶▶
Alums: see Section 22.4

$[V(O)(H_2O)_5]^{2+}$ **(23.22)** which contains a V=O double bond. Solutions containing this ion are blue. Often, $[V(O)(H_2O)_5]^{2+}$ is written simply as $[VO]^{2+}$, and it is called the oxovanadium(IV) or vanadyl ion.

Chromium(II) is also readily oxidized in aqueous solution. Solutions containing $[Cr(H_2O)_6]^{2+}$ are sky-blue and oxidation gives the violet $[Cr(H_2O)_6]^{3+}$ ion. Once again, the polarizing metal(III) ion leads to easy proton loss (equation 23.3) and the formation of a hydroxo-complex which condenses to a dinuclear complex (equation 23.4) with a structure similar to **23.21**.

$$[Cr(H_2O)_6]^{3+}(aq) + H_2O(l) \rightleftharpoons [Cr(H_2O)_5(OH)]^{2+}(aq) + [H_3O]^+(aq)$$

$$pK_a = 4.0 \quad (23.3)$$

$$2[Cr(H_2O)_5(OH)]^{2+}(aq) \rightleftharpoons [(H_2O)_4Cr(\mu\text{-}OH)_2Cr(H_2O)_4]^{4+} + 2H_2O \quad (23.4)$$

Manganese follows chromium in the first row of the *d*-block. Manganese(II) salts such as $MnCl_2 \cdot 4H_2O$, $Mn(NO_3)_2 \cdot 6H_2O$ and $MnSO_4 \cdot H_2O$ dissolve in water to give very pale pink solutions containing the octahedral $[Mn(H_2O)_6]^{2+}$ ion. Iron(II) salts such as $FeSO_4 \cdot 7H_2O$ dissolve to give pale green aqueous solutions, although atmospheric $O_2$ oxidizes iron(II) to iron(III). Solutions containing $[Fe(H_2O)_6]^{3+}$ are purple but when it is prepared, solutions usually appear yellow due to the formation of the hydroxo species $[Fe(H_2O)_5(OH)]^{2+}$ and $[Fe(H_2O)_4(OH)_2]^+$ (equations 23.5 and 23.6). These observations emphasize the polarizing power of the $Fe^{3+}$ ion. In equilibria 23.5 and 23.6, *successive* aqua ligands are deprotonated. The purple colour of $[Fe(H_2O)_6]^{3+}$ can be seen in the alum $KFe(SO_4)_2 \cdot 12H_2O$.

$$[Fe(H_2O)_6]^{3+}(aq) + H_2O(l) \rightleftharpoons [Fe(H_2O)_5(OH)]^{2+}(aq) + [H_3O]^+(aq)$$

$$pK_a = 2.0 \quad (23.5)$$

$$[Fe(H_2O)_5(OH)]^{2+}(aq) + H_2O(l) \rightleftharpoons [Fe(H_2O)_4(OH)_2]^+(aq) + [H_3O]^+(aq)$$

$$pK_a = 3.3 \quad (23.6)$$

Like the hydroxo complexes of vanadium and chromium described above, those of iron(III) undergo condensation reactions to form dinuclear compounds such as $[(H_2O)_4Fe(\mu\text{-}OH)_2Fe(H_2O)_4]^{4+}$. This type of process continues until, finally, solid state iron(III) hydroxides are produced.

Solutions containing the $[Co(H_2O)_6]^{2+}$ ion are pale pink, those of $[Ni(H_2O)_6]^{2+}$ are pale green, and solutions with the $[Cu(H_2O)_6]^{2+}$ cation are pale blue. From titanium to nickel, all the solution species that we have described have been coloured, but when zinc(II) salts dissolve in water, colourless solutions are formed. These solutions contain the aqua ion $[Zn(H_2O)_6]^{2+}$. We return to the question of colour in Section 23.10.

## Ammine complexes

The addition of $NH_3$ to an aqueous solution of *d*-block $M^{2+}$ ions *may* result in the *displacement* of aqua ligands by ammine ligands.[§] The number of ligands displaced depends on the metal and the concentrations of the aqua complex and $NH_3$ in solution. Equation 23.7 shows that the reaction between $[Ni(H_2O)_6]^{2+}$ and $NH_3$ leads to the displacement of all six ligands.

---

[§] Do not confuse the name *ammine* for a coordinated $NH_3$ ligand with *amine*, a molecule of type $RNH_2$, $R_2NH$ or $R_3N$.

Equilibrium 23.8 shows that only four of the six aqua ligands in $[Cu(H_2O)_6]^{2+}$ are readily displaced.

$$[Ni(H_2O)_6]^{2+}(aq) + 6NH_3(aq) \rightleftharpoons [Ni(NH_3)_6]^{2+}(aq) + 6H_2O(l) \tag{23.7}$$

$$[Cu(H_2O)_6]^{2+}(aq) + 4NH_3(aq) \rightleftharpoons [Cu(H_2O)_2(NH_3)_4]^{2+}(aq) + 4H_2O(l) \tag{23.8}$$

> In a ***ligand displacement reaction***, one ligand in the coordination sphere of the metal ion is replaced by another.

In Section 23.7 we consider the equilibrium constants for reactions of this type. Displacement reactions are not always the best methods of preparing ammine complexes. The reactions of metal salts with gaseous $NH_3$ may provide alternative routes. Solid iron(II) halides, $FeX_2$, react with gaseous $NH_3$ to give $[Fe(NH_3)_6]X_2$ in addition to $[Fe(NH_3)_5X]X$ and $[Fe(NH_3)_4X_2]$. The reaction must be carried out in the solid state because iron(II) ammine complexes are not stable in aqueous solution. Iron(III) does not form simple ammine complexes.

The complex cation $[Co(NH_3)_6]^{3+}$ is usually prepared by oxidizing the corresponding cobalt(II) complex (equation 23.9) which may itself be prepared by the displacement of aqua ligands in $[Co(H_2O)_6]^{2+}$ by $NH_3$ ligands. Alternatively, oxidation and complex formation can be carried out in one step as in reaction 23.10.

$$4[Co(NH_3)_6]^{2+} + O_2 + 4[NH_4]^+ \longrightarrow 4[Co(NH_3)_6]^{3+} + 2H_2O + 4NH_3 \tag{23.9}$$

$$\underset{\text{oxidation}}{\underbrace{\phantom{xxxxxxxxxxxxxxxx}}} \qquad \underset{\text{reduction}}{\underbrace{\phantom{xxxxxxxxxxxxxxxx}}}$$

$$4CoCl_2 + O_2 + 4[NH_4]Cl + 20NH_3 \xrightarrow{\underset{\text{charcoal}}{\text{Activated}}} 4[Co(NH_3)_6]Cl_3 + 2H_2O \tag{23.10}$$

## Chloro ligands

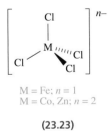

$M = Fe; n = 1$
$M = Co, Zn; n = 2$

**(23.23)**

In Chapter 22, we illustrated examples of reactions between *p*-block metal halides and excess halide ion in which the metal halide acted as a Lewis acid, e.g. the formation of $[AlCl_4]^-$ and $[SnCl_6]^{2-}$. Similar reactions occur with some of the *d*-block metal halides and equations 23.11 to 23.13 show representative reactions. Each chloro complex is tetrahedral (**23.23**).

$$FeCl_3(aq) + Cl^-(aq) \underset{}{\overset{\text{in conc HCl}}{\rightleftharpoons}} [FeCl_4]^-(aq) \tag{23.11}$$

$$ZnCl_2(aq) + 2Cl^-(aq) \underset{}{\overset{\text{in conc HCl}}{\rightleftharpoons}} [ZnCl_4]^{2-}(aq) \tag{23.12}$$

$$CoCl_2(aq) \underset{\text{in conc HCl(aq)}}{\overset{\text{in dilute HCl(aq)}}{\rightleftharpoons}} \begin{matrix} [Co(H_2O)_6]^{2+}(aq) \\ \text{Pink} \\ \\ [CoCl_4]^{2-}(aq) \\ \text{Blue} \end{matrix} \tag{23.13}$$

The chloro complexes may be isolated from solution by exchanging the cation (which is $[H_3O]^+$ in aqueous medium) for a much larger counter-ion such as tetraethylammonium, $[Et_4N]^+$. After adding $[Et_4N]Cl$, the desired salt precipitates out of solution (equation 23.14).

$$2[Et_4N]Cl(s) + [ZnCl_4]^{2-}(aq) \longrightarrow [Et_4N]_2[ZnCl_4](s) + 2Cl^-(aq) \tag{23.14}$$

$$
\left[
\begin{array}{c}
\text{Cl} \\
\text{H}_2\text{O}\,\,{}_{'''''}\,\,\underset{\displaystyle\overset{|}{\text{Ti}}}{}\,\,{}^{\backslash\backslash\backslash\backslash}\text{OH}_2 \\
\text{H}_2\text{O} \quad\quad\quad \text{OH}_2 \\
\text{OH}_2
\end{array}
\right]^{2+}
$$

(23.24)

In equation 23.13, the effects of varying the concentration of chloride ion are apparent. If a pink solution of cobalt(II) chloride in dilute hydrochloric acid is heated, the concentration of chloride ion increases as water evaporates, and the solution turns blue as $[\text{Co}(\text{H}_2\text{O})_6]^{2+}$ is converted to $[\text{CoCl}_4]^{2-}$ (equation 23.15). This process is used in experiments involving 'invisible ink' and is reversible; solutions containing the blue $[\text{CoCl}_4]^{2-}$ ion turn pale pink when diluted with water.

$$[\text{Co}(\text{H}_2\text{O})_6]^{2+}(\text{aq}) + 4\text{Cl}^-(\text{aq}) \rightleftharpoons [\text{CoCl}_4]^{2-}(\text{aq}) + 6\text{H}_2\text{O}(\text{l}) \qquad (23.15)$$

Similar effects are observed with other metal ions. For example, $[\text{Ti}(\text{H}_2\text{O})_6]^{3+}$ is stabilized in *dilute* hydrochloric acid (equation 23.1), but in *concentrated* acid, chloride ion displaces an aqua ligand and $[\text{TiCl}(\text{H}_2\text{O})_5]^{2+}$ (**23.24**) is formed.

## Cyano complexes

The cyano ligand $[\text{CN}]^-$ usually bonds to metal ions through the *carbon atom* and each $\text{M}-\text{C}\equiv\text{N}$ unit is linear. When $[\text{CN}]^-$ acts as a bridging ligand, both the C and N atoms act as donors (see Box 23.4). With chromium(II), $[\text{CN}]^-$ forms the octahedral complex anion $[\text{Cr}(\text{CN})_6]^{4-}$. Octahedral complexes are also formed with iron(II) and iron(III). Many stable salts of $[\text{Fe}(\text{CN})_6]^{4-}$ exist, for example $\text{K}_4[\text{Fe}(\text{CN})_6]$; the complex anion is stable in aqueous solution. The iron(III)-containing complex $[\text{Fe}(\text{CN})_6]^{3-}$ (Figure 23.15) is more toxic than $[\text{Fe}(\text{CN})_6]^{4-}$, because it releases cyanide ion more readily than does the iron(II) species. When $[\text{Fe}(\text{CN})_6]^{4-}$ is added to an aqueous solution containing iron(III) ions, a deep blue compound is formed and this is a qualitative test for $\text{Fe}^{3+}$ (see Box 23.4).

---

THEORETICAL AND CHEMICAL BACKGROUND

### Box 23.4 Prussian blue and Turnbull's blue

When $[\text{Fe}(\text{CN})_6]^{4-}$ is added to an aqueous solution containing $\text{Fe}^{3+}$, a deep blue compound called Prussian blue is formed. The same blue colour is observed when $[\text{Fe}(\text{CN})_6]^{3-}$ is added to an aqueous solution containing $\text{Fe}^{2+}$; this blue complex is referred to as Turnbull's blue. Both Prussian blue and Turnbull's blue are hydrated salts and possess the formula $\text{Fe}^{\text{III}}_4[\text{Fe}^{\text{II}}(\text{CN})_6]_3 \cdot x\text{H}_2\text{O}$ (where $x$ is about 14). A related compound is the potassium salt $\text{KFe}[\text{Fe}(\text{CN})_6]$ and this is referred to as 'soluble Prussian blue'. The solid state structures of these compounds contain cubic arrangements of iron centres with cyano groups bridging between the metals. One-eighth of the unit cell of $\text{KFe}[\text{Fe}(\text{CN})_6]$ is shown on the right, with the $\text{K}^+$ ions removed. Each iron centre is octahedrally sited and is in either an $\text{FeN}_6$ or an $\text{FeC}_6$ environment; the lattice must be extended to see this feature. Half of the iron sites are Fe(II) and half are Fe(III), and the $\text{K}^+$ ions occupy cavities within the cubic lattice.

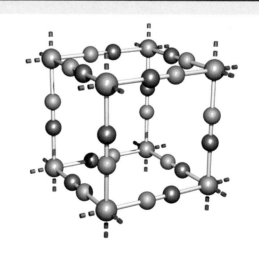

Colour code: Fe, green; N, blue; C, grey. The hashed-bonds indicate how the extended structure continues.

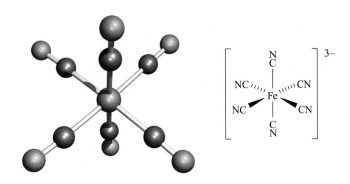

**Fig. 23.15** The octahedral structure of $[Fe(CN)_6]^{3-}$. Each of the Fe−C−N bond angles is approximately $180°$. Colour code: Fe, green; N, blue; C, grey. [X-ray diffraction data: P. A. W. Dean *et al.* (2003) *Dalton Trans.*, p. 1520.]

***Qualitative test for Fe³⁺ ions***: When $[Fe(CN)_6]^{4-}$ is added to an aqueous solution containing iron(III) ions, a deep blue complex called Prussian blue is formed.

***Qualitative test for Fe²⁺ ions***: When $[Fe(CN)_6]^{3-}$ is added to an aqueous solution containing iron(II) ions, a deep blue complex called Turnbull's blue is formed.

Although most 4-coordinate complexes of the first row *d*-block metals are tetrahedral, *those containing nickel(II) centres are* **sometimes** *square planar*. For example, $[NiCl_4]^{2-}$ and $[NiBr_4]^{2-}$ are tetrahedral, but $[Ni(CN)_4]^{2-}$ is square planar; structures **23.25** and **23.26** give two representations of $[Ni(CN)_4]^{2-}$.

**(23.25)**

**(23.26)**

### Bidentate and polydentate ligands

Table 23.2 lists examples of both symmetrical and asymmetrical bidentate ligands. The presence of three *symmetrical* bidentate ligands in an octahedral coordination sphere leads to the formation of enantiomers and, if the precursor is *achiral* (i.e. non-chiral) such as $[Co(H_2O)_6]^{2+}$ (equation 23.16), the product is formed as a *racemate* which contains equimolar amounts of the two enantiomers. These cobalt(II) complexes readily oxidize to the corresponding cobalt(III) compounds as we showed for the ammine complexes in equation 23.9.

Racemate: see Section 24.8

$$[Co(H_2O)_6]^{2+} + 3en \rightleftharpoons \left[\begin{array}{c} Co \end{array}\right]^{2+} + \left[\begin{array}{c} Co \end{array}\right]^{2+} + 6H_2O$$

$\overset{\frown}{N \quad N}$ = en = ethane-1,2-diamine

(23.16)

If the ligands are asymmetrical, additional stereoisomers are possible. For example, the reaction of $CrCl_3$ with $HOCH_2CH_2NH_2$ in water may give a mixture of isomers (Figure 23.16) of the octahedral complex $[Cr(OCH_2CH_2NH_2)_3]$ where the ligand is **23.27**.

Related complexes which can be formed by adding the ligand en to aqueous solutions of metal(II) chlorides or nitrates are $[Cr(en)_3]^{2+}$, $[Mn(en)_3]^{2+}$,

**(23.27)**

## COMMERCIAL AND LABORATORY APPLICATIONS

### Box 23.5 Anti-cancer drugs

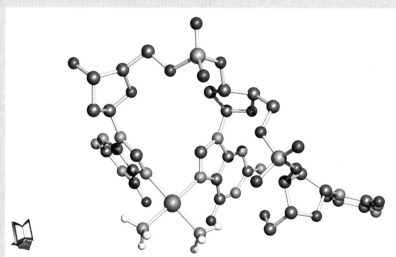

Cisplatin                    Carboplatin

The square planar complex *cis*-[PtCl$_2$(NH$_3$)$_2$] (cis-platin) has been recognized as an anti-tumour drug since the 1960s. It is used in the treatment of bladder and cervical tumours and ovarian and testicular cancers. Unfortunately, side-effects include nausea and kidney damage. An alternative drug which is clinically active but has fewer side-effects than cisplatin is carboplatin.

Research to find other active platinum-containing anti-cancer drugs is actively being pursued, and an understanding of how cisplatin and related active compounds work is an important aspect of drug development.

DNA (see Box 21.6 and Chapter 35) is composed of a sequence of nucleotide bases including guanine. It appears that when cisplatin is administered to a patient, the *cis*-Pt(NH$_3$)$_2$ units are coordinated by nitrogen donor atoms in the guanine bases of DNA:

Guanine residue in DNA showing possible sites of platinum(II) coordination

to DNA backbone

Designing 'model ligands' that mimic biological systems is commonly employed as a method of investigating otherwise complex systems. For example, a ligand incorporating a cytosine and two guanine units, connected to a phosphate-containing chain similar to the backbone of DNA has been developed. It reacts with cisplatin to give a complex in which two nitrogen atoms, one from each of the guanine residues, coordinate to the platinum(II) centre. The structure of the complex is shown below; hydrogen atoms, except those in the NH$_3$ ligands, are omitted from the diagram.

Colour code: Pt, green; N, blue; O, red; P, orange; C, grey; H, white.
[X-ray diffraction data: G. Admiraal *et al.* (1987) *J. Am. Chem. Soc.*, vol. 109, p. 592.]

Research in the area of platinum anti-cancer drugs has been described by:

Jan Reedijk (1996), *Chemical Communications*, p. 801.
T. W. Hambley (2001) *Journal of the Chemical Society, Dalton Transactions*, p. 2711.

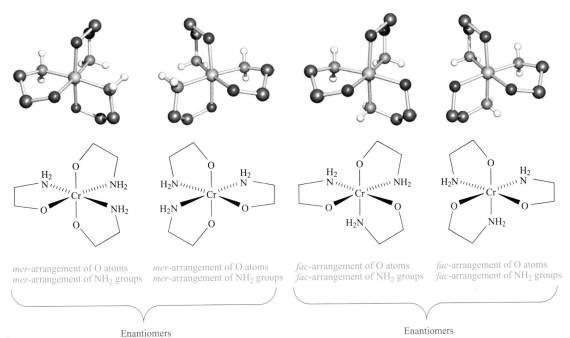

| | | | |
|---|---|---|---|
| *mer*-arrangement of O atoms | *mer*-arrangement of O atoms | *fac*-arrangement of O atoms | *fac*-arrangement of O atoms |
| *mer*-arrangement of NH₂ groups | *mer*-arrangement of NH₂ groups | *fac*-arrangement of NH₂ groups | *fac*-arrangement of NH₂ groups |

Enantiomers                                    Enantiomers

**Fig. 23.16** The complex cation $[\mathrm{Cr(OCH_2CH_2NH_2)_3}]$ possesses *mer*- and *fac*-isomers, and each forms as an enantiomeric pair. Notice that a *fac*-arrangement of NH₂ groups *necessarily* means a *fac*-arrangement of O atoms, and similarly for the *mer*-arrangements. Colour code: Cr, yellow; O, red; N, blue; C, grey; H, white. Only the H atoms of the NH₂ groups are shown.

$[\mathrm{Fe(en)_3}]^{2+}$ and $[\mathrm{Ni(en)_3}]^{2+}$, although the en ligand readily dissociates from $\mathrm{Cr^{2+}}$ in aqueous solution. Ethane-1,2-diamine may also displace $\mathrm{NH_3}$ from ammine complexes (equation 23.17).

$$[\mathrm{Ni(NH_3)_6}]^{2+}(\mathrm{aq}) + 3\mathrm{en} \rightleftharpoons [\mathrm{Ni(en)_3}]^{2+}(\mathrm{aq}) + 6\mathrm{NH_3}(\mathrm{aq}) \qquad (23.17)$$

**Entropy: see Chapter 17**

The driving force behind many of the displacements of two monodentate for one bidentate ligand may be understood in terms of entropy effects and is referred to as *the chelate effect*.[§] Similarly, the replacement of three monodentate ligands by one tridentate ligand is favourable. When a dien ligand (Table 23.2) coordinates to a metal ion in a tridentate manner, *two chelate rings* are formed (structure **23.28**). This makes dien complexes more stable than corresponding en or ammine complexes. $[\mathrm{Ni(dien)_2}]^{2+}$ is thermodynamically more stable than $[\mathrm{Ni(en)_3}]^{2+}$ which is thermodynamically more stable than $[\mathrm{Ni(NH_3)_6}]^{2+}$.

**(23.28)**

**Conformation: see Section 24.9**

Figure 23.17a shows the structure of the dien ligand in an *extended conformation*. Rotation can occur around any of the single bonds and this makes dien flexible enough to adopt either a *mer*- or *fac*-configuration in an octahedral complex. When dien reacts with $\mathrm{Co^{2+}}$ ions in air (i.e. an oxidizing environment) in aqueous solution, $[\mathrm{Co(dien)_2}]^{3+}$ is formed (equation 23.18). A mixture of three isomers (Figure 23.17b) is produced. One *mer*-arrangement of the two dien ligands is possible, but there are *two different fac*-arrangements. In the *sym,fac*-isomer, the ligands are positioned with the central NH groups *trans*- to one another, while in the

---

[§] For a more detailed introduction to the chelate effect, see: C. E. Housecroft and A. G. Sharpe (2005) *Inorganic Chemistry*, 2nd edn, Prentice Hall, Harlow, Chapter 6.

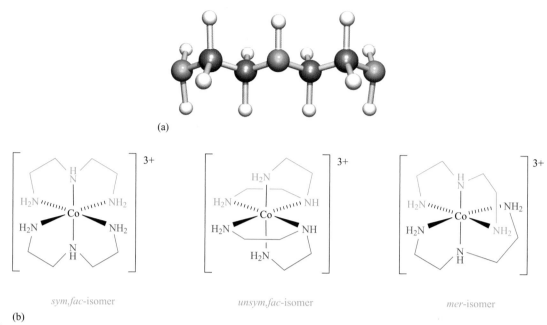

(a)

(b)

*sym,fac*-isomer          *unsym,fac*-isomer          *mer*-isomer

**Fig. 23.17** (a) The dien ligand ($H_2NCH_2CH_2NHCH_2CH_2NH_2$) in an extended conformation; N atoms are shown in blue. Free rotation can occur about any of the single bonds. (b) The complex cation $[Co(dien)_2]^{3+}$ is produced as a mixture of isomers; the *mer*- and *unsym,fac*-isomers are present as racemates (*sym-* = symmetrical; *unsym-* = unsymmetrical). The two dien ligands in each isomer are shown in different colours.

**(23.29)**

*unsym,fac*-isomer, the NH groups are *cis* to each other. These three isomers can be separated by using column chromatography (see Box 24.7). Each of the *mer*- and *unsym,fac*-isomers possesses enantiomers.

$$[Co(H_2O)_6]^{2+} + 2\,dien \underset{Co(II)}{\overset{\substack{\text{In the presence of air}\\ \text{and activated charcoal}}}{\rightleftharpoons}} [Co(dien)_2]^{3+} + 6H_2O \quad (23.18)$$

If the tridentate ligand is *not* as flexible as dien, the number of isomers of an octahedral complex may be reduced. For example, although free rotation about the C−C bonds *between* the aromatic rings in the terpyridine ligand is possible, the ligands in $[Fe(tpy)_2]^{2+}$ can only adopt a *mer*-arrangement (Figure 23.8).

Flexibility plays a vital role in the coordination behaviour of polydentate ligands, and an example is $[edta]^{4-}$ **(23.29)**; the parent carboxylic acid is tetrabasic and is represented by the abbreviation $H_4edta$. The two nitrogen atoms and the four $O^-$ groups are donors and $[edta]^{4-}$ is a potentially hexadentate ligand. The conformation of the ligand can change by free rotation

**Fig. 23.18** The solid state structure of $[Co(edta)]^-$ from the salt $[NH_4][Co(edta)]\cdot2H_2O$. The two *N*-donor atoms must be mutually *cis*, and the flexibility of the ligand allows the *O*-donor atoms to occupy the remaining four coordination sites. Colour code: Co, yellow; N, blue; O, red; C, grey. H atoms are omitted. [X-ray diffraction data: H. A. Weakliem *et al.* (1959) *J. Am. Chem. Soc.*, vol. 81, p. 549.]

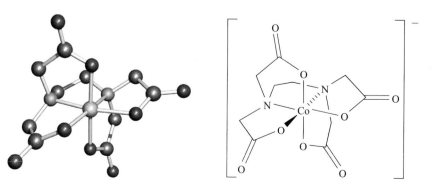

about the single C–C and C–N bonds. When $[edta]^{4-}$ coordinates to a metal ion, it may wrap around the metal centre to give an octahedral complex. Equations 23.19 and 23.20 show a two-step preparation of the ammonium salt of $[Co(edta)]^-$, the structure of which is shown in Figure 23.18.

$$Co(OH)_2 + H_4edta + 2NH_3 \rightleftharpoons [NH_4]_2[Co(edta)] + 2H_2O \qquad (23.19)$$
Cobalt(II)

$$2[NH_4]_2[Co(edta)] + H_2O_2 \longrightarrow 2[NH_4][Co(edta)] + 2H_2O + 2NH_3 \qquad (23.20)$$
Cobalt(III)

## 23.7    Ligand exchange and complex stability constants

Many of the complex formation reactions shown in Section 23.6 were written as *equilibria* and in this section we consider the thermodynamic stability of coordination complexes and values of their *stability constants*.

### Stepwise stability constant, *K*

The replacement of one aqua ligand in a hexaaqua ion by another ligand L is shown in equation 23.21. The relative stabilities of the two complex ions are given by the equilibrium constant, $K$, and this particular type of equilibrium constant is called the *stability constant of the product complex* (equation 23.22). The concentration of water is taken to be unity (see Chapter 16).

$$[M(H_2O)_6]^{2+}(aq) + L(aq) \rightleftharpoons [M(H_2O)_5L]^{2+}(aq) + H_2O(l) \qquad (23.21)$$

$$K_1 = \frac{[M(H_2O)_5L^{2+}][H_2O]}{[M(H_2O)_6^{2+}][L]} = \frac{[M(H_2O)_5L^{2+}]}{[M(H_2O)_6^{2+}][L]} \qquad (23.22)$$

The stability constant $K_1$ refers to the stepwise replacement of the first aqua ligand by L, and there are six stability constants ($K_1$, $K_2$, $K_3$, $K_4$, $K_5$ and $K_6$) associated with the conversion of $[M(H_2O)_6]^{2+}$ to $[ML_6]^{2+}$. (This is similar to the way in which we consider the stepwise dissociation of polybasic acids.) Equations 23.23–23.27 are the five steps which follow equilibrium 23.21 in the formation of $[ML_6]^{2+}$.

Polybasic acids:
see Section 16.4

$$[M(H_2O)_5L]^{2+}(aq) + L(aq) \rightleftharpoons [M(H_2O)_4L_2]^{2+}(aq) + H_2O(l) \qquad (23.23)$$

$$[M(H_2O)_4L_2]^{2+}(aq) + L(aq) \rightleftharpoons [M(H_2O)_3L_3]^{2+}(aq) + H_2O(l) \qquad (23.24)$$

$$[M(H_2O)_3L_3]^{2+}(aq) + L(aq) \rightleftharpoons [M(H_2O)_2L_4]^{2+}(aq) + H_2O(l) \qquad (23.25)$$

$$[M(H_2O)_2L_4]^{2+}(aq) + L(aq) \rightleftharpoons [M(H_2O)L_5]^{2+}(aq) + H_2O(l) \qquad (23.26)$$

$$[M(H_2O)L_5]^{2+}(aq) + L(aq) \rightleftharpoons [ML_6]^{2+}(aq) + H_2O(l) \qquad (23.27)$$

In the formation of a complex $[ML_6]^{2+}$ from $[M(H_2O)_6]^{2+}$, the ***stability constant*** $K_1$ refers to the ***stepwise replacement*** of the first aqua ligand by L. There are six stability constants ($K_1$, $K_2$, $K_3$, $K_4$, $K_5$ and $K_6$) associated with individual steps in the overall process.

The stepwise displacements of the aqua ligands in $[Ni(H_2O)_6]^{2+}$ by the bidentate ligand en are shown in equations 23.28 to 23.30; the stability constants show that the first displacement (equation 23.28) of two $H_2O$ molecules by one ethane-1,2-diamine ligand is more favourable than the second step, and this is more favourable than the third step.

$$[Ni(H_2O)_6]^{2+}(aq) + en(aq) \rightleftharpoons [Ni(H_2O)_4(en)]^{2+}(aq) + 2H_2O(l)$$

$$K_1 = 2.81 \times 10^7 \quad (23.28)$$

$$[Ni(H_2O)_4(en)]^{2+}(aq) + en(aq) \rightleftharpoons [Ni(H_2O)_2(en)_2]^{2+}(aq) + 2H_2O(l)$$

$$K_2 = 1.70 \times 10^6 \quad (23.29)$$

$$[Ni(H_2O)_2(en)_2]^{2+}(aq) + en(aq) \rightleftharpoons [Ni(en)_3]^{2+}(aq) + 2H_2O(l)$$

$$K_3 = 2.19 \times 10^4 \quad (23.30)$$

---

| Worked example 23.2 | *Stepwise formation of [Ni(NH_3)_6]^{2+}* |

The following graph shows a plot of log $K$ ($K$ is the stability constant) against the number, $n$, of $NH_3$ ligands present in the complex $[Ni(H_2O)_{6-n}(NH_3)_n]^{2+}$:

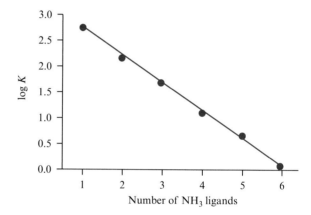

To what reactions do the values of log $K = 2.70$ and $0.65$ correspond? What does the graph tell you about the displacement of successive aqua ligands by ammine ligands?

The stability constants correspond to the stepwise displacements of $H_2O$ by $NH_3$ ligands in going from $[Ni(H_2O)_6]^{2+}$ to $[Ni(NH_3)_6]^{2+}$.
The value of log $K = 2.70$ corresponds to the first ligand exchange step:

$$[Ni(H_2O)_6]^{2+}(aq) + NH_3(aq) \rightleftharpoons [Ni(H_2O)_5(NH_3)]^{2+}(aq) + H_2O(l)$$

The value of log $K = 0.65$ corresponds to $n = 5$, and therefore to the equilibrium:

$$[Ni(H_2O)_2(NH_3)_4]^{2+}(aq) + NH_3(aq) \rightleftharpoons [Ni(H_2O)(NH_3)_5]^{2+}(aq) + H_2O(l)$$

The graph shows a decrease in the value of log $K$ as successive $H_2O$ ligands are displaced by $NH_3$ ligands. As log $K$ decreases, $K$ decreases, and this means that successive equilibria (corresponding to successive ligand replacements) lie further to the left-hand side. It is less favourable to displace $H_2O$ by $NH_3$ in $[Ni(H_2O)_5(NH_3)]^{2+}$ than in $[Ni(H_2O)_6]^{2+}$, and less favourable still in later complexes in the series.

## Overall stability constant, $\beta$

Instead of considering the *stepwise* formation of a complex such as $[ML_6]^{2+}$, it may be more appropriate or convenient to consider the formation in one step such as reaction 23.31.

$$[M(H_2O)_6]^{2+}(aq) + 6L(aq) \rightleftharpoons [ML_6]^{2+}(aq) + 6H_2O(l) \qquad (23.31)$$

We can write down an expression for the equilibrium constant for this reaction but instead of calling the equilibrium constant $K$, the symbol $\beta$ is used. $\beta_n$ is the *overall stability constant*, whereas values of $K_n$ represent *stepwise stability constants* for a series of $n$ equilibria. Equation 23.32 gives the expression for the overall stability constant for the formation of $[ML_6]^{2+}$ from $[M(H_2O)_6]^{2+}$.

$$\beta_6 = \frac{[ML_6^{2+}]}{[M(H_2O)_6^{2+}][L]^6} \qquad (23.32)$$

For the formation of a complex $[ML_6]^{2+}$ in the reaction:

$$[M(H_2O)_6]^{2+}(aq) + 6L(aq) \rightleftharpoons [ML_6]^{2+}(aq) + 6H_2O(l)$$

the ***overall stability constant*** is $\beta_6$ and is given by the equation:

$$\beta_6 = \frac{[ML_6^{2+}]}{[M(H_2O)_6^{2+}][L]^6}$$

For an $n$-step process, the ***overall stability constant*** is $\beta_n$.

## Gibbs energy changes for the formation of complexes

The reaction isotherm (equation 23.33) gives the relationship between a stability constant and the standard Gibbs energy of formation of a complex. Equation 23.33 can be used to determine thermodynamic data for the formation of complexes.

More about $\Delta G^{\circ}$ in Chapter 17; $R = 8.314\,J\,K^{-1}\,mol^{-1}$

$$\Delta G^{\circ} = -RT \ln K \qquad (23.33)$$

By applying this equation to the series of reactions given in equations 23.28 to 23.30 (for which values of $K$ are measured at 298 K), values of $\Delta G^{\circ}$ of $-42.5$, $-35.5$ and $-24.8\,kJ\,mol^{-1}$ are obtained for the successive displacements of two aqua ligands by a bidentate ethane-1,2-diamine ligand. The more en ligands that are added, the less favourable the reaction becomes. This trend is seen in many complex-forming reactions and another example is illustrated in worked example 23.3.

| Worked example 23.3 | *The thermodynamics of complex formation* |

The first five stepwise stability constants (at 298 K) corresponding to the reaction of $[Cu(H_2O)_6]^{2+}$ with ammonia are $1.58 \times 10^4$, $3.16 \times 10^3$, $7.94 \times 10^2$, $1.25 \times 10^2$ and $0.32$. Write down the equilibria to which these values correspond and determine a value of $\Delta G^{\circ}$ for each step. What can you conclude about the overall reaction between $[Cu(H_2O)_6]^{2+}$ and $NH_3$?

The relationship between $K$ and $\Delta G^{\circ}$ is:

$$\Delta G^{\circ} = -RT \ln K$$

By applying this equation, we can determine values of $\Delta G^\circ$ for the displacements of the first five aqua ligands in $[Cu(H_2O)_6]^{2+}$ by $NH_3$. For the first step:

$$[Cu(H_2O)_6]^{2+}(aq) + NH_3(aq) \rightleftharpoons [Cu(H_2O)_5(NH_3)]^{2+}(aq) + H_2O(l)$$

$$K = 1.58 \times 10^4 \qquad \ln K = 9.67$$

$$\Delta G^\circ = -[(8.314 \times 10^{-3}\,\text{kJ}\,\text{K}^{-1}\,\text{mol}^{-1}) \times (298\,\text{K}) \times 9.67]$$

$$= -24.0\,\text{kJ per mole of } [Cu(H_2O)_6]^{2+}$$

Similarly for the next four steps:

$$[Cu(H_2O)_5(NH_3)]^{2+}(aq) + NH_3(aq) \rightleftharpoons [Cu(H_2O)_4(NH_3)_2]^{2+}(aq) + H_2O(l)$$

$$\Delta G^\circ = -20.0\,\text{kJ}\,\text{mol}^{-1}$$

$$[Cu(H_2O)_4(NH_3)_2]^{2+}(aq) + NH_3(aq) \rightleftharpoons [Cu(H_2O)_3(NH_3)_3]^{2+}(aq) + H_2O(l)$$

$$\Delta G^\circ = -16.5\,\text{kJ}\,\text{mol}^{-1}$$

$$[Cu(H_2O)_3(NH_3)_3]^{2+}(aq) + NH_3(aq) \rightleftharpoons [Cu(H_2O)_2(NH_3)_4]^{2+}(aq) + H_2O(l)$$

$$\Delta G^\circ = -12.0\,\text{kJ}\,\text{mol}^{-1}$$

$$[Cu(H_2O)_2(NH_3)_4]^{2+}(aq) + NH_3(aq) \rightleftharpoons [Cu(H_2O)(NH_3)_5]^{2+}(aq) + H_2O(l)$$

$$\Delta G^\circ = +2.8\,\text{kJ}\,\text{mol}^{-1}$$

These data illustrate two points:

- at 298 K it becomes increasingly less thermodynamically favourable to replace successive $H_2O$ by $NH_3$ ligands in $[Cu(H_2O)_6]^{2+}$;
- the displacement of the fifth ligand is thermodynamically *unfavourable* because $\Delta G^\circ$ for this step is positive.

---

### 23.8    Hard and soft metals and donor atoms

We have already seen that some metal ions show a preference for particular types of donor atoms. For example, early *d*-block metals form complexes with cyclic ethers (oxygen donor atoms) but later metals prefer thioether ligands (sulfur donor ligands). In this section we describe the concept of *hard* and *soft* metals and donor atoms. This provides us with a method of 'matching' metal centres with appropriate ligand types.

### Case study 1: iron(III) halide complexes

The displacement of an aqua ligand by a halide $X^-$ in $[Fe(H_2O)_6]^{3+}$ is shown in equation 23.34; the solution must be acidic in order to stabilize the hexaaquairon(III) ion. Equation 23.35 gives an alternative form of this equation, where $Fe^{3+}(aq)$ and $[FeX]^{2+}(aq)$ signify hydrated ions. We include this because the shorthand method of writing hydrated species is so widely adopted, even though it provides less information about the identity of the aqua-species.

$$[Fe(H_2O)_6]^{3+}(aq) + X^-(aq) \rightleftharpoons [Fe(H_2O)_5X]^{2+}(aq) + H_2O(l) \qquad (23.34)$$

$$Fe^{3+}(aq) + X^-(aq) \rightleftharpoons [FeX]^{2+}(aq) \qquad (23.35)$$

**Table 23.3** Stability constants for the formation of iron(III) and mercury(II) halides $[FeX]^{2+}(aq)$ and $[HgX]^+(aq)$. The equilibria are defined in equations 23.35 and 23.36.

| Metal ion | Stability constant $K_1$ | | | |
|---|---|---|---|---|
| | $X = F$ | $X = Cl$ | $X = Br$ | $X = I$ |
| $Fe^{3+}(aq)$ | $1 \times 10^6$ | 25.1 | 3.2 | – |
| $Hg^{2+}(aq)$ | 10.0 | $5.0 \times 10^6$ | $7.9 \times 10^8$ | $7.9 \times 10^{12}$ |

Table 23.3 lists stability constants for the formation of $[FeX]^{2+}(aq)$ for $X = F$, Cl, Br. The data indicate that formation of the fluoride complex is highly favourable. Although the stability constants for $[FeCl]^{2+}(aq)$ and $[FeBr]^{2+}$ are much smaller than for $[FeF]^{2+}$, the formation of the chloro and bromo complexes is still favourable, but $[FeBr]^{2+}$ is less stable than $[FeCl]^{2+}$ with respect to the hexaaqua ion.

The trend in $[FeX]^{2+}(aq)$ complex stability is:

$$F \gg Cl > Br$$

### Case study 2: mercury(II) halide complexes

Equation 23.36 shows the reaction between a halide ion and a mercury(II) aqua ion, and Table 23.3 gives stability constants for the $[HgX]^+$ complexes with different halo ligands.

$$Hg^{2+}(aq) + X^-(aq) \rightleftharpoons [HgX]^+(aq) \tag{23.36}$$

For each halide, equilibrium 23.36 lies over to the right-hand side, and the coordination of an iodo ligand to $Hg^{2+}$ is the most favourable. The trend in $[HgX]^+(aq)$ complex stability is:

$$I > Br > Cl \gg F$$

### Hard and soft metal ions and ligands

The two case studies above illustrate that while $F^-$ is the halide preferred by aqueous $Fe^{3+}$, $I^-$ is the preferred ligand for $Hg^{2+}$. The fact that $Fe^{3+}$ and $Hg^{2+}$ exhibit opposite trends in their preferences for halo ligands is not unique to this pair of metal centres. The ions $Sc^{3+}$, $Cr^{3+}$, $Co^{3+}$ and $Zn^{2+}$ behave in a similar manner to $Fe^{3+}$, while $Cu^+$, $Ag^+$ and $Pt^{2+}$ behave similarly to $Hg^{2+}$. Metal ions are classified as being either *hard* (*class A*) or *soft* (*class B*) – $Fe^{3+}$ is hard while $Hg^{2+}$ is soft.

A hard metal ion has a high charge density. The consequence of this is that the valence electrons of the metal ion are tightly held and the metal–ligand bonding possesses a high degree of ionic character. This in turn has an effect on the type of ligands with which the metal ion prefers to interact. A hard metal ion will favour ligands containing donor atoms which are highly electronegative. Such ligands are termed *hard ligands* and include $F^-$ and $Cl^-$ ions and *O* or *N*-donor atoms.

A soft metal ion has a low charge density. Such metal ions are easily polarized and the metal–ligand bonding has a high covalent character. In

Charge density and polarization of ions: see Section 8.18

**Table 23.4** Representative hard, soft and intermediate *d*-block metal ions and ligands. Donor atoms in the ligands are indicated in pink; ligands such as nitrate and sulfate ions are oxygen donors and may be mono- or bidentate.

| | |
|---|---|
| Hard metal ions | $Li^+$, $Na^+$, $K^+$, $Be^{2+}$, $Mg^{2+}$, $Ca^{2+}$, $Mn^{2+}$, $Zn^{2+}$, $Al^{3+}$, $Sc^{3+}$, $Cr^{3+}$, $Fe^{3+}$, $Co^{3+}$, $Ti^{4+}$ |
| Hard ligands | $F^-$, $Cl^-$ |
| | $H_2O$, $[OH]^-$, $[OR]^-$, $ROH$, $R_2O$, $[RCO_2]^-$, $[SO_4]^{2-}$, $[NO_3]^-$ |
| | $NH_3$, $RNH_2$ |
| Intermediate metal ions | $Fe^{2+}$, $Co^{2+}$, $Ni^{2+}$, $Cu^{2+}$, $Ru^{3+}$, $Rh^{3+}$, $Ir^{3+}$ |
| Intermediate ligands | $Br^-$ |
| | $C_5H_5N$ (pyridine), $[NCS]^-$ |
| Soft metal ions | $Tl^+$, $Cu^+$, $Ag^+$, $Au^+$, $Hg^{2+}$, $Cd^{2+}$, $Pd^{2+}$, $Pt^{2+}$ |
| Soft ligands | $I^-$ |
| | $[CN]^-$, $CO$ |
| | $PR_3$ |
| | $[NCS]^-$, $[RS]^-$, $RSH$, $R_2S$ |

contrast to a hard metal, a soft metal ion favours ligands containing less electronegative donor atoms and these include $I^-$ and *S*-donors.

Many metal–ligand interactions can be classified as being either hard–hard (for example, iron(III)–fluoride ion) or soft–soft (for example, mercury(II)–iodide ion). However, some metal ions and ligands are intermediate in character and this permits them to be more versatile in their complex formation than the limiting hard and soft metal ions and ligands. Table 23.4 lists representative hard, soft and intermediate metal ions and ligands; the donor atoms in the ligands are highlighted in pink. Metal ions from the *s*- and *p*-blocks, as well as the *d*-block, are included.

We can now appreciate why $Sc^{3+}$ and $Y^{3+}$ ions form complexes with the oxygen donor atoms in ethers (Figures 23.9 and 23.10b), why $F^-$ and $Cl^-$ ions form more stable complexes with $Fe^{3+}$ ions than does $I^-$, and why iodo-mercury(II) complexes are more stable than fluoro-mercury(II) compounds.

While the concept of 'matching' hard metal ions with hard ligands and soft metal ions with soft ligands provides us with a means of understanding why certain complexes are more favourable than others, it is *not* a foolproof method of predicting whether a complex will form. For example, the reaction of aqueous chromium(III) acetate with potassium cyanide gives the compound $K_3[Cr(CN)_6]$ which contains $[Cr(CN)_6]^{3-}$ (**23.30**). This is composed of a hard $Cr^{3+}$ and soft cyano ligands.

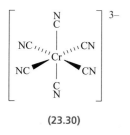

(23.30)

## 23.9    The thermodynamic stability of hexaaqua metal ions

One of the usual characteristics of a *d*-block metal is that it possesses several stable oxidation states. Exceptions include the group 12 metals zinc and cadmium, which form compounds almost exclusively in the +2 oxidation state, and the group 3 metals (scandium, yttrium and lanthanum), which predominantly form compounds in the +3 oxidation state. In this section we consider the reduction potentials of couples containing *d*-block metal ions, and quantify some of the reduction–oxidation (*redox*) reactions that are observed in aqueous solution.

## Reduction potentials for hexaaqua metal ions

Table of reduction potentials: see Appendix 12

Table 23.5 lists values of $E^o$ for selected couples involving first row $d$-block metal ions. An important point to remember is that each metal ion is of the type $M^{n+}(aq)$ and is an *aqua ion*. For example, $Co^{2+}(aq)$ tells us we have an aqueous solution of cobalt(II) ions, and indicates the presence of the complex ion $[Co(H_2O)_6]^{2+}$. Although we have made this point before, we reiterate it here because the presence of ligands other than water can significantly influence the reduction potential of a metal–ion couple. For example, the value for $E^o$ $Fe^{3+}(aq)/Fe^{2+}(aq)$ is $+0.77\,V$ and refers to half-reaction 23.37.

$$[Fe(H_2O)_6]^{3+} + e^- \rightleftharpoons [Fe(H_2O)_6]^{2+} \qquad E^o = +0.77\,V \quad (23.37)$$

$$[Fe(CN)_6]^{3-} + e^- \rightleftharpoons [Fe(CN)_6]^{4-} \qquad E^o = +0.36\,V \quad (23.38)$$

$$[Fe(bpy)_3]^{3+} + e^- \rightleftharpoons [Fe(bpy)_3]^{2+} \qquad E^o = +1.03\,V \quad (23.39)$$

The reduction potential of a couple such as $M^{n+}/M^{m+}$ is dependent upon the ligands present in the coordination shell.

When an $Fe^{3+}$ ion is surrounded by cyano or 2,2'-bipyridine ligands, their presence markedly alters the ease (or not) with which the reduction of $Fe^{3+}$ to $Fe^{2+}$ takes place (equations 23.38 and 23.39). The reasons for this difference are beyond the scope of our discussion, but it is important to

**Table 23.5** Selected standard reduction potentials (298 K). The concentration of each aqueous solution is $1\,mol\,dm^{-3}$, and the pressure of a gaseous component is 1 bar; for a half-cell involving $H^+$, $E^o$ refers to a solution at pH 0, and for a half-cell involving $[OH]^-$, $E^o$ refers to pH 14.

| Reduction half-equation | $E^o$ or $E^o_{[OH^-]=1}\,/\,V$ |
|---|---|
| $Ti^{2+}(aq) + 2e^- \rightleftharpoons Ti(s)$ | −1.63 |
| $Mn^{2+}(aq) + 2e^- \rightleftharpoons Mn(s)$ | −1.19 |
| $Zn^{2+}(aq) + 2e^- \rightleftharpoons Zn(s)$ | −0.76 |
| $Fe^{2+}(aq) + 2e^- \rightleftharpoons Fe(s)$ | −0.44 |
| $Cr^{3+}(aq) + e^- \rightleftharpoons Cr^{2+}(aq)$ | −0.41 |
| $Ti^{3+}(aq) + e^- \rightleftharpoons Ti^{2+}(aq)$ | −0.37 |
| $Co^{2+}(aq) + 2e^- \rightleftharpoons Co(s)$ | −0.28 |
| $V^{3+}(aq) + e^- \rightleftharpoons V^{2+}(aq)$ | −0.26 |
| $Ni^{2+}(aq) + 2e^- \rightleftharpoons Ni(s)$ | −0.25 |
| $Fe^{3+}(aq) + 3e^- \rightleftharpoons Fe(s)$ | −0.04 |
| $2H^+(aq) + 2e^- \rightleftharpoons H_2(g, 1\,bar)$ | 0.00 |
| $Cu^{2+}(aq) + e^- \rightleftharpoons Cu^+(aq)$ | +0.15 |
| $Cu^{2+}(aq) + 2e^- \rightleftharpoons Cu(s)$ | +0.34 |
| $O_2(g) + 2H_2O(l) + 4e^- \rightleftharpoons 4[OH]^-(aq)$ | +0.40 |
| $Cu^+(aq) + e^- \rightleftharpoons Cu(s)$ | +0.52 |
| $O_2(g) + 2H^+(aq) + 2e^- \rightleftharpoons H_2O_2(aq)$ | +0.70 |
| $Fe^{3+}(aq) + e^- \rightleftharpoons Fe^{2+}(aq)$ | +0.77 |
| $Ag^+(aq) + e^- \rightleftharpoons Ag(s)$ | +0.80 |
| $O_2(g) + 4H^+(aq) + 4e^- \rightleftharpoons 2H_2O(l)$ | +1.23 |
| $[Cr_2O_7]^{2-}(aq) + 14H^+(aq) + 6e^- \rightleftharpoons 2Cr^{3+}(aq) + 7H_2O(l)$ | +1.33 |
| $[MnO_4]^-(aq) + 8H^+(aq) + 5e^- \rightleftharpoons Mn^{2+}(aq) + 4H_2O(l)$ | +1.51 |
| $H_2O_2(aq) + 2H^+(aq) + 2e^- \rightleftharpoons 2H_2O(l)$ | +1.78 |
| $Co^{3+}(aq) + e^- \rightleftharpoons Co^{2+}(aq)$ | +1.92 |

remember that *you can only apply the reduction potentials listed in Table 23.5 to aqua ions*, and that more detailed tables should be consulted if other ligands are involved.

## The stability of $M^{2+}$(aq) ions in aqueous solutions

Predicting reactions using values of $E°_{cell}$: see Section 18.3

In Section 23.6 we considered some hexaaqua ions of the first row *d*-block metals. We stated that in aqueous solution, $Ti^{2+}$, $V^{2+}$ and $Cr^{2+}$ are readily oxidized to $Ti^{3+}$, $V^{3+}$ and $Cr^{3+}$ respectively. We also saw that atmospheric $O_2$ can oxidize $Fe^{2+}$ to $Fe^{3+}$. By considering the reduction potentials in Table 23.5, we can quantify these observations.

Each of the $Cr^{3+}$(aq)/$Cr^{2+}$(aq), $Ti^{3+}$(aq)/$Ti^{2+}$(aq) and $V^{3+}$(aq)/$V^{2+}$(aq) standard reduction potentials is *negative* and this means that the $M^{2+}$(aq) ions liberate $H_2$ from *acidic solution*; standard conditions mean that the concentration of the acid is $1\,mol\,dm^{-3}$. Equations 23.40–23.42 illustrate this for chromium(II).

$$Cr^{3+}(aq) + e^- \rightleftharpoons Cr^{2+}(aq) \qquad\qquad E° = -0.41\,V \quad (23.40)$$

$$2H^+(aq) + 2e^- \rightleftharpoons H_2(g) \qquad\qquad E° = 0\,V \quad (23.41)$$

$$2Cr^{2+}(aq) + 2H^+(aq) \longrightarrow 2Cr^{3+}(aq) + H_2(g) \qquad\qquad (23.42)$$

The value of $E°_{cell}$ for reaction 23.42 is 0.41 V and from this, the change in standard Gibbs energy can be determined (equation 23.43).

For treatment of units, see Chapter 18

$$\Delta G°(298\,K) = -zFE°_{cell} = -2 \times (96\,485\,C\,mol^{-1}) \times (0.41\,V)$$
$$= -79 \times 10^3\,J \text{ per mole of reaction}$$
$$= -79\,kJ \text{ per mole of reaction} \qquad\qquad (23.43)$$

The negative value of $\Delta G°$ indicates that $H^+$ will oxidize acidic aqueous $Cr^{2+}$ to $Cr^{3+}$ ions, i.e. the $[Cr(H_2O)_6]^{2+}$ ion is not stable in acidic aqueous solution. This means that even in the total absence of air, $Cr^{2+}$ solutions (acidic, aqueous) are unstable with respect to the evolution of $H_2$ and formation of $Cr^{3+}$. We know nothing *per se* about the *rate* of this reaction which is actually quite slow. In the presence of air, oxidation by $O_2$ occurs. Once again, this makes $Cr^{2+}$ unstable with respect to $Cr^{3+}$, but this time half-reactions 23.44 and 23.45 are relevant. The value of $E°_{cell}$ for the overall reaction 23.46 is $+1.64\,V$, and the change in standard Gibbs energy for the overall reaction is calculated in equation 23.47.

$$Cr^{3+}(aq) + e^- \rightleftharpoons Cr^{2+}(aq) \qquad\qquad E° = -0.41\,V \quad (23.44)$$

$$O_2(aq) + 4H^+(aq) + 4e^- \rightleftharpoons 2H_2O(l) \qquad\qquad E° = +1.23\,V \quad (23.45)$$

$$4Cr^{2+}(aq) + O_2(g) + 4H^+(aq) \longrightarrow 4Cr^{3+}(aq) + 2H_2O(l) \qquad\qquad (23.46)$$

$$\Delta G°(298\,K) = -zFE°_{cell} = -4 \times (96\,485\,C\,mol^{-1}) \times (1.64\,V)$$
$$= -63.3 \times 10^4\,J \text{ per mole of reaction}$$
$$= -633\,kJ \text{ per mole of reaction} \qquad\qquad (23.47)$$

As we saw in Chapter 18, reduction potentials that involve $H^+$ or $[OH]^-$ ions are dependent on the pH of the solution and the values of $\Delta G^\circ$ determined above refer to *standard conditions*, i.e. pH 0 or pH 14 (see Table 23.5). We leave it to the reader to consider the effects on $\Delta G^\circ$ of reducing the concentration of $H^+$ ions for either of reactions 23.42 and 23.46.

| 23.10 | **Colours** |

A characteristic feature of compounds of *d*-block metals is that they are often coloured, although the colour may not be intense. Although *dilute* aqueous solutions of potassium permanganate, $K[MnO_4]$, are intense purple, this depth of colour should not be considered 'normal' for a *d*-block metal compound. Blue crystals of copper(II) sulfate dissolve in water to give a *pale blue* solution; the solution must be very concentrated before the colour can be described as being intense. 'Weakly' and 'intensely' coloured solutions are quantified by molar extinction coefficients.

▰▷
Molar extinction coefficient: see Section 11.4

The colours of the hexaaqua ions of the first *d*-block row metals are listed in Table 23.6. The data illustrate that the colour depends not only on the metal ion but also on the ligands present in the coordination sphere.

The fact that solutions containing complexes of the *d*-block metals are coloured means that these species must absorb visible light. In Section 13.6, we introduced this idea by considering the electronic spectrum of an aqueous solution containing the $[Ti(H_2O)_6]^{3+}$ ion. The spectrum exhibits a band with $\lambda_{max} = 510\,nm$ (corresponding to the absorbed light), and the observed colour of aqueous $Ti^{3+}$ ions is purple.

▰▷
Complementary colours: see Table 11.2

The electronic transitions that occur and give rise to the characteristic colours of many *d*-block metal complexes are known as '*d–d*' transitions. When an ion of a *d-block element possesses partially filled d orbitals*, electronic transitions between *d* orbitals may occur. However, $Zn^{2+}$ has a filled $3d$ shell, and no electronic transitions are possible. This explains why solutions of zinc(II) compounds are usually colourless.

We now come to two major problems:

• Are not *d* orbitals with the same principal quantum number degenerate? If so, transitions between them may not be accompanied by a change in electronic energy. The consequence of this is that no electronic spectrum should be visible.

**Table 23.6** Colours of some hexaaqua ions and other complex ions of the *d*-block metals. Ligand abbreviations are given in Table 23.2.

| Complex ion | Observed colour in aqueous solution | Complex ion | Observed colour in aqueous solution |
|---|---|---|---|
| $[Ti(H_2O)_6]^{3+}$ | Violet | $[Zn(H_2O)_6]^{2+}$ | Colourless |
| $[V(H_2O)_6]^{3+}$ | Green | $[Co(CN)_6]^{3-}$ | Yellow |
| $[Cr(H_2O)_6]^{3+}$ | Violet | $[Co(en)_3]^{3+}$ | Yellow |
| $[Cr(H_2O)_6]^{2+}$ | Blue | $[Co(ox)_3]^{3-}$ | Dark green |
| $[Fe(H_2O)_6]^{2+}$ | Pale green | $[Co(NH_3)_6]^{3+}$ | Orange |
| $[Co(H_2O)_6]^{2+}$ | Pink | $[Cu(en)_3]^{2+}$ | Blue |
| $[Ni(H_2O)_6]^{2+}$ | Green | $[CoCl_4]^{2-}$ | Blue |
| $[Cu(H_2O)_6]^{2+}$ | Blue | | |

**Laporte selection rule:
see Section 3.16**

• Are not '*d–d*' transitions disallowed by the *Laporte selection rule*? This rule states that $\Delta l$ must change by $+1$ or $-1$, and therefore a $3d \longrightarrow 3d$ transition is disallowed because $\Delta l = 0$.

In the next section, we introduce a method of approaching the bonding in *d*-block metal complexes and it provides us with a partial answer to the above problems.

## 23.11      Crystal field theory

In Section 23.4, we considered the electroneutrality principle and saw that neither a wholly ionic nor a wholly covalent bonding picture provided a realistic charge distribution in the complex. Nonetheless, the assumptions that ligands may be considered to be negative point charges and that there is *no* metal–ligand covalent bonding are the bases for a model put forward to explain, among other things, the electronic spectroscopic characteristics of *d*-block metal complexes. The model is called *crystal field theory* and in this section we apply it to octahedral complexes.[§]

Throughout this section, we focus on first row *d*-block metals and, therefore, on 3*d* orbitals. The discussions can be extended to second and third row *d*-block metals involving 4*d* and 5*d* orbitals, respectively.

### Crystal field splitting

Consider an octahedral metal complex $[ML_6]^{n+}$ (Figure 23.19a) in which a metal ion $M^{n+}$ is surrounded by six ligands. We can view this complex in terms of a positively charged metal ion surrounded by six point negative charges (the electrons on the ligands) as shown in Figure 23.19b. We must remember, however, that the metal ion consists of a positively charged nucleus surrounded by negatively charged electrons.

**Point charges and
electrostatic interactions:
see Section 8.6**

There are electrostatic attractions between the metal ion and the ligands. However, as the ligands approach the *d*-block $M^{n+}$ ion, the electrons in the *d* orbitals will be *repelled* by the point charges (the ligands). This has a *destabilizing effect* upon the metal electrons. The valence electrons of a first row *d*-block metal *ion* occupy the 3*d* atomic orbitals. If the electrostatic field (termed the *crystal field*) were *spherical*, then all the metal ion 3*d*

**Fig. 23.19** (a) A structural representation of an octahedral complex $[ML_6]^{n+}$ in which a metal ion $M^{n+}$ is surrounded by six neutral ligands L. (b) The complex can be considered in terms of six negative point charges surrounding the positively charged metal ion. *Remember that the metal ion $M^{n+}$ contains a positively charged nucleus surrounded by negatively charged electrons.*

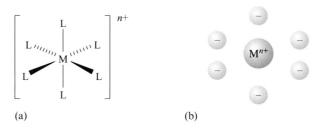

(a)                              (b)

---

[§]  Coordination geometries other than octahedral give different splittings of *d* orbitals. See for example: C. E. Housecroft and A. G. Sharpe (2005) *Inorganic Chemistry*, 2nd edn, Prentice Hall, Harlow, Chapter 20.

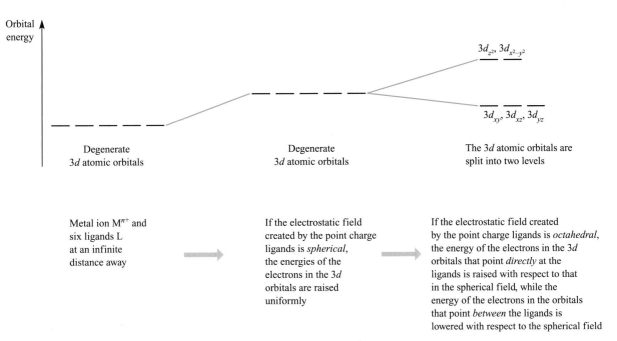

**Fig. 23.20** The changes in the energies of electrons occupying the $3d$ atomic orbitals of an ion $M^{n+}$ when the metal ion is surrounded by a spherical crystal field and an octahedral crystal field. The energy changes are shown in terms of the orbital energies.

electrons would be destabilized to the same extent as shown in Figure 23.20. In an octahedral complex $[ML_6]^{n+}$, the electrostatic field is not spherical but octahedral. Before we look at how this changes the picture, we must recognize the relationship between the shape of the $[ML_6]^{n+}$ cation and the directionalities of the five $3d$ orbitals.

Figure 23.21 shows that the M−L vectors may be defined so as to lie along a set of Cartesian axes. This means that the $3d_{z^2}$ orbital points *directly* at two of the ligands (two axial L in Figure 23.21) and the $3d_{x^2-y^2}$ orbital points *directly* at four of the ligands (the equatorial L in Figure 23.21). The three orbitals $d_{xy}$, $d_{xz}$ and $d_{yz}$ are related to one another and their lobes point *between* the Cartesian axes, that is, *between* the ligands. Now consider the effect of placing the metal ion at the centre of an octahedral crystal field. The electrons in the $3d_{z^2}$ and $3d_{x^2-y^2}$ orbitals are repelled by the point charge ligands. The electrons in the $3d_{xy}$, $3d_{xz}$ and $3d_{yz}$ orbitals are also repelled by the point charges, but since these orbitals do *not* point directly at the ligands, the repulsion is less than that experienced by the electrons in the $3d_{z^2}$ and $3d_{x^2-y^2}$ orbitals. The net effect is that the electrons in the $3d_{z^2}$ and $3d_{x^2-y^2}$ orbitals are destabilized *with respect to a spherical field* while those in the $3d_{xy}$, $3d_{xz}$ and $3d_{yz}$ orbitals are stabilized. This is represented on the right-hand side of Figure 23.20. The final result is that the $3d$ orbitals (which were originally degenerate) are split into two sets – the $3d_{z^2}$ and $3d_{x^2-y^2}$ orbitals form a higher energy set, and the $3d_{xy}$, $3d_{xz}$ and $3d_{yz}$ orbitals make up a lower energy set. This is called *crystal field splitting*.

See Box 23.6

The two sets of $d$ orbitals of the metal ion in an octahedral complex are conveniently labelled as the $e_g$ and the $t_{2g}$ sets. The $e_g$ set is doubly degenerate and consists of the $3d_{z^2}$ and $3d_{x^2-y^2}$ orbitals. The $t_{2g}$ set is triply degenerate and consists of the $3d_{xy}$, $3d_{xz}$ and $3d_{yz}$ orbitals (Figure 23.22a). The energy difference between the $e_g$ and the $t_{2g}$ orbitals is called $\Delta_{oct}$ (pronounced 'delta oct').

**Fig. 23.21** The six M−L bonds in an octahedral complex $[ML_6]^{n+}$ may be defined as lying along Cartesian axes. The directionalities of the five *d* orbitals can therefore be related to the directions of the M−L bond vectors. Although we distinguish between axial and equatorial ligands in the figure, they are identical in an octahedral $[ML_6]^{n+}$ complex.

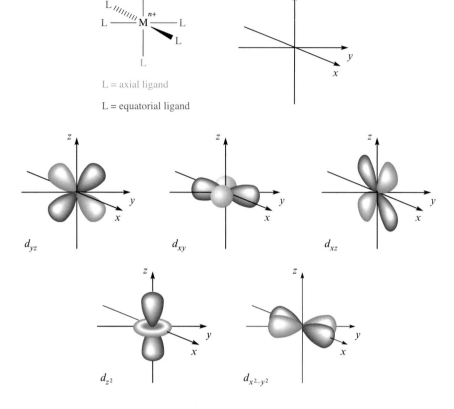

L = axial ligand

L = equatorial ligand

*Crystal field theory* is a purely *electrostatic model* which may be used to show that the *d* orbitals in a *d*-metal ion complex are no longer degenerate. In an octahedral complex, crystal field splitting results in there being two sets of *d* orbitals:

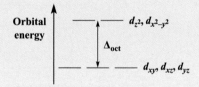

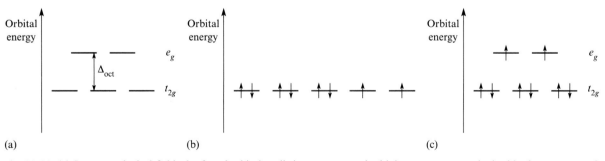

**Fig. 23.22** (a) In an octahedral field, the five *d* orbitals split into two sets: the higher energy $e_g$ set is doubly degenerate and consists of the $3d_{z^2}$ and $3d_{x^2-y^2}$ orbitals, while the lower energy $t_{2g}$ set is triply degenerate and consists of the $3d_{xy}$, $3d_{xz}$ and $3d_{yz}$ orbitals. (b) In a free, gaseous $Ni^{2+}$ ion, the 3*d* orbitals are degenerate and the eight valence electrons occupy them according to the Pauli exclusion principle. (c) In an octahedral complex such as $[Ni(H_2O)_6]^{2+}$, the 3*d* orbitals are split into the $t_{2g}$ and $e_g$ sets and the eight valence electrons of the $Ni^{2+}$ ion do not all possess the same orbital energy.

THEORETICAL AND CHEMICAL BACKGROUND

## Box 23.6 The $e_g$ and $t_{2g}$ labels

We will not delve deeply into the origins of the $t_{2g}$ and $e_g$ labels but it is useful to be aware of the following:

- A triply degenerate level is labelled $t$.
- A doubly degenerate level is labelled $e$.
- The subscript $g$ means *gerade*, German for 'even'.

- The subscript $u$ means *ungerade*, German for 'odd'.
- *Gerade* and *ungerade* designate the behaviour of the wavefunction under the operation of inversion.

### Electronic configurations

So far, we have written the ground state electronic configuration of a $d$-block metal ion in terms of occupying a *degenerate* set of five $d$ orbitals. Indeed, this is correct for the free, gaseous ion such as $Ni^{2+}$ (Figure 23.22b), but what happens in $[Ni(H_2O)_6]^{2+}$?

In an octahedral complex, the metal $d$ orbitals can be considered in terms of the $t_{2g}$ and $e_g$ sets (Figure 23.22a) and when the eight valence electrons of the Ni(II) ion in $[Ni(H_2O)_6]^{2+}$ occupy the five $3d$ orbitals, they do so such that two of the electrons are at a higher energy than the other six (Figure 23.22c). The $d^8$ configuration can be more informatively written as $t_{2g}^6 e_g^2$. Figure 23.23 shows the occupancy of the $3d$ orbitals of $Fe^{2+}$ ($d^6$) in $[Fe(CN)_6]^{4-}$ and of $Mn^{2+}$ ($d^5$) in $[Mn(CN)_6]^{4-}$; the configurations are $t_{2g}^6 e_g^0$ and $t_{2g}^5 e_g^0$ respectively.

### Strong and weak field ligands: the spectrochemical series

The energy gap $\Delta_{oct}$ is a measure of the electrostatic repulsions between the metal $d$ electrons and the point charge ligands. Some ligands cause a greater splitting of the $d$ orbitals than others. Ligands that cause a large crystal field splitting are called *strong field ligands* and $[CN]^-$ is an example. Conversely, the interaction of a metal ion with *weak field ligands* results in a smaller splitting of the $d$ orbitals. The ordering of ligands according to their ability with respect to crystal field splitting is given by the *spectrochemical series* in which $I^-$ is the weakest field ligand.

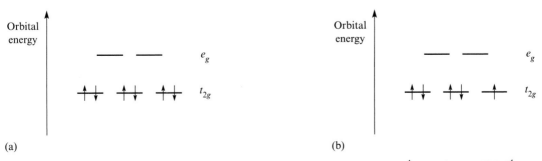

**Fig. 23.23** The arrangements of the $3d$ electrons in the octahedral complexes (a) $[Fe(CN)_6]^{4-}$ and (b) $[Mn(CN)_6]^{4-}$.

The *spectrochemical series* orders ligands according to their ability with respect to crystal field splitting:

$I^- < Br^- < [NCS]^- < Cl^- < F^- < [OH]^- < [ox]^{2-} \approx H_2O < [NCS]^- < NH_3 < en < [CN]^- < CO$

weak field ligands                                                                                    strong field ligands

In this series, note that the $[NCS]^-$ ligand has two positions depending upon whether it is *N*-bonded or *S*-bonded to the metal ion.

### High- and low-spin electronic configurations

For metal ions with $d^4$, $d^5$, $d^6$ or $d^7$ configurations, it is possible for the electrons to be arranged in the five *d* orbitals in two different ways (Figure 23.24). Consider the $d^4$ case. If the value of $\Delta_{oct}$ is relatively small, it may be energetically favourable to adopt the *high-spin* $t_{2g}^3 e_g^1$ configuration rather than the *low-spin* $t_{2g}^4$ in which there are significant electrostatic repulsions between the electrons paired in the same $t_{2g}$ orbital. The adoption of a high- or low-spin metal ion will depend upon the relative magnitudes of the *spin-pairing energies* and $\Delta_{oct}$.

It follows that strong field ligands which cause the greatest splitting of the *d* orbitals will tend to give rise to low-spin electronic configurations. Similarly, weak field ligands may lead to high-spin configurations. In Figure 23.23, the configurations for $[Fe(CN)_6]^{4-}$ and $[Mn(CN)_6]^{4-}$ are low-spin, consistent with the fact that $[CN]^-$ is a strong field ligand. We emphasize that it is not possible to predict with complete certainty which of the high- or low-spin configurations will be observed in practice, and in Section 23.13 we look at experimental data that can distinguish between them.

---

| **Worked example 23.4** | *High- and low-spin configurations* |
|---|---|

**In an octahedral complex, why can the $Ti^{3+}$ ion have only one electronic configuration in its ground state?**

Titanium is in group 4 and the electronic configuration of $Ti^{3+}$ is $[Ar]3d^1$. In an octahedral field, the $3d$ orbitals split and the single electron occupies the $t_{2g}$ level. If this electron is promoted to the $e_g$ level, it would be at the expense of completely vacating the lower energy $t_{2g}$ orbitals and therefore the configuration shown is the only one available for the $Ti^{3+}$ ion in its ground state.

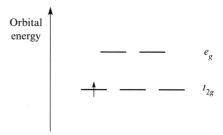

A similar argument can be applied to show that $d^2$ and $d^3$ configurations can each have only one arrangement in their ground states.

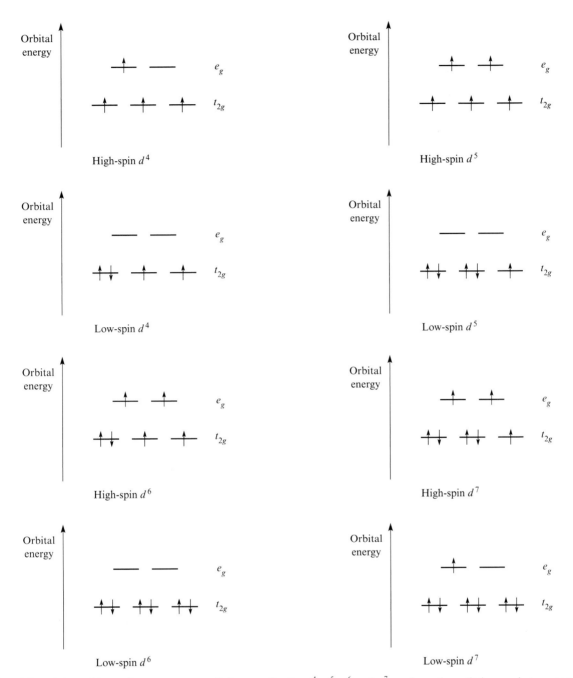

**Fig. 23.24** High- and low-spin arrangements of electrons for the $d^4$, $d^5$, $d^6$ and $d^7$ configurations. Only one electron can be promoted from the $t_{2g}$ to $e_g$ level in the $d^4$ and $d^7$ cases, but two can be promoted for $d^5$ and $d^6$. (Why is this?)

| 23.12 | **Electronic spectra** |
|---|---|

### 'd–d' Transitions

The crystal field splitting of the $d$ orbitals in an octahedral complex leads us to a qualitative understanding as to why such complexes are coloured. If the $t_{2g}$ and $e_g$ sets of five $d$ orbitals are *partially occupied* as is the case for $[Ni(H_2O)_6]^{2+}$ (Figure 23.22c), it would appear to be possible for an electron from the lower $t_{2g}$ level to undergo a transition to the $e_g$ level.

◼▷
Laporte selection rule: see
Section 3.16

There is, however, a problem – a '*d–d*' transition is forbidden by the Laporte selection rule, i.e. it is *Laporte forbidden*. The observed electronic spectra of *d*-block complexes are consistent with the fact that '*d–d*' transitions *appear to occur*. In fact the selection rule is *absolutely obeyed* and '*d–d*' transitions are *strictly forbidden*. However, the picture of the interactions that we developed in Section 23.11 is overly simplistic. In practice, spectroscopic transitions do occur.[§] In this text we shall simply relate $\Delta_{oct}$ to the spectroscopic transitions that are observed.

### The relationship between $\lambda_{max}$ and $\Delta_{oct}$

In this section, we look more closely at how the energy gap between the $t_{2g}$ and $e_g$ levels influences the value of $\lambda_{max}$ in the electronic spectrum of a complex.

The $[Ti(H_2O)_6]^{3+}$ ion absorbs light at 510 nm. The aqua ligand is relatively weak field and $\Delta_{oct}$ is correspondingly small. Exchanging $H_2O$ for even weaker field $Cl^-$ or $Br^-$ ligands gives $[TiCl_6]^{3-}$ and $[TiBr_6]^{3-}$ respectively. In these anions the energy gap between the $t_{2g}$ and $e_g$ levels is *smaller* than in $[Ti(H_2O)_6]^{3+}$. We expect therefore that the light absorbed by these halocomplexes should be of a *longer wavelength* (equation 23.48) than that absorbed by $[Ti(H_2O)_6]^{3+}$. Indeed the observed values of $\lambda_{max}$ for $[TiCl_6]^{3-}$ and $[TiBr_6]^{3-}$ are 780 and 850 nm respectively.

$$E = h\nu \qquad \text{or} \qquad E \propto \frac{1}{\lambda} \tag{23.48}$$

Although we have talked in terms of the field strength of the ligands to gain insight into the spectroscopic properties of these complexes, it is in fact the opposite arguments that allow us to construct the spectrochemical series in the first place!

---

**23.13**     ## Magnetism: the spin-only formula

### High- and low-spin configurations: how many unpaired electrons?

We have already seen that the presence of strong or weak field ligands in an octahedral complex may result in a change from low- to high-spin electronic arrangements for the $d^4$, $d^5$, $d^6$ and $d^7$ configurations. If we consider the high- and low-spin configurations of the $d^4$ ion $Cr^{2+}$, Figure 23.24 ilustrates that the difference between them is *the number of unpaired electrons*. A low-spin $Cr^{2+}$ ion possesses two unpaired electrons but a high-spin ion has four. This distinction gives rise to differences in the magnetic properties of the two ions. In this section we look at the relationship between the *effective magnetic moment* of a complex and the number of unpaired electrons present.

### The Gouy balance

Several experimental methods can be used to determine effective magnetic moments. These include the Gouy balance (Figure 23.25), the Faraday

---

[§] For a more detailed discussion of how '*d–d*' electronic spectra arise, see M. Gerloch and E. C. Constable (1994) *Transition Metal Chemistry: The Valence Shell in d-Block Chemistry*, VCH, Weinheim; C. E. Housecroft and A. G. Sharpe (2005) *Inorganic Chemistry*, 2nd edn, Prentice Hall, Harlow, Chapter 20.

## THEORETICAL AND CHEMICAL BACKGROUND

### Box 23.7 Ferromagnetic chains

The magnetic behaviour of a substance is usually described in terms of the Curie–Weiss Law which states that the *molar magnetic susceptibility* of a substance is inversely proportional to the temperature. *Ferromagnetic materials* do not conform to the Curie–Weiss Law until temperatures above the Curie temperature. *Ferromagnetism* arises through communication between metal centres, each of which has one or more unpaired electrons. Interaction between electrons on different metal centres causes their spins to become aligned but the electrons do *not* pair up so as to form a covalent bond.

The salt $[Ni(en)_2]_3[Fe(CN)_6]_2 \cdot 2H_2O$ is a beautiful example of a ferromagnetic material. In the solid state, X-ray diffraction studies have shown that the nickel(II) and iron(III) centres are connected by bridging cyano ligands and each metal is 6-coordinate. The structure consists of interconnected helical chains (see below). The ferromagnetism arises because the cyano-bridges allow the iron and nickel centres to communicate electronically with one another. This is just one example of the process of self-assembly of magnificent structures in the solid state.

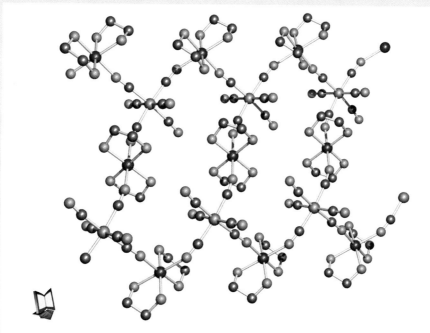

Colour code: Fe, green; Ni, red; N, blue; C, grey.
[X-ray diffraction data: M. Ohba *et al.* (1994) *J. Am. Chem. Soc.*, vol. 116, p. 11566.]

Diamagnetism and paramagnetism: see Section 4.18

balance and a more modern method that uses a SQUID (superconducting quantum interference device). The basis of the Gouy method is the interaction of unpaired electrons with a magnetic field. If a material is *diamagnetic* (all electrons paired) it is repelled by a magnetic field, but a *paramagnetic* material (possessing one or more unpaired electrons) is attracted by it.

Figure 23.25 schematically illustrates the apparatus used in the Gouy method. The compound under study is contained in a glass tube which is suspended from a balance on which the weight of the sample can be recorded. The glass tube is positioned so that one end of the sample lies at the point of maximum magnetic flux in an electromagnetic field – this is initially switched

**Fig. 23.25** Schematic representation of a Gouy balance.

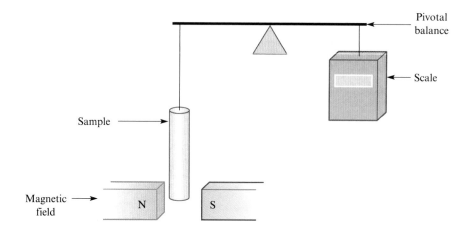

off. (An electromagnet is more convenient to use than a permanent magnet because the field can be switched on and off.) The other end of the sample lies at a point of low magnetic flux. When the magnetic field is applied, paramagnetic compounds are attracted into it, and the extent of attraction depends upon the number of unpaired electrons. The attraction causes a change in the weight of the sample and from these readings, it is possible to determine the *magnetic susceptibility*, $\chi$, of the compound. (The details of the derivation are beyond the scope of this text.) Usually, values of the *effective magnetic moment* are reported in the chemical literature rather than values of $\chi$, and the two quantities are related by equation 23.49.

$$\mu_{\text{eff}} = 2.828\sqrt{\chi_{\text{m}} \times T} \tag{23.49}$$

where:    $\mu_{\text{eff}} =$ effective magnetic moment in units of Bohr magnetons ($\mu_B$),

         $\chi_{\text{m}} =$ molar magnetic susceptibility,

         $T =$ temperature in kelvin

Equation 23.49 applies to Gaussian (non-SI) units.

We shall only be concerned with values of $\mu_{\text{eff}}$, the units of which are Bohr magnetons ($\mu_B$). Experimental values of the effective magnetic moments of selected octahedral complexes are listed in Table 23.7 and are compared with calculated values which we discuss below.

**Table 23.7** Experimentally determined values of the effective magnetic moments of selected octahedral complexes and values of $\mu$ calculated using the spin-only formula (equation 23.50). Ligand abbreviations are listed in Table 23.2.

| Compound | Experimental $\mu_{\text{eff}} / \mu_B$ | Spin only $\mu / \mu_B$ |
|---|---|---|
| $K_4[Mn(CN)_6]$ | 2.18 | 1.73 |
| $[VCl_4(bpy)]$ | 1.77 | 1.73 |
| $K_3[V(ox)_3]$ | 2.80 | 2.83 |
| $K_2[Mn(CN)_6]$ | 3.94 | 3.87 |
| $[MnCl_4(bpy)]$ | 3.82 | 3.87 |
| $[Cr(en)_3]Br_2$ | 4.75 | 4.90 |
| $[Cr(dien)_2]Cl_2$ | 4.81 | 4.90 |
| $K_3[Fe(ox)_3]$ | 5.95 | 5.92 |

**THEORETICAL AND CHEMICAL BACKGROUND**

## Box 23.8 An alternative form of the spin-only formula

In equation 23.50, we write the spin-only formula in terms of the number of unpaired electrons, $n$. An alternative form of the equation is in terms of the *total spin quantum number*, $S$.

In Section 3.10, we saw that the spin quantum number $s$ determines the magnitude of the spin angular momentum of an electron and has a value of $\frac{1}{2}$. The *total spin quantum number* $S$ is given by:

| Number of unpaired electrons: | 1 | 2 | 3 | 4 | 5 |
|---|---|---|---|---|---|
| Total spin quantum number, $S$: | $\frac{1}{2}$ | 1 | $\frac{3}{2}$ | 2 | $\frac{5}{2}$ |

Since:

$$S = n \times \frac{1}{2} = \frac{n}{2} \quad \text{or} \quad n = 2 \times S$$

we can rewrite the spin-only formula in the form:

$$\mu(\text{spin-only}) = \sqrt{2S(2S+2)} = \sqrt{4S(S+1)}$$
$$= 2\sqrt{S(S+1)}$$

### The spin-only formula

There are two effects that contribute towards the paramagnetism of a species: the electron spin and the orbital angular momentum. For octahedral complexes involving first row $d$-block metals, it is *approximately true* that the value of $\mu_{\text{eff}}$ can be estimated by assuming that the contribution made by the orbital angular momentum is negligible. Equation 23.50 is the *spin-only formula* and shows a simple dependence of the spin-only magnetic moment on the number of unpaired electrons, $n$.

$$\text{Spin-only magnetic moment} = \mu(\text{spin-only}) = \sqrt{n(n+2)} \qquad (23.50)$$

where $n$ = number of unpaired electrons

For high-spin $Cr^{2+}$ ($d^4$), there are four unpaired electrons (Figure 23.24) and the spin-only value of $\mu$ is calculated to be $4.90\,\mu_B$ (equation 23.51).

$$\mu(\text{spin-only}) = \sqrt{n(n+2)} = \sqrt{4 \times 6} = \sqrt{24} = 4.90\,\mu_B \qquad (23.51)$$

Measured effective magnetic moments are used to distinguish between high- and low-spin configurations. In a low-spin configuration, $Cr^{2+}$ ($d^4$) has only two unpaired electrons (Figure 23.24) and the observed value of $\mu_{\text{eff}}$ is expected to be close to $2.83\,\mu_B$.

---

**Worked example 23.5** | *Effective magnetic moments*

**The experimentally determined magnetic moment of the compound $[V(en)_3]Cl_2$ is $3.84\,\mu_B$. How many unpaired electrons are present?**

$[V(en)_3]Cl_2$ contains the complex cation $[V(en)_3]^{2+}$.
The effective magnetic moment can be estimated by using the spin-only formula:

$$\mu(\text{spin-only}) = \sqrt{n(n+2)}$$

Rather than substituting in the experimental value of $\mu_{\text{eff}}$, it is quicker to test values of $n$ (which can only be 1, 2, 3, 4 or 5) in the spin-only formula.

If $n = 3$:

$$\mu(\text{spin-only}) = \sqrt{3(3+2)} = \sqrt{15} = 3.87 \, \mu_B$$

This value is close to the experimental value of $\mu_{\text{eff}}$ and so there are three unpaired electrons on the vanadium centre in $[V(en)_3]^{2+}$.

This is consistent with $V^{2+}$ with electronic configuration $[Ar]3d^3$.

---

As the data in Table 23.7 illustrate, the spin-only formula is a fairly good approximation for octahedral complexes containing first row *d*-block metal ions. However, you should appreciate that the fit between the spin-only and measured effective magnetic moments varies and the values for $K_4[Mn(CN)_6]$ and $[VCl_4(bpy)]$ illustrate this. Complexes with other geometries or containing second and third row *d*-block metals usually show much larger deviations from the spin-only formula. Second and third row complexes are generally low-spin, but with magnetic moments significantly lower than the spin-only formula predicts.

## 23.14    Metal carbonyl compounds

The discussion in this chapter has focused on complexes of $M^{2+}$ and $M^{3+}$ ions with ligands that donate electrons, and we have seen that the electroneutrality principle can be used to assess reasonable charge distributions. In this final section, we introduce another group of complexes called *metal carbonyl compounds* which are representative of the much larger class of *organometallic compounds*.[§]

### The carbon monoxide ligand

Carbon monoxide is a Lewis base and we have already seen that its toxicity is due to the fact that it binds more strongly to the iron centres in haemoglobin than does $O_2$. Carbon monoxide also forms complexes with *low oxidation state* metals from the *d*-block. Its Lewis basicity arises from the properties of the highest *occupied* molecular orbital (HOMO) in the CO molecule (Figure 23.26). This MO consists largely of carbon character and may be considered to be a carbon-centred lone pair. The lowest *unoccupied* molecular orbitals (LUMO) in CO are a degenerate set of $\pi^*$ orbitals, again with more carbon than oxygen character, and one of the orbitals is shown in Figure 23.26.

Haemoglobin: see Box 5.3

Bonding in CO: see Section 5.14

### The low oxidation state metal centre

By *low oxidation state*, we usually think in terms of a metal(0) centre but we shall also include metals in oxidation states $+1$, $-1$ and $-2$. In iron pentacarbonyl, $Fe(CO)_5$, the iron centre is in an oxidation state of zero, and in $[Fe(CO)_4]^{2-}$, the oxidation state is $-2$.

---

[§] For a more detailed introduction to organometallic compounds, see: C. E. Housecroft and A. G. Sharpe (2005) *Inorganic Chemistry*, 2nd edn, Prentice Hall, Harlow, Chapter 23.

**Fig. 23.26** The highest occupied molecular orbital (HOMO) and one of the two degenerate lowest unoccupied molecular orbitals (LUMO) of carbon monoxide. More detailed diagrams are given in Section 5.14.

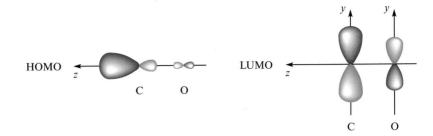

(23.31)

In a metal carbonyl compound such as $[Cr(CO)_6]$ (**23.31**), each CO ligand can donate a pair of electrons to the metal centre. However, this greatly increases the negative charge at the metal. This $\sigma$-interaction is shown for one Cr–CO interaction in Figure 23.27a; the metal orbital involved is one of the $e_g$ orbitals, exemplified here by the $d_{z^2}$ orbital.

By Pauling's electroneutrality principle, the distribution of charge in a molecule or ion is such that the charge on any single atom is ideally close to zero. Since the chromium centre in $[Cr(CO)_6]$ bears no formal charge, we appear to be saying that it is unfavourable for the chromium atom to form coordinate bonds with the six carbonyl ligands because this would give a formal charge at the Cr(0) centre of $-6$. How does the molecule cope with this situation?

### CO: a $\pi$-acceptor ligand

We used the electroneutrality principle to introduce partially ionic metal–ligand bonding with higher oxidation state metal ions. However, with the $[Cr(CO)_6]$ molecule, no degree of ionic character will remove the negative charge that appears to have built up on the electropositive Cr(0) centre. We need another means of removing this negative charge from the metal.

The key to the success of CO in stabilizing low oxidation state metal centres is its ability to act as an *acceptor ligand*. The low-lying $\pi^*$ orbitals can overlap with the occupied $t_{2g}$ orbitals of the metal centre and the result is the formation of a coordinate $\pi$-bond in which electrons *from the metal are donated to the ligand* (Figure 23.27b).

The $\pi$-interaction operates in addition to the $\sigma$-interaction shown in Figure 23.27a. The overall effect is that the negative charge that builds up on the

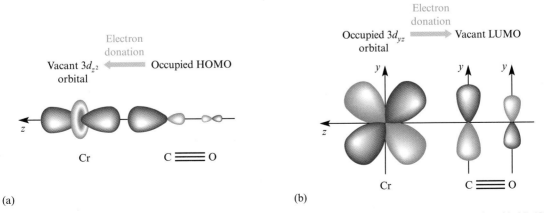

**Fig. 23.27** (a) The formation of one chromium–carbon $\sigma$-bond, and (b) the formation of a chromium–carbon $\pi$-bond in $[Cr(CO)_6]$. The electronic charge is transferred in *opposite* directions in the two interactions, and the overall effect is a *synergic interaction*.

**Table 23.8** Infrared spectroscopic data for two series of isoelectronic metal carbonyls.

| Octahedral complex | $\bar{\nu}_{CO} / cm^{-1}$ | Tetrahedral complex | $\bar{\nu}_{CO} / cm^{-1}$ |
|---|---|---|---|
| $[Mn(CO)_6]^+$ | 2090 | $[Ni(CO)_4]$ | 2060 |
| $[Cr(CO)_6]$ | 2000 | $[Co(CO)_4]^-$ | 1890 |
| $[V(CO)_6]^-$ | 1860 | $[Fe(CO)_4]^{2-}$ | 1790 |

Synergic means
'working together'

chromium centre as a result of the formation of six metal–carbon $\sigma$-bonds is redistributed *back to the ligands* as a consequence of the metal–carbon $\pi$-bonds. This is called a *synergic effect* and the bonding theory summarized in Figure 23.27 is called the Dewar–Chatt–Duncanson model.

### Infrared spectroscopic evidence for $\pi$-back donation

The back donation of electrons from a metal to CO ligand involves electrons being transferred to a carbon–oxygen $\pi^*$ MO. This orbital has C–O *antibonding character*. Occupancy of the C–O $\pi^*$ orbital weakens the C–O bond and the result is readily observed using IR spectroscopy. There is a *reduction* in the value of the C–O vibrational wavenumber in going from free CO to coordinated CO. Table 23.8 lists values of $\bar{\nu}_{CO}$ for two series of isoelectronic complexes. In the spectrum of gaseous CO, an absorption is observed at $2143 \, cm^{-1}$. The values in Table 23.8 show that, for an isoelectronic series of carbonyl complexes, the amount of back donation increases with an increase in formal negative charge of the complex. For example, back donation increases along the series $[V(CO)_6]^- > [Cr(CO)_6] > [Mn(CO)_6]^+$. This means that values of $\bar{\nu}(CO)$ follow the order $[V(CO)_6]^- < [Cr(CO)_6] < [Mn(CO)_6]^+$.

### Shapes of mononuclear metal carbonyl complexes

Kepert theory:
see Sections 6.8–6.10

The theory of Kepert can successfully be applied to mononuclear metal carbonyl complexes. Binary carbonyl complexes with four ligands are tetrahedral, with five ligands are trigonal bipyramidal and with six ligands are octahedral, and the structures of the neutral compounds $[Ni(CO)_4]$, $[Fe(CO)_5]$ and $[Cr(CO)_6]$ are shown in Figure 23.28. The anions $[Co(CO)_4]^-$ and $[Fe(CO)_4]^{2-}$ are isoelectronic and isostructural with $[Ni(CO)_4]$. Similarly $[Mn(CO)_5]^-$ is related to $[Fe(CO)_5]$. $[Mn(CO)_6]^+$ and $[V(CO)_6]^-$ are isostructural with $[Cr(CO)_6]$.

### The 18-electron rule

Low oxidation state metal carbonyl (and, more generally, organometallic) complexes tend to obey the *18-electron rule*, thereby utilizing all the valence atomic orbitals of the metal. For a first row metal, these are the $4s$, $3d$ and $4p$ orbitals. The involvement of all nine atomic orbitals arises because of the formation of both M–CO $\sigma$-bonds, and $\pi$-interactions (i.e. back donation). The bonding scheme in Figure 23.27 is an over-simplification because it considers a single M–CO interaction. In an octahedral complex such as $[Cr(CO)_6]$, the formation of six M–C $\sigma$-bonds involves the metal $4s$, $4p$, $3d_{z^2}$, $3d_{x^2-y^2}$ orbitals (six atomic orbitals) while the M–CO $\pi$-interactions involve the metal $3d_{xy}$, $3d_{xz}$ and $3d_{yz}$

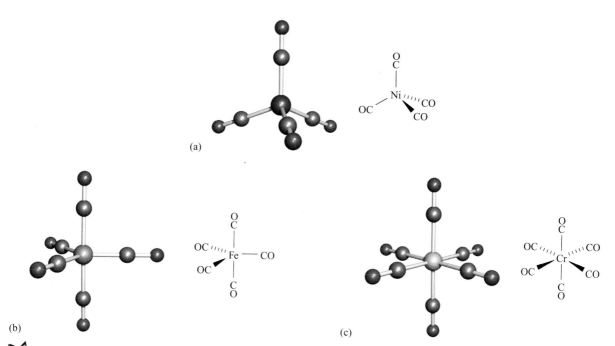

**Fig. 23.28** The structures of the binary metal carbonyl compounds (a) [Ni(CO)₄], (b) [Fe(CO)₅] and (c) [Cr(CO)₆]. Colour code: Ni, blue; Fe, green; Cr, yellow; C, grey; O, red. [X-ray diffraction data: [Ni(CO)₄], D. Braga *et al.* (1993) *Organometallics*, vol. 12, p. 1481; [Fe(CO)₅], A. W. Hanson (1962) *Acta Crystallogr.*, vol. 15, p. 930; [Cr(CO)₆], A. Whitaker *et al.* (1967) *Acta Crystallogr.*, vol. 23, p. 977.]

orbitals (three atomic orbitals). The nine molecular orbitals resulting from these $\sigma$- and $\pi$-interactions tend to be completely occupied by 18 electrons. Consider [Cr(CO)₆]. The Cr atom (group 6) is in oxidation state 0 and provides six valence electrons. Each CO ligand is a 2-electron donor, these electrons coming from the HOMO (Figure 23.26). There are therefore 18 valence electrons available in [Cr(CO)₆] (equation 23.52).

$$\text{Total number of valence electrons in } [\text{Cr(CO)}_6] = 6 + (6 \times 2) = 18 \quad (23.52)$$

The number of CO ligands with which a metal combines to give a metal carbonyl complex obeying the 18-electron rule is dictated by the number of valence electrons provided by the metal. For example, Ni(0) (group 10) has 10 valence electrons and only requires eight more electrons to satisfy the 18-electron rule; eight electrons are provided by four CO ligands and Ni(0) forms [Ni(CO)₄]. Equation 23.53 shows that [Fe(CO)₅] obeys the 18-electron rule (Fe is in group 8).

$$\text{Total number of valence electrons in } [\text{Fe(CO)}_5] = 8 + (5 \times 2) = 18 \quad (23.53)$$

| **Worked example 23.6** | *Applying the 18-electron rule* |
| --- | --- |

Show that **[Fe(CO)₄]²⁻ obeys the 18-electron rule.**

Fe is in group 8 and Fe(0) provides 8 valence electrons.
Each CO ligand provides 2 electrons.
The 2− charge provides 2 electrons.
Total number of valence electrons in $[\text{Fe(CO)}_4]^{2-} = 8 + (4 \times 2) + 2 = 18$

The 18-electron rule is not always obeyed. For example, second and third row, late *d*-block metals, e.g. Rh, Ir, often form 16-electron complexes. However, if we confine our attention to binary carbonyls of the middle first row *d*-block metals, then for the most part, complexes do obey the rule and this allows us to rationalize the number of carbonyl ligands present.

## SUMMARY

This chapter has introduced some of the basic concepts of the coordination chemistry of the *d*-block metals, with an emphasis on the elements in the first row. We have also looked briefly at the special properties of the CO ligand that make it able to stabilize *low* oxidation state metal centres and have introduced the 18-electron rule.

### Do you know what the following terms mean?

- coordinate bond
- coordination number
- ligand
- monodentate
- bidentate
- polydentate
- donor atom
- chelate
- electroneutrality principle
- stereoisomers

- enantiomers
- racemate
- hexaaqua ion
- ammine complex
- nuclearity of a complex
- hard and soft donor atoms
- hard and soft metal ions
- crystal field splitting
- $\Delta_{oct}$
- strong and weak field ligand

- spectrochemical series
- high-spin and low-spin complexes
- spin-only formula
- $\pi$-acceptor ligand
- synergic effect
- Dewar–Chatt–Duncanson model
- 18-electron rule

### You should now be able:

- to know the symbols and names of the first row *d*-block metals
- to write down the elements in sequence in the first row of the *d*-block
- to write down the ground state electronic configuration of metal atoms and ions in the first row of the *d*-block
- to recognize potential donor atoms in a ligand
- to recognize how flexible a ligand may be in terms of its ability to adopt different conformations
- to draw the abbreviations and structures of representative ligands including py, [NCS]$^-$, en, dien, bpy, phen, [acac]$^-$ and [edta]$^{4-}$
- to estimate the charge distribution in a complex of the type [ML$_6$]$^{n+}$
- to work out the possible isomers of a given complex
- to explain why some aqua ions are acidic in aqueous solution
- to give representative preparative routes to selected complexes
- to suggest possible products from complex-forming reactions, including isomers

- to write down an expression for the stability constant (stepwise or overall) of a complex
- to obtain values of $\Delta G^\circ$ from stability constant data
- to use the concept of hard and soft metal ions and ligands to suggest reasons for the preferences shown in the formation of some complexes
- to use values of $E^\circ$ to assess the stability of an aqueous ion in solution with respect to oxidation by, for example, H$^+$ or O$_2$, recognizing the restrictions that working under standard conditions imposes
- to explain why the *d*-orbitals are not degenerate in an octahedral complex
- to write down the electronic configuration of a given metal ion in an octahedral environment
- to understand the effects that the presence of strong and weak field ligands have on the electronic configuration of a metal ion in an octahedral environment
- to describe a method of measuring the effective magnetic moment of a complex
- to estimate the magnetic moment of complexes containing a first row *d*-block metal ion

- to estimate the number of unpaired electrons present given a value of $\mu_{eff}$
- to discuss the molecular orbital properties of CO that lead to its ability to act as a $\pi$-acceptor ligand
- to outline the bonding in a mononuclear metal carbonyl compound in terms of the interactions in a

single M–CO unit.

- to explain what spectroscopic evidence supports the Dewar–Chatt–Duncanson model
- to apply the 18-electron rule to mononuclear metal carbonyl complexes

## PROBLEMS

Refer to Table 23.2 for ligand abbreviations.

**23.1** Draw radial distribution curves for the $4s$ and $3d$ atomic orbitals. How is an electron in a $4s$ orbital affected by an increase in nuclear charge? How is an electron in a $3d$ orbital affected by an increase in nuclear charge?

**23.2** Write down the ground state electronic configurations of the following atoms and ions: (a) Fe; (b) $Fe^{3+}$; (c) Cu; (d) $Zn^{2+}$; (e) $V^{3+}$; (f) Ni; (g) $Cr^{3+}$; (h) $Cu^+$.

**23.3** Copy the structures of each of the following ligands. Indicate which atoms are potential donor sites. Which ligands could be chelating?

**23.4** What is the oxidation state of the metal in the following complexes?
(a) $[CoCl_4]^{2-}$
(b) $[Fe(CN)_6]^{3-}$
(c) $[Ni(gly\text{-}N,O)_3]^-$
(d) $[Fe(bpy)_3]^{3+}$
(e) $[Cr(acac)_3]$
(f) $[Co(H_2O)_6]^{2+}$
(g) $[Ni(en)_3]^{2+}$
(h) $[Ni(edta)]^{2-}$
(i) $[Cu(ox)_2]^{2-}$
(j) $[Cr(en)_2F_2]^+$

**23.5** Use the Kepert theory to suggest structures for the following complexes. Could any of these complexes possess stereoisomers? If so, suggest which isomer might be favoured (none of the 4-coordinate complexes is square planar):

(a) $[Zn(OH)_4]^{2-}$
(b) $[Ru(NH_3)_5(NCMe)]^{2+}$
(c) $[CoBr_2(PPh_3)_2]$
(d) $[WBr_4(NCMe)_2]$
(e) $[ReCl(CO)_3(py)_2]$
(f) $[Cr(NPr_2)_3]$
(g) $[CrCl_3(NMe_2)_2]$

[*Hint*: $[Pr_2N]^-$ and $[Me_2N]^-$ are *amido* ligands, isoelectronic with an ether $R_2O$.]

**23.6** Estimate the charge distribution over the metal and donor atoms in the complex ions (a) $[Fe(bpy)_3]^{2+}$ and (b) $[CrF_6]^{3-}$.

**23.7** Draw the stereoisomers of (a) octahedral $[Mn(H_2O)_2(ox)_2]^{2-}$, (b) octahedral $[Co(H_2O)(NH_3)(en)_2]^{3+}$, (c) square planar $[NiCl_2(PMe_3)_2]$ and (d) octahedral $[CrCl_3(dien)]$.

**23.8** Draw the structures of the isomers of octahedral $[Ni(gly)_3]^-$.

**23.9** Which of the following octahedral complexes possess enantiomers?
(a) $[Fe(ox)_3]^{3-}$
(b) $[Fe(en)(H_2O)_4]^{2+}$
(c) *cis*-$[Fe(en)_2(H_2O)_2]^{2+}$
(d) *trans*-$[Fe(en)_2(H_2O)_2]^{2+}$
(e) $[Fe(phen)_3]^{3+}$
(f) $[Co(bpy)(CN)_4]^-$

**23.10** For each complex, state whether there are isomers and if so, draw the structures and give them appropriate distinguishing labels.
(a) square planar $[PtCl_3(NH_3)]^-$
(b) tetrahedral $[CoBr_2(PPh_3)_2]$
(c) octahedral $[Ni(edta)]^{2-}$
(d) octahedral $[CoCl_2(en)_2]^+$

**23.11** (a) What is the oxidation state of Cr in $[CrCl_2(H_2O)_4]Cl\cdot 2H_2O$? (b) Draw the structures of the isomers of the complex cation in $[CrCl_2(H_2O)_4]Cl\cdot 2H_2O$. (c) What other types of isomerism may $[CrCl_2(H_2O)_4]Cl\cdot 2H_2O$ exhibit?

**23.12** Both tpy and dien act as tridentate ligands. Discuss possible isomerism in complexes of the type $[M(tpy)_2]^{2+}$ and $[M(dien)_2]^{2+}$.

**23.13** Suggest possible products (including isomers if appropriate) for the following reactions:
(a) $[Ni(H_2O)_6]^{2+} + 3en \rightleftharpoons$
(b) $[Co(H_2O)_6]^{2+} + 3phen \rightleftharpoons$
(c) $VCl_4 + 2py \rightleftharpoons$
(d) $[Cr(H_2O)_6]^{3+} + 3[acac]^- \rightleftharpoons$
(e) $[Fe(H_2O)_6]^{2+} + 6[CN]^- \rightleftharpoons$
(f) $FeCl_2 + 6NH_3 \rightleftharpoons$

**23.14** Figure 23.29 shows the variation in molar conductivity for the series of compounds $[Co(NH_3)_6]Cl_3$, $[Co(NO_2)(NH_3)_5]Cl_2$, $[Co(NO_2)_2(NH_3)_4]Cl$, $[Co(NO_2)_3(NH_3)_3]$ and $K[Co(NO_2)_4(NH_3)_2]$ in aqueous solutions. Suggest reasons for the observed trend.

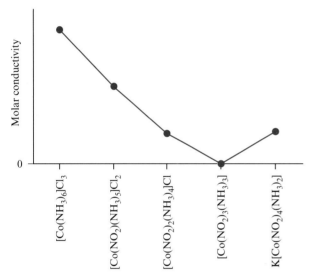

**Fig. 23.29** For problem 23.14.

**23.15** Write down expressions for the equilibrium constants for reactions 23.23–23.27. Use these expressions and that in equation 23.22 to show that the overall stability constant for $[ML_6]^{2+}$, $\beta_6$ is given by the equation:

$$\beta_6 = K_1 \times K_2 \times K_3 \times K_4 \times K_5 \times K_6$$

and that

$$\log \beta_6 = \log K_1 + \log K_2 + \log K_3 \\ + \log K_4 + \log K_5 + \log K_6$$

where $K_1$ is the equilibrium constant for the replacement of the first ligand, and $K_2$ refers to the second step, etc.

**23.16** The overall stability constant for the formation of $[Co(H_2O)_4(NH_3)_2]^{2+}$ from $[Co(H_2O)_6]^{2+}$ is $5.0 \times 10^3$, whereas for the formation of $[Co(H_2O)_4(en)]^{2+}$, $K_1 = 1.0 \times 10^6$ (both at 298 K).

Determine values of $\Delta G^\circ$ for these reactions. Is it more or less favourable to replace two aqua ligands in $[Co(H_2O)_6]^{2+}$ by two ammine ligands or one ethane-1,2-diamine ligand? Suggest a reason for this difference.

**23.17** (a) Explain briefly what is meant by the *chelate effect*. (b) How many 5-membered chelate rings are present in $[Co(en)_3]^{2+}$, $[Co(bpy)_3]^{3+}$, $[Co(edta)]^-$ and $[Co(dien)_2]^{3+}$?

**23.18** For the reaction of $Fe^{2+}$ with $[CN]^-$ in aqueous solution, $\log \beta_6 = 24$. (a) To what equilibrium does $\beta_6$ refer? (b) Write down an expression for $\beta_6$ in terms of the equilibrium concentrations of $Fe^{2+}$, $[CN]^-$ and the appropriate complex. (c) Determine a value for $\beta_6$.

**23.19** Stability constant data for the formation of $[MnF_6]^{4-}$ from $[Mn(H_2O)_6]^{2+}$ include the following: $\log K_1 = 5.52$, $\log \beta_2 = 9.02$. Determine $K_1$ and $K_2$, and comment on the relative positions of the equilibria describing the first two ligand displacements.

**23.20** (a) Draw the structure of the complex ion represented by $Co^{2+}(aq)$. (b) The ion $[Cr(H_2O)_5(OH)]^{2+}$ forms a dinuclear complex in solution. Write an equation for this reaction and draw the structure of the dichromium product. Is there a change in oxidation state during the reaction? (c) Yellow solutions of ammonium vanadate $[NH_4][VO_3]$ in dilute sulfuric acid contain $[VO_2]^+$. Draw a possible structure for the solution species represented by $[VO_2]^+$ given that the oxo ligands are *cis* to each other.

**23.21** Classify each of the following coordination complexes in terms of the concept of hard and soft metal ions and ligands, indicating which are the donor atoms in each ligand:
(a) $[AuI_2]^-$
(b) $[Ru(NH_3)_5(MeCO_2)]^{2+}$
(c) $[Y(H_2O)_9]^{3+}$
(d) $[Mn(en)_3]^{2+}$

What shape do you think the $[Y(H_2O)_9]^{3+}$ cation might adopt?

**23.22** Use the data in Table 23.4 to suggest which linkage isomer of $[Cr(NCS)_6]^{3-}$ might be favoured.

**23.23** Explain why there is only one spin arrangement for an octahedral $d^8$ or $d^9$ metal ion.

**23.24** Calculate the spin-only magnetic moment for each of the following octahedral complexes:
(a) $[Mn(en)_3]^{3+}$ (high-spin)
(b) $[Fe(en)_3]^{2+}$ (high-spin)
(c) $[Ni(HOCH_2CH_2NH_2)_2(OCH_2CH_2NH_2)]^+$

**23.25** Using the experimentally determined effective magnetic moments of the following octahedral

complexes, determine if the metal ion is in a high-
or low-spin configuration:

(a) $[Cr(en)_3]^{2+}$ $\mu_{eff} = 4.84 \mu_B$;

(b) $cis$-$[MnBr_2L]$ $\mu_{eff} = 5.87 \mu_B$.

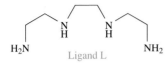

Ligand L

Draw the structure of the manganese(II) complex.

**23.26** What is the oxidation state of the metal centre in
each of the following complexes?

(a) $[Co(CO)_4]^-$

(b) $[Ni(CN)_4]^{2-}$

(c) $[Cr(CO)_6]$

(d) $[Mn(CO)_5Br]$

(e) $[Mn(CO)_3(py)_2I]$

(f) $[Mn(CO)_6]^+$

**23.27** What do you understand by the synergic effect in
metal carbonyl complexes? Suggest reasons for the
*trends* in the values listed in Table 23.8.

**23.28** Show that the following complexes obey the
18-electron rule: (a) $[Mn(CO)_5]^-$; (b) $[V(CO)_6]^-$;
(c) $[Co(CO)_4]^-$.

**23.29** Using the information in Table 23.9 and assuming
that the following complexes obey the 18-electron
rule, determine $n$ in each complex and draw
possible structures for the complexes, commenting
on possible isomerism: (a) $[Ni(CO)_n]$; (b)
$[Fe(CO)_n]$; (c) $[HMn(CO)_n]$; (d) $[Fe(CO)_n(PPh_3)_2]$;
(e) $[H_2Fe(CO)_n]$; (f) $[Cr(MeCN)_2(CO)_n]$.

**Table 23.9** Electrons provided by selected ligands.

| Ligand | $PPh_3$ | H | MeCN |
|---|---|---|---|
| Number of electrons | 2 | 1 | 2 |

**23.30** *This problem assumes a knowledge of chirality from
Chapter 24.*

Propane-1,2-diamine (1,2-pn) has the following
structure:

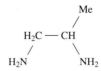

Rationalize why there are six isomers of *trans*-
$[Co(1,2\text{-}pn)_2Cl_2]^+$. Comment on relationships
between the isomers.

**23.31** Suggest identities and structures for the products
in the following reactions.

(a) The reaction of $MnCl_2$, KNCS and $Et_4NBr$
in ethanol gives a salt **A** of a complex anion
with a magnetic moment of 5.93 $\mu_B$ (295 K).
The ratio of moles of cations : anions in **A** is
4 : 1.

(b) The ligand $N(CH_2CH_2NMe_2)_3$ (L) forms a
5-coordinate complex $[CoLBr]^+$.

(c) Reaction of $[Co(NH_3)_5Br]Br_2$ with $Ag_2SO_4$
gives a complex **B** that forms a white
precipitate when treated with $BaCl_2$. **B** is a 1 : 1
electrolyte. Treatment of $[Co(NH_3)_5Br]Br_2$
with concentrated $H_2SO_4$ followed by $BaBr_2$
gives a 1 : 1 electrolyte **C**. Compound **C** does
not give a precipitate with $BaCl_2$ but does form
a white precipitate with $AgNO_3$.

# 24 Carbon compounds: an introduction

## Topics

- Structural representations
- Types of carbon compounds
- Geometry and hybridization
- Functional groups
- Hydrocarbon frameworks and nomenclature
- Isomerism
- Chiral compounds
- Conformation
- Steric energy

### 24.1  Introduction

Organic chemistry is the study of compounds of carbon. Conventionally, elemental carbon and compounds such as carbon monoxide, carbon dioxide, hydrogen cyanide and carbonates lie within the realm of the inorganic chemist. In this book, we try to show that common themes run through the chemistry of 'organic' and 'inorganic' compounds, and that divisions are artificial and often unhelpful. Nowadays, interdisciplinary research is common and divisions are not clear-cut. Real progress in science is made when researchers from different areas communicate their ideas and exchange knowledge with one another. Nonetheless, in schools, colleges and universities, the chemistry of carbon compounds is often studied separately from the chemistry of all other elements. In part, this reflects the huge numbers of organic compounds known, and the industrial, commercial and biological importance of organic chemistry. The *carbon cycle* (Box 24.1) summarizes the way in which carbon passes from one system to another in nature, although man-made events such as forest fires affect the natural balance of the cycle.

### 24.2  Drawing structural formulae

We introduced structural formulae and 'ball-and-stick' models in Section 1.17. 'Ball-and-stick' models of the type used in, for example, Figures 23.28 and 24.9, provide 3-dimensional structural information. In inorganic molecules, we deal with a range of coordination numbers and geometries, and 'ball-and-stick' models are essential in helping us to gain a full picture of the *stereochemistry* of a molecule. This is also true in organic chemistry,

The *stereochemistry* of a molecule describes the way in which the atoms are arranged in space.

## ENVIRONMENT AND BIOLOGY

## Box 24.1 The carbon cycle

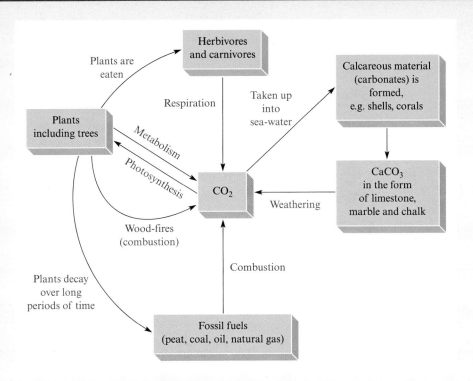

but structures are somewhat simplified by the fact that carbon has the restricted range of geometries described in Section 24.3. 'Ball-and-stick' models can be built using model kits but, nowadays, they are more often produced by a computer graphics package. Useful though this is, we also need to be able to produce schematic structural formulae of the types we have used throughout this book. For example, structure **24.1** represents $CH_4$ and has enough information to reveal that the C atom is tetrahedral. Structure **24.2** represents ethene, $C_2H_4$, with trigonal planar C atoms. The structure of acetone can be drawn as shown in **24.3**, but it is also commonly represented simply as structure **24.4**. The advantage of this representation is that it is easy to draw, but certain information is 'hidden' in the structure. The following assumptions are made:

- C atoms make up the backbone of the molecule;
- every unspecified position in the chain, including the ends, is occupied by a C atom;
- each C atom makes four bonds;
- any remaining unspecified atoms are H atoms.

By putting these points into practice, we can 'fill in the gaps' in structure **24.4** and arrive at structure **24.3** for acetone. You should study the pairs of structures below, and confirm that the lower one in each case can be derived by 'filling in the gaps' in the top structure:

**(24.1)**

**(24.2)**

**(24.3)**

**(24.4)**

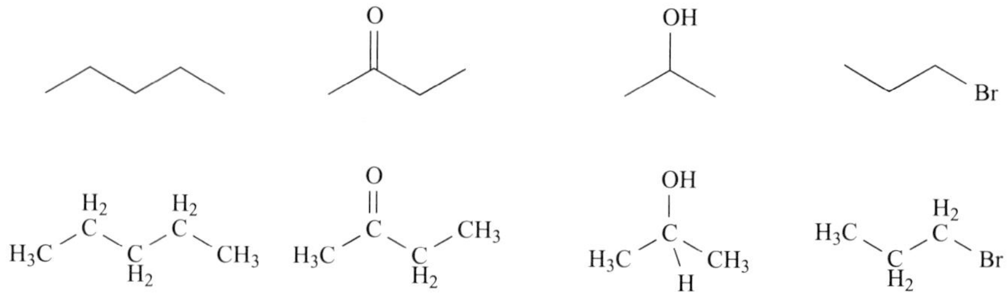

We adopt this method of representing organic structures throughout the book, and further practice in relating structural formulae to the complete formulae of molecules can be found in problems 24.5–24.7.

## 24.3     Types of carbon compounds

The natural world is one of animal, vegetable and mineral matter. We can conveniently take the *mineral* part as involving inorganic materials such as water, oxygen and silicates (see Section 22.6) while *animals and vegetables* rely on organic compounds. Nature has designed enormous numbers of complicated molecules (e.g. DNA), but it also utilizes countless relatively small ones such as those in Figure 24.1. Fundamental to our understanding of carbon compounds and how they function is an awareness of their structure and the *functional groups* that they contain. Before we embark on this classification, we need to look at the *backbone* structures of molecules. We begin with the simplest compounds, *hydrocarbons*, which contain only C and H, and then summarize the functional groups that are crucial to the discussions in much of Chapters 25–35.

DNA: see Box 21.6

### What is a hydrocarbon?

A *hydrocarbon* contains only C and H atoms.

Hydrocarbon compounds contain only C and H, and are conveniently classified according to whether they are *aliphatic* or *aromatic*, and whether the aliphatic compound is *saturated* or *unsaturated* (Figure 24.2). The word 'aliphatic' was originally used to describe fats and related compounds. It is now used to refer to hydrocarbons containing open carbon chains (*acyclic*

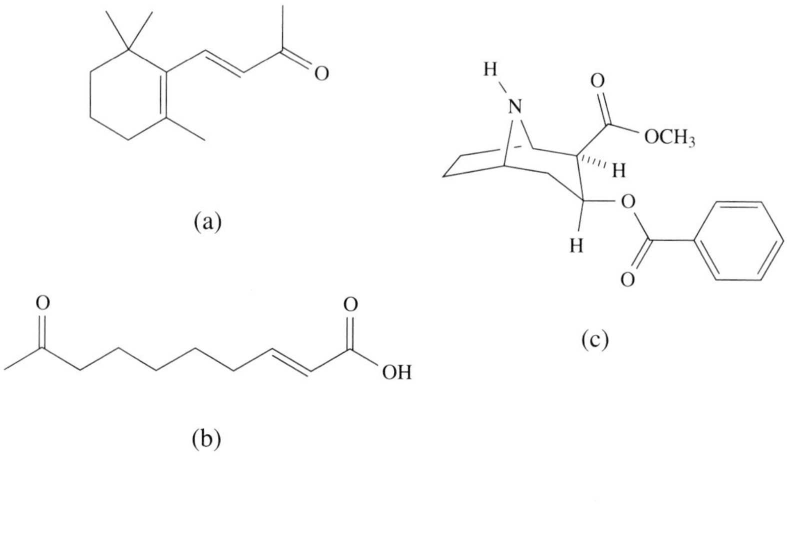

**Fig. 24.1** Examples of relatively simple molecules from nature: (a) β-ionone, an intense scent emitted by flowers such as freesias, (b) a pheromone from honeybees (pheromones are message-carrying chemicals, see Boxes 29.3 and 34.2) and (c) cocaine, a habit-forming drug obtained from coca (the dried leaves of *Erythroxylum coca* and *Erythroxylum truxillense*).

**Fig. 24.2** Classification of hydrocarbon compounds.

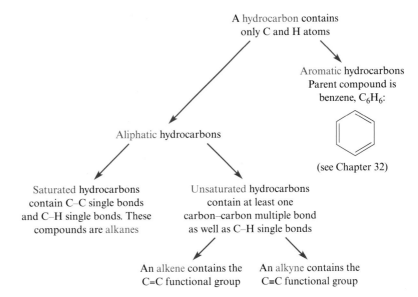

In an *acyclic* hydrocarbon, the C atoms are connected in an open chain. If they are in a ring, the compound is *cyclic*.

compounds) or compounds with cyclic structures but with properties similar to the open-chain hydrocarbons, e.g. cyclohexane.

Figure 24.2 shows two classes of aliphatic hydrocarbons. A *saturated* hydrocarbon contains only C–C and C–H bonds and has *no functional group*. The old name for a hydrocarbon was *paraffin*, derived from Latin and meaning 'little affinity'. The modern name for a saturated hydrocarbon is an *alkane*. The general formula for an acyclic alkane is $C_nH_{2n+2}$, and every C atom forms four single bonds. An *unsaturated* hydrocarbon contains at least one C=C or C≡C bond, as well as C–H bonds. Compounds that possess a C=C bond are called *alkenes*, and those with a C≡C bond are *alkynes*. The C=C and C≡C units are functional groups, and undergo characteristic reactions which we describe in Chapter 26. Despite having introduced only a handful of compound classes, we can already illustrate the common geometries in which carbon is found.

▪▷
Restricted geometries of carbon: see Section 6.14

## Geometry and hybridization

▪▷
Hybridization: see Section 7.5

A hydrocarbon may contain *tetrahedral*, *trigonal planar* or *linear* C atoms depending on the types of C–C bonds that it forms, and these three shapes are found throughout carbon compounds. The relevant hybridization schemes are $sp^3$ (tetrahedral), $sp^2$ (trigonal planar) and $sp$ (linear). Carbon obeys the octet rule and forms four localized bonds. These may be made up of:

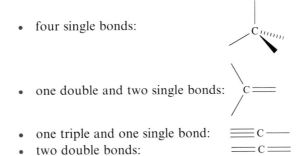

- four single bonds:

- one double and two single bonds:

- one triple and one single bond:
- two double bonds:

Within hydrocarbon chemistry, these shapes are exemplified by methane (**24.5**), ethene (**24.6**), ethyne (**24.7**) and allene (**24.8**).

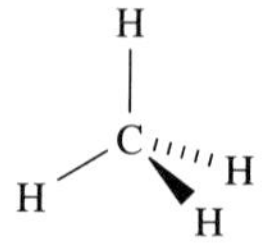

(24.5) 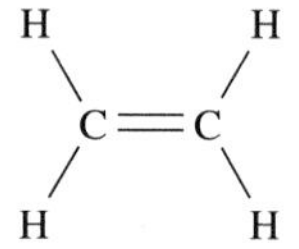          (24.6) 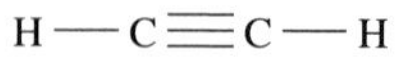          (24.7) 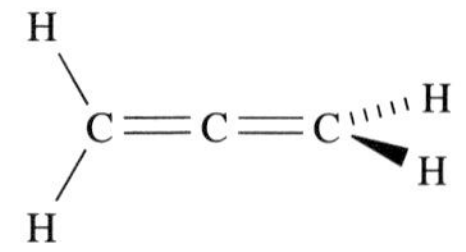          (24.8)

## Functional groups

In organic formulae, it is often useful to represent a *general organic group*. The conventional method is to use a group *R for an aliphatic group* and *Ar for an aryl (aromatic) group*. To distinguish between different R groups, use R, R' and R'', or $R^1$, $R^2$ and $R^3$, and so on.

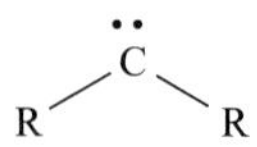

R = general organic group

(24.9)

In Table 1.9, we introduced selected functional groups in organic compounds. In this section, we expand the introduction by giving examples of compounds containing different functional groups. Examples are taken from the laboratory, medicine, the commercial world and nature, to provide insight into the applications of functionalities and the fact that many molecules contain more than one functional group. Apart from C and H, the elements that appear most often in organic compounds are N and O. Like C, both N and O are restricted in the number of bonds they can form (see Section 7.2). Carbon only rarely carries a lone pair of electrons: *carbenes* (**24.9**) provide an example of such a structure. In contrast, lone pairs of electrons play an important role in the chemistry of N and O centres, affecting both shape and reactivity. Being able to recognize the functional groups in a molecule is the key to being able to understand its chemical properties, as we discuss in later chapters. For now, you should aim to learn the names and structures of the functional groups described below. Notice also that the structures below are drawn in a form that emphasizes only the functional groups.

An *alcohol* contains an −OH group; an example is ethanol, **24.10**, which is used widely as a solvent and is the important ingredient in alcoholic drinks. Ethanol is an example of a *primary alcohol* with a terminal −CH$_2$OH group. Secondary and tertiary alcohols have the general formulae $R_2$CHOH and $R_3$COH respectively. Propane-1,3-diol (**24.11**) contains two −OH groups (hence the name *diol*) and is used commercially in the manufacture of *Corterra*, a polymer used for carpet and textile fibres. Citronellol (**24.12**) is a floral scent found in roses and contains alkene and alcohol functional groups.

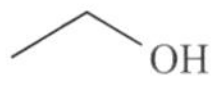

(24.10)           (24.11) 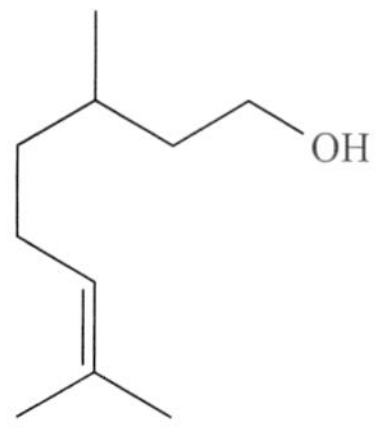          (24.12)

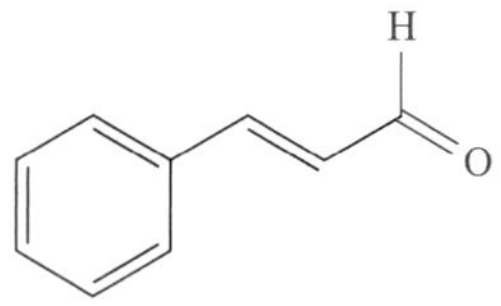

(24.13)

In an *aldehyde*, the functional group is −CHO and an example of a natural aldehyde is cinnamaldehyde (**24.13**) present in cinnamon. The functional group in a *ketone* is a C=O group, but the C atom must also be attached to two organic groups, which may be the same or different. The general formula for a *ketone* is:

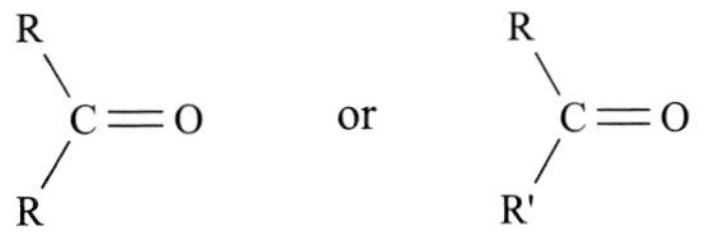

$$\begin{array}{c} R \\ \diagdown \\ C=O \\ \diagup \\ R \end{array} \quad \text{or} \quad \begin{array}{c} R \\ \diagdown \\ C=O \\ \diagup \\ R' \end{array}$$

where $R$ *and* $R' \neq H$. Acetone (**24.14**) is a common laboratory solvent and is used commercially in paints, cleaning fluids and as a solvent for coatings. Compound **24.15** (containing ketone and alkene functional groups) is responsible for the scent of magnolia flowers.

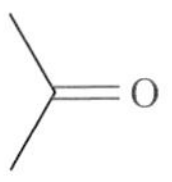

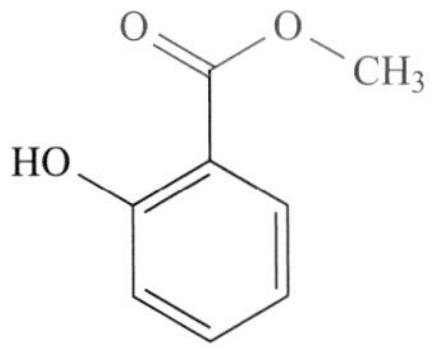

(24.14) 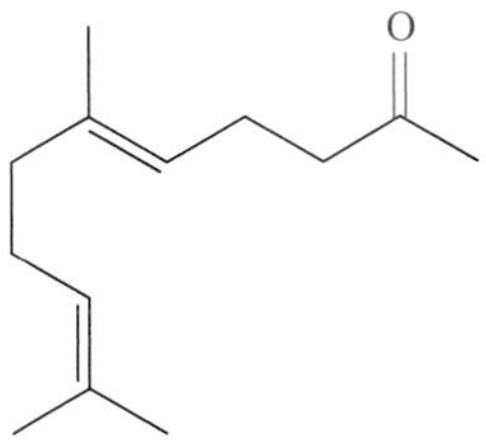                    (24.15) 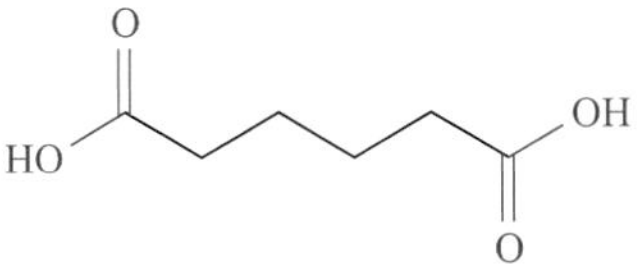                    (24.16)

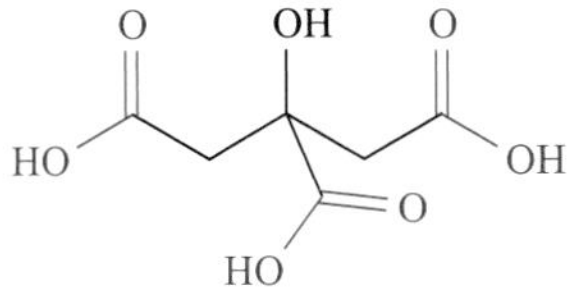

A *carboxylic acid* is characterized by a $-CO_2H$ functionality; *dicarboxylic acids* possess two such groups, and so on. Compound **24.16** is one of two precursors for Nylon 66, the other being the diamine **24.22**. Carboxylic acids play an important role in nature, for example, citric acid (**24.17**) occurs in citrus fruits as well as other fruits such as cranberries.

An *ester* is related to a carboxylic acid, but has the H atom replaced by an organic group: the functional group in an ester is $-CO_2R$. Many esters have characteristic odours, for example methyl salicylate (commonly called oil of wintergreen, **24.18**) contains an ester group in which a $CH_3$ group is the R group in the general formula. An *ether* has the general formula $R-O-R$ (symmetrical ether) or $R-O-R'$ (unsymmetrical ether). Diethyl ether (**24.19**) is a common solvent for Grignard reagents (see Section 28.4). In nature, vanillin (**24.20**), present in oil of vanilla, contains an ether functionality, as well as an aldehyde and aromatic $-OH$ (a *phenolic* group).

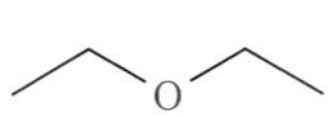

(24.17)

(24.18)

(24.19) 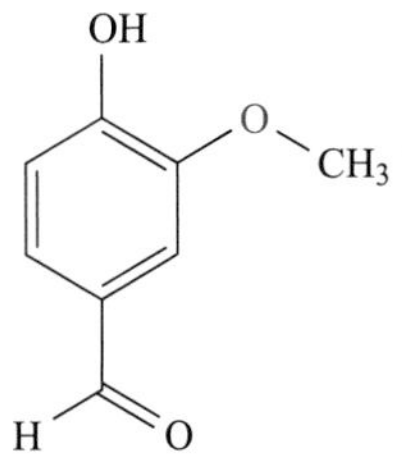                         (24.20)

A primary *amine* is characterized by having an $-NH_2$ group. Methylamine (**24.21**) is responsible for the smell of rotting fish, while diamine **24.22** is of commercial importance along with carboxylic acid **24.16** in the manufacture

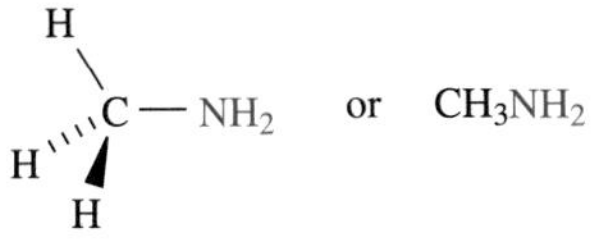

(24.21) 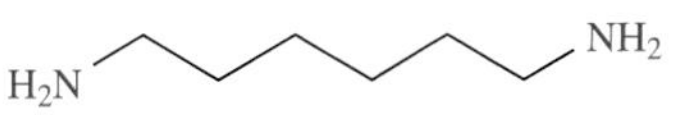                                        (24.22)

of Nylon 66. Secondary and tertiary amines possess the general formulae $R_2NH$ and $R_3N$ where the R groups may be the same or different.

The characteristic group of a simple *amide* is $-CONH_2$ and contains a carbonyl (C=O) group. The N atom may carry R substituents rather than H atoms and so the amide family also includes $RCONHR$ and $RCONR_2$:

The R groups may or may not be identical. Amides are immensely important in biology. Polypeptides (see Chapter 35) are amides and include proteins (high molecular weight polypeptides). Many polymers are polyamides, e.g. Nylon 66 and Kevlar (**24.23**). Smaller molecular amides of commercial importance include diethyltoluamide (**24.24**) which is the active component in insect repellants such as Autan (in the UK) and Off (in the US).

(24.23)  (24.24)

In *halogenated* compounds (*halo*-compounds), the functional groups are halogen atoms (F, Cl, Br or I). Halogenoalkanes are given frequent publicity because they include CFCs (chlorofluorocarbons, see Box 25.3). Chlorinated organics include common laboratory solvents such as $CH_2Cl_2$ and $CHCl_3$. Chloroethene (**24.25**, also called vinyl chloride) is of commercial importance as the precursor to the polymer PVC (polyvinyl chloride). The insecticide DDT (**24.26**) is a derivative of trichloroethane and also incorporates chlorinated aromatic substituents. DDT was first found to be an effective insecticide in 1939 but in the decades following its introduction and widespread use, many species of insect became resistant to DDT. In addition, DDT is highly toxic to fish. It is now banned as an insecticide.

The functional group in an *acid chloride* is $-COCl$ and contains a carbonyl (C=O) group. These compounds are very reactive; acetyl chloride (**24.27**) is a useful synthetic reagent in the laboratory (see Chapter 33).

In a *nitrile*, the functional group is $-C\equiv N$. Acetonitrile, $CH_3CN$, is a valuable laboratory solvent. Commercially important nitriles include acrylonitrile (**24.28**) which is the precursor to polyacrylonitrile, a synthetic fibre with trade names Orlon and Acrilan.

The *nitro*-functional group is $-NO_2$; nitromethane (**24.29**) has some uses as a laboratory solvent. In order that N obeys the octet rule, the correct

(24.25)

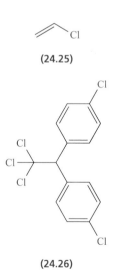

(24.26)

(24.27)      (24.28)      (24.29)

valence bond representation of the bonding in $CH_3NO_2$ is shown in the resonance pair:

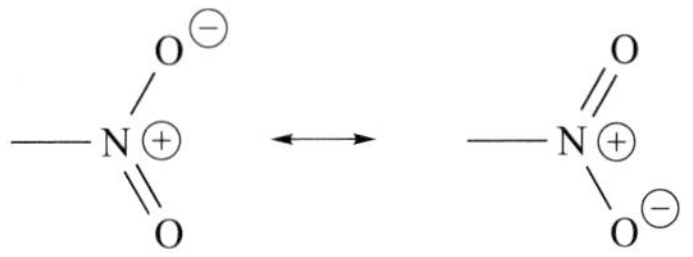

Compare this with the bonding schemes we drew for $[NO_3]^-$ in worked example 7.3. A range of compounds used as explosives possess nitro functionalities, e.g. trinitrotoluene (TNT, **24.30**).

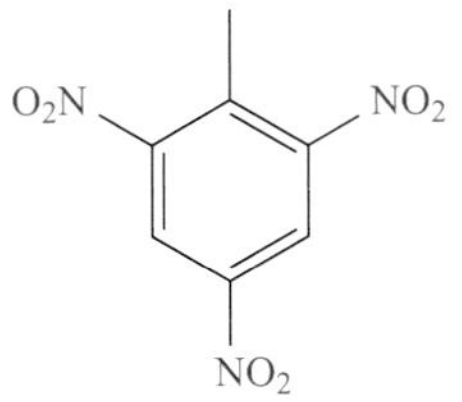

|  |  |  |
|:---:|:---:|:---:|
| **(24.30)** | **(24.31)** | **(24.32)** |

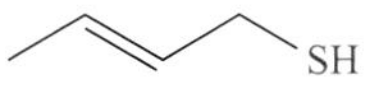

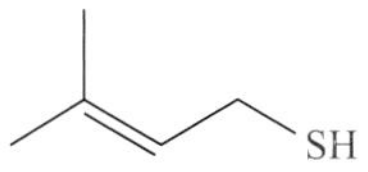

The functional group in *thiols* is −SH, and one of their characteristic properties is a foul odour. The defensive mechanism that skunks have evolved is to emit a foul-smelling spray that contains thiols **24.31** and **24.32**.

## 24.4    Hydrocarbon frameworks and nomenclature

We introduce nomenclature for different classes of organic compounds as we require it in later chapters. However, at this stage, we must become familiar with the nomenclature of *straight* and *branched* carbon chains. The names of derivative molecules are based upon those of the carbon chains that make up their backbones. Although we use recommended IUPAC nomenclature in this book, many *trivial names* are still commonly used. Names such as *acetone* and *acetic acid* are so firmly embedded in a chemist's vocabulary that it would be unrealistic to ignore them. Box 24.2 contains information on widely used laboratory chemicals for which trivial (rather than systematic) names are generally used.

IUPAC: see Section 1.2

### Aliphatic hydrocarbon families

Aliphatic hydrocarbons belong to one of three families: alkanes, alkenes or alkynes (see Figure 24.2). The name of a compound immediately tells you to which family the compound belongs:

- -ane (saturated hydrocarbon, alkane);
- -ene (unsaturated hydrocarbon with a C=C bond, alkene);
- -yne (unsaturated hydrocarbon with a C≡C bond, alkyne).

### Straight chain alkanes

The word 'straight' is somewhat misleading: see Box 24.3 and Section 24.9

The name for a *straight chain alkane* indicates the number of C atoms in the chain. Table 1.8 lists the prefixes that are used to indicate the number of C atoms in a chain. Table 24.1 lists several straight chain alkanes, giving

**THEORETICAL AND CHEMICAL BACKGROUND**

## Box 24.2 Trivial names for common laboratory chemicals

The IUPAC recognizes certain trivial names and among them are the ones given below; the abbreviation Me stands for a methyl group (see Section 24.7). The examples include many chemicals used in routine laboratory work.

Acetone    Acetic acid    Acetaldehyde    Acetyl chloride    Acetamide    Acetylene    Acetonitrile

Formic acid    Formaldehyde    Formamide    Dimethylformamide (DMF)    Ethylene glycol    18-crown-6

Toluene    Phenol    Aniline    Pyridine    Monoglyme    Diglyme

**Table 24.1** Nomenclature for selected straight chain alkanes, general formula $C_nH_{2n+2}$. The structural formulae for $n \geq 3$ are shown in abbreviated form.

| Name | Molecular formula | Structural formula |
|------|-------------------|--------------------|
| Methane | $CH_4$ | |
| Ethane | $C_2H_6$ | |
| Propane | $C_3H_8$ | |
| Butane | $C_4H_{10}$ | |
| Pentane | $C_5H_{12}$ | |
| Hexane | $C_6H_{14}$ | |
| Heptane | $C_7H_{16}$ | |
| Octane | $C_8H_{18}$ | |
| Nonane | $C_9H_{20}$ | |
| Decane | $C_{10}H_{22}$ | |

molecular and structural formulae. These hydrocarbons are members of a *homologous series* in which adjacent members differ by only a single $CH_2$ (*methylene*) group.

### Branched chain alkanes

The *root name* of a branched alkane is derived from the longest *straight* chain of carbon atoms that is present in the compound. Remember that a so-called 'straight' chain may not *look* straight (see Box 24.3). This is emphasized in Figure 24.3 where the structure of the branched alkane drawn at the top of the figure is considered. There are three possible straight chains, shown in blue in Figure 24.3. One of these chains contains seven C atoms, while the other two blue chains contain only six. Hence, the name for the branched chain shown in Figure 24.3 is based on the name for an alkane with seven

---

**THEORETICAL AND CHEMICAL BACKGROUND**

### Box 24.3 'Straight' chains in carbon chemistry

The tendency of carbon to form C–C bonds means that carbon chemistry often involves chains and rings of C atoms. The words 'straight' and 'linear' are commonly used to describe chains of atoms. A straight chain could be represented in a manner suggesting that the C atoms are joined together in a strictly linear fashion. For example, hexane might be shown as:

$$H-\overset{\overset{\displaystyle H}{|}}{\underset{\underset{\displaystyle H}{|}}{C}}-\overset{\overset{\displaystyle H}{|}}{\underset{\underset{\displaystyle H}{|}}{C}}-\overset{\overset{\displaystyle H}{|}}{\underset{\underset{\displaystyle H}{|}}{C}}-\overset{\overset{\displaystyle H}{|}}{\underset{\underset{\displaystyle H}{|}}{C}}-\overset{\overset{\displaystyle H}{|}}{\underset{\underset{\displaystyle H}{|}}{C}}-\overset{\overset{\displaystyle H}{|}}{\underset{\underset{\displaystyle H}{|}}{C}}-H$$

However, each C atom is $sp^3$ hybridized (tetrahedral) and so better representations of the 'straight' chain are:

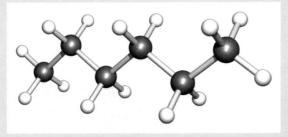

Additionally, rotation about the C–C bonds permits the chain to 'curl up':

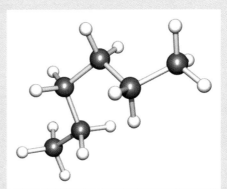

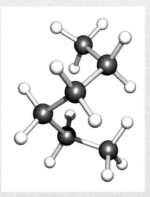

The term 'straight' means only that the C atoms are bonded in a continuous sequence, and tells us nothing about the *conformation* of the chain. When the sequence of atoms in the chain is not continuous, the chain is said to be 'branched'. We return to *conformation* in Section 24.9.

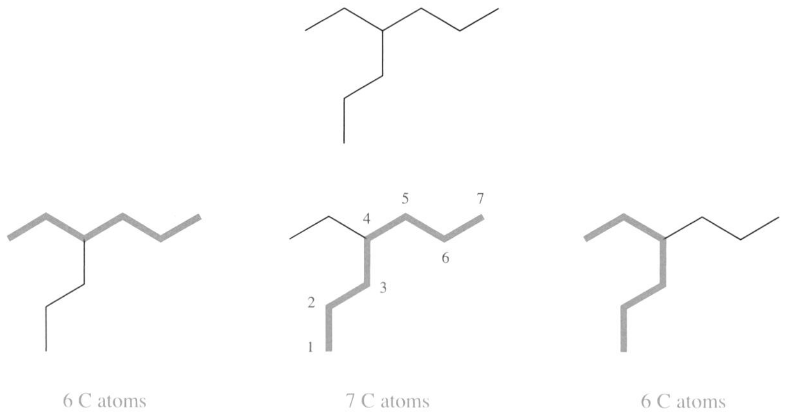

**Fig. 24.3** The IUPAC name for an organic compound is based on the longest *straight* chain in the molecule. The diagram shows a branched alkane containing nine C atoms. There are three ways to trace out a straight chain, and these are shown in blue in the lower diagrams along with the number of C atoms in the chain.

6 C atoms          7 C atoms          6 C atoms

Methyl, ethyl, propyl and butyl groups: see Section 24.7

C atoms: *heptane*. The longest chain is numbered from one end, and this is chosen to give the *lowest position numbers for the side chains*. For the alkane in Figure 24.3, the side chain is attached to atom C(4), irrespective of the end from which we begin to number the 7-membered chain. The name for a side chain (a *substituent*) is derived for the parent alkane and the general name is an *alkyl group*: *methyl* is a $CH_3$ (Me) group and *ethyl* is a $C_2H_5$ (Et) group. The correct name for the hydrocarbon in Figure 24.3 is 4-ethylheptane. Note two features about this name:

- there is a hyphen between the position number and the name of the substituent;
- there is *no* space between the substituent name and the root name.

If more than one substituent is present, their names are given in alphabetical order, each with its respective position number. The prefixes di-, tri- and tetra- are used to show the presence of more than one substituent of the same type: *dimethyl* means 'two methyl groups' and *tetraethyl* means 'four ethyl groups'. The alphabetical ordering of the alkyl groups takes priority over the multipliers, e.g. 4-ethyl-3,3-dimethyloctane, *not* 3,3-dimethyl-4-ethyloctane. Table 24.2 lists examples; remember that in the structural formulae, a single line drawn as a branch means a *methyl* substituent.

> The basic rules for ***naming a branched alkane chain*** are summarized as follows:
>
> 1. Find the longest straight chain in the molecule; this gives the root name.
> 2. The longest chain is numbered from one end, chosen to give the lowest position numbers for the substituents.
> 3. Prefix the root name with the substituent descriptors, and if more than one is present, give them in alphabetical order.

## Straight chain alkenes

When one double bond is introduced into a hydrocarbon chain, the name of the compound must indicate:

- the number of C atoms in the chain, *and*
- the position of the alkene functionality.

**Table 24.2** Structural formulae and names for selected branched chain alkanes.

| Structural formula | Compound name |
|---|---|
| | 2-methylbutane |
| | 2,2-dimethylpentane |
| | 2,4-dimethylhexane |
| | 4-ethyl-3-methylheptane (*not* 3-methyl-4-ethylheptane) |
| | 3,4-dimethylnonane |

The arrangement of the four groups attached to a C=C bond is fixed and this gives rise to (*E*)- and (*Z*)-isomers.[§] Thus, not only must the name of the alkene carry information about the position of the C=C bond in the chain and so distinguish between *constitutional isomers*, it must also indicate how the chain is arranged with respect to the C=C bond and so distinguish between *stereoisomers*. For example, placing a C=C bond in the middle of a 6-carbon chain leads to two isomers:

(*E*)-isomer            (*Z*)-isomer

Constitutional isomers: see Section 24.7

Stereoisomers: see Sections 6.11 and 24.8

The carbon chain is numbered so that the position of the double bond is described by the lowest possible site number. In structure **24.33**, the chain has five C atoms and the root name is *pent-*. The chain is numbered from the end nearest to the C=C bond, i.e. the right-hand end in structure **24.33**. The compound is pent-2-ene *not* pent-3-ene. Now we have to consider stereoisomers. The substituents attached to the C=C bond in **24.33** are arranged on *opposite* sides of the C=C unit, and therefore the full name of the compound is (*E*)-pent-2-ene. In structure **24.34**, the eight C atoms give the root name *oct-* and atom numbering begins at the left-hand end of the chain because the C=C bond is closest to this end. The compound is oct-3-ene. Now for the isomer prefix: the substituents are arranged on *opposite* sides of the C=C unit, and so **24.34** is (*E*)-oct-3-ene. Stereoisomers of pent-2-ene and oct-3-ene are shown in Figure 24.4.

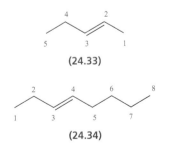

(24.33)

(24.34)

---

[§] The labels *E* and *Z* originate from the German words *entgegen* (opposite) and *zusammen* (together).

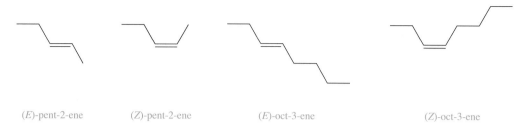

(E)-pent-2-ene          (Z)-pent-2-ene          (E)-oct-3-ene                    (Z)-oct-3-ene

**Fig. 24.4** Naming stereoisomers of alkenes. (Z)- derives from the German word *zusammen* meaning 'together' and (E)- comes from *entgegen* meaning 'opposite'.

More generally, the (Z)- or (E)-relationship between substituents attached to double bonds is determined by using a *sequence rule of preferred atoms or groups*. This rule assigns priorities to groups based on atomic number. If priorities cannot be assigned on this basis (e.g. in an isotopically labelled compound), then assignments should be made on the basis of atomic mass. Consider the alkene:

For the left-hand C atom, the priority is I > Br because the atomic number of I is greater than that of Br. For the right-hand C atom, the priority is Br > Cl. Hence, the stereochemical descriptor for the compound is based on the relationship between I and Br, i.e. a (Z)-isomer.

The following two alkenes differ only in one subsituent: $CH_3$ versus $CMe_3$:

(Z)-isomer                                    (E)-isomer

The different stereochemical descriptors arise as follows. Look first at the atoms attached *directly* to the atoms of the C=C bond. Based on atomic number, the substituent with the highest priority is Br. Now look at the substituents on the other atom of the C=C bond. In each compound, C atoms are attached directly to the C atom of the double bond. Therefore, no priorities can be made based on these atoms. Move on to the next atoms in the substituents and, based on the atomic numbers of these atoms, assign priorities to the substituents:

The assignment of (Z)- and (E)-labels for these compounds then follows.

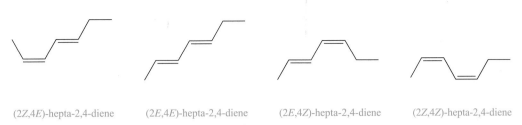

(2Z,4E)-hepta-2,4-diene      (2E,4E)-hepta-2,4-diene      (2E,4Z)-hepta-2,4-diene      (2Z,4Z)-hepta-2,4-diene

**Fig. 24.5** Stereoisomers of hepta-2,4-diene. Each carbon chain is numbered from the left-hand end. The prefixes contain both numerical *and* (Z) or (E) descriptors to ensure that there is no ambiguity.

We look further at the assignment of priorities to substituents when we discuss *R* and *S* notation for chiral compounds in Section 24.8.

### Dienes and trienes

> A *diene* contains two C=C bonds; a *triene* contains three C=C bonds.

When two C=C bonds are present in the hydrocarbon chain, the hydrocarbon is called a *diene*; this is a particular type of alkene. The rules for naming a diene follow those outlined above, but include the use of two site numbers and the prefix 'di-', e.g. buta-1,3-diene. Similarly, a *triene* contains three alkene functionalities. Going from one to two C=C bonds in a straight chain may result in a greater number of possible stereoisomers. Whereas hept-2-ene possesses two stereoisomers, hepta-2,4-diene has four (Figure 24.5).

> An *allene* contains two C=C bonds which share a common C atom (−C=C=C−).

An *allene* is a particular type of diene in which the two double bonds are attached to the same C atom. The simplest example is propadiene. Structure **24.35** illustrates that propadiene is non-planar and this is a consequence of the formation of two mutually orthogonal π-bonds (see end-of-chapter problem 24.1).

$$\begin{array}{c} H \\ \diagdown \\ C=C=C\diagup^{\text{''}H}_{H} \\ \diagup \\ H \end{array}$$

**(24.35)**

### Alkynes

The rules for naming *alkynes* are similar to those for alkenes, although no stereoisomers are possible because each C atom in the C≡C unit is linear. Numbering of the carbon chain begins at the end nearest to the alkyne functionality. The names for alkynes **24.36** and **24.37** are hex-2-yne and but-1-yne respectively.

**(24.36)**

**(24.37)**

| 24.5 | **Primary, secondary, tertiary and quaternary carbon atoms** |

A tetrahedral carbon atom is often designated as *primary*, *secondary*, *tertiary* or *quaternary* depending on the number of other C atoms to which it is connected.

> A *primary* C atom is attached to one other C atom in an organic group.
> A *secondary* C atom is attached to two other C atoms in organic groups.
> A *tertiary* C atom is attached to three other C atoms in organic groups.
> A *quaternary* C atom is attached to four other C atoms in organic groups.

For example, ethane (Table 24.1) contains two primary C atoms. Propane (**24.38**) contains one secondary and two primary C atoms. In 2-methylpropane (**24.39**), there are one tertiary C atom and three primary C atoms, and in 2,2-dimethylpropane (**24.40**), the central C atom is quaternary and the remaining four are primary.

Secondary    Primary    Tertiary    Quaternary

(24.38)      (24.39)      (24.40)

The distinction between the different types of centres is particularly important when it comes to reactivity patterns of aliphatic compounds as we shall discuss later.

## 24.6   Isomerism

Earlier in the book (Sections 6.11 and 23.5), we discussed different types of isomers in inorganic compounds. In this chapter, we have so far introduced two types of isomerism in hydrocarbons. Figure 24.6 summarizes isomerism in organic compounds in general. We can divide isomers into two types:

• *constitutional isomers* have different connections of atoms;
• *stereoisomers* are isomers that differ in the arrangement of their atoms in space.

Examples of constitutional isomers are 2-methylbutane (Table 24.2) and pentane. Further examples are shown in the left-hand side of Figure 24.6. Stereoisomers are of two types: *enantiomers* and *diastereoisomers*. Enantiomers are mirror images of each other. Diastereoisomers are *all* stereoisomers that are not enantiomers, e.g. (*Z*)- and (*E*)-but-2-ene (Figure 24.6) are diastereoisomers. We return to stereoisomerism in Section 24.8.

**Fig. 24.6** A summary of isomerism in organic compounds. The examples shown are restricted to acyclic (non-cyclic) structures.

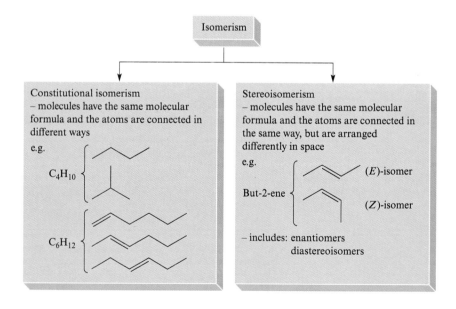

Isomerism

Constitutional isomerism
– molecules have the same molecular formula and the atoms are connected in different ways
e.g.

$C_4H_{10}$

$C_6H_{12}$

Stereoisomerism
– molecules have the same molecular formula and the atoms are connected in the same way, but are arranged differently in space
e.g.

But-2-ene    (*E*)-isomer

(*Z*)-isomer

– includes: enantiomers
         diastereoisomers

**24.7**     **Constitutional isomerism**

### Alkanes

The alkanes $CH_4$, $C_2H_6$ and $C_3H_8$ have no constitutional isomers. (Draw their structures and confirm that there can be no ambiguity about the way in which the atoms are connected.) For an acyclic alkane of formula $C_nH_{2n+2}$ with $n \geq 4$, more than one structural formula can be drawn and the number of possibilities increases dramatically with increasing $n$ (Figure 24.7). Figure 24.8 shows constitutional isomers of $C_6H_{14}$. A mixture of these isomers is used as a solvent often referred to as *hexanes*. Compositions of mixtures of this type can be analysed using gas chromatography as described in Box 24.4.

### Alkyl substituents

*Methyl* and *ethyl* substituents are derived from $CH_4$ (methane) and $C_2H_6$ (ethane) and there are no isomers of these alkyl groups:

$$CH_4 \longrightarrow -CH_3 \qquad\qquad C_2H_6 \longrightarrow -C_2H_5$$

Methane     Methyl (Me)        Ethane     Ethyl (Et)

In forming a propyl group from propane, there are two possibilities, and the two isomeric propyl groups are distinguished in their names and abbreviations:

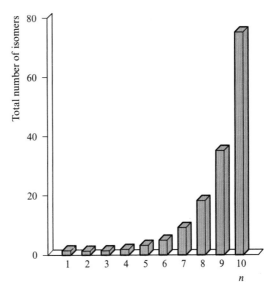

In going from butane (which has two constitutional isomers) to a butyl group, four isomers can be formed. They are distinguished in their names

**Fig. 24.7** The number of isomers (constitutional + enantiomers) for an alkane of formula $C_nH_{2n+2}$ increases dramatically with $n$. The presence of one structure of each of methane, ethane and propane is counted to be one isomer for consistency with the other data. Enantiomers are discussed in Section 24.8.

COMMERCIAL AND LABORATORY APPLICATIONS

## Box 24.4 Gas chromatography

'40–60 Petroleum ether' is not an ether at all, but is a fraction of petroleum with a boiling point range of 40–60 °C (313–333 K). The components of 40–60 petroleum ether are alkanes and the mixture can be analysed by using the technique of *gas chromatography* (GC). The trace opposite shows the result of passing a sample of 40–60 petroleum ether through a gas chromatograph. Each peak corresponds to a particular hydrocarbon and the relative areas of the peaks indicate the relative amounts of each compound in the mixture. The retention time of a fraction depends on the molecular weight and whether the chain is straight or branched. Typical components of 40–60 petroleum ether are isomers of pentane and hexane.

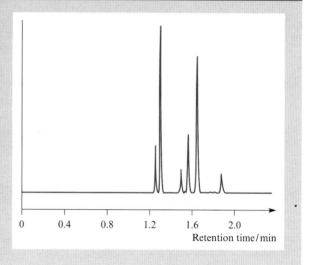

*Related information*: Box 24.7

and abbreviations:

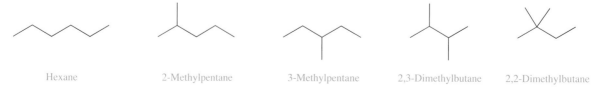

For longer chains, the situation becomes increasingly complicated.

**Fig. 24.8** Constitutional isomers of $C_6H_{14}$; only isomers with acyclic structures are included.

**24.8**

## Stereoisomerism and diastereoisomers

### Asymmetric carbon atoms and enantiomers

A compound that is **not superposable** on its mirror image is **chiral**.

A compound that is *non-superposable* on its mirror image is described as being *chiral*. We have already met chirality in Chapter 23 when we discussed enantiomers of coordination complexes. However, many chiral compounds contain an *asymmetric carbon atom*, i.e. a carbon atom with four different groups attached (Figure 24.9). The object and its mirror image are *enantiomers* and are *stereoisomers* (see Figure 24.6). Enantiomers of a compound possess *identical chemical properties* and differ *only* in their interactions with other chiral objects. Hands and gloves are chiral: your right hand will fit into a right-handed glove but not into a left-handed glove. The real test for chirality is non-superposability of the mirror image on the original object: the two structures in Figure 24.9 cannot be superimposed upon one another.

An **asymmetric carbon atom** has four different groups attached to it; in structural representations a * may be used to denote an asymmetric carbon atom:

An asymmetric carbon atom is also called a **stereogenic** or **chiral centre**.

> **Enantiomers** are mirror images of each other and are non-superposable. Enantiomers are stereoisomers.

There are some quick tests that allow you to check if a molecule is likely to be chiral, but these tests are *not infallible*:

- The presence of an asymmetric carbon atom is widely used as a test for chirality. *However*, it must be remembered that many chiral compounds are known which do not contain an asymmetric carbon atom (e.g. see Section 23.5). More importantly, many compounds containing *two or more* asymmetric carbon atoms are *not chiral*.

**Fig. 24.9** An asymmetric carbon atom has four different groups attached to it. The diagram shows a molecule that contains an asymmetric carbon atom. The molecule is 'looking' into a mirror. The mirror image is non-superposable on the original molecule. They are *enantiomers*.

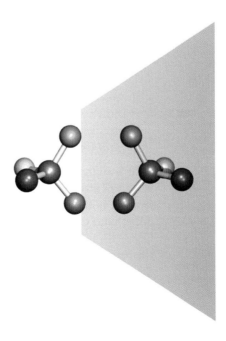

**Fig. 24.10** (a) Isomers A and B are enantiomers of 3-methylhexane. The carbon atom at the centre of the molecule is a chiral centre (an asymmetric carbon atom). The enantiomers are related to each other by reflection through the mirror plane, and are non-superposable. This is made clearer in diagram (b) where we begin with isomer B and rotate it about the vertical C−C bond indicated. The resultant structure is *not* equivalent to isomer A. Colour code: C, grey; H, white. There is no substitute for model building or using computer modelling programs to help you to confirm this point.

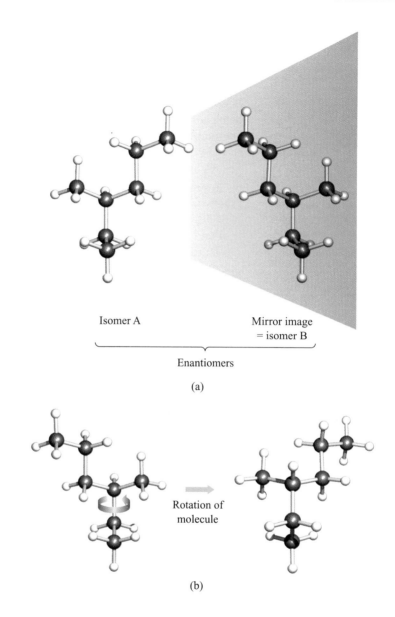

Isomer A | Mirror image = isomer B

Enantiomers

(a)

Rotation of molecule

(b)

- Look for a plane of symmetry (see Box 24.5). Compounds *with* a plane of symmetry are *not chiral*. However, the reverse is not true: some types of molecule which do not possess a plane of symmetry are, nonetheless, *achiral* for reasons of molecular symmetry.

Achiral = non-chiral

In 3-methylhexane, one of the C atoms (the central one in Figure 24.10a) has an H atom, a methyl group, an ethyl group and a propyl group attached to it. It is therefore an asymmetric carbon atom. The two enantiomers of 3-methylhexane are shown in Figure 24.10a. Figure 24.10b shows that isomer B cannot be converted back to isomer A by a simple rotation, and this confirms that the enantiomers are not superposable. Look carefully at the perspective in Figure 24.10.

If the position and nature of a substituent destroys a plane of symmetry in a molecule, then the act of substitution usually causes the molecule to become chiral. For alkanes that already possess a large number of

## THEORETICAL AND CHEMICAL BACKGROUND

## Box 24.5 Planes of symmetry

Molecular symmetry is usually described in terms of *symmetry operations* such as *axes of rotation* and *planes of symmetry*.

If a molecule is symmetrical with respect to a plane that is drawn through the molecule, then it is said to contain a plane of symmetry. A molecule may contain more than one plane of symmetry. Consider 1,1,2-tribromoethene (shown on the left below) and 1,2-dibromoethene (on the right):

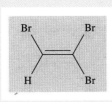

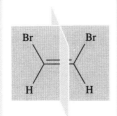

The molecule and the plane of symmetry both lie in the plane of the paper. There is only one plane of symmetry.

The molecule lies in the plane of the paper. One plane of symmetry is in the plane of the paper. The second plane is perpendicular to it.

None of the following molecules contains a plane of symmetry, and all three compounds are chiral:

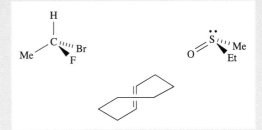

Note, however, that in only one example is there an asymmetric C atom. Compare these results with the 'tests for chirality' given in the text.

constitutional isomers, this significantly increases the total number of isomers. In Figure 24.7, *all* possible enantiomers were included in the total number of isomers.

| Worked example 24.1 | *Chiral and achiral alkanes* |

**Explain why heptane and 4-methylheptane are achiral, whereas 3-methylheptane is chiral.**

The heptane molecule contains a plane of symmetry passing through the atoms of the central methylene (CH$_2$) group. In the diagrams on the next page, the plane of symmetry is shown in yellow. The carbon atom through which the plane of symmetry passes is C(4). When a methyl group is substituted at this position, its presence does not destroy the plane of symmetry. Neither heptane nor 4-methylheptane is chiral. However, placing the Me group at the 3-position destroys the plane of symmetry and 3-methylheptane is chiral. In this chiral compound, one C atom has four different groups attached to it: an H, an Me group, an Et group and a Bu group:

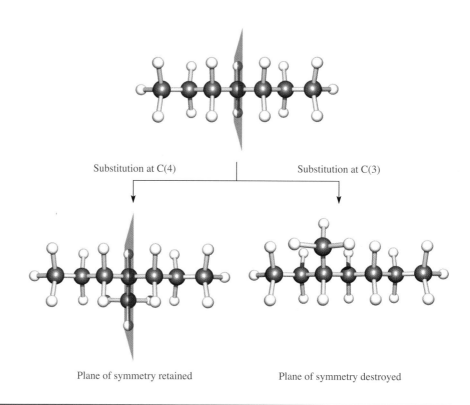

Substitution at C(4)                    Substitution at C(3)

Plane of symmetry retained              Plane of symmetry destroyed

## Specific rotation and notation for enantiomers

A material that is able to rotate the plane of polarization of a beam of transmitted plane-polarized light is said to possess *optical activity*.
The material is *optically active*.

One of the simplest ways of distinguishing between enantiomers involves their interaction with *polarized light*. Plane-polarized light is a form of light in which the electromagnetic wave lies in a single plane. When plane-polarized light interacts with a chiral molecule, the plane in which the polarized light wave lies changes. In effect, this plane rotates around the axis of propagation of the light (Figure 24.11). A convenient way of characterizing a chiral compound is by measuring the amount of rotation of this plane. The rotation, $\alpha$, may be measured in a polarimeter. In practice, the amount of rotation depends on the wavelength of the light, temperature and the

**Fig. 24.11** One enantiomer of a chiral compound rotates plane-polarized light through a characteristic angle, $\alpha^\circ$. The instrument used is a polarimeter. The direction of rotation indicated (a clockwise rotation if one views the light as it emerges from the polarimeter tube) is designated $+\alpha^\circ$. The other enantiomer of the same compound rotates the plane of polarized light in the opposite direction through an angle $-\alpha^\circ$.

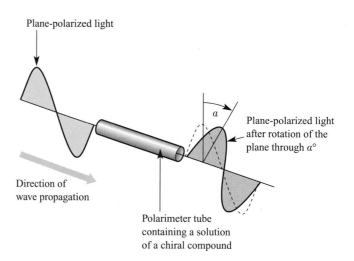

Plane-polarized light

Plane-polarized light after rotation of the plane through $a^\circ$

$a$

Direction of wave propagation

Polarimeter tube containing a solution of a chiral compound

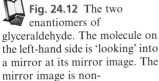 **Fig. 24.12** The two enantiomers of glyceraldehyde. The molecule on the left-hand side is 'looking' into a mirror at its mirror image. The mirror image is non-superposable on the initial molecule. Colour code: C, grey; O, red; H, white.

amount of compound in solution. The *specific rotation*, $[\alpha]$, for a chiral compound in solution is defined in equation 24.1.

$$[\alpha] = \frac{\alpha}{c \times \ell} \tag{24.1}$$

where:

$\alpha$ = observed rotation in degrees

$c$ = concentration (in $g\,cm^{-3}$)

$\ell$ = path length of the solution in the polarimeter (in dm)

*Note that these units are non-SI*

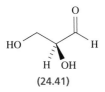

**(24.41)**

**(24.42)**

The *specific rotation* of enantiomers is equal and opposite.

It is very convenient to use light of a common wavelength for such measurements, and very often one of the lines from the atomic emission spectrum of sodium is used. This emission is known as the sodium D line, and the specific rotation at this wavelength ($\lambda = 589\,nm$) is denoted as $[\alpha]_D$. The specific rotation of enantiomers is equal and opposite. For example, the two enantiomers of glyceraldehyde (**24.41**, **24.42** and Figure 24.12) have $[\alpha]_D$ values of $+11°$ and $-11°$. This leads to a common way of distinguishing between enantiomers by stating the sign of $[\alpha]_D$. The two enantiomers of glyceraldehyde are denoted (+)-glyceraldehyde and (−)-glyceraldehyde. Sometimes (+) and (−) are denoted by the prefixes *dextro-* and *laevo-* (derived from Latin for right and left) and indicate a right-handed and left-handed rotation of the light respectively. These are often abbreviated to *d* and *l*. The $+/-$ or *d/l* notations refer to the *observed* rotation of the light and are temperature-, concentration- and wavelength-dependent. They give *no information about the absolute configuration* of the chiral compound.

The convention for labelling chiral carbon atoms and assigning the *absolute configuration* (i.e. the *exact* structure of a compound) uses *sequence rules* (also called the *Cahn–Ingold–Prelog* notation). These are the same sequence rules that we described for labelling (*Z*)- and (*E*)-isomers of alkenes. The four groups attached to a chiral carbon atom are prioritized according to atomic number, the *highest priority being assigned to the highest atomic number*. Thus for the halogens F, Cl, Br and I, I has the highest priority and F has the lowest. An H atom has the lowest priority of any atom or group of atoms. The molecule is then viewed down the C–X vector, with X facing away from you where X has the *lowest* priority. The enantiomers can now be labelled according to whether there is a clockwise (*rectus*, *R* label) or anticlockwise (*sinister*, *S* label) sequence of prioritized substituents taking the highest priority first. As an example, consider the molecule CHFClBr. The lowest priority atom is H and the priority sequence is Br > Cl > F > H.

Vladimir Prelog (1906–1998). © *The Nobel Foundation.*

Draw out the enantiomers so that the structural diagrams can be viewed along the C–H vectors:

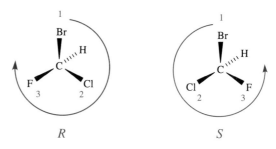

In the left-hand diagram, the clockwise sequence means that the asymmetric carbon atom is labelled $R$, and in the right-hand diagram, the anticlockwise sequence gives rise to the label $S$. These labels tell us the *absolute configuration* of the molecule. If the substituent consists of a group of atoms, e.g. $CH_3$ or OH, the priority of the group depends on the atomic number of the atom attached to the asymmetric carbon atom. Thus, OH has a higher priority than $CH_3$. If the two groups have the *same* first atom, e.g. $CH_3$ and $CO_2H$, then the priorities of the groups are determined by looking at the second atom. Therefore, based on the atomic number of the second atom in the group, $CO_2H$ has a higher priority than $CH_3$. Glyceraldehyde (**24.41** and **24.42**) illustrates a final priority problem. The asymmetric carbon atom has two C-attached groups: $CH_2OH$ and CHO. In each case, the next atom is O, but in $CH_2OH$ there is a C–O bond, and in CHO there is a C=O bond. The double-bonded O is counted as *two* O substituents and takes the higher priority. Thus, to assign an absolute configuration to enantiomer **24.41**, the steps are:

> The *absolute configuration* of a molecule specifies the *exact* arrangement of substituents around an asymmetric carbon atom.

- identify the asymmetric C atom;
- find the substituent with the *lowest* priority – in **24.41**, this is H;
- draw the structure of the molecule as viewed down the C–H bond;
- assign priorities to the OH, $CH_2OH$ and CHO groups: $OH > CHO > CH_2OH$;
- assign the $R$ or $S$ label:

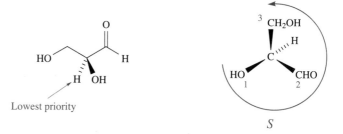

---

**Worked example 24.2**    *Assigning an absolute configuration*

Assign an absolute configuration to the enantiomer of 1-aminopropan-2-ol with the structure:

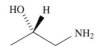

The chiral centre is the C atom with four different groups attached and these groups are H, OH, $CH_3$ and $CH_2NH_2$.

The substituent with lowest priority is H.

Redraw the structure looking down the C−H bond:

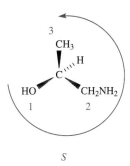

Assign priorities: OH > $CH_2NH_2$ > $CH_3$.

Assign the R or S label:

*S*

---

We stated earlier that the observed specific rotation gave no information about absolute configuration. This means that there is no connection between the (+) or (−) and R or S labels. Thus, some compounds with an S-configuration are (+) while others are (−), e.g. (S)-(−)-methylsuccinic acid but (S)-(+)-2-aminobutanol. Similarly for compounds with an R-configuration.

## Diastereoisomers

**Diastereoisomers** are stereoisomers that are not mirror images of each other.

We are now familiar with the idea that if a molecule contains a chiral centre, it possesses a special pair of stereoisomers called enantiomers. If a molecule contains *more than one chiral centre*, the situation becomes more complicated and leads us to another class of stereoisomers called *diastereoisomers*.

Consider 3-chloro-3,4-dimethylhexane (**24.43**). The molecule contains two asymmetric carbon atoms marked by asterisks in structure **24.43**. Taking structure **24.43** into three dimensions, we can draw structure **A** in Figure 24.13. Its mirror image is isomer **B**. **A** and **B** are enantiomers and are also stereoisomers. We can draw another stereoisomer of 3-chloro-3,4-dimethylhexane and this is shown as isomer **C** in Figure 24.13. The mirror image of **C** is **D**. Isomers **C** and **D** are a pair of enantiomers. Because there are two asymmetric carbon atoms in 3-chloro-3,4-dimethylhexane, there are two pairs of enantiomers: **A** and **B**, and **C** and **D**. But what of the relationship between other pairs of stereoisomers: for example, are **A** and **C** related? While **A** and **C** are non-superposable, they are *not* mirror images of one another, and so they are *not* enantiomers. These stereoisomers are called *diastereoisomers*. Similarly, pairs **A** and **D**, **B** and **C**, and **B** and **D** are diastereoisomers.

If you look carefully at the stereoisomers of 3-chloro-3,4-dimethylhexane in Figure 24.13, you will be able to see the following important points, which are key to distinguishing between enantiomers and diastereoisomers:

(24.43)

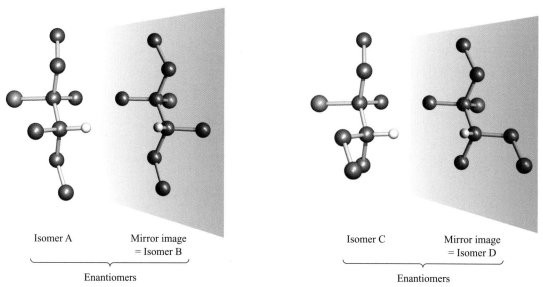

Isomer A   Mirror image = Isomer B        Isomer C   Mirror image = Isomer D

Enantiomers                                Enantiomers

**Fig. 24.13** 3-Chloro-3,4-dimethylhexane contains two asymmetric carbon atoms. There are two enantiomeric pairs (**A** and **B**, and **C** and **D**), and four diastereoisomeric pairs (**A** and **C**, **B** and **C**, **A** and **D**, and **B** and **D**). Colour code: C, grey; H, white; Cl, green. Only one H atom (that attached to an asymmetric C atom) is shown.

- **A** and **B** are mirror images (enantiomers) and have the opposite configurations at *both* asymmetric carbon centres.
- **C** and **D** are mirror images (enantiomers) and have the opposite configurations at *both* asymmetric carbon centres.
- Molecules in each of pairs **A** and **D**, **A** and **C**, **B** and **C**, **B** and **D** are not mirror images and are diastereoisomers. They have opposite configurations at *one* asymmetric carbon atom, but the *same* configuration at the other asymmetric carbon atom.

The stereoisomers of 3-chloro-3,4-dimethylhexane are distinguished by *R,S*-notation in the same way that we applied this to chiral molecules with one asymmetric carbon centre. Additional site-number labels must be added to make the name unambiguous. Take each centre in turn in Figure 24.13 and confirm the labels:

- **A** = (3*S*,4*S*)-3-chloro-3,4-dimethylhexane;
- **B** = (3*R*,4*R*)-3-chloro-3,4-dimethylhexane;
- **C** = (3*S*,4*R*)-3-chloro-3,4-dimethylhexane;
- **D** = (3*R*,4*S*)-3-chloro-3,4-dimethylhexane.

Remember than a pair of enantiomers must have opposite configurations at every chiral centre.

Although we have discussed diastereoisomers in conjunction with enantiomers in this section, we must reiterate the point we made in Section 24.6. *Diastereoisomers are **any** stereoisomers that are not enantiomers.* Examples of diastereoisomers that do not possess asymmetric carbon atoms are:

- (*Z*)- and (*E*)-oct-2-ene;
- (*Z*)- and (*E*)-1,2-dichloroethene;
- *cis*- and *trans*-$N_2F_2$ (structures **22.41** and **22.41**);
- *cis*- and *trans*-$[PtCl_2(NH_3)_2]$ (square planar complex, see Box 23.5).

## ENVIRONMENT AND BIOLOGY

### Box 24.6 Chirality and life

As you look around you in everyday life, you become aware that we live in a chiral world. On the macroscopic level, gloves and shoes have a handedness, the tendrils of vines coil in a particular direction, snail shells have a specific helical twist, and water goes down the plug-hole in a chiral, circular motion. At a molecular level, the body is optimized to use only *one* enantiomer or diastereomer of biological building blocks, and we explore this is detail in Chapter 35: *Molecules in nature*. You might think that using the 'wrong' diastereomer of a sugar (one that the body cannot metabolize) might provide an easy way to diet! However, the sweet taste of a sugar is also specific to the biologically occurring form.

image is non-superposable on the original molecule:

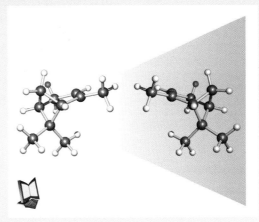

The structures below indicate the stereogenic carbon centres in each enantiomer of α-pinene:

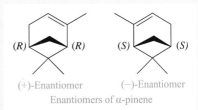

(+)-Enantiomer        (−)-Enantiomer
Enantiomers of α-pinene

The helical twist of the shell makes it chiral.
*Steve Gorton/Dorling Kindersley.*

The male gall wasp *Antistrophus rufus* provides a fascinating example of chiral recognition. The eggs of this gall wasp are laid in the stems of specific prairie plants. These plants contain a number of terpenes, including the chiral compound α-pinene. When a molecule of α-pinene 'looks' in a mirror, the mirror

The presence of the gall wasp larvae in their host plants affects the ratio of (+):(−)-enantiomers of terpenes, including α-pinene. Male gall wasps emerge in the spring from dead plant material, before the females. They are able to locate plant stems from which female wasps will emerge by recognizing plant matter with altered ratios of (+):(−)-enantiomers of specific terpenes. A 50:50 ratio indicates that no females will emerge. Terpenes are volatile compounds and possess characteristic smells. Studies show that the chiral recognition exhibited by the male gall wasps arises from olfaction, i.e. the sense of smell. [See: J. F. Tooker, W. A. Koenig and L. M. Hanks (2002) *Proceedings of the National Academy of Sciences*, vol. 99, p. 15486.]

## Racemates and resolution of enantiomers

A 50:50 mixture of two enantiomers of a compound is a **racemate** or **racemic mixture**.

A 50:50 mixture of two enantiomers of a compound is called a *racemate* or *racemic mixture*. Since the mixture contains equal amounts of the (+) and (−) enantiomers, it is designated (±). Alternatively, a *dl*-prefix can be used to indicate equal amounts of *d*- and *l*-enantiomers. A racemate is characterized by exhibiting *no specific rotation* because the (+) rotation of one enantiomer cancels out the (−) rotation of the other.

**(24.44)**

When a chiral compound is *resolved*, it is separated into its enantiomers.

Alkaloids: see Box 34.1

In some cases it is possible to separate a pair of enantiomers. The classic example was the separation of enantiomers of sodium ammonium tartrate (**24.44**) by Louis Pasteur. He observed that a carefully recrystallized sample of **24.44** consisted of two distinct crystalline forms which could be separated by hand-picking the crystals. Solutions of the two different types of crystals rotated the plane of polarized light in opposite directions, whereas a solution of the initial crystalline salt gave no rotation at all, i.e. it was optically inactive whereas the individual crystals were optically active but in different senses. These results were the start of our knowledge of chirality, and can be interpreted in terms of sodium ammonium tartrate consisting of a racemate which can be *resolved* into its component enantiomers.

The classical method of separating enantiomers involves conversion of the compound to a cation or anion, and forming a salt with a chiral counter-ion. Consider a chiral compound XA which can form a pair of enantiomeric cations $(+)$-$A^+$ and $(-)$-$A^+$. When treated with an enantiomerically pure anion $(-)$-$B^-$, a pair of *diastereoisomeric salts* is formed, $(+)$-A$(-)$-B and $(-)$-A$(-)$-B. Whereas *enantiomers* possesses *identical* physical properties because the relative arrangements of atoms in space are the same, *diastereoisomers* possess *different* physical properties because the relative arrangements of atoms in space are different (look carefully at Figure 24.13). Very often, pairs of diastereoisomers differ in their solubilities, and addition of $(-)$-$B^-$ to a racemate of $A^+$ results in the precipitation of only one of the diastereoisomeric salts, $(+)$-A$(-)$-B or $(-)$-A$(-)$-B. This process of separation of enantiomers is known as *resolution*. Chiral *resolving agents* are commercially available. Typical reagents that are used for the resolution of chiral compounds include anions such as **24.45**, or cations derived from *alkaloids* such as strychnine, **24.46**.

**(24.45)**

**(24.46)**

More modern methods of separation involve column chromatography (see Box 24.7) using a *chiral stationary phase*. The solid phase supports an enantiomerically pure species which interacts specifically with one enantiomer in a mixture of enantiomers of a compound, thereby removing it from the mixture and allowing one enantiomer to pass through the column. The second enantiomer can be removed in a later elution of the column. Once separated, enantiomers may *racemize*, i.e. revert to racemic mixtures.

The separation of enantiomers is especially important in the manufacture of drugs. A well-documented example is that of Thalidomide, prescribed in the 1960s to pregnant women as an anti-nausea drug. Thalidomide was used in the form of a racemate, but only one enantiomer provided the anti-nausea properties. The result of using the mixture of enantiomers was horrendous birth defects. Even if the correct enantiomer had been specifically administered, Thalidomide racemizes *in vitro* and side effects would again have been

## Box 24.7 Chromatography

Chromatography is a *separation* technique, and in Box 24.4 we considered *gas chromatography*. Another common technique is *column chromatography*. The components to be separated are *adsorbed* on to a *stationary phase* (a solid support) such as alumina or silica (see Section 22.6). The column is then *eluted* with a suitable solvent or solvent mixture (the *mobile phase*). The choice of solvent is made on its ability to separate the components by utilizing their differing solubilities and interactions with the stationary phase. Test separations are first carried out on a small scale using small, thin-layer chromatography plates. As the column is eluted with solvent, some components are removed from the stationary phase more readily than others, and separation of the mixture is therefore achieved.

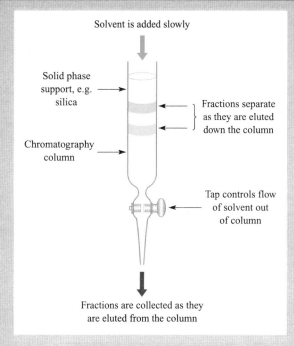

Solvent is added slowly

Solid phase support, e.g. silica

Chromatography column

Fractions separate as they are eluted down the column

Tap controls flow of solvent out of column

Fractions are collected as they are eluted from the column

*Related information*: Box 24.4.

observed. Many drugs are chiral, with only one enantiomer possessing the desired therapeutic properties. The second enantiomer may be inactive for the required therapy, active in a different way, or may produce undesired side effects. Table 24.3 gives examples of the activities of selected drugs.

### Meso compounds

We now return to a point we made near the beginning of this section: compounds with a plane of symmetry are *not chiral*. We consider the case of a molecule that contains two asymmetric carbon atoms and a plane of symmetry. The structure of tartaric acid is shown in **24.47** without any stereochemical information. Now consider the stereochemistry of the molecule. We appear to be able to draw two pairs of enantiomers: **24.48** and **24.49** make one pair, and **24.50** and **24.51** appear to make the second pair. The diagrams are simplified by omitting the carbon-attached H atoms:

HO—CH—CO$_2$H
HO—CH—CO$_2$H

**(24.47)**

HO$_2$C ⟍ OH
HO$_2$C ⁗OH

*(2S,2S)*

**(24.48)**

HO ⟍ CO$_2$H
HO⁗ CO$_2$H

*(2R,2R)*

**(24.49)**

HO$_2$C ⟍ OH
HO$_2$C OH

*(2R,2S)*

**(24.50)**

HO ⟍ CO$_2$H
HO CO$_2$H

*(2S,2R)*

**(24.51)**

However, inspection of the second pair of structures shows that these molecules are *identical*. By rotating **24.51** in the plane of the paper through

**Table 24.3** Examples of chiral drugs and their properties. The chiral centre is indicated by ∗.

| Drug | Structure | Effects of first enantiomer | Effects of second enantiomer |
|------|-----------|-----------------------------|------------------------------|
| L-Dopa | 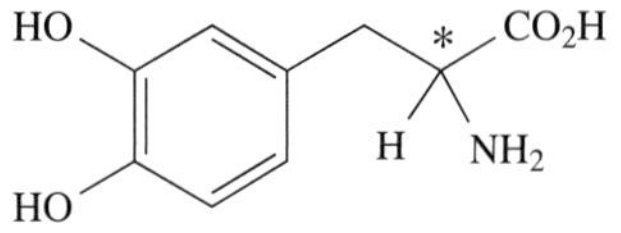 | (S)-(−)-enantiomer used to treat Parkinson's disease | (R)-(+)-enantiomer contributes to side effects |
| Indacrinone | 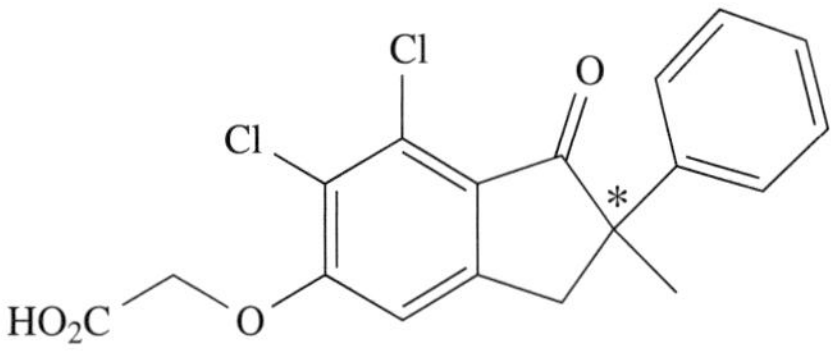 | (R)-(−)-enantiomer is a diuretic | (S)-(+)-enantiomer induces uric acid excretion |
| Penicillamine | 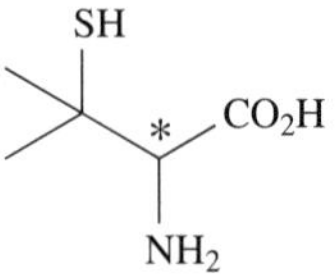 | (S)-(−)-enantiomer used to treat copper poisoning | (R)-(+)-enantiomer is toxic |
| Timolol | 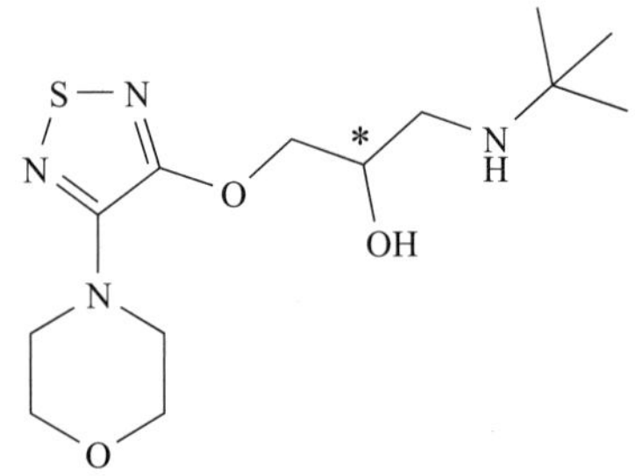 | (S)-(−)-enantiomer is used to treat angina and high blood pressure | (R)-(+)-enantiomer is used to treat glaucoma |

Data from: E. Thall (1996) *Journal of Chemical Education*, vol. 73, p. 481.

180°, we arrive at structure **24.50**:

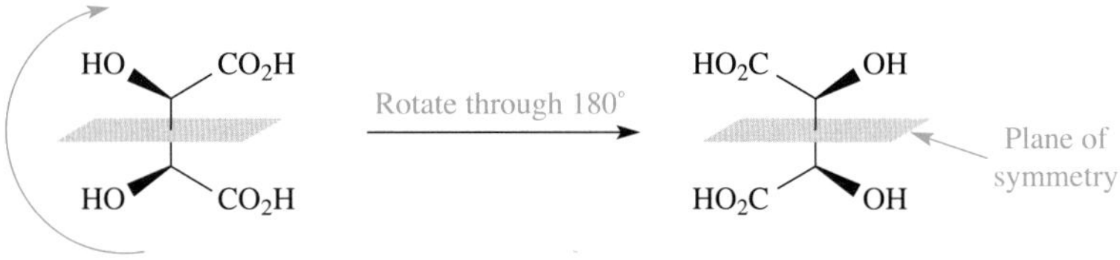

The crucial property of **24.50** (and **24.51**) is that it possesses a plane of symmetry. This makes it achiral, despite the fact that it contains two asymmetric carbon atoms. The form of tartaric acid with structure **24.50** is called a *meso form*, and tartaric acid therefore possesses one pair of enantiomers and one *meso* form. The *meso* form has different chemical and physical properties from the enantiomeric pair.

> A *meso* form of a compound contains chiral centres, but is achiral.

## 24.9 Conformation

### Staggered, eclipsed and skew conformations

> *Rotation* can usually occur about a single bond but not about a multiple bond.

Rotation about the internuclear axis of a single bond is usually possible, but rotation about multiple bonds is restricted because of the π-component. By looking back at Figure 4.18, you should be able to appreciate why this is.

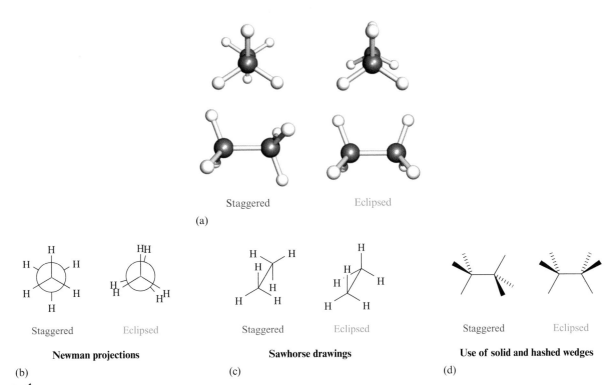

**Fig. 24.14** Conformers of $C_2H_6$. (a) 'Ball-and-stick' models of a molecule of $C_2H_6$ showing two views of the staggered and eclipsed conformers. Colour code: C, grey; H, white. Three methods are in common use to represent conformations. (b) In a *Newman projection*, atoms in front of the circle point towards you. (c) In a *sawhorse diagram*, the lower left-hand C atom is, by convention, closer to you. (d) Solid wedges and hashed lines indicate, respectively, atoms or groups that point towards you or away from you; the remaining lines are in (or close to being in) the plane of the paper.

In an acyclic alkane, rotation about *all* the single bonds in the molecule can occur. In $C_2H_6$, rotation about the C−H bonds does not affect the shape of the molecule, but rotation about the C−C bond alters the *relative* orientations of the two $CH_3$ groups. These different arrangements are called *conformers*.

The two conformers of $C_2H_6$ that represent the two extreme positions of the $CH_3$ groups with respect to one another are shown in Figure 24.14a. Between the *staggered* and *eclipsed* conformations lie an infinite number of *skew conformations*.

> *Conformers* of a molecule can be interconverted by rotation about single bonds.

### Newman projections, sawhorse drawings and the use of wedges

There are several schematic ways of representing conformations and these are shown in Figures 24.14b–d with $C_2H_6$ as the example.

In the *Newman projection* (Figure 24.14b), the stereochemistry is shown by looking along the C−C bond. Atoms attached to the C atom closest to you are drawn in front of a circle, and atoms attached to the more distant C atom lie behind the circle.

In a *sawhorse diagram* of $C_2H_6$ (Figure 24.14c), the C−C bond is drawn at an angle and, by convention, the C atom at the lower left-hand corner is nearer to you.

Stereochemistry: see Section 24.2

A third method makes use of the solid wedges and hashed lines with which we are already familiar. Solid wedges represent bonds pointing up from the paper (towards you), and hashed lines represent bonds pointing away from you.

Use of these schemes allows the stereochemistry of complicated molecules to be shown in a relatively unambiguous manner.

### Steric energy changes associated with bond rotation

Steric interactions: see Sections 6.8 and 6.9

When a $C_2H_6$ molecule is in a staggered conformation (Figure 24.14), the steric interactions between H atoms on adjacent C atoms are at a minimum. This represents the minimum *steric energy*. As rotation about the C−C bond occurs, steric interactions increase as the H atoms get closer together. Further rotation takes the H atoms further apart again and the steric interactions decrease. The steric energy varies with the degree of rotation as shown in Figure 24.15. The eclipsed conformation is the energy maximum. In a full 360° rotation about the C−C bond, the molecule passes through three identical eclipsed and three identical staggered conformations. The energy difference between the staggered and eclipsed conformations in $C_2H_6$ is $\approx 12.5 \, kJ \, mol^{-1}$. This is relatively small and means that, at 298 K, we can consider the $C_2H_6$ molecule to be freely rotating about the C−C bond.

If we replace the H atoms in Figure 24.15 by alkyl groups, a staggered conformation remains the most favourable, but the barrier to rotation increases. *Sterically demanding* substituents (i.e. bulky ones such as *tert*-butyl) hinder the rotation.

It is important to recognize that the ability of an alkane molecule to undergo C−C bond rotation means that the so-called 'straight chains' (an *extended conformation*) can in fact roll up into 'balls' (see Box 24.3). The flexibility of an alkane chain is best illustrated by using molecular models or computer molecular modelling programs.

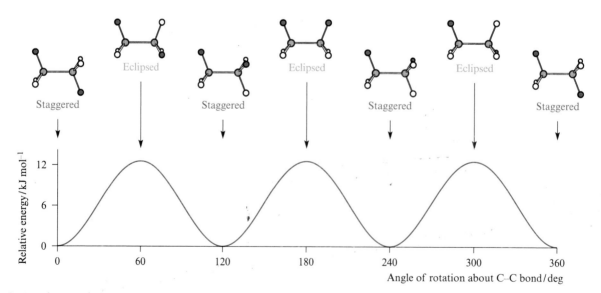

**Fig. 24.15** The relative steric energy of $C_2H_6$ changes as one $CH_3$ group rotates with respect to the other one about the C−C bond. Two of the six equivalent H atoms are marked in red to highlight their relationship throughout the rotation.

## Some general features of a reaction mechanism

Before we can discuss the reactivity of organic compounds, we must consider some general features of reaction mechanisms. Relevant topics that were introduced in Chapters 15 and 16 are:

- activation energy;
- reaction profiles;
- transition states and intermediates;
- rates of reaction;
- rate equations;
- elementary steps;
- rate-determining step;
- radical chain reactions;
- equilibrium constants;
- Brønsted acids and bases;
- $pK_a$ and $pK_b$ values.

### Bond cleavage

Bond cleavage (fission) can occur either **homolytically** or **heterolytically**.

In the *homolytic bond cleavage* (or fission) of an $X-Y$ bond, one electron is transferred to each of the two atoms and is represented by a 'half-headed' arrow. In equation 24.2, two radicals are formed as a result of homolytic bond cleavage.

$$X \overset{\frown}{-\!\!\!:\!\!\!-} Y \longrightarrow X^{\bullet} + Y^{\bullet} \tag{24.2}$$

The half-headed arrows are one type of *curly arrows* that are used throughout organic chemistry to show the movement of electrons in a reaction mechanism. The way that the arrow are drawn is very important: an arrow must start and end where the electron or electrons come from and go to. *Arrows should never be vaguely positioned; they always begin from a pair of electrons, or a single electron in the case of a radical such as $Cl^{\bullet}$.*

In *heterolytic bond cleavage*, the two electrons in the $X-Y$ bond are *both* transferred to the same atom and the transfer is represented by a 'full-headed' arrow (equation 24.3). A 2-electron transfer usually *generates* ions (equations 24.3 and 24.4), but this is not always the case. In equation 24.5, there is no change in the total number of ions.

$$X \overset{\frown}{-\!\!\!:\!\!\!-} Y \longrightarrow X^{\oplus} + Y^{\ominus} \tag{24.3}$$

$$\underset{}{\text{}} \overset{\frown}{\underset{\bullet\bullet}{:}} Cl \longrightarrow \overset{\oplus}{\underset{}{}} + Cl^{\ominus} \tag{24.4}$$

$$\underset{H}{\overset{H}{\underset{\text{}}{\overset{|}{\underset{\bullet\bullet}{:}O^{\oplus}}}}} \longrightarrow \underset{H}{\overset{H}{\underset{}{\overset{|}{O}}}} + H^{\oplus} \tag{24.5}$$

In equation 24.4, the pair of electrons from the $C-Cl$ bond becomes localized on the Cl centre. If we consider *all* the valence electrons around the Cl atom, the change is:

Similarly, in equation 24.5, the changes in the valence shell of the O atom as the pair of electrons from the H−O bond becomes localized on the O centre, leaving the H atom without electrons, can be summarized as:

These representations keep track of the electrons, and you may wish to include valence electrons in diagrams of this type. However, we shall, in general, only explicitly show pairs of electrons involved in the reaction (see the discussion of *nucleophiles* below). For example, attack by a lone pair of electrons on $H_2O$ is represented showing both lone pairs, or just the lone pair involved in attack:

If only one lone pair is shown on $H_2O$, do not forget that there are actually two! If the attacking species is an anion, lone pairs of electrons may be shown with the curly arrow starting at a lone pair. For example:

If only one of the lone pairs is shown, the representation is:

We use this representation in this book. Keep in mind that the *curly arrow indicates movement of a pair of electrons.*

One final point regards the way in which we show charges in organic mechanisms. For clarity, we draw the charge in a circle and position it by the atom that carries the charge.

*Homolytic bond cleavage (fission)* of a single bond X−Y means a 1-electron transfer to *each* atom and the formation of radicals X$^{\bullet}$ and Y$^{\bullet}$. A 1-electron transfer is represented by a 'half-headed' arrow

*Heterolytic bond cleavage (fission)* of a single bond X−Y means a 2-electron transfer to *either* X or Y, and (usually) the formation of ions. A 2-electron transfer is represented by a 'full-headed' arrow

The reactions on the previous page illustrate a single 2-electron transfer. In most reactions, a series of 2-electron transfers are coupled together in steps in the reaction mechanism. An important point to remember is that such a series of transfers *must* involve the *electron pairs* moving in the *same* direction. This may not be true for 1-electron transfers in radical reactions (look at equation 24.2). We illustrate these points in later chapters.

## Electron movement and orbitals

**(24.52)**

In equations 24.4 and 24.5, we show the electron pairs being transferred to atom centres. As we have seen in Chapters 3–5, electrons must be accommodated in orbitals and therefore an atom accepting electrons must possess a suitable, low-lying vacant molecular or atomic orbital. This concept has already been explored with respect to the formation of Lewis base adducts of $BF_3$, $BCl_3$ and $BBr_3$ (see Figure 22.6). The Lewis bases that we studied included pyridine (**24.52**) and are characterized by having a lone pair of electrons in a high-lying molecular orbital. The molecular orbital (MO) in $BX_3$ (X = halogen) that *accepts* the pair of electrons from the Lewis base is the lowest energy unoccupied MO (the LUMO) of the $BX_3$ molecule. At the end of Chapter 23, we discussed the formation of metal carbonyl complexes. We saw how a lone pair of electrons in the highest occupied molecular orbital (the HOMO) of CO is donated into a low-lying metal atomic orbital, and how the M−C bond is further stabilized by donation of electrons from a high-lying filled MO on the metal centre into the LUMO of the CO ligand (Figure 23.26). Both these examples illustrate the role of high-lying filled MOs (often the HOMO) and low-lying unoccupied MOs (often the LUMO) in bond formation.

Interactions between the HOMO and LUMO of species are crucial to the movement of electrons in organic reactions. Although we usually represent this movement by a curly arrow, you should bear in mind the orbital criteria. Electrons cannot be transferred from one atomic centre to another if there is no vacant orbital available, nor if there is a major mismatch in orbital symmetry or energy. We look further at the role of orbitals in later chapters.

HOMO = highest occupied molecular orbital

LUMO = lowest unoccupied molecular orbital

## Electrophiles and nucleophiles

An *electrophile* is electron seeking. A *nucleophile* donates electrons.

Heterolytic bond cleavage (equations 24.3–24.5) is the net consequence of a reaction with an electrophile or a nucleophile. An *electrophile* is 'electron loving' and seeks electrons. It is attracted to an electron-rich centre with negative or $\delta^-$ charge. A *nucleophile* is a species that can donate electrons and is attracted to a centre with a positive or $\delta^+$ charge. In terms of the orbital criteria discussed above:

- an electrophile accepts electrons into a low-lying vacant atomic or molecular orbital;
- a nucleophile donates electrons from a high-lying occupied atomic or molecular orbital.

Electrophiles include $H^+$, $Cl^+$, $Me^+$ and $NO_2^+$. Examples of nucleophiles are $H_2O$, $Cl^-$, and $NH_3$. More detailed discussion of electrophiles and nucleophiles comes in later chapters.

## SUMMARY

Chapter 24 begins our coverage of the chemistry of carbon compounds and we have summarized the general types of compound and important functional groups. A knowledge of basic nomenclature is needed so that we can build upon it as we encounter more complicated molecules. We have elaborated upon the introduction to isomerism given in Chapter 6, and have discussed constitutional isomers and stereoisomers, with particular emphasis on chiral compounds. It is important that you can recognize the various conformations that acyclic carbon chains can adopt; these affect properties that we discuss later. The final part of this chapter has pulled together some topics from earlier in the book that are needed for discussions of reaction mechanisms, and has introduced the concepts of 'curly arrow chemistry', electrophiles and nucleophiles. The list of terms below may look daunting, but you should become familiar with them now because we shall make constant use of them in the remaining chapters.

### Do you know what the following terms mean?

- hydrocarbon
- carbon backbone
- aliphatic
- acyclic
- cyclic
- saturated
- unsaturated
- alkane
- stereochemistry
- homologous series
- methylene group
- methyl group
- ethyl group
- propyl and isopropyl groups
- butyl, *sec*-butyl, isobutyl and *tert*-butyl groups
- primary carbon atom
- secondary carbon atom

- tertiary carbon atom
- quaternary carbon atom
- constitutional isomers
- stereoisomers
- (*E*)- and (*Z*)-isomers
- enantiomers
- asymmetric carbon atom
- plane of symmetry
- polarized light
- optically active
- specific rotation
- (+) and (−) notation
- *d* and *l* notation
- absolute configuration
- (*R*) and (*S*) notation
- diastereoisomers
- racemate (racemic mixture)

- (±)-notation
- resolution of a chiral compound
- racemization
- conformation
- staggered conformation
- eclipsed conformation
- skew conformation
- Newman projection
- sawhorse drawing
- steric energy
- extended conformation
- homolytic bond cleavage (fission)
- heterolytic bond cleavage (fission)
- HOMO
- LUMO
- electrophile
- nucleophile

### Do you know what the functional groups are in:

- an alkene?
- an allene?
- an alkyne?
- an alcohol?
- an aldehyde?

- a ketone?
- a carboxylic acid?
- a halo-compound?
- an ester?
- an ether?
- an amine?

- an amide?
- an acid chloride?
- a nitrile?
- a nitro-compound?
- a thiol?

### You should be able:

- to state what is meant by a *hydrocarbon* and to classify the different types of such compounds
- to distinguish between cyclic and acyclic compounds
- to discuss (with examples) the different geometries in which C atoms are found
- to relate suitable hybridization schemes to C atoms in different environments

- to identify functional groups given the structural formula of a compound
- to interpret a schematic structural formula (i.e. 'fill in the gaps' with atoms and interpret the geometry of each atom)
- to name straight chain alkanes up to 20 C atoms

- to name a branched alkane given the structural formula, and to interpret a name in terms of the structure
- to name a straight chain alkene or alkyne given the structural formula, and to interpret a name in terms of the structure
- to recognize primary, secondary, tertiary and quaternary carbon centres in a molecule
- to classify the types of isomerism found in organic compounds
- to discuss (with examples) constitutional isomerism among alkanes
- to discuss (Z)- and (E)-isomers among alkenes
- to identify asymmetric carbon atoms in a molecule
- to explain how enantiomers of a compound are related
- to give two 'tests' for chirality (are they infallible?)
- to discuss how specific rotation is measured and to explain the significance of (+) and (−) or *d* and *l* labels
- to state what is meant by *absolute configuration* and explain how (*R*) and (*S*) labels are assigned

- to explain what diastereoisomers are
- to describe what a racemate is and how it might be resolved
- to explain what the *meso*-form of a compound is and when it may arise
- to discuss what is realistically meant by a 'straight carbon chain'
- to describe the differences between staggered, eclipsed and skew conformations of a molecule such as ethane
- to draw Newman projections and sawhorse drawings, and to use wedges to illustrate the conformation of a given molecule
- to explain how steric energy is related to the conformation of a molecule
- to distinguish between homolytic and heterolytic bond fission
- to interpret the meanings of ⌢ and ⌢ arrows
- to understand the criteria for moving electrons along a 'curly arrow' pathway
- to distinguish between an electrophile and a nucleophile

## PROBLEMS

**24.1** (a) Give hybridization schemes appropriate for the C atoms in structures **24.5–24.8**. (b) Why is propadiene (**24.8**) not planar?

**24.2** Redraw the following compounds showing the C and H atoms, and give an appropriate hybridization scheme for each C atom:

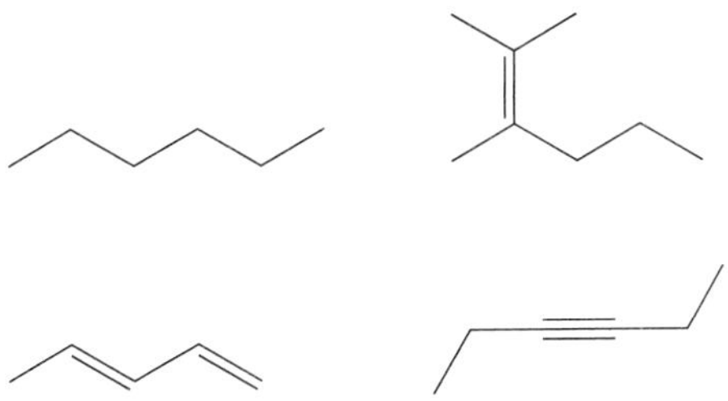

**24.3** Draw structural representations of the functional groups in the following types of compound: (a) amine; (b) alcohol; (c) carboxylic acid; (d) ether; (e) ester; (f) thiol; (g) chloroalkane; (h) nitro compound; (i) alkyne; (j) aldehyde; (k) ketone; (l) amide.

**24.4** Identify the functional groups in each of the following compounds:

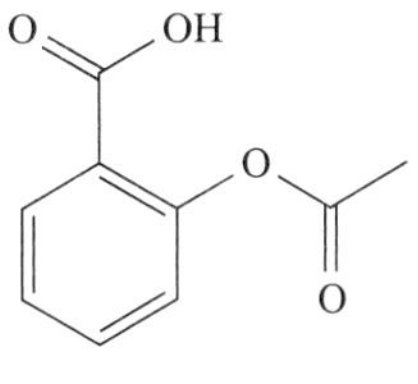

Aspirin

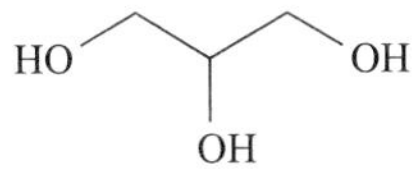

Glycerol

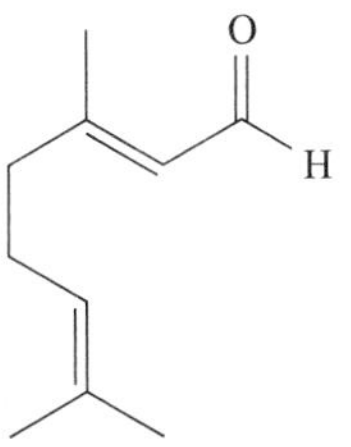

Citral-b (component in lemon grass oil)

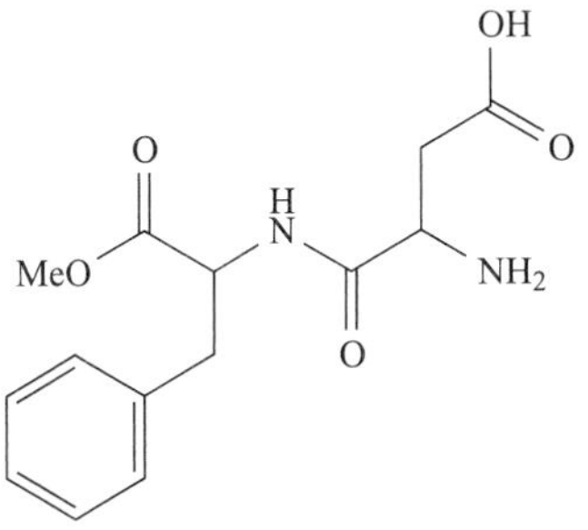

Aspartame (an artificial sweetner)

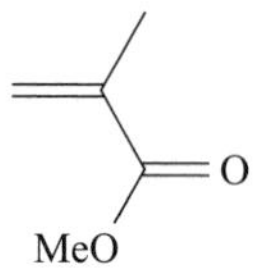

Methyl methacrylate
(used in the
manufacture of
Perspex)

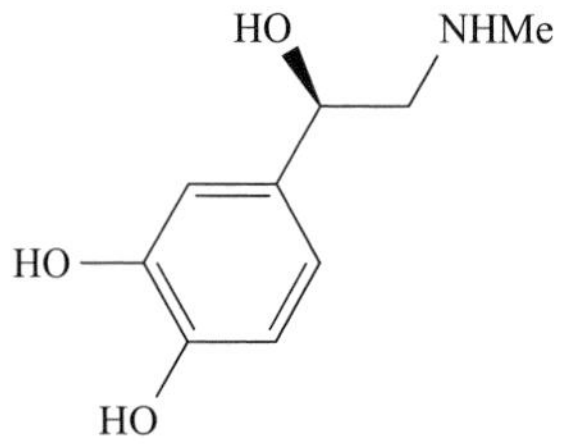

Adrenaline

**24.5** In each of the formulae in problem 24.4, 'fill in the gaps' with C and H atoms to give a full structural formula.

**24.6** In each of the formulae below, 'fill in the gaps' with C and H atoms to give a full structural formula:

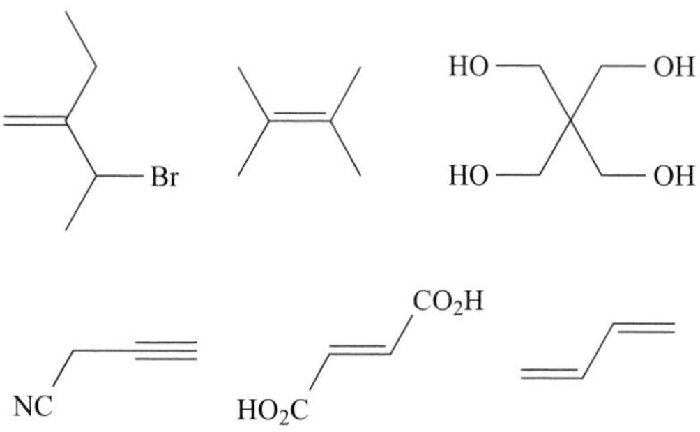

**24.7** Draw a schematic structural formula (i.e. no C and carbon-attached H atoms shown) for the following molecules:

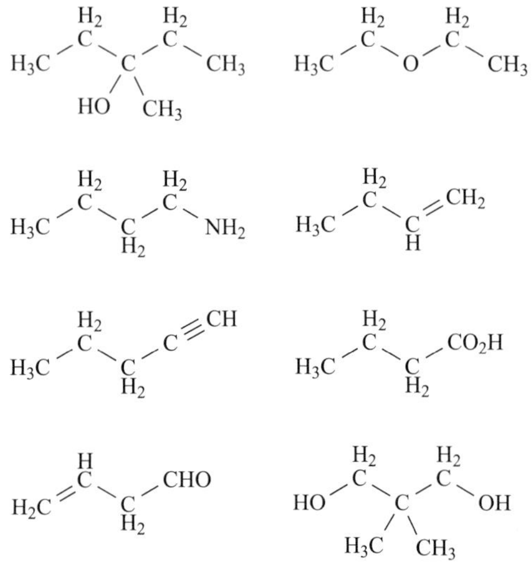

**24.8** Draw structural formulae for the following compounds:
(a) octane;
(b) 2-methylpentane;
(c) 3-ethylheptane;
(d) 2,2-dimethylpropane.

**24.9** Draw structural formulae for the following alkenes and distinguish between (*E*)- and (*Z*)-isomers where appropriate:
(a) hex-3-ene;
(b) but-2-ene;
(c) pent-1-ene;
(d) 3,4-dimethylhept-3-ene.

**24.10** Give names for the following compounds:

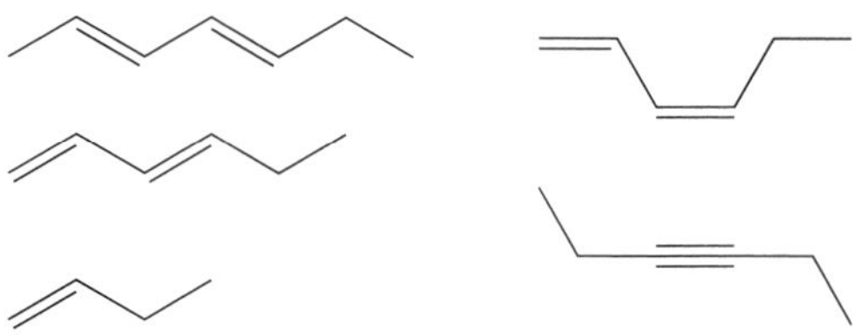

**24.11** (a) Draw structural formulae for the stereoisomers of octa-2,5-diene and name the isomers. (b) How many isomers does buta-1,3-diene possess? (c) Draw the structures of the acyclic isomers of $C_5H_{12}$.

**24.12** Draw the structures of the following compounds and identify the types of C centre (primary, secondary, tertiary or quaternary):
(a) butane;
(b) 2-methylpentane;
(c) 2,2,4,4-tetramethylhexane;
(d) 3-ethylpentane.

**24.13** Which of the following molecules contain chiral centres?

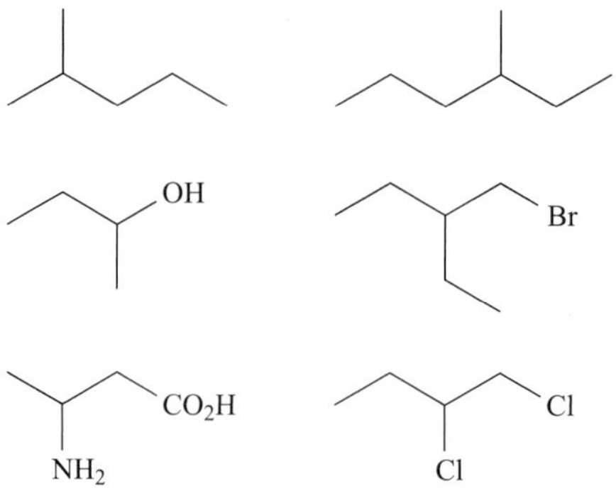

**24.14** Assign absolute configurations to the following enantiomers.

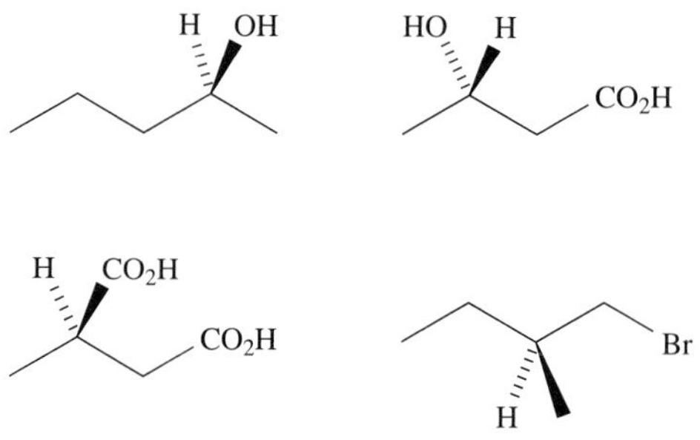

**Fig. 24.16** For problem 24.19.

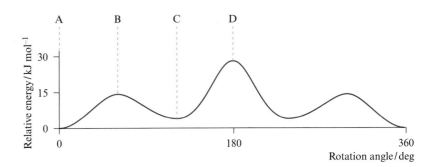

**24.15** Alanine has the structure:

H    NH₂

CO₂H

Explain what is meant by the notation (*S*)-(+)-alanine and (*R*)-(−)-alanine.

**24.16** 2,3-Dibromopentane has the structure:

Br

Br

Explain why this compound has enantiomers and diastereoisomers.

**24.17** Pentane-2,4-diol has the structure:

OH    OH

What stereoisomers does it possess?

**24.18** What isomers do the following molecules possess? Consider only acyclic structures.
(a) $C_4H_{10}$
(b) $C_4H_9Cl$
(c) $C_4H_6$
(d) $C_4H_8Cl_2$

**24.19** Figure 24.16 shows the variation in steric energy as rotation occurs around the C(2)−C(3) bond in butane. The energies are relative to a value of zero for a staggered conformation (angle = 0°). (a) Draw Newman projections of the conformation of butane at points A, B, C and D marked in the figure. (b) Explain the shape of the graph.

**24.20** Write out the products from the following steps in reaction mechanisms:

(a) Br — Br ⟶

(b) Cl ⟶

(c) Br ⟶

(d) H—O⁺—H ⟶

(e) ⁻O CN ⟶

**24.21** Write out the products from the following steps in reaction mechanisms in which bonding electron pairs are not explicitly shown:

(a) H—O⁺—H ⟶

(b) Cl⁻ ⊕ ⟶

(c) H—Br ⟶

(d) ⁻O Cl ⟶

## Additional problems

**24.22** In each of the following, choose the *most convenient and routine* experimental method appropriate.

(a) Suggest a method of distinguishing between the following isomers of $C_8H_{12}$.

(b) How might the following unsaturated hydrocarbons be distinguished?

(c) Suggest how 2,2-dimethylpropane and 2-methylbutane could be distinguished.

**24.23** The general formula of the 20 naturally occurring amino acids is drawn below.

$$H \overset{R}{\underset{NH_2}{\overset{|}{-C-}}} CO_2H$$

In glycine, R = H, in cysteine, R = $CH_2SH$, in aspartic acid, R = $CH_2CO_2H$, and in threonine, R = CH(OH)Me. (a) Which of these amino acids are chiral? (b) For each chiral molecule, draw a representation of the (R)-enantiomer. (c) What can you say about the chiral or achiral nature of the remaining 16 naturally occurring amino acids?

**24.24** A compound **A** (which has a foul smell) shows a parent ion in its mass spectrum at $m/z$ 76. The $^1H$ NMR spectrum of **A** shows signals with relative integrals as follows: $\delta$ 2.10 (singlet, 3H), 2.53 (quartet, 2H), 1.27 (triplet, 3H) ppm. The $^{13}C$ NMR spectrum exhibits three resonances. Elemental analysis for **A** gives C 47.3, H 10.6%. Compound **B** is an isomer of **A** and also has a foul smell; **B** also has three signals in its $^{13}C$ NMR spectrum. Suggest identities and structures for **A** and **B**.

# 25 Acyclic and cyclic alkanes

## Topics

- Acyclic and cyclic alkanes
- Ring strain
- Physical properties
- Industrial interconversions of hydrocarbons
- Reactivity
- Radical processes

Conformation: see Box 24.3
and Section 24.9

### 25.1 Cycloalkanes: structures and nomenclature

We are already familiar with the fact that saturated hydrocarbons are called *alkanes*, and that *acyclic alkanes* have the general formulae $C_nH_{2n+2}$. The carbon chains may be *straight* or *branched*, but these descriptions hide the fact that rotation around the C–C bonds leads to changes in conformation. Before considering the physical and chemical properties of alkanes in detail, we extend our general introduction to *cycloalkanes*.

### Structures and nomenclature of monocyclic alkanes

An *unsubstituted cycloalkane* with a single ring (*monocyclic*) has the general formula $C_nH_{2n}$. The compound name indicates the number of C atoms in the cyclic portion of the molecule. Compound **25.1** is cyclohexane and **25.2** is cyclopropane.

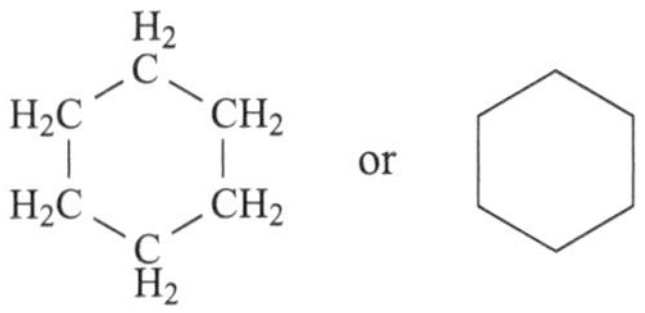

A *monocyclic compound* contains one ring.

(25.1)

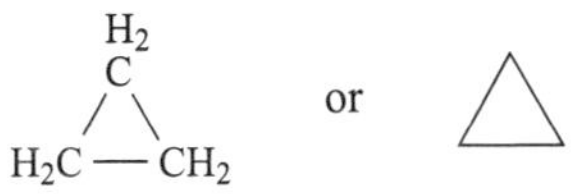

(25.2)

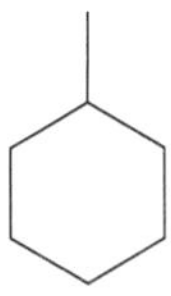

A *substituted cycloalkane* with only one substituent needs no position number. Since all the C atoms in the parent unsubstituted ring are identical, it is irrelevant which atom becomes substituted. Compound **25.3** is methylcyclohexane and **25.4** is ethylcyclopentane. When the ring has more than one alkyl substituent, it is named so that the position numbers of the substituents are *as low as possible*. This means that one substituent is taken to be at position C(1) and the ring is then numbered to place the second substituent at the lowest position number possible. For example, the

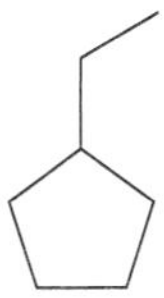

(25.3)          (25.4)

compound below is named 1,2-dimethylcyclohexane and *not* 1,6-dimethyl-cyclohexane:

and *not*

**(25.5)**

If the alkyl substituents are different, an alphabetical sequence of substituent names takes priority. Thus, **25.5** is 1-ethyl-3-methylcyclohexane (*not* 1-methyl-3-ethylcyclohexane).

## NMR spectra

NMR spectroscopy:
see Chapter 14

The use of NMR spectroscopy in structural assignment is extremely important as we have already seen. For example, one can distinguish between related acyclic and cyclic alkanes. In an *unsubstituted* cycloalkane $C_nH_{2n}$, all the C atoms are equivalent. This contrasts with their acyclic counterparts, and the $^{13}C$ NMR spectra of pairs of cyclic and acyclic alkanes allow the compounds to be readily distinguished. Figure 25.1 shows the $^{13}C$ NMR spectra of pentane and cyclopentane. Proton spectra of unsubstituted cycloalkanes are also simple: cyclopentane exhibits a singlet at $\delta$ 1.50 ppm, cyclohexane a singlet at $\delta$ 1.42 ppm and cyclooctane a singlet at $\delta$ 1.52 ppm. Even though rings with more than three C atoms are non-planar, changes in conformation (see Section 25.2) lead to the protons being equivalent on the NMR spectroscopic timescale.

## Structures of bicyclic alkanes

Although we shall not have a great deal to say about the chemistry of bicyclic compounds, their structures do appear quite often in this book, especially

**Fig. 25.1** The proton decoupled $^{13}C$ NMR spectrum of pentane exhibits three signals; the three C environments are labelled *a–c* in the structural diagram of pentane. In cyclopentane, all the C atoms are equivalent and the $^{13}C$ NMR spectrum shows one signal.

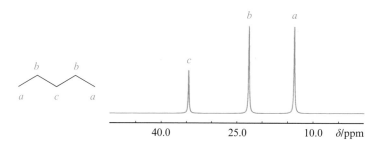

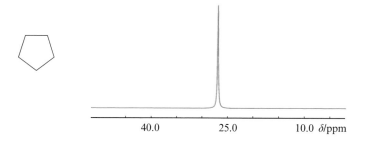

*Bicyclic* compounds contain two rings and can be *fused*, *bridged* or *spirocyclic*.

with respect to molecules that occur in nature. The framework structures can be considered in terms of three classes of bicyclic cycloalkanes. The first group consists of molecules in which two rings share a common C–C bond. These *fused bicyclics* are represented by structure **25.6**. Members of the second group of bicyclics have *bridged structures*, e.g. **25.7** and **25.8**; bridged structures share more than one atom between the rings. When ring linkage is through a single, shared tetrahedral C atom, the compound is *spirocyclic* as in **25.9**.

|  |  |  |  |
|:---:|:---:|:---:|:---:|
| (25.6) | (25.7) | (25.8) | (25.9) |

α-Pinene and β-pinene are examples of commercially important bicyclics. They are derived from turpentine and play a major role in the fragrance industry; see also Box 24.6.

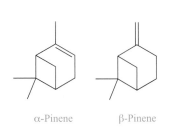

α-Pinene        β-Pinene

| 25.2 | **Cycloalkanes: ring conformation and ring strain** |

### Ring conformation

Whereas there is usually rotation about the C–C bonds in an acyclic alkane, such rotation is restricted in cycloalkanes, especially in those with small rings. In cyclopropane (Figure 25.2), there is *no* rotation possible about the C–C bonds and the $C_3$ ring is constrained to being planar. This imposes *eclipsed* conformations along each C–C bond. This unfavourable conformation contributes to cyclopropane being a *strained* ring and its chemistry reflects this. We return to ring strain in terms of orbital overlap later in the section.

As the $C_n$ ring size increases, *partial* rotation about the C–C bonds becomes possible. If the $C_n$ rings of cyclobutane and cyclopentane adopted planar conformations, adjacent $CH_2$ groups would be eclipsed. However, if partial rotation occurs about one or more C–C bonds, the rings become non-planar, non-bonded H···H interactions are reduced, and the energy of the system is lowered. In the lower energy conformations, the rings of cyclobutane and cyclopentane are *folded* (or *puckered*). The conformation of cyclopentane is also referred to as an *envelope conformation* because of its relationship to an open envelope as shown in structure **25.10**. In Figure 25.3, we compare planar and folded conformations for cyclobutane and cyclopentane. The folded conformations are close to those observed experimentally. Despite possessing some degree of flexibility, the ring in cyclobutane is strained. Larger

(25.10)

 **Fig. 25.2** Two views of the structure of cyclopropane. The ring is planar. The right-hand diagram emphasizes that the conformation along each C–C bond is eclipsed. Colour code: C, grey; H, white.

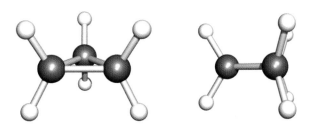

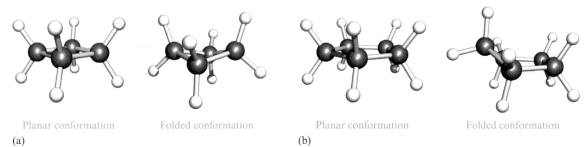

Planar conformation          Folded conformation          Planar conformation          Folded conformation

(a)                                                    (b)

**Fig. 25.3** Comparing planar and folded (experimentally favoured) conformations of (a) cyclobutane and (b) cyclopentane. In the planar forms, the conformation is eclipsed along each C–C bond. An energetically more favourable conformation is achieved if the ring is folded. This is the result of partial rotation about the C–C bonds. Colour code: C, grey; H, white.

■▷
Chair and boat conformers:
see Section 9.8

rings have greater degrees of flexibility. Partial rotation about the C–C bonds in cyclohexane results in 'flipping' (ring inversion) between the *chair, twist-boat* and *boat conformations* (Figure 25.4a). The relative energies of the conformers are $E(\text{chair}) \ll E(\text{twist-boat}) < E(\text{boat})$. Cyclohexane is a liquid at 298 K and the chair form is favoured to an extent of $\approx 10^4 : 1$. The H···H interactions are greater in the boat form than in either the twist-boat or chair conformers.

In the chair form of cyclohexane (Figure 25.4b), there are two different sites for the H atoms and these are called *axial* and *equatorial*. The origins of the names can be appreciated by looking at the Newman projection in Figure 25.4b; the projection is viewed along two of the six C–C bonds in cyclohexane. The presence of different sites leads to the possibility of constitutional isomerism, but it is important not to confuse isomers and conformers. There are two *conformers* of methylcyclohexane (Figure 25.5a) because as the $C_6$-ring flips from one chair conformation to another, the methyl group switches between axial and equatorial positions (Figure 25.6). Figure 25.5b shows the three conformers of the two isomers of 1,3-dimethylcyclohexane. The space-filling diagrams of methylcyclohexane in Figure 25.5a illustrate that the equatorial site is less sterically crowded than the axial site. The equatorially substituted conformer is energetically favoured although the energy difference is small, about 7 kJ mol$^{-1}$. Similarly,

**Fig. 25.4** (a) The chair and boat forms of cyclohexane are the two extreme conformations. The chair is energetically favoured. The twist-boat is an intermediate conformation. (b) A 'ball-and-stick' representation of the chair conformation of $C_6H_{12}$; colour code: C, grey; H, white. A schematic form of the same structure that shows the axial and equatorial positions of the H atoms. The Newman projection of the chair conformation of $C_6H_{12}$ is viewed along the two C–C bonds that are highlighted in bold in the middle diagram.

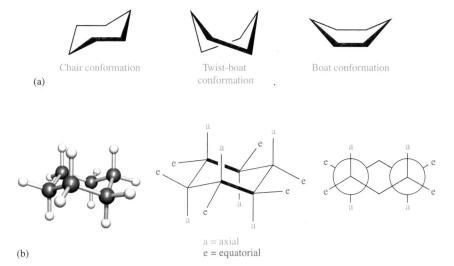

Chair conformation          Twist-boat conformation          Boat conformation

(a)

(b)

a = axial
e = equatorial

**Fig. 25.5** (a) There are two *conformations of the chair form* of methylcyclohexane: one with the Me group in an axial site, and one with it in an equatorial site. The space-filling diagrams illustrate that the Me group experiences more steric crowding when in an axial position. Colour code: C, grey; H, white. (b) There are two *isomers* of 1,3-dimethylcyclohexane in which the Me groups are either *cis* (same side of the ring) or *trans* (opposite sides of the ring). There are two *conformers* of the *cis*-isomer. The *equatorial,equatorial* conformer is sterically favoured. (The *cis* and *trans* labels are recommended in preference to the (*Z*) and (*E*) labels used for stereoisomers of alkenes.)

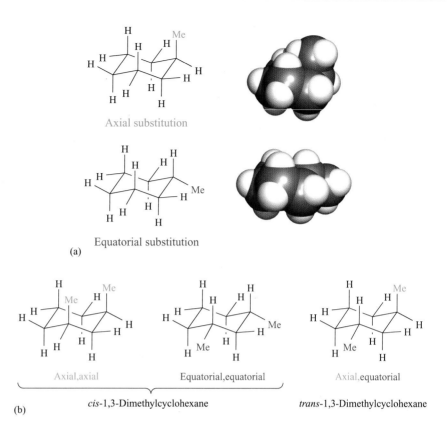

Axial substitution

Equatorial substitution

(a)

Axial,axial          Equatorial,equatorial          Axial,equatorial

*cis*-1,3-Dimethylcyclohexane          *trans*-1,3-Dimethylcyclohexane

(b)

of the structures shown in Figure 25.5b, the *equatorial,equatorial* conformer of 1,3-dimethylcyclohexane is sterically favoured.

> The chair ***conformation of cyclohexane*** is preferred over the twist-boat or boat:
>
> $$E(\text{chair}) \ll E(\text{twist-boat}) < E(\text{boat})$$
>
> Equatorial sites in the chair conformer of cyclohexane are less sterically crowded than axial sites.

### Ring strain

In a cycloalkane, each C atom forms four single bonds and can be considered to be $sp^3$ hybridized. The ideal angle for $sp^3$ hybridization is 109.5° as is found in $CH_4$. *Acyclic* alkanes have H–C–C, C–C–C and H–C–H bond

**Fig. 25.6** When the cyclohexane ring 'flips' from one chair conformation to another chair conformation, axial and equatorial sites are interchanged.

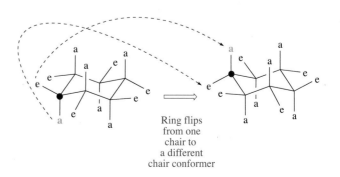

Ring flips from one chair to a different chair conformer

**Table 25.1** Internal C–C–C bond angles in selected cycloalkanes and enthalpies of combustion. Enthalpies of combustion of acyclic alkanes and the difference between successive values as $n$ increases in $C_nH_{2n+2}$ (see text). Values all refer to gaseous compounds.

| Cycloalkane | $\angle$CCC / ° | $\Delta_cH^o$(g, 298 K) / kJ mol$^{-1}$ | $\Delta_cH$(298 K) per mole of CH$_2$ units / kJ mol$^{-1}$ | Alkane | $\Delta_cH^o$(g, 298 K) / kJ mol$^{-1}$ | Difference between successive values of $\Delta_cH$(g, 298 K) / kJ mol$^{-1}$ |
|---|---|---|---|---|---|---|
| | | | | $C_2H_6$ | −1560 | |
| $C_3H_6$ | 60 | −2091 | −697 | $C_3H_8$ | −2219 | −659 |
| $C_4H_8$ | 88 | −2746 | −687 | $C_4H_{10}$ | −2877 | −658 |
| $C_5H_{10}$ | 105 | −3320 | −664 | $C_5H_{12}$ | −3536 | −659 |
| $C_6H_{12}$ | 111 | −3953 | −659 | $C_6H_{14}$ | −4195 | −659 |
| $C_7H_{14}$ | 113–115 | −4637 | −662 | $C_7H_{16}$ | −4854 | −659 |
| $C_8H_{16}$ | 115–117 | −5310 | −664 | $C_8H_{18}$ | −5512 | −658 |
| $C_9H_{18}$ | 115–118 | −5985 | −665 | $C_9H_{20}$ | −6172 | −660 |
| $C_{10}H_{20}$ | 114–118 | −6640 | −664 | $C_{10}H_{22}$ | −6830 | −658 |

angles that are close to 109.5°. For *cycloalkanes*, the picture is different. The C–C–C bond angle depends on the size of the ring. But caution is needed! Representations of the carbon frameworks of cycloalkanes (e.g. **25.1–25.4**) often make them look flat when this is not the case; the one exception is cyclopropane. The C–C–C bond angles in cycloalkanes $C_nH_{2n}$ with $3 \leq n \leq 10$ are listed in Table 25.1. In some rings, experimental angles span a small range of values. Two points stand out:

- the C–C–C bond angles increase from 60° for $n = 3$ to values of around 116° for $n \geq 7$;
- rings with $n = 3$ and 4 have C–C–C bond angles that are *significantly* smaller than the ideal angle of 109.5° for $sp^3$ hybridization.

Values of the standard enthalpies of combustion, $\Delta_cH^o$(298 K) for cyclo-alkanes provide an experimental indicator of the amount of strain that a ring exhibits. For *acyclic alkanes*, values of $\Delta_cH^o$ (defined for equation 25.1) become increasingly negative as the number of C atoms increases (Table 25.1). Because this is a comparative exercise, all data refer to gaseous molecules.

$$C_nH_{2n+2}(g) + \frac{3n+1}{2}O_2(g) \longrightarrow nCO_2(g) + \frac{2n+2}{2}H_2O(l) \qquad (25.1)$$

The right-hand column in Table 25.1 shows that the differences between successive values of $\Delta_cH^o$(g, 298 K) along the homologous series of acyclic alkanes are virtually constant ($\approx -659$ kJ mol$^{-1}$). This is consistent with the combustion of each additional CH$_2$ unit releasing the same amount of energy. This is reasonable because the environment of every CH$_2$ unit in a straight chain alkane is approximately the same. Table 25.1 also lists values of $\Delta_cH^o$ for gaseous cycloalkanes from cyclopropane to cyclodecane. Each ring consists of $n$ CH$_2$ units, and as $n$ increases, combustion becomes increasingly exothermic as expected. However, on going from $C_3H_6$ to $C_{10}H_{20}$, values of $\Delta_cH^o$ *per mole of CH$_2$ units* are not constant (Table 25.1). The data in Table 25.1 can be analysed by comparing $\Delta_cH^o$ per mole of CH$_2$ units in a cycloalkane with the typical value of $\Delta_cH^o$ per mole of CH$_2$ units ($-659$ kJ mol$^{-1}$) in an acyclic alkane. The differences between these values (e.g. $-659 - (-697) = 38$ kJ mol$^{-1}$ for $n = 3$ from Table 25.1) are expressed graphically in Figure 25.7, and are measures of

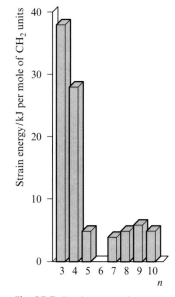

**Fig. 25.7** Strain energy in cycloalkanes $C_nH_{2n}$ for $3 \leq n \leq 10$. The strain energy is estimated from standard enthalpies of combustion as described in the text.

**Fig. 25.8** (a) $sp^3$ Hybridization is compatible with tetrahedral carbon. (b) Two $sp^3$ hybrid orbitals point directly along two bond vectors in a tetrahedral arrangement, e.g. in $CH_4$. (c) In cyclopropane, the C–C–C bond angles are 60° and the C–C bond vectors do not coincide with the directions of two $sp^3$ hybrid orbitals. (d) A similar situation arises in cyclobutane. (e) Regions of overlap between $sp^3$ hybrid orbitals on adjacent C atoms in cyclopropane lie outside the $C_3$ ring.

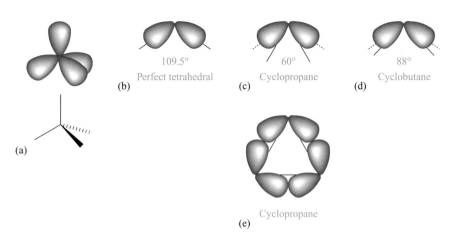

the amount of ring strain energy in the cycloalkanes. The data in Figure 25.7 illustrate that:

- $C_6H_{12}$ has no strain energy;
- $C_3H_6$ has the greatest strain energy, and $C_4H_8$ also suffers significant strain;
- $C_5H_{10}$, $C_7H_{14}$, $C_8H_{16}$, $C_9H_{18}$ and $C_{10}H_{20}$ have similar, relatively small strain energies.

If the bar chart in Figure 25.7 were continued to higher values of $n$, we would find that cycloalkanes with $n \geq 15$ suffer no ring strain.

By considering an $sp^3$ hybridization model and the structural data in Table 25.1, we can use an orbital approach to gain insight into the ring strain in $C_3H_6$ and $C_4H_8$. Figures 25.8a and 25.8b show that an $sp^3$ hybridization scheme is ideally suited to a tetrahedrally sited atom such as the C atom in $CH_4$. In cyclopropane, each C atom is 4-coordinate, but the C–C–C bond angle is only 60°. While we can adopt an $sp^3$ hybridization scheme to cope with the four bonds made by each C atom, the orientations of the hybrid orbitals are not matched to the directions of the C–C bonds (Figure 25.8c). A similar situation arises in cyclobutane (Figure 25.8d) although the situation is not as bad as in cyclopropane. Figure 25.8e illustrates that when interaction occurs between $sp^3$ hybrid orbitals on adjacent C atoms in cyclopropane, the regions of overlap lie outside the 3-membered ring. This result indicates that the C–C bonds in cyclopropane are 'bent' and the ring can be described as being 'strained'. As we emphasized in Chapter 7, all hybridization schemes are *models*. However, support for a scheme analogous to that shown in Figure 25.8e comes from experimental results: electron density maps obtained from low-temperature X-ray diffraction data reveal that the electron density associated with the C–C bonds does indeed lie outside the $C_3$ ring.

## Summary

Two factors contribute to the ring strain in cyclopropane:

- eclipsing of the $CH_2$ units (*torsional strain*);
- unfavourable orbital overlap (*angle strain*).

Similar effects, but less severe, contribute to the ring strain in cyclobutane. Rings with more than three C atoms are not constrained to being planar, and partial rotation about the C–C bonds relieves torsional ring strain. In addition, as the ring increases in size, the C–C–C bond angles are no longer constrained to the unfavourable values observed in $C_3H_6$ and $C_4H_8$. In $C_6H_{12}$, the bond angle of 111° and the staggered conformation achieved in the chair conformation lead to a situation not very different from that in straight chain alkanes and $C_6H_{12}$ has no ring strain.

The fact that cyclopropane suffers from ring strain is not a reason to assume that it and its derivatives are chemically or commercially insignificant. Pyrethrum has been used as an insecticide since the early 19th century and is extracted from the flowers of a species of chrysanthemum (*Tanacetum cinerariaefolium*). Pyrethrum is a mixture of pyrethrin I, pyrethrin II, cinerin I and cinerin II. Synthetic pyrethroids (like natural pyrethrum) cause rapid paralysis of the insects they target. Although efficient, pyrethrum suffers from being unstable to light. Synthetic analogues that are photostable include permethrin:

Pyrethrin I

Permethrin

## 25.3 Physical properties of alkanes

We could consider the physical properties of alkanes from a range of viewpoints. However, it is most useful to look for trends in behaviour and for this, it is most profitable to consider homologous series.

Homologous series: see Section 24.4

### Melting and boiling points

Pauling electronegativities and polar bonds: see Section 5.7

Intermolecular interactions: see Sections 2.10 and 3.21

The Pauling electronegativity values of C and H are 2.6 and 2.2 respectively, and the C–H bond dipole moment is small. Alkanes are essentially non-polar molecules, and in both the solid and liquid states, the intermolecular interactions are weak van der Waals forces. Figure 25.9 shows the trends in melting and boiling points along the homologous series of alkanes $C_nH_{2n+2}$ for $1 \leq n \leq 20$. $CH_4$ (see Box 25.1), $C_2H_6$, $C_3H_8$ and $C_4H_{10}$ are gases at 298 K and atmospheric pressure, while straight chain alkanes from pentane to heptadecane are liquids. Higher alkanes ($n \geq 18$) are solids at 298 K.

Figure 25.10 shows 'ball-and-stick' and space-filling models of a pentane molecule in an extended conformation. The 'space-filler' emphasizes the overall rod-like shape of the molecule, whereas the environment about each C atom and the zigzag backbone are clearer in the 'ball-and-stick' model. In the solid and liquid phases, van der Waals forces operate between

**Fig. 25.9** Trends in melting and boiling points of straight chain alkanes $C_nH_{2n+2}$ and cycloalkanes $C_nH_{2n}$.

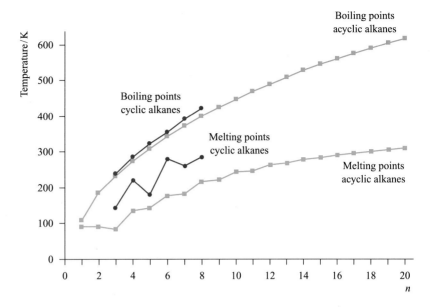

## ENVIRONMENT AND BIOLOGY

### Box 25.1 Methane: a 'greenhouse gas'

Increases in the concentrations of certain gases in the atmosphere may be leading to global warming, the so-called *greenhouse effect*. Probably the two major *greenhouse gases* are $CO_2$ (see Box 22.2) and $CH_4$.

Methane is produced by the *anaerobic* (in the absence of $O_2$) decomposition of organic material and is released into the atmosphere. Its escape as bubbles from wetlands is the origin of the name *marsh gas*. Flooded areas (e.g. rice paddy fields) are a source of large amounts of $CH_4$. Methane also originates from ruminants (e.g. cows,

sheep and goats) that feed on grass and digest their food in a particular manner. After conversion of the foodstuffs to energy, end products include $CH_4$ which passes into the atmosphere. This is a natural process, but the numbers of domestic animals have increased significantly over the last two centuries or so, and the diet of livestock influences the amount of $CH_4$ produced. The chart below summarizes the approximate emission of $CH_4$ into the atmosphere from animals and humans. Domestic cattle are responsible for 71% of the total emission.

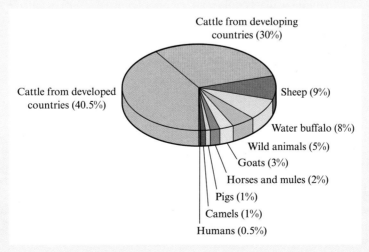

Data from: *Chemistry & Industry* (1992), p. 334.

**Fig. 25.10** Two representations of a pentane molecule shown in an extended conformation: (a) a 'ball-and-stick' model and (b) a space-filling model. Colour code: C, grey; H, white.

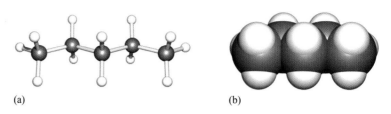

(a)  (b)

the atoms on the periphery of the molecule and those of its neighbours. Figure 25.11 illustrates the way in which molecules of icosane are packed in the solid state. The general rise in melting points with carbon chain length (Figure 25.9) reflects the fact that the intermolecular interactions increase as the surface area of the molecule increases.

Figure 25.9 also shows the trends in melting and boiling points of cycloalkanes with $3 \leq n \leq 8$. In general, values for a cycloalkane with a given number of C atoms are higher than for its straight chain counterpart. Van der Waals forces operate between molecules of cycloalkanes in the solid state, but the trend in melting points is not regular, displaying an alternating effect. This can be discussed in terms of the variation in molecular shape and molecular packing efficiencies along the series $C_3H_6$, $C_4H_8$, $C_5H_{10}$, $C_6H_{12}$, $C_7H_{14}$, $C_8H_{18}$. Contrast this with the rather similar rod-like shapes of extended conformers of straight chain alkanes.

We exemplify the effects of chain branching by considering constitutional isomers of $C_6H_{14}$, boiling points of which are listed in Table 25.2. In general, an isomer that has a branched chain boils at a lower temperature than one with a straight chain. The intermolecular forces decrease as the contact area between molecules decreases. An alkane molecule becomes more spherical in shape as the degree of branching increases.

### Densities and immiscibility with water

The densities of straight chain alkanes increase as the carbon chain lengthens, but values level off at about $0.8\,\mathrm{g\,cm^{-3}}$. All alkanes are less dense than water (density $= 1.00\,\mathrm{g\,cm^{-3}}$). The trend in densities for the liquid alkanes (at 298 K) is shown in Figure 25.12. In keeping with the fact that alkanes are non-polar and water is polar, alkanes are *immiscible* with water. When a liquid alkane is mixed with pure water, the alkane settles as the upper layer.

If two liquids are ***miscible***, they form a solution. If they are ***immiscible***, they form two layers, with the more dense liquid as the lower layer.

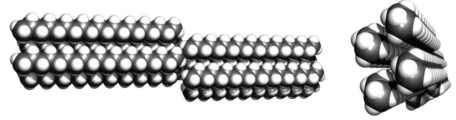

**Fig. 25.11** In the crystal lattice, molecules of icosane ($C_{20}H_{42}$) are in an extended conformation and pack parallel to one another. Colour code: C, grey; H, white. [X-ray diffraction data: S. C. Nyburg *et al.* (1992) *Acta Crystallogr.*, *Sect. B*, vol. 48, p. 103.]

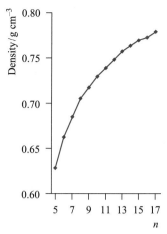

**Fig. 25.12** The trend in densities of the liquid (at 298 K) alkanes. All are less dense than water.

**Table 25.2** Boiling points (measured at 1 bar pressure) of constitutional isomers of $C_6H_{14}$.

| Compound name | Structural formula | Boiling point / K |
|---|---|---|
| Hexane | | 341.7 |
| 2-Methylpentane | | 333.3 |
| 3-Methylpentane | | 336.3 |
| 2,3-Dimethylbutane | | 331.0 |
| 2,2-Dimethylbutane | | 322.7 |

## 25.4  Industrial interconversions of hydrocarbons

Commercial demand for hydrocarbons is vast, and the major sources of alkanes (from which unsaturated hydrocarbons are derived) are natural gas and petroleum deposits. In this section, we give a brief overview of the processes involved in the hydrocarbon industry.

### Refining

After extraction from natural sources, crude oil is *refined*. This separates the oil into the main alkane components (gases, motor fuels, kerosene, diesel oil, wax, lubricants and bitumen) and involves fractional distillation and gas separation processes. Crude oil first passes through a desalter to remove water-soluble contaminants and other impurities, and is then introduced into the fractionating columns (represented schematically in Figure 25.13) at $\approx 600\,K$. The columns used for distillation each contain about 40 chambers. Vapours rise up through the column, which is cooler at the top than at the bottom. Only components with the lowest boiling points (correlating approximately to those with the lowest molecular weights) reach the highest chambers in the column. Each chamber collects fractions containing hydrocarbon components with different boiling ranges. The highest boiling fractions remain as a residue which is further separated by distillation under reduced pressure.

### Cracking

The fractions obtained from refining do not all contain a commercially viable distribution of hydrocarbons. Many higher molecular weight residues remain. Such fractions are subjected to *cracking*, a process that breaks down the long hydrocarbon chains into shorter ones and produces alkenes.

**Fig. 25.13** A schematic representation of a fractionating column used in crude oil distillation.

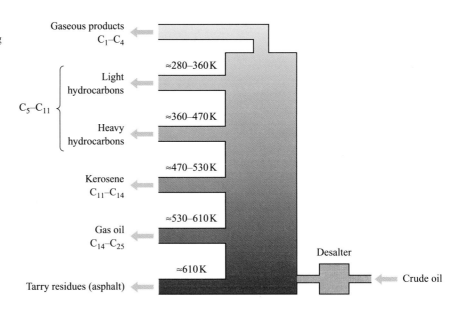

Alkenes (which are not present in crude oil) have major applications as precursors in the polymer industry. Cracking can be carried out at high temperatures and pressures (*thermal cracking*) or in the presence of a catalyst (*catalytic cracking*). Thermal cracking is used to treat both liquid fractions and solid residues, but catalytic methods are suitable only for the liquid components. Zeolite catalysts have now generally replaced the alumina and alumina–silica catalysts that used to be used in fluid catalytic cracking.

■▷
Polymers from alkenes:
see Section 26.8

■▷
Zeolites: see Figure 22.13
and discussion

## Reforming

Fractions to be used as fuels should burn as smoothly and cleanly as possible, and this requirement leads to the process of *reforming* in which unbranched alkanes are converted into branched compounds. Reforming is simply isomerization. When straight chain alkanes burn in an engine, premature ignition gives rise to 'knocking'. The octane number of a motor fuel indicates whether or not the fuel is resistant to knocking. The octane scale runs from 0 to 100, with 100 assigned to a fuel that is completely resistant. Anti-knocking agents such as tetraethyllead (**25.11**) were added to fuels. However, lead compounds are environmental pollutants (see Box 25.2) and also destroy the catalysts that are used in catalytic converters (see Box 22.9). Hence modern motor engines are designed to run on unleaded fuels.

(25.11)

---

| 25.5 | **Synthesis of cycloalkanes** |

■▷
Alkane synthesis using
Grignard reagents:
see Figure 28.2

■▷
Alkene additions are dis-
cussed in detail in Chapter 26

Whereas acyclic alkanes can be obtained from the fractionation of crude petroleum, the relative amounts of cycloalkanes depend on the source of the oil deposit. Syntheses of cycloalkanes of specific ring sizes have therefore been developed. Two general strategies are:

- to use addition reactions to alkenes;
- to cyclize an acyclic compound.

**ENVIRONMENT AND BIOLOGY**

## Box 25.2  Wine as a monitor of lead pollution

A group of researchers in France and Belgium has used the vintage wine Châteauneuf-du-Pape to monitor lead pollution from motor engine emissions. Grapevines grown at the side of French roads have been the source of grapes for this wine for many years. Wines made between 1962 and 1991 have been analysed for their lead content and the concentration of tetraethyl lead $Et_4Pb$ (the anti-knock additive) and its decomposition product $Me_4Pb$ have varied as shown in the bar chart. The initial rise follows an increase in the amount of traffic passing the vineyard. The fall after 1978 coincides with the introduction of unleaded fuels, and the more drastic fall in the late 1980s to the more common use of such fuels. At their peak, the lead levels were a possible health risk.

Vineyard in Chinon wine region, Loire valley, France.
*John Parker* © *Dorling Kindersley.*

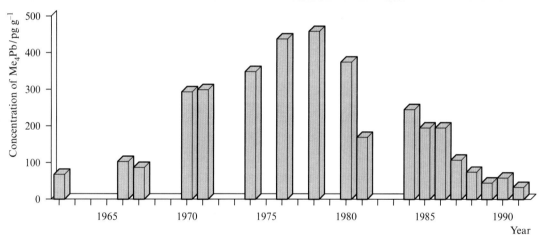

Data from: R. Lobinski *et al.* (1994) *Nature*, vol. 370, p. 24.

## Cycloadditions

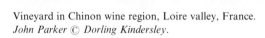

**(25.12)**

Cyclopropane can be prepared by the addition of a $CH_2$ unit to ethene. Although we have seen $CH_2$ groups as components of alkane chains, we have not considered them as independent molecules. When, for example, diazomethane is *photolysed*, $N_2$ is released and carbene (**25.12**) is formed (equation 25.2).

> A *photolysis* reaction takes place upon irradiation and is designated by the symbol $h\nu$.

$$CH_2N_2 \xrightarrow{h\nu} :CH_2 + N_2 \tag{25.2}$$

**A sextet is two electrons short of an octet: octet rule, see Section 7.2**

Carbene has a sextet of electrons and is electron deficient. It adds across the C=C bond in ethene to give cyclopropane (equation 25.3).[§]

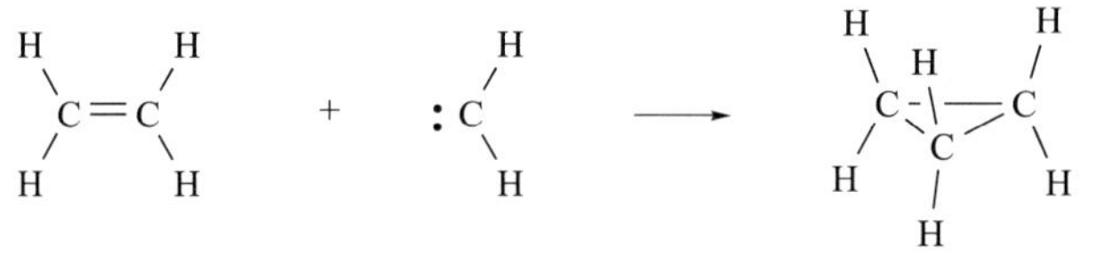

$$\tag{25.3}$$

**Cycloadditions** of unsaturated molecules such as alkenes or alkynes are their reactions to form cyclic molecules.

Substituted cyclopropanes can be prepared by starting with appropriate alkenes, e.g. reaction 25.4. Reactions 25.3 and 25.4 are examples of *cycloadditions*.

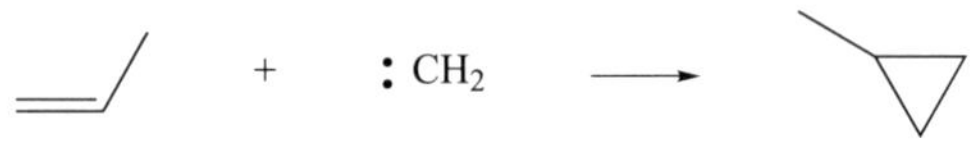

$$\tag{25.4}$$

Propene                    Methylcyclopropane

**An *organometallic* compound contains a metal–carbon bond.**

Rather than use a free carbene, a better strategy is to start with an *organometallic* reagent that produces a carbene *in situ*. The *Simmons–Smith reaction* is a common method of making cyclopropane derivatives, and involves treatment of an alkene with $ICH_2ZnI$, prepared by reaction 25.5. If this reaction is carried out in the presence of an alkene, the organometallic reagent acts as a source of $CH_2$ as in reaction 25.6.

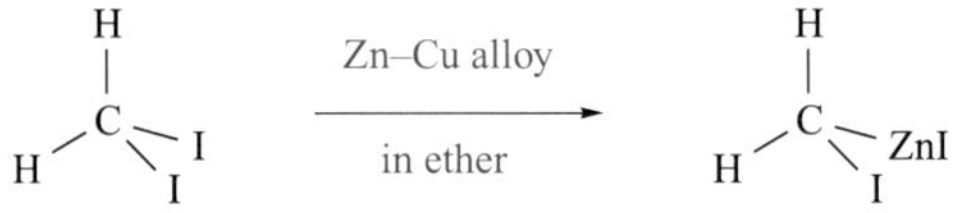

$$\tag{25.5}$$

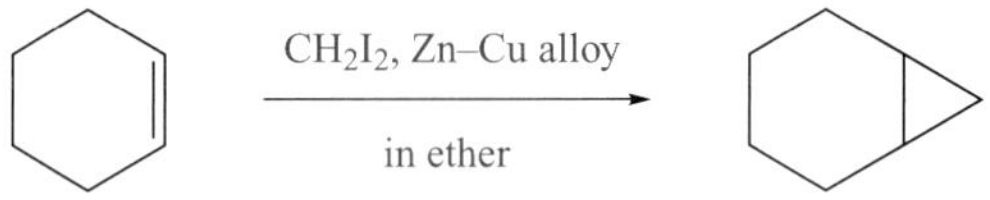

$$\tag{25.6}$$

The dimerization of ethene gives cyclobutane, but the reaction works *only* when driven *photochemically* (equation 25.7); heating ethene does *not* produce cyclobutane. Similar reactions occur between a range of alkenes and in some cases it is possible to react one alkene specifically with a different one.

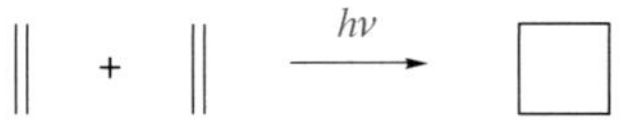

$$\tag{25.7}$$

The reason why the reagents must be photolysed can be understood in terms of the symmetry properties of the HOMO and LUMO of ethene. Each new C–C bond that forms corresponds to the movement of a pair of electrons and the reaction can be written with curly arrows as follows:

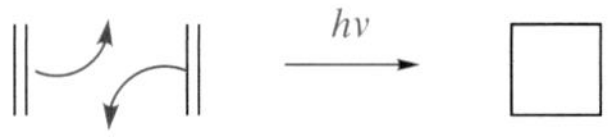

**π-Orbitals in alkenes and electronic transitions: see Section 13.5**

As we discussed in Section 24.10, electron movement requires the interaction of occupied and unoccupied MOs. Figure 25.14a shows the HOMO and LUMO of ethene in its ground state. The interaction between the HOMO of one ethene molecule with the LUMO of another cannot occur because the symmetries of the two orbitals are different. (Look back at Chapters 4

[§] For details of the reactions of carbenes, see for example: J. March (1992) *Advanced Organic Chemistry: Reactions, Mechanisms and Structure*, 4th edn, Wiley, New York.

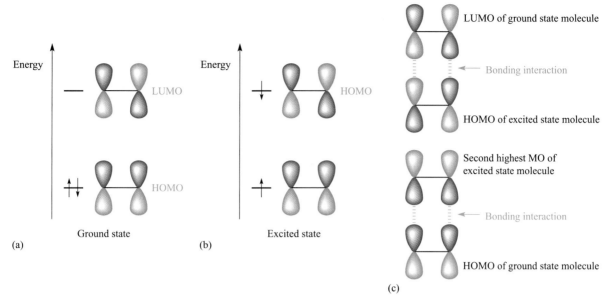

**Fig. 25.14** The $\pi$-orbitals of ethene in (a) its ground state and (b) its excited state. The excited state is produced photolytically. (c) The formation of two bonding interactions that permit the dimerization of ethene. There are two corresponding antibonding interactions.

The addition of two ethene molecules to give cyclobutane is an example of a $[2 + 2]$ *cycloaddition* and only takes place *photochemically*.

and 5 where we discussed symmetry-matching of orbitals.) However, if ethene is photolysed, it absorbs light energy and an electron is promoted from the HOMO to the LUMO. In this excited state, there are two MOs, each containing one electron (Figure 25.14b). Orbital interactions are possible between the excited state of one ethene molecule and the ground state of the other. The resultant bonding interactions are shown in Figure 25.14c. Corresponding antibonding interactions also arise. The evolution of the $C_4$ framework of cyclobutane is evident in Figure 25.14c.

### Cyclization using halogenoalkanes

The first synthesis of cyclopropane involved the reaction of sodium metal with 1,3-dibromopropane (equation 25.8).

$$Br\diagup\!\!\!\!\diagdown\!\!\!\!\diagup Br \xrightarrow[- 2NaBr]{2Na} \triangledown \tag{25.8}$$

This is an example of a *Wurtz reaction* in which a halogenoalkane is used in a C–C bond-forming process. The reactions are usually carried out in an inert ether solvent. 1,3-Dibromopropane also undergoes ring closure when treated with Zn and NaI.

### Cyclization using alkenes

When we discuss radical polymerizations in Section 26.8, we view intramolecular ring closure as an undesired complication. However, intramolecular radical reactions are also convenient methods of preparing 5- and 6-membered cycloalkanes. An example is shown in equation 25.9. We discuss radical reactions in more detail in Sections 25.7, 25.8 and 26.8.

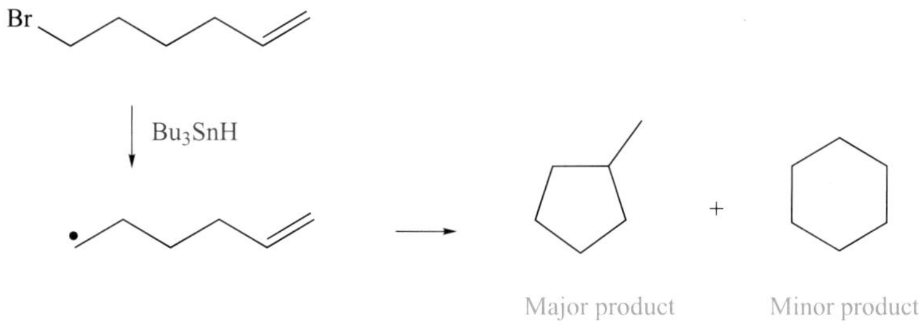

(25.9)

See problem 25.7 ▬▶

| 25.6 | **Reactions of straight chain alkanes** |

Alkanes commonly undergo the following types of reactions:

- combustion;
- cracking (see Section 25.4);
- radical substitutions.

### Combustion

Alkanes burn in $O_2$ to give $CO_2$ and $H_2O$ (equations 25.10 and 25.11) and such combustions are exothermic (see Table 25.1).

$$CH_4 + 2O_2 \longrightarrow CO_2 + 2H_2O \tag{25.10}$$

$$C_9H_{20} + 14O_2 \longrightarrow 9CO_2 + 10H_2O \tag{25.11}$$

In practice, combustion is incomplete and varying amounts of elemental carbon and other carbon compounds are formed in addition to $CO_2$. The presence of elemental carbon and large aromatic molecules results in the well-known 'sooty' flame observed when higher hydrocarbons burn. When the amount of $O_2$ is limited (e.g. burning fuels in a closed room), combustion can lead to CO. This has the potential for serious hazards: for example, CO breathed in from fuel combustion in faulty or badly vented domestic heating appliances can prove fatal.

Values of $\Delta_c H^\circ$ (standard enthalpy of combustion) can be found experimentally using a bomb calorimeter (Figure 17.5). Then $\Delta_f H^\circ$ of the compound can be determined by using a Hess cycle (see Section 2.6). Combustion is also used to determine the molecular formula of a compound as is illustrated in worked example 25.1.

| Worked example 25.1 | *Determining the formula of an alkane* |

**When it is fully combusted, an acyclic alkane X gives 211.2 g $CO_2$ and 97.2 g $H_2O$. Suggest a possible identity for X.**

The molecular formula of a hydrocarbon gives the ratio of moles of C : H atoms. For alkanes $C_nH_{2n+2}$, this can be determined from the ratio of moles of $CO_2$ and $H_2O$ formed on *complete* combustion.

First find the amount of $CO_2$ and $H_2O$ produced. Relative atomic masses (see inside the front cover of the book) are $C = 12.01$, $O = 16.00$, $H = 1.008$.

For $CO_2$:    $M_r = 12.01 + 2(16.00) = 44.01$

For $H_2O$:    $M_r = 2(1.008) + 16.00 = 18.016 = 18.02$   (to 2 dec. pl.)

$$\text{Amount of } CO_2 \text{ formed} = \frac{\text{Mass}}{M_r} = \frac{(211.2\,\text{g})}{(44.01\,\text{g mol}^{-1})} = 4.799\,\text{mol}$$

$$\text{Amount of } H_2O \text{ formed} = \frac{\text{Mass}}{M_r} = \frac{(97.2\,\text{g})}{(18.02\,\text{g mol}^{-1})} = 5.39\,\text{mol}$$

Amount of C in the quantity of compound X combusted

  $= $ Amount of $CO_2$

  $= 4.799$ mol

Amount of H in the quantity of compound X combusted

  $= $ Twice the amount of $H_2O$

  $= 10.78$ mol

Ratio of moles of $C:H = 4.799:10.78$

  $= 1:2.25$

The formula of X is therefore $C_nH_{2.25n}$. We know that X is an acyclic alkane and so has a general formula $C_nH_{2n+2}$. Therefore:

$2n + 2 = 2.25n$

$0.25n = 2$

$n = \dfrac{2}{0.25} = 8$

Compound X is $C_8H_{18}$ (octane or an isomer of octane).

*Exercise*: Why can this type of calculation not be applied to find the formula of an unsubstituted cycloalkane?

---

## Halogenation

If a mixture of $CH_4$ and $Cl_2$ is left in sunlight or is photolysed with radiation of wavelength in the range 200–800 nm, a reaction occurs. The products are HCl and members of the family of chloroalkanes $CH_{4-n}Cl_n$ where $n = 1, 2, 3$ or 4. Equations 25.12–25.15 summarize the formation of chloromethane ($CH_3Cl$) followed by sequential reactions with $Cl_2$.

$$CH_4 + Cl_2 \xrightarrow{h\nu} CH_3Cl + HCl \tag{25.12}$$

$$CH_3Cl + Cl_2 \xrightarrow{h\nu} CH_2Cl_2 + HCl \tag{25.13}$$

$$CH_2Cl_2 + Cl_2 \xrightarrow{h\nu} CHCl_3 + HCl \tag{25.14}$$

$$CHCl_3 + Cl_2 \xrightarrow{h\nu} CCl_4 + HCl \tag{25.15}$$

In a *substitution reaction*, one atom or group in a molecule is replaced by another.

These are *substitution reactions*, i.e. substitution of the H atoms in $CH_4$ by Cl atoms has taken place. This particular substitution is a *chlorination* reaction. Bromination of $CH_4$ occurs when $CH_4$ and $Br_2$ are photolysed, but no such reaction occurs with $I_2$. The reaction with $F_2$ is explosive and proceeds even in the absence of light. The relative order of reactivities of the halogens with $CH_4$ is:

$$F_2 \gg Cl_2 > Br_2 \gg I_2$$

Although there is no reaction between $CH_4$ and $I_2$, we include $I_2$ in the series because this order of reactivities is encountered in many halogenation reactions.

The fluorination, chlorination and bromination of higher alkanes lead to complex mixtures of substituted products. As the number of C atoms in the chain increases, the number of possible sites of substitution increases. As an exercise, write down the possible products of the monochlorination of $C_2H_6$ and $C_6H_{14}$ and compare the results with equation 25.12.

| 25.7 | **The chlorination of $CH_4$: a radical chain reaction** |

In this section, we discuss the mechanism by which the chlorination of $CH_4$ occurs. Why is it necessary to irradiate the mixture of gases? The mechanisms described are general for the chlorination and bromination of other alkanes. Reactions with $F_2$ may proceed by a different mechanism. The reaction between $CH_4$ and $Cl_2$ is an example of a *radical chain reaction*.

Kinetics of radical chain reactions: see Section 15.14

### General features of a radical chain reaction

A radical chain reaction involves a series of steps, classified as being one of:

- initiation, or
- propagation, or
- termination.

In the *initiation step*, radicals are *produced* from one or more of the reactants. The generation of radicals may result from heating or irradiating the reaction mixture, or by adding a *chemical initiator*. Alternatively, a 'stable' radical (a *radical initiator*) may be added to initiate the reaction. Equation 25.16 shows a common initiation step in which $X^{\bullet}$ radicals are formed from $X_2$ by homolytic bond fission on irradiation of $X_2$.

$$
\begin{array}{c}
X \overset{hv}{\cdot\vdots} X \xrightarrow{hv} 2X^{\bullet} \\
\text{or} \\
X_2 \xrightarrow{hv} 2X^{\bullet}
\end{array}
\left.\right\} \qquad \textit{Initiation step} \quad (25.16)
$$

Organic radicals are often (but not necessarily) short-lived. As they collide with other species, reactions may occur and can result in either propagation or termination. If a radical collides with a non-radical species, then a *propagation step* may take place in which a new radical is generated (equation 25.17).

$$X^{\bullet} \quad R \overset{\frown}{\vdots} Y \longrightarrow R \overset{\cdot}{\vdots} X + Y^{\bullet} \qquad \textit{Propagation step} \quad (25.17)$$

The newly generated radical can now react with a second species to produce a third radical (equation 25.18).

$$Y^{\bullet} \quad Y \overset{\frown}{\vdots} R \longrightarrow Y \overset{\cdot}{\vdots} Y + R^{\bullet} \qquad \textit{Propagation step} \quad (25.18)$$

Although reactions 25.17 and 25.18 show the radical reacting with a different species, it may combine with a like radical (equation 25.19). This removes radicals from the system without generating any new ones and is a *termination step*. It is also possible for X$^{\bullet}$ to combine with other radicals, e.g. R$^{\bullet}$, in a termination step.

$$X \overset{\bullet}{\phantom{.}} \curvearrowright \curvearrowleft \overset{\bullet}{\phantom{.}} X \longrightarrow X \div X \qquad \textit{Termination step} \quad (25.19)$$

The overall sequence of events combines to give a *chain reaction* with the propagation steps keeping the chain 'alive' (see Figure 15.24). The concentration of radicals in the reaction system at any given time is low. Thus, the interaction of a radical with a non-radical species (i.e. propagation) is far more likely than the reaction between two radicals (i.e. termination). After the initiation process, a large number of propagation steps can occur before radicals are removed in termination steps. Very often, the generation of a single radical allows the formation of many thousands of product molecules before the chain is terminated. Combustion of alkanes and other organic compounds and halogenation of alkanes are radical chain reactions.

***Radical chain reaction*** $\begin{cases} \text{1. Initiation (radicals generated)} \\ \text{2. Propagation (radicals react and more are formed)} \\ \text{3. Termination (radicals are removed from the reaction)} \end{cases}$

## The mechanism of the chlorination of CH₄

The absorption of light by $Cl_2$ causes homolytic fission of the Cl–Cl bond. This is shown in equation 25.20; compare with general reaction step 25.16.

$$Cl_2 \xrightarrow{h\nu} 2Cl^{\bullet} \qquad (25.20)$$

Collisions can occur between Cl$^{\bullet}$ radicals and $CH_4$ molecules. A sequence of propagation steps begins that includes those shown in equations 25.21–25.24 (compare with general reaction step 25.17). In each step, a radical is consumed and another radical is formed. In equations 25.21 and 25.23, a Cl$^{\bullet}$ radical *abstracts* an H atom (H$^{\bullet}$) from one of the initial reactants to give HCl.

$$Cl^{\bullet} + CH_4 \longrightarrow HCl + CH_3^{\bullet} \qquad (25.21)$$

$$CH_3^{\bullet} + Cl_2 \longrightarrow CH_3Cl + Cl^{\bullet} \qquad (25.22)$$

$$Cl^{\bullet} + CH_3Cl \longrightarrow HCl + {}^{\bullet}CH_2Cl \qquad (25.23)$$

$${}^{\bullet}CH_2Cl + Cl_2 \longrightarrow CH_2Cl_2 + Cl^{\bullet} \qquad (25.24)$$

Termination of the chain occurs when two radical species combine. Equations 25.25–25.27 show three possible steps, but others involving chloroalkane radicals such as $^{\bullet}CH_2Cl$ may also take place.

$$2Cl^{\bullet} \longrightarrow Cl_2 \qquad (25.25)$$

$$Cl^{\bullet} + CH_3^{\bullet} \longrightarrow CH_3Cl \qquad (25.26)$$

$$2CH_3^{\bullet} \longrightarrow C_2H_6 \qquad (25.27)$$

## ENVIRONMENT AND BIOLOGY

## Box 25.3 Depletion of the ozone layer

The *ozone layer* is the layer in the atmosphere that is 15–30 km above the Earth's surface. Ozone absorbs strongly in the ultraviolet (UV) region and the ozone layer shields the Earth from UV radiation from the Sun. One possible effect of UV radiation on humans is skin cancer.

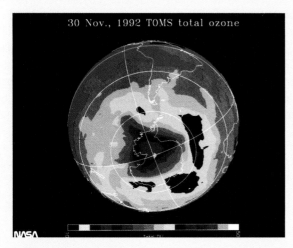

Computer image showing the ozone hole above Antarctica. *NASA/© Dorling Kindersley.*

One group of pollutants from the Earth comprises chlorofluorocarbons (CFCs). These have had extensive use as propellants in aerosols, as refrigerants, in air conditioners, as solvents and in foams for furnishings. After use, CFCs escape into the atmosphere and in the stratosphere they undergo photochemical reactions such as:

$$CCl_2F_2 \xrightarrow{h\nu} Cl^{\bullet} + {}^{\bullet}CClF_2$$

A radical chain reaction then begins, and reactions with ozone ($O_3$) include:

$$O_3 + Cl^{\bullet} \longrightarrow O_2 + ClO^{\bullet}$$

The devastating result is that the ozone layer is slowly being depleted. This fact was first recognized in the 1970s and, in some countries, action was taken

swiftly legislating against the use of CFCs. Although so-called *ozone-friendly* aerosols and other products are now increasingly available, the problem of the depletion of the ozone layer remains. Emissions of CFCs from the Earth must be drastically reduced if we are to remain protected from the intense UV radiation of the Sun. In 1987, the *Montreal Protocol for the Protection of the Ozone Layer* was established and legislation was implemented to phase out the use of CFCs. An almost complete phase-out of CFCs was required by 1996 for industrial nations, with developing nations following this by 2010. Taking the 1986 European consumption of CFCs as a standard (100%), the graph below shows the reduction in usage of CFCs from 1986 to 1993. More recently, asthma inhalers that use CFCs have been phased out.

CFCs are categorized as 'Class I' ozone-depleters. Other 'Class I' halocarbons include $CCl_4$, $CHCl_3$, $CH_2ClBr$, $CBr_2F_2$, $CF_3Br$ and $CH_3Br$. Methyl bromide has been widely used for agricultural pest control but, under the Montreal Protocol, $CH_3Br$ is banned from 2005 onwards (2015 in developing countries). Hydrochlorofluorocarbons (HCFCs) are also ozone-depleting. They appear to be less harmful to the environment than CFCs, and are classified as 'Class II' ozone-depleters. HCFCs will be phased out by 2020, but in the meantime they remain in use as refrigerants in place of CFCs. In contrast, hydrofluorocarbons seem to have little or no ozone-depleting effect and are permitted for use in refrigerants and aerosol propellants.

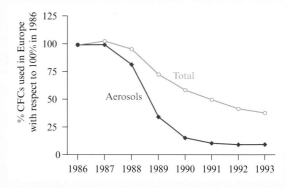

Data from: *Chemistry & Industry* (1994) p. 323.

Step 25.25 is the reverse of the initiation step 25.20, while step 25.26 illustrates the formation of a desired product. Reactions of the type shown in step 25.27 complicate the process by extending the carbon chain. The $C_2H_6$ molecules can participate in a reaction step similar to step 25.21 and this may ultimately lead to further growth of the carbon chain and production of higher alkanes. A radical chain reaction is, therefore, not specific.

**25.8**

## Competitive processes: the chlorination of propane and 2-methylpropane

For alkane chains that possess more than two C atoms, an additional complication arises during radical reactions: a radical can attack at one of several different sites. We exemplify the problem by considering the chlorination of propane and 2-methylpropane. These compounds contain different types of C atom as shown in structures **25.13** and **25.14**.

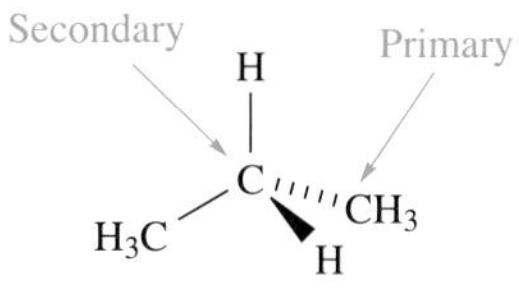

(25.13) 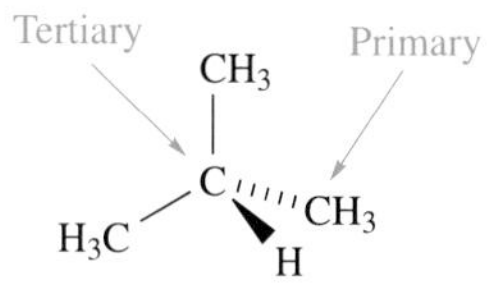                              (25.14)

### The reaction between $Cl_2$ and propane

When a mixture of $Cl_2$ and $C_3H_8$ is irradiated, the initiation is step 25.20, i.e. generation of $Cl^•$. Propagation steps then follow one of two sequences depending on whether the H involved is attached to a primary or secondary C atom. Equation 25.28 shows the abstraction of $H^•$ from a primary C atom, and in equation 25.29, $Cl^•$ abstracts $H^•$ from the secondary C atom.

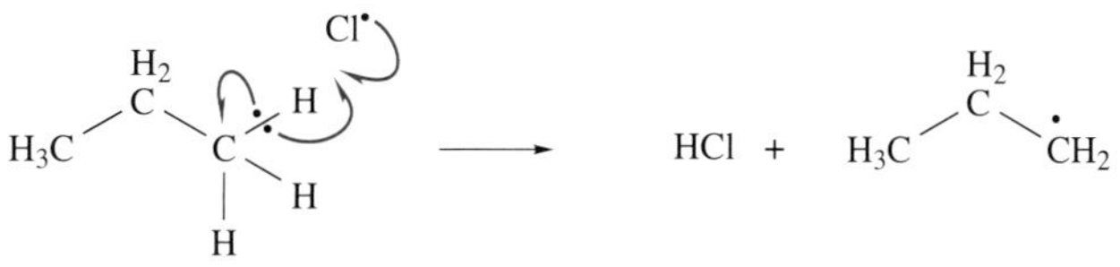

$$(25.28)$$

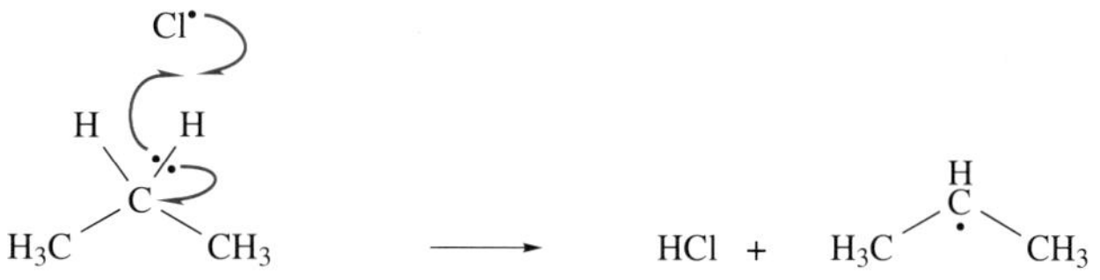

$$(25.29)$$

The competitive processes are summarized in equation 25.30.

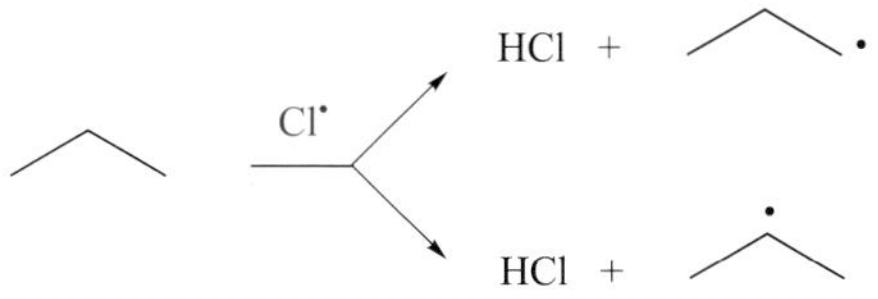

$$(25.30)$$

The abstraction of $H^•$ to give an alkyl radical with a *secondary* C atom is preferred to abstraction of $H^•$ to give an alkyl radical at a primary C atom. Thus, step 25.29 dominates over step 25.28. Reaction between an alkyl radical and $Cl_2$ now leads to either 1-chloropropane (equation 25.31) or 2-chloropropane (equation 25.32).

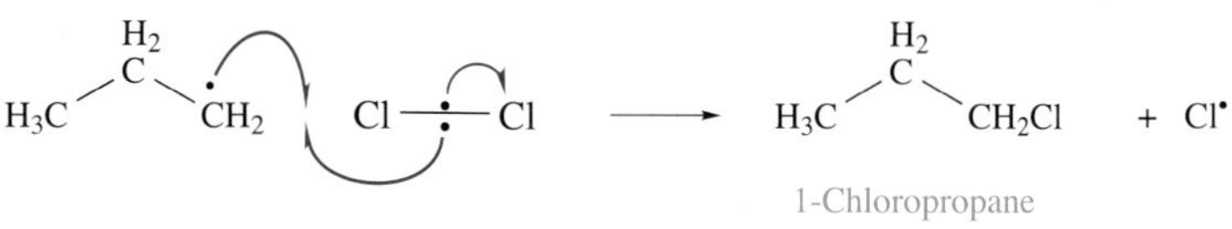

1-Chloropropane

$$(25.31)$$

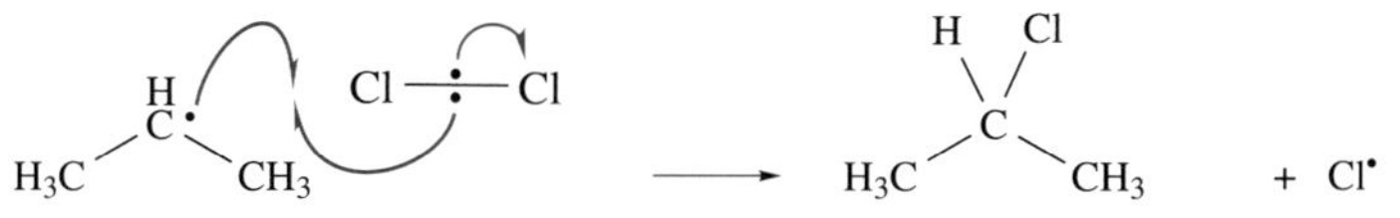

2-Chloropropane                    (25.32)

Because of the relative stabilities of the two radicals formed in scheme 25.30, the final product mixture contains more 2-chloropropane than 1-chloropropane.

## Radical stability

The reaction described above leads us to a more general, and important, result. In a radical reaction, *a radical derived from a tertiary C atom is formed in preference to one from a secondary C atom, and one derived from a secondary C atom is formed in preference to one from a primary C atom.* Thus, for the formation of radicals from propane where the competition is between secondary and primary C centres respectively, the relative stabilities of alkyl radicals are:

Experimental data indicate that the alkyl radical stability series can be generalized as:

$$R_3C^\bullet > R_2HC^\bullet > RH_2C^\bullet > H_3C^\bullet$$

The relative stabilities of alkyl radicals and the preferential formation of one over another has an effect on the final product distribution in a chain reaction as we have already seen. Hydrogen abstraction involves C−H bond cleavage and, as we saw in Chapter 4, the bond enthalpy term for a C−H bond depends on its environment. In $CH_4$, the cleavage of the first C−H bond requires $436 \, kJ \, mol^{-1}$. In contrast, typical C−H bond enthalpies along the series $RH_2C−H$ (primary C), $R_2HC−H$ (secondary C), $R_3C−H$ (tertiary C) are 420, 400 and $390 \, kJ \, mol^{-1}$. Thus, the least amount of energy is needed to break the C−H bond when the C atom is a tertiary centre. Enthalpy changes for the reactions of $Cl^\bullet$ with $RH_2C−H$, $R_2HC−H$ and $R_3C−H$ can be determined by considering appropriate bond enthalpy terms. In each case, a C−H bond is broken and an H−Cl bond is formed. Formation of the H−Cl bond *releases* $432 \, kJ \, mol^{-1}$ (see Table 4.5), and the schemes below show the approximate enthalpy changes for each reaction:

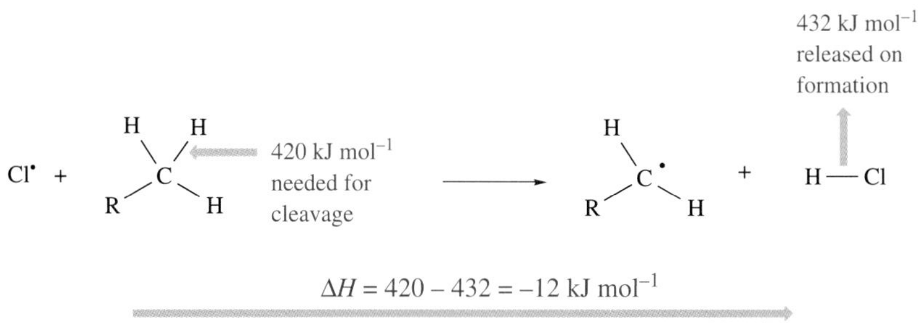

Look back at the discussion in Box 4.4

$$\Delta H = 420 - 432 = -12 \, kJ \, mol^{-1}$$

432 kJ mol$^{-1}$ released on formation

$$Cl^\bullet + \underset{R}{\overset{H\quad H}{\diagup}}\overset{|}{C}\underset{\diagdown R}{} \quad\xleftarrow{\text{400 kJ mol}^{-1}\ \text{needed for cleavage}}\quad \longrightarrow \quad \underset{R}{\overset{H}{\diagup}}\overset{\bullet}{C}\underset{\diagdown R}{} \quad + \quad H\!-\!Cl$$

$$\Delta H = 400 - 432 = -32 \text{ kJ mol}^{-1}$$

432 kJ mol$^{-1}$ released on formation

$$Cl^\bullet + \underset{R}{\overset{R\quad H}{\diagup}}\overset{|}{C}\underset{\diagdown R}{} \quad\xleftarrow{\text{390 kJ mol}^{-1}\ \text{needed for cleavage}}\quad \longrightarrow \quad \underset{R}{\overset{R}{\diagup}}\overset{\bullet}{C}\underset{\diagdown R}{} \quad + \quad H\!-\!Cl$$

$$\Delta H = 390 - 432 = -42 \text{ kJ mol}^{-1}$$

Now let us return to the reaction of $Cl^\bullet$ with propane. Figure 25.15 shows schematic reaction profiles for the competitive pathways to alkyl radical formation. Both are exothermic, but abstraction of $H^\bullet$ from the secondary C atom releases more energy (see above) and the radical intermediate is stabilized with respect to that formed by abstraction of $H^\bullet$ from the primary C atom. Figure 25.15 also shows that the activation energy for $H^\bullet$ abstraction from the secondary C atom is lower than that for primary $H^\bullet$ abstraction. As a result, the *rate* of formation of the radical from the secondary C atom is greater than that from the primary centre.

In an alkyl radical, the C atom at the radical centre has only seven electrons in its valence shell and is *electron-deficient*. Alkyl groups are better able to stabilize an electron-deficient centre than H atoms because an alkyl group is *electron-releasing* and in a C–R bond, the electrons are polarized towards the C atom in the sense $C^{\delta-}\!-\!R^{\delta+}$.

Now let us consider the competitive pathways in another example, one that involves competition between $H^\bullet$ abstraction at primary and tertiary C atoms.

◼▶
Reaction rates and activation energies: see Section 15.1

Alkyl groups are *electron-releasing* and can stabilize electron-deficient centres.

**Fig. 25.15** Reaction profiles for competitive alkyl radical formations in the reaction of propane with $Cl^\bullet$.

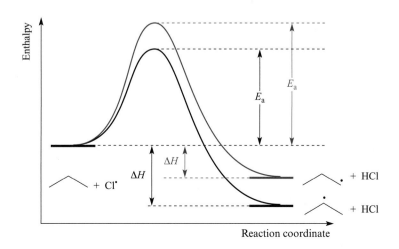

### The reaction between $Cl_2$ and 2-methylpropane

When a mixture of $Cl_2$ and 2-methylpropane (**25.14**) is irradiated, initiation step 25.20 occurs. After this, $Cl^{\bullet}$ radicals react with 2-methylpropane in competitive steps. One step involves abstraction of $H^{\bullet}$ from a primary C atom (equation 25.33), while in the other, $H^{\bullet}$ is abstracted from a tertiary C atom (equation 25.34).

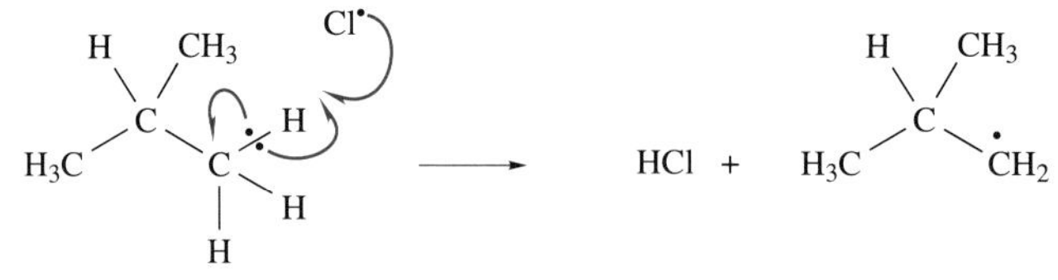

$$(25.33)$$

$$(25.34)$$

The competitive processes are summarized in equation 25.35.

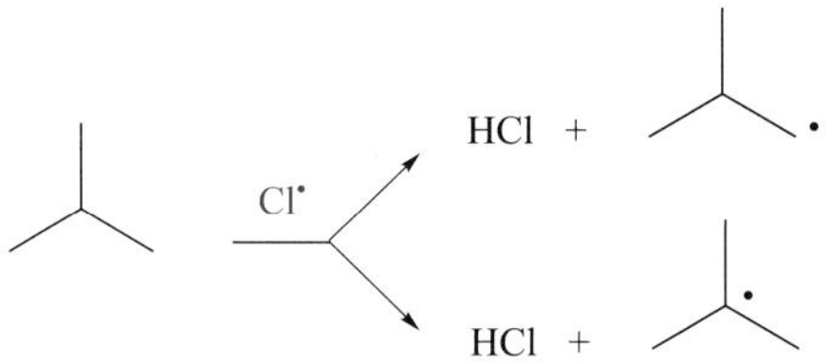

$$(25.35)$$

The alkyl radicals now go on to react with $Cl_2$ to give 1-chloro-2-methylpropane (equation 25.36) and 2-chloro-2-methylpropane (equation 25.37).

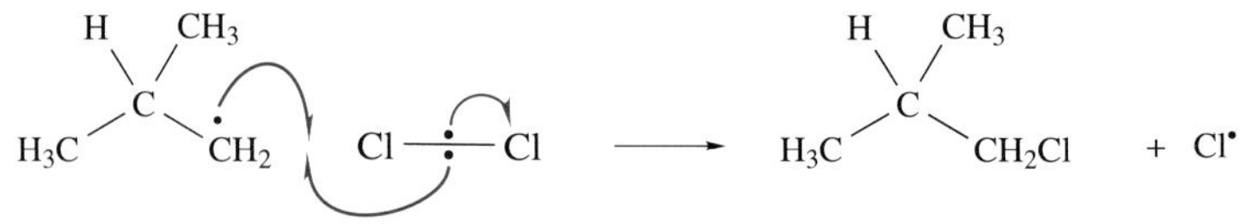

1-Chloro-2-methylpropane

$$(25.36)$$

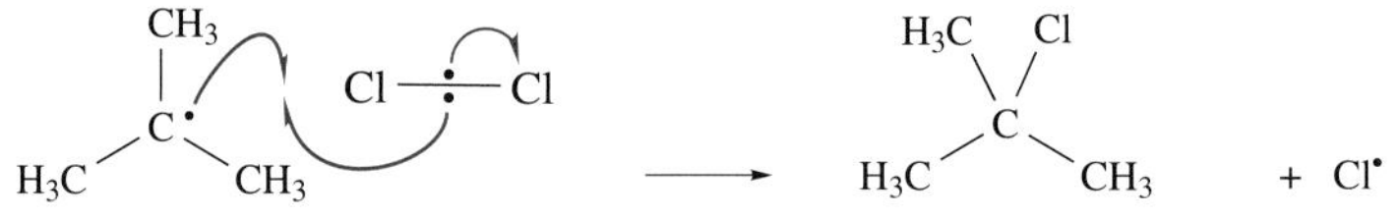

2-Chloro-2-methylpropane

$$(25.37)$$

Experimental data reveal that 1-chloro-2-methylpropane predominates over 2-chloro-2-methylpropane (65% : 35%) even though the radical intermediate formed from the abstraction of $H^{\bullet}$ from a tertiary C atom (the lower pathway in scheme 25.35) is energetically favoured over that formed from abstraction of $H^{\bullet}$ from a primary C atom (the top pathway in scheme 25.35). The experimental observations can be rationalized by considering that there are nine H atoms attached to primary C atoms in 2-methylpropane and only one H atom attached to a tertiary C atom. While abstraction of $H^{\bullet}$ from the tertiary C centre is favoured on energetic grounds and is faster than abstraction of $H^{\bullet}$ from a primary C atom, the *statistical chance* of abstracting $H^{\bullet}$ from a primary C atom is higher than that of removing $H^{\bullet}$ from a tertiary centre. Thus, there are two competing factors, and the observed product distribution confirms that abstraction of $H^{\bullet}$ from a primary C atom wins over the increased reactivity of the tertiary C atom.

## 25.9    Reactions of cycloalkanes

Cycloalkanes undergo similar reactions to straight chain alkanes: combustion (see Table 25.1) and radical substitution reactions. In addition, they undergo ring-opening reactions.

### Substitution reactions

Equation 25.38 shows the monochlorination of cyclopentane. The mechanism is similar to that outlined in Sections 25.7 and 25.8. Since all the H atoms in cyclopentane are equivalent, only one monosubstituted product is possible. Further substitution may lead to a mixture of isomers of multi-substituted products.

$$\text{(cyclopentane)} + Cl_2 \xrightarrow{h\nu} \text{(chlorocyclopentane)} + HCl \qquad (25.38)$$

Chlorocyclopentane

### Ring-opening reactions

Cyclopropane, in particular, undergoes ring-opening reactions that are *additions*. Reactions 25.39–25.41 are examples of additions that lead to the opening of the carbon ring of cyclopropane and relieve its ring strain.

$$\text{(cyclopropane)} + H_2 \xrightarrow{\Delta, \text{ Ni catalyst}} \qquad (25.39)$$

$$\text{(cyclopropane)} + H_2O \xrightarrow{\text{conc } H_2SO_4} \text{OH} \qquad (25.40)$$

$$\text{(cyclopropane)} + HBr \longrightarrow \text{Br} \qquad (25.41)$$

## SUMMARY

In this chapter, we have described the physical and chemical properties of acyclic alkanes and cycloalkanes. These hydrocarbons are saturated and do not possess functional groups. Apart from combustion, typical reactions of alkanes are radical substitutions. We have used chlorination of straight chain and branched alkanes to introduce the mechanism of radical substitution and the effects of different radical stabilities on product distribution. Reactions of cyclopropane include additions that relieve its ring strain.

### *Do you know what the following terms mean?*

- acyclic alkane
- cycloalkane
- monocyclic
- bicyclic
- spirocyclic
- ring conformation
- chair conformation
- boat conformation
- twist-boat conformation
- ring strain

- torsional strain
- angle strain
- immiscible
- miscible
- thermal cracking
- catalytic cracking
- photolysis
- photochemical
- cycloaddition
- [2 + 2] cycloaddition

- organometallic
- cyclization
- radical substitution reaction
- initiation
- propagation
- termination
- radical abstraction
- electron-releasing group

### *You should be able:*

- to draw the structures of an alkane or cycloalkane given a name or formula

- to name an alkane or cycloalkane given a structure

- to relate the number of signals in a $^{13}C$ NMR spectrum to the structure of an alkane or cycloalkane

- to relate the number of signals and coupling patterns in a $^1H$ NMR spectrum to the structure of an alkane or cycloalkane

- to discuss how partial rotation about C–C bonds in cycloalkanes alters the ring conformation

- to draw diagrams to show the chair and boat conformations of $C_6H_{12}$ and to distinguish between axial and equatorial sites in the chair conformer

- to explain how axial and equatorial sites interconvert in $C_6H_{12}$

- to distinguish between isomers and conformers of substituted cyclohexanes

- to describe what is meant by *ring strain*

- to discuss why $C_3H_6$ and $C_4H_8$ are particularly strained, but why $C_6H_{12}$ suffers no ring strain

- to illustrate trends in the physical properties of straight chain alkanes and relate these trends to a *homologous series* of compounds

- to give a brief account of the processes by which crude oil is converted to commercially viable hydrocarbons

- to give examples of syntheses of cycloalkanes

- to explain why [2 + 2] cycloadditions are carried out photochemically

- to illustrate how measurements of $\Delta_c H°(298\,K)$ are used to determine values of $\Delta_f H°(298\,K)$ for alkanes

- to determine the formula of an alkane from the amounts of $CO_2$ and $H_2O$ produced on complete combustion of the compound

- to discuss typical reactions of alkanes and cycloalkanes

- to outline the mechanism by which radical substitutions in alkanes occur

- to illustrate the competitive processes that occur in radical substitutions in alkanes with different types of C centres

- to place a series of alkyl radicals in order of stability

## PROBLEMS

**25.1** Draw structures of the following cycloalkanes:
(a) cyclobutane;
(b) cycloheptane;
(c) methylcyclopentane;
(d) 1,4-dimethylcyclohexane;
(e) 1-ethyl-3-propylcyclooctane.

**25.2** Name the following compounds:
(a)     (b)     (c)        (d)

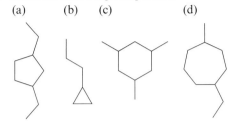

**25.3** How many signals would you expect to see in the proton decoupled $^{13}C$ NMR spectra of the compounds in problem 25.2?

**25.4** Give names and structures for isomers of dimethylcyclobutane.

**25.5** (a) Draw a diagram to show cyclohexane in a chair conformation. Label the axial and equatorial sites. (b) Distinguish between conformers and isomers of 1,3-dimethylcyclohexane.

**25.6** Discuss *ring strain* with reference to cycloalkanes $C_nH_{2n}$ for $3 \le n \le 8$.

**25.7** The standard enthalpy of combustion for $C_3H_6(g)$ at 298 K is $-2091 \, kJ \, mol^{-1}$. Determine $\Delta_f H^{\circ}(C_3H_6, g)$. Other data: $\Delta_f H^{\circ}(CO_2, g)$ and $\Delta_f H^{\circ}(H_2O, l) = -394$ and $-286 \, kJ \, mol^{-1}$ respectively.

**25.8** When fully combusted, an acyclic alkane **A** gives 13.20 g $CO_2$ and 6.306 g $H_2O$. Suggest a possible identity for **A**.

**25.9** Write out each of reaction steps 25.21–25.27 showing appropriate curly arrows.

**25.10** Distinguish between initiation, propagation and termination steps in a radical chain reaction and gives examples of each.

**25.11** Outline the steps involved in the reaction of $Cl_2$ with butane to give monochlorinated products. Comment on factors that affect the product distribution.

**25.12** How would $^1H$ NMR spectroscopy allow you to distinguish between the products of the monochlorination of 2-methylpropane?

**25.13** Give possible monochlorinated products, including isomers, of the photolysis reactions of $Cl_2$ with the following alkanes:

**25.14** The reaction of $Cl_2$ with propane gives 55% 2-chloropropane and 45% 1-chloropropane. In our discussion, we rationalized the product distribution in terms of the relative stabilities of the intermediate alkyl radicals. (a) Are the statistics of abstracting H• from a primary versus secondary C atom in propane also consistent with the observed product distribution? (b) What can you say about the relative rates of formation of the two alkyl radical intermediates?

**25.15** Why do you think that $O_2$ has a dramatic effect on many radical reactions?

**25.16** Suggest products for the following reactions. Where appropriate, comment on complications caused by non-selectivity:

(a)                    $Br_2, h\nu$

(b)                    $CH_2I_2, Zn/Cu$
                       in ether

(c)            HBr

(d)            $Cl_2, h\nu$

**25.17** $^1H$ NMR spectroscopic data for five compounds **A–E** are as follows:

| Compound | $^1H$ NMR $\delta$/ppm (coupling pattern) |
|---|---|
| A | 2.2 (quintet), 3.7 (triplet) |
| B | 0.9 (singlet) |
| C | 1.5 (triplet), 3.6 (quartet) |
| D | 2.2 (singlet), 4.0 (singlet) |
| E | 6.1 (doublet), 4.5 (triplet) |

Assign **A–E** to the following compounds:

**Fig. 25.16** Mass spectrum of heptane. See problem 25.18.

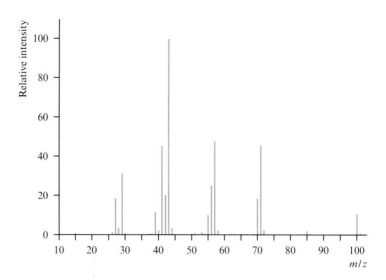

**25.18** Figure 25.16 shows the mass spectrum of heptane. Account for the observed pattern of peaks.

## Additional problems

**25.19** Rationalize the following observations.
  (a) When *n*-octane is photolysed with $Cl_2$, the product distribution is 14% 1-chlorooctane, 31% 2-chlorooctane, 28% 3-chlorooctane and 27% 4-chlorooctane.
  (b) A bicyclic isomer of $C_7H_{12}$ exhibits three resonances in its $^{13}C$ NMR spectrum.

  (c) Photolysis of $Cl_2$ and ethane is a poor method of preparing chloroethane.

**25.20** (a) An alkane **A** exhibits signals at $\delta$ 34.4, 22.6 and 14.1 ppm in its $^{13}C$ NMR spectrum. Elemental analysis shows C 83.2, H 16.8%. Identify **A** and comment on what you would expect to see in its low-resolution mass spectrum. (b) A second isomer of **A** (**B**) exhibits a singlet in its $^1H$ NMR spectrum. Suggest a structure for **B**. How might the mass spectrum of **B** differ from that of **A**? (c) Suggest (with reasons) whether **A** and **B** has the higher boiling point.

# 26 Alkenes and alkynes

## Topics

- Structure and bonding
- Synthesis of alkenes
- Reactivity of alkenes
- Mechanism of electrophilic additions
- Radical substitutions in alkenes
- Radical polymerization of alkenes
- Hydroboration of alkenes
- Synthesis of alkynes
- Reactivity of alkynes

---

**26.1**    **Structure, bonding and spectroscopy**

We have already covered a number of topics concerning the nomenclature, structures and spectroscopic characteristics of alkenes and alkynes:

- naming alkenes – see Section 24.4;
- hybridization schemes – see Sections 7.2 and 24.3;
- isomerism – see Sections 6.11 and 24.4;
- IR spectroscopy – see Section 12.6;
- electronic spectra of $\pi$-conjugated alkenes – see Section 13.5;
- $^{13}C$ NMR spectra – see Section 14.7;
- $^{1}H$ NMR spectra – see Sections 14.8 and 14.9.

The first few problems at the end of the chapter review these topics.

### Bonding schemes

Overlap of atomic orbitals and MO diagrams: see Chapter 4

Although one can develop molecular orbital schemes in terms of the interactions between atomic orbitals of C and H atoms in a hydrocarbon, it is often more convenient to use an orbital hybridization approach. In Section 7.2, we described how to choose an appropriate hybridization scheme for an atom, using the geometry of the atom as the starting point. In an alkene, the C atoms in the C=C unit are trigonal planar (**26.1**), the exception being an allene in which one C atom is involved in two C=C bonds and is in a linear environment (**26.2**). In an alkyne, C atoms in the C≡C unit are linear (**26.3**). Appropriate hybridization schemes for the C atoms are $sp^2$ (trigonal planar) or $sp$ (linear).

The formation of three $sp^2$ hybrid orbitals uses three of the four valence atomic orbitals of a C atom and one $2p$ orbital remains unhybridized. This

(26.1)                    (26.2)                    (26.3)

orbital is available for $\pi$-bond formation (see Section 7.6). Thus, for ethene, we can describe the bonding in terms of the scheme in Figure 26.1a. Each C atom forms three $\sigma$-bonds and one $\pi$-bond.

The formation of two $sp$ hybrid orbitals uses two of the four valence atomic orbitals of carbon, leaving two $2p$ orbitals unhybridized. These orbitals are orthogonal to one another and can form two $\pi$-bonds. The bonding in ethyne can therefore be described in terms of the scheme in Figure 26.1b. Each C atom forms two $\sigma$- and two $\pi$-bonds.

◼▶
Compare with the bonding scheme for $CO_2$ in Figure 7.17

### Effect of $\pi$-character on bond rotation

◼▶
$\pi$-Bond formation: see Figure 4.18

Overlap between two $p$ atomic orbitals is efficient only if the $p$ orbitals are oriented correctly and this is essential for $\pi$-bond formation. The interaction is 'turned off' if the $p$ orbitals are not properly aligned:

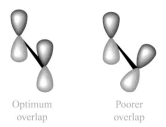

Optimum          Poorer
overlap          overlap

**Fig. 26.1** Using a hybridization approach to generate bonding schemes for (a) ethene and (b) ethyne.

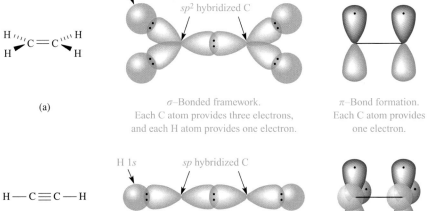

(a)

H 1$s$
$sp^2$ hybridized C

$\sigma$–Bonded framework.
Each C atom provides three electrons, and each H atom provides one electron.

$\pi$–Bond formation.
Each C atom provides one electron.

(b)

H 1$s$          $sp$ hybridized C

$\sigma$–Bonded framework.
Each C atom provides two electrons, and each H atom provides one electron.

$\pi$–Bond formation.
The two $\pi$–bonds are orthogonal.
Each C atom provides two electrons.

This has important consequences with respect to rotation about the C=C bond. For rotation to occur, the $\pi$-contribution to the carbon–carbon bond must be destroyed and then reformed. As a result, the energy barrier to rotation is high ($\approx 250\,kJ\,mol^{-1}$ in ethene) and the alkene is locked in a planar configuration. This means that, under normal conditions, ($Z$)- and ($E$)-isomers of alkenes do not interconvert. The only way to facilitate their interconversion is to temporarily convert the double bond to a single bond, thereby reducing the barrier to rotation. This can be achieved by using an acid catalyst (see Section 26.9) or by photolysis. Photolysis of an alkene promotes an electron from the HOMO to LUMO as we have already seen when we considered the [2 + 2] cycloaddition of alkenes in Section 25.5. In the excited state, $\pi^1\pi^{*1}$ configuration (Figure 25.14b) there is a net $\pi$-bond order of zero, and hence a total C–C bond order of 1 (the $\sigma$-bond is unaffected by the photolysis). The excited state of the alkene can therefore undergo C–C bond rotation and, provided that rotation takes place within the lifetime of the excited state, ($E$)–($Z$) isomerization can occur (equation 26.1).

Look back at problem 11.15 and Figure 11.9: isomerization of RN=NR by flash photolysis

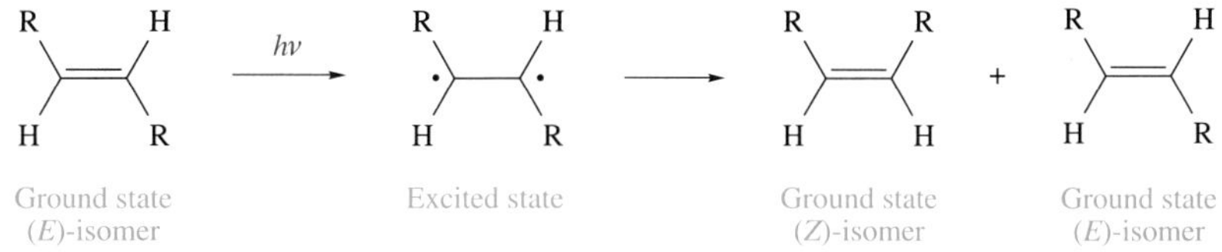

(26.1)

Steric interactions between R groups on the same side of the double bond are greater than those between an R group and H atom and so, on steric grounds, the ($E$)-isomer is favoured over the ($Z$)-isomer. For example, the steric energy of ($E$)-but-2-ene is lower than that of the ($Z$)-isomer:

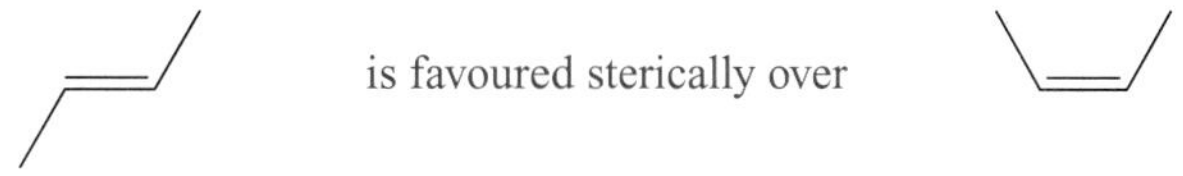

is favoured sterically over

This preference increases as the substituents become more sterically demanding.

---

## ENVIRONMENT AND BIOLOGY

## Box 26.1 Alkene isomerization in the human eye

Alkene ($Z$)–($E$) isomerization is the crucial step that triggers a nerve impulse in the human eye that provides the basis of our vision. The process involves the reversible isomerization of protein-tethered *retinal*, and this is shown below for the free retinal molecule:

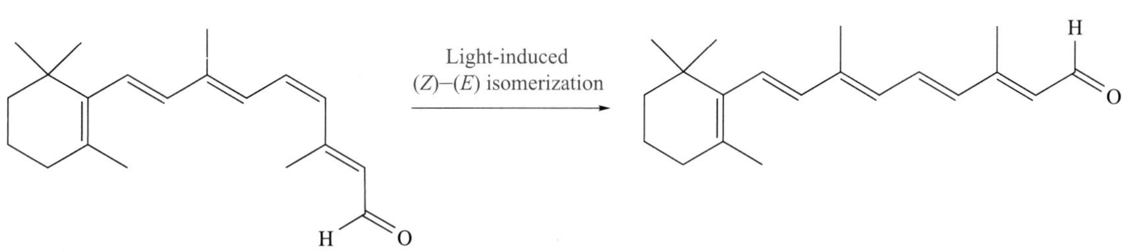

## 26.2    Cycloalkenes: structures and nomenclature

### Names and structures

The general formula for an unsubstituted monocyclic alkene containing one C=C bond is $C_nH_{2n-2}$. Its name must indicate the number of C atoms, the fact that the structure is cyclic, *and* the presence of the C=C functionality. Compound **26.4** is cyclohexene, and **26.5** is cyclobutene. No position numbers are needed.

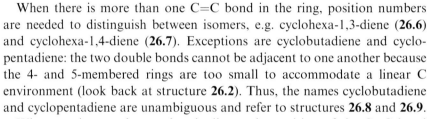

(26.4)                                                    (26.5)

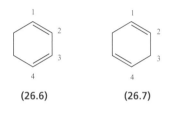

(26.6)          (26.7)

(26.8)          (26.9)

When there is more than one C=C bond in the ring, position numbers are needed to distinguish between isomers, e.g. cyclohexa-1,3-diene (**26.6**) and cyclohexa-1,4-diene (**26.7**). Exceptions are cyclobutadiene and cyclopentadiene: the two double bonds cannot be adjacent to one another because the 4- and 5-membered rings are too small to accommodate a linear C environment (look back at structure **26.2**). Thus, the names cyclobutadiene and cyclopentadiene are unambiguous and refer to structures **26.8** and **26.9**.

When naming a *substituted* cycloalkene, the position of the C=C bond takes priority over the substitution positions. Thus, the ring in compound **26.10** is numbered so as to place the C=C bond at position C(1). This information does not appear in the name because it is assumed knowledge. Cycloalkene **26.10** is therefore 3,5-dichlorocyclohexene. For a substituted cyclic diene, position numbers for the C=C bonds and the substituent are needed, and again the first C=C bond is placed at atom C(1). The direction of numbering the C atoms in **26.11** is chosen so as to give the C=C bonds the lowest position numbers, and then the ethyl substituent the lowest position number possible. Thus, **26.11** is 5-ethylcyclohexa-1,3-diene.

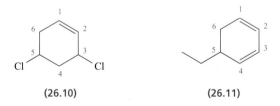

(26.10)                              (26.11)

### Ring conformation

The presence of one double bond in a cyclic hydrocarbon means that two of the C atoms are in planar environments. This makes a cycloalkene ring less flexible than its cycloalkane counterpart. Figure 26.2a shows a 'ball-and-stick' model of cyclopentene and a schematic representation of the structure. The planar part of the ring imposed by the HC=CH unit is clearly seen in the model. The part of the ring that has tetrahedral C atoms is puckered to prevent adjacent CH$_2$ units being eclipsed. When two (or more) C=C bonds are present, the flexibility of the ring is further reduced. Figure 26.2b shows

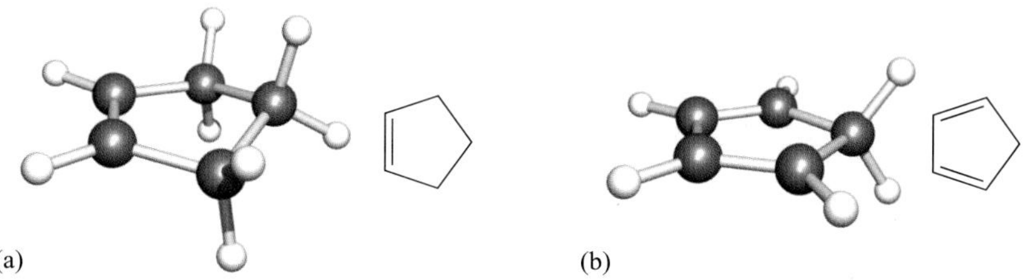
(a)
(b)

**Fig. 26.2** (a) A 'ball-and-stick' model and the structural formula of cyclopentene. There are three tetrahedral and two trigonal planar C atoms in the ring. (b) A 'ball-and-stick' model and the structural formula of cyclopentadiene. The ring contains one tetrahedral and four trigonal planar C atoms. Colour code: C, grey; H, white.

the structure of cyclopentadiene. The $C_5$-ring is planar, a consequence of there being four trigonal planar C atoms. In general:

- the ring conformation is constrained to being planar at each alkene functionality;
- partial rotation can occur around any C–C single bonds in the ring.

## 26.3 Syntheses of acyclic and cyclic alkenes

### Elimination reactions

***Dehydrogenation*** is the elimination of $H_2$.

Elimination of $H_2O$ from alcohols: see Section 30.5

The industrial formation of alkenes by *cracking* of alkanes was described in Section 25.4; the process involves *dehydrogenation* (elimination of $H_2$). Laboratory syntheses of alkenes usually employ *elimination reactions*. We discuss the details of such reactions in Chapter 28, but for now we simply illustrate their application in alkene synthesis. The elimination of HX (X = Cl, Br) from a halogenoalkane or $H_2O$ from an alcohol leads to an alkene. Elimination of HX is usually base-catalysed (equation 26.2) while elimination of $H_2O$ is catalysed by a strong acid (equations 26.3 and 26.4). The role of the base in reaction 26.2 is to remove a C–H proton and we discuss this further when we return to elimination reactions in Chapter 28.

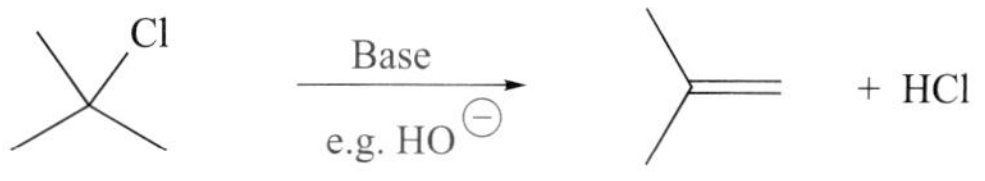
Cl
Base
e.g. HO⊖
+ HCl

(26.2)

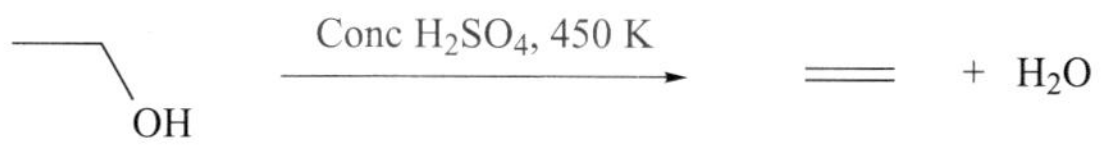
OH
Conc H2SO4, 450 K
+ H2O

(26.3)

Product distributions and isomer preferences are discussed in Chapter 30

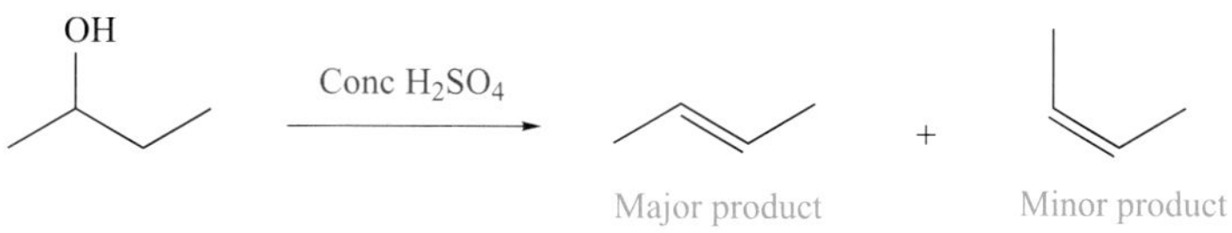
OH
Conc H2SO4
+
Major product
Minor product

(26.4)

### Diels–Alder cyclization

In a ***Diels–Alder reaction***, a double bond adds across the 1,4-positions of a 1,3-conjugated diene.

The dehydrogenation of a cycloalkane to give a *specific* cycloalkene or cyclodiene is synthetically difficult and is not usually a viable strategy. Cyclohexenes may be prepared by *[4 + 2] cycloadditions* called *Diels–Alder reactions*. The [4 + 2] notation comes from the fact that the reaction is between a $4\pi$-electron species (a diene) and a $2\pi$-electron species. In Section 25.5, we

**THEORETICAL AND CHEMICAL BACKGROUND**

## Box 26.2 *Endo* and *exo* labels in bicyclic compounds

Substituents in bicyclic compounds are often labelled *exo* or *endo* and this provides information about the relative orientation of the substituents. Consider the following structure:

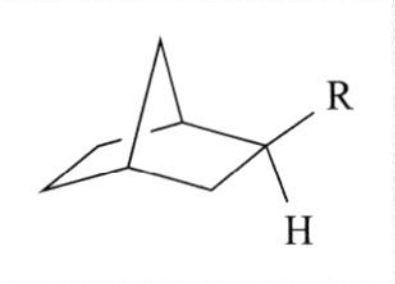

The rings are conformationally fixed with respect to each other and the R group always faces up and the H atom points down. The 'up' and 'down' descriptors are with respect to the CH$_2$ bridge. This bicyclic molecule can be described as containing three bridges, which are shown in blue, red and black below:

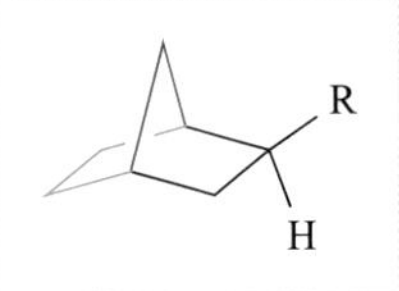

In order to decide whether the R group is *exo* or *endo*, we must look at the two *unsubstituted* bridges and decide which is the shorter bridge. In the diagram above, the red and blue bridges are the unsubstituted

ones, and of these, the red bridge is the shorter (one CH$_2$ unit as opposed to two). Now we apply the rule that *endo* refers to the substituent that lies closer to the longer unsubstituted bridge, and *exo* refers to the substituent that is closer to the shorter bridge. In the example above, therefore, the R group is *exo*. The *endo*-isomer of this compound has the structure:

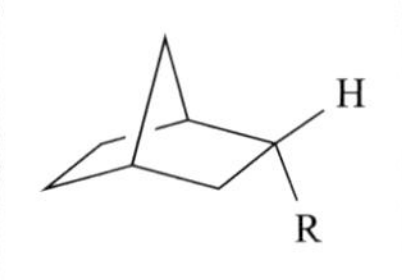

In cases where the two unsubstituted bridges are of equal length, the *exo* and *endo* labels are not required. For example in the compound:

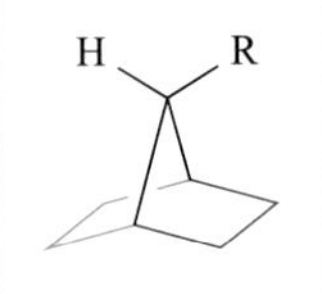

the H and R groups *both* face unsubstituted bridges that are the same length and so the H and R groups are in equivalent sites.

*[4 + 2] Cycloadditions*
(Diels–Alder reactions)
occur thermally.

stated that [2 + 2] cycloadditions only occur photochemically. In contrast, *[4 + 2] cycloadditions occur thermally*. In such a reaction, a double bond adds across the 1,4-positions of a 1,3-conjugated diene (see Section 13.5) to give a cyclohexene (equation 26.5). The reaction involves three 2-electron transfers, and the conversion of three C=C bonds and one C−C bond into one C=C and five C−C bonds. Note that your ability to draw a mechanism using curly arrows tells you nothing as to whether the reaction will proceed or not. For example, the [2 + 2] cycloaddition in equation 25.7 occurs photochemically, whereas the [4 + 2] cycloaddition in equation 26.5 occurs thermally. To understand these differences, one must investigate the behaviour of the molecular orbitals in detail, something which is beyond the scope of this book.[§]

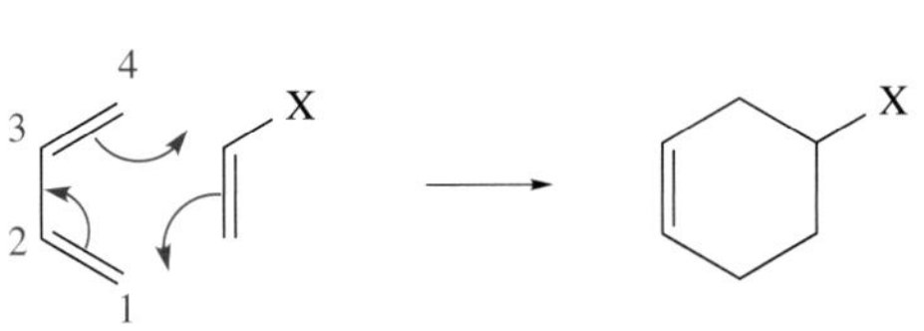

See text for identity of X

(26.5)

[§] For detailed discussion, see: J. Clayden, N. Greeves, S. Warren and P. Wothers (2001) *Organic Chemistry*, Oxford University Press, Oxford, Chapters 35 and 36.

The precursor containing *one* C=C is called a *dienophile* (it 'loves dienes'). Ethene (X=H in equation 26.5) is a poor dienophile. For a successful reaction, the dienophile should carry at least one *electron-withdrawing* substituent. For example, X could be Cl, CN, CH$_2$Cl, CH$_2$OH, CHO, CO$_2$H or CO$_2$R. Equation 26.5 illustrates a further prerequisite: for the [4 + 2] cyclo-addition, the diene must be able to undergo rotation about the central C−C bond so as to adopt the correct conformation (the so-called *s-cis* conformation) for ring closure. Dienes such as buta-1,3-diene can undergo rotation about the C−C bond (equation 26.6), and a Diels–Alder reaction is possible even though the required conformation is not the one that is sterically favoured.

Atom electronegativities: see Table 5.2

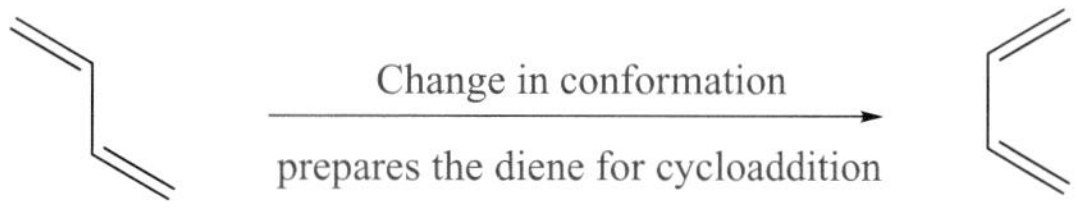

$$\text{Change in conformation} \atop \text{prepares the diene for cycloaddition} \qquad (26.6)$$

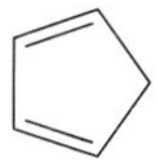

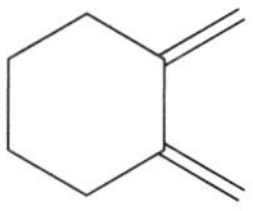

In some dienes, the *s-cis* conformation is fixed and such molecules are inherently prepared for [4 + 2] cycloadditions. Examples include cyclopenta-diene (**26.12**) and diene **26.13**. Reaction of a cyclic diene with a dienophile leads to a bridged bicyclic molecule, for example reaction 26.7. The fate of the cyclopentadiene ring in equation 26.7 is highlighted in blue.

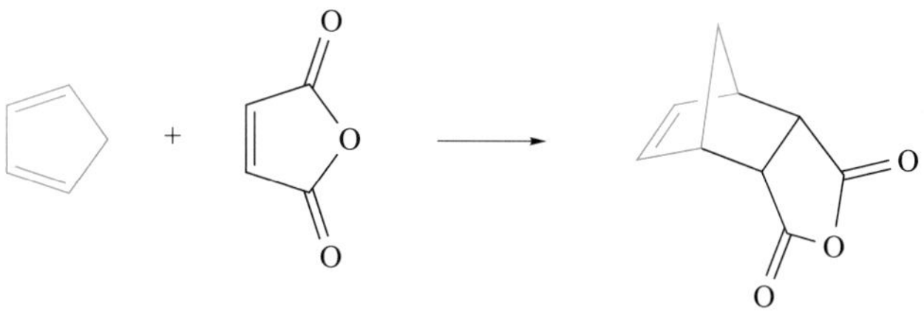

**(26.12)**    **(26.13)**

$$(26.7)$$

In equation 26.7, we have drawn the product with a specific stereochemistry. This can be better appreciated if we put in two H atoms as shown below. There are two possible products and *only* the one on the left is formed:

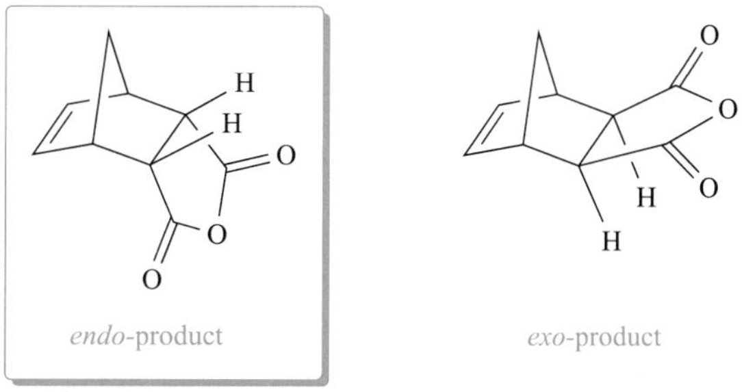

*endo*-product          *exo*-product

*Endo* and *exo* labels: see Box 26.2

In order to explain the *stereoselectivity* of the Diels–Alder reaction, we turn to molecular orbital theory and consider the interactions between the HOMO and LUMO of each reactant. The π-molecular orbitals of ethene and buta-1,3-diene were drawn in Figures 13.6a and 13.7, and these MOs are general for alkenes and 1,3-conjugated dienes. Interactions between the HOMO of the alkene and the LUMO of the conjugated diene, and between the HOMO of the diene and LUMO of the alkene, are required in order that the new C−C σ-bonds (equation 26.5) can form during the Diels–Alder cyclization. Figure 26.3 shows the HOMO and LUMO of each reactant, and illustrates that their symmetries are compatible with the evolution of σ-bonding interactions. Notice that the two reactants must approach one under the other so that σ-interactions evolve. The orbital interactions are further enhanced if the dienophile is so oriented that its electron-withdrawing

A ***stereoselective*** reaction gives, specifically, one product when, in theory, two or more products that differ in their stereochemistries are possible.

**Fig. 26.3** In a Diels–Alder reaction, the HOMO and LUMO of the diene overlap with the LUMO and HOMO of the dienophile (the alkene) to generate two C–C σ-interactions.

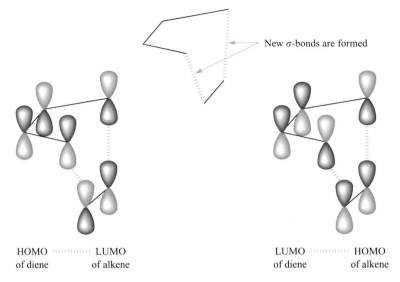

New σ-bonds are formed

HOMO ·········· LUMO
of diene    of alkene

LUMO ·········· HOMO
of diene    of alkene

substituents, X, are facing *under* the diene. The origin of this is an enhancement of orbital overlap. The crucial result of this orientation effect is that the dienophile adds to give the *endo*-product that we described earlier:

In related reactions, alkynes with electron-withdrawing substituents act as dienophiles and reaction 26.8 shows the formation of a substituted hexa-1,4-diene.

$$(26.8)$$

Aldrin

Dieldrin

The Diels–Alder reaction was applied to the manufacture of a group of extremely effective insecticides which included aldrin and its epoxy-derivative, dieldrin. These had widespread use from around 1950 to control termites. On the downside, these insecticides do not degrade in the environment and this is a major factor that led to their withdrawal from use in the 1980s.

## Cyclizations of acetylene

The trimerization and tetramerization of acetylene are also cycloadditions and occur when acetylene is heated in the presence of a nickel(0) or nickel(II) catalyst (equation 26.9).

$$\text{H}\!=\!\!=\!\!=\!\text{H} \xrightarrow[\text{catalyst}]{\text{Ni(0) or Ni(II)}} \bigcirc + \bigcirc \qquad (26.9)$$

The cycloaddition of buta-1,3-diene is also catalysed by nickel and other *d*-block metal compounds. This reaction is non-specific and yields dimers, trimers and tetramers (equation 26.10) as well as other products.

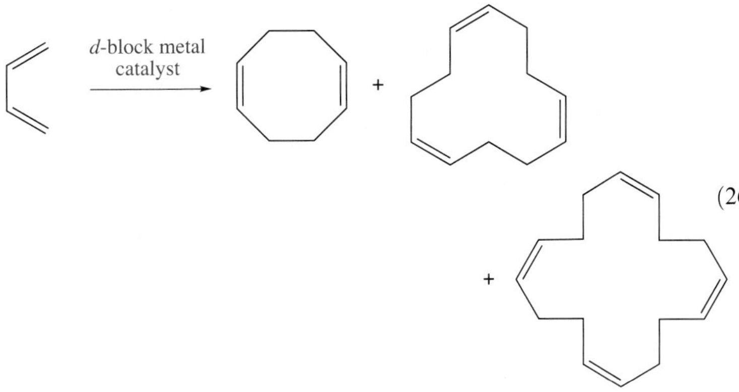

(26.10)

### 26.4    Reactions of alkenes 1: an introduction

In this section, we introduce some reactions of alkenes. In the next section, we go on to discuss the mechanism of electrophilic addition, and in Section 26.6 we continue our overview of alkene reactivity, using what we have learnt about mechanisms to explain the observed selectivity of the reactions.

Like alkanes, alkenes burn in $O_2$ to give $CO_2$ and $H_2O$ (equations 26.11 and 26.12).

$$C_4H_8 + 6O_2 \longrightarrow 4CO_2 + 4H_2O \qquad (26.11)$$
Butene

$$C_6H_{10} + \tfrac{17}{2}O_2 \longrightarrow 6CO_2 + 5H_2O \qquad (26.12)$$
Cyclohexene

Whereas alkanes undergo *substitution* reactions, the C=C functionality in an alkene undergoes *addition* reactions, a general representation of which is reaction 26.13.

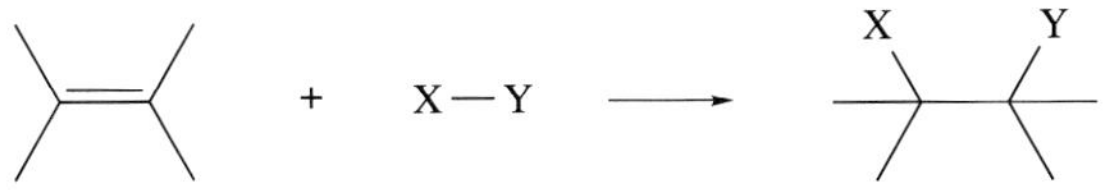

(26.13)

Addition reactions may be radical or electrophilic in nature, and can be summarized as follows:

*Typical addition reactions of alkenes* are:

- electrophilic addition;
- radical addition;
- polymerization.

### Hydrogenation

*Hydrogenation* is the addition of $H_2$.

The *hydrogenation* of an alkene involves the addition of $H_2$ and converts an unsaturated hydrocarbon into a saturated one (assuming all the C=C bonds undergo reaction). The reaction requires a catalyst, often a Ni, Pd or Pt metal surface (equation 26.14).

Catalysts: see Section 15.8

$$\text{(26.14)}$$

The metal surface is an example of a *heterogeneous catalyst*. The $H_2$ is adsorbed (equation 26.15), thereby providing a source of H atoms that can react with the alkene. This pathway has a lower activation energy than one in which $H_2$ reacts directly with the alkene.

$$\text{(26.15)}$$

Equation 26.15 can also be represented in terms of the dissociation of the $H_2$ molecule on a surface of close-packed metal atoms:

Similar hydrogenations convert cycloalkenes into cycloalkanes (reaction 26.16).

$$\text{(26.16)}$$

## Addition of $X_2$ (X = Cl or Br): formation of vicinal dihalides

The addition of $Cl_2$ or $Br_2$ to alkenes gives dichloro- or dibromo-derivatives (equations 26.17 and 26.18). In the products, the halogen atoms are attached to *adjacent* C atoms. Such compounds are known as *vicinal dihalides*.

Compare with the use of the term *vicinal coupling* between H atoms on adjacent C atoms: see Section 14.9

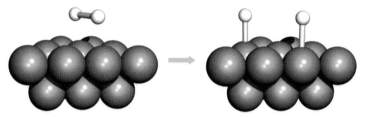

$$\text{(26.17)}$$

2,3-Dichloropentane

$$\text{(26.18)}$$

1,2-Dibromobutane

Reactions with $F_2$ are extremely violent, while those with $I_2$ are extremely slow or fail altogether. Chlorination and bromination take place at or below 298 K without the need for irradiation. This suggests that the addition does not involve radicals. However, if a source of radicals *is* available, addition still occurs but may be complicated by competing radical substitution reactions.

The decolorization of an aqueous solution of $Br_2$ (bromine water) is commonly used as a qualitative test for the presence of a C=C bond, although it is by no means definitive. Mixed products are obtained (see the discussion of the reactions of alkenes with $Br_2$ and $H_2O$ in Section 26.6).

### Addition of HX (X = Cl, Br or I): formation of halogenoalkanes

Equation 26.19 shows the reaction between an alkene and HCl, HBr or HI. The substrate in this example is a *symmetrical alkene* and when HBr is added, the product can only be 2-bromobutane.

$$(26.19)$$

2-Bromobutane

With an *unsymmetrical alkene,* two products are possible, as shown in equation 26.20 for the addition of HBr to propene. In practice, one product predominates and in reaction 26.20, 2-bromopropane is the major product. Notice that the reaction favours attachment of the halide to the *secondary (rather than primary) C atom* in the product.

$$(26.20)$$

2-Bromopropane      1-Bromopropane
Major product        Minor product

Similarly, when HCl reacts with 2-methylpropene, the major product is 2-chloro-2-methylpropane (equation 26.21). The preference is for the Cl atom to be attached to the *tertiary (rather than primary) C atom* in the product.

2-Chloro-2-methylpropane      1-Chloro-2-methylpropane
Major product          Minor product

$$(26.21)$$

The selectivity that is observed in reactions 26.20 and 26.21 can be explained if we look in detail at the mechanism of electrophilic addition.

## 26.5     The mechanism of electrophilic addition

Electrophiles, nucleophiles and 'curly arrows': see Section 24.10 for a general introduction

Additions to alkenes can occur by radical or electrophilic mechanisms. An electrophilic mechanism is favoured when the reaction is carried out in a *polar solvent,* while a radical mechanism requires a *radical initiator.*

### Addition of HBr to a symmetrical C=C bond

The HOMO of an alkene such as ethene or propene possesses C—C $\pi$-bonding character (Figure 13.6). The alkene can therefore act as a *nucleophile,* donating electrons from its HOMO to a species that has a suitably low-lying, unoccupied orbital to accept electrons. That species is an *electrophile.* Consider

**Fig. 26.4** A schematic representation of the reaction profile of the addition of HBr to $C_2H_4$. The rate-determining step is the formation of the intermediate carbenium ion. The intermediate is at a local energy minimum.

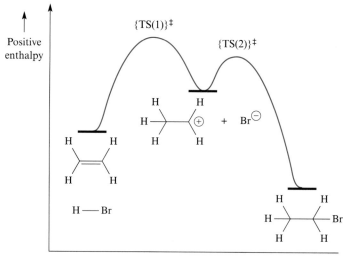

the reaction between $C_2H_4$ and HBr to give $C_2H_5Br$. The HBr molecule is polar (**26.14**), the Pauling electronegativity values being $\chi^P(H) = 2.2$ and $\chi^P(Br) = 3.0$. $H^{\delta+}$ is attracted towards the region of electron density of the C=C bond and, as a result, the H–Br bond is polarized even more. The first step in the reaction is shown in equation 26.22. Two $\pi$-electrons are transferred to the electrophile with the formation of a new C–H $\sigma$-bond and the heterolytic cleavage of the H–Br bond.

$$\text{(26.22)}$$

Carbenium ion

The C atom that does *not* form the new $\sigma$-bond is electron-deficient. It has a sextet of electrons and bears a formal positive charge. This carbon centre is called a *carbenium ion* or *carbocation*. The carbenium ion is an *intermediate* in the reaction, and is susceptible to attack by a nucleophile. In equation 26.22, the cleavage of the H–Br bond produces $Br^-$ and this acts as a nucleophile in the next step of the reaction as shown in equation 26.23.

$$\text{(26.23)}$$

Figure 26.4 shows the reaction profile for the reaction of $C_2H_4$ with HBr. The formation of the carbenium ion intermediate is the *rate-determining step*, i.e. the step in the pathway with the highest activation energy. Reaction of the carbenium ion with $Br^-$ occurs in a faster step (lower activation energy). The first transition state, $\{TS(1)\}^{\ddagger}$, may be described in terms of a species that is part way between the reactants and intermediate. The second transition state, $\{TS(2)\}^{\ddagger}$ in Figure 26.4, is considered as a species in which $Br^-$ has *begun to form a bond* with the carbenium ion, and a *stretched bond* may be drawn as in structure **26.15**. The mechanism described above can be extended to the addition of HCl and HI.

**A *carbenium ion* or *carbocation*** has the general formula $[R_3C]^+$ where R = H or an organic group.

(26.15)

Activation energy and reaction profiles: see Section 15.1

### Addition of HBr to an unsymmetrical C=C bond

When an electrophile attacks an *unsymmetrical alkene*, there are two possible pathways to carbenium ion formation. Consider the reaction of propene with HBr. The first step in the reaction is shown in scheme 26.24.

Primary carbenium ion

(26.24)

Secondary carbenium ion

We saw in Section 25.8 that carbon-centred radicals are stabilized by electron-releasing groups and, in the same way, electron-deficient carbenium ions are stabilized by electron-releasing substituents attached to the positively charged C centre. *Hyperconjugation* contributes to the stabilizing effects of having alkyl groups attached to the $C^+$ centre. Carbenium ions are planar at the central C atom and an $sp^2$ hybridization scheme is appropriate. This leaves a vacant $2p$ atomic orbital. Hyperconjugation arises because electron density from the C−H $\sigma$-bonds in the alkyl groups can be donated into the vacant $2p$ atomic orbital as shown in diagram **26.16**. (Donation of electron density into the C $2p$ orbital can also take place from C−C, C−Si and other $\sigma$-bonds of adjacent substituents.) The more alkyl groups there are attached to the central C atom, the more the positive charge can be stabilized by electron donation. As a result, the order of stabilities of carbenium ions, where R is an alkyl group, is:

$$[R_3C]^+ > [R_2CH]^+ > [RCH_2]^+ > [CH_3]^+$$

i.e. tertiary > secondary > primary > methyl.

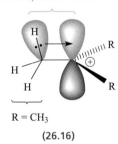
In the competitive steps in scheme 26.24, the formation of the secondary carbenium ion is therefore favoured over that of the primary carbenium ion. The next step in the reaction is nucleophilic attack by Br$^-$ on the carbenium ion intermediates to give 1-bromopropane and 2-bromopropane (equations 26.25 and 26.26).

1-Bromopropane

(26.25)

Br

2-Bromopropane

(26.26)

Because the secondary carbenium ion is more stable than the primary ion and is preferentially formed, it follows that the formation of 2-bromopropane is favoured over that of 1-bromopropane. This corresponds to the experimental result we gave for reaction 26.20. The reaction is said to be *regioselective* because, although the alkene can form two carbenium ions and hence two products, *one product is formed preferentially*.

The mechanism described for the addition of HBr to propene can be extended to the additions of HX (X = Cl, Br, I or other $\delta^-$ group) to other unsymmetrical alkenes. The general observation is that the H atom becomes attached to the less highly substituted C atom, thereby forming a tertiary carbenium ion in preference to a secondary or primary one, or a secondary carbenium ion in preference to a primary one. This is often referred to as *Markovnikov addition*.

In the ***Markovnikov addition*** of HX to an unsymmetrical alkene, the H atom attaches to the *less* highly substituted C atom, and the X atom attaches to the *more* highly substituted C atom. This follows from the carbenium ion intermediate being the most highly substituted possible.

### Addition of $X_2$ (X = Cl or Br)

Both $Cl_2$ and $Br_2$ are non-polar, but approach towards the C=C $\pi$-cloud induces a dipole, as shown for $Cl_2$ below:

Induced dipole: see Section 3.21

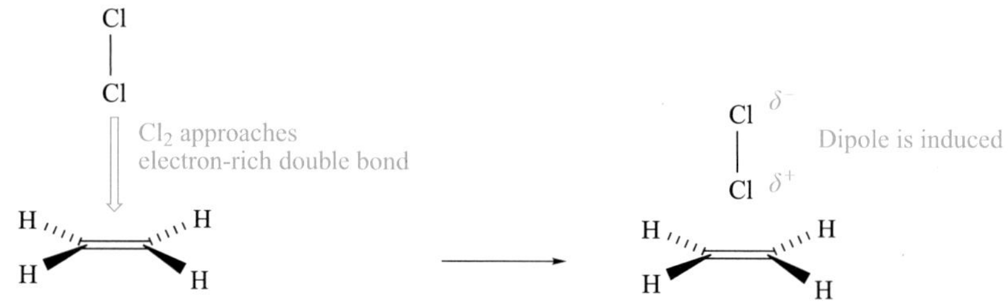

In the rate-determining step of the chlorination of an alkene, the $\pi$-electrons from the C=C bond are donated into the vacant $\sigma^*$ MO (the LUMO) of $Cl_2$. This results in the formation of a C−Cl bond and the heterolytic cleavage of the Cl−Cl bond (equation 26.27). This step generates $Cl^-$, which then acts as a nucleophile and attacks the carbenium ion in a faster step (equation 26.28).

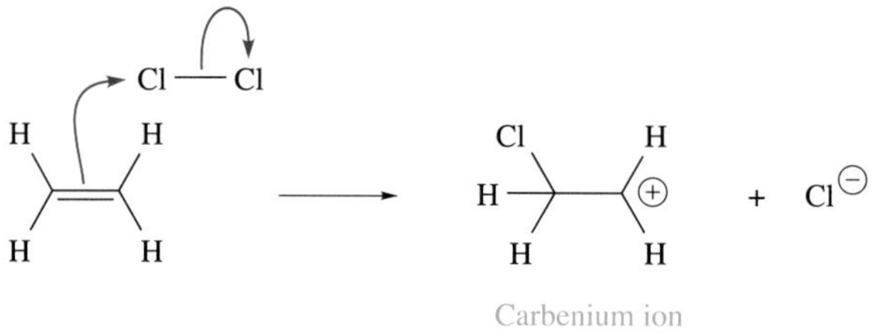

(26.27)

Carbenium ion

(26.28)

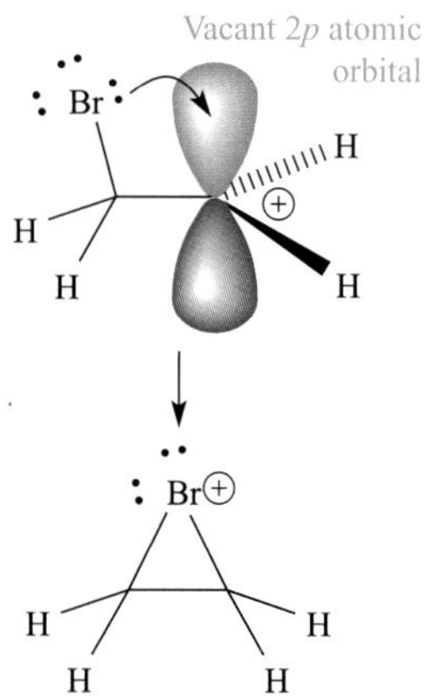

(26.17)

When the halogen is $Br_2$, there is experimental evidence that the intermediate is a *bromonium ion*. Just as we saw that carbenium ions are stabilized by hyperconjugation, we can view the stabilizing influence of the Br substituent in the intermediate by considering lone pair donation from Br to $C^+$ and evolution of the 3-membered ring. Structure **26.17** illustrates this. The reaction mechanism for the reaction of $Br_2$ with $C_2H_4$ is represented in equations 26.29 (the rate-determining step) and 26.30.

**Fig. 26.5** With extremely bulky substituents, it has been possible to isolate an example of a bromonium ion. (a) The structure of $[R_2CBrCR_2]^+$ (R = adamantyl) determined for the $[Br_3]^-$ salt by X-ray diffraction [H. Slebocka-Tilk *et al.* (1985) *J. Am. Chem. Soc.*, vol. 107, p. 4504], and (b) a scheme for its preparation. In the 3-membered ring in the cation, C–Br = 219 and 212 pm and C–C = 150 pm (in between a typical carbon–carbon single and double bond length). Colour code: C, grey; H, white; Br, brown.

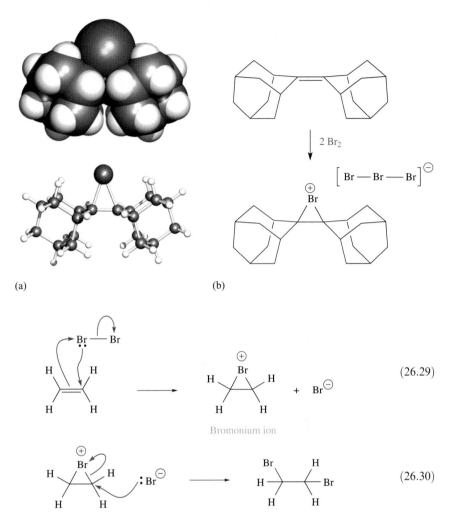

(a)                                (b)

$$(26.29)$$

Bromonium ion

$$(26.30)$$

(26.18)

Evidence for the bromonium ion comes from structural studies on an extremely sterically hindered cation (Figure 26.5) and from the stereo-selectivity shown in bromination reactions. For example, when cyclohexene is treated with $Br_2$, the reaction is completely selective as shown in equation 26.31. *None* of stereoisomer **26.18** is formed.

$$(26.31)$$

Diequatorial conformer is favoured, but diaxial is formed first

This observation is consistent with the presence of a bromonium ion intermediate, because the formation of the $BrC_2$ ring will block one side of the C–C bond from attack by the incoming nucleophile. Thus, $Br^-$ has to attack from the side opposite to the $BrC_2$ ring and when the ring opens in the second step of the reaction, the two Br substituents end up on opposite sides of the cyclohexene ring. This stereoselective addition is *anti-addition*, i.e. the two Br atoms add to opposite sides of the C–C bond. In a cyclic system, we can see this clearly. In an acyclic system, interpreting experimental data must take into account the fact that once an XY molecule has added to the C=C bond, the product can undergo rotation about the newly formed C–C bond. We now consider this scenario.

### *Syn*- and *anti*-addition

In the ***syn***-addition of XY to a C=C bond, X and Y add to the ***same*** side of the plane containing the two $sp^2$ C centres in the alkene. In the ***anti***-addition, X and Y add to ***opposite*** sides.

Let us consider the addition of HX to the alkene ABC=CAB shown in Figure 26.6. The two C atoms in the C=C bond are $sp^2$ hybridized and because of the bonding restrictions imposed by the C–C π-bond, the A and B substituents all lie in the same plane. Figure 26.6 shows *syn*- and *anti*-addition of HX to the alkene. If all the substituents had been equivalent (i.e. A = B in the figure), then the stereochemistries of the products formed from the *syn*- and *anti*-additions would have been the same. To test this out, draw two planar $C_2A_4$ molecules. To one, add HX in a *syn*-manner, and to the other, add HX in an *anti*-manner, copying the modes of additions from Figure 26.6. The products will be identical.

Now look at the case described in Figure 26.6. Each C atom in the alkene bears non-equivalent groups. *syn*-Addition of HX pushes *all* the A and B substituents on to the same side of the plane that contains the alkene molecule. Once addition has taken place, the molecule contains a C–C *single* bond, about which rotation can occur. One staggered conformer of the product is shown at the right-hand side in Figure 26.6. When *anti*-addition of HX occurs, one set of A and B substituents is pushed down and one set is pushed up. This gives an arrangement of substituents in the product that is *different* from that in the product formed from the *syn*-addition. *The stereochemistries of the products are different.*

In most cases, the nucleophile adds after the electrophile, as we have seen in the examples described earlier in this section. The nature of the intermediate cation is therefore of importance in dictating whether *syn*-, *anti*- or both types of addition occur. In an intermediate such as the bromonium ion, we have already seen that one side of the molecule is blocked with respect to attack by the nucleophile, with the result that *anti*-addition takes place. If the intermediate is a carbenium ion like that in equation 26.27, the C$^+$ centre is in a planar environment and can be attacked from

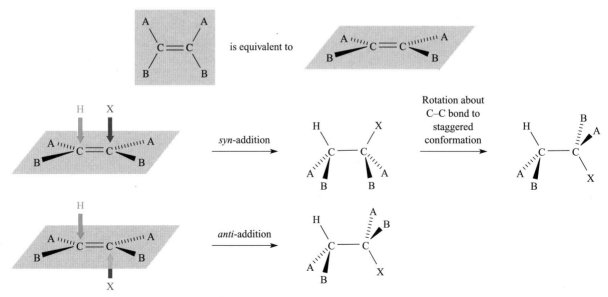

**Fig. 26.6** An alkene ABC=CAB is planar ($sp^2$ hybridized C). When H and X atoms of HX add to the C=C bond, they can do so from the same side (*syn*-addition) or from different sides (*anti*-addition). The presence of the two A and B substituents on the alkene means that the stereochemistries of the products of the *syn*- and *anti*-additions are not the same.

either side:

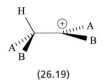

(26.19)

**26.19** is the carbenium ion that may be formed in the example in Figure 26.6 when the alkene adds $H^+$. If this carbenium ion has a relatively long lifetime, rotation about the C–C bond may occur *before* $X^-$ attacks. This will lead to non-specific addition:

or

and there will also be nucleophilic attack from the lower side. In practice, other factors such as steric demands of the substituents may favour *syn*- or *anti*-addition.

The outcome of the *syn*- and *anti*-additions shown in Figure 26.6 can be represented in terms of the Newman projections shown in equation 26.32. These projections correspond exactly to the products drawn in Figure 26.6.

> **▶▶▶**
> Newman projections and use of wedges: see Section 24.9

(26.32)

Because there are different substituents in each C atom in the alkene, the addition of HX creates *two* asymmetric C centres in the product. Each product in equation 26.32 consists of a pair of enantiomers and these are shown in Figure 26.7. The four stereoisomers in Figure 26.7 consist of two

**Fig. 26.7** The two pairs of enantiomers (drawn as Newman projections related by reflections through a mirror) formed from the *syn*- and *anti*-addition of HX to an alkene ABC=CAB (see equation 26.32). The grey planes in the figure represent mirror planes.

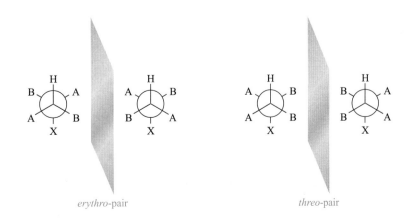

*erythro*-pair            *threo*-pair

**Look back to Figure 24.13 and review enantiomers and diastereoisomers**

enantiomeric pairs *and* four diastereoisomeric pairs, and the labels *threo-* and *erythro-* allow one to distinguish between the stereoisomers. These names originate from the aldoses threose and erythrose (structures **26.20** and **26.21**) and can be assigned as follows:

(2R,3S)-threose    (2R,3R)-erythrose

**(26.20)**       **(26.21)**

- Draw Newman projections of the enantiomers of the addition product in *staggered* conformations. View the molecule along the C−C bond that was originally the C=C bond of the alkene (Figure 26.7).
- Ensure that the H and X groups from the reagent HX point up and down in the Newman projection.
- Two of the enantiomers have *identical* groups on the *same* side of the Newman projection and these enantiomers make up the *threo*-pair.
- Two of the enantiomers have *different* groups on the *same* side of the Newman projection and these enantiomers make up the *erythro*-pair.

**26.6**     **Reactions of alkenes 2: additions and C=C oxidation and cleavage**

In this section we continue our survey of reactions of alkenes by considering more electrophilic additions in which we apply the mechanistic details from the previous section. We also consider some oxidation reactions and reactions that cleave the C=C bond.

### Addition of $H_2O$: formation of alcohols

Alkenes do not react directly with water, but if the reagents are heated in the presence of concentrated $H_2SO_4$, the alkene adds $H_2O$ and is converted into an alcohol. This reaction is used industrially to manufacture ethanol (equation 26.33).

$$ \text{(26.33)} $$

There is, however, a problem with using this as a general method for preparing alcohols from alkenes: heating an alcohol with a strong acid can dehydrate the alcohol, giving an alkene (see equation 26.3). Thus, the reaction can be driven in either direction. The acid-catalysed addition of $H_2O$ to an alkene may be useful synthetically if the intermediate is a tertiary carbenium ion. For example, 2-methylpropene undergoes the reaction shown in equation 26.34. Two products are possible, depending on whether a primary or tertiary carbenium ion is formed. The latter predominates and 2-methylpropan-2-ol is the favoured product. Equation 26.34 shows the formation of the tertiary carbenium ion and subsequent attack by $H_2O$. Water is polar (**26.22**) and acts as a nucleophile by donating a pair of electrons. The positive charge is transferred to the O atom and this triggers loss of $H^+$ to give the alcohol.

**Molecular dipole moments: see Section 6.13**

**(26.22)**

$$ \text{(26.34)} $$

An alternative and successful strategy for adding $H_2O$ to an alkene is by using mercury(II) acetate, $Hg(O_2CMe)_2$. The reaction of an alkene with $Hg(O_2CMe)_2$ gives an organometallic complex that reacts with $H_2O$ to yield an organomercury compound. On treatment with sodium borohydride, this produces the desired alcohol. In reaction 26.35, the product contains the OH substituent attached to the secondary rather than the primary C atom.

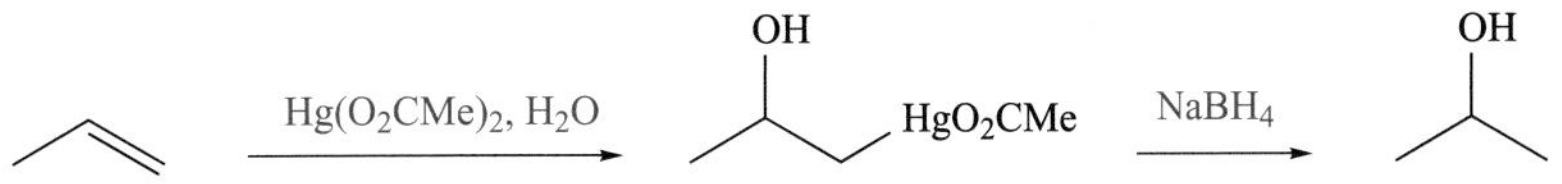

(26.35)

Mercury(II) acts as an electrophile and is attacked by the electron-rich C=C bond. This forms a cyclic intermediate as shown in the first step in scheme 26.36. Nucleophilic $H_2O$ then attacks the more highly substituted C centre in the intermediate, opening the $C_2Hg$ ring as shown in scheme 26.36. The positive charge now resides on the 3-coordinate O atom and a proton is readily lost to give the OH group.

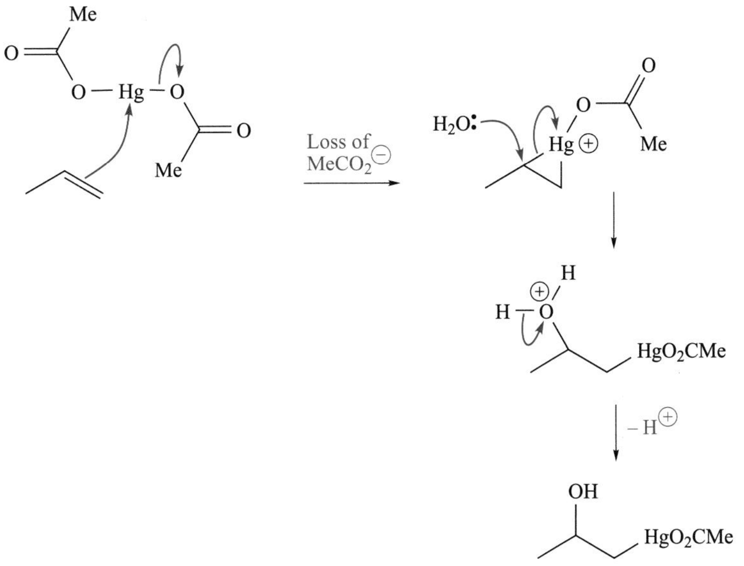

(26.36)

## Addition of $H_2O$ and $X_2$ (X = Cl or Br): formation of halohydrins

If $Cl_2$ or $Br_2$ *and* $H_2O$ react together with an alkene, the reactions proceed as illustrated in equations 26.37 and 26.38. The products are *halohydrins*: a *chlorohydrin* in reaction 26.37, and a *bromohydrin* in reaction 26.38.

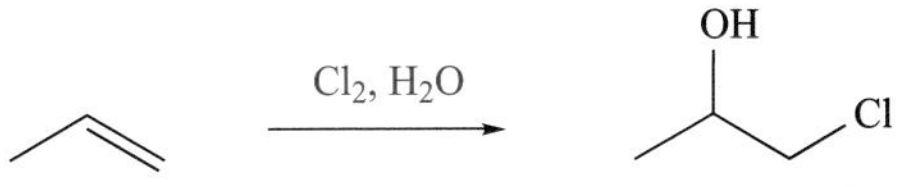

1-Chloropropan-2-ol

(26.37)

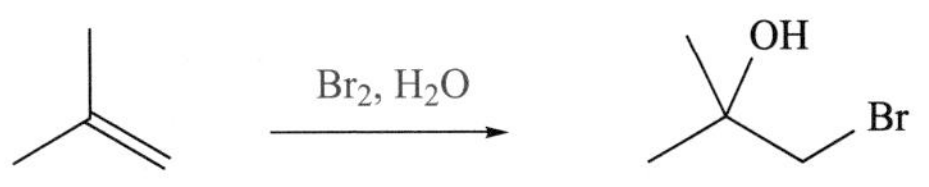

1-Bromo-2-methylpropan-2-ol

(26.38)

The OH group shows a preference to be attached to the secondary or tertiary, rather than primary, C atom. This is readily explained in terms of the relative stabilities of the intermediate carbenium ions. The first step in reaction 26.37 is attack on $Cl_2$ by the $\pi$-electrons of the C=C bond (equation 26.39). This is exactly the same as the rate-determining step in chlorination of an alkene. The secondary carbenium ion is stabilized to a greater extent than the primary carbenium ion.

$$(26.39)$$

If water is present, it competes with $Cl^-$ during the second step of the reaction. Nucleophilic attack by $Cl^-$ on either carbenium ion in equation 26.39 leads to 1,2-dichloropropane. Nucleophilic attack by $H_2O$ leads to 1-chloropropan-2-ol as the major product and 2-chloropropan-1-ol as the minor product. The mechanism for the more favoured pathway is shown in scheme 26.40.

$$(26.40)$$

### Addition of HBr to 1-methylcyclohexene

We use the reaction between HBr and 1-methylcyclohexene as an example of addition to a cycloalkene that is substituted at one of the C=C sites so as to give an unsymmetrical alkene. Equation 26.41 shows the favoured pathway of this reaction which goes through a tertiary (rather than a secondary) carbenium ion. Keep in mind that the cyclohexane ring is not planar. Structures **26.23** and **26.24** show the two conformers of the product in reaction 26.41.

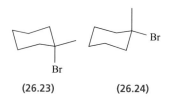

(26.23)          (26.24)

$$(26.41)$$

### Oxidation: formation of diols

The oxidation of an alkene by $OsO_4$ (tetrahedral, Os(VIII) oxide) leads to the formation of a *vicinal diol*. The *diol* part of the name indicates that there are two OH groups, and *vicinal* indicates that the OH groups are attached to adjacent C atoms. Reaction 26.42 shows the conversion of but-1-ene to butane-1,2-diol.

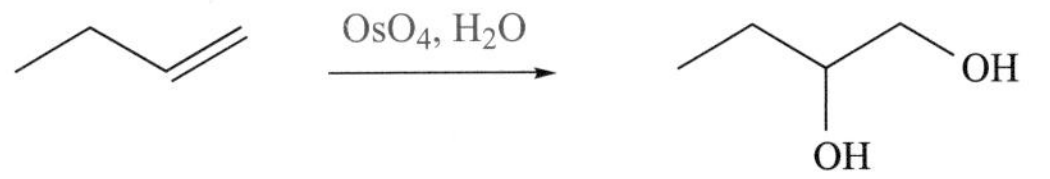

$$\text{(26.42)}$$

The proposed mechanism involves the formation of an $OsO_4$-bridged intermediate (equation 26.43) which is cleaved by $H_2O$ in the next step of the reaction. The addition of the two OH groups is specifically *syn*, i.e. the OH substituents add to the same side of the C=C bond. During the reaction Os(VIII) is reduced to Os(VI) and the alkene is oxidized.

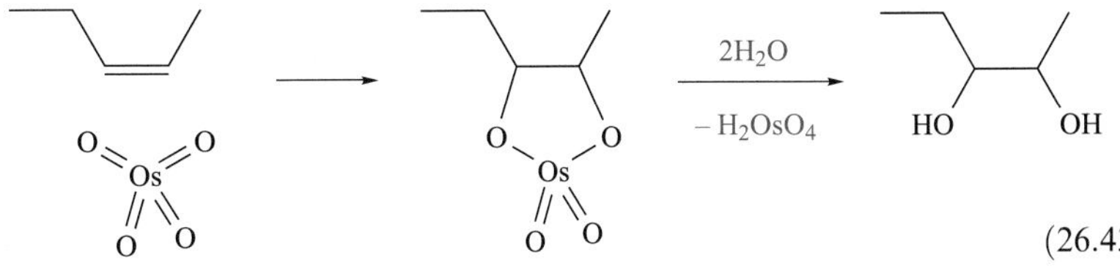

$$\text{(26.43)}$$

Now let us look more closely at the stereochemistry. Although the *addition* is *syn*, we must remember (see Figure 26.6) that once the C=C bond has been converted into a C−C bond, conformational changes occur to minimize the steric energy of the molecule. The molecule of pent-2-ene shown in equation 26.43 has been drawn in the (*Z*)-configuration, but we have not specified anything about the stereochemistry of the product. The *syn*-addition to a (*Z*)-alkene can be represented schematically as follows:

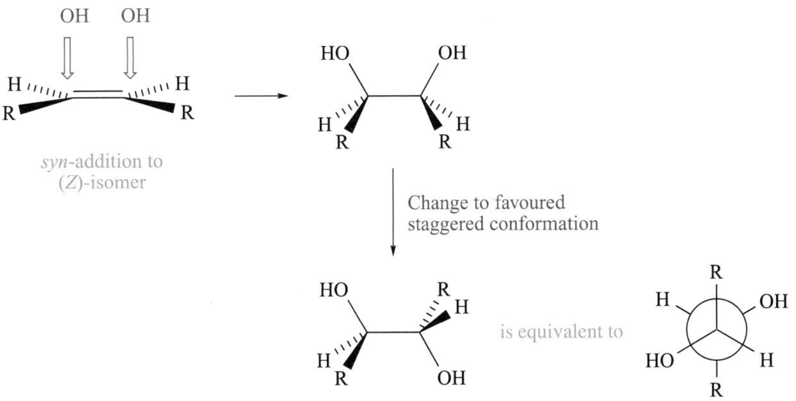

After the product has taken up a minimum energy conformation, a Newman projection drawn with the R groups from the alkene facing up and down has the two OH groups on opposite sides. Now consider *syn*-addition to an (*E*)-alkene. This is shown schematically as follows:

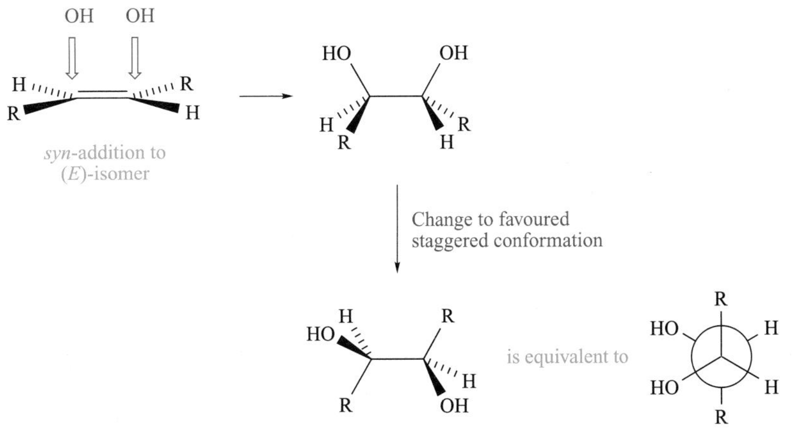

After allowing the product to adopt a favourable staggered conformation with the R groups *anti* to one another, a Newman projection confirms that the OH groups are now *syn*. Thus, although *syn*-addition occurs to *both* the (Z)- and (E)-isomers of an alkene, the stereoisomers (diastereoisomers) that are formed are not the same (unless R = H).

The conversion of a non-terminal alkene to a diol can also be carried out using cold alkaline $KMnO_4$. It is proposed that the tetrahedral $[MnO_4]^-$ ion acts in the same way as $OsO_4$, forming a similar intermediate, and *syn*-addition is again observed. During the reaction, Mn is reduced from Mn(VII) to Mn(V).

The old name for a diol is a *glycol* and this name is still used (see Box 24.2). Ethylene glycol (ethane-1,2-diol, **26.25**) is commercially important as an 'anti-freeze' additive to motor vehicle radiator water. It has a low melting point (261 K), high boiling point (471 K) and is completely miscible with water. Industrially, it is not made by the method described here, but rather by the hydration of ethylene oxide (or oxirane). Ethylene oxide is an example of an epoxide, itself a member of the family of cyclic ethers. We discuss their preparation and properties in Chapter 29.

**(26.25)**

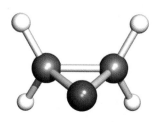

Ethylene oxide

### Alkene cleavage: reaction with $O_3$

The reaction of an alkene with $O_3$ leads to a cyclic *ozonide*. The initial addition product of the alkene and $O_3$ rapidly isomerizes to the ozonide shown in equation 26.44. The C=C bond is completely cleaved during the reaction; the two C atoms in the $C_2O_3$ ring are the original atoms of the C=C functionality. **Caution!** Ozonides are often explosive.

$$\text{(reaction scheme)} \qquad \xrightarrow{\;O_3\;} \qquad \text{(ozonide)} \tag{26.44}$$

An important application of this reaction lies not in the isolation of ozonides, but in their hydrolysis products. Reductive hydrolysis yields aldehydes or ketones depending on the substituents present in the original alkene. Reactions 26.45 and 26.46 are two examples. The $Me_2S$ or thiourea **(26.26)** acts as a reducing agent and is converted into $Me_2SO$ (dimethyl-sulfoxide, DMSO) or **26.27** respectively. Although $Me_2S$ is efficient in this role, it has the disadvantage of having an offensive smell.

**(26.26)**     **(26.27)**

$$\text{(ozonide)} \xrightarrow{\;Me_2S \text{ or thiourea}\;} \underset{\substack{\text{Butanal} \\ \text{(aldehyde)}}}{\text{(CHO product)}} + \underset{\text{Formaldehyde}}{\text{H—CHO}} \tag{26.45}$$

$$\text{(ozonide)} \xrightarrow{\;Me_2S \text{ or thiourea}\;} \underset{\substack{\text{Butanone} \\ \text{(ketone)}}}{\text{(ketone product)}} + \underset{\text{Acetaldehyde}}{\text{(CHO product)}} \tag{26.46}$$

## Radical substitution and addition in alkenes

### Radical substitution: bromination

Vinylic H atoms

Allylic H atoms

**(26.28)**

Many alkenes contain alkyl groups in addition to the C=C functionality, and radical substitutions are possible. Hydrogen atoms attached *directly* to the C atoms of the C=C bond are called *vinylic H atoms* (structure **26.28**) and are difficult to abstract. An H atom attached to the C atom *adjacent to a C=C bond* is called an *allylic H atom* (structure **26.28**) and is more easily abstracted.

After initiation (see Section 25.7), the radical reaction of propene with $Br_2$ continues with propagation steps 26.47 and 26.48. An *allylic* H atom is abstracted in step 26.47 in preference to a vinylic H atom.

$$\text{CH}_2=\text{CHCH}_3 + \text{Br}^{\bullet} \longrightarrow \text{HBr} + {}^{\bullet}\text{CH}_2\text{CH}=\text{CH}_2 \qquad (26.47)$$

See problem 26.16

$$ {}^{\bullet}\text{CH}_2\text{CH}=\text{CH}_2 + \text{Br}_2 \longrightarrow \text{Br}^{\bullet} + \text{BrCH}_2\text{CH}=\text{CH}_2 \qquad (26.48)$$

The radical formed in step 26.47 is an *allyl* radical and is *more* stable than any of the alkyl radicals that we discussed in Section 25.7. The radical stability order (where R = alkyl) is:

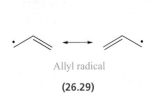

Allyl radical

**(26.29)**

$$ {}^{\bullet}\text{CH}_2\text{CH}=\text{CH}_2 \longleftrightarrow \text{CH}_2=\text{CHCH}_2^{\bullet} $$

This enhanced stability can be attributed to the contributions made by the two resonance structures drawn in **26.29**. In general, the more resonance structures that can be drawn, the more stable a species is. (Look back at Section 7.2.) In terms of a delocalized picture, the π-bonding in the allyl radical can be described in much the same way as in the allyl anion (see Figure 13.6). Both the allyl radical and allyl anion are *planar* and a π-system can therefore develop over the $C_3$ framework. Delocalizing the odd electron over the three C atoms is a stabilizing influence, and tells the same story as the combination of the two resonance structures in **26.29**.

Radical substitution in other alkenes follows the same pattern as above, with an allylic H atom being preferentially abstracted. For example, in reaction 26.49, substitution is specifically at the site *adjacent* to the C=C bond.

$$ \text{(cyclohexene)} + \text{Br}^{\bullet} \longrightarrow \text{HBr} + \text{(3-bromocyclohexene)} \qquad (26.49)$$

**(26.30)**

Specific though this H abstraction is, using $Br_2$ as the brominating agent leads to a problem: competitive *radical addition* to the C=C bond. Usually, $Br_2$ is replaced by the brominating agent *N*-bromosuccinimide (NBS, **26.30**) in $CCl_4$ solution. Initiation of the reaction requires the use of a *radical* *initiator*, the normal choice being an organic peroxide (**26.31a**), often with $R = C_6H_5C(O)$ (**26.31b**).

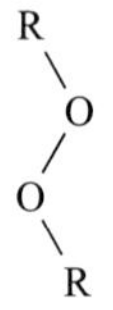

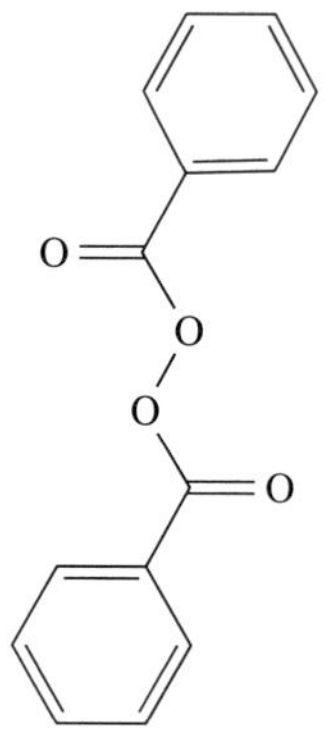

(26.31a)    (26.31b)

### Radical addition: bromination

We exemplify radical addition to a C=C bond by considering the reaction between HBr and 2-methylpropene, an *unsymmetrical* alkene. To facilitate a radical rather than electrophilic addition, a radical initiator must be present. Photolysis of the peroxide **26.31** results in homolytic fission of the O–O bond and formation of RO˙ radicals (equation 26.50). Propagation step 26.51 produces Br˙ radicals for reaction with the alkene.

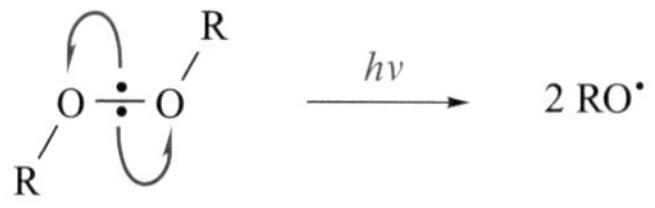

$$2\ RO^{\bullet} \tag{26.50}$$

$$RO \overset{\bullet}{-} H \ + \ Br^{\bullet} \tag{26.51}$$

The Br˙ radical now adds to the C=C bond and two pathways are possible (scheme 26.52). The more stable radical is the more highly substituted one and the more favoured pathway is highlighted in scheme 26.52. Note that reaction of Br˙ with the alkene does not abstract a vinylic H atom, but competitive abstraction of allylic H atoms will occur. Collision between an alkyl radical and HBr leads to the formation of the brominated product (equation 26.53). Termination steps in the radical chain involve reactions between two radicals.

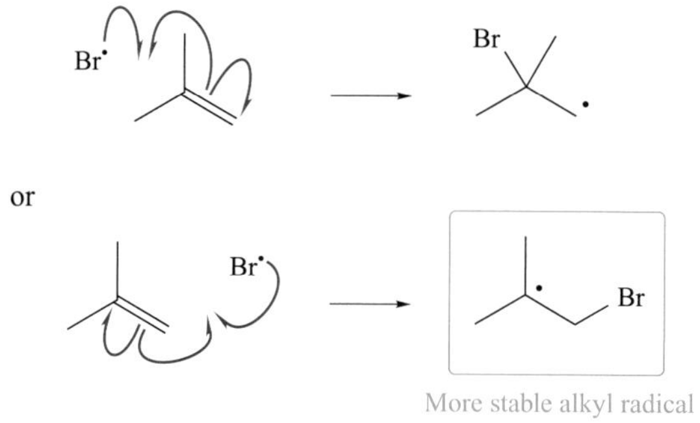

or

$$\tag{26.52}$$

More stable alkyl radical

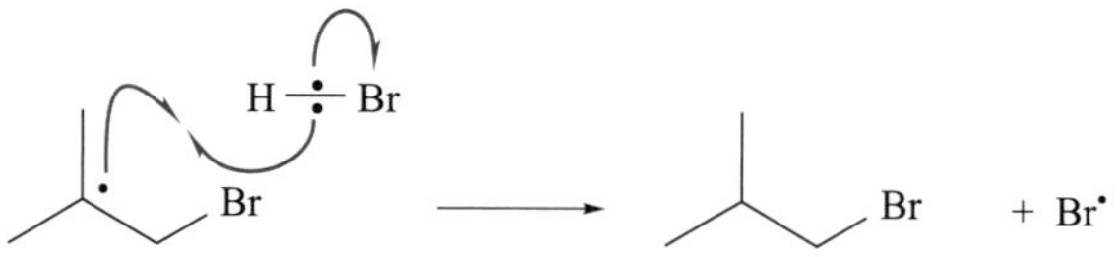

$$\tag{26.53}$$

The important point to notice in this final step is that the *radical addition* of HBr to an unsymmetrical alkene leads to a product in which the Br atom is attached to the C atom with the *lesser number of R substituents*. This is in contrast to the *electrophilic addition* of HBr which leads to a product in which the Br atom is attached to the C atom with the *greater number of R substituents*. In both cases, the product selectivity is controlled by the stability of the intermediate and the differences can be summarized as follows:

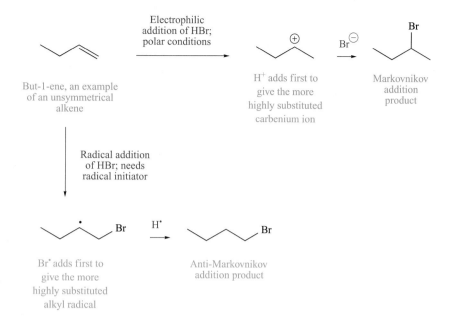

## 26.8    Polymerization of alkenes

A *polymer* is a
macromolecule containing
repeating units derived from
a *monomer*.

A *polymer* is a macromolecule that consists of repeating units. Alkenes undergo *addition polymerization*, a process in which the alkene units (*monomers*) add to one another to gives a polymer. Addition polymerization is of huge commercial importance (see Box 26.3). Figure 26.8a shows the ways in

**Fig. 26.8** (a) End uses of $C_2H_4$, estimated world consumption in 2000. [Data: www.shellchemicals.com] (b) Worldwide uses of PVC in 2001; the total demand was 26.3 million metric tons. [Data: *Chemical & Engineering News* (2002) November 25 issue, p. 12.]

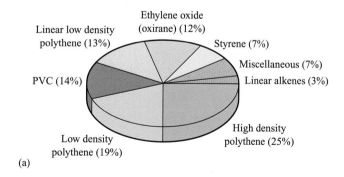

(a)

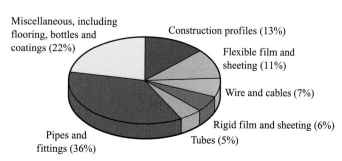

(b)

**COMMERCIAL AND LABORATORY APPLICATIONS**

## Box 26.3 Addition polymers

Some commercial polymers are listed in the table below along with the monomer from which they are made:

| Commercial name of polymer | Name of monomer | Structure of monomer |
|---|---|---|
| Polythene (polyethylene, PET) | Ethene | 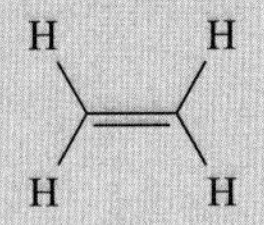 H H / H H |
| Polyvinylchloride (PVC) | Chloroethene | 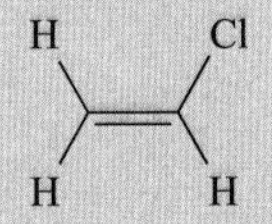 H Cl / H H |
| Polypropene (polypropylene, PP) | Propene | 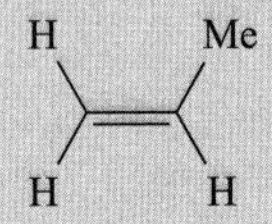 H Me / H H |
| Teflon (PTFE) | Tetrafluoroethene | 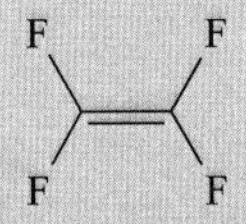 F F / F F |
| Polystyrene | Styrene (phenylethene) | 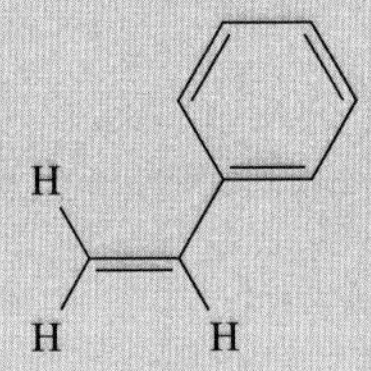 H / H H |
| Perspex | Methyl 2-methylpropenoate (methyl methylacrylate) | 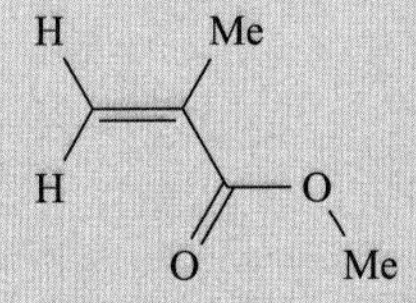 H Me / H O O Me |

### Copolymers

All the above polymers are examples of homopolymers: only one monomer is used in the manufacture of the polymer. When more than one monomer is used, the product is a *copolymer*. Copolymerization is one method of 'tuning' the properties of a polymer to meet commercial demands. Three types of copolymer in which the monomer units are represented by A and B are:

- an *alternating copolymer* in which the monomer sequence is ABABABABAB...
- a *block copolymer* which may possess a sequence such as AAABBBAAAABBB...
- a *random copolymer* in which the arrangement of monomers is irregular, e.g. ABBABAABA...

### Isotactic, syndiotactic and atactic polymers

In a polymer such as polypropene, the substituents can be arranged in several ways with respect to the carbon backbone of the polymer chain. This arrangement is important, because it affects the way that the chains pack together in the solid state.

In an *isotactic* arrangement, all the methyl groups (shown in red in the diagrams opposite) in polypropene are on the same side of the chain (shown in green):

In a *syndiotactic* arrangement, the methyl groups are regularly arranged in an alternating sequence along the carbon chain:

In an *atactic* arrangement, the methyl groups are randomly arranged along the carbon backbone:

In the solid state, molecules of both isotactic and syndiotactic polypropene pack efficiently and the material is crystalline. Atactic polypropene, however, is soft and elastic. Whereas isotactic and syndiotactic polypropene can be used, for example, as a material for plastic pipes and sheeting, atactic polypropene has no such commercial application.

### Recycling of plastics

Polymers have an enormous range of applications, e.g. uses of PVC are illustrated in Figure 26.8b. Life without polymers and plastics seems unthinkable. However, an environmentally conscious society must consider what to do with the many thousands of tonnes of waste artificial plastics that are awaiting disposal every year, and many recycling schemes are now in place. The pie chart below summarizes the 1990 attitudes towards disposal of plastics in Western Europe. The current market for degradable (e.g. photodegradable or biodegradable) plastics is steadily increasing.

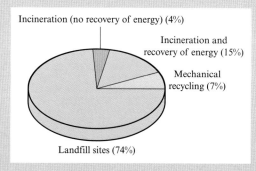

Incineration (no recovery of energy) (4%)

Incineration and recovery of energy (15%)

Mechanical recycling (7%)

Landfill sites (74%)

Data from: *Chemistry & Industry* (1992), p. 401.

which ethene is currently used, and Figure 26.8b illustrates the diversity of uses of polyvinylchloride (PVC).

## Radical addition polymerization

The polymerization of an alkene such as ethene may be a radical reaction requiring an initiator such as an organic peroxide (structure **26.31**). Equation 26.54 shows the initiation step where the initiator is represented by $Y_2$.

$$Y \overset{\cdot\cdot}{-} Y \xrightarrow{\;h\nu\;} 2\,Y^{\bullet} \tag{26.54}$$

When the radicals collide with alkene molecules (the monomers), propagation of the radical chain reaction begins. Equation 26.55 shows how attack of $Y^{\bullet}$ at a C atom of ethene leads to an alkyl radical.

$$ \tag{26.55}$$

The propagation of the polymerization reaction continues according to equations 26.56 and 26.57, and the polymer chain continues to grow in sequential radical steps of this type.

$$ \tag{26.56}$$

$$ \tag{26.57}$$

Alternatively, a non-radical product may be formed in a propagation step (equation 26.58) or termination step (equation 26.59).

$$ \tag{26.58}$$

$$ \tag{26.59}$$

The overall polymerization process with initiator $Y_2$ is summarized in reaction 26.60 where $n$ is any integer between 1 and $\infty$. The length of the polymer chain depends on the point at which termination occurs.

$$n \; \text{CH}_2=\text{CH}_2 \longrightarrow \text{Y} \cdot \left( \text{CH}_2-\text{CH}_2 \right)_n \text{Y} \qquad (26.60)$$

Monomer                    Polymer

If the monomer is the type $RHC=CH_2$ where $R \neq H$, then a propagation step analogous to, say, reaction 26.55 could lead to either a primary or secondary radical. The preference for the secondary radical intermediates controls the formation of regular polymer and end-of-chapter problem 26.18 asks you to consider this aspect of polymer growth.

An interesting complication occurs if the monomer is a diene. Consider attack by radical $Y^{\bullet}$ on one of the C=C bonds in octa-1,7-diene (equation 26.61).

$$(26.61)$$

The radical produced in this propagation step can react with another molecule of the monomer, or can undergo *intramolecular* reaction 26.62. The cyclic species formed can become involved in the polymerization process, or can react with the initiator or another organic radical in a termination step.

$$(26.62)$$

It is not difficult to envisage how many problems there are associated with the control of a radical polymerization process in which the monomer is an alkene or diene. Nonetheless, the radical addition polymerization of alkenes remains an important method of producing polymers on an industrial scale. Polymers that are made by the radical method include polyethene (polythene), polystyrene, PVC and PTFE (see Box 26.3).

## Cationic polymerization

Isoprene

**(26.32)**

Another method of polymerizing alkenes is *cationic polymerization* and this is a suitable strategy when a tertiary carbenium ion can be formed from the alkene monomer. 2-Methylpropene (isobutene) is used in the manufacture of butyl rubber. This is a copolymer of 2-methylpropene and isoprene **(26.32)** containing 1–3% of isoprene. Butyl rubber is resistant to atmospheric degradation (initiated by $O_3$) and has widespread uses in, for example, vehicle tyre inner tubes and liners for garden ponds. The polymerization of 2-methylpropene is initiated by providing a source of $H^+$. Electrophilic addition of $H^+$ to the monomer gives the tertiary carbenium ion which then undergoes nucleophilic attack by the $\pi$-bond of another monomer (scheme 26.63).

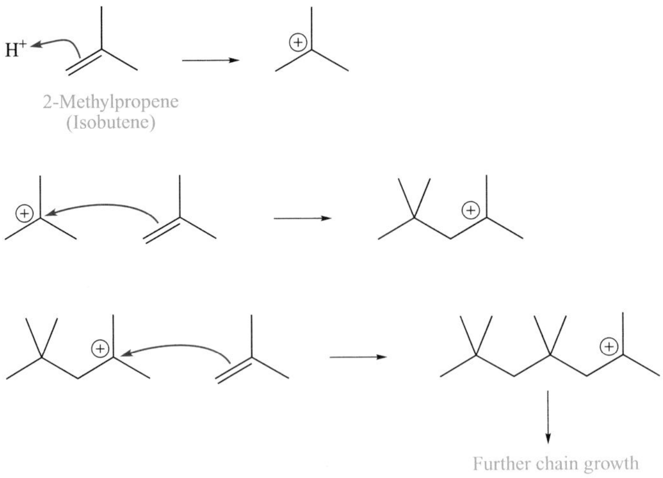

2-Methylpropene
(Isobutene)

Further chain growth          (26.63)

## Anionic polymerization

So far, we have seen alkenes *acting as nucleophiles* and attacking electrophiles. However, when an alkene has an electron-withdrawing substituent such as CN that can stabilize a negative charge, the alkene may *act as an electrophile*. Use is made of this characteristic in the polymerization of acrylonitrile to give synthetic fibres manufactured as Orlon and Acrilan. The addition polymerization is an example of *anionic polymerization*. The initiation step is the reaction of $[NH_2]^-$ with the monomer to give an active nucleophile; $[NH_2]^-$ is generated in liquid $NH_3$ by adding Na (see the end of Section 21.11). The chain reaction continues by attack on a second monomer, and so on as shown in scheme 26.64.

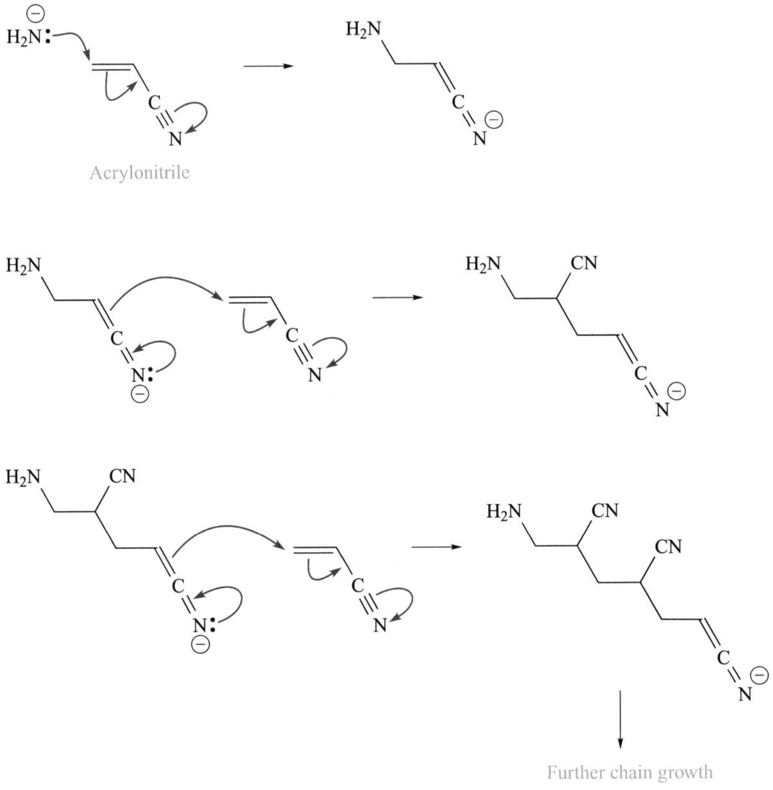

Acrylonitrile

Further chain growth          (26.64)

| 26.9 | **Double bond migration and isomerization in alkenes** |

When an alkene with four or more C atoms is treated with $H^+$, isomerization may occur by *migration of the C=C bond* along the carbon chain. A strong base such as potassium amide, $KNH_2$, can also promote double bond isomerization. The acid or base acts as a catalyst, i.e. the reactions are *acid-catalysed* or *base-catalysed* isomerizations. The C=C bond moves to a more substituted position within the carbon chain, for example but-1-ene isomerizes to but-2-ene in the presence of $H^+$ (equation 26.65).

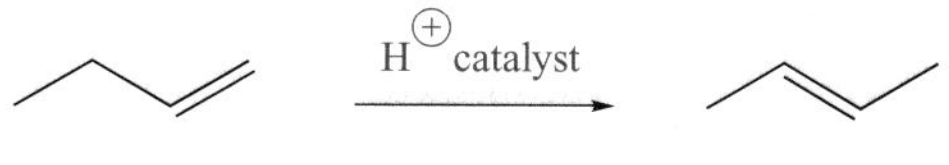

(26.65)

> The tendency is for the **C=C bond to migrate** along a carbon chain so as to increase the number of alkyl groups attached to it.

The preference for the C=C bond to be in the more highly substituted position follows from the greater electronegativity of an $sp^2$ C centre with respect to an $sp^3$ C centre. The electron-releasing alkyl groups should be attached to the $sp^2$ C atoms. (We return to the electronegativities of different C centres in Section 26.12 when we discuss the acidity of terminal alkynes.)

The longer the carbon chain, the more isomers that can be formed by migration of the C=C functionality. In practice, mixtures of isomers are usually obtained with the thermodynamically preferred isomer predominating.

### The mechanism of acid-catalysed C=C bond migration

We have already seen that the addition of $H^+$ to an alkene promotes the formation of a carbenium ion. Equation 26.66 shows that the reaction of pent-1-ene with $H^+$ gives a secondary carbenium ion in preference to a primary one.

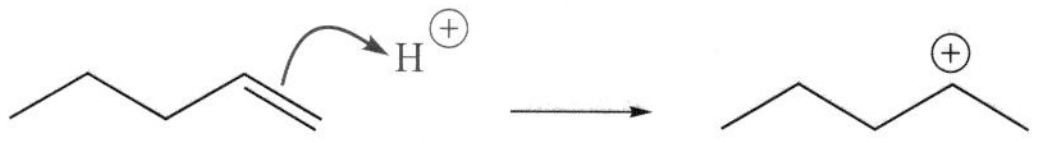

(26.66)

In previous examples (e.g. reaction 26.34), the carbenium ion formed has reacted with a nucleophile to give a final addition product. In the absence of a nucleophile, the carbenium ion intermediate can lose $H^+$. Starting from the carbenium ion in reaction 26.66, there are two competitive eliminations to give pent-1-ene or pent-2-ene (scheme 26.67). One reaction is simply the reverse of protonation step 26.66, but the competitive elimination of $H^+$ gives the isomerization product and this is the favoured pathway.

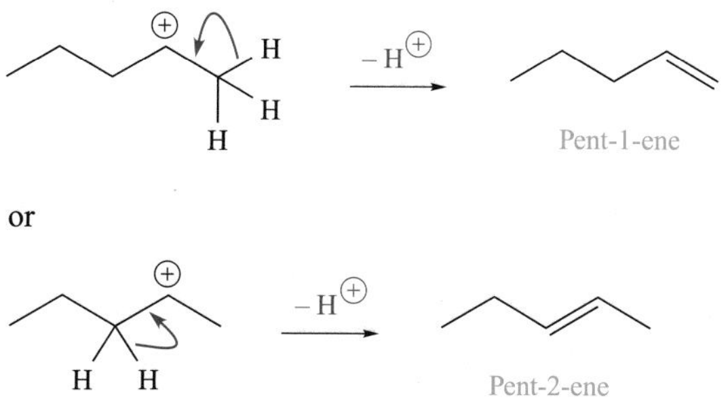

Pent-1-ene

or

(26.67)

Pent-2-ene

### The mechanism of base-catalysed C=C bond migration

An *allylic H* atom is relatively *acidic*.

Not only is an allylic H atom susceptible to radical abstraction (see equation 26.47), it is also susceptible to attack by base and removal as $H^+$. In reaction 26.68, the base $B^-$ removes $H^+$ from the C atom *adjacent* to the C=C bond (see structure **26.28**). The intermediate is a *carbanion*.

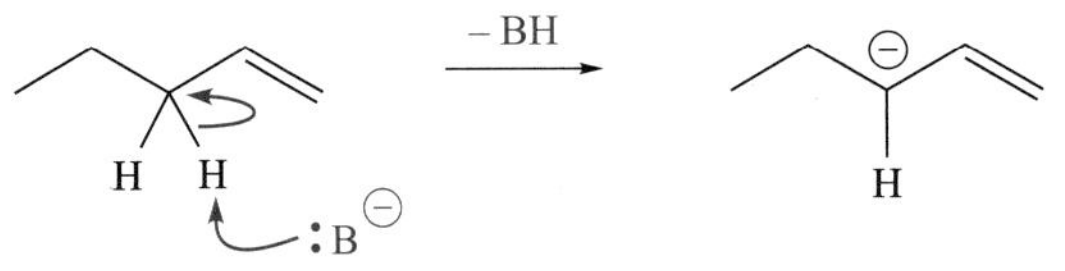

$$(26.68)$$

A *carbanion* has the general formula $[R_3C]^-$ where R = H or organic group.

The carbanion formed in step 26.68 is related to the allyl carbanion and is resonance stabilized by the contributing structures **26.33**. This allows the $\pi$-bonding to be delocalized over the three C atoms (see Figure 13.6). Structure **26.34** is an alternative way of representing resonance pair **26.33**.

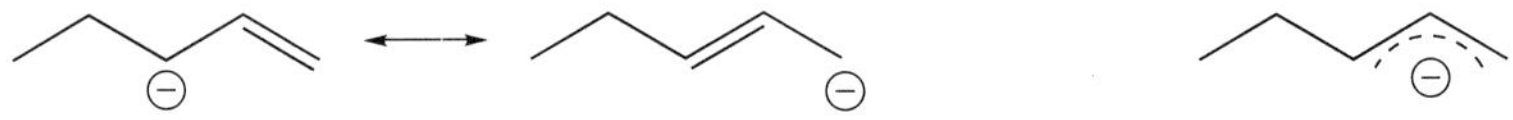

**(26.33)**                                    **(26.34)**

The carbanion can be protonated at one of two sites (equation 26.69) and the favoured product is the alkene with the greater number of R groups attached to the C=C bond. The net result of steps 26.68 and 26.69 is the migration of the C=C functionality along the carbon chain. The source of protons in reaction 26.69 is protonated base BH. The catalyst $B^-$ is regenerated in the last step of the isomerization process.

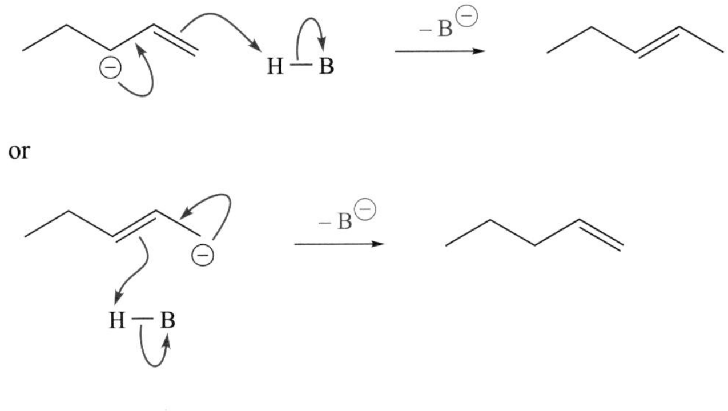

or

$$(26.69)$$

| 26.10 | **Hydroboration of alkenes** |

### Hydroboration with BH₃

When we write 'BH₃' in this section, we are referring either to B₂H₆ or to a Lewis base adduct: see Section 21.5

The reaction of an alkene with $BH_3$ leads to the *addition* of the B–H bond to the C=C double bond (equation 26.70). This reaction is called *hydroboration*.

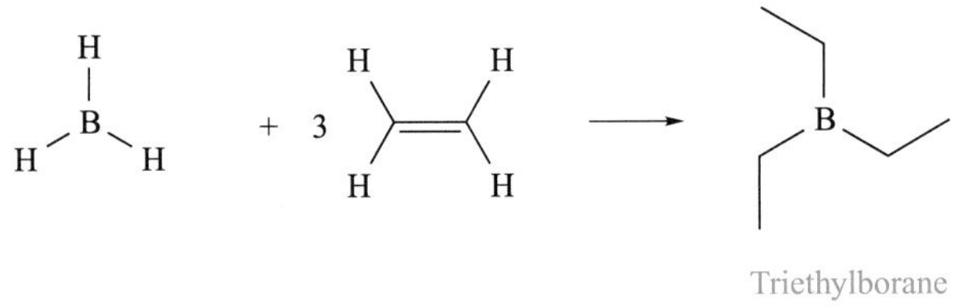

$$(26.70)$$

Triethylborane

If the alkene is sterically hindered by the presence of organic substituents, only one or two B–H bonds will be involved in the hydroboration as, for example, in reaction 26.71.

(26.71)

Hydroboration reactions can therefore be used to form organoborane compounds of the general type $RBH_2$ and $R_2BH$ (R = alkyl) and further reactions with different alkenes can yield organoboranes with different alkyl substituents. Addition of a borane to an unsymmetrical alkene is *regioselective*:

> When a B−H bond adds to a C=C bond, addition is regioselective with the B atom becoming attached to the C atom that possesses the most H atoms; this is **anti-Markovnikov addition**.

The regioselectivity arises from the fact that B, not H, is the electrophile and so the reaction differs from the electrophilic addition of, for example, HCl or HBr. The mechanism is summarized in scheme 26.72. The electrophilic nature of $BH_3$ is due to its vacant $2p$ atomic orbital which can accept a pair of electrons (see Section 21.5). Notice that in the first step in scheme 26.72, a secondary carbenium ion forms in preference to a primary one, and so the 'usual' rules are obeyed even though we end up with a so-called anti-Markovnikov product.

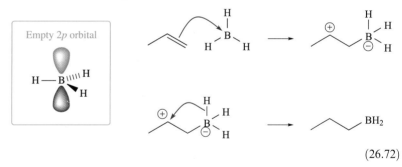

(26.72)

Synthesis of alcohols: see Section 30.2

The importance of the organoborane products is in their synthetic applications. For example, treatment with $H_2O_2$ oxidizes an organoborane to an alcohol (equation 26.73).

$$+ 3H_2O_2 \longrightarrow 3 \quad \text{OH} \quad + \quad \text{boric acid}$$

Triethylborane        Ethanol        Boric acid

(26.73)

## 26.11    Synthesis of alkynes

Naming alkynes: see Section 24.4

We now turn our attention to alkynes, in which the functional group is the C≡C bond. In this section, we look at ways to prepare alkynes and then we move on to look at their reactivity.

Vicinal dihalides:
see Section 26.4

In reaction 26.2, we showed how the elimination of HCl from a chloro-alkane yields an alkene. If we begin with a vicinal dihalide, the elimination of HX can take place twice to give an alkyne. A strong base is needed (as in reaction 26.74) to remove $H^+$ and encourage the elimination of HX. We look at elimination reactions in detail in Chapter 28.

$$(26.74)$$

This method is the most common strategy for alkyne synthesis.

## 26.12 Reactions of alkynes

In this section, we consider typical reactions of alkynes and we shall often need to distinguish between an *internal* and *terminal* alkyne:

Internal alkyne            Terminal alkyne

### Combustion

Like alkanes and alkenes, alkynes burn in $O_2$ to give $CO_2$ and $H_2O$ (equation 26.75).

$$2MeC{\equiv}CMe + 11O_2 \longrightarrow 8CO_2 + 6H_2O \qquad (26.75)$$

### Hydrogenation

The addition of $H_2$ to give alkenes takes place readily in the presence of a metal catalyst such as Ni, Pd or Pt. Activation of $H_2$ occurs as in equation 26.15. Unlike addition of $H_2$ to an alkene where only one product is possible, addition to *internal alkynes* may give (*E*)- or (*Z*)-isomers. An additional problem is the possibility of further reduction to the alkane.

(*E*)-isomer            (*Z*)-isomer

$$(26.76)$$

Na in liquid NH₃: see the end
of Section 21.11

Selectivity can be achieved by careful control of reaction conditions. Reaction of $H_2$ with an alkyne in the presence of a $Pd/BaSO_4$ catalyst produces (*Z*)-isomers. If reduction is carried out using Na in liquid $NH_3$ as the reducing agent, the products are (*E*)-isomers. (Sodium/liquid $NH_3$ cannot be used for the reduction of terminal alkynes; see later in the section.) A selective means of producing alkenes (rather than alkanes) is by using Lindlar's catalyst. This consists of Pd metal supported on a $CaCO_3/PbO$ surface which is treated with $Pb(O_2CMe)_2$. Further, Lindlar's catalyst is selective towards the production of (Z)-alkenes, i.e. $H_2$ undergoes *syn*-addition to the C≡C bond.

## Addition of $X_2$ and HX (X = Cl, Br)

The mechanisms of the electrophilic additions of $X_2$ and HX to alkynes are similar to those that we described for additions to alkenes. It is possible to halt reactions after the addition of one equivalent of $X_2$ or HX, but under appropriate conditions, the reaction continues to give a saturated product (equation 26.77). The addition of $X_2$ to the alkyne may, in principle, give rise to (E)- or (Z)-isomers of the dihalo-product. In practice, the (E)-isomer is favoured.

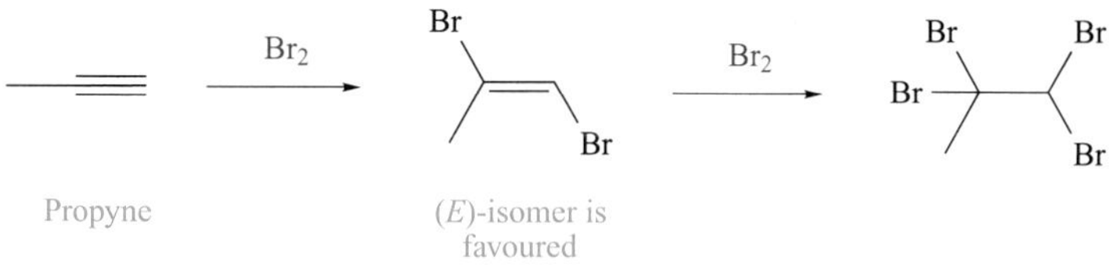

Propyne          (E)-isomer is favoured

$$(26.77)$$

The reaction of HX with an alkyne gives a *vinylic halide* and then a *gem-dihalide*; *gem* stands for *geminal* and means that the two X atoms are attached to the *same* C atom. The mechanism of addition is given in scheme 26.78, exemplified by the reaction between propyne and HCl. Only the more favoured carbenium ion is shown in each addition and this preference explains why the *gem*-dihalide is the predominant product.

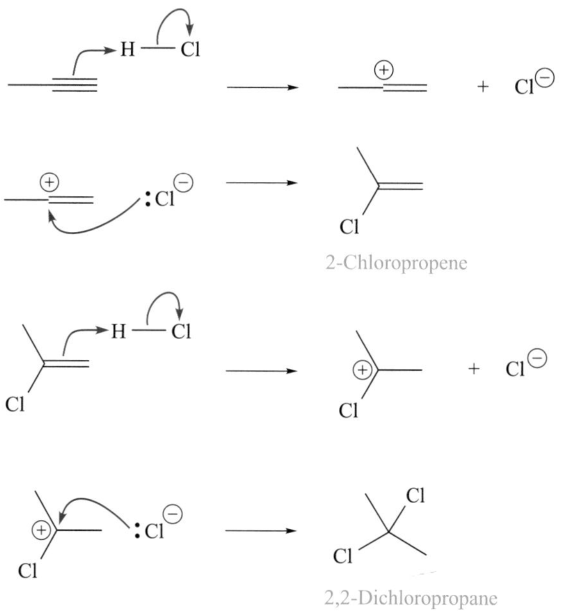

2-Chloropropene

$$(26.78)$$

2,2-Dichloropropane

The carbenium ion that forms in the first step in scheme 26.78 is *linear* at the $C^+$ centre, and is stabilized by hyperconjugation. This involves the donation of electron density from an alkyl C−H bond into the vacant $2p$ atomic orbital on the $C^+$ centre and is similar to that described in diagram **26.16**.

See problem 26.22

When HX adds to an internal alkyne, (Z)- and (E)-isomers (with respect to the carbon chain as in equation 26.76) could, in principle, be produced. In most cases, the mechanism follows scheme 26.79 with the favoured addition of HX being *anti* across the C≡C bond.

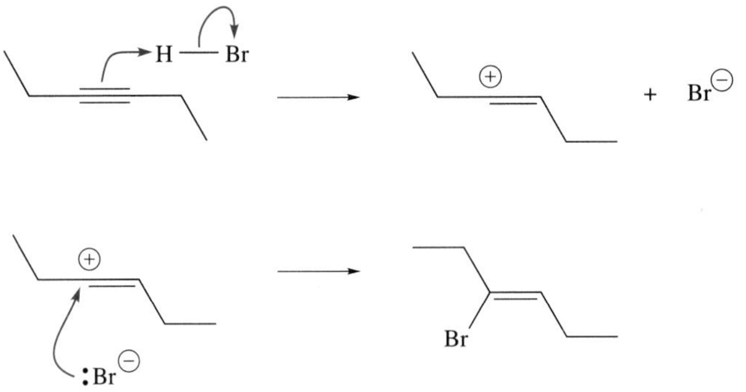

(26.79)

## Addition of H$_2$O: formation of a ketone

In reactions 26.35 and 26.36, we described the conversion of an alkene to an alcohol by using Hg(O$_2$CMe)$_2$ followed by treatment with H$_2$O and NaBH$_4$. However, the reaction between an alkyne and Hg(O$_2$CMe)$_2$ followed by treatment with H$_2$O leads to a *ketone* rather than to an alcohol. Before we look at this reaction in detail, we need to introduce the concept of *keto–enol tautomerism*. When an OH group is attached to a C=C bond, the species is unstable with respect to rearrangement to a carbonyl compound. The rearrangement involves the *transfer of a proton* and is called *keto–enol tautomerism*. The 'keto' part of the name comes from *keto*ne, and 'enol' is a contraction of alk*ene*-alcoh*ol*. The two *tautomers* are in equilibrium (equation 26.80) and the *keto*-form is usually favoured.

> Carbonyl compounds are discussed in detail in Chapter 33

> ***Tautomerism*** involves the transfer of a proton between sites within a molecule with an associated rearrangement of multiple bonds.

(26.80)

*enol*-tautomer          *keto*-tautomer

Now consider the reaction between an alkyne and Hg(O$_2$CMe)$_2$ followed by treatment with H$_2$O. The first step is nucleophilic attack by the carbon-rich C≡C bond on the Hg(II) centre and formation of an unsaturated C$_2$Hg ring. Nucleophilic attack by H$_2$O is followed by proton loss. This is shown in scheme 26.81 which should be compared with scheme 26.36.

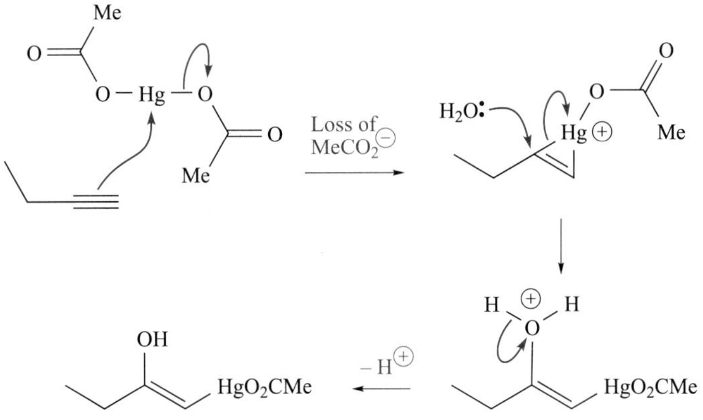

(26.81)

The product is an enol and is unstable with respect to rearrangement to the keto-tautomer (equation 26.82). The C−Hg bond in the organomercury product is converted to a C−H bond by treatment with acid, the mechanism for which also involves rearrangement from an enol to keto-tautomer (scheme 26.83).

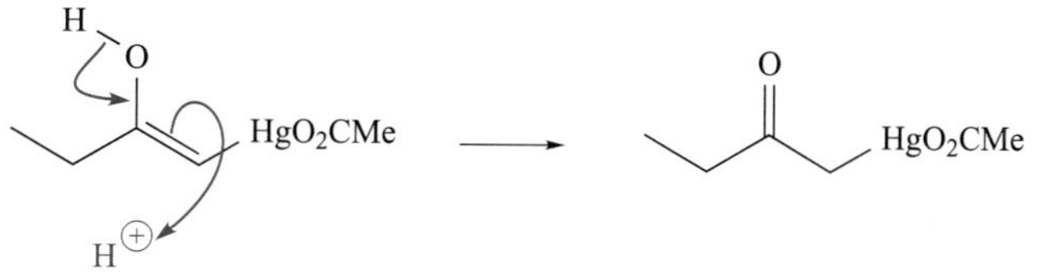

(26.82)

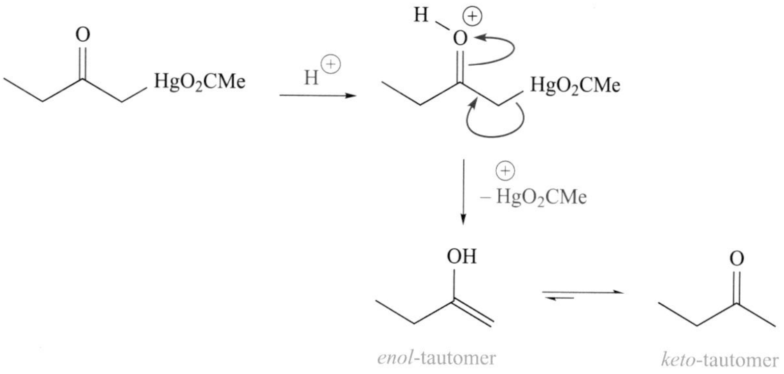

enol-tautomer          keto-tautomer

(26.83)

## Oxidation: carboxylic and dione formation

An **α-dione** has the general formula:

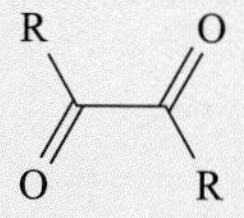

The oxidation of a terminal alkyne using $KMnO_4$ gives a carboxylic acid, whereas oxidation of an internal alkyne produces an α-dione. Reactions with $KMnO_4$ may be carried out in $CH_2Cl_2$, but a phase transfer agent is required to facilitate dissolution of $KMnO_4$ into the organic layer. Reactions 26.84 and 26.85 show examples of permanganate oxidation of terminal and internal alkynes.

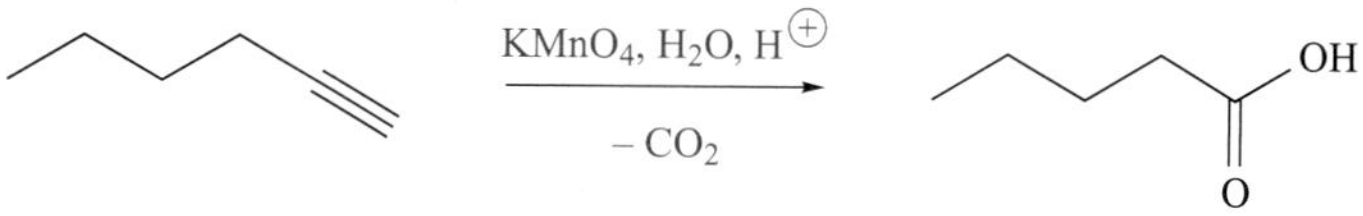

(26.84)

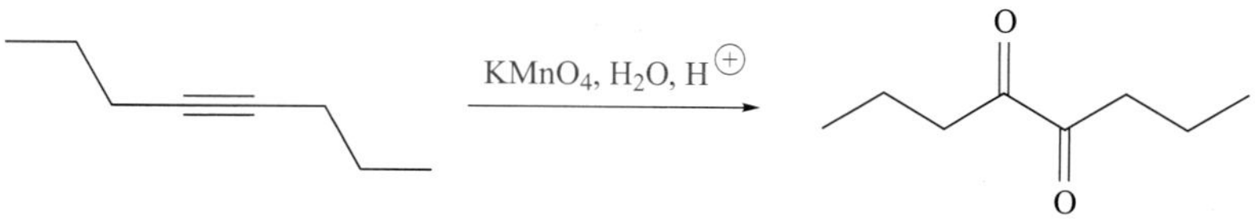

(26.85)

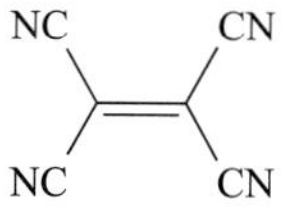

(26.35)

Ozone oxidations of terminal and internal alkynes are analogous to those of $[MnO_4]^-$ oxidations. As with the $O_3$ oxidation of alkenes (equations 26.44–26.46), a reducing agent is needed for the second stage of the reaction. Suitable reductants for the formation of α-diones include tetracyanoethene (TCNE, **26.35**) as shown in reaction 26.86.

(26.86)

## Alkynes as acids

In Table 5.2, we listed the Pauling electronegativity value of C as 2.6, and that of H as 2.2. However, we added a note of caution, stating that the electronegativity of an atom depends on the oxidation state and bond order. As the hybridization of a C atom changes from $sp^3$ to $sp^2$ to $sp$, the electronegativity changes (Table 26.1). The trend in electronegativity values means that a C–H bond in an alkyne is more polar than in an alkene, and this in turn is more polar than one in an alkane. The values also predict that an $[RC{\equiv}C]^-$ ion should be more stable than an $[R_2C{=}CR]^-$, which in turn should be more stable than an $[R_3C]^-$ ion. This is indeed the case and approximate $pK_a$ values are listed in Table 26.1. The decrease in $pK_a$ (and corresponding increase in $K_a$) shows that the terminal alkyne is the most acidic species of those listed. Note that it is *only terminal alkynes* that are acidic; internal alkynes do not possess a C–H bond. In Section 26.9, we stated that an *allylic H* atom is relatively acidic; this particular acidity arises because of the resonance stabilized allyl carbanion (**26.33** and **26.34**).

Acids and $pK_a$ values: see Chapter 16

While there is evidence for the acidity of terminal alkynes in the reactions described below, it is important to keep in mind that these are *very weak acids*; $HC{\equiv}CH$ is a weaker acid than $H_2O$. Since acetylene is such a weak acid, $[HC{\equiv}C]^-$ must be a strong base (equation 26.87).

$$H{-}{\equiv}{-}H \; \rightleftharpoons \; H{-}{\equiv}{\ominus} \; + \; H^{\oplus} \quad pK_a \approx 25 \quad (26.87)$$

Terminal alkynes react with potassium amide (equation 26.88) or sodium metal (equation 26.89) to give alkali metal salts of *acetylides*.

**Table 26.1** Electronegativity values[a] of C atoms with different hybridizations and relative acidities of C–H bonds.

| Compound | Hybridization of C atom | Electronegativity value of C | $pK_a$ |
|---|---|---|---|
| | $sp^3$ | 2.5 | 48 |
| | $sp^2$ | 2.75 | 44 |
| H—C≡C—H | $sp$ | 3.3 | 25 |

[a] These electronegativity values are calculated from ionization potential and electron affinity data and are on the Mulliken–Jaffé scale (see Section 5.10). The values are then scaled so as to be comparable with Pauling electronegativity values.

$$\text{R–C≡C–H} + K[NH_2] \longrightarrow \text{R–C≡C}^{\ominus} K^{\oplus} + NH_3$$

$$(26.88)$$

$$\text{R–C≡C–H} + Na \longrightarrow \text{R–C≡C}^{\ominus} Na^{\oplus} + {}^{1}/_{2}H_2$$

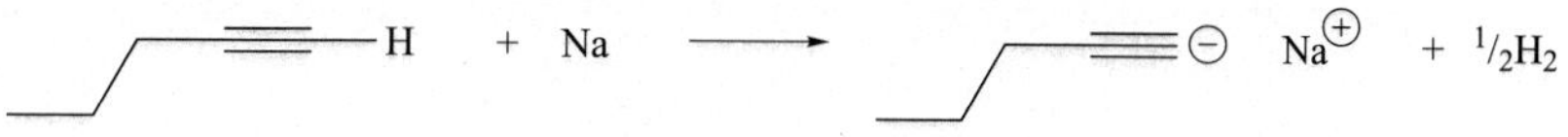

$$(26.89)$$

Reaction also occurs with $Ag^+$ ions (equation 26.90), and Cu(I) salts are formed by similar reactions. **Caution!** When dry, silver acetylides are explosive.

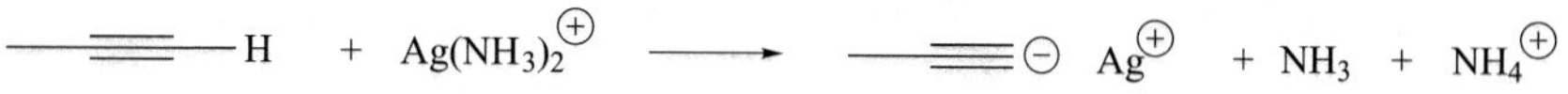

$$\text{–C≡C–H} + Ag(NH_3)_2^{\oplus} \longrightarrow \text{–C≡C}^{\ominus} Ag^{\oplus} + NH_3 + NH_4^{\oplus}$$

$$(26.90)$$

**See also Section 28.9**

An important use of alkali metal acetylides is in the preparation of alkynes with longer carbon chains. The driving force for the reactions is the elimination of an alkali metal halide (e.g. reaction 26.91). Look back at discussions of lattice energies in Chapter 8; significant energy is released when an ionic salt such as NaCl is formed.

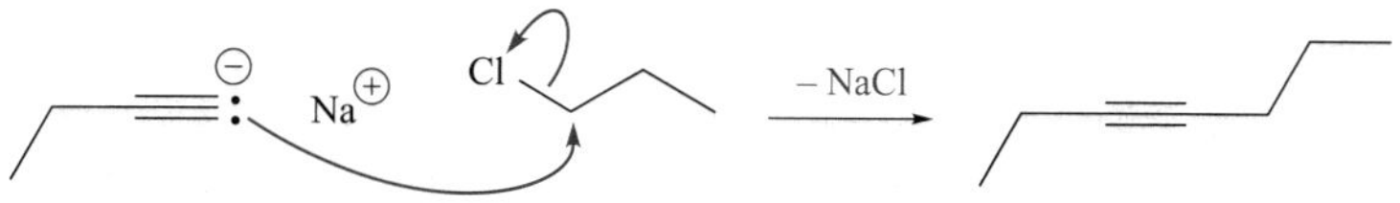

$$(26.91)$$

This strategy can be used to increase the length of a carbon chain, and the alkyne functionality then reacted further as required, e.g. reduced as in scheme 26.92.

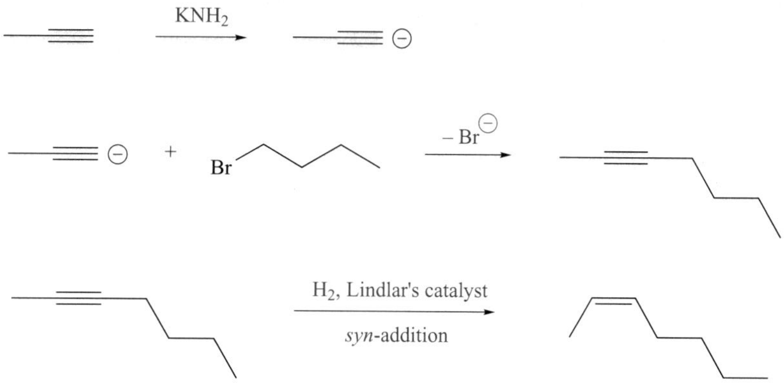

$$(26.92)$$

This last scheme is one of the first examples we have shown in which a series of reactions are put together to achieve a multi-step synthetic pathway. We shall see many more examples in later chapters, and examples are also given in the end-of-chapter problems.

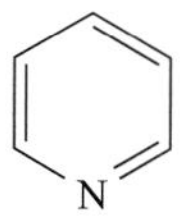

## Coupling reactions to give multi-functional alkynes

When terminal alkynes are heated with $Cu^{2+}$ salts in pyridine (py, **26.36**), a coupling reaction can take place to give a *diyne*, i.e. a molecule in which there are two C≡C functionalities. Reaction 26.93 shows an example. This is quite

**(26.36)**

a general reaction with wide application. Equation 26.94 shows how it can be used to give a cyclic hexyne starting from a terminal diyne.

$$\text{——≡——H + H——≡——} \quad \xrightarrow{\text{Cu}^{2+} \text{ in py}} \quad \text{——≡——≡——} \tag{26.93}$$

$$3 \quad \xrightarrow{\text{Cu}^{2+} \text{ in py}} \tag{26.94}$$

Aryl groups possess aromatic rings and the simplest is phenyl, $C_6H_5$: see Chapter 32

Terminal alkynes can be coupled to *aryl halides* using the *Sonogashira coupling* method. The conditions are a Pd(0) catalyst, CuI and $Et_3N$ (equation 26.95). The Pd(0) catalyst is usually $[PdCl_2(PPh_3)_2]$ (Ph = phenyl).

$$\text{Ph——I + H——≡——R} \quad \xrightarrow[- \text{HI}]{\text{Pd(0), CuI,} \atop \text{Et}_3\text{N}} \quad \text{Ph——≡——R} \tag{26.95}$$

## 26.13    Protection of a terminal alkyne: the $Me_3Si$ protecting group

Protection of –OH groups: see Section 30.6

As we have seen, the C–H group of a terminal alkyne is relatively reactive and in multi-step reactions, it may be necessary to *protect* the C–H group. This is one example of many in which a *protecting group* is introduced temporarily into a molecule and is later cleaved at the point in the multi-step synthesis when the functional group is restored. Common protecting groups are trialkyl silyl groups such as $Me_3Si$ (trimethylsilyl, abbreviated to TMS). The general strategy is to use an organometallic reagent such as $^n$BuLi to react with the terminal –C≡C–H group. This produces an organolithium compound that readily reacts with a trialkylsilyl halide. Equation 26.96 shows the general method, using an $Me_3Si$ (TMS) group as the example.

Organolithium compounds: see Section 28.4

$$R\text{——≡——}H \quad \xrightarrow[-^n\text{BuH}]{^n\text{BuLi}} \quad R\text{——≡——}Li$$

$$\downarrow \quad \text{Cl——Si(Me)}_3$$

$$R\text{——≡——Si(Me)}_3 \quad + \text{ LiCl} \tag{26.96}$$

also written as:

$$R\text{——≡——TMS}$$

Once the desired reactions have been carried out on the R group of the product in scheme 26.96, the compound can be *deprotected* using a base. If we assume that R is transformed into a new substituent R', then equation 26.97 represents the deprotection step.

$$R' \!\!\equiv\!\! TMS \xrightarrow{\text{NaOH}} R' \!\!\equiv\!\! H \tag{26.97}$$

Now let us put this protection–deprotection strategy into action in an example (scheme 26.98) where the protection is necessary because we want only *one end* of an acetylene molecule to undergo reaction with the aryl halide (relative reactivities being I > Br $\gg$ Cl) in a Sonogashira coupling (see equation 26.95). Trimethylsilylacetylene, (TMS)C$\equiv$CH, is available commercially.

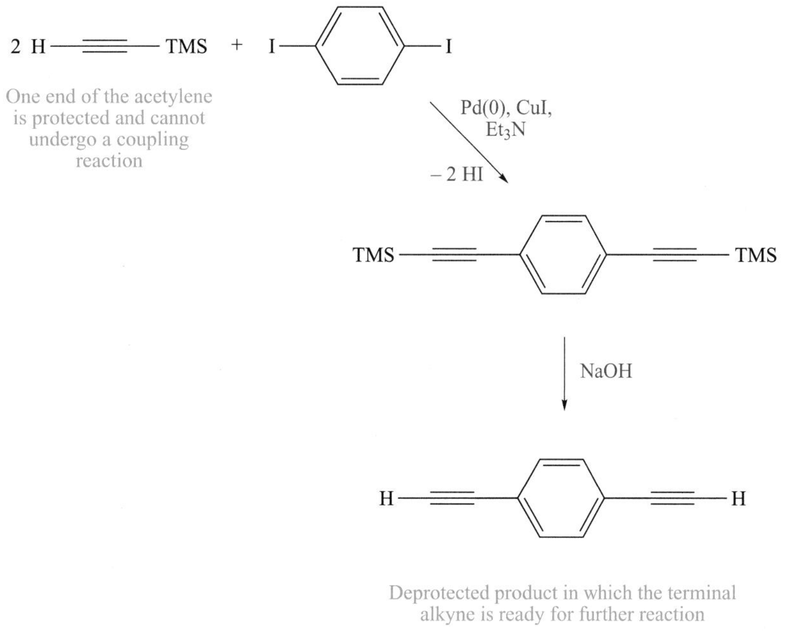

One end of the acetylene is protected and cannot undergo a coupling reaction

Deprotected product in which the terminal alkyne is ready for further reaction

$$\tag{26.98}$$

Later in the book we describe further examples of the protection of vulnerable functional groups in multi-step syntheses.

## SUMMARY

In this chapter, we have looked at the syntheses and reactivity of alkenes and alkynes and have introduced *Diels–Alder cyclizations*, *electrophilic addition* and *radical addition* reactions, oxidation reactions of C=C bonds and *addition polymerization*. We have seen that some reactions are *regioselective*. Studies of mechanisms help us to rationalize why the same reaction carried out under different conditions may lead to different products, e.g. electrophilic and radical additions of HBr to an unsymmetrical alkene give Markovnikov and anti-Markovnikov products respectively. An understanding of the stereoselectivity of some reactions is also gained by looking at reaction mechanisms. Finally, we have seen that internal and terminal alkynes have different reactivity patterns and terminal C≡C–H groups may require *protection* in multi-step syntheses.

### Do you know what the following terms mean?

- alkene
- diene
- dehydrogenation
- elimination reaction
- Diels–Alder reaction
- [4 + 2] cycloaddition
- 1,3-conjugated diene
- dienophile
- electron-withdrawing substituent
- *s-cis* conformation
- *endo-* and *exo*-configurations
- stereoselective
- hydrogenation
- vicinal
- carbenium ion
- carbocation
- primary carbenium ion
- secondary carbenium ion

- tertiary carbenium ion
- hyperconjugation
- regioselective
- Markovnikov addition
- bromonium ion
- *syn*-addition
- *anti*-addition
- halohydrin
- diol
- glycol
- ozonide
- allyl radical
- allyl anion
- allylic H atom
- vinylic H atom
- radical initiator
- anti-Markovnikov addition
- polymer

- monomer
- addition polymerization
- radical polymerization
- cationic polymerization
- anionic polymerization
- double bond migration
- acid-catalysed reaction
- base-catalysed reaction
- carbanion
- hydroboration
- terminal alkyne
- internal alkyne
- diyne
- geminal
- tautomer
- *keto–enol* tautomerism
- α-dione
- protecting group

### Some abbreviations have been used in this chapter. Do you know what they mean?

- DMSO
- NBS
- PVC
- PTFE
- TCNE
- TMS

### You should be able:

- to describe the bonding in alkenes and alkynes
- to explain how the presence of a $\pi$-bond in an alkene affects rotation about the C=C bond
- to name a cycloalkene given a structural formula
- to draw the structural formula of a cycloalkene given its name
- to compare the flexibility of a cycloalkene ring with that of the corresponding cycloalkane
- to give representative reactions to prepare alkenes and cycloalkenes
- to discuss the mechanism of a Diels–Alder cycloaddition and comment on its stereoselectivity

- to comment on how [4 + 2] cycloadditions differ from [2 + 2] cycloadditions
- to give an overview of typical reactions of alkenes
- to discuss the mechanism and selectivities of electrophilic additions using appropriate examples
- to explain the role played by hyperconjugation in stabilizing carbenium ions
- to discuss the evidence for the formation of carbenium ion intermediates
- to discuss the evidence for the formation of bromonium ion intermediates

- to explain what is meant by *syn-* and *anti-*additions and illustrate how the different additions affect the stereochemistries of addition products
- to describe ways to convert an alkene to an alcohol and to discuss relevant mechanisms
- to give an account of the competitive reactions that occur when $H_2O$ is present in reactions between alkenes and a halogen $X_2$
- to comment on the stereochemistry of the addition of HBr to a cycloalkene substituted at the C=C bond
- to describe ways of converting an alkene to a diol and to rationalize the observed stereochemistries of addition to (*E*)- and (*Z*)-isomers
- to explain (with examples) the role of a radical initiator
- to distinguish between vinylic and allylic H atoms and compare their characteristic properties
- to outline how and why the radical addition of, for example, HBr to an unsymmetrical alkene differs from its electrophilic addition
- to give a short account of radical addition polymerization of alkenes and give examples of some commercially important polymers that are made by this method

- to describe the mechanism of cationic addition polymerization and give an example of a commercial polymer manufactured by this method
- to describe the mechanism of anionic addition polymerization
- to give an account of double bond isomerization in alkenes
- to explain what is meant by hydroboration and comment on its synthetic utility
- to give examples of synthetic methods to form alkynes
- to give an overview of typical reactions of alkynes
- to describe the mechanisms of additions to alkynes and explain why the addition of HX leads to a *gem-* rather than vicinal dihalide
- to explain what is meant by *keto–enol* tautomerism and illustrate how it affects the outcome of the addition of $H_2O$ to an alkyne
- to illustrate the synthetic utility of oxidations of C=C and C≡C bonds with reagents such as $OsO_4$, $KMnO_4$ and $O_3$
- to give an account of differences between the behaviour of internal and terminal alkynes
- to illustrate the synthetic utility of alkynes as building blocks in multi-step syntheses

## PROBLEMS

**26.1** Give systematic names to the following compounds:

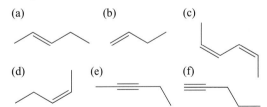

(a)          (b)          (c)

(d)          (e)          (f)

**26.2** The IR spectrum of octa-1,7-diyne is shown in Figure 26.9 (overleaf). Assign the absorptions above $1800 \, cm^{-1}$.

**26.3** How many signals would you expect to see in the proton decoupled $^{13}C$ NMR spectra of (a) penta-1,4-diene and (b) penta-2,4-diene? Could you distinguish the (*E*)- and (*Z*)-isomers of penta-2,4-diene using $^{13}C$ NMR spectroscopy?

**26.4** (a) Two isomers of cyclooctadiene exhibit two and four signals respectively in their $^{13}C$ NMR spectra. Draw the structures of the isomers and assign one of the spectra to each. (b) A third isomer of $C_8H_{12}$ has a bicyclic structure and its $^{13}C$ NMR spectrum contains three signals. Suggest a structure for this isomer. Rationalize why the $^1H$ NMR spectrum of this compound has four signals.

**26.5** Draw the structure of hexa-1,5-diyne and interpret the following spectroscopic data. (a) The IR spectrum contains strong to medium absorptions at 3665, 2922, 2851 and $2123 \, cm^{-1}$ in addition to bands in the fingerprint region. (b) The $^{13}C$ NMR spectrum has signals at $\delta$ 82.4, 69.5 and 18.6 ppm. (c) The $^1H$ NMR spectrum has singlets at $\delta$ 2.06 and 2.43 ppm.

**26.6** Develop a bonding scheme (similar to those in Figure 26.1) for allene, $H_2C=C=CH_2$.

**26.7** Draw structural formulae for the following compounds: (a) cyclohepta-1,3-diene; (b) 2-methylpentene, (c) 4-methylhexene, (d) 1,2-dibromocyclopentene and (e) cyclododeca-1,5,9-triene.

**26.8** Why is a [2 + 2] cycloaddition driven photolytically, while a [4 + 2] cycloaddition is thermally promoted?

**Fig. 26.9** IR spectrum for problem 26.2.

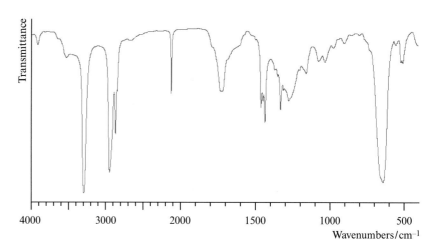

**26.9** Suggest what products are formed in the reactions of buta-1,3-diene with:

(a)

(b)

**26.10** Suggest products for the following reactions:

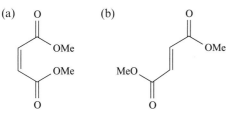

(a) $\xrightarrow{\text{Cl}_2}$

(b) $\xrightarrow{\text{H}_2;\ \text{Pt catalyst}}$

(c) $\xrightarrow{\text{HCl}}$

(d) $\xrightarrow{\text{conc } \text{H}_2\text{SO}_4;\ \Delta}$

(e) $\xrightarrow[\text{in ether}]{\text{CH}_2\text{I}_2;\ \text{Zn/Cu}}$

**26.11** Values of the bond enthalpy terms for C=C, C−C, H−H and C−H bonds are 598, 346, 436 and 416 kJ mol$^{-1}$. Determine the enthalpy change associated with the hydrogenation of $C_2H_4$. Compare your answer with that determined using values of $\Delta_f H^{\circ}$ for $C_2H_4(g)$ and $C_2H_6(g)$ of +52.5 and −83.8 kJ mol$^{-1}$.

**26.12** (a) Suggest a mechanism for the reaction of HBr with 2-methylbut-1-ene under polar conditions and indicate the relative product distribution that you might expect. (b) How does the reaction differ if it is carried out in the presence of a radical initiator?

**26.13** What happens when $Br_2$ reacts with cyclohexene in the absence of a radical initiator? Include in your answer discussion of the mechanism, evidence for proposed intermediates, and an explanation of the stereochemistry of the product.

**26.14** Consider the addition of HCl to (Z)-but-2-ene. Discuss how the stereochemistry of the product is affected by whether HCl adds in a *syn-* or *anti-* manner.

**26.15** Oxidations of alkenes include reactions with $O_3$ and $OsO_4$. Using hex-3-ene as the example, discuss the use of these reagents to produce diols and aldehydes, paying attention to the stereochemistry of the products where appropriate.

**26.16** (a) Rewrite equations 26.47–26.49 showing appropriate curly arrows. (b) Suggest the product of the reaction of hex-1-ene with NBS.

**26.17** How would you carry out the following transformations?

(a)

(b)

(c)

(d)

**26.18** The commercial polymerization of propene requires the formation of a stereoregular polymer (see Box 26.3). Suggest a mechanism for the radical polymerization of propene that is consistent with the formation of a stereoregular polymer with methyl side chains.

**26.19** Using the isomerization of hex-1-ene to hex-2-ene as an example, explain how (a) an acid and (b) a base catalyse the migration of a C=C double bond.

**26.20** 9-Borabicyclo[3.3.1]nonane (commonly called 9-BBN) is a selective hydroborating agent and has the following structure:

Starting from a suitable *cyclic diene*, suggest a method of preparing 9-BBN. What other isomers might be formed in the reaction?

**26.21** Comment on the complications that may be encountered when an alkyne such as hept-3-yne reacts with $H_2$. What methods are available to overcome the problems you describe?

**26.22** Using diagram **26.16** to help you, draw a schematic representation of the hyperconjugation that stabilizes the carbenium ion formed in the *first* step in scheme 26.78. Comment on the stereochemistry of the carbenium ion.

**26.23** Which of the following species can undergo keto–enol tautomerism? Give the keto-form where appropriate.

(a)  (b)  (c)  (d)

**26.24** Suggest products for the following reactions:

(a) Na

(b) KMnO$_4$, H$_2$O, H$^{\oplus}$

(c) 1. NaNH$_2$  2. Br

(d) Na

**26.25** Suggest suitable reagents to carry out the multi-step syntheses shown below.

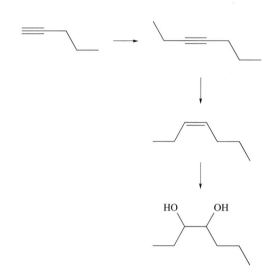

**26.26** (a) The bromination of (Z)-pent-2-ene produces a mixture of (2R, 3R)-2,3-dibromopentane and (2S, 3S)-2,3-dibromopentane. Rationalize this observation. (b) What is the stereochemistry of the product of the reaction of pentene and Br$_2$?

**26.27** Two acyclic isomers of C$_6$H$_{10}$, **A** and **B** show the following signals in their $^1$H NMR spectra:
**A** $\delta$/ppm 1.13 (doublet, 6H), 1.78 (singlet, 3H), 2.50 (septet, 1H)
**B** $\delta$/ppm 0.99 (doublet, 6H), 1.86 (multiplet, 1H), 1.96 (singlet, 1H), 2.08 (doublet, 2H)
Compound **B** reacts with KNH$_2$, but **A** does not undergo an analogous reaction. Both **A** and **B** add Br$_2$ to give compounds **C** and **D** which analyse as having 17.9% C and 2.5% H. Suggest possible structures for **A**, **B**, **C** and **D**.

# 27 Polar organic molecules: an introduction

## Topics

- Polar bonds
- Polar molecules
- Inductive and field effects
- Effects of electron-withdrawing groups on p$K_a$ values

### 27.1 Chapter aims

In Chapters 25 and 26, we were concerned with saturated and unsaturated hydrocarbons containing C−C (single, double and triple) and C−H bonds. Most of our discussions centred on the chemistry of molecules containing *non-polar bonds*. In later chapters, our attention switches to *polar* organic molecules. The aim of Chapter 27, therefore, is to remind you about electronegativities, bond polarities and molecular dipole moments, and to introduce new concepts of the *inductive and field effects*.

### 27.2 Electronegativities and polar bonds

Bond and molecular dipole moments: see Section 6.13

In Chapter 5, we introduced *electronegativity*. Of the various electronegativity scales that have been developed, that of Pauling is commonly used. Pauling electronegativity values ($\chi^P$) relevant to polar organic molecules are listed in Table 27.1; a fuller listing is given in Table 5.2. Using values of $\chi^P$, we can determine whether a heteronuclear bond is polar or non-polar. For example, values in Table 27.1 lead to the following conclusions about C−O, C−N, C−F, C−Cl and O−H bonds:

$$
\overset{\delta^+ \quad \delta^-}{\underset{\longleftarrow}{C \!-\! O}} \qquad
\overset{\delta^+ \quad \delta^-}{\underset{\longleftarrow}{C \!-\! N}} \qquad
\overset{\delta^+ \quad \delta^-}{\underset{\longleftarrow}{C \!-\! F}} \qquad
\overset{\delta^+ \quad \delta^-}{\underset{\longleftarrow}{C \!-\! Cl}} \qquad
\overset{\delta^- \quad \delta^+}{\underset{\longrightarrow}{O \!-\! H}}
$$

By SI convention, the direction of the arrow showing the bond dipole moment points from the $\delta^-$ to $\delta^+$ end of the bond. In the next few chapters, we shall be dealing with molecules containing these and related types of bonds: halogenoalkanes (e.g. **27.1**), alcohols (e.g. **27.2**), ethers (e.g. **27.3**) and amines (e.g. **27.4**).

More on naming these compounds in Chapters 28–31

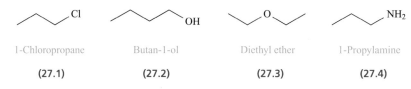

| 1-Chloropropane | Butan-1-ol | Diethyl ether | 1-Propylamine |
| (27.1) | (27.2) | (27.3) | (27.4) |

**Table 27.1** Pauling electronegativity values for selected elements, chosen because of their relevance to organic compounds.

|  | Group 14 | Group 15 | Group 16 | Group 17 |
|---|---|---|---|---|
| H 2.2 | | | | |
| | C 2.6 | N 3.0 | O 3.4 | F 4.0 |
| | | | S 2.6 | Cl 3.2 |
| | | | | Br 3.0 |
| | | | | I 2.7 |

## 27.3 Molecular dipole moments

### Polar and non-polar molecules

In a polyatomic molecule, the resultant *molecular dipole moment* depends on the magnitude and directions of the bond dipole moments, the presence of lone pairs and the shape of the molecule.

Dipole moments are vectors and a *molecular* dipole moment ($\mu$) is the resultant of the individual bond dipoles and, therefore, the presence of a polar bond does not necessarily mean that a molecule is polar. For example, tetrahedral $CF_4$ is non-polar despite possessing four polar C–F bonds because the bond dipoles (shown in blue below) cancel out:

Resultant $\mu = 0$

Many molecules contain more than one type of substituent and the various bond dipole moments act in different directions. Whether or not there is a resultant molecular dipole moment depends on:

- the shape of the molecule;
- the relative magnitudes and directions of the individual bond dipoles; and
- the presence of any lone pairs.

Consider $CH_3F$ (**27.5**). The molecule is near tetrahedral, and the only lone pairs are on the F atom. Electronegativity values from Table 27.1 show that each C–H bond has a small dipole moment in the sense $C^{\delta-}-H^{\delta+}$, and the C–F bond is polar in the sense $C^{\delta+}-F^{\delta-}$. The bond moments *reinforce* each other and the resultant molecular dipole moment is shown in pink in structure **27.5**.

$\mu = 1.86\,D$            $\mu = 0.46\,D$

(27.5)            (27.6)

Now consider $CCl_3F$ (**27.6**). The molecule is polar, but has a smaller dipole moment than $CH_3F$. Each C–Cl bond is polar in the sense $C^{\delta+}$–$Cl^{\delta-}$ (see Table 27.1) and the resultant of these three bond dipole moments *opposes* the moment of the C–F bond ($C^{\delta+}$–$F^{\delta-}$). The direction and magnitude of the molecular dipole moment in $CCl_3F$ is shown in pink in diagram **27.6**.

The effects of molecular geometry and lone pairs are not always easy to assess qualitatively. In ethanol, for example, the molecular dipole moment depends on the resultant of the C–O and O–H bond dipole moments, *and* the dipole moments associated with the two lone pairs of electrons on the O atom. The right-hand diagram below summarizes the problem schematically:

Ethanol

Qualitatively, all we can conclude is that the ethanol molecule is polar and that the OH end of the molecule will be $\delta^-$ with respect to the alkyl chain. The exact direction in which the molecular dipole moment acts cannot be deduced without the aid of quantitative vector resolution.

## Limitations of quantitative predictions

One must be cautious when drawing conclusions about the magnitude and direction of a molecular dipole moment. Although we can use the three criteria above to predict or rationalize whether a molecule is polar or not, other factors contribute to the actual magnitude of the dipole moment. To illustrate this, we look at two examples. First, consider the trend in values of molecular dipole moments for the gas phase halogenoalkanes shown below:

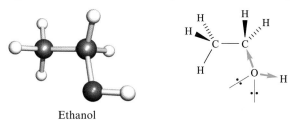

$\mu = 0$     $\mu = 1.86\,D$     $\mu = 1.89\,D$     $\mu = 1.82\,D$     $\mu = 1.62\,D$

$CH_4$      $CH_3F$      $CH_3Cl$      $CH_3Br$      $CH_3I$

Each molecule is near tetrahedral and on the basis of the electronegativity values listed in Table 27.1, we predict molecular dipole moments in the order $CH_3F > CH_3Cl > CH_3Br > CH_3I$ with $CH_4$ being non-polar. However, the experimental data for gas phase molecules reveal that the bond dipole moment of $CH_3Cl > CH_3F$, an observation that is not readily explained.

The second example concerns $CH_3F$, $CH_2F_2$ and $CHF_3$, dipole moments for which are 1.86, 1.98 and 1.65 D for gas phase molecules. At first glance, one might expect the dipole moments to follow an order that reflects the number of F atoms but this is not observed. Each molecule has bond angles close to $109.5°$. From the molecular shapes and values of $\chi^P$, we correctly predict that each molecule is polar, and resolution of vectors gives the directions of the molecular dipole moments as follows:

$\mu = 1.86$ D          $\mu = 1.98$ D          $\mu = 1.65$ D

$CH_3F$                    $CH_2F_2$                    $CHF_3$

## 27.4  Inductive and field effects

The presence of a polar $C-X$ bond can influence the chemistry of an organic molecule in many ways, and here we concentrate on two:

- the charge distribution in the molecule is altered and the C atom to which X is attached becomes a centre of reactivity;
- the charge distribution in the molecule is altered and the reactivities of C atoms other than that attached to X are modified.

The first point is straightforward. If X is $\delta^-$, the C atom to which it is attached is $\delta^+$. Thus, the C atom is likely to be susceptible to attack by nucleophiles. We look at this in detail in later chapters.

The second point concerns longer-range effects brought about by the presence of an electronegative atom. Consider chloroethane. In the $C-Cl$ bond, the electronegative Cl atom draws electron density towards itself, making the C atom attached to it partially positively charged. However, atom C(1) can partly compensate for the movement of electron density by withdrawing electron density from atom C(2) as structure **27.7** schematically illustrates. This is the *inductive effect*.

The inductive effect operates through $\sigma$-bonds and is also described as a $\sigma$-effect. Its magnitude depends on distance. The size of the induced dipole at a C centre usually decreases as the C atom becomes more remote from the electronegative atom. This is represented by $\delta^+$, $\delta\delta^+$, $\delta\delta\delta^+$ ... and therefore the charge distribution in **27.7** can be described as $Cl^{\delta-}$, $C(1)^{\delta+}$ and $C(2)^{\delta\delta+}$.

An additional effect called the *field effect* operates through space. Since experimental observables are usually the net result of both the inductive and field effects, it is generally difficult to determine the magnitude of each separate phenomenon. Thus, reference to 'the inductive effect' usually incorporates contributions from the field effect.

### Effects of electron-withdrawing groups on p$K_a$ values

The inductive effect may be used to partially rationalize the increased strength of haloacetic acids with respect to the parent carboxylic acid. For example, p$K_a$ values for $CH_3CO_2H$ and $CCl_3CO_2H$ are 4.77 and 0.70,

**(27.7)**

The *inductive effect* operates mainly through $\sigma$-bonds, and is the redistribution of charge in a molecule caused by atomic electronegativity differences.

The *field effect* is similar to the inductive effect but operates through space.

Acids and bases, p$K_a$ and $K_a$: see Chapter 16 and Section 33.12

respectively. This means that $CCl_3CO_2H$ dissociates in aqueous solution to a greater degree than $CH_3CO_2H$ (equations 27.1 and 27.2).

$$CCl_3CO_2H(aq) + H_2O(l) \rightleftharpoons H_3O^+(aq) + CCl_3CO_2^-(aq)$$

$$pK_a = 0.70 \quad (27.1)$$

$$CH_3CO_2H(aq) + H_2O(l) \rightleftharpoons H_3O^+(aq) + CH_3CO_2^-(aq)$$

$$pK_a = 4.77 \quad (27.2)$$

The greater inductive effect of the $CCl_3$ group with respect to the $CH_3$ group facilitates loss of a proton from the $CO_2H$ functional group and stabilizes the $CCl_3CO_2^-$ ion. In both $CH_3CO_2^-$ and $CCl_3CO_2^-$, the negative charge is delocalized over the two O atoms. This can be represented by resonance pair **27.8** or structure **27.9** where X = H or Cl. The negative charge is further stabilized if X is *electron-withdrawing*, e.g. X = Cl, and stabilization of $CCl_3CO_2^-$ encourages the forward reaction in equilibrium 27.1.

Compare with delocalization in the allyl anion: see structures 13.2 and 13.3

(27.8)                    (27.9)

An *electron-withdrawing* substituent attracts electrons towards itself; examples are F, Cl, CN.

The effect diminishes as more C−C bonds are introduced between the $CO_2H$ functionality and the electron-withdrawing substituent. This is illustrated by values of $pK_a$ for the carboxylic acids shown in Table 27.2. The introduction of one Cl substituent into $CH_3CO_2H$ significantly decreases the value of $pK_a$, but differences between pairs of $pK_a$ values for $CH_2Cl(CH_2)_nCO_2H$ and $CH_3(CH_2)_nCO_2H$ ($n = 1$, 2 or 3) are much smaller.

The examples above focus on the relative acidities of selected carboxylic acids. However, the introduction of electron-withdrawing groups into compounds such as $CH_4$ has dramatic effects. $CH_4$ has a $pK_a$ value of $\approx 48$ and so equilibrium 27.3 lies far to the left-hand side. The conjugate base

**Table 27.2** Values of $pK_a$ for a series of carboxylic acids and chloro-derivatives.

| Acid | $pK_a$ | Chloro-derivative | $pK_a$ |
|---|---|---|---|
| | 4.77 | | 2.81 |
| | 4.87 | | 4.09 |
| | 4.81 | | 4.52 |
| | 4.85 | | 4.72 |

$CH_3^-$ is poorly stabilized because the negative charge is localized on the C atom.

$$CH_4 \rightleftharpoons H^+ + CH_3^-$$   $pK_a \approx 48$   (27.3)

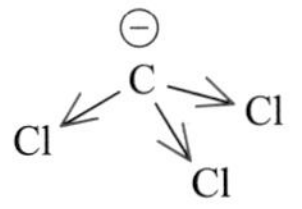

In contrast, $CHCl_3$ is considerably more acidic than $CH_4$. The negative charge in $CCl_3^-$ is spread out among the three electron-withdrawing Cl substituents (structure **27.10**).

(27.10)

---

## SUMMARY

In this short chapter, we have reviewed how to use electronegativity values to determine whether bonds are polar. Factors that contribute towards the magnitude and direction of a molecular dipole moment are bond dipoles, molecular geometry and lone pairs of electrons. The *inductive effect* operates through $\sigma$-bonds whenever electron-withdrawing groups are present; it is augmented by the through-space *field effect*. Electron-withdrawing groups such as Cl are able to stabilize negative centres and this has an effect on physical and chemical properties.

### Do you know what the following terms mean?

- electronegativity
- bond dipole moment
- molecular dipole moment
- inductive effect
- field effect
- electron-withdrawing substituent

### You should be able:

- to use electronegativity values to determine if a bond is polar
- to work out if a molecule is polar
- to interpret trends in molecular dipole moments among related compounds
- to explain how the presence of electron-withdrawing groups leads to the inductive effect
- to rationalize trends in $pK_a$ values related to the presence of electron-withdrawing groups

## PROBLEMS

**27.1** Give the directions of the dipole moments in the following bonds: (a) C−Br; (b) N−H; (c) Cl−F; (d) C−S.

**27.2** Which of the following molecules are polar? (a) $CH_3Cl$; (b) $CH_2Cl_2$; (c) $SO_3$; (d) HCHO; (e) HBr; (f) $H_2O$; (g) $CH_3OH$; (h) $CBr_4$.

**27.3** For each of the polar molecules in problem 27.2, draw a diagram to show the approximate direction of the molecular dipole moment.

**27.4** Which of the following molecules are polar? (a) $SO_2$; (b) $CO_2$; (c) $H_2O$; (d) HC≡CCl; (e) $H_2S$.

**27.5** For each of the following molecules, comment on factors that contribute to the direction and magnitude of the molecular dipole moments.

**27.6** The molecular dipole moments of the gas phase molecules $CH_2Br_2$, $CClF_3$ and HCN are 1.43, 0.50 and 2.98 D respectively. Draw the structure of each molecule and indicate the direction of the resultant dipole moment.

**27.7** Draw the structure of methanol and show the *approximate* direction in which the dipole moment of 1.70 D acts.

**27.8** How might values of molecular dipole moments help you to distinguish between (*E*)- and (*Z*)-1,2-dibromoethene?

**27.9** Explain what is meant by the inductive effect with reference to the compounds $CF_3CO_2H$ and $CH_3CO_2H$.

**27.10** Discuss the trends in values of $pK_a$ in Table 27.2.

## Additional problem

**27.11** Three compounds **A**, **B** and **C**, which are related in a series, are characterized by the data in Table 27.3. Suggest identities for the compounds.

**Table 27.3** Data for problem 27.11

|  | **A** | **B** | **C** |
|---|---|---|---|
| Dipole moment (gas phase) / D | 2.05 | 2.03 | 1.91 |
| $^1H$ NMR $\delta$/ppm | 1.48 (triplet); 3.57 (quartet) | 1.67 (triplet); 3.43 (quartet) | 1.83 (triplet); 3.20 (quartet) |
| Highest mass peaks in the mass spectrum | $m/z$ 64, 66 (rel. intensities ≈3 : 1) | $m/z$ 110, 108 (rel. intensities ≈1 : 1) | $m/z$ 156 |
| C and H analysis | C 37.2, H 7.8 % | C 22.0, H 4.6 % | C 15.4, H 3.2 % |

# 28 Halogenoalkanes

## Topics

- Structure and nomenclature
- Synthesis
- Physical properties
- Formation of Grignard reagents
- Formation of organolithium reagents
- Nucleophilic substitution reactions
- Elimination reactions
- Selected reactions of halogenoalkanes

## 28.1 Structure and nomenclature

In this section, we build on the basic rules of organic nomenclature to incorporate halogen functional groups. The general formula for a *halogenoalkane* (or *alkyl halide*) is RX where R is the alkyl group and X is the F, Cl, Br or I.

### Halogenoalkanes with one halogen atom

The alkyl group in RX can have a straight or branched carbon chain. In naming a relatively simple compound, the same basic rules apply as were outlined in Section 24.4, but with the additional point that *the longest chain contains the C atom attached to the halogen functional group*. The name of a halogenoalkane is formed by prefixing the alkane stem with *fluoro-*, *chloro-*, *bromo-* or *iodo-*. Compounds **28.1** and **28.2** are straight chain, bromo-derivatives of pentane. In each, the principal carbon chain is numbered so that the Br substituent has the lowest possible position number. In compounds in which there is another substituent in addition to the halogen atom, the carbon chain is numbered beginning at the end nearest to the first substituent, whether that is the halogen atom or the other substituent.

Four isomers of $C_6H_{13}Cl$ are shown below (see also problem 28.1). The principal chain *must contain the C atom attached to the Cl-substituent*:

1-Bromopentane

**(28.1)**

2-Bromopentane

**(28.2)**

1-Chloro-3-methylpentane     2-Chloro-3-methylpentane     3-Chloro-3-methylpentane     1-Chloro-2-ethylbutane

In three of the isomers, the principal chain contains five C atoms and the name is based on pentane. The substituents are ordered alphabetically (e.g. 1-chloro-3-methylpentane and *not* 3-methyl-1-chloropentane). In the last isomer, the principal chain (*chosen to contain the Cl atom*) contains four C atoms and so the name is based on butane. The atom numbering is shown in the diagrams.

## Halogenoalkanes with more than one halogen atom

A straight chain halogenoalkane with two like halogen substituents is named so as to give them the lowest position numbers. For example:

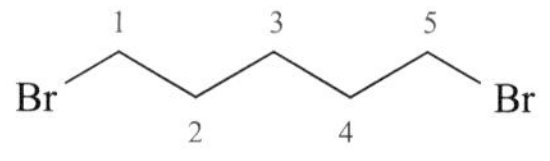

1,5-Dibromopentane 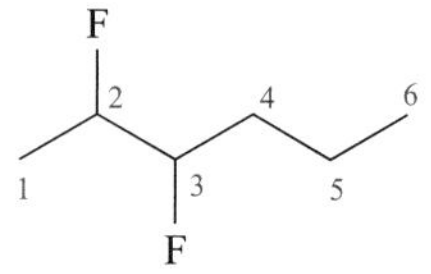            2,3-Difluorohexane

If *different halogen atoms* are present, their descriptors are ordered alphabetically, e.g. 1-bromo-2-chlorobutane.

## Primary, secondary and tertiary halogenoalkanes

Primary, secondary, tertiary C atoms: see Section 24.5

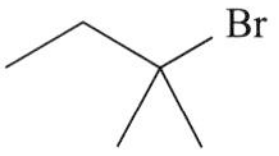

**(28.3)**

When the halogen atom is attached to a terminal C atom as in **28.1**, the compound is a *primary halogenoalkane*; the general formula is $RCH_2X$. Compound **28.2** has the general formula $R_2CHX$, and is an example of a *secondary halogenoalkane*. In a *tertiary halogenoalkane* three R groups are attached to the same C atom as the halogen substituent and the general formula is $R_3CX$. Compound **28.3** is an example of a tertiary halogenoalkane. The R groups in these general formulae may be the same or different.

---

**28.2**     ## Synthesis of halogenoalkanes

In looking at the chemistry of alkanes and alkenes, we described the formation of halogenoalkanes in a number of reactions and details can be obtained by referring to the relevant sections:

- reactions of $X_2$ with alkanes (radical substitutions): Sections 25.7 and 25.8;
- formation of vicinal dihalides from electrophilic addition of $X_2$ to an alkene: Sections 26.4 and 26.5;
- formation of a halogenoalkane from electrophilic addition of HX to an alkene: Sections 26.4 and 26.5;
- formation of a halogenoalkane from radical addition of HX to an alkene: Section 26.7;
- formation of a *gem*-dihalide by reaction of HX with an alkyne: Section 26.12.

The first of these methods is *not* a useful way of producing a specific halogenated derivative. It is difficult to control radical substitution reactions and mixtures of products are obtained. The electrophilic addition of $X_2$

($X = Cl$ or $Br$) gives vicinal dihalides in good yields (e.g. reactions 26.17 and 26.18). Monohalides are produced by addition of HX ($X = Cl$, $Br$ or $I$) to an alkene and, as we discussed in Chapter 26, the addition is regioselective with the X atom becoming attached to the more highly substituted C atom (*Markovnikov addition*). Reaction 28.1 gives an example.

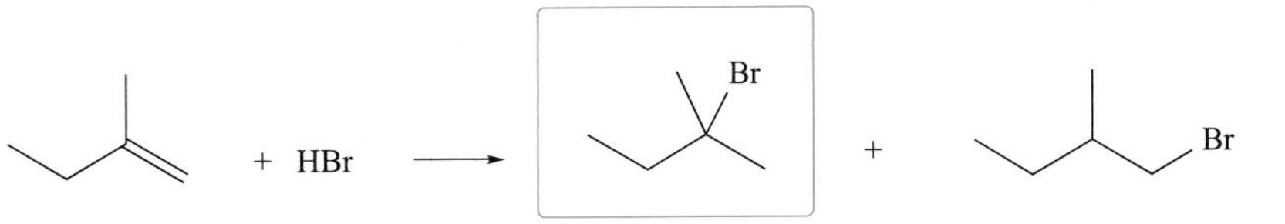

2-Bromo-2-methylbutane
Major product

1-Bromo-2-methylbutane
Minor product

(28.1)

The addition of HX under radical conditions gives an anti-Markovnikov product as in reaction 28.2.

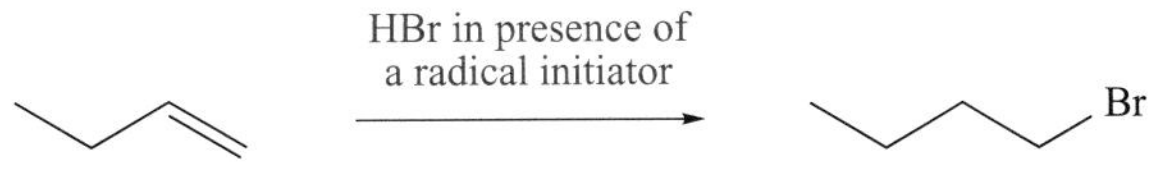

But-1-ene

HBr in presence of a radical initiator

1-Bromobutane

(28.2)

In Section 26.7, we also described *allylic bromination* using *N*-bromo-succinimide (NBS), a reaction that brominates an alkyl group adjacent to a C=C bond.

Reactions between alkanes or alkenes and $F_2$ are explosive and so fluoroalkanes are usually prepared using an alternative fluorinating agent. Equation 28.3 illustrates the use of $HF/SbF_5$ with fluorination occurring by partial halogen exchange. Fluorination occurs in a symmetrical manner, with steric crowding being relieved on each sequential halogen exchange.

Chlorofluorocarbons, CFCs: see Box 25.3

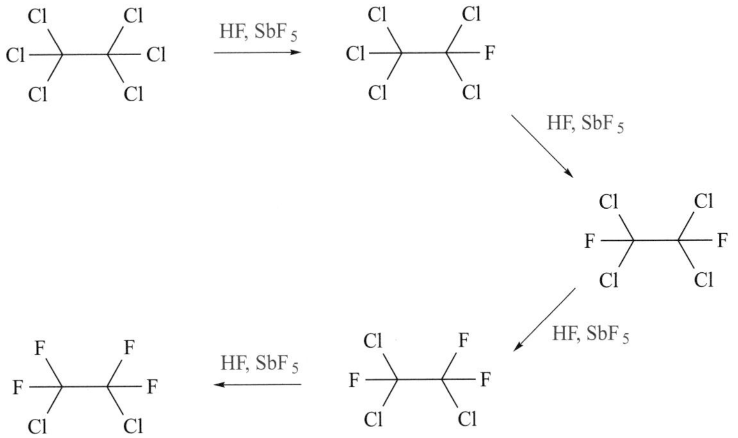

(28.3)

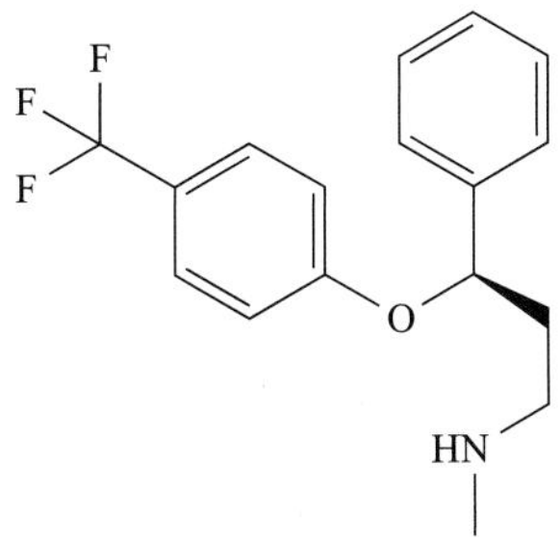

A trifluoromethyl ($CF_3$) group is present in the antidepressant Prozac which is manufactured by Lilley and ranks among the world's top-selling drugs.

A common and convenient route for the preparation of halogenoalkanes is from the corresponding alcohols as in general reaction 28.4.

Prozac

$$ROH + HX \longrightarrow RX + H_2O \qquad\qquad X = Cl, Br, I \quad (28.4)$$

Tertiary alcohols are the most reactive and the order of reactivity is:

$$R_3COH > R_2CHOH > RCH_2OH$$

Tertiary     Secondary     Primary

As a result, the reaction is used most widely for tertiary (reaction 28.5) rather than secondary and primary alcohols. We return to the mechanism of the reaction and reasons for differences in reaction rates in Chapter 30.

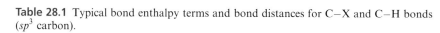

(28.5)

For the conversion of secondary or primary alcohols to halogenoalkanes, reactions with thionyl chloride ($SOCl_2$), phosphorus(III) halides ($PCl_3$, $PBr_3$, $PI_3$) or phosphorus(V) chloride ($PCl_5$) are the most useful. Reactions 28.6 and 28.7 give examples. An advantage of using $SOCl_2$ is that the side products (HCl and $SO_2$) are gases, making isolation of the organic product easy.

(28.6)

(28.7)

### 28.3     Physical properties

#### The C–X bonds

The electronegativity values in Table 27.1 show that each C–X (X = F, Cl, Br, I) bond is polar in the sense $C^{\delta+}–X^{\delta-}$. We considered the effect of this on *molecular* dipole moments in Chapter 27. The bond enthalpies of C–X bonds decrease in the order C–F > C–Cl > C–Br > C–I, and bond distances increase. Values listed in Table 28.1 refer to halogenoalkanes, i.e. $sp^3$ hybridized C atoms.

**Table 28.1** Typical bond enthalpy terms and bond distances for C–X and C–H bonds ($sp^3$ carbon).

| Bond | Bond enthalpy term / kJ mol$^{-1}$ | Bond length / pm |
|------|------|------|
| C–H | 416 | 109 |
| C–F | 485 | 132 |
| C–Cl | 327 | 177 |
| C–Br | 285 | 194 |
| C–I | 213 | 214 |

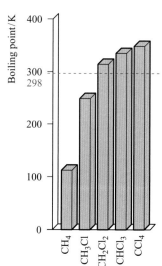

**Fig. 28.1** The trend in boiling points of members of the series $CH_{4-x}Cl_x$, $x = 0–4$.

#### Boiling points

A halogenoalkane usually has a higher boiling point than the corresponding alkane. Figure 28.1 illustrates the trend in boiling points across the series

$CH_{4-x}Cl_x$ and shows that increasing the number of Cl atoms leads to an increase in boiling point: $CH_2Cl_2$, $CHCl_3$ and $CCl_4$ are liquids at 298 K while $CH_4$ and $CH_3Cl$ are gases. The trend in boiling points follows an increase in molecular weight.

### Liquid chloro-, bromo- and iodoalkanes

The liquid chloro-, bromo- and iodoalkanes are usually denser than water and the density increases with higher degrees of halogen substitution. For example, the densities of $CH_2Cl_2$, $CHCl_3$ and $CCl_4$ are 1.33, 1.48 and 1.59 g cm$^{-3}$ respectively. Liquid halogenoalkanes are usually immiscible with water. This is consistent with the fact that water is polar, but halogenoalkanes such as $CCl_4$ and $CBr_4$ are non-polar. However, despite the fact that $CHCl_3$ and $CH_2Cl_2$ are polar as shown in structures **28.4** ($\mu = 1.60$ D) and **28.5** ($\mu = 1.04$ D), they are immiscible with $H_2O$. This can be attributed to the fact that dipole–dipole interactions between $H_2O$ and the halogenoalkane do not sufficiently compensate for the loss of hydrogen bonding between $H_2O$ molecules.

In the past, liquid halogenoalkanes were widely used as dry-cleaning solvents. However, it is now known that some of them may cause liver damage and others have been implicated as cancer-causing agents. Dichloromethane ($CH_2Cl_2$) and trichloromethane ($CHCl_3$, with the common name of chloroform) are common laboratory solvents, although health risks associated with chlorinated solvents, particularly $CCl_4$, have led to reduced usage.

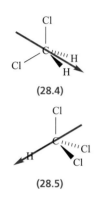

(28.4)

(28.5)

Immiscibility: see end of Section 25.3

| 28.4 | **Reactions of halogenoalkanes: formation of Grignard and organolithium reagents** |
|---|---|

Halogenoalkanes usually undergo the following general types of reactions:

- nucleophilic substitution of $X^-$;
- elimination of HX to give an unsaturated hydrocarbon;
- formation of Grignard reagents;
- formation of organolithium reagents.

In this section, we look at the use of halogenoalkanes for the formation of two types of organometallic compounds: Grignard and organolithium reagents. Later in the chapter, we move on to a detailed look at nucleophilic substitution and elimination reactions.

### Grignard reagents

A **Grignard reagent** is an organometallic compound of the type RMgX (R = alkyl, X = halogen) and is used as an alkylating reagent.

Grignard reagents are organometallic compounds of general type RMgX and are of great importance in synthetic organic chemistry as alkylating reagents. The Mg—C bond is highly polar in the sense $Mg^{\delta+}$–$C^{\delta-}$ and this contributes to the reactivity of Grignard reagents. They are air- and moisture-sensitive and are usually prepared *in situ*, i.e. the compounds are not isolated but are prepared in ether solutions as required, working under moisture- and oxygen-free conditions (*inert conditions*). Alternatively, some of the most regularly used Grignard reagents can be purchased from commercial companies. The usual solvents for Grignard reagents are *anhydrous*

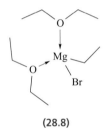

**(28.6)**      THF **(28.7)**

diethyl ether (**28.6**) or tetrahydrofuran (THF, **28.7**). Magnesium metal reacts with a halogenoalkane according to equation 28.8 and an Mg−C bond is formed.

$$R-X \;+\; Mg \quad\xrightarrow{\text{Et}_2\text{O or THF}}\quad R-Mg-X \qquad (28.8)$$

$$X = Cl,\ Br,\ I$$

The reaction can be initiated by adding $I_2$. This reacts to give a clean and highly reactive Mg surface. The rate at which reaction 28.8 proceeds follows the order RI > RBr > RCl. Primary, secondary and tertiary halogeno-alkanes all react with Mg. Equations 28.9 and 28.10 show examples. The method can be extended to aryl halides, e.g. chlorobenzene.

Victor Grignard (1871–1935).
© *The Nobel Foundation.*

$$\text{Br} \;+\; Mg \quad\xrightarrow{\text{Et}_2\text{O}}\quad \text{MgBr} \qquad (28.9)$$

$$\text{Cl} \;+\; Mg \quad\xrightarrow{\text{Et}_2\text{O}}\quad \text{MgCl} \qquad (28.10)$$

We shall see uses of Grignard reagents in later chapters, and their synthetic utility is illustrated in Figure 28.2. This includes reactions between Grignard reagents and carbonyl groups (see Chapter 33) and proton donors (for alkane synthesis).

Although we represent the structures of Grignard reagents as having 2-coordinate Mg, adduct formation with donor solvents such as THF or $Et_2O$ leads to, typically, 4-coordination. For example, EtMgBr crystallizes from diethyl ether solution as adduct **28.8** and contains approximately tetrahedral Mg.

**(28.8)**

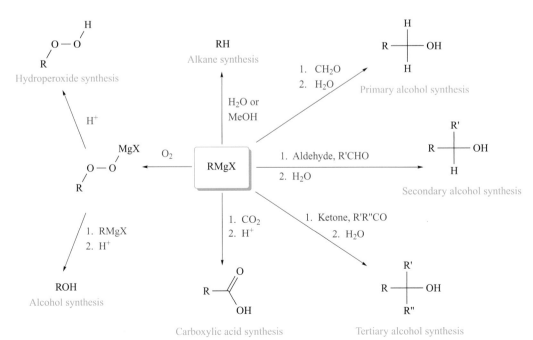

**Fig. 28.2** Representative reactions of Grignard reagents, RMgX.

## Organolithium reagents

An *organolithium reagent* is an organometallic compound of the type RLi. If R = alkyl, the compound is an *alkyllithium* compound and is an alkylating reagent.

Organolithium compounds, RLi, are organometallic compounds that behave as sources of *carbanions*, $R^-$, and have widespread synthetic uses. Alkyllithium compounds are used as alkylating reagents. Alkyllithium or aryl-lithium compounds can be prepared by treating lithium with a halogenoalkane under inert conditions (equation 28.11). The absence of air and moisture is essential since both Li metal and RLi derivatives react with $H_2O$ and $O_2$. Solvents must be anhydrous.

$$RX + 2Li \xrightarrow{\text{e.g. Et}_2\text{O or THF}} RLi + LiX \qquad (28.11)$$

A particularly important organolithium reagent is *n*-butyllithium, $^nBuLi$, which can be made by reaction 28.12.

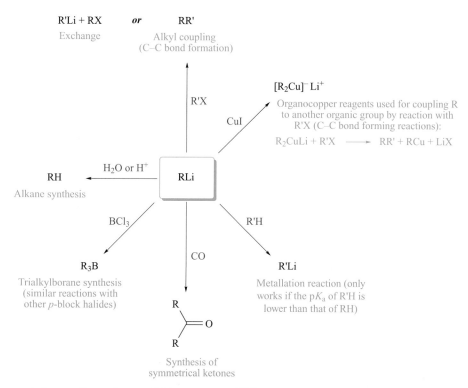

$$\xrightarrow{\text{Li, pentane}} \qquad + LiBr \qquad (28.12)$$

*n*-Butyllithium

*n*-Butyllithium is also commercially available as a solution in hexane, sealed under an atmosphere of $N_2$, and is seldom prepared on a small scale in a laboratory. Other commercially available organolithium reagents include MeLi and PhLi. Organolithium compounds can act as nucleophiles and bases. Figure 28.3 exemplifies some of the uses of RLi reagents. Methyllithium and butyllithium are commonly used as deprotonating agents. The $pK_a$ value of $CH_4$ is 48 (Table 26.1) and hence $Me^-$ is a very strong base.

Although we usually represent organolithium compounds as monomeric RLi, polymeric species may be present in solution and the solid state. For example, in hydrocarbon solvents, MeLi and $^nBuLi$ are hexameric, while

R'Li + RX  **or**  RR'

Exchange      Alkyl coupling
              (C–C bond formation)

R'X

$[R_2Cu]^- Li^+$

CuI

Organocopper reagents used for coupling R to another organic group by reaction with R'X (C–C bond forming reactions):

$R_2CuLi + R'X \longrightarrow RR' + RCu + LiX$

$H_2O$ or $H^+$

RH

Alkane synthesis

RLi

$BCl_3$

$R_3B$

Trialkylborane synthesis (similar reactions with other *p*-block halides)

CO

R'H

R'Li

Metallation reaction (only works if the $pK_a$ of R'H is lower than that of RH)

$\underset{R}{\overset{R}{>}}{=}O$

Synthesis of symmetrical ketones

**Fig. 28.3** Representative reactions of organolithium reagents, RLi.

in $Et_2O$, they are tetrameric. At low temperature and in THF, $^nBuLi$ exists as an equilibrium mixture of dimers and tetramers. Representations such as that in equation 28.12 imply covalent $C-Li$ bonding. The nature of the bonding in organolithium compounds remains a matter for debate, with both ionic and covalent models being discussed in the chemical literature.

## 28.5 Reactions of halogenoalkanes: nucleophilic substitution versus elimination

Nucleophile: see Section 24.10

The fact that the $C-X$ bond is polar in the sense $C^{\delta+}-X^{\delta-}$ means that the C atom in a halogenoalkane is susceptible to attack by nucleophiles; the $C^{\delta+}$ centre in the halogenoalkane acts as an electrophile. Two reactions may take place when a nucleophile attacks.

- the nucleophile, $Nu^-$, can become attached to the C atom of the $C-X$ bond and $X^-$ leaves (a substitution, reaction 28.13), or
- a base, B, can abstract an H atom from the C atom next to the $C-X$ bond and there is concomitant $X^-$ elimination to give an alkene (reaction 28.14).

$$R-X \ + \ \overset{\ominus}{Nu} \ \longrightarrow \ R-Nu \ + \ \overset{\ominus}{X} \tag{28.13}$$

B = base

$$(28.14)$$

In the following sections, we consider how nucleophilic substitution and elimination reactions occur, and also possible competition between these reactions.

## 28.6 Nucleophilic substitution

### Reaction kinetics: some experimental data

In Chapter 15, we discussed the relationships between reaction mechanism and observed reaction kinetics. Problem 15.6 related to the reaction of $Me_3CCl$ with $H_2O$ (equation 28.15).

$$(28.15)$$

The data from problem 15.6 are presented in Figure 28.4 and the linear relationship between $\ln[Me_3CCl]$ and time shows that the reaction is first order with respect to $Me_3CCl$. It is also found that the rate is *independent* of the concentration of $H_2O$ (the nucleophile). The overall rate equation is given by equation 28.16 where $k$ is the rate constant.

$$\text{Rate of reaction} = -\frac{d[Me_3CCl]}{dt} = k[Me_3CCl] \tag{28.16}$$

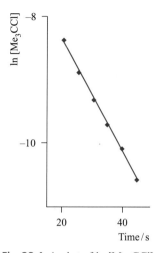

**Fig. 28.4** A plot of $\ln[Me_3CCl]$ against time for the hydrolysis of $Me_3CCl$. The linear plot shows that the reaction rate is first order with respect to $Me_3CCl$. [Data from: A. Allen *et al.* (1991) *J. Chem. Educ.*, vol. 68, p. 609.]

This rate equation is *not* general for all halogenoalkanes. Kinetics data for the nucleophilic substitution reaction between $CH_3Br$ and $Cl^-$ (equation 28.17) reveal that the reaction is first order with respect to $CH_3Br$ *and* first order with respect to $Cl^-$ (equation 28.18), i.e. second order overall.

$$\text{Rate of reaction} = -\frac{d[CH_3Br]}{dt} = k[CH_3Br][Cl^-] \tag{28.18}$$

A similar result is obtained for the reaction between 1-bromobutane and $NC^-$ (equations 28.19 and 28.20). Notice that we write $NC^-$ for cyanide ion rather than $CN^-$ or $[CN]^-$ as in earlier chapters. Writing $NC^-$ emphasizes that the attacking atom is C, and we shall use this notation for the cyanide and $HO^-$ for hydroxide throughout this and later chapters.

$$\text{Rate of reaction} = -\frac{d[RBr]}{dt} = k[RBr][NC^-]$$

$$RBr = \text{1-bromobutane} \tag{28.20}$$

The differences in kinetic data are consistent with there being two mechanisms for nucleophilic substitution. Reactions that follow first order kinetics exhibit a unimolecular rate-determining (*slow*) step involving the halogenoalkane. Reactions that follow second order kinetics exhibit a bimolecular rate-determining step involving both the halogenoalkane and nucleophile. The two mechanisms are called $S_N1$ (unimolecular) and $S_N2$ (bimolecular). Notice that the example of first order kinetics (equations 28.15 and 28.16) involves a *tertiary* halogenoalkane, while the examples of second order kinetics (equations 28.17 to 28.20) involve *primary* halogenoalkanes. We come back to this point later on.

We now look in details at $S_N1$ and $S_N2$ mechanisms and then consider the factors that favour one mechanism over the other.

Unimolecular and bimolecular mechanisms: see Section 15.10

### The $S_N1$ mechanism

Consider the reaction of $Me_3CCl$ with $H_2O$ which obeys rate law 28.16. A reaction mechanism consistent with this rate law has one step that is rate-determining involving *only* $Me_3CCl$. The nucleophile, $H_2O$, must be involved in a *fast step* (i.e. *non-rate-determining*). The steps in the $S_N1$ mechanism are shown in the scheme below:

Tetrahedral C                   Trigonal planar C

The slow step is the spontaneous, dissociative cleavage of the C–Cl bond, i.e. a unimolecular step. A chloride ion departs (Cl⁻ is the *leaving group*) and a carbenium ion is formed. This is a reaction *intermediate*, which reacts in a fast step with any available nucleophile. In the reaction above, the nucleophile is $H_2O$. As the C–O bond forms, the positive charge is transferred from C to O. This is followed by rapid loss of $H^+$ to give the product. It is, of course, also possible for the carbenium ion to be attacked by Cl⁻ to regenerate the starting material.

Figure 28.5a shows the reaction profile for the reaction of $Me_3CCl$ with $H_2O$ to give $Me_3COH$ and HCl; this profile is typical of an $S_N1$ reaction. The activation energy, $E_a$, marked on the figure corresponds to the energy barrier for the rate-determining step. The second energy barrier is lower and corresponds to that for the fast step. The carbenium ion intermediate lies at a local energy minimum along the reaction pathway. Transition state $\{TS(1)\}^{\ddagger}$ may be described in terms of a species that is part-way between $Me_3CCl$ and $[Me_3C]^+$, and can be represented as having a 'stretched' bond as in structure **28.9**.

The reaction scheme above highlights a change in geometry at the central C atom. In $Me_3CCl$ and $Me_3COH$, all C atoms are tetrahedral and $sp^3$ hybridized. In the carbenium ion intermediate, the positive charge is localized on the central C atom. This centre is electron-deficient (it has a sextet of electrons) and is trigonal planar ($sp^2$ hybridized). The incoming nucleophile donates a pair of electrons into the empty $2p$ atomic orbital on the carbon atom (diagram **28.10**), and can attack from either above or below the plane. This has stereochemical consequences that we shall consider later on.

Reaction profiles, intermediates and transition states: see Section 15.1

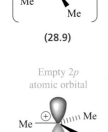

**(28.9)**

Empty $2p$ atomic orbital

**(28.10)**

### The $S_N2$ mechanism

Now let us look at a nucleophilic substitution that is second order overall, with a reaction rate that depends on both the concentration of the halogenoalkane and the nucleophile (equations 28.18 and 28.20). A mechanism that is consistent with these data involves a *one-step* process, i.e. the rate-determining step is a *bimolecular exchange* mechanism. The pathway is known as an $S_N2$ mechanism and is illustrated in the scheme below for the reaction of $CH_3Br$ with Cl⁻:

**Fig. 28.5** (a) Reaction profile for the $S_N1$ reaction between $Me_3CCl$ with $H_2O$. The intermediate is a carbenium ion and $\{TS(1)\}^{\ddagger}$ and $\{TS(2)\}^{\ddagger}$ are transition states (see text). (b) Reaction profile for the $S_N2$ reaction of $CH_3Br$ with $Cl^-$. The transition state (structure **28.11**) is $\{TS\}^{\ddagger}$.

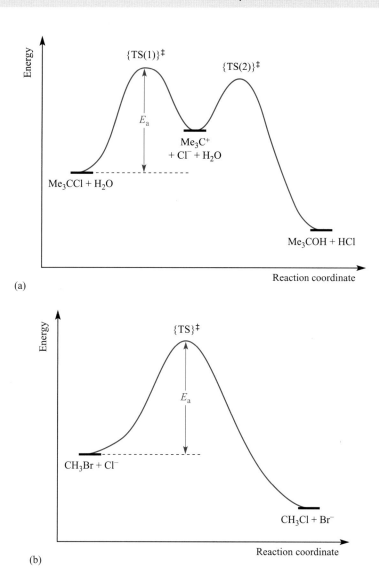

(a)

(b)

The formation of the C−Cl bond and the fission of the C−Br bond take place in a single step, and the process is said to be *concerted*. The reaction profile is shown in Figure 28.5b. There is *no intermediate*. The reaction pathway proceeds through a transition state $\{TS\}^{\ddagger}$ at an energy maximum, and this state can be considered as having two 'stretched bonds'. These are shown in structure **28.11** by the dashed lines and represent the C−Cl bond as it is being formed and the C−Br bond in the middle of being broken. The overall charge on the transition state is 1−, and this is spread out over the entering and leaving groups and is represented in **28.11** by partial charges. The $CH_3$ unit in **28.11** is planar, and the *entering* and *leaving groups* are on *opposite sides* of this plane (diagram **28.12**). The consequence of this on the stereochemistry of the product is discussed below.

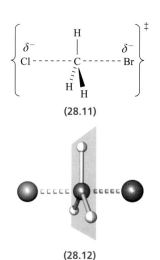

(28.11)

(28.12)

### Stereochemistries of $S_N1$ and $S_N2$ reactions

We have now outlined two limiting mechanisms whereby nucleophilic substitution may occur. Observable differences in reaction kinetics provide some evidence for the $S_N1$ and $S_N2$ mechanisms. These mechanisms are exhibited

**(28.13)**

▶▶
Chirality and absolute
configuration:
see Section 24.8

exclusively by only a few molecules. Usually there is competition between $S_N1$ and $S_N2$ mechanisms and with the eliminations that we discuss in Section 28.7. In addition to differences in rate laws, differences in the *stereochemistry of the product in relation to the reactant* distinguish $S_N1$ and $S_N2$ mechanisms.

2-Bromobutane (**28.13**) contains an asymmetric C atom (the atom to which Br is attached). When an optically pure enantiomer undergoes nucleophilic substitution with $HO^-$, the absolute configuration changes from ($R$) to ($S$) or from ($S$) to ($R$) in going from starting material to product. That is, there is a *complete inversion* of configuration as illustrated in equation 28.21. This is called a *Walden inversion* after Paul Walden who discovered the phenomenon.

$$\text{(S)-2-Bromobutane} \quad \xrightarrow[-Br^{\ominus}]{HO^{\ominus}} \quad \text{(R)-Butan-2-ol} \tag{28.21}$$

(S)-2-Bromobutane                                    (R)-Butan-2-ol

An $S_N2$ reaction is *stereospecific*, and takes place with *inversion of configuration*.

The inversion can be rationalized in terms of the $S_N2$ mechanism that we described above. The attack of the nucleophile and the release of the leaving group take place along a linear path (look at structure **28.12**). As this happens, the three substituents that are retained during the reaction are pushed 'inside out' through a plane. The process is represented below for reaction 28.21 going through a transition state like the one in structure **28.11**. Watch what happens to the absolute configuration:

(S)-2-Bromobutane                                    (R)-Butan-2-ol

The reaction is also represented in Figure 28.6 to give a better 3-dimensional picture of the stereochemistry.

We have already shown that the kinetics of an $S_N1$ reaction are consistent with an intermediate carbenium ion. Such an intermediate contains a planar $C^+$ centre which can be attacked from *either side* of the plane by the entering nucleophile. The stereochemical consequences of this can be seen if we consider the reaction of the chiral, tertiary halogenoalkane 3-chloro-3-methylhexane with $H_2O$. If we start with the optically pure ($R$) enantiomer, equation 28.22 shows that the product is a mixture of enantiomers. Conversion of ($R$)-3-chloro-3-methylhexane to ($R$)-3-methylhexan-3-ol occurs with *retention of configuration*, and ($R$)-3-chloro-3-methylhexane to ($S$)-3-methylhexan-3-ol occurs with *inversion of configuration*.

**Fig. 28.6** Inversion of configuration occurs during an $S_N2$ reaction. Here it is illustrated by the reaction of $OH^-$ with 2-bromobutane. Colour code: C, grey; H, white; O, red; Br, brown.

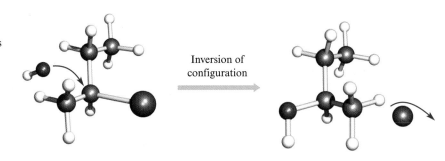

Inversion of configuration

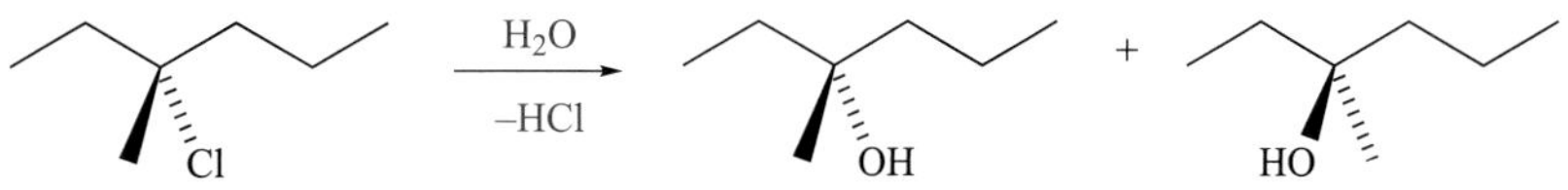

$$(R)\text{-3-Chloro-3-methylhexane} \qquad (R)\text{-3-Methylhexan-3-ol} \qquad (S)\text{-3-Methylhexan-3-ol}$$

$$(28.22)$$

This observation is consistent with the loss of Cl⁻ to form a carbenium ion (equation 28.23) followed by attack by the nucleophile as shown in Figure 28.7.

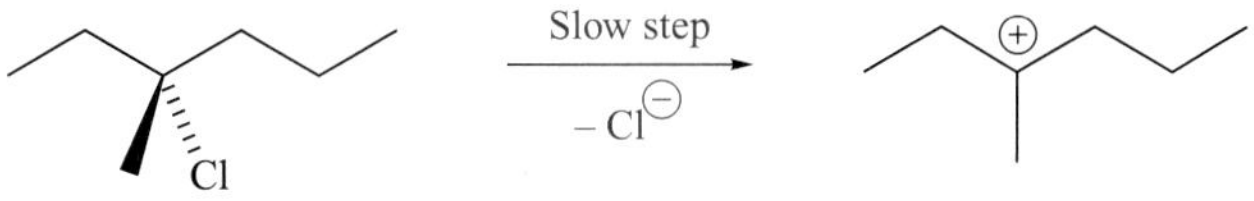

$$(28.23)$$

The top and bottom faces of the carbenium ion are identical and so *equal quantities* of the two enantiomers of the product must be formed. Thus, in equation 28.22, the product should be a racemate, i.e. 50% of $(R)$-3-methylhexan-3-ol and 50% $(S)$-3-methylhexan-3-ol. In practice, few $S_N1$ reactions give the expected 50:50 mix of enantiomers, and inversion of configuration exceeds retention to an extent of up to $\approx$20%. This probably occurs because the 'real' mechanism involves a transition state in which the carbenium ion and leaving group are still *loosely* bound by electrostatic interactions, making attack from the side opposite to the leaving group less sterically hindered than attack from the same side. In other words, the 'real' $S_N1$ mechanism involves a transition state that lies somewhere between the two extremes that we have described for the $S_N1$ and $S_N2$ pathways.

► Racemate: see Section 24.8

An $S_N1$ reaction is **stereochemically non-specific**.

## $S_N1$ or $S_N2$?

We have now outlined the details of the two extreme mechanisms of nucleophilic substitution, but what determines which pathway will be followed in a particular reaction? Do the two mechanisms compete within the same reaction? The answers to these questions are not simple because

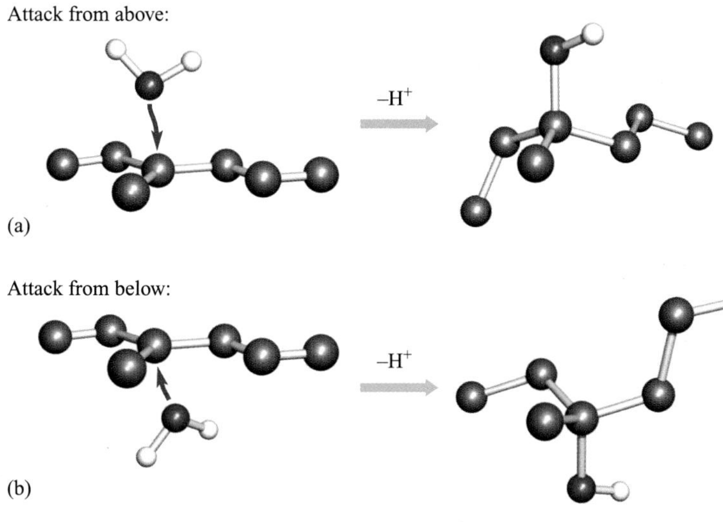

**Fig. 28.7** In an $S_N1$ reaction, the configuration at the central C atom can be retained or inverted because the nucleophile can attack from above or below the plane of the carbenium ion intermediate. The figure illustrates attack of $H_2O$ on the carbenium ion formed by loss of Cl⁻ from 3-chloro-3-methylhexane. For clarity, the H atoms are omitted from the organic backbone. Colour code: C, grey; H, white; O, red.

Attack from above:

$-H^+$

(a)

Attack from below:

$-H^+$

(b)

there are a number of contributing factors:

- the nature of the carbenium ion that is formed if the reaction goes by an $S_N1$ mechanism;
- steric effects;
- the nucleophile;
- the leaving group;
- the solvent.

We will consider each factor separately. However, it is important to keep in mind that, in practice, there is an interplay between these factors and drawing general conclusions is not at all easy or foolproof. A crucial point to keep in mind is that *for the dissociative $S_N1$ mechanism to be favoured, heterolytic cleavage of the C−X bond must occur readily.*

### The carbenium ion

■▷
Carbenium ions:
see Section 26.5

We introduced carbenium ions in Section 26.5 and described how they are stabilized by *electron-releasing* groups. *Hyperconjugation* contributes to this stabilization: electron density from the C−H $\sigma$-bonds of the alkyl groups is donated into the vacant $2p$ atomic orbital on the $C^+$ centre (see diagram **26.16**). The general order of carbenium ion stability where R is an alkyl group is:

$$[R_3C]^+ > [R_2CH]^+ > [RCH_2]^+ > [CH_3]^+$$

i.e. tertiary > secondary > primary > methyl

■▷
Compare this with
the discussion of allyl
radicals in Section 26.7

Carbenium ions formed by dissociation of $X^-$ from a simple halogeno-alkane fall into one of the categories above, but other types of cations are formed if the starting compound contains other functional groups. Two carbenium ions, the *allyl* and *benzyl* cations, have particular stabilities. If the starting halo-compound contains an X substituent attached to a C atom *adjacent to a C=C bond*, then loss of $X^-$ generates an allyl carbenium ion (equation 28.24) that is stabilized by contributions made by the two resonance structures **28.14**. Alternatively, the stabilization can be represented in terms of delocalized bonding (**28.15**, see problem 28.9).

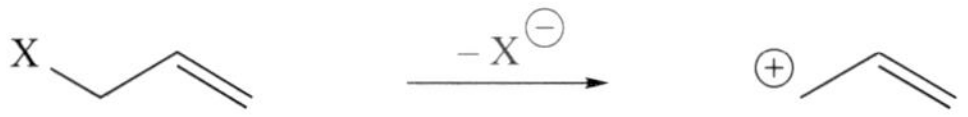

Allyl carbenium ion

**(28.14)**

**(28.15)**

$$X\diagdown\diagup \quad \xrightarrow{\;-X^{\ominus}\;} \quad \oplus\diagdown\diagup \tag{28.24}$$

When the halo-compound contains a $CH_2X$ group attached to a benzene ring, loss of $X^-$ gives a *benzyl* carbenium ion (equation 28.25). We shall discuss the chemistry of benzene and its derivatives in Chapter 32, but here we are interested in the reactivity of the $CH_2X$ substituent. The enhanced stability of the benzyl carbenium ion arises because of contributions made by the resonance structures shown in Figure 28.8. The net result is that the positive charge is delocalized around the ring. An alternative way of looking

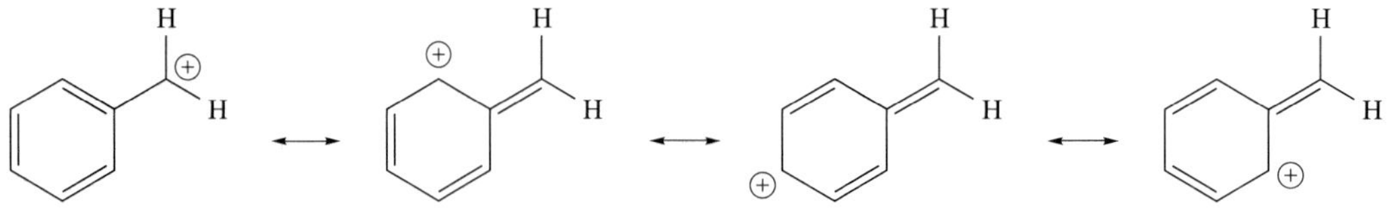

**Fig. 28.8** Resonance structures for the benzyl carbenium ion.

**THEORETICAL AND CHEMICAL BACKGROUND**

## Box 28.1 $\pi$-Delocalization in the benzyl carbenium ion

In Chapter 32, we describe the bonding in benzene in terms of a delocalized $\pi$-system. In this box, we introduce the bonding scheme briefly in order to relate it to that in the benzyl cation.

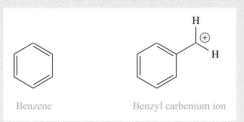

Benzene       Benzyl carbenium ion

Benzene, $C_6H_6$ is planar. Each C atom is $sp^2$ hybridized, and after the formation of the $\sigma$-bonding framework of C−C and C−H bonds, the six remaining C 2p atomic orbitals form a delocalized $\pi$-system. The

six $\pi$-electrons occupy three bonding MOs, the lowest energy one of which is shown below:

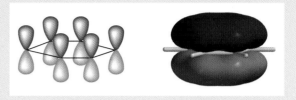

The benzyl carbenium ion is planar, and overlap between the seven C 2p atomic orbitals means that the $\pi$-system can be extended as shown below. The positive charge is therefore delocalized over the $C_7$-framework, giving extra stability to the carbenium ion.

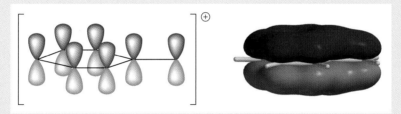

*Related information*: see Box 32.1.

at the bonding is to consider the extension of the $\pi$-system. This is considered in Box 28.1 and relates to the bonding description of benzene that we discuss in Chapter 32.

$$\text{Benzyl halide} \xrightarrow[-X^{\ominus}]{} \text{Benzyl carbenium ion} \tag{28.25}$$

The allyl and benzyl carbenium ions are types of primary carbenium ions, $[RCH_2]^+$, but their resonance stabilization means that they are more stable than primary carbenium ions where R = alkyl. Overall, the stabilities of carbenium ions can be summarized as follows, where R = alkyl:

So what do these relative stabilities mean in terms of the mechanism of nucleophilic substitution? The more stable the carbenium ion is, the more likely that loss of $X^-$ will occur as the rate-determining step of the reaction. Hence, the more likely it is that the $S_N1$ mechanism will be favoured. If you look at Figure 28.5a and apply the reaction profile in a general sense to $S_N1$ reactions, then the more stable the carbenium ion is, the lower the energy of the intermediate. Stabilization of the intermediate *also stabilizes transition state* $\{TS(1)\}^\ddagger$ because the transition state has some features in common with the intermediate that follows it in the reaction pathway. It follows, therefore, that a more stable carbenium ion means a lower activation energy, $E_a$, and hence a greater rate of reaction. Remember that $E_a$ in Figure 28.5 corresponds to the rate-determining step. Whereas the tertiary halogenoalkane $Me_3CCl$ undergoes hydrolysis *very* rapidly, the secondary and primary compounds $Me_2CHCl$ and $MeCH_2Cl$ are hydrolysed only slowly. Hydrolysis is more rapid if an allylic intermediate can be formed. Thus, under the same conditions (including same temperature and same solvent), the relative rates of hydrolysis of a series of chlorides are:

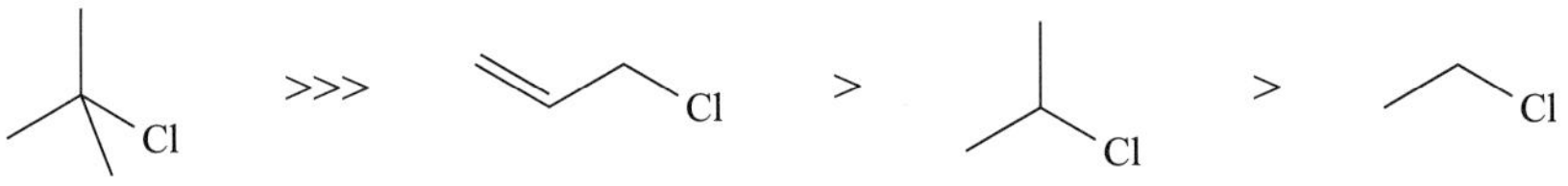

The fastest rate corresponds to the tertiary halogenoalkane undergoing an $S_N1$ reaction with $H_2O$; remember that the rate is *independent* of the nucleophile. The second fastest rate corresponds to a system that can form an allyl carbenium ion and for which an $S_N1$ mechanism is favourable. The rate of hydrolysis of the secondary chloroalkane is slow; the $S_N1$ mechanism is possible but it is not especially favourable and there is competition with an $S_N2$ mechanism. The primary chloroalkane exhibits the slowest rate of hydrolysis; the $S_N1$ mechanism is not favourable and the reaction probably takes place almost entirely by the $S_N2$ mechanism. As we have seen, the rate of an $S_N2$ reaction depends on the nucleophile as well as the halogenoalkane; the rate of hydrolysis of $MeCH_2Cl$ is very slow, and this is consistent with the fact that $H_2O$ is a poor nucleophile. We return to this later on.

In summary:

- the formation of a *tertiary* carbenium ion favours $S_N1$ over $S_N2$;
- the formation of an *allyl* or *benzyl* carbenium ion favours $S_N1$ over $S_N2$, but not to the same extent as the formation of a tertiary carbenium ion;
- the formation of a *secondary* carbenium ion is sufficiently favourable that it allows an $S_N1$ mechanism to operate, but it will compete with an $S_N2$ pathway;
- *primary* carbenium ions $[RCH_2]^+$ (R = alkyl) and the methyl cation are not favoured and spontaneous dissociation of the halide ion from the corresponding precursors does not occur.

## Steric effects

If you look back at the discussion of the $S_N2$ mechanism, then it is clear that the approach of the nucleophile and the formation of a transition state like **28.11** is sterically hindered if the central C atom carries bulky substituents. The nucleophile has to be able to attack the C atom through a 'shield' of substituents. Figure 28.9 shows 'space-filling' diagrams of

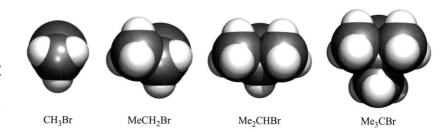

**Fig. 28.9** 'Space-filling' diagrams of $CH_3Br$, $MeCH_2Br$, $Me_2CHBr$ and $Me_3CBr$. Each is viewed down the C—Br bond. Colour code: C, grey; H, white; Br, brown; the Br atom is at the back of each diagram.

CH$_3$Br        MeCH$_2$Br        Me$_2$CHBr        Me$_3$CBr

$CH_3Br$, $MeCH_2Br$, $Me_2CHBr$ and $Me_3CBr$. Each molecule is viewed down the C—Br bond, the view that the approaching nucleophile 'sees' as it makes its way to the central C atom. The smallest possible substituents are H atoms, and nucleophilic attack at the C atom in $CH_3X$ is facile. Thus, the $S_N2$ mechanism is favourable for methyl halides. Similarly, attack by a nucleophile at the primary C atom of $MeCH_2Br$ (Figure 28.9) is sterically unhindered and the $S_N2$ mechanism can occur readily. It is more difficult for the nucleophile to approach the secondary C atom in $Me_2CHBr$, and yet more sterically hindered for the nucleophile to attack the tertiary C atom in $Me_3CBr$ (Figure 28.9). Secondary halogenoalkanes can undergo $S_N2$ reactions, but the reaction is generally slow because of steric hindrance. Tertiary halogenoalkanes do *not* undergo $S_N2$ reactions. Steric hindrance plays a part, but the rate at which the tertiary carbenium ion forms is so rapid that it precludes the $S_N2$ reaction from taking place in any case.

The two remaining types of species that we discussed in the previous section are allyl halides (**28.16**) and benzyl halides (**28.17**). We showed how both of these can undergo $S_N1$ reactions. However, both are primary halogenoalkanes and so can also undergo $S_N2$ reactions. Moreover, the $\pi$-system that is adjacent to the site of nucleophilic attack helps to stabilize the transition state. The central C atom in the transition state is 5-coordinate, although two of the interactions are 'stretched' bonds. The valence orbitals of carbon are the 2s and 2p atomic orbitals and so an appropriate hybridization scheme is to consider the C atom to be $sp^2$ hybridized with a 2p orbital used in forming the two 'stretched' bonds. Electrons in this occupied 2p orbital can be delocalized over the allyl or benzyl $\pi$-systems. Diagram **28.18** shows how we can view the bonding in the transition state formed when an allyl halide reacts with a nucleophile, Nu$^-$. Compare this with Figure 13.6b for the allyl anion.

In summary:

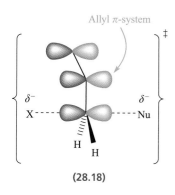

Allyl $\pi$-system

(28.16)        (28.17)

$\delta^-$        $\delta^-$
X ---- Nu

H  H

(28.18)

- $S_N2$ reactions can occur if attack by the nucleophile at the C atom is sterically unhindered;
- methyl halides ($CH_3X$) and primary halogenoalkanes are the best candidates for the $S_N2$ mechanism, and this is further favoured for allyl and benzyl halides by $\pi$-delocalization of electrons in the transition state;
- tertiary halogenoalkanes do not undergo $S_N2$ reactions.

### The nucleophile

The rate of an $S_N2$ reaction shows a first order dependence on the nucleophile (e.g. equations 28.18 and 28.20). In addition, the rate of reaction is affected by the nature of the nucleophile. Not all attacking species are as good nucleophiles as one another. The rate constant, $k$, is characteristic of a particular reaction and, if we consider an $S_N2$ reaction with a constant halogenoalkane substrate, we find that $k$ varies with the nucleophile. This observation is usually discussed in terms of the *nucleophilicity* of the incoming group, all other conditions (e.g. halogenoalkane, solvent, temperature) being constant. Unfortunately, 'nucleophilicity' is not easy to define, but it is useful to think of a good nucleophile (with respect to attack at a halogenoalkane) as a species that readily forms a bond with the $C^{\delta+}$ atom that was originally part of the C−X bond. It is also useful to make a comparison between 'nucleophilicity' and 'basicity', because we have discussed bases in detail in Chapter 16 and are familiar with the idea that a good base is a species that readily adds $H^+$.

> The *nucleophilicity* of a species is a measure of its affinity towards an electrophilic $C^{\delta+}$ atom; the *basicity* of a species is a measure of its affinity towards $H^+$. There is generally a relationship between the basicity and nucleophilicity of a species, provided the donor atom is constant in the series of nucleophiles being considered.

Conjugate acids–bases, $pK_a$ and $K_a$: see Section 16.4

The last point in the definition above needs some clarification. The reactions of $CH_3Br$ with $HO^-$, $CH_3CO_2^-$ and $H_2O$ follow $S_N2$ pathways and the order of their rates of reaction (under constant conditions) is $HO^- > CH_3CO_2^- > H_2O$. The order of the nucleophilicities of these species is therefore $HO^- > CH_3CO_2^- > H_2O$. Now consider the affinity of each nucleophile for a proton, i.e. the basicity of each species. This can be assessed by looking at the $pK_a$ values of the conjugate acids: $pK_a$ $H_2O > pK_a$ $CH_3CO_2H > pK_a$ $H_3O^+$, meaning that $H_3O^+$ donates $H^+$ the most readily and $H_2O$ accepts $H^+$ the least readily. Remember that if $pK_a$ is large, then $K_a$ for acid HA is small. Equilibrium 28.26 therefore lies towards the left-hand side, and $A^-$ is a relatively strong base. The larger the $pK_a$ of HA, the greater the basicity of $A^-$.

$$HA \rightleftharpoons H^+ + A^- \qquad\qquad pK_a = -\log K_a \quad (28.26)$$

Thus, the $pK_a$ values for $HO^-$, $CH_3CO_2^-$ and $H_2O$ show that the order of basicities is $HO^- > CH_3CO_2^- > H_2O$, an order that mimics that of their nucleophilicities. In general, in a series of nucleophiles that *contain the same donor atom*, we can use $pK_a$ values to gain an *approximate* idea of the relative strengths of the species as nucleophiles.

Where members of a series of nucleophiles contain different donor atoms (again, we are looking at attack on a constant halogenoalkane under constant reaction conditions), then trends in $pK_a$ values are not a reliable guide to trends in nucleophilicities. However, from rate data, we can derive some useful trends: nucleophiles related in a periodic group become stronger as the group is descended. For example, nucleophilicities usually follow the orders:

$$I^- > Br^- > Cl^-$$

$$HS^- > HO^-$$

$$RS^- > RO^- \qquad (R = alkyl)$$

An overview of the points that we have just made can be seen by looking at the relative rates of reactions of various nucleophiles with $CH_3Br$ (in a common solvent, ethanol). The relative rates are with respect to attack by $H_2O$:

| Nucleophile | $HS^-$ | $I^-$ | $HO^-$ | $Cl^-$ | $CH_3CO_2^-$ | $H_2O$ |
|---|---|---|---|---|---|---|
| Relative rate | ≈125 000 | ≈100 000 | ≈12 000 | ≈1 000 | ≈800 | 1 |

▣▷
Hard and soft ligands:
see Section 23.8

The terms *hard* and *soft* nucleophiles are often used, and application of these terms is similar to that used to describe ligands in coordination chemistry. Hard nucleophiles tend to have O or N as the attacking atom, e.g. $HO^-$ and $H_2N^-$, whereas soft nucleophiles feature less electronegative atoms that are lower down in the periodic groups, e.g. $RS^-$ and $I^-$. As far as $S_N2$ substitution in halogenoalkanes is concerned, soft nucleophiles are better than hard ones (look at the relative rates in the table above). But, the situation may be reversed for attack at other centres, as we shall see when we discuss nucleophilic attack on carbonyl groups (Chapter 33).

In summary:

- trends in nucleophilicity of species generally parallel trends in basicity, provided the attacking atom is constant in the series of nucleophiles;
- for species related in a periodic group, nucleophilicity generally increases down the group (e.g. $I^- > Br^- > Cl^-$).

## The leaving group

The nature of the leaving group is important in both $S_N1$ and $S_N2$ reactions. In the rate-determining step in both mechanisms, the C–X bond is broken and $X^-$ leaves. In nucleophilic substitution in halogenoalkanes, the leaving group could be $F^-$, $Cl^-$, $Br^-$ or $I^-$. The best leaving group is $I^-$, and the poorest is $F^-$:

$$I^- > Br^- > Cl^- > F^-$$

▣▷
For more bond enthalpies:
see Table 4.5

▣▷
$K_a = 10^{-pK_a}$

The ease with which the $X^-$ group leaves reflects the C–X bond enthalpy contributions: C–I = 213, C–Br = 285, C–Cl = 327 and C–F = 485 kJ mol$^{-1}$. The observations can also be correlated with the $pK_a$ values of the conjugate acids: $pK_a$ HF = +3.5, HCl = −7, HBr = −9 and HI = −11. Negative values of $pK_a$ signify strong acids, and so $Cl^-$, $Br^-$ and $I^-$ are weak bases; the ions are stable and are good leaving groups. On the other hand, HF is a weak acid, and $F^-$ is a poor leaving group.

In later chapters, we shall be concerned with nucleophilic substitution reactions in which the leaving group is something other than a halide ion, and so it is useful to include here discussion of some other groups. Poor leaving groups include $HO^-$, $H_2N^-$ and $RO^-$ and, like $F^-$, these are conjugate bases of weak acids. The poor ability of $HO^-$ to leave in a nucleophilic substitution reaction involving an alcohol substrate can be dealt with by protonation of the OH group. This means that $H_2O$ (a good leaving group) is being expelled, rather than $HO^-$ (a poor leaving group). We see this strategy at work in Chapters 29 and 30. The tosylate ion (**28.19**) is an excellent leaving group.

In summary:

(28.19)

- halide leaving groups are ranked $I^- > Br^- > Cl^- > F^-$;
- fluoroalkanes do not undergo $S_N2$ reactions because $F^-$ is such a poor leaving group.

## The solvent

■▷
Activation energy: see
Section 15.1

The choice of a suitable solvent for nucleophilic substitution reactions is important. In an $S_N1$ reaction, the rate-determining step involves heterolytic bond fission and a *polar solvent* increases the rate of ion formation. Look at the reaction profile in Figure 28.5a. The first step is rate-determining and the rate of reaction depends on the activation energy, $E_a$. If the energy of the transition state $\{TS(1)\}^{\ddagger}$ is lowered, $E_a$ decreases and the rate of reaction increases. If the reaction is carried out in a polar solvent, the carbenium ion is readily solvated in the same way that ions are solvated in water (see Figure 21.13). The solvation stabilizes the carbenium ion, lowering its energy. Transition state $\{TS(1)\}^{\ddagger}$ has some features in common with the carbenium ion intermediate (compare **28.20** and **28.21**). Although the C−X bond is stretched rather than broken, the transition state is significantly more polar than the starting halogenoalkane. Polar solvents not only stabilize the intermediate, they also stabilize the transition state, with the effect that $E_a$ is lowered. The rates of $S_N1$ reactions are therefore increased by using polar solvents such as water or ethanol.

■▷
Solvation by water: see
Section 21.7

(28.20)

> A *protic solvent* contains a Y−H bond where Y is an electronegative atom, usually O or N. The solvent can form hydrogen bonds. It is also a source of protons, although $K_a$ for the self-ionization process (see Section 16.4) may be small.

> An *aprotic solvent* does not contain H atoms that are capable of forming hydrogen bonds. The solvent is not a source of protons.

(28.21)

Solvent choice is also important for $S_N2$ reactions, but is *not* the same as for $S_N1$ reactions. Whereas the rate of an $S_N1$ reaction is independent of the nucleophile, the rate of an $S_N2$ reaction depends on it. We therefore have to consider what effect the solvent might have on the approach of the nucleophile to the halogenoalkane. The nucleophile in an $S_N2$ reaction is usually an anion (e.g. $Cl^-$, $HO^-$, $NC^-$). Anions are strongly solvated by *protic solvents* such as $H_2O$ or MeOH. These solvents form hydrogen bonds with an anionic nucleophile (see Figure 21.13) and hinder its attack on the halogenoalkane substrate. The nucleophile is stabilized by solvation and this increases the activation energy for the substitution reaction. Look at Figure 28.5b and consider what happens if the $Cl^-$ nucleophile is stabilized.

$Me-C{\equiv}N$

(28.22)

(28.23)

(28.24)

(28.25)

*Polar aprotic solvents* include acetonitrile (**28.22**), acetone (**28.23**), dimethyl sulfoxide (DMSO, **28.24**) and dimethylformamide (DMF, **28.25**). Being polar, these are good solvents for other polar molecules and for some ionic compounds. However, aprotic solvents do not contain H atoms attached to electronegative atoms and are unable to form hydrogen bonds with the nucleophile. As a consequence, the rate of an $S_N2$ reaction in a polar, aprotic solvent greatly exceeds that of the same reaction in a protic solvent.

In summary:

- the rate of an $S_N1$ reaction is enhanced by using a protic solvent, e.g. $H_2O$;
- the rate of an $S_N2$ reaction is enhanced by using a polar aprotic solvent, e.g. DMF, MeCN.

### Rearrangements accompanying S$_N$1 reactions

One complicating factor in an S$_N$1 reaction is that there may be a tendency for the carbenium ion to rearrange *before* the nucleophile has time to attack. Earlier in this section and in Section 26.5, we considered in detail the relative stabilities of different types of carbenium ions. If a carbenium ion is able to, it may undergo *methyl migration* in order to form a new carbenium ion that is more stable than the first. For example, equation 28.27 illustrates the rearrangement of a secondary carbenium ion to the more stable tertiary isomer.

Secondary carbenium ion                                    Tertiary carbenium ion                    (28.27)

Elimination of H$^+$ from the C atom *adjacent to* the C$^+$ centre in a carbenium ion is called *β-elimination*.

Thus, although loss of halide in the rate-determining step may produce, for example, a secondary carbenium ion, the product of nucleophilic substitution may not be that expected after addition of the nucleophile to this intermediate. Rather, the carbenium ion may rearrange and *then* be attacked by the nucleophile. A further complication is the possible *elimination* of H$^+$ from the carbenium ion to give an alkene:

or

Under what circumstances will this elimination compete with substitution? We answer this in the next section where we look in detail at elimination reactions and their competition with nucleophilic substitution reactions.

---

**28.7**    **Elimination reactions**

In Section 28.5, we stated that a halogenoalkane, RX, could undergo nucleophilic substitution or elimination reactions. In Section 28.6, we discussed S$_N$1 and S$_N$2 reactions, and the preference for one substitution mechanism over the other. We saw that the situation is not simple, and that a number of factors contribute to the S$_N$1 versus S$_N$2 competition. The situation is further complicated by the fact that *elimination of HX* may compete with substitution of an incoming nucleophile for X$^-$. Just as there are two substitution mechanisms, there are also two elimination mechanisms:

- E1 is unimolecular;
- E2 is bimolecular.

First, we consider the E1 and E2 mechanisms separately and then overview their competitions with S$_N$1 and S$_N$2 reactions.

### The E1 mechanism

In equation 28.15, we showed the reaction of 2-chloro-2-methylpropane with $H_2O$, giving 2-methylpropan-2-ol. In fact, minor amounts of 2-methyl-propene are also formed, and if the nucleophile is changed from $H_2O$ to $HO^-$, the predominant product is 2-methylpropene (equation 28.28).

$$(28.28)$$

Reaction 28.28 shows a first order dependence on halogenoalkane and *no* dependence on $HO^-$. Equation 28.29 gives the overall rate equation; compare this with equation 28.16.

$$\text{Rate of reaction} = -\frac{d[Me_3CCl]}{dt} = k[Me_3CCl] \qquad (28.29)$$

The rate data are consistent with the formation of a carbenium ion in the rate-determining step, i.e. a unimolecular, E1, mechanism. The first step in the E1 pathway is therefore the same as that in the $S_N1$ reaction:

Tetrahedral C          Trigonal planar C

The next step in the formation of 2-methylpropene is a fast step in which $HO^-$ abstracts $H^+$ from the carbenium ion intermediate (equation 28.30).

$$(28.30)$$

Whether or not the nucleophile acts in this manner (E1 pathway) rather than adding to the carbenium ion ($S_N1$ pathway) depends on a range of factors. The abstraction of $H^+$ is facilitated by a strong base which is a poor nucleophile, for example the amide $Li\{N(CHMe_2)_2\}$. Steric factors in the organic substrate may also swing the mechanism in favour of elimination. It may help to relieve steric strain by creating two trigonal planar in prefer-ence to tetrahedral C atoms. For example, the tertiary carbenium ion $[Bu_3C]^+$ should prefer to eliminate $H^+$ (equation 28.31) rather than react with the nucleophile at the $C^+$ centre. The latter would force the three butyl substituents closer together than in the alkene product.

$$(28.31)$$

In fact, this elimination reaction will take place with a weaker base than $HO^-$, *provided that the base is a poor nucleophile*.

E1 reactions require the formation of a stable carbenium ion and therefore are favoured for tertiary, allyl and benzyl halogenoalkanes (see Section 28.6, equations 28.24 and 28.25). Secondary halogenoalkanes may undergo E1 reactions, but primary halogenoalkanes do not because of the relatively low stability of a primary carbenium ion.

Where a choice is possible, *loss of H$^+$ will favour the formation of the more highly substituted alkene* as illustrated in the following scheme:

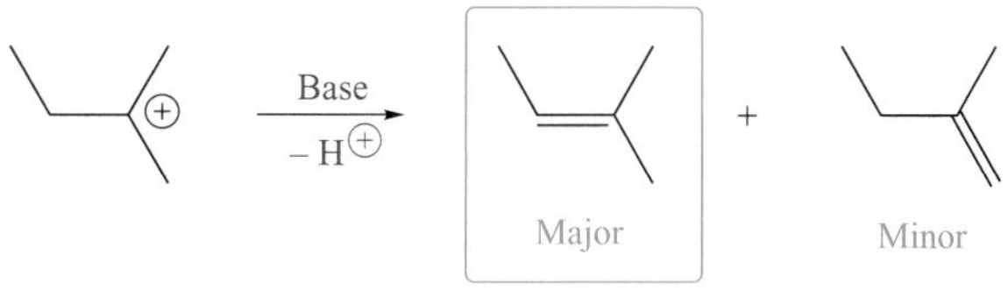

*Major*    +    *Minor*

◼▶
Regioselective:
see Section 26.5

Thus the elimination is *regioselective*.

Where the final step of the E1 mechanism could lead to (*Z*)- or (*E*)-isomers of an alkene, it will favour the formation of the less sterically crowded (*E*)-isomer, for example:

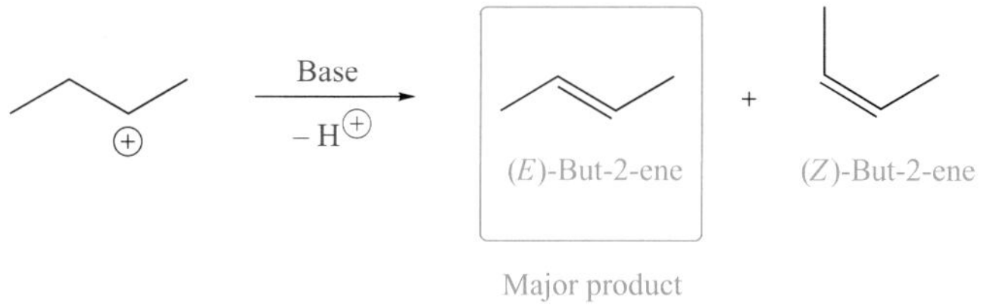

(*E*)-But-2-ene    (*Z*)-But-2-ene

Major product

## The E2 mechanism

In the presence of a base, the elimination of HX from a halogenoalkane may obey second order kinetics with the rate equation being of the form:

Rate of reaction $= k$[halogenoalkane][base]

A mechanism that is consistent with these experimental results involves a bimolecular rate-determining step (equation 28.32). This is the E2 reaction: *loss of HX promoted by base*.

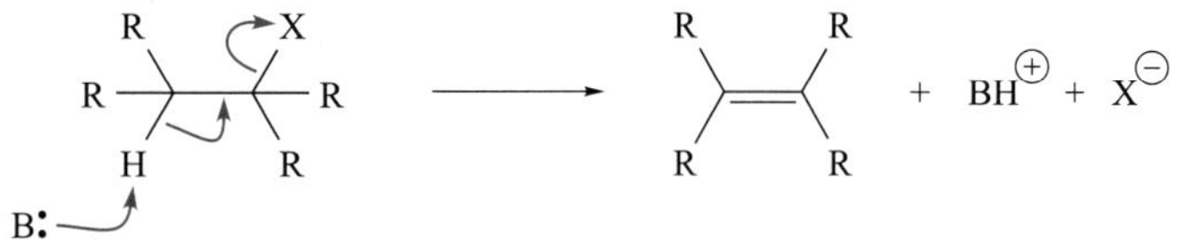

(28.32)

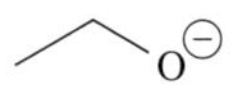

Ethoxide ion

(28.26)

The base in equation 28.32 is shown as a neutral molecule B and forms BH$^+$, but it could be negatively charged, e.g. ethoxide EtO$^-$ (**28.26**).

Figure 28.10 shows the reaction profile for the base-promoted elimination of HBr from 2-bromobutane using EtO$^-$ as base. Elimination occurs to give the most highly substituted alkene. In the transition state, the base begins to form a bond with the proton that is about to be eliminated and, *simultaneously*, the C—H bond begins to break, the π-component of the C=C double bond begins to form and the C—Br bond begins to break. These bonds are indicated by dashed lines in Figure 28.10. For this *concerted* process to occur, *all atoms in the H–C–C–X unit involved in bond making and bond breaking must lie in the same plane*. This is a *periplanar* geometry.

There are only two conformations of the halogenoalkane that allow this coplanarity in the transition state. These are shown in the

**Fig. 28.10** The reaction profile for the E2 reaction of HBr from 2-bromobutane using EtO⁻ as base. The formation of both (Z)- and (E)-isomers of the alkene is shown; see text for discussion.

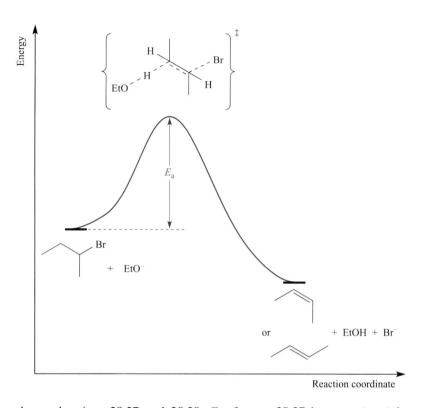

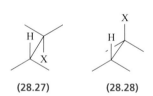

Sawhorse projections: see Section 24.9

**(28.27)**   **(28.28)**

sawhorse drawings **28.27** and **28.28**. Conformer **28.27** is an *anti-periplanar* geometry (H and X on opposite sides of the C–C bond) and **28.28** is *syn-periplanar* (H and X on the same side). Of these two conformers, the anti-periplanar arrangement is favoured: **28.27** has a staggered conformation, while **28.28** has an eclipsed conformation. Thus, **28.27** is energetically favoured over **28.28**, and the transition state in Figure 28.10 will be at lower energy if it possesses an anti-periplanar geometry. A lower-energy transition state corresponds to a lower activation barrier and a faster rate of reaction. The need for the periplanar arrangement can be understood by looking at the scheme below, and realizing that for π-bond formation, the 2p atomic orbitals on the two C atoms must be orthogonal. These orbitals evolve as the C atoms rehybridize from $sp^3$ to $sp^2$. Effectively, the $sp^3$ orbitals that were involved in C–H and C–X bond formation lose their *s* character and become pure *p* orbitals. Correct alignment of the orbitals is essential for π-bond formation:

Elimination of HX from a halogenoalkane in an *anti*-periplanar geometry is called *anti-elimination*, and from a *syn*-periplanar geometry is *syn-elimination*. The concerted nature of the E2 process means that rotation about the central C–C bond *cannot occur* during elimination of HX. This has stereochemical consequences on the elimination reaction. Starting from a particular stereoisomer, *anti-* and *syn*-eliminations lead exclusively to either (Z)- or (E)-isomers of the alkene (if the substitution pattern is such that the isomers can be recognized). This is illustrated in Figure 28.11 by

📖 **Fig. 28.11** *Anti-* and *syn-*
elimination of HCl from
one diastereoisomer of 3-chloro-
3,4-dimethylhexane. See also
problem 28.19. Colour code:
C, grey; H, white; Cl, green. The
C=C bond in each product is
shown in orange.

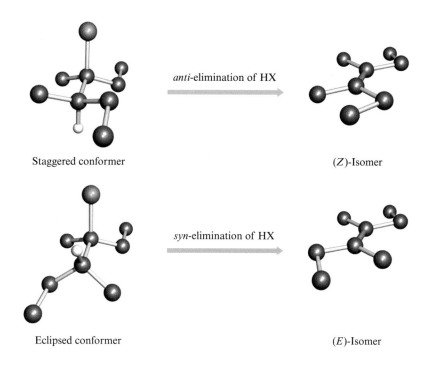

Staggered conformer          *anti*-elimination of HX          (*Z*)-Isomer

Eclipsed conformer           *syn*-elimination of HX           (*E*)-Isomer

▣▷
There are three other
stereoisomers of 3-chloro-
3,4-dimethylhexane, and in
problem 28.19 you are asked
to consider the
consequences of starting
with each isomer

the elimination of HCl from 3-chloro-3,4-dimethylhexane. The product of
*anti*-elimination is (*Z*)-3,4-dimethylhex-3-ene, and *syn*-elimination gives
(*E*)-3,4-dimethylhex-3-ene. Of the eliminations shown in Figure 28.11, we
have already seen that *anti*- is preferred over *syn*-elimination, and so the
observed product is (*Z*)-3,4-dimethylhex-3-ene.

In the example in Figure 28.11, there is only one H atom available to line
up in an *anti*-periplanar geometry with the X atom (bearing in mind that the
most highly substituted alkene is going to be formed). Now consider a
halogenoalkane in which there is a choice of H atoms, e.g. 2-bromobutane
(**28.29**). Elimination of HBr could give but-1-ene or but-2-ene, but as
we have already discussed, the reaction will be regioselective in favour of
but-2-ene. The next question is: is (*E*)- or (*Z*)-but-2-ene formed or are both
isomers produced? The halogenoalkane has two conformers with the
H−C−C−Br unit in an *anti*-periplanar arrangement. However, one is
more sterically crowded than the other. (Look at the positions of the Me
groups in each conformer below.) The elimination products reflect this
difference in conformer stabilities as the scheme below illustrates:

**(28.29)**

Me        H                    Me        H
H         Br    Rotation about   H        Br
Me        H    central C–C bond  H        Me    Favoured
                                              conformer

– HBr                          – HBr

(*Z*)-But-2-ene                (*E*)-But-2-ene

In summary:

- E2 reactions require a periplanar arrangement of the H−C−C−X unit and *anti*-periplanar is favoured over *syn*-periplanar;
- it follows that *anti*-elimination is favoured over *syn*-elimination;
- where there is a choice, elimination will favour the formation of the most highly substituted alkene (regioselectivity);
- where there is a choice, the halogenoalkane adopts the least sterically crowded conformation and the elimination product reflects this.

### 28.8 Nucleophilic substitution in competition with elimination reactions

In this section, we look at factors that influence whether a reaction follows an $S_N1$ or E1, or $S_N2$ or E2 pathway. Before working through this section, you may find it useful to review the summary points in Section 28.6 concerning carbenium ions, steric effects, nucleophiles, leaving groups and solvents.

### E1 versus $S_N1$

We have already seen that carbenium ion formation is favoured by protic solvents. This helps *both* the $S_N1$ and E1 reactions, since the rate-determining step is the same for both. Whether or not the carbenium ion is attacked by the nucleophile at the $C^+$ centre to give a substitution product, or is subject to abstraction of $H^+$ to give an elimination product, depends on the nature of the nucleophile. If the nucleophile is highly basic, the E1 mechanism is preferred. However, even a weak base can abstract a proton from the carbenium ion intermediate, and in this case there is competition between the E1 and $S_N1$ pathways as in reaction 28.33. The ratio of products depends upon reaction conditions and is affected by temperature, e.g. at 340 K and in an ethanol solvent, reaction 28.33 gives a 2:1 mixture of $S_N1$:E1 products.

$$\text{(28.33)}$$

### Tertiary halogenoalkanes

When considering substitution reactions, we saw that tertiary halogenoalkanes, $R_3CX$, undergo $S_N1$, but not $S_N2$, reactions. It is tempting, therefore, to think that for a tertiary halogenoalkane, we need only consider competition with an E1 reaction. But this is not the case. If a *strong base* is used, an E2 pathway can be followed, e.g. reaction 28.34 produces alkene almost exclusively with negligible substitution product.

$$\text{(28.34)}$$

As we saw earlier in this section, an E1 pathway for a tertiary halogenoalkane can occur with either a strong or weak base. However, the rate of an E2 reaction *depends on the concentration of base*:

Rate of reaction = $k$[halogenoalkane][base]

and is therefore favoured when the concentration of base is high.

### E2 versus $S_N2$

(28.30)

Primary halogenoalkanes, $RCH_2X$, undergo nucleophilic substitution by an $S_N2$ mechanism, and competition with an E2 reaction can occur if the nucleophile is a strong base such as $EtO^-$. The E2 reaction is especially favoured if the nucleophile is sterically demanding. Look back at Figure 28.9 where we emphasized the nucleophile's view of the $C^{\delta+}$ centre as it approaches through the R and H substituents. A nucleophile such as $Me_3CO^-$ ($^tBuO^-$, **28.30**) is very bulky as the 'space-filling' diagram in Figure 28.12 shows. It is also a strong base and reacts with 1-bromopentane to give pent-1-ene (reaction 28.35) rather than a substitution product.

**Fig. 28.12** A 'space-filling' diagram of the *tert*-butoxide ($Me_3CO^-$) ion illustrating how sterically crowded this ion is. Colour code: C, grey; H, white; O, red.

$$Br\diagdown\diagup\diagdown\diagup \xrightarrow{\ ^tBuO^{\ominus}\ } \diagup\diagdown\diagup\diagdown \qquad (28.35)$$

Competition between E2 and $S_N2$ is also observed when the substrate is a secondary halogenoalkane. We have already described how elimination of HBr from 2-bromobutane by an E2 pathway gives but-2-ene as the preferential product over but-1-ene, and that (*E*)-but-2-ene is favoured over (*Z*)-but-2-ene. Elimination is favoured over nucleophilic substitution if a base such at $EtO^-$ is used; a mixture of substitution and elimination product is generally expected (equation 28.36), but the elimination product predominates.

$$\text{(28.36)}$$

$S_N2$       E2

### Methyl halides

In the discussions of $S_N2$ reactions, we paid significant attention to MeX compounds, but we have failed to mention them in our discussions of elimination reactions. This was for a good reason. The possibility of elimination can be ignored because a $CH_3X$ molecule is unable to form an alkene since there is only one C atom.

**28.9**    ### Selected reactions of halogenoalkanes

We have considered a range of substitution and elimination reactions in the previous sections. Our focus in this final section is on selected reactions of halogenoalkanes of synthetic and biological importance.

### Ether synthesis

Reactions between alkoxide ions, $RO^-$, and halogenoalkanes are used to prepare ethers. This method is called the *Williamson synthesis*. An example was shown in reaction 28.36 and also illustrated competitive elimination. In order to reduce the chance of an E2 reaction, primary halogenoalkanes are generally used. The Williamson synthesis can be used to prepare both symmetrical (ROR) and unsymmetrical (R'OR) ethers and we return to a more detailed discussion in Chapter 29.

### Carbon chain growth

The growth of carbon chains by C–C bond formation is an important synthetic strategy. In Section 26.12, we showed the reactions between $RC{\equiv}C^-$ anions and halogenoalkanes (reaction 26.91 and scheme 26.92). Scheme 28.37 gives a further example.

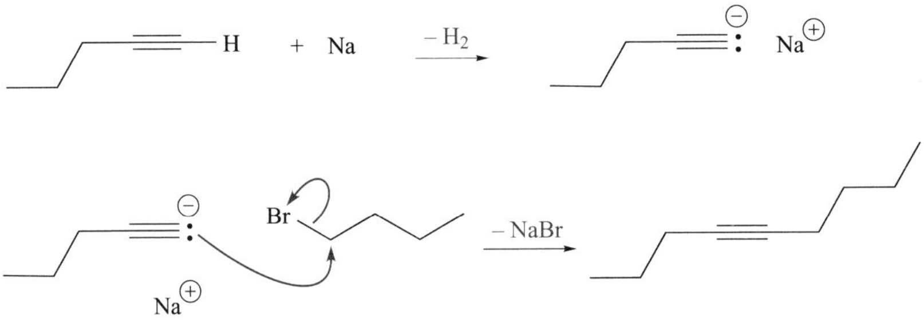

(28.37)

Competition with elimination reactions can be minimized if primary halogenoalkanes are used. Particular care has to be taken when choosing a suitable halogenoalkane substrate because $RC{\equiv}C^-$ anions are strong bases (see Table 26.1).

Another way of extending the carbon chain is by reaction of a halogeno-alkane with $NC^-$. The cyano functional group can, for example, be reduced to give an amine (equation 28.38).

Synthesis of amines: see Section 31.3

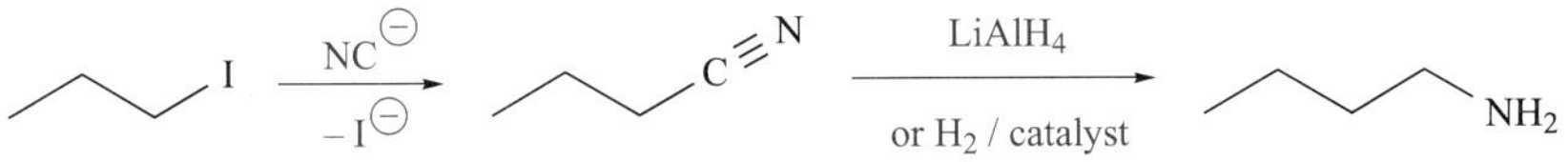

(28.38)

### Alkylating agents

DNA: see Box 21.6 and Chapter 35

The ability of RX compounds to react with nucleophiles means that many are potent *alkylating agents*. The alkylation of the organic components of DNA (the material that carries genetic information) is strongly implicated in carcinogenic (cancer-forming) and mutagenic (mutation-forming) processes. Amine ($NH_2$) and thiol (SH) groups in amino acid residues in proteins can act as nucleophiles attacking halogenoalkanes that come into contact with the protein. The resultant alkylation of the protein inhibits its natural functions, i.e. the halogenoalkane exhibits toxic effects.

## SUMMARY

This is the first in a series of chapters dealing with organic compounds containing polar bonds. Halogenoalkanes contain one or more polar C−X (X = F, Cl, Br, I) bonds, and the $C^{\delta+}$ atom is susceptible to attack by nucleophiles. We have introduced two mechanisms by which nucleophilic substitution can take place ($S_N1$ and $S_N2$), and have discussed factors that influence the choice of mechanism. The situation is further complicated by the possibility of HX elimination to give alkenes, and we have described the unimolecular E1 and bimolecular E2 reactions and the conditions under which they may occur and compete with substitution.

### *Do you know what the following terms mean?*

- halogenoalkane
- alkyl halide
- Grignard reagent
- alkyllithium compound
- nucleophilic substitution
- leaving group
- $S_N1$ and $S_N2$ mechanisms

- concerted process
- retention of configuration
- inversion of configuration
- allyl carbenium ion
- benzyl carbenium ion
- nucleophilicity
- hard and soft nucleophiles

- protic solvent
- aprotic solvent
- methyl migration
- β-elimination
- E1 and E2 reactions
- *anti-* and *syn*-periplanar
- *anti-* and *syn*-elimination

### *The following abbreviations have been used in this chapter. What do they stand for?*

- NBS
- DMF
- THF
- DMSO

### *You should be able:*

- to systematically name straight and branched chain halogenoalkanes containing one or more X atoms
- to outline methods of preparation of halogenoalkanes
- to describe typical physical properties of acyclic halogenoalkanes, and to discuss trends in boiling points
- to illustrate the use of halogenoalkanes in the preparation of Grignard reagents
- to exemplify the use of halogenoalkanes in the synthesis of alkyllithium reagents
- to give examples of nucleophilic substitution reactions
- to outline the $S_N1$ and $S_N2$ mechanisms
- to discuss the stereochemical consequences of an $S_N1$ or $S_N2$ reaction
- to discuss factors that favour one substitution mechanism over the other

- to give examples of reactions involving elimination of HX
- to outline the E1 and E2 mechanisms
- to discuss factors that favour one elimination mechanism over the other
- to comment on kinetic data that support a proposal for $S_N1$, $S_N2$, E1 or E2 mechanism
- to explain why *anti*-elimination is favoured over *syn*-elimination
- to discuss the stereochemical consequences of *anti-* and *syn*-eliminations
- to give an overview of factors that influence the outcome of competition between $S_N1$, $S_N2$, E1 and E2 pathways
- to give examples of reactions of halogenoalkanes that are synthetically useful

## PROBLEMS

**28.1** In addition to the four isomers shown in Section 28.1, draw the structures of three more isomers of $C_6H_{13}Cl$. Name these isomers.

**28.2** Indicate how $^{13}C$ NMR spectroscopy might help you to distinguish between the isomers of $C_6H_{13}Cl$ you have described in your answer to problem 28.1.

**28.3** Draw the structures of the following compounds and classify them as primary, secondary or tertiary halogenoalkanes: (a) 1-chloro-2-methylhexane; (b) 2-iodobutane; (c) 2-chloro-2,3-dimethylpentane; (d) 1-bromo-3-methylbutane.

**28.4** Using the following examples, distinguish between *bond* and *molecular* dipole moments: (*E*)-1,2-dichloroethene; 1,1-dichloroethane; 1,1,1-trichloroethane; tetrachloromethane.

**28.5** Suggest products for the following reactions:

(a)

(b)

(c)

(d)

**28.6** The boiling points of $C_2H_5X$ are 184 K for $X = H$, 235 K for $X = F$, 285 K for $X = Cl$, 311 K for $X = Br$ and 345 K for $X = I$. Comment on the trend in values and the factors that contribute to it.

**28.7** Outline the mechanism of an $S_N1$ reaction using as your example the reaction between water and 2-chloro-2-methylbutane. Include a rate equation in your answer.

**28.8** (a) If the reaction of hydroxide ion with optically pure (*S*)-3-iodo-3-methylheptane leads to a racemate, what can you deduce about the mechanism of the reaction? (b) Write a rate equation for the reaction.

**28.9** Construct an MO diagram that describes the $\pi$-bonding in the allyl cation and show that the bonding is delocalized.

**28.10** Look at Figure 28.8. Redraw the structures and add curly arrows to show how the positive charge

is pushed around the cation on going from one resonance structure to the next.

**28.11** How do the steric effects of the alkyl groups in a halogenoalkane influence the rate of $S_N2$ reactions?

**28.12** What will be the order of the rates of reactions of water with $Me_3CBr$, $Me_2CHBr$, $MeCH_2Br$ and $CH_3Br$? Rationalize the order that you propose.

**28.13** The treatment of one enantiomer of 2-iodooctane with $Na(I^*)$ (where $I^*$ is isotopically labelled iodide) leads to racemization. The rates of iodine exchange and inversion are identical. (a) Write a mechanism consistent with these observations. (b) How is the rate of racemization related to the rate of $I^-$ exchange?

**28.14** Categorize (giving reasons) the following solvents as (a) non-polar, (b) polar and protic, or (c) polar and aprotic: hexane, acetone, ethanol, DMSO, acetonitrile, water, acetic acid, dichloromethane.

**28.15** In the $S_N2$ reaction of $CH_3Br$ with various nucleophiles, the rate of substitution depends on the nucleophiles in the order $HS^- > I^- > HO^- > Cl^- > NH_3 > H_2O$. Comment on these data. Suggest where $F^-$ might come in this series.

**28.16** Discuss the role of the leaving group on the rates of $S_N1$ and $S_N2$ reactions.

**28.17** Suggest what products would result from the elimination of $H^+$ from the following carbenium ions:

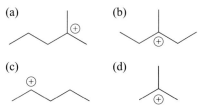

**28.18** Which of the following carbenium ions might undergo methyl migration? Give equations to show these migrations.

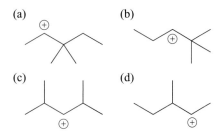

**28.19** The elimination of HCl from one stereoisomer of 3-chloro-3,4-dimethylhexane was shown in Figure 28.11. From this diastereoisomer (let this be **A**), we saw that *syn*-elimination gives exclusively (*E*)-3,4-dimethylhex-3-ene, and *anti*-elimination gives (*Z*)-3,4-dimethylhex-3-ene. (a) Draw the enantiomer of **A**, and label this **B**. What are the products of *anti*- and *syn*-elimination of HCl from **B**? (b) Draw a diastereoisomer of **A**, and label this **C**. What are the products of *anti*- and *syn*-elimination of HCl from **C**? (c) Draw the enantiomer of **C**, and label it **D**. Are the products of *anti*- and *syn*-elimination of HCl from **D** the same or different from those formed from **C**? (d) What general conclusions can you draw from your answers to parts (a)–(c)?

**28.20** If each of the following halogenoalkanes were treated with EtO⁻, what would you expect to be the predominant product? Suggest a likely mechanism in each case. How could you test that the mechanism you suggest is correct?

**28.21** Discuss briefly the structural requirements of a halogenoalkane so that it can undergo an E2 reaction, and explain why *anti*-elimination is favoured over *syn*-elimination.

**28.22** Suggest methods of converting the given precursors to products, and comment on any possible complications:

(a)

(b)

(c)

(d)

**28.23** What differences (if any) would you expect in the products of the reaction of 1-bromohexane with (a) HS⁻, (b) HO⁻, (c) ᵗBuO⁻ and (d) EtO⁻?

## Additional problems

**28.24** The following table lists four bromoalkanes and four nucleophiles. Classify the nucleophiles and suggest what types of reaction will take place between each bromoalkane and each nucleophile (16 combinations in all). Give reasons for your choices.

| Bromoalkane | Nucleophile |
| --- | --- |
| 2-Bromo-2-methylpropane | $H_2O$ |
| 2-Bromobutane | MeS⁻ |
| 1-Bromobutane | EtO⁻ |
| Bromomethane | ᵗBuO⁻ |

**28.25** Rationalize the following observations.
   (a) The following compound does not undergo $S_N1$ reactions.

   (b) Reaction between MeCl and KOᵗBu is a good route to the following ether:

   but reaction between ᵗBuCl and KOMe is a poor route.
   (c) Reaction of 2-chloro-2-methylbutane with EtO⁻ gives a mixture of two alkenes, but one predominates over the other.
   (d) Hydrolysis of (*R*)-3-chloro-3-methylheptane gives a mixture of (*R*)- and (*S*)-3-methylheptan-3-ol with %(*S*) > %(*R*).

**28.26** Rationalize the following observations.
   (a) The relative rates of reaction of $CH_3I$ with Cl⁻ in MeOH and *N,N*-dimethylformamide (**28.31**) are $1 : 8 \times 10^6$.

**(28.31)**

   (b) The rate of reaction of I⁻ with 2-bromopropane in acetone at 298 K is less than that of 1-bromobutane under the same conditions.

(c) Reaction of 1-bromo-2,2-dimethylpropane with EtOH/H$_2$O at 375 K leads to the following products:

(d) The rate of reaction of 1-bromobutane with NaOH in ethanol halves when the concentration of NaOH halves.

**28.27** Suggest how you would prepare the following product, starting with the organic compounds shown:

Starting materials                    Product

# 29 Ethers

## 29.1 Introduction

Ethers play a major role as solvents in the laboratory and we have already seen examples of their use with organolithium and Grignard reagents. Two of the most common laboratory ethers are diethyl ether (**29.1**, an example of an acyclic ether) and tetrahydrofuran (**29.2**, a cyclic ether). Each O atom carries two lone pairs as shown in structures **29.1** and **29.2**. Note that these electrons are *not* explicitly shown in structures later in the chapter.

◼▷
Organolithium and Grignard reagents: see Section 28.4

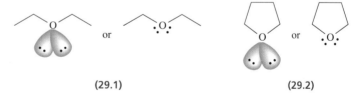

(29.1)                (29.2)

*Crown ethers* are a special group of cyclic ethers that possess cavities into which, for example, alkali metal ions can be coordinated (see Box 29.2).

Common names are still in use for some ethers and we introduce these in the next section. You should remember that *petroleum ether* is not an ether at all but is a *hydrocarbon*. 'Petroleum ether' is an incorrect usage of the term ether.

◼▷
Petroleum ether:
see Box 24.4

## 29.2 Structure and nomenclature

### Acyclic ethers

Naming acyclic ethers is not as straightforward as naming the types of organic compounds in earlier chapters. There is more than one accepted nomenclature. There are two general classes of acyclic ether:

- symmetrical ethers of general formula $R_2O$ (or ROR) in which the R groups are the same; and

- unsymmetrical ethers of general formula ROR' in which R and R' are different.

An acyclic ether may be named by prefixing the word 'ether' with the alkyl or aryl substituents, e.g. the symmetrical ethers **29.1** and **29.3** are diethyl ether (Et$_2$O) and dimethyl ether (Me$_2$O) respectively. Note that the name is written as *two words* (*not*, for example, diethylether). In an unsymmetrical ether, the two alkyl or aryl substituent names are placed in alphabetical order, so **29.4** is called ethyl methyl ether. Notice the use of spaces between the component names.

An alternative nomenclature (also recommended by the IUPAC) is to recognize the OR group as an *alkoxy substituent*. Thus, **29.3** can be named methoxymethane (i.e. an OMe substituent replacing an H in CH$_4$), and **29.1** is ethoxyethane. For an unsymmetrical ether, the parent compound is established by finding the longest carbon chain. Compound **29.4** is derived from ethane with a methoxy substituent: methoxyethane; it is incorrect to call it ethoxymethane. In this nomenclature, there are *no spaces* between the component names.

A third method of naming ethers is especially useful when more than one −O− linkage is present, e.g. compound **29.5** contains two ether functionalities. An O atom in an ether can be considered to replace an isoelectronic methylene (CH$_2$) group in the parent aliphatic chain. The name of the ether is derived by prefixing the name of the parent alkane with *oxa* (for one O atom), *dioxa* (for two O atoms), *trioxa* (for three O atoms) and so on, and with numerical descriptors to show the positions of replacement. For example:

Me—O—Me   or   (structure)

Dimethyl ether

**(29.3)**

(structure)

Ethyl methyl ether

**(29.4)**

> An **alkoxy** group has the general formula RO where R = alkyl; e.g. MeO = methoxy.

(structure)

**(29.5)**

(structure)

Parent alkane = hexane

Replace two CH$_2$ groups by two O atoms

Ether = 2,5-dioxahexane

Common names persist for a number of ethers and Table 29.1 lists several that you may encounter.

### Cyclic ethers

Systematic names for cyclic ethers have their origins in the parent cyclic alkanes, with an O-for-CH$_2$ replacement being made just as we described above for the acyclic ethers. The name is constructed as exemplified below

**Table 29.1** Selected acyclic ethers for which common names are often used.

| Common name | Systematic name | Structure |
|---|---|---|
| Ether or ethyl ether | Diethyl ether | (structure) |
| Monoglyme | 1,2-Dimethoxyethane, or 2,5-dioxahexane | (structure) |
| Diglyme | 2,5,8-Trioxanonane | (structure) |
| Triglyme | 2,5,8,11-Tetraoxadodecane | (structure) |

for 1,4,7-trioxacyclononane:

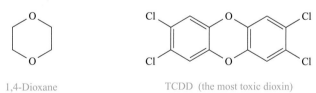

Parent alkane = cyclononane          Ether = 1,4,7-trioxacyclononane

In an organic molecule, a **heteroatom** is an atom other than C or H. A cyclic compound containing a heteroatom is called a **heterocyclic compound**.

Oxirane
(ethylene oxide)

**(29.6)**        **(29.7)**

**(29.8)**        **(29.9)**

**(29.10)**

Small rings each have a particular name, and we shall consider only 3- and 5-membered rings. Compound **29.6** is *oxirane*. The *-irane* ending is characteristic of a *saturated*, 3-atom ring and the prefix *ox-* shows that the heteroatom is oxygen. An older name for oxirane is ethylene oxide, and a general name used for compounds of the general type **29.7** (R = organic substituent) is *epoxides*. The R groups lie above and below the plane of the ring as shown in Figure 29.1a.

One of the commonest cyclic ethers that you are likely to encounter in the laboratory is *tetrahydrofuran* (THF, **29.8**). The systematic name for **29.8** is *oxolane*; the *-olane* part of the name signifies the presence of a *saturated* 5-membered ring (Figure 29.1b). The trivial name of tetrahydrofuran is always used for **29.8**, but if two O atoms are present in a saturated, 5-membered ring, a systematic name is recommended. Thus, **29.9** is called 1,3-dioxolane; the ring numbering begins at an O atom and places the second O atom at the site of lowest position number. Related to THF is the unsaturated ether **29.10**, the name of which is *furan*. The cyclic ether *1,4-dioxane* (with the older name of *dioxan*) should not be confused with the highly toxic, and much publicized, *dioxin*:

1,4-Dioxane          TCDD (the most toxic dioxin)

*p* means *para*:
see end of Section 32.5

The term dioxin is used to cover a large group of compounds, the most toxic of which is TCDD (2,3,7,8-tetrachlorobenzo-*p*-dioxin). Atmospheric pollution by TCDD arises mainly from burning chlorinated wastes. Dioxins are fat-soluble and, once in the food chain, they accumulate in body fat. They

**Fig. 29.1** Ball-and-stick models of the structures of (a) oxirane and (b) tetrahydrofuran. Colour code: C, grey; O, red; H, white.

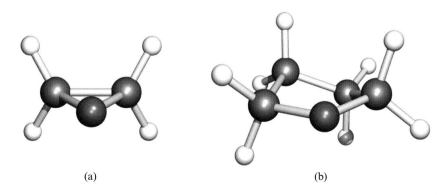

(a)                  (b)

are not excreted from the body. Women can pass dioxins on to their children, first through the placenta, and then through breast-feeding. Dioxins are carcinogenic, affect the reproductive system and damage the immune system.

Cyclic and acyclic ethers are not, of course, mutually exclusive as the structure of piperonyl butoxide illustrates. Piperonyl butoxide is used commercially as a *synergist* (a 'working partner') in pyrethrum-based insecticides. Pyrethrum paralyses insects but in the absence of the synergist, the insects quickly recover.

Piperonyl butoxide

## 29.3    Synthesis

### Acyclic ethers

Symmetrical ethers can be prepared from alcohols by heating them in the presence of acid, and this is the method used industrially for the preparation of, for example, diethyl ether (equation 29.1). Sulfuric acid acts as a catalyst.

$$2 \quad \diagdown\diagup^{OH} \quad \xrightarrow{H_2SO_4,\ heat} \quad \diagdown\diagup^{O}\diagdown\diagup \quad +\quad H_2O \qquad (29.1)$$

Ethanol                              Diethyl ether

For this method to be successful, the conditions must be carefully controlled because there is potential competition between this reaction and intramolecular dehydration of the alcohol to give an alkene. We return to this in the next chapter.

In Section 28.9, we described how reactions between alkoxide ions, $RO^-$, and halogenoalkanes, RX, can be used to prepare ethers. This is the *Williamson synthesis* and is widely used to make both symmetrical and unsymmetrical ethers (equations 29.2 and 29.3).

$$RX + RO^- \longrightarrow ROR + X^- \qquad\qquad \textit{Symmetrical ether} \quad (29.2)$$

$$R'X + RO^- \longrightarrow R'OR + X^- \qquad\qquad \textit{Unsymmetrical ether} \quad (29.3)$$

An alkoxide ion is a good nucleophile *and* a strong base, and so reactions 29.2 and 29.3 run the risk of being complicated by competing elimination reactions. In Section 28.6, we saw that *tertiary* halogenoalkanes (which tend to react by an $S_N1$ pathway) are prone to give elimination and rearrangement products when they are attacked by nucleophiles. However, when a nucleophile attacks a *primary* halogenoalkane, substitution generally predominates over elimination. These observations must be considered when choosing the starting materials for the preparation of an *unsymmetrical* ether by the Williamson synthesis. There are always two possible combinations of reagents for the formation of R'OR (i.e. RX and $R'O^-$, or R'X and $RO^-$) and one reaction route is usually better than the other in terms of maximizing the yield of ether. In scheme 29.4, the reaction of $Me_3CCl$ with methoxide ion gives elimination and rearrangement products, whereas reaction between chloromethane and *tert*-butoxide ion produces the ether in good yield.

Nucleophilic substitution:
see Section 28.6

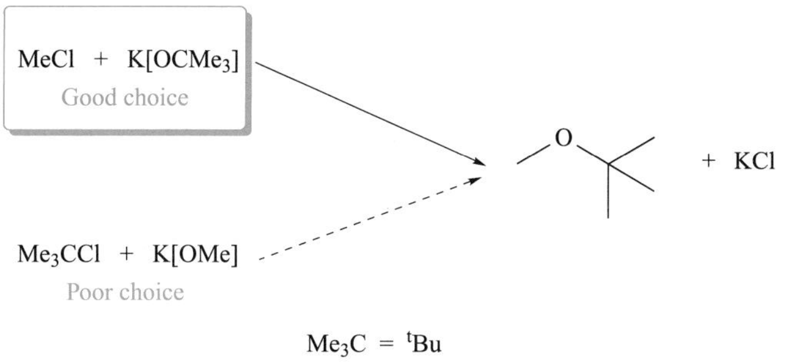

$$\boxed{\text{MeCl} + \text{K[OCMe}_3]} \quad \text{Good choice}$$

$$\text{Me}_3\text{CCl} + \text{K[OMe]} \quad \text{Poor choice}$$

$$+ \text{ KCl} \qquad (29.4)$$

$$\text{Me}_3\text{C} = {}^t\text{Bu}$$

The lower reaction in scheme 29.4 suffers from competition with elimination of HCl to give 2-methylpropene (equation 29.5).

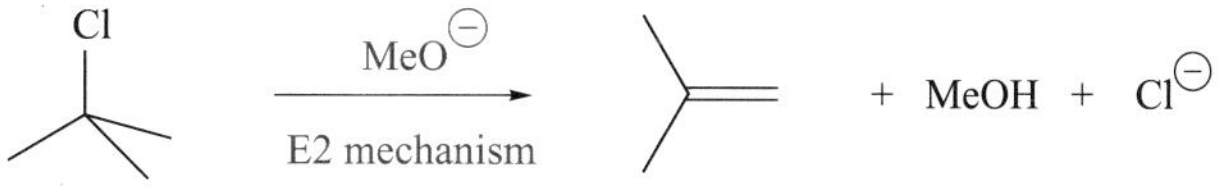

$$\xrightarrow[\text{E2 mechanism}]{\text{MeO}^{\ominus}} \qquad + \text{ MeOH} + \text{Cl}^{\ominus} \qquad (29.5)$$

## Cyclic ethers

Many cyclic ethers are available commercially, and in this section, we focus on methods of making oxirane (an epoxide that is also called *ethylene oxide*) and THF.

The catalytic oxidation of ethene gives oxirane (equation 29.6). Oxirane is manufactured on a huge scale. Applications include its use as a precursor in epoxy resins (see Box 29.1) and in the industrial preparation of ethane-1,2-diol (ethylene glycol) which is used as an 'anti-freeze' additive.

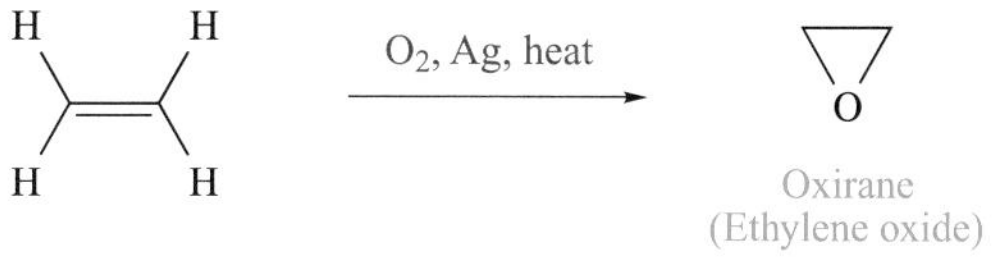

Ethane-1,2-diol:
see Section 30.2

$$\xrightarrow{\text{O}_2, \text{Ag, heat}} \qquad (29.6)$$

Oxirane
(Ethylene oxide)

Halohydrins:
see Section 26.6

On a smaller scale, oxiranes can be made from halohydrins as in reaction 29.7.

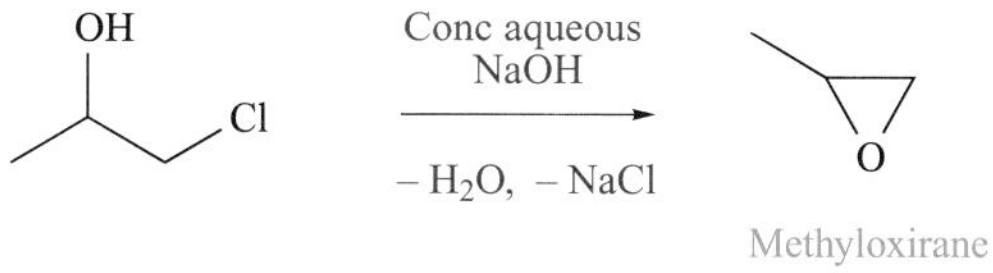

$$\xrightarrow[- \text{H}_2\text{O}, \, - \text{NaCl}]{\substack{\text{Conc aqueous} \\ \text{NaOH}}} \qquad (29.7)$$

Methyloxirane

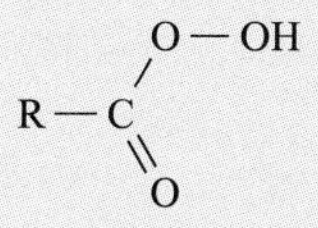

The normal laboratory synthesis of an oxirane is by the reaction of an alkene with a *peracid* such as perbenzoic acid. Reaction 29.8 shows the conversion of 2,3-dimethylbut-2-ene to tetramethyloxirane; the stereochemistry has been included in the product to emphasize the change from trigonal planar $sp^2$ to tetrahedral $sp^3$ C during the reaction.

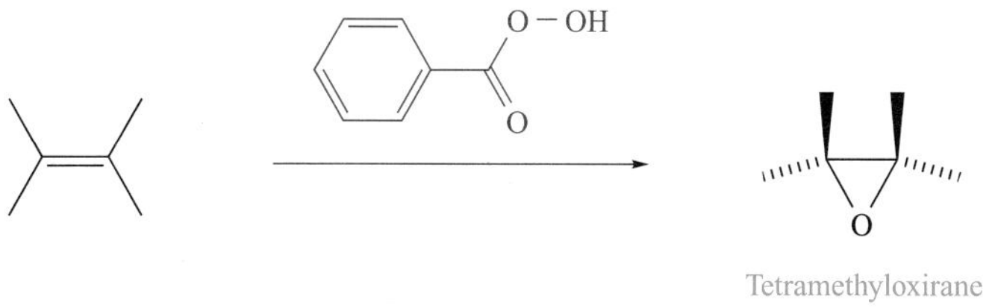

An organic ***peracid*** has the general structure:

$$\text{R} - \text{C} \begin{smallmatrix} \text{O} - \text{OH} \\ \\ \text{O} \end{smallmatrix}$$

$$\qquad (29.8)$$

Tetramethyloxirane

## Box 29.1 Epoxy resins

The glue *Araldite* is an example of a glue that is sold in two separate tubes, and when the contents of the tubes are mixed, the glue hardens rapidly. One tube contains an *epoxy resin* and the second contains a *diamine*. The reaction that occurs on mixing the components can be represented as follows:

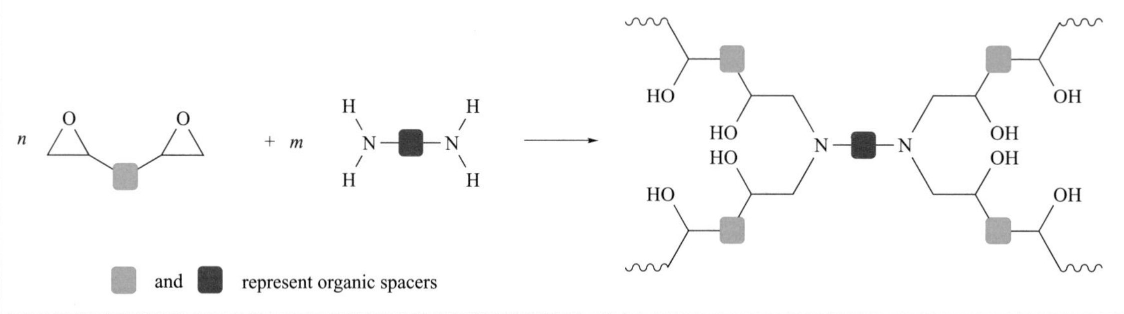

and represent organic spacers

The strained epoxide ring undergoes an instantaneous ring-opening reaction as soon as the diamine is made available as a reaction partner. The four H atoms in the amine are points from which the polymer can 'grow' and a *cross-linked* (rather than a single chain) structure results.

Reaction 29.8 is the addition of an O atom to the C=C bond, and it is the O of the OH group in the peracid that is specifically transferred. Cleavage of the O–O bond is facilitated by its weakness. The mechanism of the reaction is shown in equation 29.9, and shows that the reaction is accompanied by the conversion of the peracid into a carboxylic acid.

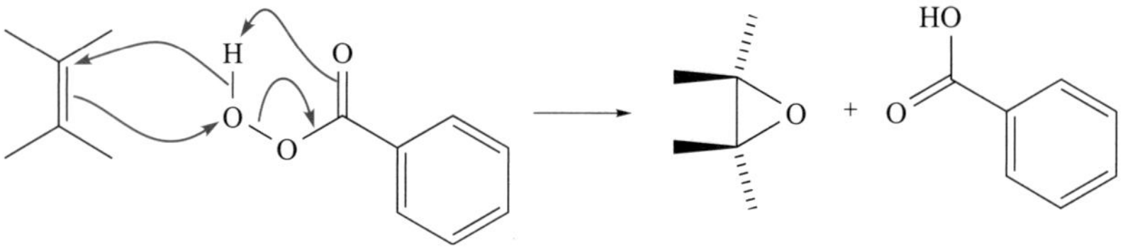

(29.9)

The formation of the epoxide from an alkene is stereospecific. The mechanism is concerted, and the alkene substituents 'bend back' as the O atom adds to the double bond. Thus, addition is always *syn*.

*Syn-* and *anti*-addition: see Section 26.5

Cyclic ethers can be prepared by ring closure starting from a suitable alcohol and treating it with lead(IV) acetate. Reaction 29.10 shows the formation of THF.

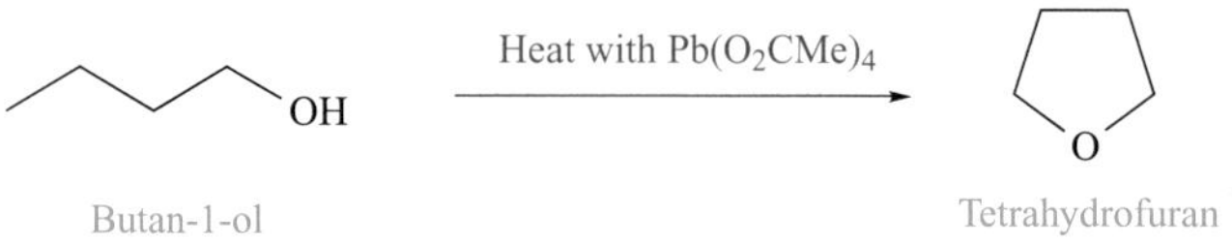

(29.10)

Butan-1-ol                Tetrahydrofuran

On a commercial scale, one method of preparing THF is by the catalytic hydrogenation of furan. The latter is made from furfural, itself manufactured by the acidic dehydration of polysaccharides in oat husks. Scheme 29.11 summarizes the last steps in the process.

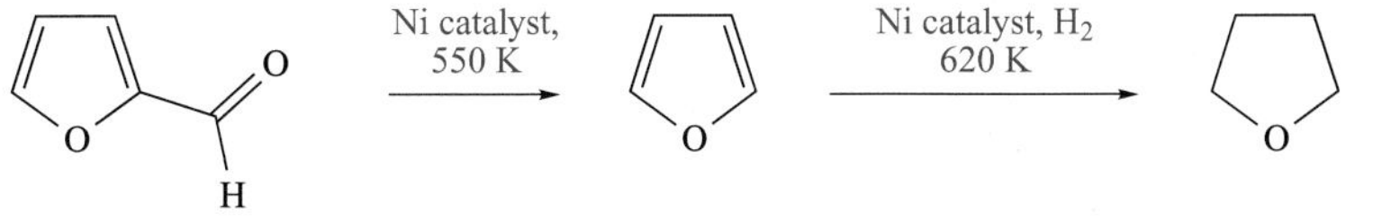

$$\text{(29.11)}$$

Furan is a building-block in one of the world's top-selling drugs, Zantac (a GlaxoSmithKline product), which is an $H_2$ antagonist and is prescribed as an anti-ulcer drug.

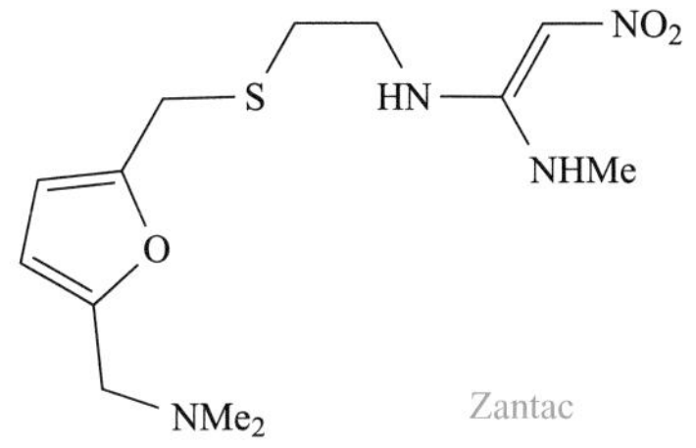

| 29.4 | **Physical properties** |

### Dipole moments, boiling points and enthalpies of vaporization

The introduction of an O atom into an alkane chain results in the presence of a small dipole moment and is exemplified below by comparing pentane and diethyl ether:

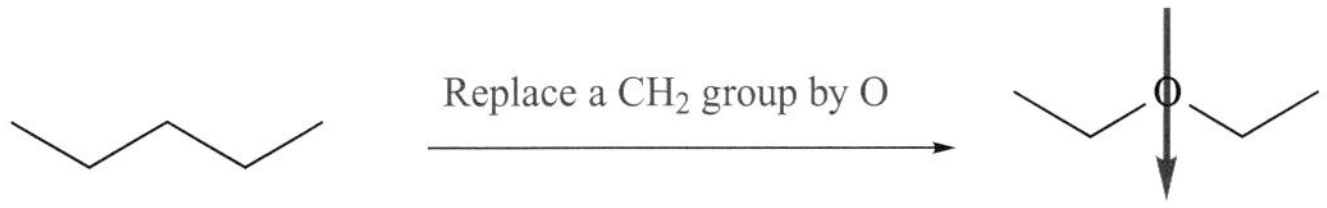

Values of dipole moments for selected ethers are listed in Table 29.2. Despite these ethers being polar, the boiling points listed in Table 29.2 show that the strengths of the dipole–dipole interactions between molecules of a liquid

**Table 29.2** Dipole moments, boiling points and enthalpies of vaporization of selected ethers and parent alkanes. The 'parent alkane' and ether are related as shown in the scheme at the beginning of Section 29.4, i.e. the replacement of the central $CH_2$ group in the alkane by an O atom.

| Ether | Relative molecular mass | Dipole moment for gas phase molecule / D | Boiling point (bp) / K | $\Delta_{vap}H^o(bp)$ / kJ mol$^{-1}$ | Parent alkane | Relative molecular mass | Dipole moment for gas phase molecule/ D | Boiling point (bp) / K | $\Delta_{vap}H^o(bp)$ / kJ mol$^{-1}$ |
|---|---|---|---|---|---|---|---|---|---|
| Dimethyl ether | 46.07 | 1.30 | 248.3 | 21.5 | Propane | 44.09 | 0 | 231.0 | 19.0 |
| Diethyl ether | 74.12 | 1.15 | 307.6 | 26.5 | Pentane | 72.15 | 0 | 309.2 | 25.8 |
| Dipropyl ether | 102.17 | 1.21 | 363.2 | 31.3 | Heptane | 100.20 | 0 | 371.6 | 31.8 |
| Dibutyl ether | 130.22 | 1.17 | 413.4 | 36.5 | Nonane | 128.25 | 0 | 424.0 | 36.9 |

**Table 29.3** Dipole moments, boiling points and enthalpies of vaporization for selected and comparable cyclic and acyclic ethers. Compounds are grouped according to molecular mass.

| Ether | Structure | Dipole moment for gas phase molecule / D | Relative molecular mass | Boiling point / K | $\Delta_{vap}H^{\circ}$(bp) / kJ mol$^{-1}$ |
|---|---|---|---|---|---|
| Oxirane (ethylene oxide) | | 1.89 | 44.05 | 283.6 | 25.5 |
| Dimethyl ether | | 1.30 | 46.07 | 248.3 | 21.5 |
| Tetrahydrofuran (THF) | | 1.75 | 72.10 | 338.0 | 29.8 |
| Diethyl ether | | 1.15 | 74.12 | 307.6 | 26.5 |
| Methyl propyl ether | | 1.11 | 74.12 | 312.1 | 26.7 |

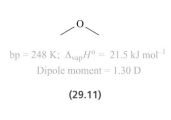

bp = 248 K;  $\Delta_{vap}H^{\circ}$ = 21.5 kJ mol$^{-1}$
Dipole moment = 1.30 D

**(29.11)**

Hydrogen bonding: see Section 21.8

bp = 351 K;  $\Delta_{vap}H^{\circ}$ = 38.6 kJ mol$^{-1}$
Dipole moment = 1.69 D

**(29.12)**

ether are not significantly different from those of the van der Waals forces between molecules of a straight chain alkane of comparable molecular mass. The low enthalpies of vaporization exhibited by low molecular mass ethers result in these compounds being volatile. If diethyl ether is spilt on your hand, it quickly evaporates as it absorbs heat from your body. Each ether has an isomeric alcohol, e.g. dimethyl ether (**29.11**) and ethanol (**29.12**) are isomers. However, the boiling points and values of $\Delta_{vap}H^{\circ}$ of Me$_2$O and EtOH are significantly different because alcohol molecules, unlike ethers, form hydrogen bonds with one another. We return to the effects of hydrogen bonding on the properties of alcohols in Chapter 30.

The physical properties of cyclic ethers follow similar trends to those of their acyclic analogues, although the selected data in Table 29.3 indicate that a cyclic ether has a higher boiling point than an acyclic ether of comparable molecular mass. This trend is similar to that observed for cyclic and acyclic alkanes (see Figure 25.9), and can be attributed to differences in packing forces associated with differences in molecular shapes. The data in Table 29.3 also allow a comparison of the dipole moments of oxirane and dimethyl ether, and of THF and acyclic ethers of similar molecular mass. In each case, the cyclic ether possesses a higher dipole moment than its acyclic analogue, and this can be rationalized in terms of the molecular shapes.

### Ethers as solvents

The importance of ethers as solvents arises from their ability to form hydrogen bonds with molecules containing H atoms attached to electronegative atoms (see below) and from their coordinating properties. Coordination through the O atom or atoms of an ether allows it to function as a ligand to cations as we discussed in Section 23.3.

Diethyl ether is widely used for the extraction of organic compounds from aqueous solutions. However, Et$_2$O is *not* completely immiscible with water, and a saturated aqueous solution contains about 8 g of Et$_2$O per

**Fig. 29.2** The use of a separating funnel to carry out an ether extraction.

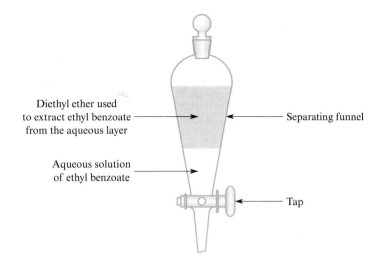

Diethyl ether used to extract ethyl benzoate from the aqueous layer

Separating funnel

Aqueous solution of ethyl benzoate

Tap

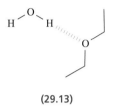

**(29.13)**

Synthesis of esters: see Section 33.9

$100 \, cm^3$. Although hydrogen bonds do not form between $Et_2O$ molecules in pure diethyl ether, they *do* form between $H_2O$ and $Et_2O$ molecules as shown in structure **29.13**. The formation of hydrogen bonds between the O atoms of ether molecules and the H atoms of X−H groups where X is an electronegative atom, or the formation of ion–dipole interactions between ions and polar ether molecules, makes ethers useful solvents. *Caution! Alkyl ethers are highly volatile and inflammable, and their vapours present a significant explosion hazard.*

Figure 29.2 illustrates how an ether extraction is typically carried out. In this example, ethyl benzoate (an ester) has been prepared from the reaction of benzoic acid and ethanol, and is in aqueous solution. The ester is extracted by pouring the aqueous solution into a separating funnel and adding $Et_2O$. The funnel is stoppered and shaken. Since $Et_2O$ is volatile, there is a build-up of vapour pressure and this can be released by inverting the funnel and opening the tap. After closing the tap again, the funnel is left in the position shown in Figure 29.2 until the two layers have separated. The stopper is then removed and the lower, aqueous layer is carefully removed through the tap. The remaining ether layer is then removed to a separate flask and the aqueous layer is returned to the funnel. The extraction is repeated several times, each with a fresh portion of $Et_2O$. The combined ether extracts are dried by adding powdered anhydrous $MgSO_4$ or $CaCl_2$ and, after filtration, the product can be recrystallized. The drying stage is necessary because the ether layer consists of a saturated solution of water in diethyl ether. In general, the solubility of $Et_2O$ in water decreases when ionic salts are dissolved in the aqueous solution. This can be rationalized in terms of the competition between $Et_2O$ molecules for hydrogen bonding with $H_2O$ molecules, and cations for solvation by $H_2O$. The enthalpies of solvation of cations are significantly greater than the enthalpies associated with hydrogen bond formation, making ion solvation the energetically more favourable process. Thus, the addition of, for example, NaCl to the aqueous phase of an ether extraction drastically reduces the solubility of $Et_2O$ in the aqueous solution, and this process is known as *salting out*.

Cyclic ethers are widely used as solvents in synthetic chemistry, for both organic and inorganic compounds, and Box 29.2 highlights an application of crown ethers.

### COMMERCIAL AND LABORATORY APPLICATIONS

## Box 29.2 Crown ethers: large cyclic ethers that encapsulate alkali metal ions

In Figure 21.13, we showed a hydrated $K^+$ ion. The O atoms of ethers are able to form similar coordinate bonds with alkali metal ions, and when the ether is a large cyclic molecule, the coordination complex that forms has the metal ion encapsulated within the cavity of the cyclic ether.

The structure of 1,4,7,10,13,16-hexaoxacycloocta-decane is shown in diagrams (a) to (c). This is one of the so-called *crown ethers* (the name arises from the crown-shaped ring) and is usually called 18-crown-6. The size of the cavity in 18-crown-6 is compatible with

a $K^+$ ion. The 18-crown-6 ring is flexible (by restricted rotation about the C–O and C–C bonds) and can alter its conformation so that it can bind the $K^+$ ion.

Compare the ring conformations in diagrams (a) and (d) which show space-filling diagrams of the free 18-crown-6 molecule and the complex [K(18-crown-6)]$^+$, respectively. (Colour code: C, grey; O, red; H, white; K, blue.) [X-ray diffraction data for figures (a), (c) and (d): E. Maverick *et al.* (1980) *Acta Crystallogr.*, *Sect. B*, vol. 36, p. 615; N. P. Rath *et al.* (1986) *J. Chem. Soc., Chem. Commun.* p. 311.]

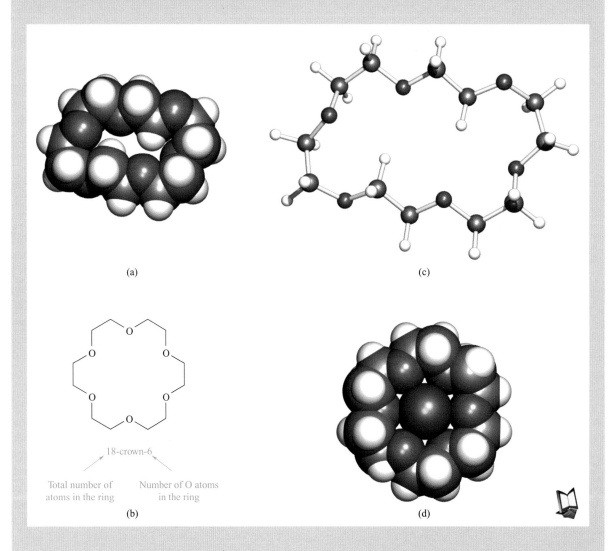

(a)

(c)

18-crown-6

Total number of       Number of O atoms
atoms in the ring         in the ring

(b)

(d)

The advantage of [K(18-crown-6)]$^+$ (and similar cations) is that they are soluble in organic solvents. For example, $KMnO_4$ is insoluble in benzene, but when 18-crown-6 is added, the salt becomes soluble,

giving a purple solution. The compound that dissolves is [K(18-crown-6)][MnO$_4$].

Different crown ethers have different cavity sizes (e.g. the radii of the cavities in 15-crown-5 and

21-crown-7 are 90 and 170 pm respectively), and selective complexation of $M^+$ ions is possible. Complexes formed by such macrocyclic ligands are appreciably more stable than those formed by closely related open-chain ligands.

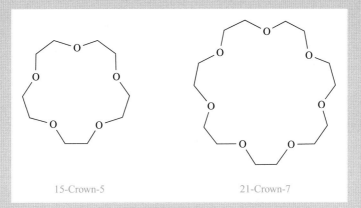

15-Crown-5          21-Crown-7

*Further reading*: C. E. Housecroft and A. G. Sharpe (2005) *Inorganic Chemistry*, 2nd edn, Prentice Hall, Harlow, Chapters 6 and 10, and references therein.

## 29.5 Identification of ethers by IR spectroscopy

(29.14)

The IR spectrum of a dialkyl ether exhibits a strong absorption in the region 1080–1150 cm$^{-1}$ assigned to the C–O stretching mode, $\nu$(C–O). Figure 29.3 shows the IR spectra of two symmetrical ethers, compared with that of an alkane of similar chain length. Absorptions assigned to $\nu$(C–H) of the aliphatic C–H groups are common to all the spectra in Figure 29.3, but the bands in the fingerprint region around 1100 cm$^{-1}$ are characteristic of the ether functionality. Cyclic ethers also show a strong absorption assigned to $\nu$(C–O) in the same range as acyclic ethers. Figure 29.4 illustrates the IR spectrum of THF (structure **29.14** and Figure 29.1). The absorptions around 3000 cm$^{-1}$ are assigned to $\nu$(C–H) of the CH$_2$ groups, and in the fingerprint region, the strong absorption at 1070 cm$^{-1}$ is assigned to $\nu$(C–O).

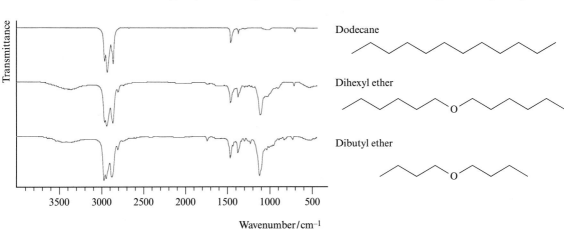

**Fig. 29.3** The IR spectra of dibutyl ether and dihexyl ether both exhibit a strong absorbance at 1120 cm$^{-1}$ assigned to $\nu$(C–O). For comparison, the figure also shows the IR spectrum of dodecane in which absorptions arising from vibrations of the C–H bonds are observed.

**Fig. 29.4** The IR spectrum of tetrahydrofuran (THF).

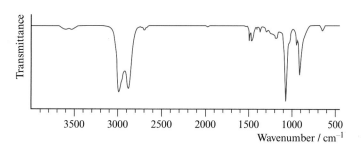

| 29.6 | **Reactivity of ethers** |
|---|---|

With the exception of epoxides, aliphatic ethers are not particularly reactive, and this is one reason why they are suitable as solvents. Acyclic ethers undergo cleavage when treated with HI or HBr, and the reactivity pattern of cyclic ethers with a ring size of four or more atoms is similar. Epoxides, on the other hand, exhibit a different reactivity pattern, and we therefore discuss them separately.

### Cleavage of ethers

Ethers (epoxides excepted) are cleaved by treatment with concentrated HI or HBr (equation 29.12).

$$\text{ROR} + \text{HX} \xrightarrow{\text{Heat}} \text{ROH} + \text{RX} \qquad (\text{X} = \text{Br or I}) \quad (29.12)$$

The first step in the reaction is the protonation of the O atom (equation 29.13), and this is followed by nucleophilic attack by $Br^-$ or $I^-$, with substitution occurring by either an $S_N1$ (scheme 29.14) or $S_N2$ mechanism (equation 29.15). The initial protonation means that the leaving group in the substitution reaction is ROH (a good leaving group). Without protonation, $RO^-$ (a strong base and a poor leaving group) would be expelled. The choice of HI or HBr over HCl as the reagent for cleavage arises from the greater nucleophilicities of $I^-$ and $Br^-$ with respect to $Cl^-$; of HI and HBr, HI is the better reagent.

▶
Leaving groups and nucleophilicity: see Section 28.6

$$(29.13)$$

$$(29.14)$$

$$(29.15)$$

The products of the reaction of HX with the unsymmetrical ether ROR' could be ROH and R'X (as shown in the equations above), or R'OH and

RX. Mixtures of products are often obtained, but if either R or R' is a methyl group, then the formation of MeX over MeOH is generally favoured. This is exemplified by reaction 29.16.

$$\text{(29.16)}$$

## Reactivity of epoxides (oxiranes)

In Section 25.2, we discussed the ring strain of small cycloalkanes, in particular cyclopropane. The 3-membered ring of oxirane is similarly strained. This imparts a unique reactivity to oxirane, and its industrial importance (see Section 29.3) stems from this characteristic. Ring opening of oxirane occurs under milder conditions than that of other cyclic ethers, and can be brought about by treatment with dilute aqueous acid. Reaction 29.17 is of industrial importance in the manufacture of ethane-1,2-diol, used as an 'anti-freeze' additive to motor vehicle radiator water.

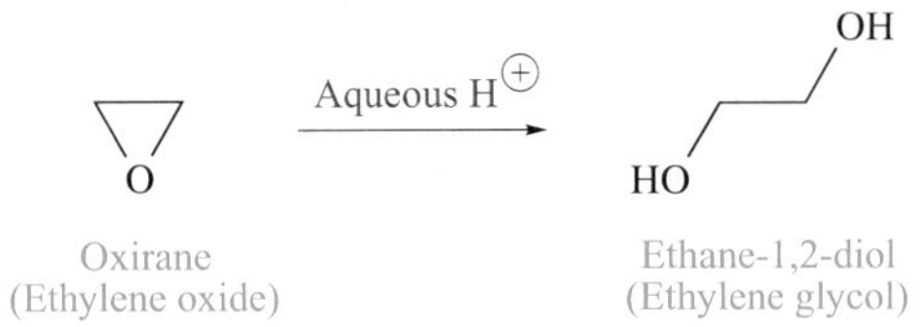

$$\text{(29.17)}$$

Oxirane
(Ethylene oxide)

Ethane-1,2-diol
(Ethylene glycol)

The mechanism of this reaction involves the protonation of the ring O atom and nucleophilic attack by $H_2O$ as shown in scheme 29.18.

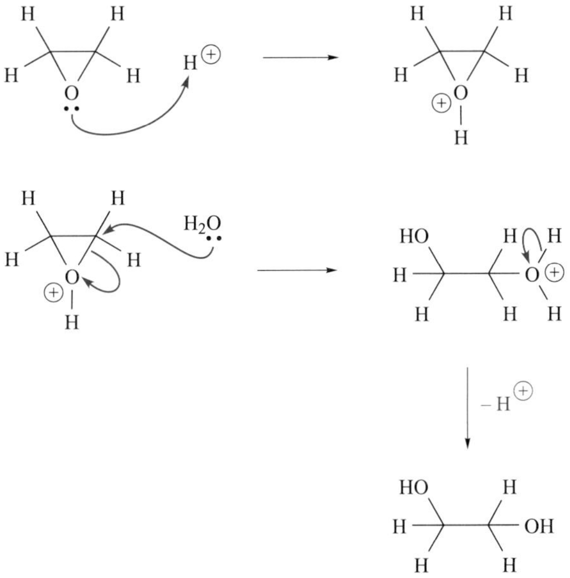

$$\text{(29.18)}$$

The stereochemistry of the reaction is specific. This cannot be appreciated in reaction 29.18 because there is rotation about the C–C bond in the product. However, in reaction 29.19 in which rotation is restricted by the cyclic structure, the product has the specific stereochemistry shown, with the two OH groups on opposite sides of the C–C bond. You can make a direct comparison between the attack of $H_2O$ on the protonated oxirane,

Look back to the discussion in Section 25.2 for a reminder about ring flipping

and the attack of $Br^-$ on a bromonium ion: compare equations 29.18 and 26.30, and equations 29.19 and 26.31. Once formed, the diaxial product flips immediately to the sterically favoured diequatorial product (equation 29.19).

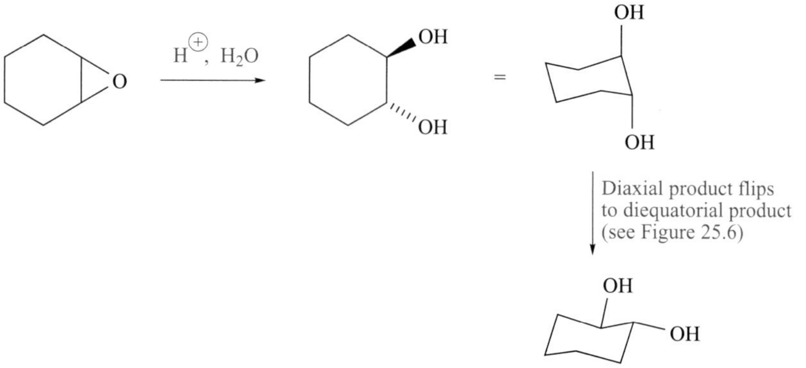

Diaxial product flips to diequatorial product (see Figure 25.6)

(29.19)

Ring opening of an oxirane also occurs on treatment with base, provided that the base is a strong nucleophile. Scheme 29.20 gives an example using methoxide ion in methanol.

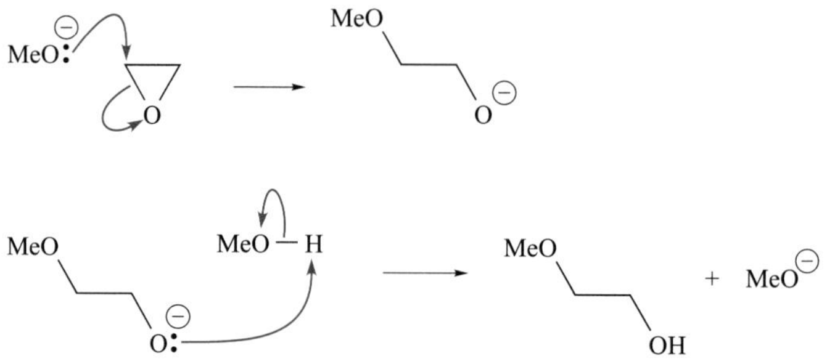

(29.20)

The choice of acidic or basic conditions for the ring opening affects the regioselectivity of the reaction. A base attacks the C atom with the *fewer number of substituents* attached to it, for example:

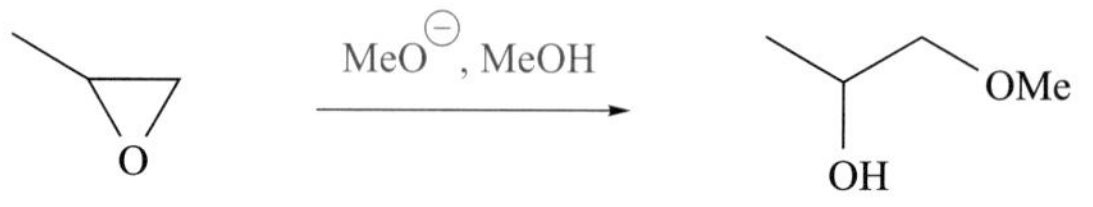

Regioselectivity: see Section 26.5

whereas when a nucleophile attacks a *protonated* oxirane, it shows a preference for the *more substituted* C atom:

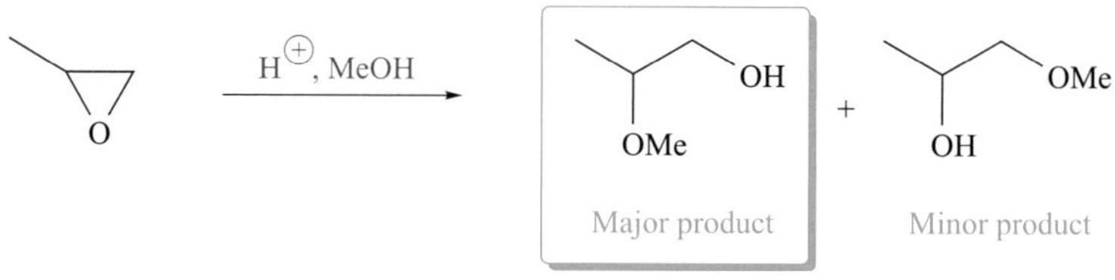

Major product    Minor product

This difference is rationalized as follows. When a base approaches the oxirane, it attacks the C atom in the 3-membered ring that is least sterically crowded. Thus, attack is at a primary C atom in preference to a secondary or tertiary centre, or at a secondary C atom in preference to a tertiary centre. In the example above, the methoxide ion has a choice of primary or secondary C atom (structure **29.15**). The mechanism proceeds by an $S_N2$ pathway, and the product is specific. In reactions where the oxirane is protonated before attack by a nucleophile, the positive charge on the intermediate can be delocalized on to the C atoms in the 3-membered ring. The stabilities of carbenium ions increase with the degree of substitution, and so the dominant product of the reaction of a nucleophile with an oxirane under acidic conditions has the nucleophile attached to the more highly substituted C atom. This description suggests the involvement of a carbenium ion and, therefore, an $S_N1$ mechanism. However, experimental evidence suggests a pathway that is in between $S_N1$ and $S_N2$, and this is represented in scheme 29.21.

**Relative stabilities of carbenium ions: see Section 26.5**

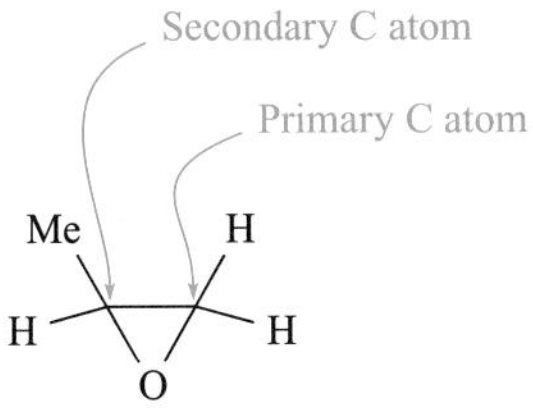

Secondary C atom

Primary C atom

**(29.15)**

Overall charge = 1+

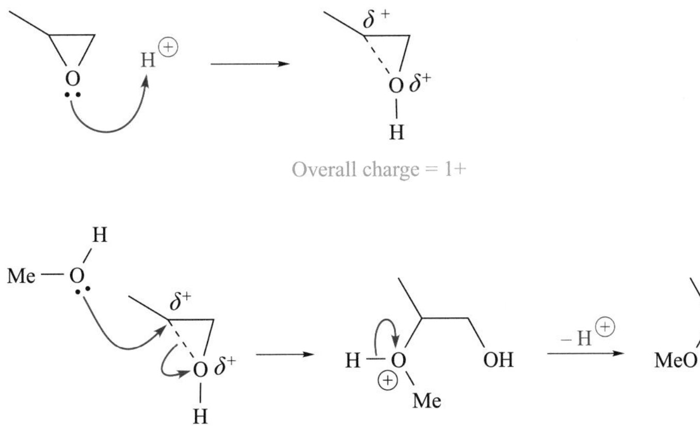

(29.21)

Oxiranes react with a variety of nucleophiles and reactions 29.22 and 29.23 give two examples.

**Applications of epoxide ring opening are in the manufacture of phenolic resins and the drug Propranolol: see Box 29.1 and Section 32.12**

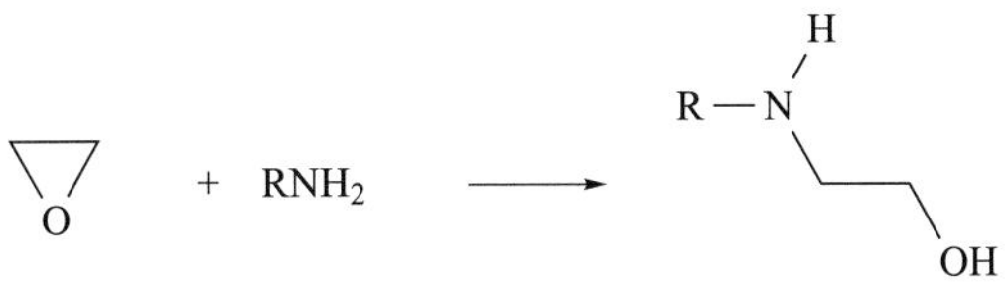

(29.22)

(29.23)

Despite its reactivity, the epoxide ring occurs in nature, for example in some pheromones (see Box 29.3) and in the juvenile-insect hormone JHIII. Juvenile hormones in insects are responsible for keeping the insect at the larval stage prior to metamorphosis.

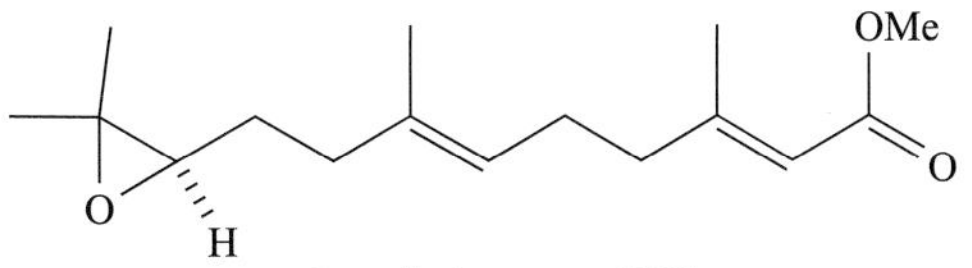

Juvenile hormone JHIII

**ENVIRONMENT AND BIOLOGY**

## Box 29.3 Pheromones 1: a female moth attracts her mate

The giant peacock moth, *Saturnia pyri*, showing its feather-like antennae.
*Claude Nuridsany & Marie Perennou / Science Photo Library.*

Pheromones are chemical messengers and nature has provided female moths with the ability to secrete pheromones to act as sexual attractants. A pheromone is more than a perfume! A particular species of moth secretes a unique pheromone, and since the 1950s, the sex pheromones of several hundreds of species have been identified. The specificity of the pheromone allows a female moth to attract males of her own species. The female of the vapourer moth is wingless. It is a remarkable sight to watch a newly emerged wingless female moth: her life is restricted to sitting on her pupal cocoon and from there she must play her part to ensure the next generation of her species. She pumps out pheromones that can be sensed by male moths some distance away. Large antennae for sensing pheromones are characteristic features of many male moths, and it is not long before the female entices her mate. Many moth pheromones consist of $C_n$ chains with a terminal functional group (e.g. alcohol or aldehyde) and also include the epoxide shown below.

Pheromone of the female silkworm moth

Pheromone of the female gypsy moth

*See also*: Box 34.2: Pheromones 2: ant trails

## SUMMARY

In this chapter, we have introduced the chemistry of acyclic and cyclic ethers. Apart from cleavage reactions, acyclic ethers and cyclic ethers with ring sizes ≥ four atoms are relatively unreactive, and are regularly used as solvents in chemical reactions. In contrast, the ring strain in 3-membered oxiranes (epoxides) provides these ethers with a high reactivity, characterized by ring-opening reactions.

### Do you know what the following terms mean?

- acyclic ether
- cyclic ether
- crown ether
- petroleum ether

- oxirane (epoxide)
- oxolane
- heteroatom
- heterocyclic
- alkoxy group

- alkoxide
- Williamson synthesis
- organic peracid
- ether extraction

### You should be able:

- to give systematic names to symmetrical and unsymmetrical ethers, and recognize the advantages of using different systems of nomenclature in different situations
- to give methods of preparing symmetrical and unsymmetrical acyclic ethers
- to outline methods of preparing oxiranes and tetrahydrofuran
- to describe typical physical properties of ethers
- to give examples of chemical reactions in which ethers are used as solvents, commenting on associated hazards

- to explain how to carry out an ether extraction, commenting on associated hazards
- to state where you expect to observe characteristic absorptions in the IR spectrum of an ether
- to describe, and give examples of, cleavage reactions of ethers
- to illustrate how the reactivity of oxiranes (epoxides) differs from that of larger cyclic ethers
- to outline the mechanisms of ring-opening reactions of oxiranes (epoxides) under acidic and basic conditions, and to discuss the regioselectivity of these reactions

## PROBLEMS

**29.1** Give names for the following ethers:

(a)     (b)

(c)     (d)

(e)

**29.2** Draw the structures for the following compounds:
(a) diethyl ether, (b) ethoxyethane,
(c) ethoxyheptane and (d) 4,7-dioxadecane.

**29.3** Draw the structures for the following compounds:
(a) THF, (b) diglyme, (c) 1,4,7-trioxacyclononane,
(d) furan and (e) oxirane.

**29.4** Draw structures for the following compounds:
(a) epoxypropane, (b) methyloxirane,
(c) trimethyloxirane, (d) 1,2-epoxycyclohexane,
(e) 3,4-epoxyheptane and (f) 2-ethyl-3,3-dimethyloxirane.

**29.5** Why would it be inefficient to prepare ethyl methyl ether by a method analogous to that shown in equation 29.1? What reaction(s) could you use to prepare this ether?

**29.6** Suggest products for the reactions of
(a) chloroethane with ethoxide, (b) iodomethane with potassium *tert*-butoxide, (c) methoxide with 2-chloro-2-methylpropane and (d) *iso*-propoxide with iodomethane. Draw the structure of each product.

**29.7** Give the products of the reactions between perbenzoic acid and (a) (*Z*)-but-2-ene and (b) (*E*)-but-2-ene. Rationalize your answer.

**29.8** (a) The text in Section 29.4 states that a 'cyclic ether possesses a higher dipole moment than its acyclic analogue'. With reference to the data in Table 29.3, suggest reasons for this observation. (b) Are the following molecules polar: 1,3-dioxalane; 1,4-dioxane; furan; dibutyl ether?

**29.9** (a) To what do you assign the broad band near 3400 cm$^{-1}$ in the spectra of the ethers in Figure 29.3? (b) How might you distinguish diethyl ether and methyl propyl ether using routine techniques?

**29.10** Comment on the following data. The $^1$H NMR spectrum of diglyme contains a singlet at $\delta$ 3.38 ppm and two multiplets at $\delta$ 3.58 and 3.64 ppm. In the $^{13}$C NMR spectrum, three signals are observed.

**29.11** A cyclic ether, **A**, shows a quintet ($\delta$ 2.72 ppm) and triplet ($\delta$ 4.73 ppm) with relative integrals of 1:2. The mass spectrum of **A** has a parent peak at $m/z = 58$. (a) Suggest a possible structure for **A**. (b) What would you expect to observe in the $^{13}$C{$^1$H} NMR spectrum of **A**?

**29.12** What would you expect to be the outcome of (a) heating Et$_2$O and HBr; (b) heating Et$_2$O with HCl; (c) adding a little Et$_2$O to water; (d) placing a few drops of Et$_2$O on your hand? Comment on any hazards involved.

**29.13** Compare scheme 29.18 with equation 26.30. Comment (with reasoning) on similarities and differences between them.

**29.14** Outline the mechanisms of the reactions of methyloxirane with (a) HBr and (b) EtO$^-$ in ethanol, and comment on the regioselectivity of the reactions.

**29.15** Suggest products for the following reactions:

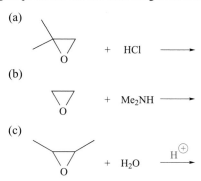

(a)

$+$ HCl $\longrightarrow$

(b)

$+$ Me$_2$NH $\longrightarrow$

(c)

$+$ H$_2$O $\xrightarrow{\text{H}^{\oplus}}$

## Additional problems

**29.16** A compound **X** has strong or very strong absorptions in its IR spectrum at 2961, 2930, 2871 and 1110 cm$^{-1}$. Elemental analysis shows that **X** contains 75.9% C, 14.0% H and 10.1% O. Reaction of **X** with HI at elevated temperature gives compounds **Y** and **Z** which analyse as containing: **Y**, 30.3% C, 5.6% H, 64.1% I, and **Z**, 68.1% C, 13.7% H and 18.2% O. The parent ion of **Y** in its mass spectrum is at $m/z$ 198. What can you deduce about the identities of **X**, **Y** and **Z**, and what other information would be useful in determining the exact structures of the compounds?

**29.17** Organic sulfides are related to ethers by exchange of S for O. (a) Draw the structure of diethyl sulfide. (b) Suggest a method of preparing methyl *iso*-propyl sulfide. (c) Me$_2$S reacts with MeI to give an ionic salt. Suggest a mechanism for the reaction and the structure of the product.

# 30 Alcohols

## Topics

- Structure and nomenclature
- Industrial manufacture and synthesis
- Physical properties
- Spectroscopic characteristics
- Reactivity
- Protection of OH groups

## 30.1 Structure and nomenclature

The functional group in an *alcohol* is an –OH (*hydroxyl*) group, and the general formulae for *primary, secondary* and *tertiary* alcohols are:

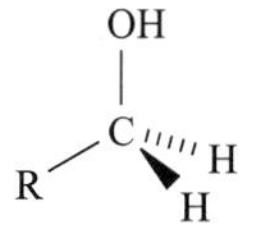

Primary alcohol 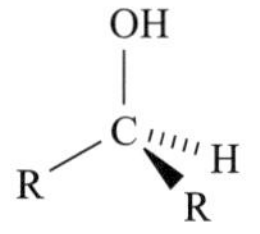      Secondary alcohol 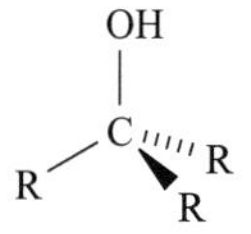      Tertiary alcohol

This introduction to nomenclature is concerned only with alcohols derived from *saturated* hydrocarbons. To name a simple alcohol, we extend the rules already described for hydrocarbons and halogenoalkanes. The name of a straight chain, primary alcohol is derived by attaching the ending *-ol* to the root name of the parent alkane, and placing an appropriate position number before the name ending, e.g. hexan-1-ol shows that the –OH group is attached to atom C(1) in a saturated, 6-membered carbon chain. The principal chain is found in the same way as was described for halogeno-alkanes (see Section 28.1). The names shown for structures **30.1–30.4** follow from these rules.

Root names: see Section 24.4 and Table 1.8

Common abbreviations:

MeOH = methanol
EtOH = ethanol
PrOH = propanol
BuOH = butanol

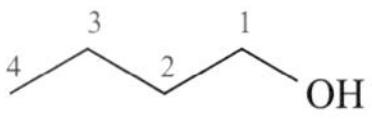

Butan-1-ol       Pentan-2-ol       Hexan-3-ol       Butan-2-ol

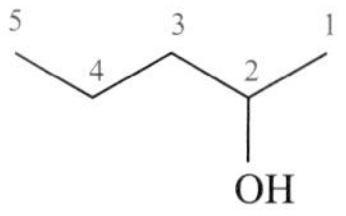

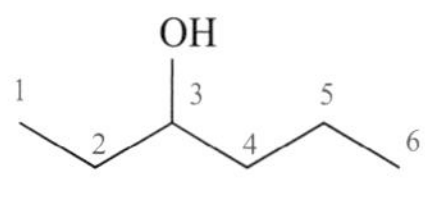

(30.1)            (30.2)            (30.3)            (30.4)

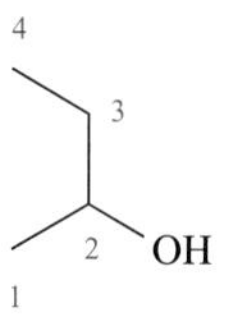

When there is more than one OH group, the name must specify the total number and positions of the functional groups. An alcohol with two OH groups is a *diol*, one with three OH groups is a *triol*, and so on. Compounds

**Table 30.1** Selected alcohols for which common names are often used.

| Common name | Systematic name | Structure |
| --- | --- | --- |
| Ethylene glycol | Ethane-1,2-diol | 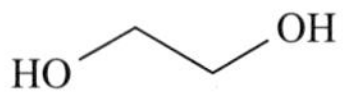 |
| Glycerol (glycerine) | Propane-1,2,3-triol | 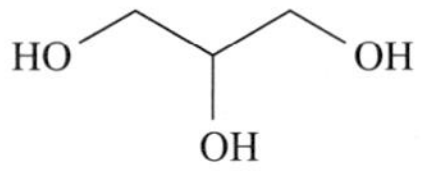 |
| *tert*-Butyl alcohol | 2-Methylpropan-2-ol |  |
| Isopropyl alcohol (isopropanol) | Propan-2-ol | 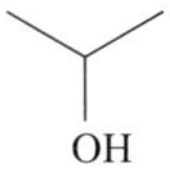 |

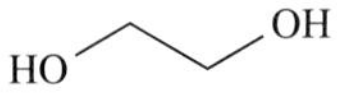

Ethane-1,2-diol

**(30.5)**

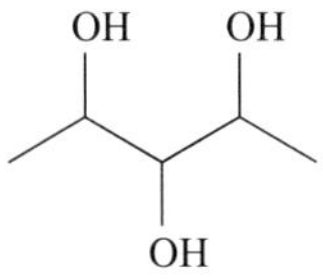

Pentane-2,3,4-triol

**(30.6)**

**30.5** and **30.6** give examples. Note that the 'e' on the end of the alkane root name is retained, in contrast to its omission in alcohols with one OH functionality. Many alcohols have trivial names and examples of ones that you may come across are listed in Table 30.1.

Examples of alcohols derived from cyclic alkanes and containing one OH group are given below. The position numbers are consistent with the OH group taking a higher priority than an alkyl substituent:

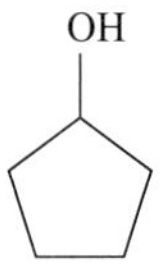

Cyclopentanol 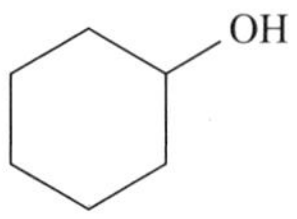          Cyclohexanol 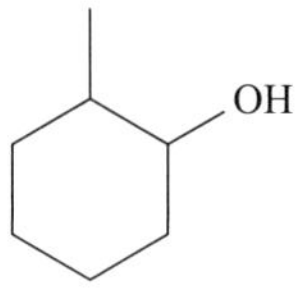          2-Methylcyclohexanol

Look back at the discussion of conformers and isomers in Section 25.2

When two OH groups are present and are attached to different C atoms in the saturated ring, stereoisomers are possible. Consider cyclohexane-1,4-diol. The OH groups may be on the same or opposite sides of the $C_6$ ring and these isomers are named *cis*- and *trans*-cyclohexane-1,4-diol respectively:

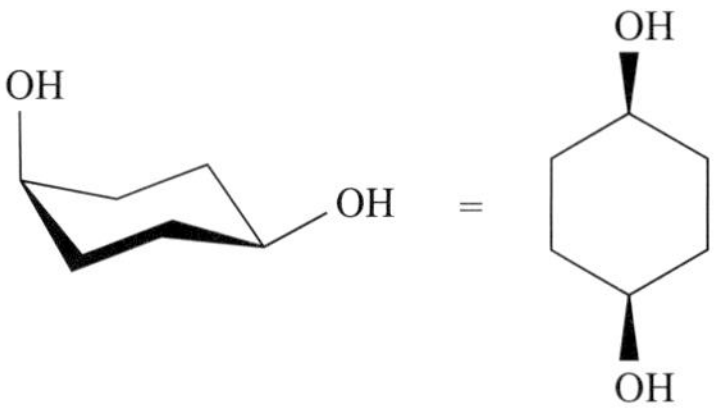

*cis*-Cyclohexane-1,4-diol 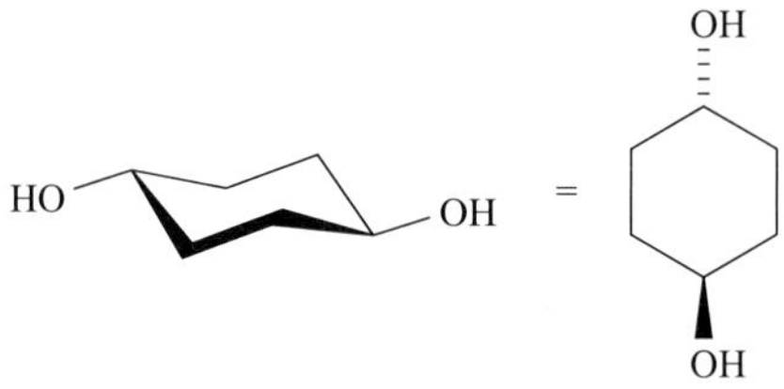          *trans*-Cyclohexane-1,4-diol

## 30.2    Industrial manufacture and synthesis

### Industrial manufacture of methanol and ethanol

The industrial manufacture of methanol and ethanol is big business, with methanol ranking among the top 30 chemicals produced commercially in the US. In the past, methanol was called *wood alcohol* because its manufacture involved heating wood in the absence of air. The modern method of production is the catalytic reduction of CO with $H_2$ (equation 30.1). Methanol is toxic, and drinking or inhaling it can lead to blindness or death.

$$CO + 2H_2 \xrightarrow[\text{ZnO catalyst}]{150\,\text{bar, }670\,\text{K}} CH_3OH \tag{30.1}$$

Ethanol is the alcohol present in alcoholic drinks, and for this purpose it is manufactured by the fermentation of glucose contained in, for example, grapes (for wine production), barley (for whisky) and cherries (for kirsch). The reaction is catalysed by a series of enzymes in yeast (equation 30.2).

$$\underset{\alpha\text{-D-Glucose}}{\text{[structure]}} \xrightarrow{\text{Yeast}} 2\ \underset{\text{Ethanol}}{\text{OH}} + 2CO_2 \tag{30.2}$$

Ethanol is widely used as a solvent and a precursor for other chemicals including acetic acid. The large-scale manufacture of EtOH is carried out by the acid-catalysed addition of $H_2O$ to ethene (equation 30.3). *Pure EtOH is often called absolute alcohol.*

$$\underset{\text{Ethene}}{\text{[structure]}} + H_2O \xrightarrow{\text{conc }H_2SO_4\text{ catalyst; 520 K}} \underset{\text{Ethanol}}{\text{[OH structure]}} \tag{30.3}$$

Ethane-1,2-diol (ethylene glycol) is commercially important as an 'anti-freeze' additive to motor vehicle radiator water; it is also used in resins and in the production of polyester fibres. The industrial synthesis involves the oxidation of ethene to form an epoxide, followed by acid-catalysed hydrolysis (scheme 30.4). The mechanism of ring opening was described in Section 29.6.

$$\underset{\text{Ethene}}{\text{[structure]}} \xrightarrow{O_2,\ \text{Ag, heat}} \underset{\substack{\text{Oxirane} \\ \text{(Ethylene oxide)}}}{\text{[structure]}} \xrightarrow{H^{\oplus},\ H_2O} \underset{\substack{\text{Ethane-1,2-diol} \\ \text{(Ethylene glycol)}}}{\text{[structure]}} \tag{30.4}$$

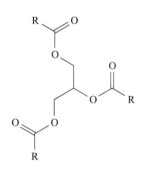

R groups may be the same or different

**(30.7)**

Propane-1,2,3-triol (glycerol, see Table 30.1) is the central building block for a wide range of fats with the general structure **30.7**. Fats are esters and we consider them in more detail in Chapter 33. Glycerol is also used as a sweetener, and as a softener in soap. Its reaction with nitric acid gives nitro-glycerine (equation 30.5). Glycerine is used in the treatment of angina, but it is also the explosive constituent of dynamite (see Box 30.3).

$$\underset{}{\text{[HO...OH structure]}} \xrightarrow[-3H_2O]{3HNO_3} \underset{}{\text{[O_2N...NO_2 structure]}} \tag{30.5}$$

Saponification: see *Hydrolysis of esters* in Section 33.9

Propane-1,2,3-triol (glycerol) is obtained commercially as a by-product of soap manufacture, and in the US, this method accounts for $\approx 40\%$ of

**ENVIRONMENT AND BIOLOGY**

## Box 30.1 Hydroxy-compounds and their derivatives in biology

The alcohol functionality occurs regularly in biology, and in this box we look at selected examples. Sugars contain a relatively large number of –OH groups and are *polyfunctional*; examples of some of the simplest

sugars (*monosaccharides*) include glucose, mannose and galactose, the structures of which (in their 6-membered cyclic *pyranose forms*) are shown below:

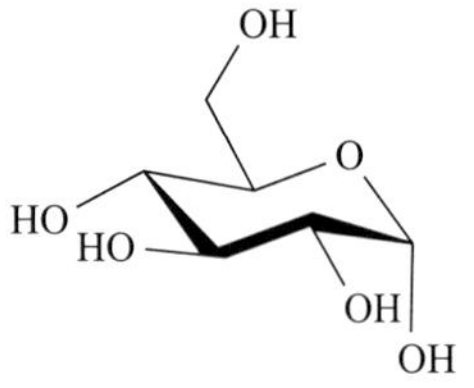

α-D-Glucose 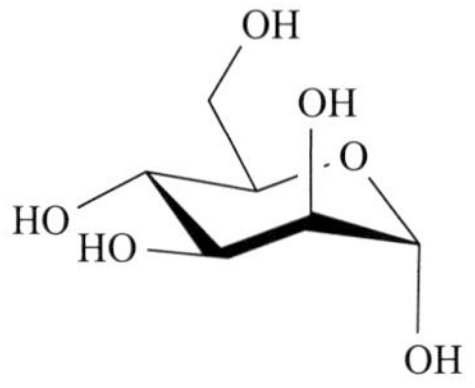          α-D-Mannose 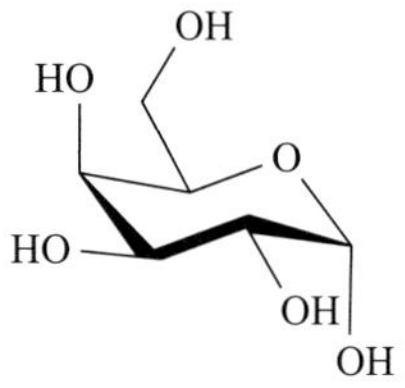          α-D-Galactose

Cholic acid occurs in bile and is involved in the digestion of fats. It is a carboxylic acid, but also contains alcohol functionalities. The central 4-ring structure is characteristic of a *steroid*. Bile acids are produced from cholesterol in the liver. Cholesterol (another

steroid) is present in the body, but increased deposits on the inner walls of arteries are associated with heart disease (see also Boxes 33.5 and 33.10). Cholesterol is the natural precursor to sex hormones such as testosterone.

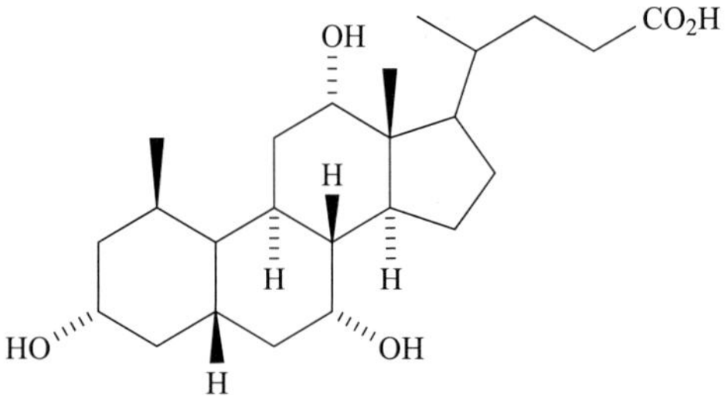

Cholic acid 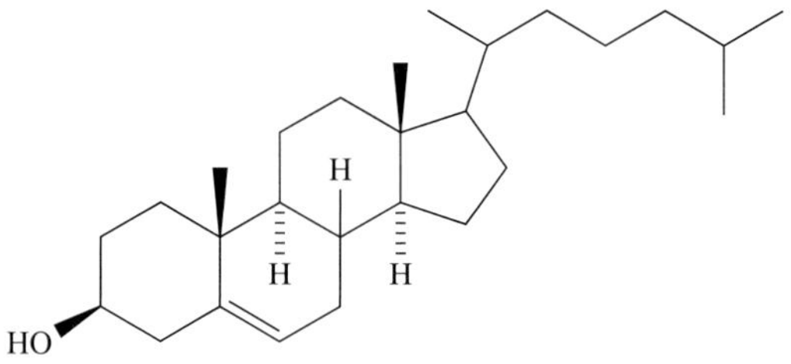          Cholesterol

Glycerol is a *triol* and is found in plant and animal oils and fats. Esterification (see Chapter 33) leads to the formation of triglycerides (fats), the form in which fatty acids are stored in the body. A related reaction

is used in the manufacture of soap (see Section 33.9). Glycerol is the building block for *phosphoglycerides*, a major group of *phospholipids*, essential to the body.

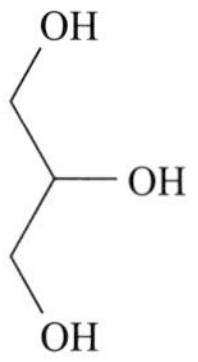

Glycerol
Propane-1,2,3-triol

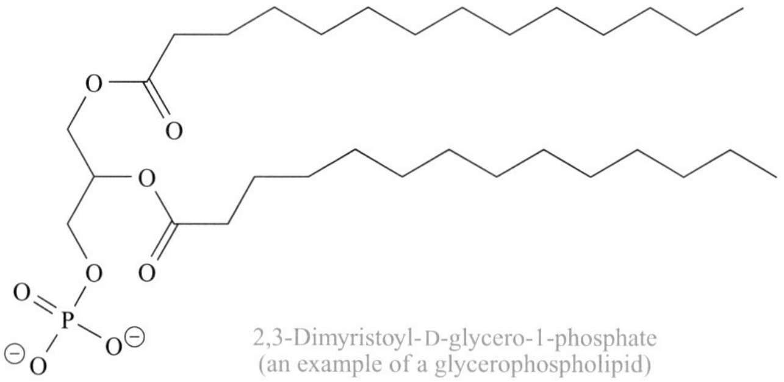

2,3-Dimyristoyl-D-glycero-1-phosphate
(an example of a glycerophospholipid)

The term *lipid* is used to cover oils, fats and related compounds that occur in living tissues. *Lipid membranes* are formed from lipids in cells and are hugely important. They act as barriers between cells and their surrounding environment, in particular, separating the aqueous

interior from the aqueous exterior. Lipid membranes are permeable but can discriminate between molecules, thus providing highly selective cell walls. Even greater specificity is provided by membrane-spanning molecules and proteins. A membrane lipid molecule is *amphiphilic*.

It possesses a polar head group which is *hydrophilic* ('water-loving') and two non-polar hydrocarbon tails which are *hydrophobic* ('water-hating') (see Figure 30.2). The double tail contrasts with the single tail found in, for example, a detergent molecule. Above a minimum concentration in water, amphiphilic molecules with single tails aggregate to form *micelles* as shown in the left-hand diagram below. (In practice, the micelle is spherical and the diagram shows only a cross section

through the micelle.) This motif is favoured because each single-tailed molecule is 'wedge-shaped' and such building blocks readily pack into a spherical domain. In contrast, double-tailed molecules resemble cylinders and molecular aggregation involves stacking of the molecules into layers. This is represented below in the right-hand diagram. The formation of a *bilayer* (rather than monolayer) means that the outer skin of the assembly is hydrophilic while the interior is hydrophobic.

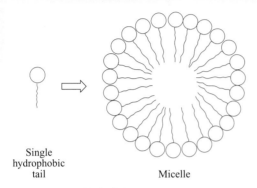

Single
hydrophobic
tail                    Micelle

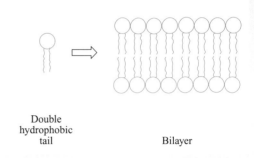

Double
hydrophobic
tail                    Bilayer

The major group of lipids involved in membrane formation are *glycerophospholipids*, an example of which is 2,3-dimyristoyl-D-glycero-1-phosphate (see opposite). A space-filling model of the 2,3-dimyristoyl-D-glycero-1-phosphate anion is shown below; the structure has been determined from an X-ray diffraction study of the sodium salt [K. Harlos *et al.* (1984) *Chem. Phys.*

*Lipids*, vol. 34, p. 115]. Features to note are the phosphate head group (polar), the two hydrocarbon tails in extended conformations (non-polar), and the overall cylindrical shape of the molecule. The diagram on the right-hand side below models how 12 anions pack together to form a small part of a bilayer.

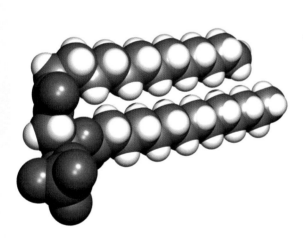

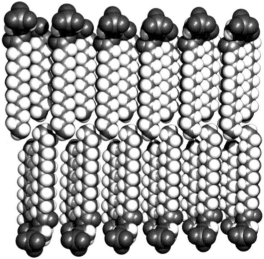

Colour code: P, brown; O, red; C, grey; H, white.

Typically, lipid bilayers are about 5–6 nm thick. The arrangement of outward-pointing polar head groups and inward-pointing hydrophobic tails is a crucial feature responsible for the bilayer being selective towards the penetration of molecules. Ionic and polar species are essentially excluded. Specific transport mechanisms involving transmembrane proteins are needed for their

uptake and removal from the cell. One of the major functions of the steroid cholesterol (see opposite) is to modify the properties of cell membranes.

*For related information*: see Figure 25.6 for a review of axial and equatorial sites in a cyclohexane ring; see Section 33.9 for saponification of glycerol; see Chapter 35 for more details of sugars.

## Box 30.2 Polyethylene glycol and the *Mary Rose*

A view of the Mary Rose, looking from the stern. *Reproduced courtesy of The Mary Rose Trust.*

The sunken Tudor warship *Mary Rose* was brought to the surface in 1982. An immediate problem (as with other salvaged wooden wrecks and wooden artefacts that have remained in the sea or bogs for long periods) was how to preserve the timbers of the ship as they began to dry out. A suitable method of conservation involves the use of polyethylene glycol:

$$\text{HO} \left( \!\!\!\diagup\!\!\!\diagdown \!\!\!\text{O} \right)_{\!n} \!\!\text{H}$$

This is a waxy compound and when it impregnates the wood, it helps to prevent the deterioration of the drying timbers.

glycerol production. Other industrial routes include conversion of propene or allyl alcohol to glycerol.

### Synthesis of alcohols: methods covered in previous chapters

Alcohols have many synthetic applications, and are often prepared as intermediate compounds in multi-step syntheses. They can be made from a range of starting materials and we have already described ways of preparing alcohols from alkenes, halogenoalkanes and ethers. A summary of alcohol-forming reactions described in previous chapters is given in Figure 30.1.

### Synthesis of alcohols: reduction of carbonyl compounds

The reduction of carbonyl compounds (e.g. aldehydes, ketones, carboxylic acids and esters) is a common method of preparing alcohols. In this section we present reactions simply as syntheses, and we return to their mechanisms in Chapter 33 when we detail the reactivity of the carbonyl (C=O) group.

Equations 30.6 and 30.7 show the reduction of an *aldehyde to a primary alcohol*, and of a *ketone to a secondary alcohol*, respectively. Reduction involves the addition of two H atoms and these are provided by reagents such as $LiAlH_4$, $LiBH_4$ or $NaBH_4$.

$$R-\overset{H}{\underset{O}{\vert\vert}} \xrightarrow{\text{Reduction}} R-\overset{H}{\underset{OH}{\vert}}-H \tag{30.6}$$

$$R-\overset{R}{\underset{O}{\vert\vert}} \xrightarrow{\text{Reduction}} R-\overset{R}{\underset{OH}{\vert}}-H \tag{30.7}$$

R groups may be the same or different

**Fig. 30.1** Methods of preparing alcohols already covered in previous chapters. Further details can be found by using the cross-references given under the equations.

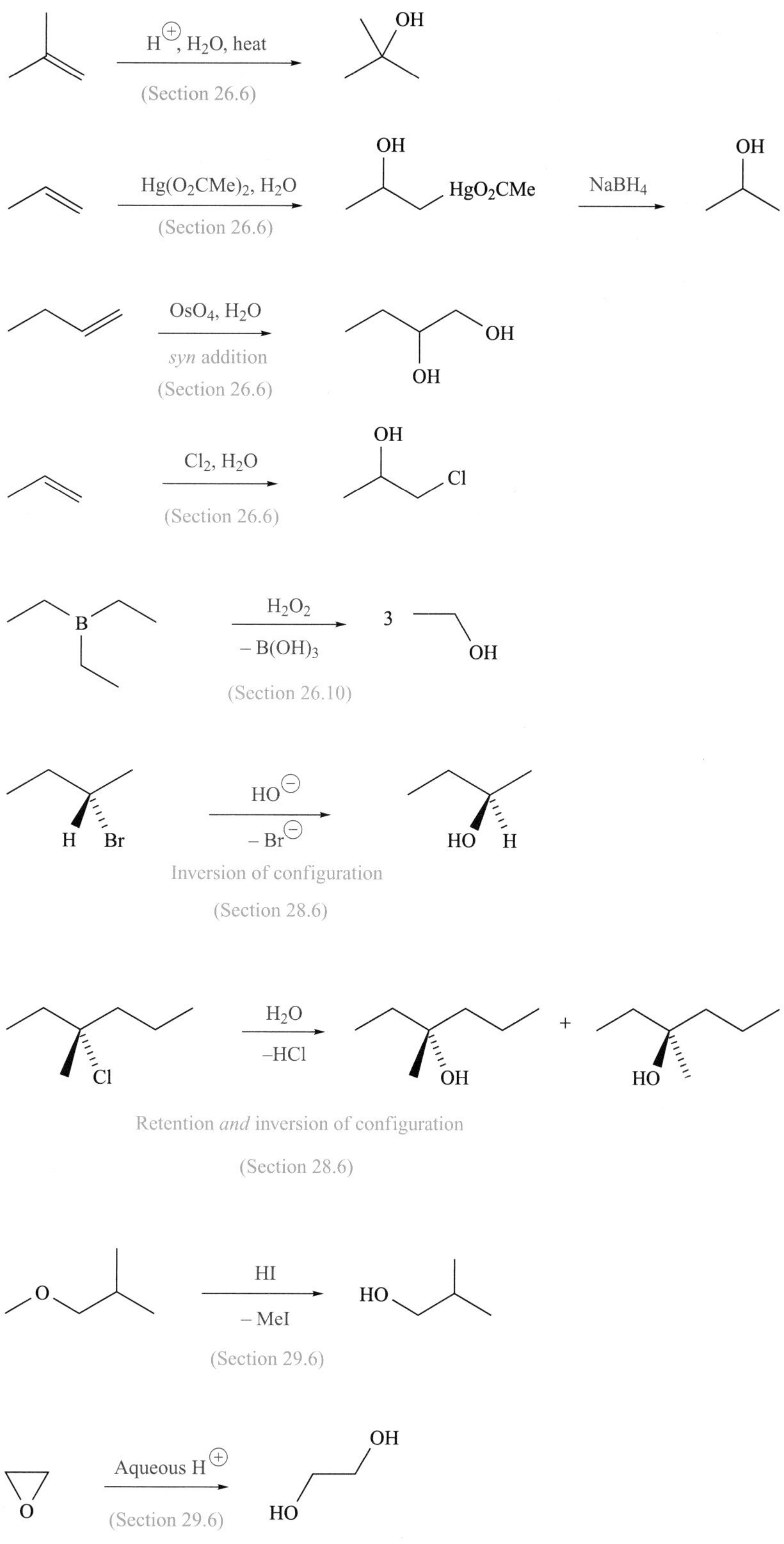

LiAlH$_4$ as a reducing agent: see Box 21.5

Reductions of aldehydes and ketones are usually carried out using sodium borohydride (NaBH$_4$), but lithium aluminium hydride (LiAlH$_4$) is also suitable. Both NaBH$_4$ and LiAlH$_4$ are crystalline solids, but whereas LiAlH$_4$ is moisture sensitive, NaBH$_4$ is reasonably air stable (see Section 21.5), making it more easily handled than LiAlH$_4$.

## COMMERCIAL AND LABORATORY APPLICATIONS

### Box 30.3 Explosives and rocket propellants

Equation 30.5 illustrated that propane-1,2,3-triol is the precursor to nitroglycerine. Nitroglycerine was discovered in 1846, and both nitroglycerine (also called nitroglycerin and trinitroglycerol) and trinitrotoluene (TNT) are well-known explosives. The common structural feature of these compounds is the nitro ($NO_2$) group. On explosion, the molecules of the nitro-compound decompose. Nitroglycerine (mp 286 K) forms *only* gaseous products making the reaction highly entropically favoured:

$$4O_2NOCH(CH_2ONO_2)_2(l) \longrightarrow$$
$$6N_2(g) + 12CO_2(g) + 10H_2O(g) + O_2(g)$$

The detonation releases 1415 kJ of energy per mole of nitroglycerine. Nitroglycerine and TNT are classed as *high explosives*. TNT had widespread application in World War I, but other explosives such as pentaerythritol tetranitrate and RDX (cyclonite) have entered the field since World War II. Pentaerythritol tetranitrate has a major application in explosive fuses in demolition charges and rock blasting.

Other nitro-compounds have been used (or studied for potential use) by the military as explosives and rocket propellants. After the end of the Cold War, more information about these compounds became publicly available. Although an explosive and a rocket propellant may appear to do the same job, the design requirements are different. For example, one aim of a propellant is to fire a missile over a given range in a controlled manner, whereas an explosive releases all its energy rapidly.

A further point to consider is the thermal and shock sensitivities of the materials. Hazards prior to the point of ignition must be minimized, e.g. picric acid is explosive, but its high shock sensitivity can result in premature detonation, and it has generally been replaced by other (slightly!) less hazardous materials. Nitroglycerine is very sensitive to shock or friction, whereas TNT can be stored moderately safely.

Nitroglycerine    Pentaerythritol tetranitrate    TNT    Picric acid    RDX

---

When we write '$BH_3$' in this section, we are referring to a Lewis base adduct such as $THF \cdot BH_3$: see Section 21.5

The reduction of an *ester to an alcohol* is achieved using $LiAlH_4$ or $LiBH_4$ (equation 30.8); $NaBH_4$ is not very efficient for ester reduction. Reduction of an ester leads to cleavage at the ester linkage; the carbonyl group evolves into one OH group and the R'O group into a second.

R and R' may be the same or different

(30.8)

Carboxylic acid reduction can be carried out using $LiAlH_4$, but $BH_3$ is also a useful reagent. In Section 21.5, we described how $BH_3$ acts as a Lewis acid, and in equation 26.72 we saw how $BH_3$ accepts a pair of electrons from an alkene and then transfers an H atom to the organic group. Similarly, $BH_3$ reduces carboxylic acids as shown in equation 30.9; the C=O group is converted into a $CH_2$ group. Despite the structural relationship between carboxylic acids and esters, $BH_3$ is a poor reagent for ester reduction because esters are poorer electron donors than carboxylic acids (see Chapter 33).

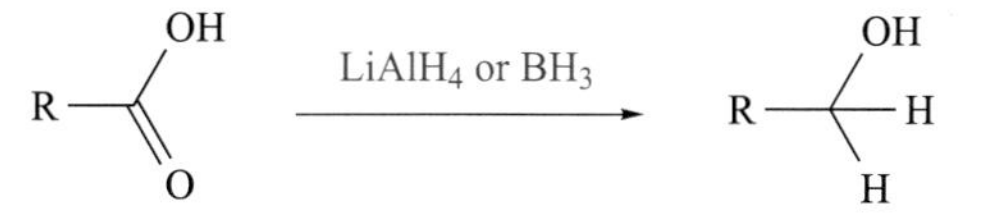

$$R-\overset{OH}{\underset{O}{C}} \xrightarrow{\text{LiAlH}_4 \text{ or BH}_3} R-\overset{OH}{\underset{H}{C}}-H \tag{30.9}$$

## Synthesis of alcohols: use of Grignard reagents

Grignard reagents: see Sections 28.4 and 33.14

Grignard reagents $RMgX$ ($X = Cl$ or $Br$) react with carbonyl compounds to give primary, secondary or tertiary alcohols. The product depends on the substituents attached to the C=O group. If the starting material is methanal, the product is a primary alcohol (equation 30.10), but if a higher aldehyde ($R'CHO$) is used, a secondary alcohol results (equation 30.11). If the precursor is a ketone, the product is a tertiary alcohol (equation 30.12).

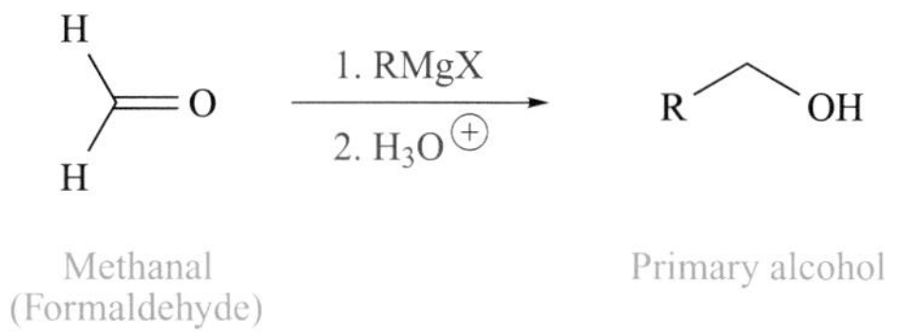

$$\underset{H}{\overset{H}{C}}=O \xrightarrow[\text{2. H}_3\text{O}^{\oplus}]{\text{1. RMgX}} R\diagup OH \tag{30.10}$$

Methanal
(Formaldehyde)                    Primary alcohol

$$\underset{H}{\overset{R'}{C}}=O \xrightarrow[\text{2. H}_3\text{O}^{\oplus}]{\text{1. RMgX}} \underset{R}{\overset{R'}{C}}\diagup OH \tag{30.11}$$

Aldehyde
$R' \neq H$                        Secondary alcohol

$$\underset{R''}{\overset{R'}{C}}=O \xrightarrow[\text{2. H}_3\text{O}^{\oplus}]{\text{1. RMgX}} \underset{R}{\overset{R' \quad R''}{C}} OH \tag{30.12}$$

Ketone
R' and R" may be
the same or
different                          Tertiary alcohol

We return to the mechanism of these reactions in Chapter 33. For now, equations 30.13–30.15 give examples of their applications in synthesis.

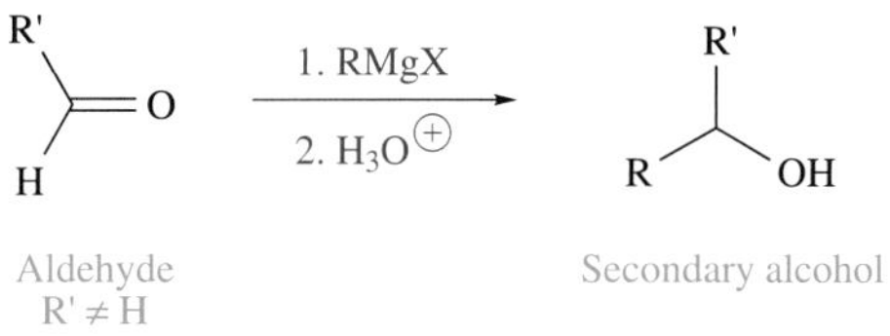

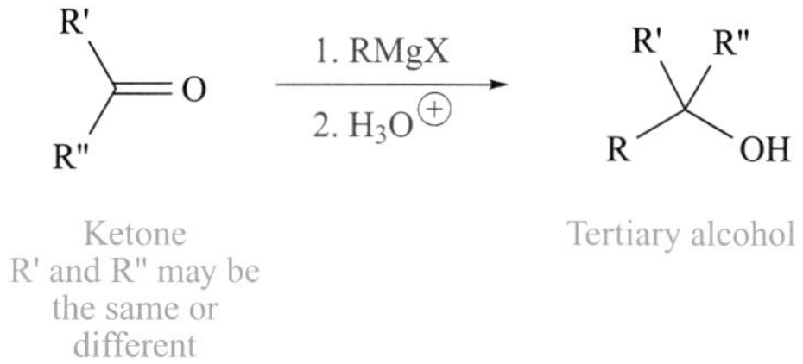

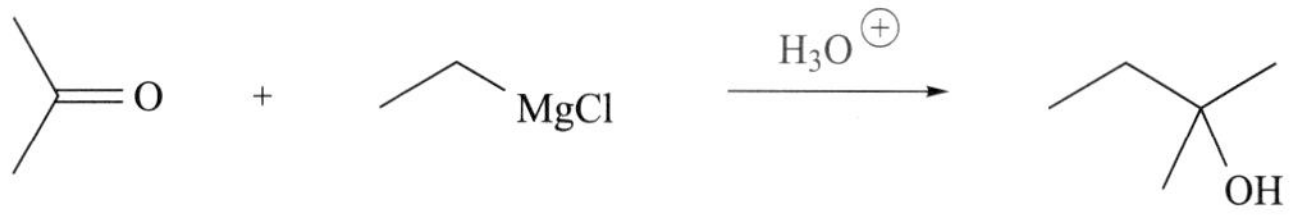

$$\text{(structure)}=O \quad + \quad \text{MgCl} \xrightarrow{\text{H}_3\text{O}^{\oplus}} \text{(structure)} OH \tag{30.13}$$

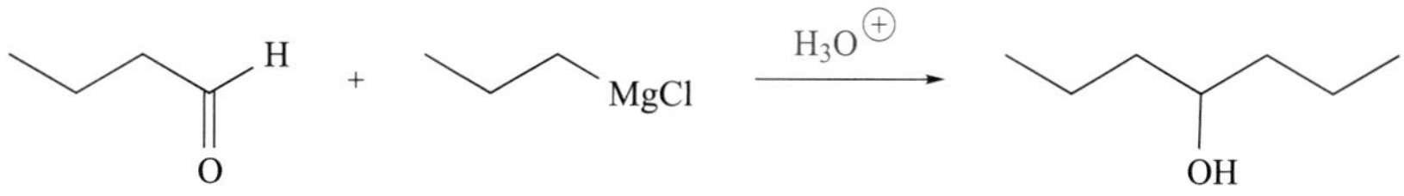

$$\text{(structure)} \quad + \quad \text{MgCl} \xrightarrow{\text{H}_3\text{O}^{\oplus}} \text{(structure)} OH \tag{30.14}$$

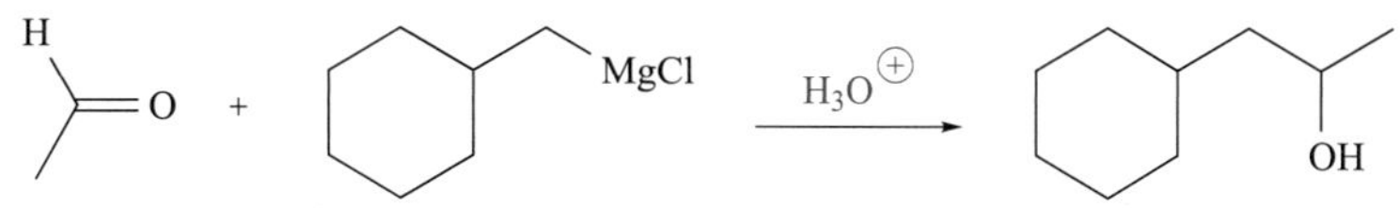

$$\text{(structure)}=O \quad + \quad \text{MgCl} \xrightarrow{\text{H}_3\text{O}^{\oplus}} \text{(structure)} OH \tag{30.15}$$

**30.3**

## Physical properties

### The effects of hydrogen bonding

Hydrogen bonding: see Section 21.8

$$\delta^- \quad \delta^+$$
$$O - H$$
$$R$$

(30.8)

The Pauling electronegativities of O and H are 3.4 and 2.2 respectively, and so an O–H bond is polar as shown in structure **30.8**. As a result of the high electronegativity of O, alcohols are involved in hydrogen bonding with each other and with other suitable molecules such as water. This is represented in diagrams **30.9** and **30.10**.

Propan-1-ol

(30.9)

Propan-1-ol and water

(30.10)

The boiling points and enthalpies of vaporization for selected alcohols are listed in Table 30.2. The relatively high values (for example, compared with those for alkanes or ethers of similar molecular weight; see Tables 25.2, 29.2 and 29.3 and Figure 25.9) reflect the intermolecular association of the alcohol molecules in the liquid.

The lower molecular weight alcohols (MeOH, EtOH, PrOH) are completely miscible with water owing to the formation of hydrogen bonds as in structure **30.10**. The solubilities in water of butanol, pentanol, hexanol and heptanol are significantly lower than those of methanol, ethanol and propanol. Higher molecular weight alcohols (octanol upwards) are essentially immiscible with water. This can be explained in terms of the increased aliphatic chain length of the molecule. The –OH group is *hydrophilic* ('water-loving'), while the hydrocarbon part of the molecule is *hydrophobic* ('water-hating'). Figure 30.2 illustrates this for heptan-1-ol. As the molecular weight of the alcohol increases, the inability of the hydrocarbon chain to form hydrogen bonds with water predominates over the effects of hydrogen bond formation by the –OH group. As a result, the solubility of the alcohol in water decreases. One application of hydrophilic molecules is highlighted in Box 30.4; see also Box 30.1.

*Hydrophilic* means 'water-loving'.
*Hydrophobic* means 'water-hating'.

### Basicity and acidity

We saw in Section 16.4 that water can act as a Brønsted acid *and* as a Brønsted base. Similarly, alcohols can both donate a proton (acts as a

**Fig. 30.2** Heptan-1-ol is an example of an alcohol with a long hydrocarbon chain. The molecule has a hydrophobic tail and a hydrophilic head. Colour code: C, grey; O, red; H, white.

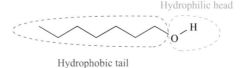

Hydrophilic head

Hydrophobic tail

**Table 30.2** Boiling points and enthalpies of vaporization of selected alcohols. Values for water are included for comparison.

| Compound | Structure | Relative molecular mass | Boiling point / K | $\Delta_{vap}H(bp) \, / \, kJ \, mol^{-1}$ |
|---|---|---|---|---|
| Water | | 18.02 | 373.0 | 40.7 |
| Methanol | | 32.04 | 337.6 | 35.2 |
| Ethanol | | 46.07 | 351.3 | 38.6 |
| Propan-1-ol | | 60.09 | 370.2 | 41.4 |
| Propan-2-ol (isopropyl alcohol) | | 60.09 | 355.3 | 39.9 |
| Butan-1-ol | | 74.12 | 390.7 | 43.3 |
| Butan-2-ol | | 74.12 | 372.5 | 40.8 |
| 2-Methylpropan-1-ol | | 74.12 | 380.9 | 41.8 |
| 2-Methylpropan-2-ol (*tert*-butyl alcohol) | | 74.12 | 355.4 | 39.1 |
| Pentan-1-ol | | 88.15 | 411.0 | 44.4 |
| Hexan-1-ol | | 102.17 | 430.6 | 44.5 |

Brønsted acid) and accept a proton (acts as Brønsted base). *Aliphatic alcohols act as weak acids and bases.* The $pK_a$ of MeOH is 15.5, a value close to that of water; compare this with a $pK_a$ value of 4.77 for acetic acid ($CH_3CO_2H$). Thus, equilibrium 30.16 lies far towards the left-hand side.

$$MeOH + H_2O \rightleftharpoons MeO^- + H_3O^+ \qquad (30.16)$$

In the presence of *strong bases* such as sodium hydride (NaH) or potassium amide ($KNH_2$), reactions to give *alkoxide* ions occur (equations 30.17 and 30.18). The $H^+$ from the alcohol combines with $H^-$ to give $H_2$, or with $NH_2^-$ to give $NH_3$.

$$EtOH + NaH \longrightarrow \underset{\text{Sodium ethoxide}}{Na^+EtO^-} + H_2 \qquad (30.17)$$

$$EtOH + KNH_2 \longrightarrow \underset{\text{Potassium ethoxide}}{K^+EtO^-} + NH_3 \qquad (30.18)$$

In Chapter 32, we discuss aromatic alcohols (*phenols*), the acidity of which is greater than that of aliphatic alcohols.

$pK_a$ and $K_a$ values: see Section 16.4

See also Section 30.5

## Box 30.4 Polymers for contact lenses

Macrophotograph of a human eye, showing contact lens.
*Argentum/Science Photo Library.*

Polymeric material for soft contact lenses must be optically transparent, chemically stable, be able to withstand being constantly wet, possess suitable mechanical properties and be permeable to $O_2$. (The cornea obtains most of its $O_2$ supply from the atmosphere rather than the bloodstream.) Such a range of requirements provides challenges to polymer chemists.

Compound **1** is a monomer used in the manufacture of polymers for contact lenses. Its radical polymerization produces a polymer that is a *hydrogel*; it absorbs water but at the same time is insoluble in water. Water retention in a contact lens is important because there is a relationship between water content and $O_2$ permeability. Designing lenses for relatively long wearing periods is a problem that the contact lens industry has had to overcome. Polymers for these lenses must possess enhanced $O_2$ permeability characteristics. This can be solved either by manufacturing thinner lenses, or by increasing the water content of the hydrogel. The latter option is preferable because reducing the thickness of a lens makes it more susceptible to tearing. Copolymerization of monomer **1** with highly *hydrophilic* monomers such as **2** or **3** has proved a successful approach to increasing the water content.

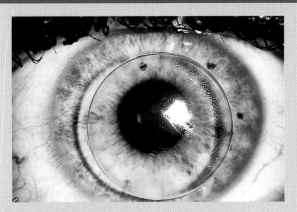

Protonation of alcohols by acids is analogous to the formation of the oxonium ion, $[H_3O]^+$, by protonation of water. Equation 30.19 shows the protonation of ethanol.

$$\text{CH}_3\text{CH}_2\overset{\cdot\cdot}{\underset{\cdot\cdot}{O}}\text{H} \;+\; \text{HX} \;\rightleftharpoons\; \text{CH}_3\text{CH}_2\overset{\cdot\cdot}{\underset{\underset{H}{|}}{\overset{\oplus}{O}}}\text{H} \;+\; \text{X}^{\ominus} \qquad (30.19)$$

We shall see in Section 30.5 that protonating an alcohol is an important means of providing the molecule with a good leaving group (i.e. $H_2O$) in nucleophilic substitution reactions.

## 30.4    Spectroscopic characteristics of alcohols

In Chapters 12 and 14, we described the applications of IR and NMR spectroscopies, and among the compounds we discussed were those

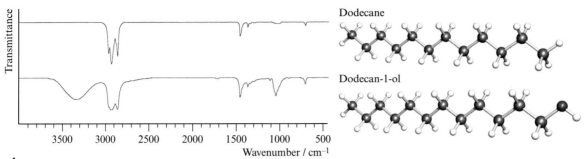

**Fig. 30.3** The IR spectra of dodecane and dodecan-1-ol. Colour code for structures: C, grey; O, red; H, white.

containing –OH groups. In this section, we review the spectroscopic characteristics of alcohols.

## IR spectroscopy

The appearance in an IR spectrum of a broad and strong absorbance near to $3300\,cm^{-1}$ is characteristic of an alcohol functional group. Figure 30.3 shows the IR spectra of dodecane and dodecan-1-ol. These compounds are structurally similar apart from the presence of the terminal OH group in the alcohol. Common to both spectra are absorbances at $\approx 2900$–$3000\,cm^{-1}$ assigned to aliphatic CH stretches, i.e. $\nu(C-H)$. There are also characteristic absorbances in the fingerprint region. The broad band centred around $3300\,cm^{-1}$ in the spectrum of dodecan-1-ol is assigned to $\nu(O-H)$. The broadness of the band arises from intermolecular hydrogen bonding (as in structure **30.9**) which leads to a range of frequencies (and therefore wavenumbers) being associated with the O−H stretch.

## NMR spectroscopy

In $^1H$ NMR spectra, a signal for an OH proton is sometimes difficult to assign. The shift range is approximately between $\delta 0.5$ and $8.0\,ppm$ (Table 14.3). The chemical shift of a signal is often solvent-dependent and may be broadened, either by the effects of hydrogen bonding or by exchange (see equation 14.3). Exchange also occurs between the H atom of an alcohol OH and the D atom of $D_2O$ (deuterated water). This exchange can be used as a means of confirming the assignment of an OH peak in a $^1H$ NMR spectrum: shaking an alcohol ROH with $D_2O$ gives a deuterated alcohol ROD, and the signal assigned to the OH proton of ROH disappears.

Figure 14.11 showed the $^1H$ NMR spectrum of ethanol dissolved in chloroform and the broad signal at $\delta 2.6\,ppm$ was assigned to the OH proton. However, the $^1H$ NMR spectrum of neat ethanol shows a triplet for the OH proton, arising from coupling between the OH and $CH_2$ protons. Figure 30.4 shows the $^1H$ NMR spectrum of 2-methylpropan-1-ol dissolved in $CDCl_3$. The alcohol contains four proton environments as shown in structure **30.11**, and the spectrum can be assigned as follows.

- Protons $a$ couple to one proton $b$ to give a doublet, relative integral = 6, and this corresponds to the signal at $\delta 0.92\,ppm$.

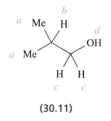

(30.11)

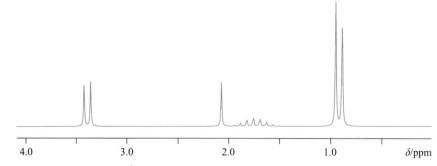

**Fig. 30.4** The 100 MHz $^1$H NMR spectrum of 2-methylpropan-1-ol dissolved in $CDCl_3$.

- Protons $c$ couple to one proton $b$ to give a doublet, relative integral = 2, and this corresponds to the signal at $\delta$ 3.39 ppm.
- The values of the coupling constants for the two doublets are equal (6.5 Hz), i.e. $J_{H_aH_b} = J_{H_bH_c}$, and therefore protons $b$ couple to both $a$ and $c$ (eight protons in all) with equal coupling constants to give a nonet. This corresponds to the signal at $\delta$ 1.75 ppm. All nine lines in this multiplet are not clearly visible in Figure 30.4, but a consideration of Pascal's triangle (see Figure 14.10) and looking at an enlargement of the signal confirm that it is a nonet.
- The singlet at $\delta$ 2.07 ppm is assigned to the OH proton.

## Mass spectrometry

Figure 10.4b showed the mass spectrum of ethanol. The parent ion appears at $m/z = 46$ and is assigned to $[C_2H_6O]^+$. Loss of H gives $[C_2H_5O]^+$. A characteristic fragmentation pathway for alcohols occurs by the cleavage of the C–C bond adjacent to the OH group. For ethanol, this gives the $[CH_3O]^+$ fragment at $m/z = 31$ (Figure 10.4a). The fragmentation is represented by the scheme:

A similar fragmentation pathway occurs with other primary alcohols. For a secondary alcohol, the analogous pathway is:

R groups may be the same or different

Loss of this first fragment in the mass spectrum is a characteristic of alcohols and assists in their identification.

| **30.5** | **Reactivity of alcohols** |
|---|---|

### Combustion

The combustion of an alcohol gives $CO_2$ and $H_2O$. Ethanol (equation 30.20) burns with a blue flame.

$$C_2H_5OH + 3O_2 \longrightarrow 2CO_2 + 3H_2O \tag{30.20}$$

Methanol ($\Delta_c H^{\circ} = -726 \, \text{kJ mol}^{-1}$) is being considered as an alternative to hydrocarbon-based motor fuels. It is especially attractive in countries where it is available in bulk from biomass (see Box 30.5) or from catalytic processes. We have given special mention of the combustion of alcohols because of their potential as fuels, but you should note that most organic compounds burn well.

### Cleaving the R–OH bond: dehydration

In Section 29.3, we described the synthesis of symmetrical ethers from alcohols by *intermolecular elimination* of water (equation 30.21), but stressed that the reaction conditions must be carefully controlled because of potential *intramolecular elimination* of water.

$$2 \quad \diagup\!\!\!\diagdown\!\!\!\text{OH} \xrightarrow{\text{H}_2\text{SO}_4, \, \text{heat}} \diagup\!\!\!\diagdown\!\!\!\text{O}\!\!\!\diagdown\!\!\!\diagup + \text{H}_2\text{O} \tag{30.21}$$

Ethanol                   Diethyl ether

---

**COMMERCIAL AND LABORATORY APPLICATIONS**

### Box 30.5 Methanol and ethanol: alternative motor fuels?

Methanol can be produced from natural gas and from *biomass* (i.e. plant and animal material) and is a possible alternative fuel for motor vehicles. Among its advantages are lower levels of $NO_x$ emissions (see Section 22.8 and Box 22.9) but disadvantages include fire and explosion risks and high toxicity. Mixed with hydrocarbon-based fuels, methanol becomes a more viable fuel.

Ethanol can also be manufactured from biomass and it possesses a high octane number (see the end of Section 25.4). 'Gasohol' (a petrol mixture containing 10% ethanol) is becoming available in the American continents. Brazil uses large amounts of ethanol-containing fuels. Cars do not have to be specially adapted, although they suffer from reduced performance. Within many European countries, the EU restricts the ethanol content of motor fuels to 5%.

Biomass makes a relatively small contribution to the energy supplies in the US as the charts below show.

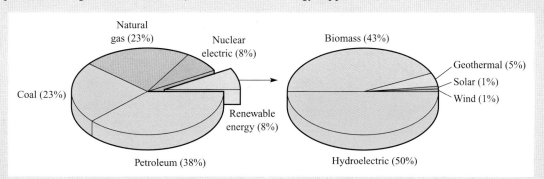

[Data from: *Chemical & Engineering News* (2000) October 9 issue, p. 45.]

*Intramolecular* elimination of $H_2O$ occurs when an alcohol is heated with concentrated $H_2SO_4$ (equation 30.22), or is passed over hot $Al_2O_3$. In Section 26.3, we described this reaction as a method of preparing alkenes.

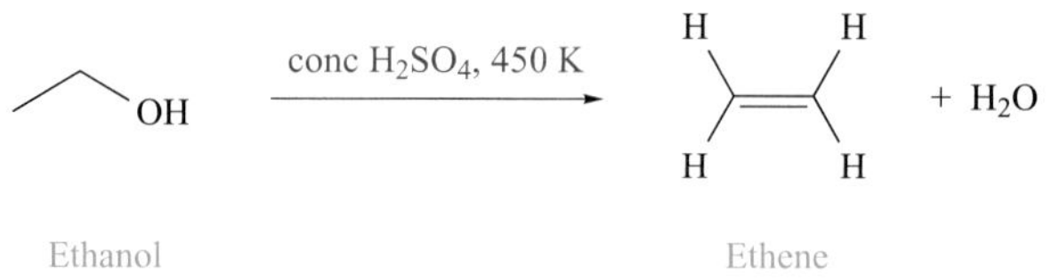

$$\text{(30.22)}$$

Ethanol (a primary alcohol) is converted to ethene only under harsh conditions (450 K). In contrast, 2-methylbutan-2-ol is dehydrated on treatment with acid at 298 K. In this reaction, two products are formed (equation 30.23), and, as we saw in Section 28.7, *loss of $H^+$ favours the formation of the more highly substituted alkene.*

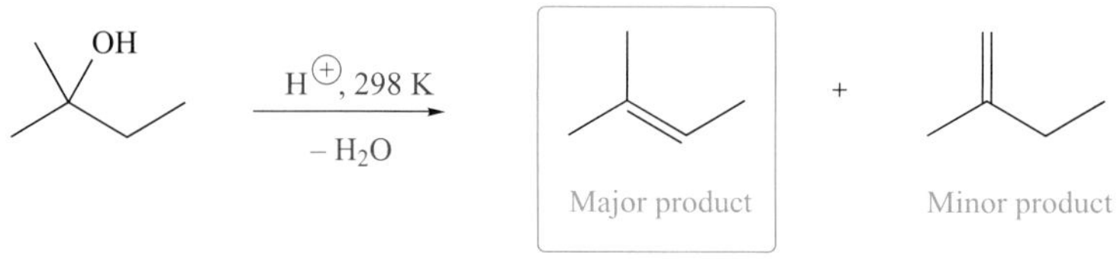

$$\text{(30.23)}$$

▪▶
E1 mechanism:
see Section 28.7

This is an E1 mechanism, and the final loss of $H^+$ is facilitated by the presence of water which acts as a base. The first step in this type of reaction is the protonation of the OH group, and the subsequent loss of $H_2O$ (a *good leaving group*). The intermediate is a tertiary carbenium ion that loses $H^+$ in a base-assisted process to give either 2-methylbut-2-ene or 2-methylbut-1-ene:

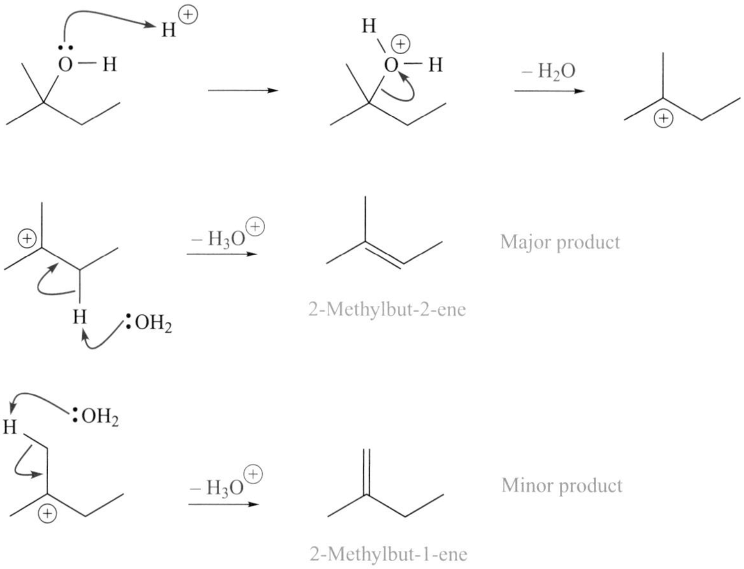

Major product
2-Methylbut-2-ene

Minor product
2-Methylbut-1-ene

▪▶
Carbenium ions:
see Section 26.5

We have previously detailed the order of stabilities of carbenium ions as $[R_3C]^+ > [R_2CH]^+ > [RCH_2]^+$ (R = alkyl substituent). It follows that tertiary alcohols undergo dehydration more readily than secondary or primary alcohols. Consider the dehydration of butan-1-ol. Protonation leads to a primary carbenium ion, but this rapidly rearranges to the more stable secondary carbenium ion as shown in equation 30.24.

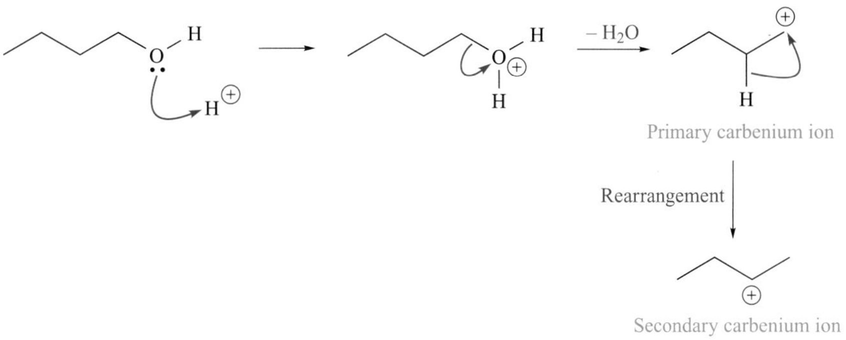

(30.24)

With $H_2O$ acting as a base, the secondary carbenium ion now loses $H^+$ and could form either but-1-ene or but-2-ene. Since the more highly substituted alkene is favoured, the predominant product is but-2-ene (equation 30.25). Of the possible (Z)- and (E)-isomers, the more favourable is the less sterically hindered (E)-isomer.

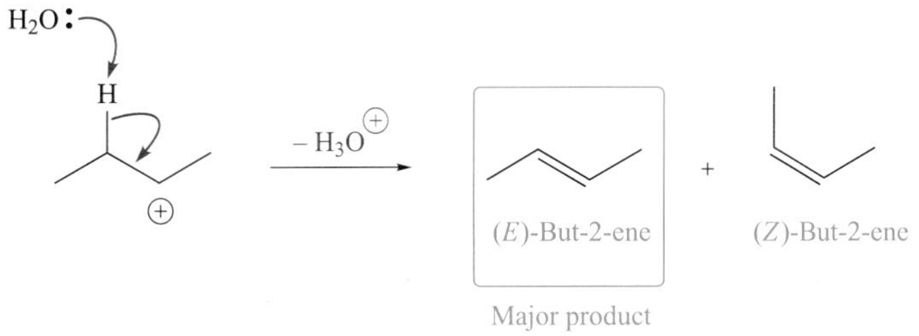

(30.25)

Equation 30.26 summarizes the overall reaction: the acid-catalysed dehydration of butan-1-ol.

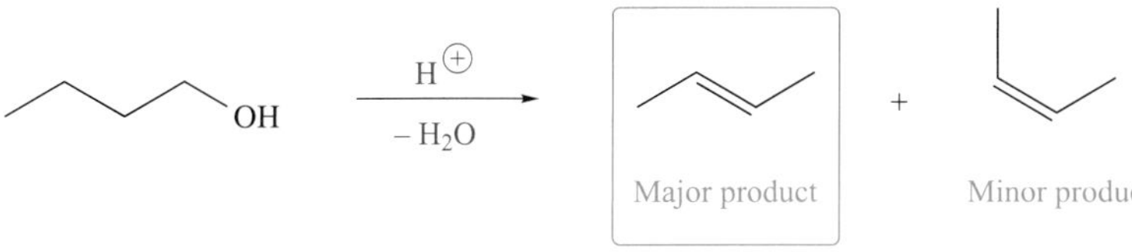

(30.26)

See problems 30.13 and 30.14

The rearrangements of carbenium ions may involve H or *Me migration* (see equation 28.27).

The dehydration of alcohols can also be carried out using $POCl_3$. This reagent converts an OH group into an $OPOCl_2$ substituent (diagram **30.12**), providing the molecule with a good leaving group: $Cl_2OPO^-$ is a far better leaving group than $HO^-$. In the presence of base, the derivatized alcohol undergoes elimination by an E2 mechanism to give an alkene. This is shown in equation 30.27 where B = base.

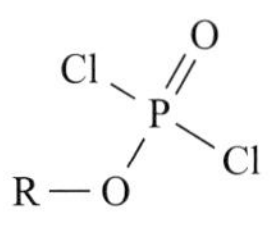

Tetrahedral P

**(30.12)**

E2 mechanism: see Section 28.7

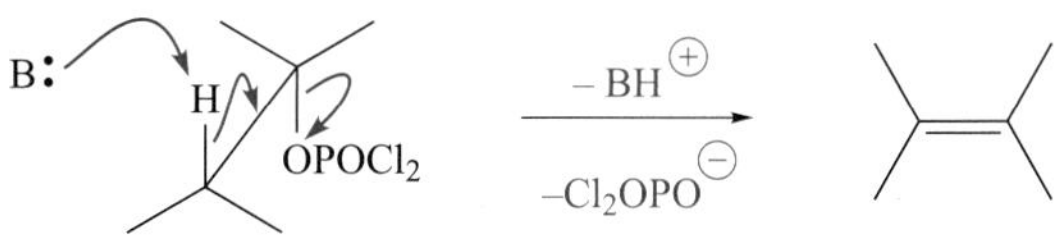

(30.27)

## Cleaving R–OH bond: conversion to halogenoalkanes

See also the end of Section 28.2

Alcohols react with phosphorus(III) halides ($PCl_3$, $PBr_3$, $PI_3$), phosphorus(V) chloride ($PCl_5$) or thionyl chloride ($SOCl_2$), and this method is used

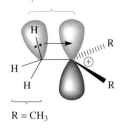

C–H $\sigma$–orbital can overlap with one lobe of the 2*p* atomic orbital

R = CH$_3$

**(30.13)**

to convert secondary or primary alcohols to halogenoalkanes, e.g. reactions 28.6 and 28.7. Reactions of alcohols with hydrogen halides (equation 30.28) also produces halogenoalkanes.

$$ROH + HX \longrightarrow RX + H_2O \qquad\qquad X = Cl, Br, I \quad (30.28)$$

The reaction is most efficient for tertiary alcohols, the order of reactivity being:

$$R_3COH > R_2CHOH > RCH_2OH$$
Tertiary    Secondary    Primary

This is consistent with an $S_N1$ mechanism, with a carbenium ion intermediate. We saw in Chapter 26 that *hyperconjugation* (diagram **30.13**) contributes to the stabilizing effects of alkyl groups attached to the $C^+$ centre, making a tertiary carbenium ion more stable than a secondary carbenium ion, and a secondary carbenium ion more stable than a primary one.

## Cleaving RO–H bond: reactions with alkali metals

When a piece of sodium is placed in ethanol, $H_2$ is evolved in a reaction that is reminiscent of, but less violent than, that of Na with water: compare equations 30.29 and 30.30. The alcohol acts as a weak acid and forms a *sodium alkoxide* salt. The $pK_a$ of ethanol is 16, slightly higher than the $pK_a$ of water.

$$2EtOH + 2Na \longrightarrow \underset{\text{Sodium ethoxide}}{2Na^+EtO^-} + H_2 \qquad\qquad (30.29)$$

$$2H_2O + 2Na \longrightarrow \underset{\text{Sodium hydroxide}}{2Na^+OH^-} + H_2 \qquad\qquad (30.30)$$

Reactions of Na and alcohols are often used to destroy excess Na,[§] for example after the metal has been used as a drying agent for a hydrocarbon solvent. However, ***care is needed!*** There is a risk of fire owing to the production of $H_2$ in an inflammable alcohol medium. Choosing higher molecular weight alcohols reduces the fire hazard, but if the alcohol is too bulky, the reaction is very slow. Propan-2-ol is a suitable compromise between safety and rate of reaction (equation 30.31).

$$+ \ Na \ \longrightarrow \qquad + \ {}^1\!/_2\,H_2 \qquad (30.31)$$

Propan-2-ol (Isopropanol)    Sodium isopropoxide

## Oxidation

Although combustion is an oxidation process, more synthetically useful oxidations are those that convert alcohols to carbonyl compounds: *oxidation to aldehydes, ketones and carboxylic acids.* (Earlier in the chapter, we described the reduction of aldehydes, ketones and carboxylic acids as a method of

---

[§] For details of a safe way to destroy Na or K using water, see: H. W. Roesky (2001) *Inorganic Chemistry*, vol. 40, p. 6855.

preparing alcohols.) Although each of primary, secondary and tertiary alcohols burns to give $CO_2$ and $H_2O$, only primary and secondary alcohols are oxidized to carbonyl compounds as the following general scheme shows:

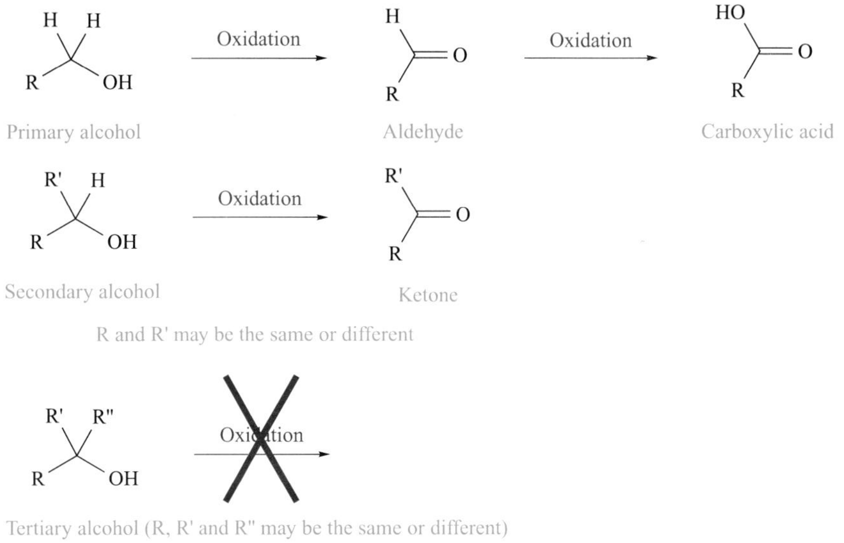

Primary alcohol      Aldehyde      Carboxylic acid

Secondary alcohol      Ketone

R and R' may be the same or different

Oxidation

Tertiary alcohol (R, R' and R" may be the same or different)

Suitable oxidizing agents for these reactions are $KMnO_4$, $Na_2Cr_2O_7$, $K_2Cr_2O_7$ and $CrO_3$, all under acidic conditions, e.g. reaction 30.32.

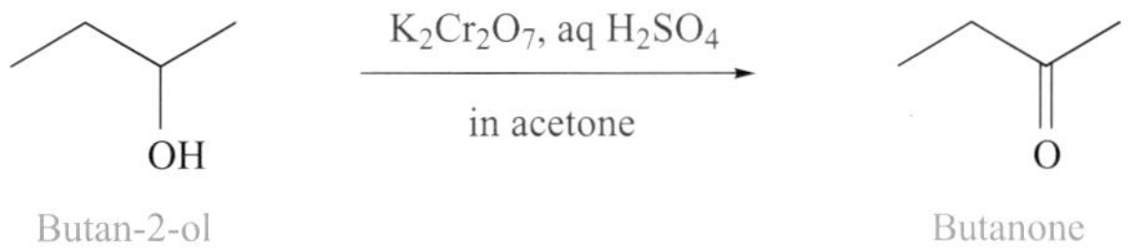

Butan-2-ol      Butanone      (30.32)

In this example, the product can only be butanone because the starting compound is a *secondary* alcohol. If the precursor had been butan-1-ol, initial oxidation would have produced the aldehyde butanal, and this has the potential for further oxidation to butanoic acid. The extent of oxidation can be controlled by careful choice of oxidizing agent. By using $KMnO_4$, $K_2Cr_2O_7$, $Na_2Cr_2O_7$ or $CrO_3$ in the presence of dilute sulfuric acid in acetone solution, a primary alcohol is converted to a carboxylic acid. Equation 30.33 shows the conversion of butan-1-ol to butanoic acid in a so-called *Jones oxidation*. During the reaction, Cr(VI) is reduced to Cr(III).

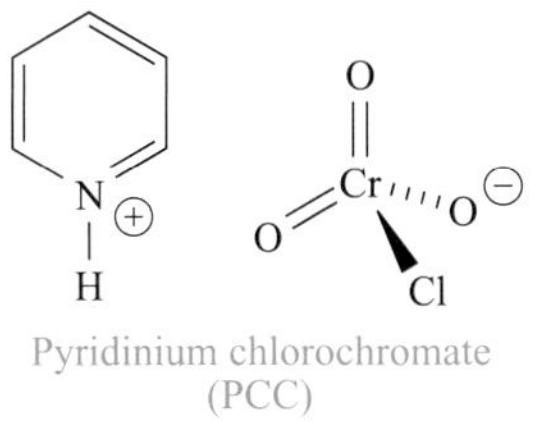

Pyridinium chlorochromate (PCC)

**(30.14)**

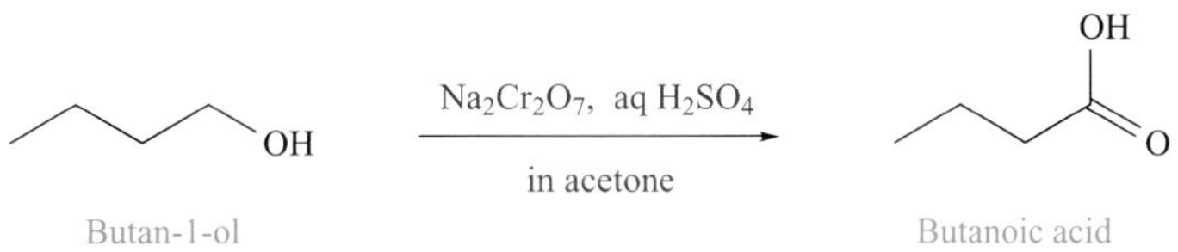

Butan-1-ol      Butanoic acid      (30.33)

If the aldehyde is the desired product, the most convenient oxidizing agent for laboratory use is pyridinium chlorochromate (PCC, **30.14**). Non-aqueous conditions must be used, and the reaction is usually carried out in $CH_2Cl_2$ solution, e.g. reaction 30.34.

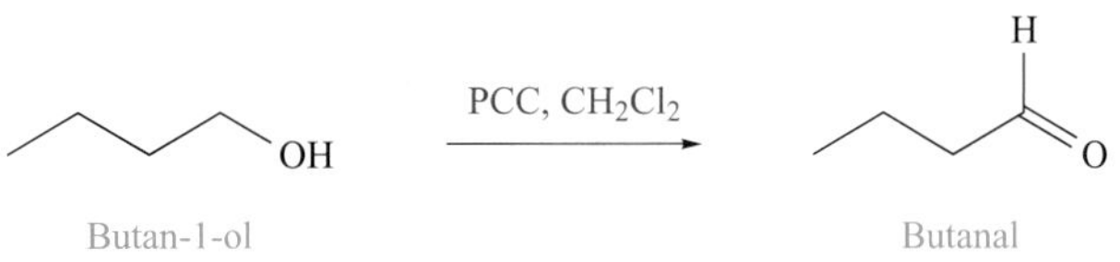

Butan-1-ol      Butanal      (30.34)

The mechanism of alcohol oxidation by $CrO_3$ and related species occurs as follows. Nucleophilic attack by the OH group at the Cr(VI) centre produces an intermediate that undergoes an elimination reaction (compare this with the E2 reactions in Section 28.7):

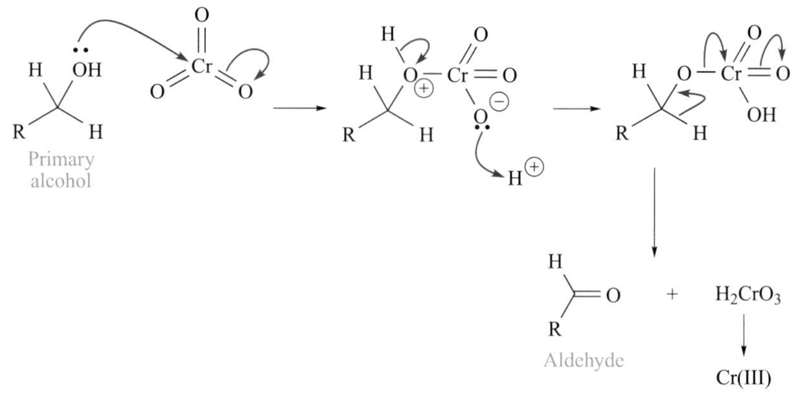

The aldehyde that forms may now be subject to further oxidation, depending on the reaction conditions. As we shall discuss in Chapter 33, the $C^{\delta+}$ atom of a carbonyl group is susceptible to attack by nucleophiles. As a consequence, an aldehyde in aqueous solution is in equilibrium with its hydrate, a *diol* (equation 30.35). The mechanism for this process is shown in scheme 30.36.

(30.35)

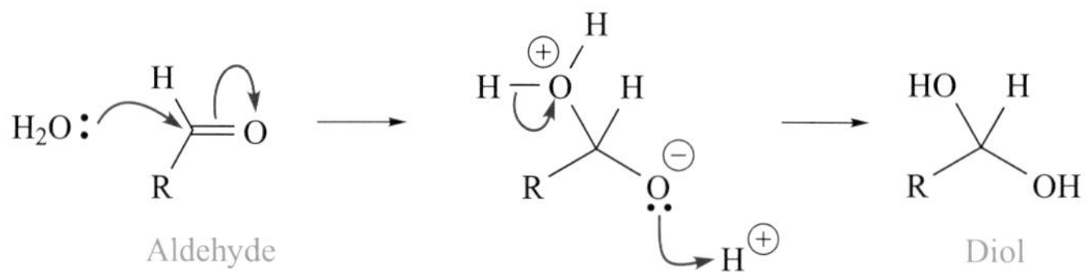

(30.36)

Thus, in aqueous solution and in the presence of a suitable oxidizing agent, the hydrate of an aldehyde undergoes oxidation to a carboxylic acid. The mechanism is analogous to that shown above for the conversion of an alcohol to an aldehyde.

## Ester formation

The reaction between an alcohol and a carboxylic acid or acyl chloride produces an *ester*. We shall look at this type of reaction in detail in Chapter 33. For now, the general reaction of an alcohol with a carboxylic acid is summarized in equation 30.37. Esters possess characteristic, often fruity, smells (see Box 33.2).

(30.37)

## 30.6 Protection of OH groups

We have already come across the idea of a *protecting group*: in Section 26.13, we described how trialkyl silyl groups such as $Me_3Si$ (TMS) can be used to protect terminal alkyne CH groups during reactions in which some other functional group in the molecule is undergoing reaction. Trialkyl silyl groups are one of several groups that are used to protect alcohol groups in cases where a reaction targeted at another functional group in the molecule would also affect the OH group, e.g. in syntheses involving sugars.

### Trialkylsilyl groups: TMS, TIPS and TBDMS

Trimethylsilyl chloride reacts with an alcohol in the presence of base as shown in equation 30.38. The base scavenges the HCl produced in the reaction.

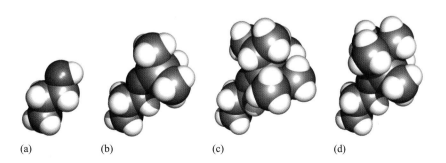

(30.38)

The trialkylsilyl derivatives are stable to base and oxidation. The steric protection offered by the TMS group can be appreciated by comparing the space-filling model of propan-1-ol (Figure 30.5a) with its TMS derivative (Figure 30.5b). The effectiveness of the trialkylsilyl protection is increased if alkyl groups that are more sterically demanding than methyl are used. Two common choices are the triisopropylsilyl (TIPS) and *tert*-butyldimethylsilyl (TBDMS) groups. The structures of triisopropylsilyl chloride (TIPSCl) and *tert*-butyldimethylsilyl chloride (TBDMSCl) are given in **30.15** and **30.16**. Figures 30.5c and 30.5d show space-filling diagrams of the TIPS and TBDMS derivatives of propan-1-ol and clearly show the 'umbrella of protection' offered by these substituents. Deprotection is achieved by treatment with $F^-$ (e.g. as $^nBu_4NF$) or $H^+$ in aqueous solution (equations 30.39 and 30.40).

TIPSCl

(30.15)

TBDMSCl

(30.16)

(30.39)

**Fig. 30.5** Space-filling diagrams of (a) propan-1-ol, $^nPrOH$, (b) $^nPrOTMS$, (c) $^nPrOTIPS$ and (d) $^nPrOTBDMS$. See text for definitions of TMS, TIPS and TBDMS. Colour code: C, grey; O, red; H, white; Si, purple.

(a)  (b)  (c)  (d)

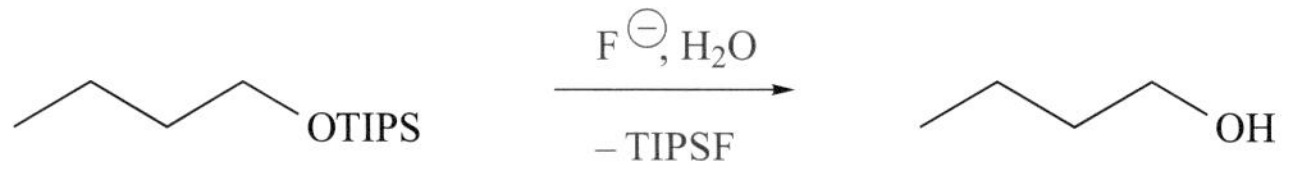

$$\text{OTIPS} \xrightarrow[\text{– TIPSF}]{F^{\ominus}, H_2O} \text{OH} \tag{30.40}$$

### Benzyl group, Bn

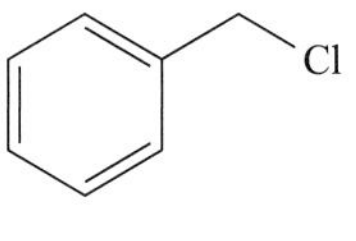

Benzyl chloride
(BnCl)

**(30.17)**

Alcohols are often protected by preparing a benzyl ether, BnOR, from the reaction of an alcohol ROH with benzyl chloride, BnCl (**30.17**), e.g. reaction 30.41.

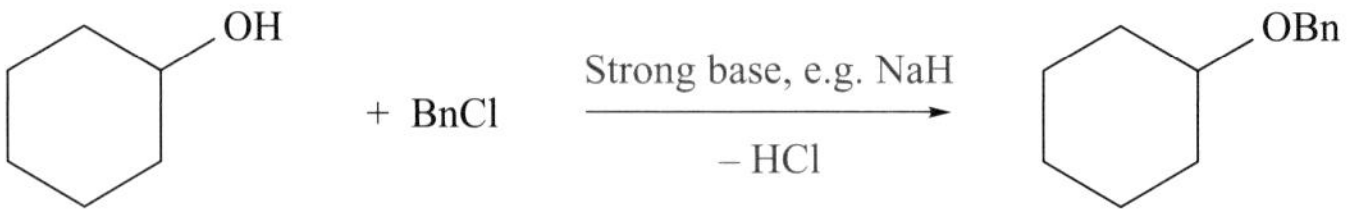

$$\text{OH} + \text{BnCl} \xrightarrow[\text{– HCl}]{\text{Strong base, e.g. NaH}} \text{OBn} \tag{30.41}$$

$$\text{OBn} = \text{(cyclohexyl-O-CH}_2\text{-phenyl)}$$

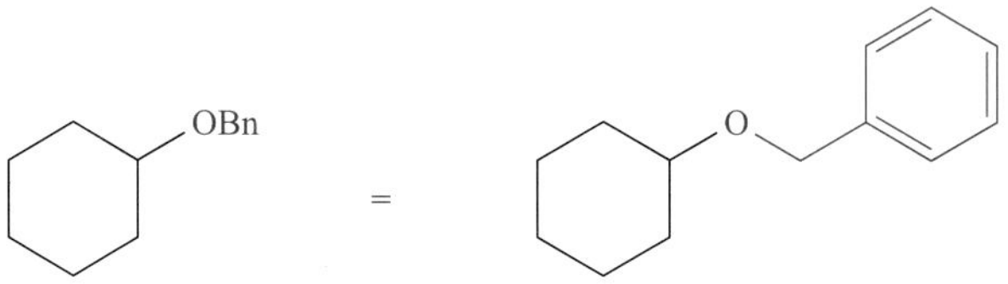

Benzyl ethers are stable to acid, base, oxidation and reduction, and the molecule can be deprotected by hydrogenation ($H_2$ with Pd catalyst) or treatment with HBr.

### Tetrahydropyranyl, THP

An *acetal* has the general formula:

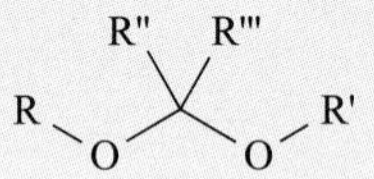

R''  R'''
   \  /
R—O—C—O—R'

The final method of protection that we mention is the use of an *acetal*. Treating an alcohol with dihydropyran and an acid catalyst gives a tetrahydropyran (THP) derivative (equation 30.42).

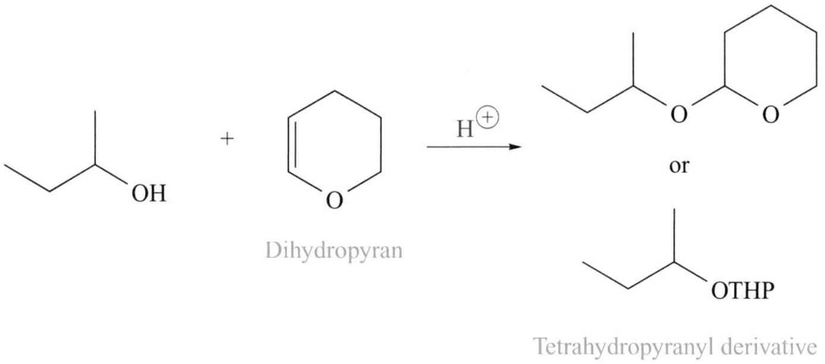

$$\text{OH} + \text{(dihydropyran)} \xrightarrow{H^{\oplus}} \text{or} \tag{30.42}$$

Dihydropyran

OTHP

Tetrahydropyranyl derivative

THP-protected alcohols are stable to attack by strong and weak bases. They are easily converted back to the respective alcohol by adding aqueous acid (equation 30.43).

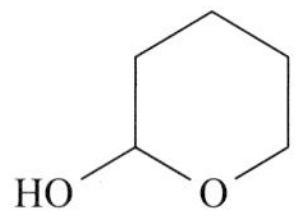

HO—O

THPOH

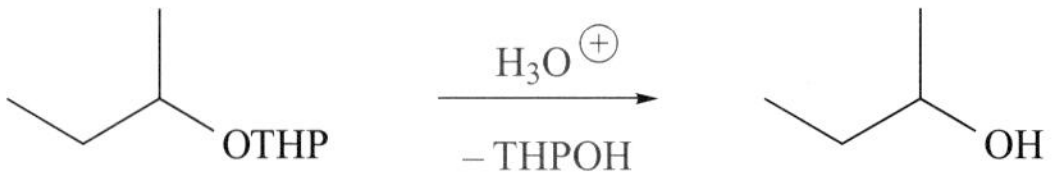

$$\text{OTHP} \xrightarrow[\text{– THPOH}]{H_3O^{\oplus}} \text{OH} \tag{30.43}$$

### Which protecting group?

The choice of an appropriate protecting group depends on the reaction conditions. For example, ROBn is stable to acids, but ROTHP is not. The differing steric demands of the alkyl groups in trialkylsilyl groups (Figure 30.5) leads to differing degrees of protection. TIPS offers the most effective protection against bases and nucleophiles.

Protecting groups play crucial roles, not only in the laboratory syntheses but also in commercial processes such as drug synthesis. An example is in the manufacture of the sweetener aspartame (see Box 35.3), the synthesis of which is a multi-step process. Aspartame contains a carboxyl OH group and a terminal $NH_2$ group, both of which must be protected during synthesis. The final deprotection step is:

Aspartame
(tradenames NutraSweet and Sanecta)

We see more protection in action during peptide synthesis in Section 35.5.

## SUMMARY

This chapter has introduced the chemistry of alcohols, ROH. Alcohols are important precursors in a range of chemical manufacturing processes, and also feature in many biological molecules, e.g. sugars. The physical properties reflect the ability of the OH group to form hydrogen bonds. Reactions of alcohols include those in which the RO−H bond is cleaved, the R−OH is broken, formation of alkoxides and oxidations to aldehydes, ketones and carboxylic acids, and they are important starting materials in the laboratory. If a molecule contains several functional groups, it may be necessary to protect the OH group by using protecting groups such as TMS, TIPS, TBDMS, THP or Bn.

### Do you know what the following terms mean?

- primary alcohol
- secondary alcohol
- tertiary alcohol
- diol

- absolute alcohol
- ethylene glycol
- glycerol
- hydrophilic

- hydrophobic
- alkoxide
- Jones oxidation
- protection of an OH group

### Some abbreviations have been used in this chapter. Do you know what they mean?

- Bn
- PCC

- TBDMS
- THP

- TIPS
- TMS

### You should be able:

- to recognize primary, secondary and tertiary alcohols
- to draw the structure of an aliphatic alcohol given its systematic name
- to give systematic names to simple, aliphatic (acyclic and cyclic) alcohols
- to give examples of alcohols that are important in commercial manufacturing processes
- to illustrate methods of making alcohols from carbonyl compounds
- to describe how alcohols can be prepared using Grignard reagents
- to discuss the effect that hydrogen bonding has on the physical properties of alcohols
- to give examples of alcohols behaving as Brønsted acids and bases
- to assign characteristic absorptions in IR spectra of alcohols

- to assign $^1$H NMR spectra of simple alcohols
- to write an equation for the combustion of a given alcohol
- to write an account of dehydration reactions of alcohols to give alkenes, and contrast them with intermolecular dehydrations to give ethers
- to illustrate the conversion of alcohols to halogenoalkanes
- to describe how alkali metals react with alcohols and comment on applications and hazards of this reaction
- to give an account of the oxidations of alcohols to aldehydes, ketones and carboxylic acids
- to write an equation for the conversion of an alcohol to an ester
- to appreciate the needs for protecting an OH group in certain reactions, and to give examples of the use of suitable protecting groups

## PROBLEMS

**30.1** Draw the structures of (a) pentan-1-ol;
(b) heptan-3-ol; (c) 2-methylpentan-2-ol;
(d) propane-1,2,3-triol.

**30.2** Name the following alcohols:

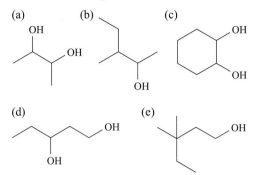

**30.3** For which of the following alcohols are
stereoisomers possible?

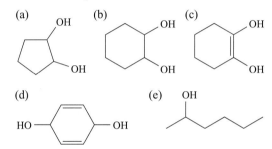

**30.4** Suggest ways of carrying out the following
transformations. Comment on any competitive
reactions that may affect the yield of alcohol, and
on the stereochemistry of the products where
appropriate.

**30.5** Write a short account of the use of hydroboration
in the synthesis of alcohols.

**30.6** Suggest starting materials for the preparation of
the following alcohols using a Grignard reagent as
one precursor. In each case, state how you might
make the Grignard reagent: (a) 2-methylpropan-2-
ol; (b) heptan-1-ol; (c) pentan-2-ol.

**30.7** What would you expect to be the organic products
of the following reactions?

(a)

(b)

(c)

**30.8** How does hydrogen bonding affect (a) the boiling
points of alcohols, (b) the solid state structure of
methanol and (c) the value of $\Delta_{vap}S$ for methanol?

**30.9** Suggest reasons for the following trend in $pK_a$
values: EtOH, $pK_a = 16.0$; Me$_3$COH, $pK_a = 18.0$;
CF$_3$CH$_2$OH, $pK_a = 12.4$; (CF$_3$)$_3$COH, $pK_a = 5.4$.

**30.10** An alcohol **X** has a composition of 64.8% C and
13.6% H. The mass spectrum shows a parent ion at
$m/z = 74$. The $^1$H NMR spectrum of **X** dissolved
in CDCl$_3$ has signals at $\delta$ 3.71 (sextet, 1H), 2.37
(singlet, 1H), 1.46 (multiplet, 2H), 1.17 (doublet,
3H), 0.93 (triplet, 3H) ppm; in the $^{13}$C NMR
spectrum, four resonances are observed. Use these
data to suggest a structure of **X** and comment on
isomer possibilities that retain the OH
functionality.

**30.11** What spectroscopic technique(s) would you choose
to distinguish between (a) propan-1-ol and
propan-2-ol; (b) diethyl ether and butan-1-ol;
(c) cyclohexanol and hexan-1-ol?

**30.12** Sodium wire can be used to dry hexane solvent.
(a) Write an equation for the reaction that occurs.
(b) How might you destroy excess Na and what
precautions should you take?

**30.13** Suggest a mechanism for the following reaction:

**30.14** Suggest the identities of the major products in the
acid-catalysed dehydrations of (a) butan-2-ol, (b) 2-
methylbutan-1-ol, (c) pentan-1-ol.

**30.15** Suggest products in the following reactions:

(a)

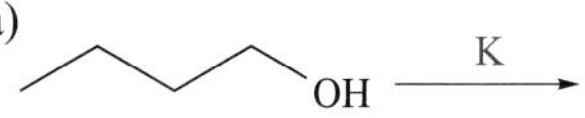

$$\text{CH}_3\text{CH}_2\text{CH}_2\text{CH}_2\text{OH} \xrightarrow{\text{K}}$$

(b)

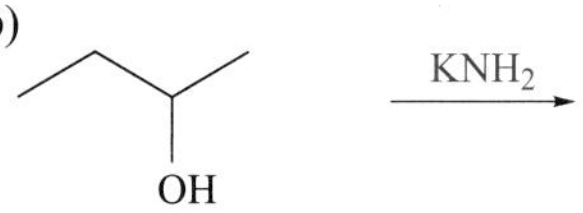

$$\xrightarrow{\text{KNH}_2}$$

(c)

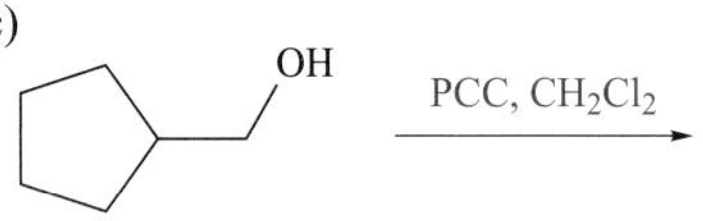

$$\xrightarrow{\text{PCC, CH}_2\text{Cl}_2}$$

(d)

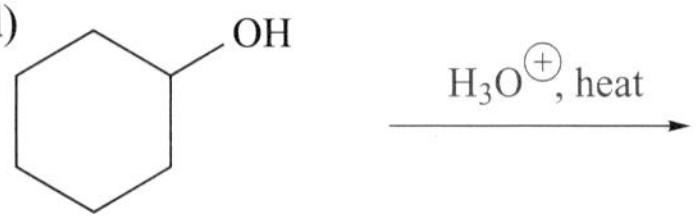

$$\xrightarrow{\text{H}_3\text{O}^{\oplus}, \text{ heat}}$$

(e)

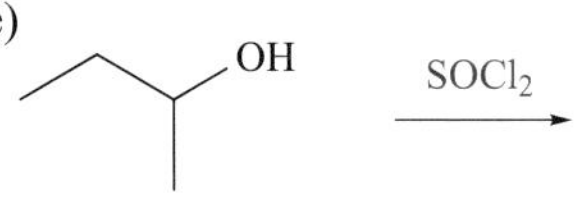

$$\xrightarrow{\text{SOCl}_2}$$

(f)

$$\xrightarrow{\text{conc HCl; 298 K}}$$

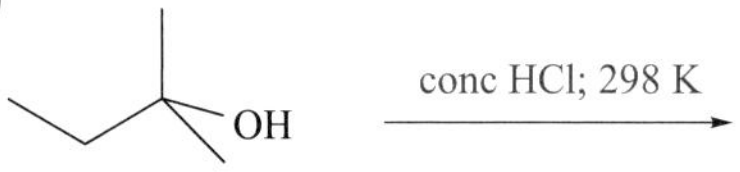

**30.16** Suggest what alcohol precursor and reaction conditions you might use to obtain the following products after oxidation:

(a)      (b) 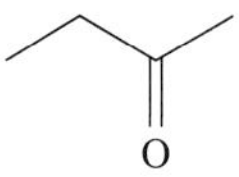      (c)

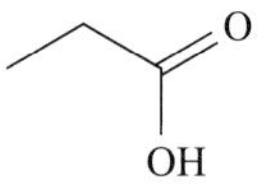

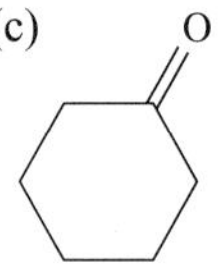

(d) 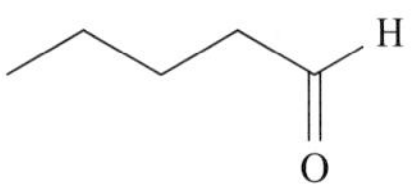      (e)

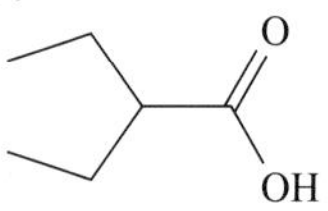

**30.17** When we discussed oxidation of alcohols in Section 30.5, we stated that the mechanism of oxidation of an aldehyde to carboxylic acid is analogous to that of the conversion of an alcohol to an aldehyde. Propose a mechanism for the oxidation of RCHO to $RCO_2H$ using $KMnO_4$ in acidic aqueous solution.

**30.18** (a) Summarize methods of protecting OH groups, stating differences in the conditions under which they work. How are the compounds deprotected?

(b) Suggest possible *disadvantages* to using protecting groups during a synthesis. (c) Write a mechanism for the deprotection of ROTMS using $^n\text{Bu}_4\text{NF}$.

## Additional problems

**30.19** Suggest explanations for the following.
(a) The $^1$H NMR spectrum of $CF_3CH_2OH$ contains a quartet ($J$ 9 Hz) at $\delta$ +3.9 ppm in addition to the signal assigned to the OH proton.
(b) The addition of $D_2O$ (D = $^2$H) to hexanol causes the disappearance of the signal assigned to the OH proton.
(c) Whereas alcohols exhibit relatively high boiling points and enthalpies of vaporization, the same is not true of thiols, RSH, e.g. propan-1-ol, bp = 370.2 K, $\Delta_{vap}H(bp) = 41.4$ kJ mol$^{-1}$; propane-1-thiol, bp = 340.8 K, $\Delta_{vap}H(bp) = 29.5$ kJ mol$^{-1}$.

**30.20** (a) Explain why the displacement of OH in an alcohol by Br is carried out under acidic conditions.
(b) Suggest products at each stage in the following reaction scheme:

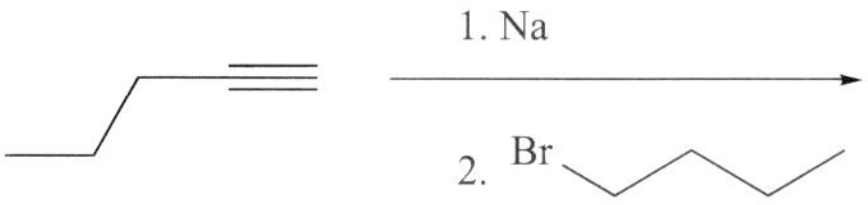

Why does the OH group need to be protected before the following reaction is carried out?

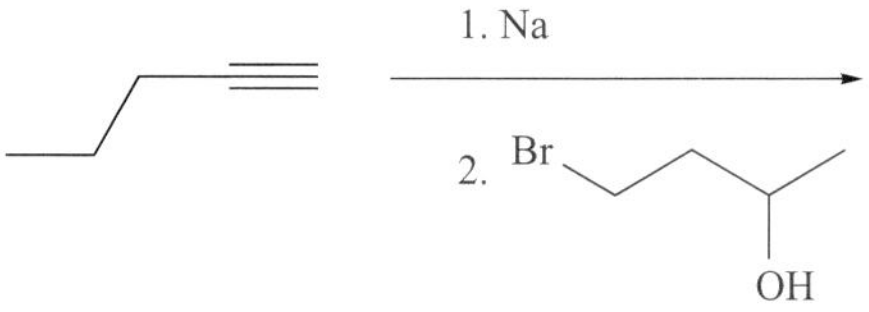

(c) Suggest how you would prepare the following $^{13}$C-labelled compound (● = $^{13}$C) starting from the precursor shown below:

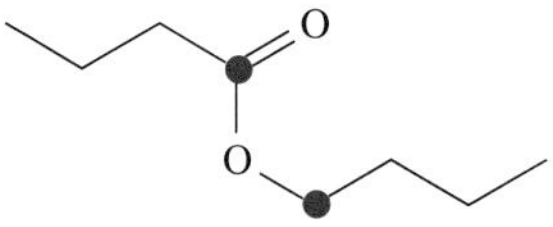

Precursor

# 31 Amines

## 31.1    Structure and nomenclature

The functional group of a primary amine is an $NH_2$ group, and the general formula of an aliphatic amine is $RNH_2$. The compound is derived from ammonia (**31.1**) by replacement of one H atom by an alkyl group, R (structure **31.2**). The trigonal pyramidal arrangement of atoms attached to the N atom in **31.1** and **31.2** is a consequence of the lone pair of electrons on the N atom (see Chapter 6). Further replacement of H atoms by R groups gives *secondary* and *tertiary amines* as in structures **31.3** and **31.4** in which the R groups may be the same or different.

The systematic name of a straight chain primary amine is derived by suffixing the root name of the alkane with 'amine'. Numerical descriptors indicate the position of the $NH_2$ group. In a primary amine, *the $NH_2$ group is not necessarily at the end of the carbon chain*. Compounds **31.5**–**31.8** are examples of primary amines. When there are alkyl substituents (e.g. methyl), the numbering of the carbon chain is such that the $NH_2$ group takes the highest priority.

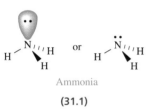

Ammonia

(31.1)

Primary amine

(31.2)

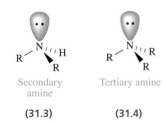

Secondary amine

(31.3)

Tertiary amine

(31.4)

1-Butylamine

(31.5)

3-Pentylamine

(31.6)

2-Butylamine

(31.7)

4-Methyl-2-pentylamine

(31.8)

Primary amine derivatives of cycloalkanes (**31.9** and **31.10**) are similarly named, again with the amine group taking priority over alkyl substituents. Primary amines containing more than one $NH_2$ group are named as in example **31.11**; note the retention of the -e in the alkane part of the name.

Secondary and tertiary amines may be symmetrical or unsymmetrical depending on whether the alkyl groups are the same or different. *Symmetrical* secondary and tertiary amines are named by adding the prefix *di-* or *tri-* before

Older names (still in use) for 1,2-ethanediamine are 1,2-diaminoethane and ethylene diamine

Cyclohexylamine
(31.9)

3,5-Dimethylcyclohexylamine
(31.10)

1,2-Ethanediamine
(31.11)

the name of the alkyl group, for example diethylamine (**31.12**, a secondary amine) and tributylamine (**31.13**, a tertiary amine). In an *unsymmetrical* secondary or tertiary amine, the root name is chosen by finding the largest substituent attached to the N atom; smaller chains are classed as *N*-bound substituents. For example, compound **31.14** has one propyl and two methyl substituents attached to the same N atom. The propyl group is the parent chain, and the methyl groups are classed as substituents on the N atom. The compound is named *N,N*-dimethylpropylamine.

Diethylamine
(31.12)

Tributylamine
(31.13)

*N,N*-Dimethylpropylamine
(31.14)

Secondary amines can also be named as *aza derivatives of alkanes*. Just as an aliphatic ether can be considered to be derived from an alkane with a $CH_2$ group replaced by an isoelectronic O atom, an amine can be derived from an alkane by replacement of $CH_2$ by an isoelectronic NH. The amine is named by prefixing the alkane root name with *aza* accompanied by numerical descriptors to show the position of the NH group or groups. This nomenclature is used particularly for compounds containing several amine functionalities, e.g. **31.15** and **31.16**.

3,5,7-Triazanonane
(31.15)

1,4,7-Triazacyclononane
(31.16)

You will often encounter *tetraalkylammonium salts* in the laboratory, especially in inorganic synthesis. These contain *quaternary ammonium ions* of general formula $[R_4N]^+$. Their names show the alkyl substituents attached to the central N atom as in **31.17**. A shorthand way of representing ion **31.17** is $Et_4N^+$. Similarly, $Et_3NH^+$ is triethylammonium ion, $^nBu_4N^+$ is tetrabutyl-ammonium ion and $Et_2NH_2^+$ is diethylammonium ion. These ions are synthetically useful because their salts are often soluble in organic solvents such as $CH_2Cl_2$.

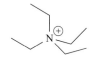

Tetraethylammonium ion

(31.17)

## ENVIRONMENT AND BIOLOGY

## Box 31.1 Amines in biology

Many biologically active compounds contain amine groups and in this box we highlight a few examples. The smell of rotting fish arises from release of $MeNH_2$, and decaying animal flesh produces a foul odour owing to release of 1,4-butanediamine (with the common name of *putrescine*) and 1,5-pentanediamine (*cadaverine*).

1,4-Butanediamine

1,5-Pentanediamine

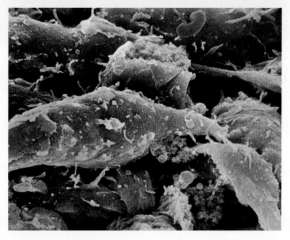

Coloured scanning electron micrograph (SEM) of ganglion nerve cells in adrenal gland.
*Prof. P. Motta/Dept. of Anatomy/University 'La Sapienza', Rome/Science Photo Library.*

The structures of *adrenaline* and *noradrenaline* also feature amine functionalities. Adrenaline and noradrenaline are secreted by the adrenal glands. The photograph opposite shows ganglion nerve cells in the medulla of the adrenal gland. The cells have terminal nerve fibres. The medulla region of the adrenal gland produces adrenaline and noradrenaline. The body uses noradrenaline as a precursor to adrenaline. Adrenaline increases the rate at which the heart beats and increases blood pressure. Noradrenaline is also secreted at sympathetic nerve endings.

R = Me    Adrenaline (epinephrine)
R = H     Noradrenaline

The drugs *amphetamine* and *amitriptyline* are primary and tertiary amines respectively. Amphetamine is a stimulant and vasoconstrictor, but it is also addictive. Amitriptyline is used in the treatment of depression.

Amphetamine          Amitriptyline

*L-Lysine* (see Table 35.1) is a naturally occurring amino acid that has an amino-functional group. Although L-lysine is essential for growth of the body, mammals cannot synthesize it and rely on dietary intake.

L-Lysine

Two further amines of biological significance are *pyridoxamine* and *glucosamine*. Pyridoxamine belongs to the vitamin $B_6$ group; the phosphate derivative acts as a coenzyme. Glucosamine is related to glucose (see Chapter 35) and is sold for the treatment of osteoarthritis; research into this application of glucosamine continues. Glucosamine building blocks are present in chitin in the skeletons of invertebrates.

Pyridoxamine          β-D-Glucosamine

Isostructural and isoelectronic: see Sections 5.12 and 7.2, and Figure 7.2

Whereas an amine contains a trigonal pyramidal N atom, a quaternary ammonium ion contains a tetrahedral N centre: $N^+$ is isoelectronic with C, and so $Me_4N^+$ is isoelectronic with $Me_4C$. The two species are isostructural.

## 31.2    Inversion at nitrogen in amines

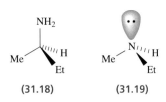

(31.18)       (31.19)

In Section 24.8, we introduced *asymmetric C atoms*. Such an atom (e.g. the central C atom in **31.18**) has four different groups attached to it. A compound containing an asymmetric C atom is chiral. A trigonal pyramidal N atom in an amine has only three groups attached to it, but it also has a lone pair. Comparison of **31.18** and **31.19** implies that **31.19** is chiral. Following from this theory, Figure 31.1 shows the two enantiomers of EtMeNH, i.e. a pair of stereoisomers, the structures of which are not superposable on their mirror images. Towards the end of Section 24.8, we discussed ways of resolving chiral compounds. *In the case of chiral amines, resolution is not usually possible because of facile inversion at the N atom that has the effect of interconverting the two enantiomers (Figure 31.2). Inversion at the N atom in most amines has a low energy barrier.*

**Fig. 31.1** When a molecule of EtMeNH 'looks' in a mirror, the mirror image is non-superposable on the original molecule. Thus, in theory, EtMeNH has enantiomers, although these *cannot* be resolved. Colour code: C, grey; N, blue; H, white.

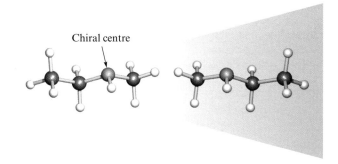

Chiral centre

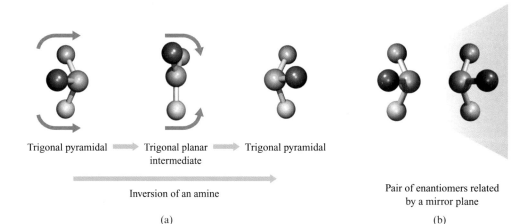

Trigonal pyramidal ⟹ Trigonal planar intermediate ⟹ Trigonal pyramidal

Inversion of an amine

Pair of enantiomers related by a mirror plane

(a)                    (b)

**Fig. 31.2** (a) Inversion of an amine; three different groups attached to the N atom (shown in blue) are represented as different coloured spheres. (b) Enantiomers of the same amine. Comparison of (a) and (b) shows that inversion leads to the interconversion of the enantiomers. Inversion is facile and the enantiomers *cannot* be resolved.

Nucleophilic substitution:
see Section 28.6

**31.3     Synthesis**

### Conversion of halogenoalkanes to amines by nucleophilic substitution

Halogenoalkanes can be converted to amines by *nucleophilic substitution* using $NH_3$. Ammonia is a good nucleophile and readily displaces a halide ion as in reaction 31.1.

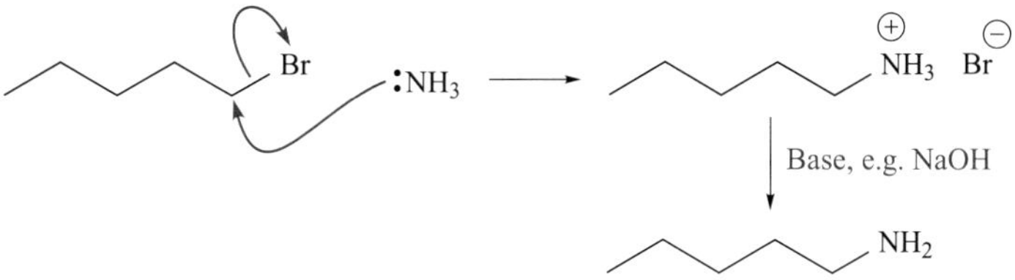

(31.1)

Although this reaction may appear to be a good method of preparing a primary amine, it has an inherent problem: the primary amine is also a good nucleophile and can attack another molecule of halogenoalkane. This leads to the formation of a secondary amine (equation 31.2).

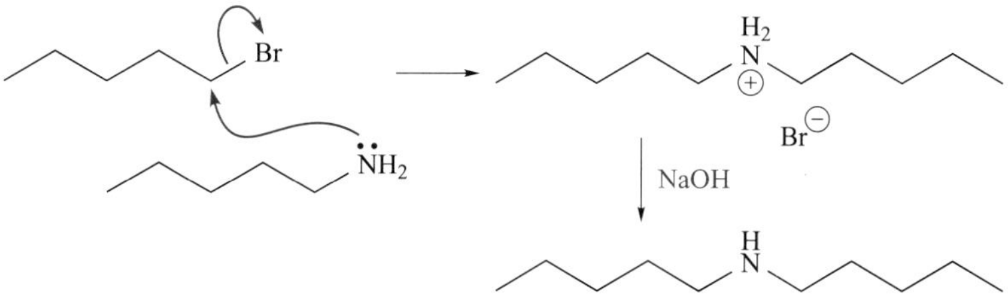

(31.2)

The reaction does not stop here. The secondary amine also behaves as a good nucleophile, and reacts with the halogenoalkane to give a tertiary amine (equation 31.3). Further reaction to give a quaternary ammonium ion then follows (equation 31.4).

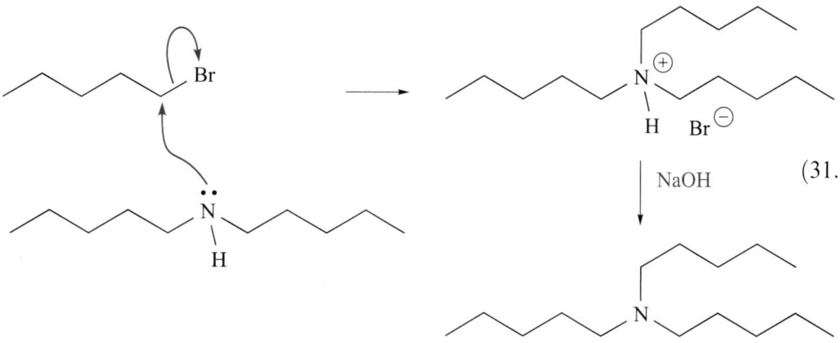

(31.3)

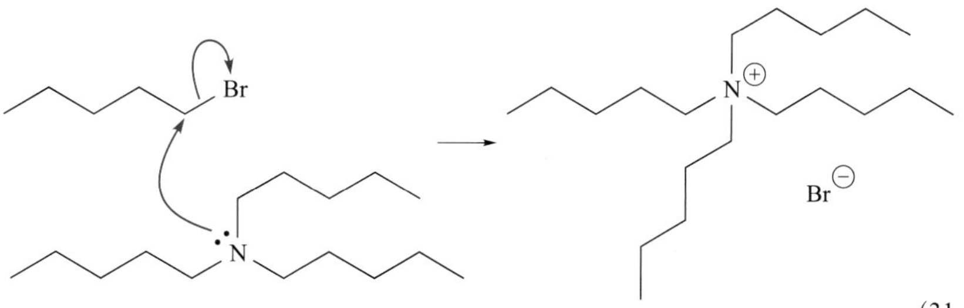

(31.4)

The method is *non-specific*: it gives mixtures of primary, secondary and tertiary amines and quaternary ammonium salts. Thus, the synthesis of a given amine from a halogenoalkane is not usually a convenient route.

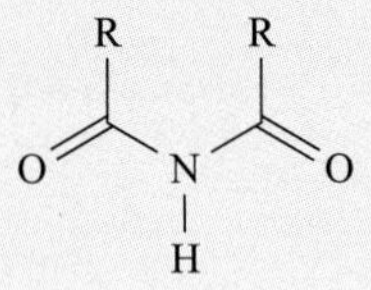

## Conversion of halogenoalkanes to amines by the Gabriel synthesis

An *imide* contains the functional group:

Instead of using $NH_3$ or an amine to introduce an amine group, an alternative method is to use potassium phthalimide in a reaction called the *Gabriel synthesis*. This *specifically yields a primary amine*. Potassium phthalimide is the salt of an imide (equation 31.5) and reacts readily with halogenoalkanes to give an alkylated derivative of the imide (equation 31.6). An alternative to hydrazine in the last step is aqueous alkali.

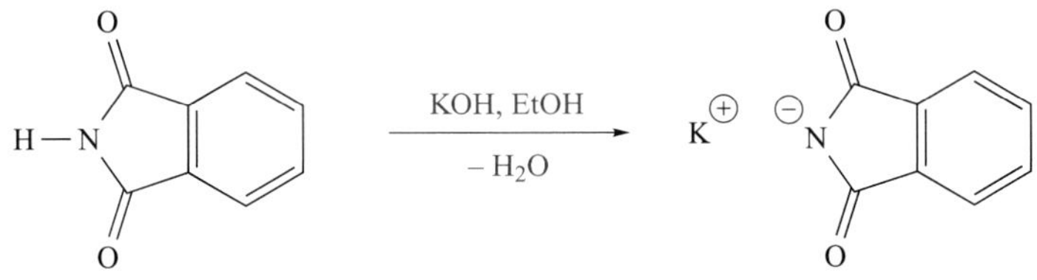

(31.5)

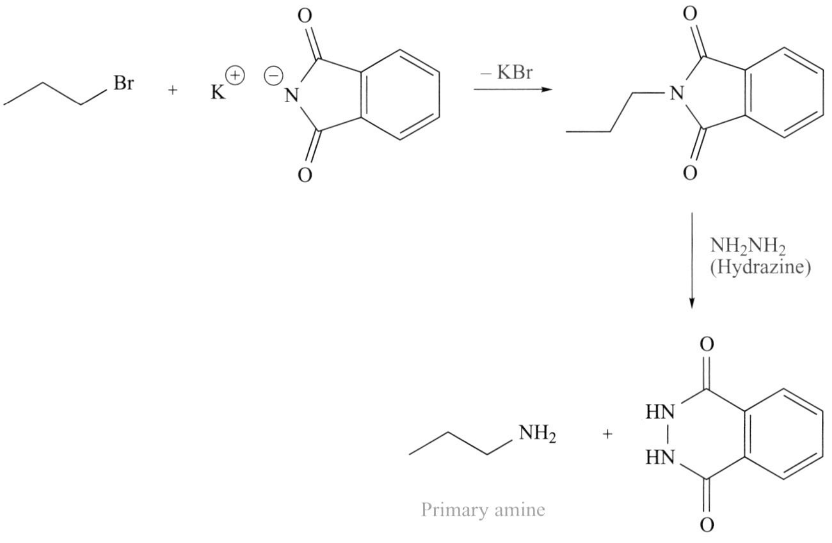

Potassium phthalimide

(31.6)

Primary amine

## Conversion of nitriles to amines

An organic *nitrile* contains the functional group:

$$R - C \equiv N$$

Primary amines can be prepared by the reduction of nitriles using lithium aluminium hydride as the reducing agent (equation 31.7). The C atom that was in the cyano functional group becomes part of the aliphatic chain in the amine. This has the effect of lengthening the carbon chain during the reaction. Equation 31.8 shows the preparation of butylamine from cyanopropane.

(31.7)

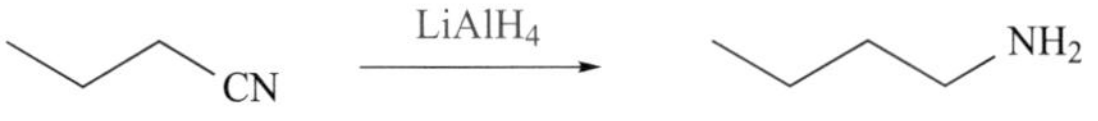

(31.8)

Since cyanide ion is a good nucleophile, reactions of type 31.7 can be coupled with the nucleophilic substitution of halide by cyanide to give a two-step synthesis of an amine from a halogenoalkane. Reaction 31.9 shows the transformation of bromoethane into propylamine.

$$(31.9)$$

The reduction of nitriles produces only primary amines. This includes the formation of $NH_2$ groups attached to terminal C atoms (e.g. equation 31.9), or C atoms in the middle of a chain (reaction 31.10).

$$(31.10)$$

*Caution!* Azides are explosive

Azide ion, $N_3^-$, can be used as an alternative to cyanide as a nucleophile. The azide derivative $RN_3$ is readily reduced by $LiAlH_4$ to the corresponding amine $RNH_2$. Note that, in contrast to the use of cyanide as the nucleophile, use of azide does *not* lead to an increase in the carbon chain.

## Reductive amination of aldehydes and ketones

*Reductive amination of an aldehyde using $NH_3$ gives a primary amine with the $NH_2$ group attached to a terminal C atom (i.e. to a primary alkyl group).*
*Reductive amination of a ketone using $NH_3$ gives a primary amine with the $NH_2$ group attached to a secondary alkyl group.*

Primary amines in which the $NH_2$ group is attached to the terminal C atom of an aliphatic chain can be prepared by the *reductive amination of aldehydes*. The aminating agent is ammonia, and a reducing agent used on a commercial scale is $H_2$ in the presence of a nickel catalyst. Reaction 31.11 shows the reductive amination of pentanal.

$$(31.11)$$

The mechanism of reductive amination involves nucleophilic attack by $NH_3$ on the $C^{\delta+}$ of the carbonyl ($C=O$) group. We look at this mechanism in detail in Section 33.14. Ketones are structurally related to aldehydes and also undergo reductive amination. Equation 31.12 shows that, starting from hexan-3-one, the product is 3-hexylamine.

$$(31.12)$$

In both reactions 31.11 and 31.12, the products are primary amines, but:

- if the precursors are $NH_3$ and an aldehyde, the product is $RCH_2NH_2$ and contains the $NH_2$ group attached to a primary alkyl group;
- if the precursors are $NH_3$ and a ketone, the product is $R_2CHNH_2$ or $RR'CHNH_2$, and the $NH_2$ group is attached to a secondary alkyl group.

The $NH_2$ group always ends up in the position originally occupied by the $C=O$ group.

When the *aminating agent is a primary amine*, the *product is a secondary amine*. Similarly, when the *aminating agent is a secondary amine*, the *product is a tertiary amine*. Examples are shown in reactions 31.13 and 31.14.

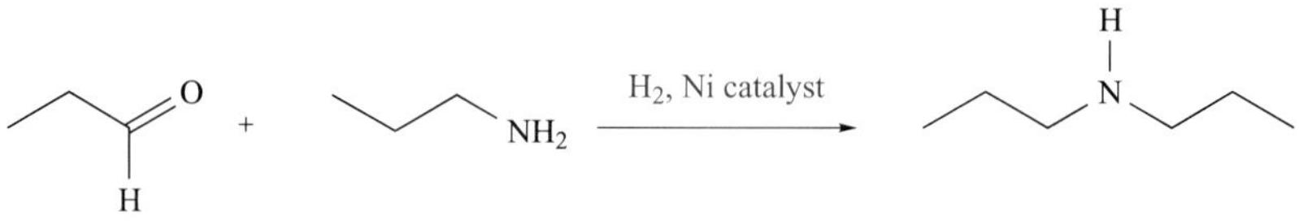

Propanal                Propylamine                              Dipropylamine
                        (Aminating agent)

$$(31.13)$$

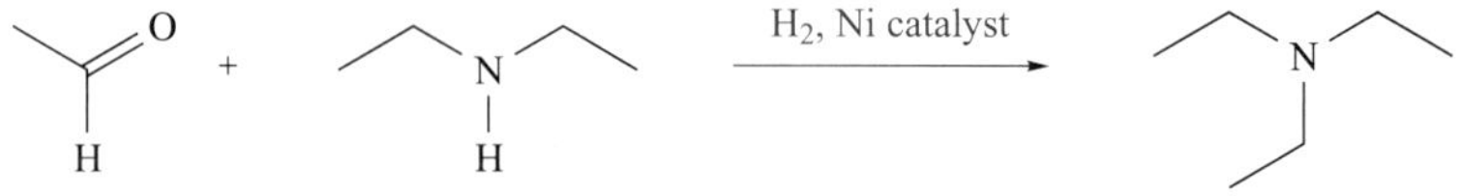

Ethanal                Diethylamine                              Triethylamine
                       (Aminating agent)

$$(31.14)$$

Each of the reductive amination reactions gives a specific product, and this method of preparing primary, secondary or tertiary amines is superior to the nucleophilic substitution reactions described in reactions 31.1–31.3.

Although high pressures of $H_2$ are used commercially, it is more convenient to use $NaBH_3CN$ (sodium cyanoborohydride) in the laboratory as in reaction 31.15. ($NaBH_4$ is not suitable as the reducing agent as it would reduce the starting aldehyde.)

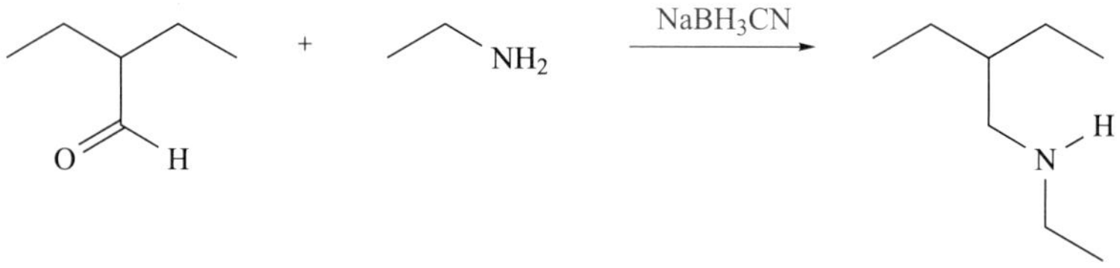

$$(31.15)$$

The drug amphetamine (see Box 31.1) provides an example of a commercial application of reductive amination of a ketone:

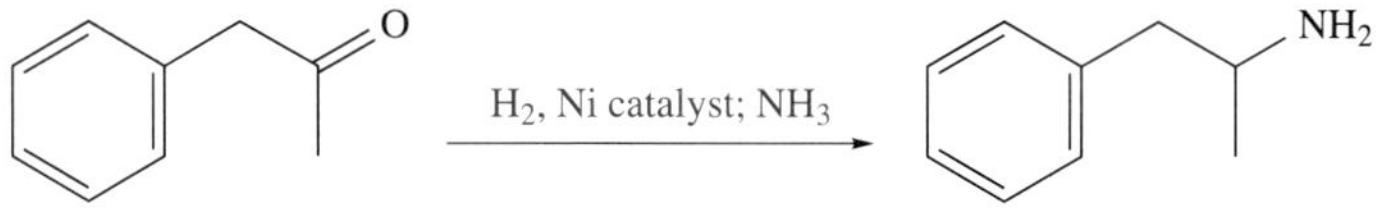

Amphetamine

Because it is so addictive, medical uses of amphetamine are limited.

## 31.4    Physical properties of amines

### Boiling points and enthalpies of vaporization

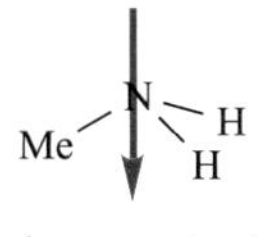

Like $NH_3$, low molecular weight amines possess characteristic smells often described as 'fishy'. Amines are polar molecules. Structure **31.20** shows the approximate direction of the dipole moment in $MeNH_2$. The presence of N–H bonds in primary and secondary amines means that some of their physical properties contrast with those of tertiary amines. In the neat

Dipole moment = 1.31 D

**(31.20)**

**Table 31.1** Boiling points and enthalpies of vaporization of selected amines. Values of $NH_3$ are included for comparison.

| Compound | Structure | Relative molecular mass | Boiling point / K | $\Delta_{vap}H(bp)$ / kJ mol$^{-1}$ |
|---|---|---|---|---|
| Ammonia | | 17.0 | 239.7 | 23.3 |
| Methylamine | | 31.1 | 266.7 | 25.6 |
| Dimethylamine | | 45.1 | 279.9 | 26.4 |
| 1-Propylamine | | 59.1 | 320.2 | 29.6 |
| 2-Propylamine (isopropylamine) | | 59.1 | 304.8 | 27.8 |
| 1-Butylamine | | 73.1 | 350.0 | 31.8 |
| 2-Butylamine (*sec*-butylamine) | | 73.1 | 335.7 | 29.9 |
| 2-Methyl-1-propylamine (isobutylamine) | | 73.1 | 340.7 | 30.6 |
| 2-Methyl-2-propylamine (*tert*-butylamine) | | 73.1 | 317.0 | 28.3 |
| Diethylamine | | 73.1 | 328.5 | 29.1 |
| 1-Hexylamine | | 101.2 | 405.8 | 36.5 |
| Triethylamine | | 101.2 | 362.0 | 31.0 |

liquid and solid state, primary and secondary amines form intermolecular hydrogen bonds, but this is not generally true of tertiary amines. Table 31.1 lists boiling points and values of $\Delta_{vap}H(bp)$ for selected amines. Some points to note are as follows.

- There is a general increase in values with increasing molecular weight for the series of straight chain primary amines ($MeNH_2$, $EtNH_2$, $^nPrNH_2 \ldots$).
- For a series of primary amines of constant molecular weight but different degrees of chain branching, the boiling point and enthalpy of vaporization are highest for the straight chain member of the family. Look at the trend for $^nBuNH_2$, $^{sec}BuNH_2$, $^{iso}BuNH_2$, $^tBuNH_2$ and consider the intermolecular interactions between the molecules.
- There is a significant difference between the boiling points and values of $\Delta_{vap}H$ for 1-hexylamine (a primary amine) and triethylamine (a tertiary amine) despite the fact that they have the same molecular weight.

Intermolecular interactions between alkane chains: see Section 25.3

The Pauling electronegativity values of N and H are 3.0 and 2.2 respectively, and intermolecular hydrogen bonds form between amine molecules containing N−H bonds. The strengths of these hydrogen bonds are less than those between alcohol molecules (see Chapter 30). Structure **31.21** shows hydrogen bonding between molecules of diethylamine. A comparison of the boiling points of hexane (341.7 K) and 1-hexylamine (405.8 K) illustrates how the introduction of the $NH_2$ group also introduces the effects of hydrogen bonding. The difference in boiling points between hexane and 1-hexylamine also reflects an increase in molecular weight. A more realistic feel for the contributions made by hydrogen bonding can be made by comparing the boiling points of hexane (bp 341.7 K, $M_r$ 86.2) and 1-pentylamine (bp 377.3 K, $M_r$ 87.2).

Hydrogen bonding: see Section 21.8

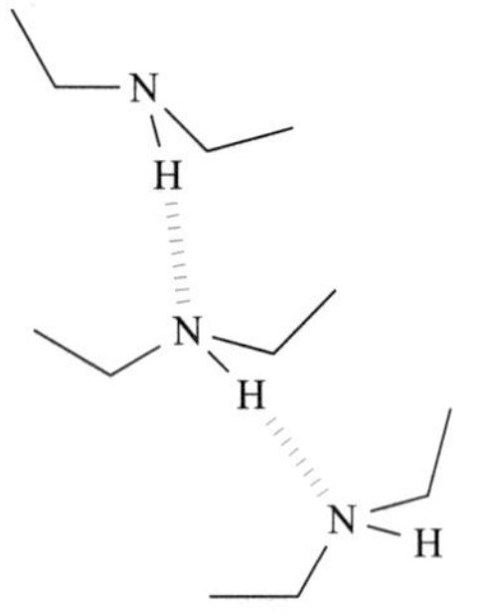

(31.21)

## Solubility in water

Amines form hydrogen bonds with water molecules. Primary, secondary and tertiary amines may interact as shown in structure **31.22**. In addition, primary and secondary amines can form hydrogen bonds of the type shown in diagram **31.23**. The structures on the right-hand side of the page remind you about the presence of lone pairs of electrons in amine and water molecules.

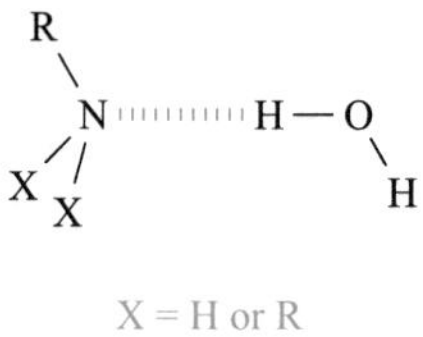

X = H or R

(31.22)

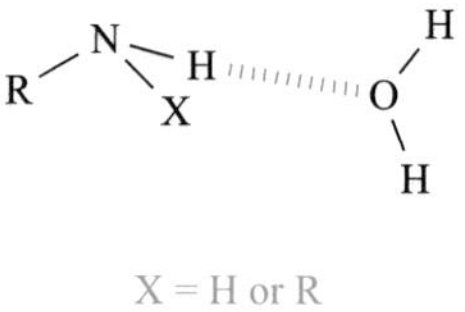

X = H or R

(31.23)

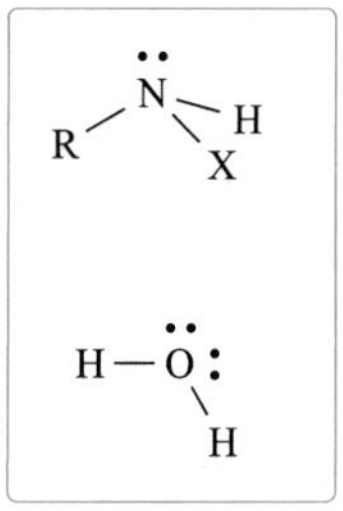

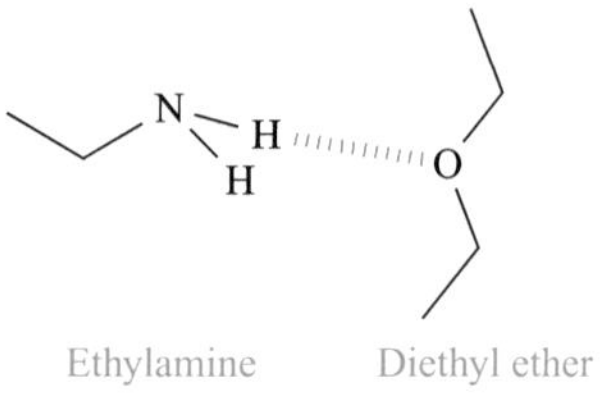

It follows from these interactions that low molecular weight amines are water soluble. They are also miscible with polar solvents such as ethers and alcohols, because of the formation of hydrogen bonds as exemplified in structure **31.24**.

Ethylamine    Diethyl ether

(31.24)

## Amines as bases

Like $NH_3$, amines behave as weak Brønsted bases and values of $pK_a$ and $pK_b$ (equation 31.16) for selected amines are listed in Table 31.2. In Chapter 16, we considered the relationship between $pK_b$ for a base B, and the value of $pK_a$ for its conjugate acid $BH^+$ in aqueous solution (equation 31.17).

$$\left. \begin{array}{l} pK_a = -\log K_a \\ pK_b = -\log K_b \end{array} \right\} \tag{31.16}$$

$$pK_a + pK_b = 14.00 \tag{31.17}$$

For $NH_3$, $K_b$ refers to equilibrium 31.18, and $K_a$ refers to equilibrium 31.19. Similarly, for an amine $RNH_2$, $K_b$ and $K_a$ apply to equilibria 31.20 and 31.21 respectively. Take care when looking up values of equilibrium constants; some tables quote $K_b$ values for amines, while others quote $K_a$

**Table 31.2** $pK_b$ values for selected amines and $pK_a$ values for their conjugate acids. Ammonia is included for comparison.

| Amine | Structure | $pK_a$ | $pK_b$ |
|---|---|---|---|
| Ammonia | | 9.25 | 4.75 |
| Methylamine | | 10.66 | 3.34 |
| Ethylamine | | 10.81 | 3.19 |
| 1-Propylamine | | 10.71 | 3.29 |
| 1-Butylamine | | 10.77 | 3.23 |
| 2-Methyl-2-propylamine (*tert*-butylamine) | | 10.83 | 3.17 |
| Diethylamine | | 10.49 | 3.51 |
| Triethylamine | | 11.01 | 2.99 |

values for their conjugate acids. In discussing the acid–base behaviour of amines in this book, we use $K_a$ or $pK_a$ values.

$$NH_3(aq) + H_2O(l) \rightleftharpoons NH_4^+(aq) + OH^-(aq) \qquad pK_b = 4.75 \quad (31.18)$$

$$NH_4^+(aq) + H_2O(l) \rightleftharpoons H_3O^+(aq) + NH_3(aq) \qquad pK_a = 9.25 \quad (31.19)$$

$$RNH_2(aq) + H_2O(l) \rightleftharpoons RNH_3^+(aq) + OH^-(aq) \qquad (31.20)$$

$$RNH_3^+(aq) + H_2O(l) \rightleftharpoons H_3O^+(aq) + RNH_2(aq) \qquad (31.21)$$

The smaller the value of $pK_a$ (and the larger the value of $K_a$) for the conjugate acid $BH^+$ of a base B, the stronger the conjugate acid and the weaker the base.

The smaller the value of $pK_a$ for the conjugate acid of a base, the stronger the conjugate acid and the weaker the base. The values in Table 31.2 illustrate that aliphatic amines are stronger bases than $H_2O$ or alcohols, but are weaker bases than hydroxide or alkoxide ions.

Alcohols and alkoxide ions: see Section 30.3

Compounds with two $NH_2$ groups (diamines) are characterized by two $pK_a$ values. We define $pK_a(1)$ to refer to the loss of a proton from the diprotonated conjugate acid of the amine. $pK_a(2)$ refers to the loss of a proton from the monoprotonated conjugate acid. Equations 31.22 and 31.23 show the appropriate equilibria for 1,2-ethanediamine. It is a general trend that the value of $pK_a(1) < pK_a(2)$, i.e. the diprotonated conjugate acid is a stronger acid than the monoprotonated conjugate acid. Conversely, the monoprotonated base is a weaker base than the neutral, parent compound.

$$pK_a(1) = 7.56 \quad (31.22)$$

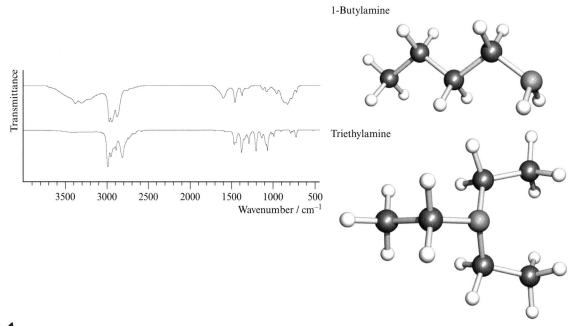

$$pK_a(2) = 10.71 \qquad (31.23)$$

We look further at the reactions of amines as bases in Section 31.5.

<table>
<tr><td>31.5</td><td>**Spectroscopic characterization of amines**</td></tr>
</table>

### IR spectroscopy

The IR spectrum of a primary or secondary amine exhibits a broad absorption band around 3000–3500 cm$^{-1}$ assigned to the N−H stretch, i.e. $\nu$(N−H). Because a tertiary amine does not possess an N−H bond, its IR spectrum does not exhibit this characteristic absorption. Figure 31.3 shows the IR spectra of 1-butylamine and triethylamine. Both spectra show absorbances between ≈3000 and 2800 cm$^{-1}$ assigned to $\nu$(C−H), and there are bands in the fingerprint region that are characteristic of each compound. The broad band centred around 3330 cm$^{-1}$ in the IR spectrum of 1-butylamine is assigned to $\nu$(N−H). No analogous absorption is present in the IR spectrum of Et$_3$N. It would not be seen either in the spectrum of an ammonium salt containing an R$_4$N$^+$ ion.

In general, bands arising from N−H stretches are less intense than those assigned to $\nu$(O−H). Compare the spectrum of 1-butylamine (Figure 31.3) with that of dodecan-1-ol (Figure 30.3). Bands arising from $\nu$(N−H) tend to be sharper than those assigned to $\nu$(O−H) because the degree of hydrogen bonding between amine molecules is less than that between alcohol molecules.

**Fig. 31.3** The IR spectra of 1-butylamine (a primary amine) and triethylamine (a tertiary amine). Colour code for structures: C, grey; N, blue; H, white.

Ammonium ions that contain $N-H$ bonds (e.g. $Et_3NH^+$) have characteristic absorptions assigned to $\nu(N-H)$ depending on the type of ion:

- $R_2NH_2^+$ and $R_3NH^+$ exhibit bands $\approx 2700-2250\,cm^{-1}$;
- $RNH_3^+$ has an absorption around $3000\,cm^{-1}$.

### $^1$H NMR spectroscopy

The $^1$H NMR spectrum of a primary or secondary amine is expected to contain a broad signal in the approximate range $\delta +1$ and $+6$ ppm (Table 14.3). However, in some cases the signal may be difficult to observe. On its own, $^1$H NMR spectroscopy is not a reliable method of confirming the presence of an NH or $NH_2$ group. Coupled with other data (e.g. IR spectroscopy), the $^1$H NMR spectrum of an unknown compound provides supporting information.

The assignment of a peak to an NH proton can be aided by adding $D_2O$ to the solution. The exchange reaction:

$$R_2NH + D_2O \rightleftharpoons R_2ND + DOH$$

leads to the disappearance of the NH proton signal in the $^1$H NMR spectrum. The same observation is made when $D_2O$ is added to a solution of an alcohol, and the signal assigned to the OH proton disappears.

### Mass spectrometry

In Chapter 10, we discussed the use of mass spectrometry in compound characterization. We focused on the identification of parent ions and fragmentation patterns. A simple organic compound containing C, H, O and N and an *odd* number of N atoms always has an odd molecular mass, e.g. the parent ion for $Me_2NH$ is at $m/z = 45$. Figure 31.4 shows the mass spectrum of ethylamine, $EtNH_2$. The parent ion comes at $m/z = 45$. Loss

**Fig. 31.4** The mass spectrum of ethylamine, $EtNH_2$.

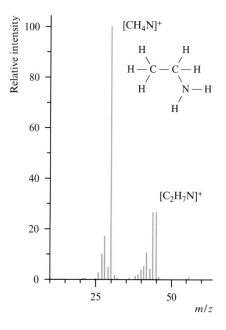

of H gives $[C_2H_6N]^+$ with $m/z = 44$. The intense peak at $m/z = 30$ arises from cleavage of the C–C bond adjacent to the $NH_2$ group. This is a characteristic cleavage, shown in equation 31.24 for ethylamine and in equation 31.25 for a general aliphatic amine.

$$m/z = 45 \xrightarrow{\text{Fragmentation}} m/z = 30 \tag{31.24}$$

$$\xrightarrow{\text{Fragmentation}} \tag{31.25}$$

R groups may be the same
or different

## 31.6 Reactivity

### Amines as weak bases

The fact that amines are weak bases means that they form salts when treated with acid. Equations 31.26–31.28 give examples, starting from primary, secondary and tertiary amines respectively.

$$\text{1-Pentylamine} + \text{HCl} \longrightarrow \text{1-Pentylammonium chloride} \tag{31.26}$$

$$\text{Dipropylamine} + \text{HBr} \longrightarrow \text{Dipropylammonium bromide} \tag{31.27}$$

$$\text{Triethylamine} + \text{HNO}_3 \longrightarrow \text{Triethylammonium nitrate} \tag{31.28}$$

Salt formation can be applied in the laboratory to the purification of amines if the amine is in a mixture of compounds in which only the amine will act as a base. The mixture of compounds is dissolved in an organic solvent such as an ether. Aqueous hydrochloric acid is then added to the solution contained in a separating funnel (see Figure 29.2). The amine forms a chloride salt (as in equation 31.26), and dissolves in the aqueous solution. The ether and aqueous solution are immiscible and the two layers are separated as described at the end of Section 29.4. Addition of base (e.g. NaOH) to the aqueous solution of the ammonium salt regenerates the amine (equation 31.29).

$$RCH_2NH_3{}^+Cl^- + NaOH \longrightarrow RCH_2NH_2 + NaCl + H_2O \qquad (31.29)$$

Soluble in an   Water<br>
organic solvent  soluble

Commercially available amines may be sold in the form of, for example, a hydrochloride, e.g. cetirizine dihydrochloride is an anti-histamine drug, and *ecstasy* (a controlled substance which has hallucinogenic effects) is a hydrochloride of a secondary amine.

Cetirizine dihydrochloride       Ecstasy

## Amines as weak acids

Just as alcohols can act as acids or bases, so too can amines. Amines are very weak acids, much weaker than alcohols, and *very* much weaker than carboxylic acids. The $pK_a$ values below illustrate this:

| Acetic acid $MeCO_2H$ | Ethanol $EtOH$ | Diisopropylamine $^iPr_2NH$ |
|---|---|---|
| $pK_a = 4.77$ | $pK_a = 16$ | $pK_a = 40$ |

We must be clear to what equilibrium this last value refers, because earlier in the chapter we discussed $pK_a$ values *of the conjugate acids of amines*. The $pK_a$ value *of an amine* refers to the loss of a proton and formation of an *amide ion*. For diisopropylamine, equilibrium 31.30 is the relevant one.

> *Caution!* The word ***amide*** is used both for an ion derived from an amine, and for a neutral compound of type $RCONH_2$:
>

Diisopropylamine       Diisopropylamide ion

$$pK_a = 40 \qquad (31.30)$$

The diisopropylamide ion is especially important in synthetic chemistry. Because it is the conjugate base of a very weak acid, $^iPr_2N^-$ is a very strong base. It is also sterically hindered. This allows $^iPr_2N^-$ to deprotonate a compound without there being competition with it acting as a nucleophile. Lithium diisopropylamide (LDA) is prepared by reaction 31.31 and is

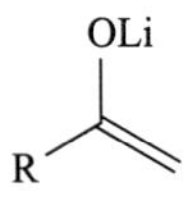

**(31.25)**

commonly used as a base in the formation of *lithium enolates* (**31.25**) from carbonyl compounds. We return to this application in Section 33.13.

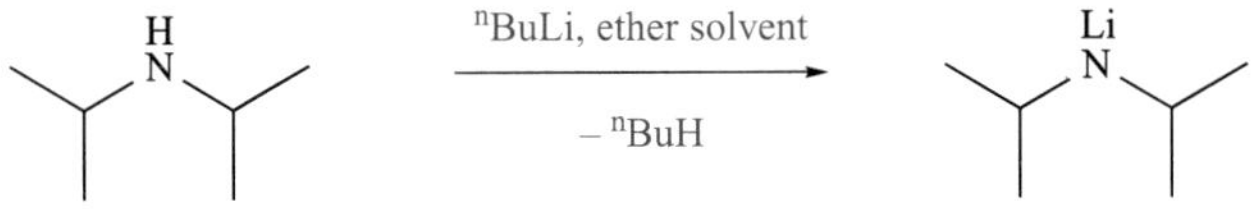

$$\text{Diisopropylamine} \xrightarrow[\text{–} ^n\text{BuH}]{^n\text{BuLi, ether solvent}} \text{Lithium diisopropylamide (LDA)} \quad (31.31)$$

### Hofmann elimination

A quaternary ammonium ion is susceptible to attack by a strong base such as hydroxide or an amide ion. The result is an elimination reaction yielding a tertiary amine and an alkene. Equation 31.32 shows an example.

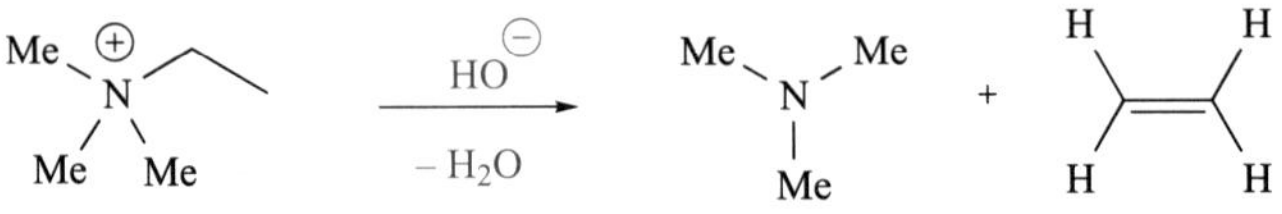

$$\xrightarrow[\text{– } H_2O]{HO^-} \qquad (31.32)$$

E2 mechanism

**E2 mechanism: see Section 28.7**

A tertiary amine, $R_3N$, is a *good leaving group* and the E2 mechanism is as follows:

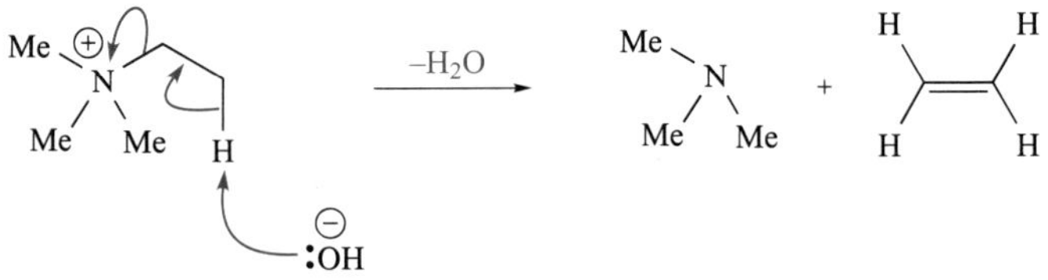

$$\xrightarrow{-H_2O}$$

Quaternary ammonium ion + strong base       Tertiary amine       Alkene

**Leaving groups: see Chapter 28**

If a similar reaction were carried out starting from an amine, the potential leaving group would be $NH_2^-$, $RNH^-$ or $R_2N^-$. All these are *poor leaving groups* and so the reaction does not occur. To overcome this problem, the amine is first methylated using methyl iodide to give a quaternary salt, e.g. reaction 31.33. Base-promoted elimination can now take place with a tertiary amine as the leaving group (equation 31.34).

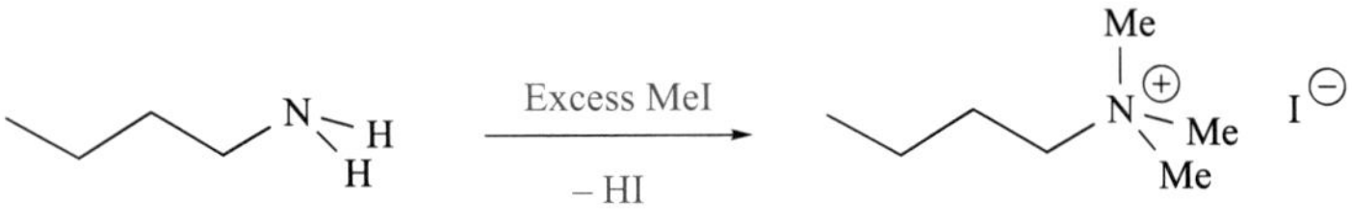

$$\xrightarrow[\text{– HI}]{\text{Excess MeI}} \qquad (31.33)$$

1-Butylamine          Butyltrimethylammonium iodide

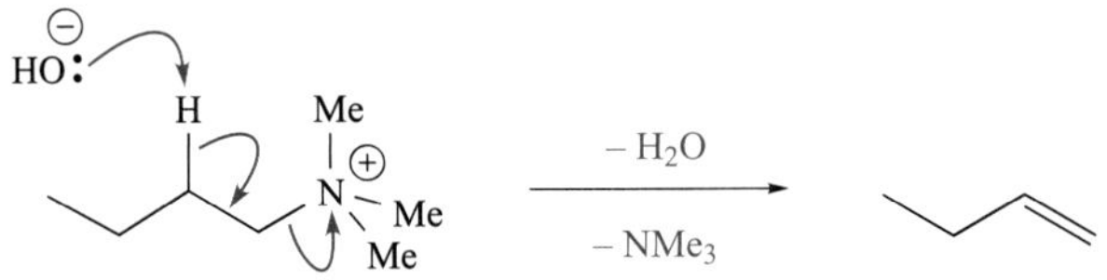

$$\xrightarrow[\text{– NMe}_3]{\text{– } H_2O} \qquad (31.34)$$

This type of elimination reaction is called a *Hofmann elimination*.

### Alkylation and acylation of amines

*Alkylation* of an amine (i.e. the conversion of an NH group to an NR group) can be achieved by reaction with a halogenoalkane, RX. However, as we saw in equations 31.2–31.4, such reactions tend to give a mixture of products as in equation 31.35. In reaction 31.33, an *excess* of MeI was used to ensure complete conversion to the ammonium salt.

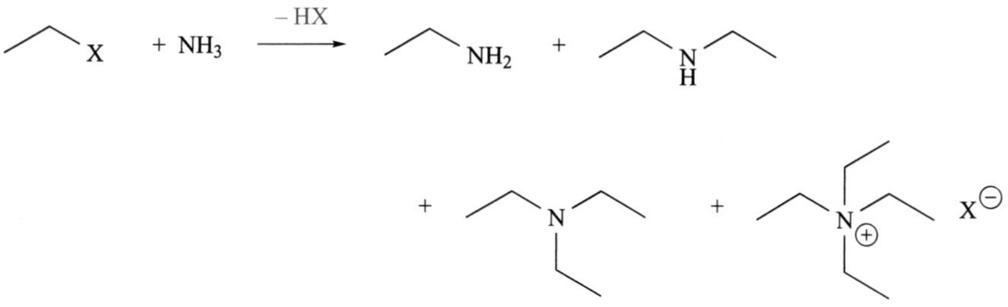

$$(31.35)$$

An *acyl chloride* has the general formula:

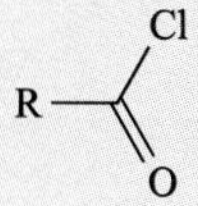

In contrast, if an amine is *acylated* (i.e. an NH group is converted to an NCOR group), the reaction is more readily controlled. Acylation takes place when ammonia or an amine reacts with an acyl chloride, and gives a specific *amide* as in reactions 31.36 and 31.37.

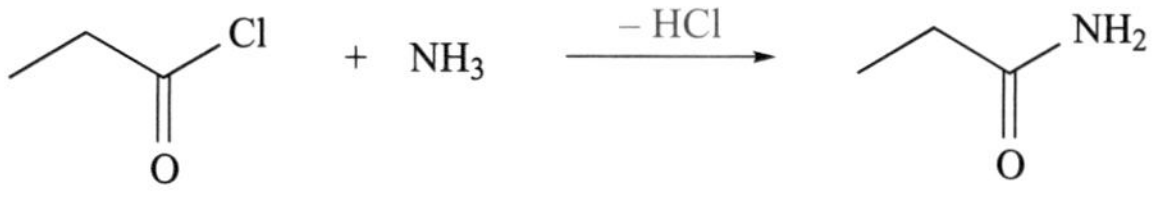

$$(31.36)$$

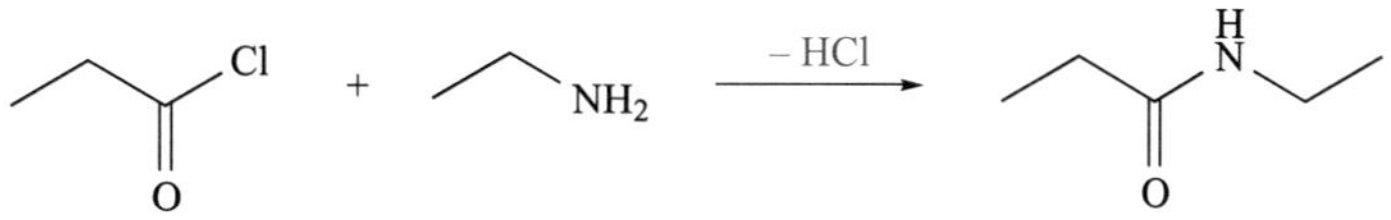

Acyl chloride                          Amide

$$(31.37)$$

Acyl chloride                          Amide

Reduction of the amide using lithium aluminium hydride gives the corresponding amine (e.g. equation 31.38).

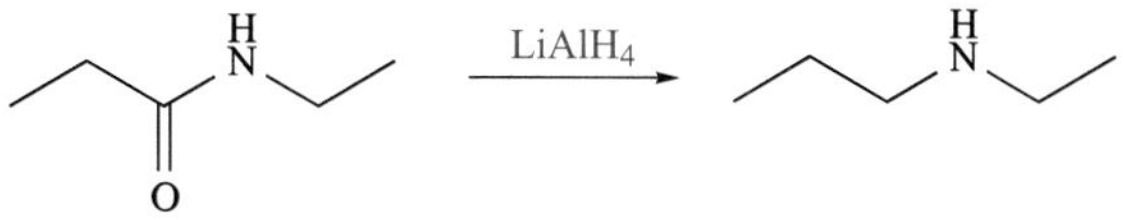

$$(31.38)$$

Combination of acylation and reduction therefore provides a means of specific alkylation, as opposed to the non-specific direct alkylation shown in equation 31.35.

## SUMMARY

This chapter has been concerned with amines: their syntheses and physical, spectroscopic and chemical properties. Hydrogen bonding involving NH groups occurs, although the strength of these hydrogen bonds is not as strong as those involving OH groups. Amines behave as weak bases; they may also act as very weak acids. The $NH_2$ group is a poor leaving group; elimination reactions involving attack by strong base are enhanced by converting amines into quaternary ammonium salts.

### Do you know what the following terms mean?

- primary amine
- secondary amine
- tertiary amine
- quaternary ammonium ion
- diamine

- aza-
- Gabriel synthesis
- imide
- reductive amination of an aldehyde or ketone
- amide functional group

- amide ion
- Hofmann elimination
- alkylation of an amine
- acylation of an amine
- acyl chloride

### The following abbreviation was introduced in this chapter. Do you know what it means?

- LDA

### You should be able:

- to give a systematic name of a simple amine given the structure
- to draw the structure of a simple amine given a systematic name
- to distinguish between primary, secondary and tertiary amines
- to know what is meant by a quaternary ammonium salt
- to recognize if an amine is chiral, and to comment on the possibility of resolving mixtures of enantiomers
- to describe methods of preparing amines, commenting on disadvantages or advantages of individual methods

- to comment on the effects of hydrogen bonding on physical properties of amines
- to relate $pK_a$ and $pK_b$ values for ammonium ions and amines to appropriate equilibria
- to assign an absorption in the IR spectrum of an amine to $\nu(N-H)$
- to interpret the $^1H$ NMR spectrum of a simple amine
- to interpret the mass spectrum of a simple amine and comment on characteristic fragmentation patterns
- to discuss characteristic reactions of amines

## PROBLEMS

**31.1** Give systematic names for the following amines:

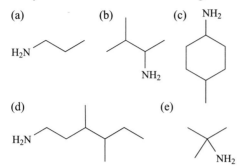

(a)    (b)    (c)

(d)    (e)

**31.2** Draw the structures of the following amines:
(a) 2-hexylamine; (b) 3,3-dimethyl-1-heptylamine;
(c) triethylamine; (d) dimethylamine;
(e) 1,4-butanediamine; (f) isopropylamine;
(g) N,N-dimethylbutylamine; (h) 1-octylamine;
(i) benzylamine. Classify each compound as a
primary, secondary or tertiary amine.

**31.3** Give structural representations of the following
salts: (a) tetraethylammonium chloride;
(b) tetrabutylammonium bromide;
(c) diethylammonium sulfate;
(d) N,N-dimethylbenzylammonium bromide.

**31.4** Explain why racemates of salts of chiral
quaternary ammonium ions can usually be
resolved whereas this is not generally true for chiral
amines in which the chiral centre is the N atom.

**31.5** (a) Suggest how the reaction of $NH_3$ with
1-bromopropane might proceed.
(b) Suggest a method of preparing 1-hexylamine in
high yield from 1-bromohexane.

**31.6** Suggest a multi-step synthesis by which ethene
might be converted to 1-propylamine.

**31.7** Suggest products in the following reactions:

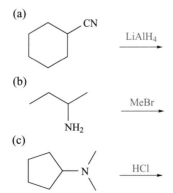

(a)

    LiAlH₄

(b)

    MeBr

(c)

    HCl

**31.8** Give the products expected from the reductive
amination using $NH_3$ of (a) butanone;
(b) propanal; (c) pentane-2,4-dione;
(d) cyclohexanone.

**31.9** How would you expect the boiling points of the
following series of compounds to vary? Rationalize
your answers.
(a) Pentane; 1-pentylamine; 2-pentylamine.
(b) 1-Butylamine; 2-methyl-1-propylamine;
2-methyl-2-propylamine.
(c) Ethane; butane; 1-propylamine;
1,2-ethanediamine.

**31.10** Write down the equilibria to which the following
$pK_a$ and $pK_b$ values refer: (a) $pK_b$ for methylamine;
(b) $pK_a$ values for the conjugate acids of 1,3-
propanediamine; (c) $pK_a$ for triethylammonium
ion; (d) $pK_b$ for benzylamine; (e) $pK_a$ for the
conjugate acid of 1-butylamine.

**31.11** (a) Match the following compounds to the
correct $^1H$ NMR spectroscopic data given in the
table. (b) What coupling patterns do you expect to
see in the spectrum of compound **B**? (c) Explain
why the signal at $\delta$ 1.15 ppm in the spectrum of **C** is
a singlet.

| Compound | $^1H$ NMR $\delta$/ppm (relative integral) |
|---|---|
| A | 2.69 (4H), 2.43 (6H), 1.24 (2H) |
| B | 2.30 (2H), 2.22 (6H), 1.06 (3H) |
| C | 1.41 (2H), 1.15 (9H) |

**31.12** (a) Match the following compounds to the correct
$^{13}C$ NMR spectroscopic data given in the table.
(b) Do any of the compounds possess stereoisomers?
Could you distinguish between them using $^{13}C$
NMR spectroscopic data? (c) The data could be
recorded as $^{13}C$ or $^{13}C\{^1H\}$ NMR spectra. What is
the difference between these spectra for a given
compound?

| Compound | $^{13}C$ NMR $\delta$/ppm |
|---|---|
| A | 49.2, 31.3, 27.2 |
| B | 44.1, 15.4 |
| C | 50.6, 46.0, 36.7, 34.5, 32.0, 25.1, 22.6 |

**31.13** Suggest likely products in the following reactions.

(a)

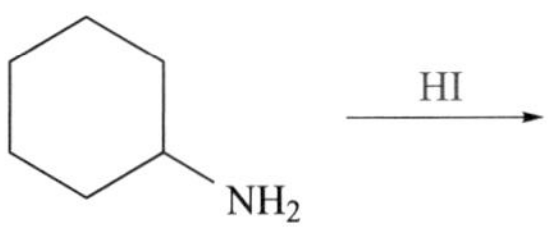

(b)

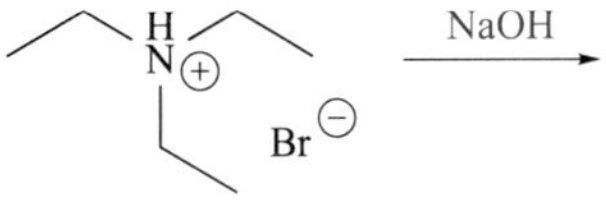

(c)

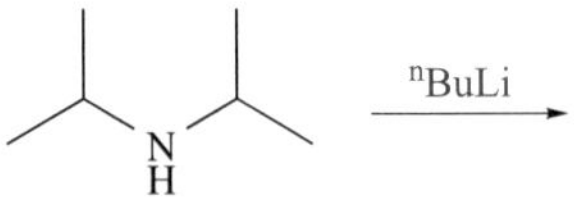

(d)

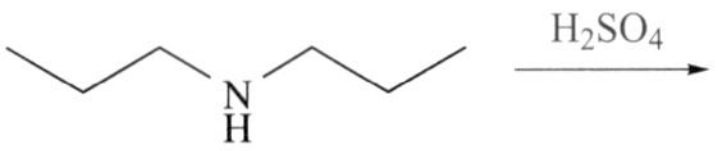

**31.14** What problems would be encountered if you used the reaction of 1-bromohexane with $NH_3$ as a route to 1-hexylamine?

**31.15** Elemental analysis for a compound **A** gives C 58.8%, N 27.4% and H 13.8%. In the mass spectrum of **A**, a parent ion is observed at $m/z = 102$ and a major fragmentation peak comes at $m/z = 30$. The $^{13}$C NMR spectrum exhibits signals at $\delta$ 42.1, 33.8 and 24.2 ppm. (a) Suggest an identity for **A**. (b) How would **A** react with $HNO_3$?

**31.16** Suggest likely products in the following reactions. For the two-step syntheses, give the intermediate compounds as well as the final products.

(a)

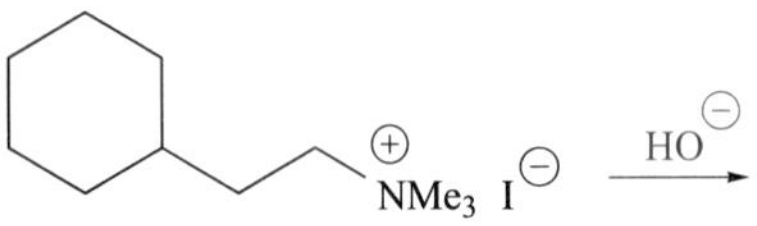

(b)

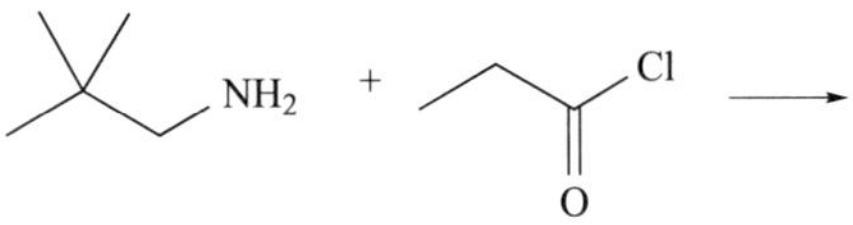

(c)

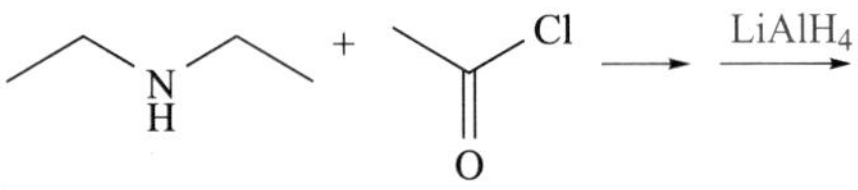

(d)

**31.17** Suggest ways of carrying out the following transformations. More than one step may be required.

(a)

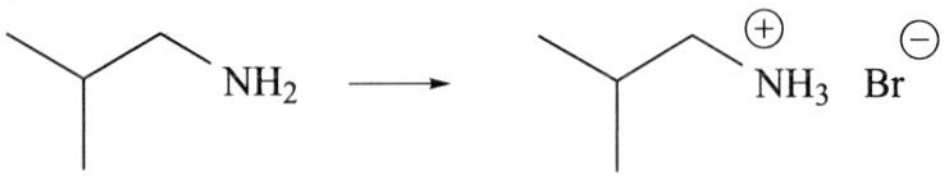

(b)

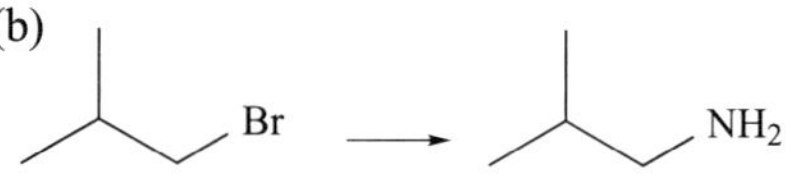

(c)

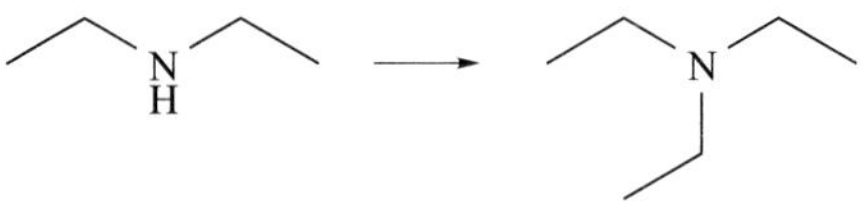

(d)

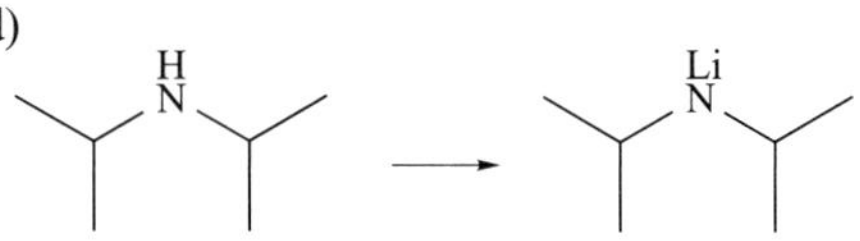

**31.18** Suggest two methods of preparing 2-aminopentane as selectively as possible and starting from a suitable halogenoalkane.

## Additional problems

**31.19** The mass spectrum of a compound **X** has a parent peak at $m/z = 69$. The $^{13}$C NMR spectrum of **X** has signals at $\delta$ 119.9, 19.3, 19.0 and 13.3 ppm. Treatment of **X** with $LiAlH_4$ gives compound **Y**, the mass spectrum of which contains major peaks at $m/z = 73$ and 31. The $^{13}$C NMR spectrum of **Y** shows signals at $\delta$ 42.0, 36.1, 20.1 and 13.9 ppm. Elemental analytical data for **X** and **Y** are: **X**, 69.6% C, 20.3% N, 10.1% H; **Y**, 65.7% C, 19.2% N, 15.1% H. Compound **Z** is an isomer of **X**; its $^1$H NMR spectrum contains signals at $\delta$ 1.3 (doublet, 6H), 2.7 (septet, 1H) ppm. Suggest identities for **X**, **Y** and **Z**. How would **Y** react with (a) $H_2SO_4$, (b) an excess of MeI and (c) butanoyl chloride, $C_3H_7COCl$?

**31.20** (a) To what equilibrium does a value of $pK_a = 40$ for $^i Pr_2 NH$ refer? Why does this value differ from a $pK_a$ value of 10.9 calculated from the equation $pK_a = 14.00 - pK_b$ where $pK_b$ refers to $^i Pr_2 NH$?
(b) Suggest how $NaN_3$ would react with 2-bromobutane. How could the product of this reaction be converted into 2-butylamine?
(c) Which of the following compounds possess stereoisomers? Draw structural diagrams that differentiate between these stereoisomers.

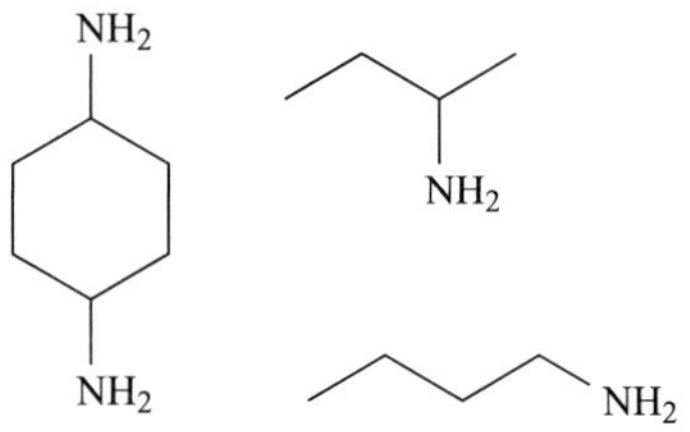

(d) The $^1$H NMR spectrum of an amine $RNH_2$ contains a broad resonance. Explain why the signal disappears when $D_2O$ is added to the solution.

# 32 Aromatic compounds

## 32.1 An introduction to aromatic hydrocarbons

Benzene  Naphthalene

(32.1)  (32.2)

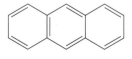

Anthracene

(32.3)

In Figure 24.2, we classified hydrocarbon compounds in two primary groups: *aliphatic* and *aromatic* compounds. Chapters 24–31 are concerned with aliphatic compounds. Now we turn our attention to *aromatic* compounds, the archetypal example of which is *benzene*, $C_6H_6$ (**32.1** and Figure 32.1).

### Sources of aromatic hydrocarbons

Figure 25.13 showed how industrial distillation of crude oil gives fractions of hydrocarbons with different molecular weight ranges. Crude oil is not naturally rich in aromatic compounds, but benzene and toluene (methylbenzene), for example, are made industrially from cyclohexane (see Section 32.6) and methylcyclohexane (see equation 32.35) respectively. The second crude material for aromatics is *coal*, although this is not of modern, commercial importance. The structure of coal is complex, but on heating at very high temperatures in the absence of $O_2$, the structure is destroyed and *coal tar* is produced. Coal tar contains a mixture of aromatic hydrocarbons and derivatives can be separated by fractional distillation. Aromatic hydrocarbons obtained from petroleum and coal tar contain at least one $C_6$ ring and include benzene, naphthalene and anthracene. Although their structures can be represented as in **32.1**–**32.3**, we shall see in the next section that they do *not* contain localized double and single C−C bonds. Derivatives of benzene obtained from coal tar include alkyl-substituted compounds such as toluene (**32.4**) and the three isomers of xylene (**32.5**–**32.7**).

| Toluene (Methylbenzene) | *ortho*-Xylene (1,2-Dimethylbenzene) | *meta*-Xylene (1,3-Dimethylbenzene) | *para*-Xylene (1,4-Dimethylbenzene) |
|---|---|---|---|
| (32.4) | (32.5) | (32.6) | (32.7) |

---

### 32.2

## The structure of benzene and its delocalized bonding

### C₆H₆: early history

Benzene was discovered in 1825 by Faraday as he studied the liquid condensed from products of gas combustion. Its formula was established as $C_6H_6$ and in the years that followed, various structures were proposed, including those shown in **32.8**–**32.10**. At this time, structural data could not be confirmed by X-ray diffraction or spectroscopic techniques. Early ideas about the structure of benzene came instead from observations about the number of isomers of the disubstitution product of the reaction of $C_6H_6$ with $Cl_2$ or $Br_2$. Consider Kekulé's structure **32.10** for benzene: for the disubstitution product 1,2-dichlorobenzene, two isomers are possible depending on whether the Cl atoms are separated by a C−C single or double bond. However, only one isomer was actually observed. Kekulé explained this observation by allowing the two theoretical isomers of 1,2-dichlorobenzene to be in *rapid equilibrium* (equation 32.1).

Landenberg structure
(32.8)

Dewar structure
(32.9)

Kekulé structure
(32.10)

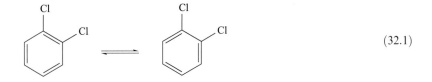

$$(32.1)$$

When looking at these early proposals, you should bear in mind that, in the mid-1800s, ideas of valency were not yet established and so Kekulé's concepts were remarkably well developed. However, although extremely valuable at the time of Kekulé, equilibrium 32.1 is an *incorrect* explanation. Within the framework of modern valence bond theory, Kekulé's ideas are represented in terms of the contributions made by the two *resonance structures* **32.11** and **32.12**. This picture gives an average bond order for each carbon–carbon bond of 1.5.

(32.11)     (32.12)

### Structural determination

The structure of benzene was determined by diffraction methods in the mid-1900s, and this confirmed the equivalence of the C−C bonds (Figure 32.1).

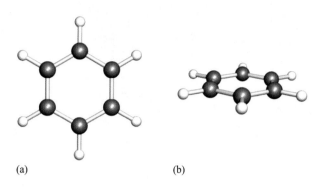

**Fig. 32.1** Two views of the structure of benzene. The C−C bond lengths are all equal (140 pm), the C−H bond lengths are equal (110 pm), and the C−C−C bond angles are all the same (120°).

(a)                                    (b)

The molecule is planar and all the C−C bonds are of length 140 pm. This value lies between typical values for C−C single (154 pm) and double (134 pm) bonds. Thus, the representation of benzene as the *triene*-like molecule shown in **32.10** is incomplete, although the resonance pair **32.11** and **32.12** does provide an explanation for the bond equivalence.

> ◼▷
> Recall from Chapter 26 that a triene contains three C=C bonds

### Thermochemical evidence for the non-triene-like nature of benzene

Thermochemical data also provide evidence that benzene should *not* be described as a triene with localized single and double C−C bonds. When cyclohexene is hydrogenated (equation 32.2), the enthalpy change accompanying the reaction is $-118 \text{ kJ mol}^{-1}$.

$$\text{Cyclohexene} + \text{H}_2 \longrightarrow \text{Cyclohexane} \qquad (32.2)$$

**(32.13)**

When cyclohexadiene (**32.13** shows one isomer) is converted to cyclohexane, the enthalpy change ($\Delta_r H$) is approximately *twice* that associated with reaction 32.2. This is consistent with a value of $-118$ kJ being a measure of the enthalpy change associated with the *addition of one mole of H$_2$ to one mole of C=C bonds*. From this, we would predict that the enthalpy change for reaction 32.3 (the hydrogenation of benzene) should be about $-(3 \times 118) = -354 \text{ kJ mol}^{-1}$. This is shown on the left-hand side of Figure 32.2.

$$\text{Benzene} + 3\text{H}_2 \longrightarrow \text{Cyclohexane} \qquad (32.3)$$

The *observed* value of $\Delta_r H$ is significantly less than the predicted value as Figure 32.2 shows. This means that benzene is *thermodynamically more stable* than the triene model suggests.

### Chemical evidence for the non-triene-like nature of benzene

> ◼▷
> Additions to alkenes: see Chapter 26

Further evidence for the failure of the triene model comes from the reactivity of benzene. Alkenes with localized C=C bonds undergo facile *addition*

**Fig. 32.2** The experimental molar enthalpy of hydrogenation ($\Delta_r H$) of benzene is smaller than that predicted by assuming a cyclic triene model.

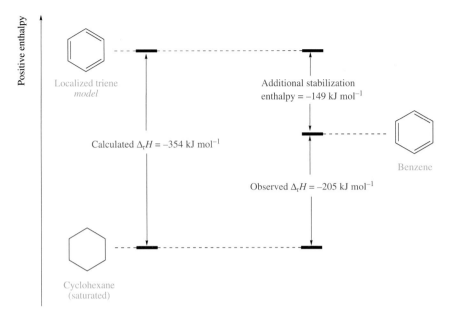

reactions. For example, when ethene is bubbled through an aqueous solution of $Br_2$, the solution is decolourized as reaction 32.4 occurs.

$$Br_2 \quad + \quad \text{Ethene} \quad \longrightarrow \quad \text{1,2-Dibromoethane} \qquad (32.4)$$

Orange, aqueous solution        Ethene        1,2-Dibromoethane (Colourless)

In contrast, benzene only reacts with $Br_2$ in the presence of a catalyst, and the reaction is a *substitution*, not an addition (equation 32.5).

$$\text{Benzene} \xrightarrow[{- HBr}]{Br_2;\ FeBr_3\ catalyst} \text{Bromobenzene} \qquad (32.5)$$

Benzene                                Bromobenzene

### A delocalized bonding model

(32.14)

The resonance pair **32.11** and **32.12** can also be represented by structure **32.14**. This indicates that the bonding is delocalized around the ring. Each C atom is in a trigonal planar environment (Figure 32.1) and so an $sp^2$ hybridization scheme is appropriate. Figure 32.3a illustrates that the C $sp^2$ hybrid orbitals and the H $1s$ orbitals overlap to give a $\sigma$-bonding framework consisting of C−C and C−H interactions. Each C atom provides three of its four valence electrons, and each H atom contributes one electron to the $\sigma$-bonding interactions. After the formation of the $\sigma$-interactions shown in Figure 32.3a, each C atom has one $2p$ orbital and one valence electron left over. The $2p$ orbitals can overlap to give a delocalized $\pi$-molecular orbital (Figure 32.3b) that is continuous around the $C_6$ ring. In Box 32.1, we consider the $\pi$-interactions in detail and show how the six $\pi$-electrons are accommodated in three $\pi$-bonding molecular orbitals.

The relationship between the Kekulé and delocalized models of bonding in benzene is similar to the one we described for the allyl anion or allyl radical in

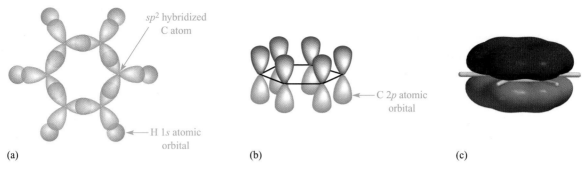

**Fig. 32.3** In benzene, each C atom provides four electrons and each H atom provides one electron. (a) The $\sigma$-bonding framework in $C_6H_6$ can be described in terms of $sp^2$ hybridized C atoms. The C–H interactions are localized, 2c–2e $\sigma$-bonds. The $\sigma$-contributions to the overall C–C bonding are localized. (b) After forming three $\sigma$-bonds, each C atom has one $2p$ orbital left over. Overlap between all six $2p$ orbitals can occur to give a delocalized $\pi$-system. (c) A computer-generated picture of the $\pi$-bonding MO in benzene (computer generation using Spartan '04, © Wavefunction Inc. 2003). Diagrams (b) and (c) are representations of the same MO.

Figure 13.6 and resonance pair **26.29**. Structure **32.14** is *exactly equivalent* to the pair of resonance structures **32.11** and **32.12**, but for the most part, we shall use the Kekulé representation of benzene. Moreover, we shall draw only one resonance structure in chemical equations. However, you should not forget that this is only part of the story so far as bonding is concerned.

### 32.3    Aromaticity and the Hückel (4n + 2) rule

The term *aromatic* originally arose because of the aromatic smells that some members of this family of compounds possess

At the beginning of the chapter, we stated that benzene is an *aromatic* compound. By aromatic, we mean that there is some special character associated with the ring system. While this is difficult to quantify in a simple definition, we can state that an aromatic species is typified by a *planar ring with a delocalized $\pi$-system*. Even so, diagrammatic representations of the ring may show alternating single and double bonds as in **32.10**. Caution is needed though: not all rings that are drawn with alternating single and double bonds are aromatic. For example, cyclobutadiene (although planar) has C–C bond lengths of approximately 159 and 134 pm (structure **32.15**).[§] Cyclobutadiene is, therefore, *not* aromatic. Most cyclobutadienes are stable for only very short periods, e.g. the lifetime of $C_4H_4$ is $<5$ s.

*Aromatic* compounds are stabilized with respect to a localized structure by delocalization of the $\pi$-electrons. For some related cyclic compounds, the delocalized structure is *less* stable than the localized one. Such compounds are described as being *anti-aromatic*. We have already seen that benzene possesses a delocalized $\pi$-system extending over the ring, but this alone does not make the compound aromatic. The *number* of $\pi$-electrons in the ring is critical, and the *Hückel rule* states that there must be $(4n + 2)$ $\pi$-electrons where $n$ is an integer. Benzene has six $\pi$-electrons and obeys the Hückel $(4n + 2)$ rule with $n = 1$. Cyclobutadiene is planar and has four $\pi$-electrons; it does not obey the Hückel rule, i.e. an integral value of $n$ cannot be found to satisfy the equation: $(4n + 2) = 4$. Cyclobutadiene is anti-aromatic. Table 32.1 lists examples of Hückel $(4n + 2)$ $\pi$-systems and some other cyclic $\pi$-systems. $C_8H_8$ does not obey the $(4n + 2)$ rule and would be anti-aromatic *if planar*. However, the non-planarity of the ring

159 pm
134 pm

**(32.15)**

The *Hückel (4n + 2) rule* states that an aromatic compound possesses a planar ring with a delocalized $\pi$-system containing $(4n + 2)$ $\pi$-electrons.

---

[§] These bond lengths are for a derivative of cyclobutadiene.

**THEORETICAL AND CHEMICAL BACKGROUND**

## Box 32.1  A molecular orbital approach to the $\pi$-bonding in benzene

Before working through this box, you may wish to review the $\pi$-bonding in the allyl anion (Figure 13.6).

The $\pi$-bonding in benzene arises from the overlap of the six carbon $2p$ orbitals. The interaction of the six $2p$ atomic orbitals must lead to six MOs. This is shown below. Each MO is represented in a diagrammatic form, and in a more realistic form generated computa-

tionally using Spartan '04, © Wavefunction Inc. 2003. The diagrams in each pair are equivalent to one another. Each $\pi$-MO possesses a nodal plane that coincides with the plane of the $C_6H_6$ molecule. Additional nodal planes which lie perpendicular to the plane of the molecule are noted in the figure shown below.

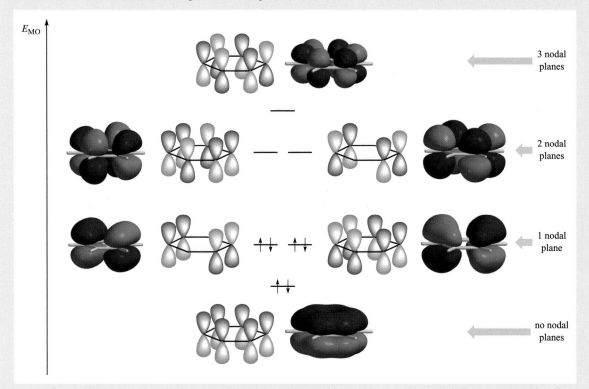

An in-phase overlap of all six $2p$ orbitals generates the lowest-energy $\pi$-bonding MO and this has $\pi$-character that extends continuously (no phase change) around the ring. This is clearly seen in the computer-generated orbital representation. The highest-energy MO results if adjacent $2p$ orbitals are out of phase with one another. This MO is antibonding between each pair of adjacent C atoms, and contains three nodal planes that lie

perpendicular to the plane of the benzene ring. The remaining four MOs consist of two degenerate pairs with either one or two nodal planes. In two of these MOs, the nodal planes pass through two C atoms and this means that there is no contribution from these $2p$ orbitals to the MO; compare this with the description of the allyl anion in Figure 13.6.

(Figure 32.4) and the alternating C—C bond lengths of 134 and 147 pm provide evidence for the localized nature of the bonding. The $^1$H NMR spectrum of $C_8H_8$ shows a signal at $\delta$ 5.7 ppm which is typical of an alkene rather than an aromatic compound. We return to NMR spectroscopy of aromatics in the next section.

Hückel species include neutral molecules, anions and cations. For example, cyclopentadiene ($C_5H_6$, **32.16**) is a diene with four $\pi$-electrons and localized bonding (see Figure 26.2), but the cyclopentadienyl anion

Cyclopentadiene

**(32.16)**

**Table 32.1** Examples of Hückel (4n + 2) aromatic systems and non-aromatic systems. Each trigonal planar C atom provides one electron to the π-system.

| Formula | Structure | Number of π-electrons | n | Comments |
|---|---|---|---|---|
| $C_6H_6$ | | 6 | 1 | Planar ring; obeys (4n + 2) rule; aromatic |
| $(C_6H_5)_2CH_2$ | | 6 (twice) | 1 | Each ring is planar and is a separate π-system; obeys (4n + 2) rule; each ring is aromatic |
| $C_{10}H_8$ | | 10 | 2 | Complete ring system is planar; aromatic (but see text) |
| $C_{18}H_{18}$ | | 18 | 4 | Planar ring; obeys (4n + 2) rule; aromatic (see Box 32.2) |
| $C_8H_8$ | | 8 | – | Does not obey (4n + 2) rule; ring is non-planar (see Figure 32.4); non-aromatic |
| $C_{10}H_{10}$ | | 10 | 2 | Obeys (4n + 2) rule, *but* the ring is non-planar; non-aromatic |

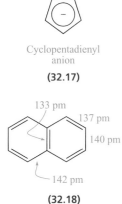

Cyclopentadienyl anion

**(32.17)**

133 pm
137 pm
140 pm
142 pm

**(32.18)**

$C_5H_5^-$ is a planar, Hückel system with six π-electrons and delocalized bonding. It is represented by structure **32.17**.

Naphthalene, $C_{10}H_8$, is listed in Table 32.1 as being aromatic, but its aromaticity deserves special mention. The question is: *should naphthalene be considered in terms of two fused benzene rings?* A structural feature of benzene that points towards the delocalization of the π-electrons is the equal lengths of the C−C bonds (140 pm). The bond lengths in naphthalene are shown in structure **32.18** and the central C−C bond is noticeably shorter than the remaining bonds. Furthermore, the remaining C−C bonds are non-equivalent. Figure 32.2 presented thermochemical evidence for the non-triene-like nature of benzene based on the enthalpy change for the reaction of $C_6H_6$ with $H_2$. A similar consideration of the enthalpy change accompanying the reaction of $C_{10}H_8$ with $H_2$ indicates that the molecule is stabilized

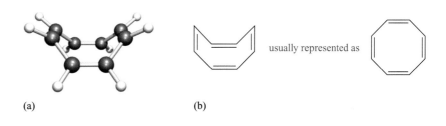

**Fig. 32.4** (a) The structure (optimized by computer modelling) of $C_8H_8$ (cyclooctatetraene). (b) The *non-planar* ring has *localized* π-bonding with alternating short (C=C) and long (C−C) bonds. Each C atom is trigonal planar and $sp^2$ hybridized.

usually represented as

(a)          (b)

by $\pi$-electron delocalization. However, the stabilization is *less* than would be expected for two benzene rings, and it is easier to hydrogenate the first ring (step 1 in scheme 32.6) than the second (step 2 in scheme 32.6).

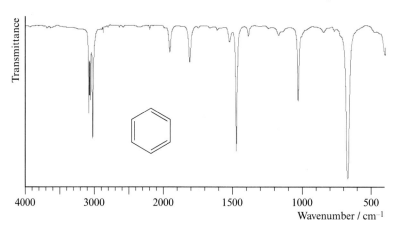

Naphthalene    1,2,3,4-Tetrahydronaphthalene    Decahydronaphthalene

(32.6)

These data support a delocalized bonding model, but one in which the aromaticity per ring is less than in benzene. Once one ring in naphthalene has been hydrogenated, the molecule contains a benzene ring, and the aromaticity increases, making hydrogenation of the second ring more difficult than that of the first. In terms of valence bond theory, the bonding in naphthalene is represented in terms of the resonance structures **A, B** and **C** below. The structural data (**32.18**) are consistent with a major contribution from resonance structure **B**:

A    B    C

### 32.4    Spectroscopy

### IR spectroscopy

The IR spectrum of benzene is shown in Figure 32.5. Absorptions in the range 3120–3000 cm$^{-1}$ are assigned to the C−H stretches. Look back at Figure 12.13, where we compared this region in the IR spectra of benzene and ethylbenzene, and drew attention to the differences in the frequencies of the vibrations of the C($sp^2$)−H and C($sp^3$)−H bonds. The IR spectra of benzene and its derivatives show a strong absorption near 1500 cm$^{-1}$. In benzene, this absorption comes at 1480 cm$^{-1}$ (Figure 32.5). It arises from the vibration of the C$_6$ ring (a *ring mode*). This band is particularly useful in aiding the identification of a 6-membered aromatic ring. The strong

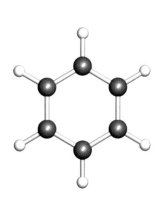

 Fig. 32.5 The IR spectrum of benzene.

absorption at $680 \, cm^{-1}$ in Figure 32.5 is assigned to the out-of-plane bending of the C−H bonds.

## $^1$H and $^{13}$C NMR spectroscopy

The $^1$H NMR spectrum of benzene is shown in Figure 32.6a. The six H atoms are equivalent and, consistent with this, the spectrum contains a singlet. The chemical shift of $\delta$ 7.2 ppm is at relatively high frequency and is typical of protons attached directly to an aromatic ring (see Box 32.2). The shift distinguishes aromatic (*aryl*) protons from those in alkenes (see Table 14.3).

Figure 32.6b shows the $^{13}$C{$^1$H} NMR spectrum of benzene. The singlet is consistent with the presence of equivalent C atoms. The chemical shift of $\delta$ 128.4 ppm is typical of $sp^2$ hybridized C atoms (see Figure 14.4 and Table 14.2). Carbon nuclei in the $C_6$ rings of derivatives of benzene resonate at similar frequencies to those in $C_6H_6$, e.g. the proton-decoupled $^{13}$C NMR spectrum of chlorobenzene is shown in Figure 32.6c. Notice that the signal integrals do *not* give a correct indication of the relative numbers of C

Signal intensities: see Section 14.7

**Fig. 32.6** (a) The $^1$H NMR spectrum of benzene, (b) the $^{13}$C{$^1$H} NMR spectrum of benzene and (c) the $^{13}$C{$^1$H} NMR spectrum of chlorobenzene. (Recall from Chapter 14 that {$^1$H} means 'proton-decoupled'.)

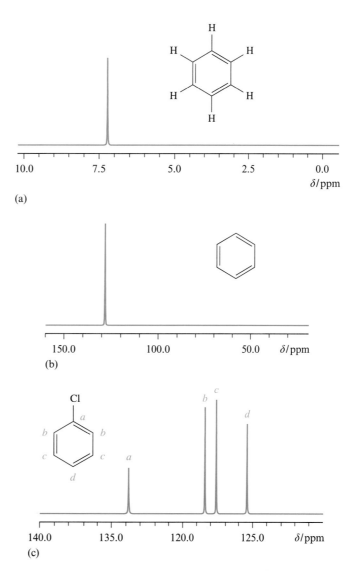

**THEORETICAL AND CHEMICAL BACKGROUND**

## Box 32.2 Deshielding of aromatic protons

The $^1$H NMR spectrum of benzene exhibits a singlet at $\delta$ 7.2 ppm. This signal comes at a higher frequency than those arising from typical alkene protons (see Table 14.3), and the significant shift in signals is further evidence for the aromatic character of benzene. The protons attached to the $C_6$ ring in derivatives of benzene also give signals around $\delta$ 7.0–7.7 ppm. Signals in the region $\delta$ 6.0 to 10.0 ppm are characteristic of many aromatic compounds. The origin of this effect is as follows.

When a benzene molecule is placed in the magnetic field of an NMR spectrometer, the $\pi$-electrons undergo circular motion (a *ring current*) that generates a secondary (*induced*) magnetic field:

In benzene, the induced field reinforces the applied field and the protons experience a greater magnetic field than do protons not affected by the induced field. The protons are *deshielded*. In the $^1$H NMR spectrum, *deshielded protons appear at higher frequency*.

In some aromatic systems, the induced magnetic field *opposes* the applied field and the aromatic protons are *shielded*. In the $^1$H NMR spectrum, *shielded protons appear at lower frequency*. The classic example of this effect is observed in the $^1$H NMR spectrum of [18]-annulene, $C_{18}H_{18}$ (see Table 32.1). The ring current *shields* the protons lying *within the ring* (the H atoms shown in red below), and *deshields* the protons lying *outside the ring* (the H atoms shown in blue below):

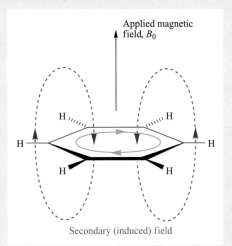

Secondary (induced) field

At low temperatures, the $^1$H NMR spectrum of [18]-annulene contains two signals at $\delta$ +9.3 and −3.0 ppm.

In a monosubstituted benzene derivative, the substituted C atom is called the *ipso*-position:

atoms in each type of environment. The signal assigned to the *substituted* or *ipso*-C atom (signal *a* in Figure 32.6c) is of low intensity and this is typical for substituted rings. In some cases, the intensity is so low that the signal is barely visible. We see more examples later in the chapter.

### Electronic spectroscopy

In Chapter 13, we considered how the effects of $\pi$-conjugation permit the $\pi^* \leftarrow \pi$ transition to be observed in the near-UV part of the electronic spectrum of conjugated polyenes. Similarly, the presence of the delocalized $\pi$-system in benzene (see Box 32.1) reduces the energy difference between the highest-lying $\pi$ MO and the lowest-lying $\pi^*$ MO, and the associated $\pi^* \leftarrow \pi$ transition is observed in the near-UV. The electronic spectrum of benzene has three absorptions, the most intense of which occurs at $\lambda_{max} = 183$ nm $(\varepsilon_{max} = 46\,000\ dm^3\,mol^{-1}\,cm^{-1})$. Selected electronic spectroscopic data for benzene and some of its derivatives are given in Table 32.2.

**Table 32.2** Observed values of $\lambda_{max}$ for benzene and some related aromatic molecules. Only the most intense band in each spectrum is listed.

| Compound | Structure | $\lambda_{max}$ / nm | $\varepsilon_{max}$ / dm$^3$ mol$^{-1}$ cm$^{-1}$ |
|---|---|---|---|
| Benzene | | 183 | 46 000 |
| Biphenyl | | 246 | 20 000 |
| Naphthalene | | 220 | 132 000 |
| Anthracene | | 256 | 182 000 |
| Styrene | | 244 | 12 000 |
| Diphenylacetylene | | 236 | 12 500 |

## 32.5 Nomenclature

### Derivatives of benzene

The general abbreviation for an **aryl** substituent is Ar. Compare the use of R for an alkyl group.

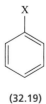

(32.19)

Before discussing the synthesis and reactivity of benzene, we introduce the basic rules for naming derivatives of benzene. An aromatic group is called an *aryl* substituent and has the abbreviation Ar, e.g. ArCl stands for a chloroarene as opposed to RCl which represents a chloroalkane. Benzene derivatives of the type shown in structure **32.19** are named in one of two ways. Group X may be taken as a substituent of the benzene ring and the compound is named by prefixing 'benzene' with the name of the functional group; e.g. for X = Cl in **32.19**, the compound is chlorobenzene. Further examples are compounds **32.20–32.23**. An exception is benzenesulfonic acid (**32.24**).

| Fluorobenzene | Bromobenzene | Ethylbenzene | Nitrobenzene | Benzenesulfonic acid |
|---|---|---|---|---|
| (32.20) | (32.21) | (32.22) | (32.23) | (32.24) |

In the second method of nomenclature, the $C_6H_5$ group is considered to be a substituent of X. $C_6H_5$ is called a *phenyl* group (abbreviated to Ph), and a compound is named by prefixing the name of the X group with 'phenyl', e.g. for $X = NH_2$ in **32.19**, the compound is phenylamine, although its common name of *aniline* is also used. Further examples are given in **32.25** and **32.26**, and abbreviating phenyl to Ph gives a shorthand way of writing these compounds as $Ph_2NH$ and $Ph_3P$.

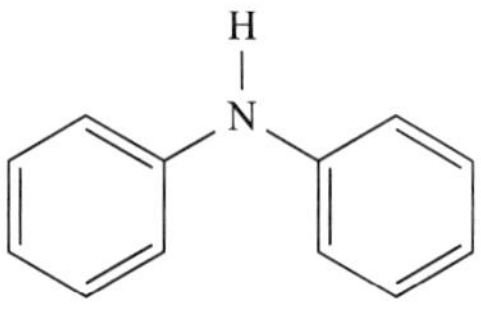

Diphenylamine

(32.25)

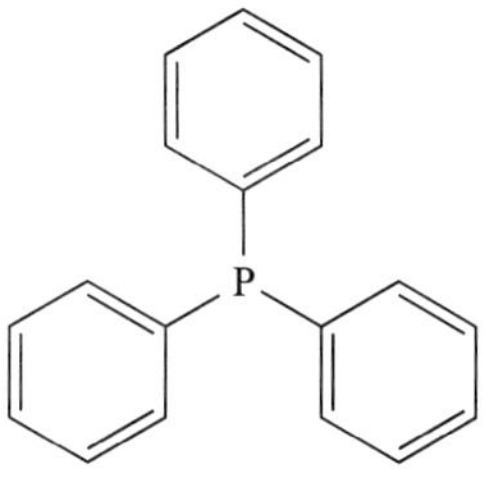

Triphenylphosphine

(32.26)

A number of important derivatives of benzene have common names that remain in use, and you should learn the names and structures of the following compounds:

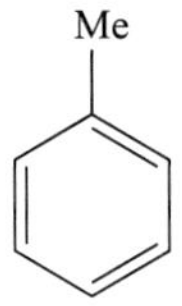

Toluene 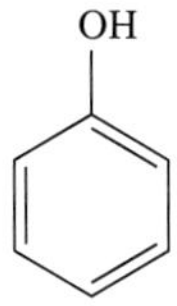    Phenol 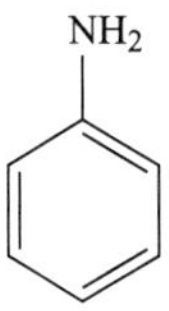    Aniline 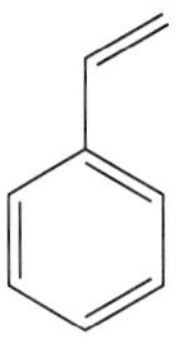    Styrene 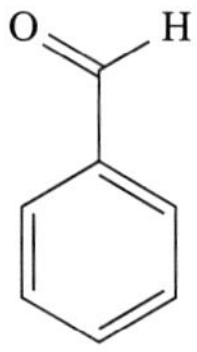    Benzaldehyde 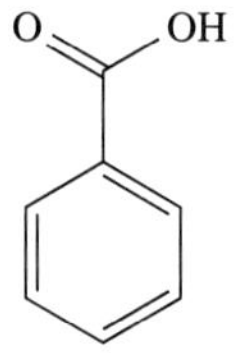    Benzoic acid 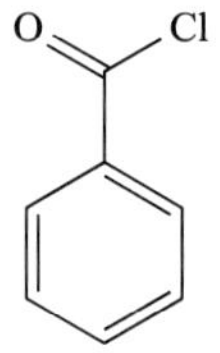    Benzoyl chloride 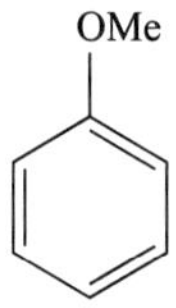    Anisole

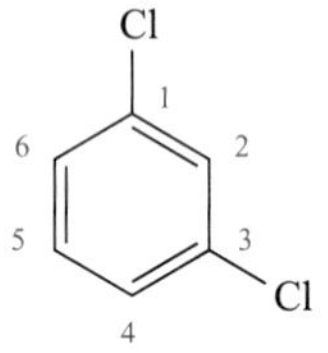

(32.27)

When there is more then one substituent on the benzene ring, numerical descriptors are used so as to place one substituent at ring atom C(1). Thus, compound **32.27** is 1,3-dichlorobenzene. Additional examples are given below:

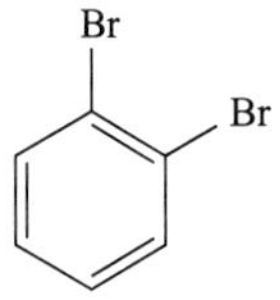

1,2-Dibromobenzene 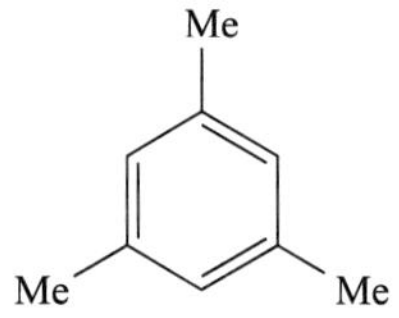    1,3,5-Trimethylbenzene    1,3-Difluorobenzene 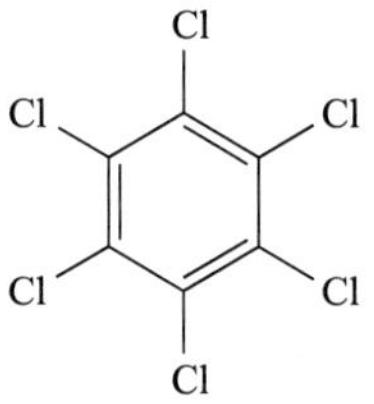    Hexachlorobenzene
(Common name =    (No numbering needed)
mesitylene)

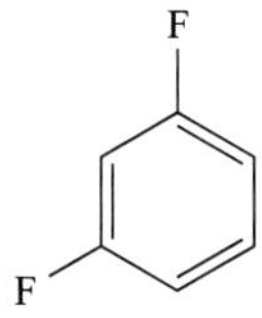

*ortho-* (*o-*) stands for 1,2-substitution.
*meta-* (*m-*) stands for 1,3-substitution.
*para-* (*p-*) stands for 1,4-substitution.

## Ortho-, meta- and para-substitution patterns

A disubstituted derivative of benzene such as dibromobenzene possesses three isomers: 1,2-dibromobenzene, 1,3-dibromobenzene and 1,4-dibromobenzene. These compounds can also be named using the prefixes *ortho-*, *meta-* and *para-* respectively. The examples below illustrate use of this nomenclature for isomers of dimethylbenzene (also called *xylene*):

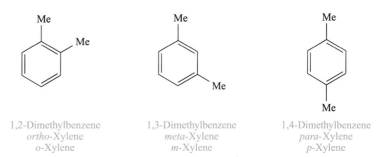

1,2-Dimethylbenzene
*ortho*-Xylene
*o*-Xylene

1,3-Dimethylbenzene
*meta*-Xylene
*m*-Xylene

1,4-Dimethylbenzene
*para*-Xylene
*p*-Xylene

### Phenyl versus benzyl

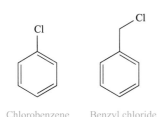

Chlorobenzene    Benzyl chloride

(32.28)         (32.29)

It is important to distinguish between a *phenyl* and a *benzyl* group:

- phenyl $= C_6H_5 =$ Ph;
- benzyl $= C_6H_5CH_2 =$ PhCH$_2$.

A compound such as *chlorobenzene* (**32.28**) has a chloro group attached *directly* to the aromatic ring. In *benzyl chloride* (**32.29**), the chloro group is part of the alkyl substituent. Thus, benzyl chloride behaves as a halogeno-alkane (see equation 28.25 and accompanying discussion). The structures of chlorobenzene and benzyl chloride are compared in Figure 32.7.

**Fig. 32.7** Ball-and-stick models of (a) chlorobenzene and (b) benzyl chloride. Colour code: C, grey; Cl, green; H, white.

(a)                    (b)

---

**32.6** | **Industrial production of benzene**

Benzene is commercially available and lies about 17th in ranking of chemicals manufactured in the US. Figure 32.8 illustrates the scale of manufacture of benzene and some of its derivatives in relation to the industrial output of ethene, propene and oxirane (see Sections 26.7 and 29.3). Industrially, $C_6H_6$ is made by the catalytic dehydrogenation of cyclohexane (equation 32.7).

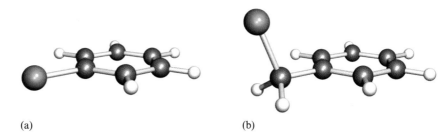

$$\text{Pt catalyst; 380 K} \quad -3H_2$$

(32.7)

---

**32.7** | **Physical properties of benzene**

*Caution!* Benzene is a carcinogen and prolonged exposure may cause leukaemia

Benzene is a colourless liquid at room temperature (mp 278.7 K, bp 353.2 K) and has a characteristic smell. It is non-polar and immiscible with water; $C_6H_6$ (density $= 0.89\,\text{g cm}^{-3}$) is less dense than $H_2O$ (density $= 1.00\,\text{g cm}^{-3}$). It is miscible with other non-polar organic solvents and alcohols. At one time, benzene was widely used as a solvent, but the toxicity of both liquid and vapour is now well recognized and its use in the laboratory is restricted.

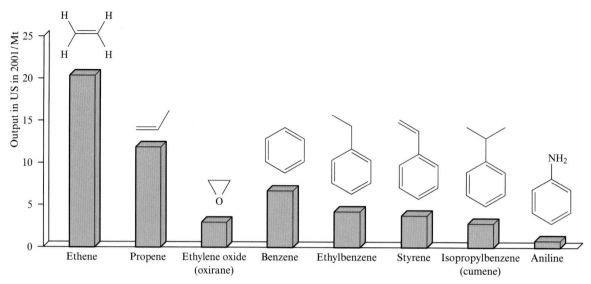

**Fig. 32.8** Industrial output in the US of benzene and important derivatives in 2001, compared to the US output of ethene, propene and oxirane. Data are in megatonnes (Mt). [Data: *Chemical & Engineering News* (2002) 24 June, p. 60.]

| 32.8 | **Reactivity of benzene** |

### Combustion

Like other hydrocarbons, $C_6H_6$ burns in air and $O_2$ to give $CO_2$ and $H_2O$ (equation 32.8).

$$2C_6H_6(l) + 15O_2(g) \longrightarrow 12CO_2(g) + 6H_2O(l) \tag{32.8}$$

$$\Delta_c H^\circ (298\,K) = -3268\,kJ \text{ per mole of } C_6H_6$$

### Hydrogenation

The hydrogenation of benzene to give cyclohexane (see Figure 32.2) can be achieved by reaction of $C_6H_6$ under a very high pressure of $H_2$ in the presence of a platinum catalyst. These drastic conditions reflect the stability of the aromatic system. Scheme 32.9 compares the conditions needed to hydrogenate a typical alkene and benzene.

$$\xrightarrow[\text{Pd or Pt catalyst}]{H_2;\ 1\ \text{bar};\ 298\ K}$$

$$(32.9)$$

$$\xrightarrow[\text{Pt catalyst}]{H_2;\ 150\ \text{bar};\ 298\ K}$$

A ***Birch reduction*** is reduction using Li, Na or K in liquid $NH_3$ in the presence of an alcohol, ROH.

Under these harsh conditions, hydrogenation is complete. Controlled reduction is achieved by a *Birch reduction*. Benzene is added to a solution of an alkali metal (Li, Na or K) in liquid $NH_3$ containing a small amount of EtOH. The addition of an alkali metal to liquid $NH_3$ generates solvated electrons (see equation 21.112). Reduction of benzene to a radical anion occurs (equation 32.10).

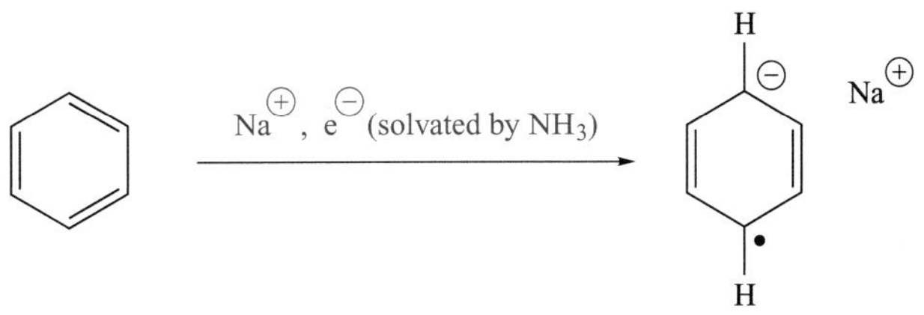

$$(32.10)$$

The alcohol acts as a source of protons, and the radical anion above is rapidly converted to a radical, which is further reduced and then protonated again as shown in equation 32.11.

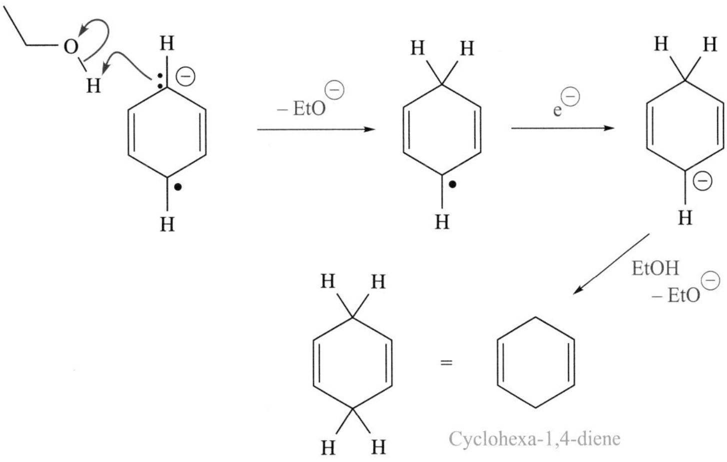

$$(32.11)$$

## Substitution reactions: an overview

The characteristic reactions of benzene are *substitution of H* by an atom or functional group and *most substitutions require a catalyst*. This introduction gives examples of substitution reactions with appropriate reaction conditions. In the next section, we look at the mechanisms of these reactions.

The reaction of $C_6H_6$ with $Cl_2$ to give chlorobenzene occurs in the presence of $AlCl_3$ (equation 32.12). Aluminium chloride is a Lewis acid and the chlorination of benzene is an example of a *Lewis acid-catalysed reaction*. An alternative Lewis acid catalyst is $FeCl_3$.

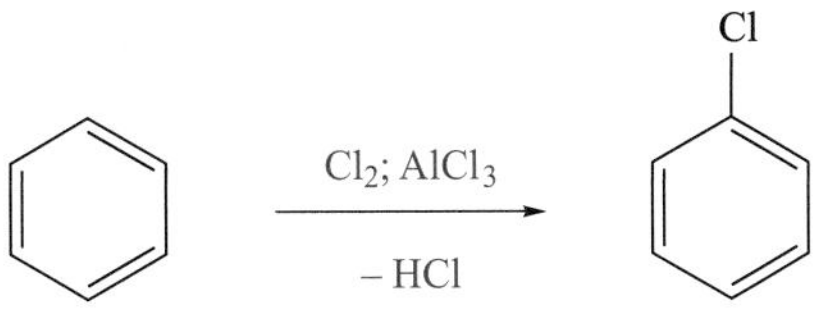

◄►
Lewis acidity of $AlCl_3$:
see discussion
accompanying
equation 22.21

$$(32.12)$$

Similarly, bromination of benzene using $Br_2$ takes place in the presence of Lewis acids such as $FeBr_3$ or $AlCl_3$. Iodination is more difficult because $I_2$ is less reactive than $Cl_2$ or $Br_2$.

Alkyl derivatives of $C_6H_6$ can be produced by *Friedel–Crafts alkylation*. This involves the reaction of a chloro- or bromoalkane, RX, with $C_6H_6$ in the presence of a Lewis acid catalyst (equation 32.13). *Polyalkylation* is a complication and rearrangements may also occur as in reaction 32.14. We return to these effects in Section 32.9.

In a *Friedel–Crafts alkylation*, an alkyl substituent R is introduced into an aromatic ring by reaction with RCl or RBr in the presence of a Lewis acid catalyst.

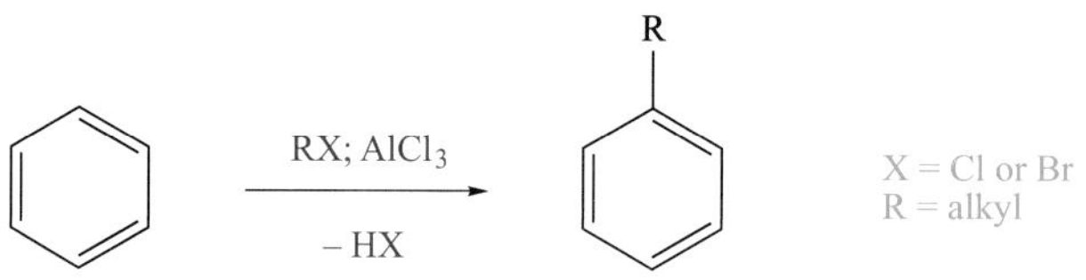

$$(32.13)$$

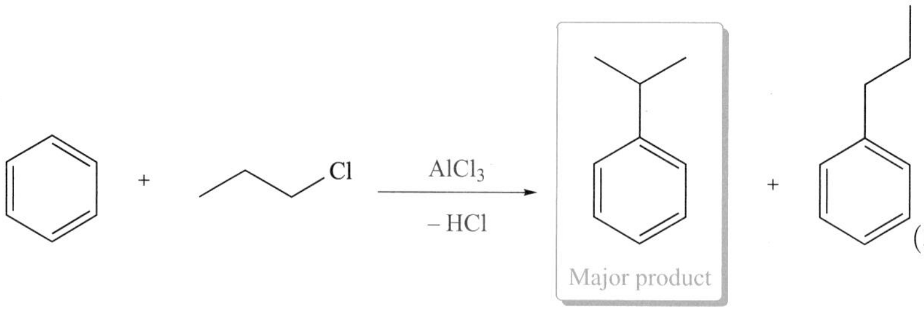

X = Cl or Br
R = alkyl

$$(32.14)$$

1-Chloropropane          Isopropylbenzene          *n*-Propylbenzene
                         (Cumene)

In a ***Friedel–Crafts acylation***, an acyl substituent RCO is introduced into an aromatic ring by reaction with RCOCl in the presence of a Lewis acid catalyst.

*Friedel–Crafts acylation* of benzene is the reaction with an acyl chloride RCOCl in the presence of a Lewis acid catalyst to give a ketone (equation 32.15). Acylations do *not* suffer from the multiple substitutions or rearrangements that are a problem with Friedel–Crafts alkylations.

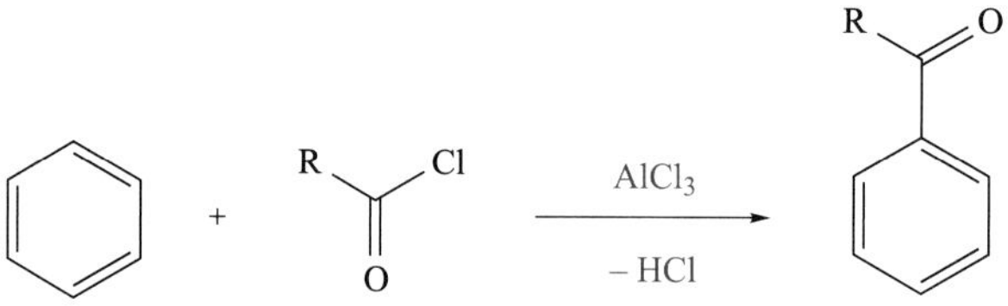

$$(32.15)$$

Nitrobenzene is a precursor to aromatic amines: see Section 32.13

The *nitration* of benzene is carried out using a mixture of concentrated $HNO_3$ and $H_2SO_4$. This acid combination generates $NO_2^+$ as the reactive species (equation 32.16); $H_2SO_4$ acts as a catalyst in the reaction. Reaction 32.17 summarizes the nitration of $C_6H_6$.

$$HNO_3 + H_2SO_4 \rightleftharpoons H_2NO_3^+ + HSO_4^-$$
$$H_2NO_3^+ \rightleftharpoons NO_2^+ + H_2O$$

$$(32.16)$$

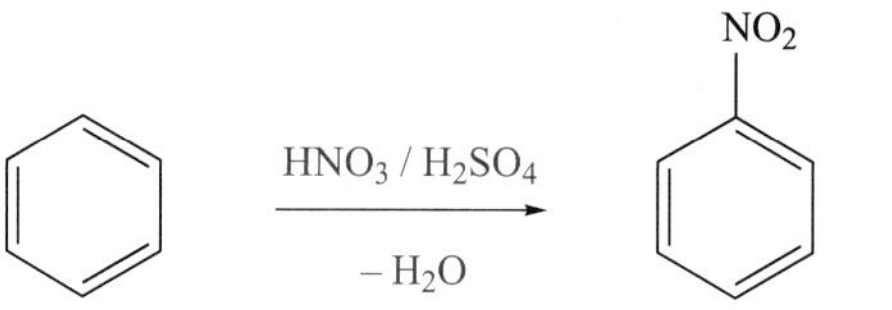

$$(32.17)$$

Nitrobenzene

The reaction between $C_6H_6$ and concentrated sulfuric acid and $SO_3$ (*fuming sulfuric acid* or *oleum*) leads to *sulfonation* of the ring. The reactive species is $HSO_3^+$ which is generated in equilibrium 32.18. Reaction 32.19 summarizes the sulfonation of $C_6H_6$.

$$H_2SO_4 + SO_3 \rightleftharpoons HSO_3^+ + HSO_4^-$$

$$(32.18)$$

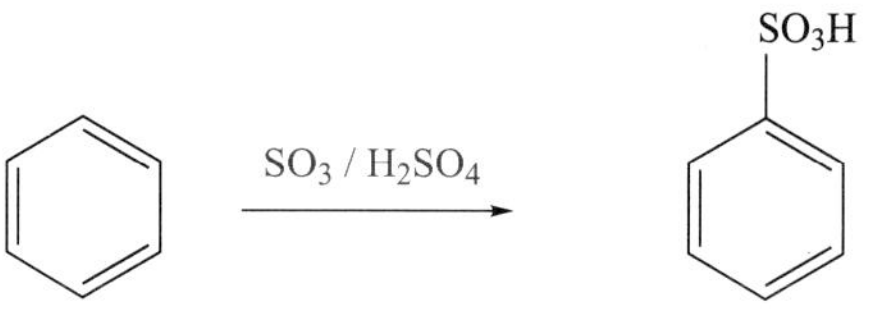

$$(32.19)$$

Benzenesulfonic acid

■▶
See also Box 5.4

Sulfonation is one step in the manufacture of the drug *Viagra* which came on to the market in 1999 to treat male erectile dysfunction. The drug inhibits the enzyme phosphodiesterase, thereby sustaining erection of the penis. Despite its success, Viagra has a number of side effects.

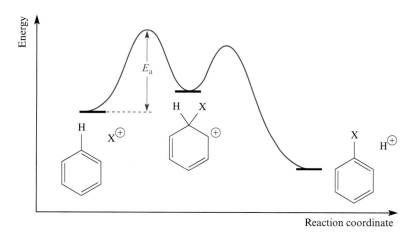

Viagra

---

## The mechanism of electrophilic substitution

### The general mechanism

The substitution reactions described in the previous section are *electrophilic substitutions*. The mechanism involves two steps:

- the reaction between the aromatic ring and an electrophile to give an intermediate carbenium ion;
- loss of the leaving group.

The first step in the reaction of benzene with an electrophile $X^+$ is shown in equation 32.20. The direction of movement of the electron pair shows that the *aromatic ring behaves as a nucleophile* and donates a pair of electrons to the electrophile. Step 32.20 is the *rate-determining step* of the electrophilic substitution.

$$ \text{(32.20)} $$

Slow step

The carbenium ion is an *intermediate* (see Figure 32.9) and possesses a 4-coordinate (tetrahedral) C atom. The six π-electron delocalized system of

**Fig. 32.9** Reaction profile for the electrophilic substitution of $H^+$ in benzene by $X^+$. One of three resonance structures for the intermediate is shown (see **32.30**).

the aromatic ring has been destroyed. Only four $\pi$-electrons remain and the Hückel $(4n + 2)$ rule is no longer obeyed. Although the positive charge in the intermediate is shown in equation 32.20 as being localized on one C atom, this structure is only one of a set of contributing resonance forms, and the charge is actually delocalized around the ring. This is represented either as the set of resonance structures shown in **32.30**, or by the single structure **32.31**. We return to electron distribution around the ring when we discuss orientation effects later in the chapter.

(32.30)                  (32.31)

Structure **32.32** emphasizes the geometries and appropriate hybridization schemes for the ring C atoms in the intermediate carbenium ion. The tetrahedral $sp^3$ C atom uses all its valence orbitals to form $\sigma$-bonds (C–C, C–H or C–X).

In the next step of the electrophilic substitution reaction, $H^+$ is lost from the intermediate by heterolytic bond cleavage of the C–H bond at the tetrahedral C atom. This is represented in equation 32.21. The step is *fast*, i.e. non-rate-determining. In this equation, we show only one of the resonance structures for the intermediate (see problem 32.14).

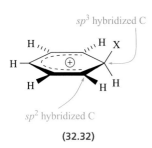

$sp^3$ hybridized C

$sp^2$ hybridized C

(32.32)

$$\text{Fast step} \longrightarrow \quad + \quad H^{\oplus} \qquad (32.21)$$

Figure 32.9 shows a reaction profile for the overall reaction of $C_6H_6$ with $X^+$. The reaction intermediate is at a local energy minimum. Each step has an associated activation barrier, but that for the first (i.e. rate-determining) step is the higher. The second step shown in Figure 32.9 is accompanied by rehybridization of one C atom from $sp^3$ to $sp^2$. As this occurs, a new $\pi$-interaction is formed and the delocalized $\pi$-system returns to the ring. The product (like the reactant benzene) is therefore stabilized by delocalization of six $\pi$-electrons. *Substitution* is energetically preferable to addition to the ring, e.g. reaction 32.22 is energetically more favourable than reaction 32.23. *In reaction 32.22, aromaticity is retained, but in reaction 32.23, aromaticity is destroyed.*

Benzene undergoes *electrophilic substitution* reactions and the *aromaticity of the ring is retained*. Addition reactions are unfavourable because the aromaticity of the $C_6$ ring is destroyed.

$$+ \ X_2 \longrightarrow \quad + \ HX$$

*Substitution is observed*     (32.22)

$$+ \ X_2 \longrightarrow$$

*Addition is **not** favoured*     (32.23)

Now we consider specific electrophilic substitution reactions, and the roles of catalysts in generating the electrophiles.

## Chlorination and bromination

Halogenation reactions require a *Lewis acid catalyst* to encourage the formation of the electrophile. Suitable catalysts are $AlCl_3$, $FeCl_3$ and $FeBr_3$. Each reacts with $Cl_2$ or $Br_2$ as illustrated in equation 32.24; the diagram in the margin illustrates how $AlCl_3$ acts as a Lewis acid during the reaction.

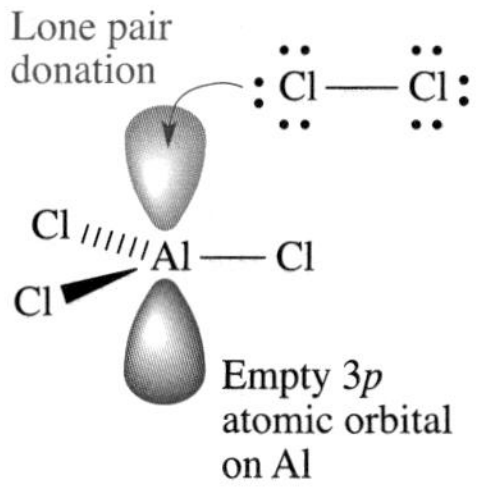

Lone pair donation

Empty $3p$ atomic orbital on Al

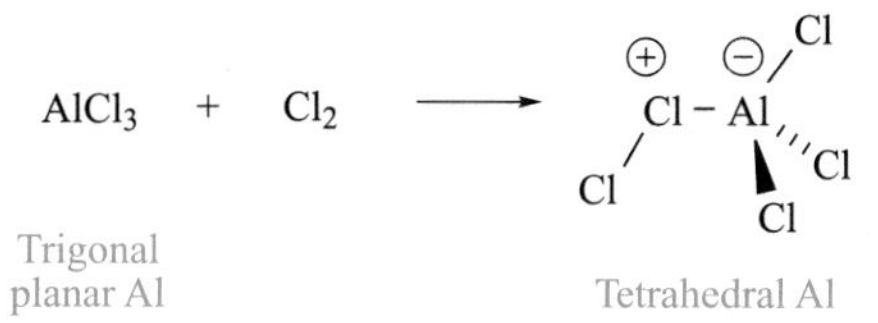

$$AlCl_3 \quad + \quad Cl_2 \quad \longrightarrow \quad \overset{\oplus}{Cl} - \overset{\ominus}{Al} \quad (32.24)$$

Trigonal planar Al

Tetrahedral Al

Scheme 32.25 shows the transfer of $Cl^+$ to $C_6H_6$ followed by release of $H^+$ to form chlorobenzene. Note how the catalyst is regenerated. A similar reaction scheme can be drawn for the conversion of benzene into bromobenzene.

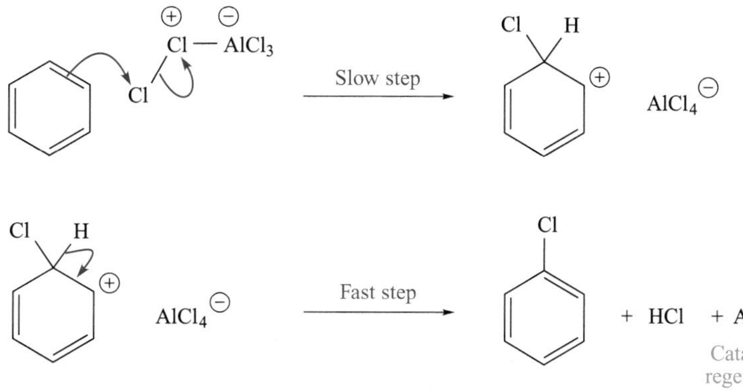

$$(32.25)$$

Catalyst is regenerated

## Friedel–Crafts alkylation

In a Friedel–Crafts alkylation (e.g. equations 32.13 and 32.14), the Lewis acid catalyst behaves in a similar way to that in the halogenations described above. Treatment of an alkyl chloride RCl with $AlCl_3$ generates $R^+$ as in equation 32.26.

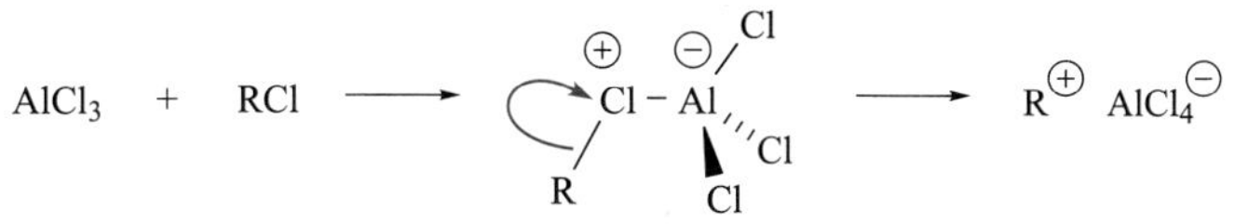

$$AlCl_3 \quad + \quad RCl \quad \longrightarrow \quad \overset{\oplus}{Cl} - \overset{\ominus}{Al} \quad \longrightarrow \quad R^{\oplus} \; AlCl_4^{\ominus} \quad (32.26)$$

The reaction of $R^+$ with benzene now proceeds through the two steps shown in scheme 32.27. Although we show $R^+$ as a separate entity, it is likely that it remains associated with $AlCl_4^-$ during the first step.

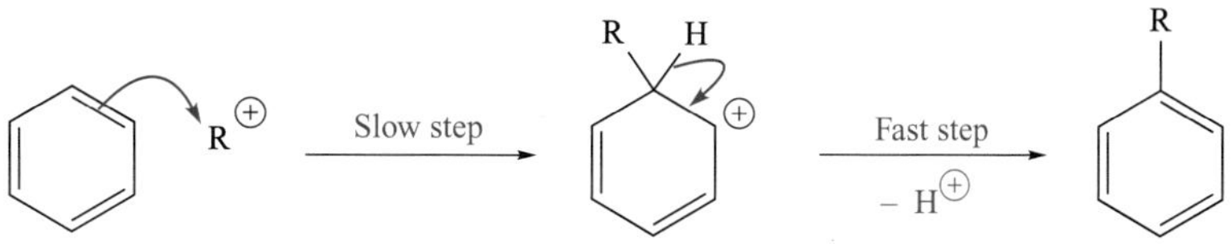

$$(32.27)$$

Relative stabilities of
carbenium ions:
see Section 26.5

In reaction 32.14, we showed the reaction of benzene with 1-chloropropane giving mixed products: isopropylbenzene (the dominant product) and *n*-propylbenzene. These products are a result of the rearrangement of the $R^+$ carbenium ion *prior* to its reaction with the aromatic ring. The relative stabilities of carbenium ions follow the sequence:

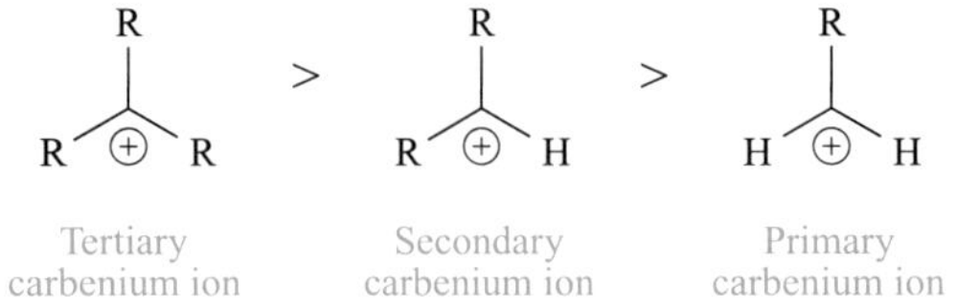

This means that when the alkylating agent is a primary halogenoalkane, the favoured product does not contain a straight chain alkyl substituent because rearrangement of the primary carbenium ion takes place as shown in equation 32.28.

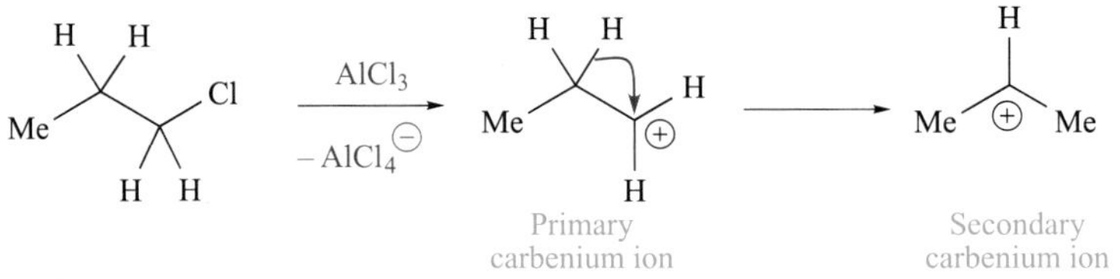

(32.28)

The reaction of benzene with 1-chloropropane gives a mixture of alkylated products that reflects the relative stabilities of the primary and secondary carbenium ions formed after loss of $Cl^-$ from 1-chloropropane. Scheme 32.29 summarizes the reaction.

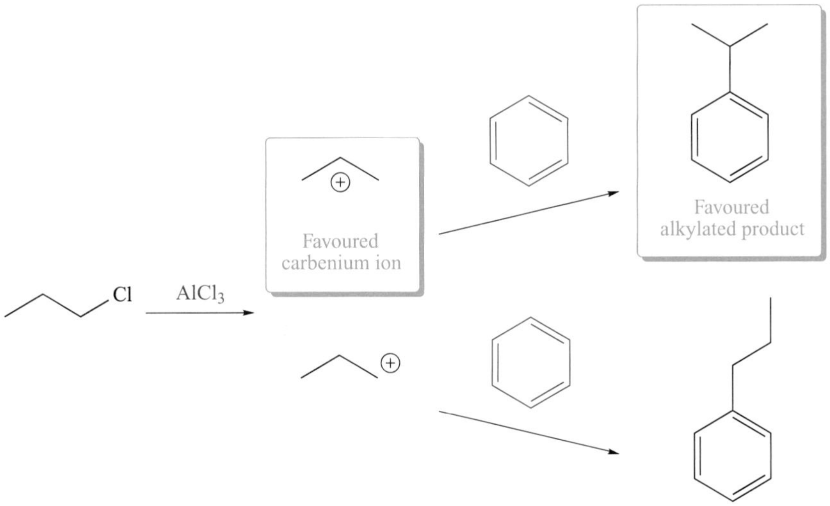

(32.29)

A further complicating factor in the alkylation of aromatic rings is that the introduction of the first alkyl substituent *activates* the ring towards further substitution at the *ortho-* and *para*-positions. In the synthesis of toluene from benzene, a mixture of products is obtained because the entry of the first Me substituent encourages further electrophilic substitution as shown in equation 32.30. In Section 32.10 we explain the origin of these observations.

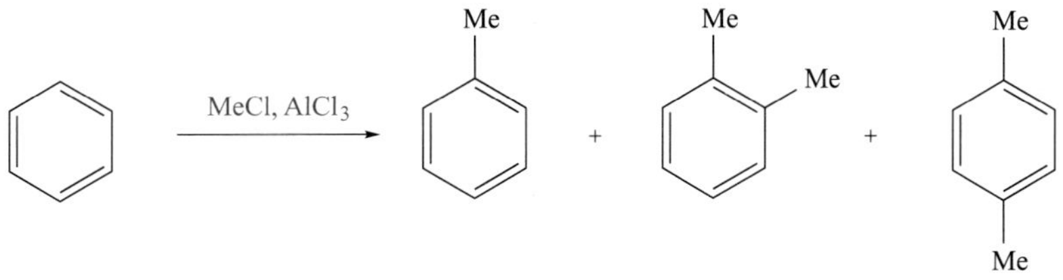

(32.30)

## Friedel–Crafts acylation

Acylation occurs in a similar manner to alkylation. The precursor is an *acyl chloride* and this reacts with a Lewis acid catalyst to produce an electrophile. Scheme 32.31 illustrates the formation of the electrophile and its subsequent reaction with benzene. The acyl group *deactivates* the ring and multiple substitution is not a problem as it is with alkylation.

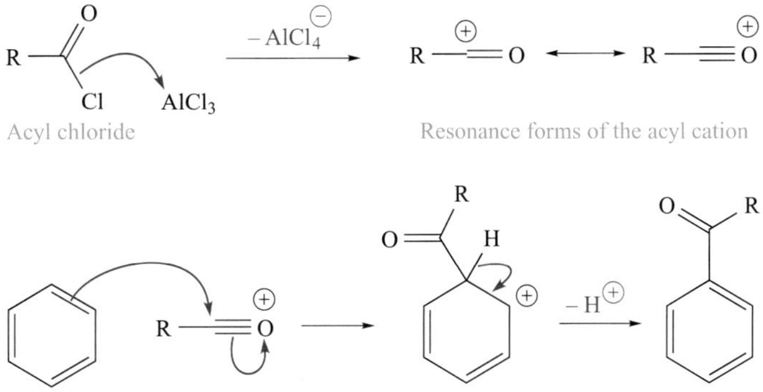

Acyl chloride          Resonance forms of the acyl cation

(32.31)

## Nitration

Nitration of benzene cannot be carried out using $HNO_3$ alone. It requires a mixture of concentrated nitric and sulfuric acids. Sulfuric acid is a stronger acid than $HNO_3$ and a proton is transferred from the stronger to the weaker acid:

$$HNO_3 \;+\; H_2SO_4 \;\rightleftharpoons\; H_2NO_3^+ + HSO_4^-$$

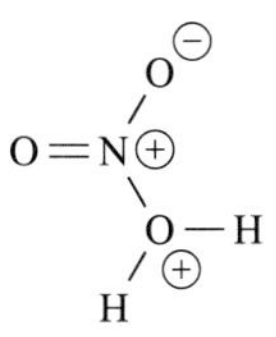

Brønsted base    Brønsted acid

The $H_2NO_3^+$ ion (**32.33**) readily loses $H_2O$ to generate the *nitryl* (or *nitronium*) ion, $NO_2^+$:

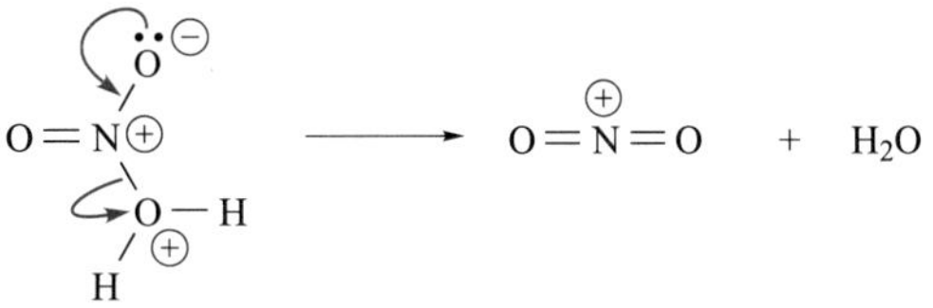

(32.33)

This is an electrophile and reacts with $C_6H_6$ to give nitrobenzene (scheme 32.32).

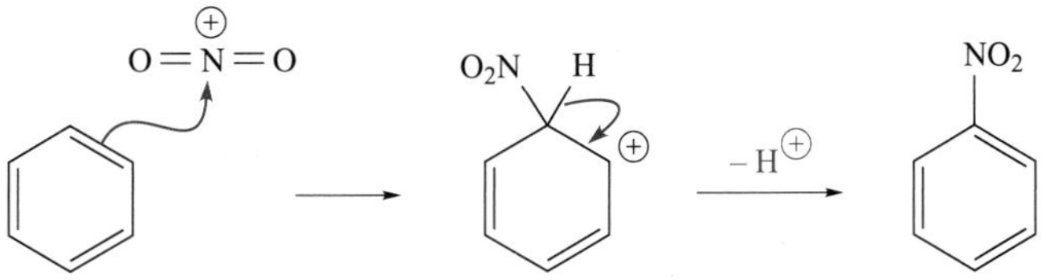

(32.32)

## Sulfonation

See problem 32.14

Benzenesulfonic acid is made by the sulfonation of $C_6H_6$. The electrophile in the reaction is either $SO_3$ or $HSO_3^+$. The latter is produced by the protonation of $SO_3$ in concentrated $H_2SO_4$. Scheme 32.33 illustrates the reaction of benzene with $HSO_3^+$.

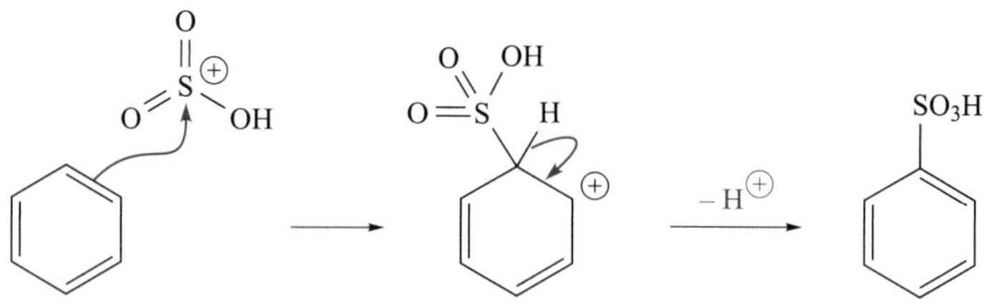

$$(32.33)$$

## 32.10 Orientation effects, and ring activation and deactivation

In the rest of this chapter, we consider the chemistry of derivatives of benzene. Electrophilic substitution reactions remain important. However, before discussing them, we must look at the effects of introducing substituents into the benzene ring. We have already mentioned that alkylations of benzene are complicated by the substitution of more than one alkyl group. For example, an Me group directs the second Me substitution into the *ortho-* and *para*-positions (equation 32.30). We have also introduced the terms *activating* and *deactivating* with respect to the effects that the alkyl and acyl groups, respectively, have on the reactivity of the $C_6$ ring towards further substitution.

Consider reaction 32.34. Although, in theory, three isomers of $C_6H_4XY$ can be formed, in practice, particular isomers are favoured.

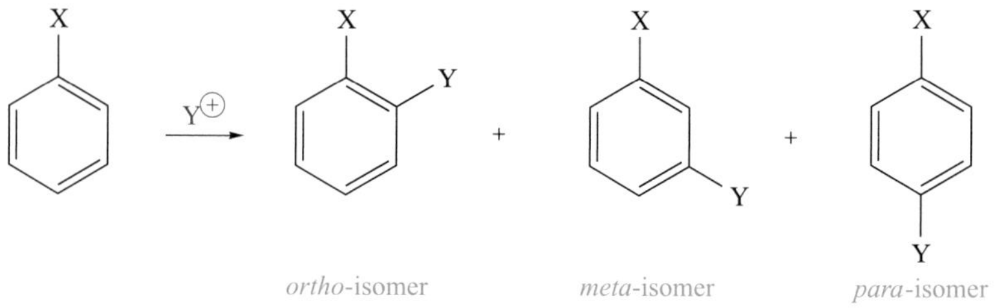

*ortho*-isomer        *meta*-isomer        *para*-isomer

$$(32.34)$$

The substituent X is:

Electron-releasing groups: see Sections 25.8 and 26.5

- either *electron-releasing* (e.g. X = alkyl) or *electron-withdrawing* (e.g. X = F, Cl or $NO_2$);
- either *activating* (rate of substitution is increased) or *deactivating* (rate of substitution is decreased).

The interplay of these two factors determines the position of the second substitution. Table 32.3 lists the effects observed for selected substituents. *Activation* of the ring tends to be associated with *ortho-* and *para*-directing groups, whereas *deactivation* is usually associated with *meta*-directing groups. Halo-substituents are the odd ones out. To understand these observations, we now look at specific examples.

**Table 32.3** The orientation and activation effects of substituents X in monosubstituted aromatic compounds $C_6H_5X$.

| Substituent X | *Ortho-*, *para-* or *meta*-directing? | Activating or deactivating? |
|---|---|---|
| Me | *ortho-* and *para*-directing | Activating |
| Et | *ortho-* and *para*-directing | Activating (less than Me) |
| OH | *ortho-* and *para*-directing | Activating |
| $NH_2$ | *ortho-* and *para*-directing | Activating |
| F | *ortho-* and *para*-directing | Deactivating |
| Cl | *ortho-* and *para*-directing | Deactivating |
| Br | *ortho-* and *para*-directing | Deactivating |
| I | *ortho-* and *para*-directing | Deactivating |
| $NO_2$ | *meta*-directing | Deactivating |
| CN | *meta*-directing | Deactivating |
| $SO_3H$ | *meta*-directing | Deactivating |
| CHO | *meta*-directing | Deactivating |
| COMe | *meta*-directing | Deactivating |
| $CO_2H$ | *meta*-directing | Deactivating |

## Aniline (phenylamine) and phenol

Aniline and phenol contain substituents with one or two lone pairs, respectively, on the atom attached to the $C_6$ ring. Resonance structures for these compounds are shown below and illustrate a distribution of negative charge onto the *ortho-* and *para-*C atoms:

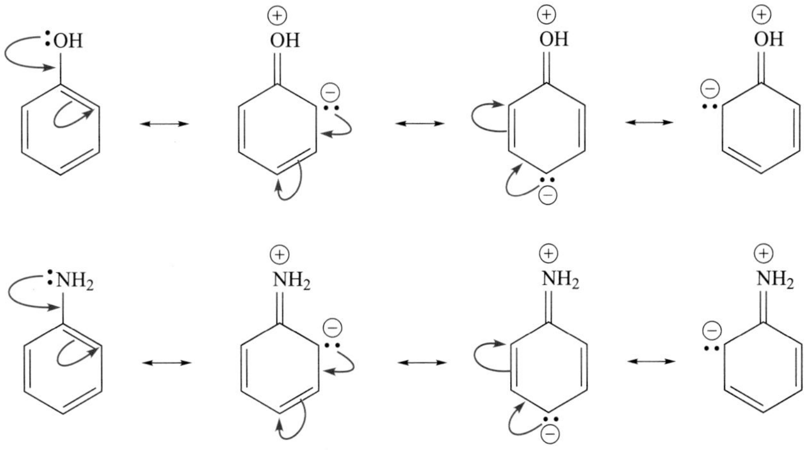

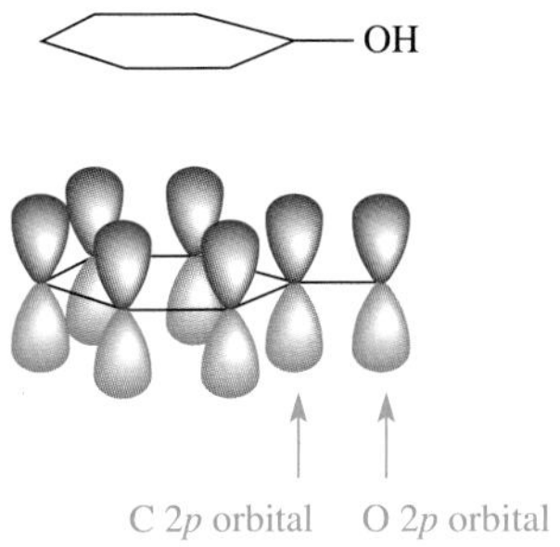

C 2*p* orbital    O 2*p* orbital

**(32.34)**

Conjugation of $\pi$-electrons in phenoxide ion: see Section 32.12

The donation of the lone pair electrons takes place by *conjugation into the $\pi$-system* as shown in diagram **32.34**. The $C^{\delta-}$ *ortho-* and *para*-positions in aniline and phenol are *activated* towards attack by electrophiles. Rates of substitution are relatively fast. Since N is less electronegative than O, the lone pair in aniline is more readily available than that in phenol, making the resonance stabilization greater for aniline. As a result, aniline is more activated than phenol towards electrophilic attack at the *ortho-* and *para*-positions.

Now consider what happens when an electrophile, $Y^+$, attacks aniline. The resonance structures for each of the three possible intermediates (*ortho-*, *meta-* and *para-*substitution) are drawn below:

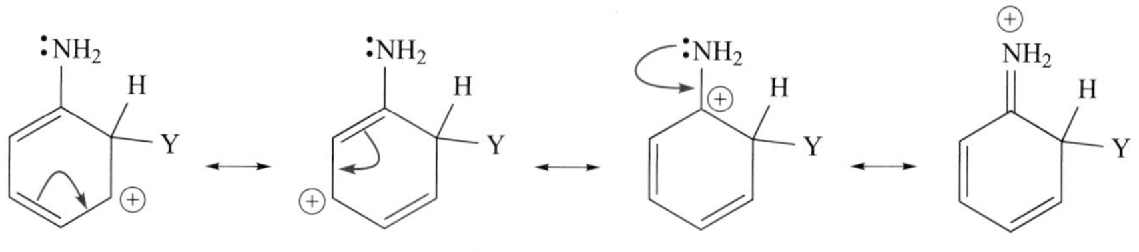

Intermediate for *ortho*-substitution

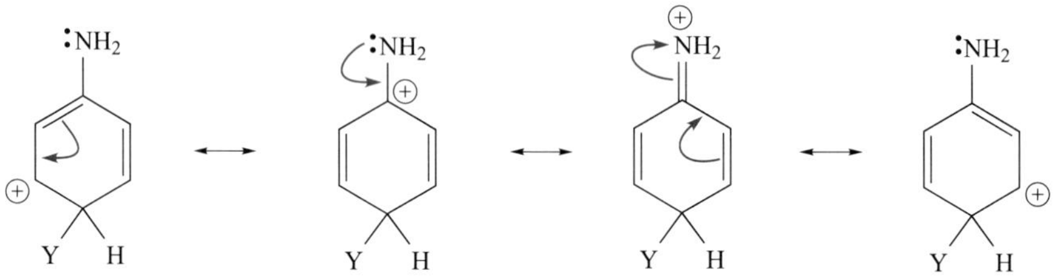

Intermediate for *para*-substitution

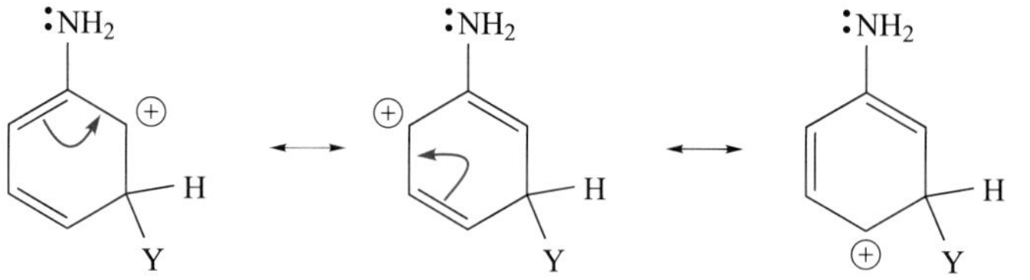

Intermediate for *meta*-substitution

For *ortho*-substitution, resonance contributions for the carbenium ion intermediate include one in which the N atom bears the positive charge. The same is true for the intermediate in *para*-substitution, but this is not the case for *meta*-substitution. The extra resonance stabilization of the intermediate in *ortho-* and *para*-substitution results in a lowering of its energy. Stabilization of the intermediate *also stabilizes the first transition state* (see Figure 32.9) because the transition state has some features in common with the intermediate that follows it in the reaction pathway. It follows that a more stable carbenium ion means a lower activation energy, $E_a$, and a greater rate of reaction. A similar argument can be put forward for faster rates of substitution at the *ortho-* and *para*-sites in phenol compared with the *meta*-positions.

In summary, resonance effects activate the *ortho-* and *para*-positions in the ground state of aniline and phenol, and are also responsible for the lowering of the activation barrier for substitution at these same sites. Thus, the OH and $NH_2$ groups are *activating* and *ortho-* and *para-directing*. But a word of caution: see the discussions following equations 32.50 and 32.63.

## Nitrobenzene

The nitro group is electron-withdrawing. The resonance structures on the following page show how charge is withdrawn from the ring in nitrobenzene, leaving the *ortho-* and *para*-positions $\delta^+$. The charge is delocalized by conjugation involving the $\pi$-system:

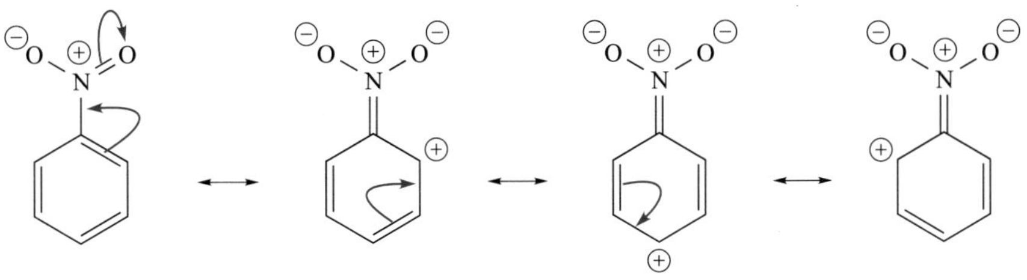

The ring is therefore *deactivated* with respect to attack by electrophiles. The resonance structures indicate that more electronic charge is removed from the *ortho*- and *para*-positions than the *meta*-positions. Thus, when nitrobenzene reacts with electrophiles, it does so at the *meta*-carbon atoms: i.e. the $NO_2$ group is *meta-directing*.

Now consider the carbenium ion intermediate during the reaction between nitrobenzene and an electrophile $Y^+$:

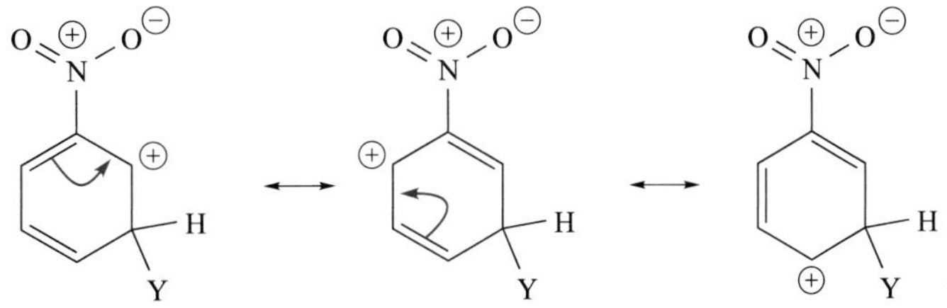

Intermediate for *meta*-substitution

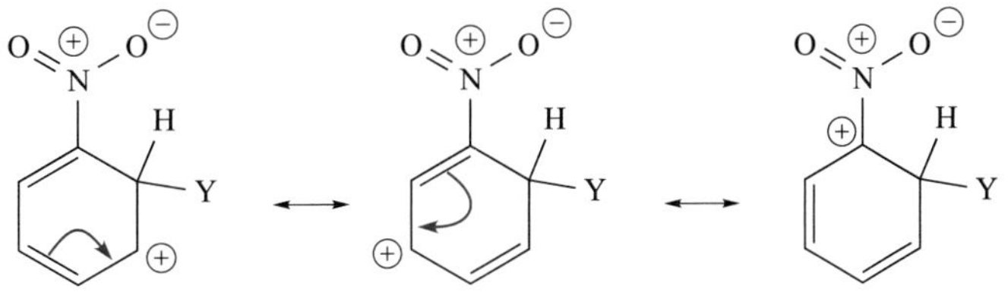

Intermediate for *ortho*-substitution

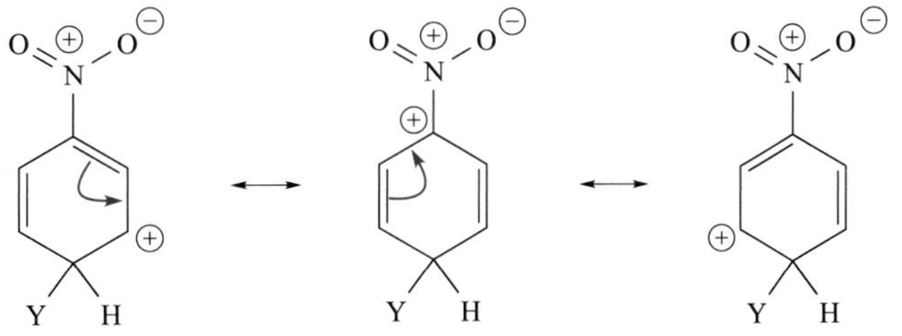

Intermediate for *para*-substitution

The resonance structures above show that when $Y^+$ enters in the *meta*-position, the positive charge is delocalized around the ring on the *ortho*- and *para*-positions. When $Y^+$ enters in the *ortho*- or *para*-position, delocalization of the charge can again occur, but there is a contribution from a resonance form in which the positive charge resides on the C atom adjacent to the positively charged N atom. This destabilizes the intermediate, raising it in energy. Destabilization of the intermediate also destabilizes the first transition state (see Figure 32.9) and raises its energy. It follows that the activation energy is increased, and as a consequence, substitutions at the *ortho-* and *para-*positions are slower than at the *meta-*site.

In summary, resonance effects deactivate the ring towards electrophilic substitution in the ground state of nitrobenzene. They are also responsible for raising the activation barrier for substitution at the *ortho-* and *para-*positions, and making substitution at the *meta-*position faster than at the *ortho-* and *para-*sites. Therefore, the $NO_2$ group is *deactivating* and *meta-directing*.

## Halobenzene derivatives

Inductive effects: see Section 27.4

Halogen substituents are electronegative and can withdraw electrons from the ring in halobenzene derivatives. This is an *inductive effect*, and is greatest for the most electronegative halogen, i.e. $F > Cl > Br > I$. There is an opposing effect: halo-groups can use their lone pair electrons to enter into *π-conjugation* just as we described for OH and $NH_2$ groups (see diagram **32.34**). We can use resonance stabilization arguments analogous to those for aniline and phenol to rationalize why a halo-group directs electrophilic substitution into the *ortho-* and *para-*positions. However, Table 32.3 lists F, Cl, Br and I as being *deactivating*, in contrast to the activating properties of $NH_2$ and OH groups. Why is there a difference? Two effects must be considered:

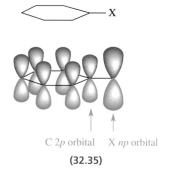

C 2p orbital     X np orbital

(32.35)

- The highly electronegative F atom withdraws negative charge from the ring, making the C atoms $\delta^+$, militating against attack by electrophiles: inductive effects win out over π-conjugation, and the ring is deactivated in fluorobenzene.
- Effective π-conjugation (diagram **32.35**) depends on there being good overlap of the C 2p orbital with the *np* orbital of halogen X. For the later halogens, overlap is poor. (This is exactly the same as the argument that we put forward in Section 22.3 to rationalize the effectiveness of π-bonding in $BF_3$ compared with that in $BCl_3$, $BBr_3$ and $BI_3$.) For Cl, Br and I, π-conjugation is not sufficiently effective to activate the rings in chlorobenzene, bromobenzene and iodobenzene towards electrophilic substitution. Substitution is therefore slower than in aniline or phenol.

## Toluene

C–H σ-orbital can overlap with one lobe of the 2p orbital

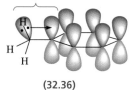

(32.36)

The methyl group in toluene releases electrons into the $C_6$ ring through *hyperconjugation* as shown in diagram **32.36**. (Compare this with the discussion in Section 26.5 of hyperconjugation in carbenium ions.) Electrons from a C–H σ-bond are conjugated into the ring π-system. This directs electrophiles to attack at the *ortho-* and *para-*positions, just as in phenol or aniline.

In the following sections, we shall see orientation effects at work as we describe the reactivity of selected derivatives of benzene. Before working through these sections, you should use the following questions to test your understanding of the first part of Chapter 32.

## MID-CHAPTER PROBLEMS

**1**   Draw a representation of the structure of benzene and comment on the bonding.

**2**   Comment on what is meant by the following diagram:

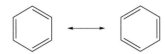

**3**   $C_5H_5^-$ is aromatic. Which of the following properties do you associate with this anion: (a) planarity; (b) non-planarity; (c) equal C−C bond lengths; (d) localized C−H bonds; (e) localized $\pi$-bonds; (f) cyclic $C_5$ framework; (g) $(4n + 2)$ $\pi$-electrons.

**4**   The IR spectrum of benzene has absorptions at 3120–3000, 1480 and 680 $cm^{-1}$. Assign these absorptions.

**5**   (a) The $^{13}C$ NMR spectrum of benzene contains one signal at $\delta$ 128.4 ppm, while that of 1,4-dimethylbenzene contains three signals at $\delta$ 129.0, 134.7 and 20.9 ppm. Comment on these observations.

(b) What spectroscopic method could you use to distinguish between 1,2-dibromobenzene and 1,4-dibromobenzene?

**6**   Draw the structure of chlorobenzene and label the *ipso*-, *ortho*-, *meta*- and *para*-carbon atoms.

**7**   Draw the structures of (a) toluene, (b) phenol, (c) benzoic acid, (d) styrene, (e) aniline and (f) diphenylamine.

**8**   Draw the structures of *o*-, *m*- and *p*-difluorobenzene and give each compound a systematic name (i.e. using position numbers).

**9**   Outline how benzene reacts with $Cl_2$ in the presence of $AlCl_3$, and describe the mechanism of the reaction.

**10**   Comment on the following observations.
(a) In the nitration of benzene, it is necessary to use a mixture of concentrated $HNO_3$ and $H_2SO_4$.
(b) The reaction of benzene with 1-bromopropane in the presence of a Lewis acid catalyst gives isopropylbenzene as the major product.
(c) The $NH_2$ group in aniline is *ortho/para*-directing and activates the ring towards attack by electrophiles.

### 32.11    Toluene

(32.37)

#### Synthesis

Toluene (**32.37**) can be prepared by a Friedel–Crafts methylation of benzene. However, the reaction suffers from multiple substitution (equation 32.30) because the first Me group activates the ring towards further electrophilic substitution at the *ortho-* and *para*-positions. On an industrial scale, toluene is manufactured by dehydrogenating methylcyclohexane (reaction 32.35).

$$\text{Methylcyclohexane} \xrightarrow[\;-3H_2\;]{Al_2O_3/Pt\ catalyst;\ heat} \text{Toluene} \qquad (32.35)$$

#### Physical properties

Toluene is a colourless liquid (mp 178.0 K, bp 383.6 K). It is less toxic than benzene and has, as far as possible, replaced benzene as a laboratory solvent. Like benzene, toluene is non-polar and immiscible with water.

#### Spectroscopy

The IR spectrum of toluene is shown in Figure 32.10. The aromatic and alkyl C−H bonds both have absorptions near to 3000 $cm^{-1}$, but bands assigned to

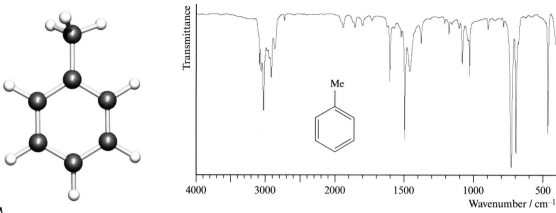

Fig. 32.10 The IR spectrum of toluene.

the alkyl C−H come at lower wavenumber than those assigned to the aromatic C−H. Compare the region around 3000 cm$^{-1}$ in Figure 32.10 with the IR spectra of benzene (Figure 32.5) and ethylbenzene (Figure 12.13). Figure 32.10 shows a strong absorption at 1496 cm$^{-1}$ assigned to the $C_6$ ring vibration. This closely resembles the absorption at 1480 cm$^{-1}$ for benzene (Figure 32.5). The strong absorptions at 730 and 695 cm$^{-1}$ in Figure 32.10 are assigned to the out-of-plane bending of the aromatic C−H bonds. This pair of absorptions is typical of a monosubstituted ring (compare with the single strong absorption at 680 cm$^{-1}$ for benzene in Figure 32.5).

Figure 32.11 shows the proton-decoupled $^{13}$C NMR spectrum of toluene and illustrates the distinction between the chemical shifts of the alkyl and aryl C atoms (see also Table 14.2). On going from benzene to toluene, there is a lowering of molecular symmetry. All the C atoms in $C_6H_6$ are equivalent and the $^{13}$C NMR spectrum exhibits one signal (Figure 32.6b), but in toluene, there are four ring C environments, labelled *b–e* in Figure 32.11. Compare the pattern of signals with that for chlorobenzene in Figure 32.6c. In both chlorobenzene and toluene, the *ipso*-C atom (the one bearing the substituent) comes at highest frequency, and this is a typical observation.

In the $^1$H NMR spectrum of toluene, the protons in the Me group give a singlet at $\delta$ 2.3 ppm. There are three aryl proton environments (*ortho*-, *meta*- and *para*-). The signals for them are at similar chemical shifts, and there is spin–spin coupling between non-equivalent proton nuclei. The final signal appears complex and is composed of overlapping multiplets centred around $\delta$ 7.1 ppm.

Refer also to Figure 32.13b

Fig. 32.11 The $^{13}$C{$^1$H} NMR spectrum of toluene showing peak assignments.

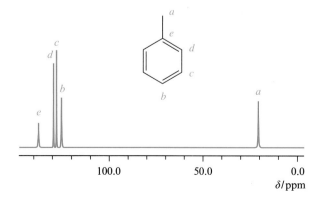

## Reactions

We have already seen that the Me group in toluene is activating and *ortho*- and *para*-directing. Electrophilic substitution in toluene occurs at the 2-, 4- and 6-positions (**32.38**); note that the 2- and 6-positions in toluene are identical. The nitration of toluene first gives a monosubstituted product, either *o*- or *p*-nitrotoluene (equation 32.36).

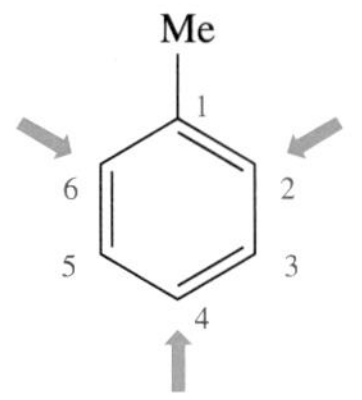

(32.38)

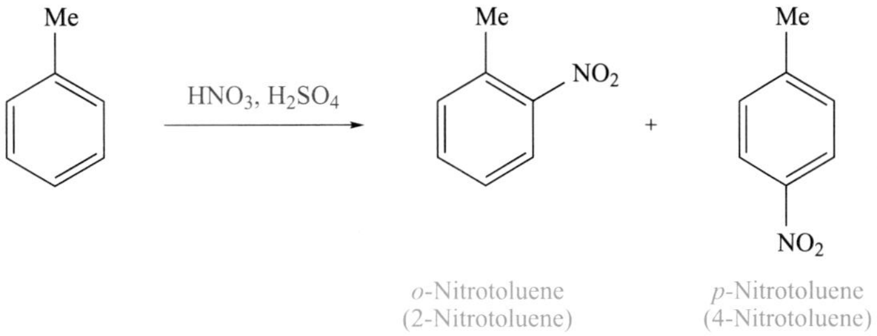

(32.36)

*o*-Nitrotoluene
(2-Nitrotoluene)

*p*-Nitrotoluene
(4-Nitrotoluene)

Further nitration can occur because, although the $NO_2$ group is de-activating, the Me group is activating. The position of substitution depends on the orientation effects of both the Me and $NO_2$ groups. The Me group is *ortho*-directing, while the $NO_2$ group is *meta*-directing. Starting from *p*-nitrotoluene, the orientation effects of the two substituents combine to direct the incoming electrophile as shown in **32.39** and a single dinitro derivative forms (equation 32.37). Starting from *o*-nitrotoluene (**32.40**), two products are possible (equation 32.38).

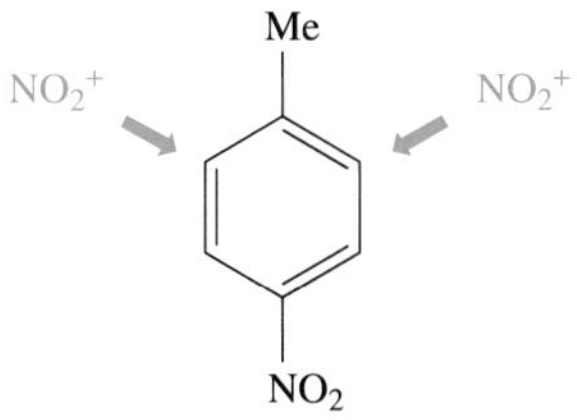

(32.39)

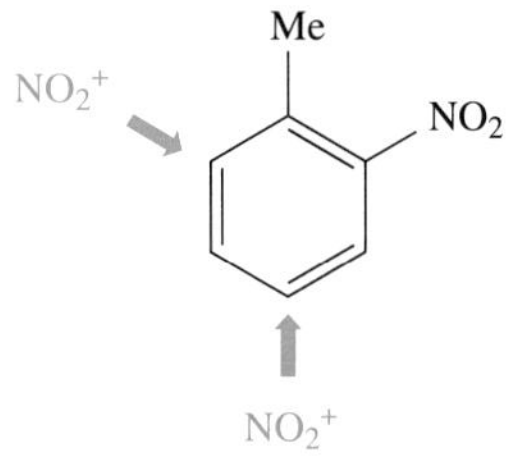

(32.37)

(32.40)

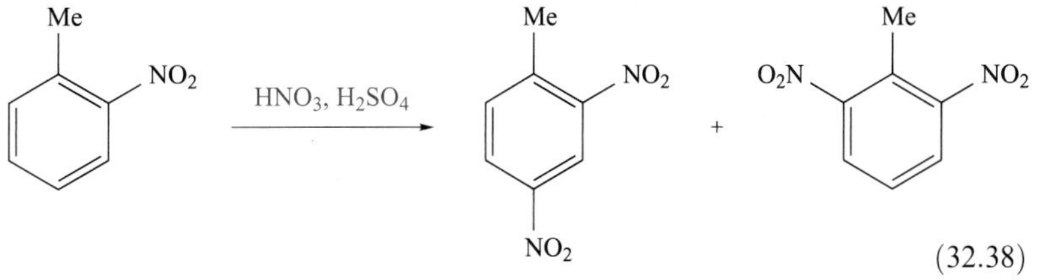

(32.38)

In the presence of $Cl_2$ or $Br_2$ and a Lewis acid catalyst, toluene is converted to *o*-chlorotoluene, *p*-chlorotoluene, *o*-bromotoluene or *p*-bromotoluene. Equation 32.39 illustrates bromination.

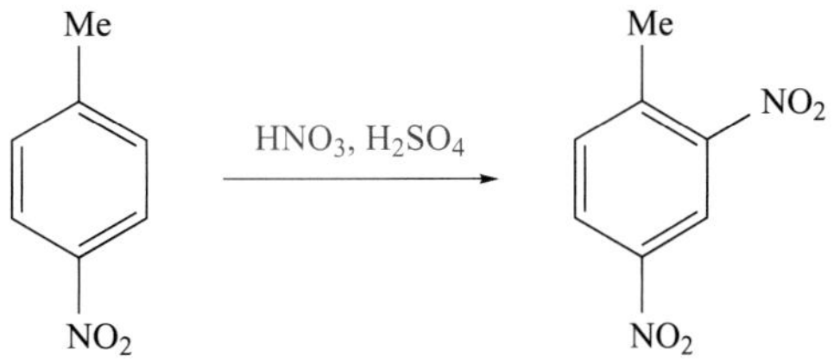

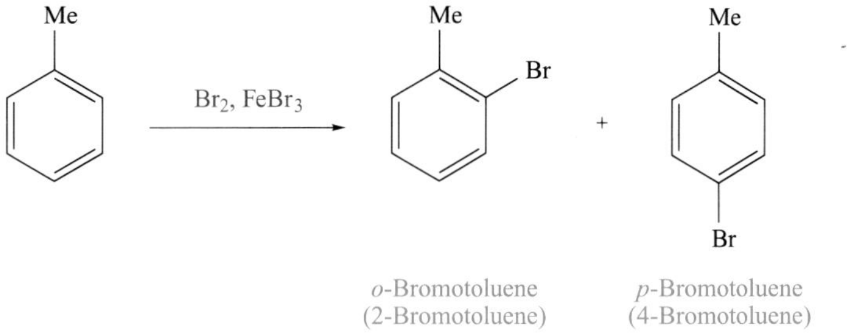

(32.39)

*o*-Bromotoluene
(2-Bromotoluene)

*p*-Bromotoluene
(4-Bromotoluene)

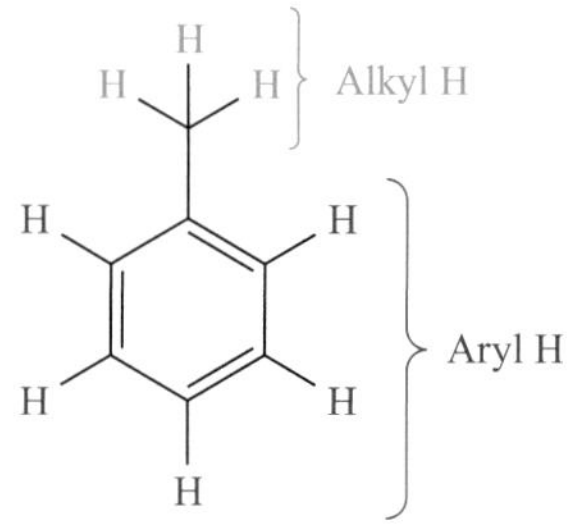

**(32.41)**

In toluene, the aryl ring undergoes electrophilic substitution by $Cl^+$ or $Br^+$, and the alkyl group undergoes radical substitution by $Cl^{\bullet}$ or $Br^{\bullet}$.

Unlike benzene, toluene contains both aryl *and* alkyl H atoms (**32.41**). Halogenation of the *aromatic ring* requires a Lewis acid catalyst. Halogenation of the *alkyl group* occurs when toluene reacts with $Cl_2$ or $Br_2$ under thermal or photolytic conditions (equation 32.40). The mechanism of this radical substitution is the same as for radical substitution in an alkane.

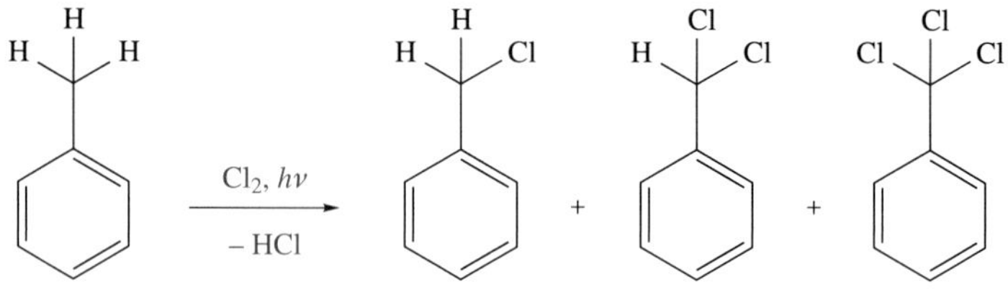

The methyl group in toluene is oxidized by $KMnO_4$ or $K_2Cr_2O_7$ to give a carboxylic acid (equation 32.41). This is in contrast to the fact that neither benzene nor alkanes are oxidized by $KMnO_4$ or $K_2Cr_2O_7$. The reaction can be extended to other alkyl derivatives, e.g. ethylbenzene. As reaction 32.42 shows, oxidation is still at the $CH_2$ group attached to the aromatic ring, giving benzoic acid irrespective of the starting alkylbenzene derivative. We return to benzoic acid and related compounds in Chapter 33.

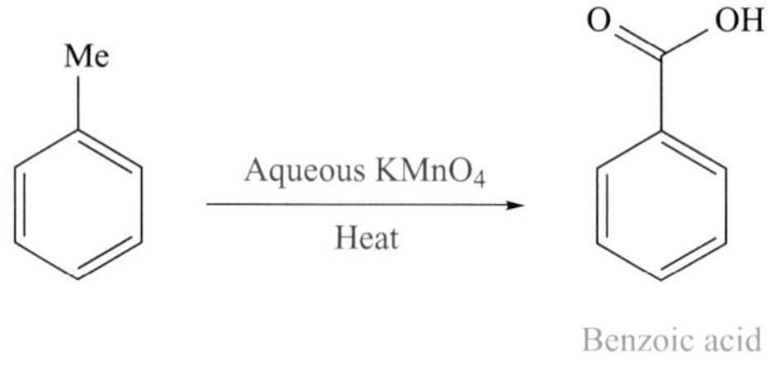

(32.41)

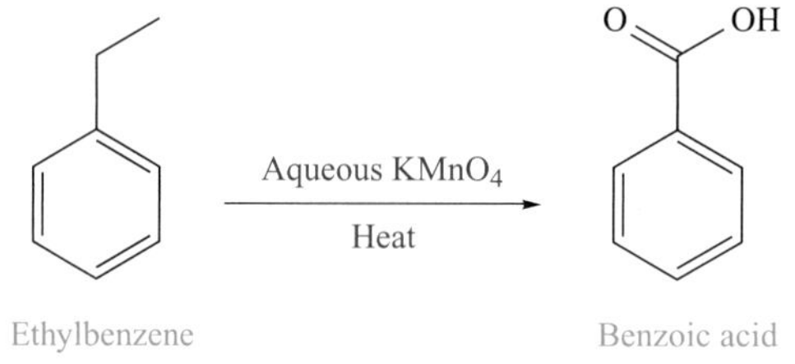

(32.42)

---

**32.12**    **Phenol**

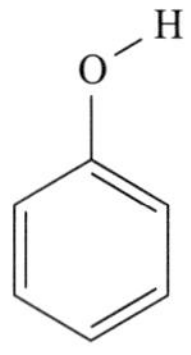

**(32.42)**

Phenols are compounds in which an OH group is attached *directly to an aromatic ring*, and the parent member of this family is called phenol (**32.42**). In Box 32.3, we highlight applications of selected phenols.

### Synthesis

Phenol (PhOH) is manufactured on a large scale and output in the US places it about 30th among industrially produced chemicals. The commercial process for production of phenol involves the oxidation of isopropylbenzene (cumene), and is also a means of obtaining acetone (scheme 32.43).

$$(32.43)$$

Commercial applications of phenol have been important for many decades, and Bakelite (see Box 32.5) was one of the first industrially manufactured polymers. Uses also include the formation of phenolic resins and bisphenol A. The latter is a precursor in the manufacture of some epoxy resins, e.g. that formed from bisphenol A and epichlorohydrin:

Epoxy resins: see Box 29.1

Bisphenol A                                    Epichlorohydrin

Two other methods of preparing phenol are the fusion of a benzene-sulfonate salt with alkali (equation 32.44) and the hydrolysis of a diazonium salt (see equation 32.68). A reaction analogous to that in equation 32.44 but starting with benzene-1,3-disulfonate is used for the manufacture of resorcinol (see Box 32.3).

Sodium                    Sodium phenoxide                    Phenol
benzenesulfonate

$$(32.44)$$

## Physical properties

Phenol is a colourless, hygroscopic, crystalline solid at room temperature (mp 313.9 K). It is polar (O $\delta^-$) and is soluble in water and a range of organic solvents. An old name for phenol is 'carbolic acid'. It is a strong disinfectant; J. J. Lister recognized this property and, in 1865, introduced phenol as the first antiseptic in surgery.

Phenol is a weak acid ($pK_a = 9.89$). Other phenols behave similarly in aqueous solution and this property makes them stand apart from aliphatic alcohols, $pK_a$ values for which are considerably larger (for ethanol, $pK_a \approx 16$). Note though, that phenols are typically weaker acids than carboxylic acids. What is so special about the direct attachment of the OH group to an aromatic ring? Consider equilibrium 32.45. The factor

Carboxylic acids: see Chapter 33

## COMMERCIAL AND LABORATORY APPLICATIONS

## Box 32.3  Phenols in action

Some commercial applications of phenols are highlighted in this box, along with examples of their biological roles. Many phenols have common names with which you should be familiar, for example:

*o*-Cresol or 2-Cresol
(also *m*- and *p*-cresols)

Catechol

Hydroquinone

Resorcinol

Pyrogallol

Commercial applications of phenols are widespread. Cresols (three isomers) are used in the manufacture of dyes and indicators, e.g. bromocresol green is an acid–base indicator (see Chapter 16). *o*-Cresol is the precursor for 4,6-dinitro-*o*-cresol (DNOC) and 2-methyl-*p*-chlorophenoxyacetic acid (MCPA). DNOC is used as a spray insecticide against, for example, locusts, and is a contact herbicide for the control of broad-leaved weeds and treatment of potatoes and leguminous crops. MCPA is also used to control broad-leaved weeds.

Bromocresol green ($\lambda_{max}$ = 423 nm)

DNOC

When fats are oxidized by atmospheric $O_2$, they become rancid. To prevent this, *antioxidants* are added to foods; these additives work by reacting preferentially with $O_2$. BHT ('butylated hydroxytoluene' or 2,6-di-$^t$butyl-*p*-cresol) and BHA ('butylated

hydroxyanisole') are used as antioxidants and can be recognized on food-packaging labels by their E-numbers (see Box 35.3) of E321 and E320 respectively. There is concern that BHT and BHA may be carcinogenic, and the Joint FAO/WHO Expert Committee on Food Additives has set an acceptable daily intake of 0–0.3 mg per kg of body weight for BHT, and 0–0.5 mg per kg for BHA.

BHT

Mixtures of phenol, the three isomers of cresol and isomers of xylenols ($C_6H_3Me_2OH$) are referred to as cresylic acids and are used as precursors to phenolic resins and disinfectants.

Quinone (paraquinone), semiquinone and hydroquinone play a vital role as mediators in the mitochondrial electron transport chain; *mitrochondria* are sites in cells where biological fuels are converted into energy:

Quinone

Semiquinone

Hydroquinone

Because it is readily oxidized, hydroquinone is used commercially as an antioxidant. One application is in creams applied to bleach melanin-hyperpigmented skin, e.g. freckles and age spots. Resorcinol and catechol are isomers of hydroquinone. Resorcinol is used as a starting material in the manufacture of resins, plastics and dyes and in the pharmaceutical industry. In Box 32.5, we look at the formation of Bakelite from phenol and formaldehyde. A related polymerization reaction between resorcinol and formaldehyde leads to a resin that is used commercially to treat nylon so that the material can be impregnated with rubber. Catechol is a precursor for various pharmaceutical drugs, e.g. L-DOPA which is used in the treatment of Parkinson's disease. Our bodies contain the hormones adrenaline and noradrenaline (see Box 31.1) which are derivatives of catechol.

L-DOPA

Pyrogallol (the name of which arises from its original source of *gallic acid*, obtained from certain tree galls, and the fact that gallic acid converts to *pyro*gallol on heating) is used widely as a photographic developer. In the laboratory, alkaline solutions of pyrogallol are used to absorb $O_2$ both qualitatively and quantitatively.

Gallic acid

*Related information*: see Box 30.3 which includes picric acid.

that determines the acidity of phenol is the stability of the *phenoxide ion* (PhO⁻).

$$\text{Phenol} + H_2O \rightleftharpoons \text{Phenoxide ion} + H_3O^{\oplus} \qquad (32.45)$$

## ENVIRONMENT AND BIOLOGY

### Box 32.4 Aromatics attract cockroaches

Surinam cockroach.
*Frank Greenaway © Dorling Kindersley.*

Cockroaches are pests in households worldwide, and chemical baits are commonly used to attract the insects to traps or poisons. Phenol and naphthol attract male cockroaches but only those of one particular species, making these aromatics too selective to be useful. However, derivatives of the partially hydrogenated derivative 1,2,3,4-tetrahydronaphthalene are irresistible to male and female cockroaches of all common species.

1-Naphthol        1,2,3,4-Tetrahydronaphthalene

**COMMERCIAL AND LABORATORY APPLICATIONS**

## Box 32.5 Thermosetting polymers

One of the early polymers to be industrially manufactured was *Bakelite*. The polymer was first prepared by the Belgian chemist Leo Baekeland, and the Bakelite Corporation was set up in about 1910. Bakelite products were at the peak of their popularity in the 1920s–1940s, but after the Second World War, new polymers and plastics dominated the market. Nowadays, Bakelite items from household utensils to jewellery are collectable pieces, and one of the largest collections was amassed by the artist Andy Warhol.

Bakelite is prepared from phenol and formaldehyde (methanal, see Chapter 33). The OH group in phenol is strongly activating and *ortho*- and *para*-directing. In the first step of polymer formation, substitution at the *ortho*-position occurs. Water is then eliminated between adjacent monomers in a *condensation reaction* to give a polymer:

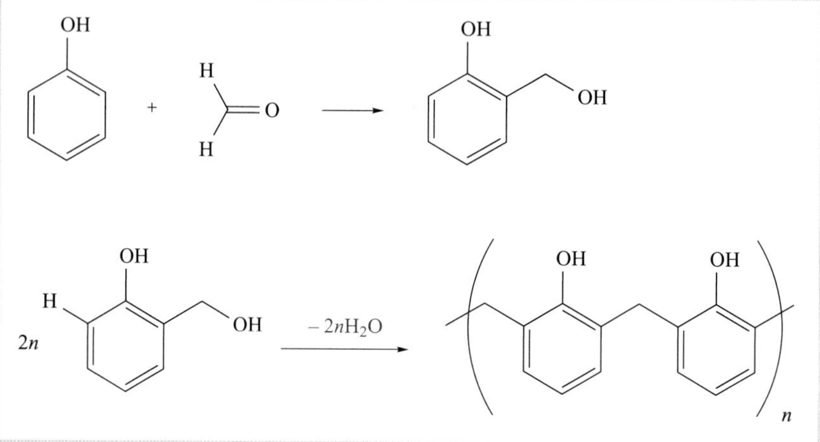

This polymer is a chain, but a *cross-linked* polymer is desirable. Cross-linking provides the polymer with added structural rigidity, making Bakelite a hard material that does not soften on heating. By controlling the stoichiometry of the reaction between phenol and methanal, *para*-substitution as well as *ortho*-substitution is obtained in some monomers. This introduces additional side-chains in the chain-polymer. Cross-linking then occurs as follows:

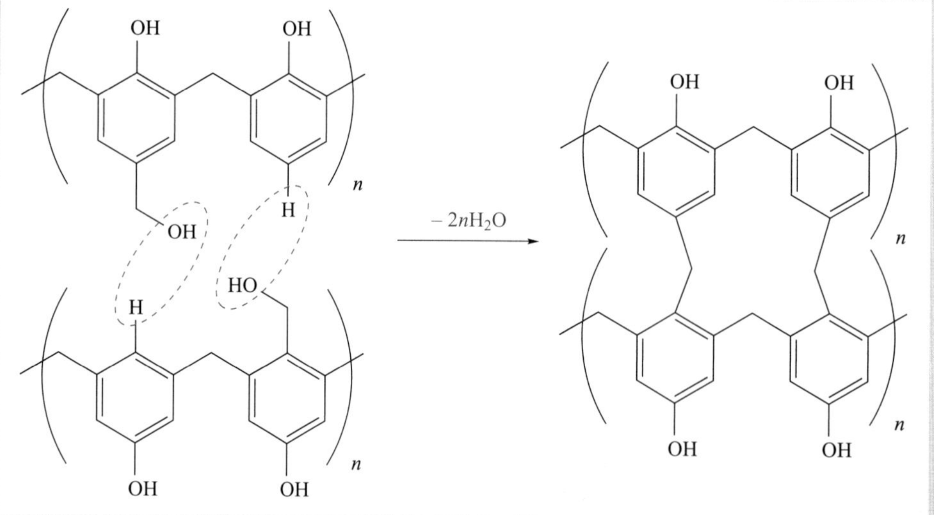

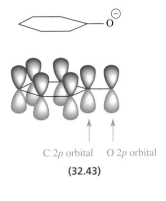

C 2*p* orbital   O 2*p* orbital

**(32.43)**

The negative charge of the phenoxide ion is not localized on the O atom, but is delocalized over the aromatic ring by conjugation into the π-system. This is represented in diagram **32.43**, in which there is overlap between an O 2*p* orbital and the adjacent C 2*p* orbital. Compare this with diagram **32.34**.

Within valence bond theory, the stabilization of the phenoxide ion can be described in terms of the following set of resonance structures:

## Spectroscopy

The IR spectrum of phenol (Figure 32.12a) exhibits the typical broad absorption around 3300 cm$^{-1}$ assigned to the OH group (compare Figure 32.12a with Figures 12.12 and 30.3). The strong absorptions at 1598, 1500 and 1475 cm$^{-1}$ are assigned to the vibrations of the ring. Lower-energy bands occur in the fingerprint region of the spectrum.

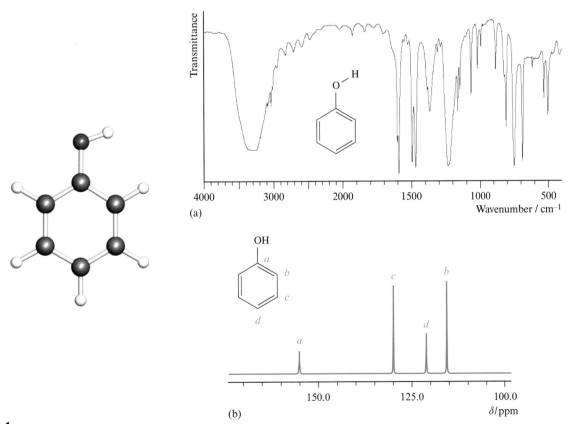

 **Fig. 32.12** (a) The IR spectrum of phenol. (b) The $^{13}$C{$^1$H} NMR spectrum of phenol (dissolved in CDCl$_3$), showing peak assignments.

The $^{13}C\{^1H\}$ NMR spectrum of phenol is shown in Figure 32.12b. The four resonances are assigned to the four C environments as shown. The *ipso*-C atom (C atom *a*) comes at highest frequency (compare the spectra of chlorobenzene, toluene and phenol in Figures 32.6, 32.11 and 32.12) and is of low intensity as is typical of *ipso*-C atoms.

The $^1H$ NMR spectrum of phenol exhibits multiplets centred at $\delta$ 7.24, 6.93 and 6.84 ppm assigned to the *meta*-, *para*- and *ortho*-H atoms respectively. A spectrum of phenol dissolved in CDCl$_3$ shows a singlet at $\delta$ 5.35 ppm assigned to the OH proton, but (as with aliphatic alcohols) the shift and lineshape of this signal are solvent-dependent.

**Refer also to Figure 32.13b**

## Reactions

The acidity of phenol discussed above means that reactions with metal hydroxides occur to give phenoxide salts, e.g. reaction 32.46. However, phenol is *not* a strong enough acid to neutralize aqueous NaHCO$_3$. This observation can be used to distinguish phenol from a carboxylic acid which does react with NaHCO$_3$ (scheme 32.47).

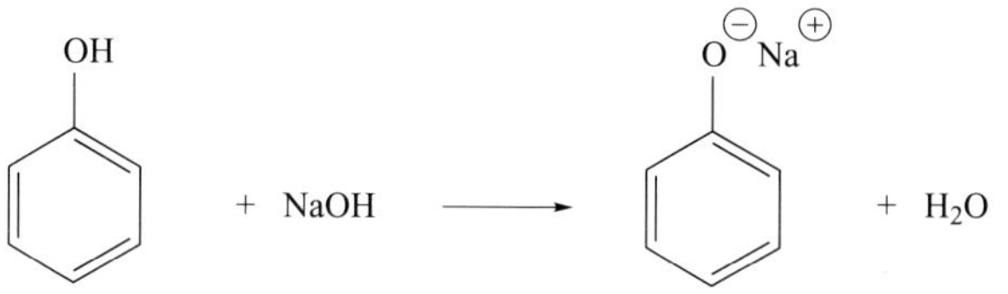

$$+ \text{ NaOH} \longrightarrow \qquad + \text{ H}_2\text{O} \tag{32.46}$$

Sodium phenoxide

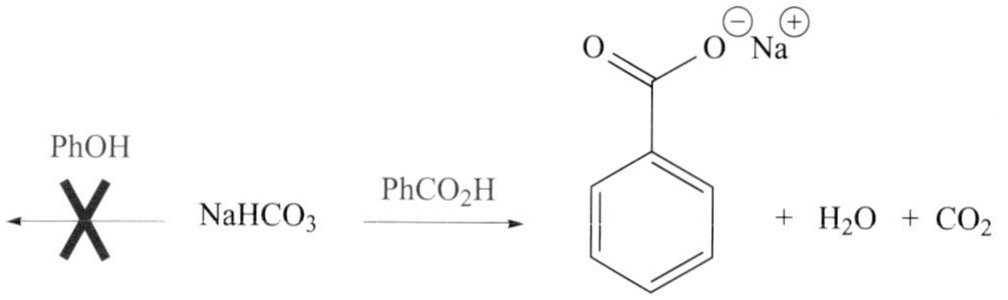

$$\text{PhOH} \quad \overset{\times}{\longleftarrow} \quad \text{NaHCO}_3 \quad \overset{\text{PhCO}_2\text{H}}{\longrightarrow} \qquad + \text{ H}_2\text{O} + \text{CO}_2 \tag{32.47}$$

An OH group attached to an aromatic C$_6$ ring is *strongly activating* and *ortho*- and *para*-directing.

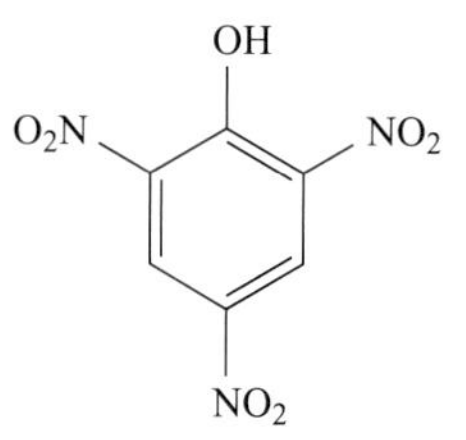

Picric acid
(see Box 30.3)

**(32.44)**

The OH group is *strongly activating* and *ortho*- and *para*-directing (Table 32.3). The conditions under which phenol undergoes electrophilic substitutions are less harsh than those required for the corresponding reactions of benzene. Nitration of phenol (equation 32.48) can be carried out using HNO$_3$ alone (compare the need for HNO$_3$/H$_2$SO$_4$ when nitrating benzene or toluene). The NO$_2$ substituent is *deactivating*, but the highly activating nature of the OH group means that nitration continues after the introduction of one and two NO$_2$ groups, the final product being 2,4,6-trinitrophenol (picric acid, **32.44**).

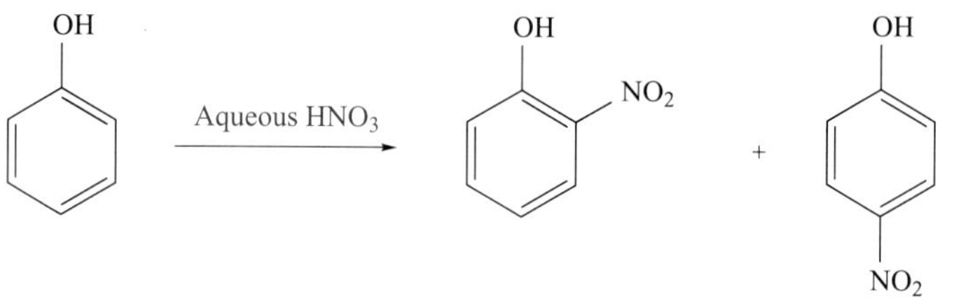

$$\overset{\text{Aqueous HNO}_3}{\longrightarrow} \qquad + \tag{32.48}$$

*o*-Nitrophenol          *p*-Nitrophenol
(2-Nitrophenol)          (4-Nitrophenol)

The activation of the aromatic ring in phenol towards electrophilic substitution means that chlorination and bromination occur on treatment with aqueous $Cl_2$ and aqueous $Br_2$ respectively (compare the use of a Lewis acid catalyst for the halogenation of benzene). Sulfonation requires only $H_2SO_4$ rather than the oleum that is required to sulfonate benzene. Bromination and sulfonation of phenol are summarized in schemes 32.49 and 32.50. The introduction of the first $SO_3H$ group deactivates the ring towards further substitution; hence the need for harsher conditions to encourage attack by the second electrophile. Halogenations are rapid and multi-substitution is common. 2,4,6-Trichlorophenol (**32.45**) is abbreviated to TCP and this is well known in the UK as the brandname of an antiseptic. This liquid, however, is an aqueous glycerol solution of various halogenated phenols and phenol; TCP itself is probably a human carcinogen and is no longer used in the US as an antiseptic or pesticide. Further chlorination under forcing conditions gives pentachlorophenol. Its use in wood preservatives, and the environmental problems associated with halogenated compounds, are described in Box 32.6.

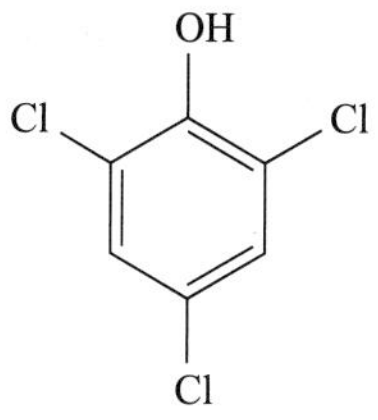

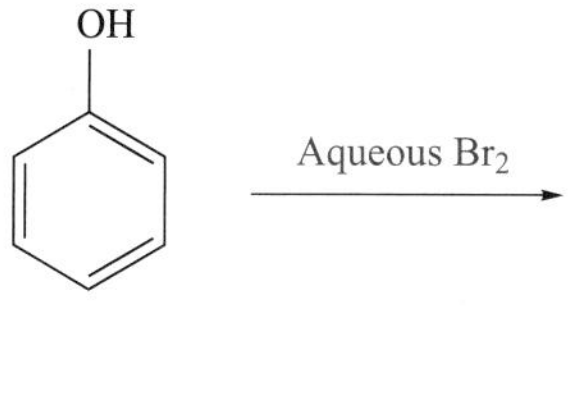

2,4,6-Trichlorophenol (TCP)

(**32.45**)

(32.49)

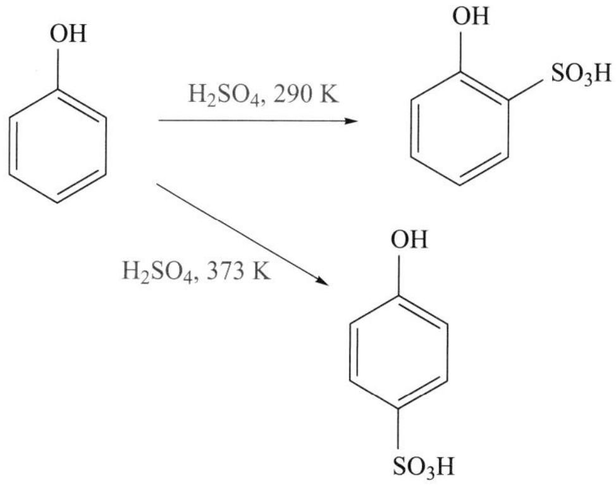

(32.50)

Friedel–Crafts alkylations and acylations of phenol are affected by the fact that the OH group can act as a base, forming an adduct with the Lewis acid catalyst. Reactions still proceed, but not as readily as one would expect. We return to this effect again in the discussion of substitutions in aniline.

Williamson synthesis of ethers: see Section 29.3

The phenoxide ion is used in the Williamson synthesis to form ethers containing the PhO group. Scheme 32.51 shows the formation of ethyl phenyl ether.

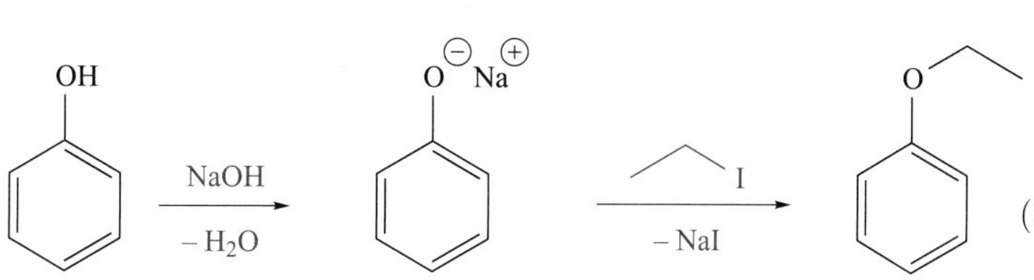

Phenol          Sodium phenoxide          Ethyl phenyl ether

(32.51)

Like aliphatic alcohols, phenols can be converted to esters. Whereas an aliphatic alcohol reacts with a carboxylic acid in the presence of an acid catalyst (see equation 30.37) or with an acyl chloride, only the acyl chloride

## ENVIRONMENT AND BIOLOGY

### Box 32.6 Timber treatment: environmental side effects

Six pipistrelle bats huddled together.
*Frank Greenaway © Dorling Kindersley.*

Pentachlorophenol (PCP) is a fungicide used to treat wood, but there are environmental side effects. Field research has shown that pipistrelle bats roosting in wooden boxes treated with PCP died within a short period after exposure to the chemical in the concentrations used in the study. Bats often roost in the roof spaces of older houses and barns, and may be in close proximity of timbers treated with PCP. Not all chemicals used as wood preservatives are as toxic to bats and other mammals, but we must be aware of, and assess, the environmental side effects that such treatments may have.

PCP is not the only chloro-substituted aromatic compound to have environmental side effects. Other examples include DDT (dichlorodiphenyltrichloro-ethane) and the family of PCB (polychlorinated biphenyl) derivatives. DDT has been used as a pesticide for over 50 years, but in 1973, it was banned in the US. Although it remains in use in other parts of the world, legislative measures are in place to control the use of DDT.

*Further information*: to read about the problems of chlorinated aromatics and related compounds and methods of disposing of them, see: M. L. Hitchman *et al.* (1995) *Chemical Society Reviews*, vol. 24, p. 423.

PCP                     DDT                     PCB

route can be applied effectively to phenols. Reaction 32.52 shows the formation of phenyl propanoate.

$$\text{Phenol} + \text{Propanoyl chloride} \xrightarrow[-\text{HCl}]{\text{NaOH (as base)}} \text{Phenyl propanoate} \quad (32.52)$$

Phenol        Propanoyl chloride                        Phenyl propanoate

Earlier, we mentioned the formation of epoxy resins from the reaction between bisphenol A and epichlorohydrin. A related reaction is used in the manufacture of the drug Propranolol, an AstraZeneca product used as a β-blocker in the treatment of high blood pressure. The starting material is 1-naphthol, a close relation of phenol:

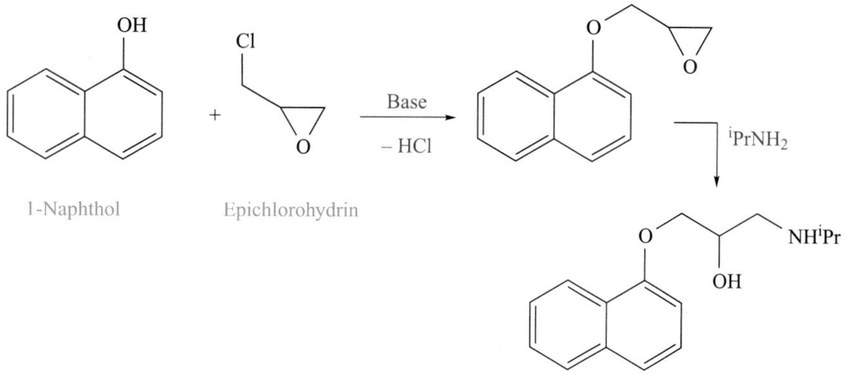

1-Naphthol          Epichlorohydrin

Propranolol

Phenols and phenoxide ions are readily oxidized, turning brown. The products of oxidation are *quinones*. Strong oxidizing agents convert phenol into *p*-benzoquinone (reaction 32.53).

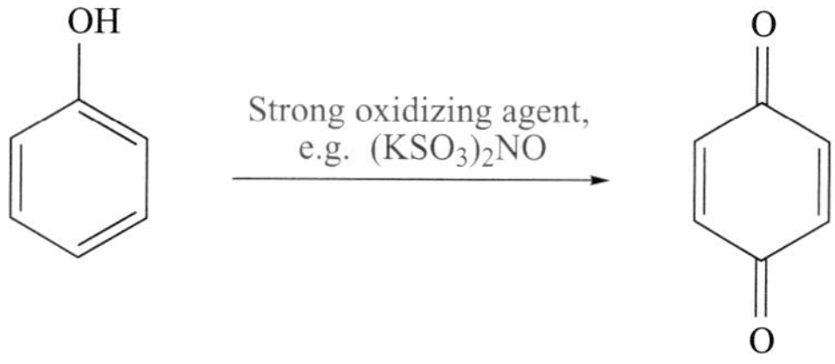

For information on *hydroquinone*: see Box 32.3

$$ (32.53) $$

*p*-Benzoquinone

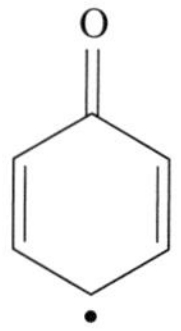

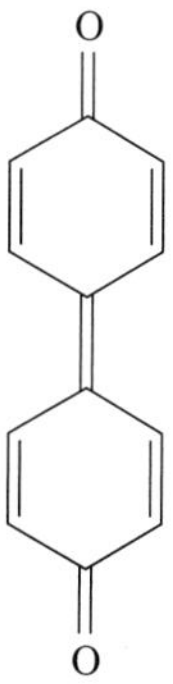

The phenoxide ion easily loses an electron (a 1-electron oxidation step) to give a phenoxide radical (equation 32.54). The odd electron is delocalized around the ring, giving **32.46** as one resonance form. Radical coupling between the *para*-C atoms occurs, followed by oxidation to paraquinone, **32.47**.

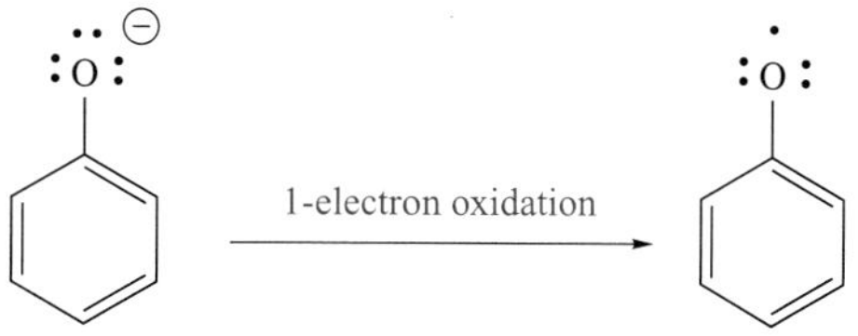

Paraquinone

**(32.46)**          **(32.47)**

1-electron oxidation

$$ (32.54) $$

## 32.13   Nitrobenzene and aniline

### Synthesis

Nitrobenzene is manufactured by the nitration of benzene (equation 32.17). It is an industrial precursor to aniline which has widespread applications in the dye-stuffs (see Box 32.7) and pharmaceutical industries. The reduction of nitrobenzene to aniline is the best way of attaching an $NH_2$ group to the aromatic ring, and can be carried out by various routes (e.g. scheme 32.55).

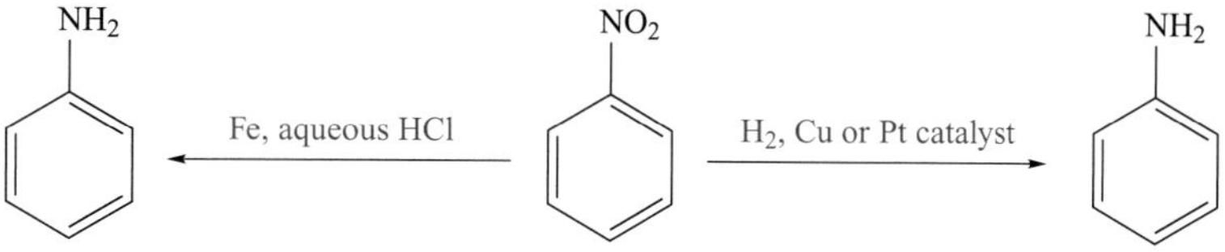

Azo dyes: see Figure 13.8

Fe, aqueous HCl          $H_2$, Cu or Pt catalyst

$$ (32.55) $$

## Box 32.7 William Henry Perkin and early synthetic dyes

The first synthetic dye, *aniline purple* or *mauveine*, was produced by William H. Perkin and commercially manufactured from 1857. The dye was used to colour a silk gown worn by Queen Victoria at the Royal Exhibition in 1862. Although mauveine has been known since 1857, its structure was only confirmed in 1994. Other early dyes produced in Perkin's factory were *Britannia violet*, *Perkin's green* and the red dye *alizarin*.

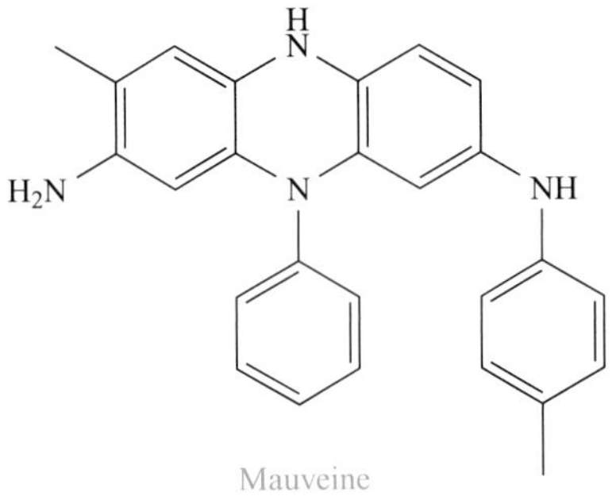

Mauveine 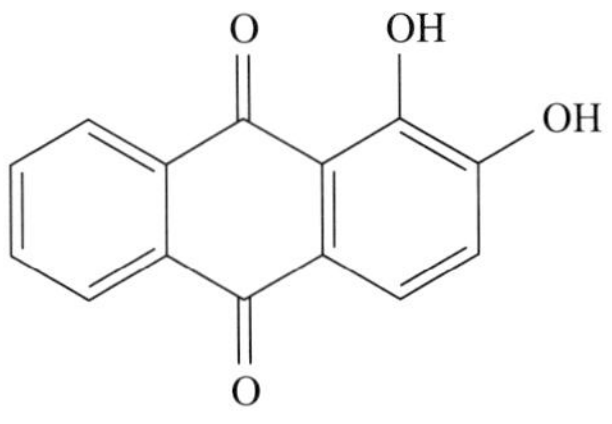                                        Alizarin

In small-scale reactions, granulated tin in hydrochloric acid is a good reducing agent. The acidic conditions mean that the initial product is an *anilinium salt* and this is neutralized with base to obtain aniline (scheme 32.56).

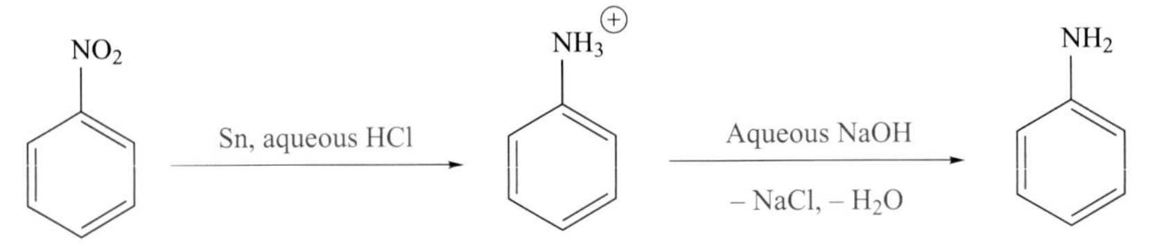

Nitrobenzene                                                                    Aniline
                                                                                (Phenylamine)

(32.56)

## Physical properties

Nitrobenzene ($PhNO_2$) is a colourless liquid with a relatively long liquid range (mp 278.7 K, bp 483.8 K). It has a characteristic marzipan-like smell and is extremely toxic. Aniline (phenylamine, $PhNH_2$) is a colourless, oily liquid (mp 267.0 K, bp 457.2 K) and is also toxic. It often appears brownish in colour owing to the presence of oxidation products. Aniline is polar (N, $\delta^-$) and is slightly soluble in water: at 298 K, 100 cm$^3$ of water dissolves 3.7 g of $PhNH_2$.

Aniline is an *aromatic primary amine* and behaves as a *very weak base* (equation 32.57). Table 32.4 lists the p$K_a$ values of the conjugate acids of aniline and some of its derivatives. Compare these values with those of selected aliphatic amines in Table 31.2. Remember that the larger the value of p$K_a$, the smaller is the value of $K_a$. The larger the value of p$K_a$, the weaker is the conjugate acid, and the stronger is the base. Thus, the data in Tables 32.4 and 31.2 show that aromatic amines are generally weaker

$pK_a = -\log K_a$

**Table 32.4** p$K_b$ values of aniline and selected derivatives of aniline. (For a reminder about p$K_a$ and p$K_b$ values, refer to Chapter 16.)

| Compound | Structure | p$K_a$ of conjugate acid |
|---|---|---|
| Aniline (phenylamine) | $NH_2$ | 4.63 |
| N-Methylaniline (N-methylphenylamine) | Me—NH | 4.85 |
| N,N-Dimethylaniline (N,N-dimethylphenylamine) | Me—N—Me | 5.15 |
| o-Chloroaniline (2-chloroaniline, 2-chlorophenylamine) | $NH_2$ Cl | 2.65 |
| p-Nitroaniline (4-nitroaniline, 4-nitrophenylamine) | $NH_2$ $NO_2$ | 1.00 |
| p-Methylaniline (4-methylaniline, 4-methylphenylamine) | $NH_2$ Me | 5.08 |

than aliphatic amines. As a Brønsted base, aniline reacts with Brønsted acids, e.g. reaction 32.58.

$$NH_2 + H_2O \rightleftharpoons NH_3^{\oplus} + HO^{\ominus} \qquad (32.57)$$

Anilinium ion

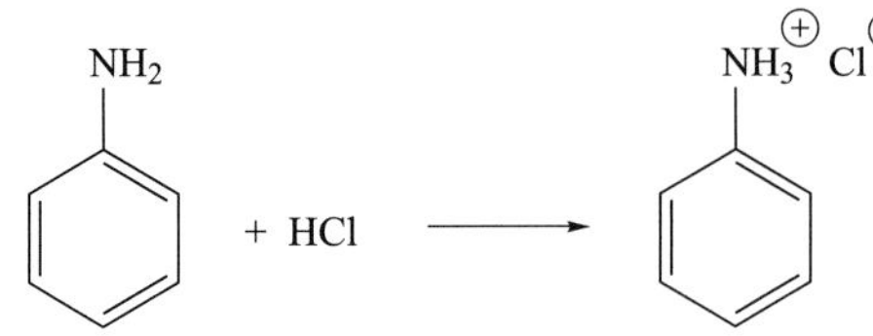

(32.58)

The weak basicity of aniline can be understood in terms of resonance stabilization. Whereas a series of resonance structures can be drawn for $PhNH_2$:

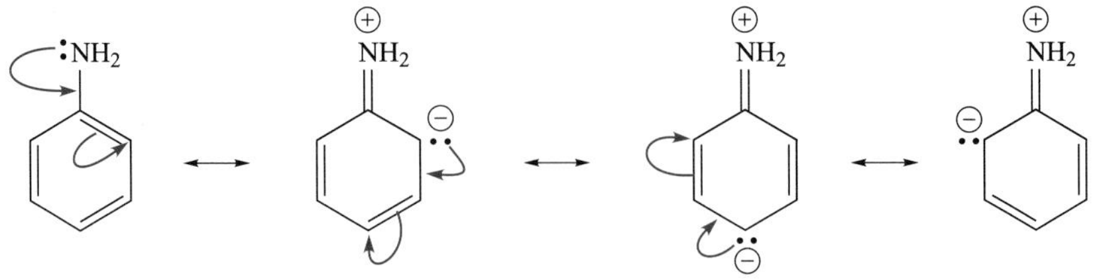

the same is not true of the anilinium ion. Therefore, equilibrium 32.57 lies towards the left-hand side. Table 32.4 includes derivatives of aniline that are substituted on the aromatic ring. The $pK_a$ values are sensitive to the nature of the substituent. Both Cl and $NO_2$ are electron-withdrawing and *destabilize* the positively charged N centre in the anilinium ion (compare this with the discussion in Section 32.10 of the deactivating effects of Cl and $NO_2$). Destabilization of the cation on the right-hand side of equilibrium 32.59 by introducing an electron-withdrawing substituent means that equilibrium 32.59 lies further towards the left-hand side than equilibrium 32.57, i.e. the substituted aniline is *less basic* than aniline itself. The relatively small values of $pK_a$ for *o*-chloroaniline and *p*-nitroaniline in Table 32.4 confirm this statement.

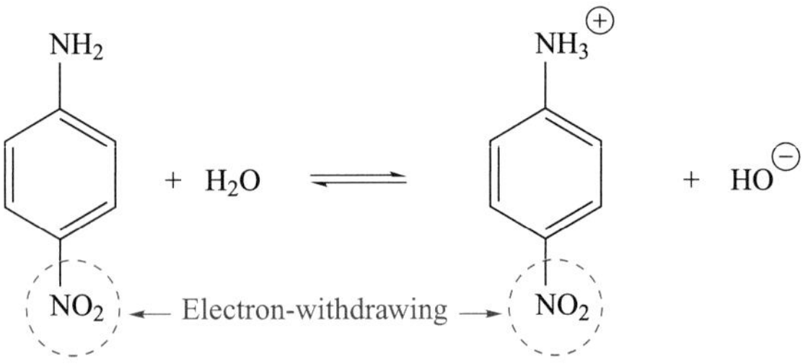

(32.59)

Table 32.4 shows that *p*-methylaniline is a *stronger base* than aniline. The Me substituent is electron-releasing and its presence *stabilizes* the *p*-methylanilinium ion. As a result, equilibrium 32.60 lies further towards the right-hand side than equilibrium 32.57.

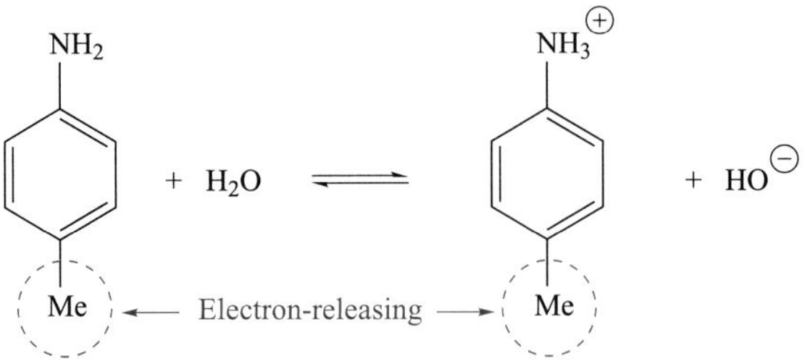

(32.60)

## Spectroscopy

Figure 32.13a shows the IR spectrum of nitrobenzene. Stretching of the aromatic C–H bonds gives rise to the absorption at $3080 \, cm^{-1}$. Strong

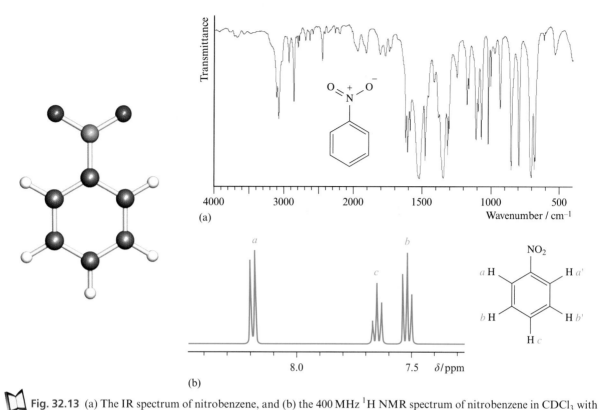

**Fig. 32.13** (a) The IR spectrum of nitrobenzene, and (b) the 400 MHz $^1$H NMR spectrum of nitrobenzene in CDCl$_3$ with peak assignments. The $^1$H NMR spectrum is simulated using experimental data.

NO$_2$
ipso $\delta$ 148.3 ppm
ortho $\delta$ 123.5 ppm
meta $\delta$ 129.4 ppm
para $\delta$ 134.7 ppm

**(32.48)**

absorptions between 1600 and 1500 cm$^{-1}$ are assigned to the C$_6$ ring vibrations and the stretching of the NO$_2$ group.

The $^{13}$C{$^1$H} NMR spectrum of PhNO$_2$ contains four signals assigned to the four C environments. The spectrum resembles that of chlorobenzene (Figure 32.6c), except that the signals are shifted, reflecting the different electron distributions in the molecules of PhCl and PhNO$_2$. Diagram **32.48** shows the chemical shifts of the signals in the $^{13}$C NMR spectrum. The signal at $\delta$ 148.3 ppm for the *ipso*-C atom is of very low intensity. The $^1$H NMR spectrum of nitrobenzene is shown in Figure 32.13b. The three multiplets are assigned to the three chemically non-equivalent proton environments. Spin–spin coupling occurs between magnetically non-equivalent $^1$H nuclei. Although the two *ortho*-H atoms are *chemically equivalent*, they are *magnetically non-equivalent* and can couple to each other. (We examined a similar example in Chapter 14: furan, structure **14.11**.) Using the atom labelling in Figure 32.13b, the coupling constants are $J(H_aH_{a'}) = 2.46$ Hz, $J(H_aH_b) = J(H_{a'}H_{b'}) = 8.35$ Hz, $J(H_aH_{b'}) = J(H_{a'}H_b) = 0.48$ Hz, $J(H_aH_c) = J(H_{a'}H_c) = 1.17$ Hz, $J(H_bH_{b'}) = 1.47$ Hz, and $J(H_bH_c) = J(H_{b'}H_c) = 7.46$ Hz. In Figure 32.13b, the largest couplings dominate, giving a spectrum that appears as a doublet for H$_a$ (*ortho*), a doublet of doublets (an apparent triplet because of the similarity in $J$ values) for H$_b$ (*meta*), and a triplet for H$_c$ (*para*).

The IR spectrum of aniline is shown in Figure 32.14. The absorptions around 3400 cm$^{-1}$ arise from the stretching of the N−H bonds, and the aromatic C−H stretches give rise to the absorptions centred at 3010 cm$^{-1}$. The strong absorptions at $\approx$1600 and 1496 cm$^{-1}$ are assigned to vibrations of the C$_6$ ring.

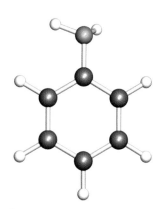

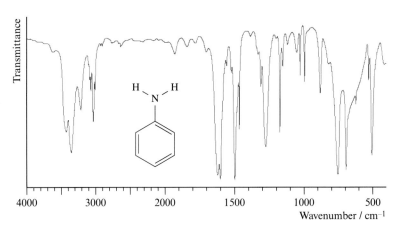

**Fig. 32.14** The IR spectrum of aniline.

$ipso$ $\delta$ 146.5 ppm

$ortho$ $\delta$ 115.1 ppm

$meta$ $\delta$ 129.3 ppm

$para$ $\delta$ 118.4 ppm

**(32.49)**

The $^{13}C$ NMR spectrum of aniline looks similar to that of phenol (Figure 32.12), with four signals assigned to the four C environments. Diagram **32.49** shows the $^{13}C$ NMR chemical shifts and their assignments. The signal at $\delta$ 146.5 ppm assigned to the *ipso*-C is of very low intensity. In the $^{1}H$ NMR spectrum of PhNH$_2$ dissolved in CDCl$_3$, a signal at $\delta$ 3.55 ppm is assigned to the NH$_2$ proton; the lineshape and chemical shift of this signal are solvent-dependent. There are three ring proton environments giving rise to three signals (*ortho* $\delta$ 6.64 ppm, *meta* $\delta$ 7.12 ppm and *para* $\delta$ 6.73 ppm). These appear as multiplets owing to $^{1}H$–$^{1}H$ coupling; look at Figure 32.13b and the accompanying discussion.

## Reactions

An NO$_2$ group attached to an aromatic C$_6$ ring is *strongly deactivating* and *meta-directing*.

NO$_2$

**(32.50)**

We have already seen that reduction of nitrobenzene gives aniline and this is an important reaction. The NO$_2$ group is *meta*-directing (Table 32.3) and substitution is directed towards the 3- and 5-positions (diagram **32.50**). However, NO$_2$ is *strongly deactivating* and electrophilic substitution in PhNO$_2$ is difficult. This explains why the synthesis of PhNO$_2$ from benzene is *not* complicated by the formation of disubstituted products. The introduction of a second NO$_2$ group into PhNO$_2$ is much more difficult than the initial nitration of benzene. Contrast this with the nitration of toluene (reactions 32.36–32.38). Since the aromatic ring in PhNO$_2$ is deactivated towards electrophilic substitution, conditions must be harsh to make such reactions occur. For example, bromination requires a Lewis acid catalyst and high temperature (equation 32.61). Compare this with the bromination of benzene which requires a Lewis acid catalyst, but not an elevated temperature, and with the bromination of phenol (OH is activating) which occurs in the absence of a catalyst.

NO$_2$    $\xrightarrow{\text{Br}_2, \text{FeBr}_3, 410 \text{ K}}$    NO$_2$ ... Br     (32.61)

*meta*-Bromonitrobenzene

The mechanism of this reaction is represented in scheme 32.62. Note the charge distribution on the intermediate carbenium ion and compare this with the resonance structures that we drew for this intermediate in Section 32.10.

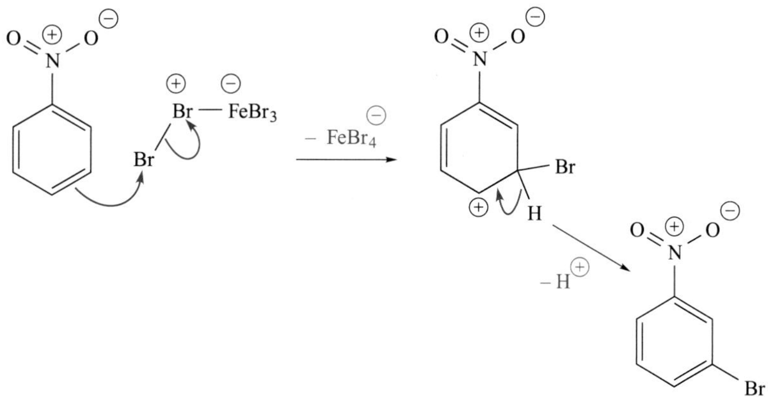

(32.62)

The nitro-group is so deactivating that some substitutions do not work. Friedel–Crafts alkylations and acylations of $PhNO_2$ are *not* readily achieved. Consequently, nitrobenzene is a suitable inert solvent for Friedel–Crafts reactions involving other substrates.

We have already described the reduction of $PhNO_2$ to $PhNH_2$. By changing the reduction conditions, it is possible to reduce nitrobenzene to *azobenzene*, although in practice, mixtures of products are generally obtained including hydroazobenzene and azoxybenzene. The scheme below shows the formation of azobenzene, and the structures of likely by-products:

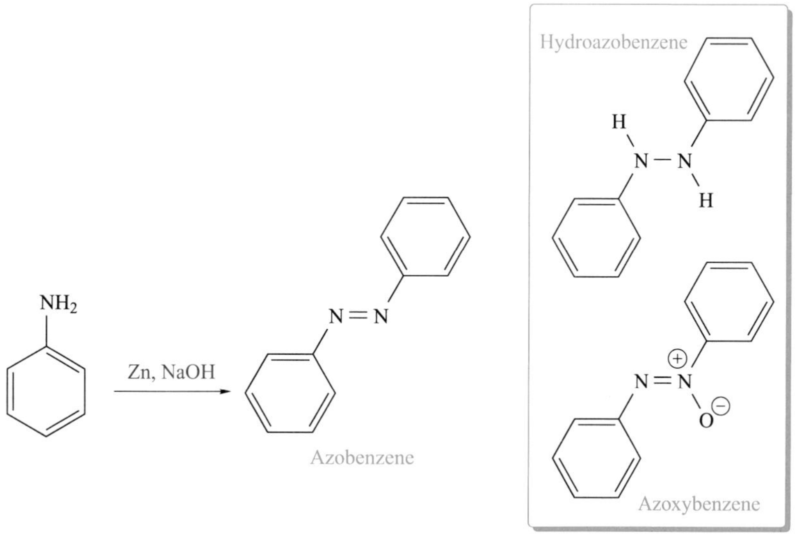

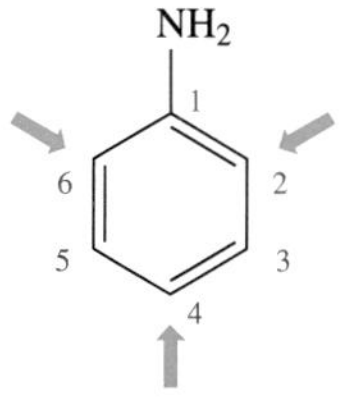

The $NH_2$ group in aniline is *highly activating*, and is *ortho-* and *para-*directing. Electrophiles are directed to the 2-, 4- and 6-positions (diagram **32.51**). Electrophilic substitutions such as halogenations occur readily and multi-substituted products are obtained, e.g. reaction 32.63.

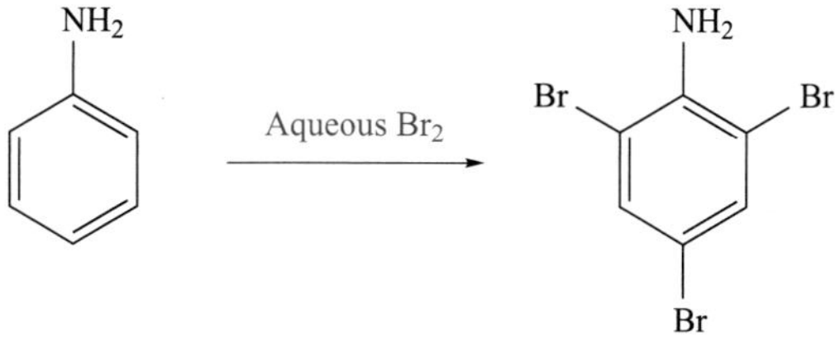

(32.51)

(32.63)

> An NH$_2$ group attached to an aromatic C$_6$ ring is ***highly activating*** and ***ortho-*** and ***para-directing***.

Lewis acid–base adducts: see 'Aluminium halides' in Section 22.4

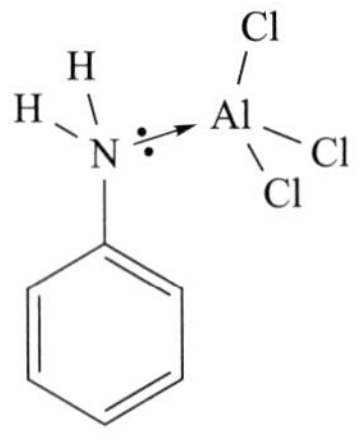

Despite the activation of the NH$_2$ group, Friedel–Crafts alkylations and acylations fail because the NH$_2$ group acts as a Lewis base and interacts with the Lewis acid catalyst (structure **32.52**). Similar problems occur in acidic conditions (e.g. during nitration) when PhNH$_3^+$ forms. A way of overcoming this problem is to convert the NH$_2$ group into an *amido group* prior to the Friedel–Crafts reaction. Reaction 32.64 shows the conversion of aniline to an amide.

**(32.52)**

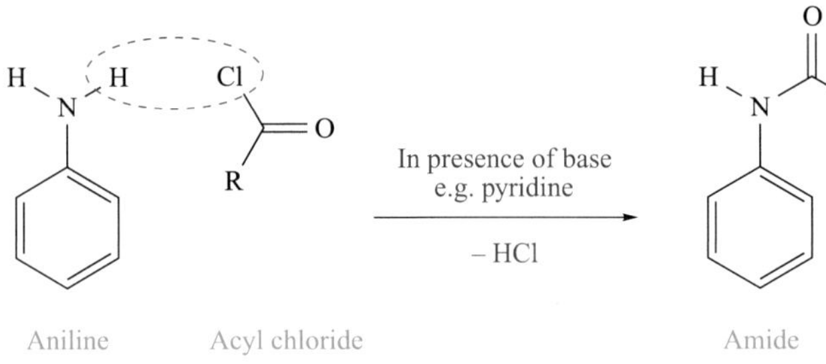

Aniline     Acyl chloride        Amide

(32.64)

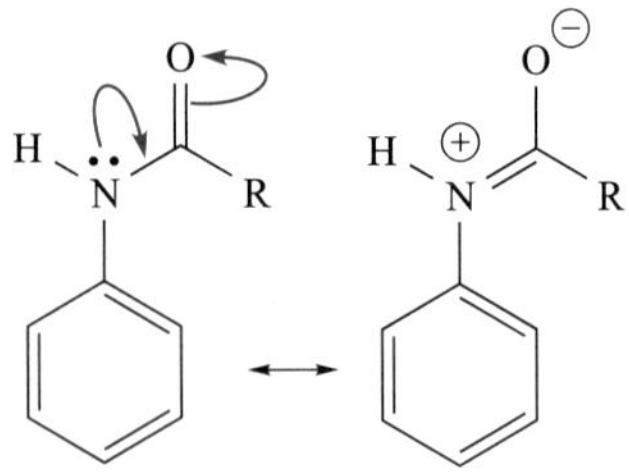

The N atom in the amide is less basic than the N in the NH$_2$ group because negative charge is delocalized on to the carbonyl O atom as shown in resonance pair **32.53**. Thus, the catalyst in the Friedel–Crafts reaction remains unhindered, and alkylations and acylations of the aromatic ring in the amide take place where they fail in the corresponding amine. The amide group is *ortho-* and *para*-directing (i.e. the same as the NH$_2$ group). The amide group can be hydrolysed back to an amine group after alkylation. Scheme 32.65 summarizes a general Friedel–Crafts alkylation of acetanilide.

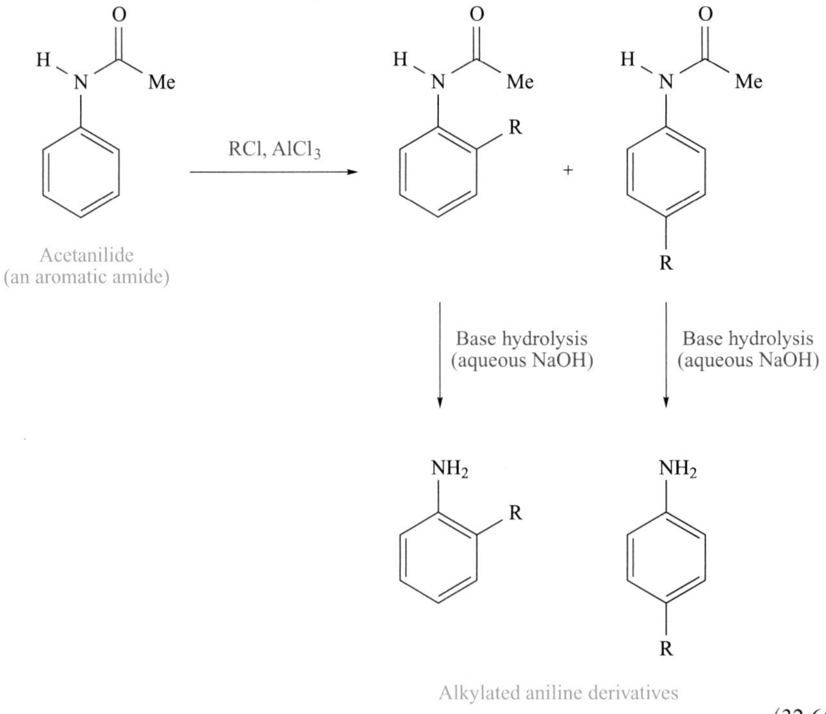

**(32.53)**

Acetanilide
(an aromatic amide)

RCl, AlCl$_3$

Base hydrolysis
(aqueous NaOH)

Base hydrolysis
(aqueous NaOH)

Alkylated aniline derivatives

(32.65)

Aniline reacts with alkyl halides to give *N*-alkyl derivatives. Scheme 32.66 shows the conversion of $PhNH_2$ to $PhNMe_2$.

$$\text{(32.66)}$$

Aniline                    *N*-Methylaniline                    *N,N*-Dimethylaniline

One of the most important reactions of aromatic primary amines is their conversion to *diazonium salt*s by treatment with nitrous acid, prepared *in situ* by the action of an acid (e.g. HCl) on sodium nitrite (equation 32.67). (The instability of nitrous acid was described in Section 22.8.) *Caution! Solid diazonium salts are explosive.*

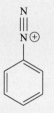

A *diazonium salt* contains the diazonium cation:

$$\xrightarrow[- NaCl, \ - 2H_2O]{NaNO_2, \ HCl(aq), \ in \ an \ ice \ bath}$$

$$\text{(32.67)}$$

The diazonium cation is resonance stabilized with the positive charge conjugated on to the aromatic ring C atoms:

Diazonium salts undergo two general types of reaction in which:

- $N_2$ is eliminated, or
- an $-N=N-$ group (an *azo group*) is formed.

Elimination of $N_2$ occurs when a diazonium ion is hydrolysed. This takes place only slowly under the aqueous conditions used in the preparation of diazonium chloride, but the reactions of diazonium salts with water are a convenient method of preparing phenols, e.g. reaction 32.68.

$$+ \ H_2O \longrightarrow \quad + \ N_2 \ + \ H^{\oplus}$$

$$\text{(32.68)}$$

In the *Sandmeyer reaction*, a diazonium ion is converted to a chloro-, bromo- or cyano-benzene derivative by reaction with CuCl, CuBr or CuCN.

Elimination of $N_2$ accompanies the conversion of $PhN_2^+$ to PhCl, PhBr, PhI and PhCN. Chlorobenzene, bromobenzene and benzonitrile (or derivatives of these compounds) are formed in the *Sandmeyer reaction* in which diazonium salts are treated with CuCl, CuBr or CuCN in the presence of an excess of the corresponding anion ($Cl^-$, $Br^-$ or $CN^-$). In reaction 32.69, the copper(I) salt acts as a catalyst. The mechanism involves radicals, with

Cu(I) being oxidized to Cu(II) as aryl radicals are generated. Copper(II) is then reduced back to Cu(I) as Ar$^{\bullet}$ is converted to ArX. The conversion of PhN$_2^+$ to iodobenzene is achieved by direct reaction with NaI. The Sandmeyer reaction is historically important as a method of preparing halobenzene compounds.

$$\text{CuX, excess } X^{\ominus}, \quad X = Cl, Br, CN \qquad + \ N_2 \qquad \textit{Sandmeyer reaction} \ (32.69)$$

The coupling of a diazonium ion with phenols or amines takes place *without* N$_2$ elimination. The product is an *azo-compound*: two aryl groups coupled by an azo (−N=N−) unit. Equation 32.70 shows the general reaction.

X = OH or NH$_2$ — Substituents are *para* to each other

$$(32.70)$$

The reaction is an electrophilic substitution (PhN$_2^+$ acts as an electrophile), and works because conjugation of the lone pairs from the OH or NH$_2$ substituents into the ring π-system makes the ring electron-rich. Electrophilic substitution takes place at the *ortho-* and *para*-positions of PhOH or PhNH$_2$; look back at the resonance structures near the beginning of Section 32.10. The mechanism is represented as follows:

Chromophores and azo-compounds: see Section 13.6

The −N=N− group is a *chromophore* and some azo-compounds absorb light in the visible region making them coloured and of commercial use as biological stains (see Box 13.2), indicators and dyes (see Figures 13.8 and 13.9).

### 32.14    Nucleophilic substitution in aromatic rings

Although halogenoalkanes undergo nucleophilic substitution reactions, halobenzenes (PhCl, PhBr and PhI) do not. However, in special cases, a halo-substituent on an aromatic ring bearing other substituents can be displaced by a nucleophile. The key to success is that the ring must contain

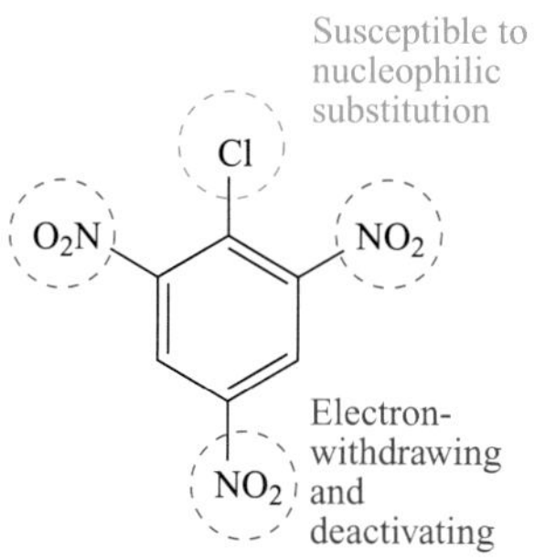

Susceptible to
nucleophilic
substitution

Electron-
withdrawing
and
deactivating

**(32.54)**

*ortho-* or *para-*substituents that are electron-withdrawing and deactivating, e.g. $NO_2$ or CN.

2,4,6-Trinitrochlorobenzene (**32.54**) fulfils these criteria, and can be converted to picric acid by treatment with hydroxide (equation 32.71) or to a derivative of benzonitrile (equation 32.72).

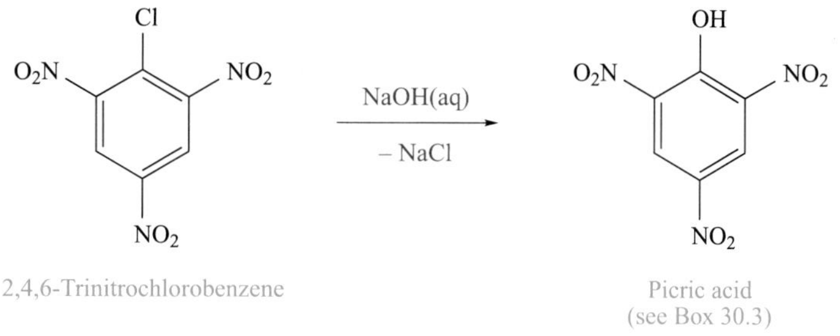

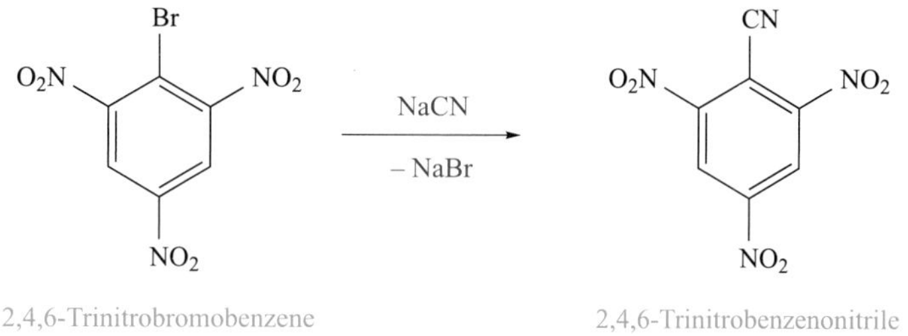

$$\xrightarrow[- NaCl]{NaOH(aq)} \tag{32.71}$$

2,4,6-Trinitrochlorobenzene

Picric acid
(see Box 30.3)

$$\xrightarrow[- NaBr]{NaCN} \tag{32.72}$$

2,4,6-Trinitrobromobenzene

2,4,6-Trinitrobenzenonitrile

The mechanism of substitution does *not* parallel that for displacement of halide from a halogenoalkane, i.e. it is not $S_N1$ or $S_N2$. The mechanism is one of *addition and elimination*:

- addition of the nucleophile to give a resonance-stabilized intermediate, followed by
- elimination of halide.

The resonance-stabilized intermediate is shown below, and the resonance forms that are dominant are those in which the negative charge is delocalized on to the nitro groups (i.e. the top three structures):

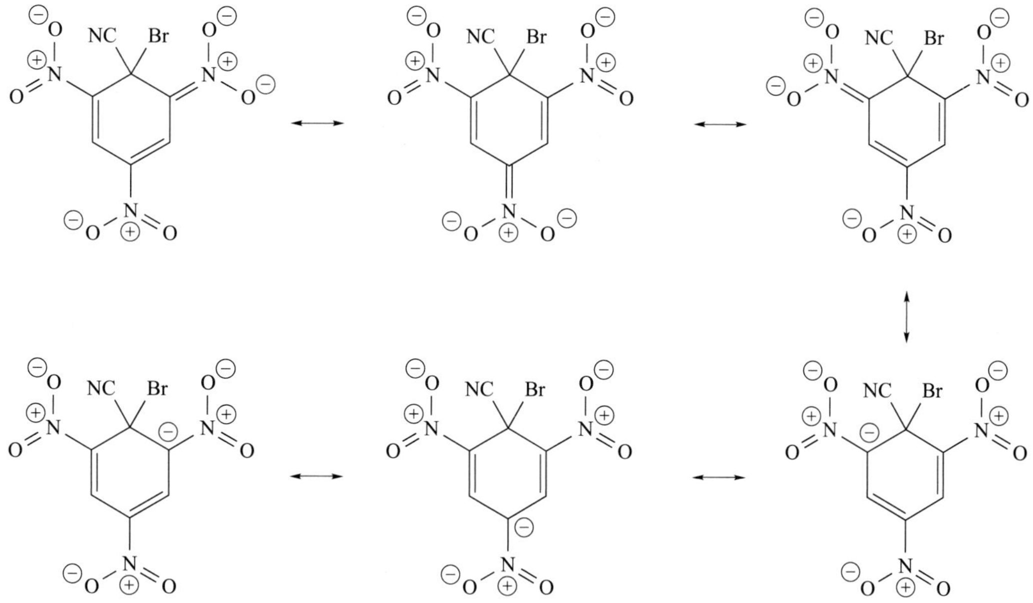

The above example highlights an important use of this type of reaction. The CN and $NO_2$ groups are both *meta*-directing, but Cl, Br and I groups are *ortho/para*-directing. Preparing the benzonitrile derivative via the halo-derivative allows the CN group to be introduced *ortho* to the $NO_2$ group.

## SUMMARY

In this chapter, we have introduced aromatic compounds. Benzene is the parent of a wide range of compounds, many of which are of commercial importance. Although representations of the structure of benzene may imply the presence of C=C bonds, the $\pi$-electrons in the ring are delocalized. As a result, benzene and its derivatives behave differently from alkenes; electrophilic substitution reactions are characteristic. The introduction of substituents such as Me, Et, OH and $NH_2$ activates the ring towards electrophilic substitution. Substituents such as Cl, Br, I, CN, $NO_2$ and $SO_3H$ deactivate the ring towards electrophilic substitution. Both the nature of the ground state and carbenium ion intermediates contribute towards the orientation effects of different substituents.

### Do you know what the following terms mean?

- benzene
- Kekulé structure
- delocalized $6\pi$-electron system in benzene
- aromaticity
- Hückel $(4n + 2)$ rule
- *ortho*-position
- *meta*-position

- *para*-position
- aryl
- phenyl
- benzyl
- Birch reduction
- electrophilic substitution
- Friedel–Crafts reaction
- Lewis acid catalyst

- electron-releasing group
- electron-withdrawing group
- activating substituent
- deactivating substituent
- orientation effects
- benzoquinone
- azobenzene
- diazonium salt

### You should be able:

- to comment on the historical development of a bonding scheme for benzene
- to rationalize why the C–C bonds in benzene are equivalent
- to apply the Hückel $(4n + 2)$ rule to test whether a compound is aromatic, and to distinguish between aromatic and anti-aromatic systems
- to give typical chemical shift values for ring C and H atoms in the $^{13}C$ and $^1H$ NMR spectra of aromatic compounds
- to interpret the $^{13}C$ and $^1H$ NMR spectra of simple aromatic compounds
- to interpret the IR spectra of simple aromatic compounds
- to name simple aromatic compounds
- to draw the structures of simple aromatic compounds given the systematic name
- to relate structures and trivial names for the following: toluene, xylene, phenol, aniline, styrene, anisole

- to discuss typical reactions of benzene
- to describe a general mechanism for electrophilic substitution in benzene and to apply this to halogenations, Friedel–Crafts alkylations, acylations, nitration and sulfonation
- to discuss the origins of activation and deactivation of a ring for selected substituents
- to describe why some substituents are *ortho/para*-directing, while others are *meta*-directing
- to give preparations of PhMe, PhOH, $PhNO_2$ and $PhNH_2$
- to describe typical reactions of PhMe, PhOH, $PhNO_2$ and $PhNH_2$
- to illustrate how diazonium salts are prepared
- to describe how to convert a diazonium salt into an azo-compound and comment on the importance of these derivatives
- to illustrate examples of nucleophilic substitution in aromatic compounds

## PROBLEMS

**32.1** Resonance structures **32.11** and **32.12** are usually drawn to rationalize the bonding in benzene in terms of the VB model. However, the set of contributing resonance structures can be extended to include Dewar benzene. Draw such an extended set that retains the equivalence of the C–C bonds.

**32.2** What evidence makes us believe that it is incorrect to consider benzene as a triene?

**32.3** Comment on the following observations.
(a) The C–C bond lengths in benzene are all 140 pm.
(b) Ethylbenzene undergoes some free radical reactions that are typical of an alkane.
(c) The electronic spectrum of benzene has an intense absorption at $\lambda_{max} = 183$ nm.

**32.4** Describe the bonding in benzene in terms of an appropriate hybridization scheme for the C atoms and account for the equivalence of the C–C bonds.

**32.5** Which of the following are aromatic:
(a) cyclopropenyl cation, $C_3H_3^+$; (b) cyclopropenyl anion, $C_3H_3^-$; (c) cyclobutadiene, $C_4H_4$; (d) cycloheptatriene, $C_7H_8$; (e) cycloheptatrienyl cation, $C_7H_7^+$; (f) cyclooctatetraene, $C_8H_8$?

**32.6** Assign appropriate hybridization schemes to the C atoms in the following molecules.

**32.7** Draw structural formulae for the following compounds: (a) PhOH, (b) $Ph_2CH_2$, (c) $Ph_2NH$, (d) $PhNO_2$ and (e) PhSMe.

**32.8** Match the following $^1$H NMR spectroscopic data to the compounds shown below; the multiplets may consist of overlapping signals.

| Compound | $^1$H NMR $\delta$/ppm |
|---|---|
| A | 7.2 (multiplet), 2.9 (septet, $J$ 8 Hz), 1.3 (doublet, $J$ 8 Hz) |
| B | 7.1 (singlet), 2.3 (singlet) |
| C | 7.1 (multiplet), 2.3 (singlet) |
| D | 7.3 (multiplet), 4.5 (singlet), 2.4 (broad) |
| E | 5.6 (multiplet), 2.4 (multiplet) |
| F | 7.2 (multiplet), 2.6 (quartet), 1.2 (triplet) |

**32.9** Consider the following compounds.

(a) What would you see in the $^1$H NMR spectrum of compound **A**? (b) Suggest a spectroscopic method that would allow you to distinguish between isomers **B** and **C**. Give the structure of a third isomer and comment on how it could be distinguished from **B** and **C**.

**32.10** Draw the structures of (a) $PhNH_2$, (b) iodobenzene, (c) isopropylbenzene, (d) PhCHO, (e) benzoic acid, (f) benzyl alcohol, (g) $PhC{\equiv}CPh$, (h) toluene, (i) *p*-xylene and (j) 1,2-diaminobenzene.

**32.11** Suggest ways of carrying out the following transformations:

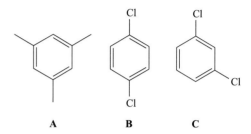

**32.12** Two isomers of tribromobenzene have been prepared. NMR spectroscopic data for them are given below. Assign a structure to each isomer.

| Isomer | $^1$H NMR $\delta$/ppm | $^{13}$C NMR $\delta$/ppm |
|---|---|---|
| **A** | 7.61 | 133.0, 123.4 |
| **B** | 7.77, 7.47, 7.29 | 136.2, 134.7, 131.7, 125.9, 123.8, 121.4 |

**32.13** Equation 32.14 shows that the Friedel–Crafts reaction of benzene with 1-chloropropane gives a mixture of products. How would you use $^1$H and $^{13}$C NMR spectroscopy to characterize the products?

**32.14** (a) Equation 32.21 shows loss of $H^+$ from a carbenium intermediate in the electrophilic substitution of $X^+$ for $H^+$ in benzene. Write two analogous equations starting from the remaining two resonance structures in **32.30**.
(b) Equation 32.33 shows the sulfonation of benzene using $HSO_3{}^+$. Write a mechanism for the conversion of benzene to benzenesulfonic acid in which the electrophile is $SO_3$.

**32.15** Suggest products for the following reactions and give reasons for your answers:

(a)
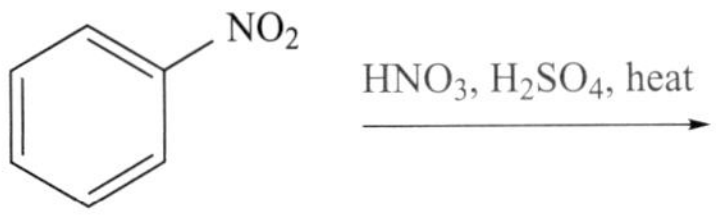
NO2
HNO3, H2SO4, heat

(b)
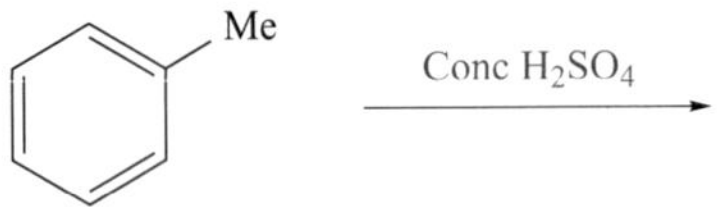
Me
Conc H2SO4

(c)
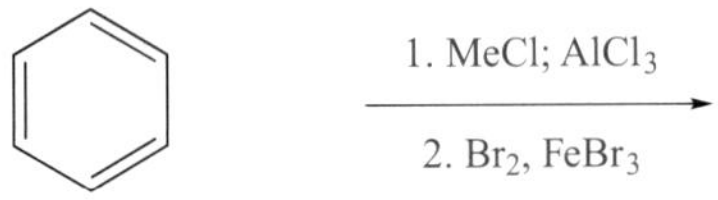
1. MeCl; AlCl3
2. Br2, FeBr3

(d)
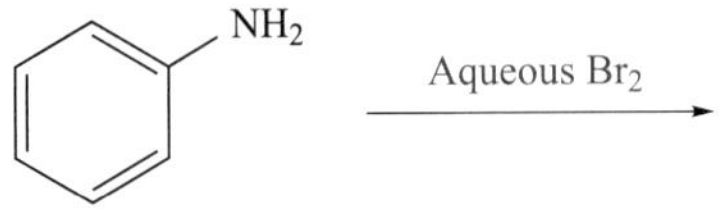
NH2
Aqueous Br2

**32.16** Use resonance stabilization arguments to describe (a) why phenol is acidic, and (b) why the OH group in phenol is *ortho*- and *para*-directing.

**32.17** Suggest ways of making the following compounds starting from phenol, toluene or nitrobenzene.

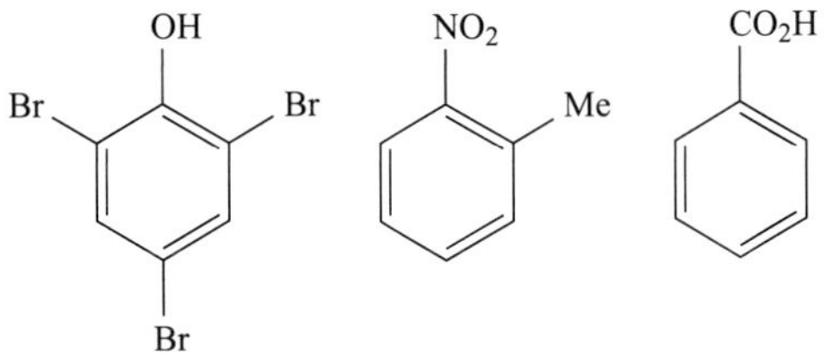
OH
Br
Br
Br
NO2
Me
CO2H

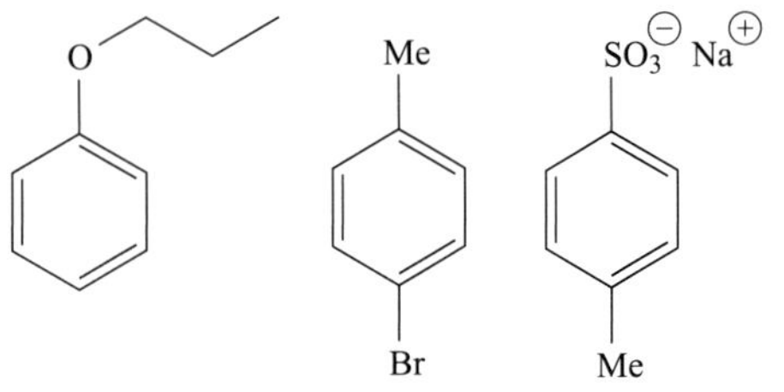
O
Me
Br
SO3 Na
Me

**32.18** How would you differentiate between the following pairs of isomers using $^{13}C\{^1H\}$ NMR spectroscopy?
(a) 1,2-dimethylbenzene and ethylbenzene
(b) 1,3-dichlorobenzene and 1,4-dichlorobenzene
(c) 1,3,5-trimethylbenzene and 1,2,3-trimethylbenzene
(d) *n*-propylbenzene and isopropylbenzene
(e) cyclooctatetraene and styrene

**32.19** Under what conditions (using one or more steps) could you convert benzene to
(a) bromobenzene, (b) aniline,
(c) *o*-chloronitrobenzene, (d) phenol?

**32.20** Suggest mechanisms for the conversions of (a) $C_6H_6$ to PhCl; (b) $C_6H_6$ to $PhSO_3H$; (c) $PhNO_2$ to *m*-bromonitrobenzene. Describe the role of catalysts where appropriate.

**32.21** Which of the following molecules are polar, and in what sense?

(a)
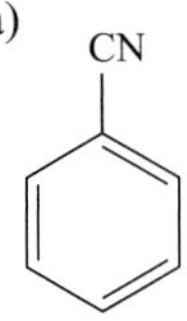
CN

(b)
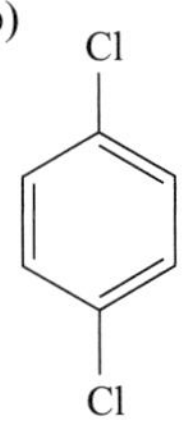
Cl
Cl

(c)
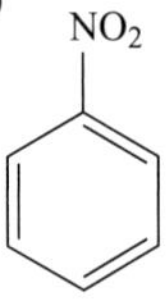
NO2

(d)
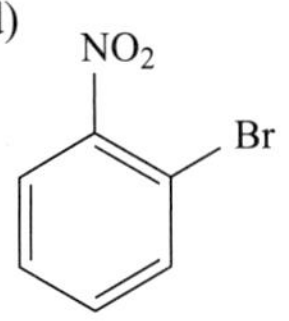
NO2
Br

**32.22** Give suitable reagents to carry out the following transformations.

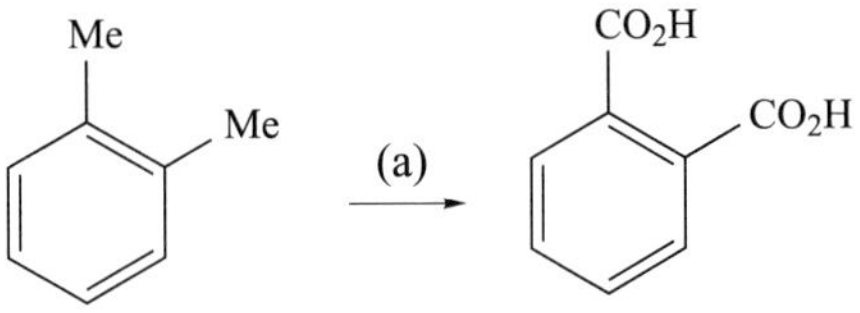
Me
Me
(a)
CO2H
CO2H

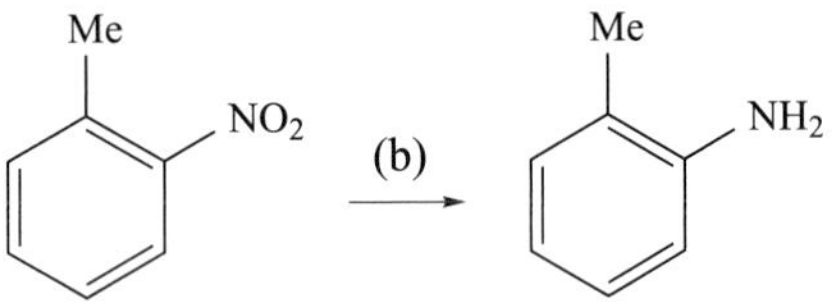
Me
NO2
(b)
Me
NH2

o-Methylacetanilide

## Additional problems

**32.23** When PhMeNH reacts with $PhN_2^+$ in aqueous solution, two azo-compounds are obtained. In one, the substituent in one aromatic ring *para* to the azo-group is MeNH, while in the second, the *para*-substituent is OH. Draw the structures of the products and suggest an explanation for these observations.

**32.24** (a) Account for the fact that the $^1H$ NMR spectrum of benzene contains a singlet at $\delta$ 7.2 ppm, but the spectrum of buta-1,3-diene shows lower frequency multiplets ($\delta$ 5.0, 5.1 and 6.3 ppm).

(b) An aromatic compound **A** shows a parent ion in its mass spectrum at $m/z = 104$. The results of an elemental analysis are: C, 92.3%; H, 7.7%. Compound **A** reacts readily with $Br_2$ to give an addition product, **B**. The $^{13}C$ NMR spectrum of **A** exhibits signals at $\delta$ 137.6, 137.0, 128.5, 127.8, 126.2 and 113.7 ppm. In the 600 MHz $^1H$ NMR spectrum, overlapping signals appear in the range $\delta$ 7.5–7.1 ppm, and there are three doublets of doublets at $\delta$ 6.7 ($J$ 18 Hz, 11 Hz), 5.7 ($J$ 18 Hz, 2 Hz) and 5.2 ($J$ 11 Hz, 2 Hz) ppm. Suggest identities for **A** and **B**.

**32.25** The conversion of benzene to *o*-methylacetanilide can be carried out in the following four steps. Suggest how each step can be carried out.

(a)

(b)

**32.26** Comment on the following observations.
(a) The $pK_a$ of phenol is 9.89.
(b) Bromination of 4-methylphenol gives 2-bromo-4-methylphenol.
(c) The $CF_3$ substituent in $PhCF_3$ is deactivating and *meta*-directing.

**32.27** (a) Nitrobenzene and nicotinic acid are isomers. How could NMR spectroscopy be used to distinguish between them?

Nicotinic acid

(b) The $^1H$ NMR spectrum of phenol contains three multiplets in the region $\delta$ 6.8–7.2 ppm. Explain the origins of the couplings in these signals.

(c) Suggest a method of preparing each of the following compounds starting from benzene. Comment on the formation of possible side-products.

**32.28** Explain why the nitration of aniline gives a mixture of *o*-, *m*- and *p*-nitroaniline.

# 33 Carbonyl compounds

## Topics

- Nomenclature
- Polarity of the C=O bond
- Structural properties
- Spectroscopy
- Tautomerism
- Syntheses of aldehydes and ketones
- Syntheses of carboxylic acids
- Syntheses of esters
- Syntheses of amides
- Syntheses of acyl chlorides
- Carbonyl compounds as acids
- Enolate ions in synthesis
- Nucleophilic attack at the C=O carbon atom

## 33.1 The family of carbonyl compounds

We have already mentioned some aspects of chemistry involving aldehydes, ketones, carboxylic acids and esters, including:

- Brønsted acid behaviour of carboxylic acids (Chapter 16);
- reduction of carbonyl compounds to prepare alcohols (equations 30.6–30.9);
- oxidation of alcohols to aldehydes, ketones and carboxylic acids (equations 30.32–30.34);
- amination of aldehydes and ketones (equations 31.11–31.15);
- formation of esters (equation 30.37).

The structural feature common to aldehydes, ketones, carboxylic acids, amides, acyl chlorides and esters is the C=O bond, a *carbonyl group*. In the formulae below, R and R' in the ketone or ester may be the same or different:

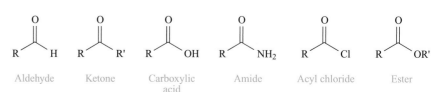

| Aldehyde | Ketone | Carboxylic acid | Amide | Acyl chloride | Ester |

## 33.2

The common name for methanal is *formaldehyde*, and for ethanal is *acetaldehyde*

## Naming carbonyl compounds

### Aldehydes

The systematic name for an aliphatic aldehyde is constructed by taking the root name for the carbon chain (e.g. methan-, ethan-, propan-) and adding the ending *-al*. Examples are:

Methanal (Formaldehyde)    Ethanal (Acetaldehyde)    Propanal    Butanal

Note two general points:

- The aldehyde functional group is written as CHO and *not* COH; this avoids confusion with an alcohol group.
- One C atom in the chain is *part of the functional group*; thus, methanal does *not* contain a methyl group, and butanal does *not* contain a butyl group.

Benzenecarbaldehyde (Benzaldehyde)

**(33.1)**

Aldehydes in which the CHO group is attached to a benzene ring are named by adding the ending *-carbaldehyde*, e.g. **33.1** is benzenecarbaldehyde, although the common name of *benzaldehyde* is usually used.

### Ketones

Hexan-2-one

**(33.2)**

Hexan-3-one

**(33.3)**

Acetone    Acetylacetone

**(33.4)**    **(33.5)**

Ketones are named using the ending *-one* (or *-dione* if there are two C=O functional groups). The carbon chain is numbered from the end *nearest* to the C=O group. In **33.2**, the six C atoms in the chain give the root name of hexan- and to this is added the ending -one as well as a position number -2- to show where the C=O group is within the chain. Compound **33.3** is an isomer of **33.2** differing only in the position of the C=O group.

One of the most common ketones in the laboratory is *acetone* (**33.4**). The common name is so well recognized that the systematic name of propanone is hardly ever used. Acetylacetone (**33.5**) is a diketone; its systematic name is pentane-2,4-dione. The conjugate base of acetylacetone is acetylacetonate (acac⁻) and we described its use as a ligand in Chapter 23 (e.g. see Figure 23.5). Ketones involving aryl groups are named by taking the aryl group as a substituent, e.g. diphenyl ketone. Substituent names are placed in alphabetical order when they are different:

Diphenyl ketone    Ethyl phenyl ketone

### Carboxylic acids

In the systematic names for carboxylic acids, the root name gives the number of C atoms in the chain and this *includes the C atom in the $CO_2H$ group*. The ending added to the name is *-oic acid*. Methanoic and ethanoic acids have the trivial names of *formic* and *acetic acids* and these are generally preferred to the systematic ones:

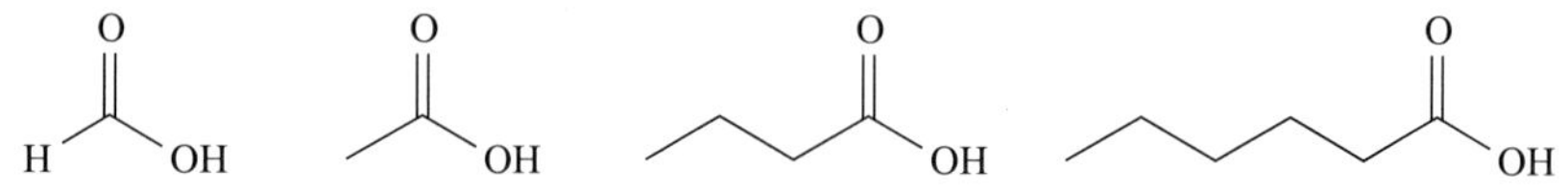

Formic acid     Acetic acid     Butanoic acid     Hexanoic acid
(Methanoic acid)  (Ethanoic acid)

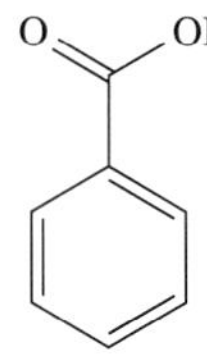

*Benzoic acid* is the name given to $PhCO_2H$ (**33.6**). Many common names for carboxylic acids are of historical interest and we highlight this aspect of nomenclature in Box 33.1.

Salts of carboxylic acids are called *carboxylates*. The names of the carboxylate anions are derived from those of their conjugate acids as follows:

| Acid | Conjugate base (carboxylate ion) |
| --- | --- |
| Formic acid | Formate ion |
| Acetic acid | Acetate ion |
| Propanoic acid | Propanoate ion |
| Butanoic acid | Butanoate ion |
| Pentanoic acid | Pentanoate ion |

The salts are named as exemplified below:

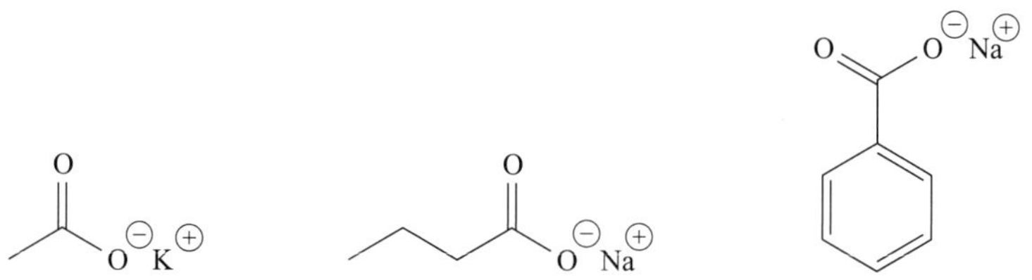

Potassium acetate       Sodium butanoate       Sodium benzoate

Although the structures above show the charge localized on one O atom, we shall see later that it is actually *delocalized*.

## Amides

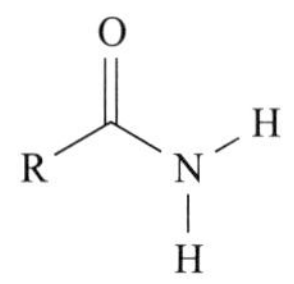

Primary amides have the general structure shown in **33.7**. Their names are constructed in a similar way to those of carboxylic acids, replacing the *-oic* of the systematic name of the acid by *-amide*:

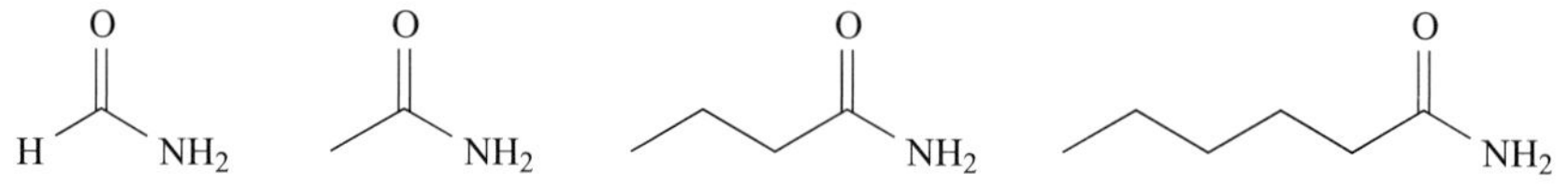

Formamide     Acetamide     Butanamide     Hexanamide
(Methanamide)  (Ethanamide)

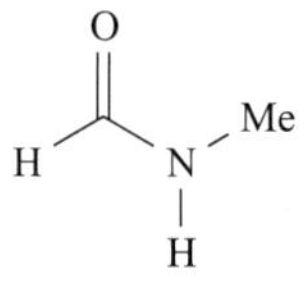

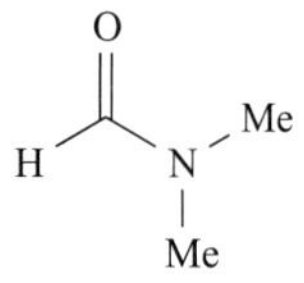

The trivial names *formamide* and *acetamide* remain in common use. If the H atoms of the $NH_2$ group are substituted by R (alkyl) or Ar (aryl) groups, the prefix *N-* is used to signify the position of the substituent, for example *N*-methylformamide (**33.8**) and *N,N*-dimethylformamide (DMF, **33.9**). The latter is a common laboratory solvent.

## HISTORY

## Box 33.1 Trivial names for carboxylic acids: names with meanings

Whereas systematic names provide structural information, many trivial names were chosen to convey some other description of the compound. *Formic acid* derives from the Latin *formica* meaning ant; formic acid is the chemical that gives an ant its sting. *Acetic acid* is the component of vinegar that provides its sour taste; the Latin word *acetum* means vinegar. *Butyric acid* is named using the Latin *butyrum* for butter; rancid butter has a characteristic smell caused by butyric acid.

Formic acid     Acetic acid     Butyric acid

Wood ant (*Formica rufa*).
*Kim Taylor* © *Dorling Kindersley.*

*Valeric acid* is the old name for pentanoic acid. *Isovaleric acid* is an isomer of valeric acid and occurs in the roots of the plant valerian.

Valeric acid     Isovaleric acid

*Caproic, caprylic* and *capric acids* are responsible for goaty smells; the names for these acids have their originates in the Latin words *capra* (a female goat) and *caper* (goat).

Caproic acid

Caprylic acid

Capric acid

*Oxalic acid* is poisonous and can cause paralysis of the nervous system. The name of the acid derives from the plant wood sorrel (*Oxalis acetosella*), the leaves of which contain potassium hydrogen oxalate ($KHC_2O_4$). *Palmitic acid* is an important 'fatty acid' (see Section 33.9) and its name originates from its main source of palm oil.

Oxalic acid     Palmitic acid

## Acyl chlorides

The name of an acyl chloride is derived from that of the corresponding carboxylic acid by replacing the *-oic acid* by *-oyl chloride*, e.g. butanoic acid becomes butanoyl chloride. Ethanoyl chloride is usually referred to by its common name of acetyl chloride. The acyl chloride derived from benzoic acid is benzoyl chloride:

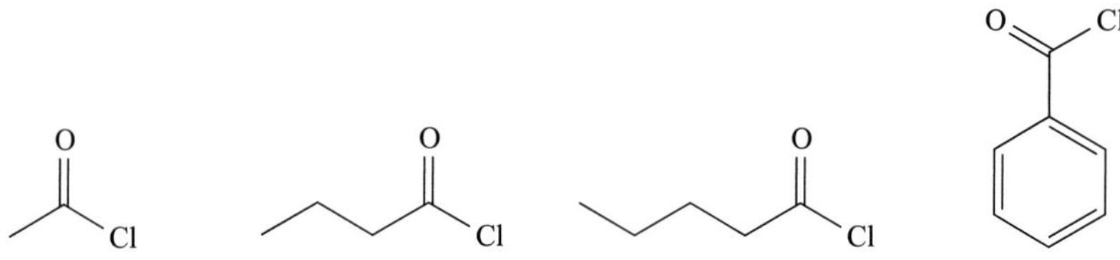

Acetyl chloride (Ethanoyl chloride)    Butanoyl chloride    Pentanoyl chloride    Benzoyl chloride

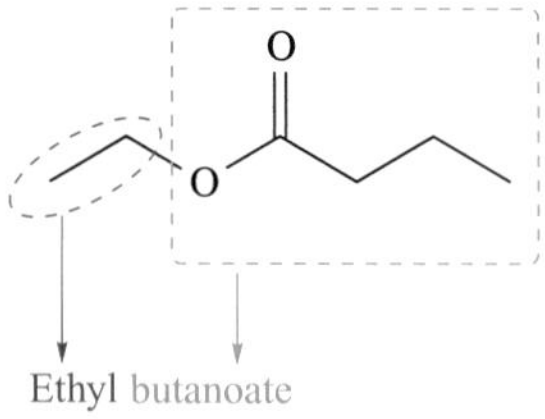

## Esters

Esters are derived from carboxylic acids by the replacement of the H of the −OH group by an alkyl (R) or aryl (Ar) substituent. The name of the ester has two parts: the first derives from the R or Ar group, and the second from the carboxylic acid. Structure **33.10** shows these derivations for ethyl butanoate, where the parent acid is butanoic acid. Further examples are shown below:

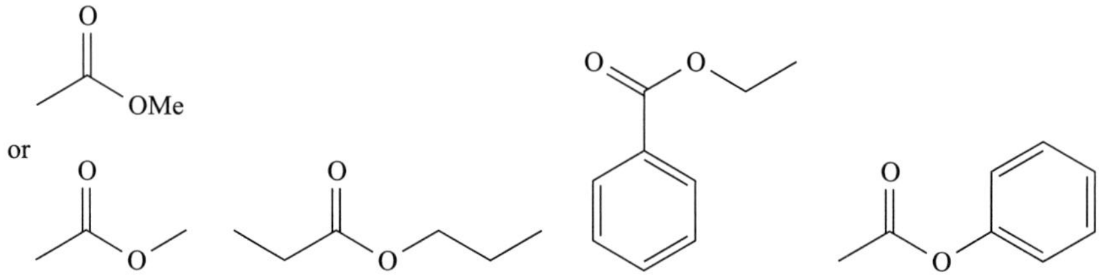

Ethyl butanoate

(33.10)

or

Methyl acetate    Propyl propanoate    Ethyl benzoate    Phenyl acetate

**33.3** ## The polar C=O bond

## Polarity and reactivity

The difference between the electronegativity values of C ($\chi^P = 2.6$) and O ($\chi^P = 3.4$) means that the C=O bond is polarized in the sense shown in diagram **33.11**. In terms of an MO bonding description (Figure 33.1), the difference in the effective nuclear charges of the C and O atoms means that both the C−O σ-bonding and π-bonding MOs contain greater contributions from the O than the C atom. The corresponding σ* and π* MOs contain more C than O character. This is similar to the situation we described for the CO molecule in Section 5.14.

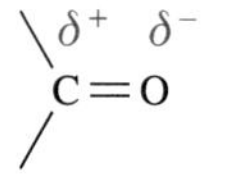

$\chi^P$ = Pauling electronegativity value: see Section 5.7

(33.11)

In a carbonyl compound, electrophiles (e.g. $H^+$) are attacked by the $\delta^-$ O atom of the C=O group; nucleophiles attack the $\delta^+$ C atom:

Attack by nucleophile at C

Electrophile is attacked by O

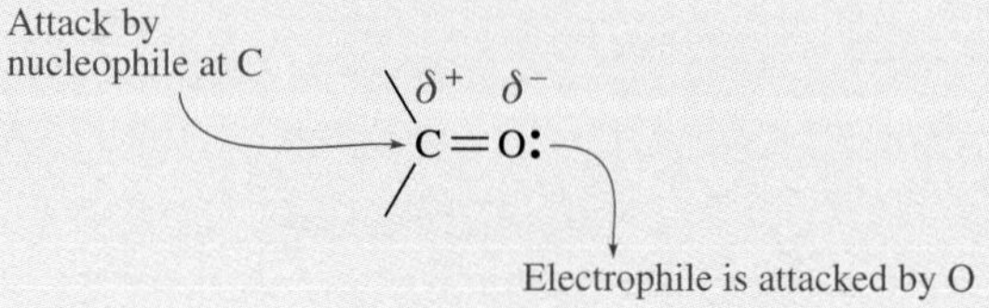

The polarization of the C=O bond has important consequences with regard to the reactivity of carbonyl compounds. Electrophiles (most commonly $H^+$) are attacked by the $\delta^-$ O atom, while nucleophiles attack the $\delta^+$ C atom.

**Fig. 33.1** Part of the MO diagram for a carbonyl compound. The diagram shows only the orbital interactions that are associated with the C=O bond. The diagrams on the right-hand side represent the characters of the four MOs in the centre of the figure. The $\sigma$- and $\pi$-bonding MOs contain more O than C character; the $\sigma^*$ and $\pi^*$ MOs contain more C than O character. See also Figure 5.14.

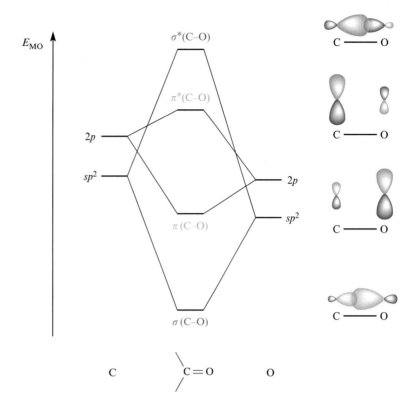

## Physical properties

The presence of the polar C=O bond in aldehydes, ketones, esters and acyl chlorides results in significant intermolecular dipole–dipole interactions and these compounds possess relatively high boiling points. Carboxylic acids and amides are able to form intermolecular hydrogen bonds. Figure 33.2 shows part of the solid state structure of 3-fluorobenzamide. The molecules are ordered into chains through the formation of intermolecular hydrogen bonds. We return to hydrogen bonding between carboxylic acid molecules in the next section. Table 33.1 lists dipole moments, melting points and boiling points for related carbonyl compounds and compares the data with those of alcohols and an ether of similar molecular weights. Alcohols also exhibit intermolecular hydrogen bonding, but simple ethers do not (see Sections 29.4 and 30.3). The effects of intermolecular association, and of hydrogen bonding in particular, are evident from the data in

Hydrogen bonding: see Section 21.8

**Fig. 33.2** Intermolecular hydrogen bonding occurs in amides between the O atoms of the C=O groups and the H atoms of the NH$_2$ groups. The solid state structure of 3-fluorobenzamide illustrates the intermolecular association resulting from hydrogen bonding. [X-ray diffraction data: T. Taniguchi *et al.* (1975) *Mem. Osaka Kyoiku Univ. Ser. 3*, vol. 24, p. 119.] Colour code: C, grey; O, red; N, blue; F, green; H, white.

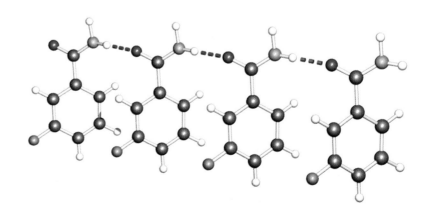

**Table 33.1** Dipole moments and melting and boiling points for selected carbonyl compounds. The values for selected polar, O-containing compounds of similar molecular weights are given for comparison.

| Compound | Structure | Dipole moment / D | Relative molecular mass | Melting point / K | Boiling point / K |
|---|---|---|---|---|---|
| Acetaldehyde | | 2.75 | 44.05 | 150.0 | 293.2 |
| Propanal | | 2.52 | 58.08 | 193.0 | 321.0 |
| Acetone | | 2.88 | 58.08 | 178.3 | 329.2 |
| Acetic acid | | 1.70 | 60.05 | 289.7 | 391.0 |
| Acetyl chloride | | 2.72 | 78.49 | 161.1 | 324.0 |
| Acetamide | | 3.76 | 59.07 | 355.0 | 494.0 |
| Methyl formate | | 1.77 | 60.05 | 174.0 | 304.8 |
| Ethanol | | 1.69 | 46.07 | 159.0 | 351.3 |
| Propan-1-ol | | 1.55 | 60.09 | 147.0 | 370.2 |
| Dimethyl ether | | 1.30 | 46.07 | 131.6 | 248.3 |

Table 33.1. Note especially the high dipole moment and high melting and boiling points of acetamide.

Many carbonyl compounds possess characteristic odours. The lower molecular weight amides smell 'mousy'. Esters often smell fruity, and some are used in artificial fruit essences (see Box 33.2). A number of carboxylic acids also have characteristic smells (see Box 33.1).

## 33.4    Structure and bonding

### Structural details

In a carbonyl compound, the C atom of the C=O group is in a trigonal planar environment. An $sp^2$ hybridization scheme for this C atom is

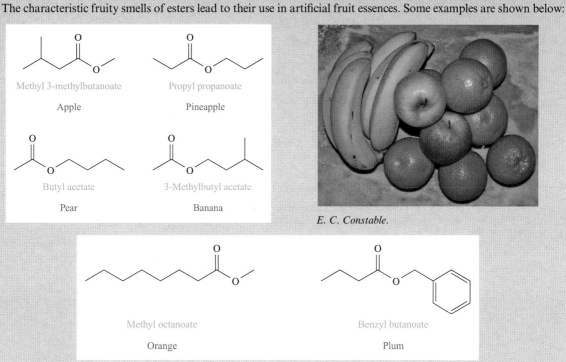

**COMMERCIAL AND LABORATORY APPLICATIONS**

**Box 33.2  Fruit essences and flavours**

The characteristic fruity smells of esters lead to their use in artificial fruit essences. Some examples are shown below:

Methyl 3-methylbutanoate

Apple

Propyl propanoate

Pineapple

Butyl acetate

Pear

3-Methylbutyl acetate

Banana

*E. C. Constable.*

Methyl octanoate

Orange

Benzyl butanoate

Plum

appropriate to allow for the formation of three $\sigma$-bonds and one $\pi$-interaction:

$\sigma + \pi$-components of C=O bond    $\sigma$-bond (C–C or C–H)

$\sigma$-bond (C–C, C–H, C–O, C–N or C–Cl)

**(33.12)**

$\angle$R–C–X ~ R–C–O ~ X–C–O ~ 120°

The $\pi$-component of the C=O bond is formed by the overlap of the C and O $2p$ orbitals. Typical lengths for C=O bonds in carbonyl compounds are around 121 pm, and the three bond angles around the trigonal planar C atom are usually close to 120° **(33.12)**.

**Dimerization of carboxylic acids**

**(33.13)**

We described earlier how carboxylic acids are able to form hydrogen-bonded intermolecular interactions. The OH and C=O groups in a carboxylic acid molecule are perfectly positioned to allow the formation of hydrogen-bonded dimers as in structure **33.13**. Figure 33.3 shows part of the solid state structure of terephthalic acid, an example of a *dicarboxylic acid*. The presence of two $CO_2H$ groups per molecule allows the formation of an extended structure in the solid state through intermolecular hydrogen bonding.

Fibres from terephthalic acid: see Box 33.3

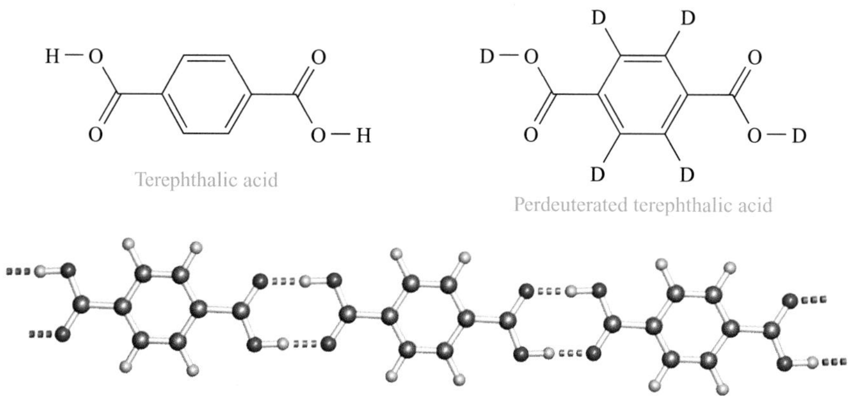

**Fig. 33.3** Terephthalic acid (see Box 33.3) is a dicarboxylic acid. Intermolecular hydrogen bonding in the solid state leads to the formation of a ribbon-like structure. The structure shown here was determined by neutron diffraction for the fully deuterated compound. This allows accurate location of the H/D positions. [Data: P. Fischer *et al.* (1986) *J. Solid State Chem.*, vol. 61, p. 109.] Colour code: O, red; C, grey; D, yellow.

Terephthalic acid

Perdeuterated terephthalic acid

In the liquid state, evidence for dimerization of carboxylic acids comes from molecular weight determinations and from the fact that the boiling points of the acids are relatively high. The boiling points of formic acid and acetic acid are significantly higher than those of aldehydes of corresponding molecular weights (Table 33.2). The ratio of $\Delta_{vap}H(bp)$ to the boiling point (in K) is equal to $\Delta_{vap}S$, and Trouton's rule states that this ratio is approximately constant (equation 33.1).

Trouton's rule: see equations 17.66 and 21.50 and associated text

$$\Delta_{vap}S = \frac{\Delta_{vap}H(bp)}{bp} \approx 88\,\mathrm{J\,K^{-1}\,mol^{-1}} \qquad \textit{Trouton's rule} \quad (33.1)$$

Both acids show anomalously low values of the molar entropy of vaporization ($\Delta_{vap}S$). That is, *they do not obey Trouton's rule*, and this can be explained in terms of the formation of hydrogen-bonded dimers which lowers the entropy of the system. The aldehydes in Table 33.2 obey Trouton's rule, because both *aldehydes are monomers*.

**Table 33.2** Values of $\Delta_{vap}H(bp)$ and boiling points for formic and acetic acids compared with aldehydes of similar molecular weights.

| Compound | Structure | Relative molecular mass | Boiling point / K | $\Delta_{vap}H(bp)$ / kJ mol$^{-1}$ | $\dfrac{\Delta_{vap}H(bp)}{bp}$ / J K$^{-1}$ mol$^{-1}$ |
|---|---|---|---|---|---|
| Formic acid | 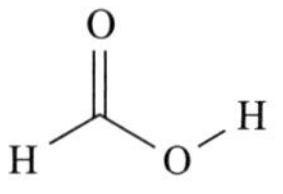 | 46.03 | 374.0 | 22.7 | 60.7 |
| Acetic acid | 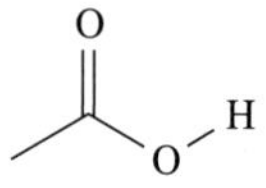 | 60.05 | 391.0 | 23.7 | 60.6 |
| Acetaldehyde | 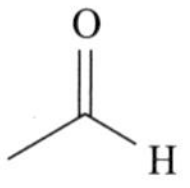 | 44.05 | 293.2 | 25.8 | 88.0 |
| Propanal | 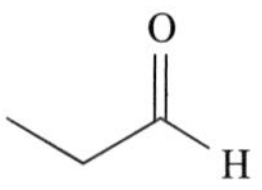 | 58.08 | 321.0 | 28.3 | 88.2 |

## 33.5    IR and NMR spectroscopy

### IR spectroscopy

In Section 12.6, we used the carbonyl group as an example of a functionality that could be detected by IR spectroscopy. Table 12.1 lists typical wavenumbers associated with the stretching of the C=O bond in different types of carbonyl compounds. Figure 33.4 shows the IR spectra of a ketone (hexan-2-one), a carboxylic acid (hexanoic acid) and an ester (methyl hexanoate). A dominant feature of each spectrum is the absorption near $1700 \, cm^{-1}$. This arises from the stretching of the C=O bond. The IR spectrum of the carboxylic acid contains a broad band centred around $3000 \, cm^{-1}$. This is assigned to the O$-$H stretch, and the broadening of the absorption arises from hydrogen bonding, exactly as it does for alcohols (see Figure 30.3). Sharper absorptions near to $3000 \, cm^{-1}$ are assigned to the alkyl C$-$H stretches, and these are observed in all the spectra in Figure 33.4. Absorptions below $1600 \, cm^{-1}$ lie in the fingerprint region.

### $^{13}$C NMR spectroscopy

Signal intensities:
see Section 14.7

$^{13}$C $\delta$ 203 ppm

H

O

Hexanal

**(33.14)**

A compound containing a carbonyl group possesses a characteristic high-frequency signal assigned to the $sp^2$ C atom of the C=O group. In addition, this signal is usually of relatively low intensity. Figure 33.5 shows the proton-decoupled $^{13}$C NMR spectrum of butanone with signal assignments. The signal at $\delta$ 209 ppm is assigned to the carbonyl C atom. This is a typical chemical shift for this type of C environment, and the signal is easily distinguished from those of alkyl or aryl C atoms (see Figure 14.4). Structures **33.14** and **33.15** give additional examples of $\delta$ $^{13}$C values for C=O groups.

$^{13}$C $\delta$ 174 ppm

Cl

O

Hexanoyl chloride

**(33.15)**

### $^1$H NMR spectroscopy

Aldehydes (RCHO), carboxylic acids (RCO$_2$H) and amides (RCONH$_2$ or RCONHR') possess protons that give characteristic signals in the $^1$H NMR spectrum (see Table 14.3).

In an aldehyde, the proton of the CHO group gives rise to a sharp signal at $\approx \delta$ 8–10 ppm. A small coupling to alkyl protons on the adjacent C atom may be observed. For example, the $^1$H NMR spectrum of acetaldehyde (**33.16**) exhibits a quartet at $\delta$ 9.8 ppm (1H, $J_{HH} = 3$ Hz) for the CHO proton, and a doublet at $\delta$ 2.2 ppm (3H, $J_{HH} = 3$ Hz) for the methyl protons.

H H O
H$-$C$-$C
H H

$J = 3$ Hz

Acetaldehyde

**(33.16)**

In a carboxylic acid, the OH proton gives a signal in the region $\delta$ 9–13 ppm in the $^1$H NMR spectrum, and the signal is typically of low intensity and broadened. Hydrogen bonding between carboxylic acid molecules typically occurs in solution. There may also be interactions between carboxylic acid and solvent molecules, making the chemical shift of the OH proton signal solvent-dependent. Just as in alcohols, the OH proton of a CO$_2$H group exchanges with deuterium from D$_2$O (equation 33.2). Such exchange does *not* occur with the proton in an aldehyde functional group.

$$RCO_2H + D_2O \rightleftharpoons RCO_2D + HOD \qquad (33.2)$$

The $^1$H NMR spectra of amides RCONH$_2$ and RCONHR' show broad signals assigned to the NH protons. The chemical shift range is variable,

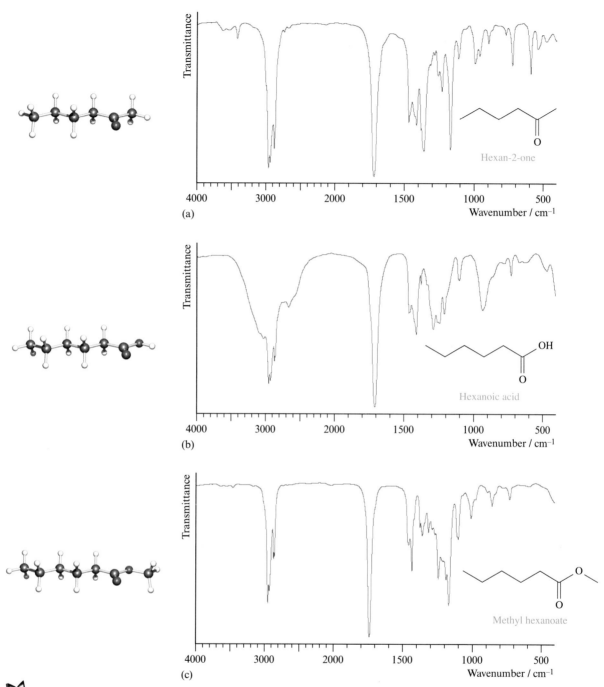

**Fig. 33.4** The IR spectra of (a) hexan-2-one (a ketone), (b) hexanoic acid (a carboxylic acid) and (c) methyl hexanoate (an ester). In each, the *strong* absorption near $1700 \, \text{cm}^{-1}$ is assigned to the stretch of the C=O bond. Colour code for the structures: C, grey; O, red; H, white.

See diagrams 33.27 and 33.28 for bonding descriptions of an amide

typically between $\delta 5$ and 12 ppm. The N atom is in a trigonal planar environment and, because $\pi$-bonding restricts rotation about the C−N bond, the two H atoms in $RCONH_2$ are magnetically non-equivalent. In hexanamide, for example, two broad signals at $\delta 6.3$ and 5.9 ppm are observed, one assigned to each $NH_2$ proton. (See end-of-chapter problem 33.26.)

**Fig. 33.5** The $^{13}C\{^1H\}$ NMR spectrum of butanone. The low-intensity peak at $\delta$ 209 ppm is assigned to the carbonyl C atom.

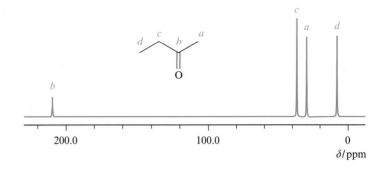

| 200.0 | 100.0 | 0 |

$\delta$/ppm

**33.6**      ## Keto–enol tautomerism

### Keto–enol tautomerism and the enolate ion

***Tautomerism*** involves the transfer of a proton between sites in a molecule with an associated rearrangement of multiple bonds.

When an OH group is attached to a C=C bond, the species is unstable with respect to rearrangement to a carbonyl compound. The rearrangement is a *keto–enol tautomerism* and the *two tautomers are in equilibrium* (equation 33.3) with the keto-form usually being favoured. The preference for the keto-form can be rationalized in terms of the sum of the bond enthalpies of the C=O ($806 \, kJ \, mol^{-1}$), C–C ($346 \, kJ \, mol^{-1}$) and C–H ($416 \, kJ \, mol^{-1}$) bonds versus the C–O ($359 \, kJ \, mol^{-1}$), O–H ($464 \, kJ \, mol^{-1}$) and C=C ($598 \, kJ \, mol^{-1}$) bonds.

$$\text{keto-tautomer} \rightleftharpoons \text{enol-tautomer} \qquad (33.3)$$

A carbon atom adjacent to the C=O carbon atom is labelled α-C, e.g.

α-C atoms

The formation of the enol from the keto-form is called *enolization* and is catalysed by acid or base. The acid-catalysed route involves protonation of the carbonyl O atom followed by loss of $H^+$ from the adjacent (or α) C atom:

keto-tautomer                    enol-tautomer

The base-catalysed route involves abstraction of the α-C–H proton by a base such as hydroxide ion, followed by protonation, e.g. by $H_2O$:

keto-tautomer                    enol-tautomer

The keto–enol pair is acidic, with the *enol-tautomer being more acidic than the keto-tautomer*. Loss of $H^+$ by treatment with base gives the conjugate base: the *enolate ion*. This anion is also the intermediate formed in the base-catalysed enolization shown above. The structure of and bonding in the enolate ion are represented by resonance pair **33.17**, or by the delocalized bonding description **33.18**. (Compare this with the allyl anion that we described in detail in Figure 13.6.) Of the two contributing resonance forms in **33.17**, the left-hand one is preferred with the negative charge on the O atom (remember that O is more electronegative than C).

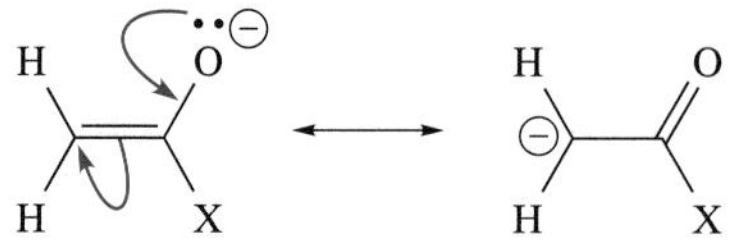

(33.17) 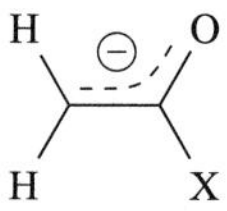                                            (33.18)

The structures that make up resonance pair **33.17** are *not separate entities*, but are *contributing forms that together describe the bonding in the enolate ion*. In contrast, the keto- and enol-forms of the conjugate acid are *discrete and different chemical compounds*. It is important to remember this distinction.

### Reactions of an enol or enolate ion with electrophiles

We described above the protonation of the O atom of the C=O group. Although $H^+$ is an electrophile, electrophiles may attack the α-C atom rather than the carbonyl O atom. This can be understood by considering the involvement of the enol-tautomer. We can draw an analogy with the reaction of a particular type of alkene with an electrophile, $E^+$:

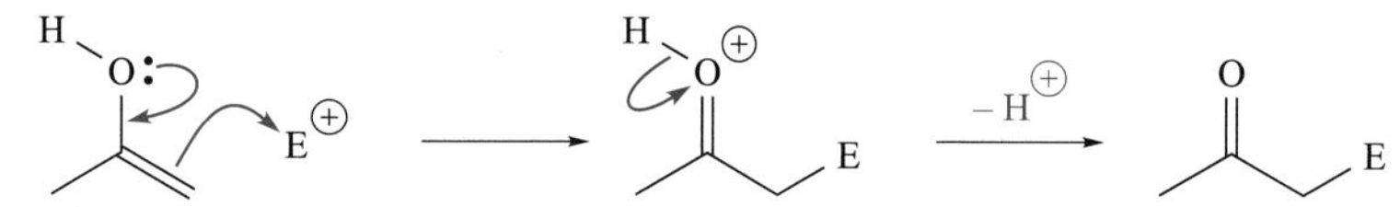

enol-tautomer

The same is also true of electrophilic attack on an enolate ion, as the scheme below shows:

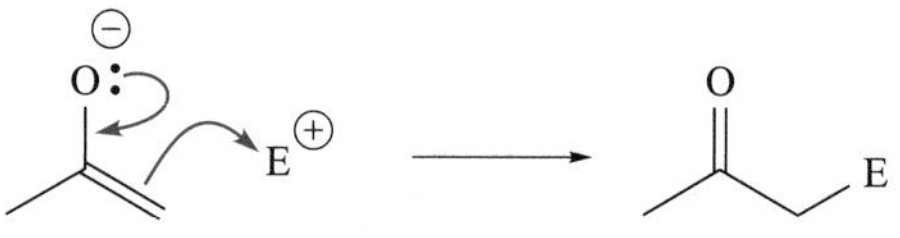

Enolate ion

We return to these mechanisms later when we apply them to a number of reactions.

| 33.7 | **Aldehydes and ketones: synthesis** |
|------|--------------------------------------|

In this and the following four sections, we turn our attention to ways of preparing carbonyl compounds. Some of the reactions interconvert different types of carbonyl compounds, e.g. making esters or acyl chlorides from carboxylic acids. Therefore the discussion necessarily includes aspects of the reactivity of carbonyl compounds.

### Oxidation of alcohols

See also under 'oxidation' in Section 30.5

The oxidation of alcohols can lead to the formation of aldehydes, ketones or carboxylic acids depending on:

- the position of the OH group (whether the alcohol is primary or secondary);
- the reaction conditions.

As we discussed in Section 30.5, primary and secondary (but *not* tertiary) alcohols are oxidized to carbonyl compounds. Suitable oxidizing agents are $KMnO_4$, $Na_2Cr_2O_7$, $K_2Cr_2O_7$ and $CrO_3$, all under acidic conditions. The oxidation of a secondary alcohol ($R_2CHOH$) gives a *ketone* (e.g. reaction 33.4) and this cannot easily be oxidized further, since oxidation of ketones involves C–C bond cleavage.

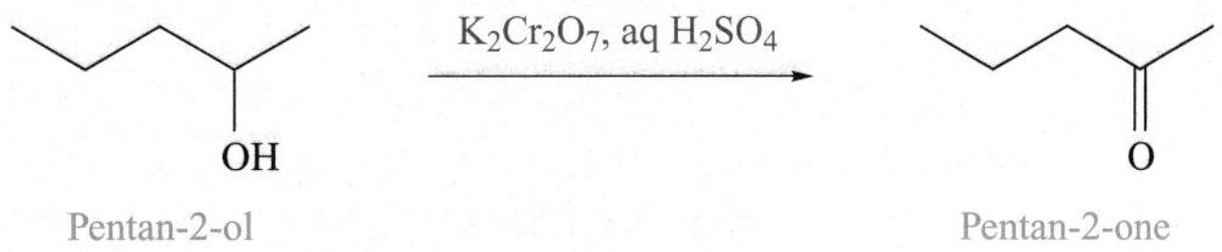

K2Cr2O7, aq H2SO4
OH
O
Pentan-2-ol
Pentan-2-one


(33.4)

Oxidation of a primary alcohol initially gives an aldehyde and this can be further oxidized to a carboxylic acid (scheme 33.5).

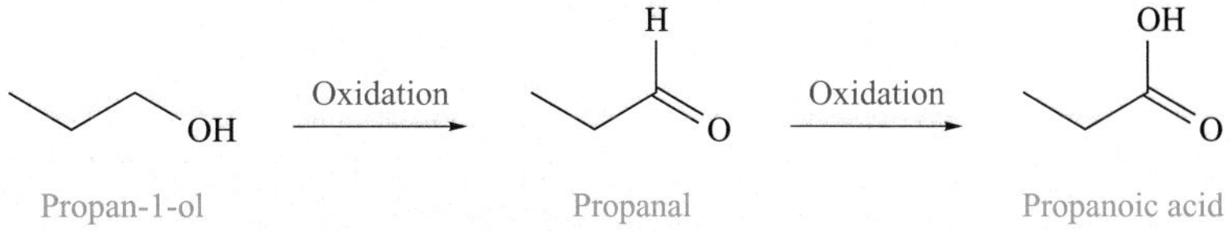

H
OH
Oxidation
Oxidation
OH
O
O
Propan-1-ol
Propanal
Propanoic acid


(33.5)

For mechanistic details: see Section 30.5

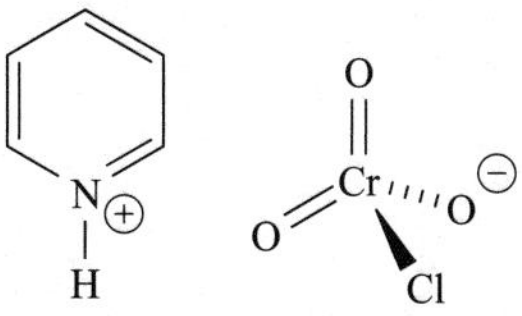

O
Cr
O
N
H
O
Cl


Pyridinium chlorochromate (PCC)

**(33.19)**

The choice of oxidizing agent affects the outcome of the oxidation of a primary alcohol. With $KMnO_4$, $K_2Cr_2O_7$, $Na_2Cr_2O_7$ or $CrO_3$ in the presence of dilute $H_2SO_4$ in acetone, oxidation to a carboxylic acid occurs (a *Jones oxidation*). To stop the reaction at the aldehyde stage, the most convenient oxidant is pyridinium chlorochromate (PCC, **33.19**) in non-aqueous (e.g. $CH_2Cl_2$) solution. Scheme 33.6 gives examples.

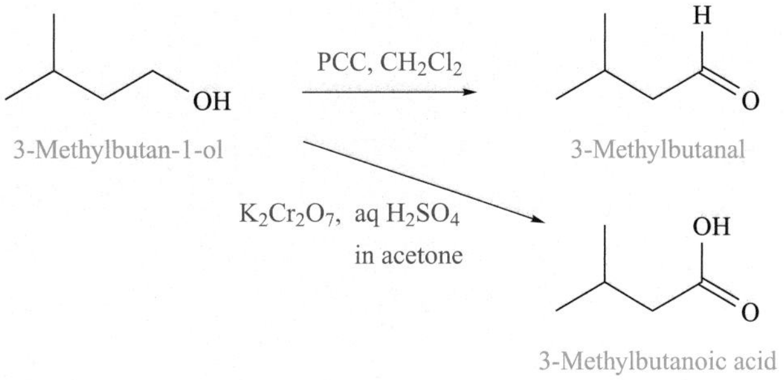

H
PCC, CH2Cl2
OH
O
3-Methylbutan-1-ol
3-Methylbutanal
K2Cr2O7, aq H2SO4
in acetone
OH
O
3-Methylbutanoic acid


(33.6)

## Reduction of acyl chlorides and esters

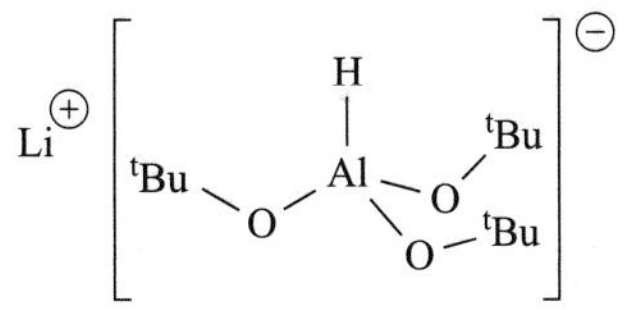

H
Li
tBu
Al
tBu
O
O
O
tBu


$LiAlH(O^tBu)_3$

**(33.20)**

The reduction of acyl chlorides or esters by $LiAlH_4$ yields alcohols (see Section 30.2) but if a milder reducing agent is used, aldehydes can be obtained. Lithium tri-*tert*-butoxyaluminium hydride (**33.20**) is related to $LiAlH_4$ by replacement of three H atoms by $^tBuO$ groups. Reaction 33.7 illustrates use of $LiAlH(O^tBu)_3$ to reduce an acyl chloride to an aldehyde. It can be applied to both alkyl and aryl derivatives.

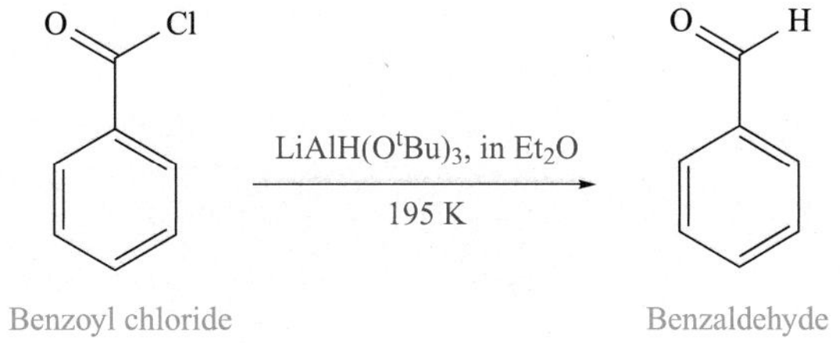

O
Cl
O
H
LiAlH(OtBu)3, in Et2O
195 K
Benzoyl chloride
Benzaldehyde


(33.7)

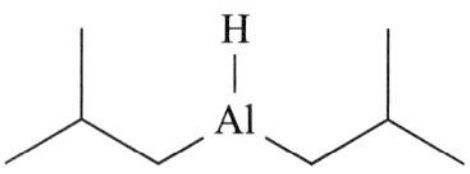
H
Al

DIBAL
**(33.21)**

Reduction of esters can be carried out using a reducing agent called DIBAL (diisobutyl aluminium hydride, **33.21**) and is a good method of preparing aldehydes (equation 33.8).

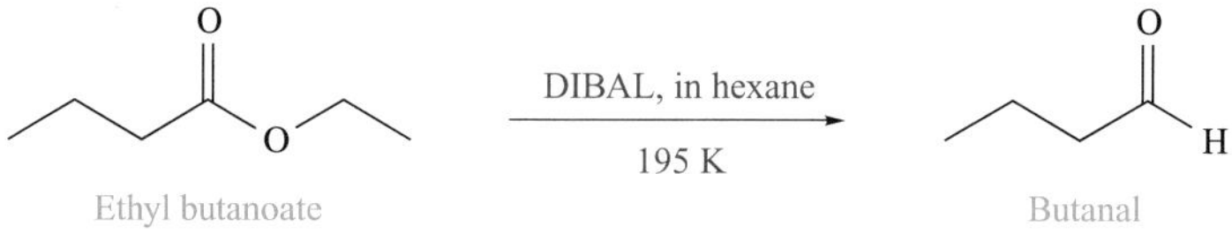
O
O
Ethyl butanoate
DIBAL, in hexane
195 K
O
H
Butanal

(33.8)

## Friedel–Crafts acylation: preparation of aryl ketones

We described the Friedel–Crafts acylation of benzene in detail in Chapter 32. This reaction is used to prepare aryl ketones, as illustrated in equations 33.9 and 33.10.

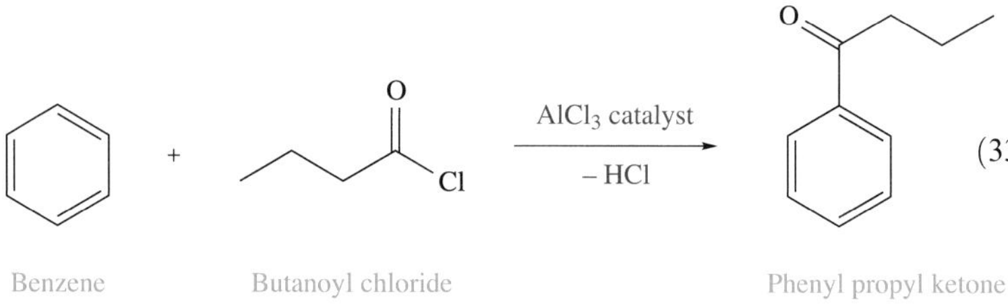
+
O
Cl
AlCl3 catalyst
– HCl
O
Benzene
Butanoyl chloride
Phenyl propyl ketone

(33.9)

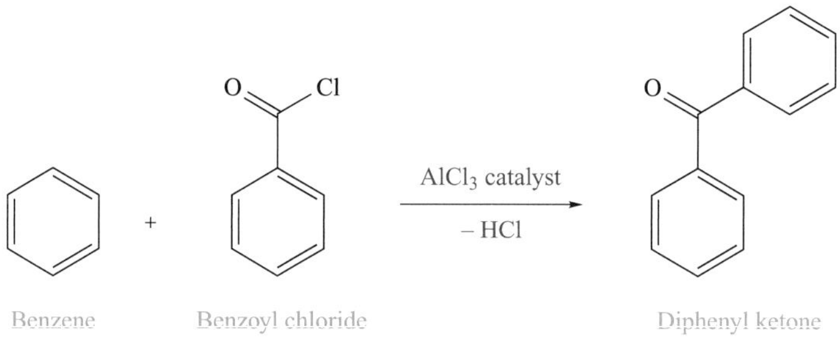
O
Cl
+
AlCl3 catalyst
– HCl
O
Benzene
Benzoyl chloride
Diphenyl ketone

(33.10)

Orientation effects, activation and deactivation: see Section 32.10

For substituted aromatic compounds, the choice of starting compounds may be critical because some substituents deactivate the ring towards electrophilic substitution. For example, consider the formation of the target molecule in scheme 33.11.

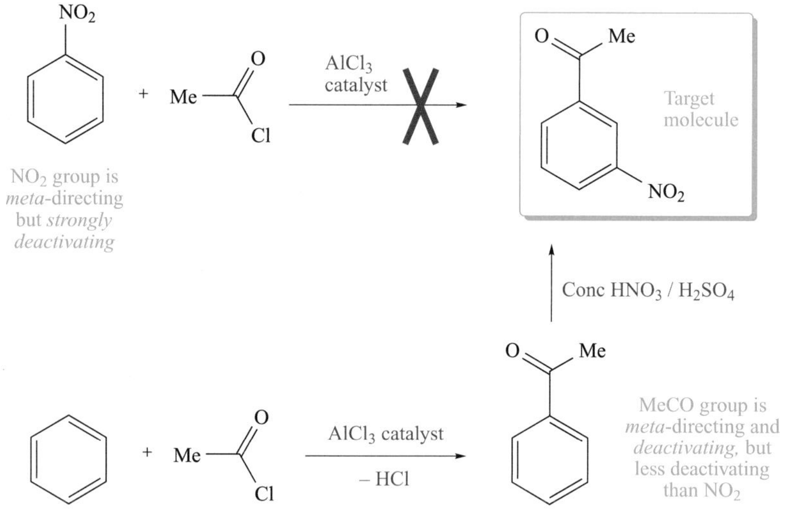
NO2
+ Me
O
Cl
AlCl3
catalyst
NO2 group is
meta-directing
but strongly
deactivating
O
Me
Target
molecule
NO2
Conc HNO3 / H2SO4
O
Me
+ Me
O
Cl
AlCl3 catalyst
– HCl
MeCO group is
meta-directing and
deactivating, but
less deactivating
than NO2

(33.11)

The $NO_2$ group in $PhNO_2$ is *meta*-directing, and so we could consider the acylation of nitrobenzene as a possible route. However, the $NO_2$ group is *strongly deactivating*, and $PhNO_2$ does *not react with acetyl chloride* in the presence of a Lewis acid catalyst. An alternative route to the target molecule is to acylate benzene *before* nitration. Like the $NO_2$ group, the MeCO group is *meta*-directing, and Table 32.3 shows that both groups are deactivating. However, the acyl substituent is less deactivating than is the nitro group, and the nitration step in the lower pathway in scheme 33.11 is successful.

## 33.8    Carboxylic acids: synthesis

### Oxidation of primary alcohols and aldehydes

We have already seen that the oxidation of primary alcohols by acidified $KMnO_4$, $K_2Cr_2O_7$, $Na_2Cr_2O_7$ or $CrO_3$ in acetone (a *Jones oxidation*) gives first an aldehyde and then a carboxylic acid (equations 30.33 and 33.5). This is a convenient method of preparing carboxylic acids and can be applied

Remember that *benzyl* alcohol is a *primary* alcohol

to aliphatic alcohols, including benzyl alcohol (equation 33.12). The same reaction conditions can be used to prepare a carboxylic acid starting from an aldehyde.

Benzyl alcohol        Acidified $K_2Cr_2O_7$ in acetone        Benzoic acid        (33.12)

### Oxidation of an alkylbenzene derivative

Oxidation of toluene and ethylbenzene: see equations 32.41 and 32.42

The oxidation of the alkyl group of an alkylbenzene derivative leads to the formation of a carboxylic acid in which the $CO_2H$ group is attached *directly* to the aryl ring. Oxidation takes place *at the $CH_2$ group attached to the ring* (equation 33.13) and oxidation of, for example, toluene, ethylbenzene, *n*-propylbenzene and *n*-butylbenzene gives the same aromatic product: benzoic acid.

Butylbenzene        Aqueous $KMnO_4$        Heat        Benzoic acid    +    Propanoic acid

(33.13)

The reaction can be extended to prepare carboxylic acids with more than one $CO_2H$ group. Phthalic acid (reaction 33.14) and its isomers are prepared from *o*-, *m*- and *p*-xylene and are of commercial importance (see Box 33.3).

$$(33.14)$$

*o*-Xylene
(1,2-Dimethylbenzene)

Aqueous KMnO$_4$

Heat

Phthalic acid
(Benzene-1,2-dicarboxylic acid)

Benzoic acid derivatives are among several compounds manufactured on a large scale as herbicides. The group consists of molecules that mimic the action of the plant hormone *auxin*. Synthetic mimics of auxin include dicamba, 2,4-D and 2,4,5-T. The hormone mimics interfere with plant growth and are selective herbicides targeting broad-leaved plants. In the Vietnam War, a mixture of 2,4-D and 2,4,5-T called *Agent Orange* was used by the US forces to defoliate forests, thereby removing cover used by the enemy. The huge quantities of agent orange used in the war and its contamination with the dioxin TCDD (see end of Section 29.2) remain controversial issues.

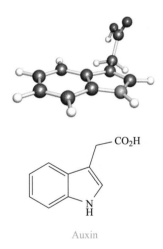

Auxin

Dicamba

2,4-D

2,4,5-T

Grignard reagents:
see Section 28.4

## Use of Grignard reagents

Halogenoalkanes, RX, can be converted to carboxylic acids by first transforming RX into the corresponding Grignard reagent, RMgX (e.g. reaction 33.15).

Et$_2$O

$$(33.15)$$

The Grignard reagent is then treated with solid CO$_2$ (*dry ice*). The reaction involves *an increase in the carbon chain length*: the precursor is RMgX and the product is RCO$_2$H (equation 33.16).

Solid CO$_2$, Et$_2$O

4-Carbon chain

5-Carbon chain

$$(33.16)$$

## COMMERCIAL AND LABORATORY APPLICATIONS

### Box 33.3 Dyes and fibres from phthalic and terephthalic acids

Benzenedicarboxylic acid has three isomers (*o*-, *m*- and *p*-benzenedicarboxylic acids) of which two are of particular commercial importance. *o*- and *p*-Benzenedicarboxylic acids are known as phthalic and terephthalic acids respectively. *Phthalic anhydride* is a derivative of phthalic acid and has applications in the dyestuffs industry and is used in the manufacture of some plasticizers.

Phthalic acid    Terephthalic acid    Phthalic anhydride

An important commercial use of phthalic acid is as a precursor to synthetic indigo dye, the major producer of which is BASF in Germany. The most well-known application of indigo is in the dyeing of blue jeans. The production process involves the following transformations:

Phthalic acid    Phthalimide

Anthranilic acid

Indigo

Terephthalic acid is high on the list of top-ranking chemicals manufactured in the US and Europe, and is used in the production of *polyester synthetic (man-made) fibres*. Polyesters are formed by the condensation of a dicarboxylic acid and a diol, as exemplified by the formation of the polyester Dacron:

$- H_2O$

Dacron

Polyester contributes significantly to the man-made fibre market; the chart below shows the relative amounts of man-made fibres manufactured in the US during 2001. Output from the US ($\approx 4.0$ Mt in 2001) makes up about 20% of the world man-made fibre market.

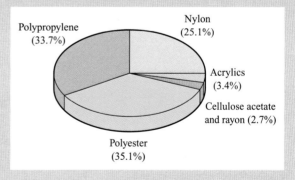

Polypropylene (33.7%)

Nylon (25.1%)

Acrylics (3.4%)

Cellulose acetate and rayon (2.7%)

Polyester (35.1%)

[Data from: *Chemical & Engineering News* (2002) June 24, p. 60.]

The $Mg-C$ bond in the Grignard reagent is polar in the sense $Mg^{\delta+}-C^{\delta-}$, and this allows it to act as a nucleophile, attacking the $C^{\delta+}$ centre of $CO_2$:

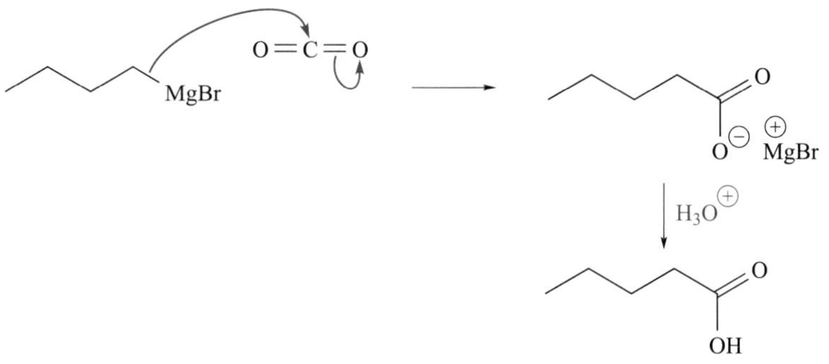

Grignard reagents can be used to prepare a wide range of carboxylic acids including alkyl and aryl derivatives.

### Acid- and base-catalysed hydrolysis of nitriles

Competition with E2 reactions may occur with secondary or tertiary halogenoalkanes: see Chapter 28

Nitriles $RC\equiv N$ can be produced by the nucleophilic substitution of a halide group in a halogenoalkane by cyanide (e.g. reaction 28.19). The $C\equiv N$ group is then readily hydrolysed in the presence of acid or base and the usual product is a carboxylic acid. Scheme 33.17 shows the conversion of 1-bromopropane to butanoic acid: *note the increase in the carbon chain length*.

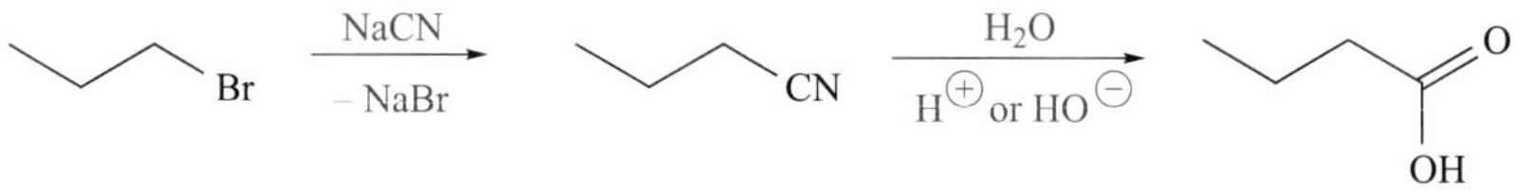

$$(33.17)$$

### 33.9     Esters: synthesis and hydrolysis

### Synthesis from a carboxylic acid and alcohol

In a ***condensation reaction***, two reactants combine with the elimination of a small molecule such as $H_2O$, $NH_3$ or HCl.

An ester can be produced by a *condensation reaction* (in which $H_2O$ is eliminated) between a carboxylic acid and an alcohol. The reaction is catalysed by acid, and is summarized in equation 33.18. The fact that the acid $O-H$, rather than the alcohol $O-H$, bond is cleaved has been confirmed by $^{18}O$ labelling studies.

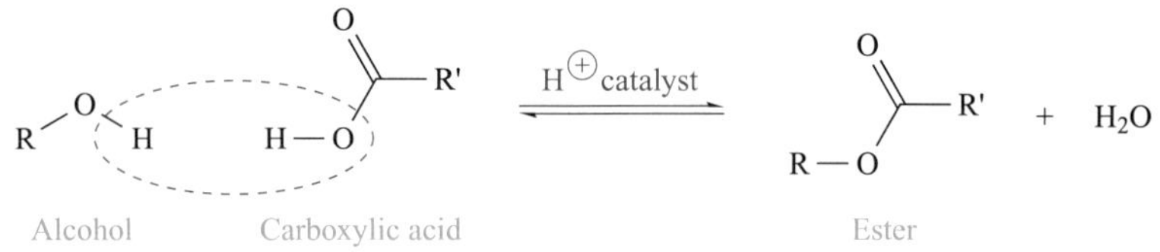

$$(33.18)$$

The reaction is general and can be applied to R and R' being alkyl or aryl groups, although usually the alcohol is an alkyl derivative. The mechanism of the reaction involves the protonation of the carbonyl group of the acid, followed by nucleophilic attack by the alcohol. Protonation of the diol that forms in the first step then facilitates conversion to the ester by loss of $H_2O$ (a good leaving group):

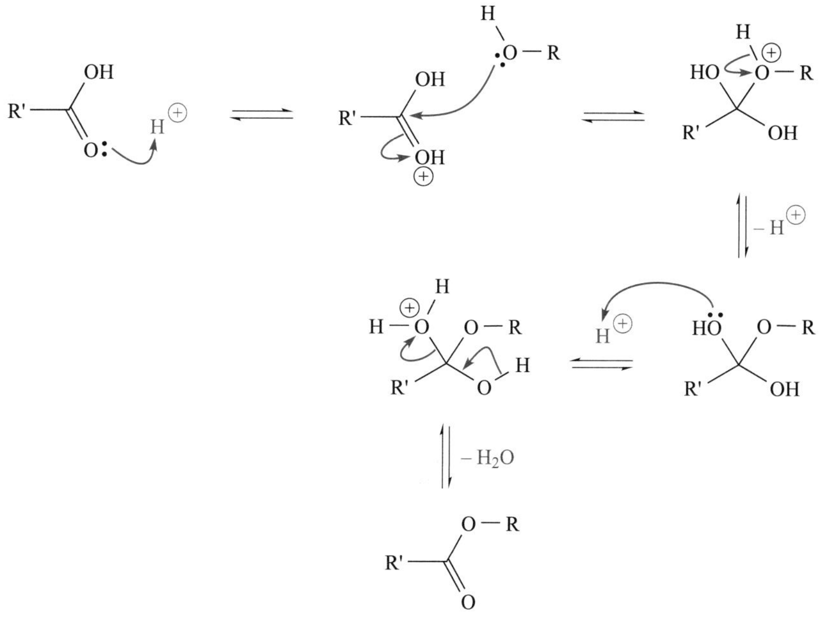

Each step in the pathway above is an equilibrium, and the general formation of the ester (equation 33.18) is an equilibrium. This has consequences for the success (or not) of the synthesis. If the ester is to be formed in good yield, equilibrium 33.18 must be driven to the right-hand side. Ways of achieving this are to continually remove the ester or water by distillation, or to add

Equilibria: see Chapter 16

## ENVIRONMENT AND BIOLOGY

### Box 33.4 Ester hydrolysis for protection from the sun

Safety in the sun (protection against skin cancer) is a topical issue, and use of sunscreen creams is recommended. High-factor sunscreens should absorb harmful UV radiation and protect the skin, but UV radiation that does reach the skin produces highly reactive radicals that damage cells irrevocably. An ingredient that is added to sunscreen lotions is the ester α-tocopheryl acetate. On the skin, the ester is hydrolysed to the corresponding phenol and this acts as a *radical quencher*. Our awareness of UV radiation has increased with publicity of ozone depletion.

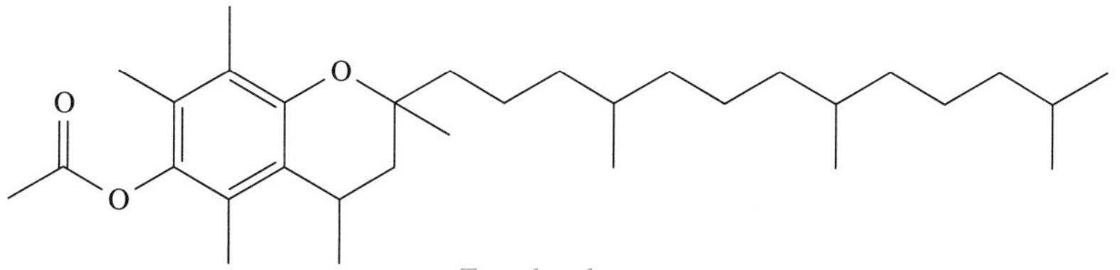

α-Tocopheryl acetate

See also: Box 25.3 'Depletion of the ozone layer'.

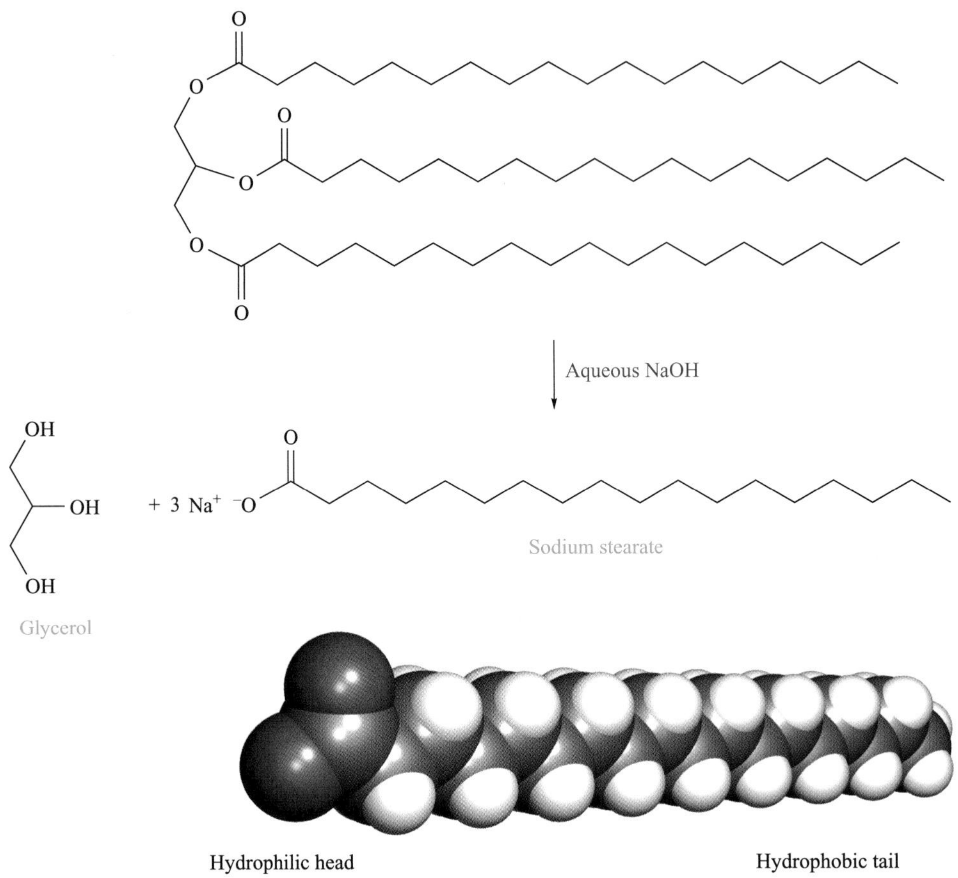

**Fig. 33.6** Saponification to produce the soap sodium stearate. The space filling model of the stearate anion shows the hydrophilic ('water-loving') and hydrophobic ('water-hating') tail that are crucial to the formation of micelles (see Box 30.1).

an excess of the alcohol and carboxylic acid. Esters are of commercial significance and in Boxes 33.2–33.5 we highlight several applications.

## Hydrolysis of esters

The hydrolysis of an ester by aqueous NaOH is called *saponification*.

Esters are hydrolysed by aqueous NaOH. This process is called *saponification* and is the key step in the manufacture of soap from fats which are esters of glycerol (propane-1,2,3-triol). The scheme in Figure 33.6 shows saponification of a fat in which the ester is derived from stearic acid. Stearic acid contains a long aliphatic chain and is an example of a *fatty acid* (see Box 33.5). The product of this saponification reaction is the soap sodium stearate.

A *fatty acid* contains a long aliphatic chain.

## Synthesis from an acyl chloride and alcohol

An alternative method of preparing esters is the condensation reaction between an acyl chloride and an alcohol in which HCl is eliminated. General reaction 33.19 can be applied to alkyl and aryl derivatives *including phenols*

**ENVIRONMENT AND BIOLOGY**

## Box 33.5 Saturated and unsaturated fats

Fats are esters of carboxylic acids (fatty acids) and play a vital role in nutrition. Those involved in the food industry fall into several categories:

- saturated (all C−C bonds in the organic chain);
- monounsaturated (one C=C in the chain);
- polyunsaturated (more than one C=C in the chain).

The relationship between blood cholesterol (see Box 30.1) and the intake of different fatty acids is the subject of research. The results are sometimes controversial, providing the consumer with confusing and often contradictory advice. Saturated fatty acids such as lauric, myristic, palmitic and stearic acids have varying effects on blood cholesterol levels. Lauric and myristic acids appear to be higher-risk components of high-fat diets than palmitic acid. Monosaturated fatty acids (usually oleic acid) and polyunsaturated (mainly linoleic acid) represent more healthy substitutes to saturated fatty acids. It is not just the presence of the C=C groups that is important in an unsaturated fat; isomers possess different nutritional characteristics. Edible oils vary in their composition, e.g. olive oil contains ≈80% oleic acid, while corn oil contains about 40% each of oleic and linoleic acids and 10% palmitic acid.

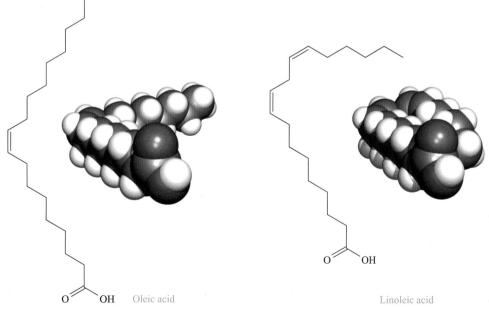

Lauric acid

Myristic acid

Palmitic acid

Stearic acid

Oleic acid

Linoleic acid

(compare with the use of reaction 33.18 in which the alcohol is usually an alkyl derivative).

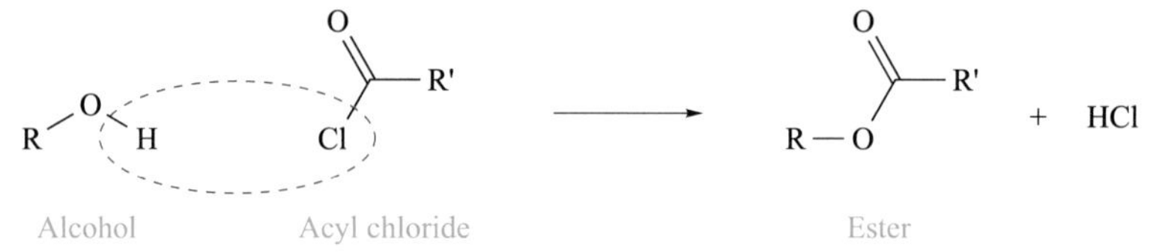

Alcohol  Acyl chloride  Ester

$$(33.19)$$

Reactions 33.20 and 33.21 give examples of ester syntheses starting from acyl chlorides.

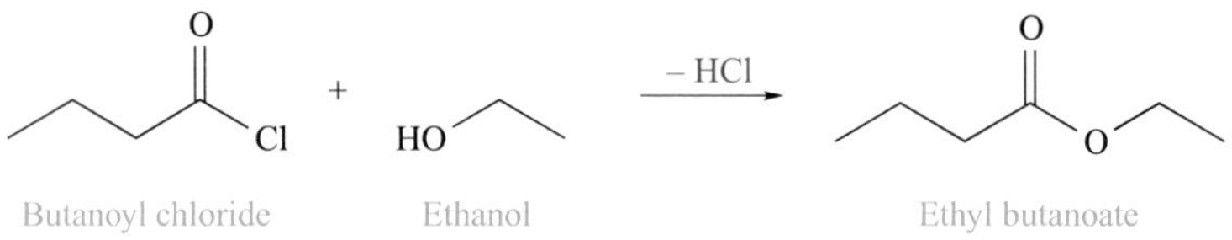

Butanoyl chloride  Ethanol  Ethyl butanoate

$$(33.20)$$

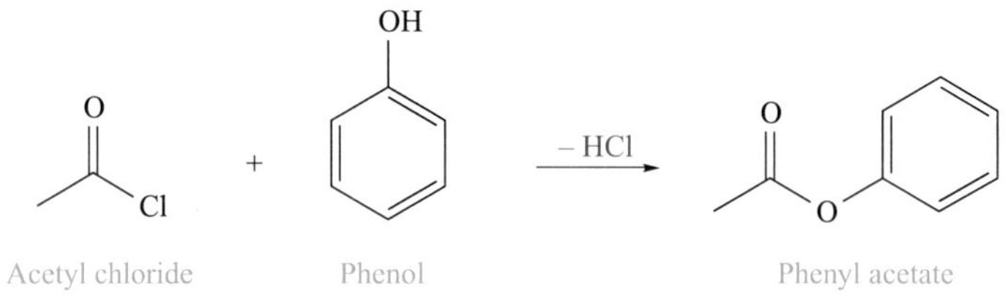

Acetyl chloride  Phenol  Phenyl acetate

$$(33.21)$$

## 33.10  Amides: synthesis

Esters react with $NH_3$ to give amides (reaction 33.22) although a better route is to start with an acyl chloride (equation 33.23). We return to the mechanisms of these reactions later in the chapter.

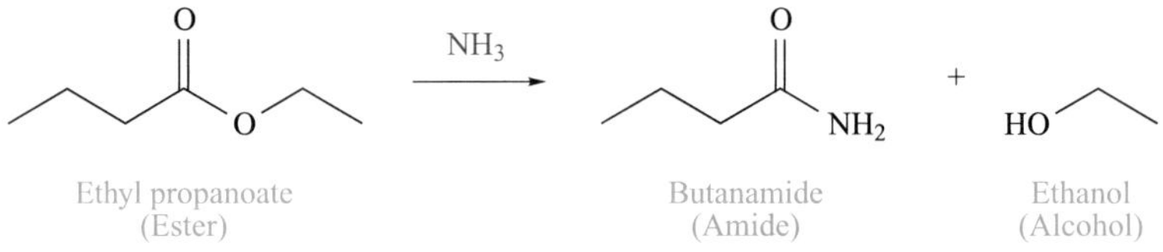

Ethyl propanoate
(Ester)  Butanamide
(Amide)  Ethanol
(Alcohol)

$$(33.22)$$

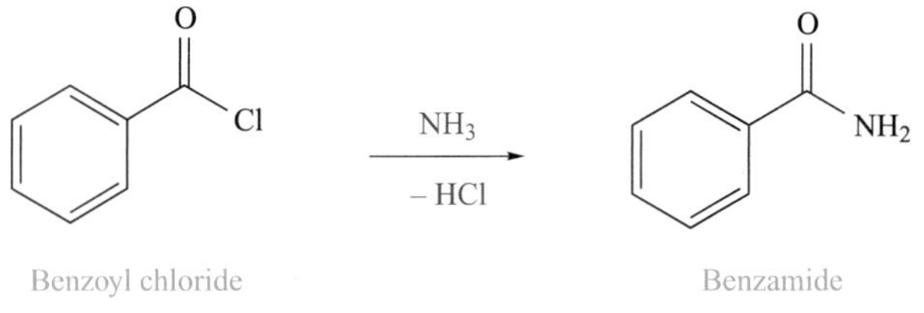

Benzoyl chloride  Benzamide

$$(33.23)$$

Reaction 33.23 is readily extended to the syntheses of N-substituted amides by using primary or secondary amines in place of $NH_3$ (scheme 33.24).

(33.24)

Urea is a unique type of amide in which the R group is $NH_2$, and has wide application as a nitrogen-based fertilizer. It was discovered in 1773 as a component of human urine, and is today manufactured on a large scale by dehydrating ammonium carbamate, itself formed from $NH_3$ and $CO_2$:

Ammonium carbamate | Heat; high pressure, $-H_2O$ | Urea

Representative amides from the pharmaceutical industry are paracetamol (used widely as a pain killer) and the AstraZeneca product Tenormin (a β-blocker prescribed for the treatment of high blood pressure). Tenormin ranks among the best-selling drugs in the world.

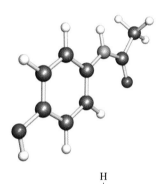

Paracetamol

Tenormin

## 33.11 Acyl chlorides: synthesis

Acyl chlorides RCOCl can be prepared from carboxylic acids $RCO_2H$ by treatment with $PCl_3$, $PCl_5$ or $SOCl_2$ (thionyl chloride). An advantage of using $SOCl_2$ is that the by-products are gases, making isolation of the acyl chloride relatively easy. Reactions 33.25–33.27 give examples of acyl chloride syntheses.

$$\xrightarrow[-H_3PO_3]{PCl_3}$$ (33.25)

$$\xrightarrow[-POCl_3]{PCl_5}$$ (33.26)

$$\xrightarrow[-SO_2,\, -HCl]{SOCl_2}$$ (33.27)

## 33.12    Carbonyl compounds as acids

pKₐ and Kₐ: see Chapter 16

In this section, we consider the $pK_a$ values of different carbonyl compounds, and illustrate that carboxylic acids (introduced as *weak acids* in Chapter 16) are not the only carbonyl compounds to behave as Brønsted acids.

**(33.22)**

### Carboxylic acids

Equilibrium 33.28 shows a carboxylic acid behaving as a weak Brønsted acid. Values of $pK_a$ for selected carboxylic acids are listed in Table 33.3.

$$RCO_2H(aq) + H_2O(l) \rightleftharpoons H_3O^+(aq) + RCO_2^-(aq) \qquad (33.28)$$

Carboxylic acids are weaker than mineral acids such as HCl, $H_2SO_4$ and $HNO_3$ but are stronger than $H_2O$ and alcohols. A carboxylic acid $RCO_2H$ has a smaller $pK_a$ value (larger $K_a$) than an aliphatic alcohol ROH because of the greater stability of the carboxylate anion $RCO_2^-$ with respect to the alkoxide ion $RO^-$. The $RCO_2^-$ anion is *resonance stabilized* and can be represented by resonance pair **33.22**. An alternative bonding description is given in **33.23** which shows delocalization of the negative charge over the O−C−O unit. The delocalization arises from overlap of the C and O $2p$ orbitals, and the $\pi$-bonding MO is represented schematically in diagram **33.24**. This system is analogous to the allyl anion (Figure 13.6) and is a four $\pi$-electron system. However, because of the differing effective nuclear charges of C and O, and the resulting different atomic orbital energies, the $\pi$-bonding MO contains more O than C character. Evidence for delocalization comes from the fact that the two C−O bonds are equivalent. In the solid state structures of salts such as $MeCO_2^-Na^+$ and $PhCO_2^-Na^+$, the C−O bond lengths are typically 126 pm. This distance is in between typical

**(33.23)**

$sp^2$ hybridized C

Overlap of one C $2p$ and two O $2p$ orbitals

**(33.24)**

---

**ENVIRONMENT AND BIOLOGY**

### Box 33.6 Enzymic browning of fruit and salads: lemon juice comes to the rescue

If you cut an apple or prepare a chicory salad, what originally appeared fresh soon turns brown and, while still edible, does not look so appetizing. The process of browning results from the enzymic oxidation of a natural substrate. The enzyme is phenolase (also called catecholase) and both it and the substrate (which varies depending on the foodstuff) are present in the fruit. However, they do not react until cell membranes are destroyed when the fruit is cut. Browning can be inhibited by the presence of reducing agents, and sulfites and $SO_2$ have been used, although their toxicity makes them far from ideal.

$$\text{Substrate} \underset{\text{Reducing agent}}{\overset{\text{Phenolase}}{\rightleftharpoons}} \text{Oxidized product}$$

A common method of preventing browning and retaining a fresh look of salad or fruit is to squeeze lemon juice over it. Lemon juice contains citric acid which is a tricarboxylic acid. The acid alters the pH of the system, and since the action of enzymes is affected by changes in pH, the lemon juice inhibits enzymic browning.

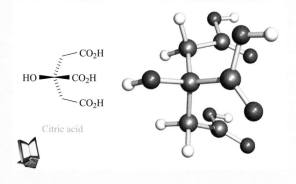

Citric acid

**Table 33.3** p$K_a$ values for selected carboxylic acids.

| Compound | Structure | p$K_a$ |
|---|---|---|
| Formic acid | | 3.75 |
| Acetic acid | | 4.77 |
| Propanoic acid | | 4.81 |
| Butanoic acid | | 4.87 |
| Chloroacetic acid | | 2.85 |
| Dichloroacetic acid | | 1.48 |
| Trichloroacetic acid | | 0.70 |
| Benzoic acid | | 4.19 |
| o-Chlorobenzoic acid | | 2.92 |
| m-Chlorobenzoic acid | | 3.82 |
| p-Chlorobenzoic acid | | 3.98 |

bond lengths for C–O (single, 135 pm) and C=O (double, 122 pm) bonds involving $sp^2$ C atoms.

Unsubstituted aliphatic acids possess similar $pK_a$ values (compare values for formic, acetic, propanoic and butanoic acids in Table 33.3). The presence of electron-withdrawing substituents such as Cl can significantly affect the equilibrium constant for equilibrium 33.28. Compare the $pK_a$ values for acetic acid and its chloro-derivatives listed in Table 33.3. We have already rationalized these trends in terms of the inductive effect (see Section 27.4).

Carboxylate salts are formed when a carboxylic acid reacts with bases as in reactions 33.29 and 33.30.

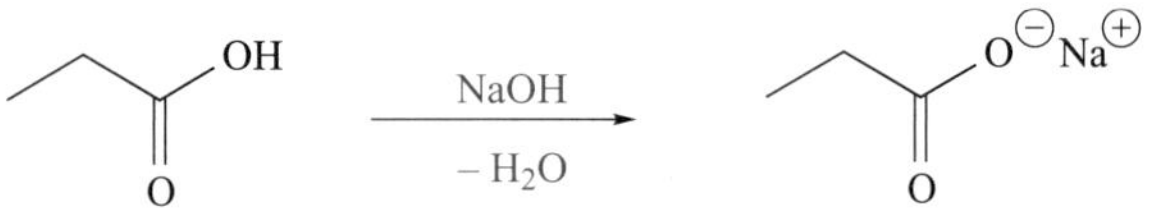

Propanoic acid — Sodium propanoate    (33.29)

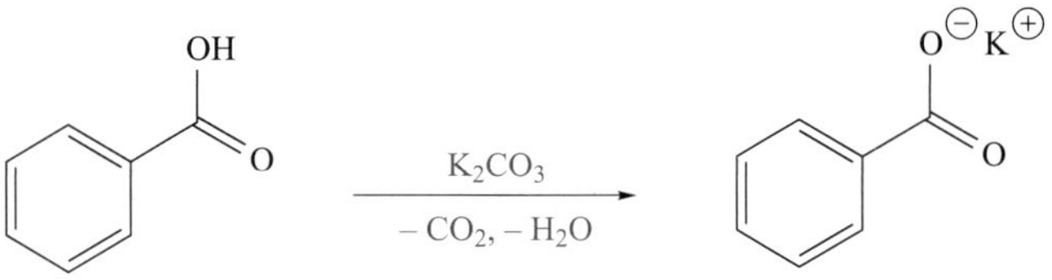

Benzoic acid — Potassium benzoate    (33.30)

See also structures 16.4–16.6 and associated discussion in Chapter 16

Two salts can be derived from dicarboxylic acids such as oxalic acid, e.g. $KHC_2O_4$ and $K_2C_2O_4$. Equilibria 33.31 and 33.32 show the first and second dissociation steps of oxalic acid. The values of $pK_a$ show the usual trend, with $pK_a(1) < pK_a(2)$, i.e. $K_a(1) > K_a(2)$.

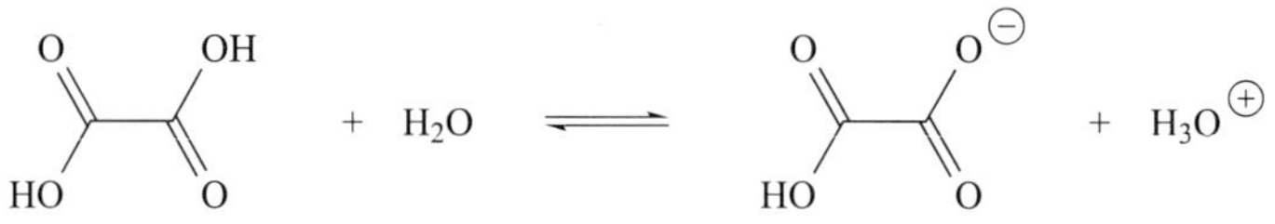

Oxalic acid

$pK_a(1) = 1.23$    (33.31)

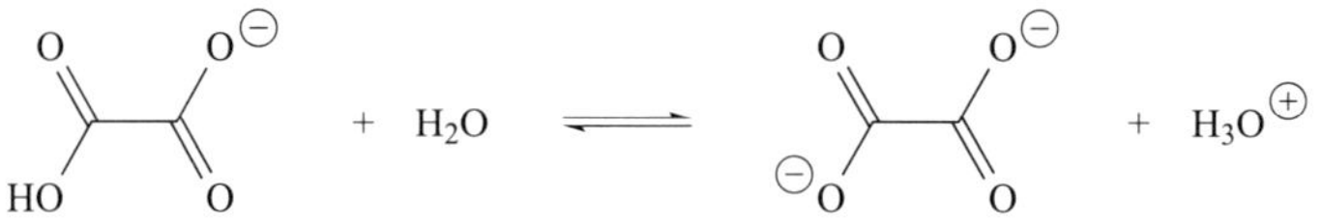

$pK_a(2) = 4.19$    (33.32)

### Amides

Just as carboxylate ions are resonance-stabilized, so too are the conjugate bases of amides. We discussed the acidity of carboxylic acids in terms of the stabilization of the conjugate base $RCO_2^-$. Similarly, the conjugate base of an amide is resonance-stabilized, but this is only sufficient to make the amide a *very weak acid* ($pK_a \approx 17$). Structure **33.25** shows the resonance pair for the conjugate base of the amide $RCONH_2$. An NH unit is isoelectronic with O, and so the bonding descriptions of the $RCO_2^-$ and $RCONH^-$ ions are related. The alternative description is a delocalized bonding scheme as in **33.26**. The MO picture (which includes $\pi$-bonding, non-bonding and antibonding MOs) is similar to that for the allyl anion (Figure 13.6) and the carboxylate ion.

## ENVIRONMENT AND BIOLOGY

### Box 33.7 Capsaicin and dihydrocapsaicin: hot amides!

The *Capsicum* family includes a wide variety of peppers ranging from the large sweet peppers to the fiery chilli peppers. Chilli peppers derive their hotness from capsaicin and dihydrocapsaicin (members of the capsaicinoid family), and restaurants specializing in 'hot' foods owe much to these molecules of nature. They work by interacting with the same receptors in your mouth that sense heat. Excessive amounts of capsaicin are toxic; capsaicin prevents the production of certain neurotransmitters and affects the function of neuroproteins in the brain. At high enough concentrations, capsaicin destroys 'substance P' in the nervous system. This effect has now been harnessed for medical use: 'substance P' is associated with the pain suffered by people with, for example, arthritis and inflammatory bowel disease, and application of a cream containing capsaicin results in pain relief.

Capsicum.
*Peter Anderson © Dorling Kindersley.*

Capsaicin

Dihydrocapsaicin

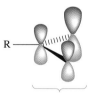

*sp²* hybridized C

*sp²* hybridized N

Overlap of C 2*p*, N 2*p* and O 2*p* orbitals

**(33.27)**

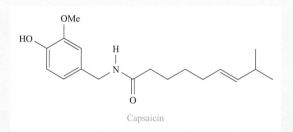

**(33.25)**          **(33.26)**

In Chapter 31, we described how amines act as Brønsted *bases*, through use of the lone pair on the N atom. Amides are almost non-basic because the lone pair is conjugated into a π-system. The N atom is trigonal planar and *sp²* hybridized, with the lone pair in a 2*p* orbital and involved in the π-system shown in diagram **33.27**. The bonding can also be represented in terms of resonance pair **33.28**. An important consequence of the π-delocalization in amides is that proteins (which are polypeptides with −CONH− units, see Sections 35.5 and 35.6) have well-defined 3-dimensional structures.

**(33.28)**

## Aldehydes, ketones and esters

Carbonyl group

α-Hydrogen atoms

**(33.29)**

> Ketones and aldehydes that have **α-hydrogen** atoms can behave as weak acids. A β-diketone has a CH₂ group between two C=O groups and loses H⁺ to give a **β-diketonate** anion.

We have seen how resonance stabilization involving the C=O group is responsible for the acidic properties of carboxylic acids and, to a lesser extent, of amides. Similar effects are responsible for the acidity of certain ketones and aldehydes. Dissociation by proton loss is associated with the *acidity of the α-hydrogen atom*, i.e. the H atom attached to the C atom *adjacent* to the C=O group (diagram **33.29**). Loss of H⁺ gives a carbanion as in equilibrium 33.33 where B is a Brønsted base. Although we show the negative charge on the C atom, the anion is actually resonance stabilized (resonance pair **33.30**) and *the negative charge resides mainly on the O atom*. This is the enolate ion that we met in Section 33.6. The extra resonance stabilization that the enolate ion possesses means that ketones and aldehydes that have α-hydrogen atoms can behave as weak acids.

(33.33)

**(33.30)**

Table 33.4 lists p$K_a$ values for selected carbonyl compounds. Of particular note is the p$K_a$ value of propanedial which has a similar acid strength to acetic acid (p$K_a = 4.77$). Diketones that have one CH₂ between the C=O groups are called *β-diketones*. An example is pentane-2,4-dione (acetylacetone) which has p$K_a \approx 9$. It is a weaker acid than acetic acid, but a stronger acid than a ketone of type RCH₂COR' (Table 33.4). Deprotonation of acetylacetone (equilibrium 33.34 in which B = base) gives an anion that is resonance stabilized through the involvement of two O atoms. This is shown in the set of resonance structures **33.31** and is a particularly favourable situation; *the negative charge resides mainly on the O atoms*. An alternative view of the bonding is to show the negative charge delocalized over five atoms (**33.32**) by π-conjugation involving 2$p$ orbitals (**33.33**). The acetylacetonate anion is an example of a *β-diketonate*.

The ligand acac⁻: see for example, Figure 23.5

(33.34)

Acetylacetone
(Pentane-2,4-dione)

Acetylacetonate ion

**(33.31)**

Each C is
$sp^2$
hybridized

Overlap of three C 2$p$ and
two O 2$p$ orbitals

**(33.33)**

**(33.32)**

## ENVIRONMENT AND BIOLOGY

### Box 33.8 Esters as environmentally friendly fuels

*Biodiesel* is an environmentally friendly substitute for diesel fuel. Diesel fuel is manufactured from crude oil, a non-renewable source of fuel. Biofuels are produced mainly from vegetable oils and are renewable sources of fuel. The major raw material for biodiesel is rapeseed, instantly recognizable as the bright yellow crop in fields throughout Europe (see Box 7.2). Biodiesel is a mixture of esters and is obtained by the base-catalysed reaction of methanol and vegetable oils, for example:

Rapeseed oil + MeOH $\xrightarrow{\text{NaOH catalyst}}$ Rapeseed oil esters + Glycerol

The component oils that are used in the manufacture of biodiesel are shown in the chart below:

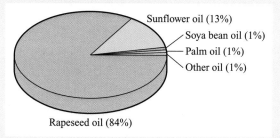

Sunflower oil (13%)
Soya bean oil (1%)
Palm oil (1%)
Other oil (1%)

Rapeseed oil (84%)

[Data from: *Chemistry & Industry* (2000) p. 530.]

After purification, biodiesel performs well with respect to regular diesel fuel, and vehicles using it require little or no engine modification. The environmental advantages of biodiesel are significant. It is non-toxic and is almost entirely biodegradable. Compared with diesel fuel, the emissions after combustion contain smaller amounts of $SO_2$, CO and hydrocarbons.

Biodiesel is a new fuel, but one for which a market is developing rapidly. It is marketed under several names throughout Europe, e.g. in France it is sold as Diester (from *diesel-ester*). Industrial-scale manufacturing plants currently operate in France (with an output of $\approx 0.12$ Mt per year), Italy ($\approx 0.08$ Mt per year) and Austria ($\approx 0.01$ Mt per year) and production levels are growing.

For more information, visit the websites:

www.biodiesel.org/
www.greenfuels.co.uk/

---

Esters such as **33.34** or **33.35** are similarly *weakly* acidic (see Table 33.4). Esters are weaker acids than ketones because the $\delta^+$ charge on the carbonyl C atom is delocalized. This can be understood in terms of the resonance pair shown below:

We shall see in the next section that the acidic nature of $\alpha$-H atoms in some carbonyl compounds produces *carbon centres* which are susceptible to attack by electrophiles.

(33.34)

(33.35)

**Table 33.4** Approximate p$K_a$ values for types of carbonyl compounds with α-H atoms; the α-H atoms in the conjugate acid are shown in blue. R and R' are alkyl substituents and may be the same or different.

| Compound type | p$K_a$ (approximate value) | Conjugate base (each is resonance stabilized; only the form derived directly by loss of the α-H proton is shown) |
|---|---|---|
| Propanedial | 5 | |
| | 9 | |
| | 11 | |
| | 13 | |
| | 16 | |
| | 19 | |
| | 24 | |

## 33.13 Enolate ions in synthesis: electrophilic substitution reactions

In Section 33.3, we showed that electrophiles such as H$^+$ can attack the O atom of a C=O group, while nucleophiles may attack the C atom. Now we describe how some *electrophilic* substitutions result in bond formation between the carbonyl C atom and the electrophile. As you work through this section, consider what makes these reactions *different* from *direct attack* by an electrophile on a carbonyl compound.

### Base-catalysed halogenation of aldehydes and ketones

When an $\alpha$-H atom (see **33.29**) is present in an aldehyde or ketone, the reaction of $Cl_2$, $Br_2$ or $I_2$ with the carbonyl compound in the presence of acid (see below) or base results in halogenation at the $\alpha$-position as in reaction 33.35.

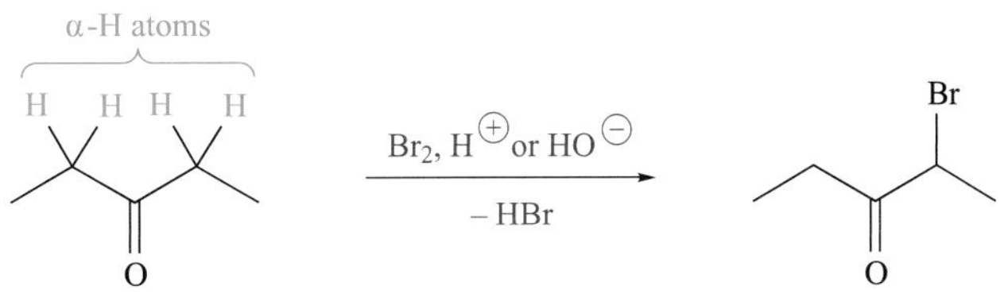

(33.35)

The base-catalysed reaction proceeds by the initial abstraction of an $\alpha$-H to give an enolate. In the scheme below, the anion is shown with the negative charge localized on the O atom; recall from **33.30** that this is the dominant resonance form. Attack on $Br_2$ then follows to give the brominated substitution product:

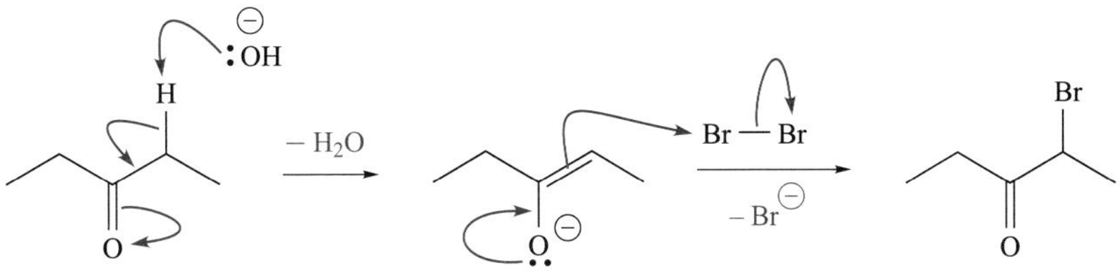

Examples of methyl ketones

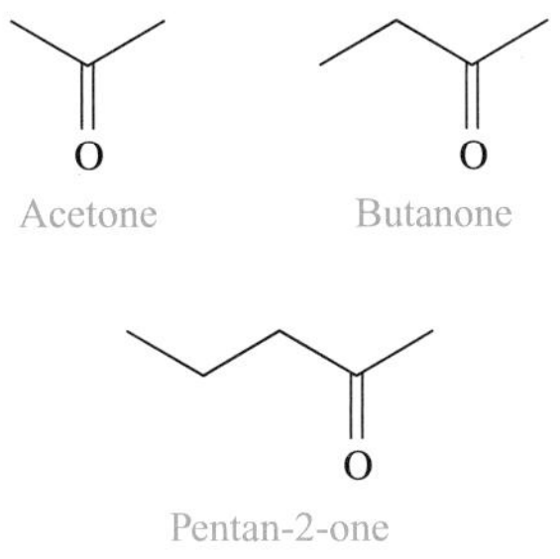

Acetone        Butanone

Pentan-2-one

In practice, mixtures of products are often observed. In reaction 33.35, for example, there are four $\alpha$-H atoms, and as substitution for Cl, Br or I occurs, the p$K_a$ of the compound decreases (Table 33.3):

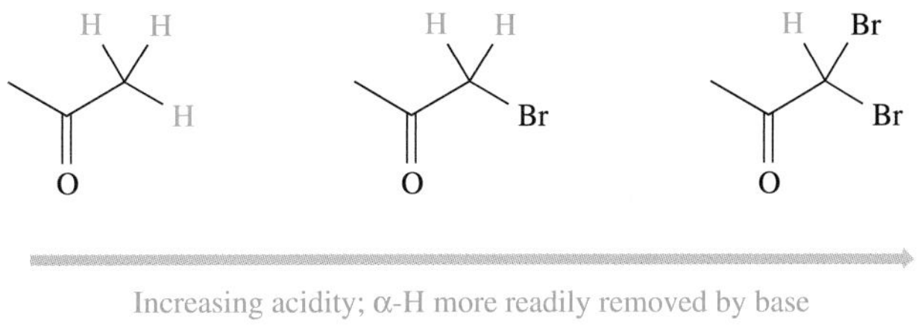

Increasing acidity; $\alpha$-H more readily removed by base

Thus, it becomes easier for the base to abstract sequential $\alpha$-protons, and multi-substitution occurs. In the bromination of acetone, for example, the electron-withdrawing effects of the Br atom in the monosubstituted product facilitate the *rapid* formation of the di- and then the trisubstituted product (equation 33.36).

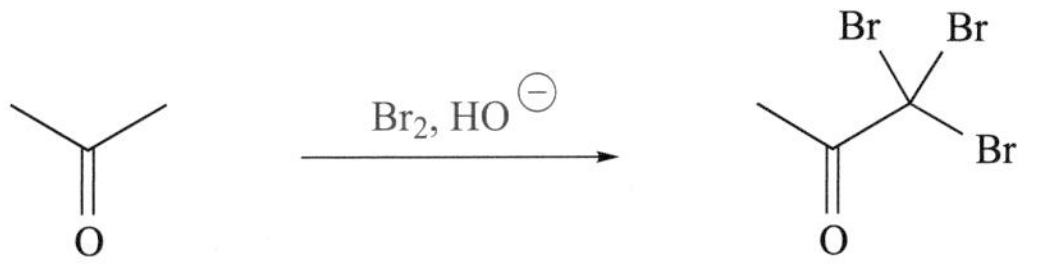

(33.36)

Once a terminal $CX_3$ (X = Cl, Br, I) unit has formed *adjacent to the C=O group*, another reaction enters into play because $CX_3^-$ is a good leaving group. In the formation of the halo-derivatives, the role of $HO^-$ is to abstract

$H^+$. However, $HO^-$ can also act as a nucleophile and attack the $C^{\delta+}$ atom of the carbonyl group. We have ignored this possibility so far because, if there is no other good leaving group, this attack leads nowhere:

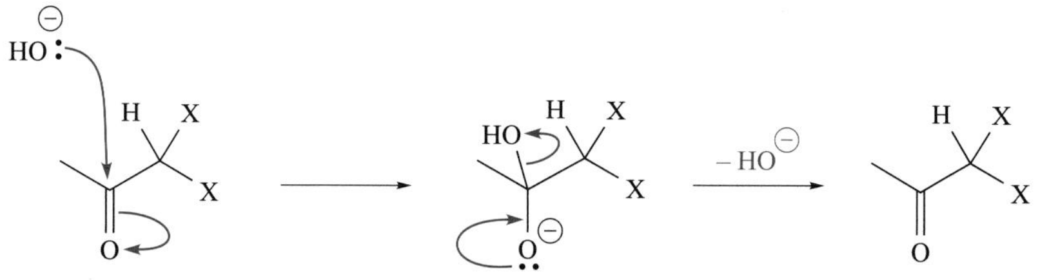

X = H, Cl, Br or I

However, when the choice of leaving group is between $CX_3^-$ and $HO^-$, loss of $CX_3^-$ is favoured:

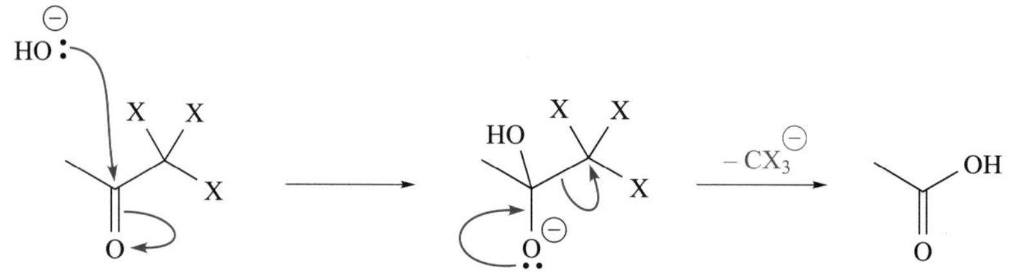

See equation 31.31 for the formation of LDA

The isolated products are $HCX_3$ and a carboxylate ion:

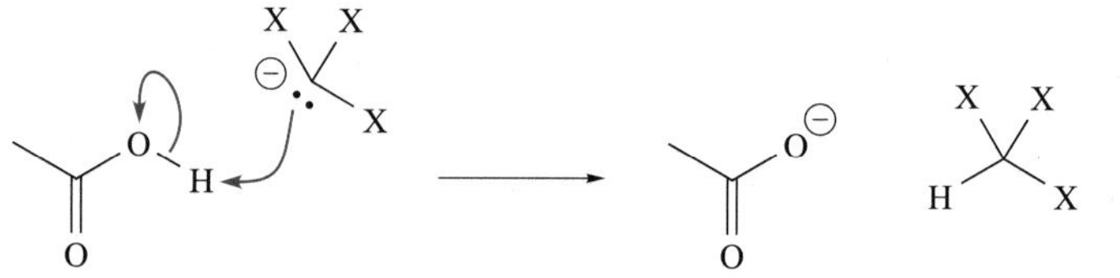

This is a general reaction of methyl ketones (e.g. acetone, butanone and pentan-2-one) and is called the *haloform reaction*.

### Acid-catalysed halogenation of aldehydes and ketones

In worked example 15.3, we considered the kinetics of the acid-catalysed bromination of acetone:

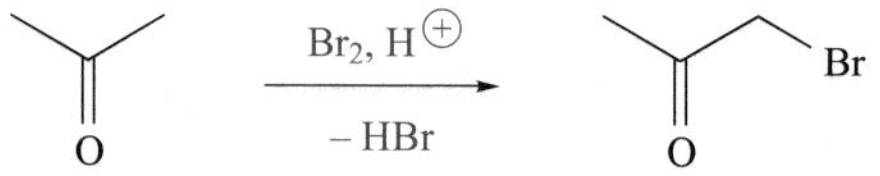

Nucleophilic substitution versus elimination: see Section 28.8

The rate equation is given below and similar expressions are found for other acid-catalysed halogenations of aldehydes and ketones:

$$-\frac{d[Br_2]}{dt} = -\frac{d[Me_2CO]}{dt} = k[Me_2CO][H^+]$$

The rate depends on the concentrations of carbonyl compound and acid catalyst, but is independent of halogen. A mechanism that is consistent with these data involves the generation of an enol. The rate-determining step is the acid-catalysed rearrangement of ketone to enol. The enol then reacts with $Br_2$ in fast steps:

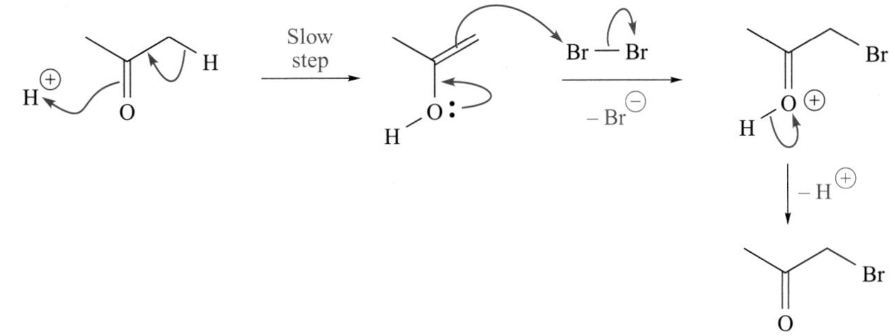

The complications of multi-substitution that occur in the base-catalysed halogenation reactions are avoided if an acid catalyst is used instead.

## Alkylation of a ketone via an enolate

If a ketone, ester, diketone or diester possesses an α-H atom, the carbonyl compound can be alkylated at the α-position, i.e. this is a *C–C bond-forming reaction*. The first step is the formation of a *lithium enolate* from the carbonyl compound. It is important that the starting material is *fully* enolized. This prevents any reaction between enolate and keto-form taking place to give secondary products. To achieve this, the carbonyl compound is treated with lithium diisopropylamide (LDA) to give a lithium enolate (reaction 33.37).

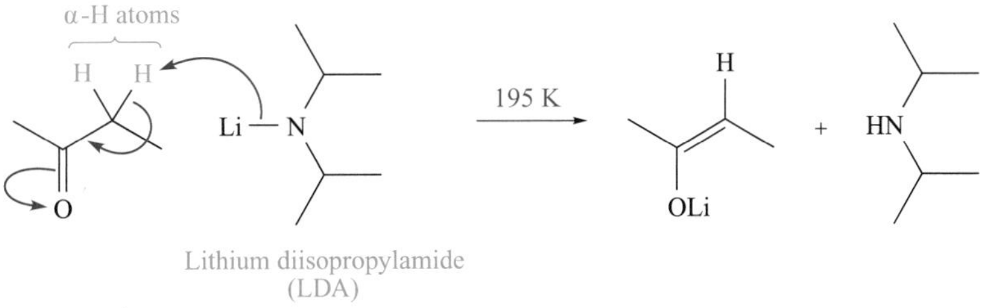

(33.37)

Treatment of the lithium enolate with a halogenoalkane leads to the formation of a C–C bond and involves the α-C atom. Reaction 33.38 converts the lithium enolate formed from acetone into pentan-2-one; remember that the C–X bond in a halogenoalkane is polar in the sense $C^{\delta+}$–$X^{\delta-}$, and that halogenoalkanes undergo nucleophilic substitution reactions.

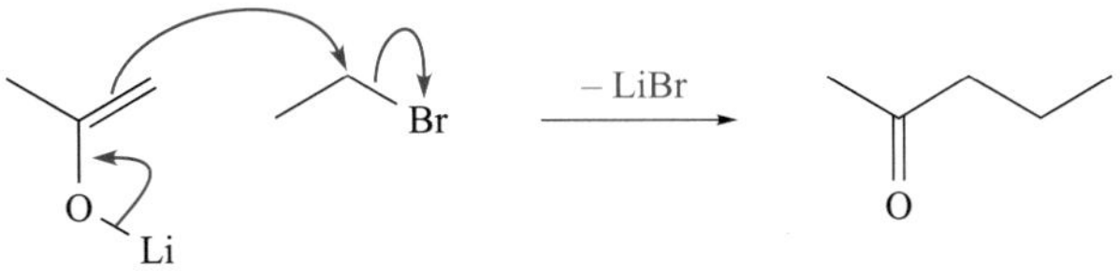

(33.38)

Such alkylation reactions are successful if the halogenoalkane is primary or secondary. Tertiary halogenoalkanes should be avoided because their reactions are complicated by competing eliminations.

## Alkylation of an ester via an enolate

Lithium enolates derived from esters react with halogenoalkanes in a similar manner to that described above. A competing reaction is that of the lithium enolate with the ester. This is the *Claisen condensation* and is considered in Section 33.16. To minimize this unwanted side-reaction, the concentration of LDA (and thus lithium enolate) should greatly exceed that of the starting ester during the initial reaction of the ester with LDA:

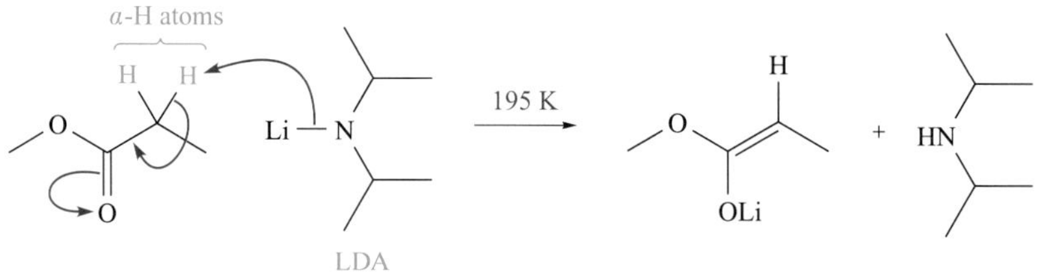

Reaction of the lithium enolate with halogenoalkane proceeds at the α-C position:

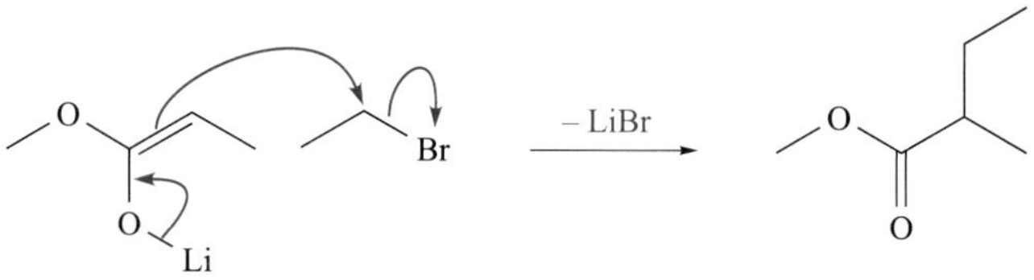

## Alkylation of β-diketones or diesters via an enolate

Data in Table 33.4 showed that dicarbonyl compounds in which the C=O groups are separated by one $CH_2$ unit (β-dicarbonyl compounds) are more acidic than ketones or esters in which the $CH_2$ group is adjacent to only one C=O group. Enolate ions of β-diketones or diesters can therefore be generated by using weaker bases than LDA. Ethoxide is a suitable base, and the enolate ion is stabilized by delocalization of the negative charge. Equilibrium 33.39 lies to the right-hand side because the β-dicarbonyl compound is a stronger acid ($pK_a \approx 13$) than ethanol ($pK_a = 16$).

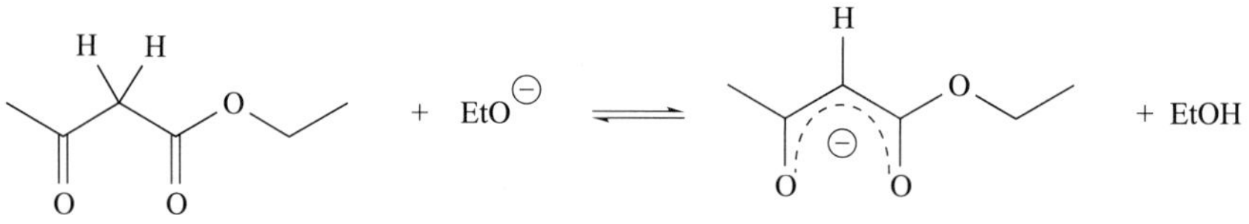

$$(33.39)$$

The reaction of the enolate ion with a primary or secondary halogeno-alkane leads to alkyl-substituted products as illustrated in reaction 33.40. The mechanism is represented starting from one of the contributing resonance forms of the enolate.

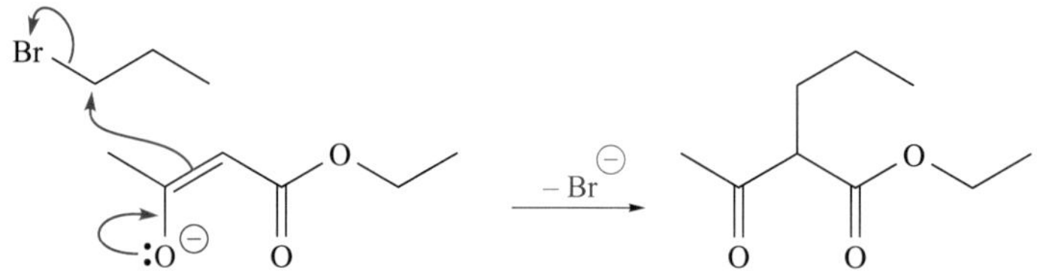

$$(33.40)$$

The product in reaction 33.40 still possesses an acidic α-H atom and so the reaction can be repeated to introduce a second alkyl substituent (the same as or different from the first substituent) at the α-position:

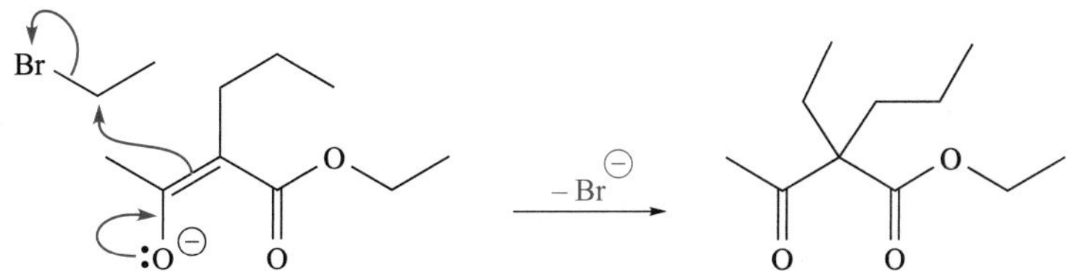

### Aldol and Claisen reactions

An ***aldol reaction*** is a C–C bond-forming reaction and occurs between two carbonyl compounds in the presence of base.

We have already seen that the addition of base (e.g. NaOH) to a carbonyl compound with an α-hydrogen atom results in *enolization*. In the enolizations described above, we used the strong bases LDA or EtO⁻ (LDA is stronger than ethoxide) to effect removal of the α-H from carbonyl compounds. In each reaction we considered, complete enolization was required to maximize the yield of the desired product, but we mentioned that competitive reactions can be problematic. If only *some* of the carbonyl compound is converted to the enolate, the reaction solution contains both enolate and the parent carbonyl compound, and reaction can occur between them. This is an *aldol reaction*. The first carbonyl compound may be an aldehyde, ketone or ester, but the final product of the aldol reaction is usually an aldehyde or ketone. The reaction involves the nucleophilic attack of the enolate ion on the $C^{\delta+}$ atom of a C=O group. A related reaction occurs between an ester and its enolate ion, formed in the presence of a strong base such as EtO⁻. This reaction is a *Claisen condensation*.

A ***Claisen condensation*** is a C–C bond-forming reaction and occurs between two ester molecules in the presence of a strong base such as EtO⁻.

We return to the details of the aldol reaction and Claisen condensation after a short introduction to nucleophilic attack at the C=O carbon atom.

---

**33.14**    ### Nucleophilic attack at the C=O carbon atom

Many reactions of carbonyl compounds involve nucleophilic attack at the $C^{\delta+}$ atom of a C=O group. We have already seen one example: the *haloform reaction* in equation 33.36. Equation 33.41 shows a general step that you will see repeatedly in this section. In the reaction, X⁻ represents a nucleophile and the carbon centre that is attacked goes from trigonal planar ($sp^2$ hybridized) to tetrahedral ($sp^3$ hybridized).

$$(33.41)$$

### Hydrolysis of acyl halides

Acyl chlorides hydrolyse readily to give the corresponding carboxylic acid (equation 33.42). Because the reaction is so facile, many acyl chlorides must be stored under water-free conditions.

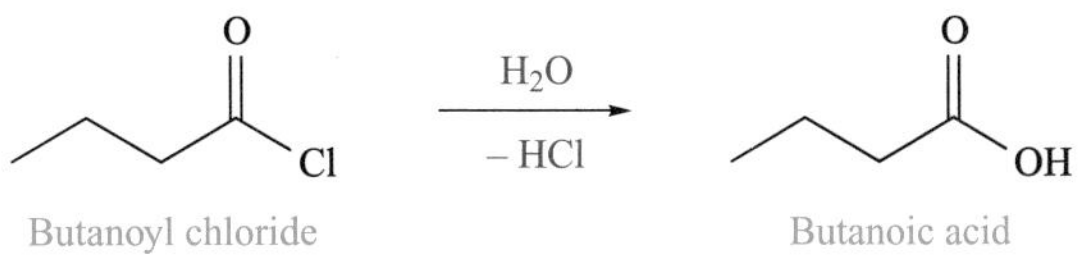

$$(33.42)$$

Butanoyl chloride                              Butanoic acid

The net result of the reaction is the substitution of $Cl^-$ by $HO^-$. However, the nucleophile is usually $H_2O$, although some acyl chlorides require aqueous base. The reaction mechanism is outlined below. Initial attack by $H_2O$ leads to O–C bond formation and opening of the C=O bond. *This is in preference to the direct loss of the leaving group* which takes place in the next step of the mechanism:

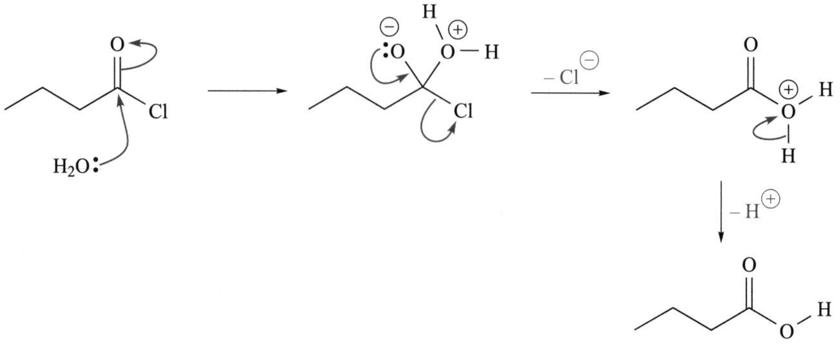

## Hydrolysis of amides

$NH_3$ is a good leaving group.

$NH_2^-$ is a very poor leaving group.

Amides may be hydrolysed to the corresponding carboxylic acid in the presence of an acid catalyst. In the hydrolysis of the acyl chloride, $Cl^-$ is the leaving group. By analogy, amide hydrolysis would involve loss of $NH_2^-$, *but $NH_2^-$ is a very poor leaving group*. In the presence of $H^+$, $NH_3$ (a good leaving group) departs from the amide. The mechanism of acid-catalysed amide hydrolysis is shown below. Notice that in the first step, the O atom is protonated in preference to the N atom. Later in the pathway, when the competition is between OH and $NH_2$ as a site of protonation, the N site is preferred on grounds of relative base strengths of ROH and $RNH_2$:

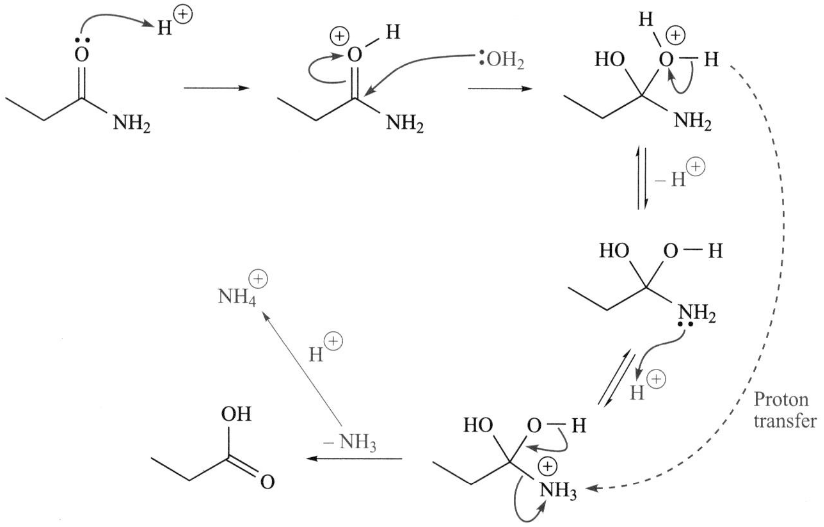

As a *rule of thumb* (but it is *not* infallible), the direction of an equilibrium of the type:

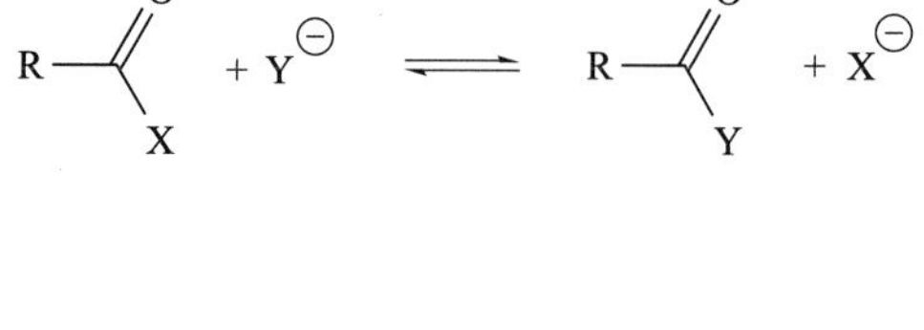

can be determined from the p$K_a$ values of HX and HY. If HX is a stronger acid than HY (i.e. p$K_a$(HX) < p$K_a$(HY)), then X$^-$ is more favoured than Y$^-$, and the equilibrium lies to the right-hand side.

## The reaction of aldehydes and ketones with CN$^-$

A *cyanohydrin* has the general structure:

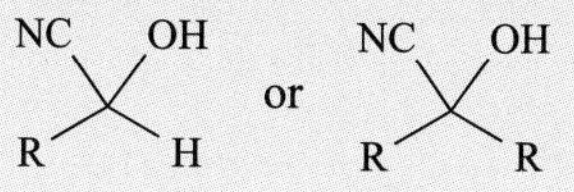

The reaction between an aldehyde or a ketone and the nucleophilic cyanide ion leads to the formation of a *cyanohydrin*. The mechanism of the reaction involves nucleophilic attack by cyanide on the C$^{\delta+}$ atom of the C=O group, followed by protonation. The source of NC$^-$ is, for example, NaCN. The mechanism is illustrated by the reaction of cyanide with acetaldehyde:

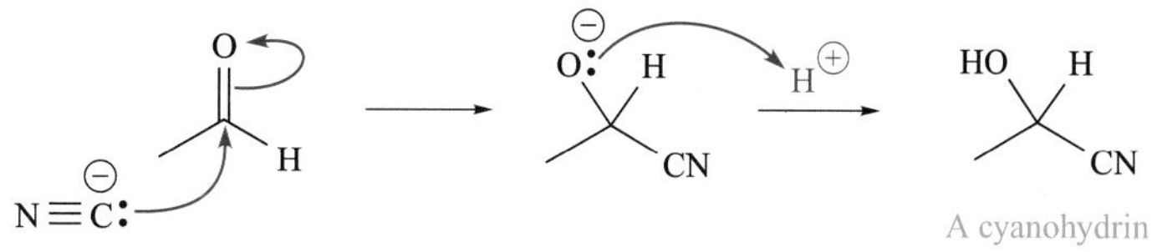

A cyanohydrin

A problem with cyanohydrin formation is that an equilibrium is established between starting materials and products. In the presence of base, NC$^-$ tends to be eliminated from the cyanohydrin (equation 33.43).

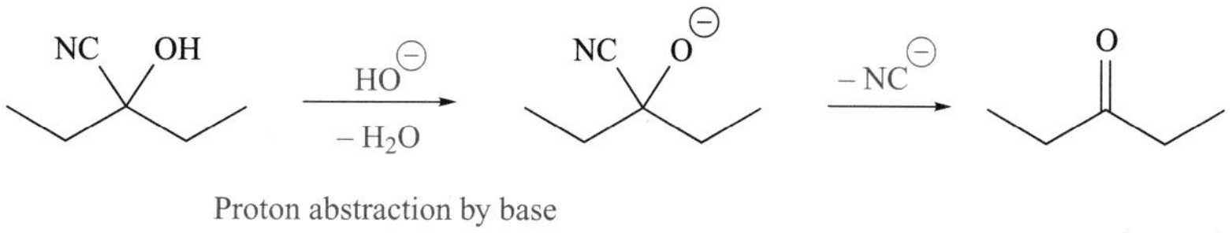

Proton abstraction by base

(33.43)

## Reactions of Grignard reagents with aldehydes and ketones

Formation of Grignard reagents: see Section 28.4

The reactions of Grignard reagents RMgX (R = alkyl, X = Cl or Br) with aldehydes and ketones are important synthetic routes to alcohols (see Section 30.2). In each of the following reactions, the solvent is usually anhydrous THF or Et$_2$O. *Formaldehyde* (methanal) reacts with Grignard reagents to give *primary alcohols* in which the terminal CH$_2$OH group is derived from the aldehyde. Reaction 33.44 shows the synthesis of butan-1-ol.

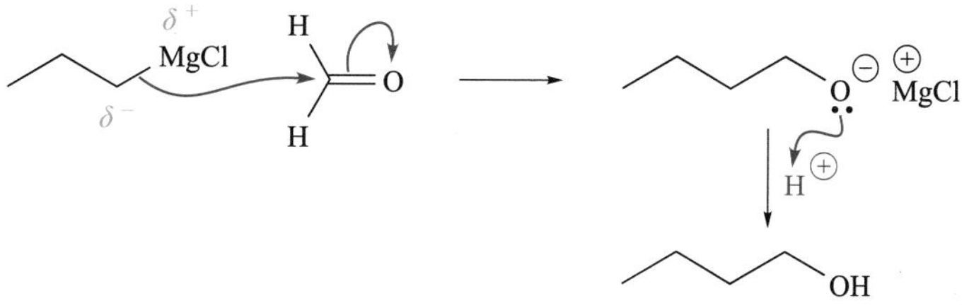

(33.44)

If the starting compound is an *aldehyde other than formaldehyde* (RCHO where R ≠ H), then reaction with a Grignard reagent gives a *secondary alcohol*. An example is the conversion of propanal to pentan-3-ol (equation 33.45).

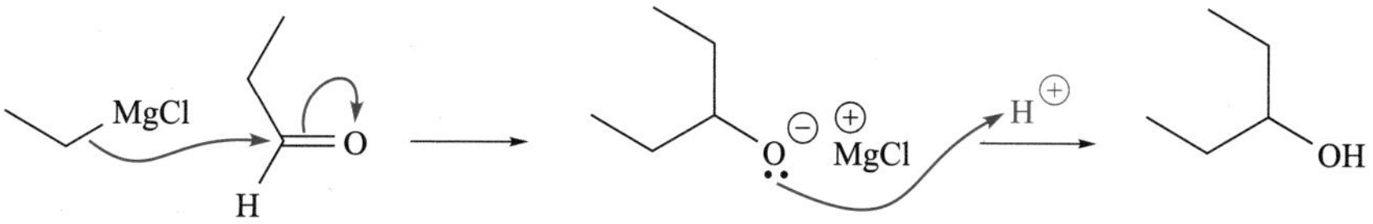

(33.45)

When a *ketone* reacts with a Grignard reagent, the product is a *tertiary alcohol*. The example in equation 33.46 shows the conversion of acetone to 2-methylpentan-2-ol.

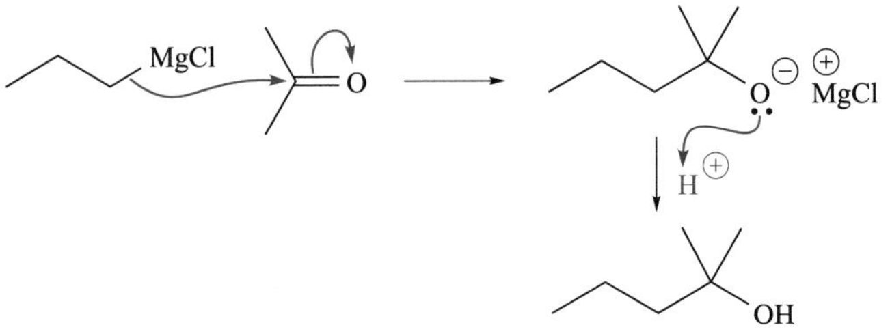


(33.46)

These reactions have widespread applications in alcohol synthesis and can be used with alkyl or aryl substituents. However, it is often found that the alcohols are dehydrated in the work-up (see Section 30.5).

## Reductive amination of aldehydes and ketones

In Section 31.3, we looked at the *reductive amination of aldehydes and ketones*. Reductive amination of an aldehyde using $NH_3$ gives a primary amine with the $NH_2$ group attached to a terminal C atom (i.e. to a primary alkyl group). Reductive amination of a ketone using $NH_3$ gives a primary amine with the $NH_2$ group attached to a secondary alkyl group. The aminating agents can also be primary or secondary amines (e.g. reactions 31.13 and 31.14). The mechanism of reductive amination involves nucleophilic attack by $NH_3$, $RNH_2$ or $R_2NH$ on the $C^{\delta+}$ atom of the C=O group of the aldehyde or ketone. Consider the reaction of acetaldehyde with $NH_3$. Attack by the nucleophile is followed by proton transfer:

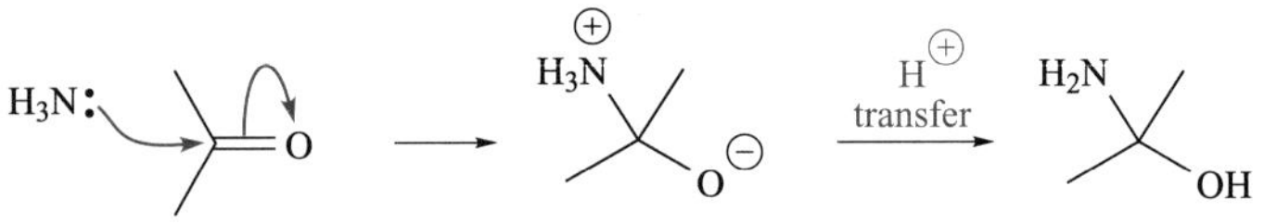


An ***imine*** has the general structure:

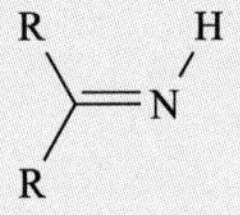


Elimination of $H_2O$ catalysed by acid then produces an *imine*:

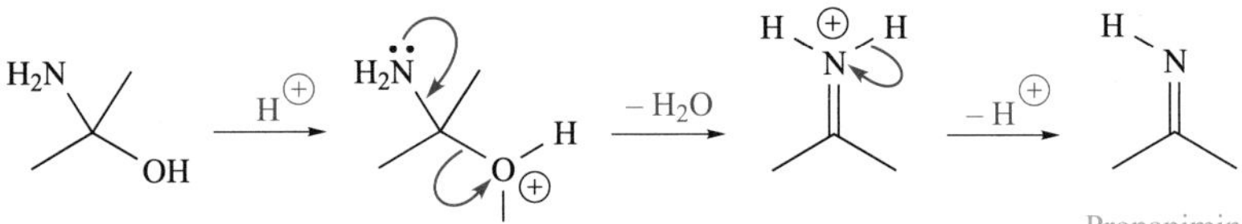


Propanimine

In the absence of a reducing agent, the reaction stops at this point. In the presence of a reducing agent (e.g. $H_2$ and a Ni catalyst on an industrial scale, or $NaBH_4$ or $LiAlH_4$ on a laboratory scale), the imine is reduced to an amine:

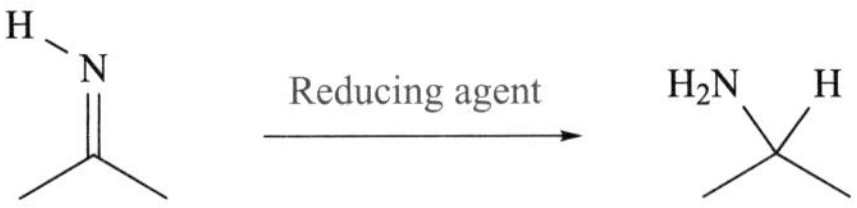


Propanimine 2-Propylamine

Analogous mechanisms can be written for the reactions of primary and secondary amines with aldehydes and ketones (see problem 33.19).

### Reactions of acyl chlorides and esters with NH₃

In Section 33.10, we described how amides can be prepared by reactions of $NH_3$ with acyl chlorides or esters (equations 33.22 and 33.23). The reaction between $NH_3$ and an acyl chloride gives $NH_4Cl$ as the second product. The mechanism (illustrated for the conversion of propanoyl chloride to propanamide) is as follows:

Notice that attack by $NH_3$ does not result in the direct loss of $Cl^-$ but rather causes the C=O bond to open up. This is a general pattern throughout examples of nucleophilic attack at the C=O carbon atom (see general reaction 33.41). In the reaction between $NH_3$ and an ester, the first step is the attack by $NH_3$ and opening of the C=O bond; this is followed by loss

---

**ENVIRONMENT AND BIOLOGY**

### Box 33.9 Feline slumbers

The authors' cats enjoy a well-earned rest.  *E. C. Constable.*

Research has shown that sleep-deprived cats accumulate (Z)-octadec-9-enamide in their spinal fluids. This unsaturated amide acts as a signalling molecule in the nervous system, inducing sleep in cats. The effect is sensitive to molecular structure. The (E)-isomer is less effective than the (Z)-isomer, and saturation of the C=C bond or lengthening of the carbon chain also alters the sleep-inducing properties of the compound. The studies suggest that amides in this family may find potential applications in drugs to treat insomnia.

(Z)-Octadec-9-enamide

of alkoxide or aryloxide depending on the ester being used in the reaction. The mechanism (illustrated with the reaction between ethyl butanoate and $NH_3$) is as follows:

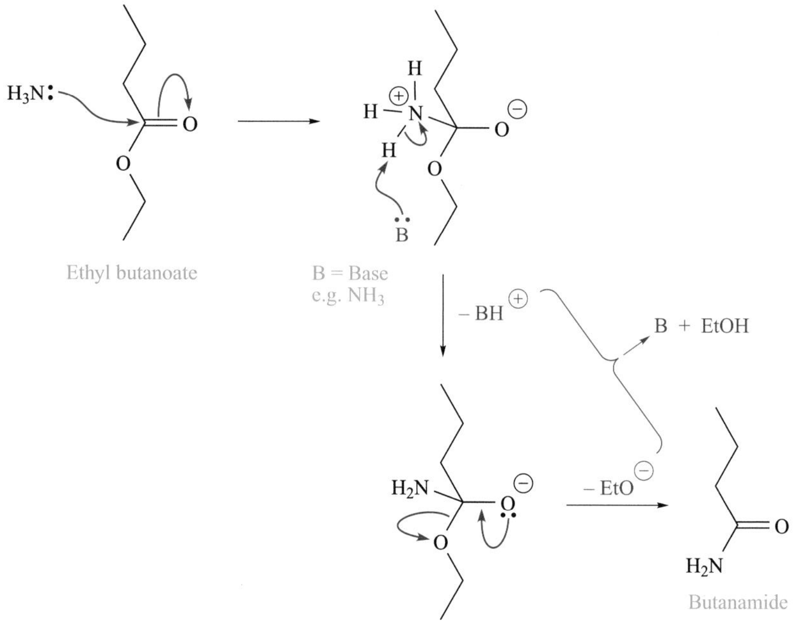

Ethyl butanoate

B = Base
e.g. $NH_3$

$- BH^{\oplus}$

$B + EtOH$

$- EtO^{\ominus}$

Butanamide

## Reactions of aldehydes with alcohols: hemiacetal and acetal formation

An *acetal* has the general structure:

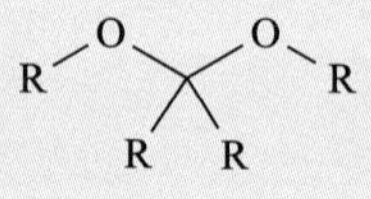

A *hemiacetal* has the general structure:

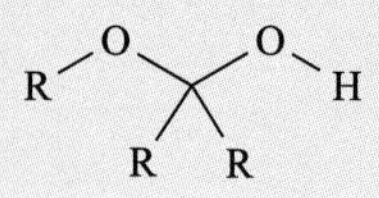

An aldehyde reacts with an alcohol in the presence of an acid catalyst to give a *hemiacetal* and then an *acetal*. The first step in the reaction is the protonation of the C=O group in the aldehyde:

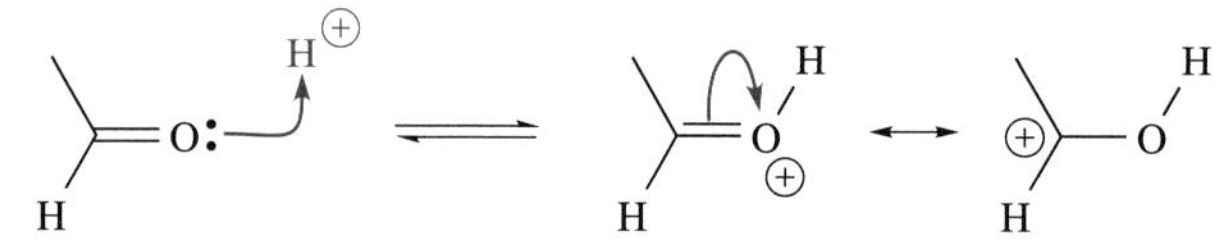

Resonance forms of protonated aldehyde

The resonance structures for the protonated aldehyde indicate that protonation makes the C atom of the carbonyl group more susceptible to attack by nucleophiles and in the next step, the alcohol (exemplified here by EtOH) acts as a nucleophile. Notice that all steps in the reaction pathway are *reversible*:

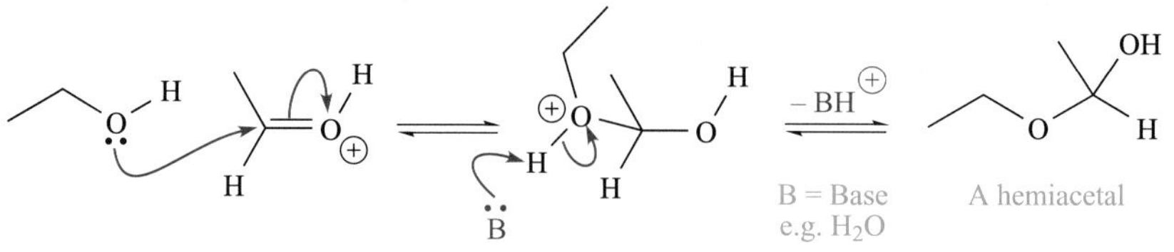

B = Base
e.g. $H_2O$

A hemiacetal

The reaction does not stop here. Protonation (remember that the acid catalyst is still in solution) of the OH group of the hemiacetal creates a good leaving group, $H_2O$, which is eliminated. This leaves a species that is susceptible to attack by the nucleophilic alcohol and the final product of the reaction is an *acetal*. The mechanism is as follows:

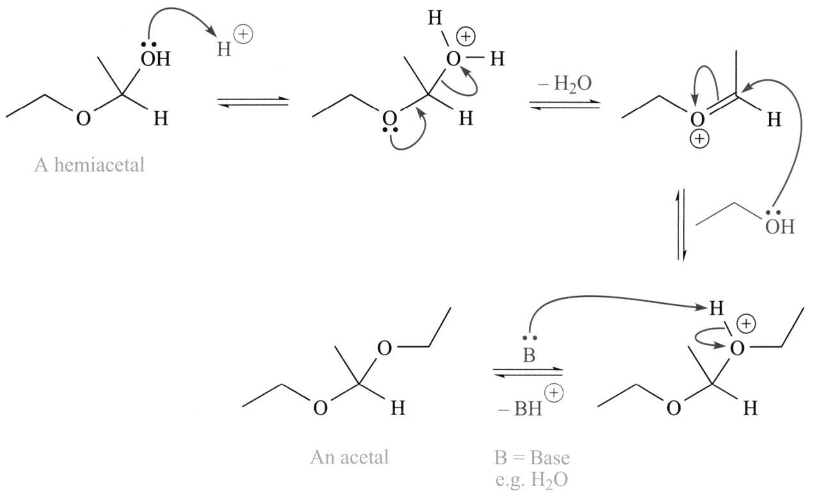

A hemiacetal

An acetal

B = Base
e.g. $H_2O$

As with the formation of the hemiacetal, notice that each step in the formation of the acetal is *reversible*. Similar reactions do occur between some ketones and alcohols but are not generally as successful as those involving aldehydes.

Acetals as protecting groups for alcohols: see Section 30.6

Acetals are very often used to *protect carbonyl compounds or alcohols* during reactions in which the C=O or OH group may undergo an unwanted side-reaction. We have already described use of the tetrahydropyranyl, THP, group to protect alcohols (equations 30.42 and 30.43). An example of the need for C=O protection is during the reduction of an ester group in a molecule that has another C=O group. Both carbonyl groups would be reduced by $LiAlH_4$ and so if, for example, *selective reduction* of the ester group is the target, the second C=O group can be temporarily converted into an acetal by reaction with a diol:

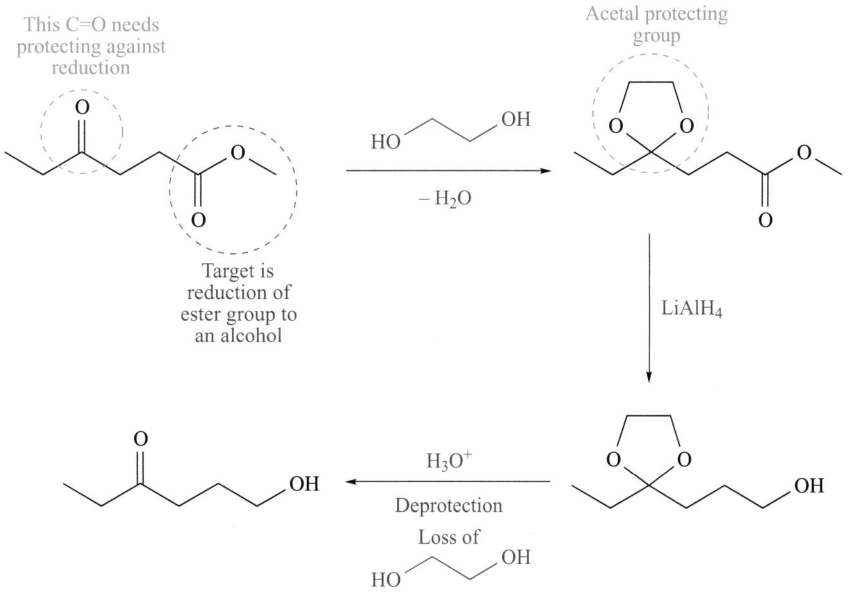

This C=O needs protecting against reduction

Target is reduction of ester group to an alcohol

Acetal protecting group

$LiAlH_4$

$H_3O^+$
Deprotection
Loss of

### Aldol reactions

We met the aldol reaction at the end of Section 33.13. Now we look in detail at the mechanism of this important C–C bond-forming reaction. The example shown is the reaction between two aldehydes in the presence of base, but similar reactions occur between two ketones, an aldehyde and a ketone, an ester and a ketone, or an ester and an aldehyde. *One member of each pair must possess an α-H atom.*

The first step in the reaction is deprotonation of the aldehyde at the α-position. This generates an enolate ion which acts as a nucleophile and attacks the second aldehyde molecule to give a β-hydroxy aldehyde:

A *β-hydroxy aldehyde* has the general structure:

R groups may be attached to the α- and β-C atoms.

A problem with such reactions is the possibility that the enolate formed in the first step of the reaction attacks a molecule of its parent aldehyde (its conjugate acid) rather than a molecule of the second aldehyde. If the reaction is a *self-condensation* (i.e. only one aldehyde is involved), the competition presents no problem since the products are the same whatever happens. However, if the aldol reaction is between *two different aldehydes*, a mixture of products results. We return to these complications later.

Now consider the reaction between two ketones, at least one of which contains an α-H atom. Enolization of the ketone is followed by nucleophilic attack on the second ketone. The product is a *β-hydroxy ketone*. In the example below, the two ketones are the same (both are acetone):

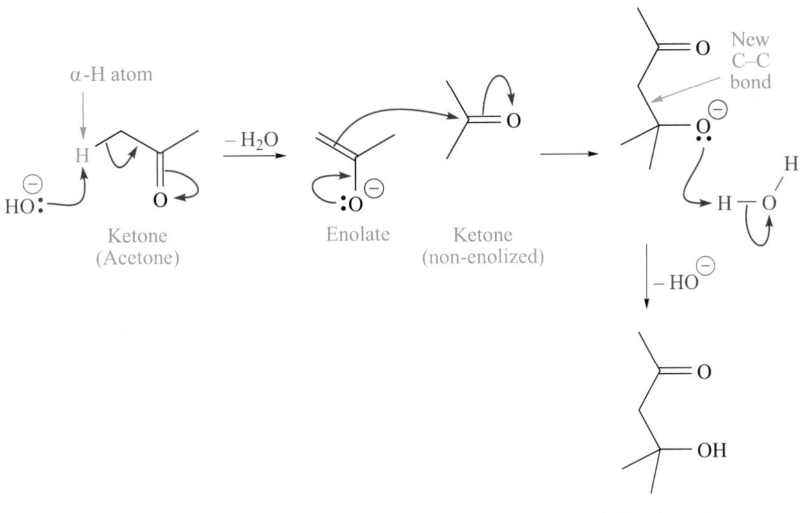


A β-hydroxy ketone

A *β-hydroxy ketone* has the general structure:

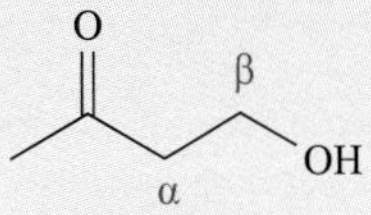


R groups may be attached to the non-carbonyl C atoms.

Although we have looked at aldol reactions between aldehydes and between ketones, we must note that aldehydes are more reactive than ketones. This affects the outcome of aldol reactions between aldehydes and ketones, because an enolate formed from an aldehyde will tend to react with a non-enolized aldehyde rather than a ketone.

## β-Eliminations accompanying aldol reactions

β-elimination: see Section 28.7

So far, in considering the aldol reaction, we have ignored the possibility that *more than one α-H atom is involved.* In reality, reactions are complicated by *β-eliminations.* A β-hydroxy aldehyde can undergo a base-promoted β-elimination, losing $H_2O$ and forming an α,β-unsaturated aldehyde. This is shown below, starting from the β-hydroxy aldehyde formed in the scheme above:

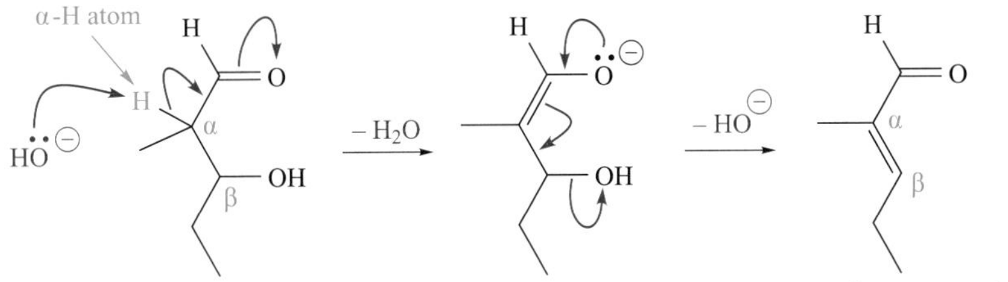


A β-hydroxy aldehyde

An α,β-unsaturated aldehyde

An *α,β-unsaturated aldehyde* has the general structure:

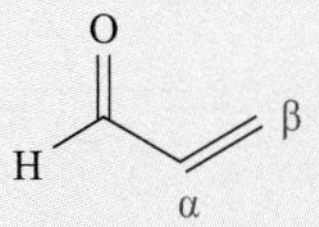


R groups may be attached to the non-carbonyl C atoms.

An *α,β-unsaturated ketone* has the general structure:

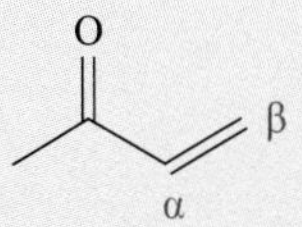


R groups may be attached to the non-carbonyl C atoms.

Similarly, *α,β-unsaturated ketones* are formed from a β-hydroxy ketone that possesses an α-H atom. In Section 13.5, we discussed the π-conjugation in α,β-unsaturated ketones and aldehydes and the effects this has on their electronic spectra.

## An overview of complications arising in aldol reactions

From the discussion above, it is clear that significant complications arise when, for example, two *different* aldehydes undergo an aldol reaction. Let us take the case of the reaction between two aldehydes, one of which has two α-H atoms. Equation 33.47 shows the formation of two possible α,β-hydroxy aldehydes.

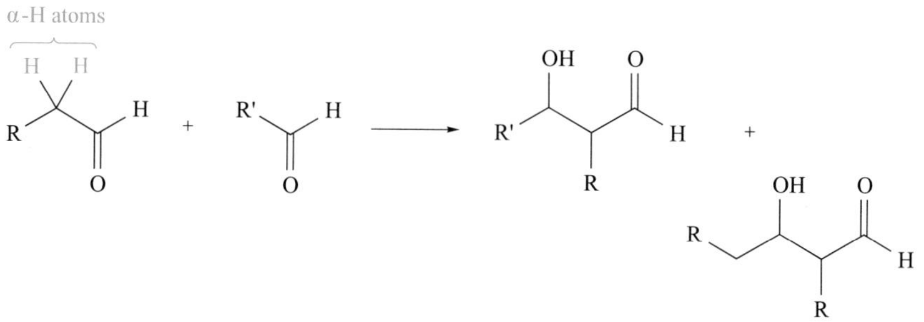

(33.47)

Both products possess α-H atoms and so can undergo β-elimination reactions. Thus the reaction is complicated by the formation of α,β-unsaturated aldehydes:

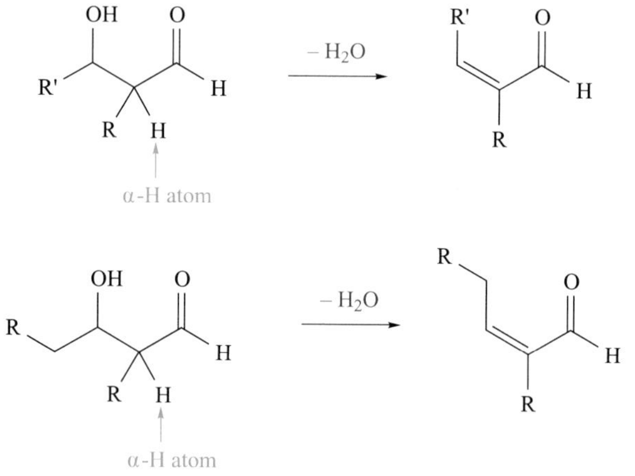

## 33.16    Nucleophilic attack at C=O: Claisen condensation

The condensation of two esters in the presence of strong base (often EtO⁻) gives a *β-keto ester* and is called a Claisen condensation. We met this reaction

---

**COMMERCIAL AND LABORATORY APPLICATIONS**

### Box 33.10 Pharmaceutical drugs: some carbonyl compounds in action

In this box, we highlight some pharmaceutical drugs. All the examples chosen feature carbonyl groups, but not always in the simple environments that we have described in this chapter.

*Aspirin* is an example of a non-steroidal anti-inflammatory drug (NSAID) and is a household name among painkillers. Taken in excess of the prescribed doses, it can prove fatal to children, and may cause stomach ulcers in adults. *Naproxen* (sold by Syntex under the name of *Naprosyn*) is a more recently developed, and now world-leading, NSAID.

*Tetracycline* is the parent compound in a major group of antibiotics and is used in the treatment of infections

that include *Streptococcus* pneumonia, cholera and typhus. Although increased resistance by organisms to tetracycline is now being encountered, the drugs remain effective against a range of diseases.

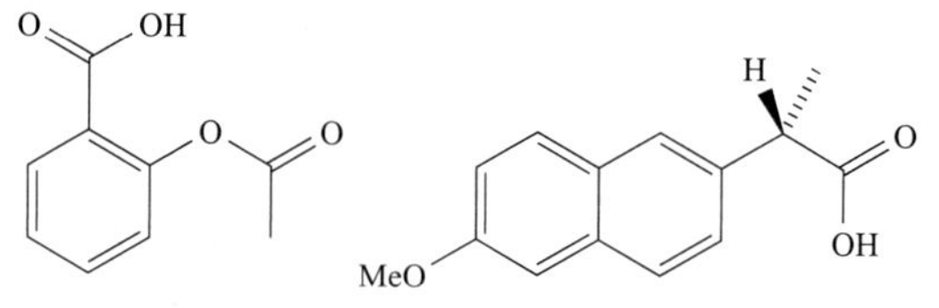

Aspirin                    Naproxen

Tetracycline

*Lovastatin* (manufactured by Merck & Co. as *Mevacor*) is among the world's best-selling drugs. It is a hypolipaemic, acting to reduce cholesterol levels and so working against arteriosclerosis (see Box 30.1).

Lovastatin (sold under the tradename Mevacor)

Taxol

Captopril (Capoten)

*Taxol* is an anti-cancer drug, first discovered as a chemical present in the bark of the Pacific yew tree and in fungus that grows on these trees. In 1994, the drug was synthesized for the first time and is now being used to treat ovarian and breast cancers.

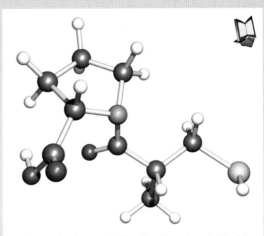

Colour code: C, grey; N, blue; S, yellow; O, red; H, white.

Drugs that are used to relieve hypertension include *captopril* (sold under the name of *Capoten*) and *enalapril* (manufactured by Merck & Co. as *Vasotec*).

The sulfur atom in captopril has a crucial role to play. It coordinates to the $Zn^{2+}$ ion present in the active site of the enzyme that the drug is designed to inhibit. The figure below shows how the anti-hypertensive drug captopril binds through the S atom to the $Zn^{2+}$ centre in human testicular angiotensin I-converting enzyme. The structure has been determined by single crystal X-ray diffraction [R. Natesh *et al.* (2004) *Biochemistry*, vol. 43, p. 8718]. The metalloenzyme is drawn in a ribbon representation with the zinc centre shown as a silver-coloured sphere. The captopril molecule is shown in a 'stick' representation, with the sulfur atom (in yellow) coordinated to the $Zn^{2+}$ centre.

at the end of Section 33.13. At least one of the esters must possess an α-H atom, and the first step in the reaction is deprotonation by ethoxide ion to give an enolate ion:

A *β-diketo ester* has the general structure:

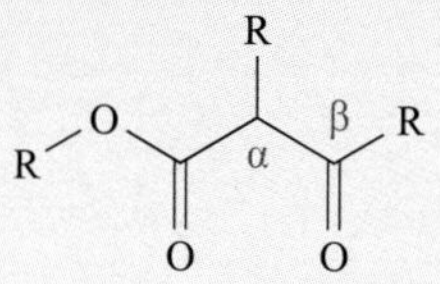

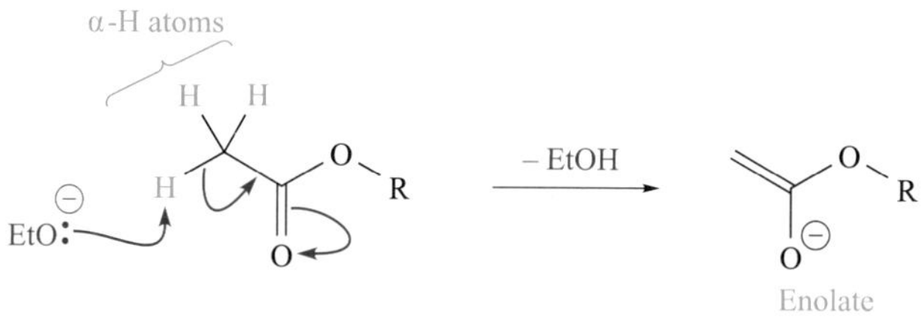

Let us take the case of *self-condensation*, i.e. reaction between the enolate and the parent ester. The reaction proceeds as follows:

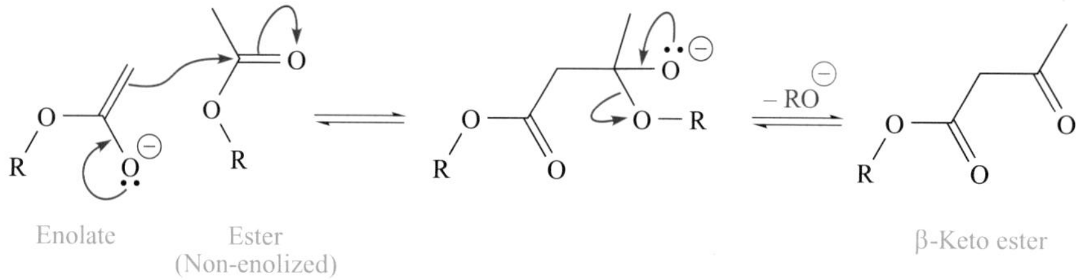

Enolate    Ester
(Non-enolized)    β-Keto ester

These equilibria lie towards the *left-hand side*, i.e. product formation is *not* favourable. However, the β-keto ester possesses a weakly acidic α-H atom and in the presence of alkoxide ion, RO⁻, it is deprotonated. Equilibrium 33.48 lies to the *right-hand side*, and it is this final step that drives the reaction to a successful end.

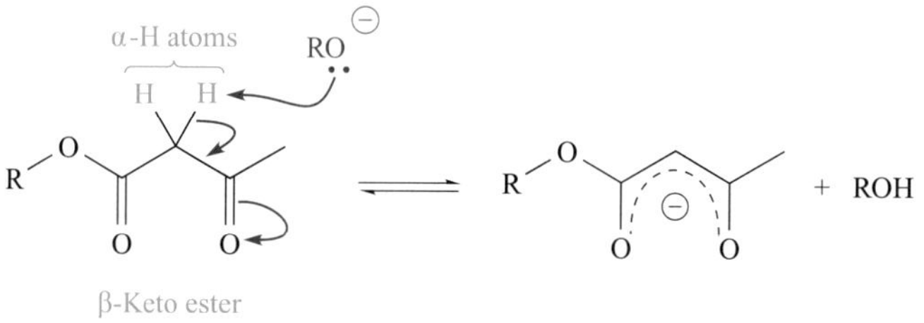

β-Keto ester

(33.48)

Why should the products in equilibrium 33.48 be favoured? Although the β-keto ester and alcohol are both weak acids, the presence of the two carbonyl groups stabilizes the conjugate base of the β-keto ester. The β-ketonate on the right-hand side of the equilibrium is shown in the form that emphasizes the delocalized bonding (see structures **33.31** and **33.32**). Consequently, the β-keto ester ($pK_a \approx 11$) on the left-hand side of equilibrium 33.48 is a stronger acid than the alcohol ($pK_a = 16$ for EtOH) on the right-hand side. As a result, the equilibrium lies towards the right. The presence of the two α-H atoms is crucial to the success of the Claisen condensation. The neutral β-keto ester can be obtained by isolating a salt of the conjugate base, and adding acid to the salt.

## SUMMARY

This chapter has been concerned with the chemistry of compounds containing a C=O bond: aldehydes, ketones, carboxylic acids, esters, acyl chlorides and amides. The polarity of the C=O bond is $C^{\delta+}-O^{\delta-}$. The O atom is susceptible to attack by electrophiles such as $H^+$, and the C atom is susceptible to attack by nucleophiles. Carbonyl compounds containing α-H atoms can be deprotonated to give enolate ions that are nucleophilic. Their reactions with other carbonyl compounds are the basis of widespread applications in C—C bond-forming reactions.

### Do you know what the following terms mean?

- aldehyde
- ketone
- carboxylic acid
- ester
- acyl chloride
- amide
- keto–enol tautomerism
- enolate ion
- β-diketonate

- condensation reaction
- fatty acid
- α-hydrogen atom (in a carbonyl compound)
- haloform reaction
- reductive amination
- imine
- cyanohydrin
- acetal

- hemiacetal
- acetal protecting group
- aldol reaction
- Claisen condensation
- β-hydroxy aldehyde
- β-hydroxy ketone
- α,β-unsaturated aldehyde
- α,β-unsaturated ketone
- β-diketo ester

### The following abbreviations have been used. Do you know what they mean?

- DIBAL
- DMF

- LDA
- PCC

- THF
- THP

### You should be able:

- to draw the functional groups of an aldehyde, ketone, carboxylic acid, ester, acyl chloride and amide
- to name simple carbonyl compounds
- to recognize trivial names such as formic acid, formate, formaldehyde, acetic acid, acetyl chloride, acetylacetone, acetylacetonate, acetamide, oxalic acid, phthalic acid
- to describe the bonding in the C=O group in terms of VB and MO theories, and to comment on the polarity of the bond
- to discuss the role of hydrogen bonding in the intermolecular association of carboxylic acids and of amides
- to comment on the effects of hydrogen bonding on some physical properties of carbonyl compounds
- to recognize IR spectroscopic absorptions characteristic of carbonyl compounds
- in $^{13}$C NMR spectra, to know typical $\delta$ ranges associated with C=O groups
- in $^1$H NMR spectra, to know typical $\delta$ ranges associated with CHO, $CO_2H$ or $CONH_2$ protons

- to understand what effect the polarity of the C=O bond has on the reactivity of carbonyl compounds
- to give methods of preparing different types of carbonyl compounds
- to describe saponification reactions and their industrial significance
- to discuss the relative acidities of different types of carbonyl compounds, and the resonance stabilization of particular conjugate bases
- to show how enolate ions are formed
- to discuss reactions that rely on the presence of α-H atoms in carbonyl compounds
- to give examples, and describe the mechanisms, of acid- and base-catalysed halogenations of aldehydes and ketones
- to give examples, and describe the mechanisms, of C—C bond forming reactions involving carbonyl compounds
- to give examples, and describe the mechanisms, of reactions involving attack by nucleophiles at the carbonyl C atom

## PROBLEMS

**33.1** Draw structures for (a) pentanal, (b) benzaldehyde, (c) butanone, (d) heptan-3-one, (e) pentanoic acid, (f) pentanoyl chloride; (g) ethyl propanoate, (h) phenyl acetate, (i) ethyl benzoate, (j) acetamide, (k) butanamide, (l) potassium propanoate.

**33.2** Name the following compounds:

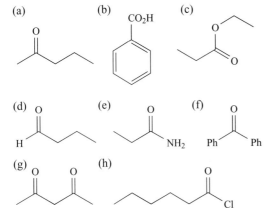

**33.3** Explain the following observations.
   (a) The $^1$H NMR spectrum of a solution of hexanoic acid contains a signal at $\delta$ 11.2 ppm, but the signal disappears when $D_2O$ is added to the solution.
   (b) The $^{13}$C NMR spectrum of octanoic acid contains seven signals in the region $\delta$ 14.1 to 34.3 ppm, and one signal at $\delta$ 181.1 ppm.
   (c) The $^1$H NMR spectrum of butanone exhibits a triplet at $\delta$ 1.06 ppm, a singlet at $\delta$ 2.14 ppm and a quartet at $\delta$ 2.45 ppm.
   (d) The $^1$H NMR spectrum of methyl formate shows signals at $\delta$ 8.08 and 3.77 ppm.

**33.4** Comment on the following statements.
   (a) Keto- and enol-forms of a compound are isomers.
   (b) The equilibrium that describes the keto–enol tautomerism for butanone lies well over to the side of the keto-form.
   (c) Treatment of pentane-2,4-dione with base results in a monoanion.

**33.5** Match the following spectroscopic data to the compounds whose structures are shown below.

(a)            (b)            (c)

(d)            (e)

| Compound | $^{13}$C NMR $\delta$/ppm |
|----------|---------------------------|
| A | 25.1, 27.1, 41.9, 211.3 |
| B | 136.4, 187.1 |
| C | 20.6, 178.1 |
| D | 13.8, 22.3, 24.9, 30.7, 47.2, 173.8 |
| E | 162.3 |

**33.6** Structure **33.21** shows DIBAL as a monomeric compound containing a 3-coordinate Al atom. DIBAL actually exists as a dimer and the bonding can be described as being similar to that in $B_2H_6$. Draw the dimeric structure of DIBAL and present a bonding scheme for the molecule.

**33.7** Suggest ways of carrying out the following transformations:

(a)

(b)

(c)

(d)

(e)

(f)

**33.8** Give methods of converting (a) *o*-bromotoluene into *o*-methylbenzoic acid, (b) toluene into benzoic acid and (c) 1-bromobutane into pentanoic acid.

**33.9** Suggest products in the following reactions:

(a)

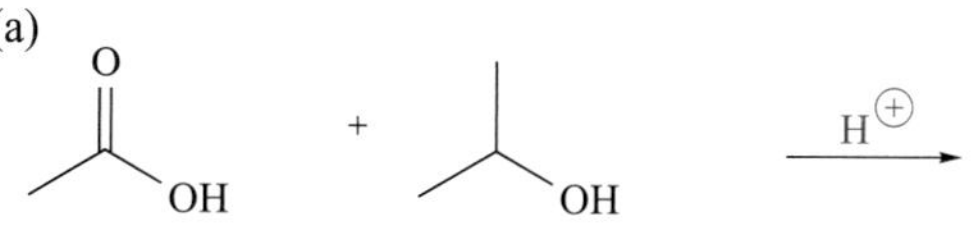

(b)

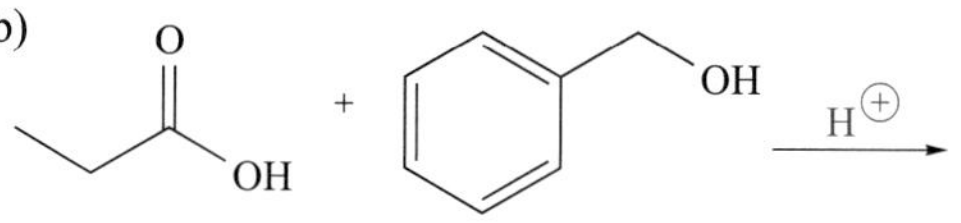

(c)

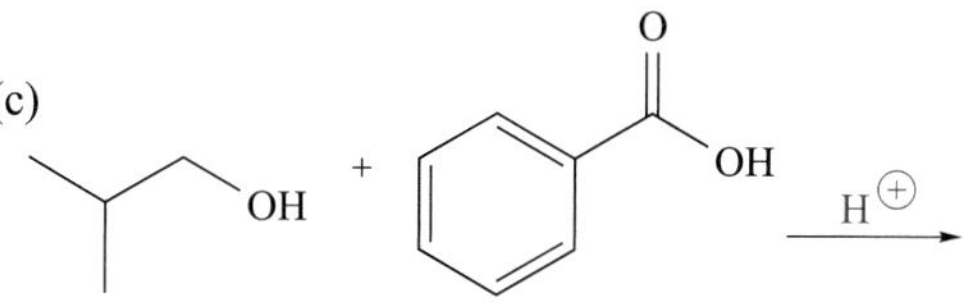

**33.10** (a) Suggest how each of the following Grignard reagents in $Et_2O$ would react with solid $CO_2$ followed by treatment with $H^+$. (b) Give a mechanism for one of the reactions.

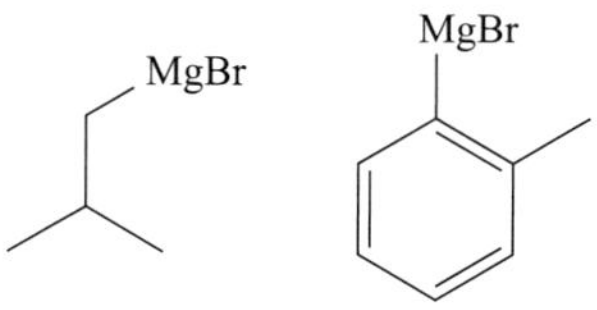

**33.11** Suggest a method of preparing each of the following compounds, starting from butanoyl chloride in each case.

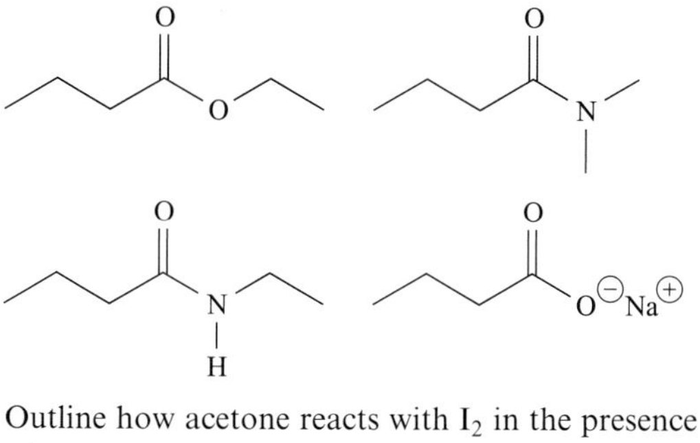

**33.12** Outline how acetone reacts with $I_2$ in the presence of (a) an acid catalyst and (b) a base catalyst. Give appropriate mechanisms for the reactions you describe.

**33.13** Which of the following would undergo a haloform reaction when treated with $Br_2$ in the presence of NaOH: (a) pentan-3-one; (b) butanone; (c) propanone; (d) hexan-2-one; (e) hexan-3-one? Give the products in each case where you state the reaction will occur.

**33.14** Arrange the following compounds in order of decreasing $pK_a$ values. In each case, give the equilibrium to which the $pK_a$ value refers.

(a)

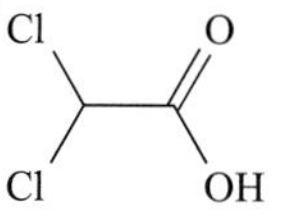

(b)

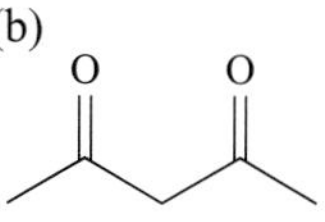

(c)

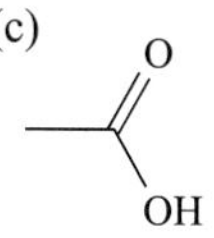

(d)

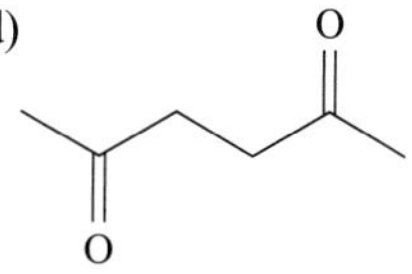

(e)

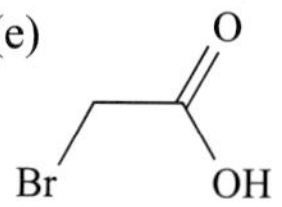

**33.15** Elemental analysis for compound **A** gives 71.9% C and 12.1% H. In the mass spectrum of **A**, the parent ion comes at $m/z = 100$, and there is an intense fragmentation peak at $m/z = 43$. The IR spectrum of **A** has strong absorptions at $2960–3000\ cm^{-1}$ and $1718\ cm^{-1}$ as well as absorptions in the fingerprint region. In the $^{13}C$ NMR spectrum, there are signals at $\delta$ 209.0, 43.5, 29.8, 26.1, 22.4 and 13.9 ppm. Compound **A** can be reduced by $NaBH_4$, but is not oxidized by $K_2Cr_2O_7$ in acidic solution. Suggest (with explanation) the identity of **A**, and give the product of its reduction.

**33.16** Give two synthetic methods by which carbon chains can be lengthened, illustrating your answer with reference to the conversion of 1-chloropentane to hexanoic acid.

**33.17** (a) Sketch a representation of the HOMO of the cyanide ion, and explain its role in the ability of $NC^-$ to act as a nucleophile.
(b) Rewrite equation 33.43 using appropriate curly arrows.

**33.18** Suggest products in the following reactions. Give appropriate mechanisms for the reactions.

(a)

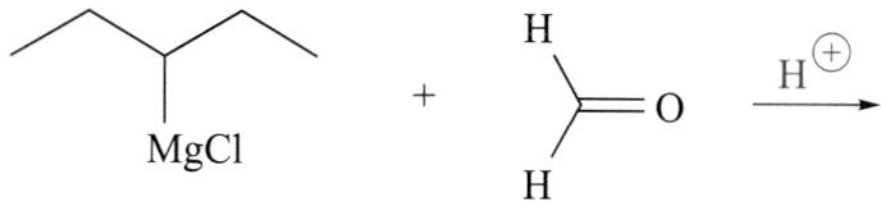

(b)

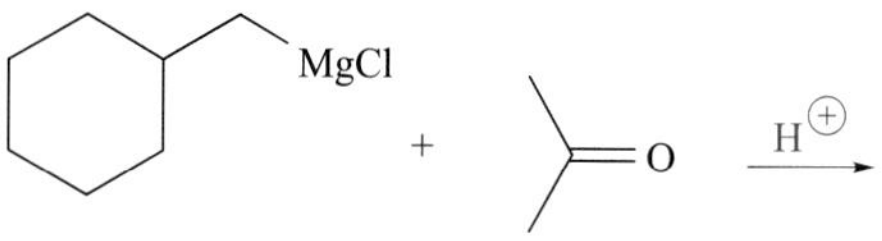

(c)

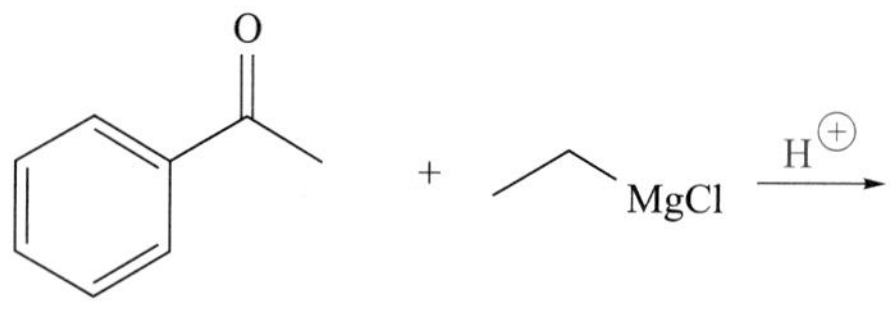

**33.19** Suggest products for the following reactions and give appropriate mechanisms; [H] represents a reducing agent.

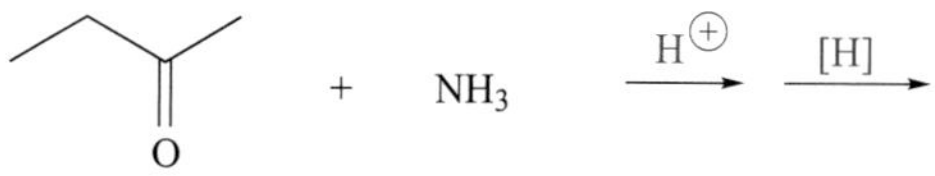

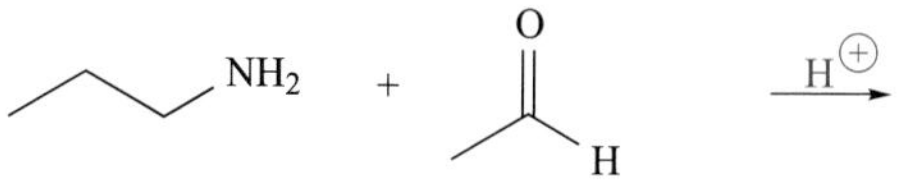

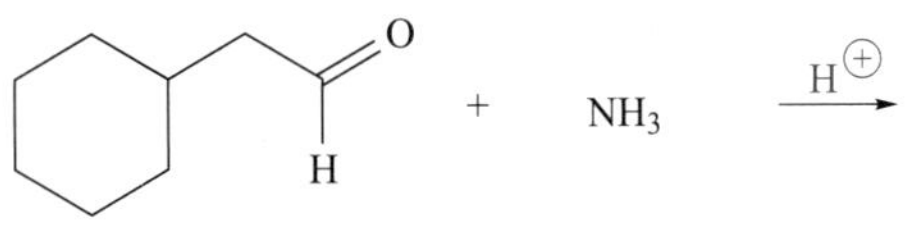

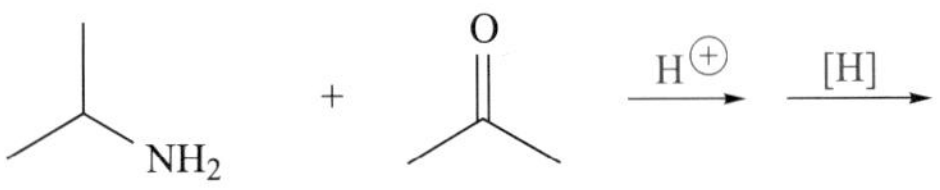

**33.20** (a) Write a mechanism for the acid-catalysed decomposition of the hemiacetal:

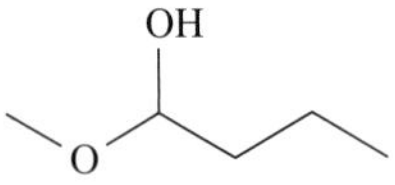

to the appropriate aldehyde and alcohol.
(b) Write a mechanism for the base-catalysed reaction between methanol and butanal to give a hemiacetal.
(c) Classify the following compounds as simple ethers, esters, hemiacetals, acetals or alcohols:

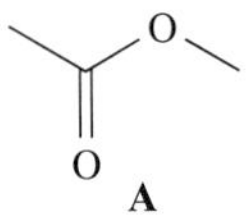

**33.21** Suggest ways of carrying out the following transformations:

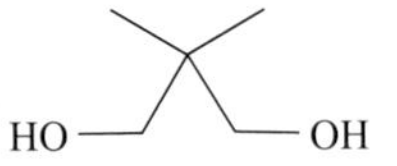

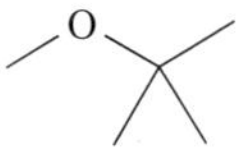

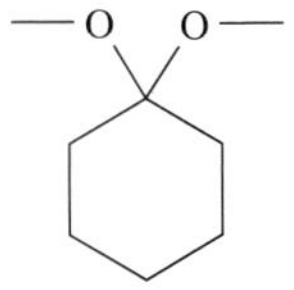

(a)

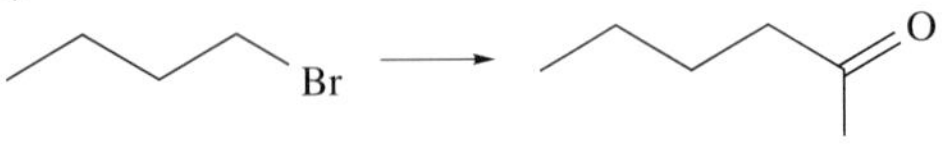

(b)

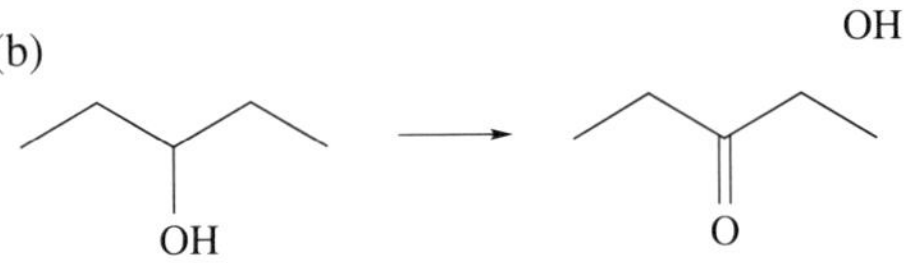

(c)

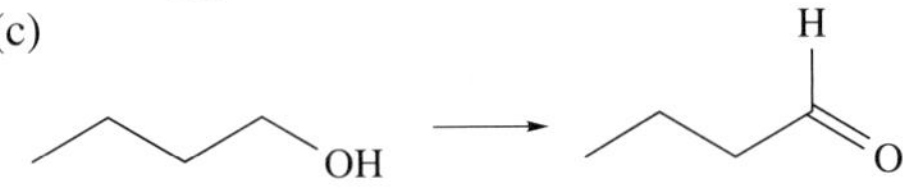

(d)

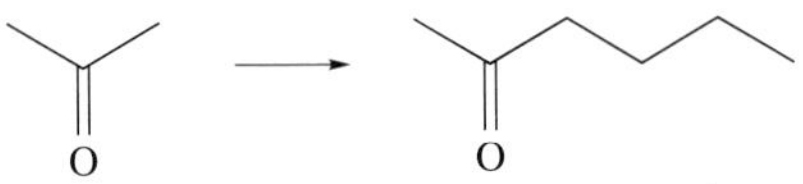

(e)

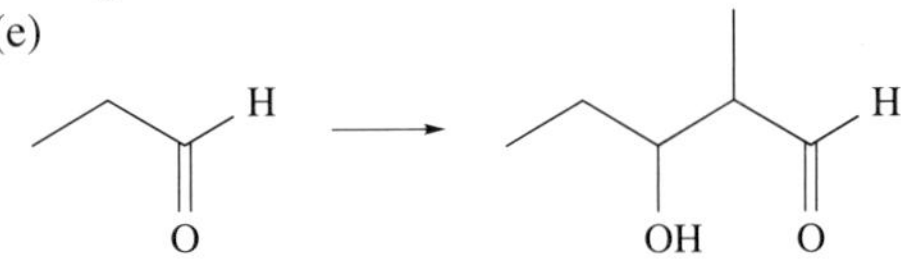

**33.22** Explain what is meant by the *aldol reaction* with reference to what happens when a small amount of NaOH is added to (a) acetaldehyde and (b) acetone.

## Additional problems

**33.23** Comment on the following observations.
(a) The $^{13}C\{^1H\}$ NMR spectrum of $CF_3CO_2H$ shows two quartets with $J$ 284 and 44 Hz. *Hint*: look at Table 14.1.
(b) $NH_2^-$ is a poor leaving group, and yet, in the presence of $H^+$, amides are hydrolysed to carboxylic acids.

**33.24** How might you carry out the following transformations?

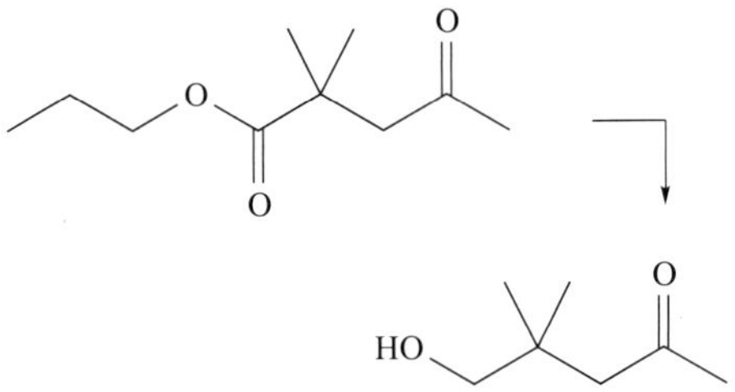

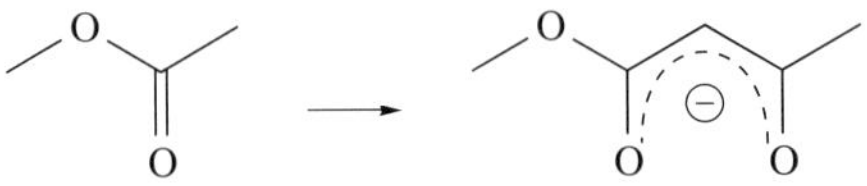

**33.25** (a) Suggest likely products from the reaction of acetaldehyde and butanal in the presence of NaOH.
(b) Explain why benzaldehyde does not form an enolate.
(c) Suggest a method for converting methyl propanoate to methyl 2-methylbutanoate.

**33.26** Dimethylformamide (DMF) is a common solvent. In its $^{13}C$ NMR spectrum, signals are observed at $\delta$ 162.4, 36.2 and 31.1 ppm. Explain why there are three, and not two, signals.

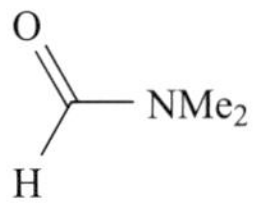

DMF

# 34 Aromatic heterocyclic compounds

## Topics

- The importance of heterocyclic compounds
- Isoelectronic relationships
- Structure and bonding in pyridine, pyrrole, furan and thiophene
- Syntheses of pyrrole, furan and thiophene
- Reactivity of pyrrole, furan and thiophene
- Syntheses of pyridine and the pyrylium cation
- Reactivity of pyridine and the pyrylium cation
- *N*-Heterocycles with more than N atom

## Why study heterocyclic compounds?

*Hetero* is a prefix meaning 'different'. In a *heterocycle*, the *heteroatom* is a different atom from the other atoms in the ring.

A heterocyclic molecule is a cyclic compound in which one or more of the atoms in the ring is a heteroatom. Within organic chemistry, the heteroatom is an atom other than carbon or hydrogen, and common heteroatoms are N, O and S. A heterocycle can be saturated (e.g. tetrahydrofuran, THF), unsaturated (e.g. pyran) or aromatic (e.g. pyridine). A large proportion of naturally occurring organic compounds are heterocyclic, and many are of biological importance, for example sugars (see Chapter 35), nucleic acids (see Chapter 35), alkaloids (see Box 34.1) and some pheromones (Box 34.2).

Figure 34.1 shows the structures of some vitamins containing heterocycles. Thiamine (vitamin $B_1$) is involved in the metabolism of carbohydrates and is converted to its active form, thiamine pyrophosphate, in the brain and liver by the enzyme thiamine diphosphotransferase. In severe cases, a carbohydrate-rich, but thiamine-deficient, diet leads to the disease beriberi. Riboflavin (vitamin $B_2$) is the precursor to flavoproteins, which are involved in the mitochondrial electron transport chain in the body. Nicotinic acid (niacin, vitamin $B_3$) is the precursor to the coenzymes nicotinamide adenine dinucleotide (NAD) and nicotinamide adenine dinucleotide phosphate (NADP), both of which are involved in hydrogen transfer and redox processes in the body. Vitamin $B_6$ consists of pyridoxal, pyridoxamine and pyridoxine. The active form of vitamin $B_6$ is pyridoxal phosphate and this

Tetrahydrofuran
(THF)

4*H*-Pyran

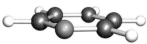

Pyridine

## ENVIRONMENT AND BIOLOGY

### Box 34.1 Alkaloids

*Alkaloid* is a term used to cover a wide range of naturally occurring organic, nitrogen-containing bases. Most alkaloids are heterocyclic compounds, and most are optically active, but are produced naturally as one enantiomer. Many alkaloids are toxic, but at the same time are active ingredients in a large number of drugs and medications. The pyridine ring is a building block in many alkaloids. Other heterocyclic groups are quinoline, isoquinoline and pyrimidine, all of which are related to pyridine and are planar aromatic systems:

Quinoline

Isoquinoline

Pyrimidine

Alkaloid drugs, many of which are addictive, include nicotine (present in tobacco), morphine (from opium poppies and highly addictive), caffeine (a stimulant in tea and coffee, and an additive to cola drinks), quinine (from cinchona bark, and used in the treatment of malaria), cocaine (found in coca, and stimulates the nervous system) and strychnine (stimulates the nervous system, causing convulsions and ultimately death even when taken in relatively small doses).

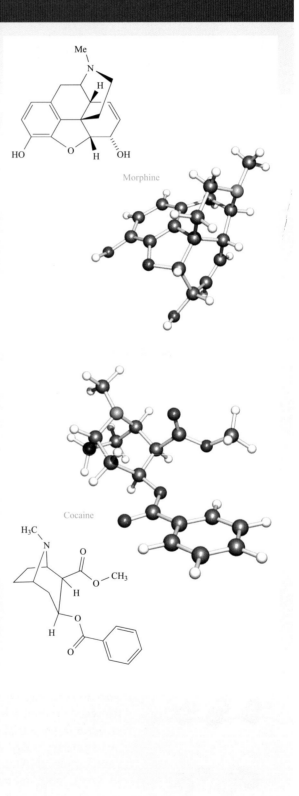

Morphine

Cocaine

Caffeine

Quinine

Nicotine

Strychnine

An especially potent natural toxin is the alkaloid batrachotoxin which is secreted by the poison dart frog *Phyllobates terribilis*. This toxin is used by the Choco Indians in Colombia as a poison for blow-pipe darts, and is far more toxic than strychnine. Batrachotoxin causes irreversible nerve and muscle damage by binding to nerve channels and, in particular, affecting sodium channels. Ultimately, the alkaloid causes death.

Batrachotoxin

Analgesic drugs such as morphine are prescribed to reduce the feeling of pain, but the addictive side effects are undesirable. The search for non-addictive analgesics is therefore an area of active interest. Epibatidine is an alkaloid that can be isolated from the skin of the Ecuadorian tree frog (*Epipedobates tricolor*). It is a powerful analgesic but has the side effect of causing paralysis. A compound that is structurally related to epibatidine, and shows analgesic effects but no paralysis, has been synthetically produced; it functions in a similar manner to epibatidine (but in a different manner from morphine and related addictive analgesics) and is a candidate to enter the pharmaceutical industry.

Epibatidine

Synthetic analogue of epibatidine

The poison dart frog *Phyllobates terribilis*. *E. C. Constable.*

is a cofactor in enzymes that are involved in, for example, decarboxylation and transamination of amino acids. Biotin acts as a coenzyme in reactions involving $CO_2$ transfer. It is required by all organisms but is produced only by bacteria, yeast and other fungi, algae and some plants. Vitamin $B_{12}$ contains an octahedral cobalt(III) centre which is coordinated in the four equatorial sites by a corrin ligand (related to a porphyrin, see Box 23.3) and in the axial positions by a ligand $X^-$ and a 5,6-dimethylbenzimidazole ribonucleotide unit. Structures **34.1** and **34.2** show benzimidazole and 5,6-dimethylbenzimidazole ribonucleotide; ribonucleotides are discussed in

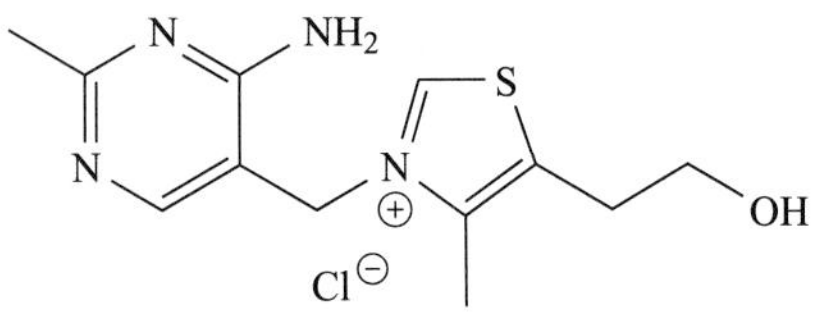

Thiamine (vitamin B$_1$)

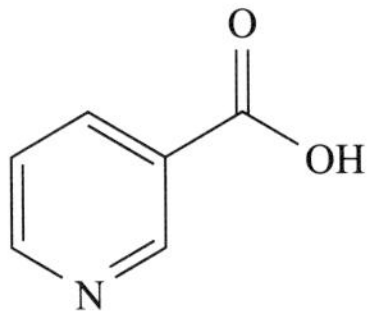

Nicotinic acid (vitamin B$_3$)

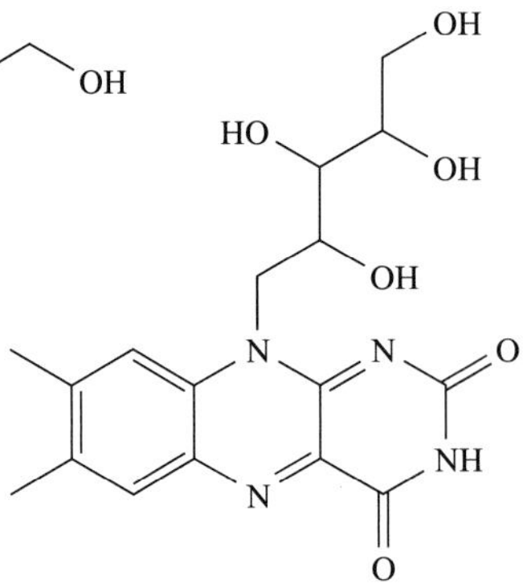

Riboflavin (vitamin B$_2$)

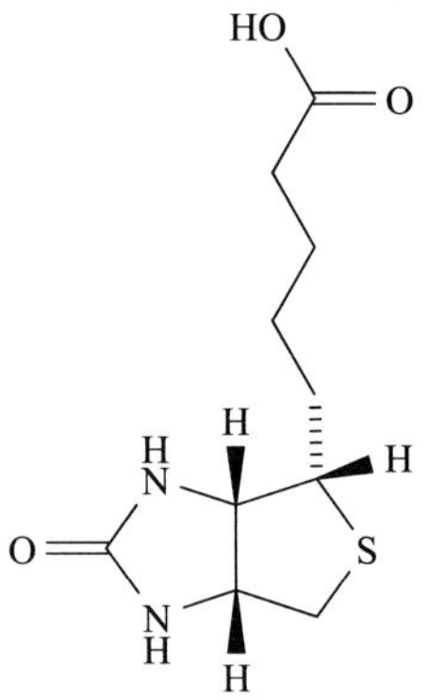

Biotin

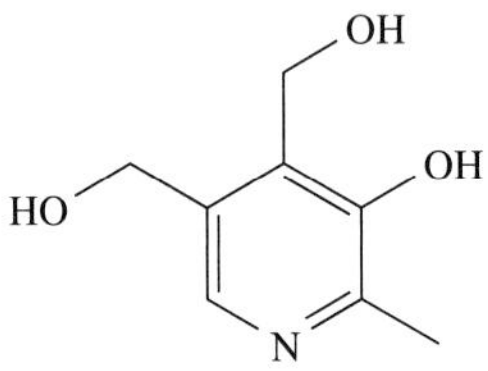

Pyridoxine

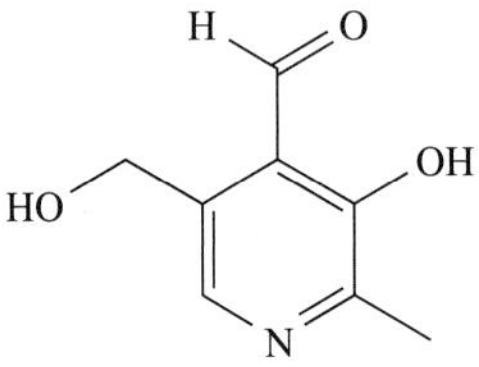

Pyridoxal

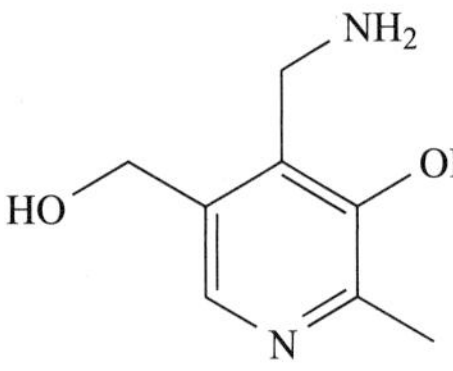

Pyridoxamine

Pyridoxine, pyridoxal and pyridoxamine (vitamin B$_6$)

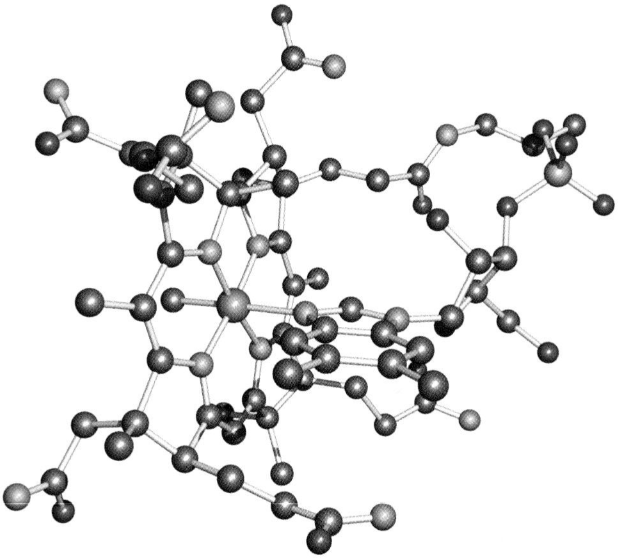

Cobalamin (vitamin B$_{12}$)
Colour code: Co, green; C, grey; N, blue; O, red; P, orange

 **Fig. 34.1** Examples of vitamins that contain heterocyclic rings.

Section 35.7. When cobalamin is isolated, X$^-$ is typically a cyano group. The complex is then referred to as cyanocobalamin. In the body, the important coenzymatic form of vitamin B$_{12}$ contains a 5'-deoxyadenosyl group as the X$^-$ ligand. This is bonded to the cobalt centre through a carbon atom as shown in structure **34.3**.

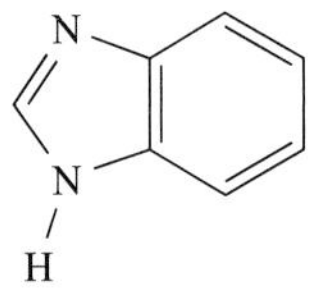

Benzimidazole

(34.1)

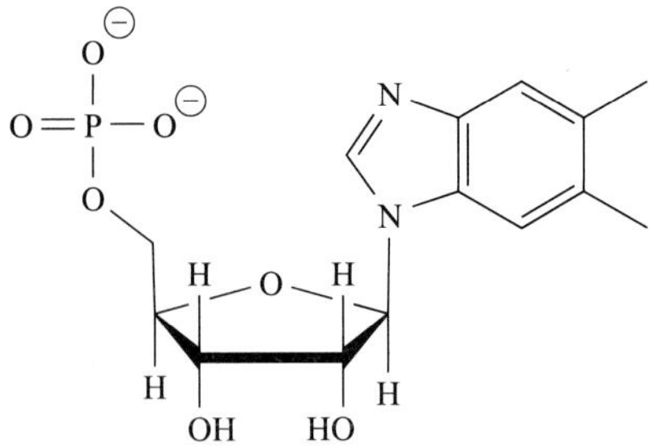

(34.2)

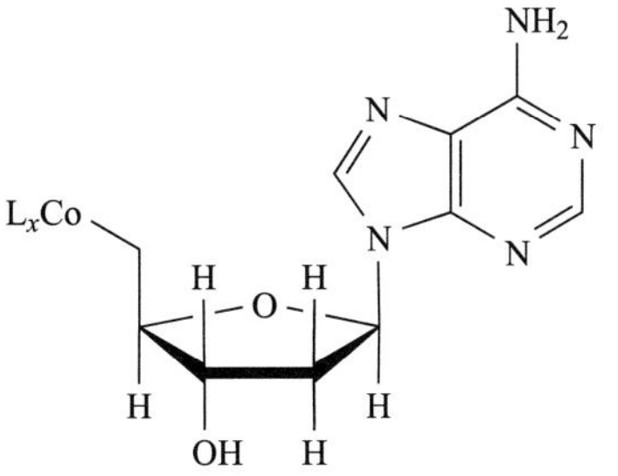

(34.3)

Heterocyclic units also feature in adenosine diphosphate (ADP) and adenosine triphosphate (ATP). In Box 22.11 and the associated text, we discussed the role of ATP (**34.4**) as an energy store.

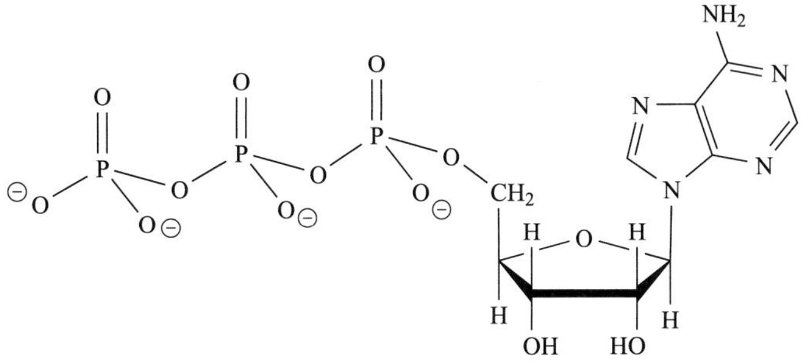

Adenosine triphosphate (ATP)

(34.4)

Structures **34.5**–**34.7** illustrate miscellaneous examples of heterocyclic compounds found in nature. In humans, small amounts of uric acid are present in urine. However, birds and reptiles metabolize nitrogen-containing compounds to give uric acid as the main end-product. Uric acid contains 33% nitrogen and, therefore, guano (the droppings of seabirds) is a commercial source of nitrogenous fertilizer. Histamine is found in body tissues and is released during allergic responses. Luciferin is responsible for the light-emission (bioluminescence) of fireflies (see Box 22.11).

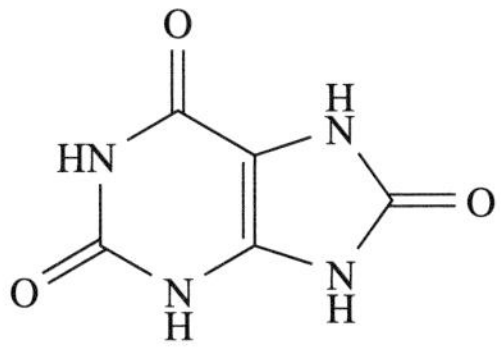

Uric acid    Histamine    Luciferin

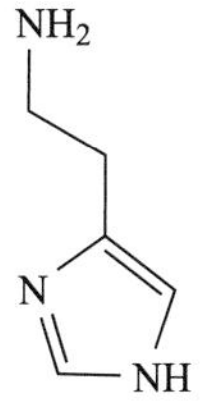

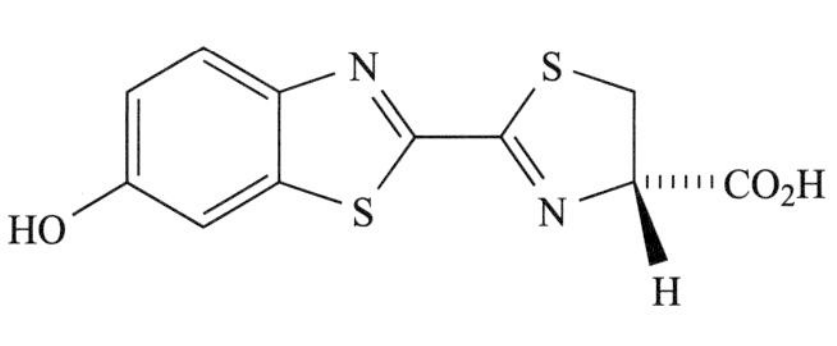

(34.5)    (34.6)    (34.7)

Oxygen-containing 6-membered aromatic heterocycles occur in a wide range of natural products in plants. Three basic building blocks are the benzopyrylium cation, coumarin and chromone:

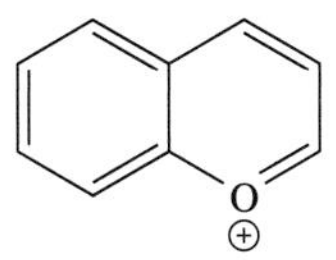

Benzopyrylium cation 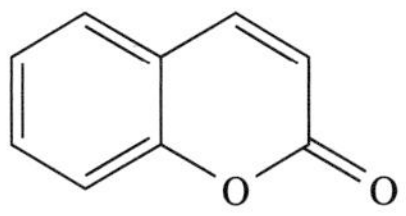    Coumarin 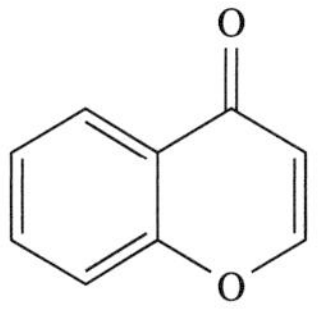    Chromone

The red, blue and purple pigments in fruit and flowers are often compounds called *anthocyanins*. These are *glycosides* of *anthocyanidins*:

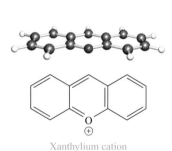

Glycosides: see Section 35.3

| Pelargonidin | R = R' = H |
| Delphinidin | R = R' = OH |
| Petunidin | R = OH, R' = OMe |
| Peonidin | R = OMe, R' = H |
| Cyanidin | R = OH, R' = H |
| Malvidin | R = R' = OMe |

The names of these anthocyanidins derive from plant genera such as pelargonium, delphinium and petunia, all of which have brightly coloured red, blue or purple flowers. Anthocyanins typically absorb light in the region 480–550 nm (see Table 11.2) with large molar extinction coefficients in the range 25 000 to 50 000 dm³ mol⁻¹ cm⁻¹. The characteristic spectroscopic properties of pyrylium derivatives can be put to use in dyes. The example below is a xanthylium derivative that is a calcium-specific fluorescent dye, used to indicate the presence of $Ca^{2+}$ in biological systems:

Xanthylium cation

Drug manufacturers rely heavily on heterocyclic compounds, and we have already mentioned a number of such compounds, for example Viagra (Box 5.4). Other commercially available pharmaceuticals include Diazepam, Zantac and Cimetidine:

Zantac (Ranitidine)
(anti-ulcer drug)

Diazepam (Valium)
(tranquillizer)

Cimetidine
(anti-ulcer drug)

Some hallucinogenic drugs are based on *indole* and two examples are psilocybin and LSD-25:

Indole

Psilocybin
(hallucinogen from 'magic mushrooms')

LSD-25
Lysergic acid diethylamide

## ENVIRONMENT AND BIOLOGY

### Box 34.2 Pheromones 2: ant trails

Nest of wood ants (Hochwald, Switzerland).
*E. C. Constable.*

Pheromones are chemical messengers and are used to communicate messages between insects of the same species. For example, when worker ants discover food, they secrete pheromones to mark a trail allowing ants from the same nest to find the food supply. The photograph shows a wood ants' nest, the centre to which food must be taken. Other pheromones are used to send alarm signals. Heterocyclic compounds feature in a range of ant pheromones. For example methyl 4-methylpyrrole-2-carboxylate is a trail and alarm pheromone of the Texas leaf-cutting ant (*Atta texana*), and 3-ethyl-2,5-dimethylpyrazine is used for the same function by red ants (*Myrmica rubra*).

*See also* Box 29.3: Pheromones 1: a female moth attracts her mate.

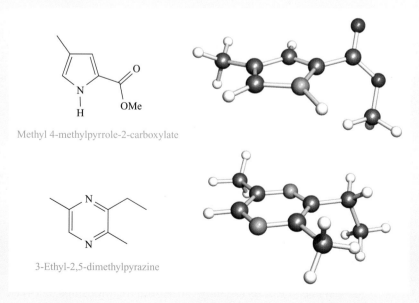

Methyl 4-methylpyrrole-2-carboxylate

3-Ethyl-2,5-dimethylpyrazine

The above examples illustrate a few of the many roles that heterocyclic compounds play in nature and the pharmaceutical industry. Despite the structural diversity of these compounds, you should be able to pick out several basic building blocks. These units include pyridine, pyrrole, imidazole and pyrimidine, all of which are aromatic and contain nitrogen as the heteroatom:

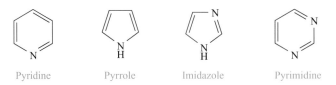

| Pyridine | Pyrrole | Imidazole | Pyrimidine |

In this chapter, we introduce some important classes of aromatic heterocycles that contain N, O or S heteroatoms.

## 34.2     Isoelectronic replacements for CH and CH₂ units

In Chapter 29, we saw that an ether is related to a parent hydrocarbon compound by replacement of a $CH_2$ group by an O atom, for example:

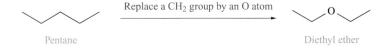

Pentane               Diethyl ether

Replace a $CH_2$ group by an O atom

■▷
Isoelectronic: see
Section 5.12

This replacement is made on the grounds that $CH_2$ and O are isoelectronic. Each possesses six valence electrons. Four electrons are involved in C−H bond formation in a $CH_2$ group, while an O atom possesses two lone pairs, leaving two electrons available for bonding.

We can extend the idea of isoelectronic units so that we have a means of relating N, O and S-containing heterocycles to parent hydrocarbons. In terms of valence electrons, and remembering that O and S are both in group 16:

• $CH_2$, NH, O and S are isoelectronic;
• CH, N, $O^+$ and $S^+$ are isoelectronic.

If we now make these substitutions in some representative cyclic hydrocarbons, we can generate series of structurally related species:

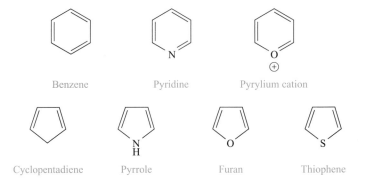

| Benzene | Pyridine | Pyrylium cation |

| Cyclopentadiene | Pyrrole | Furan | Thiophene |

While these series show structural relationships between compounds, we must be careful not to assume that the bonding in the molecules within a given series is the same. For example, whereas cyclopentadiene has two localized C=C bonds, pyrrole, furan and thiophene have delocalized bonding involving the heteroatom and are aromatic. We discuss bonding in Section 34.4.

## 34.3    Nomenclature

So far in this book, each chapter on organic compounds has begun with a section on nomenclature. One of the difficulties encountered by the beginner in heterocyclic chemistry is the large number of ring types and the corresponding array of names for parent heterocyclic rings.[§] For this reason, we shall introduce new names as we need them through the chapter. At this stage, you need to be familiar only with *pyridine*, *pyrylium cation*, *pyrrole*, *furan* and *thiophene*, the structures of which are shown on the opposite page.

### Ring numbering

In monocyclic systems containing one heteroatom, ring atom numbering begins at the heteroatom, and derivatives are named by using the position of substitution, for example:

Pyridine          2-Chloropyridine          4-Aminopyridine

Ring numberings for several other important molecules are given below for reference. Note that in fused ring systems, the heteroatom is not necessarily at position 1:

Pyrimidine          Quinoline          Isoquinoline

Imidazole          Pyrazole          Indole          Benzofuran

---

[§] For detailed guidelines on naming heterocyclic compounds, see the 1979 and 1993 recommendations in the IUPAC 'Blue Book': http://www.chem.qmul.ac.uk/iupac/

| | |
|---|---|
| 34.4 | **Structures of and bonding in aromatic heterocycles** |

### Pyridine and the pyrylium cation

(34.8)

A molecule of pyridine is a planar, 6-membered ring (Figure 34.2) with bond lengths and angles similar to those in benzene. However, the introduction of one N atom into the ring means that all the bonds in the ring are not equivalent. The bonding description that we gave for benzene in Section 32.2 is equally valid for pyridine. Two Kekulé-like resonance structures may be drawn (**34.8**). Alternatively, the bonding can be represented in terms of a delocalized $\pi$-system (Figure 34.3) consisting of six overlapping $2p$ orbitals (one from N and five from C atoms). Each C and N atom is $sp^2$ hybridized. After C$-$H and C$-$C $\sigma$-bond formation (see Figure 32.3), each C atom provides one electron to the $\pi$-system. The N atom has five valence electrons. Two electrons are used in C$-$N $\sigma$-bond formation, two remain localized outside the ring as a lone pair, and one is involved in $\pi$-bonding. Pyridine is therefore an aromatic, Hückel $6\pi$-electron system.

Hückel ($4n + 2$) rule: see Section 32.3

Because pyridine and the pyrylium cation are isoelectronic, their bonding can be described in a similar manner. However, one must take into account the fact that the O atom carries a formal positive charge.

### Pyrrole

A pyrrole molecule consists of a planar, 5-membered ring, with each C and N atom bonded to an H atom (Figure 34.4). Each of the C and N atoms is $sp^2$ hybridized and forms three $\sigma$-bonds. This leaves a $2p$ orbital on each atom in the ring available for $\pi$-bonding. In contrast to the N atom in pyridine (Figure 34.3), the pyrrole N atom contributes its lone pair of electrons to the delocalized $\pi$-system (Figure 34.4). The 5-membered ring in pyrrole therefore possesses six $\pi$-electrons and obeys the Hückel ($4n + 2$) rule. Pyrrole is therefore an aromatic heterocycle.

Within valence bond theory, the bonding in pyrrole can be represented in terms of the following set of resonance structures:

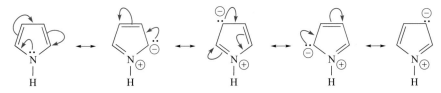

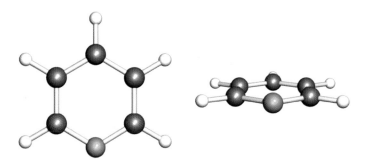

(34.9)

Throughout the rest of this chapter, we shall represent pyrrole using structure **34.9**. However, you must remember that this is only one of the

**Fig. 34.2** The planar structure of pyridine. Colour code: C, grey; N, blue; H, white.

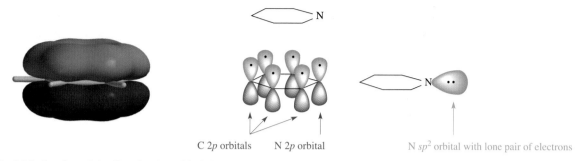

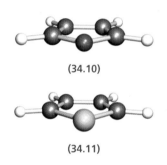

C 2p orbitals    N 2p orbital

N $sp^2$ orbital with lone pair of electrons

**Fig. 34.3** Overlap of the C and N 2p orbitals in pyridine results in delocalized bonding and a 6π-electron system. The N atom lone pair occupies an outward-pointing orbital. The diagram on the left-hand side shows a representation of the π-bonding MO generated computationally using Spartan '04, © Wavefunction Inc. 2003.

contributing resonance structures and that the lone pair on the N atom is delocalized into the π-system.

### Furan and thiophene

**(34.10)**

**(34.11)**

Furan (**34.10**) and thiophene (**34.11**) are structurally related and possess planar, 5-membered rings. Considering only valence electrons, O, S and NH are isoelectronic. The bonding in furan and thiophene can therefore be described in a similar way to that in pyrrole. The $sp^2$ hybrid orbitals of the N atom in pyrrole are involved in the formation of three σ-bonds (two N–C and one N–H). In contrast the three $sp^2$ hybrid orbitals of the O atom in furan are used to form two O–C σ-bonds and to accommodate one lone pair of electrons. The second lone pair resides in an oxygen 2p orbital and is involved in π-bonding (Figure 34.5). The 5-membered ring in furan possesses 6π-electrons, obeys the Hückel ($4n + 2$) rule, and is therefore an aromatic system. Similarly, thiophene is a 6π-aromatic species.

The bonding in furan or in thiophene can be described using valence bond theory by drawing out a set of resonance structures just as we did for pyrrole. Resonance stabilization for furan is achieved from the contributing structures shown below:

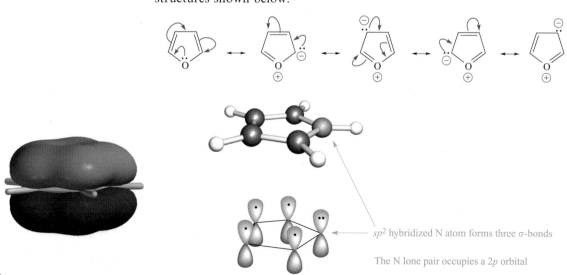

$sp^2$ hybridized N atom forms three σ-bonds

The N lone pair occupies a 2p orbital

**Fig. 34.4** Overlap of the C and N 2p orbitals in pyrrole results in delocalized bonding and a 6π-electron system. The N lone pair is involved in this delocalized bonding. The diagram on the left-hand side shows a representation of the π-bonding MO generated computationally using Spartan '04, © Wavefunction Inc. 2003.

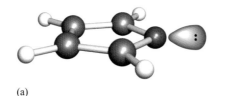

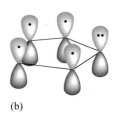

(a)    (b)

**Fig. 34.5** In furan, the O atom has two lone pairs of electrons. One lone pair occupies an outward-pointing $sp^2$ hybrid orbital. The second lone pair occupies a $2p$ orbital and is involved in the delocalized $\pi$-system. An analogous bonding scheme can be drawn for thiophene.

■▶
Deshielding of aromatic protons: see Box 32.2

### Aromatic character of pyrrole, furan and thiophene

Evidence for aromatic character in pyridine, pyrrole, furan and thiophene comes from NMR chemical shift data. The archetypal aromatic compound is benzene. In the $^1$H NMR spectrum of benzene, the six equivalent protons give rise to a signal at $\delta\,7.2$ ppm (Figure 32.6). This lies in the region $\delta\,6$–$10$ ppm that is typical for aromatic protons (Table 14.3). The $^1$H NMR spectroscopic data for the C–H protons in pyrrole, furan and thiophene are given below and in each case, the chemical shift values are indicative of aromatic character:

A comparison of the proton chemical shifts with those of the 'alkene' protons in the model dihydro-compounds shows a consistent shift to higher frequency for the aromatic species.

We have now established that the 5-membered heterocycles pyrrole, furan and thiophene possess aromatic character, but is the extent of aromaticity the same for each compound? The easiest way of measuring 'how aromatic' a compound might be is to compare it with a hypothetical non-aromatic form. In the case of benzene, it is possible to compare the experimental heat of hydrogenation with that calculated for a cyclohexatriene with isolated and alternating single and double bonds. The difference is the resonance stabilization energy, which is 149 kJ mol$^{-1}$ in the case of benzene (see Figure 32.2). A similar calculation for pyridine yields a stabilization energy of 145 kJ mol$^{-1}$, virtually the same as benzene. In the case of the 5-membered rings, this approach yields stabilization energies for the delocalized forms of 67, 92 and 122 kJ mol$^{-1}$ for furan, pyrrole and thiophene respectively. On this basis, we may say that thiophene is the most aromatic and that the aromatic characters of thiophene, pyrrole and furan follow the order:

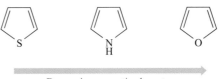

Decreasing aromatic character

We shall see later how the differences in aromaticity affect the reactivity of these compounds.

## 34.5    Physical properties and uses

### Pyridine

(34.12)

Pyridine is a colourless liquid (mp 231.3 K, bp 388.2 K) and has a characteristic smell. It is toxic, and prolonged exposure may cause sterility in men. It is polar (**34.12**, N is $\delta^-$) with a dipole moment of 2.22 D for the gas phase molecule. Pyridine has a similar density (0.98 g cm$^{-3}$) to water (1.00 g cm$^{-3}$), and is miscible with water and with a range of organic solvents including alcohols, ethers and benzene.

Pyridine is widely used as a solvent and has applications in the plastics industry and in the manufacture of some pharmaceuticals. Nicotinic acid (Figure 34.1) is a derivative of pyridine and is an essential vitamin in the diet of mammals. The name 'nicotinic acid' derives from the *alkaloid* nicotine (see Box 34.1) which may be oxidized to nicotinic acid. *Pyriproxyfen* is an example of a pyridine-based insecticide (a juvenile hormone mimic), introduced in 1996 for control of the whitefly complex that affects cotton crops in the US.

Pyriproxyfen

### Pyrrole

Pyrrole is a colourless liquid (bp 402.8 K, density 0.97 g cm$^{-3}$) when freshly distilled, but the liquid darkens in the presence of $O_2$. It is only sparingly soluble in water, but dissolves easily in alcohols, ethers and benzene. Pyrrole is polar with a gas-phase dipole moment of 1.74 D. Pyrrole is used to make polypyrrole, an electrically conducting polymer that may be used as the electrode in lithium ion batteries or capacitors. A large number of pyrrole derivatives have medicinal applications and the pyrrole ring is a key structural feature in porphyrins and related macrocycles.

### Furan

Furan is a colourless liquid (bp 304.4 K, density 0.94 g cm$^{-3}$). It is insoluble in water, but dissolves in alcohols and ether. The furan molecule is polar with a gas-phase dipole moment of 0.66 D. Furan vapour has anaesthetic properties, although the compound is highly toxic and the liquid can be absorbed through skin. There is some concern today about the occurrence of furan and furan derivatives in canned and preserved foods, where they arise from the breakdown of carbohydrates. The fully reduced compound, tetrahydrofuran, is widely used as a solvent, having properties similar to diethyl ether.

### Thiophene

Thiophene is a liquid (bp 357.2 K, density 1.06 g cm$^{-3}$) that is insoluble in water, but miscible with a range of organic solvents. Thiophene is polar with a gas-phase dipole moment of 0.55 D. Thiophene is a naturally occurring component in petroleum, often to the extent of 1–2%. There is considerable interest in the development of hydrodesulfurization catalysts that will remove the sulfur from thiophene and related sulfur-containing compounds before combustion (which would otherwise lead to the formation of sulfur dioxide and sulfuric acid). Thiophene is used in industry as a solvent and as the starting point for a wide range of derivatives. Oxidative polymerization leads to polythiophene, a conducting polymer that is under intensive investigation as a component in a wide range of electronic devices.

Acid rain: see Box 22.13

## 34.6    Pyrrole, furan and thiophene: syntheses

### The disconnection approach (retrosynthesis)

When planning the syntheses of most heterocyclic compounds (or complex molecules in general), a suitable preparative route can be designed by using the *disconnection approach*, i.e. working backwards from the desired product to determine appropriate precursors. This method is also called *retrosynthesis*. In order to apply this approach, you must have a working knowledge of basic bond-forming reactions. We have covered the principal reactions that we need for heterocyclic formation in earlier chapters and expand upon them below.

In the disconnection approach, equations are written in the following format:

$$A \Longrightarrow B$$

The type of arrow is important because it indicates that this is a retro-synthetic step, i.e. an 'exercise on paper'. Before we look at the application of the disconnection approach to the formation of pyrrole, furan and thiophene, we review a key reaction type, that of aldehydes and ketones with amines.

### Reactions of aldehydes and ketones with amines

In Section 33.14, we saw how nucleophilic attack by $NH_3$ on an aldehyde or ketone leads to the formation of a C–N bond and an imine via an intermediate aminol:

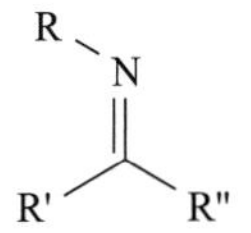

An imine

R, R' and R" may be
the same or different

**(34.13)**

Analogous reactions occur with primary amines, $RNH_2$, to yield imines of type **34.13**. When a secondary amine reacts with an aldehyde or ketone, the product is an *enamine*:

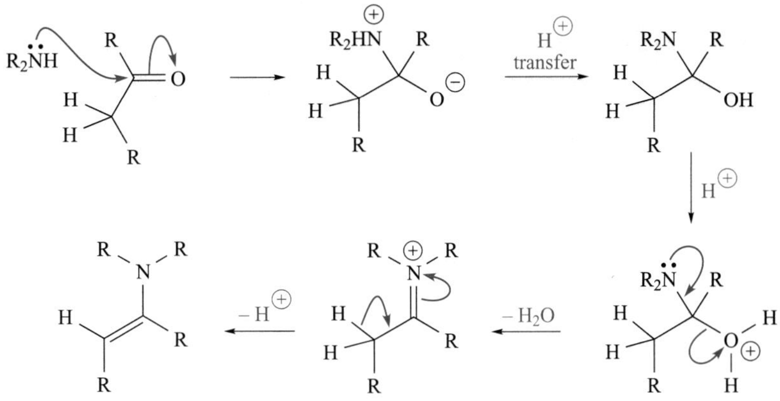

An enamine

Using the disconnection approach, we view enamine formation in reverse and represent the synthesis as follows:

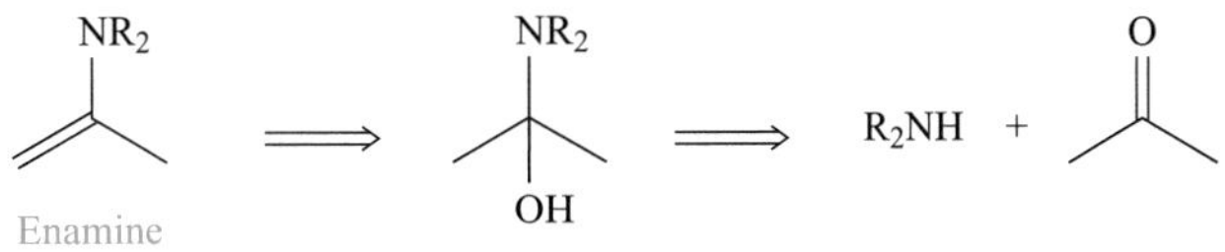

Enamine

We are now in a position to look at how we can apply the reaction of a carbonyl group and an amine or $NH_3$ to the formation of a nitrogen-containing heterocycle.

## Syntheses of pyrrole, furan and thiophene

Equation 34.1 illustrates the disconnection of a pyrrole ring and shows that a suitable precursor for pyrrole is a dicarbonyl compound. In addition to forming pyrrole itself, the reaction scheme also allows for the introduction of substituents into the ring, for example, equation 34.2.

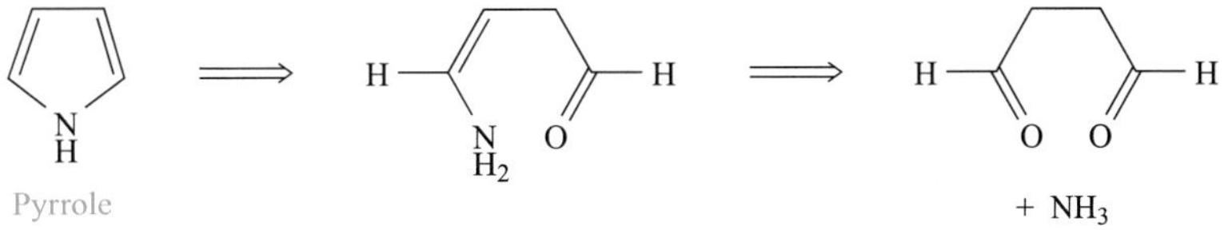

Pyrrole

$+ NH_3$

(34.1)

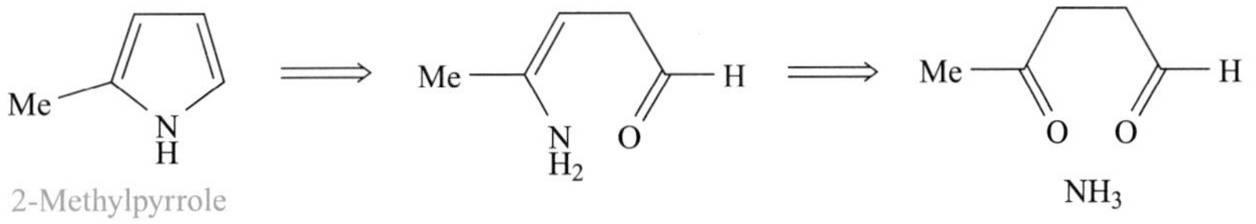

2-Methylpyrrole

$NH_3$

(34.2)

Equation 34.1 indicates that the synthesis of pyrrole can be carried out by treating butanedial with ammonia. The mechanism can be represented as follows, although the precise sequence of steps could vary:

Furans are also prepared starting from a 1,4-dione. Dehydration of butane-dial gives furan (equation 34.3).

Mechanism: see
end-of-chapter problem 34.7

$$\text{(34.3)}$$

The best way to prepare thiophene (equation 34.4) is by use of a thionating agent to convert butanedial into the corresponding thiocarbonyl compound. Lawesson's reagent or $P_2S_5$ is convenient for this purpose. Once formed, cyclization by loss of $H_2S$ is facile.

Lawesson's reagent

$$\text{(34.4)}$$

## 34.7    Pyrrole, furan and thiophene: reactivity

### Protonation of pyrrole

In Chapter 31, we described how amines ($RNH_2$, $R_2NH$ or $R_3N$) can be protonated (equation 34.5). The basic character depends on the availability

Amines as bases: see
Section 31.4

of the lone pair of electrons on the N atom. Thus, aniline, $C_6H_5NH_2$, only behaves as a weak base because resonance stabilization involving the N lone pair precludes its ready availability for protonation (equation 32.57 and discussion). As we have already seen, the relative basicities of amines can be quantified by comparing $pK_b$ values of the bases (equation 34.5) or the $pK_a$ values of their conjugate acids (equation 34.6).

$$RNH_2 + H_2O \rightleftharpoons RNH_3^+ + OH^-$$

$pK_b$ refers to protonation of base　　(34.5)

$$RNH_3^+ + H_2O \rightleftharpoons RNH_2 + H_3O^+$$

$pK_a$ refers to loss of $H^+$ from conjugate acid　　(34.6)

The $pK_a$ values for selected conjugate acids of amines are listed in Table 34.1 and are compared with the $pK_a$ value for the conjugate acid of pyrrole. This value indicates that pyrrole (**34.14**) does not behave as a typical amine and is a very weak base. In fact, pyrrole more often acts as an acid than a base, and may be deprotonated by strong bases. Pyrrole is a relatively weak acid (equation 34.7), but is a considerably stronger acid than the saturated analogue pyrrolidine ($pK_a = 44$).

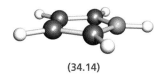

(34.14)

Pyrrolidine

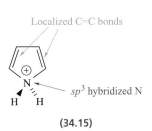

Localized C=C bonds

$sp^3$ hybridized N

(34.15)

$pK_a = 17.5$　(34.7)

The reason that pyrrole is not basic can be understood by looking at Figure 34.4. This shows that the lone pair on the N atom in pyrrole is involved in the $6\pi$-electron system and is unavailable to accept a proton.

Pyrrole can be protonated, but protonation at a carbon atom and not at nitrogen results in more stable species. If the N atom were the site of protonation, the aromatic character of the pyrrole ring would be lost (structure **34.15**). The preferred site of protonation is atom C(2) (i.e. adjacent to the N atom) and this can be rationalized in terms of resonance stabilization of the protonated species, although the C(3) protonated species is also present in the equilibrium mixture:

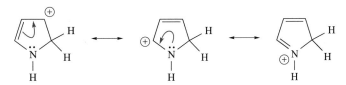

Protonation at C(2): three resonance structures

Table 34.1 $pK_a$ values for selected aliphatic amines, aniline and pyrrole. See also Tables 31.2 and 32.4.

| Compound | $pK_a$ of conjugate acid |
|---|---|
| Ethylamine (EtNH$_2$) | 10.81 |
| Diethylamine (Et$_2$NH) | 10.49 |
| Triethylamine (Et$_3$N) | 11.01 |
| Aniline (PhNH$_2$) | 4.63 |
| Pyrrole (structure **34.14**) | <0 |

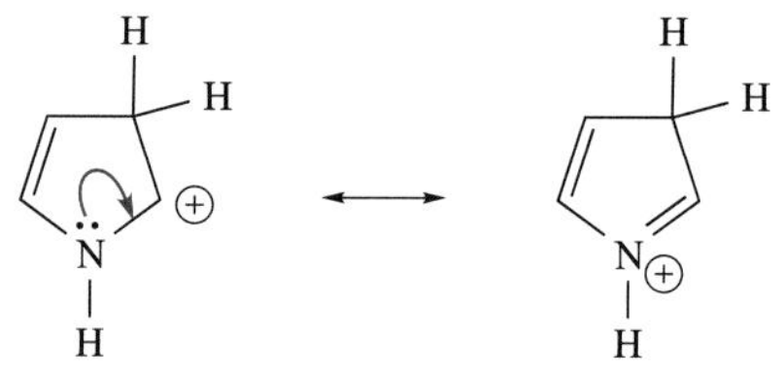

Protonation at C(3): two resonance structures

The relative stabilities of the various sites of protonation are given by the $pK_a$ values for the conjugate acids: protonation at C(2) has a $pK_a$ value of $-3.8$, at C(3), a $pK_a$ of $-5.9$, and at nitrogen the $pK_a$ is about $-10$. Interestingly, the *rate* of protonation is fastest at nitrogen, with protonation at C(2) being about twice as fast as C(3). A complication of protonation by strong acids is the reaction of the protonated species with pyrrole. The reaction continues and eventually forms a polymer:

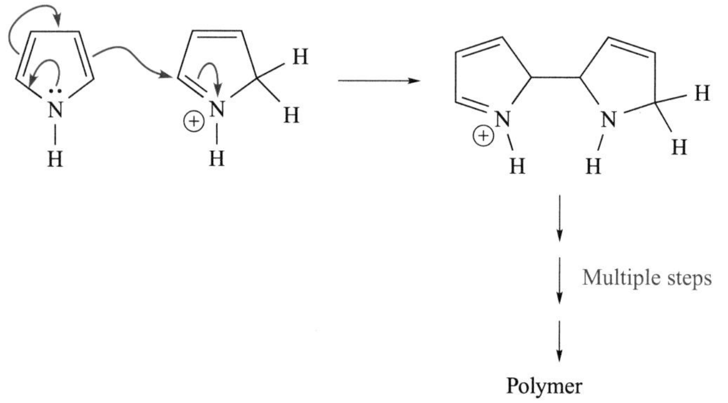

Multiple steps

Polymer

An analogous polymerization occurs when thiophene is protonated.

## Electrophilic substitution: general mechanism

Pyrrole, furan and thiophene are all susceptible to attack by electrophiles, and are generally more reactive towards electrophiles than is benzene. Furan is the most reactive of the three compounds towards electrophiles:

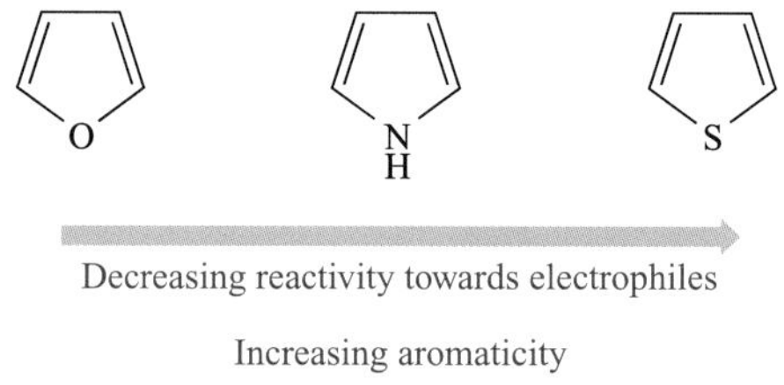

Decreasing reactivity towards electrophiles

Increasing aromaticity

Electrophilic substitution often occurs at the C(2) atom (equation 34.8 where $E^+$ is a general electrophile), although, in many cases, changes in reaction conditions or the electrophilic reagent may be used to tune the major product to that with substitution at C(2) or C(3).

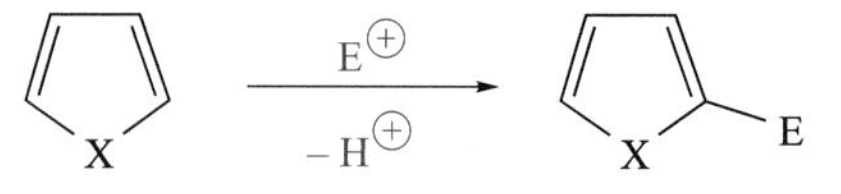

$$(34.8)$$

We can understand the reason why substitution occurs at atom C(2) rather than at C(3) by considering the relative stabilities of the two possible intermediates. For electrophilic attack at atom C(2), the first step in the mechanism is:

For attack at C(3), there is less resonance stabilization of the intermediate:

Compare this with the discussions in Section 32.10

Figure 34.6 shows a general reaction profile for a two-step reaction such as electrophilic substitution in pyrrole, furan or thiophene. The greater the resonance stabilization of the intermediate, the lower its energy. The energy of transition state $\{TS(1)\}^{\ddagger}$ is also lowered. As a consequence, the activation energy, $E_a(1)$, of the rate-determining step is lowered, and the rate of reaction is increased. The second (fast) step in each substitution pathway can be represented as follows:

For 2-substituted product:

For 3-substituted product:

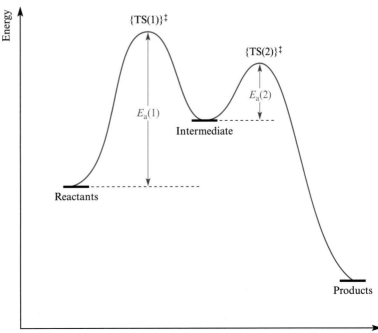

**Fig. 34.6** Reaction profile for a general two-step reaction showing the relative energy levels of the transition states (TS) and intermediate.

In each of the following examples of electrophilic substitution, notice the preference for substitution at the 2-position of the ring. The conditions for electrophilic substitution in the 5-membered aromatic heterocycles are milder than those used for analogous substitutions in benzene, and multi-substitution often occurs. Thiophene protonates primarily at the C(2) and C(3) positions.

### Electrophilic substitution: halogenation

The halogenation of pyrrole is facile and the reaction is difficult to control. Tetrahalo-derivatives tend to form, and no Lewis acid catalyst is needed (equation 34.9).

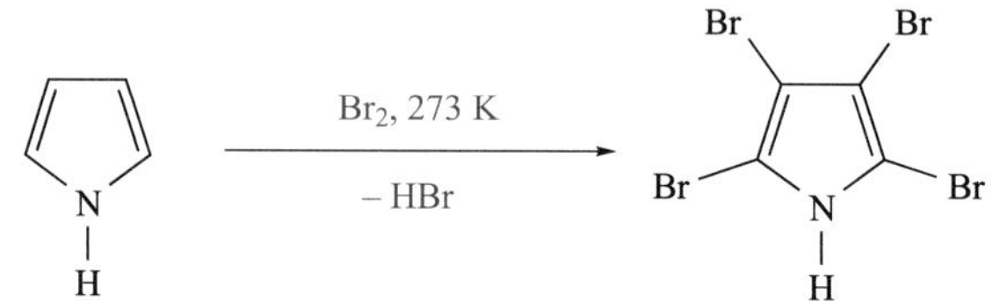

$$\text{(34.9)}$$

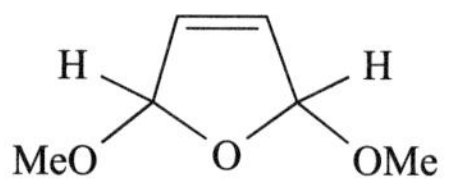

Choice of solvent can be important. For example, reaction of furan with $Br_2$ in methanol leads to compound **34.16** rather than a bromo-derivative. This is an example of the formation of a *2,5-addition* to furan occurring in preference of a substitution product, and is a complication that we shall see again later.

**(34.16)**

### Electrophilic substitution: acylation

Friedel–Crafts acylation: see Section 32.9

Whereas the acylation of benzene uses Friedel–Crafts conditions, pyrrole and furan are unstable in the presence of Lewis acids such as $AlCl_3$ or $BF_3$. These heterocycles are therefore acylated using the *Vilsmeier reaction*. In the first step, the electrophile is generated by the reaction of *N,N*-dimethyl-formamide (DMF) or a derivative, $RCONMe_2$, and $POCl_3$ (equation 34.10).

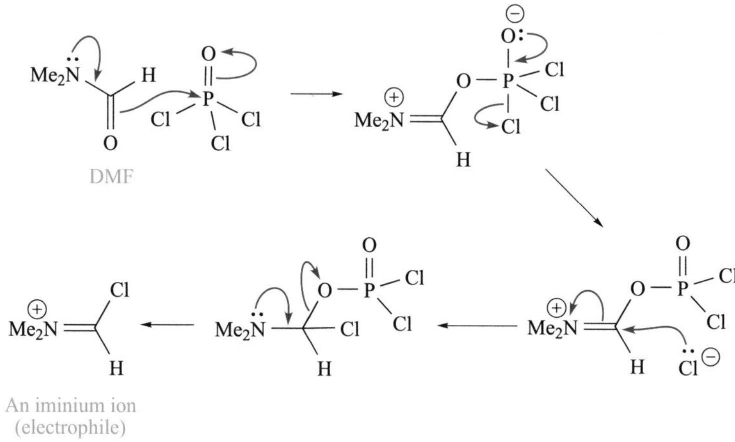

DMF

An iminium ion
(electrophile)

$$\text{(34.10)}$$

The electrophile now reacts with the aromatic heterocycle, and the product is hydrolysed in aqueous $Na_2CO_3$ to give an acyl derivative. Equation 34.11 shows the formation of pyrrole-2-carbaldehyde.

Pyrrole-2-carbaldehyde

$$(34.11)$$

Thiophene can be acylated using Friedel–Crafts conditions, for example reaction 34.12.

$$(34.12)$$

## Nitration

In contrast to the use of a mixture of concentrated nitric and sulfuric acids for the nitration of benzene, the sensitivity of pyrrole and furan to acids means that they cannot be used to nitrate these heterocycles. The nitration of pyrrole, furan and thiophene is carried out using acetyl nitrate. This is formed from the reaction of $HNO_3$ with acetic acid and acetic anhydride. Pyrrole and thiophene react with acetyl nitrate to give 2-nitropyrrole (and 3-nitropyrrole in a $5:1$ ratio) and 2-nitrothiophene (and 3-nitrothiophene in a $6:1$ ratio), respectively (equation 34.13). However, when furan reacts with acetyl nitrate, a 2,5-addition product is formed (equation 34.14). The

X = NH or S     Acetyl nitrate

$$(34.13)$$

reaction must be carried out in the presence of pyridine to ensure the formation of 2-nitrofuran (equation 34.15).

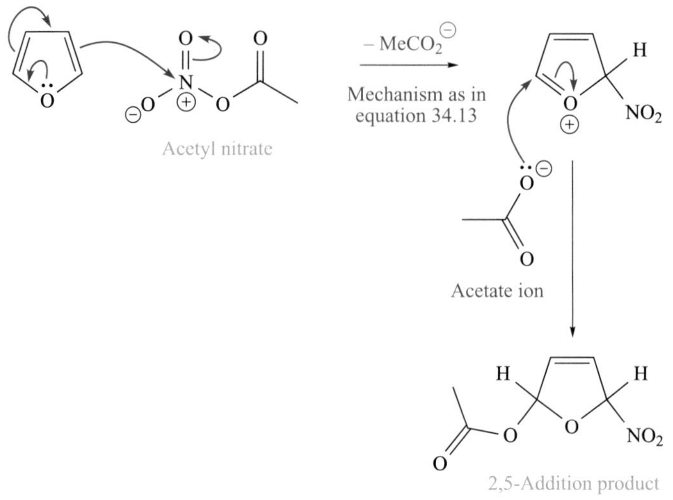

(34.14)

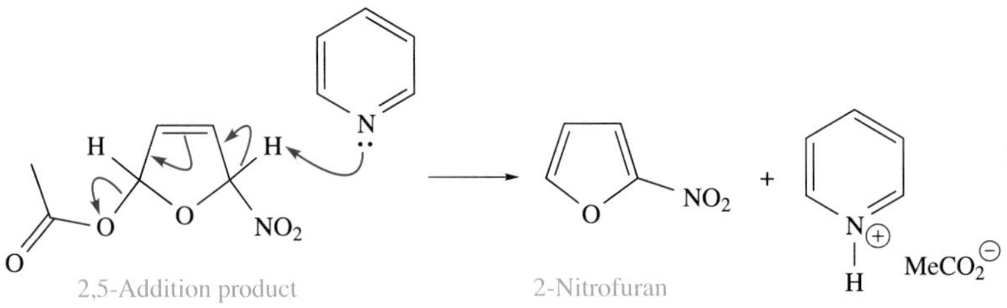

Acetyl nitrate

Acetate ion

2,5-Addition product

(34.15)

2,5-Addition product          2-Nitrofuran

## Electrophilic substitution: sulfonation

The sulfonation of pyrrole, furan and thiophene can be carried out using a pyridine adduct of $SO_3$. This acts as a source of $SO_3$, but allows the reaction to be carried out under milder conditions than if $SO_3$ itself were used (compare with equation 32.33). The sulfonation of pyrrole gives exclusively the 3-sulfonate (equation 34.16).[§] In contrast, sulfonation of thiophene and furan with pyridine $SO_3$ gives the expected 2-substituted products.

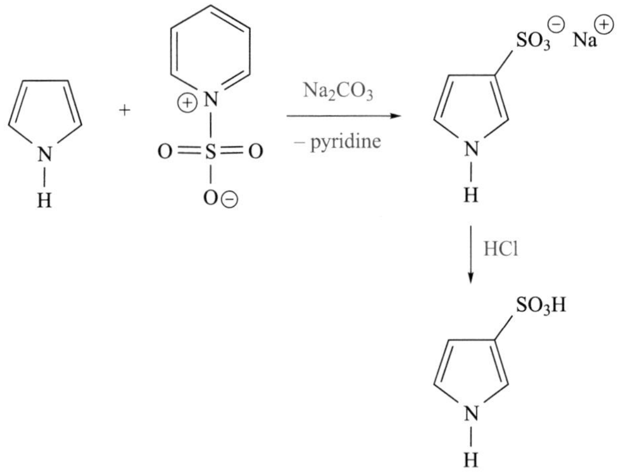

(34.16)

---

§ See: A. Mizuno, Y. Kan, H. Fukami, T. Kamei, K. Miyazaki, S. Matsuki and Y. Oyama (2000) *Tetrahedron Letters*, vol. 41, p. 6605 – 'Revisit to the sulfonation of pyrroles: is the sulfonation position correct?'

## Diels–Alder cyclizations

Diels–Alder reactions: see Section 26.3

In a Diels–Alder reaction (or [4 + 2] cycloaddition), a dienophile adds across a 1,3-diene to give a cyclic product. We have already seen that of thiophene, pyrrole and furan, the oxygen-containing heterocycle has the least aromatic character. Consistent with this is the fact that furan acts as a diene and readily undergoes Diels–Alder cyclization reactions. In equation 26.7, we showed the stereospecific formation of the *endo*-product when a dienophile reacts with cyclopentadiene. Equation 34.17 shows an analogous reaction involving furan. This time, the *exo*-product is favoured.

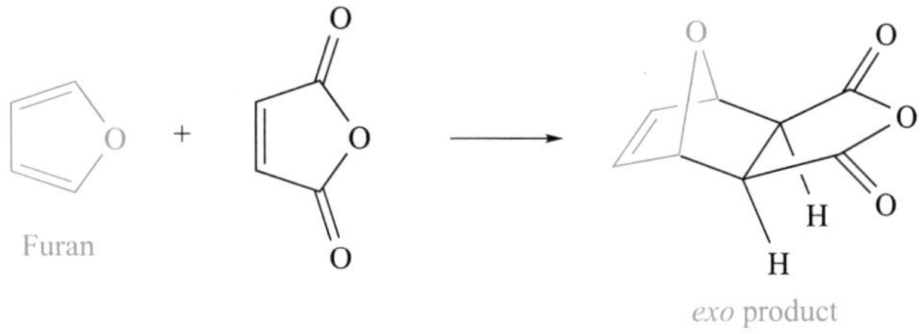

Furan     *exo* product     (34.17)

The difference between the stereospecificities of reactions 26.7 and 34.17 lies in the difference between isolating a kinetic or thermodynamic product. Diels–Alder reaction 26.7 is irreversible and the kinetic *endo*-product can be isolated. In contrast, under the conditions used for reactions of furan with dienophiles, the process is reversible and the isolated product is the more thermodynamically stable diastereoisomer, i.e. the *exo*-product.

**34.8**     **Pyridine and pyrylium ion: syntheses**

## The Hantzsch synthesis of pyridine

By using the disconnection approach, we can plan the synthesis of pyridine as follows:

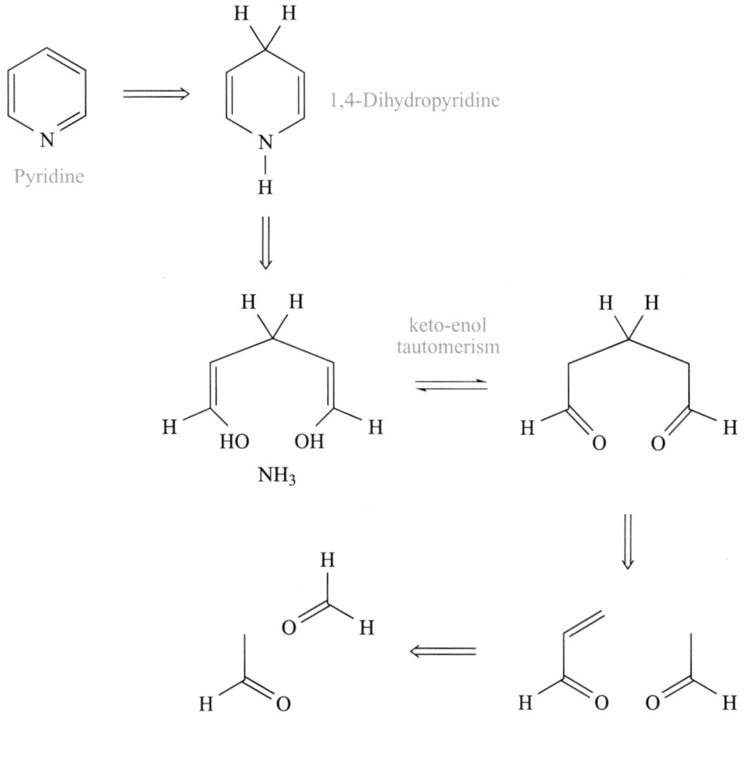

1,4-Dihydropyridine

keto-enol tautomerism

Review reactions of enolates
at the end of Chapter 33

The reasoning behind this scheme makes use of several bonding-forming reactions that we introduced in Chapter 33. Disconnecting the 1,4-dihydro-pyridine is more straightforward than disconnecting pyridine, i.e. planning the synthesis of 1,4-dihydropyridine can be achieved readily. The 1,4-dihydropyridine is easily oxidized to pyridine using an oxidizing agent such as a quinone, cerium(IV), nitric acid or even air. The synthesis of 1,4-dihydropyridine occurs most readily if electron-withdrawing substituents are present in the 3- and 5-positions of the final ring (see below). These substituents also activate and control the condensation reactions early in the reaction sequence. Working through the retrosynthesis, we can see that the final cyclization is the addition of $NH_3$ to one carbonyl group of the 1,4-dicarbonyl compound, followed by cyclization of an imine with the second carbonyl group. The 1,4-dicarbonyl compound is formed by addition to an enone, which in turn is formed by an aldol reaction. In practice, the synthesis of 1,4-dihydropyridine is carried out as a 'one-pot' reaction in which the four reagents are mixed in the correct molar ratios. The specific assembly of the 6-membered ring from four small molecules is quite remarkable: this is the Hantzsch synthesis of pyridine and can be represented in the following disconnections:

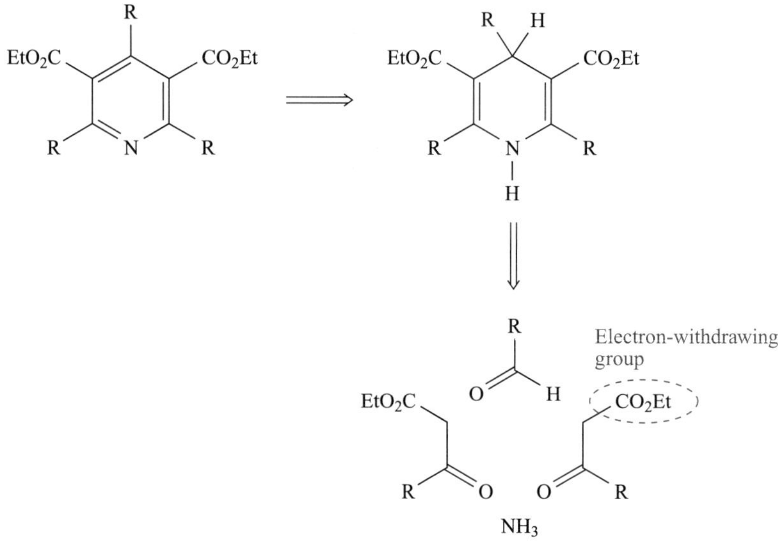

Notice the flexibility of the synthesis. Not only can it be used to prepare pyridine, but it also allows the introduction of various substituents, R, into the pyridine ring.

So how does the Hantzsch synthesis work? The precise details of the mechanism are not known, and probably vary slightly with differing substituents. However, a reasonable pathway starts with the reaction of the enolate of a keto-ester with an aldehyde. The product of this reaction is in equilibrium with its enolate form:

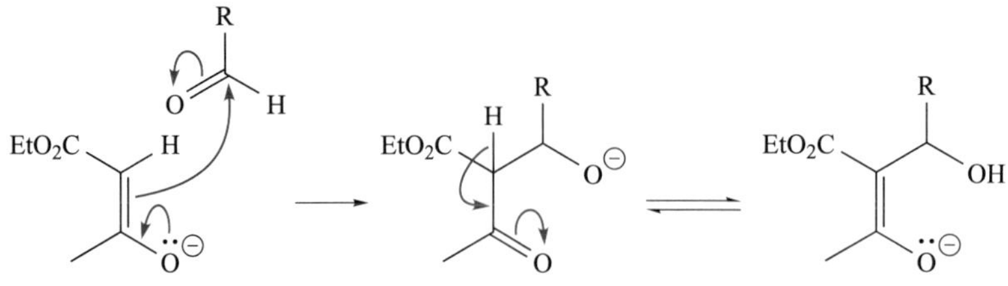

Loss of OH⁻ from the enolate (probably after initial protonation so that the leaving group is $H_2O$ rather than OH⁻) generates an unsaturated carbonyl compound. This reacts with the second equivalent of the starting keto-ester in its enolate form:

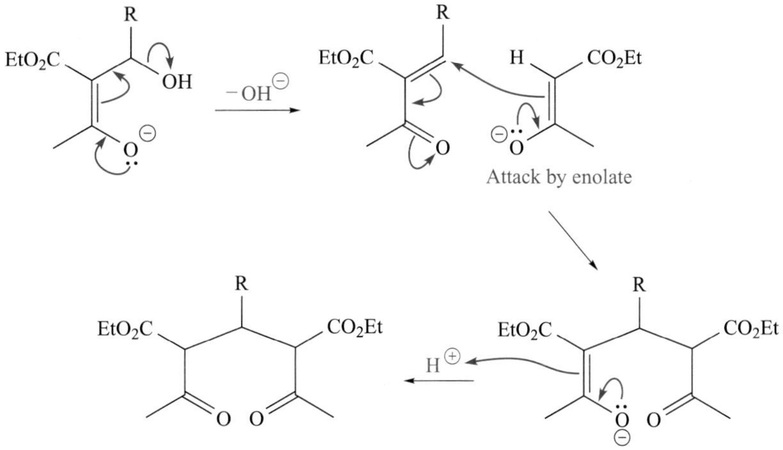

The 1,5-dicarbonyl compound now reacts with $NH_3$ to give the 1,4-dihydro-pyridine (equation 34.18). The mechanism for this reaction is the same as that for the reaction of 1,4-butanedione with $NH_3$ to give pyrrole (see equation 34.2 and the following mechanism).

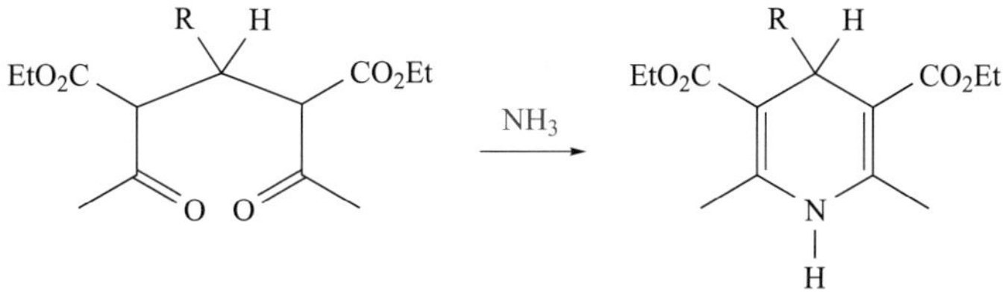

(34.18)

Finally, the 1,4-dihydropyridine must be oxidized to give an aromatic pyridine derivative (equation 34.19). A number of oxidizing agents can be used, including quinone and cerium(IV); the latter is reduced to cerium(III).

Quinones: see Box 32.3

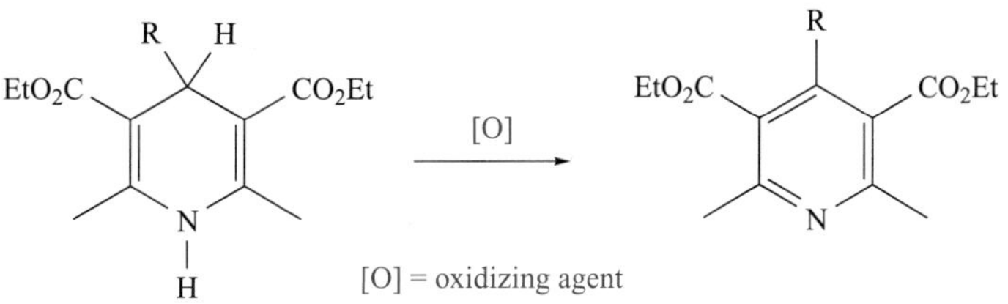

[O] = oxidizing agent

(34.19)

A direct method of preparing pyridine or a substituted derivative of pyridine is the reaction a 1,5-dicarbonyl compound with hydroxylamine, $NH_2OH$. This introduces an NOH group into the ring and the intermediate 1,4-dihydropyridine eliminates $H_2O$ (equation 34.20).

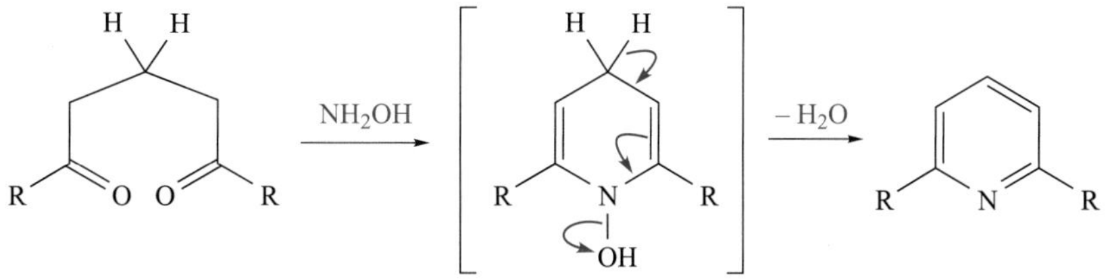

(34.20)

### Synthesis of the pyrylium cation

The synthesis of the pyrylium cation is related to the Hantzsch synthesis of pyridine, the precursor being a 1,5-dicarbonyl compound. In the case of the pyrylium cation, however, the heteroatom is already present as part of one of the carbonyl groups. Substituted pyrylium cations can be prepared in the same way as the parent compound by using a suitably substituted 1,5-dicarbonyl compound. Starting from a representative 1,5-dicarbonyl compound, a Lewis acid such as $BF_3$ is used to activate a carbonyl group. Monoenolization is then the initial step that produces a nucleophile to facilitate ring closure:

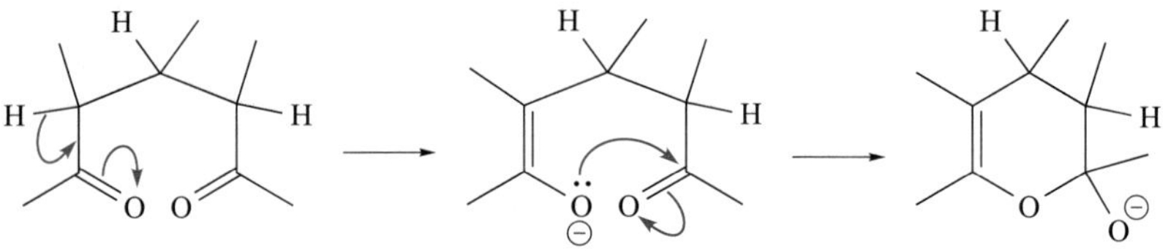

After protonation, the molecule is set up to undergo elimination of $H_2O$:

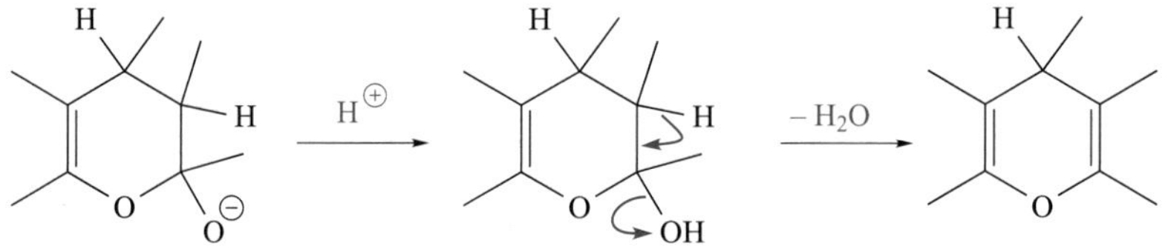

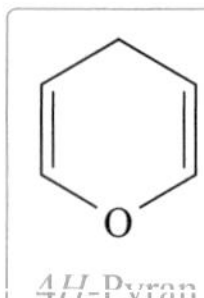

The product at this stage of the reaction is $4H$-pyran (or a derivative thereof). An oxidation step is needed to convert this to a pyrylium cation, and this step involves the abstraction of a hydride ion, $H^-$. In equation 34.21, the choice of $[Ph_3C][BF_4]$ as the hydride abstracting agent means that the pyrylium cation is isolated as a tetrafluoroborate salt.

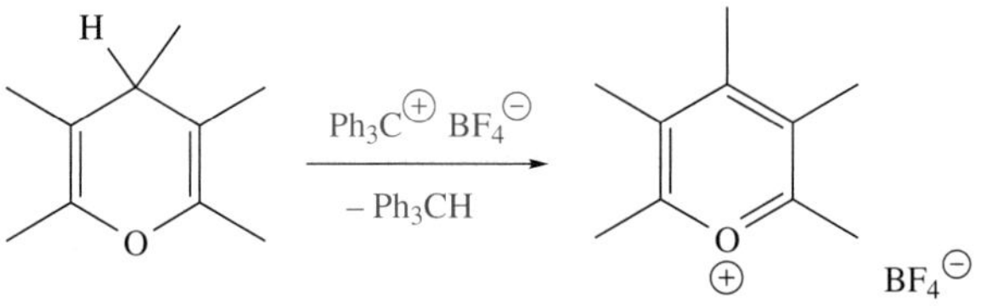

(34.21)

---

**34.9**  **Pyridine: reactivity**

### Pyridine as a base

The presence of the outward-pointing lone pair of electrons on the N atom in pyridine (Figure 34.3) means that pyridine acts as a base. Thus, pyridine reacts with acids to form pyridinium salts, e.g. equation 34.22.

$$\text{Pyridine} + \text{HCl} \longrightarrow \text{Pyridinium chloride} \quad Cl^{\ominus} \tag{34.22}$$

The pyridinium cation has a $pK_a$ value of 5.25 (equilibrium 34.23) and is therefore a slightly weaker acid than acetic acid ($pK_a = 4.77$).

$$+ H_2O \rightleftharpoons + H_3O^{\oplus} \qquad pK_a = 5.25 \tag{34.23}$$

Quaternary pyridinium salts (**34.17**) can be prepared by treating pyridine with an alkyl iodide, e.g. reaction 34.23. Salts of *paraquat* (**34.18**) are potent, non-selective contact herbicides. Paraquat works by disrupting electron transfer processes. The dication readily accepts an electron to form a radical monocation. Paraquat is extremely toxic, and now has restricted uses. In Sweden, it has been banned since 1983.

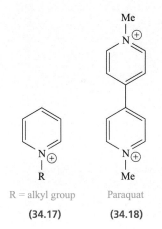

R = alkyl group

**(34.17)**

Paraquat

**(34.18)**

$$+ \text{MeI} \longrightarrow \quad I^{\ominus} \tag{34.24}$$

*N*-Methylpyridinium iodide

### Electrophilic substitution in pyridine

Compared with benzene, pyridine is *deactivated* towards electrophilic substitution and exhibits a reactivity similar to that of nitrobenzene. In part, this arises from the electron-withdrawing effects of the N atom, i.e. an inductive effect draws electronic charge towards the N atom making the C atoms $\delta^+$. Pyridine reacts with $Br_2$ but only at high temperatures (equation 34.25). Pyridine can be nitrated using concentrated $HNO_3/H_2SO_4$ (i.e. the same conditions as for the nitration of benzene), but 3-nitropyridine is isolated only in very poor yields. The reason for the lack of success in this reaction is that the pyridine N atom is protonated in acidic solution and the ring deactivated even further towards electrophilic attack. A much more successful route involves the use of $N_2O_5$ in the presence of $SO_2$ (equation 34.26). This reaction involves the initial formation of *N*-nitropyridinium ion (**34.19**).[§]

*Inductive effect: see Section 27.4*

$$\xrightarrow[{- \text{HBr}}]{Br_2, \text{ 600 K}} \tag{34.25}$$

---

[§] For proposed mechanistic details, see: J.M. Bakke (2003) *Pure and Applied Chemistry*, vol. 75, p. 1403.

**N-Nitropyridinium ion**

**(34.19)**

$$\text{(34.26)}$$

1. $N_2O_5$ in $MeNO_2$
2. $HSO_3^{\ominus} / H_2O$

3-Nitropyridine

When electrophilic substitution in pyridine does occur, it typically does so at the 3-position. This can be rationalized in terms of the possible intermediates for 2-, 3- and 4-substitution by a general electrophile $E^+$:

Resonance stabilized intermediate for 2-substitution

Resonance stabilized intermediate for 3-substitution

Resonance stabilized intermediate for 4-substitution

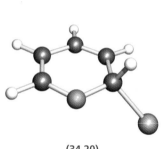

**(34.20)**

**(34.21)**

Orientation and activation effects of substituents in aromatic rings: see Table 32.3 and accompanying discussion

Remember (from Chapter 32) that the $6\pi$-aromatic system is destroyed on forming the intermediate. One C atom is $sp^3$ hybridized and tetrahedral (e.g. structure **34.20** where the orange atom is the electrophile). The intermediates in the 2- and 4-substitution pathways have resonance forms in which the positive charged is localized on the N atom. For 3-substitution, the positive charge in the intermediate is localized only on C atoms. Since N is more electronegative than C, it is less favourable to have the positive charge localized on N than C. Therefore, the lowest energy pathway from reactants to products is that for 3-substitution.

Pyridine does *not* undergo normal Friedel–Crafts reactions, because the Lewis acid catalyst (e.g. $AlCl_3$) forms an adduct with pyridine (**34.21**).

Electrophilic substitutions occur more readily in pyridine derivatives that contain electron-donating groups than in pyridine itself. For example, OR and $NR_2$ substituents each possess lone pair electrons that can be delocalized into the $\pi$-system of the pyridine ring. Electrophilic substitutions occur *ortho* or *para* to the OR or $NR_2$ groups, for example:

e.g. R = Me

## Activation of pyridine towards electrophiles: use of pyridine *N*-oxide

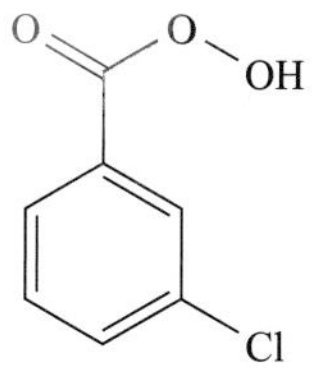

3-Chloroperbenzoic acid
(*meta*-chloroperbenzoic acid,
MCPBA)

The pyridine ring can be activated towards electrophilic substitution by first converting pyridine to pyridine *N*-oxide (equation 34.27). This oxidation may be carried out using one of a number of oxidants; both $H_2O_2$ in acetic acid and 3-chloroperbenzoic acid are convenient choices. Pyridine *N*-oxide is stabilized because a lone pair of electrons is conjugated into the $\pi$-system of the pyridine ring (Figure 34.7). Pyridine *N*-oxide is isoelectronic with the phenoxide ion, and the extension of the $\pi$-system is analogous to that described for phenol in structure **32.34**.

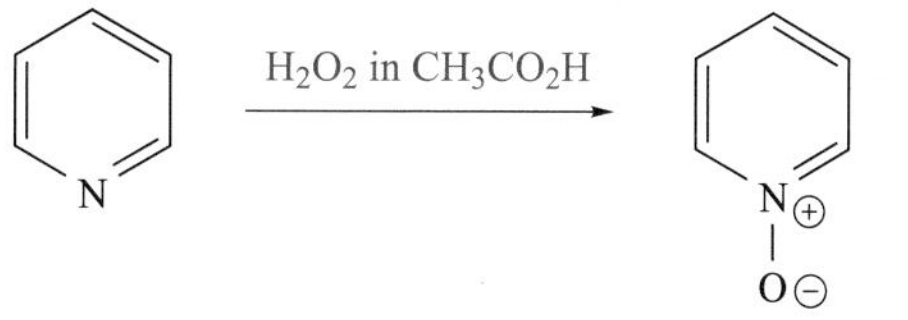

(34.27)

Pyridine *N*-oxide

Pyridine *N*-oxide now reacts with an electrophile, substitution at the 4-position being the most favoured product (scheme 34.28 in which $E^+$ is a general electrophile). The substituted pyridine *N*-oxide is finally reduced using $PCl_3$. The mechanism of this last step involves nucleophilic attack by the O atom of the pyridine *N*-oxide on $PCl_3$, and transfer of oxygen from N to P (equation 34.29). This reduces the pyridine *N*-oxide and oxidizes $PCl_3$ to $POCl_3$.

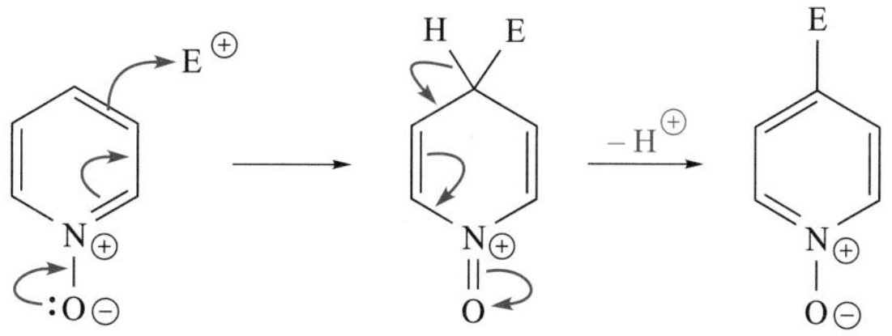

(34.28)

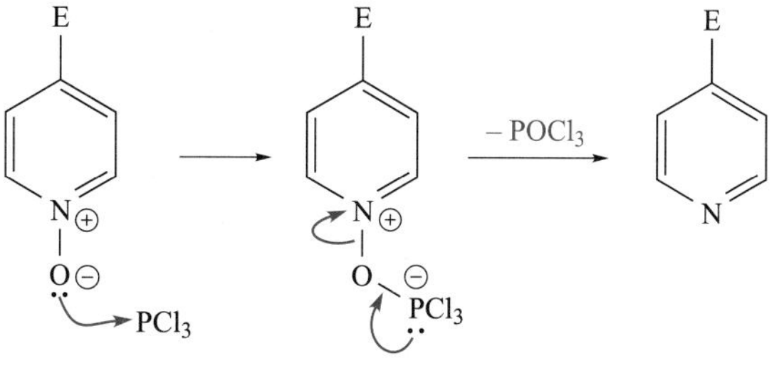

(34.29)

**Fig. 34.7** (a) Pyridine-*N*-oxide has a planar structure (colour code: N, blue; O, red; C, grey; H, white). (b) The $\pi$-system of the pyridine ring is extended to include a $2p$ orbital and a lone pair of electrons from the O atom. This representation of the $\pi$-bonding MO was generated using Spartan '04, © Wavefunction Inc. 2003.

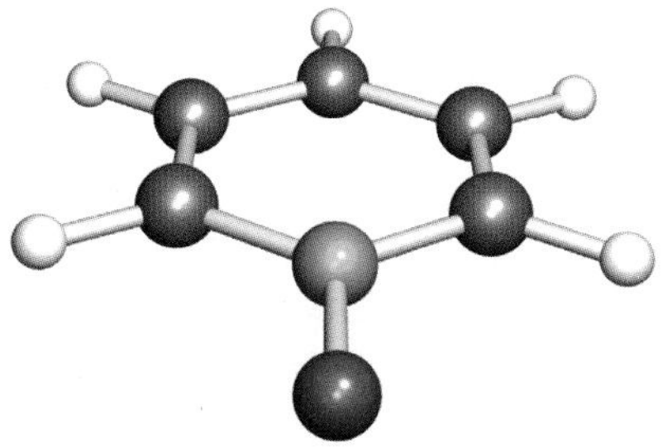

(a)

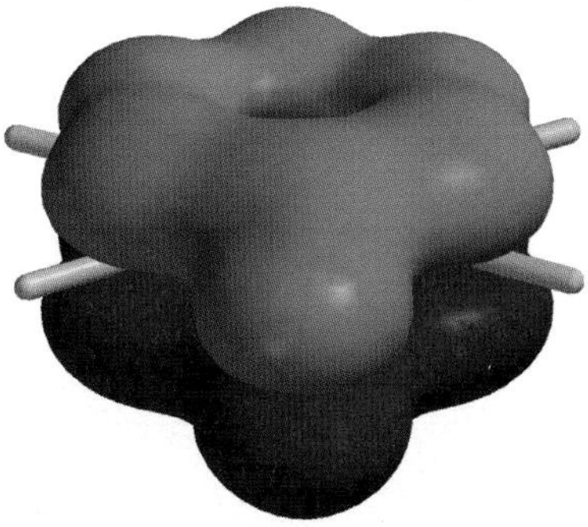

(b)

Electrophilic substitutions that are carried out this way include nitration (equation 34.30). This route gives 4-nitropyridine, compared to the formation of 3-nitropyridine when pyridine is treated with $N_2O_5$ followed by $HSO_3^-$ and $H_2O$ (equation 34.26).

$$(34.30)$$

## Activation of pyridine: use of pyridones

Keto–enol tautomerism: see Section 33.6

Another way of activating pyridine towards electrophilic attack is to use 2- or 4-hydroxypyridine as the starting compound. Each hydroxy derivative behaves similarly to an enol and, thus, there is keto-like form called a *pyridone*. Each of the tautomer equilibria 34.31 and 34.32 lies predominantly over to the right-hand side, although the position of the equilibrium is very sensitive to the solvent and substituents that are present.

$$(34.31)$$

2-Hydroxypyridine          2-Pyridone

$$(34.32)$$

4-Hydroxypyridine          4-Pyridone

The stability of 2-pyridone with respect to 2-hydroxypyridine, and of 4-pyridone with respect to 4-hydroxypyridine, is due to resonance stabilization:

Resonance stabilization of 2-pyridone          Resonance stabilization of 4-pyridone

In contrast, 3-hydroxypyridine exists solely in the enol-like form.

Pyridones react more readily with electrophiles than does pyridine. Equation 34.33 shows a general mechanism for electrophilic substitution in 4-pyridone, with the electrophile, $E^+$, entering *ortho* to the keto-group. Electrophilic substitution in 2-pyridone may occur *ortho* or *para* to the C=O (see end-of-chapter problem 34.17).

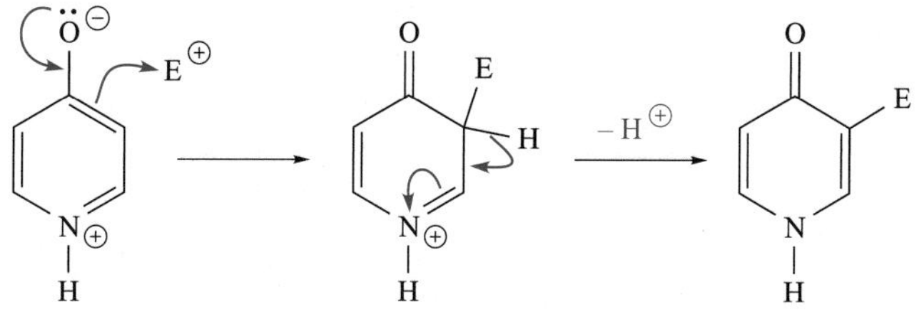

(34.33)

The reaction of a pyridone with $POCl_3$ is an important means of generating a chloropyridine derivative. For example, when 2-pyridone reacts with $POCl_3$, the product is 2-chloropyridine (scheme 34.34). Similarly, 4-pyridone can be converted into 4-chloropyridine. The crucial feature of a chloro-derivative is that $Cl^-$ is a good leaving group and this activates pyridine towards *nucleophilic substitution*.

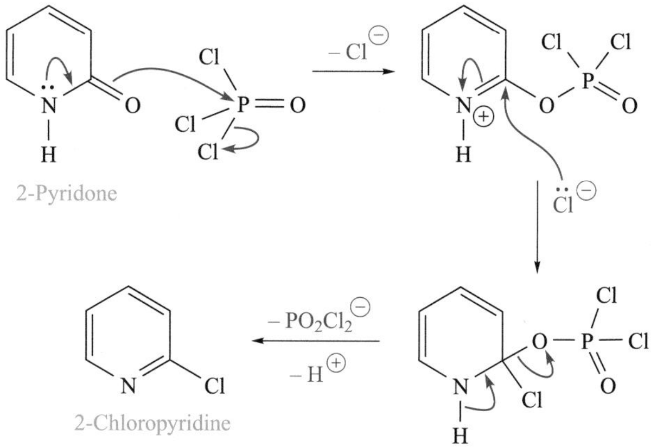

(34.34)

### Reactions of pyridine and chloropyridines with nucleophiles

We have already seen that the electronegativity of the N atom in pyridine means that the ring C atoms are $\delta^+$ and are therefore deactivated towards attack by electrophiles. On the other hand, this charge distribution makes the C atoms susceptible to attack by nucleophiles (equation 34.35 where $X^-$ is a general nucleophile).

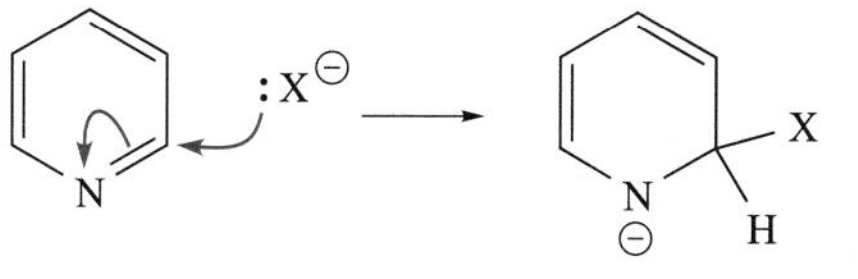

(34.35)

Attack at atoms C(2) and C(4) is more favourable than at the C(3) position. This is readily understood if we look at the resonance stabilization for the intermediate in each reaction. Because the N atom is more electronegative than a C atom, any intermediate in which the negative charge can be localized on nitrogen is favourable:

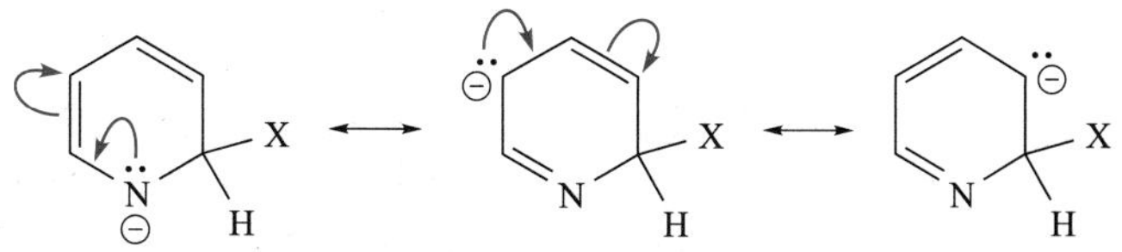

Resonance stabilized intermediate for 2-substitution

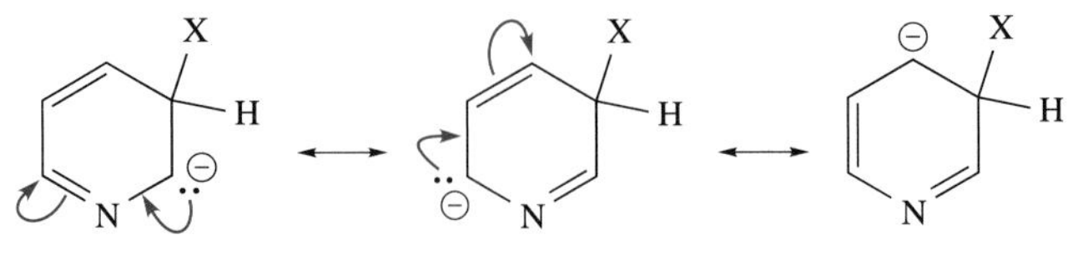

Resonance stabilized intermediate for 3-substitution

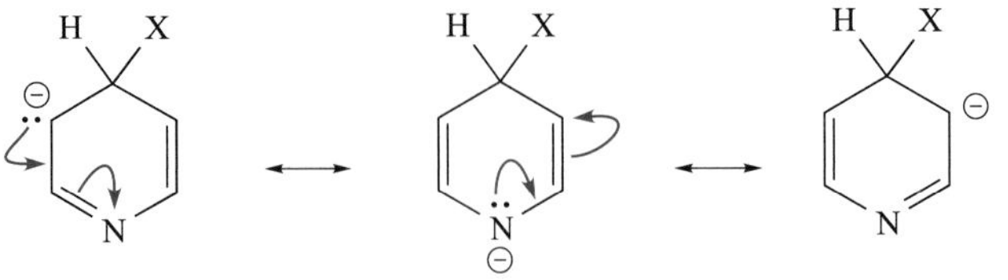

Resonance stabilized intermediate for 4-substitution

The intermediates for substitution at the 2- and 4-positions are therefore energetically more stable than that for 3-substitution. Lowering the energy of an intermediate also lowers the energy of the preceding transition state. This in turn lowers the activation energy of the step (see Figure 34.6 and accompanying discussion).

In a nucleophilic substitution reaction in pyridine, the leaving group is a hydride ion $H^-$. This is a poor leaving group, but its loss is facilitated in the presence of a hydride acceptor. An example is the reaction of pyridine with $NaNH_2$ (sodium amide or sodamide). This is shown in equation 34.36 and is called the *Chichibabin reaction*. The amino-substituted product cannot be isolated initially because it is basic and readily deprotonates under the reaction conditions.

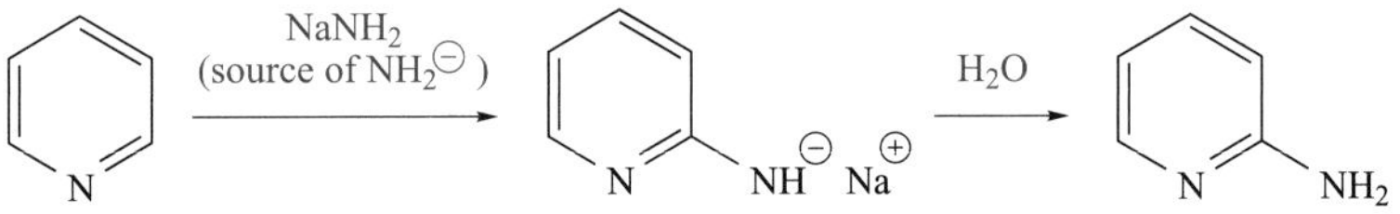

(34.36)

The mechanism of this reaction can be represented as shown below, but this is an oversimplification. Remember that the $NH_2^-$ in the first step is supplied as $NaNH_2$, and the $Na^+$ in the second step comes from the same source:

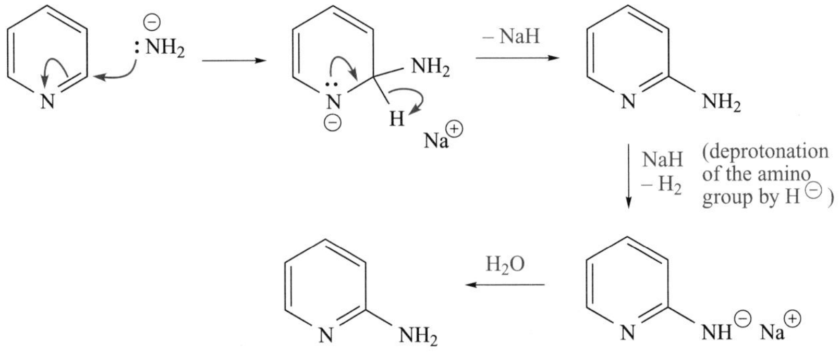

Since $H^-$ is such a poor leaving group, it is not the most useful of precursors for nucleophilic substitutions in pyridine. Chloropyridines (see equation 34.34) are far better choices because $Cl^-$ is a good leaving group. 2-Chloro- and 4-chloropyridine are more reactive with respect to nucleophilic

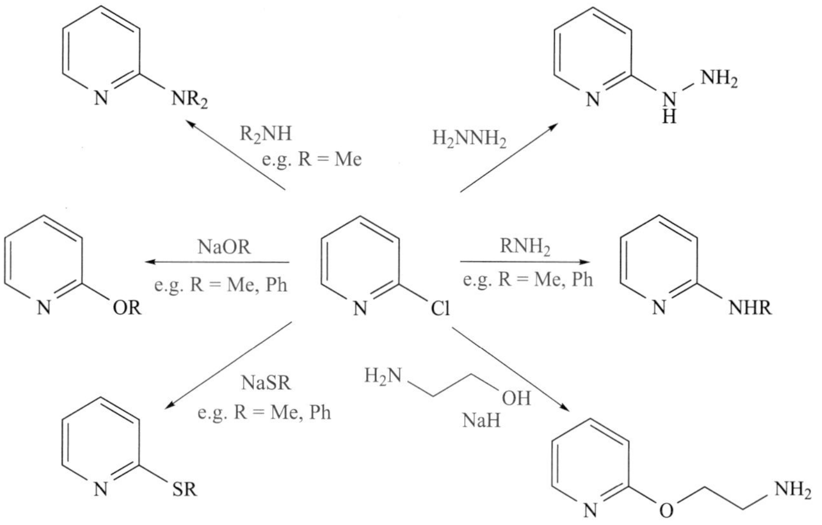

**Fig. 34.8** Representative
nucleophilic substitution
reactions of 2-chloropyridine
illustrating the general use of
chloropyridines in synthesis.

**Fig. 34.8** Representative nucleophilic substitution reactions of 2-chloropyridine illustrating the general use of chloropyridines in synthesis.

substitution than is 3-chloropyridine. Equations 34.37 and 34.38 show general mechanisms for nucleophilic substitution by a nucleophile $X^-$ in 2-chloropyridine and 4-chloropyridine respectively.

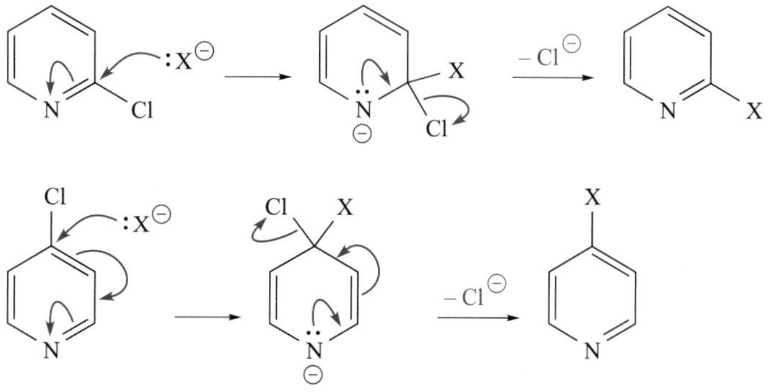

$$(34.37)$$

$$(34.38)$$

Chloropyridines are useful starting materials for synthesizing a range of derivatives of pyridine. Examples starting from 2-chloropyridine are shown in Figure 34.8. Similar reactions can readily be carried out with 4-chloropyridine as the precursor. Bromopyridines are also useful precursors for nucleophilic substitution reactions, and are also used in the lithiation reactions described below.

## Lithiation of pyridine

The lithiation of pyridine is a useful first step for the introduction of substituents at the 2-, 3- or 4-positions of the aromatic ring. Scheme 34.39 exemplifies this, starting from 2-bromopyridine. Equation 34.40 gives a specific example of synthesis involving a lithio-derivative.

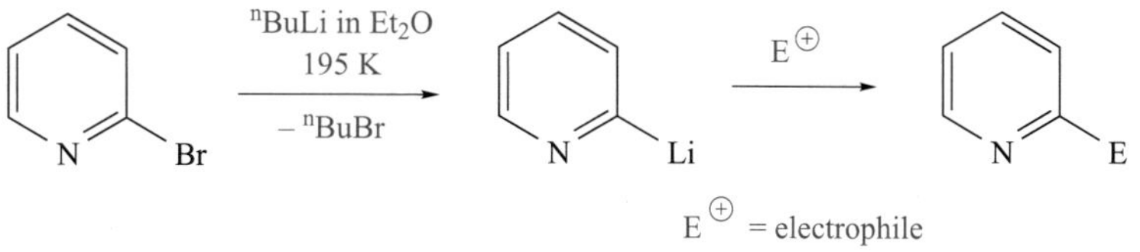

$$(34.39)$$

$E^{\oplus}$ = electrophile

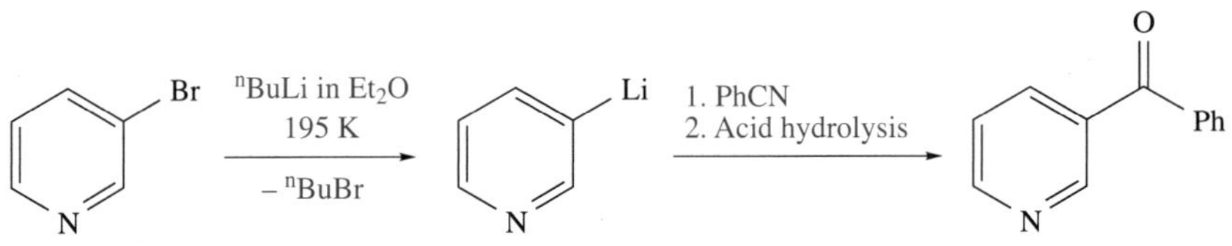

(34.40)

## 34.10 Pyrylium ion: reactivity

The reactions described below provide a selective introduction to the chemistry of the pyrylium ion. Being positively charged, the pyrylium ion is highly deactivated towards reactions with electrophiles. The pyrylium cation reacts readily with nucleophiles and one facile, but reversible, reaction is that with water, which opens the 6-membered ring:

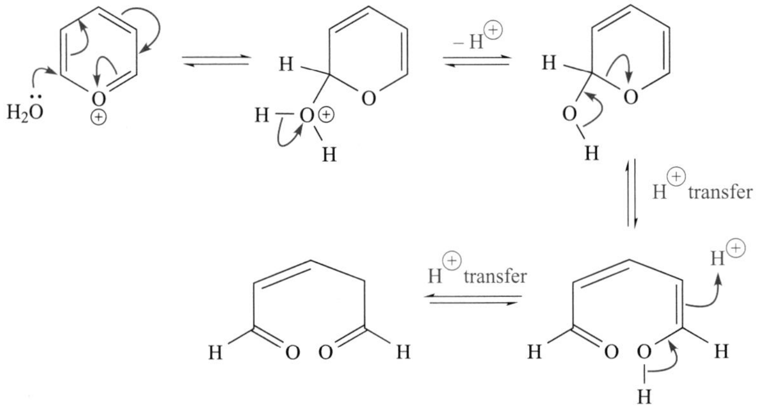

The position of equilibrium is sensitive to pH, and the pyrylium ion can be stabilized by working in acidic media.

Reaction with $NH_3$ converts the pyrylium ion to pyridine (equation 34.41).

Mechanism: see end-of-chapter problem 34.21

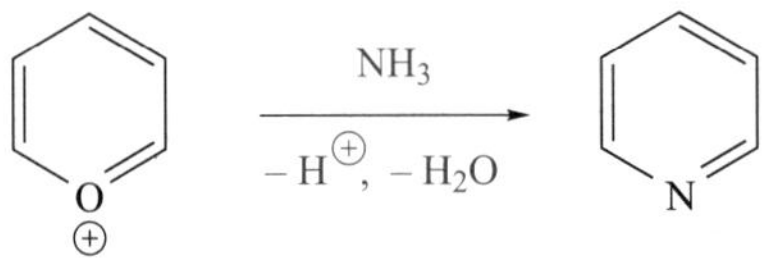

(34.41)

Analogous reactions occur with primary amines, $RNH_2$, to yield pyridinium salts.

## 34.11 Nitrogen-containing heterocycles with more than one heteroatom

In this section, we introduce a number of nitrogen-containing heterocycles that are especially important in biological systems. It is beyond the scope of this book to give more than a superficial coverage, and we focus attention only on structures and why these compounds are important. We meet more nitrogen-containing heterocycles in the next chapter.

## Imidazole

Imidazole

**(34.22)**

Using the isoelectronic principle (Section 34.2), we can replace a CH unit in pyrrole by an N atom and generate *imidazole* (**34.22**). The two N atoms in imidazole are quite different from one another in terms of bonding. One N atom bears an outward pointing lone pair of electrons and contributes one electron to the $\pi$-system. The other is bonded to an H atom and contributes two electrons to the $\pi$-system (Figure 34.9). Thus, like pyrrole, imidazole is aromatic.

In terms of acid–base behaviour, we expect the two N centres in imidazole to behave differently and this is what is observed. The first N atom behaves as a base and can be protonated (equations 34.42 and 34.43), while the NH group can be deprotonated. Imidazole is a very weak acid (equation 34.44), but not as weak as pyrrole (equation 34.7).

$$+ \; H_3O^{\oplus} \; \rightleftharpoons \; + \; H_2O \tag{34.42}$$

$$+ \; H_2O \; \rightleftharpoons \; + \; H_3O^{\oplus} \qquad pK_a = 6.95 \tag{34.43}$$

$$+ \; H_2O \; \rightleftharpoons \; + \; H_3O^{\oplus} \qquad pK_a = 14.5 \tag{34.44}$$

Imidazole can exist in two tautomeric forms but, of course, the two tautomers are identical (equation 34.45). In a derivative such as 4-methylimidazole, the tautomer equilibrium results in rapid interconversion of 4- and 5-methylimidazole (equation 34.46).

$$\rightleftharpoons \tag{34.45}$$

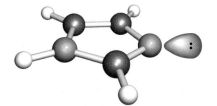

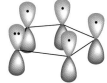

All C and N atoms are *sp²* hybridized

**Fig. 34.9** Overlap of the C and N 2*p* orbitals in imidazole results in delocalized bonding and a 6$\pi$-electron system. The NH group contributes two electrons to the $\pi$-system. The second N atom contributes one electron to the $\pi$-system and has an outward-pointing lone pair of electrons. Colour code: C, grey; N, blue; H, white.

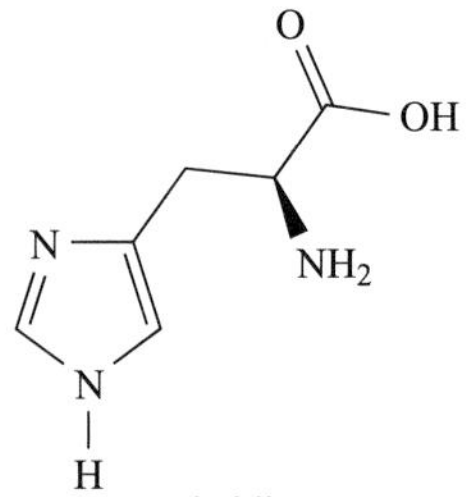

Histidine

(34.23)

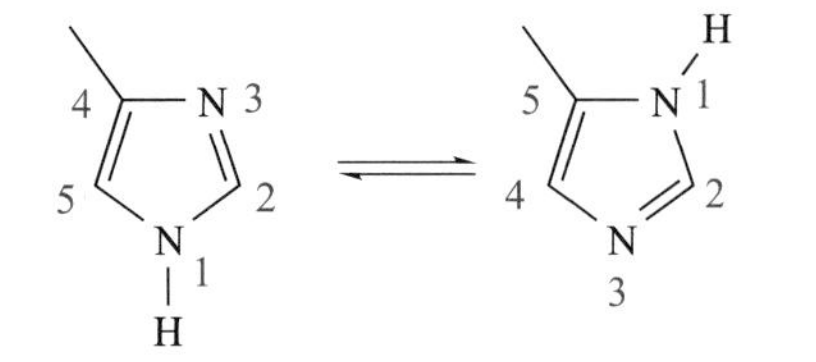

4-Methylimidazole    5-Methylimidazole

(34.46)

Imidazole is an important building block in nature. It occurs in histamine (**34.6**) and in the amino acid histidine (**34.23**) which we have encountered as a protein residue in haemoglobin (see Boxes 5.3 and 16.2). L-Carnosine (β-alanyl-L-histidine, **34.24**) is a dipeptide (see Section 35.5) that occurs in skeletal muscle and functions as a free-radical scavenger. The anti-ulcer drug Cimetidine (see Section 34.1) is also a derivative of imidazole.

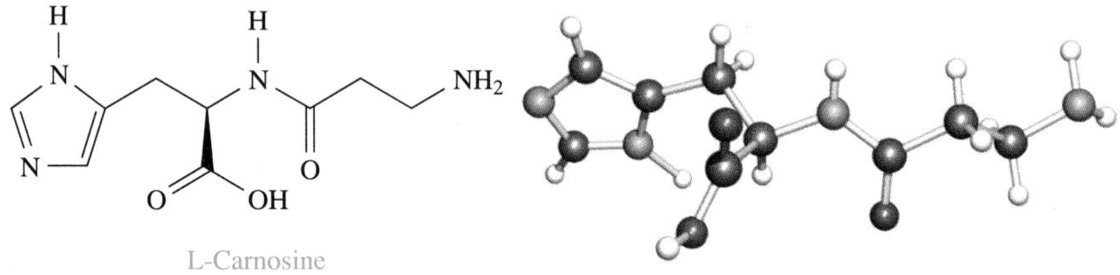

L-Carnosine

(34.24)

### Diazines: pyrimidine, pyridazine and pyrazine

Pyrimidine (Figure 34.10) is a member of the family of the aromatic *diazines* and is isoelectronic with pyridine and benzene. Each atom in the 6-membered ring contributes one electron to the π-system, and the bonding in pyrimidine can be described in an analogous manner to that in pyridine (Figure 34.3). Each N atom carries a lone pair of electrons, accommodated in an outward-pointing $sp^2$ hybrid orbital. Pyrimidine therefore behaves as a base, but it is less basic than pyridine, i.e. the conjugate acid of pyrimidine is a stronger acid than the pyridinium ion (compare equations 34.47 with equation 34.23).

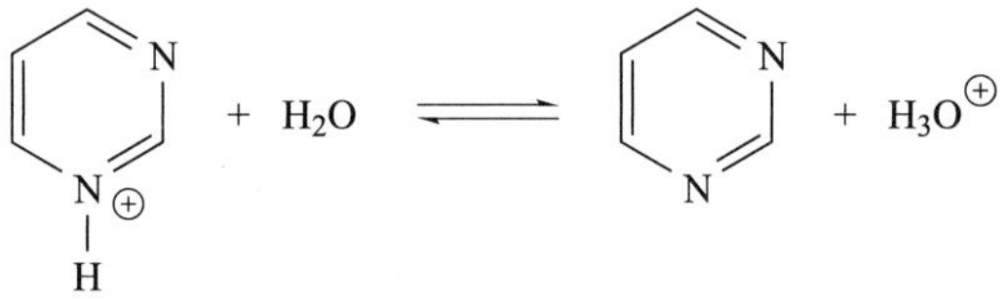

Pyridazine    Pyrazine

(34.25)      (34.26)

$$+ \; H_2O \; \rightleftharpoons \; + \; H_3O^{\oplus} \qquad pK_a = 1.23 \qquad (34.47)$$

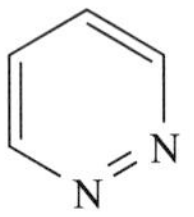

Pyridazine (**34.25**) and pyrazine (**34.26**) are isomers of pyrimidine and also contain aromatic, planar rings. Both act as bases, and the $pK_a$ of the conjugate acids of pyridazine and pyrazine are 2.24 and 0.65, respectively. In nature, pyrimidine is far more important than the other diazines. The nucleobases cytosine, uracil and thymine that we discuss in Section 35.7 are derivatives of pyrimidine and are essential to life. Fusion of pyrimidine to imidazole gives purine (**34.27**), and the other essential naturally occurring nucleobases (adenine and guanine) are derivatives of purine.

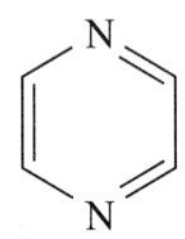

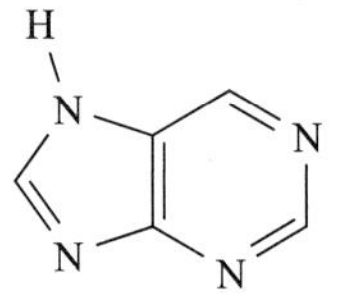

Purine

(34.27)

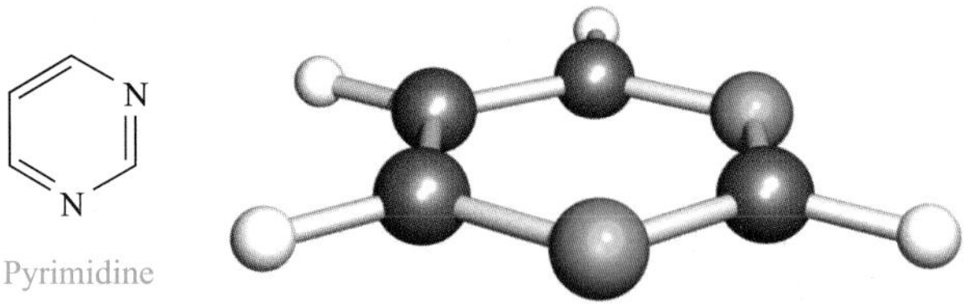

**Fig. 34.10** Pyrimidine is a planar, 6π-electron aromatic molecule. Colour code: C, grey; N, blue; H, white.

Pyrimidine

## SUMMARY

## Summary

In this chapter, we have introduced aromatic heterocyclic compounds, which are of particular significant to biology. Without nitrogen-containing heterocyclic derivatives, our bodies would not function. We have shown how the introduction of an N, O or S heteroatom into a 5- or 6-membered planar, aromatic ring alters both the bonding and reactivity when compared with all-carbon analogues.

### *Do you know what the following terms mean?*

- heteroatom
- heterocycle

- disconnection approach
- imine

- enamine
- hydroxypyridine–pyridone tautomerism

### *You should be able:*

- to give examples of biologically important heterocyclics

- to draw the structures of pyridine, the pyrylium cation, pyrrole, furan and thiophene

- to give bonding descriptions for pyridine and the pyrylium ion and explain why they are aromatic

- to give bonding descriptions for pyrrole, furan and thiophene and state which electrons contribute to their aromatic $\pi$-systems

- to compare the relative aromaticities of pyrrole, furan and thiophene

- to explain in general terms what is meant by the *disconnection approach*

- to give synthetic routes to pyrrole, furan and thiophene

- to describe the acid–base properties of pyrrole

- to give a general mechanism for electrophilic substitution in pyrrole, furan and thiophene

- to give examples of electrophilic substitution reactions in pyrrole, furan and thiophene and comment on substitution patterns

- to illustrate Diels–Alder cycloadditions to furan

- to give synthetic routes to pyridine and the pyrylium ion

- to illustrate how pyridine acts as a base

- to give examples of electrophilic substitution reactions of pyridine and to compare them with analogous reactions of benzene

- to describe ways of activating pyridine towards electrophilic attack

- to illustrate the reactions of pyridine with nucleophiles and explain why halopyridines are better precursors for these reactions than pyridine

- to describe how the pyrylium ion reacts with water and comment on how the heterocycle can be stabilized against such hydrolysis

- to give examples of heterocycles that contain more than one N atom and comment on their biological significance

## PROBLEMS

**34.1** Name the following heterocyclic compounds:

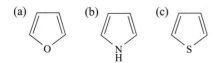

**34.2** Draw the structures of (a) pyridine, (b) indole, (c) the pyrylium cation and (d) imidazole.

**34.3** Draw the structures of (a) 4-methylpyridine, (b) 2-chloropyrrole, (c) 1,4-dihydropyridine, (d) tetrabromopyrrole, (e) nicotinic acid.

**34.4** (a) How many isomers of dimethylpyridine do you expect? Draw their structures and give each a systematic name.
(b) Draw diagrams to show the dipole moments in furan and pyrrole.

**34.5** (a) Explain how thiophene achieves a $6\pi$-aromatic system. (b) What *experimental* evidence is there for the aromatic character of thiophene? (c) Does thiophene possess more or less aromatic character than furan? Rationalize your answer.

**34.6** (a) Using a disconnection approach, suggest suitable precursors for the synthesis of 2,5-dimethylpyrrole. (b) Propose a mechanism for the formation of 2,5-dimethylpyrrole from the starting materials you have suggested.

**34.7** Suggest a mechanism for the formation of furan from butanedial (equation 34.3).

**34.8** Rationalize the following p$K_a$ values *which refer to the conjugate acids of the nitrogen-containing heterocycles* shown:

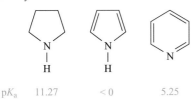

| p$K_a$ | 11.27 | < 0 | 5.25 |

**34.9** Explain why protonation of thiophene by a strong acid leads to the formation of a polymer.

**34.10** Suggest products for the following reactions, and for reaction (a), propose a mechanism.

(a)

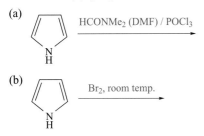

HCONMe$_2$ (DMF) / POCl$_3$

(b)

Br$_2$, room temp.

(c)

Br$_2$ in MeOH

(d)

Acetyl nitrate

**34.11** Using the disconnection approach, illustrate how you might plan a synthesis for pyridine. Rationalize your synthetic strategy.

**34.12** Illustrate the role of pyrylium ion derivatives in certain flower petals.

**34.13** In the synthesis of the pyrylium ion, the penultimate product is a 4*H*-pyran. Draw the structure of 4*H*-pyran and give the conditions under which it can be converted to the pyrylium ion.

**34.14** Suggest products in the following reactions:

(a)

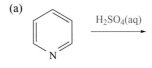

H$_2$SO$_4$(aq)

(b)

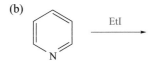

EtI

(c)

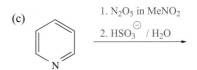

1. N$_2$O$_5$ in MeNO$_2$
2. HSO$_3^{\ominus}$ / H$_2$O

**34.15** Suggest how the following reactions may proceed, paying attention to the stereochemistry of the products.

(a)

(b)

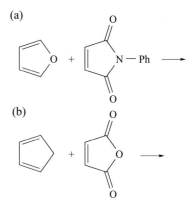

**34.16** (a) Why is the acylation of pyrrole not carried out under Friedel–Crafts conditions?

(b) With reference to the nitration of pyrrole, show how acetyl nitrate behaves as a nitrating agent. Which site in pyrrole is preferentially nitrated?

**34.17** Propose mechanisms for the reaction of 2-pyridone with a general electrophile, $E^+$, to give (a) the 3-substituted and (b) the 5-substituted products.

**34.18** Explain why nucleophilic substitution occurs more readily in 4-chloropyridine than in 3-chloropyridine.

**34.19** Suggest products for the reaction of 4-chloropyridine with (a) $NH_3$, (b) NaSPh, (c) hydrazine, and (d) NaOEt. Give a scheme that shows the mechanism of the reaction between 4-chloropyridine and $NH_3$.

**34.20** Three isomeric chloro-derivatives of pyridine (**A**, **B** and **C**) analyse as containing 40.58% C, 2.04% H and 9.46% N. The $^1$H NMR spectroscopic data for the compounds are as follows where d = doublet, dd = doublet of doublets and t = triplet:

| Compound | $^1$H NMR $\delta$ / ppm |
|---|---|
| A | 7.66 (t, $J = 7.6$ Hz), 7.31 (d, $J = 7.6$ Hz) |
| B | 8.64 (d, $J = 2.1$ Hz), 8.25 (t, $J = 2.1$ Hz) |
| C | 8.70 (dd, $J = 3.0$ and 0.3 Hz), 8.13 (dd, $J = 9.0$ and 3.0 Hz), 7.68 (dd, $J = 9.0$ and 0.3 Hz) |

In each isomer, Cl atoms are in either the 2- or 3-position with respect to the N atom. Suggest structures for **A**, **B** and **C**.

**34.21** Propose a mechanism for reaction 34.41.

## Additional problems

**34.22** Suggest how aniline might react with compound **A** to give compound **B**. Propose a mechanism for the reaction.

**A**        **B**

**34.23** (a) Suggest why 2,6-dichloropyridine is an important intermediate in the manufacturing of organic pharmaceuticals.

(b) Comment on the synthetic usefulness of the reaction of pyridine with $H_2O_2$.

**34.24** For each of the following heterocycles, comment on the ease of electrophilic substitution with respect to analogous reactions of benzene. In which of the heterocycles would the introduction of electron-donating groups aid electrophilic substitutions in the ring? Give examples of electron-donating groups and state the associated orientation effects.

# 35 Molecules in nature

## Topics

- Carbohydrates
- Amino acids
- Peptides
- Proteins
- Nucleic acids

### 35.1 The molecules of life

In this chapter, we consider some molecules that are essential to life. In the last chapter, we looked at a number of heterocyclic compounds that are used by nature as building blocks in biological molecules. We introduced DNA in Box 21.6, where we illustrated the role that hydrogen bonding plays in supporting its structure. The classes of molecule with which we are concerned in the chapter are *carbohydrates*, *amino acids*, *peptides*, *proteins* and *nucleic acids*.

Chiral molecules are everywhere in nature, and you may wish to review Section 24.8 ('Stereoisomerism and diastereoisomers') before studying Chapter 35. In particular, make sure that you are familiar with the prefixes D- and L-, and can apply and interpret sequence rules.

The classes of carbohydrate are *monosaccharides*, *disaccharides* and *polysaccharides*.

We begin with a look at *carbohydrates*. Carbohydrates contain C, H and O and were originally considered as 'hydrates of carbon', e.g. glucose has the formula $C_6H_{12}O_6$ or $(CH_2O)_6$. The family of carbohydrates consists of simple sugars (monosaccharides) and molecules in which the simple sugar units are connected together to give disaccharides and polysaccharides. Two important polysaccharides are starch (with a role in energy storage in living systems) and cellulose (the building material of cell walls in plants).

### 35.2 Monosaccharides

#### Acyclic (open) forms of monosaccharides

Monosaccharides (sugars) are a source of energy in metabolic processes, and include *glyceraldehyde*, *ribose*, *glucose*, *fructose*, *mannose* and *galactose*. Glucose is the most important. The simplest is glyceraldehyde. Structural properties common to monosaccharides are that:

*Monosaccharides* which, in their open form, possess an aldehyde functionality are called *aldoses*; those with a ketone group are *ketoses*, e.g. *glucose* is an aldose and *fructose* is a ketose.

- there is at least one asymmetric carbon atom;
- an aldehyde or a ketone group is present in the 'open form' (see later) of the sugar;
- two or more OH groups are present.

## THEORETICAL AND CHEMICAL BACKGROUND

## Box 35.1 Fischer and Haworth projections

The structures of sugars have traditionally been represented by using *Fischer projections*. A Fischer projection focuses on the open form of the sugar and draws the carbon backbone as a vertical, linear chain. The substituents are drawn on the left and right sides of the chain, and *all* the substituents project out of the plane of the paper towards you. This stereochemistry is understood in the diagram, and is not shown explicitly. For example, the left-hand diagram below shows a Fischer projection for D-glucose. On the right, the diagram is redrawn with stereochemical information about the substituents added:

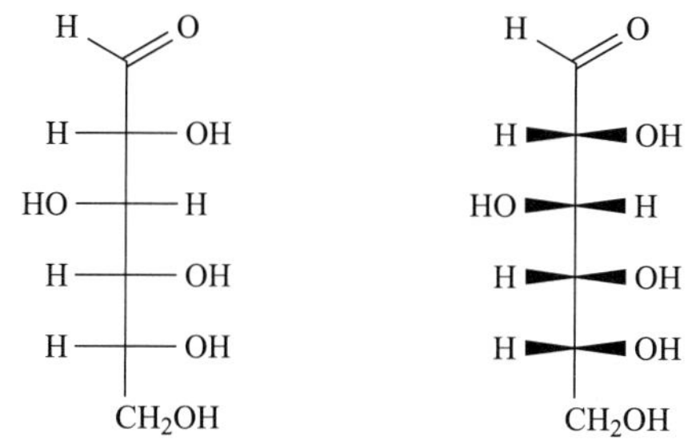

Fischer projection
of D-glucose

In reality, the environment of each C atom is tetrahedral, and so the right-hand structure above translates to the structure below:

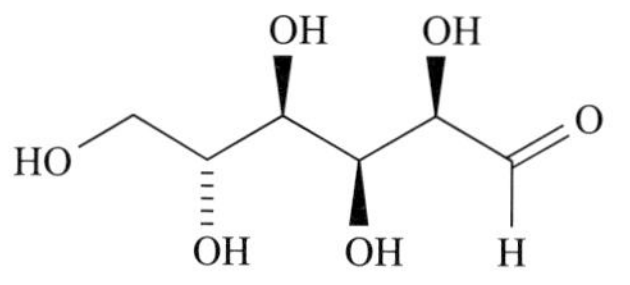

D-Glucose

A ball-and-stick representation of this structure is as follows:

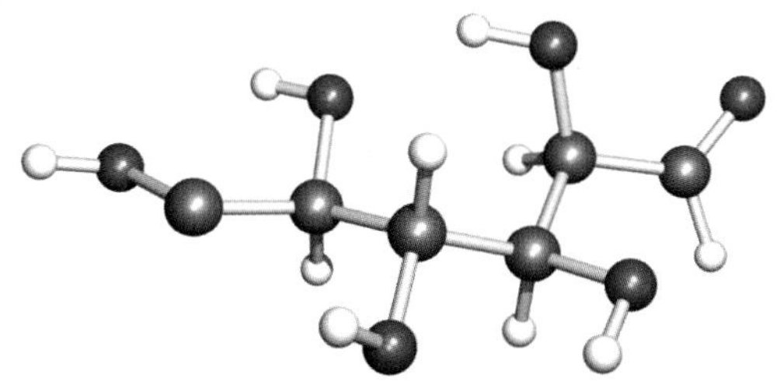

As we discuss in the text, the open form of a sugar is in equilibrium with one or more ring forms. These can be represented by *Haworth projections* in which the ring

conformation is simplified to a planar representation. D-Glucose has two ring forms: the 6-membered ring (*pyranose* form) is greatly favoured over the 5-membered ring (*furanose* form). Haworth projections of the ring forms of D-glucose are shown below (α- and β-forms are discussed in the text):

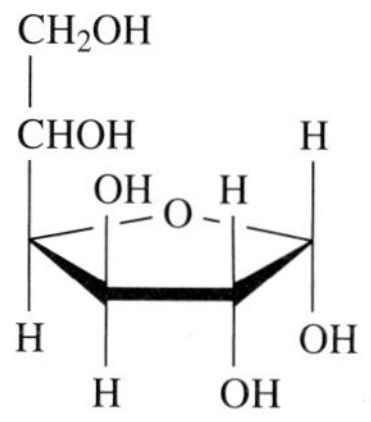

α-Furanose form of
D-glucose

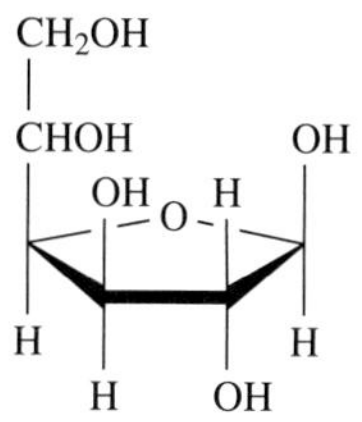

β-Furanose form of
D-glucose

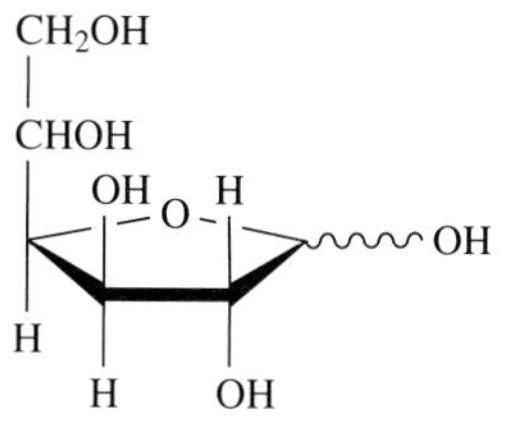

Furanose form of D-glucose: no specified sterochemistry at atom C(1)

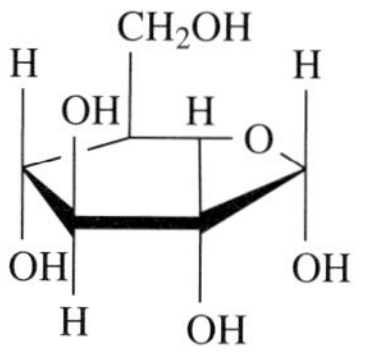

α-Pyranose form of
D-glucose

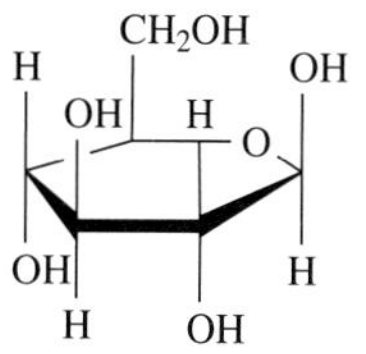

β-Pyranose form of
D-glucose

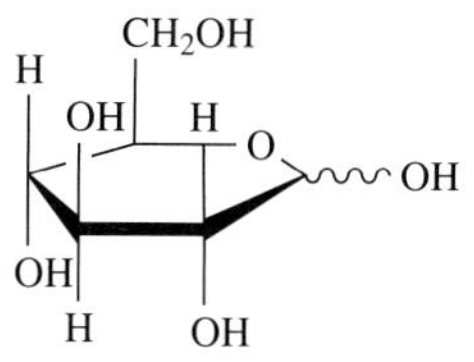

Pyranose form of D-glucose: no specified sterochemistry at atom C(1)

The rings actually adopt non-planar conformations (see Figures 25.3 and 25.4). Of the possible conformation for the 6-membered ring, the chair is energetically favoured. The Haworth projection of the pyranose form of β-D-glucose shown above translates to the following structure:

β-D-Glucose

We restrict our use of Fischer and Haworth projections to this short introduction. In the main text, we use diagrams that clearly show the chain formation in the open forms of sugars, or the chair conformation of the 6-membered rings.

Monosaccharides with an aldehyde functionality in their open form are called *aldoses* and those with a ketone group are *ketoses*.

D- and L-Glyceraldehyde, use of (+) and (−) and sequence rules: see Section 24.8

Glyceraldehyde contains one asymmetric C atom and has two enantiomers **35.1** and **35.2**. The D- and L-descriptors derive from the direction in which a given enantiomer rotates the plane of polarized light. These labels tell us nothing about the absolute configuration. This is given by using sequence rules: (R)- and (S)-glyceraldehyde. The overall information is provided by the names (R)-(+)-glyceraldehyde and (S)-(−)-glyceraldehyde. In general, nature is specific to the formation of D-sugars, and we shall mainly be concerned with these stereoisomers in this chapter.

Nature tends to be specific to the formation of D-sugars.

L-Glyceraldehyde
(S)-(−)-Glyceraldehyde

(35.1)

D-Glyceraldehyde
(R)-(+)-Glyceraldehyde

(35.2)

Extension of the carbon chain, starting from glyceraldehyde and by the introduction of CH(OH) groups, leads to families of monosaccharides called *tetroses* (four C in the chain with two asymmetric C atoms), *pentoses* (five C with three asymmetric C atoms) and *hexoses* (six C with three asymmetric C atoms). Going from glyceraldehyde to a tetrose means going from one to two asymmetric C atoms, and the combinations (R,R), (R,S), (S,R) and (S,S) are possible. Two of these stereoisomers rotate the plane of polarized light in the same direction, and two rotate it in the opposite direction. The naturally occurring (D or +) sugars possess (R,R) and (R,S) centres respectively and are called D-erythrose (**35.3**) and D-threose (**35.4**). In **35.3**, **35.4** and related structures, the carbon chain is numbered from the end carrying the aldehyde functionality.

D-Erythrose
(2R,3R)-(+)-Erythrose

(35.3)

D-Threose
(2R,3S)-(+)-Threose

(35.4)

* = asymmetric C atom

(35.5)

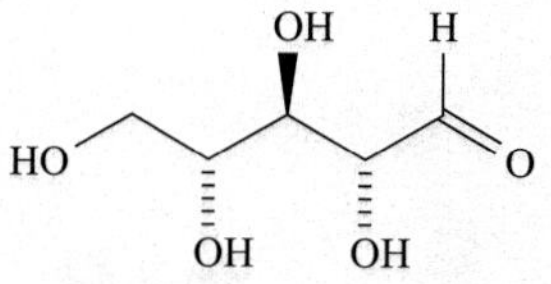

As the chain becomes longer, more stereoisomers are possible. Eight stereoisomers can be drawn for pentose (**35.5**): $(R,R,R)$, $(R,R,S)$, $(R,S,R)$, $(R,S,S)$, $(S,R,R)$, $(S,S,R)$, $(S,R,S)$ and $(S,S,S)$. Four of these are D-sugars and four are L-sugars. The naturally occurring sugars are D-ribose (**35.6**), D-arabinose, D-xylose and D-lyxose (see problem 35.1). Sixteen stereoisomers are possible for hexose (**35.7**), and the eight D-sugars occur naturally. Three of these are shown in structures **35.8**–**35.10**; the remaining monosaccharides are:

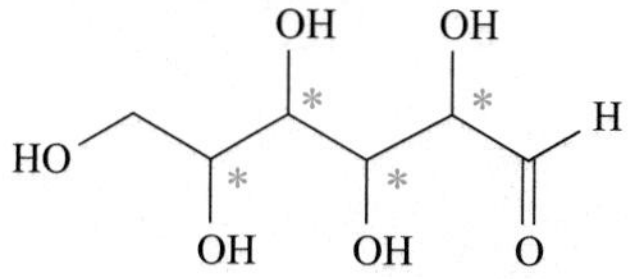

- D-allose or $(2R,3R,4R,5R)$-(+)-allose;
- D-altrose or $(2S,3R,4R,5R)$-(+)-altrose;
- D-gulose or $(2R,3R,4S,5R)$-(+)-gulose;
- D-idose or $(2S,3R,4S,5R)$-(+)-idose;
- D-talose or $(2S,3S,4S,5R)$-(+)-talose.

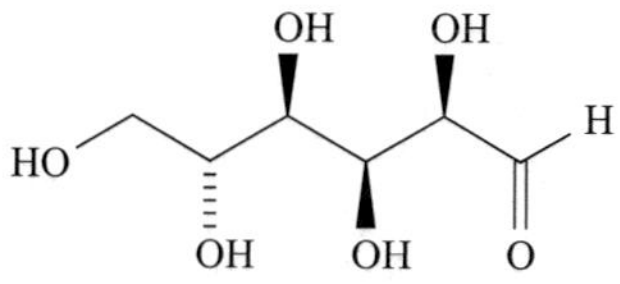

D-Ribose
$(2R,3R,4R)$-(+)-Ribose

(**35.6**)

* = asymmetric C atom

(**35.7**)

D-Glucose
$(2R,3S,4R,5R)$-(+)-Glucose

(**35.8**)

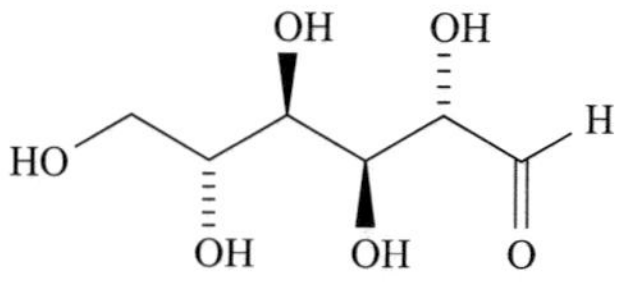

D-Mannose
$(2S,3S,4R,5R)$-(+)-Mannose

(**35.9**)

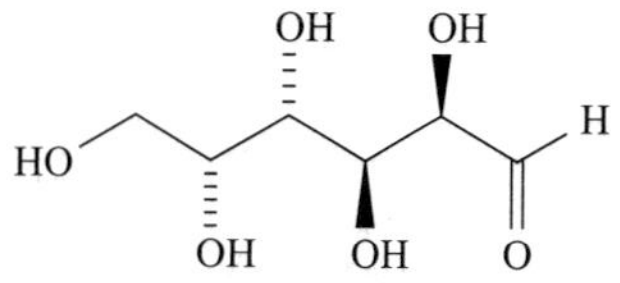

D-Galactose
$(2R,3S,4S,5R)$-(+)-Galactose

(**35.10**)

## Cyclic forms of monosaccharides

In all the structures we have drawn so far, the carbon chain of the mono-saccharide is in an acyclic (open) form. However, for each molecule, an equilibrium exists between the acyclic and one or more cyclic forms. The equilibrium arises because of nucleophilic attack by an OH group on the C atom of the carbonyl group. Cyclization of glucose to a 6-membered ring (a *pyranose form*) is shown below. Ring closure produces a new asymmetric C atom and two stereoisomers can be formed. These are called *anomers* and are labelled α- and β-forms (Figure 35.1):

Look back at Figures 25.5 and 25.6 to remind yourself about axial and equatorial sites

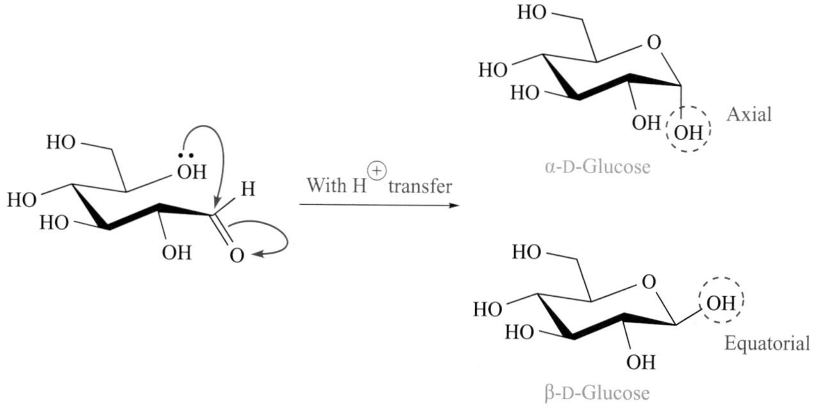

With $H^{\oplus}$ transfer

α-D-Glucose

Axial

β-D-Glucose

Equatorial

This ring closure is summarized in equation 35.1 in which the stereochemistry at the *anomeric centre* is not specified. The 'wavy line' is used to show that the OH group can be in either an axial or equatorial position.

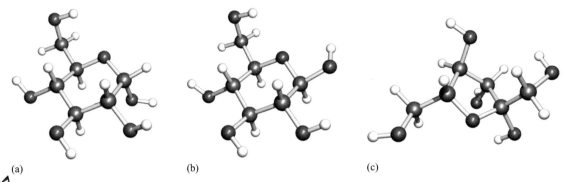

**Fig. 35.1** Ball-and-stick structures of (a) α-D-glucose, (b) β-D-glucose and (c) D-fructose. Colour code: O, red; C, grey; H, white.

$$(35.1)$$

The process is reversible and ring opening is represented in equation 35.2.

$$(35.2)$$

Steps 35.1 and 35.2 combine to give the ring closing and opening equilibrium 35.3. The cyclic form of a sugar is a *cyclic hemiacetal*, and it is the stability of the hemiacetal that pushes equilibrium 35.3 in favour of the cyclic form of glucose.

> Hemiacetal: see towards the end of Section 33.14

$$(35.3)$$

A 6-membered ring form of a sugar is the ***pyranose*** form; a 5-membered ring is the ***furanose*** form. The names derive from the heterocycles pyran and furan:

2*H*-Pyran  4*H*-Pyran

Furan

Formation of a 6-membered ring is not the only possibility. The 6-membered ring forms when the OH attached to atom C(5) of the open chain of D-glucose (look at structure **35.8**) attacks the carbonyl C atom. But if, for example, the OH attached to C(4) were to attack, a 5-membered ring (the *furanose* form of glucose) would result. However, the favoured ring closure is to the pyranose form of D-glucose, and similarly for the other hexoses. For the *pentoses*, formation of a 5-membered ring is the most favourable and equation 35.4 illustrates the formation of the furanose form of ribose:

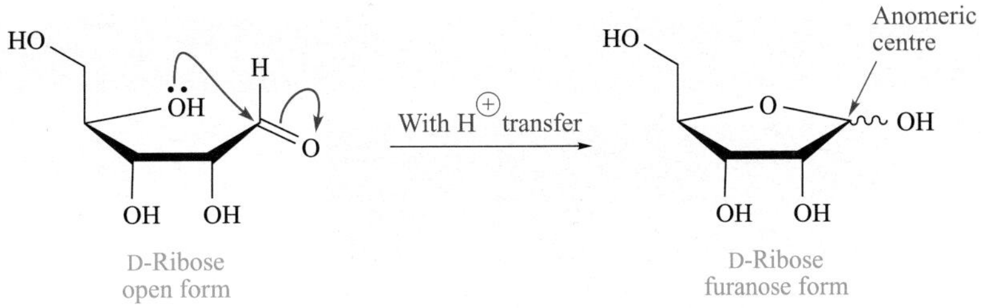

$$(35.4)$$

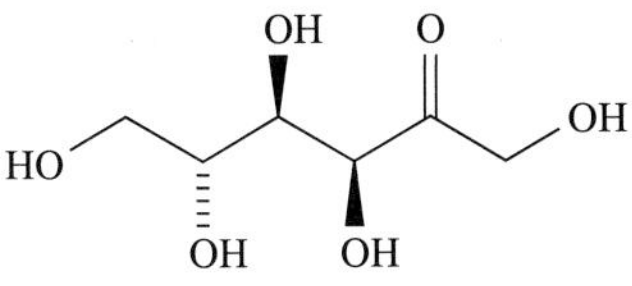

Again, this reaction is reversible, and the open form of ribose is in equilibrium with the furanose form, the cyclic form predominating.

So far, all the monosaccharides that we have considered have been *aldoses*, i.e. the sugar contains an aldehyde group in its open form. The most important *ketose* monosaccharide is *fructose*. Structure **35.11** shows the open form of D-fructose. The furanose form is shown in Figure 35.1c. Note the *puckered ring* conformation.

D-Fructose

**(35.11)**

## Glycoside formation from monosaccharides

The reaction of the cyclic form of a monosaccharide with an alcohol converts the hemiacetal into an acetal (equation 35.5). The acetal is called a *glucoside* and is wholly in the ring form, i.e. there is no equilibrium involving ring and open forms. Acid-catalysed hydrolysis of a glucoside results in the formation of the corresponding monosaccharide.

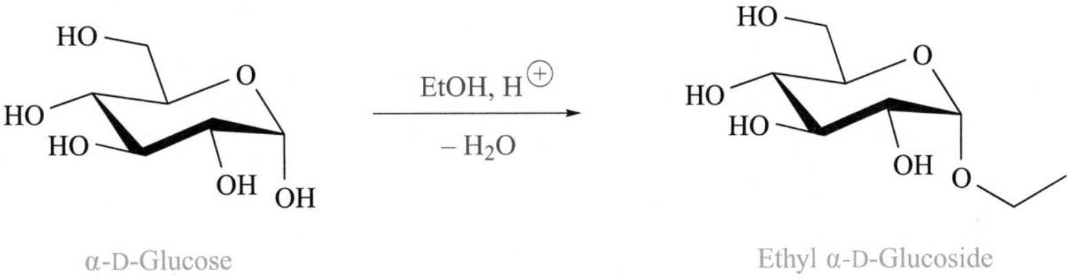

$$(35.5)$$

Glycosides are commonplace in nature (see Box 35.2), and a *glycoside link* connects sugar building blocks to a wide variety of organic units including other sugars. We return to this in Section 35.3.

## Reduction of a monosaccharide

The open form of an aldose or ketose is reduced by $NaBH_4$: the aldehyde or ketone functionality is reduced to a primary or secondary alcohol respectively:

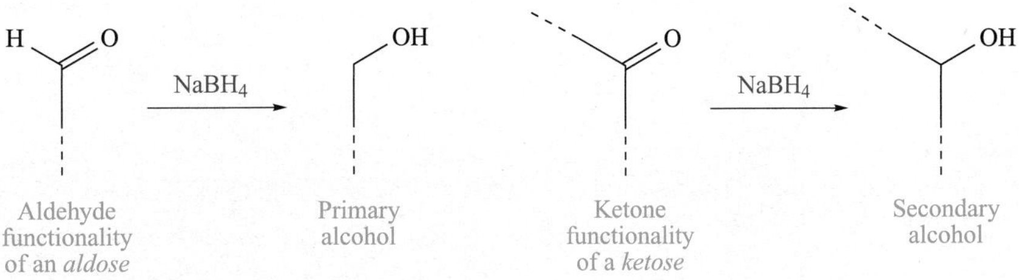

**ENVIRONMENT AND BIOLOGY**

## Box 35.2  Glycosides in nature

Purple foxglove (*Digitalis purpurea*).
© *Angus Beare*.

The term *cardiac glycoside* is given to members of a family of drugs that stimulate the heart. Cardiac glycosides are obtained from several plants including purple foxglove (*Digitalis purpurea*), woolly foxglove (*Digitalis lanata*), yellow foxglove (*Digitalis lutea*), common oleander (*Nerium oleander*), yellow oleander (*Thevetia peruviana*) and lily-of-the-valley (*Convallaria majalis*). Although cardiac glycosides can be used for treatment purposes, if misused, they have toxic effects that can be fatal.

Digitalin (or digoxin, and marketed under this and similar names) is obtained from the leaves of *Digitalis lanata* and is used in the treatment of patients who suffer from a failing heart. Its structure is shown below:

Colour code: C, grey; O, red; H, white. The structure was determined by X-ray diffraction: K. Go *et al.* (1979) *Cryst. Struct. Commun.*, vol. 8, p. 149.

Aldehyde
functionality
of an *aldose*

Carboxylic
acid

**(35.12)**

## Monosaccharides as reducing sugars

The aldehyde group of an aldose is oxidized to a carboxylic acid (diagram **35.12**) by a number of oxidizing agents. Coupled with a visual (e.g. colour) change, this provides the basis for laboratory tests for so-called *reducing sugars*. The oxidizing agents most commonly used are:

- Fehling's solution (aqueous $CuSO_4$, sodium potassium tartrate and NaOH) which turns from blue to a brick-red precipitate of $Cu_2O$;
- Benedict's solution (aqueous $CuSO_4$, $Na_2CO_3$ and sodium citrate) which turns from blue to a brick-red precipitate of $Cu_2O$;
- Tollen's reagent (a solution of $Ag_2O$ in aqueous $NH_3$) which deposits an Ag mirror on reduction from Ag(I) to Ag(0).

**35.3** | ## Disaccharides and polysaccharides

### Disaccharides: sucrose and lactose

A *disaccharide* (e.g. *lactose* and *sucrose*) contains two monosaccharide units connected by a *glycoside linkage*.

C atom numbering in glucose

**(35.13)**

Reaction 35.5 showed the reaction between a monosaccharide and an alcohol to form a glycoside. Sugars are themselves alcohols, and two or more sugars can be connected through glycoside linkages to give disaccharides, trisaccharides and so on. Glycoside bond formation in a disaccharide involves the hemiacetal OH (i.e. the one attached to atom C(1)) of the first sugar and one OH group in the second monosaccharide. In theory, a large number of isomers may form. Using the notation that the unprimed atom (see **35.13**) belongs to the first sugar (in either the α- or β-form) and the primed atom belongs to the second (in either the α- or β-form), possibilities for glycoside bond formation between two hexoses include α-C(1)–α-C(1'), α-C(1)–α-C(2'), α-C(1)–α-C(3'), α-C(1)–α-C(4'), α-C(1)–α-C(6'), β-C(1)–β-C(1'), β-C(1)–β-C(2'), β-C(1)–β-C(3'), β-C(1)–β-C(4') and β-C(1)–β-C(6'). However, in nature di- and polysaccharide assembly is carried out by enzymes and, because they work so specifically, reproducible and specific linkage formation is achieved.

*Lactose* and *sucrose* are examples of disaccharides. Lactose is the main sugar component in milk, and sucrose is the common sugar in plants. Acid-catalysed hydrolysis of disaccharides cleaves the glycoside links and produces the component sugars:

Sucrose     $\xrightarrow{H^{\oplus}, H_2O}$     α-D-Glucose    +    β-D-Fructose

Lactose     $\xrightarrow{H^{\oplus}, H_2O}$     α-D-Galactose    +    β-D-Glucose

Nature provides enzymes that carry out this task. Sucrose is hydrolysed by the enzyme *invertase*, while *lactase* catalyses the hydrolysis of lactose. In biological systems, di- and polysaccharides are a means of storing monosaccharides until they are required, with enzymic hydrolysis releasing them. On a commercial scale, glucose is produced by the acid-catalysed hydrolysis of starch (a polysaccharide, see below). In Box 35.3, we highlight natural and artificial sweeteners.

### Polysaccharides

*Starch, cellulose* and *glycogen* are polysaccharides.

Polysaccharides include *starch, cellulose* and *glycogen*. Starch is continually being made in living cells and is continually being broken down again by enzymic hydrolysis. Starch has two components: *amylose* and *amylopectin*. Amylose is water-soluble. It is composed of 200–1000 α-glucose units connected into *straight chains* by 1,4'-glycoside linkages (see **35.13**):

Part of a chain in amylose

Amylopectin has a paste-like consistency. It consists of *branched chains* of >1000 α-glucose units. 1,4'-Glycoside linkages are present within the straight parts of the chain, but points of branching involve 1,6'-glycoside linkages. Starch is a source of glucose for the body and is broken down by the action of the enzyme *amylase*, present in saliva and the pancreas. Sources of starch in our diets include potatoes, bread, pasta and cereals.

Cellulose is the main component of cell walls in plants. It consists of chains of 500–3000 β-glucose units connected by 1,4'-glycoside linkages:

Part of a chain in cellulose

Despite the similarity in structure between cellulose and amylose, humans have the necessary enzymes to digest only amylose. Herbivores such as sheep and cows, on the other hand, obtain glucose by digesting cellulose.

Glycogen is stored in the liver and is a source of glucose for the body. Its structure is complex, consisting of cross-linked chains of α-glucose units. 1,4'-Glycoside linkages are present within straight chain sections of the structure, while branching involves 1,6'-glycoside linkages. Tens of thousands of α-glucose units are present in a molecule of glycogen and its

## ENVIRONMENT AND BIOLOGY

## Box 35.3  Adding the sweetness factor

Natural sugars include glucose, sucrose (cane and beet sugar), fructose and lactose (milk sugar); honey, for example, is a mixture of glucose and fructose. The term *sweetener* is reserved for food additives that are not natural sugars. Under a European Parliament and Council Directive in June 1994, certain permitted sweeteners for use in foodstuffs have been designated and these (identified by their *E-numbers*) are:

E420  Sorbitol (both sorbitol and sorbitol syrup)
E421  Mannitol
E950  Acesulfame-K ($K^+$ salt of acesulfame)
E951  Aspartame
E952  Cyclamic acid and its $Na^+$ and $Ca^{2+}$ salts
E953  Isomalt
E954  Saccharin and its $Na^+$, $K^+$ and $Ca^{2+}$ salts
E957  Thaumatin
E959  Neohesperidine DC
E965  Maltitol (both maltitol and maltitol syrup)
E966  Lactitol
E967  Xylitol

Sweeteners are classified as either *bulk or intense sweeteners*. Sorbitol, mannitol, isomalt, maltitol, lactitol and xylitol come into the 'bulk' category, and the remaining sweeteners listed above are 'intense'. Sorbitol occurs naturally in a range of fruits and berries, e.g. berries of the mountain ash tree (*Sorbus aucuparia*).

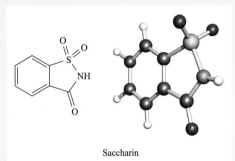

Saccharin

Aspartame is sold under several tradenames, e.g. NutraSweet and Spoonful. Although approved for use as an additive in foods and drinks, aspartame remains controversial. There is evidence that it can cause numerous side effects including headaches, nausea, depression, fatigue and insomnia, as well as more serious conditions. For updated information, see: http://www.food.gov.uk/safereating/additivesbranch/sweeteners/

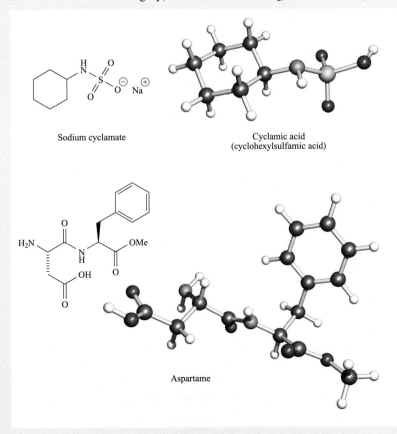

Sodium cyclamate

Cyclamic acid
(cyclohexylsulfamic acid)

Aspartame

molecular weight is $\approx 4 \times 10^6$. In the body, *glucosidases* are the enzymes responsible for breaking down glycogen into glucose.

## Amino acids

### Structure

**(35.14)**

Amino acids are the building blocks of proteins and have the general structure **35.14** where R is H or an organic substituent. The amino acid *proline* is an exception because the amino N atom is part of a heterocycle. There are 20 naturally occurring $\alpha$-amino acids and these are shown in Table 35.1; the $\alpha$-label refers to the $\alpha$-C atom to which the R group in **35.14** is attached. Except for R = H, all amino acids are chiral, because the tetrahedral C atom at the centre of structure **35.14** is an asymmetric centre. Nature is specific and uses only L-amino acids. In all but *cysteine*, the asymmetric C atom has an ($S$)-absolute configuration. Cysteine contains an SH group close to the asymmetric C atom, and the ($R$)-configuration results from the high priority of the S atom in the sequence rules. Thus, cysteine is not actually unusual among nature's amino acids.

### Zwitterionic character

A *zwitterion* carries both positive and negative charges but is neutral overall, e.g.

Table 35.1 shows that every amino acid contains at least one $NH_2$ (basic) and one $CO_2H$ (acidic) group. As with other Brønsted acids and bases, amino acids are protonated or deprotonated depending on the pH of the solution (see Chapter 16). However, at a particular pH known as the *isoelectric point*, an amino acid exists as a neutral *zwitterion* in which a proton from the $CO_2H$ group is transferred to the $NH_2$ group. The equilibrium is shown in equation 35.6 for glycine, and similar equilibria exist for all amino acids. The isoelectric points for the amino acids in Table 35.1 are all similar, and at physiological pH ($\approx$7), they exist as zwitterions. Amino acids crystallize in the zwitterionic state, and there are significant electrostatic interactions between molecules which result in the formation of a well-ordered structure. The same interactions are responsible for ordered structures in proteins.

$$H_2N \diagup CO_2H \rightleftharpoons \overset{\oplus}{H_3N} \diagup CO_2^{\ominus} \tag{35.6}$$

The zwitterion responds to a change in pH by accepting or losing $H^+$. In acidic solution, the $CO_2^-$ group of the zwitterion acts as a base and accepts $H^+$, while in basic solution, the $NH_3^+$ group acts as an acid and donates $H^+$:

$$H_2O + H_2N \diagup CO_2^{\ominus} \underset{\text{In basic solution}}{\overset{HO^{\ominus}}{\rightleftharpoons}} \overset{\oplus}{H_3N} \diagup CO_2^{\ominus} \underset{\text{In acidic solution}}{\overset{H^{\oplus}}{\rightleftharpoons}} \overset{\oplus}{H_3N} \diagup CO_2H$$

Monosodium L-glutamate

Salts of amino acids can therefore be formed, and one that is well known is the sodium salt of glutamic acid, i.e. monosodium glutamate (MSG). Sold as a food additive to intensify the flavour of foods (especially oriental cuisine), commercial MSG contains $\approx 75\%$ free glutamic acid in addition to the sodium salt. Manufactured glutamic acid is the D-enantiomer as opposed to the natural L-form. Many consumers of MSG suffer adverse reactions, and the addition of MSG to foods is a controversial issue.

### Acidic, basic and neutral amino acids

Table 35.1 classifies amino acids according to their acidic, basic or neutral character. To understand this, we need to focus on the R substituent (see diagram **35.14**).

Aspartic acid and glutamic acid are listed as being acidic, and this is also implied by their names. Both possess carboxylic acid functionalities *in addition* to the $CO_2H$ of the general amino acid **35.14**. The $pK_a$ values for the $CO_2H$ groups in the R substituents are 4.00 (aspartic acid) and 4.31 (glutamic acid); compare these values with that of 4.77 for acetic acid.

The basic amino acids in Table 35.1 are arginine, histidine and lysine. Each has at least one amino group *in addition* to that of the general amino acid **35.14**. At pH 7, the amino groups in arginine and lysine are essentially fully protonated, while histidine exists in an equilibrium between non-protonated and protonated forms in an approximately 10:1 ratio.

| 35.5 | **Peptides** |
| --- | --- |

With a library of 20 amino acids, nature can assemble an amazing variety of different proteins. The key step is the formation of a *peptide link* between two amino acids. In the remaining part of this chapter, we draw structures of amino acids in non-zwitterionic forms, but you should keep in mind how these structures change as the pH varies.

### Formation of a dipeptide

A *dipeptide* consists of two *amino acid residues* connected by a *peptide link*.

The carbonyl C atom of an amino acid is susceptible, *after suitable activation*, to attack by the $NH_2$ group of another amino acid. Reaction 35.7 summarizes the *condensation reaction* between two different amino acids to form a *peptide link*.

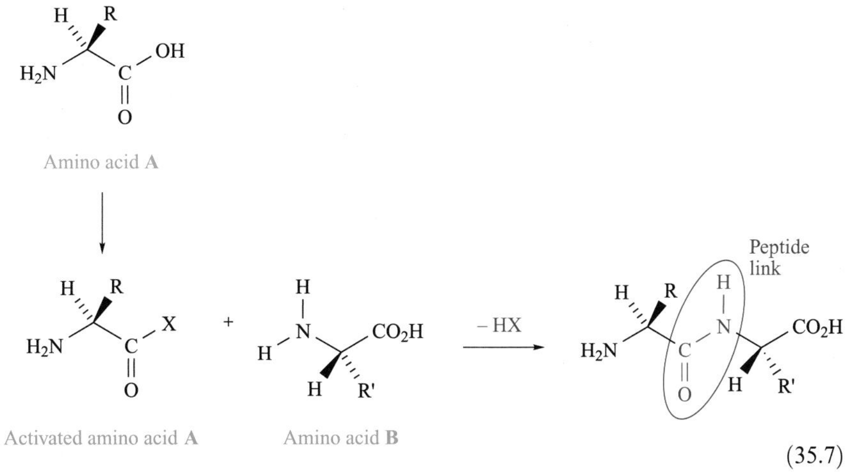

Amino acid **A**

Activated amino acid **A**     Amino acid **B**

(35.7)

**Table 35.1** The 20 naturally occurring amino acids.

| Name of amino acid | Abbreviation for amino acid residue (abbreviation used in sequence specification) | Structure | Acidic, neutral or basic |
|---|---|---|---|
| L-Alanine | Ala (A) | | Neutral |
| L-Arginine | Arg (R) | | Basic |
| L-Asparagine | Asn (N) | | Neutral |
| L-Aspartic acid | Asp (D) | | Acidic |
| L-Cysteine | Cys (C) | | Neutral |
| L-Glutamic acid | Glu (E) | | Acidic |
| L-Glutamine | Gln (Q) | | Neutral |
| Glycine | Gly (G) | | Neutral |
| L-Histidine | His (H) | | Basic |
| L-Isoleucine | Ile (I) | | Neutral |

**Table 35.1** *Continued.*

| Name of amino acid | Abbreviation for amino acid residue (abbreviation used in sequence specification) | Structure | Acidic, neutral or basic |
|---|---|---|---|
| L-Leucine | Leu (L) | | Neutral |
| L-Lysine | Lys (K) | | Basic |
| L-Methionine | Met (M) | | Neutral |
| L-Phenylalanine | Phe (F) | | Neutral |
| L-Proline | Pro (P) | | Neutral |
| L-Serine | Ser (S) | | Neutral |
| L-Threonine | Thr (T) | | Neutral |
| L-Tryptophan | Trp (W) | | Neutral |
| L-Tyrosine | Tyr (Y) | | Neutral |
| L-Valine | Val (V) | | Neutral |

(35.15)

There are several ways in which the amino acid can be activated. Examples are conversion to the acyl chloride (X = Cl in equation 35.7), or to an acid anhydride, e.g. **35.15**.

Reaction 35.7 may look straightforward, but there are problems with competitive reactions. The activated amino acid **A** could react with amino acid **B** or with another molecule of the activated amino acid **A**. To ensure that the correct nucleophile attacks the carbonyl C atom, the $NH_2$ group in amino acid **A** must be protected in such as way that it cannot act as a nucleophile. A common protecting group for the $NH_2$ group is *tert*-butoxycarbonyl (BOC). Reaction 35.8 shows the protection of L-alanine. Deprotection is achieved by treatment with acid.

(35.8)

A further complication in peptide synthesis is that the wrong carbonyl C atom may be subject to nucleophilic attack. In addition to the $CO_2H$ group in amino acid **B** in equation 35.7, some amino acids contain further groups (e.g. the additional $CO_2H$ group in glutamic acid) that may be susceptible to nucleophilic attack. To ensure attack at a specific site, one $CO_2H$ group is protected. This is done by converting it into, for example, a *tert*-butyl ester, the steric crowding in which hinders the approach of a nucleophile. Reaction 35.9 shows the protection of the $CO_2H$ group in L-valine. Deprotection is achieved by treatment with acid.

*tert*-Butyl ester protecting group

(35.9)

Two suitably protected amino acids react to give an initially protected dipeptide from which the protecting groups are then cleaved. Consider the formation of the dipeptide formed from L-alanine and L-valine. We represent this in the scheme below by showing the amine-protected and activated L-alanine reacting with the carboxylic acid-protected L-valine (see top of next page).

Dipeptides are named by a combination of the constituent amino acid names, e.g. alanylvaline, but are more often referred to by combining the abbreviations for the component amino acids, e.g. Ala-Val. The formation of a peptide link between the $NH_2$ end of one amino acid and the $CO_2H$ end of the other means that the dipeptide (and, indeed, any polypeptide)

A peptide chain has an *N-terminus* (corresponding to an $NH_2$ group) and a *C-terminus* (corresponding to a $CO_2H$ group).

A ball-and-stick model of the dipeptide alanylvaline (Ala-Val). Notice that the peptide link (CONH) is planar (see structure **33.27**)

Activated and BOC-protected L-alanine (Ala)

*tert*-Butyl ester-protected L-valine (Val)

$- HX$

Deprotection

The dipeptide Ala-Val

has a $CO_2H$ group at one end of the chain and an $NH_2$ group at the other. These are called the *C*-terminus and the *N*-terminus respectively.

## Polypeptide synthesis

By extrapolation from the dipeptide synthesis, it is easy to see that the synthesis of a polypeptide is non-trivial. Not only are there the competitive reactions described above to consider, but there is also the fact that the amino acid residues must be *sequenced* in a specific order. Automated procedures such as the Merrifield synthesis are now common practice. The Merrifield synthesis makes use of a support of polystyrene beads to which the peptide chain is anchored during growth. The method is summarized in Figure 35.2.

## Polypeptides in nature

The number of amino acid residues in naturally occurring polypeptides varies enormously. The examples below represent only a very few of the many polypeptides that operate in living cells. *Glutathione* (glutamyl-cysteinylglycine) contains only three residues: γ-Glu-Cys-Gly (**35.16**). Notice that the peptide link to the Glu residue involves the γ-$CO_2H$ and not the α-$CO_2H$ group; the latter is more usual. Glutathione is present in all cells and has an important role, reducing Cys-Cys disulfide bridges in proteins. This process can be represented as follows:

Robert B. Merrifield (1921–   ). © *The Nobel Foundation.*

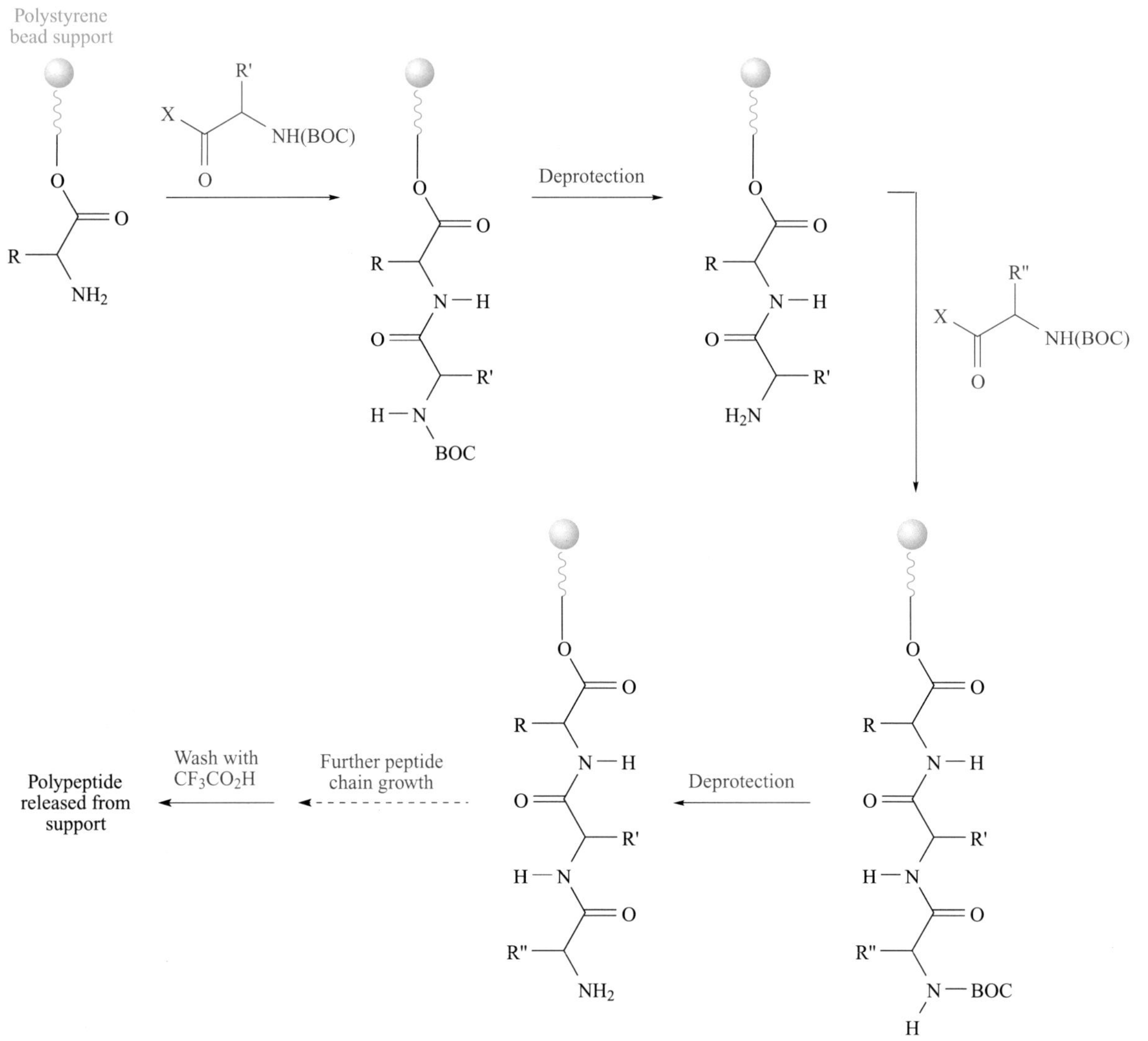

**Fig. 35.2** Schematic representation of the Merrifield synthesis of a sequenced polypeptide.

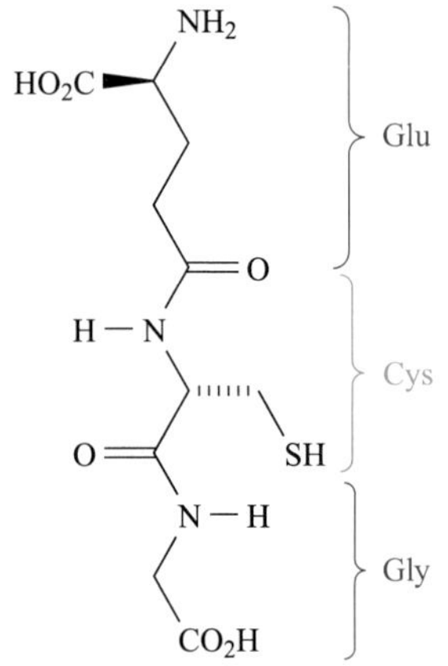

Glutathione

**(35.16)**

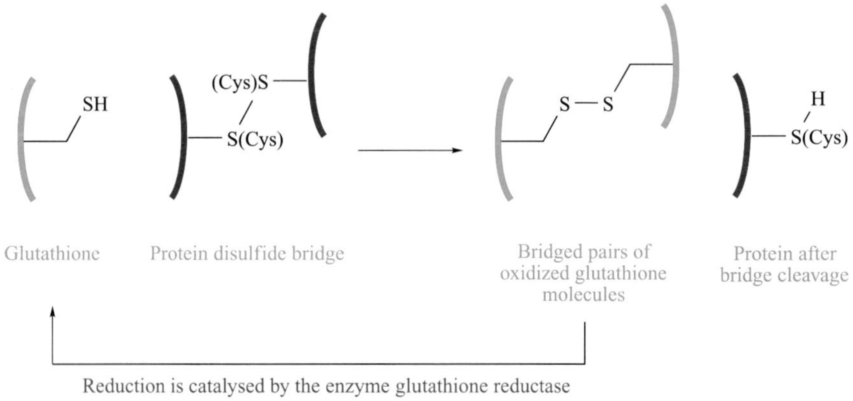

The reversible reduction of glutathione disulfide shown above is coupled to the oxidation of NADPH (nicotinamide adenine dinucleotide phosphate) and is catalysed by the enzyme glutathione reductase (Figure 35.3). The active site in the enzyme contains the coenzyme FAD (flavin adenine dinucleotide) which is involved in many redox processes in the human

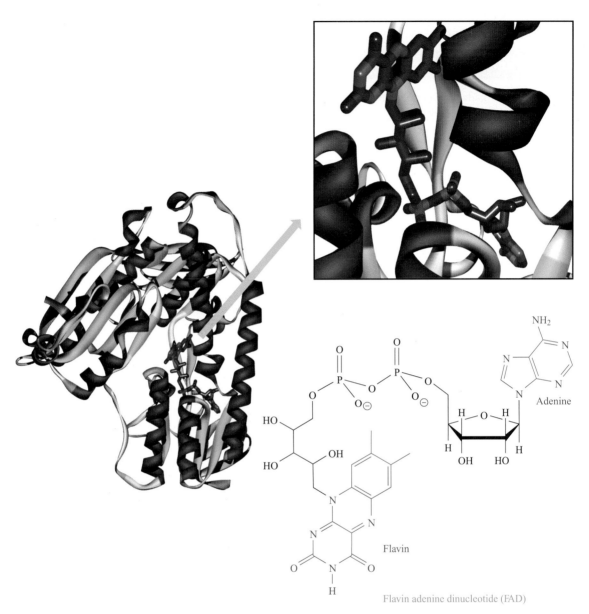

**Fig. 35.3** The structure of the enzyme glutathione reductase in its oxidized form, determined by X-ray diffraction [P. A. Karplus *et al.* (1987) *J. Mol. Biol.*, vol. 195, p. 701]. The protein is shown in a ribbon representation and the FAD molecule in the active site is shown in a stick representation. An enlargement of the active site is shown at the top right.

body. FAD is a derivative of a ribonucleotide and we return to these molecules in Section 35.7.

*Gastrin* contains 17 amino acid residues:

Tyr-Gly-Pro-Tyr-Met-Glu-Glu-Glu-Glu-Glu-Ala-Tyr-Gly-Tyr-Met-Asp-Phe

$N$-terminus

and at appropriate concentrations, it stimulates the secretion of gastric juices. *Vasopressin* is a cyclic polypeptide with the ring closure arising from the formation of a disulfide (S–S) bridge between two cysteine residues:

Cys-Tyr-Phe-Gln-Asn-Cys-Pro-Arg-Gly

$N$-terminus

## ENVIRONMENT AND BIOLOGY

## Box 35.4 Bullet-proof vests and spider silk

Kevlar is a man-made fibre (a polyamide) that is used in the manufacture of bullet-proof vests and other shock-resistant materials.

Kevlar

As is often the case, nature was there before man, producing a fibre of high tensile strength and great elasticity. The golden orb spider (*Nephila clavipes*) produces different forms of silk including dragline silk, which is used for safety lines. This is composed mainly of proteins. The relationship to Kevlar is that a protein consists of a sequence of amino acids, condensed to give a polymer:

$H_2N$ ▢ $CO_2H$    $H_2N$ ■ $CO_2H$    $H_2N$ ▢ $CO_2H$

$- H_2O$

The tensile strength of spider silk is only a quarter that of Kevlar, although it takes more energy to break the spider silk ($\approx 10\,J\,m^{-3}$) than Kevlar ($\approx 3\,J\,m^{-3}$).

Male and female golden orb spiders (*Nephila clavipes*) resting on a web; the male is much smaller than the female.  *William Ervin/Science Photo Library.*

Vasopressin is a hormone that is secreted by the pituitary gland and acts as an antidiuretic. Polypeptides that are antibiotics include *bacitracin* and *polymyxin*.

Peptide sequencing (i.e. working out the sequence of amino acid residues in a polypeptide chain) is carried out by degrading the polypeptide into the component amino acids and analysing them by chromatographic techniques.

### Electrophoresis

*Electrophoresis* is the movement of charged species through a solution under the influence of an applied electric field. The method can be used to separate polypeptides and proteins.

In a polypeptide or protein (see Section 35.6), there are often more acidic sites than the *C*-terminus, and more basic sites than the *N*-terminus. For example, whenever lysine or aspartic acid residues are present, additional amino and carboxylate groups, respectively, are incorporated into the structure. We have already discussed the isoelectric point of an amino acid, i.e. the pH at which it exists primarily as a neutral zwitterion. A similar situation exists for polypeptides and proteins, but it is more complex because there is often more than one basic and one acidic site. Whereas the isoelectric points of amino acids are very similar, this is not likely to be the case for polypeptides or proteins. It should therefore be possible to separate proteins on the basis of their isoelectric points. The technique by which this is carried out is called *electrophoresis*. A sample containing a mixture of proteins (or polypeptides) is placed on a gel that contains an electrolyte and a buffer solution. The pH is controlled by the buffer, and proteins exist with an excess of $NH_3^+$ or $CO_2^-$ sites, depending on their isoelectric points (unless the isoelectric point matches the pH exactly, in which case the protein exists as a zwitterion). A potential is then applied across the gel. Positively charged proteins migrate towards the negatively charged electrode, and negatively charged proteins move in the opposite direction towards the positive electrode. At a specific pH and potential, a given protein travels a given distance towards the anode or cathode. After protein migration, the gel is treated with a reagent that forms coloured compounds with the proteins (or the proteins are labelled in some way so that they can be visualized). This permits the protein migration patterns to be easily monitored, and makes electrophoresis a sensitive method for protein characterization.

Buffers: see Section 16.7

## 35.6   Proteins

*Proteins* are high molecular mass polypeptides with complicated structures. The transition from polypeptides to proteins can be illustrated by considering the structure of insulin. Human *insulin* is secreted by the pancreas and controls sugar levels in the body. Insulin consists of two polypeptide chains (labelled A and B) that are linked by two disulfide (S–S) bridges. Polypeptide A contains 21 amino acid residues, and chain B is made up of 30 residues:

Gly-Ile-Val-Glu-Gln-Cys- Cys-Thr-Ser-Ile-Cys-Ser-Leu-Tyr-Gln-Leu-Glu-Asn-Tyr- Cys-Asn

Phe-Val-Asn-Gln-His-Leu- Cys-Gly-Ser-His-Leu-Val-Glu-Ala-Leu-Tyr-Leu-Val- Cys-Gly-Glu-Arg-Gly-Phe-Phe-Tyr-Thr-Pro-Lys-Thr

The cross-linking of polypeptide chains, especially through disulfide bridges, is a means by which many *proteins* are assembled from peptides.

The sequence of amino acids gives the *primary structure* of the protein, while the spatial properties of the peptide chains are described by the *secondary and tertiary structures*. The secondary structure takes into account the folding of polypeptide chains into domains called α-*helices*, β-*sheets*, *turns* and *coils*. In the ribbon representations of all the protein structures illustrated in this book (excluding Box 5.3), the same colour coding is used for these features: α-helices are shown in red, β-sheets in cyan, turns in green, and coils in silver-grey. The way in which α-helices are assembled is described later in the chapter. Additional intra-chain interactions that are responsible for chain folding fall into the tertiary structure description. Disulfide (S–S) bridges and hydrogen-bonded interactions play essential roles in determining and maintaining the folded protein structure. Some proteins contain more than one polypeptide chain, and the *quaternary structure* describes how inter-chain interactions give rise to the overall protein structure. Some proteins contain a *prosthetic group*. This is an additional, non-amino acid component of the protein, which is essential for its biological activity. Where the prosthetic group involves a metal centre, the protein is called a *metalloprotein*. An example is haemoglobin which contains an iron-containing haem unit (see Box 5.3).

*Proteins* are high molecular mass polypeptides with complex structures. *Fibrous proteins* have relatively open-chain structures; in *globular proteins*, the chains are coiled into approximately spherical structures.

### Fibrous and globular proteins

Current knowledge of the structure of proteins owes much to recent developments in X-ray diffraction and NMR spectroscopic methods (see Box 14.2) as well as computer modelling studies. Proteins are classified as being *globular* or *fibrous*. In *fibrous proteins*, the chains (or *strands*) remain in more extended conformations. One of the most important fibrous proteins is *collagen* which is the natural building material of tendons, ligaments, skin, animal hide and cartilage. The polypeptide chains in collagen consist mainly of glycine and proline residues. Figure 35.4 shows how the polypeptide chains in collagen twist together to give 'hollow tubes' of fibrous protein. Hydrogen-bonded interactions between the NH and C=O groups of adjacent chains (diagram **35.17**) are important for maintaining the structure.

In *globular proteins*, the polypeptide chains are folded so that the protein has a near-spherical (or similar) structure. Among the simplest proteins are the *albumins*. These are present in living tissue and include *lactalbumin*, *ovalbumin* and *human serum albumin*. Lactalbumin is present in milk and has a molecular weight of $\approx 17\,500$. The protein is able to bind calcium ions by using O atoms in the polypeptide chains as donor atoms. The structure shown in Figure 35.5 is that of a $Ca^{2+}$ complex of lactalbumin. Ovalbumin occurs in egg-white; it has a molecular weight of $\approx 45\,000$ and consists of 385 amino acid residues. Figure 35.6 shows the structure of human serum albumin.

► ►
Hydrogen bonding: see
Section 21.8

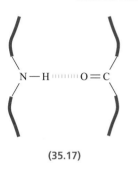

(35.17)

### Myoglobin and haemoglobin: proteins with prosthetic groups

Myoglobin and haemoglobin are *haem-iron proteins*. Myoglobin (Figure 35.7) has a molecular weight of $\approx 17\,000$ and consists of a single, coiled protein chain composed of 153 amino acid residues. Haemoglobin is a tetramer and has a molecular weight of $\approx 64\,500$; one of the four units was shown in Box 5.3.

**Fig. 35.4** The structure of collagen (from a computer modelling study). (a) 'Tube' representation of the three strands that combine to give the overall protein structure. Each strand is shown in a different colour. (b) Space-filling diagram looking down the coil formed by the three strands, with the same colour coding as diagram (a), and (c) the same view of the structure in a 'tube' representation.

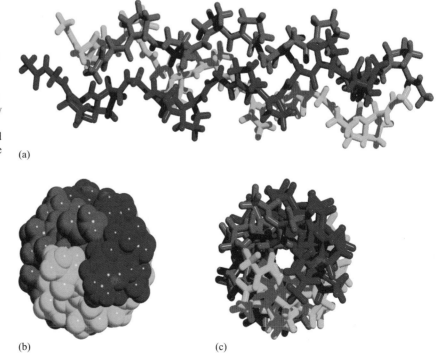

(a)

(b)                         (c)

**Fig. 35.5** The structure of human α-lactalbumin, crystallized in the presence of excess $Ca^{2+}$. The structure was determined by X-ray diffraction methods and reveals the presence of two calcium-binding sites. The protein chain is shown in a ribbon representation and the $Ca^{2+}$ ions are shown in grey. [X-ray diffraction data: N. Chandra *et al.* (1998) *Biochemistry*, vol. 37, p. 4767.]

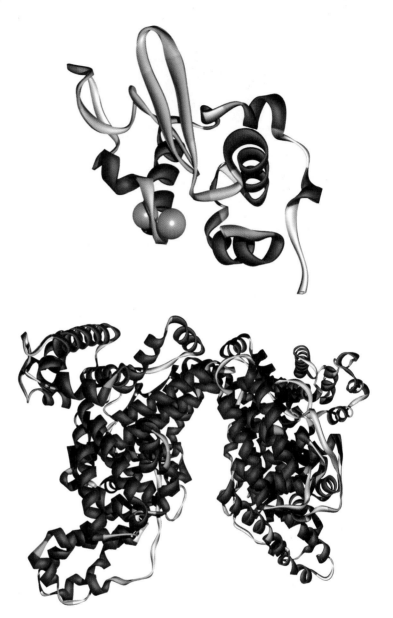

**Fig. 35.6** The structure of human serum albumin (one of several crystal forms) determined by X-ray diffraction [S. Sugio *et al.* (1999) *Protein Eng.*, vol. 12, p. 439]. The protein chain is shown in a ribbon representation.

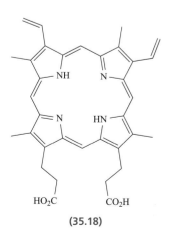

**(35.18)**

Myoglobin and each unit of haemoglobin contain a *prosthetic group* with a *haem unit*. This is the active site of the protein and is responsible for binding $O_2$; $O_2$ is transported by haemoglobin in the blood of mammals and myoglobin is the $O_2$-carrying protein in muscles. The haem unit is connected to the protein backbone through a histidine residue, and one N atom of this residue is coordinated to the Fe centre. In its $O_2$-free (*deoxy-form*), myoglobin and haemoglobin contain Fe(II) and its 5-coordinate environment is shown in Figure 35.7. Four of the coordination sites are occupied by a *protoporphyrin IX* group which is not directly attached to the protein chain. The protonated form of the protoporphyrin ligand is shown in structure **35.18**. When the protein binds $O_2$, the latter coordinates through one O atom to the Fe(II) centre giving an octahedral coordination sphere. During coordination, Fe(II) is oxidized to Fe(III), and $O_2$ is reduced to $[O_2]^-$.[§]

---

[§] For a more detailed account: see Chapter 28 in C. E. Housecroft and A. G. Sharpe (2005) *Inorganic Chemistry*, 2nd Ed, Prentice Hall, Harlow.

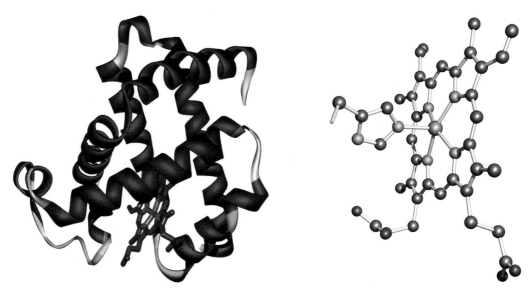

**Fig. 35.7** The structure of the deoxy-form of myoglobin determined by X-ray diffraction [J. Vojtechovsky *et al.* (1999) *Protein Data Bank code 1A6N*]. The protein chain is shown in a ribbon representation. The prosthetic group contains a haem unit, and the enlargement of this group on the right-hand side includes the histidine residue that connects the haem unit to the protein chain. The 'terminated' bond is the position of attachment to this chain. Colour code for the ball-and-stick structure: Fe, green; C, grey; O, red; N, blue.

## Enzymes and metalloenzymes

*Enzymes* are proteins that are biological catalysts. An enzyme has a specific function and operates at a specific pH.

Enzymes are proteins that act as biological catalysts and are involved in virtually all reactions taking place in living cells. In Section 15.15, we considered the kinetics of enzyme reactions, and now we comment briefly on their structures. Each enzyme has a specific role to play, and its structure reflects the way in which it must interact with the *substrate*. (Look back at the discussion of 'lock-and-key' mechanisms in Section 15.15.) After the substrate and enzyme have come together, the catalytic activity takes place at specific sites within the protein chain. Classes of enzymes are:

- *oxidases* and *reductases* that catalyse redox reactions;
- *hydrolases* that catalyse hydrolytic cleavage of bonds such as C−O and C−N;
- *transferases* that are involved in the transfer of a group such as $NH_2$, $PO_4$ or Me;
- *isomerases* that catalyse processes such as racemization, geometrical isomerization and tautomerism;
- *lysases* are involved in elimination reactions (or the reverse processes) that result in loss of e.g. $CO_2$ and formation of double bonds;
- *ligases* catalyse reactions involving bond formation.

In the discussion of polysaccharides in Section 35.3, we stated that starch is broken down by the action of the enzyme *amylase* to provide glucose for the body. Amylase is a hydrolase, and degradation of amylose in starch involves the hydrolysis of glycoside linkages. Figure 35.8 shows the globular protein structure of human pancreatic α-amylase. *Ribonuclease* is another important hydrolase and catalyses the hydrolysis of *ribonucleic acid* (RNA).

**Fig. 35.8** The structure of human pancreatic α-amylase (the R195A variant) determined by X-ray diffraction [S. Numao *et al.* (2002) *Biochemistry*, vol. 41, p. 215].

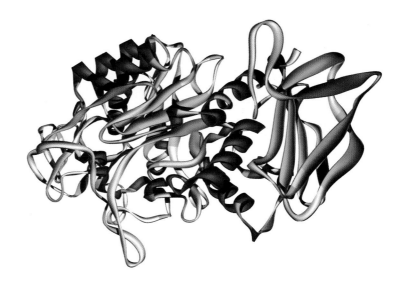

## 35.7 Nucleobases, nucleotides and nucleic acids

Although we come to nucleic acids in the last section of this book, this must not distract from the fact that they are critical to life because they are the storage molecules for genetic information. We set the scene for a discussion of nucleic acids, nucleotides and nucleobases in Chapter 34 when we introduced the structures and reactivities of some nitrogen-containing heterocycles. The nucleic acids deoxyribonucleic acid (DNA) and ribonucleic acid (RNA) are, respectively, found in cell nuclei and in cytoplasm. The building blocks of RNA and DNA are nucleotides, and these, in turn, are constructed from nucleobases, phosphate groups and a sugar.

### Nucleobases

In RNA and DNA, nature uses the purine bases *adenine* and *guanine*, and the pyrimidine bases *uracil*, *thymine* and *cytosine*.

Nature uses five nucleobases which fall into two classes: *purines* and *pyrimidines*. Structures **35.19** and **35.20** show the parent compounds purine and pyrimidine. *Adenine* and *guanine* are purines, while *uracil*, *thymine* and *cytosine* are pyrimidines:

Purine

**(35.19)**

Pyrimidine

**(35.20)**

Purine bases

Adenine

Guanine

Pyrimidine bases

Cytosine

Uracil

Thymine

The nucleobases adenine, guanine, cytosine and thymine occur in DNA, and in RNA, the component bases are adenine, guanine, cytosine and uracil.

## Nucleosides and nucleotides

A *nucleoside* is composed of a nucleobase bonded to a sugar; a *nucleotide* consists of a nucleoside covalently bonded to a phosphate group.

Connection of a nucleobase to a sugar leads to the formation of a *nucleoside*. In RNA, the sugar is ribose (**35.21**) and a nucleobase is bonded to ring-atom $C(1)$. The sugar 2-deoxyribose is a building block in DNA; again a nucleobase is connected to position $C(1)$. 2-Deoxyribose is related to ribose but carries no OH group on ring atom $C(2)$ (structure **35.22**).

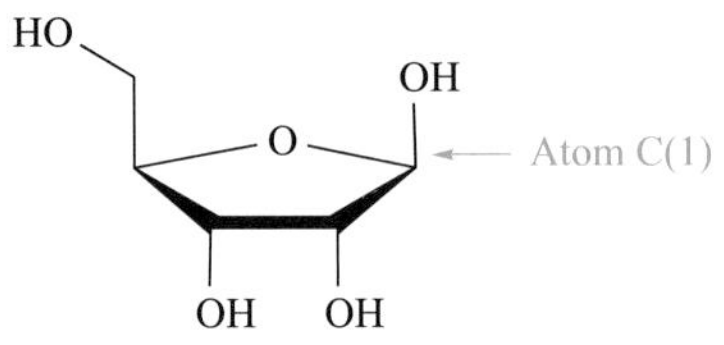

D-Ribose

**(35.21)**

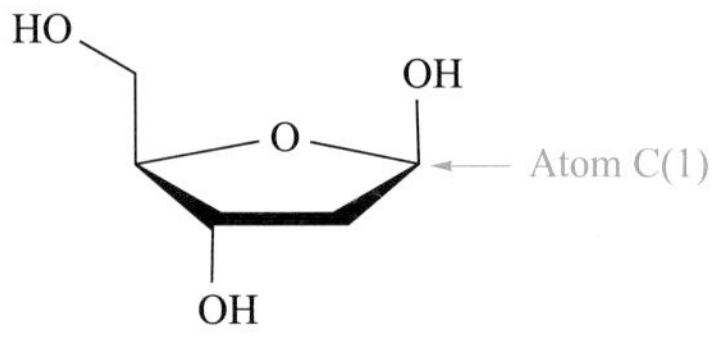

2-Deoxyribose

**(35.22)**

The condensation of a phosphate group to the $CH_2OH$ group (i.e. the $C(5)$ position) of the sugar in a nucleoside converts the latter to a *nucleotide*. Hence, we can construct the four ribonucleotides that combine to make RNA, and the four deoxyribonucleotides that are the building blocks of DNA (Figure 35.9).

## From nucleotides to nucleic acids

Using the nucleotides in Figure 35.9 as building blocks, we are now in a position to construct nucleic acids. The general scheme for this construction is the condensation of the phosphate group of one nucleotide with the OH on atom $C(3)$ of the next nucleotide. Once in the nucleic acid chain, the atom numbers change to 'primed' numbers, i.e. $C(3)$ becomes $C(3')$ and so on. Nucleic acids are macromolecules and have the general structures shown below:

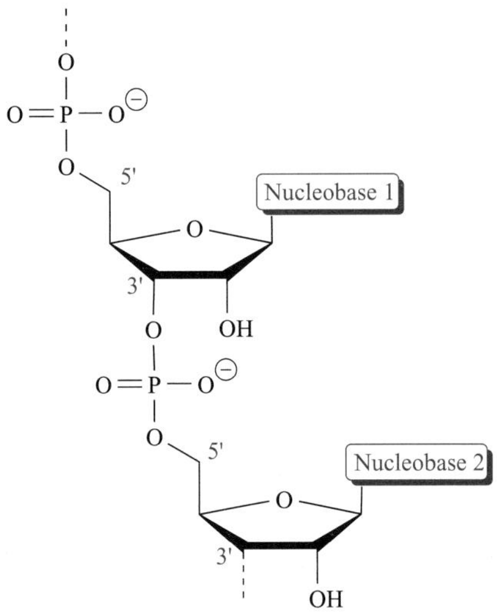

Using ribonucleotides

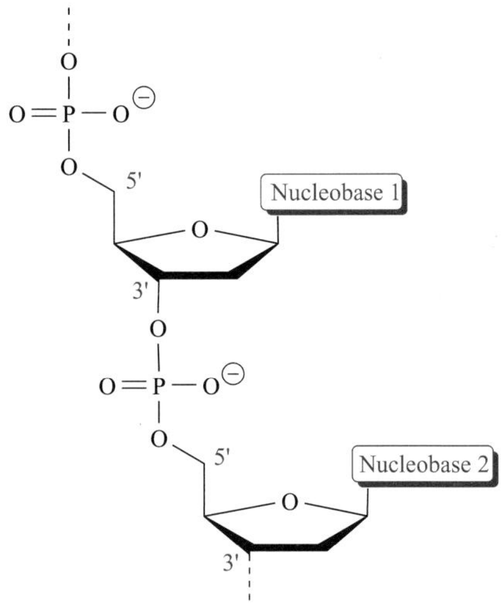

Using deoxyribonucleotides

Ribonucleotides: from adenine, guanine, cytosine and uracil

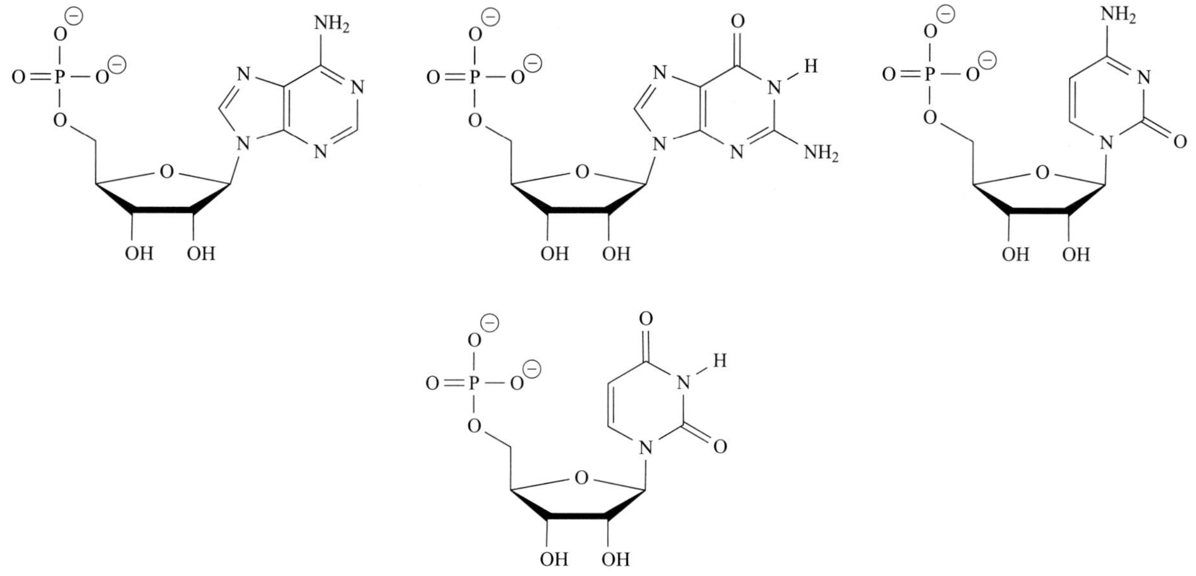

Deoxyribonucleotides: from adenine, guanine, cytosine and thymine

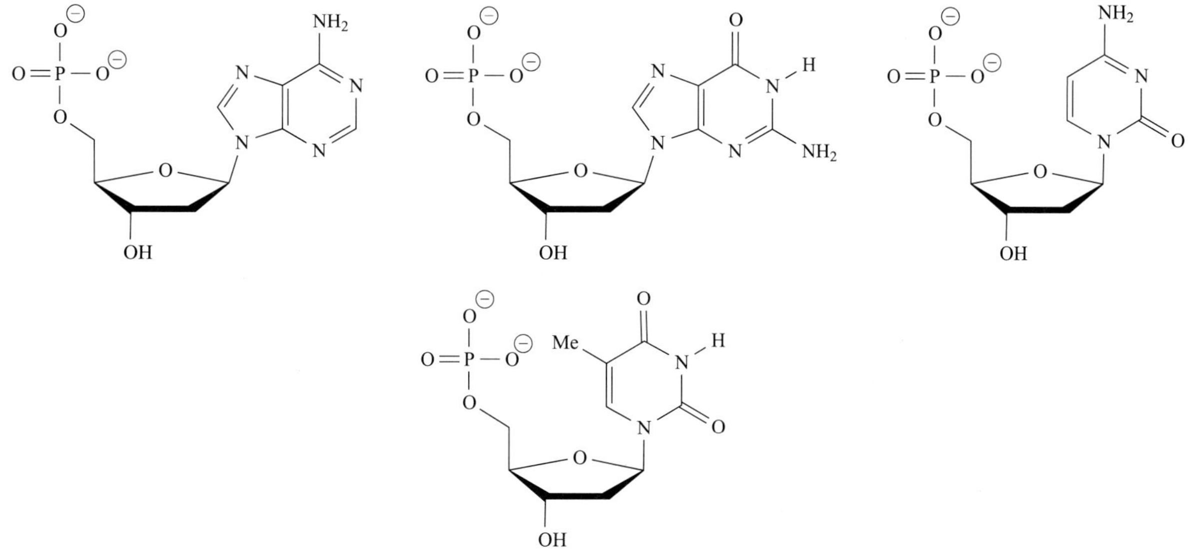

**Fig. 35.9** The four ribonucleotides that are the building blocks of RNA, and the four deoxyribonucleotides that are the construction units of DNA.

The prefix *oligo* means 'a few' and is used for compounds in which the number of repeating units is intermediate between a monomer and a high molecular weight macromolecule (polymer), e.g. *oligonucleotides*.

The sequence in which the nucleobases appear gives the nucleic acid its particular identity. The diagrams above illustrate that the direction of sequencing can be defined using 5' and 3'-nomenclature, e.g. in the left-hand chain, the sequence from 5'-end to 3'-end for the part of the chain shown is (nucleobase 1)–(nucleobase 2). The abbreviations used for the five nucleobases are adenine (A), guanine (G), thymine (T), cytosine (C) and uracil (U), and a sequence of nucleotides can be represented as, for example, 5'-AUACCUUGUCAG-3'. Short sequences of nucleotides are referred to as *oligonucleotides*.

### The structure of DNA: the double helix

Sequencing nucleotides in the manner described above generates nucleic acids composed of single strands. In DNA, the strands consist of deoxyribonucleotides. In 1953, Watson and Crick (see Box 35.5) made the important discovery that the nucleobases in one strand of DNA formed hydrogen bonds with *complementary bases* in a second strand. *Only certain interactions are favoured*: guanine and cytosine, and adenine and thymine are pairs of complementary bases. Each pair of nucleobases associates through hydrogen-bonded interactions:

Adenine–thymine (A–T) base pair

Guanine–cytosine (G–C) base pair

(35.23)

The result of these base-pairings between adjacent strands of DNA is the formation of a double chain. Moreover, the association specifically leads to a *helical assembly* with a *right-handed twist*. This is shown schematically in diagram **35.23**. Figure 35.10 shows the double helical assembly of two complementary strands of oligonucleotides composed of C, G, A and T nucleotides, i.e. the same nucleotides that are present in DNA. Figures 35.10a and b show two views of the molecular structure. Figures 35.10c and d show two ways of representing the structure so that the base-pairings and the phosphate-containing backbone are the focus of attention. A space-filling representation of part of the *double helix of DNA* is shown in Figure 35.11. The two strands of DNA can be seen in the figure, and the base pairs can be identified in the centre of the helix.

During cell replication, the double helix of DNA unwinds and each strand *templates* the construction of a new strand. By 'templating' we mean that by using the complementary base-pairing, a specific sequence of nucleobases in one strand, e.g. –ACAGC–, will control the specific sequencing in a new

## Box 35.5 Unravelling the double helix of DNA

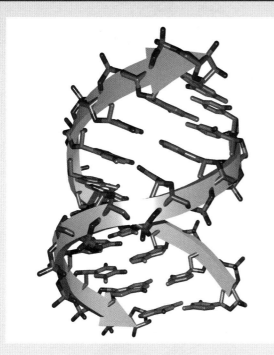

The molecular structure is shown in stick-representation. Colour code: C, grey; N, blue; O, red; P, orange. [X-ray diffraction data: M. C. Wahl (2000) *Nucleic Acids Res.*, vol. 28, p. 4356.]

Before 1950, the general nature of RNA and DNA had been established, and it was known that DNA carried genetic information. The 1950s was a period when X-ray crystallography was first being applied to biological molecules. Given the vast wealth of information about proteins that is now available in the Protein Data Bank (http://www.rcsb.org/pdb), it is perhaps difficult to appreciate how little structural information was available 50 years ago. Between 1951 and 1953, Rosalind Franklin and Maurice Wilkins were applying X-ray diffraction to try to determine the structure of DNA. Franklin discovered that DNA had a helical form, and that the phosphate groups lay on the outside of the helix. At the same time, other research groups were also working on the structure of DNA. In April 1953, James Watson and Francis Crick published a paper in the journal *Nature* outlining their proposals. Linus Pauling (see Box 5.1) had just published a separate proposal for DNA: a helical arrangement of three strands with the phosphate groups on the inside of the helical coil and the nucleobases on the outside, but it was the model of Watson and Crick that proved to be correct. This structure was made up of two strands, coiled helically around each other, each with a right-handed twist. Negatively charged phosphate groups resided on the outside of the structure within easy reach of cations. Nucleobases faced into the centre of the helix with the plane of each base perpendicular to the axis of the helix. It was proposed that the two strands ran in opposing directions and, as a consequence, bases could pair, namely adenine with thymine, and guanine with cytosine. The sequence of bases in one chain therefore determined the sequence of bases in the second strand. Watson and Crick also observed that the specific base-pairing could be the basis for a 'copying mechanism' for genetic material.

In 1962, Watson, Crick and Wilkins were awarded the Nobel Prize in Physiology or Medicine 'for their discoveries concerning the molecular structure of nucleic acids and its significance for information transfer in living material'. It is sad that Franklin did not receive credit for her contributions to the DNA story; she died of cancer in 1958, just at the start of her scientific career.

strand and this will be an exact replica of the strand with which the first strand was originally paired in the double helix:

Two strands with complementary base pairs, initially in double helix

Double helix unwinds

New strands are constructed that replicate the old ones

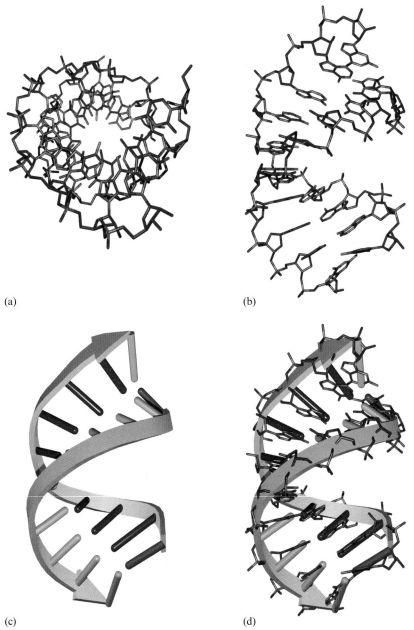

(a)　　　　　　　　　　　　　　(b)

(c)　　　　　　　　　　　　　　(d)

**Fig. 35.10** Two strands of oligonucleotides sequenced 5'-CAAAGAAAAG-3' and 5'-CTTTTCTTTG-3' assemble into a double helix. The structure has been determined by X-ray diffraction [M. L. Kopka *et al.* (1996) *J. Mol. Biol.*, vol. 334, p. 653]. The structure of the double helix (a) looking down the centre and (b) viewed from the side showing the oligonucleotides in stick representation; colour code: C, grey; N, blue; O, red; P, orange. Diagrams (c) and (d) show the backbone of each olignucleotide depicted as an arrow pointing toward the C3' end of the sequence, and the nucleobases are shown in a 'ladder' representation. Each nucleobase is colour coded (G, green; A, red; C, purple; T, cyan), making the A–T and G–C base pairs clearly visible.

## The structure of RNA

There are three classes of RNA molecules: messenger RNA (mRNA), transfer RNA (tRNA) and ribosomal RNA (rRNA). The function of mRNA is to transfer genetic coding information required for protein synthesis from chromosomes to ribosomes. Transfer RNA carries a specific amino acid and matches it to a codon (a sequence of three nucleotides) on mRNA during protein synthesis. rRNA molecules are directly involved in protein

**Fig. 35.11** Part of the right-handed, double helical structure of DNA. Colour code: C, grey; O, red; N, blue; P, orange. Hydrogen atoms are omitted for clarity.

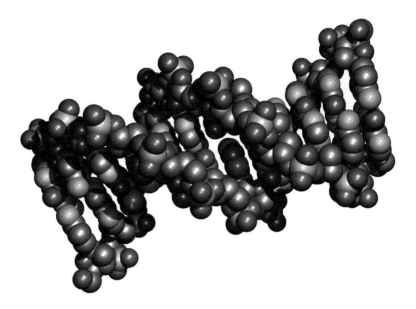

synthesis. Cellular RNA consists of *single* strands of ribonucleotides. The nucleotides present are adenine, uracil, guanine and cytosine, i.e. whereas DNA contains thymine, RNA contains uracil. Hydrogen-bonded base-pairing can occur between adenine and uracil, and between guanine and cytosine:

Adenine–uracil (A–U) base pair

Guanine–cytosine (G–C) base pair

Note that the only difference between the diagram above and that for G–C base-pairing in DNA is the presence of a ribose unit in RNA versus deoxyribose in DNA.

**Fig. 35.12** The structure of a modified hammerhead ribozyme containing two RNA strands. The backbone of each strand is depicted as an arrow pointing toward the C3' end of the sequence. The nucleobases are shown in a 'ladder' representation; each nucleobase is colour coded (G, green; C, purple; A, red; U, blue). The complementary base pairs are G–C and A–U. The structure was determined by X-ray diffraction: C. M. Dunham *et al.* (2003) *J. Mol. Biol.*, vol. 332, p. 327.

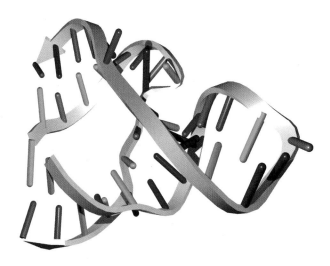

Although RNA exists as single strands, intra-strand base-pairing can occur within self-complementary sections of the strand. This gives RNA a *secondary structure* which results from folding of the chain into *loops* and *hairpins* (schematically illustrated in diagrams **35.24** and **35.25**), as well as helical twists. RNA strand lengths vary. For example, mRNA contains between ≈500 and 7000 nucleotides, whereas tRNA consists of ≈70–90 nucleotides. Figures 35.12 and 35.13 illustrate two examples of RNA structures. A *ribozyme* is an RNA segments that catalyses the cleavage of another RNA strand, and Figure 35.12 shows the structure of a modified hammerhead ribozyme containing two RNA strands. (The hammerhead ribozyme is derived from self-cleaving RNA molecules that are associated with plant RNA viruses.) These strands are relatively short, and the colour codes of the nucleotides (defined in the figure caption) allow you to see how base-pairing leads to folding of the strands. Figure 35.13 shows a 247-nucleotide ribozyme and the role of complementary base-pairing in the development of the secondary structure of the RNA strand can be seen.

**Fig. 35.13** The structure of a modified 247-nucleotide ribozyme from the ciliate *Tetrahymena thermophila*. The backbone of the RNA strand is shown as an arrow pointing toward the C3' end of the sequence; the head of the arrow is towards the lower right of the figure. The nucleobases are shown in a 'ladder' representation and are colour coded: G, green; C, purple; A, red; U, blue. The figure illustrates how G–C and A–U base-pairing in a single RNA strand leads to the complicated secondary structure. The structure was determined by X-ray diffraction: B. L. Golden *et al.* (1998) *Science*, vol. 282, p. 259.

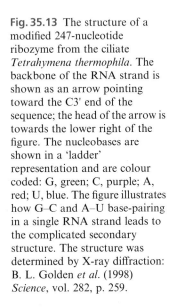

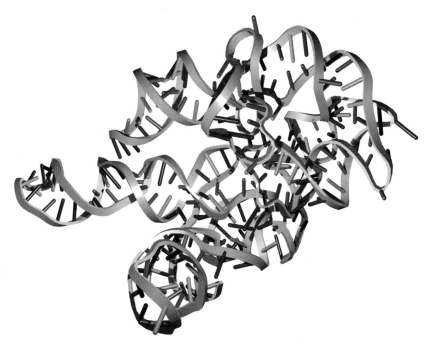

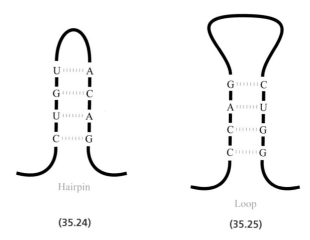

Hairpin

Loop

(35.24)                    (35.25)

By comparing the structures in Figures 35.10–35.13, you can appreciate that hydrogen-bonded interactions (which are typically of a strength of $\approx 25 \, \text{kJ mol}^{-1}$) are responsible both for the ordered double helical structure of DNA, and the complicated, folded structures of single-stranded RNA molecules.

## SUMMARY

In this chapter, we have introduced some of the molecules that are essential to life. The classes of compound that we have discussed are carbohydrates (mono-, di- and polysaccharides), amino acids, polypeptides, proteins, nucleobases, nucleotides and nucleic acids. Nature is able to carry out complicated reactions that are reproducible and specific, and enzymes (which are proteins) control the specificity of these processes. Nature is far ahead of man in terms of the chemistry that it achieves. RNA and DNA are nucleic acids that are fundamental to life and genetic coding.

### Do you know what the following terms mean?

- carbohydrate
- monosaccharide
- aldose
- ketose
- tetrose
- pentose
- hexose
- anomer
- pyranose form of a sugar
- furanose form of a sugar

- glycoside
- amino acid
- zwitterion
- isoelectric point
- peptide
- polypeptide
- $C$-terminus of a peptide
- $N$-terminus of a peptide
- protein
- electrophoresis

- protein
- enzyme
- nucleobase
- nucleoside
- nucleotide
- nucleic acid
- oligonucleotide
- DNA
- RNA

### The following abbreviations have been used. Do you know what they mean?

- Ala
- Arg
- Asn
- Asp

- Cys
- Glu
- Gln
- Gly

- His
- Ile
- Leu
- Lys

- Met
- Phe
- Pro
- Ser

- Thr
- Trp
- Tyr
- Val

- A
- C
- G
- T

- U
- BOC
- DNA
- RNA

## You should be able:

- to comment on the presence and importance of chiral compounds in nature
- to classify carbohydrates as mono-, di- and polysaccharides
- to distinguish between an aldose and a ketose and give examples of each
- to classify sugars as tetroses, pentoses and hexoses
- to describe the open and cyclic forms of sugars such as glucose, and comment on the equilibria between these forms
- to explain how different isomers of a sugar arise
- to describe typical reactions of a sugar such as glucose
- to describe how a glycoside linkage is cleaved and the role of this reaction in nature
- to give examples of mono-, di- and polysaccharides, and comment on the roles they play in nature
- to give examples of amino acids and comment on their chirality
- to discuss the Brønsted acidic and basic properties of amino acids
- to describe how a peptide linkage is formed

- to explain why amino acids need to be activated and protected before a di- or polypeptide is assembled
- to describe briefly the use of electrophoresis for the separation and characterization of polypeptides or proteins
- to distinguish between fibrous and globular proteins and give examples of each
- to comment on the role of hydrogen bonding and disulfide bridge formation in protein structure
- to give an example of a prosthetic group in a protein
- to give examples of classes of enzymes and their roles in nature
- to draw the structures of the nucleobases adenine, guanine, cytosine, thymine and uracil, and classify them as purines or pyrimidines
- to illustrate the general structures of nucleosides and nucleotides
- to describe how nucleotides assemble to give nucleic acids
- to comment on the assembly of the double helix of DNA and the importance of complementary base pairs in nature

## PROBLEMS

**35.1** The following structures show the open forms of two pentoses. Label these compounds using sequence rules.

(a)

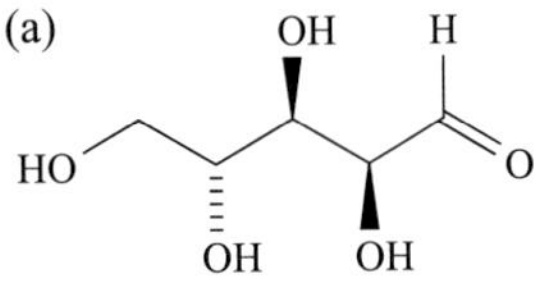

D-Arabinose

(b)

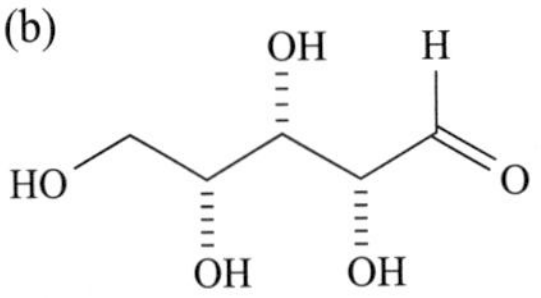

D-Xylose

(c)

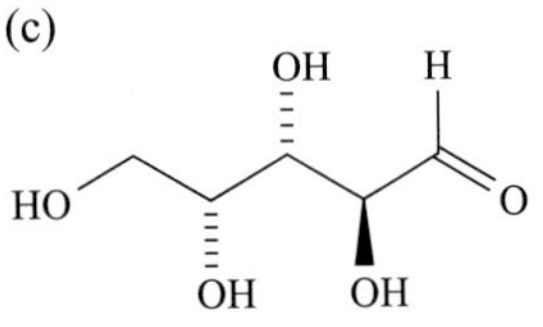

D-Lyxose

**35.2** Show how the open form of D-ribose undergoes conversion to the furanose form, and comment on why the latter is classed as a hemiacetal.

**35.3** (a) Draw the structure of the pyranose form of D-glucose and comment on what is meant by the anomeric centre. (b) Give a mechanism by which the open form of D-glucose converts to its furanose form, and suggest why this is less favourable than formation of the pyranose form.

**35.4** Suggest products in the following reactions. Give both structures and names for the products:

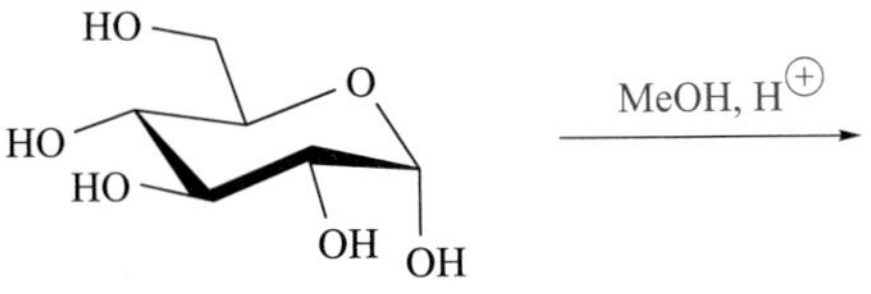

α-D-Glucose   MeOH, H⊕ →

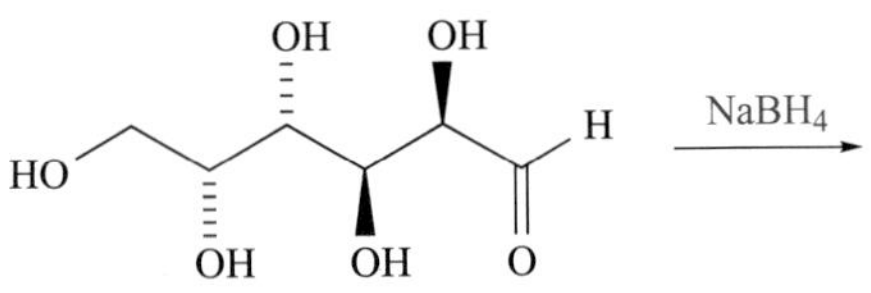

D-Galactose   NaBH₄ →

**35.5** Classify the following sugars as tetroses, pentoses or hexoses, and as aldoses or ketoses: (a) glucose, (b) fructose, (c) ribose, (d) mannose, (e) galactose, (f) threose.

**35.6** Classify the following sugars as monosaccharides, disaccharides or polysaccharides: (a) cellulose, (b) lactose, (c) galactose, (d) amylose, (e) sucrose, (f) ribose, (g) fructose.

**35.7** Explain what you understand by the following notations: (a) D- and L-forms of a particular sugar, (b) α- and β-D-glucose, (c) (R)- and (S)-glyceraldehyde.

**35.8** (a) Draw the structure of and name the only achiral amino acid. (b) Does nature tend to use D-, L- or racemic amino acids? (c) Draw the structure of (R)-cysteine. (d) Give the names of the amino acids for which the following are abbreviations: Val, Gly, Asp, Cys, Glu, His.

**35.9** (a) What is meant by the zwitterionic form of an amino acid? (b) How will the structure of L-valine respond to changes in pH if the $pK_a$ of the $CO_2H$ group is 2.32, and the $pK_b$ of the $NH_2$ group is 4.38?

**35.10** (a) In the synthesis of a dipeptide, where and why must activating and protecting groups be introduced before condensation of the two amino acids? (b) Give a scheme that shows the formation of Gly-Leu starting from suitably protected precursors, and ending with Leu at the C-terminus.

**35.11** The structure of vasopressin is given in the text as:

Cys-Tyr-Phe-Gln-Asn-Cys-Pro-Arg-Gly

N-terminus

Using Table 35.1 to help you, interpret this sequence of abbreviations and draw the full structure of vasopressin. Comment on any special features of the structure.

**35.12** Comment on the following features of proteins: (a) primary structure, (b) secondary and tertiary structure, (c) quaternary structure, (d) fibrous protein, (e) globular protein.

**35.13** Chymotrypsin is an enzyme that cleaves peptide bonds, specifically targeting the peptide bond at the C-termini of phenylamine, tyrosine and tryptophan. Interpret the sequence of amino acids Cys-Leu-Phe-GlyNH₂ into a structural form and draw a scheme to show chymotrypsin acting on this sequence.

**35.14** (a) Distinguish between a nucleoside and a nucleotide. (b) Explain what is meant by the notation 5'-AGAGAGAGAGAAAAA-3'. (c) What is the complementary nucleic acid strand to the one given in part (b)? Give reasoning for your answer.

**35.15** Discuss the role of hydrogen bonding in (a) collagen and (b) DNA.

## Additional problems

**35.16** One of the active sites of ribonuclease consists of a pocket in the folded protein chain and can be presented as follows, where the pink line is the backbone of the protein:

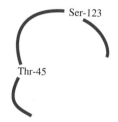

Only the two residues involved in binding the substrate are shown; the numbers refer to their positions in the chain. Suggest how this site might bind a uracil residue in RNA.

**35.17** The hormone angiotensin II plays a key role in the regulation of blood pressure. Its structure is shown in **35.26**. Rewrite the structure in an unambiguous, abbreviated form.

(35.26)

# Appendix 1  Mathematical symbols

| Symbol | Meaning |
|--------|---------|
| $>$ | is greater than |
| $\gg$ | is much greater than |
| $<$ | is less than |
| $\ll$ | is much less than |
| $\geq$ | is greater than or equal to |
| $\leq$ | is less than or equal to |
| $\approx$ | is approximately equal to |
| $=$ | is equal to |
| $\neq$ | is not equal to |
| $\rightleftharpoons$ | equilibrium |
| $\propto$ | is proportional to |
| $\times$ | multiplied by |
| $\infty$ | infinity |
| $\pm$ | plus or minus |
| $\sqrt{\phantom{x}}$ | square root of |
| $\sqrt[3]{\phantom{x}}$ | cube root of |
| $\lvert x \rvert$ | modulus of $x$ |
| $\sum$ | summation of |
| $\Delta$ | change in (for example, $\Delta H$ is 'change in enthalpy'); see also Appendix 3 |
| $\angle$ | angle |
| $\log$ | logarithm to base 10 ($\log 10$) |
| $\ln$ | natural logarithm, i.e. logarithm to base e ($\log_e$) |
| $\int$ | integral of |
| $\dfrac{\mathrm{d}}{\mathrm{d}x}$ | differential with respect to $x$ |
| $\dfrac{\partial}{\partial x}$ | partial differential with respect to $x$ |

# Appendix 2  Greek letters with pronunciations

| Upper case letter | Lower case letter | Pronounced |
|---|---|---|
| A | α | alpha |
| B | β | beta |
| Γ | γ | gamma |
| Δ | δ | delta |
| E | ε | epsilon |
| Z | ζ | zeta |
| H | η | eta |
| Θ | θ | theta |
| I | ι | iota |
| K | κ | kappa |
| Λ | λ | lambda |
| M | μ | mu |
| N | ν | nu |
| Ξ | ξ | xi |
| O | ο | omicron |
| Π | π | pi |
| P | ρ | rho |
| Σ | σ | sigma |
| T | τ | tau |
| Υ | υ | upsilon |
| Φ | φ | phi |
| X | χ | chi |
| Ψ | ψ | psi |
| Ω | ω | omega |

# Appendix 3 Abbreviations and symbols for quantities and units

For ligand structures, see Table 23.2. Where a symbol is used to mean more than one thing, the context of its use should make the meaning clear.

| | | | |
|---|---|---|---|
| A | ampere (unit of electrical current) | $C_P$ | heat capacity (at constant pressure) |
| $A$ | absorbance | | |
| $A$ | frequency factor (in Arrhenius equation) | $C_V$ | heat capacity (at constant volume) |
| $A$ | Madelung constant | $D$ | bond dissociation enthalpy |
| $A$ | mass number (of an atom) | $\bar{D}$ | average bond enthalpy term |
| $A_r$ | relative atomic mass | D | debye (non-SI unit of electric dipole moment) |
| $A(\theta,\phi)$ | angular wavefunction | | |
| Å | ångstrom (non-SI unit of length, used for bond distances) | $d$ | distance |
| | | dec. pl. | decimal places |
| $a_i$ | activity (of component i) | $dm^3$ | cubic decimetre (unit of volume) |
| AO | atomic orbital | | |
| aq | aqueous | $E$ | energy |
| Ar | general aryl group | $E_J$ | rotational energy level |
| atm | atmosphere (non-SI unit of pressure) | $E_k$ | kinetic energy |
| ax | axial | $E_v$ | vibrational energy |
| | | $e$ | charge on the electron |
| $B$ | magnetic field strength | $e^-$ | electron |
| $B$ | rotational constant | E1 | unimolecular elimination |
| bar | bar (unit of pressure) | E2 | bimolecular elimination |
| bp | boiling point | $EA$ | electron affinity |
| Bu | butyl | $E_a$ | activation energy |
| | | $E_{cell}$ | electrochemical cell potential |
| $c$ | coefficient (in wavefunctions) | $E^{\circ}$ | standard reduction potential (standard electrode potential) |
| $c$ | concentration (of solution) | | |
| $c$ | speed of light | $E^{\circ}_{[OH^-]=1}$ | standard reduction potential when $[OH^-] = 1\,mol\,dm^{-3}$ |
| $\bar{c}$ | root mean square velocity | | |
| C | coulomb (unit of charge) | EI | electron impact ionization (mass spectrometry) |
| CI | chemical ionization (mass spectrometry) | | |
| | | EPR | electron paramagnetic resonance |
| cm | centimetre (unit of length) | eq | equatorial |
| $cm^3$ | cubic centimetre (unit of volume) | ES | electrospray (mass spectrometry) |
| $cm^{-1}$ | reciprocal centimetre (wavenumber) | ESR | electron spin resonance |
| conc | concentrated | Et | ethyl |
| cr | crystal | eV | electron volt |

| | | | |
|---|---|---|---|
| $F$ | Faraday constant | $l$ | length |
| FAB | fast-atom bombardment (mass spectrometry) | $l$ | orbital quantum number |
| | | $\ell$ | path length |
| FID | free induction decay | LCAO | linear combination of atomic orbitals |
| FT | Fourier transform | | |
| | | LUMO | lowest unoccupied molecular orbital |
| $G$ | Gibbs energy | | |
| g | gas | M | molarity |
| g | gram (unit of mass) | m | molal (solution) |
| | | $m$ | mass |
| $H$ | enthalpy | m | metre (unit of length) |
| $\mathscr{H}$ | Hamiltonian operator | $m^3$ | cubic metre (unit of volume) |
| $h$ | Planck constant | $M^+$ | molecular ion (in mass spectrometry) |
| $h\nu$ | high-frequency radiation (for a photolysis reaction) | $m_e$ | electron rest mass |
| | | $m_i$ | molality (of component i) |
| HOMO | highest occupied molecular orbital | $m_s$ | magnetic spin quantum number |
| Hz | hertz (unit of frequency) | $m/z$ | mass : charge ratio |
| | | $M_r$ | relative molecular mass |
| $I$ | current | MALDI-TOF | matrix assisted laser desorption ionization time-of-flight (mass spectrometry) |
| $I$ | moment of inertia | | |
| $I$ | nuclear spin quantum number | | |
| $I$ | intensity of transmitted radiation | Me | methyl |
| $I_0$ | intensity of incident radiation | MO | molecular orbital |
| $IE$ | ionization energy | mol | mole (unit of quantity) |
| IR | infrared | mp | melting point |
| $^i$Pr | isopropyl | MS | mass spectrometry |
| J | joule (unit of energy) | N | newton (unit of force) |
| $J$ | spin–spin coupling constant | $N$ | normalization factor |
| $J$ | rotational quantum number | $n$ | Born exponent |
| | | $n$ | number of (e.g. moles) |
| $k$ | Boltzmann constant | $n$ | principal quantum number |
| $k$ | force constant | nm | nanometre (unit of length) |
| $k$ | rate constant | NMR | nuclear magnetic resonance |
| K | kelvin (unit of temperature) | | |
| $K$ | equilibrium constant | $P$ | pressure |
| $K_a$ | acid dissociation constant | $P^+$ | parent ion (in mass spectrometry) |
| $K_b$ | base dissociation constant | Pa | pascal (unit of pressure) |
| $K_c$ | equilibrium constant for a reaction in solution | PES | photoelectron spectroscopy |
| | | Ph | phenyl |
| $K_M$ | Michaelis constant | pm | picometre (unit of length) |
| $K_p$ | equilibrium constant for a gaseous reaction | Pr | propyl |
| | | $P_X$ | partial pressure of component X |
| $K_{sp}$ | solubility product constant | | |
| $K_w$ | self-ionization constant of water | $q$ | charge |
| kg | kilogram (unit of mass) | $q$ | heat |
| kJ | kilojoule (unit of energy) | | |
| kPa | kilopascal (unit of pressure) | R | general alkyl group |
| | | $R$ | molar gas constant |
| $L$ | Avogadro constant or number | $R$ | Rydberg constant |
| l | liquid | $R$ | resistance |

| | | | |
|---|---|---|---|
| $R\text{-}$ | prefix for an enantiomer using sequence rules | VB | valence bond |
| | | VIS | visible |
| $r$ | internuclear distance | VSEPR | valence-shell electron-pair repulsion |
| $r$ | radial distance | | |
| $r$ | radius | | |
| $R(r)$ | radial wavefunction | $w$ | work done |
| $r_{cov}$ | covalent radius | | |
| $r_{ion}$ | ionic radius | $x_e$ | anharmonicity constant |
| $r_v$ | van der Waals radius | $[X]$ | concentration of X |
| $r_{metal}$ | metallic radius | | |
| RDS | rate-determining step | $z$ | number of moles of electrons transferred in an electrochemical cell |
| RF | radiofrequency | | |
| | | | |
| $S$ | entropy | $Z$ | atomic number |
| $S$ | overlap integral | $Z_{eff}$ | effective nuclear charge |
| $S$ | total spin quantum number | $|z_-|$ | modulus of the negative charge |
| $S\text{-}$ | prefix for an enantiomer using sequence rules | $|z_+|$ | modulus of the positive charge |
| $s$ | second (unit of time) | $\alpha$-carbon | C atom adjacent to the one in question |
| $s$ | solid | | |
| $s$ | spin quantum number | $[\alpha]_D$ | specific rotation at $\lambda = 589$ nm (for a chiral compound in solution) |
| $S_N1$ | substitution nucleophilic unimolecular | | |
| | | $\alpha$ | angle ($\theta$ is also used) |
| $S_N2$ | substitution nucleophilic bimolecular | $\alpha$ | degree of dissociation |
| | | $\beta$ | stability constant |
| sig. fig. | significant figures | $\beta$-carbon | C atom two along from the one in question |
| soln | solution | | |
| solv | solvated | $\delta$ | chemical shift |
| | | $\delta^-$ | partial negative charge |
| $T$ | temperature | $\delta^+$ | partial positive charge |
| $T$ | transmittance | $\Delta$ | change in |
| $t$ | time | $\Delta$ | heat (in a pyrolysis reaction) |
| $t_{\frac{1}{2}}$ | half-life | $\Delta_{oct}$ | octahedral crystal field splitting energy |
| $^tBu$ | *tert*-butyl | | |
| THF | tetrahydrofuran | $\Delta_a H$ | enthalpy change of atomization |
| TMS | tetramethylsilane | $\Delta_c H$ | enthalpy change of combustion |
| TS | transition state | $\Delta_{EA} H$ | enthalpy change associated with electron attachment |
| $U$ | internal energy | $\Delta_f H$ | enthalpy change of formation |
| u | atomic mass unit | $\Delta_{fus} H$ | enthalpy change of fusion |
| UV | ultraviolet | $\Delta_{hyd} H$ | enthalpy change of hydration |
| UV–VIS | ultraviolet–visible | $\Delta_{lattice} H$ | enthalpy change for the formation of an ionic lattice |
| $V$ | potential difference | $\Delta_r H$ | enthalpy change of reaction |
| $V$ | velocity of reaction (in Michaelis–Menten kinetics) | $\Delta_{sol} H$ | enthalpy change of solution |
| | | $\Delta_{solv} H$ | enthalpy change of solvation |
| $V$ | volume | $\Delta_{vap} H$ | enthalpy change of vaporization |
| V | volt (unit of potential difference) | $\Delta_f G$ | Gibbs energy change of formation |
| v | vapour | $\Delta_r G$ | Gibbs energy change of reaction |
| v | vibrational quantum number | $\varepsilon$ | molar extinction (or absorption) coefficient |
| $v$ | velocity | | |

| | | | |
|---|---|---|---|
| $\varepsilon_{max}$ | molar extinction coefficient corresponding to an absorption maximum (in an electronic spectrum) | $\mu_B$ | bohr magneton (unit of magnetic moment) |
| | | $\mu_{eff}$ | effective magnetic moment |
| | | $\mu\text{-}$ | bridging ligand |
| $\varepsilon_0$ | permittivity of a vacuum | $\rho$ | density |
| $\gamma_i$ | activity coefficient | $\rho$ | resistivity |
| $\kappa$ | conductivity | $\nu$ | frequency |
| $\theta$ | angle ($\alpha$ is also used) | $\bar{\nu}$ | wavenumber |
| $\lambda$ | wavelength | $\nu_-$ | number of anions per formula unit |
| $\lambda_{max}$ | wavelength corresponding to an absorption maximum (in an electronic spectrum) | $\nu_+$ | number of cations per formula unit |
| $\lambda_+{}^\infty$ | cation molar conductivity at infinite dilution | $\chi$ | magnetic susceptibility |
| | | $\chi$ | electronegativity |
| | | $\chi^{AR}$ | Allred–Rochow electronegativity |
| $\lambda_-{}^\infty$ | anion molar conductivity at infinite dilution | $\chi^M$ | Mulliken electronegativity |
| | | $\chi^P$ | Pauling electronegativity |
| $\Lambda_m$ | molar conductivity | $\psi$ | wavefunction |
| $\Lambda_m{}^\infty$ | molar conductivity at infinite dilution | $\Omega$ | ohm (unit of resistance) |
| | | 2c-2e | 2-centre 2-electron |
| $\mu$ | electric dipole moment | 3c-2e | 3-centre 2-electron |
| $\mu$ | ion mobility | $^{\ominus}$ | standard state |
| $\mu$ | reduced mass | $\ddagger$ | activated complex; transition state |
| $\mu(\text{spin-only})$ | spin-only magnetic moment | $^{\circ}$ | degree |

# Appendix 4 The electromagnetic spectrum

The frequency of electromagnetic radiation is related to its wavelength by the equation:

$$\text{Wavelength } (\lambda) = \frac{\text{Speed of light } (c)}{\text{Frequency } (\nu)}$$

where $c = 2.998 \times 10^8 \, \text{m s}^{-1}$.

$$\text{Wavenumber } (\bar{\nu}) = \frac{1}{\text{Wavelength}}$$

with units in $\text{cm}^{-1}$ (pronounced 'reciprocal centimetre')

$$\text{Energy } (E) = \text{Planck's constant } (h) \times \text{Frequency } (\nu) \quad \text{where } h = 6.626 \times 10^{-34} \, \text{J s}$$

The energy given in the last column is measured per mole of photons.

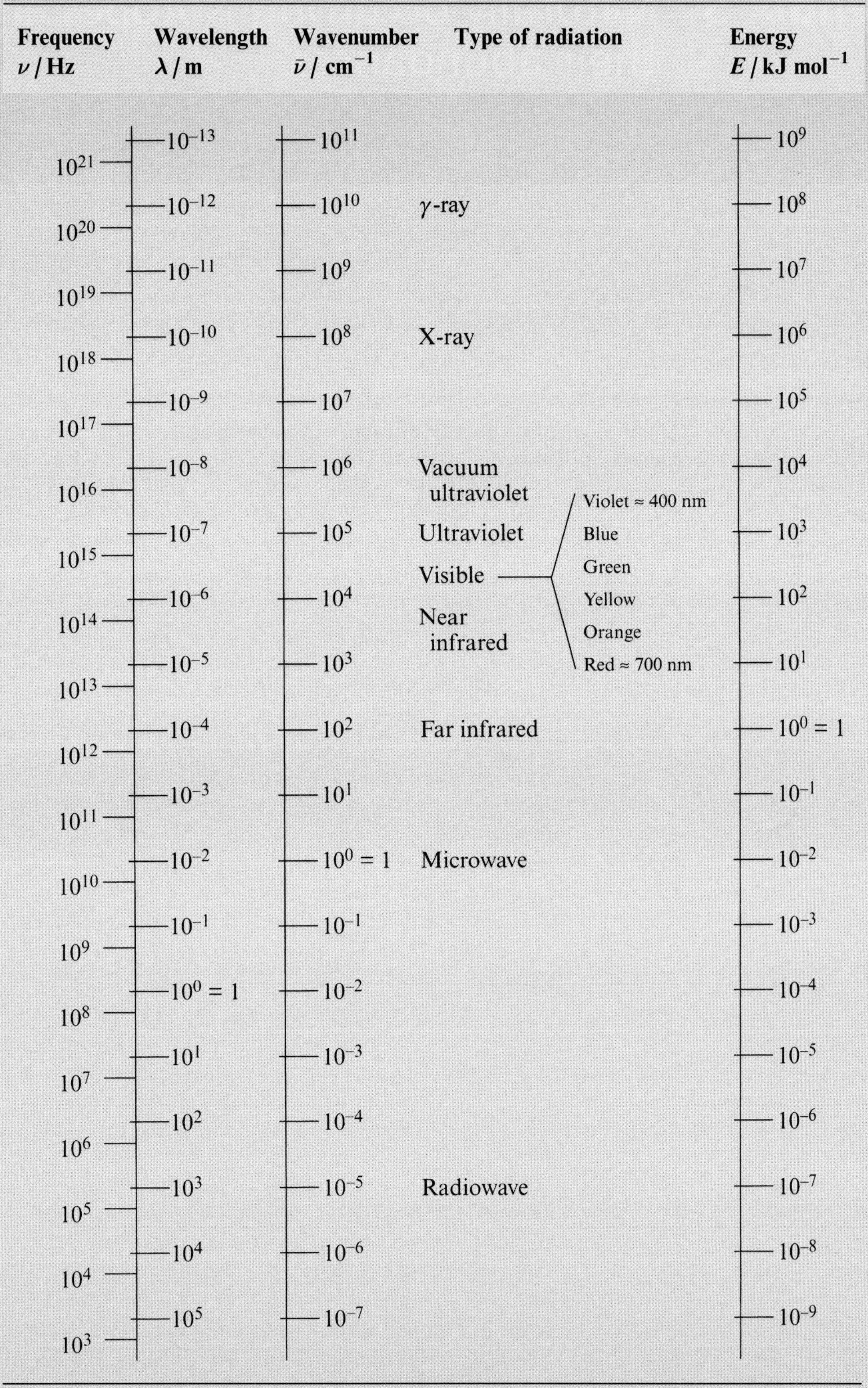

| Frequency $\nu$ / Hz | Wavelength $\lambda$ / m | Wavenumber $\bar{\nu}$ / $cm^{-1}$ | Type of radiation | Energy $E$ / kJ $mol^{-1}$ |
|---|---|---|---|---|
| | $10^{-13}$ | $10^{11}$ | | $10^9$ |
| $10^{21}$ | | | | |
| | $10^{-12}$ | $10^{10}$ | $\gamma$-ray | $10^8$ |
| $10^{20}$ | | | | |
| | $10^{-11}$ | $10^9$ | | $10^7$ |
| $10^{19}$ | | | | |
| | $10^{-10}$ | $10^8$ | X-ray | $10^6$ |
| $10^{18}$ | | | | |
| | $10^{-9}$ | $10^7$ | | $10^5$ |
| $10^{17}$ | | | | |
| | $10^{-8}$ | $10^6$ | Vacuum ultraviolet | $10^4$ |
| $10^{16}$ | | | | |
| | $10^{-7}$ | $10^5$ | Ultraviolet | $10^3$ |
| $10^{15}$ | | | Visible | |
| | $10^{-6}$ | $10^4$ | | $10^2$ |
| $10^{14}$ | | | Near infrared | |
| | $10^{-5}$ | $10^3$ | | $10^1$ |
| $10^{13}$ | | | | |
| | $10^{-4}$ | $10^2$ | Far infrared | $10^0 = 1$ |
| $10^{12}$ | | | | |
| | $10^{-3}$ | $10^1$ | | $10^{-1}$ |
| $10^{11}$ | | | | |
| | $10^{-2}$ | $10^0 = 1$ | Microwave | $10^{-2}$ |
| $10^{10}$ | | | | |
| | $10^{-1}$ | $10^{-1}$ | | $10^{-3}$ |
| $10^9$ | | | | |
| | $10^0 = 1$ | $10^{-2}$ | | $10^{-4}$ |
| $10^8$ | | | | |
| | $10^1$ | $10^{-3}$ | | $10^{-5}$ |
| $10^7$ | | | | |
| | $10^2$ | $10^{-4}$ | | $10^{-6}$ |
| $10^6$ | | | | |
| | $10^3$ | $10^{-5}$ | Radiowave | $10^{-7}$ |
| $10^5$ | | | | |
| | $10^4$ | $10^{-6}$ | | $10^{-8}$ |
| $10^4$ | | | | |
| | $10^5$ | $10^{-7}$ | | $10^{-9}$ |
| $10^3$ | | | | |

Visible breakdown:
Violet $\approx$ 400 nm
Blue
Green
Yellow
Orange
Red $\approx$ 700 nm

# Appendix 5 Naturally occurring isotopes and their abundances

Data from *WebElements* by Mark Winter. Further information on radiocative nuclides can be found using the Web link www.webelements.com

| Element | Symbol | Atomic number, $Z$ | Mass number of isotope (% abundance) |
|---------|--------|---------------------|--------------------------------------|
| Actinium | Ac | 89 | artificial isotopes only; mass number range 224–229 |
| Aluminium | Al | 13 | 27(100) |
| Americium | Am | 95 | artificial isotopes only; mass number range 237–245 |
| Antimony | Sb | 51 | 121(57.3), 123(42.7) |
| Argon | Ar | 18 | 36(0.34), 38(0.06), 40(99.6) |
| Arsenic | As | 33 | 75(100) |
| Astatine | At | 85 | artificial isotopes only; mass number range 205–211 |
| Barium | Ba | 56 | 130(0.11), 132(0.10), 134(2.42), 135(6.59), 136(7.85), 137(11.23), 138(71.70) |
| Berkelium | Bk | 97 | artificial isotopes only; mass number range 243–250 |
| Beryllium | Be | 4 | 9(100) |
| Bismuth | Bi | 83 | 209(100) |
| Boron | B | 5 | 10(19.9), 11(80.1) |
| Bromine | Br | 35 | 79(50.69), 81(49.31) |
| Cadmium | Cd | 48 | 106(1.25), 108(0.89), 110(12.49), 111(12.80), 112(24.13), 113(12.22), 114(28.73), 116(7.49) |
| Caesium | Cs | 55 | 133(100) |
| Calcium | Ca | 20 | 40(96.94), 42(0.65), 43(0.13), 44(2.09), 48(0.19) |
| Californium | Cf | 98 | artificial isotopes only; mass number range 246–255 |
| Carbon | C | 6 | 12(98.9), 13(1.1) |
| Cerium | Ce | 58 | 136(0.19), 138(0.25), 140(88.48), 142(11.08) |
| Chlorine | Cl | 17 | 35(75.77), 37(24.23) |
| Chromium | Cr | 24 | 50(4.345), 52(83.79), 53(9.50), 54(2.365) |
| Cobalt | Co | 27 | 59(100) |
| Copper | Cu | 29 | 63(69.2), 65(30.8) |
| Curium | Cm | 96 | artificial isotopes only; mass number range 240–250 |
| Dysprosium | Dy | 66 | 156(0.06), 158(0.10), 160(2.34), 161(18.9), 162(25.5), 163(24.9), 164(28.2) |
| Einsteinium | Es | 99 | artificial isotopes only; mass number range 249–256 |
| Erbium | Er | 68 | 162(0.14), 164(1.61),166(33.6), 167(22.95), 168(26.8), 170(14.9) |
| Europium | Eu | 63 | 151(47.8), 153(52.2) |
| Fermium | Fm | 100 | artificial isotopes only; mass number range 251–257 |
| Fluorine | F | 9 | 19(100) |

| Element | Symbol | Atomic number, $Z$ | Mass number of isotope (% abundance) |
|---|---|---|---|
| Francium | Fr | 87 | artificial isotopes only; mass number range 210–227 |
| Gadolinium | Gd | 64 | 152(0.20), 154(2.18), 155(14.80), 156(20.47), 157(15.65), 158(24.84), 160(21.86) |
| Gallium | Ga | 31 | 69(60.1), 71(39.9) |
| Germanium | Ge | 32 | 70(20.5), 72(27.4), 73(7.8), 74(36.5), 76(7.8) |
| Gold | Au | 79 | 197(100) |
| Hafnium | Hf | 72 | 174(0.16), 176(5.20), 177(18.61), 178(27.30), 179(13.63), 180(35.10) |
| Helium | He | 2 | 3(<0.001), 4(>99.999) |
| Holmium | Ho | 67 | 165(100) |
| Hydrogen | H | 1 | 1(99.985), 2(0.015) |
| Indium | In | 49 | 113(4.3), 115(95.7) |
| Iodine | I | 53 | 127(100) |
| Iridium | Ir | 77 | 191(37.3), 193(62.7) |
| Iron | Fe | 26 | 54(5.8), 56(91.7), 57(2.2), 58(0.3) |
| Krypton | Kr | 36 | 78(0.35), 80(2.25), 82(11.6), 83(11.5), 84(57.0), 86(17.3) |
| Lanthanum | La | 57 | 138(0.09), 139(99.91) |
| Lawrencium | Lr | 103 | artificial isotopes only; mass number range 253–262 |
| Lead | Pb | 82 | 204(1.4), 206(24.1), 207(22.1), 208(52.4) |
| Lithium | Li | 3 | 6(7.5), 7(92.5) |
| Lutetium | Lu | 71 | 175(97.41), 176(2.59) |
| Magnesium | Mg | 12 | 24(78.99), 25(10.00), 26(11.01) |
| Manganese | Mn | 25 | 55 (100) |
| Mendelevium | Md | 101 | artificial isotopes only; mass number range 247–260 |
| Mercury | Hg | 80 | 196(0.14), 198(10.02), 199(16.84), 200(23.13), 201(13.22), 202(29.80), 204(6.85) |
| Molybdenum | Mo | 42 | 92(14.84), 94(9.25), 95(15.92), 96(16.68), 97(9.55), 98(24.13), 100(9.63) |
| Neodymium | Nd | 60 | 142(27.13), 143(12.18), 144(23.80), 145(8.30), 146(17.19), 148(5.76), 150(5.64) |
| Neon | Ne | 10 | 20(90.48), 21(0.27), 22(9.25) |
| Neptunium | Np | 93 | artificial isotopes only; mass number range 234–240 |
| Nickel | Ni | 28 | 58(68.27), 60(26.10), 61(1.13), 62(3.59), 64(0.91) |
| Niobium | Nb | 41 | 93(100) |
| Nitrogen | N | 7 | 14(99.63), 15(0.37) |
| Nobelium | No | 102 | artificial isotopes only; mass number range 250–262 |
| Osmium | Os | 76 | 184(0.02), 186(1.58), 187(1.6), 188(13.3), 189(16.1), 190(26.4), 192(41.0) |
| Oxygen | O | 8 | 16(99.76), 17(0.04), 18(0.20) |
| Palladium | Pd | 46 | 102(1.02), 104(11.14), 105(22.33), 106(27.33), 108(26.46), 110(11.72) |
| Phosphorus | P | 15 | 31(100) |
| Platinum | Pt | 78 | 190(0.01), 192(0.79), 194(32.9), 195(33.8), 196(25.3), 198(7.2) |
| Plutonium | Pu | 94 | artificial isotopes only; mass number range 234–246 |
| Polonium | Po | 84 | artificial isotopes only; mass number range 204–210 |
| Potassium | K | 19 | 39(93.26), 40(0.01), 41(6.73) |
| Praseodymium | Pr | 59 | 141(100) |
| Promethium | Pm | 61 | artificial isotopes only; mass number range 141–151 |

| Element | Symbol | Atomic number, $Z$ | Mass number of isotope (% abundance) |
|---|---|---|---|
| Protactinium | Pa | 91 | artificial isotopes only; mass number range 228–234 |
| Radium | Ra | 88 | artificial isotopes only; mass number range 223–230 |
| Radon | Rn | 86 | artificial isotopes only; mass number range 208–224 |
| Rhenium | Re | 75 | 185(37.40), 187(62.60) |
| Rhodium | Rh | 45 | 103(100) |
| Rubidium | Rb | 37 | 85(72.16), 87(27.84) |
| Ruthenium | Ru | 44 | 96(5.52), 98(1.88), 99(12.7), 100(12.6), 101(17.0), 102(31.6), 104(18.7) |
| Samarium | Sm | 62 | 144(3.1), 147(15.0), 148(11.3), 149(13.8), 150(7.4), 152(26.7), 154(22.7) |
| Scandium | Sc | 21 | 45(100) |
| Selenium | Se | 34 | 74(0.9), 76(9.2), 77(7.6), 78(23.6), 80(49.7), 82(9.0) |
| Silicon | Si | 14 | 28(92.23), 29(4.67), 30(3.10) |
| Silver | Ag | 47 | 107(51.84), 109(48.16) |
| Sodium | Na | 11 | 23(100) |
| Strontium | Sr | 38 | 84(0.56), 86(9.86), 87(7.00), 88(82.58) |
| Sulfur | S | 16 | 32(95.02), 33(0.75), 34(4.21), 36(0.02) |
| Tantalum | Ta | 73 | 180(0.01), 181(99.99) |
| Technetium | Tc | 43 | artificial isotopes only; mass number range 95–99 |
| Tellurium | Te | 52 | 120(0.09), 122(2.60), 123(0.91), 124(4.82), 125(7.14), 126(18.95), 128(31.69), 130(33.80) |
| Terbium | Tb | 65 | 159(100) |
| Thallium | Tl | 81 | 203(29.52), 205(70.48) |
| Thorium | Th | 90 | 232(100) |
| Thulium | Tm | 69 | 169(100) |
| Tin | Sn | 50 | 112(0.97), 114(0.65), 115(0.36), 116(14.53), 117(7.68), 118(24.22), 119(8.58), 120(32.59), 122(4.63), 124(5.79) |
| Titanium | Ti | 22 | 46(8.0), 47(7.3), 48(73.8), 49(5.5), 50(5.4) |
| Tungsten | W | 74 | 180(0.13), 182(26.3), 183(14.3), 184(30.67), 186(28.6) |
| Uranium | U | 92 | 234(0.005), 235(0.72), 236(99.275) |
| Vanadium | V | 23 | 50(0.25), 51(99.75) |
| Xenon | Xe | 54 | 124(0.10), 126(0.09), 128(1.91), 129(26.4), 130(4.1), 131(21.2), 132(26.9), 134(10.4), 136(8.9) |
| Ytterbium | Yb | 70 | 168(0.13), 170(3.05), 171(14.3), 172(21.9), 173(16.12), 174(31.8), 176(12.7) |
| Yttrium | Y | 39 | 89(100) |
| Zinc | Zn | 30 | 64(48.6), 66(27.9), 67(4.1), 68(18.8), 70(0.6) |
| Zirconium | Zr | 40 | 90(51.45), 91(11.22), 92(17.15), 94(17.38), 96(2.8) |

# Appendix 6  Van der Waals, metallic, covalent and ionic radii for the *s*-, *p*- and first row *d*-block elements

The ionic radius varies with the charge and coordination number of the ion; a coordination number of 6 refers to octahedral coordination, and of 4 refers to tetrahedral unless otherwise specified.

| | Element | Van der Waals radius, $r_v$ / pm | Metallic radius for 12-coordinate metal, $r_{metal}$ / pm | Covalent radius, $r_{cov}$ / pm | Ionic radius, $r_{ion}$ / pm | Charge on ion | Coordination number of the ion |
|---|---|---|---|---|---|---|---|
| | | | | | | **Ionic radius** | |
| **Hydrogen** | H | 120 | | 37[‡] | | | |
| **Group 1** | Li | | 157 | | 76 | 1+ | 6 |
| | Na | | 191 | | 102 | 1+ | 6 |
| | K | | 235 | | 138 | 1+ | 6 |
| | Rb | | 250 | | 149 | 1+ | 6 |
| | Cs | | 272 | | 170 | 1+ | 6 |
| **Group 2** | Be | | 112 | | 27 | 2+ | 4 |
| | Mg | | 160 | | 72 | 2+ | 6 |
| | Ca | | 197 | | 100 | 2+ | 6 |
| | Sr | | 215 | | 126 | 2+ | 8 |
| | Ba | | 224 | | 142 | 2+ | 8 |
| **Group 13** | B | 208 | | 88 | | | |
| | Al | | 143 | 130 | 54 | 3+ | 6 |
| | Ga | | 153 | 122 | 62 | 3+ | 6 |
| | In | | 167 | 150 | 80 | 3+ | 6 |
| | Tl | | 171 | 155 | 89 | 3+ | 6 |
| | | | | | 159 | 1+ | 8 |
| **Group 14** | C | 185 | | 77 | | | |
| | Si | 210 | | 118 | | | |
| | Ge | | | 122 | 53 | 4+ | 6 |
| | Sn | | 158 | 140 | 74 | 4+ | 6 |
| | Pb | | 175 | 154 | 119 | 2+ | 6 |
| | | | | | 65 | 4+ | 4 |
| | | | | | 78 | 4+ | 6 |

[‡] Sometimes it is more appropriate to use a value of 30 pm in organic compounds.

| | Element | Van der Waals radius, $r_v$ / pm | Metallic radius for 12-coordinate metal, $r_{metal}$ / pm | Covalent radius, $r_{cov}$ / pm | Ionic radius | | |
|---|---|---|---|---|---|---|---|
| | | | | | Ionic radius, $r_{ion}$ / pm | Charge on ion | Coordination number of the ion |
| **Group 15** | N | 154 | | 75 | 171 | 3− | 6 |
| | P | 190 | | 110 | | | |
| | As | 200 | | 122 | | | |
| | Sb | 220 | | 143 | | | |
| | Bi | 240 | 182 | 152 | 103 | 3+ | 6 |
| | | | | | 76 | 5+ | 6 |
| **Group 16** | O | 140 | | 73 | 140 | 2− | 6 |
| | S | 185 | | 103 | 184 | 2− | 6 |
| | Se | 200 | | 117 | 198 | 2− | 6 |
| | Te | 220 | | 135 | 211 | 2− | 6 |
| **Group 17** | F | 135 | | 71 | 133 | 1− | 6 |
| | Cl | 180 | | 99 | 181 | 1− | 6 |
| | Br | 195 | | 114 | 196 | 1− | 6 |
| | I | 215 | | 133 | 220 | 1− | 6 |
| **Group 18** | He | 99 | | | | | |
| | Ne | 160 | | | | | |
| | Ar | 191 | | | | | |
| | Kr | 197 | | | | | |
| | Xe | 214 | | | | | |
| **First row d-block elements** | Sc | | 164 | | 75 | 3+ | 6 |
| | Ti | | 147 | | 86 | 2+ | 6 |
| | | | | | 67 | 3+ | 6 |
| | | | | | 61 | 4+ | 6 |
| | V | | 135 | | 79 | 2+ | 6 |
| | | | | | 64 | 3+ | 6 |
| | | | | | 58 | 4+ | 6 |
| | | | | | 53 | 4+ | 5 |
| | | | | | 54 | 5+ | 6 |
| | | | | | 46 | 5+ | 5 |
| | Cr | | 129 | | 73 | 2+ | 6 (low-spin) |
| | | | | | 80 | 2+ | 6 (high-spin) |
| | | | | | 62 | 3+ | 6 |
| | Mn | | 137 | | 67 | 2+ | 6 (low-spin) |
| | | | | | 83 | 2+ | 6 (high-spin) |
| | | | | | 58 | 3+ | 6 (low-spin) |
| | | | | | 65 | 3+ | 6 (high-spin) |
| | | | | | 39 | 4+ | 4 |
| | | | | | 53 | 4+ | 6 |
| | Fe | | 126 | | 61 | 2+ | 6 (low-spin) |
| | | | | | 78 | 2+ | 6 (high-spin) |
| | | | | | 55 | 3+ | 6 (low-spin) |
| | | | | | 65 | 3+ | 6 (high-spin) |

| Element | Van der Waals radius, $r_v$ / pm | Metallic radius for 12-coordinate metal, $r_{metal}$ / pm | Covalent radius, $r_{cov}$ / pm | Ionic radius | | |
|---|---|---|---|---|---|---|
| | | | | Ionic radius, $r_{ion}$ / pm | Charge on ion | Coordination number of the ion |
| Co | | 125 | | 65 | 2+ | 6 (low-spin) |
| | | | | 75 | 2+ | 6 (high-spin) |
| | | | | 55 | 3+ | 6 (low-spin) |
| | | | | 61 | 3+ | 6 (high-spin) |
| Ni | | 125 | | 55 | 2+ | 4 |
| | | | | 44 | 2+ | 4 (square planar) |
| | | | | 69 | 2+ | 6 |
| | | | | 56 | 3+ | 6 (low-spin) |
| | | | | 60 | 3+ | 6 (high-spin) |
| Cu | | 128 | | 46 | 1+ | 2 |
| | | | | 60 | 1+ | 4 |
| | | | | 57 | 2+ | 4 (square planar) |
| | | | | 73 | 2+ | 6 |
| Zn | | 137 | | 60 | 2+ | 4 |
| | | | | 74 | 2+ | 6 |

# Appendix 7  Pauling electronegativity values ($\chi^P$) for selected elements of the periodic table

Values are dependent on oxidation state.

| Group 1 | Group 2 | | Group 13 | Group 14 | Group 15 | Group 16 | Group 17 |
|---|---|---|---|---|---|---|---|
| H 2.2 | | | | | | | |
| Li 1.0 | Be 1.6 | | B 2.0 | C 2.6 | N 3.0 | O 3.4 | F 4.0 |
| Na 0.9 | Mg 1.3 | | Al(III) 1.6 | Si 1.9 | P 2.2 | S 2.6 | Cl 3.2 |
| K 0.8 | Ca 1.0 | | Ga(III) 1.8 | Ge(IV) 2.0 | As(III) 2.2 | Se 2.6 | Br 3.0 |
| Rb 0.8 | Sr 0.9 | (*d*-block elements) | In(III) 1.8 | Sn(II) 1.8 Sn(IV) 2.0 | Sb 2.1 | Te 2.1 | I 2.7 |
| Cs 0.8 | Ba 0.9 | | Tl(I) 1.6 Tl(III) 2.0 | Pb(II) 1.9 Pb(IV) 2.3 | Bi 2.0 | Po 2.0 | At 2.2 |

# Appendix 8  Ground state electronic configurations of the elements and ionization energies for the first five ionizations[‡]

$IE(n)$ in kJ mol$^{-1}$ for the processes:

| $IE(1)$ | $M(g) \longrightarrow M^+(g)$ |
|---|---|
| $IE(2)$ | $M^+(g) \longrightarrow M^{2+}(g)$ |
| $IE(3)$ | $M^{2+}(g) \longrightarrow M^{3+}(g)$ |
| $IE(4)$ | $M^{3+}(g) \longrightarrow M^{4+}(g)$ |
| $IE(5)$ | $M^{4+}(g) \longrightarrow M^{5+}(g)$ |

| Atomic number, $Z$ | Element | Ground state electronic configuration | $IE(1)$ | $IE(2)$ | $IE(3)$ | $IE(4)$ | $IE(5)$ |
|---|---|---|---|---|---|---|---|
| 1 | H | $1s^1$ | 1312 | | | | |
| 2 | He | $1s^2$ = [He] | 2372 | 5250 | | | |
| 3 | Li | [He]$2s^1$ | 520.2 | 7298 | 11820 | | |
| 4 | Be | [He]$2s^2$ | 899.5 | 1757 | 14850 | 21010 | |
| 5 | B | [He]$2s^2 2p^1$ | 800.6 | 2427 | 3660 | 25030 | 32830 |
| 6 | C | [He]$2s^2 2p^2$ | 1086 | 2353 | 4620 | 6223 | 37830 |
| 7 | N | [He]$2s^2 2p^3$ | 1402 | 2856 | 4578 | 7475 | 9445 |
| 8 | O | [He]$2s^2 2p^4$ | 1314 | 3388 | 5300 | 7469 | 10990 |
| 9 | F | [He]$2s^2 2p^5$ | 1681 | 3375 | 6050 | 8408 | 11020 |
| 10 | Ne | [He]$2s^2 2p^6$ = [Ne] | 2081 | 3952 | 6122 | 9371 | 12180 |
| 11 | Na | [Ne]$3s^1$ | 495.8 | 4562 | 6910 | 9543 | 13350 |
| 12 | Mg | [Ne]$3s^2$ | 737.7 | 1451 | 7733 | 10540 | 13630 |
| 13 | Al | [Ne]$3s^2 3p^1$ | 577.5 | 1817 | 2745 | 11580 | 14840 |
| 14 | Si | [Ne]$3s^2 3p^2$ | 786.5 | 1577 | 3232 | 4356 | 16090 |
| 15 | P | [Ne]$3s^2 3p^3$ | 1012 | 1907 | 2914 | 4964 | 6274 |
| 16 | S | [Ne]$3s^2 3p^4$ | 999.6 | 2252 | 3357 | 4556 | 7004 |
| 17 | Cl | [Ne]$3s^2 3p^5$ | 1251 | 2298 | 3822 | 5159 | 6540 |
| 18 | Ar | [Ne]$3s^2 3p^6$ = [Ar] | 1521 | 2666 | 3931 | 5771 | 7238 |
| 19 | K | [Ar]$4s^1$ | 418.8 | 3052 | 4420 | 5877 | 7975 |
| 20 | Ca | [Ar]$4s^2$ | 589.8 | 1145 | 4912 | 6491 | 8153 |
| 21 | Sc | [Ar]$4s^2 3d^1$ | 633.1 | 1235 | 2389 | 7091 | 8843 |
| 22 | Ti | [Ar]$4s^2 3d^2$ | 658.8 | 1310 | 2653 | 4175 | 9581 |
| 23 | V | [Ar]$4s^2 3d^3$ | 650.9 | 1414 | 2828 | 4507 | 6299 |
| 24 | Cr | [Ar]$4s^1 3d^5$ | 652.9 | 1591 | 2987 | 4743 | 6702 |
| 25 | Mn | [Ar]$4s^2 3d^5$ | 717.3 | 1509 | 3248 | 4940 | 6990 |

[‡] Values are from several sources, but mostly from the *Handbook of Chemistry and Physics* (1993) 74th edn, CRC Press, Boca Raton, and from the NIST Physics Laboratory, Physical Reference Data. The values in kJ mol$^{-1}$ are quoted to 4 significant figures or less depending upon the accuracy of the original data in eV. A conversion factor of $1\,\text{eV} = 96.485\,\text{kJ mol}^{-1}$ has been applied.

| Atomic number, $Z$ | Element | Ground state electronic configuration | $IE(1)$ | $IE(2)$ | $IE(3)$ | $IE(4)$ | $IE(5)$ |
|---|---|---|---|---|---|---|---|
| 26 | Fe | $[Ar]4s^2 3d^6$ | 762.5 | 1562 | 2957 | 5290 | 7240 |
| 27 | Co | $[Ar]4s^2 3d^7$ | 760.4 | 1648 | 3232 | 4950 | 7670 |
| 28 | Ni | $[Ar]4s^2 3d^8$ | 737.1 | 1753 | 3395 | 5300 | 7339 |
| 29 | Cu | $[Ar]4s^1 3d^{10}$ | 745.5 | 1958 | 3555 | 5536 | 7700 |
| 30 | Zn | $[Ar]4s^2 3d^{10}$ | 906.4 | 1733 | 3833 | 5730 | 7970 |
| 31 | Ga | $[Ar]4s^2 3d^{10} 4p^1$ | 578.8 | 1979 | 2963 | 6200 | |
| 32 | Ge | $[Ar]4s^2 3d^{10} 4p^2$ | 762.2 | 1537 | 3302 | 4411 | 9020 |
| 33 | As | $[Ar]4s^2 3d^{10} 4p^3$ | 947.0 | 1798 | 2735 | 4837 | 6043 |
| 34 | Se | $[Ar]4s^2 3d^{10} 4p^4$ | 941.0 | 2045 | 2974 | 4144 | 6590 |
| 35 | Br | $[Ar]4s^2 3d^{10} 4p^5$ | 1140 | 2100 | 3500 | 4560 | 5760 |
| 36 | Kr | $[Ar]4s^2 3d^{10} 4p^6 = [Kr]$ | 1351 | 2350 | 3565 | 5070 | 6240 |
| 37 | Rb | $[Kr]5s^1$ | 403.0 | 2633 | 3900 | 5080 | 6850 |
| 38 | Sr | $[Kr]5s^2$ | 549.5 | 1064 | 4138 | 5500 | 6910 |
| 39 | Y | $[Kr]5s^2 4d^1$ | 599.8 | 1181 | 1980 | 5847 | 7430 |
| 40 | Zr | $[Kr]5s^2 4d^2$ | 640.1 | 1267 | 2218 | 3313 | 7752 |
| 41 | Nb | $[Kr]5s^1 4d^4$ | 652.1 | 1382 | 2416 | 3700 | 4877 |
| 42 | Mo | $[Kr]5s^1 4d^5$ | 684.3 | 1559 | 2618 | 4480 | 5257 |
| 43 | Tc | $[Kr]5s^2 4d^5$ | 702 | 1472 | 2850 | | |
| 44 | Ru | $[Kr]5s^1 4d^7$ | 710.2 | 1617 | 2747 | | |
| 45 | Rh | $[Kr]5s^1 4d^8$ | 719.7 | 1744 | 2997 | | |
| 46 | Pd | $[Kr]5s^0 4d^{10}$ | 804.4 | 1875 | 3177 | | |
| 47 | Ag | $[Kr]5s^1 4d^{10}$ | 731.0 | 2073 | 3361 | | |
| 48 | Cd | $[Kr]5s^2 4d^{10}$ | 867.8 | 1631 | 3616 | | |
| 49 | In | $[Kr]5s^2 4d^{10} 5p^1$ | 558.3 | 1821 | 2704 | 5200 | |
| 50 | Sn | $[Kr]5s^2 4d^{10} 5p^2$ | 708.6 | 1412 | 2943 | 3930 | 6974 |
| 51 | Sb | $[Kr]5s^2 4d^{10} 5p^3$ | 830.6 | 1595 | 2440 | 4260 | 5400 |
| 52 | Te | $[Kr]5s^2 4d^{10} 5p^4$ | 869.3 | 1790 | 2698 | 3610 | 5668 |
| 53 | I | $[Kr]5s^2 4d^{10} 5p^5$ | 1008 | 1846 | 3200 | | |
| 54 | Xe | $[Kr]5s^2 4d^{10} 5p^6 = [Xe]$ | 1170 | 2046 | 3099 | | |
| 55 | Cs | $[Xe]6s^1$ | 375.7 | 2234 | 3400 | | |
| 56 | Ba | $[Xe]6s^2$ | 502.8 | 965.2 | 3619 | | |
| 57 | La | $[Xe]6s^2 5d^1$ | 538.1 | 1067 | 1850 | 4819 | 5940 |
| 58 | Ce | $[Xe]4f^1 6s^2 5d^1$ | 534.4 | 1047 | 1949 | 3546 | 6325 |
| 59 | Pr | $[Xe]4f^3 6s^2$ | 527.2 | 1018 | 2086 | 3761 | 5551 |
| 60 | Nd | $[Xe]4f^4 6s^2$ | 533.1 | 1035 | 2130 | 3898 | |
| 61 | Pm | $[Xe]4f^5 6s^2$ | 538.8 | 1052 | 2150 | 3970 | |
| 62 | Sm | $[Xe]4f^6 6s^2$ | 544.5 | 1068 | 2260 | 3990 | |
| 63 | Eu | $[Xe]4f^7 6s^2$ | 547.1 | 1085 | 2404 | 4120 | |
| 64 | Gd | $[Xe]4f^7 6s^2 5d^1$ | 593.4 | 1167 | 1990 | 4245 | |
| 65 | Tb | $[Xe]4f^9 6s^2$ | 565.8 | 1112 | 2114 | 3839 | |
| 66 | Dy | $[Xe]4f^{10} 6s^2$ | 573.0 | 1126 | 2200 | 3990 | |
| 67 | Ho | $[Xe]4f^{11} 6s^2$ | 581.0 | 1139 | 2204 | 4100 | |
| 68 | Er | $[Xe]4f^{12} 6s^2$ | 589.3 | 1151 | 2194 | 4120 | |
| 69 | Tm | $[Xe]4f^{13} 6s^2$ | 596.7 | 1163 | 2285 | 4120 | |
| 70 | Yb | $[Xe]4f^{14} 6s^2$ | 603.4 | 1175 | 2417 | 4203 | |
| 71 | Lu | $[Xe]4f^{14} 6s^2 5d^1$ | 523.5 | 1340 | 2022 | 4366 | |
| 72 | Hf | $[Xe]4f^{14} 6s^2 5d^2$ | 658.5 | 1440 | 2250 | 3216 | |
| 73 | Ta | $[Xe]4f^{14} 6s^2 5d^3$ | 728.4 | 1500 | 2100 | | |

| Atomic number, $Z$ | Element | Ground state electronic configuration | $IE(1)$ | $IE(2)$ | $IE(3)$ | $IE(4)$ | $IE(5)$ |
|---|---|---|---|---|---|---|---|
| 74 | W | $[Xe]4f^{14}6s^25d^4$ | 758.8 | 1700 | 2300 | | |
| 75 | Re | $[Xe]4f^{14}6s^25d^5$ | 755.8 | 1260 | 2510 | | |
| 76 | Os | $[Xe]4f^{14}6s^25d^6$ | 814.2 | 1600 | 2400 | | |
| 77 | Ir | $[Xe]4f^{14}6s^25d^7$ | 865.2 | 1680 | 2600 | | |
| 78 | Pt | $[Xe]4f^{14}6s^15d^9$ | 864.4 | 1791 | 2800 | | |
| 79 | Au | $[Xe]4f^{14}6s^15d^{10}$ | 890.1 | 1980 | 2900 | | |
| 80 | Hg | $[Xe]4f^{14}6s^25d^{10}$ | 1007 | 1810 | 3300 | | |
| 81 | Tl | $[Xe]4f^{14}6s^25d^{10}6p^1$ | 589.4 | 1971 | 2878 | 4900 | |
| 82 | Pb | $[Xe]4f^{14}6s^25d^{10}6p^2$ | 715.6 | 1450 | 3081 | 4083 | 6640 |
| 83 | Bi | $[Xe]4f^{14}6s^25d^{10}6p^3$ | 703.3 | 1610 | 2466 | 4370 | 5400 |
| 84 | Po | $[Xe]4f^{14}6s^25d^{10}6p^4$ | 812.1 | 1800 | 2700 | | |
| 85 | At | $[Xe]4f^{14}6s^25d^{10}6p^5$ | 930 | 1600 | 2900 | | |
| 86 | Rn | $[Xe]4f^{14}6s^25d^{10}6p^6 = [Rn]$ | 1037 | | | | |
| 87 | Fr | $[Rn]7s^1$ | 393.0 | 2100 | 3100 | | |
| 88 | Ra | $[Rn]7s^2$ | 509.3 | 979.0 | 3300 | | |
| 89 | Ac | $[Rn]6d^17s^2$ | 499 | 1170 | 1900 | | |
| 90 | Th | $[Rn]6d^27s^2$ | 608.5 | 1110 | 1930 | 2780 | |
| 91 | Pa | $[Rn]5f^27s^26d^1$ | 568 | 1130 | 1810 | | |
| 92 | U | $[Rn]5f^37s^26d^1$ | 597.6 | 1440 | 1840 | | |
| 93 | Np | $[Rn]5f^47s^26d^1$ | 604.5 | 1130 | 1880 | | |
| 94 | Pu | $[Rn]5f^67s^2$ | 581.4 | 1130 | 2100 | | |
| 95 | Am | $[Rn]5f^77s^2$ | 576.4 | 1160 | 2160 | | |
| 96 | Cm | $[Rn]5f^77s^26d^1$ | 578.0 | 1200 | 2050 | | |
| 97 | Bk | $[Rn]5f^97s^2$ | 598.0 | 1190 | 2150 | | |
| 98 | Cf | $[Rn]5f^{10}7s^2$ | 606.1 | 1210 | 2280 | | |
| 99 | Es | $[Rn]5f^{11}7s^2$ | 619 | 1220 | 2330 | | |
| 100 | Fm | $[Rn]5f^{12}7s^2$ | 627 | 1230 | 2350 | | |
| 101 | Md | $[Rn]5f^{13}7s^2$ | 635 | 1240 | 2450 | | |
| 102 | No | $[Rn]5f^{14}7s^2$ | 642 | 1250 | 2600 | | |
| 103 | Lr | $[Rn]5f^{14}7s^26d^1$ | 440 (?) | | | | |

# Appendix 9 Electron affinities

Approximate enthalpy changes, $\Delta_{EA}H(298\,K)$ associated with the gain of one electron by a gaseous atom or anion. A negative enthalpy ($\Delta H$), but a positive electron affinity ($EA$), corresponds to an exothermic process (see Section 8.5).

$$\Delta_{EA}H(298\,K) \approx \Delta U(0\,K) = -EA$$

|  | **Process** | $\approx \Delta_{EA}H\ /\ \text{kJ mol}^{-1}$ |
|---|---|---|
| **Hydrogen** | $H(g) + e^- \longrightarrow H^-(g)$ | $-73$ |
| **Group 1** | $Li(g) + e^- \longrightarrow Li^-(g)$ | $-60$ |
|  | $Na(g) + e^- \longrightarrow Na^-(g)$ | $-53$ |
|  | $K(g) + e^- \longrightarrow K^-(g)$ | $-48$ |
|  | $Rb(g) + e^- \longrightarrow Rb^-(g)$ | $-47$ |
|  | $Cs(g) + e^- \longrightarrow Cs^-(g)$ | $-45$ |
| **Group 15** | $N(g) + e^- \longrightarrow N^-(g)$ | $\approx 0$ |
|  | $P(g) + e^- \longrightarrow P^-(g)$ | $-72$ |
|  | $As(g) + e^- \longrightarrow As^-(g)$ | $-78$ |
|  | $Sb(g) + e^- \longrightarrow Sb^-(g)$ | $-103$ |
|  | $Bi(g) + e^- \longrightarrow Bi^-(g)$ | $-91$ |
| **Group 16** | $O(g) + e^- \longrightarrow O^-(g)$ | $-141$ |
|  | $O^-(g) + e^- \longrightarrow O^{2-}(g)$ | $+798$ |
|  | $S(g) + e^- \longrightarrow S^-(g)$ | $-201$ |
|  | $S^-(g) + e^- \longrightarrow S^{2-}(g)$ | $+640$ |
|  | $Se(g) + e^- \longrightarrow Se^-(g)$ | $-195$ |
|  | $Te(g) + e^- \longrightarrow Te^-(g)$ | $-190$ |
| **Group 17** | $F(g) + e^- \longrightarrow F^-(g)$ | $-328$ |
|  | $Cl(g) + e^- \longrightarrow Cl^-(g)$ | $-349$ |
|  | $Br(g) + e^- \longrightarrow Br^-(g)$ | $-325$ |
|  | $I(g) + e^- \longrightarrow I^-(g)$ | $-295$ |

# Appendix 10 — Standard enthalpies of atomization $(\Delta_a H^\circ)$ of the elements at 298 K

Enthalpies are given in $kJ\,mol^{-1}$ for the process:

$$\frac{1}{n}E_n(\text{standard state}) \longrightarrow E(g)$$

Elements (E) are arranged according to their position in the periodic table. The lanthanoids and actinoids are excluded. The noble gases are omitted because they are monatomic at 298 K.

| 1 | 2 | 3 | 4 | 5 | 6 | 7 | 8 | 9 | 10 | 11 | 12 | 13 | 14 | 15 | 16 | 17 |
|---|---|---|---|---|---|---|---|---|----|----|----|----|----|----|----|----|
| **H** 218 | | | | | | | | | | | | | | | | |
| **Li** 161 | **Be** 324 | | | | | | | | | | | **B** 582 | **C** 717 | **N** 473 | **O** 249 | **F** 79 |
| **Na** 108 | **Mg** 146 | | | | | | | | | | | **Al** 330 | **Si** 456 | **P** 315 | **S** 277 | **Cl** 121 |
| **K** 90 | **Ca** 178 | **Sc** 378 | **Ti** 470 | **V** 514 | **Cr** 397 | **Mn** 283 | **Fe** 418 | **Co** 428 | **Ni** 430 | **Cu** 338 | **Zn** 130 | **Ga** 277 | **Ge** 375 | **As** 302 | **Se** 227 | **Br** 112 |
| **Rb** 82 | **Sr** 164 | **Y** 423 | **Zr** 609 | **Nb** 721 | **Mo** 658 | **Tc** 677 | **Ru** 651 | **Rh** 556 | **Pd** 377 | **Ag** 285 | **Cd** 112 | **In** 243 | **Sn** 302 | **Sb** 264 | **Te** 197 | **I** 107 |
| **Cs** 78 | **Ba** 178 | **La** 423 | **Hf** 619 | **Ta** 782 | **W** 850 | **Re** 774 | **Os** 787 | **Ir** 669 | **Pt** 566 | **Au** 368 | **Hg** 61 | **Tl** 182 | **Pb** 195 | **Bi** 210 | **Po** ≈146 | **At** 92 |

The values of $\Delta_f H^\circ(298\,\text{K})$, $\Delta_f G^\circ(298\,\text{K})$ and $S^\circ$ refer to the state specified. The abbreviations for states are: s = solid, l = liquid, g = gas. The standard state pressure is 1 bar ($10^5\,\text{Pa}$). In the melting and boiling point columns: sub = sublimes, dec = decomposes.

Compounds containing carbon (with the exception of metal carbonates) are arranged alphabetically *by name*. Compounds *not* containing carbon are arranged according to the non-carbon element under which the compounds are described in the text; for example, metal hydrides were described in Chapter 21 and are listed under hydrogen. The list is necessarily selective, and not all elements are represented.

| Compound | Molecular or structural formula | State | Melting point / K | Boiling point / K | $\Delta_f H^o$(298 K) / kJ mol⁻¹ | $\Delta_f G^o$(298 K) / kJ mol⁻¹ | $S^o$ / J K⁻¹ mol⁻¹ |
|---|---|---|---|---|---|---|---|
| **Compounds containing carbon (excluding metal carbonates)** | | | | | | | |
| Acetaldehyde (ethanal) | | g | 152 | 294 | −166 | −133 | 264 |
| | | l | 152 | 294 | −192 | −128 | 160 |
| Acetamide (ethanamide) | | s | 355 | 494 | −317 | | 115 |
| Acetic acid (ethanoic acid) | | l | 290 | 391 | −484.5 | −390 | 160 |
| Acetone (propanone) | | l | 178 | 329 | −248 | 200 | 126 |
| Acetyl chloride (ethanoyl chloride) | | l | 161 | 324 | −274 | −208 | 201 |
| Aniline (phenylamine) | | l | 267 | 457 | 31 | 149 | 191 |
| Benzaldehyde | | l | 247 | 451 | −87 | | 221 |
| Benzene | | l | 278.5 | 353 | 49 | 124 | 173 |

| Compound | Molecular or structural formula | State | Melting point / K | Boiling point / K | $\Delta_f H^o$(298 K) / kJ mol$^{-1}$ | $\Delta_f G^o$(298 K) / kJ mol$^{-1}$ | $S^o$ / J K$^{-1}$ mol$^{-1}$ |
|---|---|---|---|---|---|---|---|
| Benzoic acid | | s | 395 | 522 | −385 | | 168 |
| Bromobenzene | | l | 242 | 429 | 61 | | 219 |
| Bromoethane | | l | 154 | 311 | −90 | −26 | 199 |
| (E)-But-2-ene | | g | 167.5 | 274 | −11 | | |
| (Z)-But-2-ene | | g | 134 | 277 | −7 | | |
| Buta-1,2-diene | | g | 137 | 284 | 162 | | |
| Buta-1,3-diene | | g | 164 | 269 | 110 | | |
| Butan-1-ol | | l | 183.5 | 390 | −327 | | 226 |
| Butanal | | l | 174 | 349 | −239 | | 247 |
| Butane | | g | 135 | 272.5 | −126 | | |
| Butanoic acid | | l | 268.5 | 438.5 | −534 | | 222 |

| Compound | State | | | | | |
|---|---|---|---|---|---|---|
| Butanone | l | 187 | 353 | −273 | | 239 |
| 1-Butylamine | l | 224 | 351 | −128 | | |
| Carbon dioxide ($CO_2$) | g | 195 (sub) | | −393.5 | −394 | 214 |
| Carbon disulfide ($CS_2$) | l | 161.5 | 319 | 89 | 65 | 151 |
| Carbon monoxide ($CO$) | g | 68 | 82 | −110.5 | −137 | 198 |
| Chlorobenzene | l | 227 | 405 | 11 | | |
| Chloroethane | g | 137 | 285 | −112 | −60 | 276 |
| Cyclohexane | l | 279.5 | 354 | −156 | | |
| Cyclohexene | l | 169.5 | 356 | −38.5 | | 215 |
| Cyclopentane | l | 179 | 322 | −105 | | 204.5 |
| Cyclopentene | l | 138 | 317 | 4 | | 201 |
| 1,2-Dibromoethane | l | 210 | 381 | −79 | | 223 |
| 1,2-Dichloroethane | l | 238 | 357 | −167 | | |
| (E)-1,2-Dichloroethene | l | 223 | 320.5 | −23 | 27 | 196 |

| Compound | Molecular or structural formula | State | Melting point / K | Boiling point / K | $\Delta_f H^o$(298 K) / kJ mol⁻¹ | $\Delta_f G^o$(298 K) / kJ mol⁻¹ | $S^o$ / J K⁻¹ mol⁻¹ |
|---|---|---|---|---|---|---|---|
| (Z)-1,2-Dichloroethene | | l | 192.5 | 333 | −26 | | 198 |
| 1,1-Dichloroethene | | l | 151 | 310 | −24 | 24 | 201.5 |
| Dichloromethane | | l | 178 | 313 | −124 | | 178 |
| Diethyl ether | | l | 157 | 307 | −279 | | 172 |
| Diethylamine | | l | 225 | 329 | −104 | | |
| Dimethylamine | | g | 180 | 280 | −18.5 | 68.5 | 273 |
| Ethanal (see acetaldehyde) | | | | | | | |
| Ethanamide (see acetamide) | | | | | | | |
| Ethane | $C_2H_6$ | g | 90 | 184 | −84 | −32 | 230 |
| 1,2-Ethanediamine | | l | 281.5 | 389.5 | −63 | | |
| Ethane-1,2-diol | | l | 261.5 | 471 | −455 | | 163 |
| Ethanoic acid (see acetic acid) | | | | | | | |
| Ethanol | | l | 156 | 351.5 | −278 | −175 | 161 |

| Compound | State | | | | | |
|---|---|---|---|---|---|---|
| Ethanoyl chloride (see acetyl chloride) | | | | | | |
| Ethene | g | 104 | 169 | 52.5 | 68 | 220 |
| Ethyl acetate (ethyl ethanoate) | l | 189 | 350 | −479 | | 258 |
| Ethyl benzene | l | 178 | 409 | −12 | | |
| Ethyl ethanoate (see ethyl acetate) | | | | | | |
| Ethylamine | g | 192 | 290 | −47.5 | 36 | 284 |
| Ethyne (acetylene) | g | 189 | 192 | 228 | 211 | 201 |
| Heptane | l | 182 | 371 | −224 | | |
| Hexane | l | 178 | 342 | −199 | | |
| Hexanoic acid | l | 271 | 478 | −584 | | |
| Hex-1-ene | l | 133 | 336 | −74 | | 295 |
| Hydrogen cyanide | l | 260 | 299 | 109 | 125 | 113 |
| | g | | | 135 | 125 | 202 |
| Iodobenzene | l | 242 | 461 | 117 | | 205 |
| Methane | g | 91 | 109 | −74 | −50 | 186 |
| Methanol | l | 179 | 338 | −239 | −167 | 127 |

| Compound | Molecular or structural formula | State | Melting point / K | Boiling point / K | $\Delta_f H^o$(298 K) / kJ mol$^{-1}$ | $\Delta_f G^o$(298 K) / kJ mol$^{-1}$ | $S^o$ / J K$^{-1}$ mol$^{-1}$ |
|---|---|---|---|---|---|---|---|
| 2-Methylpropan-1-ol | | l | 165 | 381 | −335 | | 215 |
| 2-Methylpropan-2-ol | | l | 298 | 355 | −359 | | 193 |
| Nitrobenzene | | l | 279 | 484 | 12.5 | | |
| Octane | | l | 216 | 399 | −250 | | |
| Oxalic acid | | s | 430 (sub) | | −822 | | 110 |
| Pentane | | l | 143 | 309 | −173.5 | | |
| Pentane-2,4-dione (Hacac) | | l | 250 | 412 | −424 | | |
| Phenol | | s | 316 | 455 | −165 | | 144 |
| Phenylamine (see aniline) | | | | | | | |
| Propan-1-ol | | l | 146.5 | 370 | −303 | | 194 |
| Propan-2-ol | | l | 183.5 | 355 | −318 | | 181 |

| Compound | State | | | | | |
|---|---|---|---|---|---|---|
| Propane | g | 83 | 231 | −105 | −23 | 270 |
| Propanoic acid | l | 252 | 414 | −511 | 191 | 153 |
| Propene | g | 88 | 226 | 20 | 63 | 267 |
| Propyne | g | 171.5 | 250 | 185 | | |
| Pyridine | l | 231 | 388.5 | 100 | | |
| Sodium acetate | s | 597 | | −709 | −607 | 123 |
| Tetrabromomethane (CBr$_4$) | s | 363 | 462 | 19 | 48 | 212.5 |
| Tetrachloromethane (CCl$_4$) | l | 250 | 349.5 | −128 | | |
| Tetrahydrofuran | l | 165 | 340 | −216 | | 204 |
| Toluene | l | 178 | 384 | 12 | 114 | 221 |
| 1,1,1-Trichloroethane | l | 243 | 347 | −177 | | 227 |
| Trichloromethane (CHCl$_3$) | l | 209.5 | 335 | −134.5 | −74 | 202 |
| Triethylamine | l | 158 | 362 | −128 | | |

| Compound | Molecular or structural formula | State | Melting point / K | Boiling point / K | $\Delta_f H^o$(298 K) / kJ mol$^{-1}$ | $\Delta_f G^o$(298 K) / kJ mol$^{-1}$ | $S^o$ / J K$^{-1}$ mol$^{-1}$ |
|---|---|---|---|---|---|---|---|
| **Compounds *not* containing carbon but including metal carbonates** | | | | | | | |
| **Aluminium** | | | | | | | |
| Aluminium | Al | s | 933 | 2793 | 0 | 0 | 28 |
| Aluminium oxide | Al$_2$O$_3$ | s | 2345 | 3253 | −1676 | −1582 | 51 |
| Aluminium(III) chloride | AlCl$_3$ or Al$_2$Cl$_6$ | s | 463 (at 2.5 bar) | 535 (dec) | −704 | −629 | 111 |
| Aluminium(III) sulfate | Al$_2$(SO$_4$)$_3$ | s | 1043 (dec) | | −3441 | | |
| **Arsenic** | | | | | | | |
| Arsenic (grey) | As (grey) | s | 887 (sub) | | 0 | 0 | 35 |
| Arsenic (yellow) | As (yellow) | s | | | 15 | | |
| Arsenic(III) chloride | AsCl$_3$ | l | 264 | 403 | −305 | −259 | 216 |
| Arsenic(III) fluoride | AsF$_3$ | l | 267 | 330 | −821 | −774 | 181 |
| Arsenic(V) fluoride | AsF$_5$ | s | 193 | 220 | | | |
| Arsenic(V) oxide | As$_2$O$_5$ | s | 588 (dec) | | −925 | −782 | 105 |
| **Barium** | | | | | | | |
| Barium | Ba | s | 1000 | 2170 | 0 | 0 | 63 |
| Barium chloride | BaCl$_2$ | s | 1236 | 1833 | −859 | −810 | 124 |
| Barium hydroxide | Ba(OH)$_2$ | s | 681 (dec) | | −945 | | |
| Barium nitrate | Ba(NO$_3$)$_2$ | s | 865 | (dec) | −992 | −797 | 214 |
| Barium oxide | BaO | s | 2191 | ≈2273 | −553.5 | −525 | 70 |
| Barium sulfate | BaSO$_4$ | s | 1853 | | −1473 | −1362 | 132 |
| **Beryllium** | | | | | | | |
| Beryllium | Be | s | 1560 | 2744 | 0 | 0 | 9.5 |
| Beryllium chloride | BeCl$_2$ | s | 678 | 793 | −490 | −446 | 83 |
| Beryllium nitrate trihydrate | Be(NO$_3$)$_2$·3H$_2$O | s | 333 | 415 | −788 | | |
| Beryllium oxide | BeO | s | 2803 | ≈4173 | −609 | −580 | 14 |
| **Bismuth** | | | | | | | |
| Bismuth | Bi | s | 544 | 1837 | 0 | 0 | 57 |
| Bismuth(III) chloride | BiCl$_3$ | s | 503 | 720 | −379 | −315 | 177 |
| Bismuth(III) hydroxide | Bi(OH)$_3$ | s | 373 (dec, −H$_2$O) | | −711 | | |
| Bismuth(III) oxide | Bi$_2$O$_3$ | s | 1098 | ≈2163 | −574 | −494 | 151 |
| **Boron** | | | | | | | |
| Boron (β-rhombohedral) | B(β-rhombohedral) | s | 2453 | 4273 | 0 | 0 | 6 |
| Borazine | [HBNH]$_3$ | l | 215 | 328 | −541 | −393 | 200 |
| Boric acid (boracic acid, orthoboric acid) | H$_3$BO$_3$ or B(OH)$_3$ | s | 442 (−H$_2$O) | | −1094 | −969 | 89 |
| Boron tribromide | BBr$_3$ | l | 227 | 364 | −240 | −238.5 | 230 |
| Boron trichloride | BCl$_3$ | l | 166 | 285 | −427 | −387 | 206 |
| Boron trifluoride | BF$_3$ | g | 144 | 172 | −1136 | −1119 | 254 |

| | | | | | | | |
|---|---|---|---|---|---|---|---|
| **Bromine** | | | | | | | |
| Bromine | Br$_2$ | l | 266 | | 0 | 0 | 152 |
| | | g | | 332 | 31 | 3 | 245.5 |
| **Caesium** | | | | | | | |
| Caesium | Cs | s | 301.5 | 978 | 0 | 0 | 85 |
| Caesium bromide | CsBr | s | 909 | 1573 | −406 | −391 | 113 |
| Caesium chloride | CsCl | s | 918 | 1563 | −442 | −414.5 | 101 |
| Caesium fluoride | CsF | s | 955 | 1524 | −553.5 | −525.5 | 93 |
| Caesium iodide | CsI | s | 899 | 1553 | −347 | −341 | 123 |
| Caesium oxide | Cs$_2$O | s | 673 (dec) | | −346 | −308 | 147 |
| **Calcium** | | | | | | | |
| Calcium | Ca | s | 1115 | 1757 | 0 | 0 | 42 |
| Calcium bromide | CaBr$_2$ | s | 1015 | 2088 | −683 | −664 | 130 |
| Calcium carbonate (aragonite) | CaCO$_3$ | s | 793 → calcite | 1098 (dec) | −1208 | −1128 | 88 |
| Calcium chloride | CaCl$_2$ | s | 1055 | >1873 | −795 | −749 | 108 |
| Calcium fluoride | CaF$_2$ | s | 1696 | ≈2773 | −1228 | −1176 | 68.5 |
| Calcium hydroxide | Ca(OH)$_2$ | s | 853 (−H$_2$O) | (dec) | −985 | −898 | 83 |
| Calcium nitrate | Ca(NO$_3$)$_2$ | s | 834 | | −938 | −743 | 193 |
| Calcium oxide | CaO | s | 2887 | 3123 | −635 | −603 | 38 |
| Calcium phosphate | Ca$_3$(PO$_4$)$_2$ | s | 1943 | | −4121 | −3885 | 236 |
| Calcium sulfate | CaSO$_4$ | s | 1723 (dec) | | −1435 | −1332 | 106.5 |
| **Chlorine** | | | | | | | |
| Chlorine | Cl$_2$ | g | 172 | 239 | 0 | 0 | 223 |
| Chlorine dioxide | ClO$_2$ | g | 213 | explodes | 102.5 | 120.5 | 257 |
| Oxygen dichloride | Cl$_2$O | g | 253 | explodes | 80 | 98 | 266 |
| **Chromium** | | | | | | | |
| Chromium | Cr | s | 2180 | 2944 | 0 | 0 | 24 |
| Chromium(II) chloride | CrCl$_2$ | s | 1097 | 1573 | −395 | −356 | 115 |
| Chromium(III) chloride | CrCl$_3$ | s | 1423 | | −556.5 | −486 | 123 |
| Chromium(III) oxide | Cr$_2$O$_3$ | s | 2539 | 4273 | −1140 | −1058 | 81 |
| **Cobalt** | | | | | | | |
| Cobalt | Co | s | 1768 | 3200 | 0 | 0 | 30 |
| Cobalt(II) chloride | CoCl$_2$ | s | 997 (under HCl gas) | 1322 | −312.5 | −270 | 109 |
| Cobalt(II) nitrate hexahydrate | Co(NO$_3$)$_2$·6H$_2$O | s | 328 (−H$_2$O) | | −2216 | | |
| Cobalt(II) oxide | CoO | s | 2068 | | −238 | −214 | 53 |
| Cobalt(II) sulfate | CoSO$_4$ | s | 1008 (dec) | | −888 | −782 | 118 |
| **Copper** | | | | | | | |
| Copper | Cu | s | 1358 | 2835 | 0 | 0 | 33 |
| Copper(I) chloride | CuCl | s | 703 | 1763 | −137 | −120 | 86 |
| Copper(II) chloride | CuCl$_2$ | s | 893 | 1266 (dec) | −220 | −176 | 108 |
| Copper(I) oxide | Cu$_2$O | s | 1508 | 2073 (dec) | −169 | −146 | 93 |
| Copper(II) oxide | CuO | s | 1599 | | −157 | −130 | 43 |
| Copper(II) sulfate (anhydrous) | CuSO$_4$ | s | 473 (dec) | | −771 | −662 | 109 |
| Copper(II) sulfate | CuSO$_4$·5H$_2$O | s | 383 (−4H$_2$O), 423 (−H$_2$O) | | | | |

| Compound | Molecular or structural formula | State | Melting point / K | Boiling point / K | $\Delta_f H^\circ$(298 K) / kJ mol⁻¹ | $\Delta_f G^\circ$(298 K) / kJ mol⁻¹ | $S^\circ$ / J K⁻¹ mol⁻¹ |
|---|---|---|---|---|---|---|---|
| **Fluorine** | | | | | | | |
| | $F_2$ | g | 53 | 85 | 0 | 0 | 203 |
| Dioxygen difluoride | $O_2F_2$ | g | 109 | 216 | 18 | | |
| Oxygen difluoride | $OF_2$ | g | 49 | 128 | 25 | 42 | 247 |
| **Gallium** | | | | | | | |
| | $Ga$ | s | 303 | 2477 | 0 | 0 | 41 |
| Gallium(III) chloride | $GaCl_3$ | s | 351 | 474 | −525 | −455 | 142 |
| Gallium(III) fluoride | $GaF_3$ | s | 1073 (sub) | | −1163 | −1085 | 84 |
| Gallium(III) hydroxide | $Ga(OH)_3$ | s | 713 (dec) | | −964 | −831 | 100 |
| **Germanium** | | | | | | | |
| | $Ge$ | s | 1211 | 3106 | 0 | 0 | 31 |
| Germanium(IV) chloride | $GeCl_4$ | l | 223 | 357 | −532 | −463 | 246 |
| Germanium(II) oxide | $GeO$ | s | 983 (sub) | | −262 | −237 | 50 |
| Germanium(IV) oxide | $GeO_2$ | s | 1359 | | −580 | −521 | 40 |
| **Hydrogen** | | | | | | | |
| | $H_2$ | g | 13.7 | 20.1 | 0 | 0 | 131 |
| Ammonia | $NH_3$ | g | 195 | 239 | −45.9 | −16 | 193 |
| Arsane (arsine) | $AsH_3$ | g | 160 | 218 | 66 | 69 | 223 |
| Calcium hydride | $CaH_2$ | s | 1089 (dec) | | −181.5 | −142.5 | 41 |
| Diborane | $B_2H_6$ | g | 108 | 180.5 | 35.5 | 87 | 232 |
| Diphosphane | $P_2H_4$ | g | 174 | 340 | 21 | | |
| Germane | $GeH_4$ | g | 108 | 184 | 91 | 113 | 217 |
| Hydrazine | $N_2H_4$ | g | 275 | 386 | 95 | 159 | 238.5 |
| | | l | | | 51 | 149 | 121 |
| Hydrogen bromide | $HBr$ | g | 185 | 206 | −36 | −53 | 199 |
| Hydrogen chloride | $HCl$ | g | 158 | 188 | −92 | −95 | 187 |
| Hydrogen fluoride | $HF$ | g | 190 | 293 | −273 | −275 | 174 |
| Hydrogen iodide | $HI$ | g | 222 | 238 | 26.5 | 2 | 207 |
| Hydrogen peroxide | $H_2O_2$ | l | 272 | 425 | −187 | −120 | 110 |
| Hydrogen sulfide | $H_2S$ | g | 187.5 | 214 | −21 | −33 | 206 |
| Lithium hydride | $LiH$ | s | 953 | | −90.5 | −68 | 20 |
| Lithium tetrahydroborate (lithium borohydride) | $Li[BH_4]$ | s | 548 (dec) | | −191 | −125 | 76 |
| Phosphane (phosphine) | $PH_3$ | g | 139.5 (dec) | 185 | 5.5 | 13.5 | 210 |
| Potassium hydride | $KH$ | s | (dec) | | −58 | | |
| Silane | $SiH_4$ | g | 88 | 161 | 34 | 57 | 205 |
| Sodium hydride | $NaH$ | s | 1073 (dec) | | −56 | −33.5 | 40 |
| Sodium tetrahydroborate (sodium borohydride) | $Na[BH_4]$ | s | 673 (dec) | | −189 | −124 | 101 |
| Stibane (stibine) | $SbH_3$ | g | 185 | 256 | 145 | 148 | 233 |
| Water | $H_2O$ | l | 273 | 373 | −286 | −237 | 70 |
| | | g | | | −242 | −229 | 189 |

| Compound | Formula | State | m.p. | b.p. | $\Delta_f H^\circ$ | $\Delta_f G^\circ$ | $S^\circ$ |
|---|---|---|---|---|---|---|---|
| **Iodine** | $I_2$ | s | 387 | 458 | 0 | 0 | 116 |
| | $I_2$ | g | | | 62 | 19 | 261 |
| **Iron** | $Fe$ | s | 1811 | 3134 | 0 | 0 | 27 |
| Iron(II) chloride | $FeCl_2$ | s | 943 | | −342 | −302 | 118 |
| Iron(III) chloride | $FeCl_3$ | s | 579 | 588 (dec) | −399.5 | −334 | 142 |
| Iron(II) oxide | $FeO$ | s | 1642 | | −272 | | |
| Iron(III) oxide (haematite) | $Fe_2O_3$ | s | 1838 | | −824 | −742 | 87 |
| Iron sulfide (pyrite) | $FeS_2$ | s | 1444 | | −178 | −167 | 53 |
| **Lead** | $Pb$ | s | 600 | 2022 | 0 | 0 | 65 |
| Lead(II) chloride | $PbCl_2$ | s | 774 | 1223 | −359 | −314 | 136 |
| Lead(II) iodide | $PbI_2$ | s | 675 | 1227 | −175.5 | −174 | 175 |
| Lead(II) oxide (litharge) | $PbO$ | s | 1159 | | −219 | −189 | 67 |
| Lead nitrate | $Pb(NO_3)_2$ | s | 743 (dec) | | −452 | | |
| Lead(IV) oxide | $PbO_2$ | s | 563 (dec) | | −277 | −217 | 69 |
| Lead sulfate | $PbSO_4$ | s | 1443 | | −920 | −813 | 148.5 |
| Lead(II) sulfide (galena) | $PbS$ | s | 1387 | | −100 | −99 | 91 |
| **Lithium** | $Li$ | s | 453.5 | 1615 | 0 | 0 | 29 |
| Lithium bromide | $LiBr$ | s | 823 | 1538 | −351 | −342 | 74 |
| Lithium carbonate | $Li_2CO_3$ | s | 996 | 1583 (dec) | −1216 | −1132 | 90 |
| Lithium chloride | $LiCl$ | s | 878 | 1598 | −409 | −384 | 59 |
| Lithium fluoride | $LiF$ | s | 1118 | 1949 | −616 | −588 | 36 |
| Lithium hydroxide | $LiOH$ | s | 723 | 1197 (dec) | −485 | −439 | 43 |
| Lithium iodide | $LiI$ | s | 722 | 1453 | −270 | −270 | 87 |
| Lithium nitrate | $LiNO_3$ | s | 537 | 873 (dec) | −483 | −381 | 90 |
| Lithium oxide | $Li_2O$ | s | >1973 | | −598 | −561 | 38 |
| **Magnesium** | $Mg$ | s | 923 | 1363 | 0 | 0 | 33 |
| Magnesium bromide | $MgBr_2$ | s | 973 | | −524 | −504 | 117 |
| Magnesium carbonate | $MgCO_3$ | s | 623 (dec) | | −1096 | −1012 | 66 |
| Magnesium chloride | $MgCl_2$ | s | 987 | 1685 | −641 | −592 | 90 |
| Magnesium fluoride | $MgF_2$ | s | 1534 | 2512 | −1124 | −1071 | 57 |
| Magnesium hydroxide | $Mg(OH)_2$ | s | 623 ($-H_2O$) | | −924.5 | −833.5 | 63 |
| Magnesium oxide | $MgO$ | s | 3125 | 3873 | −602 | −569 | 27 |
| Magnesium sulfate | $MgSO_4$ | s | 1397 (dec) | | −1285 | −1171 | 92 |
| **Manganese** | $Mn$ | s | 1519 | 2334 | 0 | 0 | 32 |
| Manganese(II) chloride | $MnCl_2$ | s | 923 | 1463 | −481 | −440.5 | 118 |
| Manganese(II) nitrate tetrahydrate | $Mn(NO_3)_2 \cdot 4H_2O$ | s | 299 | 402 | −576 (anhydrous salt) | | |
| Manganese(IV) oxide | $MnO_2$ | s | 808 (dec) | | −520 | −465 | 53 |
| Manganese(II) sulfate | $MnSO_4$ | s | 973 | 1123 (dec) | −1064 | | |

| Compound | Molecular or structural formula | State | Melting point / K | Boiling point / K | $\Delta_f H^\circ$(298 K) / kJ mol$^{-1}$ | $\Delta_f G^\circ$(298 K) / kJ mol$^{-1}$ | $S^\circ$ / J K$^{-1}$ mol$^{-1}$ |
|---|---|---|---|---|---|---|---|
| Potassium permanganate | $KMnO_4$ | s | 513 (dec) | | −837 | −738 | 172 |
| **Mercury** | | | | | | | |
| | $Hg$ | l | 234 | 630 | 0 | 0 | 76 |
| Mercury(I) chloride | $Hg_2Cl_2$ | s | 673 (sub) | | −265 | −211 | 192 |
| Mercury(II) chloride | $HgCl_2$ | s | 549 | 575 | −224 | −179 | 146 |
| Mercury(II) oxide | $HgO$ | s | 773 (dec) | | −91 | −58.5 | 70 |
| Mercury(II) sulfide | $HgS$ | s | 856 (sub) | | −58 | −51 | 82 |
| **Nickel** | | | | | | | |
| | $Ni$ | s | 1728 | 3186 | 0 | 0 | 30 |
| Nickel(II) bromide | $NiBr_2$ | s | 1236 | | −212 | | |
| Nickel(II) chloride | $NiCl_2$ | s | 1274 (sub) | 410 | −305 | −259 | 98 |
| Nickel(II) nitrate hexahydrate | $Ni(NO_3)_2 \cdot 6H_2O$ | s | 330 | | −2223 | | |
| Nickel(II) oxide | $NiO$ | s | 2263 | | −244 | −216 | 39 |
| Nickel(II) sulfate | $NiSO_4$ | s | 1121 (dec) | | −873 | −760 | 92 |
| **Nitrogen** | | | | | | | |
| | $N_2$ | g | 63 | 77 | 0 | 0 | 192 |
| Ammonia | $NH_3$ | g | 195 | 239 | −45.9 | −16 | 193 |
| Ammonium chloride | $NH_4Cl$ | s | 613 (sub) | | −314 | −203 | 95 |
| Ammonium nitrate | $NH_4NO_3$ | s | 443 | 483 | −366 | −184 | 151 |
| Ammonium nitrite | $NH_4NO_2$ | s | 333 (explodes) | | −256.5 | | |
| Ammonium sulfate | $[NH_4]_2SO_4$ | s | 508 (dec) | | −1181 | −902 | 220 |
| Dinitrogen oxide | $N_2O$ | g | 182 | 185 | 82 | 104 | 220 |
| Dinitrogen tetraoxide | $N_2O_4$ | l | 262 | 294 | −19.5 | 97.5 | 209 |
| Nitric acid | $HNO_3$ | l | 231 | 356 | −174 | −81 | 156 |
| Nitrogen dioxide | $NO_2$ | g | 262 | 294 | 33 | 51 | 240 |
| Nitrogen monoxide | $NO$ | g | 109 | 121 | 90 | 87 | 210 |
| Nitrogen trifluoride | $NF_3$ | g | 66 | 144 | −132 | −91 | 261 |
| **Oxygen** | | | | | | | |
| | $O_2$ | g | 54 | 90 | 0 | 0 | 205 |
| Ozone | $O_3$ | g | 81 | 163 | 143 | 163 | 239 |
| **Phosphorus** | | | | | | | |
| | $P_4$ (white phosphorus) | s | 317 | 550 | 0 | 0 | 41 |
| Red phosphorus | | s | | | −18 | | 23 |
| Black phosphorus | | s | | | −39 | | |
| Phosphine | $PH_3$ | g | 139.5 | 185 | 5.5 | 13.5 | 210 |
| Phosphonic acid (phosphorous acid) | $H_3PO_3$ | s | 343 | 473 (dec) | −964 | | |
| Phosphoric acid | $H_3PO_4$ | s | 315 | 486 (−$H_2O$) | −1284 | −1124 | 110.5 |
| Phosphorus(III) chloride | $PCl_3$ | l | 161 | 348 | −320 | −272 | 217 |

| Compound | Formula | State | Melting point | Boiling point | $\Delta_f H^\circ$ | $\Delta_f G^\circ$ | $S^\circ$ |
|---|---|---|---|---|---|---|---|
| Phosphorus(V) chloride | $PCl_5$ | s | 440 (dec) | | −443.5 | | |
| Phosphorus(III) fluoride | $PF_3$ | g | 122 | 171.5 | −958 | −937 | 273 |
| Phosphorus(V) fluoride | $PF_5$ | g | 190 | 198 | −1594 | −1521 | 301 |
| Phosphorus(III) oxide | $P_4O_6$ | s | 297 | | −1640 | | |
| Phosphorus(V) oxide | $P_4O_{10}$ | s | 573 (sub) | 447 (under $N_2$) | −2984 | | |
| **Potassium** | | | | | | | |
| Potassium | K | s | 336 | 1032 | 0 | 0 | 65 |
| Potassium bromide | KBr | s | 1007 | 1708 | −394 | −381 | 96 |
| Potassium carbonate | $K_2CO_3$ | s | 1164 | (dec) | −1151 | −1063.5 | 155.5 |
| Potassium chlorate | $KClO_3$ | s | 629 | 673 (dec) | −398 | −296 | 143 |
| Potassium chloride | KCl | s | 1043 | 1773 (sub) | −436.5 | −408.5 | 83 |
| Potassium cyanide | KCN | s | 907 | | −113 | −102 | 128.5 |
| Potassium fluoride | KF | s | 1131 | 1778 | −567 | −538 | 67 |
| Potassium hydroxide | KOH | s | 633 | 1593 | −425 | −379 | 79 |
| Potassium iodate | $KIO_3$ | s | 833 | (dec) | −501 | −418 | 151.5 |
| Potassium iodide | KI | s | 954 | 1603 | −328 | −325 | 53 |
| Potassium nitrate | $KNO_3$ | s | 607 | 673 (dec) | −495 | −395 | 133 |
| Potassium nitrite | $KNO_2$ | s | 713 (dec) | | −370 | −307 | 152 |
| Potassium oxide | $K_2O$ | s | 623 (dec) | | −361.5 | | |
| Potassium perchlorate | $KClO_4$ | s | 673 (dec) | | −433 | −303 | 151 |
| Potassium permanganate | $KMnO_4$ | s | 513 (dec) | | −837 | −738 | 172 |
| Potassium sulfate | $K_2SO_4$ | s | 1342 | 1962 | −1438 | 1321 | 176 |
| **Silicon** | | | | | | | |
| Silicon | Si | s | 1687 | 2638 | 0 | 0 | 19 |
| Silicon dioxide (silica, α-quartz) | $SiO_2$ | s | 1883 | 2503 | −911 | −856 | 41.5 |
| Silicon tetrachloride | $SiCl_4$ | l | 203 | 331 | −687 | −620 | 240 |
| Silicon tetrafluoride | $SiF_4$ | g | 183 | 187 | −1615 | −1573 | 283 |
| Sodium silicate | $Na_4SiO_4$ | s | 1291 | | | | |
| **Silver** | | | | | | | |
| Silver | Ag | s | 1235 | 2435 | 0 | 0 | 43 |
| Silver bromide | AgBr | s | 705 | (dec) >1573 | −100 | −97 | 107 |
| Silver chloride | AgCl | s | 728 | 1823 | −127 | −110 | 96 |
| Silver chromate | $Ag_2CrO_4$ | s | | | −732 | −642 | 218 |
| Silver fluoride | AgF | s | 708 | 1432 | −205 | | |
| Silver iodide | AgI | s | 831 | 1779 | −62 | −66 | 115.5 |
| Silver nitrate | $AgNO_3$ | s | 485 | 717 (dec) | −124 | −33 | 141 |
| **Sodium** | | | | | | | |
| Sodium | Na | s | 371 | 1156 | 0 | 0 | 51 |
| Sodium bromide | NaBr | s | 1020 | 1663 | −361 | −349 | 87 |
| Sodium carbonate | $Na_2CO_3$ | s | 1124 | | −1131 | −1044 | 135 |
| Sodium chloride | NaCl | s | 1074 | 1686 | −411 | −384 | 72 |
| Sodium cyanide | NaCN | s | 837 | 1769 | −87.5 | −76 | 116 |

| Compound | Molecular or structural formula | State | Melting point / K | Boiling point / K | $\Delta_f H^\circ$(298 K) / kJ mol$^{-1}$ | $\Delta_f G^\circ$(298 K) / kJ mol$^{-1}$ | $S^\circ$ / J K$^{-1}$ mol$^{-1}$ |
|---|---|---|---|---|---|---|---|
| Sodium fluoride | NaF | s | 1266 | 1968 | −577 | −546 | 51 |
| Sodium hydroxide | NaOH | s | 591 | 1663 | −426 | −379.5 | 64.5 |
| Sodium iodide | NaI | s | 934 | 1577 | −288 | −286 | 98.5 |
| Sodium nitrate | NaNO$_3$ | s | 580 | 653 (dec) | −468 | −367 | 116.5 |
| Sodium nitrite | NaNO$_2$ | s | 544 | 593 (dec) | −359 | −285 | 104 |
| Sodium oxide | Na$_2$O | s | 1548 (sub) | | −414 | −375.5 | 75 |
| Sodium sulfate | Na$_2$SO$_4$ | s | 1157 | | −1387 | −1270 | 150 |
| **Sulfur** | S$_8$ (orthorhombic) | s | 388 | 718 | 0 | 0 | 32 |
| | S$_8$ (monoclinic) | s | | | 0.3 | | |
| Disulfur dichloride | S$_2$Cl$_2$ | l | 193 | 409 | −59 | −300 | 248 |
| Sulfur dioxide | SO$_2$ | g | 200 | 263 | −297 | | |
| Sulfur(IV) fluoride | SF$_4$ | g | 149 | 233 | −763 | −722 | 300 |
| Sulfur(VI) fluoride | SF$_6$ | g | 222.5 | 337 | −1221 | −1116.5 | 291.5 |
| Sulfur trioxide | SO$_3$ | s | 290 | 318 | −454 | −374 | 71 |
| | | g | | | −396 | −371 | 257 |
| Sulfuric acid | H$_2$SO$_4$ | l | 283 | 603 | −814 | −690 | 157 |
| Thionyl dichloride (thionyl chloride) | SOCl$_2$ | l | 168 | 413 (dec) | −246 | | |
| **Tin** | Sn (white) | s | 505 | 2533 | 0 | 0 | 52 |
| | Sn (grey) | s | | | −2 | 0.1 | 44 |
| Tin(II) chloride | SnCl$_2$ | s | 519 | 925 | −325 | −440 | 259 |
| Tin(IV) chloride | SnCl$_4$ | l | 240 | 387 | −511 | −516 | 49 |
| Tin(IV) oxide | SnO$_2$ | s | 1903 | | −578 | | |
| **Zinc** | Zn | s | 693 | 1180 | 0 | 0 | 42 |
| Zinc bromide | ZnBr$_2$ | s | 667 | 923 | −329 | −312 | 138.5 |
| Zinc carbonate | ZnCO$_3$ | s | 573 (−CO$_2$) | −813 | −731.5 | 82 | |
| Zinc chloride | ZnCl$_2$ | s | 556 | 1005 | −415 | −369 | 111.5 |
| Zinc oxide | ZnO | s | 2248 | | −350.5 | −320.5 | 44 |
| Zinc sulfide (wurtzite) | ZnS | s | 1458 (sub) | | −193 | | |
| Zinc sulfide (zinc blende) | ZnS | s | 1293 ⟶ wurtzite | | −206 | −205 | 58 |
| Zinc sulfate | ZnSO$_4$ | s | | | −983 | −871.5 | 111 |

# Appendix 12   Selected standard reduction potentials (298 K)

The concentration of each aqueous solution is $1\,mol\,dm^{-3}$ and the pressure of a gaseous component is 1 bar ($10^5\,Pa$). (Changing the standard pressure to 1 atm (101 300 Pa) makes no difference to the values of $E^{\circ}$ at this level of accuracy.) Each half-cell listed contains the specified solution species at a concentration of $1\,mol\,dm^{-3}$. Where the half-cell contains $[OH]^-$, the value of $E^{\circ}$ refers to $[OH]^- = 1\,mol\,dm^{-3}$, hence the notation $E^{\circ}_{[OH^-]=1}$.

| Reduction half-equation | $E^{\circ}$ or $E^{\circ}_{[OH^-]=1}$ / V |
|---|---|
| $Li^+(aq) + e^- \rightleftharpoons Li(s)$ | −3.04 |
| $K^+(aq) + e^- \rightleftharpoons K(s)$ | −2.93 |
| $Ca^{2+}(aq) + 2e^- \rightleftharpoons Ca(s)$ | −2.87 |
| $Na^+(aq) + e^- \rightleftharpoons Na(s)$ | −2.71 |
| $Mg^{2+}(aq) + 2e^- \rightleftharpoons Mg(s)$ | −2.37 |
| $Al^{3+}(aq) + 3e^- \rightleftharpoons Al(s)$ | −1.66 |
| $[HPO_3]^{2-}(aq) + 2H_2O(aq) + 2e^- \rightleftharpoons [H_2PO_2]^-(aq) + 3[OH]^-(aq)$ | −1.65 |
| $Ti^{2+}(aq) + 2e^- \rightleftharpoons Ti(s)$ | −1.63 |
| $Mn^{2+}(aq) + 2e^- \rightleftharpoons Mn(s)$ | −1.19 |
| $Te(s) + 2e^- \rightleftharpoons Te^{2-}(aq)$ | −1.14 |
| $[SO_4]^{2-}(aq) + H_2O(l) + 2e^- \rightleftharpoons [SO_3]^{2-}(aq) + 2[OH]^-(aq)$ | −0.93 |
| $Se(s) + 2e^- \rightleftharpoons Se^{2-}(aq)$ | −0.92 |
| $2[NO_3]^-(aq) + 2H_2O(l) + 2e^- \rightleftharpoons N_2O_4(g) + 4[OH]^-(aq)$ | −0.85 |
| $Zn^{2+}(aq) + 2e^- \rightleftharpoons Zn(s)$ | −0.76 |
| $S(s) + 2e^- \rightleftharpoons S^{2-}(aq)$ | −0.48 |
| $[NO_2]^-(aq) + H_2O(l) + e^- \rightleftharpoons NO(g) + 2[OH]^-(aq)$ | −0.46 |
| $Fe^{2+}(aq) + 2e^- \rightleftharpoons Fe(s)$ | −0.44 |
| $Cr^{3+}(aq) + e^- \rightleftharpoons Cr^{2+}(aq)$ | −0.41 |
| $Ti^{3+}(aq) + e^- \rightleftharpoons Ti^{2+}(aq)$ | −0.37 |
| $PbSO_4(s) + 2e^- \rightleftharpoons Pb(s) + [SO_4]^{2-}(aq)$ | −0.36 |
| $Tl^+(aq) + e^- \rightleftharpoons Tl(s)$ | −0.34 |
| $Co^{2+}(aq) + 2e^- \rightleftharpoons Co(s)$ | −0.28 |
| $H_3PO_4(aq) + 2H^+(aq) + 2e^- \rightleftharpoons H_3PO_3(aq) + H_2O(l)$ | −0.28 |
| $V^{3+}(aq) + e^- \rightleftharpoons V^{2+}(aq)$ | −0.26 |
| $Ni^{2+}(aq) + 2e^- \rightleftharpoons Ni(s)$ | −0.25 |
| $Sn^{2+}(aq) + 2e^- \rightleftharpoons Sn(s)$ | −0.14 |
| $Pb^{2+}(aq) + 2e^- \rightleftharpoons Pb(s)$ | −0.13 |
| $Fe^{3+}(aq) + 3e^- \rightleftharpoons Fe(s)$ | −0.04 |
| $2H^+(aq) + 2e^- \rightleftharpoons H_2(g, 1\,bar)$ | 0.00 |
| $[NO_3]^-(aq) + H_2O(l) + 2e^- \rightleftharpoons [NO_2]^-(aq) + 2[OH]^-(aq)$ | +0.01 |
| $[S_4O_6]^{2-}(aq) + 2e^- \rightleftharpoons 2[S_2O_3]^{2-}(aq)$ | +0.08 |

| Reduction half-equation | $E^o$ or $E^o_{[OH^-]=1}$ / V |
|---|---|
| $S(s) + 2H^+(aq) + 2e^- \rightleftharpoons H_2S(aq)$ | +0.14 |
| $2[NO_2]^-(aq) + 3H_2O(l) + 4e^- \rightleftharpoons N_2O(g) + 6[OH]^-(aq)$ | +0.15 |
| $Cu^{2+}(aq) + e^- \rightleftharpoons Cu^+(aq)$ | +0.15 |
| $Sn^{4+}(aq) + 2e^- \rightleftharpoons Sn^{2+}(aq)$ | +0.15 |
| $[SO_4]^{2-}(aq) + 4H^+(aq) + 2e^- \rightleftharpoons H_2SO_3(aq) + H_2O(l)$ | +0.17 |
| $Cu^{2+}(aq) + 2e^- \rightleftharpoons Cu(s)$ | +0.34 |
| $[ClO_4]^-(aq) + H_2O(l) + 2e^- \rightleftharpoons [ClO_3]^-(aq) + 2[OH]^-(aq)$ | +0.36 |
| $O_2(g) + 2H_2O(l) + 4e^- \rightleftharpoons 4[OH]^-(aq)$ | +0.40 |
| $Cu^+(aq) + e^- \rightleftharpoons Cu(s)$ | +0.52 |
| $I_2(aq) + 2e^- \rightleftharpoons 2I^-(aq)$ | +0.54 |
| $H_3AsO_4(aq) + 2H^+(aq) + 2e^- \rightleftharpoons HAsO_2(aq) + 2H_2O(l)$ | +0.56 |
| $[MnO_4]^-(aq) + 2H_2O(aq) + 3e^- \rightleftharpoons MnO_2(s) + 4[OH]^-(aq)$ | +0.59 |
| $[BrO_3]^-(aq) + 3H_2O(l) + 6e^- \rightleftharpoons Br^-(aq) + 6[OH]^-(aq)$ | +0.61 |
| $O_2(g) + 2H^+(aq) + 2e^- \rightleftharpoons H_2O_2(aq)$ | +0.70 |
| $[BrO]^-(aq) + H_2O(l) + 2e^- \rightleftharpoons Br^-(aq) + 2[OH]^-(aq)$ | +0.76 |
| $Fe^{3+}(aq) + e^- \rightleftharpoons Fe^{2+}(aq)$ | +0.77 |
| $Ag^+(aq) + e^- \rightleftharpoons Ag(s)$ | +0.80 |
| $[ClO]^-(aq) + H_2O(l) + 2e^- \rightleftharpoons Cl^-(aq) + 2[OH]^-(aq)$ | +0.84 |
| $2HNO_2(aq) + 4H^+(aq) + 4e^- \rightleftharpoons H_2N_2O_2(aq) + 2H_2O(l)$ | +0.86 |
| $[NO_3]^-(aq) + 3H^+(aq) + 2e^- \rightleftharpoons HNO_2(aq) + H_2O(l)$ | +0.93 |
| $[NO_3]^-(aq) + 4H^+(aq) + 3e^- \rightleftharpoons NO(g) + 2H_2O(l)$ | +0.96 |
| $HNO_2(aq) + H^+(aq) + e^- \rightleftharpoons NO(g) + H_2O(l)$ | +0.98 |
| $[IO_3]^-(aq) + 6H^+(aq) + 6e^- \rightleftharpoons I^-(aq) + 3H_2O(l)$ | +1.09 |
| $Br_2(aq) + 2e^- \rightleftharpoons 2Br^-(aq)$ | +1.09 |
| $2[IO_3]^-(aq) + 12H^+(aq) + 10e^- \rightleftharpoons I_2(aq) + 6H_2O(l)$ | +1.20 |
| $O_2(g) + 4H^+(aq) + 4e^- \rightleftharpoons 2H_2O(l)$ | +1.23 |
| $Tl^{3+}(aq) + 2e^- \rightleftharpoons Tl^+(aq)$ | +1.25 |
| $2HNO_2(aq) + 4H^+(aq) + 4e^- \rightleftharpoons N_2O(g) + 3H_2O(l)$ | +1.30 |
| $[Cr_2O_7]^{2-}(aq) + 14H^+(aq) + 6e^- \rightleftharpoons 2Cr^{3+}(aq) + 7H_2O(l)$ | +1.33 |
| $Cl_2(aq) + 2e^- \rightleftharpoons 2Cl^-(aq)$ | +1.36 |
| $2[ClO_4]^-(aq) + 16H^+(aq) + 14e^- \rightleftharpoons Cl_2(aq) + 8H_2O(l)$ | +1.39 |
| $[ClO_4]^-(aq) + 8H^+(aq) + 8e^- \rightleftharpoons Cl^-(aq) + 4H_2O(l)$ | +1.39 |
| $[BrO_3]^-(aq) + 6H^+(aq) + 6e^- \rightleftharpoons Br^-(aq) + 3H_2O(l)$ | +1.42 |
| $[ClO_3]^-(aq) + 6H^+(aq) + 6e^- \rightleftharpoons Cl^-(aq) + 3H_2O(l)$ | +1.45 |
| $2[ClO_3]^-(aq) + 12H^+(aq) + 10e^- \rightleftharpoons Cl_2(aq) + 6H_2O(l)$ | +1.47 |
| $2[BrO_3]^-(aq) + 12H^+(aq) + 10e^- \rightleftharpoons Br_2(aq) + 6H_2O(l)$ | +1.48 |
| $HOCl(aq) + H^+(aq) + 2e^- \rightleftharpoons Cl^-(aq) + H_2O(l)$ | +1.48 |
| $[MnO_4]^-(aq) + 8H^+(aq) + 5e^- \rightleftharpoons Mn^{2+}(aq) + 4H_2O(l)$ | +1.51 |
| $2HOCl(aq) + 2H^+(aq) + 2e^- \rightleftharpoons Cl_2(aq) + 2H_2O(l)$ | +1.61 |
| $PbO_2(s) + 4H^+(aq) + [SO_4]^{2-}(aq) + 2e^- \rightleftharpoons PbSO_4(s) + 2H_2O(l)$ | +1.69 |
| $Ce^{4+}(aq) + e^- \rightleftharpoons Ce^{3+}(aq)$ | +1.72 |
| $H_2O_2(aq) + 2H^+(aq) + 2e^- \rightleftharpoons 2H_2O(l)$ | +1.78 |
| $Co^{3+}(aq) + e^- \rightleftharpoons Co^{2+}(aq)$ | +1.92 |
| $XeO_3(aq) + 6H^+(aq) + 6e^- \rightleftharpoons Xe(g) + 3H_2O(l)$ | +2.10 |
| $H_4XeO_6(aq) + 2H^+(aq) + 2e^- \rightleftharpoons XeO_3(aq) + 3H_2O(l)$ | +2.42 |
| $F_2(aq) + 2e^- \rightleftharpoons 2F^-(aq)$ | +2.87 |

# Answers to non-descriptive problems

Answers are given below to non-descriptive problems. Complete answers to all problems are available in the accompanying *Solutions Manual* by Catherine E. Housecroft (see www.pearsoned.co.uk/housecroft).

## Chapter 1

1.1 (a) 0.6 mm; (b) $6 \times 10^8$ pm; (c) 0.06 cm; (d) $6 \times 10^5$ nm

1.2 0.122 nm

1.3 $2.71 \times 10^{-5}$ m$^3$

1.4 J s

1.6 10.8

1.8 (a) $2.86 \times 10^{-5}$ m$^3$; (b) $4.86 \times 10^{-3}$ m$^3$; (c) 0.318 m$^3$; (d) 407 m$^3$

1.9 0.84 moles

1.10 N$_2$, 0.57 bar; Ar, 0.80 bar; 0.56 moles of one or more other gases

1.11 (a) $5.0 \times 10^{-3}$ moles; (b) $8.21 \times 10^{-2}$ moles; (c) 0.040 03 moles; (d) $4.98 \times 10^{-4}$ moles

1.12 (a) 0.166 g; (b) 1.17 g; (c) 0.710 g

1.13 (a) NaI; (b) MgCl$_2$; (c) MgO; (d) CaF$_2$; (e) Li$_3$N; (f) Ca$_3$P$_2$; (g) Na$_2$S; (h) H$_2$S

1.14 Al$_2$O$_3$; AlCl$_3$; AlF$_3$; AlH$_3$

1.18 (a) +1; (b) +2; (c) +4; (d) +3; (e) +4; (f) +5

1.20 (a) O$_2$ and O$_3$, 0; [O$_2$]$^{2-}$, $-1$; [O$_2$]$^+$, $+\frac{1}{2}$

1.22 (a) NiI$_2$; (b) NH$_4$NO$_3$; (c) Ba(OH)$_2$; (d) Fe$_2$(SO$_4$)$_3$; (e) FeSO$_3$; (f) AlH$_3$; (g) PbO$_2$; (h) SnS

1.23 (a) 8; (b) 6; (c) 3; (d) 10; (e) 4

1.26 (a) $2Fe + 3Cl_2 \longrightarrow 2FeCl_3$;
(b) $SiCl_4 + 2H_2O \longrightarrow SiO_2 + 4HCl$;
(c) $Al_2O_3 + 6NaOH + 3H_2O \longrightarrow 2Na_3Al(OH)_6$;
(d) $K_2CO_3 + 2HNO_3 \longrightarrow 2KNO_3 + H_2O + CO_2$;
(e) $Fe_2O_3 + 3CO \longrightarrow 2Fe + 3CO_2$;
(f) $H_2C_2O_4 + 2KOH \longrightarrow K_2C_2O_4 + 2H_2O$

1.27 (a) $2AgNO_3 + MgCl_2 \longrightarrow 2AgCl + Mg(NO_3)_2$;
(b) $Pb(O_2CCH_3)_2 + H_2S \longrightarrow PbS + 2CH_3CO_2H$;
(c) $BaCl_2 + K_2SO_4 \longrightarrow BaSO_4 + 2KCl$;
(d) $Pb(NO_3)_2 + 2KI \longrightarrow PbI_2 + 2KNO_3$;
(e) $Ca(HCO_3)_2 + Ca(OH)_2 \longrightarrow 2CaCO_3 + 2H_2O$

1.28 (a) $C_3H_8 + 3Cl_2 \longrightarrow C_3H_5Cl_3 + 3HCl$;
(b) $2C_6H_{14} + 19O_2 \longrightarrow 12CO_2 + 14H_2O$;
(c) $2C_2H_5OH + 2Na \longrightarrow 2C_2H_5ONa + H_2$;
(d) $C_2H_2 + 2Br_2 \longrightarrow C_2H_2Br_4$;
(e) $CaC_2 + 2H_2O \longrightarrow Ca(OH)_2 + C_2H_2$

1.29 (a) $Ag^+(aq) + Cl^-(aq) \longrightarrow AgCl(s)$;
(b) $Mg^{2+}(aq) + 2[OH]^-(aq) \longrightarrow Mg(OH)_2(s)$;
(c) $Pb^{2+}(aq) + S^{2-}(aq) \longrightarrow PbS(s)$;
(d) $Fe^{3+}(aq) + 3[OH]^-(aq) \longrightarrow Fe(OH)_3(s)$;
(e) $3Ca^{2+}(aq) + 2[PO_4]^{3-}(aq) \longrightarrow Ca_3(PO_4)_2(s)$;
(f) $2Ag^+(aq) + [SO_4]^{2-}(aq) \longrightarrow Ag_2SO_4(s)$

1.30 (a) $2Fe^{3+} + H_2 \longrightarrow 2Fe^{2+} + 2H^+$;
(b) $Cl_2 + 2Br^- \longrightarrow 2Cl^- + Br_2$;
(c) $6Fe^{2+} + [Cr_2O_7]^{2-} + 14H^+ \longrightarrow 6Fe^{3+} + 2Cr^{3+} + 7H_2O$;
(d) $2NH_2OH + 4Fe^{3+} \longrightarrow N_2O + 4Fe^{2+} + H_2O + 4H^+$;
(e) $2[S_2O_3]^{2-} + I_2 \longrightarrow [S_4O_6]^{2-} + 2I^-$;
(f) $12[MoO_4]^{2-} + [PO_4]^{3-} + 24H^+ \longrightarrow [PMo_{12}O_{40}]^{3-} + 12H_2O$;
(g) $HNO_3 + 2H_2SO_4 \longrightarrow [H_3O]^+ + [NO_2]^+ + 2[HSO_4]^-$

1.31 0.61 g

1.32 NaOH is in excess; 0.091 moles unreacted

1.33 0.415 g

1.34 10.0 cm$^3$

1.35 2.80 g; $CaO + H_2O \longrightarrow Ca(OH)_2$

1.36 0.0511 mol dm$^{-3}$

1.37 Empirical formula = molecular formula = C$_2$H$_5$NO

1.38 (a) +2; (b) A = Fe$_2$O$_3$; B = Fe$_3$O$_4$

1.39 $x = 5$

1.40 (a) Empirical formula = CH$_2$O; molecular formula = C$_6$H$_{12}$O$_6$; (b) +6

## Chapter 2

2.1 (a), (c), (d) Exothermic; (b) endothermic

2.2 (a) Cl$_2$(g); (b) N$_2$(g); (c) P$_4$(white); (d) C(graphite); (e) Br$_2$(l); (f) Na(s); (g) F$_2$(g)

2.3 $Ca(s) + \frac{1}{2}O_2(g) \longrightarrow CaO(s)$; gives out heat

2.4 $-58\,kJ\,mol^{-1}$

2.6 Exothermic; $-7.8\,kJ\,mol^{-1}$

2.7 296.5 K

2.8 (a) $-546$; (b) $-828$; (c) $-160$; (d) $-18$; (e) $+132$; (f) $-15$; (g) $+286\,kJ\,mol^{-1}$

2.9 $-5472\,kJ\,mol^{-1}$

2.10 $-2220\,kJ\,mol^{-1}$

2.11 101 kJ liberated $(\Delta H = -101\,kJ)$

2.12 (a) $S_8$(orthorhombic); (b) 3 J

2.13 $+868\,kJ$; endothermic, so suggests $LiNO_3$ stable with respect to the reaction

2.14 (a) $-157$; (b) $+124$; (c) $-3312$; (d) $-87\,kJ\,mol^{-1}$ (per mole of reaction)

2.16 (a) $+0.300\,kJ$; (b) $-0.048\,kJ$; (c) $-0.81\,kJ$

2.17 3.55 g

2.18 (a) $+0.315\,kJ$; (b) $-1.4\,kJ$; (c) $+339\,kJ$

2.19 (a) $-129$; (b) $-85$; (c) $-1532\,kJ\,mol^{-1}$ (per mole of reaction)

2.20 (a) $-2878$; (b) $-1746\,kJ$ per mole of butane

2.21 (a) $N_2(g)$ and $H_2O(l)$; (b) $-2.9\,kJ$; (c) $-85\,kJ\,mol^{-1}$ (per mole of $BCl_3$)

2.22 (a) $Ag(s) \longrightarrow Ag(l)$; endothermic; (b) $-21\,kJ\,mol^{-1}$

2.23 $-43\,kJ$ per mole of $NO_2$

2.24 (b) $-157\,kJ\,mol^{-1}$; (c) $34\,kJ\,mol^{-1}$

2.25 (a) $-99\,kJ\,mol^{-1}$; (c) $+44.5\,kJ\,mol^{-1}$

## Chapter 3

3.2 (a) $1.5 \times 10^{-5}\,m$; (b) $6.6 \times 10^{-8}\,m$; (c) $1.4 \times 10^{-9}\,m$

3.3 Seven $4f$ orbitals

3.4 (a) $n=1, l=0, m_l=0, m_s=-\frac{1}{2}$; $n=1, l=0, m_l=0, m_s=+\frac{1}{2}$; (b) $n=1, l=0, m_l=0, m_s=-\frac{1}{2}$; $n=1, l=0, m_l=0, m_s=+\frac{1}{2}$; $n=2, l=0, m_l=0, m_s=-\frac{1}{2}$; $n=2, l=0, m_l=0, m_s=+\frac{1}{2}$; $n=2, l=1, m_l=0, m_s=-\frac{1}{2}$; $n=2, l=1, m_l=0, m_s=+\frac{1}{2}$; $n=2, l=1, m_l=1, m_s=-\frac{1}{2}$; $n=2, l=1, m_l=1, m_s=+\frac{1}{2}$; $n=2, l=1, m_l=-1, m_s=-\frac{1}{2}$; $n=2, l=1, m_l=-1, m_s=+\frac{1}{2}$

3.5 (a) 9; (b) 7; (c) 3; (d) 5; (e) 5

3.6 $3s$; $m_s=\pm\frac{1}{2}$

3.9 $-1312\,kJ\,mol^{-1}$; $-328.0\,kJ\,mol^{-1}$; $-145.8\,kJ\,mol^{-1}$

3.10 Hydrogen-like: (d), (f)

3.15 (a) Lyman; (b) Balmer; (c) Lyman; (d) Lyman; (e) Balmer

3.17 $R = 3.90 \times 10^{15}\,Hz$ (or $s^{-1}$)

3.18 $1312\,kJ\,mol^{-1}$

3.19 Be, $1s^2 2s^2$; F, $1s^2 2s^2 2p^5$; P, $1s^2 2s^2 2p^6 3s^2 3p^3$; K, $1s^2 2s^2 2p^6 3s^2 3p^6 4s^1$

3.21 (a) 8; (b) 8; (c) 8; (d) 8; (e) 8; (f) 8; (g) 6; (h) 4

## Chapter 4

4.3 $F_2$, 142 pm; $Cl_2$, 198 pm; $Br_2$, 228 pm; $I_2$, 266 pm

4.5 (a) $120°$; (b) planar

4.7 (a) 1200; (b) 2128; (c) 1062; (d) $151\,kJ\,mol^{-1}$

4.8 $60\,kJ\,mol^{-1}$

4.9 $C-H$, 416; $C-C$, $331\,kJ\,mol^{-1}$

4.10 (a) 461; (b) $503\,kJ\,mol^{-1}$

4.11 (a) 321; (b) $197\,kJ\,mol^{-1}$

4.19 Both paramagnetic

4.21 $<74\,kJ\,mol^{-1}$

## Chapter 5

5.1 $Z_{eff}(A) > Z_{eff}(B)$

5.4 $262\,kJ\,mol^{-1}$

5.5 (b) $430\,kJ\,mol^{-1}$

5.6 (a) $464\,kJ\,mol^{-1}$

5.8 (a) 329; $402\,kJ\,mol^{-1}$; (b) 1316; $1608\,kJ\,mol^{-1}$

5.10 Examples: $N^{3-}$, $O^{2-}$, $F^-$, $Na^+$, $Mg^{2+}$, $Al^{3+}$

5.12 (b) Polar: HBr, IF, BrCl

5.13 CO and $N_2$; $[S_2]^{2-}$ and $Cl_2$; $[O_2]^{2-}$ and $F_2$; NO and $[O_2]^+$

5.14 (b) 1; diamagnetic

5.16 (b) 2; paramagnetic; (c) polar; O is $\delta^-$

5.18 (a) $[NO]^+$; diamagnetic; (b) shorten

## Chapter 6

6.2 Linear

6.3 (a) $CO_2$, linear; $SO_2$, non-linear; (b) $[NO_2]^+$, linear; $[NO_2]^-$, non-linear; $CS_2$, linear; $N_2O$, linear; $[NCS]^-$, linear

6.4 (a) Trigonal planar; (b) trigonal pyramidal; (c) non-linear; (d) non-linear; (e) trigonal bipyramidal; (f) octahedral; (g) tetrahedral; (h) square planar

6.5 (a) $SF_4$; (b) $AlBr_3$; (c) $[ClO_2]^+$; (d) $IF_5$

6.6 (a), (c) Linear; (b), (d) octahedral; (e), (f), (g), (h) tetrahedral

6.8 (a) Trigonal bipyramidal; (b) 3

6.9 (a) 2; (b) 2; (c) 2; (d) 0; (e) 0; (f) 0; (g) 3; (h) 2

6.10 (b), (d), (e), (f)

6.13 (a) Trigonal pyramidal; (b) trigonal planar; (c) trigonal planar; (d) non-linear; (e) trigonal planar; (f) trigonal bipyramidal; (g) trigonal planar; (a), (b), (c) and (d) are polar

6.16 (a) 2; axial and equatorial

## Chapter 7

7.1 Octet: (b), (c), (d), (e), (f), (g)

7.2 (a), (h), (i)

7.6    (a) 2; (b) 4; (c) 3

7.9    (a) $sp^3$; (b) $sp^2$; (c) $sp^2$, $sp^2$, $sp^3$; (d) $sp$

7.10    (a) $sp^3$; (b) $sp^3$; (c) $sp^3$; (d) $sp^2$; (e) $sp^3d^2$; (f) $sp^3d$; (g) $sp^3$

7.11    (a) $sp$; (b) $sp^3$, $sp^3$; (c) $sp^3$; (d) $sp^3$, $sp^3$; (e) $sp^3$, $sp^2$, $sp^2$, $sp^3$, $sp^3$; (f) $sp^2$; (g) $sp^2$

7.16    (a) $sp^3d$; (b) $sp^3$; (c) $sp^3d^2$; (d) $sp^3$; (e) $sp^3$; (f) $sp^3$

## Chapter 8

8.2    Group 2

8.4    $+575\,\text{kJ mol}^{-1}$

8.5    $-5\,\text{kJ mol}^{-1}$

8.6    $+984\,\text{kJ mol}^{-1}$; $+1608\,\text{kJ mol}^{-1}$

8.7    $+439\,\text{kJ mol}^{-1}$

8.9    Each side: $\text{kg m}^2\,\text{s}^{-2}$

8.15    (a) GaP; (b) Ce(IV) is 8-coordinate; $CaF_2$ structure

8.16    $+3$

8.17    (a) 8; (b) 7; (c) 8.5; (d) 7

8.18    $-631\,\text{kJ mol}^{-1}$

8.19    Born–Landé: $-3926\,\text{kJ mol}^{-1}$; Born–Haber: $-3843\,\text{kJ mol}^{-1}$

8.20    $6.20 \times 10^{23}\,\text{mol}^{-1}$

8.21    $-760\,\text{kJ mol}^{-1}$

8.22    250 pm

8.23    $-2961\,\text{kJ mol}^{-1}$

8.24    $-2286\,\text{kJ mol}^{-1}$

8.25    $-793\,\text{kJ mol}^{-1}$

8.26    $+2250\,\text{kJ mol}^{-1}$

## Chapter 9

9.2    ccp *and* hcp: octahedral *and* tetrahedral holes

9.4    (a) 12; (b) 12; (c) 6; (d) 8

9.8    $C_{60}\cdot C_6H_6\cdot CH_2I_2$

9.10    $r(\text{Ca}^{2+}) < r(\text{Ca})$

9.12    (a) $N_2(s) \longrightarrow N_2(l)$; (b) $N_2(l) \longrightarrow N_2(g)$; (c) $\frac{1}{2}N_2(g) \longrightarrow N(g)$

9.16    Coordination numbers: Re, 6; O, 2

## Chapter 10

10.2    $C_5H_7N_2O$

10.3    $C_5H_{14}N_2$

10.5    (b) $m/z = 192$ for $(^{32}\text{S})_6$; $m/z = 194$ for $(^{32}\text{S})_5(^{34}\text{S})$

10.8    $\mathbf{A} = N_2$; $\mathbf{B} = CO$

10.10    Unambiguous assignment not possible

10.14    $\mathbf{D} = C_5H_{12}O$

10.15    $\mathbf{E} = C_7H_8$

10.16    $\mathbf{F} = C_6H_5CH_2CH_3$; group $R = C_2H_5$

10.21    (a) $\mathbf{Y} = CH_4N_2O$; (b) proposed structure is:

10.22    (a) $\mathbf{Z} = CH_3NO_2$; (b) proposed structure is:

## Chapter 11

11.1    Energy of transitions in NMR < rotational < vibrational < electronic

11.2    (a) $20\,000\,\text{cm}^{-1}$; (b) $44\,400\,\text{cm}^{-1}$; (c) 500 nm, visible; 225 nm, UV

11.3    (a) 35.8%; (b) 0.149

11.4    700–620 nm

11.5    (a) 256 nm; (b) 256 nm; (c) $1.25 \times 10^{-3}\,\text{mol dm}^{-3}$

11.6    (a) Gradient gives $\varepsilon$; (b) $\approx 9 \times 10^{-6}\,\text{mol dm}^{-3}$

11.7    $1.5 \times 10^{-4}\,\text{mol dm}^{-3}$

11.8    $4000\,\text{dm}^3\,\text{mol}^{-1}\,\text{cm}^{-1}$

11.9    $3.5 \times 10^{-3}\,\text{mol dm}^{-3}$

11.10    2.86

11.11    (a) 0.61; (b) 0.30

11.12    *cis*:*trans* = 1.31:1.00

11.13    (a) $214\,\text{dm}^3\,\text{mol}^{-1}\,\text{cm}^{-1}$

11.14    (a) $9.16 \times 10^{-6}\,\text{mol dm}^{-3}$; (b) $4.3 \times 10^{-4}\,\text{g}$

11.15    (b) 4.3

11.17    $x = 1$; $n = 1$

## Chapter 12

12.2    (a) $1.12 \times 10^{-27}\,\text{kg}$; (b) $1.55 \times 10^{-27}\,\text{kg}$; (c) $1.19 \times 10^{-26}\,\text{kg}$

12.3    ClF > BrF > BrCl

12.4    $1593\,\text{N m}^{-1}$

12.5    (a) $H^{35}Cl$, $1.63 \times 10^{-27}\,\text{kg}$; $H^{37}Cl$, $1.63 \times 10^{-27}\,\text{kg}$

12.7    (a) Linear; all modes, IR active; (b) linear; symmetric stretch is IR inactive; (c) linear; all modes, IR active; (d) linear; symmetric stretch is IR inactive

12.9    (a) $C\equiv N$, $\approx 2200\,\text{cm}^{-1}$; (b) O–H, $\approx 3600$–$3200\,\text{cm}^{-1}$; (c) $NH_2$, $\approx 3500$–$3300\,\text{cm}^{-1}$; (d) C=O, $\approx 1700\,\text{cm}^{-1}$

12.11    (a) Toluene; (b) cyclohexane; (c) benzene; (d) phenol

12.13    (a) Trigonal planar; (c) IR inactive

12.16    $\bar{\nu} = 2047\,\text{cm}^{-1}$; change $= 736\,\text{cm}^{-1}$ to lower wavenumber

12.17    $2.63 \times 10^{-47}\,\text{kg m}^2$

12.18  $I_A = 0$; $I_B = I_C = 7.87 \ 10^{-46} \ \text{kg} \, \text{m}^2$

12.19  232 pm

12.20  HCl is polar; $H_2$ and $Cl_2$ are non-polar

12.21  (b) and (c) are allowed

12.24  $B = 0.515 \, \text{cm}^{-1}$; spacings are $2B$, $4B$ and $6B = 1.03$, 2.06 and $3.09 \, \text{cm}^{-1}$

12.26  (a) 113 pm; (b) 3.84, $7.68 \, \text{cm}^{-1}$; (c) $J = 1 \leftarrow J = 0$; $J = 2 \leftarrow J = 1$

## Chapter 13

13.1  (a) 441 nm; (b) 286 nm

13.2  near-UV and visible

13.5  (a) near-UV; (b) near-UV; (c) vacuum-UV; (d) visible; (e) near-UV

13.6  (b) Delocalized in **13.11** and **13.14**

13.8  **13.15**, $\sigma^* \leftarrow \sigma$; **3.16**, $\pi^* \leftarrow \pi$; **13.17**, $\pi^* \leftarrow \pi$; **13.18**, $\sigma^* \leftarrow n$

13.10  (b) 358, 384 nm are in near-UV; band (broad) at 403 nm tails into visible; 420 nm is just in visible

13.12  Product absorbs at 470 nm

## Chapter 14

14.1  161.9 MHz; 62.50 MHz

14.2  $\delta 178.1 \, \text{ppm}$, C=O; $\delta 20.6 \, \text{ppm}$, $CH_3$

14.3  $\delta 68.9 \, \text{ppm}$, central C; $\delta 31.2 \, \text{ppm}$, $CH_3$

14.4  $\delta 170.0 \, \text{ppm}$, C=O; $\delta 80.8 \, \text{ppm}$, C≡C; $\delta 52.0 \, \text{ppm}$, $CH_2$; $\delta 20.5 \, \text{ppm}$, $CH_3$

14.5  (a) **14.18**, 2 signals; **14.19**, 4 signals; (b) **14.18**, $\delta 74.6 \, \text{ppm}$, C≡C; $\delta 3.3 \, \text{ppm}$, $CH_3$; **14.19**, $\delta 71.9$, 86.0 ppm, C≡C; $\delta 12.3$ and 13.8 ppm, $CH_2$ and $CH_3$ (exact assignment not possible)

14.6  Isomer I = **14.20**; isomer II = **14.21**

14.7  (a) $sp$ for C≡N carbon; $sp^3$ for $CH_2$ carbon; (b) $\delta 119.3 \, \text{ppm}$, C≡N; $\delta 24.3$, 16.4 ppm, $CH_2$ (exact assignment ambiguous)

14.8  $\delta 170.3 \, \text{ppm}$, C=O; $\delta 28.7 \, \text{ppm}$, $CH_2$; $\delta 8.4 \, \text{ppm}$, $CH_3$

14.9  $CH_3$, doublet, 9H; CH, decet, 1H

14.10  (a) $\delta 1.3 \, \text{ppm}$, $CH_3$; $\delta 2.7 \, \text{ppm}$, CH; (b) $\delta 1.1 \, \text{ppm}$, $CH_3$; $\delta 2.2 \, \text{ppm}$, $CH_2$; $\delta 6.4 \, \text{ppm}$, $NH_2$; (c) $\delta 1.1 \, \text{ppm}$, $CH_3CH_2$; $\delta 2.1 \, \text{ppm}$, $CH_3CO$; $\delta 2.5 \, \text{ppm}$, $CH_2$; (d) $\delta 1.3 \, \text{ppm}$, $CH_3CH_2$; $\delta 3.7 \, \text{ppm}$, $CH_3CH_2$; $\delta 4.1 \, \text{ppm}$, $CH_2CO$; $\delta 10.9 \, \text{ppm}$, OH; (e) $\delta 2.5 \, \text{ppm}$, $CH_3$; $\delta 5.9 \, \text{ppm}$, CH; (f) $\delta 2.2 \, \text{ppm}$, middle $CH_2$; $\delta 3.7 \, \text{ppm}$, $CH_2Cl$

14.11  (a) $\delta 9.79 \, \text{ppm}$, CH; $\delta 2.21 \, \text{ppm}$, $CH_3$; (b) CH, quartet; $CH_3$, doublet

14.12  **14.18**: one singlet; **14.19**: CH, singlet; $CH_2$, quartet; $CH_3$, triplet

14.13  $CH_3$, triplet; $CH_2$, quartet

14.14  $\delta 1.7 \, \text{ppm}$, $CH_3$, triplet; $\delta 3.4 \, \text{ppm}$, $CH_2$, quartet

14.15  $CH_3$, triplet; $CH_2$, quartet

14.16  $^1H$ nuclei couple to 2 equivalent $^{19}F$

14.17  $(CH_3)_2CHBr$: CH, septet; $CH_3$, doublet; $CH_3CHBr_2$: CH, quartet; $CH_3$, doublet; $CH_2BrCH_2CH_2Br$: middle $CH_2$, quintet; terminal $CH_2$, triplet

14.18  2 inequivalent $^{13}C$; each couples to 3 equivalent $^{19}F$

14.20  (a) Trigonal bipyramidal; 2 F environments (axial and equatorial); (b) stereochemically non-rigid (Berry pseudo-rotation); (c) one doublet; (d) octahedral; (e) $^{31}P$ NMR, septet; $^{19}F$ NMR, doublet

## Chapter 15

15.1  (a) $1 \times 10^{-3} \, \text{mol} \, \text{dm}^{-3} \, \text{s}^{-1}$; (b) at $t_1$, $9 \times 10^{-4} \, \text{mol} \, \text{dm}^{-3} \, \text{s}^{-1}$; at $t_2$, $6 \times 10^{-4} \, \text{mol} \, \text{dm}^{-3} \, \text{s}^{-1}$

15.5  Rate $= k[I^-]^2[Fe^{3+}]$

15.6  (a) $n = 1$; (b) $k = 0.087 \, \text{s}^{-1}$

15.7  Reaction I: $n = 1$; $k_{\text{obs}} = 2.8 \times 10^{-4} \, \text{s}^{-1}$; reaction II: $n = 2$; $k_{\text{obs}} = 1.1 \times 10^{-2} \, \text{dm}^3 \, \text{mol}^{-1} \text{s}^{-1}$

15.8  First order

15.9  (a) Second order; (b) $4.14 \, \text{dm}^3 \, \text{mol}^{-1} \text{min}^{-1}$

15.10  (b) First order; (c) first order; (d) $\varepsilon_{\text{max}}$ for $[P]^{2-}$

15.11  First order

15.12  (a) $x = 1$; $k = 169 \, \text{dm}^3 \, \text{mol}^{-1} \, \text{min}^{-1}$; (b) second order

15.13  (a) $[C_2O_4]^{2-}$; (c) first order

15.14  (b) Rate $= k[A]^2[B]^0 = k[A]^2$; (c) $0.334 \, \text{dm}^3 \, \text{mol}^{-1} \, \text{min}^{-1}$

15.15  (a) $^{115}_{49}In \longrightarrow \, ^{115}_{50}Sn + \beta^-$; (b) $1 \times 10^{-5} \, \text{y}^{-1}$

15.16  (a) $^{211}_{84}Po \longrightarrow \, ^{207}_{82}Pb + \, ^4_2He$; (b) $1.3 \, \text{s}^{-1}$

15.17  (a) $^{241}_{95}Am \longrightarrow \, ^{237}_{93}Np + \, ^4_2He$; (b) 431.9 y

15.18  $81.0 \, \text{kJ} \, \text{mol}^{-1}$

15.19  (a), (b) and (d) are temperature dependent

15.20  $79.0 \, \text{kJ} \, \text{mol}^{-1}$

15.22  (b) Unimolecular; bimolecular; bimolecular; bimolecular; unimolecular

15.23  Rate $= k[H_2]$; Rate $= k[H^\bullet][Br_2]$; Rate $= k[Cl^\bullet]^2$; Rate $= k[Cl^\bullet][O_3]$; Rate $= k[O_3]$

15.24  (a) D = intermediate; (b) $\dfrac{d[D]}{dt} = k_1[A]$; (c) $\dfrac{d[D]}{dt} = k_1[A] - k_1[D]$

15.25  $V_{\text{max}} = 1.5 \times 10^{-4} \, \text{mmol} \, \text{dm}^{-3} \, \text{min}^{-1}$; $K_M = 8.0 \, \text{mmol} \, \text{dm}^{-3}$

15.26  (c) First order; $k = 0.20 \, \text{d}^{-1}$

15.27  (c) $k = 2.4 \times 10^{-3} \, \text{s}^{-1}$

## Chapter 16

16.1  Equilibrium moves to (a) right; (b) right

16.2  Equilibrium moves to (b) right; (c) right

16.3  Equilibrium moves to (a) right; (b) right

16.6   100

16.7   36

16.8   0.11

16.9   0.36 moles

16.10   $K_2 = \sqrt{K_1}$

16.12   HCN, $K_a = 4.9 \times 10^{-10}$; $HNO_2$, $K_a = 4.6 \times 10^{-4}$

16.13   $K_a(1) = 7.2 \times 10^{-4}$; $K_a(2) = 1.7 \times 10^{-5}$;
         $K_a(3) = 4.1 \times 10^{-7}$

16.14   (a) Acetic acid; (b) HOCl

16.15   $4.9 \times 10^{-6}$ mol dm$^{-3}$

16.16   (a) Fully dissociated; (b) 60 cm$^3$; (c) $K_b = 6.5 \times 10^{-4}$

16.18   (a) 1.0; (b) change in pH = +1 unit (to pH 2.0);
         (c) 12.70

16.19   2.68

16.20   $2.7 \times 10^{-3}$ mol dm$^{-3}$; 11.43

16.21   1.23

16.24   (b) 3.75

16.25   7.35

16.26   Initial pH = 1.30; at equivalence point, pH = 7.00;
         final pH = 12.15

16.27   5.28

16.28   (a) 4.66; 9.94; (b) 4.36; 9.08

16.30   0.08 mol $CH_3CO_2H$; 0.23 mol $C_2H_5OH$; 0.22 mol
         $CH_3CO_2C_2H_5$; 0.42 mol $H_2O$

16.31   3.96

16.32   6.80

16.33   $Na^+(aq)$, $H_3O^+(aq)$, $OH^-(aq)$, $H_2CO_3(aq)$,
         $HCO_3^-(aq)$, $CO_3^{2-}(aq)$


## Chapter 17

17.2   (a) $-496$ J per 0.200 mol; (b) $-100.7$ kJ mol$^{-1}$

17.3   $n = +3$

17.5   $-408.0$ kJ mol$^{-1}$

17.6   $+218.6$ kJ mol$^{-1}$

17.7   Valid for (a), (c), (e)

17.8   (a) $-283.0$ kJ mol$^{-1}$; (b) $-283.1$ kJ mol$^{-1}$

17.9   $-603$ kJ mol$^{-1}$

17.10   (a) $-243$; (b) $+131$; (c) $-402$ kJ mol$^{-1}$

17.12   (a) 0; (b) 0; (c) 0; (d) 0 kJ mol$^{-1}$

17.14   (a) $NO_2$, $SO_2$, $GeCl_4$; (b) CO, $SO_2$, $GeCl_4$; (c) $SO_2$;
         (e) $Ge(s) + 2Cl_2(g) \longrightarrow GeCl_4(g)$

17.15   (a) Ca, Sr, Al, Zn; (b) $\approx -35$ kJ mol$^{-1}$

17.17   $1.2 \times 10^{-12}$

17.18   (a) $5.6 \times 10^{-16}$

17.19   $\Delta_r G^\circ(298\ K) = +1.7$ kJ mol$^{-1}$;
         $\Delta_r G^\circ(298\ K) = -12.3$ kJ mol$^{-1}$

17.20   $-301$ kJ mol$^{-1}$

17.21   (b) $2.3 \times 10^9$; (c) yes

17.22   (a) Positive; (b) negative; (c) negative

17.23   (a) 148.4 J K$^{-1}$ mol$^{-1}$; (b) $-107.4$ J K$^{-1}$ mol$^{-1}$;
         (c) $-172.9$ J K$^{-1}$ mol$^{-1}$

17.24   191.6 J K$^{-1}$ mol$^{-1}$

17.25   25.5 J K$^{-1}$ mol$^{-1}$

17.27   (a) $-439$; (b) $-152$; (c) $+164$ J K$^{-1}$ mol$^{-1}$

17.28   $\Delta_{fus}S = 9$ J K$^{-1}$ mol$^{-1}$; $\Delta_{vap}S = 78$ J K$^{-1}$ mol$^{-1}$

17.29   $-83.1$ J K$^{-1}$ mol$^{-1}$; $-370.9$ kJ mol$^{-1}$

17.30   $-17.7$ kJ mol$^{-1}$

17.31   $1.0 \times 10^{-5}$ mol dm$^{-3}$

17.32   (a) $1.1 \times 10^{-12}$

17.33   (b) $K(300\ K) \approx 1.3 \times 10^{52}$; $K(600\ K) \approx 1.1 \times 10^{26}$;
         (c) $-299$ kJ mol$^{-1}$; (d) $-296$ kJ mol$^{-1}$;
         (e) 41.4 J K$^{-1}$ mol$^{-1}$

17.34   (a) $-5.4$ kJ mol$^{-1}$ (per mole of $N_2O_4$);
         (e) $-3.95$ kJ mol$^{-1}$; $+3.95$ kJ mol$^{-1}$ (per mole of
         $N_2O_4$); (f) 180 J K$^{-1}$ mol$^{-1}$


## Chapter 18

18.1   (a) 2; (b) 6; (c) 2; (d) 4; (e) 6; (f) 2

18.2   $-150$ kJ mol$^{-1}$

18.4   (b) 2.37 V; (c) $-457$ kJ mol$^{-1}$

18.5   (a) 0.13 V; $-25$ kJ mol$^{-1}$; (b) 0.13 V;
         $-12.5$ kJ mol$^{-1}$; (c) 0.56 V; $-324$ kJ mol$^{-1}$

18.7   (c) $-122$ kJ mol$^{-1}$

18.8   0.29 V

18.9   0.74 V

18.11   1.06 V

18.12   0.21 mol dm$^{-3}$

18.13   $+0.34$ V

18.15   (c) $-104$ kJ per mole of $H_2O_2$

18.16   (a) $Br_2$, anode; K, cathode; (b) $Cl_2$, anode; Ca,
         cathode; (c) $Cl_2$ or mixture of $Cl_2$ and $O_2$, anode;
         $H_2$, cathode; (d) $Cl_2$, anode; $H_2$, cathode; (e) Cu
         transferred from anode to cathode; (f) $O_2$, anode;
         $H_2$, cathode

18.17   2.3 g

18.19   0.64 dm$^3$ $H_2$; 0.32 dm$^3$ $O_2$

18.20   $8.57 \times 10^{-17}$

18.21   0.66 V

18.23   $7.0 \times 10^{13}$


## Chapter 19

19.2   0.0133 S m$^2$ mol$^{-1}$; $1.333 \times 10^{-2}$ S m$^2$ mol$^{-1}$ from
         values in Table 19.2

19.3   (a) $2.586 \times 10^{-2}$; (b) $2.752 \times 10^{-2}$; (c) $2.617 \times 10^{-2}$;
         (d) $2.602 \times 10^{-2}$ S m$^2$ mol$^{-1}$

19.4   (a), (b), (d), (e) Strong; (c) weak

19.5   (b), (d)

19.6   $3.822 \times 10^{-2}$ S m$^2$ mol$^{-1}$

19.7   $5.012 \times 10^{-2}$ S m$^2$ mol$^{-1}$; $5.020 \times 10^{-2}$ S m$^2$ mol$^{-1}$
         from values in Table 19.2

19.9   $K = 1.81 \times 10^{-5}$; $\alpha = 0.019$ or 1.9%

19.10   4.74

19.11   (a) $6.162 \times 10^{-8}$; (b) $2.052 \times 10^{-7}$; (c) $8.291 \times 10^{-8}$;
(d) $7.618 \times 10^{-8}$; (e) $7.960 \times 10^{-8}$ m$^2$ s$^{-1}$ V$^{-1}$

19.12   $1.348 \times 10^{-2}$ S m$^2$ mol$^{-1}$

19.13   $2.72 \times 10^{-2}$ S m$^2$ mol$^{-1}$

## Chapter 20

20.1   (a) 16; (b) 1; (c) 18; (d) 2; (e) 12; (f) 13

20.2   (a) 7; (b) 25; (c) $d$-block; (d) 7; (e) metal;
(f) $1s^2 2s^2 2p^6 3s^2 3p^6 4s^2 3d^6$

20.3   (a) 6; (b) $p$-block; (c) 16; (d) anion, $Z^{2-}$; (e) non-metal; (f) $1s^2 2s^2 2p^4$

20.4   (a) 4; (b) 2; (c) $p$-block; (d) 14; (e) character changes down group; (f) EH$_4$; (g) $ns^2 np^5$

20.8   (a) $\frac{1}{2}$F$_2$(g) $\longrightarrow$ F(g); (b) Rb(s) $\longrightarrow$ Rb(g);
(c) $\frac{1}{2}$Br$_2$(l) $\longrightarrow$ Br(g); (d) V(s) $\longrightarrow$ V(g);
(e) Si(s) $\longrightarrow$ Si(g)

20.9   (a) 2; (b) metal

20.16   (a) decrease; (b) increase; (c) decrease; (d) increase; (e) increase

## Chapter 21

21.1   Shifts to (a) 2281, (b) 2261, (c) 2269 cm$^{-1}$

21.3   Cu$^{2+}$; Mg, Ca, Zn, Cr

21.4   $-727$ kJ mol$^{-1}$; $-5476$ kJ mol$^{-1}$

21.7   (a) $-3$; (b) $-3$; (c) $-2$; (d) $-3$

21.8   $\Delta_{vap}H/T$: PCl$_3$, 0.0874; HI, 0.0835; CHCl$_3$, 0.0874; C$_7$H$_{16}$, 0.0857 kJ K$^{-1}$ mol$^{-1}$

21.11   (a) C$-$H, $\approx$non-polar; CH$_4$, non-polar; (b) O$-$H, polar; H$_2$O, polar; (c) B$-$H, $\approx$non-polar; BH$_3$, non-polar; (d) N$-$H, polar; NH$_3$, polar; (e) Si$-$H, $\approx$non-polar; SiH$_4$, non-polar; (f) H$-$Cl, polar; (g) Se$-$H, polar; H$_2$Se, polar

21.12   (a) K at cathode; Br$_2$ at anode; (b) H$_2$ at cathode; Cl$_2$ (or mix of Cl$_2$ and O$_2$) at anode; (c) BaSO$_4$ and H$_2$O$_2$; (d) Pb and H$_2$O; (e) H$_2$O and O$_2$; (f) I$_2$ and H$_2$O

21.16   (a) Na$_2$CO$_3$ + $\frac{1}{2}$O$_2$; (b) 2NaOH; (c) K$_2$SO$_4$ + H$_2$O; (d) KI + H$_2$O; (e) NaO$_2$; (f) 2NaCl

21.17   SrO + CO$_2$; (b) Ca(OH)$_2$ is sparingly soluble; (c) [Be(OH)$_4$]$^{2-}$; (d) MgCl$_2$ + 2H$_2$O; (e) 2HF + Ca(HSO$_4$)$_2$; (f) Ca(OH)$_2$ + 2H$_2$

## Chapter 22

22.1   (a) $+2$; (b) $+5$; (c) $+4$; (d) $+5$; (e) $+5$; (f) $+3$; (g) $+4$; (h) $+6$; (i) $+7$; (j) $+7$; (k) $-1$

22.2   $-513$ kJ mol$^{-1}$

22.3   (a) B$-$F > B$-$Cl > B$-$I

22.4   Ga$^+$[GaCl$_4$]$^-$; Ga(I) and Ga(III) both diamagnetic

22.5   (a) Octahedral; (b) trigonal bipyramidal; (c) tetrahedral

22.8   [NH$_4$]$^+$, $-3$; [NO$_2$]$^-$, $+3$; N$_2$, 0; [NH$_4$]$^+$, $-3$; [NO$_3$]$^-$, $+5$; N$_2$, 0

22.9   Linear; N$_2$O

22.10   (a) and (c)

22.11   $K \approx 5 \times 10^{-12}$

22.13   (a) [BrF$_2$]$^+$[SbF$_6$]$^-$; (b) K$^+$[AsF$_6$]$^-$; (c) [PF$_4$]$^+$[Sb$_2$F$_{11}$]$^-$

22.14   (a) $-3$ to $+1$; $+5$ to $+1$; (b) N$_2$O, $+1$; NO, $+2$; N$_2$O$_3$, $+3$; N$_2$O$_4$, $+4$; NO$_2$, $+4$; N$_2$O$_5$, $+5$

22.15   2; stereochemically non-rigid (fluxional) on NMR timescale

22.17   (a) IF, IF$_3$, IF$_5$ or IF$_7$; (b) BrCl

22.20   (a) N$_2$O$_3$; (b) N$_2$O$_5$; (c) P$_4$O$_{10}$; (d) P$_4$O$_6$; (e) Cl$_2$O; (f) I$_2$O$_5$

22.21   (b) Linear; octahedral; tetrahedral; linear

22.22   (a) BF$_3$ is a good Lewis acid; (b) effect of a lone pair of electrons; (c) to give Na[XeF$_7$] or Na$_2$[XeF$_8$]

## Chapter 23

23.2   (a) $1s^2 2s^2 2p^6 3s^2 3p^6 4s^2 3d^6$; (b) $1s^2 2s^2 2p^6 3s^2 3p^6 3d^5$; (c) $1s^2 2s^2 2p^6 3s^2 3p^6 4s^1 3d^{10}$; (d) $1s^2 2s^2 2p^6 3s^2 3p^6 3d^{10}$; (e) $1s^2 2s^2 2p^6 3s^2 3p^6 3d^2$; (f) $1s^2 2s^2 2p^6 3s^2 3p^6 4s^2 3d^8$; (g) $1s^2 2s^2 2p^6 3s^2 3p^6 3d^3$; (h) $1s^2 2s^2 2p^6 3s^2 3p^6 3d^{10}$

23.4   (a) $+2$; (b) $+3$; (c) $+2$; (d) $+3$; (e) $+3$; (f) $+2$; (g) $+2$; (h) $+2$; (i) $+2$; (j) $+3$

23.6   (a) Fe, 0; N, $+\frac{1}{3}$; (b) Cr, 0; F, $-\frac{1}{2}$

23.9   (a), (c), (e) possess enantiomers

23.10   (d) *trans*; *cis*; *cis* possesses enantiomers

23.11   (a) $+3$; (c) hydration isomers

23.13   (a) [Ni(en)$_3$]$^{2+}$, enantiomers; (b) [Co(phen)$_3$]$^{2+}$, enantiomers; (c) *trans* and/or *cis*-[VCl$_4$(py)$_2$]; (d) [Cr(acac)$_3$], enantiomers; (e) [Fe(CN)$_6$]$^{4-}$; (f) [Fe(NH$_3$)$_6$]Cl$_2$

23.16   NH$_3$ complex: $-21.1$ kJ mol$^{-1}$; en complex $-34.2$ kJ mol$^{-1}$

23.17   (b) 3; 3; 5; 4

23.18   (c) $10^{24}$

23.19   $K_1 = 3.31 \times 10^5$; $K_2 = 3.16 \times 10^3$

23.21   (a) soft–soft; (b) hard–hard; (c) hard–hard; tricapped trigonal prismatic; (d) hard–hard

23.22   [Cr(NCS-$N$)$_6$]$^{3+}$; Cr$^{3+}$ and $N$-donor are both hard

23.24   (a) 4.90 $\mu_{eff}$; (b) 4.90 $\mu_{eff}$; (c) 2.83 $\mu_{eff}$

23.25   (a) High-spin; (b) high-spin

23.26   (a) $-1$; (b) $+2$; (c) 0; (d) $+1$; (e) $+1$; (f) $+1$

23.29   Values of $n$: (a) 4; (b) 5; (c) 5; (d) 3; (e) 4; (f) 4

23.31   (a) **A** = [Et$_4$N]$_4$[Mn(NCS)$_6$];
(c) **B** = [Co(NH$_3$)$_5$Br][SO$_4$];
**C** = [Co(NH$_3$)$_5$(SO$_4$)]Br

## Chapters 24–35

For more detailed answers with structural diagrams and curly-arrow mechanisms, see the accompanying *Solutions Manual* (www.pearsoned.co.uk/housecroft).

## Chapter 24

24.1 (a) **24.5**, $sp^3$; **24.6**, $sp^2$; **24.7**, $sp$; **24.8**, $sp^2$ terminal C, $sp$, central C

24.2 All 4-coordinate C, $sp^3$; C=C atoms, $sp^2$; C≡C atoms, $sp$

24.3 See Table 1.9

24.10 (2E,4E)-Hepta-2,4-diene; (3Z)-hexa-1,3-diene; (3E)-hexa-1,3-diene; hex-3-yne; but-1-ene

24.11 (a) (2E,5E)-Octa-2,5-diene; (2Z,5E)-octa-2,5-diene; (2E,5Z)-octa-2,5-diene; (2Z,5Z)-octa-2,5-diene; (b) only one form; (c) 3 isomers

24.13 Chiral centres are marked *; the other 2 molecules in the question are achiral:

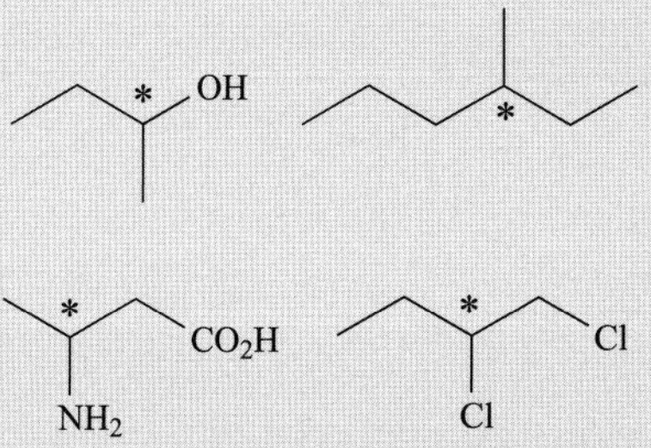

24.14

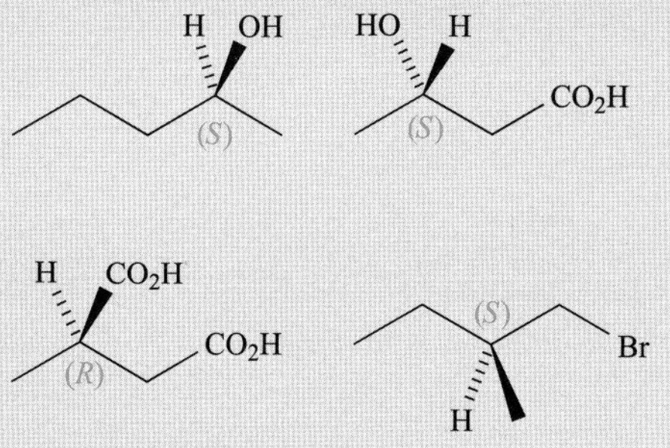

24.17 A pair of enantiomers and a *meso*-form

24.19 (a)

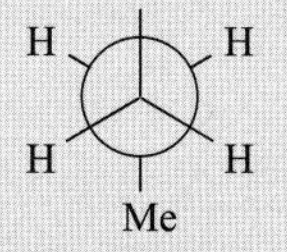

**A** (staggered)   **B** (eclipsed)

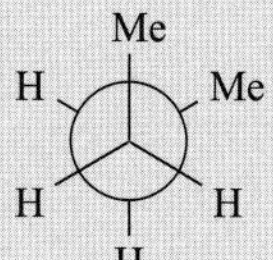

**C** (staggered)   **D** (eclipsed)

24.20 (a) $2\mathrm{Br}^{\bullet}$

(b)

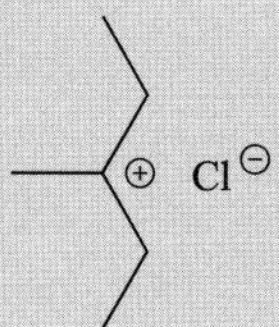

(c) $\mathrm{Et}^{\bullet} + \mathrm{Br}^{\bullet}$

(d) $+ \mathrm{H_2O}$

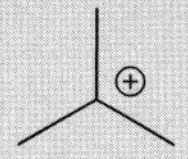

(e) $+ \mathrm{CN}^{\ominus}$

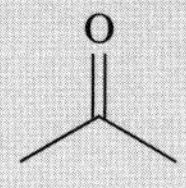

24.21 (a) $+ \mathrm{H_2O}$

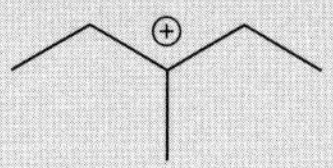

(b) $\mathrm{Me_3CCl}$

(c) $\mathrm{H}^{\bullet} + \mathrm{Br}^{\bullet}$

(d) $+ \mathrm{Cl}^{\ominus}$

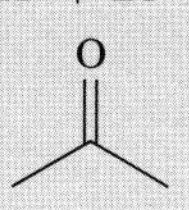

24.22 Note that the question asks for the *most convenient and routine* methods: (a) $^{13}$C NMR spectroscopy; (b) $^1$H NMR spectroscopy; (c) $^{13}$C or $^1$H NMR spectroscopy

24.23 (a) All but glycine are chiral; (c) all are chiral

24.24

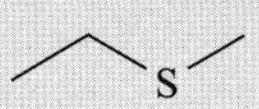

**A**   **B**

## Chapter 25

25.1 (a) (b) 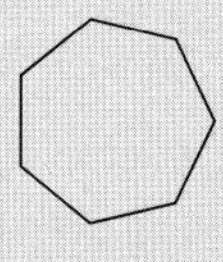 (c)

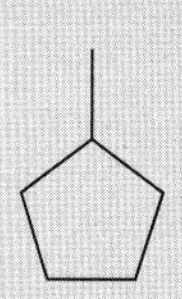

(d) 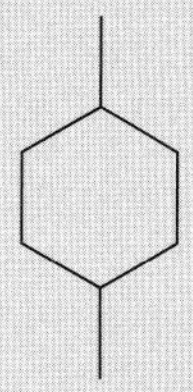 (e)

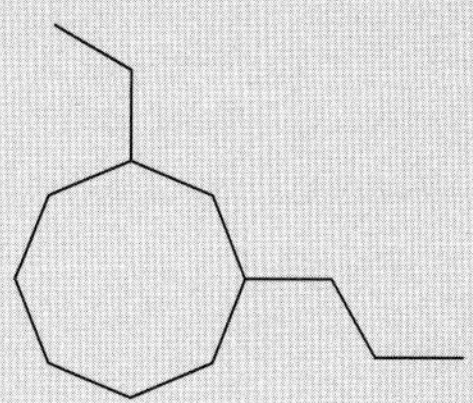

25.2 (a) 1,3-Diethylcyclopentane;
(b) propylcyclopropane;
(c) 1,3,5-trimethylcyclohexane;
(d) 1-ethyl-4-methylcycloheptane

25.3 (a) 5; (b) 5; (c) 3; (d) 10

25.4 1,1-Dimethylcyclobutane;
*cis*-1,2-dimethylcyclobutane;
*trans*-1,2-dimethylcyclobutane;
*cis*-1,3-dimethylcyclobutane;
*trans*-1,3-dimethylcyclobutane

25.7   $+51 \, \text{kJ mol}^{-1}$

25.8   $C_3H_8$

25.10  Initiation: radicals formed; propagation: radicals consumed *and* formed; termination: radicals consumed

25.12  Products are 1-chloro-2-methylpropane ($^1$H NMR spectrum = 3 multiplets) and 2-chloro-2-methylpropane (1 singlet)

25.15  $O_2$ is a diradical; it may react with radical species in the reaction mixture

25.16  (a) ... + ...   (b) ...

(c) ...   (d) ... + other substituted products

25.17  **B**   **C**   **A**

**D**   **E**

25.20  (a) **A** = pentane; $^{13}$C NMR inconsistent with branched isomers; (b) **B** = 2,2-dimethylpropane; (c) bp **A** > **B**

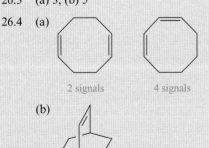

## Chapter 26

26.1   (a) (*E*)-Pent-2-ene; (b) but-1-ene;
(c) (*Z,Z*)-hexa-2,4-diene; (d) (*Z*)-pent-2-ene;
(e) pent-2-yne; (f) pent-1-yne

26.2   $3300 \, \text{cm}^{-1}$, terminal C—H (*sp* C); $\approx 3000 \, \text{cm}^{-1}$, $CH_2$ (*sp*$^3$ C); $2200 \, \text{cm}^{-1}$, C≡C

26.3   (a) 3; (b) 5

26.4   (a)

2 signals     4 signals

(b)

26.5   (a) $3665 \, \text{cm}^{-1}$, terminal C—H (*sp* C); 2922, $2851 \, \text{cm}^{-1}$, $CH_2$ (*sp*$^3$ C); $2123 \, \text{cm}^{-1}$, C≡C

(b) 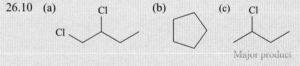 
*a* $\delta$ 69.5 ppm
*b* $\delta$ 82.4 ppm
*c* $\delta$ 18.6 ppm

(c) $\delta$ 2.06 ppm, terminal CH; $\delta$ 2.43 ppm, $CH_2$

26.7   (a)   (b)

(c)   (d)

(e)

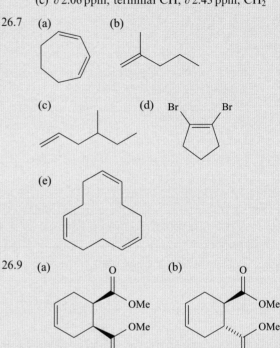

26.9   (a)   (b)

26.10  (a)   (b)   (c)

Major product

(d)   (e)

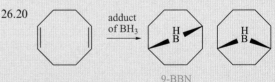

26.11  $-144 \, \text{kJ mol}^{-1}$; $-136.3 \, \text{kJ mol}^{-1}$

26.17  (a) $H_2O$, $H^+$ catalyst; (b) $H_2O$, $Cl_2$; (c) $OsO_4$, $H_2O$;
(d) HBr, radical initiator

26.20

adduct of $BH_3$ →

9-BBN

26.23  (c), (d)

26.24  (a)   (b)

+ $^1/_2 H_2$

(c)   (d)

(*Z*)-isomer

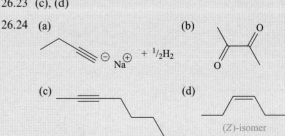

26.27

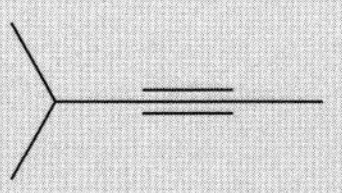

**A** 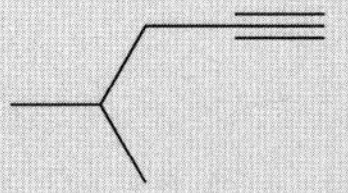          **B**

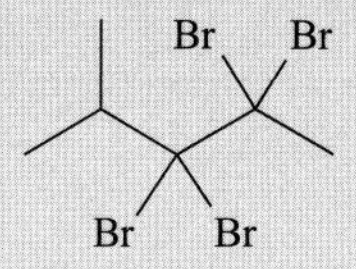

**C**

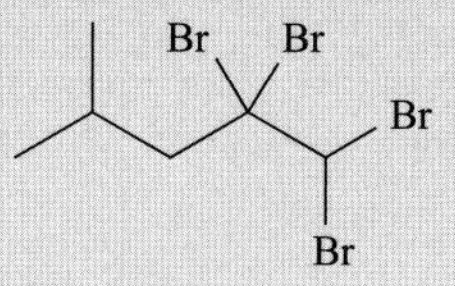

## Chapter 27

27.1 (a) $C^{\delta+}-Br^{\delta-}$; (b) $N^{\delta-}-H^{\delta+}$; (c) $Cl^{\delta+}-F^{\delta-}$; (d) approximately non-polar

27.2 Polar: (a), (b), (d), (e), (f), (g)

27.4 Polar: (a), (c), (d), (e)

27.8 (E)-isomer is non-polar; (Z)-isomer is polar:

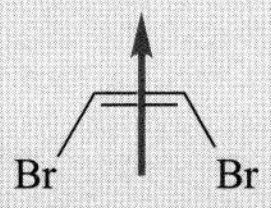

Br          Br

27.11 **A** $= CH_3CH_2Cl$; **B** $= CH_3CH_2Br$; **C** $= CH_3CH_2I$; mass spectrometric data confirm **A** contains Cl, **B** contains Br

## Chapter 28

28.3 (a)          (b)

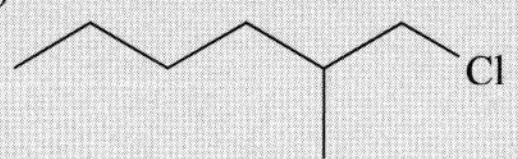

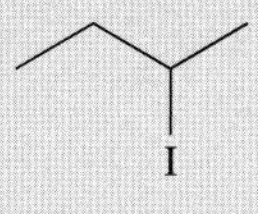

Cl

Primary          I

Secondary

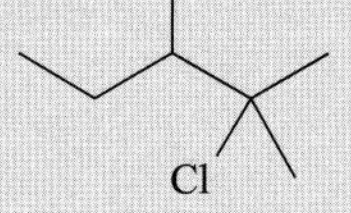

(c)          (d)

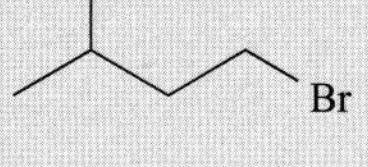

Cl          Br

Tertiary          Primary

28.5 (a)          (b)

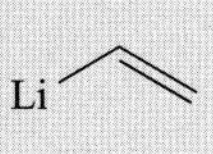

Li          Li

(c) 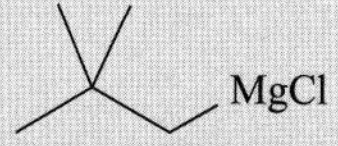          (d)

MgCl

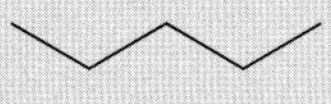

28.8 (b) If $R_3CI$ represents (S)-3-iodo-3-methylheptane, rate equation is:

$$\text{Rate} = -\frac{d[R_3CI]}{dt} = k[R_3CI]$$

28.12 $Me_3CBr \gg Me_2CHBr > MeCH_2Br > CH_3Br$

28.14 (a) Hexane is non-polar; (b) ethanol, water, acetic acid are polar and protic; (c) acetone, DMSO, acetonitrile, dichloromethane are polar and aprotic

28.17 (a)          (b)

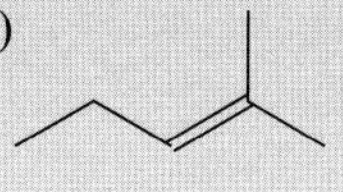

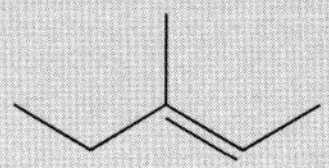

(c)           (d)

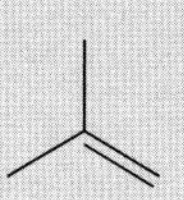

28.18 Methyl migration: (a), (b)

28.23 (a) $HS^-$, soft nucleophile, $S_N2$; (b) $HO^-$, hard nucleophile, $S_N2$; (c) $^tBuO^-$, strong/bulky base, E2; (d) $EtO^-$, strong base, E2 and $S_N2$ compete, E2 predominates

## Chapter 29

29.1 (a) Dimethyl ether; (b) dipropyl ether or propoxypropane; (c) methyl propyl ether or methoxypropane; (d) 2,4-dioxapentane; (e) 3,6-dioxaoctane

29.2 (a) and (b)

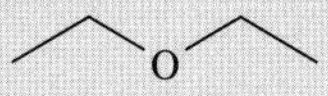

(c)

(d)

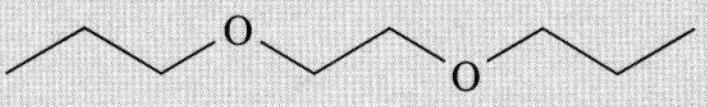

29.3 (a)

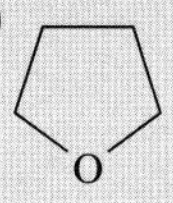

(b)

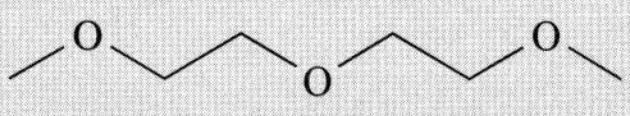

(c) 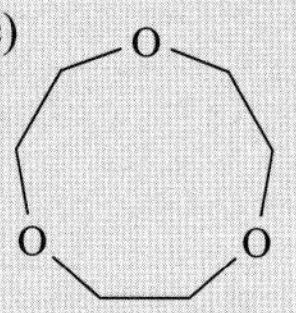          (d) 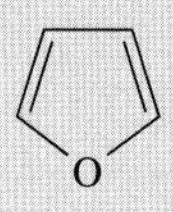          (e)

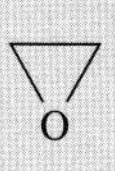

29.5 Competitions between formations of EtOMe, MeOMe and EtOEt, and also elimination of $H_2O$ from EtOH; better choice is Williamson's synthesis from EtCl and $MeO^-$

29.6 (a)          (b)

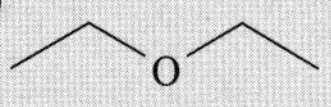

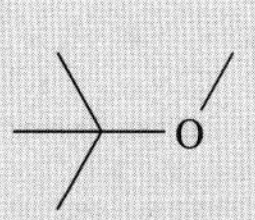

(c)          (d)

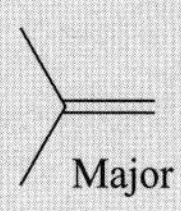

Major product

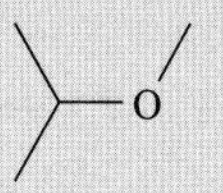

**29.7** (a)  (b)

**29.8** (b) Polar: 1,3-dioxalane, furan, dibutyl ether

**29.9** (a) Stretching mode of C−H; (b) $^1$H or $^{13}$C NMR spectra

**29.10** See structure of diglyme in answer 29.3b

**29.11** (a) ; (b) 2 signals, rel. integrals 1 : 2

**29.12** (a) Formation of EtBr and EtOH; (b) formation of EtCl and EtOH; (c) $Et_2O$ will dissolve; (d) $Et_2O$ evaporates; hand feels cold; hazards: $Et_2O$ is volatile and inflammable, and vapour is explosion hazard

**29.15** (a) (b)

(c)

**29.16** Possible compounds: **A** = Dipentyl ether; **B** = 1-iodopentane; **C** = pentan-1-ol; but there are insufficient data to confirm linear chains

**29.17** (a) ; (b) analogue of Williamson's synthesis using MeCl and $K[SCHMe_2]$; (c) $[Me_3S]^+I^-$

**Chapter 30**

**30.1** (a) (b)

(c) (d)

**30.2** (a) Butane-1,2-diol; (b) 3-methylpentan-2-ol; (c) cyclohexane-1,2-diol; (d) pentane-1,3-diol; (e) 3,3-dimethylpentan-1-ol

**30.3** Stereoisomers for (a), (b), (d), (e)

**30.6** (a) $Me_2CO$ (acetone) and MeMgCl; (b) $H_2CO$ (methanal) and $CH_3(CH_2)_5MgCl$; (c) butanal and MeMgCl, or ethanal (acetaldehyde) and $CH_3(CH_2)_2MgCl$

**30.7** (a) OH (b) OH

(c) OH

**30.10** $C_4H_9OH$; data correspond to butan-2-ol; other isomers: butan-1-ol, 2-methylpropan-1-ol, 2-methylpropan-2-ol

**30.11** (a) $^1$H or $^{13}$C NMR spectroscopy; (b) IR, $^1$H or $^{13}$C NMR spectroscopy; (c) mass spectrometry, or $^1$H or $^{13}$C NMR spectroscopy

**30.12** (a) $2Na + 2H_2O \longrightarrow H_2 + 2NaOH$; (b) reaction with propan-2-ol (isopropanol) is relatively slow and safe

**30.14** Major products:

(a) (b)

(c)

**30.15** (a) $O^{\ominus}K^{\oplus}$ (b) $O^{\ominus}K^{\oplus}$

(c) (d)

(e) Cl (f) Cl

**30.16** (a) OH (b) OH
(Jones oxidation)

(c) OH (Jones oxidation)
(Jones oxidation)

(d) OH
(PCC, $CH_2Cl_2$)

(e) OH
(Jones oxidation)

**30.19** (a) $CH_2$ protons with $J(^1H-^{19}F)$; (b) $^1$H/D exchange (D = $^2$H); (c) hydrogen bonding involving O···H stronger than that with S···H

**Chapter 31**

**31.1** (a) 1-Propylamine; (b) 3-methyl-2-butylamine; (c) 4-methylcyclohexylamine; (d) 3,4-dimethyl-1-hexylamine; (e) 2-methyl-2-propylamine

**31.2** (a) $NH_2$  Primary

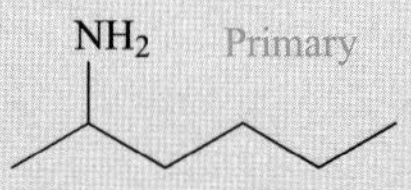

(b) $NH_2$  Primary

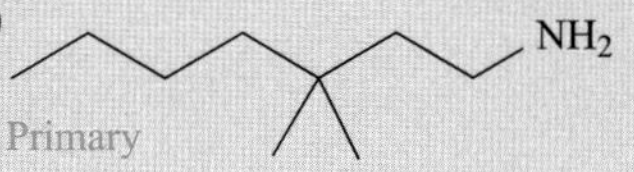

(c) N  Tertiary   (d) N H  Secondary

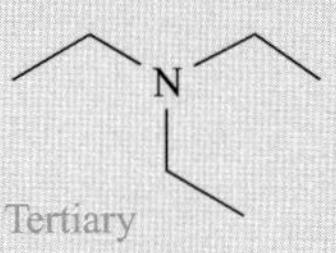

(e) $H_2N$ ... $NH_2$  Primary  Primary   (f) Primary $NH_2$

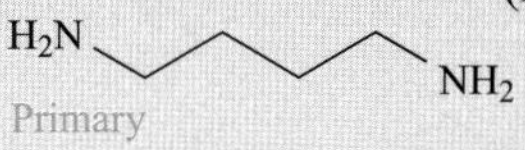

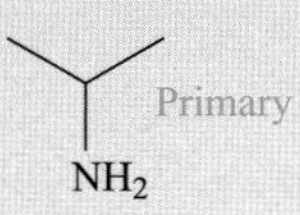

(g) N  Tertiary   (i) $NH_2$  Primary

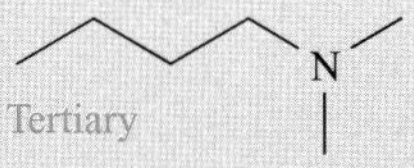

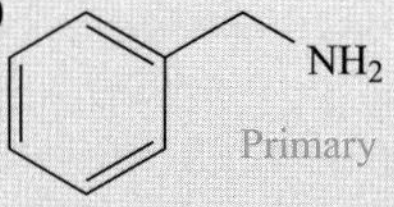

(h) $NH_2$  Primary

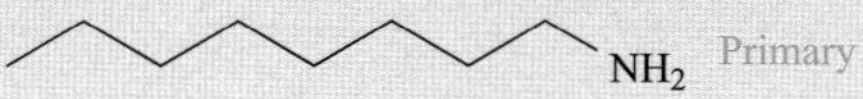

**31.3** (a) $N^+$ Et $Cl^-$   (c) $(... N^+ H_2)_2$ $SO_4^{2-}$

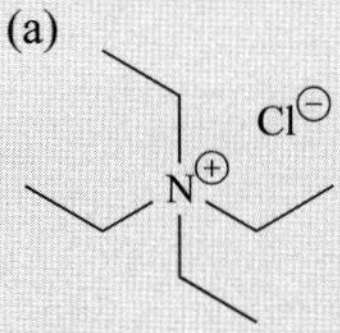

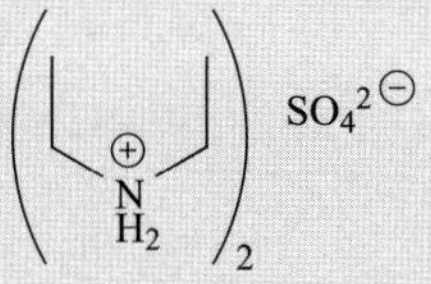

(b) $N^+$ $Br^-$

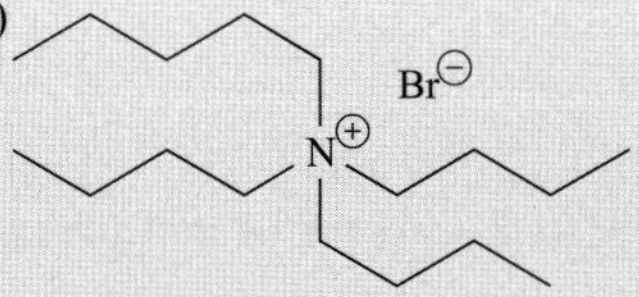

(d) benzyl $N^+$(H)(Me)(Me) $Br^-$

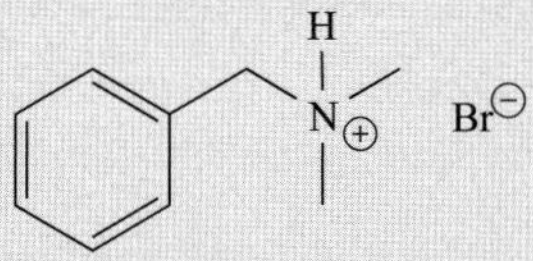

**31.4** $[RR'R''R'''N]^+$ contains tetrahedral $N^+$ centre; $RR'R''N$ undergoes facile inversion

**31.7** (a) $NH_2$   (c) $N^+H$(Me)(Me) $Cl^-$

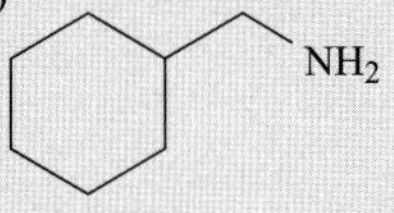

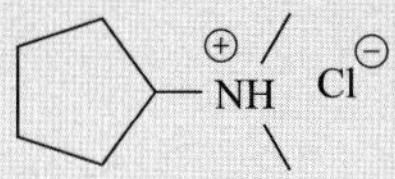

(b) $N^+Me_3$ $Br^-$ + $NMeR$  R = H, Me

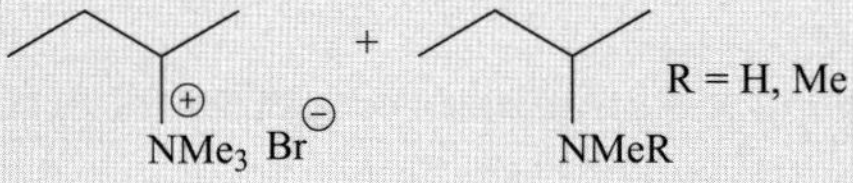

**31.8** (a) $NH_2$ / $NH_2$   (b) $NH_2$

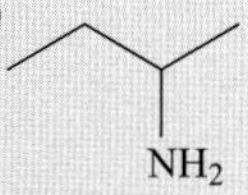

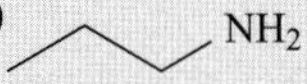

(c) $NH_2$ $NH_2$   (d) $NH_2$

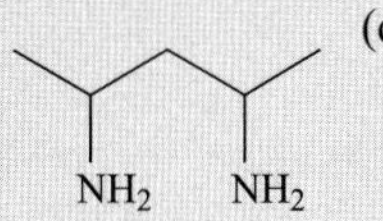

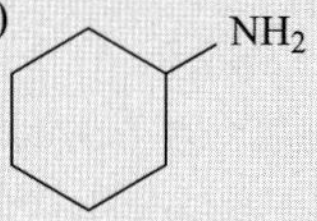

**31.10** Each $pK_b$ defined by general equilibrium:
$$RNH_2 + H_2O \rightleftharpoons RNH_3^+ + OH^-;$$
each $pK_a$ defined by general equilibrium:
$$RNH_3^+ + H_2O \rightleftharpoons RNH_2 + H_3O^+$$

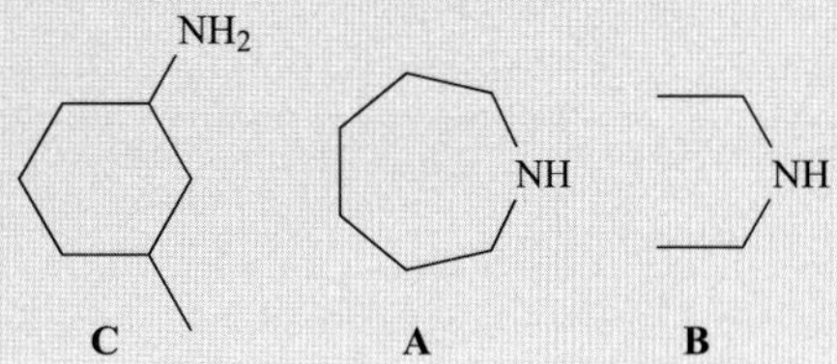

**31.11**  $NH_2$ **C**   H–N...N–H **A**   N **B**

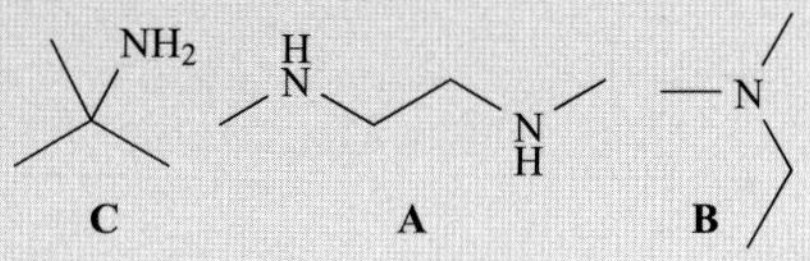

**31.12** (a) $NH_2$ **C**   NH **A**   NH **B**

(b) **C** has stereoisomers; not distinguished by $^{13}C$ NMR; (c) $^{13}C\{^1H\}$ = proton decoupled

**31.13** (a) $\overset{+}{N}H_3$ $I^-$   (b) N

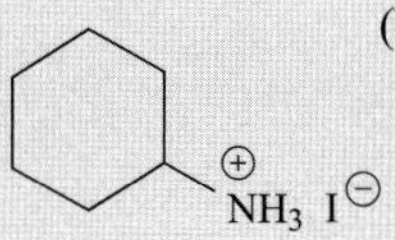

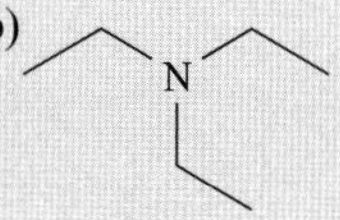

(c) N Li   (d) $\overset{+}{N}H_2$  as $SO_4^{2-}$ salt

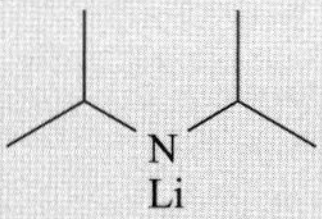

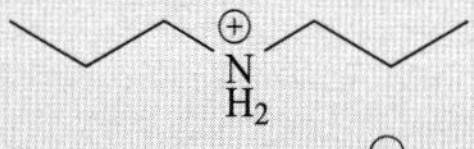

**31.15** (a) 1,3-Pentanediamine; (b) forms nitrate salt, each $NH_2$ protonated

**31.16** (a)    (b) $\overset{H}{N}$ ... O

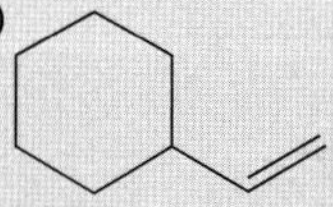

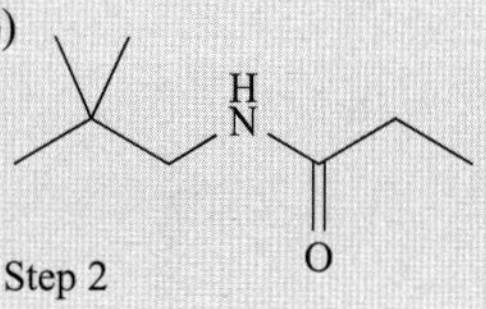

(c) Step 1    Step 2
N O    N

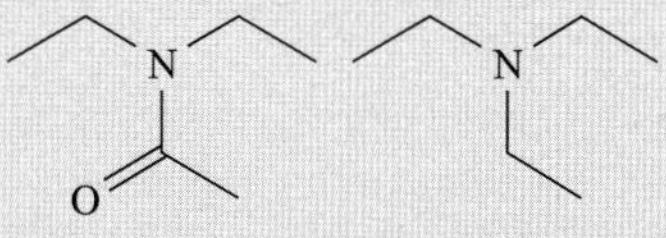

(d) Step 1    Step 2
$NH_2$    $\overset{+}{N}H_3$ $Br^-$

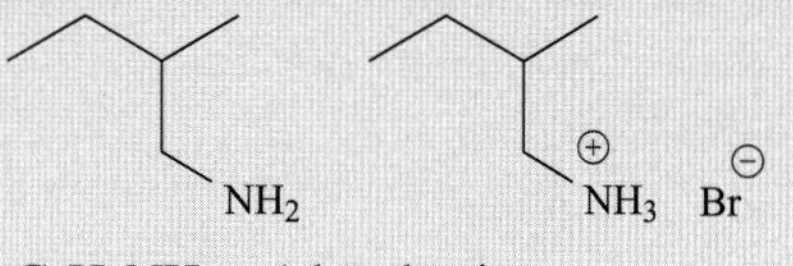

**31.19** $Y = C_4H_9NH_2 = $ 1-butylamine;
$X = $ 1-cyanopropane; $Z = $ 2-cyanopropane

## Chapter 32

### Mid-chapter problems

2   Resonance structures contribute equally to overall bonding

3   (a), (c), (d), (f), (g)

4   $3120$–$3000\ cm^{-1}$, $\nu(C-H)$; $1480\ cm^{-1}$, ring vibration; $680\ cm^{-1}$, out-of-plane C–H bending

5   (b) $^1H$ or $^{13}C$ NMR spectroscopy

**6**

**7**  (a) Me (b) OH (c) O—OH

(d) (e) NH$_2$ (f) HN

**8**  F F (ortho) / F F (meta) / F F (para)

ortho — 1,2-Difluorobenzene
meta — 1,3-Difluorobenzene
para — 1,4-Difluorobenzene

## End-of-chapter problems

**32.1** Include 3 more resonance forms:

**32.2** Equal C—C bond lengths; no addition reactions; thermochemical data (see Figure 32.2)

**32.3** (a) Delocalized bonding, not C—C and C=C alternating around ring; (b) radical substitution occurs in Et group; (c) $\pi^* \leftarrow \pi$ transition

**32.5** Aromatic: (a), (e)

**32.7** (a) OH (b) (c) N—H (d) $\overset{\ominus}{O}\overset{\oplus}{N}O$ (e) S—

**32.8** (a) **C**; (b) **A**; (c) **E**; (d) **D**; (e) **B**; (f) **F**

**32.9** (a) 2 singlets, rel. int. 1 : 3 (aromatic: alkyl); $^1$H or $^{13}$C NMR spectroscopies; isomer is 1,2-chlorobenzene

**32.11** (a) MeCl/AlCl$_3$; (b) Br$_2$/FeBr$_3$; (c) $^i$PrCl (or $^n$PrCl)/AlCl$_3$; (d) Na/liquid NH$_3$ in EtOH; (e) Cl$_2$/AlCl$_3$; other Lewis catalysts could be used in (a), (b), (c), (e)

**32.12** **A** = 1,3,5-tribromobenzene; **B** = 1,2,4-tribromobenzene

**32.15** (a) NO$_2$ ... NO$_2$  Must be *harsh* conditions
(b) Me ... SO$_3$H  + *para*-isomer
(c) Me ... Br  + *para*-isomer
(d) NH$_2$ with Br, Br, Br

**32.18** (a) 1 versus 2 signals in alkyl region; (b) 4 versus 2 signals; (c) 1 versus 2 signals in alkyl region; 2 versus 4 signals in aromatic region; (d) 3 versus 2 signals in alkyl region; (e) 1 versus 6 signals

**32.19** (a) Br$_2$/FeBr$_3$; (b) 2 steps: conc HNO$_3$/H$_2$SO$_4$; Fe/HCl(aq); (c) 2 steps: Cl$_2$/AlCl$_3$; conc HNO$_3$/H$_2$SO$_4$; (d) 4 steps: SO$_3$/conc H$_2$SO$_4$; NaOH(aq); NaOH/550 K; H$_2$SO$_4$(aq)

**32.21** Polar: (a), (c), (d)

**32.22** (a) KMnO$_4$(aq)/heat; (b) Fe/HCl(aq); (c) 2 steps: NaNO$_2$/HCl(aq)/ice bath; CuCN, excess CN$^-$; (d) 2 steps: NaNO$_2$/HCl(aq)/ice bath; PhOH

**32.24** (b)

A                B

**32.25** (a) MeCl/AlCl$_3$; (b) conc HNO$_3$/H$_2$SO$_4$; (c) Fe/HCl(aq); (d) MeCOCl

## Chapter 33

**33.2** (a) Pentan-2-one; (b) benzoic acid; (c) ethyl propanoate; (d) butanal; (e) propanamide; (f) diphenyl ketone; (g) pentane-2,4-dione (acetylacetone); (h) hexanoyl chloride

**33.5** (a) **C**; (b) **E**; (c) **A**; (d) **D**; (e) **B**

**33.6**

R = CH$_2$CHMe$_2$ ;

bonding is analogous to structure **21.3**

**33.7** (a) K$_2$Cr$_2$O$_7$, H$^+$; (b) K$_2$Cr$_2$O$_7$, H$^+$;
(c) KMnO$_4$(aq), heat; (d) LiAlH(O$^t$Bu)$_3$ in Et$_2$O,
low temp.; (e) DIBAL in hexane, low temp.;
(f) K$_2$Cr$_2$O$_7$, H$^+$

**33.9** (a) (b)

(c)

**33.13** Haloform reaction: (b), (c), (d)

**33.14** Decreasing p$K_a$ values: (d) > (b) > (c) > (e) > (a)

**33.15** **A** = hexan-2-one (or 3-methylpentan-2-one)

**33.16** NC$^-$, then hydrolysis in presence of acid or base, or
Grignard reagent, then solid CO$_2$ and H$^+$

**33.18** (a) (b)

(c)

**33.20** (c) **A**, ester; **B**, acetal; **C**, hemiacetal; **D**, alcohol
(diol); **E**, ether; **F**, acetal

**33.21** (a) NC$^-$, hydrolysis; (b) K$_2$Cr$_2$O$_7$, H$^+$; (c) PCC in
CH$_2$Cl$_2$; (d) 1-bromopropane; (e) base catalysed
condensation (aldol reaction)

**Chapter 34**

**34.1** (a) Furan; (b) pyrrole; (c) thiophene

**34.2** (a) (b)

(c) (d)

**34.3** (a) (b) (c)

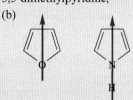

(d) (e)

**34.4** (a) 6; 2,3-, 2,4-, 2,5-, 2,6-, 3,4-,
3,5-dimethylpyridine;
(b)

**34.5** (a) 1 electron from each CH unit; 2 electrons from
O; overlap of 2$p$ orbitals; (b) $^1$H NMR chemical
shift data; (c) more

**34.8** Saturated heterocycle is typical secondary amine;
pyrrole N lone pair used in $\pi$-system; pyridinium
ion is weak acid – aromaticity stabilizes conjugate
base

**34.10** (a) (b)

(c) (d)

**34.13** ; H$^-$ abstractor, e.g. [Ph$_3$C][BF$_4$]

**34.14** (a) Pyridinium sulfate; (b) $N$-ethylpyridinium
iodide; (c) 3-nitropyridine

**34.15** Both Diels–Alder cycloadditions; (a) *exo*-product;
(b) *endo*-product

**34.16** (a) Pyrrole unstable in the presence of Lewis acid
catalyst such as AlCl$_3$

**34.19** (a) NH$_2$ (b) SPh (c) NHNH$_2$ (d) OEt

**34.20** **A** = 2,6-Dichloropyridine;
**B** = 3,5-dichloropyridine; **C** = 2,5-dichloropyridine

## Chapter 35

35.1   Starting from the aldehyde end of the chain:
       (a) $(2S,3R,4R)$; (b) $(2R,3S,4R)$; (c) $(2S,3S,4R)$

35.4

35.5   (a) Hexose, aldose; (b) hexose, ketose; (c) pentose, aldose; (d) hexose, aldose; (e) hexose, aldose; (f) tetrose, aldose

35.6   (a) Polysaccharide; (b) disaccharide; (c) monosaccharide; (d) polysaccharide; (e) disaccharide; (f) monosaccharide; (g) monosaccharide

35.8   (a) Glycine; (b) L-amino acids; (d) valine; glycine; aspartic acid; cysteine; glutamic acid; histidine

35.9   (a) Charge-separated species, neutral overall (see equation 35.6); (b) at pH <2.32, valine is protonated; isoelectric point is pH 5.97; at pH >5.97, valine is deprotonated

# Index

Note: (B) indicates text in a Box, (F) a Figure, (N) a footnote, (T) a Table, and (WE) a Worked Example.